D1218694

Encyclopedia of Agricultural, Food, and Biological Engineering

edited by

Dennis R. Heldman
Rutgers, The State University of New Jersey
New Brunswick, New Jersey, U.S.A.

MARCEL DEKKER, INC. NEW YORK • BASEL

BRADLEY UNIVERSITY LIBRARY

Although great care has been taken to provide accurate and current information, neither the author(s) nor the publisher, nor anyone else associated with this publication, shall be liable for any loss, damage, or liability directly or indirectly caused or alleged to be caused by this book. The material contained herein is not intended to provide specific advice or recommendations for any specific situation.

Trademark notice: Product or corporate names may be trademarks or registered trademarks and are used only for identification and explanation without intent to infringe.

Library of Congress Cataloging-in-Publication Data
A catalog record for this book is available from the Library of Congress.

ISBN
Print: **0-8247-0938-1**
Print/Online: **0-8247-4266-4**
Online: **0-8247-0937-3**

Main cover photograph courtesy: Malo Inc., Tulsa, Oklahoma, U.S.A.

This book is printed on acid-free paper.

Headquarters
Marcel Dekker, Inc.
270 Madison Avenue, New York, NY 10016, U.S.A.
tel: 212-696-9000; fax: 212-685-4540

Distribution and Customer Service
Marcel Dekker, Inc.
Cimarron Road, Monticello, New York 12701, U.S.A.
tel: 800-228-1160; fax: 845-796-1772

Eastern Hemisphere Distribution
Marcel Dekker AG
Hutgasse 4, Postfach 812, CH-4001 Basel, Switzerland
tel: 41-61-260-6300; fax: 41-61-260-6333

World Wide Web
http://www.dekker.com

The publisher offers discounts on this book when ordered in bulk quantities. For more information, write to Special Sales/Professional Marketing at the headquarters address above.

Copyright © 2003 by Marcel Dekker, Inc. (except as noted on the opening page of each article.) All Rights Reserved.

Neither this book nor any part may be reproduced or transmitted in any form or by any means, electronic or mechanical, including photocopying, microfilming, and recording, or by any information storage and retrieval system, without permission in writing from the publisher.

Current printing (last digit):

10 9 8 7 6 5 4 3 2 1

PRINTED IN THE UNITED STATES OF AMERICA

Ref.
TP
368.2
.E57
2003

Dennis R. Heldman, Editor
Rutgers, The State University of New Jersey
New Brunswick, New Jersey, U.S.A.

Editorial Advisory Board

Ralph P. Cavalieri — *Agricultural Research Center, Washington State University, Pullman, Washington, U.S.A.*

James C. Craig — *United States Department of Agriculture, Agricultural Research Service (USDA-ARS), Retired, Wyndmoor, Pennsylvania, U.S.A.*

Robert J. Gustafson — *Department of Food, Agricultural, and Biological Engineering, The Ohio State University, Columbus, Ohio, U.S.A.*

Edward A. Hiler — *Department of Biological and Agricultural Engineering, Texas A&M University, College Station, Texas, U.S.A.*

Glenn J. Hoffman — *Department of Biological Systems Engineering, University of Nebraska, Lincoln, Nebraska, U.S.A.*

Gerald W. Isaacs — *Agricultural and Biological Engineering Department, University of Florida, Gainesville, Florida, U.S.A.*

Yoshisuke Kishida — *Shin-Norinsha Co. Ltd., Tokyo, Japan*

Otto J. Loewer — *Department of Biological and Agricultural Engineering, University of Arkansas, Fayetteville, Arkansas, U.S.A.*

Paul B. McNulty — *Department of Agricultural and Food Engineering, University College Dublin, Dublin, Ireland*

Sidney F. Sapakie — *General Mills, Inc., Minneapolis, Minnesota, U.S.A.*

Norman R. Scott — *Department of Biological and Environmental Engineering, Cornell University, Ithaca, New York, U.S.A.*

AUG 0 2 2004

Klaas van't Riet *TNO Nutrition and Food Research Institute, Zeist, The Netherlands*

Brahm P. Verma *Department of Biological and Agricultural Engineering, The University of Georgia, Athens, Georgia, U.S.A.*

List of Contributors

Viacheslav I. Adamchuk / *University of Nebraska, Lincoln, Nebraska, U.S.A.*
Jay B. Agness / *Deere and Company, Waterloo, Iowa, U.S.A.*
José Miguel Aguilera / *Pontificia Universidad Católica de Chile, Santiago, Chile*
Louis Albright / *Cornell University, Ithaca, New York, U.S.A.*
S. Almonacid / *Oregon State University, Astoria, Oregon, U.S.A.*
Pedro M. Álvarez / *Universidad de Extremadura, Badajoz, Spain*
Alejandro Amézquita / *University of Nebraska–Lincoln, Lincoln, Nebraska, U.S.A.*
R. C. Anantheswaran / *The Pennsylvania State University, University Park, Pennsylvania, U.S.A.*
Tetsuya Araki / *The University of Tokyo, Tokyo, Japan*
J. Arul / *Université Laval, Sainte-Foy, Quebec, Canada*
Pete E. Athanasopoulos / *Agricultural University of Athens, Athens, Greece*
G. B. Awuah / *National Food Processors Association, Washington, District of Columbia, U.S.A.*
James E. Ayars / *United States Department of Agriculture (USDA), Parlier, California, U.S.A.*
Phillip C. Badger / *General Bioenergy, Inc., Florence, Alabama, U.S.A.*
Rakesh Bajpai / *University of Missouri–Columbia, Columbia, Missouri, U.S.A.*
L. Dale Baker (Retired) / *Case Corporation, Burr Ridge, Illinois, U.S.A.*
Amarjit S. Bakshi / *ConAgra Foods, Irvine, California, U.S.A.*
V. M. Balasubramaniam / *The Ohio State University, Columbus, Ohio, U.S.A.*
Gustavo V. Barbosa-Cánovas / *Washington State University, Pullman, Washington, U.S.A.*
Edward M. Barnes / *United States Department of Agriculture (USDA), Phoenix, Arizona, U.S.A.*
Sheryl Barringer / *The Ohio State University, Columbus, Ohio, U.S.A.*
Felix H. Barron / *Clemson University, Clemson, South Carolina, U.S.A.*
John W. Bartok, Jr. / *University of Connecticut, Storrs, Connecticut, U.S.A.*
K. Belkacemi / *Université Laval, Sainte-Foy, Quebec, Canada*
Fernando J. Beltrán / *Universidad de Extremadura, Badajoz, Spain*
Zeki Berk / *Technion, Israel Institute of Technology, Haifa, Israel*
D. Bermúdez / *Universidad de las Américas–Puebla, Puebla, Mexico*
Carl J. Bern / *Iowa State University, Ames, Iowa, U.S.A.*
Sandeep Bhatnagar / *Nestle Purina Petcare Co., St. Louis, Missouri, U.S.A.*
William G. Bickert / *Michigan State University, East Lansing, Michigan, U.S.A.*
Barbara Blakistone / *National Food Processors Association, Washington, District of Columbia, U.S.A.*
Waraporn Boonsupthip / *Rutgers, The State University of New Jersey, New Brunswick, New Jersey, U.S.A.*
M. K. Bothwell / *Oregon State University, Corvallis, Oregon, U.S.A.*
Joseph Boudrant / *Centre Nationale de la Recherche Scientifique (CNRS), Vandoeuvre-les-Nancy, France*
Malcolm C. Bourne / *Cornell University, Geneva, New York, U.S.A.*
Raymond A. Bourque / *Ocean Spray Cranberries, Inc., Lakeville-Middleboro, Massachusetts, U.S.A.*
C. K. Bower / *Oregon State University, Corvallis, Oregon, U.S.A.*

Timothy J. Bowser / *Oklahoma State University, Stillwater, Oklahoma, U.S.A.*

Rhonda M. Brand / *University of Nebraska, Lincoln, Nebraska, U.S.A.*

Jenni L. Briggs / *Iowa State University, Ames, Iowa, U.S.A.*

Richard W. Brinker / *Auburn University, Auburn, Alabama, U.S.A.*

Aaron L. Brody / *Packaging/Brody, Inc., Duluth, Georgia, and The University of Georgia, Athens, Georgia, U.S.A.*

Thomas Brumm / *Iowa State University, Ames, Iowa, U.S.A.*

Dennis R. Buckmaster / *The Pennsylvania State University, University Park, Pennsylvania, U.S.A.*

Roberto A. Buffo / *University of Minnesota, St. Paul, Minnesota, U.S.A.*

Joel D. Bumgardner / *Mississippi State University, Mississippi State, Mississippi, U.S.A.*

Joel D. Burcham / *Clemson University, Clemson, South Carolina, U.S.A.*

William W. Casady / *University of Missouri, Columbia, Missouri, U.S.A.*

Maria Elena Castell-Perez / *Texas A&M University, College Station, Texas, U.S.A.*

Ralph P. Cavalieri / *Washington State University, Pullman, Washington, U.S.A.*

Dong S. Cha / *The University of Georgia, Griffin, Georgia, U.S.A.*

Ellen K. Chamberlain / *Kraft Foods, Glenview, Illinois, U.S.A.*

K. Chao / *United States Department of Agriculture (USDA), Beltsville, Maryland, U.S.A.*

C. R. Chen / *McGill University, Ste Anne de Bellevue, Quebec, Canada*

Hongda Chen / *United States Department of Agriculture (USDA), Washington, District of Columbia, U.S.A.*

Munir Cheryan / *University of Illinois, Urbana, Illinois, U.S.A.*

Pavinee Chinachoti / *University of Massachusetts, Amherst, Massachusetts, U.S.A.*

Manjeet S. Chinnan / *The University of Georgia, Griffin, Georgia, U.S.A.*

Yonghee Choi / *Kyungpook National University, Taegu, South Korea*

Grace Christian / *University of Birmingham, Birmingham, United Kingdom*

Ann D. Christy / *The Ohio State University, Columbus, Ohio, U.S.A.*

J. Peter Clark III / *Consultant, Oak Park, Illinois, U.S.A.*

D. J. Cleland / *Massey University, Palmerston North, New Zealand*

Richard Cooke / *University of Illinois, Urbana, Illinois, U.S.A.*

Paul Cornillon / *Danone Vitapole, Palaiseau, France*

Francis Courtois / *UMR-GENIAL, Unité Mixte de Recherche GENie Industriel ALimentaire, ENSIA, Massy, France*

Kevin Cronin / *University College, Cork, Cork, Ireland*

Joel L. Cuello / *The University of Arizona, Tucson, Arizona, U.S.A.*

Ashim K. Datta / *Cornell University, Ithaca, New York, U.S.A.*

Christopher R. Daubert / *North Carolina State University, Raleigh, North Carolina, U.S.A.*

P. Michael Davidson / *University of Tennessee, Knoxville, Tennessee, U.S.A.*

Craig J. Davis / *State University of New York, College of Environmental Science and Forestry, Syracuse, New York, U.S.A.*

Gary DeBerg / *University of Nebraska, Lincoln, Nebraska, U.S.A.*

N. de Kruijf / *TNO Nutrition and Food Research, Zeist, The Netherlands*

Adriana E. Delgado / *University College Dublin, Dublin, Ireland*

Ali Demirci / *The Pennsylvania State University, University Park, Pennsylvania, U.S.A.*

Indrani Deo / *Rutgers, The State University of New Jersey, New Brunswick, New Jersey, U.S.A.*

Richard C. Derksen / *United States Department of Agriculture (USDA), Wooster, Ohio, U.S.A.*

Christina A. Mireles DeWitt / *Oklahoma State University, Stillwater, Oklahoma, U.S.A.*

Monte Dickson / *Deere and Company, Des Moine, Iowa, U.S.A.*

James H. Dooley / *Silverbrook Limited, Federal Way, Washington, U.S.A.*

Caye Drapcho / *Louisiana State University, Baton Rouge, Louisiana, U.S.A.*
Larry Erickson / *Kansas State University, Manhattan, Kansas, U.S.A.*
Robert G. Evans / *United States Department of Agriculture (USDA), Sidney, Montana, U.S.A.*
Robert O. Evans / *North Carolina State University, Raleigh, North Carolina, U.S.A.*
James Faller / *Mary Kay, Inc., Dallas, Texas, U.S.A.*
Qi Fang / *Banner Pharmacaps, Inc., High Point, North Carolina, U.S.A.*
Daniel F. Farkas / *Oregon State University, Corvallis, Oregon, U.S.A.*
Oladiran O. Fasina / *Auburn University, Auburn, Alabama, U.S.A.*
Hao Feng / *Washington State University, Pullman, Washington, U.S.A.*
P. Fito / *Universidad Politécnica de Valencia (UPV), Valencia, Spain*
John D. Floros / *The Pennsylvania State University, University Park, Pennsylvania, U.S.A.*
Robert D. Fox / *United States Department of Agriculture (USDA), Wooster, Ohio, U.S.A.*
Peter Fryer / *University of Birmingham, Birmingham, United Kingdom*
Takeshi Furuta / *Tottori University, Tottori, Japan*
Juan F. García-Araya / *Universidad de Extremadura, Badajoz, Spain*
Raymond M. Garcia / *LBFH Inc., Palm City, Florida, U.S.A.*
Y. Gariépy / *McGill University, Montreal, Quebec, Canada*
V. Gekas / *Technical University of Crete, Chania, Greece*
Vikram Ghosh / *The Pennsylvania State University, University Park, Pennsylvania, U.S.A.*
Thomas M. Gilmore / *International Association of Food Industry Suppliers, McLean, Virginia, U.S.A.*
James L. Glancey / *University of Delaware, Newark, Delaware, U.S.A.*
M. Marcela Góngora-Nieto / *Washington State University, Pullman, Washington, U.S.A.*
M. Gras / *Universidad Politécnica de Valencia (UPV), Valencia, Spain*
Sundaram Gunasekaran / *University of Wisconsin–Madison, Madison, Wisconsin, U.S.A.*
Yuan-Kuang Guu / *National Pingtung University of Science and Technology, Neipu, Pingtung, Taiwan*
John J. Hahn / *University of Missouri–Columbia, Columbia, Missouri, U.S.A.*
Steven G. Hall / *Louisiana State University, Baton Rouge, Louisiana, U.S.A.*
Bengt Hallström / *Lund University, Lund, Sweden*
J. H. Han / *The University of Manitoba, Winnipeg, Manitoba, Canada*
H. Mark Hanna / *Iowa State University, Ames, Iowa, U.S.A.*
Milford A. Hanna / *University of Nebraska, Lincoln, Nebraska, U.S.A.*
Arthur P. Hansen / *North Carolina State University, Raleigh, North Carolina, U.S.A.*
Pamela K. Hardt-English / *PhF Specialists, Inc., San Jose, California, U.S.A.*
Federico Harte / *Washington State University, Pullman, Washington, U.S.A.*
Iyad Hatem / *University of Missouri–Columbia, Columbia, Missouri, U.S.A.*
J. L. Hatfield / *United States Department of Agriculture (USDA), Ames, Iowa, U.S.A.*
Hiromichi Hayashi / *Tokyo University of Agriculture, Tokyo, Japan*
Dennis R. Heldman / *Rutgers, The State University of New Jersey, New Brunswick, New Jersey, U.S.A.*
Kenneth Hellevang / *North Dakota State University, Fargo, North Dakota, U.S.A.*
Martin L. Hellickson / *Oregon State University, Corvallis, Oregon, U.S.A.*
Andrew J. Hewitt / *Stewart Agricultural Research Services, Inc., Macon, Missouri, U.S.A.*
Harvey J. Hirning / *North Dakota State University, Fargo, North Dakota, U.S.A.*
Steven J. Hoff / *Iowa State University, Ames, Iowa, U.S.A.*
Joseph H. Hotchkiss / *Cornell University, Ithaca, New York, U.S.A.*
Fu-hung Hsieh / *University of Missouri–Columbia, Columbia, Missouri, U.S.A.*

Chuan-liang Hsu / *Yuanpei Institute of Science & Technology, Hsin-chu, Taiwan*
Lihan Huang / *United States Department of Agriculture (USDA), Wyndmoor, Pennsylvania, U.S.A.*
Yanbo Huang / *Texas A&M University, College Station, Texas, U.S.A.*
Kerry L. Hughes / *The Ohio State University, Columbus, Ohio, U.S.A.*
Yen-Con Hung / *The University of Georgia, Griffin, Georgia, U.S.A.*
Douglas J. Hunsaker / *United States Department of Agriculture (USDA), Phoenix, Arizona, U.S.A.*
Chang Hwan Hwang / *University of Wisconsin–Madison, Madison, Wisconsin, U.S.A.*
Héctor A. Iglesias / *University of Buenos Aires, Buenos Aires, Argentina*
Koreyoshi Imamura / *Okayama University, Okayama, Japan*
Joseph Irudayaraj / *The Pennsylvania State University, University Park, Pennsylvania, U.S.A.*
Gerald W. Isaacs / *University of Florida, Gainesville, Florida, U.S.A.*
Forrest T. Izuno / *University of Minnesota, Waseca, Minnesota, U.S.A.*
William A. Jacoby / *University of Missouri–Columbia, Columbia, Missouri, U.S.A.*
Digvir S. Jayas / *University of Manitoba, Winnipeg, Manitoba, Canada*
B. M. Jenkins / *University of California, Davis, Davis, California, U.S.A.*
Arthur T. Johnson / *University of Maryland, College Park, Maryland, U.S.A.*
Vijay K. Juneja / *United States Department of Agriculture (USDA), Wyndmoor, Pennsylvania, U.S.A.*
Gönül Kaletunç / *The Ohio State University, Columbus, Ohio, U.S.A.*
Mukund V. Karwe / *Rutgers, The State University of New Jersey, New Brunswick, New Jersey, U.S.A.*
Kevin M. Keener / *North Carolina State University, Raleigh, North Carolina, U.S.A.*
Piet J. A. M. Kerkhof / *Eindhoven University of Technology, Eindhoven, The Netherlands*
William Kerr / *The University of Georgia, Athens, Georgia, U.S.A.*
Joong Kim / *Pukyong National University, Pusan, South Korea*
Sang Hun Kim / *Kangwon National University, Kangwon-do, South Korea*
Taejo Kim / *Mississippi State University, Mississippi State, Mississippi, U.S.A.*
N. Suzan Kincal / *Middle East Technical University, Ankara, Turkey*
William S. Kisaalita / *The University of Georgia, Athens, Georgia, U.S.A.*
Richard G. Koegel / *United States Department of Agriculture (USDA), Madison, Wisconsin, U.S.A.*
Jozef L. Kokini / *Rutgers, The State University of New Jersey, New Brunswick, New Jersey, U.S.A.*
Edward Kolbe / *Oregon State University, Portland, Oregon, U.S.A.*
Heather L. Kramer / *Utah State University, Logan, Utah, U.S.A.*
John M. Krochta / *University of California, Davis, California, U.S.A.*
M. Krokida / *National Technical University of Athens, Athens, Greece*
Yubin Lan / *Fort Valley State University, Fort Valley, Georgia, U.S.A.*
Dennis L. Larson / *University of Arizona, Tucson, Arizona, U.S.A.*
Harris N. Lazarides / *Aristotle University of Thessaloniki, Thessaloniki, Greece*
Tung-Ching Lee / *Rutgers, The State University of New Jersey, New Brunswick, New Jersey, U.S.A.*
Leon Levine / *Leon Levine Associates, Albuquerque, New Mexico, U.S.A.*
Yanbin Li / *University of Arkansas, Fayetteville, Arkansas, U.S.A.*
Marybeth Lima / *Louisiana State University, Baton Rouge, Louisiana, U.S.A.*
Pekka Linko / *Helsinki University of Technology, Espoo, Finland*
Sean X. Liu / *Rutgers, The State University of New Jersey, New Brunswick, New Jersey, U.S.A.*
Otto J. Loewer / *University of Arkansas, Fayetteville, Arkansas, U.S.A.*
Jorge E. Lozano / *PLAPIQUI (UNS-CONICET), Bahía Blanca, Argentina*
Graeme Macaloney / *QSV Biologics Ltd., Edmonton, Alberta, Canada*
Diarmuid MacCarthy / *University College, Cork, Cork, Ireland*

Jay Marks / *Purdue University, West Lafayette, Indiana, U.S.A.*
Z. B. Maroulis / *National Technical University of Athens, Athens, Greece*
Takeshi Matsuura / *University of Ottawa, Ottawa, Ontario, Canada*
J. McGuire / *Oregon State University, Corvallis, Oregon, U.S.A.*
Brian M. McKenna / *University College Dublin, Dublin, Ireland*
George E. Meyer / *University of Nebraska, Lincoln, Nebraska, U.S.A.*
Christopher C. Miller / *Rose Acre Farms, Social Circle, Georgia, U.S.A.*
Robert C. Miller / *Consulting Engineer, Auburn, New York, U.S.A.*
Gauri S. Mittal / *University of Guelph, Guelph, Ontario, Canada*
Osato Miyawaki / *The University of Tokyo, Tokyo, Japan*
Gary A. Montague / *University of Newcastle, Newcastle upon Tyne, United Kingdom*
Elton F. Morales-Blancas / *Universidad Austral de Chile, Valdivia, Chile*
Rosana G. Moreira / *Texas A&M University, College Station, Texas, U.S.A.*
Michael T. Morrissey / *Oregon State University, Astoria, Oregon, U.S.A.*
Richard E. Muck / *United States Department of Agriculture (USDA), Madison, Wisconsin, U.S.A.*
H. Mújica-Paz / *Universidad Autónoma de Chihuahua, Chihuahua, Chih., Mexico*
Edgar G. Murakami / *National Center for Food Safety and Technology/Food and Drug Administration, Summit-Argo, Illinois, U.S.A.*
K. Muthukumarappan / *South Dakota State University, Brookings, South Dakota, U.S.A.*
Kazuhiro Nakanishi / *Okayama University, Okayama, Japan*
Ganesan Narsimhan / *Purdue University, West Lafayette, Indiana, U.S.A.*
Paul Nesvadba / *The Robert Gordon University, Aberdeen, Scotland, United Kingdom*
Bart M. Nicolaï / *Katholieke Universiteit Leuven, Leuven, Belgium*
Martin R. Okos / *Purdue University, West Lafayette, Indiana, U.S.A.*
Fernanda A. R. Oliveira / *University College Cork, Cork, Ireland*
Jorge C. Oliveira / *University College Cork, Cork, Ireland*
Eiichi Ono / *The University of Arizona, Tucson, Arizona, U.S.A.*
Fernando A. Osorio / *Universidad de Santiago de Chile, Santiago, Chile*
Banu F. Ozen / *Purdue University, West Lafayette, Indiana, U.S.A.*
Mahesh Padmanabhan / *Kraft Foods North America, Inc., Glenview, Illinois, U.S.A.*
John E. Parsons / *North Carolina State University, Raleigh, North Carolina, U.S.A.*
Q. T. Pham / *University of New South Wales, Sydney, New South Wales, Australia*
Raul H. Piedrahita / *University of California, Davis, Davis, California, U.S.A.*
J. H. Prueger / *United States Department of Agriculture (USDA), Ames, Iowa, U.S.A.*
Mohamed Qasim / *U.S. Army Corps of Engineer Research and Development Center, Vicksburg, Mississippi, U.S.A.*
Ximena Quintero / *Frito-Lay, Plano, Texas, U.S.A.*
G. S. V. Raghavan / *McGill University, Montreal, Quebec, Canada*
M. Shafiur Rahman / *Sultan Qaboos University, Muscat, Sultanate of Oman*
D. Raj Raman / *The University of Tennessee, Knoxville, Tennessee, U.S.A.*
H. S. Ramaswamy / *McGill University, Ste Anne de Bellevue, Quebec, Canada*
Rakesh Ranjan / *Applied Geoscience and Engineering, Reading, Pennsylvania, U.S.A.*
M. A. Rao / *Cornell University, Geneva, New York, U.S.A.*
V. N. Mohan Rao / *Frito-Lay, Plano, Texas, U.S.A.*
Randy L. Raper / *United States Department of Agriculture (USDA), Auburn, Alabama, U.S.A.*
Puntarika Ratanatriwong / *The Ohio State University, Columbus, Ohio, U.S.A.*

David S. Reid / *University of California, Davis, Davis, California, U.S.A.*
John F. Reid / *Deere and Company, Moline, Illinois, U.S.A.*
José I. Reyes De Corcuera / *Washington State University, Pullman, Washington, U.S.A.*
David Reznik / *Raztek Corporation, Sunnyvale, California, U.S.A.*
Tom L. Richard / *Iowa State University, Ames, Iowa, U.S.A.*
Mark R. Riley / *The University of Arizona, Tucson, Arizona, U.S.A.*
John S. Roberts / *Cornell University, Geneva, New York, U.S.A.*
Evangelina T. Rodrigues / *The University of Manitoba, Winnipeg, Manitoba, Canada*
Sandrine Rodriguez / *University of Birmingham, Birmingham, United Kingdom*
Yrjö H. Roos / *University College Cork, Cork, Ireland*
Shyam S. Sablani / *Sultan Qaboos University, Muscat, Sultanate of Oman*
Yasuyuki Sagara / *The University of Tokyo, Tokyo, Japan*
Takaharu Sakiyama / *Okayama University, Okayama, Japan*
Fernanda San Martín / *Washington State University, Pullman, Washington, U.S.A.*
K. P. Sandeep / *North Carolina State University, Raleigh, North Carolina, U.S.A.*
E. C. M. Sanga / *McGill University, Montreal, Quebec, Canada*
Sudhir K. Sastry / *The Ohio State University, Columbus, Ohio, U.S.A.*
Nico Scheerlinck / *Katholieke Universiteit Leuven, Leuven, Belgium*
Henry G. Schwartzberg / *University of Massachusetts, Amherst, Massachusetts, U.S.A.*
Elaine P. Scott / *Virginia Polytechnic Institute and State University, Blacksburg, Virginia, U.S.A.*
David R. Sepúlveda / *Washington State University, Pullman, Washington, U.S.A.*
Shri K. Sharma / *International Food Network, Ithaca, New York, U.S.A.*
Lon R. Shell / *Southwest Texas State University, San Marcos, Texas, U.S.A.*
Thomas H. Shellhammer / *Oregon State University, Corvallis, Oregon, U.S.A.*
Guo-Qi Shen / *Shanxi Agriculture University, Shanxi, People's Republic of China*
Nicholas Shilton / *University College Dublin, Dublin, Ireland*
Terry J. Siebenmorgen / *University of Arkansas, Fayetteville, Arkansas, U.S.A.*
Juan L. Silva / *Mississippi State University, Mississippi State, Mississippi, U.S.A.*
R. Simpson / *Universidad Técnica Federico Santa María, Valparaíso, Chile*
Rakesh K. Singh / *The University of Georgia, Athens, Georgia, U.S.A.*
Glen H. Smerage / *University of Florida, Gainesville, Florida, U.S.A.*
Shahab Sokhansanj / *University of Saskatchewan, Saskatoon, Saskatchewan, Canada*
Apinan Soottitantawat / *Tottori University, Tottori, Japan*
Benjamin C. Stark / *Illinois Institute of Technology, Chicago, Illinois, U.S.A.*
Matthew D. Steven / *Cornell University, Ithaca, New York, U.S.A.*
Nikolaos G. Stoforos / *Aristotle University of Thessaloniki, Thessaloniki, Greece*
Richard J. Straub / *University of Wisconsin–Madison, Madison, Wisconsin, U.S.A.*
Wei Wen Su / *University of Hawaii at Manoa, Honolulu, Hawaii, U.S.A.*
Da-Wen Sun / *University College Dublin, Dublin, Ireland*
Barry G. Swanson / *Washington State University, Pullman, Washington, U.S.A.*
Jinglu Tan / *University of Missouri–Columbia, Columbia, Missouri, U.S.A.*
Juming Tang / *Washington State University, Pullman, Washington, U.S.A.*
Yang Tao / *University of Maryland, College Park, Maryland, U.S.A.*
Petros S. Taoukis / *National Technical University of Athens, Athens, Greece*
Timothy A. Taylor / *Utah State University, Logan, Utah, U.S.A.*
Arthur A. Teixeira / *University of Florida, Gainesville, Florida, U.S.A.*

Donald W. Thayer (Retired) / *United States Department of Agriculture (USDA), Wyndmoor, Pennsylvania, U.S.A.*

J. Alex Thomasson / *Mississippi State University, Mississippi State, Mississippi, U.S.A.*

Allen L. Thompson / *University of Missouri, Columbia, Missouri, U.S.A.*

Graham Thorpe / *Victoria University of Technology, Melbourne, Victoria, Australia*

U. Sunday Tim / *Iowa State University, Ames, Iowa, U.S.A.*

K. C. Ting / *The Ohio State University, Columbus, Ohio, U.S.A.*

Ernest W. Tollner / *The University of Georgia, Athens, Georgia, U.S.A.*

Peggy M. Tomasula / *United States Department of Agriculture (USDA), Wyndmoor, Pennsylvania, U.S.A.*

J. Antonio Torres / *Oregon State University, Corvallis, Oregon, U.S.A.*

Gun Trägårdh / *Lund University, Lund, Sweden*

Emine Unlu / *Masterfoods USA, Vernon, California, U.S.A.*

A. Valdez-Fragoso / *Universidad Autónoma de Chihuahua, Chihuahua, Chih., Mexico*

M. D. van Beest / *TNO Nutrition and Food Research, Zeist, The Netherlands*

Rutger M. T. van Sleeuwen / *Rutgers, The State University of New Jersey, New Brunswick, New Jersey, U.S.A.*

Pieter Verboven / *Katholieke Universiteit Leuven, Leuven, Belgium*

F. Vergara / *Universidad de las Américas–Puebla, Puebla, Mexico*

Brahm P. Verma / *The University of Georgia, Athens, Georgia, U.S.A.*

D. Vidal-Brotóns / *Universidad Politécnica de Valencia (UPV), Valencia, Spain*

C. Vigneault / *Agriculture and Agri-Food Canada, St-Jean-sur-Richelieu, Quebec, Canada*

Joel T. Walker (Deceased) / *The Ohio State University, Columbus, Ohio, U.S.A.*

Peter M. Waller / *University of Arizona, Tucson, Arizona, U.S.A.*

Jaw-Kai Wang / *University of Hawaii at Manoa, Honolulu, Hawaii, U.S.A.*

Pie-Yi Wang / *ConAgra Foods, Downers Grove, Illinois, U.S.A.*

S. Wang / *Washington State University, Pullman, Washington, U.S.A.*

Zebin Wang / *Purdue University, West Lafayette, Indiana, U.S.A.*

Jochen Weiss / *University of Tennessee, Knoxville, Tennessee, U.S.A.*

Curtis L. Weller / *University of Nebraska–Lincoln, Lincoln, Nebraska, U.S.A.*

J. Welti-Chanes / *Universidad de las Américas–Puebla, Puebla, Mexico*

Fred Wheaton / *University of Maryland, College Park, Maryland, U.S.A.*

Lyman S. Willardson / *Utah State University, Logan, Utah, U.S.A.*

Shang-Tian Yang / *The Ohio State University, Columbus, Ohio, U.S.A.*

Wade Yang / *University of Arkansas, Fayetteville, Arkansas, U.S.A.*

M. Erhan Yildiz / *Rutgers, The State University of New Jersey, New Brunswick, New Jersey, U.S.A.*

R. E. Yoder / *The University of Tennessee, Knoxville, Tennessee, U.S.A.*

C. Dean Yonts / *University of Nebraska–Lincoln, Scottsbluff, Nebraska, U.S.A.*

Hidefumi Yoshii / *Tottori University, Tottori, Japan*

F. Younce / *Washington State University, Pullman, Washington, U.S.A.*

Yanyun Zhao / *Oregon State University, Corvallis, Oregon, U.S.A.*

Ying Zhu / *The Ohio State University, Columbus, Ohio, U.S.A.*

Gregory R. Ziegler / *The Pennsylvania State University, University Park, Pennsylvania, U.S.A.*

Tommy L. Zimmerman / *The Ohio State University, Wooster, Ohio, U.S.A.*

Joseph M. Zulovich / *University of Missouri, Columbia, Missouri, U.S.A.*

Contents

Preface

The *Encyclopedia of Agricultural, Food, and Biological Engineering* focuses primarily on the processes used to produce raw agricultural materials and convert the raw materials into consumer products for distribution. The conversion of raw agricultural materials into consumer products adds significant value to the outputs from the agricultural and food systems. The total expenditures for food in the U.S. in 2000 was $661.1 billion, with $390.2 spent on food consumed at home and $270.9 for food consumed away from home. The sale of foods from supermarkets increased to nearly $400 billion in 2001 and the expenditures at 858,000 restaurants in the U.S. is expected to reach $407.8 billion in 2002. These statistics do not include the significant value of many nonfood products of agricultural origin, such as forest products and a growing segment of energy products being created from renewable resources.

Many customers of products of agricultural origin have limited appreciation and understanding of production processes. In addition, the consumers of food products may not fully understand the purpose of processes used to convert raw materials into safe, convenient, and nutritious foods. Consumers of products of agricultural origin represent a significant audience for this Encyclopedia. The broad objective of the Encyclopedia is to provide readers with an improved understanding of the processes used in the production and manufacture of consumer products of agricultural origin.

The role of engineering in most industries involves the development and design of the processes and equipment for the manufacture of products associated with that industry. When considering consumer products manufactured using the raw materials from agriculture, many different types of engineering processes are involved. In this Encyclopedia, some of the articles are devoted to engineering processes associated with production of raw agricultural materials. Other articles describe the engineering processes involved in the conversion of raw materials into consumer food products. A third set of articles describes engineering processes associated with conversion of raw materials into a variety of nonfood consumer products. Historically, the development and design of processes and equipment used for production of agricultural products have been the domain of agricultural engineering. Over time, the range of raw agricultural materials has expanded to include forest products, the products of aquaculture, and biomass used for energy from renewable sources.

Over the past 50 years, food engineering has evolved as the focal point for the development and design of processes and equipment for conversion of raw agricultural materials and ingredients into consumer food products. More recently, within the past 20–25 years, biological engineering has emerged as a description for engineering processes associated with all biological materials and products. The title of this Encyclopedia reflects the interface of three engineering disciplines as well as the reader's opportunity to gain an understanding of the relationships among the three types of engineering processes. As these engineering disciplines continue to evolve, quarterly updates and future editions of the Encyclopedia will provide increased insight into these topics.

The concepts and processes described in the Encyclopedia are critical components of a system with the capacity to deliver food and other consumer products to an expanding world population. The safety and nutrition of the food supply are at the highest levels in history, and other consumer products contribute to a quality of life that continues to improve. Leading engineers and scientists from many parts of the world have contributed to the Encyclopedia. All authors were asked to present topics in a manner that will communicate to an audience with limited technical background. At the same time, a sound technical explanation has been incorporated into each contribution. All contributions have been peer-reviewed to ensure that the content is technically accurate and communicated to the anticipated audience in an effective manner. In addition, most of the articles contain reference lists directing the reader to further information.

Completion of the *Encyclopedia of Agricultural, Food, and Biological Engineering* would have been very difficult without significant input from many colleagues and professionals. The Editorial Advisory Board provided valuable input during the initial stages of development, and many members of the Board have been involved in the review of individual contributions. The opportunity to collaborate with these outstanding engineers and scientists has made this a memorable experience. The continuous support of Susan Lee, Assistant Editor at Marcel Dekker, Inc., has been appreciated by the Editor and Board throughout the development and production of the Encyclopedia.

Dennis R. Heldman

Activation Energy in Thermal Process Calculations

Elton F. Morales-Blancas
Universidad Austral de Chile, Valdivia, Chile

J. Antonio Torres
Oregon State University, Corvallis, Oregon, U.S.A.

INTRODUCTION

The Arrhenius equation is most useful to explain the nonlinear relationship temperature dependence of simple reaction rates in food and biological systems.[1–5] The literature shows a number of applications to more complex situations.[6–8] In the food science and engineering literature, both the activation energy (E_a) and the thermal resistance constant (z) are utilized to describe the influence of temperature on the reduction of microbial populations and the loss of food quality attributes.[3,9,10] This article provides a brief overview of the activation energy (E_a) and its mathematical relationship with the first-order rate constant (k_T) and the thermal resistance constant (z-value). The applications and limitations of the activation energy concept in thermal processing calculations are also discussed.

CHEMICAL REACTION KINETICS AND ARRHENIUS MODEL RELATIONSHIP

The Arrhenius equation was derived theoretically from thermodynamic laws as well as statistical mechanistic principles for reversible chemical reactions.[11] Theoretical approaches have confirmed this model for simple gaseous systems and solutions.[7] Although the Arrhenius relation was derived originally as a fundamental approach, its first application in food science, describing the temperature effect on the acid inversion of sucrose, was empirical.[12] It is now commonly used to describe the nonlinear temperature dependence of chemical reaction rates in food and biological systems, and of complex physical phenomena such as viscosity, diffusion, and sorption.[4,6–8] Its application has been extended to the evaluation of the temperature effect on the microbial inactivation during pasteurization and sterilization.[2,3,5]

Many observations and controlled studies have shown that when foods and biological systems are subjected to an elevated constant temperature, quality attributes (e.g., vitamin C concentration) and micro-organisms decrease exponentially with time.[3,9] Mathematically, a first-order kinetic model can describe this behavior as follows[2,4,13–16]:

$$\frac{dN_t}{dt} = -k_T N_t \tag{1}$$

where k_T is a first-order rate constant for the inactivation at a constant temperature T of a microbial population, N_t. Similar expressions can be defined for the degradation of a food quality attribute. Integrating Eq. 1 and using the initial condition $N = N_0$ at $t = 0$, yields:

$$\ln\frac{N_t}{N_0} = -k_T t \tag{2}$$

Although Eqs. 1 and 2 are valid only for unimolecular chemical reaction kinetics, it has become an accepted practice to use these relatively simple expressions to describe food quality changes and microbial population reductions.[2,3,6,9,10,15] The temperature dependence of the first-order rate constant (k_T) can be described by the Arrhenius equation:

$$k_T = k_0\, e^{\frac{-E_a}{RT_A}} \tag{3}$$

The activation energy (E_a) can be interpreted as resulting from collisions of reactive molecules occurring with a certain frequency k_0 and generating enough energy to provide this value. In the activated-complex theory, an activated complex formed from reactants eventually decomposes with a certain frequency k_0 to generate products. In the latter case, the activated complex is in thermodynamic equilibrium with the reactants, and complex decomposition is the limiting step.[6] As neither theory provides a means to calculate the activation energy from thermodynamic information, one has to collect kinetic data to determine the effect of temperature on reaction kinetics. Based on Eq. 3, the activation energy (E_a) can be obtained from the slope of linear $\ln(k_T)$ vs. $1/T$ plots.[3,6,7,9]

$$\ln k_T = \ln k_0 - \frac{E_a}{R}\left(\frac{1}{T_A}\right) \tag{4}$$

It should be noted that the E_a value calculated from

Encyclopedia of Agricultural, Food, and Biological Engineering
DOI: 10.1081/E-EAFE 120018364
Copyright © 2003 by Marcel Dekker, Inc. All rights reserved.

Eq. 4 does not by itself assess the reactivity of a food system and should be interpreted only as information on the temperature dependence of the reaction.[3,6] Another form of the Arrhenius equation frequently found in the literature is[2,5,7]:

$$k_T = k_{T_{\mathrm{ref}}} e^{\frac{-E_a}{R}\left[\frac{1}{T_A} - \frac{1}{T_{A_{\mathrm{ref}}}}\right]} \qquad (5)$$

where $k_{T_{\mathrm{ref}}}$ is the reaction constant at a reference absolute temperature, $T_{A_{\mathrm{ref}}}$.

ACTIVATION ENERGY (E_a) AND THERMAL RESISTANCE CONSTANT (z) RELATIONSHIP

The activation energy constant (E_a) is used to describe the influence of temperature on the first-order reaction rate constant (k_T) while the thermal resistant constant (z) is utilized to describe the influence of temperature on the decimal reduction time (D_T). As discussed in the *Thermal Resistance Constant* article of this encyclopedia, the latter constants were defined without referring to first-order chemical reaction kinetics and the Arrhenius model for its temperature dependence.[4,14,17–28] For convenience, as well as tradition, the food literature has most often reported microbial population inactivation data as thermal resistance parameters (D-value and z-value) while food quality attribute losses are presented as first-order reaction kinetic constants and activation energy parameters (k_T and E_a). However, both sets of empirical coefficients are related and conversion expressions can be obtained by comparing Eq. 2 and the definition for D-value (see Eq. 5, in the *Thermal Resistance Constant* article of this encyclopedia). This leads to the familiar relationship between D-value and k_T:

$$D_T = \frac{2.303}{k_T} \qquad (6)$$

Using the definition for thermal resistance constant (see Eq. 3, in the *Thermal Resistance Constant* article of this encyclopedia). and the Arrhenius (Eq. 5) models yields the following relationship between z and E_a values[2,5]:

$$z = \frac{2.303}{(E_a/R)} T_A T_{A_{\mathrm{ref}}} \qquad (7)$$

An alternative to Eq. 7 is the expression suggested by Ramaswamy, Van de Voort, and Ghazala[29]:

$$z = \frac{2.303}{(E_a/R)} T_{A_{\min}} T_{A_{\max}} \qquad (7a)$$

where the minimum ($T_{A_{\min}}$) and maximum temperature ($T_{A_{\max}}$) are the extreme limits of the temperature range used to obtain kinetic data. Another form of Eq. 7 commonly found in the literature is the following expression[4,7,28,30,31]:

$$z = \frac{2.303}{(E_a/R)} T^2_{A_{\mathrm{mean}}} \qquad (7b)$$

The temperature used in Eq. 7b should be chosen as a mean temperature ($T_{A_{\mathrm{mean}}}$) in the range of the original experimental data collection. It is important to emphasize that Eq. 7 and its variations 7a and 7b are valid only when the z-value and E_a cover the same temperature range.[10]

APPLICATIONS AND LIMITATIONS

The rate and extent of microbial population inactivation or degradation of quality attributes in foods subjected to thermal processing are important.[32] Availability of kinetic data is crucial to establish processing conditions optimizing food quality.[1,6,32] Fortunately, the literature provides an impressive array of kinetic parameters for microbial inactivation and food quality losses by heat.[2,6,10] These include empirical coefficients determined from log-linear kinetics (D_T and z), as well as constants based on first-order chemical reaction kinetics (k and E_a). In many sources, the temperature range used by researchers in their experimentation is narrow, or even more troubling, it is not reported. Therefore, care must be taken when published kinetic parameters are used to establish a thermal process because they may lead to severe under- or overprocessing errors.

In the lethal temperature range, the Arrhenius model can be extrapolated with relative confidence, while the z model is known to become nonlinear.[28,33] This is not surprising, because a comparison of Eq. 3 and the definition for thermal resistance constant (see Eq. 3, in the *Thermal Resistance Constant* article of this encyclopedia). reveals that the z-value is a function of temperature, whereas the activation energy, E_a, is normally independent of temperature.[4,7,34] The parameter z should be recognized as an approximate linearization of the theoretical but nonlinear relationship described by the Arrhenius equation for the temperature dependence of reaction constants.[4] From a safety point of view, the z model should be considered valid only over a limited lethal temperature range,[2] and its use should be limited to the range of lethal temperatures used to obtain experimental D-values.[5,10]

The literature makes extensive use of a reference temperature (T_{ref}) to describe the temperature dependency of reaction rates in foods. Traditionally, T_{ref} has been assumed implicit (e.g., 121°C or 250°F for microbial death in the sterilization of low-acid foods) or chosen to be around the upper processing temperature. Datta[4] developed a calculation procedure to preserve the full Arrhenius accuracy while still using the z-values available for reactions in foods. When the food temperature is constant

throughout the processing cycle (e.g., in HTST and UHT thermal processes), the reference temperature should be equal to the food temperature but when the food temperature varies during processing, as in the thermal processing of canned food, the optimum T_{ref} is typically only a few degrees below the maximum temperature achieved by the food.

Peleg et al.[8] pointed out that severe calculation errors for lower temperature processes may appear when the Arrhenius equation is used without previous identification of the temperature level at which inactivation becomes predominant; i.e., the thermal inactivation of micro-organisms is noticeable only above a certain lethal temperature. To overcome this problem, an alternative vitalistic approach was recently proposed leading to a log-logistic model based on the actual behavior of the examined system and not on any preconceived kinetics.[8] However, more experimental evidence is needed to support the general acceptance of this approach.

CONCLUSION

The mechanistic and fundamental chemical kinetic approach, involving rate constants, activation energy values, and the related mathematical models, has been a widely used tool to describe the thermal degradation kinetics of food constituents. The use of the Arrhenius equation assumes a universal analogy between complex reaction rates occurring in food and biological systems and simple chemical reactions. An examination of the recent literature suggests that it will remain widely used in chemical, biochemical, and food research.

NOMENCLATURE

D_T	Decimal reduction time, time required for 90% microbial inactivation or 90% quality attribute degradation at a constant temperature T, dimension of time (e.g., min)
E_a	Activation energy constant, dimension of energy per mole (e.g., $\mathrm{J\,mol^{-1}}$)
k_0	Arrhenius frequency constant, dimension of inverse units of time (e.g., $\mathrm{min^{-1}}$)
k_T	First-order rate constant for microbial inactivation or quality attribute degradation at a constant temperature T, dimension of inverse units of time (e.g., $\mathrm{min^{-1}}$)
N_0	Initial microbial population or initial quality attribute
N_t	Microbial population or quality attribute at any time, t
R	Universal gas constant, dimension of energy per mol and absolute temperature, dimension of energy per mole and unit temperature (i.e., $\mathrm{J\,mol^{-1}\,K^{-1}}$)
t	Time (e.g., min)
T	Relative temperature (°C or °F)
T_A	Absolute temperature (e.g., K)
$T_{A_{ref}}$	Reference absolute temperature in the range used to generate kinetic parameters (e.g., K)
T_{ref}	Reference temperature in the temperature range used to generate kinetic parameters (e.g., °C or °F)
z	Thermal resistance constant, a parameter of each micro-organism or quality attribute which measures the influence of temperature on rates of microbial population inactivation or quality losses; dimension of temperature degrees (e.g., °C or °F)

ACKNOWLEDGMENTS

Author Elton F. Morales-Blancas acknowledges Comisión Nacional de Investigación Científica y Tecnológica of Chile (CONICYT—Project No. 1970303), Universidad Austral de Chile and Ministerio de Educación of Chile (MECESUP Program—Project No. AUS9908) for supporting a professional opportunity as an Oregon State University Faculty on courtesy appointment.

REFERENCES

1. Villota, R.; Hawkes, J.G. Kinetics of Nutrient and Organoleptic Changes in Foods During Processing. In *Physical and Chemical Properties of Foods*; Okos, M.R., Ed.; American Society of Agricultural Engineering: St. Joseph, Michigan, 1986; 266–366.
2. Toledo, R.T. *Fundamentals of Food Engineering*, 2nd Ed.; Van Nostrand Reinhold: New York, 1991.
3. Texeira, A. Thermal Process Calculations. In *Handbook of Food Engineering*; Heldman, D.R., Lund, D.B., Eds.; Marcel Dekker, Inc.: New York, 1992; 563–619.
4. Datta, A.K. Error Estimates for Approximate Kinetic Parameters Used in Food Literature. J. Food Eng. **1993**, *18* (2), 181–199.
5. Ramaswamy, H.S.; Singh, R.P. Sterilization Process Engineering. In *Handbook of Food Engineering Practice*; Valentas, K.J., Rotstein, E., Singh, R.P., Eds.; CRC Press LLC: Boca Raton, FL, 1997; 37–69.
6. Villota, R.; Hawkes, J.G. Reaction Kinetics in Food Systems. In *Handbook of Food Engineering*; Heldman,

D.R., Lund, D.B., Eds.; Marcel Dekker, Inc.: New York, 1992; 39–144.

7. Taoukis, P.S.; Labuza, T.P.; Saguy, I.S. Kinetics of Food Deterioration and Shelf-Life Prediction. In *Handbook of Food Engineering Practice*; Valentas, K.J., Rotstein, E., Singh, R.P., Eds.; CRC Press LLC: Boca Raton, FL, 1997; 361–403.
8. Peleg, M.; Engel, R.; Gonzalez-Martinez, C.; Corradine, M.G. Non-Arrhenius and Non-WLF kinetics in Food Systems. J. Sci. Food Agric. **2002**, *82* (12), 1346–1355.
9. Heldman, D.R.; Hartel, R.W. *Principles of Food Processing*; Chapman & Hall-International Thomson Publishing: New York, 1997.
10. http://vm.cfsan.fda.gov/~comm/ift-over.html (accessed Aug 2002).
11. Bunker, D.L. Simple Kinetic Models from Arrhenius to Computer. Accounts Chem. Res. **1974**, *7* (6), 195–201.
12. Arrhenius, S. Über die Reaktionsgeschwindigkeit bei der Inversion von Rohrzucker durch Sauren. Z. Physik. Chem. **1889**, *4*, 226–248.
13. Stumbo, C.R. Bacteriological Considerations Relating to Process Evaluation. Food Technol. **1948**, *2* (2), 115–132.
14. Stumbo, C.R. *Thermobacteriology in Food Processing*, 2nd Ed.; Academic Press, Inc.: New York, 1973.
15. Merson, R.L.; Singh, R.P.; Carroad, P.A. An Evaluation of Ball's Formula Method of Thermal Process Calculations. Food Technol. **1978**, *32* (3), 66–72, 75.
16. Katzin, L.I.; Sandholzer, L.A.; Strong, M.E. Application of the Decimal Reduction Time Principle to a Study of the Resistance of Coliform Bacteria to Pasteurization. J. Bacteriol. **1943**, *45* (3), 265–272.
17. Bigelow, W.D.; Esty, J.R. The Thermal Death Point in Relation to Time of Typical Thermophilic Organisms. J. Infect. Dis. **1920**, *27*, 602–617.
18. Bigelow, W.D. The Logarithmic Nature of Thermal Death Time Curves. J. Infect. Dis. **1921**, *29* (5), 528–536.
19. Ball, C.O. *Thermal Process Time for Canned Foods*, Bull No. 37; Natl. Research Council: Washington, D.C., 1923; Vol. 7, Part 1.
20. Ball, C.O. Mathematical Solution of Problems on Thermal Processing of Canned Foods. Univ. Calif. (Berkeley) Publ. Public Health **1928**, *1* (2), 15–245.
21. Ball, C.O. Advancements in Sterilization Methods for Canned Foods. Food Res. **1938**, *3* (1/2), 13–55.
22. Townsend, C.T.; Esty, J.R.; Baselt, F.C. Heat-Resistance Studies on Spores of Putrefactive Anaerobes in Relation to Determination of Safe Processes for Canned Foods. Food Res. **1938**, *3* (3), 323–346.
23. Stumbo, C.R. Bacteriological Considerations Relating to Process Evaluation. Food Technol. **1948**, *2* (2), 115–132.
24. Stumbo, C.R. Further Considerations Relating to Evaluation of Thermal Processes for Foods. Food Technol. **1949**, *3* (4), 126–131.
25. Stumbo, C.R. New Procedures for Evaluating Thermal Processes for Foods in Cylindrical Containers. Food Technol. **1953**, *7* (8), 309–315.
26. Stumbo, C.R.; Murphy, J.R.; Cochran, J. Nature of Thermal Death Time Curves for P.A. 3679 and Clostridium botulinum. Food Technol. **1950**, *4* (8), 321–326.
27. Ball, C.O.; Olson, F.C.W. *Sterilization in Food Technology. Theory, Practice and Calculations*; McGraw-Hill Book Co.: New York, 1957.
28. Casolari, A. Microbial Death. In *Physiological Models in Microbiology*; Bazin, M.J., Prosser, J.I., Eds.; CRC Press Inc.: Boca Ratón, FL, 1988; Vol. II, 1–44.
29. Ramaswamy, H.S.; Van de Voort, F.R.; Ghazala, S. An Analysis of TDT and Arrhenius Methods for Handling Process and Kinetic Data. J. Food Sci. **1989**, *54* (5), 1322–1326.
30. Cerf, O. Tailing of Survival Curves of Bacterial Spores: A Review. J. Appl. Bacteriol. **1977**, *42* (1), 1–19.
31. Le Jean, G.; Abraham, G.; Debray, E.; Candau, Y.; Piar, G. Kinetics of Thermal Destruction of *Bacillus stearothermophilus* Spores Using a Two Reaction Model. Food Microbiol. **1994**, *11* (3), 229–241.
32. Ávila, I.M.L.B.; Silva, C.L.M. Methodologies to Optimize Thermal Processing Conditions: An Overview. In *Processing Foods Quality Optimization and Process Assessment*; Oliveira, F.A.R., Oliveira, J.C., Eds.; CRC Press Inc.: Boca Ratón, FL, 1999; 67–82.
33. Hayakawa, K.-I. A Critical Review of Mathematical Procedures for Determining Proper Heat Sterilization Processes. Food Technol. **1978**, *32* (3), 59–65.
34. Clark, J.P. Mathematical Modeling in Sterilization Processes. Food Technol. **1978**, *32* (3), 73–75.

Active Packaging

N. de Kruijf
M. D. van Beest
TNO Nutrition and Food Research, Zeist, The Netherlands

A

INTRODUCTION

Active packaging may be defined as packaging that changes the condition of the packaged food to extend shelf-life or to improve safety or sensory properties, while maintaining the quality of the food. In the literature other definitions are also given.[1,2] The above definition was jointly drafted by the participants of the ongoing European FAIR project "Evaluating safety, effectiveness, economic–environmental impact and consumer acceptance of active and intelligent packagings (Actipak)."[3,4]

Food condition in the definition of active packaging encompasses various aspects that may play a role in determining the shelf-life of packaged foodstuffs, such as physiological processes (e.g., respiration of fresh fruits and vegetables), chemical processes (e.g., lipid oxidation), physical processes (e.g., staling of bread, dehydration), microbiological aspects (e.g., spoilage by micro-organisms), and infestation (e.g., by insects). These conditions can be regulated in numerous manners through the application of appropriate active packaging systems. Depending on the requirements of the packaged food, food quality deterioration can be significantly reduced. In this manner the desired shelf-life extension of the packaged food can be achieved. In recent years many different packaging systems have been developed in response to increasing trends in consumer preferences towards fresh, mildly preserved, tasty, and convenient food products with a prolonged shelf-life.

Active packaging systems are sometimes compared with modified atmosphere packaging (MAP). The difference between MAP and active packaging is that MAP is a passive system, whereas active packaging plays an active role during storage and transport. Besides, active packaging systems employ a wide range of technologies, each selected to deal with specific problems, while MAP is based on one technology only.

In the United States, Japan, and Australia, active packagings are already being successfully applied to extend shelf-life while maintaining nutritional quality and ensuring microbiological safety. Examples of commercial applications include the use of oxygen scavengers for sliced processed meat, ready-to-eat meals and beer, the use of moisture absorbers for fresh meat, poultry, and fresh fish, and ethylene-scavenging bags for packaging of fruit and vegetables. In Europe, however, only few of these systems have been developed and are being applied by now. The main reasons for this are legislative restrictions and a lack of knowledge about their acceptability to European consumers, the efficacy of such systems, and the economic and environmental impact such systems may have. The European "Actipak" project will address these issues in the near future. Provided that, in particular, the safety and regulatory aspects of these systems are adequately dealt with, active packaging will no doubt increasingly be applied in Europe in the years to come.[5]

OVERVIEW OF EXISTING ACTIVE PACKAGING SYSTEMS

Active packaging systems can be distinguished into active scavenging systems (absorbers) and active releasing systems (emitters). Scavenging systems remove undesired components, such as oxygen, excessive water/moisture, ethylene, carbon dioxide, other specific food constituents, and taints. Releasing systems actively add compounds to the packaged food, such as carbon dioxide, water, antioxidants, or preservatives. Both scavenging and releasing systems are aimed at extending shelf-life and/or improving food quality.[6] Active packaging systems can be used for numerous applications. Table 1 gives an overview of current and future applications of active packaging systems.

Active Scavenging Systems

Oxygen scavengers

Oxygen can exert considerable detrimental effects on foodstuffs. Oxygen is a reactant in several chemical reactions such as lipid oxidation resulting in the formation of off-flavors in fatty foods, enzyme-induced oxidation in sliced vegetables and fruits, discoloration caused by oxidation of pigments, and oxidation of components like tocopherol (vitamin E) and ascorbic acid (vitamin C), which will result in nutritional losses. Growth of many spoilage organisms like molds and yeast relies on the availability of oxygen. Reducing the oxygen concentration in the headspace of a package will generally inhibit

Encyclopedia of Agricultural, Food, and Biological Engineering
DOI: 10.1081/E-EAFE 120007142
Copyright © 2003 by Marcel Dekker, Inc. All rights reserved.

Table 1 Overview of active packaging systems, their objectives, and possible food applications

Active packaging system	Objective	Food applications
Oxygen scavengers	Prevention of growth of micro-organisms (aerobic bacteria, yeast, and molds) Prevention of oxidation (e.g., lipid oxidation, rancidity) Prevention of discoloration Prevention of infestation Delay in ripening of climacteric fruits and vegetables	Meat, fish, poultry, prepared dishes, cheese, bakery products, pizza, pasta, nuts, fried food, milk powder, dried products, tea, herbs, spices, beans, cereals, beverages, snack-food products
Ethylene scavengers	Delay in ripening of climacteric fruits and vegetables Prevention of postharvest disorders	Fruits and vegetables and other horticultural produce
Moisture regulators	Prevention of microbial growth Removal of dripping water Prevention of fogging	Fish, meat, poultry, snack-food products, cereals, dried foods, sandwiches, bakery products, fruits, and vegetables
CO_2 scavengers	Removal of carbon dioxide after packing	Coffee, cheese
Flavor/odor absorbers	Absorption of off-flavors	Fish, fatty foods, fruit juices, biscuits, cereals
Antimicrobial packagings	Prevention of growth of micro-organisms (bacteria, molds, yeast)	Bakery products, bread, fruit, vegetables, cheese, meat, fish, snack-food products, prepared dishes, wine, flour, grain, beans
Antioxidant-releasing systems	Prevention of oxidation (e.g., lipid oxidation)	Breakfast cereals, fatty foods
Flavor-releasing systems	Release of flavor to improve the sensory quality and avoid flavor scalping	Ice-cream, orange juice

the growth of spoilage organisms and reduce several shelf-life-limiting processes. Oxygen also affects the physiological condition of respiring foods like fruits and vegetables as well as infestation. Reduction of oxygen decreases respiration rate and ethylene production and inhibits the growth of insects, for example. One possible disadvantage of the use of oxygen scavengers is a higher risk for growth of pathogenic micro-organisms. Most pathogenic micro-organisms only grow under anaerobic conditions. With the use of oxygen scavengers an anaerobic atmosphere is created in the headspace of a package.

Oxygen scavengers are the oldest and commercially most important active packaging systems. The first scavengers were marketed in Japan in 1976. Existing oxygen-scavenging technologies utilize one or more of the following mechanisms: iron powder oxidation, ascorbic acid oxidation, photosensitive dye oxidation, enzymatic oxidation, and unsaturated fatty acids or immobilized yeast on solid material. Different forms of oxygen scavengers exist. The best known scavengers are small sachets containing iron powder. They can be easily inserted into food packages. The main advantage of these sachets is than they are able to reduce the level of oxygen to less that 0.01% in a relatively short time. Their disadvantage is that there always exists a possibility, despite the label "Do not eat," that consumers ingest the content of the sachet accidentally. An alternative to sachets is the integration of the active scavenging system into the packaging materials such as films or closures. In general, the speed and capacity of oxygen scavengers incorporated in films or closures is lower than with the iron-based oxygen-scavenging sachets.[7,8]

Oxygen scavengers can be applied to various foods such as bakery products, mayonnaise, powdered milk, fruit juices, beer, fresh meat, and fish.[8–12] For example, the capacity of an oxygen scavenger to extend the shelf-life of a "Russian salad" (mixed potato salad with mayonnaise) has been evaluated. The salad was stored with and without an oxygen-scavenging sachet for 2 w at 7°C.

Figs. 1–3 show the growth of aerobic bacteria, yeasts, and lactic acid bacteria, respectively, during storage. These results show a clear difference in the growth of

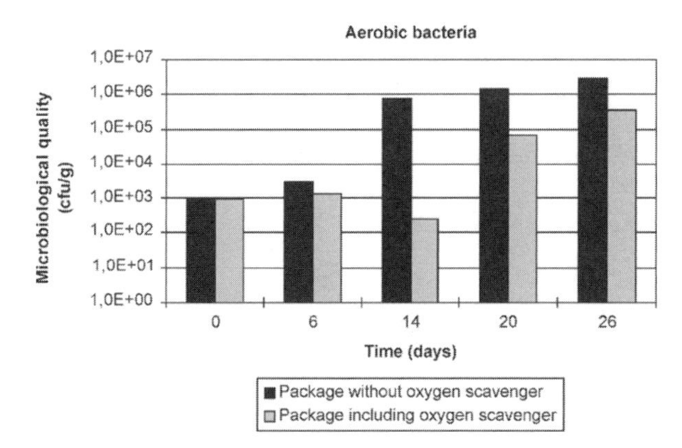

Fig. 1 Growth of aerobic bacteria in "Russian salad" packed with and without oxygen scavenger.

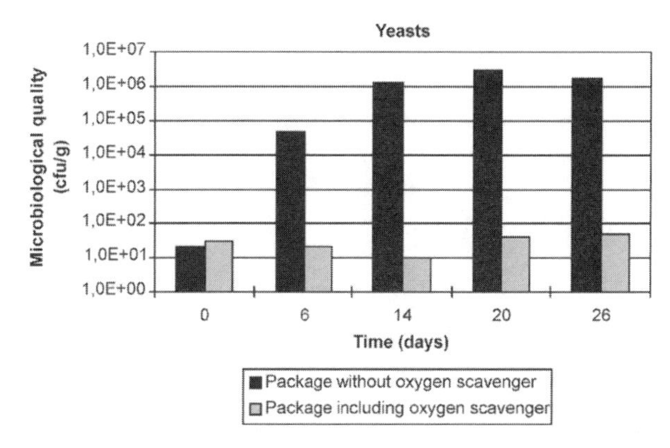

Fig. 2 Growth of yeasts in "Russian salad" packed with and without oxygen scavenger.

micro-organisms between the sample with and without oxygen scavenger. The oxygen-scavenging sachet extended the microbiological shelf-life of "Russian salad" by 15 days–20 days.

Ethylene scavengers

Ethylene is the so-called "ripening hormone" of fruits and vegetables because it has a significant impact on their physiological state. Ethylene triggers ripening, accelerates senescence, and reduces the shelf-life of climacteric fruits and vegetables. Therefore, to prolong the shelf-life and to maintain an acceptable visual and sensory quality, accumulation of ethylene in the package should be avoided.[13] Different ethylene-absorbing packaging systems currently exist, most of whom are supplied as sachets or incorporated in films. The active ingredient is often potassium permanganate on an inert mineral substrate. Minerals such as zeolite and clays, silica gel, and active carbon powder incorporated in packaging materials such as films and bags are also applied to absorb ethylene.[6,13]

Moisture regulators

Several foodstuffs require a strict control of water. For example, high water content causes softening of dry, crispy products. On the other hand, very low levels of water can decrease the shelf-life of fatty foods due to an increase of lipid oxidation. According to Rooney,[14] there are two distinct manners to regulate the moisture content of packed foods, namely liquid water control and humidity buffering. Excess water can be controlled by applying drip-absorbent sheets usually composed of a super-absorbent polymer between two microporous layers. Polyacrylate salts and copolymers of starch are suitable absorbing agents. The sheets can be used as pads, for example, under meat and fish.

Another way to control excess moisture in packed food is to regulate the relative humidity of the packed food (humidity buffering). The humectants (e.g., propylene glycol) can be placed between two plastic films. For a wide range of dry foods, desiccants such as silica gel, calcium oxide, and active clay, are being successfully used.[15,16]

Other scavengers

Other active absorbing packaging systems are developed to remove, for example, undesirable flavor constituents or carbon dioxide. Carbon dioxide is sometimes formed due to deterioration or respiration reactions. An active packaging system that scavenges both oxygen and carbon dioxide has been used to package fresh roasted coffee to avoid deterioration and/or package destruction. Generally, flavor scalping by the package is detrimental to food quality, but sometimes it is useful to selectively absorb unwanted odors or flavors such as aldehydes and amines. Amines are formed from protein breakdown in fish muscle, and aldehydes are reaction products of auto-oxidation of fats and oils. For both components packaging

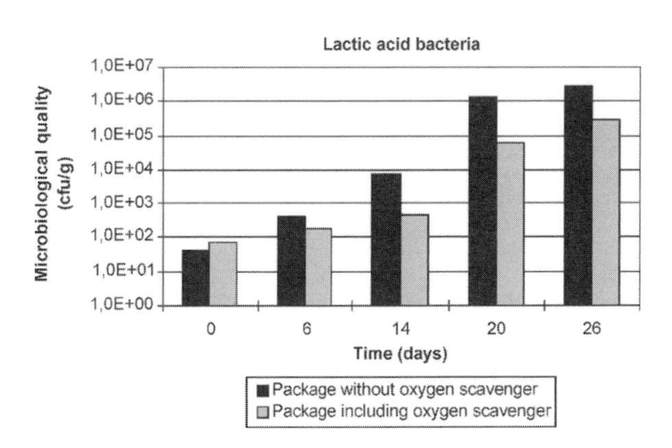

Fig. 3 Growth of lactic acid bacteria in "Russian salad" packed with and without oxygen scavenger.

systems have been developed, which are able to absorb these. Other commercial systems absorb odors due to the formation of mercaptans and hydrogen sulfide (H_2S).[17] It is important that the aforementioned technologies are not misused to mask the development of microbial off-odors.[4]

Active Releasing Systems

In active releasing systems, components migrate from the packaging material to the packaged food with the aim to extend shelf-life or to improve the quality of the packed food.

Antimicrobial packaging

Microbiological growth is one of the major modes of deterioration of fresh foods. Besides, the removal of oxygen to avoid growth of micro-organisms, the release of specific antimicrobial agents is a potential application of active packaging. When spoilage of micro-organisms is strictly due to surface growth, a film or sachet that emits an antimicrobial agent can be of value. The principal action of antimicrobial films or sachets is based on the release of antimicrobial entities. The major potential food applications include meat, fish, poultry, bread, cheese, fruits, and vegetables.[15]

Antimicrobial packaging systems with different applications and active agents are on the market. In Japan, Ag-substituted zeolite is the most common antimicrobial agent incorporated in plastics. Ag ions inhibit a range of metabolic enzymes.[18] Other active substances that can be used include ethanol and other alcohols, carbon dioxide, sorbate, benzoate, propionate, bacteriocins, fungicides, and enzymes.[19,20] Each antimicrobial agent has its own target microorganism and hence its own application. For example, ethanol is effective against surface growth of molds and can be used, for example, for bakery products.[8,21] The use of an antimicrobial packaging might lead to less-hygienic working conditions in plants. Another disadvantage could be that micro-organisms may build up resistance against the antimicrobial compound by excessive use.

Antioxidant-releasing systems

Antioxidants can be incorporated in plastic films for polymer stabilization in order to protect film from degradation. However, when the antioxidant migrates into the food it may have an additional positive effect on the shelf-life of the packed food. In the United States, release of BHA and BHT into breakfast cereals and snack products has been applied. Recently, vitamins E and C have been suggested for integration in polymer films by virtue of their antioxidant effect. Vitamin E is a safe and effective antioxidant for cereal and snack products where the development of rancid odors and flavors is often the shelf-life-limiting factor.[22]

Flavor-releasing systems

Flavor-releasing systems can be used to mask off-odors coming from the food or the packaging. Other applications of flavor-enriched packaging materials include the possibility to improve the sensory quality of the product by emitting desirable flavors into the food and to encapsulate pleasant aromas that are released upon opening. It should be mentioned that flavor-releasing systems should not be misused to mask microbial off-odors, thus introducing risks to consumers.[4,17]

Miscellaneous Active Packaging Systems

Other active packaging systems are microwave susceptors, temperature-control packaging systems, and packaging materials with foaming properties. Temperature-controlling materials include the use of innovative insulating materials, and self-heating and self-cooling packaging systems.

CONCLUSION

The industry's need to effectively and safely package foods for transport and storage while maintaining the quality, along with increasing demands from consumers for fresher, more convenient, minimally processed and safer foods with a prolonged shelf-life presents a bright future for active packaging. The interest in active packaging and the number of food-related applications is expected to increase significantly in the years to come.

REFERENCES

1. Rooney, M.L. Overview of Active Food Packaging. In *Active Food Packaging*; Rooney, M.L., Ed.; Blackie Academic & Professional: London, 1995; 1–37.
2. Ahvenainen, R.; Hurme, E. Active and Smart Packaging for Meeting Consumer Demands for Quality and Safety. Food Addit. Contam. **1997**, *14* (6–7), 753–763.
3. Kruijf de, N.; Beest van, M.; Rijk, M.A.H.; Sipiläinen-Malm, T.; Paseiro Losada, P.; Meulenaer De, B. Active and Intelligent Packaging: Applications and Regulatory Aspects. Food Addit. Contam. **2002**, *19* (supplement), 144–162.
4. Vermeiren, L.; Devlieghere, F.; Beest van, M.D.; Kruijf de, N.; Debevere, J. Developments in the Active Packaging of Foods. Trends Food Sci. Technol. **1999**, *10*, 77–86.

5. Beest van, M.D.; Kruijf de, N. Actieve en Intelligente Verpakkingen: Toepassingen en Wetgevingsaspecten. Voedingmiddelentechnologie (in Dutch) **2001**, *12*, 15–19.
6. Labuza, T.P. An Introduction to Active Packaging for Foods. Food Technol. **1996**, *50*, 68–71.
7. Day, B.P.F. Underlying Principles of Active Packaging Technology. Food Cosmet. Drug Packag. **2000**, *23*, 134–139.
8. Floros, J.D.; Dock, L.L.; Han, J.H. Active Packaging Technologies and Applications. Food Cosmet. Drug Packag. **1997**, *20*, 10–17.
9. Smith, J.P.; Ooraikul, B.; Koersen, W.J.; Jackson, E.D.; Lawrence, R.A. Novel Approach to Oxygen Control in Modified Atmosphere Packaging of Bakery Products. Food Microbiol. **1986**, *3*, 315–320.
10. Gill, C.O.; McGinnes, J.C. The Use of Oxygen Scavengers to Prevent the Transient Discoloration of Ground Beef Packaged Under Controlled Oxygen-Depleted Atmosphere. Meat Sci. **1995**, *41*, 19–27.
11. Schozen, K.; Ohshima, T.; Ushio, H.; Takiguchi, A.; Koizumi, C. Effects of Antioxidants and Packing on Cholesterol Oxidation in Processed Anchovy During Storage. Food Sci. Technol. **1997**, *30*, 2–8.
12. Berenzon, S.; Saguy, I.S. Oxygen Absorbers for Extension of Crackers Shelf-life. Food Sci. Technol. **1998**, *31*, 1–5.
13. Zagory, D. Ethylene-Removing Packaging. In *Active Food Packaging*; Rooney, M.L., Ed.; Blackie Academic & Professional: London, 1995; 38–54.
14. Rooney, M.L. Active Packaging in Polymer Films. In *Active Food Packaging*; Rooney, M.L., Ed.; Blackie Academic & Professional: London, 1995; 74–110.
15. Labuza, T.P.; Breene, W.M. Applications of Active Packaging for Improvement of Shelf life and Nutritional Quality of Fresh and Extended Shelf-life Foods. J. Food Process. Preserv. **1989**, *13*, 1–69.
16. Day, B.P.F. Active Packaging of Foods. CCFRA New Technol. Bull. **1998**, *17*, 23.
17. Nielsen, T. Active Packaging—a Literature Review. SIK-report no. 361. **1997**, 20.
18. Hotchkiss, J.H. Safety Considerations in Active Packaging. In *Active Food Packaging*; Rooney, M.L., Ed.; Blackie Academic & Professional: London, 1995; 238–255.
19. Han, J.H.; Floros, J.D. Casting Antimicrobial Packaging Films and Measuring Their Physical Properties and Antimicrobial Activity. J. Plast. Film Sheeting **1997**, *13*, 287–298.
20. Padgett, T.; Han, I.Y.; Dawson, P.L. Incorporation of Food-Grade Antimicrobial Compounds into Biodegradable Packaging Films. J. Food Prot. **1998**, *61*, 1330–1335.
21. Smith, J.P.; Hoshino, J.; Abe, Y. Interactive Packaging Involving Sachet Technology. In *Active Food Packaging*; Rooney, M.L., Ed.; Blackie Academic & Professional: London, 1995; 143–173.
22. Wessling, C.; Nielsen, T.; Leufven, A.; Jägerstad, M. Mobility of α-Tocopherol and BHT in LDPE in Contact with Fatty Food Simulants. Food Addit. Contam. **1998**, *15*, 709–715.

Aerobic Reactions

Ali Demirci
The Pennsylvania State University, University Park, Pennsylvania, U.S.A.

INTRODUCTION

Oxygen is an important ingredient for microbial growth during aerobic fermentation. Since solubility of oxygen is very low in aqueous solutions, it has to be added to a fermentation medium by aeration to keep up with the oxygen demand of growing microorganisms. The stoichiometric equations can suggest some information on how much oxygen is needed, but it may not describe true oxygen demand. Therefore, oxygen transfer rate (OTR) from gas phase to aqueous phase has to be evaluated closely to make sure that enough oxygen is provided. In order to determine OTR, it is essential to estimate volumetric oxygen transfer coefficient (k_La). OTR can be improved by adjusting air flow rates, agitation, orifice size of sparger as well as more efficient impellers, all of which can effect k_La positively.

AEROBIC MICROORGANISMS

Microorganisms are categorized based on their need or tolerance of oxygen. Some of them do not require oxygen for their metabolism and these are called anaerobic microorganisms. Anaerobes can be further categorized as: (i) facultative anaerobes that can grow in the presence of oxygen even though they do not use oxygen, and (ii) obligate anaerobes that are killed by oxygen. On the other hand, oxygen is required by many microorganisms, which are called aerobic microorganisms. Aerobes use oxygen as a substrate in the metabolism and/or as a terminal electron acceptor during respiration. Some aerobic microorganisms can adapt themselves under anaerobic conditions, which are called facultative aerobes. Some microorganisms such as *Saccharomyces cerevisiae*, are capable of switching their metabolisms from aerobic to anaerobic or vice versa depending on the availability of oxygen and, in some cases, other nutrients. *S. cerevisiae* grows anaerobically in the presence of low oxygen and high glucose concentrations and produces high amount of ethanol. On the other hand, *S. cerevisiae* grows aerobically in the presence of high oxygen and low glucose concentrations and produces mostly biomass with negligible amount of ethanol. This phenomena is known as Krebtree effect or Pasteur effect.

OXYGEN REQUIREMENT AND SOLUBILITY

Many of the commercial fermentation processes use aerobic microorganisms. Therefore, oxygen needs to be provided during fermentation process (Fig. 1). The stoichiometric equations can suggest some information on how much oxygen is needed. For example,

$$C_6H_{12}O_6(\text{Glucose}) + 6O_2 \rightarrow 6H_2O + 6CO_2$$

However, stoichiometric approach based on carbon source usually does not refer to true oxygen demand, which depends not only on carbon sources, but also on nitrogen sources in the fermentation broth. Additionally, basic stoichiometric approach may not include biomass productions in addition to product formations. Several researchers recommend equations to predict oxygen demand by considering biomass and/or product formations.[3,4,11,17] Golobic et al.[9] studied an online estimation of the specific growth rate of an aerobic fermentation based on the measurement of oxygen or carbon dioxide concentrations in the exhaust gas or dissolved oxygen concentrations in the fermentation broth.

All these clearly indicate that oxygen has to be provided to microorganisms for efficient metabolic process. However, solubility of oxygen in water or fermentation broth is very low compared to other substrates in the fermentation broth such as sugars and salts. In addition, solubility of oxygen as any other gas decreases with increase temperature (Table 1). In addition to temperature effect, the presence of other ingredients in the medium affects the solubility of oxygen. Iwai et al.[10] measured and correlated the solubility of oxygen in aqueous solutions containing salts. Eya et al.[7] performed similar experiments for aqueous solutions containing glucose, sucrose, and maltose. Both studies indicated that solubility of oxygen reduced in the presence of salts and sugars.

Dissolved oxygen concentration can be increased if partial pressure of oxygen in the gas phase is increased. Therefore, Henry's Law (Eq. 1) states that the amount of oxygen that remains dissolved in a volume of water, at

Encyclopedia of Agricultural, Food, and Biological Engineering
DOI: 10.1081/E-EAFE 120007197
Copyright © 2003 by Marcel Dekker, Inc. All rights reserved.

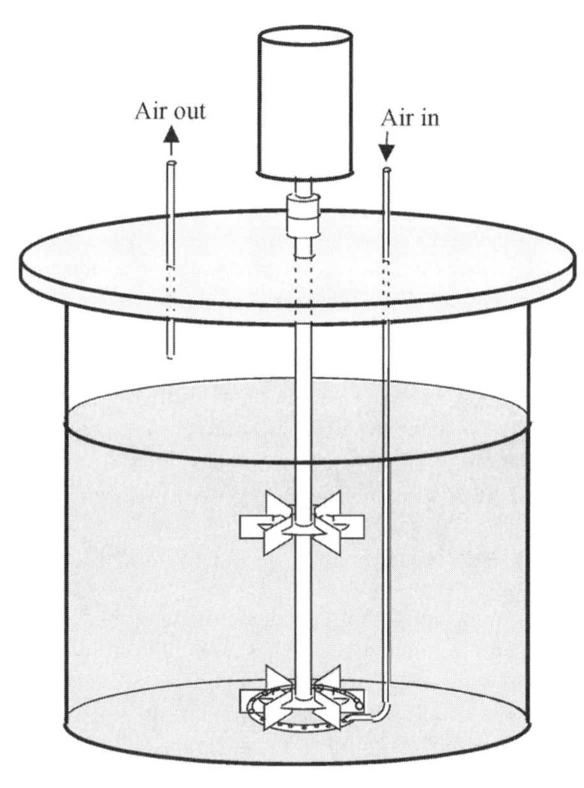

Fig. 1 Typical bioreactor design with aeration and agitation.

a constant temperature, is proportional to the ambient pressure of oxygen gas with which it is in equilibrium. Therefore, it is sometimes desirable to use concentrated oxygen gas instead of ambient air to increase dissolved oxygen concentration in a medium, but this approach will increase the fermentation operating cost.

$$P_{O2} = X_{O2}H_{O2} \tag{1}$$

where P_{O2} is the partial pressure of oxygen in gas phase, X_{O2} the mole fraction of oxygen in water, and H_{O2} the Henry's law constant for oxygen in water (3.12×10^7 mm Hg unit^{-1} mole fraction).

It is extremely difficult to maintain dissolved oxygen concentration at its saturation level as oxygen is being consumed during fermentation. However, many microorganisms do not require maximum saturation level, because oxygen uptake rate (OUR) increases as dissolved oxygen concentration increases until a certain dissolved oxygen level is achieved (critical oxygen concentrations, C_{crit}). Therefore, knowing C_{crit} value for a specific microorganism helps to determine the set dissolved oxygen level. Finn[8] has reported C_{crit} values for various microorganisms (Table 2). There is a wide range for C_{crit} values. For example, C_{crit} for *Azatobacter vinelandii* is 1.07 mg L^{-1}, whereas C_{crit} for *Pseudomonas denitrificans* is 0.29 at the same temperature.

Table 1 Solubility of oxygen (mg L^{-1}) in water at atmospheric pressure[a]

	Chlorinity (g kg^{-1})		
Temperature (°C)	**0**	**10**	**20**
0	14.62	12.89	11.36
10	11.29	10.06	8.96
20	9.09	8.17	7.35
30	7.56	6.85	6.20
40	6.41	5.84	5.32
50	5.48	5.02	4.59

[a] Summarized from Ref. 1.

OXYGEN MASS TRANSFER

In order to keep critical dissolved oxygen concentration at the desired level, oxygen has to be supplied at the rate of oxygen consumption by the growing microorganisms. In other words, OTR must be equal to or higher than OUR. OTR can be calculated by the following equation:

$$\frac{dC_L}{dt} = k_L a(C_L{*} - C_L) \tag{2}$$

where t is the time (hr), C_L the dissolved oxygen concentration at any time, C_L* the saturated dissolved oxygen concentration, k_L the oxygen transfer coefficient (cm hr^{-1}), a the gas–liquid interface area (cm^2 cm^{-3}), and k_La the volumetric oxygen transfer coefficient (hr^{-1}).

Thus k_La is a critical parameter to maintain OTR and the desired dissolved oxygen concentration in a fermentation broth during fermentation. k_La can be improved by using smaller orifices through which air is sparging or by faster agitation rates. Both play an important role in

Table 2 Typical values of critical oxygen concentrations (C_{crit}) for various microorganisms[a]

Microorganism	Temperature (°C)	C_{crit} (mg L^{-1})
Aspergillus oryzae	30	0.64
Azatobacter vinelandii	30	1.07
Escherichia coli	38	0.26
	15	0.10
Penicillum chryogenum	24	0.70
	30	0.29
Pseudomonas denitrificans	30	0.29
Saccharomyces cerevisea	35	0.15
	20	0.12
Serratia marcescens	31	0.48

[a] Reported by Finn.[8]

A

reducing bubble sizes, which directly influence k_La positively. Therefore, it is essential to determine k_La for each fermentation condition (i.e, aeration rate, agitation, and fermentation temperature). The easiest method to determine k_La is the unsteady state method in which oxygen is removed from fermentation broth first by sparging nitrogen (N_2) gas until the dissolved oxygen concentration approaches zero. Then air is sparged at the desired aeration rate, agitation, and temperature. Increase in the dissolved oxygen concentration in the fermentation broth is recorded over time by using dissolved oxygen (DO_2) probe, which provides percent dissolved oxygen (DO_2) level (i.e. 100% DO_2 is C_L*).

After rearranging (Eq. 2) as follows,

$$\int_{C_o}^{C_L*} \frac{dC_L}{(C_L* - C_L)} = k_La \int_0^t dt \qquad (3)$$

and integration yields to:

$$\ln(C_L* - C_L) = -k_La\,t + \ln(C_L* - C_o) \qquad (4)$$

Thus, plot of $\ln(C_L* - C_L)$ vs. t will provide a straight line with a slope of $-k_La$ (Fig. 2). Other methods to measure k_La such as the sulphite oxidation method and the steady state method can be found somewhere else.[2,15]

After determining k_La value, OTR can be calculated by using Eq. 1 by using set dissolved oxygen concentration (C_L) and compared with OUR. If OUR is larger than OTR, necessary adjustments have to be made such as increasing air flow rate or agitation as well as decreasing sparger orifice size. Even changing the impeller with a more efficient type can improve k_La significantly.

It should be kept in mind that every fermentation broth demonstrates different k_La. For example, Vlaev and Valeva[16] investigated the effect of various starch suspensions for α-amylase fermentation. Since higher starch concentrations of starch suspensions increased the rheological characteristics of the medium, it caused reduction in k_La value. Sometimes reological characteristics may change during fermentation, which is the case during xanthan fermentation by *Xanthomonas campestris*. OTR is significantly reduced as viscosity of medium changes due to xanthan gum production. Oxygen demand by microorganism is properly met only up to 6 hr during 72-hr fermentation. Therefore, new methods to incorporate oxygen into the medium are being investigated. For example, Sriram et al.[14] suggested the utilization of the hydrogen peroxide as a source of oxygen by which they continuously achieve 50% DO_2 level throughout the fermentation.

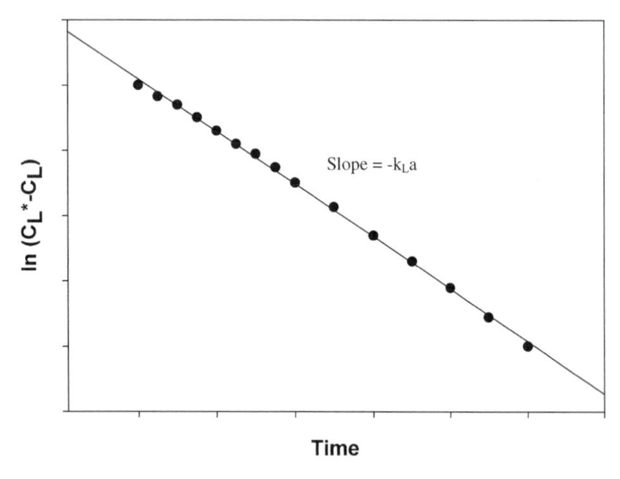

Fig. 2 Plot of $ln(C_L* - C_L)$ vs. time to calculate k_La.

CONCLUSION

Oxygen is an important ingredient for microbial growth during aerobic fermentation. Since solubility of oxygen is very low in aqueous solutions, it has to be added to a fermentation medium by aeration to keep up with the oxygen demand of growing microorganisms. Therefore, OTR from gas phase to aqueous phase has to be evaluated closely to make sure that enough oxygen is provided. If not, action has to be taken to improve OTR by adjusting air flow rates, agitation, orifice size of sparger as well as more efficient impellers. More detailed information can be found in some other references.[2,5,6,12,13,15]

REFERENCES

1. American Public Health Association (APHA). *Standard Methods for the Examination of Water and Wastewater*, 18th Ed.; American Public Health Association: Washington, D.C., 1992.
2. Baily, J.E.; Ollis, F.D. *Biochemical Engineering Fundamentals*, 2nd Ed.; McGraw Hill, Inc.: New York, 1986; 457–529.
3. Cooney, C.L. Conversion Yields in Penicillin Production: Theory versus Practice. Process Biochem. **1979**, *14*, 31–33.
4. Darlington, W.A. Aerobic Hydrocarbon Fermentation—A Practical Evaluation. Biotech. Bioeng. **1964**, *6*, 241–242.
5. Doran, P.M. *Bioprocess Engineering Principles*; Academic Press: San Diego, 1995; 190–217.
6. Madigan, M.T.; Martinko, J.M.; Parker, J. *Brock's Biology of Microbiology*, 9th Ed.; Prince-Hall: Upper Saddle River, 2000; 135–162.
7. Eya, H.; Mishima, K.; Nagatani, M.; Iwai, Y.; Arai, Y. Measurement and Correlation of Solubilities of Oxygen in Aqueous Solutions Containing Glucose, Sucrose and Maltose. Fluid Phase Equilibria **1994**, *97*, 201–209.
8. Finn, R.K. Agitation and Aeration. In *Biological Engineering Science*; Blakebrough, N., Ed.; Academic Press, Inc.: New York, 1967; Vol. 1, 69.

9. Golobic, I.; Gjerkes, H.; Bajsic, I.; Malensek, J. Software Sensor for Biomass Concentration Monitoring During Industrial Fermentation. Instrum. Sci. Technol. **2000**, *28*, 323–334.
10. Iwai, Y.; Eya, H.; Itoh, Y.; Arai, Y.; Takeuchi, K. Measurement and Correlation of Solubilities of Oxygen in Aqueous Solutions Containing Salts and Sucrose. Fluid Phase Equilibria **1993**, *83*, 271–278.
11. Krzystek, L.; Ledakowicz, S.J. Stoichiometric Analysis of *Kluyveromyces fragilis* Growth on Lactose. Chem. Technol. Biotechnol. **2000**, *75*, 1110–1118.
12. Shuler, M.L.; Kargi, F. In *Bioprocess Engineering: Basic Concepts*, 2nd Ed.; Prentice Hall: Englewood Cliffs, 2002; 285–328.
13. Soderberg, A.C. Fermentation Design. In *Fermentation and Biochemical Engineering Handbook: Principles, Process Design, and Equipment*, 2nd Ed.; Vogel, H.C., Todaro, C.L., Eds.; Noyes Publications: Westwood, 1997; 99–110.
14. Sriram, G.; Rao, Y.M.; Suresh, A.K.; Sureshkumar, G.K. Oxygen Supply Without Gas–Liquid Film Resistance to *Xanthomonas campestris* Cultivation. Biotechnol. Bioeng. **1998**, *59*, 714–723.
15. Stanbury, P.F.; Whitaker, A.; Hall, S.J. *Principles of Fermentation Technology*, 2nd Ed.; Elsevier Science, Inc.: Tarrytown, 1995; 243–272.
16. Vlaev, S.D.; Valeva, M. Rheology and Oxygen Transfer in Starch Suspensions During High Temperature (35°C) α-Amylase Fermentation. Chem. Eng. Biochem. Eng. **1990**, *44*, B51–B55.
17. Wincure, B.M.; Cooper, D.G.; Rey, A. Mathematical Model of Self-Cycling Fermentation. Biotech. Bioeng. **1995**, *46*, 180–183.

Agricultural Engineering History

Gerald W. Isaacs
University of Florida, Gainesville, Florida, U.S.A.

INTRODUCTION

Agricultural engineering was first recognized as an engineering discipline in 1907 with the formation of the American Society of Agricultural Engineers (ASAE).[1] The new constitution of this professional organization stated "the object of this society shall be to promote the art of science of engineering as applied to agriculture." Agricultural engineers have since then developed a myriad of labor-saving farm machines, farm buildings, irrigation and drainage systems, and processes for preserving and converting agricultural products to useful food, feed, and fiber products.

Agricultural engineers have been major contributors towards building a highly efficient agriculture in the United States today. In 1870, over half the labor force of the United States was devoted to agricultural production. In 2001, about 2% of the population was directly involved in agricultural production. A major portion of the income of American households used to be spent on food consumed in home. In 2001, less than 15% of U.S. family income was used for food.

ENGINEERING FOR AGRICULTURE IN THE 19TH CENTURY AND BEFORE

Engineering for agriculture prior to the formalization of the agricultural engineering profession in 1907 was done primarily by practitioners of other engineering disciplines. Many mechanization developments were made by mechanical engineers. Civil engineers designed farm buildings, drainage and irrigation systems, and waste disposal systems. Electrical engineers were responsible for some of the modest applications of electric power to agriculture in these early days.

Due credit must be given to many developers of mechanization for agriculture who were not engineers but rather blacksmiths or very talented farmer inventors. These early developers worked without the sophisticated technologies used routinely by today's agricultural engineers, such as engineering analysis, computer simulation, and computer graphics. The old time inventors were equally creative for their time and in many cases had the added advantage of first hand knowledge of the need for their inventions. Spending many hours in the drudgery of such activities as hoeing or picking corn by hand gave plenty of time for reflection about better ways to do the job.

There were great numbers of important developments in the 19th century that increased the labor efficiency of agricultural production and served as a prelude to the explosion of new accomplishments in the 20th century.

Early in the 19th century, nearly all motive power for agriculture was animal power, mostly horses and oxen. By the beginning of the last quarter of the 19th century, steam traction engines were beginning to replace animal power for high-energy demand operations like tillage and grain threshing. Steam engines for agriculture were adapted by mechanical engineers from steam railroad engine designs. The major challenge was to make an engine capable of pulling a load under field conditions that were highly variable.

The resulting steam traction engines were powerful under the right conditions, but they were cumbersome to fire up before using and required the operator to have some special technical skills. They were used well into the 20th century for pulling large plows and for running grain threshing machines.

Internal combustion engines utilizing liquid petroleum fuels like kerosene and eventually gasoline began to replace the steam engines at the beginning of the 20th century. Traction engines with internal combustion engines became known as the agricultural tractor, a machine that would serve as a basis for many farm machinery developments in the new century. The first tractors looked very much like the steam engines, being not very maneuverable to drive and useful mostly in plowing and other heavy tillage operations.

There was a great need felt by farmers on both sides of the Atlantic for smaller tractors that could do the plowing on small farms and could also be used for planting, cultivating, and pulling the farm wagon for materials handling chores. Smaller tractors were produced in the 1920s, e.g., the Lanz Bulldog in Germany and the Fordson in America. These and many other smaller tractors were a step in the right direction for the small farmers, but these tractors still lacked adaptability to many farm chores, and as a result many farmers kept their horses for some farm operations.

Encyclopedia of Agricultural, Food, and Biological Engineering
DOI: 10.1081/E-EAFE 120006945
Copyright © 2003 by Marcel Dekker, Inc. All rights reserved.

In the 1930s, Henry Ford and Harry Ferguson collaborated to produce the Ford 9N tractor that was highly adaptable to many farm tasks. It had a three-point hitch with automatic draft control that improved traction for plowing with a small 23 hp tractor. The hydraulic system made the Ford 9N adaptable to a variety of attached implements for chores like manure handling, grading, cultivating, and planting. The horses could at last be relegated to recreational activities.

Great progress was made in the 19th century in the development of the plow. The wooden plow with its obvious shortcomings was replaced gradually with steel. In 1835, the Wurttembergischer Hagen (hook) had two steel knives to cut the sod and soil ahead of a steel point, and a wooden moldboard to turn over the soil. John Deere, an inventive blacksmith from Illinois, saw the need to replace the wooden moldboard because the heavy prairie soils adhered to it and made the desired scouring action difficult to attain. He made the first steel moldboard plow in 1837. It worked well in a wide variety of soils, and Deere's basic design is still widely used today.

Harvesting grain by hand with a sythe was very hard slow work, so the reaper was welcomed by farmers of the early 19th century. Cyrus Hall McCormick invented the first successful reaping machine in Rockbridge County, Virginia in 1831. It was pulled through the field by horses to cut a swath of grain and place it in a windrow for hand tying. The bunches were later hand tied and placed in a shock to dry before threshing.

AGRICULTURAL ENGINEERING DEVELOPMENTS IN THE 20TH CENTURY

In 2000, ASAE published a study by its various divisions that identified the most significant developments in agricultural engineering in the 20th century. In addition, during the year 2000, the National Academy of Engineering listed the 20 most significant development by engineers in general during the 20th century.[2] Seventh on the list was the mechanization of agriculture, standing alongside such great engineering accomplishments as television, computers, and a safe and abundant water supply. At the top of the list was electrification of U.S. homes, businesses, and industries including farms.

The ASAE Power and Machinery Division listed the most significant developments of the 20th century in their area of interest.[3] The agricultural tractor was recognized for having changed the way food is grown and produced, and the Ford 9N tractor mentioned herein previously was cited. Certainly tractors did greatly increase the productivity of agricultural workers and reduced the need for producing feed for draft animals.

Many state universities in the United States and Europe offered course work in internal combustion engines and farm machinery design. The Nebraska tractor Testing Program at the University of Nebraska provided performance data on tractors that was invaluable to designers of tractors and farmers making purchase decisions.

The self-propelled combine harvester was another major farm machinery development of the 20th century. It replaced the grain reaper and binder previously used to cut and bundle grain to place grain in shocks to dry in the field before hauling to a stationary threshing machine. The combine harvester cuts, threshes the grain, spreads the straw or places it in windrows, and transports the grain to a truck or wagon for transport to storage or market. The combine harvester later replaced the ear corn picker, bringing shelled corn to the edge of the field. The modern combine harvester with a single operator can harvest grain at a rate 100 times of what was achievable with 19th century technology.

Like many agricultural machinery developments, there was a long time between invention of the combine and its general use. The first combine harvester was built by Hiram Moore and John Hascall in 1836 but commercial versions first became available nearly 100 years later.

Harvesters were developed for crops other than grain that were formerly harvested by hand. The first attempt at building a cotton harvester was made in 1850 by S. S. Rembert and Jebediah Prescott of Memphis, Tennessee. Several experimental machines were built later, but the first commercial machines did not become available until 1943 when introduced by International Harvester Company. Mechanical cotton pickers have mostly replaced hand harvesting today in the industrialized nations.

Other specialized harvesting machines have been developed for a wide variety of fruit and vegetable crops like potatoes, tomatoes, carrots, peas, and beans. Some mechanical harvesting of tree fruits is done by mechanical tree shakers with catching frames under the trees. Sugar beet harvesters are widely used and have partly replaced hand labor for this crop.

Advancements in tillage equipment in the 20th century were also recognized by the ASAE Power and Machinery Division. After a century of using the moldboard plow as a primary tillage tool, interest in the chisel plow began to grow on the Oklahoma farm of Fred Hoeme in 1933. He was concerned about the great difficulty in plowing drought-parched prairie soils at that time and had seen road-building scarifiers that could rip up old roadbeds. He based his chisel plow design on these scarifiers and it became a common tillage tool for soils difficult to till. It had the added advantage of being able to leave crop residue and clods on the soil surface to prevent soil erosion from wind and water. The chisel plow was the forerunner of later generations of conservation tillage implements sometimes combined with planters for tilling and planting in a single operation.

It was not surprising that the ASAE Soil and Water Division[4] cited conservation tillage as their first nomination for the top achievements of the 20th Century. The chisel plow, till planter, and no-till planter were key machines in implementing conservation tillage, which is used in many U.S. farms today. Use of agricultural chemicals for weed and insect control has made conservation tillage feasible and has resulted in major reductions in labor and energy input for crop production.

Devising methods of controlling soil erosion became a major activity of agricultural engineers and soil scientists in the 20th century. Development of the Universal Soil Loss Equation to predict soil erosion was a major early accomplishment of Walter Wischmeier, a USDA soil scientist working at Purdue University in the 1950s. This equation became an early basis for computer simulations used to pre-evaluate management methods for controlling erosion and generally improving the management of soil and water resources.

Advances in irrigation technology have made possible efficient crop production on soils lacking in natural rainfall throughout the world. Finding ways to apply water to crops efficiently has been a major issue for agricultural engineers working in irrigation. Irrigation competes heavily with other water uses for our limited water resources. The first impact sprinkler was developed by Arton Englehart in 1933. A higher proportion of land is irrigated today by piped sprinkler systems than by flood irrigation.

Increasing concerns about conserving water resources throughout the world prompted the first development of drip irrigation system in Israel in the 1960s. Drip and the later microsprinkler systems are designed to apply only the water needed by individual plants and are usually automatically controlled by a small computer.

The Structures and Environment Division of ASAE[4] identified several outstanding achievements by agricultural engineers in the 20th century. They recognized the contribution of an organization of agricultural engineers called the Midwest Plan Service, founded in 1929. This organization has developed farm building plans and equipment recommendations for a wide variety of farmstead operations. Some of these agricultural engineers were directly involved in the design of prefabricated wood roof trusses that made the construction of large clear-span buildings possible. Many of these same truss designs are widely used for home construction in the United States today.

Other 20th century agricultural building developments recognized by ASAE included advancements in the ventilation of farm buildings being used for intensive animal production. Without such ventilation, a healthful environment for animals and workers could not be maintained in these buildings. Also recognized as a major agricultural engineering achievement was the cyclone air pollution abatement system used to reduce dust emissions from agricultural processing operations like feed milling, cotton ginning, and wood products manufacturing.

Agricultural engineers have given much attention to the design of greenhouses to make them more energy efficient and lower in construction cost than the traditional glass and frame structures. Today about 64% of commercial greenhouses in the United States use double polyethylene film construction to house winter production of ornamental plants and vegetables. These buildings frequently use computer controlled heating, ventilation, and irrigation to create a productive plant environment.

The responsibilities of the agricultural engineer did not end with the harvesting of food and feed crops. The quality of harvested crops is maintained sometimes for months or even years until they are marketed. Several important developments that made preservation of grain crops possible after harvest were identified as major achievements of the 20th century by the ASAE Information and Electrical Technology Division.[5]

With use of the grain combine, it became necessary to remove the field heat and sometimes excessive moisture from grain by aerating it in storage. Aeration was later used in the 1940s to prevent the migration of moisture in large grain masses during cold weather. This moisture transport phenomenon caused crusting and spoilage of the upper layers of grain and could be prevented by equalizing the temperature of the grain mass by passing a small amount of air through the grain.

As grain combines were applied to the harvesting of corn in the early 1950s, mechanical drying of shelled corn became a necessity in most harvest years to remove excess moisture before storing. Drying with high-temperature air has been used in order to keep up with the high harvesting rate of the combine harvesters.

High-temperature drying and rapid cooling in the dryer were shown to cause stress cracks that caused kernel breakage in subsequent handling and reduced value for some uses. USDA research agricultural engineers George H. Foster and Ralph Thompson at Purdue University developed a slow bulk cooling process, which they called dryeration. The grain was conveyed from the dryer while hot to an intermediate storage bin equipped with aeration. Low rates of air were then passed through the grain to cool it and finish the drying process before conveying to long-term storage. The result is improved grain quality through reduction of stress cracks, higher dryer output, and higher energy efficiency of the drying process.

Agricultural engineers working with rice were concerned about stress cracks even before the advent of

dryeration. E. B. Copeland of the University of the Philippines observed stress cracks in most sun-dried rice in 1924. Otto Kunze of Texas A&M University in the early 1960s developed improved harvesting and drying procedures that were used worldwide to improve the availability and quality of vital food crops.

Bringing electrical power to the farms became a strong movement in the 1930.[6] This new energy source made many new machines and processes possible that saved labor and greatly enhanced the quality of food produced. Farm homes benefited from electrical power, which made many conveniences of city living available on the farm. Electrical power on the farm made possible the mechanical grain handling and feed-processing systems that allowed the farmer to hold his grain for a better price and use it to make feed for his livestock. Milking machines replaced milking cows by hand. Refrigeration for cooling newly harvested milk greatly extends the shelf life of today's dairy products. Refrigeration also cooled fruits and vegetables to maintain their quality after harvesting. Electric power made automatic control of many farm operations possible.

Assessing the quality of agricultural products as they are grown, harvested, processed, and marketed has been an important issue for agricultural engineers. Rapid measurement of critical quality factors is frequently desirable as our food products move from production to consumption. Electronic moisture meters for grain were developed over the past 70 yr to guide the harvesting, processing, and storage of grains for food and feed. In the early 1970s, agricultural engineer Karl Norris and his coworkers in the USDA laboratories in Beltsville, Maryland pioneered the use of spectroscopic methods to measure the protein, oil, and moisture contents of grain.

Agricultural engineers along with food technologists and other engineers and scientists have played an important role in the development of the food processing industry that converts raw farm products into endless variety of food products in our supermarkets. Among the first packaged food products made available to consumers were the grain cereals. Physicians W. K. Kellogg, his brother John Kellogg, and C.W. Post operated a health sanitarium in Battle Creek, Michigan at the turn of the 20th century. Recognizing the nutritional benefits of grain products and the need for a convenient ready-to-eat breakfast product, they developed the first process for making cereal flakes from grain in1895. W. K. Kellogg started his company in 1906 to make and sell corn flakes and eventually a host of other cereal products. C. W. Post soon after formed his own company to sell Grape Nuts, a nonflaked ready-to-eat cereal product made from hard baked loaves milled to a crunchy cereal product.

The achievements of agricultural engineers serving agriculture and the consuming public are numerous and highly varied. Many technical developments of agricultural engineers have had enormous social and economic impacts. Many other developments, though proven technically feasible, have not yet shown significant impact, presumably for economic reasons. Although many inventions and designs may never achieve impact, we should remember that many of the most important technical developments in agriculture were initially developed 20 yr–50 yr before they were widely adapted.

REFERENCES

1. Stewart, R.E. *7 Decades that Changed America*; American Society of Agricultural Engineers (ASAE): St. Joseph, MI, 1979.
2. Anonymous. Agricultural Mechanization Named Among Top 20 Engineering Achievements. *Resource Magazine*; ASAE: St. Joseph, MI, April 2000; Vol. 7, No. 4, 17.
3. Coello, J.L.; Huggins, L.F. Outstanding Agricultural Engineering Achievements of the 20th Century. *Resource Magazine*; ASAE: St. Joseph, MI, January 2000; Vol. 7, No. 1, 17–19.
4. Coello, J.L.; Huggins, L.F. Outstanding Agricultural Engineering Achievements of the 20th Century. *Resource Magazine*; ASAE: St. Joseph, MI, March 2000; Vol. 7, No. 3, 18–19.
5. Coello, J.L.; Huggins, L.F. Outstanding Agricultural Engineering Achievements of the 20th Century. *Resource Magazine*; ASAE: St. Joseph, MI, April 2000; Vol. 7, No. 4, 18–19.
6. Coello, J.L.; Huggins, L.F. Outstanding Agricultural Engineering Achievements of the 20th Century. *Resource Magazine*; ASAE: St. Joseph, MI, May 2000; Vol. 7, No. 5, 18–19.

Agricultural Implements

Jay B. Agness
Deere and Company, Waterloo, Iowa, U.S.A.

INTRODUCTION

Agricultural implements are as varied as the variety of plant and animal tissue we call agricultural products. While the commodities are diverse (meat, cotton, peanuts, tomatoes, pulpwood, and honey illustrate only some of the range of materials), the basic elements of the implement design process are common. Understanding the task and the materials to be processed is the key to develop an agricultural implement that performs the desired task. Perfecting the implement into a robust, durable, safe, efficient, and esthetically pleasing form is the key to widespread acceptance and usage.

The term agriculture is used to describe the cultivation and husbandry of plant and animal products used for food and fiber. The need for implements to aid this process predates even the desire to control the food supply rather than simply hunt and gather. The first agricultural implements were simple extensions of the human bodywork aids designed to enhance performance and reduce wear and tear on the body. The tilling stick and scraping flint are examples. The prime attributes of good implements are functionality and durability. The complexity of the tools advanced with the vision of doing multiple and complex tasks and doing them faster with less effort. Significant advances in civilization, like the industrial and informational revolutions, have provided materials, mechanisms, and control means to expand both the complexity and capacity of implements.

With the definition of agriculture presented above, agricultural implements include those associated with the production of grain, oilseeds, nuts, forage, cotton, eggs, pulpwood, sugar beets and cane, wool, citrus and other fruits, fresh vegetables, milk, meat, coffee, root crops, forage crops, comb honey, and many more. Agricultural implements work in the soil—stirring, turning, planting, incorporating, loosening, digging, and burying; orchard—transplanting, thinning, grafting, pruning, spraying, and harvesting; field—cultivating, weeding, spraying, cutting, beating, picking, separating, cleaning, and accumulating; livestock buildings—gathering, cleaning, holding, shearing, cooling, heating, protecting, milking, ventilating, vaccinating, and inseminating; processing sheds—cleaning, conveying, aerating, lifting, cutting, stirring, icing, polishing, packing, and wrapping.

AGRICULTURAL IMPLEMENT DESIGN

The variety in task and crop combinations is mind-boggling. What can be common between the hoe and the egg washer? However, even with great diversity in form and function between agricultural implements, some common elements apply to agricultural implement design. As in any design activity, there are some tried and true principles to be followed. The rigor to which they are implemented will vary, but the principles remain.

Define the Task

The common place to begin the design process is to completely understand the task at hand. The task may be as simple as moving material from point A to point B or as complex as selectively removing ripe tomatoes from tangled vines, washing them, and separating them by color and size and packing them into supermarket-ready, sealed packages.

A key part of task definition involves visualizing the attributes of a successfully completed task. Criteria must be developed in terms of general or specific end conditions, and they may include one or more very controlled intermediate states. They must be detailed enough to include all the important characteristics of the end product like position and orientation. As in most human activity, a fundamental understanding of the problem and the characteristics of a solution are essential to a successful process. Attempts to shortcut this part of the process usually results in having to recycle through it again when the proposed design just is not quite right.

Understand the Materials Involved

The common thread in the design of implements of production agriculture is the interface with nonhomogeneous animal and vegetable material. Biological material

Encyclopedia of Agricultural, Food, and Biological Engineering
DOI: 10.1081/E-EAFE 120006858
Copyright © 2003 by Marcel Dekker, Inc. All rights reserved.

is the common element. Whether it is actively growing, dead and decaying, or in a dormant condition, the variety is great. Plant and animal cells may be turgid and compliant or dry and tough. They may be loosely associated or joined in a tight matrix. Biological material may be resilient and pliable or hard and strong.

The agricultural materials will need to be defined and understood in terms of properties that are pertinent to the task. These properties may include biological, physical, thermal, mechanical, rheological, chemical, mass, photo, and even magnetic information. Pertinent properties are those that directly relate to the details of the task. For example, if an egg must slide, the friction properties of its shell must be understood. If it is to be rolled, the surface texture, curvature, mass, and center of gravity are important. If it is to be picked up, its mass, size, shape, and shell strength become relevant.

Sometimes, we want the implement to preserve the basic properties of the material such as maintaining the succulent quality of fresh fruit. At other times, we may want to change the properties such as shape, moisture content, or aggregate size. Therefore, we must understand the material properties in both static and dynamic terms.

Develop a Functional Implement

After the successful task has been visualized and the biological material characterized, you can begin to develop the solution. For simple tasks, the methodology may be trial-and-error. Complex tasks will probably not yield so easily. Complex implements may perform a variety of tasks or have multiple end products. Dividing the tasks into segments is often a good way to attack the problem, but at some point in time the entire task must be addressed. The simple implements capable of doing many simple tasks may not perform as desired when integrated into a complex implement.

It is in this phase that the material properties of implement components come into play. Implement components that interact with biological materials must be chosen to effectively perform the desired functions. While soft, smooth surfaces may be a prime attribute in handling fresh fruits and vegetables, sharp, abrasive actions may be required to perform some harvesting and separating operations with other crops. Sometimes the best tools are sacrificial, and erode away rather than cause damage to a delicate crop.

Perfect the Implement

The ability to appropriately perfect the implement often marks the difference between success and failure. Refining the functioning tool into a robust, durable, efficient, and esthetic implement usually takes more time and energy than developing the first working model. The success in the perfecting process is often the difference between a "nice idea" and a "historically significant implement." Simple, elegant designs successfully combine the following in just the right proportion:

- Robust designs provide strength where strength is needed to perform the desired functions. Design analysis methods and extensive testing are key to successfully implementing the right structure and materials.
- Durable designs remain effective for an appropriate implement life. They must be both strong and functional over long periods of time.
- Efficient designs make appropriate use of materials, processes, energy, and time. The balancing of these factors is often more an art than a science.
- Esthetic designs look appropriate for the task to be done. Coconut processing machines should have a different appearance than cherry pitters and egg handling equipment.

Almost all the standard measures of a good design are violated at one time or another in the design of agricultural implements. Functionality, robustness, durability, efficiency, and esthetics all will appear to be ignored in some cases. Sometimes it may be appropriate that the implement break rather than injure or destroy a valuable animal. Durable is not the prime attribute of a gentle, degradable plant transfer tube. Efficiency may be much less important than gentleness in the harvesting and handling of perishable fruits, vegetables, eggs, or meat. Sometimes, an esthetically pleasing curved surface may also be the most efficient form to move and place material. At other times sharp, saw-toothed edges are required. The key is to thoroughly understand the task at hand and use appropriate materials and actions to complete it.

EVOLUTION IN IMPLEMENT DESIGN

No matter how complete and perfect an agricultural implement may seem at the time, some person, knowledge, or technology will come along to improve it. The tilling stick became the hoe and then the plow and still later the residue incorporating chisel. The field of genetic engineering opens more doors. Now we may be able to genetically change the biological material to improve the end product or aid in the processing.

The information age offers more detailed knowledge of the exact product needed, access to the precise values of the material properties involved, instantaneous analysis of the forces and motions involved, and graphics to visualize the implement in action even before it is built. Agricultural implement designs will continue to evolve with agriculture and the surrounding world as long as food and fiber remain vital building blocks of civilization.

FURTHER READING

Goering, C.E. *Engine & Tractor Power*; American Society of Agricultural Engineers: St. Joseph, MI, 1992; 533.

Krutz, G.; Thompson, L.; Claar, P. *Design of Agricultural Machinery*; John Wiley & Sons: New York, 1984; 472.

Srivastava, A.K.; Goering, C.E.; Rohrbach, R.P. *Engineering Principles of Agricultural Machines*; American Society of Agricultural Engineers: St. Joseph, MI, 1993; 616.

Airflow Measurement

Robert D. Fox
Richard C. Derksen
United States Department of Agriculture (USDA), Wooster, Ohio, U.S.A.

A

INTRODUCTION

There are many agricultural and food engineering applications where measuring airflow is a necessity. Typical processes requiring airflow information include: pipe flow for food processing; flow in ducts for grain, fruit, and vegetable drying and storage; flow in open areas such as room ventilation, machinery, and wind velocities. In this article, some operating characteristics of several common techniques used to measure airflow are discussed.

BASIC CONCEPTS

An understanding of flow measurement requires basic knowledge of the physical properties of air, such as temperature, pressure, density, specific gravity, viscosity, sonic conductivity, and compressibility. The density of air changes with changes in temperature and pressure. Because of this, airflow rate is usually defined at "standard" conditions, i.e., 20°C temperature and 101.3 kPa pressure. Mass flow depends on volume and air density. The Ideal Gas Law provides a relationship between pressure, temperature, and density. Methods are available to correct for compressibility due to high velocity flows.

Basic concepts that form the basis for many flow measurement devices include the equation of continuity, Bernoulli's theorem, and Reynolds number. The equation of continuity states that the volume of flow passing every position in a conduit must be equal. Thus, if the cross-sectional area in one section of a conduit is smaller than the other, flow velocity must be greater in the smaller area to permit the same volume of air to pass. Bernoulli's theorem states that total energy in the flow must be constant. Thus, if the flow velocity increases in one section of a flow system, pressure must decrease. This principle is used in many flow-measuring devices. Reynolds number (Re) is a ratio of inertial to viscous forces in the flow. Below a critical Re_C (about 2300 for pipe flow), flow will be laminar while above the critical value, flow will be turbulent. These two flow regimes have quite different velocity profiles across the pipe diameter that may influence how flow measurements are made (Fig. 1).

The profile of air velocities within a conduit (or pipe) where flow is measured is an important factor in measurement accuracy. In addition to laminar/turbulent effect on flow profile, air flowing through a turn in the pipe or through an increase or decrease in pipe diameter can significantly change air velocity profiles across the pipe. Thus, many flow-measuring devices require a straight section of pipe, several pipe diameters in length, both up- and downstream from the position of the device (see, e.g., Buzzard, Koop, and Burgess, 2000, p. 34).

Desired information about the airflow condition may be volume, mass, or velocity. Some types of flow meters measure volume or mass flow rate directly, others measure velocity and calculate volume or mass flow rate from dimensions of the meter, and others determine flow by measuring differential pressure and the density of the air.

Differential pressure flow meters (orifice, venturi, etc.) require measurement of pressure differences between pressure taps upstream and downstream from the pressure drop device. For low flow rates, the pressure difference is usually small, and sensitive manometers or pressure transducers are required.

Table 1 lists several common types of flow meters used to measure air flow in pipes or conduits. Some application parameters and limitations are given in the table for each type of meter. There are over 100 different types of flow meters commercially available. Picking the best meter for a specific task requires experience and careful investigation. Important factors to consider in selecting a flow meter include flow range, sensitivity, stability, response time, cleanliness of the air, reliability, calibration frequency required, and price (Table 1).

MEASUREMENT METHODS

Orifice meters are composed of a thin, flat plate with a circular hole in the center, mounted perpendicular to the flow stream in a section of the pipe. Pressure differences between taps located up- and downstream from the orifice are calibrated to flow rate through a pipe. These widely used meters can be used in pipes from 2 cm to 20 cm in diameter.

Encyclopedia of Agricultural, Food, and Biological Engineering
DOI: 10.1081/E-EAFE 120006897
Copyright © 2003 by Marcel Dekker, Inc. All rights reserved.

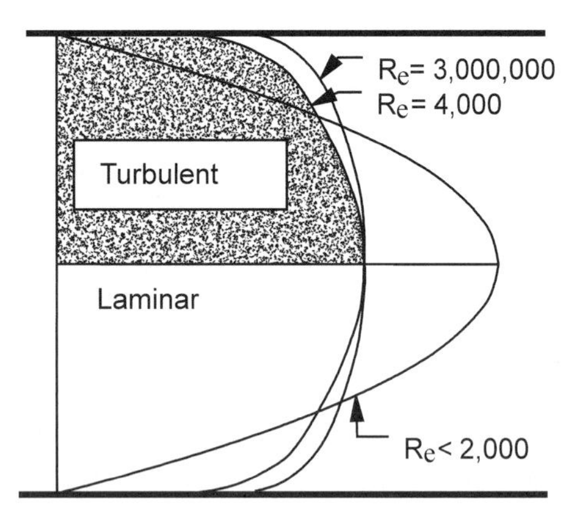

Fig. 1 Velocity profiles in pipe flow. (From Miller, 1996; copyright McGraw-Hill.)

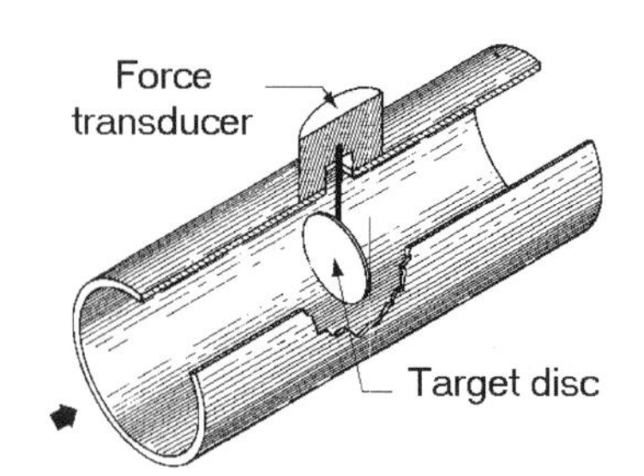

Fig. 2 Drag meter. (From Galley, 1969; copyright ISA.)

Table 1 Flow meters for measuring airflow

	Pipe size (mm)	Clean	Dirty	Temperature (°C)	Pressure (kPa)		Reynolds number/other limitations
Flow meters for pipe flow							
Flow proportional to pressure drop squared; maximum single range about 4:1						Accuracy, uncalibrated	
Orifice	12–1800	■	■	Process temperatures to 540°C, transmitters limited to − 30 to 120°C	$<$ 41,000	± 2% URV[a]	*Re* $>$ 2000
Target	12–100	■	■			± 1% URV	*Re* $>$ 10,000
Venturi	$>$ 50	■	■			± 1%– ± 2% URV	*Re* $>$ 75,000
Flow nozzle	$>$ 50	■	✤			± 1%– ± 2% URV	*Re* $>$ 10,000
Pitot[b]	$>$ 75	■	□			± 5% URV	$>$ 5 m sec^{-1}
Multiport averaging	$>$ 25	■	✤			± 1.25% URV	*Re* $>$ 10,000
Flow directly proportional to measured value; typical range 10:1						Typical calibrated accuracy	
Thermal mass[b]		■	□	$<$ 480	$<$ 21,000	± 1% URV	No limit
Turbine	6–600	■	□	− 268–260	$<$ 21,000	± 0.5% URV	No limit
Ultrasonic, time of flight	$>$ 12	■	□	− 180–260	Pipe rating	± 1% of rate to ± 5% URV	No limit
Variable area	$<$ 75	■	□	Glass: $<$ 200 Metal: $<$ 540	Glass: 2,400 Metal: 5,000	± 0.5% of rate to ± 1% URV	$<$ 15,000 L min^{-1}
Vortex	12–400	■	✤	$<$ 400	$<$ 10,500	± 0.5%– ± 1.5% of rate	*Re* $>$ 10,000
Laser-Doppler[b]		■	✤	System	System	± 1% URV	Transparent walls, requires small particles in flow
Flow meters for measuring in open flow							
Imaging system		■	✤	0–40	50–150	± 1% URV	Requires small particles in flow
Propellers and cups		■	□	− 20–40	Atmospheric	± 2% URV	$<$ 40 m sec^{-1}

■ = designed for this application; ✤ = normally applicable; □ = not designed for this application.

[a] URV = upper range value of the flow rate.

[b] Flow meters can be used in open flow also.

(From Miller, 1996; Copyright McGraw-Hill.)

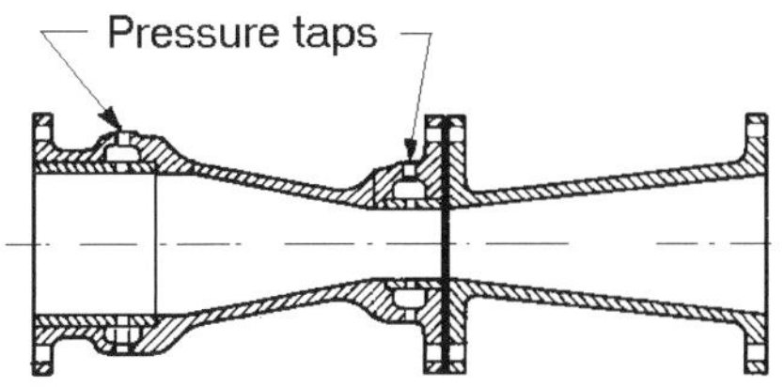

Fig. 3 Venturi flow meter.

Target meters use the force on the support arm of a disk placed perpendicular to the air stream to determine flow speed. A transducer measures the force on the support arm (Fig. 2).

Venturi meters consist of a short length of straight tubing connected at both ends to the flow pipe. Velocity is determined by measuring the difference in pressure upstream and within the venturi. The converging entrance and diverging exit sections of the venturi meter are designed to reduce disturbance in the flow. This reduces the pressure loss across the flow meter compared to an orifice meter (Fig. 3).

Flow nozzles operate on the same principle as orifices, but with the inlet section streamlined. This allows higher flow rates to be measured. These nozzles can be used for direct discharge to the atmosphere (Fig. 4).

Pitot tubes provide velocity by measuring the difference between impact (total) and static pressures. Manufacturers produce multiple pitot-tube fixtures for measuring flow at several points in a pipe for accurate total flow. These instruments often include cleaning systems that use high-pressure gases to blow contaminants from the impact and static ports (Fig. 5).

Thermal mass flow meters can use heated thermocouples or hot wire/films. Air velocity is determined by measuring the cooling effect of the gas flowing over a heated sensing element. Constant temperature, hot-wire/film anemometers adjust electrical current flow to maintain sensors at a constant temperature. The electrical current required to maintain film temperature is proportional to the air velocity over the sensor. Multiple

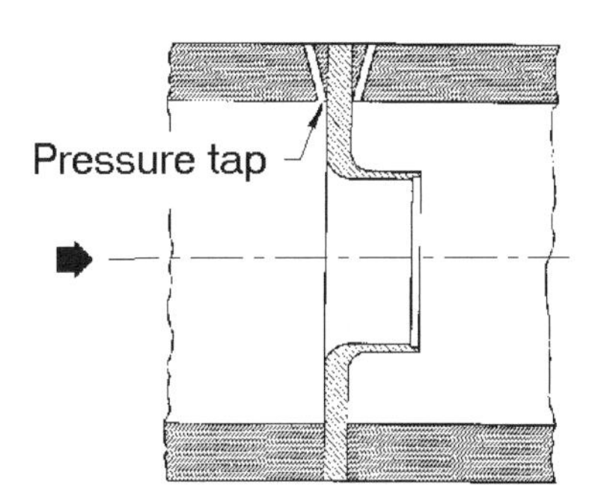

Fig. 4 Flow nozzle. (From Galley, 1969; copyright ISA.)

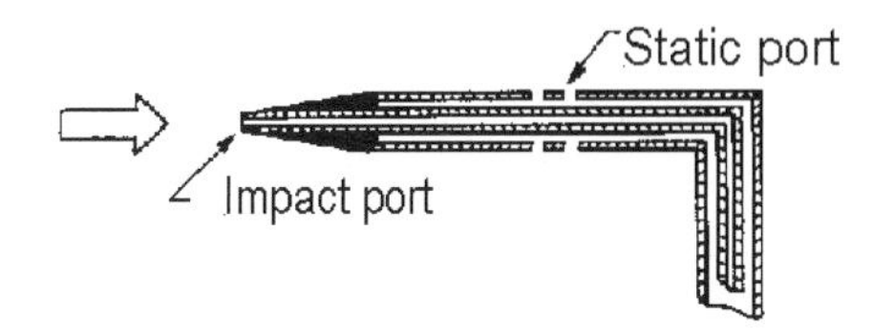

Fig. 5 Pitot tube. (From Miller, 1996; copyright McGraw-Hill.)

sensors can be used to measure flow in two or three directions at a point (Fig. 6).

Turbine flow meters contain a turbine mounted on bearings in the center of a pipe section. Magnetic detectors are used to measure turbine speed, which is directly proportional to the volume flow rate of the air (Fig. 7).

Ultrasonic flow meters transmit energy pulses in both directions between a pair of sensors. The difference in travel time for the ultrasonic pulses moving in opposite directions between the two sensors provides a measure of the wind vector in the plane of the sensor axis. These sensors are robust, linear, and maintain calibration. Three pairs of ultrasonic sensors can be combined to measure air velocity components in three directions at a location. Typical sensor pair spacing (sound wave path length) is from 5 cm to 10 cm. Ultrasonic sensors are used to measure atmospheric flow or flow in a conduit.

Area meters (float sensors, rotameters) consist of a float, suspended in a slightly tapered tube. The float is free to move up or down the tube to reach the position where the open area around the float accommodates the flow volume. Remote sensors can be used to determine float position (Fig. 8).

Vortex-shedding flow meters measure the number of vortices shed by a blunt body inserted into the flow stream. The rate of vortex shedding is directly proportional to the volumetric flow rate. Thermisters, movable metal balls, vanes, and other methods are used to detect the vortex-shedding rate, so there are no moving parts in the air conduit.

Laser-Doppler anemometry is a nonintrusive method of measuring flow velocity at a point in an air stream. A laser is used to provide the light source. Velocity is determined from the change in frequency of the radiation between the source and the a receiver due to the scattering of light by particles in the beam path. A sufficient number of particles that accurately track the airflow velocity must be present. The fluid must be transparent to permit light transmission.

Imaging systems store multiple images of particles in the flow stream in frames captured only a few

Fig. 6 Hot-film sensor. (Copyright TSI.)

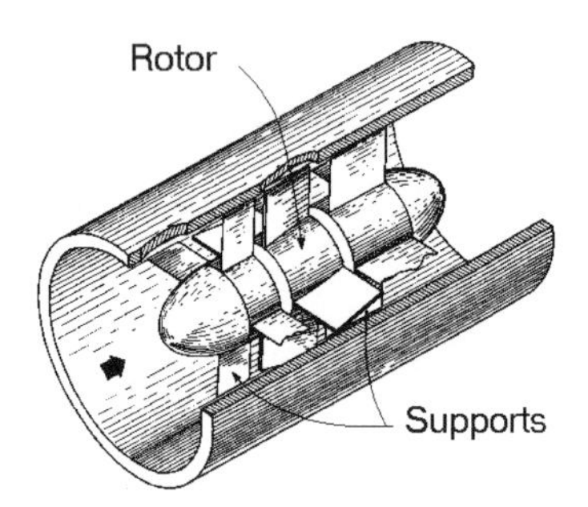

Fig. 7 Turbine flow meter. (From Galley, 1969; copyright ISA.)

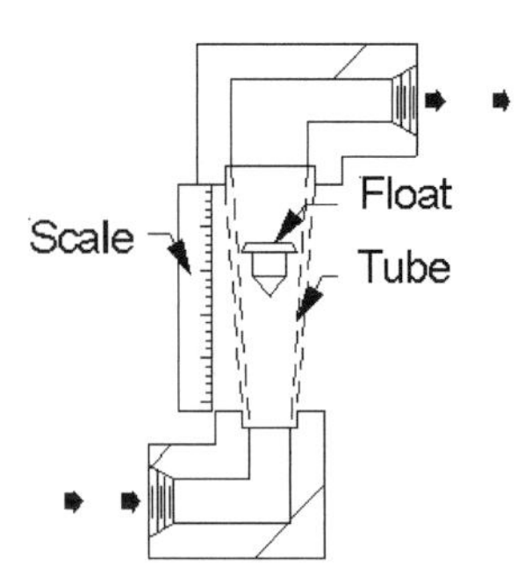

Fig. 8 Area meter. (From Galley, 1969; copyright ISA.)

microseconds apart. Then, special software is used to track particles from one frame to the next. In this way, the velocity of the particles in the flow stream can be determined. Stereoscopic systems can measure 3-D flow fields.

Propeller anemometers use carefully shaped blades to provide a cosine response to wind from any angle. Generators or pulse-counting systems are used to measure propeller rotational speed. Three orthogonal propellers can measure the three components of atmospheric flow at one location. One propeller mounted on a two-axis vane can be used to measure the total wind vector.

Cup anemometers are used to measure wind speed and are usually used together with direction vanes that provide wind direction. Usually three or more cups are mounted on a horizontal support bracket and turned about vertical axes. Cups are made as light as possible for low speed measurements or very rugged for measuring wind in storms. Generators or pulse-counting systems are used to measure propeller rotational speed that is proportional to the wind speed.

There are many manufacturers for the many types of flow-measuring equipment. Many have web sites and can be located by a simple search. The Instrument Society of America (ISA) web site, www.isa.org, lists over 100 suppliers under the heading: Flowmeters, Air and Gas.

FURTHER READING

ASME. *Fluid meters: Their Theory and Application*, 5th Ed.; American Society of Mechanical Engineers: New York, 1959; 43–108.

Buzzard, B.; Koop, J.G.; Burgess, T.H. Fundamentals of Flow Measurement. *ISA Encyclopedia of Measurement and Control (EMC)*; ISA: Research Triangle Park, NC, 2000; Vol. EMC 13.01, 25–47.

Galley, R.L. Flow. In *ISA Transducer Compendium, Second Edition—Part 1*; Harvey, G., Ed.; IFI/Plemum: New York, 1969; 157–203.

Miller, R.W. *Flow Measurement Engineering Handbook*, 3rd Ed.; McGraw-Hill: New York, 1996; 5.1–6.54.

Ower, E.; Pankhurst, R.C. The Measurement of Air Flow, 5th Ed.; Pergamon Press: New York, 148–257.

Anaerobic Reactions

Shang-Tian Yang
Ying Zhu
The Ohio State University, Columbus, Ohio, U.S.A.

A

INTRODUCTION

This article provides a brief review of the biochemistry of anaerobic micro-organisms and their applications in industrial fermentation. Anaerobic reactions are biological processes by which complex organic materials are decomposed partially in the absence of air to yield a variety of reduced organic compounds (e.g., ethanol, butanol, acetic acid, methane, sulfide), instead of complete aerobic combustion to CO_2 and H_2O. These reactions occur everywhere in nature, ranging from aquatic environments to the gastrointestinal tract of animals.[1–3] Anaerobic metabolism also plays an important role in the global sulfur[4,5] and nitrogen cycle.[6,7] In industry, anaerobic fermentations have been applied in wastewater treatment,[8–11] food processing including alcoholic beverages[12] and lactic acid-fermented foods,[13] and bulk chemical and fuel production.[14] Metabolic pathways are usually very complex in anaerobic micro-organisms. Within the total group of anaerobic micro-organisms, four general kinds of reactions are usually observed. They include alcohol and ketone production by solventogens, acid production by acidogens, methane production by methanogens, and hydrogen sulfide production by sulfate-reducing bacteria (SRB). The anaerobic metabolisms of nitrate-reducing bacteria and facultative anaerobic bacteria that are found in many anaerobic wastewater treatment processes are not discussed here.

PRODUCTION OF ALCOHOLS AND KETONES

Ethanol and 2,3-Butanediol Fermentations

Ethanol has received by far the most attention to date as a fuel and chemical produced by fermentation. Yeast (*Saccharomyces cerevisiae*) has traditionally been utilized to achieve such a conversion. Under anaerobic conditions, yeast metabolizes glucose to ethanol primarily by the Embden-Meyerhof pathway. The overall net reaction produces 2 mol each of ethanol, CO_2, and ATP per mol of glucose fermented. The metabolic reactions involved are as follows:

$$\text{Glucose} \rightarrow 2\,\text{Pyruvate} + 2\,\text{ATP} + 2\,\text{NADH}$$

$$\text{Pyruvate} \rightarrow \text{Acetaldehyde} + CO_2$$

$$\text{Acetaldehyde} \rightarrow \text{Ethanol}$$

A great number of bacteria are capable of ethanol formation. Those bacteria producing ethanol as the main fermentation product are listed in Table 1. In recent years, the gram-negative bacterium *Zymomonas mobilis* has attracted increased interest as a strict ethanol producer. It is one of the few facultative anaerobic bacteria, which metabolize glucose and fructose via the Entner-Doudoroff pathway, which is usually present in aerobic micro-organisms.[15] Glucose is phosphorylated and then oxidized to 6-phosphogluconate. Then 2-keto-3-deoxy-6-phosphogluconate (KDPG) is formed after dehydration and further cleaved by KDPG-aldolase to yield pyruvate. *Z. mobilis* has a higher ethanol production capability, which in the presence of oxygen gives a decreased ethanol yield and increased biomass production. The highest level of ethanol that *Z. mobilis* can tolerate is equivalent to or higher than that of *S. cerevisiae*.[16] Future progress in ethanol tolerance and range of possible substrates used in these fermentation systems can be made through genetic recombination techniques.

Most of ethanologenic micro-organisms, such as the "enteric" group of facultative anaerobic bacteria also produce other end products, including other alcohols (e.g., 2,3-butanediol), organic acids (e.g., acetic, lactic, and formic), and gases (H_2 and CO_2). They metabolize glucose by the Embden-Meyerhof pathway and then the phosphoenolpyruvate is further broken down to diverse products (Fig. 1). Several organisms are known to accumulate 2,3-butanediol in reasonable quantities. These organisms include *Klebsiella oxytoca*, *Aeromonas hydrophila*, *Bacillus polymyxa*, *B. subtilis*, *B. amyloliquefaciens*, and *Seratia marcescens*. They are gram-positive, facultatively anaerobic rods, and produce butanediol in different stereoisomers.

Encyclopedia of Agricultural, Food, and Biological Engineering
DOI: 10.1081/E-EAFE 120007198
Copyright © 2003 by Marcel Dekker, Inc. All rights reserved.

Table 1 List of bacterial species producing ethanol, lactate, acetate, or butyrate as a major fermentation product

Ethanol[15]	Lactate[19]	Acetate[21,22]	Butyrate[24]
Mesophilic	Homofermentative	*Clostridium*	*Clostridium*
Clostridium	*Lactobacillus*	*aceticum*	*butyricum*
sporogenes	*thermobacterium*	*formicoaceticum*	*tyrobutyricum*
indolis	*streptobacterium*	*thermoaceticum*	*beijerinckii*
sphenoides	*sporolactobacillus*	*thermoautotrophicum*	*pasteurianum*
sordelli		*magnum*	*barkeri*
Zymomonas	*Enterococcus*		*acetobutylicum*
mobilis	*Lactococcus*	*Acetobacterium*	*thermobutyricum*
mobilis ssp pomaceae	*Pediococcus*	*woodii*	*thermopalmarium*
Spirochaeta	*Streptococcus*	*carbinolicum*	
aurantia			*Butyribacterium*
stenostrepta	Heterofermentative	*Acetogenium kivui*	*methylotrophicum*
litoralis	*Lactobacillus betabacterium*	*Moorella thermoautotrophica*	*Pseudobutyrivibrio ruminis*
Erwinia amylovara	*Leuconostoc*		
Leuconostoc mesenteroides			
Streptococcus lactis			
Sarcina ventriculi			
Thermophilic			
Thermoanaerobacter ethanolicus			
Bacillus stearothermophilus			
Thermoanaerobium brockit			
Clostridium			
thermohydrosulfuricum			
thermosaccharolyticum			
thermocelllum			

Acetone and Butanol Fermentations

The acetone–butanol fermentations are of great industrial significance. It became in volume, the second largest fermentation process in the world in the first half of last century, only exceeded by ethanol fermentation of yeast, and up until the 1960s was able to compete successfully with synthetic processes.

Solvents such as acetone and *n*-butanol are natural products of only a few bacterial species. Most of them belong to the genus *Clostridium*. However, *Butyribacterium methylotrophicum* and *Hyperthermus butylicus* also produce *n*-butanol as a major product.[17] The most widely known producer of acetone and butanol in industry probably is *C. acetobutylicum* because it produces and tolerates the highest solvent concentration. In addition, the species *C. aurantibutyricum*, *C. beijerinckii*, *C. butyricum*, *C. cadaveris*, *C. chauvoei*, *C. felsineum*, *C. pasteurianum*, *C. puniceum*, *C. roseum*, *C. sporogenes*, *C. tetani*, *C. tetanomorphum*, and *C. thermosaccharolyticum* are also butanol producers. All butanol-producing anaerobes display a mixed-solvent fermentation including butanol, acetone, ethanol, and isopropanol. Solvent yields are dependent on type and concentration of substrates and the specific strain utilized.

Clostridium species generally employ the Embden-Meyerhof pathway for hexose metabolism and multiple end products are formed (Fig. 1). In addition to the two main products acetone and butanol, major byproducts including ethanol, acetate, and butyrate, and small amounts of isopropanol and lactate can also be produced. Most anaerobic bacteria employ the Warburg-Dickerns pathway for pentose metabolism (i.e., xylolysis). In this pathway, pentoses are converted into ribose-5-phosphate and xylulose-5-phosphate, which are fed into the pentose phosphate cycle, the resulting hexose phosphates are subsequently catabolized by the Embden-Meyerhof pathway.

During exponential growth of *C. acetobutylicum* on a variety of substrates when pH is greater than 5.6, the major fermentation products are butyrate, acetate, CO_2, and H_2. Lactate production is detected under very slow growth conditions. This phase is so called acidogenesis. As acids accumulate in a batch culture, growth becomes linear and gradually stops when the pH goes down to 4.5–4.0, a shift from acidogenesis to solventogenesis occurs, and solvent

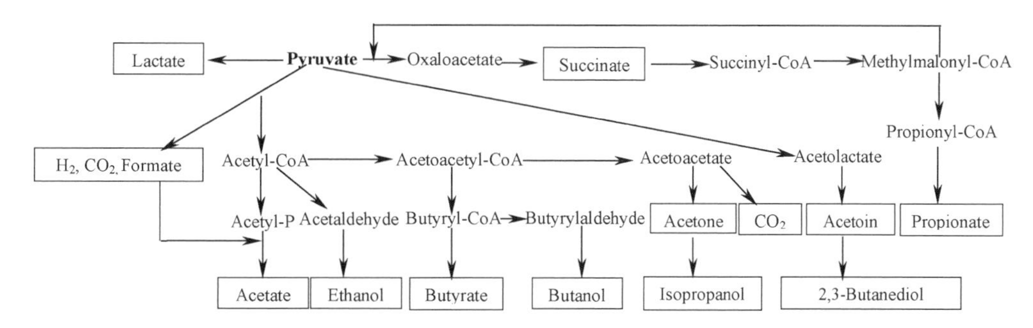

Fig. 1 Generalized metabolic pathways for pyruvate degradation to various anaerobic fermentation products.

formation is triggered. Solvent formation requires the induction of new enzyme pathways in these cells catalyzing the formation of acetone, butanol, and ethanol.[18] Physiological control of solvent production in acetone–butanol fermentation is very complex. Enormous progress has been made in elucidation of gene structure, mRNA regulation, and intracellular, molecular signaling systems in *C. acetobutylicum* fermentation.[18] In addition, various research activities are devoted to the improvement of the micro-organisms. The inhibitory effect of solvents on cell growth contributes greatly to the shut-down of the fermentation. Chemical mutagenesis has been used to obtain butanol-tolerant strains. Genetic manipulations to alter product formation (e.g., less acetone production) and improve substrate utilization (e.g., broader substrate range) are now available and might lead to the development of this fermentation at an industrial scale.

PRODUCTION OF ACIDS

Many organic acids are produced by anaerobic micro-organisms in sufficient yields, which can be manufactured economically by fermentation. Certain species can produce lactic or acetic acid with nearly 100% of substrate-to-product conversion yield. These organic acids have a variety of applications ranging from large-volume industrial chemicals (i.e., acetic acid) to food chemicals (i.e., lactic acid and propionic acid) or feedstocks for production of other chemicals (i.e., succinic acid and butyric acid).

Lactic Acid

Lactic acid occurs widely in nature, being found in man, animals, plants, and micro-organisms. Commercial manufacture of lactic acid by lactic acid bacterial fermentation of sugars has been carried out for more than a century. Lactic acid bacteria are facultative anaerobic or microaerophilic gram-positive bacteria.

Lactic acid bacteria can be classified as homofermentative (produce lactic acid only) and heterofermentative (produce lactic acid and other metabolic products such as acetic acid, ethanol, and CO_2) (Table 1). Homofermentative lactic acid bacteria utilize the Embden-Meyerhof-Parnas pathway of glucose metabolism to produce lactic acid as the only end product. No gas is generated. Heterofermentative bacteria use pathways of pentose metabolism because they do not have aldolase, one of the key enzymes in glycolysis. Instead, the presence of phosphoketolase in pentose metabolism results in lactic acid and other products such as acetic acid, ethanol, and CO_2 (Fig. 1). The reactions for these two different kinds of bacteria are summarized as follows:

$$\text{Homofermentative: Glucose} \rightarrow 2\,\text{Lactate}$$

$$\text{Heterofermentative: Glucose} \rightarrow \text{Lactate} + \text{Ethanol} + CO_2$$

The homofermentative ones are of greatest interest for commercial lactic acid production for economic reasons. They generally give $> 90\%$ of the theoretical yield in fermentation. However, cultivation conditions can influence the activity of enzymes involved in the metabolism and control the yield of lactic acid. For certain strains, homolactic acid fermentation can be converted to heterolactic acid fermentation under glucose-limited condition. Similarly, heterofermentative behavior may change to homofermentative at high glucose concentration.[14,20]

Acetic Acid

Homoacetogenic fermentation is of interest from a biotechnological standpoint because nearly all the substrate carbon can be recovered in the product acetate without any byproduct, as compared with 67% of theoretical yield in aerobic vinegar process. Many anaerobic bacteria can convert carbohydrates to acetic acid via homofermentation (Table 1). The best characterized homoacetogenic bacteria are the gram-positive spore-forming bacteria *C. formicoaceticum* and *C. thermoaceticum*, and the gram-negative nonspore-forming *Acetobacterium woodii*. None of these species ferment biomass polymers, but the homoacetogens can form acetic acid from a variety of hexoses, pentoses, and lactic acid. In addition, some species can grow on formic acid and H_2/CO_2.

Homoacetogenic fermentation has a theoretical yield of 3 mol of acetate per mol of metabolized glucose. Homoacetogens ferment glucose via glycolysis converting glucose to 2 mol of pyruvate. From this point, two molecules of acetate are produced. The third acetate comes from the reduction of two molecules of CO_2, using the four electrons generated from glycolysis plus the four electrons produced during oxidation of two pyruvates to two acetates. This reaction is so called acetyl-CoA pathway, in which CO_2 is utilized as a terminal electron acceptor.[23] The acetogenic Clostridia are very pH sensitive. At pH 5.0 or below, acetic acid functions as an uncoupler, and growth and acetogenesis is no longer possible.[18] Therefore, the homoacetic fermentation can be used to produce acetate only if pH is properly controlled.

Butyric Acid

There are several bacterial strains producing butyric acid, which belong to the genera *Clostridium*, *Butyribacterium*, *Butyrivibrio*, *Sarcina*, *Eubacterium*, *Fusobacterium*, and *Megasphera*. The genera *Clostridium*, *Butyribacterium*, and *Butyrivibrio* are the mostly used micro-organisms (Table 1). The preferred strains are in the genus *Clostridium*, regarding potential commercial uses. They are gram-positive, chemo-organotrophic, strict anaerobes, and spore formers. Most butyric acid-producing bacteria ferment glucose, hexose, pentose, and oligo- and polysaccharides, and form acetic acid in addition to butyric acid as their major fermentation products. Several species produce additional products such as ethanol and lactate, but *C. butyricum* and *C. tyrobutyricum* are purely acidogenic, and they are of interest for the production of butyrate. *Butyribacterium methylotrophicum* is very interesting because it can ferment methanol in addition to hexose, lactic acid, and H_2/CO_2. In batch culture, butyric acid is the only product during metabolism of methanol by *B. methylotrophicum*, whereas acetate is the major product on H_2/CO_2.[14]

The Embden-Meyerhof pathway of sugar breakdown in these micro-organisms is similar to the one used by homoacetogens or solvent producers. These bacteria do not contain alcohol dehydrogenase and acetaldehyde dehydrogenase so that only acids are produced. The maximal theoretical yields of 1 mol of butyrate and 2 mol of hydrogen and CO_2 are obtainable via the butyric acid fermentation pathway. Acetate is also formed as an intermediate, and is required in high concentrations to drive the butyryl CoA–acetyl-CoA transfer reaction involved in butyric acid synthesis (Fig. 1). Butyric acid fermentation is controlled at the level of pyridine nucleotides and by intermediary product and proton concentrations. Under a low pressure of H_2, the butyrate/acetate ratio decreases and more acetate is produced, accompanied by an increase in the ATP yield.

Propionic Acid

Homoacidogenic fermentation for propionic acid and succinic acid has not been reported. Propionibacteria, such as *Propionibacterium freudenreichii ss. shermanii* and *P. acidipropionici* are the most commonly used bacteria in propionic acid fermentation. They are gram-positive, nonspore-forming, and anaerobic or facultative bacteria. Propionibacteria can ferment lactic acid, carbohydrates, and polyhydroxy alcohols, producing propionic acid, succinic acid, acetic acid, and CO_2. Their nutritional requirements are complex, and they usually grow rather slowly.

The enzymatic reactions in the dicarboxylic pathway yield 2 mol of propionic acid and 1 mol of acetic acid and CO_2 per 1.5 mol of glucose fermented (Fig. 1):

$$1.5\,\text{Glucose} \rightarrow 2\,\text{Propionate} + \text{Acetate} + CO_2$$

The carbon flow in this pathway is unique. Pyruvate accepts a carboxyl group from methylmalonyl-CoA by the transcarboxylase reaction, leading to the formation of oxalacetate and propionyl-CoA. The latter reacts with succinate, producing succinyl-CoA and propionate. The succinyl-CoA is then isomerized to methylmalonyl-CoA, and the cycle is complete.[23]

Another bacterium, *Propionigenium*, also produces propionic acid fermentatively. It is a gram-negative, strictly anaerobic bacterium that ferments succinate to propionate and CO_2:

$$\text{Succinate} \rightarrow \text{Propionate} + CO_2$$

In the fermentation, Na^+ plays an important role in establishing an ion gradient that ultimately drives ATP synthesis. It can also grow on fumarate, malate, aspartate, oxalacetate, and pyruvate, but does not ferment sugars.[23]

Succinic Acid

Succinic acid is a common intermediate in the metabolic pathway of several anaerobic and facultative micro-organisms. For example, succinate is formed from carbohydrates or amino acids by rumen bacteria such as *Fibrobacter succinogenes* (previously named *Bacteroides succinogenes*), *Ruminococcus flavefaciens*, *Ruminobacter amylophilus*, *Succinivibrio dextrinisolvens*, *Succinimonas amylolytica*, *Prevotella ruminocola*, *Wolinella succinogenes*, and *Cytophaga succinican*, and nonrumen bacteria such as *Anaerobiospirillum succiniproducens*, *Clostridium* species, *Propionibacterium*, *Lactobacilli*, and *Bacteroides*.[25,26] A few anaerobic fungi *Neocallimastix* and yeast are able to produce succinate as well. Succinic acid bacteria such as *F. succinogenes* and *R. flavefaciens* are the only cellulose-fermenting species that produce high yields of acids (i.e., succinate, acetate, and formate). A newly discovered rumen organism *Actinobacillus succinogenes* also produces succinate in very high concentrations.[26] The biochemical pathways for succinate production by rumen bacteria are quite similar. It appears that equal molar quantities of succinate, acetate, and formate are formed per mole of glucose and CO_2 fermented (Fig. 1):

$$\text{Glucose} + CO_2 \rightarrow \text{Succinate} + \text{Acetate} + \text{Formate}$$

SULFATE-REDUCING BACTERIA (SRB)

SRB are obligate anaerobes, employing a respiratory mechanism with sulfate as the terminal electron acceptor for the dissimilation of organic compounds, which consequently gives rise to hydrogen sulfide as the major metabolic end product[5]:

$$SO_4^{2-} + \text{organic compound} \rightarrow HS^- + H_2O + HCO_3^-$$

Most SRB can also utilize other oxidized sulfur compounds as terminal electron acceptor, such as sulfate or thiosulfate.[5] Many other reductants have been recognized in certain species, such as nitrate[27] and metal ferric.[28]

A relatively wide range of genera of SRB has been identified. The two well-known genera of SRB are spore-forming *Desulfotomaculum* containing straight or curved rods and the nonsporing genus *Desulfovibrio* containing curved, motile vibrios or rods.[29] The type of carbon source utilized for the reduction of electron acceptor varies from organic acids (e.g., acetate, lactate, pyruvate, formate, and malate) to alcohols (e.g., ethanol, propanol, methanol, and butanol). The individual metabolic reactions were summarized by Gibson[29] and detailed carbon and energy metabolism of SRB has been studied by Hansen.[30]

Sulfate reduction by SRB is important not only in the fulfillment of sulfur cycle in nature, but also in anaerobic degradation in wastewater treatment systems to remove sulfate and heavy metals. Furthermore, SRB are of considerable technological interest because of their role in anaerobic corrosion, marine and estuarine pollution, and hydrocarbon reservoir souring.[31,32]

METHANE PRODUCTION

Methane production is carried out by a group of Archaea, the methanogens, which are obligate anaerobes

Table 2 Substrates and reactions of methanogenesis by various methanogenic Archaea

Substrates	Reactions	Genus
CO_2-type		
CO_2	$CO_2 + 4H_2 \rightarrow CH_4 + 2H_2O$	*Methanobacterium, Methanolbrevibacter, Methanothermus, Methanococcus, Methanomicrobium, Methanogenium, Methanospirillum, Methanoplanus, Methanosarcina, Methanoculleus, Methanopyrus, Methanocorpusculum*
Formate	$4HCOO^- + 4H^+ \rightarrow CH_4 + 3CO_2 + 2H_2O$	
CO	$4CO + 2H_2O \rightarrow CH_4 + 3CO_2$	
Methyl		
Methanol	$4CH_3OH \rightarrow 3CH_4 + CO_2 + 2H_2O$	*Methanosphaera, Methanosarcina, Methanoobus, Methanoculleus, Methanohalobium, Methanococcoides, Methanohalophilus, Methanocorpusculum*
	$CH_3OH + H_2 \rightarrow CH_4 + H_2O$	
Methylamine	$4CH_3NH_3Cl + 2H_2O \rightarrow 3CH_4 + CO_2 + 4NH_4Cl$	
Dimethylamine	$2(CH_3)_2NH_2Cl + 2H_2O \rightarrow 3CH_4 + CO_2 + 2NH_4Cl$	
Trimethylamine	$4(CH_3)_3NHCl + 6H_2O \rightarrow 9CH_4 + 3CO_2 + 4NH_4Cl$	
Methylmercaptan	$4CH_3SH + 2H_2O \rightarrow 3CH_4 + CO_2 + 4H_2S$	
Dimethylsulfide	$2(CH_3)_2S + 2H_2O \rightarrow 3CH_4 + CO_2 + 2H_2S$	
Acetate	$CH_3COO^- + H_2O \rightarrow CH_4 + HCO_3^-$	*Methanosarcina, Methanothrix (Methanosaeta), Methanococcus*

(From Ref. 34.)

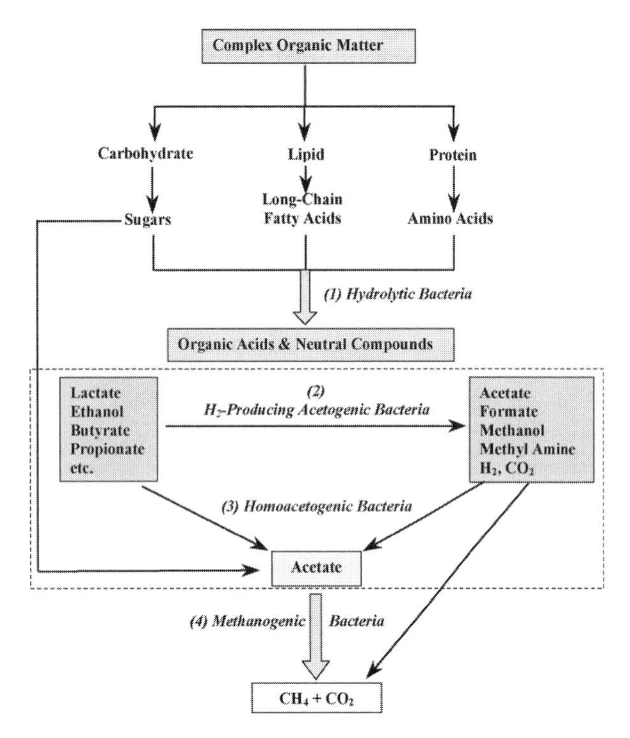

Fig. 2 Overall process of anaerobic digestion for methane production from organic matter. (From Ref. 35.)

and share the ability to make methane as a means of obtaining energy.[33] A variety of morphological types of methanogenic bacteria have been isolated and contain a total of 18 genera.[34] Methanogens catabolize a very restricted range of substrates that include a variety of CO_2-type substrates, methyl substrates, and acetotrophic substrates (Table 2).

In anaerobic methanogenic fermentation, various organisms and biological pathways are involved. High molecular weight substances such as polysaccharides, proteins, and fats, are converted to CH_4 by cooperative interactions of several physiological groups of prokaryotes. In anaerobic digestion of a typical organic matter, as many as four major physiological groups of fermentative anaerobes may act together in the conversion of complex polymer ultimately to methane and CO_2 (Fig. 2). At present, methane fermentations are economically feasible for energy recovery and treatment of concentrated organic materials in industrial wastes, municipal solids and sludge, and animal manure. Its application in agricultural residues has also been addressed.[36]

CONCLUSION

Anaerobic bacteria have potential for industrial scale chemical production because they offer the promise of simple efficient processes not common to previous aerobic industrial fermentations.[14] Because of the wide range of products formed and substrates fermented, high substrate-to-product conversion yields and production rates, and the potential for enhanced process stability and product recovery by use of thermophilic organisms, anaerobic fermentations are of interest to develop new biotechnology.

REFERENCES

1. Lin, K.W.; Patterson, J.A.; Ladisch, M.R. Anaerobic Fermentation: Microbes from Ruminants. Enzyme Microb. Technol. **1985**, *3*, 98–107.
2. Wallace, R.J. Rumen Microbiology, Biotechnology and Ruminant Nutrition: The Application of Research Findings to a Complex Microbial Ecosystem. FEMS Microbiol. Lett. **1992**, *100*, 529–534.
3. Wallace, R.J. Rumen Microbiology, Biotechnology and Ruminant Nutrition: Progress and Problems. J. Anim. Sci. **1994**, *72*, 2992–3003.
4. Hines, M.E.; Visscher, P.T.; Devereux, R. Sulfur Cycling. In *Manual of Environmental Microbiology*; Hurst, C.J., Knudsen, G.R., McInerney, M.J., Stetzenbach, L.D., Walter, M.V., Eds.; American Society for Microbiology: Washington, D.C., 1996; 324–333.
5. Hao, O.J.; Chen, J.M.; Huang, L.; Buglass, R.L. Sulfate-Reducing Bacteria. Crit. Rev. Environ. Sci. Technol. **1996**, *26* (1), 155–187.
6. Capone, D.G. Microbial Nitrogen Cycling. In *Manual of Environmental Microbiology*; Hurst, C.J., Knudsen, G.R., McInerney, M.J., Stetzenbach, L.D., Walter, M.V., Eds.; American Society for Microbiology: Washington, D.C., 1996; 334–342.
7. Ye, R.W.; Thomas, S.M. Microbial Nitrogen Cycles: Physiology, Genomics and Applications. Curr. Opin. Microbiol. **2001**, *4*, 307–312.
8. de Beer, D.; Schramm, A.; Santegoeds, C.M.; Nielsen, H.K. Anaerobic Processes in Activated Sludge. Water Sci. Technol. **1998**, *37* (4–5), 605–608.
9. Lettinga, G.; Field, J.; van Lier, J.; Zeeman, G.; Hulshoff Pol, L.W. Advanced Anaerobic Wastewater Treatment in the Near Future. Water Sci. Technol. **1997**, *35* (10), 5–12.
10. Pavlostathis, S.G.; Beydilli, I.; Misra, G.; Prytula, M.; Yeh, D. Anaerobic Processes. Water Environ. Res. **1997**, *69* (4), 500–521.
11. Riffat, R.; Dararat, S.; Krongthamchat, K. Anaerobic Processes. Water Environ. Res. **1999**, *71* (5), 656–676.
12. Rose, A.H. History and Scientific Basis of Alcoholic Beverage Production. In *Alcoholic Beverages*; Rose, A.H., Ed.; Academic Press: New York, 1977; 1–41.
13. Steinkraus, K.H. Lactic Acid Fermentation in the Production of Foods from Vegetables, Cereals and Legumes. Antonie Leeuwenhoek **1983**, *49* (3), 337–348.

14. Zeikus, J.G. Chemical and Fuel Production by Anaerobic Bacteria. Annu. Rev. Microbiol. **1980**, *34*, 423–464.
15. Kosaric, N. Ethanol—Potential Source of Energy and Chemical Products. In *Biotechnology*, 2nd Ed.; Rehm, H.-J., Reed, G., Pühler, A., Stadler, P., Eds.; VCH Verlagsgesellschaft mbH: Weinheim, Federal Republic of Germany, 1996; Vol. 6, 121–203.
16. Gunasekaran, P.; Chandra Raj, K. Ethanol Fermentation Technology—*Zymomonas mobilis*. Curr. Sci. **1999**, *77* (1), 56–68.
17. Dürre, P.; Bahl, H. Microbial Production of Acetone/Butanol/Isopropanol. In *Biotechnology*, 2nd Ed.; Rehm, H.-J., Reed, G., Pühler, A., Stadler, P., Eds.; VCH Verlagsgesellschaft mbH: Weinheim, Federal Republic of Germany, 1996; Vol. 6, 230–268.
18. Rogers, P.; Gottschalk, G. Biochemistry and Regulation of Acid and Solvent Production in Clostridia. In *The Clostridia and Biotechnology*; Woods, D.R., Ed.; Butterworth-Heinemann: Stoneham, MA, 1993; 25–50.
19. Litchfield, J.H. Lactic Acid, Microbially Produced. In *Encyclopedia of Microbiology*, 2nd Ed.; Lederberg, J., Alexander, M., Bloom, B., Eds.; Academic Press: San Diego, CA, 2000; Vol. 3, 9–17.
20. Kašcák, J.S.; Komínek, J.; Roehr, M. Lactic Acid. In *Biotechnology*, 2nd Ed.; Rehm, H.-J., Reed, G., Pühler, A., Stadler, P., Eds.; VCH Verlagsgesellschaft mbH: Weinheim, Federal Republic of Germany, 1996; Vol. 6, 293–306.
21. Yang, S.-T.; Huang, Y.L.; Jin, Z.; Huang, Y.; Zhu, H.; Qin, W. *Calcium Magnesium Acetate at Lower-Production Cost: Production of CMA Deicer from Cheese Whey*; FHWA-RD-98-174; U.S. Department of Transportation: Washington, D.C., 1999; 9–10.
22. Talabardon, M.; Schwitzguébel, J.-P.; Péringer, P. Anae robic Thermophilic Fermentation for Acetic Acid Production from Milk Permeate. J. Biotechnol. **2000**, *76*, 83–92.
23. Madigan, M.T.; Martinko, J.M.; Parker, J. Prokaryotic Diversity: Bacteria. *Brock Biology of Microorganisms*, 8th Ed.; Prentice-Hall, Inc.: Upper Saddle River, NJ, 1997; 635–740.
24. Zigová, J.; Šturdlk, E. Advances in Biotechnological Production of Butyric Acid. J. Ind. Microbiol. Biotechnol. **2000**, *24*, 153–160.
25. Gokarn, R.R.; Eiteman, M.A.; Sridhar, J. Production of Succinate by Anaerobic Microorganisms. In *ACS Symposium Series 666, Fuels and Chemicals from Biomass*; Saha, B.C., Woodward, J., Eds.; American Chemical Society: Washington, D.C., 1997; 237–263.
26. Zeikus, J.G.; Jain, M.K.; Elankovan, P. Biotechnology of Succinic Acid Production and Markets for Derived Industrial Products. Appl. Microbiol. Biotechnol. **1999**, *51*, 545–552.
27. Moura, I.; Bursakov, S.; Costa, C.; Moura, J.J.G. Nitrate and Nitrite Utilization in Sulfate-Reducing Bacteria. Anaerobe **1997**, *3*, 279–290.
28. Lovley, D.R. Microbial Reduction of Iron, Manganese, and Other Metals. Adv. Agron. **1995**, *54*, 175–231.
29. Gibson, G.R. Physiology and Ecology of the Sulfate-Reducing Bacteria. J. Appl. Bacteriol. **1990**, *69*, 769–797.
30. Hansen, T.A. Metabolism of Sulfate-Reducing Prokaryotes. Antonie Leeuwenhoek **1994**, *66*, 165–185.
31. Hamilton, W.A. Sulphate-Reducing Bacteria and Anaerobic Corrosion. Annu. Rev. Microbiol. **1985**, *39*, 195–217.
32. Hamilton, W.A. Bioenergetics of Sulphate-Reducing Bacteria in Relation to Their Environmental Impact. Biodegradation **1998**, *9*, 201–212.
33. Daniels, L. Biotechnological Potential of Methanogens. Biochem. Soc. Symp. **1992**, *58*, 181–193.
34. Madigan, M.T.; Martinko, J.M.; Parker, J. Prokaryotic Diversity: Archaea. *Brock Biology of Microorganisms*, 8th Ed.; Prentice-Hill, Inc.: Upper Saddle River, NJ, 1997; 741–768.
35. Yang, S.-T. Kinetic Study of Defined Methane Fermentation of Lactose. Ph.D. Thesis, Purdue University, West Lafayette, IN, 1984; 13.
36. Hashimoto, A.G.; Chen, Y.R.; Varel, V.H.; Prior, R.L. Anaerobic Fermentation of Agricultural Residue. In *Utilization and Recycle of Agricultural Wastes and Residues*; Shuler, M.L., Ed.; CRC Press, Inc.: Boca Raton, FL, 1980; 136–196.

Aquacultural Harvesting Systems

Fred Wheaton
University of Maryland, College Park, Maryland, U.S.A.

INTRODUCTION

Most people view aquatic organisms as finfish, shellfish, or aquatic plants. However, there are at least 20,000 species of finfish ranging from sharks and tuna fish to tropical fish used in home aquariums. Shellfish have a similar diversity; crabs are very different from clams. Aquatic plants range from single celled phytoplankton to kelp that may be 50 or more feet in length. Obviously, harvesting techniques for these widely diverse organisms must also vary. Thus, harvesting techniques discussed in this article will concentrate on aquaculture crops of economic importance. Given the space limitations for this entry, only the more commonly used harvesting techniques will be discussed. No attempt will be made to be comprehensive in coverage and most harvesting systems for juveniles fish will be neglected.

DIP NETS

Dip nets are simple units comprised of a handle of appropriate length with a hoop attached at one end (Fig. 1). The hoop is made of rigid material such as steel so the opening will always stay open. A net or other mesh material is attached to the hoop such that it extends below the hoop and is formed into a closed bag. The dip net is dipped into the water where fish are crowded together and the fish enter the net. As the hoop is raised above the water surface, the water drains out through the mesh material leaving the fish trapped in the dip net.

PUMPS

Pumps are one of the most efficient methods of harvesting fish. There are several manufactures that have developed and commercially sell fish pumps. There are several types of pumps used including partial vacuum, centrifugal, a variation of Archimedes Screw, and several types of elevator systems. Pump capacity depends on the species of fish, fish size, lift needed, and other variables. Pumps tend to have the highest capacity (although capacity is a function of pump size and operating conditions) of the various harvesting systems, and use of a pump minimizes the labor requirement for harvesting.

Centrifugal pumps (Fig. 2) with large internal clearances and rounded internal components (to reduce damage to the fish) have been used.[1] The pump must be sized for the fish size and species to be pumped. Fish up to at least 40 cm in length have been pumped with these pumps without significant damage. Capacities are available up to several 1000 $kg\,hr^{-1}$. These pumps generally are not self-priming, and thus once started are best allowed to continuously pump water with some mechanism on the intake side of the pump to prevent fish from entering the pump except when the operator wishes to pump fish. Because these pumps are not self-priming it is best, if possible, to operate them with a positive inlet pressure. Positive inlet pressure allows the pumps to self-prime. To operate these pumps, a flexible inlet pipe is placed into the tank or pond where the fish are crowded together. The pump is primed and allowed to pump water for a short time. The fish exclusion device is opened allowing fish to be sucked into the pump inlet. The fish and water pass through the pump and out of the discharge to a grader, transport vehicle, or other receiving device. The lift of these pumps is limited by the pressure change the fish can tolerate as they move through the pump.

Another type of pump uses a partial vacuum to pull fish from the source into a tank in the pump. An automated valve first seals the outlet, and a partial vacuum is drawn in the tank, which is part of the pump. Water and fish are drawn into the pump tank by the partial vacuum. A series of automated valves then closes the inlet to the pump and opens the pump discharge. The water and fish drain from the pump tank by gravity or the pump tank is pressurized forcing the water and fish out of the pump. This cycle is repeated and the systems functions as a fish pump that does little or no damage to the fish. The fish and water can be separated using a mesh that allows passage of the water but not the fish, or the fish and water can be discharged into another tank.

Another type of fish pump (Fig. 3) is based on Archimedes screw. This pump consists of an auger that fits snugly inside a semi-rigid tube. The lower end of the tube and auger are placed into a tank or pond where fish have been crowded together. Rotation of the auger causes water and fish to be trapped by the auger flights. Because

Encyclopedia of Agricultural, Food, and Biological Engineering
DOI: 10.1081/E-EAFE 120006879
Copyright © 2003 by Marcel Dekker, Inc. All rights reserved.

Fig. 1 Dip net.

Fig. 2 Centrifugal type fish pump. (Courtesy Dr. Andrew Lazur.)

Fig. 3 Fish pump using Archimedes principle. (Courtesy Dr. Andy Lazur.)

the auger is angled upward, auger rotation causes the fish and water to travel upward as they move along the length of the tube. By sloping the tube upward, the water and fish are lifted (i.e. pumped) upward. These units have a limited lift, primarily because of the mechanical strength requirements of the unit, but they handle fish gently.

Various kinds of conveyors are also used as fish pumping devices. However, these units are essentially flat conveyors with flat plates attached at right angles to the conveyor belt or chain. The flat plates stick up and act as paddles to move the fish along with the conveyor belt. The lower end of the conveyor is placed in the tank or pond where the fish have been crowded together and as the conveyor moves along it traps fish with the paddles and carries them along the conveyor belt. By setting the conveyor at an angle, the fish can be conveyed from a pond up and into a transport vehicle. Although these systems are wet they do not convey the fish in water. Thus, they tend to cause more damage to the fish.

SEINING

Pond culturists use seines to harvest fish from ponds (Fig. 4). The seine may be constructed from a variety of materials ranging from bamboo stems formed into a long flexible wall such as used in many parts of the far East to plastic rope seines commonly used in the United States. The seine is placed across one of the narrower ends of a pond, usually the deeper end, and then pulled along the length of the pond. The seine mesh is chosen to accomplish the desired function. If all of the fish are to be removed from the pond, the mesh size is relatively small so even the small fish will be trapped. If it is desirable to remove only the larger fish the mesh size is larger, just large enough to contain the smallest fish the culturist wishes to harvest. As the seine is pulled through the pond the fish are forced to

Fig. 4 Pond seining. (Courtesy Don Webster.)

move ahead of the seine. In some cases, the seine is used to pull the fish out of the water and haul them up on the bank of the pond where they are picked up by hand. In other cases, the primary seine is used to crowd the fish into a small area, and they are removed using dip nets or pumps or they are placed in live cars, a net that attaches to the back of the primary seine and allows the fish to move into it. In large ponds, several live cars may be used to contain the fish from the pond. As each live car becomes filled it is sealed off, removed from the seine and another live car is attached to the seine. The fish become confined in the live cars that take up a small proportion of the pond. Dip nets or other methods are then used to remove the fish from the live cars or from the primary seine. When dip nets are used the fish are dipped out of the seine or live car and placed into the transport vehicle. In small ponds, manual dip netting is used while in large ponds hydraulically or mechanically powered dip nets may be used.

CROWDING GATES

Crowding gates are used in raceway and tank culture to force fish into a smaller volume and make them easier to remove from the growing systems. Crowding gates are manufactured from some type of mesh material, or a series of bars spaced slightly apart, of sufficient opening size that the fish of a desired size and larger cannot pass through the net but the water can. The crowding device is designed to completely fill the cross section of a raceway or the radius of a tank such that the fish cannot swim around the crowding gate. As the crowding gate is moved through the water, usually manually, the large fish are forced to crowd together ahead of the crowding gate. In a raceway or large rectangular tank, the crowding gate is moved linearly through the raceway causing the fish to be concentrated in one end of the raceway. In a round tank, two crowding gates are used, one is placed in a stationary position across the tank radius while the other is moved around the tank causing the fish to be concentrated between the stationary and moving crowding gates. The crowding gate only concentrates the fish in a small volume of water and makes it easier to harvest them with a dip net or pump.

Crowding gates are usually manufactured to fit a specific system because the gates have to fit the entire width of a raceway or radius of a tank. The mesh size or the bar spacing on the crowding gates are selected to fit the fish size the culturist wishes to harvest. If a complete harvest is desired, the mesh size or bar spacing must be set small enough to prevent the smallest fish from passing through the gate. Larger mesh sizes or bar spacing will allow partial harvest as the small fish can pass through the crowding gate.

DRAINING

One of the simplest harvesting techniques is to drain the water out of the fish production structures. Pond draining must be slow and uniform enough to allow the fish to move with the falling level of water so they do not get caught in small pool caused by bottom irregularities.[2] Draining concentrates the fish, but the actual capturing of the fish can be done in a variety of ways. The simplest way is to wait until the culture facility is completely drained and then pick up the dead fish from the bottom. This will not produce high quality fish in most cases as the fish die slowly and struggle for considerable time. This can lead to bruising and/or changes in the fish muscle due to stress. A better harvesting method is to wait until the water gets down to the harvesting level and then dip the fish out of the confined area. Oxygen concentration in the harvesting area must be maintained by aeration or the flow of fresh water. If the topography of the area permits, the fish can be allowed to exit the pond with the water. By passing the water and fish through a mesh material the fish can be separated out and the water allowed to escape. Although this latter method is efficient it requires all of the water in a pond to be dumped, a practice that usually is not very environmentally friendly.

Draining is an effective method of concentrating fish in tank culture systems, particularly if the water can be stored in another tank while harvesting is occurring. Tank harvesting is often accomplished by partially draining the tank to concentrate the fish, and then removing the fish by dip net or pump. In high intensity systems, the fish are already highly concentrated and draining is only necessary to remove the last group of fish.

TRAPS

Traps are widely used to capture crabs, finfish, and many other species, especially in wild fisheries. In aquaculture systems, traps are used for harvesting finfish, some crustaceans (e.g., shrimp and crawfish), and other species.[2] Traps are devices that lead fish or another species into a maize or some device that is easy to enter but very difficult to get out of when following the same path by which they came in. The trap is set in a pond or other body of water where the species is grown and the crop is enticed (by bait or some other means) to enter the trap. Once in the trap the animal cannot find their way back out. The trap is periodically lifted out of the water and the trapped animals are removed by hand. Traps are not widely used in aquacultural applications because there are more efficient harvesting methods.

DREDGES

Dredges are widely used to harvest shellfish such as oysters, clams, and scallops. There are two principle types of dredges: hydraulic and mechanical. Mechanical dredges (Fig. 5) vary widely in size and design depending on the species, size of the power unit, and legal limitations. Mechanical dredges are typically made of a steel rectangle that is the desired width and anywhere from 15 cm to 45 cm high. In operation the rectangle is in a vertical plane. The bottom bar of the rectangle may have teeth 10 cm–15 cm in length attached to it so they can pull the clams or other organisms out of the bottom. A bag is attached to one side of the rectangle to catch the crop as it is forced through the rectangular frame. Rope or chains are attached to the other side of the rectangle. These are usually extended to a boat or other power source and used to pull the dredge along the bottom.

Fig. 5 Oyster dredge. (Courtesy Don Merritt.)

Hydraulic dredges may be a suction type or they may be similar to a soft clam escalator dredge used in Maryland. Suction dredges, as their name implies, work like a household vacuum cleaner except they draw in water instead of air. Suction is generated using a pump, either on the suction side of a pump or by pumping water through a venturi that creates the suction. The inlet of the suction dredge is passed over the harvest area and the rapid inflow of water entrains the desired crop, as well as mud and other debris. The crop (shellfish) are separated from the water using screens or mesh bags. The water passes through the screens and the crop is retained on the screen. The water is returned to the source and the crop is harvested.

Escalator dredges consist of a conveyor, a large pump, a power source, and a flotation device (i.e., boat). The top end of the conveyor is attached to the flotation device so the top end of the conveyor is above the water surface. The lower end of the conveyor is fitted with a dredge head. The dredge head consists of a series of short pipes pointed at the bottom just ahead of the bottom end of the conveyor. Water is pumped through the short pipes at high velocity. This high velocity water erodes the bottom and causes any clams, or other partially buried organisms, to be washed free of the bottom mud and blown up onto the conveyor. The conveyor carries the crop (clams) to the surface and deposits them in the harvesting containers. If there is lot of debris, the crop is manually picked off the conveyor by the operator and placed in the harvest containers. The debris goes off the top end of the conveyor and back to the bottom. The floatation device is power driven and continuously moves the dredge head and conveyor forward permitting continuous harvesting.

BEHAVIORAL CONTROL HARVESTING METHODS

Behavior control harvesting methods take advantage of natural tendencies of fish to aid the harvesting process. It is well known that many species of fish will swim against the current if given the opportunity.[3] Other species, especially the young, often are attracted to light. The effects of electric fields on fish is somewhat understood. At certain electric field strengths, fish will orient themselves parallel to the lines of force in the field, while at higher field strengths some fish will be forced to move toward one of the field electrodes. If the electric field strength is very high it will stun or kill the fish.

Stunning fish with electric shock is widely used by biologist to make population assessments in streams and small bodies of water.

Some aquaculture facilities are beginning to exploit fish behavior to reduce stress associated with handling and harvesting. For example, fish stocking densities need to be reduced as the fish gets larger. One way to do this without much stress is to have large pipes connecting tanks. This pipe needs to have a valve to close it off and a bar grader sized to fit over the inlet end of the pipe. Fish are initially stocked in tank 1. As they grow they need to expand into tank 1 and 2. Thus, the operator opens the valve between the two tanks, and adjusts the water levels such that there is flow from tank 2 to tank 1. The fish try to swim against the current. The bar grader in the pipe connecting the two tanks prevents the large fish from entering the pipe. The small fish swim through the bar grader and into tank 2. The same number of fish now have the culture volume of two tanks with the small fish in tank 2 and the larger fish in tank 1. The fish experienced little if any stress during the move. Another example is to use a light to attract oyster larva. If the tank, light, and associated equipment is set up correctly, the strongest larva can be separated from the weak or dying larva without touching the larva.

The use of fish behavior to assist in handling and harvesting aquatic animals is just beginning. This area is, however, a fertile area for research.

AQUATIC PLANT HARVESTING

Aquatic plants may be water lilies where the blossoms are harvested, kelp where the leaves and stalk are harvested, or some other species. Harvesting methods vary widely because the plants are so different. Various algae species are commercially valuable and are harvested from rock, robes, or net surfaces, usually manually by use of the hands or simple hand tolls such as rakes. Other plants such as giant kelp, water hyacinth, and others are harvested by mowing devices that cut the plants a few centimeters to more than a meter below the water surface with a mowing system similar to an agricultural hay mower. The plants are then conveyed to a barge for temporary storage. The barge is towed to the processing plant or disposal site where it is unloaded with automated equipment.

REFERENCES

1. Leitrizt, E.; Lewis, R.C. *Tout and Salmon Culture*; California Fish Bulletin Number 64; California Department of Fish and Game: Berkeley, California, 1980.
2. Huet, M. *Textbook of Fish Culture Breeding and Cultivation of Fish* (English Translation by Henry Kahn); Fishing News (Books) Ltd.: London, 1970.
3. Billard, R. In *Culture of Salmonids in Fresh Water. Aquaculture*; Gilbert, B., Ed.; Ellis Horwood: New York, 1989; Vol. 2, 551–592.

Aquacultural Product Storage Systems

Jaw-Kai Wang
University of Hawaii at Manoa, Honolulu, Hawaii, U.S.A.

A

INTRODUCTION

The U.S. annual per capita consumption of seafood, including shellfish, rose steadily from 12.5 lb in 1980[1] to peak at 16.2 lb in 1987,[2] then declined to 14.9 lb in 1998.[3] Consumption of all seafood has declined since 1987, however, per capita consumption of fresh and frozen seafood has been increasing since 1990 and seafood prices have also risen steadily.[2] While per capita seafood consumption has not regained the peak of 1987, its value has risen due to the increase in demand for high-quality products.

As a result, seafood storage has become increasingly important to government regulatory agencies, wholesalers, retailers, and consumers. Consumers expect seafood to make a positive contribution to their diets as a low-fat and low-calorie source of high-quality proteins, fatty acids, and minerals. General consumers are also becoming increasingly aware of the importance of seafood quality, a fact long recognized by oriental consumers, and are starting to pay higher prices for better quality. As this happens, the storage and transport of high-quality fresh seafood assume heightened importance.

There is a long existing tradition that heavily favors fresh seafood over the frozen products. In many instances, however, the word "fresh" has simply been taken to mean "never frozen." But the quality of seafood is not entirely dependent upon whether or not it has been frozen. Frozen seafood loses its quality much slower than nonfrozen seafood. It is rare that the quality of fresh or nonfrozen seafood can survive a week even with the best handling, including stable storage temperature and the frequent replacement of ice. On the other hand, live seafood is an entirely different matter altogether.

FACTORS INFLUENCING SEAFOOD QUALITY CHANGES DURING STORAGE

Color, flavor, and texture of seafood change during storage. Colby, Enriquez-Ibarra, and Flick[4] give an excellent review of the physical, bacterial, biochemical, and enzymatic changes that occur during the storage of seafood. Bacterial and enzymatic degradations and the oxidation of unsaturated lipids are the major factors that affect seafood storage life.

One can readily judge the quality of seafood by observing its appearance. Fresh seafood has a shimmering opalescent appearance that changes to increasing opaqueness due primarily to the disintegration of myofibrils, which are made up of thick and thin myofilaments that help give the muscle its striped appearance. The thick filaments are composed of myosin and the thin filaments are predominantly actin. The degradation of myofibrils also causes seafood texture to change from firm to soft during storage.

The red color that one finds in fresh fish and shellfish mostly comes from the presence of carotenoid pigments such as astaxanthin, which is often included in fish and shrimp feeds, and heme proteins such as myoglobin and hemoglobin.[5] The auto-oxidation process during storage causes the red color to fade. As storage temperature and the concentration of oxygen affect the oxidation process, they are both important in controlling the physical changes that occur during storage.

Bacterial activities are among the primary underlying reasons for quality degradation of fresh seafood during storage. While micro-organisms can be found only in the gills, skins, and intestines of healthy fish and mollusks, they can be found everywhere on dead animals. Widespread micro-organisms and endogenous enzymes degrade trimethylamine oxide (TMAO) to trimethylamine (TMA) and formaldehyde (FA), which produces the smell that is commonly associated with spoiled fish. Magno-Orejana, Juliano, and Banasihan[6] and Sen Gupta, Mondal, and Mitra[7] used TMA concentration as an index of freshness.

According to Ikeda,[8] more TMA is produced from TMAO by bacterial action than by fish tissue enzymes. This correlation of TMA to bacterial counts in fresh fish, particularly with psychrotrophic plate counts, has led to the adoption of TMA as a standard of quality in fresh shrimp in Australia and Japan.[9]

Frozen Storage

High-quality frozen seafood begins with high-quality seafood. Freezing never improves the quality of

Encyclopedia of Agricultural, Food, and Biological Engineering
DOI: 10.1081/E-EAFE 120006906
Copyright © 2003 by Marcel Dekker, Inc. All rights reserved.

seafood. It merely reduces the rate of quality deterioration. For a first class freezing operation, fish must be fresh and kept very cold and well iced before it is frozen, and after freezing it must be held at low, stable temperatures.

The best example of high-quality frozen seafood is raw fish for sushi and sashimi. Sushi and sashimi have become the rage of the world's gourmets. In Japan, where sushi and sashimi originated, a great deal of the fish used for raw consumption is first frozen. Freezing removes the problem of parasites and allows a large variety of fishes to be available in many markets. In the United States, where litigation has become a common occurrence, restaurants have taken to serving only prefrozen seafood if it is to be eaten raw.[10]

The most important factor governing the storage life of frozen fish is the storage temperature. When fish is frozen and stored at subfreezing temperatures, bacterial growth is retarded and enzyme and chemical actions are slowed. A general rule of thumb regarding the effect of temperature is that for every 10° of temperature reduction the rate of chemical reaction is halved. As seafood quality deterioration is a result of chemical reaction, enzymatic or otherwise, it is therefore important to maintain the storage temperature as low as is economically feasible.

At subfreezing temperatures, most of the free water becomes ice, and bacteria require free water to survive and grow. The lowered temperature itself can destroy from 50% to 90% of the bacteria found on fish. During frozen storage, there is a continued slow, steady die-off of the bacteria, with the rate of decrease depending on the temperature and bacterial species.[11] However, complete elimination of bacteria by freezing is not possible.

Quick cooling to and storage at −3°C–0°C is sometimes referred to as superchilling. At these temperatures, seafood is partially frozen but retains the sensory quality associated with the fresh state.[4] Tomlinson et al.[12] reported that superchilling was effective in preventing the darkening as well as the proteolysis of the flesh adjacent to the visceral cavity, and odors generally associated with spoilage developed in iced fish but were greatly reduced or absent in superchilled fish (− 2°C).

Frozen seafood needs to be carefully packaged to reduce moisture loss, avoid bacterial contamination, prevent the absorption of refrigeration odors, and facilitate labeling. The loss of moisture from the seafood surface to the surrounding air through sublimation is the cause for "freezer burn." The loss of moisture will cause the surface to become wrinkled and the dried flesh is very likely to be tough after cooking. To reduce moisture loss, tight packaging using material impermeable to moisture is necessary.

Live Storage

Seafood quality suffers from the moment of the animal's death. Hence, the best way to preserve seafood quality is to preserve the life of the animal until it is consumed.

Live seafood has always commanded surprisingly high value in the Orient. *Penaeus japonicus* (Kuruma ebi) gained a predominant place in Japanese cuisine because in the days before refrigeration it was among the few species of aquatic animals that could be shipped live on trips that lasted many days. Today, glass tanks containing live marine animals are becoming a common sight in Chinese restaurants in Asia as well as in the United States.

In the United States and Europe, oysters, clams, crabs, and lobsters are often marketed live for the gourmet trade. "Marketing and Shipping Live Aquatic Products"[13] contains many useful articles on the shipment and storage of live animals.

Smoking, Curing, and Drying

Crance[14] and Paparella[15] believe that smoking and open-air drying of fish long preceded civilization and the present day method of hot smoking evolved from these early practices. The development of controlled smokehouses in recent history has begun to change the art of smoking into a science. The primary curing agent is still salt, with other ingredients such as sugar and spices, and artificial preservatives such as sodium nitrite also becoming widely used.

The percentage of salted and smoked seafood sold in the United States is small when compared to the total consumption of seafood products. However, cultural, culinary, and other preferences mean that the demand for such products, while small, will remain persistent and constant.

Irradiation

In the 1960s, the Atomic Energy Commission (AEC) funded a national program on the low-dose irradiation of fishery products. However, the program was phased out when the U.S. Food and Drug Administration (FDA) decided to adopt strict guidelines for testing the wholesomeness of irradiated foods.[4] In 1963, FDA approved the use of irradiation for wheat and wheat flour to eliminate insects. Over the years, irradiation has been approved in the United States for pork, spices, fruits, vegetables, and poultry. In 1997, red meat joined the list

and a petition to irradiate seafood is pending.[16] One of the major arguments raised against commercial irradiation of seafood is that under low-dose irradiation, while spoiling bacteria are being destroyed, some human pathogens may survive.

Low-dose irradiation extends the storage life of seafood by lowering the bacterial counts of micro-organisms, particularly *Pseudomonas*. The applied dose must be below 5 kGy, as higher dosages will induce an off-flavor in seafood products.[4]

There is research to show that vibrios, particularly *Vibrio vulnificus*, *V. parahemolyticus*, and *V. cholerae*, and other pathogens, such as *Aeromonas hydrophila*, *Salmonella typhimurium*, *Streptococcus faecalis*, and *Staphylococcus aureus*, are radiation sensitive. Even *Clostridium botulinum* type E has been reported to stop toxin production in shrimp after a dose of 1.5 kGy and ice storage at 0°C for 31 days.[17]

It is anticipated that irradiation of seafood will gain acceptance in the next 10 years as scientific data will increasingly show the commercial benefits of such application. However, it is equally true that it has not yet been clearly established that both human pathogens and the bacteria that cause spoilage can be eliminated from seafood with equal efficacy under irradiation treatment.

Controlled Atmosphere Storage

Controlled atmosphere storage has been widely and successfully used in agriculture, particularly for fresh vegetables and fruits. However, the National Academy of Sciences does not recommend the use of controlled atmosphere packaging for fresh fishery products.

By controlling the percentage of carbon dioxide, oxygen, and inert gases, such as nitrogen, in the storage environment, the life process of harvested vegetables and fruits can be retarded and the process of spoilage reduced. For seafood, the growth of gram-negative bacteria, such as *Pseudomonas*, *Actinobacter*, *Flavobacterium*, *Micrococcus*, and *Staphylococcus*, can be retarded by enriching the carbon dioxide content in the surrounding atmosphere. Unfortunately, the growth of lactic acid bacteria such as *Streptococcus* and *Lactobacillus* is not as affected by the increase in carbon dioxide in the storage atmosphere. In addition, the increase in nitrogen content in the storage environment does not seem to have much effect on the shelf life of seafood.

When combined with low storage temperature, however, the effectiveness of controlled atmospheric storage of seafood can be greatly enhanced.

REFERENCES

1. Wilkes, A.P. Food Product Design: Hooking onto Seafood Safety. http://www.foodproductdesign.com/archive/1991/0891QA.html (accessed Nov 2000).
2. National Marine Fisheries Service. Overview of the U.S. Fishing Industry. *Our Living Oceans, The Economic Status of U.S. Fisheries 1996*; U.S. Department of Commerce, National Oceanic and Atmospheric Administration, National Marine Fisheries Service: Washington, D.C., 1996; Chap. 2, http://www.st.nmfs.gov/st1/econ/oleo/oleo.html (accessed Dec 2000).
3. Fisheries Statistics and Economics Division. *Fisheries of the United States, 1998*; Current Fishery Statistics No. 9800; U.S. Department of the Interior, Fish and Wildlife Service Bureau of Commercial Fisheries: Washington, D.C., 1999.
4. Colby, J.-W.; Enriquez-Ibarra, L.G.; Flick, G.J., Jr. Shelf Life of Fish and Shellfish. In *Shelf Life Studies of Foods and Beverages, Chemical, Biological, Physical and Nutritional Aspects*; Charalambous, G., Ed.; Elsevier Science Publishers B.V.: Amsterdam, The Netherlands, 1993; 85–143.
5. Chen, H.M.; Myers, S.P.; Hardy, R.W.; Biede, S.L. Color Stability of Astaxanthin Pigmented Rainbow Trout Under Various Packaging Conditions. J. Food Sci. **1984**, *49* (5), 1337–1340.
6. Magno-Orejana, F.; Juliano, R.O.; Banasihan, E.T. Trimethylamine and Volatile Reducing Substances in Frigate Mackerel (Auxis Thasard Lacepede). Philipp. J. Sci. **1971**, *100* (3–4), 209–226.
7. Sen Gupta, P.; Mondal, A.; Mitra, S.N. Separation and Quantitative Estimation of Dimethylamine and Trimethylamine in Fish by Paper Chromatography. J. Inst. Chem. (India) **1972**, *44*, 49–50.
8. Ikeda, S. Other Organic Components and Inorganic Components. In *Advances in Fish Science and Technology*, the Jubilee Conference of the Torry Research Station, Aberdeen, Scotland, July 23–27, 1979; Connell, J.J., Ed.; Fishing News Books, Ltd.: Farnham, England, 1980; 111–124.
9. Montgomery, W.A.; Sidhu, G.S.; Vale, G.L. The Australian Prawn Industry. I. Natural Resources and Quality Aspects of Whole Cooked Fresh Prawn and Frozen Prawn Meat. CSIRO Food Preserv. Q. **1970**, *30* (2), 21–27.
10. Dore, I. *The New Frozen Seafood Handbook*; Osprey Books: Huntington, New York, 1989; 360.
11. Licciardello, J.J. Freezing. In *The Seafood Industry*; Martin, R.E., Flick, G.J., Eds.; Osprey Book Published by Van Nostrand Reinhold: New York, NY, 1990; 205–218.
12. Tomlinson, N.; Geiger, S.E.; Kay, W.W.; Uthe, J.; Roach, S.W. Partial Freezing as a Means of Preserving Pacific Salmon Intended for Canning. J. Fish. Res. Board Can. **1965**, *22* (4), 955–968.
13. Paust, B.; Peters, J.B., Eds. In *Marketing and Shipping Live Aquatic Products*, Proceedings from Marketing Live Products '96, Seattle, Washington, Oct 13–15, 1996; Northwest Regional Agricultural

Engineering Service, Cooperative Extension: Ithaca, NY, 1997; 288.

14. Crance, J.H. *Smoked Fish*; Fact Sheet #L-1043; Texas A&M University: Texas, 1955.
15. Paparella, M. *Information Tips*; University of Maryland Sea Grant: Maryland, 1979.
16. Knehr, E. Food Product Design: Making Progress in Food Preservation. http://foodproductdesign.com/archive/1998/0398NT.html (accessed Nov 2000).
17. Grodner, R.M.; Andrew, L.S. Irradiation. In *Microbiology of Marine Food Products*; Ward, D.R., Hackney, C.R., Eds.; Van Nostrand Reinhold: New York, NY, 1991; 429–440.

Aquacultural Production Systems

Raul H. Piedrahita
University of California, Davis, Davis, California, U.S.A.

INTRODUCTION

Aquaculture is the fastest growing sector of U.S. agriculture and one of the fastest growing sectors of agriculture worldwide. In 1996, aquaculture production was over 26 million t, constituting almost 22% of total worldwide fish production (including fresh and seawater fisheries, excluding seaweeds). Worldwide aquaculture production of Atlantic salmon accounts for approximately 98% of the total production. The corresponding aquaculture contributions are 27%, 95%, and 61%, for shrimp, oysters, and tilapias, respectively.[1] The prospects for continued growth are excellent given the increasing demand for fish, the limitations on further increases in wild fisheries production, and the constant improvements in production systems.

Aquaculture is extraordinarily diverse in terms of the number of species cultured and the types of culture systems used. Fish, crustaceans, and molluscs are cultured in fresh and salt water, warm and cool water, oligotrophic and eutrophic waters. Aquacultured species consume oxygen from the water and release waste products such as ammonia, carbon dioxide, and particulate organics. To ensure good health and fast growth of cultured organisms, a production system must be managed to provide the required temperature, water velocity, light, and overall water quality and environmental characteristics.

A number of systems are used for aquaculture production, and the main types are described here. They include ponds, flow-through, recirculation, and cage systems. Ponds typically have low water exchange rates, resulting in complex ecosystems in which the animals cultured are but one of the components determining water quality. Flow-through systems are the second type of production system described. In these, water typically has a residence time of about one hour, and water quality is determined almost exclusively by the metabolic activity of the cultured animals. The third type of production system described here are recirculation systems. In these, water is reused after being treated to remove waste products and replenish oxygen used by the cultured animals. Cages constitute a unique type of production system and will also be described.

AQUACULTURE PONDS

Earthen aquaculture ponds continue to be the most common type of production system worldwide.[2] Ponds range in size from under 0.01 ha to over 25 ha and usually have an average water depth of less than 1.5 m. The most common pond construction technique is the "cut and fill" method in which a thin layer of soil excavated to form the pond bottom is used to form the embankments.

Preferred soils for aquaculture ponds have very low permeabilities, low organic matter content, and lack potentially deleterious compounds. Low permeability is desirable to ensure adequate compaction of pond banks during construction and to minimize water losses to infiltration. Low soil organic matter content is important because it permits good soil compaction during pond construction and reduces settling and weakening of embankments as organic matter decomposes over time. Acid sulfate soils, which are common in some coastal areas where aquaculture is practiced, are to be avoided. When these soils are exposed to oxygen, the iron pyrite present oxidizes to produce sulfuric acid that leaches into the pond water.[3] In addition to the low pH and high acidity in ponds constructed on acid sulfate soils, there is a risk that potentially toxic ions may be released from the soil.[3]

Although water use in aquaculture ponds varies greatly, environmental concerns, government regulations, and variable supply have resulted in a trend towards lower water usage. Typically, replacement water to freshwater ponds equals losses to evaporation and infiltration. Until recently, salt and brackishwater shrimp production ponds generally were operated with water exchanges as high as 25% of pond volume per day. However, potential water quality degradation in receiving waters, high pumping energy costs, better understanding of pond ecosystems, and increased environmental awareness on the part of producers have resulted in substantial reductions in the amount of water that is exchanged. A number of pond construction and management practices have been developed in which very high production intensities can be maintained with systems in which no effluents are released from the ponds. These production systems rely on mechanical aeration to provide the oxygen necessary for the cultured organisms and for the aerobic decomposition of wastes released into the water.

Encyclopedia of Agicultural, Food, and Biological Engineering
DOI: 10.1081/E-EAFE 120006866
Copyright © 2003 by Marcel Dekker, Inc. All rights reserved.

Although the category boundaries are blurred, the quantity and quality of inputs and the level of management control exerted normally determine the intensity of production in ponds. Three levels are often recognized[4]: extensive (only organic and inorganic fertilizers are used, cultured animal biomass density is relatively low); semi-intensive (fertilizers may be used but prepared diets are also used, animal biomass densities are intermediate); and intensive (fertilizers may be used, feeds and mechanical aeration are used, density of target animal biomass is relatively high).

FLOW-THROUGH SYSTEMS

Water quality in flow-through systems is maintained primarily by the continuous inflow of new water to replenish oxygen and remove wastes produced by the fish. Biomass, species, and age of the animals in the system, the amount of feed offered daily, and the quality of the influent water determine the flow rate needed. Two basic types of containers are used to hold the animals: raceways and tanks. These are described below.

Raceways

Raceways are commonly used for the production of salmonids, especially trout. A typical raceway consists of a concrete channel approximately 30 m long, 3 m wide and 1 m deep. Water flows along the long axis of a raceway creating conditions similar to those of a plug flow reactor. Average velocities of approximately $0.03\,m\,s^{-1}$ are maintained to prevent excessive settling and accumulation of solids.[5] Maintenance of this velocity in a typical raceway requires a flow of about $0.09\,m^3\,s^{-1}$, which is substantially higher than the flow needed to provide oxygen to the fish and remove their waste products. As a result, raceways often are constructed in series in an effort to increase the biomass produced with a given flow, and up to five raceways can be in a series receiving the $0.09\,m^3\,s^{-1}$ mentioned above. Some form of aeration, such as splashboards, must be used between successive raceways. Modern raceways also include a "quiescent zone" used for solids removal. The quiescent zone is an area of approximately 6 m at the downstream end of a raceway from which fish are excluded by the use of a screen. In the absence of fish, solids settle and are then removed by vacuuming or pumping every one to two weeks and are treated before disposal.

Tanks

Whereas raceways resemble plug flow reactors with definite water quality gradients, tanks approximate fully mixed. Tank shapes used commercially include round, square, hexagonal, or octagonal. Tanks with straight sides may have sharp or rounded corners. The surface area of tanks varies widely; and commonly ranges from less than $1\,m^2$ to $180\,m^2$ (corresponding to a 15 m diameter). Water depth usually is maintained under 4 m, even for the largest tanks. Bottoms may be flat or have a mild slope towards the center.

Water is normally introduced along the edge of the tank. A number of inlet configurations are used to impart a tangential velocity to the flow and to try to create a uniform velocity distribution throughout the tank. The most common inlets consist of perforated pipes placed either horizontally at the water surface or submerged vertically close to the tank wall. In both cases the perforations are directed tangentially. In some cases a combination of the two types of inlets is used.

Tank effluent often is removed through a center drain. A relatively new development in tank design is the use of dual drains in which a small fraction of the water flow is removed through a center drain located close to the tank bottom. The majority of the water flow is removed through a second drain located either at the tank center and further from the bottom than the previous drain, or along the edge of the tank.[6] The dual drains are used to concentrate and collect the solids produced in the tank (uneaten feed and fish feces), with the majority of these being removed through the bottom center drain (85%–95% of the solids in 5%–10% of the water flow), and a small fraction being carried out with the majority of the water through the second drain (5%–15% of the solids in 90%–95% of the flow). The low solids fraction can be treated for reuse in the system, or can be discharged after minimal treatment. The high solids fraction must be treated prior to discharge or disposal, minimizing the risks of negative environmental impacts.

RECIRCULATION SYSTEMS

Recirculation systems have been introduced in an attempt to reduce water use, improve water quality, increase flexibility in site selection, and reduce the potential impact of discharges from aquaculture.[7] In a recirculation system, a number of water treatment operations are used to treat the water and make it suitable for reintroduction into the culture container (usually a tank). Normally, solids removal is the first step in a water treatment sequence, and a variety of systems are used. These range from screen filters with openings of about 100 μm to particulate media filters to gravitational separators. The screens are usually configured such that automatic cleaning can take place, and "drum," "disk," and "belt" filters are available. Particulate media filters range from the traditional swimming pool sand filters to upflow filters through a bed of plastic beads. Gravitational separators include systems such as settling basins and foam fractionators.

Ammonia, the primary nitrogenous fish waste product, is often removed in biofilters by nitrification, a two-step process in which bacteria convert ammonia to nitrate. While unionized ammonia may be toxic to aquatic animals at very low concentrations (typically well under $1\,mg\,NH_3\text{-}N\,L^{-1}$), nitrate is not toxic even at concentrations exceeding $1000\,mg\,NO_3\text{-}N\,L^{-1}$. A variety of basic biofilter configurations are available, and no biofilter type has emerged as the preferred system for all conditions. Biofilter types being used in aquaculture include trickling filters, rotating biological contactors (RBC), packed bed filters (e.g., "bead filters"), and fluidized bed filters.

Aeration may serve the dual purpose of replenishing the oxygen used by the fish and removing the carbon dioxide produced. However, pure oxygen is commonly used as an oxygen source instead of air. If pure oxygen is used, carbon dioxide removal is achieved either by a separate aeration system or by the addition of strong bases. These cause the carbon dioxide chemical equilibrium to shift, resulting in decreased carbon dioxide concentration and increased bicarbonate and carbonate concentrations.[7]

After years of being confined to research settings, water reuse systems are now being used for commercial aquaculture production of species such as salmon (smolt production), seabass, seabream, striped bass, sturgeon, tilapia, and turbot. The degree of water reuse in these facilities ranges widely, and depends on economic, technical, and biological considerations.

CAGES

Cages are enclosures used to restrict the movement of fish while allowing for the flow of water. Cages may be placed in natural or man-made water bodies, such as the ocean, reservoirs, lakes, rivers, or irrigation canals. Cages have been used for hundreds of years, and some of the traditional types of structures are still being used.[8] However, significant developments in aquaculture cages have been brought about by the availability of new materials and construction techniques, and by environmental considerations. The growth of the salmon industry along coastal areas of Scandinavia, the British Isles, Chile, and North America has occurred around the use of advanced cage designs. Materials used for cage construction should be, among other things, strong, light, and corrosion, weather, and fouling resistant.[8] New alloys and plastics offer considerable improvements over traditional materials. As a result of these material developments, advances in design and construction techniques, and environmental pressures, the tendency over the last few years has been towards the development of larger cages that can be moved further off-shore. Modern cages may be as large as $800\,m^2$ in surface area (corresponding to about a 32 m diameter), and have depths of up to 35 m.[8]

Continued expansion and growth of cage aquaculture has been offset by concerns over the possible environmental impact. The issues center on the release of nutrients from uneaten feed and fecal matter. In response to these concerns, new technologies are being developed as well as guidelines for siting of cages to minimize negative water quality impacts. Some of the technologies include the production of new cages for open ocean deployment; new feeds that maximize assimilation of nitrogen and phosphorus by the fish, and the use of collection systems to remove waste products from under cages.

REFERENCES

1. New, M.B. Global Aquaculture: Current Trends and Challenges for the 21st Century. World Aquac. **1999**, *30* (1), 8–13.
2. Wheaton, F.W.; Singh, S. Aquacultural Systems. In *CIGR Handbook of Agricultural Engineering. Volume II Animal Production & Aquacultural Engineering*; Bartali, E.H., Jongebreur, A., Moffitt, D., Wheaton, F.W., Eds.; American Society of Agricultural Engineers: St. Joseph, MI, 1999; 211–217.
3. Boyd, C.E.; Tucker, C.S. *Pond Aquaculture Water Quality Management*; Kluwer Academic: Boston, 1998; 700 pp.
4. Egna, H.S.; Boyd, C.E.; Eds., *Dynamics of Pond Aquaculture*; CRC Press: Boca Raton, FL, 1997; 437 pp.
5. Timmons, M.B.; Riley, J.; Brune, D.; Lekang, O.-I. Facilities Design. In *CIGR Handbook of Agricultural Engineering. Volume II Animal Production & Aquacultural Engineering*; Bartali, E.H., Jongebreur, A., Moffitt, D., Wheaton, F.W., Eds.; American Society of Agricultural Engineers: St. Joseph, MI, 1999; 245–280.
6. Timmons, M.B.; Summerfelt, S.T.; Vinci, B.J. Review of Circular Tank Technology and Management. Aquac. Eng. **1998**, *18* (1), 51–69.
7. Timmons, M.B.; Losordo, T.M., Eds. *Aquaculture Water Reuse Systems: Engineering Design and Management*; Elsevier: Amsterdam, 1994; 333 pp.
8. Beveridge, M.C.M. *Cage Aquaculture*, 2nd Ed.; Fishing News Books: Surrey, 1996; 346 pp.

Batch Process Control

Graeme Macaloney
QSV Biologics Ltd., Edmonton, Alberta, Canada

Gary A. Montague
University of Newcastle, Newcastle upon Tyne, United Kingdom

INTRODUCTION

The aim of process control in any production system is to achieve, in a safe and environmentally acceptable way, the desired quality, cost, and productivity targets. The fundamental principles covered in this article, therefore, apply to a wide variety of agricultural and food processing systems. Control is an essential tool that promotes reproducible process operation and facilitates process optimization. This article discusses the basics of batch control and explore approaches to address technological constraints in process monitoring that limit batch control.

MEASUREMENT AND SIGNAL PROCESSING

Measurement for Control

The availability of accurate and reliable measurements is essential if control is to be effective. Unfortunately when batch systems are considered, control quality tends to be constrained by the lack of available on-line measurement methods. Notably, many important biochemicals can only be measured off-line, often producing results after the batch is finished hence making corrective actions ineffective. Recent progress in measurement technology is beginning to overcome some of the limitations. A notable example is Near infrared spectrometry (NIRS), which is a potentially valuable analytical technique, since it lends itself to an extensive range of processes including agricultural, baked, dairy, beverages, polymers, and pharmaceuticals.[1] As a process control tool, NIRS offers rapid (<1 min), multicomponent analysis with instrumentation that is robust, precise, and easily operated by technically unskilled personnel.[2] As an at-line[3] methodology (Fig. 1), NIRS may be used in open-loop control.[4] When equipped with fibre-optic probes it can be used for on-line analysis and closed-loop control.

Signal Processing

Even if instrumentation is available to provide information on key process variables, problems can still remain. Output signals from process instrumentation often exhibit excessive signal noise and spurious data spikes that mask the underlying true process condition and that can be passed on to the control system and lead to unacceptable actions being taken. Such corruption of the signal can arise from instrument recalibration, nonuniform conditions in the vessel leading to variation at the point of measurement, electrical interference etc. Excessive noise in these signals can be removed using algorithms known as "data filters." First order filters exponentially weight past information to produce a moving average that places decreasing emphasis on older values. By increasing a time constant in these filters, greater weight is given to older values and therefore greater data smoothing or noise reduction is achieved. A side effect of heavy filtering can be the creation of a lag or offset in the filtered data. In essence noise is removed but at the expense of delaying observation of process change and therefore the ability to compensate for deviations. More sophisticated filters are available to deal with such situations.[5] As the frequency of signal information decreases (i.e., reducing the observations per unit time) the ability of filters to effectively reduce noise and avoid phase lag becomes more difficult. If data frequency is an issue, then the preferred method of filtering is via local "analogue" filters on the data monitoring equipment. Data spikes or "outliers" are a common occurrence in industrial environments and it is recommended that they be removed prior to filtering because they have a disproportionate effect upon the smoothed data over a significant period of time. Methods to detect and remove outliers from on-line data are available.[6]

Modeling and Estimation

Critical variables that cannot be measured on-line may be estimated using other available related on-line data combined with a model called a software sensor. The software sensor relates the measured data to the variable to be estimated. Software models based upon a mechanistic understanding of bioprocesses include mass balances, time varying parametric structures, kinetic

Encyclopedia of Agricultural, Food, and Biological Engineering
DOI: 10.1081/E-EAFE 120007213
Copyright © 2003 by Marcel Dekker, Inc. All rights reserved.

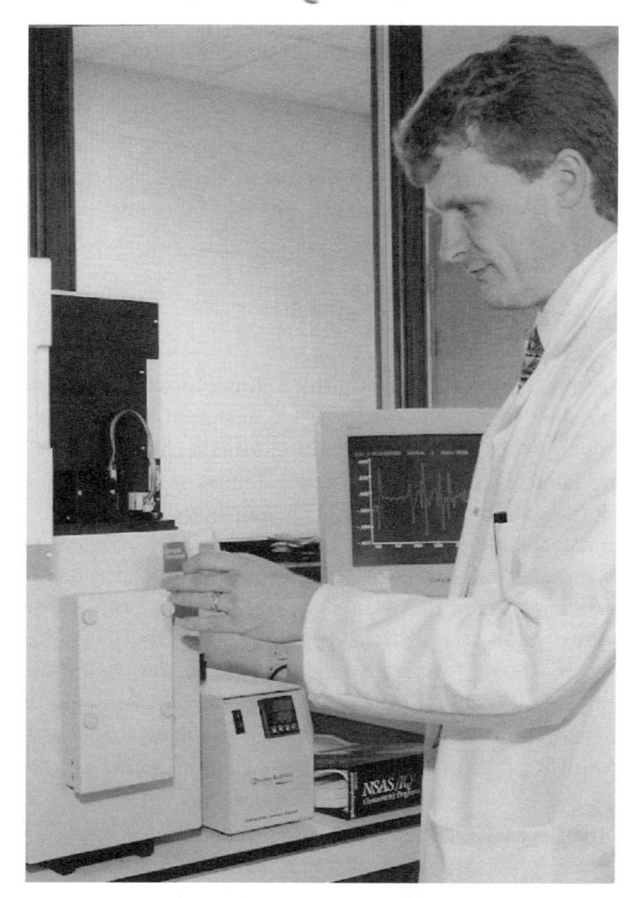

Fig. 1 One of the authors operating an at-line Foss-NIRSystems near infrared spectrometer for the monitoring and control of a bioprocess.

models, and physiological models.[7] These models involve the use of extended Kalman filters for nonlinear systems whereby a knowledge of noise level in the system is used to provide optimal estimates.[8] Unfortunately, they have not been widely adopted by industry due to the difficulty in developing sufficiently accurate models of complex industrial processes. Alternatively, data-based modeling techniques including linear or nonlinear/linear in the parameters time series and neural network models[8] may be employed. Neural networks have been extensively studied by workers at the University of Newcastle in the UK.[9–11] Data-based models do not provide an understanding of bioprocess behavior and may not be sufficiently robust if required to extrapolate beyond their original data range. Hybrid or fused mathematical/mechanistic models offer considerable promise in the future. Here, errors in approximate mechanistic models are corrected by data-based models. This approach constrains the model behavior and hence increases robustness.[12]

SEQUENCE CONTROL

Batch control may be considered to consist of sequence control and regulatory control. Sequence control incorporates algorithms designed to control a defined sequence of events or batch operations in a hierarchical manner. Commercial systems for defining procedures are available lately (Fig. 2) following the Instrumentation, Systems, and Automation Society (ISAS) SP88 standard.[13] SP88 provides standard terminology and definitions that facilitate the design and operation of manufacturing plant control systems. See the ISA web site: http://www.isa.org/ for more information.

REGULATORY CONTROL

Regulatory control uses algorithms to control a process or equipment at a desired predetermined level or "set-point". Feedback control (Fig. 3) occurs when a measurement of process output is compared with the desired output value. The resulting error is used to calculate changes in a

Fig. 2 Automated 20 L Applikon fed-batch fermenter, with SCADA system incorporating SP88 batch recipe standard.

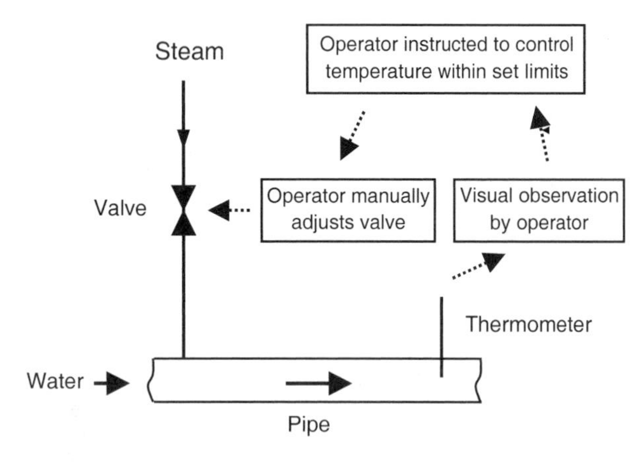

Fig. 3 Open loop (manual) feed back for temperature control.

manipulated variable that will effect the desired change in the process output. For example, in a heating system the temperature (process output) may be controlled by a valve (manipulated variable) on a cooling line or perhaps by the speed of a fan for an air-cooled system.

Open-Loop (Manual) Control

In batch control there is usually a requirement for operator intervention. This may be to initiate the next phase of a sequence or to manually adjust a flow-control valve as part of a regulatory control loop (Fig. 3). This is referred to as manual or "open loop" control. Manual control requires skill on the part of operators in knowing when and how much adjustment to make.

Closed-Loop (Automatic) Feedback Control

Critical control applications or those requiring frequent adjustments are usually automated first. This enables

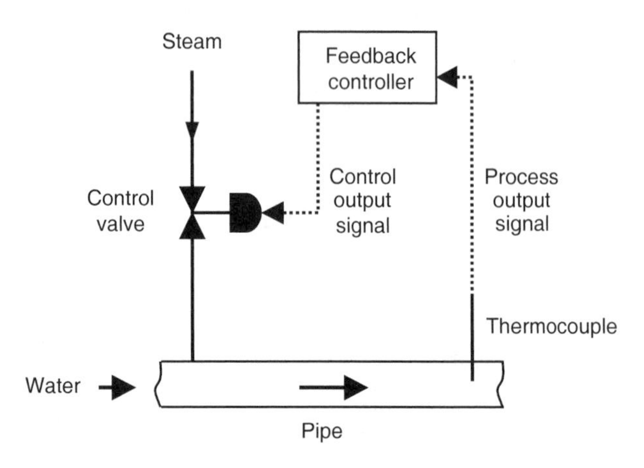

Fig. 4 Closed loop (automatic) feed back for temperature control.

closed-loop control leading to benefits including reduced operator error, improve safety, and improve process reproducibility. When automated, the manual control scheme illustrated in Fig. 3 would transform to that shown in Fig. 4. Commercially available control algorithms are mostly limited to variations on the following control strategies.

On/Off Control

In this simplest of algorithms, the control action is switched full on in response to a deviation in the output signal. When the system returns to the desired set point, the control action is switched off. The resulting output oscillates around the desired set point, which may be acceptable for certain applications (e.g., tank level). A variation on simple on/off control is low/off/high action found in pH controllers. In this situation, if the pH is within an acceptable range defined by a high and low set point, then no control action is required. However, if the pH drops below the lower set point, the alkali pump is switched full on. Similarly, if the pH is high then an acid pump is switched on.

Three-Term or PID Control

Three-term controllers, or PID (proportional, integral, derivative) controllers are used in the majority of industrial situations. Comprehensive coverage of their function can be found in many control texts (e.g. Refs. [14] and [15]). The basic PID algorithm is:

$$u = K(\text{error} + 1/T_\text{i}) \int \text{error} + T_\text{d}\text{d}(\text{error})/\text{d}t)$$

where each of the three additive terms within the parentheses corresponds to proportional, integral, and derivative control elements respectively, when multiplied by the controller gain (K). Where u is the manipulated variable, "error" is the difference between the measured value and the desired output/set point, T_i is integral time and T_d is derivative time. Transformations of this basic form, which provide improved robustness, may be encountered in practice.

Various combinations of P, I, and D control may be encountered. If P-control is employed alone, the resulting dynamic response shown in Fig. 5 will be observed. As can be seen, once the control action reaches steady state there is an offset between the desired value and the steady-state output measure. Given that an offset is often undesirable, P and I controller action may be combined. The effect of the I-control is to force the steady-state error to zero thereby eliminating the offset. The resulting superior dynamic response is

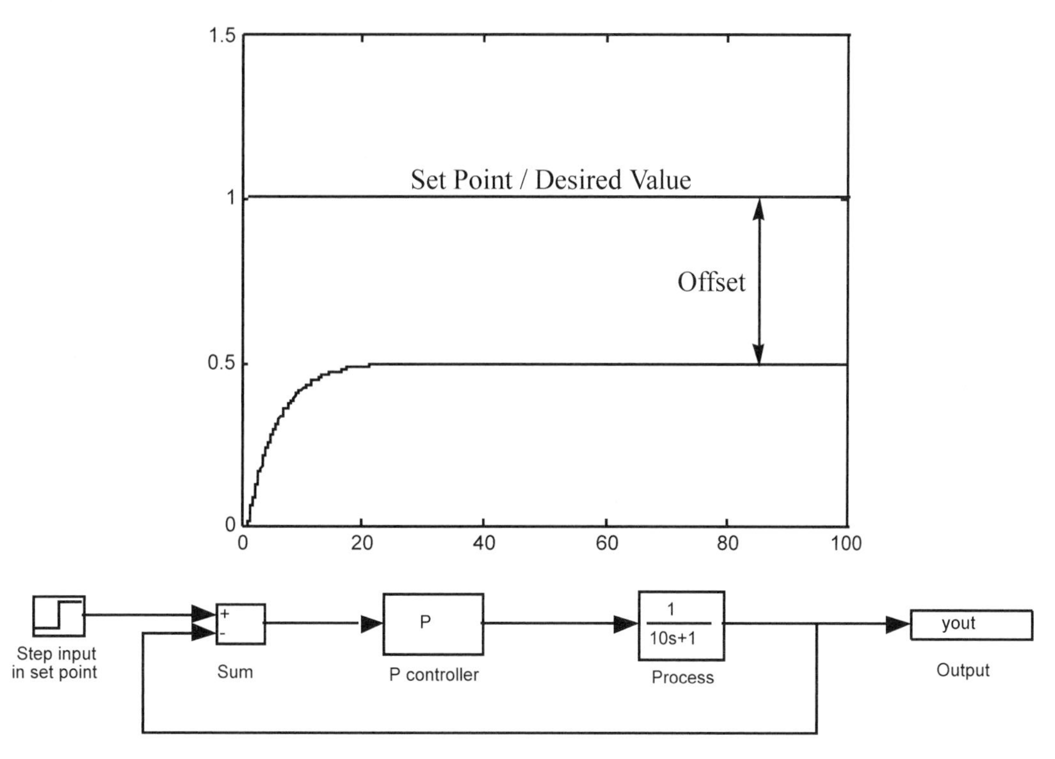

Fig. 5 P-control response (showing offset) and control system block diagram. Note: Scenario is based upon a first order system with a gain of 1, a time constant of 10 s and involves a Laplace transformation (used in the manipulation of the differential equations).

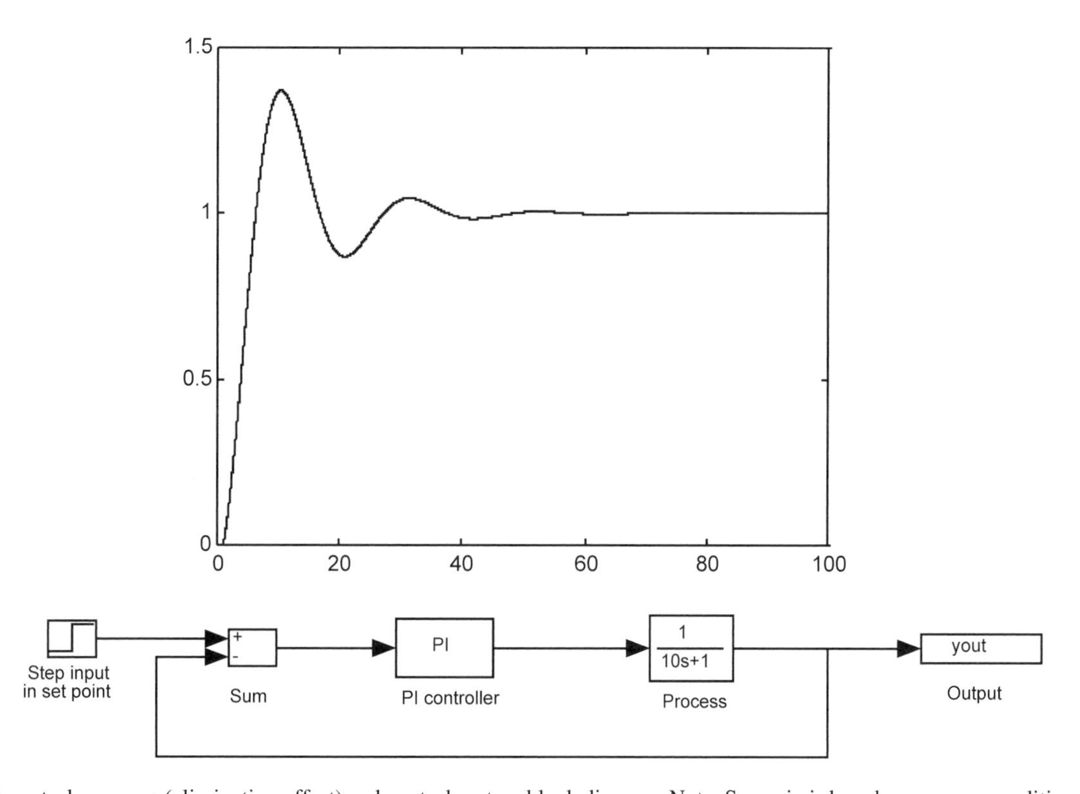

Fig. 6 PI-control response (eliminating offset) and control system block diagram. Note: Scenario is based upon same conditions detailed in Fig. 5.

shown in Fig. 6. PI-control is the most common form employed in industry.

Derivative control is used as part of full PID control for applications where sudden changes in error require rapid changes in control signal to compensate. D-control in effect magnifies the change in controller output signal. However, in industrial environments where signal noise levels are significant, the derivative action results in an extremely noisy control output. This may be mitigated by reducing the derivative time, but usually the derivative control is removed altogether.

Tuning of controllers is an important consideration and may be accomplished manually by skilled staff, or automatically using commercially available packages. Difficulties arise when controllers are operated with factory default settings instead of being tuned for the process, or when inadequately tuned leading to increased process variability (relative to open loop control). Automatic tuners help avoid improperly tuned controllers that may result from tuning by unskilled staff, and their ease of use can also facilitate timely retuning relative to process changes. Tuning is considered in detail by many authors.[14,15]

Advanced Control

In systems where large disturbances are anticipated, feedback control can be improved with the addition of a feed forward component. Feed forward controllers eliminate known, anticipated disturbances thereby providing tighter, more precise control than feedback control alone. Where the effects of a disturbance can be regulated with a local loop but the variable controlled is required to regulate a desired process output, cascade control may be employed. With cascade control, the output of the basic (primary) feedback controller now becomes the input for the new secondary controller. The secondary controller in turn takes over actuation of the final control element. Each controller has an input signal provided by different measured process output. Flow control loops are typical examples where cascade loops may be commonly found. Advanced control is an area of extensive research. Feed forward, cascade, and other forms of advanced control including flow ratio, adaptive, multivariate, fuzzy, and neural network controllers are discussed in detail elsewhere.[16,17]

IMPLEMENTATION ISSUES

The safe design and implementation of control systems is a crucial issue reviewed by Sawyer[18] and Andow.[19] The operator interface in Supervisory Control and Data Acquisition (SCADA) systems requires careful consideration for the concise and effective presentation of large amounts of data. Albert and Coggan[20] discuss the practical considerations of interface design. For those processes that are regulated by the FDA, validation of batch control systems can be an important consideration. Useful reference papers on validation are available[21,22] with up to date information available on the FDA web site (www.fda.gov).

REFERENCES

1. Burns, D.A.; Cuirczak, E.W. *Handbook of Near Infrared Analysis*; Marcel Dekker, Inc.: New York, 1992.
2. Macaloney, G. NIR. Chem. Eng. **1996**, *619* (3), 5–37.
3. Callis, J.B.; Illman, D.L.; Kowalski, B.R. Process Analytical Chemistry. Anal. Chem. **1987**, *59* (9), 624A–637A.
4. Macaloney, G.; Draper, I.; Preston, J.; Anderson, K.B.; Rollins, M.J.; Thompson, B.G.; Hall, J.W.; McNeil, B. At-line Control and Fault Analysis in an Industrial *E. coli* Fermentation Using NIR Spectrometry. Trans. IChemE **1996**, *74* (C4), 212–221.
5. Gardner, E.S. Exponential Smoothing: The State of the Art. J. Forecasting **1985**, 1–28.
6. Tham, M.T.; Parr, A. Succeed at On-line Validation and Reconstruction of Data. Chem. Eng. Prog. **1994**, *90* (5), 46–56.
7. Thornhill, N.F.; Royce, P.N.C. Modeling Fermenters for Control. In *Measurement and Control in Bioprocessing*; Carr-Brion, K.G., Ed.; Elsevier Science Publishers: London, 1991; 67–107.
8. Montague, G.A. *Monitoring and Control of Fermenters*; Institution of Chemical Engineers: Rugby, U.K., 1997.
9. Glassey, J.; Ignova, M.; Ward, A.C.; Montague, G.A.; Morris, A.J. Bioprocess Supervision: Neural Networks and Knowledge Based Systems. J. Biotechnol. **1997**, *52* (3), 201–205.
10. Warnes, M.R.; Glassey, J.; Montague, G.A.; Kara, B. Application of Radial Basis Function and Feedforward Artificial Neural Networks to the *Escherichia coli* Fermentation Process. Neurocomputing **1998**, *20* (1–3), 67–82.
11. Ignova, M.; Montague, G.A.; Ward, A.C.; Glassey, J. Fermentation Seed Quality Analysis with Self-organising Neural Networks. Biotechnol. Bioeng. **1999**, *64* (1), 82–91.
12. Zorzetto, L.F.M.; Maciel-Filho, R.; Wolf-Maciel, M.R. Process Modeling Development Through Artificial Neural Networks and Hybrid Models. Comput. Chem. Eng. **2000**, *24*, 1355–1360.
13. *ISA-SP88.01, Batch Control, Part 1: Models and terminology*; International Society for Measurement and Control, 1995.

14. Stephanopoulos, G. *Chemical Process Control: An Introduction to Theory and Practice*; Prentice Hall, Englewood Cliffs, NJ, 1994.
15. Dorf, R.C.; Bishop, R.H. *Modern Control Systems*; Prentice Hall, Englewood Cliffs, NJ, 2001.
16. Edgar, T.F. Control of Unconventional Processes. J. Proc. Control **1996**, *6* (2/3), 99–110.
17. Haley, T.A.; Mulvaney, S.J. Advanced Process Control Techniques for the Food Industry. Trends Food Sci. Technol. **1995**, Apr*il*, 103–110.
18. Sawyer, P. *Computer-Controlled Batch Processing*; Institution of Chemical Engineers: Rugby, U.K., 1993.
19. Andow, P. Guidance for HAZOP procedures for computer controlled plants; HSE Contract Research Report No 26; HMSO, 1991.
20. Albert, C.L.; Coggan, D.A. *Fundamentals of Industrial Control*; Instrument Society of America, 1992.
21. Grigonis, G.J.; Wyrick, M.L. Computer System Validation: Auditing Computer Systems for Quality. Pharm. Technol. Eur. **1994**.
22. Kuzel, N. Fundamentals of Computer System Validation and Documentation in the Pharmaceutical Industry. Pharm. Technol. **1985**, *19* (9), 60–76.

Bernouilli Equation

Fernando A. Osorio
Universidad de Santiago de Chile, Santiago, Chile

INTRODUCTION

The Bernouilli equation (Eq. 11, Daniel Bernouilli [1700–1782]) is of great importance in solving many food flow situations such as pipeline design calculations and design of the holding tube for sterilization of continuously flowing fluids. To most fluid flow situations found in food engineering, it is necessary to make modifications such as correction of the kinetic energy term for the variation of local velocity with radial position within a pipe. This velocity profile is determined by the rheological properties of the fluid and the correction for the presence of fluid friction in the boundary layer.

MATHEMATICAL DEVELOPMENT

To develop a mathematical expression of this equation the following assumptions are stated:

- Steady state.
- Incompressible fluid, i.e., constant density.
- Inviscid flow along a streamline (a curve in the flow field in which the velocity vector is tangent to the curve) in an inertial reference frame. Inviscid flow means that the flow outside the thin viscous boundary layer (where the viscous effects are confined) is insensitive to viscosity, this means to assume an ideal fluid with zero viscosity.[1] This assumption originates little error in many practical applications, and allows the use of the control volume technique, where there is no need to consider what happens within the control volume as information can be obtained by fluid properties as it enters and leaves through the control surface.[2]
- Negligible viscous effects, shearing stresses due to velocity gradients are not considered because they are small when compared to pressure differences in the flow field. This is not valid for long distances or in regions with high velocity gradients, because the shearing stresses in this case could affect flow conditions and viscous effects should be considered.[1]

Consider a fluid element occupying an infinitesimal volume (differential area dA and differential length ds) moving along a streamline at an instant t as shown in Fig. 1 Applying Newton's second law ($\sum\vec{F} = m\vec{a}$), the mathematical expression in the $\hat{s}$—direction is obtained[1] as:

$$pdA - \left(p + \frac{\partial p}{\partial s}ds\right)dA - \rho g ds dA \cos\theta = \rho ds dA a_s \quad (1)$$

where:

$\rho dsdA$ = mass of the fluid element
p = pressure of the surrounding fluid
pdA = force acting at $ds = 0$ in the $\hat{s}$—direction
$\left(p + \frac{\partial p}{\partial s}ds\right)dA$ = force acting at ds in the $\hat{s}$—direction
$\rho g dsdA \cos\theta$ = weight component in the $\hat{s}$—direction of the fluid element
$\rho dsdAa_s$ = net force acting on the fluid element in the $\hat{s}$—direction
a_s = the fluid element acceleration in the $\hat{s}$—direction
Eq. 1 yields

$$-\frac{\partial p}{\partial s}dsdA - \rho g dsdA \cos\theta = \rho dsdAa_s$$

dividing both sides by ds dA, gives

$$-\frac{\partial p}{\partial s} - \rho g \cos\theta = \rho a_s \quad (2)$$

The acceleration a along a streamline is[1]:

$$a = \frac{\partial(V\hat{s})}{\partial t} + V\frac{\partial(V\hat{s})}{\partial s} - \frac{V^2}{R}\hat{n} \quad (3)$$

where:

$\frac{\partial(V\hat{s})}{\partial t}$ = variation of velocity with time along a streamline
$V\frac{\partial(V\hat{s})}{\partial s}$ = variation of velocity with coordinate $\hat{s}$ along a streamline
$\frac{V^2}{R}\hat{n}$ = variation of velocity with coordinate $\hat{n}$ along a streamline

The fluid element acceleration a_s in the $\hat{s}$—direction [i.e., $(V^2/R)\hat{n} = 0$, meaning that the fluid element velocity is zero along $\hat{n}$] for steady state flow [$(\partial V/\partial t) = 0$ meaning that the fluid element velocity is constant with time) is obtained from Eq. 3 as:

$$a_s = V\frac{\partial V}{\partial s} \quad (4)$$

Encyclopedia of Agricultural, Food, and Biological Engineering
DOI: 10.1081/E-EAFE 120006960
Copyright © 2003 by Marcel Dekker, Inc. All rights reserved.

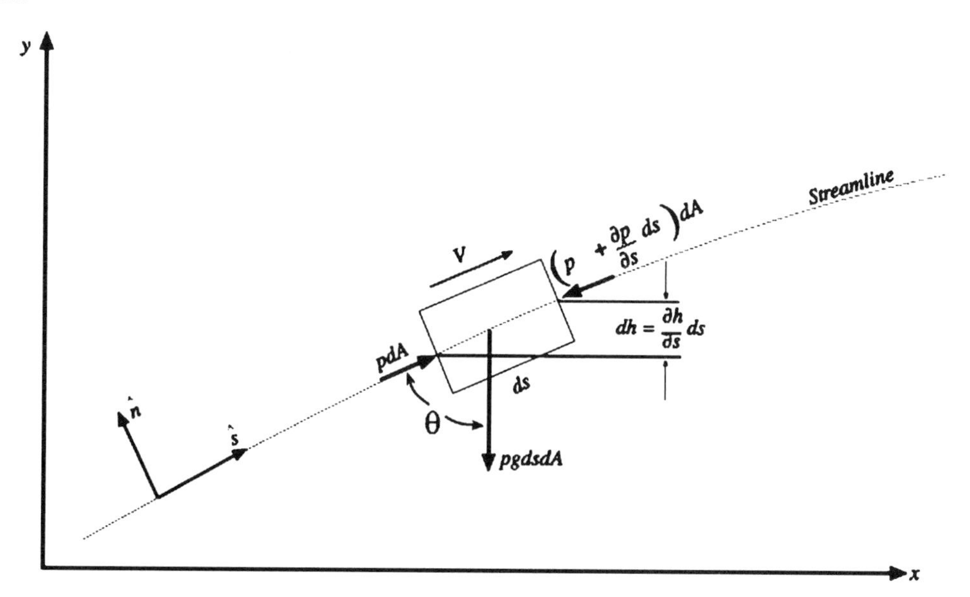

Fig. 1 Flow of a fluid element along a streamline.

and from derivative properties the following expression holds

$$V\frac{\partial V}{\partial s} = \frac{\partial}{\partial s}\left(\frac{V^2}{2}\right) \tag{5}$$

replacing Eq. 5 into Newton's second law (Eq. 2) gives,

$$-\frac{\partial p}{\partial s} - \rho g \cos\theta = \rho\frac{\partial}{\partial s}\left(\frac{V^2}{2}\right) \tag{6}$$

Also, from Fig. 1,

$$dh = \frac{\partial h}{\partial s} ds \tag{7}$$

Where h = vertical length in Fig. 1, taking a value zero for $ds = 0$ and dh when $ds = ds$ along the streamline.

From trigonometry:

$$\sin(90 - \theta) = \cos\theta = \frac{\partial h}{\partial s} \tag{8}$$

Replacing Eq. 8 into Eq. 7 and assuming constant density Eq. 6 gives

$$\frac{\partial}{\partial s}\left(\frac{V^2}{2} + \frac{p}{\rho} + gh\right) = 0 \tag{9}$$

Eq. 9 holds if along the streamline the expression

$$\left(\frac{V^2}{2} + \frac{p}{\rho} + gh\right) = \text{constant} \tag{10}$$

Applying Eq. 10 between two points along a streamline yields:

$$\frac{V_1^2}{2} + \frac{p_1}{\rho} + gh_1 = \frac{V_2^2}{2} + \frac{p_2}{\rho} + gh_2 \tag{11}$$

Eq. 11 is the known Bernouilli equation where:

$\frac{V_1^2}{2}$ = kinetic energy per unit of mass of stream reaching point 1
$\frac{p_1}{\rho}$ = work done by external forces (surroundings) on the fluid in pushing it into the tube (point 1)
gh_1 = potential energy per unit of mass of stream reaching point 1
$\frac{V_2^2}{2}$ = kinetic energy per unit of mass of stream leaving point 2
$\frac{p_2}{\rho}$ = work done by the system on the fluid in pushing it out of the tube (point 2)
gh_2 = potential energy per unit of mass of stream leaving point 2

The Bernouilli equation (Eq. 11) is a special form of a mechanical energy balance; all the terms in Eq. 11 are scalar and have dimensions of energy per unit of mass.

APPLICATIONS

There are many applications for the Bernouilli equation. Eq. 11 shows that in the absence of friction, when the velocity V is reduced, either the height above datum h or the pressure p, or both, must increase. If velocity increases, h or p or both must decrease. To apply Eq. 11 to a specific problem, it is necessary to identify the streamline or tube and to choose upstream and downstream locations (points 1 and 2).

As the two pressure terms in Bernouilli equation (Eq. 11) are divided by the same density, it can either be used as genuine thermodynamic pressure, known as absolute pressure, or measure the pressure above an arbitrary datum. When measured above atmospheric pressure, the difference is known as the gauge pressure.

Pitot Tube

This tube measures the local velocity along a streamline by measuring the difference between impact (also called total pressure) and static pressure.[1] Fig. 2 shows a Pitot tube with a tube whose opening is normal to the flow which impinges into it and creates a pressure, the stagnation pressure; and a sidewall tap to measure static pressure. The two tubes are connected to the legs of a manometer or equivalent device for measuring small pressure differences.

Applying the Bernouilli equation (Eq. 11) between these two points, and considering that at the stagnation pressure the velocity is zero, the fluid velocity at the position where the Pitot tube is located is obtained as:

$$V = C\sqrt{\frac{2(p_i - p_e)}{\rho}} \tag{12}$$

The pressure difference $(p_i - p_e)$ is measured by a manometer with a manometer fluid with a density ρ_m; therefore the following expression is obtained,

$$p_i - p_e = (\rho_m - \rho)g\Delta h \tag{13}$$

Replacing Eq. 13 into Eq. 12, gives the expression for the local velocity as:

$$V = C\sqrt{\frac{2(\rho_m - \rho)g\Delta h}{\rho}} \tag{14}$$

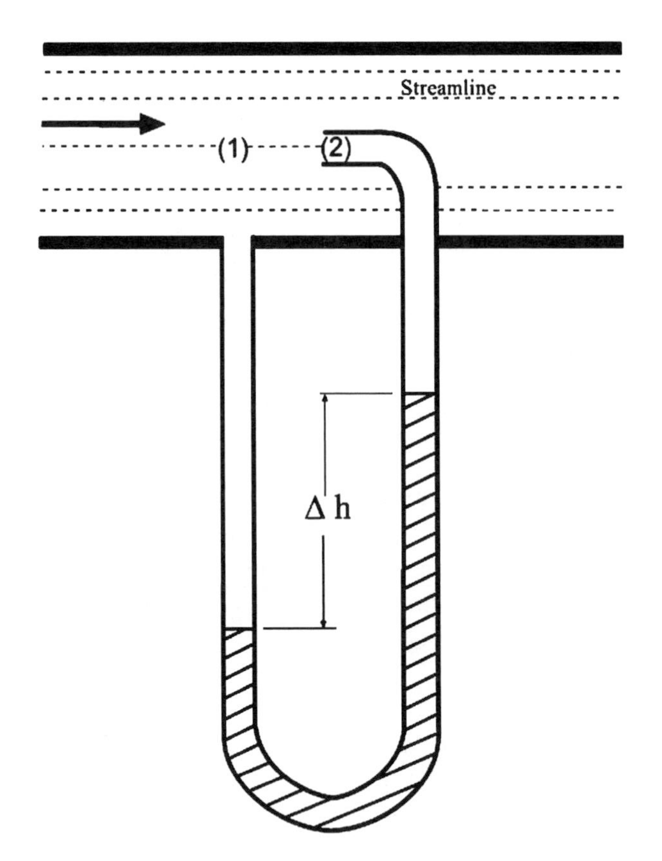

Fig. 2 Pitot tube.

The Venturi Meter

As indicated in Fig. 3, the Venturi is simply a smooth conical contraction in the cross section of a tube followed by a gradual expansion to the original area. The Venturi gives a measure of the average velocity over the entire cross section.[2]

Applying Bernouilli equation between points 1 and 2 as shown in Fig. 3 gives:

$$\frac{V_1^2}{2} + \frac{p_1}{\rho_1} = \frac{V_2^2}{2} + \frac{p_2}{\rho_2}$$

and using continuity equation gives:

$$\rho_1 A_1 V_1 = \rho_2 A_2 V_2$$

considering incompressible fluid (i.e., $\rho_1 = \rho_2$), substitution for V_1 in terms of V_2 results in

$$V_1 = \left(\frac{A_2}{A_1}\right)V_2 = \left(\frac{D_2}{D_1}\right)^2 V_2 = \beta^2 V_2$$

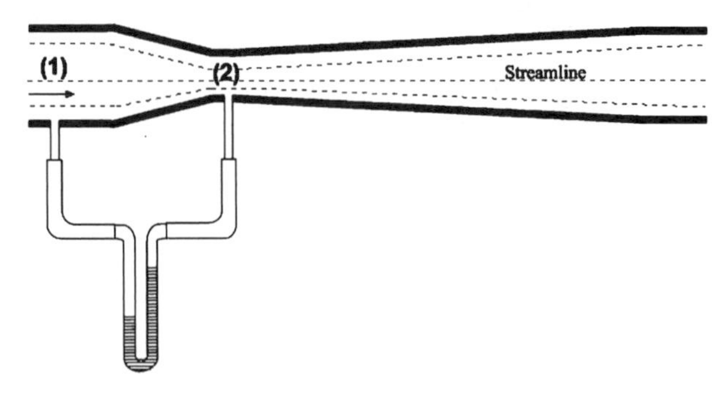

Fig. 3 Venturi meter.

where β is the ratio of the diameters. Replacing V_1 in Bernouilli equation gives the mean velocity at the throat:

$$V_2 = \mathrm{Cd}\sqrt{\frac{2(p_1 - p_2)}{\rho(1 - \beta^4)}} \tag{15}$$

The value of Cd, coefficient of discharge, depends on the Reynolds number.[3] For a well-designed Venturi, Cd usually is between 0.95 and 0.99.

Bernouilli Equation Applications in Food Engineering

In order to apply Bernouilli equation to most fluid flow situations found in food engineering it is necessary to make two modifications: one is a correction of the kinetic energy term for the variation of local velocity with radial position within a pipe, this velocity profile is determined by the rheological properties of the fluid; and the second is the correction for the presence of fluid friction in the boundary layer.[2,4]

Kinetic energy corrections for fluids in circular tubes

The average kinetic energy per unit mass (KE) of any fluid stream moving in a circular tube is[4]:

$$\mathrm{KE} = \frac{1}{R^2\bar{V}}\int_0^R V^3 r\,\mathrm{d}r \tag{16}$$

The KE in laminar flow can be expressed in terms of a kinetic energy correction factor as:

$$\mathrm{KE} = \frac{(\bar{V})^2}{\alpha} \tag{17}$$

where α is the kinetic energy correction factor.

The kinetic energy term can be evaluated if the kinetic energy correction factor α is known. In turbulent flow of any fluid the value of α is 2.[5] Expressions to evaluate α for some common fluid food models during laminar flow are given in Table 1.

Values of the ratio of actual kinetic energy to turbulent or plug flow were obtained for the Heschel-Bulkley rheological model[6] which considers Newtonian, power law, and Bingham plastic models as particular cases.

Introducing the kinetic energy correction factor in Eq. 11, the Bernouilli equation becomes:

$$\frac{V_1^2}{\alpha_1} + \frac{p_1}{\rho} + gh_1 = \frac{V_2^2}{\alpha_2} + \frac{p_2}{\rho} + gh_2 \tag{18}$$

Correction of Bernouilli equation for fluid friction

Fluid friction is any conversion of mechanical energy into heat in a flowing stream. In frictional flow the expression $\left(\frac{V^2}{2g} + \frac{p}{\rho g} + h\right)$ decreases along a streamline, and due to the principle of energy conservation, an equivalent amount of mechanical energy is lost when heat is generated due to friction.

For incompressible fluids, the Bernouilli equation is corrected for friction by adding a term h_f to the right-hand side of Eq. 18. After introducing the kinetic energy correction factor and fluid friction correction, Bernouilli equation becomes:

$$\frac{V_1^2}{\alpha_1} + \frac{p_1}{\rho} + gh_1 = \frac{V_2^2}{\alpha_2} + \frac{p_2}{\rho} + gh_2 + h_f \tag{19}$$

the units of all the terms in Eq. 19 are energy per unit of mass. The term h_f represents all the friction generated per unit of mass of fluid that occurs in the fluid between points 1 and 2.

Table 1 Kinetic energy correction factor for laminar flow in tubes

Fluid	**α, Dimensionless**
Newtonian $\sigma = \mu\dot{\gamma}$	1.0
Power law $\sigma = K(\dot{\gamma})^n$	$\alpha = \frac{2(2n+1)(5n+3)}{3(3n+1)^2}$
Herschel Bulkley $\sigma = \sigma_0 + K(\dot{\gamma})^n$	$21\alpha = \frac{L}{[M(N+Q+S)]}$
	where:
	$L = 2(1 + 3n + 2n^2 + 2n^2c + 2nc + 2n^2c^2)^3(2 + 3n)(3 + 5n)(3 + 4n)$
	$M = (1 + 2n)^2(1 + 3n)^2$
	$N = 18 + n(105 + 66c) + n^2(243 + 306c + 85c^2)$
	$Q = n^3(279 + 522c + 350c^2) + n^4(159 + 390c + 477c^2)$
	$S = n^5(36 + 108c + 216c^2)$
	$c = \frac{\sigma_0}{\sigma_w}$

The final form of Bernouilli equation (Eq. 19) is very useful in food engineering applications such as pipeline design calculations and design of the holding tube for sterilization of continuously flowing fluids.

NOMENCLATURE

a_s	fluid element acceleration in the $\hat{s}$—direction, m sec^{-2}
A	area, m^2
C	correction factor for Pitot tube, dimensionless
Cd	coefficient of discharge for Venturi meter, Eq. 15, dimensionless
$c = \frac{\sigma_0}{\sigma_w}$	dimensionless
ds	differential length along a streamline
g	acceleration due to gravity, m sec^{-2}.
h	height above a reference, datum, and plane, m
h_f	friction generated in the fluid between points 1 and 2, J kg^{-1}
n	flow behavior index, dimensionless
p	pressure of the surrounding fluid, Pa
p_i	impact pressure for Pitot tube, Eq. 12, Pa
p_e	local static pressure for Pitot tube, Eq. 12, Pa
p_1	static pressure at upstream pressure tap for Venturi meter, Eq. 15, Pa
p_2	static pressure at downstream pressure tap for Venturi meter, Eq. 15, Pa
R	radius, m
V	local velocity, m sec^{-1}
$\bar{V}$	average velocity, m sec^{-1}
α	kinetic energy correction factor, dimensionless
β	diameter ratio in Eq. 15, dimensionless
$\dot{\gamma}$	shear rate, sec^{-1}
Δh	length of the manometer column, m
K	consistency coefficient, Pa sec^n
μ	Newtonian viscosity, Pa sec
ρ	fluid density, kg m^{-3}
ρ_m	density of manometer fluid, kg m^{-3}
σ	shear stress, Pa
σ_0	yield stress, Pa

REFERENCES

1. Potter, M.; Foss, J. The Bernoulli Equation. *Fluid Mechanics*, 1st Ed.; Great Lakes, Inc.: Michigan, 1982; 49–51.
2. Nedderman, R. Newtonian Fluid Mechanics. In *Chemical Engineering for the Food Industry*, 1st Ed.; Fryer, P., Pyle, D., Rielly, C., Eds.; Blackie Academic and Professional: London, 1997; 68–77.
3. Sakiadis, B. Fluid and Particle Mechanics. In *Perry's Chemical Engineers' Handbook*, 6th Ed.; Perry, R., Gree, D., Maloney, J., Eds.; McGraw-Hill, Inc.: New York, 1984; 5-6–5-12.
4. Heldman, D.R.; Singh, R.P. Rheology of Processed Foods. *Food Process Engineering*, 2nd Ed.; AVI Publishing Company, Inc.: Wesport, CT, 1981; 46–48.
5. Steffe, J.F. Tube Viscometry. *Rheological Methods in Food Process Engineering*, 2nd Ed.; Freeman Press: East Lansing, MI, 1996; 128–138.
6. Osorio, F.A.; Steffe, J.F. Kinetic Energy Corrections for Non-Newtonian Fluids in Circular Tubes. J. Food Sci. **1984**, *49* (5), 1295–1296, 1315.

Biodegradation Waste Treatment

Pedro M. Álvarez
Fernando J. Beltrán
Juan F. García-Araya
Universidad de Extremadura, Badajoz, Spain

B

INTRODUCTION

Many food processing plant effluents constitute a growing environmental problem and their disposal and management are of great interest today. In this category, production of ethanol from fermentation of fruit and vegetables and subsequent distillation usually generates large volumes of high strength wastewater (i.e., spentwash) with high BOD, COD, solid content, acidic pH, and variable amounts of inorganic materials.[1] The treatment of spentwash from distilleries is, therefore, necessary to be applied before its disposal in the environment. In many cases, among the treatment choices, anaerobic digestion after neutralization seems to be the preferred method because of the production of biogas which can be further used to recover energy.[2] However, depending on the distillery type, anaerobic treatment of spentwash sometimes does not show suitable performance because of the presence of toxic or inhibitory substances towards biological oxidation.[3] In the specific case of distilleries located in urban areas, the best management option to deal with spentwash is to apply an anaerobic digestion as pretreatment on-site and discharge the partially degraded effluent to the nearest municipal wastewater treatment plant (MWTP) to be treated as a mixture with domestic sewage.

In MWTP, receiving municipal and industrial wastewaters a system comprising solids and greases separation, neutralization, mixing and dilution, activated sludge biodegradation and tertiary treatment (i.e., nutrient removal and chemical oxidation) may represent an efficient way to obtain an effluent suitable to be released to water courses or reused. The activated sludge system is the key factor in achieving efficient organic matter removal. However, when dealing with industrial wastewater the presence of toxic substances for aerobic population in activated sludge (i.e., high concentration of phenols which are often present in distillery spentwash) is a common event. This reduces or even completely inhibits the efficiency of such a biodegradation technology. The use of ozone to reduce the level of toxics, like phenolic compounds, in distillery wastewater has been shown to be a promising technology.[4]

The aim of this work is to compare the single biodegradation of a combined distillery wastewater–domestic sewage by activated sludge with other systems in which ozonation is integrated with biodegradation.

MATERIALS AND METHODS

Wastewater and Sludge Source and Preparation

Wastewater was taken from a wastewater treatment plant located in Almendralejo (Badajoz province, Spain) that receives separately wastewater from the municipal sewerage system and various food-processing industries (i.e., manufacturers of table olives and wine and spirit distilleries). Wastewater coming from distilleries only could be collected separately due to the seasonal nature of the food-processing activities. This industrial effluent was received in the MWTP after being subjected to partial degradation by anaerobic digestion in their own factories so that the average characteristics of collected wastewater were those presented in Table 1. Domestic sewage was also taken and used for dilution purposes. For this work typically industrial wastewater was neutralized with sodium hydroxide and mixed with domestic sewage with a dilution ratio close to five by volume, resulting in a wastewater with about $2.5\,g\,L^{-1}$ COD. Hereafter, this combined distillery–domestic wastewater will be referred simply as wastewater.

Sludge from the full-scale wastewater treatment plant was used as inoculum to start the lab-scale activated sludge process. Micro-organisms of the inoculum were first acclimated to the specific wastewater of this work in a batch digester operating in fill and drawn mode by increasing the content of industrial wastewater step by step until it reaches a $2.5\,g\,L^{-1}$ influent COD. After this acclimation period (over 1 wk) the sludge showed good metabolic and settling characteristics and maximum biological activity was attained.

Continuous Wastewater Treatment System

Wastewater was continuously treated in the experimental apparatus whose schematic is shown in Fig. 1. For

Encyclopedia of Agricultural, Food, and Biological Engineering
DOI: 10.1081/E-EAFE 120007217
Copyright © 2003 by Marcel Dekker, Inc. All rights reserved.

Table 1 Main features of the industrial wastewater and domestic sewage used in this work (average values)

Parameter of wastewater	Industrial wastewater	Domestic sewage
COD ($g L^{-1}$)	10.8	0.28
BOD_5 ($g L^{-1}$)	8.4	0.19
TOC ($g L^{-1}$)	3.8	0.93
TKN ($g L^{-1}$)	0.8	0.04
Total phenols[a] ($g L^{-1}$)	0.4	—
A_{254}[b]	1.2[c]	0.9
PH	5–6	7.6

[a] As gallic acid.
[b] Measured with 1 cm path length quartz cell.
[c] Sample diluted 5 times for measure.

the treatment of wastewater, the overall process combined preozonation, activated sludge system, and postozonation. The system also comprised ozonation of the recycled sludge. Details of each individual process and apparatus can be found elsewhere.[5] Experimental conditions used for each treatment stage of this work are summarized in Table 2. The system was operated up to reach the steady state. Then, samples of influent and effluent gases, water, and sludge were withdrawn and analyzed.

Analyses

Liquid effluents from both biological and ozonation systems were analyzed for pH, COD, BOD, ultraviolet

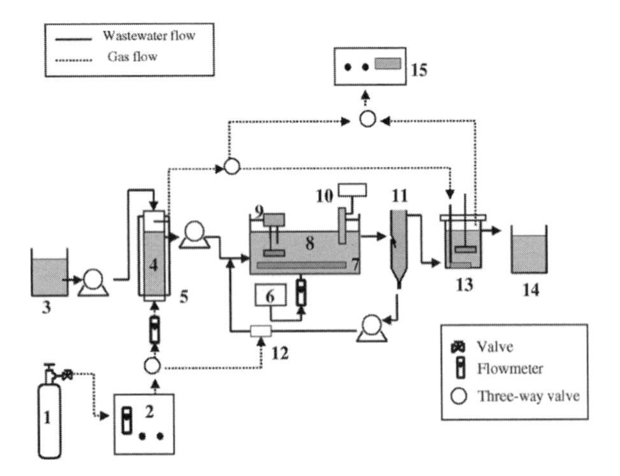

Fig. 1 Schematic diagram of the lab-scale wastewater treatment system: 1. Oxygen; 2. Ozone generator; 3. Wastewater feed tank; 4. Ozonation bubble column; 5. Thermoregulated jacket; 6. Air pump; 7. Diffuser; 8. Aeration tank; 9. Temperature and agitation speed controller; 10. Dissolved oxygen meter; 11. Clarifier; 12. Sludge ozonation chamber; 13. Postozonation tank; 14. Effluent collection tank; 15. Ozone analyzer.

absorbance (UV_{254}), and total phenols. UV_{254} was measured with a Hitachi 2000 spectrophotometer using a 1-cm path length quartz cell, and total phenols were analyzed by the Folin Ciocalteau method using gallic acid as the standard.[6] The rest of the analyses were carried out according to Standard Methods.[7] pH and dissolved oxygen (DO) in the aeration tank were continuously monitored by a pH meter (Crison 507) and an oxygen meter (YSI 58), respectively. Total and volatile suspended solids (TSS and VSS, respectively), and sludge volumetric index (SVI) were also measured in the mixed liquor of the aeration tank according to standard procedures. Ozone was measured in the gas phase by means of an Anseros Ozomat GM19 analyzer.

RESULTS AND DISCUSSION

The wastewater treatment system outlined above was used to investigate various operating modes, which could include various ozonation steps.

Single Biodegradation

The activated sludge system was operated at a hydraulic retention time (HRT) between 9 hr and 36 hr, average MLVSS of about $2.5 g L^{-1}$ and 100% recycle ratio. The operation of the activated sludge system could be arranged to provide maximum BOD and COD removals of 95% and 72%, respectively. However, at these conditions, corresponding to a food to micro-organism ratio (F/M) of 0.67 kg COD applied/kg MLVSS day, sludge settling was poor ($SVI > 160 mL g^{-1}$). Better sludge settleability (i.e., $SVI < 130 mL g^{-1}$) while keeping suitable biodegradation performance (i.e., 84% and 65% of BOD and COD removals, respectively) was attained when the system was operated at F/M of 1.3 kg BOD applied/kg MLVSS day which meant a HRT of 18 hr. At these conditions, removals of total phenols and UV_{254} were found to be just 42% and 37%, respectively. Therefore for the case-study single biodegradation was shown unsuitable to achieve desirable degradation levels.

Preozonation Followed by Biodegradation

In order to improve results from single biodegradation, wastewater was continuously preozonated by applying an ozone dose of about 0.45 g of ozone per liter of wastewater. Ozone mass transfer efficiency was slightly lower than 80% and COD degradation level after ozonation was found to be 15.8% (i.e., 1.1 g COD removed/g ozone absorbed). In spite of moderate COD removal, ozonation also led to significant total phenols and UV_{254} removals (36% and 29%, respectively). Biodegradation after ozonation, operating

Table 2 Experimental conditions applied in this work at each treatment stage

System	Effluent preozonation	Activated sludge system	Effluent postozonation	Sludge ozonation
Reactor	Bubble column	Aeration tank	Stirred tank	Chamber
Working volume (L)	1	18	4	0.1
Wastewater flow ($L\,hr^{-1}$)	0.5–2	0.5–2	0.5–2	0.5–2
pH	5–7	6.5–8.0	7–8	—
DO ($mg\,L^{-1}$)	—	> 2	—	—
Temperature (°C)	20	20	20	20
VSS concentration ($g\,L^{-1}$)	—	2.5	—	4–6
Gas flow ($L\,hr^{-1}$)	20–30	200		10–30
Inlet ozone concentration ($mg\,L^{-1}$)	15	—	2–8	15

at 18 hr HRT was shown to be more effective than single biodegradation, leading to an effluent with 540 $mg\,L^{-1}$ COD (i.e., 74% removal) and very low phenol content (i.e., 22.3 $mg\,L^{-1}$). In addition to having a higher specific oxygen uptake rate, the sludge from aeration tank, where biological oxidation was carried out after wastewater preozonation, showed good settling ability without being observed bulking or foaming problems.

Preozonation—Biodegradation—Postozonation

Postozonation was aimed at the removal of biorefractory material and to take advantage of the residual ozone from preozonation. Thus, the gas leaving the preozonation bubble column entered the postozonation tank. 95% of Ozone was consumed there, leading to a 14.2% COD removal which represented 0.78 g COD removed per g of ozone supplied.

Recycle Sludge Ozonation

Ozone has been shown as a potential chemical agent to stabilize and mineralize excess sludge produced from the metabolism of activated sludge during biodegradation.[8] Also, ozone can prevent bulking problems in activated sludge systems.[9] With both purposes sludge was ozonated in the sludge return line. For the specific conditions of this work, the rate of production of excess of activated sludge from continuous experiments was measured to be about 0.7 $g\,hr^{-1}$, regardless of whether preozonation was applied or not. Preliminary experiments performed in a batch reactor operating with high concentration of suspended solids (SS = 5 $g\,L^{-1}$) showed that 1 g of ozone can degrade up to 0.26 g of sludge (as suspended solids), depending on the ozonation efficiency. Thus, taking into account the aforesaid conditions, 2.7 g $O_3\,hr^{-1}$ should be supplied to the ozonation chamber to achieve zero sludge production within the system. However, such an ozone requirement does not seem recommended because it would kill microorganisms of activated sludge. Substantially lower ozone mass rates of 0.45 $g\,hr^{-1}$ and 0.15 $g\,hr^{-1}$ were used. For the former, the excess of sludge was reduced by a 73%. However this effect was not only due to solubilization and mineralization but also to activated sludge inactivation, the overall COD removal being lower than 50%. When the system was operated with 0.15 g of ozone per hour supplied to the ozonation chamber, less than 10% of excess of sludge was removed, but the biological activity was kept going (i.e., overall COD removal over 80%) and the sludge showed very good settling characteristics with SVI in the range 80 $mL\,g^{-1}$–100 $mL\,g^{-1}$.

Comparison of Treatment Efficiencies

Fig. 2 presents a comparison of efficiencies of the various treatment sequences used in this work. From this figure it is observed that ozonation technologies enhance the overall process performance, especially total phenols removal. Fig. 3 shows the advantages of applying ozone to the sludge return line, in order to avoid settling problems and slightly reduce the excess of sludge.

Economic Analysis

Final decision about the recommended treatment process, or combination of process, that will best clean up a given industrial wastewater should be accomplished after all specifications are established and detailed cost estimates

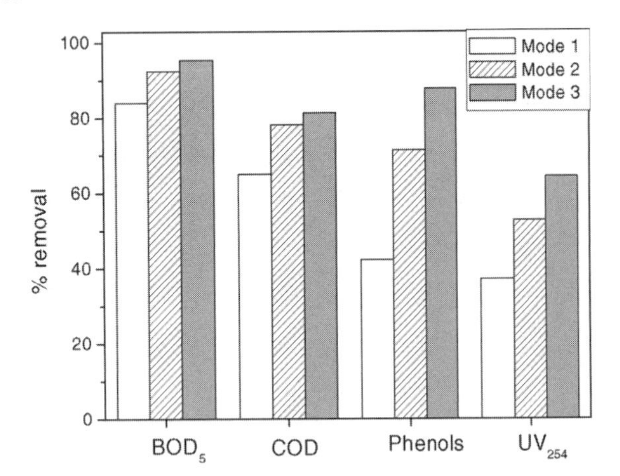

Fig. 2 Comparison of BOD_5, COD, total phenols, and UV_{254} removals achieved by the various treatment modes: 1. Single biodegradation; 2. Preozonation–biodegradation; 3. Preozonation—Biodegradation—Postozonation.

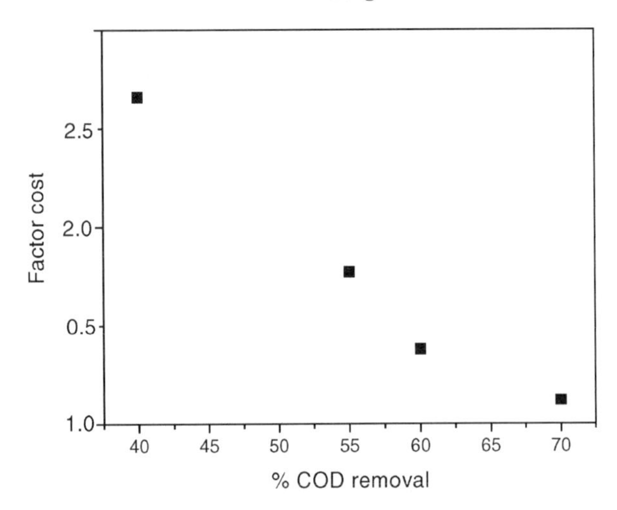

Fig. 4 Factor cost of total cost of integrated activated sludge-ozonation plant for the treatment of distillery wastewater. Influence of the % of COD removal.

are developed for any specific application. In this case-study, a sounding economic analysis of the ozonation phases is difficult to assess since to our knowledge there is no plants using such technology in the way, as was investigated here. Therefore no bids of capital costs (i.e., ozone generation equipment, ozone contactor, ozone destruction equipment, piping, valves, instrumentation, etc.) are available. Also operating costs are highly dependent on the price of oxygen (or air) and electricity, which in turns depend on local markets. However, before the preliminary design stage an order of magnitude estimate is considered useful for initial planning purpose.

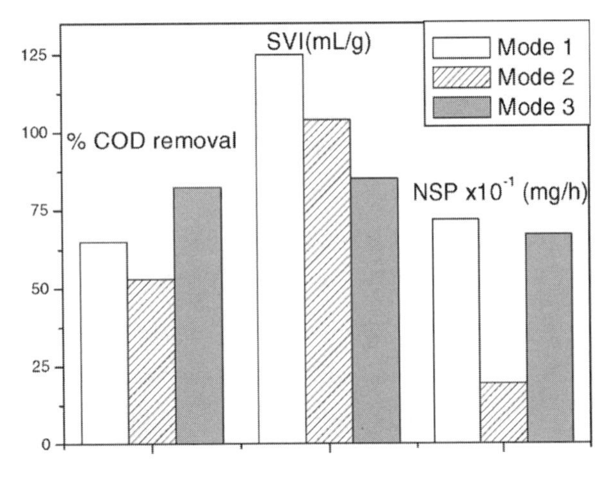

Fig. 3 Comparison of COD removal, SVI and net sludge production and (NSP) in treatments with and without ozonation of the sludge: Mode 1. Single biodegradation; Mode 2. Biodegradation with ozonated sludge (Ozone dose: $0.45\,g\,L^{-1}$); Mode 3. Biodegradation with ozonated sludge (Ozone dose: $0.15\,g\,L^{-1}$).

Cost estimates have been performed according to the methods outlined by Bellamy et al.[10] For doing this, estimated sizing of main process equipment has also been carried out previously according to known procedures.[11,12] Major variables affecting overall cost were organic load of wastewater, level of purification required, and energy cost. As exemplary cases Fig. 4 shows the factor costs of the integrated activated sludge-ozonation plant as here proposed when referred to a base case: a conventional activated sludge plant treating wastewater of a 60,000 equivalent population.

$$\text{Cost of integrated plant} = \text{Factor cost X Cost of base case} \qquad (1)$$

As can be observed in Fig. 4, economic results suppose a drawback of ozonation systems as factor cost was estimated to be higher than unity. However if high quality effluent discharge is taken in mind, single biological systems cannot cope with the required purification demand (i.e., 72% maximum COD removal) and should be accompanied with the aid of a physical–chemical technology. Then, ozonation can be considered as a sounding alternative choice.

REFERENCES

1. Sheehan, G.J.; Greenfield, P. Utilisation, Treatment and Disposal of Distillery Wastewater. Water Res. **1980**, *14*, 257–277.
2. Romero, L.I.; Nebot, E.; De la Ossa, E.M.; Sales, D. Microbiological Purification Kinetics of Wine-Distillery Wastewaters. J. Chem. Technol. Biotechnol. **1993**, *58*, 141–149.

3. Borja, R.; Martin, A.; Maestro, R.; Luque, M.; Duran, M.M. Improvement of the Kinetics of Anaerobic Digestion of Molasses by the Removal of Phenolics Compounds. Biotech. Lett. **1993**, *15*, 311–316.
4. Beltrán, F.J.; García-Araya, J.F.; Rivas, F.J.; Álvarez, P.; Rodríguez, E. Kinetics of Competitive Ozonation Of Some Phenolic Compounds Present in Wastewater From Food Processing Industries. Ozone Sci. Eng. **2000**, *22*, 167–183.
5. Beltrán, F.J.; García-Araya, J.F.; Álvarez, P.M. Integration of Continuous Biological and Chemical (Ozone) Treatment of Domestic Wastewater. 1. Biodegradation and Post-ozonation. 2. Ozonation Followed by Biological Oxidation. J. Chem. Technol. Biotehnol. **1999**, *74*, 884–890.
6. García-García, I.; Bonilla-Venceslada, J.L.; Jiménez-Peña, P.R.; Ramos-Gómez, E. Biodegradation of Phenol Compounds in Vinasse Using Aspergillus Terreus and Geotrichum Candidum. Water Res. **1997**, *31*, 2005–2011.
7. APHA, AWWA, WPCT, *Standard Methods for the Examination of Water and Wastewater*, 16th Ed.; American Public Health Association: Washington D.C., 1985.
8. Délereis, S.; Paul, E.; Audic, J.M.; Roustan, M.; Debellefontaine, H. Effect of Ozonation on Activated Sludge Solubilization and Mineralization. Ozone Sci. Eng. **2000**, *22*, 461–472.
9. Van Leeuwen, H. Domestic and Industrial Wastewater Treatment with Ozonated Activated Sludge. Ozone Sci. Eng. **1988**, *10*, 291–308.
10. Bellamy, W.D.; Langlais, B.; Lykins, B.; Rakness, K.L.; Robson, C.M.; Schulhof, P. Economics of Ozone Systems: New Instalations and Retrofits. In *Ozone in Water Treatment. Application and Engineering*; Langlais, B., Reckhow, D.A., Brink, D.R., Eds.; Lewis Publishers Inc.: USA, 1991.
11. Bellamy, W.D.; Damez, F.; Langlais, B.; Montiel, A.; Rakness, K.L.; Reckhow, D.A.; Robson, C.M. Engineering Aspects. In *Ozone in Water Treatment. Application and Engineering*; Langlais, B., Reckhow, D.A., Brink, D.R., Eds.; Lewis Publishers Inc.: USA, 1991.
12. Metcalf & Eddy, Inc., *Wastewater Engineering. Treatment, Disposal and Reuse*, 3rd Ed.; McGraw-Hill: New York, 1991.

Biological Engineering Definition

James H. Dooley
Silverbrook Limited, Federal Way, Washington, U.S.A.

INTRODUCTION

The engineering profession is among the many important and enduring roles supported by society. The role of technical designer, with decisions based on observations about nature, has been traced to at least 5000 B.C. when the pyramids and major irrigation canal systems were designed and constructed along the Nile river.[1,2] The ancient role of technical designer as engineer is disputed by Florman,[3] who holds the belief that science is necessary for there to be engineering. Florman asserts that the modern profession of engineering had its start in the eighteenth century when science was first applied to the disciplined solution of technical problems.[3] Now, as then a substantial cadre of engineers hold a strong belief that science is necessary for there to be engineering.[4] Engineering disciplines have evolved along two lines—scientific basis and field of application. Examples of science-based engineering disciplines include electrical, mechanical, chemical, and most recently biological engineering. Examples of application-based engineering disciplines include biomedical, aerospace, agricultural, and environmental.

EMERGENCE OF ENGINEERING DISCIPLINES

Engineering disciplines, like other defined roles within society, emerge from the ether of events, ideas, and ever-increasing knowledge. The catalysts for emergence of a defined societal role, such as that of biological engineer, include a critical mass of skilled practitioners and an increasing societal need. Social systems are an integral part of the human condition. Bertrand (Ref. 5, p. 25) defined a social system as "... two or more people in interaction directed toward attaining a goal and guided by patterns of structured and shared symbols and expectations." Humans from the earliest times have banded together into organized groups, communities, and societies to achieve goals that are not attainable by singular individuals.[6]

Continued existence of a social system is dependent on the formation and maintenance of systems of cooperation among inter-linked social units. Predictable and ordered patterns of interaction among members of a social unit constitute an important part of the culture of the unit.[7] According to Bertrand,[8] the structural elements of any social system include: 1) norms; 2) roles; 3) status–position; 4) situs; 5) station; and 6) subsystems.

Norms constitute patterns of behavior expected of members of a particular social system (Ref. 9, p. 161). Norms for a particular group may include scientific literacy, mathematical competency, cultural awareness, and social sensitivity, as well as those more traditionally identified such as respect for the individual. In the case of biological engineering, there are increasingly well-defined norms for education,[10] methods for problem solving,[11] and the lexicon for describing biological systems. The societal roles for those who are identified as biological engineers, including the applied fields dependent upon biological engineering sciences, are broadly accepted by academia, research institutions, and industry. Biological engineers are rapidly being included on multidisciplinary teams and respected for their bio-technical and biosystems expertise.

The discipline of biological engineering has itself evolved to the point where subsystems and defined subroles exist. Role is defined as a labeled set of behaviors that one who holds a certain status or position in a social unit is expected to exhibit.[12] Individuals in a group develop one or more identities associated with the roles they occupy. Roles are the dynamic action elements of social systems. Roles provide conduits for communication within and across social unit boundaries.[8] Functional interdependence of the members of a social system is a naturally accepted part of existence.[13] An outgrowth of interdependence is further specialization of roles to improve group efficiency, reinforce the mutual interdependence, and reduce interpersonal conflict.[8] Those with the skills, attitudes, and desire become the farmers, cobblers, builders, historians, and philosophers of a community—provided the community places sufficient value on these roles to endorse and support individuals to become specialists in one or more roles. Biological engineering examples include educators,[14] engineering scientists, systems modelers, process designers, and compliance engineers. The applied bio-based engineering fields such as food engineering, agricultural, forest, bioresources, biomedical, and biochemical are recognized by society for their combined competence in the collective set of biological engineering science, biosystems

Encyclopedia of Agricultural, Food, and Biological Engineering
DOI: 10.1081/E-EAFE 120007160
Copyright © 2003 by Marcel Dekker, Inc. All rights reserved.

design/problem solving, and the particular economic enterprise or field of application.

ROLE OF THE ENGINEER IN SOCIETY

There are a number of published sources describing normative traits of those who perform the role of engineer. Morrison[15] proposes the following relevant norms for engineering behavior:

1. (Engineers) are logical, methodical, objective, and make unemotional decisions based on facts.
2. (Engineers) use their technical knowledge to check the validity of information.
3. (Engineers) can analyze problems thoroughly, look beyond the immediate ones, and ask good questions to explore alternative solutions to technical problems.
4. (Engineers) can engage in future planning with appropriate consideration for technology and its relationship to cost effectiveness.

The Accreditation Board for Engineering Technology[16] defines the profession of engineering as follows:

> "Engineering is that profession in which the knowledge of the mathematical and natural sciences gained by study, experience, and practice is applied with judgment to develop ways to utilize, economically, the materials and forces of nature for the benefit of mankind."

Based on a review of many historic and current monographs, we can synthesize a set of conclusions that society expects engineers to:

1. Act in service to other members of society.
2. Be knowledgeable about mathematics and basic sciences.
3. Be competent in problem analysis and getting at the core technical issues.
4. Provide creative technical solutions to societal needs.
5. Protect the health and safety of all members of the society as affected by technical works.
6. Ensure that technological works are sensitive to the cultural mores of the society within which they are created.

Asimow[17] developed a philosophy of engineering design that is generally consistent with the norms we have just stated. Asimow asserts that "Design must be a response to individual or social needs which can be satisfied by the technological factors of culture" (Ref. 17, p. 5). The first of the norms listed above suggest that societies are likely to support activities that meet understood needs, and have an expected net benefit to the community. Societies are unlikely to share their limited resources with members who are pursuing apparently counterproductive or frivolous design objectives. They are likely to withdraw support from activities where the costs exceed the perceived benefits. The second through fourth norms suggest that other members of the social unit expect the work products from designers to be the best solution to the problem or need. A trust is placed upon the designer to make compromising trade-offs between the inevitable conflicts in values held by various stakeholders. The fifth norm suggests that society places a trust on engineers to make decisions from a basis of knowledge and foresight to preclude harm to those who do not have scientific and technical understanding. A principal facet of engineering life is the willingness to accept responsibility for the consequences of design decisions.[3] The sixth and final norm suggests that engineers and their sponsors appreciate that they are members of a broader social group. It is axiomatic that the current beliefs, risk tolerance, and collective wisdom of the broader social group will influence the design of technical works.

DEFINITION OF BIOLOGICAL ENGINEERING

The current definition(s) of biological engineering are as much social constructs as is the societal role of biological engineering itself. As is common with many terms, a number of alternative definitions exist to help readers properly apply the term in various contexts. As noted earlier, there are two dominant worldviews with respect to what it means to be a biological engineer, practice the profession of biological engineering, or do biological engineering independently of training or credentialing.[18]

(1) Biological engineering is a biological sciences-based, application independent engineering discipline.

Johnson and Davis[19] assert that the inseparability of biological engineers from the science of biology is parallel to those of chemical engineering with chemistry, mechanical engineering with mechanics, etc. While many other applied and science-based engineering disciplines work with biological materials and systems, Johnson and Phillips[4] suggest that the practice of biological engineering is founded upon a "substantial and intrinsic knowledge base in the biological sciences."

(2) Biological engineering is an applied engineering discipline with special knowledge and interest in problems related to biological organisms, materials, processes and systems.

Modern biology and biological systems modeling techniques enable designers to apply engineering thinking and disciplined mathematical rigor to the engineering practitioner arena. Cherry[20] suggests that biological engineering encompasses a set of "skills and knowledge" that enables commercial practice of current biological discoveries. Verma[21] further observes that the convergence of computer technologies and biotechnologies enable rigorous practice of engineering with and for inherently complex biological systems.

Many current members of the emergent profession of biological engineering have their academic and practice roots in longstanding disciplines such as chemical, electrical, or mechanical engineering. Both converted and newly minted biological engineers tend to be competent in a set of core concepts of biological engineering that help differentiate it from other disciplines, including:

- Understanding the fundamental life processes (genetics, biochemistry, cellular to organismal to ecosystems systems, etc.) sufficiently to make informed judgments during the design of technical works that are based on the first principles of biology.
- Understanding the interrelationships between living organisms and their physical environment.
- Sensitivity to the physical properties, environmental and life-process needs associated with living organisms when designing bio-physical systems.
- Understanding the unique chemical and physical properties, including temporal properties, of living and organic materials.
- Knowing how to make disciplined design decisions when faced with highly variable, uncertain, and unknown properties.

Dr. Norman Scott, 2001 president of the Institute of Biological Engineering (IBE), worked with members of the Institute to develop a working definition of the profession and discipline of biological engineering.[22]

(3) Biological engineering is the biology-based engineering discipline that integrates life sciences with engineering principles in the advancement of fundamental concepts of biological systems from molecular to ecosystem levels.

(4) Biological Engineering is the discipline of engineering that integrates biology, physics, chemistry and mathematics with engineering principles in the design of biologically based products and processes from the biomolecular to organisms to the ecosystem.

The IBE definitions reflect membership interest in biological sciences and their application at all scales, from molecular to ecosystem. The Institute's effort to refine and codify a coherent definition is ongoing as of this writing. The working definitions reflect both a biological science-based application-independent dimension to the profession and a biologically competent practitioner dimension.

CONCLUSION

We have explored relevant literature from the social sciences and engineering in order to synthesize a definition of biological engineering consistent with the modern view of the role of engineering in society. Engineering is a socially endorsed and supported role with the dual expectations that engineers will provide solutions to society's technical problems and will participate in societal deliberations with a perspective that is grounded in science, mathematics, and other technical matters. Engineering is, thus, a social construct that has grown and developed its own institutions, subdisciplines, roles, and norms, just like any other segment of society.

The alternate definitions of biological engineering reflect the continuing parallel worldviews of engineer as designer and engineer as holder of a unique set of application-independent skills and knowledge that may or applied through many avenues of communication beyond the design and creation of technical works. What is constant across all worldviews is that the discipline of biological engineering necessarily includes competencies in both biological sciences and engineering.

REFERENCES

1. Howell, E.B. *Different by Design*; American Management Association: New York, 1996.
2. Mendelssohn, K. *The Riddle of the Pyramids*; Praeger: New York, 1974.
3. Florman, S.C. *The Civilized Engineer*; St. Martin's Press: New York, 1987.
4. Johnson, A.T.; Phillips, W.M. Philosophical Foundations of Biological Engineering. J. Eng. Educ. **1995**, *October*, 311–318.
5. Bertrand, A.L. *Basic Sociology*; Appleton-Century-Crofts: New York, 1967.
6. Sorokin, P.A. *Society, Culture, and Personality: Their Structure and Dynamics*; Harper & Bros.: New York, 1947.
7. Chinoy, E., *Sociological Perspective*, 2nd Ed.; Random House: New York, 1968.
8. Bertrand, A.L. *Social Organization: A General Systems and Role Theory Perspective*; F.A. Davis Co.: Philadelphia, 1972.
9. Parsons, T. *Social Systems and the Evolution of Action Theory*; The Free Press: New York, 1977.
10. Garrett, R.E. *Core Engineering Courses for Biological Engineering. 1990 ASEE Annual Conference*, American Society for Engineering Education, 1990.
11. Dooley, J.H. Paradigms for the 21st Century 'Living World' Engineer—Redefining the 'System'. In *International Symposium II—The Culture of Engineering in a Rapidly Changing World*, University of California at Berkeley, 1993.

B

12. Cohen, B.J.; Orbuch, T.L. *Introduction to Sociology*; McGraw-Hill: New York, 1979.
13. Warriner, C.K. *The Emergence of Society*; The Dorsey Press: Homewood, IL, 1970.
14. Wells, J.H.; Taylor, T.A. A Contemporary Vision of Biological Engineering Education. Resource **1995**, *2* (7), 13–15.
15. Morrison, P. Making Managers of Engineers. J. Manag. Eng. **1986**, *2* (4), 259–264.
16. ABET. *Criteria for Accrediting Programs in Engineering in the United States: Effective for Evaluations During the 1996–97 Accreditation Cycle*, 1996; Accreditation Board for Engineering and Technology: Baltimore, MD.
17. Asimow, M. *Introduction to Design*; Prentice-Hall, Inc.: Englewood Cliffs, NJ, 1962.
18. Dooley, J.H. Biological Engineering Core Concepts. In *Proceedings of the IBE 2001 Workshop "DNA of Biological Engineering: Defining the Body of Knowledge for the Discipline"*, 2001; Institute of Biological Engineering: Cortland, NY, 8–19.
19. Johnson, A.T.; Davis, D.C. Biological Engineering: A Discipline Whose Time Has Come. Eng. Educ. **1990**, *80* (1), 15–18.
20. Cherry, R.S. It Is Time to Recognize Biological Engineering. Biotechnol. Prog. **1988**, *4* (1), M3.
21. Verma, B.P. An Emerging New Order. Biological Engineering Is Evolving as a Discipline and a Profession. Resource **1995**, *2* (11), 8–10.
22. IBE. Proceedings of the IBE Workshop "DNA of Biological Engineering: Defining the Body of Knowledge for the Discipline". In *IBE Annual Meeting. 2001: An Engineering BIOdyssey*, 2001; Institute of Biological Engineering: Sacramento, CA.

Biological Engineering Evolution

Brahm P. Verma
The University of Georgia, Athens, Georgia, U.S.A.

INTRODUCTION

Engineering is a practical art learned from scientific knowledge, mathematical logic, and experience from practice. Over the past 5000 years, engineering has evolved into the means by which ideas are turned into useful products, process, and systems, thus contributing to the development of technologies that have helped the society to advance and prosper. In other words, engineering meets societal needs. Perhaps engineering, as an attempt to design methods for the use and control of natural materials, began with the beginning of civilization itself.

This article presents a brief history of engineering and the evolution of several engineering disciplines. Biological engineering, the newest engineering discipline, is in a nascent stage. The growth and potential of biological engineering in the age of information and biology is discussed.

EARLY ENGINEERS

As early as 6000 B.C.E., ancient Mesopotamians (inhabitants of land between the Tigris and Euphrates rivers in present-day Iraq) built temples, water canals, city walls, and significant irrigation and flood control systems. Ancient Egyptians, whose work in building large structures is legendary, also developed extensive water transporting canals and drainage systems. Similarly, engineering feats of the Greeks and Romans include extraordinary temples, extensive networks of roads and bridges for transportation, aqueducts, and the development of construction methods and machines (such as pile drivers, hoists, and bucket wheels for lifting water) and building materials. Without the present-day knowledge of the natural sciences, these "engineers" must have relied upon intuition developed from an acute sense of observation rather than on a critical understanding of the laws of nature and mathematical logic.

Advances in science and mathematics in the Middle Ages (1300–1700 A.D.) spurred the development of science-based engineering principles and practices. These principles and practices constitute a body of knowledge known as engineering science. Engineering science gives a fundamental basis for engineering work, provides an objective basis for evaluating the functionality of design concepts and enables the prediction of behavior in designed systems under anticipated "operational" conditions. The science of engineering evolved with the use of mathematical logic to interpret natural laws and to provide objective principles for engineering design. The discoveries of Copernicus, Galileo, Boyle, Hooke, and Newton were useful for the advancement of engineering science as they contributed to the development of subjects such as statics, dynamics, fluid mechanics, and heat transfer—all of which are part of the canon of engineering science.

CIVIL ENGINEERING

Over the centuries, many branches of engineering have evolved. Initially the primary focus of engineering was on designing machines for the military and for the construction of structures. Englishman John Smeaton is credited for coining the term "civil engineering" in 1761 to describe a type of engineering undertaken for the development of civil society, thus distinguishing it from military engineering with the primary focus on war and destruction. The oldest of all the branches of engineering science, civil engineering continues its original focus today, encompassing a wide range of engineering activities related to structures, roads, bridges, water systems, flood control, protection from adverse environmental conditions, and sanitation. The British established the Institution of Civil Engineers in 1818. The American Society of Civil Engineers was established in 1852.

MECHANICAL ENGINEERING

Advances in the design of the English blast furnace played a key role in the beginnings of mechanical engineering. The blast furnace was used to process iron ore, an essential element in the production of high quality wrought iron and steel. The greater availability of high quality steel provided materials for designing new machines and manufactured products. The profession of mechanical engineering experienced its fastest growth during the Industrial revolution of the 18th and 19th centuries. A wide

Encyclopedia of Agricultural, Food, and Biological Engineering
DOI: 10.1081/E-EAFE 120007161
Copyright © 2003 by Marcel Dekker, Inc. All rights reserved.

variety of machines were needed to manufacture goods and provide power to fuel emerging industries. Today, mechanical engineering is a broad-based engineering discipline with a focus on mechanisms, machines, and power generation. The fundamental principles of classical mechanics and thermodynamics from physics and mathematics provide the basis for mechanical engineering. The British established the Institution of Mechanical Engineers in 1847. The American Society of Mechanical Engineers was established in 1880.

ELECTRICAL ENGINEERING

The beginning of electrical engineering is associated with the need for electricity to power machines, devices, and industries that were being developed by mechanical engineers. The fundamental principles of electro-magnetic phenomena and mathematical logic are the basis for electrical engineering. With a wide range of applications for advanced technologies in power generation and transmission, communication, instrumentation and measurement, electronics, computers, and controls, it is not surprising that electrical engineering has become the largest branch of engineering. One of the fastest growing areas in electrical engineering is electronics engineering that exploits emission, behavior, and effects of electrons for designing efficient electrical devices. The Institute of Electrical and Electronics Engineers was established in 1884, four years after Thomas Edison's invention of a practical incandescent bulb and two years after the first electric generating station on Edison's Pearl Street in New York.

CHEMICAL ENGINEERING

Advancements in manufacturing during the late 1800s precipitated the need to understand chemical changes in materials during industrial operations. Such changes became an important consideration in the design of processes in industrial plants. Thus, the principles of chemistry and a quantitative understanding of distillation, heat and mass transfer, reaction kinetics, and extraction gave rise to the discipline, of chemical engineering. Today chemical engineers are involved in the design of biochemical processes for the production of useful products such as food, fuels, and pharmaceuticals. In 1908 the American Institute of Chemical Engineers (AIChE) was established.

Within a 100-year period, starting at the beginning of the nineteenth century, four engineering disciplines (civil, mechanical, electrical, and chemical) made major advances and established professional societies to further the disciplines and profession of engineering. These engineering fields started with an application-focused orientation in order to fulfill the societal needs of the time. At the same time, advancements in physics, chemistry, and mathematics provided careers in engineering research that began formulating engineering principles and practices for the foundation of engineering science. With the advent of engineering science, these engineering fields began to transform from application-focused studies to science-based disciplines. This transformation liberated these engineering disciplines from the confining perspective derived from the application needs of limited areas. Thus, these engineering disciplines are now viewed as fundamental fields of engineering that are application-independent, i.e., their work is applied to all areas where designed products, processes or systems are useful. (Note that in the evolution of science-based engineering disciplines the two components of natural sciences, physics, and chemistry, have been the knowledge-pool for developing engineering practices and principles. However, biology has contributed a very limited input or perspective. Interaction between engineering and biology has been limited at best. This has been dealt with in the later part of the article.)

MINING, METALLURGICAL, AND PETROLEUM ENGINEERING

The nineteenth century saw the rise of several new areas of engineering focused on meeting societal needs. These areas included designing systems for the extraction of raw materials deep inside the earth (mining); designing systems for producing materials (mostly metals) with properties desired for the many products of industrial production (metallurgy); and designing systems for extracting and refining petroleum to satisfy the increasing energy appetite of industry and the growing demands of society for machines and appliances of convenience. These engineering needs were so important and widespread that in 1871, even before the establishment of the American Society of Mechanical Engineers (in 1880), the American Institute of Mining, Metallurgical and Petroleum Engineers was established and became the second engineering professional society in America to be formed after the American Society for Civil Engineers.

AGRICULTURAL ENGINEERING

One of the most important engineering challenges at the beginning of the 20th century was to meet the needs of agricultural mechanization. There were two powerful forces driving this need: 1) more than 95% of Americans lived in rural farm communities and technologies that relieved farmers from hard work in extreme outdoor conditions became a primary national need and 2) the establishment of new industries was dependent on a large supply of workers and the only additional indigenous source of labor was found in rural and farm communities. Thus, in addition to their work producing food, farm laborers were needed to help support the industrial revolution. The only solution was to increase the productivity of farmers by the use of technology, thereby freeing them to work in new industries. The field of agricultural engineering began evolving at the end of the 1800s and in 1907, one year before chemical engineers organized their professional society, the American Society of Agricultural Engineers (ASAE) was established by pioneers who were devoted to designing systems for reducing the time and drudgery involved in agricultural production. Today, one farmer produces enough food for more than 100 people and more than 97% of Americans are involved with nonfarm related work. (The 20 "Greatest Engineering Achievements of the 20th Century," a project launched by the National Academy of Engineering in 2000, named Electrification as the greatest engineering achievement of the past century and Agricultural Mechanization the seventh—agricultural engineering played a key role in both.)

An important observation in the evolution of agricultural engineering as a discipline is that unlike mechanical, chemical, and electrical engineering (that transformed from their initial application-focus to science-based engineering disciplines), agricultural engineering has largely remained unchanged and continues to be focused on application to agricultural systems. By the late 1960s, with the mechanization of agriculture nearly complete, there was a sharp decline of interest in agricultural engineering.

Attempts began in the early 1970s to transform the agricultural engineering discipline by enlarging its application domain to include new but related areas of postharvest, food, natural resources, and the environment. In many ways, these new areas are logical extensions of agricultural engineering as they are inherent components of "agriculture" itself. However, other emerging engineering disciplines focused singularly on each of these areas. Food engineering and environmental engineering, e.g., attracted engineers from the science-based disciplines of chemical, mechanical, and electrical engineering to these agriculturally-related application fields.

Unlike the agricultural engineering of the past when the discipline had sole dominion over agricultural mechanization, the changes of the 1970s brought several other players to the agricultural field of application. The growth of agricultural engineering in new application areas was slower than the attrition due to lack of interest and low national priorities placed on farm efficiency. By the mid-1980s the profession began to exhibit serious decline and by the later part of the decade the rate of decline became alarming. Several academic programs had unacceptably low enrollments and were on university administrators' short list for elimination. Similarly research programs in agricultural engineering were barely recognizable in the USDA's national priorities. Membership in ASAE declined precipitately.

BIOLOGY AND ENGINEERING

A revolution in the field of biology began in the mid-1950s. Starting with the discovery of double helix, an incredible understanding of genetics and the functions that govern life processes at the molecular level has been gained. In the 1980s, scientists were able to successfully engineer living systems that express desired behaviors at the genetic level. Many call this type of work genetic engineering. The approach used by genetic engineers to design living systems is similar to that used by mechanical engineers, as the goal of both disciplines is to conceptualize and build systems that satisfy a specific need. The difference is that mechanical engineers, e.g., employ their knowledge of mechanics and other principles from physics in their design work, where genetic engineers use their understanding of genes and gene functions.

For the past several years, genetic engineering techniques have been successfully used to develop pest-resistant plant varieties. Today, genetic engineers have developed sophisticated cloning methods that enable the creation of complex life forms such as sheep, cows, and pigs. In addition to bioengineered plants and animals, the new discipline of biological engineering is also being used to develop pharmaceuticals and other useful products and processes designed to meet the needs of society.

Advances in biology are leading to the creation of useful methods for designing life forms. For example, biologists are gaining a new understanding of living systems and employing this new knowledge (through trial and error methods) to design systems at the molecular level. While the work of these biologists shows great potential, there is an equally fervent effort to develop the discipline of biological engineering by using an application-focused approach. The two primary areas of application are agriculture and medical and health systems. On this front, biomedical engineers are focusing their work on the design and creation of useful products, such as biomaterials and artificial organs that are able to perform the functions of living tissues and organs.

To a great degree, the trial and error methods of biologists are similar to those of the early designers in

Mesopotamia who created complex systems by intuition and experience using limited scientific and mathematical (objective) logic. However, engineers designing systems for agriculture and medicine are employing greater mathematical logic in their designs. These engineers have been greatly handicapped, as the scientific understanding of biology has been largely unamenable to classical mathematical formalism. Today, engineers, biologists, mathematicians, computer scientists, and statisticians are actively engaged in developing computational methods that describe the processes of living systems. These computational methods are allowing "mathematical" (objective) logic to be used in biological engineering design.

SCIENCE AND ENGINEERING

Fig. 1 is a schematic illustration of the perspectives of science and engineering, and their relationship in the design of products, processes, and systems. Although the role of cultural values in design was not discussed earlier, it plays a very important role in the work of engineers and is included in the illustration.

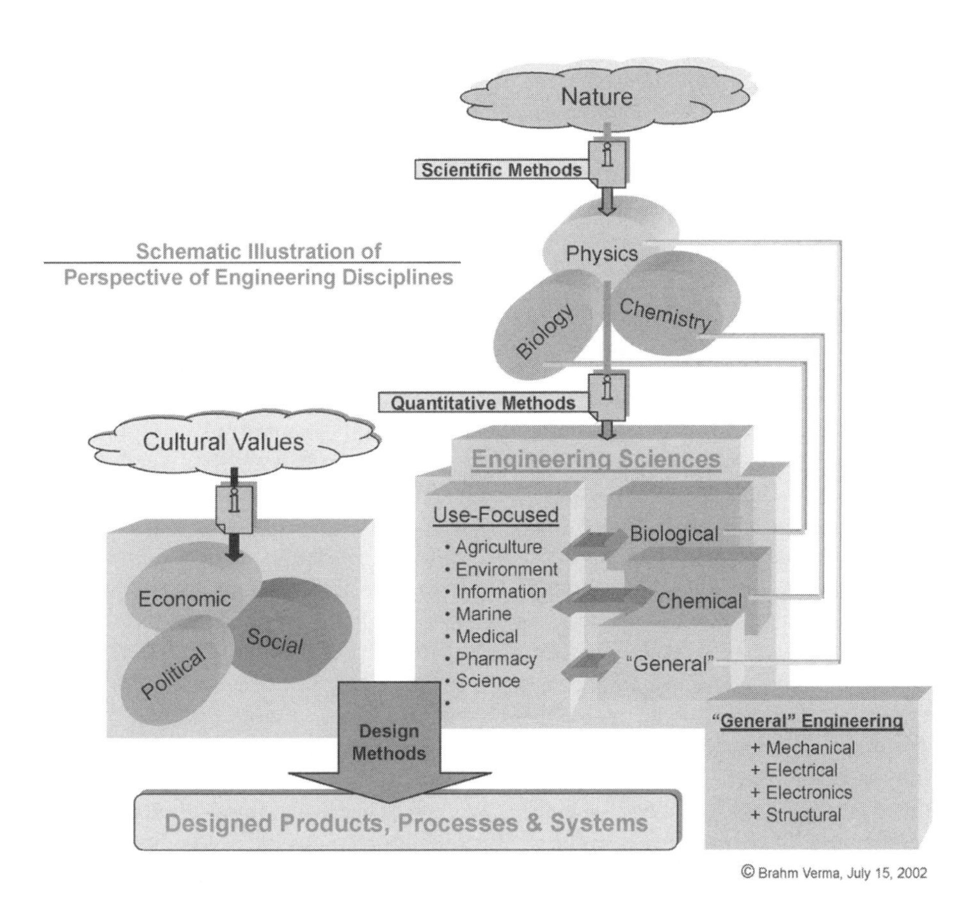

Fig. 1 Schematic illustration of interrelationship between science and engineering. The knowledge of the three components of natural sciences (Physics, Chemistry, and Biology) is discovered by critical inquiry (shown with i) of nature by using the Scientific Methods. Similarly, principles of Engineering Sciences are gained by critical investigations (shown with i) of natural sciences using Quantitative Methods of mathematical and computational sciences. The direct connections between Physics to General engineering, Chemistry to Chemical engineering, and Biology to Biological engineering show the fundamental sources knowledge of the three Engineering Sciences. The knowledge of the three engineering sciences is used to develop principles useful to the application (Use-Focused) areas, such as Agriculture, Environment and others. (See bidirection arrows in the engineering science box indicating this relationship.) Engineering Sciences for the Use-Focused areas often have customized principles and practices that evolve from experience and intuitions gained from the designing experience of engineers. The knowledge of Cultural Values (Social, Economical and Political) are critical to create successful and useful designs. Design Methods a systematic process that employs knowledge from sciences and cultural values for Designing Products, Processes and Systems beneficial to the prosperity of society.

Fig. 1 shows that the study of nature (with the use of proper research methods) leads to the discovery of knowledge (by revealing the laws and governing observations on how natural systems work). For simplicity, "nature" was reduced to three components: physical, chemical, and biological (reflective of the natural sciences). In reality, however, nature is an integrated whole and cannot be so reduced.

Engineering sciences provide the principles and practices necessary for creating successful designs. Knowledge of the natural sciences is fundamental to the development of the engineering sciences, and objective logic (mathematical and computational methods) is critical to transforming scientific knowledge into engineering principles and practices. This connection is shown in Fig. 1. For example, the earliest scientific understanding of physics gave rise to the "general" engineering sciences that provide the overall perspective and framework for engineering science itself. Similarly, the discipline of chemical engineering is developed from chemistry. The inability to understand complex living biological systems and the absence of developed objective logic has been impediments in the evolution of biological engineering. With recent advances, however, biological engineering is developing but is still in a nascent stage.

All engineering work is undertaken for the central objective of creating new and useful designs. Fig. 1 shows that the knowledge of general, chemical, and biological engineering sciences has to be applied for designing useful solutions to societal needs. Agriculture, pharmacy, and medicine are just a few examples of application-focused engineering disciplines that are developing from both the knowledge of science and the intuition of practitioners. The bidirection arrows between the science-based and application-focused areas are important. Without this interaction, engineering science is mute to society's needs.

Finally, Fig. 1 illustrates that cultural values are an integral part of engineering design. A perfectly viable technical design is useless if it does not fit into a society's value system and fulfill a societal need. Fig. 1 shows that the successful design of products, processes, and systems is a result of the proper use of scientific and cultural knowledge in engineering design methodology.

In summary, biological engineering is evolving from both directions in terms of its engineering science knowledge-base and design methodologies for diverse applications. Scientists, e.g., are advancing the understanding of biology using trial and error methods, engineers in several engineering disciplines are designing application-specific systems and computational scientists are developing quantitative methods for biology. Furthermore, there is growing evidence that new engineering design methodologies may emerge from the new understanding of living systems. Advances on all of these fronts are needed to contribute to the development of the biology-based engineering science.

ENGINEERING DISCIPLINES IN THE AGE OF BIOLOGY

In the decade of the 1990s, engineers from all disciplines have carefully assessed the role of biology and have made visible attempts to incorporate biology into the work of their respective engineering fields.

Biological Engineering and Agricultural Engineering

To the agricultural engineering discipline, the debate on the role of biology is an old one. As early as the 1930s, a small group of visionaries in ASAE led by Ohio State University Professor C. O. Reed, argued vehemently but unsuccessfully that the discipline of agricultural engineering is based not on the application of civil, mechanical, and electrical engineering to the industry of agriculture but on the science of biology; "it is the engineering of biology." This profound shift in the perspective of agricultural engineering in the 1930s from application-focused to science-based engineering would have led the discipline to a different course. With the death of Professor Reed in 1940 this argument became dormant and would not surface again until the beginning of 1960s.

In the 1960s, Wilson B. Bell, an agronomist and administrator at Virginia Agricultural Experiment Station awakened agricultural engineers by stating that, "Your sphere of activity is more closely intertwined with the life sciences ... You cannot escape, even if you wished, the world of living things." Professor Wallace Giles of North Carolina (N.C.) State University led this discussion and spoke widely for a biology-based engineering discipline. An ASAE Committee appointed in 1962 headed by another professor from Ohio State University, Robert Stewart, did not support the advancement of agricultural engineering as a biology-based engineering discipline, but instead recommended the formation of a division of biological engineering in ASAE to heighten the use of biology in the work of agricultural engineers. Despite the committee's recommendation, ASAE formed a bioengineering technical committee. N.C. State University changed its name to Biological and Agricultural Engineering in 1965 and became the only department in the country with a name other than agricultural engineering. It was only a year later that Mississippi State University took heed to these changes and became the second department in the country to change its name to Agricultural and Biological Engineering reflecting the added biology-focus in the department's work. In 1969,

Mississippi State University became the first and the only department to offer a separate undergraduate degree program in biological engineering based on the science of biology. The pioneering program educated students to apply biological engineering to a range of areas including medicine, agriculture, environment, and veterinary. Other than N.C. State University and Mississippi State University, there was little interest shown by other universities, and industries were not supportive to modify the professional content of agricultural engineering.

For the next decade and a half, the issue of biological engineering was not a national debate among agricultural engineers. However, there were several departments actively experimenting with adding biology and biological science perspective to agricultural engineering. In 1987, at the Conference for Administrative Heads of North American Agricultural Engineering Departments in Columbus, Ohio (the home of the late Professor E. O. Reed who had in the 1930s argued unsuccessfully that agricultural engineering should be viewed as "the engineering of biology"), the potential of biological engineering was extensively discussed. Three years later in 1990, the ASAE Academic Program Administrators Committee developed a "Vision of the Future" with four recommendations: 1) offer a biological science based, biological applications focused engineering curriculum that defines our uniqueness among engineering disciplines; 2) have a core curriculum designed to define the biological science base of our discipline; 3) provide areas of emphasis within the curriculum that focus upon applications involving biological systems; and 4) adopt "Biological Engineering" as the name of our curriculum.

Work throughout the past decade provided a much greater understanding of the role of biological sciences in the discipline of agricultural engineering. The 1990s was a decade of profound debates in ASAE and in academic departments. From the mid-1980s, Professor Norm Scott of Cornell University championed the argument for biological engineering in the ASAE. During his term as the President of ASAE in 1993–1994, Scott called for an active debate and asked how ASAE should change to capture the promise of biological engineering. This debate led to the establishment of the Institute of Biological Engineering (IBE) in 1995 with the objective "to encourage inquiry, application, and interest in biological engineering in the broadest and most liberal manner and to promote the professional development of its members." ASAE agreed that IBE should have autonomy in membership, programs, and services normally available to an independent professional society. They saw that they could benefit from the new class of IBE members who would otherwise not be attracted to their organization (which was a great incentive during the rapidly declining membership) and IBE founders recognized the opportunity to provide a scientific and professional forum for the rapidly growing discipline of biological engineering. Although the leadership of ASAE was very supportive of this arrangement, many prominent and active members of ASAE were uncomfortable with the ASAE/IBE relationship. ASAE found the new arrangement difficult to implement. On December 28, 1999, IBE separated from ASAE and became an independent professional society.

The 1990s were also a period of great advancement for the ASAE. Many departments began to appreciate the value of incorporating biology (some say agro-biology) and a biological science perspective to the discipline of agricultural engineering. Others moved to this position because of the alarming decline in student enrollment. Major efforts were undertaken both locally and nationally to reevaluate curriculum. By the end of the decade, all agricultural engineering departments in the United States, with the exception of two, added "biosystems," "bioresource," "biological," "food," and/or "environment" with "agricultural" to their names. A few departments went as far as completely dropping the word "agricultural" from their name, citing the argument that new titles such as "biosystems" inherently include agricultural systems. By the mid-1990s, even the ASAE, the professional society that had been home for "agricultural" engineers, changed its marketing strategy at the recommendation of a presidential commission and started promoting itself by its abbreviations "ASAE." The organization added a byline to read "The society for agricultural, food and biological systems." When IBE became a separate professional society, ASAE membership excitedly formed a division of biological engineering at the 2000 Annual Meeting. Today, the ASAE division of biological engineering has undertaken the application perspective for the areas of agricultural, food, and biological systems, and the IBE continues its vision to be the "headquarters" of the application-neutral biological-science based engineering discipline evolving from the advances of biology.

Biological Engineering and Biomedical Engineering

The two engineering societies, active in advancing biological engineering from the application in medicine perspective are the Biomedical Engineering Society (BMES) and the Institute of Electrical and Electronics Engineering–Engineering in Medicine and Biology Society (IEEE–EMBS). The stated purpose of BMES is "to promote the increase in biomedical engineering knowledge and its utilization." BMES addresses only a single but a critical application area of medicine that is essential for the prosperity of society. It was established in 1968. BMES has seen a phenomenal growth in the nineties with the advances in biology and medical technologies. It has also benefited from the Whitaker Foundation's generous support for developing biomedical engineering and the increasing priority in the United States and increase in the research budget of the National Institutes of Health (NIH).

IEEE–EMBS is similarly focused on medicine but with a more interest in biology as well. IEEE is the primary home of electrical engineers. EMBS being a part of IEEE has evolved from the perspectives of electrical engineering. Its stated interest is "The field of interest of the IEEE-EMBS is the application of the concepts and methods of the physical and engineering sciences in biology and medicine. This covers a very broad spectrum ranging from formalized mathematical theory through experimental science and technological development to practical clinical applications. It includes support of scientific, technological, and educational activities."

BIOLOGICAL ENGINEERING AND CHEMICAL ENGINEERING

There are many academic departments adding biology to the chemical engineering program. Perhaps the recent change of MIT is indicative to the evolution of biological engineering in the field of chemical engineering. MIT states that biology is not only amenable to engineering but by requiring biology-based engineering analysis and synthesis for design has formed a division of biological engineering. American Institute of Chemical Engineers has been organizing many biologically engineering related technical sessions in their annual meetings. Now there is an active move to form a biological engineering division in the AIChE.

BIOLOGICAL ENGINEERING AND MECHANICAL ENGINEERING

Mechanical engineers have been active in the development of devices for the medical and health industries. They are contributing to engineering science knowledge in such fields as biomechanics, biofluid mechanics, heat and mass transfer in biotechnology and cell and tissue engineering. The American Society of Mechanical Engineers has a Bioengineering Division which is "focused on the application of mechanical engineering knowledge, skills, and principles from conceptions to the design, development, analysis, and operation of biomechanical systems."

CONCLUSION

Engineering is a practical art learned from scientific knowledge, objective logic, and experience from practice. Over the past 5000 years, the science of engineering has evolved to meet the continuously changing needs of society. Many disciplines of engineering have evolved by first focusing on designing products for a clearly identifiable need of society. A few of these engineering disciplines (mechanical, electrical, and chemical) have evolved into science-based disciplines drawing its fundamental engineering understanding from the knowledge of natural sciences. Other disciplines have remained primarily application focused (e.g., agricultural engineering to the industry of agriculture).

Biological engineering is evolving from both directions. Many scientists are intimately involved in designing new living systems by using advancing knowledge of biology at the molecular and genetic levels and some objective logic. These designs are results of trial and error approaches. They are contributing to new engineering principles of biological engineering. Simultaneously, those engineers who are primarily focused on application and are designing new systems for the agriculture, medical, and health industries are developing practices of biological engineering. Together biologists and engineers are evolving a new science-based engineering discipline—the biological engineering. The 21st century will make incredible advances through the evolving discipline of biological engineering which will make the advances of the 20th century look elementary.

FURTHER READING

Cuello, J.L. *Faces of Change, Part II. RESOURCE*; ASAE: St. Joseph, MI, 1995.

Cuello, J.L. An Operational Approach to Systemic Development of a Proposed Framework for Biological Engineering Curriculum. Proceedings of the Institute of Biological Engineering; IBE Publications: Athens, GA, 1998.

Cuello, J.L. Mechanism and Provincialism: Two Philosophical Movements Shaping Biological Engineering. Proceedings of the Institute of Biological Engineering; IBE Publications: Athens, GA, 1998.

DNA of Biological Engineering: Defining the Body of Knowledge for the Discipline. Proceedings; The Institute of Biological Engineering: Sacramento, CA, 2001; 80.

Final Report of Project 2001—Engineering for the 21st Century. International Conference for Administrative Heads of North American Agricultural Engineering Departments, Columbus, Ohio, October 26–28, 1987; 17.

Garrett, R.E. Core Engineering Courses for Biological Engineering. ASEE Annual Conference Proceedings, Session 2508, 1990; 1248–1251.

Undergraduate Biological Engineering Curriculum: VISION for the FUTURE. Report of the Academic Program Administrators Committee of the American Society of Agricultural Engineers. St. Joseph, MI, 1990; 11.

Garrett, R.E.; Davis, D.C.; Edwards, D.M.; Johnson, A.T.; Rehkugler, G.E. Development of Biological Engineering Courses and Faculty Enhancement. Report of CSRS USDA Agreement No. 91-38411-6842; University of California: Davis, CA, 1994; 95616-5294.

Giles, G.W. Biological Engineering: Reaching for our Destiny: 1950–1990; Department of Biological and Agricultural Engineering, North Carolina State University: Raleigh, N.C., 1993–1994.

Johnson, A.T. Hierarchical Competencies in Biological Engineers. ASEE Annual Conference Proceedings, Session 2508, 1990; 1252–1253.

Johnson, A.T.; Phillips, W.M. Philosophical Foundations of Biological Engineering. J. Eng. Ed. **1995**.

Oaks, W.C.; Leone, L.L.; Gunn, C.J.; Dilworht, J.B.; Pottre, M.C.; Young, M.F.; Diefes, H.A. ; Flori, R.E. Engineering Your Future. Great Lakes Press, Inc. ISBN 1-881018-26-1, 1999.

Scott, N.R. Promise of Biological Engineering. Presented at the ASEE, Reno, Nevada. 1987.

Stewart, R.E. Seven Decades That Changed America: A History of the American Society of Agricultural Engineers 1907–1977. The American Society of Agricultural Engineers, 2950 Niles Rd, St. Joseph, MI. ISBN 0-916-150-15-18, 1979.

Verma, B.P. An Emerging New Order. RESOURCE, November, 8–10, 1995.

Wright, P.H. Introduction to Engineering. Second Edition. John Wiley and Sons, Inc. ISBN 0-471-57930-0, 1994.

SOME USEFUL WEBSITES

The Institute of Biological Engineering: http://www.ibeweb.org

American Society of Agricultural Engineers: http://www.asae.org/about.html

American Institute of Chemical Engineers: http://www.aiche.org/

American Society of Mechanical Engineers: http://www.asme.org/divisions/bed/

Biomedical Engineering Society: http://www.bmes.org/about.asp

IEEE Engineering in Medicine and Biology Society: http://www.eng.unsw.edu.au/embs/index.html

Biological Engineering History

Otto J. Loewer
University of Arkansas, Fayetteville, Arkansas, U.S.A.

INTRODUCTION

Biological engineering has often been a point of dissention within the agricultural engineering profession since the founding of the American Society of Agricultural Engineering (ASAE) in 1907. Over the years, the advent of computers and the decrease in farm population greatly influenced the agricultural engineering academic programs, thus changing the perspective of the agricultural engineering profession. This change is still very much at the forefront of discussion among those with an interest in the interface of biology, agriculture, and engineering.

IN THE BEGINNING

The American Society of Agricultural Engineers (ASAE) was formed in 1907, a time when mules still provided 75% of the total horsepower used by American agriculture but clearly when "agriculture and engineering were on a collision course."[1] From the beginning, the name and identity of this new professional society was debated. A major philosophical question was whether agricultural engineering was a branch of agriculture or a branch of engineering, or both. The constitution read: "The object of this Society shall be to promote the art and science of engineering as applied to agriculture." In the second meeting of ASAE in 1908, a committee was formed to prepare resolutions stating that "agricultural engineering" was preferable to all other titles. It was 30 yr from ASAE's founding before there was a formal effort to alter the original philosophical base of the organization.

THE 1937 INITIATIVE

In 1937, ASAE headquarters identified their view of the "Big Problems for Agricultural Engineers" in an editorial. One of these was the concept of agricultural engineering as "the engineering of agricultural biology."

This concern was championed by C. O. Reed, a professor at Ohio State University. Reed had been a longtime ASAE member, having served as treasurer in 1912. Reed argued that the factor that conferred distinction upon agricultural engineering was that "it is the engineering of biology... This unique kind of engineering should be based on the energy transformations and transfer conducted by living cells; a methodology and efficiency concept so based would open a new world to the agricultural engineer."

ASAE headquarters agreed with Reed and stated that this concept "... would not step on the toes of older branches of engineering... Every agricultural engineer may be able to see new light and new opportunity in his particular job and abilities by looking at them in relation to the engineering of agricultural biology."

Few ASAE members agreed with Reed and ASAE headquarters. In 1938, a three-member committee was appointed, entitled the ASAE Committee on the Energetics of the Biology in Agriculture (this group also called itself the "Horse-feathers Club"!). Reed died in 1940, and the committee died with him.

THE 1960s INITIATIVE

In 1960, G. W. Giles made a speech to the Farm Equipment Institute (FEI) as part of the Winter ASAE meeting held in Memphis. Giles was head of the Agricultural Engineering Department at North Carolina State College. Giles said, "... the fact remains that the mathematical relationships of the physical to the biological processes are basic to developing superior engineering systems. Our profession needs some fundamental law on which to base our judgments and guide our direction and pattern of growth for engineering the biological system. The core of our profession should be built on engineering laws governing the intricate complex processes of plants and animals. This is the thing that distinguishes agricultural engineering from other engineering professions."

This time the idea of "biological engineering" was received with more enthusiasm than 20 yr before, especially among the ASAE members employed in higher education. However, those in the private sector did not look upon this approach with great favor.

The discussion about "biological engineering" went on for another 2 yr. One of the ASAE directors was quoted as saying that "modern" agricultural engineering was essentially biological engineering and that perhaps

Encyclopedia of Agricultural, Food, and Biological Engineering
DOI: 10.1081/E-EAFE 120007159
Copyright © 2003 by Marcel Dekker, Inc. All rights reserved.

ASAE should consider "broadening its base to include all aspects of biological engineering." In fact there was the hint that ASAE should change its name to the "American Society of Biological Engineering."

An ASAE committee on the "Relationship of Biological Engineering to Agricultural Engineering" was appointed which concluded that ASAE would be a doubtful nucleus for a "Society of Biological Engineers." They reported that under no circumstances should ASAE change its name. However, they did suggest changing technical division names to describe the functional aspects of engineering the biological systems of agriculture. The committee's report was not well received, and it became clear that the enthusiasm for biological engineering centered largely in the academic institutions. In the end, ASAE even rejected the creation of a "bioengineering division" in 1966, although a bioengineering committee was formed.

The visible mark of this effort, however, was seen at a few academic institutions where "biological" was included as part of the department name at North Carolina State University in 1965, and a little later at Mississippi State University, and Rutgers University. In retrospect, the seeds that were planted in 1937 had begun to germinate in the classrooms of agricultural engineering departments in the 1960s.

COMPUTERS, BIOLOGICAL ENGINEERING, AND THE 1970s

Digital computers helped in bringing biological engineering to the forefront in the 1970s by making it possible to quantitatively describe biological processes over time. Many agricultural engineers became experts in mathematical modeling. They led interdisciplinary teams who integrated agricultural and biological knowledge in a way that was used to develop new insights into biological systems. It could be argued that those agricultural engineers who developed computer models were the first to make the transition to become biological engineers. In retrospect, perhaps the computer became to biological engineers what the internal combustion engine had been to the first agricultural engineers.

THE 1987 DEPARTMENT CHAIRS MEETING

The economic downturn in U.S. agriculture in the 1980s adversely impacted agricultural engineering academic programs. The mid-1980s saw a steady decline in agricultural engineering undergraduates whose numbers had peaked a few years before. Graduation rates were about half of what they had been a few years earlier. There had been an exodus from the youth of the farm who had populated the agricultural engineering programs in years gone by. In addition, the "computer oriented" group of academic ASAE members had begun to have considerable influence within the organization and their departments.

By 1987, the agricultural engineering department heads were greatly concerned about the continued vitality of their departments, especially at the undergraduate engineering level.[2] A meeting was called to gather at Ohio State University on October 26–28, 1987, to discuss this serious situation and to develop a unified approach for solving the problems of low enrollment. Thirty-seven departments in the United States and Canada were represented. There were strong philosophical differences with some believing that enrollments would increase if the general public were better informed about agricultural engineering while others argued that this approach had been tried for years with little success, and that a fundamental change was in order.

A formal report (Project 2001—Engineering for the 21st Century) was generated that tried to accommodate a wide range of viewpoints. Little consensus was reached, and the prevailing opinion after the workshop seemed to be that each department was on its own.

The next 3 yr would bring considerable turbulence to agricultural engineering programs across the country as each department attempted to market its own program. It became increasingly clear to most that fundamental change was needed quickly if the engineering undergraduate program was to survive.

THE 1990 DEPARTMENT CHAIRS MEETING

At the 1989 Winter meeting of ASAE in New Orleans, the mood of the department chairs was similar to that of the Ohio meeting 2 yr earlier. Enrollments were still low, and there was some feeling of despair about the future of agricultural engineering undergraduate programs. However, there had been considerable turnover in departmental leadership, and many departments were now ready to try something new to revitalize their programs.

Many were convinced that curriculum and associated name changes were required, and that a move was needed towards a science-focused engineering program (biological engineering) away from an industry-focused engineering program (agricultural engineering). The decision was made to have a department heads workshop in St. Louis early the next year, and a planning committee, headed by Dick Hegg of Clemson University, was given the task of developing the workshop program. This meeting was to be devoted exclusively to a discussion of the undergraduate engineering program as compared to the Ohio State

University meeting that had addressed a wide range of issues.

The St. Louis meeting was held just 4 mo later, on April 24–25, 1990. Forty-three attended representing 39 academic programs, a large percentage of the total number of programs in the United States and Canada. Less than half of those in attendance were present at the Ohio State University meeting 3 yr before.

The St. Louis group was committed to being proactive and, to their surprise, found that there was considerable agreement as to what needed to be done. In fact, many departments, acting independently, had taken the first steps required for development of biological engineering programs. This approach had considerable support among the faculty, especially among the midsized to smaller departments where survival supplanted tradition as the order of the day.

At the end of the meeting, the attendees made a bold statement as to the path that they believed their academic units should follow and endorsed the concept that agricultural engineering was a subset of biological engineering.

A follow-up workshop was held on June 26, 1990, at the 1990 summer ASAE meeting. Forty-six were in attendance representing 36 academic units. A report ("A Vision for the Future") was formally approved by the group with only a single negative vote. The report was printed in ASAE's Agricultural Engineering magazine and endorsed offering a "biological science based, biological applications focused engineering curriculum that defines our uniqueness among engineering disciplines" and "... adopt 'Biological Engineering' as the name of our curriculum." This group of administrators was preparing to launch out in a new direction with or without the blessing of ASAE.

THE 1990s AMERICAN SOCIETY FOR ENGINEERING EDUCATION (ASEE) INITIATIVE

The department heads were not alone in their belief that changes in curriculum and name were in order. A significant number of agricultural engineering faculties were also pushing in the same direction. They began by creating the Biological and Agricultural Engineering Division within the ASEE in the late 1980s. From this base, they held a workshop in Atlanta on January 10–13, 1991, to develop a core curriculum in biological engineering. A second workshop was held as part of the 1991 Winter ASAE meeting where "core" biological engineering courses were identified. This ASEE group, with initial leadership provided by Roger Garrett (UC Davis) was also influential in getting USDA to fund the development of biological engineering core courses through its Challenge Grant Program.

ASAE AND BIOLOGICAL ENGINEERING IN THE 1990s

The plight of ASAE followed that of academic programs. Hard economic times in agriculture took a heavy toll on ASAE membership. ASAE found it caught in an internal struggle with a significant part of its leadership having one view of biological engineering and its academic leadership having another. During this time, ASAE had a series of presidents who worked hard to accommodate differences of opinion. In the early 1990s, a commission chaired by former ASAE president John Walker of the University of Kentucky, examined a name change for the society, the result being the decision to refer to the American Society of Agricultural Engineers as *ASAE: The Society for Engineering in Agricultural, Food, and Biological Systems*. A series of formal ASAE discussions, promoted by ASAE presidents Doug Bosworth of John Deere and Norm Scott of Cornell University, did much to bring the two schools of thought together. The creation of the Institute of Biological Engineering (IBE) in 1995 with close connections to ASAE was indicative of the rather rapid evolution of ASAE to a greater acceptance of biological engineering. In 1999, IBE became an independent professional society.

REFERENCES

1. Stewart, R.E. *Seven Decades that Changed America—A History of the American Society of Agricultural Engineers 1907–1977*; American Society of Agricultural Engineers: St. Joseph, MI, U.S.A., 1979.
2. Loewer, O.J. Notes and Recollections of the meetings he attended as a Biological/Agricultural Engineering department head, 1985–1996.

Biological Reaction Kinetics

B

Marybeth Lima
Louisiana State University, Baton Rouge, Louisiana, U.S.A.

INTRODUCTION

Kinetics is an important subject in biological engineering because it allows for the measurement and prediction of how fast organisms and their biological products are degraded and produced. Kinetics is used as the basis for the design of biological reactors, to predict the shelf life of foods and drugs, to estimate how fast microorganisms degrade toxins in the environment, and to assess the rate of human population expansion. Kinetic models regarding enzyme production and catalysis can be seen in future sections of this chapter.

The purpose of this section is to give a general overview of rate equations as they relate to the kinetics of biological reactions. Simple kinetic models based on these rate equations will be derived, methods for determining kinetic parameters will be discussed, and several examples will be presented.

ZERO, FIRST, AND SECOND ORDER DEGRADATION REACTIONS

The kinetics of most biological reactions are zero or first order, though second order reactions are also possible. It is also common for a reaction to be a combination of two orders, for example, zero and first, depending on environmental conditions. These reactions are termed biphasic. Monod growth kinetics and Michaelis–Menten kinetics are kinetic models that are biphasic; during low substrate concentration, reactions are first order, while at high substrate concentrations they are zero order.

The kinetic models developed to describe zero, first, and second order degradation reactions are presented below. These models involve one reactant, A, which is "reacting" to form product B. We can model kinetic phenomenon by the rate at which A depletes according to the generic model

$$-r_A = -\frac{dC_A}{dt} = kC_A^x$$

where r_A is the rate of reaction, dC_A/dt, the rate at which the concentration of reactant A is changing with time, k, the reaction rate constant or rate constant, C_A, the concentration of reactant A, and x is the order of the reaction. There is a negative sign in the equation because reactant A is disappearing or decreasing. The rate constant is unchanged for the entire reaction (provided the temperature does not change), though the units of k change according to the order of the reaction.

Zero order: For a zero order reaction, $x = 0$ and the generic kinetics model reduces to

$$\frac{-dC_A}{dt} = k$$

This means that the rate of depletion of reactant A is a constant k. One can separate the variables and integrate this equation with the boundary conditions $C_A = C_{A0}$ (initial concentration) at time $t = 0$ (initial time), and $C_A = C_A$ at any time $t = t$.

The resulting equation for a zero order reaction is

$$C_A = C_{A0} - kt$$

The units of the reaction rate constant for the zero order reaction are $ML^{-3}t^{-1}$.

One parameter of interest for researchers studying kinetics is the half time of a reaction, which is defined as the time at which the concentration of component A is at half its initial value. To derive the half time for a zero order reaction, substitute $t = t_{1/2}$ and $C_A = C_{A0}/2$ into the equation for a zero order reaction. The resulting equation for half time for a zero order reaction is $t_{1/2} = C_{A0}/2k$.

First order: For a first order reaction, $x = 1$ and the generic kinetics model reduces to

$$\frac{-dC_A}{dt} = kC_A$$

This equation states that the rate of depletion of reactant A is proportional to the concentration of A. By separating the variables and integrating this equation using the same boundary conditions as in the zero order case, the equation for first order degradation kinetics is $C_A = C_{A0}e^{-kt}$. The units of k for a first order reaction are t^{-1}. The half time for a first order reaction is $t_{1/2} = \ln 2/k$.

Example: Degradation of the antibiotic gentamicin (used in the treatment of pneumonia) follows first order kinetics in the human bloodstream. If the initial concentration of gentamicin in the bloodstream is $9.4\ \mathrm{mg\,l^{-1}}$ and $k = 0.45\ \mathrm{hr^{-1}}$, what is the concentration

DOI: 10.1081/E-EAFE 120007179
Copyright © 2003 by Marcel Dekker, Inc. All rights reserved.

of gentamicin in the bloodstream 3 hr later? When does this system go to steady state?

Solution: To determine the concentration of gentamicin in the bloodstream after 3 hr, use the equation

$$C_A = C_{A0}e^{-kt}$$

Substituting into this equation yields $C_A = 9.4\,\mathrm{mg\,l^{-1}}\ e^{-(0.45\,\mathrm{hr^{-1}})(3\,\mathrm{hr})}$ and $C_A = 2.45\,\mathrm{mg\,l^{-1}}$.

Steady state is achieved in five half lives (or half times) for any first order degradation reaction.[1] Find the half-time of this reaction, and multiply by five. This yields the time to steady state.

$$t_{1/2} = \frac{\ln 2}{k} = \frac{0.693}{0.45\,\mathrm{hr^{-1}}} = 1.54\,\mathrm{hr}$$

time to *steady state* $= 5 \times 1.54\,hr = 7.7\,hr$.

Second order: For a second order reaction, $x = 2$ and the generic kinetics model becomes

$$\frac{-dC_A}{dt} = kC_A^2$$

This equation states that the rate of depletion of reactant A is proportional to the square of the concentration of A. Integration of the differential equation yields a second order reaction equation of $1/C_A = (1/C_{A0}) + kt$. The units of k are $\mathrm{L^3\,M^{-1}\,t^{-1}}$. The half time of a second order reaction is $t_{1/2} = 1/C_{A0}k$.

A graph of C_A vs. time is shown in Fig. 1 for zero, first, and second order reactions. The first order reaction is also referred to as exponential decay.

Researchers are often interested in determining the rate constant experimentally, and determining if a reaction is zero, first, or second order. This information will allow an investigator to predict the concentration of reactant A at any time t. The rate constant and order of reaction can be found experimentally as follows:

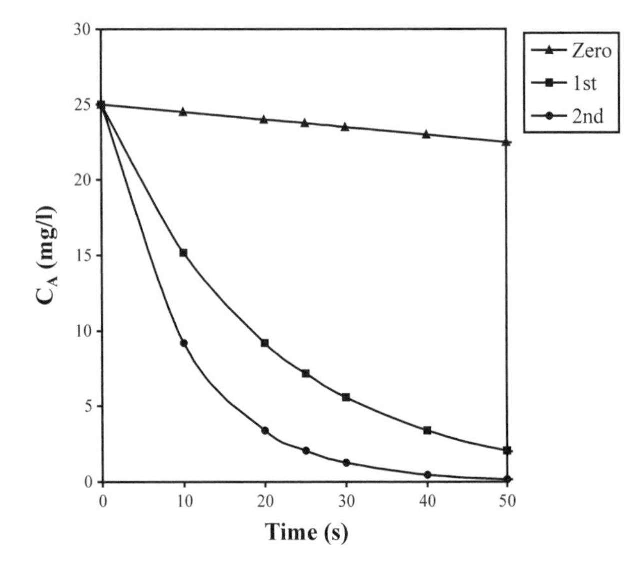

Fig. 1 Zero, 1st, and 2nd order reactions.

1. Put reactant A into a batch reactor and measure the concentration of A initially (C_{A0}).
2. Measure the concentration of A over time.
3. Analyze the data as follows:

- Plot C_A vs. time; if you have a linear line then it's a zero order reaction. The rate constant is the negative slope of the best fit line.
- Plot $\ln C_A$ vs. time; if you have a linear line then it's a first order reaction. The rate constant is the negative slope of the best fit line.
- Plot $1/C_A$ vs. time; if you have a linear line then it's a second order reaction. The rate constant is the slope of the best fit line. See Fig. 2 for details of this analysis.

Two things should be considered when experimentally determining the rate constant. All three graphs with regression analysis should be executed for each set of experimental data because occasionally it is difficult to distinguish among reaction orders. In addition, the reaction should be allowed to proceed approximately 50% to completion. It is often difficult to distinguish between reaction orders during initial stages of the reaction, thus, significant degradation should occur before analysis. Other methods exist for determining rate constant, but are beyond the scope of this chapter. See Ref.[2] for details.

Example: The following data for a zero order reaction has been tabulated. Find the rate constant for this reaction.

Concentration ($\mathrm{mg\,l^{-1}}$)	Time (min)
15.0	0
12.1	5
9.2	10
6.0	15

Solution: Plot concentration vs. time and take the best fit line of this data (see Fig. 3). The negative slope of this graph is the rate constant. Thus, $k = 0.598\,\mathrm{mg\,l^{-1}\,min^{-1}}$).

EFFECT OF TEMPERATURE ON REACTION RATE

The rate constant k is a strong function of temperature. In general, as the temperature at which a reaction occurs is increased, the rate of reaction is increased exponentially (at critically high temperatures, this generality does not hold). Investigators studying kinetics usually quantify

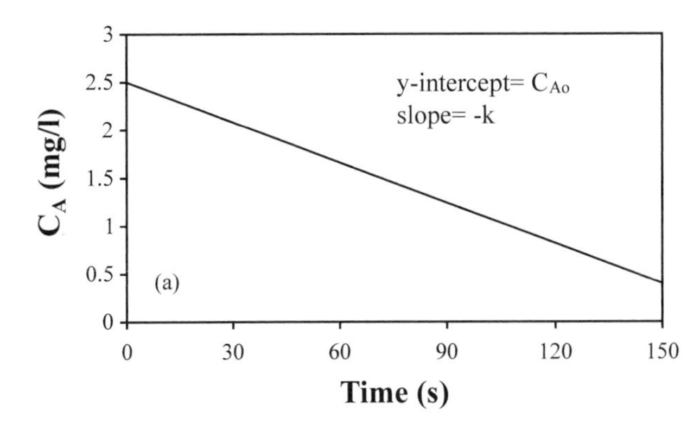

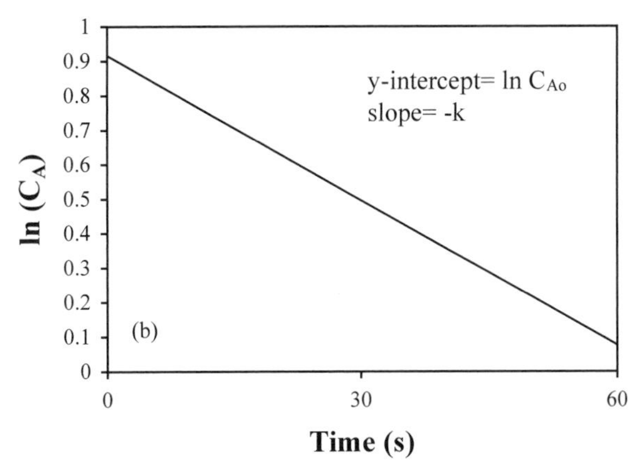

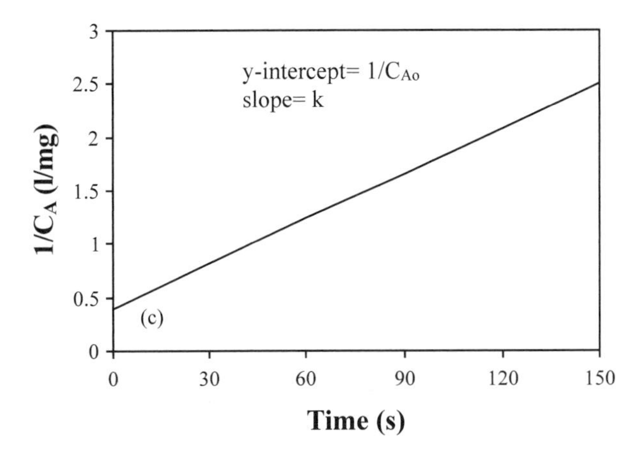

Fig. 2 (a) Zero order reaction, (b) first order reaction, and (c) second order reaction.

the effect of temperature on reaction rate. This is done using the Arrhenius equation:

$$k = Ae^{-\frac{E_a}{RT}}$$

where k is the rate constant, A, the frequency factor (also referred to as the pre-exponential factor or Arrhenius constant), E_a, the activation energy, R, the universal gas constant, and T is the absolute temperature. See Ref.[2] for the theoretical development of this model. Investigators

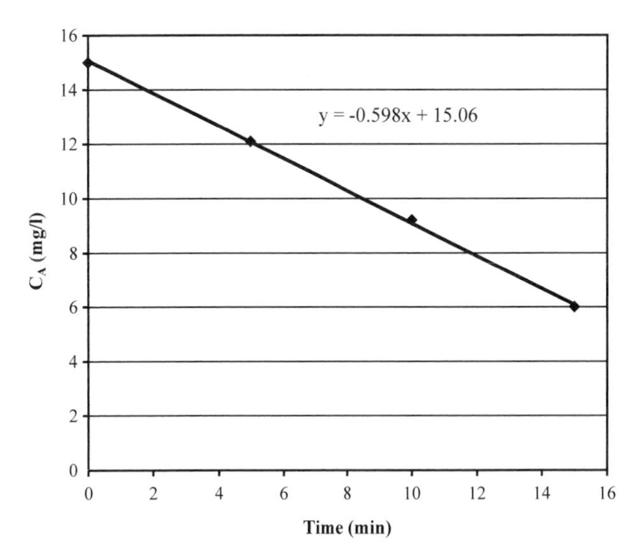

Fig. 3 Determination of rate constant using regression analysis.

report activation energy and sometimes the frequency factor; this information, in conjunction with the value of the rate constant at a given temperature, allows the rate constant to be determined at other intermediate temperatures.

To establish the Arrhenius relationship with a kinetic reaction, determine the rate constant at different temperatures. Then plot the natural log of k vs. the reciprocal of the absolute temperature. The best fit line yields a y-intercept of the natural log of the frequency factor, and a slope of $-E_a/R$. See Fig. 4 for a sample regression.

GROWTH KINETICS

This subject area is important because it allows investigators to determine production rates of microorganisms and their products, and how fast organisms will

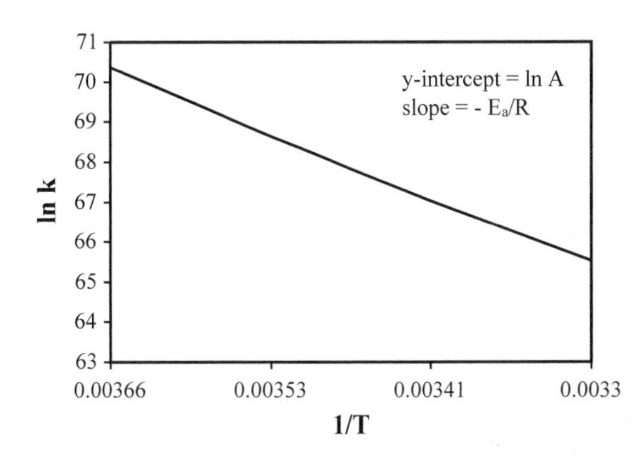

Fig. 4 Regression of Arrhenius relationship.

uptake substrates or gases. Common rate equations pertaining to this area include first order growth kinetics, and the Monod equation:

$$\mu = \frac{\mu_{\max} S}{K_S + S}$$

First order growth kinetics are commonly used to describe microorganism and/or cell growth in batch culture. Microbes placed into a batch reactor start growing at an exponential rate when they are unconstrained by substrate concentration, toxic by-product accumulation, or volume limitations. During this so-called exponential growth phase, the rate of cell growth is described as follows:

$$r_X = \frac{dX}{dt} = \mu X$$

where r_x is the rate of microorganism (or biomass) production, X, the cell concentration, and μ is the specific growth rate. One can see that this equation is very similar to first order degradation kinetics; there is no negative sign in this equation, indicating that the growth rate is positive. Separating the variables and integrating this equation with the boundary condition that cell concentration is at an initial value X_0 at time zero, the following equation is obtained:

$$X = X_0 e^{\mu t}$$

The doubling time τ is also analogous to the half time equation in first order kinetics, and is derived the same way. It is defined as the amount of time it takes for an organism population to double. *Clostridium perfringens* is the microorganism with the fastest known doubling time at 6 min–8 min. The doubling time is determined using the following equation:

$$\tau = \frac{\ln 2}{\mu}$$

Problem: The number of microorganisms in a batch reactor at 12:00 noon is 25,000. How many microorganisms will be in the batch reactor at 2:00 if the doubling time is 15 min? Assume that the microorganisms are in exponential growth phase at noon and at 2:00.

Solution: Use $X = X_0 e^{\mu t}$. Before substituting into this equation to solve for X, the specific growth rate μ must be found. Use $\tau = \ln 2/\mu$ to get μ, then solve for X.

$$\mu = \frac{0.693}{15\,\text{min}} = 0.0462\,\text{min}^{-1}\left(\frac{60\,\text{hr}^{-1}}{\text{min}^{-1}}\right) = 2.77\,\text{hr}^{-1}$$

$$X = 25,000\, e^{(2.77\,\text{hr}^{-1})(2\,\text{hr})} = 6,400,000$$

CONCLUSION

In this section, rate equations for zero, first, and second order degradation kinetics were derived, related parameters of interest were discussed, and illustrative examples were given. Models for growth kinetics were also provided. Reaction rates are a strong function of temperature, and this dependence can be modeled using the Arrhenius equation. The rate equations for kinetics of a single reactant were discussed in this section. Rate equations have also been developed for multiple reactants. See Refs.[2,3] for models involving multiple reactants.

LIST OF SYMBOLS

A	frequency factor or pre-exponential factor
C_A	concentration of reactant A
C_{A0}	initial concentration of reactant A
$\frac{dC_A}{dt}$	rate of change of the concentration of reactant A
$\frac{dX}{dt}$	rate of change of microorganism concentration
e	exponent (exp)
E_a	activation energy
k	rate constant or reaction rate constant
l	liter
L	length
ln	natural logarithm
M	mass
mg	milligrams
r_A	rate of reaction
R	universal gas constant
S	substrate concentration
t	time
$t_{1/2}$	half time of reaction
T	absolute temperature
τ	doubling time
μ	specific growth rate
$\mu_{\max}$	maximum specific growth rate
X	cell concentration or cell number
X_0	initial cell concentration or cell number

ACKNOWLEDGMENTS

Support for this research was provided in part by the Louisiana Agricultural Experimental Station and the Louisiana State University Agricultural & Mechanical College. Julianne Forman created the graphics for this section and assisted with editing and revisions.

REFERENCES

1. Gibaldi, M.; Perrier, D. *Pharmacokinetics*, 2nd Ed.; Marcel Dekker: New York, NY, 1982; 504 pp.
2. Doran, P. *Bioprocess Engineering Principles*, Academic Press: London, 1995; 439 pp.
3. Fogler, H. *Elements of Chemical Reaction Engineering*, Prentice-Hall: Englewood Cliffs, NJ, 1992; 838 pp.

B

Biomass Harvesting Systems

Dennis L. Larson
University of Arizona, Tucson, Arizona, U.S.A.

INTRODUCTION

Harvesting implies in-field gathering and removal of selected crop materials or products and their delivery to a storage site or transport vehicle. Harvest may include sizing and densification and is interrelated with other activities required to meet user needs, activities that may include additional transport, drying, and storage. The desired type and sequence of these activities based on operational costs, energy consumption, environmental effects, and personnel and management considerations depends on biomass form, conversion process, scale of operation, and harvest/utilization schedules.

Most plant biomass residues currently used for energy production are after-processing residues that accumulate at processing plants, principally sugarcane bagasse and wood residues. These byproducts of sugar and wood processing fuel cogeneration to meet industry energy needs and produce surplus electricity for sale to other users. High harvesting and transport costs relative to value have limited the specific harvest of residues. Biomass harvest cost estimates vary greatly depending on custom or owner operation, harvest area, yield, and management variables. Estimates range from \$25–35 Mg^{-1} to rake, bale, pick-up, and transport large bales of rice straw[1] to \$25 Mg^{-1} to cut, rake, and big bale switchgrass[2] and \$35 Mg^{-1} to bale and transport corn stover to roadside.[3] Some of these estimates include storage under plastic or tarp, but none include cost of replacement nutrients or other lost benefits of residues to soil potential or crop productivity, as described later.

HARVEST ISSUES

Environmental restrictions have renewed interest in harvesting residues that had commonly been burned or buried for crop or pest management reasons, such as rice straw,[4] cotton stalks,[5] and sugar cane residues.[6] And potential economic benefits from new markets are motivating the study and development of systems for harvesting high yield crops, such as corn stover, crops harvestable with conventional equipment, such as other cereal residues, and for the harvest of woody energy and other specialty crops.[7]

Current residue harvesting systems commonly utilize available forage or timber harvesting equipment, the forage harvesting systems consisting of swather or windrower, rake, baler and bale stackwagon, and tree harvest utilizing timber cutting, collection, perhaps chipping, and loading equipment. Desired residue harvest equipment modifications may include improved mobility and gathering devices for harvest in rough, muddy field conditions, such as rice straw and sugar cane bagasse collection.[1,6] Biomass shape and size and longer annual operating schedules per year also make it desirable that machines be stronger, more durable, and configured for larger stems and material than conventional grinders, balers, and other equipment.[2]

Because crop residues and woody biomass may have low bulk densities, be large and irregularly shaped, and have high moisture content and be shipped long distances, a systems approach is suggested to determine the most appropriate set, sequence, and location of operations for a given application considering harvest separation, drying, sizing and/or densification, transport, and storage needs. Different harvesting approaches may be most beneficial for dry and small sized, dry and large, wet and small, and wet and large sized energy crops and depend on the relative importance of cost, CO_2 emission, energy input, and other management objectives.[2]

IN-FIELD OR CENTRALLY

Several of the individual operations could be done either at the harvest site or centrally, the choice depending on economic, social, and other factors. For example, on-site drying and size reduction could reduce transport costs and prior size reduction might facilitate drying, but large scale processing may be much more cost effective. A corn stover collection project in Iowa collects stover and cob after grain harvest, then separates higher value cobs at the processor.[3] On the West Coast, rectangular bales are transported to a central location, compressed to double the baling density and wrapped in plastic for overseas shipment. Alternatively, bales are chopped and cubed for long shipments at large, centralized processing plants.[1]

Encyclopedia of Agricultural, Food, and Biological Engineering
DOI: 10.1081/E-EAFE 120006880
Copyright © 2003 by Marcel Dekker, Inc. All rights reserved.

RESIDUE MANAGEMENT

Residue removal could degrade crop production potential or require alternative soil amendments, crop nutrients or erosion mitigation measures to make up for residue losses. The value of nutrients was estimated to be \$9–15 Mg^{-1} of rice straw in one study.[1] Erosion control value may be more nebulous, but Midwest corn stover recommendations include leaving a 10 cm anchored stubble to prevent wind erosion and 30%–40% of the stover to minimize erosion caused by rainfall.[8] On the other hand, residue removal offsets costs associated with such alternative residue management operations as chopping and/or soil incorporation to facilitate machine operations, improve soil properties, or reduce insect survival. For example, residue harvest can replace the chopping and incorporation of cotton residues to reduce insect overwintering sites or substitute for prohibited burning of rice stalks and sugar cane leaves. A multicriteria approach to harvesting system selection may be useful to indicate the best methods for meeting conflicting economic, environmental, and other sustainable harvest goals.

HARVEST OPERATIONS

Separation

The first harvesting step usually is separation of desired biomass from plant stalk, root, or soil by cutting, felling, pulling, or windrowing. For grain harvested by cutting the stem near the grain head or pulling the ear from the stalk, this involves cutting, or recutting the standing stem or stalk at the desired height near the ground. Where stems are not vertical, the harvester may need to reorient stems for effective separation from retaining roots. Raking then may be needed to pick-up, turn, and windrow the material to facilitate drying and collection.

Residue removal height and pick-up from the ground may be constrained by the need to minimize the collection of soil, rocks, and other foreign material that can damage equipment or reduce processing efficiency and residue value. Cane harvest compromises the desire to maximize sugar harvest by cutting cane at the soil interface with the additional separation and cleansing necessitated to remove collected foreign materials.[9] Insect control in some locales mandates removal or burial of all above ground cotton plant materials after lint and seed harvest. An experimental cotton harvester cuts the stalk below the soil surface and pulls the plant from the soil.[10] The plant then is subjected to physical agitation by the conveying system to remove adhering soil. Gravity separation using a blower mechanism similar to that in cotton harvester lint cleaners has been evaluated to separate undesired fines.

Drying

Forage harvesting operations subsequent to separation of plant material from stalk, roots, or soil may include drying, collection, sizing, densification, and transport. Crop moisture content and bulk density have been found to influence transport costs more highly than shipping distance in an analysis of energy crop harvesting.[2] Drying and sizing are often pretreatment or preparation for biomass conversion or combustion. However, the high energy requirements and cost of artificial drying, and moisture content limitations for package stability and storage life of bales and cubes dictate in-field drying whenever time and weather permit for all applications favoring dry input materials. Artificial in situ drying of woody biomass from 50% to 25% moisture content was found to require more energy than the gain in energy potential of dried wood used for combustion.[2] The drying methods used cold air ventilation, hot air (natural gas) ventilation, and solar collectors. And though transport of 50% moisture product may be 50% more costly per unit dry matter than transport of 25% material, transport savings may not offset on-farm artificial drying costs.[2]

Collection, Sizing, and Densification

Cereal residues, corn stover, sugar cane residue, hemp, miscanthus, and cotton plants are among the crops for which baling has been accepted as the most efficient, cost effective harvesting method.[1,2,10] Forage choppers and forage wagons can efficiently remove the material from the field, but subsequent handling is difficult and long-distance transport expensive. Cubes and double density bales are commonly used to ship forage long distances, and though commercial portable machines have been used to a limited extent, at present, these are made from standard density bales at centralized processing facilities with less energy at a lower cost. Some packaging properties are summarized in Table 1.

Three common commercial machines, small and large rectangular bales, and large round bales, have been used successfully to gather residues from the ground and densify and package them in bale form. A variety of forage handling equipment is also available for in-field collection and transport of bales, such as stackwagons, and for loading bales onto semitrailers or other long-distance transport vehicles, including modified forklifts. Round balers may have increased capabilities for packaging biomass with tough stems and spread-out configuration and round bales may better preserve quality in uncovered storage, but capacity, handling, and transport challenges favor rectangular bales in many applications. The Iowa corn stover project uses both large round and large rectangular balers, but the round baler required use of

Table 1 Relative size, density, and energy input of alternative sizing or densification systems

Operation	Package size	Density ($kg\,m^{-3}$)[1–3]	Energy input ($MJ\,t^{-1}$)[2]
Baling	0.4 m × 0.6 m × 1.2 m, 32 kg	110	1,445
	1.2 m × 1.2 m × 2.4 m, 600 kg	150–175	
	1 m dia round bales, 500 kg	100–150	
Chipping	Fines, < 3 mm	80	400–3,000
	Chips, < 50 mm	100	200–800
	Chunks, 51 mm–250 mm	150	100–450
Mobile chipping and pelleting	3 cm	500	5,800
Central chipping and pelleting	3 cm	500	3,060

a shredder attachment to achieve the desired density.[3] Bale handling and cost advantages may favor large over smaller rectangular bales. For example, a rice straw harvesting analysis found in-field raking, swathing, baling, and roadsiding costs averaged $11 Mg^{-1} for large bales, $23 Mg^{-1} for small rectangular bales.[1]

Energy crops, such as poplar and willow, are harvested as whole stems or chopped or chipped material, using commercial timber cutting, gathering, chipping, and transporting equipment.[2] The principle "the smaller and wetter the material, the greater the decline in quality" favors material sizing just before use. Handling may offset this principle and transport cost advantages of chipped material. A variety of commercial mobile and stationary machines are available for cutting, chipping, or crushing woody biomass to the desired material size. Financial analysis of energy crop harvest in the Netherlands found harvest and transport costs generally ranged from lowest with delivery of dry, compressed chips, becoming progressively higher with delivery as dry, loose chips; wet chips; and wet stems. However, the report recognized some methods may require more, higher cost, in-plant preprocessing.[2]

Transport

Transportation requirements vary with crop yield, areal density, and harvest season length as well as scale and type of conversion process.[11] Trucking, often via semitrailers, has become the favored form of transport due to harvest site dispersion and access to processing locations via convenient road systems. Conventional or slightly modified forklifts, baleloaders, and timber handling equipment have been used for loading and unloading. Chips, cubes, and other small materials can be handled with augers and other conventional conveyance equipment.

Models have been utilized to estimate transport distances required to meet the capacity of different sized power plants for a range of energy crop land production factors.[2] These models have then been used to compare cost and energy requirements of alternative transport systems based on load, load density, and loading and unloading operations. Costs were more sensitive to density and moisture content than distance; semitrailers often were found to be least cost, containers highest cost in a study of energy crop transport.[2] For miscanthus, transport costs increased progressively with shipping density from lowest with pressed chips (312 $kg\,m^{-3}$), large round bales (300 $kg\,m^{-3}$), big square bales (153 $kg\,m^{-3}$), and chips (118 $kg\,m^{-3}$), to highest cost with bundles (82 $kg\,m^{-3}$) for average shipping distances of 13.5 km–38.5 km. The miscanthus study also found container transport to be more costly than trucks or semitrailers.

The most appropriate transportation may depend on whether harvested biomass will be converted to energy via a thermochemical or wet bioconversion process, when transport of wet product could be advantageous. But the view that transporting water for thermoconversion is uneconomical is not a truism. A Swedish processor transports wet chipped energy trees to a power plant, delivering the material just-in-time for use due to exorbitant drying and storage costs.[2]

Storage

Storage is required whenever biomass harvest and utilization schedules do not match. Storage options are limited by and product losses related to product size and moisture content. Woody biomass studies of chopped and chipped material found the wetter and smaller the material, the greater the quality degradation in storage.[2] Other storage goals are low cost handling, reduction in land use and fire risk, including arson as well as spontaneous combustion. Provision for pressurized ventilation or industrial drying is needed to preserve high, perhaps above 25%, moisture biomass. The common storage options, open stacks, covered with plastic or tarp, or under a shed roof, compromise least cost storage with minimum loss of biomass value. Uncovered storage of bales is justified only when loss is confined to the uppermost bales in the stack.[1] Tarps and plastic are inexpensive covers, but may hinder ventilation and must physically be placed over and taken off the stored product. Storage in pole buildings or under other shelters is more expensive, but permits observation and may facilitate handling.[1]

WHOLE-CROP HARVESTING

The harvest of biomass residues, the harvest of remaining plant materials following the selective harvest of grain or fiber, typically is a separate operation. The development of one-pass machines for collection of all product constituents has promising potential for more cost effective harvest of the total crop.[11] Whole crop harvesting machines developed for simultaneous collection of cereal grain and straw have not been adopted, but might gain favor with increased residue value. A whole-crop approach is being applied to the development of green sugar cane harvesting equipment that harvests cane, leaves, and tops for separation and processing at the sugar mill.[9] Research is also advancing the possibilities for whole crop harvest of corn stover, cobs, and grain. The McLeod Harvester provides an intermediate approach to the harvest of cereal grains, collecting grain, chaff, and weed seed for separation at a central mill.[12] While whole crop harvesters offer hope for reducing residue harvesting costs, development is interdependent with development of markets and increased value for the residues.

SYSTEM ASPECTS

Harvesting system considerations include economic, energy, time, and other management objectives. One analysis of energy crop harvesting alternatives reported values for three factors for management consideration: harvesting and transport costs, CO_2 emissions, and environmental effects.[2] But comparison of harvest alternatives can be confusing since published harvesting evaluations may define harvest differently. One study defined harvest as collection and roadsiding of residues or biomass while another reported all operations required to gather and deliver biomass to the user in the form desired for input to the biomass conversion process, e.g., fines or pellets at a specified moisture content. An evaluation of rice straw feedstock collection, storage and delivery in California that anticipated development of a standardized approach to the design of preferred systems was frustrated by variability among production and utilization conditions and data available for objective analysis.[4] This problem is multiplied with different products, uses, and locations. The definition of harvesting as all operations required to collect, process, transport, and store biomass in preparation for its intended use is advocated, a definition which permits clear comparison of in-field and centralized drying, sizing, and densification. This definition might also provide a better measure of cost/value to potential users of the biomass.

SUMMARY AND CONCLUSIONS

Conventional forage and timber harvest and transport systems have been applied to harvest of crop residues and energy crops with minimal modification. However, this use has indicated the need for increased mobility and capability for pick-up and separation of materials in rough or wet terrain, better configuration for collection and processing of plants with larger stems and different configurations, and tougher, more durable equipment to accommodate greater annual utilization.

Harvesting system evaluation should include energy, time, and other management objectives in addition to financial aspects. Economic considerations might include deferred costs of alternative residue management practices, value of residues for erosion control, to soil properties and as nutrients for crop growth, and other environmental factors.

Whole crop harvesting systems show promise for reducing total harvesting costs. Separation of plant constituents to maximize value is another goal. The optimal combination of in-field and centrally located processes will evolve with development of equipment to achieve sometimes conflicting grower/user objectives.

REFERENCES

1. Jenkins, B.M.; Bakker-Dhaliqal, R.; Summers, M.D.; Bernheim, L.G.; Lee, H.; Huisman, W.; Yan, L. Equipment Performance, Costs and Constraints in the Commercial Harvesting of Rice Straw for Industrial Applications. Paper 006035, American Society of Agricultural Engineers Annual Int. Mtg., Milwaukee, WI, July 27–30, 2000; American Society of Agricultural Engineers: St. Joseph, MI, 2000.
2. van den Heuvel, E. *Pretreatment Technologies for Energy Crops*. Technical Report to the Netherlands Agency for Energy and the Environment and Nat. Research Program on Global Air Pollution and Climate Change; MHP Mgmt & Sec. Services: The Netherlands, 1995.
3. Glassner, D.A.; Hettenhaus, J.R.; Schechinger, T.M. Corn Stover Collection Project. Proc. BioEnergy '98: Expanding BioEnergy Partnerships **1998**, *2*, 1100–1110.
4. Fife, L.; Miller, W. *Rice Straw Feedstock Supply Study for Colusa County California*. Report Prepared for the USDOE Western Energy Program; Rice Straw Joint Venture: Woodland, CA, March 2000; www.bioenergy.org/ricestraw (accessed July 1999).

5. Gemtos, T.A.; Tsiricoglou, T. Harvesting of Cotton Residue for Energy Production. Biomass Bioenergy **1999**, *6* (1), 51–59.
6. Molina, W.F., Jr.; Ripoli, T.C. Sugarcane Harvesting Trash Baling: Operational and Economical Performance Evaluation. Paper 016082, American Society of Agricultural Engineers Annual Int. Mtg., Sacramento, CA, July 27–30, 2001; American Society of Agricultural Engineers: St. Joseph, MI, 2001.
7. Lindley, J.A.; Baker, L.F. *Agricultural Residue Harvest and Collection*; Report for the USDOE Western Regional Energy Program; Agricultural Engineering Dept., North Dakota State University: Fargo, ND, 1994.
8. Nelson, R.G. Resource Assessment and Removal Analysis for Corn Stover and Wheat Straw in the Eastern and Midwestern United States—Rainfall and Wind-Induced Soil Erosion Methodology. Biomass Bioenergy **2002**, *22* (5), 349–363.
9. Braunbeck, O.; Bauen, A.; Rosillo-Calle, F.; Cortez, L. Prospects for Green Cane Harvesting and Cane Residue Use in Brazil. Biomass Bioenergy **1999**, *7*, 495–506.
10. Coates, W.E. Harvesting Systems for Cotton Plant Residue. Am. Soc. Agric. Eng. Appl. Eng. in Agric. **1996**, *12* (6), 639–644.
11. Larson, D.L.; Turner, A.K. Possible Costs for Harvesting Cereal Residues for Energy Production in Victoria. J. Aust. Inst. Agric. Sci. **1986**, *52* (1), 45–51.
12. Badger, P.J. The McLeod Harvester. Summary in Bioenergy Update, **2002**, *4* (11), 1–4; www.bioenergyupdate.com (accessed November 2002).

Biomass Production Systems

B. M. Jenkins
University of California, Davis, Davis, California, U.S.A.

INTRODUCTION

Biomass production systems constitute a broad array of technologies and concepts, some familiar, some only emerging. They span the range from precisely controlled intensive reactor culture systems to the extensive agricultural and managed and natural forest and aquatic systems. Biomass production is an integral consideration in regenerative life support systems for extended human space travel.

Production of biomass is intended for a number of purposes, including traditional uses as food and feeds, fiber, and structural materials. The historical use of biomass for energy, primarily as heat for cooking and space heating, continues in many parts of the world, but has largely been supplanted by fossil fuels in power generation, industrial production, and transportation. Biomass production has over the last few decades received increasing attention for its role in again providing renewable sources of energy, fuels, chemicals, and other industrial products, but at higher efficiencies and greater selectivity than previously possible. Motivating much of this renewed emphasis are issues of energy cost, supply, and security, global environmental benefits stemming from reductions in fossil fuel use and greenhouse gas emissions, local environmental benefits due to changes in disposal practices for crop residues and other biomass wastes or to biomass production in bioremediation and phytoremediation applications, and rural economic improvement. Biomass production is the first step in a chain of activities that results in product for final demand. Biomass is produced either specifically for the intended application, as is the case of energy crops grown for fuel, or as a residue of some other enterprise, such as agriculture, but for which some useful applications can still be found. Following production are harvesting, handling, processing, storage, transportation, conversion, and product distribution that must be properly integrated into an economically feasible biomass utilization system. Methods to optimize the overall system have been the subject of much research, and the production step remains a key element especially in terms of the quantity, distribution, quality, and cost of raw material available. For example, power generation systems using biomass fuels have specific constraints as to fuel moisture and composition depending on whether they employ combustion, thermal gasification, or anaerobic digestion technologies, and similar considerations apply to biofuels production, such as ethanol and biodiesel, composite materials, biopolymers, pharmaceuticals, and the wide array of other industrial products that can be manufactured from biomass.

Biomass is biologically derived matter of all types, including that from plants, animals, protists, and the monera. The production system design depends on the type of organism and on the nature of the energy and nutrients required. The autotrophic organisms, such as higher plants and certain bacteria, require only inorganic nutrients for growth and production, whereas the heterotrophs also require organic materials for their nutrition. The heterotrophs include all animals, fungi, and most bacteria. Phototrophs are photosynthetic organisms that utilize light energy for conversion to chemical energy (see later discussion of photosynthetic pathways). Chemotrophs, such as the animals, use organic and inorganic (e.g., oxygen) compounds for energy conversion rather than light. Lithotrophs use inorganic compounds in cellular synthesis (e.g., water as the electron donor for photosynthesis), whereas organic compounds such as carbohydrates are used by the organotrophs. An organism can by classified according to all three categories. Green plants are lithotrophic in addition to being photoautotrophic. Bacteria may be photoheterotrophic, photoautotrophic, or chemolithotropic, and production systems must necessarily account for these differences.

Biomass is produced in natural ecosystems and in human managed systems although most natural ecosystems evidence some influence of human management or impact, e.g., fire suppression in forests, deforestation in land conversion to agriculture, and vegetative damage due to acid rain from industrial pollution. Managed or cultured systems include agriculture and aquaculture (and its marine constituent, mariculture). Agriculture comprises the growing of terrestrial crops as well as forestry. Aquaculture and mariculture include the production of biomass in aquatic systems and the oceans. Energy crop production, as in short-rotation woody crops grown for fuelwood, is a form of agriculture. The managed systems also include many intensive operations such as the aerobic

Encyclopedia of Agricultural, Food, and Biological Engineering
DOI: 10.1081/E-EAFE 120006867
Copyright © 2003 by Marcel Dekker, Inc. All rights reserved.

and anaerobic processes used in waste treatment and the fermentation systems employed in biotechnology.

In more popular use, the word biomass has come to refer primarily to materials obtained through photosynthesis that are used for fuels, chemicals, and other industrial and manufactured products. By this terminology, biomass includes matter produced principally by green plants, algae, and photosynthetic bacteria. This definition includes primary or virgin materials, such as wood, as well as wastes and secondary materials, such as the organic fraction of municipal solid waste and animal manures. The discussion that follows is largely directed at the biomass included under this definition, particularly the green plants.

GLOBAL BIOMASS PRODUCTION

Total global annual net photosynthetic biomass production is variously estimated at around 100 Gt. Of the incident solar power received at the top of the atmosphere (173×10^{15} W), an estimated 0.02% is used in biomass production through photosynthesis.[1] At a mean higher heating value (heat released by combustion) of 16 MJ kg^{-1}, this net photosynthetic yield translates to 70 Gt yr^{-1} dry matter. More detailed analyses of biomass production by type of ecosystem yield estimates of 170 Gt yr^{-1}–220 Gt yr^{-1}.[2,3] Total energy content of the earth's annual biomass production is of the order of 10^{21} J. By comparison, the total annual commercial energy consumption by humans including fossil, nuclear, geothermal, hydro, biomass, and other solar energy is currently one-third of this value, but expected to grow rapidly over the next 50 yr. Biomass provides 15% of world energy needs, but in developing countries constitutes a much higher fraction of energy supply: 35% overall, and in excess of 80% in many rural areas.[3,4] Total global plant biomass currently accumulated in all ecosystems is estimated roughly at 10^{12} t dry matter, or 10 times annual production.[2,5]

PHOTOSYNTHETIC EFFICIENCIES AND BIOMASS YIELDS

Plants utilize three principal pathways in assimilating atmospheric carbon dioxide and synthesizing carbohydrate structures and other compounds through photosynthesis. The biological or dry matter yield is dependent in part on the pathway used. Light energy is absorbed in two pigment systems called photosystem I (PSI) and photosystem II (PSII). In both systems, absorption of light by chlorophyll and accessory pigments leads to the emission and transport of electrons against an adverse voltage gradient. The electrons are derived from the photolysis of water, mediated by a manganese-containing enzyme in PSII.[5,6] Electrons are transferred from PSII through the Z-scheme to PSI, storing energy in the carriers adenosine triphosphate (ATP) and the reduced form of nicotinamide adenine dinucleotide phosphate (NADPH) for later use in CO_2 reduction and compound synthesis. Mineral nutrients are directly involved in the electron transport, and in other processes of the plant. The mineral or ash concentration and composition of the plant are often quite important in the subsequent use of the biomass, and can influence the design of the production and utilization system.

The light reactions store energy in NADPH and ATP. In so called C3 plants, CO_2 and water react with ribulose-1,5-diphosphate to produce 3-phosphoglyceric acid as part of the Calvin-Benson cycle. The glyceric acid is subsequently converted using NADPH and ATP to 3-phosphoglyceraldehyde and then to hexose-phosphate and ribulose-5-phosphate. The latter reacts with ATP to regenerate ribulose-1,5-diphosphate, whereas the hexose is used in the synthesis of the primary storage products, sucrose and starch. The C3 pathway takes its name from the 3-carbon intermediates produced during the cycle. C3 plants include the cereals barley, oats, rice, and wheat, alfalfa (lucerne), cotton, *Eucalyptus*, sunflower, soybeans, sugar beets, potatoes, tobacco, *Chlorella*, and others. Gymnosperms (with a few exceptions), bryophytes, and algae are C3 plants, as are most trees and shrubs.[5]

Along the C4 pathway, CO_2 combines with phosphoenol pyruvate (PEP) via a PEP-carboxylase catalyzed reaction to form oxaloacetate, which is reduced to malic acid (malate) or aspartic acid (aspartate), 4-carbon intermediates giving the pathway its name. These are translocated from the mesophyll cells, where the primary CO_2 assimilation occurs, to the bundle sheath cells, where CO_2 is released for subsequent fixation through reactions of the Calvin-Benson cycle as in C3 plants. Decarboxylation of the acids regenerates PEP. C4 plants are usually of tropical origin and under higher intensity solar radiation have higher photosynthesis rates and higher biological yields (dry matter production) compared with C3 plants. C4 plants include sugarcane, sorghum, maize, and Bermuda grass.[2,6] *Euphorbia* species are mostly C3, but a few have evolved to use the C4 pathway.[5]

The third primary pathway is that of crassulacean acid metabolism (CAM) used by many succulents. The CAM pathway also fixes CO_2 via PEP-carboxylase, but in CAM species the stomata are open at night rather than during the day in order to conserve water.[5,6] Malate is stored in the vacuoles during the night, then released during the day when the stomata are closed. CAM plants have lower growth rates than C4 species but have high water use efficiency due to their adaptation to low-water environments, including semiarid and salt-marsh regions, and epiphytic sites, such as the orchids use.

The dominant structural compounds making up plant biomass are cellulose (C6 polymers) and hemicellulose (predominantly C5 polymers) produced via condensation polymerization of the monosaccharides. The other primary structural components are lignins, aromatic polymers of variable structure derived in one proposed pathway from coniferyl, sinapyl, and *p*-coumaryl alcohols. The alcohols arise through the shikimic acid pathway, and are polymerized into lignin via free-radical reactions.[5] Organic compounds in biomass also include proteins, triglycerides (fats and oils), terpenes (including isoprenes), waxes, cutin, suberin, phenolics, phytoalexins (antimicrobial compounds produced by the plant), flavonoids, betalains, alkaloids, and other secondary compounds as well as sugars and starch.[2,5] Plants also accumulate inorganic materials (ash), sometimes in concentrations exceeding those of hemicellulose or lignin. Structural and elemental composition of several types of biomass are listed in Table 1.

Biological yield is a function of the net of photosynthesis and respiration, the latter including photorespiration in C3 plants. Respiration provides energy through the oxidation of organic compounds, but also provides substrates for the synthesis of other plant products. Maximum theoretical photosynthetic efficiencies can be derived based on the minimum requirement of 8 photons of photosynthetically active radiation (PAR, 400 nm–700 nm wavelength band) per molecule CO_2 used to produce hexose (glucose).[2,3,7] About 43% of the energy in sunlight at the earth's surface is PAR, and of this a maximum of about 80% is actively absorbed in photosynthesis. Only about 28% of the absorbed energy is captured in hexose products. In C4 plants, respiration consumes somewhere between 25% and 40% of the energy in the sugar. The maximum net efficiency of photosynthesis based on incident sunlight is therefore 6%–7%. Photorespiration in C3 plants generally leads to efficiencies of around 3%, substantially lower than C4 plants.[3] Photosynthetic efficiencies can be translated to biomass yields using site-specific insolation data and biomass heating values. At maximum efficiency and high insolation, theoretical yields can exceed $400\,Mg\,ha^{-1}\,yr^{-1}$.[2,7]

Actual biological yields are less than maximum yields due to nonoptimal temperature and crop conditions (e.g., incomplete canopy closure or lodging of crop), limited supplies of water and nutrients, and losses due to diseases and pests. Commercial yields are lower still because not all biomass is or can be harvested. Efficiencies for commercial agricultural crops typically are of the order of 1%, although tropical crops such as sugar cane and high yielding grasses can produce at 2%–3% efficiency with dry matter yields of $50\,Mg\,ha^{-1}\,yr^{-1}$–$100\,Mg\,ha^{-1}$-yr^{-1}.[2] Intensive production of green algae can approach 5% efficiency, similar to the best efficiencies with C4 crops under research conditions. Seasonal efficiency for many C3 crops when given sufficient water and nutrients with adequate light and temperature is approximately 2%.[8]

B

BIOMASS RESOURCES AND PRODUCTION SYSTEMS

Primary or virgin sources of biomass include residues from agriculture and forestry (e.g., cereal straws, orchard and vineyard prunings, forest slash, sugarcane trash), biomass from forests and forest stand management activities (thinnings from stand improvement operations, brush and other biomass from fire hazard reduction operations, increased forest harvesting), byproducts of food, fiber, wood, and feed processing and manufacturing operations (sawdust, sander dust, nut shells, fruit pits, pomace, rice husks, cotton gin trash, sugarcane bagasse, substandard grain), and industrial and energy crops grown for biomass (plantation trees, energy grasses, kelp, algae). Waste biomass generally includes processed materials such as animal manures, biosolids from wastewater treatment, and paper and other organic materials in municipal solid wastes. Black liquor is a lignin-containing product from wood or other biomass pulping and papermaking, but is most commonly burned in recovery boilers at the pulp-plant in order to recover pulping chemicals and generate steam and power. Classification of these materials as waste is inappropriate when they are used as resources for energy and products. When this occurs, they can be classified with other industrial byproducts.

Residues

Agricultural and forest residues are produced as byproducts of the primary commodity production system. Most equipment operations involving residues are therefore associated with harvesting, handling, transportation, storage, and utilization (see related articles discussing these aspects for biomass), and not with the actual biomass production that is accessory to the primary economic crop component. In some cases, the intention to use residue biomass can lead to changes in the overall crop production system, such as increased fertilization to replace nutrients exported with the residue biomass when harvested, or alternatively, the application of recycled ash or other nutrient containing byproducts of the biomass utilization system; changes in the fertilizer composition in order to reduce the uptake of undesirable constituents in

Table 1 Composition of selected biomass materials

Type	Alfalfa straw	Rice straw	Wheat straw	Miscanthus	Switch grass	Jose Tall wheat grass	Hybrid poplar	Willow	Water hyacinth	Nonrecyclable waste paper	Sugarcane bagasse	Municipal digester sludge (class B biosolids)
Typical harvest moisture (% wet basis)	14	14	10	14	14	10	45	45	85	6	50	75
Proximate composition (% dry matter)												
Ash	4.88	18.67	14.48	4.90	6.53	12.34	1.6	0.95	22.40	8.21	3.61	37.91
Organic fraction	95.12	81.33	85.52	95.10	93.47	87.66	98.40	99.05	77.60	91.79	96.39	62.09
Volatiles	76.48	65.47	69.94	78.20	77.03	72.18	86.14	85.23		82.50	84.51	53.68
Fixed carbon	18.64	15.86	15.58	16.90	16.44	15.48	12.26	13.82		9.29	11.88	8.41
Higher heating value ($MJ\,kg^{-1}$)												
Moisture free (dry)	18.16	15.09	16.33	18.05	18.90	17.86	18.93	19.38	16.02	21.52	18.50	15.38
Moisture and ash free	19.09	18.55	19.09	18.98	20.22	20.38	19.24	19.56	20.64	23.44	19.19	24.77
Wet	15.62	12.97	14.69	15.52	16.26	16.08	10.41	10.66	2.40	20.22	9.25	3.85
Structural composition (% dry matter)												
Cellulose	29	34	40	45			41	49	16		36	
Hemicellulose	12	28	29	30		33	28	56		30		
Lignin		9	14	21			26	22	6		19	
Ultimate elemental composition (% moisture and ash free)												
Carbon	48.40	47.02	48.08	53.31	50.69	52.86	51.65	49.56	52.96	53.66	49.99	58.29
Hydrogen	6.15	6.39	6.08	4.63	6.08	5.05	5.99	5.95	6.82	7.60	5.86	7.18
Oxygen (by difference)	43.05	44.77	43.99	41.59	42.57	39.59	41.75	44.11	37.16	38.18	43.94	24.17
Nitrogen	1.19	1.07	1.19	0.21	0.60	2.00	0.60	0.35	2.53	0.38	0.15	9.08
Sulfur	0.19	0.22	0.28	0.11	0.06	0.37	0.02	0.03	0.53	0.22	0.08	1.72
Chlorine	1.17	0.71	0.71	0.21	0.10	2.21		0.01		0.04		0.16
Ash analysis (% ash)												
SiO_2	7.04	74.67	54.64	70.60	66.53	46.71	1.17	8.08		19.44	41.87	47.11
Al_2O_3	1.12	1.04	5.73	1.10	6.98	3.09	0.41	1.39		63.97	22.25	17.9
TiO_2	0.05	0.09	0.23	0.06	0.34	0.09	0.21	0.06		3.81	3.87	1.22
Fe_2O_3	0.41	0.85	6.16	1.00	3.56	1.00	0.76	0.84		0.42	20.90	5.64
CaO	21.37	3.01	5.02	7.50	7.14	4.59	59.16	45.62		8.37	3.50	8.65
MgO	5.83	1.75	2.45	2.50	3.17	2.03	5.76	1.16		1.68	1.45	2.98
Na_2O	11.2	0.96	2.16	0.17	1.03	8.60	0.31	2.47		0.83	0.26	1.33
K_2O	22.9	12.3	14.09	12.80	7.00	15.68	26.76	13.2		0.23	2.59	1.32
P_2O_5	6.32	1.41	2.43	2.00	2.80	2.70	0.20	10.04		0.10	1.13	14.65
SO_3	4.27	1.24	3.03	1.70	2.00	2.05	5.26	1.15		1.14	0.90	1.38
Cl						13.43						0.01
CO_2						0.05						0.21

Structural data representative only and not necessarily from same sample used for elemental analysis and heating value. Oxygen in ultimate analysis is by difference, including Cl, S, and C in ash.
(Adapted from Refs. [2], [9], [18], [23], [24].)

the biomass such as chlorides; modified tillage strategies to protect against soil erosion when residue cover is reduced or to take advantage of reduction in surface biomass to be incorporated into the soil; changes to irrigation practices to manage soil and crop moisture for enhanced biomass harvesting; and changes in chemical applications due to changes in weed, pest, and disease pressure resulting from residue removal. Other changes to the production system may occur when modifications are desired in the properties of the biomass. One example is the delayed harvesting of cereal straws to take advantage of natural precipitation in the leaching of salts (alkali metals and chloride) to improve the combustion properties for biomass power plants and to reduce the export of nutrients.[9] In such cases, the equipment used for, and the timeliness of tillage and other soil preparation activities may be adjusted as a direct result of the use of the residue biomass. Residue removal also reduces air pollution when substituted for traditional open burning disposal practices with some crops.[10] When substituted for fossil fuels in controlled combustion processes, residues or other biomass generally generate lower SO_x emissions due to typically lower sulfur contents. A decision to utilize the residue component of a crop can also lead to changes in the variety selected, and in the location of the production site in order to optimize the overall system.[11]

B

Forest Biomass

In addition to residues, biomass can be removed from existing forests in the form of thinnings comprising low-quality stock that is often unsuitable for traditional markets but which contributes to poor forest health, or as increased production from more intensively managed regrowth forests.[3] In many existing temperate forests, mean annual growth far exceeds mean annual harvest, and overall stand quality can be diminished. Understory vegetation, small trees, and other biomass can contribute to high fuel loadings in forests, with increased danger of catastrophic wildfire, higher incidence of crown fires with greater damage to mature trees, and high costs of fire suppression if practiced. Changes to the biomass production system in these cases are largely associated with harvesting equipment (see related article titled *Biomass Harvesting Systems*), but could also include greater use of prescribed fire as a management tool and better monitoring of forest conditions. The sustainability of these practices requires careful consideration of forest biodiversity, habitat, soil erosion, forest preservation, riparian quality, air quality, and global atmospheric impacts including impacts on soil carbon fluxes and greenhouse gas emissions associated with disturbances to the forest floor.

Industrial and Energy Crops

Prior to the development of coal, petroleum, and natural gas as primary fuel and chemical resources, biomass supplied most energy and chemical needs, and continues to do so in many areas of the world. Although substantial hydrocarbon resources still exist,[12] the use of biomass as fuel in substitution of fossil resources results in almost no net atmospheric carbon emissions when the biomass is renewed at a rate equal to consumption, and hence serves to mitigate greenhouse gas and global climate change impacts. Biomass production also serves to store solar energy, thus allowing continuous power generation. This is in contrast to solar photovoltaic and solar thermal power systems that operate only during the day and that require other storage mechanisms to supply continuous power if another primary source, such as grid-power from other generators, is not available. Like other solar technologies, biomass is a distributed resource, and its use to supply large quantities of energy requires large amounts of land. The overall conversion efficiency from solar energy to final energy product is low as well, due to the inherently low efficiency of photosynthesis. The current overall efficiency of electricity, e.g., using biomass produced at 2% photosynthetic efficiency as fuel in biomass-dedicated power plants operating at 20% average thermal efficiency is 0.4% compared with solar photovoltaic systems generating at efficiencies of 6%–10% during the day. To meet current world energy demands of 380 EJ in heating value would require 380×10^6 ha of land (1 TJ $ha^{-1} yr^{-1}$) in terrestrial crops continuously producing at 2% photosynthetic efficiency (about 6 kg $m^{-2} yr^{-1}$ dry matter). This is roughly 25% of the world's cultivated land area and 10% of forest lands. It is 20% of the area of degraded tropical lands, and half the area of these lands considered suitable for reforestation.[3] The actual land requirement to meet world energy demand would be much larger, especially as world population and energy demand increases. This estimate does not include aquatic or marine species that could also contribute. Generation of all commercial energy from biomass is not necessarily desirable, but the energy potential of biomass is large, and its use for energy, chemicals, and other industrial products is likely to continue to increase.

Terrestrial plants, including industrial and energy crops, can be classified as woody or herbaceous. Woody crops are predominantly trees, frequently grown in plantations using short-rotation intervals of 1 yr–20 yr. Cultural practices for these crops have been well established for roundwood and papermaking, and have been more recently extended to the production of fuel wood. Herbaceous crops include annual and perennial grasses and other nonwood plants. Production practices for these crops are in most cases similar to other agricultural crops. A number of more commonly considered industrial and energy crops are listed in Table 2. The design of the production system considers climate, soil preparation and preservation, species and variety selection, planting, weed and pest control, nutrients and fertilization, water and irrigation, harvesting, and postharvesting operations.

Along with temperature and light, water availability and cost are key constraints in biomass production. Water requirements are typically 300 kg kg^{-1}–1000 kg kg^{-1} dry matter produced.[3] Due to soil and water limitations, arid or semiarid regions of the world are not anticipated to produce substantial quantities of biomass even though good light and temperature conditions often exist. However, biomass production can occur in such areas where waste or degraded water supplies are available (e.g., treated municipal waste water) and in regions of irrigated agriculture. In the latter case, biomass can play an integral role in managing salts and remediating other undesirable impacts of agriculture, such as in the integrated farm drainage management (IFDM) systems now developing in California and where water is sequentially reused on crops of increasing salt tolerance, including biomass crops.[13] Biomass crops can in general be used for restoration of degraded or deforested lands,[3] and for economic and environmental reasons are frequently proposed for planting in marginal areas.

Short-rotation woody crops are typically planted in plantations with stand densities ranging up to 10,000 trees per ha depending on the size of tree desired at harvest, harvesting technique employed, and end use for the biomass.[14] Currently 100×10^6 ha are in industrial tree plantations,[3] most in longer rotations for roundwood and pulp production. Production site selection depends on a number of factors, including soil properties, water availability, slope, climate, and distance from market. As with agricultural crops, high yields are associated with better soil types, although tree species can take advantage of higher ground water tables.[15] Soil preparation involves land clearing to remove any existing vegetation and eliminate weeds; soil amelioration to adjust pH and improve tilth, drainage, and nutrient concentrations; leveling, and in some cases mulching. The plant bed is prepared in a manner similar to agricultural crops, although tillage depth is typically deeper than for cereals. Following soil preparation, cuttings, slips, or seedlings are planted either manually or by machine or machine assist. Cuttings are commonly cut from nursery grown whips. Whips are harvested, bundled, and machine cut to about 25 cm lengths.[16] For machine planting, cuttings are placed in magazines after sorting for straightness and size. Automatic planters singulate cuttings by picking them individually from the magazines, in some designs using pneumatic heads on robotic arms. Cuttings are transferred to a planting conveyor and deposited in the furrow generated by a disc opener. Machines plant 4–6 rows simultaneously. Slips and cuttings are typically planted manually by simply pushing them into the prepared soil bed. Seedlings can be planted bare root or in plant pots. Soil moisture management following planting is critical as deeper tillage increases drying rate, potentially leading to inadequate water availability without rain or irrigation. Tending after planting is important for proper stand establishment, and involves cultivation, chemical application, or mulching (including plastic sheet mulching) to control weeds, and in some cases, pruning to improve shoot vigor. Some crop species, such as *Salix* (willow), are

Table 2 Selected crops considered for industrial and energy biomass production

Herbaceous species—biomass/fiber/energy grain	Woody species—biomass/fiber/pulp
Alfalfa (*Medicago sativa*)	Alder (*Alnus* spp.)
Flax (*Linum usitatissimum*)	Australian pine (*Casuarina*)
Hemp (*Cannabis sativa*)	Birch (*Onopordum nervosum*)
Jose Tall wheatgrass (*Agropyrum elongata*)	Black locust (*Robinia pseudoacacia*)
Kenaf (*Hibiscus cannabinus*)	Eucalyptus (*Eucalyptus* spp.)
Miscanthus (*Miscanthus* spp.)	Lucaena (*Lucaena leucocephala*)
Napier grass/Banagrass (*Pennisetum purpureum*)	Poplar (*Populus* spp.)
Spanish thistle or Cardoon (*Cynara cardunculus*)	Willow (*Salix* spp.)
Spring barley (*Hordeum vulgare*)	
Switchgrass (*Panicum virgatum* L.)	
Triticale (*Triticosecale*)	
Winter rye (*Secale cereale*)	
Winter wheat (*Triticum aestivum*)	
Herbaceous species—sugar/starch/biomass	Wetland species—biomass/fiber
Buffalo gourd (*Curcurbita foetidissima*)	Cattail (*Typha* sp.)
Cassava (*Manihot esculenta*)	Cordgrass (*Spartina* spp.)
Jerusalem artichoke (*Helianthus tuberosus*)	Giant reed (*Arundo* spp.)
Maize/corn (*Zea mays*)	Giant reed (*Phragmites* spp.)
Sugar/energy cane (*Saccharum* spp.)	Reed canary grass (*Phalaris arundinacea*)
Sugar/fodder beet (*Beta vulgaris*)	
Sweet sorghum (*Sorghum bicolor*)	
Herbaceous species—seed/oilseed/terpenes	Aquatic species—biomass/lipids/chemicals
Amaranth (*Amaranthus* spp.)	Brown algae (*Sargassum* spp.)
Castor (*Ricinus communis*)	Giant kelp (*Macrocystis pyrifera*)
Crambe (*Crambe abyssinica*)	Microalgae (*Botryoccus braunii*)
Euphorbia (*Euphorbia* spp.)	Red algae (*Gracilaria tikvahiae*)
Jojoba (*Simmondsia chinensis*)	Unicellular algae (*Chlorella, Scenedesmus*)
Linseed (*Linum usitatissimum*)	Water hyacinth (*Eichhornia crassipes*)
Oilseed rape (*Brassica* spp.)	
Safflower (*Carthamus tinctorius*)	
Soybean (*Glycine max*)	
Sunflower (*Helianthus annuus* L.)	

(Adapted from Refs. [2], [13], [14], [25].)

B

highly susceptible to common herbicides. Biological weed control is sometimes practiced, usually by planting weed-competitive crops such as *Trifolium*, although these can also compete for nutrients and water with the primary crop until good canopy cover is achieved. Fencing to control grazing by animals may also be needed. Irrigation and fertilization can be accomplished in the same manner as for agricultural crops. Drip irrigation systems can be combined with chemigation (injection of nutrients and other chemicals into the irrigation system for distribution to the plants). Frost protection is not commonly practiced, but frost damage is a concern in many locales. Where frosts are frequent, more tolerant species or varieties must be selected. In dry areas, fire suppression may be needed. Varietal or species selection is also important in reducing damage from insects and diseases. Plantation design can include set aside areas harboring native predators of pests, and division of the plantation into blocks of different clones or species to make the overall plantation less susceptible.[3] Due to the lower frequency of planting and other operations in tree plantations, soil erosion rates are generally reduced compared with agricultural crops. Survival rates are typically above 85% for trees under good management without frost or other catastrophic natural events. Coppice crops are harvested on 3 yr–10 yr cycles, with up to 6 cycles before replanting. Mean annual dry matter increments in practice under good conditions are $10\,Mg\,ha^{-1}\,yr^{-1}$–$20\,Mg\,ha^{-1}\,yr^{-1}$, generally declining at higher latitudes for reasons of climate including temperature and insolation. Crop improvement is anticipated to extend these yields to $15\,Mg\,ha^{-1}\,yr^{-1}$–$30\,Mg\,ha^{-1}\,yr^{-1}$ or higher.[2]

Production of herbaceous crops in many respects bears greater resemblance to conventional forage and other agricultural crops, although many of the same considerations apply to the production system design as for woody crops.[14,17] A large number of species are suitable for industrial and energy crop production.[2,14] Regional mixed cultivation of perennial and annual species, as opposed to large-scale monoculture, is seen in many cases to be of benefit both in terms of environmental and economic performance. Depending on location, the regional mix could include legumes and warm and cool season grasses. Total dry matter yields for herbaceous species under good management are typically in the range of $10\,Mg\,ha^{-1}\,yr^{-1}$–$30\,Mg\,ha^{-1}\,yr^{-1}$, although in practice yields below $10\,Mg\,ha^{-1}\,yr^{-1}$ are still common. Sugar cane, a C4 species, is one of the most productive crops, with world average yields around $35\,Mg\,ha^{-1}\,yr^{-1}$ dry matter.[3] The C4 grasses are generally associated with high yields. Where proposed as energy crops, particular attention must be given to the composition of the plant. Herbaceous species, especially grasses, generally contain more ash than wood, with the ash containing undesirable proportions of alkali metals and silica leading to problems of ash slagging and fouling in combustion and other thermal conversion systems. Leaching can remove most of the alkali and chloride, and delayed or spring harvest to take advantage of rain-washing can improve the combustion properties in the same manner as for residues.[18,19] Reducing chloride in fertilizers is advantageous, as is proper management of sulfur. In all cases, the composition of the feedstock needs to be considered in terms of the requirements of the manufacturing or conversion process and any accompanying waste disposal. Consideration also needs to be given to the introduction of exotic species that may grow well initially but lack biological disease and pest controls in new environments and are therefore more susceptible to later damage,[3] or that may prove adaptive and invasive to the detriment of native species or other crops.

Except for phytoplankton, aquatic species tend to demonstrate higher yields than terrestrial crops.[2] Aquatic species include both unicellular (e.g., *Chlorella*, *Scenedesmus*) and macroscopic multicellular (e.g., *Macrocystis*, *Gracilaria*, *Sargassum*) algae (seaweeds), and a number of water and salt-marsh plants such as cordgrass (*Spartina*), reed (*Arundo*, *Phragmites*), bulrush (*Scirpus*), and water hyacinth (*Eichhornia*).[2,13,14] Dry matter yields range between $5\,Mg\,ha^{-1}\,yr^{-1}$ and $75\,Mg\,ha^{-1}\,yr^{-1}$. The high water content of these species implies their use in biochemical conversion systems when considered for energy or fuels, as dewatering and drying are not generally economically feasible. Controlled production of marine species in the open ocean is also difficult and expensive, but total production potential is large. Water hyacinth is considered a nuisance plant in many inland waterways and is difficult to control, but is hardy and disease-resistant.[2] Many of the aquatic species are currently produced under much more intensive conditions, and generally for more valuable products than energy and fuels.

Industrial Byproducts and Wastes

Production systems for byproduct and waste biomass generally involve processing operations to improve handling, mitigate hazards, and reduce transportation, storage, and conversion costs. These operations include mechanical dewatering, drying, separation, sorting and classification, decontamination, sterilization, densification (e.g., briquetting, pelleting), blending, and size reduction. Such operations can often apply equally well to other forms of biomass, but are generally considered part of the harvesting and processing system rather than the production system.

COSTS OF BIOMASS PRODUCTION

Exclusive of harvesting and downstream processing and conversion operations, production costs for agricultural and other biomass residues are typically allocated to the primary crop production system and not separately accounted. Byproduct or waste biomass may be available at no cost, or in some cases, tipping (disposal) fees are applied to cover the costs of handling. In contrast, industrial and energy crops grown for biomass assume full allocation of production costs. These costs are quite variable depending on species, production location, level of management, and resulting yield. Total average delivered costs, including harvesting and transportation (85 km), for plantation *Eucalyptus* in northeast Brazil producing at 12.5 Mg ha^{-1} yr^{-1} have been estimated at \$1.90–2.60 per GJ, or about \$40–52 per Mg dry matter (2001 U.S. dollars, CPI adjusted from 1990 dollars).[3] Of the total, 40% is associated with stand establishment including nursery production, land, planting, and administration, and another 10% is associated with plantation maintenance including management, cultivation, and research. Half the delivered cost is in production of the biomass. Under Canadian conditions, total delivered cost including harvesting, chipping, and transportation for a typical 5 yr rotation, 4 rotation cycle forestry crop with a yield of 12 Mg ha^{-1} yr^{-1}, was similarly estimated at \$45 per Mg (2001 U.S. dollars, CPI adjusted from 1985 dollars), of which 40% was allocated to the production system including land, nursery, planting, and tending.[16]

Cost estimates for energy and industrial crops in the United States range between \$28 and \$115 per Mg, or \$1.60–5.80 per GJ.[2,3,20,21] These costs depend on the scale of production, the crop planted, management level, soil type, geographical region, and contributions of various governmental incentives and restrictions. An analysis for the southeastern United States estimates costs for short-rotation woody crops with yields of 5 Mg ha^{-1} yr^{-1}–10 Mg ha^{-1} yr^{-1} at \$32–51 per Mg on crop land, and \$48–70 per Mg on pasture land. Half of the cost is in production exclusive of harvesting.[20] For switchgrass (*Panicum virgatum*), with yields of 10 Mg ha^{-1} yr^{-1}–20 Mg ha^{-1} yr^{-1}, costs are \$30–70 per Mg, of which 40% is attributed to production prior to harvesting, but this fraction is sensitive to yield. Biomass production costs before harvesting are therefore in the range of \$16–35 per Mg. A study of farm gate prices needed to increase biomass production in the United States estimated that 6×10^6 ha would be brought into production by 2008 with prices at \$33–37 per Mg, and 16×10^6 ha would be brought into production with prices at \$44–48 per Mg.[21] Current costs of producing and harvesting biomass feedstocks were seen as being limiting to meeting U.S. objectives of tripling bioenergy production by 2010. Fossil energy costs influence such conclusions. Natural gas prices in the United States in the period 1999–2001, e.g., fluctuated between \$2 and 15 per GJ. Research along with genetic and cultural improvements are projected to reduce biomass production costs by 20%–40% over the next 10 yr–30 yr.[2,3] Future biomass production levels will also be influenced by direct and indirect environmental and socioeconomic consequences that are in many cases external to the direct costs of production.[25–30] Costs of biomass produced under intensive culture reactor systems such as used in biotechnology can greatly exceed the costs just described for energy crops grown in what are largely agricultural settings, but the products of these intensive systems have much higher market values than energy crops.

ENERGY BALANCES

Where biomass is grown for energy, the quantity of energy invested in the production system should be less than the energy yield in biomass when expressed in equivalent forms. In general this implies that the fossil energy input to the production system should be less than the fossil energy equivalent of the biomass. Net energy yields from biomass have been of particular concern for the production of ethanol, especially from grain; less so where biomass is grown for thermal conversion (e.g., combustion in power plants).[22] The fossil energy investment, however, requires consideration in the design of biomass production systems. Energy balances for hybrid poplar, sorghum, and switchgrass show that production energy inputs amount to 2%–5% of crop heating value, with harvesting and transportation energy adding another 3%–5%.[3] The energy content of biomass is generally much higher than the fossil energy used in its production. The overall energy yield of a biomass energy system, however, depends on the type and efficiency of the conversion system employed, as well as the energy costs of production.

REFERENCES

1. Hubbert, M.K. The Energy Resources of the Earth. Sci. Am. **1971**, *Sept*, 31–40.
2. Klass, D.L. *Biomass for Renewable Energy, Fuels, and Chemicals*; Academic Press: San Diego, CA, 1998.
3. Hall, D.O.; Rosillo-Calle, F.; Williams, R.H.; Woods, J. Biomass for Energy: Supply Prospects. In *Renewable Energy: Sources for Fuels and Electricity*; Johansson, T.B., Kelley, H., Reddy, A.K.N., Williams, R.H., Eds.; Island Press: Washington, D.C., 1993; 593–651.

4. Bain, R.L.; Overend, R.P.; Craig, K.R. Biomass-Fired Power Generation. Fuel Proc. Tech. **1998**, *54*, 1–16.
5. Salisbury, F.B.; Ross, C.W. *Plant Physiology*; Wadsworth Publishing Co.: Belmont, CA, 1992; 225–248.
6. Marschner, H. *Mineral Nutrition of Higher Plants*; Academic Press: London, 1986; 115–125.
7. Loomis, R.S.; Williams, W.A. Maximum Crop Productivity: An Estimate. Crop Sci. **1963**, *3*, 67–72.
8. Monteith, J.L. Climate and the Efficiency of Crop Production in Britain. Trans. R. Soc. Lond. B. **1977**, *281*, 277–294.
9. Jenkins, B.M.; Bakker, R.R.; Wei, J.B. On the Properties of Washed Straw. Biomass Bioenergy **1996**, *10* (4), 177–200.
10. Jenkins, B.M.; Turn, S.Q.; Williams, R.B. Atmospheric Emissions from Agricultural Burning in California: Determination of Burn Fractions, Distribution Factors, and Crop-Specific Contributions. Agric. Ecosyst. Environ. **1992**, *38*, 313–330.
11. Jorgensen, U.; Sander, B. Biomass Requirements for Power Production: How to Optimize the Quality by Agricultural Management. Biomass Bioenergy **1997**, *12* (3), 145–147.
12. Rogner, H.H. An Assessment of World Hydrocarbon Resources. Annu. Rev. Energy Environ. **1997**, *22*, 217–262.
13. Cervinka, V.; Finch, C.; Martin, M.; Menezes, F.; Peters, D.; Buchnoff, K. *Drainwater, Salt and Selenium Management Utilizing IFDM/Agroforestry Systems*; Final Report, Grant Number 4-FG-20-11640; Bureau of Reclamation, US Department of the Interior: Fresno, CA, 2001.
14. Smith, N.O.; Maclean, I.; Miller, F.A.; Carruthers, S.P. *Crops for Industry and Energy in Europe*; European Commission: Luxembourg, 1997.
15. Siren, G.; Sennerby-Forsse, L.; Ledin, S. Energy Plantations—Short Rotation Forestry in Sweden. In *Biomass: Regenerable Energy*; Hall, D.O., Overend, R.P., Eds.; John Wiley & Sons: Chichester, U.K., 1987; 119–143.
16. Golob, T.B. Machinery for Short Rotation Forestry. In *Biomass: Regenerable Energy*; Hall, D.O., Overend, R.P., Eds.; John Wiley & Sons: Chichester, U.K., 1987; 145–173.
17. Gosse, G. *Lignocellulosic Energy Crops in Different Agricultural Scenarios*; European Commission: Luxembourg, 1996.
18. Jenkins, B.M.; Baxter, L.L.; Miles, T.R., Jr.; Miles, T.R. Combustion Properties of Biomass. Fuel Proc. Tech. **1998**, *54*, 17–46.
19. Huisman, W. Harvesting and Handling of Miscanthus giganteus, Phalaris arundanicea and Arundo donax in Europe. In *Biomass: A Growth Opportunity in Green Energy and Value-Added Products*; Overend, R.P., Chornet, E., Eds.; Pergammon: Oxford, 1999; 327–333.
20. Graham, R.L.; Downing, M.E. *Potential Supply and Cost of Biomass from Energy Crops in the TVA Region*; Oak Ridge National Laboratory (ORNL-6858): Oak Ridge, TN, 1995.
21. Wright, L.L.; Walsh, M.; Downing, M.; Kszos, L.; Cushman, J.; Tuskan, G.; McLaughlin, S.; Tolbert, V.; Scurlock, J.; Erhenshaft, A. Biomass Feedstock Research and Development for Multiple Products in the United States. In Proceedings First World Conference and Exhibition on Biomass for Energy and Industry, Sevilla, Spain, 2000.
22. Jenkins, B.M.; Knutson, G. *Energy Balances in Biomass Handling Systems: Net Energy Analysis of Electricity from Straw*; ASAE Paper No. 843593; American Society of Agricultural Engineers: St. Joseph, Michigan, 1984.
23. Broder, J.D.; Badger, P.C. Acid Hydrolysis and Subsequent Ethanol Production from Cellulosic Resources. In *Renewable Energy in Agriculture and Forestry*; Vanstone, B.J., Ed.; University of Toronto: Toronto, Canada, 1995; 189–214.
24. Jenkins, B.M.; Way, Z. Properties of Biomass. In *Biomass Energy Fundamentals*; Wiltsee, G.A., Jr., Ed.; Electric Power Research Institute (EPRI TR-102107): Palo Alto, California, 1993; Vol. 2, Appendices, 58–59.
25. Venendaal, R.; Jorgensen, U.; Foster, C.A. European Energy Crops: A Synthesis. Biomass Bioenergy **1997**, *13* (3), 147–185.
26. Braunstein, H.M.; Kanciruk, P.; Roop, R.D.; Sharples, F.E.; Tatum, J.S.; Oakes, K.M. *Biomass Energy Systems and the Environment*; Pergammon Press: New York, 1981; 182.
27. Hanegraaf, M.C.; Biewinga, E.E.; van der Bijl, G. Assessing the Ecological and Economic Sustainability of Energy Crops. Biomass Bioenergy **1998**, *15* (4/5), 345–355.
28. Scholes, H. Can Energy Crops Become a Realistic CO_2 Mitigation Option in South West England? Biomass Bioenergy **1998**, *15* (4/5), 333–344.
29. Rafaschieri, A.; Rapaccini, M.; Manfrida, G. Life Cycle Assessment of Electricity Production from Poplar Energy Crops Compared with Fossil Fuels. Energy Convers. Manag. **1999**, *40*, 1477–1493.
30. Chiaramonti, D.; Grimm, H.-P.; El Bassam, N.; Cendagorta, M. Energy Crops and Bioenergy for Rescuing Deserting Coastal Area by Desalination: Feasibility Study. Bioresource Technol. **2000**, *72*, 131–146.

Biomass Transport Systems

Phillip C. Badger
General Bioenergy, Inc., Florence, Alabama, U.S.A.

INTRODUCTION

Biomass is land and water vegetation and materials derived from this vegetation, including agricultural and forestry crops and their harvesting and processing residues. Although biomass covers all vegetation, the term is usually used in the context of biomass energy or biobased product applications, noting that many systems used for energy or biobased products have been adopted from existing agricultural and forestry equipment and systems. Biomass resources typically have low bulk densities, making the costs for their transportation and handling a significant factor in their use. Innovative methods to reduce transportation costs are the focus of this article.

TECHNOLOGY REVIEW

In addition to cost, the type of transportation system used depends on the type of biomass, the form that it is in, whether the transportation will be off-road or over-the-road or both, and the method of transportation. The loading and unloading are part of the overall transport systems, since these operations can constitute a major portion of the transportation time and expense and depend on transportation method and biomass form. Table 1 summarizes the current transportation practices for various common biomass resources. These practices serve a baseline from which improvements and new technologies are developed. Raw biomass used for energy and biobased products can be represented by all the forms shown in Table 1, and variations of the methods of transportation shown in Table 1 are frequently used for transporting biomass materials.

Whole Tree Systems

Harvesting, handling, and conversion of biomass feedstocks must be performed as integrated systems, as each operation impacts the other to some degree. Conventional forestry harvesting and handling systems use feller-bunchers that shear trees off at the stump and haul them to a landing for chipping and loading onto transport trucks, and the chips are burned in a conventional boiler. The Whole Tree Energy™ system takes advantage of fast growing trees such as hybrid poplars that can be harvested every three or four years and coppice (resprout) after harvest. These trees tend to be long and slender with slender small-diameter branches. The Whole Tree Energy™ system shears trees off at the stump using standard feller-bunchers, loads the whole trees onto a "pole-trailer" (commonly used to haul tree stems in the pulpwood industry), and transports the whole trees to the power plant, where they are stacked inside an air-inflated dome. Inside the dome, waste heat from the power plant is used to reduce the moisture content by half over a 30-day period. The trees are then loaded onto a conveyor, which feeds the whole trees to the furnace with a cut-off saw at the furnace door to sever the trees into 20-ft pieces to fit into the furnace. Bringing in whole trees reduces handling costs by about 35% over conventional practices, allows for air drying, and the large "particles" require less combustion air, thus reducing NO_x formation and sensible heat loss.[1]

Bale-Based Systems

Biomass, by nature, tends to be bulky, which increases transportation costs. Thus, many transportation systems include methods to densify biomass and increase the bulk density of the material. Baling is one method of densification commonly used for grass-type materials, and standard enclosed semitrailer vans are frequently used for over-the-road transportation of baled biomass materials. Older vans have standard internal dimensions of 96-in. wide by 48-ft long, and newer vans have dimensions of 102-in. wide by 53-ft long. Both have an internal height of 110 in. To optimize transportation of bales using these vans, John Deere has designed a baler (100 Intermediate Rectangular Baler) that produces hay or straw bales with end dimensions of 31.5 in. by 31.5 in. By adjusting the length of a bale through adjustments to the baler, a bale can be produced that will fit snuggly inside a semitrailer van. For loading such vans, bales are stacked three high on a forklift and slid into the van from the rear.[2]

The John Deere Company has offered tractor pull-type balers that produce high density, small square bales by doubling the plunger crank cycle revolutions. These balers

Encyclopedia of Agricultural, Food, and Biological Engineering
DOI: 10.1081/E-EAFE 120006887
Copyright © 2003 by Marcel Dekker, Inc. All rights reserved.

B

Table 1 A summary of current transportation practices for selected agricultural and forestry resources

Type of biomass	Form	Activity	Method	Loading method	Unloading method
Silage (hay or corn)	Bulk loose chop	Off road	Silage wagon	From field chopper	Self, live bottom
		Off/on road	Stakeside truck	From field chopper	Self dumping
Hay/straw	Bales, small square	On road	Flat bed truck or trailer	Front-end loader	Front-end loader
	Bales, large square	Off/on road	Flat bed truck	Front-end loader	Front-end loader
		On road	Enclosed truck	Fork lift	Fork lift
	Bales, large round	On road	Flat bed trucks and trailers	Front-end loader	Front-end loader
	Loose stack (Stackhand)	Off road	Live bottom trailers	Self	Self
Grain	Bulk loose	Off road	Hopper wagons	From combine	Self, gravity
			Stakeside trucks	From combine	Self dumping
		On road	Stakeside trucks	From combine, elevator, or conveyor	Self dumping or with truck dumper
			Semitrailer	From combine, elevator, or conveyor	Self dumping or with truck dumper
			Train hopper cars	From elevator or conveyor	Gravity dump
Cotton	Loose	Off road	Oversized stakeside wagons	From picker	Pneumatic conveyor
	Loose, modularized	Off/on road	Live bottom, tilt bed trucks	Self, live bottom	Self, live bottom
	Baled	On road	Flat bed trucks	Front end loader	Front end loader
Root crops (sugar beets, potatoes, vegetables)	Bulk	Off/on road	Stakeside semitrailers, wagons	From harvester	Self dumping or truck dumping
Sugar cane	Bulk, whole stalks	Off/on road	Caged, flat-bed wagons	From cane harvester	Knuckleboom with grapple hooks
Pulpwood	Whole tree stems	On road	Pole-side truck	Front end loader or knuckleboom with grapple hook	Front end loader or knuckleboom with grapple hook
	6-ft lengths	Off/on road	Pole-side truck	Knuckleboom with grapple hook	Knuckleboom with grapple hook
	Bulk, whole tree chips	On road	Open top hopper semitrailer vans	From chipper	Truck or semitrailer dumper
Fuel	Bulk, hog fuel, sawdust, shavings, etc.	On road	Open top hopper semitrailer vans	Conveyor or gravity bin	Truck or semitrailer dumper or self, live-bottom semitrailer

were discontinued after about a year of production because the increased densities facilitated molding in the bales, unless additional hay drying was performed prior to baling.[3]

The hay and straw industries have also developed presses for large square and round bales to double their density. These presses are being used in the U.S. and Canadian West Coasts primarily to facilitate shipping hay to the orient in standard shipping containers.[4,5] For straw in the Willamette Valley of Oregon, it costs about $45 per ton to harvest, store, and deliver straw to the press; $30 per ton to compress it, and another $15 to deliver it to the docks in Portland.[6]

Picking up bales, even automatically by machine, is a time consuming process. Preventing the need to pick up bales requires a trailing unit to aggregate the bales. Such equipment has been available for years for small square bales but not for large round or square bales. A new bale accumulator for large round bales, made by Kelly Manufacturing Company in Tifton, Georgia, is in the form of a trailer towed behind the baler. The trailer has a deck that tilts backward to facilitate unloading and, depending on the model, can haul two or three bales at a time. For loading, a loading arm lying on the deck catches bales as the baler ejects them, and assists the bales to roll by gravity as far back on the trailer as possible. For transport, the front gate on the trailer is raised to prevent the bales from rolling off the front. For unloading, the rear gate on the trailer is dropped and the bed tilted for the bales to roll off.[7]

Along the same theme, several small companies have developed a variety of machines to mechanically pick up large square or round hay or straw bales from the ground. (Such machines have been available for small square bales for some time.) One such system, built by Martin Welding and Repair in Lititz, Pennsylvania, is a custom hay stacker that allows a single person to load large round bales lying on the ground, haul, and unload bales quickly. The pull-type unit is towed and powered by a tractor, and is designed to haul 15 bales of 3 ft by 4 ft by 8 ft or 8 square bales of 4 ft by 4 ft by 8 ft. The unit is towed with an offset hitch so that a pair of hydraulically operated lift forks can be slid under bales as they lie on the ground. In operation, the operator drives parallel to the bale to be loaded, and the lift forks are slid beneath the bale. An arm on the loading forks rotates the bale 90°, the forks are tilted to slide the bale back, and the forks lowered to pick up additional bales.[8]

The U.S. Department of Energy has recently become interested in using large quantities of corn stover for energy purposes at off-farm locations. This interest has necessitated the development of equipment capable of efficiently harvesting and transporting large quantities of stover over-the-road. The wet, muddy field conditions for harvesting corn stover make the task for harvesting and transporting bales from the field much more difficult than for hay or straw fields. One result of this work has been the development of specialized "load and go" trailers that allow one person to quickly pick up and stack bales onto a tractor-pulled trailer. The tractor cab contains controls that allow the driver to operate a hydraulically actuated arm that can pick up bales regardless of their orientation, rotate them to the correct orientation for loading, and then stack them on the trailer.[9]

Forestry harvesting residues can be an important source of biomass feedstocks; however, they also have quite low density. One solution to this problem has been baling the residues into large round bales using equipment similar to large round hay balers, but designed for heavier duty. The balers are designed for stationary operation and are mounted on a self-propelled heavy-duty frame to allow the unit to be taken to landing sites in the forest. The baler has a knuckleboom loader, which the operator uses to feed material into the baler. Material up to 8 in. in diameter is forced into the bale chamber by feed rollers that also serve to crush the material, which facilitates drying and bending. Inside the chamber, the material is further crushed as it is rolled under pressure into a cylindrical shape. When the chamber is full, 3 or 4 layers of polyethylene netting are wrapped around the bale to hold its shape, the chamber is opened, and the bale is ejected. Baling not only almost doubles the bulk density, but also converts the residues into a form that is easier to handle and store. The porous netting allows the bales to breathe, and since the bales are not very dense and most material in the bale is not touching the ground, baling allows the material to readily air dry. Bale dimensions are roughly 44 in. by 49 in. with weights in the range of 1200 lb–1300 lb at 40% moisture content.[10]

Pellet-Based Systems

Converting biomass into pellets is another way to increase density and convert biomass feedstocks into a form that is easier to handle and transport. The sooner in the production process that the biomass is converted into pellets, the sooner the advantages associated with pellets can be realized. Thus the Germans have developed a mobile, self-propelled biomass harvester (Biotruck, 2000) for grasses that cuts, chops, and dries the grasses with waste heat from the engine, compresses the grass into pellets having a density of about 75 lb ft^{-3}, and then transfers the pellets to a storage hopper on the harvester.[11]

Several plants in the United States and Canada make fuel pellets (0.25-in. diameter by 0.5-in. long) from sawdust and other forms of clean wood waste. To date, most of the markets for this fuel have been for residential pellet stoves, with the pellets marketed in 40-lb bags. Methods for handling bulk quantities of pellets for commercial and industrial users require further development. One-ton canvas bags with built-in canvas spouts have been used for bulk transportation of pellets by Spokane Pres-to-Log, Spokane, Washington. The bag bottom dimensions conform to the dimensions of a standard wooden pallet and sit on a standard pallet, making it roughly cubical in shape. The pallet allows the bag to be moved with a forklift. At the use point for the pellets, the pellets can be stored in the bag and fed directly from the bag to the feeding device.

A second method of handling fuel pellets in bulk by Spokane Pres-to-Log has been hopper trucks commonly used by the feed industry for bulk delivery of feed. These trucks have a built-in auger spout that can swivel and elevate to reach bulk bins at the customer's site. Load cells on the truck record the amount of delivery.

The Finns have developed a special truck bed for bulk delivery of wood fuel pellets. The enclosed special truck bed, which can be interchanged with a standard bed for other uses, is divided into three 212-ft^3 sections that can be unloaded separately using the truck's hydraulics to open the gates to the individual bins and power the blower for the pneumatic conveyor built into the truck. Upon arrival at a customer's location, the gate to a bin is opened and the truck bed tilted to permit the pellets to flow by gravity to the feeder of the pneumatic conveyor. The pellets flow through a flexible tube at distances up to 20 m to the customer's storage silo.[12]

Pipeline Transportation Systems

One of the biomass transportation systems that is more innovative and still very experimental is the capsule pipeline concept, under development by the University of Missouri. In this system, biomass is compacted into large (5 to 6 in. diameter) solid cylinders (logs) that can be inserted into a pipeline of somewhat larger diameter and transported through the pipe using water. Another version uses small-wheeled capsules that are transported through the pipeline using compressed air. The pipelines are located underground, thus providing safety and environmental benefits by reducing truck traffic.[13]

Bio-Oil Based Systems

Another form of biomass densification is to convert biomass materials into bio-oil at the field edge or forest-landing site using mobile equipment and thermo-chemical processes. The biomass is dried as part of the process, and part of the biomass is consumed during the conversion process, providing the energy for the process and decreasing the weight to be transported. Conversion into bio-oil provides a high energy density liquid material that, relative to unprocessed biomass, can be more easily and more cost-effectively stored in tanks and moved with pumps and piping, or over longer distances, with tank-trucks.[14]

CONCLUSION

New crops and new methods of processing, along with new raw feedstock requirements, require new methods for handling and transporting biomass materials. The design of new systems has been based on experiences and existing equipment from the agricultural and forestry industry.

Harvesting, handling, and conversion of biomass feedstocks must be performed as integrated systems, as each operation impacts the other to some degree. Minimizing the effects of the low bulk density of biomass materials and loading and unloading methods are important factors in systems design, as well as design for heavier duty service and flexibility to reduce operating costs. New systems will continue to be developed as the demand for bioenergy and biobased products continues to grow and new crops, conversion processes, and products are developed.

REFERENCES

1. Ragland, K.W.; Ostlie, L.D. *High Efficiency Bioenergy Steam Power Plant*, Proceedings of BioEnergy '98: Expanding BioEnergy Partnerships, Madison, WI, Oct 4–8, 1998; U.S. Department of Energy Great Lakes Regional Biomass Energy Program, Coalition of Great Lakes Governors: Chicago, IL; 772–782.
2. http://products.deere.com/webapp/commerce/command/ProductDisplay (accessed April 5 2002).
3. Badger, D. Personal communication. John Deere Company, April 5, 2002.
4. http://www.agfiber.org/ (accessed April 7 2002).
5. http://www.doublepress.net/specs.thm (accessed April 7 2002).
6. Miles; Thomas, T.R. Personal communication. Miles Consultants, Portland, Oregon, April 5, 2002.
7. Round Bale Accumulator Saves Loading Time. Farm Show **2001**, *25* (6), 19.
8. New-Style Big Square Bale "Stacker." Farm Show **2002**, *25* (6), 8.
9. Glassner, D.A.; Hettenhaus, J.R.; Schechinger, T.M. In *Corn Stover Collection Project*, Proceedings of BioEnergy '98: Expanding BioEnergy Partnerships, Madison, WI, Oct 4–8, 1998; Wichert, D., Ed.; U.S. Department of Energy Great Lakes Regional Biomass Energy Program, Coalition of Great Lakes Governors: Chicago, IL; 1100–1111.
10. Bala Press AB Stallgatan, Company Literature, Nossebro, Sweden, 1997.
11. Sutor, P. In *A New System of Harvesting Biomass Is Born*, Proceedings of the Second Biomass Conference of the Americas, Portland, OR, Aug 21–24, 1995; Klass, D.L., Ed.; National Renewable Energy Laboratory: Golden, CO; 1228–1235.

12. http://www.tekes.fi/opet/pdf/Kiihtelysvaara-e.pdf (accessed April 5 2002).
13. Graham, J.; Kiesler, J.; Morgan, A.; Liu, H.; Marrero, T.R. *Compacting Biomass Waste Materials from Co-firing with Coal*, Proceedings of the 24th International Technical Conference on Coal Utilization and Fuel Handling Systems, Clearwater, FL, March 8–11, 1999.
14. http://www.renewableoil.com (accessed April 5 2002).

Bioprocess Residence Time Distribution

B

Wei Wen Su
University of Hawaii at Manoa, Honolulu, Hawaii, U.S.A.

INTRODUCTION

Residence time distribution (RTD) describes the distribution of times required for fluid elements to pass through a continuous-flow system. RTD function and associated flow models have been widely used to characterize hydrodynamic behavior of a variety of systems encountered in agricultural, food, and biological engineering. RTD alone tells us how long the various fluid elements have been in the system, and it yields distinctive clues to the type of macromixing (distribution of matters by bulk circulation currents) occurring within the system, but it does not tell us anything about the degree of intermixing between fluid elements of different ages (i.e., micromixing).[1] In addition to the RTD, an adequate model of the flow pattern and knowledge of the extent of micromixing or degree of segregation[2] are required to fully characterize the hydraulic characteristics of a system.[1] In the article *Fermentation Residence Time Distributions*,[3] the basic concept, measurement techniques, and associated mathematical models of RTD have been discussed. The scope of this article is to review recent literature on bioprocess applications of the RTD analysis. Additional applications of the RTD analysis in the biomedical, food process, and environmental fields are reviewed in the article, *Residence Time Distribution—Biomedical, Food, and Environmental Applications.*

Industrial bioprocessing typically consists of upstream (production) and downstream (separation/purification) operations. RTD analysis has been applied in both operations. Most studies have focused on bioreactors and protein-purification systems for upstream and downstream operations, respectively.

CHARACTERIZATION OF BIOLOGICAL REACTORS

Gas-Phase Hydrodynamics

In the study of Tescione, Ramakrishnan, and Curtis,[4] an *Agrobacterium*-transformed root culture of *Solanum tuberosum* was grown in a 15-L bubble column, and the gas-phase hydrodynamics was investigated using RTD analysis. The second moment, or variance, of the gas-phase RTD was used to quantify gas dispersion in the reactor. Gas dispersion measured from argon tracer RTDs increased fourfold due to increased stagnation and channeling of gas through the bed of growing roots; however, introduction of an antifoam surfactant into the reactor greatly reduced dispersion with no accompanying change in respiration.

Gas-phase hydrodynamics was also investigated using RTD analysis in a rotating drum bioreactor for solid-state fermentation.[5] The RTD studies of gas through the bioreactor were performed using carbon monoxide as a tracer gas. Carbon monoxide behaves similarly like air, with comparable molecular weight, diffusivity, critical temperature, and critical volume,[6] and it is readily detectable at very low concentrations. Further, CO is only sparingly soluble in water and hence is not absorbed in the water associated with the bran. Hardin, Howes, and Mitchell[5] showed that the tracer response of the rotating drum bioreactor differed considerably from profiles expected for plug flow, plug flow with axial dispersion, and continuous stirred tank reactor (CSTR) models. The RTD data were then fitted by least-squares analysis to mathematical models describing a central plug-flow region surrounded by either one dead region (a three-parameter model) or two dead regions (a five-parameter model). Model parameters were the dispersion coefficient in the central plug-flow region, the volumes of the dead regions, and the exchange rates between the different regions. The three-parameter model (Fig. 1A) is discussed here. The model allows for two subsystems within the drum, a plug-flow region and a dead region (Fig. 1B). Hardin, Howes, and Mitchell[5] suggested that the central plug-flow region is likely to be located near the central axis of the drum, due to the entry and removal of air at the centers of the ends of the drum. The dead region is made up of the stagnant parts of the headspace surrounding the central plug-flow core plus the interparticle gas within the bran bed. The gas superficial velocity was shown to significantly affect the model parameter values.[5] Increased superficial velocity tends to decrease dead region volumes, exchange rates, and axial dispersion. The amount of dispersion of gas entering the drum was found sensitive to the geometry of the entry port and the geometry of the drum. Changing the pattern of inlet flow, e.g., by installing a diffuser plate at the gas inlet, could very easily alter the volume and

Encyclopedia of Agricultural, Food, and Biological Engineering
DOI: 10.1081/E-EAFE 120007202
Copyright © 2003 by Marcel Dekker, Inc. All rights reserved.

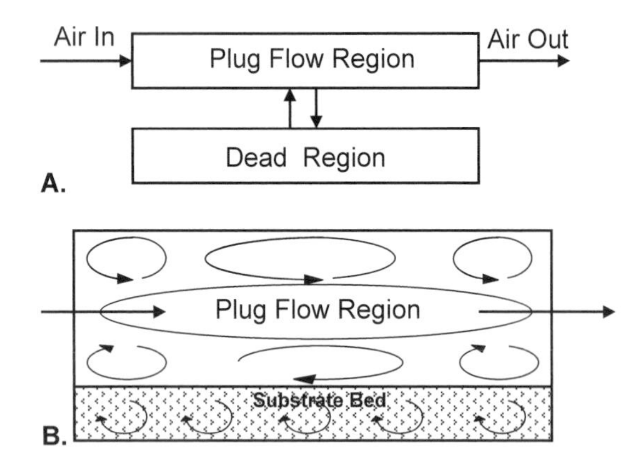

Fig. 1 (A) A three-parameter mixing model for a rotating drum bioreactor. (B) Hypothesized flow pattern within the drum. (From Ref. 5.)

velocity of the jet and the degree of backmixing. This study illustrated the needs for a reasonable flow model in order to analyze the RTD data and how such analysis could be used to understand gas-flow hydrodynamics and estimate mass and heat transfer parameters, from which an optimal rotary drum bioreactor can be designed and scaled up.

Liquid-Phase Hydrodynamics

Another important application of RTD analysis has been in characterizing liquid-phase hydrodynamics in bioreactors. Knowledge of such characteristics is highly valuable since liquid-phase hydrodynamics could greatly affect the overall conversion in the bioreactor. It is well known that except for first-order reactions, the product yield is a function of the system segregation, which in turn is governed by both macro- and micromixing.[1,7] Such segregation can result in the formation of dissolved oxygen, substrate, and pH gradients, rendering a decrease in reactor productivity. This is especially problematic at large scales, as with increasing bioreactor dimensions, circulation times increase and variation in the micro-environment experienced by the cells becomes augmented.[7] A useful strategy to examine such phenomenon is to conduct scale-down studies. The RTD analysis in this case assures the proper setting of the flow regime in the scaled-down model to mimic the hydrodynamics in a large-scale bioreactor with imperfect mixing. Amanullah et al.[7] devised a scaled-down model to simulate spatial pH variations in large-scale bioreactors. This model consisted of a stirred tank reactor (STR) and a recycle loop containing a plug-flow reactor (PFR, simulated using a very long plastic tubing), with the pH in the STR being maintained by addition of alkali in the loop. Tracer experiments and RTD analysis were applied to probe and compare the liquid-phase hydrodynamics under different recirculating flow rates. Amanullah et al.[7] found that the RTD in the tubing section could not be well represented by a common dispersion model.[8] However, vessel dispersion numbers (i.e., N_D, the dimensionless dispersion number equivalent to the inverse of the Peclet number) could still be used to indicate whether the reactor is closer to plug flow or mixed flow. N_D was calculated based on the mean residence time and the variance of the experimentally determined RTD according to Levenspiel.[8] These dispersion numbers were found to have very low values and increased with increasing flow rate. The low values indicating that dispersion is small and the long tubing section gave a good approximation to plug flow. The scaled-down STR-PFR system could be a useful tool to optimize culture performance with respect to pH fluctuations in the large-scale bioreactors.

CHARACTERIZATION OF SEPARATION SYSTEMS

Expanded Beds

The RTD analysis has been used to characterize expanded bed adsorption systems in several recent studies. Expanded bed adsorption has become popular for bioseparations in recent years. In such system, the adsorbent particles are slightly fluidized, and hence offers reduced flow resistance compared with conventional packed-beds. This system simplifies sample-processing requirements and could lead to improved processing throughput and better recovery. However, it is necessary to optimize the hydrodynamics in the expanded beds to allow optimal and reliable performance. Willoughby, Hjorth, and Titchener-Hooker[9] used RTD studies to estimate bed voidage in an expanded bed adsorption system, which contains semifluidized adsorbent packing. Voidages within a given section in the expanded beds were calculated using the following equation:

$$\varepsilon = \frac{V_l}{V_c} = \frac{Qt_d}{\pi R^2(\Delta h)} \tag{1}$$

where V_c is column volume in a given section, which is equal to the liquid volume V_l, plus that occupied by the particles in that section, Q the volumetric flow rate, t_d the tracer residence time in the given section (which can be obtained from RTD measurements), R the bed radius, and Δh the axial height of the section. In using Eq. 1, it was assumed that the fluid in the column behaves as plug flow and the tracer sees only the voidage that is exterior to the particles. To determine the extent the flow deviates from plug flow, Willoughby, Hjorth, and Titchener-Hooker[9] characterized the axial dispersion in the bed. The extent of axial dispersion was estimated from the RTD curves obtained at the sampling points for the section under consideration using the dimensionless dispersion number, N_D, according to Levenspiel[8]:

$$\sigma_\theta^2 = 2N_D - 2N_D^2(1 - e^{-1/N_D}) \tag{2}$$

where σ_θ is the dimensionless standard deviation of the RTD curve. Eq. 2 was solved for N_D where σ_θ was input from the RTD curve. From this calculation, it was possible to determine the operating condition under which Eq. 1 could be used to estimate the voidage in the expanded bed. As mentioned earlier, the dimensionless dispersion number was also used to characterize the PFR region of a scaled-down bioreactor with nonideal mixing.[7] In addition, this approach (i.e., calculation of dimensionless dispersion numbers) was employed by Mullick and coworkers[10,11] in characterizing the effect of bed expansion and height of the settled bed to the column diameter ratio on the degree of dispersion in an expanded bed adsorption system packed with fluoride-modified zirconia particles.

In expanded bed adsorption, whole culture broth that contains biological particles may be processed in the adsorption bed directly without prior separation of cell particles from the broth. However, the presence of the biological particles could potentially interfere with the adsorption efficiency. Fernandez-Lahore et al.[12] studied the effect of biomass loading on the hydrodynamic stability of expanded bed adsorption systems by RTD analysis and a hydrodynamic model. Here, fluorescent tracer (tryptophan or fluorescein) was used in lieu of UV/Vis absorbing or salt tracers to avoid problems associated with the presence of biomass suspended solids and changes in ionic strength from using salt pulses.[12] Analysis of the RTD data was done according to the PDE model (axially dispersed plug-flow exchanging mass with stagnant zones[13]), which contains three parameters: the number of mass transfer units between the dynamic (mobile) and stagnant zones (N), the Peclet number that accounts for the degree of axial dispersion (Pe), and the fraction of liquid in the dynamic zone (φ)[12] (Fig. 2). When the expanded bed was tested using particle-free buffer, the normalized response signal (upon perfect input pulse) was symmetric (N: 0; Pe: 50–100; φ: 1), indicating a homogeneous fluidized (expanded) bed. After adding suspended biomass (yeast cells, yeast cell homogenate, or *Escherichia coli* homogenate), the RTD became skewed. Depending on the adsorbent used and the type and level of biomass present in the sample, three groups of hydrodynamic behaviors (represented by different values of N, Pe, and φ) were noted in the expanded bed[12]: 1) a well-fluidized system (N: ≥ 7–10; Pe: ≥ 40; φ: 0.80–0.90); 2) a system exhibiting bottom channeling (N: < 1–2; Pe: ≥ 40; φ: 0.5–0.7); and 3) a system where extensive agglomeration develops (N: 4–7; Pe: 20–40; φ: < 0.5).[12] Fernandez-Lahore et al.[12] noted that changes in the hydrodynamics of expanded bed adsorption system could occur even in the presence of moderate biomass concentrations. Interactions between biomass and adsorbent led to changes in the stagnant zone fraction, $1 - \varphi$. The presence of the biomass also changed the hydrodynamics of the system as reflected by the parameters Pe and N. This is a clear example where by using RTD analysis along with a suitable flow model could lead to useful information for designing bioprocesses.

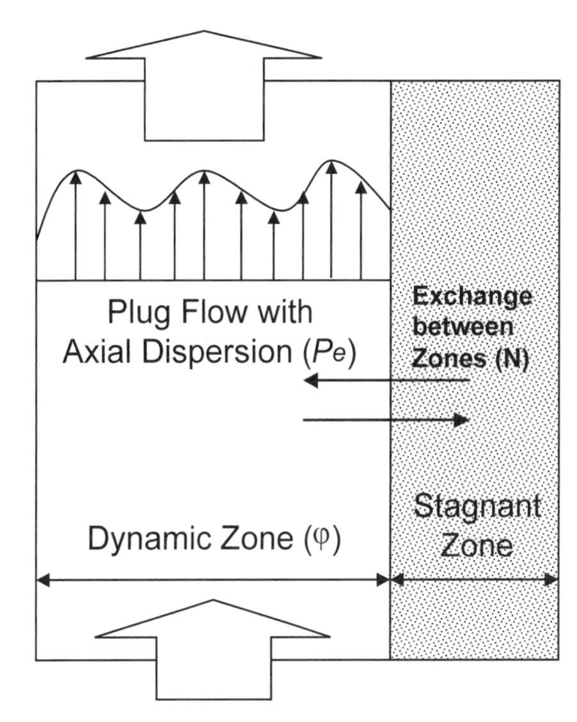

Fig. 2 PDE model applied to an expanded bed adsorption column. (From Ref. 12.)

Voute et al.[14] also employed the RTD analysis to determine the liquid-phase hydrodynamic characteristics in an expanded bed by measuring the exit stream age distribution following an upstream step input. The fluidizing liquid and tracer were, respectively, a $3\,g\,L^{-1}$ NaCl solution and a 1% v/v acetone solution. The acetone concentration was followed at the column outlet by UV absorbance at 280 nm. The dimensionless output tracer concentrations were plotted against the dimensionless volume. These normalized results were fitted to the tank-in-series model using the number of mixing plate as model parameter[15]:

$$F = 1 - \exp(n\psi)\sum_{j=0}^{n-1}\frac{(n\psi)^n}{j!} \tag{3}$$

where F is the dimensionless outlet concentration, ψ the dimensionless column effluent volume, and n the number of mixing plates. Based on the RTD analysis, Voute et al.[14] showed that dense mineral oxide gel composites could be used as effective expanded bed adsorbents.

Membrane Filtration Processes

In addition to expanded bed chromatographic systems, RTD techniques have been used to study other types of separation systems, such as the membrane filtration processes. One of the key technical challenges in membrane separation of biological molecules is the tendency of membrane fouling. To combat this problem, cross-flow tubular membrane separators are commonly employed. In such systems, it is desirable to augment mixing in the radial direction of the tubular membrane separator without raising shear that might damage the biomolecules.[16] This could be achieved by applying oscillatory flow and operating the flow in the laminar regime. The boundary-layer theory suggests that pulsing the flow might increase both heat and mass transfers from the wall to the fluid.[17] To this end, Najarian and Bellhouse[18] demonstrated enhanced capturing of target molecules in a microfiltration affinity membrane separator by increasing the radial mixing via a suitable combination of piston stroke length and pulsation frequency. Najarian and Bellhouse[16] further examined the use of such strategy in improving the performance of a cross-flow tubular affinity membrane separator by means of RTD studies. Two different methods of analyzing the RTD data were employed by Najarian and Bellhouse.[16] The first method was based on the evaluation of the variance of the normalized RTD functions (the E-curves),[8] and the second was based on the use of Fourier transform analysis. Under certain experimental conditions, tailing of the tracer curves was so long that the measurement of the variance was badly influenced by the baseline noise encountered in the tracer measurement.[16] Consequently another method of analysis, which was based on frequency response of the system, was used to determine the Peclet numbers of the system. Here, the experimental E-curves were transferred to the frequency domain, through the use of Fourier transforms.[19,20] The study of Najarian and Bellhouse[16] showed that appropriate combination of pulsation frequency, piston stroke length, and geometry resulted in an optimum value for the Peclet number. It also illustrated how tracer response data could be analyzed in the frequency domain.

CONCLUSION

Selected recent RTD studies related to industrial bioprocessing are reviewed in this article. These studies illustrated how RTD data could be collected and analyzed for a variety of bioprocess systems. Collectively, these studies also point to how reasonable flow models could be established to aid interpreting the RTD data. RTD studies could lead to substantial improvements in gas/liquid hydrodynamics and/or heat/mass transfer performances, and hence are of great value to design and scale-up of bioprocess systems.

REFERENCES

1. Fogler, H.S. *Elements of Chemical Reaction Engineering*, 2nd Ed.; Prentice-Hall: Englewood, 1992.
2. Danckwerts, P.V. Continuous-Flow System (Distribution of Residence Times). Chem. Eng. Sci. **1953**, *1*, 1–13.
3. Cuello, J. Fermentation Residence Time Distribution. In *Encyclopedia of Agricultural, Food, and Biological Engineering*; Heldman, D., Ed.; Marcel Dekker: New York, 2003.
4. Tescione, L.D.; Ramakrishnan, D.; Curtis, W.R. Role of Liquid Mixing and Gas-Phase Dispersion in a Submerged, Sparged Root Reactor. Enzyme Microb. Technol. **1997**, *20* (3), 207–213.
5. Hardin, M.T.; Howes, T.; Mitchell, D.A. Residence Time Distributions of Gas Flowing Through Rotating Drum Bioreactors. Biotechnol. Bioeng. **2001**, *74*, 145–153.
6. Treybal, R.E. *Mass Transfer Operations*, 3rd Ed.; McGraw-Hill: New York, 1981.
7. Amanullah, A.; McFarlane, C.M.; Emery, A.N.; Nienow, A.W. Scale-Down Model to Simulate Spatial pH Variations in Large-Scale Bioreactors. Biotechnol. Bioeng. **2001**, *73*, 390–399.
8. Levenspiel, O. *Chemical Reaction Engineering*, 3rd Ed.; Wiley: New York, 1999.
9. Willoughby, N.A.; Hjorth, R.; Titchener-Hooker, N.J. Experimental Measurement of Particle Size Distribution and Voidage in an Expanded Bed Adsorption System. Biotechnol. Bioeng. **2000**, *69* (6), 648–653.
10. Mullick, A.; Flickinger, M. Expanded Bed Adsorption of Human Serum Albumin from Very Dense *Saccharomyces cerevesiae* Suspensions on Fluoride-Modified Zirconia. Biotechnol. Bioeng. **1999**, *65* (3), 282–290.
11. Mullick, A.; Griffith, C.M.; Flickinger, M.C. Expanded and Packed Bed Albumin Adsorption on Fluoride Modified Zirconia. Biotechnol. Bioeng. **1998**, *60* (3), 333–340.
12. Fernandez-Lahore, H.M.; Kleef, R.; Kula, M.-R.; Thoemmes, J. Influence of Complex Biological Feedstock on the Fluidization and Bed Stability in Expanded Bed Adsorption. Biotechnol. Bioeng. **1999**, *64* (4), 484–496.
13. Villermaux, J.; van Swaaij, W.P.M. Modele Representativ de la Distribution des Temps de Sejour dans un Reacteur Semi-infini a Dispersion Axiale Avec Zones Stagnantes. Application a L'ecoulement Ruisselant dans des Colonnes D'anneaux Raschig. Chem. Eng. Sci. **1969**, *24*, 1097–1111.
14. Voute, N.; Bataille, D.; Girot, P.; Boschetti, E. Characterization of Very Dense Mineral Oxide-Gel Composites for

Fluidized-Bed Adsorption of Biomolecules. Bioseparation **1999**, *8* (1–5), 121–129.
15. Walas, S. Chemical Reactors. In *Perry's Chemical Engineers' Handbook*; Perry, P., Ed.; McGraw-Hill: New York, 1997.
16. Najarian, S.; Bellhouse, B.J. Residence Time Distribution Studies in a Simulated Tubular Affinity Membrane Separator. Chem. Eng. J. **1999**, *75* (2), 105–111.
17. Schlichting, H. *Boundary-Layer Theory*, 7th Ed.; McGraw-Hill: New York, 1968.
18. Najarian, S.; Bellhouse, B.J. Effect of Oscillatory Flow on the Performance of a Novel Cross-Flow Affinity Membrane Device. Biotechnol. Prog. **1997**, *13*, 113–116.
19. Hays, J.R.; Clements, W.C.; Harris, T.R. The Frequency Domain Evaluation of Mathematical Models for Dynamic Systems. AICHE J. **1967**, *13*, 374–378.
20. Nigam, K.D.R.; Vasudeva, K. Studies on Tubular Flow Reactor with Motionless Mixing Elements. Ind. Eng. Chem. Process Des. Dev. **1976**, *15*, 473–476.

Bioreactor Landfills

Kerry L. Hughes
Ann D. Christy
The Ohio State University, Columbus, Ohio, U.S.A.

INTRODUCTION

The production of solid waste is an inevitable consequence of human life, and since the industrial revolution the disposal of these wastes has become an increasingly pressing problem. As an improvement over open dumps, sanitary landfills were first constructed during the 1930s and 1940s.[1] Today, landfills are highly engineered containment systems and the primary method of disposing of refuse, or municipal solid waste (MSW), in the United States. Waste deposited in the landfill is isolated from ground water by landfill liners, and a cap prevents the infiltration of rainwater, thus minimizing the production of leachate (water that passes through the MSW transporting soluble contaminants). This "dry-tomb" concept of landfilling fails to take advantage of the potential to utilize the landfill as a treatment system for MSW. In such low-moisture environments, microorganisms are unable to degrade components of MSW. Eventually, without ongoing maintenance, the final caps of these landfills can fail, allowing the infiltration of water and initiating decomposition of the refuse. Under these conditions, the degradation of refuse is uncontrolled and usually results in release of leachate and gas into the environment posing a potentially serious threat to human and environmental health.

More recently, the concept of bioreactor landfills has emerged.[2,3] Bioreactor landfills are solid-state reactors where microorganisms degrade many of the components of MSW resulting in the treatment of solid waste in situ.[3] Several different methods have been proposed to create bioreactor landfills in which refuse degradation and landfill stabilization are accelerated. The efficacy of the bioreactor landfill is dependent on the growth and metabolism of microorganisms in the refuse mass. Efficacy is also influenced by the characteristics of the MSW such as carbon-to-nitrogen ratio, nutrient limitations, and relative quantities of less degradable, recalcitrant materials (e.g., lignin and cellulose). Proposed methods to create landfill bioreactors include leachate recirculation and the addition of nutrients, buffering capacity, or inoculum to the refuse mass (Fig. 1). All of these methods would alter the landfill's interior environment and create more favorable conditions for microbial growth. The most widely investigated method of creating a landfill bioreactor is using leachate recirculation to increase the moisture content of the refuse mass and, thus, promote microbial growth and degradation of MSW components. Leachate recirculation optimizes environmental conditions within the landfill by reapplying moisture that has percolated through the landfill to the refuse mass. Sometimes, the leachate will be supplemented with water if the leachate alone is insufficient to raise the moisture content of the refuse to the required level (40%–60%).

ADVANTAGES OF BIOREACTOR LANDFILLS

The suitability of bioreactor landfills for the treatment of MSW has been demonstrated in several full-scale trials,[2–4] and many of the potential benefits of bioreactor landfills have been realized in these full-scale trials. These benefits include rapid waste degradation, controlled gas production rates, improved leachate quality, and minimization of the environmental impact of landfills.

The use of bioreactor landfills decreases the time required to degrade and stabilize the MSW, from 30 to 50 years or more in a "dry-tomb" landfill, to a projection of less than 10 years using bioreactor technology. The shortened lifecycle of a bioreactor landfill has many advantages. The more rapid and extensive biodegradation of refuse that occurs in a bioreactor system results in accelerated stabilization of the refuse mass. Landfill settlement after refuse emplacement is due to the release of pore water and gas from void spaces, mechanical compression, physicochemical action, and biological decay processes.[5] The enhanced biological degradation in a bioreactor landfill therefore results in accelerated settlement. Once the landfill has settled, reuse of the land is possible, or more refuse may be disposed of in the landfill, extending its useful life.

Leachate is partially remediated by microorganisms within the refuse mass in a bioreactor landfill. Recirculated leachate from stabilized bioreactor landfills has reduced concentrations of both organic and inorganic components. Most metals are mobilized at the low pH values that characterize leachate from landfills in the acidogenic

Encyclopedia of Agricultural, Food, and Biological Engineering
DOI: 10.1081/E-EAFE 120007240
Copyright © 2003 by Marcel Dekker, Inc. All rights reserved.

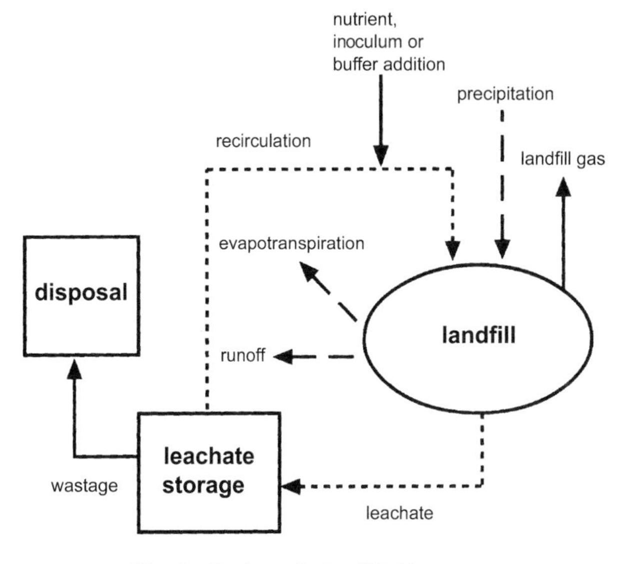

Fig. 1 Design of a landfill bioreactor.

phase of degradation. The pH of the leachate increases under the anaerobic conditions that develop during the methanogenic phase. Under these alkaline conditions, many metals form insoluble metal sulfides or metal hydroxides and subsequently precipitate out of the leachate. The stabilized landfill may also act as storage for the leachate after closure.

Increased conversion of organic carbon to methane and carbon dioxide results in greater gas production over a shorter time period in bioreactor landfills. Landfill gas recovery and utilization, thus, becomes more technically and economically feasible.

REFUSE DEGRADATION

Refuse in a landfill degrades in a series of well-characterized phases; each stage of the degradation is mediated by a distinct group of microorganisms (Fig. 2). During the initial phase, any oxygen in the refuse is utilized in aerobic respiration of bacteria in the landfill, and carbon dioxide is the main product. During anaerobic acidogenesis (Phase II), volatile fatty acids are the main products, and there is a resultant decrease in the pH of the refuse mass. Methanogens begin to proliferate during this phase, and their growth rate increases during the next phase of accelerated methane production (Phase III). Finally, decelerated methane production occurs as the biodegradable portion of the refuse decreases, and the refuse mass stabilizes.[6,7]

A different profile of gases is evolved and vented from the landfill during each of these distinct phases of refuse degradation reflecting the activity of the microorganisms that mediate each process (Fig. 3). There is much theoretical knowledge about the broad categories of microorganisms that mediate each phase of refuse degradation, but it remains difficult to identify the exact genera of organisms active in the refuse mass at any time. For example, microbiologists are able to describe a broad group of cellulolytic organisms (including *Cellulomonas, Clostridium* and *Eubacterium* species) that have been isolated from landfills. These organisms degrade cellulose but, from the available information, it remains impossible to determine which organisms actively mediate refuse degradation in a landfill. Most of the data available have been derived from bench- and small-scale lysimeter studies and extrapolated to full-scale landfills.[8] The microbiological characterization of landfills is hindered by the chemical, physical, and biological heterogeneity of these systems.[9] The advent of molecular identification techniques for bacteria and fungi may provide tools for identifying the essential microbial species and interactions during each distinct degradation phase.

DESIGN OF BIOREACTOR LANDFILLS

Much information has been published describing the construction, design, and operation of traditional and modern nonbioreactor landfills.[1,10–12] The design and construction of bioreactor landfills is, to a large extent, based on knowledge from the operation of nonbioreactor landfills. U.S. federal regulations, promulgated under Subtitle D of the Resource Conservation and Recovery Act (RCRA), 40 CFR 258.28, allow leachate recirculation if a composite liner and leachate collection system are in place. In addition, landfills to be operated as bioreactor landfills must meet the requirements of the Clean Air Act, the Clean Water Act, and many other federal, state and local regulations. It is possible to retrofit existing landfills as bioreactor landfills, but some recently designed landfills have been constructed and operated as bioreactor landfills.

A schematic cross section of a bioreactor landfill is shown in Fig. 4. The liner and leachate collection system must be carefully designed and installed to handle the potentially large volumes of leachate generated during refuse decomposition.[13] The liner usually consists of geosynthetic materials and clay, or soil, components. The drainage system is located immediately above the liner system and consists of permeable materials and a filter zone to prevent clogging of the drainage system by particulate matter. Leachate may be stored in concrete tanks, underground storage tanks, or lined ponds. The wetting of the refuse mass to achieve the minimum moisture content (40%) required for microbial growth is an important step in the creation of a bioreactor landfill.

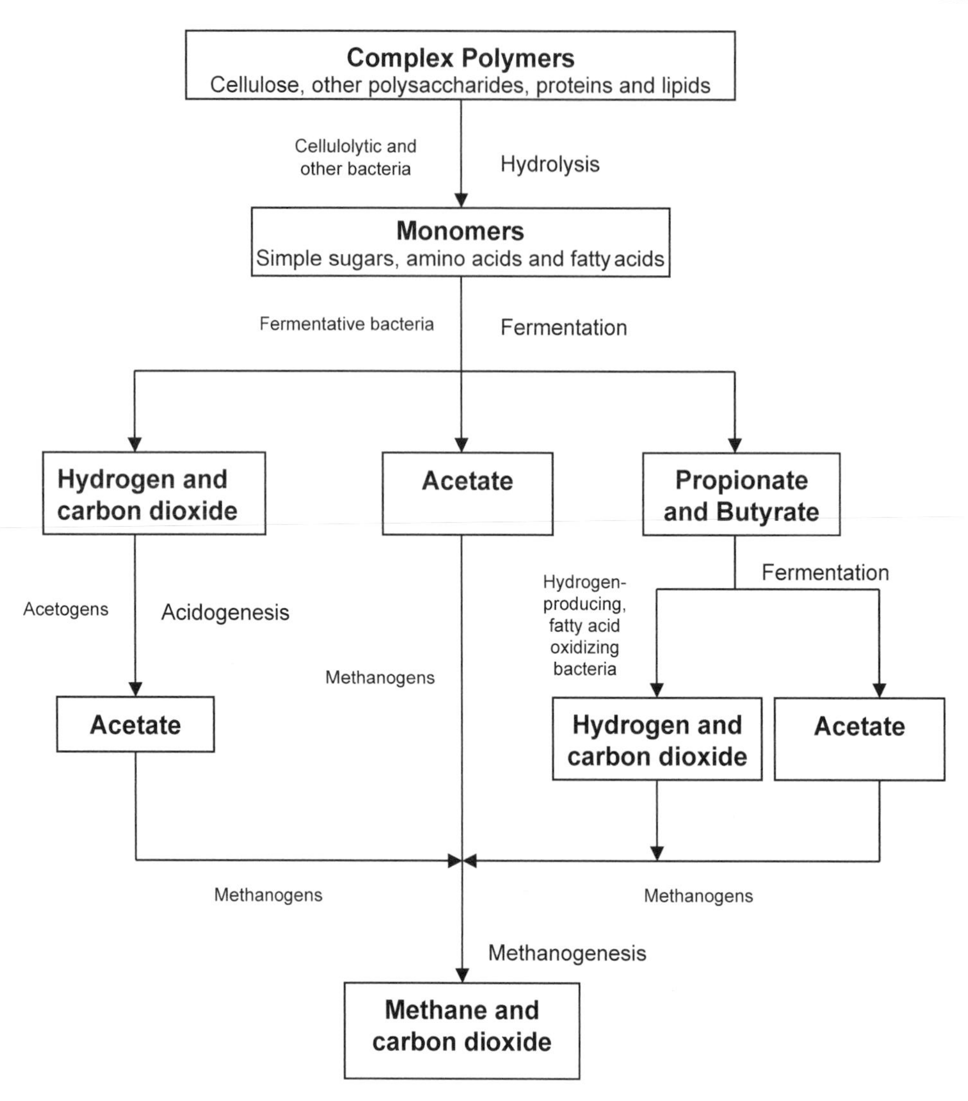

Fig. 2 Carbon breakdown pathway within a landfill. (Adapted from Ref. [6].) The microorganisms that mediate each process are labeled on the left and the process is identified to the right of each arrow.

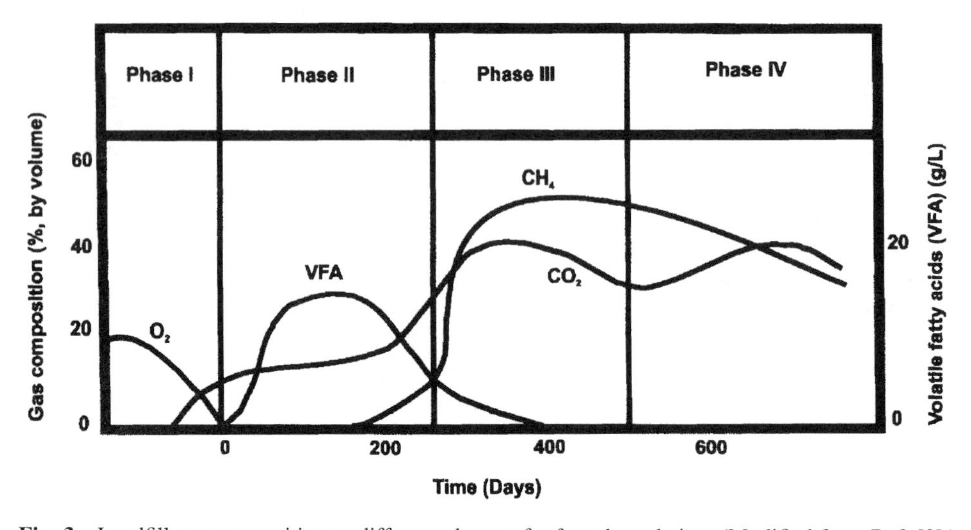

Fig. 3 Landfill gas composition at different phases of refuse degradation. (Modified from Ref. [2].)

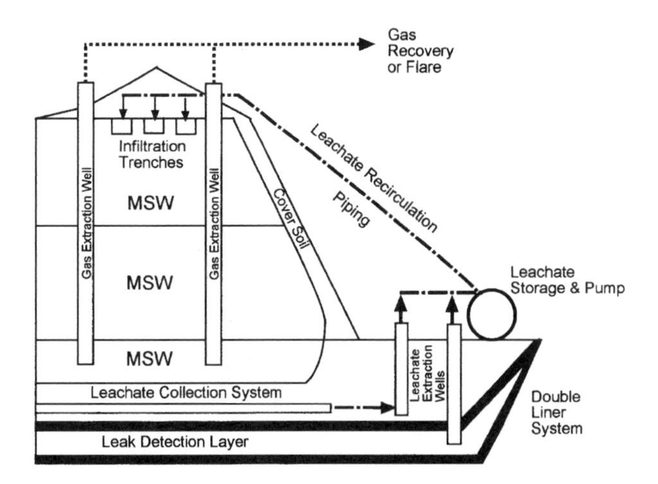

Fig. 4 Design of a landfill bioreactor showing a schematic cross-section. (Adapted from Ref. [13].)

Channeling of water and leachate through preferential flow routes, due to MSW heterogeneity and unsaturated flow characteristics, results in nonuniform wetting of the refuse.[2] Many different methods of applying leachate have been attempted in full-scale trials including surface spraying, infiltration ponds, vertical recharge wells, horizontal trenches with infiltrators, and vertical well clusters. No consensus has emerged yet as to which of these techniques is optimal. Increased moisture content of the MSW can lead to slope instability. The designer, therefore, needs to include appropriate geotechnical engineering measures to prevent side slope failures. When the landfill is closed, a final cap is placed over the landfill surface to minimize infiltration of rainwater, reduce the negative impacts of subsidence, enhance site aesthetics, and facilitate long-term maintenance. The cap usually consists of a gas control system, a hydraulic barrier (usually a geosynthetic fabric layer), a filter and drainage layer, soil, and vegetation.

CONCLUSIONS

Bioreactor landfill technology, which optimizes the microbial degradation of solid waste, has been studied and applied to many projects. Research challenges and regulatory issues still exist and have limited the widespread adoption of this new technology. Managing landfills using bioreactor technology would increase the capacity and working life of existing landfills, lessen the risk of off-site contaminant transport, and thereby reduce the health risk presented by landfills to neighboring communities.

ACKNOWLEDGMENT

The authors would like to thank Diane Yagich, graphic artist.

REFERENCES

1. Tchobanoglous, G.; Theisen, H.; Vigil, S. *Integrated Solid Waste Management: Engineering Principles and Management Issues*; McGraw Hill: Boston, Massachusetts, 1993.
2. Reinhart, D.R.; Townsend, T.G. *Landfill Bioreactor Design and Operation*; Lewis Publishers: Boca Raton, Florida, 1998.
3. Pacey, J.; Augenstein, D.; Morck, R.; Reinhart, D.; Yazdani, R. The Bioreactive Landfill. MSW Manag. **1999**, *9* (5), 53–60.
4. www.bioreactor.org, (accessed January 2002).
5. Wall, D.K.; Zeiss, C. Municipal Landfill Biodegradation and Settlement. J. Environ. Eng. **1995**, *121* (3), 214–223.
6. Barlaz, M.A. Microbiology of Solid Waste Landfills. In *Microbiology of Solid Waste*; Palmisano, A.C., Barlaz, M.A., Eds.; CRC Press: Boca Raton, Florida, 1996; 31–69.
7. Senior, E.; Balba, M.T.M. Refuse Decomposition. In *Microbiology of Landfill Sites*; Senior, E., Ed.; CRC Press: Boca Raton, Florida, 1989; 17–57.
8. Pourcher, A.-M.; Sutra, L.; Hebe, I.; Moguedet, G.; Bollet, C.; Simoneau, P.; Gardan, L. Enumeration and Characterization of Cellulolytic Bacteria from Refuse in a Landfill. FEMS Microbiol. Ecol. **2001**, *34*, 229–241.
9. Grainger, J.M.; Jones, K.L.; Hotten, P.M.; Rees, J.F. Estimation and Control of Microbial Activity in Landfill. In *Microbiological Methods for Environmental Biotechnology*; Grainger, J.M., Lynch, J.M., Eds.; Academic Press: Orlando, Florida, 1984; 259–273.
10. Bagchi, A. *Construction and Monitoring of Sanitary Landfill*; Wiley-Interscience: New York, 1990.
11. McBean, E.A.; Rovers, F.A.; Farquhar, G.J. *Solid Waste Landfill Engineering and Design*; Prentice-Hall: Englewood Cliffs, New Jersey, 1995.
12. Vesilind, P.A.; Worrell, W.A.; Reinhart, D.R. *Solid Waste Engineering*; Brooks/Cole: Pacific Grove, California, 2002.
13. Pohland, F.G. Landfill Bioreactors: Fundamentals and Practice. Water Qual. Int. **1996**, *Sept–Oct*, 18–22.

Bioremediation

Rakesh Bajpai
University of Missouri–Columbia, Columbia, Missouri, U.S.A.

Joong Kim
Pukyong National University, Pusan, South Korea

Mohamed Qasim
U.S. Army Corps of Engineer Research and Development Center, Vicksburg, Mississippi, U.S.A.

INTRODUCTION

Chemicals toxic to receptors (plants and animals) and/or having potential for causing birth defects (teratogenic) or cancer (carcinogenic) are categorized as "hazardous." Chemicals are extensively produced and used in the civilian, agricultural, and manufacturing sectors, and this has resulted in widespread environmental contaminations in industrial world. The hazardous nature of the contaminants and the severe environmental stress generated by them have spurred intense activity in clean up and restoration of contaminated sites and streams. The treatment technologies include physical, chemical, thermal, as well as biological processes. Of these, the biological methods combine advantages of mild conditions of operation, transformation of chemicals into environmentally less harmful and potentially nontoxic compounds, in-place treatment, and a likelihood of significant savings in treatment and restoration costs compared to other methods.[1] The term "bioremediation" is generally reserved for biological processes in which micro-organisms or enzymes thereof are used for clean up of sites contaminated with hazardous pollutants. Although bioremediation has been used for organic as well as inorganic pollutants, the present review would be limited to organic pollutants.

The ability of enzymes present in the micro-organisms (the microscopic or submicroscopic single- or multicellular basic units of life) to act upon and transform the organic chemicals, including the hazardous chemicals, is the basis of bioremediation. The micro-organisms are highly adaptive and utilize the organic and inorganic resources in their milieu to derive the raw materials and energy needed for their survival and procreation. In a number of cases, the micro-organisms use energy derived from metabolization of one organic compound to drive the transformation of another compound (cometabolism). Even when the chemicals are totally "man-made" and thus released into the environment relatively recently (compared to evolutionary time scale), micro-organisms have shown a remarkable adaptability towards them. Indigenous microbial activity exists in most environments to attack the different chemicals present in them. In case micro-organisms capable of degrading a specific chemical do not exist at a given location, external nonindigenous biodegrading micro-organisms (bioaugmentation) may be delivered to stimulate bioremediation.

FACTORS INVOLVED IN EFFICIENT BIOREMEDIATION

In bioremediation terminology, "transformation" refers to the conversion of one compound into another compound, "degradation" is transformation into compounds of lower molecular weight, "detoxification" is production of compounds with less or no toxicity compared to the parent chemical, and "mineralization" is the conversion of contaminant carbon into carbon dioxide. "Intrinsic bioremediation rate" refers to the rate of biodegradation of the contaminant by indigenous microbial population at any site under naturally existing physical–chemical conditions. Success of bioremediation at a given site may be defined in terms of reduction in concentration of the parent compound to a specified level (clean-up target) and/or reduction in toxicity of the contaminated media to specified receptors. Defining target clean-up levels is a critical activity in carrying out efficient bioremediation of any site.[2] The receptors present at/around a contaminated site, toxicity of the contaminants to the receptors, principal responsible parties, regulatory bodies, and the concerned public are factored in defining the target clean-up levels at any site.

The rate and the extent of bioremediation at a specific site depends on: 1) presence of microbial population able to degrade the contaminants at the site and 2) compatibility

Encyclopedia of Agricultural, Food, and Biological Engineering
DOI: 10.1081/E-EAFE 120007218
Copyright © 2003 by Marcel Dekker, Inc. All rights reserved.

of physical–chemical and biological conditions at the site for transformation of the contaminants into nontoxic innocuous compounds. The physical–chemical conditions include moisture content, pH, temperature, and concentrations of nutrients, carbon source, electron donor (energy source), electron acceptor, and availability of contaminants to the micro-organisms. Biological conditions include toxicity of available contaminants to the micro-organisms and presence of synergistic, competing, and preying organisms.

The very fact that a contaminant persists in a given environment is indicative of the fact that either the microbial population at the site is incapable of degrading the contaminant or the physical–chemical conditions are not conducive to rapid biodegradation. Success of bioremediation at any site depends on a good understanding of biodegradability of contaminants, microbial metabolism (pathway and intermediates), microbial ecology (nature and composition of microbial population) at the site, microbial physiology (effect of nutrients, pH, temperature, concentrations, etc. on microbial growth and biodegradation), local geochemistry (soil and/or groundwater composition, pH, redox potential, etc.), and management and control of physical–chemical conditions during bioremediation. The same factors would also determine if the bioremediation of subsurface contamination should be conducted in-situ or aboveground on-site.

CHEMICAL NATURE OF CONTAMINANTS AND THEIR SURFACE AND SUBSURFACE DISTRIBUTIONS

Hazardous chemicals found at major contaminated sites can be roughly divided into four categories depending on the frequency, complexity, source, and medium of contamination. These are: 1) hydrocarbons including polynuclear aromatic hydrocarbons (PAHs); 2) aliphatic and aromatic halogenated compounds; 3) pesticides and herbicides; and 4) nitroaromatics.

Petroleum hydrocarbons and creosote form the largest group of hazardous organic contaminants found in nature. Contaminations by petroleum hydrocarbons stem from accidental spills and discharges of petroleum products during their refining, transportation, storage, filling, and use. Widespread creosote contaminations were caused by extensive coal-gas operations and use of creosote in wood treatment. Both are complex mixtures of a large number of volatile and nonvolatile compounds. The most prominent hazardous contaminants in this category are the BTEX compounds (benzene, toluene, ethyl benzene, and xylene) and the PAHs. Petroleum hydrocarbons are lighter than water. In case of a spill, these compounds seep through the unsaturated soil (vadose zone) to the capillary fringe and finally to the top of the saturated layer where the light nonaqueous phase liquids (LNAPL) spread on top of the aquifer. This LNAPL layer acts as a source of dissolved contaminants in the aquifer. The contamination spreads further into the capillary and unsaturated zones as the aquifer water level rises and drops seasonally. The net result is that spills of petroleum hydrocarbons contaminate large areas of saturated and unsaturated soils. Due to significant differences in the sorption coefficients of different components of the petroleum hydrocarbons on soils, a chromatographic separation of the components is expected in the soil column with more strongly sorbed PAHs concentrating at the top and less strongly sorbed aliphatic hydrocarbons moving down significantly faster. Because of differences in volatility, sorptivity, and biodegradability of the different components in this category, contaminant composition changes as the weathering process occurs. Weathering may also result in selective deposition (precipitation and caking) of higher molecular weight hydrocarbons on solid surfaces, thus reducing the bioavailability of several components. Lack of bioavailability of the strongly sorbed and deposited chemicals is a major reason for their persistence in nature.

Aliphatic and aromatic halogenated organic compounds have been extensively used as solvents, cleaning solutions, intermediates in chemical synthesis operations, and insulating agents. The main contaminants from this category include halogenated aliphatic hydrocarbons, chlorophenols, and polychlorobiphenyls (PCBs). PCBs are generally found in the top layers of soil whereas the chlorophenols and the halogenated aliphatics contaminate deeper layers of soil also. Halogenated organic chemicals are heavier than water. Hence, their spills result in downward flow of the spilled material through the unsaturated zone, the capillary zone, and even the aquifer until the contamination encounters an impervious surface. It then spreads along the surface. This causes significant problems in bioremediation of contamination by dense nonaqueous phase liquids (DNAPL). When spilled or discharged in streams, PCBs sink to the sediment/detritus layer where these bind strongly with the natural organic matter. PCBs are a mixture of chlorinated biphenyls with differing degree of biodegradability under aerobic and anaerobic conditions. As a result, PCB contaminations undergo weathering process whereby aerobic conditions (i.e., in surface soils) cause selective utilization of less chlorinated PCBs leaving highly chlorinated PCBs behind.[3] Under anaerobic conditions (i.e., in sediments), highly chlorinated PCBs undergo selective dehalogenation leaving behind a mixture of lower chlorinated PCBs and products. Hence, the bioremediation strategies of the two types of weathered PCB contaminations would differ widely.

Contaminations of soil and water by *pesticides and herbicides* result from their production, distribution, and agricultural use. Compounds in this category include halogenated aliphatic, cyclialiphatic, aromatic, and phenoxyalkyl compounds. Some additional pesticides belong to halogenated anilines and cyclodienes. Their sorption characteristics on soil range from very strong to weak.[4] Their solubility in water also ranges from very low to mild.[4] Due to the nature of application of these compounds, pesticides and herbicides contaminate soils and surface waters in a nonpoint source manner.

Nitroaromatics and cyclonitramines are energetic chemicals and intermediates for the production of explosives and several industrial products. Several ammunition plants and depots in the United States and Europe, where these compounds were either produced or processed, have extensive soil and groundwater contaminations by energetic chemicals.[5] Nitroaromatic and nitramine explosives do not absorb strongly on the soil[6] and hence, move quickly through the soil columns and reach aquifer easily.[7]

SUSCEPTIBILITY OF DIFFERENT CONTAMINANTS TO BIOREMEDIATION

Petroleum Hydrocarbons

Micro-organisms capable of degrading and mineralizing hydrocarbons are abundantly found in nature. These include both bacteria and fungi. Aerobic conditions are required for biodegradations of hydrocarbons. Most micro-organisms are able to use aliphatic and cyclic hydrocarbons as carbon and energy sources under aerobic conditions. However, microbial susceptibility of different hydrocarbons depends upon the structure and number of carbon units in the hydrocarbons. The most common mechanism of aerobic degradation of linear chain saturated hydrocarbons (alkanes) is an oxygenase-mediated oxidation of a terminal carbon, resulting in formation of a fatty acid, followed by removal of a two-carbon compound by oxidation of β-methylene carbon.[8] Higher molecular weight alkanes are more easily degraded compared to lower molecular weight compounds. Although methane is rapidly used by "methanotrophs" as sole source of carbon and energy, ethane, propane, and butane are not. These are transformed essentially by cometabolism.[9] Other higher chain alkanes can serve as carbon and energy sources for micro-organisms, but compounds containing up to 10 carbon atoms exhibit toxicity to cells; the toxicity decreases with increasing chain length. Presence of unsaturated bonds and branch chains in hydrocarbons decrease their biodegradability.[4] Quaternary carbon compounds often form the terminal end products in biodegradation of several branched-chain hydrocarbons. Aerobic biodegradation of cyclic hydrocarbons also proceeds with hydroxylation of one of the carbons to form cycloalkanols and cycloalkanones, which are readily metabolized by a large number of naturally occurring micro-organisms by β-oxidation and ring cleavage to aliphatic acids.[10] The nature of substitutions in the cycloalkanes strongly influences their biodegradability. Presence of hydroxyl, oxygen, or carboxylic groups enhances biodegradability, but halogens make the compounds resistant to biodegradation.

Aromatic hydrocarbons such as benzene, toluene, ethyl benzene, and xylene are degraded by a number of soil micro-organisms under aerobic as well as anaerobic conditions. Initiation of biodegradation under aerobic conditions generally involves an attack by a dioxygenase to form dihydrodiols that further oxidize to form catechols. Catechols then undergo ring cleavage either by ortho-fission (opening of the C—C bond between two hydroxyl groups) or by meta-fission (opening of the C—C bond adjacent to the carbon containing hydroxyl groups) forming acids and aldehydes that undergo rapid metabolism. Several bacteria can grow on *m*- and *p*-xylene as sole carbon and energy source, but *o*-xylene does not serve as carbon and energy source for cells.[11] Under anaerobic conditions, nitrate, sulfate, and carbon dioxide may act as the terminal electron acceptors. The rates of biodegradation decrease in the following order of electron acceptors: oxygen > nitrate > sulfate > CO_2. The organic compound may itself act as terminal electron acceptor under fermentative conditions. Under denitrifying conditions, the biodegradation rate is 50% of that under aerobic conditions.[12] Under anaerobic conditions, the transformations of aromatic compounds may occur by a reductive pathway forming cyclic hydrocarbons (which may be easy or difficult to biodegrade depending on the nature of substituted groups) or by oxidation of the ring or of the substituted groups incorporating oxygen from water molecules.[13] The incorporation of oxygen may be the rate controlling step.[1] The terminal electron acceptors mentioned above are utilized to maintain a balance of reductive equivalents. Once oxygenated, further metabolism proceeds by reduction of the ring and biotransformation of the substituted cycloalkanes by β-oxidation and ring cleavage.[14]

PAHs are the compounds containing two or more fused aromatic rings. Several soil bacteria and fungi, isolated from contaminated sites, have been found to degrade PAHs. Biodegradation of PAHs takes place under aerobic as well as anaerobic (denitrifying and methanogenic) conditions. As considerable amount of PAH contamination is found in surface soils, aerobic biodegradation has received most attention. Biodegradability of PAHs depends on the number of fused benzene rings and

B

the nature, location, and number of substitutions. Bacteria of genus *Pseudomonas*, *Nocardia*, *Flavobacterium*, *Arthrobacter*, and several others have been reported to use low molecular weight PAHs (those with two and three fused benzene rings) as sole carbon and energy sources.[15–17] PAHs with four fused benzene rings (benzo(a)anthracene, pyrene, chrysene, and fluoranthene) are considerably more recalcitrant, and micro-organisms capable of biodegrading these compounds are found only in well acclimatized soils. Most of these organisms belong to *Sphingomonas* and *Mycobacterium* genera. Several fungi are also able to partially degrade high molecular weight PAHs.[18] Only recently, organisms able to grow on fluoranthene and pyrene have been reported.[19–21] A number of micro-organisms biodegrade the higher PAHs under cometabolic conditions[22–24] using an external source of primary carbon. This introduces a major complication in design and operation of the bioremediation process. Excessive amounts of the primary carbon source inhibit biodegradation of the contaminants and shortage results in insufficient energy in the micro-organisms to cometabolically degrade the pollutants. The presence of appropriate growth and nongrowth substances has been found to enhance the biodegradation of high molecular weight PAHs.[25,26] Use of surfactants enhances biodegradation of PAHs by increasing their bioavailability. Bioaugmentation with micro-organisms capable of producing surfactants has also been found to increase biodegradation rates of the highly sorbed PAHs. It is critical that the bioremediation of PAH-contaminated sites be properly managed for supply of oxygen, nutrients, microbial capability, carbon and energy source, and inducing agent. Absence of appropriate management often results in a rapid loss of low molecular weight PAHs from the contaminated sites followed by very slow or no change in the concentration of more toxic high molecular weight PAHs.[27–29] Exactly how this should be done remains an active area of research and development.

Pathways utilized by micro-organisms for degradation of PAHs are diverse. Oxidation of PAHs with two and three benzene rings takes place by the action of *dioxygenase* to form catechol via dihydroxydiol, followed by β-oxidation to cause ring cleavage.[30] In biodegradation of naphthalene by *Pseudomonas putida* or by *Beijerinckia* sp. Strain B-836, this dioxygenase attack occurs at carbons 1,2-positions by extraction of two more electrons and meta-cleavage of the ring. Biodegradation of pyrene by *Mycobacterium* and *Rhodococcus* sp. takes place by 1,2-dioxygenation followed by meta-cleavage of the ring,[31] and the 4,5-dioxygenation followed by ortho-cleavage of the ring.[32] Both of these strains are able to grow on pyrene. On the other hand, *Sphingomonas* strains do not grow on pyrene. Several fungi degrade high molecular weight PAHs. Of these, white rot fungi *Phanerochaete* is especially active against five and six ring structures. The biodegradation activity is catalyzed by a peroxidase forming epoxide that hydrolyses into dihydroxydiols. Although the epoxides readily hydrolyze, they are mutagenic and their presence in the system is a cause for concern.

Halogenated Organics

Several aerobic bacteria capable of degrading methylene chloride, dichloromethane, chloroform (trichloromethane), vinyl chloride, dichloroethylene, trichloroethylene, dichloroethane, trichloroethane, and trichloropropane have been isolated. These belong to genera *Alkaligenes*, *Mycobacterium*, *Pseudomonas*, *Nitrosomonas*, *Rhodococcus*, *Xanthanomonas*, and *Ancylobacter*.[33,34] Although some micro-organisms utilize monochlorinated hydrocarbons as sole energy source,[35] cometabolism is often the primary mode of microbial attack on halogenated compounds. Toluene, phenol, propane, and methane are common energy sources. The rate of aerobic biodegradation decreases as the number of halogen atoms in the molecule increases. Dichloroethylene and chloroethylene are readily biodegraded under aerobic conditions. Trichloroethylene also undergoes biodegradation under both oxidative conditions but carbon tetrachloride and tetrachloroethylene do not.

Oxidation state of the halogenated compounds increases as the degree of halogenation increases, making them prone to reductive dehalogenation. Redox state of the system and the oxidation state of the compound determine the rate of reductive dehalogenation. Dehalogenations of carbon tetrachloride, tetrachloroethylene, and trichloroethylene are strongly inhibited under denitrifying conditions, but take place rapidly under methanogenic conditions.[1] Acetate, hydrogen, or even ferrous ions can be electron donors. Presence of sulfate slows down the dehalogenation of tetrachloroethylene but blocks that of trichloroethylene.[36] During reductive dehalogenation of trichloroethylene, 1,2-dichloroethylene and chloroethylene (vinyl chloride) might accumulate under moderately negative redox conditions but no accumulation occurs under strongly negative redox status of the system. The rate of dehalogenation depends on the micro-organisms also. Glucose, sucrose, formate, acetate, lactate, methanol, and ethanol are excellent electron donors that support reductive dehalogenation. The micro-organisms able to do these transformations belong to genus *Methanogens*, *Acetogens*, and *Clostridia*.

Chlorophenols are often utilized by *Pseudomonas*, *Nocardia*, and *Arthrobacter* as carbon and energy sources under aerobic conditions. The rate of biodegradation decreases as the degree of chlorination increases. This is both due to toxicity of the chemicals to the micro-organisms and to the increasing oxidation state of the contaminant. Thus pentachlorophenol is highly toxic to *Pseudomonas cepacia* in aqueous solutions with an upper limit of $120\,mg\,L^{-1}$.[37] At the same time, successful landfarming treatments of PCP-contaminated soils have been reported for PCP concentrations ranging up to several thousand mg/kg soil. The mode of toxicity of chlorophenols is suggested to be partitioning into lipid bilayers and disruption of ATP production.[38] Cometabolism plays a role in biodegradation of chlorophenols also. The biodegradation takes place by formation of chlorinated catechol whereby ortho-cleavage results in destruction of ring followed by elimination of chlorine.[39] In some cases, as for 3-chlorobenzoate, chloride elimination may take place by nucleophilic substitution by a hydroxyl group. The nonhalogenated compound is then degraded as substituted aromatic hydrocarbons discussed above. Due to the oxidative nature of the chlorinated aromatics, removal of chloride moiety from the aromatic ring may also occur under anaerobic conditions. Here, the oxidized chloro-aromatic compounds would act as terminal electron acceptors. Another carbon source would be an electron donor. The rate and extent of dehalogenation would depend on the redox system and the presence of other electron acceptors. Presence of a consortium of micro-organisms could be beneficial for complete mineralization.[40]

Polychlorinated biphenyls (PCBs) consist of two nonfused benzene rings connected by a covalent C—C bond (biphenyl base) and one or more hydrogen atoms on the remaining carbons substituted by chlorine. The commercial PCB preparations consist of a mixture of compounds with differing degrees of chlorination. Hence, the preparations are characterized in terms of weight percent chlorine in the preparation. The weight percent typically ranges between 42 and 60. These are extremely hydrophobic compounds with very low solubility in aqueous phase.[4] PCBs are soluble in lipids, which results in their concentration in tissues amplifying the carcinogenic effects to tissues exposed even to aqueous phase containing very low PCB concentrations. Biodegradability of PCBs decreases as the number of substitutions on the biphenyl base increases. Long-term persistence of PCBs occurs in nature because of their low solubility in the aqueous phase, absence of molecular capability for PCB biodegradation at contaminated sites, lack of sufficiently induced enzyme system, and formation of toxic end-products. Addition of biphenyl enhanced biodegradation in aerobic Arochlor 1242[41] but not of anaerobic Arochlor 1254 under anaerobic conditions.[42] Biodegradation of PCBs is carried out by micro-organisms of genera *Pseudomonas*, *Acinetobacter*, *Alcaligenes*, *Arthrobacter*, *Rhodococcus*, *Comamonas*, *Burkholderia*, and *Corynebacterium*. Aerobic biotransformation of PCBs is limited to congeners containing five or fewer chlorine atoms.[43] Under anaerobic conditions, higher chlorinated PCBs are dehalogenated (due to their higher oxidative state) resulting in an increase in the concentration of lower chlorinated PCBs. The biodegradation of dichlorobiphenyl occurs by formation of catechol via dihydrodiol, followed by meta-cleavage of the ring.[3,39,44] No naturally occurring micro-organisms capable of growing on important PCB congeners has been discovered. It is suspected that this is due to the toxicity and inability of PCB degraders to metabolize chlorobenzoic acid. Rodrigues et al.[45] have synthesized chlorobenzoic acid utilizing PCB degraders by genetic engineering. These genetically engineered strains are able to grow on PCBs. Addition of surfactants and ferrous sulfate to the medium also enhanced biodegradation of Arochlor 1242.

Organic Pesticides

These compounds belong to the categories of linear chain and cyclic halogenated aliphatics, substituted aromatics (phenylalkanoates, halogenated aniline base, organophosphates), cyclodienes, carbamates, and pyrethroids. Their biodegradation follows the same basic routes suggested earlier for the compounds mentioned above, modulated by the specific character of the substitutions. Bioavailability of pesticides often is a critical factor in determining their biodegradability.[46] The bioavailability depends strongly on the soil moisture content, and biodegradation rates are reduced by 80%–90% when soil moisture content goes below 25% field moisture capacity.[47] Depending on the oxidation state of the pesticide, the anaerobic dehalogenation may be a more viable option compared to aerobic biodegradation in reducing environmental toxicity of the pesticides. Biodegradation may take place by substitutions on the substituted groups or by nucleophilic attack on the ring structure. In some cases, the substituted products bind strongly with the natural organic matter and thus is eliminated from the vicinity of exposed receptors.

Nitroaromatics

Nitrophenols, nitrobenzenes, and nitrotoluenes are often used in chemical industry and in manufacturing and formulation of explosives. Most of these compounds have been found to be susceptible to biodegradation under aerobic and anaerobic conditions. Mono- and dinitrotoluenes are used by *Burkholderia* species as sole source of carbon and energy.[48] Trinitrotoluene is not used as carbon and energy source and its biodegradation occurs cometabolically in presence of a primary carbon source. Biodegradation of mono- and dinitrotoluenes takes place by nucleophilic attack on the benzene ring forming epoxide as intermediate (mono-oxygenation), dihydrodiol intermediate (dioxygenation), or a hydride-Meisenheimer complex. This is followed by release of a nitro group as nitrite in solution. In some micro-organisms, electrophilic attack on the nitro group takes place forming hydroxylamino compound. The hydroxylamino compound either gives rise to catechol with simultaneous release of ammonia or undergoes rearrangement to form aminophenols. These compounds then undergo ring fission followed by further metabolization. For mono- and dinitrotoluenes, complete reduction of hydroxylamino group to amine has not been reported. Trinitrotoluene undergoes successive reduction of the nitrogroups to form monoaminodinitro-, diaminomononitro-, and then triaminotoluenes before ring fission occurs. The aminonitro intermediates are highly reactive and form strong polymeric bonds with soil surfaces.[49] As a result, most of TNT might disappear without any significant mineralization of the parent compound. Mononitrophenols undergo hydroxylation, to release nitrite, forming dihydroxy benzene. Only in some bacteria, biodegradation of 4-nitrophenol would take place by hydroxylation to produce nitrocatechol. Di- and trinitrophenol biodegradation takes place via formation of a hydride-Meisenheimer complex before elimination of nitrite in solution. Under anaerobic conditions, the mono- and dinitro-aromatics also undergo amination of the nitrogroups. However, significant biodegradations under anaerobic conditions have not been reported. Generally, microbial ability to biodegrade nitroaromatics is ubiquitous at sites contaminated with such compounds. Extensive biodegradations of contaminants have been reported at these sites by manipulation of the environmental conditions.

BIOREMEDIATION'S ESSENTIALS

Successful bioremediation requires not only ensuring the presence of microbial ability for biodegradation, but also providing the conditions promoting the biodegradation activity. These conditions include making the contaminant available to the micro-organisms and providing suitable electron donor–acceptor system, carbon and energy source, and the necessary macro- and micronutrients. As indicated in the previous section, the contaminant itself might serve one or more of these purposes. Appropriate selection of the missing ingredients can be made once the potential bioremediation route is identified.

An electron acceptor provides for an oxidation–reduction balance within a cell by oxidizing the reduced dinucleotides back to their oxidized states. In this process, energy-rich phosphate bonds (ATP) are regenerated for use in the energy-demanding metabolic reactions. An oxidation–reduction imbalance in a cell would result in highly reductive intracellular environment, which either shuts down the cellular metabolism or drives reactions towards fermentative situations. A common deficiency resulting in persistence of an otherwise easily degradable contaminant is the unavailability of a suitable electron acceptor. For aerobic micro-organisms, dissolved molecular oxygen is the desired electron acceptor. In the absence of molecular oxygen, nitrate, sulfate, iron, manganese, and carbon dioxide can also be electron acceptors if appropriate enzymatic machinery exists. Depending on the redox status of the system, oxidized substrates (such as chlorinated aromatics and hydrocarbons) can accept electrons and get dehalogenated.

The contaminant present in the system forms the basis of the total amount of electron acceptor needed. If the contaminant is hydrocarbon, stoichiometry suggests that 3.1 g oxygen is needed to mineralize 1 g of hydrocarbon.[50] When cell growth is considered on hydrocarbons under aerobic conditions with ammonia as nitrogen source, the oxygen requirements are 1 g per g of benzene, 1.4 g per g of toluene, and 1.7 g per g of toluene. In an actual system, however, natural sinks of electron acceptor (especially molecular oxygen) can increase the electron acceptor demand considerably.[51] In case of cometabolic biodegradation of contaminant, the primary carbon source would present its own requirement for electron acceptor (oxygen). Thus, use of methane as a primary carbon source would require 3.32 g oxygen per g methane and result in formation of 0.5 g of cell mass.[52]

The rate of supply of the electron acceptor depends on the microbial activity. Thus if the microbial activity amounts to mineralization rate of 1 g contaminant per day in a section of contaminated area, then the supply of electron acceptor would be 3 g per day in the contaminated section. As oxygen solubility at groundwater temperature is around $10\,\text{mg}\,\text{L}^{-1}$, it is likely that oxygen cannot be delivered in the form of oxygen saturated nutrient solution at a rate enough to match the oxygen demand of aerobic metabolism. In such a case, the alternatives could include delivery of oxygen in the form of hydrogen or some other peroxide (can deliver as high as $100\,\text{mg}\,\text{L}^{-1}$ dissolved oxygen) or by use of pure oxygen (up to $40\,\text{mg}\,\text{L}^{-1}$ dissolved oxygen). Alternative electron acceptors may also be employed provided the microbial ecology permits their use. When using alternative electron acceptors, care must be taken to control their spread as there may be regulatory limits (such as $10\,\text{mg}\,\text{L}^{-1}$ of nitrate in water). In general, it is not easy to change the redox status of in-situ systems and it is desirable to design bioremediation system around the existing redox status. In this case, one also takes advantage of local microbial flora that might have gotten adapted to the existing conditions.

Macro and micronutrients at specific location may also be a limiting factor in biodegradation. Nitrogen and phosphorous are two major nutrients that can often be in short supply. The accepted range of C:N:P for good microbial activity ranges from 100:10:1 to 100:1:0.5.[50] McCarty[53] suggests using ammonia or nitrate nitrogen at a rate of 2 kg–8 kg per 100 kg of contaminant, and P at one-fifth the rate for nitrogen. C:N:P ratio of 800:1:0.08 was used by Bossert and Bartha.[54] High concentrations of N and P may be inhibitory to cellular activity. Hence, a pilot study is always desirable. Other micronutrients, K, Mg, Ca, Fe, Na, Co, Zn, Mo, Cu, Mn, are generally not limiting and are not added.

Moisture levels below 50% field moisture capacity appear to inhibit degradation activity in hydrocarbon contaminated soils. Moisture content under 40% seriously hinders biological activity. On the up side, > 70% field capacity interferes with oxygen transfer and thus hinders aerobic activity.[55] Fungi can grow well even at 20% field moisture capacity.[56]

pH is critical for proper microbial action. Typically, it should be maintained around neutral for bacteria and somewhat acidic for fungi (4.0–4.5 for *Phanerochaete chrysosporium*[57]). pH above 8.5 also limits hydrocarbon biodegradation. Similarly, temperature also profoundly influences the microbial activity and hence the rates of bioremediation. For each micro-organism, there is an optimum temperature and pH range that is often established by conducting feasibility studies.

As indicated earlier for redox status, it is not easy to change the temperature and pH of subsurface either. The ex-situ operation offers a potential for enhancing the rates of bioremediation by manipulation of environmental conditions. However, excavation, cost, and disposal issues may overwhelm a desire to conduct a rapid biorestoration of contaminated sites using ex-situ means. A summary of bioremediation options has been presented by Bandyopadhyay, Bhattacharya, and Majumdar[58] and by Banerji.[59]

EX-SITU BIOREMEDIATION

Ex-situ bioremediation involves excavation of contaminated material followed by biodegradation of contaminants in bioreactors. The bioreactors may be conventional reactors as in slurry and packed-bed bioremediation, or unconventional as in landfarming and composting. Major advantages of slurry operation include creation of significantly homogeneous contaminated matrix, improved contact between the micro-organisms, contaminants, and nutrients, and efficient control of environmental factors (pH, temperature, nutrients, etc.). As a result, fast biodegradation rates have been reported in such systems.[60] Off gases can also be treated before discharge.[61] Slurry reactors are used for treatment of contaminated soil after coarse particles are removed from the contaminated matrix by screening. Most are operated in a batch mode. Coarse materials such as gravel and sand particles have little capacity to adsorb organic contaminants and simple washing of this solid fraction often gives rise to clean disposable solids; the liquid stream containing the washed contaminants is treated in suspended cell or immobilized cell reactors. The fine solid fraction of contaminated matrix (generally the silt and clay portions) is suspended in aerated nutrient medium, which may be inoculated with appropriate microbial population, if necessary. Solid slurry of 10%–40% is used typically; Castaldi[62] reported 40% to be optimal. Brox[63] has presented a number of designs of bioslurry reactors. As the contaminants in nature are rarely pure components, preferential metabolization of easily degradable components results in a rapid initial decline in the total contaminant concentration. This is then followed by a very slow biodegradation process that appears to be stalled. Often, a number of toxic and/or carcinogenic chemicals belonging to the category of slowly degradable compounds might still be left in the system if the treatment is terminated too soon. Thus, extended treatment times might be necessary, making this process quite costly.[64] Tobajas, Siegel, and Apitz[65] have proposed multistaged slurry reactors as means of enhancing the overall biodegradation rates. Such operations, however, have not found much acceptance due to complexity of multistage operations involving solid phase. Attempts have been made to use in-ground ditches and lagoons as bioslurry reactors with different degrees of success.[1]

Immobilized cell bioreactors have been proposed for continuous treatment of contaminated liquid and gaseous streams. One such reactor involves fluidized sand/activated carbon on which the cells are immobilized.[66,67] Oxygen transfer to the cells may be done either by sparging air directly into the reactor or by spraying the particle-free liquid through an air chamber. The latter avoids the complications of more than two phases in the main body of the reactor and eases design. Another mode of operation is a trickle-bed reactor in which cells are immobilized on solid support over which the contaminated liquid or gas flows. This type of reactor (biofilter) is commonly used for treatment of volatile contaminants in gases.[68] A trickling liquid stream may still be used in biofilters to ensure adequate nutrients and moisture for the immobilized cells.[69]

Solid phase bioremediation (landfarming, composting) has been conducted at a large number of sites[70,71] for several different types of contaminants (petroleum hydrocarbons, chlorophenols, pesticides, trinitrotoluene, PCB). It has been used for aerobic as well as anaerobic processes. Landfarming[1] is mostly conducted in lined or unlined soil beds that are regularly tilled to ensure enough gas–solid mass transfer. Depths of beds are limited by the tilling devices used (generally 0.45 m although some deep tillers capable of going up to 1.8 m have been used). If the soil is high in silt or clay content, bulking agents might be added to increase moisture drainage and bed porosity.[72] Nutrients are mixed in the soil at the start and additional supplementation depends on the residual levels at different times during the bioremediation cycle. In dry season, moisture is also controlled regularly. Berms are used to control undesirable flowing-in of liquid streams into the bed and vice versa. Drain-tiles are also installed in some cases to remove excess moisture in the beds. Composting is conducted either in static piles or in windrow systems.[1,59] Bulking agents are invariably added in order to increase soil permeability, improve soil texture, and provide organic carbon source for microbial activity. Air might be supplied to static piles using distribution pipes in static piles that may be as high as 6 m.

IN-SITU BIOREMEDIATION

In-situ bioremediation deals with in-place biodegradation of contaminants in the subsurface without excavating any soil. Here, activities of indigenous micro-organisms are stimulated by circulating solutions of nutrients, electron acceptor, and/or electron donors as necessary through contaminated subsurface. Absence of any excavation is a major advantage of in-situ treatment and in many cases, it is the only method of treatment possible. Its major disadvantages are lack of control of distribution of electron donor/acceptor systems, heterogeneity, and presence of any salts that may interfere with biological activity. In-situ treatment can effectively treat unsaturated or vadose zone contaminations, as well as saturated or aquifer contaminations. Air sparging, a technique to remove volatile organic compounds (VOCs) from unsaturated and saturated zones, can be combined with biodegradation of the VOCs in the unsaturated zone to minimize VOC release in environment.

One of the key considerations in operation of in-situ bioremediation is hydraulic control of contamination in the zone of treatment.[1] It is necessary to ensure containment of contamination, delivery of nutrients, and electron acceptor/donor through injection wells. Recovery wells can be used to control the bioremediation zone and any undesired compounds in the recovery stream might be removed ex situ before returning the stream to the injection wells. One might even consider using recovery wells to lower water table so that bioventing (air sparging + bioremediation) may be conduced in otherwise saturated zone. Alternatively, injection system might be used to raise the water table so that the unsaturated zone can be treated as saturated zone. In the case of bioventing, the nutrients can be delivered by gravity feed using either sprinklers at the surface or injection wells/injection galleries. If necessary, a system of injection and recovery wells may be employed to ensure the desired rate of delivery of nutrients to the biologically active zone.

Selection of in-situ bioremediation mechanism depends to a great extent on the nature and concentration of contaminants and the existing redox conditions. A spectrum of in-situ technologies is provided in a table of innovative biotreatment technologies in North America.[73] Most of the project handled by in-situ bioremediation dealt with volatile and semivolatile, halogenated and nonhalogenated organic compounds. Aerobic treatment technologies abounded, although several cases of anaerobic treatments with nitrate as terminal electron acceptors have also been attempted. The reason for this appears to be lack of sufficient understanding of reductive processes and hesitation to introduce nitrate in subsurface. Cometabolic biodegradation of chlorinated solvents using methane as electron donor has also been demonstrated at the field scale. In one case, microaerobic degradation of chlorinated solvents in presence of nitrate and nutrients has also been attempted. So far, the reductive dehalogenation under denitrifying conditions has been considerably more costly when compared to aerobic processes. Recently, an effort has been made to use reductive dehalogenation as an after effect of solvent flushing operation to remove separate phase DNAPL.[74,75] It was found that the residual ethanol in the solvent-flushed subsurface could be utilized by the indigenous micro-organisms to carry out reductive dehalogenation of PCE in the passively managed subsurface.[76] When coupled with another technology, natural in-situ bioremediation processes appear to supplement and assist in achieving the target levels of contaminant reduction at a lower total cost. In this respect, one should not overlook the potential of natural attenuation whereby indigenous microbial populations carryout the biodegradation without active intervention. This is especially attractive where risk assessment analysis suggests a long-time span available for clean-up processes.

CONCLUSION

Over the past decade, bioremediation has been used at a large number of sites and is now an established means of site restoration. It has overcome some serious doubts in the minds of technology vendors and regulatory agencies regarding its efficacy in degrading hazardous contaminants. Still, it should be recognized as one of the tools among many for site cleanup. Recent trends are to couple this technology with others in order to make the total process faster and more economical than either alone can be. The principles presented here provide general guidance concerning where bioremediation may be effective and what type of techniques may be used.

REFERENCES

1. Cookson, J.T. *Bioremediation Engineering*; McGraw Hill, Inc.: New York, NY, 1995.
2. Washburn, S.T.; Warnasch, J.; Harris, R.H. Risk Assessment in the Remediation of Hazardous Waste Sites. In *Remediation of Hazardous Waste Contaminated Sites*; Wise, D.L., Trantolo, D.J., Eds.; Marcel Dekker: New York, NY, 1994; Chap. 2, 9–37.
3. Unterman, R. A History of PCB Biodegradation. In *Bioremediations—Principles and Applications*; Crawford, R.L., Crawford, D.L., Eds.; Cambridge University Press: New York, NY, 1996; 209–253.
4. Dragun, J. *The Soil Chemistry of Hazardous Materials*; Hazardous Materials Control Research Institute: Silver Springs, MD, 1988.
5. Spain, J.C. Introduction. In *Biodegradation of Nitroaromatic Compounds and Explosives*; Spain, J.C., Hughes, J.B., Knackmuss, H.J., Eds.; CRC Press: Boca Raton, FL, 2000; Chap. 1, 1–5.
6. Brannon, J.M.; Pennington, J.C. *Environmental Fate and Transport Process Descriptors for Explosives*; ERDC/EL TR-02-10; U.S. Army Engineer Research and Development Center: Vicksburg, MS, 2002.
7. Waisner, S.; Hansen, L.; Fredrickson, H.; Nestler, C.; Zappi, M.; Bajpai, R. Biodegradation of RDX in Contaminated Water. J. Hazard. Mater. **2002**, *B95*, 91–106.
8. Britton, L.N. Microbial Degradation of Aliphatic Hydrocarbons. In *Microbial Degradation of Organic Compounds*; Gibson, D.T., Ed.; Marcel Dekker: New York, NY, 1984; 89–130.
9. Horvath, R.S. Microbial Cometabolism and the Degradation of Organic Compounds in Nature. Bacteriol. Rev. **1972**, *36* (2), 146–155.
10. Trudgill, P.W. Microbial Degradation of the Alicyclic Ring: Structural Relationships and Metabolic Pathways. In *Microbial Degradation of Organic Compounds*; Gibson, D.T., Ed.; Marcel Dekker: New York, NY, 1984; 130–180.
11. Gibson, D.T.; Subramanian, V. Microbial Degradation of Organic Hydrocarbons. In *Microbial Degradation of Organic Compounds*; Gibson, D.T., Ed.; Marcel Dekker: New York, NY, 1984; 181–252.
12. Hutchins, S.R.; Wilson, J.T. Laboratory and Field Studies on BTEX Biodegradation in a Fuel-Contaminated Aquifer Under Denitrifying Conditions. In *In-Situ Bioremediation Applications and Investigations for Hydrocarbon and Contaminated Site Remediation*; Hinchee, R.E., Olfenbutter, R.F., Eds.; Butterworth Heinemann: Boston, MA, 1991; 157–172.
13. Vogel, T.N.; Grbic-Galic, D. Incorporation of Oxygen from Water into Toluene and Benzene During Anaerobic Fermentative Transformation. Appl. Environ. Microbiol. **1986**, *52*, 200–202.
14. Young, L.Y. Anaerobic Degradation of Aromatic Compounds. In *Microbial Degradation of Organic Compounds*; Gibson, D.T., Ed.; Marcel Dekker: New York, NY, 1984; 487–523.
15. Grifoll, M.; Casellas, M.; Bayona, J.M.; Solanas, A.M. Isolation and Characterization of a Fluorine-Degrading Bacterium: Identification of Ring Oxidation and Ring Fission Products. Appl. Environ. Microbiol. **1992**, *58*, 2910–2917.
16. Hogan, J.A.; Tiffoli, G.R.; Miller, F.C.; Hunter, J.V.; Einstein, M.S. Composting Physical Model Demonstration: Mass Balance of Hydrocarbons and PCBs. In the *International Conference on Physicochemical and Biological Detoxification of Hazardous Wastes*; Wu, X.C., Ed.; Technomic Publication: Lancaster, PA, 1988; Vol. II, 742–758.
17. McKenna, E. Biodegradation of Polynuclear Aromatic Hydrocarbon Pollutants by Soil and Water Microorganisms. Final Report, Project Number A-078-ILL; University of Illinois Water Resources Center: Urbana, IL, 1976.
18. Baldrian, P.; Wiesche, C.; Gabriel, J.; Nerud, F.; Zadrazil, F. Influence of Cadmium and Mercury on Activities of Ligninolytic Enzymes and Degradation of Polycyclic Aromatic Hydrocarbons by *Pleuotus ostreatus* in Soil. Appl. Environ. Microbiol. **2000**, *66*, 2471–2478.
19. Boldrin, B.; Tiehm, A.; Fritzsche, C. Degradation of Phenanthrene, Fluorene, Fluoranthene, and Pyrene by a *Mycobacterium sp.* Appl. Environ. Microbiol. **1993**, *59*, 1927–1930.
20. Dean-Ross, D.; Cerniglia, C. Degradation of Pyrene by *Mycobacterium flavescens*. Appl. Microbiol. Biotechnol. **1996**, *46*, 307–312.
21. Thibault, S.L.; Anderson, M.; Frankenberger, W.T. Influence of Surfactants on Pyrene Desorption and Degradation in Soil. Appl. Environ. Microbiol. **1996**, *62*, 283–287.
22. Dagher, F.; Deziel, E.; Lirette, P.; Paquette, G.; Bisaillon, J.; Villermur, R. Comparative Study of Five Polycyclic Aromatic Hydrocarbon Degrading Bacterial Strains Isolated from Contaminated Soil. Can. J. Microbiol. **1996**, *43*, 368–377.
23. Mueller, J.; Devereux, R.; Santavy, D.; Lantz, S.; Willis, S.; Pritchard, P. Phylogenetic and Physiological Comparisons of PAH-Degrading Bacteria from Geographically Diverse Soils. Anionic Leeuwenhoek **1997**, *71*, 329–343.
24. Schocken, M.; Gibson, D. Bacterial Oxidation of Polycyclic Aromatic Hydrocarbons Acenaphthene and Acenaphthylene. Appl. Environ. Microbiol. **1984**, *48*, 10–16.
25. Aitken, M.D.; Stringfellow, W.T.; Nagel, R.D.; Kazunga, C.; Chen, C. Characteristics of Phenanthrene-Degrading Bacteria Isolated from Soils Contaminated with Polycyclic Aromatic Hydrocarbons. Can. J. Microbiol. **1998**, *44*, 743–749.
26. Stringfellow, W.T.; Chen, S.; Aitken, M.D. Induction of PAH Degradation in a Phenanthrene-Degrading *Pseudomonad*. In *Microbial Processes for Bioremediation*;

Hinchee, R.E., Brockman, F.J., Vogel, C.M., Eds.; Battelle Press, Columbus, OH, 1995; 83–89.
27. Thoma, G. Summary of the Workshop on Contaminated Sediment Handling, Treatment Technologies, and Associated Costs, April 21–22; Committee on Contaminated Sediments, Marine Board, National Research Council: Washington, D.C., 1994.
28. Luthy, R.G.; Dzombak, D.A.; Peters, C.A.; Roy, S.B.; Ramaswami, A.; Nakles, D.V.; Nott, B.R. Remediating Tar-Contaminated Soils at Manufactured Gas Plant Sites. Environ. Sci. Technol. **1994**, *28* (6), 266A–277A.
29. Loehr, R.C.; Webster, M.T. Effects of Treatment on Contaminant Availability, Mobility, and Toxicity. In *Environmentally Acceptable Endpoints in Soil*; Linz, D.G., Nakles, D.V., Eds.; American Academy of Environmental Engineers: Annapolis, MD, 1997; Chap. 2.
30. Gibson, D.T. *Microbial Degradation of Organic Compounds*; Marcel Dekker: New York, NY, 1984.
31. Walter, U.; Beyer, M.; Klein, J.; Rehm, H.-J. Degradation of Pyrene by *Rhodococcus sp. UW 1*. Appl. Microbiol. Biotechnol. **1991**, *34*, 671–676.
32. Rehmann, K.; Noll, H.; Steinberg, C.; Kettrup, A. Pyrene Degradation by *Mycobacterium sp.* Strain KR2. Chemosphere **1998**, *36*, 2977–2992.
33. Vanelli, T.; Logan, M.; Arciero, D.M.; Hooper, A.B. Degradation of Halogenated Aliphatic Compounds by the Ammonia-Oxidizing Bacterium *Nitrosomonas europaea*. Appl. Environ. Microbiol. **1990**, *56* (4), 1169–1171.
34. Shields, M. Treatment of TCE and Degradation Products Using *Pseudomonas cepacia*. In the *Symposium on Bioremediation of Hazardous Wastes: EPA's Biosystems Technology Development Program Abstracts*, Falls Church, VA, April 16–18, 1991.
35. Ensign, S.A.; Hyman, M.R.; Arp, D.J. Cometabolic Degradation of Chlorinated Alkenes by Alkene Monooxygenase in a Propylene Grown *Xanthanobacter* Strain. Appl. Environ. Microbiol. **1992**, *53*, 3038–3046.
36. Bouwer, E.J.; Wright, J.P. Transformations of Trace Halogenated Aliphatics in Anoxic Biofilm Columns. J. Contam. Hydrol. **1988**, *2*, 155–169.
37. Hsieh, C.-Y. Biodegradation of Pentachlorophenol by *Phanerochaete chrysosporium* in Soil. MS Thesis, University of Missouri–Columbia, Columbia, MO, 1991.
38. Smejtek, P. The Physiological Basis of the Membrane Toxicity of Pentachlorophenol: An Overview. J. Membr. Sci. **1987**, *33*, 249–268.
39. Reineke, W. Microbial Degradation of Halogenated Organic Compounds. In *Microbial Degradation of Organic Compounds*; Gibson, D.T., Ed.; Marcel Dekker: New York, NY, 1984; 319–360.
40. Zhang, X.; Wiegel, J. Sequential Anaerobic Degradation of 2,5-Dichlorophenol in Freshwater Sediments. Appl. Environ. Microbiol. **1990**, *56*, 1119–1127.
41. Brunner, W.; Sutherland, F.H.; Focht, D.D. Enhanced Biodegradation of Polychlorinated Biphenyls in Soil by Analog Enrichment and Bacterial Inoculation. J. Environ. Qual. **1985**, *14*, 324–328.
42. Rhee, G.Y.; Sokul, R.C.; Bush, B.; Bethoney, C.M. Long Term Study of the Anaerobic Dechlorination of Arochlor 1254 with and without Biphenyl Enrichment. Env. Environ. Env. Sci. Technol. **1993**, *27*, 714–719.
43. Quensen, J.F.; Boyd, F.A.; Tiedje, J.M. Dechlorination of Four Commercial Polychlorinated Biphenyl Mixtures (Aroclors) by Anaerobic Microorganisms from Sediments. Appl. Environ. Microbiol. **1990**, *56*, 2360–2369.
44. Adriaens, P.; Kohler, H.P.E.; Kohler-Staub, D.; Focht, D.D. Bacterial Dehalogenation of Chlorobenziates and Coculture Biodegradation of 4,4-Dichlorobiphenyl. Appl. Environ. Microbiol. **1989**, *55* (4), 887–892.
45. Rodrigues, J.L.M.; Maltseva, O.V.; Tsoi, T.V.; Helton, R.R.; Quensen, J.F., III.; Fukuda, M.; Tiedje, J.M. Development of a *Rhodococcus* Recombinant Strain for Degradation of Products from Anaerobic Dechlorination of Aroclor 1242. Environ. Sci. Technol. **2001**, *35*, 663–668.
46. Anderson, J.P.E. Herbicide Degradation in Soil: Influence of Microbial Biomass. Soil Biol. Biochem. **1984**, *16*, 483–489.
47. Skipper, H.D. Enhanced Biodegradation of Carbanothioate Herbicides in South Carolina. In *Enhanced Biodegradation of Pesticides in the Environment*; Racke, K.D., Coats, J.R., Eds.; ACS: Washington, D.C., 1990; 37–53.
48. Shirley, F.N.; Spain, J.C.; He, Z. Strategies for Aerobic Degradation of Nitroaromatic Compounds by Bacteria: Process Discovery to Field Application. In *Biodegradation of Nitroaromatic Compounds and Explosives*; Spain, J.C., Hughes, J.B., Knackmuss, H.-J., Eds.; CRC Press: Boca Raton, FL, 2000; 7–61.
49. Lenke, H.; Achmich, C.; Knackmuss, H.-J. Perspectives of Bioelimination of Polynitroaromatic Compounds. In *Biodegradation of Nitroaromatic Compounds and Explosives*; Spain, J.C., Hughes, J.B., Knackmuss, H.-J., Eds.; CRC Press: Boca Raton, FL, 2000; 91–126.
50. Mohammed, N.; Allayla, R.L.; Nakhia, G.F.; Farooq, S.; Husain, T. State of the Art Review of Bioremediation Studies. J. Environ. Sci. Health **1996**, A*31* (7), 1547–1574.
51. Barker, K.H.; Berson, D.S. In-Situ Bioremediation of Contaminated Aquifers and Subsurface Soils. Geomicrobiol. J. **1990**, *8*, 133–146.
52. Strand, S.E.; Bjelland, M.D.; Stensel, H.D. Kinetics of Chlorinated Hydrocarbon Degradation by Suspended Culture of Methane Oxidizing Bacteria. Res. J. Water Pollut. Control Fed. **1990**, *62* (2).
53. McCarty, P.L. Engineering Concepts for In-Situ Bioremediation. J. Hazard. Mater. **1991**, *28*, 1–11.
54. Bossert, I.D.; Bartha, R. The Fate of Petroleum in Soil Ecosystems. In *Petroleum Microbiology*; Atlas, R.M., Ed.; McMillan Publishing Co. Inc.: New York, NY, 1984.
55. Cookson, J.T.; Leszczynski, J.E. Restoration of a Contaminated Drinking Water Aquifer. *Proceedings of the 4th National Outdoor Action Conference on Aquifer Restoration, Groundwater Monitoring, and Geophysical Methods*; National Groundwater Association: Las Vegas, NV.
56. Lamar, R.T.; Evans, J.W.; Glaser, J.A. Solid-Phase Treatment of a Pentachlorophenol-Contaminated Soil

Using Lignin-Degrading Fungi. Environ. Sci. Technol. **1993**, *27*, 2566–2571.
57. Lewandowski, G.A.; Armenante, P.M.; Pak, D. Reactor Design for Hazardous Waste Treatment Using a White Rot Fungus. Water Res. **1990**, *24*, 75–82.
58. Bandyopadhyay, S.; Bhattacharya, S.K.; Majumdar, P. Engineering Aspects of Bioremediation. In *Remediation of Hazardous Waste Contaminated Sites*; Wise, D.L., Trantolo, D.J., Eds.; Marcel Dekker: New York, NY, 1994; Chap. 4, 55–75.
59. Banerji, S.K. Bioreactors for Soil and Sediment Remediation. Ann. N. Y. Acad. Sci. **1996**, *829*, 302–312.
60. Zappi, M.E.; Rogers, B.A.; Teeter, C.L.; Gunnison, D.; Bajpai, R.K. Bioslurry Treatment of a Soil Contaminated with Low Concentrations of Total Petroleum Hydrocarbons. J. Hazard. Mater. **1996**, *46*, 1–12.
61. Christodoulatos, C.; Agamemnon, K. Bioslurry Reactors. In *Biological Treatment of Hazardous Waste*; Lewandowski, G.A., DeFilippi, L.J., Eds.; John Wiley & Sons Inc.: New York, NY, 1998.
62. Castasldi, F.J. Slurry Bioremediation of Polycyclic Aromatic Hydrocarbons in Soil Wash Concentrates. In *Applied Biotechnology for Site Remediation*; Hinchee, R.E., Anderson, D.B., Metting, F.B., Jr., Sayles, G.D., Eds.; CRC Press: Boca Raton, FL, 1994.
63. Brox, G. Bioslurry Treatment. In *Proc. Appl. Bioremediation*, Fairfield, NJ, October 25–26, 1993; Referenced by Cookson, J.T. *Bioremediation Engineering*; McGraw Hill, Inc.: New York, NY, 1995.
64. Exner, J.H. *Bioremediation Field Experience*; CRC Press: Boca Raton, FL, 1994.
65. Tobajas, M.; Siegel, M.H.; Apitz, S.E. Influence of Geometry and Solids Concentration on the Hydrodynamics and Mass Transfer of a Rectangular Airlift Reactor for Marine Sediment and Soil Bioremediation. Can. J. Chem. Eng. **1999**, *77* (4), 660–669.
66. Sutton, P.M.; Mishra, P.N. Activated Carbon Based Biological Fluidized Bed for Contaminated Water and Wastewater Treatment: A State-of-the-Art Review. Water Sci. Technol. **1994**, *29* (10–11), 309–317.
67. Lendenman, U.; Spain, J.C.; Smets, B.F. Simultaneous Biodegradation of 2,4-Dinitrotoluene and 2,6-Dinitrotoluene in a Fluidized Bed Reactor. Environ. Sci. Technol. **1998**, *32*, 82–87.
68. Marek, J.; Paca, J.; Koutsky, B.; Gerrard, A.M. Determination of Local Elimination Capacities and Moisture Contents in Different Biofilters Treating Toluene and Xylene. Biodegradation **1999**, *10* (5), 307–313.
69. Barton, J.W.; Davison, B.H.; Klasson, K.T.; Gable, C.C., III. Estimation of Mass Transfer and Kinetics in Operating Trickle-Bed Bioreactors for Removal of VOCs. Environ. Prog. **1999**, *18* (2), 87–92.
70. Block, D. Military Wins with Bioremediation Through Composting. Biocycle **2001**, *42* (3), 52–54.
71. Cole, E.S.; Mark, S. E & P Waste: Manage It Cost Effectively by Landfarming. World Oil **2000**, *August*, 132–134.
72. Hicks, R.J. Above Ground Bioremediation: Practical Approaches and Field Experiences. In *Proceedings Applied Bioremediation*, Fairfield, NJ, October 25–26, 1993.
73. EPA. Innovative Remediation Technologies: Field Scale Demonstration Projects in North America, 2nd Edition: Year 2000 Report, Report Number EPA-542-B-00-004, June 2000.
74. Sewell, G.W.; Mravik, S.C.; Wood, A.L. Field Evaluation of Solvent Extraction Residual Biotreatment (SERB). *Proc. of the 7th International FZK/TNO Conference on Contaminated Soil (ConSoil 2000)*, September 18–22; *Contaminated Soil 2000*; Thomas Telford Publishing, 2000; Vol. 2, 982–988.
75. Sewell, G.W.; Mravik, S.C.; Wood, L. Field Application of the Solvent Extraction Residual Biotreatment (SERB) Technology. Submitted to the Federal Integrated Biotreatment Research Consortium, SERDP, as FINAL REPORT in Chlorinated Solvent Thrust Area, 2001.
76. Helton, R.R.; Löffler, F.E.; Sewell, G.W.; Tiedje, J.M. Characterization of the Microbiota in a Tetrachloroethene (PCE)-Contaminated Aquifer Following Solvent Extraction Remediation Biotreatment. In *Abstracts of the 100th General Meeting of the American Society for Microbiology*, Los Angeles, CA, 2000.

Biosensors

José I. Reyes De Corcuera
Ralph P. Cavalieri
Washington State University, Pullman, Washington, U.S.A.

B

INTRODUCTION

Industrial instrumentation for analysis is scarce and often limited to pH and conductivity. There exist on-line optical instruments such as refractometers that may be used to assess composition. However, their applicability to biological material is often limited by the presence of interfering compounds in variable concentration that interfere with the measurement. In most cases, accurate analyses of biological materials are expensive and need to be performed in external laboratories equipped with more sophisticated instrumentation. Most of these analyses require previous purification that require too much time relative to the processing time, making their on-line implementation impossible for control purposes. However, in living organisms, biological components like antibodies and enzymes work as natural sensing and controlling "devices." The ability of isolating and purifying these proteins and other biological elements such as cells or organelles has allowed their integration with physicochemical transduction devices to produce biosensors. The most widely accepted definition of a biosensors is: "a self-contained analytical device that incorporates a biologically active material in intimate contact with an appropriate transduction element for the purpose of detecting (reversibly and selectively) the concentration or activity of chemical species in any type of sample."[1] The first biosensor, an enzyme-based glucose sensor, was developed by Clark and Lyons.[2] Since then, hundreds of biosensors have been developed in many research laboratories around the world. Over 200 research papers about biosensors have been published each year for the past three years.

The objective of this article is to review the principles of biosensor fabrication and operation, their existing and potential applications in the food and agricultural industries, and to briefly discuss recent research and future trends. For more comprehensive discussion on the topic, the reader is referred to several excellent books and reviews on which most of this article is based.[3–9]

TYPES OF BIOSENSORS

Biosensors can be grouped according to their biological element or their transduction element. Biological elements include enzymes, antibodies, micro-organisms, biological tissue, and organelles. Antibody-based biosensors are also called immunosensors. When the binding of the sensing element and the analyte is the detected event, the instrument is described as an affinity sensor. When the interaction between the biological element and the analyte is accompanied or followed by a chemical change in which the concentration of one of the substrates or products is measured the instrument is described as a metabolism sensor. Finally, when the signal is produced after binding the analyte without chemically changing it but by converting an auxiliary substrate, the biosensor is called a catalytic sensor.[10] The method of transduction depends on the type of physicochemical change resulting from the sensing event. Often, an important ancillary part of a biosensor is a membrane that covers the biological sensing element and has the main functions of selective permeation and diffusion control of analyte, protection against mechanical stresses, and support for the biological element. The most commonly used sensing elements and transducers are described below.

Sensing Elements

Enzymes

Enzymes are proteins with high catalytic activity and selectivity towards substrates (see the article *Enzyme Kinetics*). They have been used for decades to assay the concentration of diverse analytes.[11] Their commercial availability at high purity levels makes them very attractive for mass production of enzyme sensors. Their main limitations are that pH, ionic strength, chemical inhibitors, and temperature affect their activity. Most enzymes lose their activity when exposed to temperatures above 60°C. Most of the enzymes used in biosensor fabrication are oxidases that consume dissolved oxygen and produce hydrogen peroxide [see Fig. 1(a)]. Enzymes have been immobilized at the surface of the transducer by adsorption, covalent attachment, entrapment in a gel or an electrochemically generated polymer, in bilipid membranes or in solution behind a selective membrane. Several reviews of enzyme immobilization have been published.[12–19] Enzymes are commonly coupled to electrochemical and fiber optic transducers.

Encyclopedia of Agricultural, Food, and Biological Engineering
DOI: 10.1081/E-EAFE 120007212
Copyright © 2003 by Marcel Dekker, Inc. All rights reserved.

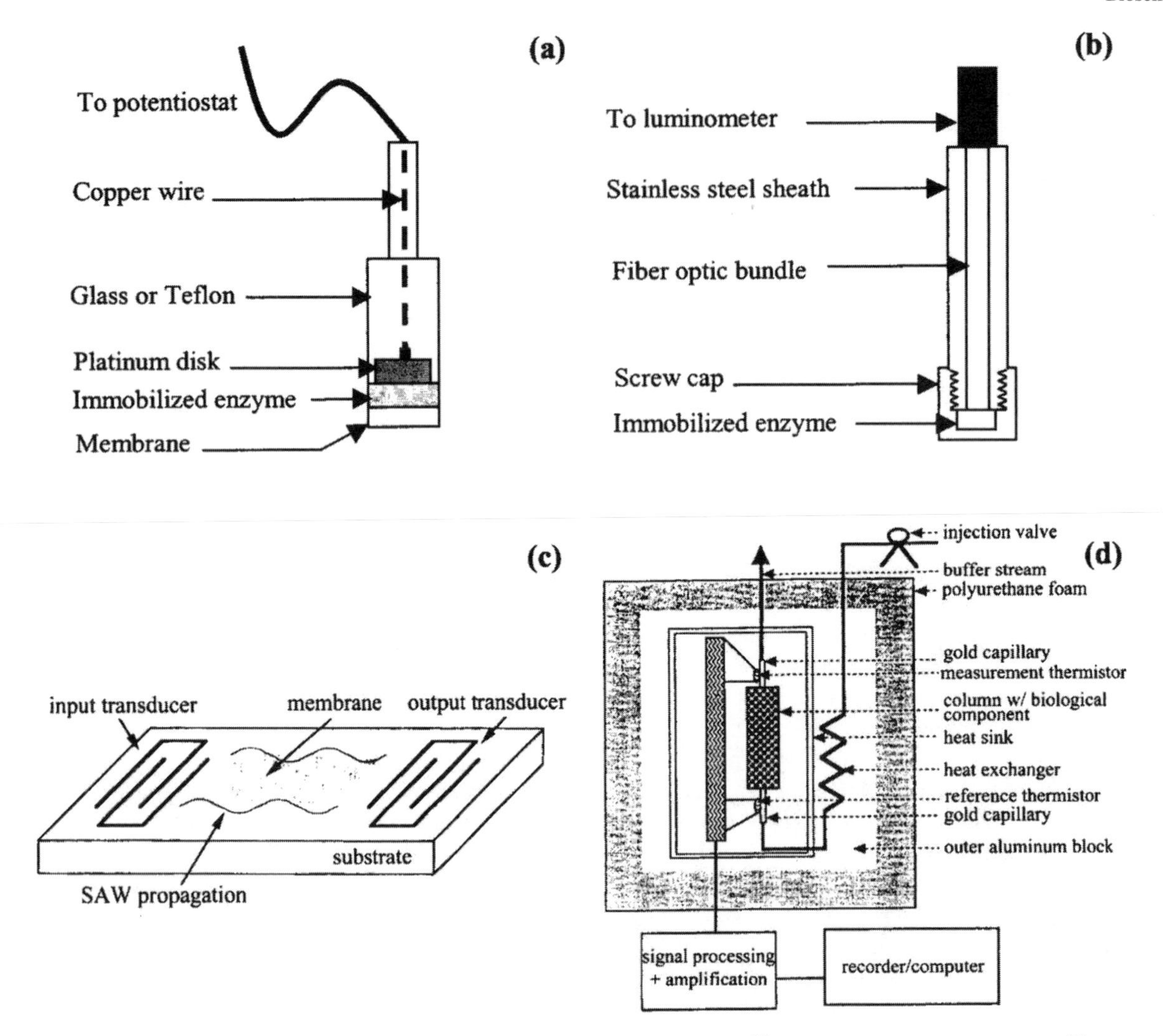

Fig. 1 Schematic representation of (a) amperometric enzyme membrane electrode[7]; (b) fiber optic enzyme sensor[7]; (c) surface acoustic wave propagation sensor[3]; and (d) enzyme thermistor.[3]

Antibodies

Antibodies are proteins that show outstanding selectivity. They are produced by β-lymphocytes in response to antigenic structures, that is, substances foreign to the organism. Molecules larger than about 10 kDa can stimulate an immune response. Smaller molecules like vitamins or steroids can be antigenic (also called haptens) but they do not cause an immune response unless they are conjugated to larger ones like bovine serum albumin. Many antibodies are commercially available and commonly used in immunoassays. Antibodies are usually immobilized on the surface of the transducer by covalent attachment by conjugation of amino, carboxyl, aldehyde, or sulfhydryl groups. The surface of the transducer must be previously functionalized with an amino, carboxyl, hydroxyl, or other group. A review of conjugation techniques can be found elsewhere.[20] Antibodies share similar limitations with enzymes. Furthermore, binding may not be reversible and regeneration of the surface may require drastic changes in conditions like low pH, high ionic strength, detergents, etc. Therefore, efforts are being made to produce low cost, single use sensors. Probably the main potential advantage of immunosensors over traditional immunoassays is that they could allow faster and in-field measurements. Immunosensors usually employ optical or acoustic transducers.

Microbes

The use micro-organisms as biological elements in biosensors is based on the measurement of their metabolism, in many cases accompanied by the consumption of oxygen or carbon dioxide, and is, in most cases, measured electrochemically.[21] Microbial cells have the advantage of being cheaper than enzymes or antibodies,

B

can be more stable, and can carry out several complex reactions involving enzymes and cofactors. Conversely, they are less selective than enzymes, they have longer response and recovery times,[22] and may require more frequent calibration. Micro-organisms have been immobilized, for example, in nylon nets,[21] cellulose nitrate membranes,[23] or acetyl cellulose.[24]

Other biological elements such as animal of vegetable tissue and membranes as well as organelles and nucleic acids have been researched but are out of the scope of this article. A summary of some biological elements and transducers used in the fabrication of biosensors is presented in Table 1.

Transducer elements

Electrochemical. Amperometric and potentiometric transducers are the most commonly used electrochemical transducers. In amperometric transducers, the potential between the two electrodes is set and the current produced by the oxidation or reduction of electroactive species is measured and correlated to the concentration of the analyte of interest. Most electrodes are made of metals like platinum, gold, sliver, and stainless steel, or carbon-based materials that are inert at the potentials at which the electrochemical reaction takes place. However, because some species react at potentials where other species are present, either a selective membrane is used or an electron mediator that reacts at lower potential is incorporated into the immobilization matrix or to the sample containing the analyte. Potentiometric transducers measure the potential of electrochemical cells with very low current. Field effect transistors (FET) are potentiometric devices based on the measurement of potential at an insulator–electrolyte interface. The metal gate of a FET can be substituted by an ion selective membrane to make a pH transducer (pH ISFET). Enzymes have been immobilized on the surface of such pH ISFET to produce enzyme-sensitized field effect transistors (ENFET). A complete description of such sensors can be found elsewhere.[25]

Table 1 Biological elements and transducers commonly used in the fabrication of biosensors

Biological elements	Transducers
Enzymes	Electrochemical
Antibodies	Amperometric
Receptors	Potentiometric
Cells	Ion selective
Membranes	Field effect transistors
Tissues	Conductimetric
Organisms	Optical
Organelles	Fiber optic (optrode)
Nucleic acids	Surface plasmon resonance (SPR)
Organic molecules	Fiber optic SPR
	Calorimetric
	Heat conduction
	Isothermal
	Isoperibol
	Acoustic
	Surface acoustic wave
	Piezocrystal microbalance

(Adapted from Refs. 4,5.)

Optical. Fiber optic probes on the tip of which enzymes and dyes (often fluorescent) have been co-immobilized are used. These probes consist of at least two fibers. One is connected to a light source of a given wave length range that produces the excitation wave. The other, connected to a photodiode, detects the change in optical density at the appropriate wavelength [see Fig. 1(b)]. Surface plasmon resonance transducers, which measure minute changes in refractive index at and near the surface of the sensing element, have been proposed. Surface plasmon resonance (SPR) transducers have been proposed. SPR measurement is based on the detection of the attenuated total reflection of light in a prism with one side coated with a metal. When a p-polarized incident light passes through the prism and strikes the metal at an adequate angle, it induces a resonant charge wave at the metal/dielectric interface that propagates a few microns. The total reflection is measured with a photodetector, as a function of the incident angle. For example, when an antigen binds to an antibody that is immobilized on the exposed surface of the metal the measured reflectivity increases. This increase in reflectivity can then be correlated to the concentration of antigen. The basic theory of SPR excitation and some examples of its application to biosensors are presented elsewhere.[26,27] A few SPR biosensors have been commercialized but no compact inexpensive portable device is available yet.

Acoustic. Electroacoustic devices used in biosensors are based on the detection of a change of mass density, elastic, viscoelastic, electric, or dielectric properties of a membrane made of chemically interactive materials in contact with a piezoelectric material. Bulk acoustic wave (BAW) and surface acoustic wave (SAW) propagation transducers are commonly used. In the first, a crystal resonator, usually quartz, is connected to an amplifier to form an oscillator whose resonant frequency is a function of the properties of two membranes attached to it. The latter is based on the propagation of SAWs along a layer of a substrate covered by the membrane whose properties affect the propagation loss and phase velocity of the wave. SAWs are produced and measured by metal interdigital transducers deposited on the piezoelectric substrate as shown in Fig. 1(c).[28]

Calorimetric. Calorimetric transducers measure the heat of a biochemical reaction at the sensing element. These devices can be classified according to the way heat is transferred. Isothermal calorimeters maintain the reaction cell at constant temperature using Joule heating or Peltier cooling and the amount of energy required is measured. Heat conduction calorimeters measure the temperature difference between the reaction vessel and an isothermal heat sink surrounding it. Using highly conducting materials ensure quick heat transferred between the reaction cell and the heat sink. Finally, the most commonly used is the isoperibol calorimeter that also measures the temperature difference between the reaction cell and an isothermal jacket surrounding it. However, in this case the reaction cell is thermally insulated (adiabatic). This calorimeter has the advantage of being easily coupled to flow injection analysis systems[29] [see Fig. 1(d)].

APPLICATIONS

One of the major driving forces for the development of biosensors is biomedical diagnosis. The most popular example is glucose oxidase-based sensor used by individuals suffering from diabetes to monitor glucose levels in blood. Biosensors have found also potential applications in the agricultural and food industries. However, very few biosensors have been commercialized.

Agricultural Industry

Enzyme biosensors based on the inhibition of cholinesterases have been used to detect traces of organophosphates and carbamates from pesticides. Selective and sensitive microbial sensors for measurement of ammonia and methane have been studied.[30] However, the only commercially available biosensors for wastewater quality control are biological oxygen demand (BOD) analyzers based on micro-organisms like the bacteria *Rhodococcus erythropolis* immobilized in collagen or polyacrylamide. Standard BOD5 measurements in which the effluent is pretreated and exposed to bacteria and protozoa require incubation at 20°C for 5 day. In contrast, BOD biosensors have throughputs of 2 to 20 samples per hour and can measure $0\,mg\,L^{-1}$ to $500\,mg\,L^{-1}$ BOD. When coupled with automatic sampling they can be implemented on-line.[30]

Food Industry

Biosensors for the measurement of carbohydrates, alcohols, and acids are commercially available. These instruments are mostly used in quality assurance laboratories or at best, on-line coupled to the processing line through a flow injection analysis system. Their implementation in-line is limited by the need of sterility, frequent calibration, analyte dilution, etc. Potential applications of enzyme based biosensors to food quality control include measurement of amino acids, amines, amides, heterocyclic compounds, carbohydrates, carboxylic acids, gases, cofactors, inorganic ions, alcohols, and phenols.[31] Biosensors can be used in industries such as wine beer, yogurt, soft drinks producers. Immunosensors have important potential in ensuring food safety by detecting pathogenic organisms in fresh meat, poultry, or fish.

CURRENT RESEARCH AND TRENDS

Because in many cases the transduction technology is well established, most of the research is focused on improving immobilization techniques of the biological element to increase sensitivity, selectivity, and stability. While critical, the latter has received relatively little attention probably in part because there is a tendency to design disposable devices that are most useful in quality assurance laboratories but do not allow on-line implementation for process control. Another dynamic area of research is miniaturization of sensors and flow systems. Development of these technologies is mainly driven by the need for in vivo applications for medical diagnosis and may not find immediate use in the agricultural and food industries. After almost 40 yr of research in biosensors, a wide gap between research and application is evident. The lack of validation, standardization, and certification of biosensors has resulted in a very slow transfer of technology. With faster computers and automated systems this process should accelerate in the future.

REFERENCES

1. Arnold, M.A.; Meyerhoff, M.E. Recent Advances in the Development and Analytical Applications of Biosensing Probes. Crit. Rev. Anal. Chem. **1988**, *20*, 149–196.
2. Clark, L.C.; Lyons, C. Electrode systems for continuous monitoring cardiovascular surgery. Ann. N. Y. Acad. Sci. **1962**, *102*, 29–45.
3. Kress-Rogers, E. *Handbook of Biosensors and Electronic Noses*; CRC Press Inc.: New York, 1997.
4. Kress-Rogers, E. *Instrumentation and Sensors for the Food Industry*; Woodhead Publishing Lmtd.: Cambridge, England, 1998.
5. Turner, A.P.F.; Karube, I.; Wilson, G.S. *Biosensors Fundamentals and Applications*; Oxford Univeristy Press: Oxford, 1987.

B

6. Schmid, R.D.; Scheller, F. *Biosensors, Applications in Medicine, Environmental Protection and Process Control*; GBF Monographs; VCH Publishers: New York, 1989.
7. Scott, A.O. *Biosensors for Food Analysis*; The Royal Society of Chemistry: Cambridge, UK, 1998.
8. Ramsay, G. *Commercial Biosensors*; in the series Chemical Analysis; Winefordner, J.D., Ed.; John Wiley & Sons Inc.: New York, 1998.
9. Wise, D.L. *Bioinstrumentation and Biosensors*; Marcel Dekker: New York, 1991.
10. Schubert, F.; Wollenberger, U.; Scheller, F.W.; Müller, H.G. Artificially Coupled Reactions with Immobilized Enzymes: Biological Analogs and Technical Consequences. In *Bioinstrumentation and Biosensors*; Wise, D.L., Ed.; Marcel Dekker, Inc.: New York, 1991; 19.
11. Guibault, G.G. Analysis of Substrates. In *Handbook of Enzymatic Methods of Analysis*; Schwartz, M.K., Ed.; Clinical and Biochemical Analysis; Marcel Dekker, Inc.: New York, 1976; 189–344.
12. Cosnier, S. Biomolecule Immobilization on Electrode Surfaces by Entrapment or Attachment to Electrochemically Polymerized Films. A Review. Biosens. Biolelectron. **1999**, *14*, 443–456.
13. Tien, H.T.; Wurster, S.H.; Ottova, A.L. Electrochemistry of Supported Bilayer Lipid Membranes Background and Techniques for Biosensor Development. Bioelectrochem. Bioenerg. **1997**, *42*, 77–94.
14. Bartlett, P.N.; Cooper, J.M. A Review of the Immobilization of Enzymes in Electropolymerized Films. J. Electroanal. Chem. **1993**, *362*, 1–12.
15. Scouten, W.H. A Survey of Enzyme Coupling Techniques. Methods Enzymol. **1987**, *135*, 30–65.
16. Tien, H.T.; Wurster, S.H.; Ottova, A.L. Electrochemistry of Supported Bilayer Lipid Membranes Background and Techniques for Biosensor Development. Bioelectrochem. Bioenerg. **1997**, *42*, 77–94.
17. Bartlett, P.N.; Whitaker, R.G. Electrochemical Immobilization of Enzymes. Part I. Theory. J. Electroanal. Chem. **1987**, *224*, 27–35.
18. Bidan, G. Electroconducting Polymers: New Sensitive Matrices to Build Up Chemical or Electrochemical Sensors. A Review. Sens. Actuators, B **1992**, *6*, 45–56.
19. Guibault, G.G. Immobilized Enzyme Electrode Probes. In *Solid Phase Biochemistry. Analytical and Synthetic Aspects*; in the series Chemical Analysis; Elving, P.J., Winefordner, J.D., Kolthoff, I.M., Scouten, W.H., Eds.; John Wiley & Sons: New York, 1983; 479–505.
20. Hermanson, G.T. *Bioconjugute Techniques*; Academic Press: San Diego, 1996.
21. Karube, I. Micro-organism Based Biosensors. In *Biosensors Fundamentals and Applications*; Turner, A.P.F., Karube, I., Wilson, G.S., Eds.; Oxford Science Publications: Oxford, 1987; 13–29.
22. White, S.F.; Turner, A.P.F. Mediated Amperometric Biosensors. In *Handbook of Biosensors and Electronic Noses. Medicine, Food and the Environment*; Kress-Rogers, E., Ed.; CRC Press: New York, 1997; 227–244.
23. Watanabe, E.; Tanaka, M. Determination of Fish Freshness with a Biosensor System. In *Bioinstrumentation and Biosensors*; Wise, D.L., Ed.; Marcel Dekker, Inc.: New York, 1991; 39–73.
24. Karube, I.; Sode, K. Microbial Sensors for Process and Environmental Control. In *Bioinstrumentation and Biosensors*; Wise, D.L., Ed.; Marcel Dekker, Inc.: New York, 1991; 1–18.
25. Kress-Rogers, E. Chemosensors, Biosensors and Immunosensors. In *Instrumentation and Sensors for the Food Industry*; Kress-Rogers, E., Ed.; Woodhead Publishing Lmtd.: Cambridge, England, 1998; 599–611, 636–639.
26. Lawrence Chris, R.; Geddes, N.J. Surface Plasmon Resonance (SPR) for Biosensing. In *Handbook of Biosensors and Electronic Noses. Medicine, Food and the Environment*; Kress-Rogers, E., Ed.; CRC Press: New York, 1997; 149–168.
27. Tao, N.J.; Boussaad, S.; Huang, W.L.; Arechabaleta, R.A.; D'Agnese, J. High Resolution Surface Plasmon Resonance Spectroscopy. Rev. Sci. Instrum. **1999**, *70* (12), 4656–4660.
28. D'Amico, A.; Di Natale, C.; Verona, E. Acoustic Devices. In *Handbook of Biosensors and Electronic Noses. Medicine, Food and the Environment*; Kress-Rogers, E., Ed.; CRC Press: New York, 1997; 197–223.
29. Kröger, S.; Danielsson, B. Calorimetric Biosensors. In *Handbook of Biosensors and Electronic Noses. Medicine, Food and the Environment*; Kress-Rogers, E., Ed.; CRC Press Oxford Science Publications: New York, 1997; 279–298.
30. Wittman, C.; Riedel, K.; Schmid, R.D. Microbial and Enzyme Sensors for Environmental Monitoring. In *Handbook of Biosensors and Electronic Noses. Medicine, Food and the Environment*; Kress-Rogers, E., Ed.; CRC Press: New York, 1997; 305–306.
31. Despande, S.S.; Rocco, R.M. Biosensors and Their Potential Use in Food Quality Control. Food Technol. **1994**, *48* (6), 146–150.

Biot Number

Chuan-liang Hsu
Yuanpei Institute of Science & Technology, Hsin-chu, Taiwan

INTRODUCTION

Biot number is defined as the ratio of the internal resistance to heat transfer in the solid to the external resistance to heat transfer in the fluid (liquid or gas) during transient heat transfer. The Biot number is a dimensionless parameter and it compares the relative magnitudes of internal conduction resistances and surface convection to heat transfer.

BIOT NUMBER

The Biot (pronounced Bee-oh) number or modulus, usually abbreviated as N_{Bi}, is associated with B. J. Biot. He proposed the problem of external convection in heat conduction analyses in 1804. J. B. Fourier (1768–1830) read Biot's work and by 1807 had determined how to analyze the problem.[1] Biot number is defined as the ratio of the internal resistance to heat transfer in the solid to the external resistance to heat transfer in the fluid (liquid or gas) during transient heat transfer. It can be expressed as:

$$N_{Bi} = \frac{D/k}{1/h} \tag{1}$$

or

$$N_{Bi} = \frac{hD}{k} \tag{2}$$

where h is the heat transfer coefficient or convection coefficient at the surface of the body ($W\,m^{-2}K^{-1}$); k is the thermal conductivity of material inside the object ($W\,m^{-1}K^{-1}$); and D (m) is a characteristic dimension of the body involved in the heat-transfer computation—this is taken to be the shortest distance between the surface of the body and the thermal center (i.e., the location that heats or cools slowest). For a sphere or infinite cylinder, this distance is the radius. For an infinite slab it is half the thickness. The Biot number is a dimensionless parameter and it compares the relative magnitudes of internal conduction resistances and surface convection to heat transfer. The N_{Bi} provides a measure of the temperature drop in the solid relative to the temperature difference between the surface and the bulk fluid.[2]

The governing equation describing transient heat transfer is:

$$\frac{\partial T}{\partial t} = \alpha\left(\frac{\partial^2 T}{\partial x^2} + \frac{\partial^2 T}{\partial y^2} + \frac{\partial^2 T}{\partial z^2}\right) \tag{3}$$

According to the properties of the body and of the ambient fluid as well as the conditions of fluid flow surrounding the body, the ratio may be discussed as the following three conditions.

Negligible Internal Resistance ($N_{Bi} < 0.1$)

When the Biot number is less than 0.1 ($N_{Bi} < 0.1$), there is negligible internal resistance to heat transfer, indicating that the value of k is much larger than the value of h, the internal conduction resistance to heat transfer is considerably less than the surface resistance and means that the internal conduction resistance is negligible in comparison with surface-convection resistance.

Consider an object at low uniform temperature T_0 (K) at time $t = 0$, immersed in a hot fluid at temperature T_∞ which is held constant with time. Assume that the heat transfer coefficient h is constant with time. Making a heat balance on the solid object for a small interval of time dt (sec), the heat transfer from the fluid to the object must equal the change in internal energy of the object.

$$hA(T_\infty - T)\,dt = c_p \rho V\,dT \tag{4}$$

where A is the surface area of the object in m^2, T the average temperature of the object at time t in sec, ρ the density of the object in $kg\,m^{-3}$, and V the volume in m^3. By proper separation of the parameters in Eq. 4 followed by integration, the following expression is obtained:

$$\frac{T - T_\infty}{T_0 - T_\infty} = \exp\left(-hA/c_p\rho V\right)t \tag{5}$$

which describe the temperature history within the object. The term $c_p\rho V$ is often called the lumped thermal capacitance of the system. This type of analysis is often called the lumped capacity method or Newtonian heating or cooling method.

Encyclopedia of Agricultural, Food, and Biological Engineering
DOI: 10.1081/E-EAFE 120007000
Copyright © 2003 by Marcel Dekker, Inc. All rights reserved.

After some rearrangement, Eq. 5 can be written in completely dimensionless form in the following manner:

$$\frac{T - T_\infty}{T_0 - T_\infty} = \exp[-(N_{Bi})(N_{Fo})] \quad (6)$$

where N_{Fo} is the Fourier Modulus defined in the following way.

$$N_{Fo} = \frac{\alpha t}{D^2} \quad (7)$$

where $\alpha = k/c_p\rho$ is referred to as the thermal diffusivity in $m^2 sec^{-1}$.

This in turn implies that the temperature will be nearly uniform throughout the solid and can be utilized to describe the heating or cooling characteristics of the body.[3] In bodies whose shape resembles a plate, a cylinder, or a sphere, the error introduced by the assumption that the temperature at any instant is uniform within the solid will be less than 5% when the internal resistance is less than 10% of the external surface resistance ($N_{Bi} < 0.1$). A typical example of this type of transient heat transfer flow is the cooling of a small metal casting or a billet in a quenching bath after its removal from a hot furnace.[4] Such a condition will usually not occur with solid foods, since the thermal conductivity of a solid food is relatively small. Negligible internal resistance to heat transfer also means that the temperature is uniform throughout the interior of the object. This condition is accomplished in objects with high thermal conductivity where heat is transferred instantaneously through the object, thus avoiding temperature gradients with location. Another way to obtain such a condition is a well-stirred liquid food in a container. For this case, there will be no spacious temperature gradient. For $N_{Bi} \ll 1$, it is reasonable to assume a uniform temperature distribution across a solid at any time during a transient process. The temperature gradient in the solid is small and all the temperature difference is between the solid and the bulk fluid.[2]

Finite Internal and Surface Resistance ($0.1 < N_{Bi} < 40$)

Between a Biot number of 0.1 and of 40 there is a finite resistance to heat transfer both internally and at the surface of the object undergoing heating or cooling.[5] Under such conditions neither the temperature gradient from the surface to the surrounding nor from the surface to the center is small enough to be neglected. For these moderate values of the Biot number, the temperature gradients within the solid are significant. Hence, the temperature difference across the solid is now much larger than that between the surface and the fluid. Again, the important ratio to consider involves the thermal conduction resistance inside and the thermal convection resistance outside.

Negligible Surface Resistance ($N_{Bi} > 40$)

For Biot numbers greater than 40, there is negligible surface resistance to heat transfer, in other words, the h value is considerably higher than k/D. A high Biot number (greater than 40) implies that external resistance to heat transfer is small (e.g., steam as a heating medium for foods) indicating that the temperature difference between the surface and the surrounding is negligible and that the major temperature change takes place from the surface of the body to the center. In this case the internal resistance cannot be neglected, and internal temperature distribution must be considered. An example of this case is a large object with low thermal conductivity in a bath of well-stirred liquid.

In the food processing, most processes involve heat transfer operations. Heating, cooling, freezing, evaporation, and drying or dehydration are examples of those processes involving heat transfer. In order to analyze heat transfer in food processing, thermal properties of the food materials are necessary, so that energy efficient processes and equipment designs can be attained. The Biot number is an important parameter which is essential in the calculations involving the transient temperature response of solids.

CONCLUSION

The Biot number is an important parameter which is essential in the calculations involving the transient temperature response of solids. According to the properties of the body and of the ambient fluid as well as the conditions of fluid flow surrounding the body, the ratio may be discussed as the following three conditions: negligible internal resistance ($N_{Bi} < 0.1$), finite internal and surface resistance ($0.1 < N_{Bi} < 40$), and negligible surface resistance ($N_{Bi} > 40$).

REFERENCES

1. Fourier, J. *The Analytical Theory of Heat*; Dover Publications, Inc.: New York, 1955.

2. Incropera, F.P. Transient Conduction. In *Introduction to Heat Transfer*, 3rd Ed.; John & Wiley Sons, Inc.: New York, 1996; 211–282.
3. Singh, R.P.; Heldman, D.R. Heating and Cooling Process. In *Food Process Engineering*, 2nd Ed; AVI Publishing Co. Inc.: Westport, CT, 1981; 87–157.
4. Kreith, F. Conduction of Heat in the Unsteady State. In *Principles of Heat Transfer*, 3rd printing; International Textbook Co.: Scranton, PA, 1960; 116–174.
5. Singh, R.P. Heating and Cooling Processes for Foods. In *Handbook of Food Engineering*; Heldman, D.R., Lund, D.B., Eds.; Marcel Dekker, Inc.: New York, 1992; 247–276.

Centrifugation

Sandeep Bhatnagar
Nestle Purina Petcare Co., St. Louis, Missouri, U.S.A.

Milford A. Hanna
University of Nebraska, Lincoln, Nebraska, U.S.A.

INTRODUCTION

Centrifugation is one of the major mechanical–physical separation processes used in the food industry. The mechanical–physical forces include gravitational and centrifugal, actual mechanical, and kinetic forces arising from flow. Particles and/or fluid streams are separated because of the different effects produced on them by these forces.

Centrifugation, for the separation of solids from liquids, is of two general types[1]:

1. *Centrifugal settling and sedimentation*: In centrifugal separations, the particles are separated from the fluid by centrifugal forces acting on the various size and density particles.
2. *Centrifugal filtration*: In filtration, a pressure difference is set up that causes the fluid to flow through small holes of a screen or cloth, which block the passage of the large solid particles, which in turn build up on the cloth as a porous cake. Centrifugal filtration is similar to ordinary filtration where a bed or cake of solids builds up on a screen, except that centrifugal force is used to cause the flow instead of a pump or other flow-producing means.

FORCES DEVELOPED IN CENTRIFUGAL SEPARATION

Centrifugal separators make use of the common principle that an object whirled about an axis at a constant radial distance from the axis is acted on by a force. The object being whirled about an axis is constantly changing direction and is thus accelerating even though the rotational speed is constant. This centripetal force acts in a direction toward the center of rotation.[1]

If the object being rotated is a cylindrical container, the contents of fluid and solids exert a force equal and opposite to the centripetal force, called centrifugal force, outward to the walls of the container. This is the force that causes the settling or sedimentation of particles through a layer of liquid or filtration of a liquid through a bed of filter cake held inside a perforated rotating chamber.

Fig. 1 shows a cylindrical bowl rotating, with slurry feed of solid particles and liquid being admitted at the center. The feed that enters is immediately thrown outward to the walls of the container. The liquid and solids are now acted upon by the vertical gravitational force and the horizontal centrifugal force. The centrifugal force is usually so large that the force of gravity may be neglected. The liquid layer then assumes the equilibrium position with the surface almost vertical. The particles settle horizontally outward and press against the vertical bowl wall.

The centrifugal force on a particle of mass m is given by $F = mr\omega^2$ where r is the radial distance from center of rotation and ω is the angular speed.

If two liquids of different densities are being separated by the centrifuge, the more dense fluid will occupy the outer periphery as the centrifugal force is greater on the denser fluid.

CENTRIFUGE EQUIPMENT FOR THE FOOD INDUSTRY

In the food industry, centrifugation has been used for clarification and separation of food products, ever since the centrifugal separator was invented by De Laval more than 100 yr ago.[2,3] Designed originally as a farm cream separator (Fig. 2) for separating butter fat from milk, it has been modified to meet requirements for increased capacity and handling a variety of products. It is impossible to list all the applications of centrifugation in the food industry but some of the most common and few of the more unusual applications are listed in Table 1.

Centrifugal Clarifier and Separators

The centrifuges are classified as clarifiers (which remove suspended solids from a liquid) and separators (which separate two liquids of different densities).

DOI: 10.1081/E-EAFE 120007078
Copyright © 2003 by Marcel Dekker, Inc. All rights reserved.

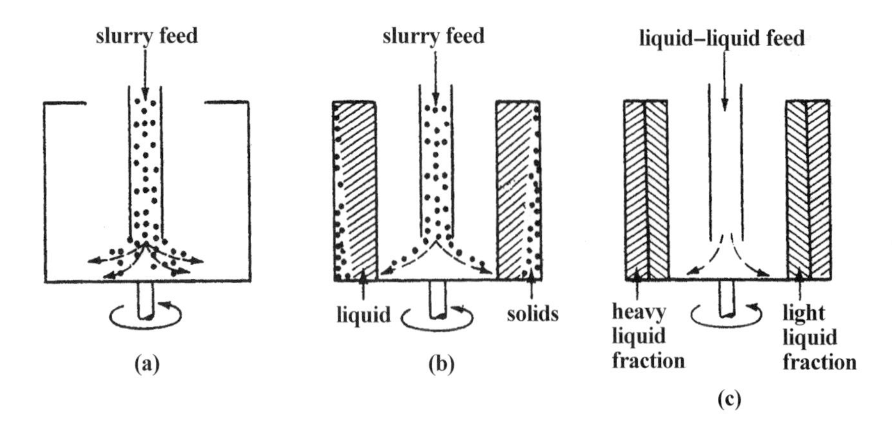

Fig. 1 Principle of centrifugal separation: (a) initial slurry feed entering, (b) settling of solids from a liquid, and (c) separation of two liquid fractions. (From Ref. 1.)

In a clarifier (Fig. 3), the liquid and solids enter along the axis of rotation near the bottom of the bowl.[4] They are pumped up by the rotary motion to the outer edge of the bowl where the solids remain while the liquid flows to the exit ports at the top of the bowl near the axis. The rotary motion imparts a very high energy to the fluid. Some of this energy can be recovered as the fluid flows back toward the axis by using disks within the bowl (Fig. 3). The disks act as turbine blades as the fluid flows past. At the periphery, the velocity of the fluid is in the circumferential direction. Left unrestricted, the fluid would tend to accelerate tangentially as it moves toward the axis. The disks in a centrifuge are fitted with ribs that form channels, which force the fluid to flow only radially inward. The tangential inertia is imparted to the ribs, thus reducing the electrical energy needed to turn the bowl. Eliminating the tangential motion also reduces back mixing of the fluid with the separated suspension of solids. A third function of the disks is to reduce the settling length in case agglomeration of the particles is important to their separation. In this case, the particles need to settle only a short distance before they come in contact with a disk surface, which aids agglomeration. The agglomerated particles are then rapidly centrifuged to the periphery of the bowl. This effect is particularly important with liquid droplets, which coalesce rapidly at a surface. The solids are deposited at the outer edge of the bowl (Fig. 4) and build up over a period of time. Once the layer of solids reaches the ends of the disks, the centrifuge must be cleaned.[5,6]

An alternate answer is to use self-cleaning centrifuges and several types are commercially available. The nozzle discharge is used in the oil and starch industries. A small fraction of the fluid is continuously discharged at the periphery of the bowl, sweeping out the solids. In the dairy and oil industries, an intermittent discharge of solids is accomplished with a periodic opening of the bowl itself.

Centrifugal Filters

Centrifuges, which filter—i.e., cause the liquid to flow through a bed of solids held on a screen—are commonly called centrifugals or centrifugal filters (Fig. 5). The continuous decanter[7] provides the transition between the sedimentation type centrifuge, on the principle of which it is dependent for its primary separation action, and the centrifugal filter, in which the solids are supported on a permeable membrane such as a screen through which the liquid phase is free to pass by the action of centrifugal force, which is the driving force for the separation. The opposing forces are essentially the interfacial tension between the solids and liquid phases, the surface area of the solids phase, the density and viscosity of liquid phases. They include both batch machines and continuous machines. They differ in whether the feed is batch, intermittent, or continuous and the way in which the solids are removed from the basket. While centrifugal sedimentation is most commonly used in the food industry, centrifugal filters have application in filtration of frying oil as in snack food manufacturing industry,[8] dewatering and removal of starch from the fibrous solids in the corn wet milling industry, dewatering and removal of starch from potato fiber, and removal of pulp skins and seeds from various fruit and vegetable products such as tomato juice, citrus juices, and fruit purees.

GAS SOLID CYCLONE SEPARATORS

Cyclone separators are used to separate small solid particles from gases.[9] In applications such as spray drying (where milk or coffee particles need to be separated from air), flour milling industry, or pneumatic conveying of powdery materials, the most commonly used equipment

Table 1 Centrifugation in the food industry applications (compiled from Ref. 2)

Food product	Application	Type of centrifugal equipment	Comments
Milk	Separation of fat from milk	Cream separator	Separate fat from skim milk up to 0.01%
Whey	Separation of cheese from whey	Solid ejecting or solid retaining separator is used	
Milk albumin	Separation of protein from whey	A solid ejecting type of centrifuge	Cheese whey contains approximately 0.9%–1% protein. This can be recovered using a centrifuge after heat and acid treatment of the whey. The protein discharge is in slurry form with approximately 20% dry solids
Liquid eggs	Clarification of liquid eggs prior to spray drying to prevent plugging of spray nozzles	Air tight clarifiers	De Laval clarifiers are used to continuously remove eggshell dust and shell fragments, chalazae, and protein strands and fibers from liquid eggs
Vegetable oils	Clarification of crude vegetable oils	Airtight solids retaining centrifuges	Crude vegetable oils containing 0.8%–10% impurities in the form of fatty acids, phosphatides, and color bodies are pretreated with heat and caustic prior to centrifugation
Fruit juices	Pulp control		Controlling the amount of pulp in citrus juices
	Peel oil recovery		Recover peel oil or essential oils from peels
Yeast manufacturing	Recovery of yeast cultures from fermenters		Typical are baker's yeast or Troula grown on molasses or spent sulfate liquor from paper mills
Starch processing	Separates starch from the lighter gluten	High-speed centrifugal centrifuges	In starch processing, stream of starch and gluten following separation from the fiber and bran is separated into starch and gluten in a one or two-step separation process. In the one-step process, the stream is directly fed into a primary centrifuge that produces two streams, one primarily starch with some gluten and the other primarily gluten and some starch. The two-step process involves the addition of a second centrifuge, which removes a portion of the water from the starch–gluten mixture prior to separation in the primary centrifuge

Fig. 2 Cutaway diagram of a centrifugal separator with disks. (From Ref. 4.)

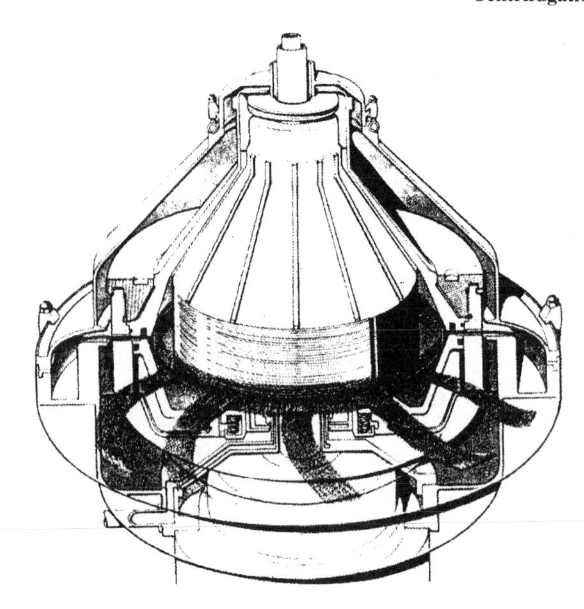

Fig. 4 Bowl opening type of self-cleaning centrifuge. (From Ref. 4.)

is the cyclone separator. The cyclone consists of a vertical cylinder with a conical bottom (Fig. 6). The gas–solid particle mixture enters the cyclone in a tangential inlet near the top. In the cyclone, the gas path involves a double vortex. On entering, the gas in the cyclone flows downward in a spiral vortex adjacent to the wall. The vortex thus formed develops a centrifugal force, which throws the solid particles radially toward the wall where they slide down toward the bottom of the cone. When the air reaches near the bottom of the cone, it spirals upward in a smaller spiral in the center of the cone and cylinder. The upward and downward spirals are in the same direction.

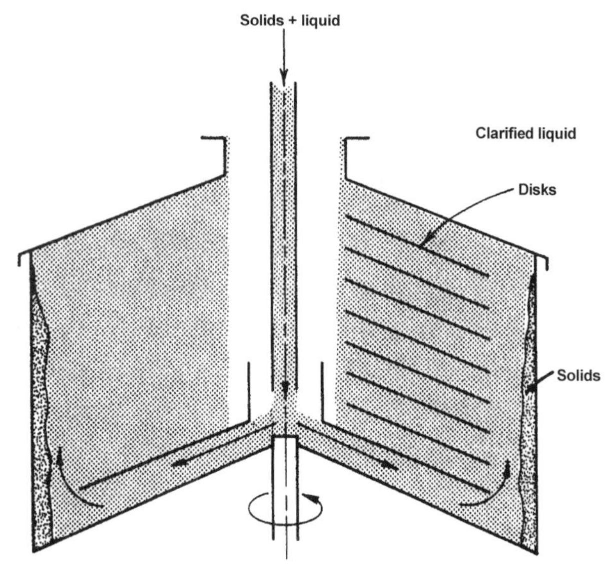

Fig. 3 Cross-section of a centrifugal clarifier. Left side is a clarifier without disks; right side shows half of a clarifier with disks. (From Ref. 4.)

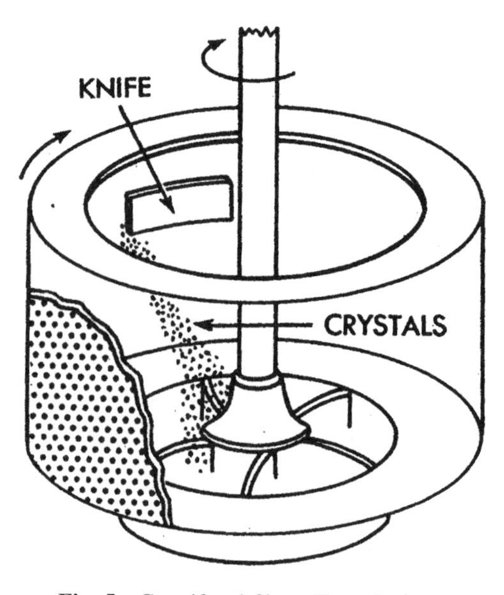

Fig. 5 Centrifugal filter. (From Ref. 7.)

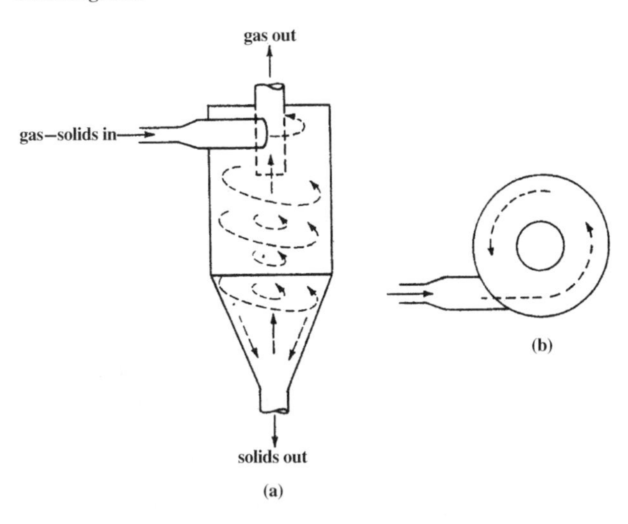

Fig. 6 Gas–solid cyclone separator: (a) side view and (b) top view. (From Ref. 1.)

A cyclone thus, is essentially a settling chamber in which gravitational acceleration is replaced by centrifugal acceleration. Cyclones offer one of the least expensive means of gas–particle separation. They are generally applicable in removing particles over 5 μ in diameter. Below 5-μ size, their efficiency is low. For particles above 200 μ, gravity settling chambers are generally satisfactory.[1]

C

REFERENCES

1. Geankoplis, C.J. *Transport Processes and Unit Operations*; Prentice Hall: Englewood Cliffs, NJ, 1983.
2. Johnson, A.H.; Peterson, M.S. *Encyclopedia of Food Technology*; The AVI Publishing Co. Inc.: Westport, CT, 1974.
3. Peterson, M.S.; Johnson, A.H. *Encyclopedia of Food Science*; The AVI Publishing Co. Inc.: Westport, CT, 1978.
4. Loncin, M.; Merson, R.L. *Food Engineering: Principles & Selected Applications*; Academic Press: New York, 1979.
5. Cheremisinoff, N.P. *Handbook of Chemical Processing Equipment*; Butterworth-Heinemann: Boston, 2000; 334.
6. McCabe, W.L.; Smith, S.C. *Unit Operations in Chemical Engineering*; McGraw Hill: New York, 1993; 994.
7. Mead, W.J. *The Encyclopedia of Chemical Process Equipment*; Reinhold Publishing Corporation: New York, 1964.
8. Simms, R.L. Corn Wet Milling. In *Starch Conversion Technology*; Van Beynum, G.M.A., Roels, J.A., Eds.; Marcel Dekker, Inc.: New York, NY, 1985; 47–72.
9. Green, D.W.; Maloney, J.O. *Perry's Chemical Engineers' Handbook*; McGraw-Hill: New York, 1997.

Chemigation

Peter M. Waller
University of Arizona, Tucson, Arizona, U.S.A.

INTRODUCTION

Chemigation is the application of fertilizers, pesticides, and other agricultural chemicals through irrigation systems. There are several advantages of chemigation:

- Low operator exposure to chemicals.
- Reduced tractor traffic, energy use, and compaction of soils.
- Incorporation of chemicals within the root zone.
- Reduced wind drift.
- The capacity to apply small doses of fertilizer throughout the growing season.
- Timeliness of applications relative to crop requirements.
- Potentially less chemical use.

Disadvantages of chemigation include the following:

- Some chemicals are not labeled for chemigation.
- Possible leaching of chemicals to the ground water.
- Low uniformity of chemical application if irrigation water application is nonuniform.
- Need for safety equipment on the irrigation system.
- High level of management needed by producer.

OVERVIEW

Chemigation has the potential to cause environmental contamination. Examples of safety equipment include backflow prevention on both the irrigation pipeline and the chemical injection line, interlocked electrical or mechanical power for the irrigation pump and chemical pump: if one pump is not supplied with power, then the other pump is automatically shut off. Pressure switches are installed that turn off both pumps when water pressure drops below a preset pressure.

The type of backflow prevention device that is required depends on the type of chemical pumped and the source of the irrigation water. The most hazardous situation is the application of pesticides through an irrigation system that is connected to a potable water supply. In this case, a reduced pressure backflow prevention device is required. The least hazardous case is the application of fertilizers through an irrigation system that is connected to an agricultural water source. In this case, a simple spring-loaded check valve is required.

Chemicals are labeled for chemigation by the Environmental Protection Agency. Criteria such as the half-life of the chemical, adsorption of the chemical to clays and organic matter, toxicity of the chemical, and toxicity of metabolites are used to determine whether a chemical is suitable for chemigation. Many agriculture chemicals that are labeled for application by conventional pesticide or fertilization application units (tractor or airplane) are not labeled for application by irrigation systems.

Chemicals are stored in tanks near the irrigation system pumping station. Secondary containment tanks are often required in order to contain chemical leaks from the primary chemical tank. Chemical tanks should be translucent so that the operator can observe the rate of chemical injection. If tanks are opaque, then a clear plastic tube is mounted on the outside of the tank to allow visual inspection of volume of chemical injected. Many tanks are constructed of high-density polyethylene in order to resist UV degradation and to provide durability and mechanical strength. Some tanks are equipped with propeller stirring devices.

Chemigation injection devices include Venturi injectors and positive displacement pumps. As water velocity is increased in the throat of the Venturi, pressure drops to near absolute zero pressure. Thus, the chemical mixture is forced into the Venturi throat from the chemical tank that is at atmospheric pressure. Generally, a centrifugal booster pump is used to remove water from the irrigation pipeline, and increase water velocity in the Venturi. The chemical and water mixture is then injected back into the irrigation system. The advantage of Venturi injectors is that there is very little contact between any moving parts and chemicals. The disadvantage of Venturi injectors is that flow rate can vary based on chemical temperature and viscosity and irrigation mainline pressure. Thus, standard Venturi injectors should not be used when precise control of chemical application rate is required. However, they are suitable for systems where chemicals are injected during an irrigation event until a given volume of chemical is injected. If precise flow rate control is required, then a Venturi injection system that is equipped with a sensor and

Encyclopedia of Agricultural, Food, and Biological Engineering
DOI: 10.1081/E-EAFE 120006937
Copyright © 2003 by Marcel Dekker, Inc. All rights reserved.

control system, which precisely controls the chemical injection rate, should be used.

The most popular types of positive displacement pumps used in chemigation are diaphragm pumps and piston pumps. Diaphragm pumps consist of a flexible diaphragm that is oscillated by a rotating device behind the diaphragm. Check valves on the inlet and outlet side of the pump work together with the diaphragm to inject chemical into the irrigation system. The advantages of diaphragm pumps include minimal contact of chemical with moving parts and the ability to adjust the injection rate while the pump is operating. Piston pumps draw chemical into a cylinder and utilize check valves on the inlet and outlet side of the pump to inject chemical into the irrigation water. Piston pump injection rate cannot be adjusted as the pump is operating and piston materials are more subject to corrosion than diaphragm pumps. Both piston and diaphragm pumps provide precise control of chemical injection flow rate over a range of chemical viscosity and irrigation mainline pressure; thus, there is very little change in injection rate with variation of chemical temperature or of the irrigation mainline pressure. Some positive displacement pumps are powered by irrigation water pressure; these systems vary chemical injection rate with water flow rate and are popular injection devices for field and greenhouse pressurized irrigation systems.

Fertilizers are often spoon-fed to the crop by continuous injection into irrigation systems. For example, a crop that requires nitrogen during the early part of the growing season, nitrogen and phosphorous during the middle part of the growing season, and calcium and nitrate during the latter part of the growing season might have the following fertigation schedule: 80 kg ha^{-1} of urea ammonium nitrate during the first-third of the growing season, 50 kg ha^{-1} of urea ammonium nitrate and 50 kg ha^{-1} of monoammonium phosphate during the second-third, and 80 kg ha^{-1} of calcium ammonium nitrate during the latter third. Fertilizer injection rate is calculated based on depth of irrigation water applied and irrigation water flow rate. The most dramatic example of spoon-feeding nutrients to crops is the case of greenhouse crops grown in rock wool media. All fertilizers required by the crop are added to irrigation water whenever irrigation water is applied (up to 50 irrigation events per day). In general, a computer control system is used to apply the precise amount of nutrients that are required by the crop.

If multiple chemicals such as fertilizers, acids, and pesticides are injected into irrigation water, then there is a danger of chemical reaction and precipitation within the irrigation pipeline. When this occurs, plugging of emitters occurs. In addition, some chemicals cause precipitation of calcium or magnesium carbonate. Growers can avoid precipitation problems by paying attention of chemical manufacturer's guidelines, testing mixtures in containers before injection into the irrigation system, and by observing the past experience of other growers in the vicinity.

One major use of chemigation is the injection of chemicals that are used for maintenance of drip irrigation systems. Chlorine is required to prevent growth of bacteria in irrigation systems when the water source is a shallow well, river water, or canal water: waters with organic matter. Typical injection concentrations range between 1 ppm and 4 ppm. Acid is injected into irrigation water in order to lower pH to 6.5 and prevent calcium carbonate and magnesium carbonate precipitation and clogging of drip emitters. Dropping pH to the range of 6.5 is very important when chlorine is injected into the irrigation system because the form of chlorine that is present at lower pH (hypochlorous acid) is effective at killing bacteria, but the form of chlorine that is present at high pH (hypochlorite) is not an effective bactericide.

Many chemicals are applied through center pivot irrigation systems. Injection rates are calculated based on the acres per minute covered by the center pivot and required chemical application rate. Chemigation of preplant herbicides is effective because the irrigation water supplies sufficient moisture for activation, and the irrigation water carries the herbicide past surface litter that can tie up the herbicide and into the soil. Depth of soil incorporation is regulated by depth of irrigation water applied. The primary form of fertilizer applied through center pivots is nitrogen and this is most often in the form of urea ammonium nitrate. Anion forms of fertilizers are thought to be best for center pivot application because they are not adsorbed by soil particles and will move into the soil with the irrigation water. Fungicides have been applied effectively through center pivots.

Application of foliar insecticides such as chlorpyrifos by center pivot sprinkler systems is a common practice. These chemicals are often formulated as oils or with a phytophyllic sticker that causes the chemical to adhere to the plant and not wash off the plant with the irrigation water. However, if oil drops are large, then they tend to float upward in the irrigation pipeline, and most of the chemical is discharged from the sprinklers at the beginning of the pipeline. If oil drops are small, then the chemical tends to be washed off of the plant. The optimal size oil drip is in the range of 100 μm diameter. This size drop does not float up in the irrigation pipeline (drop buoyancy force is less then the forces of turbulent dispersion) and also tends to stick to the plant foliage. One method that is used to control droplet size is a spring-loaded injection check valve that reduces oil droplet size within the irrigation pipeline. However, the level of control of droplet size with the spring-loaded check valve is not precise.

Surface irrigation systems are frequently used for fertigation: fertilizer is dripped into the canal upstream from the field and allowed to flow with the irrigation water onto the field. Advantages of this approach include low cost of application and incorporation of fertilizer into the soil matrix. Disadvantages include the possibilities of leaching and low application uniformity.

The fact that chemigation is used to apply many types of chemicals through many types of irrigation systems to many types of crops means that each chemigation system must be tailored to the specific application. Design of chemigation systems requires consideration of durability, required precision, cost of the system, type of chemicals, safety, and volumetric flow rate.

FURTHER READING

Bar-Yosef, B. Advances in Fertigation. In *Advances in Agronomy*; Academic Press: New York, 1999; Vol. 65, 2–77.

Burt, C.M.; O'Connor, K.; Ruehr, T. Fertigation, ITRC, San Luis Obispo, California, 1995.

Solomon, K.H.; Zoldoske, D.F. Backflow Prevention and Safety Devices for Chemigation. Research Notes—Center for Irrigation Technology, CATI Publication #981201; 1998.

Wright, J.; Bergsrud, F.; Peckham, J. Chemigation Safety Measures. University of Minnesota Extension Service Publication, FO-6122-GO, 1993.

Cleaning Chemistry and Physics

C

Peter Fryer
Grace Christian
University of Birmingham, Birmingham, United Kingdom

INTRODUCTION

Cleaning in the food industry is carried out for a range of reasons: to remove food product left on process plant surfaces, to destroy microbial growth within the system, and to maintain process efficiency by returning the equipment to a clean and hygienic state. Deposit layers that form on the surface of equipment are termed "foulants"; those produced upon heating food are particularly difficult to remove. Heat treatment is used in practice to kill microorganisms, and to create flavor and texture. However, these treatments will often degrade the food material. Heat treatment systems can become heavily fouled, and due to the large, closed nature of the processing equipment, manual cleaning is unsuitable.

Fouling is undesirable as it reduces the efficiency of the process and allows contamination of products. The former arises due to reduction of heat transfer and increased pressure drops. The layer of deposit acts as an insulator; thereby reducing heat transfer, so process temperatures have to be increased to compensate. This layer also causes large pressure drop increase across the plant, resulting in increased pumping power being required. Contamination of the product can arise from food deposits and the growth of microorganisms (biofilm). It is possible to have contamination of one product by another at changeover and after: this is especially important in cases where trace amounts of allergens (such as nuts) may be left on the surface of the processing plant. Often plants have to incorporate separate process lines for nut-free products to avoid risk.

Heat transfer systems are not the only ones to become fouled; evaporators and filtration membranes are also highly susceptible. Filtration membranes are widely used (for cold sterilization) in the processing of fluids such as milk, beer, and fruit juices[1] and become fouled due to solid particles within the fluid. Evaporators are often used in food processing, particularly in the sugar refinery[2] and dairy industry.[3]

One serious question is: "what does clean mean?" There has been some debate as to when a surface is termed clean; depending on whether the surface being considered needs to be (1) optically, (2) physically, (3) chemically, or (4) microbially clean. In the food process industry "cleaning" often describes returning the system to a state indistinguishable from the original, before fouling commenced.

CLEANING OF FOOD PLANT

Milk fouling, due to heat treatment, has been most widely studied due to the tenacious nature of the deposit formed (not removed by water alone) and also the industrial relevance of this process. Fouling from milk-based fluids is so severe that daily cleaning procedures are often required. Cleaning-in-place (CIP) procedures are commonly used, to enable the cleaning of equipment without dismantling and without manual cleaning. These systems are highly automated and employ complex cleaning chemicals. There are two kinds of procedures continuously utilized in the dairy industry: (1) single-stage, (2) two-stage procedures. Traditionally two-stage cleaners were used, and these typically consist of 5–7 steps[4]:

1. prerinse, to remove any loose fouling deposit
2. alkaline rinse (commonly sodium hydroxide), removes protein and fats and some of the mineral
3. intermediate rinse, flushes any loose deposit out of the system
4. acid rinse (commonly nitric acid), removes the remaining minerals (usually calcium phosphate)
5. postrinse, removes any detergent from the system.

Sometimes an extra sanitization step is included to destroy microorganisms on the surface. The acid and alkaline steps are reversed[4] in some cases. However, two-stage cleaners have been shown not to totally clean the system, sometimes leaving residual minerals on the surface. Also, where the acid and alkaline rinses directly follow one another the chemicals are inefficiently used. Single-stage cleaners have been proven to be more cost effective and thorough in their action.[5] Alkaline single stage cleaners compose of a range of materials,[6] shown in Table 1.

Cleaning is often a function of process history. Both the composition of the deposit and the amount of deposit formed depend on the temperature and composition of the fouling fluid, and the process time. All of these will affect the rate of cleaning, and thus the time required.

Encyclopedia of Agricultural, Food, and Biological Engineering
DOI: 10.1081/E-EAFE 120007148
Copyright © 2003 by Marcel Dekker, Inc. All rights reserved.

Table 1 Components of single stage alkali cleaning agents

Component	Action
Strong alkalines	Degradation of organic soils
Weak alkalines	Keeping removed soils dissolved
Sequestrating agents	Binding water hardeners and preventing the formation of scale and insoluble precipitates
Surface active agents	Lowering the interfacial tension, emulsifying fatty soils, and controlling the foam properties
Stabilizers	For solutions stored in a hot condition
Oxidizers	Intensify the cleaning effect
Corrosion inhibitors	
Solubilizers	For stabilizing the concentrates

The engineering effect of fouling within heat treatment plants is monitored in terms of heat transfer and pressure drop, therefore, cleaning can be reported in terms of heat transfer recovery and reduction in pressure drop.

Effect of heat transfer are modeled in terms of the fouling resistance, R_F whose change with time is found from:

$$\frac{1}{U_t} = \frac{1}{U_0} + R_F(t)$$

where U_t is the overall heat transfer coefficient at time t and U_0 the heat transfer coefficient of the clean surface. The changes in pressure drop are often modeled in terms of the pressure drop ratio:

$$\frac{\Delta P - \Delta P_0}{\Delta P_0}$$

in which ΔP and ΔP_0 are the fouled and clean pressure drops, respectively.

The cleanliness of a food processing plant is often monitored using swabbing techniques. Dairy plant vessels and pipe work that have been cleaned efficiently will give a total viable count result of less than 1000 colonies per 1000 cm^2 and will give a negative result in the presumptive coliform test.[7] However, the test may not be totally accurate as the swabs tend to be taken from areas that are easily cleaned. A possible gauge for directly monitoring thickness of soft deposits is under development[8]; this might enable on line monitoring of fouling layer thickness rather than the secondary effects of the deposit described previously.

PARAMETERS AFFECTING CLEANING

Many authors have identified the parameters important to cleaning as: (i) temperature, (ii) flow rate, and (iii) concentration of cleaning chemical. In general, if the cleaning solution flow rate is increased, the time to clean is reduced and an increased maximum cleaning rate is observed.

Flow Rate

Flow rate can influence the cleaning process both by increasing the mass transfer of cleaning solution to the cleaning interface and by increasing the shear forces on the deposit. Many of the rules governing cleaning processes in industry are anecdotal; for example 1.5 m sec^{-1} is widely quoted as being necessary to allow cleaning, however, this value has been dismissed by a number of researchers as cleaning has been observed below this value.[9]

Temperature

The temperature of the cleaning solution affects the cleaning rate. Temperature may be most important in the early stages of cleaning when the foulant layer comes into thermal equilibrium with the cleaning fluid. Cleaning requires energy to remove the deposits. Energy is obtained from the heat and mechanical force of the cleaning solution. In addition to reaction and diffusion, mass transfer and viscosity will be Arrhenius functions of the temperature.[10] Researchers[4,11] have also reported a marked increase in cleaning rate above 60°C.

Cleaning Solution Concentration

Milk and other food deposits are not removed by water alone: physical fluid shear is not enough. In order to optimize cleaning efficiency and reduce environmental impact, the concentration of chemical required has been investigated. Bird and Fryer[9] identified an optimum concentration value of 0.5 wt% NaOH for cleaning of tubes fouled by whey proteins. They observed that above this value the deposit became sticky and was not removed, i.e., was not susceptible to shear from the cleaning solution. Low concentrations also result in long cleaning times. Cleaning of pilot scale plate heat exchangers have also shown the existence of optimum cleaning chemical concentrations.[5,12]

MECHANISMS OF DEPOSIT REMOVAL

Cleaning processes are controlled by a combination of mass transfer, diffusion, or reaction. Visualization of the cleaning of a dairy deposit from a flat surface[6,13] shows that the process is nonuniform; once the deposit layer has been treated with sodium hydroxide lumps are more readily removed by fluid shear. Cleaning has often

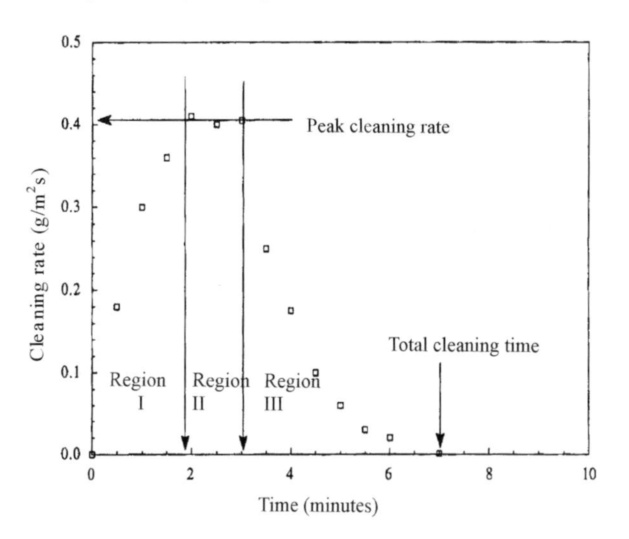

Fig. 1 Typical cleaning curve. (From Ref. 14.)

been reported in terms of the amount of deposit to be removed from the surface per second (g sec^{-1}). Fig. 1 is a cleaning curve for a whey protein deposit in a laboratory experiment[13] showing that the (i) time to clean, (ii) maximum cleaning rate, and (iii) time to reach maximum cleaning rate are measurable. The three regions of the cleaning curve in Fig. 1 correspond to (I) the time needed for cleaning solution to penetrate the deposit, followed by (II) swelling of the deposit and rapid deposit removal then (III) reduced cleaning rates where final chunks are removed from the surface. This relates to results seen in large-scale equipment, where pressure drop values are monitored. When the plant fouls the pressure drop across the plant increases, cleaning results in a reduction of this pressure drop. During the first stages of cleaning the pressure drop increases, as the deposit swells; then the pressure drop is reduced as the deposit is washed away from the surface and dissolved into the cleaning solution. Fig. 2 shows a schematic of deposit removal (modified from Ref.[14]). Christian and Fryer,[15] investigating whey protein removal, showed that chemicals were needed throughout cleaning, both for the swelling stage and the subsequent removal of deposit.

INDUSTRIAL SPRAY CLEANING

Storage and preheat vessels are subject to fouling. Cleaning is carried out using spray balls; fluids under high pressure cause bust spraying and an intermittent action is achieved by turning the pump on and off. Cleaning policies vary between fixed time and fixed volume control. The advantages of the later policy were demonstrated: (a) it is very easy to implement and tune, (b) any dynamic changes of the flow are inherently taken into account, (c) the cleaning solution level/volume in the tank can be strictly controlled between high and low limits—therefore accumulation of cleaning solution in the tank and emptying of the tank are avoided, and (d) a lower average level of fluid in the tank being cleaned can be safely achieved. Initial trials have been carried out on the use of high pressure jet spray lances[16] to be used when deposits, such as those from chocolate milk processing, are not removed by conventional CIP processes. Important parameters and processes involved with sprayball cleaning (e.g., flow rate, temperature, hole size, and spacing) have been investigated.[17]

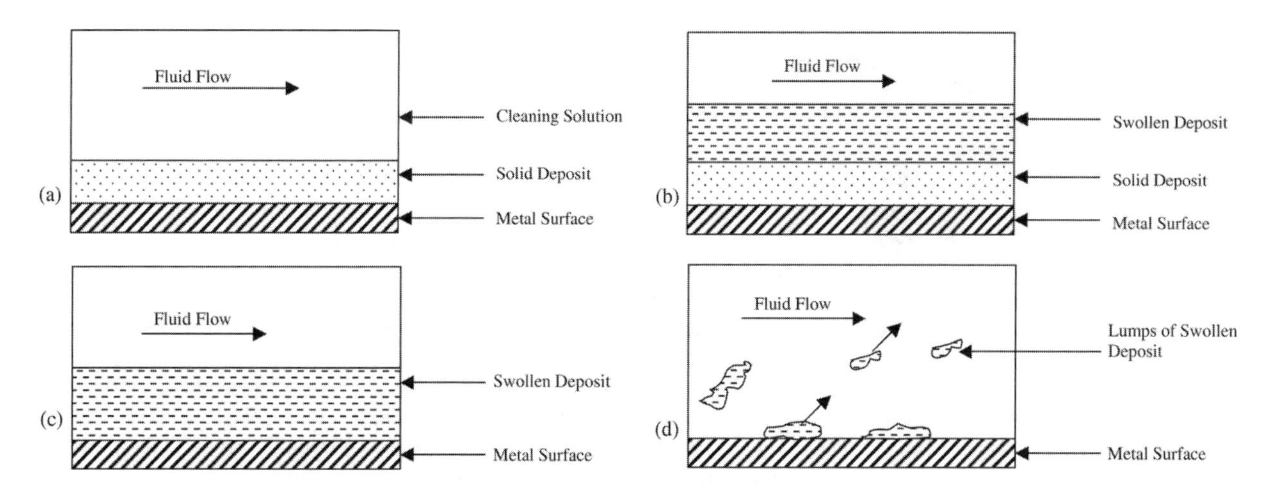

Fig. 2 (a–d): Schematic representation of the cleaning of protein deposits. (a) shows the layers involved; (b) the deposit swells and is uniformly removed; (c) all the remaining deposit is swollen; (d) decay phase, random removal of "chunks" from the surface.

THE COST OF CLEANING

Predictions have been made for the costs of fouling and the resulting cleaning processes; there are many factors that add to the cost and these include:

1. increased capital costs, e.g., hygienic design of the plant,[18,19] cost of cleaning solution;
2. increased energy cost during operation, to compensate for the reduction in process efficiency;
3. maintenance (treatment of effluent);
4. production losses associated with plant down time for cleaning;
5. energy losses, due to heating and circulation of the cleaning agent;
6. labor, although CIP is highly automated some manual cleaning may still be required.

Typical "down" times can be up to 11 hr, during which time no products are produced, also costing the producers money. There is also an environmental consideration; effluent from the plant must be treated before it can be passed out into the streams. Increased capital expenditure arises as plants are being forced to have more complex recovery and effluent treatment systems, the awareness of environmental hazards from these cleaning processes has increased over the last 10 yr.

CONCLUSION

The principles underpinning cleaning processes in the food industry have been briefly reviewed. Cleaning processes depend critically on the type of deposit to be removed, and cleaning rates vary with both the flow velocity, temperature and composition of the cleaning fluid. The mechanisms of cleaning are understood in some cases, but the processes by which equipment becomes clean are not fully quantified.

REFERENCES

1. Klein, G.M.; Meier, J.; Kottke, V. *Fouling and Cleaning Mechanisms in Membrane Apparatus During Crossflow Microfiltration. 1998; 87–94. Fouling and Cleaning in Food Processing '98, EUR 18804*; Office for the European Communities: Luxembourg, 1999.
2. Smaili, F.; Angadi, D.K.; Hatch, C.M.; Herbert, O.; Vassiliadis, V.S.; Wilson, D.I. Optimization of Scheduling of Cleaning in Heat Exchanger Networks Subject to Fouling: Sugar Industry Case Study. Trans. Inst. Chem. Eng. **1999**, *77* (C), 159–164.
3. Jeurnink, Th.J.M.; Brinkman, D.W. The Cleaning of Heat Exchangers and Evaporators After Processing Milk or Whey. Int. Dairy J. **1994**, *4*, 347–368.
4. Timperley, D.A.; Smeulders, C.N.M. Cleaning of Dairy HTST Plate Heat Exchangers: Comparison of Single- and Two-Stage Procedures. J. Soc. Dairy Technol. **1987**, *40* (1), 4–7.
5. Timperley, D.A.; Smeulders, C.N.M. Cleaning of Dairy HTST Plate Heat Exchangers: Optimization of the Single-Stage Procedure. J. Soc. Dairy Technol. **1988**, *41* (1), 4–7.
6. Grasshoff, A. Cleaning of Heat Treatment Equipment. Bulletin of the International Dairy Federation No. 328/1997. *Fouling and Cleaning of Heat Transfer Equipment*; 1997; 32–44.
7. Romney, A. *CIP: Cleaning in Place*, 2nd Ed.; Society of Dairy Technology: London, 1990.
8. Tuladhar, T.R.; Paterson, W.R.; Macleod, N.; Wilson, D.I. Development of a Novel Non-contact Proximity Gauge for Thickness Measurement of Soft Deposits and Its Application in Fouling Studies. Can. J. Chem. Eng. **2000**, *78*, 935–947.
9. Bird, M.R.; Fryer, P.J. An Experimental Study of the Cleaning of Surfaces Fouled by Whey Proteins. Trans. Inst. Chem. Eng. **1991**, *69* (C), 13–21.
10. Jennings, W.G.; McKillop, A.A.; Luick, J.R. Circulation Cleaning. J. Dairy Sci. **1957**, *40*, 1471.
11. Fryer, P.J.; Bird, M.R. Factors Which Affect the Kinetics of Cleaning Dairy Soils. Food Sci. Technol. Today **1994**, *8* (1), 36–42.
12. Changani, S.D. An Investigation into the Fouling and Cleaning Behavior of Dairy Deposits. Ph.D. Thesis, University of Birmingham, U.K., 2000.
13. Bird, M.R. Cleaning of Food Process Plant. Ph.D. Thesis, University of Cambridge, U.K., 1993.
14. Gillham, C.R. Enhanced Cleaning of Surfaces Fouled by Whey Protein. Ph.D. Thesis, University of Cambridge, U.K., 1997.
15. Christian, G.K.; Fryer, P.J. Pulsed Cleaning: Physical and Chemical Effects on Cleaning of a Dairy Fouled Pilot Scale Plat Heat Exchanger (PHE). In *Fouling, Cleaning and Disinfection in Food Processing*; Wilson, D.I., Hasting, A.P.M., Fryer, P.J., Eds.; Dept. of Chemical Engineering: Cambridge, 2002; 135–142.
16. Grasshoff, A. *Fundamental Trials to Clean UHT Tubular Modules. Fouling and Cleaning in Food Processing '98, EUR 18804*; Office for the European Communities: Luxembourg, 1999; 238–245.
17. Morison, K.R.; Thorpe, R.J. Cleaning Characteristics of CIP Sprayballs. In *Fouling, Cleaning and Disinfection in Food Processing*; Wilson, D.I., Hasting, A.P.M., Fryer, P.J., Eds.; Dept. of Chemical Engineering: Cambridge, 2002; 205–212.
18. EHEDG (European Hygienic Equipment Design Group). A Method for Assessing the In-place Cleanability of Food-Processing Equipment. Trends Food Sci. Technol. **1992**, *3*, 325–328.
19. EHEDG (European Hygienic Equipment Design Group). A Series of Articles on the Basis of Hygienic Design. Trends Food Sci. Technol. **1993**, *4*, 21–25, 52–55, 80–82, 190–192, 225–229, 306–310.

Commercial Sterilization Systems

Pamela K. Hardt-English
PhF Specialists, Inc., San Jose, California, U.S.A.

C

INTRODUCTION

Differing packaging requirements and manufacturing conditions have necessitated the development of several different types of retorting systems to safely and consistently sterilize products in sealed containers. Products that need gentle agitation and a short process to maintain product quality can be sterilized by water immersion retorts. Products in flexible or nonrigid packages such as pouches, which need overpressure to maintain package integrity, can be sterilized in cascading water retorts, water spray retorts, water immersion, or steam/air retorts. Hydrostatic retorts and continuous rotary sterilizers handle containers continuously and decrease labor requirements. Several of these types of retorts, as well as steam and water retorts, give manufacturers the flexibility they need to quickly change the container material, dimensions, etc., in order to satisfy their customers.

EARLY WATER AND STEAM RETORTS

The early commercial sterilization systems were batch retorts, which used water or steam to sterilize the food in glass jars or metal cans. These are still commonly used today.

For glass containers, water cooks were typically used in vertical retorts, where baskets were loaded on top of each other. Water and steam are usually fed in from the bottom of the retort and re-circulated to assure even water temperature. Because the internal pressure builds up in the jar as the product heats, overriding air pressure can be added to the retort during processing and before cold water is added for cooling, to keep the lids from popping off the jars. Thermal shock to the glass must also be considered, and therefore the initial water added during the cooling cycle may not be as cold as the water used for the final steps of cooling.

Simple steam retorts are used to process cans. The retort in Fig. 1 is horizontal in design and the baskets are oriented horizontally. In vertical steam retorts, baskets are loaded one on top of the other. The steam is injected into the bottom of the retort and the air is vented from the retort through bleeders, vents, or other pipes at the top. Adequate venting is critical in steam retorts because steam has much more heat capacity than steam and air combined. Air pockets can insulate the cans and result in under processing of those cans surrounded by an air pocket.

Air may be forced into the retort at the end of the cook to maintain some positive pressure in the retort during cooling, which can minimize can buckling or peaking problems. In addition, in large industrial cans with viscous products, such as cream style corn and pureed pumpkin, the overpressure at the start of the cool keeps the can volume constant, so that the product continues to heat and accumulate lethality during the first minutes of cooling. If the can expands due to high internal pressure, the steam in the headspace will collapse and the product will start cooling quickly. The early systems usually could not be operated above 121°C.

Although the unloading and reloading operations were labor intensive, a well-managed cook room can have up to 100 retorts operating at full production. Although each retort is operated as a batch operation, the whole cook room can operate as a continuous production system in that filled and sealed unsterilized cans enter the cook room continuously from the filling line operations, and fully processed sterilized cans leave the cook room continuously in route to case packing and warehousing. Within the cook room, teams of workers move from retort to retort to carry out the loading and unloading operations and retort operators are responsible for a given number or bank of retorts.

CONTINUOUS ROTARY STERILIZERS

These steam sterilizers were developed in order to handle containers continuously and process large quantities of cans automatically. Cans are held within a turning reel and are conveyed through several shells (Fig. 1) by a spiral pathway. Cans are fed individually by a conveyer into the first shell. The shell can be a sterilization or a preheater shell. The cans are then transferred into the next shell, which might be another cooking shell or a pressure or atmospheric cooling shell. The unique transfer value, which connects one shell to the next allows the cans to move from one shell to another without losing steam pressure. These sterilizers were designed to operate at temperatures up to 130°C.

In addition to being a continuous operation, the cans are agitated in two ways. The cans are first carried over the top

Encyclopedia of Agricultural, Food, and Biological Engineering
DOI: 10.1081/E-EAFE 120007108
Copyright © 2003 by Marcel Dekker, Inc. All rights reserved.

Fig. 1 Continuous rotary sterilizers. (Courtesy of FMC Corp.)

two-thirds of the reel, then fall off the reel, and roll around the bottom third of the shell. To maximize the effectiveness of the agitation, headspace is considered when developing a thermal process. If there is some headspace, e.g., 6 mm or 1/4 in. in a retail size can, the product may heat more quickly than if there is no headspace. This agitation causes products to heat much more rapidly than they would in a nonagitating sterilizer. Continuous rotary sterilizers are common in the United States, where there are standard size containers, labor costs are high, and many canned food items are considered commodities and are produced in very large quantities. The disadvantages are that only one can size can be processed in each sterilizer and the cost per system is high.

HYDROSTATIC STERILIZERS

These systems (Fig. 2) rely on the weight of columns of water to maintain steam pressure in the sterilization dome. The columns of water may be 20 m–25 m high, depending on the processing temperature. Containers, usually cans, are conveyed in rows of cans situated end-to-end or "sticks." Alternatively, some systems move pallets of containers. Containers are carried by a conveyor chain down the feed leg and up into the sterilization dome. Because hydrostatic pressure is used there are no transfer valves required when the cans move into the sterilization dome. As the cans are lowered down the feed leg, the pressure gradually increases until the cans are at the sterilization pressure. In the steam dome, the cans are carried up and down several passes by the chain. The cans then move into the discharge leg, which starts at the sterilization pressure. Gradually, the external pressure is reduced as the cans move up the leg until they are at atmospheric conditions. There may be a shower cooling section followed by a final cooling canal to complete cooling before discharge.

Fig. 2 Hydrostatic cooker. (Courtesy of FMC Corp.)

Hydrostatic sterilizers may be equipped with several chains to convey the containers, allowing for several different process times or different container diameters. These systems can also be designed to sterilize jars. Fully automatic loading, processing, cooling and unloading, large capacity and efficient use of floor space are just a few of the advantages of hydrostatic sterilizers. The disadvantages are that the initial cost of the machine is high and changes in container dimensions can be limited in some systems.

WATER IMMERSION RETORTS

Water immersion retorts usually have two tanks, an upper water storage tank and a lower processing retort. The water is preheated in the upper storage tank to a temperature greater than that required for the thermal process. At the same time, baskets of containers are loaded into the lower retort. Once the retort is loaded and the water in the upper tank is preheated and the water is pumped down into the bottom retort and then is re-circulated within the retort. Steam is injected into the water to maintain process temperature. During cooling, the hot water is pumped back to the upper storage tank and cold water is pumped into the retort. This reuse of the processing water is an advantage of this type of retort.

For cans or jars, water immersion retorts are usually operated in an agitating mode to minimize the process time. The containers are held firmly in place by a clamping system so that they are not damaged as the baskets are

Fig. 3 Batch rotary retort.

turned end-over-end. In addition, water immersion retorts are capable of maintaining overpressure, and so offer the processor the advantage of being able to use any type of packaging material, including pouches, trays, cans, or jars.

CASCADING AND WATER SPRAY BATCH RETORTS

These systems (Fig. 3) are horizontal retorts, which may or may not be operated in an agitating mode. Water is pumped quickly through sprays or a water spreader. Any type of container can be processed because air can be added to the retort to maintain independent overpressure. Because there is a layer of water next to the containers, the heat is transferred from hot water to the container. This system uses significantly less water than the water immersion system and can handle any type of metal, glass or flexible packaging.

STEAM/AIR BATCH RETORTS

This horizontal batch retort system can be operated in a stationary or agitating mode, and because air can be added to the machine to maintain overpressure, this type of machine can accommodate several types of containers. Steam and air are mixed into uniformity by a fan located at the back of the retort. Because heat is transferred to the containers by a mixture of steam and air, heating of the containers may be less efficient than when pure steam or water is used.

CRATELESS RETORTS

Crateless retorts (Fig. 4) do not use baskets, hence the name. Crateless retorts consist of a large tank, which is jumble-loaded with cans through a door at the top and

Fig. 4 Crateless retort. (Courtesy of Malo Inc.)

discharged through a door at the bottom. In order to cushion the fall of the cans, the retort is first filled with water. Once the retort is full, the top door is closed and the water is pushed out of the retort by the processing medium, steam. Once the water is completely displaced, the retort is brought up to processing temperature. During cooling, water is brought into the retort and air can be added if overpressure is needed. The cans are then discharged from the retort onto a conveyor in a cooling canal. The bottom door of the retort can be submerged in the cooling canal. This creates a vacuum within the retort, which slows the emptying of the retort and minimizes damage to the containers. Typically, there are several crateless retorts in a line which all empty into the same cooling canal in sequence, creating a continuous discharge of containers. If the installation is properly designed, crateless retorts can accommodate a continuous flow of cans to and from the retort room.

Crateless retorts offer the advantage that any size can may be sterilized, however, usually only cans may be sterilized because other types of containers may suffer significant damage during the loading and unloading of the retort. Another advantage is that the simple design of this type of retort allows easy maintenance.

AXIALLY AGITATING BATCH STERILIZERS

Similar to retorts, which allow containers to roll across a surface, this type of sterilizer provides axial agitation to containers, usually cans. Containers are conveyed into the shell until the sterilizer is full. At that point, the feed and discharge ports are closed and the containers are clamped onto a central rotating reel. The reel then turns at up to 36 rpm during processing. This system is used primarily for institutional size cans of viscous products. The headspace within the cans combined with the rapid rotation of the reel allow the products to heat much more rapidly than they would in a typical nonagitating batch sterilizer.

MODERN CONTROL SYSTEMS

Most of the retorts sold now have sophisticated controls which can optimize the time, temperature, and pressure of the process during preheating, processing, and cooling. Some control systems can even calculate thermal processes, and therefore can make process time adjustments in the event of an unplanned decrease in process temperature.

CONCLUSION

Steam and water are the heating media for retorts, but there are many types of retorts. When choosing a retort, consider present and future packaging materials, sizes and shapes; production requirements, equipment costs and the costs related to utilities, labor and maintenance.

FURTHER READING

Staff Members of The Food Processors Institute. Thermal Processing Systems. *Canned Foods—Principles of Thermal Process Control, Acidification and Container Closure Evaluation*, 5th Ed.; The Food Processors Institute: Washington, DC, 1989; 67–105.

Eisner, M. Rotary Sterilization. *Technology and Techniques in Rotary Sterilization* (*Einfuhrung in die Technik und Technologie der Rotationssterilisation*), 2nd Ed.; Verlag Gunter Hempel: Wolfsburg, Germany, 1985; 9–44.

Kimball, R.; Heyliger, T, et al. Sterilization Methods: Principles of Operation and Temperature Distribution Studies. Food Technol. **1990**, *44* (12), 99–118.

Continuous Process Control

Yanbin Li
University of Arkansas, Fayetteville, Arkansas, U.S.A.

C

INTRODUCTION

Continuous control is commonly applied to biological and food process control, such as bioreactor, thermal process, and minienvironmental control. In a process control system, the variables associated with any process are monitored and subsequent control actions are implemented to maintain variables within predetermined process constraints.[3] The biological or food processing system to be controlled is a group of physical components combined to perform a specific function. Variables to be controlled may include temperature, pressure, pH, oxygen, flow rate, liquid level, density, speed, voltage, current, position, among others. Analog or digital techniques can be employed to individually or simultaneously implement the desired control action.

A control system is usually designed to provide the best response of the complete system to external, time-dependent disturbances. A control system can be basically established as an interconnection among operational components of the system. These individual components may be electrical, mechanical, hydraulic, pneumatic, thermal, or chemical in nature. The major functions of a control system are: 1) to minimize errors between the actual and desired output and 2) to minimize the time response to changes in system load.[4]

All control systems have three basic components: sensor/transmitter, controller, and final control element, and perform the three basic operations, measurement, decision, and action.[11] An agitated and heated reactor with a pump is illustrated in Fig. 1 as an example of continuous process control. The control actions required for operating the reactor are to turn on/off an agitator and a pump based on the level of liquid materials in the reactor and to turn on/off a heater according to a temperature set point. The measurements for control of the reactor are the temperature and level of reaction materials through a thermocouple and a level sensor.

CONTROL SYSTEMS

Objectives of Control Systems

The economic goal of process control is to achieve maximum productivity or efficiency while maintaining a satisfactory level of product quality.[3] Important reasons for process control are to prevent injury to plant personnel, to protect the environment, to reduce human labor demand, and to maintain product quality and plant production rate on a continuous basis and at minimum cost. The objective of an automatic process control system is to adjust the manipulated variables to maintain the controlled variables at their set points in spite of disturbances.[10,11] Control systems, equipment design, and operating conditions are the three critical components for achieving excellent performance in biological and food processing.

Linear and Nonlinear Control Systems

A control system is linear when it satisfies both the amplitude proportionality criteria and the principle of superposition.[4] If a system output at a given time is $Y(t)$ for a given input $X(t)$, an input of $KX(t)$ must produce an output $KY(t)$ to satisfy the amplitude proportionality. In similar manner, if an input of $X_1(t)$ produces an output of $Y_1(t)$, while an input of $X_2(t)$ produces an output of $Y_2(t)$, an input of $X_1(t) + X_2(t)$ should produce an output of $Y_1(t) + Y_2(t)$ to satisfy the superposition principle. A system is nonlinear when it does not satisfy both of the criteria. In most cases, these nonlinear systems are compensated or linearized so that their behavior approaches an equivalent linear system.[4]

Open- and Closed-Loop Control

A typical open-loop system is shown in Fig. 2. The input provides the information on the desired value of the controlled variable, X. This information is then converted to an action on the controller to alter the output, Y, to the load. External disturbances may be fed in as shown in Fig. 1 and result in the variation of the output from desired values.

For the closed-loop control system, two more elements need to be added to the open-loop system. They are: 1) a monitoring element to measure the output, Y and 2) a comparing element, to measure the difference between the actual output, Y, and the desired value, X. The monitoring and comparing elements are connected through a feedback link as illustrated in Fig. 3.

Encyclopedia of Agricultural, Food, and Biological Engineering
DOI: 10.1081/E-EAFE 120007214
Copyright © 2003 by Marcel Dekker, Inc. All rights reserved.

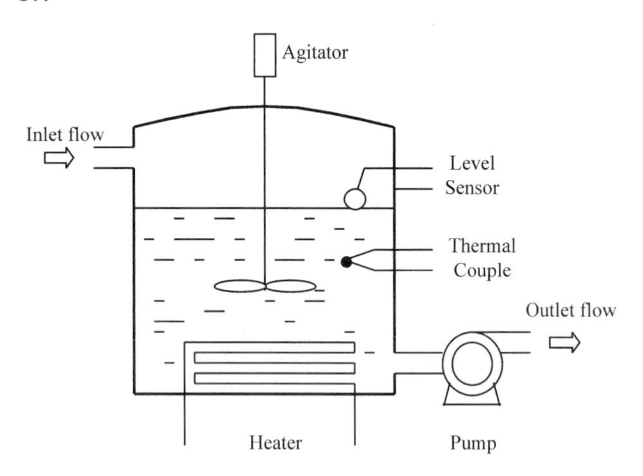

Fig. 1 An agitated and heated tank reactor with a pump, a thermocouple, and a level sensor.

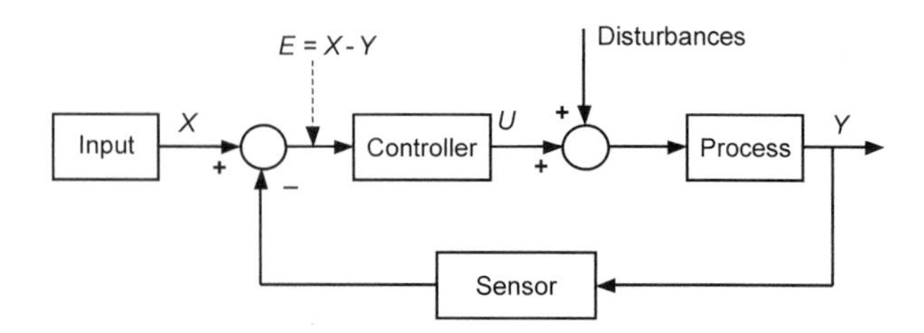

Fig. 3 Block diagram of a closed-loop feedback control system.

Characteristics of Control Systems

A control system can be characterized based on the output behavior of the system in response to any given input. The parameters used to define control system characteristics include stability, accuracy, speed of response, and sensitivity.[10] The system is said to be stable when the output attains a certain value in a finite interval after the input has undergone a change. When the output reaches a constant value the system is said to be in steady state. The system is unstable if the output increases or decreases with time.

The accuracy of a system is a measure of the deviation of the actual controlled value in relation to its desired value. Accuracy is synonymous with the steady-state error. The stability is closely associated with the time response of the system. Accuracy and stability are therefore interactive in the same conflicting manner as steady-state error and time response. The accuracy of a system may be improved, but this will be at the expense of reduced stability. The speed of response is a measure of how quickly the output attains a steady-state value after the input has been altered. The speed of the response is also constrained by stability limitations. Sensitivity is a measure of how the system output responds to external environmental conditions or the change in system parameters. The desired output should be only a function of the input, not influenced by any undesirable noise signals. Accuracy and response speed can be analyzed with time-domain methods, whereas stability and sensitivity can be evaluated with frequency-domain methods.

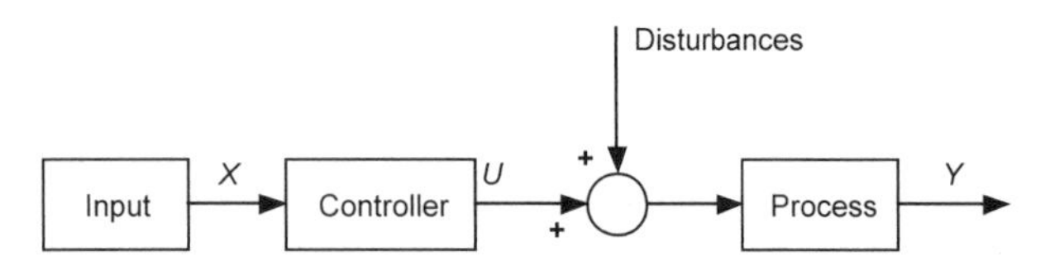

Fig. 2 Block diagram of an open-loop control system.

Dynamic System Response

The dynamic response of a control system can be evaluated by mathematical modeling or by experimental measurement of the output of the system in response to a particular set of test input conditions.[7] The step input is the most important test input as a system which is stable to a step input will also be stable under other forms of input. The step input is applied to gage the transient response of the system and give a measure of how the system can cope with a sudden change in the input. A ramp input is used to indicate the steady-state error in a system attempting to follow linearly increasing input. Finally, a sinusoidal input over a varying range of input frequencies, is a standard test used to determine the frequency dependent characteristics of the system. The three standard test inputs may not be able to represent actual inputs to which the system will be subjected to, but they do cover a broad range of the inputs in biological and food processing. In general, a system that performs satisfactorily under these idealized test inputs will perform well under a more natural range of inputs.

MATHEMATICAL MODELING

Time and Frequency Response Analyses

The time-domain model of a system results in an output $Y(t)$ with respect to time, t, for an input $X(t)$. The time-domain system model is expressed as a differential equation, the solution of which is displayed as a function of output against time. In contrast, a frequency domain model describes the system in terms of the effect that the system has on the amplitude and phase of sinusoidal inputs. The differential equation that describes a first-order system is expressed as the follows.

$$\tau \frac{dy}{dt} = -y + Kx \qquad (1)$$

where y is the output, x is the input, K is the process gain, and τ is the time constant to determine the speed of response for the system. For analysis in the frequency domain it is customary to write the differential equation in terms of the Laplace operator, s. The advantage in using the Laplace transform method is that differential equations can be expressed as equivalent algebraic relationships in s. The mathematical definition of the Laplace transform is defined as

$$L(f(t)) = f(s) = \int_0^{\infty} f(t)\, e^{-st}\, \mathrm{d}t \tag{2}$$

And then the Laplace transforms of the differential equation in Eq. 1 can be obtained as the follows.

$$\tau s\, Y(s) + Y(s) = KX(s) \tag{3}$$

or in a general form

$$Y(s) = G(s)X(s) \tag{4}$$

where $Y(s)$ is transform of the output variable, $G(s)$ is the transfer function, and $X(s)$ is the input variable.

This gives rise to the system transfer function which is formed by replacing the input and output, X and Y, respectively, with their corresponding Laplace transforms $X(s)$ and $Y(s)$. The method applies only to linear differential equations. In practice, many systems contain some degree of nonlinearity and various assumptions need to be made to simplify and approximately linearize the governing equations. The Laplace transforms of some common time functions are listed in Table 1.

Table 1 Laplace transforms of some common time functions

Time domain $f(t)$, $t > 0$	Frequency domain $f(s) = L(f(t))$
$\delta(t)$, unit impulse	1
$u(t)$, unit step or constant	$1/s$
t	$1/s^2$
t^n	$n!/s^{n+1}$
e^{-at}	$1/(s+a)$
te^{-at}	$1/(s+a)^2$
$t^n e^{-at}$	$n!/(s+a)^{n+1}$
$\sin(\omega t)$	$\omega/(s^2+\omega^2)$
$\cos(\omega t)$	$s/(s^2+\omega^2)$
$e^{-at}\sin(\omega t)$	$\omega/[(s+a)^2+\omega^2]$
$e^{-at}\cos(\omega t)$	$(s+a)/[(s+a)^2+\omega^2]$
$1-(e^{-t/\tau})$	$1/[s(\tau s+1)]$

System Transfer Functions

A system transfer function is defined as the ratio of the output to the input and can be determined by taking the Laplace transform of the ordinary differential equation that describes the system. The transfer function, $G(s)$ shown in Eq. 4 describes the dynamic characteristics of the process. For linear system it is independent of the input variable and it applies for any time-dependent input signal. Transfer functions can also be used to describe the dynamic behavior of instruments, controllers, and valves. The order of the transfer function is the power of the polynomial in s in the denominator of $G(s)$ and is equivalent to the order of the differential equation describing the system. For the first-order system described in Eq. 1, the transfer function is expressed as

$$G(s) = \frac{K}{\tau s + 1} \tag{5}$$

The transfer function for the second-order system

$$\frac{\mathrm{d}^2 Y(t)}{\mathrm{d}t^2} + 2\xi\omega_n \frac{\mathrm{d}Y(t)}{\mathrm{d}t} + \omega_n^2 Y(t) = KX(t) \tag{6}$$

can be obtained as

$$G(s) = \frac{K}{s^2 + 2\xi\omega_n s + \omega_n^2} \tag{7}$$

where ξ is the constant of damping ratio, and ω_n is the constant of natural frequency.[5]

Linearization of Nonlinear Systems

Most control systems are nonlinear and general methods for developing analytical solutions for nonlinear models are not available. Thus, linearization technique is used to approximate the response of nonlinear systems with linear differential equations so that Laplace transforms can be applied.[10,11] The linear approximation to the nonlinear equations is valid for a region near some base points around which the linearization is made. The transfer functions of the linearized equations need to be developed first and then response characteristics can be related to function parameters. An important feature of nonlinear systems is that their response depends on the operating point, i.e., the parameters of the linearized system are valid at the base point instead of a region of values. Procedures for the linearization of one or more variables and differential equations are discussed in the literature.[7,10,11]

FEEDBACK CONTROL

On–Off Control

On–off control is widely used to control the output variable within preset limits. The on–off control action results in either full power or zero power being applied to the process under control. In on–off control, there are only two possible values for the output,

$$p(t) = \begin{cases} p_{\max} & \text{if} \quad e > 0 \\ p_{\min} & \text{if} \quad e < 0 \end{cases} \tag{8}$$

where $p_{\max}$ and $p_{\min}$ are the on and off values, respectively, and e is the error or difference between the controlled variable and its set point.

Three-Term (PID) Control

As complicated transfer functions can be very difficult to model, the most common strategy used to define the controller transfer function is the three-term, or PID (proportional, integral, and derivative) controller as shown in Fig. 4. The three elements of controller action based on the evaluated error are discussed as the follows.

Proportional action can be expressed as

$$U = KE \tag{9}$$

where U is the controller output, K is the controller gain, and E is the evaluated error.

Manufacturers of three-term controllers tend to use the proportional band, PB, in preference to the gain, K. The proportional band represents the range of input over which the output is proportional to the input. The PB is usually expressed as a percentage of the input normalized between 0% and 100%.

$$PB\% = \frac{100}{K} \tag{10}$$

The limitation of proportional control is that very high gain may result in system unstable oscillation. This can be partly improved by adding a controller action that gives an output contribution related to the integral of the error value with respect to time, i.e.,

$$U = K_{\mathrm{i}} \int E \, \mathrm{d}t \tag{11}$$

where K_{i} is the controller integral gain which equals K/T_{i} and T_{i} is the controller integral time.

The nature of integral action suggests that the controller output will increase monotonically as long as an error exits. As the error tends to zero the controller output tends towards a steady value.

By adding derivative action, the stability of a system can be improved and any tendency to overshoot can be reduced. Derivative action is based on the rate of change in the error and implements a contribution to the controller output based on the assumption that the process variable will continue to change at a rate currently measured. In this manner derivative action acts as a process lead function, predicting the effects of process changes, and compensating to a certain extent for lags in the process dynamics, i.e.,

$$U = K_{\mathrm{d}} \frac{\mathrm{d}E}{\mathrm{d}t} \tag{12}$$

where K_{d} is the controller derivative gain which equals KT_{d}, and T_{d} is the derivative time, or rate of the controller. The proportional action governs the speed of the response, the integral action improves the accuracy of the final steady state, and the derivative action enhances the stability.

The PID control action is expressed in Eq. 13, and its three-term controller transfer function is shown in Fig. 4.

$$U = K\left(E + \frac{1}{T_{\mathrm{i}}}\int E \, \mathrm{d}t + T_{\mathrm{d}}\frac{\mathrm{d}E}{\mathrm{d}t}\right) \tag{13}$$

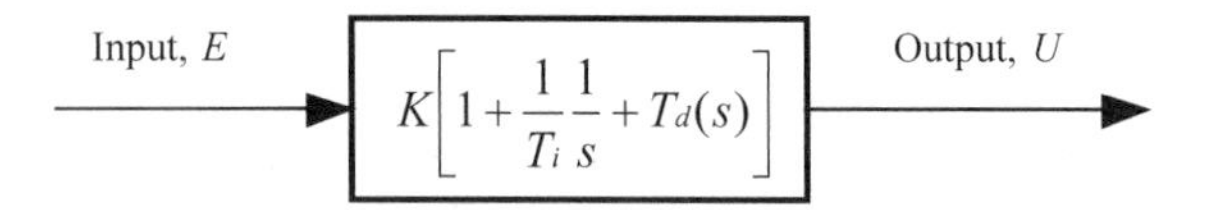

Fig. 4 Three-term, or PID control transfer function.

Stability Analysis

A control system is stable if all output variables are bounded when all input variables are bounded. The possibility of sustained oscillations always exists in closed-loop systems. The step response of a second-order system is dependent on the damping ratio, ξ, defined in Eqs. 6 and 7. As shown in Fig. 5, when the damping ratio is greater than 1, the response is overdamped, and when the damping ratio is less than one, the response is underdamped. The two regions are separated by the critically damped response when the damp ratio is equal to one. Three parameters, rise time, overshoot, and settling time, are used in the characterization of step response. Rise time is defined as the time that elapses while the step response rises from 10% to 90% of its final value. Overshoot occurs in case of underdamped response, and is

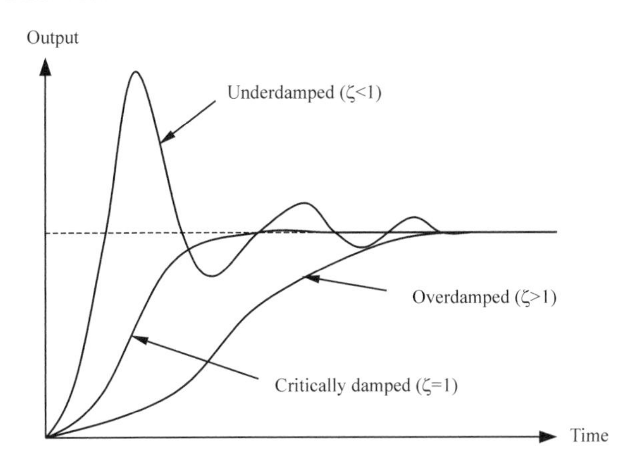

Fig. 5 Step-function response of various second-order control systems.

defined as the percentage by which the peak of the step-response exceeds the steady-state value. Setting time is described as the time required for the response to settle within a tolerance (5% in most cases) of its final value. If the oscillations damp out quickly, the control system is generally acceptable. In some cases, however, the oscillation may persist or their amplitudes may increase with time until a physical bound is reached. In the latter situation, the closed-loop system is considered to be unstable. The conditions under which a feedback control system becomes unstable can be determined theoretically using a number of different methods including the root of the characteristic equation, Bode and Nyquist methods.[1,7,10,11]

In on/off control, input noise may cause the final control element frequently turning on and off. To keep outputs stable, one method to address this problem is to low-pass filter input signals to remove the bulk of the noise, but with disadvantage of delaying the system response to rapidly changing input at the same time.[8] And another method is hysteresis that adds positive feedback to the control system for building up a "guard band" around the set point.

Control System Tuning

Control system tuning is to determine the values of constants in the control equations in the design and operation of a reliable, predictable control system.[8] Several methods are used to tune PID controllers, including model-based correlation response specification, in-circuit emulator, and frequency response.[2,3] The basic procedure for tuning a control system is to set up the system and operate it with variable parameters under a step change or disturbance. Usually, the first step is to check whether the response of the control equation is slower than that of the system being controlled for ensuring stability. After the controller speed is determined, proportional gain, then integral, and finally derivative gains can be set by testing P, PI, and PID controls. In general, PID control fits in most systems. However, it is uncertain that all three terms would be needed in a specific system. If the optimum gain for any of the control terms is less than 10% of the others, it can be omitted.[8] When the gain of the proportional term is high a tolerable error, integral term may not be needed. Optimized control system is a compromise among oscillation, overshoot, convergence, and others. For example, lowering the integral gain would decrease the oscillation at the expense of slower rise to the set point.

Advanced control techniques are needed when a continuous process has features of slow dynamics, time delays, frequencies, and/or multivariable interactions. Those techniques include multivariable control, model predictive control, feedforward control, adaptive control, statistical process control, digital control, and neural network control as described in the literature.[3,6,9]

REFERENCES

1. Bateson, R.N. *Introduction to Control System Technology*, 3rd Ed.; Merrill Publishing: Columbus, OH, 1989.
2. Coughanowr, D.R. *Process Systems Analysis and Control*; McGraw-Hill: New York, NY, 1991.
3. Edgar, T.F. Process Dynamics and Control. In *The Electronics Handbook*; Whitaker, J.C., Ed.; CRC Press: Boca, FL, 1996, 1823–1839.
4. Fraser, C.; Milne, J. Principles of Continuous Control. In *Electro-Mechanical Engineering*; IEEE Press: Piscataway, NJ, 1994, 365–419.
5. Kuo, B.C. *Automatic Control Systems*, 6th Ed.; Prentice-Hall: Englewood Cliffs, NJ, 1991.
6. Joseph, B. Expert Systems—-Artificial Intelligence and Neural Networks. In *Instrument Engineers' Handbook: Process Control*, 3rd Ed.; Liptak, B.G., Ed.; Chilton Book: Radnor, PA, 1995, 44–54.
7. Marlin, T. *Process Control: Designing Processes and Control Systems for Dynamic Performance*; McGraw-Hill: New York, NY, 1995.
8. Mosteller, T. Control Theory for Embedded Controllers—An Introduction to the Basics of Computerized Control. Circuit Cellar **1990**, *17*, 58–66.
9. Downs, J.J. and Boss, J.E. Present Status and Future Needs—A Review from North American Industry. In *Chemical Process Control—CPC IV*, Ray, W.H. and Arkun, Y., Eds.; CACHE/AIChE: New York, NY, 1991, 15–21.
10. Shinskey, F.G. *Process Control Systems*, 4th Ed.; McGraw-Hill: New York, NY, 1996.
11. Smith, C.A.; Corripio, A.B. *Principles and Practice of Automatic Process Control*, 2nd Ed.; John Wiley & Sons: New York, NY, 1997.

Controlled Atmosphere Storage

G. S. V. Raghavan
Y. Gariépy
McGill University, Montreal, Quebec, Canada

C. Vigneault
Agriculture and Agri-Food Canada, St-Jean-sur-Richelieu, Quebec, Canada

INTRODUCTION

Freshly harvested fruits and vegetables are living plant materials that are highly perishable. Their quality and consequently their storage life are reduced by the loss of moisture, decay, and physiological breakdown. It is possible to slow the deterioration process by controlling the environment to which they are exposed. Low temperature, high humidity, and modification of the surrounding gas composition are the most common techniques utilized to keep these commodities in their most usable form for consumers and processing industries.

In conjunction with low temperature and high humidity levels, controlled and modified atmosphere (CA and MA, respectively) imply the addition or removal of gases involved in the metabolism of the stored commodity. The oxygen (O_2) is generally reduced and the carbon dioxide (CO_2) level increased. The difference between CA and MA is in the way the storage atmosphere is achieved and maintained. Controlled atmosphere storage involves the active control of the gas levels, while it is done passively in MA storage.[1]

CONTROLLED ATMOSPHERE STORAGE SYSTEMS

In addition to refrigeration, cost efficient control of the storage atmosphere requires an airtight storage enclosure and equipment designed to regulate the gas levels. A controlled air vent or a pressure compensation bag must also be added to the storage room to prevent excessive air pressure differences that could compromise the storage structure integrity. There are several ways to achieve and maintain the desired atmospheric composition inside a storage room. Depending on their operating principle, CA systems are usually designed to control either O_2 or CO_2. Systems such as the Produce Package System and the Marcellin System have the ability to control both gases simultaneously.

External Gas Generators

External gas generators use either open flame or catalytic burners to control the O_2 level in the storage room. In the open flame system (Fig. 1), air is mixed with a hydrocarbon fuel and burned. The exhaust gas is cooled with a water spray, and blown into the storage room. A CO_2-scrubber is required to absorb the excess CO_2 resulting from the combustion process and the respiration activity of the stored commodity. In the catalytic system, air is also mixed with a hydrocarbon fuel but burned with a catalyst. Heat is removed with a condenser, and the CO_2 absorbed with a scrubber. Both systems can operate with the air intake located outside the storage room, as in the flushing system,[2] or inside the storage room, as in the recirculating system.[3] Flushing systems work best with open flame burners, as they are less expensive and easier to maintain. Recirculation systems are more expensive to install, but have lower operating costs. With catalytic burners, the possibility of contaminating the storage CA with undesirable hydrocarbon volatiles resulting from incomplete fuel combustion is also greatly reduced.

Liquid Nitrogen Atmospheric Generators

Liquid nitrogen (N_2) atmospheric generators are mostly utilized to achieve a rapid cooling of the stored product and to lower the O_2 level in the CA storage room. During the cooling phase, liquid N_2 is pulverized in the cold room by spraying headers located in front of the evaporator blowers. During that phase, a thermostat (Fig. 2) controls the N_2 flow. When the desired level of O_2 is reached, an O_2 sensor regulates the N_2 flow. Sometimes, they are also used to maintain the desired CO_2 level during the entire storage period. A regulated supply of outside air is then used to maintain the required O_2 level.

Encyclopedia of Agricultural, Food, and Biological Engineering
DOI: 10.1081/E-EAFE 120007246
Copyright © 2003 by Marcel Dekker, Inc. All rights reserved.

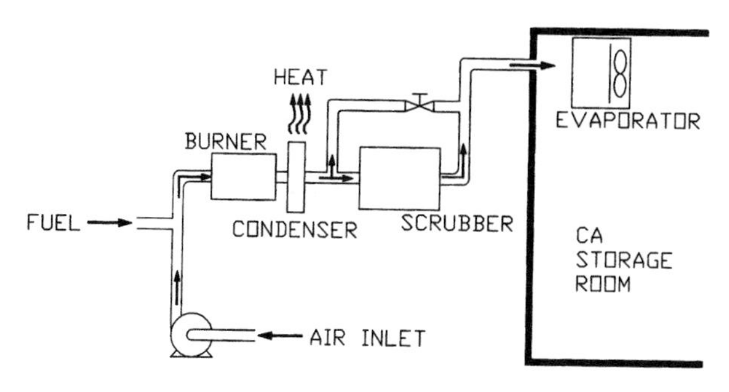

Fig. 1 External gas generator.

Gas Separators

Gas separators use either solid absorbent or membrane modules to create a N_2 enriched gas stream. They are used primarily for the rapid establishment of CA in the storage room.[4] Once the desired O_2 level is reached, they can be used to control CO_2 levels during storage. Two types of gas separators are available commercially: the pressure swing absorption (PSA) system and the hollow-fiber membrane (HFM) system.[5] Bartsssch and Blanpied[6] have looked at the energy consumption of these systems.

The PSA make use of the physical adsorption properties of a specially treated carbon molecular sieve. The enriched N_2 stream is generated through a combination of filtration and pressure cycle during which O_2 and other contaminants are selectively adsorbed by the molecular sieve. Then, the enriched N_2 stream is directed towards the storage enclosure. The depressurization phase releases the adsorbed O_2 and contaminants to the outside air.

The HFM separates N_2 from air by selective permeation. When air is supplied to the membrane cartridge, O_2, CO_2, and water vapor permeate rapidly through the membrane fiber. N_2, which does not readily permeate the membrane, goes directly into the distribution system and is used to flush the storage room.

Hypobaric Storage

A hypobaric storage consists of a reinforced, airtight, and refrigerated room in which the air is continuously removed by a vacuum pump (Fig. 3). Depending on the temperature and the stored commodity, the system will operate at a constant pressure, usually in the range of 10 mm Hg–80 mm Hg. When the desired subatmospheric pressure is obtained, fresh humid air is admitted into the cold room at a controlled rate. The main advantages of this system are: 1) Ease of manipulation of the O_2 and the relative humidity levels; 2) Removal of by-products of metabolism such as CO_2, C_2H_4, and other volatiles; 3) Its use as a vacuum cooler; and 4) Ability to store normally noncompatible products. However, this system has major technical difficulties such as: 1) The cost of the structure required to sustain the desired vacuum; and 2) The impossibility to operate at functional levels of CO_2.

CO_2-Scrubbers

In a storage room equipped with a CO_2-scrubber, the desired CO_2 levels are maintained by controlling the airflow to the scrubber. There are four main reagents that are commercially used for CO_2 absorption: hydrated lime, water, activated charcoal, and molecular sieves. With these systems, the O_2 levels are usually maintained by regulating the supply of outside air into the storage room.

Lime CO_2 -Scrubbers

One of the simplest methods used to regulate the CO_2 level in a CA room is with hydrated lime [$Ca(OH)_2$]. As shown in Fig. 4, the lime scrubber is an airtight enclosure located adjacent to the CA room. It is usually designed to hold half the total amount of lime required for the entire storage period. This is, for example, 12 kg of lime per ton of stored apple. This amount needs to be replaced with fresh lime,

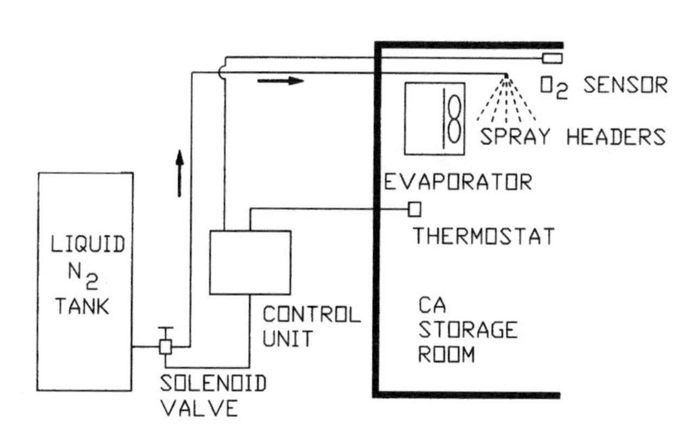

Fig. 2 Liquid nitrogen atmospheric generator.

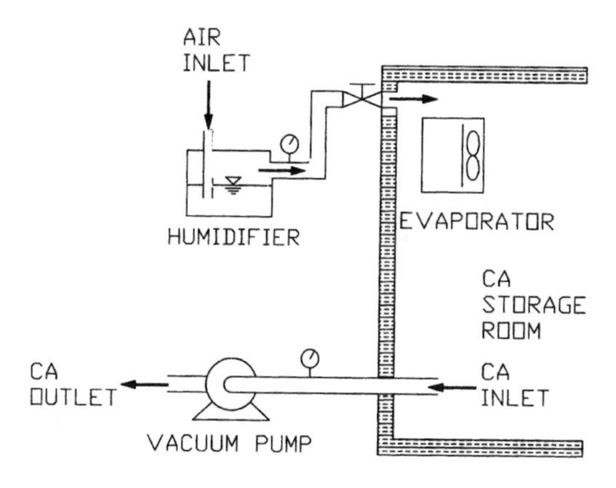

Fig. 3 Hypobaric storage

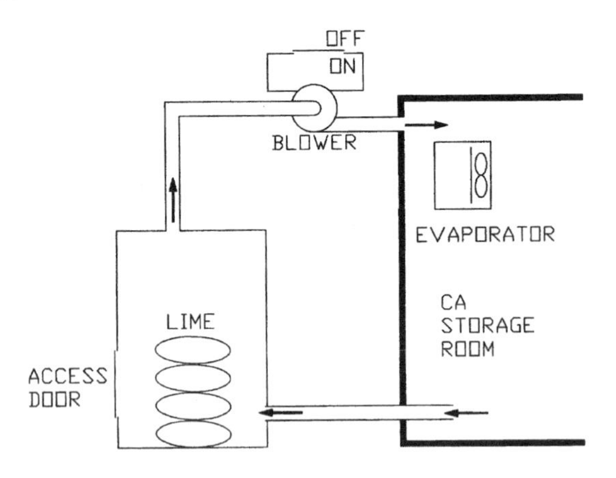

Fig. 4 Lime CO_2-scrubber.

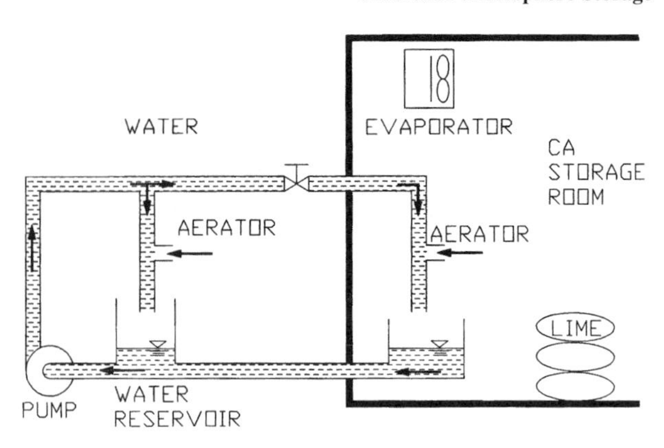

Fig. 6 Water CO_2-scrubber with two aerators.

somewhere halfway during the storage period. It can either work by natural convection or, if better control is required, the air is fed to the scrubber with blowers and dampers. Efforts have been made to decrease costs and more fully use the hydrated lime in a system that accepts bulk lime directly from trucks.

Water CO_2-Scrubbers

There are two types of CO_2 scrubbers that use water as the active absorber. In the brine CO_2-scrubber (Fig. 5), brine-water is pumped over the evaporator coils, picks up CO_2, and is gravity fed to the water reservoir located outside the CA room. The water is then pumped from the reservoir to the aerator where CO_2 is released to the outside air. Corrosion problems resulting from the use of salt water can be minimized if dry evaporator coils are used. The second system (Fig. 6) makes use of two aerators: one located outside and the other located inside the storage room. These two types of scrubbers are very efficient in controlling the CO_2 levels and in helping to maintain high relative humidity levels inside the CA room. Pflug[7] recommended a circulation rate of $100\,\mathrm{L\,hr^{-1}}$ per ton of apple stored at 1°C in a CA of 5% CO_2 and 2% O_2. It is sometimes recommended to place additional hydrated lime bags inside the CA room to absorb the excess CO_2 produced during the establishment of CA.

Activated Charcoal and Molecular Sieve CO_2-Scrubbers

These units (Fig. 7) mainly consist of a container filled with the absorbent, blowers, and solenoid valves controlled by a timer or an electronic controller. Their operations require two consecutive steps. At first the CA is circulated into the scrubber where CO_2 is absorbed and returned with a lower CO_2 level into the CA room. When this step is completed, circulating outside air through the scrubber regenerates the absorbent. Systems that utilize molecular sieve require a heater to increase temperature of the absorbent during the reactivation process. Both systems have low operating costs, and

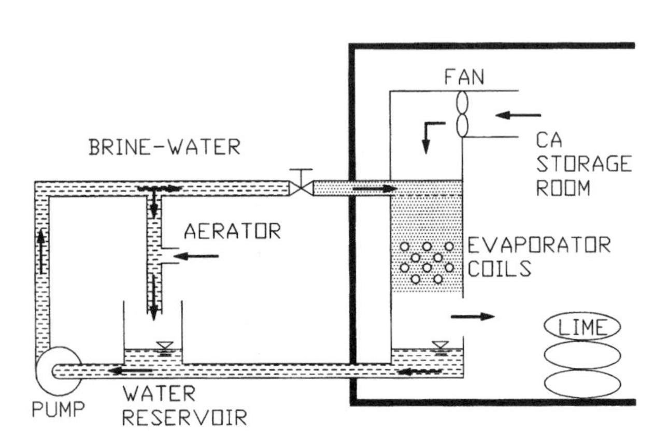

Fig. 5 Brine-water CO_2-scrubber.

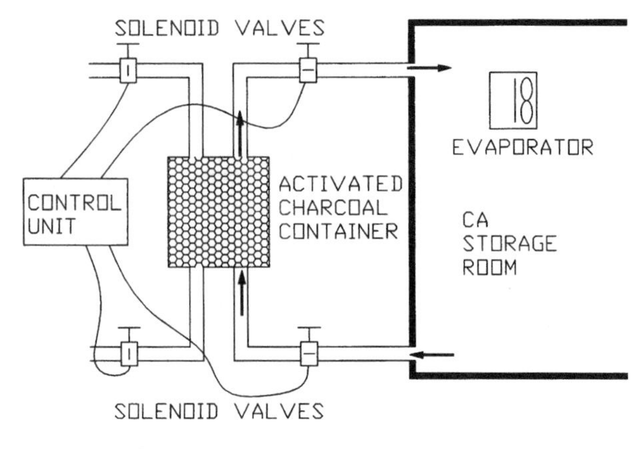

Fig. 7 Activated charcoal CO_2-scrubber.

replacement of the absorbent is usually done once every five years.

Membrane Systems

Marcellin and Leteinturier[8] have developed two MA/CA systems using a silicone based semipermeable membrane. With these systems, MA/CA is achieved and maintained by the metabolic activity of the stored produce and by gas exchanges across the silicone membrane. Because of the selectivity ratio O_2/CO_2, this system is able to regulate both O_2 and CO_2 levels in the enclosure. The first system consists of a pallet box wrapped in a heavy gauge polyethylene bag on which a silicone membrane window is installed. The advantages of this system are ease of manipulation and possibility to market the produce progressively without affecting the MA within the remaining pallet boxes. To work properly, the plastic covered pallets need to be well spaced which reduces storage capacity. In addition, one must take the time to analyze the atmosphere in individual pallet loads. Procedures required to determine the surface area of the silicone membrane are well documented by Raghavan et al.[9]

The second system (Fig. 8) creates the CA condition in storage rooms by using a series of rectangular bags of silicone rubber connected in parallel. The number of bags depends on the size of the cold room and the types of produce stored. These units can either be installed inside or outside the storage room. When the unit is located outside the storage room (exposed to ambient air), the CA circulates inside the bags. It is the analysis of the CA composition that indicates whether or not more bags should be put into use. As a general rule, a storage atmosphere of 3% O_2 and 3% CO_2 requires the installation of 50 m^2 of silicone membrane per 100 ton of fruits at a bulk density of 200 kg m^{-3}–250 kg m^{-3}.

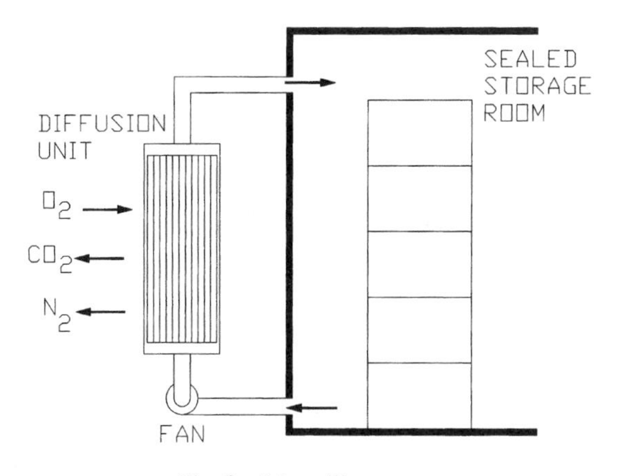

Fig. 8 Marcellin system.

AUTOMATED CONTROL SYSTEMS

The difficulty in controlling the CA storage environment is due to the dynamic nature of the system. Physiological response of the stored produce, CO_2 and O_2 regulating equipment, building airtightness and integrity, cyclical operation of the refrigeration system, and changes in weather are all major elements affecting the gas balance. The automated control system must therefore compensate for these elements in achieving and stabilizing the O_2 and CO_2 levels to the desired concentrations. At regular time intervals, the O_2 and CO_2 levels are measured and compared to their respective desired levels. The control system attempts to reduce the differences between the preset and the actual levels and decides to or not to operate the gas regulation equipment.[1]

Common devices utilized to control the operation of the gas regulating equipment include ON/OFF switches, and proportional (P), proportional-integral (PI), or proportional-integral-derivative (PID) controllers. Personal computer (PC) based automated control systems offer great flexibility since in addition to controlling the gas levels, they can be programmed to control other parameters/operation, to manage sensitive interaction, and to log valuable information for later retrieval and analysis. Dedicated systems for monitoring and control of the environment in CA storage facilities are now available commercially. They usually consist of a microprocessor-based controller, gas analyzers, temperature sensors, and devices to control the operation of the gas-regulating equipment.

ACKNOWLEDGMENTS

The assistance of Mr. Timothy J. Rennie in the preparation of this paper is gratefully acknowledged. Financial support from the Natural Sciences and Engineering Council of Canada is gratefully appreciated.

REFERENCES

1. Raghavan, G.S.V.; Vigneault, C.; Gariépy, Y.; Markarian, N.R.; Alvo, P. Refrigerated and Controlled/Modified Atmosphere Storage. In *Processing fruits: Science and Technology, Biology, Principles and Applications*, 2nd Ed.; Bennett, H., Ed.; CRC Press: Boca Raton, 2003; Vol. 1.
2. Pflung, L.L.; Guerevitz, D. Externally Generated Atmosphere for CA Storage. *Storage of Fruits and Vegetables in CA Cold Storage*; Int. Inst. Refrig., 1966; 674.
3. Thomas, O. *The Generator Concept*, Hort. Report. 9; Michigan State Univ.: East Lansing, MI, USA, 1969; 32–37.

4. Waelti, H.; Cavalieri, R.P. Matching Nitrogen Equipment to Your Needs. Tree Fruit Postharvest J. **1990**, *1* (2), 3–13.
5. Anonymous. On-site Nitrogen Generation. Airdyne Inc. Houston, TX, USA; http://www.airdyne.net (accessed May 2002).
6. Bartsch, J.A.; Blanpied, G.D. *Air Separator Technology for Controlled Atmosphere Storage*; Cornell Cooperative Extension. Cornell University: Ithaca, NY, USA, 1988; 13.
7. Pflug, I.J. Oxygen Reduction in CA Storages: A Comparison of Water Versus Caustic Soda Absorbers. Mich. Agric. Expt. Stat. Q. Bull. **1960**, *43* (2), 455–466.
8. Marcellin, P.; Leteinturier, J. *Premières Applications des Membranes de Caoutchouc de Silicone à l'entreposage des Pommes en Atmosphère Controlée*; Int. Inst. Refrig.: Paris, France, 1967; 1–9.
9. Raghavan, G.S.V.; Gariépy, Y.; Thériault, R.; Phan, C.T.; Lanson, A. System for Controlled Atmosphere Long-term Cabbage Storage. Int. J. Refrig. **1984**, *7* (1), 66–71.

Convection Heat Transfer in Foods

Brian M. McKenna
University College Dublin, Dublin, Ireland

C

INTRODUCTION

In theory, convection heating occurs within a fluid by the redistribution of thermal energy between the sections of the fluid by the bulk movement of molecules of the fluid. However, it is most commonly encountered when a fluid and solid of different temperatures come in contact with each other and exchange heat. Consequently, it becomes difficult to consider convection in isolation from other forms of heat exchange since in solid–fluid heat exchange, heat must move by conduction to or from the solid surface. In addition, many examples of convection heating in the food industry also incorporate other forms of heat exchange (e.g., radiant heat transfer in baking ovens). This short summary of convection heating will therefore consider both the fundamental theory of convection heating and also combination heat transfer mechanisms in which convection plays an important role.

CONVECTION

The bulk movement of the molecules of the fluid is induced in two different ways. First, in *free convection* the movement of the fluid molecules is induced by density differences caused by temperature differences within the fluid. In a limited number of cases, this is supplemented by movement in the fluid induced by a boiling or condensation process on the solid surface involved in the heat transfer. On the other hand, *forced convection* uses an artificial movement of the fluid induced by mechanical pumping.

A measure of the effectiveness of convection heat transfer is the convective heat transfer coefficient (sometimes called the surface or film heat transfer coefficient). This is included in the simple Newton law of convective heat transfer:

$$q = hA(T_s - T_\infty)$$

where q is the overall rate of heat transfer between the fluid and the solid surface, h is the convective heat transfer coefficient, A is the area of the solid surface, T_s is the surface temperature, and T_∞ is the temperature of the bulk of the fluid far from the surface. The units of h are normally $\mathrm{W\,m^{-2}\,K^{-1}}$.

The two variables in the above equation of greatest concern to the processor are h and T_∞. The first can be partially controlled by selection of the heating fluid and the operating conditions; the second is normally totally under the control of the processor. Obviously, the value of h influences the magnitude of the heat transfer. A selection of typical values are given in Table 1.

Deposits formed on the solid surfaces from liquid foods can also adversely affect values of h and reduce the efficiency of either heating or cooling operations (see the article titled *Cleaning Chemistry and Physics*).

Determination of the h value is quite a complex measurement process and is complicated in the case of foods that are rarely heated or cooled by convection in the absence of either evaporation or condensation of water on the surface. Several measurement systems rely on sensors that measure the flow of heat to the food surface to effectively measure the q term in the above equation. Then, by measuring the area involved and the temperatures involved, h can be calculated. However, this can give erroneous results since the heat used in causing any evaporation is ignored and can be quite considerable. A more common approach is to estimate the value of h by using a wide range of dimensionless equations that are published in the heat transfer literature. These normally have the form (forced convection):

$$Nu = \text{constant}(Re)^a(Pr)^b$$

where Nu is the Nusselt number and is equal to hD/k (h is the convective heat transfer coefficient, D is a characteristic dimension of the solid surface, e.g., the diameter of a cylindrical pipe or the length of a flat surface, and k is the thermal conductivity of the solid surface). On the right hand side of this equation, we have two further dimensionless numbers, Re—the Reynolds number and Pr—the Prandtl number. The Reynolds number is a good indicator of the flow characteristics of the fluid over the surface and the Prandtl number takes account of some of

Encyclopedia of Agricultural, Food, and Biological Engineering
DOI: 10.1081/E-EAFE 120006954
Copyright © 2003 by Marcel Dekker, Inc. All rights reserved.

Table 1 Typical values of h in W m^{-2} K^{-1}

Fluid	h
Gases (including air)	
Free convection	6–25
Forced convection	10–150
Liquids (including water)	
Free convection	20–500[a]
Forced convection	500–10,000[a]
Boiling water	3,000–50,000
Condensing steam	1,000–100,000

[a] All values are dependent on viscosity of the liquid in question.

the factors affecting heat transfer to the fluid.

$$Re = (\text{fluid density } \rho)(\text{fluid velocity } v) \times (\text{characteristic dimension } D)/(\text{fluid viscosity } \mu)$$

$$Pr = (\text{fluid specific heat } c_p) \times (\text{fluid viscosity } \mu)/(\text{thermal conductivity } k)$$

Typical values of the exponents are $a = 0.8$ and $b = 0.33$.

It is obvious from an examination of this equation that the fluid velocity v has a significant influence of the magnitude of h, the convective heat transfer coefficient. An increase in fluid velocity, will increase the magnitude of the Reynolds number, thereby increasing the Nusselt number and hence the value of h. The reason for this is not simply a mathematical expediency. As fluid velocity increases, the fluid may become more turbulent. Increased turbulence will increase the movement of eddies of the fluid to or from the solid surface. This effectively increases the rate of heat transfer since it is by fluid movement that all convective heat exchange takes place.

Free convection is not as commonly used as forced convection in food processing. Dimensionless correlations are also available for free convection but with the Reynolds number replaced by the Grasshof number Gr. This incorporates elements relating to buoyancy and the free movement of the fluid.

It should be stressed that the above correlations apply only to Newtonian liquids. Non-Newtonian liquids, in particular power-law fluids, are difficult to incorporate into dimensionless numbers. For such fluids, the reader is directed to the reading list at the end of this section.

While water is one of the most popular media for convective heat transfer, examination of the ranges of heat transfer coefficients in Table 1 highlights the fact that, in the form of condensing steam, it is a significantly more effective heat transfer medium for foods. It is widely used in heat exchangers for pasteurization and sterilization of liquid foods. In such cases, it is really a combination heat transfer process with convection from the condensing steam to either the metal surface of a heat exchanger or the wall of a metal can, conduction through the metal, and convection again within the food as heat passes from the metal wall to the interior of the liquid food. However, high pressure superheated steam is often used so that the variable T_∞ in the Newton equation can be increased. This will increase the rate of heat transfer but not as much as might be expected since superheated steam normally has a lower h value than condensing steam. Of course, a condensation effect will be obtained from even superheated steam when in contact with the cold solid surfaces.

Steam is also used in some industrial catering ovens where cooking times can be significantly improved over the use of alternative heating fluids such as air. In all such cases, care must be taken to ensure that air is eliminated from the heating medium as it will both reduce the heat transfer coefficient and lower the temperature at the selected operating pressure.

However, some forms of food processing (e.g., drying of both liquid and solid foods) require hot air to be the heat transfer medium. Most dryers use hot air, almost exclusively under a forced convection regime where the hot air is blown over the surface of the drying food. The processor is faced with the conflicts of wanting to maximize the airflow velocity so as to maximize the heat transfer rate and, consequently, the drying rate. This, however, requires excessive fan sizes and may require the costly exercise of heating overly large volumes of hot air. Costs may also constrain the processor from increasing heat transfer rates through increasing the bulk air temperature T_∞. High values of T_∞ may also cause heat damage to the food in the later stages of drying.

Another classical use of convective heat transfer is in baking ovens. Here, hot air is the medium of choice. The air is normally heated by either gas or electrical sources and the food is heated by a combination of convective heat transfer from the hot air and radiant heat transfer from the original heat source. While this is beneficial to the process, it does lead to complications in process calculations.

All the examples quoted thus far have referred to heating of foods by convection from a hot fluid. However, cooling of foods by cold fluids is equally important in the food industry. The major process fluids involved here are cold air and cold or refrigerated water. Most process operations that utilize heating require a corresponding cooling process either before or after packaging and before storage and distribution of the food. Good examples of this are sterilized products, either processed in plate heat exchangers or in metal cans. Both processes require heating to sterilization temperatures in excess of

120°C. Cold and chilled water are then used to either cool the sterile cans to room temperature or to cool a flowing sterile food liquid to around 4°C before aseptic packaging. For baked goods, cold air would be the normal heat removal medium. Here again the conflict arises for the processor between maximizing the airflow rate and yet not incurring the high costs of cooling excessive quantities of air.

Of course, convective cooling may be required in process operations that do not involve any heating. Most meat processing operations require cooling of the carcasses to 4°C or lower as rapidly as possible so that microbial growth and spoilage is minimized. In this case, the processor may be constrained from maximizing the convective heat transfer removal rates as "cold-shortening" or toughening of the product may be induced. A second constraint on the processor is the achieving of high convective heat transfer by the use of very low cooling air temperatures T_{∞}. Use of air temperatures of less than −2°C can give rise to undesirable freezing of the outer surfaces of the product. These are particular problems with beef processing.

The opposite problem is encountered in food freezing—a process that normally requires very rapid heat removal by convection. In this case, slow removal of heat and slow freezing can lead to product damage caused by both the development of large ice crystals within the cells of the food structure and internal migration of water and other food constituents before the food structural matrix assumes the rigidity of a fully frozen product. This suggests that the processor should use air velocities as high as are economically possible to carry out freezing processes together with air temperatures T_{∞} that are as low as can be economically justified. However, there is a risk inherent in such a use. High air velocities and their high heat transfer coefficients, while ensuring high convective heat removal rates and rapid freezing, can lead to surface dehydration or "freezer burn" even at the low air temperatures normally used for freezing.

C

SUGGESTED FURTHER READING

General Convection

Becker, M. *Heat Transfer: A Modern Approach*; Plenum: New York, London, 1986.

Hallström, B.; Skjöldebrand, C. *Heat Transfer and Food Products*; Elsevier Applied Science: London, 1988.

Lewis, M.; Heppell, N. *Continuous Thermal Processing of Foods*; Aspen Publishers: Gaithersburg, 2000; 19–25.

Lindon, C.T. *Heat Transfer*; Capstone Publishing Co.: Tulsa, 1999.

Macrae, R.; Robinson, R.K.; Sadler, M.J. *Encyclopaedia of Food Science, Food Technology and Nutrition*; Academic Press: London & San Diego, 1993; 2300–2305.

Singh, R.P.; Heldman, D. *Introduction to Food Engineering*; Academic Press: London & San Diego, 2001; 225–227.

Heat Flux Sensors

Danielsson, U. Convective Heat Transfer Measured Directly with a Heat Flux Sensor. J. Appl. Physiol. **1985**, *68* (3), 1275–1281.

Langley, L.W.; Barnes, A.; Matijasevic, G.; Gandhi, P. High-Sensitivity, Surface-Attached Heat Flux Sensors. Microelectronics J. **1999**, *30* (11), 1163–1168.

Convection to Non-Newtonian fluids

Metzner, A.B.; Gluck, D.F. Heat Transfer to Non-Newtonian Fluids Under Laminar-Flow Conditions. Chem. Eng. Sci. **1960**, *12* (3), 185–190.

Naccache, M.F.; Souza Mendes, P.R. Heat Transfer to Non-Newtonian Fluids in Laminar Flow Through Rectangular Ducts. Int. J. Heat Fluid Flow **1996**, *17* (6), 613–620.

Ruckenstein, E. Turbulent Heat Transfer to Non-Newtonian fluids. Chem. Eng. Sci. **1972**, *27* (5), 947–957.

Convective Heat Transfer Coefficients

Adriana E. Delgado
Da-Wen Sun
University College Dublin, Dublin, Ireland

INTRODUCTION

The transfer of heat by convection implies the transfer of heat by bulk transport and mixing of macroscopic elements of warmer portions with cooler portions of a gas or a liquid.[1] The process also often involves energy exchange between a solid surface and a fluid. However, there are many practical situations (e.g., in chilling, freezing, thawing, cooking, and drying) in which heat transfer to or from the product surface (e.g., unpackaged foods) is combined with a certain degree of mass transfer, which in turn modifies the heat transfer.

The surface heat transfer coefficient h, defined by Newton's law of cooling, is an important parameter to describe convection thermal processes. Examples of heat transfer by convection are freezing of hamburgers and pizzas, cooling of harvested fruits and vegetables in cold storage, cooking of foods in a vessel being stirred, etc. Certainly, in these processes, heat conduction takes place within the solid foods. There is a great availability and variability of the h values reported in the literature, which can be either measured or calculated through correlations of dimensionless numbers.

MEASUREMENT METHODS

A study of the literature has indicated that only a small number of experimental values of convective transfer coefficients for food products have been obtained.[2] One of the main problems found when comparing the literature values is that it is not clearly defined if a local or average, convective or effective heat transfer coefficient was reported. The "effective" value generally includes the heat transfer by conduction, radiation, or by phase change at the surface as depicted in Fig. 1, which illustrates the situation when a chilled food product is put in a convection oven.

Rahman[3] presented detailed information on the techniques available for measuring h. The techniques can be divided into steady state, quasi-steady, transient, and surface heat flux methods. Among them, the transient technique is the most commonly used one. Heat flux sensors can be used to find an approximate value of the effective heat transfer coefficient since the measuring device can change the surface properties.[4] Despite the disadvantages of the steady-state methods, Harris, Lovatt, and Willix[5] developed a sensor based on a steady-state technique for measuring local convective heat transfer coefficients on carcass shaped objects, and the measured values were within 7% of the true value.

The psychometric method used by Kondjoyan, Daudin, and Bimbenet[6] is very helpful because heat and mass transfer coefficients can be simultaneously determined using the Lewis analogy.[7] This method has been successfully adapted to bodies of complex shapes.

The determination of fluid to particle convective heat transfer coefficients (h_{fp}) is very important in aseptic processes. Recently, Sablani and Ramaswamy[8] developed a method to measure h_{fp} that provides more realistic h_{fp} values. They used a flexible fine-wire thermocouple to measure h_{fp} in cans so that sufficient particle motion during agitation processing could be achieved.

When the heat transfer medium is air, three kinds of variables should be controlled and measured to determine experimentally h values. These variables are the thermal properties of the air, the characteristics of the object, and the characteristics of the airflow (e.g., velocity, turbulence intensity).[6] It is worth noticing that the turbulence intensity (Tu), which characterizes the velocity variation around its means over a short period of time, is a parameter that was rarely considered in food engineering studies with the exception of a few works.[6,9,10] Tu can vary from 17% to 60% in a chilling room and a dryer.[6] Kondjoyan and Daudin[9] measured the effect of free stream turbulence intensity ranging from 1.5% to 40% and of air velocity ranging from 0.5 m sec^{-1} to 5.0 m sec^{-1} on heat and mass transfer coefficients at the surface of circular and elliptical cylinders. The results showed that the effect of Tu was as important as the influence of velocity and was greater than that of body shape characteristics. Lind[4] observed that turbulence could influence the results during thawing in humid air. In general, h values obtained when Tu was taken into account were higher than those evaluated from chemical engineering literature. Thus, the turbulence intensity is a parameter that should be known.

Encyclopedia of Agricultural, Food, and Biological Engineering
DOI: 10.1081/E-EAFE 120007248
Copyright © 2003 by Marcel Dekker, Inc. All rights reserved.

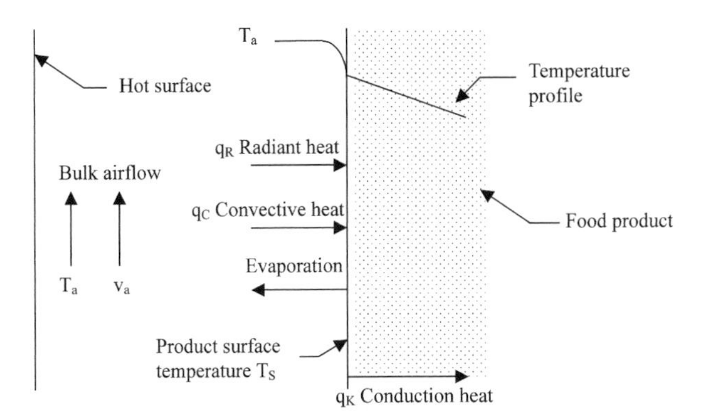

Fig. 1 Heat and mass transfer in a solid food from the surface.

PREDICTIVE MODELS

Since only under simple flow situations the convective coefficients can be obtained by solving the boundary layer equations, the more practical approach involves calculating h by correlating measured convection heat and mass transfer results in terms of appropriate dimensionless groups.[7] The correlations often take the following format[10]:

$$\overline{Nu_{\mathrm{D}}} = (C_1 + C_2 \mathrm{Tu}^a Re_{\mathrm{D}}^{\mathrm{b}}) C Re_{\mathrm{D}}^m Pr^n$$

where the dimensionless number are the average Nusselt number (Nu_{D}), the Reynolds number (Re_{D}), the Prandtl number (Pr), and Tu is the turbulence intensity. The constants C, C_1, C_2, a, b, m, and n are determined by regressing from experimental data and vary with the nature of geometry, surface conditions, and type of flow.

A distinction must be made between forced-convection heat transfer, where a fluid is forced to flow passing a solid surface by mechanical means; and natural or free convection, where warmer or cooler fluid next to the solid surface causes a circulation because of a density difference resulting from the temperature differences in the fluid.[1] In free convection, the Grashof number (Gr) plays the same role as the Reynolds number in forced convection. However, there are a number of situations where heat is transferred by the combination of free and forced convection. The combined free and force convection regime is generally the one for which $(Gr/Re^2) \approx 1$.[7] Compared to either extreme forced or natural convection, mixed convection processes are not well understood.[3] Correlations for h under mixed and forced convection regimes in aseptic processing can be found in Ref. 11.

An alternative approach for obtaining h values was developed by Ansari and Khan for air blast cooling.[12] The authors first developed a correlation, which predicted the effective heat transfer coefficient by including the effects of moisture evaporation. Then they regressed the effective h with the produce temperature, and used the linear regression obtained to solve the transient heat conduction equation. The proposed method is simple to study transient thermal analysis with variable h and when simultaneous heat and moisture transfers occur.

A new method for calculating h_{fp} was proposed by Hulbert, Litchfield, and Schmidt.[13] The method uses a finite element program (ANSYS 5.0) in combination with two-dimensional magnetic resonance imaging (2D MRI) temperature mapping. The application of the method could be useful to study heat transfer in irregular shapes or where the fluid is not uniform around the sample.

As accurate experimental data are difficult to obtain, it is common to use numerical methods as a complimentary tool. The recent advances in computer-aided simulation have allowed the use of computational fluid dynamics (CFD) to simulate food process.[10] However, many authors have reported that CFD could cause maximum error of up to 25% in such applications.[10] The main difficulty seems to be related to properly model the boundary layers near the food surface.

Overall, a compilation of convection correlations for different processes can be found in published literature.[1,3,7,14,15]

CONCLUSIONS

Mathematical modeling, design, control, and optimization of different thermal processes require data on heat transfer coefficients. Therefore, the development of methods for measuring or predicting accurate heat transfer coefficients is important. As many situations involve coupled transport processes, information on whether a true convective or effective coefficient and on the conditions of applicability should be clearly provided when reporting these values. Further studies on heat transfer at low Reynolds number, the influence of turbulence intensity, and the application of CFD would be very useful.

REFERENCES

1. Geankoplis, C.J. Principles of Steady-State Heat Transfer. *Transport Processes and Unit Operation*, 3rd Ed.; Prentice-Hall International, Inc.: New Jersey, 1993; 214–329.
2. Kondjoyan, A.; Daudin, J.D. Heat and Mass Transfer Coefficients at the Surface of a Pork Hindquarter. J. Food Eng. **1997**, *32*, 225–240.
3. Rahman, S. Surface Heat Transfer Coefficient in Food Processing. *Food Properties Handbook*; CRC Press, Inc.: Florida, 1995; 393–456.
4. Lind, I. Surface Heat Transfer in Thawing by Forced Air Convection. J. Food Eng. **1988**, *7*, 19–39.

5. Harris, M.B.; Lovatt, S.J.; Willix, J. Development of a Sensor for Measuring Local Heat Transfer Coefficients on Carcass-Shaped Objects. In 20th International Congress of Refrigeration, Sidney, Sept 19–24, 1999.
6. Kondjoyan, J.D.; Daudin, J.D.; Bimbenet, J.J. Heat and Mass Transfer Coefficients at the Surface of Elliptical Cylinders Placed in a Turbulent Air Flow. J. Food Eng. **1993**, *20*, 339–367.
7. Incropera, F.P.; DeWitt, D.P. *Fundamentals of Heat and Mass Transfer*, 4th Ed.; John Wiley & Sons: New York, 1996.
8. Sablani, S.S.; Ramaswamy, H.S. Fluid/Particle Heat Transfer Coefficient in Cans During End-Over-End Processing. Lebensm. Wiss. Technol. **1995**, *28* (1), 56–61.
9. Kondjoyan, A.; Daudin, J.D. Effects of Free Stream Turbulence Intensity on Heat and Mass Transfers at the Surface of a Circular Cylinder and an Elliptical Cylinder, Axis Ratio 4. Int. J. Heat Mass Transfer **1995**, *38* (10), 1735–1749.
10. Hu, Z.; Sun, D.-W. Predicting Local Surface Heat Transfer Coefficients During Air-Blast Chilling by Different Turbulent k-ε Models. Int. J. Refrig. **2001**, *24* (7), 702–717.
11. Awuah, G.B.; Ramaswamy, H.S. Dimensionless Correlations for Mixed and Forced Convection Heat Transfer to Spherical and Finite Cylindrical Particles in an Aseptic Processing Holding Tube Simulator. J. Food Process. Eng. **1996**, *19* (3), 269–287.
12. Ansari, F.A.; Khan, S.Y. Application Concept of Variable Effective Surface Film Conductance for Simultaneous Heat and Mass Transfer Analysis During Air Blast Cooling of Food. Energy Convers. Manag. **1999**, *40*, 567–574.
13. Hulbert, G.J.; Litchfield, J.B.; Schmidt, S.J. Determination of Convective Heat Transfer Coefficients Using 2D MRI Temperature Mapping and Finite Element Modeling. J. Food Eng. **1997**, *34*, 193–201.
14. Rao, M.A.; Anantheswaran, R.C. Convective Heat Transfer to Fluid Foods in Cans. Adv. Food Res. **1988**, *32*, 39–84.
15. ASHRAE. *ASHRAE Handbook: Fundamentals*, SI Ed.; American Society of Heating, Refrigeration and Air-Conditioning Engineers, Inc.: Atlanta, GA, 1993.

Cooling Tunnels

Timothy A. Taylor
Heather L. Kramer
Utah State University, Logan, Utah, U.S.A.

C

INTRODUCTION

Cooling tunnels are utilized in many industries to cool product from unit operations, such as ovens, fryers, and dryers, and enrobes to packaging temperatures. The common objective of cooling tunnels is to cool the product without inducing detrimental changes to product quality. The cooling rate may need to be controlled to ensure proper hardening of coatings such as chocolate without inducing bloom. Too cold of cooling air on baked goods may cause rapid staling of the product. In baked goods, the humidity of the cooling air will affect the surface properties and final moisture content of the product.[1] Cooling tunnels can use ambient temperature air or refrigerated air depending on the cooling requirements and the ambient conditions available.

COOLING TUNNEL DESIGN

Cooling tunnels utilize a combination of convective and conductive heat transfer to remove heat from the product. The heat is transferred within the product by conduction and then removed from the surface by convective transfer to the cooling air.

The critical parameters involved in the design process include the total cooling load required, estimated convective heat transfer coefficient, thermal conductivity of the product, and the desired operational conditions of the tunnel.

Total Cooling Load

The total cooling load is the amount of heat that needs to be removed from the product per given time. The cooling load can be calculated from:

$$\dot{q} = \dot{m}c_{\text{p}}\Delta T \tag{1}$$

where $\dot{q}$ is the heat removed per time, $\dot{m}$ is the mass flow rate of material to be cooled per time, c_{p} is the heat capacity of the product, and ΔT is the desired temperature change from entering to exiting the cooling tunnel.

The heat capacity ($\text{kJ kg}^{-1}\text{K}^{-1}$) can be estimated using the composition of the product:

$$c_{\text{p}} = 1.424X_{\text{c}} + 1.549X_{\text{p}} + 1.675X_{\text{f}} + 0.837X_{\text{a}} + 4.187X_{\text{m}} \tag{2}$$

where X is the mass fraction of carbohydrate (c), protein (p), fat (f), ash (a), and moisture (m).[2]

Cooling Time

The time required to cool the product to the desired temperature is an important parameter in the design of a cooling tunnel. It is best to run experiments under conditions that duplicate production conditions as closely as possible to determine the cooling rate of the product. It is important to remember that the interactions between products will affect the airflow characteristics when designing these experiments. Products should be placed in standard density and oriented as they would be in the tunnel. If experimental data is not available, mathematical estimates of cooling time can be developed from empirical equations for the surface convective properties and the internal conduction properties. These transient heat transfer equations will predict the temperature of the product with both time and location. These equations will provide an estimate of the cooling time but can have a significant margin of error that must be considered in the final design phase of the project.

The contribution of internal resistance to heat flow (conduction) and the external resistance (convection) are both significant in tunnel cooling of food products. As both internal and external resistance to heat transfer must be considered, the Biot number will be greater than 0.1:

$$N_{\text{Bi}} = \frac{hD}{k} \tag{3}$$

where h is the convective heat transfer coefficient ($\text{W m}^{-2}\text{C}^{-1}$), D is the characteristic dimension (m), and k is the thermal conductivity of the product ($\text{W m}^{-1}\text{C}^{-1}$). Heisler charts can be utilized to estimate the cooling time (see entry titled *Transient Heat Transfer Charts*), however, for their use, both the convective heat transfer

Encyclopedia of Agricultural, Food, and Biological Engineering
DOI: 10.1081/E-EAFE 120007015
Copyright © 2003 by Marcel Dekker, Inc. All rights reserved.

Table 1 Equations for calculation of convective heat transfer coefficient

Nusselt number $N_{Nu} = CN_{Re}^{m}N_{Pr}^{1/3} = \frac{hL}{k}$	N_{Nu}	Nusselt number (at T_f)
	C	Constant based on product geometry
Reynolds number $N_{Re,L} = \frac{Lv\rho}{\mu}$	N_{Re}	Reynolds number (utilizing properties of air at T_f)
	m	Constant based on product geometry
Prandtl number $N_{Pr} = \frac{c_p\mu}{k}$	N_{Pr}	Prandlt number (at T_f)
	h	Convective heat transfer coefficient ($W\,m^{-2}C^{-1}$)
Film temperature $T_f = \frac{T_w+T_b}{2}$	L	Product dimension exposed to airflow (m)
	k	Thermal conductivity of air ($W\,m^{-1}C^{-1}$)
	v	Velocity of air ($m\,sec^{-1}$)
	ρ	Density of air ($kg\,m^{-3}$)
	μ	Viscosity of air (Pa sec)
	c_p	Heat capacity of air ($kJ\,kg^{-1}C^{-1}$)
	T_f	Film temperature (°C)
	T_w	Wall temperature (°C)
	T_b	Bulk fluid temperature (°C)

(From Ref. [3].)

coefficient and the thermal conductivity of the product must be estimated.

Convective Heat Transfer Coefficient Estimation

The convective heat transfer coefficient represents the rate that heat can be removed from the surface of the product to the cooling air. An estimate of the average heat transfer coefficient for an immersed body can be estimated from the Nusselt number evaluated at the film temperature (T_f) utilizing the equations given in Table 1.[3]

Values for the constants m and C for a cylindrical geometry at various Reynolds numbers are given in Table 2.

Table 2 Constants m and c (cylindrical geometry) for Nusselt number calculation

N_{Re}	m	C
1–4	0.330	0.989
4–40	0.385	0.911
$40–4 \times 10^3$	0.466	0.683
$4 \times 10^3–4 \times 10^4$	0.618	0.193
$4 \times 10^4–4 \times 10^5$	0.805	0.0266

(From Ref. [3].)

Table 3 Thermal conductivity component contributions

Component	k_i ($W\,m^{-1}K^{-1}$)
Fat	0.180
Protein	0.200
Carbohydrate	0.245
Air	0.025
Water	0.600
Ice	2.240

(From Ref. [6].)

Thermal Conductivity Estimation

The thermal conductivity of the product gives the rate of heat flow through the product. Ranges include thermal conductivity values from 0.06 to 0.2 for grain products, oils, and bread, and values of 0.4–1.4 for fruits, vegetables, meat and dairy products.[4,5] An estimation of thermal conductivity values can be calculated from the product composition using a weighted average of the component conductivities:

$$K = \sum_{i=1}^{n} V_i k_i \tag{4}$$

where V_i is the volume fraction of the component and k_i is the thermal conductivity contribution for the component.[4] Thermal conductivity values for various product components are given in Table 3.[6]

Tunnel Configuration

The cooling time determines the length of time the product must be in the cooling tunnel, however the way in which this is achieved varies depending upon the tunnel configuration and conveyor operating speed. Factors such as space/location constraints and initial capital investment can affect the tunnel configuration and the required tunnel operating temperature to achieve the required cooling time. The total length required for the tunnel is determined by the production rate ($kg\,hr^{-1}$, $lb\,hr^{-1}$), the product load on the belt ($kg\,m^{-1}$, $lb\,ft^{-1}$), and the cooling time (hr):

$$\text{Length} = \frac{\text{production rate}}{\text{product load}} \times \text{cooling time} \tag{5}$$

Air Supply Considerations

The choice of operating temperature depends upon how quickly the cooling must be achieved. Ambient air may provide the cooling necessary for the product but for a greater cooling requirement, refrigerated air should be used, which will increase operating costs. Certain production lines may utilize the cooling stage to maintain the surface characteristics of the product. Humidity variations can cause changes in the surface structure and appearance. The suggested humidity for tunnel cooling supply air for chocolate is 78%, while the suggested humidity for bread is 83%.[7] Other products may require lower humidity conditions and this must be considered in the design and operation of the cooling tunnel. Products may need lower humidity and thorough cooling to prevent condensation inside the package, which can be detrimental to product quality. Dependent upon the surrounding air quality, an air filtration system may be required to remove particulates and mold spores from the air before using it as a cooling medium in the tunnel.

Other Considerations

Ease of cleaning should be an important consideration for design and operation of a cooling tunnel. Belts for conveyance of product should be appropriately selected for the product(s) being moved through the tunnel as well as allow for easy cleanup. The design should allow for ease of cleaning of all areas and provide accessible service locations along the length of the tunnel. The overall construction of the unit should withstand the environmental conditions within the plant and provide expandability if necessary. The cooling tunnel will probably be used in conjunction with other unit operations, therefore product transfer from preceding and orientation for subsequent unit operations should also be considered.

CONCLUSION

Cooling tunnel design requires consideration of both the product requirements to maintain maximum quality as well as the physical properties to establish engineering criteria. By evaluating and quantifying the operating parameters that affect final product quality, the final design can maximize quality and operational effciency.

REFERENCES

1. Taylor, T.A.; Heldman, D.R.; Chao, R.R.; Kramer, H.L. Simulation of Evaporative Cooling Process for Tortillas. J. Food Process. Eng. **1998**, *21* (5), 407–425.
2. Charm, S.E. *The Fundamentals of Food Engineering*, 3rd Ed.; AVI Publishing Co.: Westport, CT, 1978.
3. Geankopolis, C.J. *Transport Processes and Unit Operations*, 3rd Ed.; PTR Prentice Hall: Englewood Cliffs, NJ, 1993.
4. Stroshine, R. *Physical Properties of Agricultural Materials and Food Products*; Purdue University: West Lafayette, IN, 1999.
5. Goedeken, D.L. Microwave Baking of Bread Dough with Simultaneous Heat and Mass Transfer. UMI Dissertation Services, Ann Arbor, MI, 1995.
6. Miles, C.A.; Van Beek, G.; Veerkamp, C.H. Calculation of Thermophysical Properties of Foods. In *Physical Properties of Foods*; Jowitt, R., Escher, F., Hallstrom, B., Meffert, H.F.Th., Spiess, W.E.L., Voss, G., Eds.; Applied Science Publishers: Essex, England, 1983.
7. Carel USA. What Is Humidity? www.dghsys.com/what_is_humidity.htm (accessed May 2002).

Dairy Production Systems

William G. Bickert
Michigan State University, East Lansing, Michigan, U.S.A.

INTRODUCTION

Buildings and equipment on a dairy farm facilitate the job of caring for the animals, allowing essential tasks prescribed by the management plan to be carried out on a regular basis. Labor requirements, flow of animals and materials, pollution control, future expansion, management requirements, and animal environment are important considerations in design. Providing an environment that meets the needs of the animals being housed permits calves, heifers, and cows to grow, mature, reproduce, and maintain health. If the basic needs of the animals are not met, no amount of management can guarantee success. Facility decisions are difficult, facilities are expensive, and the required capital investments do not occur uniformly over time. The farmer must live with the consequences of decisions about facilities for many years. Long-range planning is important to all aspects of a dairy operation.

FACTORS AFFECTING CHOICE OF HOUSING TYPE

Climate impacted the type of housing that evolved in a particular area. In northern climates, barns provided protection for animals and their caretakers against long, cold winters. Pastures or other outside areas were used during milder times of the year. Where freezing temperatures and snow were not problems, shelter was less necessary. Year-round pasture systems often provided the source of energy for the cattle. Or, corrals or dry lots where feed is brought to the animals may have been used.

In colder climates, stanchion and tie stall barns served well for herds ranging up to 50 or 60 milk cows. However, stall barns are labor intensive, both for milking and feeding, an important reason for dairy farmers with larger herds to consider loose housing and milking parlor systems.

Changing economics has led to an increase in herd size, up to 3000 cows or more. Increasing management intensity leads to cows being milked three or four times daily on some dairies. Freestall barns are common for housing lactating cows in both cold and warm climates.

Keeping abreast of the increasingly superior animals from the standpoint of their genetic potential for milk production has been an important reason for changes in housing and equipment. The more traditional facilities and methods often become an obstacle to higher milk production by limiting the ability to implement appropriate management recommendations. In addition, existing facilities may not meet the environmental needs of animals thus placing restrictions on productivity, health, and general welfare.

Evaporative cooling systems, based on rapidly moving air and controlled use of water, enable producers in warmer areas to successfully produce milk and manage breeding programs. Even in northern climates, high yielding dairy cows benefit from circulation fans that help dissipate metabolic heat from the body of the cow to the ambient air during periods of hot weather.

HOUSING FOR MILKING COWS

Stall Barns

Early stanchion barns were typically two-story buildings with two rows of stanchions in a face-out arrangement and a few pens for calves and freshening cows on the first floor. The mow on the second floor was devoted to hay storage. The typical northern stall barn of today is a well-insulated, single-story structure with mechanical ventilation. Barn space is primarily for milking cows, with calves and other animals housed in other facilities. Comfort or tie stalls are more common than stanchions. A trend to year-round housing and feeding in the barn has required more ventilation. Few new stall barns are built today.

Freestall Barns

Freestall barns constructed today generally are based on using a mobile feeding system with a truck- or trailer-mounted feed mixer equipped with electronic scales. Horizontal silos for corn silage and possibly grass silage and bulk storage for dry grains and supplement are compatible with this system. A front-end loader tractor loads feed components into the mobile scale-mixer. The scale-mixer is used to deliver the mixed rations to the cows. Groups of cows then can be fed differently according to their particular nutritional requirements.

Encyclopedia of Agricultural, Food, and Biological Engineering
DOI: 10.1081/E-EAFE 120006914
Copyright © 2003 by Marcel Dekker, Inc. All rights reserved.

This has led to barn arrangements that permit division of milking herds into groups, usually by production.

Freestalls are used to some extent in combination with outside lots or corrals, particularly where periods of excessive rain make concrete in the housing and feeding area desirable. In the north, these open-lot systems, where feeding was done in an exposed concrete feed alley or lot and a feed bunk, are gradually being abandoned or converted in favor of a completely covered facility. In new construction, joining the milking center to the freestall barn allows all animal traffic areas to be covered reducing problems from frozen manure and snow and the need for runoff control.

Corral Systems

In warmer areas where freezing temperature and snow are not problems, loose housing systems usually are based on outside corrals or lots. However, freestall barns are being used with increasing frequency, particularly where periods of excessive rain make concrete surfaces in outside housing and feeding areas desirable. A dirt lot may be provided as well.

Open dry-lot corral systems without freestalls remain popular where rain does not pose a problem during most of the year. These systems are based on dirt lots, sloped for drainage, with feed mangers along one end or a side. Drives are provided for the mobile scale-mixers usually used for feeding. In many cases, evaporative coolers are incorporated into shade structures to reduce stress further during hot weather. Cow traffic patterns to and from the milking center are an important part of the overall design.

DRY COW HOUSING

The early dry cow management category begins with a successful transition from a lactating state to a nonlactating (dry) state. Early dry cows are housed as one group for approximately 40 day on pasture, on a bedded pack, in free stalls, or in comfort stalls.

Approximately two to three weeks prior to expected calving date, dry cows and heifers should be moved to a precalving group. Precalving management calls for more careful attention to nutrition and sanitation as cows and heifers near, during, and after giving birth to offspring are at greater risk for infectious diseases and noninfectious disorders.

Immediately prior to calving time, cows and heifers should be moved to individual pens that are separate from other animals, especially younger calves. The location of this area and assignment of responsibilities to personnel should encourage frequent observation.

CALF AND HEIFER HOUSING

The calves and heifers on a dairy farm represent the future of the milking herd, genetically superior to the older animals. This results from a well-managed breeding program; taking advantage of the genetic superiority requires implementing a sound management program based on current recommendations.

From an environmental standpoint, raising healthy, potentially high producing replacement animals requires adequate space for water, feed, resting, and exercise. High humidity is especially detrimental to animal health. Adequate ventilation removes moisture and other air pollutants so as to maintain suitable air quality.

Calf Housing

From birth to age of weaning (2 mo), calves are housed in individual hutches, pens, or stalls. Naturally ventilated housing where inside conditions follow outside conditions closely is preferred. A calf hutch that is clean and dry with ample amounts of dry bedding has proven to be a desirable option. From an environmental view, the benefits of a hutch are no contact between calves, good ventilation, adequate space, and a clean spot for each calf.

The calf hutch is the gold standard for calf housing in terms of the environment it provides. Moving calves to pens inside a building may compromise ventilation or the ability to provide a clean, fresh place for the newborn calf.

Transition Housing

An important management category is comprised of calves from weaning up to 5 mo or 6 mo of age—the transition group. From an environmental standpoint, these calves have special needs that differ from older heifers. Maintaining small groups to assure uniform size of calves in a group is essential. For a 100-cow dairy, 4 to 6 calves per group will sufficiently minimize the difference in size among calves in the group.

Transition housing for calves that have been raised in individual hutches or other cold housing can be provided with portable calf shelters or super hutches, group pens in an existing building that is well ventilated, or a transition barn, a building constructed specifically for the transition stage.

The environment should be similar to the environment to which the calves have already become acclimated with sufficient ventilation for moisture control and protection from the elements in winter. In cold weather, a thick, dry, erect hair coat is essential to reduce heat loss from the calf and accompanying environmental stress.

Heifer Housing

In colder climates, heifers are usually housed on a bedded pack or in free stalls, at least in winter. Pasture and other outside lot systems are used too. Regardless of housing type, animals should be grouped according to a management plan. Variations in size and nutritional health and reproductive needs may then be accommodated.

SORTING, HANDLING, AND RESTRAINT FOR TREATMENT

Handling, restraint, and treatment facilities are essential to a successful dairy facility, permitting implementation of many management recommendations. One person should be able to observe, identify, separate, and restrain animals. Choice of method will vary with age or size and the particular housing facility. Treatments and examinations include vaccinations, dehorning, weighing, artificial insemination, estrus synchronization, and pregnancy checking. Adequate consideration must be given to safety of persons handling, examining, and treating animals and safety of the animals themselves.

MANURE MANAGEMENT

Manure management is a major issue facing animal agriculture. The traditional view of a manure system has resulted in manure system designs that meet the needs of the dairy farm and generally satisfy the interests of the rest of the society. However, future planning must go beyond the traditional view. Manure management is a complex issue and decisions are closely intertwined with decisions about many different aspects of the dairy production system, including the buildings and other equipment selected for managing the herd and the cropping program.

The traditional view of manure system planning centers on viewing the system as components—collection in the barn alleys, transfer to storage or a spreading device, storage, possible treatment, and nutrient utilization. Good design assures that all components of the manure system will operate together to economically and effectively manage the manure until it reaches its final destination, typically being applied to the land.

Ideally, manure nutrients are considered in the cropping program, especially to offset the use of commercial fertilizers. Because excess nutrients are primarily why manure is an environmental concern, animal scientists are now examining ration formulation, realizing that a high percentage of the various nutrients fed to animals are excreted in the manure.

Treatment options for dairy manure include composting, anaerobic digestion, and for sand-laden manure, sand separation. In addition, biological and chemical methods for separating phosphorus from the manure stream, common in municipal and industrial wastewater treatment, may find application in animal manure treatment.

Manure management is different from other areas of farm management. Problems are more social in nature than experts in the physical sciences are accustomed to addressing, solutions to manure management problems require knowledge from several different subject matter areas, and manure management has the attention of several segments of society besides farmers.

Decisions about manure system design are among the most complex decisions on the dairy farm and are related to the requirements of a well-thought-out management program and a clear description of the environment to be provided for every animal on the farm.

CONCLUSION

Providing an environment for animals that has a positive influence on the animals' health, welfare, and productivity and enhancing the farmer's ability to fulfill the requirements of a sound management program are goals of housing design. The ultimate goal is to permit the genetically superior cows in dairy herds today to produce at a level commensurate with their potential and the management style of the dairy farmer.

FURTHER READING

Bickert, W.G. Designing Dairy Facilities to Assist in Management and to Enhance Animal Environment (Keynote Paper). *Proceedings of the Third International Dairy Housing Conference*; ASAE: St. Joseph, MI, 1994.

Bickert, W.G. Manure Management: Challenges and Opportunities (Keynote Paper). *Proceedings of the Fifth International Dairy Housing Conference*; ASAE: St. Joseph, MI, 2003.

Bickert, W.G.; Holmes, B.; Janni, K.; Kmmel, D.; Stowell, R.; Zulovich, J. *Dairy Freestall Housing and Equipment*, 7th Ed.; MWPS-7; MidWest Plan Service: Ames, IA, 2000.

Palmer, R.W. *Dairy Modernization Planning Guide*; MidWest Plan Service: Ames, IA, 2001.

Decimal Reduction Times

P. Michael Davidson
Jochen Weiss
University of Tennessee, Knoxville, Tennessee, U.S.A.

D

INTRODUCTION

At temperatures above the optimum for growth of a microorganism, growth rate declines until, at some elevated temperature, the microorganism begins to die. Heat in the form of steam or hot water is used to produce shelf-stable food products, such as low acid canned foods by killing pathogenic and spoilage microorganisms that could reproduce under normal storage conditions. Products that have been processed in such a way are termed "commercially sterile." To design thermal processes for foods that will inactivate microorganisms causing foodborne illness or spoilage, it is necessary to know the heating rate of a food product and the heat resistance of the target microorganisms in the food. The characteristic unit of microbial heat resistance is the "decimal reduction time" or "*D*-value." By definition, a *D*-value is the time required to kill 90% of a population of microorganisms at a constant temperature and under specified conditions and it is measured by constructing a "survivor curve." Factors influencing heat resistance of microorganisms include the microbial population itself, the food product, and the heating environment. While survivor curves are often linear, there are factors that may cause deviations in the linearity of these curves. Potential implications of using a log-linear model with nonlinear survivor curve data include overestimation of the *D*-value of a target microorganism leading to a thermal process design that overcooks a food or underestimation of a *D*-value creating the potential for survival of target pathogenic or spoilage microorganisms. Several models have been developed to describe nonlinear survival curves.

DEATH OF MICROORGANISMS

At temperatures above the optimum for growth of a microorganism, the growth rate declines until, at some elevated temperature, the microorganism begins to die. Death is generally defined as a failure to reproduce.

Purpose of Thermal Preservation

Utilization of moist heat to inactivate microorganisms is one of the most efficient methods of preservation of many products, including foods. Heat in the form of steam or hot water is used to produce shelf-stable products, such as low acid canned foods, by killing pathogenic and spoilage microorganisms and their spores that could reproduce under normal storage conditions. Products that have been processed in such a way are termed "commercially sterile." This term indicates that viable microorganisms, such as thermophilic sporeforming bacteria with extremely high heat resistance, could be present in a thermally processed low acid food, but that they are not able to reproduce at normal storage temperatures. Lower temperature thermal processes are also used to eliminate vegetative (non-sporeforming) pathogenic microorganisms from certain foods, such as raw milk or eggs, to make them safe for consumption. These are called "pasteurization" treatments. Pasteurization also reduces the concentration of vegetative spoilage microorganisms to improve shelflife. While the goal of thermal processing is to eliminate populations of certain microorganisms from food products, the nature of their inactivation kinetics dictates that these processes actually only reduce the probability of survival to an acceptable level.[1] This is because of the limited sensitivity of available techniques for enumeration of surviving microorganisms.

Purpose for Determining Decimal Reduction Times

To design thermal processes for foods that will inactivate microorganisms causing foodborne illness or spoilage, it is necessary to know the heating rate of a food product and the heat resistance of the target microorganisms in the food. The characteristic unit of microbial heat resistance is the "decimal reduction time" or "*D*-value." By definition, a *D*-value is the time required to kill 90% of a population of microorganisms at a constant temperature and under specified conditions. Another important heat resistance characteristic associated with microorganisms is the thermal death time (TDT) or "*z*-value." The *z*-value is the change in *D*-value associated with a change in

Encyclopedia of Agricultural, Food, and Biological Engineering
DOI: 10.1081/E-EAFE 120006977
Copyright © 2003 by Marcel Dekker, Inc. All rights reserved.

temperature (See the article *Thermal Death Time*). In addition to moist heat, decimal reduction times are sometimes used to describe the resistance of microorganisms or spores to dry heat, irradiation, chemical inactivation, or other inactivation processes.

ORDER OF DEATH

Survivor Curves

To measure the *D*-value of a microbial population or population of bacterial spores, a survivor curve must be developed. To construct a survivor curve, a population of microorganisms or a bacterial spore crop of known number is subjected to a lethal temperature. The microorganisms or spores are suspended in a heating medium (buffer or food) in various types of containers. Acceptable containers to be used for heat resistance testing include capillary tubes or TDT cans, tubes, or pouches. Each container has advantages and disadvantages relating to convenience of use and heating come-up time. Common heating vessels include water baths, oil baths, miniature retorts, and the thermoresistometer.[1,2] Temperature monitoring is usually done with thermocouples. In all cases, heating and cooling lag times must be taken into account when determining total heating time. During exposure to the lethal temperature, samples are taken from the population at several times and the number of surviving microorganisms is determined. Survivors may be enumerated using a plate count method or a most-probable number technique. Nonselective microbiological medium and optimal incubation conditions are used for maximal recovery. Following incubation of the samples, the log number of survivors is plotted (*y*-axis) vs. heating time (*x*-axis) to construct a "survivor curve" (Fig. 1).

Measurement of *D*-Values

To obtain the *D*-value for a microorganism, it should be first determined whether a straight line can represent the survivor data. If so, a line can be constructed using linear regression. The *D*-value may then be determined by calculating the time required for the survivor curve to cross one logarithm (cotangent[1]) or from the negative reciprocal of the slope (− 1/slope) of the curve (Fig. 1). The *D*-value may also be determined using the equation:

$$D_T = \frac{t}{\log N_0 - \log N_t}$$

where N_0 is the initial number, N_t the final number of survivors after a total treatment time t.

Since the *D*-value is specific for a temperature, this should always be specified and/or added as a subscript to the "D" term, e.g., $D_{110°C}$. For sporeforming bacteria, survivor curve data from homogenous populations will generally form straight line curves. In contrast, vegetative bacteria often do not have straight line survivor curves, especially at lower temperatures.[1]

Rahn[3] was probably the first to show that moist heat at a constant temperature resulted in logarithmic death of vegetative microorganisms and bacterial spores. According to Pflug and Gould,[1] if a homogenous culture of microorganisms is heated at a constant temperature under constant conditions of heating and handling, the number of survivors decreases geometrically and a plot of log number of survivors vs. time is generally a straight line. There are exceptions as noted below. The most common hypothesis to account for the logarithmic order of death for microorganisms has a thermodynamic basis.[1] This is based on the fact that there is a distribution of energy levels among molecules at any given time in a system above a temperature of absolute zero. Therefore, at

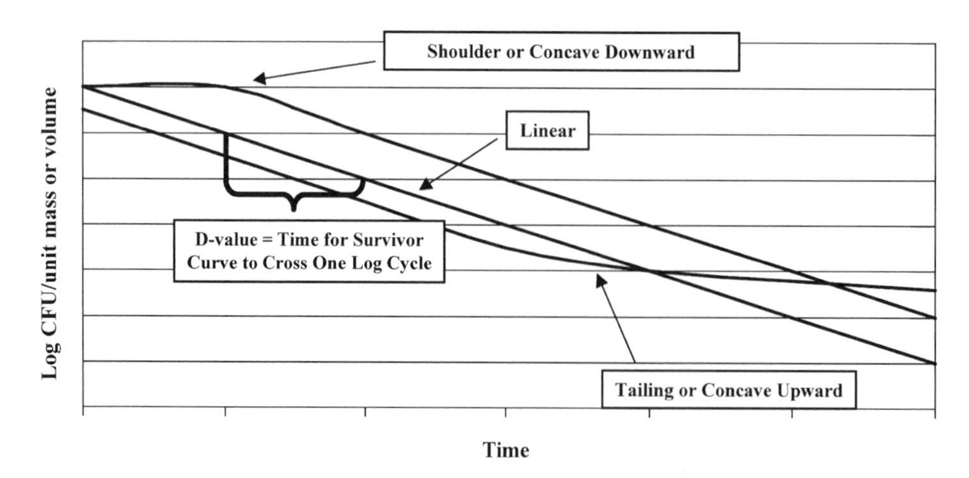

Fig. 1 Examples of Microbial Survivor Curves (CFU = colony forming unit).

D

Table 1 Examples of *D*-values of vegetative bacterial pathogens in various foods

Microorganism (strain/species)	Temp (°C)	*D*-value (min)	Food
Aeromonas hydrophila			
	48	3.3–6.2	Raw milk
Campylobacter jejuni			
	55	0.74–1.0	Skim milk
	56	0.62–0.96	Ground beef
	55	0.96–1.26	Lamb
	55	2.12–2.25	Cooked chicken
Escherichia coli			
ATCC 9637	57.2	1.3	Raw milk
O111:B4	55	5.5	Skim milk
	55	6.6	Whole milk
O157:H7	57.2	4.5–6.4	Ground beef
	60	1.63	Chicken
	60	1.89	Turkey
	60	2.01	Pork
	60	3.2	Beef
Listeria monocytogenes			
	57.8	3.97–8.17	Raw milk, raw skim milk, raw whole milk, cream
	63.3	0.22–0.58	
	60	1.6–4.5	Beef, roast beef
	60	2.3	Weiner batter
	60	3.12	Ground meat
	60	5.02–5.29	Chicken homogenate
	60	5.02–7.76	Carrot homogenate
	60	5.6	Chicken leg
	60	6.27–8.32	Beef homogenate
	60	8.7	Chicken breast
	60	9.2–11	Fermented sausage
	60	16.7	Ground meat, cured
Salmonella			
Anatum	65.5	1.4	Milk
Binza	65.5	1.5	Milk
Cubana	65.5	1.8	Milk
Eastbourne	71	270	Milk chocolate
Meleagridis	65.5	1.1	Milk
Newbrunswick	65.5	1.3	Milk
Senftenberg	65.5	0.66	Beef boullion
	65.5	1.11	Pea soup
	65.5	1.11	Skim milk
	65.5	34.0	Milk
	71	276	Milk chocolate
Tennessee	65.5	1.4	Milk
Typhimurium	57	2.13–2.67	Ground beef
	71	396	Milk chocolate
Staphylococcus aureus			
	50	10	Milk
	55	3	Milk
	60	0.9	Milk
	60	3	Pasta
Vibrio parahaemolyticus	48	10–16	Fish homogenate
	55	0.02–0.29	Clam homogenate

(Continued)

Table 1 Examples of *D*-values of vegetative bacterial pathogens in various foods (*Continued*)

Microorganism (strain/species)	Temp (°C)	*D*-value (min)	Food
	55	2.5	Crab homogenate
cholerae	60	0.35	Oyster homogenate
	60	2.65	Crab homogenate
Yersinia enterocolitica			
	51.7	23.4–29.9	Milk
	55	1.8–2.2	Milk
	60	0.067–0.51	Milk

(From Refs. [4, 9, 10].)

a constant temperature, target molecules in a population will all be subjected to the same chance of inactivation per unit time and the kinetics of inactivation will be logarithmic.[1] Assumptions associated with the thermodynamic approach include: 1) there is a distribution of energies among molecules surrounding the target molecule(s); 2) all target molecules have the same inactivation or activation energy; 3) there is only one target molecule per cell and; 4) damage repair is not possible under the exposure conditions.[1]

FACTORS AFFECTING *D*-VALUES

The *D*-value for a given population of microorganisms or spores is specific for the conditions under which the population was produced, the heating conditions utilized, and the conditions of recovery. Factors influencing heat resistance may be divided into inherent, environmental, and intrinsic. Inherent factors are those associated with the microorganism. Obviously the type of microorganism will have a significant influence on thermal resistance. Bacterial spores, including those of the species *Clostridium* and *Bacillus*, are by far the most heat resistant microbial forms. Next most resistant are certain heat resistant spores of the fungi *Byssochlamys* and *Neosartorya*. Most vegetative bacteria, molds, and yeasts are inactivated by pasteurization temperatures (ca. 63°C). Enteroviruses are similarly susceptible to heat while hepatitis A virus may require slightly higher temperatures (80°C).[4] Heat resistance can even vary among species and strains of microorganisms. Heat resistance, as measured by the *D*-value, and/or changes in the shape of the survivor curve may be influenced by treatment of microorganisms prior to heating such as incubation temperature, stage of growth, growth media, and homogeneity of the population.

Heat resistance may vary depending upon environmental and intrinsic (food or suspending medium) conditions to which the vegetative microorganism or spore is exposed. During exposure to lethal thermal stress, pH, water activity (a_w), and composition of the suspending medium or food (presence of proteins, carbohydrates, fats, salts) used in the heating process may influence *D*-value. For example, increased heat resistance is generally associated with increased pH, reduced a_w, and the presence of proteins, carbohydrates, or fats. Following exposure to heat, composition of the microbial recovery or subculture medium is vitally important in determining the actual number of survivors. Since the general definition of microbial death is failure to reproduce, it is conceivable that the apparent *D*-value of a microorganism grown in an inappropriate medium will be lower than that for the same microorganism grown in an appropriate medium. Similarly, a microbiological medium that contains selective agents may not recover microorganisms that have received thermal injury. Certain chemical agents that have been shown to enhance recovery of thermally treated cells include starch, charcoal, pyruvate, and lysozyme.[2,5] Optimum incubation conditions (e.g., temperature and atmosphere) and incubation time are important in determining actual *D*-values.

DEVIATIONS FROM *D*-VALUES

Nonlinearity of Survivor Curves

While survivor curves are often linear, there are many factors that may cause deviations in the linearity of these curves. Survivor curves may be concave downward or upward, i.e., they may have shoulders and/or tails, respectively (Fig. 1). Concave downward curves or those with shoulders may be caused by microbial populations that consist of several subpopulations, each with its own inactivation kinetics, multiple targets within a cell, clumping of cells or poor heat transfer through the heating medium.[5,6] When cells initially occur in clumps, each colony may in fact contain two or more cells. Therefore, logarithmic death will not be apparent until the final cell in each clump is inactivated.[2] Tailing can be the result of small

D

Table 2 Examples of *D*-values of bacterial spores in various heating media

Microorganism	Temperature (°C)	*D*-value (min)	Food/heating medium
Bacillus cereus			
	95	1.8–19.1	Milk
	95	2.7–15.3	Infant formula, pH 6.3
	95	2.9–36.2	Bread
Bacillus stearothermophilus			
	121	2.1–3.4	Phosphate buffer, pH 7.0
Clostridium botulinum Type A			
62A	110	0.61–1.22	Peas, canned
	110	0.61–1.22	Asparagus, canned
	110	0.61–1.74	Spinach, canned
	110	0.92–0.98	Tomato juice, pH 4.2
A16037	110	1.50–1.59	Tomato juice, pH 4.2
62A	110	1.89	Corn, canned
	110	1.98	Peas, puree
	110	2.01	Squash
	110	2.37	Spanish rice, pH 7.0
	110	2.48	Macaroni creole, pH 7.0
Clostridium botulinum Proteolytic Type B			
7 strains	110	0.49–0.99	Mushrooms, puree
213B	110	0.86	Beans, snap
	110	0.94	Carrots, fresh
	110	1.03	Corn
	110	1.06–1.09	Asparagus, canned
	110	1.17	Beets
	110	1.19–1.54	Spinach, canned
	110	1.52–3.07	Peas, canned
	110	1.75	Spinach, fresh
	110	2.14–12.42	Peas, puree
	110	2.15	Corn, canned
	110	2.88	Corn, puree
A35	110	2.97–3.33	Rock lobster, liquor
Clostridium botulinum Nonproteolytic Type B			
	85	100	Phosphate buffer
	90	18.7	Phosphate buffer
	90	0.80	Turkey
Clostridium botulinum Nonproteolytic Type E			
	70	72	Oyster homogenate
	70	100	Oyster homogenate + 1% NaCl
	70	72	Oyster homogenate + 0.13% K sorbate
	70	79	Oyster homogenate + NaCl + K sorbate
	73.9	8.66	Menhaden surimi
	74	6.8–13.0	Blue crabmeat
Clostridium perfringens			
	98.9	31.4	Beef gravy, pH 7.0
	104.4	6.6	Beef gravy, pH 7.0
	110	0.5	Beef gravy, pH 7.0

(From Refs. [4, 5, 11].)

numbers of large clumps of cells in the population, actual cell to cell differences in heat resistance, variation in life cycle (e.g., stationary phase cells are more resistant), or possibly heat adaptation.[1,5] According to Stumbo,[2] solids in foods and bacteria suspended in salt solutions will often flocculate during heating. These clumps are difficult to disperse during dilution and agitation can influence counts obtained when survivor curves are developed. Flocculation during heating will cause counts to level off or tail. The converse may also occur if the suspending medium has flocculated to any extent before heating. Nonlinear survivor curves may also be artifacts of nonhomogenous populations, mixed cultures, use of improper heating techniques, or limitations of enumeration techniques utilized.

The potential implications of using a log-linear model with nonlinear survivor curve data are twofold.[5] Overestimation of the *D*-value of a target microorganism may lead to a thermal process design that overcooks a food resulting in sensory quality deterioration and nutritional losses. In contrast, underestimation of a *D*-value is much more serious in that it creates the potential for survival of a target microorganism. If the target microorganism is a pathogen, survival could lead to foodborne illness.

Novel Kinetic Thermal Inactivation Models

The concept of *D*-values clearly becomes problematic when nonlinearity in survival curves is observed.[7] There have been many models developed to describe nonlinear survival curves.[1,5,7,8] A new but simple model to describe semilogarithmic survival curves that have either upwards or downwards concavity was suggested by Peleg and Cole.[8] In this model, the probability of a microorganism dying due to a thermal event is distributed. This means that not all microorganisms have the same probability of surviving, which is the underlying assumption in the conventional *D*-value model. Survival curves are therefore an expression that is based on a distribution of inactivation times.[6] The Weibull probability function, which is used extensively in reliability engineering, has proven to be especially suitable to describe the survival probability. The survivor curve based on this function is given by:

$$\log \frac{N_t}{N_0} = -b_T t^{n_T}$$

where b_T and n_T are temperature dependent constants.

If $n_T < 1$, the semilogarithmic curves will have an upward concavity while at $n_T > 1$ curves will be downward concave. If $n_T = 1$, the regular *D*-value model re-emerges as a special case with $b_T = D_T$. While the Weibull model is strictly empirical, it can be linked to physiological adaptation if $n_T < 1$ and accumulated damage if $n_T > 1$ and may thus signal if the thermal resistance of microorganisms is changing due to, for example, growth conditions.

EXAMPLES AND USE OF *D*-VALUES

As was stated previously, microorganisms have vastly different *D*-values. Examples of *D*-values of vegetative cells of pathogenic bacteria are presented in Table 1 while bacterial spore *D*-values are shown in Table 2.

Regulatory agencies are increasingly defining heat processes for foods as a function of the *D*-value of a particular pathogen and a target log reduction (*n*). For example, a 12D thermal process for *Clostridium botulinum* at 121.1°C indicates a process time capable of reducing viable *C. botulinum* spores by 12 log cycles. This has been used to describe heat or other inactivation processes for pathogens such as *Salmonella*, *Escherichia coli* O157:H7 and *Listeria monocytogenes* in meats and other foods (e.g., apple cider). There are several weaknesses of such applications. First, *nD* does not specify an endpoint, i.e., the number of viable microorganisms remaining or the risk of viable pathogens being present. This is because it is not possible to determine an endpoint without an initial number (N_o).[1] In general, the initial number of pathogens on a food product is unknown or only estimated based upon incomplete data. Secondly, since *D*-values are specific for application conditions and microorganism, they must be determined empirically for every process. To overcome these weaknesses, studies on the level of contamination by pathogens or bioburden should be expanded for potentially hazardous foods. This, in conjunction with expanded use of risk assessment by regulatory agencies to determine safe processing endpoints, should make target log reductions more scientific and therefore more reliable for food processors and consumers.

CONCLUSION

Decimal reduction times of *D*-values are useful for indicating the relative heat resistance of microorganisms. They allow the food scientist or technologist to calculate appropriate heat processes for inactivating spoilage or pathogenic microorganisms and thereby increase shelflife or improve safety. However, as was discussed, many factors contribute to the *D*-value of a microorganism and the user must be aware of these limitations to design valid thermal processes.

REFERENCES

1. Pflug, I.J.; Gould, G.W. Heat Treatment. In *The Microbiological Safety and Quality of Food*; Lund, B.M., Baird-Parker, T.C., Gould, G.W., Eds.; Aspen Publ.: Gaithersburg, MD, 2000; 36–64.
2. Stumbo, C.R. *Thermobacteriology in Food Processing*; Academic Press: New York, 1965.
3. Rahn, O. Physical Methods of Sterilization of Microorganisms. Bact. Rev. **1945**, *9*, 1–45.
4. ICMSF (International Commission on the Microbiological Specifications for Foods). *Microorganisms in Foods 5: Microbiological Specifications of Food Pathogens*; Blackie Academic & Professional: London, 1996.
5. Juneja, V.K. Thermal inactivation of microorganisms. In *Control of Foodborne Microorganisms*; Juneja, V.K., Sofos, J.N., Eds.; Marcel Dekker, Inc.: New York, 2002, 13–53.
6. Van Boekel, M.A.J.S. On the Use of the Weibull Model to Describe Thermal Inactivation of Microbial Vegetative Cells. Int. J. Food Micr. **2002**, *74*, 139–159.
7. Peleg, M.; Penchina, C.M. Modeling Microbial Survival During Exposure to a Lethal Agent with Varying Intensity. Crit. Rev. Food Sci. Nutr. **2000**, *40* (2), 159–172.
8. Peleg, M.; Cole, M.B. Estimating the Survival of *Clostridium botulinum* Spores During Heat Treatment. J. Food Prot. **2000**, *63*, 190–196.
9. Doyle, M.P.; Schoeni, J.L. Survival and Growth Characteristics of *Escherichia coli* Associated with Hemorrhagic Colitis. Appl. Environ. Microbiol. **1984**, *48*, 855–856.
10. Line, J.E.; Fain, A.R.; Moran, A.B.; Martin, L.M.; Lechowich, R.V.; Carosella, J.M.; Brown, W.L. Lethality of heat to *Escherichia coli* O157:H7: D-value and z-value determinations in ground beef. J. Food Prot. **1991**, *54*, 762–766.
11. Russell, A.D. Destruction of bacterial spores by thermal methods. In *Principles and Practices of Disinfection, Preservation and Sterilization*, 3rd Ed.; Russell, A.D., Hugo, W.B., Ayliffe, G.A.J., Eds.; Blackwell Science: Osney Mead, Oxford, UK, 640–656.

Deep-Fat Frying

Rosana G. Moreira
Texas A&M University, College Station, Texas, U.S.A.

INTRODUCTION

Frying is the process of cooking foods using oil as the heating medium and can be classified as pan-frying and deep-fat frying. In pan-frying, the food is moistened with fat, but not soaked. A little amount of oil is used in pan-frying so that the food is cooked until golden brown and crisp. In deep-fat frying, the food is immersed in the oil, which should be enough to cover the food by at least 2 cm. The objective of deep-fat frying is to seal the food by immersing it in hot oil so that all the flavors and juices are retained in a crisp crust. The frying technology is important to many sectors of the food industry, including suppliers of oils and ingredients, food service operators, and manufacturers of frying equipment. The United States produces more than 2.5 million tons of snack food per year, the majority of which are fried. Commonly fried products in the United States include: potato chips, French fries, doughnuts, extruded snacks, fish sticks, and the traditional fried chicken products. The term frying is interpreted broadly in this article and includes only the process of deep-fat frying.

THE FRYING PROCESS

The processes used to fry food products can be divided into two broad categories: 1) those that are static and smaller (whose capacity can range from 8 L–28 L of oil), classified as batch fryers used in the catering restaurants; and 2) those that fry large amounts of products in a moving bed, used in the food industry, classified as continuous fryers (having a throughput that varies from 250 kg product hr^{-1}–25,000 kg product hr^{-1}). These fryers can operate at atmospheric and low/high pressure conditions.

Deep-fat frying is a process of cooking and drying through contact with hot oil and it involves simultaneous heat and mass transfer. The frying process of a single product such as French fries can been divided into four stages that are characterized by Ref. [1]: a) *Initial heating*—short period, submersion of the food in the hot oil, product's surface heated to the boiling temperature of water, natural convection between oil and product's surface, and no evaporation; b) *Surface boiling*—the initial of evaporation, turbulence around the oil surrounding the product, forced convection between product's surface and hot oil, and the initial of crust formation at the product's surface; c) *Falling rate*—more internal removal of moisture from the food, rise in internal core temperature to the boiling point, increase of crust layer in thickness, and decrease in vapor transfer to the surface; and d) *Bubble end-point*—reduction of the rate of moisture removal, decrease in the bubbles leaving the product's surface, and a continued increase in the crust layer thickness.

During frying, oil is absorbed into the product via capillary pressure difference as water is removed from the product. Crust formation limits the oil absorption along with the rise in the product temperature and thus pressure. The oil remains mostly along the outer edge (crust) of the product, but does penetrates toward the center of the product.[2] The majority of the oil content in fried foods results from the absorption of oil when the product is removed from the oil and it is assumed to be a function of capillary pressure.[3] As the product is taken out from the oil, its temperature and the pressure inside the pore spaces decrease resulting in an increase in oil absorption.

Changes taking place during frying, such as shrinkage and expansion, are difficult to predict mathematically due to the number of interrelated factors that have to be taken into consideration. Shrinkage occurs mostly due to the loss of the least bound water during the first seconds of frying and product's expansion begins at low-water saturation (< 0.20).[2] The porosity of the product formed during the process plays an important role in the subsequent oil uptake. When a crust begins to form at the product's surface, there is an excessive pressure buildup by the water and air trapped inside of the pores, as the temperature increases, leading to the expansion and puffing of the product.[4] Therefore, a better understanding of the transport processes and their relationships to various parameters should provide ways to optimize the frying process, and thus control oil pickup.

FACTORS THAT AFFECT OIL ABSORPTION

Several factors affect oil absorption in fried foods, including process conditions (temperature and residence

Encyclopedia of Agricultural, Food, and Biological Engineering
DOI: 10.1081/E-EAFE 120007045
Copyright © 2003 by Marcel Dekker, Inc. All rights reserved.

time), initial moisture content of product, raw material composition, slice thickness, prefrying treatment, degree of starch gelatinization prior to frying, and oil quality.

During frying, the rapid drying is critical for ensuring the desirable quality of the final product. However, that loss of water results in a substantial absorption of oil by the product. Higher oil temperatures lead to a faster crust formation thus favoring the conditions for oil absorption. Oil content is temperature- independent when potato chips are fried at a temperature range of 145–185°C.[5] As the moisture is reduced with frying time, the ratio of oil content to the amount of water removed becomes independent of the oil temperature indicating that the oil content is not directly related to the oil temperature but to the remaining moisture present in the product.

Potatoes with high specific gravity (> 1.10) and dry matter (> 24.0%) produce chips with lower oil content.[6] French fries made from high dry matter content raw potatoes (around 24.0%) contain 9% less oil content than those made from potatoes containing lower dry matter content (around 19.5%).[7] Thickness of potato slices is another factor that affects oil uptake in chips. The thickness of the potato slice is found to be nonlinearly related to the chip's oil content.[8] The thicker is the potato slice, and thus the smaller surface area relative to the chip volume, the lower will be the chip's volume oil content.[9]

Oil uptake varies with the product pretreatment before frying. In producing French fries, blanching alone does not reduce significantly oil absorption in these products. Blanching makes the fries color more uniform, reduces oil uptake by gelatinizing the surface starch, shortens frying time, and improves the fries' texture. However, by blanching and then drying the potatoes for 15 min with hot air at 80°C and 2% relative humidity prior to frying, the oil content in French fries was reduced by 40% less than the blanched chips.[10]

The amount of water in the raw material before frying can affect the oil content of the fried product. By removing some of the water prior to frying can help reduce oil uptake after frying. However, depending on the drying process used, the final product will develop a pore structure that will affect how the absorbed oil will be distributed after frying. Microwave dried potato slices can produce chips with lower oil content than chips that are only fried. This drying process provides chips with heterogeneous moisture distribution structure and with large oil-free zones. In hot-air dried chips, the structure of the fried product shows oil more uniformly distributed around the surface than the fried only chips (which have most of the oil concentrated at the edges). Hot-air dried chips also have less oil content than the microwaved potato chips. On the other hand, freeze-drying will produce oily potato chips after frying. Freeze-dried potato slices has shown no major change in the product's structure during freeze-drying thus resulting in an even distribution of oil all over its surface after frying.[11]

Baking is used in the manufacturing of tortilla chips, for example. The objective of baking is to reduce the raw tortilla moisture content from about 55% wet basis (w.b.) to around 40%–38% w.b., to strengthen the product structure, and to develop its flavor. During baking, the raw tortilla is exposed to high temperatures (320–420°C) resulting in severe starch gelatinization (> 40%).[12] Baking time can produce tortilla with different moisture contents and consequently different degree of starch gelatinization before frying. Unbaked tortillas ($MC = 55\%\ w.b.$) yielded chips with 31% w.b. oil content after frying, compared to tortillas that were baked for 70 sec ($MC = 44\%\ w.b.$) and baked for 140 sec ($MC = 28\%\ w.b.$), which produced chips with 26% w.b. and 22% w.b. oil content, respectively.[13]

Another important factor in oil uptake in fried products is the level of particle size distribution in the dry mass flouer. Tortilla chips prepared from finely ground masa showed excessive puffing and pillowing, resulting in higher oil absorption and lower porosity; tortilla chips made from coarse particles presented the lowest oil content and no puffing; and intermediate masa produced tortilla chips with final oil content closer to the control sample.[13] The function of the coarse particles is to produce fissures in the product that allow water to escape during frying and reduce oil absorption, and to reduce the extent of pillowing during baking and frying.[14]

The degree of starch gelatinization in the product before frying can significantly affect oil absoprtion after frying. Tortillas containing high degree of starch gelatinized (80%) before frying yielded chips with lower oil content (14% w.b.) than tortilla chips made with tortillas containing 45% degree of starch gelatinized (23% w.b.) and those made with tortillas having 5% degree of starch gelatinized (35% w.b.). The degree of starch gelatinization affects the oil distribution. The higher the degree of starch gelatinized the more oil will be at the chip's surface than at the core. Starch gelatinization and subsequent swelling of the starch granules inhibits oil uptake in these products.[15]

Used or poor quality of the frying oil will make the product look more oily although the total oil content will be the same as those fried in fresh oil. Most of the oil absorbed by chips fried in degraded oil is concentrated at the surface. The higher viscosity of the degraded oil could cause the oil to adhere to the product's surfaces thus making it more difficult for the oil to be drained off from the chip's surface during cooling.[16]

QUALITY PROPERTIES OF FRIED PRODUCTS

The changes that occur in the composition (increasing oil content) of a food during frying are affected by frying oil composition; texture, size and shape of the food; and frying operating conditions (temperature, residence time, etc.). As most of the water is evaporated from the food, the product surface temperature rises to that of the frying oil causing the food to turn golden yellow (Maillard-type reactions) and to form a porous and crispy surface (a crust). These changes make the product more palatable to the consumer.

Water Loss During Frying

During the frying process, the moisture evaporates at the product surface and leaves the product due to the partial vapor pressure difference between the product and the frying oil. Oil temperature affects the drying rate of potato chips, being larger at a temperature raise from 150–160°C than from 170–180°C.[17] It takes about 140 sec for the product to reach a final moisture content of about 2% (w.b.) when fried at 150°C in comparison to only 80 sec when potato chips are fried at 180°C, i.e., a reduction in drying time of 60 sec.

Color

Color is among the major factors influencing consumer acceptability of a fried product. It can indicate high-quality product (such as the golden yellow of a potato chip) and can also influence flavor recognition. Panel evaluation and comparison to standards is the most common approach for determining color consistencies or differences in fried foods in the food industry. Color development in fried product during frying is due to the *Maillard browning* reaction that occurs between a carbonyl (aldehyde or ketone group) found in a reducing sugar and an amine found in proteins and amino acids. The results of the reaction are the flavor and color development of potato chips, for example. Potatoes containing low reducing sugars usually develop the desired golden yellow color when fried.[18] Poor quality frying oil can also affect the color of the final product by producing a dark and uncooked product when fried.[19]

Texture

Texture is a very important quality characteristic of a fried product. Crispness is a main desired textural characteristic for snack foods. It is a highly valued and universally liked textural characteristic, and its presence signifies freshness and high quality.[20]

The Universal testing machines such as the Instron and the Texture Analyzer are typically used for force-deformation studies to determine texture in food products. Fracturability (or hardness), defined as the first peak of the force vs. distance curve obtained using any of these machines have been used by many scientists to study texture characteristics in fried products.[21] When frying tortilla chips, for example, fracturability values change as frying time increases.[4] Tortilla chips become tougher as frying time increases to a maximum value at 30 sec of frying. After that time, the values of fractuarability tend to decrease slightly as the product becomes crunchier toward the end of frying (60 sec when fried in fresh oil at 190°C).

Raw material characteristics, prefrying treatment as well as processing conditions all affect the texture of fried products. Generally, instrumental methods are correlated with sensory evaluation of texture properties when determining best textural quality of a fried product.

Shelf-Life Flavor Stability

In the food industry, the most important test of product quality is shelf-stability, i.e., how the product tastes and smells after few months storage.

Rancidity resulting from oxidation of lipids is a primary concern during storage of fried products. The amount of fat a food must have to develop oxidation off-flavor depends not only on the type of frying oil but also the type of food and storage conditions. Snack products should be stored at temperature around 21°C[22] to keep their flavor quality.

Nutrition

Manufacturing of fried products requires complex mechanical and thermal operations that can result in enormous losses of nutritional components of the material. On the other hand, when foods are fried they become fat enriched, thus increasing their energy content. This fat can also help the transport of liposoluble components such as unsaturated fatty acids and liposoluble vitamins.[23]

Lipids influence flavor through their effect on flavor perception (mouthfeel, taste, and aroma), flavor stability, and flavor generation. A serious flavor defect of low-fat food is its quick disappearance in the mouth.[24] A reduction in fat content will result in higher flavor loss during processing and storage due to the increase in flavor volatility; it can also result in a decrease of the chemical stability of the flavor.

Potato chips contain 50%–70% of the nutrients present in raw potatoes. In French fries, losses of total nitrogen is in the order of 29%–43%, nonprotein nitrogen, 20%–35% and amino-acids, 45%.[25] As long as the oil used to frying the foods is not abused, fried foods pose no danger

to health. The negative aspects of frying are due to overheating. If frying is done gently, frying does not produce toxic agents which depress intake and growth.[26]

REFERENCES

1. Farkas, B.E.; Singh, R.P.; Rumsey, T.R. Modeling Heat and Mass Transfer in Immersion Frying. I, Model Development. J. Food Eng. **1996**, *29*, 211–226.
2. Yamsaengsung, R.; Moreira, R.G. Modeling the Structural Change During Deep-Fat Frying of Foods. Part I: Model Development. J. Food Eng. **2003**, *53* (1), 1–10.
3. Moreira, R.G.; Barrufet, M.A. A New Approach to Describe Oil Absorption in Fried Foods: A Simulation Study. J. Food Eng. **1998**, *35*, 1–22.
4. Kawas, M.L.; Moreira, R.G. Characterization of Quality Attributes of Tortilla Chips During Deep-Fat Frying. J. Food Eng. **2000**, *47* (2), 97–107.
5. Gamble, M.H.; Rice, P.; Selman, J.D. Distribution and morphology of oil deposits in some deep fried products. J. Food Sci. **1987**, *52* (6), 1742–1745.
6. Lulai, E.C.; Orr, P.H. Influence of potato specific gravity on yield and oil content of chips. Am. Potato J. **1979**, *56*, 379–390.
7. Lesinska, G.; Leszczynski, W. *Potato Science and Technology*; Elsevier Applied Science: New York, NY, 1989.
8. Baumann, B.; Escher, F. Mass and Heat Transfer During Deep-Fat Frying of Potato Slices. Rate of Drying and Oil Uptake. Lebensm.-Wiss.u.-Technol. **1995**, *28*, 395–403.
9. Gamble, M.H.; Rice, P. The Effect of Slice Thickness on Potato Crisp Yield and Composition. J. Food Eng. **1988**, *8*, 31–46.
10. Lamberg, I.; Halstrom, B.; Olsson, H. Fat Uptake in a Potato Drying/Frying Process. Lebensm.-Wiss.u.-Technol. **1990**, *23*, 295–300.
11. Gamble, M.H.; Rice, P. Effect of Pre-frying Drying of Oil Uptake and Distribution in Potato Crisp Manufacture. Int. J. Food Sci. Technol. **1987**, *22*, 535–548.
12. Gomez, M.H.; Lee, J.K.; McDonough, C.M.; Waniska, R.D.; Rooney, L.W. Corn Starch Changes During Tortilla and Tortilla Chip Processing. Cereal Chem. **1992**, *69*, 275–279.
13. Moreira, R.G.; Sun, X.; Chen, Y. Factors Affecting Oil Uptake in Tortilla Chips in Deep-Fat Frying. J. Food Eng. **1997**, *31* (4), 485–498.
14. Gomez, M.H.; Rooney, L.W.; Waniska, R.D. Dry corn masa flours for tortilla and snack foods. Cereal World **1987**, *32* (5), 370–377.
15. Kawas, M.L.; Moreira, R.G. Effect of starch gelatinization on quality of tortilla chips. J. Food Sci. **2001**, *66* (2), 195–210.
16. Tseng;, Y.; Moreira, R.G.; Sun, X. Total frying-use time effects on soybean oil deterioration and on tortilla chip quality. Int. J. Food Sci. Technol. **1996**, *31*, 287–294.
17. Baumann, B.; Escher, F. Mass and heat transfer during deep-fat frying of potato slices. Rate of drying and oil uptake. Lebensm.-Wiss.u.-Technol. **1995**, *28*, 395–403.
18. Ranken, M.D.; Kill, R.C. *Food Industries Manual*; Blackie Academic & Professional: New York, NY, 1993.
19. Blumenthal, M.M.; Stier, R.F. Optimization of deep-fat frying operations. Trends Food Sci. Technol. **1991**, 144–148.
20. Bourne, M.C. *Food Texture and Viscosity*; Academic Press: New York, NY, 1982.
21. Moreira, R.G.; Castell-Perez, M.E.; Barrufet, M.A. *Deep-Fat Frying: Fundamental and Applications*; Aspen Publisher: Boston, MA, 1999.
22. Paradis, A. Nitrogen in Total Quality for Snack Food. INFORM **1993**, *4* (12), 1378–1382.
23. Moreiras-Varela, O.; Ruiz-Roso, B.; Varela, G. Effects of frying on the nutritive value of food. In *Frying of Foods: Principles, Changes, New Approaches*; Varela, B., Morton, Eds.; VCH Publishers: New York, NY, 1988.
24. Roos, K.B. How lipids influence food flavor. Food Technol. **1997**, *51* (1), 60–62.
25. Lisinska, G.; Leszczynski, W. *Potato Science and Technology*; Elsevier Applied Science: New York, NY, 1989.
26. Cuesta, G.; Sanchez-Muniz, J.; Varela, G. Nutritive value of frying fats. In *Frying of Foods: Principles, Changes, New Approaches*; Varela, B., Morton, Eds.; VCH Publishers: New York, NY, 1988.

Dehydration Process Energy Balances

Yuan-Kuang Guu
National Pingtung University of Science and Technology, Neipu, Pingtung, Taiwan

INTRODUCTION

Drying is an efficient way to reduce moisture content and water activity for food preservation, but is an energy consuming process. Most fruits, vegetables, and food materials are dried to different extents in terms of product identification. Therefore, when dealing with drying, quality standards and energy costs must both be taken into account. During the drying process, two kinds of heat are involved in moisture removal, which are sensible and latent heat. Sensible heat is the heat needed to change the temperature, and is normally related to the specific heat of a substance. Latent heat is the heat needed to change the phase of a substance, e.g., changing from liquid to gas phase. In energy auditing, the percentage of latent heat is significantly greater than that of sensible heat, and is responsible for most of the energy cost for drying. This is especially the case for freeze drying.

However, a drying process must be evaluated from the standpoint of both energy and time efficiencies. Sometimes, retaining qualities of the original food materials are also key factors for choosing an adequate process. Therefore, many drying processes and equipments have been developed to meet the demands of the food industry. Drying processes can be categorized as: 1) hot-air drying, such as tunnel or kiln drying, spray drying, and fluidized-bed drying, etc. where hot air is the thermal medium and moisture carrier; 2) direct-contact drying, such as drum or pan drying, where thermal conduction is the mechanism to convey heat for drying; 3) freeze drying, where food materials are first frozen, and then followed by sublimation under negative pressure; and 4) electromagnetic or radiation drying such as microwave and infrared drying, etc.

Different drying processes adopt different devices and mechanisms, but all involve sensible and latent heats for removing moisture. Generally speaking, water removal is accomplished by phase change from liquid to vapor, except in freeze drying, where solid sublimates into vapor directly. Whether vaporization or sublimation is involved, the temperature at which phase change occurs is the decisive factor in energy consumption.

GENERAL CONSIDERATIONS FOR ENERGY BALANCES OF DRYING

Consider a block diagram as shown in Fig. 1, which represents any drying process. Calculations of energy balances are illustrated based on the First Law of Thermodynamics as stated by Eq. 1, where $\sum(\text{Energy})_{\text{in}}$ and $\sum(\text{Energy})_{\text{out}}$ are comprised of several enthalpy terms entering and leaving the drying unit, depending on what kind of process and drying mechanism are employed. *Accumulation* is the enthalpy stored within the unit, and is normally zero when steady-state operation is reached. Included in the term of $\sum(\text{Energy})_{\text{out}}$, heat loss due to improper insulation or operations has to be seriously considered for the sake of energy savings.

$$\sum(\text{Energy})_{\text{in}} = \sum(\text{Energy})_{\text{out}} + \text{Accumulation}. \quad (1)$$

ENERGY BALANCES IN HOT-AIR DRYING

Hot-air drying employs a heater to raise the temperature and increase the energy content of the drying air. Saturated or superheated steam and electrical heating devices are always the major heat sources. Hot air plays two roles in drying: as an energy carrier and as the vehicle to take away moisture from the dryer. Dry bulb and wet bulb temperatures of the drying air are two very important operating parameters and the larger the difference between these two temperatures, the faster the drying rate. Thermal calculations in this section are applicable to tunnel or kiln drying, spray drying, fluidized-bed drying, or rotary-louvre drying, etc. whether in batch or continuous operations.

A hot-air drying process can be illustrated by Fig. 2, and $\sum(\text{Energy})_{\text{in}}$ and $\sum(\text{Energy})_{\text{out}}$ in Eq. 1 are described by Eqs. 2 and 3. $H_{\text{mass in}}$ and $H_{\text{mass out}}$ represent the enthalpies carried by the mass; H_{source} is the heat provided to the heater, and H_{loss} the heat loss. Enthalpies of drying air entering and leaving the drying unit, $H_{\text{air in}}$ and $H_{\text{air out}}$, can be divided into two parts: enthalpies of bone dry air $H_{\text{bone dry air}}$ and vapor in bone dry air H_{g}. They are mathematically expressed by Eqs. 4 and 5, where $\dot{m}_{\text{air}}$ is

Encyclopedia of Agricultural, Food, and Biological Engineering
DOI: 10.1081/E-EAFE 120007086
Copyright © 2003 by Marcel Dekker, Inc. All rights reserved.

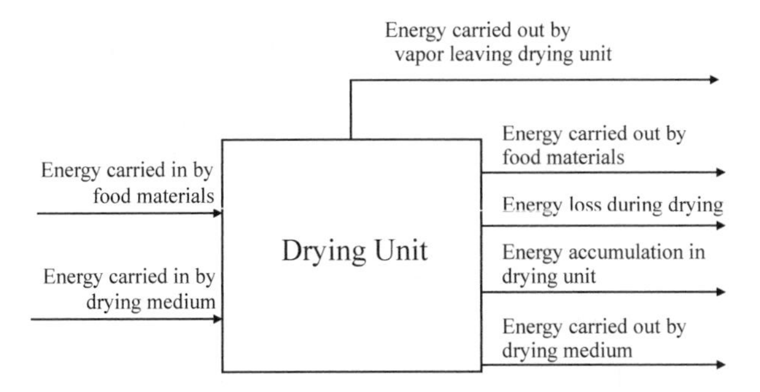

Fig. 1 Block diagram for energy balance considerations.

the mass flow rate of bone dry air, χ the absolute humidity, Cp_{air} the specific heat of bone dry air and available from any handbook of chemical engineering, and T_{ref} the reference temperature in thermal calculations and normally taken as zero.

$$\sum(\text{Energy})_{in} = H_{air\,in} + H_{mass\,in} + H_{source} \quad (2)$$

$$\sum(\text{Energy})_{out} = H_{air\,out} + H_{mass\,out} + H_{loss} \quad (3)$$

$$H_{air} = H_{bone\,dry\,air} + H_g \quad (4)$$

$$H_{air} = \dot{m}_{air}\left[\int_{T_{ref}}^{T} Cp_{air}\,dT + xH_{g\,at\,T}\right] \quad (5)$$

Enthalpies of food materials, $H_{mass\,in}$ and $H_{mass\,out}$, are calculated by Eq. 6, where $\dot{m}_{mass}$ is the mass flow rate of food materials. Cp_{mass} is the specific heat of food material, and can be evaluated by Sieble's empirical equations,[1] Choi and Oko's correlations,[2] or any database available.[3] Accumulation is normally regarded as zero for a steady state process. Therefore, Eq. 1 after rewriting is shown as Eq. 7.

$$H_{mass} = \dot{m}_{mass}\int_{T_{ref}}^{T} Cp_{mass}\,dT \quad (6)$$

$$H_{air\,in} + H_{mass\,in} + H_{source} = H_{air\,out} + H_{mass\,out} + H_{loss} \quad (7)$$

When dealing with H_{loss}, thermal insulation and operation should be taken into account, which normally vary from case to case in the range of 10%–40% of the total energy in the industrial practice. If the heat source is provided by condensation of steam, H_{source} is expressed as Eq. 8, where $\dot{m}_{steam}$ is the mass flow rate of steam and H_g and H_l are enthalpies of steam and condensate, respectively.

$$H_{source} = \dot{m}_{steam}[H_g - H_l] \quad (8)$$

ENERGY BALANCES IN DIRECT-CONTACT DRYING

Direct-contact drying equipment, such as a drum (single or duo) or a pan dryer is normally used to dry liquid materials into film or flake. In this category, a large surface with a controlled high temperature directly contacts food materials for a period of time depending on the degree of drying needed. Thermal conduction is the mechanism of heat transfer and is the sole heat source for heating food materials. During rotation of the drum or retention on the pan, moisture is heated and vaporized, and dried material is collected using a scraper to peel off the film or flake.

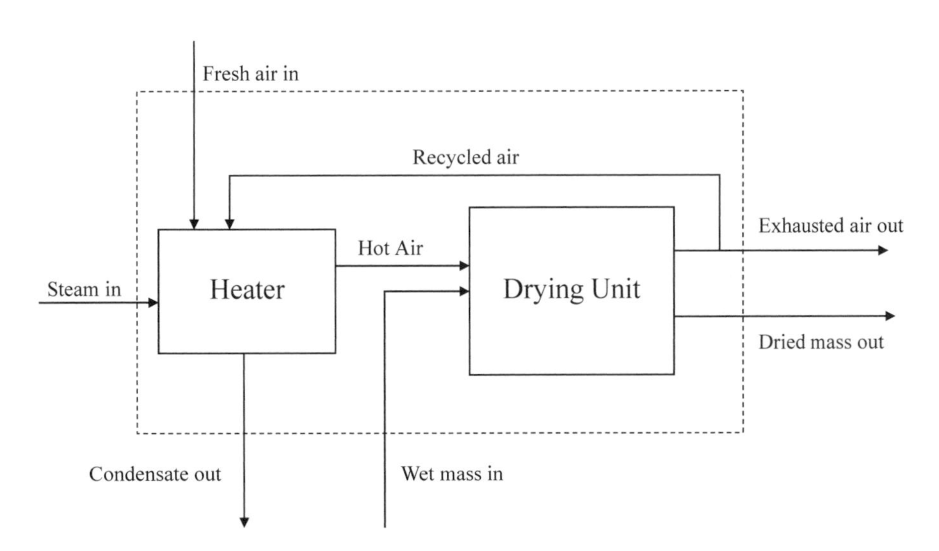

Fig. 2 Illustration of a hot-air drying process.

Energy balances for direct-contact drying can be expressed by Eq. 1. For simplicity, accumulation is taken as zero for steady state operation, and Eq. 1 is rewritten as Eq. 9. Analysis of energy terms in this equation leads to subsequent equations as shown by Eqs. 10–13.

$$\sum(\text{Energy})_{\text{in}} = \sum(\text{Energy})_{\text{out}} \tag{9}$$

$$\sum(\text{Energy})_{\text{in}} = \dot{m}_{\text{wet mass}} \int_{T_{\text{ref}}}^{T_{\text{in}}} Cp_{\text{wet mass}}\, dT + Q \tag{10}$$

$$\sum(\text{Energy})_{\text{out}} = \dot{m}_{\text{dry mass}} \int_{T_{\text{ref}}}^{T_{\text{out}}} Cp_{\text{dry mass}}\, dT + \dot{m}_{\text{vapor}} H_{\text{lg}} \tag{11}$$

$$\dot{m}_{\text{wet mass}} = \dot{m}_{\text{dry mass}} + \dot{m}_{\text{vapor}} \tag{12}$$

$$Q = UA(\Delta T)_{\text{lm}} \tag{13}$$

Eq. 13 represents the heat Q transferred from the drum or the pan to the food materials during contact drying, where U is the overall heat transfer coefficient, and normally consists of convective and conductive heat transfer mechanisms; $(\Delta T)_{\text{lm}}$ is the log mean temperature difference between the contact surface and the food materials, and A is the contact area.

ENERGY BALANCES IN FREEZE DRYING

Freeze drying always involves a three-stage process. Normally, food material is first frozen from ambient temperature T_a to its freezing point T_f, then from T_f to the frozen state T_{fs}, and then allowed to sublimate at a designated temperature T_b under vacuum. Therefore, energy balances have to be calculated in three parts. These three parts are all energy consuming processes, whether heat is removed from, or provided to, the food material. A schematic diagram is shown in Fig. 3 to depict the three parts of energy described above.

When freezing from T_a to T_f, the amount of heat ΔH_1 removed from the food materials can be calculated by Eq. 6, except that the temperature range of integration is from T_a to T_f. When freezing from T_f to T_{fs}, a commonly used procedure developed by Chang and Tao[4] for calculating the amount of heat removal ΔH_2 is adopted. Two assumptions are made: first, moisture content M of food materials is better in the range of 73%–94%; second, moisture in foods is completely frozen at T_{cf}, 227.6K (−50°F). The procedure is described as follows:

1. Calculations of T_f[K] for different food materials:

Meats : $T_f = 271.18 + 1.47M$ (14)

Fruits and vegetables :

$$T_f = 287.56 - 49.19M + 37.07M^2 \tag{15}$$

Juices : $T_f = 120.47 + 327.35M - 176.49M^2$ (16)

2. Definition of a reduced temperature T_r:

$$T_r = \frac{T_{fs} - 227.6}{T_f - 227.6} \tag{17}$$

3. Calculations of parameters, a and b:
Meats:

$$a = 0.316 - 0.247(M - 0.73) - 0.688(M - 0.73)^2 \tag{18}$$

$$b = 22.95 + 54.68(a - 0.28) - 5589.03(a - 0.28)^2 \tag{19}$$

Vegetables, fruits, and juices:

$$a = 0.362 + 0.0498(M - 0.73) - 3.465(M - 0.73)^2 \tag{20}$$

$$b = 27.2 - 129.04(a - 0.23) - 481.46(a - 0.23)^2 \tag{21}$$

4. Calculation of enthalpy H_f [J kg^{-1}] at the freezing point T_f:

$$H_f = 9792.46 + 405,096M \tag{22}$$

5. Calculation of enthalpy H_{fs} [J kg^{-1}] at the frozen state temperature T_{fs}:

$$H_{fs} = H_f[aT_r + (1 - a)T_r^b] \tag{23}$$

6. Calculation of heat removal ΔH_2:

$$\Delta H_2 = H_{fs} - H_f \tag{24}$$

When sublimating, the amount of heat ΔH_3 needed to

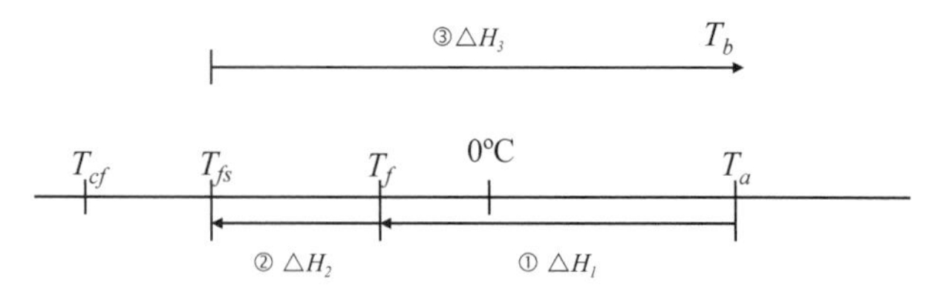

Fig. 3 Schematic diagram for the three-stage freeze-drying process, where ① is freezing from T_a to T_f; ② freezing from T_f to T_{fs}; and ③ sublimating directly from T_{fs} at a designated temperature, T_b.

supply to the food materials is estimated by Eq. 25, where H_{sg} is the heat of sublimation. Total energy needed by a freeze-drying process is the sum of the absolute values of ΔH_1, ΔH_2, and ΔH_3.

$$\Delta H_3 = |\Delta H_2| + H_{sg} + \int_{T_f}^{T_b} \text{Cp}\, dT \tag{25}$$

ENERGY BALANCES IN RADIATION DRYING

Sun drying is the oldest way of using solar radiation energy to dry food materials, and is still being used for many agricultural products. Recently, two forms of electromagnetic energy, microwave and infrared, have begun to be applied for drying purposes. High capital investment for the radiation equipment is discouraging, but better food quality and drying efficiency are anticipated. Microwave and infrared energy is used for heating and vaporizing moisture from the food materials. Thus, Eq. 6 can be used except that the energy source from microwave is expressed by Eq. 26, where P [W m^{-3}] is the power absorbed by food materials, f is the frequency of microwave [Hz], E is the electrical field strength [V m^{-1}], and ε'' is the dielectric loss factor.[5] For energy from infrared, the Stefan-Boltzmann equation for radiation energy can be used as shown in Eq. 27, where σ is the Stefan-Boltzmann constant, ε is the emissivity of food materials, A is the projected area, and T_1 and T_2 are the temperatures of the emitter and the absorber, respectively.

$$P = 55.61 + 10^{-14} f E^2 \varepsilon'' \tag{26}$$

$$Q = \sigma \varepsilon A (T_1^4 - T_2^4) \tag{27}$$

REFERENCES

1. Toledo, R.T. Energy Balances. In *Fundamentals of Food Process Engineering*; Van Nostrand Reinhold: New York, 1991; 132–159.
2. Choi, Y. Food Thermal Property Prediction as Effected by Temperature and Composition. Doctoral Thesis, Purdue University, West Lafayette, 1985.
3. Singh, R.P. *Food Properties: Database*; CRC: New York, 1995.
4. Chang, H.D.; Tao, L.C. Correlation of Enthalpy of Food Systems. J. Food Sci. **1981**, *46*, 1493.
5. Fellows, P.J. Microwave and Infrared Radiation. In *Food Processing Technology: Principles and Practices*; Ellis Horwood: New York, 1990; 343–355.

SUGGESTED READING

Barbosa-Canovas, G.V.; Vega-Mercado, H. Fundamentals of Air–Water Mixtures and Ideal Dryers. In *Dehydration of Foods*; Chapman and Hall: New York, 1996; 9–28.

Brennan, J.G. Dictionary of Food Dehydration. In *Food Dehydration: A Dictionary and Guide*; Butterworth-Heinemann: Oxford, 1994; E45–E47.

Earle, R.L. Drying. In *Unit Operations in Food Processing*; Pergamon Press: Oxford, 1983; 85–104.

Keey, R.B. The Industrial Drying of Foods: An Overview. In *Industrial Drying of Foods*; Baker, C.G.J., Ed.; Chapman and Hall: London, 1997; 1–6.

McCabe, W.L.; Smith, J.C.; Harriott, P. Drying of Solids. In *Unit Operations of Chemical Engineering*; McGraw-Hill, Inc.: New York, 1993; 767–809.

McKeon, J. Market Trends in Dehydrated Foods. In *Concentration and Drying of Foods*; MacCarthy, E., Ed.; Elsevier Science Publishing: Essex, 1986; 1–10.

Mujumdar, A.S. Drying Fundamentals. In *Industrial Drying of Foods*; Baker, C.G.J., Ed.; Chapman and Hall: London, 1997; 7–30.

Opila, R.L. Energy Use in Process Design. In *Food Process Engineering*; Linko, P., Malkki, Y., Olkku, J., Larinkari, J., Eds.; Applied Science Publishing Co.: Essex, 1979; 187–193.

Pan, Y.K.; Gu, F.J. Basic Calculations During Drying Processes. In *Modern Drying Technology*; Pan, Y.K., Wang, X.J., Eds.; Chemical Industrial Publishing Co.: Beijing, 1998; 48–67 (all in Chinese).

Singh, R.P.; Heldman, D.R. Food Dehydration. In *Introduction to Food Engineering*, 2nd Ed.; Academic Press: San Diego, 1993; 415–442.

Toledo, R.T. Dehydration. In *Fundamentals of Food Process Engineering*; Van Nostrand Reinhold: New York, 1991; 456–506.

Dehydration System Design

Harris N. Lazarides
Aristotle University of Thessaloniki, Thessaloniki, Greece

INTRODUCTION

Thermal dehydration of solids or semisolids is a high-energy demand process, which may cause severe deterioration of quality characteristics including aroma, taste, color, and nutrient losses. It is therefore a challenging task to design dehydration systems with the dual aim of maximizing process efficiency (i.e., maximizing product throughput, minimizing energy and processing cost) and maximizing quality retention.

The main objective of this article is to discuss the most important process and product parameters, which affect the design of a food dehydration system. Although emphasis is placed on thermal dehydration systems (being the most popular ones), alternative, mild dehydration methods (i.e., freeze-, dielectric- and osmotic-dehydration) are also discussed, wherever it is needed.

KINETICS IN THE DESIGN OF FOOD DEHYDRATION SYSTEMS

General Remarks

For an extensive discussion of various aspects of food dehydration systems, the reader can look into several valuable reference books.[1–4]

The successful design of a food dehydration system must take into consideration feed and product characteristics in order to define desirable properties of the drying environment (dryer type, operating conditions). On the feed side, it is very important to know its water binding characteristics, which are adequately described by its moisture isotherm. We also need to know its structural characteristics and all possible mechanisms of heat and mass (water) transfer both inside and outside the product.

Water Binding Properties

Water is present in two distinct situations. *Free or unbound* water is the part of moisture, which behaves as pure water at the same temperature. *Bound* water is moisture, which is physically or chemically bound to solid matrix or solutes, exhibiting a vapor pressure less than that of pure water at same temperature.

Free water is the first fraction of food moisture to be removed. It is water filling the pores and the capillaries. A fraction of bound moisture is loosely adsorbed to food solids, while higher binding forces characterize water trapped in colloidal gels. The last fraction of moisture to be removed is water chemically bound in sugar or salt hydrates.[5]

Relative distribution of bound and free water, as well as the degree of binding, define the hygroscopicity of a specific food. In turn, hygroscopicity strongly affects product behavior (i.e., rates of drying) during dehydration. The shape of a sorption isotherm offers a gross estimate of a product's hygroscopicity. A highly hygroscopic material gives a sorption isotherm of type I, while a low hygroscopicity material is characterized by a sorption isotherm of type II (Fig. 1).

Most food materials exhibit average (medium) hygroscopicity and they are described by a typical sigmoid sorption isotherm (type III). The relative distance of a food's isotherm from types I and III, is characteristic of its level of hygroscopicity.

It is obvious that highly hygroscopic materials are more difficult to dry and require much more energy compared to their less hygroscopic counterparts. Therefore, accurate mathematical (model) description of the moisture isotherm for a given food is an absolute prerequisite for satisfactory design of any dehydration process.

Among a large number of empirical or semiempirical models, which have been proposed, the Guggenheim-Anderson-de Boer (GAB) model has been most successful in closely describing the sorption isotherm of most foods for the widest range of a_w-values.

Osmotic dehydration is associated with substantial changes of the soluble solids profile due to some exchange of soluble solids between the product and the osmotic medium.[6] Such changes result in a shift of the isotherm curve and corresponding changes in drying behavior, when a complimentary drying step is applied.

Water Transport Mechanisms

During thermal dehydration, heat is provided to a wet product (feed) and water (in liquid or vapor form) moves from different locations within the feed towards its surface.

Encyclopedia of Agricultural, Food, and Biological Engineering
DOI: 10.1081/E-EAFE 120007087
Copyright © 2003 by Marcel Dekker, Inc. All rights reserved.

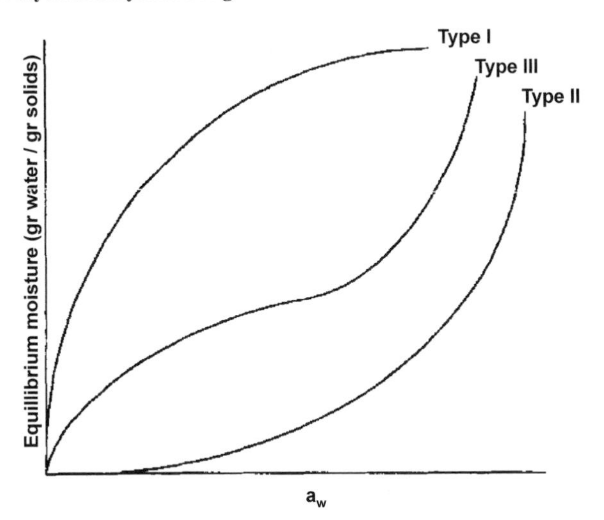

Fig. 1 Types of moisture sorption isotherms.

Movement of water within porous solids (such as food) takes place by different mechanisms, depending on structure and composition of the raw material, process conditions, and the stage of dehydration (residual moisture content). After a short initial period, where moisture is removed at a constant rate, it is the predominant mechanism of water movement, which mainly controls the rate of water removal and the shape of the drying curve.

Depending on the properties of the feed material and the progress in dehydration, water transfer may happen by one or more of the following major mechanisms[7]:

- Capillary flow due to gradients of capillary suction pressure.
- Liquid diffusion due to moisture concentration gradients.
- Vapor diffusion due to partial vapor-pressure gradients.
- Viscous vapor flow due to total pressure gradients caused by external pressure or high temperatures.

Minor water transfer mechanisms such as thermal or surface diffusion, flow induced by shrinkage pressure or gravity forces, are not considered large enough for modeling purposes since they may only result in unnecessarily complex models.

As different water movement mechanisms are associated with substantially different water transfer resistance, the designer of a food dehydration system ought to seriously consider design alternatives that promote the prevalence of most efficient moisture transfer mechanisms. The situation is different with *freeze-*, *dielectric*, and *osmotic drying* of solids.

In *freeze-drying*, all moisture moves in the form of water vapor. Progressive sublimation of ice gives rise to a so-called "*receding moisture front*" and a dry material zone of increasing thickness and heat resistance. Low process pressures and relatively high porosities in the dry zone allow for substantial contribution of *Knudsen diffusion* to water vapor transfer. A thorough review of mass transfer models for freeze-drying operations has been presented by Liapis and Marchello.[8]

In *dielectric* (RF or microwave) drying, volumetric dissipation of energy takes place throughout the wet product, often leading to automatic (self) leveling out of moisture variation.

During *osmotic dehydration*, food solids are immersed in highly concentrated sugar or salt solutions. Due to a chemical potential gradient (osmotic pressure difference) between the material and the osmotic solution, liquid water is transferred to the food surface to be taken away by the osmotic medium. There have been several reviews covering recent developments on the osmotic process.[9–11] In the case of plant tissue, water movement may follow three different pathways to reach the osmotic solution:

- *Trans-membrane transport*, i.e., transport from a vacuole to cytoplasm or from cytoplasm to the intercellular space.
- Transport through small channels between neighboring cells (*sym-plasmatic transport*).
- Transport within intercellular space (*apo-plasmatic transport*).

In case of a destructive pretreatment (i.e., blanching, freeze/thawing), the cell organization is disrupted and the above mass transfer pathways are strongly modified.

Heat and Mass Transfer Considerations

Convective or vacuum drying are unit operations where two transport phenomena take place simultaneously in a counter-current fashion. Heat is transferred from the drying chamber to the product interior. At the same time water moves from the interior to product surface and escapes to the environment.

In atmospheric (convective) air-drying, the capacity of water removal from the food's surface strongly depends on drying air characteristics, namely air temperature, humidity, and flow velocity (relative to food). Besides, the way of air-feed contacting plays a crucial role in achieving satisfactory drying rates. Typical convective heat transfer coefficients for various common, drying situations can be found in literature.[12,13]

Conduction heating takes place in contact (vacuum) dryers as well as freeze-drying. In vacuum (contact) drying, for a given feed, the pressure and temperature of the drying chamber define the rate of water removal.

Internal heat transfer is carried mainly by conduction with only a small fraction transferred by convection (through moving water). Normally, internal heat transfer resistance is much greater than the external resistance, especially as dehydration proceeds and a dry (low-conductivity) layer of increasing thickness develops.

In terms of mass transfer calculations, although we usually assume Fickian diffusion, in reality we have complex, multimechanism water transfer in capillary-cellular shrinking material, often undergoing drastic physiochemical changes and transformations (i.e., crystallization, glass transition). This is the main cause of deviations between projected and observed (actual) values in mass transfer studies.

Typical Drying Curves

Food dehydration is a process very unique to the specific product and dehydration system under consideration. A large number of parameters play significant role in the kinetics of dehydration yielding a slightly different drying curve, even when the same product is dehydrated under slightly modified conditions, or when exactly the same conditions are applied to a slightly different product. A typical thermal dehydration curve is shown in Fig. 2.

A slight initial increase of rate (section AB) represents a very short period of surface temperature equilibration to reach the hot air wet-bulb temperature. Provided that there is enough free water on product surface and that evaporated water can be rapidly substituted by new water coming from the interior, a relatively short constant rate period may exist (BC). This is a period of most efficient (fast) dehydration, where heat and mass transfer is controlled by external parameters; i.e., hot air properties and flow rate, air-product contacting, dryer design. A large fraction of product's moisture can be removed within a short period of constant rate dehydration. Process design should target at maximizing the duration of this drying period. This is equivalent to minimizing the *critical moisture* content (i.e., the average product moisture) when the falling rate period begins.

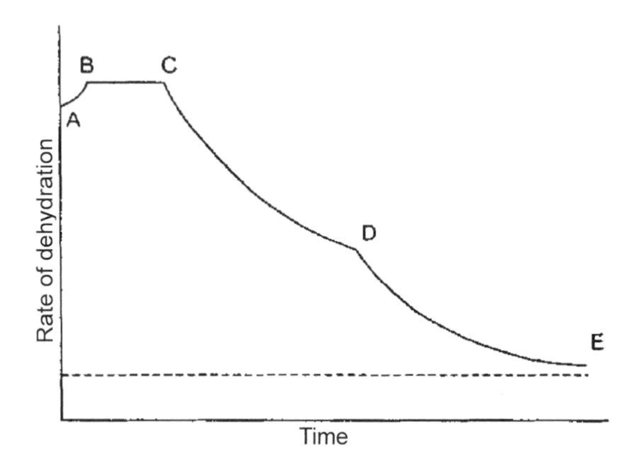

Fig. 2 Typical drying curve for solid food material.

It is important to notice that critical moisture strongly depends on product properties, including porosity, hygroscopicity, specific surface (surface to volume ratio). In order to minimize the critical moisture content of a certain solid feed, we need to minimize its size (thickness) within the limits of commercial acceptance.

In convective food dehydration a noticeable constant rate period is a relatively rare situation. Most often, following the short equilibration period, dehydration enters a long falling rate period, which may be subdivided into two (or more) sections (CD, DE). Curve breaks in drying curve correspond to changes in the predominant mechanism of moisture transfer due to physicochemical changes, depletion of moisture, and changes in heat/mass transfer gradients throughout a hygroscopic shrinking material. Point D corresponds to second critical moisture content. In certain cases the process may go through a third inflection, corresponding to third critical moisture content. Final product moisture asymptotically tends to reach an equilibrium moisture content (dotted line in Fig. 2).

The combined effects of several product-specific parameters on both the shape and length of sections of a drying curve make modeling of the drying phenomenon a rather challenging task.

PROCESS DESIGN CONSIDERATIONS

General Remarks

For a successful system design, the food engineer needs a detailed process definition; i.e., detailed information describing the properties of the feed, the target properties of the end product, the required drying capacity (throughput), and the position of the drying process in the framework of a complete processing line. A detailed checklist for selecting a drying process has been presented by Baker.[14]

Feed and Product Characteristics

The first, coarse classification of feed materials is based on their form. Feeds may be liquids, pastes, foams, gels, or

solids. The specific form of a feed material dictates the range of equipment that can be used to achieve a certain dehydration task.

A second important criterion is the desired extent of dehydration. This can vary from partial dehydration to any desired moisture level, up to complete dehydration (bone-dry products). Evaporative (or membrane-) concentration of liquid feed materials is an essential step before their complete dehydration.

Solid food is often a complex heterogeneous material with substantial local differences both in structure and composition. Structural properties like porosity, tortuosity, matter orientation (i.e., tissue fibers) affect heat and mass transfer characteristics in a dynamic way, as these properties are under continuous change during dehydration. Partial loss of porosity may cause a slow down in dehydration although it may enhance heat transfer. Techniques of maintaining or even increasing porosity (i.e., puffing) may be used to optimize the drying effect, balancing between faster mass transfer and slower heat transfer of a relatively open (porous) structure. Along with structural properties, chemical composition, and (especially) solute concentration define the water binding characteristics and the ease of drying.

Besides internal properties, shape and size of the dehydrated material can have a substantial impact on the rate of dehydration as they define the *specific surface* of the material and affect the (first) *critical moisture content*.

All of the above properties refer to the feed material. Equally or more important to dryer system design should be the desired properties of the end (dry) product. Such properties include: shape and size, bulk density, rehydration ability, color, appearance, aroma, taste, nutritional value, etc. In the case of powder products, solubility, wettability, sinkability, dispersibility, and stickiness can also play a crucial role in system design.

Required Dryer Capacity (Throughput)

Although dryer capacity is often reported in terms of product throughput (wet of dry basis), for design purposes it is preferable to express it in terms of moisture removal rates (i.e., $kg\,hr^{-1}$) since small changes in initial and/or final moisture contents can drastically change the dryer capacity in terms of product throughput.

As a general guideline, the demand for relatively small dryer capacities should lead to selection of simple, relatively cheaper, batch dryers; while large throughputs can pay for investment in more sophisticated, highly efficient, continuous drying systems.

In every case, the required production flexibility and the projected production dynamics should be taken into consideration when defining the dryer capacity.

EQUIPMENT AND PROCESS DESIGN

Largely variable material characteristics and dehydration needs have called for the development of a large variety of drying equipment. Each type is nearly ideal for certain applications, suitable or unsuitable for other.

A classification of common dryer types used for liquid and solid food has been presented by Potter and Hotchkiss.[5] Mujumdar[2] has presented a thorough classification of dryer types according to eleven design or operation criteria (i.e., mode of operation, physical form of feed, operating pressure, flow direction, etc).

Overall, process and dryer design should carefully consider the following requirements[2]:

- Ability to handle a feed at the required rate and to deliver a product of specified quality.
- Ease of hygienic operation (without contamination).
- Turndown ratio.
- Flexibility (ability to process other similar products).
- Safe operation (no fire/explosion or toxicological risk).
- Ease of control.
- Energy consumption.
- Low cost, small space requirements.
- Multiprocessing capability (e.g., cooling, granulation/instantizing, in the same unit).

Fig. 3 is a simplified scheme describing the step-wise iterative procedure of equipment/process design.

Required information may come from actual measurements, previous experience, or literature. In any case, thorough pilot plant testing is an absolute requirement for confirming suitability of dryer type and process conditions, actual measurement of processing time, and assessment of final product quality. Major (unexpected) problems in pilot testing may lead to modification or redefinition of appropriate process conditions. If this is not enough to correct the problem, the designer may need to consider a different, more suitable dryer type(s) and go through the whole exercise once again. For certain, especially demanding dehydration jobs the iterative procedure of defining process conditions and pilot plant testing may point to the need of using a combination of dryer types, instead of carrying the entire dehydration in a single dryer. Finally, existing types of dryers can be modified to better serve specific product needs.

Suggested applications, performance characteristics and detailed description for selected types of dryers can be found in Crapiste and Rotstein.[7]

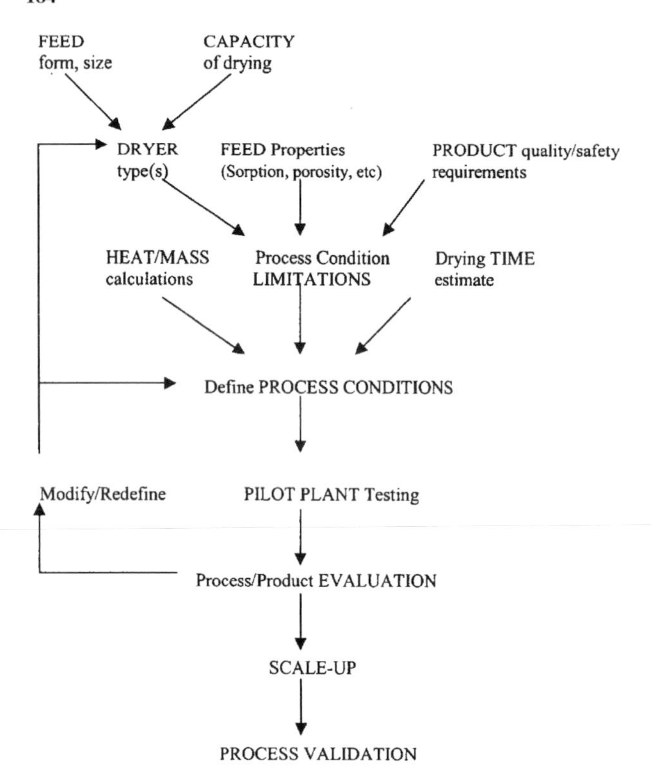

Fig. 3 Flow diagram for a typical dehydration process design.

QUALITY PARAMETERS AFFECTING DESIGN

A major goal of dryer design is to accelerate water removal up to a certain limit, which is defined by the internal mass transfer resistance for the specific material. If this limit is largely exceeded, there is a risk of fast, excessive surface drying with drastic superficial shrinkage. This leads to the undesirable situation of "*case-hardening,*" where moisture is trapped inside the food by a highly compact (hard) surface layer. The trapped moisture is very difficult or practically impossible to remove. The end result is a poor quality or unacceptable product (waste). Food with high concentration of sugars and other soluble solids are especially sensitive to case-hardening.

Uneven loss of moisture and collapse of voids may give rise to uneven shrinking and shape distortion, with serious impact on drying behavior, appearance, and functional properties of the final product (i.e., rehydration).

As any other food, dry food is expected to have some desirable properties at a satisfactory level. Such properties include appearance (i.e., shape, color, surface look), functionality (i.e., rehydration), taste, aroma, nutritional value. Thermal dehydration poses a significant thermal stress on thermo-sensitive ingredients, such as aroma, color, nutrients. Proper process design can minimize adverse effects, on the cost of lower dehydration rates.

Decreased process temperature can raise serious questions of food safety, as prolonged processing at such temperatures can promote the growth of pathogens, especially in nutrient-rich products.

Based on maximum acceptable loss in quality, process design may seek proper dryer and process conditions for a maximum throughput. Freeze drying and osmotic dehydration help to better maintain heat-sensitive quality parameters.

ECONOMIC AND ENVIRONMENTAL CONSIDERATIONS

The above iterative procedure used to establish the appropriate type of dryer(s) and the suitable process conditions does not always lead to a unique choice. Several alternatives may be possible, each offering certain advantages and disadvantages over the other ones. In such cases, economics of the designed operation will "say" the last word. Capital investment and operational costs are the two major elements of process economics. A small capital investment is often tempting and can lead to wrong decisions, if the operational cost is not taken into proper consideration. Besides fuel and electricity, maintenance, cleaning, manpower needs are some key elements of the operational cost.

Finally, the public demand for "clean" processing facilities raises environmental considerations, especially with respect to fuel choice. Natural gas is definitely the choice whenever it is offered as a fuel alternative.

RESEARCH DYNAMICS

Progressive access to cell membrane permeability data (from cell level experiments) and the use of universal modeling parameters (i.e., porosity), are expected to greatly help future attempts for successful modeling of dehydration processes.

REFERENCES

1. Baker, C.G.J. *Industrial Drying of Foods*; Chapman and Hall: New York, 1997; 309.
2. Mujumdar, A.S., Ed. *Handbook of Industrial Drying*, 2nd Ed.; Marcel Dekker: New York, 1995.
3. Barbosa-Canovas, G.V.; Vega-Mercado, H. *Dehydration of Foods*; Chapman and Hall: New York, 1996; 330.
4. Bruin, S., Ed. *Preconcentration and Drying of Food Materials*; Elsevier: New York, 1988; 353.

5. Potter, N.; Hotchkiss, J.H. Food Dehydration and Concentration. *In Food Science*, 5th Ed.; Chapman and Hall: New York, 1996; 200–244.
6. Lazarides, H.N.; Nicolaidis, A.; Katsanidis, E. Sorption Behavior Changes Induced by Osmotic Preconcentration of Apple Slices in Different Osmotic Media. J. Food Sci. **1995**, *60* (2), 348–350, 359.
7. Crapiste, G.H.; Rotstein, E. Design and Performance Evaluation of Dryers. In *Handbook of Food Engineering Practice*; Valentas, K.J., Rotstein, E., Singh, R.P., Eds.; CRC Press: New York, 1997; 125–166.
8. Liapis, A.; Marchello, J.M. Advances in Modelling and Control of Freeze-Drying. In *Advances in Drying*; Mujumdar, A.S., Ed.; Hemisphere: NY, 1984; Vol. 3, 217–244.
9. Fito, P.; Chiralt, A. Osmotic Dehydration: An Approach to the Modeling of Solid Food—Liquid Operations. In *Food Engineering 2000*; Fito, P., Ortega-Rodríguez, E., Barvosa-Cánovas, G., Eds.; Chapman and Hall: New York, 1996; 231–252.
10. Spiess, W.E.L.; Behsnilian, D. Osmotic Treatments in Food Processing. Current State and Future Needs. In *Drying '98*; Akritidis, C.B., Marinos-Kouris, D., Saravakos, G.D., Eds.; Ziti Editions: Thessaloniki, 1998; Vol. A, 47–56.
11. Lazarides, H.N.; Fito, P.; Chiralt, A.; Gekas, V.; Lenart, A. Advances in Osmotic Dehydration. In *Processing Foods: Quality Optimization and Process Assessment*; Oliveira, F.A.R., Oliveira, J.C., Eds.; CRC Press: New York, 1999; 175–199.
12. Whitaker, S. Forced Convection Heat Transfer Correlations for Flow in Pipes, Past Flat Plates, Single Cylinders and for Flow in Packed Beds and Tube Bundles. AIChE J. **1972**, *18* (2), 361–371.
13. Perry, R.H.; Chilton, C.H. *Chemical Engineer's Handbook*, 5th Ed.; McGraw-Hill: Tokyo, 1973.
14. Baker, C.G.J. Dryer Selection. In *Industrial Drying of Foods*; Baker, C.G.J., Ed.; Chapman and Hall: New York, 1997; 242–271.

Dew Point Temperature

John S. Roberts
Cornell University, Geneva, New York, U.S.A.

INTRODUCTION

Dew point temperature is an aspect of moist air below which partial condensation of the water vapor present in the air can occur. At the dew point temperature the relative humidity of the moist air is 100%, and further cooling of the air would result the water vapor losing its energy that it gained at evaporation and thus begin during condensation. Some examples of this dew formation are observed on windowpanes, on grass in the evening, and on cooling pipes. When the dew point is below freezing temperature, 0°C, sublimation will occur where the water vapor will convert directly to frost. When this condition applies, the term frost point temperature is used. The objectives of this article are to further explain the dew point temperature and related concepts, and to provide several ways in which to determine dew point temperature.

DEFINITION AND CONCEPTS

Dew point temperature is the temperature to which moist air must be cooled at constant pressure and constant humidity ratio for the partial pressure of water vapor in the air to equal the equilibrium water vapor pressure at that same temperature and pressure.[1–3] Some definitions state that it is the temperature where moist air must be cooled under constant pressure and humidity ratio to reach saturation.[4,5] The term saturation can be misleading because it indicates that the air cannot accept more water vapor molecules physically, where in fact it can if temperature increases. Several published texts have stated concerns in using saturation in terms of moist air.[1,6,7] This misunderstanding of saturation of air has lead to the misconception of how much water vapor air can "hold." The limitation of water vapor in air is not due to a lack of space in the air or due to any binding of water vapor to oxygen or nitrogen gas molecules, which does not occur. Rather, the limitation of water vapor in air is solely on the amount of available thermal energy to maintain water in the vapor state in the air.[1,8] This is shown in the relationship of the equilibrium (saturation) water vapor pressure with temperature. As temperature increases, the equilibrium water vapor pressure increases. In maintaining the use of saturation with respect to air, several works have defined saturated air as an equilibrium between water vapor in the air and the condensed water phase at the existing temperature and pressure.[4,6] This concept of saturation of air is with respect to an equilibrium and should be kept in mind when studying water vapor in air. In addition, moist air is a term typically used when describing the water vapor in air in the amount between complete dryness and saturation.[4]

The composition of moist air can be described by the humidity ratio, or specific humidity, and the relative humidity. The specific humidity, W, is the ratio of the mass of water vapor, m_v, to the mass of the dry air, m_a:

$$W = \frac{m_v}{m_a} \tag{1}$$

The humidity ratio can also be expressed in terms of partial pressures:

$$W = \frac{m_v}{m_a} = \frac{M_v p_v V/RT}{M_a p_a V/RT} = \frac{M_v p_v}{M_a p_a} \tag{2}$$

where M is molecular weight, V is volume, p is partial pressure, R is gas constant, T is temperature, subscript v represents water vapor, and subscript a represents dry air. The ratio of the molecular weight of water vapor to the molecular weight of dry air is approximately 0.62198. The vapor pressure of air, p_a, can be expressed as $p - p_v$ from the relationship $p = p_a + p_v$, where p is the total barometric pressure of the moist air. Thus, the equation for humidity ratio can be reduced to:

$$W = 0.62198 \frac{p_v}{p - p_v} \tag{3}$$

Dew point temperature, T_d, is defined mathematically as the solution $T_d(p, W)$ of the equation:

$$W_s(p, T_d) = W \tag{4}$$

W_s is the humidity ratio of air at saturation (equilibrium) and can be expressed similar to Eq. 1:

$$W_s = 0.62198 \frac{p_{vs}}{p - p_{vs}} \tag{5}$$

where p_{vs} is the saturation vapor pressure at the dew point

Encyclopedia of Agricultural, Food, and Biological Engineering
DOI: 10.1081/E-EAFE 120007052
Copyright © 2003 by Marcel Dekker, Inc. All rights reserved.

temperature. Eq. 5 reduces to:

$$p_{\mathrm{vs}}(T_{\mathrm{d}}) = p_{\mathrm{v}} = \frac{pW}{0.62198 + W} \qquad (6)$$

An important aspect either in determining the dew point or as moist air approaches the dew point is that both the pressure and the composition of moist air (humidity ratio or absolute humidity) remain constant while the relative humidity increases. The relative humidity is defined as the mole fraction of water vapor in moist air, n_{v}, to the mole fraction of equilibrium water vapor in air, $n_{\mathrm{v,sat}}$, at the same temperature and pressure. As partial vapor pressure, p_{v}, equals $n_{\mathrm{v}}p$ and saturation vapor pressure, p_{s}, equals $n_{\mathrm{v,sat}}\,p$, the relative humidity can be expressed in terms of pressure:

$$\phi = \left.\frac{p_{\mathrm{v}}}{p_{\mathrm{s}}}\right)_{T,p} \qquad (7)$$

In words, the relative humidity is the ratio of the amount of moisture in the air to the equilibrium moisture in air at the same temperature and pressure. When the pressure and humidity ratio remain constant, the relative humidity increases as temperature decreases. So as the air cools, the partial pressure of the water vapor remains constant (constant moisture content in the air) and the equilibrium vapor pressure of water decreases as the temperature decreases. Eventually the air temperature will have decreased such that the equilibrium vapor pressure equals the partial vapor pressure. Thus as the air cools, the numerator remains constant while the denominator decreases until unity is reached in Eq. 7. In the case where pressure remains constant, the relative humidity would be a function of both temperature and humidity ratio. As the dew point temperature is solely a function of the humidity ratio (at constant pressure), the dew point temperature is a better indicator of the amount of moisture in the air than relative humidity.[1,6,7]

DETERMINATION OF DEW POINT TEMPERATURE

Dew point temperature can be determined by the following means when other properties of moist air are known: from a psychrometric chart, a table on thermodynamic properties of water at saturation, equations, or physical measurements using hygrometers. This section describes how each method is used to determine dew point temperature.

Pyschrometric Chart

The easiest means to determine dew point temperature is from a psychrometric chart. Fig. 1 is a reduced version of the psychrometric chart showing only the dry bulb temperature, humidity ratio, and the saturation (equilibrium curve). The condition of the moist air can be located on

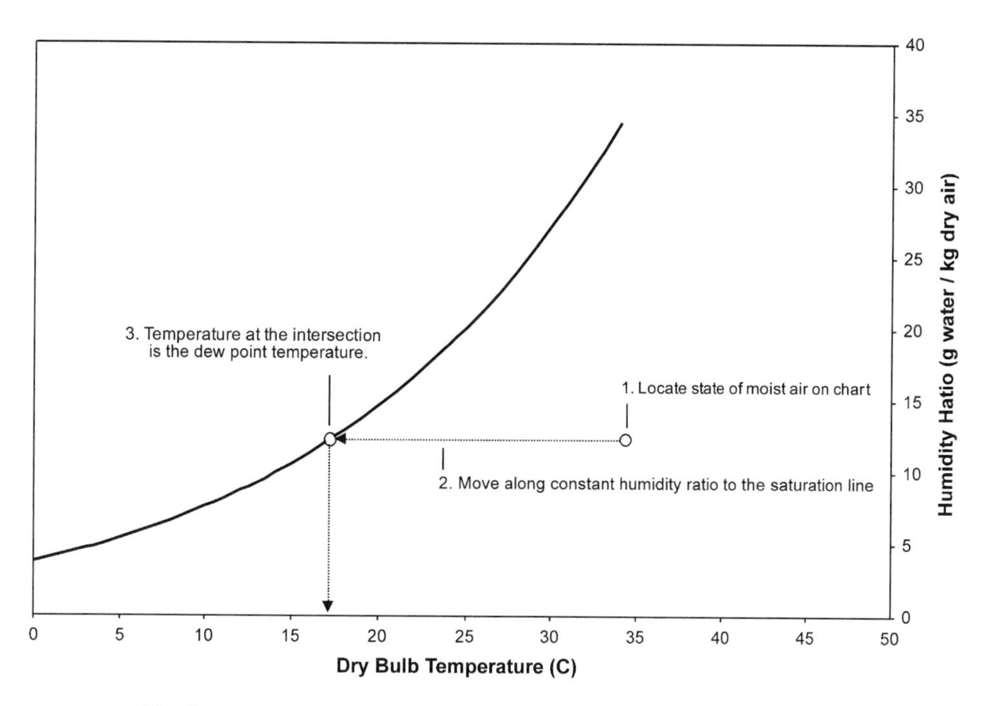

Fig. 1 Dew point temperature determination using a psychrometric chart.

the chart by knowing any two of the following properties: dry bulb temperature, humidity ratio or relative humidity, and wet bulb temperature. Once the state of the moist air has been established on the chart, the dew point temperature of this moist air can be determined by following along constant humidity ratio to the saturation curve. The temperature at this intersection is the dew point temperature.

Tables and Equations

When a psychrometric chart is not available, the dew point temperature can be determined by tables or by equations when the dry bulb temperature is given along with either the relative humidity or the humidity ratio. The state of moist air at the dew point temperature is saturated, 100% relative humidity, thus the partial vapor pressure equals the saturation vapor pressure. The corresponding temperature is the dew point temperature from the saturation vapor pressure–temperature relationship, which can be determined from either tables or from equations. If humidity ratio is given, as expressed in Eq. 3, the partial pressure of water vapor is equal to the saturation vapor pressure of the moist air and can be directly calculated from Eq. 6. This direct determination is valid since the humidity ratio remains constant when the dew point temperature is determined, as illustrated in Fig. 1. When relative humidity is given and is below 100%, the process of determining the partial pressure of the water is more involved since the relative humidity of the moist air is less than the relative humidity of the moist air at the dew point temperature. Once the partial vapor pressure is determined, the partial vapor pressure can be set equal to the saturation vapor pressure since that is the condition of the moist air at the dew point temperature.

As shown in the expression for relative humidity in Eq. 7, the saturation vapor pressure–temperature relationship must be known to determine the partial pressure of water vapor. This relationship is also needed to calculate the dew point temperature, as shown in Eq. 4. The saturation vapor pressure can be found in a table on thermodynamic properties of water at saturation or can be calculated from equations. A table on thermodynamic properties of water at saturation consists of temperature and corresponding saturation water vapor pressure, specific volume, enthalpy, and entropy.[4] Several equations are widely used to express the saturation vapor pressure of water with respect to temperature. One set of equations widely used is given below.[4]

For the temperature range − 100–0°C:

$$\begin{aligned}\ln(p_{vs}) &= (-5674.5329/T) + (6.3925247) \\ &\quad + (-0.009677843)T + (6.22115701 \\ &\quad \times 10^{-7})T^2 + (2.0747825 \times 10^{-9})T^3 \\ &\quad + (-9.484204 \times 10^{-13})T^4 \\ &\quad + (4.1635019)\ln(T)\end{aligned} \tag{8}$$

For the temperature range 0–200°C:

$$\begin{aligned}\ln(p_{vs}) &= (-5,8002206/T) + (1.3914993) \\ &\quad + (-0.048640239)T + (4.1764768 \\ &\quad \times 10^{-5})T^2 + (1.4452093 \times 10^{-8})T^3 \\ &\quad + (6.5459673)\ln(T)\end{aligned} \tag{9}$$

where p_{vs} is the saturation vapor pressure, Pa; and T is absolute temperature, K. Another widely used equation to calculate variation of saturation vapor pressure with temperature is the Magnus-Tetens equation[9]:

$$p_{vs} = 0.6105\, e^{(aT/(b+T))} \tag{10}$$

where $a = 17.27$, $b = 237.7$, and T is in °C. This equation is an approximation of the variation of saturation vapor pressure with temperature but its advantage is in its simplicity. The saturation vapor pressure of water can then be used to determine the corresponding dew point temperature either by tables or through equations.

The following equations are expressed to solve for dew point temperature. Eqs. 11 and 12 have been commonly used to calculate dew point temperature from saturated vapor pressure.[4]

For the dew point temperatures below 0°C:

$$T_d = 6.09 + (12.608)\ln(p_v) + 0.4959(\ln(p_v))^2 \tag{11}$$

and for dew point temperature in the range 0–93°C:

$$\begin{aligned}T_d &= 6.54 + (14.526)\ln(p_v) + 0.7389(\ln(p_v))^2 \\ &\quad + 0.09486(\ln(p_v))^3 + 0.4569(p_v)^{0.1984}\end{aligned} \tag{12}$$

where T_d is in °C and p_v is in kPa.

The dew point temperature is often calculated from the Magnus-Tetens equation, Eq. 10, which can be expressed to solve for dew point temperature as shown:

$$T_d = \frac{\ln\left(\frac{p_v}{0.6105}\right)b}{a - \ln\left(\frac{p_v}{0.6105}\right)} \tag{13}$$

A commonly used equation to calculate dew point temperature is also given[10]:

$$T_d = T - \Big[(14.55 + 0.114T)(1 - 0.01\phi) + ((2.5 + 0.007T)(1 - 0.01\phi))^3 + ((15.9 + 0.117T)(1 - 0.01\phi)^{14})\Big] \quad (14)$$

A new set of empirical equations were introduced to calculate dew point.[11]

For $-20 < T < 10°C$:

$$T_d = 193.03 + 28.633 p_v^{0.1609} \quad (15)$$

For $10 < T < 180°C$:

$$T_d = \frac{3816.44}{23.197 - \ln(p_v)} + 46.13 \quad (16)$$

The dew point temperature calculated in Eqs. 15 and 16 are in Kelvin.

Table 1 shows the comparison of dew point temperatures determined from the different methods described above with dry bulb temperature and relative humidity known. Eqs. 12, 13, and 14 produced very similar dew point temperatures and were close to the dew point temperature determined from a table on thermodynamic properties of water at saturation.[4] The difficulty in determining dew point temperature from the table is that the saturation vapor pressures are given to whole number dry bulb temperatures, and therefore interpolation is needed when the saturation vapor pressure lies between two dry bulb temperature entries. The psychrometric chart is useful in accessing approximate dew point temperatures, usually in whole numbers, with reasonable accuracy. But as illustrated in Fig. 1, the chart's advantage is in the ease of determining the dew point temperature. The results of using Eq. 16, which was recently offered as an alternative empirical equation in calculating dew point temperature,[11] is also shown in Table 1. The dew point temperatures were predicted slightly higher in all calculations than the other methods, more at lower dry bulb temperature conditions.

Methods of Measurement

The gravimetric hygrometer is regarded as a primary standard of measurement of moisture in air because it determines the humidity ratio directly.[4,5] As shown by the psychrometric chart in Fig. 1, the humidity ratio directly correlates with the dew point temperature. This standard method of measurement consists of passing moist air through a series of three U-tubes filled with desiccant. The amount of water vapor absorbed and the volume of dry gas can be measured precisely under controlled temperature and pressure conditions. This instrument is not used in industrial applications because the equipment is cumbersome, requires special care, and measurement times are long (up to 30 hr for low dew point temperature measurements).[4,5] There are several more practical devices to measure dew point temperature, and the three most commonly used in industry and research are capacitive (aluminum oxide) sensors, saturated salt (lithium chloride) sensors, and optical surface condensation [chilled mirror (CM)] hygrometers.[12,13] These instruments vary in the method of measurement, range of measurement, level of accuracy, and cost. The choice of measurement depends on the environment it will be used and the level of accuracy required. The following sections will describe the operation of these devices along with their advantages and disadvantages.

Capacitive (aluminum oxide) sensor

Aluminum oxide sensors are the most widely used, especially in HVAC applications, where cost is more

Table 1 Knowing the dry bulb temperature, T_{db}, and relative humidity, ϕ, dew point temperatures, T_d, determined from a table, a psychrometric chart, and various equations are compared

T_{db} (°C)	ϕ (× 100) (%)	p_{vs} (table) (kPa)	p_v (= ϕp_{vs}) (kPa)	T_d (table) (°C)	T_d (psych. chart) (°C)	T_d (Eq. 12) (°C)	T_d (Eq. 13) (°C)	T_d (Eq. 14) (°C)	T_d (Eq. 16) (°C)
9	70	1.1481	0.80367	3.823	4	3.8369	3.8450	3.8726	4.1699
29	20	4.0083	0.80166	3.787	4	3.8011	3.8094	3.7554	4.1349
19	40	2.1978	0.87912	5.105	5	5.1260	5.1271	5.0134	5.4335
19.5	70	2.2683	1.58781	14.431	14	13.9243	13.927	13.973	14.116
34	30	5.3239	1.59717	14.476	14	14.015	14.017	13.927	14.206
29.5	50	4.1272	2.0636	17.994	18	18.015	18.035	18.064	18.177
46	29	10.0976	2.928304	23.675	24	23.683	23.735	23.727	23.817
35	80	5.6278	4.50224	31.024	31	31.007	31.099	31.127	31.115

critical than accuracy. A capacitor consists of two electrodes with a dielectric in between. These capacitive sensors consist of an aluminum substrate, followed by a layer of a hygroscopic material, aluminum oxide, then followed by a coating of gold film.[12] The aluminum substrate and gold film are the electrodes of the capacitor, and the aluminum oxide acts as the dielectric. Some recent capacitive sensors have a polymer as their dielectric. Water vapor from moist air absorbs into aluminum oxide, and the amount of absorbed water vapor directly correlates with the capacitance of the sensor. The instrument determines the dew point based on a series of correlations, first on the capacitance–water vapor relationship and then from the temperature–vapor pressure relationship.

Capacitive sensors have the advantage of measuring dew points in extreme temperature conditions. These sensors can measure low frost points, down to − 100°C, and are able to withstand very high temperatures as found in pressurized heating applications. Capacitive sensors have good sensitivity at low humidity levels, so these sensors are frequently used in petrochemical and power industries.[13] Capacitive sensors are small and portable, so they are often mounted in ducts or walls in industrial processing stream applications. These sensors also have a fast response time thus making them suitable in applications where a quick read is critical in saving time and product, such as spot testing of gas cylinders.[14] However, capacitive sensors can become saturated in environments where the relative humidity is above 85%.[13] Also, these sensors are based on a secondary measurement, capacitance, so periodic recalibration is necessary.

Saturated salt (lithium chloride) sensor

Lithium chloride saturated salt sensors consist of a thin wall metal tube covered with a glass sleeve impregnated with saturated lithium chloride solution. Lithium chloride salt solution is the most desired because at saturation it has a low equilibrium relative humidity of 11% and maintains this humidity over wide range of temperatures, 11.2% at 0°C and 10.8% at 70°C.[5] This sensor bobbin is wrapped spirally with two wire electrodes, or bifilar, which are connected to an alternating current voltage source.[4] The ionic solution completes the electrical circuit between the wires. There is a direct relationship between the amount of condensation and the amount of current drawn between the two electrodes.

The operation of this sensor is best described by explaining how the dew point is determined of the following two conditions: initial moist air with a partial vapor pressure greater that the saturated salt, and initial moist air with a partial vapor pressure less than the saturated salt. In the first condition, when the initial moist air has a partial vapor pressure greater than that of the saturated salt solution, water vapor from the air will condense onto the sensor due to the vapor pressure gradient. As more water vapor condenses, the salt solution becomes more conductive. The current creates resistive heating, which increases the vapor pressure of the salt solution. The rate of condensation and rate of heating decrease until equilibrium is reached between the partial vapor pressure of the solution and the surrounding ambient moist air. A temperature probe is embedded into the sensor to measure the solution temperature needed to achieve this equilibrium vapor pressure. At this increased sensor temperature, the increased partial vapor pressure of the solution is proportional to the saturated vapor pressure of water at that same temperature to maintain the 11% equilibrium relative humidity of the lithium chloride saturated solution. So with the temperature reading, the saturated water vapor pressure can be determined. The partial vapor pressure can then be determined from this saturated water vapor pressure using the correlation shown in Eq. 7 and based on 11% equilibrium relative humidity of the saturated salt solution. This partial pressure of the solution is the same partial vapor pressure of the ambient moist air. As dew point temperature directly correlates with vapor pressure, the dew point can be determined based on the temperature–vapor pressure relationship. The hygrometer internally calculates all of these relationships to determine dew point. If the initial moist air has a partial vapor pressure less than the saturated salt solution initially, evaporation would occur until the sensor is at equilibrium with the air. As the solution losses some water to its environment, less current is drawn, and the temperature sensor would record the evaporative cooling.

The limits of this sensor are that moist air conditions below 11% RH cannot be measured since this is the equilibrium vapor pressure of saturated lithium chloride, and dew point above 70°C cannot be measured. As equilibrium has to be reached, the measurement times are longer than with most sensors. Also, exposure to liquid water can wash away the salt crystals in the sensor. These low cost sensors are typically used in applications where an inexpensive, slow, and moderate accuracy is needed, such as in refrigeration controls, dryers, dehumidifiers, air line monitoring, and pill coaters.[13]

Optical surface condensation (CM) hygrometer

The most direct method to determine dew point temperature is the CM hygrometer and is the most widely used for precise dew points.[5] This method of measurement involves a metallic mirrored surface, a mechanism to cool this mirrored surface, a light-emitting device (LED), and a photodetector. The principle of measurement is

based on the reflectance of light from the LED by the mirror. The gas (i.e., moist air) is passed over the mirror, and when the temperature of the mirror is above the dew point of the moist air, the photodector reads the direct reflectance of the light. However, when the mirror is cooled to the dew point, water vapor begins to condense on the mirror surface. The light is then scattered, and the decrease of reflected light intensity is picked up by the detector. The key factors in accurate measurement of the CM hygrometers are controlling the rate of cooling of the mirror, thus controlling the amount of condensing water vapor on the mirror surface, preventing contaminants on the mirror surface, and precise reading of the temperature of the mirror.

The mirrored surface is made of a good conductor, normally of copper or silver, and coated with an inert metal to prevent tarnishing and oxidation. As water vapor can condense at or below the dew point temperature, precise control of the mirror temperature is important not to allow significant dew formation. If significant dew has formed, the humidity ratio will reduce slightly, and the reading will not be the precise dew point temperature of the initial air condition. Modern CM hygrometers cool the mirror using a solid-state heating element, such as a Peltier device, to control cooling of the mirror. High quality CM hygrometers have a feedback control mechanism where the photodetector communicates with the Peltier device to maintain the mirror temperature such that the dew drop level is at a minimum and remains constant, thus the rates of evaporation and condensation are equal.

Contaminants, either insoluble or water-soluble, on the mirror surface can scatter light and give false readings of dew formation. When insoluble contamination occurs, the mirror surface has to be cleaned manually. Water-soluble contamination is normally in the form of dissolved salts, and there are several methods to correct this type of contamination. One method developed by General Eastern is known as the Programmable Automatic Contaminant Error Reduction (PACER) system.[13,14] This method chills the mirror for an extended time, so excessive amounts of dew form to dissolve the salts on the surface. Then the system rapidly heats the mirror to evaporate the water. This results in localized regions of salt crystals on the mirror surface while the majority of the surface is clean. The other method that optical CM hygrometers use to correct for contamination is the Cycling Chilled Mirror (CCM) hygrometer. In this method of measurement the mirror is chilled to the dew point for a short time (5% of the measurement time) and then heated above the dew point.[12] This process is repeated to give a reproducible reading of dew point. By having the mirror at dew point for a limited time reduces the chance of contamination. In both methods, eventually the surface will have to be manually cleaned.

Moisture on the mirror is at equilibrium with the water vapor in the air at one unique temperature, so the measurement of the mirror temperature is a fundamental and primary measurement of dew point temperature.[12] A precise temperature sensor, such as a platinum resistance thermometer, measures the mirror temperature at the dew point. CM hygrometers have a dew point measurement accuracy of $\pm 0.2°C$.

The advantages of CM hygrometers is in its accuracy, reliability, fast response time, ability for longer continuous and unattended operation due to its self-correcting capabilities, and can measure dew points ranging from -70 to 95°C. However, these instruments can be expensive. Typical applications of the CM hygrometers are for humidity calibration standards, for critical environmental monitoring (i.e., clean rooms, pharmaceutical labs, areas where sensitive computers and electronics are located, and environmental chambers), and for monitoring extreme environmental conditions (i.e., heat treating furnaces, engine test beds).[13]

REFERENCES

1. Bohren, C.F.; Albrecht, B.A. *Atmospheric Thermodynamics*; Oxford University Press, Inc.: New York, 1998.
2. Mohen, M.J.; Shapiro, H.N. *Fundamentals of Engineering Thermodynamics,* 4th Ed.; John Wiley and Sons, Inc.: New York, 2000.
3. http://www.natmus.min.dk/CONS/tp/atmcalc/atmoclcl.htm (accessed July 2001).
4. *ASHRAE Handbook, Fundamentals Volume*; SI Edition; American Society of Heating, Refrigeration, and Air Conditioning Engineers, Inc.: Atlanta, Georgia, 1989.
5. Kuehn, T.H.; Ramsey, J.W.; Threlkeld, J.L. *Thermal Environmental Engineering*, 3rd Ed.; Prentice Hall: Saddle River, NJ, 1998.
6. Williams, J. *The Weather Book*, 2nd Ed.; Vintage Books: New York, 1997.
7. http://www.shorstmeyer.com/wxfaqs/humidity/humidity.html (accessed July 2001).
8. http://www.owt.com/seltech/apps/supdpair.htm (accessed July 2001).
9. http://www.paroscientific.com/dewpoint.htm (accessed Apr 2000).
10. http://weatherwise.about.com/science/weather/library/bldefine.htm#dewpoint (acessed July 2001).
11. Li, S.Q.; Gong, Z.X. Calculations of the State Variables of Moist Air. In *Drying '92*; Mujumdar, A.J., Ed.; Elsevier Science Publishers B.V.: The Netherlands, 1992.
12. http://analyzer.com/Theory/moisture/rhdewpoint/rh.htm#Dewpointhygrometers (acessed Feb 2002).
13. http://iceweb.com.au/Analyzer/humidity_sensors.html (accessed Mar 2002).
14. Larson, K. Humidity Measurement Update. Control **1990**, *June*, 49–59.

Distillation System Design

M. Krokida
National Technical University of Athens, Athens, Greece

V. Gekas
Technical University of Crete, Chania, Greece

Z. B. Maroulis
National Technical University of Athens, Athens, Greece

INTRODUCTION

Distillation is the recovery of a valuable component from a liquid phase(s), by vaporization and condensation, usually in several stages. Distillation usually combines stripping (removal) with rectifying (enrichment) of volatile components in a column, consisting of both stripping and rectifying sections.

Distillation systems, used in separating complex mixtures in the chemical and petrochemical industries, consist of a number of continuous distillation columns, arranged in complex configurations. Instead, the food processing industry uses limited distillation processes of medium to small size, some of them batch-operated.

The most common applications of distillation in food and agricultural industry are the ethanol recovery from fermentation solution and the recovery of aroma or volatile components from processing fruit juices.

A variety of design procedures for distillation columns, ranging from shortcut methods to rigorous tray-by-tray calculations, are described in numerous books. Space limitation preclude a detailed review of these various techniques. Instead, a systematic step-by-step design procedure, based on the Fenske–Underwood–Gilliland (FUG) analytical method, is presented.

The proposed procedure is suitable for spreadsheet calculations and it is applied, as an example, to design an ethanol recovery distillation unit from a fermentation solution.

PROCESS DESCRIPTION

The separation of volatile components of liquid mixtures is usually achieved in a series of equilibrium stages, operated countercurrently in distillation columns. Single stage separators can separate partially a component, because of equilibrium limitations. Single stages or flash units are used to separate some components from food liquids, e.g., off-flavors from milk in a high temperature short time (HTST, i.e., aseptic) sterilization process.

Most stripping (removal) and rectifying (enrichment) of volatile components in a mixture is carried out in columns, using various types of trays (plates), each one of which acts as a vapor/liquid equilibrium stage. As thermodynamic equilibrium is not possible to be reached in a tray liquid/vapor contactor, the number of trays, for a given separation, is always greater than the number of theoretical stages.

Most of the industrial distillation columns are operated as continuous units, although there are some batch columns, used in small-scale operations. Fig. 1 shows diagrammatically a simple continuous distillation column. The unit consists of a long vertical column, containing the required number of trays, made up of the stripping (lower) and the fractionating (upper) sections. The trays (perforated, bubble cups or valves) allow the counter flow of liquid and vapor, after thorough mixing to approach equilibrium. The column is equipped with a reboiler at the bottom, which produces the required vapor flow upward, and a condenser at the top, which supplies the required liquid flow downward.

Feed F is introduced near the middle of the column, while a distillate D is received from the top and a residue B is obtained from the bottom. Steam S is used to heat the liquid in the reboiler and cooling water W is used in the condenser. The column is designed to separate a component from the feed of concentration X_F to a distillate X_D and a residue X_B.

In most distillation columns, a total condenser is used, i.e., all the vapors coming out of the first (top) tray are condensed, and the liquid condensate is split into two streams, the distillate product and the reflux, which is returned to the column.

Encyclopedia of Agricultural, Food, and Biological Engineering
DOI: 10.1081/E-EAFE 120007047
Copyright © 2003 by Marcel Dekker, Inc. All rights reserved.

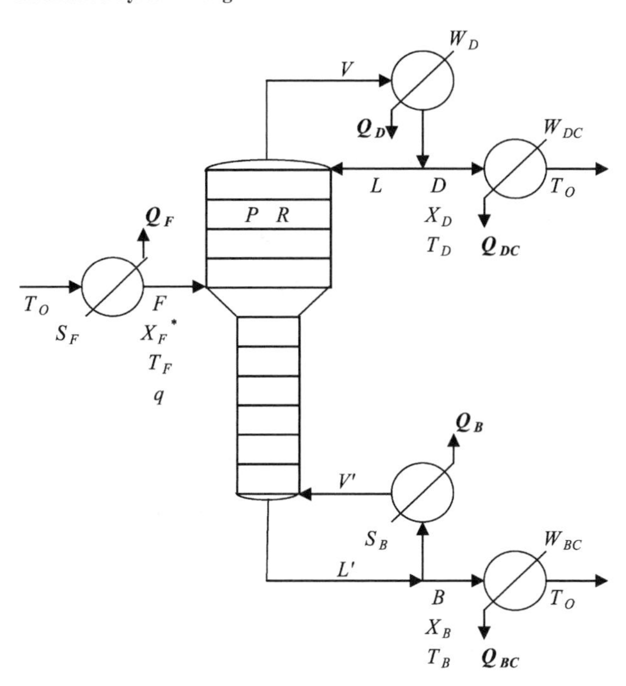

Fig. 1 Flow sheet of distillation process.

PROCESS MODEL

Vapor–Liquid Equilibrium

Vapor–liquid equilibrium data are required for the design of the distillation columns. The basic laws and definitions concerning vapor–liquid equilibrium are summarized in Table 1.

Aqueous food mixtures (solutions) of volatile components are usually nonideal, and the partial pressure is given by the Eq. 1, where, γ_i is the activity coefficient of the *i*th component in the mixture, x_i is the mole fraction,

Table 1 Vapor–liquid equilibrium

Raoult law	
$P_i = \gamma_i x_i P_i^o$	(1)
Antoine equation	
$P_i^o = A_{1i} - \frac{A_{2i}}{A_{3i}+T}$	(2)
Dalton law	
$P_i = y_i P$	(3)
Equilibrium	
$y_i P = \gamma_i x_i P_i^o$	(4)
Partition coefficient	
$K_i = \frac{y_i}{x_i}$	(5)
$K_i = \frac{\gamma_i P_i^o}{P}$	(6)
Activity Coefficient	
$\gamma_i = \begin{cases} 1 \text{ for ideal mixtures} \\ \gamma_i(P,T,\underline{x}) \text{ for nonideal} \end{cases}$	(7)

Table 2 Bubble and dew point calculations

Top	
$X_D K_{LC}(P,T_D) + (1 - X_D) K_{HC}(P,T_D) = 1$	(8)
$X_D / K_{LC}(P,T'_D) + (1 - X_D)/K_{HC}(P,T'_D) = 1$	(9)
Bottom	
$X_F K_{LC}(P,T_B) + (1 - X_B) K_{HC}(P,T_B) = 1$	(10)
$X_F / K_{LC}(P,T'_B) + (1 - X_B)/K_{HC}(P,T'_B) = 1$	(11)
Feed	
$X_F K_{LC}(P,T_F) + (1 - X_F) K_{HC}(P,T_F) = 1$	(12)
$X_F / K_{LC}(P,T'_F) + (1 - X_F)/K_{HC}(P,T'_F) = 1$	(13)

and P_i^o is the vapor pressure at the given temperature T. The vapor pressure P_i^o is estimated from the Antoine Equation as a function of the temperature T (Eq. 2).

The vapor phase in food systems can be considered as ideal and the Dalton law is applicable (Eq. 3), where y_i is the mole fraction in the vapor phase and P is the total pressure.

The assumption of an ideal vapor phase is reasonable, since most food processing operations are carried out at the atmospheric pressure or in vacuum. Nonideal gas phases characterize high-pressure operations ($P > 10$ bar).

At equilibrium, the partial pressure of a component i is the same in both phases, as shown by Eq. 4.

The partition coefficient K_i of component i between the two phases is defined by Eq. 5. Taking into account the equilibrium condition of Eq. 4, Eq. 5 is transformed to Eq. 6. It should be pointed out that, for a given system, K_i is directly proportional to the activity coefficient γ_i.

For ideal solutions $\gamma_i = 1$ (Eq. 7).

The activity coefficients for nonideal mixtures are functions of concentration of the liquid mixtures (Eq. 7). Empirical correlations, used to correlate the activity coefficients, include the Margules (two-parameter), the van Laar (two-parameter), and the Wilson (n-parameter) equations. The Wilson equation and its modifications are suited for computer calculations. A generalized correlation for multicomponent equilibrium data is the universal quasi-chemical (UNIQUAC) equation. The universal function activity contribution (UNIFAC) method is based on the contributions of certain structural parameters

Table 3 Relative volatility calculation

$a = \min(a_D, a_F, a_B)$	(14)
$a_D = \dfrac{K_{LC}(P,T_D)}{K_{HC}(P,T_D)}$	(15)
$a_F = \dfrac{K_{LC}(P,T_F)}{K_{HC}(P,T_F)}$	(16)
$a_B = \dfrac{K_{LC}(P,T_B)}{K_{HC}(P,T_B)}$	(17)

Table 4 Feed quality (flash calculation)

$$\frac{X_F}{(1-q)+\frac{1}{K_{LC}(P,T_F^*)-1}}+\frac{1-X_F}{(1-q)+\frac{1}{K_{HC}(P,T_F^*)-1}}=0 \quad (18)$$

$$X_{FL}=\frac{X_F}{(1-q)(K_{LK}(P,T_F^*)-1)+1} \quad (19)$$

$$X_{FV}=K_{LK}(P,T_F^*)X_{FL} \quad (20)$$

of the components of the liquid mixture, like relative volume and surface area, which are given in thermodynamic tables.[1]

The main conclusion from Table 1 is that the partition coefficients describe the liquid–vapor equilibrium. They are function of pressure, temperature, and eventually (for nonideal mixtures) of liquid concentration.

$$K_i(P,T,\underline{x})=\frac{\gamma_i(P,T,\underline{x})P_i^o(T)}{P}$$

In distillation design, partition coefficients are used to calculate:

- The bubble and dew point temperatures.
- The relative volatility.
- The feed quality.

Bubble and dew point temperatures at feed, top, and bottom are useful in distillation column design. They are calculated using the equations summarized in Table 2.

Relative volatility is crucial in theoretical plates calculation. Once we have estimated the bubble points of the feed, overhead and bottom streams at the operating pressure of the column, we can calculate the relative volatility of the components with respect to the heavy key component at feed, top, and bottom. It is common practice to use as a design value of the relative volatility, the geometric mean of the feed, top, and bottom values. Alternative for safe design, the smallest of these values, is proposed (Table 3).

Feed quality affects strongly the distillation design. It can be obtained by applying the well-known flash-equations presented in Table 4 to the feed stream.

Table 5 Mass balances

$F=B+D$	(21)
$FX_F=BX_B+DX_D$	(22)
$R=L/D$	(23)
$V=L+D$	(24)
$q=(L'-L)/F$	(25)
$L'=B+V'$	(26)

Table 6 Heat balances

Feed heater

$$Q_F=F(H_F-H_{Fo}) \quad (27)$$

$$H_{Fo}=X_F C_{PL}^{LC}T_o+(1-X_F)C_{PL}^{HC}T_o \quad (28)$$

$$H_F=qH_{FL}+(1-q)H_{FV} \quad (29)$$

$$H_{FL}=X_{FL}C_{PL}^{LC}T_F^*+(1-X_{FL})C_{PL}^{HC}T_F^* \quad (30)$$

$$H_{FV}=X_{FV}\left(dH_o^{LC}+C_{PV}^{LC}T_F^*\right)+(1-X_{FV})X\left(dH_o^{HC}+C_{PV}^{HC}T_F^*\right) \quad (31)$$

Bottom reboiler

$$Q_B=V'(H_{B'}-H_B) \quad (32)$$

$$H_{B'}=X_B\left(dH_o^{LC}+C_{PV}^{LC}T'_B\right)+(1-X_B)\times\left(dH_o^{HC}+C_{PV}^{HC}T'_B\right) \quad (33)$$

$$H_B=X_B C_{PL}^{LC}T_B+(1-X_B)C_{PL}^{HC}T_B \quad (34)$$

Bottom product cooler

$$Q_{BC}=B(H_B-H_{Bo}) \quad (35)$$

$$H_{Bo}=X_B C_{PL}^{LC}T_o+(1-X_B)C_{PL}^{HC}T_o \quad (36)$$

Overhead condenser

$$Q_D=V(H_{D'}-H_D) \quad (37)$$

$$H_{D'}=X_D\left(dH_o^{LC}+C_{PV}^{LC}T_{D'}\right)+(1-X_D)\times\left(dH_o^{HC}+C_{PV}^{HC}T_{D'}\right) \quad (38)$$

$$H_D=X_D C_{PL}^{LC}T_D+(1-X_D)C_{PL}^{HC}T_D \quad (39)$$

Overhead product cooler

$$Q_{DC}=D(H_D-H_{Do}) \quad (40)$$

$$H_{Do}=X_D C_{PL}^{LC}T_o+(1-X_D)C_{PL}^{HC}T_o \quad (41)$$

Eq. 18 is used to calculate the feed quality when the feed temperature is known or reversely to calculate the feed temperature when the feed quality is known. Eqs. 19 and 20 estimate the concentrations of the liquid and vapor phases, which are needed in enthalpy calculations.

Mass and Heat Balances

The mass balances of the distillation column system are summarized in Table 5. Eq. 21 is the total mass balance for the column unit and Eq. 22 the balance of the volatile

Table 7 The FUG theoretical number of trays calculation procedure

Fenske

$$SF=\frac{X_D}{1-X_D}\frac{1-X_B}{X_B} \quad (42)$$

$$N_{min}=\frac{\ln(SF)}{\ln(a)} \quad (43)$$

Underwood

$$\frac{aX_F}{a-\theta}+\frac{1-X_F}{1-\theta}=1-q \quad (44)$$

$$R_{min}=\frac{aX_D}{a-\theta}+\frac{1-X_D}{1-\theta}-1 \quad (45)$$

Gilliland

$$\frac{N-N_{min}}{N+1}=0.75\left[1-\left(\frac{R-R_{min}}{R+1}\right)^{0.57}\right] \quad (46)$$

Table 8 Column sizing

Column height	
$E_o = \frac{0.50}{(0.30a)^{0.25}}$	(47)
$N_{act} = \frac{N}{E_o}$	(48)
$H = H_0 + N_{act}H_1$	(49)
Column diameter (rectifying section)	
$\rho_V = \frac{P}{RT_D}$	(50)
$u_F\sqrt{\rho_V} = 0.47H_1^{0.74}$	(51)
$\frac{\pi D^2}{4} = \frac{V}{0.6\rho_V u_F}$	(52)
Column diameter (stripping section)	
$\rho'_V = \frac{P}{RT_F}$	(53)
$u'_F\sqrt{\rho'_V} = 0.47H_1^{0.74}$	(54)
$\frac{\pi D^2}{4} = \frac{V}{0.6\rho'_V u'_F}$	(55)
Feed-point location	
$\ln\left(\frac{N_D}{N_B}\right) = 0.206\ln\left[\frac{B}{D}\left(\frac{1-X_F}{X_F}\right)\left(\frac{X_B}{1-X_D}\right)^2\right]$	(56)
$N_{act} = N_D + N_B$	(57)

component at the total column unit. Eq. 23 is the definition of the reflux ratio R and Eq. 25 the definition of the feed quality q. Eq. 24 is the total mass balance at the top splitter and Eq. 26 at the bottom splitter.

The heat balances at the distillation column are summarized in Table 6. They are used in the calculation of thermal duties of the auxiliary heat exchangers. The stream enthalpies are estimated by suggesting that the heat of mixing is negligible and the enthalpy of any component is described by using three constant characteristic quantities: 1) the latent heat of vaporization at 0°C; 2) the average specific heat of liquid component; and 3) the average specific heat of vapor component.

Table 9 Heat exchangers sizing

Feed heater	
$Q_F = U_F A_F[(T_S - T_{Fo}) - (T_S - T_F)]/\ln[(T_S - T_{Fo})/(T_S - T_F)]$	(58)
Bottom reboiler	
$Q_B = U_B A_B(T_S - T_B)$	(59)
Bottom product cooler	
$Q_{BC} = U_{BC}A_{BC}[(T_B - T_{w2}) - (T_o - T_{w1})]/\ln[(T_B - T_{w2})/(T_o - T_{w1})]$	(60)
Overhead condenser	
$Q_D = U_D A_D[(T_D - T_{w2}) - (T_D - T_{w1})]/\ln[(T_D - T_{w2})/(T_D - T_{w1})]$	(61)
Overhead product cooler	
$Q_{DC} = U_{DC}A_{DC}[(T_D - T_{w2}) - (T_o - T_{w1})]/\ln[(T_D - T_{w2})/(T_o - T_{w1})]$	(62)

Table 10 10 Utilities flow rates

Feed heater	
$Q_F = S_F dH_S$	(63)
Bottom reboiler	
$Q_B = S_B dH_S$	(64)
Bottom product cooler	
$Q_{BC} = W_B Cp_w(T_{w2} - T_{w1})$	(65)
Overhead condenser	
$Q_D = W_D Cp_w(T_{w2} - T_{w1})$	(66)
Overhead product cooler	
$Q_{DC} = W_D Cp_w(T_{w2} - T_{w1})$	(67)

Column Size

One of the most commonly used procedures for obtaining quick estimates of the number of theoretical trays required for a distillation separation is called the FUG procedure.

Gilliland developed an empirical graphical correlation for the number of theoretical trays N in terms of the minimum number of trays at total reflux N_{min}, the minimum reflux ratio R_{min}, and the actual reflux ratio R. A simple equation for Gilliland's data was developed by

Table 11 Cost estimation

Equipment cost	
$C_{eq} = C_{Sh} + C_{trs} + C_F + C_B + C_D + C_{BC} + C_{DC}$	(68)
Column	
$C_{Sh} = fC_{1Sh}\left(DH^{0.8} + D'H'0.8\right)$	(69)
$C_{trs} = fC_{1trs}\left(D^{1.5}H + D'1.5H'\right)$	(70)
Exchangers	
$C_F = fC_{1exc}A_F^{0.65}$	(71)
$C_B = fC_{1exc}A_B^{0.65}$	(72)
$C_D = fC_{1exc}A_D^{0.65}$	(73)
$C_{BC} = fC_{1exc}A_{BC}^{0.65}$	(74)
$C_{DC} = fC_{1exc}A_{DC}^{0.65}$	(75)
Correction factors	
$f = f_T f_P\left(\frac{CEP}{500}\right)$	(76)
$f_T = 0.85\exp\left(\frac{T}{1000}\right)$	(77)
$f_P = 0.98\exp\left(\frac{P}{50}\right)$	(78)
$\left(\frac{CEP}{500}\right) = \exp[0.02(t - 2000)]$	(79)
Annual operating cost	
$C_{op} = C_S(S_F + S_B) + C_W(W_D + W_{DC} + W_{BC})$	(80)
Total annualized cost (objective function)	
$TAC = \text{crf}\,C_{eq} + C_{op}$	(81)
Capital recovery factor	
$\text{crf} = \frac{i_r(1+i_r)^{l_f}}{(1+i_r)^{l_f} - 1}$	(82)

Table 12 Process specifications

Process streams	
F	Feed flow rate
X_F	Feed composition
X_D	Overhead composition
X_B	Bottom composition
Utilities	
T_S	Steam temperature
T_{w1}	Cooling water inlet temperature
T_{w2}	Cooling water outlet temperature
T_o	Ambient temperature

Eduljee (Table 7, Eq. 46). Hence, we can calculate N as a function of R after we have estimated N_{min} and R_{min}.

For constant a systems, Fenske derived an expression for the minimum number of theoretical plates at total reflux. The result is Eq. 43 in Table 7, where the separation factor SF is defined by Eq. 42. The resulting N_{min} is sensitive to relative volatility and consequently good estimates are needed. Smaller values of a overestimate the column size but afford safety results.

Underwood equations (Eqs. 44 and 45) are the most widely used for estimating the R_{min}. The parameter θ is calculated from Eq. 44 and used in Eq. 45. Its value must be in the range $(1,a)$.

The overall plate efficiency is defined by Eq. 48 in Table 8. A simple, but accurate, technique for estimating overall plate efficiency is to use O'Connell's correlation (Table 8, Eq. 47). The column height is estimated by the geometric equation Eq. 49.

The estimation of the column diameter is based on the assumption that the vapor velocity in the column should be about 60% of the flooding velocity, i.e., Eq. 52 for the rectifying section and Eq. 55 for the stripping section. The flooding velocity can be estimated from the Fair data, to which Eq. 51 has been fitted. Eq. 50 calculates the vapor molar density using the state equation of perfect gases.

Finally the feed-point location can be estimated from the Kirkbridge empirical equation (Eq. 56).

Table 13 Degrees of freedom analysis

Process variables	73
Process equations	60
Degrees of freedom	13
Degrees of freedom	13
Specifications	8
Design variables	5

Table 14 Design variables

P	Operating pressure
R	Reflux ratio
q	Feed quality
H_1	Plate spacing
H_0	Space at the ends of the column

Distillation Column Auxiliaries Sizes

All the distillation column auxiliaries are essentially heat exchanger and their sizing is based on the well-known equation for heat flow in the heat exchangers. The logarithmic mean temperature difference is considered as the driving force. These equations are summarized in Table 9.

The corresponding utilities flows are calculated in Table 10, using heat balances for the utilities streams. These flows could be decreased by energy integration methods.

Costing

The total annualized cost of the process, TAC, is selected as the objective function to be optimized. Eq. 81 in Table 11 defines the total annualized cost, TAC, which is a weighted sum between the equipment C_{eq} and operating cost C_{op}. The weighting factor is the capital recovery factor, crf, which is defined by a well-known equation (Eq. 82). Eq. 68 calculates

Table 15 A systematic design procedure for distillation columns

Step 1
Clarify process specifications presented in Table 12
Step 2
Gather from the literature the appropriate technical and costing data according to the Tables 16 and 17
Step 3
Assign initial values for the design variables according to the guidelines in Table 18
Step 4
Solve the distillation column model presented in Tables 2–10 according to the solution algorithm presented in Table 19
Step 5
Estimate the process equipment and operating cost according to the cost model in Table 11 and the algorithm in Table 20
Step 6
Estimate the optimal values for the design variables using an optimization algorithm
Step 7
Analyze the sensitivity of the results with respect to process specifications and process data (technical and costing)

D

Table 16 Technical data

Latent heat of vaporization at reference temperature (0°C)	
dH_{os}	Water
dH_{oLC}	Light component
dH_{oHC}	Heavy component
Average specific heat of liquids	
Cp_{LW}	Water
Cp_{LLC}	Light component
Cp_{LHC}	Heavy component
Average specific heat of vapors	
Cp_{VW}	Water
Cp_{VLC}	Light component
Cp_{VHC}	Heavy component
Antoine constants	
A_{1LC}	Light component
A_{2LC}	
A_{3LC}	
A_{1HC}	Heavy component
A_{2HC}	
A_{3HC}	
Heat transfer coefficients	
U_F	Feed heater
U_B	Bottom reboiler
U_{BC}	Bottom product cooler
U_D	Overhead condenser
U_{DC}	Overhead product cooler

the equipment cost, C_{eq}, and Eq. 80 the annual operating cost C_{op}.

The column cost is estimated by using the Guthrie's equations. Eq. 69 calculates the cost of the column shell and Eq. 70 the cost of the plates. Eqs. 71–75 calculate the cost of the auxiliary exchangers. The correction factors for pressure (Eq. 78) and for temperature (Eq. 77) presented in Table 11 come from fitting exponential equations to data presented by Biegler, Grossmann, and Westerberg.[2]

Finally, Eq. 79 updates the chemical engineering plant index (CEP).

Table 17 Cost data

Utility cost	
C_W	Cost of cooling water ($/kWh)
C_S	Cost of steam ($/kWh)
Equipment unit cost	
C_{1shell}	Column shell (k$/m^2)
C_{1trays}	Column trays (k$/m^2)
C_{1exc}	Heat exchanger (k$/m^2)
Other	
t_y	Annual operating time (hr yr^{-1})
i_r	Interest rate (—)
l_f	Life time (yr)
t	Calendar year for CEP estimation

Table 18 Guidelines for assigning initial values to design variables

Column pressure (P)

The operating pressure P for a distillation column normally is fixed by the economic desirability of using a condenser supplied with the available cooling water and a reboiler supplied with the available heating steam

(1) The bubble point at the condenser T_D should be greater than the operating range of cooling water (T_{W1}, T_{w2}) plus the minimum acceptable temperature difference dT_{min} for operating the condenser. Thus, $T_D - T_{w2} > dT_{min}$

(2) The available steam temperature T_S should be greater than the dew point at the reboiler T'_B plus the minimum acceptable temperature difference dT_{min} for operating the reboiler. Thus, $T_S - T'_B > dT_{min}$

Feed quality (q)

Feed quality q is defined as the fraction of liquid phase in feed stream

$q > 1$	Subcooled liquid
$q = 1$	Saturated liquid
$0 < q < 1$	Vapor–liquid mixture
$q = 0$	Saturated vapor
$q < 0$	Superheated vapor

Feed quality affects strongly the operating curves of column, the thermal load distribution between the reboiler and the feed heater, the column diameter at stripping and rectifying section, etc.

Reflux ratio (R)

As the reflux ratio is increased, the number of trays required for a given separation decreases, so that the capital cost of the column decreases. However, increasing the reflux ratio will increase the vapor rate in the column, which corresponds to more expensive condensers and reboilers, along with higher cooling water and heating steam costs. Therefore, there is an optimum reflux ratio for any specified separation

However, experience has shown that the value of the optimum reflux ratio normally falls in the range $1.05 < R/R_{min} < 1.50$. The slope of the cost curve is very steep below the optimum, but relatively flat above the optimum. Hence, it is common practice to use the rule-of-thumb value $R/R_{min} = 1.20$

Tray spacing (H_1, H_0)

It is common practice to use $\frac{1}{4}$ m–1 m (1 ft–3 ft) space for each tray plus $1\frac{1}{2}$ m–3 m (5 ft–10 ft) to include additional space at the top and the bottom of the column.

It is undesirable to build very tall and skinny columns because they will bend. A design guideline often used is that the column height should be less than about 60 m (175 ft), but a better design guideline is that the height to diameter ratio should be less than 20 to 30.

Table 19 Solution algorithm for process model

Bubble and dew point temperature (Table 2)
Eq. 8 → T_D
Eq. 9 → T'_D
Eq. 10 → T_B
Eq. 11 → T'_B
Eq. 12 → T_F
Eq. 13 → T'_F
Design value for relative volatility (Table 3)
Eq. 17 → a_B
Eq. 16 → a_F
Eq. 15 → a_D
Eq. 14 → a
Feed Quality (Table 4)
Eq. 18 → T^*_F
Eq. 19 → X_{FL}
Eq. 20 → X_{FV}
Mass balances (Table 5)
Eqs. 21 and 22 → D
Eq. 21 → B
Eq. 23 → L
Eq. 24 → V
Eq. 25 → L'
Eq. 26 → V'
Heat balances (Table 6)
Eq. 31 → H_{FV}
Eq. 30 → H_{FL}
Eq. 29 → H_F
Eq. 28 → H_{Fo}
Eq. 27 → Q_F
Eq. 34 → H_B
Eq. 33 → H'_B
Eq. 32 → Q_B
Eq. 36 → H_{Bo}
Eq. 35 → Q_{BC}
Eq. 39 → H_D
Eq. 38 → H'_D
Eq. 37 → Q_D
Eq. 41 → H_{Do}
Eq. 40 → Q_{Do}
Number of theoretical plates estimation. the FUG method (Table 7)
Eq. 42 → SF
Eq. 43 → N_{min}
Eq. 44 → θ
Eq. 45 → R_{min}
Eq. 46 → N
Column sizing (Table 8)
Eq. 47 → E_o
Eq. 48 → N_{act}
Eq. 49 → H
Eq. 50 → ρ_V
Eq. 51 → u_F
Eq. 52 → D
Eq. 53 → ρ'_V
Eq. 54 → u'_F
Eq. 55 → D'
Eqs. 56 and 57 → N_D
Eq. 57 → N_B
Heat exchangers sizing (Table 9)
Eq. 58 → A_F
Eq. 59 → A_B
Eq. 60 → A_{BC}
Eq. 61 → A_D
Eq. 62 → A_{DC}
Utilities flow rates estimation (Table 10)
Eq. 63 → S_F
Eq. 64 → S_B
Eq. 65 → W_B
Eq. 66 → W_D
Eq. 67 → W_{DC}

PROCESS DESIGN

In a typical distillation design problem, the feed stream is known and the concentrations of top and bottom streams are fixed by product purity specifications. Usually, the utilities characteristics, i.e., the temperature of the heating steam and the temperature operating range of the cooling water are also known. These process design specifications are summarized in Table 12.

A degrees of freedom analysis is presented in Table 13. The mathematical model described in Tables 2–10 consists of 60 equations, in which 73 variables are incorporated resulting in 13 degrees of freedom. Process

Table 20 Solution algorithm for the cost model in Table 11

Eq. 82 → crf
Eq. 79 → CEP
Eq. 78 → f_P
Eq. 77 → f_T
Eq. 76 → f
Eq. 69 → C_{Sh}
Eq. 70 → C_{trs}
Eq. 71 → C_F
Eq. 72 → C_B
Eq. 73 → C_D
Eq. 74 → C_{BC}
Eq. 75 → C_{DC}
Eq. 68 → C_{eq}
Eq. 80 → C_{op}
Eq. 81 → TAC

D

Table 21 Process specifications for ethanol distillation

Process streams	
$F = 0.143\,\text{kmol}\,\text{sec}^{-1}$	Feed flow rate
$X_F = 0.05\,\text{kmol}\,\text{kmol}^{-1}$	Feed composition
$X_D = 0.75\,\text{kmol}\,\text{kmol}^{-1}$	Overhead composition
$X_B = 0.01\,\text{kmol}\,\text{kmol}$	Bottom composition
Utilities	
$T_S = 160°\text{C}$	Steam temperature
$T_{w1} = 15°\text{C}$	Cooling water inlet temperature
$T_{w2} = 45°\text{C}$	Cooling water outlet temperature
$T_o = 20°\text{C}$	Ambient temperature

specifications fix 8 variables and consequently 5 design variables are available. The variables summarized in Table 14 are usually selected as design variables.

Table 15 summarizes a systematic design procedure for distillation columns. Table 16 provides the technical data and Table 17 provides the cost data. Table 18 summarizes the guidelines for assigning initial values to design variables. Solution algorithm for process model and cost model are given in Tables 19 and 20, respectively.

APPLICATIONS IN THE FOOD INDUSTRY—A CASE STUDY OF ETHANOL DISTILLATION

Historically, distillation has been applied, for the first time, in the binary system of water/ethanol. In practice, fermentation aqueous mixtures of ethanol contain other volatile components as well, although in low concentrations, but the system is not exactly a binary one. The same happens with the second important application of distillation in the food and beverage industry, i.e., the production of concentrated fruit juices. Typical volatile components are higher alcohols, esters, and aldehydes. In both cases, the design is based on water and a "key" component, which is ethanol in the first case and the less volatile of the components in the fruit juice case. Thus, the design allows for the removal and/or recovery of the more volatile compounds as well.

Table 22 Design variables for ethanol distillation

$P = 1\,\text{bar}$	Operating pressure
$R/R_{min} = 1.20$	Reflux ratio
$q = 0.50$	Feed quality
$H_1 = 0.75\,\text{m}$	Plate spacing
$H_0/H_1 = 3$	Space at column ends

Table 23 Technical data for ethanol distillation

Latent heat of vaporization at reference temperature (0°C)	
$dH_{os} = 44.8\,\text{MJ}\,\text{kmol}^{-1}\,\text{K}^{-1}$	Water
$dH_{oLC} = 40.5\,\text{MJ}\,\text{kmol}^{-1}\,\text{K}^{-1}$	Light component
$dH_{oHC} = 44.8\,\text{MJ}\,\text{kmol}^{-1}\,\text{K}^{-1}$	Heavy component
Average specific heat of liquids	
$Cp_{LW} = 75.3\,\text{KJ}\,\text{kmol}^{-1}\,\text{K}^{-1}$	Water
$Cp_{LLC} = 96.7\,\text{KJ}\,\text{kmol}^{-1}\,\text{K}^{-1}$	Light component
$Cp_{LHC} = 75.3\,\text{KJ}\,\text{kmol}^{-1}\,\text{K}^{-1}$	Heavy component
Average specific heat of vapors	
$Cp_{VW} = 34.2\,\text{KJ}\,\text{kmol}^{-1}\,\text{K}^{-1}$	Water
$Cp_{VLC} = 73.9\,\text{KJ}\,\text{kmol}^{-1}\,\text{K}^{-1}$	Light component
$Cp_{VHC} = 34.2\,\text{KJ}\,\text{kmol}^{-1}\,\text{K}^{-1}$	Heavy component
Antoine constants	
$A_{1LC} = 1.23\text{E} + 01$	Light component
$A_{2LC} = 3.80\text{E} + 03$	
$A_{3LC} = 2.31\text{E} + 02$	
$A_{1HC} = 1.19\text{E} + 01$	Heavy component
$A_{2HC} = 3.99\text{E} + 03$	
$A_{3HC} = 2.34\text{E} + 02$	
Molecular weight	
$M_{LC} = 46.1\,\text{kg}\,\text{kmol}^{-1}$	Light component
$M_{HC} = 18.0\,\text{kg}\,\text{kmol}^{-1}$	Heavy component
Heat transfer coefficients	
$U_B = 1.50\,\text{kW}\,\text{m}^{-2}\text{K}^{-1}$	Bottom reboiler (evaporation/ evaporation)
$U_F = 1.25\,\text{kW}\,\text{m}^{-2}\text{K}^{-1}$	Feed heater (evaporation/heating–evaporation)
$U_D = 1.00\,\text{kW}\,\text{m}^{-2}\text{K}^{-1}$	Overhead condenser (condensation/heating)
$U_{BC} = 0.75\,\text{kW}\,\text{m}^{-2}\text{K}^{-1}$	Bottom product cooler (cooling/heating)
$U_{DC} = 0.75\,\text{kW}\,\text{m}^{-2}\text{K}^{-1}$	Overhead product cooler (cooling/heating)

Table 24 Cost data for ethanol distillation

Utility cost	
$C_W = 0.004\,\$/\text{kWh}$	Cost of cooling water
$C_S = 0.040\,\$/\text{kWh}$	Cost of steam
Equipment unit cost	
$C_{1Sh} = 10\,\text{k}\$/\text{m}^2$	Column shell
$C_{1trs} = 1\,\text{k}\$/\text{m}^2$	Column trays
$C_{1exc} = 5\,\text{k}\$/\text{m}^2$	Heat exchanger
Other	
$t_y = 2000\,\text{hr}\,\text{yr}^{-1}$	Annual operating time
$i_r = 0.08$	Interest rate
$l_f = 5\,\text{yr}$	Life time
$t = 2005$	Calendar year for CEP estimation

Table 25 Results of model solution

Flow rate (kmol sec^{-1})		
$F = 0.143$		Feed
$D = 0.008$		Top product
$B = 0.135$		Bottom product
$V = 0.114$	0.247	Vapor at rectifying section
$V' = 0.114$	0.104	Vapor at striping section
$L = 0.107$	0.239	Liquid at rectifying section
$L' = 0.250$	0.239	Liquid at striping section
Temperature (°C)		
$T_F^* = 98.2$	99.1	Feed
$T_F = 98.2$		Feed bubble point
$T_F' = 99.1$		Feed dew point
$T_D = 81.9$		Top bubble point
$T_D' = 85.2$		Top dew point
$T_B = 99.5$		Bottom bubble point
$T_B' = 99.7$		Bottom dew point
Composition (kmol kmol^{-1})		
$X_F = 0.05$		Feed
$X_D = 0.75$		Top product
$X_B = 0.01$		Bottom product
$X_{FV} = 0.11$	0.05	Feed
$X_{FL} = 0.05$	0.02	Feed
Thermal load (MW)		
$Q_F = 0.85$	6.67	Feed heater
$Q_B = 4.66$	4.22	Bottom reboiler
$Q_{BC} = 0.81$		Bottom product cooler
$Q_D = 4.51$	9.71	Overhead condenser
$Q_{DC} = 0.04$		Overhead product cooler
Heat transfer area (m^2)		
$A_F = 11$	88	Feed heater
$A_B = 51$	46	Bottom reboiler
$A_{BC} = 31$		Bottom product cooler
$A_D = 89$	193	Overhead condenser
$A_{DC} = 4$		Overhead product cooler
Column geometric characteristics		
$N = 15.0$	14.8	Total number of theoretical plates
$N_{min} = 7.01$		Minimum number of theoretical plates
$N_{act} = 27$		Actual number of plates
$N_D = 13$		Number of plates at rectifying section
$N_B = 14$		Number of plates at rectifying section
$D = 2.5$	3.7	Tower diameter at rectifying section
$D' = 2.5$	2.4	Tower diameter at stripping section
$H = 23$	23	Column height
Column operating characteristics		
$P = 1$ bar		Operating pressure
$q = 1$	0	Feed quality
$R = 13.8$	30.9	Reflux ratio
$R_{min} = 11.5$	25.8	Minimum reflux ratio
$u_f = 1.1$ m sec^{-1}		Flooding velocity at the rectifying section
$u_f' = 1.1$ m sec^{-1}		Flooding velocity at the stripping section
SF = 297		Separation factor
$\alpha = 2.25$		Design relative volatility
$\alpha_F = 2.26$	2.25	Relative volatility at feed
$\alpha_D = 2.30$		Relative volatility at top
$\alpha_B = 2.25$		Relative volatility at bottom
$E_o = 0.55$		Overall plate efficiency

Distillation is used in the production of brandy and other alcoholic beverages from wine and other ethanol-fermented liquids. Traditional brass (copper) stills of small capacity are used to separate the ethanol and other volatiles in simple one-stage distillations. Partial fractionation of ethanol is obtained by partial reflux in the piping of the still, obtaining distillates of about 50% ethanol by volume.

Fractional distillation of alcoholic beverages in medium to large-scale operations is carried out in batch or continuous distillation columns made of stainless steel, producing ethanol of high concentration, about 95% by volume.

Ethanol distillation for the production of alcoholic beverages and spirits is governed by strict regulations, concerned with taxation, like the U.S. Bureau of Alcohol and Cigarettes.

The step-by-step design procedure described in Table 15 is applied to design a distillation column to recover the ethanol from a fermentation solution,

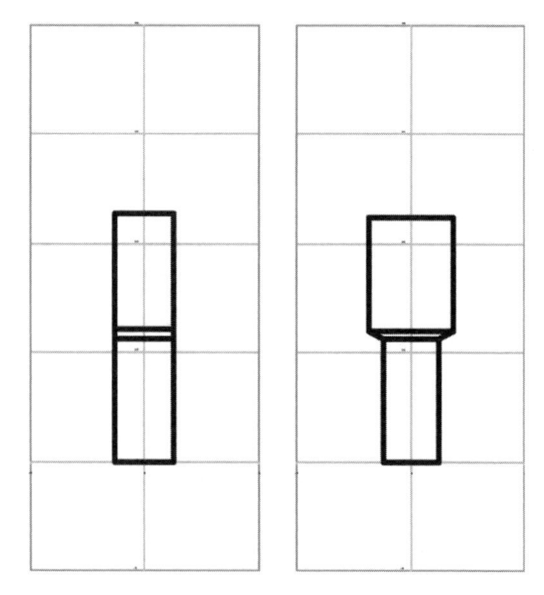

Fig. 2 Column shape in scale for $q = 1$ (left) and $q = 0$ (right), respectively. The side of the square is 10 m.

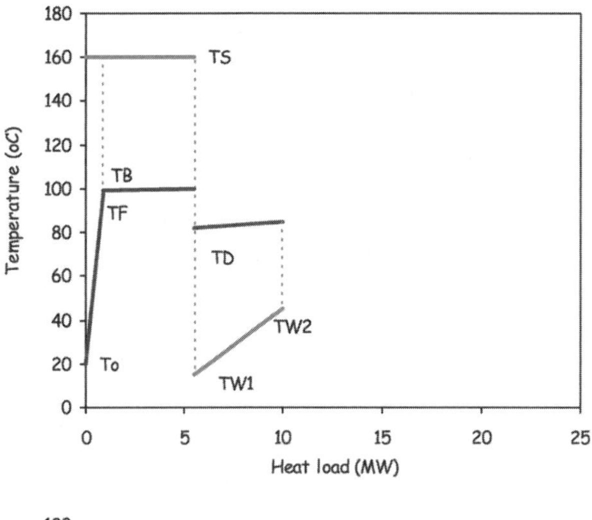

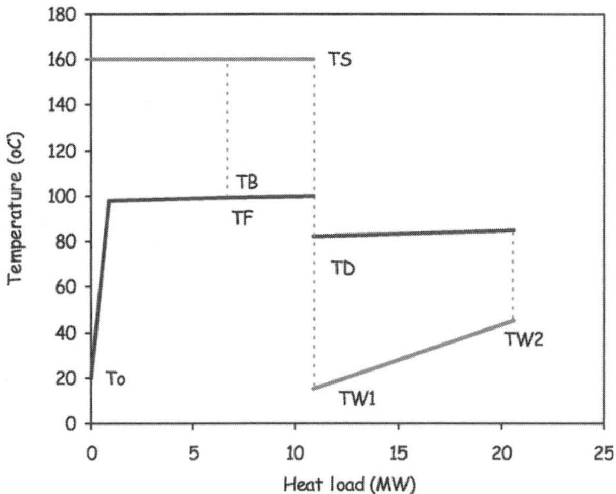

Fig. 3 Enthalpy–temperature diagram for feed heater, reboiler, and condenser, for $q = 1$ (upper) and $q = 0$ (lower), respectively.

containing 5% of ethanol. The composition of the distillate and the bottom products will be 75% and 1% ethanol, respectively (concentrations in molar basis). The feed to the column will be 0.143 kmol sec^{-1}, i.e., 10 tn hr^{-1}.

The results of steps 1–3 (Table 15) are summarized in Tables 21–23. Table 21 shows the process specifications of both process streams and utilities. Table 22 contains the applied design parameters. Table 23 contains the technical data (physicochemical property values). Table 24 exhibits the cost data taken into account for costing the process. The most significant results from the solution of the model are presented in Table 25 and Figs. 2–6. In Table 25, the first column refers to saturated vapor (feed quality $q = 1$) and the second column to saturated liquid (feed quality $q = 0$) In Figs. 2–6, a comparison between the two cases of feed quality can be made. Thus, Fig. 2 shows that a bigger size rectifying part is needed for $q = 0$, whereas

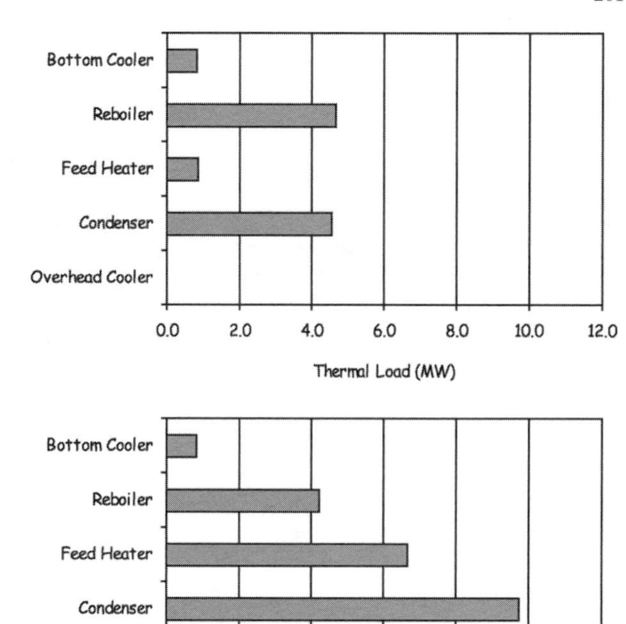

Fig. 4 Thermal duties of the distillation unit, for $q = 1$ (upper) and $q = 0$ (lower), respectively.

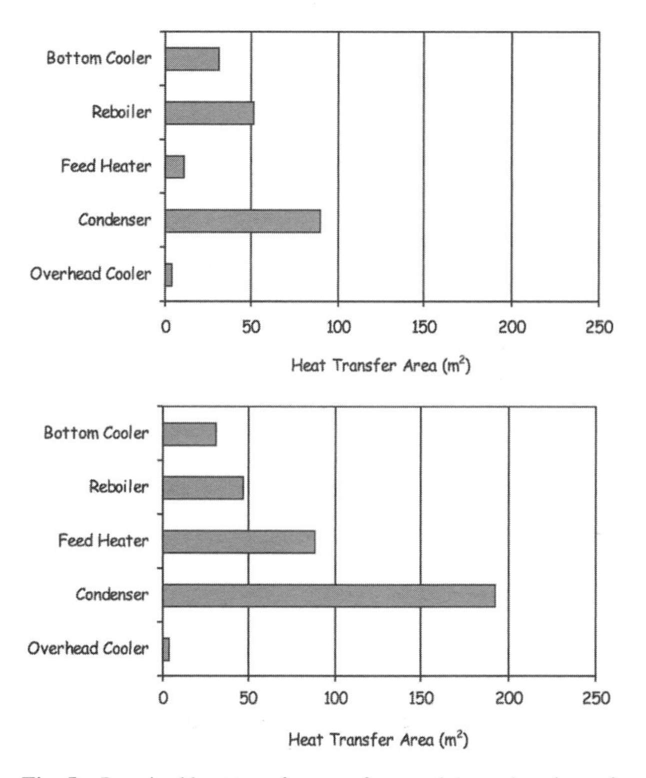

Fig. 5 Required heat transfer areas for $q = 1$ (upper) and $q = 0$ (lower), respectively.

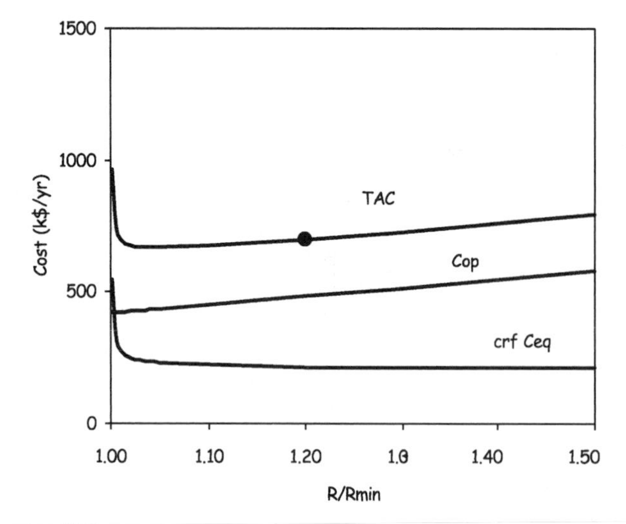

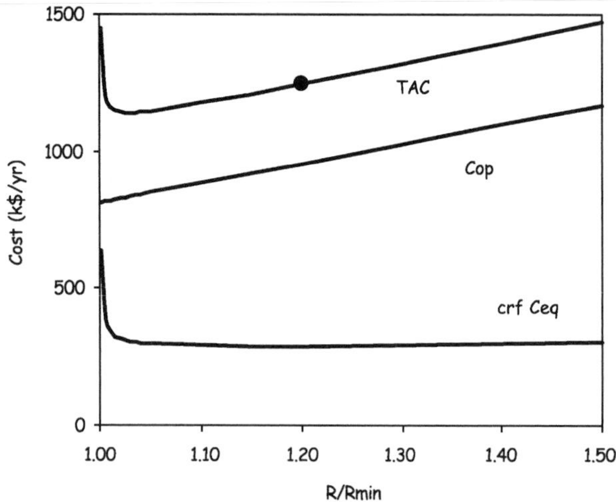

Fig. 6 Effect of R/R_{min} on the total annualized cost TAC, equipment cost C_{eq}, and operating cost C_{op} for $q = 1$ (upper) and $q = 0$ (lower), respectively.

the stripping part is the same for both cases. Consequently Figs. 3–5 show in a quantitative way the enthalpy, thermal load, heat transfer area requirements and finally Fig. 6 gives the results of the cost analysis for both high and low feed quality applications of ethanol distillation.

NOMENCLATURE

A Transfer area
- A_B Bottom reboiler transfer area
- A_{BC} Bottom product cooler transfer area
- A_D Overhead condenser transfer area
- A_{DC} Overhead product cooler transfer area
- A_F Feed heater transfer area

B Bottom product or residual flow rate
C Cost
D Top product flow rate
D Tower diameter at rectifying section
D' Tower diameter at stripping section
E_o Overall plate efficiency
F Feed flow rate
H Column height
H_0 Space at the ends of the column for vapor disengagement and liquid sump
H_1 Plate spacing
H Enthalpy
- H_B Bottom liquid before the bottom cooler stream enthalpy
- $H_{B'}$ Vapor after the bottom reboiler stream enthalpy
- H_{Bo} Bottom liquid after the bottom cooler stream enthalpy
- H_D Overhead liquid before the overhead cooler stream enthalpy
- $H_{D'}$ Vapor before the overhead condenser stream enthalpy
- H_{Do} Overhead liquid after the overhead cooler stream enthalpy
- H_F Vapor–liquid in equilibrium after the feed heater stream enthalpy
- H_{Fo} Liquid before the feed heater stream enthalpy

K Partition coefficient
L Liquid at rectifying section flow rate
L' Liquid at striping section flow rate
N Total number of theoretical plates
N_{act} Actual number of plates
N_B Number of plates at rectifying section
N_D Number of plates at rectifying section
N_{min} Minimum number of theoretical plates
P Operating pressure
q Feed quality
Q Thermal load or heat load
- Q_B Bottom reboiler thermal load
- Q_{BC} Bottom product cooler thermal load
- Q_D Overhead condenser thermal load
- Q_{DC} Overhead product cooler thermal load
- Q_F Feed heater thermal load

R Reflux ratio
R_{min} Minimum reflux ratio
S_B Steam at bottom reboiler flow rate
SF Separation factor
S_F Steam at feed heater flow rate
t Time
T Temperature
- T'_B Bottom dew point temperature
- T'_D Top dew point temperature
- T'_F Feed dew point temperature
- T_B Bottom bubble point temperature
- T_D Top bubble point temperature

	T_F Feed bubble point temperature
	T_F^* Feed temperature
	T_o Ambient temperature
	T_S Heating steam temperature
	T_{w1} Cooling water inlet temperature
	T_{w2} Cooling water outlet temperature
u_f	Flooding velocity at the rectifying section
u_f'	Flooding velocity at the stripping section
U	Overall heat transfer coefficient
V	Vapor at rectifying section flow rate
V'	Vapor at striping section flow rate
W_{BC}	Water at bottom product cooler flow rate
W_D	Water at overhead condenser flow rate
W_{DC}	Water at overhead product cooler flow rate
X	Molar fraction of a component in the liquid phase
	X_B Bottom product molar fraction
	X_D Top product molar fraction
	X_F Feed molar fraction
	X_{FL} Feed molar fraction in liquid phase
	X_{FV} Feed molar fraction in vapor phase

Greek

α	Design relative volatility
α_B	Relative volatility at bottom
α_D	Relative volatility at top
α_F	Relative volatility at feed
γ	Activity coefficient
θ	Parameter in Underwood equation
ρ_V	Vapor molar density at the rectifying section
ρ_V'	Vapor molar density at the stripping section

Abbreviations

CEP	Chemical engineering plant index
crf	Capital recovery factor
HTST	High temperature short time
TAC	Total annualized cost

REFERENCES

1. Reid, R.C.; Prausnitz, J.M.; Poling, B.E. *The Properties of Gases and Liquids*, 4th Ed.; McGraw-Hill: New York, 1987.
2. Biegler, L.T.; Grossmann, I.E.; Westerberg, A.W. *Systematic Methods of Chemical Process Design*; Prentice Hall: New York, 1997.
3. Douglas, J.M. *Conceptual Design of Chemical Processes*; McGraw-Hill: New York, 1988.
4. Perry, R.H.; Green, D. *Chemical Engineers' Handbook*, 7th Ed. McGraw-Hill: New York, 1997.
5. Sinnott, R.K. *Coulson and Richardson's Chemical Engineering, Vol 6, Design*; Butterworth-Heinemann: Oxford, 1996.

Drainage Materials

Tommy L. Zimmerman
The Ohio State University, Wooster, Ohio, U.S.A.

INTRODUCTION

Much of the world's land has a problem with wet soils that need surface drainage, subsurface drainage, or a combination of both. Wet soils cause problems such as reduced crop yields, poor crop quality, difficulty with many different field operations, and loss of income for growers. Previous articles in this series have discussed drainage in general, soil properties, water erosion and wind erosion control, and surface drainage and subsurface drainage systems. This article will discuss the materials used in subsurface drainage systems. The discussion will include the materials used as the pipe or conduit for collecting and transporting the excess water away and the various envelope materials used to surround the pipe or as a base for supporting the pipe.

For the purpose of this article, the term, tile, will mean pipe made from clay or concrete materials. The term, tubing, will mean plastic pipe.

DRAINAGE MATERIALS

Clay and Concrete Tile

The first subsurface drainage tile was made of clay or concrete. These tiles were manufactured in various diameters and wall thicknesses. Usually, they were manufactured in 1-ft lengths and placed end-to-end in a downward sloping trench. Water entered the tile where the individual tiles were "butted" together. Various accessories were also manufactured to accommodate connections where one or more lateral tile lines needed to be hooked to a main line or to accommodate connections of different lateral and main line diameters.

Clay tiles were manufactured from locally derived clay soils that were molded into the desired shape and then fired in a kiln. The firing hardened the clay into a durable material. Concrete is a mixture of water, cement, sand, and coarse aggregate. As with clay tiles, concrete tiles were molded in the desired shape and then allowed to harden and cure. Various accessory connectors were also manufactured for concrete tile.

Sometimes, clay and concrete tile lie too tightly together in the trench where they are butted together. This would seriously slow down the rate at which water entered the tile. To allow water to enter the tiles more effectively, tiles were sometimes manufactured with holes or wavy edges.

One advantage of both clay and concrete tiles are their durability. On the other hand, they are heavy and have to be placed in the trench with near-perfect alignment. This means that the individual tiles must be butted together on the same plane. If not, then problems with the flow and soil falling into the tile may occur.

Concrete tile are less subject to freezing and thawing deterioration than clay tile. Very acid or very alkaline soils can cause chemical deterioration of concrete tile, while clay tile is usually not affected.[1]

Plastic Tubing

During the 1960s, plastic materials began to be used in drainage systems.[2] Most drainage pipe is corrugated plastic tubing (CPT). This tubing is fairly thin walled and is corrugated for increased strength. It has holes or slots to allow water to enter. It is available in various diameters and mostly available in rolls that are several hundred to several thousand feet in length. Accessory fittings are available for making connections.

CPT is manufactured from high-density polyethylene (HDPE) or polyvinyl chloride (PVC). HPDE is common in the United States and PVC is common in Europe.[3] Some PVCs are manufactured as smooth wall pipes with holes for water entry, but these are more commonly used for drain fields for septic systems. CPT is available in many different colors and may be formulated with chemicals that retard UV light degradation (a problem with outdoor storage) or to prevent iron ochre formation. (Iron ochre is a complex material of iron bacteria, organic matter, and other materials that forms as a sludge in the pipe and can seriously impede flow and capacity of the pipe.)

The main advantage of the plastic pipes is their light weight. They are durable and easy to handle. Because this pipe is not in short lengths like clay and concrete tile, the alignment requirement for installation is not as important. Ideally, they should be installed in a smooth-bottomed trench, but small elevation variations in the trench bottom are not going to be problematic as in the case of the clay and concrete materials.

Encyclopedia of Agricultural, Food, and Biological Engineering
DOI: 10.1081/E-EAFE 120006925
Copyright © 2003 by Marcel Dekker, Inc. All rights reserved.

Although the corrugations provide strength to the pipe, it is important that a 120°-angle groove be cut in the bottom of the trench as the pipe is being installed. This groove is cut by the trenching equipment, and provides side support against bulging and helps provide proper alignment. If this groove is not provided, then an envelope of gravel is recommended for proper support.[4]

The plastic materials can be installed by any suitable trenching method. CPT can also be installed by a "trenchless" method using drainage plows. This involves using a large subsoiler type chisel plow point that allows the flexible CPT to be fed or "pulled" into the soil without actually opening up a trench. The main advantage is that the pipe can be installed at a higher rate of speed ($80\,\mathrm{ft\,min^{-1}} - 150\,\mathrm{ft\,min^{-1}}$ of ground speed) as compared to trenching methods. In addition, no backfilling of trenches is necessary.

Envelope Materials

Envelope materials, sometimes called filters, prevent soil particles from entering and plugging the pipe. This is not a problem with well-aggregated soil. Most plugging problems occur because of poorly structured sandy soils, especially soils with fine sand.

The first material used as envelopes was gravel. Sometimes gravel was not available locally or it was too heavy for economical transporting or required in too large a volume to be convenient. For these reasons, synthetic materials were developed.

These synthetic materials are made from various materials such as nylon, polyester, polypropylene, and others that may be woven, knitted, or spun-bonded. Weaving the material into a "sock" enables the envelope to encase the pipe. The envelope may be installed at the time of pipe manufacture or may be installed when the pipe is being installed. The main advantages of these synthetics are that they are relatively low-cost, lightweight, and long lasting.

Envelopes may eventually plug. This is because they are designed to allow some fine soil particles to pass through them so as not to overly restrict the flow of water. After some time, these particles could plug the pipe.[1]

Miscellaneous Materials

Metal pipes are often used for drainage outlets and in other situations where heavy loads would breakdown the other materials. A typical use might be as a culvert pipe that would cross beneath a driveway, lane, or road. Normally, they would not be used for field drainage. These pipes may be corrugated or smooth. As in the case with plastic pipe, smooth metal pipes are thicker walled and corrugated metal pipes are thinner walled. They do not have holes in them, because they are usually used for water transport and not for water entry.

CONCLUSION

Subsurface drainage materials include clay tile, concrete tile, plastic pipes or tubing, envelope (filter) materials like gravel and synthetic fabrics, and thick-walled smooth metal or plastic pipes or thinner-walled corrugated metal or plastic pipes used for outlets or for culverts. The use of these materials for draining wet agricultural soils is one important reason why crop yields, crop production, and crop quality have improved over time.

REFERENCES

1. Schwab, G.O.; Fangmeier, D.D.; Elliot, W.J.; Frevert, R.K. Subsurface Drainage Design. *Soil and Water Conservation Engineering*, 4th Ed.; John Wiley and Sons, Inc.: New York, 1993; 312–313.
2. Fouss, J.L.; Reeve, R.C. Advances in Drainage Technology: 1955–85. In *Farm Drainage in the United States. History, Status, and Prospects*; Pavelis, G.A., Ed.; Misc. Publ. No. 1455; Economic Research Service, U.S. Department of Agriculture: Washington, D.C., 1987; 33–43.
3. Schwab, G.O.; Fangmeier, D.D.; Elliot, W.J. Subsurface Drainage. *Soil and Water Management Systems*, 4th Ed.; John Wiley & Sons, Inc.: New York, 1996; 245–247.
4. Pira, E. Tile Drainage Systems. *Golf Course Irrigation System Design and Drainage*; Ann Arbor Press, Inc.: Chelsea, MI, 1997; 310–315.

Drip Irrigation

Allen L. Thompson
University of Missouri, Columbia, Missouri, U.S.A.

INTRODUCTION

Drip irrigation is the slow application of water to the soil in discrete droplets or small streams.[1] Its development resulted from the need to provide adequate water to meet plant requirements using limited water supplies. Drip irrigation uses small "emitters" to deliver water to a specified point. Various terminologies have been used to describe this low flow, discharge rate. The term "trickle" was frequently used during early development in the 1960s and 70s. Today, this technology is most often referred to as "drip" irrigation. Subsurface drip irrigation (SDI) describes the placement of drip irrigation laterals below the soil surface within the root zone. Microirrigation encompasses a number of different application methods including bubbler, drip, trickle, mist, or spray.[1]

OPERATING PARAMETERS

Drip systems have been successfully installed on many agronomic crops. Permanently installed systems are most often used on tree and vine crops, and SDI on rowcrops. Portable drip systems may also be found on row crops. For SDI, laterals are typically buried below the tillage zone at depths between 20 cm and 60 cm (0.6 ft–2 ft).

A typical drip irrigation system layout is shown in Fig. 1. It consists of a pump or other pressurized water source, air relief/vacuum devices, equipment for chemical injection (venturi meter, by-pass fertilizer tank, or hydraulic injection pump), filters, mainline, manifold and submanifold lines, and laterals on which the emitters are attached. Check valves may be required to prevent backflow to the water supply. Pressure relief valves may be installed downstream of check valves and at the end of the pipeline to protect from surge pressures.[2] Flow meters and pressure gauges can be included to assist in system management. If there is significant elevation change or undulation within the field, pressure regulators and/or pressure-compensating emitters can also be used to improve application uniformity. Vacuum devices should be included at high points in the system to facilitate system drainage and reduce the potential of drawing foreign material into emitters when the system is shut off.

Various emitter types are commercially available, including long path, short orifice, vortex, pressure-compensating, tortuous, and porous pipe or micro-tube.[3] Emitters are classified as inline (Fig. 2), which are molded into the lateral during manufacturing, and online (Fig. 3), which are placed on the lateral by the installer through holes punched through the sides of thicker-walled tubing. Emitter flow rate can be related to water pressure through the following empirical relationship:

$$q = kP^x$$

where q = discharge rate (volume/time), P = pressure (force/area), and k and x are empirical coefficients. The value of x indicates the flow regime. For laminar flow, x approaches 1.0 and discharge rate varies directly with pressure. For fully turbulent, orifice flow, $x = 0.5$. Pressure-compensating emitters have an x-value between 0.0 and 0.1. For spiral long-path emitters, $x = 0.7$, and for vortex emitters, $x = 0.4$.[4] Flow rates for point-source emitters typically range from 1.5 lph to 8.0 lph (0.4 gph–2.0 gph), while line-source rates may reach 12 lph per meter (1 gph per ft).[5] Water pressures are typically between 50 kPa and 240 kPa (7 psi and 35 psi). Flow rates for point-source bubblers may reach 225 lph (60 gph),[5] but often require soil basins to prevent water runoff.[6] Flow rates for micro-sprayers are generally less than 60 lph (15 gph),[7] but may be as great as 175 lph (45 gph).[5]

Drip lateral tubing is usually made from polyethylene (PE) and contains carbon black to resist deterioration by sunlight. Polyethylene used for drip laterals should meet standards for hydrostatic design stress, and failure due to bursting, cracking, splitting, or weeping.[8] Lateral wall thickness varies from 0.1 mm to 1.3 mm (0.004 in.–0.05 in.).[9] Tubing thickness less than 0.4 mm is often termed "drip tape" where discharge occurs through small holes or slits in bi-wall chambers pre-stamped into the PE tubing. Because of the thin walls, drip tape is normally used only one or two seasons before requiring replacement. Flexibility of thicker walled tubing varies with temperature, which may influence application performance. Outside tubing diameters vary from 6 mm to 35 mm (0.25 in.–1.4 in.), with commonly used diameters in the 14 mm–22 mm range (0.55 in.–0.87 in.). Tube fittings for

Encyclopedia of Agricultural, Food, and Biological Engineering
DOI: 10.1081/E-EAFE 120006935
Copyright © 2003 by Marcel Dekker, Inc. All rights reserved.

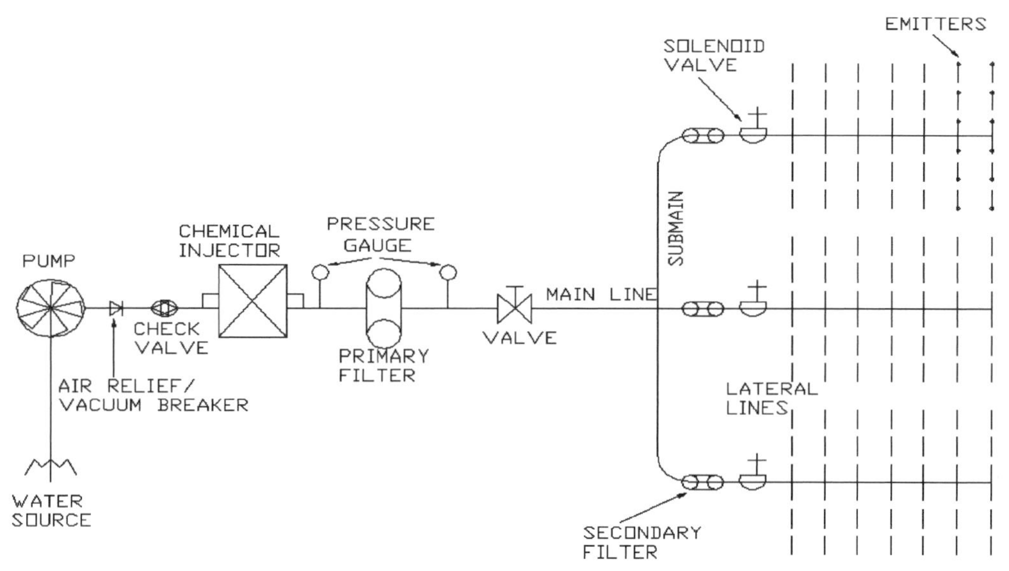

Fig. 1 A typical layout of a drip irrigation system.

items such as elbows and tees are either inside-barbed for a compression fit (Fig. 4) or outside barbed where the tubing is slipped over the fitting and secured with hose clamps, twistlocks, or similar fittings. Distribution tubing is often made from poly-vinylchloride (PVC), which may be rigid or flexible, PE, which may be either high or low density, or vinyl. Mainline pipe diameters for drip systems vary from 2.5 cm to 15 cm (1 in.–6 in.) or larger depending on total system capacity, and are commonly buried. Submains and manifolds will vary in diameter depending on the number of plants to be watered within a given irrigation zone.

Unlike other types of irrigation, drip systems are designed to replace crop water needs on a daily basis. The area wetted as a percent of the total cropped area typically ranges from 20% for widely spaced crops to over 75% for row crops.[5] Wetted patterns vary with soil type. For sandy soils, the wetted "bulb" of soil moves downward quickly with reduced horizontal wetting. For heavier clayey soils, the wetted bulb is more uniformly shaped (Fig. 5). For line-source applications, such as row crops, emitters on a given lateral should be spaced close enough for wetted patterns to merge on the surface. Because soil texture influences this pattern, it should be taken into account during design. Laterals with inline emitters can be selected with spacings between 15 cm and 120 cm (6 in. and 48 in.). Online emitters are most often selected where plants are widely spaced. Laterals are typically spaced along rather than between crop rows because salts tend to buildup along the outer boundaries of the wetted bulb, which is more likely to occur nearer the plant base when laterals are placed midway between rows. Normally one-third to three-fourths of the root zone should be supplied with water.[3] Because each zone is watered daily, the number of irrigation sets will depend on system capacity and degree of automation. A daily irrigation schedule of 22 hr is often used to provide adequate downtime for system repair and maintenance.[3]

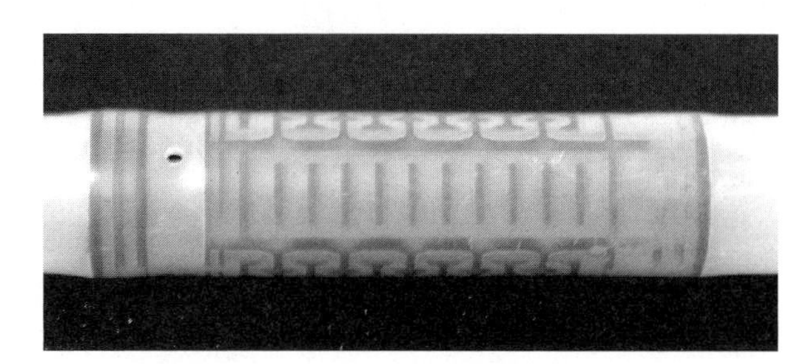

Fig. 2 An inline emitter placed within the lateral (clear tubing for visual purposes only).

Fig. 3 An online emitter inserted along the lateral.

Fig. 4 Inside compression fitting for a drip lateral.

Drip irrigation has both advantages and disadvantages compared to sprinkler or surface irrigation. These include:

Advantages

- Higher application efficiencies because water can be applied directly to the base of the plant, minimizing runoff and evaporation, and improving fertigation.
- Potential reduction in disease problems because foliage is not wetted.
- Reduced weed competition when the soil surface is not completely wetted.
- More saline water supplies can be used because plants are wetted on a daily basis.
- Improved plant growth for many plants because soil water content can be maintained near an optimum level.
- Lower operating pressures and flow rates allow for smaller pipes, pumps, and reduced energy costs.
- Adaptable to problem soils with low intake rates and low soil water storage.
- Field work can continue while the irrigation system is in operation.
- The system design adapts easily to automated control using time clocks and automatic valves (electric or hydraulic).

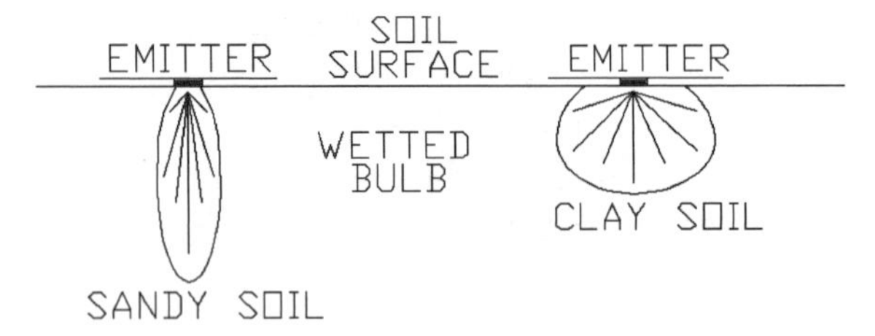

Fig. 5 Typical wetted patterns for sandy and clay soils.

Disadvantages

- Emitters are susceptible to plugging, therefore water filtration and maintenance are very important.
- Laterals are subject to rodent damage.
- Salt problems may develop in the root zone over time because irrigation system capacity may be insufficient to provide deep percolation.
- Because water pressure is low, nonuniform application may result on soils with undulating topography unless laterals and mainlines are properly designed.
- Equipment cost on a per unit area basis may be higher than surface or some sprinkler systems.
- Increased monitoring of equipment is needed; this may be compounded for SDI systems and inline emitters because these components are not readily visible.

MAINTENANCE REQUIREMENTS

The key to successful operation of a drip system is proper maintenance. Because emitter openings are small, proper filtering is required to maintain system performance. Filter selection should match the type and size of particles to be removed and the emitter orifice size and design. This will vary with the water source, which typically includes municipal, ground wells, and reservoirs. In general, filtering should reduce the smallest particle size to one-tenth the emitter opening. Additionally chemical reactions may cause precipitates that will require filtering. Chemical clogging effects of water high in calcium may be reduced by lowering pH levels below 6.[4]

Particles can be removed through settling basins, sand filters, large flow disk filters, and centrifugal sand separators. These primary filter systems should be placed just downstream of the pump and any chemical injection equipment (Fig. 1). Primary filters are often placed in parallel so that one can be back-flushed while the second is simultaneously used for irrigation. Flow rates through media filters should fall between 10 lps and 18 lps per square meter (14 gpm and 26 gpm per square foot) of filtration surface.[5] Back-flushing of primary filters may be on a time schedule or based on pressure drop across the filter not to exceed 70 kPa (10 psi).[5] Screen, disk, and cartridge filters are often used downstream of primary filters before water enters the laterals. Screen and disk filters can be designed for automatic or manual cleaning. Screen flow capacity should not exceed 135 lps (200 gpm per ft) of screen opening.[5] In addition, flushing of laterals, mains, submains, and manifolds should be provided, either manually or with automatic valves, at velocities of at least 0.3 mps (1 fps).[5]

Where bacterial slimes and algae are to be controlled, chemical or acid treatments are often necessary. Chlorine injections can successfully control these organics, and may provide added protection from dissolved iron. Adequate chlorination is reached when residual chlorine at the end of the lateral is at least 1 ppm.[4] Dissolved iron can also be precipitated out prior to entering the system using oxidants such as sodium hyperchlorite. Copper sulfate may be used to control algae. Acids can be used to control water pH in excess of 8.0 when calcium and magnesium exceed 50 ppm.[10] Special attention is needed to prevent emitter clogging and root intrusion on SDI systems. Root intrusion can be controlled using a herbicide (treflane, which may be impregnated in the emitters), dilute acid, or frequent irrigation to keep saturated conditions near emitters.[4]

SYSTEM EVALUATION

Drip systems can be evaluated based on emission uniformity (EU) and application efficiency. For properly designed and maintained drip systems, uniformity and application efficiency are often greater than for other types of irrigation. Because of the low flow rates, the variation in discharge of individual emitters can have a significant influence on overall system uniformity. This is quantified for new emitters by determining the coefficient of manufacturing variation (C_v). Pressure variations will also be reflected in nonuniform water application among emitters.

The concept of EU can be used to evaluate the performance of a subunit system. This is determined by

$$\mathrm{EU} = 100\{1 - 1.27C_v/n^{0.5}\}q_{\min}/q_{\mathrm{avg}}$$

where n = the number of emitters per plant, and $q_{\min}$ and q_{avg} are the respective minimum and average flow rates within a subunit. When classifying point-source emitters, a C_v less than 0.05 is considered excellent, 0.05–0.07 is average, 0.07–0.11 is marginal, 0.11–0.15 is poor, and greater than 0.15 is unacceptable.[5] For line-source emitters, this can be relaxed such that less than 0.10 is good, 0.10–0.20 is average, and greater than 0.20 is marginal to unacceptable.[5] Although C_v-values will increase with emitter age, proper maintenance should maintain acceptable performance. The EU for new systems may exceed 90%, but because of differences in installation and maintenance, actual values are often less. Plugged emitters are frequently the cause of a decrease in EU over time. Depending on field slope, topography, and crop type, acceptable EU-values for point-source and line-source emitters will vary from 70%–95%.[5]

CONCLUSION

Drip irrigation has been successfully adapted to many agronomic crops. The benefit of high application efficiencies with limited water supplies have made irrigation successful in regions that would not otherwise have been possible. The wide variety of emitter devices having potentially low energy requirements continue to make it an attractive alternative. However, to be successful, drip systems must be both properly designed and maintained. This is more critical for drip irrigation than for sprinkler or surface, because of small diameter laterals and emitters, and low design pressures. Although, soil properties are important to the proper design and management of any irrigation system, they are especially important for drip systems. The list of advantages and disadvantages of drip irrigation should carefully be considered before making a final irrigation system selection. Improvements in drip components and reliability will continue to make it an attractive alternative. Understanding its strengths and limitations will help guide future improvements in drip irrigation management and its successful implementation.

REFERENCES

1. ASAE S526.2. *Soil and Water Terminology*; American Society of Agricultural Engineers Standards: St. Joseph. MI, 2001; 971–990.
2. ASAE S376.1. *Design, Installation and Performance of Underground, Thermoplastic Irrigation Pipelines*; American Society of Agricultural Engineers Standards: St. Joseph. MI, 2001; 849–863.
3. Howell, T.A.; Stevenson, D.S.; Aljibury, F.K.; Gitlin, H.M.; Wu, I-Pai.; Warrick, A.W.; Ratts, P.A.C. Design and Operation of Trickle (Drip) Systems. In *Design and Operation of Farm Irrigation Systems*; 1st Ed.; Jensen, M.E., Ed.; ASAE monograph 3, American Society of Agricultural Engineers: St. Joseph. MI, 1980; 663–717.
4. Dasberg, S.; Or, D. Drip System Components. In *Drip Irrigation*, 1st Ed.; McNeal, B.L., Tarkieu, F., Van Keulen, H., Van Vleck, D., Yaron, B., Eds.; Applied Agriculture. Springer: New York, 1999; 15–35.
5. ASAE EP405.1. *Design and Installation of Microirrigation Systems*; American Society of Agricultural Engineers Standards: St. Joseph. MI, 2002; 903–907.
6. Heermann, D.F.; Wallender, W.W.; Bos, M.G. Irrigation Efficiency and Uniformity. In *Management of Farm Irrigation Systems*, 1st Ed.; Hoffman, G.J., Howell, T.A., Solomon, K.H., Eds.; ASAE monograph, American Society of Agricultural Engineers: St. Joseph. MI, 1990; 125–149.
7. Burt, B.M.; Styles, S.W. *Drip and Micro Irrigation for Trees, Vines, and Row Crops*; California Polytechnic State University: San Luis Obispo, CA, 1999; 292.

8. ASAE S435. *Polyethylene Pipe Used for Microirrigation Laterals*; American Society of Agricultural Engineers Standards: St. Joseph. MI, 2002; 931–933.
9. Pereira, L.S.; Trout, T.J. Irrigation Methods. In *CIGR Handbook of Agricultural Engineering Volume I Land and Water Engineering*; van Lier, H.N., Pereira, L.S., Steiner, F.R., Eds.; American Society of Agricultural Engineers: St. Joseph, MI, 1999; 297–371.
10. James, L.G. *Principles of Farm Irrigation System Design*, 1st Ed.; Krieger Publishing Co.: Malabar, FL, 1988; 543.

Drum Drying

Juming Tang
Hao Feng
Washington State University, Pullman, Washington, U.S.A.

Guo-Qi Shen
Shanxi Agriculture University, Shanxi,
People's Republic of China

D

INTRODUCTION

Drum dryers were developed in early 1900s. They were used in drying almost all liquid food materials before spray drying came into use. Nowadays, drum dryers are used in the food industry for drying a variety of products, such as milk product, baby foods, breakfast cereal, fruit and vegetable pulp, mashed potatoes, cooked starch, and spent yeast.[1] In a drying operation, liquid, slurry, or puree material is applied as a thin layer onto the outer surface of revolving drums that are internally heated by steam. After about three-quarters of a revolution from the point of feeding, the product is dried and removed with a static scraper. The dried product is then ground into flakes or powder. Drum drying is one of the most energy efficient drying methods and is particularly effective for drying high viscous liquid or pureed foods.

SYSTEM DESCRIPTION

A drum dryer consists of one or two horizontally mounted hollow cylinder(s) made of high-grade cast iron or stainless steel, a supporting frame, a product feeding system, a scraper, and auxiliaries. Typical structures of single and double drum dryers are shown in Fig. 1. The diameter of typical drums ranges from 0.5 m to 6 m and the length from 1 m to 6 m.

In operation, steam at temperature up to 200°C heats the inner surface of the drum. The moist material is uniformly applied in a thin layer (0.5 mm–2 mm) onto the outer drum surface. Most of the moisture is removed at water boiling temperature. The residence time of the product on the drum ranges from a few seconds to dozens of seconds to reach final moisture contents of often less than 5% (wet basis). The energy consumption in a drum dryer may range between 1.1 kg steam per kg of evaporated water and 1.6 kg steam per kg of evaporated water, corresponding to energy efficiencies of about 60%–90%.[2–4] Under ideal conditions, the maximum evaporation capacity of a drum dryer can be as high as 80 kg H_2O/hr m^2.[4] A drum dryer can produce products at a rate between 5 kg hr^{-1} m^{-2} and 50 kg hr^{-1} m^{-2}, depending upon type of foods, initial and final moisture content, and other operation conditions.[2]

Drum dryers are classified into single drum dryer [Fig. 1(a)], double drum dryer [Fig. 1(b)], and twin drum dryer.[3] A double drum dryer has two drums that revolve toward each other at the top. The spacing between the two drums controls the thickness of the feed layer applied to the drum surfaces. A twin drum dryer also has two drums, but they rotate away from each other at the top. Among the three types, single and double drum dryers are most commonly used for fruits and vegetables. For example, large quantities of mashed potato flakes are produced using single drum dryers with specially designed roll feeding system. Double drum dryers are used in California to dry tomato paste. Twin drum dryers are used only for drying materials yielding dusty products.

For materials sensitive to heat damage, a vacuum drum dryer may be used to reduce drying temperature. A vacuum drum dryer is similar to other drum dryers except that the drums are enclosed in a vacuum chamber. In continuous vacuum drum dryers, receivers and air locks are designed to provide appropriate seal. Equipment and operation of vacuum drum dryers are relatively expensive, which limits vacuum drum drying to only high-value products or products that cannot be produced more economically by other means.

Perforated (suction) drum dryer is another variation from ordinary drum dryers. It utilizes heated air to heat the drum inside surface and the product is sucked to the perforated drum surface during drying. Several perforated drums can be linked together so that product can be transferred from one drum surface to another to achieve high production rates.

The advantages of drum drying include:

- The products have good porosity and hence good rehydration due to boiling evaporation.

Encyclopedia of Agricultural, Food, and Biological Engineering
DOI: 10.1081/E-EAFE 120007091
Copyright © 2003 by Marcel Dekker, Inc. All rights reserved.

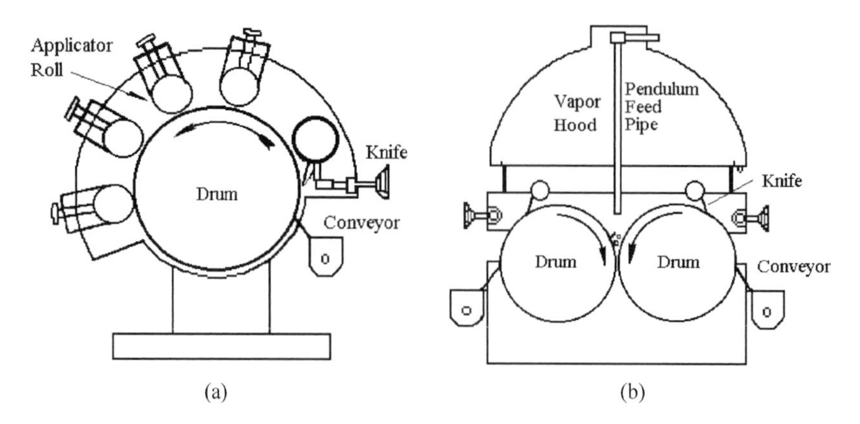

Fig. 1 (a) Single drum dryer and (b) double drum dryer.

- Drum dryers can dry very viscous foods, such as pastes and gelatinized or cooked starch, which cannot be easily dried with other methods.
- Drum dryers normally have high energy efficiency.
- Drum drying can be clean and hygienic.
- Drum dryers are easy to operate and maintain.
- The dryers are flexible and suitable for multiple but small quantity production.

The disadvantages of drum drying are the following:

- Some products may not form a good film on the drum surface and are not suitable for drum drying.
- Some products, especially those with high sugar content, may not be easily scrapped off from the drum.
- Relatively low throughput compared to spray drying.
- High cost of changing drum surface because of the precision machining that is required.
- Possible scorching of the product to impart cooked flavor and off-color due to direct contact with high temperature drum surface.
- Not able to process salty or other corrosive materials due to potential pitting of drum surface.

FEEDING METHODS

The method of applying product onto the drum surface differs, depending on the drum arrangement, the solid concentration, viscosity, and wetting ability of the product. Industrial drum dryers use five basic feeding methods, namely, roll feeding, nip feeding, dipping, spraying, and splashing, as shown in Fig. 2.[2–5]

Roll feeding [Fig. 2(a)] is used in both single and twin drum dryers. It is particularly effective for viscous and glutinous materials. Multirolls are often used to increase the film thickness, and hence the throughput. The gap between the rolls and the drum can be adjusted individually and the peripheral velocity of the rolls may or may not be the same as the drum. Roll feeding is sometimes used in combination with other feeding methods to meet the needs in the drying of certain product.

Nip feeding [Fig. 2(b)] is the simplest feeding method solely used in double drum dryers. It is suitable for drying of thin solutions, such as milk and whey. Nip feeding utilizes the adjustable gap between the two drums as a means to control the film thickness. The uniform distribution of the feed over the length of the drums is essential. Pendulum feed or perforated pipe is used to supply material into a pool in the space between the two drums for most applications.

Dipping is used in both single and twin drum dryers [Fig. 2(c)]. With this method, the dryer is partially submerged in a tray and product in the tray adheres to the surface as the drum rotates. It is good for certain suspensions of solids and used usually with a recirculation of material to prevent setting of the solid in the tray. For materials that cannot stand prolonged exposure to heat, a small tray may be used with constantly supplied fresh material.

In *spray* feeding [Fig. 2(d)], the material is atomized by nozzle onto the drum surface. Spray nozzles can be located at the bottom of the drum or other locations. The quantity of the product applied is controlled by the nozzle system and independent of other operational parameters.

Splashing [Fig. 2(e)] is a method especially suitable for products with a high rate of sedimentation. It can be used in single and twin drum dryers.

As the quality of the product film will directly affect the quality of the dried product, the control of the system, the throughput of the dryer, the proper selection of feeding method is very important. Usually the decision to select a proper feeding method relies on previous experience and/or by conducting tests with pilot units.

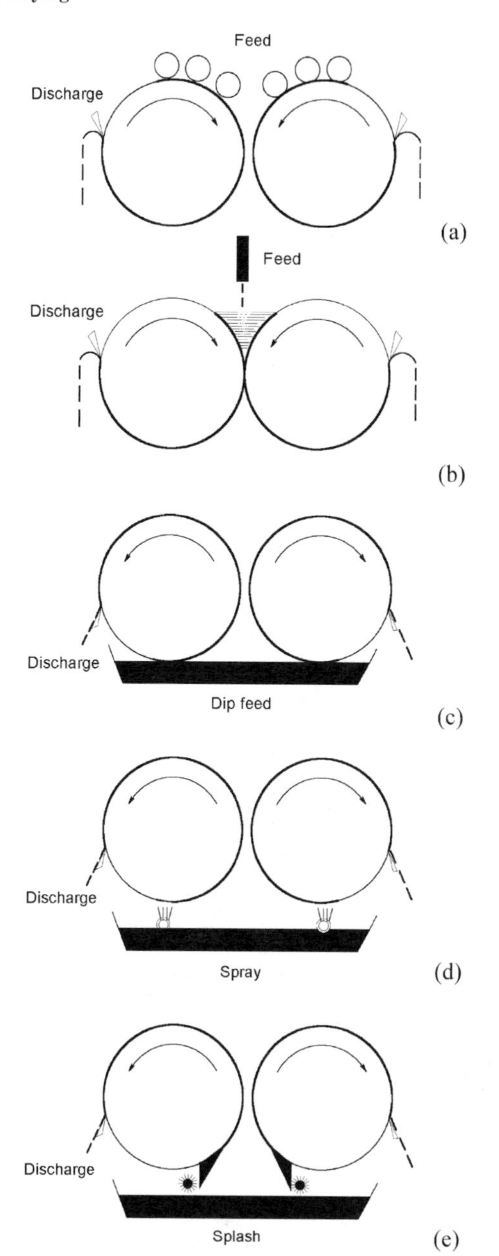

Fig. 2 Feeding methods used in drum dryers: (a) roll feed; (b) nip feed; (c) dip feed; (d) spray; and (e) splash.

PRINCIPLE AND DESIGN EQUATIONS

In drum drying, a large amount of thermal energy is released by the condensing steam in the drum and conducted through drum wall to the product. During drying, a product may go through three general periods. *Initial heating period*: after wet material is applied onto the drum surface in a thin layer, intensive heat transfer takes place due to a great temperature difference between the drum surface and the wet product. Product temperature increases rapidly to reach the boiling point of free water (Fig. 3).[5] *Constant product temperature period*: after reaching the boiling temperature, a large amount of free water evaporates and product temperature remains constant. The drum surface temperature, however, decreases due to an intense evaporative cooling. *Rising product temperature period*: after removing most of the free water, the amount of moisture for evaporation is dramatically reduced. The heat transferred from the steam gradually exceeds the energy used for evaporation. As a result, drum surface temperature increases. The bound water starts to play a major role in controlling the rate of evaporation. As bound water has a higher boiling temperature, product temperature gradually increases as drying proceeds. This trend continues till it reaches the knife where the dried product is scraped off. After the product is removed from the dryer, drum surface temperature continues to increase until new wet material is applied.

The evaporation rate of free water can be estimated by the following relationship[6]:

$$\frac{dM}{dt} = 30.94V^{0.8}\Delta P \qquad (1)$$

where dM/dt is the rate of moisture removal per unit drum surface (kg H_2O/hr m^2), V is the velocity of ambient air (m sec^{-1}), and $\Delta P = p_s - p_a$ (atm) is the difference between the vapor pressure at product surface p_s and the vapor pressure in the ambient air p_a. In period III, the drying rate is controlled by moisture diffusion as well as heat transfer.

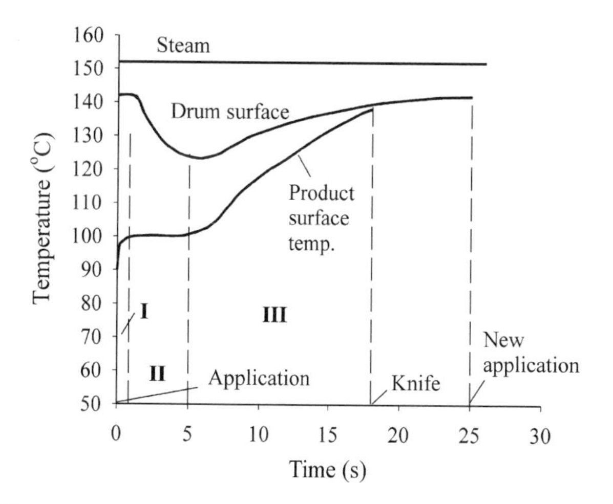

Fig. 3 Product and drum surface temperatures during one revolution of drum drying in three different drying periods. (From Ref. 5.)

Alternatively, water removal rate may be estimated from an energy balance equation:

$$\frac{dM}{dt} = 3.6\frac{h(T_w - T_{evp})}{L} \quad (2)$$

where T_w is temperature of the drum surface (°C), T_{evp} is temperature of evaporating surface (°C), L is latent heat (kJ/kg H_2O), and h is an overall heat transfer coefficient (W m^{-2}°C^{-1}). The value of h varies between 200 and 2,000 (W m^{-2}°C^{-1}), depending on the type and the thickness of the film being dried.

OPERATION

In the operation of a drum dryer, a delicate balance needs to be established among feed rate, steam pressure, roll speed, and thickness of the product film. It is desirable to maintain a uniform film on the drum surface to ensure maximized throughput and consistent final moisture content. Problems, however, are often encountered due to fluctuations in the moisture content and thickness of the feed. Accumulation of noncondensable gases in the drum also influences drying uniformity. Drum surface temperature may vary along the drum width as much as 20°C. All these factors may result in inconsistent drying performance and nonuniform final moisture content in the dried product. Means have been developed to automatically detect the moisture content and temperature, integrated with automated feedback control to minimize the fluctuations.[1,7]

Products containing high sugar contents, such as tomato puree, may be difficult to remove from the drums at high temperatures due to the thermoplasticity of those materials. A cooling mechanism (e.g., a jet of cold air) may be used at the location just before the product reaches the scraper. The purpose of the cooling is to bring the product from a rubbery state into a glassy state to facilitate separation of the product from the drum surface. An understanding of the glass transition temperature and its relation to moisture content is beneficial.

REFERENCES

1. Rodriguez, G.; Vasseur, J.; Courtois, F. Design and Control of Drum Dryers for the Food Industry. Part 1. Set-Up of a Moisture Sensor and an Inductive Heater. J. Food Eng. **1996**, *28*, 271–282.
2. Hall, C.W.; Farrall, A.W.; Roppen, A.L. Drum drier. In *Encyclopedia of Food Engineering*, 2nd Ed.; Hall, C.W., Farrall, A.W., Roppen, A.L., Eds.; AVI Publishing Company, Inc.: Westport, Connecticut, 1986; 264–266.
3. APV Crepaco Inc. Dryers: Technology and Engineering. In *Encyclopedia of Food Science and Technology*, 2nd Ed.; Francis, J., Ed.; John Wiley & Sons, Inc.: New York, 2000; 542–578.
4. Moore, J.G. Drum Dryers. In *Handbook of Industrial Drying*, 2nd Ed.; Mujumdar, A.S., Ed.; Marcel Dekker, Inc.: New York, 1995; Vol. 1, 249–262.
5. Bonazzi, C.; Dumoulin, E.; Raoult-Wack, A.; Berk, Z.; Bimbenet, J.J.; Courtois, F.; Trytram, G.; Vasseur, J. Food Drying and Dewatering. Drying Technol. **1996**, *14*, 2135–2170.
6. Okos, M.R.; Narsimhan, G.; Singh, R.K.; Weinauer, A.C. Food Dehydration. In *Handbook of Food Engineering*; Heldman, D.R., Lund, D.B., Eds.; Marcel Dekker, Inc.: New York, 1992; 507–516.
7. Rodriguez, G.; Vasseur, J.; Courtois, F. Design and Control of Drum Dryers for the Food Industry. Part 2. Automatic Control. J. Food Eng. **1996**, *30*, 171–183.

Dry Air Properties

Nicholas Shilton
University College Dublin, Dublin, Ireland

D

INTRODUCTION

Knowledge of the properties of dry air is required for many engineering applications in food processing, in particular drying. For example, these properties allow the engineer to calculate the mass of air required in order to provide sufficient energy to the product in order to achieve the desired drying parameters. The physical and thermodynamic properties of dry air are described, with values given where appropriate.

The temperature referred to in this section is at all times the dry bulb temperature. This is the temperature as measured by an ordinary, unmodified thermometer or thermocouple.

OVERVIEW

Composition of Dry Air

Dry air contains a number of component gases. The standard composition of these gases in dry air is given in Table 1. For most applications involving dry air, the reference temperature is adopted as being 273.15 K and the atmospheric pressure 101.325 kPa.

Atmospheric Pressure

The barometric pressure of dry air varies considerably with decreasing altitude. Standard atmospheric pressure is considered to be 101.325 kPa at sea level and 15°C. As altitude increases in the lower atmosphere (the troposphere), the temperature decreases according to the linear relationship

$$T = 15 - 0.0065Z \tag{1}$$

where T is the temperature (K) and Z is the altitude (m). In the stratosphere, above 10,000 m the temperature is found to be constant at −55°C. The pressure also decreases with increasing altitude, according to the following relationship

$$P = 101.325(1 - 2.25577 \times 10^{-5}Z)^{5.2559} \tag{2}$$

where P is the atmospheric pressure (kPa). For altitudes up to 10,000 m these properties are summarized in Table 2; however, pressure continues to decrease as altitude through the stratosphere increases. The effect of increasing altitude, and so decreasing pressure on the physical properties of dry air such as density, dynamic viscosity, and kinematic viscosity, is also given in Table 2.

Physical Properties of Dry Air

In conditions of the lower atmosphere, as described earlier, dry air can be considered to behave as an ideal gas. As such it is governed by the ideal gas law

$$PV = \mathrm{n}RT \tag{3}$$

where P (kPa) is the pressure of the gas, V, the volume (m^3), n, number of moles of gas, R, the universal gas constant (8.314 kPa m^3 kmol^{-1} K^{-1}), and T (K) is the absolute temperature. This can also be considered in the form of a virial equation of state

$$\frac{Pv_{\mathrm{m}}}{RT} = 1 + \frac{A_2(T)}{v_{\mathrm{m}}} + \frac{A_3(T)}{v_{\mathrm{m}}^2} + \cdots = Z \tag{4}$$

where v_{m} (m^3 mol^{-1}) is the molar volume of dry air, and the sum of the series, Z, is called the compressibility factor.[4] The coefficients A_2 and A_3 are called the virial coefficients. They represent the departure of the system under consideration from the ideal gas law. For dry air, the coefficients can be defined as follows:

$$A_2 = 0.349568 \times 10^{-4} - \frac{0.668772 \times 10^{-2}}{T} - \frac{2.10141}{T^2} - \frac{0.924746 \times 10^{2}}{T^3}\ \mathrm{m^3\,mol^{-1}} \tag{5}$$

$$A_3 = 0.125975 \times 10^{-8} - \frac{0.190905 \times 10^{-6}}{T} - \frac{0.632467 \times 10^{-4}}{T^2}\ \mathrm{m^6\,mol^{-2}} \tag{6}$$

Other terms in the series can be considered negligible.[5]

Encyclopedia of Agricultural, Food, and Biological Engineering
DOI: 10.1081/E-EAFE 120007048
Copyright © 2003 by Marcel Dekker, Inc. All rights reserved.

Table 1 Standard composition of air by volume

Gas	% By volume
Nitrogen	78.0840
Oxygen	20.9476
Argon	0.9340
Carbon dioxide	0.0314
Neon	0.0018
Other trace gases	0.0018

(From Ref. [1].)

Molecular weight

The molecular weight of dry air can be calculated to be 28.9645, based on the composition given in Table 1.[1]

Volume of dry air

Molar Volume. The molar volume of dry air can be calculated using the ideal gas law, as follows:

$$v_m = \frac{RT}{P} \tag{7}$$

where P (kPa) is the partial pressure of dry air. If the virial coefficients are taken into account, then Eq. 7 becomes

$$v_m = \frac{RT}{P}\left(1 + \frac{A_2}{v_m} + \frac{A_3}{v_m^2}\right) \text{m}^3\,\text{mol}^{-1} \tag{8}$$

Specific Volume. The specific volume of dry air can be calculated using:

$$v_a = \frac{v_m}{C_m} \tag{9}$$

where v_a ($m^3\,kg^{-1}$) is the specific volume of air and C_m ($kg\,mol^{-1}$) is the molar concentration.

Density

At 20°C and standard atmospheric pressure the density of air can be taken to be 1.2059 kg m^{-3}. Density will change with decreasing pressure (Table 2) and increasing temperature (Table 3).

Dynamic viscosity

At 20°C and standard atmospheric pressure the dynamic viscosity of air can be taken to be 1.816×10^{-6} Pa sec. Dynamic viscosity will change with decreasing pressure (Table 2) and increasing temperature (Table 3).

Kinematic viscosity

At 20°C and standard atmospheric pressure the kinematic viscosity of air can be taken to be $1.5061 \times 10^{-6}\,m^2\,sec^{-1}$. Kinematic viscosity will change with decreasing pressure (Table 2) and increasing temperature (Table 3).

Thermodynamic Properties of Dry Air

Enthalpy

Determination of the enthalpy of dry air is dependent on the selection of a standard reference temperature and

Table 2 Effect of increasing altitude on the physical properties of dry air

Altitude (m)	Temperature (°C)	Pressure (kPa)	Density ($kg\,m^{-3}$)	Dynamic viscosity (Pa sec)	Kinematic viscosity ($m^2\,sec^{-1}$)
0	15.0	101.325	1.2255	0.00001796	0.00001466
500	11.8	95.461	—	—	—
1,000	8.5	89.875	1.1120	0.00001765	0.00001587
1,500	5.2	84.556	—	—	—
2,000	2.0	79.495	1.0068	0.00001733	0.00001721
2,500	− 1.2	74.682	—	—	—
3,000	− 4.5	70.108	0.9094	0.00001700	0.00001869
4,000	− 11.0	61.640	0.8193	0.00001668	0.00002036
5,000	− 17.5	54.020	0.7363	0.00001635	0.00002221
6,000	− 24.0	47.181	0.6598	0.00001602	0.00002428
7,000	− 30.5	41.061	0.5896	0.00001568	0.00002659
8,000	− 37.0	35.600	0.5252	0.00001534	0.00002921
9,000	− 43.5	30.742	0.4664	0.00001499	0.00003214
10,000	− 50.0	26.436	0.4127	0.00001464	0.00003547

(From Refs. [2,3].)

D

Table 3 Physical and thermodynamic properties of dry air at atmospheric pressure (101.325 kPa) and 20°C

Temperature T (°C)	Density ρ ($kg\,m^{-3}$)	Dynamic viscosity μ (Pa sec)	Kinematic viscosity ν ($m^2\,sec^{-1}$)	Specific heat Cp ($kJ\,kg^{-1}\,K^{-1}$)	Thermal conductivity k ($W\,m^{-1}\,K^{-1}$)	Thermal diffusivity α ($m^2\,sec^{-1}$)
− 60	1.6573	0.00001409	0.000008501	1.003	0.01956	0.00001115
− 40	1.5153	0.00001513	0.000009987	1.003	0.02111	0.00001388
− 30	1.4534	0.00001565	0.000010777	1.004	0.02189	0.00001502
− 20	1.3965	0.00001617	0.000011587	1.004	0.02267	0.00001618
− 10	1.3436	0.00001669	0.000012436	1.004	0.02345	0.00001740
0	1.2946	0.00001719	0.000013295	1.004	0.02413	0.00001857
10	1.2486	0.00001767	0.000014168	1.005	0.02404	0.00001981
20	1.2059	0.00001816	0.000015061	1.005	0.02561	0.00002114
30	1.1663	0.00001864	0.000015990	1.006	0.02634	0.00002246
40	1.1293	0.00001912	0.000016965	1.006	0.02700	0.00002378
50	1.0936	0.00001960	0.000017941	1.007	0.02778	0.00002523
60	1.0604	0.00002005	0.000018916	1.007	0.02856	0.00002674
70	1.0345	0.00002050	0.000019892	1.008	0.02918	0.00002811
80	1.0021	0.00002091	0.000020888	1.009	0.02980	0.00002952
90	0.9734	0.00002135	0.000021941	1.010	0.03055	0.00003108
100	0.9470	0.00002179	0.000023024	1.011	0.03122	0.00003262
110	0.9227	0.00002222	0.000024090	1.012	0.03185	0.00003412
120	0.8996	0.00002263	0.000025166	1.013	0.03254	0.00003572
130	0.8770	0.00002304	0.000026281	1.014	0.03326	0.00003741
140	0.8557	0.00002345	0.000027401	1.016	0.03389	0.00003901
160	0.8153	0.00002431	0.000029806	1.018	0.03513	0.00004232
180	0.7807	0.00002513	0.000032191	1.022	0.03638	0.00004561
200	0.7487	0.00002590	0.000034606	1.025	0.03763	0.00004905
220	0.7176	0.00002666	0.000037161	1.029	0.03887	0.00005266

(From Ref. [6].)

pressure. The standard reference temperature is taken to be 273.15 K (0°C), while the reference pressure is taken as atmospheric pressure, 101.325 kPa. Thus, enthalpy can be calculated by the following equation:

$$H_a = 1.005(T_a - T_r) \tag{10}$$

where H_a is the enthalpy of dry air ($kJ\,kg^{-1}$), T_a is the temperature of dry air (either K or °C), and T_r is the reference temperature (either K or °C). The molar enthalpy of dry air can be calculated by the expression

$$h_a = \sum_{i=0}^{5} b_i T^i + RT\left[\left(A_2 - T\frac{dA_2}{dT}\right)\frac{1}{V} + \left(A_3 - \frac{T\,dA_3}{2\,dT}\right)\frac{1}{V^2}\right] \tag{11}$$

where the values for the constants A_2 and A_3 are as defined in Eqs. 5 and 6, and values for series of constants b are summarized in Table 4.

Entropy

The molar entropy of dry air can be calculated by the expression

$$S_a = \sum_{i=0}^{4} \ell_i T^i + \ell_5 \ell n T - R\ell n\left(\frac{P}{101325}\right) + R\ell n\left(\frac{PV}{RT}\right) - R\left[\left(A_2 + T\frac{dA_2}{dT}\right)\frac{1}{V} + \frac{1}{2}\left(A_3 - \frac{T\,dA_3}{2\,dT}\right)\frac{1}{V^2}\right] \tag{12}$$

where the values for the constants A_2 and A_3 are as defined in Eqs. 5 and 6, and values for series of constants ℓ are summarized in Table 4.

Specific heat

At 20°C and standard atmospheric pressure the specific heat of air can be taken to be 1.005 $kJ\,kg^{-1}\,K^{-1}$. Specific heat changes with increasing temperature is given in Table 3.

Table 4 Values for the constants b and ℓ that are used in Eqs. 11 and 12[5]

Subscript	b	ℓ
0	$-0.79078691 \times 10^{4}$	$-0.16175159 \times 10^{3}$
1	0.28709015×10^{2}	$0.52863609 \times 10^{-2}$
2	$0.26431805 \times 10^{-2}$	$-0.15608795 \times 10^{-4}$
3	$-0.10405863 \times 10^{-4}$	$0.24880547 \times 10^{-7}$
4	$0.18660410 \times 10^{-7}$	$-0.12230416 \times 10^{-10}$
5	$-0.97843331 \times 10^{-11}$	0.28709015×10^{2}

Thermal conductivity

At 20°C and standard atmospheric pressure the thermal conductivity of air can be taken to be $0.02561\ \mathrm{W\,m^{-1}\,K^{-1}}$. Thermal conductivity changes with increasing temperature is given in Table 3.

Thermal diffusivity

At 20°C and standard atmospheric pressure the thermal diffusivity of air can be taken to be $2.114 \times 10^{-5}\ \mathrm{m^{2}\,sec^{-1}}$. Thermal diffusivity changes with increasing temperature is given in Table 3.

CONCLUSIONS

In this article, the physical and thermodynamic properties of dry air have been described for use in food engineering calculations. These properties are dependent on both temperature and pressure and these relationships have been described earlier.

REFERENCES

1. Singh, R.P.; Heldman, D.R. *Introduction to Food Engineering*, 3rd Ed.; Academic Press: London, 2001.
2. ASHRAE. *ASHRAE Fundamentals*; American Society of Heating, Refrigerating, and Air-Conditioning Engineers: Atlanta, GA, 1997.
3. Keenan, J.H.; Chao, J.; Kaye, J. *Gas Tables International Version, Second Edition (SI Units)*; John Wiley: New York, 1983.
4. Threlkeld, J.L. *Thermal Environmental Engineering*; Prentice Hall: Englewood, NJ, 1962.
5. Hyland, R.W.; Wexler, A. Formulations for the Thermodynamic Properties of Dry Air from 173.15 K to 473.15 K, and of Saturated Moist Air from 173.15 K to 473.15 K, at Pressures to 5 MPa. ASHRAE Trans. **1983**, *89*, 520–535.
6. Henderson, S.M.; Perry, R.L.; Young, J.H. *Principles of Process Engineering*, 4th Ed.; American Society of Agricultural Engineers: St. Joseph, MI 1997.

Dry Bulb Temperature

Shahab Sokhansanj
University of Saskatchewan, Saskatoon, Saskatchewan, Canada

INTRODUCTION

Dry bulb temperature is commonly termed as "temperature." The dry bulb temperature is a specific term where the temperature of a medium is measured without the influence of evaporation or condensation.

In modern times, temperature is measured by a wide range of techniques and instruments. The oldest, yet the most widely used, method is to place a heat sensitive fluid such as alcohol or mercury in a spherical reservoir (bulb) of a glass thermometer (Fig. 1). As the bulb heats or cools, the fluid inside the bulb expands or contracts accordingly. This is termed as thermal expansion or contraction of the fluid.

SYSTEM OF UNITS

The SI base unit for thermodynamic temperature is Kelvin (K). Because of the wide usage of the degree Celsius, particularly in engineering and nonscientific areas (formerly called centigrade scale), it may be used when expressing temperature.[1] The Celsius scale is related to the Kelvin scale as follows:

$$°\mathrm{C} = \mathrm{K} - 273.15 \tag{1}$$

In English units, temperature has the units of degrees Fahrenheit (°F) and Rankin (°R).

$$°\mathrm{F} = °\mathrm{R} - 459.69 \tag{2}$$

The conversion between SI and English units are

$$°\mathrm{F} = 1.8°\mathrm{C} + 32 \tag{3}$$

$$°\mathrm{R} = 1.8\mathrm{K} \tag{4}$$

Temperature scales are based on the International Practical Temperature Scale (IPTS). This scale which was last revised in 1990 (IPTS-90) identifies specific conditions of substances as reference points.[2] For example, the triple point of water at 273.16K, triple point of hydrogen at 13.8033K, and freezing point of tin at 505.078K are among 16 reference points. For food engineering purposes, we use the 100 divisions between the freezing point (273.15K) and the boiling point of water (373.15K).

RESPONSE TIME–TIME CONSTANT

The dry bulb temperature measured represents the heat exchange between the sensor (thermometer bulb) and the medium. In the case of convection heat transfer, a good stream of flowing fluid past the sensor is needed. In the case of conduction, a good physical contact between the sensor and the medium is needed. For radiation measurement, the sensor should see the entire radiating surface unobstructed. The following characterizes the response time of a temperature sensor:

1. The response of thermometers is characterized by a time constant.
2. The time constant is a measure of how long it takes the thermometer to respond to a change in a medium temperature.
3. It is normal (but not always preferable) for the thermometer to have a time constant as small as possible.
4. The time constant is a function of the sensor size and its material of construction (Table 1).
5. To ensure having a small time constant, sensors are normally built with a small volume to area ratio and a large thermal conductivity.

TEMPERATURE MEASUREMENT

Hand-held thermometers discussed in previous sections are versatile, easy to use, and often low cost. Their use, however, in industry for on line data taking is limited. Numerous types of sensors including thermocouples, resistance temperature devices (RTD), infrared sensing devices, and fiber optics are available that lend themselves easier to interfacing with electronic devices. Some of these techniques mostly used in food industry are reviewed.

Thermocouple

Thermocouples work on the basis that when two wires of different materials are joined at one end and subjected to

Encyclopedia of Agricultural, Food, and Biological Engineering
DOI: 10.1081/E-EAFE 120007050
Copyright © 2003 by Marcel Dekker, Inc. All rights reserved.

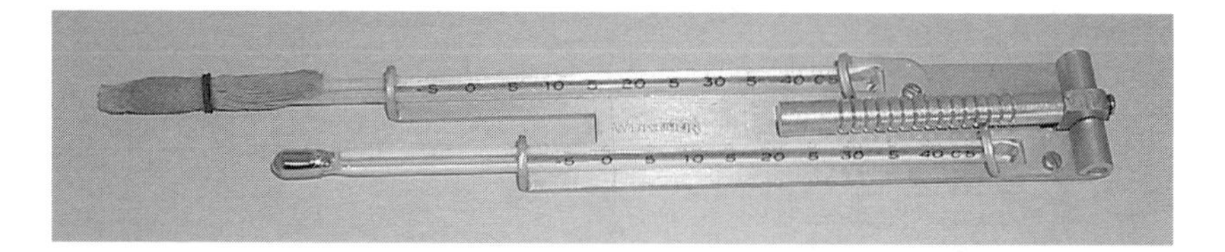

Fig. 1 Dry bulb (lower) and wet bulb (upper) thermometers placed in a hand operated sychrometer.

a temperature gradient, a thermoelectric voltage (*V*) will be observed at the other end. Fig. 2 shows a schematic of thermocouple where one of the two ends is called the reference junction. It is a common practice to call the junction at the known temperature, the reference junction and to call the junction to be measured as the measured junction. The standard calibrations (temperature vs. millivolts) are available with respect to the reference junction placed at 0°C (a mixture of ice and water). The modern measuring instruments compensate electronically for the absence of 0°C conditions at the instrument terminals.

A thermocouple circuit can be interrupted by a third or more types of conducting wires as long as the temperatures at the point of new junctions are all the same (Fig. 2). This property allows remote sensing of temperatures eliminating for the use of high grade (and expensive) thermocouple wire for the entire length. It also allows to place thermocouple switches and connectors along the conducting wires.

The conducting wires are made of at least eight combination of alloys. The most widely used ones are types T, J, and K listed in Table 2 where the maximum temperature range and the useful temperature ranges for each of thermocouples are listed. The uncertainty in measured temperatures[4] ranges from 1.1 to 2.2°C depending upon the grade of the thermocouple for type J and K. Type T has a more tighter tolerance in the range of 0.5–1.0°C.[8]

Resistant Thermometer

The resistant thermometers are based on the fact that the electrical resistance of a conducting material changes with temperature. The thermometer is composed of a sensing element and the electronic circuit (e.g., Wheatstone bridge). The sensors are made of either a metallic conductor or a ceramic semiconductor. Metallic resistance thermometer is called resistance temperature device (RTD). Platinum makes a good metallic conductor with a reasonable constant slope (resistance vs. temperature). A common type platinum RTD is called Pt100 because it is constructed to have 100 ohm resistance at 0°C. The semiconductor sensors are termed thermistor. The general slope of temperature vs. resistance for thermistors is negative and nonlinear. In spite of their nonlinearity, thermistors are stable, produce a strong signal, and for these reasons are popular. The nonlinearity is dealt with using matched thermistors and other electronic compensation techniques.

Infrared Thermometer

Infrared thermometer are optic devices that sense the temperature of a material from distance. The transfer of energy from the heated surface to the sensor is by electromagnetic radiation. The sensor is a photo sensitive material that upon "seeing" a source of thermal heat produces an electrical signal. The signal (voltage)

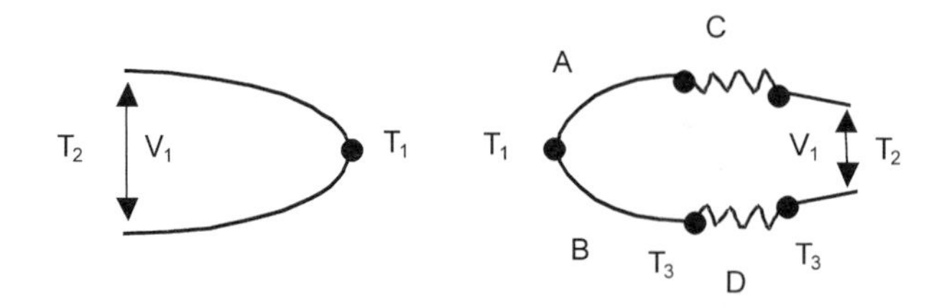

Fig. 2 Temperature T_1 measurement using thermocouples. Left diagram shows the principle of measurement. Right diagram shows that a thermocouple can be extended provided the junction temperatures at B and D remain the same.

D

Table 1 Thermal properties of some material used in construction of thermometers and thermocouples. Numbers in the parenthesis in column 1 are source of data

Material	Density ($kg\,m^{-3}$)	Specific heat ($kJ\,kg^{-1}K^{-1}$)	Thermal conductivity ($kJ\,m^{-1}K^{-1}$)
Alcohol[6]	0.79	2.30	0.18
Constantan[3]	8922	0.41	22.70
Copper[3]	8954	0.38	386
Glass[3]	2700	0.84	0.78
Mercury[3]	13579	0.14	8.69
Tin[3]	7304	0.23	64

is magnified and conditioned for calibration against temperature.

Infrared thermometers are produced for a wide range of temperature applications (0–2800°C). It is important to select a thermometer for the correct temperature range application. It is also important to pay attention to the temperature of environment in which the optical sensor is located as the sensor's performance is affected by spurious heat as well. The correct field of view, the target area, is also an important parameter. One manufacturer[5] recommends that the target area must be 50% larger than the field of view.

Infrared thermometers are graded by their optical resolution, which is defined as the ratio of distance to the target (*D*) divided by the size of the view area (spot size) (*S*). The *D*/*S* ratio may range from 2 to 300. Larger *D*/*S* ratio is the higher resolution infrared device and thus more expensive.

For the correct selection of the infrared sensors, it is important to know the expected temperature ranges, the size of target area, distance from target area, and the emissivity of the material whose temperature is to be measured. Emissivity is the ratio of the heat radiation emitted from a material to the heat radiation that is emitted from a perfect black body at the same temperature.

Fiber Optics Thermometry

Fiber optics are high quality glass strands that transmit electromagnetic waves including light and heat. One strand or many strands can be used to sense temperature in an environment where ferric-metallic material cannot be used.[7] Microwave ovens are an example where thermocouples and thermistors may cause sparks and interference in measurements. Fiber optics can sense the temperature of an object by channeling the radiated heat through the fiber to a remote optical sensor similar to infrared thermometer. In another technique, a thin film of heat sensitive material (such as gallium arsenide (GaAs) crystal) is embedded on one end of the fiber optic. The other end is connected to an instrument that can produce pulsed light or laser. The pulse light travels the length of the fiber and interacts optically with the end material. The fraction of the light, which has not been absorbed or transmitted through the end material, is reflected back and its properties are measured on a spectrometer. The spectral

Table 2 Three most commonly used thermocouples with temperature ranges and expected error

Type	Alloys	Temperature range °C		Error[a] (°C)
		Maximum	Useful	
J	Iron—Constantan	0–750	0–200	1.1–2.2
K	Chromel—Alumel	− 200–1250	0–200	1.1–2.2
T	Copper—Constantan	− 200–350	− 60–350	0.5–1.0

[a] Error range is for special grade to standard grade.
(From Ref. 4.)

properties of the reflected light depends upon the temperature of the end material.

REFERENCES

1. ASAE. *ASAE Standards. Use of SI (Metric) Units*; American Society of Agricultural Engineers: St. Joseph, MI, 2001; 26–33.
2. Preston-Thomas, H. *The International Temperature Scale of 1990. Technical References*; http://www.omega.com (acccessed July 2001) Omega Engineering, Inc.: Stamford, CT, 2000; 14.
3. Holman, J.P. *Heat Transfer*; McGraw Hill Publishing Company: New York, 1990.
4. Omega. *Practical Guidelines for Temperature Measurement*; Omega Engineering, Inc.: Stamford, CT, 2000; http://www.omega.com (accessed July 2001).
5. Raytek. *Selecting the Right Infrared Temperature Sensor for Your Application*; Raytek: Santa Cruz, CA, 2001; http://www.raytek.com/ (accessed July 2001).
6. Earle, R.L. *Unit Operations in Food Processing*; Pergamon press: New York, 1983.
7. Sowers, H. *Temperature Monitoring Using Fiber Optics. Appllication Notes*; SteamTech Environmental Services: Bakersfield, CA, 2000; http://www.Steamtech.com/ (accessed July 2001) .
8. ASHRAE. *Fundamentals Handbook*; ASHRAE American Society of Heating, Refrigeration, and Air Conditioning Engineers, Inc.: Atlanta, 2001.

Dry Food Transport

K. Muthukumarappan
South Dakota State University, Brookings, South Dakota, U.S.A.

D

INTRODUCTION

There are various types of mechanical devices, which are in use for the transport of materials from one location to another. The transport devices may be anything depending on the material to be transported, e.g., for transporting powder and ground products we may use screw conveyors, for dry foods we may use chain, belt, roller, continuous-flow, pneumatic, and bucket conveyors, and for liquid foods pumps are generally used. The development of mechanical devices for transport of food materials dates back to 1930s. During the last seven decades the technology has improved in many aspects such as flexibility, versatility, cost, and efficiency. This was possible because of development of new materials and improved design. In this article, we will specifically discuss about the different transport devices and dry food transport properties necessary to design a transport system.

DRY FOODS

Dry foods are categorized as the food products with low moisture content (2%–6% dry basis) and low water activity (0.3 –0.5 a_w). The various dry foods, which are generally used in the food processing industries, are wheat, soybeans, rice, rapeseed, oats, cowpeas, cottonseed, corn, castor beans, barley, etc. Various snack items, ready to eat foods, and other finished products use these materials as their raw ingredients for their preparation. So for the bulk handling of these ingredients, it is necessary to design transport systems that reduce the manual labor and increase the production capacity.

When designing a system both physical and chemical properties of foods should be considered. Friction and flowability are the most important properties; however, abrasiveness, friability, and lump size are also important. Chemical properties may dictate the structural materials out of which conveyor parts are fabricated. For example, the effect of oil on rubber is an important design parameter. Friction plays an important and in numerous cases a decisive role in transportation of dry foods. The rating of screw conveyors and behavior of granular materials and cereals depend on friction coefficient.

The ratio of friction force, F, and the force normal to the surface of contact, W, is given by the relationship:

$$f = F/W$$

where f is the coefficient of friction.

Factors that affect friction are sliding velocity, water film, and surface roughness. In general, at low velocities, the coefficient of friction increases with velocity, and at high velocities, friction either remains constant or decreases. There are considerable data available on the coefficient of friction of agricultural and food materials and they can be found in Mohsenin[1] and in ASAE Standards.[2]

Flowability of a granular solid is a function of several forces namely gravitational forces, friction, cohesion (interparticle attraction), and adhesion (particle–surface attraction). Gravity is the natural driving force of unaided flow. Noncohesive granular materials are those in which interparticle attractions are negligible and the major resistant to flow is its internal friction. Jenike[3] developed fundamental methods for determining flowability properties of granular solids and has been successfully used for various food granular materials by Peleg.[4] The best mode of handling the dry foods within the manufacturing unit is the conveyor.

CONVEYORS

Conveyors are mechanical devices used to assist in the movement of materials or products from one location to another without using a vehicle such as truck, forklift, cart, or similar devices. The chief advantage attributed to a conveyor is greater production realized in a manufacturing plant or warehouse because of the large reduction in labor, heavy lifting, and carrying. Conveyors of one type or another can be found in virtually every food processing plant today.

Conveyors are often manually fed at some point in the system, and in some operations the products at the discharge end of the conveyor are removed manually. A mechanical device or transfer is installed on the end of a conveyor section in certain situations. The flexibility of conveyors and their versatility in handling a wide variety of products make them extremely useful in moving

Encyclopedia of Agricultural, Food, and Biological Engineering
DOI: 10.1081/E-EAFE 120006969
Copyright © 2003 by Marcel Dekker, Inc. All rights reserved.

products from one operation to another including transfer to different floor levels. Selection of a transport system for a specific material in a specific situation is complicated and large number of interrelated parameters must be considered. Perry and Green[5] have covered well these parameters in the seventh edition of Perry's Chemical Engineer's Handbook.

Processing operations in a processing plant are frequently linked together with a conveyor. The conveyor, as a result, becomes an integral part of the total materials handling systems. Rather complicated control systems, which may employ cybernetics, can be utilized in an automated conveyorized material handling systems. Many types and adaptations have been made to utilize conveyors for dry foods. As a result, some of the types of conveyors in existence are: metal or rubber belt, metal screw, bucket, pneumatic, flexible, spiral, air, and vacuum conveyors. The various conveyor transport systems, which are most preferred for the handling of dry foods, are discussed below.

SCREW CONVEYORS

Screw conveyors (Fig. 1) are very efficient for the transport of dry foods and porous materials, which are very small in size. It is one of the oldest and versatile conveyor types. A typical screw conveyor consists of a helicoid flight (helix rolled from flat steel bar) or a sectional flight, mounted on a pipe or shaft and turning in a trough. In addition to transport of dry foods, almost any degree of mixing can be achieved with screw conveyor flights cut, cut and folded, or replaced by a series of paddles. Moreover, heating and cooling of dry foods for further processing can be achieved using hollow screws and pipes for circulating hot or cold fluids. Different designs (variable-pitch, tapered-flight, or stepped-flight) can be used when precise control of the transport rate is required. On the other hand short-pitch units are used for inclined and vertical transport of dry foods.

Other advantages are:

1. It is relatively easy to seal a screw conveyor to operate in its own atmosphere and maintain internal temperature and humidity in areas of high or low ambient temperature and humidity.
2. The casing can be designed with a drop bottom for easy cleaning and avoid contamination with different materials.

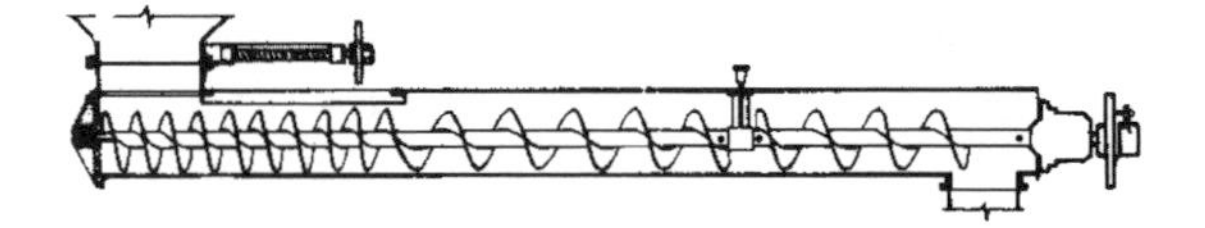

Fig. 1 Screw conveyor.[6]

BELT CONVEYORS

Belt conveyors (Fig. 2) are used principally in dry storage warehouses and for unloading warehouse items from delivery trucks. It can be used for any size of foods or food products without any major physical damage to foods. In fact, the belt conveyor is almost universal in application. Belt conveyors made of super performance elastomers such as neoprene and hyear have been used in the food plant. This type of conveyor is good for dry foods because of material not sticking to the belt and chutes. On the other hand, temperature and humidity extremes may dictate total enclosure of the belts; surroundings which involve conditions such as high temperature and humidity.

BUCKET CONVEYORS

Bucket elevators (Fig. 3) are primarily used for vertical transportation of dry foods. It is the simplest and most dependable unit. They are available in a wide range of capacities both as an open or closed system. Among different versions of the design, spaced-bucket centrifugal-discharge and spaced-bucket positive-discharge elevators are more suitable for dry food transportation. The centrifugal-discharge elevators are suitable for handling

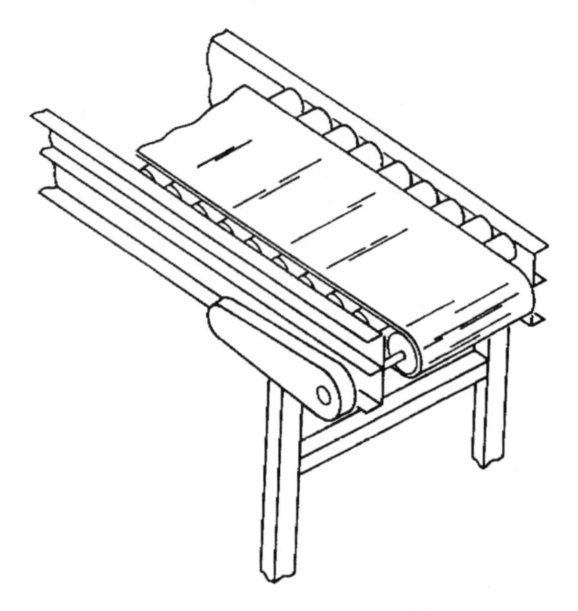

Fig. 2 Flat belt conveyor.[6]

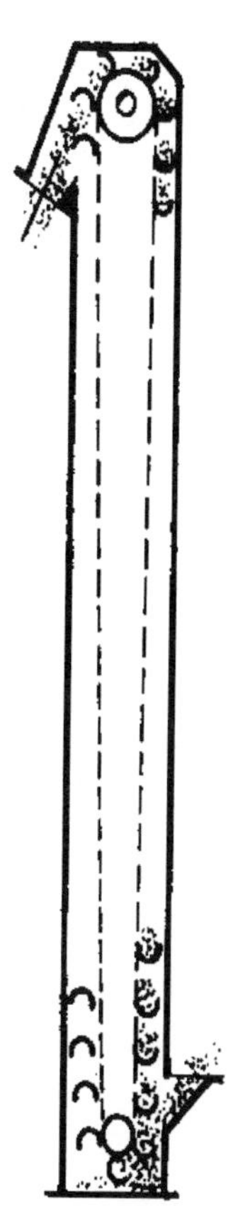

Fig. 3 Bucket conveyor.[6]

almost any free-flowing fine or small-lump dry foods such as grains, powders, etc. Buckets are loaded partly by material flowing directly into them and partly by scooping the material. The positive-discharge elevators are same as centrifugal-discharge units except that the buckets are mounted on two strands of chain and are snubbed back under the head sprocket to invert them for positive discharge. These units are specifically designed for materials that are sticky such as dry foods with high oil content.

PNEUMATIC CONVEYORS

Pneumatic transport is frequently used to convey granular materials. Aerodynamic resistance supplies force required for pneumatic conveying (Fig. 4). If this force exceeds the force required to displace the material, the latter starts to move. Two basic types of pneumatic transport are flying and plug transport. In the case of low particle concentrations the flying transport occurs, the individual grains move freely in the tube and also have a velocity component normal to the main direction of motion. In the case of plug transport, the material fills the tube and the

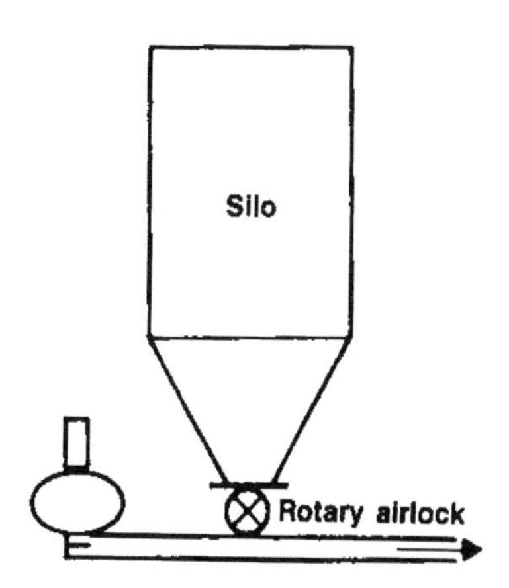

Fig. 4 Pneumatic conveyor.[6]

material moves in common. The two transport types differ with respect to velocity and static pressure. Generally, an air velocity of $20\,\mathrm{m\,sec^{-1}}$–$25\,\mathrm{m\,sec^{-1}}$ with lower static pressure is required for flying transport and for plug transport an air velocity of $0.5\,\mathrm{m\,sec^{-1}}$–$2.0\,\mathrm{m\,sec^{-1}}$ with relatively high static pressure is required.[7]

FLUIDIZED TRANSPORT

A special case of pneumatic transport is fluid-bed conveying. In this conveying method, the food material is brought into the fluidized state by an appropriate airflow. The fluidized transport system is designed primarily for transporting fine-grained (floury) materials, since the quantity of transporting air required is relatively low.

VACUUM CONVEYORS

Dry food products in the form of powders and granulates can be conveyed gently and efficiently by means of vacuum. To convey such small particles in a duct, the pressure at one end of the duct is lowered. The pressure differential induces the material to flow to the low-pressure end, where the air and material are separated. This is relatively a new technology.[8]

REFERENCES

1. Mohsenin, N.N. *Physical Properties of Plant and Animal Materials*; Gordon and Breach: New York, NY, 1986.
2. ASAE. *Properties Standards*; American Society of Agricultural Engineers: St. Joseph, MI, 1999; 538–540.
3. Jenike, A.W. *Storage and Flow of Solids*; Bulletin 123 of the Utah Engineering Experiment Station, University of Utah: Salt Lake City, Utah, 1970.

4. Peleg, M. Flowability of Food Powders and Methods for Its Evaluation. J. Food Process. Eng. **1977**, *1*, 303–328.
5. Perry, R.H.; Green, D. Handling of Bulk Solids and Packaging of Solids and Liquids. *Perry's Chemical Engineers' Handbook*; McGraw-Hill: New York, NY, 1997; Vol. 7, 1–63.
6. http://www.ie.ncsu.edu/kay/mhetax/TransEq/Conv/index.htm#Pneumatic%20conveyor.
7. Sitkei, G. Mechanics of Agricultural Materials. *Developments in Agricultural Engineering*; Elsevier Science Publishing Co: New York, NY, 1986; 284–347.
8. http://www.piab.com/.

Drying of Air

Francis Courtois
UMR-GENIAL, Unité Mixte de Recherche GENie Industriel ALimentaire, ENSIA, Massy, France

D

INTRODUCTION

It may seem unusual to dry the air, the most common drying medium, but it is not so. There are several occurrences where air needs to be dried prior to any product contact. Since the drying capacity of air is related to its relative humidity (RH), lowering it will improve the overall drying performance. Starting from psychrometric considerations, there are two obvious ways for decreasing the air RH:

1. By increasing its temperature, for a given absolute humidity (i.e., moisture content), the RH will diminish accordingly.
2. By decreasing its pressure, for a given absolute humidity (i.e., moisture content), the RH will diminish accordingly.

These methods are not exactly for "drying of air" since they do not remove any water from it. They just change its drying capacity. Thus they will not be discussed here.

When using an air-conditioning system, the air will pass through a heat exchanger where it will be in contact with a cold surface. If the surface temperature is cold enough, e.g., below the air dew point, part of the water vapor contained in the air will condense.

There is another way of removing the extra water vapor from the air. It consists of the use of water desiccants. Through contact with such a product the air will transfer part of its water vapor. This method implies the regeneration of the desiccant, loaded with high moisture content. There are mainly two industrial methods to remove water vapor from the air using desiccants.

1. Spray drying of a liquid desiccant combined with a spray rewetting regeneration method, both at different temperatures.
2. Vapor adsorption on a solid desiccant, on a rotating wheel, in parallel with a regeneration counter current system.

In any cases, the air drying system will modify the air temperature. An additional air heater or cooler may be needed for proper temperature adjustment, to fulfill the process requirements.

DEFINITION

Drying of air is the action of removing some of the water (in gas phase) contained in wet air. Furthermore, lowering the air RH should not be considered as a drying action since the overall moisture content is kept constant. It should be noted that it is possible to diminish the RH of a given air while increasing its water content and its temperature.

Some ambiguity can arise when considering the drying of air as a way to increase its drying capacity. Clearly, to increase the drying capacity of a given air there is no need to dry it. Increasing its temperature or decreasing its pressure will decrease its RH and increase its maximum water uptake capability. Drying air is thus a specific action allowing an increase of the drying capacity at atmospheric pressure without the need for any temperature increase.

OBJECTIVES

The general purpose of such an operation is to increase its drying capacity. The drier the air, the more effective the dryer will be. With an air being dried, the dryer will allow either a lower air flow rate for the same drying capacity or a higher product flow rate for the same air flow rate. In any case, one can expect a reduction in:

- Energy cost.
- Dust emission.
- Dryer size (including fan and burners).

The drying capacity of a given air is highly reduced at near-ambient or even colder temperatures. There are many cases when heating the air is not an option (e.g., product can be damaged or dryer is located in a temperature-controlled environment). In those cases, drying the air prior to any product drying or ventilation may be the right solution.

USE OF CONDENSATION

Through the use of heat exchanger with a cooling fluid, the air temperature will drop down to its dew point temperature where an important condensation will occur.

Encyclopedia of Agricultural, Food, and Biological Engineering
DOI: 10.1081/E-EAFE 120007057
Copyright © 2003 by Marcel Dekker, Inc. All rights reserved.

In most warm and wet climates, it may be useful to cool down the drying air through the use of a heat pump for instance. Depending on the temperature of the cooling fluid, vapor in excess in the treated air will either condense or freeze. In the former case, liquid water will be continuously recycled in a proper wastewater treatment system. In the latter case, ice water will be accumulated in the exchanger leading to more complex cleaning cycles. An additional heater will be turned on when the ice layer becomes too thick, diminishing the system performance. The unit is therefore switched from its production cycle to its cleaning cycle: melting the ice out of the device.

USE OF DESICCANTS

Liquid Desiccants

A hygroscopic salt solution is pumped and sprayed into a dehumidifier chamber (Fig. 1). The humid air will pass and come into contact with droplets. The hygroscopic solution will absorb the moisture present in the air. Dry air will leave the chamber by the top while the solution, including the absorbed moisture, will be collected in the lower part of the chamber. A cyclone or any similar device will ensure that the air stream does not contain any particles.

Cooling the solution prior to its spraying is an option to both dry and cool down the air in a single unit, with some energy savings.

The absorbed moisture has to be evaporated to maintain a constant concentration of the solution in the dehumidifier. Therefore, the diluted solution, or part of it, is heated up and pumped into a regenerator chamber and sprayed again. A minor secondary air stream will pass through the regenerator chamber and carry the evaporated vapor outside of the unit. This concentrated solution will then be able to return to the dehumidifier chamber.

It is possible to add a heat exchanger. It will preheat the cold diluted solution with the warm regenerated solution. Some energy savings can be obtained this way.

Dehumidifying and regenerating chambers can be located remote from each other for design flexibility. Additionally, several dehumidifying units can be used in connection with one single central regenerator.

When choosing this technology, one should carefully check the corrosion-proof and microbiological decontamination performance.

Solid Desiccants

A compound made of an organic fiber and an adsorbent is used in the form of a rotor. Common adsorbents are silica gels, sodium, calcium, or lithium chloride salts. The wheel turns slowly by means of a drive motor. Two fans are used in combination, in opposite direction to each other, creating two separate airflows: the process and the regeneration air. The process air flows through the rotor, sees its water adsorbed on the rotor, and leaves it as dry air. The regeneration air works in opposite direction. Heated by an electrical or a steam heater, it will pass through the regeneration section of the rotor, evaporating its moisture content to evacuate it outside of the unit (Fig. 2). Thus the rotor is heated during the regeneration process leading to some temperature increase of the dried air. An additional air cooler may be necessary.

Several options are available for better energy recovery. For instance, the rotor can be divided into three separate air zones: a process sector, a heat recovery sector, and a regeneration sector. Air to be dried is passed through the process sector where the desiccant adsorbs its moisture vapor. A separate smaller quantity of air recovers

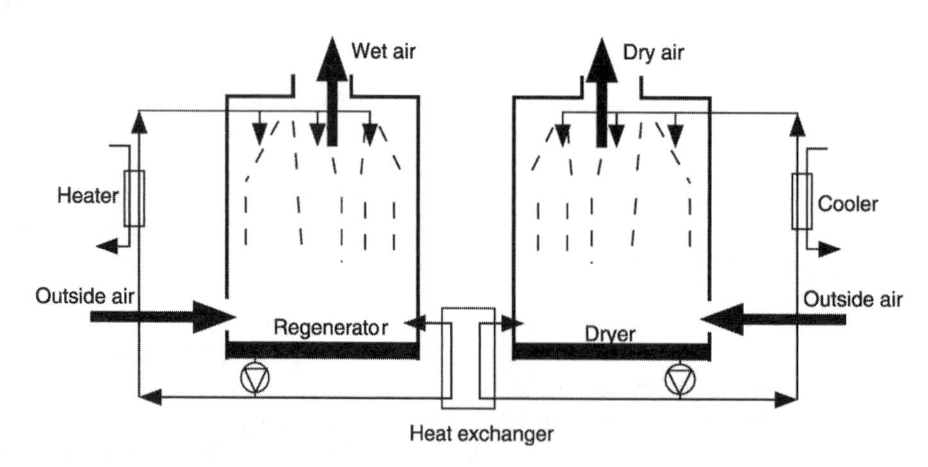

Fig. 1 Working principle of a liquid desiccant dehumidifier.

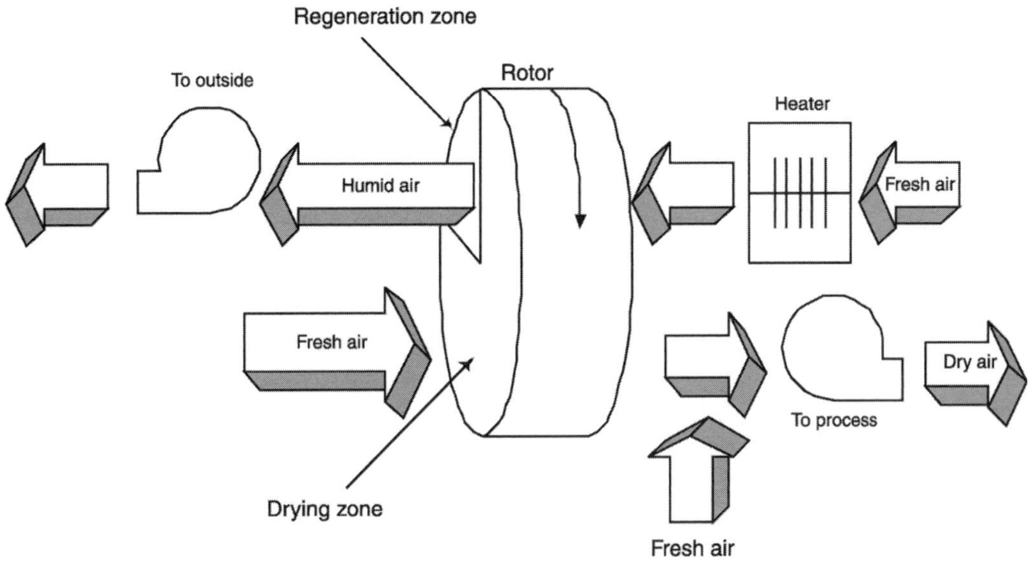

Fig. 2 Working principle of a dry desiccant dehumidifier.

heat from the rotor. Then the air is heated and passed through the regeneration sector where the adsorbed moisture is taken from the desiccant.

For a correct moisture control of the treated air, two possibilities can be considered: adjusting the rotation speed of the rotor or the temperature setpoint of the regeneration heater.

This system is simple and reliable for continuous dehumidification. The desiccant can be reactivated easily using heat generated from electricity, steam, hot water, gas, or oil. It is compatible with temperatures below 0°C, enabling humidity control in various climatic conditions. Air filters are required to avoid any possible accumulation of particles on the rotor. Eventually the rotor can be washed for complete reactivation.

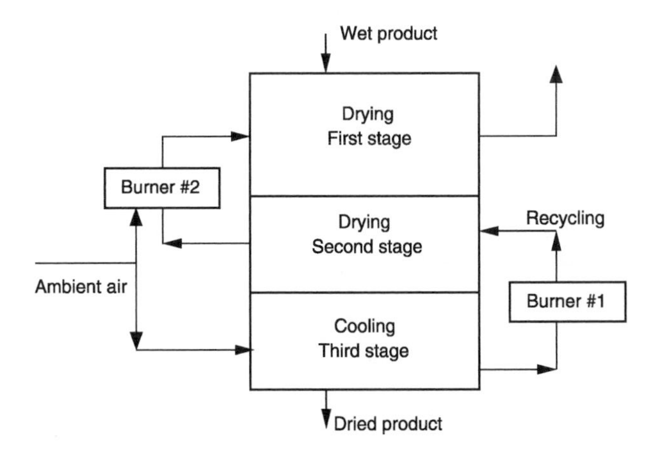

Fig. 3 Working principle of a dryer with air recycling system with additional mixing with fresh air.

MIXING OF AIRS

Many industrial dryers are composed of two or three drying stages. A stage is defined as a drying unit where inlet air is at a given temperature, moisture content, and flow rate. Usually air is recycled from the latter drying stages to the first one. Counter current flow configuration is preferred for obvious reasons of higher efficiency. This energy-efficient design implies the used air, somehow wetter, to be reused for product drying. Furthermore, drying capacity is reduced while energy is saved. In order to maintain a minimum drying capacity in the early stages, two different methods are combined.

- Increase of the air temperature with an additional heater.
- Mixing (warm) used and (cold) fresh air with an additional fan.

This air recycling design with additional air input flow and partial reheating is generally very efficient in terms of energy performance, investment, and maintenance costs.

Fig. 3 shows a typical trajectory for the air flowing from the last to the first stage.

CONCLUSION

Drying of air is an energy-costing operation with a possible need for an additional cooling or heating unit. When the dried air is later used for drying purpose, it is generally more cost effective in increasing simply the drying temperature to obtain a higher drying capacity or to reconsider the dryer design to include air recycling. Drying of air is generally of higher interest for storage preservation where continuous ventilation of dry air is required. Typically, very dry air is required when the product is highly hygroscopic and/or the air very humid and/or in near ambient air temperature.

FURTHER READING

For further Reading and Calculation

American Society of Heating, Refrigerating and Air-Conditioning Engineers, http://www.ashrae.org

Online Psychrometric Calculator, http://www.psychro.com

For further Information on Dehumidifier Design

Dehumidifier Corporation of America (Dehumidification by dew point method), http://www.dehumidifiercorp.com/industry.html

Kathabar Systems Europe (Liquid Dessicant Dehumidification), http://www.kathabar.com

HB Dehumidification Systems (Sorption Wheel), http://www.hb-almelo.nl

Drying Theory

Wade Yang
Terry J. Siebenmorgen
University of Arkansas, Fayetteville, Arkansas, U.S.A.

D

INTRODUCTION

Moisture removal is a vital step to achieve safe storage and quality retention of agri-food materials. Artificial drying is commonly used to remove moisture and thus improve storability and quality. In the past two decades, drying theory and technology have advanced greatly, which has led to the development of more energy-efficient, high-throughput, and automated dryers for producers and food processors. A good number of published treatises on drying theory and practice have contributed to the core knowledge base on drying fundamentals, mathematical modeling, equipment design, and drying operations. In this article, a description of drying fundamentals and mathematical modeling, especially as pertains to cereal grains, is presented. Coverage will also be extended to some recent developments in moisture relationships including sorption hysteresis modeling. The drying theory discussed here is limited to solids and emphasizes particulate and loose materials like grains. For a more complete and in-depth discussion of specialized topics in drying, such as properties of moist air,[1,2] industry-specific drying processes,[3,4] nonconventional drying methods,[3] types of dryers and their applications,[3,5] and drying simulation and numerical computation,[4,6–10] the reader is referred to the above-listed references.

DRYING FUNDAMENTALS

Drying, in the context used in this chapter, is a process of moisture removal in which the internal water of a hygroscopic material moves, in the form of liquid, vapor, or both, to the surface of the material and is evaporated or transferred by convection to air passing over the material, with or without the supplement of heat. The process of drying usually involves simultaneous heat and moisture transfer. The fundamental driving force for drying is the chemical potential associated with the water in the material, i.e., moisture moves from inside a moist material where the chemical potential is high to outside the material where it is low. Some fundamental principles involved in a drying process are briefly described below.

The States of Water and Types of Moisture in Solids

Under different pressures and temperatures, water can assume one or a combination of the three states: solid, liquid, or vapor as depicted in the phase diagram of water (Fig. 1), where OA, OB, and OC represent the equilibrium states between solid and liquid, liquid and vapor, and solid and vapor, respectively. From Fig. 1, it can be seen that water can become vapor from either a liquid state (evaporation, OB) or a solid state (sublimation, OC), depending on the pressure and temperature. Most drying processes for agri-food materials involve changing water from a liquid to a vapor state and removing the vapor by passing air over the material surface. This can take place at atmospheric pressure or in vacuum conditions. The process of changing from a solid state to a vapor state is typically called freeze drying.[38–40]

Moisture in solids can be broadly divided into the following three categories:

1. Surface moisture—Water retained on the surface of solids because of surface tension. Its behavior is similar to that of free water.
2. Unbound moisture—Water that exists within solids in pores and interstitial voids. It behaves in a manner similar to that of free water.
3. Bound or nonfreezable moisture—Water that attaches to solids by physical and/or chemical binding. Bound moisture acts quite differently from free water. For example, it requires much more energy to drive bound moisture out of solids than free or unbound water.

Water in materials such as grains can exist in all three of the above forms. In addition, the degree of binding associated with the water can vary. The relative degree of binding ultimately determines the drying characteristics of the material, including parameters such as the critical MC and the drying rate constant.

Moisture Representations

Moisture content can be expressed on either a dry or a wet basis (Eqs. 1 and 2) and can be numerically expressed in

Encyclopedia of Agricultural, Food, and Biological Engineering
DOI: 10.1081/E-EAFE 120006893
Copyright © 2003 by Marcel Dekker, Inc. All rights reserved.

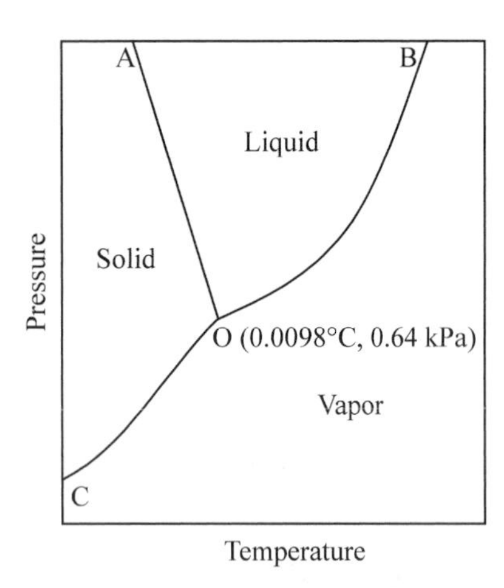

Fig. 1 A phase diagram of water.

either a percentage or decimal (fraction) format.

Wet basis moisture content, M_w

$$= \frac{\text{mass of water}}{\text{mass of wet material}} \quad (1)$$

Dry basis moisture content, M_d

$$= \frac{\text{mass of water}}{\text{mass of dry matter}} \quad (2)$$

Eqs. 3 and 4 give the conversion formulas between dry and wet basis.

$$M_w = \frac{M_d}{100 + M_d} \times 100 \quad (3)$$

$$M_d = \frac{M_w}{100 - M_w} \times 100 \quad (4)$$

where M_w and M_d are in percent basis.

Each expression of moisture content has advantages in certain applications. For example, M_d is typically used in engineering calculations in which mass balances are performed on processes involving moisture content changes; having the denominator remain constant in Eq. 2 can be an advantage in such calculations. Furthermore, during heat and mass transfer modeling and computation, M_d rather than M_w should be used to allow moisture content to be added or subtracted due to a common base. Conversely, the wet basis expression represents a readily interpretable way of quantifying the amount of water in a material and is thus almost exclusively used in grain trading and processing.

Heat and Moisture Movement During a Drying Process

Heat and moisture transfer occurs when a drying medium, typically heated air, contacts a hygroscopic material. Heat moves from the medium into the material, and moisture moves from inside the material to the surface of the material and evaporates to the medium. Heat can be transferred to the material surface by conduction, convection, and/or radiation, but convective heat transfer dominates on the surface of the material as air moves over the surface of the material, while conduction is the predominant mode of heat transfer inside the material. For moisture transfer within a material, it is believed that moisture moves from inside the material to the surface predominantly by diffusion.[2,5] Moisture is transferred into the drying medium by, as mentioned earlier, evaporation and/or convective mass transfer.

Drying of a bulk of solids can theoretically be considered an adiabatic process. In other words, heat exchange takes place entirely between the solids and the drying medium in a closed system where no heat is transferred from the solid–air system to the environment or vice versa. The energy necessary to evaporate moisture

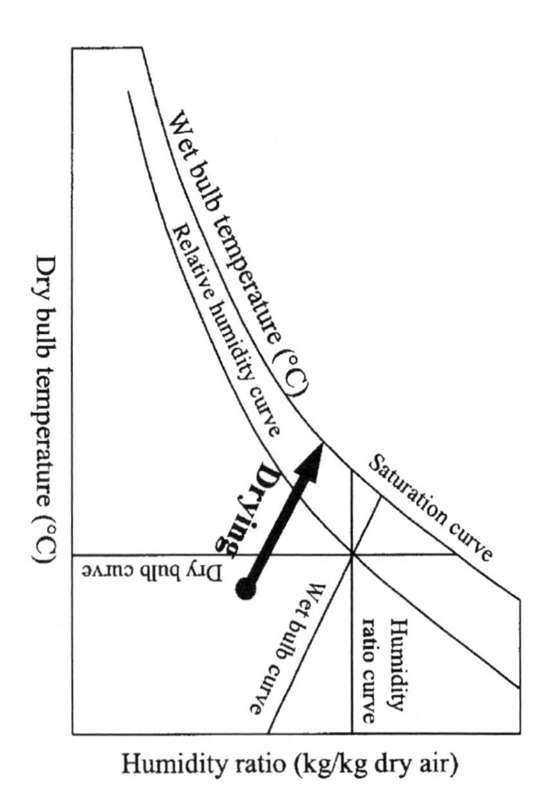

Fig. 2 A psychrometric representation of the drying process of particulate and loose materials like grains.

is supplied by sensible heat from the drying medium and, to a limited extent in certain situations, the solid. Psychrometrically, a drying process more or less follows a constant wet bulb temperature curve (Fig. 2). As drying air is passed through solids to be dried, the air dry bulb temperature decreases, and relative humidity and humidity ratio increase. For specific information on psychrometric charts and their application to drying, readers are referred to Refs. [1] and [2].

Drying Curves and Drying Rates

Drying kinetics, often associated with drying curves given a constant drying air temperature, describes the relationship between moisture content and drying duration at a given temperature. A typical drying curve at a constant drying temperature is illustrated in Fig. 3 (upper graph). It is obtained by measuring the weight loss of solids at different drying durations over the course of a drying period. In Fig. 3, the drying rates corresponding to the drying curve are also illustrated (lower graph). From the onset of drying to point A in Fig. 3, the material is gradually heated to the drying medium temperature and the drying rate increases dramatically. From A to B, drying proceeds at a constant rate and is characterized by a linear decline in the drying curve and a horizontal segment in the drying rate curve. This period represents the removal of moisture existing on the surface of the solid; it is referred to as the "constant rate" period and is typically of a short duration. From B to C, drying rates progressively decrease and the material gradually approaches an equilibrium state with the drying medium. During this "falling rate" period, the drying curve assumes a concave shape. Beyond C, the material is basically in an equilibrium state with the drying medium. For most agricultural products, particularly grains, there is usually no observable constant rate period as depicted by AB in Fig. 3. Most drying occurs in the falling rate period, except for situations in which liquid exists on the surface of solids at the beginning of drying or some deep-bed drying cases where an overall constant rate period may exist in the initial stages of drying.[6,33]

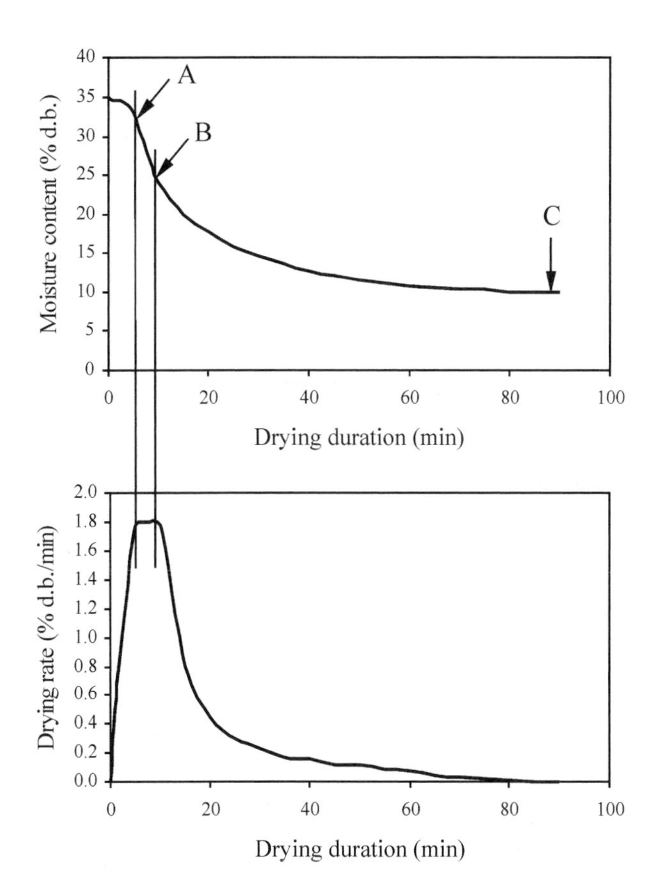

Fig. 3 Illustration of a drying curve and the drying rates associated with it.

Equilibrium Moisture Content (EMC)

EMC and moisture sorption hysteresis are two major aspects of drying and conditioning of hygroscopic materials. EMC refers to the moisture content a hygroscopic material will reach after it is exposed to an environment comprising a certain relative humidity and temperature for an infinite duration. As an example, Fig. 4 shows the EMC curves (isotherms) of ground alfalfa at 25°C[34] and cowpea at 30 and 50°C.[11] As can be seen, given a relative humidity, EMC decreases with increased temperature following the Clausius-Clapeyron equation.[35] The isotherms of most agri-food products exhibit a sigmoid shape as depicted in the upper graph of Fig. 4.

The American Society of Agricultural Engineers (ASAE) publishes EMC data for plant-based products in ASAE Standard D245.5.[11] A comprehensive bibliography, which readers can refer to for more information on moisture sorption isotherms of specific agri-food materials, has been compiled by Sokhansanj and Yang.[12] The isotherms in the current ASAE D245.5 are categorized into four parts to reflect the various physical and chemical characteristics of agri-food materials. These categories are starchy materials (e.g., wheat and corn), fibrous materials and selected feeds (e.g., forage and grain hulls), high oil and protein materials (e.g., canola and soybean), and specialty agricultural products (e.g., tobacco and sugar beets). In addition to EMC data, ASAE D245.5 also gives other pertinent information such as temperature, isotherm

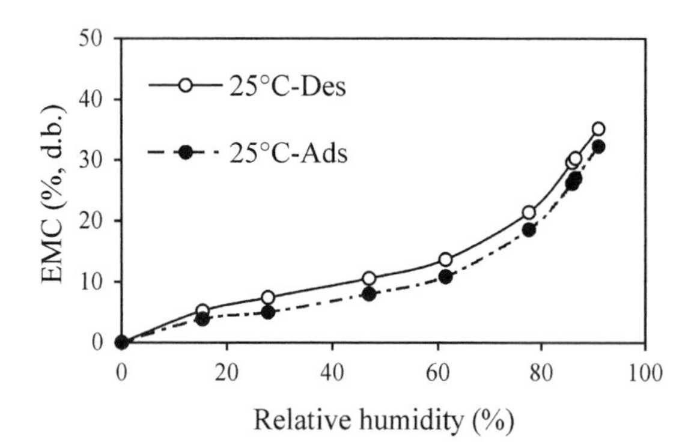

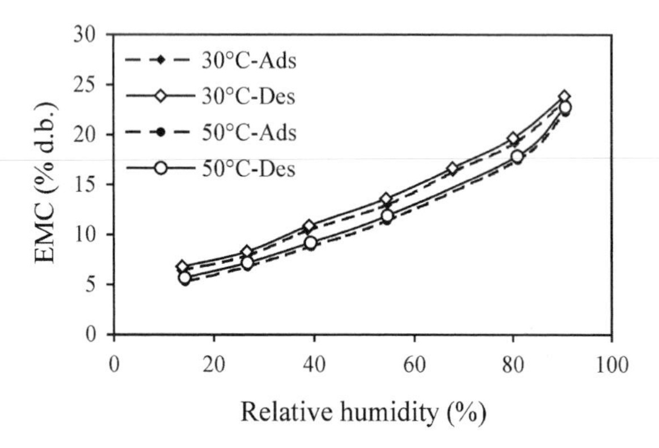

Fig. 4 Moisture sorption isotherms illustrating the hysteresis effect for ground alfalfa at 25°C (upper graph)[34] and Nigerian cowpea at 30 and 50°C (lower graph).[11] Ads and Des stand for adsorption and desorption, respectively.

path (e.g., adsorption, desorption, or mixture of both), data sources, and related references where more information about a given product can be found. ASAE Standard D245.5 recommends five isotherm equations for various agri-food materials:

1. Modified Henderson equation

$$\mathrm{RH} = 1 - \exp\left[-A(T + C)m^{B}\right] \quad \text{or}$$

$$M = \left[-\frac{\ln(1 - \mathrm{RH})}{a + bT}\right]^{\frac{1}{c}} \tag{5}$$

2. Modified Chung-Pfost equation

$$\mathrm{RH} = \exp\left[-\frac{A}{T + C}\exp(-B \times M)\right] \quad \text{or}$$

$$M = -a\ln\{(b + cT)[-\ln(\mathrm{RH})]\} \tag{6}$$

3. Modified Halsey equation

$$\mathrm{RH} = \left[-\frac{\exp(A + B \times T)}{M^{C}}\right] \quad \text{or}$$

$$M = \left[-\frac{\exp(a + bT)}{\mathrm{RH}}\right]^{\frac{1}{c}} \tag{7}$$

4. Modified Oswin equation

$$\mathrm{RH} = \left[\left(\frac{A + B \times T}{M}\right)^{C} + 1\right]^{-1} \quad \text{or}$$

$$M = (a + bT)\left(\frac{1}{\mathrm{RH}} - 1\right)^{\frac{1}{c}} \tag{8}$$

5. Guggenheim–Anderson–de Boer (GAB) equation

$$M = \frac{a \times b \times c \times \mathrm{RH}}{(1 - b \times \mathrm{RH})(1 - b \times \mathrm{RH} + b \times c \times \mathrm{RH})} \tag{9}$$

where RH is relative humidity, T the temperature, M the moisture content, and A, B, C, a, b, and c constants. A, B, C and a, b, c are used to distinguish the constants in the two separate equations with M or RH being the dependent variable, respectively.

Among the five equations, the Modified Henderson and the Modified Chung-Pfost equations generally describe the isotherms of starchy and fibrous materials better.[13] The Modified Halsey and the Modified Oswin equations tend to better approximate the isotherms of oil-rich materials,[13,14] although some exceptions exist. The GAB equation has been reported to approximate well the isotherms of most agricultural and food materials.[15]

Moisture Sorption Hysteresis

Moisture sorption hysteresis refers to the phenomenon in which the isotherms reached through adsorption do not overlap those reached through desorption, leaving a gap between the two curves (see Fig. 4). This phenomenon has been extensively studied since it was discovered over a century ago. The importance of moisture sorption hysteresis lies in both its theoretical implications for biological and food materials, such as the thermodynamic irreversibility of sorption, and its practical implications such as the effects on chemical and microbiological deterioration, quality changes, storage stability, and engineering properties of biological and food materials.[16–19]

Yang et al.[20] developed an analytical model to describe equilibrium hysteresis based on a hypothesis that recognizes the thermodynamic nature of a sorption process. This hypothesis proposes that the internal instantaneous generation of heat during adsorption causes a change in local temperature around sorption sites. It was hypothesized that altered local temperatures would induce a temperature gradient between the sorption site and the environment. The temperature gradient, in turn, affects the sorption process in such a way that moisture redistribution would occur within sorption sites and in the regions between the sorption site and the outer surface of the material. The outcome of the adjustment of moisture at the end of a sorption process is believed to give rise to sorption hysteresis.

The total hysteresis in the entire material, H, can be calculated as,[20]

$$H = \frac{m_0 g_0(r_0, t_0)}{L_{T_{\mathrm{ave}}}} \frac{1}{2\pi\rho_{\mathrm{w}} r_0^3} \times \left\{ 2\mathrm{erf}\left[\frac{r_0}{\sqrt{4\alpha_{\mathrm{w}}(t - t_0)}}\right] - \mathrm{erf}\left[\frac{2r_0}{\sqrt{4\alpha_{\mathrm{w}}(t - t_0)}}\right]\right\} + \frac{g_0(r_0, t_0)}{L_\Gamma}\frac{\sqrt{\alpha_{\mathrm{s}}(t - t_0)}}{\sqrt{\pi} r_0}\left\{1 - \exp\left[-\frac{r_0^2}{\alpha_{\mathrm{s}}(t - t_0)}\right] + \frac{2r_0 \mathrm{erfc}(r_0)}{\sqrt{\alpha_{\mathrm{s}}(t - t_0)}}\right\} \quad (10)$$

where m_0 is the mass of the spherical source with a radius r_0 that instantaneously releases a pulse of energy $g_0(r_0, t_0)$ in adsorption at time t_0. $L_{T\mathrm{ave}}$ and L_Γ are latent heats of vaporization, ρ_{w} and α_{w} the density and thermal diffusivity of liquid water, respectively, α_{s} the thermal diffusivity of dry matter, and t the time. For details on the derivation of Eq. 10, readers are referred to Yang et al.[20]

As isotherms are typically obtained by sealing a material in an environment of constant temperature for a sufficiently long duration for equilibrium to establish, the isotherm process can be approximately regarded as the situation when $(t - t_0) \rightarrow \infty$. Taking the limit of Eq. 10 as $(t - t_0) \rightarrow \infty$ and considering the surface tension effect and the net enthalpy change,[20] we have,

$$H = \frac{2m_0 \Delta E}{\sqrt{\pi} L_\Gamma} \mathrm{erfc}\left[\frac{-2\sigma V \cos\theta}{RT \ln(p/p_0)}\right] \quad (11)$$

where p/p_0 is the relative pressure of the sorbate (i.e., relative humidity) at temperature T, σ the surface tension, V the molar volume of the sorbate, θ the angle of contact, R the universal gas constant, and m_0 now becomes the EMC after infinite duration and can be expressed by an isotherm equation, such as Eqs. 5–9 shown earlier. Eq. 11 has been verified with sorption data over a wide variety of organic and inorganic materials.[20,21]

MATHEMATICAL MODELING

General Heat and Mass Transfer Models

As mentioned earlier, in normal drying conditions, diffusion and conduction can be regarded as the predominant modes for moisture and heat transfer, respectively, inside solids. For particulate and loose materials like grains, heat and moisture transfer can be modeled separately during drying, because it was found that the coupled effects of heat and moisture transfer are minimal on the drying process.[22] Therefore, the following moisture diffusion and heat conduction equations apply:

$$\frac{\partial m}{\partial t} = \nabla^2 D m \quad (12)$$

$$\frac{\partial T}{\partial t} = \nabla^2 \alpha T \quad (13)$$

where m is moisture content on a dry basis, T the temperature in °C, t the time in sec, and D and α the moisture diffusivity and thermal diffusivity, respectively, in $\mathrm{m^2\,sec^{-1}}$. Eqs. 12 and 13 can be applied to the drying of solids from three different perspectives: 1) single-particle drying, where moisture and temperature distributions inside a single particle during drying can be obtained; 2) thin-layer drying, where instead of looking into the intraparticle distributions, the entire thin layer is examined as a whole, the temperature of the particle is assumed to be constant across the thin layer, and the kinetic relationship of average moisture content can be obtained; and 3) deep-bed drying, where the drying is examined in a bulk of particles and the moisture and temperature distribution inside the bulk can be obtained. Of course, for deep-bed drying, there are, in addition to Eqs. 12 and 13, other theoretical and/or semiempirical equations as described by Brooker, Bakker-Arkema, and Hall.[2]

Single-Particle Drying Models

Although most agricultural and food materials are usually dried in bulk, it is actually the individual particles inside a bulk that interact with the drying medium to cause a change in moisture content in response to different drying conditions. The drying behavior of individual kernels is therefore of primary importance in describing a drying process.

For most grains, an axisymmetric geometrical relation can be assumed to approximate kernel shape. Thus, the single-particle drying problem can be represented by a 2-D coordinate system. Eqs. 14 and 15 are the diffusion equation and heat conduction equation, respectively, in

2-D cylindrical coordinates.

$$\frac{\partial m}{\partial t} = \left(\frac{1}{r}\frac{\partial}{\partial r}\left(rD\frac{\partial m}{\partial r}\right) + \frac{\partial}{\partial z}\left(D\frac{\partial m}{\partial z}\right)\right) \tag{14}$$

$$\left(\frac{\partial T}{\partial t}\right) = \left(\frac{1}{r}\frac{\partial}{\partial r}\left(r\alpha\frac{\partial T}{\partial r}\right) + \frac{\partial}{\partial z}\left(\alpha\frac{\partial T}{\partial z}\right)\right) \tag{15}$$

where r and z are the axes in a cylindrical coordinate system. On the interface of the material and drying medium, convective heat and mass transfers prevail. The corresponding boundary and initial conditions are:

$$-D\frac{\partial m}{\partial n} = h_m(m - m_e) \tag{16}$$

$$-k\frac{\partial T}{\partial n} = h_t(T - T_a) \tag{17}$$

$$t = 0, \quad m = m_0, \quad T = T_0 \tag{18}$$

where h_m and h_t are convective mass and heat transfer coefficients, respectively, m_e the EMC, T_a the air temperature, and n flux in normal direction. It is usually complicated to solve Eqs. 14 and 15 analytically at the given boundary and initial conditions, so numerical methods, such as the finite element method (FEM) or finite difference method, are often used to solve the equations.

Solutions to these equations allow for predicting temperature and moisture content inside particles, from which insight into the drying process can be generated. As an example, Figs. 5 and 6 show the FEM-predicted temperature and moisture content profiles, respectively, inside a long-grain rice kernel at 22% initial moisture content during drying at 60°C and 17% relative humidity; for numerical solution techniques, as well as the physical parameters and property formulas used, the reader is referred to Yang et al.[23] and Chen et al.[24] for details. From Fig. 5, it can be seen that temperatures at nodes 1, 2, and 3 increased differently as drying proceeded, but they all converged and reached the drying air temperature within 2.5 min. From Fig. 6, it can be seen that the moisture content at node 2 decreased much faster than that at nodes 1 and 3 and the center moisture content (node 1) decreased most slowly. While the numerical complexity increases greatly, additional modeling can be used to build on such simulation to predict stress distribution inside kernels.[36,37] This extended modeling adds tremendously to the ability to predict kernel integrity and overall physical quality.

Thin-Layer Drying Models

A thin layer is usually referred to as a layer thickness no more than three times the average particle dimension of the material.[25] As with this modeling approach, temperature across a particle is assumed to be constant, Eq. 13 is no longer relevant. Thin-layer drying represents a

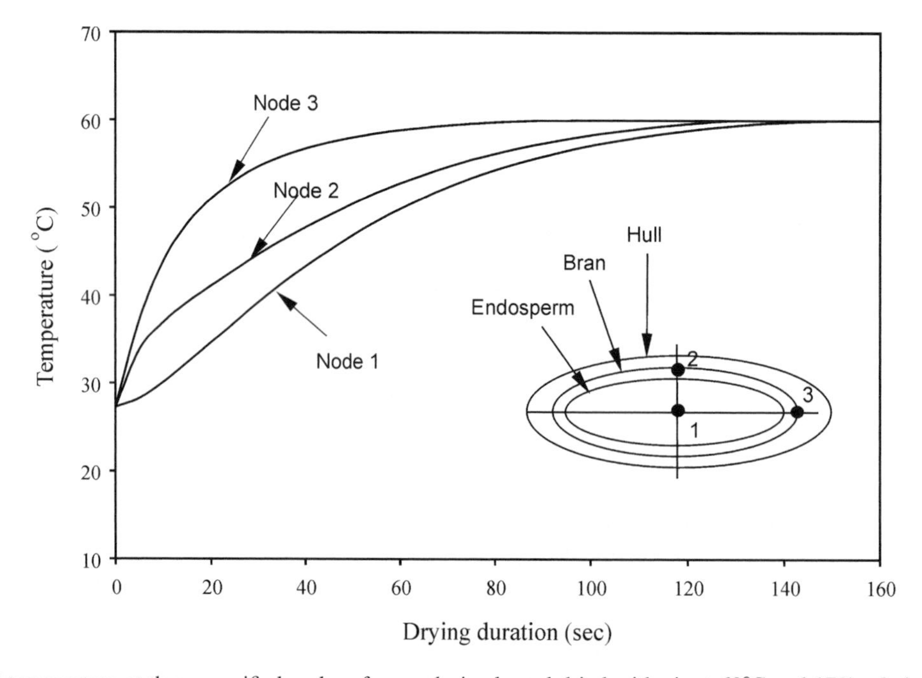

Fig. 5 Predicted temperature at three specified nodes of a rough rice kernel dried with air at 60°C and 17% relative humidity. (From Refs. [23] and [24].)

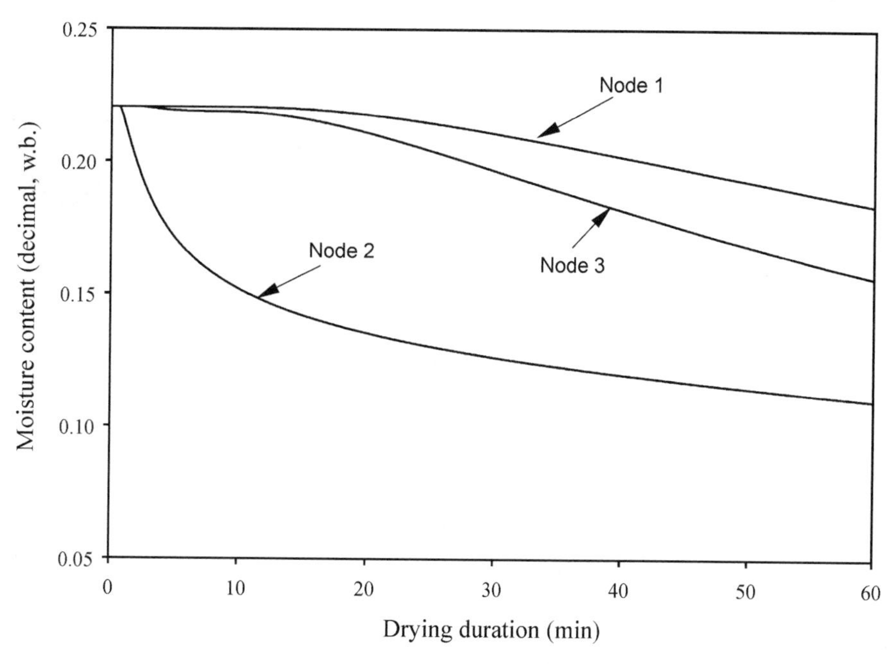

Fig. 6 Predicted moisture content at three specified nodes of a rough rice kernel dried with air at 60°C and 17% relative humidity. Refer to Fig. 5 for the specific locations of the three nodes. (From Refs. [23] and [24].)

single dimension problem, so Eq. 12 becomes:

$$\frac{\partial M}{\partial t} = D\frac{\partial^2 M}{\partial x^2} \tag{19}$$

$$M = M_0, \quad t = 0 \tag{20}$$

$$M = M_e, \quad x = \delta \tag{21}$$

where M is the average moisture content of the particles in the layer (fraction, dry basis) as opposed to the distributed moisture content, m, as used previously, M_0 the initial moisture content (fraction, dry basis), M_e the EMC (fraction, dry basis), and δ the half-thickness of the thin layer. The moisture diffusivity D defined above, which is assumed constant, refers to the overall effective diffusion coefficient that accounts for the liquid and vapor diffusion through the voids of the sample bulk and the pores in the individual particles.

The solution to Eq. 19 at the initial and boundary conditions specified in Eqs. 20 and 21 is shown in Eq. 22[26]:

$$\mathrm{MR} = \frac{M - M_e}{M_0 - M_e} = \sum_{n=1}^{\infty} \frac{8}{(2n-1)^2\pi^2} \exp\left[-(2n-1)^2\pi^2 Dt/(4\delta^2)\right] \tag{22}$$

where MR is the moisture ratio.

The series in Eq. 22 can be represented by two components, i.e., the component corresponding to $n = 1$ and the component comprising the sum of terms for $n = 2$ to $n = \infty$, as shown in Eq. 23.

$$\mathrm{MR} = \frac{8}{\pi^2}\exp\left[-\pi^2 Dt/(4\delta^2)\right] + \sum_{n=2}^{\infty} \frac{8}{(2n-1)^2\pi^2}\exp\left[-(2n-1)^2\pi^2 Dt/(4\delta^2)\right] \tag{23}$$

where $(8/\pi^2)\exp[-\pi^2 Dt/(4\delta^2)]$ is the first term of the series in Eq. 22. If it is assumed that the other terms in Eq. 23 can be approximately expressed in terms of the first term multiplied by a factor, Eq. 23 can be rewritten as:

$$\mathrm{MR} = \lambda\frac{8}{\pi^2}\exp\left[-\pi^2 Dt/(4\delta^2)\right] \tag{24}$$

where λ is a constant. Using the initial condition $M = M_0$ at $t = 0$, λ is determined to be $\pi^2/8$. Therefore,

$$\mathrm{MR} = \exp(-kt) \tag{25}$$

where k is a drying constant (s^{-1}) equal to $\frac{\pi^2 D}{4\delta^2}$.

Eq. 25 resembles the standardized first-order thin-layer drying model given in ASAE Standard S448,[25] although the theoretical basis for the derivation of the equations is different. The ASAE thin-layer equation was based on moisture diffusion in a single kernel, while Eq. 25 was based on moisture diffusion through a thin layer of

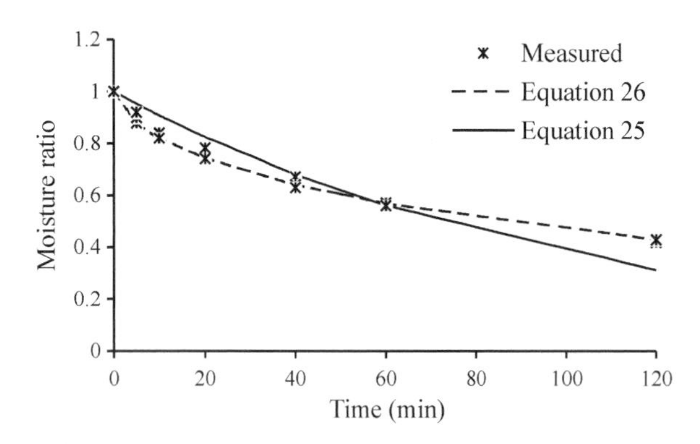

Fig. 7 Goodness-of-fit of Eqs. 25 and 26 as applied to thin-layer drying data of a long-grain rice cultivar (Cypress) using drying air at 50°C and 16% relative humidity.

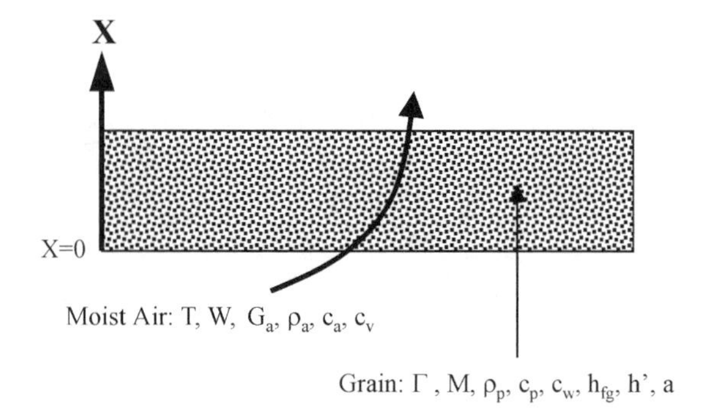

Fig. 8 Schematic indicating the conditions of moist air and grain conditions in deep-bed modeling.

thickness 2δ. By varying the power of t in Eq. 25, we have Page's equation[25]:

$$\mathrm{MR} = \exp(-kt^n) \tag{26}$$

where n is an empirical constant.

Because of the simplification and approximation during the derivation of Eq. 25, it often poorly describes thin-layer drying data of most grains. Fig. 7 shows an example of applying Eqs. 25 and 26 to thin-layer drying data for rice. It can be seen that the goodness-of-fit of Eq. 25 is poor ($R^2 = 0.84$), while Eq. 26 represents the measured values closely ($R^2 = 0.99$).

Deep-Bed Drying Models

There has been a great deal of research conducted in deep-bed drying of particulate and loose materials like grains. Parry[6] and Brooker et al.[2] give a critical review and in-depth analysis of deep-bed drying models.

Deep-bed drying can generally be categorized into two types: 1-D and 2-D. In 1-D deep-bed drying, both material flow and airflow progress along the same axis or dimension (the flows can be in different directions). Examples of this category are fixed-bed drying (e.g., in-bin drying), concurrent-flow drying, or counter-flow drying. In 2-D deep-bed drying, material flow and airflow are at right angles, e.g., cross-flow drying. For each deep-bed drying type, the drying models differ. Nevertheless, deep-bed drying models fall into three types depending on the method used to calculate moisture content throughout the bed: logarithmic, heat and mass transfer balances, and partial differential equations.[6,27] These models are usually complicated and necessitate computer-aided numerical techniques[7,9,10] to solve. In this chapter, emphasis is placed on mathematical models for fixed (stationary) bed drying.

Brooker et al.[2] present the fixed-bed drying equations as follows:

Air enthalpy balance :

$$\frac{\partial T}{\partial x} = \frac{-h'\alpha}{G_a c_a + G_a c_v W}(T - \Gamma) \tag{27}$$

Grain enthalpy balance :

$$\frac{\partial \Gamma}{\partial t} = \frac{h'\alpha}{\rho_p c_p + \rho_p c_w M}(T - \Gamma) + \frac{h_{fg} + c_v(T - \Gamma)}{\rho_p c_p + \rho_p c_w M} G_a \frac{\partial W}{\partial x} \tag{28}$$

Air moisture balance : $$\frac{\partial W}{\partial x} = -\frac{\rho_p}{G_a}\frac{\partial M}{\partial t} \tag{29}$$

Grain moisture balance (diffusion) : $$\frac{\partial M}{\partial t} = D\frac{\partial^2 M}{\partial x^2} \tag{30}$$

where T and Γ are the air and grain temperatures, respectively, h' the convective heat transfer coefficient of the grain bulk, α the particle surface area per unit bed volume, G_a the airflow rate, W the air humidity ratio, c_a, c_v, and c_w the specific heats of dry air, water vapor, and liquid water, respectively, ρ_p the bulk density of the grain, and h_{fg} the latent heat of vaporization. A schematic for the deep-bed model, including the conditions of the moisture air and the grain, is shown in Fig. 8. The boundary and initial conditions are:

$$M(x,0) = M_0 \tag{31}$$

$$\Gamma(x,0) = \Gamma_0 \tag{32}$$

$$T(0,t) = T_{\text{inlet}} \tag{33}$$

$$W(0,t) = W_{\text{inlet}} \quad (34)$$

For an in-depth analysis of deep-bed drying models, formulas expressing physical and thermal properties, numerical solutions and validations, and calculation examples, readers are referred to Brooker et al.[2]

Stochastic Drying Models

Most moisture contents referred to in grain drying are the average moisture content of a sample of particles, except for the case of intrakernel moisture distribution, where moisture content is given for different locations inside a particle. The models that use average moisture contents, such as those previously presented, are of a deterministic type in which it is assumed that the moisture content of individual grain kernels equals the bulk average moisture content. Such an assumption is, more often than not, oversimplified, because as research has suggested,[28,29] a significant variance exists in moisture content from kernel to kernel even within a panicle of rice or a cob of corn, let alone the bulk of grain entering a dryer. Fig. 9 shows the variation of kernel moisture content within a panicle for rice cultivar Cypress at two mean harvest moisture contents. To consider the probabilistic nature of moisture content distribution of individual kernels, stochastic models are developed to account for average moisture content and its variance. When a stochastic model is used, the average moisture content of a certain quantity of grain will be the statistical mean of the moisture content of the individual kernels comprising the grain bulk, which is more realistic and accurate, rather than the bulk average of the moisture content in whole.

Liu and Bakker-Arkema[30] and Ryniecki et al.[31] derived stochastic models for thin-layer and deep-bed drying and used them in corn and wheat drying, respectively. In the former case, the variation in initial kernel moisture content was considered, while in the latter case, the variation of ambient temperature and relative humidity was accounted for. In this chapter, emphasis is placed on the stochastic models related only to the variation of kernel moisture contents.

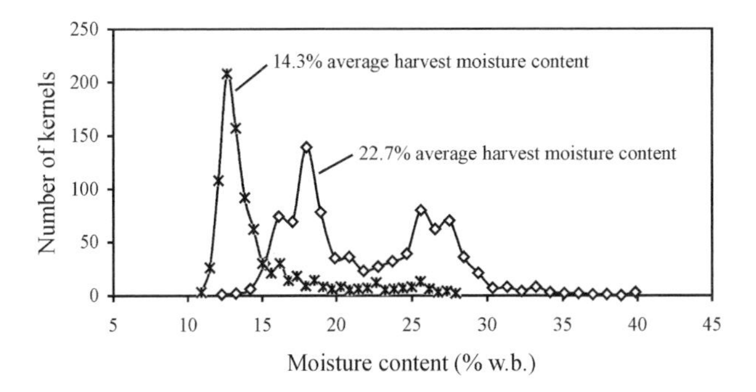

Fig. 9 Individual kernel moisture content distributions in a panicle of a long-grain rice cultivar Cypress at two mean harvest moisture contents from Stuttgart, Arkansas in 1998.

For thin-layer drying, the following stochastic models were used[30]:

$$E(M_{\text{f}}) = A\,E(M_0) + B \quad (35)$$

and

$$\sigma_{\text{f}} = A\sigma_0 \quad (36)$$

where $E(M_{\text{f}})$ is the expected value of moisture content distribution (i.e., mean individual kernel moisture content with a standard deviation of σ_{f}), $E(M_0)$ the expected value of initial moisture distribution of individual kernels with a standard deviation of σ_0, and A and B are model constants from the deterministic thin-layer drying model as specified in Eq. 25.

For deep-bed drying, both 1-D (material and air flows are in the same dimension, e.g., concurrent or counter-flow drying) and 2-D (material and air flows are in different dimensions, e.g., cross-flow) stochastic models were developed by Liu and Bakker-Arkema.[30] For concurrent or counter-flow drying situations, the stochastic models are:

$$E(M_{\text{f}}) = C\,E(M_0) + D \quad (37)$$

and

$$\sigma_{\text{f}} = C\sigma_0 \quad (38)$$

where C and D are parameters determined by the air temperature and relative humidity in the grain bed. In the derivation of Eqs. 35 and 36, it was assumed that all individual kernels in the dryer were exposed to the same drying air conditions and both initial and final moisture contents followed a normal distribution. It is noted that Eqs. 37 and 38 resemble Eqs. 35 and 36 in format and the model parameters A, B, C, and D are all dependent on temperature and relative humidity of the drying air. Parameters C and D are determined based on drying conditions and the drying rates for each of the thin layers divided over the thickness of the deep bed.[30]

Cross-flow drying stochastic models are[30]:

$$E(M_{\text{f}}) = \frac{1}{n}\sum_{i=1}^{n} E(M_{\text{fi}})$$

and

$$\sigma_{\mathrm{f}} = \sqrt{\frac{1}{n}\sum_{i=1}^{n}\sigma_{fi}^{2} + \frac{1}{m}\sum_{i=1}^{n}[E(M_{\mathrm{fi}})]^{2} - [E(M_{\mathrm{f}})]^{2}} \qquad (39)$$

where n is the number of thin layers comprising the deep-bed column (i.e., the deep-bed column is divided into n subcolumns or thin layers), $E(M_{\mathrm{fi}})$ the expected value of moisture distribution at a subcolumn with a standard deviation of σ_{fi}, and $E(M_{\mathrm{f}})$ and σ_{f} are the expected value and standard deviation of moisture content exiting the dryer, respectively. For model validations, readers are referred to Liu and Bakker-Arkema[30] and Bakker-Arkema and Liu.[32]

CONCLUSION

Both the fundamental aspects and theoretical modeling of drying agri-food materials have been discussed in this chapter. Recent advances in drying theory pertaining to quantitative modeling of sorption hysteresis and stochastic modeling to account for kernel moisture content variations are introduced. In contrast to the conventional drying theories that deal with a bulk of grains, the current trend of theoretical development of grain drying is to describe drying behavior of single grain kernels, because during drying, it is actually individual kernels inside a bulk that interact with the drying medium and undergo quality changes.

REFERENCES

1. Stanislaw, P.; Jayas, D.S.; Cenkowski, S. *Grain Drying: Theory and Practice*; John Wiley: New York, 1998; 303.
2. Brooker, D.B.; Bakker-Arkema, F.W.; Hall, C.W. *Drying and Storage of Grains and Oilseeds*; Van Nostrand Reinhold: New York, NY, 1992; 450.
3. Mujumdar, A.S. *Handbook of Industrial Drying*, 2nd Ed.; Marcel Dekker: New York, 2001; 1471.
4. Turner, I.W.; Mujumdar, A.S. *Mathematical Modeling and Numerical Techniques in Drying Technology*; Marcel Dekker: New York, 1996; 688.
5. Strumillo, C.; Kudra, T. *Drying: Principles, Applications and Design*; Gordon and Breach Science Publishers: New York, 1986; 448.
6. Parry, J.L. Mathematical Modeling and Computer Simulation of Heat and Mass Transfer in Agricultural Grain Drying: A Review. J. Agric. Eng. Res. **1985**, *32*, 1–29.
7. Irudayaraj, J.; Haghighi, K.; Stroshine, R.L. Finite Element Analysis of Drying with Applications to Cereal Grains. J. Agric. Eng. Res. **1992**, *53*, 209–229.
8. Oliveira, L.S.; Haghighi, K. Finite Element Modeling of Grain Drying. In *Mathematical Modeling and Numerical Techniques in Drying Technology*; Turner, I.W., Mujumdar, A.S., Eds.; Marcel Dekker: New York, 1996.
9. Sandeep, K.P. *Theory and Application of the Finite Difference Method in Food Processing*; ASAE Paper No. 993063; ASAE: St. Joseph, MI, 1999.
10. Sandeep, K.P.; Irudayaraj, J. Introduction to Modeling and Numerical Simulation. In *Food Processing Operations Modeling. Design and Analysis*; Irudayaraj, J., Ed.; Marcel Dekker: New York, 2001; 25–36.
11. *ASAE Standard D245.5: Moisture Relationships of Plant-Based Agricultural Products*; ASAE: St. Joseph, MI, 2002.
12. Sokhansanj, S.; Yang, W. Revision of the ASAE Standard D245.4: Moisture Relationships of Grains. Trans. ASAE **1996**, *39*, 639–642.
13. Chen, C.C.; Morey, R.V. Comparison of Four EMC/ERH Equations. Trans. ASAE **1989**, *32*, 983–990.
14. Yang, W.H.; Cenkowski, S. Enhancement of the Halsey Equation for Canola Isotherms. Can. Agric. Eng. **1995**, *37*, 169–182.
15. Sokhansanj, S.; Yang, W. *Enhancing ASAE D245.4: Moisture Relationships of Grains*; ASAE Paper No. 956103; ASAE: St. Joseph, MI, 1995.
16. Rizvi, S.S.H.; Benado, A.L. Thermodynamic Properties of Dehydrated Foods. Food Technol. **1984**, *38*, 83–92.
17. Kapsalis, J.G. Influence of Hysteresis and Temperature on Moisture Sorption Isotherms. In *Water Activity: Theory and Applications to Food*; Rockland, L.B., Beuchat, L.R., Eds.; Marcel Dekker: New York, 1987; 173–213.
18. Multon, J.L.; Bizot, H.; Doublier, J.L.; Lefebvre, J.; Abbott, D.C. Effect of Water Activity and Sorption Hysteresis on Rheological Behavior of Wheat Kernels. In *Water Activity: Influences on Food Quality*; Rockland, L.B., Stewart, G.F., Eds.; Academic Press: New York, 1981; 179–197.
19. Yang, W.H.; Cenkowski, S. Effect of Multiple Adsorption and Desorption Cycles and Drying Temperatures on the Hygroscopic Equilibrium of Canola. Can. Agric. Eng. **1993**, *35*, 119–126.
20. Yang, W.; Sokhansanj, S.; Cenkowski, S.; Tang, J.; Wu, Y. A General Model for Sorption Hysteresis in Food Materials. J. Food Eng. **1996**, *33*, 421–444.
21. Yang, W.; Sokhansanj, S.; Wu, Y.; Tang, J. *Quantification of Sorption Hysteresis Loops*; CSAE Paper No. 96-312; CSAE: Saskatoon, Canada, 1996.
22. Husain, A.; Chen, C.S.; Clayton, J.T.; Whitney, L.F. Mathematical Simulation of Mass and Heat Transfer in High Moisture Foods. Trans. ASAE **1972**, *15*, 732–736.
23. Yang, W.; Jia, C.-C.; Siebenmorgen, T.J.; Howell, T.A.; Cnossen, A.G. Intra-kernel Moisture Response of Rice to Drying and Tempering Treatments by Finite Element Simulation. Trans. ASAE **2002**, *45*, 1037–1044.
24. Chen, H.; Siebenmorgen, T.J.; Yang, W. *Finite Element Simulation to Relate Head Rice Yield Reduction During Drying to Internal Kernel Moisture Gradient and Rice

State Transition; ASAE Paper No. 996156; ASAE: St. Joseph, MI, 1999.

25. *ASAE Standard S448: Thin-Layer Drying of Grains and Crops*; ASAE: St. Joseph, MI, 2002.
26. Crank, J. *The Mathematics of Diffusion*; Clarendon Press: Oxford, England, 1975.
27. Morey, R.V.; Keener, H.M.; Thompson, T.L.; White, G.M.; Bakker-Arkema, F.W. *The Present Status of Grain Drying Simulation*; ASAE Paper No. 783009; ASAE: St. Joseph, MI, 1978.
28. Liu, Q.; Montross, M.D.; Bakker-Arkema, F.W. Stochastic Modeling of Grain Drying: Part 1: Experimental Investigation. J. Agric. Eng. Res. **1997**, *66*, 267–273.
29. Bautista, R.; Siebenmorgen, T.J. *Characterization of Rice Individual Kernel Moisture Content and Size Distributions at Harvest and During Drying*; ASAE Paper No. 996053; ASAE: St. Joseph, MI, 1999.
30. Liu, Q.; Bakker-Arkema, F.W. Stochastic Modeling of Grain Drying: Part 2: Model Development. J. Agric. Eng. Res. **1997**, *66*, 275–280.
31. Ryniecki, A.; Molinska, A.; Jayas, D.S. *Stochastic Modeling of Grain Moisture Content for Near-Ambient Drying*; ASAE Paper No. 946594; ASAE: St. Joseph, MI, 1994.
32. Bakker-Arkema, F.W.; Liu, Q. Stochastic Modeling of Grain Drying: Part 3: Analysis of Crossflow Drying. J. Agric. Eng. Res. **1997**, *66*, 281–286.
33. Simmonds, W.H.C.; Ward, G.T.; McEwen, E. The Drying of Wheat Grain. I. The Mechanisms of Wheat Drying. II. Through Drying of Deep Beds. Trans. Ind. Chem. Eng. **1953**, *31*, 265–288.
34. Yang, W. Characteristics of Ground Alfalfa in Relation to Steam Conditioning. Ph.D. Dissertation, Dept. of Agricultural and Bioresource Engineering, University of Saskatchewan, Saskatoon, SK, Canada, 1998.
35. Bell, L.N.; Labuza, T.P. Moisture Sorption. *Practical Aspects of Isotherm Measurement and Use,* 2nd Ed.; American Association of Cereal Chemists, Inc.: St. Paul, MN, 2000; 122.
36. Jia, C.-C.; Sun, D.-W.; Cao, C.-W. Mathematical Simulation of Stresses Within a Corn Kernel During Drying. Drying Technol. **2000**, *18*, 887–906.
37. Irudayaraj, J.; Haghighi, K. Stress Analysis of Viscoelastic Material During Drying. Part 2: Application to Grain Kernels. Drying Technol. **1993**, *11*, 929–959.
38. King, C.J. *CRC Freeze-Drying of Foods*; CRC Press: Cleveland, 1971; 86.
39. Schoen, M.P.; Jefferis, R.P., III. Simulation of a Controlled Freeze-Drying Process. Int. J. Modelling Simulation **2000**, *20* (3), 255–263.
40. Corridon, G.A. *Freeze-Drying of Foods: A List of Selected References*; National Agricultural Library, USDA: Washington, D.C., 1963; 79.

Drying Time Prediction

Piet J. A. M. Kerkhof
Eindhoven University of Technology, Eindhoven, The Netherlands

INTRODUCTION

In this article a modern treatment is given to the prediction of drying times, for a large part based on considerations of the diffusion of water inside drying products. As is stated in the paragraph on "Classification of Methodology," this is the most reliable and versatile approach. It comes however at a cost, namely the understanding and use of diffusion theory, of process models, the collection of physical data, and the numerical implementation. Some readers will be familiar with this; however for some it may at first sight pose a large barrier. The author has attempted to provide also the physical phenomena clearly. As in many other areas of food processing, optimization and design has evermore become an interdisciplinary effort. The author hopes that readers working in practice will not hesitate to seek assistance from experienced modeling engineers to make the theory applicable and profitable for their situation.

CLASSIFICATION OF METHODOLOGY

Foods to be dried range from liquid solutions through dispersions to entire solids and pieces. The generally complex composition and the variety of drying techniques necessitate prudence when attempting a generalized treatment. The methods for drying time prediction in literature are very diverse. A classification can be made according to the level of detailed description of the transport and exchange phenomena taking place, as follows: unstructured methods, characteristic drying curve (CDC), curve-fitting, structured models with more or less lumping of parameters.

Unstructured methods consist of the measurement of drying times under a number of circumstances, and estimate the drying time of interest from that. This can be very successful, provided the drying circumstances in the experiments are very close to those encountered in the real plant. They have, however, no predictive power outside this region. As an example, a rule of thumb is that for spray drying skimmed milk one needs an air residence time of about 40 sec, at air outlet temperature of 80°C.

A classic method is the use of the *characteristic drying curve*.[1] Here it is assumed that the falling-rate period under various external circumstances, e.g., temperatures and humidities, can be normalized into one single characteristic curve. From diffusion theory it follows that this can be a good approach when the diffusion coefficient is constant; for foods, however, this is generally not valid. In an analysis of the drying of *L. Plantarum* preparations it turned out that the CDC-method gave errors in drying time prediction up to over 100%[2] (see Fig. 1).

One of the more primitive methods is *curve-fitting*. The curves used may vary from a simple exponential to two exponentials, of which various coefficients are then again considered to be functions of initial moisture content, temperature, layer thickness, etc. These methods can only be used for interpolation when the condition of interest lies within the range investigated experimentally. In spite of claims sometimes made, that they represent e.g., Fick's law, they lack both physical basis and predictive power, and so it is strongly advised not to use them.[3]

Based on the solutions of the diffusion equation several *short-cut calculation methods* have been devised, which make use of some integrated form of the diffusion coefficient over the particle coordinate.[4–7] Predictive power has been shown to be high for situations in which the surface water concentration becomes low rapidly such as in spray drying. In situations in which the surface concentration decreases much more slowly the predictions still deviate considerably from the solution of the diffusion equation.

The most advanced and reliable are *structured models*. Here the physics of water and heat transport inside and around the product are taken into account, together with considerations of product structure. The surroundings of the product are related to larger-scale process variables by means of equipment models. Both product and equipment models may have more than one level of consideration, and so the total description requires the solution of a large number of mathematical (sub-)models, with a multitude of physical parameters to be introduced. The predictive power of such models is high; however the efforts to set them up and determine the parameters are considerable. An example is the modeling of spray drying by computational fluid dynamics-modeling (CFD) of the air and particle flow pattern, coupled to the internal diffusion in the hollow particles.[8]

For the prediction of drying times often, models with a certain *lumping of parameters* may give sufficient

Encyclopedia of Agricultural, Food, and Biological Engineering
DOI: 10.1081/E-EAFE 120007085
Copyright © 2003 by Marcel Dekker, Inc. All rights reserved.

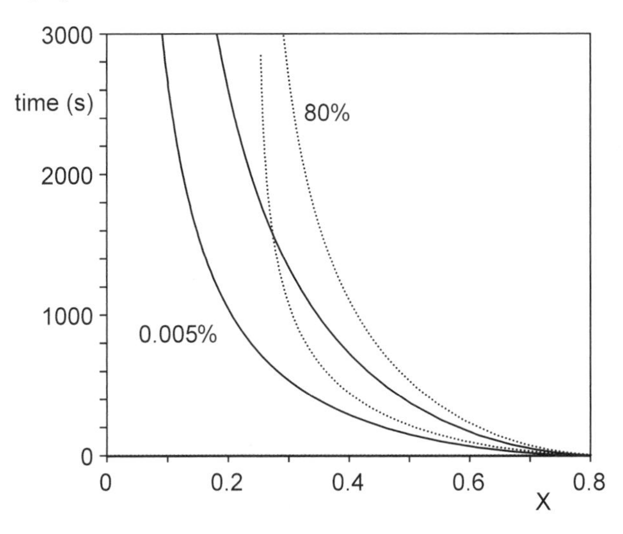

Fig. 1 Predicted drying times for bioproduct granules at 40°C in a fluidized bed, at two different air humidity levels. Drawn lines obtained from solution of diffusion model, dotted lines obtained by extrapolation from 30°C, with method based on CDC.[2]

accuracy on one hand, and reduce the efforts on the mathematical modeling and parameter estimation on the other hand. For foods in many situations the internal water transport can be described by means of an effective diffusion coefficient, depending on water concentration and temperature, and on effective geometry e.g., sphere, long cylinder, plate. This model will be treated in more detail in "Single-Particle Kinetics." Although numerical solutions can be obtained readily nowadays, they still require a considerable effort when the equipment model is of a distributed nature, such as in multiparticle models.

SINGLE-PARTICLE KINETICS

The Diffusion Model

In this paragraph the model is treated, in which water transport in the drying product takes place by diffusion, both for nonshrinking and for shrinking products, and accounting for the change in temperature during drying, which influences the transport properties. The internal transport is coupled to the external transport of water and heat, and so to the conditions in the drier. First the theory itself is presented, followed by a mathematical adaptation for shrinking systems, after which some example results are shown.

The internal transport can then be described with the diffusion equation:

$$\frac{\partial \rho_{\mathrm{w}}}{\partial t} = \frac{1}{r^{\nu}}\frac{\partial}{\partial r}\left(r^{\nu} D_{\mathrm{w}} \frac{\partial \rho_{\mathrm{w}}}{\partial r}\right) \tag{1}$$

Here ν is a geometry factor (0 for an infinite slab, 1 for an infinite cylinder, and 2 for a sphere), and D_{w} is the effective diffusion coefficient of water. In general, this is strongly dependent on water concentration and temperature, as shown in Fig. 2 for skim milk.

Initial and boundary conditions are given by

$$\begin{aligned} &t = 0 \quad 0 < r < R_0 && \rho_{\mathrm{w}} = \rho_{\mathrm{w}0} \\ &t > 0 \quad r = 0 && \frac{\partial \rho_{\mathrm{w}}}{\partial r} = 0 \\ &\qquad\quad r = R(t) && j_{\mathrm{w}}^{\mathrm{i}} = k_{\mathrm{f}}\rho_{\mathrm{f}}(Y^{\mathrm{i}} - Y^{\mathrm{b}}) \end{aligned} \tag{2}$$

In the outer boundary condition possible shrinkage is taken into account. For nonshrinking products $R = R_0$. The symbol $j_{\mathrm{w}}^{\mathrm{i}}$ stands for the water flux with respect to the (moving) boundary; it is related to the internal diffusion by:

$$\begin{aligned} &\textit{nonshrinking} && j_{\mathrm{w}}^{i} = -D_{\mathrm{w}} \frac{\partial \rho_{\mathrm{w}}}{\partial r} \\ &\textit{maximum shrinkage} && j_{\mathrm{w}}^{i} = -D_{\mathrm{w}} \frac{D_{\mathrm{w}}}{1-\rho_{\mathrm{w}}\overline{V}_{\mathrm{w}}} \frac{\partial \rho_{\mathrm{w}}}{\partial r} \end{aligned} \tag{3}$$

The set of water transport equations is completed by the relation between the equilibrium humidity at the interface and the internal water concentration there:

$$Y_{\mathrm{i}} = \frac{M_{\mathrm{w}}}{M_{\mathrm{g}}} \frac{p_{\mathrm{w}}^{\mathrm{i}}}{p_{\mathrm{t}} - p_{\mathrm{w}}^{\mathrm{i}}} \tag{4}$$

$$\begin{aligned} p_{\mathrm{w}}^{\mathrm{i}} &= a_{\mathrm{w}}^{\mathrm{i}} p_{\mathrm{w,s}} \\ a_{\mathrm{w}}^{\mathrm{i}} &= f\left(\rho_{\mathrm{w}}^{\mathrm{i}}, T\right) \end{aligned} \tag{5}$$

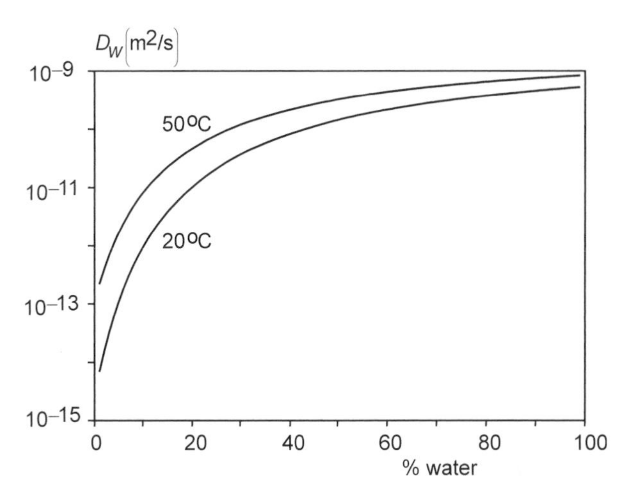

Fig. 2 Water diffusion coefficient in skim milk in dependence on water concentration for two temperatures. Data have been derived from droplet drying experiments through RR analysis.

An example of the sorption isotherm is given for skim milk at 25°C in Fig. 3.

Although in laboratory circumstances product samples can sometimes be dried isothermally by adjustment of external circumstances, in general the temperature will be variable, and equations are needed for heat transfer. In many drying situations the internal gradient in temperature is small, and the system can be considered to be at uniform temperature. For the heat transfer by convection holds:

$$V_p \rho c_p \frac{dT_p}{dt} = \left[-j_w^i\left(-c_{pv}T_p + \Delta h_0\right) + \alpha_f\left(T_{air} - T_p\right)\right]A_p \tag{6}$$

The set of equations can be solved numerically by means of finite-difference methods, such as modified Crank-Nicholson schemes. Recently we have embedded the NAGLIB-procedure D03PCF[10] in a DELPHI program to perform such calculations (Borland DELPHI is a PASCAL-based programming language of Inprise Corporation). For shrinking systems a coordinate transformation to solids-based coordinates as developed by van der Lijn[11] can be useful.

Coordinate Transformation for Shrinking and Expanding Systems

Here the treatment of van der Lijn[11] is followed, defining the following coordinate on water-free substance (solids, but also possibly dry air). For a system with an enclosed air bubble, such as a hollow droplet, the following transformation is applied:

$$\begin{aligned} &0 < r < R_b \qquad z = \textstyle\int_0^r \rho_a r^\nu \,dr \\ &r > R_b \qquad z = z_b + \textstyle\int_{R_b}^r \rho_s r^\nu \,dr \end{aligned} \tag{7}$$

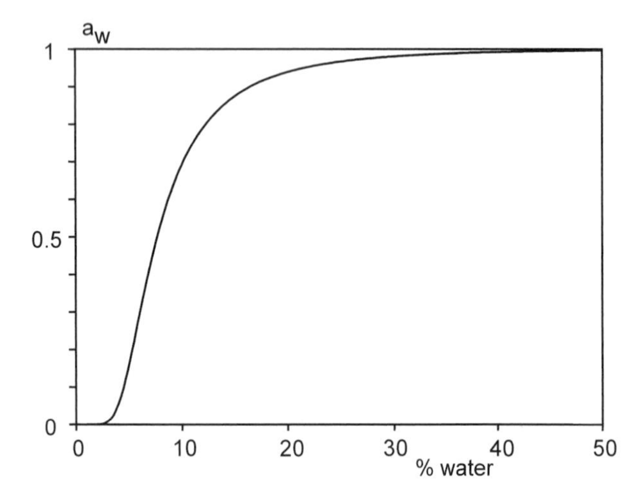

Fig. 3 Water vapor sorption isotherm of skim milk at 25°C.[9]

in which ν is a geometry factor: 0, 1, 2 for slab, infinite cylinder, sphere, respectively. For a nonhollow system $R_b = 0$, and only the second transformation of Eq. 7 is used.

Taking the water content per unit of solids:

$$X = \rho_w / \rho_s \tag{8}$$

the transformed continuity equation reads:

$$\frac{\partial X}{\partial t} = \frac{\partial}{\partial z}\left(r^{2\nu} s \frac{\partial X}{\partial z}\right) \tag{9}$$

In this equation s is a modified diffusion coefficient:

$$s = D_{ws}\rho_s^2 = \frac{D_{ws}}{\left(X\overline{V}_w + \overline{V}_s\right)^2} \tag{10}$$

The initial and boundary conditions read:

$$\begin{aligned} &t = 0 \quad 0 < z < Z \qquad X = X_0 \\ &t > 0 \qquad z = 0 \qquad \tfrac{\partial X}{\partial z} = 0 \\ &\qquad\qquad z = Z \qquad -s r^\nu \tfrac{\partial X}{\partial z} = j_w^i \end{aligned} \tag{11}$$

in which r follows from:

$$r^{\nu+1} = (\nu + 1)\int_0^z (X\overline{V}_w + \overline{V}_s)\,dz \tag{12}$$

In addition, the heat balance, assuming a flat temperature profile and nonhollow particle, reads:

$$\rho c_p \frac{dT}{dt} = \frac{\nu + 1}{R}\left[\alpha_f(T_b - T) - j_w^i(c_{pv}T + \Delta h_0)\right] \tag{13}$$

In the case of a hollow sphere holds:

$$t > 0 \quad z = Z_b \quad \tfrac{\partial X}{\partial z} = 0 \tag{14}$$

$$\frac{1}{3}R_b^3 = Z_b \frac{1}{M_a}\frac{RT}{P_t - P_w} \tag{15}$$

in which equal pressures in- and outside the particle are assumed.

$$r = R_b + \left[3\int_{Z_b}^z (X\overline{V}_w + \overline{V}_s)\,dz\right]^{1/3} \tag{16}$$

In the above it is assumed that in case of a bubble, the internal diffusion is so rapid that virtually a flat humidity profile is present.

D

In case of a spherical particle with an enclosed bubble the heat balance follows as:

$$\rho c_{\mathrm{p}} \frac{\mathrm{d}T}{\mathrm{d}t} = \frac{3}{R} \frac{1}{1 - (R_{\mathrm{b}}/R)^3} \left[\alpha_{\mathrm{f}}(T_{\mathrm{b}} - T) - j_{\mathrm{w}}^{\mathrm{i}}(c_{\mathrm{pv}}T + \Delta h_0) \right] \tag{17}$$

in which it is assumed that the internal heat of evaporation is negligible with respect to other terms.

Example Results

From the calculations the concentration profiles are obtained and by means of integration also the average moisture content and temperature are obtained as they develop in time. An example of results is given in Figs. 4 and 5, for the spray drying of a hollow skim milk particle with an initial external diameter of 100 μm, an initial gas bubble diameter of 25 μm, an initial solids content of 50 wt% in ideally mixed air of 100°C and $Y = 0.03\,\mathrm{kg\,kg^{-1}}$. In Fig. 4 the development of the concentration profiles is presented. In the course of time the development of steep concentration profiles, and the shrinkage of the outer diameter, can be seen. Due to the increase in temperature an increase in the internal bubble diameter is observed, when the internal surface still has a high enough water concentration, causing a nearly 100% water activity in the bubble. In the later stages of drying the water concentration becomes so low that the bubble water activity decreases, leading to decreased water vapor pressure and shrinkage of the bubble.

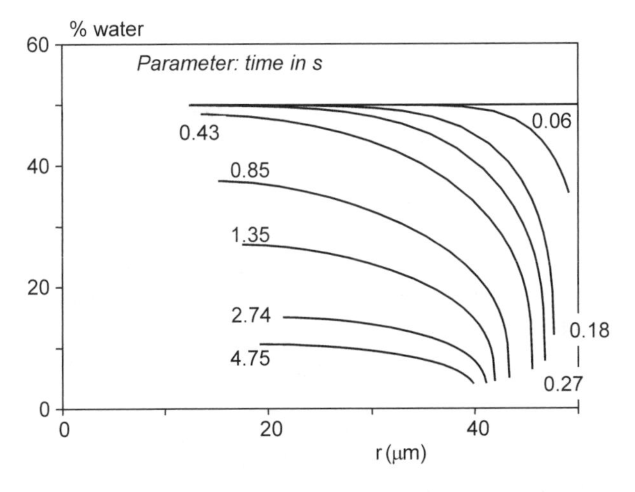

Fig. 4 Development of water concentration profiles inside a hollow skim milk droplet drying in air of 100°C and $Y = 0.03\,\mathrm{kg\,kg^{-1}}$. Initial droplet temperature 50°C, initial external diameter 100 μm, initial internal gas bubble diameter 25 μm, initial slids content 50 wt%. Parameter is time in sec.

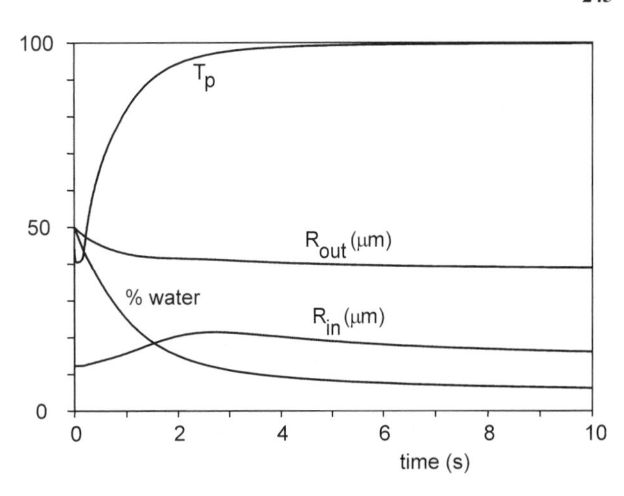

Fig. 5 Water content, temperature, and particle inner and outer radius developing during drying of a hollow particle of skim milk. Same conditions as in Fig. 4.

Fig. 5 shows the initial decrease from feed temperature towards wet-bulb temperature, at which the particle remains for some time. The water activity at the surface remains more or less constant until the surface water content decreases below about 10%. As soon as the water activity decreases the particle heats up due to the decreased evaporation rate. The average water content decreases with time, thus giving the drying curve. Similar results can be obtained for nonhollow particles and other geometries, and can be combined with simplified trajectory calculations and quality degradation.[7,9,12]

MORE COMPLEX SYSTEMS

For many dryers the description by the single particle model is not sufficient. In addition, a model is needed for the heat and mass flows through the drier and the local transfer between air and particles. For some dryers such as a fixed-bed dryer the drying equations need to be solved for many particles simultaneously. Although much effort has been dedicated to simplify such calculations in the past, e.g., by applying the CDC concept, there is no general criterion for the validity of such methods. In contrast to this the diffusion model on the particle level can be used quite successfully in more complex situations; the recent increase in computing power enables such calculations within seconds or a few minutes.

Well-Mixed Fluidized Bed

Here it is assumed that the air flows in plug flow through the bed, and the particle phase is ideally mixed. There may be a bypass flow of the air, causing a less than optimal usage of the drying air, as in the two-phase bypass model of Subramanian, Martin, and Schlünder[13] (Fig. 6). Upon passage through the bed, the air looses heat and gains humidity. As each particle performs a complex and chaotic passage through the bed, some simplification is needed. For the drying process the time-averaging hypothesis is assumed to hold, which means that the drying process of a particle in rapidly fluctuating conditions follows in good approximation the same drying curve as for constant external conditions equal to the time-averaged values.[14] This results in a model in which the local mass and heat transfer to the particle are replaced by effective values, in which both the time-averaging for the particles and the change of air conditions over the bed height are accounted for. So the same diffusion equation applies, but the boundary condition at the surface and the heat balance are now coupled to the inlet humidity and temperature:

$$t > 0 \quad r = R(t) \quad j_w^{\mathrm{i}} = k_{\mathrm{eff}}\rho_{\mathrm{f}}(Y^{\mathrm{i}} - Y^{\mathrm{in}}) \tag{18}$$

$$\rho c_{\mathrm{p}}\frac{\mathrm{d}T}{\mathrm{d}t} = \frac{\nu+1}{R}\left[\alpha_{\mathrm{eff}}(T^{\mathrm{in}} - T) - j_{\mathrm{w}}^{\mathrm{i}}(c_{\mathrm{pv}}T + \Delta h_0)\right] \tag{19}$$

with:

$$\begin{aligned} k_{\mathrm{eff}} &= k_{\mathrm{f}}\tfrac{1}{N_{\mathrm{te}}}\left[1 - \exp(-N_{\mathrm{te}})\right] \\ \alpha_{\mathrm{eff}} &\approx \alpha_{\mathrm{f}}\tfrac{1}{N_{\mathrm{the}}}\left[1 - \exp(-N_{\mathrm{the}})\right] \end{aligned} \tag{20}$$

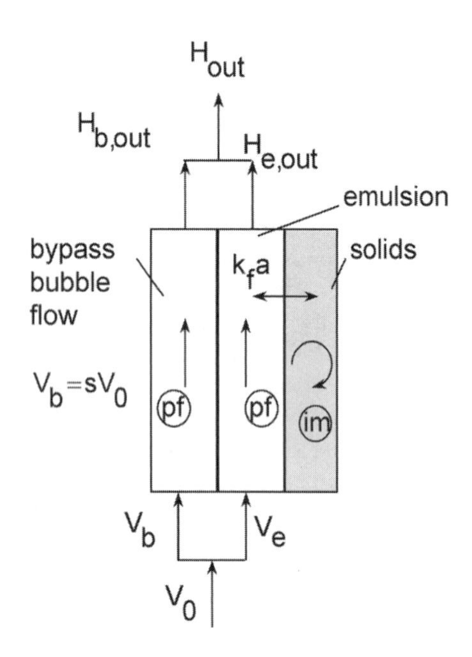

Fig. 6 Model of Subramanian, Martin, and Schlünder[13] for fluidized bed with bypass. Solids are assumed ideally mixed, the air is split up in a bypass and an emulsion flow. Mass is exchanged between solids and emulsion gas.

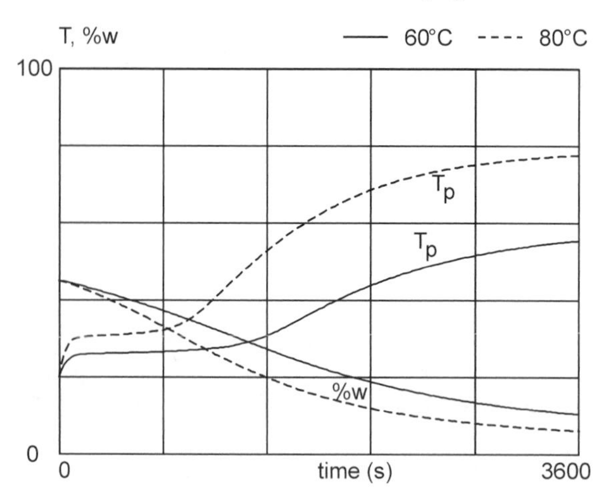

Fig. 7 Typical drying curves and temperature histories for bioproduct in a batch fluidized bed, for two air inlet temperatures. Particles are cylindrical with a diameter of 1 mm, and a length of 10 mm, initial moisture content is 45 wt%, initial temperature is 20°C. Initial product mass 1000 kg, air flow rate 50,000 kg dry air hr^{-1}, bed diameter 3 m. For more details, see Ref. [14].

where the number of heat and mass transfer units are given by:

$$\begin{aligned} N_{\mathrm{te}} &= \frac{k_{\mathrm{f}} a L_{\mathrm{bed}}}{u_0(1-s)} \\ N_{\mathrm{the}} &= \frac{\alpha_{\mathrm{f}} a L_{\mathrm{bed}}}{u_0(1-s)\rho_{\mathrm{a}} c_{\mathrm{pa}}} \end{aligned} \tag{21}$$

From literature correlations the film coefficients can be found, enabling the integration. For more details see Ref. [14]. In Fig. 7 a typical drying history is shown, for two inlet air temperatures. Also the effect of a bypass can be modeled this way.

Fixed-Bed Drying

Here layers of particles are stacked, and since the air flowing along the particles changes in conditions depending on the local drying rates, which vary in time and space, a multiparticle model is necessary. The approach taken here is to divide the layer in segments, assuming the particles in each segment to be ideally mixed, with the air in plug flow through each segment, as depicted in Fig. 8. This can be viewed as a cascade of well-mixed fluid-bed dryers with respect to the air flow. For the film coefficients now fixed-bed correlations are taken, but the rest of the model for each segment remains the same as for the fluidized bed. In the calculations the drying history for a particle in each segment is solved for a small time step, starting with the lowest segment. The outlet conditions of this segment are the inlet conditions for the next one. After all segments have been calculated, the procedure is repeated for the next time step. In this way

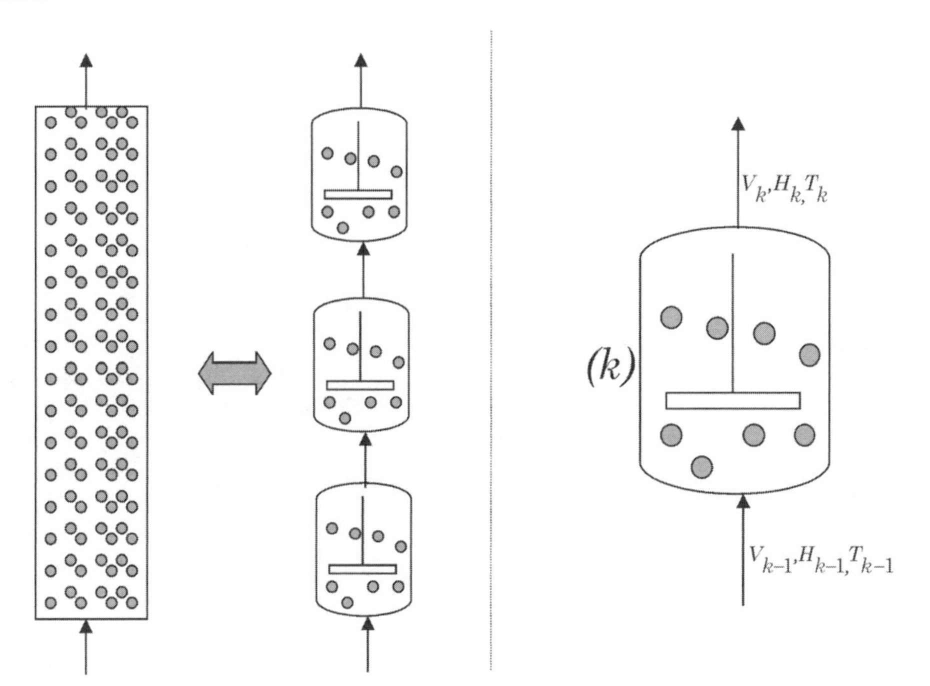

Fig. 8 Model representation of approximation of fixed-bed dryer by a cascade of ideally mixed dryers.

the dynamic profiles of moisture content and temperature over the bed are obtained, and by integration the drying curve of the whole bed.

In Fig. 9 the drying history of two particles is presented, one in the bottom and one in the top layer. In Fig. 10 a snapshot is given for the temperature and moisture profile in the product for a bed divided into 10 segments.

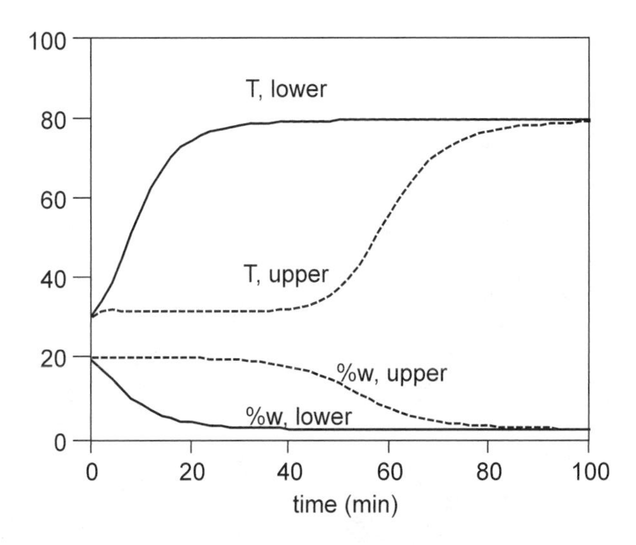

Fig. 9 Drying curves and temperature histories for a typical extrudate in a fixed bed, as modeled by the cascade model of Fig. 8. Dotted lines are for particles near air inlet, drawn lines for particles near air outlet. Air flow 600 kg dry air $m^{-2} hr^{-1}$, air inlet temperature 80°C, bed load 50 kg particles m^{-2}, initial moisture content 20 wt%, particle diameter 7.5 mm, bed height 150 mm.

Moving Beds

Examples of moving beds are the cross-flow belt dryers, and the vertical grain dryers. For a cross-flow belt dryer a model can be set up in for elements of the bed moving with the belt, which can be solved as in the previous paragraph. By storing the outlet air conditions after various times, one can integrate these outlet conditions over the area of

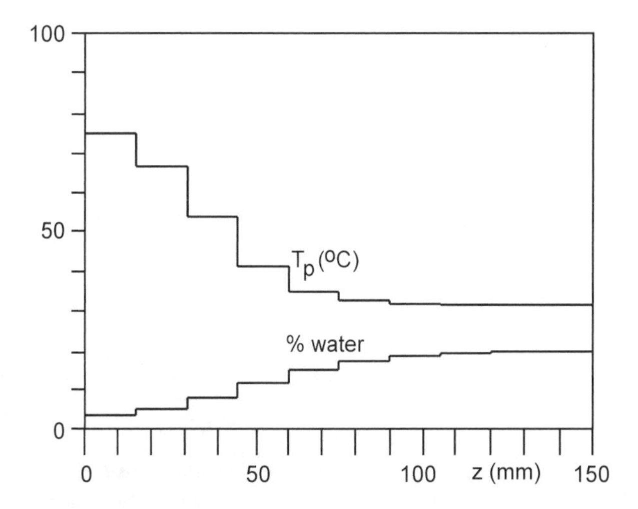

Fig. 10 Calculated temperature and moisture distribution in fixed bed after 20 min of drying time. Conditions as in Fig. 9.

the section of the bed dryer. In one industrial application we modeled a vertical grain dryer in which the product flows downward and in a number of sections is dried in cross-flow, alternated by tempering sections.[15] This means that in the active drying sections concentration profiles build up, which flatten out again in the tempering sections. The diffusion coefficient of water as a function of water content and temperature was obtained from so-called regular regime (RR) methods, as discussed in the next paragraph. A typical set of results is shown in Fig. 11. The deviation between actual and modeled moisture content at the dryer outlet was less than 0.2% wt%, which was acceptable for the application. The simulation program has been used in the design of a new dryer, as well as for daily operator use to adjust for changes in weather and feed conditions.

Versatility of the Diffusion Model

Here the diffusion model has been chosen and several applications have been shown. A number of classic methods and correlations have been discarded. Although the implementation is not straightforward, and the physical parameters have to be determined accurately, the diffusion model has the advantage that it can be put to work for a wide span of variable external conditions. In addition to the above-mentioned examples the cooling on a belt at different relative humidities and air temperatures has been modeled, showing rehydration after cooling down the product.

REGULAR REGIME METHODS AND THE DIFFUSION COEFFICIENT

Here a methodology is presented which enables the estimation of drying times for a limited region of application, but is also fit to determine the diffusion coefficient from drying experiments.

Based on a large-scale study of solutions of the diffusion model the so-called short-cut methods were developed first by Schoeber[4] for a variety of dependence of the diffusion coefficient on concentration. This was based on his RR observation. For a given set of boundary

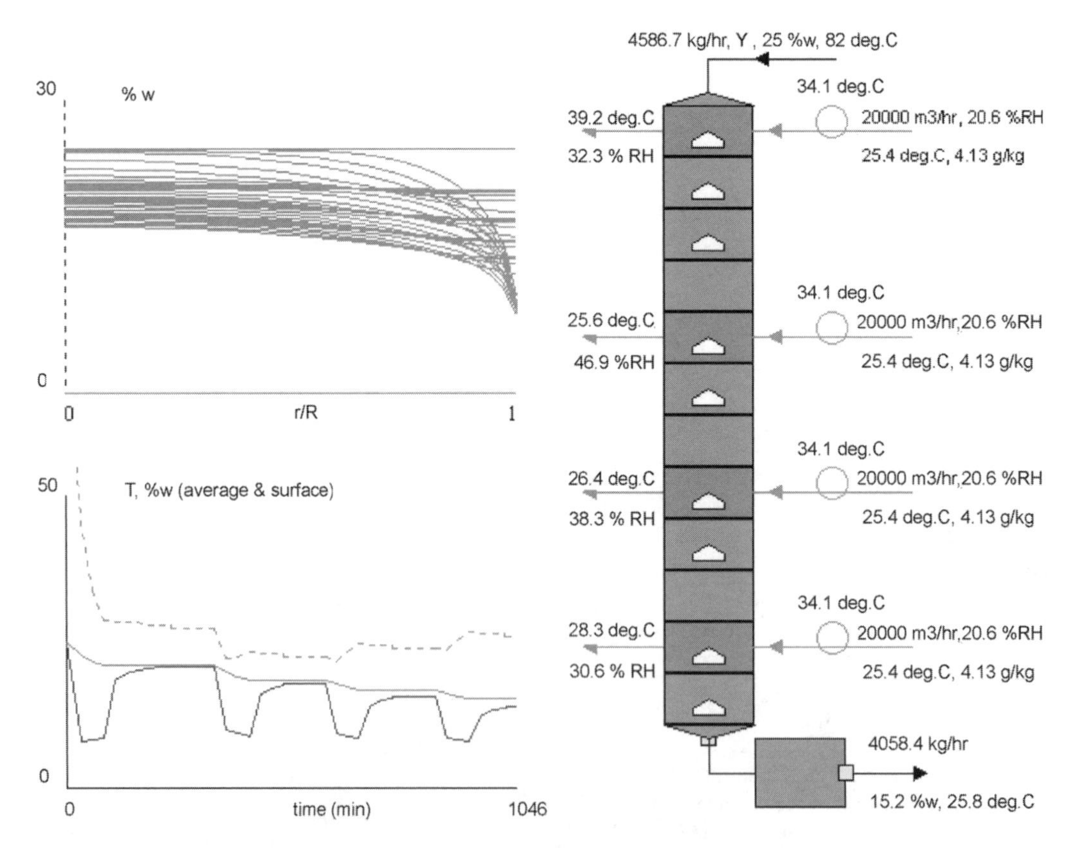

Fig. 11 Output screen of simulation program of continuous industrial grain dryer. Upper left water concentration profiles, lower left product temperature, interfacial and averaged moisture content as a function of time. Right scheme of dryer, with various active drying sections and tempering sections. Air outlet and product outlet conditions follow from simulation, and are in good agreement with plant data.

conditions the drying rate will after some time only depend on the actual moisture content, and the influence of the initial condition decreases virtually to zero. To describe a drying process he identified three stages: the *constant activity period* (CA), the *penetration period*, and the *RR*, and methods to calculate transition points between these regions. Furthermore he derived relationships between the concentration dependence of the diffusion coefficient and the RR part of the drying curve, thereby opening the way of determination of the diffusion coefficient from relatively simple drying experiments. The advantage of his methods was the generality of the treatment, which held for all kinds of concentration dependencies.

In further studies Liou and Coumans developed methods of calculating the concentration profiles inside the material by short-cut methods, for the special case of a concentration dependence given by the power-law relation of the type $D = D_0X^a$. Also they provided several simplified calculation methods for various geometries.[5,6] An overview of further development in short-cut calculations is given by Coumans.[16]

In this methodology the drying curve is represented by the so-called flux parameter F vs. moisture content. The flux parameter has been defined in several ways in the literature; here the simplest one is used:

$$F = j_w R_s \tag{22}$$

in which j_w is the water flux, and R_s is the so-called solid thickness. For nonshrinking systems this is equal to the thickness or radius of the particle. For a slab R_s is the thickness to be reached upon total shrinkage after complete drying. For a shrinking solid sphere or cylinder this would be the final radius, and for a hollow sphere it would be the thickness of the solid part of the shell at a given bubble size.

In Fig. 12 the principle of the method is illustrated, for the drying of a layer of skimmed milk at 30°C layer temperature, with air of zero humidity. Starting with a given initial moisture content, first a period occurs in which the interfacial concentration is so high that the water activity at the interface remains virtually constant: the constant activity period (CAP). For a slab the flux parameter remains constant since the flux is determined by the external conditions for a given slab temperature. Inside the layer concentration profiles build up, and thus at a certain point the interfacial water concentration decreases below the point of inflection of the sorption isotherm, and so the flux will also start to decrease. As the diffusion coefficient in that region is strongly decreasing with decreasing water concentration, the interfacial water concentration will rapidly drop to a value close to equilibrium with the drying air. If the initial flux was high, the water concentration at the nonevaporating side of the slab will still have the initial value; in that case the profiles will start to *penetrate*, until the heart is reached. This is called as penetration period. Once the center concentration starts to decrease the third period is entered, the RR. The interesting observation made by Schoeber is that independent of initial conditions all drying curves, for the given layer temperature and air humidity, merge into the same RR curve. Also starting from a given moisture content at high F-values, the penetration parts of the drying curves merge into the same penetration period curves. When the initial flux, or rather F, is low, during the CA-period the concentration profiles will be shallow, and there is no penetration period, but the CA-period goes over directly into the regular regime.

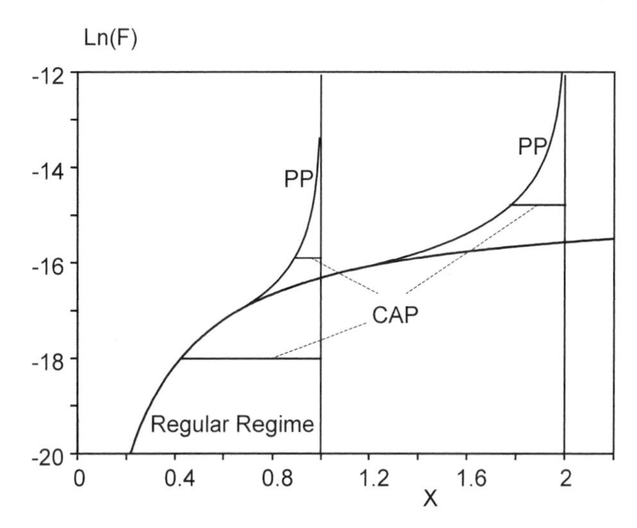

Fig. 12 Isothermal RR drying curve for a layer of skim milk at 30°C.

In the following the RR base curve (RR*) is defined as the curve for isothermal drying at zero surface moisture content, on which all drying curves would end, irrespective of their initial moisture content or initial flux.

The RR base curve for a shrinking layer is related to the concentration-dependent diffusion coefficient by:

$$F^*_{\mathrm{RR,l}} = \frac{Sh_{\mathrm{d,l}}}{2} d_s \int_0^X \frac{D}{\left(1 + Xd_s/d_w\right)^2} \mathrm{d}X \tag{23}$$

with the Sherwood number for the drying phase given by:

$$Sh_{\mathrm{d},l} = \frac{2.3 + 7.4 \frac{\mathrm{d}\left(\ln F^*_{\mathrm{RR,l}}\right)}{\mathrm{d}\ln X}}{1 + \frac{\mathrm{d}\left(\ln F^*_{\mathrm{RR,l}}\right)}{\mathrm{d}\ln X}} \tag{24}$$

Here d_s and d_w are the densities of solids and water, respectively.

For a nonshrinking layer the relation between F^*_{RR} and D is given by:

$$F^*_{RR,l} = \frac{Sh_{d,l}}{2} d_s \int_0^X D \, dX \tag{25}$$

Thus the RR* curve for a slab may be regarded as a material property.

The dependence on the temperature is described by means of an Arrhenius equation:

$$F^*_{RR,l}(T,X) = F^0_{RR,l}(X) \exp\left(-\frac{E_F(X)}{RT} \right) \tag{26}$$

in which the frequency factor $F^0_{RR}(X)$ and the activation energy $E_F(X)$ both depend on the average moisture content.

Construction of the Drying Curve

The basic principle here is that the curve of the flux parameter F in dependence of the moisture content X is constructed, for variable conditions. The drying time is then found from:

$$\tau = d_s R_s^2 \int_X^{X_0} \frac{dX}{F} \tag{27}$$

Let us first consider the hypothetical case of a slab drying isothermally, at zero bulk humidity. The basic RR-curve is then denoted by F^*_{RR}. Also the penetration period is that based on zero surface concentration.

From the theoretical work first of all follows the determination method of the transition point (X_T,F_T) between RR and PP:

$$\left. \frac{d \ln F_{RR}}{dX} \right|_{X_T} = \frac{1}{X_0 - X} \tag{28}$$

The equation for the penetration period reads:

$$F_{PP} = \frac{E_T F_T}{E} \tag{29}$$

where the efficiency E is given by:

$$E = \frac{X_0 - X}{X_0 - X_i} \tag{30}$$

with X_i the interfacial moisture content (equal to zero for this first case).

From the external conditions follows the flux parameter in the CA period, as given by Eqs. 2–5, with $a^i_w = 1$. If this value, F_{CA}, is higher than F_T, the intersection point between CA and PP is calculated, otherwise the transition between CA and RR is evaluated.

Intersection of the corresponding F-value with either the PP-curve or the RR-curve gives the transition of CA to the other periods, and thus the whole drying curve is known.

In the considerations now let a nonzero bulk humidity be assumed. The RR-curve for a given geometry depends on X_i, the interfacial concentration in equilibrium with the bulk humidity, as follows from the sorption isotherm. Formally written, it was found that:

$$\begin{aligned} F_{RR}(T,X,X_i) &\approx F_{RR}(T,X,0) - F_{RR}(T,X_i,0) \\ &= F^*_{RR}(T,X) - F^*_{RR}(T,X_i) \end{aligned} \tag{31}$$

in which F^*_{RR} represents the RR flux parameter for drying with zero surface concentration. Again in this case the penetration period by definition is based on a constant surface concentration, now equal to X_i. So now first the RR-curve is determined, corresponding to X_i, from Eq. 31. The transition point (X_T,F_T) is now again found by intersection of the RR- and PP-curves, through Eq. 28, and the CA-curve follows again from the external conditions through Eq. 2.

The foregoing illustrates that for a given particle temperature, initial and average moisture content, and bulk humidity, the full drying curve can be constructed. For the general case of variable temperature, one of the hypotheses of the RR short-cut method is that upon change of temperature the actual drying curve will follow the corresponding points on the isothermal drying curves. Starting with a given initial temperature, at that temperature the F-value is calculated according to the external flux, from the CA-curve. For a small time step then the temperature change is calculated with the enthalpy balance,[13] and the change in average moisture content X. A new value of X_i is estimated; the corresponding RR- and PP-curves at the new temperature and the transition point to the CA are calculated. If the drying curve is still in the CA-region, the F- value follows directly. When the transition point has been passed, from the guessed X_i follows a guessed F-value either from the PP or the RR. This can be checked against the one following from the external flux, through Eq. 2, and so iteratively the correct value of X_i can be found. Finally then by the integration of Eq. 27 follows the drying time. The foregoing illustrates the methodology of the RR method for layers; for construction of the drying curve for other geometries the reader is referred to the literature.[4]

Determination of the Diffusion Coefficient from Drying Experiments

Here the focus is on the RR method. Recently MRI-profile measurements have been developed which enable directly the determination of the diffusion coefficient in dependence of X.[17] Methods are being developed in which the diffusion coefficient is determined iteratively by numerical solution of the diffusion equation and matching it with experimental drying data.[18,19]

As follows from the foregoing, only part of the drying curve constitutes the RR, and so only in that part the diffusion coefficient can be found. As is clear from Fig. 12 preferably one would start with as high an initial moisture content as possible, and a high enough value of F, so that there is a penetration period. For solutions this may be carried out by drying layers with a small amount of gelling agent such as agar–agar to prevent turbulence in the sample. External mass transfer can be promoted by lowering the pressure. For several food substances there is a limit to the amount of water that can be accommodated, for others the shape is fixed. Also for granular materials it is unpractical to perform drying experiments at constant particle temperature. In the following a stepwise treatment will be given.

Isothermal drying of a shrinking layer

A sample is dried with air with close to zero humidity. The weight of the sample is recorded in the course of time. From that the flux is calculated, and from the initial moisture content the solid thickness R_s, leading to a table of the flux parameter F in dependence on X. If there is a reasonable CAP, the mass transfer coefficient can be evaluated from the flux in that period through Eq. 2. Once the mass transfer coefficient is known, for each value of F the interfacial water activity and moisture content X_i follow. For many food systems one will find that for most of the drying curve the interfacial concentration will be very low, approaching zero. The RR part of this initial curve represents $F_{RR}(T,X,X_i)$. For each value of X_i in good approximation the value of $F^*_{RR}(T,X_i) = F_{RR}(T,X_i,0)$ can be estimated directly from this initial curve. This leads then to the RR base curve through

$$F^*_{RR}(T,X) = F_{RR}(T,X,X_i) + F^*_{RR}(T,X_i) \tag{32}$$

Now the transition point to the PP is determined, according to Eq. 28, for zero interfacial concentration. If the initial F-value was high enough there will have been a PP, otherwise there has been a direct transition from CAP to RR. In both cases it is now clear what the maximum moisture content is for which the diffusion coefficient can be known from this experiment. From Eqs. 23 and 25,

$$\int_0^X S\,dX = \frac{2F^*_{RR,l}}{d_s Sh_{d,l}} \tag{33}$$

with:

$$S = D \quad \textit{nonshrinking}$$
$$S = \frac{D}{(1+Xd_s/d_w)^2} \quad \textit{shrinking} \tag{34}$$

and the Sherwood number following from Eq. 24. By applying numerical differentiation, if needed with some smoothing, one then obtains the values of

$$S = \frac{d}{dX}\left(\frac{2F^*_{RR}}{d_s Sh_{d,l}}\right) \tag{35}$$

and from those the values of the diffusion coefficient D. Finally a suitable expression can be chosen to fit the diffusion coefficient as a function of X. By doing experiments at different temperatures, the activation energy can be determined as a function of X. For some liquid foods a good fit was obtained with the Antoine-type expression[7]:

$$\ln D = a_1 - b_1\left(\frac{273.15}{T}\right) - \frac{X}{c_1+X}\left[d_1 - e_1\frac{273.15}{T}\right] \tag{36}$$

Data for skim milk and maltodextrin are given in the reference.

Isothermal drying of cylinders and spheres

Although experiments like this can be performed for some samples in practice, the main issue here is to illustrate the effect of geometry. In the theory one needs the hypothetical RR base curve for a slab, and conversions have been worked out by Schoeber, Liou and Coumans.[4–6] Let us first regard the analysis of the drying experiment. Again it is assumed that the air is dry, and that there exists a CAP. In case of nonshrinking samples the mass transfer coefficient is again determined, and similarly to the procedure for layers the table of $F(T,X,X_i)$ is made up. In case of shrinking systems, however, the mass transfer coefficient will depend on the changing particle diameter. In general one can approach the external mass transfer coefficient through a relation of the type:

$$Sh_c = \frac{k_f d_p}{D_c} = f(Re_c, Sc)$$
$$Re_c = \frac{\rho_c d_p u}{\eta_c} \tag{37}$$

Assuming the shrinkage proportional to the water loss, the change in radius for a sphere or an infinite cylinder follows:

$$\left(\frac{R}{R_0}\right)^{\nu+1} = \frac{1 + Xd_s/d_w}{1 + X_0 d_s/d_w} \tag{38}$$

From the initial flux, air velocity, and conditions, the Re- and Sc-number and the initial mass transfer coefficient can be evaluated. Subsequently for every value of X the diameter follows, leading to the mass transfer coefficient with Eq. 37. The value of R_s follows as:

$$\left(\frac{R_s}{R_0}\right)^{\nu+1} = \frac{1}{1 + X_0 d_s/d_w} \tag{39}$$

Also here the effect of the interfacial concentration is eliminated by estimation of $F^*_{\mathrm{RR},\nu}(T, X_i)$ from the actual drying curve, and addition:

$$F^*_\nu(T, X) = F_v(T, X, X_i) + F^*_{\mathrm{RR},\nu}(T, X_i) \tag{40}$$

Note that at this stage the intersection point with the PP and CAP have not yet been determined, so it is not yet known which part of the drying curve is actually the RR.

For the geometry effect, Schoeber gave the expression:

$$\frac{F^*_{\mathrm{RR},\nu}}{F^*_{\mathrm{RR},l}} = \theta = \frac{Sh_{d,\nu}}{Sh_{d,l}} \tag{41}$$

Coumans[6] derived the Sh_d-relationships for power-law dependence, which are assumed to hold for other relationships between D and X:

$$\begin{aligned} Sh_{d,cyl} &= 2.07 Sh_{d,l} - 4.45 + \Delta Sh_{d,shr,cyl} \\ Sh_{d,sph} &= 3.18 Sh_{d,l} - 9.13 + \Delta Sh_{d,shr,sph} \end{aligned} \tag{42}$$

in which the shrinkage contribution $\Delta Sh_{d,shr}$ are zero for nonshrinking systems, and for radial shrinkage are given by:

$$\begin{aligned} \Delta Sh_{d,shr,cyl} &= 10.87\left[\left(1 + Xd_s/d_w\right)^{1/2} - 1\right] \\ \Delta Sh_{d,shr,sph} &= 14.39\left[\left(1 + Xd_s/d_w\right)^{1/3} - 1\right] \end{aligned} \tag{43}$$

The process now goes in an iterative fashion. The derivative, $\mathrm{d}\ln F^*_{\mathrm{RR},l}/\mathrm{d}\ln X$, is initially not known; for a first estimate it is evaluated from that of the cylinder or sphere curve. This provides a first estimate of $Sh_{d,l}$ in dependence on X from Eq. 24, and with Eq. 42 the $Sh_{d,\nu}$-number follows. This provides the first estimate of θ, from which with Eq. 41 the first estimated RR base curve for the layer is found. In the next iteration, $\mathrm{d}\ln F^*_{\mathrm{RR},l}/\mathrm{d}\ln X$ is determined from the estimated RR base curve, the Sh_d-values are determined again, and a new estimate for the slab curve results. This process converges in 1 or 2 iterations. Then the transition point with the PP-curve for the layer is determined, giving the maximum moisture content for the validity of the diffusion coefficient, and through the differentiation procedure of the previous paragraph the diffusion coefficient is found in dependence of X. Analysis of drying experiments at different temperatures again enables the determination of the activation energy.

Nonisothermal drying experiments

We performed drying experiments by attaching gelled droplets at the end of a thin wire, suspending this on a sensitive balance, and drying them by air of constant temperature. We did not measure the temperature directly, but from the drying curve and the heat balance, Eq. 6 we estimated the particle temperature in the course of time. Such experiments were performed at different air temperatures. The next step was to estimate hypothetical isothermal drying curves. From the experimental F-curves we deduced the activation energy for the drying curves in dependence of the moisture content:

$$\ln\left(\frac{F(T_{p1})}{F(T_{p2})}\right)_X = -\frac{E_d}{R}\left(\frac{1}{T_1} - \frac{1}{T_2}\right) \tag{44}$$

where the T_p-values were the momentary values of the temperature at a given moisture content, and applied the same equation to construct the estimated isothermal drying curve. The rest of the analysis was performed as in the previous paragraph. The results for aqueous maltodextrin solutions were in good agreement with those obtained through other methods; the data presented in Fig. 2 for skim milk have been derived with the same method.

Additional Remarks

For some granular products the drying in a small spouted bed can give good results, as shown by Lievense et al.[20] By taking small samples they obtained the drying curve. In the extruded preparations of a bioproduct, they hardly found any constant-rate period; this may also be the case for other food products with a low diffusion coefficient and a reasonable size, such as beans. Experimentally the temperature of the particles will very rapidly rise toward the air temperature. In such a case the process has entered the penetration period very rapidly, and the process for the derivation of the diffusion coefficient can start directly from the isothermal drying curves as discussed in the previous sections.

NOTATION

A	area (m^2)
a	thermodynamic activity
a	specific area (m^{-1})
a	power in power law
a_1	constant
b_1	constant
c_1	constant
c_p	specific heat ($kJ\,kg^{-1}\,K^{-1}$)
D	diffusion coefficient ($m^2\,sec^{-1}$)
d	diameter (m)
d_1	constant
d_s, d_w	pure component densities of solids, water ($kg\,m^{-3}$)
E	activation energy
E	efficiency
e_1	constant
F	flux parameter ($kg\,m^{-1}\,sec^{-1}$)
R	particle radius (m)
Δh_0	heat of evaporation at 0°C
j	flux ($kg\,m^{-2}\,sec^{-1}$)
k	mass transfer coefficient ($m\,sec^{-1}$)
L_{bed}	bed height (m)
M	molar mass ($kg\,kmol^{-1}$)
N_{te}	number of mass transfer units
N_{the}	number of heat transfer units
p	pressure (Pa)
r, R	spatial coordinate (m)
R	gas constant ($kJ\,kmol^{-1}\,K^{-1}$)
Re	Reynolds number
s	modified diffusion coefficient ($kg^2\,m^{-4}\,s^{-1}$)
s	bypass fraction of air flow
S	diffusion coefficient, or modified, see Eq. 34
Sc	Schmidt number
Sh	Sherwood number
T	temperature (°C, K)
t	time (sec)
u_0	superficial gas velocity ($m\,sec^{-1}$)
V	volume (m^3)
$\overline{V}$	specific volume ($m^3\,kg^{-1}$)
X	moisture content on dry basis (kg water/kg dry solids)
Y	humidity (kg water/kg dry air)
z,Z	solids based coordinate, see Eq. 7 ($kg\,m^{\nu-2}$)
α	heat transfer coefficient ($kJ\,m^{-2}\,K^{-1}$)
∂	partial derivative
ν	shape factor
η	dynamic viscosity (Pa sec)
θ	ratio
ρ	concentration ($kg\,m^{-3}$)

Subscripts

0	at time zero
0	at reference state
a, air	air
b	bubble
c	continuous phase
CA	constant activity period
cyl	cylinder
d	dispersed phase
F	with respect to flux parameter
i	interfacial
l	layer
p	particle
PP	penetration period
RR	regular regime
s	saturation, solids
shr	shrinking system
sph	sphere
T	transition
t	total, at time t
v	vapor
w	water
ν	for cylinder or sphere

Superscripts

b	bulk
eff	effective
f	film
i	with respect to interface
in	at equipment inlet
*	at zero interfacial concentration

REFERENCES

1. Keey, R.B. *Drying: Principles and Practice*; Pergamon Press: Oxford, 1972.
2. Kerkhof, P.J.A.M. A Test of Lumped-Parameter Methods for the Drying Rate in Fluidized Bed Driers for Bioproducts. Drying Technol. **1995**, *13* (5–7), 1099–1111.
3. Kerkhof, P.J.A.M. Drying: Growth Toward a Unit Operation. Drying Technol. **2001**, *19* (8), 1505–1541.
4. Schoeber, W.J.A.H. A Short-Cut Method for the Calculation of Drying Rates in Case of a Concentration Dependent Diffusion Coefficient, Proceedings of the 1st International Drying Symposium, Montreal, Aug 3–5 1978; Mujumdar, A.S., Ed.; Science Press: Princeton, 1978.
5. Liou, J.K.; Bruin, S. An Approximate Method for the Nonlinear Diffusion Problem with the Power Relation Between Diffusion Coefficient and Concentration (Part I and II). Int. J. Heat Mass Transfer **1982**, *25*, 1209–1229.

6. Coumans, W.J. Power Law Diffusion in Drying Processes. Ph.D. Thesis, Eindhoven University of Technology, 1987.
7. Kerkhof, P.J.A.M. The Role of Theoretical and Mathematical Modeling in Scale-Up. Drying Technol. **1994**, *12* (1&2), 1–46.
8. Kieviet, F.G. Modelling Quality in Spray Drying. Ph.D. Thesis, Eindhoven University of Technology, 1997.
9. Wijlhuizen, A.B.; Kerkhof, P.J.A.M.; Bruin, S. Theoretical Study of the Inactivation of Phosphatase During Spray Drying of Skim Milk. Chem. Eng. Sci. **1979**, *34*, 651–660.
10. The NAG Fortran Library Manual-Mark 18, Software edition 2.1, D03PCF, July 27, 1998.
11. van der Lijn, J. Simulation of Mass and Heat Transfer in Spray Drying. Ph.D. Thesis, Wageningen University, 1976.
12. Kerkhof, P.J.A.M.; Schoeber, W.J.A.H. Theoretical Modeling of the Drying Behaviour of Droplets in Spray Dryers. In *Advances in Preconcentration and Dehydration of Foods*; Spicer, A., Ed.; Applied Science Publishers: London, 1974; 349–397.
13. Subramanian, D.; Martin, H.; Schlünder, E.U. Stoff übertragung zwischen Gas und Feststoff in Wiberlschichten. Vt-Verfahrenstechnik **1997**, *11*, 748–750.
14. Kerkhof, P.J.A.M. Some Modeling Aspects of (Batch) Fluid-Bed Drying of Life-Science Products. Chem. Eng. Proc. **2000**, *39*, 69–80.
15. Boeff, F. de; Kerkhof, P.J.A.M.; Deckers, P.; Houben, R.; Jacops, L. Optimisation of a Column Dryer Using Computer Modelling, Proceedings of the 13th International Drying Symposium, (IDS '2002), Beijing, China, Aug 27–31, 2002.
16. Coumans, W.J. Models for Drying Kinetics Based on Drying Curves of Slabs. Chem. Eng. Proc. **2000**, *39*, 53–68.
17. Kroes, B.; Coumans, W.J.; Pel, L.; Kerkhof, P.J.A.M. *Validation of a Physically Based Drying Model with the Aid of NMR-Imaging. Drying '98*; Akritidis, C.B., Marinos-Kouris, D., Saravakos, G.D., Eds.; 1998; Vol. A, 264–271.
18. van der Zanden, A.J.J. An Iterative Procedure to Obtain the Concentration Dependency of the Diffusion Coefficient from the Space Averaged Concentration Vs Time. Chem. Eng. Sci. **1998**, *53* (7), 1397–1404.
19. Räderer, M.; Besson, A.; Sommer, K. A Thin Film Dryer Approach for the Determination of Water Diffusion Coefficients in Viscous Products. Chem. Eng. J. **2002**, *86*, 185–191.
20. Lievense, L.C.; Verbeek, M.A.M.; Meerdink, G.; van 't Riet, K. Inactivation of *Lactobacillus Plantarum* During Drying. I. Measurement and Modeling of the Drying Process. Bioseparation **1990**, *1*, 149–159.

Electrodialysis

D. Vidal-Brotóns
P. Fito
M. Gras
Universidad Politécnica de Valencia (UPV), Valencia, Spain

E

INTRODUCTION

Electrodialysis (ED) is an electrically driven membrane process.[1,2] If an electrical potential difference is applied to a salt solution, the cations migrate to the cathode (negative electrode) whereas the anions migrate to the anode (positive electrode). Uncharged molecules are not affected by this driving force, and hence can be separated from charged components. Electrically charged ion-exchange membranes are used to control the migration of the ions.

The principal application of ED is the desalting of brackish ground water, and, in the food industry, the deionization of cheese whey.

ED MEMBRANES

The ED membranes can be subdivided into cation-exchange membranes and anion-exchange membranes. In ion-exchange membranes, charged groups are attached to the polymer backbone of the membrane material. These fixed charged groups partially or completely exclude ions of the same charge (co-ions) from the membrane, by Donnan exclusion mechanism. This means that an anionic membrane, with fixed positive groups (for e.g., those derived from quaternary ammonium salts), is preferentially permeable to negative ions, whereas a cation-exchange membrane, with fixed negative groups (primarily sulfonic or carboxylic acid groups), is preferentially permeable to cations.

The basic parameters for a good ED membrane are: high selectivity (the membrane should be permeable to counter ions only), high electrical conductivity (high counter ion permeability), good mechanical and form stabilities (moderate degree of swelling in dilute solutions), and high chemical stability (over the entire pH range and in the presence of oxidizing agents and organic solvents). The base polymer determines the mechanical, chemical, and thermal stability of the membrane. The type and concentration of the fixed ions determine the selectivity and the electrical resistance.

Two different types of ion-exchange membrane can be distinguished: homogeneous and heterogeneous. Homogeneous membranes are obtained by the introduction of an ionic group into a polymer film. The charge is distributed uniformly over the membrane, so they swell relatively uniformly when exposed to water, the degree of swelling depending on their crosslinking density. Heterogeneous membranes are prepared by combining ion-exchange resins with a film forming polymer. The ion-exchange groups are contained in small domains distributed throughout the inert support matrix, which provides mechanical strength. The difference in the degree of swelling may give rise to leaks at the boundary between both materials.

Current ion-exchange membranes contain a high concentration of fixed ionic groups, typically $3\,\text{meq}\,\text{g}^{-1}$–$4\,\text{meq}\,\text{g}^{-1}$ or more. These ionic groups tend to absorb water and their charge repulsion then cause the membrane to swell. Most ion-exchange membranes are crosslinked to limit swelling. High crosslinking densities make polymers brittle, so the membranes are usually stored and handled wet to allow absorbed water to plasticize the membrane. The diffusion coefficient of the ions inside the membrane may vary from $10^{-6}\,\text{cm}^2\,\text{sec}^{-1}$ for a highly swollen system to $10^{-10}\,\text{cm}^2\,\text{sec}^{-1}$ for a highly crosslinked one, and the electrical resistance lies in the range $2\,\Omega\,\text{cm}^2$–$10\,\Omega\,\text{cm}^2$.

ED PROCESSES

In a typical ED system, anionic and cationic membranes are arranged in an alternating pattern between an anode and a cathode. Each set of anion and cation membranes forms a cell pair. Up to several hundreds of cell pairs may be assembled in a stack. When an electrolyte solution is pumped through these cells and an electrical potential is applied between the electrodes, the cations migrate towards the cathode and the anions towards the anode. The cations can permeate the cationic membranes but not the anionic membranes. Likewise, the anions can permeate the anionic membranes but not the cationic membranes. The overall effect is an increase in the ion

Encyclopedia of Agricultural, Food, and Biological Engineering
DOI: 10.1081/E-EAFE 120016145
Copyright © 2003 by Marcel Dekker, Inc. All rights reserved.

concentration in alternate compartments while the other compartments simultaneously become depleted. This means that alternate concentrate and dilute solutions are formed.

A breakthrough in ED system design was made in the mid-1970s: the electrodialysis polarity reversal. The flow streams and the polarity of the d.c. power applied to the ED stack are reversed 2–4 times per hour. By switching cells and reversing current direction, freshly precipitated scale is flushed from the membrane before it can solidify, and colloids do not form a film on the membrane. Fouling and scaling are greatly reduced, and costly membrane cleaning procedures are unnecessary, but there is each time a brief period when the concentration of the deionized solution does not meet the product quality specification, and a certain amount of the product is lost to the waste stream.

CONCENTRATION POLARIZATION (CP) AND LIMITING CURRENT DENSITY

In an ideal well-stirred cell, the flux of ions across the membranes, and hence the productivity of the ED system, can be increased without limit by increasing the current across the stack. In practice, the resistance of the membrane is often small in proportion to the resistance of the water-filled compartments, particularly in the dilute compartment where the concentration of ions carrying the current is low. In this compartment, the formation of ion-depleted regions next to the membrane places an additional limit on the current and hence the flux of ions through the membranes.

Because ions selectively permeate the membrane, the concentration of some ions in the solution immediately adjacent to the membrane surface becomes significantly depleted compared to the bulk solution concentration. As the voltage across the stack is increased, the solution next to the membrane surface becomes increasingly depleted of the permeating ions, its conductivity decreases, and an increasing fraction of the voltage drop is dissipated in transporting ions across the boundary layer rather than through the membrane. The energy consumption per unit of salt transported increases significantly. A point can be reached at which the ion concentration at the membrane surface is zero. The current through the membrane at this point is called the limiting current density. Any further increase in voltage difference across the membrane will not increase ion transport or current through the membrane.

The CP can be partially controlled by circulating the salt solutions at high flow rates through the cell chambers. But even when very turbulent flow is maintained in the cells, significant CP occurs.

ED APPLICATIONS

The most important application of ED is the production of potable water from brackish water. For water with a relatively low salt concentration (less than 5000 ppm), ED is generally considered to be the cheapest process. The use of new ED membranes and different operating conditions are improving the performance of the process.[3]

In Japan, ED is used on a large scale to concentrate seawater to about 18%–20% solids for the production of table salt. After ED, the brine is further concentrated by evaporation, and the salt recovered by crystallization.

Several applications of ED in the food industries, such as the demineralization of whey,[4] or the production of boiler feed water, are yet well established. Others are at a preindustrial scale, or still in an experimental stage. The applications of bipolar membrane ED in the food industry have been recently reviewed.[5] Bipolar membranes (laminates of anionic and cationic membranes) are used to deacidify apple juice.[6] Other applications are: citric acid recovery from fermentation broths[7]; denitrification of drinking water, placing a nitrate-selective anion-exchange resin in the desalination compartment,[8] or in combination with a membrane bioreactor[9]; perchlorate removal from groundwater[10]; separation of amino acids from each other.[11]

TRANSPORT IN ELECTRICALLY DRIVEN MEMBRANE PROCESSES

Mass transfer in electrolyte solutions is determined by the driving forces acting on the individual ions of the solution and by the friction of the ions with other components in the solution. The driving forces can be expressed by gradients in the electrochemical potential of individual components, and the friction that has to be overcome by the driving force can be expressed by the ion mobility or diffusivity. Ref. 2 provides a very good overview of mathematical relationships, which can be applied to study energy and mass transfer phenomena that occur in ED processes.

CONCLUSION

Electrodialysis has a long and proven history in the desalination of brackish waters. However, new applications in waste-water treatments as well as in the food and the chemical industry are becoming more and more important. There are still a multitude of problems to be solved. Some are related to the properties of the membranes and the process design, while others are caused by the lack of application know-how and practical experience.

ACKNOWLEDGMENTS

The description of the CP phenomena in ED is summarized from Ref. 1, Baker, R.W., *Membrane Technology and Applications*, 385–390, copyright (2000), with permission from the author.

The conclusion is reprinted from Ref. 2, Strathmann, H. Electrodialysis. In *Encyclopedia of Separation Science*; Wilson, I.D., Ed., 2000; 1707–1717, copyright with permission from Elsevier Science.

REFERENCES

1. Baker, R.W. *Membrane Technology and Applications*; McGraw-Hill: New York, 2000.
2. Strathmann, H. Electrodialysis. In *Encyclopedia of Separation Science*; Wilson, I.D., Ed.; Academic Press: London, 2000; 1707–1717.
3. Amor, Z.; Malki, S.; Taky, M.; Bariou, B.; Mameri, N.; Elmidaoui, A. Optimization of fluoride removal from brackish water by electrodialysis. Desalination **1998**, *120* (3), 263–271.
4. Biondi, A.; D'Ascenzo, F.; Tantini, C.; Vinci, G.; Chiaccierini, E. Milk waste treatment: innovative process. Ind. Aliment. **1999**, *38* (379), 239–243.
5. Bazinet, L.; Lamarche, F.; Ippersiel, D. Bipolar-membrane electrodialysis: applications of electrodialysis in the food industry. Trends Food Sci. Technol. **1998**, *9* (3), 107–113.
6. Quoc, A.L.; Lamarche, F.; Makhlouf, J. Acceleration of pH variation in cloudy apple juice using electrodialysis with bipolar membranes. J. Agric. Food Chem. **2000**, *48* (6), 2160–2166.
7. Moresi, M.; Sappino, F. Economic feasibility study of citrate recovery by electrodialysis. J. Food Eng. **1998**, *35* (1), 75–90.
8. Kerose, K.; Janowski, F.; Shaposhnik, V.A. Highly effective electrodialysis for selective elimination of nitrates from drinking water. J. Membr. Sci. **1997**, *127* (1), 17–24.
9. Wisniewski, C.; Persin, F.; Cherif, T.; Sandeaux, R.; Grasmick, A.; Gavach, C. Denitrification of drinking water by the association of an electrodialysis process and a membrane bioreactor: feasibility and application. Desalination **2001**, *139* (1/3), 199–205.
10. Roquebert, V.; Booth, S.; Cushing, R.S.; Crozes, G.; Hansen, E. Electrodialysis reversal (EDR) and ion exchange as polishing treatment for perchlorate treatment. Desalination **2000**, *131* (1/3), 285–291.
11. Mulder, M. *Basic Principles of Membrane Technology*, 2nd Ed.; Kluwer Academic Publishers: Dordrecht, The Netherlands, 1996.

Electroheating

David Reznik
Raztek Corporation, Sunnyvale, California, U.S.A.

INTRODUCTION

Electroheating™, also called ohmic heating, resistance heating, or Joule heating, is based on the passage of alternating electrical current through a biological fluid that serves as an electrical resistance. The electrical current is passed along or across a flowing fluid, and the electrical power introduced into the product is translated into heat.[1,2]

The advantage of Electroheating over conventional heating is the departure from the limiting heat transfer coefficient and the need for high-wall temperatures. Electroheating is a novel technology for heating fluids in the food, drug, and biotechnology industries for the purpose of sterilization and pasteurization, providing fast heating rates, more uniform heating, and improved product safety and quality.[3] It overcomes problems of fouling and formation of off flavor and color formed in heat exchangers. The electroheating technology enables unprecedented rapid heating to very high temperatures, permitting very short holding time.

DESCRIPTION

Electroheating is based on passing electrical current through a food or biotech fluid by application of a voltage source across electrodes, which are placed in contact with the product. As a fluid will present an electrical resistance to the current, it will be rapidly heated in proportion to the square of the magnitude of the current. This well-known law of electrical engineering is depicted in Fig. 1. The patented Electroheater, depicted in its pilot plant configuration in Fig. 2, is the fundamental building block for Electroheating. It enables a fluid in a continuous flow system to be rapidly and accurately heated. The Electroheater is constructed of nonconductive FDA-approved materials such as ceramic and plastic. Specially treated, pure carbon (rather than metal) electrodes are employed to avoid metal dissolution by electrolysis.

Electroheating provides several unprecedented advantages over conventional heat exchangers that inherently develop temperature gradients between a heated surface and a product. In this system, electrical current flow through the product generates heat instantaneously. Consequently, the food or biotech product is heated very rapidly—typically a rise of 100°F/55°C can be achieved in less than 0.1 sec.

Uniform heating is achieved due to a uniform current flow through the bulk of the homogeneous product with no temperature gradient perpendicular to the fluid flow. In addition, there is no residual heat in the system when the current is shut off. This is not the case in a conventional heat exchanger, which has relatively large thermal mass and, thus, retains heat even when the operation has stopped.

FACTORS TO BE CONSIDERED

As there is very limited experience in electroheating on industrial scale, all the parameters that could be involved must be considered in the process design. Many of these parameters have parallels in conventional heating, but their magnitudes and the specific effects may be very different. These include the following:

1. Electrolysis may occur at alternating current at low frequency. The major effect is the dissolution of the metallic electrodes, which may contaminate the product. Utilizing high frequency or insoluble, specially treated, pure carbon electrodes that enable using the more readily available electrical power at a frequency of 50 or 60 cycles can solve this problem.

2. Electrical resistance of the product and its change with temperature is perhaps the most important parameter. The specific resistance is the electrical resistance of 1 cm^3 with units of ohms $cm^{-2} cm^{-1}$. This specific resistance decreases with temperature by a factor of 2–3 over a 120°C temperature rise. The actual resistance of the electroheating device is a function of the specific resistance of the product and the geometry of the device. To make the best use of the available power, the resistivity of the product has to be studied and carefully considered in the design.

3. Power consideration: The electrical power is the multiplication of the voltage by the current:

$$P = VI = RI^2$$

All the power is converted into thermal energy and every kW-hr equals 860 kcal.

Encyclopedia of Agricultural, Food, and Biological Engineering
DOI: 10.1081/E-EAFE 120007119
Copyright © 2003 by Marcel Dekker, Inc. All rights reserved.

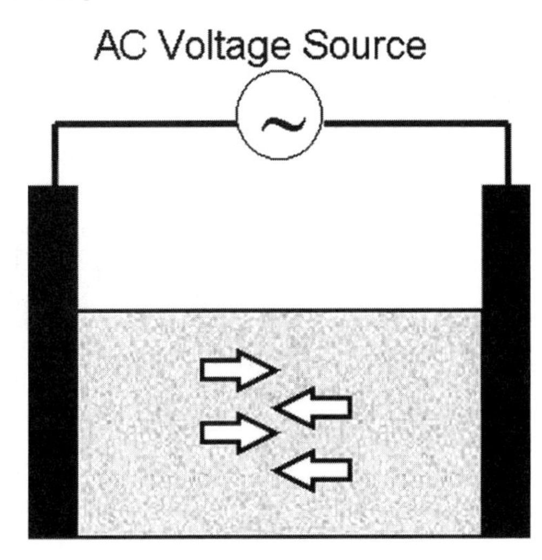

Fig. 1 Scientific basis of electroheating.

4. Voltage: Due to maximum current density limitations, it is prudent to raise the voltage, using transformers, to enable the use of low currents. The power specification of the transformers should be about 30% higher than the power requirement. This will compensate for minor changes in power demand, eliminate the need to use the maximum voltage and current, and allow some flexibility in the design of the electroheater.

5. Current density: This is the most critical parameter. It is the current divided by the area of the electrode. Every product has a specific critical current density above which arcing is likely to occur. So, once the limiting current density is known and the total current has been derived from the already known power and voltage, the minimum area of the electrode is dictated.

The geometry of the electroheater determines the resistance, which determines the current. Ideally,

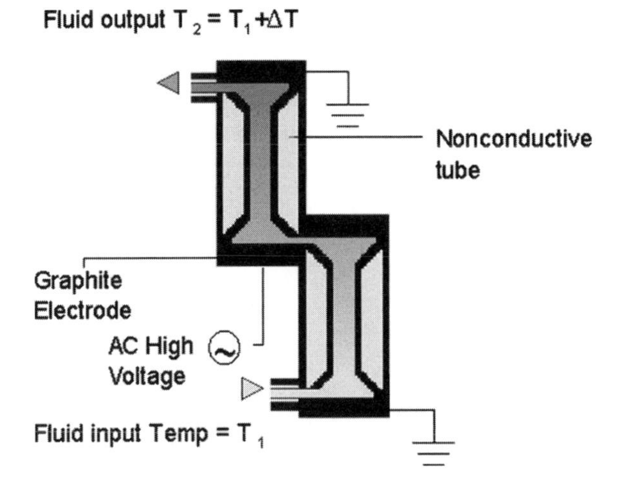

Fig. 2 Pilot scale—single phase unit.

a low total current should be utilized. This requires a high resistance, which dictates a small cross-sectional area of the tube and/or a long distance between the electrodes.

In order to obtain a high resistance and large electrode area, the tube is designed to have two cones connected by a narrow tube. Practically, the heating takes place in the narrow tube only, where the area is smaller than that of the electrode and therefore does not affect the critical current density.

The burden is put now on the pumps, since the narrow tubes create a high-pressure drop. This shifting of the burden is similar to shifting the burden from the current to the voltage to reach the required power.

The actual dimensions of the electroheater range from a 6-ft long tube with an external diameter of 3 in. for heating thousands of gal/hr to less than 1-ft long tube with an external diameter of 2 in. for heating hundreds of L/hr. Typical internal diameters of the tubes are 1 in. for the large unit and 1/8 in. for the small one. In all cases, the electroheating device becomes a part of the piping.

6. Velocity and heating rate: The velocity of the product in the electroheater is critical for applications that require high temperature rise, especially for proteinaceous products where some coagulation of proteins may occur. Arcing may occur when the material or part of it changes phase to a solid or gas. When material solidifies and remains in the electroheater, it overheats, turns into a small carbon electrode, and this leads to arcing. It is therefore important in some applications to induce a turbulent flow and keep the pressure well above that of the boiling point.

Compared to the velocity of the electrical current, the velocity of the product is negligible and the current flows as if the product is still. However, when the velocity of the product is not uniform in the cross-section, the dwelling time of the slower moving fluid in the electroheater is longer. With electroheating systems that heat at a rate of 500°C sec^{-1}, a very small delay would lead to a very high difference in temperature. It is, therefore, important to avoid even small differences in velocity in the cross section. With extremely high heating rates and when proteinaceous products such as liquid egg are electrically heated, it is important that the velocity of the fluid be maintained uniform along the tube.

7. Holding time: Electroheating is usually used for pasteurization and sterilization of food and other biological products. The product has to be held at the peak temperature for a certain period of time to ensure the desired level of bacterial kill. Electroheating the product to the same temperature as in conventional heat transfer technology will require about the same holding time. Elevating the temperature to unprecedented high temperatures will enable shorter, also unprecedented, holding times. Industries using electroheating systems claim that

higher temperature and shorter holding times are beneficial to achieve higher quality.

APPLICATIONS

The electroheating technology has been developed for applications that were not feasible by heat exchangers. Heat treatment is known to be the cleanest and safest way to kill micro-organisms and inhibit enzymes in biological fluids. The destruction of micro-organisms, viruses, and enzymes is a function of time and temperature, so is the damaging effect to the functionality of the biological medium. The electroheating system enables extreme rapid heating[3] and relatively rapid cooling,[4,5] which enables destruction of micro-organisms and viruses before damaging the biochemical functionality of the product.

The electroheating system enables, e.g., to heat proteinaceous products such as liquid egg beyond the coagulation point without significant coagulation. Similarly, animal blood could be pasteurized to destroy viruses and pathogenic bacteria at about 70°C without adverse effects on the functionality of the blood.

Electroheating could be used to heat viscous products that could not be heated effectively by conventional heat exchangers. Pasteurized fruit juices retain its fresh flavor by electroheating. Similarly, milk electroheated to 155°C yields shelf stable milk without off flavor and color while retaining the nutritional value of fresh milk.

Recent developments of the electroheating system enables vary rapid heating particles such as diced tomatoes and fruit slices.[6] In most cases, the electroheated product has a significantly better quality than conventionally heated products. The new development also enables to heat products such as dough or liquid egg, which solidify during the heating process.

REFERENCES

1. Reznik, D. Ohmic Heating of Fluid Foods. Food Technol. **1996**, *50* (5), 250–251.
2. Kinetics of Microbial Inactivation for Alternative Food Processing Technologies: Ohmic and Inductive Heating. U.S. Food and Drug Administration, Center for Food Safety and Applied Nutrition, June 2, 2000. http://vm.cfsan.fda.gov/~comm/ift-ohm.html (accessed June 2001).
3. Reznik, D. Electroheating Apparatus and Methods. US Patent 5,636,317, June 3, 1997.
4. Reznik, D. Apparatus and Method for Rapid Cooling of Liquids. US Patent 5,928,699, July 27, 1999.
5. Reznik, D. Rapid Cooling Apparatus. US Patent 6,158,504, December 12, 2000.
6. Reznik, D. Conical Shaped Electrolyte Electrode for Electroheating. US Patent 6,088,509, July 11, 2000.

Enzyme Kinetics

Mark R. Riley
The University of Arizona, Tucson, Arizona, U.S.A.

E

INTRODUCTION

Enzymes are catalysts, which speed the rate of a reaction without being altered themselves. Typically, enzymes are proteins, although enzymatic activity by nucleic acids has been observed. Enzymes are employed in a wide range of processing steps in agricultural, food, environmental, and biotechnological industries. Examples range from the use of amylase to cleave starch into simple sugars to the use of restriction endonucleases to cleave DNA. The most remarkable properties of enzymes are their high catalytic power, their specificity (each enzyme catalyzes only one type of reaction), and their high degree of regulation. Their catalytic power can be tremendous, with some enzymes reported to increase the rate of a reaction by as much as 10^{14}-fold.[1]

ASSAYS OF ENZYMATIC ACTIVITY

An enzyme is most easily characterized by observation of the rate with which it catalyzes a specific chemical reaction. For stoichiometric conversion of substrate to product, the reaction rate can be determined by either monitoring the loss of substrate being consumed in the reaction or following the formation of product. The latter is preferred as it involves the increase of a signal from zero, rather than a decrease of a substrate concentration from a high value.[2]

Continuous assays in which product concentrations are frequently sampled provide an ideal approach for quantifying the rate at which an enzymatic reaction proceeds, also called the kinetics. These methods can be automated to provide substantial information in a short period of time. Direct, continuous assays often involve nondestructive methods such as spectroscopic measurement of alterations in light absorbance brought about by changing concentrations of a reactant or product. An example includes the increase in sample absorbance at 340 nm due to the production of nicotinamide adenine dinucleotide phosphate (NADPH), a common enzymatic byproduct.[2]

A discontinuous assay usually involves removing a sample, which has been acted upon by the enzyme, stopping the reaction, and quantifying the concentration of products by methods such as HPLC. Indirect assays are employed when the reactants and products cannot be easily monitored through change in a conventional signal. Oftentimes, the assay includes incorporation of a secondary reagent, which reacts with one of the products to generate a detectable signal. One must ensure that this secondary reagent does not impact the activity of the enzyme or bind to any of the substrates. Such assays not only require more work than continuous methods, but they invariably introduce larger measurement errors. However, such methods do provide tremendous flexibility in the types of compounds that may be quantified.

As preparations of enzymes may vary in the rate at which they catalyze their reaction, quantities of enzymes are often reported as "enzyme units" or "units of enzyme," rather than as mass or concentration of enzyme. Typically, a unit represents some amount of enzyme, which is able to catalyze conversion of 1 μg of substrate to product in 1 min at a defined pH and temperature. Enzymatic preparations may have widely varying activities per mass, also called the specific activity. For example, the enzyme lipase may have specific activities from 700 unit per mg to 400,000 unit per mg depending on the source of the enzyme [Sigma Catalog]. In most applications, maintaining a prescribed level of enzymatic activity is more important than maintaining similar mass of enzyme.

MATHEMATICAL RELATIONS

The kinetics of a reaction catalyzed by an enzyme typically follows the form of:

$$E + S \underset{k_b}{\overset{k_f}{\rightleftharpoons}} ES \overset{k_2}{\rightarrow} E + P \quad (1)$$

where E is the enzyme, S the substrate(s), ES the intermediate component in which the substrate is transiently bound to the enzyme, and P the product(s). Constants k_f, k_b, and k_2 characterize the rate at which individual reactions occur. It is a reasonable assumption that the formation of the ES complex and its reversible separation back into E and S is a rapid process, which cycles as long as sufficient substrate is available. Therefore, the rate-limiting step in the enzymatic reaction

Encyclopedia of Agricultural, Food, and Biological Engineering
DOI: 10.1081/E-EAFE 120007182
Copyright © 2003 by Marcel Dekker, Inc. All rights reserved.

typically is the formation of product from the ES complex.[3] Mathematically, the rate of this last reaction is:

$$v = k_2[\mathrm{ES}] \tag{2}$$

where v is termed the "reaction velocity" (with units typically of mass per volume per time) and [ES] represents the concentration of the ES complex in solution. While this concentration is not readily quantifiable, it can be derived through application of a mass balance:

$$\frac{d[\mathrm{ES}]}{dt} = k_\mathrm{f}[\mathrm{E}][\mathrm{S}] - k_2[\mathrm{ES}] - k_\mathrm{b}[\mathrm{ES}] \tag{3}$$

At steady state, [ES] will be fairly constant. Incorporating the steady-state value of [ES] into Eq. 2 above, the reaction velocity becomes:

$$v = \frac{k_\mathrm{f}k_2[\mathrm{E}][\mathrm{S}]}{k_\mathrm{b} + k_2 + k_\mathrm{f}[\mathrm{S}]} \tag{4}$$

where [S] is the concentration of unreacted S. Following a substitution and grouping of constants, one can obtain the common Michaelis-Menten relation for enzyme kinetics:

$$v = \frac{v_\mathrm{max}[\mathrm{S}]}{K_\mathrm{m} + [\mathrm{S}]} \tag{5}$$

where v_max is the maximal reaction velocity achievable and

$$K_\mathrm{m} = (k_\mathrm{b} + k_2)/k_\mathrm{f} \tag{6}$$

K_m is commonly called the Michaelis constant, a parameter with some interesting properties. The value of K_m is the concentration of S, which yields a reaction velocity that is one-half of the maximal velocity achieved in this enzymatically catalyzed reaction. The value of K_m relative to the concentration of the substrate has an impact on the sensitivity of the reaction velocity to the substrate concentration. For example, if $[\mathrm{S}] > 10 \times K_\mathrm{m}$, then the reaction follows zero order kinetics (in which the reaction velocity is insensitive to the changing values of [S]). However, if $[\mathrm{S}] < K_\mathrm{m}/10$, then the reaction follows first order kinetics (in which the reaction velocity is linearly proportional to [S]). Fig. 1 demonstrates these varying reaction conditions. The steeply rising slope on the left-hand-side represents the first order reaction kinetic regime. The invariant rate on the right-hand-side represents the zero order kinetic regime.

A more fundamental parameter of enzymatic activity is the turnover number or $k_\mathrm{cat} = v_\mathrm{max}/[\mathrm{E}]_0$, where $[\mathrm{E}]_0$ is the total amount of enzyme initially in solution. It can be shown that Eq. 5 may be rearranged to

$$v = \frac{k_\mathrm{cat}}{K_\mathrm{m}}[\mathrm{E}][\mathrm{S}] \tag{7}$$

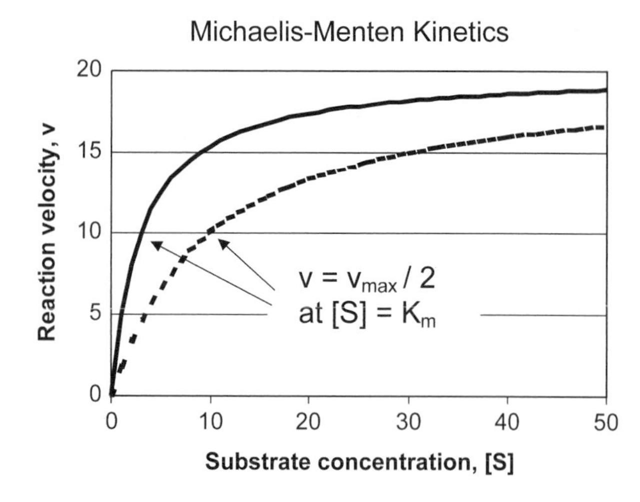

Fig. 1 Demonstration of Michaelis-Menten enzyme kinetics. The solid curve represents an enzymatically catalyzed reaction in which $v_\mathrm{max} = 20$ and $K_\mathrm{m} = 3$. The dotted curve represents a second enzymatic reaction in which v_max again is equal to 20, but K_m is 10. In both cases, the reaction velocity is highly dependent on the substrate concentration.

where [E] is the concentration of free enzyme. $k_\mathrm{cat}/K_\mathrm{m}$ is then an apparent second-order rate constant that describes the rate in terms of concentrations of the free enzyme and free substrate.[1] The ratio $k_\mathrm{cat}/K_\mathrm{m}$ can thus be used as a measure of the specificity of an enzyme for a substrate. Such comparisons are particularly useful when multiple substrates or enzymes are available in solution.

DETERMINATION OF KINETIC CONSTANTS

Despite the apparent complexity of the kinetic behavior of enzymatically catalyzed reactions as shown in Fig. 1, the Michaelis-Menten relation relies upon only two kinetic constants, v_max and K_m, which are fairly straightforward to specify. Several approaches exist to manipulate the Michaelis-Menten relationship such that experimental data may easily be used to characterize these constants. The relationship may be inverted to reveal the Lineweaver-Burk equation:

$$\frac{1}{v} = \frac{K_\mathrm{m}}{v_\mathrm{max}[\mathrm{S}]} + \frac{1}{v_\mathrm{max}} \tag{8}$$

A plot of $1/v$ vs. $1/[\mathrm{S}]$ will yield a straight line with a slope of $K_\mathrm{m}/v_\mathrm{max}$ and an intercept of $1/v_\mathrm{max}$. Unfortunately, this approach amplifies errors in the measurement of v, particularly for low values of [S] (where typically analytical error will be at its maximum). An alternative manipulation of the Michaelis-Menten relationship yields

$$\frac{[S]}{v} = \frac{K_m}{v_{max}} + \frac{[S]}{v_{max}} \quad (9)$$

often called the Langmuir relation. A plot of [S]/v vs. [S] yields a straight line with a slope of $1/v_{max}$ and an intercept of K_m/v_{max}. This approach typically reduces the impact of measurement error in [S] and v on the determined values of K_m and v_{max}. Other similar transformations are available, however, the Langmuir relation will in most cases give reliable parameter estimation.

E

REFERENCES

1. Price, N.C.; Stevens, L. Enzyme Turnover. *Fundamentals of Enzymology*; Oxford Univ. Press: Oxford, UK, 1989.
2. Gul, S.; Sreedharan, S.K.; Brocklehurst, K. Enzyme Assays, an Overview. *Enzyme Assays, Essential Data*; Wiley and Sons: Chichester, West Sussex, UK, 1998.
3. Shuler, M.L.; Kargi, F. Enzymes. *Bioprocess Engineering*; Prentice Hall: Englewood Cliffs, NJ, 1992.

Equilibrium Moisture Content

Digvir S. Jayas
University of Manitoba, Winnipeg, Manitoba, Canada

INTRODUCTION

An average annual world production of cereals, oilseeds, and legumes (hereafter referred to as grains or products) is 2 Gt (billion tonnes).[1] These grains when stored in bulk create a man-made ecosystem where interactions between abiotic (temperature, moisture content, intergranular gas composition) and biotic (grain, insects, mites, molds, bacteria) factors can cause their deterioration. By reducing temperature (to 10–20°C) and moisture content (to 12%–14% for cereals), these products can be stored for long durations because the activities of biological agents are significantly reduced at low temperatures. The concept of equilibrium moisture content (EMC) plays a significant role in understanding the drying process of grains and their potential to deteriorate.

DEFINITION

When at constant temperature, moisture content of grains comes into equilibrium with the surrounding air if exposed to it for a long enough time. The time to reach equilibrium can be reduced significantly if air is forced around the product. The moisture content of the product in equilibrium with air is known as the EMC and the relative humidity of air as the equilibrium relative humidity (ERH). The EMC–ERH data for several grains are given in Table 1. A generalized plot of the EMC on the ordinate and ERH on the abscissa at a constant temperature is known as an isotherm (Fig. 1). The product reaches equilibrium by losing moisture content if the water vapor pressure of the product is greater than the partial vapor pressure of air, or by gaining moisture content when the vapor pressures are reversed. The former is known as the desorption process and the latter as the sorption process. The sorption process includes both adsorption, physical adhesion of water molecules by the product molecules, and absorption, the movement of water molecules in the interspaces of the product and then adhesion by the product molecules.

HYSTERESIS

At constant temperature and relative humidity air, products reach higher EMC by desorption than by sorption (Fig. 2). The difference between the desorption and sorption EMCs at a given relative humidity is known as hysteresis. The product hysteresis can thus be defined by the distance between the two curves. Hysteresis decreases with an increase in temperature and with repeated wetting and drying cycles.

EFFECT OF TEMPERATURE

As the temperature of the product increases, the molecules require less energy to detach from the product molecules, thus reducing the capacity of the product to hold water. Therefore, an increase in temperature causes a reduction in the EMC at constant ERH (Fig. 3, Table 1).

SOURCES OF VARIATION IN EMC OF GRAINS

Different researchers have reported slightly different EMC values for the same grain. Such variations may be due to differences in grain variety, grain maturity, and grain history. Different EMC–ERH measuring techniques may also contribute to variation in EMC values of the same grain.

EMC AND GRAIN DETERIORATION

The survival and reproduction of biological agents in grain are dependent largely on the temperature and moisture levels.[2] Stored-product insects can live at temperatures from 8 to 41°C and ERH from 1% to 99%. Usually development and multiplication are optimum near 30°C and 50%–70% ERH. Mites can live at temperatures from 3 to 41°C and ERH from 42% to 99% with the optimum for development and multiplication near 25°C and 70%–90% ERH. Fungi can develop at temperatures from 2 to 55°C and ERH from 70% to 90% with the optimum temperature near 30°C and ERH around 80%.[3] There is a considerable variation in optimum conditions for different species.

Encyclopedia of Agricultural, Food, and Biological Engineering
DOI: 10.1081/E-EAFE 120006890
Copyright © 2003 by Marcel Dekker, Inc. All rights reserved.

Table 1 EMC of grains and seeds (percent wet basis)

	Temperature (°C)	Relative humidity (%) 10	20	30	40	50	60	70	80	90	100
Barley	25	4.7	6.9	8.4	9.6	10.6	11.9	13.4	15.7	19.2	26.5
Buckwheat	25	5.6	7.7	9.2	10.2	11.2	12.4	13.9	15.9	19.1	24.1
Cottonseed	25				6.9	7.8	9.1	10.1	12.9	19.6	
Dry beans, Michelle	4						12.8	14.4	17.0		
	10						13.6	15.3	18.1		
	25	5.5	7.4	8.5	9.5	11.0	12.6	14.9	18.2[a]		
	38						12.0	14.2	17.1		
	54						12.5	14.3	18.6		
Flaxseed	25	3.8	5.0	5.5	6.1	6.7	7.7	9.2	11.2	14.9	21.1
Oats	25	4.5	6.6	8.2	9.4	10.3	11.4	12.8	15.0	18.2	23.9
Rice, whole grain	25	5.9	8.0	9.5	10.9	12.2	13.3	14.1	15.2	19.1	
	38	4.9	7.0	8.4	9.8	11.1	12.3	13.3	14.8	19.1	
Rice, milled	25	4.9	7.7	9.5	10.3	11.0	12.0	13.4	15.3	18.3	23.3
Rice, rough	0		8.2	9.9	11.1	12.3	13.3	14.5	16.6	19.2	
	20		7.5	9.1	10.4	11.4	12.5	13.7	15.2	17.6	
	25	4.6	6.5	7.9	9.4	10.8	12.2	13.4	14.8	16.7	
	30		7.1	8.5	10.0	10.9	11.9	13.1	14.7	17.1	
Rye	25	5.3	7.4	8.8	9.8	10.8	12.2	13.9	16.3	19.6	25.7
Shelled corn, YD	4	6.4	8.6	9.9	11.2	12.6	13.9	15.6	17.7	21.4	
	16	5.6	7.8	9.3	10.5	11.6	12.6	14.2	16.2	19.8	
	27	4.2	6.4	7.9	9.2	10.3	11.5	12.9	14.8	17.5	
	38	4.2	6.2	7.5	8.5	9.8	11.3	12.5	14.4	16.9	
	50	3.6	5.7	7.0	8.1	9.3	10.5	11.9	13.8	16.3	
	60	3.0	5.0	6.0	7.0	7.9	8.8	10.3	12.1	14.6	
Shelled corn, WD	25	5.2	7.4	8.9	10.1	11.0	12.2	13.7	15.9	19.1	24.5
Shelled popcorn	25	5.8	7.5	8.4	9.2	10.2	11.4	13.1	15.1	18.2	22.7
Sorghum	− 1	6.1	8.3	10.0	11.3	12.4	13.4	14.6	15.8		
	16	5.4	7.7	9.5	10.7	11.9	13.0	14.1	15.2		
	32	4.7	7.1	8.8	10.1	11.3	12.4	13.5	14.7		
	49		6.5	8.2	9.5	10.7	11.7	12.9	14.1		
Sorghum, kafir	4	6.8	8.5	9.7	11.0	12.3	13.7	15.3	17.3		
	21	6.0	7.7	9.1	10.4	11.5	12.8	14.2	16.0		
	32	5.0	7.0	8.4	9.6	10.8	12.0	13.2	14.7		
Soybeans	5	5.2	6.3	6.9	7.7	8.6	10.4	12.9	16.9	22.4	
	15	4.3	5.7	6.5	7.2	8.1	10.1	12.4	16.1	21.9	
	25	3.8	5.3	6.1	6.9	7.8	9.7	12.1	15.8	21.3	
	35	3.5	4.8	5.7	6.4	7.6	9.3	11.7	15.4	20.6	
	45	2.9	4.0	5.0	6.0	7.1	8.7	11.1	14.9		
	55	2.7	3.6	4.2	5.4	6.5	8.0	10.6			
Wheat	− 1		7.1	9.1	10.6	12.1	13.5	14.7	16.5		
	16		6.0	8.2	9.7	11.3	12.6	13.9	15.6		
	20	5.5	7.0	8.2	9.6	10.9	12.0	13.4	14.8	17.1	
	32		5.1	7.1	8.8	10.4	11.7	13.0	14.7		
	40	5.3	6.0	7.4	8.6	9.7	11.0	12.3	14.0	16.3	
	49			6.2	7.9	9.5	10.8	12.1	13.8		
	80	2.4	3.6	4.5	5.5	6.7	7.8	9.6	11.4	13.9	

[a] Unreliable because of mold growth.
(From Ref. [5].)

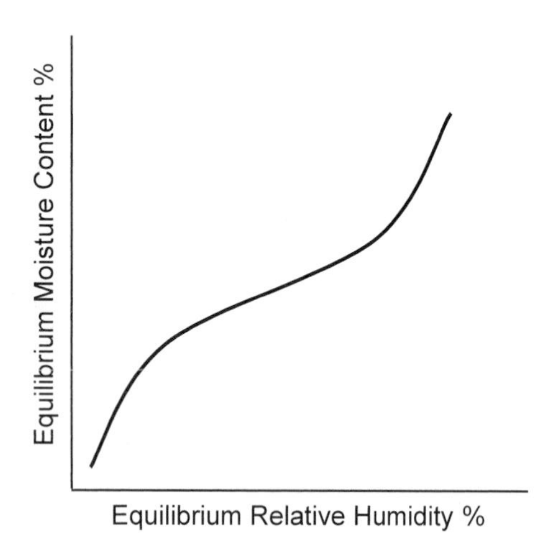

Fig. 1 Typical sorption isotherm of grains, oilseeds, and legumes.

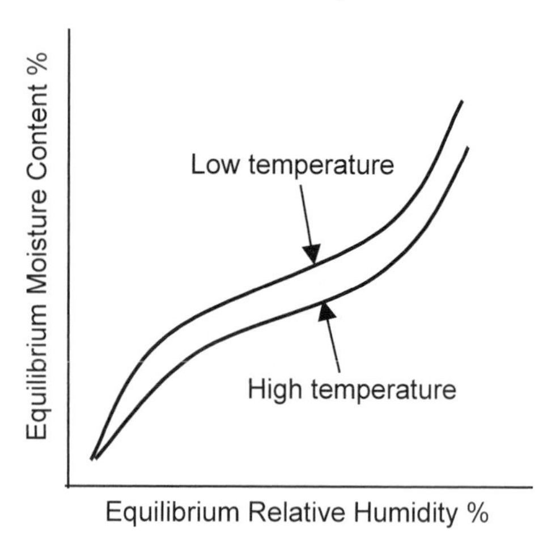

Fig. 3 Effect of temperature on sorption isotherms of grains, oilseeds, and legumes.

Localized regions may occur in stored-grain ecosystems for optimum development and multiplication of insects, mites, and fungi even when the average conditions of the bulk would prevent pest infestation.

MEASUREMENT OF EMC

The EMC is commonly measured by two methods (static and dynamic). In the static method, a sample of known mass is allowed to reach equilibrium with air maintained at a constant relative humidity and temperature. The constant relative humidity environments are usually created using saturated salt solutions in containers. A container with a 10 g–15 g sample suspended in the environment above the saturated salt solution is kept at a constant temperature. The experiment is repeated at several temperatures and relative humidities. The sample is weighed at a regular interval of 3 hr–12 hr until the change in sample mass between two successive readings is less than 0.01 g (at this stage it is assumed that the sample has reached equilibrium). The time for the samples to reach equilibrium may vary from 1 wk to 5 wk depending on the relative humidity and temperature. Therefore, mold usually develops on samples in high humidity environments and treatment of the sample with a mold inhibitor such as propionic acid is required.

In the dynamic method, a small amount of air is brought into equilibrium with a 0.5 kg–1.0 kg sample of known moisture content by recirculating the air in a sealed unit that is housed in a room at a constant temperature within $\pm$ 0.1°C. The relative humidity of the recirculating air is monitored until it becomes constant at which stage it is assumed that equilibrium has been attained, and

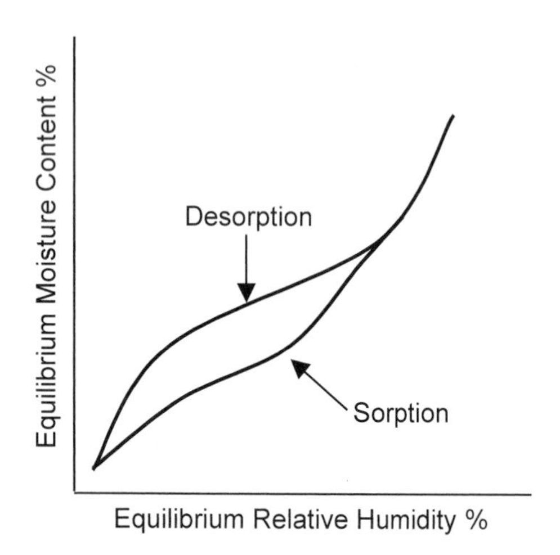

Fig. 2 Schematic of hysteresis effect for grains, oilseeds, and legumes.

Table 2 Commonly used EMC–ERH relations to analyze EMC–ERH data of agricultural products

Modified Henderson equation (Fig. 1)	$1 - \mathrm{RH} = \exp[-A(T + C)M^{B}]$
Modified Chung-Pfost equation	$\mathrm{RH} = \exp[\{-A/(T + C)\}\exp(-BM/100)]$
Modified Halsey equation	$\mathrm{RH} = \exp[\{-\exp(A + BT)\}/M^{-C}]$
Modified Oswin equation	$\mathrm{RH} = 1/[\{A + BT)/M\}^{C} + 1]$

A, *B*, *C* are constants, *M* is percent water content dry basis, RH is equilibrium relative humidity, decimal, and *T* is temperature, °C.
(From Ref. [6].)

E

Table 3 Constants of selected equations[a] for the isotherm of various grains

Seed	Equation[b]	Isotherm equation constants		
		A	*B*	*C*
Barley	PF	475.12	0.14843	71.996
Corn (shelled corn)	HE	6.6612E − 05	1.9677	42.143
Oats (cv. *Dumont*)	PF	433.157	21.581	41.439
Rough rice				
Long grain (Australia)	HE	4.1276E − 05	2.1191	49.828
Medium grain (California)	HE	3.5502E − 05	2.31	27.396
Short grain (Japan)	HE	4.8524E − 05	2.0794	45.646
Wheat durum (*Wakooma*)	OS	13.101	− 0.052626	2.9987
Wheat hard red (*Waldron*)	OS	15.868	− 0.10378	3.0842
Wheat hard red (*Napayo*)	OS	14.736	− 0.05459	3.3357
Rapeseed (*Candle*)	HL	3.0026	− 0.004897	1.7607
Canola (*Tobin*)	HL	3.489	− 0.010553	1.86
Flaxseed (*Linnot*)	HE	0.000176	1.9054	56.228
Peanut kernel	HL	3.9916	− 0.017856	2.2375
Safflower seed	HE	0.000203	1.8883	57.401
Sunflower seed	HE	0.00031	1.7459	66.603

[a] Equations are given in Table 2.
[b] HE—modified Henderson; PF—modified Chung-Pfost; HL—modified Halsey; OS—modified Oswin.
(From Ref. [7].)

the measured relative humidity is the ERH. The moisture content of the sample is measured again. The average of the initial and final moisture contents is taken as the EMC. Because the amount of recirculating air is small, the change in moisture content of the sample usually is within the error limits of the method of moisture measurement and thus the initial moisture content of the sample can be taken as the EMC. The time to reach equilibrium is reduced to 6 hr–24 hr depending on the conditions of the sample. At low temperatures, this time may be considerably higher.[4] When determining a desorption isotherm by the dynamic method, the initial relative humidity of air must be well below the expected ERH so the grain loses moisture to the air; and the reverse must be guaranteed when determining the sorption isotherm.

ANALYSIS OF EMC–ERH DATA

The EMC–ERH data of grains are analyzed by fitting various equations to the data using nonlinear regression. The commonly used equations are: the modified Henderson, Chung-Pfost, Halsey, Oswin, and G.A.B (Table 2). The constants of the most appropriate equations for common seeds are given in Table 3.

REFERENCES

1. Anonymous. *Statistical Handbook*; Canada Grains Council: Winnipeg, MB, Canada, 2000.
2. Jayas, D.S. Mathematical Modeling of Heat, Moisture, and Gas Transfer in Stored-Grain Ecosystems. In *Stored-Grain Ecosystems*; Jayas, D.S., White, N.D.G., Muir, W.E., Eds.; Marcel Dekker, Inc.: New York, 1995; 527–567.
3. White, N.D.G. Insects, Mites, and Insecticides in Stored-Grain Ecosystems. In *Stored-Grain Ecosystems*; Jayas, D.S., White, N.D.G., Muir, W.E., Eds.; Marcel Dekker, Inc.: New York, 1995; 123–167.
4. Hulasare, R.B.; Habok, M.N.N.; Jayas, D.S.; White, N.D.G. Near Equilibrium Moisture Content Values for Hull-Less Oats. Appl. Eng. Agric. **2001**, *17* (3), 325–328.
5. ASAE. D245.4. Moisture Relationships of Grains. In *Standards, Engineering Practices, and Data*, 42nd Ed.; Am. Soc. Agric. Eng.: St. Joseph, MI, 1995.
6. Mazza, G.; Jayas, D.S.; Oomah, B.D.; Mills, J.T. Comparison of Five Three-Parameter Equations for the Description of Moisture Sorption Data of Mustards. Int. J. Food Sci. Technol. **1994**, *29*, 71–81.
7. ASAE. D245.5. Moisture Relationships of Plant-Based Agricultural Products. In *Standards, Engineering Practices, and Data*, 48th Ed.; Am. Soc. Agric. Eng.: St. Joseph, MI, 2001.

Equilibrium Moisture Contents of Food

Héctor A. Iglesias
University of Buenos Aires, Buenos Aires, Argentina

INTRODUCTION

It is generally accepted that life started at the sea. This fact makes water a common substance to all living creatures and it is one of the most important components of foods. Water will be present as either solvent, product, or reactant in chemical reactions (hydrolysis, oxidation, browning); in enzymatic activity, as a plasticizer conferring foods a particular texture; and in controlling the growth of microorganisms. It also plays an important role in food spoilage.

Drying is one way of food preservation involving water removal. Perhaps in the early stages of man evolution, it was observed that a probably sun-dried food lasted longer than a fresh one. Later on it was realized that different foods with the same amount of water—moisture content—did not behave alike regarding their microbial shelf life, some will perish while others will be stable.

It was not until the middle of last century that the concept of water activity arrived in the food engineering area. This was primarily the result of the work of Scott,[1] an Australian microbiologist. Equilibrium moisture content and water activity are related through the food isotherm.

The importance of equilibrium moisture content values becomes apparent when dealing with: a) drying processes; b) chemical, physical and/or microbiological stability of foods; and c) the prediction of the shelf life of packaged foods.

DEFINITIONS

Moisture content is the amount of water that is held by a food. It can be expressed on a wet basis, i.e.,

$$\text{mass of water}/\text{mass of sample}$$
$$= \text{mass of water}/(\text{mass of water}$$
$$+ \text{mass of dry solids})$$

This basis is more often used commercially. On a dry basis, it is:

$$\text{mass of water}/\text{mass of dry solids}$$

and normally it is expressed in percentage as:

$$\text{moisture content (g of water/g dry matter} \times 100)$$

This basis is used in technical papers as well as in drying calculations. The dry matter offers the advantage of remaining constant throughout the drying process, and therefore, is an ideal stream for basing calculus.

The activity coefficient of a substance "j" in a gas mixture is defined as:

$$a_j = f_j/f_j^0$$

where f is the fugacity and f^0 is the fugacity at a reference state. Without going into thermodynamic details, which will be covered in water activity "Equilibrium Moisture Contents," the activity may be defined in terms of partial pressures when dealing with ideal gases or a mixture of real gases at low pressures as:

$$a_j = p_j/p_j^0$$

and if the component j is water, it becomes:

$$a_w = p_w/P_w$$

where p_w is the partial pressure of water vapor and P_w the vapor pressure of pure water at the same temperature.

Therefore, if a food substance is in equilibrium with the environment, the water activity of the food is the same as that of the environment, which is known as the relative humidity. Consequently, water activity of the food can be evaluated through the equilibrium relative humidity:

$$a_w = p_w/P_w = \%\text{ERH}/100$$

where %ERH is the percent equilibrium relative humidity. Moisture content values of some fresh foods are shown in Table 1.

EQUILIBRIUM MOISTURE CONTENTS

One way of measuring the equilibrium moisture content of foods is by placing the food in a container with a constant relative humidity. The food will adsorb or desorb water depending on whether its water activity is lower or higher than that of the environment, until equilibrium is reached.

Encyclopedia of Agricultural, Food, and Biological Engineering
DOI: 10.1081/E-EAFE 120007080
Copyright © 2003 by Marcel Dekker, Inc. All rights reserved.

Table 1 Moisture contents of fresh foods and monolayer values in % dry basis and net isosteric heats in kJ/g^{-1}mol^{-1}

Food	Fresh[b]	Monolayer values[a]		Net isosteric heats[c]	Reference
		BET	GAB		
Fruits					
Banana	280	4.0	—	8.4 (A)	39
Grapefruit	830	6.5	—	29.3 (A)	39
Pineapple	650	19.5	—	4.2 (A)	39
Protein foods					
Cheese (emmental)	60	3.3	3.5	40.6 (A)	39
Chicken, cooked	200	4.7*	5.5*	54.4 (D)	39
Chicken, raw	280	5.0*	5.8*	38.9 (D)	39
Fish protein concentrate	400	5.3	6.1	14.2 (A)	40
Trout, cooked	200	4.3*	4.7*	51.1 (D)	39
Trout, raw	400	4.3*	5.5*	16.7 (A)	39
Spices					
Coriander	620	5.4	6.2	30.1 (D)	39
Ginger	700	7.0	7.8	30.6 (D)	39
Laurel	100	4.5	5.1	24.3 (A)	39
Vegetables					
Avocado	270	3.2	3.9	69.9 (A)	39
Celery	1400	6.2	7.9	18.8 (A)	39
Eggplant	1100	6.7	8.2	46.9 (A)	39
Salsify	340	5.2*	5.6*	49.0 (A)	39

[a] BET (Brunauer, Emmett, and Teller[38]) monolayer values correspond to adsorption at 25°C, with the exception of (*) which are at 45°C. (From Ref. 2.) GAB (Guggenheim,[42] Anderson,[43] and DeBoer[44]) monolayer values are unpublished data.

[b] Values of fresh foods are only illustrative and obtained from different sources.

[c] Highest net isosteric heats (A) values from adsorption isotherms and (D) values from desorption isotherms. (From Ref. 37.)

Placing the substance in different environments, with constant relative humidity and temperature, and allowing equilibrium to be reached enables the food technologist to obtain the water sorption isotherm.

Factors Affecting the Equilibrium Moisture Content

As can be seen in the literature[2] there is no unique isotherm for each food product. Several factors will alter the equilibrium moisture content. Among them, it is worth mentioning the following.

Variety

Three different varieties of rapeseed were studied by Pixton and Warburton.[3] The sorption values of five different varieties of Canadian wheat were determined by Pixton and Henderson.[4] Different varieties of rough rice were analyzed by Juliano[5] while wheat and corn varieties were compared by Hubbard, Earle, and Senti.[6]

Drying method

The way drying (air, puff, freeze-drying) influences the equilibrium moisture was analyzed by Saravacos[7] for apples and potatoes; more water was sorbed by freeze-dried products. The adsorption values for carrots depended on the drying method.[8] The effect of drying temperature of precooked beef was studied by Iglesias and Chirife[9] who found that as drying temperature increases, equilibrium moisture content decreases for the same water activity. Lewicki and Lenart[10] studied the effect of osmotic dehydration on the sorption of air-dried apples. The desorption branch of lactose-hydrolyzed milk was different if freeze-dried or spray-dried.[11] These two drying methods also affected the adsorption of dried coffee products, according to Hayakawa, Matas, and Hwang.[12] In the case of milk baby food, the equilibrium moisture values for spray-dried samples were higher than those for roller-dried ones.[13] In high-sugar foods, the drying method may lead to crystalline or amorphous sugars in the dried product, which in turn will determine the adsorption behavior as well as changes during the shelf life of the product.[14,15] Mazza[16] found that freeze-dried potato slices adsorbed more water than the vacuum-dried ones. Carrillo, Gilbert, and Daun[17] studied the effect of freeze drying on the moisture sorption values of low and high-amylose corn starches, with and without sucrose added. They found an increase in equilibrium moisture content values for freeze-dried samples compared with untreated samples. Conversely, the presence of sucrose decreased sorption capacity of freeze-dried samples. Paakkonen and Roos[18] also found an increase in water sorption of horseradish roots when the surface temperature during freeze drying was increased from 20 to 60°C.

Preliminary treatments

Heldman, Hall, and Hedrick[19] showed that increasing preheat treatment of milk before drying caused higher equilibrium moisture contents at all temperatures. Iglesias and Chirife[20] found a significant reduction of equilibrium moisture values for precooked beef heated at 80 and 95°C. The same effect was reported by Pilosof et al.[21] for flour and protein isolates from bean (*Phaseolus vulgaris*) and by Gerschenson, Boquet, and Bartholomai[22] for protein isolate, starch, and flour from chickpea. The same was observed by Henderson and Pixton[23] for wheat flour heated to 70°C. Lewicki and Lenart[8] determined the influence of blanching and freezing on the adsorption values of carrots. Blanching effect on adsorption was investigated by Paakkonen and Kurkela[24] for Northern milk cap mushrooms. Cal-Vidal and Falcone[25] found lower moisture content values for freeze-dried passion fruit juice, which was previously frozen in liquid nitrogen (−195°C), than for samples conventionally freeze-dried (−45°C). Kim and Bhowmik[26] reported that concentration affected the equilibrium moisture content of concentrated yogurt, as lower levels were detected when compared with yogurt. In addition, microwave vacuum-dried concentrated yogurt powder showed lower levels compared with freeze and spray-dried yogurts.

Composition

During the shelf life of a product it suffers temperature cycles, different storage conditions, and as time passes, the food may undergo some changes in its chemical composition that will influence its sorption values and finally its stability and quality. Bolin[27] determined the influence of maturity on the sorption capacity of fresh raisins, because of the modification of the sugar content. Malthlouthi, Michel, and Maitenaz[28] found that the degree of proteolysis of gruyere cheese affected the water sorption values. An increase in proteolysis resulted in an increase of equilibrium moisture content values. Riganokos, Demertzis, and Kontominas[29] reported a decrease in equilibrium moisture content of wheat flour with an increase of crystalline sugar added. Fat is another food constituent that alters the sorption capacity. A higher fat content always results in a lower moisture content value.[19] However, Iglesias and Chirife[30] obtained the same adsorption isotherm for beef samples with different fat contents, when expressed on a fat-free basis. Cooking is another way of altering composition, as well as giving the cooked sample an additional heat treatment. This alters the adsorption values as shown by Palnitkar and Heldman[31] for precooked freeze-dried beef vs. raw freeze-dried beef.

Moisture content modification

The way moisture is modified will affect the equilibrium relative humidity values. This was observed by Henderson and Pixton[32] for white and blue lupins. Added water gave higher water activity values than adsorbed water vapor. Berlin and Anderson[33] observed that differences in the driving force for the water uptake of cottage-cheese whey solids resulted in different isotherms upon desorption.

HYSTERESIS

On performing the isotherm, some peculiar feature may appear. This is shown in Fig. 1. What is observed is that when starting with the dry substance, performing the adsorption branch, the equilibrium moisture content is lower at a given water activity than the corresponding one when starting with the fresh substance, i.e., performing the desorption branch. Moreover, Labuza et al.,[35] studying the kinetics of lipid oxidation in chicken and pork meat,

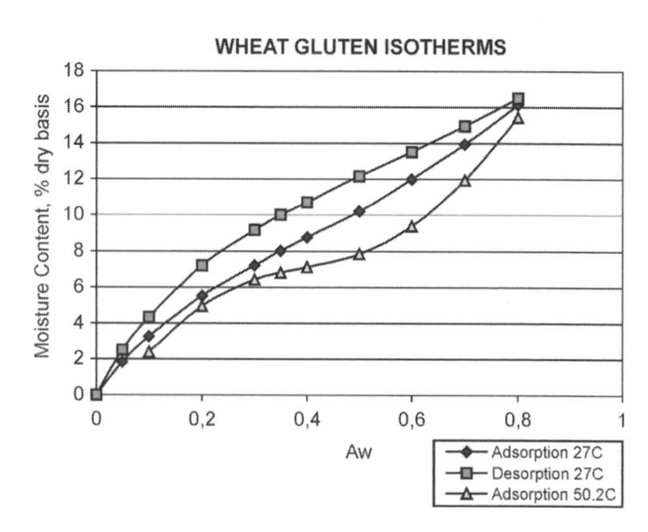

Fig. 1 Water vapor sorption for wheat gluten isotherms. (From Ref. 34.)

found that the reaction rate was greater during desorption than on adsorption. Labuza, Cassil, and Sinskey[36] also noticed that micro-organisms may grow more quickly on the desorption branch than on the adsorption one at the same value of water activity. This should be kept in mind because drying is a desorption process. The way drying is performed, temperature, rate of water removal, characteristics of food substance, sugar content, and so on, will affect the shape of the hysteresis.

It has been argued whether the moisture content is a true thermodynamic equilibrium value. Thermodynamic equilibrium comprises mechanical, chemical, and thermal equilibrium. When measuring moisture contents, true equilibrium values are always challenged by minor thermal and/or mass variations. Therefore, equilibrium moisture content is defined as constant weight when measuring food isotherms. This means that apparent equilibrium is achieved, when no variation in sample weight is detected. The appearance of hysteresis in most foods also indicates that perhaps pseudo-equilibrium values would be a better description of what happens.

EFFECT OF TEMPERATURE

As shown in Fig. 1, an increase in temperature will produce a decrease in the equilibrium moisture content for a given value of water activity. This is so because water sorption is an exothermic, spontaneous process and therefore it will not be favored by an increase in temperature. This temperature dependence is used to evaluate the isosteric heat of sorption through the application of Clausius–Clayperon equation:

$$\left.\frac{\partial \ln a_{\mathrm{w}}}{\partial T}\right|_{\mathrm{na}} = \frac{Q_{\mathrm{na}}^{\mathrm{st}}}{RT^2}$$

where $Q_{\mathrm{na}}^{\mathrm{st}}$ is the net isosteric heat of sorption and na the amount of water sorbed.

By plotting $\ln a_{\mathrm{w}}$ vs. $1/T$, the net isosteric heat of sorption may be calculated. The highest values observed, those at very low moisture content, are given in Table 1. The appearance of heat of sorption different from the latent heat of vaporization of water is in the range of water activity: 0.70–0.80 depending on the specific food. The corresponding equilibrium moisture content values are in the range: 26%–32% d.b. for fruits; 16%–22% d.b. for protein foods; 14%–18% d.b. for spices; and 13%–24% d.b. for vegetables. The differential heat of sorption is a useful quantity because it yields information on the energetics of water sorption processes in foods.

APPLICATIONS

Equilibrium moisture content values are a very important tool when solving drying problems because knowing the temperature and relative humidity specified for the process, the final moisture content and water activity of the food can be evaluated. During a drying process, the substance being dried follows the S-shaped curve of the isotherm up to the point of the equilibrium relative humidity of the drying medium. However, if the dried food adsorbs water from the environment or from any other food, it will follow the adsorption branch. This is why hysteresis must be taken into consideration. Besides, the amount of water to be removed can be calculated. BET[38] monolayer values are generally accepted as those of maximum stability regarding physical and chemical deterioration of dried foods. Monolayer values are obtained through the application of BET equation to sorption data:

$$\frac{a_{\mathrm{w}}}{(1 - a_{\mathrm{w}})M} = \frac{1}{M_{\mathrm{M}}C} + \frac{a_{\mathrm{w}}(C - 1)}{M_{\mathrm{M}}C}$$

where M is the moisture content (g of water/g of dry matter), M_{M} the monolayer moisture content (same units), and C the energy related constant.

When plotting $a_{\mathrm{w}}/(1 - a_{\mathrm{w}})M$ vs. a_{w}, a straight line is obtained up to water activity values of 0.35–0.40, where the straight line starts to curve. Monolayer values of most foods are around water activity values of 0.20–0.35. As shown by Labuza[41] and Labuza et al.[35] in their well-known stability map, the above mentioned water activity range corresponds to the lowest rate of deteriorative reactions. This is why the monolayer value is of paramount importance in food processing. Although slightly different, it has been shown that the GAB (Guggenheim,[42] Anderson,[43] and DeBoer[44]) monolayer has more physical meaning than the BET,[45] and the GAB equation is recommended by the European Project Group COST 90 on Physical Properties of Foods.[46] Both values are given in Table 1. In addition, in Table 1 are shown the extra energy requirements, isosteric heat of sorption, to be considered when performing drying cost analysis. In any theoretical analysis of the drying process, there is a need for the interfacial conditions of the material being dried and the drying medium. These conditions are given by the corresponding isotherm. When solving Fick's second law of diffusion for the first falling rate period, it is common practice to plot experimental results as:

$$\frac{M - M_{\mathrm{e}}}{M_0 - M_{\mathrm{e}}} \quad \text{vs.} \quad \text{time}$$

where M is the average moisture content at any time, M_0 the initial moisture content, and M_{e} the equilibrium moisture content. Once again the equilibrium moisture content is necessary to model the drying process.

If the mechanism of moisture transport when drying below saturation is vapor phase diffusion,[47,48] the effective diffusion coefficient, D_{eff}, is related to the environmental conditions as well as to several properties of the food as:

$$D_{eff} = \frac{M_w b}{\rho_s} \frac{\partial a_w}{\partial M_e}\bigg|_T P_w \frac{\alpha}{1+\alpha}$$

where $\alpha = (RT^2 k)/(b a_w p_0 \Delta H_s^2)$, b is the vapor space permeability, k the thermal conductivity, M_w the molecular weight of water, M_e the equilibrium moisture content, P_w the vapor pressure of pure water at temperature T, ΔH_s the heat of sorption, and ρ_s the bulk density. As can be seen, this effective diffusion coefficient is a function of the inverse slope of the sorption isotherm.[49]

Regarding packaging, it must be realized that if different food ingredients are put together as in a dry soup or sauce, e.g., there will be a moisture transfer between the components of the mixture and with the environment until equilibrium is reached. This moisture transfer will affect each component in different ways, depending on their characteristic isotherm.

The prediction of storage life of foods packaged in flexible films is based on the rate of water vapor transport through the film and the sorption isotherm of the food. Any food can gain or lose moisture up to a point where it becomes unacceptable to the consumer. This defines a limiting value for its water activity and consequently its moisture content.

The rate of water vapor transport through a flexible film is given by Labuza, Mizrahi, and Karel[50]:

$$\frac{dW}{d\theta} = \frac{PA}{e}(p_e - p)$$

where W is the weight of water transferred across the film (g), θ the time (sec), P the permeability of the film (g of water-cm/cm^2-sec-cm Hg), e the film thickness (cm), A the area of film (cm^2), p_e the vapor pressure of water outside the film (cm Hg), and p the vapor pressure of water inside the film (cm Hg).

As pointed out by Mizrahi, Labuza, and Karel,[51] the major resistance to water transport is in the film, i.e., water entering the package rapidly equilibrates with the food. Consequently, the value of p, the internal vapor pressure, is given by the sorption isotherm of the food. The above equation may be written as:

$$\frac{dW}{d\theta} = \frac{d(Mm_s)}{d\theta} = \frac{PA}{e}(p_e - p) = \frac{PA}{e}(p_e - a_w P_w)$$

where M is the moisture content (g of water/g of dry solids × 100) and m_s the weight of dry solids.

Integrating:

$$\int_{M_i}^{M_\theta} \frac{dM}{(p_e - a_w P_w)} = \int_0^\theta \frac{PA}{em_s} d\theta = P_f \int_0^\theta d\theta = P_f \theta$$

where P_f is known as the packaging factor (g of water/sec-g of solids-cm Hg), M_i the initial moisture content (g of water/g of dry solids × 100), and M_θ the moisture content at time θ (same units). The left term of the equation may be numerically evaluated, for the water activity range of interest, obtaining the equilibrium moisture content values from the corresponding isotherm.

The temperature cycles suffered by the food through the commercial chain will also affect its stability. As is shown in Fig. 1, an increase in temperature results in an increase of water activity for a particular equilibrium moisture content. Some deleterious effects may appear, which were absent when the food was packed at room temperature.

CONCLUSION

Equilibrium moisture contents of foods are essential for efficient design and modeling of drying operations in the selection of the most adequate packaging material as well as in the definition of storage conditions.

REFERENCES

1. Scott, N.F. Water Relations of Food Spoilage Microorganisms. Adv. Food Res. **1957**, *7*, 83–127.
2. Iglesias, H.A.; Chirife, J. *Handbook of Food Isotherms*; Academic Press: New York, 1982.
3. Pixton, S.W.; Warburton, S. The Moisture Content/Equilibrium Relative Humidity Relationship and Oil Composition of Rapeseed. J. Stored Prod. Res. **1977**, 77–81.
4. Pixton, S.W.; Henderson, S. The Moisture Content–Equilibrium Relative Humidity Relationship of Five Varieties of Canadian Wheat and of Candle Rapeseed at Different Temperatures. J. Stored Prod. Res. **1981**, *17*, 187–190.
5. Juliano, B.O. Hygroscopic Equilibria of Rough Rice. Cereal Chem. **1964**, *41* (3), 191–197.
6. Hubbard, J.E.; Earle, F.R.; Senti, F.R. Moisture Relations in Wheat and Corn. Cereal Chem. **1957**, *34* (6), 422–433.
7. Saravacos, G.D. Effect of the Drying Method on the Water Sorption of Dehydrated Apple and Potato. J. Food Sci. **1967**, *32*, 81–84.
8. Lewicki, P.P.; Lenart, A. Wptyw Procesu Technologicznego na Wlasciwosci Adsorocyjne Marchwi y Porow Suszonych. Prezmysl Spozywczy **1975**, *29*, 73–75.

9. Iglesias, H.A.; Chirife, J. Equilibrium Moisture Content of Air-Dried Beef. Dependence on Drying Temperature. J. Food Technol. **1976**, *11*, 565–573.
10. Lewicki, P.P.; Lenart, A. Wptyw Wstepnego Odwadniania Osmotycznego na Wlasci Adsorpcyjne Jablek Suszonych Owiewowo. Prezmysl Spozywczy **1977**, *31*, 394–397.
11. San Jose, C.; Asp, N.G.; Burvall, A.; Dahlqvist, A. Water Sorption in Lactose Hydrolyzed Dry Milk. J. Dairy Sci. **1977**, *60* (10), 1539–1543.
12. Hayakawa, K.I.; Matas, J.; Hwang, P. Moisture Sorption Isotherms of Coffee Products. J. Food Sci. **1978**, *43*, 1026–1027.
13. Varshney, N.N.; Ojha, T.P. Water Vapour Sorption Properties of Dried Milk Baby Foods. J. Dairy Res. **1977**, *44*, 93–101.
14. Strolle, E.O.; Cording, J., Jr.; McDowell, P.E.; Eskew, R.K. Effect of Sucrose on Crispiness of Explosion-Puffed Apple Pieces Exposed to High Humidities. J. Food Sci. **1970**, *35*, 338–342.
15. Iglesias, H.A.; Chirife, J.; Lombardi, J.L. Water Sorption Isotherms in Sugar Beet Root. J. Food. Technol. **1975**, *10*, 299–308.
16. Mazza, G. Moisture Sorption Isotherms of Potato Slices. J. Food. Technol. **1982**, *17*, 47–54.
17. Carrillo, P.J.; Gilbert, S.G.; Daun, H. Starch/Sucrose Interactions by Organic Probe Analysis: An Inverse Gas Chomatography Study. J. Food Sci. **1989**, *54* (1), 162–165.
18. Paakkonen, K.; Roos, Y.H. Effects of Drying Conditions on Water Sorption and Phase Transitions of Freeze-Dried Horseradish Roots. J. Food Sci. **1990**, *55* (1), 206–209.
19. Heldman, D.R.; Hall, C.W.; Hedrick, T.I. Equilibrium Moisture of Dry Milk at High Temperatures. Trans. ASAE **1965**, *8*, 535–537, 541.
20. Iglesias, H.A.; Chirife, J. Effect of Heating in the Dried State on the Moisture Sorption Isotherm of Beef. Lebensm.-Wiss. Technol. **1977**, *10*, 249–250.
21. Pilosof, A.M.R.; Bartholomai, G.B.; Chirife, J.; Boquet, R. Effect of Heat Treatment on Sorption Isotherms and Solubility of Flour and Protein Isolates from Bean *Phaseolus vulgaris*. J. Food Sci. **1982**, *47* (4), 1288–1290.
22. Gerschenson, L.N.; Boquet, R.; Bartholomai, G.B. Effect of Thermal Treatments on the Moisture Sorption Isotherms of Protein Isolate, Starch and Flour from Chickpea. Lebensm.-Wiss. Technol. **1983**, *16*, 43–47.
23. Henderson, S.; Pixton, S.W. The Relationship Between Moisture Content and Equilibrium Relative Humidity of Five Types of Wheat Flour. J. Stored Prod. Res. **1982**, *18*, 27–30.
24. Paakkonen, K.; Kurkela, R. Effect of Drying Method on the Water Sorption of Blanched Northern Milk Cap Mushrooms (*Lactarius trivialis*) at Different Temperatures. Lebensm.-Wiss. Technol. **1987**, *20*, 158–161.
25. Cal-Vidal, J.; Falcone, M. Processing Conditions Affecting the Hygroscopic Behaviour of Freeze-Dried Passion Fruit Juice. J. Food Sci. **1985**, *50* (5), 1238–1241, 1253.
26. Kim, S.S.; Bhowmik, S.R. Moisture Sorption Isotherms of Concentrated Yogurt and Microwave Vacuum-Dried Yogurt Powder. J. Food Eng. **1994**, *21*, 157–175.
27. Bolin, H.R. Relation of Moisture to Water Activity in Prunes and Raisins. J. Food Sci. **1980**, *45*, 1190–1192.
28. Malthlouthi, M.; Michel, J.F.; Maitenaz, P.C. Study of Some Factors Affecting Water Vapor Sorption of Gruyere Cheese. I. Proteolysis. Lebensm.-Wiss. Technol. **1981**, *14*, 163–165.
29. Riganokos, K.A.; Demertzis, P.G.; Kontominas, M.G. Effect of Crystalline Sucrose on the Water Sorption Behaviour of Wheat Flour as Studied by Inverse Gas Chromatography. Lebensm.-Wiss. Technol. **1992**, *25* (4), 389–394.
30. Iglesias, H.A.; Chirife, J. Effect of Fat Content on the Water Sorption Isotherm of Air Dried Minced Beef. Lebensm.-Wiss. Technol. **1977**, *10*, 151–152.
31. Palnitkar, M.P.; Heldman, D.R. Equilibrium Moisture Characteristics of Freeze-Dried Beef Components. J. Food Sci. **1971**, *36*, 1015–1018.
32. Henderson, S.; Pixton, S.W. The Relationship Between Moisture Content and Equilibrium Relative Humidity of White Lupins. J. Stored Prod. Res. **1980**, *16*, 45–46.
33. Berlin, E.; Anderson, B.A. Reversibility of Water Vapor Sorption by Cottage Cheese Whey Solids. J. Dairy Sci. **1975**, *58*, 25–29.
34. Bushuk, W.; Winkler, C.A. Sorption of Water Vapour on Wheat Flour, Starch, and Gluten. Cereal Chem. **1957**, *34* (2), 73–86.
35. Labuza, T.P.; McNally, L.; Gallagher, D.; Hawkes, J.; Hurtado, F. Stability of Intermediate Foods. I. Lipid Oxidation. J. Food Sci. **1972**, *37*, 154–159.
36. Labuza, T.P.; Cassil, S.; Sinskey, A.J. Stability of Intermediate Moisture Foods. 2. Microbiology. J. Food Sci. **1972**, *37*, 160–162.
37. Iglesias, H.A.; Chirife, J. Isosteric Heats of Water Vapor Sorption on Dehydrated Foods. Part I. Analysis of the Differential Heat Curves. Lebensm.-Wiss. Technol. **1976**, *9*, 116–122.
38. Brunauer, S.; Emmett, H.P.; Teller, E. Adsorption of Gases in Multi-molecular Layers. J. Am. Chem. Soc. **1938**, *60*, 309–319.
39. Wolf, W.; Spiess, W.E.L.; Jung, G. Die Wasserdampfsorptionsisothermen Einiger, in der Literatur Bislang Wening Berucksichtigter Lebensmittel. Lebensm.-Wiss. Technol. **1973**, *6*, 94–96.
40. Rasekh, J.G.; Stilling, B.R.; Dubrow, D.L. Moisture Adsorption of Fish Protein Concentrate at Various Relative Humidities and Temperatures. J. Food Sci. **1971**, *36*, 705–707.
41. Labuza, T.P. Proceedings of the 3rd International Congress of Food Science and Technology, 1970, Washington, D.C., 1971; 618
42. Guggenheim, E.A. *Applications of Statistical Mechanics*; Clarendon Press: Oxford. U.K., 1966.
43. Anderson, R.B. Modifications of the Brunauer, Emmett and Teller Equation. J. Am. Chem. Soc. **1946**, *68*, 686–691.
44. DeBoer, J.H. *The Dynamical Character of Adsorption*; Clarendon Press: Oxford, U.K., 1968.

45. Timmermann, E.O.; Chirife, J.; Iglesias, H.A. Water Sorption Isotherms of Food and Foodstuffs: BET or GAB Parameters? J. Food Eng. **2001**, *48*, 19–31.
46. Bizot, H. Using the GAB Model to Construct Sorption Isotherms. In *Physical Properties of Foods*; Jowit, R., Escher, F., Hallstrom, B., Meffert, H.F.T., Spiess, W.E.L., and Voss, G., Eds.; Applied Science Publishers: London, 1983; 43–54.
47. King, C.J. Rates of Moisture Sorption and Desorption in Porous, Dried Foodstuffs. Food Technol. **1968**, *22*, 509–515.
48. Karel, M. *Fundamentals of Dehydration of Foods*; De Spicer, A., Ed.; Applied Sci. Publishers Ltd.: London, 1974.
49. Viollaz, P.; Chirife, J.; Iglesias, H.A. Slopes of Moisture Sorption Isotherms of Foods as a Function of Moisture Content. J. Food Sci. **1978**, *43*, 606–608.
50. Labuza, T.P.; Mizrahi, S.; Karel, M. Mathematical Models for Optimization of Flexible Film Packaging for Foods for Storage. Trans. ASAE **1972**, *15*, 150–155.
51. Mizrahi, S.; Labuza, T.P.; Karel, M. Computer-Aides Predictions of Extent of Browning in Dehydrated Cabbage. J. Food Sci. **1970**, *35*, 799–803.

Evaporators

Oladiran O. Fasina
Auburn University, Auburn, Alabama, U.S.A.

E

INTRODUCTION

Evaporators are used in food and agricultural processing to reduce the moisture content of a dilute liquid product to a more concentrated product. Since evaporation involves the boiling of excess moisture, most evaporation equipment consists of a heat exchanger, and a separation vessel to allow the disentrainment of liquid droplets from the vapor. The efficiency of the evaporation process and product quality is highly dependent on the type of evaporator. This manuscript is a concise review of the types of evaporators commonly employed in the food industry. The design, operating characteristics, advantages, and disadvantages of the evaporators are discussed. Examples of the types of food products suitable for the different evaporator types are also given.

EVAPORATOR TYPES

Several types of evaporators are used in the food industry. The choice of an evaporator depends on factors such as physical characteristics of material to be concentrated, heat transfer properties, energy, and cost. Irrespective of the type, evaporators may be classified based on the means by which the solvent (water in food applications) is removed from the liquid product. Four general classes are:

1. Heating medium separated from evaporating liquid by tubular heating surfaces.
2. Heating medium confined by coils, jackets, double walls, flat plates, etc.
3. Heating medium brought into direct contact with evaporating liquid.
4. Heating by solar radiation.

Heating by solar radiation has become obsolete in the food industry because of the long time required to achieve the desired concentration levels. In food products, this was frequent after the onset of spoilage. In addition, it is impossible to design a continuous system with this method. The use of solar radiation for evaporation is now limited to man-made lagoons designed for the sole purpose of concentrating salt solutions.

Most of the evaporators used in the food industry employ the tubular heating surfaces where circulation of liquid past the heating surface is induced by boiling or by mechanical means. Some of the common types of evaporators in the food industry are described below.

Batch Type or Open Pan Evaporator

This is one of the oldest means of concentration in the food industry [Fig. 1(a)]. Due to its low throughput, the use of open pan evaporator is limited to few applications such as concentration of jams and jellies. The low throughput is primarily due to the low heat transfer area per unit volume and the low heat transfer coefficient resulting from movement by natural convection. Evaporation of liquid in open pan evaporator is achieved by use of internal coils, direct heaters, or by means of jackets. When heat sensitive products are to be processed, the heating vessel is connected to a vacuum system so that boiling of the product occur at lower temperatures.

Short-Tube or Calandria Evaporator

This evaporator type consists of a vessel (or shell) which contains a bundle of short vertical tubes [Fig. 1(b)] through which the product passes by natural circulation. Liquid in the tube is heated by steam outside of the tubes. The liquid rises through the tubes and recirculated through a central well at the middle of the evaporator. Product is removed from the evaporator when the desired concentration has been attained. The vapor produced escapes to the top of the dome. The short tube evaporator is mainly used for concentrating cane sugar solutions, syrups, salt, and fruit juices.

Long-Tube Vertical (LTV) Evaporator

LTV evaporators are one of the most efficient evaporators because more moisture per unit steam consumed is removed in comparison to other evaporator types. The evaporators consist of simple one-pass vertical shell and tube heat exchanger discharging into a relatively small vapor head. The tubes are typically 5 m–12 m long and 25 mm–35 mm in internal diameter. There are three types of LTV—rising (or climbing) film, falling film, and combination rising–falling film.

In the rising film LTV evaporator [Fig. 1(c)], the feed, preheated to near boiling, is introduced at the bottom

Encyclopedia of Agricultural, Food, and Biological Engineering
DOI: 10.1081/E-EAFE 120007062
Copyright © 2003 by Marcel Dekker, Inc. All rights reserved.

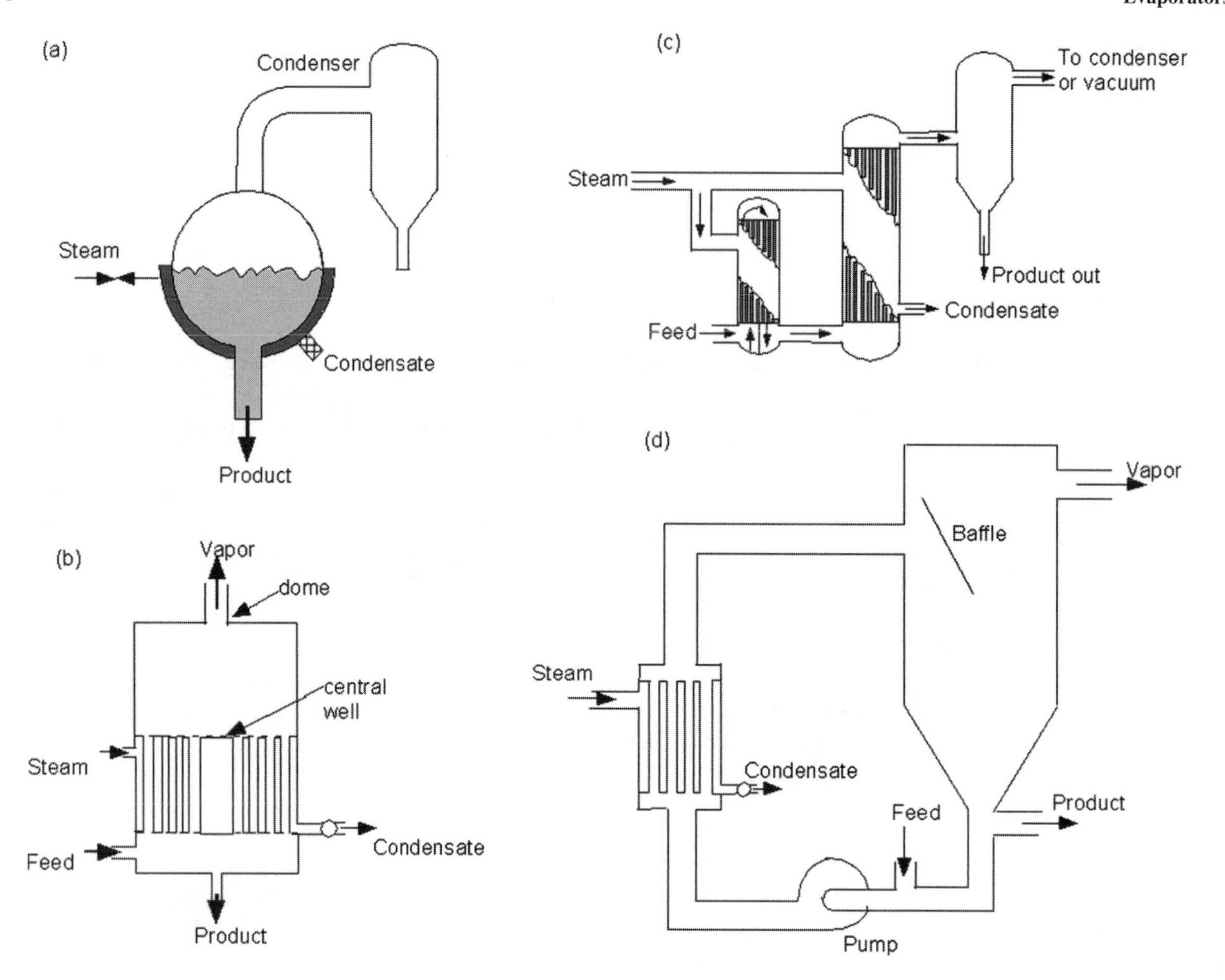

Fig. 1 Schematic diagram of some evaporator types: (a) batch type; (b) short tube; (c) rising film long tube; and (d) forced circulation evaporators.

of the tube assembly. It is further heated inside the tubes. Boiling commences a short distance up the tubes and evaporation proceeds as the liquid ascends. The upward movement of vapors result in a thin film of liquid to be formed up the walls of the tubes. The concentrate is separated from the vapor and removed from the evaporator or recirculated until the desired solid concentration is attained. Rising film LTV evaporator is especially good for low viscosity liquids such as milk.

In the falling film LTV evaporator, the feed is introduced at the top of the tube bundle and flows down the wall as a film. Vapor and liquid separation is usually carried out at the bottom of the evaporator. This evaporator type is suited to more viscous foods or those that are heat sensitive. The film flows with gravity rather than against it. This result in thinner, faster moving film and gives rise to shorter product contact time and a further improvement in the value of heat transfer coefficient. The problem with the falling film evaporator is that it is difficult to achieve even distribution of liquid to each of the tubes. This is sometimes overcome by use of specially designed sprayers.

As the name implies, the rising–falling film evaporator is a combination of the rising and falling film LTV evaporators. The dilute feed is partially concentrated in the rising film section and the more viscous material is concentrated to the final level at the falling film section of the evaporator. The advantage of this evaporator type is that the tube bundle is approximately half the height of either the rising or falling film evaporator. In addition, the vapor–liquid separator is positioned at the bottom of the calandria. Rising/falling film evaporators are commonly used in the food, juice, and diary industries where low residence time and temperatures lower than 90°C are crucial for the production of quality concentrate.

Forced Circulation Evaporator

The forced circulation evaporator [Fig. 1(d)] is best suited for processing products with suspended solids and with high viscosity (e.g., tomato products), and for products susceptible to scaling, fouling, and crystalizing. The liquid is pumped through the heating chamber at a constant rate regardless of the evaporation rate. This makes it easier to analyze and control the system. The system is designed such that no boiling takes in the heating area until the product reaches the separation chamber where flashing occurs. The capital and operating costs of this evaporator type are higher than others because of energy associated with the use of the recirculating pipe network and pump.

Plate Evaporator

Plate evaporator was developed by APV in the late 50s mainly for heat sensitive products that require low residence time. This evaporator is similar in construction to the heat exchangers used for pasteurization and sterilization of fluid food products except that the spacing between the plates and appropriate passages are increased so that much larger volume of vapors can be accommodated. The conventional plate evaporator uses the rising and falling film principle to concentrate liquids in the spaces between the plates. Feed enters at the base of each climbing film section, boils, and rises to the top of the plates. It then enters a falling film section where boiling continues. The mixture of vapor and concentrate exiting from the evaporator is separated outside the evaporator. Some of the advantages of the plate evaporator include flexible capacity by adding more plate units, more compact design with low headroom, high energy efficiencies, and high rates of heat transfer. Products that are currently concentrated with plate evaporators include fruit juices, beef and chicken broth, fruit purees, caragenan, sugars, pectin, whole milk, liquid egg, and coffee.

Other Evaporator Types

Several lesser known evaporators are used in the food industry today to cater for specific processing needs. Three of these are the agitated thin film evaporator, the expanding flow evaporator, and the APV paraflash evaporating system. The agitated film evaporator is used for concentrating liquids that are heat sensitive, viscous and when foaming of the evaporating liquid can occur. The expanding flow evaporator is used in situations where the sensory and nutritional qualities of the product are to be preserved while the APV paraflash evaporation system is designed for the evaporation of liquids containing high concentrations of solids.

CONCLUSION

A review of the types of evaporators used in food and agricultural processing is provided in this chapter. It was shown that several factors affect the choice of an evaporator type. As evaporation is an energy intensive process, efficient and economic energy utilization will continue to play a prominent role in the design and manufacturing of new evaporation systems. This requirement for energy utilization will have to be balanced with other factors such as process efficiency, improved product quality, and ease of cleaning.

FURTHER READING

Rubin, F.L.; Moak, H.A.; Holt, A.D.; Standiford, F.C.; Stuhlbarg, D. Heat-Transfer Equipment. In *Perry's Chemical Engineering Handbook*, 5th Ed.; Perry, R.H., Green, D.W., Maloney, J.O., Eds.; McGraw-Hill Book Company: New York, 1984; 11–40.

Singh, R.P.; Heldman, D.R. *Introduction to Food Engineering*, 2nd Ed.; Academic Press: New York, 1993; 499 pp.

Fellows, P. *Food Processing Technology: Principles and Practice*; VCH: New York, 1988; 505 pp.

Lavis, G. Evaporation. In *Handbook of Separation Techniques for Chemical Engineers*, 3rd Ed.; Schweitzer, P.A., Ed.; McGraw-Hill Book Company: New York, 1997; 2–140.

Evapotranspiration

J. L. Hatfield
J. H. Prueger
United States Department of Agriculture (USDA), Ames, Iowa, U.S.A.

INTRODUCTION

Evaporation of water is a critical environment process. Evaporation is the physical process of converting liquid water into vapor and the release of large amounts of energy during this conversion process. The cooling that results from the energy released during evaporation benefits animals, humans, and plants. This benefit includes maintaining temperatures within critical limits for optimum physiological function. The application of the evaporation process to plants and plant communities has some added dimensions compared to animals or humans. Plants have roots that extend into the soil and extract water from deeper regions of the soil profile. This water is transported through the vascular tissue contained in roots, stems, and leaves and is released to the atmosphere through stomata. These stomata serve as the sites for the evaporation process to occur within the leaf with water moving through the stomata into the atmosphere. This special process of evaporation from plants is referred to as transpiration.

COMPONENTS OF EVAPOTRANSPIRATION

Plant communities can be characterized as having two components that release water vapor to the atmosphere. Evaporation of water from either soil or plant surfaces uses free water that is at that surface. Transpiration is defined as the evaporation process occurring through the leaf stomata. Since both processes occur in plant communities, evapotranspiration has been defined as the sum of evaporation from a surface including both soil and plants. It is expressed as $\mathrm{ET} = E_s + E_P$, where ET is evapotranspiration, E_s is the soil water evaporation, and E_p is evaporation from the plant. Evaporation of free water on plant surfaces, e.g., dew or rainfall can be accounted for by expanding the equation to include this term as E_{ps}, where the notation expresses plant surfaces. The combination of evaporation from free water and via the stomata has been referred to as evapotranspiration for the past 50 yr since Moneith[1] described this process and some of its limitations in describing environmental physics. Values of ET can be expressed as depth of water equivalent ($\mathrm{mm\,day^{-1}}$) of in energy units ($\mathrm{MJ\,m^{-2}\,day^{-1}}$). In these expressions the unit expressions vary by application.

QUANTIFYING EVAPOTRANSPIRATION FOR AGRICULTURAL APPLICATIONS

Understanding and knowing the amount of ET is important for a number of agricultural purposes. Crop water use and crop water demand have been used to determine the amount of irrigation water supplies needed to achieve maximum crop yield. Irrigation water supplies are often required in many areas of the world in order to control salt content in the soil profile. Water is a critical component of world crop production that the Food and Agriculture Organization (FAO) and others have published several reports on methods of estimating crop water requirements.[2] Other important needs to understand ET are associated with soil water content for estimating tillage operations and trafficability of machinery across fields. Knowing ET is important for a number of within field decisions, however, across multiple fields a more regional scale knowledge or water use is important for understanding water supplies and impacts on the water resources. For example, water use in recharge areas for municipal water supplies or amount of water used by different cropping systems relative to river flow or water quality is important for decision-making processes about future management practices.

Measurement of ET

Evapotranspiration can be determined through either direct or indirect measurements. The methods for obtaining ET values are described by Hatfield.[3] Direct measurements of ET have been made with lysimeters or other techniques that measure the amount of water lost from a given soil surface. Lysimeters are mechanical devices that contain a given volume of soil that has some type of recording system to be able to measure the mass of water lost. The accuracy of these systems vary; however, a well-maintained lysimeter can have an accuracy of greater than 0.5 $\mathrm{mm\,day^{-1}}$. Lysimeters have been used to evaluate different methods of estimating ET and because of their

Encyclopedia of Agricultural, Food, and Biological Engineering
DOI: 10.1081/E-EAFE 120006931
Copyright © 2003 by Marcel Dekker, Inc. All rights reserved.

permanency to a given site there are few of these available structures in the world.

Indirect methods of estimating ET are based on the conservation of energy and mass. The premise that underlies this approach is given in the following simple expression,

$$\mathrm{ET} = R_n - G - H \quad (1)$$

where R_n is the net radiation ($\mathrm{MJ\,m^{-2}\,hr^{-1}}$), G the soil heat flux ($\mathrm{MJ\,m^{-2}\,hr^{-1}}$), and H the sensible heat flux (($\mathrm{MJ\,m^{-2}\,hr^{-1}}$). The balance of energy exchanges in the soil–plant–atmosphere continuum is best shown in the following figure (Fig. 1). Estimation methods are based on utilization of the relationships shown in Eq. 1. There have been several methods that have been used that incorporate measurements of these parameters over an agricultural field. These fall under three primary categories: Bowen ratio, Flux gradient, or eddy correlation. Bowen ratio methods use measurements or air temperature and water vapor at two heights above the crop canopy along with R_n and G to estimate ET. Flux gradient methods use multiple height measurements of windspeed, water vapor, and air temperature to estimate the fluxes of water vapor from the surface. Eddy correlation uses high-frequency measurements of 3-D windspeed components, water vapor density, and temperature (10 Hz) at one height above the surface to estimate water vapor fluxes. The time scale of the Bowen ratio and Flux gradient methods are typically on the 30 min averaging interval while eddy correlation is averaged over 10 min–15 min intervals. Over the years, there have been several different approaches developed to estimate ET from routine meteorological observations. These relationships were developed after observations of ET were obtained from lysimeters or indirect methods and some type of statistical approach used to determine the relationships. These methods are detailed in Hatfield.[3]

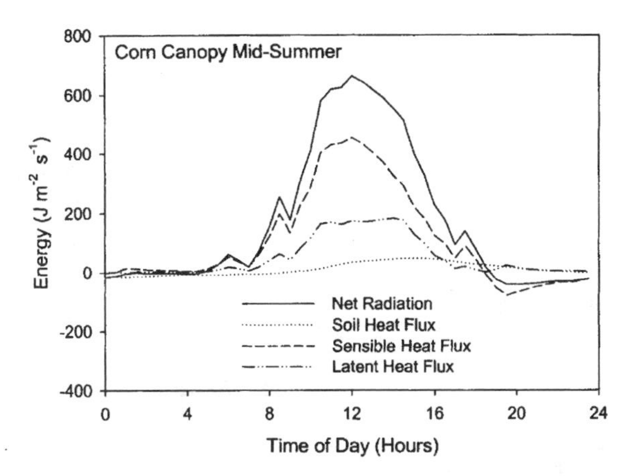

Fig. 1 Energy balance for an agricultural surface throughout a day.

Factors Influencing ET Rates

The ET amount from a given agricultural surface is controlled by four major factors: the amount of energy available, the gradient of water vapor between the surface and atmosphere, the rate of transport of water vapor mass from the surface to atmosphere (windspeed gradient), and the amount of soil water stored in the soil profile. Canopy characteristics, e.g., amount of leaf area, height, rooting depth, or foliage wetness, are included in these factors. A series of research studies conducted by Ritchie defined the importance and role of each of these factors in the ET rates from agricultural crops.[4–7] These interrelationships are best shown through the following diagrams. In Fig. 2a and b two days with different net radiation have different partitioning of energy into ET because the latent heat component is the same between the two days. There are many potential reasons for this response but illustrates the point that a single value of any of the energy balance terms (Eq. 1) is not possible. ET varies throughout within a day,

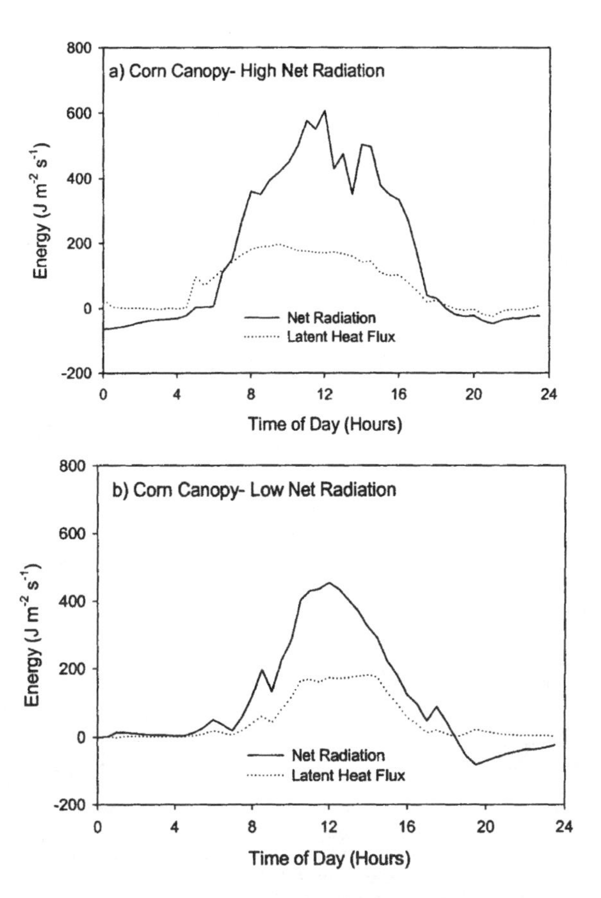

Fig. 2 Evapotranspiration from two days (a, high net radiation conditions; b, low net radiation conditions) with adequate soil water supplies. ET rates are similar between the two days because of changes in water vapor and temperature conditions to offset the R_n changes.

among days, and throughout the year because R_n is related to amount of energy from solar and longwave radiation and except at the equator the amount of solar radiation varies from month to month. Other critical factors affecting ET rates are humidity, windspeed, and cloud cover. One of the better ways to explain the effect of humidity and windspeed is to use an illustration relative to human comfort. On a very humid day with low windspeed, although evaporation occurs at the skin there is little gradient to move the water vapor away and there is a large feeling of discomfort. While on a day with high windspeeds and even high humidity, there is a greater feeling of comfort because of the effect of moving water vapor away from the skin. Finally, on a day with high windspeeds and low humidity, there is little feeling of water vapor on the skin and you hardly notice the process of evaporation because it occurs so quickly. Gradients of temperature, windspeed, and water vapor are controlling factors in the evaporation process and their effect on evaporation is included in measurement methods. One special case for windspeed and temperature gradients is advection. Advection is the horizontal gradient of water vapor and temperature often found in semiarid regions and results from warm, dry air moving across vegetated surfaces. This source of energy causes the ET rate to be increased above that expected from the energy available from R_n (Eq. 1).

The effect of soil water supply on ET rate is one of the most important factors in agriculture since we rely on soil water as the source of water for ET. Evaporation from the soil surface is affected by water availability on the soil surface, however, as soon as the soil surface dries there is little soil water evaporation and ET is dominated by transpiration. As an illustration, two days with similar R_n are shown in Fig. 3. These data were observed from a corn field in central Iowa after a period without rainfall and the soil water supply was low (Fig. 3a). A few days later, there was a large amount of rainfall on the field and the ET rate was nearly all of the available R_n (Fig. 3b). These changes in ET throughout a growing season on crops are not unusual in agricultural areas that depend on rainfall as their source of water. The process of ET is critical to crop production and for maximum production levels to be achieved, ET has to be as large as possible. Soil water availability affects the ET rate and as soil water decreases the ET rates begin to decrease (Fig. 4). Over the period of a few days that soil water is constantly being removed from the soil profile without being replenished from rainfall, the ET rate would begin to decrease. When the available soil water reaches 50% then the ET rates are severely impacted. Reduction in ET rates would cause air temperature to increase since there is no cooling effect from evaporation. We see the plant begins to change during this time, e.g., 10 days–14 days, and

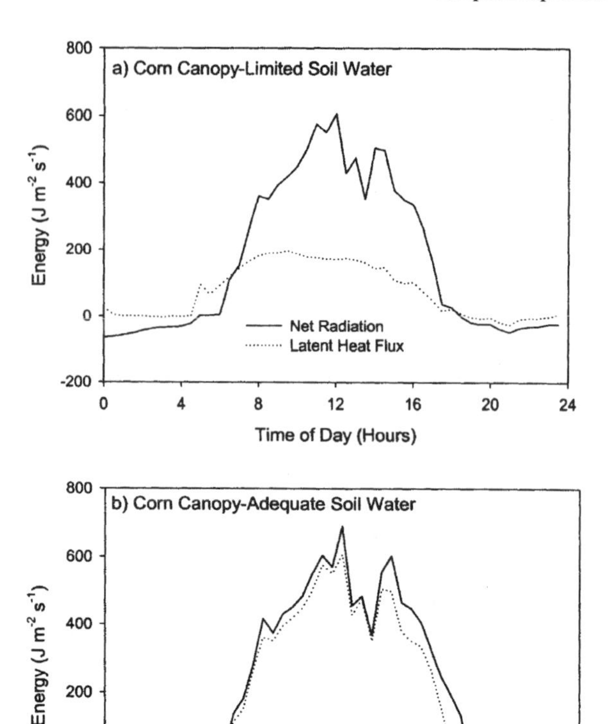

Fig. 3 Evaportanspiration from two days with similar net radiation regimes but differences in soil water availability (a, limited soil water, and b, adequate soil water). These conditions would be typical of a sequence of days before and after a rainfall event.

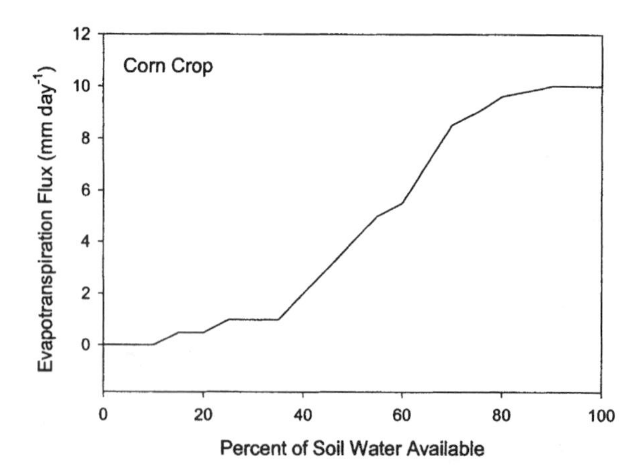

Fig. 4 Evapotranspiration losses from a cropped surface as the soil water supply decreases from a wet to dry soil profile. These relationships represent typical conditions for corn in the Midwest.

the stomata react to the decreased soil water supply and the ability of the plant as a conduit of soil water to the atmosphere decreases. Differences among soils would determine the amount of soil water that is available within the soil profile to crops. Crops also vary among species or varieties in their ability to extract soil water through differences in rooting depth.

CONCLUSION

Evapotranspiration as a physical process is critical to the well being of plants. The cooling action provided by the evaporation of water allows the plant temperatures to be maintained within limits that permit optimum physiological function. Plant canopies are unique because they are comprised of evaporation amounts that occur from the soil surface and the plant surface, however, plants through roots that extend throughout the soil profile are not limited to only soil water supplies at the soil surface. There is a dynamic nature of ET across time and space scales that needs to be considered in interpreting ET values or designing studies that measures or estimate ET.

REFERENCES

1. Montieth, J.L. Evaporation and Environment. In *The state and Movement of Water in Living Organisms*, 19th Symp. Soc. Exp. Biol., 1964; 205–234.
2. Allen, R.G.; Periera, L.S.; Raes, D.; Smith, M. *Crop Evapotranspiration. Guidelines for Computing Crop Water Requirements*; FAO Irrigation and Drainage Paper No. 56 (FAO-56); FAO (Food and Agriculture Organization of the United Nations): Rome, Italy, 1998.
3. Hatfield, J.L. Methods of Estimating Evapotranspiration. In *Irrigation of Agricultural Crops*; Stewart, B.A., Nielsen, D.R., Eds.; Am. Soc. Agron.: Madison, WI, 1990; 435–474.
4. Ritchie, J.T. Dryland evaporative flux in a subhumid climate:I. Micrometeorological influences. Agron. J. **1971**, *63*, 1–55.
5. Ritchie, J.T.; Burnett, E. Dryland evaporative flux in a subhumid climate: II. Plant influences. Agron. J. **1971**, *63*, 56–62.
6. Ritchie, J.T.; Burnett, E.; Henderson, R.C. Dryland evaporative flux in a subhumid climate: III. Soil water influences. Agron. J. **1972**, *64*, 168–173.
7. Ritchie, J.T.; Jordan, W.R. Dryland evaporative flux in a subhumid climate: IV. Relation to plant water status. Agron. J. **1972**, *64*, 173–176.

Extensional Rheology

Mahesh Padmanabhan
Kraft Foods North America, Inc., Glenview, Illinois, U.S.A.

INTRODUCTION

The mechanics of extensional flows are different from shear flows. Extensional flows are an important part of many food-processing operations and in consumer perception and quality of food products. Liquid foods are structurally complex, composed of polymers in various stages of dissolution, suspended particles, and incompatible phases; consequently, their flow behavior is complex. For complex liquids, the nature of extensional flow kinematics result in a behavior that cannot be understood or predicted from its shear rheology. Thus, the design considerations for formulation and process of many foods should include the characterization and understanding of its extensional rheology. The viscous and elastic characteristics of a viscoelastic liquid in extensional flows are measured through a single material property, viz., the extensional viscosity. Due to the difficulty of measuring extensional viscosity, several ingenious techniques have been developed. Many of these techniques are increasingly being used to characterize foods. A greater understanding and exploitation of the extensional flow behavior of foods remain to be achieved.

BACKGROUND

Considering many foods to be dispersions, Giesekus[1] illustrated, using examples of model suspensions, that the rheological behavior is dependent on flow type for complex liquids. In most instances, the structural components in liquid foods have a different and sometimes, a more dramatic effect on its extensional rather than shear rheology. Thus, measuring and understanding the extensional rheological behavior is very important to food manufacture.

Extensional flows are prevalent and sometimes dominant in processing, quality, and consumer perception of foods.[2] Operations such as mixing, homogenization, sheeting, extrusion, extrudate expansion, baking expansion, flow in and out of constrictions, and coating/enrobing are all examples of processes with significant extensional component. Phenomena such as "cold flow" (e.g., soft confections) and settling of particles (e.g., salad dressings), important to the shelf stability of the product, are partly governed by extensional rheology. From the consumer's standpoint, spreadability, pourability, mouthfeel, stringiness, and stretchability are important attributes where extensional rheology dominates the textural perception. Thus, to the product developer, choosing the right ingredient(s) at the right level(s) exhibiting the desired extensional rheological behavior is part of the equation in developing and commercializing a successful food product. A thorough discussion of various aspects of extensional flows was given by Petrie.[3]

FUNDAMENTALS

A simplified illustration of the difference in the kinematics between shear and extensional deformations is given in Fig. 1. Extensional flows also go by the names of Elongational, Stretching, or more correctly, Shearfree flows[4]; the term extensional flows will be used throughout this article. Following the notation of Meissner et al.,[5] the velocity field for homogeneous extensional flows can be written as

$$v_x = \dot{\varepsilon}(t)x \tag{1}$$

$$v_y = m\dot{\varepsilon}(t)y \tag{2}$$

$$v_z = -(1+m)\dot{\varepsilon}(t)z \tag{3}$$

where v_x, v_y, and v_z are the velocities in the x, y, and z coordinate directions, $\dot{\varepsilon}(t)$ the extension rate (velocity gradient), and m a parameter representing extensional flow type, taking on the values of $-1/2$ for uniaxial, 1 for biaxial, and 0 for planar extensional flows. Deformations achieved for the three types of extensional flows for an incompressible material are illustrated in Fig. 2.[4]

For steady extensional flows, the extension rate is independent of time ($\dot{\varepsilon}_0$—constant). Unlike shear flows, where the separation of two points within the fluid is a linear function of the elapsed time Δt $(= t - t_0)$, in extensional flows the separation ℓ (at any time t) is an exponential function of time:

$$\ell = \ell_0 e^{\dot{\varepsilon}_0 \Delta t} \tag{4}$$

where ℓ_0 is the initial separation between the points.

Encyclopedia of Agricultural, Food, and Biological Engineering
DOI: 10.1081/E-EAFE 120005744
Copyright © 2003 by Marcel Dekker, Inc. All rights reserved.

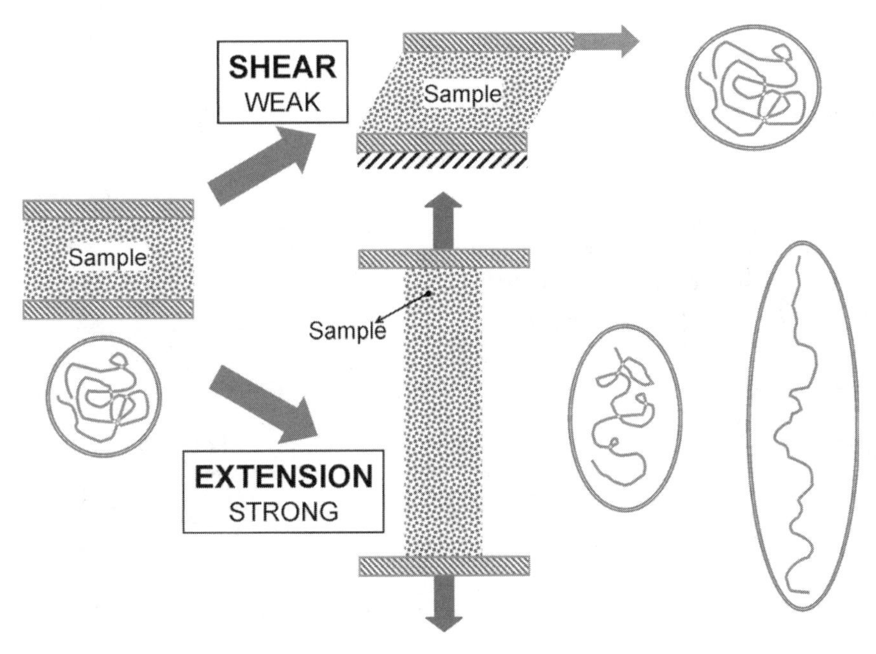

Fig. 1 Simplified illustration of the differences in the kinematics between shear and extensional deformations for a sample held between two flat plates. In shear, the top plate slides parallel to the bottom plate, which may be held fixed. In extension, the two plates move away from each other. The corresponding deformation of a single molecule is also illustrated for shear and extension.

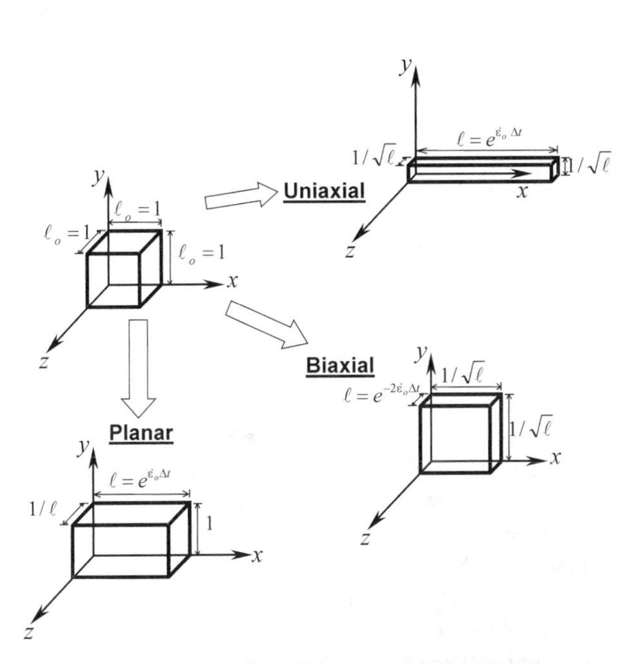

Fig. 2 Illustration of the deformation of a unit cube of the material element subjected to uniaxial, biaxial, and planar extensional flows after a certain elapsed time of Δt. Note the change in the dimensions and shape of the material element, depending on the extensional flow type. (Adapted from Fig. 3.1-3 in Ref. 4.)

The exponential quantity in Eq. 4 is the strain ε_0, known as Hencky strain, and defined as follows:

$$\varepsilon_0 = \dot{\varepsilon}_0 \Delta t = \ln \frac{\ell}{\ell_0} \tag{5}$$

For infinitesimally small strains, Eq. 5 approaches the familiar "engineering" or Cauchy strain e given by

$$e = \frac{\ell - \ell_0}{\ell_0} \tag{6}$$

The rheological material function (a fundamental material property) that governs the flow behavior in extensional flows is the extensional viscosity η_E defined as

$$\eta_E^+(t) = \frac{\sigma_E^+(t)}{\dot{\varepsilon}(t)} \tag{7}$$

where the extensional stress σ_E is a function of the force required to sustain the flow; the superscript sign " + " indicates the transient nature and may be deleted for the steady-state case.

There are two types of measurements needed to understand the extensional flow characteristics of a given liquid. One is the transient behavior, exemplified by Eq. 7, which may start from an initial time (e.g., from rest) and continue on, if possible, until a steady state is reached for a fixed extension rate. The second is the effect of extension rate.

Fig. 3 illustrates typical transient extensional flow behavior and compares with shear behavior. At short times (or low strains, as the deformation rate is fixed for each curve), the viscosity increases; such an increase in viscosity with increasing strain is known as strain hardening. In shear, after a maximum is reached, the viscosity decreases (strain softening) and reaches a constant, steady-state value (curves 2 and 3); for a shear-thinning liquid, curve 2 is for a lower shear rate than curve 3. In extension, the strain-hardening behavior may continue for longer times, and the extensional viscosity could increase by several decades over this time.

Fig. 4 illustrates steady-state extensional flow behavior and compares it with shear behavior. Box 1 shows the viscosity in the linear viscoelastic (LVE) limit, where the viscosity is independent of stress (or deformation rate). Box 2 shows a region of increasing extensional viscosity with increasing stress (or extension rate). Such a behavior is known as extension thickening; if this occurs in shear, it is known as shear-thickening behavior. Box 3 shows regions of decreasing viscosity with increasing stress (or deformation rate) known as extension thinning or shear thinning for extensional and shear flows, respectively.

Trouton ratio, defined as the ratio of extensional to shear viscosity, is a convenient quantity used to represent the rheological behavior of liquids. Following the convention proposed by Jones, Walters, and Williams,[6]

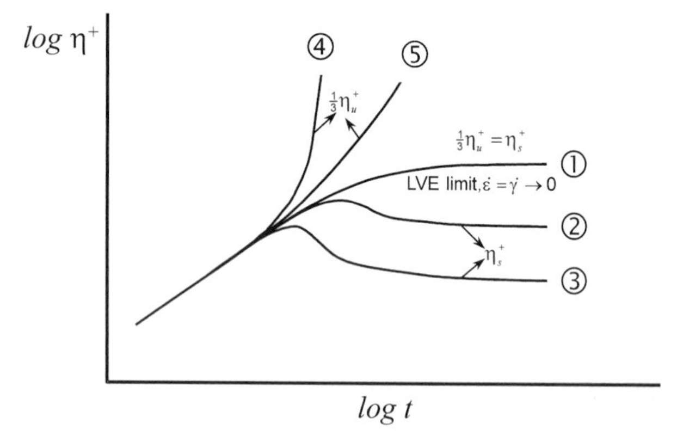

Fig. 3 Typical transient uniaxial extensional and shear viscosity curves for polymer solutions, shown schematically. Each curve is obtained at a constant deformation rate (shear or extension, as the case may be). In the LVE limit of small strain rates, the normalized uniaxial extensional viscosity is the same as the shear viscosity (curve 1); the LVE limit also applies at short times, irrespective of the deformation rate. Curves 2 and 3 are for shear viscosity, which may be contrasted with curves 4 and 5 for the normalized uniaxial extensional viscosity. (Adapted from Fig. 2a in Ref. 2.)

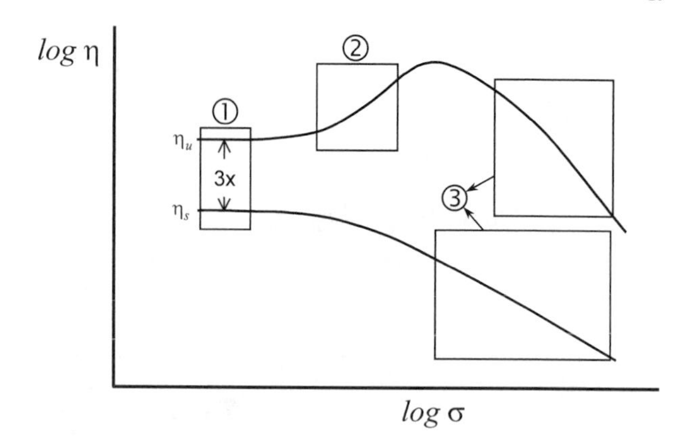

Fig. 4 Typical schematic steady-state (or pseudosteady-state) uniaxial extensional and shear viscosity curves for polymer solutions. (Adapted from Fig. 2b in Ref. 2.)

the Trouton ratio, T_r, is defined as

$$T_r = \frac{\eta_E(\dot{\varepsilon})}{\eta_s(c\dot{\varepsilon})} \tag{8a}$$

where the shear viscosity η_s is evaluated at a shear rate, $\dot{\gamma} = c\dot{\varepsilon}$, and the magnitude of constant c is based on the type of extensional flow as follows:

$$c = \begin{cases} \sqrt{3} & \text{for uniaxial} \\ 2 & \text{for planar} \\ \sqrt{6} & \text{for biaxial} \end{cases} \tag{8b}$$

MEASUREMENT TECHNIQUES

Realization of a simple, well-defined, and controlled flow is critical for rheometrical purposes. This is more difficult for complex liquids. Further experimental complication arises to generate extensional rheometrical flows due to the need, in principle, to "hold" the ends of a "cylinder" of the liquid and "pull" it lengthwise at an exponential speed. Consequently, several ingenious and sophisticated techniques have been developed to measure the extensional viscosity of non-Newtonian liquids.[7] In practice, measurement techniques subscribing to the philosophy espoused by Walters and coworkers[6] serves well: "*...it may often be sufficient to seek simple, but scientifically acceptable, techniques which yield extensional viscosity levels and are able to indicate whether the liquids are tension stiffening or tension thinning, i.e., whether the extensional viscosity increases or decreases with strain rate.*" Thus, although generation of uniform extensional flows is difficult with all the existing techniques, the results obtained are of great practical value.

In practice, uniaxial and biaxial extensional flows are encountered more commonly. Experimentally, uniaxial and biaxial extensional flows are more easily achieved than planar extensional flows.[7] From a rheological standpoint, uniaxial extensional flows are the strongest, while shear flows are the weakest—these two flows together provide the range of nonlinear behavior exhibited by the material.

In simple or uniaxial extension, a rod-shaped sample (cylindrical or rectangular cross-section) is stretched at a constant extension rate and the force is measured [Fig. 5(a)]. Sample clamping problems can introduce significant errors in to the measurement.[7] Instrument size can limit the maximum strains that may be achieved (Eq. 4); e.g., a sample of initial length 1 cm will have a final length of 1100 cm when stretched to a strain, $\varepsilon = 7$. de Bruijne, de Loof, and van Eulem[8] applied this technique to study the rheology of bread dough; they programmed the tensile testing machine to achieve constant extension rates. Extensional tests used for doughs (e.g., Brabender Extensigraph[9]) and those developed for cheese[10] also apply this principle; however, these tests do not generate uniform extensional flows. Ring-shaped samples [Fig. 5(b)] can be used to overcome clamping problems and increase the magnitude of the measured force. Meissner[11] overcame the clamping and instrument size problems through the development of the rotary-clamp technique [Fig. 5(c)]. A commercial version of this rheometer has been used to study the extensional rheology of bread dough.[12] For sufficiently high-viscosity liquids, the fiber-windup technique can be used on conventional rotational shear rheometers.[2] For lower viscosity liquids such as polymer solutions, the filament stretching rheometer was recently developed[13]; application of this technique to food systems would be of great value and interest.

When a liquid flows through a contraction, a mixed extensional and shear flow exists at the contraction (Fig. 6). Due to large extensional viscosities of polymeric liquids, the extensional component constitutes a large portion of the pressure drop at the contraction. This pressure drop at the contraction, known as the entrance pressure drop, can be used to estimate the extensional rheological characteristics of liquids of wide range of viscosities.[2] When the cross-sectional geometry is circular, the extensional flow in the contraction region is uniaxial in nature. Contraction flow analyses have been used to estimate extensional viscosities of potato powder and corn meal during extrusion cooking,[14,15] wheat dough,[2,16] and spreads.[17]

The opposed nozzles device consists of two nozzles placed opposite to each other and immersed in a sea of liquid (Fig. 7). Extensional flow is generated between circular cross-section nozzles when the liquid is sucked (uniaxial) or expelled (biaxial), with the nozzles separated by a known distance apart. This technique has gained popularity to characterize the extensional flow behavior of low-viscosity liquids. Among foods, such measurements

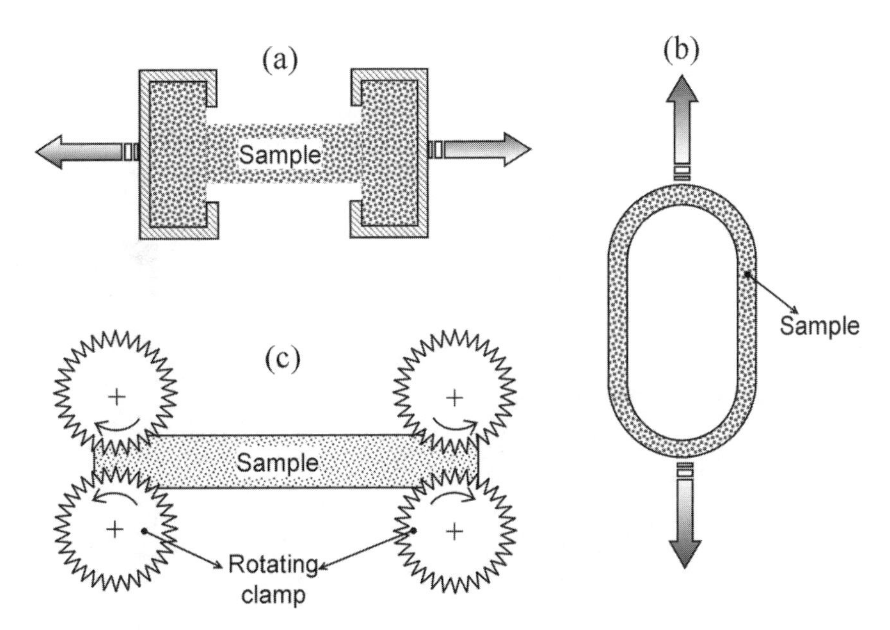

Fig. 5 Various approaches to overcome sample clamping and instrument size problems to generate uniaxial extensional flows for high-viscosity liquids such as doughs and melts. (a) Conventional approach of stretching a dumb-bell shaped sample that has sample clamping and instrument size problems; (b) use of ring-shaped sample to overcome the clamping problems but still has the instrument size problem; and (c) Meissner's rotary-clamp technique that avoids clamping and instrument size problems.

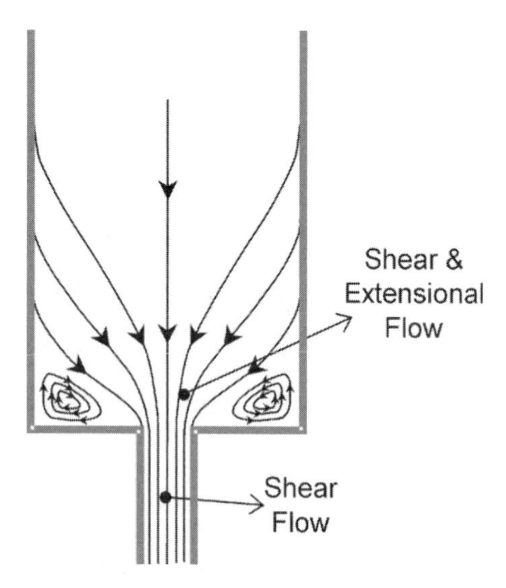

Fig. 6 Illustration of the region of mixed shear and extensional flow in an abrupt contraction geometry.

have been reported for food gums[18,19] and starch suspensions.[20]

By extending the filament extruded from a die, known as fiber spinning or fiber drawing, the liquid can be subjected to an extensional flow. For low-viscosity liquids, the filament may be stretched with a suction device while high-viscosity liquids may be stretched using a windup device or toothed wheels.[7] By the very nature of this technique, a uniaxial extensional flow is generated. Experimental uncertainties allow the measurement of only an apparent extensional viscosity.[7] Although not very common, fiber spinning is gaining prominence as a process to create various food textures[21–23]; an extensional thickening behavior of the liquid will contribute to the stability of the process.

The two most common techniques used to generate biaxial extensional flows are lubricated squeezing (or uniaxial compression) and sheet (or bubble) inflation. When a cylindrical piece of the sample is compressed between two circular disks, the adhesion between the sample and the disk introduces a strong shearing component. A low-viscosity lubricating liquid placed between the sample and the disk can greatly reduce the friction facilitating a biaxial extensional flow (Fig. 8). The relative simplicity of this technique and its ability to be executed with a universal testing machine, which is available in most laboratories, have made this a very popular technique for foods.[24] The complex nature of the flow demands careful experimentation and interpretation of the data. A number of studies can be found in the literature looking at a wide range of food products.[24]

When a sheet of a high-viscosity material is inflated, the bubble that expands (Fig. 9) is subjected to a biaxial extensional flow.[25] When properly instrumented, the technique can be used to characterize the biaxial extensional rheology of food systems. The Chopin Alveograph is an example of an instrument used to characterize wheat flour doughs by sheet inflation.[26] Although the deformation achieved is not uniform in nature, the data obtained has found extensive industrial use to characterize wheat flours for use in baking applications. Recently this technique has gained renewed attention.[27]

APPLICATIONS

The importance of extensional rheology for baking leavened breads has been recognized for nearly a century.[28] Extensional flows are prominent during the stages of mixing, fermentation, and baking. Magnitude of extension rates during the fermentation and baking stages have been estimated[29]; to be practically relevant, the rheological measurements should encompass the kinematic conditions actually present. Recent work has also demonstrated that the strain-hardening behavior during

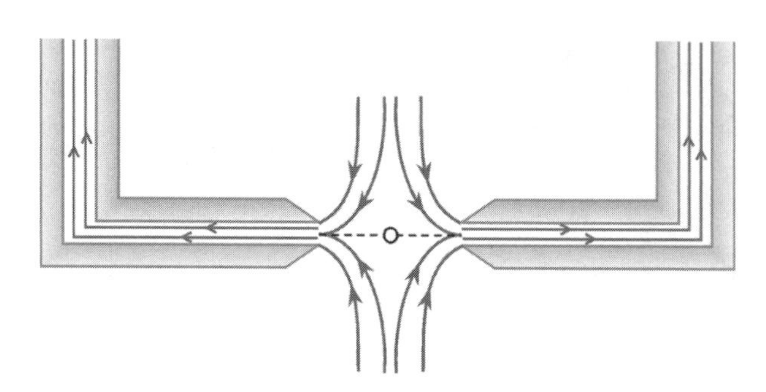

Fig. 7 Opposed nozzles device for generating a stagnation point extensional flow.

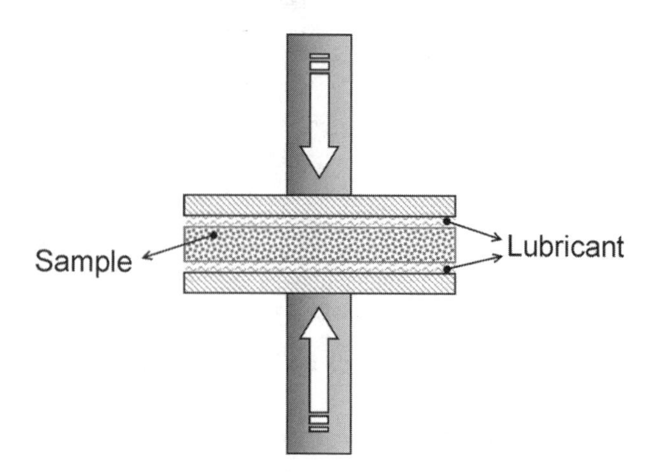

Fig. 8 Lubricated squeezing (or uniaxial compression) technique to generate biaxial extensional flows.

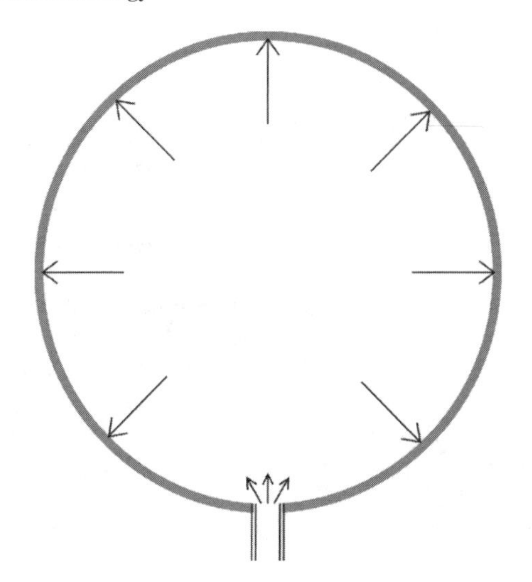

Fig. 9 Generation of biaxial extensional flow through the expansion of the bubble formed from the inflation of a sheet of the sample.

bubble growth is important to gas retention and hence the final structure and texture of the baked product.[30] The importance of extensional deformation to mixing, in general, has been illustrated[31]; extension of its application to foods remains to be demonstrated.

The ability of extensional deformation to orient elements within the sample enables its use to create various types of structures and textures.[32] Recently, the use of extensional flows to shape drops of gelling hydrocolloids was demonstrated.[33]

Various hydrocolloids are used in the formulation of a large number of food products. In aqueous solutions, the semirigid molecules of xanthan gum have been shown to display extension-thinning behavior,[18,19,34] while flexible chain molecules of guar[34] and carboxymethyl cellulose[19] gums show extension-thickening behavior. Additives such as salt or change in pH can have a dramatic effect on the polymer conformation and hence its rheology. As shown in Fig. 10, addition of salt resulted in a large decrease in the extensional viscosity magnitude of a dilute xanthan gum solution, while the shear viscosity showed a considerably reduced shear-thinning behavior, with a much smaller drop in shear viscosity magnitude at higher stresses.[18] Taking pouring of syrup (such as pancake syrup) as an example, an extension-thinning behavior will allow a clean break when pouring is stopped, in contrast to an extension-thickening behavior.

A number of sensory attributes such as creaminess, mouthfeel, and mastication (structure breakdown and consequent in-mouth perception), all have important contributions from extensional rheology. Similarly, stretchability of cheese,[10] spreadability, pouring (see above), and swallowing are all examples with significant extensional deformation presence.[2] However, a systematic study of this important area will require

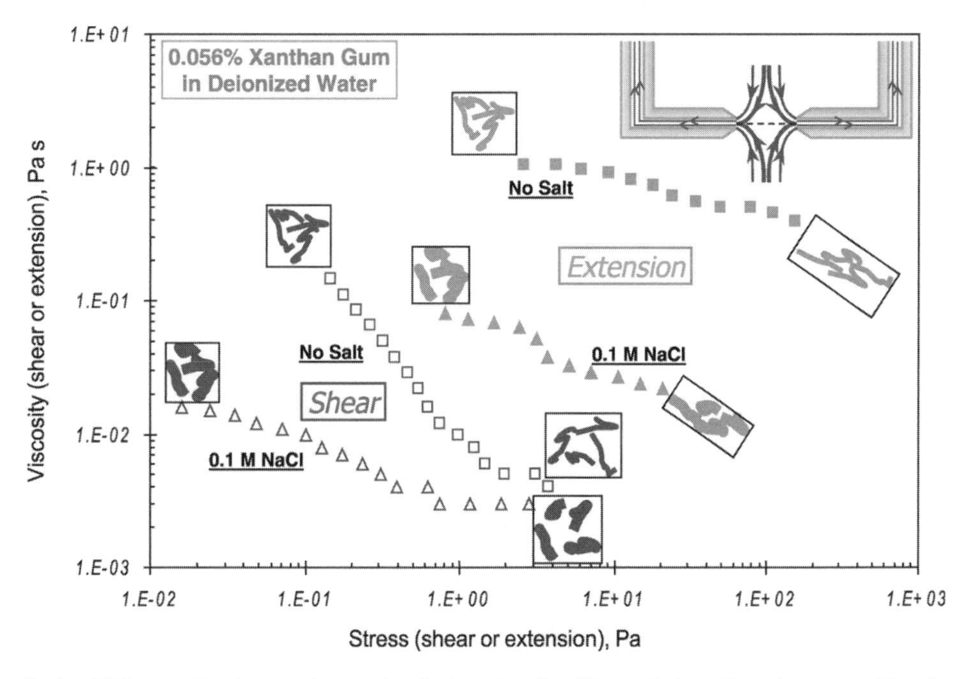

Fig. 10 Effect of salt addition on the shear and extensional viscosity of a dilute solution of xanthan gum. Also shown are schematic diagrams of molecular conformations that would exist for the various situations (Data replotted from Ref. 18.)

developing correlations between the rheological measurements and human perception; such studies can be expensive and daunting.

NOMENCLATURE

c	Extensional flow dependent constant for Trouton ratio calculation
e	Cauchy or engineering strain
ℓ	Separation distance between two points in the fluid
ℓ_0	Initial separation distance between two points in the fluid
m	Extensional flow parameter
t	Time
t_0	Initial time
T_r	Trouton ratio
Δt	Elapsed time
v_x	Velocity in the x-coordinate direction
v_y	Velocity in the y-coordinate direction
v_z	Velocity in the z-coordinate direction
x	Position in the x-coordinate direction
y	Position in the y-coordinate direction
z	Position in the z-coordinate direction
$\dot{\gamma}$	Shear rate
ε_0	Hencky strain
$\dot{\varepsilon}(t)$	Extension rate, as a function of time
$\dot{\varepsilon}_0$	Constant extension rate
η_E	Steady-state extensional viscosity
η_E^+	Transient extensional viscosity
η_s	Shear viscosity
η_s^+	Transient shear viscosity
η_u	Uniaxial extensional viscosity
σ_E	Steady-state extensional stress
σ_E^+	Transient extensional stress

ACKNOWLEDGMENTS

The author is grateful to the management of Kraft Foods for their support and permission to publish this article. Helpful comments provided by Dr. Mary Amini (Kraft Foods) and Dr. T. van Vliet (Wageningen Agricultural University) are greatly appreciated.

REFERENCES

1. Giesekus, H. Disperse Systems: Dependence of Rheological Properties on the Type of Flow with Implications for Food Rheology. In *Physical Properties of Foods*; Jowitt, R., Escher, F., Hallström, B., Meffert, H., Speiss, W.E.L., Vos, G., Eds.; Applied Science: London, 1983; 205–220.
2. Padmanabhan, M. Extensional Viscosity of Viscoelastic Liquid Foods. J. Food Eng. **1995**, *25*, 311–327.
3. Petrie, C.J.S. *Elongational Flows: Aspects of the Behaviour of Model Elasticoviscous Fluids*; Pitman: San Francisco, 1979.
4. Bird, R.B.; Armstrong, R.C.; Hassager, O. *Dynamics of Polymeric Liquids. Volume 1: Fluid Mechanics*; Wiley Interscience: New York, 1987; 100–103.
5. Meissner, J.; Stephenson, S.E.; Demarmels, A.; Portmann, P. Multiaxial Elongational Flows of Polymer Melts—Classification and Experimental Realization. J. Non-Newtonian Fluid Mech. **1982**, *11*, 221–237.
6. Jones, D.M.; Walters, K.; Williams, P.R. On the Extensional Viscosity of Mobile Polymer Solutions. Rheol. Acta **1987**, *26*, 20–30.
7. Macosko, C.W. *Rheology: Principles, Measurements, and Applications*; Wiley-VCH: New York, 1994; 285–336.
8. de Bruijne, D.W.; de Loof, J.; van Eulem, A. The Rheological Properties of Bread Dough and Their Relation to Baking. In *Rheology of Food, Pharmaceutical and Biological Materials with General Rheology*; Carter, R.E., Ed.; Elsevier: London, 1990; 269–283.
9. Keiffer, R.; Wieser, H.; Henderson, M.H.; Graveland, A. Correlations of the Breadmaking Performance of Wheat Flour with Rheological Measurements on a Micro-scale. J. Cereal Sci. **1998**, *27*, 53–60.
10. Ak, M.M.; Bogenrief, D.; Gunasekaran, S.; Olson, N.F. Rheological Evaluation of Mozzarella Cheese by Uniaxial Horizontal Extension. J. Texture Stud. **1993**, *24*, 437–453.
11. Meissner, J. Polymer Melt Elongation–Methods, Results, and Recent Developments. Polym. Eng. Sci. **1987**, *27*, 537–546.
12. Schweizer, T.; Conde-Petit, B. Bread Dough Elongation. In *Proceedings of 1st International Symposium on Food Rheology and Structure*; Windhab, E.J., Wolf, B., Eds.; Vincentz: Hannover, 1997; 391–394.
13. McKinley, G.H.; Sridhar, T. Filament-Stretching Rheometry of Complex Fluids. Annu. Rev. Fluid Mech. **2002**, *34*, 375–415.
14. Senouci, A.; Smith, A.C. An Experimental Study of Food Melt Rheology. II. End Pressure Effects. Rheol. Acta **1988**, *27*, 649–655.
15. Padmanabhan, M. In-Line Measurement of Shear-Elastic and Extensional Rheological Properties of Food Doughs. Ph.D. Dissertation, University of Minnesota, St. Paul., 1992.
16. Wikström, K. Rheology of Wheat Flour Dough at Large Deformations and the Relation to Baking Quality and Physical Structure. Ph.D. Dissertation, Lund University, Lund, 1997.
17. Stading, M.; Bohlin, L. Measurements of Extensional Flow Properties of Semi-solid Foods in Contraction Flow. In *Proceedings of the 2nd International Symposium on Food Rheology and Structure*; Fischer, P., Marti, I., Windhab, E.J., Eds.; Laboratory of Food Process Engineering, ETH: Zürich, 2000; 117–120.
18. Clark, R.C. Extensional Viscosity of Some Food Hydrocolloids. In *Gums and Stabilizers for the Food Industry 6*;

E

Phillips, G.O., Williams, P.A., Wedlock, D.J., Eds.; IRL Press: New York, 1992; 73–85.

19. Clark, R. Evaluating Syrups Using Extensional Viscosity. Food Technol. **1997**, *51*, 49–52.
20. Kapoor, B.; Bhattacharya, M. Dynamic and Extensional Properties of Starch in Aqueous Dimethylsulfoxide. Carbohydr. Polym. **2000**, *42*, 323–335.
21. Luyten, J.M.J.G. The Creation of New Food Structures by Spinneretless Spinning of Protein–Polysaccharide Mixtures. In *Proceedings of the 2nd International Symposium on Food Rheology and Structure*; Fischer, P., Marti, I., Windhab, E.J., Eds.; Laboratory of Food Process Engineering, ETH: Zürich, 2000; 377.
22. Tolstoguzov, V.B. Creation of Fibrous Structures by Spinneretless Spinning. In *Food Structure: Its Creation and Evaluation*; Blanshard, J.M.V., Mitchell, J.R., Eds.; Butterworths: London, 1988; 181–196.
23. Visser, J. Dry Spinning of Milk Proteins. In *Food Structure: Its Creation and Evaluation*; Blanshard, J.M.V., Mitchell, J.R., Eds.; Butterworths: London, 1988; 197–218.
24. Campanella, O.H.; Peleg, M. Squeezing Flow Viscometry for Nonelastic Semiliquid Foods—Theory and Applications. Crit. Rev. Food Sci. Nutr. **2002**, *42*, 241–264.
25. Dealy, J.M. Uniform Extensional Flows. *Rheometers for Molten Plastics*; Van Nostrand Reinhold: New York, 1982; 161–166.
26. Launay, B. Theoretical Aspects of the Alveograph. In *The Alveograph Handbook*; Faridi, H., Rasper, V.F., Eds.; AACC: St. Paul, 1987; Chap. II., 10–16.
27. Dobraszczyk, B.J.; Roberts, C.A. Strain Hardening and Dough Gas Cell-Wall Failure in Biaxial Extension. J. Cereal Sci. **1994**, *20*, 265–274.
28. Faridi, H.; Rasper, V.F. *The Alveograph Handbook*; AACC: St. Paul, 1987; Chap. I., 1–9.
29. Bloksma, A.H. Rheology of the Breadmaking Process. Cereal Foods World **1990**, *35*, 228–236.
30. Van Vliet, T.; Janssen, A.M.; Bloksma, A.H.; Walstra, P. Strain Hardening of Dough as a Requirement for Gas Retention. J. Texture Stud. **1992**, *23*, 439–460.
31. Utracki, L.A.; Luciani, A. Mixing in Extensional Flow Field. Appl. Rheol. **2000**, *10*, 10–21.
32. Richmond, P.; Smith, A.C. Rheology, Structure, and Food Processing. In *Food Structure and Behavior*; Blanshard, J.M.V., Lillford, P., Eds.; Academic Press: London, 1987; 259–283.
33. Hamberg, L.; Walkenström, P.; Hermansson, A.-M. Shaping of Gelling Biopolymer Drops in an Elongation Flow. J. Colloid Interface Sci. **2002**, *252*, 297–308.
34. Secor, R.B.; Schunk, P.R.; Hunter, T.B.; Stitt, T.F.; Macosko, C.W.; Scriven, L.E. Experimental Uncertainties in Extensional Rheometry of Liquids by Fiber Drawing. J. Rheol. **1989**, *33*, 1329–1358.

Extraction System Design

Larry Erickson
Kansas State University, Manhattan, Kansas, U.S.A.

INTRODUCTION

The terms leaching and extraction are used to describe the transfer of a solute from one phase to another. When the term leaching is used, the source of the solute or product to be recovered is usually a solid. Extraction may refer to a solute being removed from either a solid or a liquid. This treatment of the design of extraction systems includes both solids and liquids as sources of solutes to be extracted. There are many books that provide additional information on the design of extraction systems.[1–5]

OVERVIEW

The market for extraction systems is very large, and extraction operations are performed at many different locations. Whenever, a cup of tea is prepared by extracting the solutes from tea leaves or a pot of coffee is made by extracting the solutes of interest from coffee beans, extraction systems are employed. Thus, extraction systems include those for making tea and coffee at home, restaurant, and institutional food service establishments. Extraction operations range in size from the individual serving size such as that used to make a cup or small pot of tea to large scale commercial activities where tea leaves are used to produce large quantities of instant tea. Decaffeinated coffee is produced by extracting caffeine from coffee beans. Cooking oils are often extracted from raw materials such as soybeans and cottonseeds using hexane as solvent. Vanilla can be extracted from vanilla beans using a 65% ethanol solution.[4] Soy protein isolates are extracted from defatted soy flour using pH 9 aqueous NAOH.[4]

Leaching and extraction have many biological applications where specialty products are manufactured. Blanch and Clark[6] provide a list of enzymes that are extracted from cell homogenates. Belter et al.[7] provide a list of biochemical solutes and the solvent that is frequently used for product recovery. Table 1 is a representative list from these sources. Enzymes that are biologically active proteins are often purified using aqueous two-phase extraction processes in which both phases are aqueous phases.[6,8]

Leaching and extraction are common separation processes in the processing of agriculture raw materials into industrial products such as oils that are used for fuels and other industrial purposes. Most plants that process corn, milo, sunflowers, soybeans, and other grains have one or more extraction processes. Extraction may be used with microbial fermentation bioconversion processes to extract continuously a product from the broth.[9]

The market for extraction systems is competitive, and there is a need to understand the fundamentals of extraction when designing the process and equipment. The phase equilibrium relationship for the solute of interest, the mass transfer process associated with the extraction of the solute, and the phase separation process should be considered in any design. Materials of construction, corrosion, capital and operating cost, and safety are other considerations.

Process Design Fundamentals

The phase equilibrium relationship influences the extraction process. At any given temperature and pressure, there is a phase equilibrium condition that relates the concentration of the solute in the solvent or extract phase to the concentration of the solute in the solid or raffinate phase. Temperature is an important variable because the phase equilibrium relationship is a function of temperature. For example, hot water is usually used in making tea and coffee. The solubility of the solute in the solvent can influence the phase equilibrium if the amount of solute is sufficient to saturate the solvent at a given temperature. In most cases, the solubility of a solute increases with temperature. The viscosity of the solution is generally lower at higher temperature and the diffusivity is higher.[1] With food and biological products, there may be undesirable chemical transformations at higher temperatures which limit the practical temperature that should be used. For liquid–liquid extraction systems, there are many sets of equilibrium data in the published literature.[1–5] For solid–liquid systems, the phase equilibrium may be quite simple if the solute does not sorb strongly to the solid phase. If all of the solute dissolves in the liquid, then the solute retained with the solid is related to the amount of solvent retained with the solid. In many cases the solute associated with the solvent that is retained by the solids accounts for the majority of the retained solute.

Table 2 contains equilibrium data taken from Ref. 5 for the two-phase system in which furfural (F) is used to

Encyclopedia of Agricultural, Food, and Biological Engineering
DOI: 10.1081/E-EAFE 120007044
Copyright © 2003 by Marcel Dekker, Inc. All rights reserved.

Table 1 Examples of specialty products and solvents from Refs. [7,6]

Solute product	Solvents	Reference
Amino acids (glycine, alanine, lysine, and glutamic acid)	*N*-butanol	Ref. [7]
Antibotics (Penicillin K. and Erythromycin)	Amyl acetate	Ref. [7]
Proteins (glucose isomerase, fumarase)	Polyethylene glycol and potassium phosphate	Refs. [5,6][a]

[a] Aqueous two-phase extraction.

extract ethylene glycol (EG) from water (W). Furfural and water separate into two liquid phases. The ethylene glycol has greater affinity for furfural than water as shown by the equilibrium data. For example, when the mass fraction of ethylene glycol is 0.13 in the water-rich phase, the equilibrium concentration is 0.42 in the furfural rich phase. Seader and Henley[5] show how this data can be presented using a triangular graph.

The solute diffuses out of the solid particles because of a difference in concentration. As the equilibrium condition is approached the concentration difference decreases. The design of extraction systems is often based on the concept that sufficient time is allowed for the solute to approach a phase equilibrium condition. The rate of mass transfer of solute (its movement by diffusion and convection) depends on mixing, surface area, and particle size. Size reduction of solids is common in order to increase surface area and reduce particle size. The time required to approach the equilibrium condition increases with particle size. This is why coffee beans are ground, for example.

Phase separation must be considered in the design of extraction systems. The commonly used tea bag serves to contain the tea leaves and allows simple and easy separation of the liquid phase from the solid phase. In some cases, the solid phase is contained as a packed bed and the solvent is passed through the bed. In some operations, solids have greater density and they settle to the bottom. Where solids are

Table 2 Equilibrium data[a] for the system ethylene glycol (EG), furfural (F) and water (W)

Furfural rich phase			Water rich phase		
EG	F	W	EG	F	W
0.50	0.36	0.14	0.32	0.09	0.59
0.42	0.49	0.09	0.13	0.08	0.79
0.30	0.64	0.06	0.09	0.08	0.83
0.22	0.74	0.04	0.08	0.08	0.84
0.15	0.81	0.04	0.6	0.08	0.86
0.09	0.87	0.04	0.02	0.08	0.90
0	0.96	0.04	0	0.08	0.92

[a] Data from Ref. 5.

0.1 mm or smaller, auxiliary separation devices such as centrifuges or filtration systems are often required. In liquid–liquid extraction, the solvent is selected such that the two phases can be separated based on difference in density.

The choice of solvent may be simple such as selecting water for making tea; however, in extraction of oils from seeds, organic solvents are regularly selected because of the greater solubility of the oil in the organic phase. The solvent should be one which is effective as well as one that is safe. Food safety is often based on the ability to either remove the solvent or have it as an accepted food grade material. Processing safety must also be considered because of the danger of fire and explosions when working with organic solvents. It is desirable to have solvents that are noncorrosive and nonflammable such as carbon dioxide. Often the solvent needs to be recovered and recycled for the process to be competitive. The choice of solvent may greatly influence the separation and recovery costs associated with solvent removal and recycle. The cost of the solvent is also an important consideration. Commonly used solvents in the food industry include water, aqueous solutions of acids and bases, hexane, ethanol, carbon dioxide, and vegetable oils. In liquid–liquid extraction, the solubility of the solvent in the raffinate should be small to reduce the loss of solvent.

Fig. 1 shows the different parts of an extraction process to separate the oil from ground soybeans. In the extractor a solvent such as hexane is used to extract the oil from the soybeans. In the separator, the liquid phase containing oil and hexane is separated from the solid phase by filtration, gravity separation, or centrifugation. The solvent is recovered by distillation and recycled to the extractor. In practice a multistage extractor is commonly used.

The process analysis required to design a solvent extraction system is based on mass balances and the phase equilibrium relationship. The processes can be batch or continuous. Multiple stages are needed where the value of the products is sufficient to justify the additional capital and operating cost. Perry and Green,[1] Treybal,[2] King,[3] Rousseau[4] and Seader and Henley[5] provide detailed information on the design of a variety of extraction systems. For a single completely mixed stage that approaches equilibrium, the equilibrium relationship is used to relate

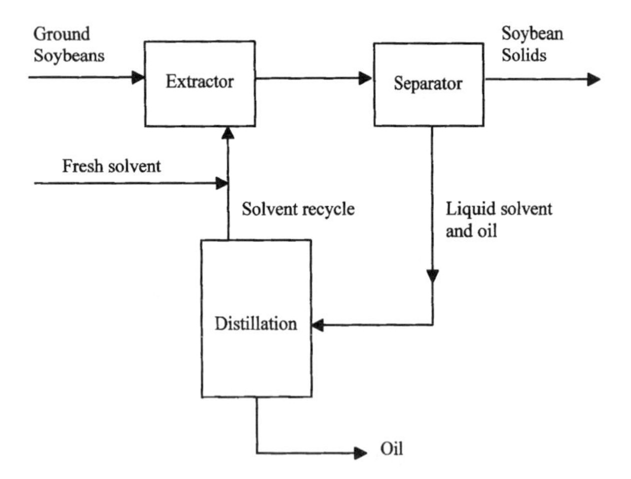

Fig. 1 Flow diagram showing the extractor, separator, and solvent recovery process for extraction of oil from soybeans.

the solute concentration in the exiting underflow or solids stream to that in the exiting extract stream. The amount of liquid retained with the solids needs to be known; however, this is easily determined from an experiment if it is not available.

Contacting Methods

There are many different contacting methods that have been used to extract solutes from solids. In Chapter 10 of Ref. 4, Schwartzberg has described several different methods, including batch extraction, differential extraction (continuous passage of extract through a well-mixed slurry that contains the solids), fixed bed extraction (passage of extract through a fixed bed of solids), countercurrent percolation (continuous countercurrent flow of the extract phase through a moving bed of solids), and countercurrent slurry extraction (the dispersed solids pass through a flowing extract phase).

Other Considerations

The design of extraction systems must be integrated into the process design for the entire plant. Sometimes, there are issues such as solubilization of the solutes so that they can be extracted. This often involves chemical transformations that may be designed to take place in the extractor or in a separate reactor. Solubilization times may control the duration of the extraction process. In other cases, both the solids and the extract have significant value, and the choice of solvent are limited by the final product requirements which must be met.

Capital and Operating Costs

The cost of the raw materials are significant and they must be considered in any economic analysis. Capital equipment costs may range from very modest for simple extractions to moderate for more difficult separations. Where the solvent is recovered, solvent costs and solvent recovery costs are often substantial. Waste treatment costs can be substantial if there are streams that need to be treated. Energy costs often are important as well. The capital equipment and economics are considered in detail in several of the references.[1,3,10]

CONCLUSION

Extraction systems design is based on phase equilibrium data, product purity requirements, and the chemical characteristics of the material to be processed. The solvent is selected based on affinity for the solute, phase separation, cost, ease of recovery, and safety.

REFERENCES

1. Perry, R.H.; Green, D.W. *Perry's Chemical Engineering Handbook*, 7th Ed.; McGraw-Hill: New York, 1997.
2. Treybal, R.E. *Mass-Transfer Operations*, 3rd Ed.; McGraw-Hill: New York, 1980.
3. King, C.J. *Separation Processes*, 2nd Ed.; McGraw-Hill: New York, 1980.
4. Rousseau, R.W. *Handbook of Separation Process Technology*; Wiley: New York, 1987.
5. Seader, J.D.; Henley, E.J. *Separation Process Principles*; Wiley: New York, 1988.
6. Blanch, H.W.; Clark, D.S. *Biochemical Engineering*; Marcel Dekker: New York, 1996.
7. Belter, P.A.; Cussler, E.L.; Hu, W.S. *Bioseparations*; Wiley: New York, 1988.
8. Hatti-Kuhl, R., Ed. *Aqueous Two Phase Systems*; Humana Press: Totowa, New Jersey, 2000.
9. Mattiason, B.; Holste, O.; Eds. *Extractive Bioconversions*; Marcel Dekker: New York, 1991.
10. Asenjo, J.A., Ed. *Separation Process in Biotechnology*; Marcel Dekker: New York, 1990.

Extrudate Rheology

James Faller
Mary Kay, Inc., Dallas, Texas, U.S.A.

Emine Unlu
Masterfoods USA, Vernon, California, U.S.A.

E

INTRODUCTION

Extrusion processing in the food industry is popular for a number of food products: snacks, breakfast cereals, pasta and other noodles, confectioneries, pet foods, and modified starches. Most of these products are cooked but some are simply formed by the process. The extrusion process derives its popularity from its ability to perform multiple unit operations in a single unit requiring little floor space. The extruder (see Fig. 1) converts agricultural commodities, usually in a granular or powdered form, into fully cooked and formed, near finished, products through the action of shearing friction generated between the extruder screws and extruder barrel. Agricultural materials are processed at relatively low moisture contents and are easily dried to their final forms by frying, baking, gun-puffing, or drying.

RHEOLOGY OF EXTRUDES

In all food extrusion processes, whether cooking, in the sense of starch gelatinization and protein denaturation, occurs or not, agricultural materials are converted from a solid, granular state into a fluid, viscoelastic state through the introduction of heat, shear, and moisture. Most agricultural materials contain starch and/or protein, high molecular weight polymers, which are converted to an amorphous (noncrystalline) state exhibiting a glass transition during the extrusion process. As the agricultural materials are fluid during the process, characterization and measurement of the fluid rheological properties aid in the design and development of food extrusion systems and new extruded food products.

The rheological properties of extruded foods have been investigated either off-line or on-line. Off-line measurement is useful in characterizing the polymeric fraction of the agricultural material in its raw or preprocessed state. The postextrusion state is also characterized off-line for the purpose of analyzing the effects of the extrusion process on the polymeric materials.

The most common methods for characterizing rheological behavior of pre- and postextruded products are the Viscoamylograph[1] and the Rapid Visco Analyzer (RVA).[2] The basic procedure is to ramp the temperature of a suspension of the agricultural material in a concentric cylinder rheometer to a given set point (usually above the starch gelatinization temperature of the material), hold that temperature while continuing the shearing action within the rheometer, and then to ramp the temperature down to a final temperature point. The resulting graph yields information on the gelatinization temperature of the agricultural material, its ability to maintain a viscous state, and its retrogradation upon cooling (Fig. 2).

While the off-line methods give valuable information about the character of the agricultural materials, the information cannot be used directly in the estimation of pressure drops, flow rates, or shear stresses within an extrusion system. Only on-line methods will allow rheological characterization of agricultural materials under the conditions of the extrusion environment. This is critical for extrusion cooking since the objective is to convert the materials from the raw to the cooked state. Starch gelatinization and dextrinization,[3,4] protein denaturation and hydrolysis,[5] and other chemical modifications[6] occur during the extrusion process, and therefore on-line methods should offer superior information for extrusion system design considerations. A number of on-line rheometers have been developed for food extrusion systems. The essence of these rheometers is to modify the extruder die to allow the introduction of pressure transducers along a length of die. Knowing the pressure drop over a given length at a given flow rate is used to characterize the melt rheology.

RHEOLOGICAL CHARACTERISTICS OF AGRICULTURAL MATERIALS

Due to their high molecular weight polymer constituents, many agricultural and food materials exhibit viscoelastic properties, meaning that they have flow properties similar to purely viscous fluids (like water),

Encyclopedia of Agricultural, Food, and Biological Engineering
DOI: 10.1081/E-EAFE 120007120
Copyright © 2003 by Marcel Dekker, Inc. All rights reserved.

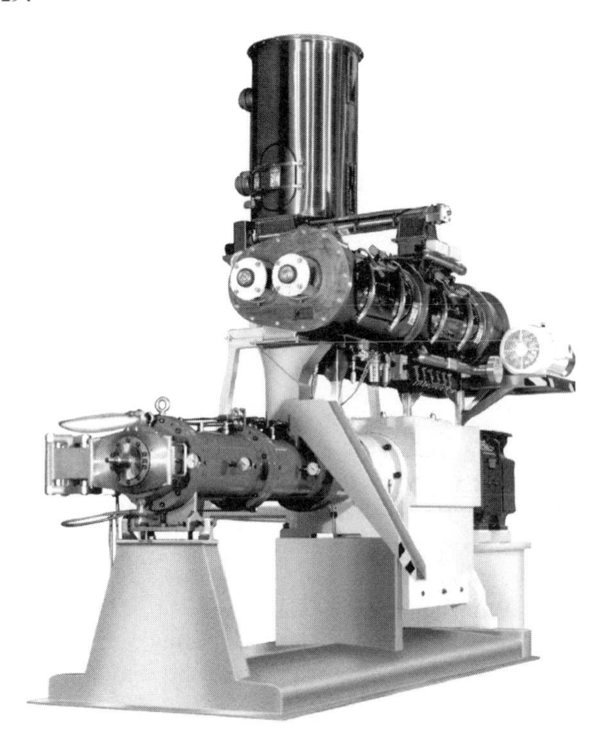

Fig. 1 Wenger® TX Magnum Extruder. (Courtesy of Wenger®, Sabetha, Kansas.)

but also have elastic behavior like elastic solids (e.g., rubber bands).[7] Viscoelastic fluids flow under a pressure gradient, but have memory effects and attempt to recoil after the pressure gradient is removed. The elastic nature of these fluids presents itself in extrusion as die swell as the product exits the die. For example, pasta products will have a greater diameter than the die opening. Die swell can also occur in direct expanded products.[8] Viscoelastic fluids also are known to produce vortices in the entrance region of extruder dies.[9] This complicates the on-line measurement of rheological properties. High molecular weight polymers also tend to be shear thinning, meaning that the viscosity of the fluid decreases as the shear rate increases. This behavior is characterized by the power law model, which is the model that has been predominantly used in on-line extruder rheology studies:

$$\tau = m\gamma^n \tag{1}$$

where τ is the shear stress, γ is the shear rate, n is the flow behavior index, and m is the consistency coefficient. This model has also been applied for elongational flow in the entrance region of the die.[10]

The power law model has also been modified to account for changes in temperature, moisture content, and processing history.[11,12] The assumption is made that changes in these parameters only affect the consistency coefficient, m. A correction factor, usually in the form of an Arrhenius relationship, is added to the power law. For example, the Arrhenius correction to consistency coefficient, m, for temperature would be:

$$m = m_0 \exp\left[\frac{E_a}{R}\left(\frac{1}{T} - \frac{1}{T_0}\right)\right] \tag{2}$$

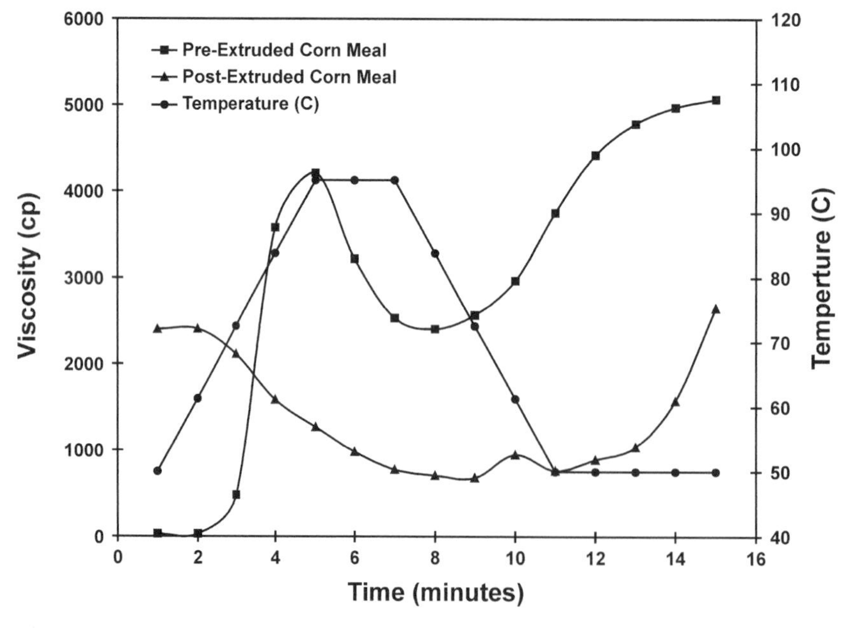

Fig. 2 RVA graphs for pre- and postextruded corn meal. (Data courtesy of the University of Illinois College of ACES Office of Research, Urbana, IL.)

where E_a is the activation energy, R is the universal gas constant, T is the absolute temperature, and T_0 is the reference temperature at which m_0 is determined.

OFF-LINE RHEOLOGICAL CHARACTERIZATION METHODS

The off-line rheological measurements have been used commonly for the characterization of extruded materials. Even though these methods can easily be performed under controlled conditions with sensitive instruments, they fail to represent the shear rates and moisture conditions achieved during a typical extrusion operation.[13] However, they still provide valuable information for the rheological characterization of the finished product as a function of processing conditions during process. Two common instruments used off-line are the Viscoamylograph (also the RVA) and capillary rheometer. The Viscoamylograph and RVA measure and chart the viscosity changes of starch–water dispersions upon exposure to a preset heating period followed by a cooling period. The typical response for uncooked starchy biomaterials is a low viscosity while uncooked starch particles are in suspension, and a rapid increase in viscosity at the gelatinization temperature (Fig. 2). After a peak is reached, the viscosity reduces due to breakdown of the starch gel matrix, and then increases as the temperature is cooled. Postextruded starchy materials have no peak if they are fully cooked during extrusion.

The moisture contents used for Viscoamylograph and RVA studies are much greater than the normal operating range for food extrusion. Therefore, these off-line method results are difficult to interpret for in-process concerns. As the a_w is much lower in-process, the gelatinization temperature of the starch is higher. Also, the Viscoamylograph and RVA units are operated only at atmospheric pressure, and therefore are limited in temperature range by the boiling point of water. Food extruders are operated well above atmospheric pressure (25 atm would not be uncommon), and therefore can develop temperatures well above 100°C, where the denaturation of some proteins becomes important. For example, textured soy protein requires temperatures in excess of 150°C[5] to form a gel under limited moisture conditions.

Capillary rheometry has been the most widely used off-line method to study rheological characteristics of extrudates. In this method, the test material is forced through a small capillary die equipped with a piston or plunger to produce back-pressure. Pressure drop across the capillary (ΔP), volumetric flow rate (Q), and characteristic geometry of the capillary (length, L, and radius, R) are used to calculate the shear stress and shear rate at the wall after end-effect corrections are made.[13]

Studies by Harper, Rhodes, and Wanniger[14]; Harper[15]; and Mackey et al.,[11] are some of the examples on the food dough rheology measured by capillary viscometers.

Vergnes and Villemaire[12] used a different type of off-line viscometer called the Rheoplast to study rheological behavior of molten maize starch. The viscometer consists of a shearing chamber, configured like a Couette rotational viscometer, and a capillary die. Test material is sheared and heated in the shearing chamber until melted, and the molten material is then forced through the capillary die where the melt viscosity is determined from the pressure drop across the die. This operational principle, therefore, resembles an extruder.

ON-LINE RHEOMETERS—SLIT AND CAPILLARY DIES

To determine the power law model coefficients using on-line rheometry, the relationship between pressure drop along a given length and the flow rate is required under several different flow conditions. The pressure drop is related to the shear stress that is achieved under the flow condition and the flow rate combined with the rheometer geometry defines the shear rate. For slit dies, the equation that defines this relationship using the power law is (derived from Steffe[16]):

$$\Delta P = \left[\frac{2mL}{h}\left(\frac{2(2+\frac{1}{n})}{wh^2}\right)^n\right]Q^n \tag{3}$$

where ΔP is the pressure drop over length L, Q is the flow rate, w is the width of the slit die, h is the height of the slit die, and m and n are the power law coefficients.

The slit die rheometer is used to measure the pressure drop over the length L at a known flow rate. As Eq. 3 contains two unknown parameters (m and n), the pressure drop/flow rate relationship is commonly measured at three or more conditions to determine m and n by regression. The regression analysis is performed by plotting $\ln(\Delta P)$ vs. $\ln(Q)$, with the slope resulting in the flow behavior index, n, and the intercept being used to determine consistency coefficient, m.

Rheological properties for similar agricultural materials as reported in the literature vary widely. Differences may be attributable to the different extruder systems and conditions and also to rheometer designs and measurement procedures. One of the main difficulties with on-line measurements is maintaining consistent fluid properties over the three or more flow conditions needed to determine flow behavior index and consistency coefficient by regression. This has led to widely varying values for flow behavior index for the same agricultural materials, and even to negative values in some cases. Clearly, the influence of extrusion processing on rheological state is profound. Chemical analyses have shown a relationship between molecular weight distribution of starch and extrusion processing parameters. Attempts have been made with some success to alleviate this problem by using a flow diversion valve to alter the flow conditions within the on-line rheometer, thus allowing the extruder to maintain screw speed and feed rate. However, the influence of die restriction on rheological properties is also a concern.

ALTERNATIVE ON-LINE RHEOMETRIC METHODS

The difficulty of determining power law rheological properties using different extruder conditions can be overcome by using methods to measure the properties at a single extruder condition. This is accomplished by recognizing that the flow behavior index uniquely defines the velocity profile within the rheometer (Fig. 3). Possibilities include using nuclear magnetic resonance technology to measure the velocity profile within the die,[17] which will determine n for any given condition, and then using pressure drop analysis to determine the flow behavior index. Another approach would be to bifurcate the flow at the die into two channels having different shear rates. The agricultural material flowing through each channel would have seen identical extrusion environment. Drozdek and Faller[18] used this principle to determine flow behavior index from the ratio of flow rates emanating from a bifurcated capillary die. Also, Della Valle[19] used a split stream, slit die rheometer to measure power law properties, achieving results with good consistency.

GLASS TRANSITION CHARACTERIZATION

An important consideration in the rheological analysis of a fluid within an extruder is its glass transition temperature (T_g). Below this temperature, amorphous materials are glassy, hard, and brittle. Near T_g but above, amorphous materials are rubbery or leathery, but still are not fluid. Above a characteristic melting temperature (T_m), these materials behave like fluids. Techniques that are used to measure T_g and T_m are differential scanning calorimetry (DSC), dynamic mechanical thermal analysis (DMTA), and thermal mechanical analysis (TMA).

An instrument that measures T_g and T_m specifically for extrusion applications is a phase transition analyzer (PTA).[20] In contrast to the traditional capillary rheometers, the PTA consists of two sealed chambers separated by an interchangeable capillary die. Product

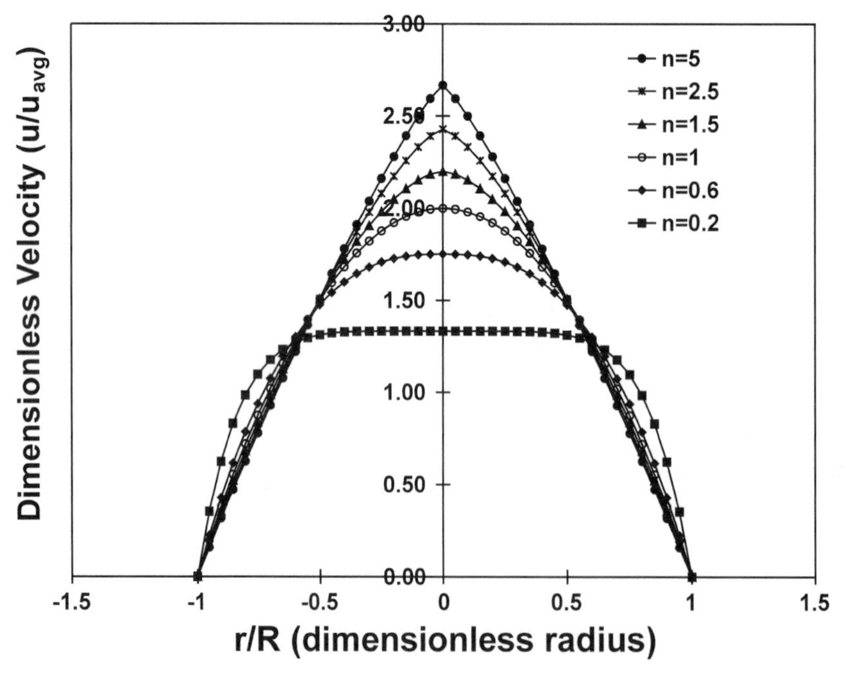

Fig. 3 Dimensionless velocity profile within a capillary extruder die as a function of power law flow behavior index (n). R is the capillary radius.

extrudes from one closed chamber into the other, preventing moisture loss and thus allowing the test to be conducted at temperatures higher than 100°C for samples containing water. Load cells measure the applied pressure while a linear displacement transducer measures the compaction, deformation, and flow relative to the initial sample height. T_g and T_m are determined from the temperatures at which compaction and flow, respectively, occur as measured by the displacement transducers.

REFERENCES

1. Kim, J.C. Effect of Some Extrusion Parameters on the Solubility and Viscograms of Extruded Wheat Flour. In *Thermal Processing and Quality of Foods*; Zeuthen, P., Cheftel, J.C., Eriksson, C., Jul, M., Leniger, H., Linko, P., Varela, G., Vos, G., Eds.; Elsevier Applied Science Publishers: London, 1984; 251–255.
2. Whalen, P.J.; Bason, M.L.; Booth, R.I.; Walker, C.E.; Williams, P.J. Measurement of Extrusion Effects by Viscosity Profile Using the Rapid Visco Analyser. Cereal Foods World **1997**, *42* (6), 469–475.
3. Chiang, B.Y.; Johnson, J.A. Gelatinization of Starch in Extruded Products. Cereal Chem. **1977**, *54*, 436–443.
4. Gomez, M.H.; Aguilera, J.M. Changes in Starch Fraction During Extrusion Cooking of Corn. J. Food Sci. **1983**, *48*, 378–381.
5. Kitabatake, N.; Doi, E. Denaturation and Texturization of Food Protein by Extrusion Cooking. In *Food Extrusion Science and Technology*; Kokini, J.L., Ho, C.T., Karwe, M.V., Eds.; Marcel Dekker, Inc.: New York, 1992; 361–371.
6. Camire, M.E.; Camire, A.; Krumhar, K. Chemical and Nutritional Changes in Foods During Extrusion. Crit. Rev. Food Sci. Nutr. **1990**, *29* (1), 35–57.
7. Murakami, E.G.; Kim, M.H.; Okos, M.R. Viscoelastic Properties of Extrudates, Grains and Seeds. In *Viscoelastic Properties of Foods*; Rao, M.A., Steffe, J.F., Eds.; Elsevier Applied Science: London, 1992; 103–155.
8. Faller, J.F.; Huff, H.E.; Hsieh, F. Evaluation of Die Swell and Volumetric Expansion in Corn Meal Extrudates. J. Food Proc. Eng. **1995**, *18*, 287–306.
9. Bird, R.B.; Armstrong, R.C.; Hassager, O. *Dynamics of Polymeric Liquids*, 2nd Ed.; John Wiley and Sons: New York, 1987.
10. Padmanabhan, M.; Bhattacharya, M. Planar Extensional Viscosity of Corn Meal Dough. J. Food Eng. **1993**, *18*, 389–411.
11. Mackey, K.L.; Ofoli, R.Y.; Morgan, R.G.; Steffe, J.F. Rheological Modeling of Potato Flour During Extrusion Cooking. J. Food Process. Eng. **1989**, *12*, 1–11.
12. Vergnes, B.; Villemaire, J.P. Rheological Behaviour of Low Moisture Molten Maize Starch. Rheol. Acta **1987**, *26*, 570–576.
13. Davidson, V.J. The Rheology of Starch-Based Materials in Extrusion Processes. In *Food Extrusion Science and Technology*; Kokini, J.L., Ho, C., Karwe, M.V., Eds.; Marcel Dekker: New York, 1992; 263–275.
14. Harper, J.M.; Rhodes, T.P.; Wanniger, L.A. Viscosity Model for Cooked Cereal Doughs. Am. Inst. Chem. Eng. (AIChE) Symp. Ser. **1971**, *67*, 40–43.
15. Harper, J.M. Dough Rheology. In *Extrusion of Foods*; Harper, J.M., Ed.; CRC Press: Boca Raton, FL, 1981; 21–46.
16. Steffe, J.F. *Rheological Methods in Food Process Engineering*, 2nd Ed.; Freeman Press: East Lansing, MI, 1996; 418.
17. McCarthy, K.L.; Kerr, W.L. Rheological Characterization of a Model Suspension During Pipe Flow Using MRI. J. Food Eng. **1998**, *37* (1), 11–23.
18. Drozdek, K.; Faller, J.F. Use of a Dual Orifice Die for On-line Extruder Measurement of Flow Behavior Index in Starchy Foods. Journal of Food Engineering **2002**, *55*, 79–88.
19. Della Valle, G.; Colonna, P.; Patria, A. Influence of Amylose Content on the Viscous Behavior of Low Hydrated Molten Starches. J. Rheol. **1996**, *40*, 347–362.
20. Strahm, B.; Plattner, B.; Huber, G.; Rokey, G. Application of Food Polymer Science and Capillary Rheometry in Evaluating Complex Extruded Products. Cereal Foods World **2000**, *45* (7), 300–302.

E

Extruder Power Requirements

Fu-hung Hsieh
University of Missouri–Columbia, Columbia, Missouri, U.S.A.

INTRODUCTION

Extrusion is a process that combines several unit operations, which include mixing, kneading, shearing, heating, cooling, shaping, and forming. This process also involves compressing and working a material to form a semisolid mass under a variety of controlled conditions and then forcing (or pushing) it to pass through a restricted opening such as a hole or a slot at a predetermined rate.[1] All these operations of mixing, kneading, shearing, heating, cooling, shaping, forming, and pushing require power. While power to the extruder screws for mixing, kneading, shearing, shaping, forming, and pushing is usually delivered by the main drive motor through a gear box, power for heating and cooling is supplied by electric heaters or circulating heating and cooling media such as steam, hot oil, or cold water through the extruder barrel.

EXTRUDER MECHANICAL POWER REQUIREMENTS

Extruder mechanical power consumption is a complex function of the properties of processed material, design of extruders, type of motor drive, and extrusion conditions. For example, once the screws are designed and fixed, the mechanical power required for single-screw extruders and both counter-rotating and corotating, intermeshing extruders is expressed as follows[2,3]:

$$P_m = A\bar{\mu}Nn^2 + \frac{Q^2\mu}{K_f} = A\bar{\mu}Nn^2 + Q\Delta P \quad (1)$$

where P_m is the extruder mechanical power consumption, A a coefficient dependent on the screw design, μ the average viscosity, N the number of filled flights, n the screw speed, Q the volumetric flow rate, K_f the die conductance, and ΔP the net pressure rise in the extruder.

The first term on the right-hand side of Eq. 1 is the mechanical power consumed to shear the feed material in the extruder screw channel and convert the mechanical energy into heat energy. For single-screw extruders, shearing develops between flight tips and barrel. A similar expression for the power consumption by the extruder screw as well as methods for estimating apparent viscosity and the filled length of the screw element for single-screw extruders have been suggested by Levine.[4] For both counter-rotating and corotating, intermeshing extruders, in addition to shearing between the flight tips and the barrel, shearing also develops between 1) the flight tips of one screw and the screw roots of the other screw and 2) the flanks of the flights in the intermeshing region.[3] The second term on the right-hand side of Eq. 1 is the mechanical power consumed to forcibly push semisolid mass through a die. According to Eq. 1, it may be concluded that the two most important factors influencing the extruder mechanical power consumption are the screw speed and the volumetric flow rate or Q.[3]

EXTRUDER POWER MEASUREMENT

The most direct method of measuring mechanical power input, P_m, to the extruder screws involves measuring both torque and speed simultaneously. Torque measurement may be conducted by a torque transducer, which measures the torsion of the drive shaft using strain gauges. Speed may be obtained by a tachometer. Instead of measuring torque and speed to calculate mechanical power input, another approach is to measure electrical power input to the extruder's drive motor such as amperage or wattage using ammeters or wattmeters.[5] However, motor efficiency may vary with load, and both transmission losses and electrical losses in the drive motor should also be accounted for in this method. These may be obtained either from the data supplied by the manufacturer or from direct calibration against a mechanical break mounted in place of the screws. For AC induction motors and DC shunt motors, a single efficiency curve is sufficient. For variable-speed AC motors, the efficiency is also dependent on the nominal speed setting.[6]

Barrel heater power input can be determined by measuring voltage and current to the heaters or a wattmeter with a power integration function. Another simple and easy method is to use a multichannel event

Encyclopedia of Agricultural, Food, and Biological Engineering
DOI: 10.1081/E-EAFE 120007123
Copyright © 2003 by Marcel Dekker, Inc. All rights reserved.

recorder to record the ON and OFF periods of each heater and give an average percentage ON time for each zone. This should be taken over 20 min–30 min to allow for fluctuations due to interaction between various zones. This percentage ON time is then multiplied by the rated wattage of each barrel heater to provide the heater power input. If the extruder barrel is heated by steam, the heating power input can be estimated by the steam consumption rate. For most low-moisture food extrusion applications, barrel heater power input, in general, is not significant when compared with mechanical power input.[6,7]

ENERGY BALANCE

In an extrusion system, there are two major energy input sources: the drive motor and the barrel heaters. The mechanical energy from the drive motor, which is mainly dissipated in the form of heat, plus the energy from barrel heating elements cause temperature, chemical, and perhaps phase (latent heat) changes in the food product.[8] The total energy consumed by the system consists of the energy used for driving various loads, energy for heating the barrel, and energy losses within the system. The other areas of energy dissipation and losses, except the work done on feed material, include:

1. Magnetic core, heat, and friction losses from the drive motor and silicon-controlled rectifier (SCR);
2. Gear-box losses;
3. Barrel-cooling losses;
4. Convective and radiative heat losses to the surroundings from hot barrel and die, conduction loss to the barrel and die supports and to the feed pocket; and
5. Energy to drive feeder, pumps, cooling fans, and instrument panel.

An extruder is a thermodynamic unit. Under a steady-state operation, all energy that is introduced into the system must also come out again or must be accountable. In such an energy balance system, several different terms can be distinguished. The first is the mechanical energy added by the rotation of the screws and the second is the heat transferred by the barrel heaters. Furthermore, the mechanical energy is partly used to increase the pressure of the material and partly converted into heat by viscous dissipation. The thermal energy (generated by viscous dissipation and transferred through the barrel wall) results in an increase of the temperature and chemical reactions such as starch gelatinization and protein denaturation.[8,9] The last is the energy loss in the system. Thus,

$$P_{in} = P_{out} \tag{2}$$

$$P_{in} = P_m + P_h \tag{3}$$

$$P_{out} = P_{th} + P_{loss} \tag{4}$$

where P_{in} and P_{out} are power input and output, P_m and P_h the mechanical power applied to the screw shafts and heat power from the barrel heaters, P_{th} and P_{loss} the theoretical power required in an extruder and power losses, respectively.

The theoretical power required in an extruder, P_{th}, can be expressed as:

$$P_{th} = Q\int_{T_i}^{T_o} C_p \, dT + \frac{Q\Delta P}{\rho} + Q\Delta H \tag{5}$$

where Q is the mass flow rate, T_i and T_o the feed and product temperatures, respectively, C_p the heat capacity of the feed, ΔP the discharge pressure, ρ the density of melted feed, and ΔH the enthalpy changes associated with starch gelatinization and protein denaturation per unit mass of extrudate.

In Eq. 5, the first term on the right-hand side represents the power required to heat the feed material to a certain temperature. The second term is the power required for conveying and pushing the material through the die. The third term is the enthalpy change associated with starch gelatinization and protein denaturation. This equation does not include melting of fat and browning occurring within the ingredient because their contribution to overall energy needs is usually very small.[10] In addition, the power requirements for the external auxiliary devices, such as the feeder, water pump, and control panel are not considered in this equation.

The heat capacity, C_p (kJ kg^{-1}K^{-1}), and enthalpy changes associated with starch gelatinization and protein denaturation, ΔH (kJ kg^{-1}), may be estimated by the following two equations[10,11]:

$$C_p = 1.424m_c + 1.549m_p + 1.675m_f + 0.837m_a + 4.187m_m \tag{6}$$

$$\Delta H = 14m_c + 95m_p \tag{7}$$

where m is the mass fraction, and the subscripts c, p, f, a, m denote carbohydrate, protein, fat, ash, and moisture, respectively, in feed.

Once power inputs, including the mechanical power, P_m, the barrel heater power, P_h, and the theoretical power required in the extruder, P_{th}, are known, all power losses in the extruder, P_{loss}, can then be easily obtained using the energy balance according to Eqs. 2–4. Using a corotating twin-screw extruder, Karwe and Godavarti[12] measured the surface heat flux and found that the heat loss from the barrel bottom surface was the highest, followed by top surface and side surface for the same surface temperature. In addition, total heat loss from the surface varied from 2% to 19% of the total heat input. Levine[13] suggested that the heat transfer coefficient, h, of the extruder barrel surface may be assumed to be $10\,W\,m^{-2}K^{-1}$–$15\,W\,m^{-2}K^{-1}$.

In summary, extrusion process comprises many unit operations that consume power. The total power requirements for extrusion, including the electrical power input to the extruder's drive motor and the electrical power input to the extruder's barrel heaters, can be measured by using ammeters or wattmeters. If the extruder barrel is heated by steam, the power input can be determined by the steam consumption rate. All power losses in the extrusion process can be estimated using the energy balance approach.

REFERENCES

1. Dziezak, J.D. Single- and Twin-Screw Extruders in Food Processing. Food Technol. **1989**, *43* (4), 164–174.
2. Middleman, S. *Fundamentals of Polymer Processing*; McGraw-Hill, Inc.: New York, 1977; 525.
3. Martelli, F.G. *Twin-Screw Extruders*; Van Nostrand Reinhold: New York, 1983; 137.
4. Levine, L. Extruder Screw Performance. Part V. Cereal Foods World **2001**, *46* (4), 169.
5. Harper, J.M. Instrumentation for Extrusion Processes. In *Extrusion Cooking*; Mercier, C., Linko, P., Harper, J.M., Eds.; Am. Assoc. Cereal Chem.: St. Paul, 1989; 39–55.
6. Stevens, M.J. *Extruder Principles and Operation*; Elsevier Applied Sci.: New York, 1985; 339.
7. Rauwendall, C. *Polymer Extrusion*; Hanser Publishers: New York, 1986; 567.
8. Harper, J.M. *Extrusion of Foods*; CRC Press: Boca Raton, 1981; Vol. I, 212.
9. Harper, J.M. *Extrusion of Foods*; CRC Press: Boca Raton, 1981; Vol. II, 174.
10. Caldwell, E.F.; Fast, R.B.; Ievolella, J.; Lauhoff, C.; Levine, H.; Miller, R.C.; Slade, L.; Strahm, B.S.; Whalen, P.J. Unit Operation and Equipment I. Blending and Cooking. In *Breakfast Cereals—and How They Are Made*; Fast, R.B., Caldwell, E.F., Eds.; Am. Assoc. Cereal Chem.: St. Paul, 2000; 55–131.
11. Singh, R.P.; Heldman, D.R. *Introduction to Food Engineering*; Academic Press: New York, 1984; 499.
12. Karwe, M.V.; Godavarti, S. Accurate Measurement of Extrudate Temperature and Heat Loss on a Twin-Screw Extruder. J. Food Sci. **1997**, *62* (2), 367–372.
13. Levine, L. More on Extruder Energy Balances. Cereal Foods World **1997**, *42* (9), 772.

Extrusion System Components

Qi Fang
Banner Pharmacaps, Inc., High Point, North Carolina, U.S.A.

Milford A. Hanna
University of Nebraska, Lincoln, Nebraska, U.S.A.

Yubin Lan
Fort Valley State University, Fort Valley, Georgia, U.S.A.

INTRODUCTION

Extrusion refers to the forming of products to the desired shape and size by forcing the material through a die opening under pressure. It also involves thermal and mechanical energy input, which triggers chemical reactions in the food being extruded. Extrusion has been used extensively in the food industry to process a variety of raw materials into foods, including breakfast cereals, snacks, pastas, texturized vegetable protein, flat bread, meat products, and pet foods.

Depending on the product to be processed, a complete extrusion system generally consists of storage bins, dry mix feeders, liquid pumps and meters, a preconditioner, an extruder assembly, a die, and a cutter (Fig. 1).

STORAGE BIN

A storage bin is used for the storage of dry ingredients. It provides a buffer of raw material so that an extruder has a continuous and stable supply of feed ingredients. This bin is usually equipped with rotating blades to prevent bridging.

DRY INGREDIENT FEEDERS

Two types of dry feeders are used to feed extruders: volumetric feeders and gravimetric feeders. A volumetric feeder provides a constant volume of dry ingredients, but cannot guarantee a constant mass flow rate due to changes in feed material density. Gravimetric feeders, on the other hand, control feed flow rate based on the mass delivered and are, therefore, more accurate feeding devices.

Volumetric Feeding Devices

Volumetric feeders deliver dry ingredients on a volume basis. Several types of volumetric feeders are available. The most common ones are the single screw feeders. The volumetric feed rate is proportional to the screw speed. Use of a twin screw instead of single screw improves the feeding accuracy, but with higher manufacturing cost.

Gravimetric Feeding Devices

Gravimetric feeders meter the weight of the dry ingredients, which gives better precision than the volume metering used with the volumetric devices. Gravimetric feeders are used mostly in large-scale extrusion systems. They include weigh-belt and loss-in weight feeders. Both feeders monitor the quantity of dry ingredients, and adjust the feed rate accordingly if any change is detected.

LIQUID FEEDERS

Common liquid raw materials used in extrusion include water, fat, syrup, and some minor ingredients. Metering of liquid ingredients is critical for successful extrusion. Metering of liquid feeds can be achieved either by volume or by mass. Mass flow meters are more accurate than volumetric meters. Devices include the rotameter, differential pressure meter, fluid displacement meter, velocity flow meter, and mass flow meter.

PRECONDITIONER

The most important functions of a preconditioner are moisture adjustment and precooking of the raw materials prior to extrusion. During preconditioning, raw materials are held in a warm, moist environment where they are

Encyclopedia of Agricultural, Food, and Biological Engineering
DOI: 10.1081/E-EAFE 120007125
Copyright © 2003 by Marcel Dekker, Inc. All rights reserved.

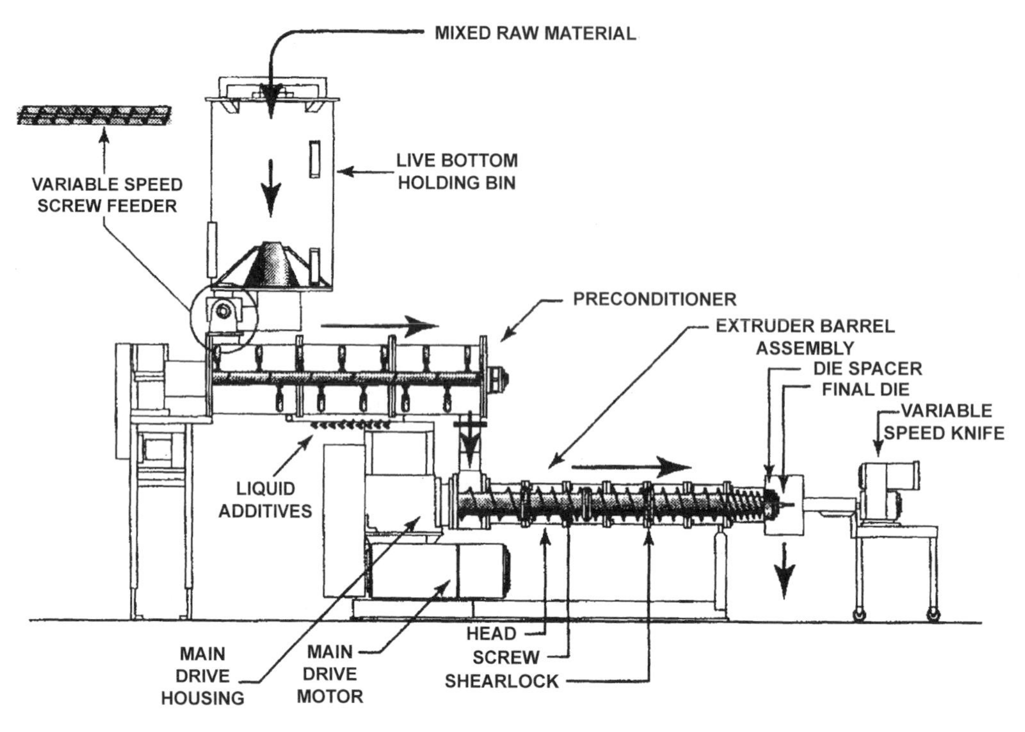

Fig. 1 General view of an extruder system. (Reprinted from Ref. 3.)

mixed for a given time, and then discharged into the extruder. Preconditioning provides several benefits, including improved product quality, reduced extruder wear, increased extruder capacity, and reduced power consumption.

Preconditioners are mounted between the feeding device and the extruder. Earlier preconditioners had a single-shaft design. The shaft, having mixing elements, ran at relatively high speeds, giving a retention time of 30 sec or less. This was insufficient. Most modern preconditioners have a double-shaft design as shown in Fig. 2. The two shafts have different dimensions, and rotate at different speeds, which results in better mixing, and retention times of between 2 and 4 min.[2]

Preconditioners can be operated at both atmospheric pressure and elevated pressure. Elevated pressure preconditioning increases cooking temperature to above 100°C which is sometimes advantageous. However, they are more complex, and cost more to purchase and maintain. During preconditioning, steam and/or water are supplied to increase the temperature and moisture content of the raw materials. Steam is added from the bottom of the preconditioner; hot water (80–90°C) is added from the top through spray nozzles for uniform distribution.

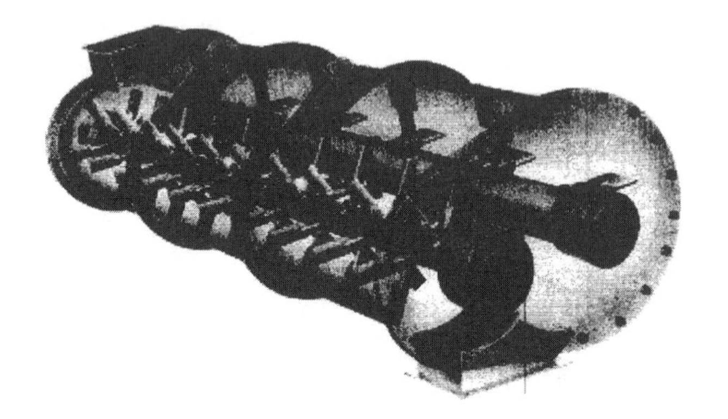

Fig. 2 Double-shaft preconditioner.

EXTRUDER ASSEMBLY

Extruders can be classified by the number of screws used. There are two basic designs: single screw extruders and twin screw extruders. Compared to single screw extruders, twin screw extruders are more versatile in that they can handle raw materials with a wider range of characteristics. For large-scale extruders, the screws are made into segments so that various screw configurations can be achieved to meet the processing requirements of different products. When needed, various screw segments with different processing functions, such as conveying, kneading, mixing, shearing, and compression can be configured to meet the processing requirement for a specific product.

Single Screw Extruder Assembly

A single screw extruder assembly consists of a screw and a barrel. The conventional practice is to rotate the screw within a stationary barrel, but the converse is being explored. Depending on the processing requirements, several types of screw configurations are available as shown (Fig. 3). Different screw configurations provide different shearing conditions and pressure profiles, which are uniquely suited for the production of various products. According to Harper,[11] the extruder barrel for a single screw extruder can be divided into three sections: feeding, compression, and final metering zones, as shown in Fig. 4. Along the length of the screw, compression increases from the feeding zone to the final metering zone. The main function of the feeding zone is to push the incoming raw materials to the subsequent metering zone, forming a continuous flow of mass within the barrel. The screw conveying capability is determined by the screw geometry, i.e., the pitch, flight angle, and flight depth. The flight of a screw is the helical conveying surface of the screw to push the product forward. The pitch of a screw is the axial distance of a full flight circle. The flight angle is defined as the angle of the flight relative to the axis of the screw shaft. Generally, screw feeding zones have relatively deep flights, large pitches, and large flight angles to enhance conveying capacity. As the raw material is conveyed to the metering zone, it starts to melt due to the thermal energy from the barrel wall as well as from mechanical energy dissipation. Within the compression zone, additional mixing can be accomplished by means of special screw elements, such as kneading blocks and interrupted flights. At the end of the compression zone, the material is fully melted, exhibiting a rubbery behavior similar to flour dough. In the final metering zone, the dough is further compressed by the shallow flights and short pitches of the screw. The dough is fully cooked before exiting the extruder die.

In many extruders a heating/cooling medium flows through a jacket surrounding the extruder barrel. Heating is typically accomplished with overheated steam or hot oil. Cooling is achieved with tap water. The insider surface of the barrel can be smooth or grooved to prevent the dough from rotating with the screw. The barrels of many commercial extruders are segmented and individually

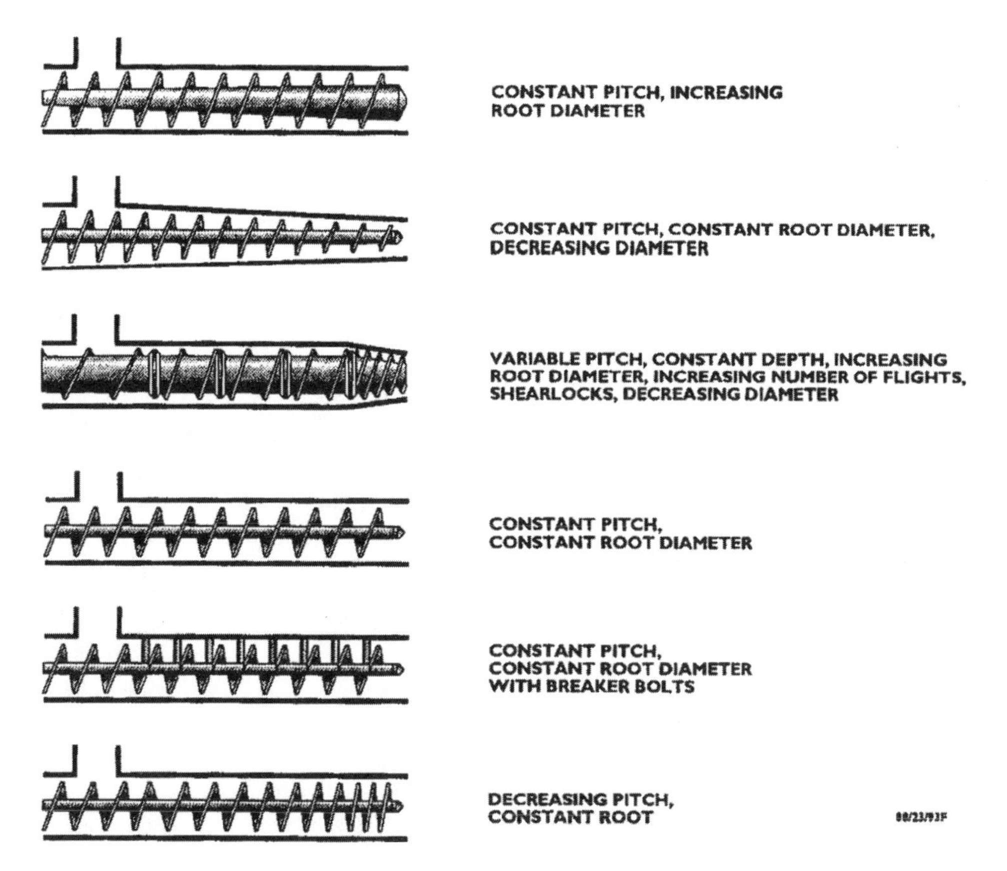

Fig. 3 Common screw configurations for a single screw extruder. (Reprinted from Ref. 3.)

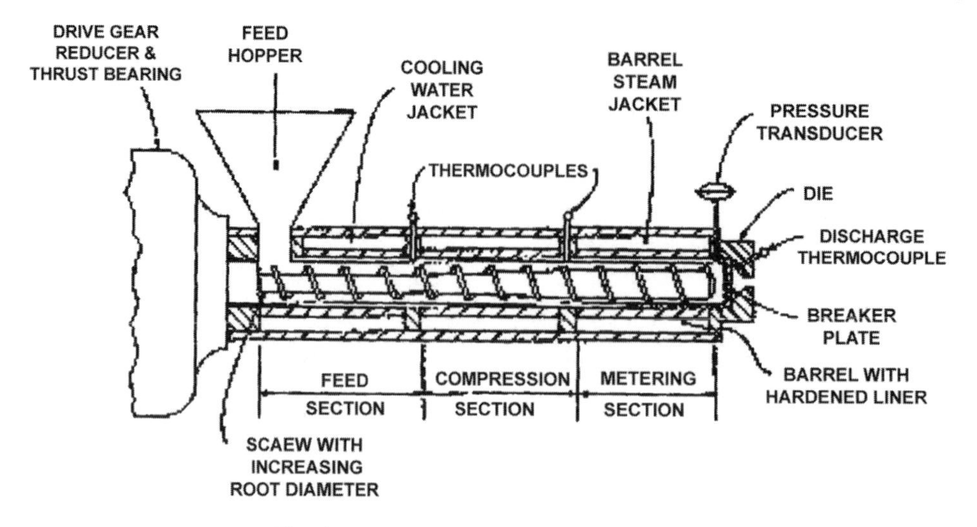

Fig. 4 Extruder barrel zones. (From Ref. 1.)

jacketed so that various temperature profiles can be achieved.

Twin Screw Extruder Assembly

Twin screw extruders are becoming popular in the food industry because they can process a wider variety of materials, including high moisture and sticky materials than the single screw extruders. The versatility of the twin screw extruder makes it suitable for manufacturing a wider range of products. On the basis of the relative positions of the screws and their directions of rotation, twin screw extruders can have four possible screw configurations (Fig. 5), i.e., corotating intermeshing, corotating non-intermeshing, counterrotating intermeshing and counter-rotating nonintermeshing.

The food industry most often uses twin screw extruders with a corotating intermeshing configuration because they are self-cleaning, achieve better mixing, produce moderate shearing forces, and have higher capacities. The barrel of twin screw extruder, like the barrel of the single screw extruder, is divided into three zones: feeding, kneading, and final cooking. These zones have processing functions similar to those of a single screw extruder.

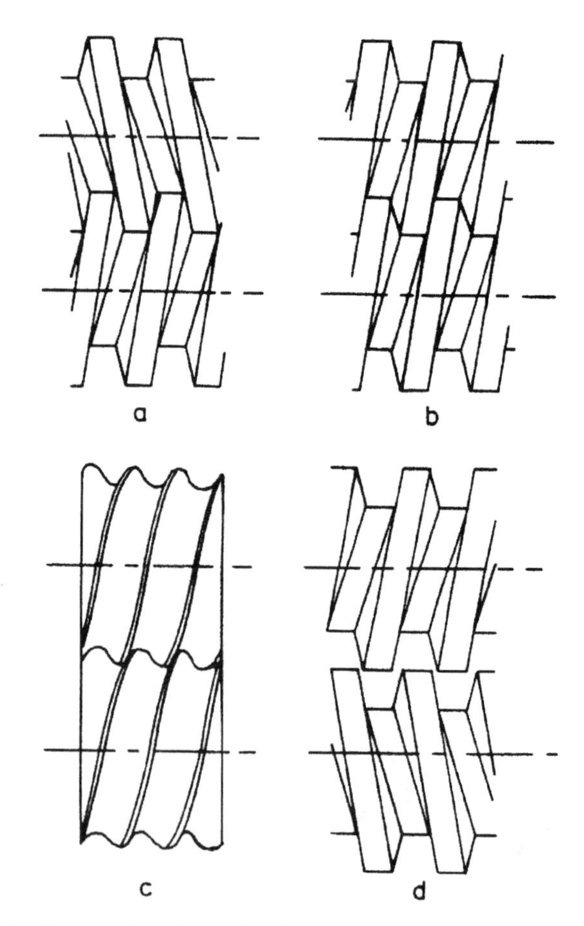

Fig. 5 Classifications of twin screw extruder. (a) Counter-corotating with intermeshing; (b) Corotating with intermeshing; (c) Self-wiping; (d) Nonintermeshing. (From American Association of Cereal Chemists, Inc., St. Paul, MN, 1989.)

VENTING PORTS

Openings can be provided in the barrels of both single screw and twin screw extruders to allow moisture and other volatile gases to be removed from the dough. The venting ports are located in the middle or end sections of the barrel. The removal of moisture from the dough reduces product expansion, which is beneficial for nonexpanded foods and for adjusting the density of aquatic foods that must be sink or suspended beneath the surface when placed in water.

DIE ASSEMBLY

The die plate functions as a restriction and forming device mounted at the end of the barrel. By adjusting the die opening, pressure, and retention time, the dimensions and shape of the final products can be controlled. For small-scale extruders, the die assembly has only one opening. Multiple openings are available for large or commercial scale extruders.

CUTTER

A cutter consists of a group of rotating knives mounted in front of the die plate. Cutters are used to cut the extrudate into finite lengths. The rotational speed of the knives is adjustable. The length of extrudate is determined by the number of knives and the rotational speed of the knives.

DRIVING UNIT

The main function of the driving units is to provide power to rotate the extruder screws. Part of the mechanical energy from the driving unit is converted into heat due to friction within the extruder barrel. Depending on the degree of shearing, friction heating can be significant. The extruder-driving unit usually consists of an electric motor, V-belts, and a gearbox. The V-belts and gearbox achieve speed reduction. The power rating of the electric motor is determined by the size of the extruder and the product to be extruded. Low shearing extruders may require as little as $10\,\mathrm{kWh\,tn}^{-1}$ of throughput, while high shearing extruders may require as much as $120\,\mathrm{kWh\,tn}^{-1}$ of throughput.[3] The gearbox of the single screw extruder requires only one output shaft to drive the screw, which makes the construction of the gearbox simple. In the twin screw extruder, the gearbox has two output shafts, rotating at the same speed, to drive the two screws. The limited space in the radial direction makes the bearing arrangement difficult since the bearings have to support both the radial load and the thrust load from the extruder screws. Therefore, a combination of various types of bearings is usually used in the gearbox.

CONCLUSION

Extrusion systems are complex systems in which the screws, barrel, and the dies are the main components to accomplish the food processing tasks. On the other hand, extruders can combine multiple unit operations (mixing, melting, degassing, cooking, and forming) into a single processing step, which provides enormous benefits in terms of saving overall equipment cost, energy and spaces. The segmented screw and barrel designs allow unlimited number of configurations to satisfy processing requirements of a wide variety of products. Finally with the incorporation of computer automation, extrusion processing can ensure the highest and consistant product quality with minimum human labor cost.

REFERENCES

1. Harper, J.M. The Food Extruder. In *Extrusion of Foods*; Harper, J.M., Ed.; CRC Press, Inc.: Boca Raton, FL, 1981; 8.
2. Hauck, B.W. Preconditioning apparatus for extruder. US Patent 4,752,139, June 21, 1988.
3. Rokey, G.J. Single Screw Extruder. In *Extruders in Food Applications*; Riaz, M.N., Ed.; Technomic Publishing Co.: Lancaster, PA, 2000; 30–31.

Extrusion System Design

Qi Fang
Banner Pharmacaps, Inc., High Point, North Carolina, U.S.A.

Milford A. Hanna
University of Nebraska, Lincoln, Nebraska, U.S.A.

Yubin Lan
Fort Valley State University, Fort Valley, Georgia, U.S.A.

INTRODUCTION

Extrusion technology has been widely applied to the food industry for manufacturing numerous products such as pasta, ready-to-eat cereal, flat bread, pepperoni, and pet food. Extrusion can be used to perform various types of tasks, including mixing, kneading, dispersing, plasticizing, gelatinizing, texturizing, cooking, melting, roasting, caramelizing, sterilizing, drying, crystallizing, reacting, and shaping.[1] Extrusion technology has advantages over other conventional processing methods such as adaptability, productivity, quality, and energy efficiency. Each food product requires a different setup of the extrusion system to achieve the desired degree of processing. A complete extrusion system generally consists of raw material receiving and storage, mixing, preconditioning, extrusion, drying, and packaging (Fig. 1).

RAW MATERIAL RECEIVING AND STORAGE

Raw materials are received usually in bulk or bags. They are first checked for quality control purposes. Then, they are stored in various bins for subsequent use.

MIXING

Blending of various ingredients can be performed either before extrusion or inside the extruder. Mixing prior to extrusion can be done in a mixer by preblending all dry and liquid ingredients together according to a predetermined formulation. In order for moisture to be distributed more evenly, the mixed ingredients are to be held for a certain time in a closed container. Ingredients also can be mixed in the preconditioner of the extruder. Liquid ingredients can be injected through ports located in the preconditioner or in the extruder barrel.

EXTRUSION SYSTEM

Fig. 2 shows a complete extruder commonly used for food processing. It consists mainly of a preconditoner, a main drive unit, a barrel, one or two screws, and dies. Extruders are typically classified into two categories: single-screw extruder and twin-screw extruder. Historically, single-screw extruders have been used extensively in the food industry. But, twin-screw extruders have gained popularity because of their versatility and other advantages over single-screw extruders.

PRECONDITIONER

Preconditioning adjusts the moisture content of the feed materials and can be used to precook the feed materials. The incorporation of a preconditioner in the system results in reduced wear to the extruder, lower overall energy usage, and higher extruder throughput. Preconditioners can be classified into single-shaft and double-shaft designs. Double-shaft preconditioners have gained popularity in recent years for their ability to uniformly mix and effectively transfer heat.

SINGLE-SCREW EXTRUDER

A single-screw extruder has only one screw rotating within the barrel. The screw can be built in a solid piece or assembled of screw segments. The latter allows changes in screw configurations. The rotating action of the screw mixes and transports the feed materials from the feeding zone to the die exit. A forward motion takes place with the assistance of raised flights having a helical profile (Fig. 3). Similar to the screw, the barrel in which the screw rotates can be manufactured into a single piece (for smaller laboratory extruders) or segmented sections being put together (for commercial production). A single-screw extruder can be modeled as a drag flow pump. The material within the barrel is transported in the form of

Encyclopedia of Agricultural, Food, and Biological Engineering
DOI: 10.1081/E-EAFE 120007122
Copyright © 2003 by Marcel Dekker, Inc. All rights reserved.

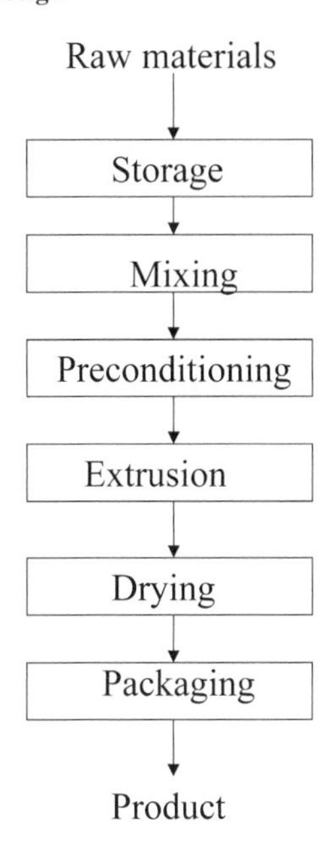

Fig. 1 Components of an extruder system.

a plug. Mixing between the inner layer next to the screw and the outside layer next to the barrel is limited. By adding interrupted flights and kneading blocks, the mixing action can be improved but with a reduced throughput.

Depending on the extent of shearing action, single-screw extruders can perform low, intermediate, and high shear extrusions. Low shear extruders with smooth barrels, deep flights, and low screw speeds are used for the production of pastas, meats, cereals, and fried snacks. Moderate shear extruders with grooved barrels and high compression ratio screws are used for cooking extrusions, such as bacteria pasteurization, enzyme inactivation, protein denaturation, and starch gelatinization.[2] The high shear extruders with grooved barrels, variable pitch and flight depth screws, and interrupted screw flights are used for high temperature/short time (HTST) cooking of ready-to-eat cereals, snacks, candies, expanded pet foods, soup mixes, and texturized soy proteins.[3]

TWIN-SCREW EXTRUDER

According to the relative positions of the two screws, there are four types of screw configurations available: corotating intermeshing, corotating nonintermeshing, counter-rotating intermeshing, and counter-rotating nonintermeshing. For food processing, corotating, intermeshing twin-screw extruders are most widely used. This type of twin-screw extruder works as a positive

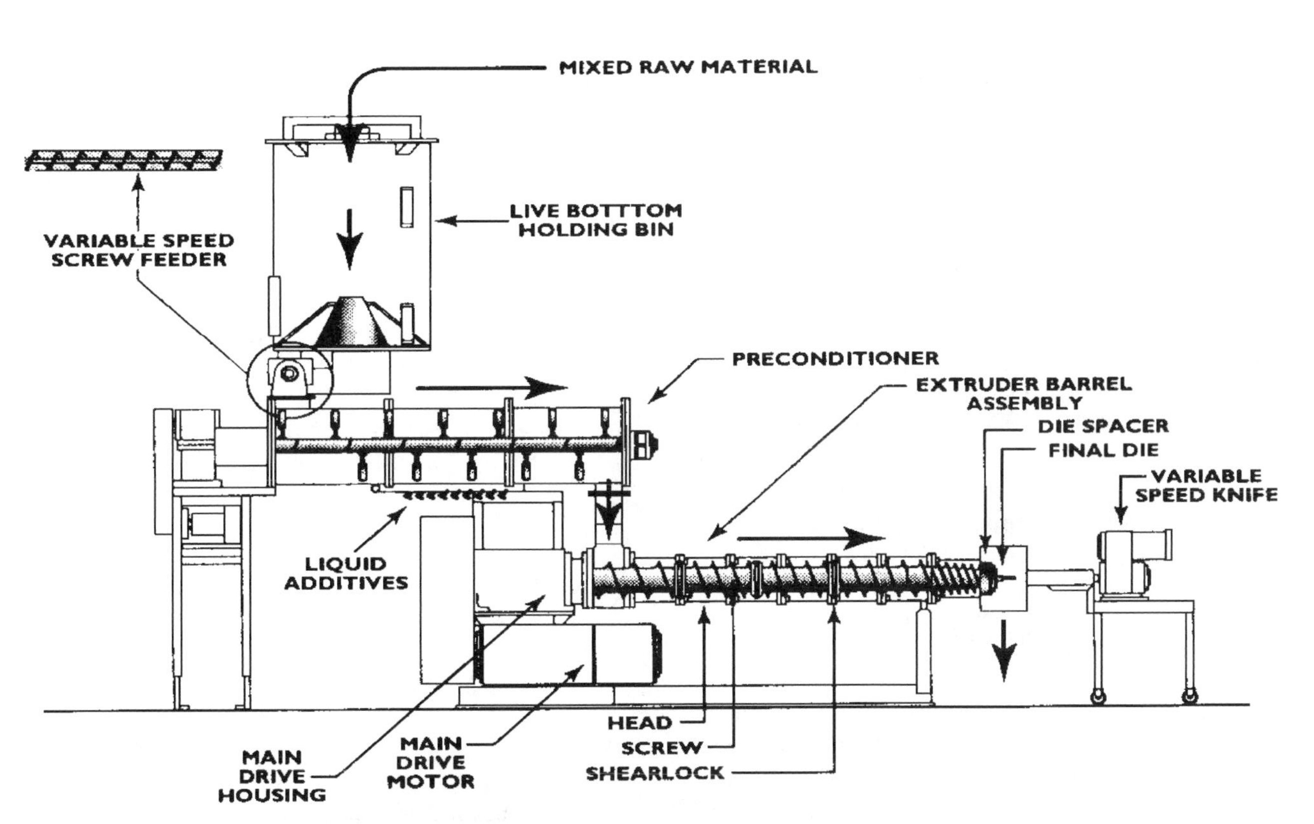

Fig. 2 A single-screw extruder. (From Ref. [2].)

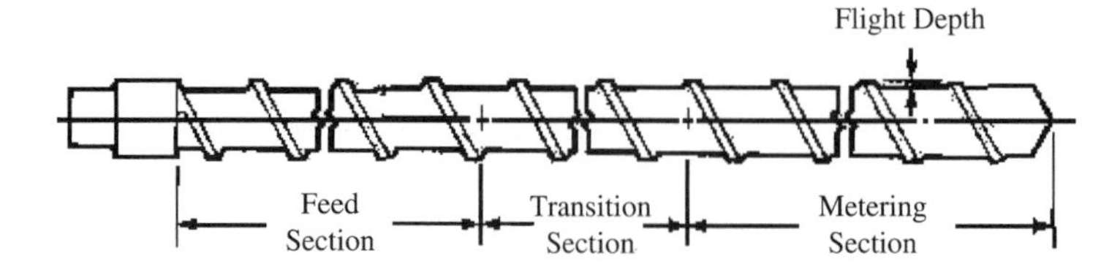

Fig. 3 A typical screw configuration.

displacement pump in transporting the materials inside the extruder barrel that allows the twin-screw extruder to convey materials with very high or very low viscosities,[4] which means a wider range of moisture contents is allowed in the raw materials. Other distinct advantages of the twin-screw extruders over the single-screw extruders are the excellent material mixing and heat transfer, which take place mostly at the intermeshing region of the two screws. Good mixing and heat transfer are critical for the manufacturing of certain products. Except small-scale machines, most twin-screw extruders have segmented screws and barrels for flexible screw configurations and easy manufacturing. For example, the inclusion of kneading blocks provides excellent mixing action for processing of materials containing ingredients having different characteristics.

FACTORS AFFECTING EXTRUSION

Feed Moisture Content

Moisture in the raw materials has multiple functions during extrusion. First, moisture functions as a plasticizer to soften the dough and achieve certain rheological characteristics. Secondly, moisture is needed to achieve starch gelatinization and protein denaturation. Thirdly, moisture acts as a lubricant within the barrel. Moisture reduces friction, lowering mechanical power input. Finally, moisture is used as a blowing agent in expanded food extrusions. The effect of moisture on expansion is twofold (Fig. 4). At lower moisture contents, expansion increases as the moisture increases. There is a maximum expansion at optimum moisture content. Above that, expansion actually decreases as more moisture is added.

Barrel Temperature

Temperature is an indication of the thermal energy input to the system. For the raw materials to be fully cooked, heat and moisture are both needed. Generally, increasing temperature has a positive effect on the characteristics of the extruded products, such as degree of starch gelatinization and extrudate expansion. However, product color is negatively affected by barrel temperature. Higher temperature results in darker product color.

Screw Rotational Speed

Screw rotational speed determines the shearing action acting upon the dough, which in turn affects the heat dissipation from the mechanical energy input. Higher screw speeds introduce greater shear action upon the extrudate. Screw speed also influences the residence time of the dough. Slower screw speed results in a longer residence time, which is sometimes critical for the completion of texture formation and chemical reactions.

Screw Configuration

Screw configuration refers to the selection of screw elements to achieve the desired processing condition. As most of the commercial extruders use segmented screws,

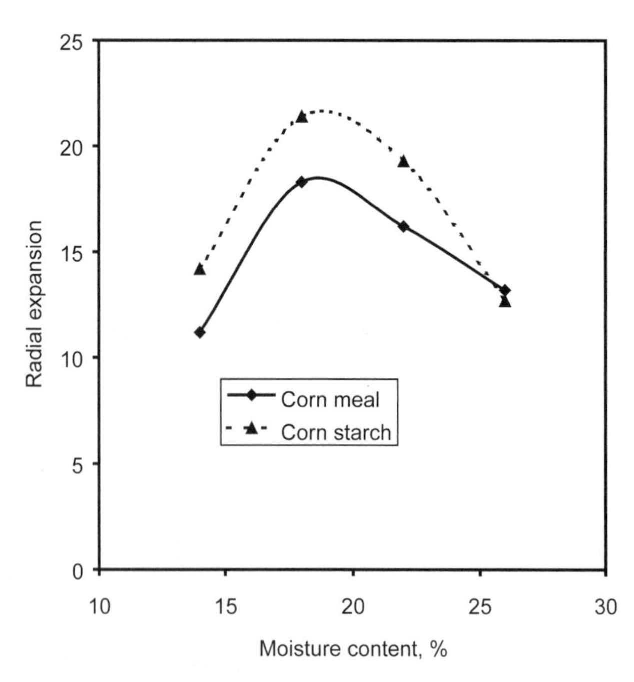

Fig. 4 Relationship between moisture content and extrudate expansion.

E

unlimited numbers of screw configurations can be achieved. The setup of screw configuration depends mostly on past experiences. For example, one can use double-flighted elements to speed up movement of materials, kneading blocks to increase mixing action on the dough, and reverse pitched elements to increase retention time.

Feed Rate

Feed rate has an overall effect on extrusion. Feed rate affects the residence time, torque load, barrel pressure, and dough temperature. Generally, an extruder is operated at the maximum feed rate to achieve maximum output. However, starve feeding is sometimes employed to increase the residence time and achieve the desired extrudate characteristics.

DRYING OPERATION

Some freshly extruded products contain too much moisture for safe storage. The excessive moisture needs to be removed by drying operation. Drying operations for semimoist extruded products are carried out mostly on fluidized bed dryers. This type of dryer is operated continuously with high heat and mass transfer efficiencies and self-cleaning.

EXTRUSION MODELING AND AUTOMATIC CONTROL

As extrusion involves multiple variables as mentioned above, it is difficult for an extruder operator to manually adjust all the variables to the optimum levels. Therefore, it is desirable to apply automatic controls to an extrusion system. Automation means that the operating parameters are controlled and regulated to operate at the preset values by a control system.[5] Extrusion automation improves product quality, maximizes extruder throughput, and saves operating costs.

In order to apply an automation technique to an extrusion system, a basic understanding of the extrusion process is essential. In order to eliminate the dependency of a process model on a special machine, secondary or process variables are developed, including dough viscosity, residence time, and energy input. Product characteristics are directly correlated with the above mentioned process variables and not the operating variables. The developed models are not machine-dependent. With the development of computer technology, new methods of modeling, neural network and fuzzy logic have been employed to model the extrusion process. Compared to the traditional mathematical and statistical models, neural network and fuzzy logic models have the advantages of handling multiple variables, predicting more accurately, and being implemented more easily.

REFERENCES

1. Meuser, F.; Wiedmann, W. Extrusion Plant Design. In *Extrusion Cooking*; Mercier, C., Linco, P., Harper, J., Eds.; American Association of Cereal Chemists: St. Paul, MN, 1989.
2. Riaz, M.N. Introduction to Extruders and Their Principles. In *Extruders in Food Applications*; Riaz, M.N., Ed.; Technomic Publishing: Lancaster, PA, 2000.
3. Linko, P.; Colonna, P.; Mercier, C. High Temperature Short Time Extrusion. In *Advances in Cereal Science and Technology*; Pomeranz, Y., Ed.; American Association of Cereal Chemists: St. Paul, MN, 1981; 145–235.
4. Starer, M.S. When to Consider a Twin Screw Extruder for Making Pet Food and the Use of High Speed. In *Advances in Extrusion Technology*; Chang, Y.K., Wang, S.S., Eds.; Technomic Publishing Co., Inc.: Lancaster, PA, 1998.
5. Millauer, C. The Monitoring and Control of an Extrusion System. In *Extrusion Technology for the Food Industry*; O'Connor, C., Ed.; Elsevier Applied Science Pub.: London, 1987; 54–70.

Extrusion System Residence Time Distribution

Leon Levine
Leon Levine Associates, Albuquerque, New Mexico, U.S.A.

Robert C. Miller
Consulting Engineer, Auburn, New York, U.S.A.

INTRODUCTION

Flow in screw extruders is created by shear between the barrel and screw surfaces. Laminar flow in shear creates areas of differing velocity with a resulting spectrum of retention time within the process. This is the residence time distribution (RTD). Although inherently deterministic, prediction of the RTD can be complicated by the presence of various flow components from drag, pressure, leakage, and calendering, and by variations in rheological properties at various points in the equipment. Descriptions of these various flow components may be found in a number of texts.[1-4]

For several reasons, there is significant interest in the RTDs of extrusion systems. A number of different physical and chemical phenomena take place within the extruder including: gelatinization of starch; melting of starches and proteins; hydration of particles; chemical destruction of materials such as vitamins, micro-organisms, and antinutritional factors; depolymerization of biopolymers; and a number of browning reactions. All these have rate equations (kinetics) associated with them. If different elements of the extrudate have different retention times, the degree of conversion associated with the chemico-physical changes will vary. There is also interest in auxiliary equipment in extrusion systems, particularly preconditioners, in which some of the same physical and chemical reactions occur, as well as simple heat transfer, a time dependent phenomenon, controlled chiefly by conduction.

RTD, or change in the RTD, also provides us with information about the hidden flow components in the extruder. For example, an increase in leakage flow, because of screw or barrel wear, is often exposed by a change in the RTD. Furthermore, many, if not most, extruders operate in a starved condition: the extruder screw is incompletely filled. Since heat transfer and viscous energy dissipation, important determinants of product quality, take place primarily in regions where the screw is filled, the measurement of RTD tells us the degree of fill in the extruder. This is an important factor in establishing screw configurations.

MEASUREMENT AND ANALYSIS OF RTDs

RTDs are measured via introduction of a tracer material into the extruder feed and analysis of tracer concentration in the exit stream. Several approaches have been used: introduction of a pulse of tracer, making a step change in the concentration of the tracer in the feed, or introduction of a sinusoidal variation of tracer concentration in the feed. The pulse method is the easiest and most common. Sinusoidal variation is, by far, the least common and will not be mentioned again. Tracer selection is important. Many types of tracers have been used, such as colorants, various salts, proteins, suspended particles, and radioactive tracers. The characteristics of a suitable tracer are:

1. Concentration in the extrudate should be easily and accurately measured, preferably at infinitesimal levels.
2. Introduction of a tracer should not alter extrudate rheology. Since flow patterns in the extruder are a consequence of the laminar flow fields, changes in rheology will change them and the RTD. Since most additives affect rheology, one reason that tracers should be detectable at infinitesimal levels is that these effects can be minimized. Examples of this kind of distortion may be found in the literature.[5]
3. The tracer should be completely miscible. If not, the tracer may flow in a different pattern and not reflect the bulk product RTD accurately. When measuring RTDs where the product is granular, as in the feed zone or in a preconditioner, this can cause a particular problem because flow properties depend on particle size.

Intense colorants, which are inert, detectable at very low levels, and easily analyzed, seem to be a natural choice. However, care must be taken to analyze the actual concentration of the colorant and not simply the color intensity—an error (usually small) in some published studies. The problem is that concentration and color intensity are not proportional. The relationship between color and concentration is actually nonlinear (as shown in Fig. 1).[6] To avoid this error, one should have a calibration curve for concentration as a function of color.

Encyclopedia of Agricultural, Food, and Biological Engineering
DOI: 10.1081/E-EAFE 120007124
Copyright © 2003 by Marcel Dekker, Inc. All rights reserved.

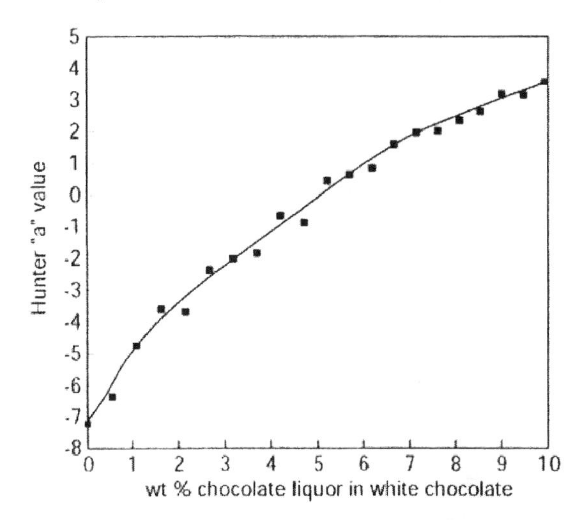

Fig. 1 Relationship between color and concentration.

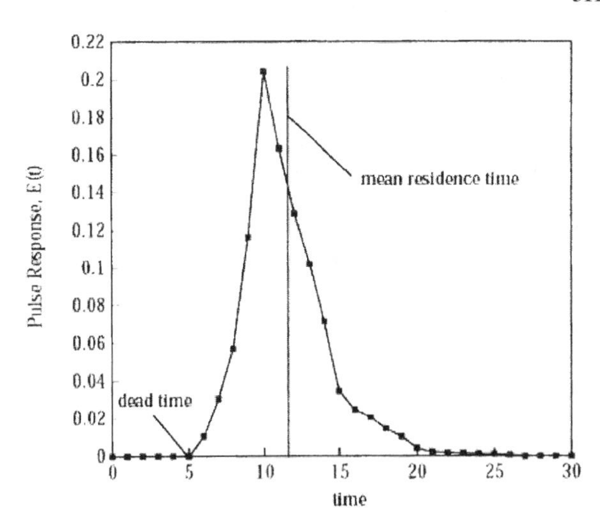

Fig. 2 Typical pulse response.

Another problem that appears in the literature is omission of the extrusion die RTD (a particular issue when using small laboratory extruders with relatively large dies). The die (with a different shape RTD) may have a residence time, which is a significant portion of that in the system. Or worse, a poorly designed die may have significant stagnant regions. These factors introduce artifacts resulting in distortions of the RTD.

ANALYSIS OF DATA: MODELS

Pulse Tracer Input

Analysis of RTDs is very well covered in a number of texts (Refs. [7] and [8] and others) as summarized below.

Response of the extruder output to a pulse input of tracer (Fig. 2) is $E(t)$, the exit age time response. This is determined by measuring the concentration, c, at the outlet of the system at time, t, after introduction of the pulse at the inlet. $E(t)$ is given by:

$$E(t) = \frac{c(t)}{\int_0^{\infty} c(t)\mathrm{d}t} \tag{1}$$

The area under the exit age distribution from time 0 to time t is the fraction of material residing within the extruder for less than time t. Average (mean) residence time, τ, is found by taking the first moment of the exit age distribution, $E(t)$.

$$\tau = \int_0^{\infty} tE(t)\mathrm{d}t \tag{2}$$

Eq. 1 should be more properly expressed in dimensionless form. This scales the distribution with average residence time and allows one to make more direct comparisons between different sets of data. Defining dimensionless time as:

$$\Theta = \frac{t}{\tau} \tag{3}$$

The dimensionless age distribution becomes:

$$E(\Theta) = \tau E(t) \tag{4}$$

As stated above, plotting dimensionless exit age distribution, $E(\Theta)$, vs. dimensionless time, Θ, allows one to directly compare different situations. Not doing this is a common error. If one simply changes the feed rate to a system, say by reducing it, the average residence time gets longer and the spread of $E(t)$ distribution gets wider. Many people have interpreted the wider distribution to mean that there is more mixing or a wider spread in the residence times. However, when the data are looked at in dimensionless form, the different distributions look identical, meaning that the spread is proportional to the median residence time.

Step Change in Tracer Concentration

The integral of $E(t)$ or $E(\Theta)$ allows easy determination of the median residence time, and also shows the response to a step change of tracer concentration in the system, in either the real or dimensionless domain. This curve, called $F(t)$ or $F(\Theta)$, is illustrated in Fig. 3.

$$F(t) = \int_0^{\infty} E(t)\mathrm{d}t \tag{5a}$$

$$F(\Theta) = \int_0^{\infty} E(\Theta)\mathrm{d}t \tag{5b}$$

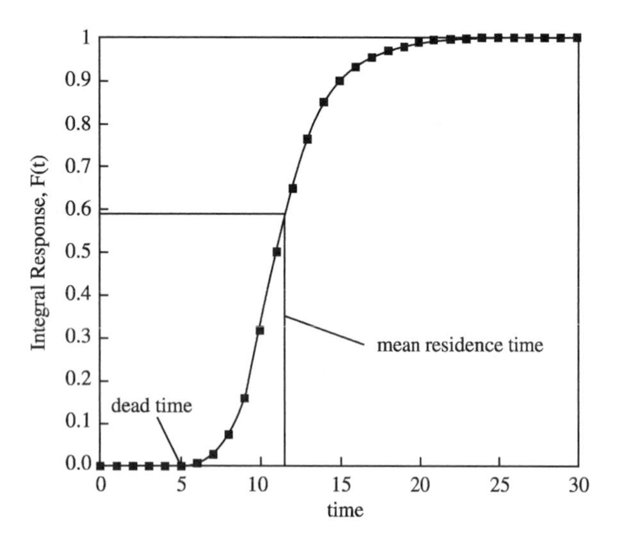

Fig. 3 Typical step response.

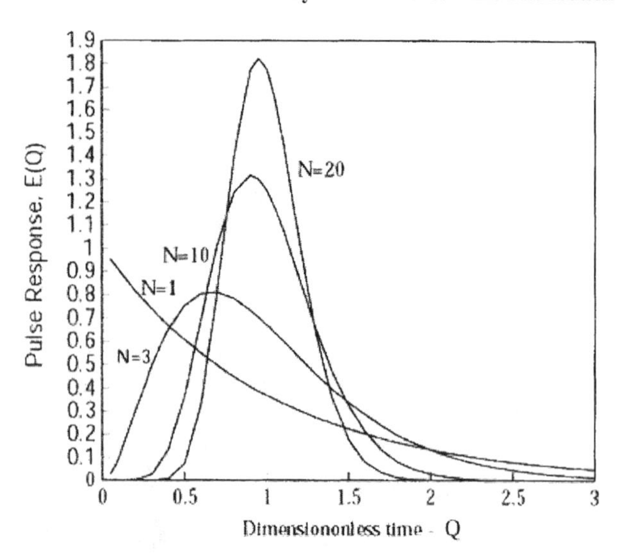

Fig. 4 Pulse response for various numbers of CSTRs.

$F(t)$ and $F(\Theta)$ may be interpreted as fractional responses to step changes in concentration of tracer in the feed.

RTD, like any other distribution, has not only an average (the mean residence time) but also a variance. The variance is given by the second moment of the exit age distribution. In dimensionless form,

$$\sigma^2 = \int_0^\infty (\Theta - 1)^2 E(\Theta) \mathrm{d}\Theta \tag{6}$$

Physical Model Equivalents

The variance, σ^2, has been interpreted in a number of ways.

Stirred tank model

We can imagine a series of equal residence continuous stirred tank reactors (CSTRs). As we add more tanks in series, the average residence time in the system increases, and the distribution narrows, giving us a series of RTD curves, one of which will most closely resemble actual data from extrusion measurements. Fig. 4 is an illustration of the pulse response for various numbers of CSTRs in series.

One way of interpreting the variance of a system is to find the equivalent number, N, of CSTRs it is equivalent to.

$$\frac{1}{N} = \sigma^2 \tag{7}$$

There is some suggestion[9] that the multiple stirred tank model is not really a good representation of the extruder's RTD, although it has been used to describe both extruders and preconditioners. One serious flaw is that it does not allow for dead time—in real extruders, a significant portion of the total residence time elapses before any tracer emerges. In the stirred tank model, a tiny portion of tracer should emerge immediately after it is injected into the system.

Diffusion model

Another interpretation of the variance, σ^2, is in terms of an axial dispersion coefficient, D_L, which is similar to apparent diffusivity. Here, we can imagine that the product moves through the extruder in plug flow, with simultaneous axial diffusion causing the residence times of different elements to diverge. An explanation of the axial dispersion coefficient may be found in Ref. [8]. Axial dispersion is related to the variance by

$$\sigma^2 = 2\frac{D_L}{vL} - 2\left(\frac{D_L}{vL}\right)^2 \left(1 - \mathrm{e}^{\left(\frac{-vL}{D_L}\right)}\right) \tag{8}$$

Here v and L represent the average channel velocity and length, respectively. This model is based on the analogy of dispersion with diffusivity. The model is based on the deviation from plug flow in a pipe. The model is based on the concept of a "closed" vessel, with no dispersion of the tracer occurring at the exit and entrance to the vessel. Dispersion is assumed to be uniform at all points within the vessel.

Since the diffusion model is simply related to the stirred tank model via Eqs. 7 and 8, it is also flawed by its prediction of some instant change in exit tracer concentration after change at the inlet.

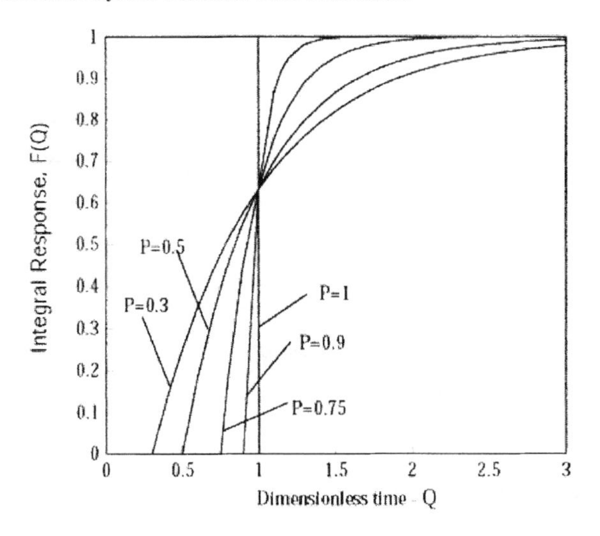

Fig. 5 Wolf–White models for various fractions of dead time P.

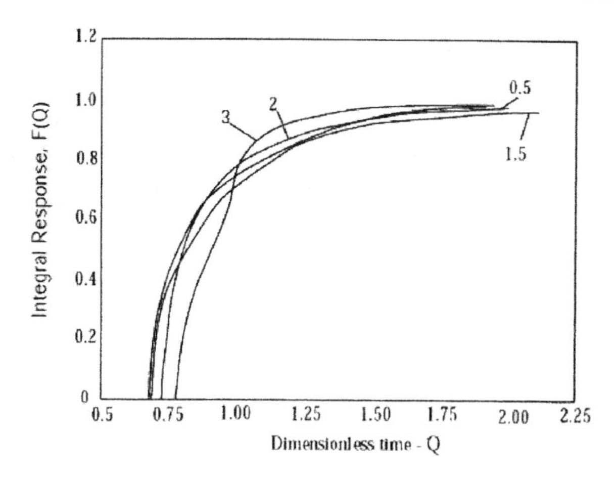

Fig. 6 Step response for single-screw extruders (parameter dimensionless pressure, $n = 0.4$).

Wolf–White model

In general, extruder RTD data can be adequately described by the Wolf–White model,[10] which approximates the system as a combination of a pure dead time in series with a CSTR. Some investigators[2,11,12] have used more than one CSTR to improve the description of the "tail" of the distribution. The Wolf–White model is given by

$$F(\Theta) = 0 \quad \Theta \leq P \tag{9a}$$

$$F(\Theta) = 1 - e^{-\left(\frac{1}{1-P}\right)(\Theta - P)} \quad \Theta \geq P \tag{9b}$$

Here, the constant P represents the fraction of the average residence time that is seen as pure dead time. Because of the compactness of the Wolf–White form, wherever possible, we will try to summarize the data in this form. Fig. 5 illustrates the Wolf–White model for various values of P.

EXTRUSION RTD STUDIES AND DATA

RTD Theory

In single-screw extruders (where most work has been focused because of their relative simplicity), a dead time of 75% of the medical residence time ($P = 0.75$) is predicted for Newtonian flow and when the pressure flow component is very small.[13] Shear-thinning behavior makes the shape of the RTD curve a function of screw geometry (e.g., pitch), flow index (n), and magnitude of pressure flow, as shown in Fig. 6 for $n = 0.4$.

The value of P (Fig. 7) is predicted to increase (for more plug-flow-like behavior) with high-pressure flow and low-flow index, but in all cases should lie between 0.62 and 0.84, the dead time for intermediate pressure flows of shear-thinning materials.

One attempt to create an RTD theory for the more complicated corotating twin-screw extruders[14,15] predicts behavior of simple screw profiles using forward and reverse-pitch elements (including those with slotted flights that increase leakage flows). This model predicts that dead time is reduced at high-screw speed, but not affected by feed rate, screw pitch, or the size of axial slots in the screw. Backmixing, i.e., the length of the RTD tail, is predicted to be less at higher feed rates and with shorter screw pitch. The general belief that these machines have a narrower RTD than do single-screw extruders, because their

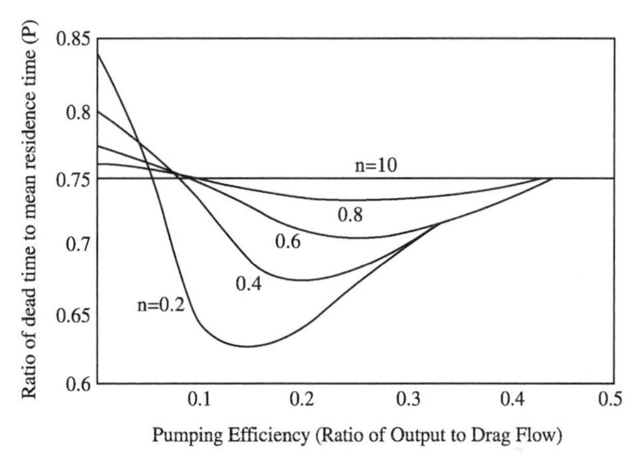

Fig. 7 Single-screw extruder dead time for various flow indices, n, and outputs.

self-wiping feature should eliminate dead flow zones, is not necessarily the case.

Counter-rotating twin-screw extruders do have a narrow RTD because there is no pressure flow component (except from small leakage flows), giving them plug-flow behavior in comparison to other machine styles. Some attempts at modeling of the RTDs of these extruders have been reported.[2] The literature suggests that they are best modeled as a series of two CSTRs plus a pure dead time.[2]

Extruder Performance

Single-screw extruders

There is generally good agreement between single-screw extruder performance and the Biggs and Middleman theory from their own measurements and others. P has been found to be about 0.75 for $n = 0.5$[16] and between 0.80 and 0.89 (possibly skewed by using color as a response)[17] over a wide range of conditions. Some improvement in the data representation has been found by using the two-CSTR-plus-dead-time model.[11]

Corotating twin-screw extruders

Smaller values for P have been determined for corotating twin-screw extruders: about 0.5 over a range of screw speeds, moisture, feed rate, and barrel temperature[18]; 0.55[19]; and between 0.4 and 0.6, depending on screw geometry, with more constrictive elements increasing the extruder dead time.[20] This change in P with constriction is not large and difficult to quantify (mainly affecting the tails of the F distribution), but have been confirmed in other studies as well, with ranges of 0.45–0.7[19] and 0.43 and 0.66.[9] Altomare also reports that the dispersion model did not agree well with RTD data. The Teledyne-Readco style mixer (a corotating extruder with short screws having very deep flights) has a median residence time and dispersion both inversely proportional to flow rate, so the RTD curves are geometrically similar.[21]

Counter-rotating twin-screw extruders

Limited studies of counter-rotating twin-screw extruders confirm their theoretical description, with one study showing a P range of 0.91–0.93[22] and another with a qualitative description suggesting plug flow.[23]

Counter-rotating, nonintermeshing twin-screw extruders,[3,4] rarely used in the food industry, have very open channels resulting in a lower percentage of dead time than do single-screw extruders.[24]

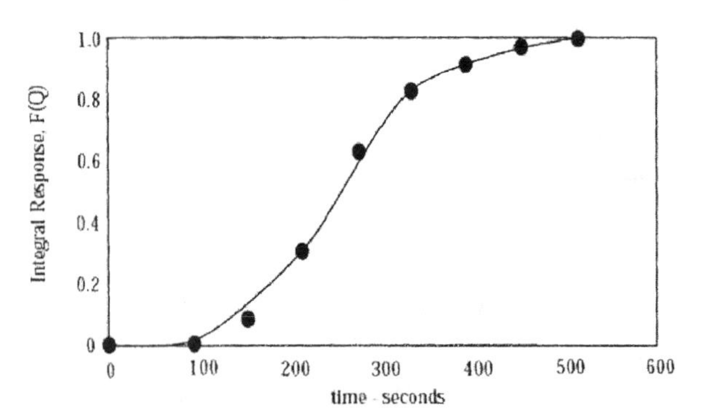

Fig. 8 Typical preconditioner RTD.

"Kneading" extruders

Another style is the kneading extruder (like the Buss machine) encountered occasionally in the food industry and in the plastics industry as a compounder (e.g., the Co-Kneader). These machines have a single interrupted-flight screw that oscillates in the axial direction as it rotates, intermeshing with a series of stators. For starch extrusion,[25] the RTD is similar to that of an intermeshing corotating twin-screw extruder with a P of about 0.6. In addition (for plastics), constriction decreases the percentage seen as a pure dead time.[26]

Preconditioners

The product within a preconditioner is a fluidized bed of granules and is physically more similar to a CSTR than is an extruder, with a generally broader RTD. One preconditioner study[27] indicates that RTD data are well modeled with the multiple CSTR model consisting of 5–15 tanks in series and that the degree of backmixing increases with shaft speed. A typical RTD is illustrated in Fig. 8. Feed rate variations do not affect degree of dispersion significantly if the rates are changed in proportion to shaft speed. The fluidized, granular nature of the conveyed material can create variations of RTD between components of the mix if they have different flow characteristics,[28] a particular concern in selecting a tracer material. This work agrees with the Bouvier study, showing a broader RTD at higher shaft speed, unaffected by other variables. In addition the average residence time appears to decrease at higher feed rates, higher shaft speeds, and in the presence of moisture (water and steam) in the preconditioner.

REFERENCES

1. Harper, J.M. *Extrusion of Foods*; CRC Press: Boca Raton, FL, 1981; Vol. I.

2. Janssen, L.P.B.M. *Twin Screw Extrusion*; Elsevier: New York, 1978.
3. White, J.L. *Twin Screw Extrusion, Technology and Principles*; Hanser Pub.: New York, 1991.
4. Rauwendaal, C. *Polymer Extrusion*; Hanser Pub.: New York, 1990.
5. Levine, L.; Symes, S.; Weimer, J. Automatic Control of Moisture in Food Extruders. J. Food Process. Eng. **1986**, *8* (2), 97–116.
6. Zeigler, G. Private Correspondence from Dept. of Food Science, Penn State University, University Park, PA, 2001.
7. Levenspiel, O. *Chemical Reaction Engineering*; John Wiley & Sons: New York, 1972.
8. Himmelblau, D.M.; Bischoff, K.B. *Process Analysis and Simulation, Deterministic Systems*; John Wiley & Sons: New York, 1968.
9. Altomare, R.E. The Effect of Screw Element Selection on the Residence Time Distribution in a Twin Screw Cooking Extruder. In the National Meeting of the AIChE, Washington, D.C., 1988.
10. Wolf, D.; White, D.H. Experimental Study of the Residence Time Distribution in Plasticating Screw Extruder. AIChE J. **1976**, *22* (1), 22.
11. Eerikaninen, T.; Linko, P. *Extrusion Cooking Modeling Control and Optimization in Extrusion Cooking*; Mercier, C., Linko, P., Harper, J.M., Eds.; American Association of Cereal Chemists: St. Paul, MN, 1989.
12. Davidson, V.J.; Paton, D.; Disady, L.L.; Spratt, W.A. Residence Time Distributions for Wheat Starch in a Single Screw Extruder. J. Food. Sci. **1983**, *48* (4), 1157–1161.
13. Bigg, D.; Middleman, S. Mixing in a Screw Extruder. A Model for Residence Time Distribution and Strain. Ind. Eng. Chem. Fundam. **1974**, *3* (1), 66.
14. Tayeb, J.; Vernes, B.; Della Valle, G. Theoretical Computation of the Isothermal Flow Through Reverse Screw Element of a Twin Screw Extrusion Cooker. J. Food Sci. **1988**, *53* (2), 616–624.
15. Tayeb, J.; Vernes, B.; Della Valle, G. A Basic Model for a Twin-Screw Extruder. J. Food Sci. **1988**, *53* (4), 1047–1056.
16. Bruin, S.; Van Zuilicham, D.J.; Stolp, W.J. Fundamental and Engineering Aspects of Extrusion of Biopolymers in a Single Screw Extruder. J. Food Process. Eng. **1973**, *2* (1), 1.
17. Likimani, T.A.; Sofos, J.N.; Maga, J.A.; Harper, J.M. Methodology to Determine Destruction of Bacterial Spores During Extrusion Cooking. J. Food. Sci. **1990**, *55* (5), 1388–1393.
18. Altomare, R.E.; Ghossi, P. An Analysis of Residence Time Distribution Patterns in a Twin Screw Cooking Extruder. Biotechnol. Prog. **1986**, *2* (3), 157–163.
19. Curry, J.; Kiani, A.; Dreiblatt, A. Feed Variance Limitations for Corotating Intermeshing Twin Screw Extruders. Polym. Process. **1991**, *6* (2), 148–155.
20. Todd, D.B. Residence Time Distributions in Twin Screw Extruder. Polym. Eng. Sci. **1975**, *15* (6), 437.
21. Aquilar, C.A.; Ziegler, G.R. Residence Time Distributions of Chocolate Coating Co-rotating Twin Screw Continuous Mixer. J. Food Process. Eng., Manuscript #223H.
22. Lin, J.K.; Armstrong, D.J. Process Variables Affecting Residence Time Distributions of Cereals in an Intermeshing, Counter Rotating Twin Screw Extruder. Trans. ASAE. **1990**, *33* (6), 1971–1978.
23. Ilo, S.; Berhofer, E. Kinetics of Thermomechanical Destruction of Thiamine During Extrusion Cooking. J. Food Sci. **1998**, *63* (2), 312–316.
24. Bash, T.F. *Welding Engineers' CRNI Twin-Screw Extruders in Plastics Compounding, Equipment and Processing*; Todd, D.B., Ed.; Hanser Pub.: Cincinnati, OH, 1998.
25. Jager, T.; Santulte, P.; van Zuilichem, D.J. Residence Time Distribution in Kneading Extruders. J. Food Eng. **1995**, *24*, 284–294.
26. Elemans, P.H.M. *Modeling of the Cokneader in Mixing and Compounding of Polymers, Theory and Practice*; Manas-Zloczower, I., Tadmor, Z., Eds.; Hanser Pub.: Cincinnati, OH, 1994.
27. Bouvier, J.M. Preconditioning in the Extrusion Process. Int. Milling Flour Feed **1995**, Dec, 34–37.
28. Morales, J.; Twombly, W.; Brent, J.; Glaser, B.; Strahm, B. Examination of the Residence Time Distribution in a Differential Diameter Preconditioner (Poster Presentation). In the National Meeting of the AACC, Baltimore, MD, September 1996.
29. Choudhury, G.S.; Gautam, A. On-Line Measurement of Residence Time Distribution in a Food Extruder. J. Food Sci. **1998**, *63* (8), 529–534.
30. Tadmor, Z.; Gogos, C.G. *Principles of Polymer Processing*; John Wiley & Sons: New York, 1979.

Fermentation Process Material and Energy Balances

Joseph Boudrant
Centre Nationale de la Recherche Scientifique (CNRS), Vandoeuvre-les-Nancy, France

INTRODUCTION

Fermentations concern biological transformations of products using micro-organisms. These bioprocesses lead to biomass and biosynthesis of metabolites due to displacements of atoms and radicals (molecular rearrangements). The obtained products are generally solubilized, at least during a minimum period of time, some of them being volatile (gas or organic compounds). The corresponding reactions are under the control of a large number of enzymes. Indeed, each degradation and biosynthesis is controlled by a set of different enzymatic reactions. Besides, heat is generated by catabolism. Living micro-organisms permanently evolve by themselves, whereas raw materials and bioproducts generally do not.

The raw materials, which can be transformed, are a mixture of organic compounds or molecules made of carbon, nitrogen, and basic elements such as oxygen, hydrogen, and minerals. In aerobic conditions, oxygen mainly comes from air whereas the other elements are mainly from raw materials (Fig. 1).

The micro-organisms (whose most important use in industry is as chemotroph) are mainly made of elements such as carbon, nitrogen, oxygen, hydrogen, and a few others that are present in their composition but in a lower content. The cells are in structurized form, whereas the organization of the raw material is not. Indeed, the latter might be even more efficient after having been destructurized, which can be critical. The metabolites are most of the time molecules with specific composition and defined molecular structure.

Fermentations are multiparametric. Globally, fermentations can be evaluated in terms of raw material or substance transformation and availability, metabolite production, energy requirement, production, or elimination. So there are different manners of evaluating fermentation performances, depending on the objective of the chosen fermentation process. In fact, fermentations can be analyzed with objectives of either to increase the knowledge in order to know what is possible in terms of potential performances (study or improvement at laboratory or pilot levels) or to control some of its process parameters (optimization, scale-up).

Material and energy balances imply stoichiometry, which is the science concerned with the quantitative composition of chemical compounds and conversion in chemical reactions. It is one of the fundamentals of reaction engineering together with thermodynamics and kinetics. Stoichiometric calculations require knowledge of the mechanisms and the balance of major compounds. Normally, it could be applied to all compounds. But generally for conservating properties, only macrobalances are used. Indeed it is difficult to apply it to fermentation as several hundred components are involved in the whole metabolism (Figs. 2 and 3). But specific microbalances can be used. The main problem in applying stoichiometry to fermentation arises from the complex network of metabolism. Due to high complexity, the approach is made simpler by using the concept of yield. But this macroscopic variable cannot be considered as a biological constant.

The main elements of this chapter are based on the exhaustive, clear, and well-written review of Nagaï,[1] published in 1977. Also mentioned are the interesting articles of Bailey and Ollis,[2] Luong and Volesky,[3] Parisi,[4] Krämer,[5] and the well-known book of Pirt.[6]

ENERGETIC

The Growth Yields

The yield factor was defined by Monod in mass units by the quotient of the increase in biomass or metabolites in regard to the amount of utilized substrate. This mass ratio is fully sufficient for many cases. But it is purely macroscopic and descriptive and can be applied to all compounds. Several growth yields can be defined as follows.

$Y_{X/S}$

Growth yield from substrate is expressed as grams of dry cells produced per gram of substrate consumed or as grams

Encyclopedia of Agricultural, Food, and Biological Engineering
DOI: 10.1081/E-EAFE 120007196
Copyright © 2003 by Marcel Dekker, Inc. All rights reserved.

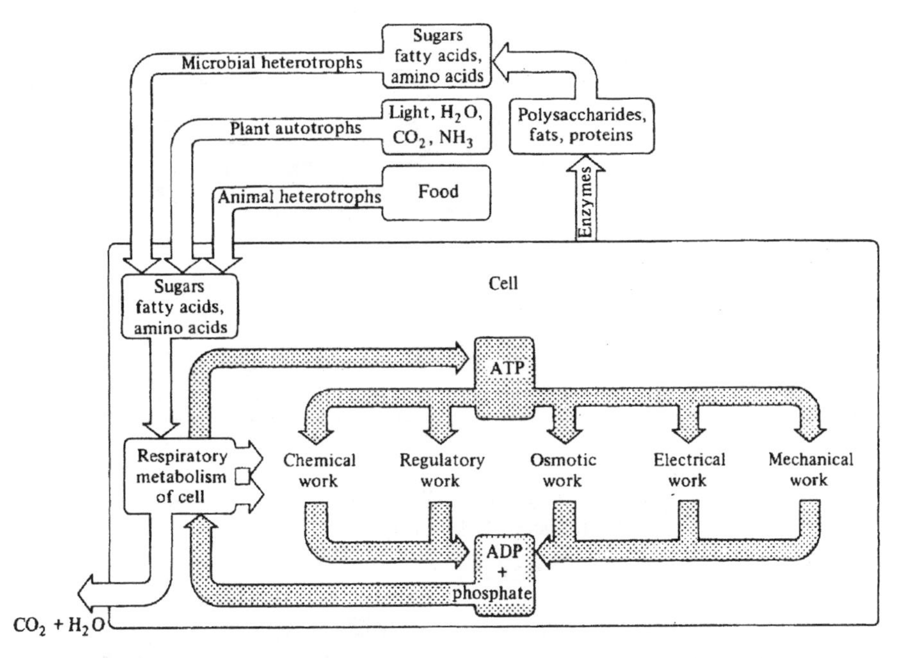

Fig. 1 Schematic diagram showing synthesis of biological macromolecules from simple nutrients in a cell (bacteria). (From Mandelstan, J., McKillen, K., Eds; *Biochemistry of Bacterial Growth*, 2nd Ed.; Blackwell Scientific Publications: Oxford, 1973; 4.)

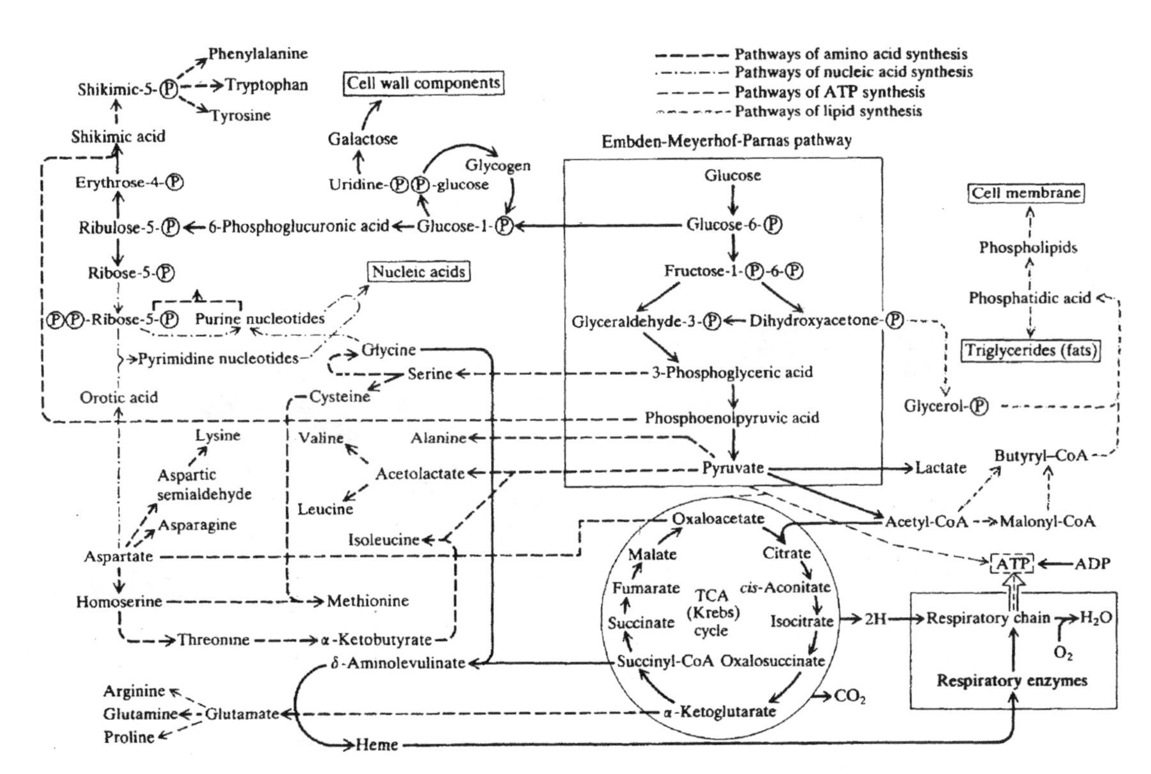

Fig. 2 A summary of a few of the major metabolic pathways in a cell (the bacteria *Escherichia coli*). (From Watson, J.D. *Molecular Biology of the Gene*, 2nd Ed.; W. A. Benjamin, Inc.: New York, 1970; 96–97.)

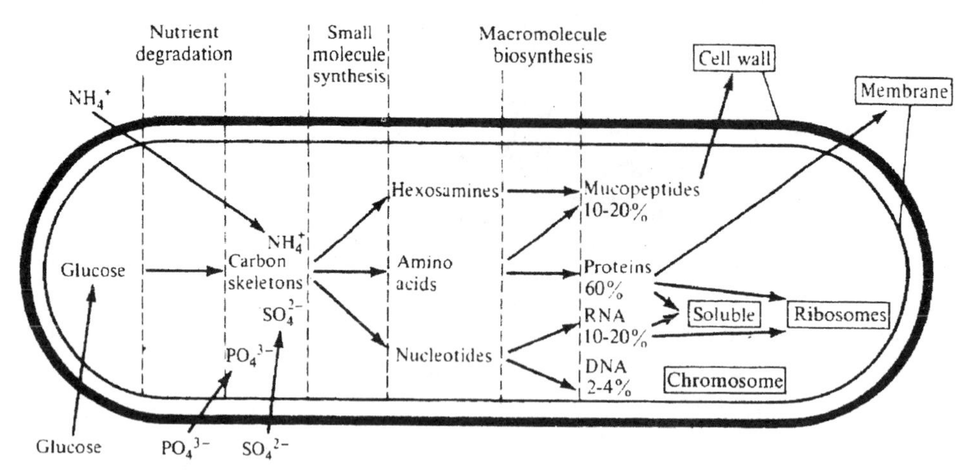

Fig. 3 Overall schematic diagram of energy flows in the cell and in the environment. (From Loewy, A.G.; Siekevitz, P. *Cell Structure and Function*, 2nd Ed.; ©Holt, Rinehart and Winston, Inc., 1963 and 1969; 26.)

of dry cells produced per mole substrate consumed.

$$Y_{X/S} = \Delta X / -\Delta S$$

where $Y_{X/S}$ is the growth yield from substrate, g g^{-1} or g mol^{-1}, X the biomass concentration, g L^{-1}, and S the substrate concentration, g L^{-1} or mol L^{-1}.

The same concept of yield can be applied to quantify the relation between consumed and produced amounts. The unit of yield depends on the unit used for quantification. For biomass yield evaluation, an empirical formula for cell must be used (see later), but indeed, even for one micro-organism, there is no unique empirical formula: the elementary composition of cell or biomass varies with growth and limiting nutrient.

Y_{ave}

To evaluate growth yield from substrate, the dimension of dry cell produced per gram carbon of substrate consumed is used without taking into account the other constituent elements of the substrate. Corresponding to this, another growth yield defined by the amount of dry cell per electron equivalent initially available from the substrate was proposed:

$$Y_{ave} = Y_{X/S} / Y_{ave/S}$$

where Y_{ave} is the growth yield based on electron available from substrate, g (av e)$^{-1}$ and $Y_{ave/S}$ the electron available from substrate, av e mol^{-1}.

$Y_{X/O}$

Growth yield based on oxygen, $Y_{X/O} = \Delta X/\Delta O_2$, gram cell produced per mole oxygen consumed, means the efficiency of biomass produced to catabolic energy expended since ΔO_2 corresponds to the representative value of overall catabolic activity when energy yielding reactions are via the oxidative phosphorylation pathway without depending on the glycolytic pathway.

$Y_{X/C}$

The growth yield based on catabolic activity $Y_{X/C}$ can be written as:

$$Y_{X/C} = \Delta X/\Delta H_C = \Delta X/\left[\Delta H_S(-\Delta S) - \sum \Delta H_P \cdot \Delta C_P\right]$$

$$= Y_{X/S}/\left[\Delta H_S \cdot - \sum \Delta H_P \cdot Y_{P/S}\right]$$

where ΔH_C is the heat generation by catabolism, Cal L^{-1}, ΔH_S the heat of combustion of substrate, Cal mol^{-1}, ΔH_P the heat of combustion of end-product, Cal mol^{-1}, C_P the product concentration, mol L^{-1}, and $Y_{P/S}$ the product yield from substrate, mol mol^{-1}.

When a micro-organism grows in a complex medium without producing products, $Y_{X/S}$ is identical with $Y_{X/C}$, however, when significant extracellular products are discharged from the cells, the former becomes a pretented yield differing completely from the latter.

Growth Yields Based on Total Energy

A growth yield based on total energy available, g Cal^{-1}, can generally be written as

$$Y_{Cal} = \Delta X/[\Delta H_a \Delta X + \Delta H_C]$$

where Y_{Cal} is the growth yield based on total energy available, g Cal^{-1}, ΔH_a the heat of combustion of dry cell,

F

Cal g^{-1}, and ΔH_C the heat generated by catabolism, Cal L^{-1}.

In complex media, since a carbon source is completely dissimilated by catabolism, Y_{Cal} can be expressed by the following formulas:

$$Y_{Cal} = 1/[\Delta H_a + \Delta H_S/Y_{X/S}] \quad \text{or}$$

$$Y_{Cal} = 1/[\Delta H_a + \Delta H_O/Y_{X/O}]$$

where ΔH_O is the heat generated based on oxygen consumed, Cal mol^{-1}.

In minimal media, the sole carbon source is naturally metabolized partly via both biosynthetic and catabolic pathways, and Y_{Cal} is given by the following formula:

$$Y_{Cal} = Y_{X/S}/[\Delta H_a \cdot Y_{X/S} + \Delta H_S[1 - (\alpha_2/\alpha_1) \cdot Y_{X/S}] - \sum \Delta H_P \cdot Y_{P/S}$$

where α_1 is the carbon content of substrate, g mol and α_2 the carbon content of cell, g g^{-1}.

Growth Yield Based on ATP Generation, Y_{ATP}

Y_{ATP} in energy coupled growth

Amounts of growth of cells (case of yeast) growing anaerobically in complex media are directly proportional to the moles of ATP produced by catabolism. Thus Y_{ATP} can be defined as

$$Y_{ATP} = \Delta X/\text{ATP} = Y_{X/S}/Y_{A/S} \ \text{g mol}^{-1}$$

where Y_{ATP} is the growth yield based on ATP generation, g mol^{-1} and $Y_{A/S}$ the ATP yield from energy source catabolized, mol mol^{-1}.

For the estimation of Y_{ATP}, $Y_{A/S}$ must be assessed on the basis of the established pathways accompanied by ATP formation, and calculated by the amount of ATP on the basis of the experimental data with respect to energy source utilized and/or end-products formed. Estimation of *P/O* ratio, the efficiency of ATP formation in relation to oxygen consumed via the oxidative phosphorylation pathway, has always been attempted using crude cell-free extracts. The prediction of *P/O* ratio has been carried out on the basis of Y_{ATP} concept.

Y_{ATP} in energy uncoupled growth

If energy yielding metabolism is fully coupled with macromolecular synthesis, i.e., the limitation of growth is at the level of energy production, Y_{ATP} could be counted to be more or less 10 in many organisms. However, this is contrary if the formation rate of essential compounds required for cell biosyntheses is rate limiting rather than that of ATP in energy yielding processes. Energy would be excess and wasted as heat without coupling the growth, i.e., the so-called energy uncoupled growth.

MASS AND ENERGY BALANCES DURING MICROBIAL GROWTH

Stoichiometry

A stoichiometry equation with respect to growth, carbon, noncellular product formation, oxygen uptake, carbon dioxide evolution can be regarded as the basis of the law of conservation of substrate metabolized by a microorganism. The following expression can be written as:

$$aC_xH_yO_z + bO_2 + cNH_3 = dC_\alpha H_\beta O_\gamma N_\xi + eC_{\alpha'}H_{\beta'}O_{\gamma'}N_{\xi'} + fH_2O + gCO_2$$

x, y, z, α, β, γ, ξ, α', β', γ', and ξ' can be fixed if the respective molecular composition of material is known. So the remaining seven [or six, if one (e.g., a) is taken as 1] unknowns of the stoichiometric coefficients, $a(1)$, b, c, d, e, f, and g, must be decided either by material balance or in other ways. Based on material balance, four independent equations can theoretically be written with respect to carbon, hydrogen, oxygen, and nitrogen. But most of the time, it is possible only with carbon, whereas it is nearly impossible with hydrogen and oxygen due to water, for nitrogen because of analytical difficulties and variation of nitrogen content as a function of cell cultivation conditions.

Carbon and Oxygen Balances

Instead of stoichiometric equation, the overall expression on the growth reaction in minimal media will be written as:

$$\nu + Q_{O2} \rightarrow \mu' + Q_{CO2} + Q_P$$

where ν is the specific rate of substrate consumption, $\text{mol g}^{-1}\text{hr}^{-1}$, Q_{O2} the specific rate of oxygen consumption, $\text{mol g}^{-1}\text{hr}^{-1}$, μ' the specific growth rate here in $\text{mol g}^{-1}\text{hr}^{-1}$, equivalent to μ, hr^{-1}, if biomass is expressed in g, Q_{CO2} the specific rate of carbon dioxide evolution, $\mathrm{mol\,g^{-1}hr^{-1}}$, and Q_P the specific rate of noncellular product formation, $\mathrm{mol\,g^{-1}hr^{-1}}$.

Thus the carbon balance is

$$\alpha_1\nu = \alpha_2\mu' + \alpha_3 Q_{CO2} + \alpha_4 Q_P$$

where α_1 is the carbon content of substrate, $\mathrm{g\,mol^{-1}}$, α_2 the carbon content of cells, $\mathrm{g\,mol^{-1}}$, α_3 the carbon content of carbon dioxide, $\mathrm{g\,mol^{-1}}$, and α_4 the carbon content of product, $\mathrm{g\,mol^{-1}}$.

Another mass balance of oxygen on the growth reaction can be written as:

$$A\nu = B\mu' + Q_{CO2} + CQ_P$$

where A is the amount of oxygen required for the combustion of substrate to CO_2, H_2O, and NH_3, $\mathrm{mol\,mol^{-1}}$, B the amount of oxygen required for the combustion of dry cells to CO_2, H_2O, and NH_3, $\mathrm{mol\,mol^{-1}}$, and C the amount of oxygen required for the combustion of noncellular product to CO_2, H_2O, and NH_3, $\mathrm{mol\,mol^{-1}}$.

ATP Generation During Growth

It is supposed that ATP formed is immediately utilized mainly for cellular biosyntheses and partly for maintenance metabolism. For the assessment of a maintenance coefficient, mass balance on ATP formed during a small interval can be written as:

$$(\Delta\mathrm{ATP})_F = (\Delta\mathrm{ATP})_M + (\Delta\mathrm{ATP})_G$$

where subscripts are F: formed, M: maintenance, and G: growth.

Assuming that the amount of ATP utilized for maintenance metabolism is proportional to the cell density, and that ATP requirement for the growth is proportional to the amount of biomass produced, the following equations will be established:

$$(\Delta\mathrm{ATP})_M = m_A X\Delta t$$

$$(\Delta\mathrm{ATP})_G = \Delta X / Y_{\mathrm{ATP}}^{\mathrm{MAX}}$$

where m_A is the maintenance coefficient for ATP, $\mathrm{mol\,g^{-1}hr^{-1}}$ and $Y_{\mathrm{ATP}}^{\mathrm{MAX}}$ the maximum growth yield for ATP, $\mathrm{g\,mol^{-1}}$. Here $Y_{\mathrm{ATP}}^{\mathrm{MAX}}$ differs from Y_{ATP} at the point in which Y_{ATP} is variable as a function of μ' as defined by μ'/Q_{ATP} whereas $Y_{\mathrm{ATP}}^{\mathrm{MAX}}$ is constant.

Relationship Between Substrate Consumption, Growth, Respiration, and Noncellular Products

Carbon source in minimal media

When the carbon source is mainly metabolized for cellular synthesis without discharging any noncellular products in the medium, the carbon source consumed might approximately be balanced by the following equation based on the heat of consumption of the substrate:

$$\begin{aligned}\Delta H_S(-\Delta S) &= \Delta H_S(-\Delta S_M) + \Delta H_S(-\Delta S_W) \\ &\quad + \Delta S_S(-\Delta S_C) \\ &= m'X\Delta t + \Delta H_S \cdot Y_W \cdot \Delta X\end{aligned}$$

where ΔH_S is the heat of combustion of substrate, $\mathrm{Cal\,mol^{-1}}$, $\Delta H_S(-\Delta S)$ the total substrate consumed, $\Delta H_S(-\Delta S_M)$ the substrate expended for maintenance metabolism, $\Delta H_S(-\Delta S_W)$ the substrate expended for biosynthetic activity, $\Delta H_S(-\Delta S_C)$ the substrate incorporated into cellular components, m' the maintenance coefficient based on heat of combustion, $\mathrm{Cal\,g^{-1}hr^{-1}}$, and Y_W the substrate catabolized for true biosynthetic activity to build up 1-g cell, $\mathrm{mol\,g^{-1}}$.

Carbon source in complex media

Carbon source in complex media would then be considered almost completely dissimilated in energy yielding processes. Within this context, the energy balance in terms of carbohydrate (energy source) consumed can be represented by

$$\Delta H_S(-\Delta S) = m'X\Delta t + \Delta H_S \cdot Y_w \cdot \Delta X$$

Respiration

Assuming that energy yielding processes are mostly dominated by the oxidative phosphorylation rather than the substrate level phosphorylation, and that most of the oxygen consumed is oxidized via oxidase systems, the amount of oxygen consumed would be generally regarded as a representative value in catabolism and consequently this could be subdivided into two parts

$$Q_{O2} = m_O + 1/Y_{GO}\mu$$

where m_O is the maintenance coefficient for oxygen, $\mathrm{mol\,g^{-1}hr^{-1}}$ and Y_{GO} the true growth yield for oxygen, $\mathrm{g\,mol^{-1}}$.

Heat Evolution During Growth

Heat evolution during cultivation is a real industrial problem since the heat must be removed during cultivation so as to maintain an optimum temperature for growth or metabolism. Heat evolution measurement has been carried out successfully by calorimetric analysis and by heat balance based on heat losses and gains. A calorimetric method gives data, however, this can be applicable under very limited culture conditions. Similarly the heat balance methods are only applicable in the case where a small reactor is used. The heat evolution during cultivation can be calculated on the basis of the difference between the heat of combustion of substrate consumed and that for the sum of the products formed during growth. When no particular products other than biomass and carbon dioxide are observed during the growth in minimal media, the heat evolution during cultivation can be calculated from the difference between the heat of combustion of substrate and for biomass produced (e.g., for yeast it is 3.63 Cal g^{-1}). But only few estimations of heat production have been reported when noncellular products accompanied growth. Note that the mean for the estimation of heat evolution based on the amount of oxygen consumption can be used ($108{\cdot}10^3$ Cal per mole oxygen consumed).

Estimation of heat generation based on substrate consumed and products formed

Different stoichiometric equations have been established for different cells growing in anaerobic or aerobic conditions on different substrates. Indirect estimation of heat evolution can be calculated from the difference between the heat of combustion of substrate consumed and that for the products formed:

$$\Delta H_C = \left(\Delta H_S(-\Delta S) - \sum \Delta H_P\right) - \Delta H_a \cdot \Delta X$$

where ΔH_C is the heat production accompanying growth, Cal L^{-1}.

Estimation of heat generation based on respiration

The oxygen balance can be established, and it is possible to estimate the quantity of heat production based on the amount of oxygen consumed during aerobic cultivation.

CONCLUSION

During fermentation the cellular reactions can be subdivided into three classes: degradation of nutrients, biosynthesis of small molecules, biosynthesis of large molecules (besides transport of ionic and neutral substances, mechanical work required for cell division and motion). The number of different chemical reactions necessary for sustenance of cell life is of the order of 1000 or more. Each reaction is catalyzed by at least one enzyme. Many reacting substances within the cell can be attacked simultaneously by several different enzymes. Thus the sequences of reactions occurring in the cell intersect and overlap in complex ways.

Biosynthesis work is performed with relatively high efficiency of free-energy utilization typically about 20%. The transport work also involves ATP consumption in a process unique to living systems. Small molecules and ions can be moved through membrane against gradient concentration to achieve a ratio of concentration on the two sides of the membrane as great as 10^3. Mechanical work exists during cell division and bacterial movement. All these processes are by themselves nonspontaneous and result in an increase of free energy of the cell. Consequently they occur when simultaneously coupled to another process, which has a negative free-energy change of greater magnitude. The resulting direct conversion of chemical free-energy into mechanical work is also unique for life. Losses during chemical energy conversions in cell result in heat generation, which must be considered in an engineering process point of view.

Many released substances or metabolic end products from the cell are not necessary or useless for the cell function: organic molecules, antibiotics, extracellular enzymes. But many of them are useful for the mankind.

As the elemental composition of a particular species of micro-organism is not extremely variable, certain stoichiometric and thermodynamic constraints can be applied to metabolic activities. The synthesis of a certain amount of cellular material implies the utilization of amounts of carbon, nitrogen, and oxygen from the cell environment simply to account for the new cell mass. If the chemical form in which these different elements are supplied in the medium is known, it is possible to establish additional constraints relating the amount of substrate consumed, the amount of cell material produced, and the amount of certain products formed. In addition, as there is an energy requirement associated with synthesizing all the necessary components required to produce additional cells, a particular amount of energy generation activity is implied. Thus, a particular reaction pathway for obtaining energy in the form of ATP implies corresponding requirements with respect to the amount of chemical energy utilized by the cell.

In conclusion, material balance constraints and thermodynamic requirements give useful relationship with very simplified view of cellular activity, which is entirely macroscopic and which does not utilize any information about the internal chemical workings of the cell.

REFERENCES

1. Nagai, S. Mass and Energy Balances for Microbial Growth Kinetics. *Advance in Biochemical Engineering*; Springer-Verlag: Berlin, 1979; Vol. 11, 49–83.
2. Bailey, J.E.; Ollis, D.F. *Biochemical Engineering Fundamentals*; Chemical Engineering Studies, 2nd Ed.; Mc-Graw Hill International Editions: New York, 1986; 228–306.
3. Luong, J.H.; Volesky, B. Heat Evolution During the Microbial Process: Estimation, Measurement, and Applications. *Advance in Biochemical Engineering*; Springer-Verlag: Berlin, 1983; Vol. 28, 1–40.
4. Parisi, F. Energy Balance for Ethanol as a Fuel. *Advance in Biochemical Engineering*; Springer-Verlag: Berlin, 1983; Vol. 28, 41–68.
5. Krämer, R. Analysis and Modelling of Substate Uptake and Product Releases by Prokaryotic and Eukaryotic Cells. *Advance in Biochemical Engineering*; Springer-Verlag: Berlin, 1996; Vol. 54, 31–74.
6. Pirt, S. *Principles of Microbes and Cell Cultivation*; Blackwell Scientific Publications: Oxford, 1975.

Fermentation Residence Time Distributions

Joel L. Cuello
Eiichi Ono
The University of Arizona, Tucson, Arizona, U.S.A.

F

INTRODUCTION

Except in the case of the ideal plug flow reactor, fluid elements flowing through real (nonideal) reactors spend different amounts of time traversing the reactors from the feed stream to the effluent stream. Nonideality of flow conditions in reactors arise from such factors as fluid channeling, by-passing, or short-circuiting of certain fluid elements through the reactors; longitudinal mixing caused by vortices and turbulence; the presence of stagnant regions; the failure of mixing devices to provide perfect mixing, etc. All these result in flow patterns in real reactors deviating from those in the ideal plug flow or the continuous stirred tank reactors. Residence time distribution (RTD), which measures the duration of stay of different elements of a fluid that is continuously moving through a reactor, is a practical method for determining the magnitude of deviation of the flow patterns in real reactors from those in ideal reactors. The magnitude of such deviation is a critical consideration in scale up, as this magnitude varies widely between large and small reactors. Thus, failure to account for such deviation could lead to significant errors in reactor design and scale up.

DETERMINATION OF RESIDENCE TIME DISTRIBUTION

Stimulus–Response Technique

The RTD of a fluid flowing through a reactor is determined by changing, as a function of time, some property of the feed stream and then measuring the response at the effluent stream. The most common approach employed in this form of stimulus–response technique is that of varying the concentration of one of the nonreactive components (tracers) of the feed stream. The choice of a particular tracer depends on the convenience of its use, facility and accuracy of its measurement (e.g., via electrical conductivity, absorbance, optical density, etc.), its chemical inertness, its not disturbing the flow pattern, and its not disappearing during the course of the experimental measurement possibly through surface adsorption or settling out.[2]

The E Curve

When elements of a fluid passing through a reactor take different routes on their way to the effluent stream, they will take different lengths of time to reach the effluent stream. The RTD of a fluid passing through a reactor is simply the distribution of times the various elements of that fluid take to reach the outlet stream from the inlet stream. Hence, RTD is an exit age distribution for the fluid elements, with the "age" of an element defined as the time spent by that element within the reactor.

A normalized exit age distribution or RTD is the **E** curve, with the total area under the curve being equal to 1 (Fig. 1), and is represented by Eq. 1. In Eq. 1, $\mathbf{E}\,dt$ is the fraction of the effluent stream whose age is between t and $t + dt$. The fraction that is younger than age t_1 is represented by Eq. 2 while the fraction that is older than age t_1 is given by Eq. 3.

$$\int_0^{\infty} \mathbf{E}\,dt = 1 \tag{1}$$

$$\int_0^{t_1} \mathbf{E}\,dt \tag{2}$$

$$\int_{t_1}^{\infty} \mathbf{E}\,dt = 1 - \int_0^{t_1} \mathbf{E}\,dt \tag{3}$$

The F and C Curves

In conducting a stimulus–response experiment to determine the RTD of a given reactor, there are four ways by which the input tracer signal can be introduced into the feed stream, each resulting in a distinctive time distribution of the tracer as it comes out in the effluent stream. These include: 1) step input; 2) pulse input; 3) cyclic input; and 4) random input. Only the first two are considered here as they are the simplest to treat.

In using the step input, no tracer is initially present in the stream entering the reactor. Then, a constant concentration of the tracer C_0 is imposed on the input stream continuously over time (Fig. 2). The time record of

Encyclopedia of Agricultural, Food, and Biological Engineering
DOI: 10.1081/E-EAFE 120007203
Copyright © 2003 by Marcel Dekker, Inc. All rights reserved.

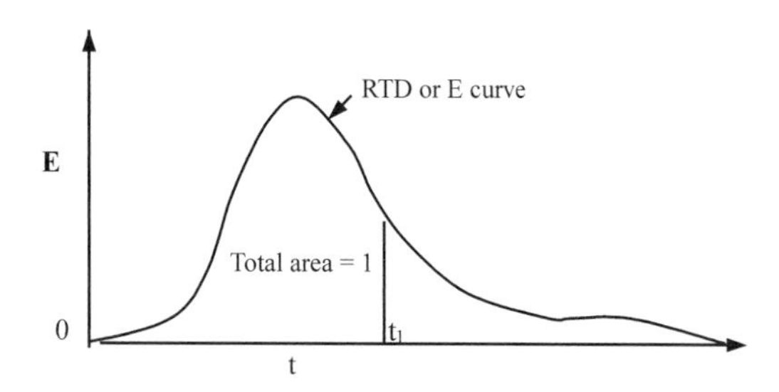

Fig. 1 The **E** curve or the normalized exit age distribution for a fluid flowing through a reactor. RTD = residence time distribution.

the tracer as it comes out in the effluent stream of the reactor, expressed as C/C_0, is called the **F** curve (Fig. 2). The **F** curve ascends over time from 0 to 1.

In using the pulse input, no tracer is again initially present in the stream entering the reactor. Then, an idealized instantaneous pulse of tracer—also known as an impulse, a Dirac delta function, or simply a delta function—is imposed on the input stream (Fig. 3). The normalized time record of the tracer as it comes out in the effluent stream of the reactor is called the **C** curve (Fig. 3). The normalization is performed (Eqs. 4 and 5) by dividing the measured concentration C by Q, which is the area under the concentration–time curve.

$$\int_0^\infty \mathrm{C}\,dt = \int_0^\infty \mathrm{C}/Q\,dt = 1 \tag{4}$$

$$Q = \int_0^\infty \mathrm{C}\,dt \tag{5}$$

The **F** and **C** curves are related to the **E** curve as described by Eqs. 6–8. Note, however, that these relationships are true only with the following assumptions: 1) single fluid flowing through the reactor; 2) steady-state flow; 3) constant fluid density; and 4) closed reactor vessel. In a closed vessel, the flow enters and leaves solely by plug flow, though varying velocities, back diffusion, swirls, and

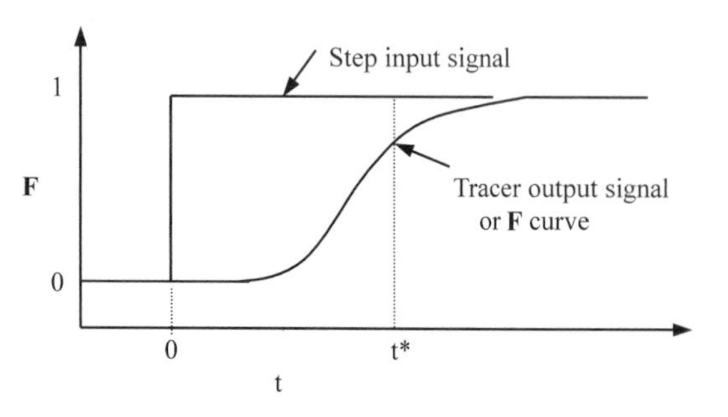

Fig. 2 A step input signal and a resulting output signal, called the **F** curve.

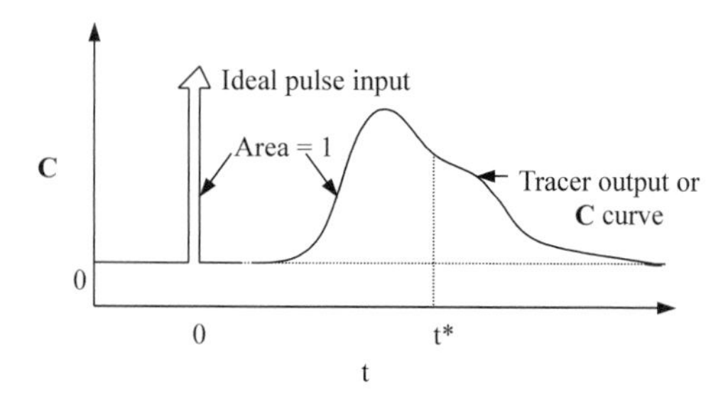

Fig. 3 A pulse input signal and a resulting output signal, called the **C** curve.

eddies may be assumed to occur within the vessel itself. Real vessels often reasonably satisfy this assumption.

$$\mathbf{C} = \mathbf{E} \tag{6}$$

$$\frac{d\mathbf{F}}{dt} = \mathbf{E} \tag{7}$$

$$\mathbf{F} = \int_0^t \mathbf{E}\,dt = \int_0^t \mathbf{C}\,dt \tag{8}$$

Mean Residence Time and Variance

The mean residence time t^*, which is the centroid, of a C vs. t curve is given by Eq. 9. In terms of discrete time values t_i, t^* is given by Eq. 10. Under the same assumptions enumerated in the previous section, Eq. 11 holds true.

$$t^* = \frac{\int_0^\infty t\mathrm{C}\,dt}{\int_0^\infty \mathrm{C}\,dt} \tag{9}$$

$$t^* \cong \frac{\sum t_i \mathrm{C}_i \Delta t_i}{\sum \mathrm{C}_i \Delta t_i} \tag{10}$$

$$t^*_{\mathrm{C}} = t^*_{\mathrm{E}} = t^* \tag{11}$$

The variance σ^2, which is the spread, of a C vs. t curve is given by Eq. 12, and in discrete form is given by Eq. 13.

$$\sigma^2 = \frac{\int_0^\infty (t - t^*)^2 \mathrm{C}\,dt}{\int_0^\infty \mathrm{C}\,dt} = \frac{\int_0^\infty t^2 \mathrm{C}\,dt}{\int_0^\infty \mathrm{C}\,dt} - t^{*2} \tag{12}$$

$$\sigma^2 \cong \frac{\sum (t_i - t^*)^2 \mathrm{C}_i \Delta t_i}{\sum \mathrm{C}_i \Delta t_i} = \frac{\sum t_i^2 \mathrm{C}_i \Delta t_i}{\sum \mathrm{C}_i \Delta t_i} - t^{*2} \tag{13}$$

When the distribution is normalized, the foregoing equations are simplified for a closed vessel. Thus for a continuous curve or for discrete values at equal time increments, the mean residence time is given by Eq. 14

while the variance is represented by Eqs. 15 and 16.

$$t^* = \int_0^\infty t\mathbf{E}\,dt \cong \frac{\sum t_i \mathbf{E}_i}{\sum \mathbf{E}_i} = \sum t_i \mathbf{E}_i \Delta t \tag{14}$$

$$\sigma^2 = \int_0^\infty (t - t^*)^2 \mathbf{E}\,dt = \int_0^\infty t^2 \mathbf{E}\,dt - t^{*2} \tag{15}$$

$$\sigma^2 = \frac{\sum t_i^2 \mathbf{E}_i}{\sum \mathbf{E}_i} - t^{*2} = \sum t_i^2 \mathbf{E}_i \Delta t - t^{*2} \tag{16}$$

USING RTD IN MODELS OF NONIDEAL FLOW

Two one-parameter models of nonideal flow in reactors are described here. These models sufficiently represent packed beds and tubular vessels.

Dispersion Model

In ideal plug flow, no back mixing or intermixing occurs along the longitudinal or axial direction. In real flow approximating plug flow, some degree of back mixing or intermixing takes place. The latter condition is represented by the plug flow dispersion model, or simply the dispersion model, and is represented by a differential equation (Eq. 17), which is analogous to that of Fick's Law for molecular diffusion. In Eq. 17 that describes flow in the x direction, D is the longitudinal or axial dispersion coefficient, which characterizes the degree of back mixing during flow. Note that in Fick's Law, the coefficient is called the coefficient of molecular diffusion, which accounts alone for molecular diffusion.

$$\frac{\partial C}{\partial t} = D \frac{\partial^2 C}{\partial x^2} \tag{17}$$

In dimensionless form where $z = x/L$ and $\theta = t/t^* = tu/L$, the differential equation representing the dispersion model is given by Eq. 18, wherein the dimensionless parameter D/uL, called the vessel dispersion number, is the parameter which measures the degree of axial dispersion. Thus, $(D/uL) \rightarrow 0$ when dispersion is negligible, hence plug flow, and $(D/uL) \rightarrow \infty$ when dispersion is large, hence mixed flow.

$$\frac{\partial C}{\partial \theta} = \left(\frac{D}{uL}\right) \frac{\partial^2 C}{\partial z^2} - \frac{\partial C}{\partial z} \tag{18}$$

Small Extent of Dispersion

For small extents of dispersion, meaning small (D/uL), the solution to Eq. 18 is given by Eq. 19, which represents a family of normal curves. The mean and variance are given by Eqs. 20 and 21, respectively.

$$\mathbf{C}_\theta = \frac{1}{2(\pi(D/uL))^{1/2}} \exp\left[-\frac{(1-\theta)^2}{4(D/uL)}\right] \tag{19}$$

$$\theta^*_C = \frac{t^*_C}{t^*} = 1 \tag{20}$$

$$\sigma_\theta^2 = \frac{\sigma^2}{t^{*2}} = 2\left(\frac{D}{uL}\right) \tag{21}$$

Fig. 4 shows various ways of calculating (D/uL) from an experimental curve, such as through calculation of

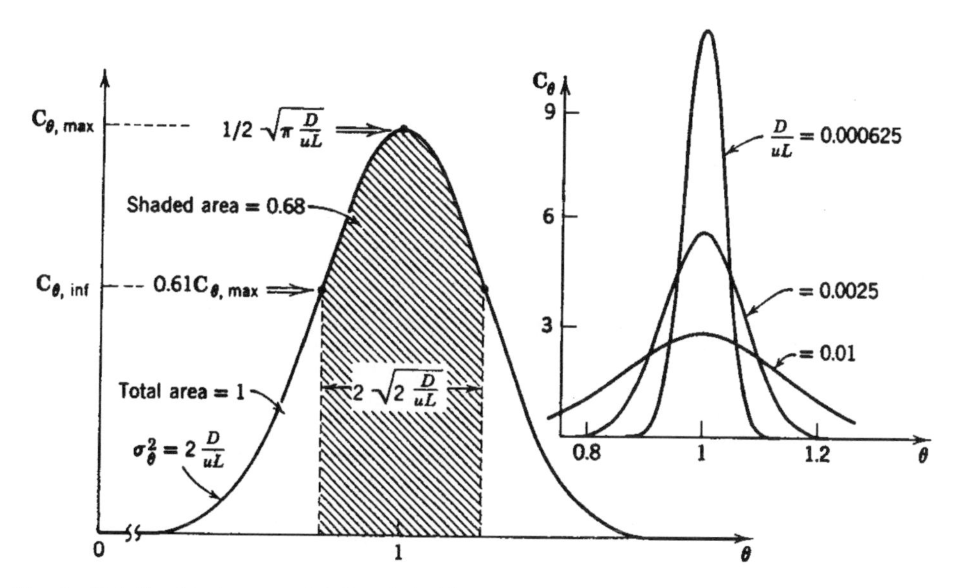

Fig. 4 Relationship between D/uL and the **C** curve for small extents of dispersion. (From Ref. 1.)

the variance, measurement of the maximum height of the curve, measurement of the width of the curve at the point of inflection, etc. The error involved in estimating (D/uL) is less than 5% when $(D/uL) < 0.01$ and is less than 0.5% when $(D/uL) < 0.001$.

Large Extent of Dispersion

In the case of large extent of axial dispersion in a closed vessel, the mean and variance are given by Eqs. 22 and 23, respectively. The **C** curves for closed vessels are illustrated in Fig. 5.

$$\theta^*_C = \frac{t^*_C}{t^*} = 1 \tag{22}$$

$$\sigma^2_\theta = \frac{\sigma^2}{t^{*2}} = 2\frac{D}{uL} - 2\left(\frac{D}{uL}\right)^2\left(1 - e^{-uL/D}\right) \tag{23}$$

Tanks-in-Series Model

The tanks-in-series model assumes the fluid to flow through a series of ideal stirred tanks connected in series, and the one parameter in the model is the number of tanks N in the series. For N tanks, the various forms of the **C** or **E** curve obtained are represented by Eqs. 24–26, where t^*_i is the mean residence time in one tank, t^* the mean residence time in the N tank system, $\theta_i = t/t^*_i = Nt/t^*$, and $\theta = t/t^* = t/Nt^*_i$. As shown in Fig. 6, the curves have their mean and variance given by Eqs. 27–29. Note that

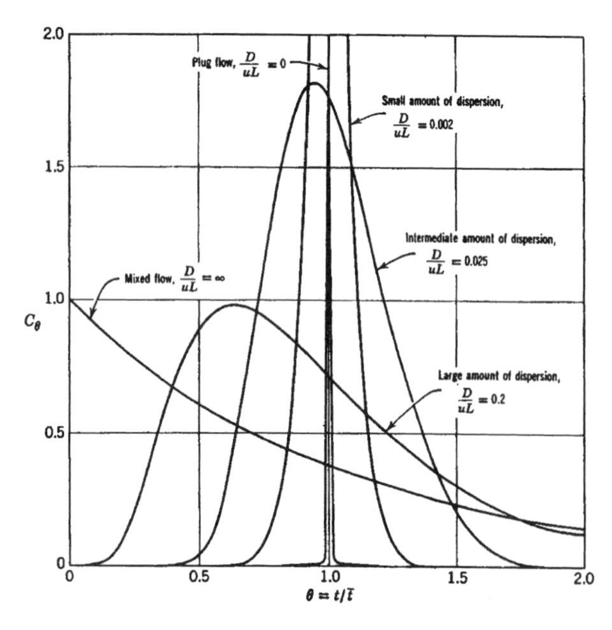

Fig. 5 **C** curves in closed vessels for various extents of back mixing. (From Ref. 1.)

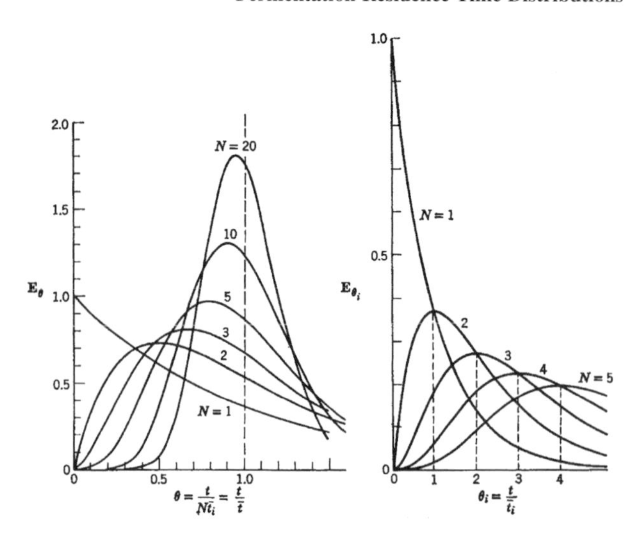

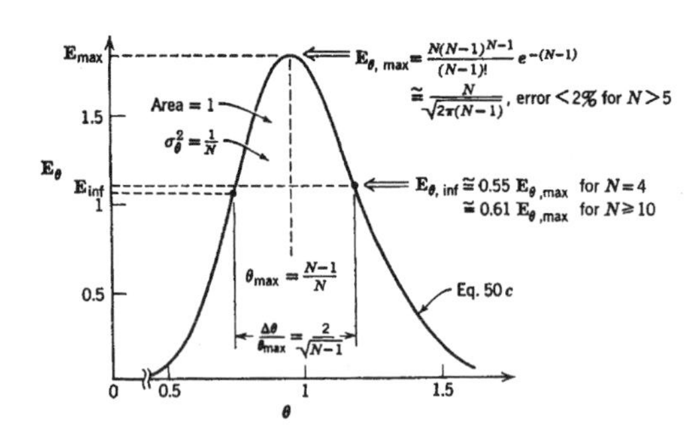

Fig. 6 RTD curves (top) and a curve's properties (bottom) for the tank-in-series model. (From Ref. 1.)

increasing N leads to an approximation of plug flow.

$$t^*_i\,\mathbf{E} = \left(\frac{t}{t^*_i}\right)^{N-1}\frac{1}{(N-1)!}\exp[-t/t^*_i] \tag{24}$$

$$\mathbf{E}_{\theta i} = t^*_i\mathbf{E} = \frac{\theta_i^{N-1}}{(N-1)!}\exp[-\theta_i] \tag{25}$$

$$\mathbf{E}_\theta = (Nt^*_i)\mathbf{E} = \frac{N(N\theta)^{N-1}}{(N-1)!}\exp[-N\theta] \tag{26}$$

$$t^* = Nt^*_i \quad \sigma^2 = Nt^{*2}_i = \frac{t^{*2}}{N} \tag{27}$$

$$t^*_{\theta i} = N \quad \sigma^2_{\theta i} = N \tag{28}$$

$$t^*_\theta = 1 \quad \sigma^2_\theta = \frac{1}{N} \tag{29}$$

Fig. 6 shows some properties of the curves that can be used to estimate the value of N. For instance, N can be estimated

from the maximum height of the curve, its width at the point of inflection, or the variance of the measured **C** curve.

CONCLUSION

RTD, which measures the duration of stay of different elements of a fluid that is continuously moving through a reactor, is a practical method for determining the magnitude of deviation of the flow patterns in real reactors from those in ideal reactors. The magnitude of such deviation is a critical consideration in scale up, as this magnitude varies widely between large and small reactors. This magnitude of deviation could reasonably be determined for packed beds and tubular vessels using the one-parameter dispersion model and tanks-in-series model.

REFERENCES

1. Levenspiel, O. *Chemical Reaction Engineering*, 2nd Ed.; John Wiley & Sons: New York, 1972.
2. Hill, C.G. *An Introduction to Chemical Engineering Kinetics & Reaction Design*; John Wiley & Sons: New York, 1977.

Filtration

G. S. V. Raghavan
E. C. M. Sanga
McGill University, Montreal, Quebec, Canada

INTRODUCTION

Solid–liquid separation is a major unit operation that exists in almost every flow scheme related to the chemical process industries, mining/ore industries, pharmaceuticals, food industries, and water and waste treatment. Filtration is one of the solid–liquid separation methods in which the solid–liquid mixture is directed towards a medium (screen, paper, woven cloth, membrane, etc.) for separation purposes. The liquid phase or filtrate flows through the medium while solids are retained, either on the surface or within the medium. For filtration of coarse materials a filter medium can be a woven wire mesh, which can retain particulates on the surface of the screen. As the size of particulate decreases, other screens may be required such as woven cloth and cellulose membranes.[1,2] For the filtration process various techniques and modes of operations may be employed to aid in separation such as gravity, vacuum, pressure and centrifugal. The separation techniques are diverse and the objective of this article is to highlight filtration mechanisms, equipment, operational models, and techniques for plant and design engineers, research personnel, and students.

CLASSIFICATION OF FILTRATION

The filtration process can either be intermittent or continuous depending on the choice of the equipment. In a batch filtration process the sequence is alternated between filling and discharge. The available filtration techniques are gravity, vacuum, pressure, and centrifugal. Each technique has its advantages and limitations. Fig. 1 gives a general summary and classification of filtration equipment.

Vacuum Filtration

This type of operational mode has a wide number of applications. The absolute downstream pressure of the filter medium is kept low using vacuum pumps. This creates suspension flow to the filter medium. There are various types of filtration machines that operate in this mode (Fig. 1).

Advantages and disadvantages of vacuum filtration

The advantages and disadvantages of vacuum filtration when compared to other filtration methods are:

Advantages

- Intensive soluble recovery or removal of contaminants from the cake by counter-current washing.
- Producing relatively clean filtrates by using a sedimentation basin (on Horizontal Belt, Tilting Pan, and Table Filters).
- Convenient access to the cake for sampling or operator's activities.
- Easy control of operating parameters such as cake thickness.

Disadvantages

- Higher residual moisture in the cake.
- Difficult to clean (mainly as required for food-grade applications).
- High power consumption by the vacuum pump.

Pressure Filtration

Filtration medium can be fitted to various types of equipment and can be pressure-operated. Raising the pressure by pumping creates the fluid flow through the medium. The pressure-operated filtration is done above atmospheric pressure by creating pressure differential across the medium. Equipment can be operated at either constant pressure differential or constant flow rate in a wash and cake discharge mode at the end of the filtration cycles.[3] The filtration cycle may extend from 5 min to 10 min on cake filtration applications and up to 8 or even more hours for polishing of liquids. Since the operation is in batches, usually fed from and discharged to a continuous process, a surge tank is required upstream to the filter and batch collection of cake downstream. The collection rate of

Encyclopedia of Agricultural, Food, and Biological Engineering
DOI: 10.1081/E-EAFE 120007204
Copyright © 2003 by Marcel Dekker, Inc. All rights reserved.

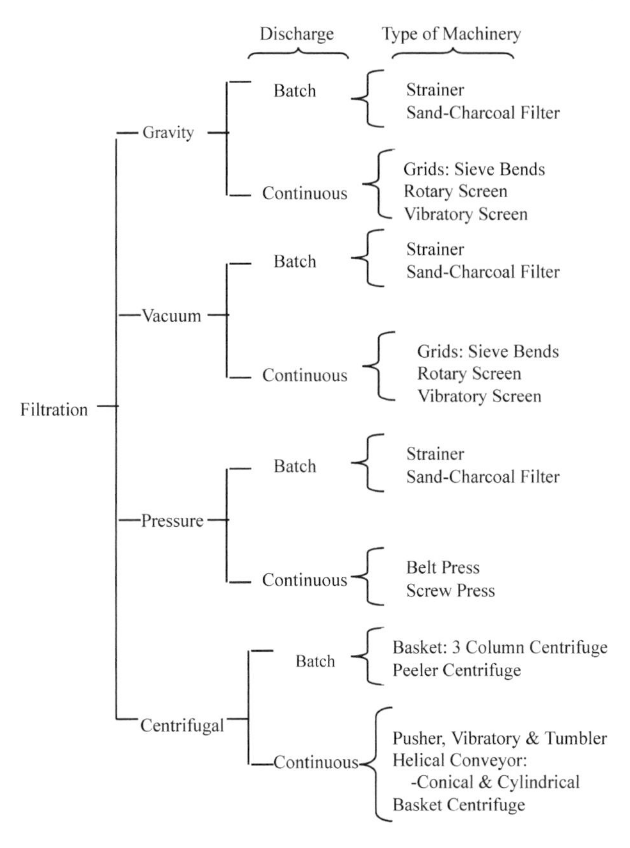

Fig. 1 Classification of filtration equipment. (From Ref. [3].)

the filtrate depends on the operating mode of the filter, which can be constant flow rate, constant pressure, or both with pressure rising and flow rate reduced as dictated by a centrifugal pump.

Advantages and disadvantages of pressure filtration

The advantages and disadvantages of pressure filtration compared to other separation methods are[3,4]:

Advantages

- Cakes are obtained with very low moisture content.
- Cakes may be disposed and flattened in layers provided.
- Intensive soluble recovery or removal of contaminants from the cake may be achieved.
- Clean filtrates may be produced by re-circulating the filtrate for 1 min–2 min or by precoating if a clear filtrate is required right from the start.
- Solutions may be polished to a high degree of clarity.
- The filter bodies and internals may be constructed from a wide variety of alloys including synthetic materials.
- Pressure filters are available in a wide range of automation category starting from labor-intensive operator controlled to fully automatic machines.

Disadvantages

- Washing the filter medium is difficult and especially when the solid cake is sticky.
- The internals are difficult to clean and this may be a problem with food-grade applications.

Centrifugal Filtration

Spinning the suspension can also create pressure differential across the medium. This creates a centrifugal force in the system, thus fluid flow through the medium. This type of operational modes requires centrifugal filters with a suitable filter media to withstand the centrifugal force. Filtration machines operated in this mode are commonly found in food industries, beverage, and pharmaceutical industries.

FILTRATION MECHANISMS

For proper selection and operation of the filtration equipment, the physical mechanism has to be understood. Understanding the quantitative relationships between operational variables such as liquid flow rate, viscosity, particle size and concentration, and filter medium size rating is an important aspect for proper operation of the filtration process. The deposition of the particulate on the surface of filter medium is due to the fluid flow resulting from:

1. The resistance of the filter medium R_m.
2. The resistance of the particulate layer or cake R_c.

The velocity v_0 through the clean filter medium is proportional to the pressure differential ΔP imposed over the medium, inversely proportional to the viscosity of the flowing fluid μ and the resistance of the medium. Mathematically the relationship can be expressed as[5]:

$$v_0 = \frac{\Delta P}{\mu R_m} \tag{1}$$

The filtrate velocity after the deposition of particles on the surface of the filtration medium under the same overall pressure differential decreases to v_f where[5]:

$$v_f = \frac{\Delta P}{\mu (R_c + R_m)} \tag{2}$$

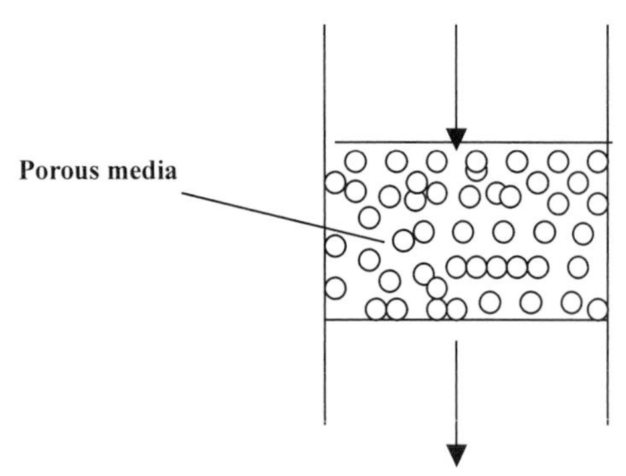

Fig. 2 Schematic diagram of porous media.

The above expressions are simple and do not take into consideration particle concentration effects, cake compressions, pore sizes, and permeability changes.[6] Equations are based on the assumption that the medium resistance does not change during filtration process. The equations were developed based on Darcy's law that describes the relationship between the pressure drop and the flow rate of fluid passing through a packed bed of solids. As the liquid passes between the particles of a packed bed (Fig. 2), the frictional losses lead to a pressure drop.

Darcy's law states that the pressure drop is directly proportional to the flow rates of fluid flowing through fixed porous media. Fig. 3 shows the relationship between the pressure drop and liquid flow rates through the porous media. This relationship can be expressed mathematically as[5]:

$$\frac{\Delta P}{L} = \frac{\mu}{k}\frac{\mathrm{d}V}{\mathrm{d}t}\frac{1}{A} \tag{3}$$

where ΔP is the pressure drop, L, the porous bed depth, μ, the fluid viscosity, k, the porous medium permeability, $\mathrm{d}V$, the volume flowing in $\mathrm{d}t$, and A, the cross-sectional area of the bed.

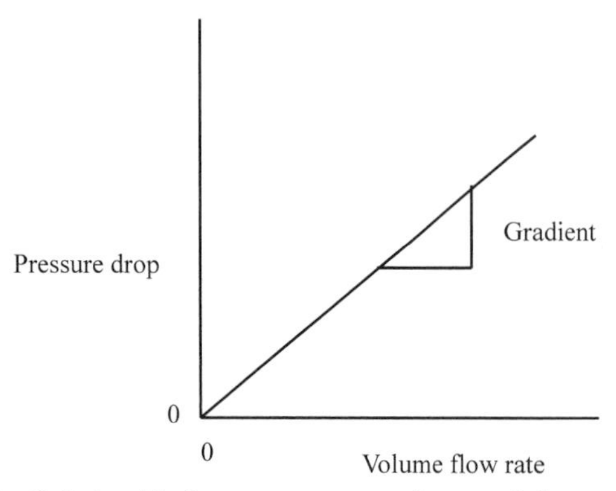

Fig. 3 Relationship between pressure drops and flow rates of liquid flowing through fixed porous media.

The slope of the curve from the Fig. 3 can be found in Eq. 3 above as being equal to:

$$\text{Slope} = \frac{\mu}{k}\frac{L}{A} \tag{4}$$

FILTRATION EQUATIONS

During filtration, solids are deposited on the filter medium resulting in a filter cake. The filter cake formation can be described from Darcy's law (Eq. 3), which relates the flow rate and the pressure drop. During filtration the cake depth increases due to solid deposits and changes both the flow rate and the pressure differential.[5] In incompressible filtration the filter cake volume increases uniformly throughout filtration process. Rearranging Darcy's equation we get a cake filtration equation as:

$$\frac{\mathrm{d}V}{\mathrm{d}t} = \frac{A^2 \Delta P_\mathrm{c}}{\mu c V \alpha} \tag{5}$$

where c is the dry mass of solids per unit filtrate volume, V, the filtrate volume flow rate, α, the local specific cake resistance.

Eq. 5 has three variables: time, filtrate volume, and pressure. The four remaining, area, viscosity, dry mass of solids per unit filtrate volume, and specific resistance, are all constant. For Eq. 5 to be solved, one of the three variables has to be kept constant. Thus, the operational mode of filtration dictates the variable to be kept constant. Mathematically we can express vacuum filtration operated at constant pressure difference and pressure filtration that is operated under constant flow rate.

The filter medium resistance R_m is constant throughout the filtration process. The resistance does not change. Darcy's equation can be rearranged to take into account the constant filter medium resistance as:

$$\frac{\mathrm{d}V}{\mathrm{d}t} = \frac{A}{\mu R_\mathrm{m}}\Delta P_\mathrm{m} \tag{6}$$

The filter medium R_m is defined in terms of filter medium depth L_m and the filter permeability k_m as:

$$R_\mathrm{m} = \frac{L_\mathrm{m}}{k_\mathrm{m}}$$

Constant Pressure Filtration

For constant pressure filtration, pressure drop across the filter case and the filter medium is considered to give the overall pressure drop as:

$$\Delta P = \Delta P_c + \Delta P_m \tag{7}$$

Rearranging Eq. 7 in terms of the flow rate through the filter cake and filter medium we get:

$$\Delta P = \frac{\mu c \alpha}{A^2} V \frac{dV}{dt} + \frac{\mu}{A} R_m \frac{dV}{dt} \tag{8}$$

Under constant pressure Eq. 9 can be rearranged and integrated under with the following limits.

At time $t = 0$ filtrate volume is zero;

Volume V at t

$$\int_0^t dt = \frac{\mu c \alpha}{A^2 \Delta P} \int_0^V V dV + \frac{\mu R_m}{A \Delta P} \int_0^V dV \tag{9}$$

Integrating Eq. 9 with limits and rearranging we get:

$$\frac{t}{V} = \frac{\mu c \alpha}{2A^2 \Delta P} V + \frac{\mu R_m}{A \Delta P} \tag{10}$$

In Eq. 10, t/V is the dependent and V is the independent variable. Experimental data points of t/V and V can be used to plot a graph. From the graph of the experimental data the gradient and intercept of Eq. 10 can be found.

$$\text{Gradient} = \frac{\mu c \alpha}{2A^2 \Delta P}$$

$$\text{Intercept} = \frac{\mu R_m}{A \Delta P}$$

Constant Rate Filtration

Constant rate filtration can be observed on a plot of filtrate volume against time. For this type of filtration, cake filtration, Eq. 8 is rearranged using the set conditions of:

$$\frac{dV}{dt} = \frac{V}{t} = \text{Constant}$$

Eq. 8 then becomes:

$$\Delta P = \left(\frac{\mu c \alpha V}{A^2 t}\right) V + \left(\frac{\mu R_m}{A} \frac{V}{t}\right) \tag{11}$$

A plot of Eq. 11 (filtration pressure against filtrate volume) is a straight line with gradient and intercept. The intercept and gradient from the plot can be used to calculate the specific resistance and filter medium resistance provided the liquid viscosity, filter area, and mass of dry cake per unit volume of filtrate are known.

Table 1 Filter medium classification

Type	Example	Minimum trapped particles (μm)
Edge filters	Wire-wound tubes; Scalloped washers	5–25
Metallic sheets	Perforated plates	100
	Woven wire	5
Woven fabrics	Woven cloths	10
	Natural and synthetic fibres	
Cartridges	Spools of yarns or fibre	2

(From Ref. [7].)

CLASSIFICATION OF FILTER MEDIA

Filter medium plays a key role of separating particulate solids from flowing fluid. Resistance of filter medium needs to be of minimum value to reduce energy consumption during filtration process. The criteria for the selection of filter medium are[7]:

1. The permeability of new filter medium.
2. The stopping power of the medium.
3. The permeability of used filter medium.

The classification of filter medium is based on the rigidity of the media. Table 1 shows the filter media and minimum sizes of particulate that can be trapped.

The classification gives a wide guide to the types of filter medium available. However, individual manufacturer and supplier may provide a detail guide to the types of filter medium available and the sizes of particulate that can be retained.

CONCLUSION

In this article, the basic fundamental principles underlying solid–liquid filtration have been presented. Filtration techniques, operational modes, and equipment have also been highlighted and discussed. Fundamental equations describing the filtration process have been given to guide the design, selection, and proper operation of filtration equipment. Equations presented enable the prediction of equipment performance provided the resistance of the filter medium, the anticipated pressure differential, operation of equipment, and the viscosity of flowing fluid are known. However, for in-depth understanding of the filtration process and other aspects of modeling, readers are referred to the literature cited in this article.

NOMENCLATURE

A	Area (m^2)
c	Dry mass of solids per unit filtrate (—)
C	Solid concentration by volume fraction (m^2)
k	Filter cake permeability (—)
L	Filter cake or bed thickness (m)
ΔP	Pressure differential ($N\,m^{-2}$)
R_c	Filter cake resistance (m^{-1})
R_m	Filter medium resistance (m^{-1})
t	Time (sec)
V	Volume of filtrate (m^3)
v_f	Filtrate velocity after cake formation ($m\,sec^{-1}$)
v_0	Filtrate velocity before cake formation ($m\,sec^{-1}$)

Greek Symbols

α	Local specific cake resistance ($1/kC\rho_s$) ($m\,kg^{-1}$)
μ	Liquid viscosity ($kg\,m^{-1}sec^{-1}$)
ρ_s	Solid density ($kg\,m^{-3}$)

Subscripts

c	Filter cake
m	Filter medium

REFERENCES

1. Matsuura, T.; Sourirajan, S. Physicochemical and Engineering Properties of Food in Membrane Separation Processes. In *Engineering Properties of Foods*; Rao, M.A., Rizvi, S.S.H., Eds.; Marcel Dekker, Inc.: New York, 1995; 311–388.
2. Rushton, A.; Griffiths, P.V.R. Filter Media. In *Filtration: Principles and Practices Part I*; Orr, C., Ed.; Marcel Dekker, Inc.: New York, 1977; 251–308.
3. Rushton, A. Solid Liquid Separation Technology. In *Solid–Liquid Filtration and Separation Technology*; Rushton, A., Ward, A.S., Holdrich, R.G., Eds.; Wiley-VCH: Weinheim, 1996; 1–32.
4. Akers, R.J.; Ward, A.S. Liquid Filtration Theory and Filtration Pretreatment. In *Filtration: Principles and Practices Part I*; Orr, C., Ed.; Marcel Dekker, Inc.: New York, 1977; 169–250.
5. Holdich, R.G. Filtration Fundamentals. In *Solid–Liquid Filtration and Separation Technology*; Rushton, A., Ward, A.S., Holdrich, R.G., Eds.; Wiley-VCH: Weinheim, 1996; 33–84.
6. Espadel, M.S.; Fasano, A.; Mikelic, A. *Filtration in Porous Media and Industrial Application*; Springer-Verlag: New York, 1998.
7. Loff, L.G. Filter Media, Filter Rating. In *Solid–Liquid Separation*; Svarovsky, L., Ed.; Butterworth and Co., Ltd.: London, 1990.

Flexible Packaging

Thomas H. Shellhammer
Oregon State University, Corvallis, Oregon, U.S.A.

F

INTRODUCTION

Flexible packaging is evolving to become the largest segment of the food packaging industry. Its dominance is driven by factors such as packaging material source reduction, convenience for the end user, and visual and handling appeal. The tremendous growth in flexible packaging in the last century is linked directly to the growth of the plastics industry. This category of packaging is defined by its material's properties, i.e., the flexural strength of the packaging material is sufficiently weak that when empty, the package generally does not retain a specific shape. Although it may be formed into a definite shape when empty, this form of packaging generally relies on the product being contained to provide structure. Developments in polymer synthesis and film manufacturing and converting during the latter half of the 20th century resulted in a wide array of film material choices with improved barrier and mechanical properties. Selection of appropriate materials in combination with modified atmosphere packaging (MAP) resulted in substantial improvement in product shelf life and quality. Compared to the other two major forms of packaging, semirigid and rigid, flexible packaging is arguably the most diverse and largest segment of the food packaging industry, one that is an $18 billion industry in the United States at the turn of the 21st century.[1]

HISTORY

Until the early 20th century, package design was dictated as form following function. That function was containing, protecting, and dispensing material. Following the industrial revolution, the types and availability of materials for packaging increased, and the package function evolved to include informing consumers of product contents as well as marketing the product or brand.

In its earliest cases, flexible packaging was based on natural, flexible materials, such as parchment (animal skins), vellum (skins of newborn calves, kids, or lambs), papyrus, paper, or woven plant fibers (baskets).[2] Paper wrapping served as flexible packaging material for centuries. The earliest surviving printed package was a wrapper used by a German papermaker from the 1550s.[3] The mid-nineteenth century saw the introduction of thermal processing for preserving foods, and thus the beginning of a large growth in rigid packaging made of metal or glass. Not until the 1930s did polymeric plastics become a significant contributor to the packaging industry. The tremendous growth in the flexible packaging in the last century is linked directly to the growth of the plastics industry. Techniques in synthesizing new synthetic polymers and developments in film manufacturing and converting have lead to a wide range of versatile materials suitable for flexible packaging. Since World War II, flexible packaging has grown into the most varied industry in the field of packaging.

ADVANTAGES OF FLEXIBLE PACKAGING

Material Source Reduction

In the last quarter of the 20th century, the food industry was driven by needs for convenience, portion control, and waste reduction. Flexible packaging assisted in delivering these needs by lightweight, easy to open, minimal packaging systems. The main advantage of flexible packaging is the source reduction of the material used to create the package. For example, replacing a steel soup can with a flexible pouch reduces the product weight by 93% and the packaging material by 97%. Similarly, packaging juice in a flexible box, instead of a glass bottle, reduces the weight of the product by 90% and amount of packaging material by 70%.[1] The fact that the materials used in this type of packaging are flexible means the final package need not be limited to regular shapes such as cylinders or rectangular solids. For example, high-quality printing techniques and the convenience of resealability produced an exciting and new packaging option, the flexible, self-standing, resealable pouch.

While pouches have been used for packaging dry snack goods for years, they have often lacked resealability. A wide range of closures can be used with pouch systems for dry goods and perishables, with the largest entry being interlocking ridge or zippered closures. The use of resealable flexible packaging has provided greater consumer convenience in the refrigerated fresh-cut and frozen produce categories, as well as broadening the snack

Encyclopedia of Agricultural, Food, and Biological Engineering
DOI: 10.1081/E-EAFE 120007136
Copyright © 2003 by Marcel Dekker, Inc. All rights reserved.

food category beyond savory baked/fried food and baked sweets. The self-standing pouch, with or without resealability, has opened an entirely new category or style for liquid packaging that benefits greatly from source material reduction and improved printability of the film stock.

Modified Atmosphere Packaging with Flexible Films

The shelf-life extension achieved by flexible packaging systems is due in many cases to using MAP. Strictly speaking, MAP is packaging items in an atmosphere other than air (see the article *Modified Atmosphere Packaging*). However, the requirements for MAP depend largely upon the material being packaged. Oxygen sensitive, dry goods such as snack chips require an environment that is free of oxygen and light and is able to prevent the ingress of oxygen over time. A high barrier material, such as vacuum metallized polypropylene, works well at achieving this goal, as it is a great barrier to oxygen (Table 2) and light. Fresh produce on the other hand requires a system to reduce but not prevent the exchange of gases by the polymer film. Produce respires, both consuming oxygen and producing carbon dioxide. Reduction in available oxygen and increases in environmental carbon dioxide slow the respiration process and extend the life of the product. However, excessively high levels of carbon dioxide or the absence of oxygen will cause the product to respire anaerobically, thereby producing off-flavors and odors. The key to MAP of fresh produce is the selection of the appropriate flexible film to achieve the proper restriction of oxygen ingress and carbon dioxide egress, such that the package environment causes a reduction in the produce respiration rate and in turn, an extension in shelf life.

ISSUES FACING FLEXIBLE PACKAGING

Barrier Properties

One of the largest issues that flexible packaging is faced with is due to the barrier properties of polymeric materials. Unlike glass and metal, polymeric packaging possesses some degree of permeation to moisture, permanent gases, flavors, and light. While semirigid, and in particular, rigid packaging made from polymeric materials benefit from having greater material thickness than flexible packaging, the barrier property issue remains. The issue is further complicated by the fact that due to their small size, many flexible packaging systems result in a high film surface area:volume ratio. In terms of shelf-life extension by preventing migration of moisture or gases through a package, a high surface area:volume ratio increases the demand placed on the barrier properties of the packaging materials. High barrier needs are accommodated by first determining the tolerable level of permeant ingress or egress, and then selecting the appropriate material or layers of materials.

The physical (glass transition temperature, degree of cystallinity) and chemical (monomer type, degree and type of chemical substitution, copolymer ratio) properties of the polymer system determine its resistance to permeation of moisture and gases of varying polarity and molecular weight (Table 1). For example, low-density polyethylene (LDPE) is a highly economical moisture barrier, but a relatively poor oxygen barrier. Furthermore, hydrophobic molecules, typical of many food flavors, either permeate freely through LDPE or are selectively bound by the polymer. The latter case is known as flavor scalping. Contrast this with an ethylene vinyl alcohol copolymer (EVOH) in which the polymer system is highly resistant to oxygen permeation (when dry) and to flavor permeation or scalping, but is not resistant to moisture permeation (see also the article *Package Permeability*). In many cases, the former is laminated to or coextruded with the latter, such that the LDPE offers moisture resistance while the EVOH offers gas and flavor resistance for the final multilayered polymeric system. Additionally, these systems, single material or multilayered, may be coated with a high barrier resin or vacuum metallized with aluminum to improve the barrier properties (Table 2). The barrier properties of metallized films depend little on the substrate being coated and almost solely on the amount of aluminum being deposited on the substrate. While metallizing does increase the cost of the final system, it produces a film with a significant barrier to moisture, gases, and UV light and produces a mirror-like appearance, qualities that are desirable to both the producer and the consumer, respectively.

Solid Waste Issues

An issue some users have with flexible packaging involves material use. While the weight of the package is reduced by using polymeric films, the combined barrier, structural, sealing, and printing properties necessitate using multilayered films. Recycling of metal and glass is rather straightforward, but recycling laminated polymeric structures is not yet a reality. Nevertheless, the flexible packaging industry continues to make a strong case about the net reduction in material and energy costs derived from using less material at the outset vs. the cost and energy consumption of using recyclable packaging.

F

Table 1 Attributes and properties of commonly used flexible packaging films

Material	Principle advantages[a]	Principle disadvantages[a]	Oxygen[b]	Moisture[c]	Reference[d]
LDPE	Moisture barrier Heat-seal adhesive Flexible, tough	Poor clarity when thick Absorbs odors and flavors	1934	83	5
Polyvinyl chloride (PVC)	Tough, flexible, clear transparent, oil resistant	Limited hot fill capacity	1321	671	4
Polystyrene, oriented (OPS)	High rigidity, clear, high luster		1304	546	
Polyethylene, high density (HDPE)	Moisture barrier Heat-seal adhesive Flexible, tough	Poor clarity when thick Absorbs odors and flavors	677	28	5
Polypropylene, cast (PP)	Tough, clear Tolerates retorting	Lowered stiffness at elevated temperatures Hazy in thick sections	677	44	5
Polypropylene, oriented (OPP)	Tough, clear Orientation improves barrier properties		501	22	5
Polyester, oriented (OPET)	Clear transparency Good printability Good gas barrier		21	98	6
Polyamide, Nylon 6 (Nylon)	Excellent gas barrier when dry Good aroma barrier Slick, slippery	Moisture sensitive barrier properties	10.1	1355	4
Polyamide, biaxially oriented (Biax Nylon)	Improved flex-crack resistance, mechanical and barrier properties over nonoriented polyamide		7	765	5
Cellophane (Cello)	Stiff, strong, clear Good printability	Moisture sensitive barrier properties	0.7	7.27	7,8
Polyvinylidene chloride (PVdC)	Excellent gas and moisture barrier Unaffected by moisture Retortable	Subject to thermal degradation during extrusion	0.57	3	4
EVOH 30% PE	Stiff, strong, clear Excellent gas barrier when dry Good aroma barrier	Moisture sensitive barrier properties	0.1	385	7,9
Ethylene vinyl acetate (EVA)	Tough, used as sealant Adheres to many substrates	Low stiffness Low barrier properties			
Ionomer	Tough, strong, excellent sealant, high gloss oil resistant	Limited service temperature		135	7,9

[a] Adapted from Ref. [9], p. 2139.
[b] 0% RH, 23–25°C ($cm^3\ \mu m\ m^{-2}\ d^{-1}\ kPa^{-1}$).
[c] 90% RH, 38°C ($g\ \mu m\ m^{-2}\ d^{-1}\ kPa^{-1}$).
[d] Barrier property data adapted from these sources for oxygen and moisture permeability, respectively, or both.

MANUFACTURERS OF FLEXIBLE PACKAGING

Flexible packaging is produced through the coordinated efforts of the plastic film extruding and converting industries. Generally speaking, extruders produce flexible film materials in roll stock quantities, while converters take roll stock, or the resin itself, of base materials and convert them into a multilayered flexible package. The converting operations include laminating, coextruding, printing, and finishing. As the overall barrier and functional properties of a flexible package are often

Table 2 Barrier properties are significantly improved by metallizing with aluminum or coating with high barrier resins

Film type	Oxygen[a]	Moisture[b]	Reference[c]
Vacuum metallized (aluminum) films			
OPET	0.23	7	4,5
Cello	0.14	5	4
Biax Nylon 6	0.20	88	4,5
OPP	4.62	0.7	5
LDPE	143	10	4
HDPE		2.0	5
Coated films			
PVOH coated OPP	0.06	20.6	5
PVdC coated OPP	0.83	7.7	5
PVdC coated Nylon 6	1.93	39	6
PVdC coated OPET	2.01	35	6

[a]0% RH, 23–25°C ($cm^3\ \mu m\ m^{-2}\ d^{-1}\ kPa^{-1}$).
[b]90% RH, 38°C ($g\ \mu m\ m^{-2}\ d^{-1}\ kPa^{-1}$).
[c]Barrier property data adapted from these sources for oxygen and moisture permeability, respectively, or both.

the result of combining properties of several different materials, converters must either join two or more distinct polymeric films or coextrude them together. In the former case, an adhesive is applied to one substrate, allowed to dry, and then the two webs are joined together using heat and pressure. This process is complicated by having to deal with solvent emissions and issues related to unreacted residuals in the joined films. Alternatively, the two substrates may be joined by extruding a thin layer of polymer melt (typically LDPE) between them. The greatest flexibility in producing multilayered films comes through coextrusion. By simultaneously extruding several layers of polymer melt, a very thin multilayered structure can be produced in a single step.

Flexible packaging is printed using flexographic (images on rubber plates that stamp ink onto a substrate) and rotogravure (images etched into metal cylinders that release ink to a substrate from small engraved cells) printing processes.[4] If printing is performed on the outside surface, an additional protective coating is necessary to prevent scuffing of the image and to produce a glossy appearance. Alternatively, the printing can be performed on an internal layer and then covered with an external film, such as polyethylene, polyester, or nylon. The final step in the converting process is called finishing, which consists of preparing the final product for the end-user's packaging equipment. This may entail slitting the printed web to the appropriate size for a particular packaging machine, such as a vertical or horizontal form, fill, and seal or it may entail performing pouches that the end-user fills and seals at a later date.

REFERENCES

1. Anonymous. The Flexible Packaging Association, 2001; http://www.flexpack.org (accessed Sep 2001).
2. Sacharow, S.; Griffin, R.C. *Principles of Food Packaging*, 2nd Ed.; AVI Publishing Company, Inc.: Westport, CT, 1980; 484.
3. Griffin, R.C.; Sacharow, S.; Brody, A.L. *Principles of Package Development*, 2nd Ed.; AVI Publishing Company, Inc.: Westport, CT, 1985; 378.
4. Brody, A., Marsh, K., Eds. *The Wiley Encyclopedia of Packaging Technology*, 2nd Ed.; John Wiley & Sons, Inc.: New York, 1997.
5. ExxonMobil. Packaging Film Property Data, 2001; http://www.exxonchemical.com/chemical/customer/products/families/oppfilms/americas/index.html (accessed Sep 2001).
6. DuPont. Packaging Materials Product Property Data, 2001; http://www.dupont.com/packaging/products/index.html (accessed Sep 2001).
7. Salame, M. Barrier Polymers. In *The Wiley Encyclopedia of Packaging Technology*; Bakker, M., Ed.; John Wiley & Sons: New York, 1986; 48–54.
8. Taylor, C.C. Cellophane. In *The Wiley Encyclopedia of Packaging Technology*; Bakker, M., Ed.; John Wiley & Sons: New York, 1986; 159–163.
9. Brown, W.E. *Plastics in Food Packaging, Properties, Design and Fabrication*; Marcel Dekker, Inc.: New York, 1992.

Flow Measurement

George E. Meyer
Rhonda M. Brand
Gary DeBerg
University of Nebraska, Lincoln, Nebraska, U.S.A.

INTRODUCTION

Flows in biological systems involve primarily air, water, or water-based solutions through pores and capillary systems. Fluid flow involves blood movement for animals, transpiration for plants, and exchange of gases for both plants and animals. Measurement of these flow processes can be used to assess physiological condition. All these flow processes take place in small tubes or capillaries making nonintrusive measurement of flow rates difficult, but basic principles are the same as for larger systems.

OVERVIEW

Fluid flow processes were originally addressed using concepts of particle mechanics first developed by Joseph-Louis Lagrange (1731–1831), but later, a more convenient approach specifying the density and the velocity at a point in space and instant of time was proposed by Leonard Euler (1707–1783). Fluid flow measurement can be addressed by the First Law of thermodynamics, considering heat exchange, work, kinetic energy, and potential energy, as:

$$\dot{Q} - \dot{W} = \Delta\dot{U} + \Delta\dot{\mathrm{KE}} + \Delta\dot{\mathrm{PE}} \tag{1}$$

where $\dot{Q}$ is the net heat transfer in system (W), $\dot{W}$ the net work in system (W), $\Delta\dot{U}$ the change in internal energy (W), $\Delta\dot{\mathrm{KE}}$ the change in kinetic energy (W), and $\Delta\dot{\mathrm{PE}}$ the change in potential energy (W).

The First Law applies to any closed or open system and steady or nonsteady flow condition. The internal energy, $\Delta\dot{U}$, can be replaced by $\Delta\dot{H}$ (enthalpy) when the flow energy or boundary work is included with internal energy. The mass flow rate of the fluid is included, and given for a steady flow system as:

$$\dot{Q}_{\mathrm{in}} + \dot{W}_{\mathrm{in}} + \dot{m}\left(h_{\mathrm{in}} + \frac{V_{\mathrm{in}}^2}{2} + gz\right) = \dot{Q}_{\mathrm{out}} + \dot{W}_{\mathrm{out}} + \dot{m}\left(h_{\mathrm{out}} + \frac{V_{\mathrm{out}}^2}{2} + gz_{\mathrm{out}}\right) \tag{2}$$

where h_{in} is the inlet enthalpy (kJ kg^{-1}), h_{out} the outlet enthalpy (kJ kg^{-1}), V_{in} the inlet average fluid velocity (m sec^{-1}), V_{out} the outlet average fluid velocity (m sec^{-1}), Z_{in} the inlet elevation (m), Z_{out} the outlet elevation (m), g the acceleration gravity (9.807 m sec^{-2}), $\dot{W}_{\mathrm{in}}$ and $\dot{W}_{\mathrm{out}}$ the nonboundary forms of work, which could include electrical induction or capacitance, $\dot{Q}_{\mathrm{in}}$ the heat transfer in (kJ sec^{-1}), and $\dot{Q}_{\mathrm{out}}$ the heat transfer out (kJ sec^{-1}).

Eq. 3 can be simply rewritten as:

$$\dot{Q} - \dot{W} = \dot{m}\left(\Delta h + \frac{\Delta V^2}{2} + g\Delta z\right) \tag{3}$$

However, the First Law must also be accompanied by a mass balance, which takes into account changes in fluid density, velocities, and flow inlet and exit opening areas. The mass balance, well known as the continuity equation, is given as:

$$\sum \int_{A_{\mathrm{i}}} (\rho V_{\mathrm{n}}\, \mathrm{d}A)_{\mathrm{i}} - \sum \int_{A_{\mathrm{e}}} (\rho V_{\mathrm{n}}\, \mathrm{d}A)_{\mathrm{e}} = \frac{\mathrm{d}}{\mathrm{d}t}\int_{V} (\rho\, \mathrm{d}V)_{\mathrm{cv}} \tag{4}$$

where A is the cross-sectional area of pipe or duct (m^2), ρ the fluid density (kg m^{-3}), V_{n} the velocity component normal to dA (m sec^{-1}), and i the inlet, e the exit, and cv the control volume.

The continuity equation is the mass balance, which takes into account the variation of fluid and channel properties.

Most flow measurements are made across boundaries of open systems, so work is required to push mass into or out of the control volume. This work is known as flow work or flow energy and is needed to maintain continuous flow. Thus, the work required to push a fluid element across the control volume boundary is given as:

$$W_{\mathrm{flow}} = v\Delta P \tag{5}$$

where W_{flow} is the flow work (J), ΔP the pressure of system (Pa), and v the specific volume (m^3 kg^{-1}).

That work, a form of energy, is given by 1 joule per Pascal m^3, but in most biomedical literature as mm of mercury.

Encyclopedia of Agricultural, Food, and Biological Engineering
DOI: 10.1081/E-EAFE 120007166
Copyright © 2003 by Marcel Dekker, Inc. All rights reserved.

MEASUREMENT PRINCIPLES

There are a variety of different sensing technologies for measuring porous and capillary fluid flow. These are often categorized as micro differential-pressure flow sensors and use principles of the First Law; mechanical flow sensors resulting from the displacement of vanes and rotary impellers; thermal flow sensors, which look at the effects of convecting heat flow of moving fluids; magnetic flow sensors, which look at inductive and capacitive effects of electrical charge fields and fluid movement; ultrasonic flow sensors, which use the fluid media for transmission of differential sound propagation using either transit time or doppler frequency shifts; and photonic sensors that consider doppler shifts of laser light through moving fluids. These sensing methods are described by Gardner[1] and Webster.[2] There are also methods of counting pulses of fluid movement, which are often important in biological systolic and diastolic capillary flow.

THERMAL FLOW MICROSENSORS

A thermal flowmeter is based on the immersion of a heating element into a moving fluid. From the First Law (Eq. 1), heat $\dot{Q}_{in}$ is applied to a point within the fluid rejoin. The internal energy $\Delta\dot{U}$ plus the flow energy of the fluid can be represented by the enthalpy change $\Delta\dot{H}$. If we assume no changes in kinetic $\Delta\dot{KE}$ or potential $\Delta\dot{PE}$ energies, then:

$$\dot{Q}_{in} = \Delta\dot{H} = \dot{m}\Delta h \tag{6}$$

Eq. 6 can be rewritten as a difference in temperatures:

$$Q_{in} = \dot{m}C(t_2 - t_1) \tag{7}$$

where C is the specific heat of fluid (kJ kg^{-1} °C^{-1}) and t_2, t_1 are the upstream and downstream temperatures of the fluid (°C), respectively.

Solving for the mass flow rate gives:

$$\dot{m} = \dot{Q}_{in}/C(t_2 - t_1) \tag{8}$$

Eq. 8 carries the assumption of a complete immersion of the heat source in the fluid. Heat sources near the pipe or duct walls need to take into account boundary layer resistances to heat flow and other heat losses. Radiant heat loss is usually small, less than 0.5°C accuracy loss in measurement. Temperature measurements need to be rather precise using thermocouples. The placement of these thermocouples is important. One disadvantage is that the measurement process is somewhat intrusive, reducing the accuracy of the mass flow rate obtained.

The measurement of microcapillary sap flow or transpiration in intact stems using a stem-surface heat balance measurement technique has become widespread. These have resulted from the works of Sakuratani[3,4] and Baker and VanBavel.[5] Cohen et al.[6–8] demonstrated a heat pulse method for sap flow based on the rate of heat transfer, stem area, and a coefficient based on the structure of the xylem conductive area. Such a system is shown by Fig. 1. For low flow rates (0–0.22 mm sec^{-1}), the convective heat velocity was given as:

$$V = \frac{x_1 - x_2}{2t_0} \tag{9}$$

where V is the flow rate (mm sec^{-1}), x_1, x_2 the distances from the heater to the thermocouple junctions (mm), respectively, and t_0 the time elapsed from the pulse emission to first recurrence of the initial temperature difference (sec).

Above 0.22 mm sec^{-1}, the flow rate is given as:

$$V = \frac{(x_1 - 4kt_m)^{0.5}}{t_m} \tag{10}$$

where k is the thermal diffusivity of the stem (mm^2 sec^{-1}) and t_m the time at which the temperature at the upper sensor reaches a maximum (sec).

The constant heat balance method of Baker and VanBavel[5] applies a known amount of heat to a stem segment from a thin, flexible heater encircling the stem. Cohen found that the heat pulse method was limited for transpiration rates less than 7 grams H_2O per hour.

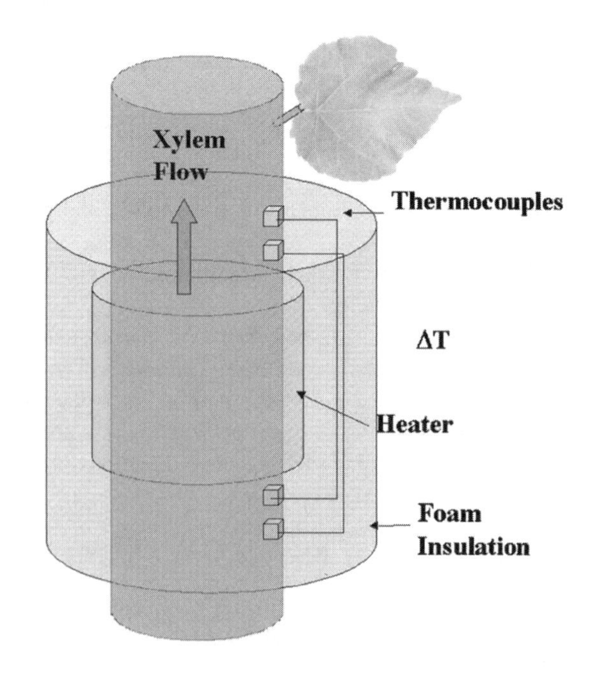

Fig. 1 Heat flow system for transpiration measurements.

However, the heat balance method was satisfactory when transpiration rates were less than 100 grams per hour. Ishida, Campbell, and Calissendorff[9] found that the continuous heat balance method came within 8% of lysimeter weight losses for potted maize, sunflower, and potato.

LASER-DOPPLER FLOW SENSORS

The laser-doppler flow sensor attempts to obtain a measurement of the fluid velocity using the principle of kinetic energy $\Delta\dot{K}E$. These methods fall into a broader class of laser velocimetry or anemometry. The most common approach is the use of a dual laser beam system. An optical beam splitter is used to create two parallel laser beams of the same luminance. Optics is used to focus and cross the two beams onto a target area, the intersection of which becomes the region of measurement. Particles in the fluid intercept and scatter the laser light. Receiving optics and a photodetector collect some of the scattered light. The photodetector converts the scattered light to an analog electric signal, which is a pulsed signal whose frequency is proportional to the fluid particle velocity. Frequency information is provided by a digital signal processor or counter.

The dual-beam system provides laser light beams of a known wavelength λ and an angle δ between them, based on the focusing optics. The optics also focuses the laser beams to a small measurement region. At the point the beams cross or intersect, the wave fronts of each beam interfere with one another, creating what is called a fringe pattern. These are alternating light and dark regions of light, equally spaced. As particles in the fluid traverse the fringe pattern, the velocity of the fluid particle is proportional to the fringe spacing (d_{fs}) and doppler frequency (f_{D}) or time ($1/f_{\mathrm{D}}$) to cross a pair of fringes. The fringe spacing is given as:

$$d_{\mathrm{fs}} = \frac{\lambda}{2\sin\delta} \tag{11}$$

and the apparent fluid velocity is given as:

$$V = f_{\mathrm{D}} d_{\mathrm{fs}} \tag{12}$$

Laser-doppler flow systems are quite straightforward in that the fringe spacing is only dependent on the wavelength of light and the optics and the angle δ between the two laser beams. Laser-doppler systems are very accurate for turbulent flows and for very small fluid particle sizes.

Lu et al.[10] described the use of laser-doppler flowmetry for measuring blood flow rates in conscious rats. Optical fiber probes (0.5-mm diameter) were inserted into the kidneys of rats. Cortical and medullary fibers were inserted and connected to a master probe of a commercial laser-doppler flowmeter. The use of acutely implanted optical fibers for laser-doppler measurement of regional blood flow correlated very closely to similar measurements with electromagnetic flow probes.

F

ULTRASONIC FLOW SENSORS

A flowmeter that uses ultrasound is based on principles of doppler sonic frequency shift or transit time for sound to propagate through a moving fluid. Generally, ultrasound waves are used and include frequencies higher than the range of the human ear, i.e., greater than 18 kHz. The advantages of an ultrasound signal system are that the shorter wavelengths are more easily controlled and can be focused for a more noninvasive measurement. Ultrasound signals can be launched through metal walls or any material with a high conductance for sound transmission. The sound propagation speed is much slower than light in fluids, e.g., $10^6\,\mathrm{m\,sec^{-1}}$ vs. $1000\,\mathrm{m\,sec^{-1}}$ water. A disadvantage is that ultrasound signals cannot be as well focused or as specific as optical laser flow measurements. Ultrasound emitters and earphones need to be tuned to specific sound frequencies. As with light, suspended particles or bubbles interrupt the ultrasound propagation, resulting in a frequency shift that is proportional to the velocity of the particles and fluid flow rate.

The transmitting crystal or emitter sends an ultrasonic wave with a frequency of 1 mHz and a sound propagation angle θ in the direction of the flow. An earphone or receiver in-line downstream will observe an apparent sound wave velocity given as:

$$c + V\cos\theta \tag{13}$$

where V is the mean velocity of the fluid say, $10\,\mathrm{m\,sec^{-1}}$ and c is the sound propagation velocity, e.g., $1000\,\mathrm{m\,sec^{-1}}$ for water.

The apparent frequency f^1_{us} and its ratio to the original frequency is proportional to the apparent sound and original sound velocities as:

$$\frac{f^1_{us}}{f_{\mathrm{us}}} = c + \frac{V\cos\theta}{c} \tag{14}$$

However, the velocity of sound waves relative to source is:

$$c - V\cos\theta \tag{15}$$

Therefore, the corresponding frequency ratio of the apparent sound velocity at the source to the sound velocity at the receiver is given as:

$$f_{\text{ratio}} = \frac{c + V\cos\theta}{c - V\cos\theta} \quad (16)$$

which can be reduced to a frequency difference and sound and particle velocity ratio as:

$$\Delta f_{\text{us}} = \frac{2f_{\text{us}}V}{c}\cos\theta \quad (17)$$

For example, when $\theta = 30°$, $\Delta f_{\text{us}} = 17.3\,\text{kHz}$.

Variations of the ultrasound systems include the cross-correlation flowmeter and the transit time flowmeter. The transit time is used to measure pulse–echo time differences, which are dependent on pipe diameter as well as on the angle of sound transmission into the fluid stream.

ELECTROMAGNETIC FLOWMETERS

Electromagnetic flowmeters or flow probes are noninvasive devices used for measuring blood flow in vessels through animal circulatory systems. Consider a conductor shown in Fig. 2 consisting of a fluid in motion with a velocity V and subjected to a uniform electric charge across the flow field. Every charged particle in the fluid will experience an electric induction force given as:

$$F = qVB \quad (18)$$

where F is the induction force (N), q the charge (C), V the flow velocity (m sec^{-1}), and B the electromagnetic field strength (W m^{-2}).

Induction simply means the transfer of energy from one electromagnetic field to another field using electrical coils. The induction rate is governed by basic electrical permittivity and dimensions of the induction path. The induced voltage in the receiving coil is given by:

$$\text{EMF} = BLV \times 10^{-8} \quad (19)$$

where EMF is the induced voltage (V), B the magnetic field strength (G), L the diameter of the tube (m), and V the velocity of liquid (m sec^{-1}).

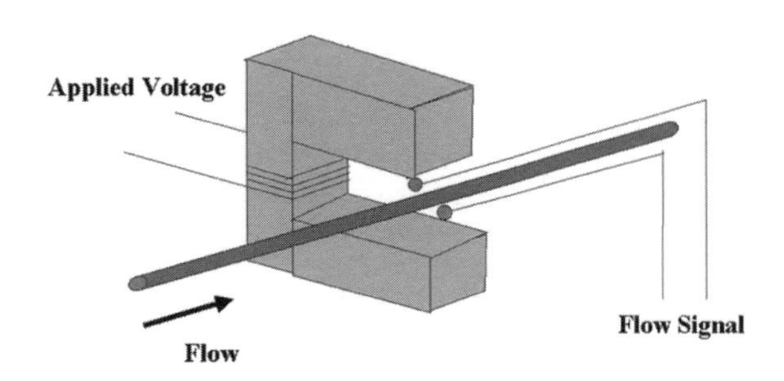

Fig. 2 Electromagnetic flow measurement systems..

Scott and Sandler[11] discuss uses and practical considerations of electromagnetic flowmeters for blood flow measurements: 1) there are polarity considerations in mounting these probes around or within blood vessels; 2) while probes may be precalibrated, variations in actual and measured flows will occur due to variation in conductivity or vessel walls, changes in electrolytes and ionic composition; 3) the use of pulse or continuous electromagnetic wave forms; 4) probe size and contact area; 5) orientation of electromagnetic devices; and 6) careful cleaning and calibration in saline solutions are necessary for accurate readings. Neuwirth and Rawlings[12] discuss calibration of electromagnetic flowmeters. Durand et al.[13] measured mean portal blood flow rates in calves.

POSITIVE DISPLACEMENT FLOWMETERS

The positive displacement flowmeter called the PD meter provides a volume flow rate through entrapment of a small quantity of fluid in a chamber of known volume and release back to a continuous stream of fluid. Plant transpiration rates can be measured using a computer control system and water potential-controlled incremental volumetric (V) monolithic lysimeter, shown by Fig. 3. This lysimeter can supply both water and nutrients (microirrigation) to each plant and provides measured transpiration data for feedback. Various localized heating and cooling treatments at various plant temperatures and transpiration were studied for crops of New Guinea Impatiens by Al-Faraj, Meyer, and Fitzgerald.[14] High daily water use accuracy can be obtained with large metering values of ΔV. However, to keep the daily error rate within an acceptable range, ΔV should be less than or

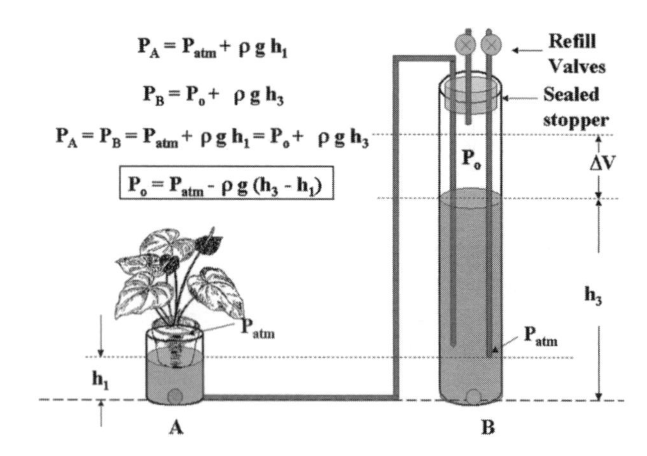

Fig. 3 Mariotte equilibrium system for transpiration measurements

equal to one-tenth of the total daily plant water use on an individual pot basis. However, for dynamic diurnal water use fluctuations, ΔV must be even smaller to track these changes.

FINAL COMMENTS

A brief survey of flow measurement methods for microcapillary systems in animals and plants has been presented. As with any measurement system, calibration to known standards is important. Emphasis should be given to nonintrusive measurement techniques for greater accuracy.

REFERENCES

1. Gardner, J.W. *Microsensors Principles and Applications*; John Wiley and Sons: New York, 1994; 331.
2. Webster, J.G. *The Measurement, Instrumentation and Sensors Handbook*; CRC and IEEE Press: Boca Raton, FL, 1999; 1026.
3. Sakuratani, T. A Heat Balance Method for Measuring Water Flux in the Stem of Intact Plants. J. Agric. Meteorol. **1981**, *37*, 9–17.
4. Sakuratani, T. Improvement of the Probe for Measuring Water Flow Rate in Intact Plants with the Stem Heat Balance Method. J. Agric. Meteorol. **1984**, *40*, 273–277.
5. Baker, J.M.; VanBavel, C.H.M. Measurement of Mass Flow of Water in the Stems of Herbaceous Plants. Plant Cell Environ. **1987**, *10*, 777–782.
6. Cohen, Y.; Fuchs, M.; Falkenflug, V.; Moreshet, S. Calibrated Heat Pulse Method for Determining Water Uptake in Cotton. Agron. J. **1988**, *80*, 398–403.
7. Cohen, Y.; Huck, M.G.; Hesketh, J.D.; Frederick, J.R. Sap Flow in Stem of Water Stressed Soybeans and Maize Plants. Irrig. Sci. **1990**, *11*, 45–50.
8. Cohen, Y.; Takeuchi, S.; Nozaka, J.; Yano, T. Accuracy of Sap Flow Measurement Using Heat Balance and Heat Pulse Methods. Agron. J. **1993**, *85*, 1080–1086.
9. Ishida, T.; Campbell, G.S.; Calissendorff, C. Improved Heat Balance Method for Determining Sap Flow Rate. Agric. Forest Meteorol. **1991**, *56*, 35–48.
10. Lu, S.; Mattson, D.L.; Roman, R.J.; Becker, C.G.; Cowley, A.W., Jr. Assessment of Changes in Intrarenal Blood Flow in Conscious Rats Using Laser-Doppler Flowmetry. Am. J. Physiol. **1993**, *264* (33), F956–F962.
11. Scott, E.A.; Sandler, G.A. Electromagnetic Blood Flowmeters and Flow Probes: Theoretic and Practical Considerations. Am. J. Vet. Res. **1978**, *39* (9), 1567–1571.
12. Neuwirth, J.G.; Rawlings, C.A. Calibration and Use of Electromagnetic Flow Transducers. Am. J. Vet. Res. **1979**, *40* (1), 1659–1661.
13. Durand, D.; Bauchart, D.; LeFaivre, J.; Donnat, J.P. Method for Continuous Measurement of Blood Metabolite Hepatitic Balance in Conscious Prereminant Calves. J. Dairy Sci. **1988**, *71* (6), 1632–1637.
14. Al-Faraj, A.; Meyer, G.E.; Fitzgerald, J.B. Simulated Water Use and Canopy Resistance of New Guinea Impatiens (*Impatiens X hb.*) in Single Pots Using Infrared Heating. Trans. ASAE **1994**, *37* (6), 1973–1980.

Food Engineering Education

Juan L. Silva
Taejo Kim
Mississippi State University, Mississippi State, Mississippi, U.S.A.

INTRODUCTION

Food engineering may be defined as the sum of those engineering activities associated with the processing, packaging, and delivery of food products from the farm to the consumer.

BACKGROUND

The food engineer is expected not only to have food engineering background (mathematics, physics, chemistry, and engineering) but also to know how to deal with biological systems (microbiology, biochemistry, etc.). Thus, students should be able to understand and resolve problems dealing with factors that affect the biological systems in question. These may deal with simple formulation and interaction of ingredients to such complex problems as modeling of food product spoilage and safety under various conditions.

TEACHING AND LEARNING METHODS

The conventional process of developing a food engineering curriculum is based on the instructive paradigm, i.e., a list of courses and their contents to meet a minimum number of credit hours needed to graduate. Evaluation of the students was usually based on a set of exams and questions that may not evaluate to what extent the students had learned. A shift from the "instructive paradigm" to the "learning paradigm" is needed.[1,2] This experience should be supported by a strong web-based education (to afford enough practical experience) and practicing experience (internships).

The Institute of Food Technologists (IFT) has developed education standards for degrees in food science based on outcome-based measures of learning.[3] This replaced the 1992 minimum standards. The new standards are based on five core areas, one being processing and engineering. The areas covered (in classes, assignments, experiments, etc.) are: characteristics of raw food material, principles of food preservation, engineering principles, principles of food processing techniques, packaging materials and methods, cleaning and sanitation, and water and wastewater management. The engineering principles include mass and energy balances, thermodynamics, fluid flow, and heat and mass transfer.

CHALLENGES IN TEACHING AND LEARNING

There are various challenges in providing the best education in food engineering. The first is recruiting and instructing the necessary background knowledge to the students. The second is educating instructors in providing the correct information at the level needed. Thirdly, the diversity of departments where the course(s) is taught: food science (agricultural/biological engineering or chemical engineering) makes it difficult to have uniform course content. Finally, correct counseling, advising, and monitoring of students due to their diverse background is key to a successful program. The first hurdle can be overcome by introducing and teaching mandatory basic science and math at middle and high school education, by directing freshmen in agriculture to take math and science courses, and by including a review, a basics of pre-engineering, in the introduction course in food engineering. The second barrier is overcome by providing the instructor with needed information and assistance to provide the students with. The third barrier is difficult to overcome, but a clear curriculum with defined learning objectives and outcomes should aid in the teaching of the course(s). The final hurdle is mediated through work-study sessions, matching engineering with nonengineering students for homework and other assignments, and providing ample tutoring and mentoring. The instructor is also challenged to teach logic and problem solving ability to nonengineering students, who are used to memorizing.

PROPOSED COURSE(S) CONTENTS

Food engineering course or course(s) should include theoretical background, problem solving ability,

Encyclopedia of Agricultural, Food, and Biological Engineering
DOI: 10.1081/E-EAFE 120006947
Copyright © 2003 by Marcel Dekker, Inc. All rights reserved.

Table 1 Suggested content for a food engineering course

Units and dimensions
Review: math, physics, thermodynamics, food processing
Mass and energy balance
Fluid flow
Heat transfer and thermal processing
Mass transfer
Unit operations: mixing, filtration, packaging, extraction, evaporation, refrigeration, freezing, dehydration, distillation
Process control: automation and sensors

(From Refs. [3], [9], [12], [13].)

practical experience, and some type of design. The common understanding in academia is that a food engineering course should cover the following areas: mass and energy balance, fluid flow, and heat and mass transfer (Table 1). The course(s) should be taught in such a manner as to introduce the students to the theory for each section accompanied by theoretical and practical problem solving. A hands-on plus experimental laboratory should complement the class. Experiments on viscosity, fluid flow, heat transfer, thermal processing, refrigeration, freezing, psychometrics and dehydration, extraction, sensors, and/or others are suggested. Work-study sessions to direct problem solving and a project on a design (theoretical or practical) should complete course(s) requirements and prepare the student.

LEARNING OBJECTIVES AND OUTCOMES

Following IFT 2001 standards[4] and the principles of learning outcomes,[5,6] a summary is shown in Table 2. This should guide academicians and others in providing students of food engineering with the necessary tools to learn the subject. It describes learning objectives and outcomes in the core subjects in food engineering, plus various unit operations and other subjects such as process control. These should be covered in at least two courses. It is understood that students should have a background not only in mathematics and physics but also in food microbiology, biochemistry, and food processing and preservation principles. In order to be effective, these learning outcomes should be evaluated by the use of written quantitative measures, behavioral quantitative measures, and qualitative measures. The first refers to exams and homework (individual, team), the second to attendance rates, contact hours, participation, etc. and the third one to design projects, real world problem solving, focus groups, etc. Finally, the learning outcomes should be assessed, not only by evaluation of the instructor, but also by exit interviews, alumni surveys, employer focus groups, and comparison with other similar programs.[7]

RESOURCES

In order to accomplish the goals of a good food engineering learning experience, a combination of resources are needed. These include textbook(s) and journals, web-based information, laboratory experiments, real world problems, and industry internships. The available funds to accomplish these are dwindling very quickly, including funding for instructors. Thus, the professor(s) have to rely on available resources internally and externally. There should be a minimum of contact hours for lectures and problem solving, minimum number of laboratory experience/demonstrations (at least one per unit/module), and time and resources for plant design and research. The web offers some very good opportunities these days to complement learning good engineering. Two sites, the IFT Food Engineering Division[8] and Dr. R. Paul Singh[9] sites, offer a large and substantial amount of information. The latter offers information, problem solving, laboratories, and much more. There are other resources that are readily available or accessible.

FOOD (PROCESS) ENGINEERING CURRICULUM

Although not the aim of this work, there are many food engineering programs available in the United States and abroad. These vary in content, depending on their emphasis and where they are located. There are two ABET accredited programs (Purdue, Ohio State) and 35 other programs in food engineering, agricultural, bioprocess or biosystem engineering in the United States. There are many other universities across the globe that offer these programs. Welti-Chanes et al.[10] reported a total of 150 FE programs in Latin America, with Brazil, Mexico, and Argentina having 76 of them. Welti-Chanes, Parada-Arias, and Ondorica-Vargas[11] reported a total of six undergraduate and one graduate program in food engineering in Mexico, with 15% and 2% of the total "Food Science and Technology" undergraduate and graduate students, respectively, in Mexico.

Table 2 Food processing, engineering course contents, and learning objectives and outcomes

Module, content (% of total load)	Learning objectives	Learning outcomes
1. Units and dimensions (5%)	Understand the different systems of measurement, conversion factors, and dimensional analysis	The student will be able to recognize the units used in a problem and how to convert them to the proper units
2. Material and energy balances(15%)	Understand the processes needed to solve mass and energy balances, including the setup of the unit operations and equation to solve the problems. Understand properties of materials, thermodynamics and systems, and mixing and formulation	The student will be able to formulate mass or energy balances from a statement or a problem, and to solve for the unknown(s) using system(s) of equations and needed assumptions
3. Fluid flow (15%)	Understand a fluid transport system, properties of fluids including viscosity, fluid flow regimes, Bernoullis equation, pump selection, and flow/viscosity measurement	The student will be able to calculate/determine fluid properties, calculate/design piping and pumping requirements to move a fluid and identify the possible effects on the fluid
4. Heat transfer and thermal processing (20%)	Understand energy/power generation and types of heat, systems and heat transfer, thermal properties of foods, steady and in-steady heat transfer, thermal processing, microwave, ohmic, and dielectric heating	The student should be able to identify and determine/estimate thermal food properties, design a heat exchanger, and calculate process depending on source of heat
5. Mass transfer (10%)	Understand principles of diffusion of fluids through solids, influence of fluid flow regime on diffusion, steady- and unsteady-state diffusion	The student should be able to estimate mass transfer/diffusion coefficients
6. Unit operations		
Refrigeration (5%)	Understand components in a refrigeration system. Enthalpy diagrams and other factors in refrigerated storage, calculate cooling load and cleaning out of place (COP)	The student should be able to calculate refrigeration needs or estimate cooling capacity available from an existing system
Freezing (5%)	Understand various freezing systems, influence of freezing on food properties, calculate freezing time, frozen storage	The student should be able to design the best conditions for freezing and maintaining frozen foods. The student should be able to design and determine a thermal process required
Evaporation and concentration (5%)	Understand the components of an evaporation process, calculate single vs. multiple effect evaporator	The student should be able to calculate energy needed to accomplish evaporation requirements, calculate evaporator size
Package and membrane filtration (10%)	Understand the principles of gas diffusion through a membrane, osmotic design systems	The student should be able to estimate gas permeability or film thickness, membrane size
Extraction (5%)	Understand the principles of liquid–liquid extraction, limitation	The student should be able to design a liquid–liquid extraction system
Dehydration (5%)	Understand psychrometrics and the psychrometric chart, moisture and water activity, dehydration systems, dehydration calculation	The student should be able to combine heat and mass transfer balances, calculate dehydration times and design dehydration systems
7. Other topics		
Process control (5%)	Understand logic control	The student should be able to understand a control system and sensors

(*Continued*)

F

Table 2 Food processing, engineering course contents, and learning objectives and outcomes (*Continued*)

Module, content (% of total load)	Learning objectives	Learning outcomes
Water and wastewater (5%)	Understand the principles of water conservation and wastewater engineering, including primary, secondary, and tertiary treatment, organic load	The student should be able to calculate water requirements and design wastewater treatment systems
Cleaning and sanitation	Understand the principles of cleaning and sanitation, COP vs. cleaning in place (CIP) systems	The student should be able to design a sanitation system where possible
Food plant design (10%)	Understand the logical sequence in a food plant operation, conveying systems, scale-up	The student should be able to draw a layout of the plant, flow diagram, and process components

(From Refs. [3], [4], [7], [14], [15].)

REFERENCES

1. Iwaoka, W.T.; Britten, P.; Dong, F.M. The Changing Face of Food Science Education. Trends Food Sci. Technol. **1996**, *7*, 105–112.
2. Barr, R.B.; Tagg, J. From Teaching to Learning—A New Paradigm for Undergraduate Education. Change **1995**, *27* (6), 13–25.
3. Hartel, R.W. IFT Revises Its Education Standards. Food Technol. **2001**, *55* (10), 53–59.
4. www.ift.org/education/standards.shtml#standards (accessed Jan 2003).
5. Frye, R. Assessment, Accountability, and Student Learning Outcomes. Dialogue **1999**, *2*, 1–12 (Western Washington University).
6. Jacob, J.E. *A Simple and Effective Student Learning Outcomes Assessment Plan That Works*; California State University: Chicago 1998. www. csuchico.edu (accessed Dec 2002).
7. Silva, J.L. Steps in Responding to the Revised Curriculum Guidelines. Abstract 95-4, IFT Annual Meeting, Anaheim, CA, June 10–16, 2002; Institute of Food Technologists: Chicago, IL.
8. www.ift.org/divisions/food-eng/shtml (accessed Jan 2003).
9. www.rpaulsingh.com (accessed Jan 2003).
10. Welti-Chanes, J.; Vergara-Balderas, F.; Palou, E.; Alzadora, S.; Aguilera, J.M.; Barbosa-Canovas, G.V.; Tapia, M.S.; Parada-Arias, E. Food Engineering Education in Mexico, Central America, and South America. J. Food Sci. Educ. **2002**, *1*, 59–65.
11. Welti-Chanes, J.; Parada-Arias, E.; Ondorica-Vargas, C. Past, Present, and Future of Food Engineering Education in Mexico. World Food Sci. **2000**, *1* (3): www.worldfoodscience.org/vol1_3/focus/-3.html (accessed Jan 2003).
12. Hartel, R.W. Food Engineering in Food Science Programs: Course Material and Teaching Techniques. Abstract 142, 1993 IFT Annual Meeting Book of Abstracts, Chicago, IL; Institute of Food Technologists: Chicago, IL.
13. Silva, J.L.; Chamul, R.S. Challenges in Teaching an Introduction Food Engineering Class to Non-engineering Students. Processing 6th Conference of Food Engineering (COFE 99), Dallas, TX, Oct 31–Nov 5, 1999; Barbosa-Canovas, G.V., Lombardo, S.P., Eds.; AICHE: New York, 1999.
14. www.after.edu.au/mschtml (accessed June 2001).
15. Singh, R.P.; Heldman, D.R. *Introduction to Food Engineering*, 3rd Ed.; Academic Press: New York, 2001.

Food Engineering History

Daniel F. Farkas
Oregon State University, Corvallis, Oregon, U.S.A.

INTRODUCTION

The field of food engineering has developed over the past 100 yr to help food processors apply scientific food processing findings to the cost effective manufacturing of safe, nutritious foods with extended shelf lives. The growth of cities during the late 1800s and the concentration of production agriculture in geographic areas removed from consuming populations, such as Australia, created an incentive for processed foods capable of being transported around the world. Because processed foods can provide a safe, year-round, nutritious food supply, the food engineer has assumed greater and greater responsibility for developing processing methods that deliver these qualities. Increasing disposable income and opportunities for education, travel, and leisure have fostered the increasing use of packaged, convenient, prepared foods. The food processing industry has responded with convenient packaging, automation, high-speed materials handling, and by pioneering statistical quality control. Food Engineers are now deeply involved in the hazard analysis of critical control points (HACCP) and understanding the kinetics of microbial growth and inactivation by heat, radiation, chemical preservatives, high pressure, and other nonthermal technologies.

The low unit cost of foods and the tonnage needed to feed millions of people has challenged the food engineer to develop high-speed processing lines and highly reliable, automated machinery. The need to conserve water and energy has resulted in new minimum water use equipment. The need to reduce waste generated during the cleaning, peeling, separation, size reduction, and heating and cooling unit operations has been met by detailed studies of existing unit operations and processes so that losses can be reduced.

While heat (canning) has been a major preservation tool of the food process industry for 200 yr, there is a continuing demand for convenient and safe prepared foods preserved with a minimum of heat. This need has lead to the dependence on mechanical refrigeration equipment to complement minimal thermal and nonthermal commercial preservation processes such as high pressure, pulsed electric fields, modified/controlled atmospheres, filtration, and radiation. The history of food engineering can be traced conveniently by following the development of heat, refrigeration, drying, and chemical preservation processes. Their effectiveness against microbes, enzymes, and chemical changes must be known. Modern food processing and preservation combines several preservation methods with functional packaging. The historical development of the engineering of preservation by heat, cold, chemicals, drying, and mechanical operations are summarized. Individual pioneering food engineers are identified and references are provided for further reading.

ENGINEERING THE PRESERVATION OF FOODS BY HEAT

Preservation of foods by heat may be considered as the first invented food preservation technology. The origins of drying and chemical preservation by alcohol or acid (generated by fermentation) have been lost in history as is the use of ice. Nicholas Appert, in the late 1700s, applied heat to acid and low acid foods sealed in bottles and eventually received a prize of 12,000 French francs from the French government for inventing a method for safely preserving foods for long term storage. Appert established a food preservation business in 1812. Peter Durand in England, in the early 1800s, adopted Appert's process to foods packed in tin coated steel canisters. While the canisters were hand made, heating of foods in hermetically sealed, tinned steel containers, formed the basis for the first true manufacturing of acid and low acid convenience food products.

The tin can made heat preservation practical as metal containers were more compatible with package filling, sealing, heat processing, and the subsequent rigors of storage and distribution.

Early heat preservation technology was constrained by the misconception that air caused food spoilage and by the lack of safe, reliable, and properly instrumented steam pressure retorts. Appert attributed his success at preserving foods by heat to the excellent packaging he developed using glass bottles of his own specification and his carefully prepared stoppers made from hand-cut and glued cork. He assumed that air was responsible for the spoilage of food. His meticulous procedures for filling and sealing his glass bottles reflect this belief. Appert was limited to

Encyclopedia of Agricultural, Food, and Biological Engineering
DOI: 10.1081/E-EAFE 120006946
Copyright © 2003 by Marcel Dekker, Inc. All rights reserved.

boiling water baths for preservation and heating times were in the order of hours for some products. Appert's successor, his son, Raymond Chevallier-Appert, adapted the steam autoclave so that packaged foods could be heated above 100°C. The higher temperature allowed shorter heat treatments. Raymond Chevallier-Appert could be considered among the first food engineers as he invented a manometer, which allowed him to control steam pressures to provide temperatures accurate to 1°C. Prior to his invention steam temperatures could vary 10–20°C during the heat treatment of bottled foods.

Research on heat preservation using temperatures above the boiling point of water, from the early to late 1800s addressed technologies to safely generate elevated temperatures. Because steam boilers were so unreliable, and often dangerous to operate due to the lack of pressure controls, alternate methods were sought. Food processors used salt, and later calcium chloride, in the process water to increase its boiling point. Calcium chloride gave a higher boiling point than salt and calcium chloride solutions became accepted for elevated temperature processing to allow the safe and accurate heat treatment of canned and bottled low acid foods. Advancing steam boiler technology coupled with advancements in the science of microbiology finally allowed food processors to develop reproducible empirical procedures for the sterilization of canned foods at temperatures above 100°C. However, the times at various temperatures needed to prevent spoilage and the growth of *Clostridium botulinum* were the trade secrets of each canner. In the 1860s Pasteur showed that heat inactivated food-borne pathogens and spoilage microbes.

The first scientific studies to link the time and temperature combinations needed to "commercially sterilize" low acid foods in various commercial containers were conducted in 1895. Prof. Samuel C. Prescott worked with William Underwood at the Underwood cannery in the Boston, Massachusetts area. Cans containing low acid food were inoculated with microbes isolated from spoiled cans. They were fitted with maximum reading thermometers and placed in steam retorts for various time periods. By changing the pressure (and hence temperature) of the steam, Prescott was able to construct tables showing the time at each temperature needed to obtain commercially sterile food placed in the can. The work of Prescott and Underwood determined the quantitative flow of heat into cans of food and compared this data to the rate of inactivation of food spoilage microbes. These studies help put the preservation of foods by heat on a sound mathematical basis. By the 1920s sufficient microbial heat inactivation kinetic data had been accumulated to allow food engineers to develop graphic models of the time and temperature needed to obtain commercial sterility in a wide range of steel containers. C. Olin Ball, in 1925, was among the first to use computers to model the heat inactivation kinetics of *Clostridium botulinum*. The availability of thermocouples and pH meters in the 1930s provided the basis for food engineers to predict the rate of inactivation of heat resistant microbes isolated for spoiled canned foods. The mathematical models developed by C. Olin Ball and later Stumbo form the basis of heat preservation today.

ENGINEERING THE PRESERVATION OF FOODS BY REFRIGERATION AND FREEZING

Ice harvested in winter and held in insulated houses was an article of commerce in the American colonies. Ice was shipped south in trade for cotton, rum, and other warm climate goods. The addition of salt to a water and ice mixture to lower its freezing point was the basis of commercial food freezing (ice cream, meat, and poultry) as early as the 1850s. The value of ice as a food preservative was a powerful incentive for scientists and inventors to develop methods for the production of ice by mechanical means. Experiments with evaporating ether and other volatile liquids showed that ice would form on the surface of containers as the liquid evaporated. Ferdinand Carré patented the first ammonia compression refrigeration machine in France in 1864. The ammonia refrigeration system was refined by Dr. Carl Linde in Germany and by David Boyle in the United States in the 1870s. Safe, reliable ammonia compression systems were rapidly adapted for the production of ice, refrigerated, and frozen food storage, and the cooling of beer in the 1880s. Frozen meat shipments from Australia and New Zealand began in the 1880s using shipboard mechanical refrigeration. The ability to machine steels to accurate tolerances enabled the construction of reliable, efficient, ammonia compressors. In this period a quantitative understanding of the laws of thermodynamics was being developed. This provided a sound scientific basis for understanding the flow of energy, fluids, and work in refrigeration machinery.

While foods were frozen commercially for nonretail distribution, the birth of the retail frozen food business was pioneered by Clarence Birdseye with the founding of his company in 1923. As a food engineer, Birdseye developed consumer size packages for the distribution of frozen foods and commercial equipment for rapidly freezing these packages. His research showed the need for rapid cooling and freezing of fruits, vegetables, and meats to preserve their structure after thawing. The continuous plate freezer and belt freezer were developed by Birdseye. Additionally Birdseye saw the need for low temperature storage of his products in retail stores. This need was met by mechanically refrigerated retail display cases of his design in 1928.

Since 1950 frozen foods have become a staple in the market place. Food engineers have developed thermodynamic data on the heat transfer rates, heat capacities, and heat transfer coefficients for various freezing methods, including cryogenic freezing by liquid nitrogen and solid carbon dioxide. Food processors can now predict the temperature distributions in foods being frozen using computer models.

ENGINEERING THE PRESERVATION OF FOODS BY CHEMICAL TREATMENT AND OTHER NONTHERMAL METHODS

Salt, sugar, organic acids such as vinegar and lactic acid, and ethanol have been used to preserve foods before recorded history. These compounds were often used in conjunction with drying. Chemical preservation is dependant on delivering the preserving compound uniformly throughout the food. Understanding the rate of diffusion of preservation chemicals as a function of temperature, concentration, and food structure has occupied the interest of the food engineer. Major drawbacks of preservation by chemical treatment are the changes in the flavor, color, and structure of foods.

While chemical preservation probably ranks as the oldest method of nonthermal preservation, the food process industry has continued to search for chemical additives that leave no residue and that do not alter the quality of the food. Ionizing radiation is an example of a process that can pasteurize and sterilize selected foods and meet these requirements. Soon after the discovery of ionizing radiation, from radium and x-rays, in the early 1900s, these radiations were shown to inactivate microbes. However the quantities of radiation needed to treat foods commercially did not become available until after World War II. Large electron beam machines were built and isotopes, such as cobalt 60, and cesium 137, could be produced in quantities sufficient to treat food products at thousands of pounds per hour.

Proctor and Goldblith at the Massachusetts Institute of Technology initiated studies on the effect of ionizing radiation on a variety of foods using Van de Graff electron beam accelerators and later cobalt 60 radiation to determine the effect of dose rate, presence of oxygen, and temperature on the rate of inactivation of vegetative and spore forming bacteria, yeasts, and molds. The Office of Naval Research sponsored initial research. Later the U.S. Army Quartermaster Corps undertook the development of packaging, dosimetry, and products on a scale that allowed troop feeding. Studies focused on nutritional value, sensory qualities, and microbiological safety. Detailed animal and human feeding studies were performed to facilitate approval of the technology by the Food and Drug Administration. The economic value of ionizing radiation sterilization of medical devices was recognized in the 1950s.

High hydrostatic pressure was first applied to food preservation by Bert Heit at the West Virginia Agricultural Experiment Station in the early 1900s. His research showed that pressures in the range of 60,000 psi–100,000 psi (400 MPa–680 MPa) could inactivate a wide range of vegetative microbes, yeasts, and molds. The first successful commercial application of high pressure in the United States was with guacamole using equipment developed by the Flow Corporation in the 1990s. The product was packed in plastic bags, pressure treated at 600 MPa, and distributed under refrigeration. A 1-mo refrigerated shelf life was obtained.

ENGINEERING THE PRESERVATION OF FOODS BY DRYING

Grains preserved by natural drying have been used as foods before recorded history. Modern drying technologies and related engineering developments started in the 1920s with research on diffusion of water through porous media and the chemical binding of water by starch, sugar, protein, and lipid domains in foods. The 1950s brought an understanding of the relationship between water activity and moisture content. The storage stability of a dried food was found to be determined by its water activity rather than its water content.

Freeze drying emerged as a commercial process for drying foods in the 1950s and 1960s. The high quality of foods preserved by freeze drying provided a strong incentive for food engineers to understand and optimize the process. Coffee, diced meats, vegetables, and whole fruits were freeze dried in tonnage amounts.

Bacteria were preserved by freeze drying in 1921 and blood, sera, and biological specimens were routinely freeze dried for extended preservation of their biological activity without refrigeration. World War II accelerated the use of freeze drying for blood plasma on a commercial basis and equipment became available to speed the drying of frozen biological materials from the frozen state. The understanding of the material and energy balance in the drier, the need for conduction or radiant heating in the high vacuum surrounding the food, and the selection of an appropriate vacuum for the product being dried resulted in shortened batch drying cycles. High quality was achieved as shown by rapid and complete rehydration, good flavor and color retention, and consistent particle integrity.

CONCLUSION

Food engineering as a discipline has provided the food processing industry with the knowledge to produce safe, high-quality processed foods at lower and lower operating costs. Research findings have provided guidelines for improved equipment, packaging, and materials handling while establishing operating conditions that insure a safe product and guaranteed shelf life at prescribed storage conditions.

FURTHER READING

American Can Co. Research Division. *Canned Food Reference Manual*, 3rd Ed.; American Can Co.: New York, NY, 1947.

Appert, N. *The Art of Preserving Animal and Vegetable Substances for Many Years*; Translated by K.G. Bitting; Published by the translator, 1920.

Collins, J.H. *The Story of Canned Foods*; E. P. Dutton and Co. : NewYork, NY, 1924.

Cruess, W.V. *Commercial Fruit and Vegetable Products*, 4th Ed.; McGraw-Hill Book Co.: New York, NY, 1948.

Desrosier, N.W., Tressler, D.K., Eds. *Fundamentals of Food Freezing*; AVI Publishing Co.: Westport, CN, 1977.

Flosdorf, E.W. *Freeze-Drying*; Reinhold Publishing Co.: New York, NY, 1949.

Prescott, S.C.; Underwood, W.L. Micro-organisms and Sterilizing Processes in the Canning Industries. Technol. Quart. **1897**, *X* (1), 183–199.

Von Loesecke, H.W. *Drying and Dehydration of Foods*; Reinhold Publishing Co.: New York, NY, 1943.

Food Freezing History

Dennis R. Heldman
Rutgers, The State University of New Jersey, New Brunswick, New Jersey, U.S.A.

Paul Nesvadba
The Robert Gordon University, Aberdeen, Scotland, United Kingdom

INTRODUCTION

Like most food preservation methods, the freezing process for foods has evolved over a significant period of time. As an extension of cooling, the primary goal of food freezing has been the reduction of microbial growth and/or control of reactions causing spoilage of the food, as well as the corresponding extension of product shelf-life. In addition, the process has extended the availability of many food commodities for consumption at any time during the year. Frozen foods can be transported for longer distances and the process has contributed to making foods available on a worldwide basis. For most applications, the food products are held in storage for some period of time before thawing, and prepared for consumption. More recently, the process has evolved to include a variety of products that are consumed in a frozen state. The purpose of this contribution is to review the historical evolution of food freezing and the frozen foods.

THE FIRST 1000 YR—BEFORE MECHANICAL REFRIGERATION

During the early evolution of food freezing, the preservation of the food was the primary motivation. There is evidence that natural ice or cold air (in appropriate geographic regions) was used to create a cold environment for food products as early as 500 B.C., or possibly as early as 1000 B.C. Freezing as a method of preserving perishable foods such as meat may have been used by prehistoric man in temperate and cold climates. Although specific references to lowering of product temperatures to below the initial freezing temperature are not described, the use of water as a medium to freeze products may have occurred as early as 1550, when salt and other additives were used to reduce the freezing temperature of liquid water.

THE IMPACT OF MECHANICAL REFRIGERATION

Large-scale use of freezing was not realized until development of mechanical refrigeration in the 1830s. The first method of refrigeration (cooling air by the evaporation of liquids in vacuo) was invented in 1748 by William Cullen of the University of Glasgow, Scotland. Cullen did not apply his discovery to any practical purposes. Michael Faraday, an English physicist liquefied ammonia to cause cooling (in the 1800s). Faraday's idea would eventually lead to the development of compressors, which compress gas to liquid, a process that absorbs heat. Jacob Perkins, an American working in England, invented the first refrigeration machine, based on discoveries in 1834. In 1844, John Gorrie, an American doctor from Florida made a device that would make ice in order to cool the air for feverish yellow fever patients.

An ammonia compression machine was patented by Ferdinand Carre in France in 1864. Following this event, developments associated with frozen foods increased in frequency. Around 1870, key contributions were the development of ammonia compressor applications by Carl Linde in Germany and David Boyle in the United States. The first commercial refrigerator designed to keep food cold was sold in 1911 (by the General Electric Company) and in 1913 (invented by Fred W. Wolf of Fort Wayne, Indiana). These refrigerators consisted of a unit that was mounted on top of an icebox. A self-contained refrigerator (with a compressor on the bottom of the cabinet) was invented by Alfred Mellowes in 1915. Mellowes produced this refrigerator commercially (each unit was hand made), but was bought out by W. C. Durant (the president of General Motors) in 1918, who started the Frigidaire Company in order to mass-produce refrigerators in the United States.

Encyclopedia of Agricultural, Food, and Biological Engineering
DOI: 10.1081/E-EAFE 120007022
Copyright © 2003 by Marcel Dekker, Inc. All rights reserved.

THE EVOLUTION OF FOOD FREEZING

The first patent (British Patent # 9240) related to food freezing was for rapid freezing of a product by immersion in cold medium and was granted to H. Benjamin in 1842. Although there are references to ice cream being served at social events as early as 1700, the first evidence of commercial ice cream freezing was by Jacob Russell (Baltimore, Maryland) in 1851. Other examples of the use of ice to refrigerate and partially freeze food commodities include the shipment of fish to Newfoundland by Captain H. O. Smith in 1853–1854, and a U.S. Patent on fish freezing granted to Enoch Piper in Camden, Maine in 1861. During the same period, the first cold storage facility for frozen meat was opened in Australia. An innovative freezing system for fish was patented by William Davis in 1868; the fish were placed in metal containers and sealed before immersion in an ice/salt mixture.

The early developments in food freezing and mechanical refrigeration opened the way for large-scale freezing of raw materials for later use, such as seasonal fruit and vegetables and fluctuating fish supplies. Developments in mechanical refrigeration triggered a series of food applications that continued through the early 1900s. Mechanical refrigeration systems were used in combination with insulated rooms for storage of frozen foods from about 1880. Frozen meats were shipped from Australia to England using mechanical refrigeration, and an ammonia compressor system was used to freeze fish in 1892. The applications of mechanical refrigeration to freezing of other foods continued through 1905, when the first applications to fruits occurred.

Commercially packed fish products appeared in 1929, under the Birds Eye label, (the divided name of the founder), and the first retail-styled frozen food packs were launched in 1939. During the second-world war vast quantities of produce were frozen in the United States for the war effort. During the postwar years the frozen food industry in the western world continued to evolve due to demand driven by demographic trends and lifestyle and emergence of a middle class with less time available for preparing food at home. In 1953, ready meals began to appear in a meat and two vegetables format. The ready meals frozen food sector continued to grow through the prosperous 1960s (for example "boil in the bag" products by Birds Eye and Findus) onto the consumer boom in the 1980s and the food safety and health conscious 1990s.

THE BEGINNINGS OF RESEARCH ON FROZEN FOODS

Throughout the early history of the food freezing and frozen food storage, the process was viewed narrowly as a preservation step, and the quality of the product was accepted as being inferior to fresh counterpart products. An interest in addressing these concerns resulted in a period of significant emphasis on the freezing process and the storage of the frozen food products. The research began around 1910 and resulted in several key developments within a relatively short period. An early pioneer was Clarence Birdseye, who was directly responsible for the early versions of "quick freezing" and the corresponding improvements in the quality of many frozen foods. An early British patent to M. T. Zarotschenzeff was granted for a device to spray food products with cold brine to increase rate of freezing. By 1930, Birdseye was granted three U.S. patents for plate freezing systems.

The significant contributions from research led by Clarence Birdseye deserve special attention. Over 20 yr, beginning in 1923, Birdseye provided the energy and leadership for the following developments for frozen foods:

- Demonstrated that "quick freezing" prevented loss of quality attributes in many foods, including most fruits and vegetables.
- The marketing of packaged frozen foods through retail outlets.
- Invention and commercialization of double- and multiple-plate freezing systems to accomplish rapid freezing of foods in retail type packages.
- The integration of production, harvesting, handling, preparation, and freezing of several fruits and vegetables, for retail markets.
- Demonstration of quality attributes in frozen foods that are superior to canned food counterparts after preparation.
- Assembly of early evidence on the importance of maintaining uniform temperature during storage of frozen foods at below −12°C.
- Development of processes for packaging and freezing of precooked foods.

It appears that much of the evidence that led to the standard recommendation on commercial frozen food storage at −18°C have origins in the reports from Birdseye's research efforts.

During the time of Birdseye's contributions, several other significant developments were occurring. During the early 1930s, M. A. Joslyn and W. V. Cruess, at the University of California, were reporting important results on the need for blanching of vegetables prior to freezing, and resulted in expanded use of freezing for preservation of vegetables, as well as fruits. During the late 1920s and early 1930s, there were a series of developments in the ice cream industry, including the continuous ice cream freezer by Henry Vogt in 1927.

RECENT RESEARCH HISTORY

A significant contribution to the understanding of the food freezing process emerged from research by L. Riedel at the Food Research Institute in Karlsruhe, Germany. The research involved the calorimetric measurement of the thermal energy content or enthalpy of foods at temperatures as low as $-40°C$. The results of this research have provided a firm basis for estimating refrigeration requirements for freezing of foods, and have stimulated research on prediction of changes in thermal energy content of foods during freezing.

An important component of the frozen food industry is the storage of the product following the freezing process. Although the importance of maintaining the temperature of the frozen food at levels well below initial freezing temperature were obvious during early research, the quantification of the impact of adverse temperatures on frozen food quality was not addressed until the 1950s. Beginning with the relationships proposed by Schwimmer et al., quantitative approaches to estimating the influence of fluctuating storage temperatures and other thermal abuse conditions on the quality of the frozen food product have been explored. Beginning in the mid-1950s, Van Arsdel and other researchers at the Western Regional Research Laboratory of USDA completed an extensive series of frozen food storage investigations, in an effort to quantify the influence of storage temperature on the quality of frozen foods. The results of these efforts were expressed in terms of time–temperature-tolerance (TTT) impacts on frozen food quality.

During more recent periods of time, the World Food Logistics Organization (WFLO) and The Refrigeration Research and Education Foundation have supported research efforts to improve the quality of frozen foods reaching the consuming public. These efforts have included investigations on the influence of freezing rates, storage temperatures, and thawing conditions on the quality of the frozen food. The WFLO has sponsored and distributed a computer-based program for prediction of freezing times, as a function of process parameters. The organization has encouraged research of the optimum storage temperatures for frozen foods, with the sponsorship of research on the relationships of molecular mobility to shelf-life of the frozen products.

The future of frozen foods clearly depends on steps needed to reduce the impact of the freezing process on product quality attributes, minimizing the influence of storage conditions on quality, and the development of methods for control of the thawing process. As these developments occur, the quality of the frozen foods will approach the quality attributes of the fresh food counterparts.

FURTHER READING

Arbuckle, W.S.; Robert, T.M. *Ice Cream*, 5th Ed.; Kluwer Academic/Plenum Publishers: New York, NY, 2000; 349 pp.

Desrosier, N.W.; Donald, K.T. *Fundamentals of Food Freezing*; The AVI Publishing Co, Inc.: Westport, CT, 1977; 629 pp.

Mallet, C.P. *Frozen Food Technology*; Blackie Academic & Professional: Glasgow, Scotland, 1993.

Riedel, L. The Refrigeration Required to Freeze Fruits and Vegetables. Refrig. Eng. **1951**, *59*, 670–673.

Schwimmer, S.; Ingraham, L.L.; Hughes, H.M. Temperature Tolerance in Frozen Food Processing. Effective Temperatures in Thermally Fluctuating Systems. Ind. Eng. Chem. **1955**, *47*(6), 1149–1151.

Van Arsdel, W.B. The Time–Temperature Tolerance of Frozen Foods. I. Introduction—the Problem and the Attack. Food Technol. **1957**, *11*, 28.

WFLO. *Successful Refrigerated Warehousing*, 3rd Ed.; World Food Logistics Organization: Bethesda, MD. 2001; 588 pp.

Food Texture

Malcolm C. Bourne
Cornell University, Geneva, New York, U.S.A.

F

INTRODUCTION

Rheology is an important component but not the only component of the texture perception of liquid foods, semiliquid foods, and solid foods. Many texture notes such as the feeling of fracture, particle size and shape, juiciness and oiliness are unrelated to rheology. Also, the human senses seem to be "blind" to some rheological properties. We conclude that "texture" is composed partly of rheological properties and partly of nonrheological properties. Although food texture and food rheology have much in common they should never be used as synonymous terms.

DEFINITIONS

The previous articles in this section have already explained that "rheology" means the study of the deformation and flow of matter. This definition was coined in 1929. Therefore, it is not necessary to spend time explaining rheology.

In contrast, the word "texture" is not as clearly defined as rheology which makes it necessary to begin with a short review of concepts of texture that have been proposed over the years and the general consensus of the measuring of texture that has developed over the last three decades.

Most of the older definitions of texture were commodity oriented, referring to a single property of one type of food that was not applicable to other foods. For example, in ice cream grading, "texture" refers only to its smoothness; while in bread grading, it means the uniformity of the crumb and even distribution in size of the gas bubbles.

These "one food, one property" definitions have been largely supplanted by definitions that encompass all properties of all foods as the following examples demonstrate.

Kramer. "Texture is one of the three primary sensory properties of foods that relates entirely to the sense of touch or feel and is, therefore, potentially capable of precise measurement objectively by mechanical means in fundamental units of mass or force."[1]

Sherman. "Texture is the composite of those properties (attributes) which arise from the structural elements of food and the manner in which it registers with the physiological senses."[2]

Jowitt. "Texture is the attribute of a substance resulting from a combination of physical properties and perceived by the senses of touch (including kinesthesis and mouthfeel), sight, and hearing."[3]

Szczesniak. "Texture can be defined as the sensory manifestation of the structure of the food and the manner in which this structure reacts to applied forces, the specific senses involved being vision, kinesthetics, and hearing."[4]

The International Organization for Standardization. "Texture (noun): All the mechanical (geometrical and surface) attributes of a food product perceptible by means of mechanical, tactile and, where appropriate, visual and auditory receptors" (Standard 5492). It is interesting to note that the first version of this definition in 1979 used the words "rheological and structural attributes" but these were replaced by "mechanical attributes" in the 1992 version given above.

All these definitions have several common features:

1. Texture is a group of properties, not a single property.
2. Texture is sensed by humans, therefore, it is a sensory property.
3. Texture is primarily perceived by the tactile sense, but the senses of sight and sound are sometimes involved. The chemical senses (flavor and odor) are not used to perceive texture.

Foods have many different physical properties: mechanical, rheological, electrical, thermal, optical, acoustical, magnetic, etc. Some of these are related to texture, but many of them are not. All textural properties are physical properties, but not all physical properties are textural properties. The way to determine whether or not a physical property is a textural property is to present several foods that possess the property of interest at several levels of intensity to a panel of people. If the property is detected and people call it "texture," and if it is quantified in the same sequence as the physical measurement, then it is a textural property. If not, it is not a textural property. The principle is explained in Fig. 1.[5]

An analogy can be made with the electromagnetic spectrum. The visible wavelengths of light are in the range 0.4 μm–0.7 μm while the invisible range extends from ultraviolet to x-rays at shorter wavelengths (0.4 μm–10^{-14} m), and from infrared to radio waves (0.7 μm–10^{4} m) at longer wavelengths than the visible region.

Encyclopedia of Agricultural, Food, and Biological Engineering
DOI: 10.1081/E-EAFE 120006974
Copyright © 2003 by Marcel Dekker, Inc. All rights reserved.

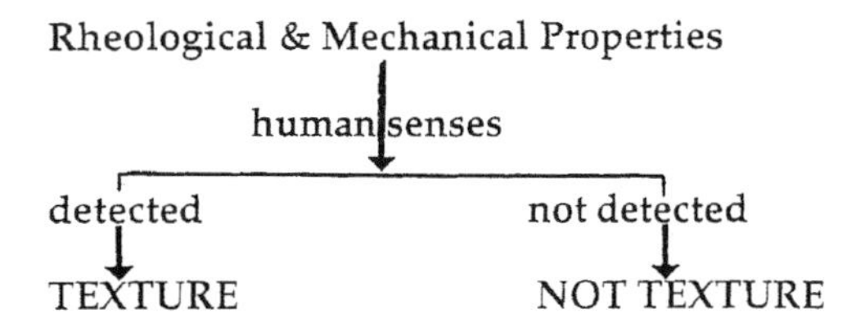

Fig. 1 Rheological properties may, or may not be textural properties. (From Ref. [4].)

Those wavelengths of the electromagnetic spectrum that are detected by the eye are called "light" and those not detected by the eye are not light.

In a similar manner we need to understand that just as humans are blind to most of the electromagnetic spectrum they are also "blind" to many physical properties. There may be good reasons for measuring these physical properties, but it is a waste of time to measure them for textural purposes. Since texture is a sensory attribute the ultimate calibration of any rheological test that is intended to measure a textural property has to be against humans.

Food rheology has been defined as "the study of deformation and flow of the raw materials, the intermediate products, and the final products of the food industry."[6]

Foods exhibit rheological properties that range from ideal Newtonian fluids to ideal Hookean solids, with almost every combination of plasticity, elasticity, viscoelasticity, recoverability, and time effects in between. Few foods are ideal Hookean or Newtonian systems and the majority are very complex rheologically.

A number of food processing operations depend heavily on the rheological properties of the product at an intermediate stage of manufacture because control at this point has a profound effect upon the quality of the finished product. For example, the rheology of bread dough, milk curd, and meat emulsions are important aspects in the manufacture of high quality bread, cheese, and sausage.

RHEOLOGICAL COMPONENTS OF FLUID AND SEMIFLUID FOODS

Viscosity is an important quality factor in most fluid and semifluid foods. Cutler, Morris, and Taylor[7] studied the oral perception of "thickness" in fluid foods over a range of almost five orders of magnitude of viscosity. For Newtonian fluids a plot of the logarithm of perceived thickness (sensory assessment) vs. the logarithm of viscosity (instrument measure) was rectilinear with a correlation coefficient $r = 0.995$. For non-Newtonian fluids made from aqueous solutions of alginate, pectin, guar, and xanthan gums the correlation coefficient between log apparent viscosity measured at $50\,sec^{-1}$ and perceived thickness was $r = 0.933$. Fig. 2a shows the relationship between viscosity and thickness of Newtonian fluids while Fig. 2b is for apparent viscosity of non-Newtonian fluid.[8]

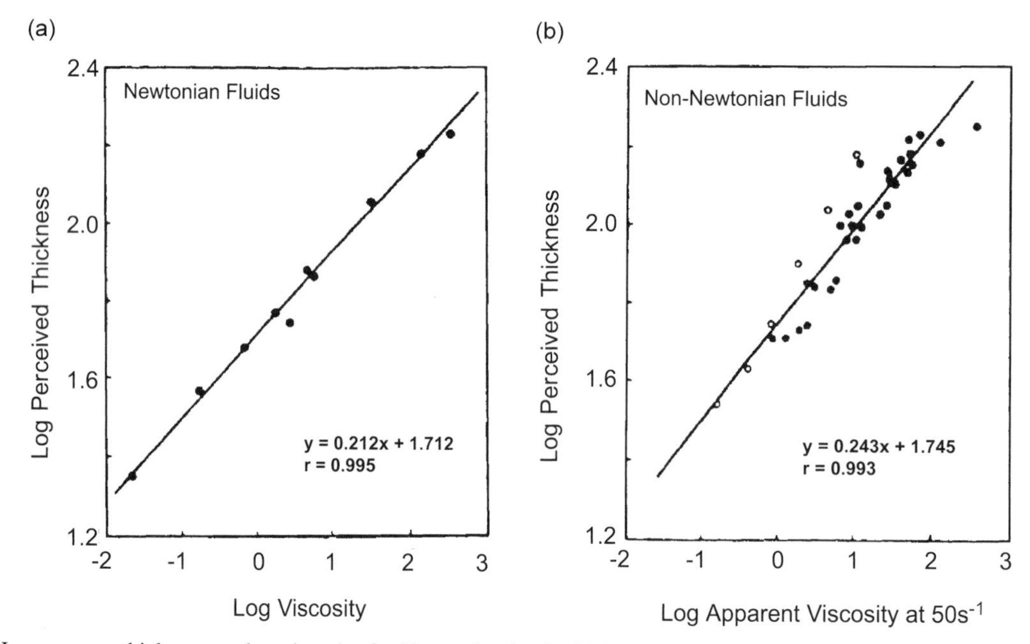

Fig. 2 (a) Log sensory thickness vs. log viscosity for Newtonian foods. (b) Log sensory thickness vs. log apparent viscosity at a shear rate of $50\,sec^{-1}$ for non-Newtonian foods. (From Ref. [7].)

Yield stress is another important rheological textural property of many foods. The ease with which tomato ketchup flows out from the bottle, but then "stays put" on the food shows that good quality ketchup must have a yield stress. It must be low enough to initiate flow under a gentle force, but high enough that it will not flow under the force of gravity. In contrast, butter, margarine, and cheese spreads have a higher yield stress because the force exerted by a knife is greater than what can be achieved by shaking a bottle, but it should not be so high that it tears the bread as it is spread.

The property of shear thinning is necessary for many foods. Whatever form a food is in when introduced into the mouth, the combined process of mastication and saliva secretion must convert it into a liquid or slurry that can be easily swallowed. Dilution with saliva causes the needed reduction in viscosity for some foods. For other foods, such as custard, pudding, and yogurt, the shear thinning caused by the stirring of the food by the tongue is sufficient to bring the viscosity down to a level where the food can be swallowed comfortably.

The shear rate generated in the mouth depends upon the viscosity of the food. Shama and Sherman[9] determined this by offering pairs of non-Newtonian fluid foods to a sensory panel and asking them to determine which was the "thicker" of the pair. The criteria used to select these foods was that the shear stress–shear rate curve of one sample intersected and crossed over the curve for the other sample. Whether the first sample has a lower, equal, or greater sensory thickness showed approximately the shear rate operating in the mouth for that pair of foods. In this manner they were able to construct the graph shown in Fig. 3. The pair of equidistant lines that curve downward from the top LHS to the lower RHS of the graph represent the shear rates operating in the mouth and it shows that shear rates vary over about two decades depending on the viscosity of the sample. For samples of low viscosity, the stimulus is the shear rate developed at a constant shear stress of about 10 Pa and the shear rate ranges from about $100\,\text{sec}^{-1}$ to almost $1000\,\text{sec}^{-1}$. For highly viscous samples the shear rate is approximately constant at about $10\,\text{sec}^{-1}$ and the mouth measures the shear stress developed to maintain this shear rate. Since most semifluid foods are non-Newtonian, any viscosity measurement intended to measure "thickness" should be made at a shear rate comparable to that generated in the mouth.

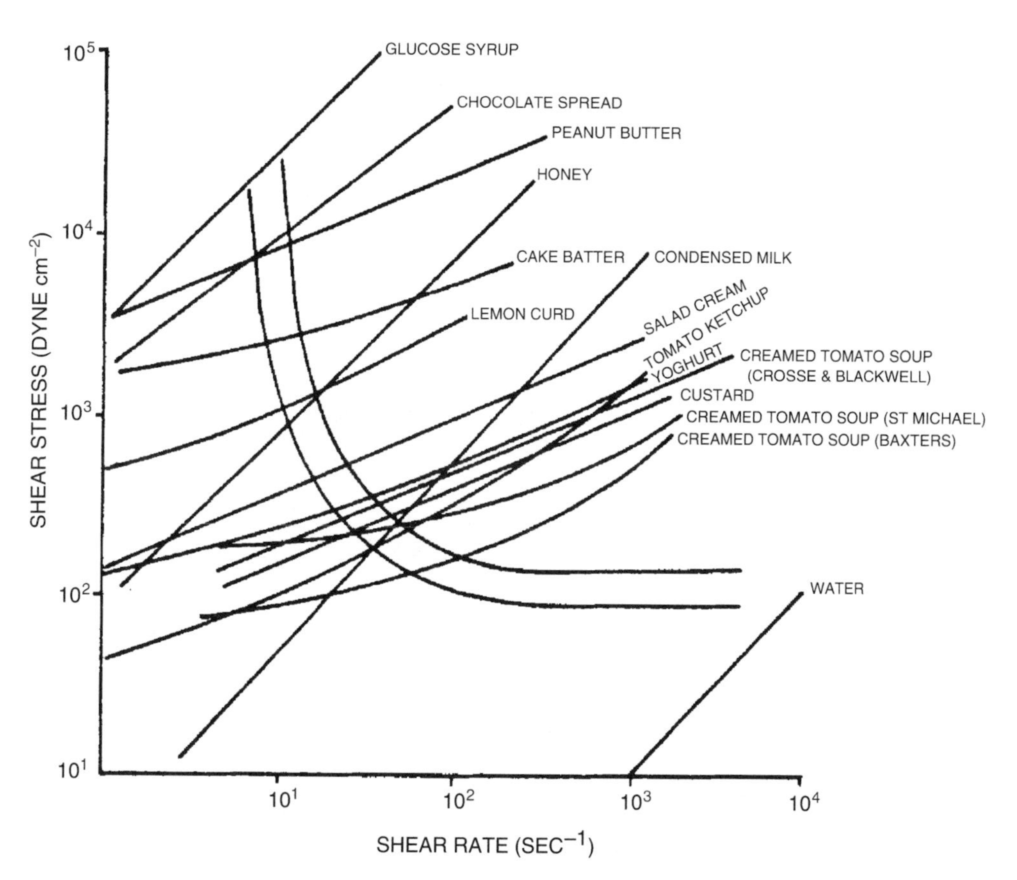

Fig. 3 Bounds for shear stress and shear rate associated with oral evaluation of viscosity. (From Ref. [8].)

RHEOLOGICAL COMPONENTS OF SOLID FOODS

The major rheological component in measuring texture of solid foods is their deformability. A widely used practice is to hold an article of food in the hand and measure the force needed to compress it a short distance. The manner in which the compressed food recovers its shape is also sensed at times. A food that recovers its shape quickly is said to have good "fight back." This simple hand squeeze gives the customer valuable information about the textural quality of bread, many fruits, cheeses, sausage, and other foods with a similar deformability. It is a nondestructive test.

Peleg[10] studied the sensitivity of the human tissue in squeeze tests and pointed out that in these types of tests there can be significant deformation of the human tissues (e.g., the balls of the fingers) in addition to the deformation of the specimen. Peleg pointed out that the combined mechanical resistance in a squeezing test is given by the equation:

$$M_c = \frac{M_1 M_x}{M_1 + M_x} \tag{1}$$

where M_c is the combined mechanical resistance of the sample and the fingers; M_1, the resistance of the human tissue; and M_x, the resistance to deformation of the test specimen. This equation provides an explanation as to why there are differences in the sensing range between the fingers and the jaws and why the human senses are practically insensitive to sample hardness beyond certain levels as detailed below in Case 3.

There are three different types of responses that can be drawn from this equation:

Case No. 1: $M_1 \gg M_x$. This case occurs when a soft material is deformed between hard contact surfaces (e.g., a soft food is deformed between the teeth). Under these conditions Eq. 1 becomes $M_c = M_x$ (since $M_1 + M_x \simeq M_1$).

Case No. 2: M_x and M_1 are of comparable magnitude. In this case the response is regulated by both the properties of the test material and the tissue applying the stress, as given in Eq. 1.

Case No. 3: $M_x \gg M_1$. This case occurs when a very firm product is compressed between soft tissues. For example, pressing a nut or hard cookie between the fingers. Under these conditions the equation becomes $M_c = M_1$ (because $M_1 + M_x \simeq M_1$). In this situation the response is due to the deformation of the tissue and is insensitive to the hardness of the specimen. This is interpreted as "too hard to detect differences" or "out of range" for human perception.

The increase in deformation or strain as a function of time (creep compliance) is of rheological interest for some foods.[11] The creep and recovery of a food under a stress applied for a long time is relevant for some foods (for example, see the flat surfaces on whole grapefruit that have been tightly packed in a carton). However, the sensory evaluation of firmness is usually rapid because people simply give a food item a quick squeeze and let go. Hence, creep is not of great interest for many foods.

NONRHEOLOGICAL TEXTURE NOTES

Sensory texture analysis reveals that most foods possess from 15 to 35 texture notes.[12] These texture notes appear all through the masticatory cycle from the first bite to the final swallow of the bolus of food and saliva. The flow of liquid and semiliquid foods, and the deformability of solid foods are texture notes that use rheological principles. However, the cutting, crushing, and grinding that occur between the teeth impart texture sensations that are not related to rheological properties. The sensations of particles on the tongue and gums, their shape, size, smoothness, and hardness are also texture properties that are not rheological in nature. The expression of juice, oil, or fat during mastication are textural properties and are not related to rheology.

Another problem in relating rheological measurements to food texture is that some of the assumptions that underlie the rheology do not hold up for foods. For example, Young's modulus of elasticity is a well-known rheological property that is widely used for materials of construction and is sometimes used for foods. There are four assumptions underlying Young's modulus of elasticity:

1. The material is homogeneous, isotropic, and continuous.
2. The material is elastic.
3. The strain is small.
4. The sample has a regular shape, e.g., cylinder.

While these assumptions are generally valid for structural materials, most foods fail to comply with one or more of them. Many foods are not elastic and are irregular in shape and structure. Uniaxial compression tests on foods generally use large strains. Hence, the use of Young's modulus should not be used in most food testing.

Rheological measurements will continue to have an increasingly large impact in the quest for better quality and more uniform quality of foods. However, it must be remembered that many texture notes have no relationship with rheology and that while there is considerable overlap between food rheology and food texture they are not synonyms.

REFERENCES

1. Kramer, A. Food Texture—Definition, Measurement and Relation to Other Food Quality Attributes. In *Texture Measurements of Foods*; Kramer, A., Szczesniak, A.S., Eds.; Reidel Publishing Co.: Dordrecht, Netherlands, 1973; 1–9.
2. Sherman, S. *Industrial Rheology*; Academic Press: New York, 1970.
3. Jowitt, R. The Terminology of Food Texture. J. Texture Stud. **1974**, *5*, 351–358.
4. Szczesniak, A.S. Texture: Is It still an Overlooked Attribute? Food Technol. **1990**, *44* (9), 86, 88, 90, 92, 95.
5. Bourne, M.C. Why So Many Tests to Measure Texture? In *Hydrocolloids Part 2*; Nishinari, K., Ed.; Elsevier: New York, 2000; 425–430.
6. White, G.W. Rheology in Food Research. J. Food Technol. **1970**, *5*, 1–32.
7. Cutler, A.N.; Morris, E.R.; Taylor, L.J. Oral Perception of Viscosity in Fluid Foods and Model Systems. J. Texture Stud. **1983**, *14*, 377–395.
8. Bourne, M.C. Calibration of Rheological Techniques Used for Foods. J. Food Eng. **1992**, *16*, 151–163.
9. Shama, F.; Sherman, P. Identification of Stimuli Controlling the Sensory Evaluation of Viscosity. II. Oral Methods. J. Texture Stud. **1973**, *4*, 111–118.
10. Peleg, M. A Note on the Sensitivity of Fingers, Tongue and Jaws as Mechanical Testing Instruments. J. Texture Stud. **1980**, *10*, 245–251.
11. Sherman, P. Structure and Textural Properties of Food. In *Texture Measurement of Foods*; Kramer, A., Szczesniak, A.S., Eds.; Reidel Publishing Co.: Dordrecht, Netherlands, 1973; Chap. 5, 52–70.
12. Bourne, M.C. *Food Texture and Viscosity: Concept and Measurement,* 2nd Ed.; Academic Press: New York and London, 2002; Chap. 7.

Forage Harvesting Systems

Dennis R. Buckmaster
The Pennsylvania State University, University Park, Pennsylvania, U.S.A.

INTRODUCTION

Common forage crops used for livestock feed and biomass-energy conversion include perennial grasses (e.g., bromegrass, orchardgrass, timothy), legumes (e.g., alfalfa, clover), and annual grasses (e.g., sudangrass, corn for silage). All these crops may be harvested as silage (moisture 45%–75%); the perennial grasses and legumes are also commonly dried and packaged to be stored as hay (moisture less than 30%) (Ref. 1 and see the article *Forage Storage Systems*). Standing crops typically vary from 60% to 85% moisture, so wilting is needed to get these crops to a proper moisture level for stable, long term, aerobic (hay) or anaerobic (silage) storage. As corn dries as it matures, it can be harvested in a single direct-cut operation with properly timed harvest. Other harvest methods include grazing, green chopping, pelletizing, and direct-cut ensiling.

Geographic location has a large effect on the forage harvest system because of growing and drying conditions and markets. Capacity needs (and other farm operations) dictate whether pull-type (PT) or large self-propelled (SP) machinery is better suited[2] (Table 1). The needed capacity, labor requirements, and economics of the harvest system are affected by the number and timing of harvests within the year. Multiple harvests of hay crops (perennial grasses and legumes) within the year result in lower per-harvest yields, but better utilization of machinery and more distributed labor requirements.[2] Comparatively, corn-silage harvest, a once and done operation, may require large capacity for timely harvest (see the article *Forage Transport Systems*).

HAY HARVEST SYSTEMS

Fig. 1 illustrates the primary operations in forage harvest systems. During harvest, hay crops must be cut and conditioned for wilting, optionally manipulated for drying and windrow management, and eventually packaged for removal from the field and subsequent storage. Moisture of hay crops should be about 20% or less at harvest unless preservatives are applied (Ref. 2 and see the article *Forage Storage Systems*). As moisture of harvest increases, loss of nutrients during harvest decreases, but losses during storage increase.[3]

Cutting and Conditioning

Cutting of forage crops at a height of 5 cm–10 cm is most often accomplished with sickle (shear) or disc (impact) cutting.[3] Sickle cutting requires less processing power (1.2 kW per meter of machine operating width) and can yield a cleaner cut—particularly with lodged crops; however, travel speed is typically limited to less than 10 km hr^{-1}.[4] Disc cutting yields higher capacity (speeds up to 19 km hr^{-1}) and lowered tendency for plugging, but power requirement is about 5.0 kW per meter of width. As with other operations (Fig. 1), SP units can offer higher capacity; the higher investment of SP units must be offset with additional use.

To facilitate field drying, most mowers accomplish mechanical conditioning[3] with flails (impact) or rolls (crushing), which require approximately 3 kW of power per meter of width.[4] Breaking and abrasion of the plant stem is needed to facilitate water movement from the plant to the ambient air. Combined with mechanical conditioning, spray treatment with a chemical conditioner (such as a potassium or sodium carbonate solution) can improve drying of legume crops by removing the waxy layer on stems. Maceration (severe crushing and shredding), followed by pressing into mats (to minimize subsequent losses by holding all plant parts together) can improve forage-nutrient value due to cell rupture[5]; drying rate is also much higher.

There is a tradeoff between drying rate and induced loss of dry matter (hence nutrients); mechanical treatments that improve drying rate the most also cause the most crop loss.[6] Optimal machine selection depends upon the crop and climatic conditions; greatest improvements in drying rate are needed where drying conditions are poor and likelihood of rain damage is high.

Swath and Windrow Manipulation

Mowers and mower-conditioners can lay cut crops into swaths (wide strips) or windrows (narrow strips); the setting choice is based on the subsequent need for drying and subsequent machine operations. As with conditioning treatments, ideal swath and windrow manipulation strikes a balance between drying improvement and machine-induced loss. Inverters simply turn the crop over, putting

Encyclopedia of Agricultural, Food, and Biological Engineering
DOI: 10.1081/E-EAFE 120006875
Copyright © 2003 by Marcel Dekker, Inc. All rights reserved.

Table 1 Harvest capacities, labor requirements, and costs of typical forage harvest systems

System description	Capacity[a] (Mg DM/yr)	Labor[b] (hr/Mg DM)	Cost[c] ($/Mg DM)
Hay in small rectangular bales (3.7 m mower, tandem rake, medium baler, 3 wagons)	200–400	1.6–1.1	57–40
Hay in large round bales (3.7 m mower, tandem rake, medium baler, 2 wagons)	300–500	1.2–1.0	47–40
Hay in large rectangular bales (4.9 m SP mower, tandem rake, midsize baler, truck)	500–3000	0.8–0.5	43–20
Silage in large round bales (2.8 m mower, rake, medium baler, wagon, bale wrapper)	200–500	1.9–1.5	100–85
Wilted silage in tower silo (2.8 m mower, rake, small chopper, 2 wagons, blower)	200–500	1.3–1.0	65–45
Wilted silage in bunker silo (3.7 m mower, tandem rake, medium chopper, 3 wagons, bunker packing)	400–800	1.1–0.9	57–40
Wilted silage in bunker silo (4.9 m SP mower, tandem rake, large SP chopper, trucks, bunker packing)	1000–3000	0.7–0.6	54–32
Corn silage in tower or bunker silo (2-row chopper, 3 wagons, blower or bunker packing)	400–1200	0.7–0.5	32–22
Corn silage (6-row SP chopper, 3 trucks, bunker packing)	2000–5000	0.3–0.2	20–15

[a] Total annual production of silage; DM = dry matter.
[b] Total labor requirement in person-hours/Mg DM of forage produced.
[c] Total production cost including equipment depreciation, interest on equipment investment, insurance, shelter, repairs, maintenance, fuel, labor, and material (twine and plastic) costs.
(From Ref. 3.)

the top (drier) material underneath and the bottom (wetter) material on top for improved drying exposure. Tedders spread the crop thinner over the field to maximize exposure to solar energy for drying. Rakes pull wide swaths into narrow windrows, expose the wetter material for improved drying, and gather tedded material or multiple swaths into one windrow. Rakes are used after tedding, after rain, and many times simply to form a well-structured windrow for harvest by a baler or chopper. Typical sequences of swath-manipulation treatments are shown in Fig. 1.

Swath-manipulation machines come in various configurations that may drag, tumble, roll, or pick up, carry, and lay down the crop.[3] Good machines handle the crop gently; this is particularly important with legumes that have nutritionally valuable leaves that are easily lost. Losses during swath manipulation can vary from 2% to 20%.[8]

Baling

Tremendous variability in labor availability and feeding facilities of livestock enterprises requires hay to be packaged in many package sizes. Rectangular bales range in size from 20 kg to 700 kg allowing for manual or mechanized handling.[7] Large round bales allow one-person hay making with a reasonably low investment; they range in size from 150 kg to 600 kg. Optimal package shape and size depends primarily on the market (fed on farm or sold), transportation distance, and labor availability.

At the time of baling, there may be a need or desire to apply chemical preservatives. Organic acids, occasionally buffered, are commonly spray applied at rates of 0.5%–2.0% to increase the allowable baling moisture. These acids minimize storage loss, expand the harvest period, and can reduce baler-induced losses.

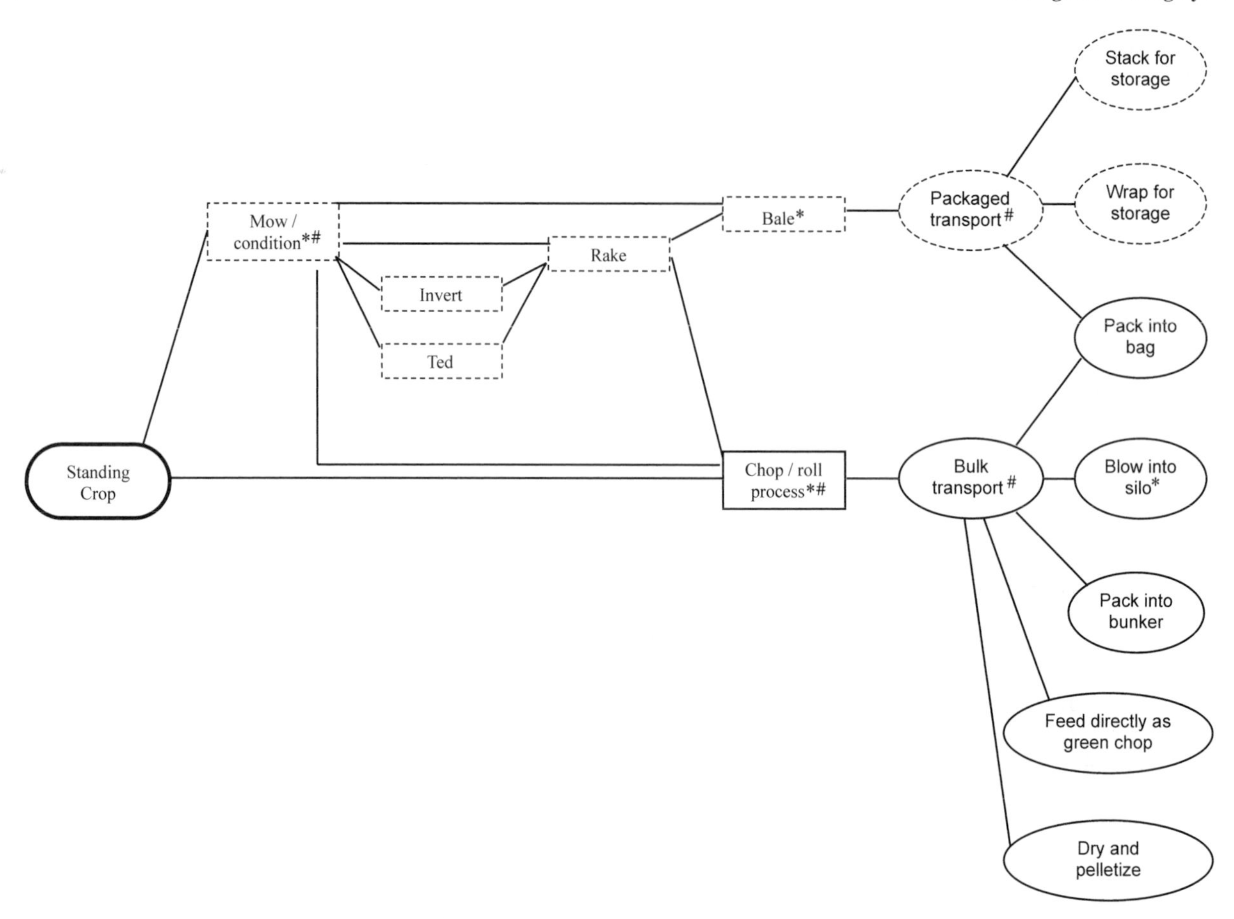

Fig. 1 Primary operations of forage harvest systems sequenced left to right (rectangles—harvest operations; ovals—nonharvest operations; dashed enclosures—wilted crops only; solid enclosures—wilted or direct-cut crops; *common points of chemical application; #sometimes SP).

Balers require 1.1 kW–1.8 kW of power per Mg of harvested hay.[4] Balers making small rectangular bales have theoretical capacities from 6 Mg to 15 Mg of dry matter (DM) per hour. Round baler theoretical capacities can reach 40 Mg DM hr^{-1}, but their field efficiencies are lower because they must stop to wrap or tie bales. Large rectangular baler capacities can reach 45 Mg DM hr^{-1} in good conditions.

SILAGE HARVEST SYSTEMS

Crop Preparation

Grasses and legumes harvested as wilted silage are cut and conditioned with the same machines used with hay harvest. Generally, less emphasis is placed on drying rate as the crop must only dry from 75%–80% moisture to 45%–70% moisture. Regardless of silo type or moisture level, bacterial inoculants are commonly applied to most silage crops during harvest to promote better fermentation (see the article *Forage Storage Systems*). Windrows of hay crops often yield sufficient drying rate, and swath manipulation is generally avoided to reduce stone pickup into the chopper. However, with large capacity harvesters, multiple windrows or swaths must be merged to improve harvest efficiency.

Large Packages

Many balers, used to package hay in large packages, can be used to pack wetter forage for subsequent ensiling. The round or rectangular packages must be sealed (wrapped individually or stuffed in tubes). An advantage of large package silage systems is that the same equipment used for hay can be used for silage; this reduces machinery investment. As packed silage—particularly if frozen—can be difficult to break apart, cutting mechanisms are incorporated on some balers so the bales are easier to feed. The cutting also increases bale density.

Chopping Bulk Silage

Bulk silage is harvested in self-loading wagons or choppers. PT self-loading/unloading wagons reduce particle size through slicing or flail action; with integral hauling capacity, they reduce the number of machines for harvest, but limit capacity when transport distances are large. Choppers generate ensilable material using either precision cutting (shear) mechanisms or flail (impact) cutting. Precision cutting with a cutter head and shear bar is more efficient (3 kW/Mg DM–6 kW/Mg DM) with smaller theoretical lengths of cut (TLC), but flail chopping is more tolerant of foreign material and requires less expensive equipment. The ideal TLC depends on many factors including the silo structure, moisture content, the animal to be fed, other ration ingredients, and the crop type. Smaller TLC improves packed density and reduces storage loss; however, it takes more power, and in the extreme, may not be beneficial to ruminant consumers (i.e., dairy cattle, beef cattle, or sheep). Hay crops harvested as silage should have a TLC between 6 mm and 12 mm.[1]

PT forage chopper capacities range from 20 Mg DM hr^{-1} to 30 Mg DM hr^{-1}; SP units can reach capacities of 70 Mg DM hr^{-1}. Power requirement (in kW) for harvest can be estimated by multiplying capacity in good harvest conditions (in Mg wet silage) by a constant (2.7 for hay crops, 1.7 for corn silage).[8]

Corn is a composite silage crop consisting of fodder (leaves and stalk), cob, and kernel. For proper storage and good feed value with chopping alone, TLC must be approximately 9 mm.[1] Serrated, narrow clearance, and differential speed rolls that crack kernels, break cobs, and shred fodder are sometimes used to improve the nutritive value of corn silage by increasing animal digestibility and utilization. Because these roll processors reduce particle size, the TLC setting of the cutter head should be increased to compensate for this subsequent particle size reduction.

OTHER HARVEST SYSTEMS

Grazing uses livestock to harvest the forage directly from the field. Green chopping (sometimes called zero grazing) is direct-cut harvest without storage; the harvested crop is transported immediately to livestock for feeding. Post-harvest drying and pelletizing can yield a premium feed but has high processing costs; generally, forages to be pelletized only wilt for a short period in the field before harvest by chopping because nutrients are lost during field respiration. Direct-cut ensiling of forages with over 70% moisture requires preservatives to prevent clostridial fermentation[2]; these preservatives are only approved in some countries, so this is not a widespread technique.

REFERENCES

1. Ishler, V.A.; Heinrichs, A.J.; Buckmaster, D.R.; Adams, R.S.; Graves, R.E. *Circular No.* 396: *Harvesting and Utilizing Silage*; Pennsylvania Cooperative Extension Service: University Park, PA, 1993; 32, Also available at: www-das.cas.psu.edu/dcn/catforg/396/index.html.
2. Rotz, C.A. Mechanization: Planning and Selection of Equipment. In *Proc. XIX International Grassland Congress*, Sao Pedro, Sao Paulo, Brazil, 2001; 763–768.
3. Miller, D.A.; Rotz, C.A. Harvesting and Storage. In *Forages: An Introduction to Grassland Agriculture*, 5th Ed.; Barnes, R.F., Miller, D.A., Nelson, C.J., Eds.; Iowa State University Press: Ames, IA, 1995; Vol. 1, 163–174.
4. ASAE. D497.4: Agricultural Machinery Management Data. *Standards*; ASAE: St. Joseph, MI, 2000.
5. Koegel, R.G. Emerging Technologies: Forage Harvesting Systems. In *Forages: The Science of Grassland Agriculture*, 5th Ed.; Barnes, R.F., Miller, D.A., Nelson, C.J., Eds.; Iowa State University Press: Ames, IA, 1995; Vol. 2, 137–146.
6. Rotz, C.A.; Muck, R.E. Changes in Forage Quality During Harvest and Storage. In *Forage Quality, Evaluation and Utilization*; Fahey, G.C., Collins, M., Mertens, D.R., Moser, L.E., Eds.; American Society of Agronomy: Madison, WI, 1994; 828–868.
7. Srivastava, A.K.; Goering, C.E.; Rohrbach, R.P. Hay and Forage Harvesting. *Engineering Principles of Agricultural Machines*; American Society of Agricultural Engineers: St. Joseph, MI, 1993; 325–398.
8. Example: A SP harvester with a capacity of 30 Mg DM/hour, harvesting grass silage at 65% moisture would be harvesting 30/(1 − 0.65) or 86 Mg/hour. This would require approximately (86)(2.7) or 230 kW.

Forage Production Systems—Machine Design

Richard G. Koegel
United States Department of Agriculture (USDA), Madison, Wisconsin, U.S.A.

INTRODUCTION

In general, forage production systems rely on equipment also used for the production of other crops. Exceptions to this or modifications of standard equipment will be noted. Maize (*Zea mays*) is also considered a forage crop when used for whole plant silage. However, equipment for production does not differ from that used for grain production.

STAND ESTABLISHMENT

Two major goals of stand establishment are attaining desired plant density and control of weeds. The former is made more challenging by the small size of forage crop seeds and the consequent shallow depth of planting relative to many other field crops. For this reason special seeding equipment has been developed.

Preplanting

Two important aids to stand establishment prior to seeding are bringing soil fertility and pH to the desired level and elimination of perennial weeds. Fertilizer and soil amendments (e.g., lime to increase pH of acid soils) can be spread using variety of standard fertilizer distributors. Lime must be incorporated into the soil using a tillage implement such as a disk harrow. Alternately, fertilizer may be applied in properly placed bands concurrent with seeding using a seeding drill. Perennial weeds, such as quack grass (*Elytrigia repens*) can be largely eliminated using herbicides such as glyphosate applied with standard field sprayers. Alternately, repeated cultivation during the summer prior to seeding using a spring-tooth harrow can be effective.

Seed for forage crops is generally produced by specialized farms where desirable growing and harvesting conditions can be provided. Harvest of seed is normally carried out using standard grain combines adapted with properly sized screens. Harvesting is made more difficult by the small size of seed, tendency of seed to shatter, and nonuniformity of ripening. To mitigate these problems, the crop can be cut and swathed before total ripeness. After thorough drying in the swath, the crop can be picked up by a combine equipped with a belt pick-up attachment and threshed with minimal loss.

Before planting, the seeds of leguminous crops should be treated with commercially available inoculum of the proper Rhizobium bacteria to assure symbiotic nitrogen fixing from the atmosphere. Treatment can be carried out in a rotating drum mixer. Seeds of certain leguminous forages may have very hard seed coats, which can delay or prevent germination. Scarification of the seed coat by methods such as tumbling with abrasive material, soaking in dilute acid, or heating and cooling may be used to overcome this problem.

Soil Preparation

Traditionally, soil in which forage crops are seeded is very thoroughly tilled by sequential use of plow, disk harrow, and spike-tooth harrow. Reducing the soil to relatively small aggregates is necessary to get intimate contact between seed and soil to promote germination and seedling survival. More recently, however, "no-till" seeding equipment has come into use. This equipment allows seeding in the residue of the previous crop without the need for soil preparation.

Seeding

The earliest and most rudimentary form of seeding is broadcasting. This can be accomplished by hand or by spinning impellers either hand-cranked and carried or mounted on the back of a tractor, wagon, or other vehicle. As this process leaves the seed on the surface, it is generally followed by a spike-tooth harrow or some form of drag to shallowly incorporate the seed into the soil. Following with some type of roller to firm the soil for more intimate contact with the seed is helpful.

Because forage seed can be relatively expensive and drilling (seeding in closely spaced rows or bands) generally leads to higher germination, this method is

Encyclopedia of Agricultural, Food, and Biological Engineering
DOI: 10.1081/E-EAFE 120006862
Published 2003 by Marcel Dekker, Inc. All rights reserved.

usually preferred. The seed drilling machine has at least three functions: 1) meter out the seed uniformly at the desired rate; 2) place the seed into the soil at the desired depth; and 3) compact the soil around the seed.

As forages may be seeded with companion (weed suppressing) crops such as cereal grains, the drill is also equipped with a separate set of equipment for concurrently drilling the larger companion crop seed. In addition, the drill typically has the capability of placing fertilizer. As the forage seed is placed shallower than either the companion crop seed or the fertilizer, the placing of these latter precedes that of the forage seed. The mechanism for drilling the forage seed is thus rear most on the drill machine. The sequence of events at a given point in the row is furrow opening, seed metering and deposition, and furrow closing and firming. Metering may be done by fluted wheels or disks rotating in the bottom of the seed hopper. Rate of seeding may be controlled by rotational speed of the mechanism and/or the extent of the mechanism exposed to the seed. Metering mechanisms for forage seed are typically miniaturizations of the mechanisms used for larger seeds such as grains. Alternately, seed may be metered by using differential air pressure to hold seeds to orifices on disks or drums rotating in the seed hopper.

In the case of conventional drills, furrow openers frequently consist of two vertical plates attached together at the front and diverging from front to rear with seeds being dropped between the plates. Furrow closers can be weighted wheels with concave faces or pairs of wheels canted in such a way as to force the furrow shut. The furrow closers are usually equipped with some form of scraper to minimize soil build-up on them. In the case of no-till drills, (Fig. 1) furrow openers consist of two sharpened diverging disks capable of cutting through crop debris and forcing the soil apart to create a planting slot. More downward force is required for the furrow openers and furrow closers of no-till drills, supplied either by the weight of the drill and/or the tractor. Use of herbicides is a necessity with no-till planting. Depth of seed placement is critical. Individual furrow openers should "float" on a suitable linkage with their depth controlled by gage wheels. Depth of fertilizer placement is likewise important and should be 2 cm–3 cm deeper than the seed.

A popular forage seeder for conventionally tilled soil is known as the culti-packer seeder or the "Brillion" seeder after the original manufacturer (Fig. 2). It consists of two corrugated rollers with the seed metering mechanism dropping the seed between them. The first roller firms and corrugates the soil. Seed falls onto the corrugations. The second roller splits the ridges, covers the seed shallowly, and firms the soil. Performance on wet or finer textured soils is less satisfactory than on lighter, more friable soils.

Crusting on finer textured soils after heavy rains can create an impediment to seedling emergence. Judicious use of a spike-tooth harrow with teeth oblique, culti-packer or rotary weeder for crust breaking can improve emergence.

Some forage species, such as Bermuda grass (*Cynodon dactylon*) are more successfully propagated vegetatively using sprigs or shoots. Sprigs can be generated by running a spring-tooth harrow repeatedly over a nursery area. A side-delivery rake can be used for gathering and soil removal. Sprigs can be manually pushed into the pretilled soil using a stick. Transplanters developed for tobacco or vegetable crops can be used. Alternately sprigs can be scattered using a manure spreader and covered using a tandem disk harrow followed by a roller. A sprig depth of around 6 cm with part of the sprig emerging is recommended.

Fig. 1 No-till drill, upper—front view showing disk-type furrow openers and lower—rear view showing seed drop tubes and furrow closers.

Fig. 2 Brillion seeder showing both corrugated rollers and seed and fertilizer hoppers.

PEST CONTROL

Weed Control

Because of being broadcast or drilled in narrowly spaced rows or bands, interrow cultivation for weed control is rarely used when establishing forage crops. Herbicides used for weed control typically target either broad-leafed weeds or grasses, but not both. Therefore certain herbicides can be used on pure grass stands and other herbicides on pure legume stands, but use of herbicides on mixed stands of grasses and legumes is problematic.

Based on time of application, herbicides may be classed as: 1) preplant incorporated (PPI); 2) pre-emergent; or 3) postemergent. Field sprayers (Fig. 3) can be used to apply all types with immediate incorporation of PPI into the soil using a disk harrow. Sprayers may be tractor-mounted with saddle tanks, trailed behind a towing vehicle, or mounted on special purpose vehicles with floatation tires; the last used mainly by contract operators. Timing, soil moisture, and temperature are all important to successful use of herbicides.

Fig. 3 Field sprayer, trailed sprayer with folding spray booms.

Insect Control

Insects, which damage forage foliage, may be controlled by spraying with selected insecticides. The same spraying equipment used for herbicides may be used for insecticides. A recommended technique for insect control is to periodically assess the insect population using a sweep net. Only when the population exceeds a recommended economic threshold is spraying carried out. An alternative to spraying may be to mow the crop if it is approaching harvesting maturity.

FORAGE PRODUCTION FOR GRAZING

Rationale

Because harvesting, transporting, and storing forage crops are a major part of total production cost, utilization of at least part of the herbage by grazing may be economically attractive. Grazing also reduces the cost of manure handling and transportation. Proponents of grazing also cite improved animal health as an advantage. Forages on grazing land may be established in the same way as those for machine harvesting, in which case the forage may be part of a rotation with other crops. However, under certain difficult topographical or climatic conditions, land is permanently grazed, and thus is used only for perennial forage production. The productivity of permanent grazing land may be enhanced by management practices including fertilization, interseeding, and weed and brush control.

Rotational Grazing

For best forage utilization by grazing, animals are moved frequently onto new, relatively small areas. The time periods between moves typically range from less than a day to several days. This allows herbage in a given area to be intensively grazed, minimizes damage to plants by animal traffic, and then provides a period for the plants to regrow to a desirable stage before animals are rotated onto the area again. In some cases, animals with the highest nutritional requirements, such as lactating dairy cows, are given first access to an area and are then followed by animals with lesser nutritional requirements such as dry cows and young stock.

Fencing systems, usually electrified, play an important part in rotational grazing. Typically the grazing area is divided into sectors by permanent or semipermanent fences. The sectors are then subdivided into grazing areas by movable cross fences. These two types of fencing might be pictured as the uprights and rungs of a ladder, respectively. Electric fence controllers may be powered by line current, batteries, or solar cells. Tractors or other vehicles can provide significant help in the fencing process by auguring post holes, driving posts, and carrying multiple spools of fencing wire.

Provision of drinking water to the individual grazing areas is another challenging requirement. This can be accomplished by a combination of permanent and temporary pipelines from various water sources or by water tanker(s). Sometimes the geometry of sectors can be arranged to converge on a common watering point. The resulting heavy traffic, however, may destroy the plants around the drinking area. After grazing, regrowth and weed control can be enhanced by mowing of ungrazed herbage, by distributing localized manure deposits by dragging a spreading device behind a vehicle, and by fertilizing where needed.

Milking of grazing cattle is accomplished either by a mobile milking unit brought to the cows or by driving the cows back to the barn for each milking. Where allowed by regulations, the mobile unit saves time and energy. Mobile milking equipment typically includes an engine to power the vacuum pump for the milking machines, means for tethering and feeding supplement to the cattle during milking, and water and sanitary supplies for preparing the cattle.

While grazing is normally associated with ruminant animals, it has also been used to advantage for poultry and swine where improvement in animal health and substantial decreases in feed costs have been claimed. In recent years, rotational grazing of poultry by means of movable pens has been reported. The increased labor involved is offset by premium prices paid by consumers desiring poultry produced under nonconfinement conditions, often referred to as "free range" poultry.

IMPROVEMENT OF PERMANENT GRAZING LAND

Measures to improve permanent grazing land include: fertilization, pH adjustment (liming), weed control, and interseeding. As the topography of permanent grazing land frequently makes it prone to erosion, the existing sod is maintained. No-till equipment may be used for fertilizing and seeding. Alternately, fertilizer, lime, and seed may be broadcast and worked into the soil with a disk harrow. Herbicides may be sprayed prior to seeding to kill broad-leafed weeds or may be used at a sublethal dose to retard the existing grass while the seedlings become established. Interseeding the existing grass sod with forage legumes may have particular value, since they fix nitrogen from the atmosphere, provide additional protein to the grazing animals, and may contribute to the uniformity of the yield by providing forage when grass growth is low due to high temperatures and/or low soil moisture.

IRRIGATION OF FORAGE CROPS

As animal products are in demand in arid regions as well as in humid regions, forage crops may be grown under irrigation. For surface methods of water distribution, fields must be planed, using grading equipment, to provide uniform slope. Forage crops are surface irrigated most frequently by means of borders or corrugations. The former are strips usually 10 m–15 m wide running down slope with low berms running along both edges to confine the flow. Corrugations are small furrows running down slope. In either case, water is introduced from a ditch at the upper end of the field and flowed to the lower end in a uniform sheet. Berms and corrugations must be small enough to allow harvesting operations. Because of the cost of land planing and the difficulty of uniform water application using surface methods, application of water by overhead sprinklers has become widespread. These are frequently mounted on wheeled center pivot systems 400 m long with the pivot located near the water source. On certain low-lying terrain with high water table, subirrigation may be practiced. In this system, artificial drainage is used to lower the water table. When irrigation is desired, drain outlets are closed to bring the water table up into the forage root zone.

FURTHER READING

Barnes, R.F.; Miller, D.A.; Nelson, C.J., Eds. *Forages, Vol. II*; The Science of Grassland Agriculture; Iowa State University Press: Ames, IA, U.S.A., 1995; 357.

Hanson, A.A.; Barnes, D.K.; Hill, R.R., Jr., Eds. *Alfalfa and Alfalfa Improvement*; Am. Soc. of Agronomy, Crop Sci. Soc. of America, Soil Sci. Soc. of America: Madison, WI, U.S.A, 1988; 1084.

Srivastava, A.K.; Goering, C.E.; Rohrbach, R.P. *Engineering Principles of Agricultural Machines*; American Society of Agricultural Engineers: St Joseph, MI, U.S.A., 1993; 601.

Stout, B.A., Ed. *CIGR Handbook of Agricultural Engineering, Vol.III*; Plant Production Engineering; American Society of Agricultural Engineers: St Joseph, MI, U.S.A., 1999; 632.

Undersander, D.; Martin, N.; Cosgrove, D.; Kelling, K.; Schmitt, M.; Wedburg, J.; Becker, R.; Grau, C.; Doll, J.; Rice, M. *Alfalfa Management Guide*; NRC 547; Am. Soc. of Agronomy, Inc., Crop Sci. Soc. of Am., Inc., Soil Sci. Soc. of Am., Inc.: Madison, WI, U.S.A., 2000; 53.

F

Forage Storage Systems

Richard E. Muck
United States Department of Agriculture (USDA), Madison, Wisconsin, U.S.A.

INTRODUCTION

Forages (primarily grasses and legumes) are a major component of the diet of cattle, sheep, and goats. Livestock may harvest forages directly by grazing, but normally farmers need to harvest and store some forage to provide feed when forage production is low.

Forages are stored in one of two general ways: hay or silage. In making hay, the crop is dried to approximately 15% moisture, wet basis. At that moisture content, the crop is typically too dry for spoilage microorganisms to grow. In making silage (i.e., ensiling), the crop is stored at much higher moisture contents, 40%–85%. Two conditions preserve forage in ensiling. First, the crop is sealed to keep out oxygen. The lack of oxygen keeps many spoilage microorganisms from growing. This may be accomplished in a structure (silo), or it may be done simply by covering a pile of forage with plastic. Second, a low pH inhibits undesirable microorganisms that grow anaerobically (i.e., without oxygen). A low pH is normally achieved by the fermentation of crop sugars to lactic acid by lactic acid bacteria naturally present on the crop.

The relative amounts of hay and silage produced vary considerably by region (Table 1). This variation is due to a number of factors including climate, feed, and herd management options available, aversion to silages for milk to be made into certain cheeses, and sometimes tradition.

Hay has several key advantages over silage. It is stable in storage for long periods if kept dry. Harvesting and storage costs are generally lower for hay. The low moisture content of hay makes transport over long distances economically feasible, and thus hay is the primary means of buying and selling forages. Hay can be kept separate in storage (by field, cutting, etc.) and then fed by quality, maximizing utilization by animals.

On the other hand, hay production requires longer wilting times that make it more susceptible to harvest losses from rainfall, particularly in humid climates. Also, shattering and loss of nutritious leaves during harvest are greater in a dry crop. As a result, a crop preserved as silage generally has a higher feed value than if preserved as hay. Ensiled crops are normally chopped into small pieces, easing mechanization of handling and feeding relative to hay.

Overall, silage tends to dominate in regions where it is difficult to make hay without rainfall damage such as northern Europe, eastern Canada, and the north central and northeastern United States. In contrast, hay is dominant in the dry climates found over much of the western United States and Canada.

HAY SYSTEMS

Hay is harvested in a number of forms. In the most processed form, hay is formed into pellets and cubes. On the opposite extreme, hay may be stored loose. However, hay is most often harvested in bales of varying dimensions [small rectangular (46 cm–56 cm wide), large rectangular (80 cm–120 cm wide), and large round (120 cm–150 cm diameter) bales].

In humid climates, hay bales are often stacked and stored indoors in low-cost pole barns or sheds (Fig. 1). Often the shed is open on one side to facilitate storage and removal. More traditionally, hay was stored in mows or upper stories of barns and then dropped down chutes to the animal feeding area below. This system worked well for loose hay as well as small rectangular bales. However, such systems are more costly, more labor intensive, and not well suited to the larger bales more common today. Forage cubes and pellets are also likely to be stored indoors or in covered commodity bins.

Particularly in dry climates, hay bales may be stored outside, either in a stack or individually. They may be laid on the soil surface or on various pads (concrete, gravel, etc.) to minimize water absorption by the bottom layer of bales. In wetter climates, covering stacks of bales with a plastic tarp minimizes losses from water absorption and subsequent respiration and heating of the hay by the growth of spoilage microorganisms. Large round bales are often stored individually on their sides (Fig. 2), and the top half of the circumference of the bale may be covered with polyethylene film to help shed rainfall.

Losses in storage are primarily from the growth of spoilage microorganisms (molds, yeasts, and bacteria). Activity is greatest in the first month of storage when bales may enter storage at higher moisture contents (20%–30%). With time, bales dry to an equilibrium moisture content of approximately 15% or less. Once the bales reach such

Encyclopedia of Agricultural, Food, and Biological Engineering
DOI: 10.1081/E-EAFE 120006898
Published 2003 by Marcel Dekker, Inc. All rights reserved.

Table 1 Estimated quantities (million t dry matter) of hay and silage produced in Europe and North America in 1994[1]

	Hay	Silage	Total
Europe			
W. Europe[a]	60.1	91.6	152.6
E. Europe[b]	32.3	15.4	47.7
Russian Federation	59.7	45.0	104.7
Total	152.1	152.0	305.0
North America			
USA	123.0	39.3	162.3
Canada	40.9	7.4	48.3
Total	163.9	46.7	210.6

[a] European Union + Norway and Sweden.
[b] 15 countries.

Fig. 2 Large round bales stored outdoors without protection. Bales in the foreground tied with plastic net wrap; bales in the background with sisal twine.

moisture contents, microbial growth is essentially stopped. The time to reach equilibrium is greater for denser bales and bigger stacks. Some farms may reduce that time by drying the hay with heated or unheated air. This adds to capital and energy costs. Losses increase as the moisture content of the bale going into storage increases. At 15% moisture, approximately 1% of dry matter will be lost in the first month whereas at 30% moisture a bale will lose approximately 8% of dry matter. These losses are typically of the most digestible fractions of the crop.

A slow loss of dry matter occurs throughout storage. Under indoor conditions, bales will lose approximately 0.5% of dry matter per month. Losses in bales stored outdoors are much more variable, 0.5%–3.0% per month, dependent on the amount of rainfall and protection from rainfall absorption. Rainfall will typically wet the outer 10 cm–20 cm of an exposed bale, allowing microbial growth in that outer layer until the bale dries again. In dry climates, the differences in losses between outdoor and indoor storage are small. In wet climates, covering of bales is needed to minimize spoilage losses, and this may justify the expense of indoor storage.

SILAGE SYSTEMS

Farmers can choose from a wide variety of silo types. Options range from expensive tower silos to covering a pile on the ground.

Fig. 1 Open front hay shed being filled with small rectangular bales.

Tower silos were the dominant silo type in North America 40 yr–50 yr ago. Tower silos are fabricated from a number of materials. The most common are concrete stave (Fig. 3), poured concrete, and glass-lined steel (Fig. 4). In the concrete stave silo, rows of interlocking concrete slabs or staves are held together with steel hoops. All tower silos are filled with a blower that carries the forage through a pipe outside the silo to the top of the silo. On entering the silo, the forage may hit a device called a distributor that spreads the forage more uniformly within the silo. Concrete stave silos are typically unloaded from the top, using an augur-type device that draws silage to the middle of the silo and blows the silage through doors on one side of the silo. Steel and poured concrete silos are usually emptied from the bottom with a chain saw-like device that cuts the silage and drops it onto a conveyor.

Some tower silos, particularly steel silos, are called oxygen-limiting. These silos are sealed at the top and bottom to keep out air. A breather bag at the top of the silo maintains a constant gas pressure in the silo by allowing the silo gases to expand and contract due to diurnal heating and cooling without drawing in air. This limits the amount of oxygen entering these silos.

Fig. 3 Concrete stave tower silos with a blower (bottom center) and feed table (left) for filling.

Fig. 4 Oxygen-limiting, steel tower silos.

Today, horizontal silos have become the most dominant, largely because they require less capital investment. There are three general types: the drive-over pile or clamp silo, the bunker (Fig. 5) or trench silo, and the pressed bag (Fig. 6). The main difference between drive-over piles and bunker silos is the presence of walls in the bunker silo. The walls in bunker silos are typically large concrete slabs whereas trench silos are built into hills so that the walls are undisturbed soil. In both pile and bunker silos, forage brought to the silo area is spread and packed by tractors or bulldozers. After filling is complete, the crop is covered with polyethylene plastic. The plastic is held tightly to the crop by used tires, soil, netting, or other means. Some farmers do not cover the crop, but this leads to substantial spoilage in the upper layers of silage.

The pressed bag is one of the newest silo types. Chopped forage is pressed into a tube of approximately 0.22 mm thick polyethylene. Typical diameters are 1.8 m–3.6 m, and the most common lengths are 30 m, 60 m, and 90 m. Bags can be laid on the ground, but feed out losses are minimized by placing the bags on a concrete pad or other hard, smooth surface.

Fig. 5 Bunker silo with concrete slab walls.

Fig. 6 Filling of a pressed bag silo.

A final silo type is the wrapped bale. Wrapping machines put multiple layers of 0.025 mm thick stretch polyethylene film around either large round (Fig. 7) or large rectangular bales. A minimum of 4–6 layers is needed for good preservation. The early machines wrapped only individual bales. Now machines are available that wrap large round or rectangular bales placed end-to-end to form a line of bales. In addition, machines are available that place large round bales end-to-end in tubes of plastic-like that used in pressed bag silos. These latter two machines reduce the amount of plastic needed to ensile baled forage.

Losses during silo storage come from several sources. At ensiling, plant respiration consumes sugars and oxygen and produces carbon dioxide, water, and heat. Even with

Fig. 7 Large round bales being wrapped with stretch polyethylene film. (From K. J. Shinners.)

rapid filling and sealing of the silo, plant respiration will cause a 1%–2% loss of dry matter. Once the crop is anaerobic, lactic acid bacteria ferment crop sugars primarily to lactic acid. Fermentation, dependent on the sugars fermented and the dominant species of microorganisms, produces more dry matter loss, typically 1%–3%. In addition, some protein is broken down to amino acids and ammonia. Seepage of oxygen into the silo, during storage and while the silo is emptied, allows spoilage microorganisms to grow. The loss from spoilage microorganisms is highly variable and affected by silo type and management as well as storage time.

Overall losses of dry matter for well-managed silos are 4%–8% for oxygen-limiting tower silos, 6%–10% for concrete stave tower silos, 10%–16% for bunker and pile silos, 6%–10% for pressed bag silos, and 6%–12% for wrapped bale silages. Poor management increases losses, particularly for the nontower silos that depend on the integrity of plastic to minimize spoilage.

FURTHER READING

Cavalchini, A.G. Forage Crops. In *CIGR Handbook of Agricultural Engineering, Vol. III Plant Production Engineering*; Stout, B.A., Cheze, B., Eds.; Amer. Soc. of Agric. Engrs.: St. Joseph, Michigan, 1999; 348–380.

McDonald, P.; Henderson, A.R.; Heron, S.J.E. *The Biochemistry of Silage*, 2nd Ed.; Chalcombe Publications: Marlow, Bucks, UK, 1991.

Rotz, C.A.; Muck, R.E. Changes in Forage Quality During Harvest and Storage. In *Forage Quality, Evaluation, and Utilization*; Fahey, G.C., Jr., Michael, M., David, R., Moser, L.E., Eds.; Amer. Soc. of Agronomy: Madison, Wisconsin, 1994; 828–868.

Wilkinson, J.M.; Bolsen, K.K. Production of Silage and Hay in Europe and North America. In *Proceedings of the XIth International Silage Conference*, Univ. of Wales, Aberystwyth, Sept 8–11, 1996; Jones, D.I.H., Jones, R., Dewhurst, R., Merry, R., Haugh, P.M., Eds.; IGER: Aberystwyth, 1996; 42–43.

Forage Transport Systems

Richard J. Straub
University of Wisconsin–Madison, Madison, Wisconsin, U.S.A.

F

INTRODUCTION

Forage transport from field to storage has increasingly become mechanized over the past 50 yr. While basic harvesting processes of baling dry or near dry forage for hay and of chopping forage for silage have not changed in principle, both processes have seen significant changes in the size of handling units, rate of harvest, and speed of transport and handling.

CHOPPED FORAGES

Forage harvesting for silage has traditionally used some sort of trailed wagon to accumulate the chopped forage as it leaves the discharge spout of the forage harvester. When the wagons are filled, they are detached and hauled to the storage site using a tractor. These wagons are usually called "self-unloading wagons" because they have the ability to unload mechanically from the front or rear (Fig. 1). Some self-unloading wagons are built to allow either mode of discharge. The unloading mechanisms of these wagons are usually powered by the power take-off (PTO) of the tractor towing them for transport, although some are now hydraulically powered by the tractor. These wagons utilize a live bed using a chain and slat conveyor to move forage to the front or rear of the wagon. On front-unload wagons during unloading, the live bed pulls forage material into a set of two or three beaters that tear or mill the material from the pile. Material pulled loose by the beaters drops onto a cross conveyor, either a chain and slat type or a large auger, that moves it to the side of the wagon for discharge into a forage blower or bagging unit. On rear-unload units, a false end-gate is used. This swings upward to allow forage to drop off the rear of the wagon into a pile or bunker silo or into a special receiver for use with a forage blower or bagger used to fill tower and bag silos, respectively. If these wagons are to be towed behind the forage harvester, they may have a roof to help retain the forage as it is blown into the wagon. If the unit is to be towed along side the forage harvester, a roof cannot be used. These wagons have lengths of 4.3 m–7.0 m (14 ft–24 ft) with capacities of $14.2\,m^3$–$28.3\,m^3$ ($500\,ft^3$–$1000\,ft^3$).

As forage harvesters have increased in capacity, these towed units, which require frequent hitching and unhitching to change the transport units as they are filled, are being replaced either: 1) direct discharge of the forage harvester into open dump trucks (multiaxle straight trucks or semis); 2) into the longer versions of the rear-discharging, live bed wagon; or 3) by side-dump receivers (Fig. 2), which are dumped sideways into a dump truck when full. These side-dump units hydraulically lift and dump the receiver into a truck for transport without having to hitch and unhitch units. Use of such units increases the field efficiency and effective fill capacity of the harvesting operation by minimizing time lost due to changing wagons. It also allows for larger haul units to be used because traction and power of the harvesting unit do not limit the size of transport unit used. These units are usually used to transport the forage to bunkers or drive-over-piles for packing and ensiling.

If these larger rear-discharging units are used to fill a tower silo, an intermediate receiver is used to accumulate the chopped forage as it is discharged. This receiver is configured similar to a side-discharger, self-unloading box with a live chain and slat bed, a beater to evenly meter the material, and a side-discharge conveyer to move the metered forage into the blower.

Forage blower is a straight-bladed material fan that pneumatically conveys chopped forage into a tower silo (Fig. 3). In spite of high power demands, blowers are used because they require a small amount of space, they are easily moved from one silo to another, and they have good capacity when ample power is available. The blower has a receiver to accept forage from a side-discharge transport or metering unit. A short auger is usually located in the receiver to feed the fan, which blows the forage into the silo. Fan diameters of 1.2 m–1.5 m (48 in.–60 in.) are common, with a fan tip speed of about $35\,m\,sec^{-1}$ ($115\,ft\,sec^{-1}$). The fan blows the material vertically through a conveying pipe to the top of the silo for discharge and distribution into the silo. Capacities of more than $100\,tn\,hr^{-1}$ are attainable.

Encyclopedia of Agricultural, Food, and Biological Engineering
DOI: 10.1081/E-EAFE 120006882
Copyright © 2003 by Marcel Dekker, Inc. All rights reserved.

Fig. 1 Self-unloading forage wagon. (Courtesy of H&S Manufacturing.)

HANDLING SMALL RECTANGULAR BALES

If forage is not harvested by chopping for silage, it is probably harvested by baling. Baling traditionally was done at crop moistures below 20% to be stored as dry hay. As the size of balers has increased and handling has been mechanized, it is now possible to bale at moistures of 60% or even higher. This high-moisture hay must be wrapped to exclude oxygen and to allow the forage to ferment for silage.

Fig. 2 Side-dump wagon. (Courtesy of Miller-St. Nazianz.)

Fig. 3 Forage blower. (Courtesy of Miller-St. Nazianz.)

Small rectangular balers produce bales containing 18 kg–60 kg (40 lb–125 lb) of material. Rectangular bales may be dropped in the field as they emerge from the bale chamber. This practice improves the field efficiency of the baling operation since it eliminates the chore of hitching and unhitching wagons. Bales must be picked up later, either by manually lifting them onto a wagon or truck or by using an automatic, self-loading bale wagon. Mechanical lifters are also available to lift bales from the ground onto a moving wagon or truck where they are piled by hand.

Balers may be equipped with long chutes, which guide emerging bales onto a trailing wagon. As the bales are pushed back toward the wagon, a person must stack each one into position. More commonly, a bale-throwing device eliminates the need for a person on the training wagon. These devices are designed to toss short 76 cm–100 cm (30 in.–40 in.) bales into a wagon, which must be equipped with high sides and a rear-end gate. Field speeds may be increased somewhat when using bale throwers as the driver need not be concerned about overworking someone on the trailing wagon.

Bale throwers are made in several designs. One common design uses a pair of high-speed flat belts mounted at an angle at the rear of the baler. As the bale emerges from the bale chamber, it is caught between the belts and is tossed into the trailing wagon. Distance of throw is changed by adjustable variable pitch V-sheaves in the bale thrower drive. On turns, the entire bale thrower turns with the trailing wagon to avoid misses.

Another bale thrower design consists of a hydraulically operated "pan" mounted on arms pivoted from the rear of the baler. The bale emerging from the baler trips a valve that allows oil under pressure to operate a hydraulic cylinder, which causes the pan to move upward and rearward suddenly. This motion tosses the bale toward the trailing wagon. The entire thrower is pivoted horizontally, and its position can be altered by a tractor-operated hydraulic cylinder to compensate for side hills and curves. Control of the thrower hydraulic pressure provides adjustment for throwing distance.

Bale throwers convey bales of hay into a large "cage" type rack on a trailed wagon. These bale cages or racks can hold 100–200 bales, depending on their length (Fig. 4). They are usually unloaded manually into storage or onto a chain conveyor for elevation into a barn or onto a pile. The chain conveyor or elevator may use a single chain with fingers that grab the bale and pull it along an elevating surface, usually of tubular steel construction, or it may use a double chain and slat design running in a sheet metal pan. The latter type of conveyor can also be used for chopped forage or grain if desired.

Automatic bale wagons are made to pick bales from the ground and stack them mechanically onto a platform as shown in Fig. 5. Bales are turned to stand on their edge by means of a special quarter-turn chute attached to the rear of the baler. As they are picked up by the bale wagon, they are grouped onto a table at the front of the wagon. When enough bales have been accumulated to form a tier, the entire group is tilted up onto the platform where they are held by stabilizing bars. When the automatic bale wagon has been completely loaded, the operator may drive the entire load into the storage site. There the entire load may be tipped intact and added to an existing stack. Another alternative is to tip the stack off at an elevator, leaving another person to load the bales onto the elevator while the wagon returns to the field for another load. Yet another alternative for an automatic bale wagon is to unload in reverse of the loading operation. In this arrangement, single bales are unloaded from the front of the wagon onto an elevator or into a feed bunk. Automatic bale wagons relieve much of the drudgery involved with harvesting rectangular bales. A careful operator may transport many hundreds of bales without handling them manually. These self-loading units may be tractor drawn or self-propelled.

Fig. 5 Self-loading bale wagon. (Courtesy of New Holland North America.)

Fig. 4 Throw bale rack. (Courtesy of New Holland North America.)

Several types of bale accumulators are also available to aid in bale handling. An accumulator is a device attached to the rear of a baler, which accumulates groups of bales and leaves them in the field for subsequent pickup by a special front-end loader. They are then loaded onto a wagon or truck for transport into storage.

MOVING LARGE ROUND BALES

A variety of devices are available to move large round bales from the field to the storage site. These range from simple bale forks to rather complex self-loading trailers.

Bale forks consist of one, two, or more large tines fastened to a frame arranged to mount on the three-point hitch of a tractor. The tines are built to lift and support the bale. Those with one or three tines are intended to "spear" or penetrate the bale. Those with two tines are intended to slide under the bale. By operating the three-point hitch, the bale is then lifted for transport.

Tractor front-end loaders or skid steer loaders are also used for lifting, loading, and sometimes transporting large bales. A special tined fork is sometimes used, but a special grapple offers more control of the bale.

A front-end loader with an ordinary materials bucket should not be used to lift bales. Serious accidents have resulted from falling bales. Large bales may roll back onto the tractor driver if they are lifted too high with a loader that is not equipped with a safety frame.

Towed bale carriers are long trailers onto which several large bales are loaded in a single row for transport. Some are simply loaded with a tractor-mounted loader. Others are equipped with a mechanical or hydraulic loader mounted directly onto the mover.

MOVING LARGE RECTANGULAR BALES

Midsize and large rectangular bales have bale cross-sections, which are approximately $0.8\,m^2$ (32 in.) and $1.2\,m^2$ (48 in.), respectively. Bale lengths are variable but generally range between 1.8 m (72 in.) and 2.5 m (100 in.). Because of the weight of these bales, they must be handled mechanically, much like round bales. These bales are discharged from the baler directly onto the ground, or sometimes a mechanical accumulator is used to collect them for grouping prior to discharging them to the ground. The bales are then loaded mechanically using a hydraulic loader with a multitined bale fork or grapple. Tractor mounted or skid steer loaders are both commonly used to load the bales on trucks or wagons for transport. The same type of loader is used to unload the bales and stack them into storage and/or feed them. Telescoping boom fork lifts (Fig. 6) are also used to load and unload these large rectangular bales. This type of fork lift offers greater height and mobility for stacking.

Fig. 6 Telescoping fork lift. (Courtesy of Gehl Company.)

SILO BAGS AND BALE WRAPPING

Over the last decade, there has been increased use of machines that move forage into plastic covering for ensiling or preservation. Bagging equipment typically consists of a conveyor that moves the forage from a truck or wagon into a hopper where a rotating shaft or cylinder with backward-curved fingers pushes the chopped forage into an attached folded bag. The bagging unit pushes itself away from the packed forage while more of the folded bag is released to receive more storage. The unit is restrained by a cable and braking system, providing resistance to motion and thereby increasing the packing density of the forage. The bags typically are 8 mil–10 mil plastic with UV inhibitor. Their diameters range from 2.1 m to 3 m (7 ft–10 ft), with lengths ranging up to approximately 60 m (200 ft). Ensiled material is generally removed from the bags with a skid steer or hydraulic loader for feeding.

Bale wrappers apply stretchable plastic (approximately 2 mil) in overlapping layers to large bales to preserve them. Some units can accommodate both rectangular and round bales. Others will only handle round bales. These units apply the plastic by rotating an individual bale about two of its axes simultaneously, allowing plastic to be applied in overlapping layers to cover the entire bale surface. Other wrapping mechanisms wrap the bales by applying plastic circumferentially while advancing along the longitudinal axis of the bale. The ends of the bales are not covered up completely, but are butted together end to end to attempt to form a seal between bales. Some units can wrap multiple bales to form a continuous, sausage-like tube of wrapped bales. Individually wrapped bales can be handled with a hydraulic loader equipped with a tong-like mechanism to squeeze and lift the bales.

FURTHER READING

Finner, M.F.; Straub, R.J. *Farm Machinery Fundamentals*, 2nd Ed.; American Publishing Company: Madison, WI, 1985.

Srivastava, A.K.; Goering, C.E.; Rohrback, R.P. *Engineering Principles of Agricultural Machines*; ASAE Textbook No. 6; American Society of Agricultural Engineers: St. Joseph, MI, 1993.

Forest Harvesting Systems

F

Richard W. Brinker
Auburn University, Auburn, Alabama, U.S.A.

INTRODUCTION

Forest harvesting systems serve a complex function in the toolkit of the forest manager. These systems are used not only to harvest wood used as raw material for the multitude of consumer products derived from the forest, but are also used to implement the silvicultural harvesting prescription in a cost effective, environmentally acceptable manner. The harvester must satisfy the landowner's needs, meet the quality requirements of the timber conversion mill, and provide a margin of profit to the harvesting company operator to maintain the financial health of the company. These systems range from simple engineered systems that are low cost and more labor intensive to those that are mechanically complex and technologically advanced.

The complete system of harvesting entails seven distinct elements in the actual harvesting process. These elements include: 1) road design, layout, and construction; 2) felling; 3) bucking, limbing, and topping; 4) skidding or forwarding; 5) loading; 6) transportation; and 7) unloading. Elements 1–5 will be discussed in this section; transportation and unloading at the mill are detailed in other chapters. Additionally, harvest planning and layout should occur prior to the physical process of harvesting to ensure a more effective harvesting operation.

ROAD CONSTRUCTION

Road construction is an initial and crucial step. Roads that are improperly planned, located, and constructed can result in irregular and costly access to the forest, and can have an adverse effect on water quality due to sedimentation. Road construction for harvesting purposes may be for one-time use, or for continued future access for other purposes. Adequate time should be allocated for properly trained and experienced personnel to plan and construct harvesting roads. Greater details concerning forest road construction can be found on several web sites (e.g., www.pfmt.org).

FELLING

Felling of timber entails severing a standing tree from its stump. Felling may be manual using hand tools or chainsaws, or mechanical through the use of felling heads on mobile carriers. Chainsaws continue to be used in many parts of the world, but are becoming less common due to safety and labor concerns. Mechanical felling machines vary from simple scissors-type, hydraulic directional shears to various designs of bar saw heads and high-speed disk saws. These machines are frequently designed as feller-bunchers, which can accumulate several stems in the cutting head before laying down the bunch of stems. High-speed disk saws have become common on most U.S. harvesting systems (Fig. 1). These felling machines have gained widespread use due to increased productivity and reduced degradation of wood quality.

The carrier, a self-propelled power source used to operate these felling devices can be propelled via tracks or rubber-tires. They can also be classified by the method of cutting. The tree-to-tree machine moves after each felling cut, contrasted to the limited area machine that allows a boom mounted felling head to cut a number of stems from one setup position. The carrier type is selected based on cost, terrain, timber type, and stand conditions.

BUCKING, LIMBING, AND TOPPING

Bucking, limbing, and topping occurs after felling and involves severing the stem into product length segments, removal of limbs from the bole, and cutting the top to a size or quality specification. Motor-manual, primarily chainsaw, or mechanical methods may be used to accomplish these activities. Mechanical methods include slideboom delimbers and pull-through types of delimbers for trees that have a large number of size of limbs to remove. This phase may be accomplished in the woods or at the landing. More recently, chain-flail delimbers have been developed for use on smaller diameter stems. Frequently seen in the southern United States, are gate or grid delimbers. These are fabricated from heavy steel beams or pipe welded in a grid pattern with 2 ft^2 openings through which several trees are simultaneously backed through the grid. This method works relatively well in southern pines.

If the stems are not bucked into product lengths, they are frequently transported as tree-length. The determination of

Encyclopedia of Agricultural, Food, and Biological Engineering
DOI: 10.1081/E-EAFE 120006881
Copyright © 2003 by Marcel Dekker, Inc. All rights reserved.

Fig. 1 High-speed disk saw mounted on a tree-to-tree rubber-tired carrier.

tree-length or product length is dependent on timber size, product specifications at the mill, and local custom.

Felling, bucking, limbing, and topping may be combined into a single machine operation through the use of a mechanical harvester (Fig. 2). Mounted on a carrier, the harvester head can accomplish all four of these operations simultaneously. Machines of the most advanced design have optically scanned length and diameter measuring capability and computer control of product lengths and diameter during the bucking and topping process. The felling mechanism most frequently used on these heads is a hydraulically operated chainsaw bar. Harvesters can be categorized as single-grip, which has one head to fell, limb, and buck; or as double-grip, which has one head to fell, and a second attachment to limb and buck. Harvesters are frequently mounted on a boom arm as a limited area machine, but may also be mounted on a tree-to-tree carrier.

FORWARDING, SKIDDING, OR YARDING

Forwarding, skidding, or yarding is the process of moving logs or tree-length stems from the stump to the landing. Forwarding entails moving product length bolts in a carrier that keeps the stems off of the ground. The forwarder is usually equipped with a self-contained loader to pick up logs or bolts from the ground and place them into a rack designed to carry from 5 tn to 15 tn of wood (Fig. 3). The forwarder is often used in conjunction with a harvesting machine.

Use of a harvester/forwarder combination is widely used in some locales due to the benefits of this system. During the delimbing operation, the harvester lays the limbs in a mat or bed on which the harvester and forwarder operate. This allows the limbs to remain in the forest stand and provides for reduced soil compaction, increased aesthetic appearance of the harvesting operation, and decreased environmental impact. Other advantages of this type of harvesting system include load sizes that are as much as five times larger than those of a skidder, resulting in fewer trips through the forest stand and longer forwarding distances. However, use of this system is frequently limited due to higher operational costs than other ground-based systems, and handling systems at the final unloading point may not be designed to accommodate the lengths produced.

Skidding is probably the most widely used form of moving stems from the woods to the log deck. Skidders are specialized rubber-tired or tracked power sources that drag the stems along the ground, using either a winch-and-cable system or a mechanical grapple to secure single stems or multiple stems laid down in bunches by a feller-buncher (Fig. 4). Skidders vary greatly in size from less than 100 hp

Fig. 2 Limited-area harvester mounted on a rubber-tired carrier.

Fig. 3 Twelve-ton capacity forwarder.

Fig. 4 Grapple skidder.

to over 260 hp and haul load capacities up to approximately 2.5 tn. Specialized clam-bunk skidders are designed similar to a forwarder, and can skid loads up to 15 tn. These are more complex skidding machines that usually have a self-contained loader.

Cable yarding is a method used to move logs from the stump to a landing using steel cable haul lines and multiple drums and winches. Using this system, yarders can drag the logs along the ground as a ground-lead system, or use a spar tree or tower as a high-lead system to elevate the logs partially or totally off of the ground. These systems can operate on terrain too steep or rocky for a ground-based system. However, these systems require larger crews to operate than most tractor systems and offer less flexibility to successfully accomplish partial cutting.

LOADING

Loading of the harvested stems is any means of transferring the logs onto a means of transport to a final destination. Mechanical loading devices may be mobile machines such as a front-end log lift or stationary, such as a crane or hydraulically powered knuckleboom loader. The knuckleboom loader is the most commonly used approach, as it is relatively fast, and can operate on a variety of terrain and ground conditions. Additionally, the bucking and delimbing process can be included in this step as pull-through delimbers and hydraulic bucking saws can be incorporated onto this type of loader (Fig. 5).

Fig. 5 Knuckleboom loader equipped with a hydraulic bucking saw.

Fig. 6 Whole-tree chipper for producing undebarked or rough chips.

A variation that incorporates several steps in the timber conversion process involves whole-tree chipping. This system involves commutating the tree to include the main bole and limbs through a high-speed, wood-chipping disk (Fig. 6). This type of machine can produce relatively low bark content chips used for pulp and other reconstituted wood products when the stems are processed through a chain-flail debarking system prior to chipping. Chips are usually blown directly into vans for transport. This system requires a high amount of energy to operate, but material utilization is increased and chips produced are ready for final processing.

TRANSPORT AND UNLOADING

Transport from the woods to a mill or concentration point is commonly accomplished by water, railroad, or highway truck. Unloading is the final step in the harvesting process and takes place at the mill. Both of these elements are discussed in another section of this encyclopedia.

CONCLUSION

The machines used to accomplish the logging process can be combined in various configurations to comprise the forest harvesting system. The goal of an efficient harvester is to combine machines to provide a system that is balanced so that each phase of the process achieves a relatively equal rate of productivity at the lowest cost of production, and meets environmental and productivity requirements. Much greater detail describing harvesting systems can be found in textbooks and forest engineering publications, such as the proceedings of the Council on Forest Engineering (www.cofe.org).

FURTHER READING

http://www.cofe.org/page20.html

http://www.pfmt.org

Stenzel, G.; Walbridge, T.A., Jr.; Pearce, J.K. *Logging and Pulpwood Production*; John Wiley & Sons: New York, 1985; 358pp.

Forest Products Transport Systems

Craig J. Davis
State University of New York, College of Environmental Science and Forestry, Syracuse, New York, U.S.A.

INTRODUCTION

The objective in the primary transport[1] phase of timber harvesting systems is to move either the tree or segments of the tree from the stump location to a collection point (i.e., the landing). Three general approaches have been used to accomplish this objective. In ground-based systems, a machine moves to the stump area, the cut trees are attached to the machine, and then the combination of the machine and cut trees moves to the landing while the machine remains in contact with the ground. In cable systems, the machine is stationary (usually at the landing) and a series of cables are used to transport the cut trees from the stump area to the landing. The aerial systems approach is very similar to ground-based systems with the exception that the machine does not remain in contact with the ground and the cut trees remain fully suspended in the air.

GROUND-BASED SYSTEMS

Ground-based systems are characterized by the use of animals or machines that move from the stump to the landing, carrying or dragging the cut trees. All ground-based systems can be divided into two categories: skidding systems and forwarding systems. The difference between the two is whether the load is dragged behind or carried on the animal or machine.

Skidding

Skidding is the process of dragging the load behind some source of tractive power. If the load is in full contact with the ground, frictional forces are maximized and the load may shift and turn when encountering obstacles. If one end of the load is lifted free of the ground, frictional forces and soil surface disturbance are reduced and the load is easier to control. The most commonly used systems include: animals, crawler tractors, and 4-wheel drive rubber-tired skidders.

Animals

Although common in the past, horses are still used for special harvesting situations in North America. Belgian, Clydesdale, or Suffolks are the preferred species due to their larger size. Usually, a horse can pull only up to 80% of its own weight and can skid up to 100 m on level ground. Skid distances of up to 150 m are possible when skidding downhill.[1] The use of horses has been most successful in partial cuts and thinnings involving relatively small timber and moderate slopes with little underbrush or slash.

Crawler Tractors

Tracked crawler tractors with a rigid track frame were introduced in logging operations in 1919,[2] as a replacement for animals due to their ability to handle rougher terrain and larger volumes. John Deere, Case, Komatsu, and Caterpillar are the current major manufacturers of logging crawlers. Crawlers range in power from 20 hp to 230 hp (drawbar) and weigh between 5 tn and 30 tn. Skidding distances[1] of 100 m–200 m are common. Crawlers are able to operate where the terrain is difficult, such as on steep slopes (due to their low center of gravity) and rocky terrain. Elevated levels of soil disturbance are common due to the scraping that is caused by their rigid steel tracks. Crawlers are most commonly employed in thinnings and partial cuts and can be used for some clearcuts.

Rubber-Tired Skidders

The term skidder is usually associated with a design introduced in 1951[2] that incorporated 4-wheel drive, rubber tires, an articulated frame (used for steering), a built-in arch, and a diesel engine. The articulated frame and the 4-wheel drive made these skidders more stable and maneuverable on rugged terrain than 2-wheel drive agricultural tractors. The modern skidder maintains many of the elements of the original design; however, engines in modern skidders range from 70 hp to over 250 hp. John Deere, Timberjack, Caterpillar, Tigercat, Tree Farmer, and Franklin are the major manufacturers of

Encyclopedia of Agricultural, Food, and Biological Engineering
DOI: 10.1081/E-EAFE 120006888
Copyright © 2003 by Marcel Dekker, Inc. All rights reserved.

Fig. 1 Skidding hardwood saw logs with a Caterpillar 528 cable skidder in New York.

rubber-tired skidders in North America. Two types of rubber-tired skidders are common today: cable skidders and grapple skidders.

Cable skidders (Fig. 1) are commonly used in areas with large timber or difficult terrain that prevent the machine from driving to the stump. In addition to the characteristics mentioned above, these skidders are equipped with a blade (used for positioning logs), a winch containing a wire-rope mainline (1.5 cm in diameter, 30+ m in length), and 6 to 8 chokers (used for attaching logs to the mainline). In operation, the skidder operator attaches the chokers to the logs, then winches the logs to the back of the skidder. Upon arrival at the landing, the logs must be unchoked by hand. While skids of up to 1500 m are possible, skidding distances of 300 m–600 m are more common.[1] Generally, skidders are able to operate on 20%–30% side-slopes. Cable skidders are best suited for partial cuts.

Grapple skidders (Fig. 2) were developed to take advantage of the development of mechanized felling (feller-bunchers).[2] On a grapple skidder, the winch assembly is replaced by a hydraulically powered grapple to assemble and hold the logs during skidding, eliminating the time-consuming step of attaching and detaching chokers to logs. A consequence of this change is that grapple skidders lose their ability to drop their load and recover it whenever they get stuck. Grapple skidders operate under the same skidding distances and terrain conditions as cable skidders. Grapple skidders are best suited for clearcuts.

Forwarding

Forwarding is defined as the in-woods transportation operation in which the material being moved is wholly supported on or by the carry vehicle.[2] Modern forwarders (Fig. 3) are 4- or 6-wheel drive machines, usually mounted on an articulated frame, equipped with a knuckleboom loader and log bunks to hold the logs. Forwarders are commonly used in areas with single species stands, where road building costs are high, and the distance from the stump to the landing is long. Forwarding distances in the range of 600 m–900 m are very common.[1] Forwarders generally can be used in partial cuts or thinnings. Forwarders are limited to side-slopes less than 20% and are unsuitable for soft soils and wet sites.

CABLE YARDING SYSTEMS

Cable yarding can be defined as the movement of logs, tree lengths, or whole trees from the stump area to the landing using a stationary machine with a single or multiple winches and a system of flexible cables.[1] Cable systems

Fig. 2 Skidding tree length spruce and fir with a Timberjack 460 grapple skidder in Maine.

Fig. 3 Partially loaded Timberjack 910 forwarder in Quebec's spruce–fir region.

usually operate on terrain conditions that are not suited to the operation of ground-based systems (e.g., steep slopes, soft soils, wet sites). There are two major types of cable yarding systems in use today: highlead and skyline.

Highlead Cable Systems

The primary characteristic of the highlead system is that one "end" of the cable is elevated on a single "spar tree" or tower to provide lift for the logs being yarded. Highlead systems consist of a yarder with a two-drum winch; one drum holds a heavier mainline, the other holds a lighter haulback line. The haulback line is fastened to the mainline at the point (i.e., butt-rigging) where the chokers are attached. From the butt-rigging, the haulback line is run through anchored sheaves back to the yarder. With this configuration, one end of the choked logs is raised by applying tension to the mainline and braking pressure to the haulback or vice versa. Yarding distances from 200 m to 250 m are common, with 370 m considered to be a maximum. Highlead systems are primarily used in clearcuts and are best suited for yarding uphill (constant or concave slopes). Because the logs are partially dragged on the ground, considerable soil disturbance is possible with highlead systems.

Skyline Yarding Systems

Skyline yarding systems are cable systems where the cable is suspended between two or more spar trees or towers and a carriage travels along the skyline. Either chokers or a grapple is used to attach the load to the carriage. In comparison to highlead systems, the skyline yarder usually has a larger engine and the winch system has at least three drums, the third of which controls the tension in the skyline. Skylines can be rigged in several different ways: the "running skyline" configuration where the carriage runs on the haulback cable; the "live skyline" and the "slackline" configurations where the skyline can be raised and lowered during yarding; and the "standing skyline" configuration where the skyline is fixed in position during yarding. The maximum yarding distance with skyline systems is limited by the amount of cable carried by the yarder; a typical yarding distance is between 200 m and 400 m, although some specialized machines can yard up to 2000 m or more. Skyline systems are used in both clearcuts and partial harvests. As the logs are usually fully suspended above the ground, soil disturbance can be less than with highlead systems.

AERIAL YARDING

Aerial logging systems are unique in that they can fly the timber from one location to another without regard to intervening obstacles. They are most effective when used in environmentally sensitive or inoperable areas that contain high-value timber. These systems are extremely sensitive to weather conditions; rain, snow, and moderate winds can make them inoperable. Two forms of aerial systems have been employed operationally: balloon systems and helicopter systems.

Balloon systems are two-drum cable yarding systems with a lighter-than-air balloon (4500 kg–9000 kg lift capacity) tethered to the butt-rigging to lift the logs into the air. The lift provided by the balloon allows yarding distances of up to 1500 m, depending upon the cable capacity of the yarder.

Helicopter systems employ a helicopter (9000 kg–15,000 kg load capacity) to lift logs directly from the stump area and transport them to a landing. While economics determines the maximum flying distance, distances of up to 2 km are common.[1] Helicopter systems are the most expensive systems to own and operate. The cost of the helicopter is, in itself, enormous and fuel requirements can be extremely high (up to 2000 L of high-octane aviation fuel per hour). Because of these high costs, helicopter systems are usually employed only on sites that would be inaccessible with any other system.

REFERENCES

1. MacDonald, A.J. *Harvesting Systems and Equipment in British Columbia*; FERIC Handbook No. HB-12; Forest Engineering Research Institute of Canada: Vancouver, BC, 1999; 197.
2. Drushka, K.; Konttinen, H. *Tracks in the Forest: The Evolution of Logging Machinery*; Timberjack Group: Helsinki, Finland, 1997; 254.

Fossil Fuel Energy

Joel T. Walker (Deceased)
The Ohio State University, Columbus, Ohio, U.S.A.

INTRODUCTION

Fossil fuels, in the various forms, make possible the mechanized agriculture that feeds the world. Direct inputs of fossil fuel energy are easily observed—fuel to run engines in tractors, combines, irrigation pumps, and trucks and as heat source for crop dryers, buildings, and animal confinement. Indirectly, fossil fuel energy is converted to electricity to provide for pumping, conveying, cooking, and otherwise processing and packaging of agricultural products, production of fertilizers and agricultural chemicals, and manufacturing of farm equipment. Current technology offers no cost-effective replacement for the portable, convenient source of energy so vital to modern agriculture.

FOSSIL FUELS

Fossil fuels are so called because they are derived from biological materials trapped in the earth and transformed over millennia into concentrated forms of energy. Fossil fuels occur as petroleum (and the associated natural gas) and mineral forms, primarily coal. Liquid and gaseous forms derived from petroleum deposits are convenient for compact, portable applications. Mineral forms require more bulky conversion equipment and therefore are better suited to transformation into electricity for distribution by wire to stationary uses.

Petroleum is obtained from underground pools through drilling wells and pumping. The earliest sources were near the surface and relatively easy to obtain. More recent wells are deep in the earth and the petroleum may be trapped in porous formations that require special treatment to release the liquid. In addition, many petroleum formations are deep below ocean areas, compounding the cost of producing and containing the material. Petroleum is separated into components by processes of fractionation and distillation. This results in a wide range of products from liquid/gaseous fuels (such as gasoline, diesel, kerosene, propane, heating oil, lubricating oils) to materials with nonenergy uses (asphalt and raw materials for plastics and agrochemicals). Molecules of "petroleum" are primarily hydrocarbon chains of various lengths. Each of the petroleum fuels results from separating a range of molecule lengths from the petroleum stock. Fuels with shorter molecule length (methane, propane) are gaseous, highly volatile and easily ignited, but contain less energy density. Fuels with greater molecule length (diesel, kerosene) are liquid and have relatively low volatility, but contain a greater energy density. Gasoline falls between these two conditions, having a significant volatile component that makes for significant risk of accidental ignition. The proportion of specific molecules obtained from raw petroleum can be adjusted to some degree by the processes of cracking or polymerization. Cracking breaks longer molecules to produce greater number of lighter ones. Polymerization can join shorter molecules into fewer, heavier ones. (Table 1)

Coal occurs in underground deposits and is obtained by mining. Surface mines are used when the deposits are very near the surface. Shaft mines are used when the deposits are deeper underground. Coal is primarily a form of carbon with various minerals and compounds present in varying amounts. Because coal is not as easily or efficiently converted in mobile applications, it is mainly used in steam production for electrical generation or heating applications. Grinding of coal to a fine powder allows better mixing with oxygen to promote efficient combustion. Byproducts of combustion often include mineral ash and compounds of sulfur. The process of gasification uses coal to produce combustible gas that may in turn be burned as a cleaner source of energy.

COMBUSTION

Release of fossil energy is accomplished primarily through the combustion process of oxidizing carbon and hydrogen, which are the primary components of the fuel. This is a chemical process that may occur with varying speed and effectiveness. Volatile gaseous components and even finely ground coal may burn with explosive speed, releasing high quantities of heat in a short time. When confined and controlled, such explosive release of energy is used to power engines.

Combustion requires the presence of oxygen in proper amounts and a source of heat for ignition. In ideal conditions, carbon will oxidize to carbon dioxide and hydrogen will oxidize to water. Widespread burning of

Encyclopedia of Agricultural, Food, and Biological Engineering
DOI: 10.1081/E-EAFE 120006842
Copyright © 2003 by Marcel Dekker, Inc. All rights reserved.

F

Table 1 Properties of fossil fuels

Fossil fuel	Chemical formula[a]	Form at room temperature	Density ($kg\,m^{-3}$)	Energy content ($MJ\,kg^{-1}$)	Ideal air/fuel ratio[b]
Methane (natural gas)	CH_4	Gas	0.67	55.7	9.5 v/v
Propane[c]	C_3H_8	Gas	509	50.3	23.8 v/v
Gasoline	C_6H_{18}	Liquid	734	47.6	15:1 m/m
Kerosene (jet fuel)	—	Liquid	810	46.0	15:1 m/m
Diesel	$C_{16}H_{34}$	Liquid	840	45.5	15:1 m/m
Heating oil	—	Liquid	860	44.0	15:1 m/m
Coal	C	Solid	1400	29.0	—

[a] Gasoline and diesel formulas approximate the blended properties.
[b] v/v is volume/volume; m/m is mass/mass ratio.
[c] Properties are given for liquid propane (in pressurized tank).

fossil fuels (and other organic sources) is being blamed for rise of CO_2 concentration in the atmosphere, which is said to cause atmospheric heat retention or "global warming." If the amount of oxygen available during combustion is insufficient to completely oxidize all hydrogen and carbon in the fuel, the release of energy will be incomplete (inefficient), thus reducing the amount of energy released. Incomplete combustion of carbon may produce carbon monoxide, an odorless, colorless, and toxic gas or carbon particulates (soot). Therefore, efficient combustion requires precise control of amounts of fuel and air as well as effective mixing to ensure oxygen is available to all fuel molecules. In gasoline engines, the volume of air entering the engine is controlled and the amount of fuel to produce an ignitable but efficient-burning mixture is metered into the air. This is done either by a carburetor or by computerized fuel injection. Some degree of vaporization of gasoline in this process is necessary to produce a mixture that can be ignited by an electrical spark (across a gap in a spark plug). In diesel engines, the amount of air is unregulated, but varies with engine speed. Fuel is metered as appropriate for the power and speed desired and is sprayed directly into the combustion chamber where heat of compression ignites the swirling mixture.

Because of high temperatures during combustion, other elements present in the fuel or air may become involved chemically in the process and produce some undesirable compounds. Sulfur, present in many fossil fuels, may form sulfur dioxide, recognized in the formation of "acid rain." Nitrogen, present in air, will form oxides that are also considered pollutants.

THERMODYNAMICS

Conversion of fossil fuels to other energy forms must follow the "laws" of thermodynamics. The law of conservation of energy provides that an energy conversion process can never result in more energy than was put into the process. The "second law" of thermodynamics says an energy conversion will never break even—less energy is always available for output than was put into the system. The ratio of energy output to energy input is called efficiency. Thermodynamic theory defines the maximum efficiency that can be obtained from current engine processes. In addition, practical design constraints (such as the temperature at which engine materials begin to fail, the speed with which valves can conduct gases through the engine, and some processes used to control formation of pollutants) will also lower efficiency. It is common for modern tractors to provide around one-third or less of the original fuel energy for useful work.

All modern engines use thermal expansion of air to drive mechanical components. Energy from the burning fuel heats combustion gases and creates pressure that forces motion of pistons or turbine wheels. Gasoline engines (Otto cycle) and diesel engines(diesel cycle) are the most common in agriculture. However, turbine engines are used on modern agricultural aircraft.

FUTURE

Fossil fuels are a limited, nonrenewable resource. While there is much disagreement as to how long supplies will continue to meet the demands, there will come a time when supplies become inadequate to support the needs of all energy-using sectors. Geographic distribution of fossil fuels and political policy (local and global) will play a critical role in the availability of energy for agriculture. Difficult choices will be necessary. Now is the time to seek novel ways to power our food supply with renewable energy sources (ultimately, the sun, wind, waves, or hydroelectric). All of these are easily converted to electricity. Practical means of powering food production with electricity are needed.

Recent concern for energy cost and supply has led to sporadic research on extending liquid fossil fuels by blending with liquid fuels from renewable sources. Mixtures of ethanol with gasoline (gasohol) and mixtures of vegetable oil with diesel (biodiesel) are workable fuels for spark ignition and diesel engines, respectively. However, so much energy is sequestered in the production of ethanol or vegetable oil that these methods provide limited benefit to the overall energy supply.

FURTHER READING

Fluck, R.C.; Baird, C.D. *Agricultural Energetics*; AVI Publishing Company, Inc.: Westport, CT, 1980.

Goering, C.E., *Engine and Tractor Power*; Breton Publishers: Boston, MA, 1986.

Liljedahl, J.B.; Turnquist, P.K.; Smith, D.W.; Hoki, M. *Tractors and Their Power Units*, 4th Ed.; AVI, Van Nostrand Reinhold: New York, NY, 1989.

Pimentel, D. *Handbook of Energy Utilization in Agriculture*; CRC Press, Inc.: Boca Raton, FL, 1980.

Srivastava, A.K., Goering, C.E., Rohrbach, R.P. *Engineering Principles of Agricultural Machines*; American Society of Agricultural Engineers: St. Joseph, MI, 1993.

Stout, B.A. *Energy Use and Management in Agriculture*; Breton Publishers: North Scituate, MA, 1984.

Freeze Concentration

Osato Miyawaki
The University of Tokyo, Tokyo, Japan

F

INTRODUCTION

Concentration of liquid foods is important to reduce volume of liquid to be handled and stored, to reduce cost of shipping, to increase solute concentration, etc.[1] Freeze concentration is a technique for concentrating solutions that gives the best quality with a good retention of flavors and fragile compounds although the cost of the process is relatively high. In this article, freeze concentration is compared with other methods of concentration; principle of freeze concentration has been reviewed, and two representative methods of freeze concentration, suspension crystallization and progressive freeze-concentration, have been compared.

OVERVIEW

There are three methods for the concentration of liquid foods: 1) evaporation; 2) membrane concentration (reverse osmosis); and 3) freeze concentration. Table 1 compares these three methods for concentration. Evaporation effectively uses the gas–liquid phase separation and is the most convenient method due to its lowest capital cost and highest maximum concentration obtainable in spite of the lowest quality and the highest energy consumption in a single-effect use. Freeze concentration is based on the solid–liquid phase separation. The process temperature in this method is low so that a good retention of flavors and thermally fragile components is expected. Freeze concentration has been known to give the best quality among the three although it is the most expensive.[2] Reverse osmosis is based on molecular-sieve mechanism of a membrane. Reverse osmosis requires least energy for separation among the three methods for concentration because of no necessity for phase change but the cost of membrane plays an important role and the maximum concentration obtainable is relatively low. In reverse osmosis, some flavor components could be lost by permeation through the membrane and adsorption.

As for energy requirement in the concentration process, the vapor recompression techniques including multiple-effect operation are available for evaporation and some thermal energy is required even for reverse osmosis when aroma recovery is necessary. Table 2 compares the energy consumption for processes with aroma recovery. In this case, the difference in energy consumption among the three methods can be reduced substantially.[3]

Separation by freeze concentration is based on solid–liquid phase separation. A simple phase diagram of a binary system is shown in Fig. 1.[4] If a binary mixture at a composition of W_A with freezing point T_A is cooled down to temperature T_B, the concentration of the solution increases to W_B, separating out ice crystals. This concentration process goes along the line (W_A,T_A)–(W_B,T_B) until the point (W_E,T_E), eutectic point. The concentration process cannot proceed beyond W_E because the crystallizing solid has the same composition as the mother solution to be concentrated. Liquid foods are mostly considered to be a pseudo-binary system so that the same principle for a binary system is generally applicable in practical freeze-concentration.

Freeze concentration has been applied to concentration of fruit juices, concentration of dairy products, preconcentration of coffee extract for freeze drying, preconcentration of solutes for analytical purposes, desalination, and wastewater treatment.[1,6] A freeze concentration system includes three processes: ice crystallization; ice crystal growth; and ice crystal separation from mother solution. In ice crystallization, direct and indirect crystallizers are available. The former cools the solution by direct contact with the refrigerant; so it is very simple with no necessity for heat exchanger and the heat transfer coefficient is very high. This is mostly used for nonfood applications because direct mixing of refrigerant with food products is not desirable. In indirect contact crystallizer, the solution is cooled by refrigerant indirectly by use of heat exchanger.

For ice crystal growth process, a recrystallizer with Ostwald ripening mechanism is available.[5] In this process, the difference in freezing point, ΔT, between a huge ice crystal and a small crystal with a diameter d is given by the Gibbs–Thomson equation as follows:

$$\Delta T = (4\sigma T_i)/(\Delta H \rho d) \quad (1)$$

where σ is the surface free energy, T_i is the freezing point of the solution, ΔH is the heat of ice fusion, and ρ is the specific weight of ice. When the bulk temperature in the recrystallizer is kept between equilibrium temperature of the largest and that of the smallest crystal determined

Encyclopedia of Agricultural, Food, and Biological Engineering
DOI: 10.1081/E-EAFE 120007066
Copyright © 2003 by Marcel Dekker, Inc. All rights reserved.

Table 1 Comparison among methods for the concentration of liquid food

Method	Principle	Energy consumption (kJ/g water)	Max. conc. obtainable (Brix)	Quality	Cost
Evaporation	Gas–liquid phase separation	2.26	> 50	Low	Low
Reverse osmosis	Molecular sieve of membrane	~ 0	~ 30	Medium	Medium
Freeze concentration	Solid–liquid phase separation	0.33	~ 50	High	High

by Eq. 1, the larger crystals grow while the smaller crystals melt.

For ice crystal separation, filter press, centrifuge, and wash column are available. Among these, filter press is less effective for crystal separation and centrifuge has a serious disadvantage of aroma loss. On the contrary, wash column provides a perfect separation of ice crystal with an aroma loss virtually equal to zero.[3] In freeze concentration, maximum concentration obtainable is around 50 Brix depending on the viscosity and the insoluble solid content of the sample solution. Solid loss due to freeze concentration can be higher than those due to evaporation and reverse osmosis because of the incorporation of solute components into ice crystals. By using wash column, however, solid loss can be substantially reduced to less than 0.01%.[2]

In the practical application of freeze concentration to food, suspension crystallization[5] is most commonly used. Fig. 2 shows the schematic diagram of this method, in which small ice crystals are first formed on the surface scraping heat exchanger and transferred to recrystallizer to grow larger by the Ostwald ripening mechanism. After recrystallization, ice crystals are transferred to washing tower to separate ice crystals from the concentrated mother solution. When the necessitated concentration is very high, multi-stage operation is employed. Suspension crystallization requires a complex system with a delicate control. For this reason, the capital investment of this process is high so that the application of this process is still limited.

As an alternative to suspension crystallization, progressive freeze-concentration or layer freezing has been developed. This method was originally proposed as a method for preconcentration of dilute solutes for analytical purpose.[7] In this method, a single ice crystal is formed and grown on the cooling surface in a concentration vessel (Fig. 3). In this method, all of the three processes of ice crystallization, ice crystal growth, and ice crystal separation are carried out in one vessel so that the system is much simpler as compared with the suspension crystallization. Therefore, a substantial cost reduction will be expected in this method. Ice crystallization and ice crystal growth proceed simultaneously on the cooling plate. Ice crystal separation from the mother solution is easily done, typically by gravitation because the ice crystal formed continues to stay on the cooling plate. During the separation process, the contaminated ice surface can be washed to increase the solid recovery.

In the progressive freeze-concentration, the growth rate of ice crystal and the mass transfer at the ice–liquid interface are two important operating parameters in consideration of the concentration polarization at the ice–liquid interface.[8] Concentration polarization is a sharp concentration distribution of solute near the ice–liquid interface where the rejected-out solute components from the ice phase is accumulated. This causes an increase in the partition coefficient of solute in the ice phase to increase the solid loss in the process. Therefore, the slower ice crystal growth rate and the larger mass transfer rate at

Table 2 Energy consumption in various concentration processes

Process	Steam equivalents	
	Without aroma recovery	With aroma recovery
Evaporation		
One effect	1.20–1.28	1.25–1.32
Two effects	0.36	0.53
Four effects	0.18	0.45
Freeze concentration	—	0.25–0.50
Reverse osmosis	0.01–0.02	0.48

(From Ref. [3].)

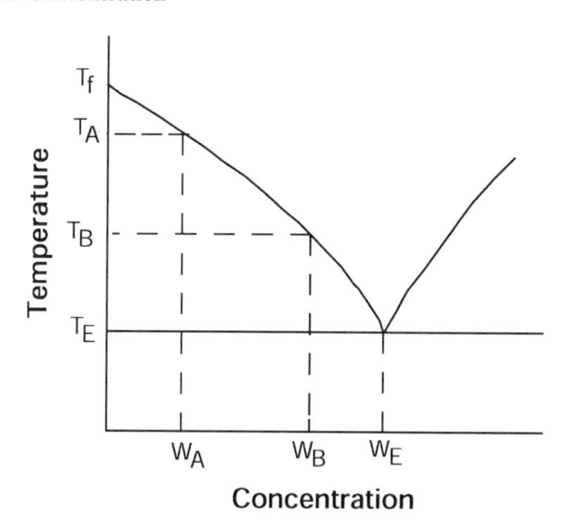

Fig. 1 Phase diagram for a simple binary system. (From Ref. [4].) (T_f, freezing point of pure solvent; T_A, initial freezing temperature of solution; T_B, final freezing temperature of solution; T_E, eutectic temperature; W_A, initial concentration of solution; W_B, final concentration of solution; W_E, eutectic concentration.)

the ice–liquid interface are recommended because the effect of the concentration polarization is less at these conditions.[8] At the appropriately chosen operating conditions, a good solid recovery of more than 99% is reported for a small test apparatus of progressive freeze-concentration.[9] To improve solid recovery more, multi-stage operation is effective.

For the scale-up of the progressive freeze-concentration, a tubular ice system is effective to produce an ice crystal at the inside surface of a cooling tube with circulating flow.[10] A drawback for progressive freeze-concentration is the smaller area of ice–liquid interface, which limits productivity. This drawback, however, can be compensated by the larger driving force (temperature difference) in the ice-crystal growth process so that high

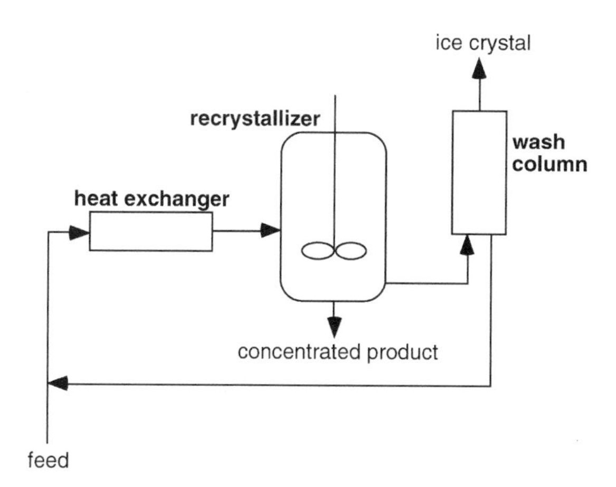

Fig. 2 Schematic diagram of suspension crystallization process for freeze concentration.

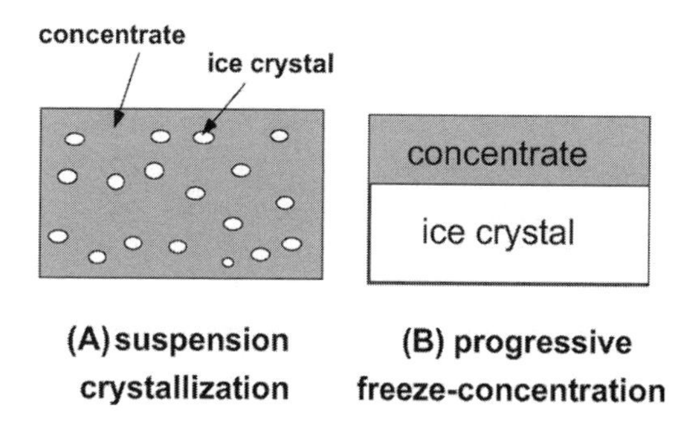

Fig. 3 Principle of two methods for freeze concentration.

productivity can be obtained. The operation mode of suspension crystallization is continuous so that this is suitable only for a large-scale production. On the contrary, progressive freeze-concentration is operated in a repeated batch mode so this method is flexibly applicable from small- to large-scale production. Progressive freeze-concentration is applicable even to a suspension and a highly viscous solution, which are not suitable for suspension crystallization.

REFERENCES

1. Muller, J.G. Freeze Concentration of Food Liquids: Theory, Practice, and Economics. Food Technol. **1967**, *21*, 49–61.
2. Deshpande, S.S.; Bolin, H.R.; Salunkhe, D.K. Freeze Concentration of Fruit Juices. Food Technol. **1982**, May, 68–82.
3. Ramteke, R.S.; Singh, N.I.; Rekha, M.J.N.; Eipeson, W.E. Methods for Concentration of Fruit Juices. J. Food Sci. Technol. **1993**, *30* (6), 391–402.
4. Karel, M. Concentration of Food. In *Principles of Food Science, Part 2*; Karel, M., Fennema, O.R., Lund, D.B., Eds.; Marcel Dekker, Inc.: New York, 1975; 265–308.
5. Huige, N.J.J.; Thijssen, H.A.C. Production of Large Crystals by Continuous Ripening in a Stirred Tank. J. Cryst. Growth **1972**, *13/14*, 483–487.
6. Muller, M.; Sekoulov, I. Waste Water Reuse by Freeze Concentration with a Falling Film Reactor. Water Sci. Technol. **1992**, *26* (7–8), 1475–1482.
7. Matthews, J.S.; Coggeshall, N.D. Concentration of Impurities from Organic Compounds by Progressive Freezing. Anal. Chem. **1959**, *31* (6), 1124–1125.
8. Miyawaki, O.; Liu, L.; Nakamura, K. Effective Partition Constant of Solute Between Ice and Liquid Phases in Progressive Freeze-Concentration. J. Food Sci. **1998**, *63* (5), 756–758.
9. Liu, L.; Miyawaki, O.; Hayakawa, K. Progressive Freeze-Concentration of Tomato Juice. Food Sci. Technol. Res. **1999**, *5* (1), 108–112.
10. Wakisaka, M.; Shirai, Y.; Sakashita, S. Ice Crystallization in a Pilot-Scale Freeze Wastewater Treatment System. Chem. Eng. Process. **2001**, *40*, 201–208.

Freeze Drying

Tetsuya Araki
Yasuyuki Sagara
The University of Tokyo, Tokyo, Japan

INTRODUCTION

Freeze drying is accomplished by reducing the product temperature so that most of the product moisture is in a solid state, and by decreasing the pressure around the product, sublimation of ice can be achieved.[1] Developments of freeze drying owe much to advances in vacuum pumping and refrigeration machines since the late 19th century. Yet freeze drying had been regarded as an occasional scientific tool until Flosdorf started his study on its industrial applications to food drying[2] and pharmaceutical drying[3] in 1930s. Since then, research and development in freeze drying of foods were promoted for military and commercial purposes in the United States and several European countries, and then in the beginning of 1960s, these results, including the process of accelerated freeze drying (AFD) method,[4] were released to industry for commercial exploitation.[5]

This article presents a brief explanation of the freeze drying process, freeze-dryer, and applications to food drying, together with key historical references of research and development needs in freeze drying of foods.

PROCESS, EQUIPMENT, AND COSTS

Freeze Drying Process

Fig. 1 shows a typical freeze drying process for manufacturing an instant coffee, together with temperature profiles of the product during the period of main freeze drying operations: Prefreezing, Primary drying, and Secondary drying. It should be noted that all of them are not mechanical, but thermal unit operations.

As shown in the figure, raw materials are prepared and pretreated, and then frozen to reduce the temperature under the eutectic point, at which all free water in the materials will become ice. In commercial operations of food drying, the final freezing temperature is about − 40°C.

After freezing, the process of primary drying should be controlled at the maximum rate of drying, as long as its superior quality is maintained without any deterioration such as collapse as well as scorch in the dried layer and melting in the frozen layer. In principle, freeze drying can proceed under the conditions of temperature and pressure below the triple point of water (0°C, 633.3 Pa).

Secondary drying can be regarded as the moisture-desorbing process from the surface of porous materials, and the drying time depends on the end-point of drying. The final moisture content of the product would be 1%–2% in wet basis.

Furthermore, industrial freeze drying has several important processes to be considered: the optimization of freeze-concentration, packaging and storage, as well as the rehydration ratio of freeze-dried products.

Freeze-dryer

Fig. 2 illustrates a laboratory-scale freeze-dryer, comprised of four main components (a drying chamber, a vacuum pumping, condensers for subliming vapors, and a refrigerator to cool condensers), together with measurement devices. The surface of a food sample is usually heated by radiation and/or conduction, and then operating temperature and pressure are controlled by the method, which is similar to that of accelerated freeze-dryer. The weight loss of the sample is followed by supporting a sample holder on an electronic balance located in the center of the drying chamber. This can be accomplished by separating a part of load cell from an indicator containing electric circuits. Absolute pressure in the drying chamber is monitored by a diaphragm-type analyzer. For industrial-scale freeze-dryers, recent books and reports have more.[6]

Costs

Freeze drying process consists of four main operations, and shares the total energetic consumption as follows: freezing (4%), vacuum (26%), sublimation (45%), and condensing (25%).[7]

Fig. 3 shows a comparison of freeze drying costs associated to the processing of two types of raw materials (high and low value foods) sold in similar consumer-type packages. As can be seen, the energy spent in the freeze drying process itself becomes insignificant when dealing with high-value raw materials.[7] For further information, you can refer to several economic studies on freeze drying of foods.[8]

Encyclopedia of Agricultural, Food, and Biological Engineering
DOI: 10.1081/E-EAFE 120007090
Copyright © 2003 by Marcel Dekker, Inc. All rights reserved.

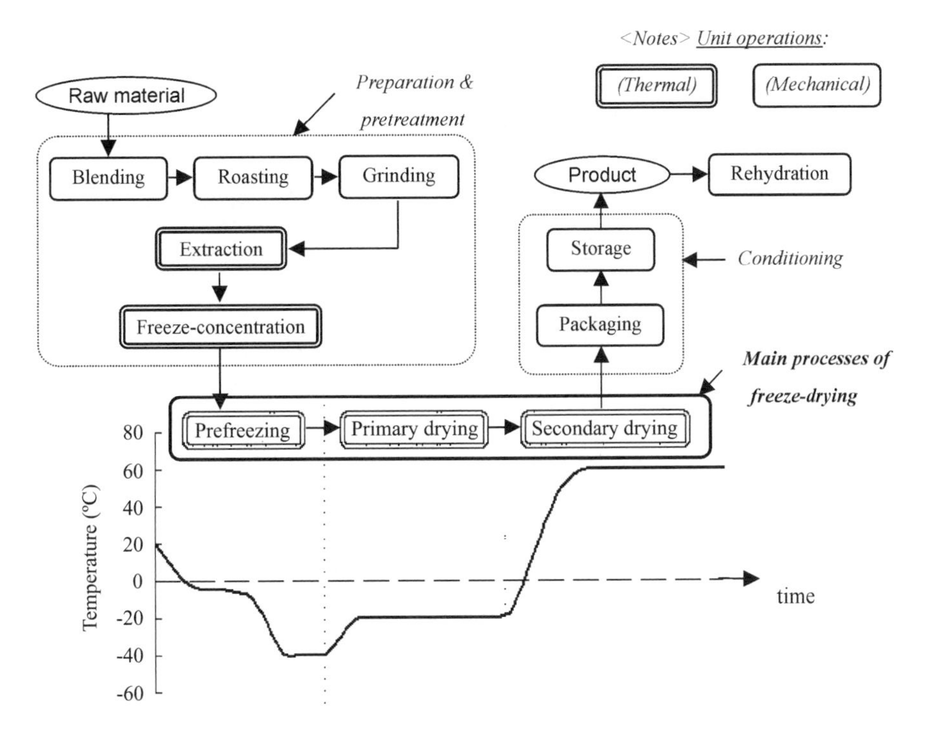

Fig. 1 Manufacturing processes and product temperatures for freeze-dried coffee.

APPLICATIONS TO FOOD DRYING

Freeze drying has a wide range of applications in food drying: instant coffee, meat products, ready-to-eat meals, vegetables, powders from liquid food extracts,[9] and miscellaneous processed foods. The largest end-use market of freeze-dried foods is instant coffee, which reached to 30% of the total production of coffee products in the United States in 1982.

On the other hand, the end-use market of freeze-dried food in Japan has developed in a different way from western countries, mainly because of the extended consumption of instant noodles and ready-to-eat meals, especially freeze-dried soups. Recently, original AFD method[4] was improved to manufacture instant egg-soup products.[10]

RESEARCH AND DEVELOPMENT NEEDS

Prefreezing

The performance of the overall freeze drying process depends significantly on this stage because the shape of the pores, the pore size distribution, and pore connectivity of the porous network of the dried layer formed by the sublimation of frozen water during the primary drying stage,[11] depend on the ice crystals that are formed during the freezing stage. However, the appropriate measurements of these parameters are still difficult challenges, although several types of structural models have been proposed.[5,12]

Another R&D issue of this stage is glass transition temperature, which can be defined as the temperature at which an amorphous system changes from the glassy to the rubbery state.[13] However, quantifiable expressions between quality parameters and glass transitions have not been found yet.[7]

Primary and Secondary Drying

Since the freeze drying rate is limited by the rates of heat and mass transfer across the dried layer, the values of its effective thermal conductivity and permeability of water vapor are indispensable to determine the drying rate. In many previous works, these transport properties have been measured by the steady state[5,14] and transient method.[15]

Fig. 4 shows a uniformly-retreating-ice front (URIF) model, originally proposed by Dyer and Sunderland,[16] to determine the transport properties for the dried layer of the material undergoing primary drying stage of freeze drying.[12] In the model, the material is assumed to have the geometry of a semi-infinite slab and the dried layer is separated from the frozen layer by the sublimation front. The insulated bottom can be regarded as the center plane

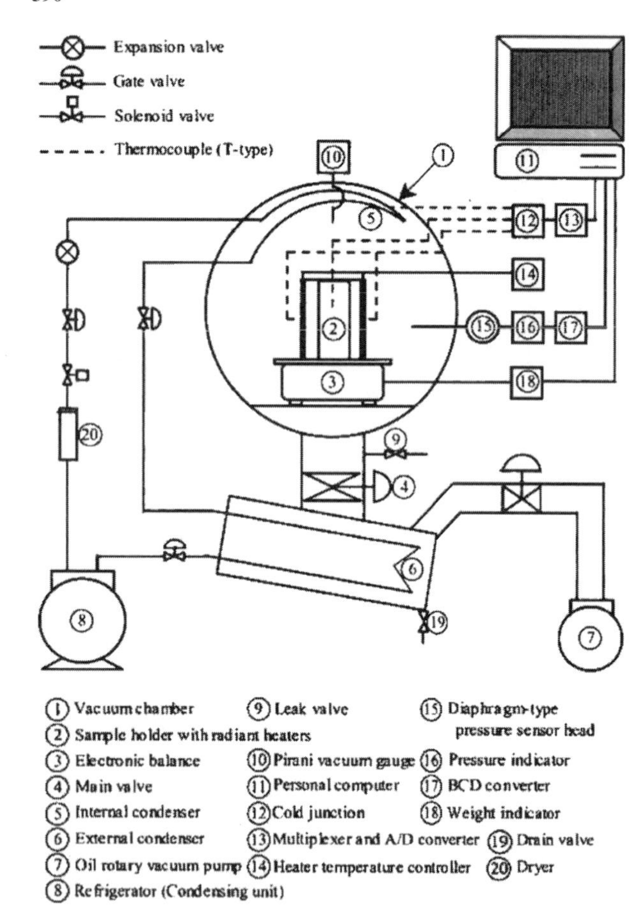

Fig. 2 Experimental freeze-dryer and measurement system.

of the material heated by radiation from both surfaces of the material.

On the other hand, secondary drying stage is rarely investigated except some previous works,[17,18] because of the difficulty in measurement of weight loss during this stage.

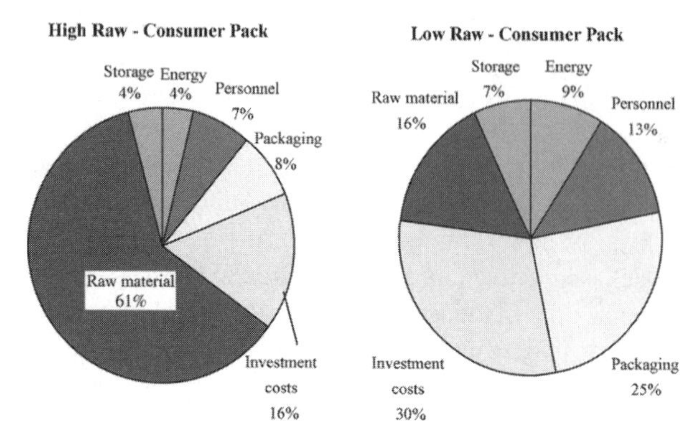

Fig. 3 Cost breakdown in two freeze-drying plants, processing high and low value foods. (From Ref. 7.)

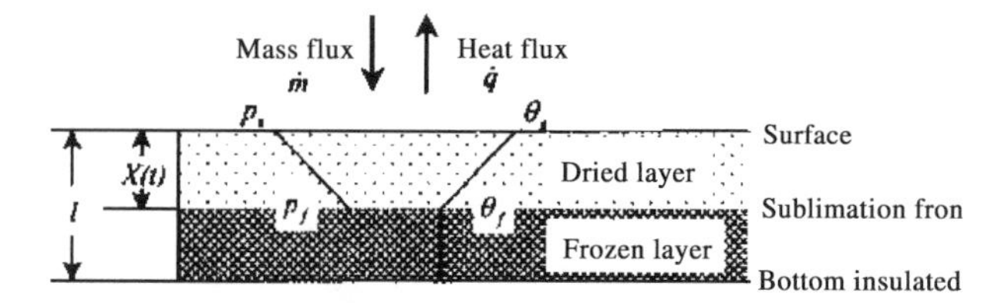

Fig. 4 Freeze drying model for transport properties analysis.

Quality Assessment of the Product

Main quality parameters of freeze-dried foods are rehydration, color, and volume. Rehydration ratio of freeze-dried foods in general 4–6 times higher than air-dried foods, making freeze-dried products excellent for ready-to-eat instant meals or soups.[7] The deterioration of volume and color can be assessed from the appearance and by the measurement of structural natures. In addition to it, Flink[19] points out that the quality of freeze-dried coffee should be assessed from the viewpoint of color, package size, uniformity of granule size, aroma, and flavor.

CONCLUSION

Freeze drying process promises continued expansion of the number of applications in the food industry because of the superior quality of the product obtained. However, it would be only feasible if the cost of production can be lowered by optimum plant operations.[20]

To sustain R&D in freeze drying of foods, there are two main challenges: 1) quantitative measurement and control of the structural nature for the dried layer; and 2) the investigation of the mutual relationships among factors influencing the final quality of the product.

REFERENCES

1. Singh, R.P.; Heldman, D.R. *Introduction to Food Engineering*, 3rd Ed.; Academic Press: New York, 2001; 567.
2. Flosdorf, E.W. *Freezing and Drying*; Reinhold: New York, 1949; 1–280.
3. Rey, L.R. Glimpses into the Realm of Freeze-Drying: Classical Issues and New Ventures. In *Freeze-Drying/Lyophilization of Pharmaceutical and Biological Products*; Rey, L.R., May, J.C., Eds.; Marcel Dekker, Inc.: New York, 1999; 1–30.
4. Ministry of Agriculture, Fisheries and Food. *Accelerated Freeze Drying Method of Food Preservation*, UK, 1961.
5. Mellor, J.D. *Fundamentals of Freeze-drying*; Academic Press: New York, 1978.
6. Global Industry Analysts, Inc., *Freeze Drying Equipment*; Global Industry Analysts, Inc.: CA, 2002.

7. Ratti, C. Hot Air and Freeze-Drying of High-Value Foods: A Review. J. Food Eng. **2001**, *49*, 311–319.
8. Lorentzen, J. Freeze-Drying of Foodstuffs, Quality and Economics in Freeze-Drying. Chem. Ind. **1979**, *14*, 465–468.
9. Flink, J.M. Applications of Freeze Drying for Preparation of Dehydrated Powders from Liquid Food Extracts. In *Freeze Drying and Advanced Food Technology*; Goldblith, S.A., Rey, L., Rothmayr, W.W., Eds.; Academic Press: New York, 1975; 309–329.
10. Sagara, S.; Yoshizawa, S.; Nakayama, H.; Okutani, Y.; Tsujimoto, S.; Nagashima, K.; Nishinomiya, T. Manufacturing Method for Instant Egg-soup Products. Japan Patent P2000-139427A, May 23, 2000.
11. Liapis, A.I.; Pikal, M.J.; Bruttini, R. Research and Development Needs and Opportunities in Freeze-Drying. Drying Technol. **1996**, *14* (6), 1265–1300.
12. Sagara, Y. Structural Models Related to Transport Properties for the Dried Layer of Food Materials Undergoing Freeze-Drying. Drying Technol. **2001**, *19* (2), 281–296.
13. Roos, Y.; Karel, M. Plasticizing Effect of Water on Thermal Behavior and Crystallization of Amorphous Food Models. J. Food Sci. **1991**, *56*, 38–43.
14. Harper, J.C. Transport Properties of Gases in Porous Media at Reduced Pressures With Reference to Freeze-Drying. AIChE J. **1962**, *8* (3), 298–302.
15. Sandall, O.C.; King, C.J.; Wilke, C.R. The Relationship Between Transport Properties and Rates of Freeze-Drying of Poultry Meat. AIChE J. **1967**, *13* (3), 428–438.
16. Dyer, D.F.; Sunderland, J.E. Heat and Mass Transfer Mechanisms in Sublimation Dehydration. Trans. ASME, J. Heat Transfer **1968**, 379–384.
17. King, C.J. *Freeze-drying of Foods*; CRC Press, 1971; 1–54.
18. Sadikoglu, H.; Liapis, A.I.; Crosser, O.K. Optimal Control of the Primary and Secondary Drying Stages of Bulk Solution Freeze Drying in Trays. Drying Technol. **1998**, *16* (3–5), 399–431.
19. Flink, J.M. The Influence of Freezing Conditions on the Properties of Freeze Dried Coffee. In *Freeze Drying and Advanced Food Technology*; Goldblith, S.A., Rey, L., Rothmayr, W.W., Eds.; Academic Press: New York, 1975; 143–160.
20. Sagara, Y.; Ichiba, J. Measurement of Transport Properties for the Dried Layer of Coffee Solution Undergoing Freeze-Drying. Drying Technol. **1994**, *12* (5), 1081–1103.

F

Freezing System Design

Yen-Con Hung
The University of Georgia, Griffin, Georgia, U.S.A.

INTRODUCTION

The freezing of foods slows down the chemical and biochemical reactions and hence prevents the loss of sensory and nutritional qualities. Although any freezing system can be used for food freezing, the key for selecting an appropriate freezing system is its ability to freeze the intended raw materials properly. In general, in order for the freezing operation to preserve the quality of food, the freezing process itself needs to be as rapidly as possible. However, fast freezing may cause freeze cracking[1] and costs are usually high for fast freezing operation. Usually the balance is to select a freezing rate that can maintain the desirable quality of the frozen food and the cost of freezing operations. The freezing equipment selected to produce the desirable freezing rate should also be integrated into the overall process and space.

FOOD FREEZING SYSTEM

Food freezing rate is governed by the properties of food products to be frozen (how fast can heat be removed from the inside of the product) and the rate of heat transfer from food surface to the freezing medium. Food freezing systems can be characterized according to the rate of heat transfer from food to the freezing medium ($h = 5\,\mathrm{W\,m^{-2}\,K^{-1}} - 10\,\mathrm{W\,m^{-2}\,K^{-1}}$ for slow freezing to above $100\,\mathrm{W\,m^{-2}\,K^{-1}}$ for ultrafast freezing), by the type of the contact between refrigerant and food (with or without direct contact), by whether the refrigerant is recyclable (mechanical vs. cryogenic freezing). Following is a brief description of different freezing systems currently used by the food industry. Detailed description of different systems can be found in Refs. [2–4].

Batch-type cabinet freezer: This method involves simply placing products to be frozen in a cabinet and then circulating either cold air or cryogenic liquid or vapor over the products to remove heat from food surface. Products selected for freezing by this method are usually not sensitive to the freezing rate. The cost of this freezing method is usually the lowest and has a low maintenance cost. It is good for small production operations. However, this method has a slow freezing rate ($h = 5\mathrm{W\,m^{-2}\,K^{-1}} - 10\mathrm{W\,m^{-2}\,K^{-1}}$) and requires considerable handling of products.

Tunnel freezer: This is a very popular freezing system for the food industry. The most common type of this freezing system is a straight through, single-belt tunnel. Food product is carried through the tunnel on a conveyor from one end of the freezer and freezing media (cold air or cryogenic refrigerant) to the other end. The rate of freezing depends on the temperature and speed of the freezing media in contact with the foods. The tunnel freezer is best for products that have a large surface area-to-volume ratio (like pizza). There are many variations of the tunnel freezer like the multideck tunnel freezer to reduce the floor space requirement, or the flighted freezer to turn over the product during freezing and prevent clumping and provide even exposure of product surface to the freezing media. Tunnel freezer has the advantages of flexibility, ease of operation, and economy of operation for large production volume.

Spiral freezer: This type of the freezer has a circular conveyor carrying the product from the bottom to the top of the freezer. Cold air usually comes in contact with the product at the outlet and then continues to travel through the freezer toward the inlet while cryogenic media can be sprayed on the product at the inlet to produce crust freezing or from the top of the freezer. This type of freezer is very economical on the floor space requirement and lowers the energy consumption than a straight through tunnel freezer due to efficient usage of all heat capacity of freezing media.

Fluidized-bed freezer: This system is defined as a method to keep solid particles floating in an upward-directed flow of cooling media (this could be air or cryogenic vapor). Because of the upward lifting motion, effective contact between cooling media and product is achieved and hence produces very high heat transfer rate. This method has the advantage of uniform freezing but is only good for small granular or diced products.

Plate freezer: For this method, the product is firmly pressed between two plates with refrigerant circulated in channels housed inside the plates. Because of the physical contact, fast freezing ($h = 50\,\mathrm{W\,m^{-2}\,K^{-1}} - 100\,\mathrm{W\,m^{-2}\,K^{-1}}$) can be achieved through conductive heat transfer. Because products are held under pressure during freezing, products bulging due to volume expansion from water to ice can be prevented. However, this

Encyclopedia of Agricultural, Food, and Biological Engineering
DOI: 10.1081/E-EAFE 120007028
Copyright © 2003 by Marcel Dekker, Inc. All rights reserved.

method is only suited for rectangular-shaped product and product with thickness less than 50 mm.

Immersion freezer: Freezing is achieved by direct immersion of food products in low temperature brine or cryogenic liquids. The direct immersion can be applied to packaged food products or food directly. However, a refrigerating medium for immersion freezing of food directly must be food grade, not affect the flavor or quality of the product, and leave no residue on the product after freezing. Immersion freezers are most commonly used for crust (surface) freezing while using other means (like tunnel freezer) finishes the freezing. This method is not limited by product shape and has a very fast freezing rate ($h > 100\,W\,m^{-2}K^{-1}$). Because of the crust freezing effect, product dehydration can be reduced and achieve individually quick frozen (IQF) products. The IQF products also have the benefit of easily portioning, ease of handling, and stock control.

Impingement freezing: This new technique increase the freezing rate by increasing the surface heat transfer coefficient using very high, localized velocity jets through radii holes or v-shaped slots. The jets are either cold air or atomized liquid nitrogen at temperatures as low as $-120°C$ and has an effect of breaking down the insulation layer of air or gas surrounding the products and hence increasing the heat transfer coefficient. For some products, impingement freezing achieves freezing rate comparable to cryogenic freezing.

Cryogenic freezer: Freezing is achieved at very low temperatures like below $-60°C$ using expendable, liquefied gases such as nitrogen or carbon dioxide. It can achieve very fast freezing rate, produce high quality frozen products, and the equipment required is simple and usually not costly. Most cryogenic freezers also require less floor space than traditional mechanical food freezing equipment. Freezing by liquid nitrogen can be applied to many different system discussed above (Cabinet, tunnel, spiral, immersion, and fluidized-bed freezers). Carbon dioxide is another cryogenic medium and one method is to spray liquid CO_2 onto the food. A CO_2 snow is formed as the carbon dioxide liquid under high pressure expands to dry ice snow at the spray nozzles because at atmospheric pressure carbon dioxide does not exist as a liquid. Cryogenic freezers can usually be installed quickly, are small in size, low in maintenance cost, and have great flexibility in the range of products to be frozen and production rate. Detailed discussion of cryogenic freezing can be found in Ref. [2].

High pressure assisted freezer: Freezing point of water decreases with increase in pressure. When temperature of the product is decreased under pressure, water inside the product is kept in the liquid state at subzero temperature. When the pressure is released, rapid ice nucleation occurred due to the large supercooling effect.[5] This quick freezing process resulted in many small ice crystals. For large volume product, a thermal gradient exists between an interior point and the surface of the product using traditional freezing method. This thermal gradient contributes to the larger ice crystal toward the center than the surface of the product. High pressure assisted freezing can help eliminate this problem and produce a product with less structural damage. However, this system is still at research/development stage. Table 1 summarizes several different freezing systems discussed above.

DESIGN AND SELECTION OF FOOD FREEZING SYSTEM

Before designing a food freezing equipment, it is beneficial to compare the available systems and evaluate its use for the products to be frozen. There are many other factors than just the price of the freezer that need to be considered when selecting a freezing system. Following is a list of the factors that should be taken into consideration before the selection of a particular freezing system for a particular operation.

- Capital and operating costs—Mechanical freezers usually have high capital costs. Cryogenic freezers usually have high operating costs and sometime high operating costs can offset the advantage of a low initial investment cost. Operating costs are factors of energy consumption, labor cost, and maintenance (this should include frequency and degree of ease to maintain the equipment). Amortization of the system should also be considered.
- Throughput and refrigeration load—The total refrigeration load (transmission load, product load, internal load, infiltration air load) as described in Ref. 6 and throughput of the system should be calculated before selecting a system.
- Flexibility and reliability—The system should be reliable with minimal downtime for maintenance. For mechanical freezers, use air-defrost system to prevent buildup of frost and ice on the evaporators to facilitate longer production runs. A system that is easy to modify and expand will make it possible to respond quickly to the market place.
- Cleanability and good hygienic design—Surfaces in direct contact with the product should be easily cleaned. Time for defrosting and removing debris and product particles should be minimized. For example, a spiral freezer, because of the mesh belt and structural support for the belt, can trap product spillage and is difficult to clean. A self-stacking belt has been developed to eliminate the supportive steel structure and the peripheral cladding. This design leaves

Table 1 Summary of various freezing options

Batch freezer	Tunnel or spiral freezer	Fluidized-bed freezer	Cryogenic	Plate freezer	High pressure assisted freezer
Ease of handling	Economical to operate	Fast freezing rate	Low product dehydration	Relative high freezing rate and efficient heat transfer	High product quality
Wide range of product can be frozen	Continuous, in-ine processing	Uniform and IQF product	High quality and IQF product	Improved bulk product handling	Suitable for large volume product
Cabinets and equipment can be easily cleaned	Flexible production line	High operation costs	Low capital investment	Can prevent product bulging	High costs
High labor cost	Low maintenance	Only suited for small granular or diced products	Low maintenance	Only suited for regular-shaped products and with less than 50 mm thickness	Still at research/ development stage
Slow freezing rate	Reasonable freezing rate		Variable production rate		
			Module design allow for future expansion		
			Refrigerants are not recyclable		
			High operation costs		

the floor of the equipment almost clear and easy accessible for cleaning.

- Floor space—Particularly for replacing existing systems or adding a new freezing line to an existing facility, availability of floor space should be carefully checked.
- Type and number of different products to be run—Consideration needs to be given to the variety, size, shape of the product to be processed to ensure that the system can accommodate them. If the product is sticky or prone to dehydration then a step to first crust freeze the product will be appropriate. Product surface to volume ratio, thermal diffusivity of the food product to be frozen should also be considered.
- Number of days of operation per month and year—Depending on whether it is a year round operation or a seasonal production, cryogenic freezing will have the advantage for seasonal operation due to ease of startup and low maintenance during the off season.
- Product quality—Some products are more sensitive to the freezing rate (like fruits and seafood) and some are less sensitive to the freezing rate (like beef). Weight loss due to dehydration should also be considered. Crust freezing or fast freezing rate can reduce product dehydration.

Ultimately, the design or selection of freezing system for a food freezing operation is a compromise among all the factors discussed above.

FUTURE TRENDS OF FOOD FREEZING SYSTEMS

The rapid changes of the food market place reflect the consumers' desire for more variety of new products; this translates to a shorter market life of existing products and calls for flexibility of freezing systems. New systems will also have to be easily modified to respond quickly to produce the products to meet the need of market place. In the past several years, much advancement has been achieved on freezing systems. More energy efficient enclosures and drive motors have resulted in less energy consumption. Overall systems, spiral belts and drive motors are more reliable. Improvements on defrost systems enable production systems to run continuous. New designs for impingement freezing system achieve fast freezing rate by injecting cold air or LN2 from both top and bottom of the food. Design of the freezer has also improved to allow ease of cleaning while minimizing labor costs. New material like fiberglass has been used to lower

F

the thermal mass and allow rapid cool down for cryogenic freezers. Timers have also been installed on the LN2 storage tank to lower the set pressure during nonproduction times to reduce loss, and vacuum-insulated piping have replaced urethane tubing to improve insulation and reduce loss.

In the future, improvements will continue to come in the areas of mechanical handling and process control to ensure the optimal freezing rate of the food product and reductions in costs. We will see the development of using an old technology but environmental friendly ammonia in the refrigeration system with improved safety. Ammonia is the most efficient and most cost effective refrigerant and has the minimal impact to the environment. Ammonia also has a strong odor, so a leak can be found and repaired quickly. Safe handling of ammonia is a major concern and OSHA and EPA guidelines should be followed.

Combinations of different freezers to take the advantage of different freezing systems will help ensure easy handling of difficult to handle products (like wet, sticky, or product with delicate structure) and at the same time control the cost for the process. Because of the required reduction of using chlorofluorocarbons (CFCs) and hydrofluorocarbons (HCFCs) for the freezing systems, we will also see the development of new nontoxic and nonflammable refrigerant liquids that exhibit appropriate thermodynamic efficiency and minimal environmental impact.

REFERENCES

1. Hung, Y.-C.; Kim, N.-K. Fundamental Aspect of Freeze-Cracking. Food Technol. **1996**, *50* (12), 59–61.
2. Hung, Y.-C. Cryogenic Refrigeration. In *The Advances in Food Refrigerations*; Sun, D.-W., Ed.; Leatherhead Food RS: Surrey, UK, 2001: 305–325.
3. George, R.M. Freezing Systems. In *The Quality in Frozen Food*; Erickson, M.C., Hung, Y.-C., Eds.; Chapman and Hall: New York, 1997: 3–9.
4. ASRAE. *Refrigeration Systems and Applications*; American Society of Heating, Refrigerating and Air-conditioning Engineers, Inc.: Atlanta, 1998.
5. Denys, S.; Van Loey, A.M.; Hendrickx, M.E.; Tobback, P.P. Modeling Heat Transfer During High-Pressure Freezing and Thawing. Biotechnol. Process. **1997**, *13*, 416–423.
6. Hung, Y.-C. Food Freezing. In *The Encyclopedia of Food Science and Technology*; Hui, Y.H., Ed.; John Wiley & Sons: New York, 1991: 1041–1050.

Freezing Time Calculations

D. J. Cleland
Massey University, Palmerston North, New Zealand

INTRODUCTION

Food freezers should completely freeze the product with minimal deterioration of quality for low cost. Accurate prediction of food freezing time allows the designer or operator of food freezers to:

- Define the required product residence time and hence the freezer size or production rate.
- Specify appropriate freezer operating conditions (e.g., air temperature and velocity).
- Assess how changes in operating conditions affect the extent of freezing, and
- Minimize expensive freezing trials.

This article briefly overviews freezing time calculation methods, presents a recommended method, and provides guidance for its use.

THE FREEZING PROCESS

Most foods that are frozen have high moisture contents and are either a solid when unfrozen and/or confined in packaging. Freezing involves exposing the product surface to a cooling medium so that heat transfers out of the product. As the surface cools below the initial freezing point (θ_{if}), pure ice starts forming in the aqueous phase and solutes in the food become more concentrated in the remaining water, further depressing the freezing point below that for pure water. Thus, there is no sharp freezing point as for pure water, and latent heat is released over a range of temperatures. Continued heat removal from the food's interior is by conduction through the outer frozen shell and a notional "freezing front" moves inwards.

Food thermo-physical properties change markedly with temperature particularly as the water changes phase. Fig. 1gives enthalpy, volumetric heat capacity, and thermal conductivity as a function of temperature for a typical high-moisture food based on data measured by Willix and Amos.[1]

FREEZING MODEL

Food freezing is most commonly modeled by assuming that:

- Changes in size and shape of the product during freezing are negligible (i.e., constant density).
- Freezing is heat transfer controlled (ice crystal nucleation and growth are instantaneous).
- Heat transfers within the product by conduction only.
- Thermal properties are homogeneous throughout the product, and
- Mass transfer within the product can be ignored.

The general formulation for a 3-D product becomes heat conduction with temperature-variable thermal properties:

$$C(\theta)\frac{\partial \theta}{\partial t} = \frac{\partial}{\partial x}\left(\lambda(\theta)\frac{\partial \theta}{\partial x}\right) + \frac{\partial}{\partial y}\left(\lambda(\theta)\frac{\partial \theta}{\partial y}\right) + \frac{\partial}{\partial z}\left(\lambda(\theta)\frac{\partial \theta}{\partial z}\right) \quad \text{for } t > 0 \text{ within the product} \tag{1}$$

subject to the combined convection, radiation, and evaporation boundary condition:

$$\lambda(\theta)\left(l_x\frac{\partial \theta}{\partial x} + l_y\frac{\partial \theta}{\partial y} + l_z\frac{\partial \theta}{\partial z}\right) = \alpha(\theta_a - \theta) + F\varepsilon\sigma((\theta_a + 273)^4 - (\theta + 273)^4) + h_{fg}k(p_a - p_s(\theta)) \quad \text{for } t > 0 \text{ on the surface of the product} \tag{2}$$

and the initial condition:

$$\theta = \theta_i \quad \text{for} \quad t = 0 \tag{3}$$

To simplify predictions it is also often assumed that:

- Heat transfer to product surfaces is by convection only, with radiation, evaporation, and/or conduction to surfaces being represented using an effective α value.
- θ_i is uniform.
- θ_a and α are constant with both position and time.
- Thermal properties in the fully unfrozen and frozen states are constant, and
- Product geometry can be approximated by a regular shape.

Encyclopedia of Agricultural, Food, and Biological Engineering
DOI: 10.1081/E-EAFE 120007027
Copyright © 2003 by Marcel Dekker, Inc. All rights reserved.

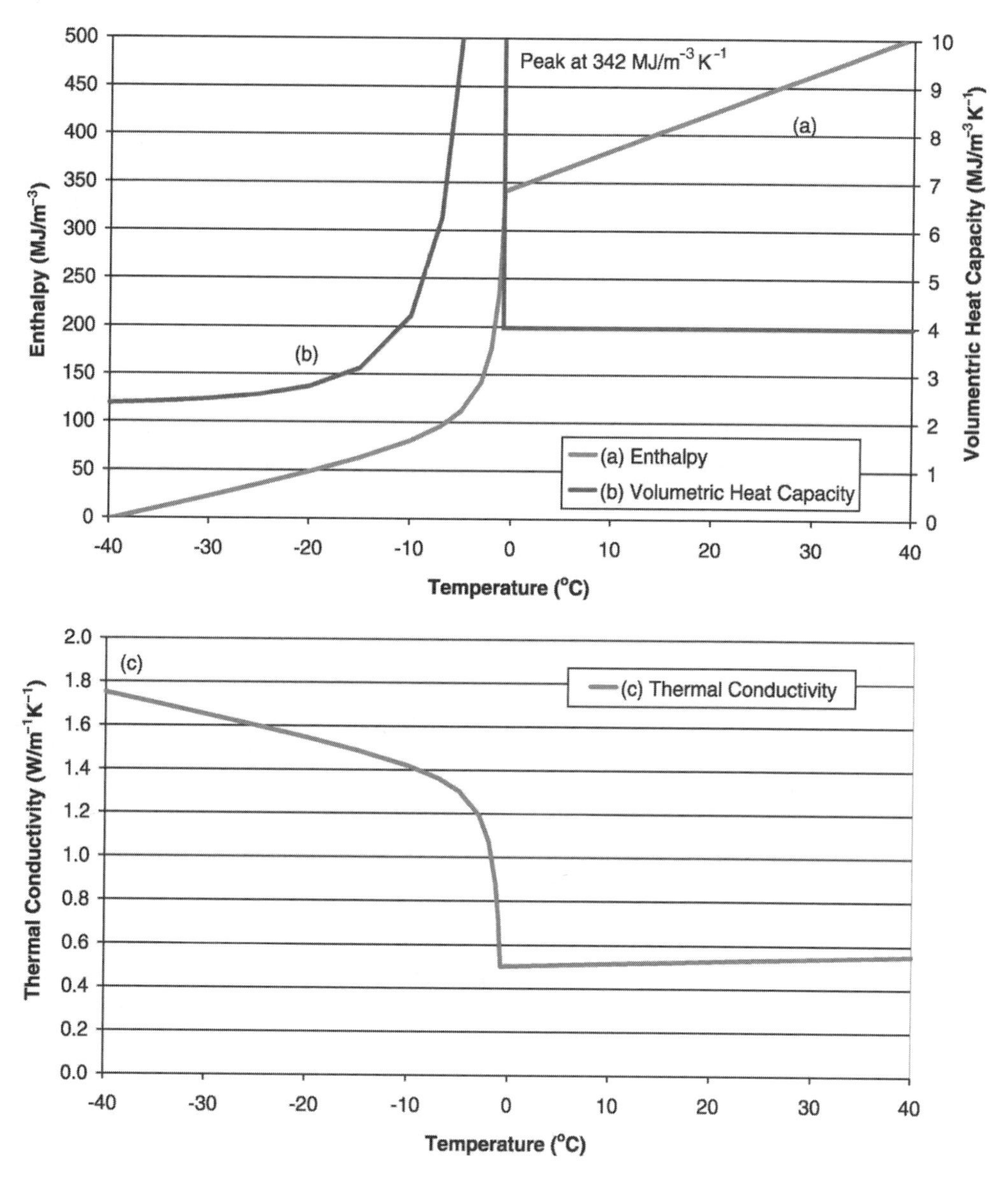

Fig. 1 Thermal properties for fish fillets based on data measured by Willix and Amos[1]: (a) Enthalpy (datum of 0 J m^{-3} at -40°C), (b) volumetric heat capacity, (c) thermal conductivity.

PREDICTION METHODS

Delgado and Sun[2] and Cleland and Ozilgen[3] provide recent reviews of food freezing time prediction literature. Prediction methods are of two main types—numerical methods and simple formula.

Numerical Methods

Overviews of numerical methods used for food freezing time prediction are given by Cleland[4] and Singh and Mannapperuma.[5] The best numerical methods use space and time discretization techniques, such as finite differences and finite elements, to approximate Eqs. 1–3. The release of latent heat across a range of temperature mean that numerical methods that track a moving phase change front are not appropriate for food freezing.[6] The main generic advantages of numerical methods are their ability to make predictions where:

- Mechanisms other than convection occur at the surface.
- Boundary conditions vary with position and time.
- Product composition is heterogeneous, and
- Product geometry is complex.

In addition, they can predict full temperature–time profiles, surface heat transfer rates, and evaporative weight loss during freezing.

Many engineering software packages provide the above functionality with convenient user-friendly interfaces. However, the very rapid changes in thermal properties with temperature, particularly just below θ_{if} (Fig. 1), mean that the discretization methods used by many generic software will not be accurate for freezing problems represented by Eqs. 1–3.

Special three time-level implicit time-discretization schemes have proved the most successful for freezing time predictions based on Eqs. 1–3, being both accurate and unconditionally stable and convergent.[4] Even with such schemes, the latent heat peak in the heat capacity vs. temperature profile can be partially missed ("jumped") leading to under-prediction of freezing time. Calculating the difference between the change in product internal heat content and the cumulative surface heat loss detects this problem that can be minimized by reducing the time interval used for calculations.[2,4]

The enthalpy transformation, where Eq. 1 is replaced by Eq. 4, overcomes peak jumping because H is a smoother function of temperature than C, but can only be used as part of an explicit numerical scheme because temperature must be evaluated from predicted enthalpy values.[5] Hybrid schemes that combine the best features of implicit and enthalpy schemes have also been shown to be effective.[7]

$$\frac{\partial H}{\partial t} = \frac{\partial}{\partial x}\left(\lambda(\theta)\frac{\partial \theta}{\partial x}\right) + \frac{\partial}{\partial y}\left(\lambda(\theta)\frac{\partial \theta}{\partial y}\right) + \frac{\partial}{\partial z}\left(\lambda(\theta)\frac{\partial \theta}{\partial z}\right)$$

$$\text{for } t > 0 \text{ within the product} \quad (4)$$

Commercial software employing such specialized techniques are available, but are often expensive, may not have the full functionality of generic packages particularly with respect to product geometry and user interface, and are often not accessible by industrial practitioners. These reasons, and the requirement for thermal property data across the full temperature range, are the main disadvantages of numerical methods.

Combining the enthalpy transformation and a similar smoothing of the thermal conductivity vs. temperature function such as the Kirchoff transformation,[6] may allow the simpler time discretization schemes employed by generic packages to be used without significant loss of accuracy. However, the cost of generic packages and the lack of familiarity with such transformations remain significant barriers to their widespread use by users who infrequently undertake freezing time predictions.

Simple Formula

In many situations, control of conditions such as air velocity and air temperature and knowledge about product size, shape, and composition are imprecise. In such cases, data uncertainty becomes more limiting than the calculation precision of the method being used, and low cost, easy-to-use prediction formula will often be as accurate as more sophisticated numerical methods.

A large number of simple prediction formula have been proposed that combine sound engineering judgement to simplify the physical problem with empirical modifications.[2,3] Most are based on Plank's equation which is an analytical solution where, in addition to the simplifying assumptions listed above, it is assumed that:

- Latent heat is released at a unique phase change temperature, and
- Frozen and unfrozen sensible heat capacities are negligible compared with latent heat.

A number of such formula have been shown to be of similar accuracy to each other and to numerical methods when assessed against a large set of experimental data.[2,3]

The effect of geometry is virtually independent of other factors affecting freezing time so generalized shape factors can be used to apply simple formula based on Plank's equation to a wider and more realistic range of product geometries.[3,8] Proposed shape factor approaches include defining an "equivalent other shape,"[9,10] a mean conducting path length from the thermal center to the surface,[11] and the equivalent heat transfer dimensionality.[12,13] Of these, the last is the best developed, with formulae available that both have an analytical basis and cover most product geometries.

Recommended Simple Prediction Method

The method recommended on the basis of being quite physically realistic, and having good accuracy for a wide range of freezing conditions with low calculation complexity is that proposed by Pham[14] combined with the generalized shape factor proposed by Hossain, Cleland, and Cleland[12]:

$$t_f = \frac{1}{E}\left[\frac{\Delta H_1}{\Delta \theta_1} + \frac{\Delta H_2}{\Delta \theta_2}\right]\left[\frac{R}{\alpha} + \frac{R^2}{2\lambda_f}\right] \quad (5)$$

where: $\Delta H_1 = C_u(\theta_i - \theta_{fm})$

$$\Delta H_2 = L + C_f(\theta_{fm} - \theta_{fin})$$

$$\Delta \theta_1 = 0.5(\theta_i + \theta_{fm}) - \theta_a$$

$$\Delta \theta_2 = \theta_{fm} - \theta_a$$

$$\theta_{fm} = 1.8 + 0.263\theta_{fin} + 0.105\theta_a$$

Eq. 5 was derived from data for high-moisture foods where:

$0.02 < Bi < 11 \quad 0.11 < Ste < 0.36$

$0.03 < Pk < 0.61 \quad -20 < \theta_{fin} < -10°C$

No strict physical significance should be assigned to the value of θ_{fm} because it is an empirical function of freezing conditions and is not a function of the product composition.

For all product geometries, the shape factor can be estimated by the formulae for an equivalent ellipsoid:

$$E = 1 + \frac{2 + Bi}{\beta_1(2 + \beta_1 Bi)} + \frac{2 + Bi}{\beta_2(2 + \beta_2 Bi)} \tag{6}$$

where:

$$\beta_1 = \frac{A}{\pi R^2} \tag{7}$$

$$\beta_2 = \frac{3V}{4AR} \tag{8}$$

$$Bi = \frac{\alpha R}{\lambda_f} \tag{9}$$

More complex formulae available for multidimensional regular geometries are only marginally more accurate than Eq. 6 for practical food freezing conditions.[13]

The accuracy of this approach has been proven by comparison with over 300 high quality experimental freezing data from various sources for a wide range of freezing conditions and product geometries.[8,13,14] Although maximum differences were as large as ± 30%, the 95% confidence limits were ± 15%–20%. Most of the prediction error could be attributed to experimental uncertainty so the inherent uncertainty in the method is probably less than ± 10% when it is used within the range of conditions for which it was developed. Given that in practice product characteristics and freezing conditions are seldom known accurately or tightly controlled, freezing time estimates should be treated as being accurate to within ± 20% at best.

Guidelines for Use

Guidelines for the use of Eqs. 5 and 6 are available.[8,15,16] Some of the main considerations are summarized below:

- *Cryogenic freezing and thawing*: The accuracy has not been proven for the extremely low θ_a values achieved with cryogenics (less than − 50°C) so predictions for such applications should be treated cautiously. For thawing, Eq. 5 must be replaced by different semiempirical formula.[8] Eq. 5 cannot be used to predict partial (crust) freezing.
- *Product dimensions*: The dimensions used should be for the smallest item that the cooling medium cannot penetrate. For example, for freezing of fish fillets the product item could be an individual fillet (if a belt freezer is used), a single carton of fillets (if cartons are sealed and placed on separate racks in an air-blast freezer), or a stack of cartons (if air cannot penetrate between individual cartons). Use of frozen product dimensions provides a safety margin if there is any expansion during freezing.
- *Thermal properties*: If available, thermal properties measured using reliable techniques for the actual food product should be used. Otherwise, simple models that allow properties to be estimated from product composition are available.[15] Measurement of θ_{if} is always advised because it is difficult to predict and an accurate value is critical for both precise thermal property and freezing time predictions.
- *Heterogenous products*: If the product is heterogeneous, then effective thermal properties must be used. Measurement or prediction of effective L, C_u, and C_f values is relatively straightforward using differential scanning calorimetry or volume-weighted sums of the product components, respectively.[15] However, effective λ_f is more difficult to measure and is structure dependent. If the components are randomly distributed then combined (weighted) series and parallel models are appropriate.[15] Layers of components with low conductivity orientated across the direction of heat flow (e.g., fat or trapped air layers) can have a particularly large effect, because series conductivity models are generally appropriate. Component layers orientated parallel to the direction of heat flow are uncommon.
- *Heat transfer coefficients*: Radiation and evaporation are only significant if air is the cooling medium. A number of correlations are available for effective α values including radiative and evaporative contributions for many common air freezers.[16,17] For air freezers, if air velocity is less than about 0.5 m sec^{-1} then radiation and evaporation often become dominant over convection. A minimum effective α value under such circumstances without packaging is 5 W m^{-2}K^{-1}.
- *Variable freezing conditions*: If variations with time or position are small then use of average values of α, θ_a, and θ_i is usually adequate (e.g., a pull-down in θ_a at the start of a batch freezing process). Differential forms of Eq. 5 or full numerical methods should be used if time-variability is significant (e.g., if the desired θ_{fin} is close to a varying θ_a).[2]

- *Packaged products*: Packaging acts to increase the effective size of the product, trap air voids within the product and influence heat transfer properties directly. The heat capacity of packaging and trapped air is usually negligible compared with the food itself so volumetric properties can be adjusted in proportion to the voids fraction. Packaging and trapped air layers located at the surface of the product item reduce α so their thermal resistance should be added in series to the convective resistance to heat transfer.[15,16] Packaging and air voids inside the product item affect the effective λ_f value.
- *Sensitivity analysis*: Where data uncertainty is high, undertaking a sensitivity analysis by repeating predictions for slightly different values of key variables is recommended. It can also provide guidance as to which variables will provide greater benefits in terms of process efficiency. For example, if Bi is high then both uncertainty in and improvements to α are relatively unimportant, whereas the effect of air voids on λ_f could be highly influential.

CONCLUSION

Carefully formulated and implemented numerical prediction methods can be very accurate and flexible freezing time prediction methods. However, they tend to be expensive, require specialist expertise often not available to industrial practitioners, and their accuracy is limited by uncertainty about product characteristics and freezing conditions. For many industrial situations, semiempirical prediction formulae are just as accurate, yet are simpler to use and have lower data requirements. A simple prediction method and guidelines for its use are presented. The method is accurate to within $\pm$ 10% for most practical freezing conditions if data uncertainty is low.

NOMENCLATURE

A	area of the smallest product cross-section (m^2)
Bi	Biot number
C	food volumetric heat capacity ($J\,m^{-3}\,K^{-1}$)
E	equivalent heat transfer dimensionality (shape factor)
F	radiation view factor
h_{fg}	latent heat of vaporization for water ($J\,kg^{-1}$)
H	food enthalpy ($J\,m^{-3}$)
k	mass transfer coefficient ($kg\,sec^{-1}\,m^{-2}\,Pa^{-1}$)
l	cosine of outward normal
L	food latent heat of freezing; water latent heat × volume fraction of freezable water ($J\,m^{-3}$)
p	partial pressure of water vapor (Pa)
Pk	Plank number $= C_u(\theta_{in} - \theta_{if})/[L + C_f(\theta_{if} + 10)]$
R	minimum distance from product thermal center to surface (m)
Ste	Stefan number $= C_f(\theta_{if} - \theta_a)/[L + C_f(\theta_{if} + 10)]$
t	time (sec)
V	volume of the product (m^3)
x, y, z	distance in x, y, or z directions (m)
α	surface heat transfer coefficient ($W\,m^{-2}\,K^{-1}$)
β	ratio of axes length for equivalent ellipsoid
ε	emissivity
λ	food thermal conductivity ($W\,m^{-1}\,K^{-1}$)
θ	temperature (°C)
σ	Stefan–Boltzmann constant ($W\,m^{-2}\,K^{-4}$)

Subscripts

a	cooling medium
f	fully frozen
fin	final condition at the thermal center
fm	mean freezing condition
i	initial condition
if	initial freezing point
s	in equilibrium with the product surface
u	unfrozen
x, y, z	in x, y, or z direction

REFERENCES

1. Willix, J.; Amos, N.D. *Thermal Property Data for Foods*; MIRINZ Technical Report 976; Meat Industry Research Institute of New Zealand: Hamilton, New Zealand, 1998.
2. Delgado, A.E.; Sun, D.W. Heat and Mass Transfer Models for Predicting Freezing Processes—A Review. J. Food Eng. **2001**, *47*, 157–174.
3. Cleland, A.C.; Ozilgen, S. Thermal Design Calculations for Food Freezing Equipment—Past, Present and Future. Int. J. Refrig. **1998**, *21*, 359–371.
4. Cleland, A.C. *Food Refrigeration Processes—Analysis, Design and Simulation*; Elsevier Applied Science: London, 1990.
5. Singh, R.P.; Mannapperuma, J.D. Developments in Food Freezing. In *Biotechnology and Food Process Engineering*; Schwartzberg, H.G., Rao, M.A., Eds.; Marcel Dekker: New York, 1990; 309–358.
6. Nicolai, B.M.; Verboven, P.; Scheerlinck, N.; Hoang, M.L.; Haddish, N. Modelling of Cooling and Freezing Operations. In *Modelling of Food Chilling and Freezing*, Preprints of the International Institute of Refrigeration

Rapid Cooling of Foods Conference, Bristol, UK, March 28–30, 2001.

7. Pham, Q.T. A Fast Unconditionally Stable Finite-Differences Scheme for Heat Conduction with Phase Change. Int. J. Heat Mass Transfer **1985**, *28*, 2079–2084.
8. Cleland, D.J. A Generally Applicable Simple Method for Prediction of Food Freezing and Thawing Times. Proc. 18th Int. Congr. Refrig. **1991**, *4*, 1874–1877.
9. Arroyo, J.G.; Mascheroni, R.H. A Generalized Method for the Prediction of Freezing Times of Regular or Irregular Foods. Refrig. Sci. Technol. **1990**, *4*, 643–649.
10. Ilicali, C.; Hocalar, M. A Simplified Approach for Predicting the Freezing Times of Foodstuffs of Anomalous Shape. In *Engineering and Food, Volume II—Preservation Processes and Related Techniques*; Spiess, W.E.L., Schubert, H., Eds.; Elsevier Applied Science: London, 1990; 418–425.
11. Pham, Q.T. Analytical Methods for Predicting Freezing Times for Rectangular Blocks of Foodstuffs. Int. J. Refrig. **1985**, *8*, 43–47.
12. Hossain, Md.M.; Cleland, D.J.; Cleland, A.C. Prediction of Freezing and Thawing Times for Foods of Three-Dimensional Irregular Shape by Using a Semi-analytical Geometric Factor. Int. J. Refrig. **1992**, *15*, 241–246.
13. Cleland, A.C.; Cleland, D.J.; Davey, L.M. Accuracy of a Simplified Approach for Finding the Geometric Factor of Regular Geometric Shapes During Freezing and Thawing Time Prediction. Proc. 19th Int. Congr. Refrig. **1995**, *2*, 57–64.
14. Pham, Q.T. Simplified Equation for Predicting the Freezing Time of Foodstuffs. J. Food Technol. **1986**, *21*, 209–219.
15. Cleland, D.J.; Valentas, K.J. Prediction of Freezing Time and Design of Food Freezers. In *Handbook of Food Engineering Practice*; Rotstein, E., Singh, R.P., Valentas, K.J., Eds.; CRC Press: Boca Raton, FL, 1997; 71–124.
16. Cleland, A.C.; Cleland, D.J.; White, S.D. *Cost-Effective Refrigeration*; Massey University: Palmerston North, New Zealand, 2000.
17. ASHRAE. *ASHRAE Handbook—Refrigeration*; American Society of Heating, Refrigerating and Air-Conditioning Engineers: Atlanta, 1998.

Frozen Food Enthalpy

Rutger M. T. van Sleeuwen
Dennis R. Heldman
Rutgers, The State University of New Jersey, New Brunswick, New Jersey, U.S.A.

INTRODUCTION

Food freezing is a preservation method that retains many of the important quality attributes of fresh foods, including nutritional value, flavor, and texture. The principal objective of freezing operations is to preserve foods with minimal quality losses, and current advances in freezing systems aim to realize this goal.[1] In designing and optimizing freezing processes, it is imperative to have accurate information on the thermophysical properties of food products. The enthalpy, specific heat capacity, thermal conductivity, and density of the food govern to a large extent the total amount of thermal energy that needs to be removed, as well as the rate of removal.[2]

Freezing of water that contains a solute (representative of food) differs from freezing of pure water in the sense that the initial freezing point is depressed below the freezing point of pure water. As solidification of this solution proceeds and pure ice crystals are formed, the residual solution becomes increasingly concentrated. This results in a further depression of the freezing point, and in contrast to freezing of pure water, which is completed at a single temperature, freezing of a solution involves a gradual drop in temperature throughout phase change.[3]

The enthalpy–temperature relationship for a food, in the temperature region where freezing normally occurs, reflects both the variance of enthalpy with temperature as well as the state (frozen or unfrozen) of the water in the product. This relationship contains information on the thermal energy that has to be removed to freeze the product, and it can provide insight into how much unfrozen water is present at different temperatures.[4] Finally, this correlation between enthalpy and temperature is used to simulate the temperature distribution in food products. Such models are used to predict freezing and thawing times under various conditions.[5,6] In this section, attention will be given to the measurement and prediction of enthalpy of foods as a function of temperature during freezing and how this relationship can be used in freezing research and commercial operations.

ENTHALPY BASICS

Charm[7] describes that the enthalpy H of a substance is the sum of the specific internal energy U and the product of absolute pressure P and specific volume V:

$$H = U + PV \tag{1}$$

The author[7] further explains that when the pressure is constant, the derivative of enthalpy in respect to temperature is known as the specific heat capacity C_P:

$$C_P = \left(\frac{\partial H}{\partial T}\right)_P \tag{2}$$

where C_P is the specific heat capacity, H the enthalpy, T the temperature, and P the absolute pressure.

Wunderlich[8] describes that when the composition of a substance remains constant, the enthalpy–temperature relationship is described by:

$$H(T) = H(T_0) + \int_{T_0}^{T} C_P \mathrm{d}T \tag{3}$$

where T_0 is the reference temperature.

Eq. 3 describes the enthalpy as a result of changes in sensible heat of the substance, and this term is related to changes in temperature (molecular motion). When water changes phase within a substance, there is a second contribution to the heat content, the latent heat of fusion that is released during phase change, which is related to the organization and internal bonding of the substance.[9]

Heldman[10] explains that when the composition of the substance is primarily solids, unfrozen water, and ice, the total enthalpy of the substance is the sum of the sensible heat portions of these components and the latent heat:

$$H = H_U + H_S + H_I + H_L \tag{4}$$

where U stands for unfrozen water, S for solids, I for ice, and H_L is the latent heat.

The latent heat of fusion is released from a food over a range of temperatures rather than at a constant temperature.[11] In nearly all biological materials,

Encyclopedia of Agricultural, Food, and Biological Engineering
DOI: 10.1081/E-EAFE 120007025
Copyright © 2003 by Marcel Dekker, Inc. All rights reserved.

the majority of this phase change occurs over the region between −1 and −8°C.[12] In addition, foods experience major changes in their thermophysical properties when water is converted to ice,[13,59] and as most foods contain a large water fraction, the food properties are dramatically affected during phase change.[10,14]

Heldman[10] adds that in order to predict the enthalpy of a product, the individual components of thermal energy are integrated over an appropriate region, using −40°C as the reference temperature, where the enthalpy is set to zero:

$$H = m'_S C_{PS} \int_{-40}^{T_i} dT + m'_U C_{PU} \int_{T_f}^{T_i} dT + \int_{-40}^{T_f} m'_U(T) C_{PU}(T) dT + m'_U(T) L + \int_{-40}^{T_f} m'_I(T) C_{PI}(T) dT \quad (5)$$

where m' is the mass fraction, T_i the initial temperature, T_f the initial freezing temperature, and L the latent heat.

A typical enthalpy–temperature relationship for a food material is presented as a continuous curve in Fig. 1. The change in enthalpy with temperature above freezing is modest and this is related to the specific heat capacity of the unfrozen food.[10] A sharp decrease in enthalpy is observed just below the freezing point, which is related to the removal of most of the latent heat of fusion.[15]

MEASUREMENT OF ENTHALPY OF FROZEN FOODS

According to Lind,[16] common methods for measuring the relative enthalpy content or the specific heat capacity of foods are: 1) mixing methods; 2) differential scanning calorimetry (DSC); and 3) adiabatic calorimetry. These are described below in more detail. Other methods have also been employed and they include: the guarded hot-plate method, thermal analysis, differential thermal analysis,[16] and differential compensated calorimetry.[4] Nuclear Magnetic Resonance (NMR) has also been used successfully to measure the fraction of frozen water as a function of temperature in model foods, and thermal properties were predicted from this relationship.[17] Comprehensive overviews of measurement techniques are given by Ohlsson,[18] Lind,[16] and Mohsenin.[20] In addition, an extensive collection of thermophysical properties of foods has been compiled by the American Society of Heating, Refrigerating and Air-Conditioning Engineers.[19]

Method of Mixtures

Lind[16] describes that this method measures the specific heat capacity of a sample $C_{p,samp}$ with a given initial temperature ($T_{i,samp}$) by mixing it with water, held in a calorimeter at a particular temperature ($T_{i,cal}$). The author[16] explains that the specific heat capacity of the sample is obtained from the heat balance:

$$C_{p,samp} m_{samp} (T_{i,samp} - T_e) + C_{p,cal} m_{cal} (T_{i,cal} - T_e) = C_{p,w} m_w (T_e - T_{i,w}) + Q_{loss} \quad (6)$$

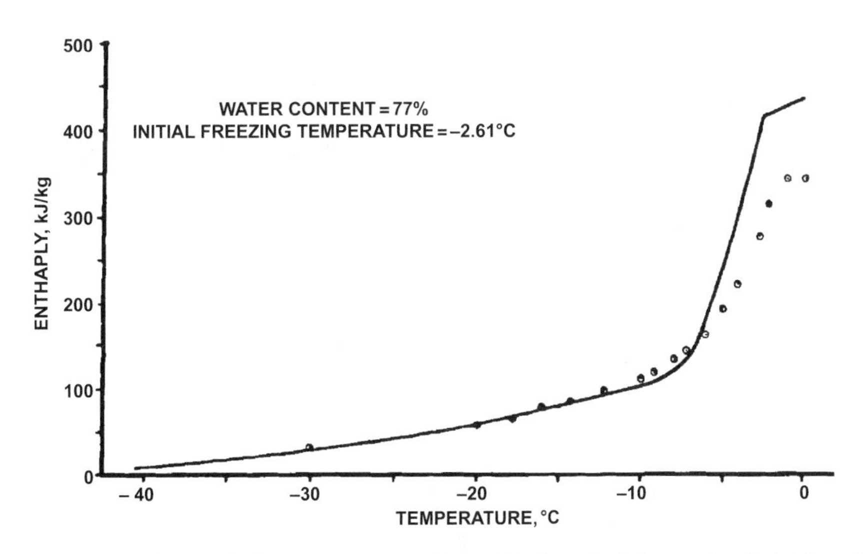

Fig. 1 General picture of the enthalpy–temperature relationship for a food (sweet cherries). (From Ref. [10].)

where m is the mass, e equilibrium, samp sample, cal calorimeter, i initial, w water, and Q_{loss} the energy loss.

In order to ensure accurate results, thermal leakage must be minimized during measurements.[20] This method is easy to conduct, and relatively large samples can be analyzed. However, the measurements are time-consuming and not suitable at temperatures where $C_{p,samp}$ is a strong function of temperature.[16] This factor may limit the application to frozen foods.

Differential Scanning Calorimetry

DSC has been used to measure the enthalpy or specific heat capacity of different foods in the temperature region where freezing typically occurs.[2,21,22] It involves measurement of the heat flow required to maintain the same temperature in a small sample of interest and a reference material,[16] while the machine is scanning a particular temperature region at a specified rate.[20] The advantage of DSC is that it gives quick results on homogeneous samples. However, accuracy is limited and precision may have to be enhanced by measuring multiple replicates. This increases the total measuring time, and precision is still inferior to the more labor-intensive adiabatic calorimetry method.[18]

Adiabatic Calorimetry

The specific heat capacity of large food samples can be measured using an adiabatic calorimeter, which is internally heated by an electrical heater. This calorimeter is surrounded by an adiabatic jacket, which can be heated or cooled to follow the temperature of the calorimeter and maintain adiabatic conditions (a zero net heat flux) between the calorimeter and its environment. The electrical heat input, sample temperature, and heat absorbed when the calorimeter is empty are all used to calculate the specific heat capacity.[8] Adiabatic calorimetry provides a measurement of specific heat capacity with high precision, but long measuring times and proper insulation of the measuring cell are required.[18] The enthalpy–temperature relationships for various foods have been measured by different researchers.[23–27] A schematic picture of the adiabatic calorimeter used by Pham et al.[23] is shown in Fig. 2. It features a vacuum flask (1.0 L) that holds a special calorimetric fluid in which the samples will be placed. The fluid is mixed and heated by an assembly comprising of a fully immersed stirrer and heater. Thermopiles are used to measure the temperature of the calorimeter against an ice reference (not shown) as well as the difference in temperature between the calorimeter and the alcohol in the jacket. The temperature of the alcohol

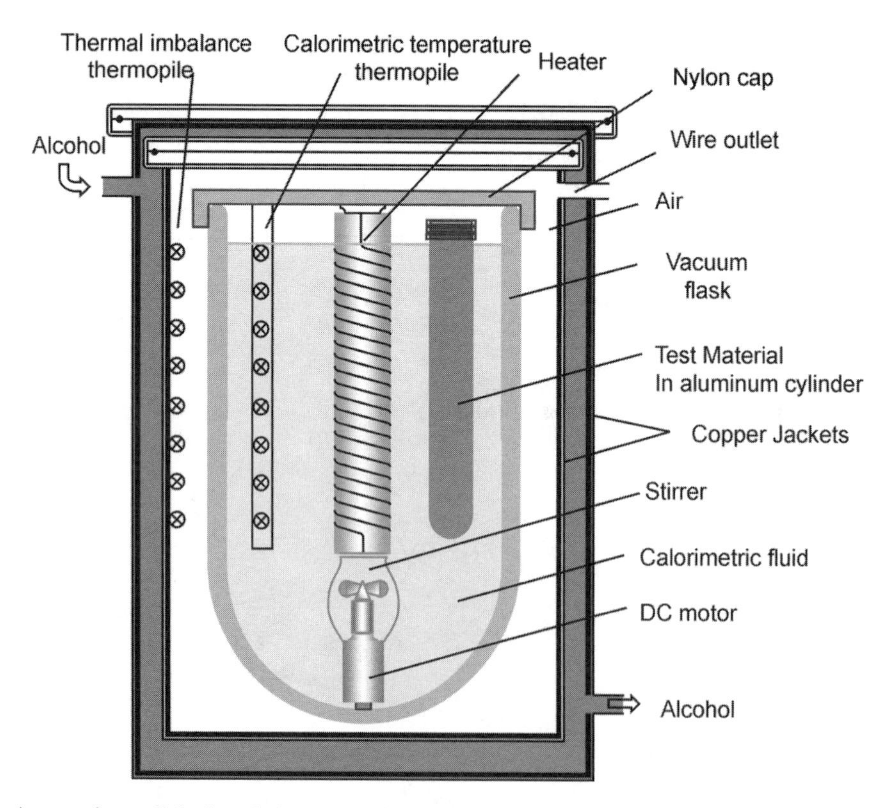

Fig. 2 Schematic picture of an adiabatic calorimeter. (Adapted / reprinted from Ref. [23]; copyright 1993, with permission from Elsevier Science)

can be controlled by means of a temperature bath and an in-line heater (not shown) to preserve adiabatic conditions. The calorimeter and jacket are positioned in a container that is packed with expanded polystyrene as insulation (not shown).[23]

PREDICTION OF ENTHALPY OF FROZEN FOODS

Because of the wide variety of food products and their composition, it is not feasible to provide experimental data on thermal properties for all types of foods, under any conceivable condition. Due to the significant influence of composition and temperature on the properties of food products, mathematical models have been developed that estimate thermophysical properties as a function of these factors.[28]

Heldman[29] used a relationship describing the freezing-point depression to create a method for estimating the amount of unfrozen water as a function of temperature in a food. This approach[30] can be described by:

$$\frac{\lambda'}{R_g}\left[\frac{1}{T_{A0}} - \frac{1}{T_A}\right] = \ln X_A \tag{7}$$

where λ' is the latent heat of fusion per mole, R_g the gas constant, A the component, T_{A0} the freezing temperature of pure liquid A, and X the mole fraction.

Heldman[31] describes that once the absolute initial freezing temperature (T_A) for the food is known, the mole fraction of water (X_A) can be calculated using Eq. 7, and this mole fraction can also be expressed as:

$$X_A = \frac{m'_A/M_A}{m'_A/M_A + m'_S/M_S} \tag{8}$$

where m' is the mass fraction and M the molecular weight.

From Eq. 8, an "effective molecular weight" M_S can be estimated. Once this value is known, Eqs. 7 and 8 can be used again by replacing X_A, m'_A, and M_A by X_U, m'_U, and M_U, respectively, to predict the fraction of unfrozen water. At a temperature of interest T_A (e.g., 260K), the "apparent mole fraction" of unfrozen water X_U can be calculated using Eq. 7. When this value is substituted into Eq. 8, the fraction of unfrozen water (m'_U) can be computed as M_S is known. An example of a typical relationship between unfrozen water and temperature is given as a continuous curve in Fig. 3. After the unfrozen and frozen fractions are known as a function of temperature, these values can be employed in Eq. 5 to obtain an enthalpy–temperature relationship.[31]

Many other researchers have developed or summarized mathematical equations or empirical correlations to predict the thermal properties of a great variety of foods.[32–42] Miles, van Beek, and Veerkamp[41] compiled many of such predictive equations and standardized them to equivalent units and terminology for ease of comparison.

APPLICATION OF THE ENTHALPY–TEMPERATURE RELATIONSHIP

Enthalpy–temperature relationships of food products are used in the following applications.

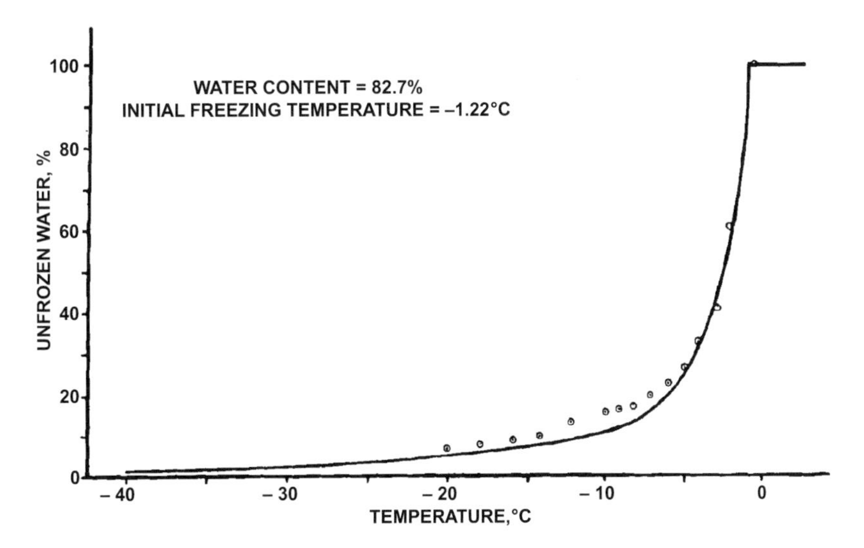

Fig. 3 Percentage of unfrozen water in raspberries as a function of temperature. (From Ref. [29].)

Simulations of the Freezing Process

Freezing time is often defined as the time needed to cool the slowest cooling location from the initial temperature to a defined final temperature. Accurate predictions of freezing time are needed in the design of a freezing process, as this establishes the minimum residence time of the product in a continuous freezer. This residence time is related to the size of the freezer and the speed at which the product is moving through this freezing unit.[31]

In order to simulate the temperature change during freezing of foods, analytical, empirical, and numerical solutions have been developed and compiled by various researchers.[43–49]

Analytical solutions are usually simple and can provide quick and reliable estimates of freezing time.[50] Numerical simulations can incorporate variable thermophysical properties during freezing and can be applied to a range of product shapes with various boundary conditions. Most analytical models are more restricted as not all of them account for the gradual variation in thermophysical properties during freezing.[51] Numerical solutions, such as finite difference or finite element methods are recognized as the most accurate procedures to simulate food freezing when implemented correctly.[43,46]

Most numerical food freezing and thawing models describing heat conduction and solidification use either the so-called apparent specific heat formulation or the enthalpy formulation to account for the gradual phase change.[5]

Mannapperuma and Singh[5] have illustrated that the enthalpy–temperature relationship (Fig. 4a), which is used in the enthalpy formulation, does not include the strong discontinuity shown by the apparent specific heat relationship (Fig. 4b). This diminishes some of the problems observed in numerical modeling using the apparent specific heat formulation (e.g., "peak jumping" or "stable oscillations").[5]

In summary, accurate enthalpy–temperature relationships can be used directly in certain numerical food freezing and thawing models.

Unfrozen Fraction

The enthalpy–temperature relationship can be used to estimate the unfrozen water fraction or ice fraction in foods as a function of temperature.[37] These predictions on the state of water are important, due to the significant effect on the thermophysical properties of the food.[10] Thermal conductivity increases dramatically just below the initial freezing point of a food, due to the higher thermal conductivity of ice as compared to unfrozen water.[52] Information on the fraction of frozen water can also help diminish detrimental effects of ice on the product.[33] The quality of meat, for instance, may be compromised by slow reactions that occur in the unfrozen phase of the product.[27] Quantification of the amount of frozen water may help in assessing the extent to which solutes have been concentrated in the remaining unfrozen water.[53] It is plausible that food product quality may be negatively impacted by this freeze concentration effect, although the contribution of freeze concentration to freezing damage has not been clearly confirmed.[54]

Cooling/Refrigeration Loads

The cooling load describes the rate at which thermal energy is removed from a certain object or space from an initial temperature to a required final temperature.[55]

The evaporator in a freezing unit must be capable of removing efficiently the thermal energy from the food material. The refrigeration requirement equals the change from the initial product enthalpy to the enthalpy of the

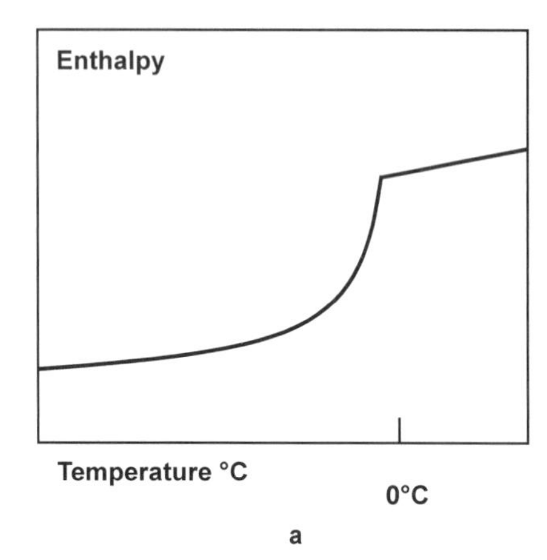

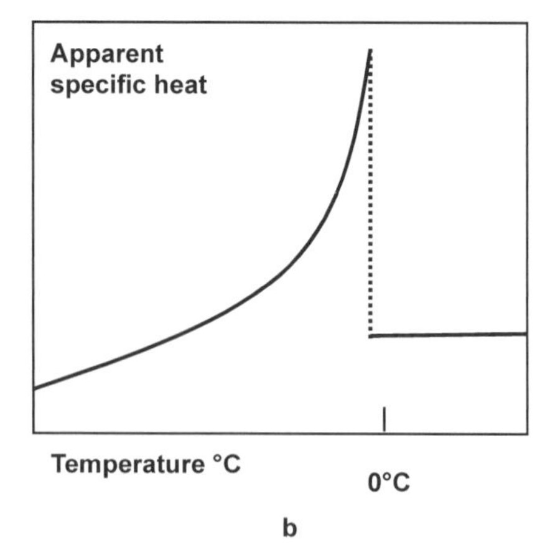

Fig. 4 Comparison of (a) enthalpy and (b) apparent specific heat as a function of temperature. (From Ref. [5].)

frozen product. This heat needs to be removed within the freezing time or residence time of the food within the freezer, and this determines the capacity or the refrigeration load on the freezing unit.[31] The direct relationship of the product enthalpy to the total refrigeration load for product freezing illustrates the importance of accurate enthalpy–temperature relationships.[56] The contribution of the product enthalpy to the total energy requirements of a freezing process is significant when compared to other factors, such as transmission, fans, and pumps. For example, the removal of thermal energy from the product in air-blast freezing demands about 57%–66% of the total refrigeration load[57] according to Ref. 56. Another example is the calculation of liquid nitrogen required for cryogenic freezing of food products, where the total amount of heat to be removed from the product plays an important role in establishing the overall cost of liquid nitrogen per kg of product.[58]

SUMMARY

The enthalpy–temperature relationship for a food product provides general information about the changes in thermal energy during the freezing process. The applications of the enthalpy–temperature relationship are most evident when predicting cooling load, unfrozen water fraction, and freezing times. These types of information are typically needed by engineers involved in designing food freezing equipment and in establishing the capacity of refrigeration systems.

LIST OF SYMBOLS

H	Enthalpy, $\mathrm{kJ\,kg^{-1}}$
U	Specific internal energy, $\mathrm{kJ\,kg^{-1}}$
P, p	Pressure, Pa
V	Specific volume, $\mathrm{m^3\,kg^{-1}}$
T	Temperature, °C
T_{A0}	Equilibrium freezing point of liquid A, K
T_{f}	Initial freezing temperature, °C
m	Mass, kg
m'	Mass fraction
M	Molecular weight, $\mathrm{g\,mol^{-1}}$
L	Latent heat of fusion/solidification, $\mathrm{kJ\,kg^{-1}}$
Q_{loss}	Energy loss, kJ
R_{g}	Gas constant, $\mathrm{J\,mol^{-1}\,K^{-1}}$
C_{P}	Specific heat capacity, $\mathrm{kJ\,kg^{-1}\,°C^{-1}}$
λ'	Latent heat of fusion/solidification, $\mathrm{kJ\,mol^{-1}}$
X	Mole fraction

Subscripts

S	Solids
U	Unfrozen
I	Ice
w	Water
A	Identifies component in solution
L	Latent
loss	Loss
0	Reference (e.g., − 40°C)
e	Equilibrium
samp	Sample
cal	Calorimeter
i	Initial

REFERENCES

1. George, R.M. Freezing Systems. In *Quality in Frozen Food*; Erickson, M.C., Hung, Y.-C., Eds.; Chapman & Hall, International Thomson Publishing: New York, NY, 1997; 3–9.
2. Ramaswamy, H.S.; Tung, M.A. Thermophysical Properties of Apples in Relation to Freezing. J. Food Sci. **1981**, *46*, 724–728.
3. Fennema, O.; Powrie, W.D. Fundamentals of Low-Temperature Food Preservation. In *Advances in Food Research*; Chichester, C.O., Mrak, E.M., Stewart, G.F., Eds.; Academic Press: New York, 1964; Vol. 13, 219–347.
4. Kerr, W.L.; Ju, J.; Reid, D.S. Enthalpy of Frozen Foods Determined by Differential Compensated Calorimetry. J. Food Sci. **1993**, *58* (3), 675–679.
5. Mannapperuma, J.D.; Singh, R.P. Prediction of Freezing and Thawing of Foods Using a Numerical Method Based on Enthalpy Formulation. J. Food Sci. **1988**, *53* (2), 626–630.
6. Mannapperuma, J.D.; Singh, R.P. A Computer-Aided Method for Prediction of Properties and Freezing/Thawing Times of Foods. J. Food Eng. **1989**, *9*, 275–304.
7. Charm, S.E. *The Fundamentals of Food Engineering*, 3rd Ed.; AVI Publishing Company, Inc.: Westport, CT, 1978.
8. Wunderlich, B. *Thermal Analysis*; Academic Press: San Diego, CA, 1990.
9. Reid, D.S. Overview of Physical/Chemical Aspects of Freezing. In *Quality in Frozen Food*; Erickson, M.C., Hung, Y.-C., Eds.; Chapman & Hall, International Thomson Publishing: New York, NY, 1997; 10–28.
10. Heldman, D.R. Food Properties During Freezing. Food Technol. **1982**, *36*, 92–96.
11. Dickerson, R.W., Jr. Thermal Properties of Foods. In *The Freezing Preservation of Foods*; Tressler, D.K., van Arsdel, W.B., Copley, M.J., Eds.; AVI Publishing Company, Inc.: Westport, CT, 1981; Vol. 2, 26–51.
12. Rebellato, L.; Del Giudice, S.; Comini, G. Finite Element Analysis of Freezing Processes in Foodstuffs. J. Food Sci. **1978**, *43*, 239–244.
13. Sastry, S.K. Freezing Time Prediction: An Enthalpy-Based Approach. J. Food Sci. **1984**, *49*, 1121–1127.

14. Saad, Z.; Scott, E.P. Estimation of Temperature Dependent Thermal Properties of Basic Food Solutions During Freezing. J. Food Eng. **1996**, *28*, 1–19.
15. Heldman, D.R.; Taylor, T.A. Modeling of Food Freezing. In *Quality in Frozen Food*; Erickson, M.C., Hung, Y.-C., Eds.; Chapman & Hall, International Thomson Publishing: New York, NY, 1997; 51–64.
16. Lind, I. The Measurement and Prediction of Thermal Properties of Food During Freezing and Thawing—A Review with Particular Reference to Meat and Dough. J. Food Eng. **1991**, *13*, 285–319.
17. Cornillon, P.; Andrieu, J.; Duplan, J.-C.; Laurent, M. Use of Nuclear Magnetic Resonance to Model Thermophysical Properties of Frozen and Unfrozen Model Food Gels. J. Food Eng. **1995**, *25*, 1–19.
18. Ohlsson, T. The Measurement of Thermal Properties. In *Physical Properties of Foods*; Jowitt, R., Escher, F., Hallström, B., Meffert, H.F.Th., Spiess, W.E.L., Vos, G., Eds.; Applied Science Publishers: London, 1983; 313–328.
19. ASHRAE. *ASHRAE Handbook: Fundamentals*; American Society of Heating, Refrigerating and Air-Conditioning Engineers, Inc.: Atlanta, GA, 1981.
20. Mohsenin, N.N. *Thermal Properties of Foods and Agricultural Materials*; Gordon and Breach Science Publishers: New York, 1980.
21. Tocci, A.M.; Flores, E.S.E.; Mascheroni, R.H. Enthalpy, Heat Capacity and Thermal Conductivity of Boneless Mutton Between − 40 and + 40°C. Lebensm. Wiss. Technol. **1997**, *30* (2), 184–191.
22. Tocci, A.M.; Mascheroni, R.H. Characteristics of Differential Scanning Calorimetry Determination of Thermophysical Properties of Meats. Lebensm. Wiss. Technol. **1998**, *31* (5), 418–426.
23. Pham, Q.T.; Wee, H.K.; Kemp, R.M.; Lindsay, D.T. Determination of the Enthalpy of Foods by an Adiabatic Calorimeter. J. Food Eng. **1994**, *21*, 137–156.
24. Riedel, L. Kalorimetrische Untersuchungen über das Schmelzverhalten von Fetten und Ölen. Fette, Seifen, Anstrichm. **1955**, *57* (10), 771–782.
25. Riedel, L. Kalorimetrische Untersuchungen über das Gefrieren von Seefischen. Kältetechnik **1956**, *8* (12), 374–377.
26. Riedel, L. Kalorimetrischen Untersuchungen über das Gefrieren von Fleisch. Kältetechnik **1957**, *9* (2), 38–40.
27. Fleming, A.K. Calorimetric Properties of Lamb. J. Food Technol. **1969**, *4*, 199–215.
28. Becker, B.R.; Fricke, B.A. Food Thermophysical Property Models. Int. J. Heat Mass Transfer **1999**, *26* (5), 627–636.
29. Heldman, D.R. Predicting the Relationship Between Unfrozen Water Fraction and Temperature During Food Freezing Using Freezing Point Depression. Trans. ASAE **1974**, *17*, 63–66.
30. Heldman, D.R.; Singh, R.P. *Food Process Engineering*; 2nd Ed.; AVI Publishing Company, Inc.: Westport, CT, 1981.
31. Heldman, D.R. Food Freezing. In *Handbook of Food Engineering*; Heldman, D.R., Lund, D.B., Eds.; Marcel Dekker, Inc.: New York, NY, 1992; 277–315.
32. Chen, C.S. Thermodynamic Analysis of the Freezing and Thawing of Foods: Enthalpy and Apparent Specific Heat. J. Food Sci. **1985**, *50*, 1158–1162.
33. Chen, C.S. Thermodynamic Analysis of the Freezing and Thawing of Foods: Ice Content and Mollier Diagram. J. Food Sci. **1985**, *50*, 1163–1167.
34. Schwartzberg, H.G. Effective Heat Capacities for the Freezing and Thawing of Food. J. Food Sci. **1976**, *41*, 152–156.
35. Bartlett, L.H. A Thermodynamic Examination of the "Latent Heat" of Food. Refrig. Eng. **1944**, *47*, 377–380.
36. Riedel, L. Eine Formel zur Berechnung der Enthalphie Fettarmer Lebensmittel in Abhängigkeit von Wassergehalt und Temperatur. Chem. Mikrobiol. Technol. Lebensm. **1978**, *5*, 129–133.
37. Chang, H.D.; Tao, L.C. Correlations of Enthalpies of Food Systems. J. Food Sci. **1981**, *46*, 1493–1497.
38. Succar, J.; Hayakawa, K. Empirical Formulae for Predicting Thermal Physical Properties of Food at Freezing or Defrosting Temperatures. Lebensm. Wiss. Technol. **1983**, *16* (6), 326–331.
39. Miki, H.; Hayakawa, K. An Empirical Equation for Estimating Food Enthalpy in a Freezing Temperature Range. Lebensm. Wiss. Technol. **1996**, *29* (7), 659–663.
40. Miles, C.A. The Thermophysical Properties of Frozen Foods. In *Food Freezing: Today and Tomorrow*; Bald, W.B., Ed.; Springer Verlag London Limited: London, 1991; 45–65.
41. Miles, C.A.; van Beek, G.; Veerkamp, C.H. Calculation of Thermophysical Properties of Foods. In *Physical Properties of Foods*; Jowitt, R., Escher, F., Hallström, B., Meffert, H.F.Th., Spiess, W.E.L., Vos, G., Eds.; Applied Science Publishers: London, 1983; 269–312.
42. Fikiin, K.A.; Fikiin, A.G. Predictive Equations for Thermophysical Properties and Enthalpy During Cooling and Freezing of Food Materials. J. Food Eng. **1999**, *40*, 1–6.
43. Cleland, A.C. *Food Refrigeration Processes: Analysis, Design and Simulation*; Elsevier Applied Science: London, 1990.
44. Hayakawa, K. Estimation of Heat Transfer During Freezing or Defrosting of Food. I.I.F-I.I.R-Commissions C1, C2 **1977**, *1*, 293–301.
45. Becker, B.R.; Fricke, B.A. Freezing Times of Regularly Shaped Food Items. Int. J. Heat Mass Transfer. **1999**, *26* (5), 617–626.
46. Succar, J. Heat Transfer During Freezing and Thawing of Foods. In *Developments in Food Preservation*; Throne, S., Ed.; Elsevier Applied Science Publishers: London, 1989; Vol. 5, 253–304.
47. Delgado, A.E.; Sun, D.-W. Heat and Mass Transfer Models for Predicting Freezing Processes—A Review. J. Food Eng. **2001**, *47*, 157–174.
48. Cleland, A.C.; Earle, R.L. Assessment of Freezing Time Prediction Methods. J. Food Sci. **1984**, *49*, 1034–1042.
49. Bakal, A.; Hayakawa, K. Heat Transfer During Freezing and Thawing of Foods. Adv. Food Res. **1973**, *20*, 217–256.

50. Cleland, A.C.; Earle, R.L. A Comparison of Analytical and Numerical Methods of Predicting the Freezing Times of Foods. J. Food Sci. **1977**, *42* (5), 1390–1395.
51. Wilson, H.A.; Singh, R.P. Numerical Simulation of Individual Quick Freezing of Spherical Foods. Int. J. Refrig. **1987**, *10*, 149–155.
52. Hsieh, R.C.; Lerew, L.E.; Heldman, D.R. Prediction of Freezing Times for Foods as Influenced by Product Properties. J. Food Process Eng. **1977**, *1*, 183–197.
53. Meryman, H.T. Review of Biological Freezing. In *Cryobiology*; Meryman, H.T., Ed.; Academic Press: London, 1966; 1–114.
54. Fennema, O.R. Freezing Preservation. In *Principles of Food Preservation*; Karel, M., Fennema, O.R., Lund, D.B., Eds.; Marcel Dekker, Inc.: New York, 1975; 173–215.
55. Singh, R.P.; Heldman, D.R. *Introduction to Food Engineering*, 2nd Ed.; Academic Press: San Diego, CA, 1993.
56. Meffert, H.F.Th. History, Aims, Results and Future of Thermophysical Properties Work Within COST 90. In *Physical Properties of Foods*; Jowitt, R., Escher, F., Hallström, B., Meffert, H.F.Th., Spiess, W.E.L., Vos, G., Eds.; Applied Science Publishers: London, 1983; 229–267.
57. Maake, W.; Eckert, H.J. *Pohlmann, Taschenbuch für Kältetechniker*, 15th Ed.; Müller, C.F., Ed.; Karlsruhe, 1971.
58. Woolrich, W.R.; Novak, A.F. Refrigeration Technology. In *Fundamentals of Food Freezing*; Desrosier, N.W., Tressler, D.K., Eds.; AVI Publishing Company, Inc.: Westport, CT, 1977; 23–80.
59. Heldman, D.R. Computer Simulation of Food Freezing Processes. Proc. IV Int. Congress Food Sci. and Technol.,1974; Vol. IV, 397–406.

Frozen Food Properties

Paul Nesvadba
The Robert Gordon University, Aberdeen, Scotland, United Kingdom

INTRODUCTION

The chain of events in the "life" of a frozen food may include the following processes: pretreatment (blanching for vegetables, osmotic impregnation, and partial dehydration for fruits), *freezing*, *storage* and display in the frozen state, tempering, cutting in the tempered state, *thawing*. The words in *italics* represent the processes that are always present, the other processes are optional.

All of these processes involve in some way the properties of frozen foods. The most important properties for frozen foods tend to be the physical properties because, ideally, freezing stops all biochemical changes. The biochemical changes do, certainly, take place and cause deterioration of the quality of the foods after a long time of storage, e.g., by oxidation of fats in the food. Many processes in frozen foods depend on one or more of these properties in an interactive way; for e.g., the growth and aggregation of ice crystals in frozen foods depend not only on thermal properties but also on mass diffusion, surface energy, and other physico-chemical properties. This article focuses on the physical properties of foods in the following categories: thermal, mechanical, mass diffusion, electrical, and optical. Thermal properties of foods are important for predicting the energies and speeds of freezing and thawing (and heat transfer in general). Similarly, all the other properties play a role in food processing.

THERMAL PROPERTIES OF FOODS

The Major Influence of Ice on the Thermal Properties of Foods

The thermal properties of frozen foods are dependent on their water content and temperature. This is because of the following reasons:

- Water and ice having large specific heat capacity and thermal conductivity in comparison with the remaining components of the food (protein, fat, and carbohydrate).
- The proportion of ice in the food increases as its temperature is lowered from the initial freezing point down to -30°C and lower.

In water-containing foods, the main features of the variation of the ice fraction with temperature are as follows:

- There is no ice present above the initial freezing point of the food (T_f).
- Below T_f the ice fraction gradually increases as temperature is lowered to -30°C.
- Below about -40°C the increase of the ice fraction levels off.
- As the temperature of a water-containing food is lowered from chill temperature (above 0°C) through the initial freezing point temperature, T_f, the first ice crystals appear only below T_f, due to supercooling. Water can be present as a liquid in a metastable state below T_f. When ice nucleation and growth occur (as discussed in section "Ice Crystal Kinetics") the temperature shoots up to T_f due to the release of the latent heat of crystallization.
- T_f is lower than 0°C (pure water) because in foods there are dissolved substances in the water, lowering the freezing point according to the Raoult's law.
- The crystals consist of pure water, therefore, the solution left behind becomes more concentrated and its freezing point is lowered still more (according to the Raoult's law). Therefore, more ice can appear only after a further lowering of temperature.
- Lowering the temperature to very low values does not convert all the water in the food to ice. About 5% of the water appears unfreezable. This may be partly due to the last water molecules being bound to the proteins in the food (a thermodynamic reason). However, a more likely explanation is that the mobility of the water molecules becomes so low that they move extremely slowly and do not reach the ice crystals within practical time frames (a kinetic reason). The unfrozen water solution appears to be in a nonequilibrium, glassy state.

Encyclopedia of Agricultural, Food, and Biological Engineering
DOI: 10.1081/E-EAFE 120007024
Copyright © 2003 by Marcel Dekker, Inc. All rights reserved.

For these reasons the general dependence of the ice fraction is as shown in Fig. 1.

The ice fraction sharply increases from zero at T_f, then increases more slowly, and reaches an asymptote given by the amount of unfreezable water.

When water starts to freeze at the initial freezing temperature (− 1°C in this example), the ice that is formed is pure (does not contain solutes). This makes the unfrozen solution more concentrated and the freezing stops if the temperature is at the new lower melting point. When the temperature is lowered further, more ice is formed, leaving behind a progressively more concentrated unfrozen water solution. Thus, unlike in pure water, the ice is formed over a temperature range that extends from the initial freezing point down to about − 40°C (in meat or fish). Even at these low temperatures about 5% of water still remains unfrozen. This is due to the speed of conversion of water into ice being almost zero because of vastly increased viscosity and decreased mobility of molecules. The unfrozen solution at very low temperatures is essentially in a glassy state, thermodynamically unstable, but for practical purposes kinetically stable.

Enthalpy of Frozen Foods

The enthalpy of frozen foods is related to the energy (heat) that a freezer has to remove in order to lower the temperature of the food from the unfrozen to the frozen state (Fig. 2). This amount of heat can be thought of as consisting of two parts: the sensible and the latent heat. The sensible heat can be "sensed" by measuring temperature because it is related to the energy stored in a substance by increasing the speed of its molecules. On the other hand the latent heat is related to the conversion of water into ice (and vice versa). In pure water at atmospheric pressure this phase transition occurs at a single fixed temperature (0°C) and requires 333 kJ/kg. In foods, the freezing point is depressed by typically 1°C, however, this depends on the concentration and molecular weight of solutes according to the Raoult's law.

Water, compared with other food components, has large latent heat and sensible heat (the specific heat capacity of water is about 4.2 kJ/kg/°C and of ice is 2.1 kJ/kg/°C). Thus the amount of water in a food has a large influence on its enthalpy. For foods such as fish or meat, containing typically 70–80% water (wet basis), the total enthalpy change when freezing from + 10°C to − 30°C is about 300 kJ/kg^{-1}. Fig. 2 shows that 200 kJ kg^{-1} of this change occurs between the initial freezing point and − 8°C. The knowledge of the dependence of enthalpy on the water content data is important for the design and operation (refrigeration load) of industrial freezers.

Thermal Conductivity

Ice is about four times more thermally conductive than water. Therefore the thermal conductivity of foods in the frozen state increases with decreasing temperature (as the ice content of the food increases). Fig. 3 shows this dependence in meat.

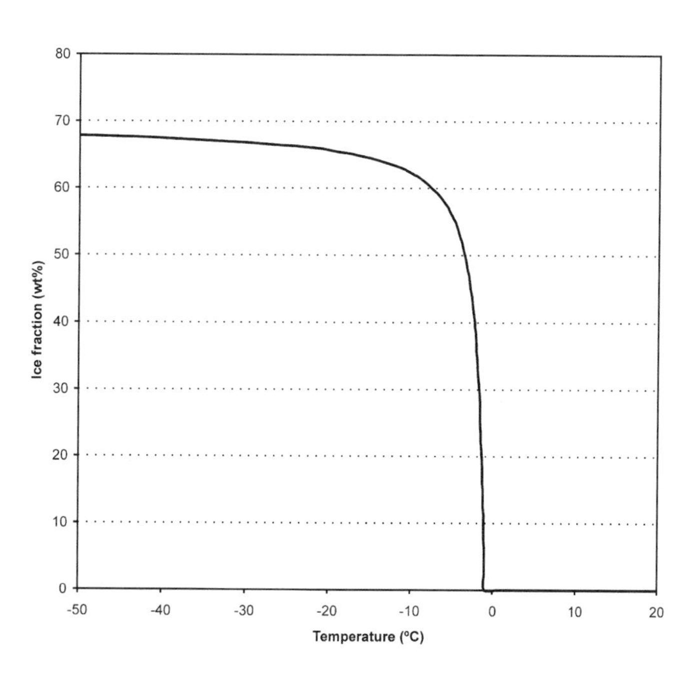

Fig. 1 Temperature dependence of the ice fraction in meat containing 74 wt% water.

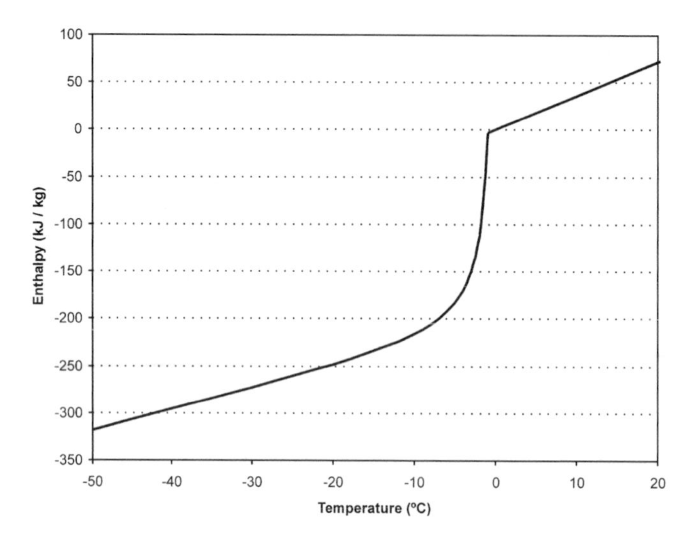

Fig. 2 Temperature dependence of enthalpy of meat containing 74 wt% water.

Lipids are poor conductors of heat and therefore fatty foods tend to have lower thermal conductivity than nonfatty foods of the same water content. This has to be taken into account during freezing of meat of fish. Especially in pelagic fatty fish there is a seasonal variation of the fat level and fish with high fat content may be incompletely frozen if the freezing is controlled by a constant set time of freezing.

Thermal conductivity, unlike the heat capacity, depends on anisotropy of the food. For example, the thermal conductivity of beef is about 15% higher when the heat flows across (rather than along) the muscle fibers.

Thermal conductivity also depends crucially on porosity. Air present in pores and cavities of foods (e.g. bread) reduces the thermal conductivity. This shows the crucial importance of structure for thermal conductivity of foods.

Measuring and Predicting Thermal Properties of Foods

The measurement of thermal properties, especially of thermal conductivity of foods is difficult. This is because of the strong temperature dependence of the thermal properties of frozen foods, their inhomogeneous and anisotropic nature and their perishability. For these reasons new measurement methods are being developed that are transient and therefore rapid and use small temperature gradients thereby minimizing the problems with temperature dependence of the properties and the migration of moisture induced by thermal gradients.[1]

Because of the difficulties of measurement, food engineers seek to predict the thermal properties whenever possible. A computer program COSTHERM uses theoretical and semi-empirical thermodynamic equations to predict the thermal properties of foods from their composition. The accuracy of these predictions is about 10%, sufficient for most food engineering applications. The early version of the software was developed under the EU collaborative project COST90[7] and later improved in another EU project (CT94-0240), by adding new models for predicting thermal conductivity and the freezing point.[2]

MECHANICAL PROPERTIES OF FROZEN FOODS

There are many more mechanical properties than thermal properties[3] that are important in frozen foods. The most basic of these is density, but there are many other mechanical properties, some of them difficult to quantify, such as the force of adhesion.

The Force of Adhesion

Aggregating small pieces of fish into larger fish “fillets” relies on the adhesion between frozen pieces and strengthening by enrobing in breadcrumbs or batter. The adhesion of frozen foods to molds (ice lollies) and conveyor belts is another technologically important example of a mechanical property. Within a certain temperature range (approximately from $-15°C$

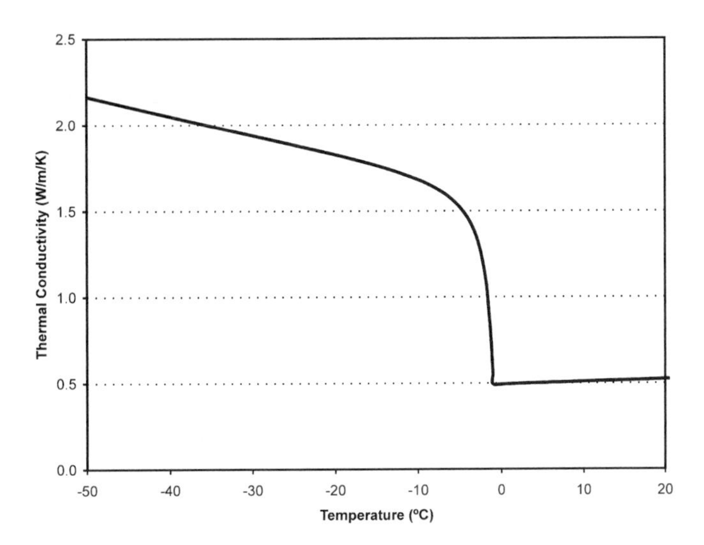

Fig. 3 Temperature dependence of thermal conductivity of meat containing 74 wt% of water.

to −50°C) the adhesion is very strong and the food (ice lolly) cannot be released from the mold. However, lowering the temperature to −60°C releases the ice lolly.

The yield strength is important for producing frozen blocks by cutting, sawing, and guillotining of frozen foods into smaller portions (for e.g. guillotining frozen blocks of fish into fish fingers). In high-speed slicing of ham the temperature of the meat is of importance. If it is too low, there is too much ice in the meat and the cutting blade gets blunt quickly because or breaks. If the temperature is too high, the blade tears the meat, pulling bits of it out of the slices. The correct temperature is obtained by tempering.

These are just some examples of the interaction between the mechanical and thermal properties.

Density

Density is a basic physics/engineering property that enters the design and control of both thermal and mechanical processes. Again, as for other properties, the ice fraction in foods plays a dominant role in variation of density with temperature. This is because ice has a larger specific volume than water and the proportion of ice in the food increases with decreasing temperature. Therefore the density of frozen foods strongly increases with decreasing temperature and the foods expand as temperature decreases. Fig. 4 shows the variation of density with temperature of frozen meat.

The increase in density of frozen food with decreasing temperature has implications for situations where the frozen food is constrained in one direction, as in between the plates of a plate freezer. The food must be allowed to expand in order to avoid damage. Nonuniform expansion during rapid freezing (in liquid nitrogen) leads to thermal stresses that can crack the food.

Ultrasonic Attenuation

Frozen meat absorbs ultrasound more strongly than unfrozen meat. Ultrasound is therefore better suited for thawing than microwaves. The attenuation of ultrasound increases with temperature, reaching a maximum near the initial freezing point of the food.[4] The ultrasonic energy is dissipated predominantly at the frozen/thawed boundary (the mushy region) because of the compressibility of ice and the friction between the crystals and the food matrix. This offers the possibility of ultrasonic thawing of foods[9] if the frequency is chosen in an optimum range, around 500 kHz. Frequencies lower than 430 kHz cause cavitation and loss of efficiency on transmitting the ultrasonic energy into the food. Ultrasonic waves with frequencies higher than 740 kHz are strongly absorbed in the unfrozen meat, again reducing the efficiency of thawing.

Apart from the application of ultrasonic power to thawing of foods, ultrasound can be used in sensors for estimating the ice fraction in foods and tracking the position of the frozen/unfrozen boundary.

ELECTRICAL PROPERTIES OF FROZEN FOODS

Electrical properties of frozen foods are important in Ohmic and microwave thawing. The d.c. and low-frequency electrical conductivity depend mainly on

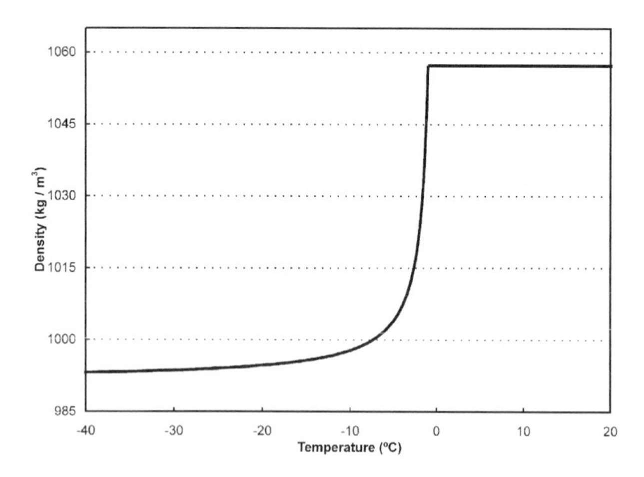

Fig. 4 Temperature dependence of the density of meat containing 74 wt% water.

the salt content (ionic conduction). At high frequencies (915 and 2450 MHz used in industrial and domestic microwave ovens) the dipole relaxation of water increases. The dielectric losses can thus exceed those to ionic conduction. The dielectric loss in ice is lower than that in water. This causes undesirable nonuniform heating during microwave thawing. The thawed layer absorbs more energy than the frozen part, necessitating a reduced rate and a close control of the process. From this point of view ultrasonic thawing offers a decisive advantage.

ICE CRYSTAL KINETICS

Ice Nucleation and Growth

Ice crystals are formed by nucleation and growth. The nucleation is always heterogeneous (rather than homogeneous), starting at sites at which the molecules of water are most easily arranged into the crystal structure of ice. The nucleation can be assisted by providing sites that closely resemble the structure of ice, most notably ice nucleation active (INA) bacteria such as *Pseudomona syringae* present on lilac plants and affording a degree of protection against frost damage. On the other hand ice nucleation can be suppressed using anti-freeze proteins, first observed in sea animals in the arctic waters.

The speed of nucleation and growth depends on the temperature difference between the food and the external cooling medium. Generally the faster the cooling the finer are the ice crystals formed. This has implications for the structural damage of the animal or plant tissues. Ice has a larger specific volume than water and therefore freezing causes internal stresses in the food. Large crystals distend the tissue locally more severely and cause a greater damage resulting in increased drip loss on thawing.

Small ice crystals tend to aggregate into larger ones during frozen storage (Ostwald ripening), especially if the temperature is not maintained below $-18°C$ or is fluctuating. For example, the size of 25 μm ice crystals may double in 24 days at $-18°C$ and in 24 hr at $-5°C$. Such growth can be arrested using anti-freeze proteins. They prevent the formation of ice by binding to specific sites within the ice crystal lattice. This lowers the freezing point by only 1 or 2 degrees, however, it strongly reduces the surface energy and the rate of growth at the surface of an ice crystal. These proteins occur in arctic fish in some plants and genetic engineering could provide further recrystallization resistant cultivars.

It has been known since the early 1900s that the application of sound waves to a supercooled solution will cause the nucleation of ice crystals. The explanation of this phenomenon is that any small bubbles within the solution are forced to oscillate by the sound wave.[5,6] During the compression stage of the oscillation pressures of several kilobars are generated in a small shell at the edge of the bubble leading to nucleation of ice crystals. Such a technique, therefore, offers the potential of growing small ice crystals throughout the food.

Ice-nucleating lipoproteins offer another possible way to control nucleation and growth of ice crystals. These proteins have been isolated in bacteria expressing these proteins (ice nucleation active bacteria—INA) such as *Xanthomonas campestris* and also from cold-blooded animals such as frogs. In plants the lipoproteins act as seeds for the formation of ice crystals and allow organisms to determine where ice crystal formation will occur. This limits the damage and allows healing when the plant thaws out. It is possible that genetically modified crops will be found, which express the proteins that offer increased frost resistance. The nucleating agents have also found commercial use in the freeze concentration of liquid foods.

Glass Transition Temperature

Ice cream is a major food product in the developed world and its structure and texture critically depend on the physical chemistry of freezing such as colloid stability, emulsification, foaming, amount of trapped air, and ice crystal size. The stability of the ice phase in frozen foods (ice crystallization/recrystallization/Ostwald ripening) influences the storage life of frozen foods. In the 1980s the concept of glass transitions started to be used in understanding the storage stability of frozen foods. The changes in frozen foods during storage are thus governed by kinetic as well as thermodynamic properties and processes. Various additives such as polysaccharides influence the storage stability. In the 1990s, interest also arose in anti-freeze proteins, initially found in arctic fish and having the property of protecting the animal from freezing even at -3 or -4°C.

Pressure Shift Freezing and Thawing

The properties of the ice–water system are unusual in that the melting temperature initially decreases when the pressure is raised. This means that it is possible to increase the pressure on a food and then cool it to around -20°C without the transition from water to ice occurring. If the pressure is then rapidly removed, a fraction of the water changes to ice in the form of very small crystals in the food. If the remaining latent heat is removed rapidly the resultant ice crystal structure is a matrix of much smaller crystals than those obtaining in traditional freezing techniques. The process is known as pressure shift freezing.

CONCLUSIONS

On the macroscale, the physical and engineering properties of frozen foods play an important role in designing and controlling technological processes. An online database at *http://www.nelfood.com* aims to facilitate ready availability of data.

On the meso- and microscale, understanding of the kinetics of the formation and growth of ice crystals as influenced by newly emerging organic molecules gives great possibilities for improving the yield of agricultural crops and quality and variety of food products.

REFERENCES

1. Nesvadba, P. Methods for the Measurement of Thermal Conductivity and Diffusivity of Foodstuffs. J. Food Eng. **1982**, *1*, 92–113.
2. Miles, C.A.; Mayer, Z.; Morley, M.J.; Houška, M. Estimating the Initial Freezing Point of Foods from Composition Data. Int. J. Food Sci. Technol. **1997**, *32*, 389–400.
3. Houška, M.; Nesvadba, P.; Mayer, Z. Database of Mechanical Properties of Foods: Subgroup of Mechanical and Rheological Properties. J. Texture Stud. **2001**, *32* (2), 155–160.
4. Shore, D.; Woods, M.O.; Miles, C.A. Attenuation of Ultrasound in Post Rigor Bovine Skeletal Muscle. Ultrasonics **1986**, *24*, 81–87.
5. Hickling, R. Nucleation of Freezing by Cavity Collapse and Its Relation to Cavitation Damage. Nature **1965**, *206*, 915–917.
6. Kennedy, C.J. Formation of Ice in Frozen Foods and Its Control by Physical Stimuli. In *The Properties of Water in Foods ISOPOW 6*; Reid, D.S., Ed.; Blackie A & P: London, 1998.
7. Miles, C.A.; van Beek, G.; Veerkamp, C.H. Calculation of Thermophysical Properties of Foods. In *Physical Properties of Foods*; Jowitt, R., Escher, F., Hallstrom, B., Meffet, H. Th., Spiess, W.E.L., Vos, G., Eds.; Applied Science Publishers: London, 1983; 269–312.
8. Watanabe, M.; Arai, S. In *Application of Bacterial Ice Nucleation in Food Processing* in *Biological Ice Nucleation and Its Applications*; Lee, R.E., Warren, G.J., Gusta, L.V., Eds.; The American Phytopathological Society: St Paul, U.S.A., 1995; 299–313.
9. Miles, C.A.; Morley, M.J.; Rendell, M. High Power Ultrasonic Thawing of Frozen Foods. J. Food Eng. **1999**, *39*, 151–159.

Frozen Food Thawing

Edward Kolbe
Oregon State University, Portland, Oregon, U.S.A.

INTRODUCTION

Once foods are packaged, quickly frozen, and maintained at a low uniform temperature, they will retain an excellent level of quality for a long time. There are many links along the "cold chain"; however, that can diminish not only the food's quality, but also its safety as well. One of these is the process of thawing, which often proceeds at a rate markedly slower than freezing. For most conventional thawing processes, heat flows in through outer surfaces. Emerging technologies are producing faster results through various methods of internal heating. All seek to speed up process rates while maintaining high levels of quality and safety.

THE THAW PROCESS

The problem: time required to thaw can be far longer than that required to freeze. This is because the thermal conductivity, and thus the rate of heat transfer, through an unfrozen layer can be less than one third of the conductivity and heat flow rate through an equivalent frozen layer, given the same temperature difference.

Fig. 1 describes this situation for an example block of frozen food. For the freezing block (a) whose vertical surfaces are depicted in contact with a cold freezer plate, the heat flux *Q/A* (heat flow rate per unit area) through the outer surface is described by the equation

$$Q/A = -k\frac{\mathrm{d}T}{\mathrm{d}x} \approx -\frac{k\Delta T}{\ell} \qquad (1)$$

where ℓ is the thickness of the outer layer; k is the thermal conductivity; and ΔT is the temperature difference between surfaces of the outer layer.

At this moment in the freezing process, the temperature difference driving the heat flow through the outer layer is 13°C, roughly the difference between the initial freezing point (−2°C) and the surface temperature, currently − 15°C (Fig. 1a). Meanwhile, the same momentary heat flux ($\mathrm{W\,m^{-2}}$) occurs at the external boundary, as heat flows to the cold sink according to the equation:

$$Q/A = h(T_{\mathrm{surf}} - T_{\mathrm{amb}}) \qquad (2)$$

where h is the surface heat transfer coefficient; and T_{surf} is the surface temperature; T_{amb} is the ambient temperature.

Fig. 1b depicts the same block undergoing thawing in warm (26°C) air. At the moment shown, all of the same temperature differences are in effect to drive the heat flux. But note that now the flux through the outer layer will be substantially lower because the thermal conductivity of the thawed layer will be close to one-third that of the frozen layer (Table 1). Thus, given the same driving temperature differences, thawing will be far slower than freezing.

The 11°C (and warming) surface temperature depicted in this thawing example highlights another potential problem beyond that of a slow production rate: surface damage by high temperature microbial or enzymatic spoilage. Conventional thawing technologies which transfer heat in from the product's outer surface, face a dilemma. Raising environment temperature or heat transfer coefficient will speed up thawing (and production) rates with increased heat flux, as Eq. 2 shows. But such increases will then raise surface temperature (Eq. 1) and worsen degradation of quality and safety.

One consequence of prolonged exposure of the food to a warm environment is *quality damage*. Examples include color changes in red meats, surface desiccation (drying) of fruits and vegetables, loss of nutrients, and flavor components in all foods. The *rate* of thawing, as well, can influence quality attributes. Although rapid thaw rates may be good for berries and pastries, they may have detrimental effects on meat, fish, and poultry, due to drip loss or toughened texture.[6] One explanation for this: as water crystals thaw under slow conditions, there will be more time for melt water to be reabsorbed by the muscle fibers.[7]

Because the growth rate of micro-organisms increases with time and temperature, *food safety* also becomes a concern during the thaw process. General thawing guidelines suggest that surrounding air or water temperature be limited to 18 or 20°C.[8,9] Some chilled storage guidelines, however, instruct that surface temperatures be held at lower values such as 7.2°C.[10] All food safety issues must be evaluated in view of the food product, time–temperature exposure, pathogen hazards, local regulations, and other factors.

Encyclopedia of Agricultural, Food, and Biological Engineering
DOI: 10.1081/E-EAFE 120007031
Copyright © 2003 by Marcel Dekker, Inc. All rights reserved.

F

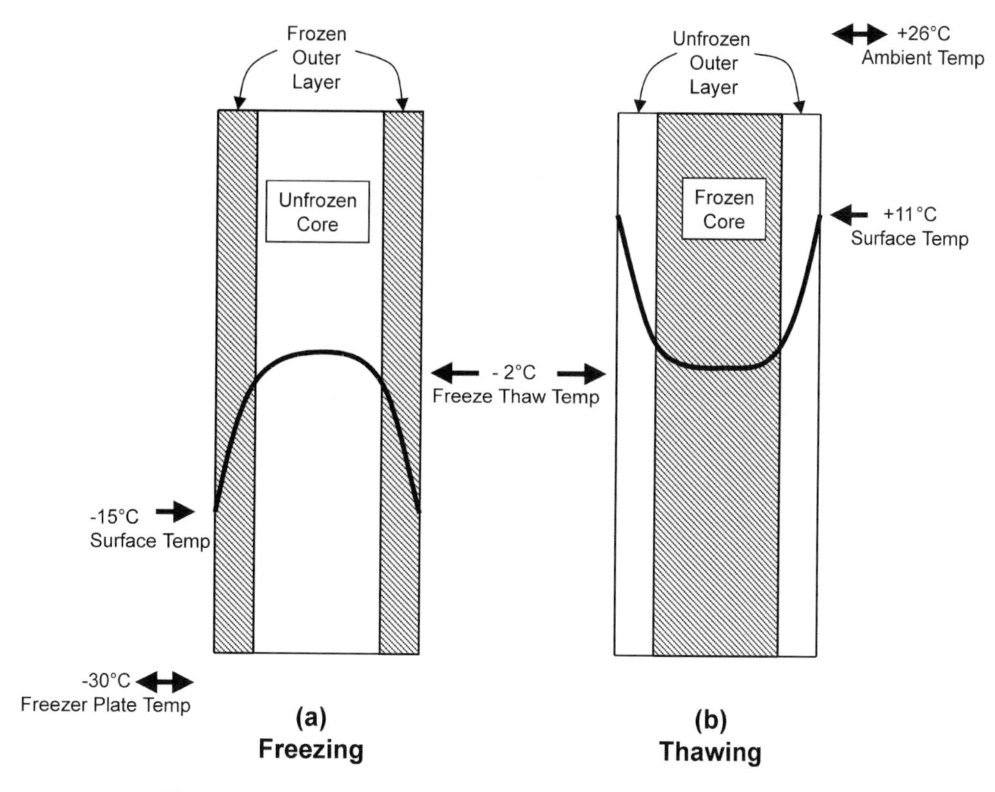

Fig. 1 Simulated temperature profiles for freezing and thawing blocks.

CONVENTIONAL TECHNOLOGIES—SURFACE HEATING

Common industrial thawing systems use air, water, or plates to supply heat to the product's surface. Fig. 2 describes the basic components of a batch thawer for fish blocks. Heat is supplied from electric or steam-heated grids, as well as from the fans which can account for as much as 30% of the heat energy.[8] Air velocity and direction, flow uniformity, temperature, and relative humidity can all be controlled in this system. One alternative to the air-flow system shown employs a vacuum chamber into which moisture or steam is injected. Without much air, this vapor can flow into spaces in and around the frozen food, releasing a very high "heat of vaporization" as it condenses on the food surface.

Thawing with flowing water is a second common method. It requires a bath or spray of clean or treated circulated water; a high heat transfer coefficient will result from water agitated by pumps, sprayers, or bubbled air. Rapid rates of thaw can occur with water temperatures to 20°C, depending upon product quality and safety issues.

A third method of surface heating is by contact plates, a system suitable for frozen blocks or packages of regular shape. Hollow plates containing controlled-temperature circulating liquid, contact parallel surfaces of the frozen samples. Contact can be uneven, making it difficult to

Table 1 Some thermal conductivity values for frozen and thawed foods

	Thermal conductivity (W [mC] $^{-1}$)		
Product	**Thawed**	**Frozen**	**Reference**
Cod (83% moisture)	0.53	1.46 @ − 15°C	1
White fish	0.43–0.58	1.87 @ − 30°C	2
Orange juice (89% moisture)	0.55	1.15 @ − 25°C	3,4
Lean beef (79% moisture)	0.43	1.43 @ − 15°C	1
Chicken white meat (74% moisture)	0.47	1.37 @ − 25°C	5

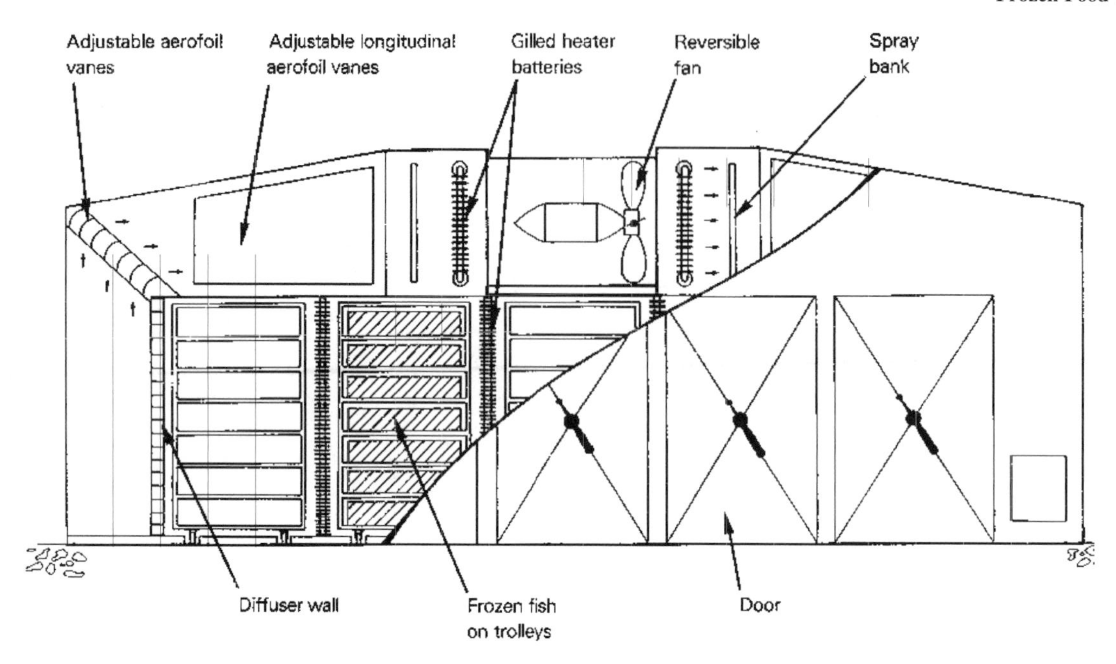

Fig. 2 Diagram of a recirculated humid air thawer for frozen fish blocks.[22] [Reprinted by permission from Her Majesty's Stationary Office, UK.]

control product temperature. These units appear to be relatively less common commercially.

EMERGING TECHNOLOGIES—VOLUMETRIC EFFECTS

The processes noted above all conduct heat from the outer surfaces; this can take a long time. To address this problem and to seek ways to conserve energy and, in some cases, fresh water, other *volumetric* heating methods have been explored. Some are used commercially.

In *ohmic heating*, electrodes directly contacting the frozen food conduct electrical current which generates heat by electrical resistance (I^2R) heating. The electrical conductivity of water is far greater than that of ice, making "thermal runaway" a potential problem. One research group has addressed this by using a grid of electrodes, each with its own controller to regulate local voltage.[11] AC current at 60 Hz is common with ohmic systems, although frequencies above 5 kHz appear to virtually eliminate corrosion of metal electrodes.[12]

With radio frequencies (*RF heating*) in the 20 MHz–80 MHz range, it is possible to internally thaw frozen foods without direct electrode contact. Thus packaged foods held between plate electrodes could be thawed about four times faster than when thawing in water.[13] Absorption of heat is by both resistance heating and by capacitive dielectric heating of the dipolar molecules. Thermal runaway of thawed sectors can be a problem that could be minimized with RF field control, a subject of current research.

Microwaves can penetrate and internally thaw frozen foods at the higher commercial frequencies of 915 MHz and 2450 MHz.[14] In these systems it is the water molecule that resonates and internally generates heat. As with household microwave thawing, penetration (non-uniformity) and thermal runaway are common problems, also being addressed by current research on cavity and control system design. Industrial microwave ovens are commonly used to "temper" frozen foods, i.e., heat them to a temperature just below freezing, when a fraction of the water just begins to change phase.

Past research has investigated the absorption of acoustic energy as well to accelerate thawing. Both low frequencies of 1500 Hz[15] and higher frequencies of 500 kHz[16] have effectively accelerated thawing.

In addition, researchers are now exploring the use of high hydrostatic pressures.[17] When applied at the 2000 atm–3000 atm level, ice turns immediately to water at a low temperature approaching − 20°C. Such a rapid thaw has the advantage of minimizing thaw damage[18] and reducing the heat load needed to bring already-thawed food to final chilled-stored temperature.

MODELING AND PREDICTION

Simulation models enable process designers to evaluate expected thawing without performing a series of expensive experiments. The literature describes many numerical models that can simulate both freezing and surface thawing of foods. The accuracy depends in large part on the knowledge of food properties and heat transfer coefficients. The thaw models of Cleland and others[19] are formulas that have proven most useful for those somewhat familiar with engineering parameters. A more user-oriented program, developed at UC Davis,[20] can be purchased from the World Food Logistics Organization (Bethesda, Maryland; email@wflo.org).

Other mathematical models have been under development in recent years to simulate the volumetric heating by ohmic, RF, and microwave energy. Commercial options for microwave heating are currently under review by the Industrial Microwave Modeling Group, Worcester Polytechnic Institute.[21]

ACKNOWLEDGMENTS

This publication was funded by the National Sea Grant College Program of the U.S. Department of Commerce's National Oceanic and Atmospheric Administration under NOAA grant number NA76RG0476, and by appropriations made by the Oregon State legislature.

REFERENCES

1. ASHRAE. *Fundamentals Handbook*; American Society of Heating, Refrigerating, and Air Conditioning Engineers: ASHRAE, Atlanta, 1985.
2. Johnston, W.A.; Nicholson, F.J.; Roger, R.; Stroud, G.D. Freezing and Refrigerated Storage in Fisheries. FAO Fisheries Technical Paper 340; Food and Agriculture Organization of the United Nations: Rome, 1994.
3. Heldman, D.R.; Singh, R.P. Thermal Properties of Frozen Foods. In *Physical and Chemical Properties of Food*; Okos, M.R., Ed.; American Society of Agricultural Engineers: St. Joseph, MI, 1986; 120–137.
4. Choi, Y.; Okos, M.R. Thermal Properties of Liquid Foods—Review. In *Physical and Chemical Properties of Food*; Okos, M.R., Ed.; American Society of Agricultural Engineers: St. Joseph, MI, 1986; 35–77.
5. Sweat, V.E.; Haugh, C.G.; Stadelman, W.J. Thermal Conductivity of Chicken Meat at Temperatures Between − 75 and 20°C. J. Food Sci. **1973**, *38* (1973), 158–160.
6. Jul, M. *The Quality of Frozen Foods*; Academic Press: London, 1984.
7. Calvello, A. Recent Studies on Meat Freezing. In *Developments in Meat Science–2*; Lawrie, R., Ed.; Applied Science Publishers: U.K., 1981; Chap. 5, 125–158.
8. Merritt, J. Guidelines for Industrial Thawing of Ground Fish in Air and in Water. Report IDD 109; Nova Scotia Department of Fisheries: Halifax, 1993.
9. FAO (Food and Agriculture Organization). Proposed Draft Code of Practice for Fish and Fishery Products. Agenda Item 4, CX/FFP 004; Codex Alimentarious Commission, United Nations: Rome, 2000; 81–82.
10. CFR (Code of Federal Regulations). Food and Drug Administration. Title 21. Part 110. Current Good Manufacturing Practice in Manufacturing, Packing, or Holding Human Food. Revised. 2001.
11. Roberts, J.S.; Balaban, M.O.; Zimmerman, R.; Luzuriaga, D. Design and Testing of a Prototype Ohmic Thawing Unit. Comput. Electronics Agric. **1998**, *19* (2), 211–222.
12. Zhao, Y.; Kolbe, E. A Method to Characterize Electrode Corrosion During Ohmic Heating. J. Food Process. Eng. **1999**, *22*, 81–89.
13. Jason, A.C. Thawing Frozen Fish. Torry Advisory Note No. 25 (Revised); Torry Research Station, Ministry of Agriculture, Fisheries and Food: Aberdeen, 1974.
14. Ohlsson, T. Low Power Microwave Thawing of Animal Foods. In *Thermal Processing and Quality of Foods*; Zeuthen P., Cheftel, Eriksson, Jul, M., Leniger, Linko, Varela, Vos, Eds.; Elsevier: New York, 1984; 579–584.
15. Kissam, A.D.; Nelson, R.W.; Ngao, R.; Hunter, P. Water Thawing of Fish Using Low Frequency Acoustics. J. Food Sci. **1981**, *47* (1), 71–75.
16. Miles, C.A.; Morley, M.J.; Randell, M. High Power Ultrasonic Thawing of Frozen Foods. J. Food Eng. **1999**, *39* (2), 151–159.
17. Knorr, D.; Schlueter, O.; Heinz, V. Impact of High Hydrostatic Pressure on Phase Transitions of Foods. Food Technol. **1998**, *52* (9), 42–45.
18. Zhao, Y.; Flores, R.A.; Olson, D.G. High Hydrostatic Pressure Effects on Rapid Thawing of Frozen Beef. J. Food Sci. **1998**, *63* (2), 272–275.
19. Cleland, A.C. Thawing Time Prediction. *Food Refrigeration Processes*; Elsevier Applied Science: London, 1990; Chap. 7, 137–152.
20. Mannapperuma, J.D.; Singh, R.P. A Computer-Aided Method for the Prediction of Properties and Freezing/Thawing Times of Foods. J. Food Eng. **1989**, *9* (4), 275–304.
21. Yakovlev, V.V. Examination of contemporary electromagnetic software capable of modeling problems of microwave heating. *Proc. 8th AMPERE Conference on Microwave and RF Heating (Bayreyth, Germany, September 2001)*, pp. 19-20; [full version (http://users.wpi.edu/~vadim/8th-AMPERE-Paper_Full-version.pdf) in: Advances in Microwave and High-Frequency Heating, Springer Verlag, 2002 (to be published)].
22. Jason, A.C. Thawing. In *Fish Handling and Processing*; Aitken, A., Mackie, I.M., Merritt, J.M., Windsor, M.L. Eds.; Her Majesty's Stationary Office: Edinburg, 1982; Chap. 9.

Frozen Foods Shelf Life

David S. Reid
University of California, Davis, Davis, California, U.S.A.

INTRODUCTION

What is meant by the term "shelf life"? The shelf life of a food can be defined as the time during which all of its primary characteristics remain acceptable for consumption. Many food items are highly perishable materials, and are subject to continual change with time. Some of these changes are accompanied by a reduction in the acceptability of the material for consumption. However what is meant by "acceptability for consumption" is a complex question, to which most intuitively know at least a partial answer, and yet which is a challenging problem if one is asked for a comprehensive definition.

OVERVIEW

Clearly, the concept of acceptability for consumption encompasses safety and nutritional characteristics of the food, factors which are often subject to legislative definition. The concept also encompasses sensory characteristics of the food, such as taste, odor, texture, and color. Additional factors may also be of importance for particular food products. For the same food product, the factors, which define suitability for consumption, may vary from culture to culture and from geographic area to geographic area. When some critical character fails to meet the acceptability criterion, then the food is deemed to have reached the end of its shelf life. In order to predict shelf life, one must first identify the critical factors. Identification of the critical factors is the essence of the challenge. Realize that for any given product, the critical factors that define acceptable storage life may change dependent upon the exact conditions of storage.

The changing characteristics of foods with time are primarily the consequences of combinations of chemical and physical processes. It has long been known that the rates of chemical and physical processes are influenced by temperature. It has been observed that most processes proceed more rapidly at higher temperatures, and more slowly at lower temperatures. Other factors, too, have been found to influence the rapidity of a process. For example, in order for certain changes to take place, the molecules that generate the change must come into close proximity. This may be more likely if the concentrations of these molecules are higher.

Given the influence of temperature on rates of change, it is not surprising that it has been found that, the application of refrigerated or frozen storage can extend the shelf life of foods. Most of us are aware of this, and we utilize refrigeration for food storage every day, in the form of domestic refrigerators. In what we term refrigerated storage, while the temperature of storage has been lowered from ambient, there has been no significant change in the state of the food. However, if temperature is lowered too much, freezing may occur. We then enter the realm of frozen foods as distinct from refrigerated foods. The phase change from water to ice within the food has important consequences.

If we consider refrigerated foods, it is possible often to estimate the potential shelf life at any refrigerated storage temperature of interest through knowing its shelf life at higher temperatures. A simple extrapolation suffices. This is not necessarily the case for frozen foods. The presence of ice in a frozen food changes the characteristics of the food, and this may significantly influence the storage characteristics. At the extreme, some foods are found to be so damaged, they are entirely unsuited for frozen storage, and hence essentially have a zero shelf life. An example of this is lettuce. While lettuce stays for a longer time at refrigerator temperatures than it does at higher temperatures, if it becomes frozen, the characteristics change so markedly that on thawing it becomes immediately unacceptable. The crisp textures of the leaf are lost. The texture loss is a consequence of the formation of ice.

For frozen foods it is therefore necessary to determine that the freezing process of itself does not lead to unacceptable quality change. For some materials, it has been found that, special treatment is required to prevent such change. An example of this is the blanching heat treatment of vegetables, which prevents the color and flavor changes that can result from the uncontrolled enzyme action associated with the freezing of many types of unblanched vegetable tissues. The thermal effect of blanching inactivates the enzymes. The other changes associated with the partial heat treatment of blanching are found to be more acceptable than the changes associated with the uncontrolled enzymatic processes. Blanching, however, is not an acceptable treatment if the tissue is required to retain the characteristics of the fresh tissue. There are many discussions of the factors which influence quality change as a consequence of the freezing process.

Encyclopedia of Agricultural, Food, and Biological Engineering
DOI: 10.1081/E-EAFE 120007034
Copyright © 2003 by Marcel Dekker, Inc. All rights reserved.

This article will assume that any frozen food product has been produced according to good manufacturing practices, and is suited for freezing and frozen storage.

Assuming that the freezing process per se does not lead to an unacceptable product, the shelf life of a frozen food usually increases as the temperature is lowered, so long as the storage temperature is below about 20°F. This parallels the behavior of refrigerated foods, and reflects the same dependence of reaction rates upon temperature. At 0°F product shelf-lives are attained that are sufficiently long to make this an effective and economic storage temperature, though temperatures below 0°F lead to even longer storage lives, and to maintenance of high quality for a significantly longer time.

A primary purpose of low temperature (0°F or less) storage is to achieve extended shelf life. As with refrigerated (38–42°F) storage, it would be useful to be able to predict storage life at lower temperatures from measurements made at higher temperatures. Unfortunately, the procedure mentioned earlier that is so successful in refrigerated storage, fails if applied to frozen storage. There is not a simple, extrapolatable relationship between shelf lives at higher (frozen) temperatures and lower (frozen) temperatures. Though the same relationships of reaction rates with temperature still hold, another factor has also to be considered, namely the formation of ice. As ice forms, the concentration of the unfrozen aqueous phase increases. Reaction rates depend upon both temperature and concentration. Above freezing, concentration does not change, and so only temperature need be considered. Below about 20°F the relative change in concentration of the unfrozen aqueous phase is small, and so temperature is again the only significant factor. In the temperature range between the freezing point (close to 32°F) and 20°F, the change, with temperature, of relative concentrations of the unfrozen aqueous phase is large. The result is that the overall observed rate of reaction in this range of temperature may increase, stay relatively constant, or may decrease, depending on the specific system. As a result of this, there is no generally applicable system to estimate low temperature storage life from measurements at higher temperatures. For frozen foods, the only sure way to estimate shelf life at low temperatures is to perform a storage experiment at the temperature of interest. On the basis of many such storage life determinations, an important lesson has been learned. It is important that frozen storage temperatures not be allowed to rise above 0°F for any length of time, as the result is a marked reduction in quality, and therefore significant shortening of shelf life.

Another concept that must be kept in mind when considering the shelf life of a frozen stored product takes us back to the original discussion of the factors that define acceptable. Assume that there are two factors that, if changed, might define the end of acceptability. Each factor will have a characteristic rate of change, and also a characteristic temperature dependence of this rate of change. It is entirely possible that at one temperature the change in factor A defines the end of acceptable life, and at another temperature it is the change in factor B that defines the shelf life. In the final reckoning, shelf life is defined by consumer acceptance.

CONCLUSION

Frozen foods have an extended shelf life as long as storage is maintained at 0°F or below. Shelf lives in excess of one year are not uncommon. Storage at temperatures higher than 0°F leads to a significant reduction in shelf life. In this respect, it is important to realize that the simple act of exposing a frozen food to temperatures above 32°F for even a short time leads to internal temperatures around 20–26°F due to the high thermal diffusivity of ice compared to liquid water.

FURTHER READING

Erickson, M.C., Hung, Y.C., Eds.; *Quality of Frozen Foods*; Chapman, Hall: New York, 1997.

Jul, M. *The Quality of Frozen Food*; Academic Press: London, 1984.

Mallett, C.P., Ed.; *Frozen Food Technology*; Blackie: London, 1993.

Taub, I.M., Singh, R.P., Eds. , *Food Storage Stability*; CRC Press: Boca Raton, FL, 1998.

Fruit Storage Systems

Martin L. Hellickson
Oregon State University, Corvallis, Oregon, U.S.A.

INTRODUCTION

The global objective of fruit storage is to provide conditions that extend the postharvest period products can be marketed while minimizing losses in quality and quantity. Research has shown that specific fruits respond best to storage in environments optimized to each variety's physiological needs. Some fruit varieties can be successfully stored for only a few hours or days after harvest, while others can retain high quality for many months.

Modern refrigeration systems provide the ability to rapidly establish and precisely maintain proper temperature control plus create air circulation and humidity conditions appropriate for the stored product. Fruit varieties that have short postharvest viability are most commonly held in refrigerated air or "regular atmosphere" storages. Fruit varieties that exhibit relatively long postharvest viability are frequently stored in "controlled atmosphere" facilities. Controlled atmosphere storages combine sophisticated refrigeration system and room atmosphere control technologies to establish environmental conditions that can maximize both fruit quality retention and energy conservation. Accomplishing these two objectives requires careful evaluation of initial fruit quality, rapid cooling of the fruit to storage temperature, and efficient establishment and precise maintenance of controlled atmosphere storage conditions. Depending upon the product being stored, controlled atmosphere conditions typically include reduced oxygen and increased carbon dioxide levels plus high relative humidity levels that are precisely maintained throughout the storage period.

The type and capacity of refrigeration system present in a storage facility has significant effect on fruit cooling rates and room humidity development. Management of the refrigeration system directly affects fruit quality, temperature management throughout the room, and maintenance of humidity, oxygen and carbon dioxide levels, room pressure, and energy use.

REFRIGERATION SYSTEMS

Refrigeration systems most frequently used in modern fruit storage warehouses are Freon, ammonia, and indirect glycol systems. Freon systems use either R-12 or R-22 (or their replacements) as the primary refrigerant. These refrigerants typically completely change from a liquid/gas mixture entering the evaporator unit to all vapor at the exit. An evaporator pressure regulator controls the refrigerant pressure and temperature inside the evaporator unit. These direct expansion systems operate with a greater temperature difference between the entering liquid/gas mixture and the exiting vapor than either the ammonia or glycol systems.

Ammonia systems operate such that the refrigerant enters the evaporator unit as a boiling liquid/gas mixture and exits as a liquid/gas mixture as well. The excess liquid exiting the evaporator unit flows into a low-pressure liquid receiver. The evaporators are termed flooded units and can develop and maintain very narrow temperature differences between entering and exiting refrigerant. Pressure and refrigerant temperature is controlled by a thermostatically controlled pressure-modulating regulator (back pressure regulator or BPR). Refrigerant temperatures within the evaporator are typically only 1.1 – 1.5°C below the room air temperature. The difference in air temperature entering and exiting the evaporator coil is 0.3°C or less.

Indirect glycol refrigeration systems are ones in which a secondary coolant (commonly glycol) is chilled in the primary refrigeration unit, typically one that uses Freon. The chilled glycol is then circulated through the evaporator unit located in the storage room. This system also allows very narrow temperature differences to be maintained between entering and exiting refrigerant and air entering and exiting the cooling unit. Rooms equipped with either ammonia or glycol systems typically can be maintained at higher relative humidity levels than Freon equipped rooms.

REFRIGERATION SYSTEM MANAGEMENT

Several refrigeration system management options have been successfully tested and adopted by warehouses in Pacific Northwest region of the United States. Storage companies adopting one or more of these options have saved significant amounts of electrical energy and improved fruit quality by reducing mass lost during storage. These options include:

Encyclopedia of Agricultural, Food, and Biological Engineering
DOI: 10.1081/E-EAFE 120006903
Copyright © 2003 by Marcel Dekker, Inc. All rights reserved.

- Time controlled evaporator fan cycling. Evaporator fans are turned on and off on a time interval set by the system manager. Typical on/off periods are 2 hr on followed by 2 hr off or 4 hr on followed by 4 hr off. Evaporator fan cycling should not be initiated until fruit in the room has cooled to storage temperature (typically 5–7 days). Both of these options achieve 50% savings of the electrical energy the fans would use if no cycling were introduced. Data recorded in a 1000 bin room filled with D'anjou pears showed that 988 kWh per week were used when the four evaporator fans operated continuously.[1] Cycling the fans (2 hr on/2 hr off) reduced their energy use to 501 kWh per week. Additional energy is also saved as the rest of the refrigeration system experiences less operational requirement. Experience has shown that when fans are cycled off for times greater than fan on periods, slight increases in fruit temperature in some areas of the storeroom occur. Evaporator fan cycling can also reduce moisture loss from the fruit. Some warehouses have achieved approximately 1% less mass loss in fruit stored for mid-year or late-year marketing when evaporator fan cycling was used.
- Evaporator fan cycling controlled by fruit temperature. This technology requires that additional temperature sensors be placed among the fruit in a room. Sensor placement should be such that both vertical and horizontal locations throughout the room are included. Research has shown that controlling this system to automatically turn on the evaporator unit when the weighted average temperature of the sensors increased 0.1°C and allowing the system to operate until the room set-point temperature was reached, was beneficial in both energy savings and reduction of fruit mass loss.[2]

  Rooms that use lime bags to control carbon dioxide may or may not be able to adopt fan cycling. Research has shown that CO_2 levels increase in lime controlled rooms during fan off periods. When the fans come back on, CO_2 levels come back down as the air more readily impinges the lime bags. Rooms filled with fruit that are not highly sensitive to moderate increases in CO_2 (0.02% increase) most likely can benefit from fan cycling.
- Turning off evaporator fans. Simply turning off evaporator fans does not require any capital investment. Once the fruit in a room has cooled to the set-point temperature, half of the evaporator fans can be deactivated. This management practice does create several changes in how the evaporator performs. Research in a 1000 bin room filled with D'anjou pears[1] has shown that when two of four evaporator fans were deactivated, air returning to the evaporator immediately increased approximately 0.28°C (0.5°F). Refrigerant temperature decreased approximately 0.28°C (0.5°F). Air temperature returning to the cooling coil increased because only half of the previous volume was now being circulated in the room. The refrigerant temperature decreased because approximately half of the coil surface areas were providing the necessary cooling. This research also found that turning off two of the four fans decreased the time the unit could operate between necessary defrost periods. Although these minor differences in refrigerant and air temperatures were recorded, fruit temperatures and humidity levels remained stable. No reductions in fruit quality were detected.
- Variable frequency drive fan motor speed controllers. Incorporation of this technology is the most capital extensive, however, potential energy savings are maximized. Because electrical energy use of a fan motor is proportional to revolutions per minute (RPM) raised to the third power (power = RPM^3), moderate reductions in motor speed result in large energy savings. Reducing motor speed to 50% results in an 87.5% reduction in motor energy use. One VFD controller is required for each storeroom. Depending upon the room, a VFD may cost from $2500 to $3000. The cost to install VFD control of a 112 kW (150 HP) compressor motor may be from $15,000 to $18,000. Experience has shown that specific electrical filtering is required for these systems to prevent premature motor burnout.

  Reducing motor speed in storerooms reduces the volumetric flow rate by the same amount. For example, reducing motor speed by 50% reduces airflow by 50%. Based on studies of air distribution in storerooms, significant changes in how uniformly air moves around and through the bin stacks may result. Caution must be exercised to insure that local areas in the room do not become warmer than others, thereby adversely affecting fruit quality.

ROOM MANAGEMENT

Bin placement in the storeroom affects capacity and air distribution. Recent studies have shown that more bins can be placed in many rooms and that the uniformity of air distribution can be improved by careful placement of the bins (Fig. 1).

- Tight stacking bins. Tight-stacking bins in existing storerooms by eliminating spaces between rows is a viable alternative to new construction provided that the room has sufficient evaporator coil capacity and floor dimensions to accommodate the additional bins.

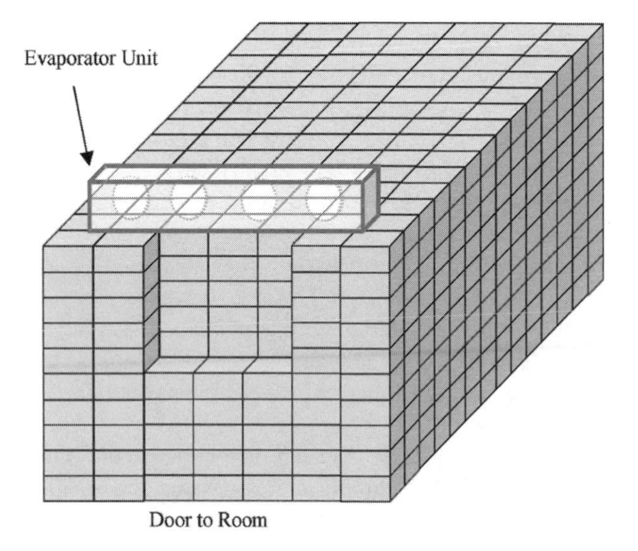

Fig. 1 General stacking pattern in a room with tight-stacked bins with the evaporator unit above the door.

Room dimensions will determine if one or more additional rows of bins can be added. Careful attention must be paid to insure that sufficient space is maintained between the wall opposite the evaporator unit (end wall) and the bin stacks. A minimum of 18 in.–24 in. (46 cm–60 cm) is recommended between the bins and the end wall to insure proper air circulation. Space between bin stacks and the sidewalls should be less than at the end walls. Spaces of 4 in.–6 in. (10 cm–15 cm) are recommended. The sidewall space should allow some air movement toward the evaporator unit, but forcing the majority of the air to pass through the bin runner spaces is mandatory.

- Bin placement near evaporator units. Space between evaporator units and the top layer of bins should be minimized to reduce the amount of air that bypasses the cooling surfaces. Air exits the evaporator at a high velocity, which creates a low pressure area near the face of the unit. Restricting the open space between bins and the cooling unit helps reduce the amount of air that does not circulate through the evaporator coil, and thus is not cooled before flowing back to the fruit bins.
- Bin placement near doors. Careful placement of bins to eliminate blocking air movement through the runner spaces is mandatory. The number of bins cross-stacked as a result of required forklift maneuvering space should be minimized. Research has shown that improved uniformity of air distribution throughout the entire room is closely linked to bin placement that eliminates spaces between stackes.[3]

CONCLUSION

Modern fruit storage systems are capable of lengthening the period many high quality fresh commodities can be supplied to world markets. As world population increases and predicted increases in per capita consumption of fresh fruits become realities, additional storage capacity will be needed. Future building designs and room environmental control systems will evolve to meet these requirements. Unchanged in future systems will be steadfast committment to fruit quality retention and the need for universal implementation of appropriate energy conservation systems and practices.

REFERENCES

1. Hellickson, M.L.; Baskins, R.A. Reducing Energy Costs in Fruit Storage Warehouses. Or. Ag. Exp. Stn. Tech. Paper No. 11772. Proceedings of the 17th Annual Washington Tree Fruit Postharvest Conference, Wash. State Hort. Assn. P.O. Box 136, Wenatchee, WA 98807, March, 2001.
2. Hellickson, M.L.; Koca, R.W.; Staples, J. Computer Control of Ammonia Refrigeration Systems in Apple Storages. Or. AES Tech. Paper No. 10245. Proceedings of the Sixth International Controlled Atmosphere Research Conference; Cornell University, June, 1993.
3. Hellickson, M.L.; Baskins, R.A. Visualization of Airflow Patterns in a Controlled Atmosphere Storage. Or. AES Tech. Paper No. 11788. Atca Horticulturae. Accepted for publication, 2001.

Fruit Transport System

Yang Tao
University of Maryland, College Park, Maryland, U.S.A.

F

INTRODUCTION

Fruit transport systems involve the transportation of fruit from harvest to storage, packing, and shipping. Fruit transport systems vary depending on kinds of fruit; however, many fruit handling operations from harvest to packing are common. This section describes a typical fruit transport system from picking and storage to packing and shipping for a few major fruit types.

FRUIT PICKING

Most fruits are still harvested through hand-picking, especially for the fresh market. Although certain fruits can be harvested mechanically, many popular fruits such as apples, pears, oranges, and stonefruit must be handled carefully to prevent bruising or other damage. Shipping and handling damage is one of the biggest problems that the fruit industry must deal with.

During the harvest season, fruit pickers pick the fruit off the tree and put them into a holding container, typically a bag or basket. When the container is full, the picker unloads the fruit into a nearby collecting bin. For apples and pears, the standard bin size is about 40 ft^3, which typically holds about 1000 lb of apples or about 20 bu. Bins are made from wood or plastic and their size and construction varies from region to region and the type of fruit it is used for.

Trucks or tractor-trailers are often used to transport fruit bins to a packing house. In small operations, the fruit can be packed and sold in the local market. Some types of fruit can be stored in cold storage to extend its shelf life. In large operations, for year-around marketing, some types of fruit are stored in a controlled atmosphere (CA) storage room, which extends its shelf life even longer. Based on the market demand, fruit is taken from the CA storage for processing and packing. Within the packing house, the bins are handled with forklifts, roller conveyors, chain conveyors, automatic equipment for unstacking, emptying, and restacking.

FRUIT CLEANING AND WAXING

Typically in the United States, fruit from the field or CA storage must go through a series of processes including cleaning, coating and treating, and sorting before it can be packed and shipped to the consumer. Some fruits need only minimal cleaning, however, fruits such as apples and pears must be cleaned thoroughly. This not only enhances the appearance, but also prolongs the shelf life.

In the packing house, the fruit must be removed from the bin or other shipping container gently to reduce damage. There are many methods to accomplish this. The container can be submerged in water so the fruit can float out, it can be dumped into water, or it can be gradually dumped onto a slow moving conveyor. In a water tank, the water acts as an accumulator and the water flow carries the fruit onto a conveyer for subsequent processing. See Fig. 1. The volume of fruit fed to the packing line is controlled at this stage in the operation.

The fruit moves through a cleaning process, usually using chlorine or detergent to remove dirt, spray residue, and natural wax. Rotating brushes are often used to do both cleaning and conveying of fruit. In many cases, the fruit passes over drying and polishing brushes after the cleaning stage. Some types of fruits are coated with a food-grade wax or other material to improve its appearance and shelf life. This coating is usually sprayed on the fruit as they rotate on waxing brushes. In most cases, the fruit needs to be clean and dry before a coating is applied. Some of the coatings require a drying operation to set the coating. For a large apple packing line, a dryer can be about 40 ft in length and it takes about 2 min to dry the fruit. Citrus dryers are often in two or three stages. See Fig. 2 for a commonly used apple waxer and drier.

FRUIT SORTING

After the fruit has gone through the cleaning, coating, and drying process, it is carried to a sorting area for the inspection, as shown in Fig. 3. Along the sorting conveyer, workers inspect each fruit, remove defective fruit or unwanted items, and redirect different grades of fruits as required. Commonly, a roller conveyer is used to rotate the fruit for overall surface inspection. At this stage, workers

Encyclopedia of Agricultural, Food, and Biological Engineering
DOI: 10.1081/E-EAFE 120006884
Copyright © 2003 by Marcel Dekker, Inc. All rights reserved.

Fig. 1 Apples are loaded to a water tank by a hydrofeeder (far left) and carried by water flow to avoid bruises.

perform sorting operations that machines are not capable of doing.

In large packing house operations, sorting machines using machine vision are installed to sort fruits into different grades based on weight, dimensional size, color, shape, limited defects, and other parameters. A typical machine vision fruit sorting system[1] is shown in Fig. 4. On the roller conveyers, fruits are singulated into pockets so they can be handled individually. Usually the fruit is rotated on rollers while passing through a vision chamber. Inside the chamber, video cameras take one or more images of each fruit. Linked to each camera, a computer embedded with a digital image processor instantly processes the images. This information is combined with information from other sensors, in many cases weight, and each individual fruit is assigned a grade based on this information compared against the sorting parameters specified by the operator. A typical user control console of a vision system is shown in Fig. 5. Research and development in nondestructive inspection and automated sorting according to quality factors such as internal and external defects, acidity, soluble solids, firmness, and other parameters are still continued efforts in the scientific and engineering community.[2–34]

After grading, the fruit is transported by a mechanical sorting machine that discharges the fruit at many different locations along its length. The "drop" location for each individual fruit is determined by the operator based on its

Fig. 2 After the cleaning process, the waxer and drier make the fruit shining for prolonged shelf life.

Fig. 3 Apples are inspected at the sorting table (conveyer), where workers remove seriously defected and other unwanted items. The sorting table consists of rollers on which apples are rotated for surface inspection.

assigned grade and where in the packing house it is to be packed, as shown in Fig. 4. When a fruit reaches its assigned location, it is removed from the sorting machine by some methods. Generally the fruit is carried in a pocket or holder, which can hinge down backwards, hinge down to the side, or tilt up to cause the fruit fall out. In some cases, an ejector is used to knock the fruit out. It then falls onto a conveyer, which carries the fruit to its packing location. The mechanisms for releasing are carefully engineered to minimize damage to the fruit. Typically, soft pads, brushes, and/or ramps are also used to reduce the impact of fruit to each other and to the collection conveyer. The computer-controlled sorting machines give the packing house the flexibility to sort fruits to meet many

Fig. 4 A typical machine vision fruit inspection system. The fruits are rotated on the singulator for full surface views by the cameras inside the vision chamber.

Fig. 5 A typical user control console of the machine vision system, where an operator can program and adjust the sorting criteria for sorting.

different customer requirements. Large apple packing lines sometimes have the capability of separating fruit into 64 or more classes.

FRUIT PACKING

Fruit is packed into containers for shipping after being graded. The type of packing is determined by the type of fruit, how easy it is damaged, its shelf life, distance it is to be shipped, and its value. Many high-value, easily damaged fruits are placed in trays or cells. These have pockets that hold individual fruits and are designed such that when stacked into a carton, the fruits are prevented from hitting one another. This can be done by a semiautomatic tray packing machine, which inserts fruits into the trays, filling most of the pockets. Operators check the tray, fill any empty pockets, remove any incorrectly graded fruit, and orient the fruit to give the best appearance, as shown in Fig. 6. The individual trays are conveyed to a carton loading station where an operator places the trays into a carton.

Hand-packing is a common packing method but is very labor intensive and slower than the tray packing. It allows the operator flexibility in the way the fruit is packed. A typical hand-packing line is shown in Fig. 7, where workers are taking individual pears from a bin, wrapping each in a paper, and placing them in a carton.

Fruit that is not so easily damaged can be hand-placed into a carton based on a pattern. This is sometimes called a pattern pack. The pattern used is determined by the carton size and fruit size and is designed so that the fruit fits in the carton with no room to move. This prevents fruit from fruit damage during shipping. This type of packing is usually

Fig. 6 Fruits are sorted by dropping from the main line to the semiautomatic tray-pack lines, where workers place full trays of fruits into boxes before transporting them by conveyers to the CA storage.

Fig. 7 Fruits are hand-packed by using wrapping papers to prevent the impact between fruits.

done by hand but there are few machines available for citrus.

Fruits can also be bulk-filled into cartons or bags based on weight or number of pieces. This is generally done by machine and has the advantage of being fast and less labor intensive than other packing methods. It can cause damage to the fruit and is generally used on lower grade of fruit or fruit that is less prone to bruising.

FRUIT SHIPPING

The cartons are conveyed by roller, belt, or chain conveyor to a closing or lidding area after being filled. A pad is sometimes placed on top of the fruit before the carton is closed. The cartons with open flaps pass through a machine where the flaps are glued, taped, or stapled shut. For cartons requiring separate lids, they are placed by hand or machine. In some cases, the cartons are individually strapped to hold the lid on securely and coded with size, grade, packer number, and other information. The packed cartons are usually stacked onto pallets or slip-sheets for handling with forklifts. Some large packing houses have automatic palletizers to stack the cartons. The stacked cartons are then strapped, banded, or wrapped together on the pallet to hold them securely for shipping.

FRUIT SIZES AND GRADES

Many fruits have size designations based on how many of a certain size range it takes to make a bushel. For example, a bushel of apples weighs 42 lb and has 16 commonly used sizes by USDA standard of grading of apples[35]: 36, 48, 56, 64, 72, 80, 88, 100, 113, 125, 138, 150, 163, 180, 198, and 215. Typical grapefruit sizes are 18, 23, 27, 32, 36, 40, 48, 56, and 64; typical peach sizes are 98, 127, 159, 215, 293, and 388. Tomatoes are sized according to the pattern in which they are packed: 4×4, 4×5, 5×5, 5×6, 6×6, 6×7, 7×7, and 7×8. Grades can be determined by the percentage of a certain color, the shade of a color, shape, and number and/or size of defects.

ACKNOWLEDGMENTS

The author is grateful to Mr. Lynn Chance for his peer review and verification of the content.

REFERENCES

1. Tao, Y.; Chance, L.; Liu, B. Full-Scale Fruit Sorting System Design—Factors and Considerations. Proc. of the Food Processing Automation IV, Chicago, IL, 1995; 14–22.
2. Crowe, T.G.; Delwiche, M.J. Real-Time Defect Detection in Fruit—Part I: Design Concepts and Development of Prototype Hardware. Trans. ASAE **1996**, *39* (6), 2299–2308.
3. Crowe, T.G.; Delwiche, M.J. Real-Time Defect Detection in Fruit—Part II: An Algorithm and Performance of a Prototype System. Trans. ASAE **1996**, *39* (6), 2309–2317.
4. Donahue, D.W.; Bushway, A.A.; Moore, K.E.; LaGasse, B.J. Evaluation of Current Winnowing Systems for Maine Wild Blueberries. Trans. ASAE **1999**, *15* (5), 423–427.
5. Guyer, D.E.; Yang, X. Use of Genetic Neural Networks and Spectral Imaging for Defect Detection on Cherries. Comput. Electron. Agric. **2000**, *29*, 179–194.
6. Heinemann, P.H.; Pathare, N.P.; Morrow, C.T. Automated Inspection Station for Machine Vision Grading of Potatoes. J. Mach. Vis. Appl. **1996**, *9* (1), 14–19.
7. Li, M.; Slaughter, D.C.; Thompson, J.F. Optical Chlorophyll Sensing System for Banana Ripening. Postharvest Biol. Technol. **1997**, *12* (3), 273–283.
8. Lu, R. Predicting Firmness and Sugar Content of Sweet Cherries Using Near-Infrared Diffuse Reference Spectrometry. Trans. ASAE **2001**, *44* (5), 1265–1271.
9. Miller, B.K.; Delwiche, M.J. Peach Defect Detection with Machine Vision. Trans. ASAE **1991**, *34* (6), 2588–2597.
10. Peiris, K.H.S.; Dull, G.G.; Leffler, R.G.; Burns, J.K.; Thai, C.N.; Kays, S.J. Nondestructive Detection of Section Drying, an Internal Disorder in Tangerine. HortScience **1998**, *33* (2), 310–312.
11. Schatzki, T.F.; Haff, R.P.; Young, R.; Can, I.; Le, L.-C.; Toyofuku, N. Defect Detection in Apples by Means of X-Ray Imaging. Proc. from the Sensors for Nondestructive Testing Int. Conf., Orlando, FL, 1997; 161–171.

12. Slaughter, D.C.; Barrett, D.; Boersig, M. Nondestructive Determination of Soluble Solids in Tomatoes Using Near Infrared Spectroscopy. J. Food Sci. **1996**, *61* (4), 695–697.
13. Tao, Y. Method and Apparatus for Sorting Objects by Color. US Patent 5,339,963, 1994.
14. Tao, Y. Methods and Apparatus for Sorting Objects Including Stable Color Transformation. US Patent 5,533,628, 1996.
15. Tao, Y. Spherical Transform of Fruit Images for On-Line Defect Extraction of Mass Objects. Optical Eng. **1996**, *35* (2), 344–350.
16. Tao, Y. Defective Object Inspection and Separation System Using Image Analysis and Curvature Transformation. US Patent 5,732,147, 1998.
17. Tao, Y. Closed-Loop Search Method for On-Line Automatic Calibrations of Multi-camera Inspection Systems. Trans. ASAE **1998**, *41* (5), 1549–1555.
18. Tao, Y. Method for Calibrating a Color Sorting Apparatus. US Patent 5,799,105, 1998.
19. Tao, Y. Defective Object Inspection and Separation System Using Image Analysis and Curvature Transformation. US Patent 5,732,147, 1998.
20. Tao, Y.; Wen, Z. An Adaptive Spherical Image Transform for High-Speed Fruit Defect Detection. Trans. ASAE **1999**, *42* (1), 241–246.
21. Tao, Y.; Morrow, C.T.; Heinemann, P.; Sommer, H.J., III. Fourier-Based Separation Technique for Shape Grading of Potatoes Using Machine Vision. Trans. ASAE **1995**, *38* (3), 949–957.
22. Throop, J.A.; Aneshansley, D.J.; Upchurch, B.L. Apple Orientation on Automatic Sorting Equipment. Proc. Sensors for Nondestructive Testing, Orlando, FL, 1997; 328–342.
23. Throop, J.A.; Aneshansley, D.J.; Upchurch, B.L.; Anger, B. Apple Orientation on Two Conveyors: Performance and Predictability Based on Fruit Shape Characteristics. Trans. ASAE **2001**, *44* (1), 99–109.
24. Tollner, W.; Hung, E.Y.; Maw, B.; Sumner, D.; Gitaitis, R. Nondestructive Testing for Identifying Poor-Quality Onions. SPIE **1995**, 2345, 392–402.
25. Upchurch, B.L.; Throop, J.A. Effects of Storage Duration on Detecting Watercore in Apples Using Machine Vision. Trans. ASAE **1994**, *37* (2), 483–486.
26. Upchurch, B.L.; Affeldt, H.A.; Hruschka, W.R.; Norris, K.H.; Throop, J.A. Spectrophotometric Study of Bruises on Whole, 'Red Delicious' Apples. Trans. ASAE **1990**, *33* (2), 585–589.
27. Wen, Z.; Tao, Y. Dual-Camera NIR/MIR Imaging for Stem-End/Calyx Identification in Apple Defect Sorting. Trans. ASAE **2000**, *43* (2), 446–452.
28. Wen, Z.; Tao, Y. *Adaptive Spherical Transform of Fruit Images for High-Speed Defect Detection*; ASAE Paper No. 973076; ASAE: St. Joseph, MI, 1997.
29. Wen, Z.; Tao, Y. *Method of Dual-Camera NIR/MIR Imaging for Fruit Sorting*; ASAE Paper No. 983040; ASAE: St. Joseph, MI, 1998.
30. Wen, Z.; Tao, Y. Brightness-Invariant Image Segmentation for On-Line Fruit Defect Detection. Optical Eng. **1998**, *37* (11), 2948–2952.
31. Wen, Z.; Tao, Y. *Method of Dual-Camera NIR/MIR Imaging for Fruit Sorting*; ASAE Paper No. 983040; ASAE: St. Joseph, MI, 1998.
32. Wen, Z.; Tao, Y. Dual-Wavelength Imaging for Online Identification of Stem Ends and Calyxes. SPIE **1998**, 3460, 249–253.
33. Wen, Z.; Tao, Y. Fuzzy-Based Determination of Model and Parameters of Dual-Wavelength Vision System for On-Line Apple Sorting. Optical Eng. **1998**, *37* (1), 293–299.
34. Wen, Z.; Tao, Y. Building a Rule-Based Machine Vision System for Defect Inspection on Apple Sorting and Packing Lines. Expert Syst. Appl. **1999**, *16* (1999), 307–313.
35. USDA. *Standard for Grading of Apples*; USDA: Washington, D.C., 1972.

Gas Phase Kinetics

Tom L. Richard
Iowa State University, Ames, Iowa, U.S.A.

G

INTRODUCTION

Gas exchange is among the most important interactions between organisms and their environment. Oxygen (O_2) and carbon dioxide (CO_2) production and consumption are fundamental to photosynthesis and aerobic respiration, while anaerobic respirations produce CO_2 and methane (CH_4). Sulfate reduction produces hydrogen sulfide (H_2S) and other reduced gaseous sulfur compounds, while denitrification generates di-nitrogen (N_2) as well as nitrous oxide (N_2O) and nitric oxide (NO). These and other nitrogen gases, including other oxides of nitrogen (NO_x) and ammonia (NH_3), leave and enter the gas phase through such biological reactions as nitrogen fixation and deamination. Water vapor is another gas of critical importance in biological systems, with losses through evaporation, transpiration, perspiration, and/or respiration causing dehydration, loss of turgidity, and in extreme cases death.

The biological reactions discussed here do not actually occur in the gas phase, but instead in the liquid phase or within living cells. However, the compounds indicated above are gases under standard temperature and pressure, and as these particular biological reactions proceed, they affect the gas composition immediately around the organism. The production, consumption, and exchange of all these gases is a function of the types and rates of biological activity that occur, as well as the physical and chemical properties of the biological system and its surrounding environment.

GAS REACTIONS

Photosynthesis is among the most crucial biological reactions on the planet, affecting the global atmosphere and climate by generating the oxygen needed for aerobic organisms, consuming CO_2, and sequestering carbon in organic compounds. In addition to CO_2, photosynthesis requires solar energy and water, as well as various macro and micronutrients needed for plant metabolism. If water and nutrients are adequate, the rate of photosynthesis will depend largely on the plant surface area and solar energy available, although there is significant variation among plants that have evolved a variety of leaf architectures and different photosynthetic pathways.[1] Photosynthesis rates increase with increase in CO_2 concentrations, reducing (but not eliminating) the global warming impacts of fossil fuel consumption.

Micro-organisms liberate the energy needed for metabolism and growth by combining electron donors (carbohydrates, proteins, lipids, etc.) with electron acceptors, a process commonly referred to as respiration.[2,3] Most of the respiration reactions produce or consume gases, and many do both. Oxygen is the most energetically efficient electron acceptor, providing a competitive advantage to aerobic organisms, with nitrate (NO_3^-) a close second and thus readily consumed under anoxic conditions. Sulfate (SO_4^{-2}), and CO_2 will also serve as electron acceptors under anaerobic conditions, but with significantly lower amounts of energy released. When NO_3^- serves as the electron acceptor the process is called denitrification. The NO_3^- is sequentially reduced to nitrite (NO_2^-), NO, N_2O, and finally N_2, the latter three products being only slightly soluble and thus rapidly leaving the system as gases. Sulfate reduction similarly forms the gas hydrogen sulfide (H_2S) and leads to other odorous sulfide gases including dimethyl sulfide (DMS) and dimethyl disulfide (DMDS). Methanogenic reactions recycle part of the CO_2 liberated during anaerobic degradation as an electron acceptor, producing a biogas that includes both CO_2 and CH_4. Autotrophic organisms can also consume gases; both as electron acceptors (O_2 and CO_2) and as the inorganic electron donors they oxidize for energy. Gaseous electron donors for autotrophs include NH_3 and hydrogen gas (H_2).

Although photosynthesis and respiration reactions produce and consume the largest quantities of biologically mediated gases, other reactions can also be important to gas exchange. In systems where mineral nitrogen sources are scarce, blue–green algae and symbiotic nitrifying organisms can fix organic nitrogen from atmospheric N_2. Protein degradation involves deamination, producing NH_3 under both aerobic and anaerobic conditions. Under aerobic conditions nitrification consumes oxygen, converting NH_3 to NO_3^-. When tracking nitrogen flows through biological systems the possible impacts of these other reactions must always be considered.

Finally, because of the fundamental importance of water to all living processes, the consumption and

Encyclopedia of Agricultural, Food, and Biological Engineering
DOI: 10.1081/E-EAFE 120007181
Copyright © 2003 by Marcel Dekker, Inc. All rights reserved.

formation of water must also be considered. Photosynthesis and anaerobic respiration consume water, while aerobic respiration produces water. Although this water is generally a liquid at the cellular level, systems that include gas flow will transfer water vapor in and out of the system. This water vapor transfer is influenced by wetted surface area, airflow velocity, and the temperature and relative humidity of the ambient air, and can be analyzed using the psychrometric equations and mass transfer analysis.[4] Additional physical characteristics such as surface area and other system characteristics are further discussed below.

INFLUENCES OF ENVIRONMENTAL FACTORS

Temperature, pH, oxygen concentration, and moisture are key environmental factors that have particularly important effects on the biological reactions that produce and consume gases. Temperature will be discussed first, as it has a significant effect on all biological organisms and reactions. Biological reaction rates generally increase with temperature up to an optimum, above which rates decrease or ultimately cease when temperature are so hot that proteins denature, cellular processes stop, and the organism dies. While a few multicellular organisms like mammals and birds have regulatory mechanisms that maintain a nearly constant temperature, most organisms must adapt to the fluctuating temperatures of their ambient environment. Each individual species has an optimum temperature, while in more complex ecosystems there is an ecological optimum. In multicellular ecosystems these optima are generally near the peak ambient temperature for that environment, and significant disruptions can occur when those peak temperatures change. Mixed microbial ecosystems also experience disruptions as temperatures change, but because of the diversity and complexity of these systems the response curves are relatively smooth. Both individual microbial species and mixed cultures generally experience a near linear increase in rate with temperature up to a plateau zone near the optimum temperature, followed by a steep decline in reaction rates above the optimum. In mixed aerobic cultures this optimum is near 60°C[5,6] (see Fig. 1), while in anaerobic cultures there are two optima, a mesophilic peak at about 30–35°C and a thermophilic peak at about 50–55°C.[7]

As with temperature, there is an optimum pH for most biological processes that varies with species and ecosystem. While most multicellular organisms can moderate or control their own pH, micro-organisms are much more at the mercy of ambient conditions. Optimum pH for many bacteria is between 6.5 and 7.5,[8] while many fungi can tolerate a much lower pH, and thus dominate under those conditions.[9] Of particular interest to engineers interested in biological gas formation is anaerobic digestion. The methanogenic bacteria that are crucial to this process area have little tolerance for low pH, and methane digesters must therefore be maintained at a pH of 6.2 or higher.[7]

In acidic environments ammonia converts to the highly soluble ammonium ion (NH_4^+), while in basic environments the equilibrium shifts to gaseous NH_3, as illustrated in Fig. 2. Acidification can be intentionally used to reduce

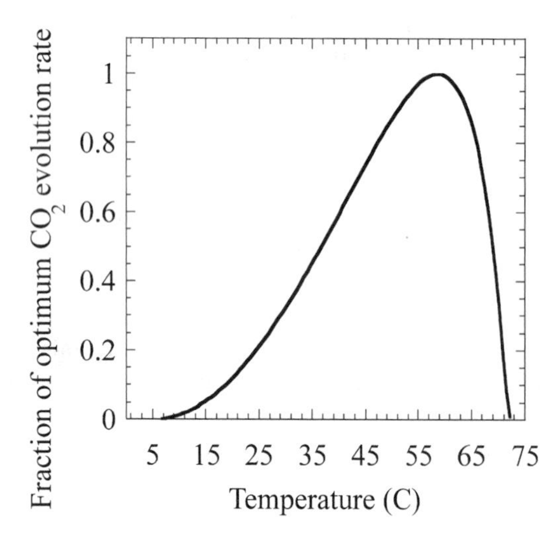

Fig. 1 Effect of temperature of kinetics of mixed aerobic biodegradation. (Adapted from Richard, T.L. The Kinetics of Aerobic Solid Biodegradation. Ph.D. Thesis, Cornell University, 1997.)

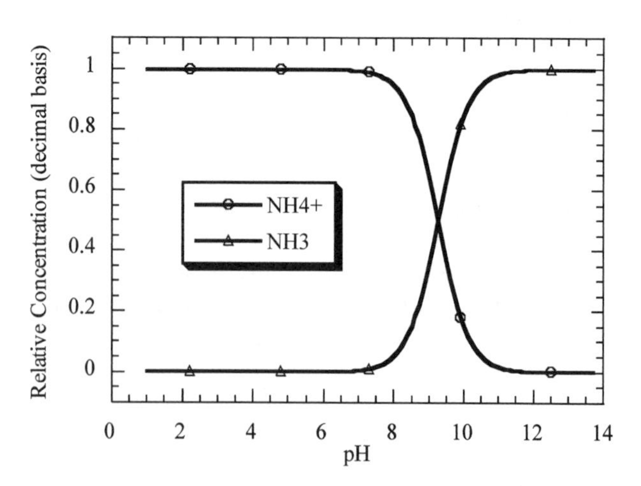

Fig. 2 Relative concentrations of NH_3 and NH_4^+ as a function of pH. (Adapted from Richard, T.L. Ammonia Odors, 1996. http://www.cfe.cornell.edu/compost/odors/ammonia.html [accessed Sept 2002].)

volatilization losses of NH_3, as has been demonstrated in applications ranging from pure culture fermentations to manure storage facilities.

Two other biologically produced gases, H_2S and CO_2, also have chemical equilibria with dissolved forms that are strongly affected by pH, but in the opposite direction to ammonia gas. Both these compounds are in their gaseous form under acidic conditions, and can be retained in solution at high pH. Applications of this principal range from wet scrubbing towers for removal of reduced sulfur compounds from industrial exhausts to the hydroxide traps used by several wet-chemistry methods to measure CO_2 respiration rates. The dissolved forms of carbon dioxide (CO_2, H_2CO_3, HCO_3^-, and CO_3^{-2}) collectively form the carbonate system, playing a major role in buffering the pH of aquatic systems through exchange with atmospheric CO_2.[10]

Oxygen availability remains one of the critical factors in all biological processes. The presence or absence of oxygen determines whether the respiration reactions previously discussed are aerobic or anaerobic. Anaerobic conditions are generally slower than aerobic reactions, in part because of reduced energy capture by the micro-organisms, and for lignicellulosic substrates because the enzymatic hydrolysis of lignin sheaths appears to be an exclusively aerobic reaction.[11] In dilute liquid systems such as lakes, rivers, and oceans oxygen uptake is relatively slow, and is normally replaced by surface diffusion and mixing and photosynthetic sources to maintain moderate to high levels of dissolved O_2. However, in high strength organic particulate systems, whether submerged or solid-state, completely aerobic conditions are difficult to achieve. Biofilm models of readily degradable substrates predict anaerobic conditions in the interior of particles greater than 0.25 mm in diameter, even with high oxygen concentrations in the gas or liquid phase surrounding the particles.[12]

In solid-state microbial ecosystems, oxygen limitations can significantly reduce substrate degradation. Fig. 3 illustrates the effect of oxygen on carbon dioxide evolution in composting systems. Oxygen concentration is measured in the bulk gas of pores within the compost matrix, with the maximum CO_2 evolution at normal atmospheric concentration of 20.9% O_2. Although this dataset did not include oxygen concentrations below 1%, the CO_2 evolution rate would not fall to zero at low oxygen levels, since CO_2 would continue to be produced by anaerobic respiration. Interestingly, under thermophilic temperatures around 65°C the relationship in Fig. 3 can become flat or inverse, with higher CO_2 evolution rates observed under low oxygen conditions.[13]

Moisture is the other environmental factor that has a significant impact on gas phase reactions. In liquid systems moisture provides a medium for gas and substrate exchange, while helping dilute any toxic byproducts that might result in substrate inhibition. In solid-state systems moisture effects biodegradation kinetics through both physical factors such as porosity, air permeability, thermal conductivity and specific heat, and biological factors such as microbial species distribution and growth rates.[14,15] These relationships are made more complex by the dynamic nature of moisture in solid-state systems, with changing particle size and structure, metabolic water production, and biological drying. Mass transfer processes in the gas and liquid phases are important in managing moisture and the other key environmental factors such as oxygen, temperature and pH, as described below.

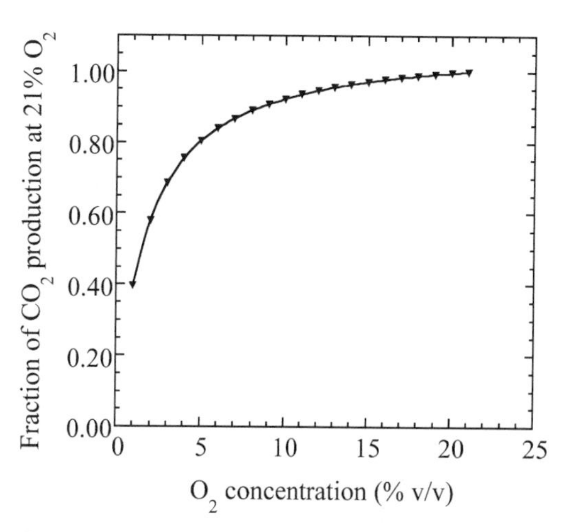

Fig. 3 Effect of oxygen on kinetics of mixed aerobic biodegradation for temperatures less than 55°C. (Adapted from Richard, T.L. The Kinetics of Aerobic Solid State Biodegradation. Ph.D. Thesis, Cornell University, 1997.)

GAS TRANSFER PROCESSES

The exchange of gases between a biological organism and a surrounding atmosphere is rarely simple. Even when cells are in direct physical contact with the atmosphere, gases must still pass through the cell wall or membrane where various active and passive transport mechanisms will affect the exchange. In more complex systems there are additional biological regulatory mechanisms, such as stomata in plants, lungs in vertebrates, and the exoskeletons of insects. Many systems include an aqueous phase, which can range in thickness from a few microns or millimeters surrounding micro-organisms or roots in soil, to many meters in streams, lakes, and oceans. Aqueous phases frequently occur in reactor systems, ranging from thin layers in the solid-state fermentations used in pharmaceutical, biochemical, mushroom and compost production to the fully submerged reactors used in food, beverage, and biochemical

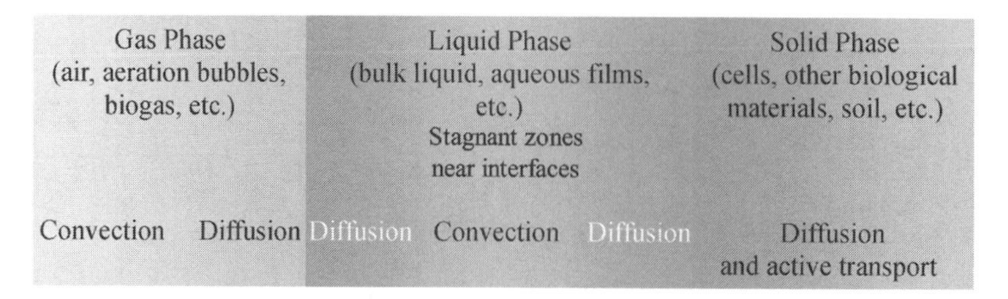

Fig. 4 Gas transfer mechanisms in a three-phase system.

production as well as wastewater treatment.[3,16–18] While the gas phase is not always obvious in liquid systems, gas transfer remains critical to management of O_2 supply, CH_4 removal, and CO_2 impacts on pH.

Gas transfer occurs both within and between the solid, liquid, and gas phases.

Within a gaseous or liquid phase this transfer results from diffusion and/or convection, while cells aggregated in and on solids can use active transport mechanisms (see Fig. 4). Pure diffusion is a function of concentration gradients and fluid properties, increasing with greater concentration gradients. Diffusion is much more rapid in the gas phase, with the oxygen diffusion coefficient about 10,000 times higher in air than in water.[19] Convective transfer rates depend on the velocity of fluid movement. In practice many systems are either convection dominated (mixed reactors, streams, and aerated solid-state or liquid systems) or diffusion dominated (soil, landfill bioreactors, anaerobic lagoons). In aerobic solid-state systems convection may dominate in the gas phase due to aeration, while diffusion may dominate in the relatively static liquid phase. Living organisms can influence or even control convection and diffusion within or near their organisms, through physical mechanisms such as breathing and pumping blood. Living cells also have a range of active transport and exclusion mechanisms for gases, ranging from stomata and various semipermeable cell membrane structures to carrier molecules such as hemoglobin.

Gas transfer can generally be neglected within a pure solid, although biological materials that are commonly referred to as solids (wood, fruit, grain, etc.) actually consist of multiple phases, ranging from solid and gas if fully dry to solid and liquid if fully saturated. Intermediate moisture levels contain three phases, as is usually the case in soil, solid-state fermentations, and similar systems (see Fig. 5). These systems are generally designated porous media, and gas transfer within them depends on the distribution of air and water filled pores and solid particles, the resulting gas pore structure and aqueous film characteristics, as well as the diffusion and convection within the gas and liquid phases.[20] In porous media that are biologically active, there are also sources and sinks of gases from the biological reactions previously described.

Gas transfer rates between phases are generally a function of gas concentrations on either side of the interface and the properties of the interface itself. In simple unmixed liquid–gas systems the interphase transfer of dilute concentrations can be quantified using the Henry's

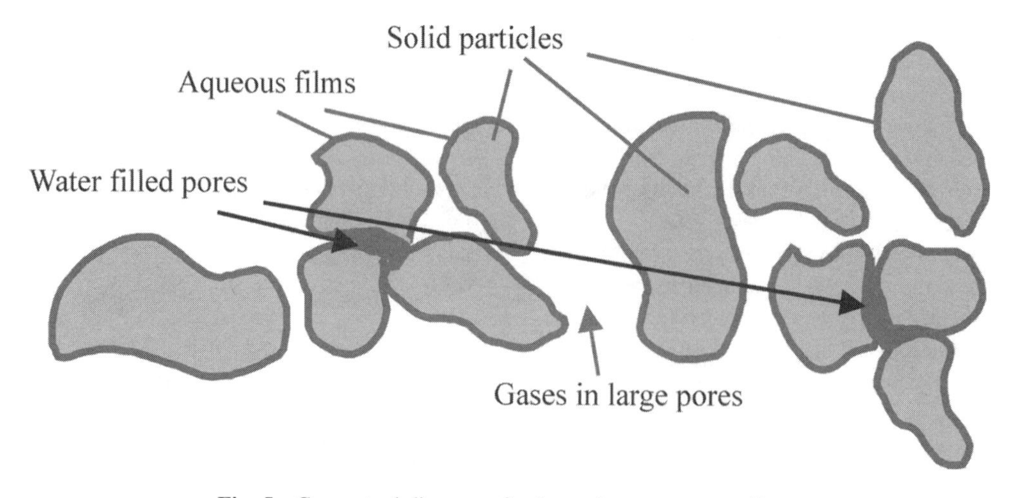

Fig. 5 Conceptual diagram of a three-phase porous medium.

law coefficient, which relates the gas concentrations in bulk fluids on either side of the interface, and varies for different gases and fluid interfaces.[4] Where mixing or aeration occurs, the added complexity usually requires the use of an empirically derived mass transfer coefficient, usually expressed using dimensionless groups to facilitate analysis and design at multiple scales.[3,17,18]

CONCLUSION

Gas phase interactions are fundamental to many living processes, ranging from respiration and photosynthesis to denitrification and methanogenesis. The specific gas forming or consuming reactions and their kinetic rates are a function of the living organism(s), available substrate(s), and several key environmental factors including temperature, oxygen, moisture, and pH. The physical arrangement of the system also has a major influence on reaction kinetics, since mass transfer limitations are very common in high rate systems, particularly where there is a solid particulate organic phase. Understanding and, where appropriate, managing these gas phase reactions provide a wealth of opportunities for biological, food, and agriculture engineers to enhance the interaction of organisms with their environments.

REFERENCES

1. Merva, G.E. *Plant Biosystems*; ASAE: St. Joseph, MI, 1995.
2. Sawyer, C.N.; McCarty, P.L. *Chemistry for Environmental Engineering*., 3rd Ed.; McGraw Hill Book Co.: New York, 1978.
3. Bailey, J.E.; Ollis, D.F. *Biochemical Engineering Fundamentals*, 2nd Ed.; McGraw Hill Book Co.: New York, 1986.
4. Edwards, W.M. Mass Transfer and Gas Absorption. In *Perry's Chemical Engineering Handbook*, 6th Ed.; Perry, R.H., Green, D.W., Malony, J.O., Eds.; McGraw-Hill, Inc.: New York, 1984; 14-1–40-40.
5. Haug, R.T. *The Practical Handbook of Compost Engineering*; Lewis Publishers: Boca Ratan, FL, 1993.
6. Richard, T.L.; Walker, L.P. Temperature Kinetics of Aerobic Solid-State Biodegradation. In *Proceedings of the Inst. of Biological Engineering*, Orlando, FL, July 10–12, 1998; Eiteman, M., Ed.; IBE Publications: Athens, GA, 1998; Vol. 1, A22–A39.
7. McCarty, P.L. Anaerobic Waste Treatment Fundamentals; Part Two, Environmental Requirements and Control. Public Works **1964**, *Oct*,123–126.
8. Lim, D.V. *Microbiology*; West Publishing Co.: St. Paul, MN, 1989.
9. Alexander, M. *Introduction to Soil Microbiology*, 2nd Ed.; John Wiley & Sons: New York, 1977.
10. Stumm, W.; Morgan, J.J. *Aquatic Chemistry: Chemical Equilibria and Rates in Natural Waters*; Wiley Interscience: New York, 1996.
11. Richard, T.L. The Effect of Lignin on Biodegradability, 1996. http://www.cfe.cornell.edu/compost/calc/lignin.html (accessed Sept 2002).
12. Hamelers, H.V.M. A Theoretical Model of Composting Kinetics. In *Science and Engineering of Composting*; Hoitink, H.A.J., Keener, H., Eds.; Renaissance Publications: Columbus, OH, 1992; 36–58.
13. Richard, T.L.; Walker, L.P.; Gossett, J.M. The Effects of Oxygen on Solid-State Biodegradation Kinetics. In *Proceedings of the Inst. of Biological Engineering*, Charlotte, NC, June 18–20, 1999; Eiteman, M., Ed.; IBE Publications: Athens, GA, 1999, Vol. 2, A10–A30.
14. Griffin, D.M. Water and Microbial Stress. In *Advances in Microbial Ecology*; Alexander, M., Ed.; Plenum Press: New York/London, 1981; 91–136.
15. Richard, T.L.; Hamelers, H.V.M.; Veeken, A.H.M.; Silva, T. Moisture Relationships in Composting processes. Compost Sci. Utilization **2002**, *10* (4), 286–302.
16. Shuler, J.L.; Kargi, F. *Bioprocess Engineering: Basic Concepts*; Prentice Hall, Inc.: Englewood Cliffs, NJ, 1992.
17. Lee, J.M. *Biochemical Engineering*; Prentice Hall, Inc.: Englewood Cliffs, NJ, 1992.
18. Grady, C.P.L.; Daiger, G.T.; Lim, H.C. *Biological Wastewater Treatment*, 2nd Ed.; Marcel Dekker: New York, 1999.
19. Richard, T.L. Oxygen Transport, 1996. http://www.cfe.cornell.edu/compost/oxygen/oxygen.transport.html (accessed Sept 2002).
20. Nield, D.A.; Bejan, A. *Convection in Porous Media*, 2nd Ed.; Springer Verlag: New York, 1998.

Glass Containers

Felix H. Barron
Joel D. Burcham
Clemson University, Clemson, South Carolina, U.S.A.

INTRODUCTION

Glass containers are still very important in food packaging despite the continuous switching by food processors to polymer (plastic) containers such as jars (Fig. 1) for peanut butter and bottles for catsup. By analyzing the following main advantages and disadvantages of glass containers, the food processor could make a decision about selecting the right package: *Advantages*: chemically resistant to all foods, no internal coating is necessary, only one seal (at the closure) is present, impermeable to gases, and is available in different colors. *Disadvantages*: it is breakable, has a limited thermal shock resistance, and is heavy.

GLASS CONTAINERS: SHAPES AND SIZES

Many shapes and sizes of glass containers can be found in the market. Their selection by the food processor depends on many factors including the type of food, the processing and closing method, and the marketing issues. Many manufacturers are now offering their products electronically by presenting detailed information about their containers. For e.g., Pont Packaging BV[1] provides several tables with specifications that are useful to users of glass containers. A summary of information for some of their products is presented here:

	Wide jars	Round jars
Capacity range	40 g–100 g	30 ml–750 ml
Color	Amber and white	Amber and white
Height range	11.3 cm–17.1 cm	3.76 cm–15.1 cm
Diameter range	6.03 cm–9.18 cm	4.42 cm–9.83 cm
Some typical uses	Dry products: coffee, sweeteners	Preserves and pickles

BASICS OF GLASS MANUFACTURING

The manufacturing of glass containers is energy intensive. The basic unit operations involved in this process include mixing, melting, forming, and annealing. A brief description[2–4] is as follows:

Mixing

Soda-lime glass makes up about 90% of all glass produced and is by far the most common glass used in making food containers. A typical composition of soda-lime glass (white or clear) consists of 72% silica (SiO_2), 13% lime (Na_2O), 11% calcium oxide (CaO), 2% aluminum oxide (Al_2O_3), 0.6% ferric oxide (Fe_2O_3), 0.6% barium oxide (BaO), 0.5% sodium oxide (K_2O), 0.3% sulfur trioxide (SO_3), and 0.2% magnesia (MgO).

Glass is produced in a wide variety of colors. Amber, green, and opal are much more prominent than others. Table 1 shows a list of some of the materials added to provide the colors.

Melting

Cullet is added to the raw materials in a ratio making up 15%–50% of the total batch to be melted. Cullet is recycled glass and glass recovered from breakages in the plant. The percentage of cullet in the mixture depends on the quality and availability of recycled glass.

Forming

The most widely used method of glass container formation uses an individual section (IS) machine with one of two general processes: "blow and blow" (B and B) or "press and blow" (P and B). The two-step B and B process drops a gob into an inverted blank mold and forces the glass down by compressed air into a neck ring mold, which shapes the appropriate finish. Then, air blows from the bottom of the mold to make the interior of the container.

The initial preformed glass is called a parison and the mold is air-cooled, so the parison can be handled somewhat and turned upright in the final mold. This upright position stretches the glass and air is blown into the parison to expand it to the final glass shape. The P and B process also has two steps and is generally used for wide-mouthed containers such as jars. After the gob is dropped into the initial mold, a plunger presses up from

Encyclopedia of Agricultural, Food, and Biological Engineering
DOI: 10.1081/E-EAFE 120007138
Copyright © 2003 by Marcel Dekker, Inc. All rights reserved.

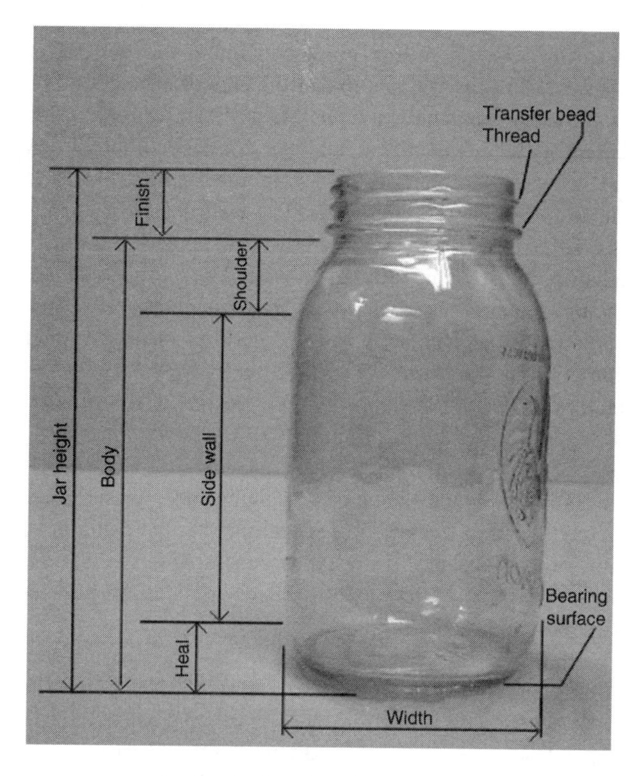

Fig. 1 Profile of a glass container.

the bottom and presses the glass until it is forced to fill the neck ring mold. The parison now follows a process similar to that of the B and B method.

Annealing

From the forming machine, the container is passed onto the annealing lehr, which is an oven where a heating and cooling process removes stress in the glass that is caused by unequal cooling rates for different parts of the container. The container with reduced stress is significantly stronger.

Table 1 Glass colors and the coloring agents

Colors	Materials
Green	Ferrous sulfide, chromic oxide
Amber	Sodium sulfide
Opal	Calcium fluoride
Blue	Cobalt oxide
Red	Cupric oxide, cadmium sulfide
Yellow	Ferric oxide, antimony oxide
Purple	Manganese
Black	Iron oxides

SURFACE TREATMENTS

Abrasions and wear to glass containers cause minute fractures that propagate under stress and can destroy the container. Exterior surface treatments strengthen the glass and reduce friction between the container and other contact surfaces.

Interior surface treatments are generally used to improve the chemical properties of glass. White glass is very resistant to chemicals; it may be still susceptible to some chemical attack. Using an interior surface treatment can greatly enhance the chemical resistance of glass. A common interior surface treatment is applied as a fluorocarbon gas at elevated temperatures typically between 550 and 600°C.

GLASS DEFECTS

Glass defects can occur in every glass manufacturing plant.[5] The quality of glass containers that leave the plants is controlled to a large extent by the process of detecting these defects, categorizing the type of defect, and determining the source of the problem.

There are three main categories of defects that are generally recognized by food processors and glass manufacturers alike. They are critical, major, and minor defects. Critical defects are imperfections that pose a danger for the user and render the product totally unusable. Some examples are spikes and cord. These are thin strands of glass on the interior of the glass that could break-off when the container is filled and a strain causing weakness in the glass that was not relieved by annealing. Major defects are problems that limit the life of the container and/or its contents. A chipped finish can allow an imperfect seal with the closure resulting in premature spoilage of the product. Minor defects reduce the appearance of the container and can limit the acceptability by the consumer.

There are a number of inspection tests used by glass manufacturers to detect defects in the glass.[4] Squeeze testers apply a fixed pressure with rollers to the sides of the container. If the glass does not have a reasonable tensile strength, then the container will shatter. Plug gauges measure the height and finish opening size of the container. Light is used to inspect the container for a number of different visible defects. As the container is rotated in the beam of light, any defects will reduce the amount of light passing through the container and cause it to be rejected. Depending on the manufacturer and glass requirements, additional tests are performed, such as glass thickness, surface stress, and container weight.

Ideal shipping and abuse testing of empty or filled glass containers determine the degree of physical protection required to keep the integrity of the containers especially during handling and transportation practices.[6] Among the most important factors causing transportation damages to glass packages are vibration, jolting, and dropping. A vibration tester is used to simulate vibration of trucks and trains, inclined impact testers are used to determine the resistance to jolting on all sides of an already vibrated packaged product, and a drop tester is used to determine the resistance to dropping of a packaged product from given height.

CLOSURES

The glass container and its closure provide the integrity of glass packages. The safety of the product is maintained as long as the hermetic seal is effective. In addition to safety, the ease of opening is a desired property by consumers. It is necessary for the contents in food containers to be readily accessible without sacrificing any safety factors. Furthermore, containers that hold more than a single serving require a closure with the ability to form an acceptable seal at re-closing by the consumer.

The quality and safety of many food products depend on controlling the environmental factors within the container. Since glass is highly impermeable, the critical region is the seal between the closure and the container, especially when a specific internal pressure has to be maintained, whether low (vacuum) or high.

Vacuum closures typically have a safety button that depresses when the container is holding a sufficient vacuum. This feature benefits both the consumer and the packager by letting them know whether a container contains vacuum. A cold-water vacuum test needs[7] to be performed before filling glass containers that will have vacuum closures. This is to test the capper efficiency.

Many closures have a liner[2] to assist in forming a seal between the environment and the container contents. A liner typically consists of a facing material and a cushioning material (the wad). The wad deforms when it is compressed to conform better to the shape of the finish, resulting in a better container seal. The purpose of the facing material is to protect the wad from potential contamination. Rubber or elastic gaskets used as a liner can combine the function of both the wad and the facing material. Plastic closures can be designed to function appropriately without a liner and still produce a hermetic seal.

Additional factors such as vacuum and adhesives are frequently used to assist in the sealing of some containers. Since the different closure types generally satisfy similar objectives, their functionality and application process separates the major types.

Common closures available in the market include: continuous thread (CT), press-on/twist-off (PT) caps, and twist caps. The selection of a particular closure will depend on type of glass container, food product, capping or closing speeds, hand or machine capping, easy opening, and other desired features. Filling processes such as hot fill, cold fill, and thermal processes such as pasteurization or commercial sterilization should also be considered for the selection of closures.

Glass closures may be made from plastic, steel, tin plate, or aluminum. They may also have a liner to obtain a higher degree of sealing if needed. The PT closure does not have lugs or threads. It has a liner, typically made of plastisols, which is deformable under high heat, but holds

Table 2 Selected closure types and uses

Closure type	Commercial benefits	Technical benefits	Closure application	Common uses
Continuous thread (CT)	Traditional old fashion image; sealing and re-closing; suitable for vacuum or nonvacuum packaging	Fast and easy closing; hand or machine closing; hot or cold fill; pasteurization or sterilization	Average speeds of 600 closures per minute	Preserves, spreads, and pates; fish and seafood; vegetables; baby foods
Press twist (PT)	Easy to open	Suitable for extreme processing conditions; hot or cold fill; pasteurization or sterilization	High speeds: 1500 closures per minute	Preserves, spreads, and pates; fish and seafood; vegetables; baby foods; juices and drinks
Twist cap	Suitable for most foods and beverages; hot, cold, or aseptic filling; re-close.	Vacuum sealing	Applied by many closing machine; speeds up to 800 closures per minute	Dressings and condiments; preserves, spreads, and pates; fish and seafood; vegetables; pickles

Courtesy of Crown Cork and Seal Co.

its shape when cooled. This lid is pressed on while hot and as it cools the liner permanently assumes the shape of the finish. The twist cap is a lug closure commonly used for vacuum glass packaging.

Table 2 shows some examples of closure types, their benefits, and common uses.

REFERENCES

1. Pont Packaging BV. http://www.pont.nl (accessed March 2001).
2. Robertson, G.L. Glass Packaging Materials. *Food Packaging, Principles and Practice*; Marcel Dekker, Inc.: New York, 1993; 232–251.
3. Crosby, N.T. Glass. *Food Packaging Materials. Aspects of Analysis and Migration of Contaminants*; Applied Science Publishers, Ltd: London, 1981; 167–169.
4. Cavanagh, J. Glass Container Manufacturing. In *The Wiley Encyclopedia of Packaging Technology*, 2nd Ed.; Brody, A.L., Marsh, K.S., Eds.; John Wiley & Sons, Inc.: New York, 1997; 483 pp.
5. Hanlon, J.F. Glassware. *Handbook of Package Engineering*, 2nd Ed.; McGraw-Hill Book Company: New York, 1984; 9-1–9-28.
6. Barron, F.H. *Food Packaging and Shelf Life: Practical Guidelines for Food Processors*; Extension Publication EC 686; Cooperative Extension Service, Clemson University: Clemson, South Carolina, January 1995.
7. Code of Federal Regulations, *Title 9, Part 3/8*; Office of the Federal Register National Archives and Records Administration, U.S. Government Printing Office: Washington, 1992.

Glass Transition Temperatures

Yrjö H. Roos
University College Cork, Cork, Ireland

INTRODUCTION

Phase transitions in food systems include a number of changes in the physical state, e.g., evaporation, condensation, freezing, and melting of water, crystallization of sugars, melting and crystallization of polymers (starch components), and melting and crystallization of fats and oils.[1] The glass transition, as the transition between various equilibrium states, is not a thermodynamically well-defined phase transition, but rather a state transition and a change in the physical state of an amorphous, noncrystalline material. The glass transition occurs over a temperature range and it is accompanied with a change between a brittle, glass-like solid state of the material and a supercooled, viscous liquid state with leathery, rubbery, or syrup-like properties. Various material properties change over the glass transition, which has also some of the thermodynamic characteristics of a second-order phase transition. Hence, there is a change in the heat capacity and thermal expansion coefficient over the glass transition. Furthermore, the transition can be observed from changes in dielectric and mechanical properties and changes at a molecular level, which can be detected with various spectroscopic techniques.[1,2]

THE GLASSY STATE AND THE GLASS TRANSITION

Foods are generally nonequilibrium systems and nonfat food solids tend to form amorphous, nonequilibrium structures in dehydrated and frozen foods.[3] Glassy structures of food solids, and particularly of sugars, carbohydrate polymers, and proteins are often formed in baking, dehydration, freezing, and extrusion. These processes involve formation of a melt of the solids that may vitrify during cooling or removal of the solvent water.[3] For example, the rapid removal of water in the spray-drying process results in formation of vitrified particles of the remaining solids. In freezing, the separation of water as ice also results in supersaturation of dissolved substances, which vitrify at low-temperatures.[4] Such glassy structures of food solids can be retained in the freeze-drying process.[5]

The glass transition of supercooled liquids occurs at about 100°C below the equilibrium melting temperature (Fig. 1). As a noncrystalline material is cooled to well below its equilibrium melting temperature, its viscosity increases until the molecules of the material become frozen in a solid, glassy state. The properties of a glassy material may differ depending on the cooling rate and thermal history. Generally, rapid cooling produces glasses with a high free volume and more "dense" glasses are formed with slow cooling rates or by annealing of the materials. The free volume of glassy materials may also decrease with physical ageing in the glassy state. Glassy materials exhibit molecular vibrations and rotations, while translational motions are attained at the glass transition. The glass transition is often accompanied with enthalpy and volume relaxations that reflect the enthalpy and volume state of the glass and the differences between the cooling and heating rates and relaxation to the supercooled liquid state (Fig. 1). Therefore, the glass transition temperature cannot be exactly defined as, e.g., the melting temperature of a pure substance.

OBSERVATION OF THE GLASS TRANSITION

The glass transition behavior of food solids is related to the glass transition of the main constituents.[1] Therefore, glass transition temperatures have been measured for a number of carbohydrates and proteins.[4,6,7] These studies often use differential scanning calorimetry, which allows determination of the change in heat capacity occurring over the glass transition. In such measurements, the glass transition temperature, T_g, is often taken from the onset or midpoint of the change in heat capacity observed during heating of an amorphous sample (Fig. 2). Depending on the glass formation kinetics and packing of the molecules in the glassy state, the glass transition may be accompanied with an endotherm or exotherm[1,2] referred to as an enthalpy relaxation.

The glassy state is a highly viscous state with restricted molecular mobility. The relaxation times of various

Encyclopedia of Agricultural, Food, and Biological Engineering
DOI: 10.1081/E-EAFE 120006983
Copyright © 2003 by Marcel Dekker, Inc. All rights reserved.

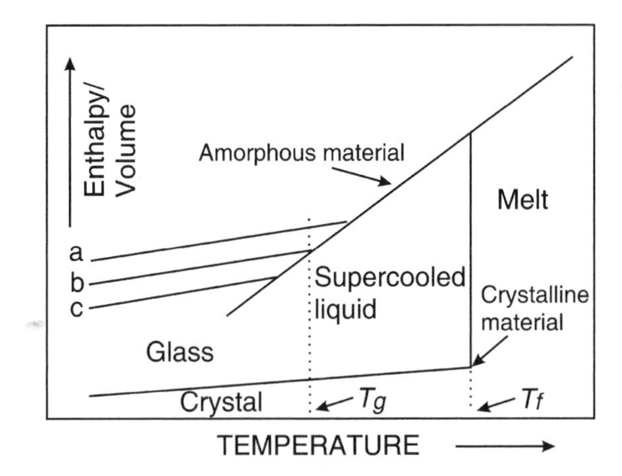

Fig. 1 Schematic representation of enthalpy and volume changes in amorphous and crystalline materials. The crystalline and liquid states are equilibrium states with the phase transition temperature, T_f. The supercooled, amorphous materials may vitrify in the glass transition, T_g, to different enthalpy/volume states. Depending on the transformation kinetics, the glass transition may be accompanied by enthalpy and volume relaxations.

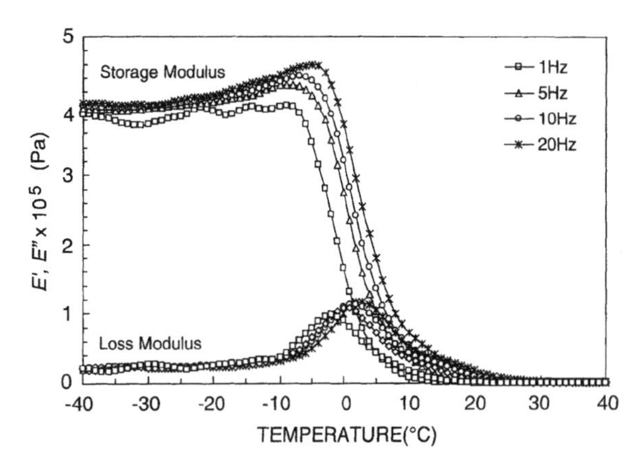

Fig. 3 Frequency-dependence of storage, E', and loss, E'', modulus of the α-relaxation of amorphous sorbitol, as determined using DMA.

changes, including flow, decrease dramatically over the glass transition. The viscosity of glassy materials is $> 10^{12}$ Pa sec, but the viscosity decreases rapidly over and above the glass transition. Hence, the glass transition can be observed from changes in mechanical properties and the "α-relaxation" corresponding to the calorimetric glass transition can be detected with dynamic mechanical analysis (DMA). The DMA is used to measure the storage, E', and loss modulus, E'', and the $\tan\delta = E''/E'$ as a function of frequency and temperature. The changes in the mechanical properties are highly frequency-dependent, because molecular mobility and relaxations in the nonequilibrium material are time-dependent. However, the glass transition temperature is often taken from the tan δ peak at a given frequency (Fig. 3). The advantage of the DMA is that it is much more sensitive in observing the glass transition than the Differential Scanning Calorimetry (DSC). Unfortunately, sample preparation and the exact control of temperature and water content of biological materials in the DMA are more difficult than in the DSC.

Dielectric analysis (DEA) may be used to determine dielectric relaxations as a function of frequency or temperature (Fig. 4). The changes occurring in the dielectric properties are analogous with mechanical property changes, i.e., the α-relaxation can be observed from changes in permittivity, ε', loss factor, ε'', and

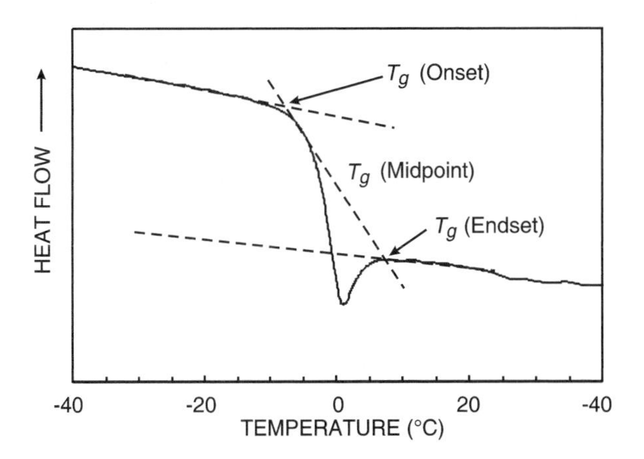

Fig. 2 DSC analysis of glass transition of amorphous sorbitol. The glass transition is observed from a step change in heat flow as the heat capacity changes over the transition temperature range. The glass transition temperature, T_g, is often reported as the onset or midpoint temperature of the change in heat capacity.

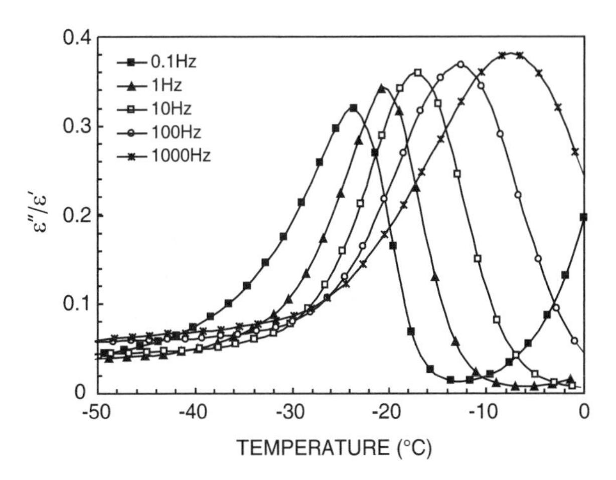

Fig. 4 Frequency-dependence of the tan δ (ϵ''/ϵ') of the α-relaxation of amorphous xylitol, as determined using DEA.

the ratio of loss factor to permittivity, $\varepsilon''/\varepsilon'$ (dissipation factor, $\tan \delta$). A maximum in the $\varepsilon''/\varepsilon'$ peak at a given frequency may be taken as the glass transition temperature, T_g. In general, DEA is more sensitive in observing the glass transition than the DMA, but the possible contribution of ions to the dielectric data should be taken into account.

GLASS TRANSITION OF FOOD COMPONENTS

Glass transition temperatures are available for a number of food components, mainly carbohydrates (Table 1) and proteins.[4,6,8] The glass transition temperature of polymers with the same repeating unit increases with increase in molecular weight. For example, the glass transition temperatures of glucose polymers increase gradually from that of glucose at 32°C to those of maltodextrins and starch components.[9] The reported glass transition temperatures of food components vary because of the sensitivity of the transition to water (water plasticization decreases the glass transition temperature).[1,4] Furthermore, there are difficulties in the determination of the transition of biopolymers (small changes in heat capacity and decomposition of the polymers at high temperatures), the materials exhibit nonequilibrium and time-dependent behavior, and the transition may be obtained using a number of methods that detect changes occurring at different temperatures due to differences in relaxation times.[1,2] For example, the glass transition temperatures of food proteins and carbohydrate polymers vary greatly because they are either measured using different techniques or predicted on the basis of other material properties.

GLASS TRANSITION IN FROZEN FOODS

Ice formation in food systems results in freeze-concentration of solids to a temperature-dependent extent.[1,4] A concurrent decrease in temperature and increase in viscosity of freeze-concentrated solutes and food solids decrease diffusion and the rate of ice formation. At a material specific temperature, the freeze-concentrated solutes and solids with some unfrozen water vitrify and ice formation ceases. Such materials are maximally freeze-concentrated and they show an initial concentration independent glass transition of the maximally freeze-concentrated phase. In most carbohydrate solutions, the maximally freeze-concentrated solute phase contains about 80% solids and 20% unfrozen water.[6] The formation of the maximally freeze-concentrated state and vitrification has been proposed to provide maximum stability to frozen foods.[4,10]

GLASS TRANSITION AND FOOD PROPERTIES

Several time-dependent changes typical of low-moisture and frozen food materials are related to changes in material properties occurring over the glass transition. These changes often result from decreasing viscosity and increasing molecular mobility above the glass transition

Table 1 Glass transition temperature, T_g, (range of typical values), change in heat capacity over the glass transition, ΔC_p, and the glass transition temperature of the maximally freeze-concentrated solution, T'_g, for common carbohydrates and sugars

Material	T_g (°C)		ΔC_p (J g^{-1}K^{-1})	T'_g (°C)
Fructose	5	(+ 5– + 100)	0.75	− 57
Galactose	30	(+ 30– + 11)	0.50	− 56
Glucose	31	(+ 21– + 41)	0.63	− 57
Lactose	101	(+ 100– + 106)		− 41
Maltitol	39	(+ 39– + 44)	0.56	− 47
Maltose	87	(+ 43– + 100)	0.61	− 42
Mannose	25	(+ 25– + 36)	0.72	− 58
Sorbitol	− 9	(− 9–0)	0.96	− 63
Starch	250			− 6
Sucrose	62	(+ 52– + 70)	0.60	− 46
Trehalose	100	(+ 79– + 107)	0.55	− 40
Xylitol	− 29	(− 39– − 18.5)	1.02	− 72

Table 2 Glass transition related time-dependent changes/properties typical of low-moisture and frozen foods

Change/property	Mechanism	Occurrence
Caking	Decrease in surface viscosity and agglomeration of particles	Amorphous powders
Collapse	Loss of structure due to decreasing viscosity and increasing flow	Collapse in dehydration
		Collapse of dried foods
Crystallization	Increasing translational mobility and diffusion	Sugar crystallization in low-moisture and frozen foods
		Retrogradation of starch
Loss of crispness	Plasticization	Low-moisture foods, e.g., cereals
Reaction kinetics	Increasing molecular mobility	Diffusion-controlled reactions (Maillard reaction; oxidation) in low-moisture and frozen foods
Recrystallization	Increasing translational mobility and diffusion	Ice recrystallization
Stickiness	Decrease in surface viscosity and adhesion of particles	Amorphous powders

(Table 2). Therefore, several mechanical properties of amorphous food materials are related to glass transition and they may dramatically affect stability and textural characteristics.[1,4,7] Typical glass transition related changes in food and related systems are stickiness and caking in many hygroscopic powders and loss of crispness of many cereal foods, such as breakfast cereals and snacks. Rates of such changes are defined by relaxation times, which decrease rapidly above the glass transition as a result of thermal or water plasticization. The changes in the physicochemical properties are also related to other properties of amorphous food solids.[10] For example, retention of flavors in dehydrated foods is often a result of flavor encapsulation in amorphous, glassy carbohydrates and sugars. Structural changes above the glass transition often enhance oxidation of such compounds and encapsulated lipids. Furthermore, component crystallization may result in full release of encapsulated substances, which can be observed from substantial flavor losses and rapid oxidation of lipids. Crystallization of amorphous compounds often occurs above the glass transition as translational diffusion of molecules allows molecular rearrangements in the supercooled liquid state and formation of the equilibrium, crystalline structure. For example, crystallization of amorphous lactose is a well-known phenomenon in dairy powders.

REFERENCES

1. Roos, Y.H. *Phase Transitions in Foods*; Academic Press: San Diego, CA, 1995; 360.
2. Sperling, L.H. *Introduction to Physical Polymer Science*; John Wiley & Sons: New York, NY, 1986; 439.
3. Roos, Y.; Karel, M. Applying State Diagrams to Food Processing and Development. Food Technol. **1991**, *45* (12), 66, 68–71, 107.
4. Slade, L.; Levine, H. Glass Transitions and Water–Food Structure Interactions. Adv. Food Nutr. Res. **1995**, *38*, 103–269.
5. Roos, Y.H. Frozen State Transitions in Relation to Freeze Drying. J. Thermal Anal. **1997**, *48*, 535–544.
6. Roos, Y. Melting and Glass Transitions of Low Molecular Weight Carbohydrates. Carbohydr. Res. **1993**, *238*, 39–48.
7. Tolstoguzov, V.B. The Importance of Glassy Biopolymer Components in Food. Nahrung **2000**, *44* (2), 76–84.
8. Matveev, Y.I.; Grinberg, V.Ya.; Tolstoguzov, V.B. The Plasticizing Effect of Water on Proteins, Polysaccharides and Their Mixtures. Glassy State of Biopolymers, Food and Seeds. Food Hydrocolloids **2000**, *14*, 425–437.
9. Roos, Y.; Karel, M. Water and Molecular Weight Effects on Glass Transitions in Amorphous Carbohydrates and Carbohydrate Solutions. J. Food Sci. **1991**, *56*, 1676–1681.
10. Roos, Y.H.; Karel, M.; Kokini, J.L. Glass Transitions in Low Moisture and Frozen Foods: Effects on Shelf Life and Quality. Food Technol. **1996**, *50* (11), 95–108.

Global Positioning Systems

William W. Casady
University of Missouri, Columbia, Missouri, U.S.A.

Viacheslav I. Adamchuk
University of Nebraska, Lincoln, Nebraska, U.S.A.

INTRODUCTION

The global positioning system (GPS) is a space-based radio-navigation system that provides the capability to determine geographic location anywhere on Earth. Developed by the U.S. Department of Defense (DoD), the GPS consists of three segments: a space segment, a control segment, and a user segment. The space segment includes a constellation of 24 operational NAVigation by Satellite Timing And Ranging (NAVSTAR) satellites orbiting our planet at an altitude of 10,900 nautical miles. These satellites follow six orbital planes (four satellites per plane) and circle the globe every 12 hr. The control segment is comprised of a network of ground stations that monitor and correct system attributes. Every satellite continuously transmits its own pseudorandom signal at two designated frequencies: 1575.42 MHz and 1227.60 MHz. The user segment is comprised of GPS receivers that determine position by calculating distances to several satellites simultaneously. The distance to a satellite is calculated through multiplication of the measured transit times by the speed of radio waves. A minimum of four visible satellites is required to compute position and time coordinates.

Geographical coordinates (latitude and longitude) and altitude are used to express location of a GPS receiver in space. Both geographical longitude and latitude represent angular measures of a position on the Earth's surface (Fig. 1). The geographic longitude defines east–west position with respect to the prime (Greenwich) meridian from 0° to 180°, while the geographic latitude indicates north–south position with respect to the equator from 0° to 90°. Several models have been developed in the past to represent our planet. For GPS technology, the World Geodetic System 1984 (WGS-84) has been adopted. This model assumes the Earth as an ellipsoid with a semimajor axis (equatorial radius) $a = 6,378,137$ m, and a semiminor axis (polar radius) $b = 6,356,752.3142$ m [defined as $1/f = 1/298.257223563$, where $f = (a - b)/a$]. Various map projection methods are used to transform geographic coordinates into linear measures used for practical applications.

Russia's Global Navigation Satellite System GLObalnaya NAvigatsionnaya Sputnikovaya Sistema (GLONASS) is similar to the U.S. GPS and also allows determination of geographic location while operating 24 satellites in three orbit planes. In addition, a European system is planned that will incorporate GPS, GLONASS, and another segment to support navigation.

GPS RECEIVERS AND DIFFERENTIAL CORRECTION SERVICES

Every GPS receiver uses the same satellites and similar methods to determine position based on the principle diagram of Fig. 2. However, the complexity and the number of user functions built into the receiver determine both accuracy and cost. Some handheld units may cost less then $100; others may cost far more than $5000. Some receivers may have an LCD screen and operational keys for easy use; others are designed only for connection to another properly equipped electronic device such as a laptop or pocket size computer, various data loggers, or other types of equipment.

By definition, each GPS receiver calculates geographic location (latitude and longitude) of its antenna. Other useful information that is available from GPS includes universal time coordinate (UTC), sea level altitude and/or height above WGS-84 ellipsoid, travel velocity and direction, position of viewed satellites, quality of signal, and other system information. All these parameters can be transferred via serial (RS 232) communication. The National Marine Electronic Association has defined a standard protocol (NMEA-0183) for GPS data transfer using a series of comma-delimited expressions following a defined pattern. Although most hardware and software products follow NMEA-0183, some manufacturers may implement their own data code, which is usually in a binary format.

Even though GPS implements very precise time measurement technology, other factors such as location of viewed satellites, atmospheric delays, multipathing of radio signals, and various sources of noise reduce accuracy. Environmentally crowded areas where trees, roofs, or other tall structures often block GPS signals can cause GPS receivers to function poorly or to become disabled. Furthermore, a security measure known as selective

Encyclopedia of Agricultural, Food, and Biological Engineering
DOI: 10.1081/E-EAFE 120006939
Copyright © 2003 by Marcel Dekker, Inc. All rights reserved.

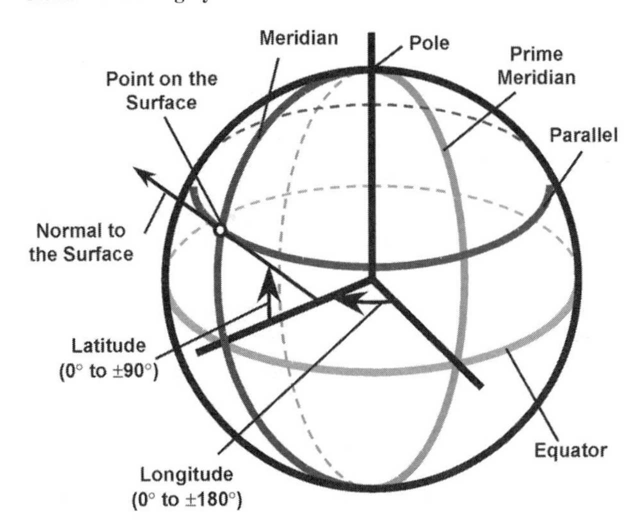

Fig. 1 Definition of geographical location (longitude and latitude).

availability (SA) artificially limits GPS accuracy by broadcasting deliberate errors in timing signals and position data prior to May 1, 2000. Even in the absence of SA, miscalculation of geographic location due to all other errors can be as large as 15 m.

Most agricultural applications of GPS receivers require much higher accuracy than 15 m. Differential correction of GPS data (DGPS) is used to reduce positioning error to a more acceptable level of about 1 m. Differential correction can be performed in real-time or as a postprocessing technique. In either procedure, errors calculated by stationary receivers with known geographic location are transmitted to users in the surrounding area. The most common source of differential correction for the first several years of operation has been a network of the U.S. Coast Guard stations transmitting amplitude modulated (AM) radio signals along coastal lines and major rivers. In certain locations, frequency modulated (FM) differential correction signals have been available as well. Several private geo-stationary satellite services covering much wider areas have also been used; these satellite-based DGPS services usually require a subscription fee. The Federal Aviation Administration (FAA) is also developing a geo-referenced satellite based wide area augmentation system (WAAS) to broadcast correction messages using a GPS frequency. A level of accuracy often referred to as "centimeter accuracy" can be achieved when using survey-grade GPS receivers that use their own on-site portable differential correction base station.

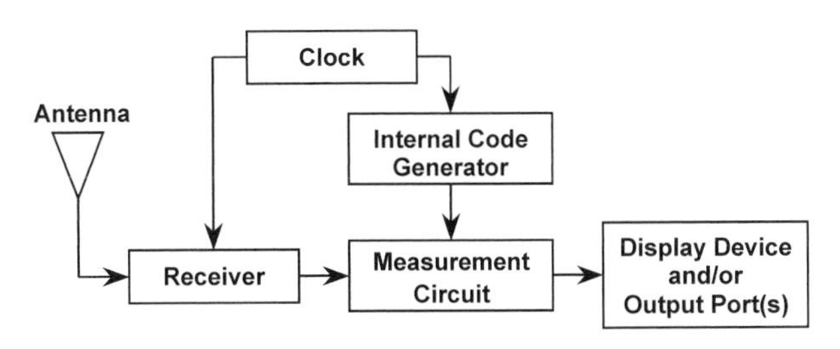

Fig. 2 Principle diagram of a GPS receiver.

AGRICULTURAL APPLICATIONS OF GPS

Agriculture has become a large user of GPS with applications that range from field mapping and navigation to vehicle guidance. The availability of GPS to civilian users made practical the development of site-specific management (SSM) of crop production (also known as precision farming) and motivated development of new technologies that would allow geo-referenced measurement of production-related parameters and distribution of agricultural inputs according to local needs. Global positioning system has become an essential component of both spatial data acquisition and variable rate technology (VRT).

In order to prescribe proper management strategies, various geo-referenced data are combined using geographic information systems (GIS). The use of GPS technology has become nearly indispensable for yield mapping, soil sampling, field surveying, and other data collection practices. For example, yield-monitoring systems periodically record the geographical position of harvesting equipment while measuring the rate of flow of grain or other crops as well as properties such as moisture content. Other GPS data such as travel velocity and UTC are also sometimes used for data processing. Similarly, geographical coordinates are used for point data collection such as soil sampling or crop scouting.

Global positioning system is also used to determine the location of machinery equipped to apply variable rates of seeds, fertilizers, pesticides, lime, and other agricultural inputs. The controller on such equipment changes machine settings according to prescribed application maps that are typically developed through analysis of previously gathered information and knowledge. Global positioning system vehicle guidance can be accomplished with either a passive system that allows the operator to guide the equipment following a visual indicator (light bar), or an active system that incorporates mechanical and/or hydraulic assistance. Completely automatic guidance of agricultural vehicles is also under development.

CONCLUSION

The availability of GPS to civilian users has been largely responsible for the development of SSM practices in agriculture. This new level of management has the potential to increase profitability of farm production and to reduce negative environmental impacts through efficient use of resources. The knowledge gained from SSM can also provide good records that can improve overall farm management. Continued research will lead to many new GPS-based practices that will change agriculture in new and exciting ways.

FURTHER READING

Morgan, M.; Parsons, S.; Ess, D. Global. Global Positioning Systems. In *Precision Farming Profitability*; Lowenberg-DeBoer, J., Ed.; Purdue University: West Lafayette, IN, 2000; 56–61.

Morgan, M.T.; Ess, D.R. In *The Precision-Farming Guide for Agriculturists. An Agriculture Primer;* Kuhar, J.E., Ed.; John Deere Publishing: Moline, IL, 1997.

Sonka, S.T.; Bauer, M.E.; Cherry, E.T.; Colburn, J.W.; Heimlich, R.E.; Joseph, D.A.; Leboeuf, J.B.; Lichtenberg, E.; Mortensen, D.A.; Searcy, S.W.; Ustin, S.L.; Ventura S.J. In *Precision Agriculture in the 21st Century. Geospatial and Information Technologies in Crop Management;* Dixon, J., McCann, M., Eds.; Committee on Assessing Crop Yield: Site-Specific Farming, Information Systems, and Research Opportunities, Board of Agriculture, National Research Council. National Academy Press: Washington, DC, 1997.

Strang, G.; Borre, K. *Linear Algebra, Geodesy, and GPS.* Wellesley-Cambridge Press: Wellesley, MA, 1997.

Tyler, D.A.; Roberts, D.W.; Nielsen, G.A. Location and Guidance for Site-Specific Management. In *The State of Site-Specific Management for Agriculture*; Pierce, F.T., Sadler, E.J., Eds.; ASA-CSSA-SSSA: Madison, WI, 1997; Chap. 10, 183–210.

Grain Harvesting Systems

William W. Casady
University of Missouri, Columbia, Missouri, U.S.A.

G

INTRODUCTION TO GRAIN HARVESTING SYSTEMS

Grain is harvested in most farms in the United States with a harvesting machine called a "combine" (pron. 'käm-, bīn). The chassis and drive train of a typical combine is reversed from that of a tractor. The large primary drive wheels are located on the front, while the engine and smaller wheels that help turn the combine are located on or near the rear.

Combines throughout the world may vary somewhat in appearance and in name, sometimes referred to as headers, etc., but nearly all are designed to harvest clean grain from a cultivated crop leaving most of the rest of the plant material in the field. To accomplish this task the combine performs five primary functions including cutting and feeding, threshing, separating, cleaning, and materials handling.

CUTTING AND FEEDING

The cutting and feeding mechanisms of combines are called "headers" and are often interchangeable from one machine to another. Although several variations exist, the two primary types of headers in wide use are row-type and platform headers. The most common row-type header is a corn header, which usually has an odd number of long snouts that are guided through an even number of rows. The most common platform header has a cylindrical bat reel or pickup reel spanning the front and is designed for general purposes and to cut many different crops. Many farms, especially those in the Corn Belt of the United States, will have both types of headers available for a single combine. In either case, the primary function of the header is to remove an appropriate amount of the crop plant to be fed into the combine for threshing.

Corn Headers

A corn header is specifically designed for the special harvesting characteristics of corn. By removing just the ear of the corn plant, the corn head significantly reduces the amount of plant material that would otherwise pass through the combine. As the combine travels through the field with a corn head, a pair of parallel snapping rolls on each row quickly pulls the corn stalk downward. The ear, which is too large to pass through the opening, is snapped off. Gathering chains convey the ears up into a trough, and a large auger carries them to the center of the header where they are fed into the threshing system of the combine.

Platform Headers

The more general design of a platform header is suitable for a wide variety of crops. Setting the cutting height of a platform header is somewhat more important than it is for a corn header. When harvesting a crop such as soybean, which may set pods along the entire stem, it is important to set the platform very low to the ground to gather as much of the crop as possible. When harvesting crops such as wheat or rice, which produce a single head at the top of the stem, the platform is set high to cut only enough of the plant to gather the head. This minimizes the amount of plant material that must pass through the combine.

The cutting mechanism on a platform header is a reciprocating bar with many serrated knife sections called a sickle bar. A sickle bar may move back and forth as many as ten times per second, cutting everything in its path. Sickle bars are also used on various types of mowers to cut grass or forage crops such as alfalfa.

Most platform headers also use a reel that helps to feed the crop into the header. The height, forward location, and speed of the reel on a platform header can be adjusted according to crop conditions to gently lay a crop into the header or to provide a more aggressive lifting action if needed. For standing crops in good condition, a simple bat reel is sufficient to lay the crop down into the platform. A pickup reel, which includes a set of tines on each bat, is the most common form of reel and can help pick up a crop that has been blown over by wind or is otherwise not standing well. Reel speed should be adjusted so that the peripheral speed of the reel is somewhat faster than the groundspeed of the combine.

Feeder House

The center of the header is attached to the feeder house (Figs. 1a and 2a) or throat of the combine. The plant material gathered by the header is fed into the threshing

Encyclopedia of Agricultural, Food, and Biological Engineering
DOI: 10.1081/E-EAFE 120006878
Copyright © 2003 by Marcel Dekker, Inc. All rights reserved.

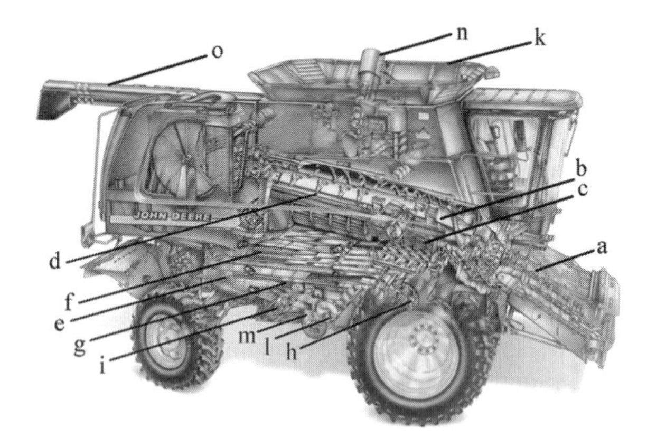

Fig. 1 Cutaway of a combine equipped with a rotor revealing internal and external components: a) feederhouse, b) threshing stage of rotor, c) concaves, d) separation stage of rotor, e) cleaning shoe, f) chaffer, g) cleaning sieve, h) fan, i) tailings auger, j) tailings elevator—not visible; see Fig. 2, k) grain tank, l) clean grain auger, m) clean grain elevator—only partially visible, n) grain-tank loading auger, o) unloading auger.

area of the combine through the feeder house. The feeder house is a rectangular section with rotating drums on each end that drive a floating feeder chain. The feeder chain grabs the straw, stalks, or ears gathered by the header and drags them up into the main part of the combine where the grain is threshed from the rest of the plant material.

THRESHING

The threshing mechanism of a combine is either a rotor (Fig. 1b) or a cylinder (Fig. 2b) that turns at high speeds and with narrow clearances inside a "set of concaves" (Figs. 1c and 2c). The term "concaves" is taken from the concave shape of this metallic grid located below the cylinder or rotor. The concaves confine material to a small space and allow threshed seeds and other small debris to pass through where they are collected below on the grain pan.

The primary threshing action for combines equipped with a cylinder is from high velocity impacts created as the cylinder spins at several hundred rpm. Threshing also takes place from the rubbing action against the plant material especially for combines equipped with a rotor. Cylinders have traditionally been equipped with either spike teeth or rasp bars to provide an aggressive action for threshing. Rasp bars (Fig. 3) typically span the entire width of a cylinder and are often mounted in a spiral pattern along the axis of a rotor. Rasp bars are designed with a set of diagonal grooves that assist in propelling and rubbing plant material to cause a threshing action. Although rasp bars are used on most machines, spike teeth are still used for some crops such as rice.

Seeds tend to have a higher density than the rest of the plant material. Most of the seeds are separated from the bulk of the material during threshing as they fall down through the small openings of the concaves. The rest of the material is discharged near the top of the cylinder onto the straw walkers. Hence, the material leaving the threshing step is divided into two components. The lower component consists primarily of grain with lots of chaff and small bits of plant material. The upper component is primarily comprised of straw, leaves, stalks, and only a small amount of grain.

The aggressive action of threshing can cause damage to seeds. Seed damage is highest for overly dry seeds, which become cracked or broken during threshing. Excessive damage can also be caused to the seed coat of very wet seeds when harvest begins too early. Seed damage is minimized by harvesting seed with moderately high

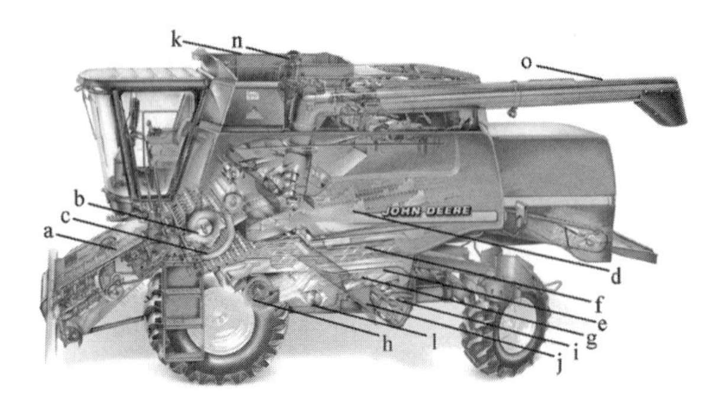

Fig. 2 Cutaway of combine equipped with a cylinder and straw walkers revealing internal and external components: a) feederhouse, b) cylinder, c) concaves, d) straw walkers, e) cleaning shoe, f) chaffer, g) cleaning sieve, h) fan, i) tailings auger, j) tailings elevator, k) grain tank, l) clean grain auger, m) clean grain elevator—not visible; see Fig. 1, n) grain-tank loading auger, o) unloading auger.

Fig. 3 Rasp bar.

moisture contents and by reducing cylinder or rotor speed to the minimum level necessary for good threshing. Cylinder or rotor speeds and clearances should be changed throughout the day to match the requirements of the crop.

SEPARATING

Although most of the grain is separated during threshing in the cylinder or rotor, further processing with either straw walkers (Fig. 2d) or the final stage of the rotor (Fig. 1d) will remove the remaining grain from the straw in the upper portion.

Combines that use a cylinder and concave threshing system typically employ a set of parallel straw walkers that alternately accelerate material upward and toward the rear of the combine. This agitating action further separates the high-density particles, such as seed from the straw, leaves, or stalks. The seed falls down through the straw walkers to a pan or conveyor that returns it to the front of the combine where it is added to the uncleaned grain in the lower portion. The remaining larger material is expelled by the straw walkers out of the back of the combine.

Combines equipped with a rotor separate the remaining grain near the end of the rotor as an extension to the threshing portion of the rotor. Because the rotor spins at high speeds, the resulting force on the grain can be many times the force of gravity. The denser grain is propelled by helical fins through the openings in the rear of the rotor. Lighter and larger material such as straw, leaves, and stalks are propelled rearward until they leave the rotor and are expelled out of the back of the combine.

CLEANING

Grain and other plant materials deposited in the grain pan from the threshing and separation stages finally reach a set of sieves in the lowest level of a combine known as the cleaning shoe (Figs. 1e and 2e). The cleaning shoe expels lightweight material known as chaff as well as material with intermediate aerodynamic properties, such as straw, from the back of the combine. The shoe also separates unthreshed heads or pods and returns them to the cylinder to complete the threshing process leaving only clean grain at the bottom of the shoe. The clean grain is then elevated to the clean grain tank.

At the shoe, grain deposited in the grain pan from the threshing and separation stages passes over two or more oscillating sieves. The top sieve is known as the chaffer (Figs. 1f and 2f), while the lower sieve is known as the cleaning sieve (Figs. 1g and 2g). A fan (Figs. 1h and 2h) blows air through the chaffer to remove lightweight material known as chaff. Grain and small, unthreshed pods or seed heads fall through the chaffer sieve down to the cleaning sieve. Large unthreshed pods or heads are propelled to the rear of the chaffer where they may fall through the chaffer extension and drop into the tailings auger (Figs. 1i and 2i). Any material that does not pass through the cleaning sieve is propelled to the rear of the sieve where it too is collected in the tailings auger. Materials with intermediate aerodynamic properties, such as straw, are both floated and mechanically propelled over the chaffer and the chaffer extension where they are expelled.

MATERIALS HANDLING

Combines are equipped with extensive materials handling components to return partially threshed materials to the threshing area, to move clean grain to the holding tank, and to move grain from the combine for transport from the field in a vehicle such as a truck or grain cart.

Tailings, the partially threshed material from the chaffer and sieve, must be returned to the cylinder or rotor where threshing is completed. Tailings are usually returned in a two-step process with a transversely mounted tailings auger (Figs. 1i and 2i) that empties into a tailings elevator (Fig. 2j). The tailings elevator is typically a paddle chain conveyor that slopes upward from the bottom and rear of the combine to the front of the combine and empties into the cylinder or rotor.

Clean grain that passes through the sieve is collected in a grain tank (Figs. 1k and 2k). Clean grain is usually delivered to the grain tank in three steps with a transversely mounted clean grain auger (Figs. 1l and 2l), a clean grain elevator (Fig. 1m), and a grain-tank loading auger (Figs. 1n and 2n). The grain tank loading auger is often oriented transversely across the grain tank to help distribute grain, or diagonally to lift grain to the top and center of the tank.

Grain is typically moved from the grain tank with a system of two or three augers. A transverse auger at the bottom of the tank moves grain to one side of the tank. A second auger may elevate the grain to a final unloading auger (Figs. 1o and 2o). The unloading auger is positioned perpendicularly to the direction of travel for unloading and can usually be rotated toward the rear of the combine when not in use as shown in both Figs. 1 and 2. The second and third augers are sometimes replaced with a single auger that both lifts and conveys grain away from the bottom of the tank diagonally.

ACCESSORIES

Combines are typically equipped with many other accessory systems such as straw or stalk choppers and spreaders, monitoring systems and various automatic controls. It is especially important in crops with heavy residue to distribute the threshed and separated plant materials evenly across the entire swath of the header. The redistributed plant residues help cushion the impact of rain and reduce soil erosion. Automatic controls are often used to regulate platform header height so that the platform can operate as closely to the ground as possible for crops that produce seed along the entire stalk such as soybeans.

Many of the systems within the combine are electronically monitored to provide valuable information to the operator that may be used to improve the performance of the threshing, separating, and cleaning systems. Combines may also be equipped with yield monitors and other grain sensors that measure properties such as moisture content. When collected together with position data from a global positioning system (GPS) receiver, the yield data and other grain data can be assembled into the form of a map that depicts the variability of productivity and grain properties across the field.

CONCLUSION

The combine is a complex piece of equipment that combines the five primary functions of cutting and feeding, threshing, separating, cleaning, and materials handling into one harvesting machine. Although the design, appearance, and accessories available on combines may vary from one manufacturer to another, all combines are designed so that the individual systems work together to harvest clean grain from fields of cultivated crops leaving most of the rest of the plant material in the field. Modifications to current designs will continue to improve data gathering capabilities that will provide traceable information about the crops harvested.

REFERENCES

1. Kepner, R.A.; Bainer, R.; Barger, E.L. Grain and Seed Harvesting. *Principles of Farm Machinery*, 3rd Ed.; AVI Publishing Company: Westport, CT, 1978; 392–431.
2. Kutzbach, H.D.; Quick, G.R. Harvesters and Threshers—Grain. In *CIGR Handbook of Agricultural Engineering*; ASAE: St. Joseph, MI, 1999; Vol. III, 311–347.

Grain Production Systems

H. Mark Hanna
Iowa State University, Ames, Iowa, U.S.A.

G

INTRODUCTION

In the industrialized world, grain production systems are dependent on machines for tillage, planting, and chemical application. Tillage modifies soil conditions. Planting affects seed germination and population. Application of crop inputs such as fertilizer and pesticide require calibrated, uniform application equipment. This section covers functional and performance requirements and examples of the range of equipment used before harvest.

TILLAGE

Tillage has developed as an art throughout most of agriculture's history. Tillage is the manipulation of soil to meet a desired objective. Because of the resources (time, equipment, labor, and energy) required for tillage operations, it is important to determine whether tillage is required and if so, identify tillage objectives for a specific grain production scenario. Weed control was the primary reason for tillage before the advent of chemical herbicides and continues to be a common objective today. Shallow tillage in the top 5 cm–10 cm of soil is used in drier areas to control weeds during fallow periods or is used preplant to produce an early flush of weed growth that is destroyed by subsequent tillage prior to planting. Tillage loosens compacted soil or soil sealed on the surface and alters soil structure. Benefits may be temporary, however, if further compaction occurs from field traffic. Tillage can be detrimental if soil structure was initially desirable. Tillage is used to mechanically incorporate fertilizer, lime, herbicide, or other plant amendments into the soil.

Tillage is used to change soil temperature and moisture. Surface soil is generally warmed by tillage from burial of surface plant residues or loosening of the soil. Such actions tend to allow the soil surface to absorb more solar energy without transferring it to deeper depths. Soil moisture can be affected with tillage by allowing soil to more rapidly dry within the tilled layer. In drier conditions, a surface-tilled layer may interrupt capillary flow of soil moisture to the surface and serve as a form of mulch to inhibit subsurface moisture loss.

Because tillage manipulates soil, fertilizer, and pesticide and impacts soil physical structure, tillage performance is usually evaluated by changes in physical or mechanical properties of soil. Measurements may include soil moisture, dry soil bulk density, soil mechanical strength (with a penetrometer), or distribution of soil aggregate sizes.

Tillage for grain production is divided into primary, secondary, and cultivating tillage operations. Primary tillage is initial tilling of soil and is usually associated with loosening soil and tilling at depths greater than 10 cm. Examples of primary tillage equipment include the chisel plow, subsoiler (Fig. 1), moldboard plow, offset-disk harrow, disk plow, and wide-sweep plow.[1] Secondary tillage is shallower and associated with soil mixing of pesticides, soil leveling for subsequent operations, or weed control. Examples of secondary tillage equipment include the field cultivator, tandem disk harrow, and rod weeder.[1] Postplant cultivating tillage with a row crop cultivator is used for weed control. Soil-engaging tools mounted on implement shanks may be considered narrow- or wide-tools.[2] Narrow tools such as a chisel spike or subsoiler point till at least as deep as the width of the tool. Soil fractures on either side of the tool so that effective width of soil fracture is significantly greater than the width of the tool itself when soil is dry enough to be tillable. Wide tools such as a sweep or moldboard plow till only about one-half as deep or less as the width of the tool. Most soil fracture occurs directly in front of the tool. Draft of tillage implements is a function of operating depth and speed as well as soil parameters.[3]

Soil erosion potential from water or wind is usually increased by soil detachment from tillage. Because of erosion, crop input expense, and in drier regions soil moisture loss, some grain producers avoid tillage. Eliminating all tillage before planting (except perhaps fertilizer injection into the soil) is referred to as a no-till system. Other systems that reduce tillage operations include conservation, mulch, and minimum tillage.

PLANTING

Planting equipment design critically affects seed germination, plant emergence, and final plant population. Seed depth, soil coverage, and seed-to-soil contact are important criteria for most grains. Uniform horizontal seed spacing is an additional factor for coarse grain, such as corn. Planting methods may include broadcast

Encyclopedia of Agricultural, Food, and Biological Engineering
DOI: 10.1081/E-EAFE 120006865
Copyright © 2003 by Marcel Dekker, Inc. All rights reserved.

Fig. 1 Primary tillage with a subsoiler.

Fig. 2 Planter operating in crop residue with row cleaner and coulter.

application of seed by random scatter across or into the soil surface, drilling seed by random spacing of seeds within a seed furrow, or precision planting with more uniform spacing of seeds within the furrow.[4]

Because small grain does not require precise horizontal seed spacing, drills or broadcast seeders are commonly used. Broadcast seeders typically use a variable orifice or gate and fluted wheel to meter seed onto one or two spinning disks. Seed is spread through the air by ballistic action and may be affected by size and density of seeds and wind. Seed may be broadcast by aerial application with ram-air in wet soil conditions (e.g., rice). Seed drills, on the other hand, use individual seed furrow openers to plant seed in rows. Seed exits a central hopper and is metered by a fluted wheel or internal double-run seed cup from a variable orifice or gate. Air drills are also used. An air drill transports metered seed pneumatically for release behind individual shanks of a field cultivator. Metering rate determines how many seeds are planted in a given area.

Precision planters use more sophisticated mechanisms for single-seed metering (singulation) of individual seeds than drills and achieve more uniform spacing along seed furrows. An early mechanism used to singulate seed was gravity feeding of seeds into individual cells in a horizontal rotating plate. During the 1960s a "plateless" planter was developed. The finger pickup planter meter uses mechanical fingers rotating in a vertical plane to pick up individual seed; then drop it onto individual seed cells on conveyor belt to the seed drop tube. Later a system was developed using air pressure to singulate and hold seeds from a central seed hopper in a rotating drum prior to release through a seed discharge tube. Subsequent air metering systems use pressure or vacuum to singulate and hold seeds onto cells in a rotating seed plate before release into the discharge tube. Metering system performance is evaluated by both total seeds planted per area and seed spacing uniformity as measured by frequency of skips or multiple seed drops in smaller areas where an individual seed drop is desired.

After seed is metered, furrow opening- and closing-devices are used to place seed at the correct depth with adequate soil contact and coverage.[5] Drills may use single- or double-disk or hoe-type furrow openers. A press wheel or drag chain may be used for soil coverage. Precision planters typically use a double-disk seed furrow opener to place seed at deeper depths (e.g., 7 cm–10 cm) in variable soil conditions. A runner or shoe-type opener may be used in tilled, uniform soil. Single- or double-press wheels are used for closing the seed furrow. If crop residue is present on the surface, row cleaners or a coulter may be required ahead of the furrow opener to improve seed placement and germination as well as planter performance (Fig. 2). Depth of seed placement is controlled by position of the bottom of the seed furrow opener in relation to the depth-gauging wheel. Performance of planter soil-engaging components is evaluated by measuring seed depth (should be uniform and at desired depth) and observing adequate contact between soil and seed without creating excessive compacted soil strength. Percentage seed germination, emergence rate, and final plant stand are further measures of planter performance.

CHEMICAL APPLICATION

Application equipment is used for other crop inputs such as nutrients, pesticides, or growth regulators in grain production systems. Input materials may be in liquid or dry form and be applied by ground or aerial applicators. Application may be pre- or postplant or during planting. It may be broadcast or applied in concentrated bands either on the surface or incorporated into the soil.

Dry granular fertilizer or pesticide is metered through a variable orifice and broadcast onto the soil by ballistic action of spinner spreaders, transported pneumatically by tubes to individual delivery points along a boom, or simply dropped by gravity from the spreader. Uniform particle size and density are important with a spinner spreader to avoid ballistic separation across the swath. Individual drops on a planter or applicator can apply concentrated bands on the surface. Banded material may be injected into the soil behind a knife. Broadcast material may be incorporated by subsequent tillage.

Liquid materials are applied with low-pressure sprayers. Liquid fertilizers may also be applied with ground-driven squeeze pumps. If anhydrous ammonia is used for fertilizer nitrogen, the liquid/gas mixture is injected behind knives below the soil surface. Other liquid products can be incorporated by knife-injection or tillage after surface application. Important components of liquid application systems are pumps, tanks, nozzles, and agitation systems as well as valves, strainers, pressure gauges, and connecting hoses.

Performance of chemical application systems is typically evaluated by a calibrated overall application rate. Calibration is an important separate step prior to any type of application. Calculation of a coefficient of variation from discrete points across the application swath may also be used to gauge application uniformity.[6] For liquid materials, off-target drift concerns may include using the sizes of spray droplets produced to evaluate drift potential and target coverage.

CONCLUSION

Equipment affects soil conditions, seed environment, and the distribution of fertilizer and pesticide. Engineering design and grower selection, operation, and management of equipment are a key to establishing a healthy growing crop. Knowledge of how machines interact with the soil and crop advance modern grain production.

REFERENCES

1. ASAE Standards. *S414.1 Terminology and Definitions for Agricultural and Tillage Implements*, 47th Ed.; ASAE: St. Joseph, MI; 2000.
2. Koolen, A.J.; Kuipers, H. *Agricultural Soil Mechanics*; Springer: New York, 1983; 197 pp.
3. ASAE Standards. *S497.4 Agricultural Machinery Management Data*, 47th Ed.; ASAE: St. Joseph, MI, 2000.
4. Srivastava, A.K.; Goering, C.E.; Rohrbach, R.P. *Engineering Principles of Agricultural Machines*; ASAE: St. Joseph, MI, 1993; 601 pp.
5. ASAE Standards. *S477 Terminology for Soil-Engaging Components for Conservation-Tillage Planters, Drills, and Seeders*, 47th Ed.; ASAE: St. Joseph, MI; 2000.
6. Hofstee, J.W.; Speelman, L.; Scheufler, B. In *CIGR Handbook of Agricultural Engineering, Vol. iii, Plant Production Engineering*; Stout, B., Cheze, B., Eds.; ASAE: St. Joseph, MI, 1999; 240–268.

Grashof Number

Mukund V. Karwe
Indrani Deo
Rutgers, The State University of New Jersey, New Brunswick, New Jersey, U.S.A.

INTRODUCTION

Grashof number (*Gr*) is a dimensionless number used in heat transfer studies involving free or natural convection. When fluids are heated/cooled, they expand/shrink causing the density of the fluid to decrease/increase. Thus, if a hot fluid is surrounded by cooler fluid, the hot fluid will rise up due to the upward buoyancy force in presence of gravity. Similarly, cold fluid surrounded by warmer fluid will flow downward. These flows occur naturally or freely, without any external driving force. The induced flow is opposed by viscous drag, usually at the solid walls of the system. The Grashof number compares the buoyancy force to viscous drag.

Grashof number is named after Franz Grashof (1826–1893), born in Dusseldorf, Germany. He was a professor of applied mechanics and mechanical engineering at Karlsruhe University. He was the founder and editor of Verbandes Deutscher Ingenieure (VDI), an association of German engineers. However, it was H. Groeber[1] who first defined the Grashof number in 1921, well after Grashof's death in 1893.

Even though the origin of Grashof number is in the area of heat transfer, it is also used in mass transfer where flow can occur due to concentration differences. The Grashof number is used to describe natural convection in laminar and turbulent flows in heat and mass transports. In heat transfer, the Grashof number is defined as:

$$Gr = \frac{g\beta\Delta TL^3\rho^2}{\mu^2} = \frac{g\beta\Delta TL^3}{\nu^2} \tag{1}$$

where L is the characteristic length, ρ the density, g the acceleration due to gravity, β the thermal expansion coefficient, and ΔT the temperature difference. μ and ν are the dynamic and kinematic viscosities of the fluid, respectively, with $\nu = \mu/\rho$. The expansion coefficient β is a measure of the rate at which the volume V of the fluid changes with temperature at a given pressure p:

$$\beta = \frac{1}{V}\frac{\partial V}{\partial T}\bigg|_p \tag{2}$$

For a perfect gas, β equals the reciprocal of temperature in kelvin. β for water varies from $-6.8 \times 10^{-5}\text{K}^{-1}$ to $75 \times 10^{-5}\text{K}^{-1}$ over the range 0–100°C at 1 atm.[2] The negative value of β is due to the density inversion of water near the freezing point.

For a given length L and temperature difference ΔT, the fluid buoyancy parameter, $g\beta/\nu^2$ determines the magnitude of the Grashof number. For example, glycerin has low buoyancy ($g\beta/\nu^2 = 3200\,\text{m}^{-3}\text{K}^{-1}$), gases such as air have moderate buoyancy ($g\beta/\nu^2 = 1.5 \times 10^8\,\text{m}^{-3}\text{K}^{-1}$) while lighter liquids such as water ($g\beta/\nu^2 = 2 \times 10^9\,\text{m}^{-3}\text{K}^{-1}$) and liquid metals such as mercury ($g\beta/\nu^2 = 1.4 \times 10^{11}\,\text{m}^{-3}\text{K}^{-1}$) have high buoyancy parameters.[3] During scale-up of systems with natural convection, it is desired to maintain the same Grashof numbers in order for the two systems to be dynamically similar.

In buoyancy driven flows, the value of Grashof number (*Gr*) indicates when the flow undergoes transition from laminar to turbulent, whereas in forced convection flows, the Reynolds number ($Re = \rho VL/\mu$, where V is the characteristic velocity) indicates this transition. Under both circumstances, the Prandtl number defined as $Pr = \nu/\alpha$, where α is the thermal diffusivity, has an additional influence on the flow. Natural convection or buoyancy driven flows are self-driven, whereas forced convection flows are driven by an external source. The relative magnitudes of the Grashof number and the Reynolds number indicate which mode of convection (free or forced) is more dominant. When $Gr \gg Re^2$, forced convection effects are negligible as compared to free or natural convection effects. Both free and forced convections are important (mixed convection) when Gr and Re^2 are of the same order of magnitude.

EXAMPLES OF FLOWS CHARACTERIZED BY GRASHOF NUMBER

Free Convection Near a Semi-Infinite Vertical Plate

Natural convection flows near the surface of isothermal hot or cold, vertical, semi-infinite plates are characterized

Encyclopedia of Agricultural, Food, and Biological Engineering
DOI: 10.1081/E-EAFE 120007001
Copyright © 2003 by Marcel Dekker, Inc. All rights reserved.

by an induced velocity boundary layer near the surface. Far away from the surface, the effects of buoyancy are not felt. The velocity boundary layer begins to become unstable at $Gr_x \approx 400$, and the flow is fully turbulent at $Gr_x \approx 10^9$.[3] Gr_x is the local Grashof number obtained by substituting x for L in Eq. 1, where x is the distance from the leading edge of the semi-infinite plate. For a given fluid, the boundary layer thickness is proportional to $Gr^{-1/4}$.[2]

For natural convection flow near a vertical, semi-infinite plate with constant heat flux q'', the Grashof number is defined[3] as:

$$Gr = \frac{g\beta q'' L^4}{k\nu^2} \tag{3}$$

where k is the thermal conductivity of the fluid.

The surface heat transfer coefficient h for the vertical plate is governed by the product of the Grashof and Prandl numbers. It is often expressed in nondimensional form in terms of Nusselt number $Nu = hL/k$.[4] For an isothermal vertical plate, it is given as:

$$Nu = 0.59(Gr\,Pr)^{1/4} \quad \text{for} \quad 10^4 < Gr\,Pr < 10^9 \tag{4}$$

$$= 0.10(Gr\,Pr)^{1/3} \quad \text{for} \quad 10^9 < Gr\,Pr < 10^{13} \tag{5}$$

Free Convection Near a Horizontal Plate with Hot Upper Surface or Cold Lower Surface

$$Nu = 0.54(Gr\,Pr)^{1/4} \quad \text{for} \quad 10^4 < Gr\,Pr < 10^7 \tag{6}$$

$$= 0.15(Gr\,Pr)^{1/3} \quad \text{for} \quad 10^7 < Gr\,Pr < 10^{11} \tag{7}$$

Free Convection Near Horizontal Plate with Hot Lower Surface or Cold Upper Surface

$$Nu = 0.27(Gr\,Pr)^{1/4} \quad \text{for} \quad 10^5 < Gr\,Pr < 10^{11} \tag{8}$$

Free Convection Around a Sphere of Diameter *d*

$$Nu_d = 2 + \frac{0.589(Gr\,Pr)^{1/4}}{\left[1 + \left(\frac{0.469}{Pr}\right)^{9/16}\right]^{4/9}} \quad \text{for}$$

$$Gr\,Pr < 10^{11} \quad \text{for} \quad Pr > 0.7 \tag{9}$$

It should be noted that the Grashof number as described by Eq. 1 does not have a velocity term (unlike the Reynolds number). This is because the velocities associated with natural convection are typically small, and there is no characteristic "free stream" velocity as in forced convection flows. However, if the velocities are large, a natural convection velocity V_c can be estimated, using force balance and dimensional analysis, as follows[5]:

$$V_c = \sqrt{Lg\beta\Delta T} \tag{10}$$

and the Grashof number can be defined in terms of the natural convection velocity V_c as:

$$Gr = \left[\frac{V_c L}{\nu}\right]^2 \tag{11}$$

The term in the square brackets is the same as Reynolds number.

GRASHOF NUMBER IN MASS TRANSFER

In mass transfer, the Grashof number is formulated in terms of $\beta^*\Delta c$ instead of $\beta\Delta T$, where $\beta*$ is the concentration coefficient of volumetric expansion and Δc is the difference in defined concentration.[2] Gr_m is given as:

$$Gr_m = \frac{L^3\rho^2 g\beta^*\Delta c}{\mu^2} \tag{12}$$

and $\beta*$ is given as

$$\beta* = -\frac{1}{\rho}\frac{\partial\rho}{\partial c}\bigg|_{p,T} \tag{13}$$

The value of $\rho\beta*$ for diffusion of sodium chloride in water is 0.7, for diffusion of ethyl alcohol in air is −0.37, and for diffusion of hydrogen in air is 13.7.

GRASHOF NUMBER IN WAVE MECHANICS

A specialized Grashof number is used in wave mechanics, where it is defined as the ratio of the square of the dissipation time to the internal wave period[6]:

$$Gr_w = \frac{N^2 H^4}{\nu^2} \tag{14}$$

where N is the internal wave frequency, H the characteristic depth, and ν the kinematic viscosity. A value of $Gr > 1$ indicates a slowly decaying wave field while a $Gr < 1$ indicates that the waves are damped by viscous dissipation as soon as they are formed.

REFERENCES

1. Jakob, M. *Heat Transfer*; John Wiley and Sons: New York, 1949; Vol. 1.
2. Gebhart, B.; Jaluria, Y.; Mahajan, R.L.; Sammakia, B. *Buoyancy-Induced Flows and Transport*; Hemisphere Publishing Co.: New York, 1988.
3. White, M. *Viscous Fluid Flow*, 2nd Ed.; McGraw-Hill Publishing Co.: New York, 1991.
4. Singh, R.P.; Heldman, D.R. *Introduction to Food Engineering*, 3rd Ed.; Academic Press: Orlando, FL, 2001.
5. Jaluria, Y. *Natural Convection Heat and Mass Transfer*; Pergamon Press: New York, 1980.
6. Fisher, H.B.; List, E.J.; Koh, R.C.Y.; Imberger, J.; Brooks, N.H. *Mixing in Inland and Coastal Waters*; Academic Press: New York, 1979.

Heat Exchangers for Liquid Foods

Pete E. Athanasopoulos
Agricultural University of Athens, Athens, Greece

H

INTRODUCTION

Heating and cooling of liquid foods is a common practice in handling of these products. Heat is transferred to and from process fluids by using appropriate designed heat transfer equipment. A considerable number of books have been published on the design of heat exchangers and a number of construction companies have been developed. In order to face special practical needs of the food industry, the construction companies follow their own design methods, which are of a proprietary nature and are not widely available.

All the equipment used in heating and cooling are called "heat exchangers." This term is often used for equipment specially designed, in which heat is exchanged between two fluids through a metal sheet.

The most common types of heat exchangers used in heating and cooling of liquid foods are the following:

1. Double-pipe exchanger: the simplest type.
2. Triple-pipe exchanger: mostly for aseptic filling.
3. Shell and tube exchanger: the most common type.
4. Plate heat exchangers: common in milk and juice industry.
5. Spiral heat exchanger: for special processing.
6. Scraped surface heat exchanger: for viscous foods.

BASIC DESIGN EQUATIONS

The heat transfer in a heat exchanger can be described by the general equation:

$$q = UA\Delta T_{\mathrm{L}} \tag{1}$$

ΔT_{L} in Eq. 1 is the logarithmic temperature difference, °C, expressed as:

$$\Delta T_{\mathrm{L}} = \frac{\Delta T_2 - \Delta T_1}{\ln(\Delta T_2/\Delta T_1)} \tag{2}$$

The ln mean temperature difference is evaluated in the inlet and exit ΔT.[1] The value of (ΔT_{L}) is affected by the heat exchanger configuration. Correction factors can be obtained from special charts.[2] If the ratio $(\Delta T_{\mathrm{maximum}})$ over $(\Delta T_{\mathrm{minimum}})$ is less than 1.5, the arithmetic mean temperature (ΔT_{m}) can be used instead of (ΔT_{L}). In this case, an error of 1% or less should be expected.

In the design of heat exchangers, the primary objective is to determine the heat exchange area (A) required for the specific amount of heat (q) to be transferred. Since the temperature on both sides of the exchange surface can be measured, or estimated, calculation of ΔT_{L} is not a problem. The most difficult design parameter is the overall heat transfer coefficient. For tubular heat exchanger, it can be calculated form the equation:

$$\frac{1}{U} = \frac{1}{h_o} + \frac{1}{f_o} + \frac{d_o \ln(d_o/d_i)}{2kw} + \frac{d_o}{d_\mathrm{i}} \times \frac{1}{f_\mathrm{i}} + \frac{d_o}{d_\mathrm{i}} \times \frac{1}{h_\mathrm{i}} \tag{3}$$

The U calculated is based on the outside area of the tube.[3] Values of h_o and h_i are affected by the thermophysical properties of the fluids: density, viscosity, thermal conductivity, and the fluids velocity:

$$Nu = \frac{h_o(\mu_\mathrm{b}/\mu_\mathrm{s})}{D} \quad \text{and} \quad Nu = \mathrm{Gr\ Re} \tag{4}$$

For shell and tube heat exchangers, typical values of heat-transfer coefficient, for liquid foods is 600 W/m^2 °C–1200 W/m^2 °C.

Fouling factors are usually quoted as heat-transfer resistance, rather than coefficients. Fouling of the heat-transfer surfaces in an exchanger is common in processing liquid foods. It is expected this resistance to be more important on the food side surface. Conventional thermal processing calculation method cannot be employed to the establishment of processing of low-acid heterogeneous liquid foods containing discrete particulate because of the difficulties associated with gathering experimental time–temperature data at the particle center as it travels through the system.[4]

COMMON TYPES OF HEAT EXCHANGERS

A number of heat exchanger configurations have been adapted by the food industry. The most common types will be described in brief.

Encyclopedia of Agricultural, Food, and Biological Engineering
DOI: 10.1081/E-EAFE 120007003
Copyright © 2003 by Marcel Dekker, Inc. All rights reserved.

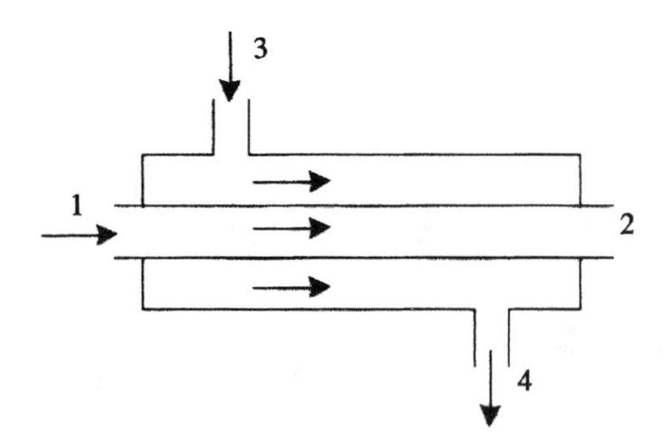

Fig. 1 Tubular heat exchanger. 1: product in, 2: product out, 3: heating medium in, 4: heating medium out.

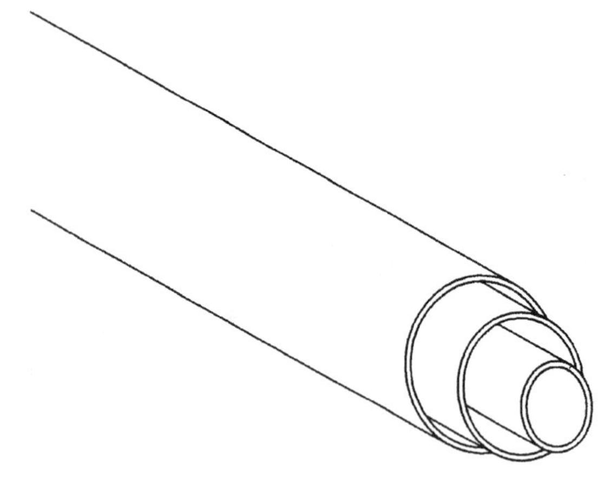

Fig. 2 Triple pipe heat exchanger.

Tubular Heat Exchanger

The concentric tube heat exchanger is the most common type of the tubular heat exchangers (Fig. 1). The hot liquid, or steam, flows through the outer jacket and the cold liquid flows through the inner tube.

Depending on the flow direction, the heat exchanger is characterised as concurrent (if the fluids directions are the same) and as counter current (in the case of opposite directions). The heat flax in this type of heat exchangers can be described by Eq. 1. The temperature difference Eq. 2 is calculated from ΔTs.[5] The presence of particles in a liquid food has a strong positive effect on heat transfer.[6]

$$\Delta T_1 = T_{ho} - T_{ci} \quad \text{and} \quad \Delta T_2 = T_{hi} - T_{\infty} \tag{5}$$

Triple Tube Heat Exchanger

This type of heat exchanger is the most common in the food industries that have adapted the aseptic filling of liquid foods such as tomato juice, orange juice, and other fruit juices. A triple tube heat exchanger consists of three concentric tubes (Fig. 2).

The heating or cooling medium flows through the outer and inner tubes and the liquid food through the annulus. The convective heat transfer coefficient, in this arrangement, is expressed somewhat different, both for laminar and turbulent flow.

Shell and Tube Heat Exchanger

The shell and tube heat exchanger is the most common type of heat-transfer equipment used in food industry today. This configuration has a number of advantages, the principal of which is that it gives a large surface area in a small volume. Heat exchangers of heavy duty are of this type.

Essentially, a shell and tube heat exchanger consists of a bundle of tubes enclosed in a cylindrical shell (Fig. 3). The type of head arrangement allows for one tube pass or multitube pass for the product. In a one-pass arrangement, the product enters at one end and exits at the opposite end. In a multipass arrangement, the product may travel back and front, through different tubes with each pass, before finally leaving the heat exchanger. The heating or cooling medium flows through the annulus with one pass and the liquid food through the tubes. Depending on the number of passes of food, heat exchangers are characterised as single-pass or 1-1, double-pass or 1-2, and triple-pass or 1-3.[7]

In most shell and tube exchangers, the flow will be a mixture of co-current counter current and cross flow. The usual practice in the design of shell and tube exchangers is estimate the "true temperature difference" from the logarithmic mean (Eq. 2) by applying a correction factor.[3]

Shell: It is constructed from stainless steel sheet for better hygienic conditions. It comes in contact with heating or cooling medium such as water, steam, refrigerants, or other media.

Tubes: They are seamless constructed from high-quality stainless steel. This material has unfortunately, very low-heat conductivity, compared to common metals. Since liquid food is passed through the tubes, their construction should be according to hygienic rules. The geometry of the tube arrangement is the square and the triangle one. For the same shell diameter, the triangular arrangement allows higher number of tubes than the square one. The last arrangement leaves small cleaning lanes, for brass cleaning of the outside surface of the tubes, and leads to higher pressure drop on the shell side.

According to TEMA standards,[8] the least distance between the tube centers should be 1.25 times of the outside tube diameter, for the triangular arrangement, and least cleaning lane of 0.6 cm.

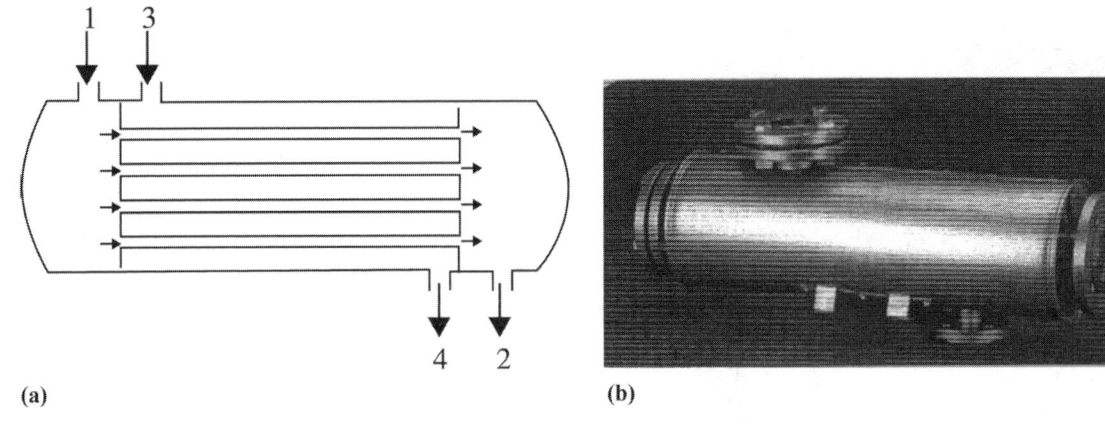

Fig. 3 a) Single pass shell and tube heat exchanger. 1: product in, 2: product out, 3: heating medium in, 4: heating medium out. b) Single pass shell and tube heat exchanger (courtesy GEA).

Plates: Metal discs, on both ends of the heat exchanger separate the heating and cooling streams and support the tubes. The tubes are stuck or mandreled on the plates which, in turn are supported by the tie rods.

Baffles: The baffles are used to increase convective heat transfer coefficient. They direct the shell-side fluid to a vertical to the tubes turbulent flow (Fig. 3). Baffles are metallic discs supported by spacers located on the tie rods. The baffle pitch is recommended to be equal or higher to one-fifth of the shell diameter.

Heat transfer in a shell and tube heat exchanger: Calculations of the convective heat transfer coefficient (h) are complicated in this type of equipment. The tube inside heat transfer coefficient can be calculated from empirical equations.[3] The calculations of the outside of the tube's convective heat transfer coefficient (h_o) are difficult due to the vertical flow of the heating or cooling medium, to the leaks through the baffles and shell and baffles and tubes connection. A number of diagrams are available for the design purposes for heat exchangers of two, three and four fluid passes.[7]

Plate Heat Exchanger

The gasketed type plate heat exchanger is commonly used in heating and cooling of liquid foods such as milk, beverages, and fruit juices. It consists of a frame supporting a stack of closely spaced thin stainless steel plates. A thin gasket fixed with adhesive seals the plates round their edges. The plates of the same type and gasket are compiled in packs with the alternate plate rotated by 180°. The gap between the plates is normally in the range of 3 mm–7 mm. The plates are embossed by high pressure on a special mold with a pattern of ridges. The ridgetity of the plates minimizes the channelling of the fluid, increases the mixing, and leads to a higher convective heat transfer coefficient (Fig. 4). The corrugated plates are compressed in a frame by two deflecture resistance and plates and tightened with bolts. Different plate packs can be separated in sections by intermediate plates that allow the performance of several independent thermal process steps in one frame. Generally, heat transfer coefficients are higher in plate heat exchangers than those in shell and tube heat exchangers.

Plate heat exchangers are more compact than shell and tube exchangers, they exhibit small pressure drop, are easily cleaned by CIP system, and they are expandable. They offer a wide range of plate surfaces, number of plates, and shape of ridges depending on the manufacturer. The operating pressure of plate heat exchanger is much higher compared to that of a shell and tube heat exchanger for the same follow rate. Standard pressure stages in a range of 16 but up to 25 bar or higher for narrow gap heat exchanger and 4–6 bar for wide gap design. The calculation of the pressure drop is very difficult but it

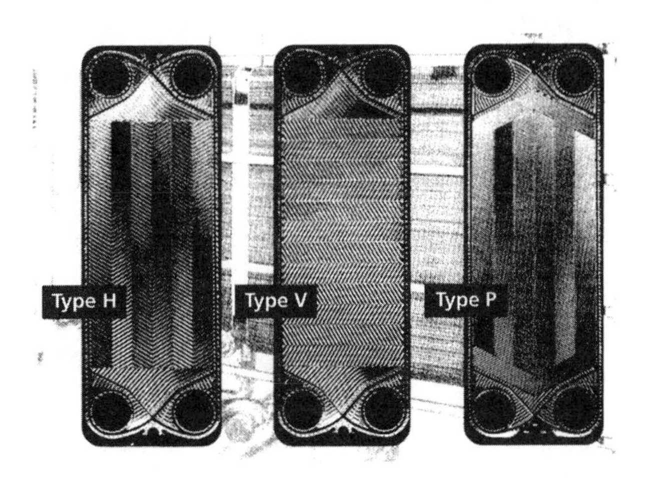

Fig. 4 Three different plates, with horizontal (H), with vertical (V), and with V-shaped (P) profile (courtesy GEA).

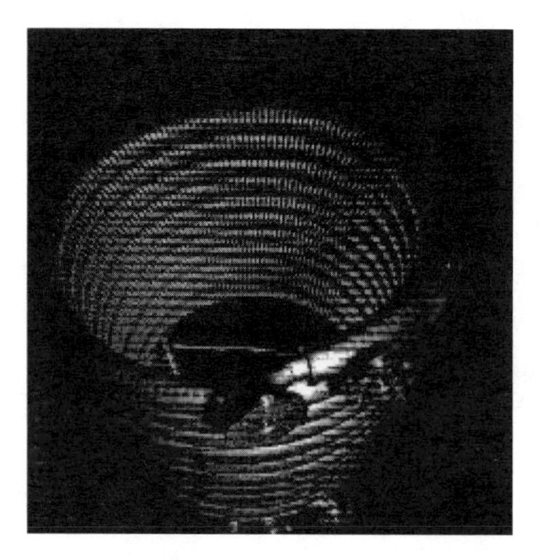

Fig. 5 Tube type spiral heat exchanger (courtesy HRS).

can be estimated using Eq. (6)

$$\Delta P_p = 8j_f\left(\frac{L_p}{d_e}\right)\frac{p(G_p/p)^2}{2} \quad (6)$$

In Eq. 6, the friction factor will depend on the design of the corrugated plate surface. For turbulent flow, the following relationship can be used.

$$j_f = 1.25\text{Re}^{-0.3} \quad (7)$$

With some plate heat exchangers designs, turbulent flow can be achieved at very low Reynolds numbers.

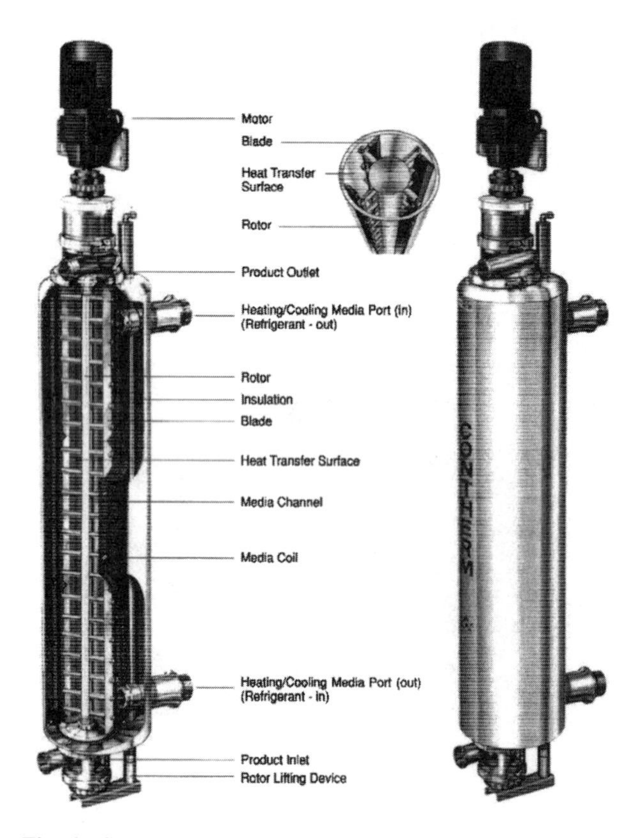

Fig. 6 Scraped-surface heat exchanger (courtesy Tetra Pack).

Spiral Heat Exchanger

This type of heat exchanger rarely is used in food processing. They are compact units. The exchanger is made up of two long metal strips wound around a center to form two spiral passages separated by leak-proof welds.[3] Covers secured by hook bolts retain the pressure. This configuration has been adapted also for spiral tubular heat exchanger that can be used for sterilization or pasteurization of milk, puddings, custards, yoghurt, salad dressings, ice cream mix etc. (Fig. 5).

Scraped Surface Heat Exchanger

Conventional heat exchangers are not suitable for viscous, sticky, and crystalline products. Scrapped surface heat exchangers (Fig. 6) are ideal for heating and cooling of such products. A rotor and a blade assembly continually removes the product from the heat surface and simultaneous it is blended and the temperature distribution ensured.[5]

Typical application of this heat exchanger is in cooking and cooling of sauces, gravies, soups, marmalade, preserves, jams, and dressings.

NOMENCLATURE

A	the area through which the heat is transferred, m^2
ΔP_p	pressure drop
d_e	twice the gap between the plates, m
d_i	tube inside diameter, m
d_o	tube outside diameter, m
f_i	inside fouling factor, W/m^2 °C
f_o	outside fouling factor, W/m^2 °C
G_p	mass flow rate per unit cross-sectional area, kg sec m^{-2}
h_i	inside surface coefficient, W/m^2 °C
h_o	outside surface coefficient, W/m^2 °C
J_f	friction factor, dimensionless
k_w	thermal conductivity of the tube wall material, W/m^2 °C
L_p	the length path, m

P	fluid density, kg m^{-3}
q	heat transferred across the surface, W
U	the overall coefficient, W/m^2 °C
u	the overall transfer coefficient, W/m^2 °C
x	tube wall thickness, m

REFERENCES

1. Toledo, R.T. Heat Transfer. In *Fundamentals of Food Process Engineering*, Toledo, R.T., Ed.; AVI Publishing Company: Westport, Conn, 1980; 228–232.
2. Perry, H.R.; Green, W.D. Heat Transfer Equipment. *Perry's Chemical Engineers' Handbook*, 7th Ed.; McGraw-Hill: Singapore, 1998; 11.4–11.7.
3. Sinnott, K.R. Heat-Transfer Equipment. *Chemical Engineering*, 2nd Ed.; Betterwoth and Heinemann: New York, USA, 1991; Vol. 6, 584–698.
4. Ramaswamy, H.S.; Abdelrahim, A.K.; Simpson, K.B.; Smith, P.J. Residence Time Distribution (RTD) in Aseptic Processing of Particulate Foods: A Review. J. Food Eng. **1995**, *28*, 291–310.
5. Heldman, D.R.; Singh, P.R. Heating and Cooling Processor. *Food Process Engineering*, 2nd Ed.; AVI Publishing Company, Inc.: Westport, Connecticut, 1981; 116–124.
6. Sannarvik, J.; Bolmstedt, U.; Tragardh Heat Transfer in Tubular Heat Exchangers for Particulate Containing Liquid Foods. J. Food Eng. **1996**, *29*, 63–74.
7. Mc Cabe, L.W.; Smith, C.J. Heat Transfer Equipment. *Unit Operations of Chemical Engineering*, 2nd Ed.; Greek version Mc Graw-Hill Book Company, 1971; 500–512.
8. TEMA. *Standards of the Tubular Heat Exchanger Manufactures Association*, 7th Ed.; Tubular Heat Exchanger Manufactures Association: New York, 1998.

Heat Transfer

Kevin M. Keener
North Carolina State University, Raleigh, North Carolina, U.S.A.

INTRODUCTION

Heat transfer is a fundamental phenomenon involved in agricultural production and food processing from controlling temperature of animal production facilities to hydrocooling of fruits and vegetables. In many applications, heat transfer occurs simultaneously with mass transfer. One example would be high-temperature drying where grain is harvested and then placed in a high-temperature air stream. The hot air heats the grain and at the same time increases the rate of water loss (rate of mass transfer). The higher the temperature, the greater the mass transfer rate. Heat transfer processes are also used to enhance biological processes such as fermentation. For example, in the production of beer, yeast, malt, and barley are placed in a large kettle and heat is added to increase yeast activity and fermentation rate. There are numerous other examples of heat transfer that occur in processing and drying and storage systems. The purpose of this article is to provide an introductory overview of these processes and the mechanisms of heat transfer, which predominates during the process.

Heat is energy transferred between materials due to temperature differences (e.g., different amounts of internal energy). There are three mechanisms that are responsible for thermal energy transfer: convection, conduction, and radiation. These mechanisms can occur individually or in combination. An example of a situation where all three mechanisms occur is heat being conducted through a double pane window, Fig. 1. On the outside of the window, air transfers heat by convection to the window glass. Conduction transfers the energy through the glass. On the inside surface of the glass, energy (heat) is radiated between the two glass panes and is also conducted through the air filled gap. Conduction occurs instead of convection because the air gap is small and convection currents cannot develop. The energy is then conducted through this window glass to the inside of the room. Convection then transfers the energy away from the glass warming the room air. The direction of energy transfer (heat) is always from high temperature (warm) to low temperature (cold).

CONDUCTION

Conduction is the process of heat transfer by molecular transport and microscopic interactions. Conduction can occur in solids, liquids, and gases, but usually is only a major contributor in heat transfer through solids. Heat loss through the wall of a building is an example. There are three primary mechanisms of heat transfer by conduction: direct contact (collisions) between neighboring molecules, lattice vibrations, and in metals, "electron" transfer. Electron transfer refers to the ability of free electrons to transfer heat, analogous to conducting electricity. It is possible to quantify the rate of heat transfer through a solid by conduction. This equation is called Fourier's law,

$$q = \frac{kA(T_i - T_o)}{d} \quad (1)$$

Fourier's law states that the heat flux (q) is equal to the thermal conductivity (k) multiplied by the area (A) and the temperature difference between two surfaces divided by the distance (d) between them. The thermal conductivity, k, is a proportionality constant that is experimentally measured, and describes the combined effect of all three conduction mechanisms. The larger the thermal conductivity, the greater the heat transfer by conduction. Thermal conductivity is dependent on composition and temperature for most materials. For materials with large water contents, such as foods, thermal conductivity can be approximated by empirical equations such as

$$k = 0.26 + 0.0034M \quad (2)$$

for unfrozen materials,[1] where M is the percent moisture content on a wet basis. Table 1 summarizes thermal conductivity values for some common materials.

Example 1

Determine the rate of heat transfer by conduction through a single-pane glass window. Assume an outside glass surface temperature of 38°C (100°F) and an inside surface temperature of 32°C (90°F). The window pane is 0.6096 m

Encyclopedia of Agricultural, Food, and Biological Engineering
DOI: 10.1081/E-EAFE 120006891
Copyright © 2003 by Marcel Dekker, Inc. All rights reserved.

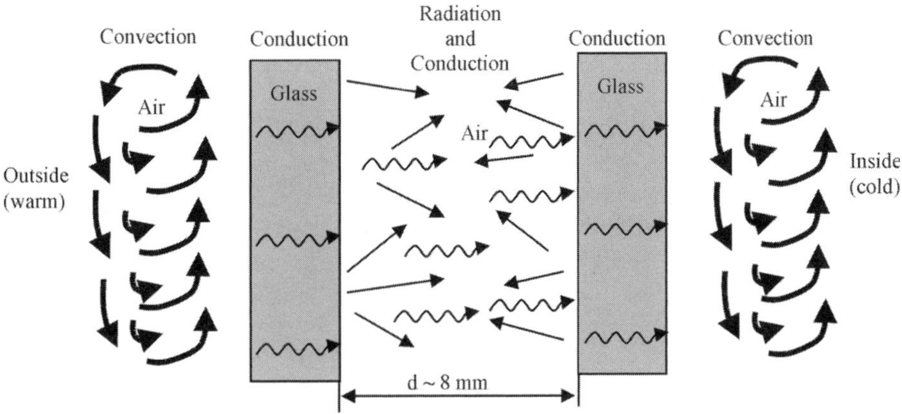

Fig. 1 Energy transfer by heat through a double pane window.

(24 in.) × 0.9144 m (36 in.) and 0.635 cm (0.25 in.) thick.

$$q = \frac{kA(T_i - T_o)}{d}$$

$$= \frac{(0.84\,\mathrm{W\,m^{-1}\,K^{-1}})(0.6096\,\mathrm{m})(0.9144\,\mathrm{m})(305\mathrm{K} - 311\mathrm{K})}{0.00635\,\mathrm{m}}$$

$$q = -442\,\mathrm{W}$$

This indicates that approximately 442 W of energy is being conducted through this window. The negative sign means that the energy moves from the outside (warm) to inside (cold).

Example 2

Using the same data as those in "Example 1" calculate the energy loss for fiberglass insulation ($k = 0.0318\,\mathrm{W\,m^{-1}\,K^{-1}}$) instead of a glass windowpane.

$$q = -16.8\,\mathrm{W}$$

This is why it is advantageous to limit the size of window in an oven or the number of windows in a heated/cooled building.

CONVECTION

Convection is the process of heat transfer by macroscopic movement of molecules. There are two mechanisms responsible for convection—molecular diffusion and macroscopic motion. Molecular diffusion (self-diffusion) involves random motion of molecules due to their internal energy content. "Hotter" molecules have higher internal energy. When molecules collide, some energy is transferred from the high-energy molecules to the low-energy molecules. Macroscopic motion typically occurs because of either forced movement of the fluid (stirring, pumping, etc.) or natural convection that results from density changes due to temperature differences in the fluid. Convection is the mechanism for heat transfer between

Table 1 Thermal conductivity values for some common materials

Material	Temperature (K)	Thermal conductivity ($\mathrm{W\,m^{-1}\,K^{-1}}$)	Reference
Still air (101.3 kPa)	300	0.026	2
Insulation (glass fibers)	300	0.0318	2
Wood (pine, across grain)	300	0.151	2
Water	293	0.602	2
Ice	273	1.88	2
Frozen chicken (74.4% m.c.)	198	1.60	2
Meat (unfrozen)		0.460	3
Shelled corn (9.8% m.c.)	300	0.152	4
Concrete or stone		0.84	5
Glass	300	0.84	5
Steel	300	40	5
Aluminum	300	200	5

a liquid or gas (fluid) and a solid. At the interface between a fluid and a solid, most energy transfer occurs by molecular diffusion because the fluid velocity is near zero. For example, a cold fluid flowing through a heated pipe would have a thin, stagnant layer of fluid at the pipe wall where molecular diffusion would be the dominant heat transfer mechanism. The combined effect from convection processes is quantified using Newton's Law of Cooling,

$$q = hA(T_s - T_b) \tag{3}$$

Heat transfer (q) is proportional to the heat transfer coefficient (h) multiplied by the area of heat transfer (A) and the temperature difference between the average surface temperature (T_s) and the average (bulk) fluid temperature (T_b).

There are two types of convection encountered: natural convection and forced convection. In natural convection, fluid movement occurs due to fluid density differences. A warm fluid has a lower density than a cold fluid. Fluid density differences create natural circulation patterns in the fluid, an example being the steam rising from a cup of hot coffee. Forced convection occurs when mechanical energy is expended to enhance the convection process. An example is the pumping of milk through a plate heat exchanger for pasteurization. In this situation, mechanical motion supplies momentum to the fluid, which increases its velocity and circulation.

In convection processes, the heat transfer coefficient (h) is impacted by the flow regime that develops. Usually in forced convection, the goal is to develop turbulent flow. Turbulent flow is where a fluid is well mixed and little temperature gradient exists through the bulk fluid. In turbulent flow, a large number of eddies (localized circulation patterns) develop that enhance fluid mixing. If the fluid has a consistent flow pattern (e.g., streamlines) then the fluid movement is characterized as laminar. The type of flow conditions that exist can be found by calculating the Reynolds (Re) number. Re is a dimensionless number that compares the inertial force to the viscous force,

$$Re = \frac{\rho V L}{\mu} \tag{4}$$

For pipe flow, if $Re > 10{,}000$ then the flow is turbulent, less than 2,100 the flow is laminar, and between 2,100 and 10,000 the flow is in transition, a combination of both.[6]

Determination of the convective heat transfer coefficient (h) for a process is complex and depends on a number of factors including fluid properties, geometry, temperature difference, and forced or natural convection. Because of the difficulty in quantifying all these factors, the heat transfer coefficient is usually determined experimentally; however, generalized equations have been developed for some common situations based on dimensionless numbers: Re, Prandtl (Pr) number, Grashof (Gr) number, and Nusselt (Nu) number. The Pr number compares the momentum diffusivity to thermal diffusivity,

$$Pr = (C_p)(\mu)/k \tag{5}$$

Gr relates the inertial force and buoyant force to (viscous force)^2 as

$$Gr = \frac{\rho^2 g L^3 \beta \Delta T}{\mu^2} \tag{6}$$

The Gr number is an important factor in natural convection processes. The Nu number relates the convective heat transfer to conductive heat transfer as

$$Nu = (h)(L)/k \tag{7}$$

For natural convection processes,

$$Nu = a(Pr)^m(Gr)^m \tag{8}$$

where $m = 0.33$ for turbulent flow and 0.25 for laminar flow.[6] For forced convection,[6]

$$Nu = b(Re)^n(Pr)^p \tag{9}$$

Once Nu number is determined a convective heat transfer coefficient (h) can be calculated using Eq. 7. Tables 2 and 3 summarize the convective heat transfer equations for natural and forced convection, respectively. All properties information is determined at the film temperature, T_f, which is the average temperature between the surface and bulk fluid. Simplified equations for water and air have been included since these are the two most common fluids. Values for convective coefficient range from $20\,\mathrm{W\,m^{-2}\,K^{-1}}$ for air cooling a baked cake to $30{,}000\,\mathrm{W\,m^{-2}\,K^{-1}}$ for French fries immersed in hot oil.

Example 3

Determine the convective heat transfer coefficient for fresh picked cherries (1.5 cm diameter) being cooled in a recirculating water tank at 7°C. The tank is 2 m in diameter and 2 m deep. The relative velocity between a cherry and water is $0.020\,\mathrm{m\,sec^{-1}}$. Assume the tank is full of cherries and the water moves around each individual cherry. The initial temperature of the cherry is 30°C with a moisture content of 84%.

For forced convection, for spheres

$$Nu - 2 = b(Re)^n(Pr)^p$$

where

$$Re = (\Delta VL)/\mu, \quad Pr = (C_p)(\mu)/k$$

Table 2 Natural convection

Surface type	Characteristic length (m)	$Gr \times Pr$	a	m	h_{air} at 1 atm ($W m^{-2} K^{-1}$)	h_{water} at 294K ($W m^{-2} K^{-1}$)
Vertical surface—plates and pipes	L = vertical length ($L < 1.0$ m)	$< 10^4$	1.36	0.20		
		10^4–10^9	0.59	0.25	$1.37(?T/L)^{0.25}$	$127(?T/L)^{0.25}$
		$> 10^9$	0.13	0.33	$1.24(?T)^{0.33}$	
Horizontal cylinder	L = cylinder diameter ($L < 0.20$ m)	$< 10^{-5}$	0.49	0		
		10^{-5}–10^{-3}	0.71	0.04		
		10^{-3}–1	1.09	0.10		
		1–10^4	1.09	0.20		
		10^4–10^9	0.53	0.25	$1.32(T/L)^{0.25}$	
		$> 10^9$	0.13	0.33	$1.24(T)^{0.33}$	
Horizontal plate—facing upward—heating top or cooling bottom	L = plate length	10^5–2×10^7	0.54	0.25	$1.32(?T/L)^{0.25}$	
		2×10^7–3×10^{10}	0.14	0.33	$1.52(?T)^{0.33}$	
Horizontal plate—facing downward	L = plate length	3×10^5–3×10^{10}	0.27	0.25	$0.59(T/L)^{0.25}$	

(From Ref. 6.)

Here, μ(water) = 0.00142 Pa-sec, $\Delta = 1000\ \text{kg m}^{-3}$, $C_p = 4.20\ \text{kJ kg}^{-1}\text{K}^{-1}$, $k = 0.574\ \text{W m}^{-1}\text{K}^{-1}$, L (cherry diameter = 0.015 m), and $V = 0.02\ \text{m sec}^{-1}$.

From Eq. 2

$$k(\text{cherry}) = 0.26 + 0.0034(84) = 0.5456\ \text{W m}^{-1}\text{K}^{-1}$$

Hence

$$Re = (1000\ \text{kg m}^{-3})(0.02\ \text{m sec}^{-1}) \times (0.015\ \text{m})/(0.00142\ \text{Pa-sec}) = 211.2$$

$$Pr = (4.20\ \text{kJ kg}^{-1}\text{K}^{-1}) \times (0.00142\ \text{Pa-sec})/(0.5456\ \text{W m}^{-1}\text{K}^{-1}) = 10.93$$

From Table 3, for forced convection, $b = 0.6$, $n = 0.5$, $p = 0.33$ for a sphere

$$Nu = 2 + (0.6)(2112)^{0.5}(10.93)^{0.33} = 21.35$$

In addition,

$$Nu = hL/k$$

As $L = 0.015$ m,

$$h = 21.35(0.5456\ \text{W m}^{-1}\text{K}^{-1})/0.015\ \text{m} = 777\ \text{W m}^{-2}\text{K}^{-1}$$

RADIATION

In most systems, radiation is seldom a significant contributor to the total heat flux; however, when vacuum is present or extremely large temperature differences, greater than 100°C, exist, radiation effects should be considered. There are many systems where radiation from the sun is the only contributor, such as solar drying of grass, fruit, or lumber.

Radiation is the energy transfer by electromagnetic waves (e.g., energy from the sun). This energy transfer can include both visible and invisible (ultraviolet, infrared, gamma radiation, etc.). All materials emit energy (heat) in the form of radiation and the amount of energy emitted increases with temperature. The net energy transfer is from the hotter material to the colder material. Radiation does not require a medium, such as fluid or solid, to transfer energy. Radiation can be reflected, absorbed, or transmitted between surfaces. Materials vary in their ability to absorb radiation. For example, a windowpane is fairly transparent to sunlight (short wavelengths), but reflects radiation from objects in a room (long wavelengths). This is why on a sunny day a room will warm up when the sun is shining in the window. A black body is defined as a material that absorbs 100% of incident radiation and reflects none. It also emits radiation at the maximum rate for a given temperature. There are no true black bodies, thus a correction factor known as emissivity, γ, is used to account for differences in radiation between real materials ($\gamma < 1.0$) and a black body ($\gamma = 1.0$). Emissivity is the ratio of emitted energy by a material to that of a black body at the same temperature.

Table 3 Forced convection

Surface type	Characteristic length (m)	Re	Pr	b	n	p	h_{air} at 1 atm ($W m^{-2} K^{-1}$)	h_{water} at 294K ($W m^{-2} K^{-1}$)
Vertical plate ζ flow	L = vertical length	4×10^3–1.5×10^4	0.7–500	0.228	0.731	0.33		
Sphere (replace Nu by $Nu - 2.0$)	L = diameter	1–7×10^4	0.6–400	0.6	0.5	0.33		
Horizontal cylinder ζ flow	L = cylinder diameter	0.4–4	> 0.7	0.989	0.330	0.33		
		4–40	> 0.7	0.911	0.385	0.33		
		40–4000	> 0.7	0.683	0.466	0.33		
		4×10^3–4×10^4	> 0.7	0.193	0.618	0.33		
		4×10^4–4×10^5	> 0.7	0.027	0.805	0.33		
Horizontal plate	L = plate length	$< 5\times10^5$ (laminar)	> 0.6	0.664	0.5	0.33		
Through horizontal circular tubes[a,b]	L = tube diameter	< 2100 (laminar)	0.5–16,700	1.86	0.33	0.33		
Through horizontal circular tubes[b]	L = tube diameter	$> 10{,}000$ ($L/D > 10$)	0.7–16,700	0.027	0.8	0.33	$3.52(V^{0.8}/D^{0.2})$	$1057(0.02T-4.108)(V^{0.8}/D^{0.2})$

[a] Must multiply equation by $(D/L)^{0.33}$.
[b] Must multiply equation by (μ_b/μ_w).
(From Ref. 6.)

Emissivity varies with temperature and wavelength of incident radiation, but for most nonmetallic materials γ ranges from 0.90–0.95. This result is what allows infrared thermometers to accurately measure surface temperatures of most materials. For metals, γ varies from 0.02 to 0.9 depending on surface finish and material composition. Radiation is not a single wavelength of light emitted, but a distribution described by Planck's law,

$$E_\lambda = \frac{3.7418\times10^{-16}}{\lambda^5(e^{1.438\times10^{-2}/\lambda T} - 1)} \tag{10}$$

Planck's law states that emitted radiation varies continuously with wavelength and at any wavelength emissive power increases with increasing temperature. In addition, as temperature increases, the wavelength of maximum spectral emissive power decreases. To obtain total emissive power per unit area emitted from a blackbody, one can integrate Planck's law (Eq. 10) and obtain the Stefan-Boltzmann Law,

$$E_B = \int_0^\infty E_\lambda \, d\lambda = \sigma T^4 \tag{11}$$

Stefan-Boltzmann Law provides the total emissive power per unit area for a black body at a specific temperature.

As shown by the Stefan-Boltzmann Law (Eq. 11), emissive power is a function of T^4; however, the emissivity and geometry effects need to be accounted for in order to calculate the net energy transfer between two objects by radiation. A generalized equation can be developed describing energy transfer by radiation between two objects,

$$q = \sigma F_A F_R A (T_1^4 - T_2^4) \tag{12}$$

Table 4 has information on correction factors for common geometries, and Table 5 lists some emissivities for some common materials at specific temperatures.

Example 4

Determine the heat transfer to a 3 m × 3 m flat roof (black shingles, $\gamma = 1.0$) on a cold ($T = 0°C$) sunny day. The sun is directly overhead.

Assume $T_{roof} = 0°C$, $T_{sun} = 5800K$, $r = 697{,}000$ km, $d_s = 150{,}000{,}000$ km, $N = 0°$

$$F_A = (697{,}000/150{,}000{,}000)^2 \cos(0 \text{ rad}) = 2.159\times10^{-5}$$

$$F_R = \gamma_1 = 1.0$$

$$A_1 = 3\,m \times 3\,m = 9\,m^2$$

$$q = (9\,m^2)(2.159\times10^{-5})(1.0)(5.676\times10^{-8}\,W\,m^{-2}\,K^{-4}) \times [(5800K)^4 - (274K)^4]$$

$$q = 12{,}482\,W$$

H

Table 4 Radiation between solids

Radiating surface	Area, A	F_A	F_R
Infinite parallel planes	A_1 or A_2	1	$1/(1/\gamma_1 + 1/\gamma_2 - 1)$
Completely enclosed body, 1, small compared with enclosing body	A_1	1	γ_1
Completely enclosed body, 1, large compared with enclosing body	A_1	1	$1/(1/\gamma_1 + 1/\gamma_2 - 1)$
Flat surface, 1, and point source (e.g., sun)	A_1	$(r/d_s)^2 \cos N$	γ_1

(From Ref. 8.)

Example 5

Assume an aluminum roof ($\gamma = 0.04$) and the sun is at 60° from the horizon (30° from normal).

$$F_A = (697,000/150,000,000)^2 \cos(0.523\,\text{rad})$$

$$= 1.870 \times 10^{-5}$$

$$F_R = \gamma_1 = 0.04$$

$$q = (9\,m^2)(1.870 \times 10^{-5})(0.04)$$

$$\times (5.676 \times 10^{-8}\,W\,m^{-2}\,K^{-4})[(5800K)^4 - (274K)^4]$$

$$q = 432.5\,\text{W}$$

LIST OF ABBREVIATIONS

A Area, m^2
A_1 Surface area of material 1, m^2
A_2 Surface area of material 2, m^2
∃ Thermal coefficient of volumetric expansion, m^3 m^{-3} K^{-1}
C_p Specific heat of a material, kJ kg^{-1} K^{-1}
D Diameter of pipe, m
E_λ Emissive power at specific wavelength, W m^{-3}
E_b Total emissive power, W m^{-2}
γ Emissivity, $0 < \gamma < 1$, dimensionless
γ_1 Emissivity of surface 1, dimensionless
γ_2 Emissivity of surface 2, dimensionless
d Distance between temperature T_i and T_o, m
d_s Distance between source and surface, m
F_A Radiation area factor, dimensionless
F_R Reradiation factor, dimensionless
g Gravitational constant, 9.81 m sec^{-2}
Gr Grashof number, dimensionless
h Convective heat transfer coefficient, W m^{-2} K^{-1}
h_{air} Convective heat transfer coefficient for air, W m^{-2} K^{-1}
h_{water} Convective heat transfer coefficient for water, W m^{-2} K^{-1}
k Thermal conductivity, W m^{-1} K^{-1}
L Characteristic length, m

Table 5 Emissivity of common materials

Surface	Temperature (K)	Emissivity	Reference
Polished aluminum	300	0.04	2
Polished stainless steel	300	0.17	2
Rolled sheet steel	293	0.82	7
Cast iron, oxidized	313	0.95	7
Window glass	300	0.90–0.95	2
Tungsten filament	6,000	0.39	6
Red brick	300	0.93–0.96	2
Planed oak	293	0.895	6
Lube oil (infinitely deep)	273	0.82	6
Paints (nonmetallic)	274–373	0.80–0.96	6
Water	300	0.96	2
Ice	273	0.95–0.98	2
Snow	273	0.82–0.90	2
Human skin	300	0.95	2

λ	Wavelength, m
M	Moisture content wet basis, %
μ_b	Absolute viscosity at bulk fluid temperature, $kg\,m^{-1}\,sec^{-1}$
μ_w	Absolute viscosity at wall temperature, $kg\,m^{-1}\,sec^{-1}$
Nu	Nusselt number, dimensionless
N	Angle between source and the normal to the surface, rad
Δ	Density, $kg\,m^{-3}$
N	Angle between point source and surface, rad
q	Heat transfer, W
Pr	Prandtl number, dimensionless
r	Equivalent radius of point source, m
Re	Reynolds number, dimensionless
Φ	Boltzmann's constant, $5.676 \times 10^{-8}\,W\,m^{-2}\,K^{-4}$
ΔT	Temperature difference between bulk fluid and surface, K
T	Absolute temperature, K
T_b	Bulk temperature of fluid, K
T_f	Film temperature, average of T_b and T_s, K
T_i	Inside surface temperature, K
T_o	Outside surface temperature, K
T_s	Surface temperature, K
V	Fluid velocity, $m\,sec^{-1}$

REFERENCES

1. Sweat, V.E. Thermal Properties of Food. In *Engineering Properties of Foods*, 2nd Ed.; Rao, M.A., Rizvi, S.H., Eds.; Marcel Dekker, Inc.: New York, 1995; 99–138.
2. Incropera, F.P.; Dewitt, D.E. *Introduction to Heat Transfer*; John Wiley and Sons: New York, 1985; 711.
3. Sweat, V.E. Modelling Thermal Conductivity of Meats. Trans. ASAE **1975**, *18* (3), 564–568.
4. ASAE. D243.3 Thermal Properties of Grains and Grain Products. *ASAE Standards*; ASAE: St. Joseph, MI, 1998; 495.
5. Giancoli, D.C. Heat. In *Physics for Scientists and Engineers with Modern Physics*, 2nd Ed.; Prentice Hall: Englewood Cliffs, NJ, 1989; Vol. 1, 449–464.
6. Knudsen, J.G. Heat Transfer. In *Perry's Chemical Engineers' Handbook*, 6th Ed.; Perry, R.H., Green, D., Eds.; McGraw-Hill Book Co.: New York, 1984; 10.1–10.68.
7. Weast, R.C. *Handbook of Chemistry and Physics*, 67th Ed.; CRC Press, Inc.: Boca Raton, FL, 1986; E393–E395.
8. Midwest Plan Service. *Structures and Environment Handbook*, 11th Ed.; Midwest Plan Service, Iowa State University: Ames, IA, 1987; 601.3.

Heating of Air

M. Marcela Góngora-Nieto
Gustavo V. Barbosa-Cánovas
Washington State University, Pullman, Washington, U.S.A.

H

INTRODUCTION

Hot air is used in most drying operations, mainly in air dryers that are widely used around the world. The utilization of hot air has a relevant place in food dehydration, in which direct dryers such as cabinet, band, fluid bed, spray, and rotary dryers are used. These systems are commonly found in the manufacture of cookies and cereals, drying of grains, fruits, and vegetables, and processing of instant and other food powder products.[1]

The principle of a direct air dryer relies on the passing of hot gas, commonly air, over or through a food product; the hot air carries off the evaporated moisture. The higher the temperature and lower moisture content of the air with respect to the food being dried the better the driving forces favoring an efficient dehydration process.

When a high moisture food product is exposed to a sufficiently hot and dry air current, a gradient in temperature and water pressure is immediately established. This generates simultaneous heat and mass transfer between the hot air and the food. Since the food product has a lower temperature with respect to the air, the heat is transferred from the air to the product, while the higher moisture content in the product (also higher partial water pressure) drives the mass transfer from the product to the air. Thus, the drying phenomenon depends on the heat and mass transfer characteristics of both the drying air and the food product, while the evaporation rate and air temperature mainly influence the production rate. The airflow is determined by the required evaporation rate and the dryer's inlet and outlet air temperatures, which govern how much moisture the air can accept. Ideally, air should be fed at the minimum moisture content and maximum temperature the product can accept; if the air is too hot and dry, it may be prone to the formation of crust or other undesirable physicochemical changes in the product. However, careful selection of the maximum permissible air temperature, for each product, can lead to long run energy savings (less energy expenditure and shortening of drying time) with little or no capital expenditure.[3]

Generally, in direct dryers the air is heated while entering the dryer by means of heat exchangers (indirect heaters) or direct mixture with combustion exhaust gases (direct heaters). Direct-fired heating is cheaper and more efficient than indirect heating systems, but the high temperature and chemical nature of combustion gases may damage some food products. The maximum air temperature in a direct heater ranges from 300 to 600°C, while in indirect heaters the hot air temperature ranges from 200 to 400°C.[2,3] Although food dehydration processes at atmospheric conditions take place at temperatures below 100°C (whenever free moisture needs to be removed), the use of super hot air can damage some heat sensitive products. Thus, the study of air heating processes and their effect on the psychrometric properties of the drying air is a must in food dehydration.

AIR HEATING IN THE PSYCHROMETRIC CHART

Hot air is a mixture of air and water, thus its basic properties, such as humid volume, enthalpy, and humidity, (which define the efficacy of the hot air to dry a product), can be located in the psychrometric chart. This chart can also easily illustrate the air heating process. There are several forms of the psychrometric chart commonly used in drying calculations, all use the same data but are plotted differently. The two main forms are those that plot moisture against temperature, and those that plot moisture against enthalpy. Since, it is beyond the scope of this article to explain in detail the construction and equations generating these plots, as they can be found elsewhere,[4] this section focuses on the practical application and use of both types of plots in air heating.

To identify the two different air heating processes in the psychrometric charts, it is important to recall that in direct air heating the combustion process adds water vapor to the air due to the hydrogen molecules present in the fuel.[3] The moisture added depends on the fuel's molecular weight, the number of hydrogen atoms in its chemical formula, and its gross heating value; such data are available in fuel properties charts. Direct air heating is also considered more energy efficient than indirect air heating, because in the direct combustion of fuel the heat load can be raised about 12% (with high thermal efficiency) and the dew point about 2.2°C.[2,3] In contrast, indirect air heating

Encyclopedia of Agricultural, Food, and Biological Engineering
DOI: 10.1081/E-EAFE 120007055
Copyright © 2003 by Marcel Dekker, Inc. All rights reserved.

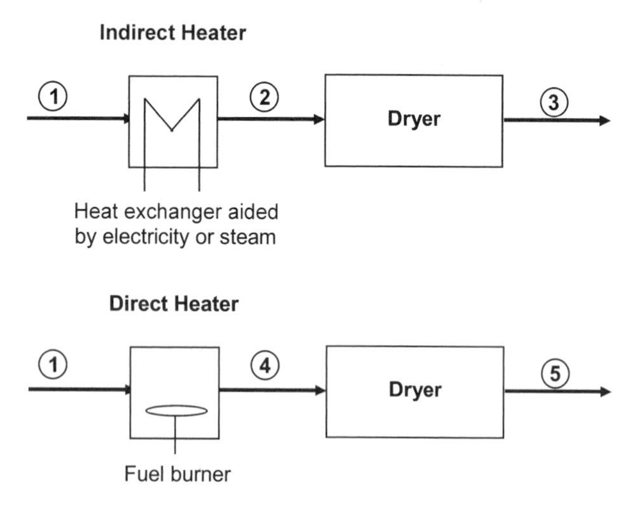

Fig. 1 Schematic representation of two different air-heating systems. Points 1 through 5 are represented in Figs. 2 and 3.

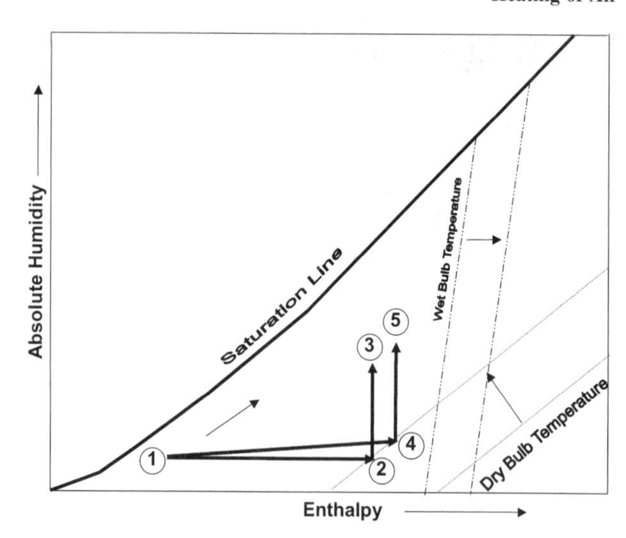

Fig. 3 Heating of air represented in a moisture vs. enthalpy psychrometric chart. Operating points 1 through 5 are represented in Fig. 1.

processes usually have lower thermal efficiencies due to heat losses and low heat transfer rates. Indirect air heating processes can have very good thermal efficiencies when economizers are used on modern steam boilers and when good insulation and steam heat recovery practices are used. With direct air heaters, heating takes place at constant air moisture content, by raising the dry bulb temperature and increasing the saturation vapor pressure and, hence, lowering the relative humidity of the air.[5] Fig. 1 illustrates the direct and indirect air heating processes in a generalized form.

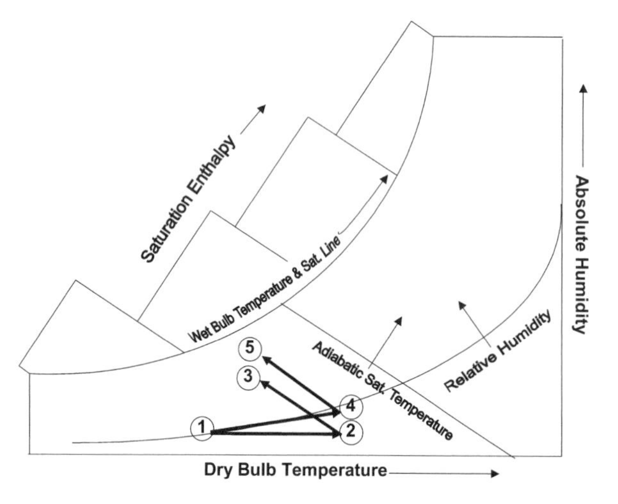

Fig. 2 Heating of air represented in a temperature vs. moisture psychrometric chart. Operating points 1 through 5 are represented in Fig. 1.

Representations of the operating points in Fig. 1 in both types of psychrometric charts is illustrated in Figs. 2 and 3. Point 1 corresponds to fresh air, from point 1 to point 2 a heat exchanger is used for indirect air heating, while from point 1 to 4 a fuel burner is used for direct air heating; in this case, observe the increase in moisture content of the air as it is heated, due to the combustion of the fuel. The adiabatic food or product drying operation takes place from point 2 to 3 and from point 4 to 5, depending on the air heating system used. In both cases, the air heats up and removes moisture from the product, with no heat loss or gain (assuming ideal drying conditions).

In Fig. 3, the vertical drying lines extending from point 2 to 3 and from point 4 to 5 may become angled to the left with heat losses, due to poor insulation of the dryer, or to the right with heat gains, due to use of an internal heat exchanger.

Effect of Air Heating on the Psychrometric Properties of Air

The psychrometric properties of moist air react differently depending on the type of heating. When indirect heaters are used, the only two properties remaining the same as the air is heated, are the absolute humidity (moisture content) and the dew point or saturation temperature, while an increase occurs in the enthalpy, specific volume, and wet and dry bulb temperatures, and a decrease in the relative humidity of the heated air. In the case of direct heaters, there is a gain in moisture content, thus a rise in the dew point of the heated air, as in the previous case (indirect heating) where an increase in enthalpy, specific

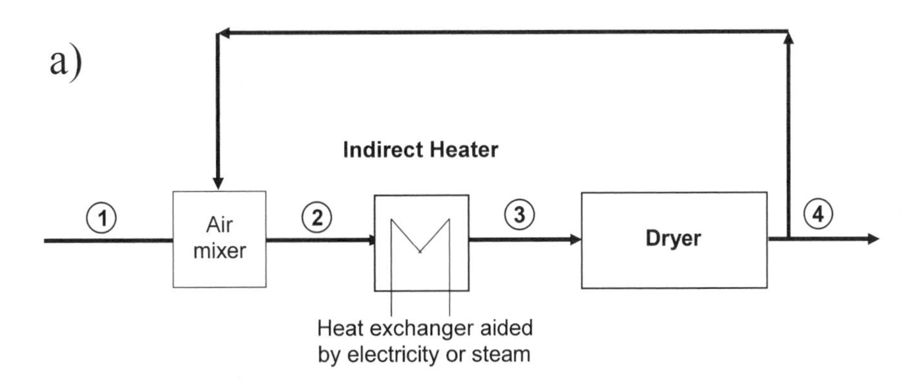

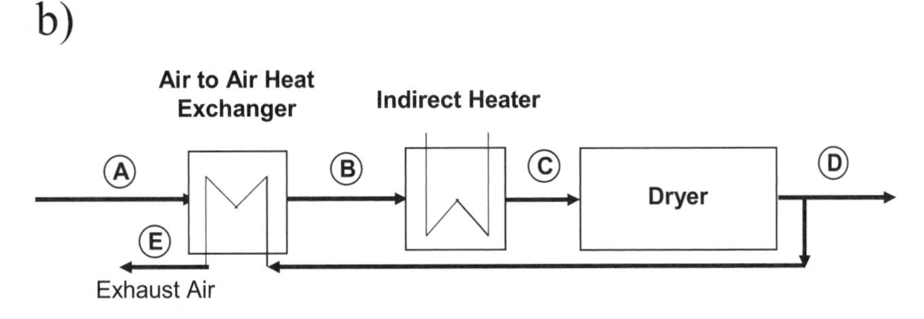

Fig. 4 Schematic representation of heat recovery during food drying.

volume, wet and dry bulb temperatures also exists, as well as a decrease in relative humidity.[6]

EXHAUST HEAT RECOVERY IN AIR HEATING

Air heating, in hot air drying applications, is the main factor that makes drying expensive, thus heat losses must be kept to a minimum. In direct dryers, the exhaust air carries out much of this heat. From 50% to 70% of the available sensible heat of the hot exhaust air can be recovered using an "exhaust to supply air-heat exchanger."[2,4] The four types of heat exchangers most commonly used for this purpose are circulating recuperators, plate heat exchangers, heat pipes and coils, and scrubbers with two section exchangers.[3,4] Fig. 4a illustrates the reutilization of exhaust air in a drying operation with an indirect air heating process. This air-heating situation is commonly found in tray dryers, where fresh air is mixed with part of the exhaust air from the dryer, and then indirectly heated. Fig. 4b illustrates the use of an air-to-air heat exchanger as an exhaust heat recovery device.

The operating points of Fig. 4 are represented in the moisture vs. temperature psychrometric chart in Figs. 5 and 6. To identify the operating point for the atmospheric and exhaust air mixture, in Fig. 4a, consider the mixing point as being somewhere in the straight line that links points 1 and 5, and the distance from point 1 to 2 as proportional to the air mass and flow rate of the streams coming from points 1 and 5. When reuse of exhaust air takes place it is not uncommon that up to a 60% of this stream is recycled to recover most of the energy carried out by the hot air.[3] It is worth mentioning that in Figs. 4 and 6 the energy recovery from points D to E must be conducted carefully, because there is a risk of condensation in the heat exchanger or other parts of the drying system.

Among food drying processes, spray drying is one of the most energy intensive, largely due to the limited heating time to achieve the evaporation of the product's water content, the relatively small allowable temperature difference between air and product, and above all, due to

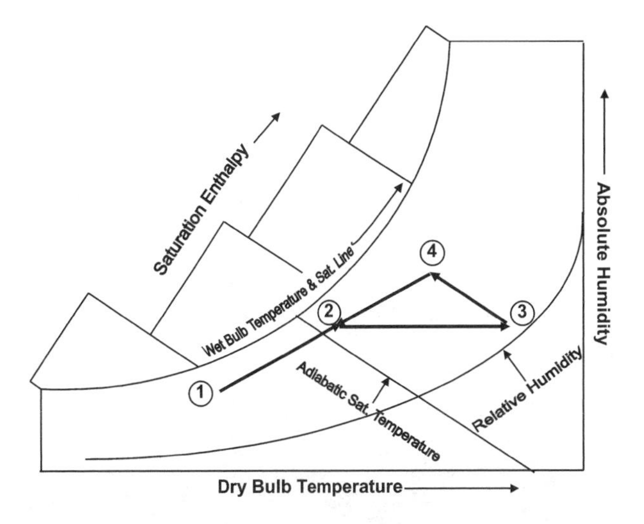

Fig. 5 Preheating of air by recovered exhaust heat. Operating points 1 through 4 are represented in Fig. 4a.

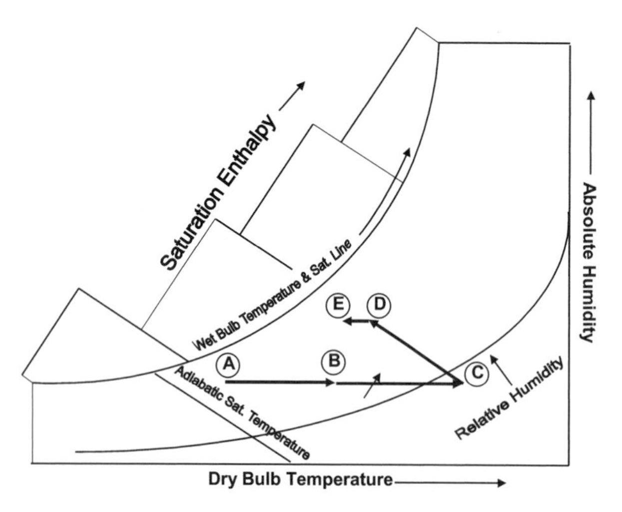

Fig. 6 Preheating of air by recovered exhaust heat and indirect heater. Operating points A through E are represented in Fig. 4b.

the amount of heat lost in the exhaust air. Thus, in this drying operation the use of multiple stages is a common practice. The thermal efficiency of multiple stage dryers is better than that of single units, with an energy savings of about 20%, mainly because the temperature of the outlet air is lower than that achieved with a single unit.[4] However, it is worth mentioning that energy savings related to heat efficient-utilization start with proper insulation of the heater, and drying system.

CONCLUSION

The properties of hot air, defined according to how it was heated, have an important impact on the heat and mass transfer processes taking place during food dehydration. The principles of air heating and the use of psychrometric charts that identify the heating process, reviewed herein, set the basis for a better understanding of dehydration processes, as well as for a better identification of the driving forces needed for an efficient drying operation. Heating of air, although seemingly a simple operation, is certainly one area of food dehydration that offers opportunities for innovation, improvement, and optimization of energy utilization. Two excellent examples on how energy consumption can be appreciably reduced are the reutilization (recycle) of exhaust air as a heating source, and the use of multiple drying stages.

REFERENCES

1. Barbosa-Cánovas, G.V.; Vega-Mercado, H. *Dehydration of Foods*, 1st Ed.; Chapman & Hall: New York, 1996; 157–227, 306–313.
2. Cook, E.E.; DuMont, H.D. *Process Drying Practice*; McGraw-Hill: New York, 1991; 20, 21, 64, 86–95.
3. Azbel, D. *Heat Transfer Applications in Process Engineering*; Noyes Publications: Park Ridge, NJ, 1984; 4–5, 501–503, 506–534.
4. Mujumdar, A.S. *Handbook of Industrial Drying*, 2nd Ed.; Marcel Dekker, Inc.: New York, 1995; Vol. 1, 85, 296–300, 536; Vol. 2, 1241–1248.
5. Brundrett, G.W. *Handbook of Dehumidification Technology*; Butterwords: London, 1987; 11–12.
6. Carrier Air Conditioning Company. *Handbook of Air Conditioning Systems Design*; McGraw-Hill: New York, 1965; I115–I141.

Heating and Cooling in Agitated Vessels

Gauri S. Mittal
University of Guelph, Guelph, Ontario, Canada

H

INTRODUCTION

Cylindrical vessels are generally used to heat or cool liquids and are agitated with an impeller mounted on a shaft, which is driven by an electrical motor. Typical vessels and agitators are shown in Figs. 1 and 2.[1] The heat is transferred through jackets in the vessel wall or immersed coiled pipes in the liquid. In most cases, baffles are also used inside the vessel for better mixing and hence, uniform heat transfer. Jackets are preferred due to inexpensive construction material, lower fouling, larger surface area, and easier maintenance and cleaning. If jacket's inside pressure is $> 1034\,\text{kPa}$ gauge, or high temperature vacuum processing is needed, a coil should be used.

BASICS ON HEAT TRANSFER

For unsteady heat transfer, conservation of energy provides:

$$\text{heat, in} - \text{heat, out} = \text{rate of heat accumulation in the vessel, or}$$

$$m_P c_{PP} \frac{dT_P}{d\theta} = UA(T_m - T_P)$$

Integrating between $\theta = 0$, $T_P = T_{Pi}$ and $\theta = \theta$, $T_P = T_P$

$$\ln\left[\frac{T_m - T_{Pi}}{T_m - T_P}\right] = \frac{UA}{m_P c_{PP}}\theta$$

Equations were summarized[2] for heat transfer in agitated vessels with the assumptions of negligible heat loss, no phase change, and constant values of U, c_P, flow rates, and temperatures. The above equation is for counter flow, isothermal heating/cooling medium in jacketed vessels or coil in tank. For other situations, the equations are given below.

Counter flow, nonisothermal heating/cooling medium (T_m at the inlet):

$$\ln\left[\frac{T_m - T_{Pi}}{T_m - T_P}\right] = \frac{(K_1 - 1)}{K_1}\frac{m_m c_{Pm}}{m_P c_{PP}}\theta \text{ for heating}$$

$$\text{or} = \frac{(K_2 - 1)}{K_2}\frac{m_{mc} c_{Pmc}}{m_P c_{PP}}\theta \text{ for cooling}$$

$$K_1 = \exp(UA/m_m c_{Pm}), \quad K_2 = \exp(UA/m_{mc} c_{Pmc})$$

For counter flow, external heat exchanger, isothermal heating/cooling medium:

$$\ln\left[\frac{T_m - T_{Pi}}{T_m - T_P}\right] = \frac{(K_2 - 1)}{K_2}\frac{m_{mc} c_{Pmc}}{m_P c_{PP}}\theta \text{ for heating}$$

$$\text{or} = \frac{(K_1 - 1)}{K_1}\frac{m_m c_{Pm}}{m_P c_{PP}}\theta \text{ for cooling}$$

For counter flow, external heat exchanger, nonisothermal heating/cooling medium (T_m at the inlet):

$$\ln\left[\frac{T_m - T_{Pi}}{T_m - T_P}\right] = \frac{(K_3 - 1)}{m_P}\frac{m_{mc} m_m c_{Pm}}{K_3 m_m c_{Pm} - m_{mc} c_{Pmc}}\theta$$

$$K_3 = \exp\left(UA\left[\frac{1}{m_{mc} c_{Pmc}} - \frac{1}{m_m c_{Pm}}\right]\right) \text{ for heating}$$

$$\ln\left[\frac{T_m - T_{Pi}}{T_m - T_P}\right] = \frac{(K_4 - 1)}{m_P}\frac{m_m m_{mc} c_{Pmc}}{K_3 m_m c_{Pm} - m_{mc} c_{Pmc}}\theta$$

$$K_4 = \exp\left(UA\left[\frac{1}{m_m c_{Pm}} - \frac{1}{m_{mc} c_{Pmc}}\right]\right) \text{ for cooling}$$

Equations for other cases are given in Ref. 2.

When a condensing vapor is used in the agitated jacketed vessels to heat or cool a food product, the controlling thermal resistance is usually that of the liquid in the vessel. On the other hand, when a fluid is used for heat transfer without phase change, the jacket-side resistance may be controlling. The heat transfer in the vessel is a function of the properties of fluid and heat transfer medium, vessel dimensions and material of construction, and agitation level. U is given by:

$$\frac{1}{U} = \frac{1}{h_i} + R_{Fi} + \frac{\Delta x}{k} + R_{Fo} + \frac{1}{h_o}$$

Encyclopedia of Agricultural, Food, and Biological Engineering
DOI: 10.1081/E-EAFE 120007009
Copyright © 2003 by Marcel Dekker, Inc. All rights reserved.

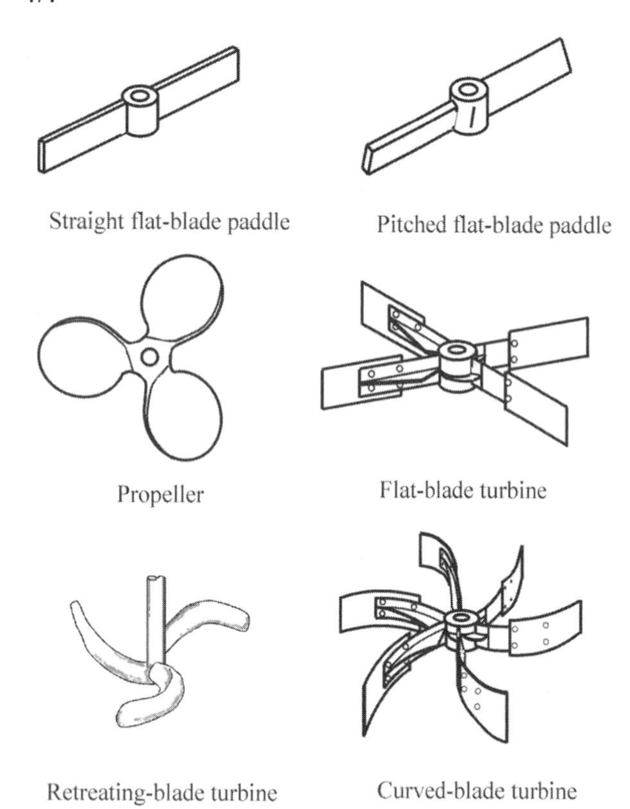

Fig. 1 Different types of impellers/agitators used in heating/cooling vessels/tanks.

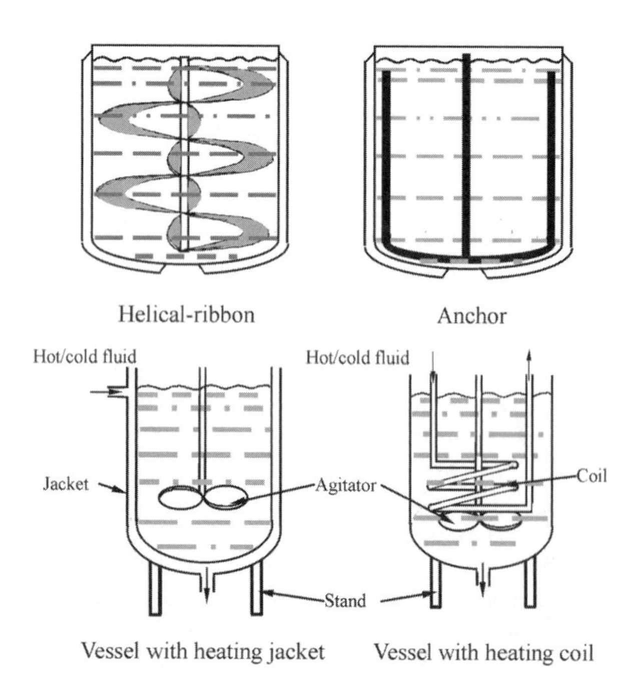

Fig. 2 Helical ribbon and anchor used for agitation in vessels. Vessel with heating/cooling jacket and vessel with heating/cooling coil.

For internal coils, U based on outer surface (U_o) is:

$$\frac{1}{U_o} = \frac{1}{h_i} + R_{Fi} + \frac{\Delta x}{k} \frac{D_{co}}{D_{cm}} + \frac{1}{h_{ci}} \frac{D_{co}}{D_{ci}} + R_{Fci}$$

Fouling is the accumulation of solids on the boundary separating two heat transfer fluids. This reduces the rate of heat transfer across the boundary. Fouling can be removed in some cases by weak acid, solvents, etc. Some fouling deposits can be removed by scraping the surface. Table 1 provides typical U for jacketed vessels.[3]

IMPELLERS

In proximity agitators, the blades are in close proximity to the vessels. Proximity types are suitable for viscous, non-Newtonian, semisolid, or granular materials,[4] e.g., anchor, wall scraper, straight ribbon, helical screw. These provide a high agitation and lower energy dissipated by shear. Nonproximity type agitators operate in the bulk liquid well removed from the vessel wall. Baffles near the vessel wall are used with nonproximity agitation to improve the turbulence by preventing swirl and vortexing.[5] These are suitable for fluids of medium viscosity (0.1 cP–1000 cP), e.g., propellers, turbines, paddles. These provide a high agitation and lower energy dissipated by shear.[4] Flat blade turbine or marine type impeller is recommended for low to medium apparent viscosity fluid. A marine propeller has three blades with a blade pitch equal to the propeller diameter.[6]

Radial and mixed type impellers are used for agitation. Fluid is discharged horizontally from blades to the tank wall in radial impellers, while axial flow impellers create vertical flow up or down the blades. Axial flow impellers include propellers, turbines—variable pitch curved blade and fixed pitch flat blade (Figs. 1 and 2). These are suitable for low to moderate apparent viscosity fluids. *Flat bladed radial flow* or *Rushton turbine* provides higher shear, turbulence, and balanced flow. The *anchor* or *horseshoe* impeller consists of contoured two-bladed radial flow turbine for high apparent viscosity (up to 50 Pa sec) fluids, and runs at slow speeds and high torque.[3] Wipers are attached to the outside of the anchor in high apparent viscosity fluids to avoid fouling as wipers scrap the inside tank wall surface during rotation. *Pitched blade turbines* are generally used in larger tanks requiring high power for mixing. These are relatively inexpensive to fabricate. The *constant pitch, variable blade angle, axial flow turbine* (3 to 6 blades) results in maximum flow and minimum shear rate. Three-bladed turbines are suitable for low apparent viscosity (< 2 Pa sec) fluids. Four to six blades are needed for higher apparent viscosity (2 Pa sec–5 Pa sec) fluids. These are more effective at low Re (< 1000) applications.

Table 1 Overall heat transfer coefficient (U, $W\,m^{-2}K^{-1}$) for jacketed vessels

Fluid inside jacket	Fluid in vessel	Wall material	Agitation	U ($W\,m^{-2}K^{-1}$)
Steam	Water	Enameled C.I.	0 rpm–400 rpm	545–682
Steam	Milk		None	1,136
Steam	Milk		Stirring	1,704
Steam	Milk boiling		None	2,840
Steam	Milk		200 rpm	490
Steam	Fruit slurry		None	188–510
Steam	Fruit slurry		Stirring	875
Steam	Solution	Cast iron	Double scraper	994–1,192
Steam	Slurry			910–994
Steam	Paste			710–852

(Modified from Ref. 3.)

Fluid foil or *hydrofoil* impeller consists of constant pitch axial flow turbine. The blades are curved and twisted, and taper more in width from hub to tip. These provide the highest flow and lowest shear rate. These are used in $Re > 1000$ applications. Newtonian fluids are mixed efficiently with narrow-bladed hydrofoils unless $Re < 1000$. Hydrofoils are more expensive than conventional fixed pitch turbines. Double spiral axial flow impeller is used in high apparent viscosity applications.

Thus, for low apparent viscosity liquids, conventional propeller, impeller, and hydrofoils are preferred. For higher apparent viscosity liquids, larger diameter, slower speed impellers are generally used. Table 2 summarizes the characteristics of agitators to be used in vessels.[7] Anchor type agitator is more economical to manufacture than the helical ribbon agitator. For $Re < 50$, the helical ribbon impeller is superior in terms of mixing and hence heat transfer.[5] Thermal gradients is much greater in an anchor agitated vessel than in a helical ribbon agitated vessel at low Re. For $Re < 50$, helical ribbon impeller is a better choice due to its superior mixing performance. Vessel geometry has little effect when used in the fully turbulent regime ($Re > 500$). For this, the nonproximity impellers (turbine, propeller, etc.) should be used. From Re

Table 2 Characteristics of agitators to be used in vessels

Impeller type	Applicability Re	μ_a (Pa sec)	Others
Nonproximity type			
Propeller	> 300	< 2	Weight of large propeller makes them more expensive than large turbines
Disk flat blade turbine	> 50	< 20	—
Flat blade turbine	> 50	< 20	For applications when impeller is located $< D_a/2$ off the vessel bottom
Pitched blade turbine	> 100	< 10	Liquid–liquid and solid–solid agitation
Glass coated turbine	> 50	< 20	Corrosion protection
Proximity type			
Anchors	> 50	20–100	—
Helical ribbons	< 50	100–1,000	Not for highly non-Newtonian liquids
Double motion	—	High	For non-Newtonian liquids
Votator	—	< 100	For concentric floating blades
	—	100–300	For eccentric floating blades
	—	300–1,000	For oval-tube floating blades
Scraped wall pipe	—	< 100	—
	—	100–1,000	With special designs

(From Ref. 7.)

500 to 50, vessel geometry becomes progressively more significant; for $Re < 50$, vessel geometry affects the performance significantly.[7]

The following equation is recommended to determine minimum impeller speed (N) to ensure forced convection in a vessel.[3]

$$N \geq 43[D_T/D_a]^{2.5}[W_B/2]^{0.3}[N\nu]^{0.2}$$

COILS AND OTHERS

When placing helical coils in vessels, a spacing of 2–4 tube diameters (centerline to centerline) is needed for uniform agitation.[3] Vertical plates or panel coils are easier to install and maintain than helical coils, and baffling effect is better than with vertical tubes. For uniform fluid flow and heat transfer, panels should be installed at equal spacing around the inside tank wall, but no closer than 10° apart.[3]

DESIGN

To design heating and cooling foods in agitated vessels, the following information will be required[3]:

1. Food properties: Solid content, initial and final temperatures, μ and ρ at desired temperatures.
2. Process time for the heating/cooling required, heating or cooling, temperature uniformity requirement.
3. Tank: Specific heats and thermal conductivities of tank and vessel walls; tank dimensions, jacket/coil types, and dimensions, etc.
4. Heat transfer medium: Properties of the medium, temperatures.
5. Fouling characteristics of heat transfer medium and food.
6. Impeller: Type, dimensions, speed of rotation, etc.

HEAT TRANSFER COEFFICIENTS (h_i AND h_o)

The general relationship is:

$$Nu = h_i D/k = f[Re, Pr, \mu/\mu_w, D_a/D_T, D_c/D_T]$$

Characteristic length for Nu is D_T for vessels, D_c for coils or pipes, and L_{pc}/number of internal passes for plate coils. For heated coil in baffled tank, h_i can be written as[3]:

$$h_i \propto \mu^{-0.30} D_a^{1.44} N^{0.67} D_T^{-0.6} k^{0.63} D_c^{-0.5} \rho^{0.67} c_{PP}^{0.37} (\mu/\mu_w)^M$$

This provides effect of various processes and equipment variables on h_i. As for impeller, power number is $\text{hp}/(\rho N^3 D_a^5)$, at constant D_a, $h_o \propto \text{hp}^{0.22}$, and at constant speed, $h_i \propto \text{hp}^{0.29}$. Hence, increasing h_i by increasing mixing level through greater energy consumption is uneconomical. Other factors are more effective. Mixing will add $2690\,\text{kJ}\,\text{hp}^{-1}$ into the product due to mixing. In moderately viscous fluid mixing, the heat load due to impeller rotation may be a significant factor.

Generally, flat blade radial turbine, pitched-bladed axial turbine, hydrofoil, and propeller provide approximately the same h_i. Thus, other application criteria should be considered when selecting the impeller type. For heat transfer, axial flow turbines are better compared to others. Among these turbines, hydrofoil impeller provides the highest flow efficiency. In limited space, a higher power number flat bladed radial turbine is more suitable. Impeller should be located preferably at mid depth in the vessel.[3]

Jacketed Walls

Table 3 summarizes some typical relationships to calculate h_i. The general h_i equation is $Nu = a\,Re^{2/3} Pr^{1/3} (\mu/\mu_w)^{0.25}$. "a" varies in the range of 0.36–0.74.[4] Pursell (1954), quoted by Ref. 4, provided the following equation:

$$Nu = 0.112 Re^{0.75} Pr^{0.44} (\mu/\mu_w)^{0.25} (D_T/D_a)^{0.40} (w/D_a)^{0.13}$$

Turbines provided h_i about 30% higher than the paddle of the same D_a. Similarly two curved blade turbines mounted on a simple shaft provided 15% greater h_i than single turbine. Central location was optimum for pitched blade or disk type flat blade turbines for heat transfer.[4]

The h_i is related to the agitator speed by $h_i \propto N^{2/3}$.[6] The h_i with the fan blade turbine was about 10% lower than with the curved blade turbine.

Outside Heat Transfer Coefficient (h_o)

For jacketed vessels[1]:

1. Annular jacket with spiral baffling

$$Nu = 0.027 Re^{0.8} Pr^{0.33} (\mu/\mu_w)^{0.14} [1 + 3.5 D_e/D_c]$$

for $Re > 10{,}000$

$$Nu = 1.86 [Re\,Pr\,D_e/L]^{0.33} (\mu/\mu_w)^{0.14}$$

for $Re < 2100$

Table 3 Inside heat transfer coefficient (h_i)—jacketed vessels $Nu = a\,Re^b Pr^c (\mu/\mu_w)^d E$

Impeller type	a	b	c	d	E	Conditions
Flat 6-blade turbine	0.74	0.67	0.33	0.14	$(5L_i/D_a)^{0.15}(n_p/6)^{0.15}(\sin\alpha)^{0.5}$	$Re > 4{,}000$, baffled, $H_L/D_T = 1$, $D_a/D_T = 0.33$
	0.54	0.67	0.33	0.14	$(5L_i/D_a)^{0.2}(n_p/6)^{0.2}(\sin\alpha)^{0.5}$	$Re < 400$, baffled or unbaffled, $D_a/D_T = 0.33$
	0.85	0.66	0.33	0.14	$(H_L/D_T)^{-0.56}(D_a/D_T)^{0.13}$	$Re > 400$, nonstandard geometry
Retreading 6-blade turbine	0.68	0.67	0.33	0.14	1	Unbaffled
Retreading 3-blade glassed steel turbine	0.33	0.67	0.33	0.14	1	Baffled
Retreading 3-blade alloy impeller	0.37	0.67	0.33	0.14	1	Baffled
45° pitched 4-blade propeller	0.54	0.67	0.25	0.14	1	Divide h_i by 1.3
3-blade propeller	0.37	2/3	1/3	0.14	$(D_T/D_a)^{0.25}(H_i/H_L)^{0.15}$	Unbaffled
3-blade propeller	0.50	2/3	1/3	0.14	$1.29P/D_a/(0.29 + P/D_a)$	Baffled
Paddle	0.36	0.67	0.33	0.14	1	$Re > 4{,}000$, baffled or unbaffled
	0.415	0.67	0.33	0.24	1	$20 < Re < 4{,}000$
Anchor	1.00	0.67	0.33	0.18	1	$30 < Re < 300$, anchor to wall clearance < 2.5 cm
	0.38	0.67	0.33	0.18	1	$300 < Re < 4{,}000$, anchor to wall clearance < 2.5 cm
	0.55	0.67	0.25	0.14	1	$4000 < Re < 37{,}000$, anchor to wall clearance = 2.5 cm–14 cm
Helical ribbon	0.248	0.50	0.33	0.14	$(e/D_a)^{-0.22}(i/D_a)^{-0.28}$	$Re < 130$
	0.238	0.67	0.33	0.14	$(i/D_a)^{-0.25}$	$Re > 130$
Nonstandard impeller height and diameter	1.01	0.66	0.33	0.14	$(H_i/D_T)^{0.12}(D_a/D_T)^{0.13}$	
Baffled flat bottom	0.73	0.65	0.33	0.24	1	Standard configuration
	1.15	0.65	0.33	0.24	$(H_i/D_T)^{0.4}(H_L/D_T)^{-0.56}$	Nonstandard configuration

(Compiled from Refs. 1, 6, and 7.)

For transition region, these equations should be used with care.

2. Annular jacket with no baffle
   For steam condensation, $h_o = 5678\ \mathrm{W\,m^{-2}\cdot K^{-1}}$
   Use above equations for baffled vessels with coil correction factor $1 + 3.5D_e/D_c$

$$D_e = (D_{jo}^2 - D_{ji}^2)/D_i, \quad A = \pi(D_{jo}^2 - D_{ji}^2)/4$$

For very low Re and water in annuli

$$Nu = 1.02Re^{0.45}Pr^{0.33}(D_e/L)^{0.4}(D_{jo}/D_{ji})^{0.8} \times (\mu/\mu_w)^{0.14}Gr^{0.05}$$

Gr is based on food properties at T_P, $D_e = D_{jo} - D_{ji}$

3. Half pipe coil jacket
   From equations given in (1) above, we get:
   For 180° central angle: $D_e = \pi/2D_{ci}$, $A = \pi/8D_{ci}^2$
   For 120° central angle: $D_e = 0.708\,D_{ci}$, $A = 0.154D_{ci}^2$
4. Dimple jacket
   Use equations given above in (1) without coil correction factor.

For internal coils, use equations given above in (1); use D_{ci} for D_e in the first equation and $(D_{ci}/D_c)^{0.5}$ for D_e/L in the second equation.

Table 4 provides h_i for anchors and helical ribbons. For vertical tubes and plate coils, the following relationships are given[7]:

Table 4 h_i for anchors and helical ribbons, $Nu = a\,Re^b Pr^{1/3}(\mu/\mu_w)^{0.14}$

Anchor	$Re < 12$	$12 < Re < 100$	$Re > 100$
a	1.05	0.69	0.32
b	1/3	0.5	2/3
Helical ribbon ($P/D_T = 0.25$)	$Re < 9$	$9 < Re < 135$	$Re > 135$
a	0.98	0.68	0.3
b	1/3	0.5	2/3
Helical ribbon ($P/D_T = 0.5$)	$Re < 13$	$13 < Re < 210$	$Re > 210$
a	0.94	0.61	0.25
b	1/3	0.5	2/3

Agitator manufacturers can provide more accurate and reliable correlations for their equipment.
(Compiled from Ref. 7.)

For 3-vertical tube baffles, 4-blade disk turbine:

$$Nu = 0.04 Re^{0.65} Pr^{0.3} (\mu/\mu_w)^{0.4} (5L_i/D_a)^{0.2} \times (D_p/D_T/0.04)^{0.5} (2/\#\,\text{baffles})^{0.2} (D_T/H_L)^{0.15}$$

For 4-vertical baffle coils at 45° to radius, two 6-flat blade turbine:

$$Nu = 0.021 Re^{0.67} Pr^{0.4} (\mu/\mu_w)^{0.27} (8L_i/D_a)^{0.2} \times (D_p/D_T/0.04)^{0.5}$$

For baffled, plate coils, two 6-flat blade turbine:

$$Nu = 0.031 Re^{0.66} Pr^{0.33} (\mu/\mu_w)^{0.5} (8L_i/D_a)^{0.2}$$

Immersed or Internal Coils

Table 5 lists typical relationships to calculate h_i for internal or immersed coils. The following equation was provided after investigating 25 different configurations (Pratt, 1947 quoted by Ref. 6) for coils in square vessels:

$$\frac{h_i L_v}{k} = 39 Re^{0.5} Pr^{0.3} \left(\frac{d_c}{L_c}\right)^{0.8} \left(\frac{w_L}{D_c}\right)^{0.25} \left(\frac{D_a^2 L_v}{D_{CT}^3}\right)^{0.1}$$

For cylindrical vessels, 39 was replaced with 34 and L_v with D_T. The following correlation is given to relate h for coils and pipes[5]:

$$h_{ci} = h_{ip}[1 + 3.5 D_p/D_c]$$

For heat transfer inside coiled tubes[8] for fully developed laminar flow with uniform wall temperature:

$$Nu = 3.65 + 0.08\left[1 + 0.08(D_{ci}/D_c)^{0.9}\right] Re^{\phi} Pr^{1/3}$$

$$\phi = 0.5 + 0.290 \exp(D_{ci}/D_c)^{0.914}$$

Vertical coils and natural convection[9]:

$$Nu = \left[4.36 + 2.84\left(\frac{Gr}{Re^2}\right)^{3.94}\right]\left[1 + 0.027 D_n^{0.75} Pr^{0.197}\right] \times \left[1 + 0.934\left(\frac{Gr}{D_n^2}\right)^{2.78} \exp\left(-1.33\frac{Gr}{D_n^2}\right)\right]\left[\frac{\mu}{\mu_w}\right]^{0.14}$$

$$D_n = Re\left(\frac{D_{ci}}{D_c}\right)^{0.5}$$

For transition and turbulent regions[8]

$$Re_{cr} = 2(10^4)(D_{ci}/D_c)^{0.32}$$

For $Re_{cr} < Re < 2.2(10^4)$

$$Nu = 0.023[1 + 14.8(1 + D_{ci}/D_c)](D_{ci}/D_c)^{1/3} Re^k Pr^{1/3}$$

$$k = 0.8 - 0.22(D_{ci}/D_c)^{0.1}$$

For $2(10^4) < Re < 1.5(10^5)$

$$Nu = 0.023[1 + 3.6(1 + D_{ci}/D_c)](D_{ci}/D_c)^{0.8} Re^{0.8} Pr^{1/3}$$

The following corrections were suggested[3] when h_i was calculated for jacketed tank for a particular ρ, c_{PP}, k, and μ:

1. For adding tubes or helical coils to a jacketed tank

   $$h_{i2}/h_{i1} = 0.2 D_a^{-0.5} D_T^{0.5} \text{ when } D_a/D_T \le 0.04$$

2. For product c_{PP} or k factors

   $$h_{i2}/h_{i1} = [c_{PP2}/c_{PP1}]^{0.37}[k_2/k_1]^{0.63}$$

3. The following correction factors for various surfaces: vertical tube (1.54); helical coils—first bank (1.54); helical coils—second bank (1.31); helical coils—third bank (1.08); and plate or panel coils (1.12).

Vertical baffle type coil provided h 13% greater than for a helical coil in a baffled tank at similar power input. A pitched blade turbine provided h about 10% lower than a curved blade turbine at the same conditions. Pitched blade turbine provided 20% greater h in an intermediate than in a low position.[4]

Table 5 Inside heat transfer coefficient (h_i)—internal helical coils $Nu = a\,Re^b Pr^c(\mu/\mu_w)^d E$; $Nu = h_i D_{co}/k$ if not mentioned

Impeller type	*a*	*b*	*c*	*d*	*E*	Conditions
Flat 6-blade turbine	0.17	0.67	0.37	b[a]	$(D_a/D_T)^{0.1}(D_{co}/D_T)^{0.5}$	$400 < Re < 1.5 \times 10^6$, all tanks, $0.018 < D_{co}/D_T < 0.036$, $\mu \leq 10{,}000\,cP$
Flat 4-blade turbine	0.09	0.65	0.30	0.14	$(D_a/D_T)^{0.33}(2/B)^{0.2}$	Baffled; $1{,}300 < Re < 2 \times 10^6$
Flat blade turbine, 4 baffles	0.0118	0.684	0.40	0.20	1	Heating, $D_a/D_T = 0.33$
	$0.0583Re^{0.5} + 0.000115Re$		0.40	0.20	1	Cooling, $12{,}900 < Re < 475{,}000$, $4.6 < Pr < 49$
Retreading 6-blade turbine, $Nu = h_i D_T/k$	1.40	0.62	0.33	0.14	1	
Pitched blade turbine, 45° baffles	0.00752	0.722	0.40	0.20	1	Heating, $D_a/D_T = 1/3$
	$0.0543Re^{0.722} + 0.000127Re$		0.40	0.20	1	Cooling, $D_a/D_T = 1/3$, $22{,}400 < Re < 521,600$, $4.2 < Pr < 41$
	0.013	0.684	0.40	0.20	1	Heating, $D_a/D_T = 0.5$
	$0.073Re^{0.5} + 0.000103Re$		0.40	0.20	1	Cooling, $D_a/D_T = 0.5$
Disk, flat and pitched 6-blade turbines	0.08	0.56	1/3	0.14	$(5L_i/D_a)^{0.15}(D_p/D_T/0.064)^{0.5}$	Unbaffled
	0.03	0.67	1/3	0.14	$(5L_i/D_a)^{0.2}(D_p/D_T/0.04)^{0.5}(D_T/H_L)^{0.15}(n_P/6)^{0.2}(\sin\alpha)^{0.5}$	Baffled
3-blade propeller	0.016	0.67	0.37	0.14	$(3D_a/D_T)^{0.1}(D_p/D_T/0.04)^{0.5}$	Baffled
Propeller	0.078	0.62	0.33	0.14	$(D_p/D_T/0.03)^{0.5}(3D_T/D_a)^{0.2}$	Unbaffled, divide h_i by 1.3 for design
Propeller, 4 baffles	0.091	0.67	0.37	0.20	$(D_a/D_T)^{0.1}(D_{co}/D_T)^{0.5}$	$400 < Re < 1E6$, heating/cooling, $2 < Pr < 6{,}330$, $\mu = 4E-4$ Pa sec–0.4 Pa sec
Paddle $Nu = h_i D_T/k$	0.87	0.62	0.33	0.14	1	
Vertical plate or panel	0.1788	0.448	0.33	0.50	1	$Re < 1{,}400$
coil $Nu = h_i L_e/k$	0.0317	0.667	0.33	0.50	1	$Re > 1{,}400$
Rushton disc turbine, 4 baffles	0.17	0.67	0.37	0.20	$(D_a/D_T)^{0.1}(D_{co}/D_T)^{0.5}$	Heating/cooling, $400 < Re < 1.5E6$, $2 < Pr < 6{,}330$, $D_a/D_T = 0.25–0.58$

[a] $b = 0.1738\mu^{-0.2}$.

(Compiled from Refs. 1, 3, and 10, vessel geometries are in Refs. 6 and 10.)

Non-Newtonian Fluids

The viscous fluids are agitated at lower rate, and thus provide lower h_i. Table 6 provides the heat transfer data in viscous liquids in a jacketed flat bottom vessel using various impellers.

For internal coils, various methods and correlations are summarized. For laminar region, the average μ_a is the μ of Newtonian fluid that needs the same power requirement as the non-Newtonian fluid in the same vessel and at the same impeller speed. The μ_a can then be used to calculate average shear rate and h by using the correlations developed for Newtonian fluids. Table 7 summarizes few correlations for non-Newtonian fluids. Re, Pr, and μ/μ_w are calculated using the method of Metzner-Otto or Calderbank-Moo Young.[10] There is no data on dilatant and time dependent fluids.

For jacketed vessels, pseudoplastic fluid, $n_1 = 0.343 - 0.633$, $Re = 100 - 5000$, $Pr = 100 - 800$, using 4-blade pitched turbine, 4 baffles[11]:

For heating: $Nu = 3.41Re^{2/3}Pr^{1/3}$ (mean deviation = 11.8%)

For cooling: $Nu = 1.43Re^{2/3}Pr^{1/3}$ (mean deviation = 14.0%)

Combined for heating/cooling: $Nu = 1.474\,Re^{2/3}Pr^{1/3}(\mu/\mu_w)^{0.24/n}$ (mean deviation = 19.3%)

Table 6 Heat transfer data for jacketed flat bottom vessels for viscous liquids $Nu = a\,Re^{0.5} Pr^{c} (\mu/\mu_w)^{0.14}$

	Vessel characteristics					
Impeller type	D_T/D_a	H_L/D_a	H_i/D_a	a	c	**Comment**
2-blade flat paddle	2.00	2.27	0.67	1.60	0.24	
	2.00	2.27	0.21	1.45	0.24	
2-blade paddle with blade angle < 45°	2.00	2.27	0.69	1.30	0.24	Bladewidth = $0.194D_a$
	2.00	2.27	0.23	1.20	0.24	
	2.82	3.18	1.04	0.82	0.33	Bladewidth = $0.182D_a$
	2.82	3.18	0.31	0.74	0.33	
Gate anchor	1.12	1.27	0.15	0.80	0.33	
Horseshoe anchor	1.15	1.30	0.12	1.38	0.28	
3-blade marine propeller	2.21	2.50	0.59	0.85	0.33	

(Data of Kapustin (1963) quoted by Ref. 6.)

Gas sparged vessels: Hart's correlation is recommended[7]:

$$\frac{h_i}{\rho v_s c_{Pg}} = 0.125 \left[\frac{v_s \rho}{\mu g}\right]^{-0.25} Pr^{-0.6}$$

Height to diameter ratio is > 5 for open pipe sparger at the column bottom, and < 5 for ring or finger style sparger.

HEAT TRANSFER AREA[1]

1. Annular jacket: The area wetted by both the vessel's contents and the heat transfer fluid.
2. Half pipe coil jacket: Ratio of effective heat transfer area to total heat transfer area for 5.1-cm pipe diameter = 0.90; for 7.6-cm diameter = 0.93; and for 10.2-cm pipe diameter = 0.94.
3. Dimple jacket: The ratio of areas is generally 0.92.

Table 7 h_i for internal coils, non-Newtonian fluids $Nu = a\,Re^{b} Pr^{c} (\mu/\mu_w)^{d} E$

Impeller type	a	b	c	d	E	**Conditions**
Propeller[12]	0.258	0.62	0.32	0.2	$(D_a/D_T)^{0.1}(D_{co}/D_T)^{0.5}$	$330 < Re < 2.6E5$
Pitched 4-blade turbine[13]	0.0196	0.71	0.33	0.17	1	$83 < Re < 4{,}855$
Pitched 4-blade, flat blade, Rushton 4-blade turbines[14]	$36.7n_1^{1.2}$	0.50	0.30	0.14	$(D_{co}/D_T)^{1.7}(n_P w \sin\alpha/H_c)^{0.15}$	$0.2 < Re < 1{,}000$
	36.7	0.50	0.30	0.14	Same as above	$10^4 < Re < 8E4$
Rushton 4-blade turbine[15]	0.067	0.65	0.33	—	1	$100 < Re < 10^4$
Rushton 4-blade turbine[16]	0.21	0.66	0.33	—	$(D_a/D_T)^{0.17}(D_{co}/D_T)^{0.55}(H_i/D_T)^{0.13}(D_c/D_T)^{-0.29}$	$200 < Re < 21{,}700$
Rushton 6-blade turbine[17,18]	0.0675	0.607	0.345	0.20	1	$400 < Re < 9.2E5$
	0.036	0.641	0.353	0.20	$(D_c/D_T)^{-0.375}$	Same as above
	0.236	0.641	0.353	0.20	$(D_a/D_T)^{0.2}(D_{co}/D_T)^{0.5}(D_c/D_T)^{-0.375}$	Same as above
Anchor[19]	0.077	0.66	0.33	0.14	$(D_{co}/D_T)^{0.52}(D_c/D_T)^{-0.27}$	$200 < Re < 6E5$
Anchor[20]	$0.0313\delta^{9.5}$	0.66	0.30	0.18	1	$100 < Re < 5.5E5$, $\delta = (3n+1)/4n$
Anchor[21]	0.084	0.55	0.33	0.14	$(D_c/D_T)^{0.74}$	$30 < Re < 2.8E4$

Correlations are less precise when $Re < 1000$. More correlations and data are given in Ref. 10 for various situations.
(Modified from Ref. 10. Details on geometries and experimental conditions are in the reference.)

4. Internal coils: Total wetted area based on coil's outside surface (A_{co}).

$$A_{co} = \pi D_{co} H_c n[(\pi D_c)^2 + n^{-2}]^{0.5}$$

SCALE UP OF VESSELS

In standard configurated agitated vessels for the heating/-cooling of liquids, h_i for a vessel of any size is[6]:

$$\frac{h_i D_T}{k} = 0.73 Re^{0.65} Pr^{0.33} \left[\frac{\mu_w}{\mu}\right]^{0.24}$$

As many properties (k, μ, ρ, c_P) are independent of vessel size, taking 1 for pilot and 2 for industrial scale vessels, and $D_T = 3D_a$, the ratio of h_i is:

$$\frac{h_{i2}}{h_{i1}} = \left[\frac{N_2}{N_1}\right]^{0.65} \left[\frac{D_{a2}}{D_{a1}}\right]^{0.30}$$

The degree of agitation in baffled vessels can be classified based on impeller tip speed[6] as: low agitation (2.54 m sec^{-1}–3.30 m sec^{-1}); medium agitation (3.30 m sec^{-1}–4.06 m sec^{-1}); and high agitation (4.06 m sec^{-1}–5.59 m sec^{-1}). The tip speed (V_T) is $\pi D_a N$ or $N \propto V_T/D_a$, or the ratio can be written after including tip speed:

$$\frac{h_{i2}}{h_{i1}} = \left[\frac{V_{T2}}{V_{T1}}\right]^{0.65} \left[\frac{D_{a1}}{D_{a2}}\right]^{0.35}$$

The ratio of heat loads (Q) for the same temperature differences and taking $U \approx h$ is:

$$Q_2/Q_1 \approx h_{i2}A_{i2}/(h_{i1}A_{i1})$$

For standard configurated vessels, the liquid height is the tank diameter[6] and heat transfer area is πD_T^2, hence $Q_2/Q_1 \approx h_{i2}D_{a2}^2/(h_{i1}D_{a1}^2)$, or substituting h_i:

$$Q_2/Q_1 \approx (V_{T2}/V_{T1})^{0.65}(D_{a2}/D_{a1})^{1.65}$$

Taking q as the heat transfer rate per unit volume of liquid:

$$q_2/q_1 \approx (V_{T2}/V_{T1})^{0.65}(D_{a1}/D_{a2})^{1.35}$$

When keeping the degree of agitation same on the two scales, the vessels cannot be scaled up on the basis of same heat transfer and temperature considerations.[6] Doubling the tip speed will slightly improve the heat transfer. Similar heat transfer per unit mass in the two scales can be achieved by pumping liquid through an external heat exchanger.

Constant power per unit volume can be used to get about the same h for the same type of impeller and constant D_a/D_T. Heat transfer surface area per unit volume is better in scaling up than power. Other variables, e.g., temperature driving force (ΔT), may have to be adjusted to compensate smaller heat transfer area.[3]

Effects of geometric parameters upon h are discussed in detail.[10] For scale-up, keep Re constant for any impeller geometry. Some equations for transition region, natural convection, and mixed convection were also summarized.[10]

CONCLUSION

Considerable data and relationships were reported on heat transfer coefficients for agitated vessels using various mixing devices and mechanisms. However, for some situations, heat transfer coefficient data or relationships are not available particularly for non-Newtonian foods and/or turbulent mixing conditions.

NOMENCLATURE

A	Surface area for heat transfer, m^2
A_{co}	Coil surface area, m^2
B	Number of vertical tubes acting as baffles or number of baffles
c_{Pg}	Specific heat of gas, J kg^{-1} K^{-1}
c_{Pm}	Specific heat of heating medium, J kg^{-1} K^{-1}
c_{Pmc}	Specific heat of cooling medium, J kg^{-1} K^{-1}
c_{PP}	Specific heat of the product, J kg^{-1} K^{-1}
D_a	Agitator or impeller diameter, m
D_c	Mean or centerline diameter of internal coil helix, m
d_c	Space between turns of a helical coil, m
D_{ci}	Coil inner diameter, m
D_{cm}	Log mean coil diameter, m
D_{co}	Coil outer diameter, m
D_{CT}	Diameter of coil tube, m
D_e	Equivalent heat transfer diameter; for rectangular cross section, $D_e = 4W$ where W is width of the annular space, m
D_{ji}	Inner diameter of annular jacket, m
D_{jo}	Outer diameter of annular jacket, m
D_p	Outside pipe diameter, m
D_T	Inside diameter of the tank or vessel, m
D_n	Dean number
e	Clearance, $(D_T - D_a)/2$, m

g	Acceleration due to gravity, m s^{-2}
Gr	Grashof number, $D_e^3 \rho^2 g \beta \Delta T_G / \mu^2$
H_c	Coil height, m
h_{ci}	h on coil side referred to inside coil area, W m^{-2} K^{-1}
h_i	Inside heat transfer coefficient for jacket or outside h for coil or process side h, W m^{-2} K^{-1}
H_i	Impeller height from vessel bottom, m
H_L	Liquid height in the tank, m
h_o	Outer heat transfer coefficient for jacket or inside h for coil, W m^{-2} K^{-1}
hp	Horse power
i	Agitator ribbon pitch, m
k	Thermal conductivity, W m^{-1} K^{-1}
L_c	Length of coil or jacket passage or plate width per panel coil or plate, m
L_i	Height of impeller blade, m
L_{pc}	Length of plate coils, m
L_v	Length of outside of a square vessel, m
m_m	Medium mass flow rate through jacket or coil or external heat exchanger, kg hr^{-1}
m_{mc}	Cold medium mass flow rate through external heat exchanger, kg hr^{-1}
m_P	Product mass, kg
n	Number of coil turns per m of coil height = 1/coil pitch
n_1	Flow behavior index
n_P	Impeller blades
N	Rotational impeller speed, rps
Nu	Nusselt number, $h_i D_T / k$, $h_o D_e / k$
P	Pitch of impeller or helical ribbon impeller, m
Pr	Prandtl number, $c_{PP} \mu / k$
Q	Heat transfer rate, W
Re	Reynolds number, $D_e V \rho / \mu$
Re_{cr}	Critical Re for transition from laminar to turbulent flow
R_{Fi}	Fouling factor, inside vessel, m^2 K W^{-1}
R_{Fo}	Fouling factor, inside jacket or coil, m^2 K W^{-1}
R_{Fci}	Fouling factor on coil side referred to inside coil area, m^2 K W^{-1}
T_m	Medium temperature of the jacket or coil, °C
T_{mi}	Jacket/coil inlet temperature, °C
T_P	Product temperature, °C
T_{Pi}	Initial product temperature, °C
U	Overall heat transfer coefficient, W m^{-2} K^{-1}
U_o	Overall h in an internal coil vessel, W m^{-2} K^{-1}
V	Velocity in spiral coil or jacket, m sec^{-1}
v_s	Sparged gas superficial velocity, m sec^{-1}
W	Width of annular space, m
W_B	Baffle width, m
w	Blade width, m
w_L	Blade width, m
α	Blade pitch angle
β	Coefficient of volumetric expansion, °C^{-1}
ΔT_G	Difference between temperature of bulk liquid in jacket and vessel wall, °C
Δx	Wall thickness of vessel or coil, m
θ	Process time, sec
μ	Liquid viscosity, Pa sec
μ_a	Apparent liquid viscosity, Pa sec
μ_w	Liquid viscosity at the wall temperature, Pa sec
ρ	Fluid density, kg m^{-3}
ν	Kinematic viscosity, μ/ρ, m^2 sec^{-1}

REFERENCES

1. Bondy, F.; Lippa, S. Heat Transfer in Agitated Vessels. Chem. Eng. **1983**, 62–71, Ap. 4.
2. Kern, D.R. *Process Heat Transfer*; McGraw Hill Book Co.: New York, 1950; 626–633, 716–723.
3. McDonough, R.J. Fundamentals of Mixing. *Mixing for the Process Industries*; Van Nostrand Reinhold: New York, 1992; 9–151.
4. Uhl, V.W. Mechanically Aided Heat Transfer. In *Mixing: Theory and Practice*; Uhl, V.W., Gray, J.B., Eds.; Academic Press, Inc.: New York, 1966; Vol. 1, 279–327.
5. Hewitt, G.F.; Shires, G.L.; Bott, T.R. Heat Transfer in Agitated Vessels. *Process Heat Transfer*; CRC Press, Inc.: Boca Raton, FL, 1994; 937–954.
6. Holland, F.A.; Chapman, F.S. *Liquid Mixing and Processing in Stirred Tanks*; Reinhold Pub. Corp., Chapman and Hall: London, 1966; 3–205.
7. Penney, W.R. Agitated Vessels. In *Hemisphere Handbook of Heat Exchanger Design*; Hewitt, G.F., Ed.; Hemisphere Pub. Corp.: New York, 1990; 3.14.1-1–3.14.1-8.
8. Schmidt, E.F. Waermeuebergang und Druckverlustin Rohrschlangen. Chem. Ing. Tech. **1967**, *39*, 781–787.
9. Abul-Hamoyel, M.A.; Bell, K.J. *Heat Transfer in Helically-Coiled Tubes with Laminar Flow*; ASME Paper 79-WA/HT-11; Am. Soc. Mech. Eng.: New York, 1979.
10. Bruxelmane, M.; Desplanches, H. Design and Scale-Up for Heat Transfer to Coils in Agitated Vessels. In *Heat Transfer Handbook of Heat and Mass Transfer*; Cheremisinoff, N.P., Ed.; Gulf Pub. Co.: Houston, TX, 1986; Vol. 1: Operations, 965–1007.
11. Carreau, P.; Charest, G.; Corneille, J.L. Heat Transfer to Agitated Non-Newtonian Fluids. Can. Chem. Eng. **1966**, *44* (1), 3–8.
12. Skelland, A.H.P.; Dimmick, G.R. Heat Transfer Between Colloids and Non-Newtonian Fluids with Propeller

Agitation. Ind. Eng. Chem. Process Design Dev. **1969**, *8* (2), 267–274.

13. Desplanches, H.; Llinas, J.R.; Chevalier, J.L. Heat Transfer from a Helical Coil to Newtonian and Non-Newtonian Fluids Agitated by a Pitched-Blade Turbine. Can. J. Chem. Eng. **1980**, *58* (4), 160–170.
14. Mizushina, T. Experimental Study of the Heat Transfer to the Cooling Coil in an Agitated Vessel. Kagaku Kogaku **1967**, *5* (1), 93–97.
15. Pandian, J.R.; Raja Rao, M. Heat Transfer to Non-Newtonian Liquids in Agitated Vessels. Ind. Chem. Eng. **1970**, *12*, 29–33.
16. Suryanarayanan, S.; Mujawar, B.A.; Raja Rao, M. Heat Transfer to Pseudo-plastic Fluids in an Agitated Vessel. Ind. Eng. Chem. Process Design Dev. **1976**, *15* (4), 564–569.
17. Edney, H.G.S.; Edwards, M.F.; Marshall, V.C. Heat Transfer to a Cooling Coil in an Agitated Vessel. Trans. Inst. Chem. Eng. **1973**, *51*, 4–9.
18. Edney, H.G.S.; Edwards, M.F. Heat Transfer to Non-Newtonian and Aerated Fluids in Stirred Tanks. Trans. Inst. Chem. Eng. **1976**, *54*, 160–166.
19. Pollard, J.; Kantyka, T.A. Heat Transfer to Agitated Non-Newtonian Fluids. Trans. Inst. Chem. Eng. **1969**, *47*, T21–T27.
20. Krishnan, R.M.; Pandya, S.B. Heat Transfer to Non-Newtonian Fluids in Jacketed Agitated Vessels. Indian Chem. Eng. **1970**, *12*, 29–33.
21. Blasinki, H.; Heim, A.; Kuncewicz, C. Heat Transfer to a Spiral Coil During the Mixing of a Non-Newtonian Liquid by Anchor Mixers. Int. Chem. Eng. **1977**, *17* (3), 548–553.

Heating and Cooling Lag Constants

Nikolaos G. Stoforos
Aristotle University of Thessaloniki, Thessaloniki, Greece

INTRODUCTION

When a product at an initial temperature T_{IT} is suddenly exposed to a heating or cooling environment, a significant immediate change in product temperature, from its initial value, is not observed due to hysteresis. The magnitude of this hysteresis depends on a variety of parameters, as discussed in this article. It is customary to describe this temperature lag hysteresis by means of a parameter, j, which is defined as the heating or cooling lag constant.

DEFINITION, ORIGIN, DETERMINATION, AND USE OF THE *j* VALUE

In developing his formula method for thermal process calculations, Ball[1] introduced two parameters, the f and j values, to describe product temperature evolution (at the cold spot) during the heating cycle of a thermal process as indicated in the following equation (all symbols are defined in the "Nomenclature"):

$$\frac{T_{RT} - T}{T_{RT} - T_{IT}} = j_h 10^{-t/f_h} \tag{1}$$

From Eq. 1 "Ball's" process time, B, may be calculated as given here, in its original notation, by Eq. 2.

$$B = f_h(\log(j_h I) - \log(g)) \tag{2}$$

Both the f_h and j_h values can be determined from the heating curve, an inverted, semilogarithmic plot of the difference between retort and product temperature (at the critical point) vs. time, as shown in Fig. 1. The j_h value, a dimensionless temperature lag correction factor, is defined as:

$$j_h = \frac{T_{RT} - T_A}{T_{RT} - T_{IT}} \tag{3}$$

where T_A is an extrapolated pseudoinitial product temperature at the beginning of heating defined as the intercept, with the temperature axis at time zero, of the linear portion of the experimental heating curve (Fig. 1).

While the f_h value is related to the slope of the heating curve, the j_h value, as Ball[1] defined it, is "a factor, which, when multiplied by I, designates the point of intersection of the vertical line representing the beginning of a process with the extension of the straight portion of the semilog heating curve, when no time is consumed in bringing the retort to holding, or processing, temperature." Despite its initial role as a descriptor of the straight line intercept, the j_h value is often referred to as the "lag factor" since it is a measure of the lag in establishing a uniform heating rate.[2]

Graphical determination of the j_h value is directly affected by the duration of the curvilinear portion of the heating curve, which, by its turn, is strongly influenced by the retort temperature profile. For cases when the retort is not instantaneously brought to the processing temperature, the duration of the retort coming-up-time (CUT) as well as the shape of the retort temperature profile during CUT must be considered for the correct determination of the j_h value. Several authors investigated the influence of the retort CUT on the j_h value.[3–8] Ball[1] suggested evaluation of the j_h value at a corrected axis obtained by shifting (increasing) the zero heating time by 0.58 CUT. As he pointed out, this correction is not always applicable. Nevertheless, it still represents the most commonly used practice.

On top of the uncertainties introduced due to retort CUT, graphical determination of the j_h (and the f_h) value is subject to errors, basically resulting from the limited amount of experimental data used for the parameter estimation. This is related to the well-documented issue of using tangents instead of the true asymptotes to experimental heating curves[2,9–11] coupled to the fact that the relationship between the j_h and f_h values and a heating curve is not unique.[12] As an alternative to the graphical j_h (and f_h) value determination, when an appropriate model for product temperature prediction exists, one can use a statistical fitting procedure to select the values of j_h and f_h that best describe the experimental data.[13]

Several researchers, following Ball's example, used the empirical f_h and j_h heat penetration parameters in various formulae or in developing new approaches for describing single point product temperatures during thermal processing (key contributions include Refs. 9,14–24). Although empirical in nature, the use of these parameters resulted in simple and flexible methodologies, independent of product

Encyclopedia of Agricultural, Food, and Biological Engineering
DOI: 10.1081/E-EAFE 120007100
Copyright © 2003 by Marcel Dekker, Inc. All rights reserved.

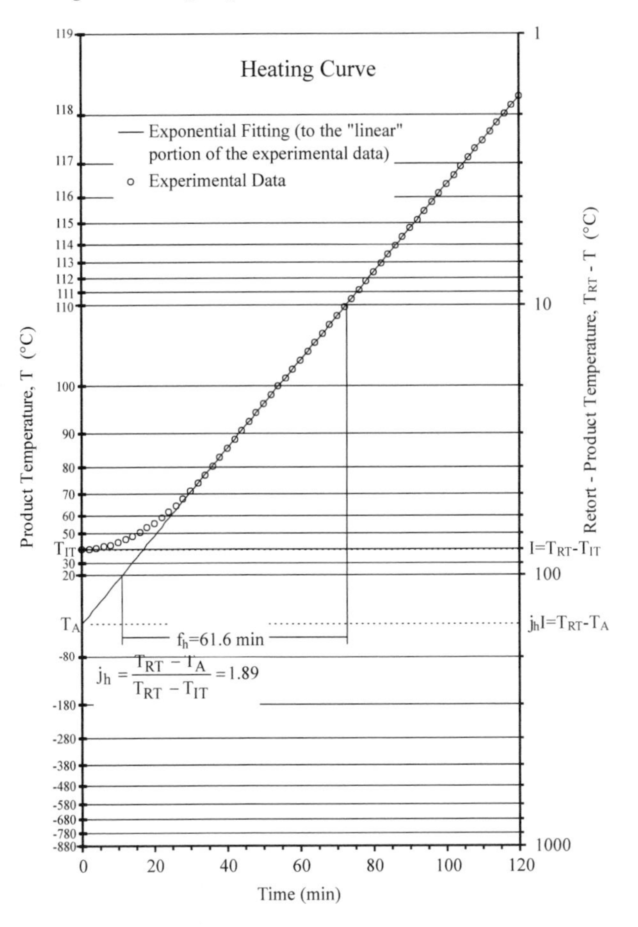

Fig. 1 Semilogarithmic heating curve, as traditionally plotted in thermal processing literature, used for parameter estimation.

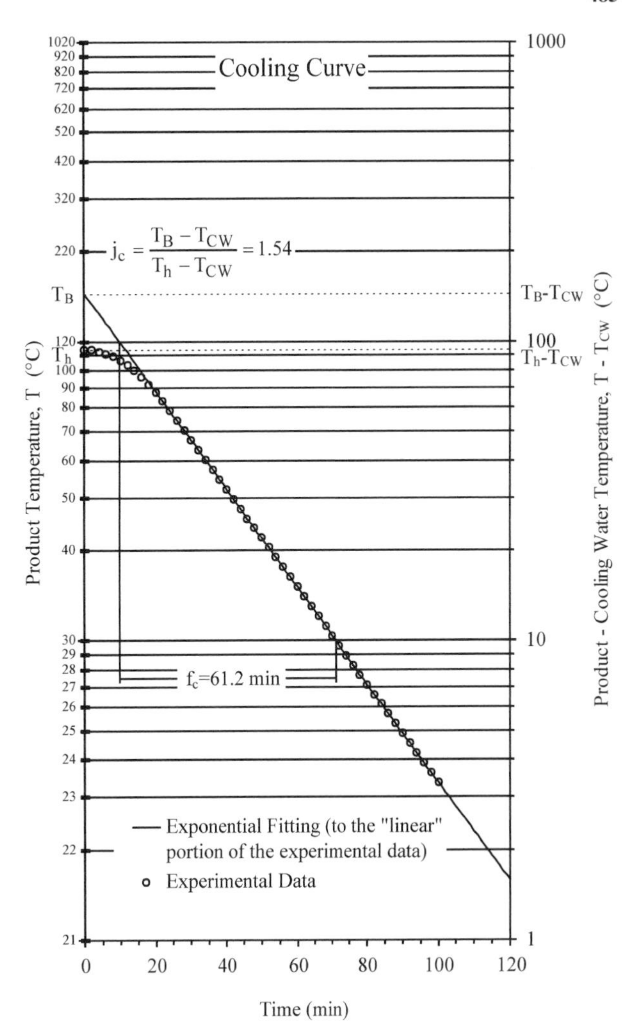

Fig. 2 Semilogarithmic cooling curve used for parameter estimation.

geometry and mode of heat transfer, for thermal process evaluation, design, and control.

In a similar way, as for the j_h value of the heating curve, the j_c value of the cooling curve is defined as:

$$j_c = \frac{T_B - T_{CW}}{T_h - T_{CW}} \tag{4}$$

for T_B being the extrapolated product temperature at the beginning of cooling defined as the intercept, with the temperature axis at zero cooling time, of the linear portion of the experimental cooling curve (Fig. 2).

For all methods for thermal process calculations that rely on the f and j parameters, accurate knowledge of the j_c value of the cooling curve is essential for proper processing time or process lethality calculations. Errors associated with j values affect mainly predictions of product temperature at the beginning of the heating or cooling cycles. Discrepancies between actual and predicted product temperature data at the beginning of the heating cycle result in negligible error when predicted product temperatures are used for calculation of process lethality or required processing time for commercial sterility, since at the beginning of the heating cycle product temperature is generally well below lethal temperature values. On the contrary, at the onset and the early stage of cooling, the product is in the range of the highest temperatures attained through out the whole process (including both heating and cooling cycles). Therefore, errors in product temperature predictions, resulting from erroneous j_c values, are critical. Difference of one unit in the j_c value can cause up to 40%–50% differences in the sterilizing values in the normal processing range, while variations in the j_h values do not cause significant variations on sterilizing values.[25] Therefore, contrary to Ball[1] who developed tables and graphs associated with his formula method for thermal process calculations using a fixed value of 1.41 for the j_c value, a number of subsequent methods[26] have been developed allowing for variable j_c values.

THEORETICAL VALUES AND FACTORS AFFECTING *j* VALUES

By comparing Eq. 1 to analytical equations for product temperature at a given point in a container as a function of time during a heating or a cooling process, theoretical expressions for the j values can be obtained. For example, the temperature evolution at any point of a conduction heating product in a can of uniform initial temperature, T_{IT}, suddenly immersed in a constant temperature, T_{RT}, environment, with infinite heat transfer coefficient between heating medium and product surface, is given by[27]:

$$\frac{T_{RT} - T}{T_{RT} - T_{IT}} = 4\sum_{m=1}^{\infty}\sum_{n=1}^{\infty} \frac{(-1)^{m+1}}{\lambda_m} \cos\left(\lambda_m \frac{2x}{L}\right) \times \frac{J_0(\beta_n \frac{r}{R})}{\beta_n J_1(\beta_n)} e^{-\left(\frac{4\lambda_m^2}{L^2} + \frac{\beta_n^2}{R^2}\right)\alpha t} \tag{5}$$

where

$$\lambda_m = (2m - 1)\frac{\pi}{2} \tag{6}$$

and β_n is the nth positive root of

$$J_0(\beta_n) = 0 \tag{7}$$

The one term approximation of Eq. 5, i.e., the equation obtained when considering only the term with $m = n = 1$ (for which, $\lambda_1 = \pi/2$, $\beta_1 = 2.4048$, $J_1(\beta_1) = 0.5191$) is valid for long times, represents the asymptote to the heating curve, and, by comparing it to Eq. 1 gives theoretical expressions for the j value at any axial and radial location. In particular, for the center of the can, i.e., for $x = r = 0$ (for which $J_0(0) = 1.0$) the approximation of Eq. 5 for long times becomes[28]:

$$\frac{T_{RT} - T}{T_{RT} - T_{IT}} = 2.0397\, e^{-\left(\frac{\pi^2}{L^2} + \frac{2.4048^2}{R^2}\right)\alpha t} \tag{8}$$

By comparing Eq. 8 with Eq. 1, the theoretical value for j_h at the can center, under the conditions used in obtaining Eq. 5, is determined as $j_h = 2.0397$. In a similar way, theoretical values (or expressions) for j_h and j_c for a number of different cases can be obtained. Furthermore, the above analysis can be used to determine factors affecting the j values.

The mode of heat transfer affects the magnitude of the j value. For perfectly mixed, forced convection, heating products, the theoretical value for both j_h and j_c (for any geometry and at any point) is 1.0.[28] Experimentally obtained j values of less than unity are attributed to variations in the overall heat transfer coefficient during the process.[9] For pure conduction heated foods (of uniform initial temperature, processed in a constant medium temperature, and for infinite heat transfer coefficient between heating medium and product), product shape influences j values as illustrated in Table 1 for selected product geometries.[2] The position inside the product where the temperature is evaluated affects also the j value as shown in Table 1.[2]

If the initial product temperature is not uniform, the existing temperature distribution must be known as it influences the j value. Differences in j values between heating and cooling curves of a thermal process are mainly due to the temperature distribution that usually exists at the beginning of the cooling cycle, contrary to heating where initial product temperature is usually uniform. If the heating cycle is long enough so that the product, at any point, has practically reached the heating medium temperature, then the j_c value of the cooling curve should be the same as the j_h value of the heating curve. With this in mind, the j values presented in Table 1 are equally good for heating and cooling curves. One can derive theoretical expressions for j values for a given initial product temperature distribution. For thermal process calculations, it is of particular interest to have an expression for the j_c value. This represents the case where the product is cooled (or heated) in a constant temperature medium, having a particular temperature distribution originated from a previous heating (or cooling)

Table 1 Theoretical j values for conduction heating products as a function of product geometry (for the case of uniform initial product temperature, constant heating medium temperature, infinite heat transfer coefficient)

Product geometry	j Value at geometrical center	j Value at any point
Infinite slab	1.27324	$1.27324\cos\left(\frac{\pi x}{L}\right)$
Rectangular rod	1.62114	$1.62114\cos\left(\frac{\pi x}{L}\right)\cos\left(\frac{\pi y}{W}\right)$
Brick	2.06410	$2.06410\cos\left(\frac{\pi x}{L}\right)\cos\left(\frac{\pi y}{W}\right)\cos\left(\frac{\pi z}{H}\right)$
Infinite cylinder	1.60218	$1.60218 J_0\left(\beta_n \frac{r}{R}\right)$
Finite cylinder	2.03970	$2.03970\cos\left(\frac{\pi x}{L}\right) J_0\left(\beta_n \frac{r}{R}\right)$
Sphere	2.00000	$0.63662\frac{R}{r}\sin\left(\frac{\pi r}{R}\right)$

step. This latter step involves heating (or cooling) of the product, initially being at a uniform temperature, at a medium of constant temperature. For this case and for a finite cylinder geometry, the following empirical equation for calculation of the j value at the center of a can (of T_0 temperature just before cooling or heating starts) based on the knowledge of the temperature of the product at a point about 0.254 cm from the can wall (T_w) just before cooling (or heating) starts, has been proposed[2]:

$$j = 1.27 + 0.77(T_{CW} - T_w)/(T_{CW} - T_0) \tag{9}$$

Based on an analytical solution of heat conduction (for the case of finite cylinder, uniform initial product temperature at the beginning of heating, constant heating and cooling medium temperature, infinite heat transfer coefficient during heating and cooling), the following equation for j_c has been proposed[29,30]:

$$j_c = 2.0397\frac{T_{RT} - T_{CW}}{T_h - T_{CW}} - 2.0397\frac{T_{RT} - T_{IT}}{T_h - T_{CW}} e^{-\left(\frac{\pi^2}{L^2} + \frac{2.4048^2}{R^2}\right)\alpha t_h} \tag{10}$$

When heating time is long enough, the second term of the right hand side of Eq. 10 becomes negligible, and the expression proposed by Hicks[31] is obtained. Furthermore, as the product temperature, T_h, approaches the retort temperature, T_{RT}, j_c approaches the value of 2.0397. Theoretical expressions for the j value for several retort temperature profiles (including both heating and cooling cycles) have been presented.[30]

Differences in j values between heating and cooling curves of a thermal process can also be attributed to differences in the heat transfer coefficients during the heating and the cooling phase. Indeed, for foods heated by conduction, the magnitude of the heat transfer coefficient (h) between the heating (or cooling) medium and the product affects the j value. For the conditions used in deriving Eq. 5, but allowing for finite heat transfer coefficient, the following expression for the j value at the geometric center can be obtained:

$$j = \frac{4\sin(\xi_1)}{\xi_1 + \sin(\xi_1)\cos(\xi_1)} \frac{Bi_R}{(\zeta_1^2 + Bi_R^2)J_0(\zeta_1)} \tag{11}$$

where ξ_1 is the first positive root of

$$\xi_n \tan(\xi_n) = \frac{hL}{2k} \tag{12}$$

and ζ_1 is the first positive root of

$$\zeta_n J_1(\zeta_n) = Bi_R J_0(\zeta_n) \tag{13}$$

Using Eq. 11, the effect of the heat transfer coefficient on the j value is illustrated in Table 2 for a particular product (with $k = 0.25\,\mathrm{W\,m^{-1}\,K^{-1}}$, $\alpha = 1.1\cdot10^{-7}\,\mathrm{m^2\,sec^{-1}}$, $R = 0.0326\,\mathrm{m}$, and $L = 0.0950\,\mathrm{m}$).

Table 2 Theoretical j values for a conduction heating product as a function of the heat transfer coefficient between the heating (or cooling) medium and the product (for the case of finite cylinder, uniform initial product temperature, constant medium temperature, $R = 0.0326\,\mathrm{m}$, $L = 0.095\,\mathrm{m}$, $k = 0.25\,\mathrm{W\,m^{-1}\,K^{-1}}$ $\alpha = 1.1\cdot10^{-7}\,\mathrm{m^2\,sec^{-1}}$)

h (W m^{-2}K^{-1})	j Value at geometrical center
5	1.28
20	1.68
200	1.95
5000	2.04
∞	2.04

CONCLUDING REMARKS

We should emphasize that, in addition to experimental variation and the error due to the limited amount of experimental product temperature data, differences between experimentally determined and theoretical j values may be due to variations in the following parameters: (nonconstant) heating or cooling medium temperature profile, initial product temperature distribution, magnitude of heat transfer coefficient during heating and cooling, position in the container where product temperature is monitored, nonconstant heat transfer coefficient throughout the whole process. All these parameters, together with product shape and product and process characteristics (as they determine the mode of heat transfer) influence the magnitude of the j values, the heating or cooling lag constants.

NOMENCLATURE

Latin Letters

B — "Ball's" process time, heating time required for commercial sterilization, measured from the "corrected" zero time, i.e., from the time where the temperature axis, where the j value is defined, is located, sec (unless otherwise explicitly stated)

Bi_R — Biot number, $Bi_R = hR/k$, dimensionless

f — Time required for the difference between the medium and the product temperature to change by a factor of 10, sec (unless otherwise explicitly stated)

g — Difference between heating medium and product temperature at the cold spot at steam-off time (note, that in defining g, Ball, 1923 used "the maximum temperature attained by the center of the can during its processing"), °C

H — Height of brick, m

h — Convective heat transfer coefficient between heating medium and product surface, $W\,m^{-2}K^{-1}$

I — Difference between heating medium and initial product temperature, $I = T_{RT} - T_{IT}$, °C

J_0 — Bessel function of the first kind of order zero

J_1 — Bessel function of the first kind of order one

j — A temperature lag correction factor defined by Eqs. 3 and 4 for the heating and cooling curve, respectively, based on the intercept, with the temperature axis at time zero, of the linear portion of the experimental heating or cooling curve plotted in a semilogarithmic temperature difference scale as shown in Figs. 1 and 2, respectively, dimensionless

k — Product thermal conductivity, $W\,m^{-1}K^{-1}$

L — Can (finite cylinder) length, brick length, thickness of rectangular rod, thickness of infinite slab, m

R — Can (finite cylinder) or sphere radius, m

r — Can (finite cylinder) or sphere radial distance, m

T — Product temperature at a particular point, °C

T_A — Extrapolated pseudoinitial product temperature at the beginning of heating defined as the intercept, with the temperature axis at time zero, of the linear portion of the experimental heating curve plotted as shown in Fig. 1, °C

T_B — Extrapolated pseudoinitial product temperature at the beginning of cooling defined as the intercept, with the temperature axis at zero cooling time, of the linear portion of the experimental cooling curve plotted as shown in Fig. 2, °C

T_{CW} — Cooling (water) medium temperature, °C

T_h — Product temperature at the beginning of the cooling cycle, °C

T_{IT} — Initial product temperature, °C

T_0 — Product center temperature just before cooling (or heating) starts, °C

T_{RT} — Heating medium (retort) temperature, °C

T_w — Temperature of the product at a point about 0.254 cm from the can just before cooling (or heating) starts, °C

t — Processing time, sec (unless otherwise explicitly stated)

t_h — Total heating time, sec (unless otherwise explicitly stated)

W — Brick width, width of rectangular rod, m

x — Can (finite cylinder) axial distance, m

x, y, z — Rectangular cartesian coordinates, m

Greek Letters

α — Product thermal diffusivity, $m^2\,sec^{-1}$

β_n — nth eigenvalue given by Eq. 7

ζ_n — nth eigenvalue given by Eq. 13

λ_n — nth eigenvalue given by Eq. 6

ξ_n — nth eigenvalue given by Eq. 12

Subscripts

c — Cooling phase

h — Heating phase

REFERENCES

1. Ball, C.O. *Thermal Process Time for Canned Food*; Bulletin of the National Research Council No. 37, Natl. Res. Council: Washington, D.C., 1923; Vol. 7, Part 1; 76.
2. Olson, F.C.W.; Jackson, J.M. Heating Curves. Theory and Practical Application. Ind. Eng. Chem. **1942**, *34* (3), 337–341.
3. Alstrand, D.V.; Benjamin, H.A. Thermal Processing of Canned Foods in Tin Containers. V. Effect of Retorting Procedures on Sterilization Values in Canned Foods. Food Res. **1949**, *14*, 253–260.
4. Townsend, C.T.; Reed, J.M.; McConnell, J.; Powers, M.J.; Esselen, W.B., Jr.; Somers, I.I.; Dwyer, J.J.; Ball, C.O. Comparative Heat Penetration Studies on Jars and Cans. Food Technol. **1949**, *3* (6), 213–226.
5. Uno, J.; Hayakawa, K. Correction Factor of Come-Up Heating Based on Critical Point in a Cylindrical Can of Heat Conduction Food. J. Food Sci. **1980**, *45* (4), 853–859.
6. Succar, J.; Hayakawa, K. Prediction of Time Correction Factor for Come-Up Heating of Packaged Liquid Food. J. Food Sci. **1982**, *47*, 614–618.
7. Berry, M.R., Jr. Prediction of Come-Up Time Correction Factors for Batch-Type Agitating and Still Retorts and the Influence on Thermal Process Calculations. J. Food Sci. **1983**, *48*, 1293–1299.
8. Ramaswamy, H.S. Come-Up Time Effectiveness for Process Calculations Involving Thin-Profile Packages. J. Food Eng. **1993**, *19*, 109–117.
9. Ball, C.O.; Olson, F.C.W. *Sterilization in Food Technology. Theory, Practice and Calculations*; McGraw-Hill Book Co.: New York, 1957; 654.
10. Cowell, N.D.; Evans, H.L. Studies in Canning Processes. IV. Lag Factors and Slopes of Tangents to Heat Penetration

Curves for Canned Foods Heating by Conduction. Food Technol. **1961**, *15*, 407–409.
11. Hayakawa, K.; Ball, C.O. A Note on Theoretical Heating Curve of a Cylindrical Can of Thermally Conductive Food. Can. Inst. Food Technol. J. **1968**, *1* (2), 54–60.
12. Stoforos, N.G.; Noronha, J.; Denys, S.; Hendrickx, M.; Tobback, P. Progress in Procedures for Designing Thermal Processes of Foods. In *Cooperation on Science and Technology in Europe (COST Project 93 and COSEMI 93/7) Seminar*, Copenhagen, Denmark, May 8–10, 1995.
13. Noronha, J.; Van Loey, A.; Hendrickx, M.; Tobback, P. Alternative Algorithms for Evaluation of Process Deviations in Batch-Type Sterilisation Processes. *Food Processing Automation III, Proceedings of the FPAC III Conference*; ASAE: St. Joseph, MI, 1994; 516–526.
14. Gillespy, T.G. Estimation of Sterilizing Values of Processes as Applied to Canned Foods. I.—Packs Heating by Conduction. J. Sci. Food Agric. **1951**, *2* (3), 107–125.
15. Hayakawa, K. Experimental Formulas for Accurate Estimation of Transient Temperature of Food and Their Application to Thermal Process Evaluation. Food Technol. **1970**, *24* (12), 1407–1418.
16. Griffin, R.C., Jr.; Herndon, D.H.; Ball, C.O. Use of Computer-Derived Tables to Calculate Sterilizing Processes for Packaged Foods. 3. Application to Cooling Curves. Food Technol. **1971**, *25* (2), 134–138, 140, 143.
17. Stumbo, C.R. *Thermobacteriology in Food Processing*, 2nd Ed.; Academic Press, Inc.: New York, 1973; 329.
18. Pham, Q.T. Calculation of Thermal Process Lethality for Conduction-Heated Canned Foods. J. Food Sci. **1987**, *52* (4), 967–974.
19. Pham, Q.T. Lethality Calculation for Thermal Processes with Different Heating and Cooling Rates. Int. J. Food Sci. Technol. **1990**, *25* (2), 148–156.
20. Larkin, J.W.; Berry, R.B. Estimating Cooling Process Lethality for Different Cooling j Values. J. Food Sci. **1991**, *56* (4), 1063–1067.
21. Teixeira, A.A.; Tucker, G.S.; Balaban, M.O.; Bichier, J. Innovations in Conduction-Heating Models for On-line Retort Control of Canned Foods with Any j-Value. In *Advances in Food Engineering*; Singh, R.P., Wirakartakusumah, M.A., Eds.; CRC Press: Boca Raton, FL, 1992; Chap. 10, 293–308.
22. Noronha, J.; Hendrickx, M.; Van Loey, A.; Tobback, P. New Semi-empirical Approach to Handle Time-Variable Boundary Conditions During Sterilisation of Nonconductive Heating Foods. J. Food Eng. **1995**, *24*, 249–268.
23. Kim, K.H.; Teixeira, A.A. Predicting Internal Temperature Response to Conduction-Heating of Odd-Shaped Solids. J. Food Process. Eng. **1997**, *20* (1), 51–64.
24. Teixeira, A.A.; Balaban, M.O.; Germer, S.P.M.; Sadahira, M.S.; Teixeira-Neto, R.O.; Vitali, A.A. Heat Transfer Model Performance in Simulation of Process Deviations. J. Food Sci. **1999**, *64* (3), 488–493.
25. Stumbo, C.R.; Longley, R.E. New Parameters for Process Calculation. Food Technol. **1966**, *20* (3), 341–345.
26. Stoforos, N.G.; Noronha, J.; Hendrickx, M.; Tobback, P.A. Critical Analysis of Mathematical Procedures for the Evaluation and Design of In-Container Thermal Processes of Foods. Crit. Rev. Food Sci. Nutr. **1997**, *37* (5), 411–441.
27. Carslaw, H.S.; Jaeger, J.C. *Conduction of Heat in Solids*, 2nd Ed.; Clarendon Press: Oxford, Great Britain, 1959; 510.
28. Merson, R.L.; Singh, R.P.; Carroad, P.A. An Evaluation of Ball's Formula Method of Thermal Process Calculations. Food Technol. **1978**, *32* (3), 66–72, 75.
29. Hayakawa, K.; Ball, C.O. A Note on Theoretical Cooling Curve of a Cylindrical Can of Thermally Conductive Food. Can. Inst. Food Technol. J. **1969**, *2* (3), 115–119.
30. Hayakawa, K.; Ball, C.O. Theoretical Formulas for Temperatures in Cans of Solid Food and for Evaluating Various Heat Processes. J. Food Sci. **1971**, *36*, 306–310.
31. Hicks, E.W. On the Evaluation of Canning Processes. Food Technol. **1951**, *5*, 134–142.

High Pressure Food Preservation

V. M. Balasubramaniam
The Ohio State University, Columbus, Ohio, U.S.A.

INTRODUCTION

High pressure processing (HPP) is a novel method of food processing wherein the food is subjected to elevated pressures (pressures up to 900 MPa or approximately 9000 atm) with or without the addition of heat to achieve microbial inactivation or to alter the food attributes in order to achieve desired qualities. Pressures used in HPP are almost ten times greater than in the deepest oceans on earth. HPP-processed foods are reported to have better flavor, texture, nutrient retention, and color compared to thermally processed foods. Pressure can be applied at ambient temperatures, thereby eliminating thermally induced cooked off-flavors. This benefits heat sensitive products such as citrus juices. The effectiveness of HPP in combination with temperature is also being studied, providing an extended opportunity for the HPP from the freezing and thawing of foods to the processing of low-acid foods. High pressure processing may also be described as high hydrostatic pressure processing (HHP) or ultra high pressure processing (UHP). High pressure processing can be used to process both liquid and solid foods. High pressure processing retains food quality and natural freshness, and extends microbiological shelf life. The technology has its roots in the material and process-engineering industry where it has been commercially used in sheet metal forming and isostatic pressing of advanced materials such as turbine components and ceramics. The effect of HPP on food microorganisms was demonstrated by early 1900.[1] Only over the last two decades has the food industry started to exploit the commercial potential of this technology. At present, HPP is used mainly for processing high-value or novel products of superior quality. Some of the HPP processed products commercially available in Japanese and European markets include jams, jellies, fish, meat products, sliced ham, salad dressing, rice cakes, juices, and yogurt. HPP processed guacamole (traditional Mexican sauce made with avocado puree, onion salt, and other ingredients) and oysters are commercially available in the United States. Other potential applications include processing shelf-stable products, blanching, and pressure assisted freezing and thawing. Equipment and processing costs are typically estimated to be less than \$0.10 per kg of the food processed.[2]

PROCESS ENGINEERING PRINCIPLES

The basic principles governing the HPP process[3,4] are: 1) Le Chatelier's principle which states that any phenomenon (phase change, change in molecular configuration, chemical reaction) that is accompanied by a decrease in volume is enhanced by pressure (and vice versa); and 2) the isostatic principle which indicates that pressure is transmitted in a uniform and quasi-instantaneous manner throughout the whole sample. Therefore, the process is independent of volume and geometry of the product. Once the desired pressure is reached, it can be maintained for an extended period of time without the need for further energy input.

Pressure creates no shear force to distort food particles. Most foods contain water, which prevents the food from being crushed. Materials containing air (e.g., marshmallow) are likely to be crushed by pressure because of the difference in material compressibility. Traditionally, high pressure has been used in the food industry mostly for homogenization. Using a pressure pump, the product is increased in pressure and then allowed to discharge through a very narrow gap. High product velocity and high shear characterize the homogenization process. On the other hand, during HPP, the product pressure is simply increased, held, and then lowered without product discharge. The product will not experience any velocity, shear, or homogenization effects.[5,6] Commonly used pressure units and conversion factors are given in Table 1.

PROCESS EQUIPMENT

High pressure processing equipment consist of: 1) pressure vessel; 2) two end closures; 3) yoke (structure for restraining end closures); 4) high pressure pump and intensifier for generating desired target pressures; and 5) system control and instrumentation. The treatment can be designed as a batch process or semicontinuous process. Batch HPP systems (Fig. 1A) are similar in operation to batch thermal processing retort systems. Batch vessel volumes ranges from 40 L to 950 L. A typical process cycle consists of loading the vessel with the prepackaged product, and then fills the reminder of the vessel with water, which acts as the pressure-transmitting fluid. The vessel is closed and the desired process pressure is achieved through compression of

Encyclopedia of Agricultural, Food, and Biological Engineering
DOI: 10.1081/E-EAFE 120007111
Copyright © 2003 by Marcel Dekker, Inc. All rights reserved.

Table 1 Commonly used pressure conversion factors

	Atmosphere	Bars	Mega Pascals	Pounds inch^{-2}
Atmosphere	1.000	0.987	9.901	0.068
Bars	1.013	1.000	10.000	0.069
Mega Pascal	0.101	0.100	1.000	0.00689
Pounds inch^{-2}	14.696	14.504	145.038	1.000

pressure transmitting fluid. After holding the product for desired time at target pressure, the vessel is depressurized and the product is unloaded.[5,7] Pressure holding times of 10 min or less may be required to develop a commercially viable process. Batch processes can be used for any kind of food in a flexible package, often the final consumer package.

Semicontinuous systems (Fig. 1B) for treating liquid foods use two or more pressure vessels containing a free-floating piston to compress liquid foods. A low-pressure transfer pump is used to fill the pressure vessel. As the vessel is filled, the free piston is displaced. When filled, the inlet valve is closed and pressure-transmitting fluid (usually water) is introduced behind the free piston to compress the liquid food. After an appropriate holding time, releasing the pressure on the pressure transmitting fluid decompresses the system. A pump is used to move the free piston towards the discharge port. The treated liquid food can be filled aseptically into presterilized containers.[5] The batch vessels in a semicontinuous system are connected such that, when one vessel discharges the product, the second system pressurizes, while the third system gets loaded. In this way, the output is maintained in a continuous fashion. Currently, continuous high-pressure process equipment is not available. Table 2 summarizes the list of vendors providing commercial scale high-pressure food equipment.

TYPICAL PROCESS

During HPP, the pressure is uniformly distributed at all points within the pressure chamber. Unlike traditional

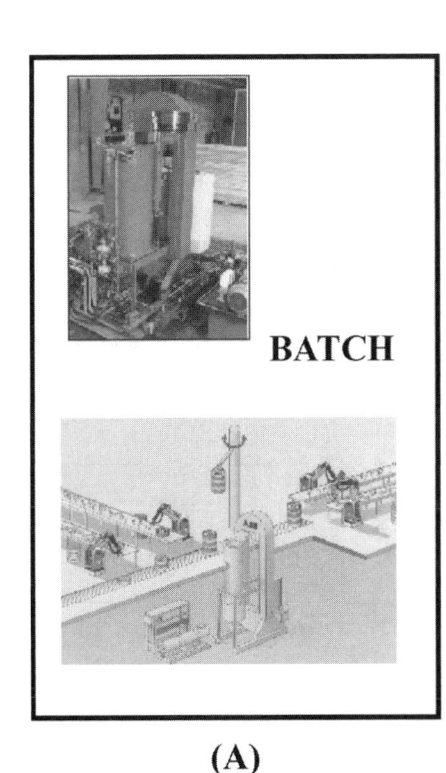

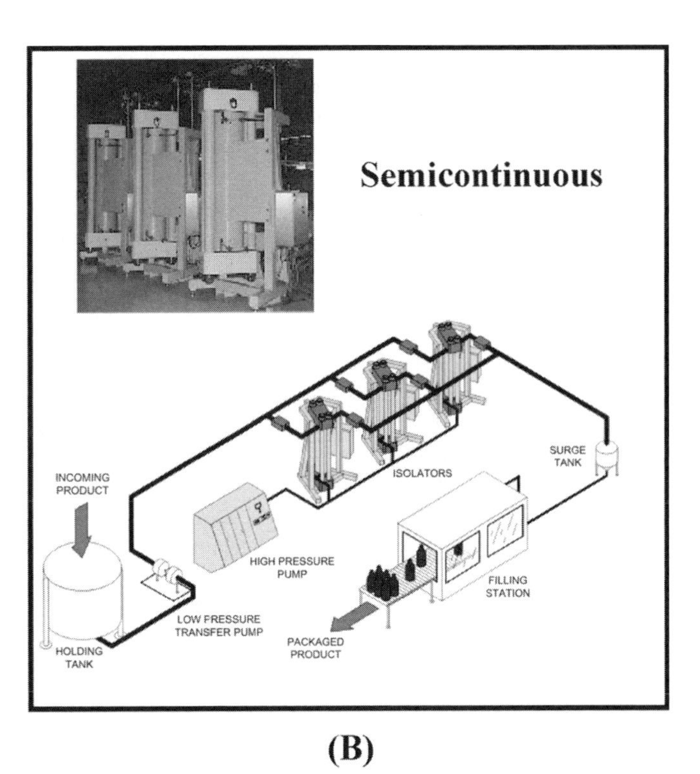

(A) (B)

Fig. 1 Both batch (A) and semicontinuous (B) high pressure food processors are commercially available (Source: http://www.flowcorp.com; accessed: 11/15/01).

Table 2 Supplier of high pressure processing equipment

Vendor	Address
Alstom	Boulevard Prairie au Duc, 44945 Nantes Cedex 9, France; http://www.alstom.com
Flow International Corp.	23500 64th Ave., S. Kent, Washington 98032; http://www.flowcorp.com
Kobe Steel Ltd.	9–12, Kita-Shinagawa 5-chrome, Shinagawa-ku, Tokyo, 141-8688, Japan; http://www.kobelco.co.jp
Stork Food & Dairy Systems	Ketelstraat 2, 1021 JX, Amsterdam, Netherlands; http://www.storkgroup.com

thermal process, pressure is "instantaneously" transmitted to all points within the HPP processor. Typical pressure–temperature curve for a batch HPP treatment is shown in Fig. 2. The pressure is generated by direct compression, indirect compression, or by the heating medium. Pressure come-up time (AB) is the time required to increase the pressure of the sample from atmospheric pressure to the process pressure. The rate of compression is proportional to the power of the pump used. Pressure holding time (BC) is the time interval between the end of compression and the beginning of decompression. A process time of 10 min or less is desired for commercially viable process. Once the pressure is released, expansion occurs. Decompression time is the time required to bring a food sample from process pressure to atmospheric pressure.

Use of pulsed pressure or oscillatory pressure treatments has been shown to be generally more effective than an equivalent single pulse of equal time.[2,8] The difference in effectiveness varies, and the measure of improved inactivation by pulsed pressurization must be weighed against the design capabilities of the pressure unit, added wear on the pressure unit, possible detrimental effect to the sensory quality of the product, and possible additional time required for cycling.[5]

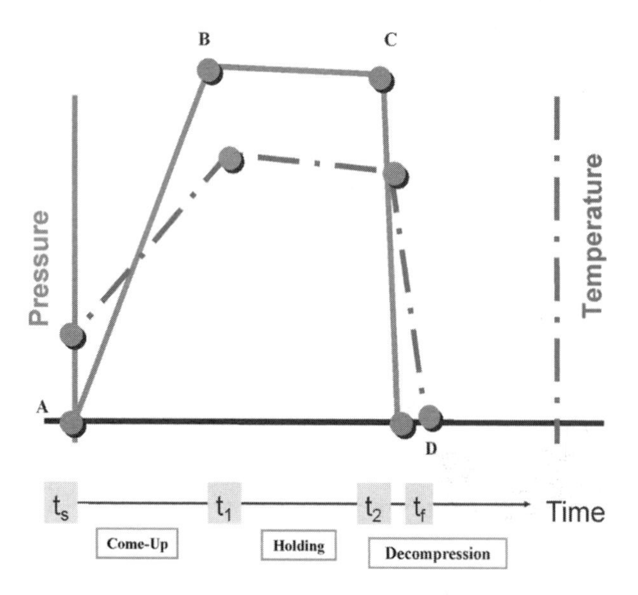

Fig. 2 Typical pressure–temperature curves for high pressure processing of foods.

During pressurization (Fig. 2), the sample temperature increases (Table 3) as a result of compression (AB). Sample temperature curve followed the pressure curve during pressure come-up time and exhibited slight time lag (up to 30 sec) to reach maximum temperature during pressure holding time.[9] This could be attributed to material phase change under pressure and likely influence the microbial inactivation kinetics. The degree of temperature increase and time lag to reach maximum temperature under pressure depends upon the composition of product being processed. Upon pressure release, the temperature drops back to or below the initial temperature. For example, compression heating of water is about 3.0°C/100 MPa pressure. Thus, high pressure offers unique way to increase temperature of the product only during the treatment. Product temperature at process pressure is independent of compression rate as long as heat transfer to the surroundings is negligible. Compression heating of materials is predictable.[10,11] To date, water appears to have the slowest compression heating values, while fats and oil seems have the highest compression heating value.[9] For processing refrigerated foods, moderate pressure (up to 600 MPa) treatment at ambient temperature is sufficient. On the other hand, for low-acid shelf stable type foods, combination of elevated pressures (up to 900 MPa) and temperatures (up to 110°C) are required. Under such process conditions, potential nonuniformity issues generally arise because of variations in temperature that result from compression heating and subsequent heat transfer. Temperature during processing may not be uniform because of compression heating differences between the product and the pressure transmitting fluid, and heat loss or heat gain between the food product, pressure transmitting fluid, and the pressure vessel. To minimize heat loss to the surroundings, the pressure vessel should be held at a temperature equal to the final process temperature (initial temperature of the food plus the apparent temperature increase due to compression heating).

Table 3 Temperature change due to adiabatic compression for selected food substances

Material	Temperature increase (°C) per 100 MPa
Water, mashed potato, orange juice, tomato salsa, 2% milk	Approximately 3.0
Salmon	Approximately 3.2
Chicken fat	Approximately 4.5
Water/glycol mix (50:50)[a]	Between 3.7 and 4.8
Beef fat	Approximately 6.3
Olive oil[a]	Between 6.3 and 8.7
Soy oil[a]	Between 6.2 and 9.1

[a] Substances exhibited decreasing ΔT as pressure increased.
(From Ref. [9].)

EFFECT OF HIGH PRESSURE ON MICROBIAL INACTIVATION

Vegetative Bacteria

Research has demonstrated microbial inactivation in foods by HPP.[4,5,12] HPP treatments are effective in inactivating most of the vegetative microorganisms, with pressures between 200 MPa and 600 MPa at ambient temperatures adequate to inactivate most of the vegetative microorganisms. The extent of inactivation also depends on type of microorganism, pH, and water activity. Microorganisms in exponential growth phase are more pressure sensitive than those in stationary phase. Gram-positive organisms are more resistant than Gram-negatives. There can be significant variations in pressure resistances between strains.

Often food composition can have a protective effect and it is important to evaluate HPP resistance of microorganism in foods rather than in traditional buffer solutions.[12] It should be noted that pressure resistance of microorganisms is usually higher at ambient temperatures and this resistance is reduced at higher or lower temperatures. Combined pressure–temperature processes could be used to achieve similar microbial inactivation as traditional thermal processes, but at significantly lower temperatures. It has long been recognized that pressure will sublethally stress or injure bacteria. Care should be taken to use nonselective microbiological media to allow detection of all viable organisms of concern.

Yeast, Molds, and Viruses

Most yeasts are inactivated by exposure to 300 MPa at 25°C within a few minutes; however, yeast ascospores (fungal spores formed within ascus) require treatment at higher pressures. High pressure processing inactivation of molds has not been studied as thoroughly as inactivation of bacteria.[5] Among viruses, there is a high degree of structural diversity and this is reflected in a wide range of pressure resistances.[12] Human viruses appears to be more pressure sensitive than the Tobacco Mosaic viruses. Overall, these results suggest that most human viruses should be eliminated by pressure treatments designed for elimination of bacteria of concern (for example, 400 MPa); however, this requires further study before such broad range conclusions can be drawn.[5]

Bacterial Spores

There are only limited studies on the efficacy of HPP on spore inactivation.[2,13–18] Bacterial spore inactivation requires combination of elevated pressures and moderate temperatures.[19] For each spore, there exists a threshold pressure and temperature below which increasing processing time may not produce any significant microbial reductions. To date, only limited number of *C. botulinum* strains (types E, A, and B) have been tested.[17,20] Nonproteolytic type B spores have been identified as the most pressure resistant spore-forming pathogens found to date. Further studies are necessary to identify the most resistant strain of *C. botulinum* and a suitable surrogate spore for industrial process validation studies. Microbial criteria to produce safe, high-pressure processed foods needs to be established. Additional research is needed to understand the mechanism of combined action of the pressure and heat on the spores, spore injury, and the uniformity of the treatment delivered.

Influence of pH

Most food constituents contain water. The pressurization of food constituents is thought to cause the ionic dissociation of water molecules with a corresponding decrease in pH.[4,21] The product pH is likely to shift as a function of pressure. A pressure of 100 MPa at 25°C, causes a 0.73 pH unit decrease in the pH of water.[4] With neutral pH phosphate buffer a pressure of 68 MPa causes a 0.4 unit decrease in pH.[22] pH changes likely influence protein properties (gel formation, enzyme activity) and microbial inactivation kinetics. As pH is lowered, most microorganisms become more susceptible to HPP inactivation, and sublethally injured cells fail to repair.[21] At present, reliable instrumentation for monitoring pH during high pressure processing is not readily available. Further research is necessary to define the exact role of pH during high pressure processing.

Influence of Water Activity

Few studies are available on the relationship between HHP treatment and water activity (a_w) of foods on microbiological inactivation. Oxen and Knorr[23] reported a reduction of water activity at ambient pressure from 0.98–1.0 to 0.94–0.96 that resulted in a marked reduction in inactivation rates for microbes suspended in a food. Rodriguez et al.,[24] observed that the interaction effect between pressure and water activity to be significant on inactivation of *E. coli*. Reducing the water activity appears to protect microbes against inactivation by HPP; however, it is to be expected that bacteria may be sublethally injured by pressure, and recovery of sublethally injured cells can be inhibited by low water activity. Consequently, the net effect of water activity may be difficult to predict.[5]

EFFECTS OF PRESSURE ON FOOD CONSTITUENTS

High pressure processed foods, in general, will not undergo chemical transformation due to pressure treatment itself. Pressure effects on the structure and texture of foods are more marked and variable and the mechanisms are not fully understood.[25] During HPP, only the noncovalent bonds (hydrogen bonds, ionic bonds, and hydrophobic bonds) in the food component are broken or formed depending on the volume decrease of the system; covalent bonds are not affected during the pressurization.[4,14] This unique characteristic suggests that reactions like Maillard reaction and production of cooking flavors (off flavors) that commonly occur in thermal treatments will not occur during HPP treatment of foods.

Most biochemical reactions are influenced by high pressure since they often involve a change in volume.[6] Volume-increasing reactions will tend to be inhibited by pressure while volume-decreasing reactions will tend to be promoted. The melting point of triglycerides increases with increasing pressure. Lipids present in the liquid state at ambient conditions crystallize under pressure.[25] High pressure enhances formation of the denser and more stable crystals.

Enzymes present in foods if not inactivated could cause degradation of the physical, chemical, and nutritional characteristics of the foods during their shelf life. During thermal processing, inactivation of the microbial flora and enzyme inactivation occur simultaneously. In contrast, pressure levels of 600 MPa at 25°C are sufficient to inactivate the vegetative microorganisms present, but enzymes may not be inactivated. To inactivate enzymes such as polyphenol oxidase and pectinase, a combination of high pressure and temperature treatments is necessary.[14] The pressure treatment to inactivate enzymes depends on a various factors like the type of the enzyme, the pressure level, the process time, the pH value, and the temperature.

COMBINATION PROCESS

Although HPP is effective in microbial inactivation, researchers are attempting to combine HPP with other interventions (CO_2, ultrasound, UV, irradiation, heat) to further reduce the pressure requirement. For example, researchers have reported that a combination of HPP and CO_2 may reduce the time and temperature needed for microbial inactivation.[26–31] Reduction of time and temperature of treatment would result in substantial cost-savings to the producer and allow greater industrial adaptation of these methods.[32]

CONCLUSION

High pressure processing of foods offers unique opportunities and challenges to the food industry. High pressure processing is likely to be used commercially before its scientific basis is well understood and its potential fully exploited. The technology enables processors to make novel, minimally-processed, extended shelf-life convenient food items with fresh like attributes and natural colors. Identification of commercially viable products would be a major challenge. Pressure in combination with moderate temperature seems to be a promising approach for processing low-acid foods. Compression heating could reduce the severity of the process requirement. Criteria for safe processing of various foods need to be developed.

REFERENCES

1. Hite, B.H. The Effect of Pressure in Preservation of Milk. Morgantown Bull. W.V. Univ. Agric. Exp. Sta. **1899**, 58, 15–35.
2. Meyer, R.; Cooper, K.L.; Knorr, D.; Lelieveld, H.L.M. High-Pressure Sterilization of Foods. Food Technol. **2000**, *54* (11), 67–72.
3. Balny, C.; Masson, P. Effects of High Pressure on Proteins. Food Rev. Int. **1993**, *9* (4), 611–628.
4. Cheftel, J.C. Review: High-Pressure, Microbial Inactivation and Food Preservation. Food Sci. Technol. Int. **1995**, *1* (2–3), 75–90.
5. Farkas, D.; Hoover, D. High Pressure Processing. In Special Supplement: Kinetics of Microbial Inactivation for Alternative Food Processing Technologies. J. Food Sci. Suppl. **2000**, 47–64.
6. Ting, E.; Tremoulett, S.; Hopkins, J.; Many, R. A Comparison Between UHP Hydrostatic Exposure and UHP Discharge Production Methods. Conference on High Pressure Bioscience and Biotechnology, Heidelberg, Germany, Aug 30–Sept 3, 1998.
7. Träff, A.; Bergman, C. High Pressure Equipment for Food Processing. In *High Pressure and Biotechnology*; Balny, C., Hayashi, R., Heremans, K., Masson, P., Eds.; Colloque INSERM/J. Libbey Eurotext Ltd: Paris, France, 1992; 224, 509–514.
8. Hayakawa, I.; Kanno, T.; Yoshiyama, K.; Fujio, Y. Oscillatory Compared with Continuous High Pressure Sterilization on *Bacillus stearothermophilus* Spores. J. Food Sci. **1994**, *59* (1), 164–167.
9. Rasanayagam, V.; Balasubramaniam, V.M.; Ting, E.; Sizer, C.E.; Bush, C.; Anderson, C. Compression Heating of Selected Food Substances During High Pressure Processing. In 2001 Annual Meeting of The Institute of Food Technologists, New Orleans, LA., June 23–27, 2001; Abstract No. 28-3.
10. Bridgman, P.W. Water, in the Liquid and Three Solid Forms, Under Pressure. Proc. Am. Acad. Arts Sci. **1912**, *47*, 439–558.
11. Harvey, A.H.; Peskin, A.P.; Klein, S.A. *NIST Standard Reference Steam Database 10 version 2.2*; U.S. Department of Commerce, National Institute of Standards and Technology: Boulder, CO, 1996.
12. Smelt, J.P.P.M. Recent Advances in the Microbiology of High Pressure Processing. Trends Food Sci. Technol. **1998**, *9*, 152–158.
13. Patterson, M.F.; Kilpatrick, D.J. The Combined Effect of High Hydrostatic Pressure and Mild Heat on Inactivation of Pathogens in Milk and Poultry. J. Food Prot. **1998**, *61* (4), 432–436.
14. Gola, S.; Foman, C.; Carpi, G.; Maggi, A.; Cassara; Rovere, P. Inactivation of Bacterial Spores in Phosphate Buffer and in Vegetable Cream Treated with High Pressures. In *High Pressure Bioscience and Biotechnology*; Hayashi, R., Balny, C., Eds.; Elsevier Science B.V.: Amsterdam, 1996; 253–259.
15. Rovere, P. The Third Dimension of Food Technology. Technol. Aliment., Tetra Pak Rep. **1995**, *4*, 2–8.
16. Rovere, P.; Carpi, G.; Dall'Aglio, G.; Gola, S.; Maggi, A.; Miglioli, L.; Scaramuzza, N. High-Pressure Heat Treatments: Evaluation of the Sterilizing Effect and of Thermal Damage. Ind. Cons. **1996**, *71* (4), 473–484.
17. Reddy, N.R.; Solomon, H.M.; Fingerhut, G.A.; Rhodehamel, E.J.; Balasubramaniam, V.M.; Palaniappan, S. Inactivation of *Clostridium botulinum* type E Spores by High Pressure Processing. J. Food Saf. **1999**, *19*, 277–288.
18. Okazaki, T.; Kakugawa, K.; Yoneda, T.; Suzuki, K. Inactivation Behavior of Heat-Resistant Bacterial Spores by Thermal Treatments Combined with High Hydrostatic Pressure. Food Sci. Technol. Res. **2000**, *6* (3), 204–207.
19. Balasubramaniam, V.M. High Pressure Processing of Foods. Review of Basic Principles. Proceedings of Short Course on Principles of High Pressure Processing, National Center for Food Safety and Technology, Summit-Argo, IL, Aug 14–18, 2000; Balasubramaniam, V.M., Sizer, C.E., Knutson, K. Eds.
20. Reddy, N.R.; Solomon, H.M.; Tetzloff, R.C.; Balasubramaniam, V.M.; Rhodehamel, E.J.; Ting, E.Y. Inactivation of *Clostridium botulinum* Spores by High Pressure Processing. 2001 Annual Report, National Center for Food Safety and Technology: Summit-Argo, IL, 2001.
21. Hoover, D.; Metrick, C.; Papineau, A.M.; Farkas, D.F.; Knorr, D. Biological Effects of High Hydrostatic Pressure on Food Microorganisms. Food Technol. **1989**, *43* (3), 99–107.
22. Linton, M.; McClements, J.M.J.; Patterson, M.F. Inactivation of *Escherichia coli O157:H7* in Orange Juice Using a Combination of High Pressure and Mild Heat. J. Food Prot. **1999**, *62* (3), 277–279.
23. Oxen, P.; Knorr, D. Baroprotective Effects of High Solute Concentrations Against Inactivation of *Rhodotorula rubra*. Lebensm.-Wiss. Technol. **1993**, *26*, 220–223.
24. Rodriguez, J.J.; Sepulveda, D.R.; Barbosa-Canovas, G.V.; Swanson, B.G. Combined Effect of High Hydrostatic Pressure and Water Activity on *Escherichia coli* Inhibition. 2000 Meeting of The Institute of Food Technologists, Dallas, TX, June 10–14, 2000; Abstract No. 86H-5.
25. Cheftel, J.C. Effects of High Hydrostatic Pressure on Food Constituents: An Overview. In *High Pressure and Biotechnology*, 1st Ed.; Balny, C., Hayashi, R., Heremans, K., Masson, P., Eds.; INSERM and John Libbey: Paris, 1992; 195–209.
26. Farr, D. High Pressure Technology in the Food Industry. Trends Food Sci. Technol. **1990**, *4*, 14–16.
27. Balaban, M.O.; Arreola, A.G. Supercritical Carbon Dioxide Applied to Citrus Processing. Trans. Citrus Eng. Conf. **1991**, *37*, 16–37.
28. Ballestra, P.; Cuq, J.L. Influence of Pressurized Carbon Dioxide on the Thermal Inactivation of Bacterial and Fungal Spores. Lebensm.-Wiss. Technol. **1998**, *31* (1), 84–88.
29. Hong, S.I.; Park, W.S.; Pyun, Y.R. Non-Thermal Inactivation of *Lactobacilllus plantarum* as Influenced

by Pressure and Temperature of Pressurized Carbon Dioxide. Int. J. Food Sci. Technol. **1999**, *34* (2), 125–130.

30. Isenschmid, A.; Marison, I.W.; Stockar, U.V. The Influence of Pressure and Temperature of Compressed CO_2 on the Survival of Yeast Cells. J. Biotechnol. **1995**, *39*, 229–237.
31. Lin, H.M.; Cao, N.; Chen, L.F. Antimicrobial Effect of Pressurized Carbon Dioxide on *Listeria monocytogenes*. J. Food Sci. **1994**, *59* (3), 657–659.
32. Mermelstein, N.H. High-Pressure Pasteurization of Juice. Food Technol. **1999**, *53* (4), 86–90.

Homeostasis in Animal Environment

Louis Albright
Cornell University, Ithaca, New York, U.S.A.

H

INTRODUCTION

Homeostasis is the maintenance of deep body temperature within a narrow range over a wide range of environmental conditions. Heat is produced by metabolic processes and retained during cold weather by vasoconstriction and various behavioral adaptations. Heat loss is facilitated in hot weather by vasodilation, water evaporation from skin surfaces, and various behavioral adaptations. A homeothermic animal automatically adjusts to environmental conditions in an effort to remain within its comfort zone. Excursions outside the comfort zone can lead to stress and, in extreme conditions, death. Metabolic heat production can be predicted mathematically and linked to heat loss models and environmental control algorithms to provide optimized aerial conditions to animals in confined animal housing.

BODY

Plants and some animal species maintain internal temperatures close to those of their (convective plus radiant) environments; other animal species do not. Mammals and birds, in particular, maintain their body temperatures within narrow ranges unless environmental conditions are extreme. This is homeostasis—maintaining constant deep body temperature over a wide range of environmental conditions. Various terms characterize how animals produce heat and interact with their thermal environments. Some more commonly-used terms related to animal temperature biology are in Table 1.

When homeothermic animals are exposed to ambient temperatures that are either too high or too low, they must adjust by changing the rate at which heat is lost from their body surfaces, the rate at which heat is produced by metabolism within their bodies, or both. Heat loss rates can be reduced by mechanisms such as vasoconstriction (movement of blood vessels away from the skin surface), feather fluffing, huddling, piloerection, or seeking different environments. Housed animals have few options to seek different environments, making environmental control crucial in agricultural building design and operation.

The metabolic responses of (adapted) animals depend on surrounding temperatures. Extensive data have shown homeothermic animals gradually develop higher metabolic rates when subjected to continued cold temperatures. If cold exposure is too limited to permit adaptation, increased metabolic rates may occur by involuntary physical activity such as shivering or increased activity such as more rapid or frequent voluntary movement. Metabolism must increase to provide the additional heat required to balance the greater heat loss to the cold environment if hypothermia and eventual death are to be avoided.

Conversely, when the environment becomes too warm and heat loss cannot balance the metabolic rate, metabolic heat production must decrease. Physical inactivity and reduced feed intake are two means by which metabolic rates can be decreased. People as well as other animals become lethargic when exposed to continued hot weather, for example. If metabolic rates cannot be reduced sufficiently, or sufficiently rapidly, death by hyperthermia is inevitable.

A region where no special metabolic adaptation is required to maintain body temperature lies between these two extremes. This region is defined at its cold end by the onset of full vasoconstriction. That is, greater temperature differences due to colder environments are compensated by increasing thermal resistances. The warm end of the region is defined by onset of full vasodilation (movement of blood vessels toward the surface of the skin), beyond which the animal can do no more to lose heat (except to seek another environment or increase evaporative heat loss by sweating or wallowing in a wet material, as examples.)

Varying the degree of vasoconstriction/vasodilation requires essentially no effort on the part of the animal. Thus, another method to identify this region of ready adaptation is to define it as the zone of minimum thermoregulation effort, termed the "thermoneutral zone" (TNZ). The TNZ is bounded at its low temperature end by the "lower critical temperature," (LCT), and at the high temperature end by the "upper critical temperature," (UCT). Note: LCT and UCT define ambient temperatures, not body temperatures, because body temperature should remain unchanged in an animal that exhibits homeostasis. Reactions to temperature environments described previously are frequently expressed graphically (Fig. 1).

Conceptually, deep body temperature remains constant over a wide ambient temperature range. In practice, there may be small changes not greater than diurnal variations of

Encyclopedia of Agricultural, Food, and Biological Engineering
DOI: 10.1081/E-EAFE 120006909
Copyright © 2003 by Marcel Dekker, Inc. All rights reserved.

Table 1 Some terms and definitions related to animal temperature biology

Term	Definition
Bradymetabolic	Characterized by a low level of resting (basal) metabolism
Tachymetabolic	Characterized by a high level of resting (basal) metabolism
Poikilotherm	Variable body temperature ("cold-blooded")
Homoiotherm	Relatively constant body temperature ("warm-blooded")
Ectotherm	Deep-body temperature is determined by external heat sources
Heliotherm	Relies on the sun's heating for warming
Thigmotherm	Relies on conduction from environment for warming (e.g., soil or water)
Endotherm	Deep-body temperature is determined by cellular metabolism
Homeotherm	Relatively constant deep-body temperature ("warm-blooded")

body temperature such as, for example, lower body temperature during sleep. However, if the surroundings are too cold (temperature at or below T_1 shown in Fig. 1), hypothermia results, possibly followed by death. Metabolism slows, accelerating the decline of deep body temperature in a positive feedback sequence. In contrast, if the surroundings are too warm (temperature at or above T_2), hyperthermia results as the body works harder to rid itself of heat. This, too, is a positive feedback sequence that can hasten the onset of heat stroke and possibly death.

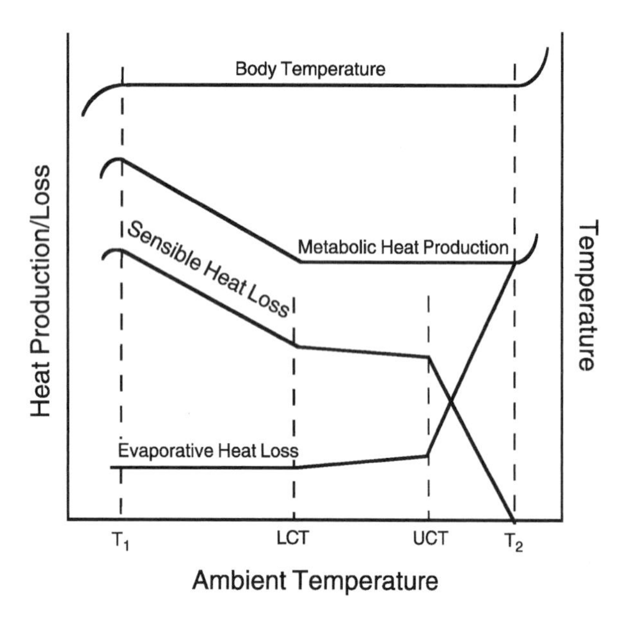

Fig. 1 Diagram of the relationship between the body temperature of homeothermic animals and ambient temperature, metabolic heat production, and partitioning between evaporative and nonevaporative heat loss mechanisms.

Within the ambient temperature range where body temperature remains relatively constant, metabolic heat production can fall into one of two zones. When ambient temperature lowers below the LCT, metabolism increases to compensate for the unavoidable and large heat loss rate. This is the sloped region of the "Metabolic Heat Production" line in Fig. 1. As stated previously, metabolism may increase in part through slow adaptation to the cold or may be the result of increased short-term physical activity, such as shivering. A third method of compensation may occur in certain animals. This is the metabolism of brown adipose tissue (BAT). The BAT is a form of body fat with very high energy density, perfused with many blood vessels, and able to metabolize rapidly in response to cold. This added heat source is found in young calves, for example, when housed (immediately after birth) outdoors in hutches during winter in cold climates. Once a calf loses its ability to create, store, and use BAT, that ability is unlikely to be regained by re-exposure to cold. The ability to store and use BAT permits the practice of housing calves in outdoor hutches during the winter in cold climates, where they are less likely to be exposed to disease organisms.

Evaporative heat loss is only by respiration and insensible perspiration when ambient conditions are below the LCT. Data have shown insensible evaporative heat loss is a relatively constant fraction of the total metabolic rate. Although, one might surmise that evaporative heat loss will vary as the ambient relative humidity varies, that effect is relatively small unless air temperature is near body temperature. When ambient temperatures are low, the humidity ratio of air (kg of water vapor per kg of dry air) is small whether the air is dry or near saturation. Respired air will be near deep body temperature and near saturation. In this condition, its humidity ratio is large and relatively constant. The numerical difference between the two humidity ratio values determines the mass of water and, thus, the heat lost by respiration. The difference between a large, stable number and a small varying number is relatively constant.

Relative constancy of evaporative heat loss below the LCT is the reason sensible heat loss must be higher at lower temperatures and must rise in parallel with metabolic heat production. (In reality, sensible loss does not rise due to metabolic heat production but, rather, heat production rises to compensate the greater heat loss.) The slope of the sensible heat loss line is the idealized thermal conductance of the animal between deep body and

ambient temperatures. Vasoconstriction is maximized. Thus, the resistance remains (more or less) constant.

Between the LCT and the UCT, evaporative heat loss may increase and sensible heat loss may decrease slightly. However, as a first approximation, the change is assumed to be small and may be ignored in a basic description of the process. A homeothermic animal is able to lose sensible heat at a constant rate through the processes of vasodilation and vasoconstriction. Thermal resistance between deep body and ambient conditions varies to compensate for the changing temperature difference.

Above maximum vasodilation, corresponding to the UCT, body temperature can no longer be maintained with minimum effort. More effort must be expended to compensate for the manner in which sensible heat loss must decrease as ambient temperature rises. The slope of the sensible heat loss line is greater than the slope below the LCT because of the smaller value of thermal resistance between deep body and ambient conditions. Evaporative losses must rise to compensate for the reduced capacity to lose heat by sensible means. Although many animals do not "sweat," insensible perspiration may increase somewhat and behavioral changes such as shallow panting, wallowing (for example, swine), or licking fur or skin is used to increase evaporative losses. If greater evaporative heat loss is not an option for the animal and a cooler environment cannot be found, temperature T_2, the temperature of extreme heat stress, will be close to the UCT.

The concept of "basal metabolism" is precisely defined. It is a measure of the rate of heat production (through oxygen uptake or carbon dioxide expiration) while in a fasting and resting state in a thermally neutral environment. Maintenance metabolism is a measure of the averaged rate of heat production expected from an animal involved in normal activities.

Basal metabolism, M_B, for homeotherms, from birds to rodents to people to elephants, can be approximated by a single relationship that depends only on body mass, B.

$$M_B = k_m B^n$$

Specific basal metabolism (metabolism per unit of body mass) is

$$(M_B)_s = k_m B^{n-1}$$

where k_m is a coefficient often treated as a constant, B is body mass (kg), and M_B is heat production measured in watts. The exponent n has been suggested to range from 2/3 to 3/4; a value of 0.75 is often assumed. As a perspective on the exponent, one may interpret the metabolic size of an animal as proportional to its body mass raised to the 0.75 power. A value for k_m equal to 3.4 W kg$^{-0.75}$ is suggested by Monteith and Unsworth,[1] corresponding to a daily heat production of 70 kcal per kg$^{0.75}$.

As an approximation for many homeotherms, basal metabolism normalized to a unit area of body surface is approximately 30 W m^{-2}–50 W m^{-2}. Maintenance metabolism is, of course, somewhat greater than this value because it encompasses a higher level of activity (for example, active digestion in progress) while basal metabolism is referenced to resting periods in the fasting state. At the other extreme, a rough rule is that an animal can sustain a maximum metabolic rate approximately ten times its basal rate.

REFERENCE

1. Monteith, J.L.; Unsworth, M.H. *Environmental Physics*; Edward Arnold, a division of Hodder & Stoughton: London, 1990.

FURTHER READING

Cossins, A.R.; Bowler, K. *Temperature Biology of Animals*; Chapman and Hall: London, 1987.

Kleiber, M. *The Fire of Life*, Robert E.; Kreiger Publishing Co.: Malabar, FL, 1975.

Kleiber, M. *Effect of Environment on Nutrient Requirements of Domestic Animals*; National Academy Press, Washington, DC, 1981.

Human Factors and Ergonomics

L. Dale Baker (Retired)
Case Corporation, Burr Ridge, Illinois, U.S.A.

INTRODUCTION

Human Factors and Ergonomics is a collection of knowledge "to achieve compatibility in the design of interactive systems of people, machines, and environments to ensure their effectiveness, safety, and ease of performance."[1] Historically, from the invention of the first tools, tool design and tool usage has gradually improved, but the science and art of human factors and ergonomics was born during World War (WW) II. Numerous projects evolved during this war, but only three will be briefly described here.

OVERVIEW

For the first time, aircrafts were built in large quantities. Engineers had questions regarding the size of the operator seat and the location of operator controls relative to that seat. What is the optimum width of the pilot's seat? How far forward of that seat should the controls be placed? To answer questions of that nature, the military asked volunteers to pose nude for photographs in front of a screen marked in squares. These photographs were then used to determine the height, width, arm length, leg length, etc. for this population sample.[2] The summaries of these types of measurements are called anthropometric data.

Airplane crashes were a major issue in pilot training. In a lecture at the University of Michigan, Dr. Alphonse Chapanis described one investigation of dozens of reports of pilots of B-17s retracting the landing gear instead of the landing flaps after landing.[3] In the cockpit were two identical toggle switches, side by side, for the landing gear and the flaps. Dr. Chapanis calls this a "designer error" and not "pilot error."

One altimeter design for WW II training aircraft had an altimeter with hands like a clock. The little hand sweeps out thousands of feet and the large hand sweeps out hundreds of feet. Trainee pilots found this altimeter confusing and a height of 4500 ft might be misread as 5400 ft, as the pilot experienced a "controlled flight into terrain." A linear altimeter resolved this problem.

These WW II examples illustrate the beginning of the discipline of human factors. For many years, people in the field have generally described human factors as relating to the man–machine interface. At this same time in Europe, a major area of investigation concerned measuring the capacity of people to exert physical effort. Examples include measuring the maximum effort a person can exert during a specific task and the ability to work at a constant physical level, such as measuring the kilograms of effort a person can sustain in a repeated task during a day of work. The European term for this area of investigation was "Ergonomics." In 1992, the Human Factors Society changed its name to the Human Factors and Ergonomics Society and the terms are becoming synonymous. In addition, note that personal safety is a component in human factors evaluations and human factors is a component of a "safety analysis" of a product.

The family automobile experienced significant human factors related development in the last 30 yr. Parallel developments in the evolution of agricultural equipment occurred during the same period. A few examples of tractor evolution:

- Ingress/egress of the cab changed from typically vertical steps to steps, which approximate steps in the home. The operator is able to safely and comfortably walk forward into and out of the cab, as opposed to the requirement to turn around and back out of the cab.
- Visual obstructions have been removed. The engine exhaust and muffler were moved to align with the cab post on the right front corner of the cab. Visually, very little of the exhaust system is seen behind the shadow of the cab corner post. The side post on the right side of the cab was designed out. Visual blockage is reduced to the four cab corner posts and the left-hand doorpost.
- The seat suspension uses air and oil systems to dampen the ride.
- The effort required to steer and apply the brakes of new machines parallels the efforts required for automobile systems.
- In recent years, some of the larger machines have the machine function controls mounted in the right-hand armrest. The physical effort to operate these controls is minimal.
- The noise level in the cab is lower than a new cab built 30-yr ago.

Encyclopedia of Agricultural, Food, and Biological Engineering
DOI: 10.1081/E-EAFE 120006869
Copyright © 2003 by Marcel Dekker, Inc. All rights reserved.

- The lighting for operating at night is greatly improved in both the intensity of the light and the ability to focus the light to where it is desired.

Newer equipment is easier to operate and the stress of operation of this newer equipment is typically much less than equipment built 30-yr ago. The operator is likely to be less tired, either mentally or physically, at the end of the same length of day.

Moving most of the controls in the armrest of these new machines sends an electrical input to electronic controllers. This controller then produces the desired result, either electrically or hydraulically. There are many human factor issues in the design of these controls. In the United States, moving a light switch "up" will turn on the light. In England, the switch is moved down to turn on the light. The operator will "transfer" this expectation that moving the light switch will turn on the light. The practical application for tractors is that light switches move up to turn on lights on equipment in the United States but the switch is rotated 180° for the European market. The new controls in the armrest need to be designed to anticipate the "transference" from the older style controls that are replaced. The PTO control was typically a lever that was moved forward to engage the PTO. Now, the new toggle or rocker switch in the armrest moves forward to engage the PTO. Some features in this switch should minimize the probability of inadvertent engagement of this switch from a casual hand movement or catching shirtsleeve.

The use of electronic controllers created the opportunity to provide additional safety benefits for the operators. It is now possible and practical to implement methods to minimize the probability of any inadvertent or unintended movement or engagement of these machines. For example, if the tractor PTO switch is left in the "ON" position, the PTO will not engage when the engine is started. Another system uses switches to insure that when the engine is started, the tractor will not move forward or backward, unless the operator is in the seat and intentionally engages the transmission in the normal manner. (If the transmission control is not in the "neutral" position when the operator sits down, the control must be moved into "neutral" and then "into gear" before the tractor will move.) The challenge to the designer and the "human factors/safety" person is to identify these risks of unexpected movement/engagement that require attention and include software and sensors to address these risks.

An additional opportunity exists in designing controls and controller systems that are easy to use in routine applications and are easy to calibrate when new and after maintenance on the machine. Many of these machines are used intensively for a few weeks and then are stored until next year. A well-designed control and controller will minimize the amount of relearning necessary to operate the machine during subsequent years of operation.

In the years after WW II, many tables of anthropometric data were published. Much of the data was based on the military population, typically young people who were physically fit. Very short and very tall people are not inducted into the military. This data is not a good fit to the general population. Some companies collected their own proprietary data. For many years, General Motors collected proprietary anthropometric data from volunteer visitors to the GM Design Center.[4] More recently, SAE coordinated a project called CAESAR (Civilian American and European Surface Anthropometry Research project). Anthropometric data was collected from civilians, both men and women between 18 yr and 65 yr of age. The sample is approximately 2500 residents in Europe and 2500 residents in the United States. Data is available from SAE after July 2002.

One of the issues for most users of this data is the need to select the percentile of population to be included in the calculation or design. The human population is a normal distribution about the 50th percentile. Then, selecting dimensions to include the 5th percentile to the 95th percentile will mean that 5 people out of 100 will be smaller than the design criteria and 5 people will be larger than the design criteria. Using 1st to 99th percentile data will include 98% of the population. A careful analysis of the anthropometric data and an analysis of the risks of this design for those outside these percentile limits are necessary in selecting the percentile to be used in this calculation or design.

Most machines used in production agriculture are used for a few weeks and then are stored until next year. Repetitive stress injuries are not common from using these large machines. However, repetitive stress injuries may be an issue for the food processing industry. Management of this risk may include carefully selecting the workers, carefully training the workers, and in some instances, redesigning the tasks to minimize the stress or minimize the time of exposure to the stress.

CONCLUSION

This discussion of "Human Factors and Ergonomics" is a brief discussion of the practical application of human factors concepts to a segment of agricultural equipment. Human factors and ergonomics is a broad discipline with many more aspects than presented here and this broad discipline should be explored by the reader.

REFERENCES

1. Definition of Human Factors and Ergonomics from Human Factors and Ergonomics Society Website (www.hfes.org).
2. Christensen, J.M. An Introduction to Physical Anthropology. In the *Human Factors Engineering Summer Conference*, Wright-Patterson Airforce Base, Dayton, OH, July 9–20, 1973.
3. Chapanis, A. Visual Information Displays I and II. In the *Human Factors Engineering Summer Conference*, Wright-Patterson Airforce Base, Dayton, OH, July 9–20, 1973.
4. Personal communications. Staff at GM Laboratories, Detroit, MI, 1974.

FURTHER READING

Chapanis, A. *The Chapanis Chronicles: 50 Years of Human Factors Research, Education, and Design*; Aegean Publishing Company: Santa Barbara, CA, 1999.

Murphy, D.J. Safety and Health for Production Agriculture; American Society of Agriculture Engineers: St. Joseph, Michigan, 1992; 256.

Hydrology in Drainage Systems

John E. Parsons
North Carolina State University, Raleigh, North Carolina, U.S.A.

INTRODUCTION

In general, drainage is done in areas that are too wet to support some activity of man. The areas may be too wet because of seasonal high water tables and/or standing surface waters. In agriculture, these conditions can limit or prevent field preparation activities, limit oxygen availability for plants, and limit or prevent harvesting operations. In arid areas, drainage systems are often required in conjunction with irrigation systems to control root zone salinity.

Agricultural drainage systems can be defined as systems that improve the removal of surface and subsurface water. These systems function to improve field trafficability, remove excess water from crop root zones, and provide drainage for controlling salinity. Design of these systems requires considerations of both surface and subsurface water movement. Eastern North Carolina is typical of humid areas in that rainfall exceeds evapotranspiration. In this region, annual rainfall ranges from 1100 mm to 1500 mm with corresponding annual evapotranspiration from a low of 800 mm to a high of 1000 mm.[1] This leaves an excess water ranging from 300 mm to 500 mm that is lost via surface runoff, shallow subsurface drainage, and deep seepage to recharge confined groundwater aquifers. Evans et al.[1] estimated that recharge to deep aquifers accounted for 25 mm–50 mm of the annual excess leaving from 250 mm to 475 mm in shallow subsurface drainage and surface runoff.

In arid and semiarid regions, low rainfall amounts are not sufficient to remove excess salts from the root zone. Irrigation water in these areas tends to be saline. With agronomic irrigation levels, salts tend to build up in the crop root zone in these areas. Additional supplemental irrigation along with adequate subsurface drainage leaches the salts from the root zone.[2]

Hydrology of agricultural drainage systems involves the study of surface and subsurface water distribution and movement. This involves an understanding of all processes affecting water movement and distribution under natural and artificial surface and subsurface drainage systems. The intent of this chapter is to examine procedures and methods to quantify the processes to describe the hydrology of agricultural drainage systems.

HYDROLOGIC CYCLE

Fig. 1 depicts the hydrologic cycle in a typical agricultural drainage system. Typical agricultural drainage systems can include subsurface artificial drainage such as tile drains and open ditches, and surface drainage improvements such as land surface modifications to improve surface water flows to the field outlets. Drainage requirements are usually implemented to either remove excess water or provide leaching to control salinity. In natural drainage systems where these drainage improvements have not been made, surface and subsurface water movement is also important in describing water dynamics and many of the same concepts can be used. Thus, the hydrologic cycle in Fig. 1 can be equally important for describing both natural and artificial drainage systems.

Potential evapotranspiration (PET) and rainfall are the sources and sinks for the system. These are dictated by the weather parameters for the site. In the case of PET, this is the atmospheric demand, which may or may not be available depending on the status of soil water. In general, actual evapotranspiration (ET) is used to reflect what can be delivered from soil via evaporation and transpiration. One of the easiest ways to visualize the hydrologic cycle is a water balance. For Fig. 1, the water balance for a given time period with all quantities as a depth of water per unit area can be written as:

$$\text{Rain} - \text{ET} = \text{Runoff} + \text{Drainage} + \text{DeepSeepage} \pm \Delta\text{SoilWater}$$

where:

Rain = rainfall over the period of interest
ET = actual evapotranspiration
Runoff = surface runoff
Drainage = subsurface drainage
DeepSeepage = deep seepage or recharge to the deep groundwater aquifer, and
ΔSoilWater = change in soil water in the profile.

The hydrologic balance of the water in the soil can be split into two water balances for illustration of the drainage process. The first is the water balance in the unsaturated zone and the second is the balance in the saturated zone.

Encyclopedia of Agricultural, Food, and Biological Engineering
DOI: 10.1081/E-EAFE 120007235
Copyright © 2003 by Marcel Dekker, Inc. All rights reserved.

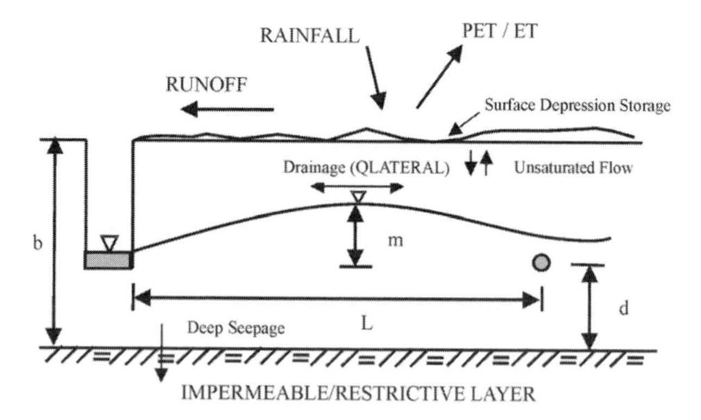

Fig. 1 Schematic depicting the hydrologic cycle for agricultural drainage systems.

These two balances are:

$$\Delta \text{UNSAT} = \text{INFIL} + \text{UPFLUX} - \text{ET} - \text{SEEPUNS}$$

$$\Delta \text{SAT} = \text{SEEPUNS} - \text{UPFLUX} - \text{DEEPSEEP} \pm \text{QLATERAL}$$

where

ΔUNSAT = change in water in the unsaturated zone
INFIL = amount of infiltration
UPFLUX = amount of upward flux from the water table
ET = actual evapotranspiration
SEEPUNS = amount of water flow from the unsaturated zone to the water table
ΔSAT = change in the amount of water in the saturated zone
DEEPSEEP = amount of water leaving the saturated zone through the restrictive (impermeable) layer
QLATERAL = amount of lateral subsurface drainage
Again all quantities are depth per unit area.

These balances describe the overall hydrology for agricultural drainage systems. The first balance describes the overall linkage between weather parameters and the soil water in both the unsaturated and saturated soil zones. Actual evapotranspiration is computed using PET estimates based on atmospheric demand. The PET can be estimated based on any number of methods[3] and the method is generally selected based on availability of weather parameters and location. Actual evapotranspiration is computed based on the ability of the soil to supply this amount of water. The ET is split between soil evaporation and plant transpiration. The ability of the soil to supply this water is based on availability and the transmission characteristics to move the water to surface and plant roots.

Surface Runoff and Infiltration

Rainfall is balanced at the surface between runoff, surface storage, and infiltration. Infiltration is determined based on the soil water status. In agricultural drainage systems, infiltration is directly related to the water table depth. Subsurface drainage improvements generally increase infiltration. Infiltration rates are influenced by both the soil hydraulic properties and the water table. In many of these situations, runoff occurs due to the water table rising to the surface, whereas, in a well-drained situation runoff is caused by rainfall intensities exceeding the infiltration rate of the soil. Theoretical methods such as Richards equation[4] can be used to describe the infiltration process in drained and poorly drained soils for small time steps. Richard's equation in two dimensions based on the hydraulic head formation is given by

$$c(h)\frac{\partial H}{\partial t} = \frac{\partial}{\partial x}\left(k(h)\frac{\partial H}{\partial x}\right) + \frac{\partial}{\partial y}\left(k(h)\frac{\partial H}{\partial y}\right) \pm S$$

where $c(h) = \mathrm{d}\theta/\mathrm{d}h$ the soil water capacity function (m^{-1}), θ is the volumetric soil water content ($\mathrm{m}^3\,\mathrm{m}^{-3}$), $H = h + z$ the total hydraulic head (m), h is the soil water pressure head (m), x and y are the space coordinates (m), S is a source or sink term (per unit time), and t is time. For the infiltration problem, one specifies the initial conditions in the profile (H at every solution point at time = 0) and two boundary conditions (one at the surface and one at the bottom of the profile). Typical surface boundary conditions are water fluxes corresponding to rainfall rates and a constant head for ponded conditions. At the bottom boundary, a constant flux is often selected. Procedurally, the equation is usually solved numerically, requires fairly extensive soil hydraulic properties, and is generally computationally intensive for short periods of time. These conditions make the equation most effective for evaluating a single rainfall event.

Simplified approaches such as the Green-Ampt method have been modified for use in poorly drained soils.[5] The equation for the Green-Ampt method is:

$$f = K_s + \frac{K_s S_{av} M}{F}$$

where f is the infiltration rate ($\mathrm{mm\,hr}^{-1}$), K_s, the vertical saturated hydraulic conductivity ($\mathrm{mm\,hr}^{-1}$), S_{av}, the average suction at the wetting front (mm) estimated using the soil water characteristic, M, the available storage for infiltration in soil profile ($\mathrm{m}^3\,\mathrm{m}^{-3}$), and F is the cumulative infiltration (mm). Skaggs' approach uses the vertical saturated conductivity of the topsoil layers and estimates $S_{av}{*}M$ based on the soil water characteristic and the depth of the water table at the start of the rainfall event.

The runoff component consists of the amount that fills surface depressions and that amount which can move to other portions of the area. In areas with surface slopes greater than 1%, runoff routing can be accomplished by methods developed for handling routing in upland conditions. These include methods based on theoretical approaches such as describing the overland flow using kinematic wave equations coupled with a surface flow equation such as the Chezy Manning equation to describe the relationship between flow rate and depth, surface roughness, and slope (see for example, Refs.[6–9]). Another approach that is often used is empirical approximations based on the relationship between the source area characteristics such as surface roughness, slope, and land use to relate rainfall and runoff outflow hydrographs. This approach is used in a number of hydrology models such as ANSWERS,[10] CREAMS,[11] EPIC,[12] etc.

In situations where land surface slopes are less than 1%, surface depressions and crop row orientations play an important role in overland flow. Slope plays a less dominant role in surface hydrology and the theoretical approaches discussed above do not do as well describing overland flow. Often surface drainage improvements such as land leveling eliminate or minimize periodic ponding problems. In this case, runoff is determined as the amount in excess of depression storage and is routed to the drainage outlets.

Unsaturated Flow

Water movement through the unsaturated zone is often assumed to be primarily in the vertical direction and negligible in the horizontal directions. Theoretical approaches such as 1-D formulations of Richards equation can be used[13,14] to describe the flow. An alternative approach, which assumes the unsaturated zone is in hydrostatic equilibrium with the water table (sometimes referred to as drained to equilibrium[5]), can also be used to balance water flows. With this approach, infiltration and water extractions add or deplete water from the profile. Any excess water (above the drained to equilibrium water contents) is assumed to be recharge to the water table. Depletion of water from the unsaturated zone decreases the drained to equilibrium contents and a dry zone is developed. The dry zone depth influences upward movement of water from the water table by effectively lowering the apparent water table. This approach has been further refined to estimate unsaturated fluxes.[15,16] In areas with water tables near the surface, this method has been shown to be an effective water balance procedure.[5]

Saturated Flow

The hydrology and water dynamics in the saturated zone is linked to the hydraulic head gradients between the water table and the drainage outlets. Upward flux from the saturated zone along with recharge from infiltration provides vertical gradients to the water table. For steady flow situations, constant recharge or upward flux and drainage outlet heads, the Laplace partial differential equation can be used to describe the flow.

$$\frac{\partial^2 H}{\partial x^2} + \frac{\partial^2 H}{\partial y^2} = 0$$

where H is the hydraulic head (m) and x and y are the spatial coordinates (m).[17] Fig. 2 shows the flow lines for a steady state analysis.

For nonsteady flow, formulations of the Richards equation can be used to integrate water flow in the saturated and unsaturated zones.[13,14,18] This computationally intensive method requires extensive soil and water input data. For short-term problems, this method tends to accurately represent the hydrology. However, perturbations to the input parameters or boundary conditions can lead to instabilities and poor performance.

As the ratio of the distance between the drains to the depth to the impermeable layer becomes large, the flow tends to obey the Dupuit-Forcheimer assumptions. In this case, the flow lines are assumed to be parallel to the impermeable layer. Water flow in this situation can be described by the Boussinesq equation.[19,20] For a 3-D area, the 2-D formulation is:

$$f\frac{\partial h}{\partial t} = \frac{\partial}{\partial x}\left(Kh\frac{\partial h}{\partial x}\right) + \frac{\partial}{\partial y}\left(Kh\frac{\partial h}{\partial y}\right) + R$$

where f is the drainable porosity (m m^{-1}), h, the water table height above the impermeable (zone of minimal vertical flow) layer (m), K, the lateral saturated hydraulic conductivity (m d^{-1}), x and y are the spatial coordinates (m), and R is the vertical recharge to the water table (m d^{-1}) with positive values indicate vertical flow to

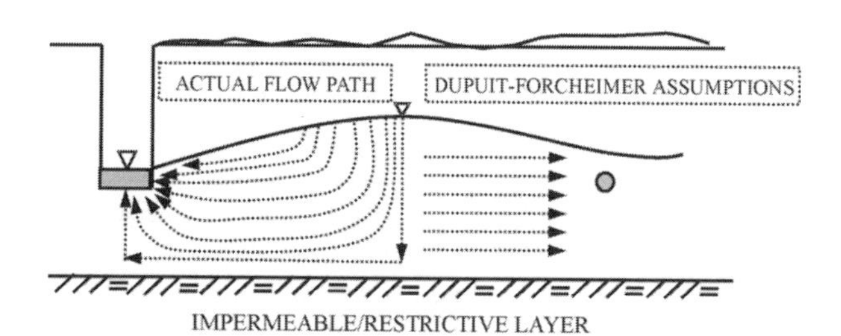

Fig. 2 Flow path comparison for agricultural drainage systems assuming Dupuit-Forcheimer.

the water table and negative values flow upward from the water table into the unsaturated zone. For movement between parallel drains, the 1-D Boussinesq equation is often used.

Further simplifications of the Dupuit-Forcheimer assumptions can be made for steady state conditions. These include equations such as Houghoudt's[19,20] that predicts the flow from parallel drains based on midpoint water table depths, drain spacing, and lateral soil hydraulic conductivity. Houghoudts' equation is:

$$q = \frac{8Kd_e m + 4Km^2}{L^2}$$

where q is the drainage flux (m d^{-1}), K, the lateral saturated hydraulic conductivity (m d^{-1}), m, the height of the water table above the drains (m), L, the distance between drains (m), and d_e is the equivalent depth (m) which corrects for convergence of the flow lines near the drains. The equivalent depth, d_e is found using Moody's equations:

For $0 < \frac{d}{L} < 0.3$

$$d_e = \frac{d}{1 + \frac{d}{L}\left(\frac{8}{\pi}\ln\left(\frac{d}{L}\right) - \alpha\right)} \quad \text{where}$$

$$\alpha = 3.55 - \frac{1.6d}{L} + 2\left(\frac{d}{L}\right)^2$$

and for $\frac{d}{L} \geq 0.3$

$$d_e = \frac{L\pi}{8\left(\ln\left(\frac{L}{r}\right) - 1.15\right)}$$

where r is the radius of the drain (m) and d is the actual depth from the drains to impermeable layer (m).

EXAMPLES OF COMBINED APPROACHES

Comprehensive water management models have been developed to simulate the hydrology of poorly drained soils with natural and artificial drainage. Some examples of these models include DRAINMOD,[5] WATRCOM,[21] EPIC-WT,[22] CREAMS-WT,[23,24] GLEAMS-WT,[25,26] ADAPT,[27,28] and SWACROP.[29] Each model takes different approaches to simulating the hydrology.

For this discussion, the DRAINMOD model is briefly described as an example of a field-scale model since it has been applied extensively and continues to be developed.[30] This model is based on simplified approaches enabling long-term (multiyear) simulations to evaluate performance of water management system designs. The model performs a 1-D water balance at the midpoint between parallel drains based on hourly rainfall and daily PET data. The water balance includes routines to simulate surface and subsurface drainage, infiltration, and evapotranspiration. For days with rainfall, infiltration is determined using the Green-Ampt equation with excess rainfall divided between storage in surface depressions and runoff. Infiltration fills the unsaturated zone to the hydrostatic equilibrium water contents with any excess amounts as recharge to the water table. For days without rainfall, ET is computed based on the amount of water available in the root zone and the amount of upward flux based on the water table depth. Other routine modify available unsaturated water based on deviations from the hydrostatic equilibrium water contents. Subsurface drainage is computed based using Houghoudt's equation. A number of output options including daily water balance components, wet and dry crop stresses, and the number of trafficable periods for planting and harvest. Examples of the water balance components for a 20 m and 50 m spacing of tile drains are shown in Fig. 3 for a poorly drained soil.

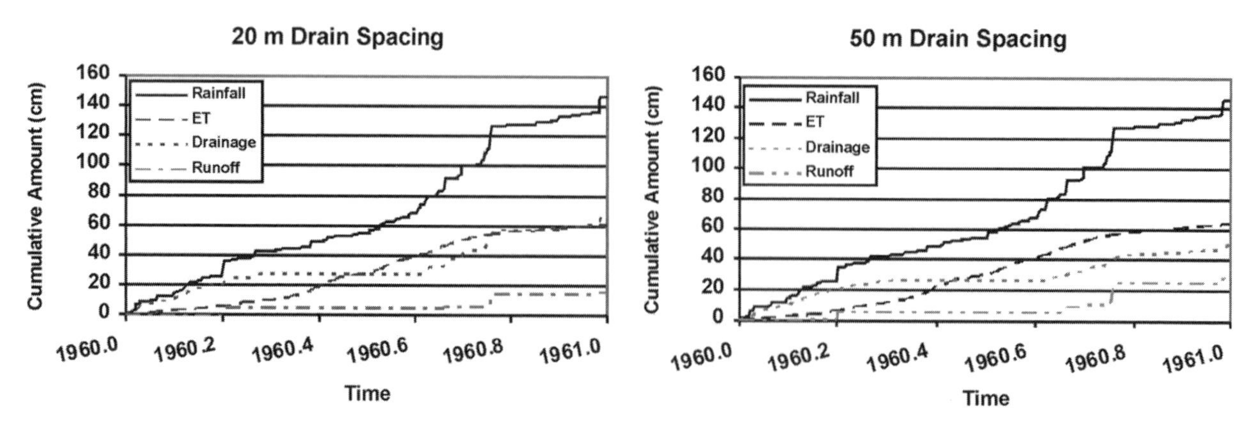

Fig. 3 Water balances for a tile drainage system with 20 m and 50 m drain spacing simulated with DRAINMOD.

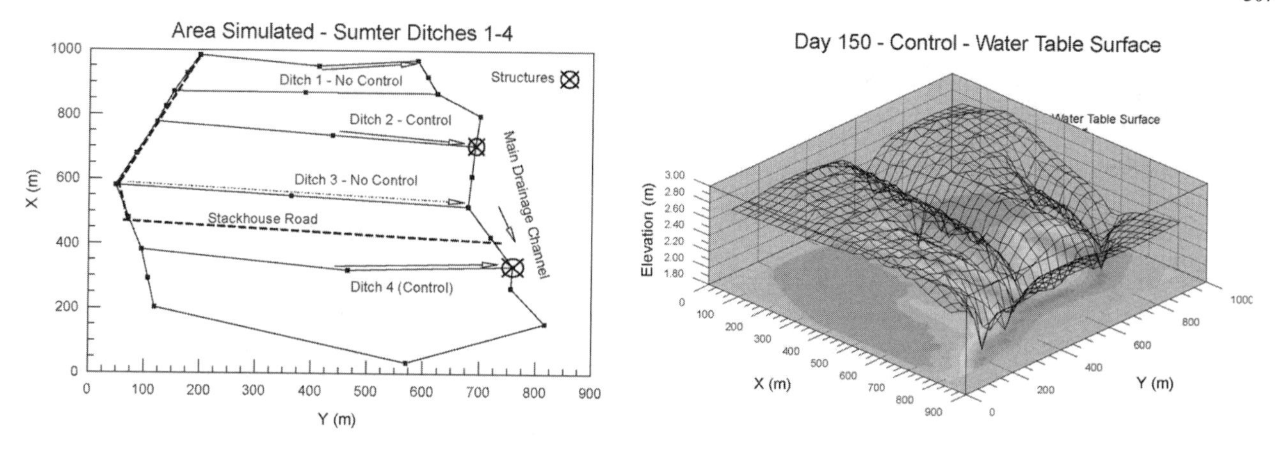

Fig. 4 Layout of the WATRCOM simulation area and a snapshot of the water table elevation on day 150 with control structures on the outlets of catchments 2 and 4.

In more complex drainage situations, such as irregular drainage networks, the Boussinesq equation can be used to describe water flows to and from the drains. Parsons, Skaggs, and Doty[21] developed the model WATRCOM using the 2-D version of the Boussinesq equation to describe saturated flows. The model uses linkages to a 1-D unsaturated balance at each solution point similar to DRAINMOD. A finite element solution procedure is used to numerically solve the Boussinesq equation. Water balances on the surface and in the unsaturated zone are performed at each time step and provide the boundary conditions for the finite element solution of the subsurface saturated flow problem. The model inputs include: rainfall and PET data, soil hydraulic properties, surface topography, and channel water levels or estimates of upstream channel inflow. Soil hydraulic inputs include lateral saturated hydraulic conductivity and soil water characteristics. Other soils inputs that are computed or estimated from these inputs include: Green Ampt infiltration parameters, volume drained vs. water table levels, and upward flux vs. water table levels.

Model setup requires a finite element grid of the project area. At each node of the grid, soils data and surface elevations are assigned. Simulation time steps range from hourly to daily depending on the soil water status at each node. Outputs from the model include surface and subsurface drainage to each drainage ditch and available saturated and unsaturated water at each node. Other outputs related to water status can be summarized over daily and yearly time periods.

An example aerial view of a typical simulations area is given in Fig. 4. This area is a forested watershed located near Sumter, SC. A snapshot of the simulated water tables at day 150 is shown in Fig. 4. The influence of the main and laterals on the drainage pattern is shown. Outlets 2 and 4 were simulated with weirs at 0.5 m below the surface.

REFERENCES

1. Evans, R.O.; Lilly, J.P.; Skaggs, R.W.; Gilliam, J.W. *Rural Land Use, Water Movement, Coastal Water Quality*; North Carolina Cooperative Extension Service Bulletin AG-605; North Carolina Cooperative Extension Service, CALS, NC State University: Raleigh, NC, 2000; 14.
2. Bernstein, L. Crop Growth and Salinity. In *Drainage for Agriculture. Agronomy 17*; van Schilfgaarde, J., Ed.; ASA: Madison, WI, 1974; 39–54.
3. Jensen, M.E.; Burman, R.D.; Allen, R.G. (Eds.) *Evapotranspiration and Irrigation Water Requirements*; ASCE Manuals and Reports on Engineering Practice No. 70; ASCE: New York, NY, 1990; 332.
4. Richards, L.A. Capillary Conduction of Liquids Through Porous Media. Physics **1931**, *1*, 318–324.
5. Skaggs, R.W. *A Water Management Model for Artificially Drained Soils*; Tech. Bull. No. 267; NC Agricultural Research Service, NC State University: Raleigh, NC, 1980.
6. Lighthill, M.J.; Whitham, C.B. On Kinematic Waves: Flood Movement in Long Rivers. Proc. R. Soc. London, Ser. A. **1955**, *22*, 281–316.
7. Woolhiser, D.A.; Liggett, J.A. Unsteady, One-Dimensional Flow over a Plane, the Rising Hydrograph. Water Resour. Res. **1967**, *3* (3), 752–771.
8. Lane, L.J.; Woolhiser, D.A. Simplifications of Watershed Geometry Affecting Simulation of Surface Runoff. J. Hydrol. **1977**, *35*, 173–190.
9. Ross, R.B.; Contractor, D.N.; Shanholtz, V.O. A Finite Element Model of Overland and Channel Flow for Assessing Hydrologic Impact of Land-Use Change. J. Hydrol. **1979**, *41*, 11–30.
10. Beasley, D.B.; Huggins, L.F.; Monke, E.J. ANSWERS: A Model for Watershed Planning. Trans. ASAE **1980**, *23* (4), 938–944.
11. Knisel, W.G. (Ed.) *CREAMS: A Field-Scale Model for Chemicals, Runoff, and Erosion from Agricultural Management Systems*; Conservation Research Report No. 26; U. S.

Department of Agriculture, Science and Education Administration: Washington, DC, 1980.
12. Sharpley, A.N.; Williams, J.R. (Eds.) *EPIC—Erosion/Productivity Impact Calculator 1. Model Documentation*; Technical Bulletin; U. S. Department of Agriculture, Agricultural Research Service, Washington, DC, 1990.
13. Feddes, R.A.; Kowalik, P.J.; Zaradny, H. *Simulation of Water Use and Crop Yield*; Simulation Monograph; PUDOC: Wageningen, 1978.
14. Feddes, R.A.; Kabat, P.; van Bakel, P.J.T.; Bronswijk, J.J.B.; Halbertsma, J. Modeling Soil Water Dynamics in the Unsaturated Zone—State of the Art. J. Hydrol. **1988**, *1000*, 69–111.
15. Skaggs, R.W.; Karvonen, T.; Kandil, H.M. *Predicting Water Flux in Drained Lands*; ASAE Meeting Paper No. 912090; ASAE: St. Joseph, MI, 1991.
16. Karvonen, T.; Skaggs, R.W. Comparison of Different Methods for Computing Drainage Water Quantity and Quality. In *15th International Congress of ICID, The Hague, Workshop on Subsurface Drainage Simulation Models, ICID-CIID, Cemagref*; 1993, 201–216.
17. van der Ploeg, R.R.; Horton, R.; Kirkham, D. Steady Flow to Drains and Wells. In *Agricultural Drainage. Agronomy 38*; Skaggs, R.W., van Schilfgaarde, J., Eds.; American Society of Agronomy: Madison, WI, 1999; 213–264.
18. Nieber, J.L.; Feddes, R.A. Solutions for Combined Saturated and Unsaturated Flow. In *Agricultural Drainage. Agronomy 38*; Skaggs, R.W., van Schilfgaarde, J., Eds.; American Society of Agronomy: Madison, WI, 1999; 145–212.
19. van Schilfgaarde, J. Nonsteady Flow to Drains. In *Drainage for Agriculture. Agronomy 17*; van Schilfgaarde, J., Ed.; ASA: Madison, WI, 1974; 245–270.
20. Youngs, E.G. Non-Steady Flow to Drains. In *Agricultural Drainage. Agronomy 38*; Skaggs, R.W., van Schilfgaarde, J., Eds.; American Society of Agronomy: Madison, WI, 1999; 265–298.
21. Parsons, J.E.; Skaggs, R.W.; Doty, C.W. Development and Testing of a Three-Dimensional Water Management Model (WATRCOM): Development. Trans. ASAE **1991**, *34* (1), 120–128.
22. Sabbagh, G.J.; Bengtson, R.L.; Fouss, J.L. Modification of EPIC to Incorporate Drainage Systems. Trans. ASAE **1991**, *34* (2), 467–472.
23. Heatwole, C.D.; Campbell, K.L.; Bottcher, A.B. Modified CREAMS Hydrology Model for Coastal Plain flatwoods. Trans. ASAE **1987**, *30*, 1014–1022.
24. Stone, K.C.; Campbell, K.L.; Baldwin, L.B. A Microcomputer Model for Design of Agricultural Stormwater Management Systems in Florida's Flatwoods. Trans. ASAE **1989**, *32* (2), 545–550.
25. Reyes, M.R.; Bengtson, R.L.; Fouss, J.L.; Rogers, J.S. GLEAMS Hydrology Submodel Modified for Shallow Water Table Conditions. Trans. ASAE **1993**, *36* (6), 1771–1778.
26. Reyes, M.R.; Bengtson, R.L.; Fouss, J.L. GLEAMS-WT Hydrology Submodel Modified to Include Subsurface Drainage. Trans. ASAE **1994**, *37* (4), 1115–1120.
27. Chung, S.O.; Ward, A.D.; Schalk, C.W. Evaluation of the Hydrologic Component of the ADAPT Water Table Management Model. Trans. ASAE **1992**, *35* (2), 571–579.
28. Ward, A.D.; Desmond, E.; Fausey, N.R.; Logan, T.J.; Knisel, W.G. Development Studies with the ADAPT Water Table Management Model. In *15th International Congress of ICID, The Hague, Workshop on Subsurface Drainage Simulation Models, ICID-CIID, Cemagref*; 1993, 235–245.
29. Belmans, C.; Wesseling, J.G.; Feddes, R.A. Simulation Model of the Water Balance of a Cropped Soil: SWATRE. J. Hydrol. **1983**, *63*, 271–286.
30. Skaggs, R.W. Drainage Simulation Models. In *Agricultural Drainage. Agronomy 38*; Skaggs, R.W., van Schilfgaarde, J., Eds.; American Society of Agronomy: Madison, WI, 1999; 469–500.

Ice Crystal Kinetics

Waraporn Boonsupthip
Tung-Ching Lee
Rutgers, The State University of New Jersey, New Brunswick, New Jersey, U.S.A.

I

INTRODUCTION

The kinetics of ice crystallization refers to the rate of transformation from the disordered liquid phase to an organized solid phase. The phase transformation develops through two steps, ice crystal nucleation and propagation. Therefore, the overall kinetics of ice crystallization is the result of the combination of both nucleation and propagation.

The nucleation process is the initiating step of ice crystallization. Typically, a few hundred water molecules align into an organized and stable ice nucleus (ice-like cluster). Nucleation is governed by three different mechanisms: homogeneous, heterogeneous, and secondary nucleation. In homogeneous nucleation, an ice nucleus spontaneously forms without any facilitation. In heterogeneous nucleation, a foreign particle or container surface catalyzes the formation of nuclei—in secondary nucleation, a small ice crystal acts as a seed. Once the ice nucleus is formed, it can serve as an ice-anchoring base for further growth during the propagation process.

It should be noted that although a single water molecule (H_2O) is structurally simple, its liquid state and phase transition behavior are very complex and not yet well understood.[1,2] A general overview of the topic is given here; however, more details can be found in literatures.[1,3–5]

GENERAL ICE CRYSTAL NUCLEATION

During freezing, water in the system cools down as sensible heat is removed. When the temperature falls below the equilibrium freezing temperature (θ_E), water does not freeze but remains supercooled (undercooled) (Fig. 1). The classical theory of nucleation suggests that in this supercooled region, ice-like clusters may occasionally form by statistical and continuous fluctuations of density and structure of water.[1] However, the clusters are in an energetically unstable stage.

The stability of a forming cluster of radius (r) is governed by the net Gibbs free energy (ΔG) of the water–ice phase transformation (Fig. 2). ΔG is the summation of the changes of volume (W_v) and interfacial surface energy (W_s) terms:

$$\Delta G = W_v + W_s = \frac{4}{3}\pi r^3 \Delta G_v + 4\pi r^2 \gamma_{ls} \tag{1}$$

where ΔG_v is the difference in volumetric free energy between the solid and liquid phases per volume and γ_{ls} is the liquid/solid interfacial surface energy per unit area.[6,7] At the beginning of the nucleation process, each cluster is small. Its surface-to-volume ratio is large, resulting in large interfacial barrier energy (γ_{ls}) (Fig. 2). As the cluster grows, the magnitude of the (negative) volume energy term increases dramatically. The cluster size at which the interfacial barrier energy is overcome is expressed as:

$$r^* = \frac{2\gamma_{ls}\theta_E^2}{\Delta H_f \Delta\theta} = -\frac{\gamma_{ls}^{5/2}}{0.12\Delta G^{*1/2}} \tag{2}$$

where r^* is the minimum size of a stable cluster and θ_E the absolute equilibrium freezing temperature. $\Delta\theta$ ($= \theta_E - \theta$) is the degree of supercooling at a particular absolute temperature (θ) and ΔH_f the latent heat of fusion. At radius r^*, the probability of further growth or disintegration of the forming cluster is equal (Fig. 2). The addition of one molecule to this cluster reduces ΔG, resulting in a stable cluster.

A kinetic approach considers nucleation to be a reversible chemical reaction.[1]

$$C_i + C_1 \Leftrightarrow C_{i+1} \tag{3}$$

A cluster C_i containing i molecules grows by an addition of a free molecule C_1. In reverse reaction, the free molecule C_1 disassociates back into the liquid. The kinetic constraint of this reaction is the activation energy of C_{i+1} formation (Δg_1) (Fig. 3). At the steady state, the net growth of the reaction gives a rate (kinetics) of nucleation, $J(\theta)$. Since this is a strongly temperature dependent reaction, the general form of the nucleation rate can be expressed using the temperature dependent Arrhenius equation[1,8]

$$J(\theta) = A\exp(-Bf(\theta)), \quad f(\theta) = [(\Delta\theta)^2\theta^3]^{-1} \tag{4}$$

where A and B are positive constants, which represent the physical properties of ice and free water, respectively, and

Encyclopedia of Agricultural, Food, and Biological Engineering
DOI: 10.1081/E-EAFE 120007026
Copyright © 2003 by Marcel Dekker, Inc. All rights reserved.

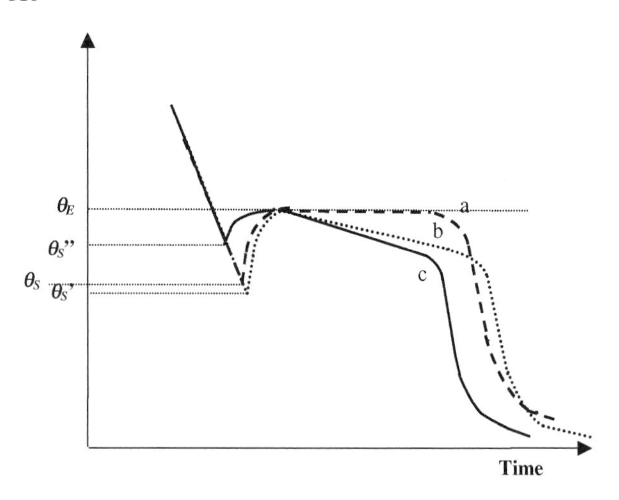

Fig. 1 Typical freezing curves of a water-containing system (a) without a solute, (b) with a solute, and (c) with a nucleator and a solute. θ_S, θ'_S, and θ''_S are nucleating temperatures of the system a, b, and c, respectively. θ_E is equilibrium freezing temperature.

are characteristics of the nucleator (if present). $J(\theta)$ represents the concentration of nuclei formed over time at temperature θ. Ultimately, one nucleus can freeze the entire body of water. The temperature at which $J(\theta) = 1$ nucleus cm^{-3} sec^{-1} is customarily defined as nucleating threshold temperature.[9]

Homogeneous Nucleation

In most practical situations, homogeneous nucleation is of no concern because a nucleator such as dust or irregular surface is always present. However, the kinetics of homogeneous nucleation has been the subject of a great deal of research.

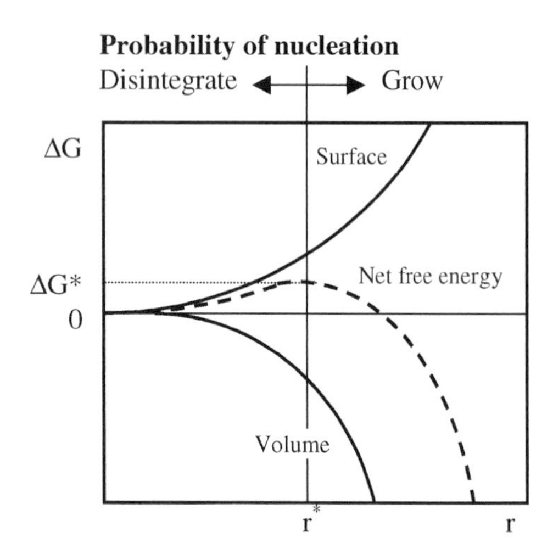

Fig. 2 Probability of nucleation as a function of net free energy and radius sizes of forming ice cluster. (Modified from Ref. [3].)

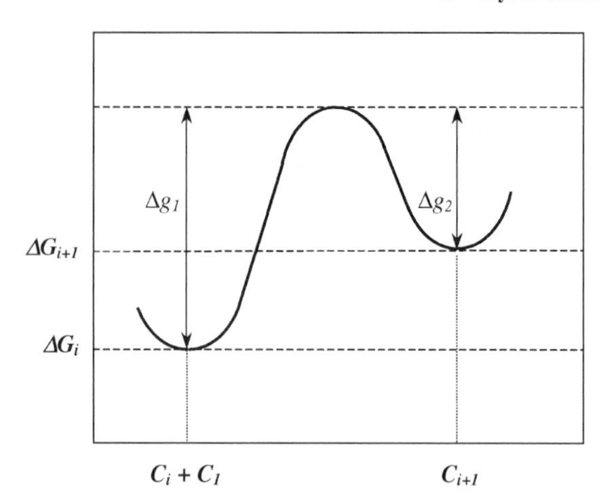

Fig. 3 Change of the net free energy of an ice cluster C_i when it incorporates another water molecule C_1 to form a larger ice cluster C_{i+1}. (Modified from Ref. [1].)

The general form of the nucleation kinetics (Eq. 4) can be reduced to describe the kinetics of homogeneous nucleation:[1,8]

$$J'(\theta) = A' \exp^* (-\Delta G^* / k\theta) \quad (5)$$

where $J'(\theta)$ is the rate of nuclei formed in a body of liquid, A' a constant dependent on the physical properties of ice and free water, and k the Boltzmann gas constant. Using this equation, several studies found the nucleating temperature of pure water to be around $-40°C$.[1,6,7,10]

Heterogeneous Nucleation

The mechanism of heterogeneous nucleation is similar to that of homogeneous nucleation in the sense that the presence, in contact with the liquid, of an energetically stable form of ice template is required. As a result of the presence of the nucleator, the degree of supercooling of the liquid is lower than that in homogeneous nucleation (Fig. 1—line c). The mechanism is based on formation of a nucleus on the surface of a nucleator instead of in the bulk. The barrier energy, Δg_I or ΔG^* is lower for heterogeneous nucleation, as evidenced by the degree of supercooling.[1,11] The barrier energy reduction depends on the geometry and surface properties of the nucleator.[1]

The kinetics of heterogeneous nucleation can be specified from the general equation (Eq. 4) as:

$$J''(\theta) = A'' \exp(-\Delta G^*_H / k\theta) \quad (6)$$

where $J''(\theta)$ is the nuclei formation rate per unit volume per nucleator, A'' a constant, and ΔG^*_H the free energy barrier to the ice cluster formation. ΔG^*_H is defined as:

$$\Delta G^*_H = \Delta G^* f(m, R) \quad (7)$$

ΔG_H^* is modified from that of homogeneous nucleation (ΔG^*) by incorporating the characteristics of the nucleator in the form of a function $f(m, R)$.[1,6,11–13] $f(m, R)$ is always less than unity. m represents the degree of surface compatibility of the nucleator with ice. It ranges from -1 for the least compatibility to $+1$ for a perfect match. R is the physical dimension of the nucleator.

Several geometries of nucleators have been evaluated, ranging from disks to needles.[1,11,12,14,15] A spherical geometry can be applied to represent the general behavior of the other geometries. Its relationship to nucleating temperatures is demonstrated in Fig. 4. A nucleator with a larger radius (R) and higher surface compatibility (larger m) is more effective in the catalysis of nucleation; therefore, water freezes at a higher temperature.

An interesting theory considers the contact angle (α) between the nucleator and the nucleus as a major factor.[2,16] ΔG_H^* is represented as a function of the contact angle and the free energy of homogeneous nucleation:

$$\Delta G_H^* = \Delta G^* f(\alpha) \tag{8}$$

The contact angle is governed by the interfacial surface energy (γ) between the interfaces of solution–particle–cluster, and is small when γ is low. A small contact angle gives a low value of $f(\alpha)$ (Fig. 5), leading to rapid nucleation.

Reported organic ice nucleators include urea,[17] steroids,[18,19] substituted fluorenes,[20] benzenoid compounds,[21] diazines,[22] amino acids,[23] metaldehyde,[24] phloroglucinol,[25–27] ice nucleation active (Ina^+) bacteria[1,28,29] and their products, and extracellular ice nucleators (ECINs).[30] These last two nucleators are very effective so that their potential usages in food products have recently been the subject of a great deal of research.[31–38] Studies are hindered by the lack of purity and low-level expression of the nucleators.[39–43]

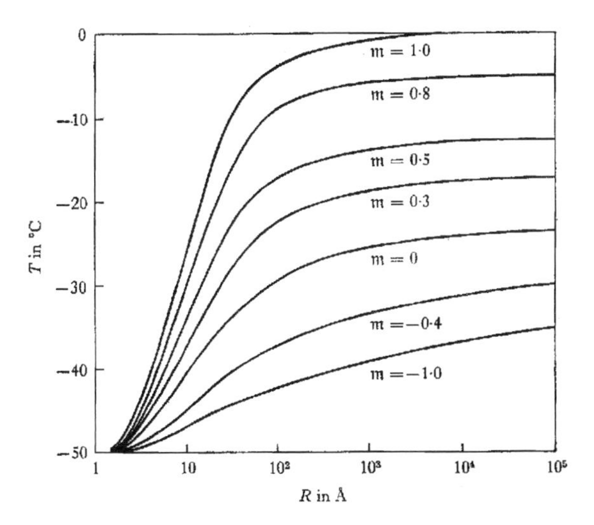

Fig. 4 Computed nucleating temperatures of the systems catalyzed with different spherical nucleators (various radius sizes R and values of the surface parameter m). The computation is based on the assumption that freezing occurs in less than a second. (From Fletcher, 1958.)

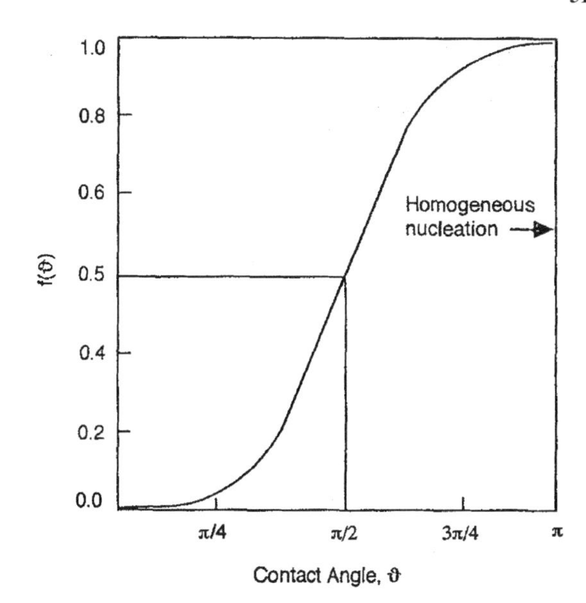

Fig. 5 Heterogeneous nucleation factor $f(\alpha)$ as a function of contact angle, α. (From Ref. [16].)

Secondary Nucleation

Secondary nucleation is an essential element in freeze-concentration of fluid foods.[44] Potential nuclei, finer ice debris, are formed from existing ice crystals during collision with the impeller, baffler, and other crystals.[2,8,45] Ice crystals nucleate freezing within the range of 0.05–0.2°C, depending on concentration of the solutes.[46,50] The rate of nucleation depends on crystal form, the removal rate of potential nuclei from the impeller, and solute type and concentration.[45–49]

CONTROLLING NUCLEATION

The factors that control ice nucleation are food volume, food composition, heat removal rate, and nucleator.[2,5,51] A larger volume provides more prospective regions for nucleation; therefore, a higher nucleation rate is likely.[1] However, the volume increase causes a corresponding reduced heat removal rate. Slower heat removal (lower supercooling or higher nucleating temperature) results in slower nucleation[10] (Fig. 6). This will increase the size of ice crystals at the end of the crystallization process. The nucleation rate can be increased by increasing the heat removal rate (due to a decrease in nucleating temperature)

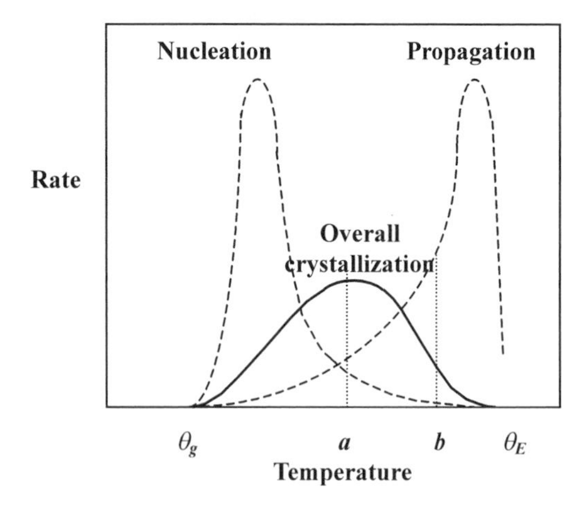

Fig. 6 A typical kinetics of ice crystallization. (Modified from Sahagian and Goff, 1996)

(Fig. 6). However, as the temperature approaches the glass transition temperature (θ_g) water mobility limits nucleation and so the rate drops sharply.

The influence of food composition is preferably examined using simple models containing one or a small number of ingredients rather than food, in order to simplify the problem. The model studies show that small solutes (i.e., sugars and salts) decrease nucleation rate[52,53] and temperature (θ_S).[2,54–58] For homogeneous nucleation, the extent of the effect on θ_S is related to the ability of the compound to depress θ_E of water. The relationship is:

$$\Delta\theta_S = K\Delta\theta_E \tag{9}$$

where K is a constant in the reported range 1.75–2.0,[54] and $\Delta\theta_S$ and $\Delta\theta_E$ are the nucleating and freezing depression temperatures, respectively. $\Delta\theta_E$ is a colligative property and a function of water activity, which varies with solute molar concentration.[2,51] For heterogeneous nucleation, $\Delta\theta_S$ depends on the type of nucleators and type and concentration of the solute.[2] Our knowledge of the effects of nucleators has been gained from studies using inorganic substances, especially AgI. Different nucleators nucleate water at different temperatures, such as ~ -4°C for AgI, ~ -6°C for PbI_2, and ~ -7°C for CuS.[1] Different types and concentrations of solutes showed different effects on the nucleating activity of nucleators.[2,55–58]

Macromolecules [i.e., polyethyleneglycol (PEG)[52,53,59]; polyvinyl pyrrolidone (PVP)[60]; hydroxyethyl-starch (HES)[52,53,59]; and antifreeze proteins (AFPs)[61, 62]] inhibit ice nucleation in proportion to their mass rather than molecular number concentrations. AFPs are hundreds of times more effective than small molecules on a mole basis.[62–64]

There is no clear explanation of the different abilities of small molecules and macromolecules to inhibit nucleation. Some researchers have hypothesized that it results from the different reduction in the diffusion of supercooled water.[52,53] Others have suggested that it might be due to differences in the volume fraction of water[52] or in γ_{ls}.[52,65]

CRYSTAL PROPAGATION

Once ice nuclei exist in a system, ice propagation proceeds spontaneously. The liquid temperature rises from the nucleation temperature to the equilibrium freezing temperature (θ_E) (Fig. 1). As the latent heat is being removed, if the system contains no substance, which reduces the water activity of the system, the temperature remains constant (Fig. 1—line a). Otherwise, temperature drops progressively (Fig. 1—line b) corresponding to the solute concentration effect on freezing point depression.[6,51,66,67]

As ice crystals propagate, free water molecules move from the bulk volume to the ice growing sites. In the last stage of freezing, the majority of water molecules are already frozen. The unfrozen portion becomes increasingly concentrated, viscous, and less mobile as the frozen fraction increases. The freezing rate is ultimately restricted by the rate of mass transfer.[68,69] Ideally, ice crystallization is complete when latent heat is minimal. Sensible heat is released while the system is cooling down. Temperature is progressively reduced until it equilibrates with the external freezing temperature. Kinetics of crystal propagation is complicated and different from case to case.[70] However, any condition can be developed based either on heat or mass transfer constraints as described above.

In general, the kinetics of propagation can be experimentally defined for two different situations.[67] One is when ice crystals grow suspended in liquid. Each individual ice crystal can grow freely. The mechanism and the rate of size change are governed by the properties of the growing crystal surface, including the morphology and concentration of the growing sites.[2] The basic theory has been discussed in literatures.[2,67,70] The other situation is when ice crystals form in a small volume of water. This can be water or solutions in a small tube or confined space such as inter- or intra-cellular sections of living tissue. Growth of one crystal in such a condition impacts another. The propagation rate of the growing front of the formed ice pack is a better way to model the process.[51] In this case, propagation rate can also be characterized as linear crystallization velocity.[71]

CONTROLLING PROPAGATION

The main factors influencing propagation are temperature (heat removal rate) and composition. Propagation is sharply accelerated when the temperature is reduced slightly below θ_E (low heat removal rate) (Fig. 6). However, it slows down if the temperature reduces further. When propagation occurs at temperatures close to θ_g, the limited mobility of free water molecules dramatically reduces the propagation rate.

Types and concentrations of solutes affect propagation rate. Solutions have linear reductions in propagation rate with increasing concentrations of sucrose and glucose at constant $\Delta\theta$.[2,72–74] The rate is also reduced with the increase in the initial concentration of the solute.[68] The reduction effect is supposedly associated with the reduction in the diffusion coefficient of water (moving toward the growing interface) and those of the solutes (moving away from the growing interface).[68,69]

Hydrocolloids or macromolecules are normally applied to inhibit crystal growth. The inhibition mechanism appears to involve a mechanical hindrance exerted by the macromolecular solutes.[51] It is not due to the limitation of the diffusion of water. The extent of this mechanical effect may be augmented by increasing the concentration or the length of the macromolecule. The latter factor is more effective due to the large mechanical hindrance of the entangled chains of long molecules.[51,75] Gelatinization and rigidity of the macromolecular networks may amplify the reduction of crystal propagation rates.[51]

ICE CRYSTALLIZATION AND SIZES OF ICE CRYSTALS

An ice crystal size control strategy can be made based on the notion of nucleation and propagation. Rapid nucleation combined with slow propagation will result in a large number of small ice crystals (Fig. 6). For example, this situation can take place where the freezing condition is at point a in Fig. 6. Overall crystallization rate[76–79] reaches a maximum at a temperature approximately in the middle of θ_E and θ_g. It approaches zero at θ_E and θ_g where no nucleation or propagation occurs. The effect of glass phase transformation on mobility of water can be found in literatures.[80,81]

Within a food matrix, ice crystal size is not uniform. A size distribution is found in correspondence with the temperature variation from surface to center. The temperature profile depends on external temperature, food geometry, and initial temperature.[82] Ice size distribution can be estimated by superimposing the effect of the temperature dependent kinetics of ice crystallization on the temperatures within the food matrix.

Once the ice crystals are fully developed in the crystallization process, they are subject to further change due to ice recrystallization. Ice recrystallization occurs when temperature fluctuates during freezing, storage, and transportation,[83] and is usually detrimental to food quality. Several mechanisms and kinetics of recrystallization are discussed in literatures.[30,84] Recently discovered AFPs have captured the attention of scientists due to their inhibition of ice recrystallization in several systems including food.[85–92] However, use of AFPs with foods is still limited by problems such as unclear functionality, application conditions (time, temperature, concentration, and food composition), and health concerns.[62,63]

CONCLUSION

Ice crystallization rate is determined by both nucleation and propagation. They are affected mainly by temperature (heat removal rate or cooling rate), food composition, and additives. We do not fully understand the combined effects of these influencing factors in real food systems.

REFERENCES

1. Fletcher, N.H. Liquid Water and Freezing. In *The Chemical Physics of Ice*; Fletcher, N.H., Ed.; Cambridge University Press: Great Britain, 1970; 73–103.
2. Sahagian, M.E.; Goff, H.D. Fundamental Aspects of the Freezing Process. In *Freezing Effects on Food Quality*; Jeremiah, L.E., Ed.; Marcel Dekker, Inc.: New York, 1996; 1–50.
3. Franks, F. *Water: A Compensive Treatise: Water and Aqueous Solutions at Subzero Temperature*, 1982; 484.
4. Lewis, B. Nucleation and Growth Theory. In *Crystal Growth*; Pamplin, B.R., Ed.; Pergamon Press: Oxford, 1980; 23–64.
5. Reid, D.S. Basic Physical Phenomena in the Freezing and Thawing of Plant and Animal Tissues. In *Frozen Food Technology*; Mallett, C.P., Ed.; Blackie Academic & Professional: New York, 1993; 1–19.
6. Franks, F. The Properties of Aqueous Solutions at Sub-zero Temperatures. In *Water: A Comprehensive Treatise*; Franks, F., Ed.; Plenum Press: New York, 1979; Vol. 7, 215–334.
7. Franks, F. *Biophysics and Biochemistry at Low Temperatures*; Cambridge University Press: Cambridge, 1985; 210.
8. Michel, B. Formation and Types of Ice. In *Ice Mechanics*; Michel, B., Ed.; Les Presses De L'Universite Laval: Quebec, 1978; 1–80.
9. Vali, G. Principles of Ice Nucleation. In *Biological Ice Nucleation and Its Applications*; Lee, R.E., Jr., Warren, G.J., Gusta, L.V., Eds.; The American Phytopathological Society Press: Minnesota, 1995; 1–28.
10. Reid, D.S. Fundamental Physicochemical Aspects of Freezing. Food Technol. **1983**, *37*, 110–113.
11. Fletcher, N.H. Size Effects in Heterogeneous Nucleation. J. Chem. Phys. **1959**, *29*, 572–579.

12. Fletcher, N.H. Nucleation by Crystalline Particles. J. Chem. Phys. **1963**, *38*, 237–240.
13. Wilson, P. Physical Basis of Action of Biological Ice Nucleating Agents. Cryo-Letters **1994**, *15*, 119–124.
14. Burke, M.J.; Lindow, S.E. Surface Properties and Size of the Ice Nucleation Site in Ice Nucleation Active Bacteria: Theoretical Considerations. Cryobiology **1990**, *27*, 80–84.
15. Turnbull, D. Kinetics of Heterogeneous Nucleation. J. Chem. Phys. **1950**, *18*, 198–203.
16. Toner, M.; Cravalho, E.G.; Karel, M. Thermodynamics and Kinetics of Intracellular Ice Formation During Freezing of Biological Cells. J. Appl. Phys. **1990**, *67*, 1582–1593.
17. Sano, L.; Fujitani, Y.; Maena, Y. The Ice-Nucleating Property of Some Substances and Its Dependence on Particle Size. Mem. Kobe Mar. Obs. **1960**, *14*, 107–118.
18. Head, R.B. Steroids as Ice Nucleators. Nature **1961**, *191*, 1058–1059.
19. Kinneberg, B.I. Sterol Ice Nucleation Catalysts. US Patent 5,239,819, March 6, 1992.
20. Head, R.B. Ice Nucleation by Some Cyclic Compounds. J. Phys. Chem. Solids **1962**, *23*, 1371–1378.
21. Komabayasi, M.; Ikebe, Y. Organic Ice Nuclei: Ice-Forming Properties of Some Aromatic Compounds. J. Meteorol. Soc. Jpn. **1961**, *39*, 82–95.
22. Head, R.B. Ice Nucleation by α-Phenazine. Nature **1962**, *198*, 736–737.
23. Power, B.A.; Power, R.F. Some Amino Acids as Ice Nucleators. Nature **1962**, *194*, 1170–1171.
24. Fukuta, N. Ice Nucleation by Metaldehyde. Nature **1963**, *199*, 475–476.
25. Bashkirova, G.M.; Krasikov, P.N. Experiments with Certain Substances as Crystallization Agents for Supercooled Fogs. Tr. Gl. Geofiz. Obs. **1957**, *72*, 118–126.
26. Langer, G.; Rosinski, J.; Bemsen, S. Organic Crystals as Icing Nuclei. J. Atmos. Sci. **1963**, *20*, 557–562.
27. Braham, R.R., Jr. Phloroglucinol Seeding of Undercooled Clouds. J. Atmos. Sci. **1963**, *20*, 563–568.
28. Gurian-Sherman, D.; Lindow, S.E. Bacterial Ice Nucleation: Significance and Molecular Basis. FASEB J. **1993**, *7*, 1338–1343.
29. Hirano, S.S.; Upper, C.D. Ecology of Ice Nucleation—Active Bacteria. In *Biological Ice Nucleation and Its Applications*; Lee, R.E., Jr., Warren, G.J., Gusta, L.V., Eds.; APS Press: St. Paul, 1995; 41–61.
30. Fennema, O.R. Freezing Preservation. In *Principles of Food Science: Part II Physical Principles of Food Preservation*; Fennema, O.R., Ed.; Marcel Dekker, Inc.: New York, 1975; Vol. 4, 173–218.
31. Arai, S.; Watanabe, M. Freeze Texturing of Food Materials by Ice Nucleation with the Bacterium *Erwinia ananas*. Agric. Biol. Chem. **1986**, *50*, 169–175.
32. Li, J.; Izquierdo, M.P.; Lee, T.-C. Effects of Ice-Nucleation Active Bacteria on the Freezing of Some Model Food Systems. Int. J. Food Sci. Technol. **1997**, *32*, 41–49.
33. Li, J.; Lee, T.-C. Bacterial Ice Nucleation and Its Potential Application in Food Industry. Trends Food Sci. Technol. **1995**, *6*, 259–265.
34. Li, J.; Lee, T.-C. Bacterial Extracellular Ice Nucleator Effects on Freezing of Food. J. Food Sci. **1998**, *63*, 375–381.
35. Watanabe, M.; Arai, S. Freezing of Eater in the Presence of the Ice Nucleation Active Bacterium, *Erwinia ananas*, and Its Application for Efficient Freeze-Drying of Foods. Agric. Biol. Chem. **1987**, *51*, 557–563.
36. Watanabe, M.; Arai, S. Applications of Bacterial Ice Nucleation Activity in Food Processing. In *Biological Ice Nucleation and Its Applications*; Lee, R.L., Warren, G.J., Gusta, L.V., Eds.; APS Press: New York, 2001; 299–313.
37. Watanabe, M.; Kumeno, K.; Nakahama, K.; Arai, S. Heat-Induced Gel Properties of Freeze-Concentrated Egg White Produced Using Bacterial Ice Nuclei. Agric. Biol. Chem. **1990**, *54*, 2055–2059.
38. Watanabe, M.; Watanabe, J.; Kumeno, K.; Nakahama, K.; Arai, S. Freeze Concentration of Some Foodstuffs Using Ice Nucleation—Active Bacterial Cells Entrapped in Calcium Alginate Gel. Agric. Biol. Chem. **1989**, *53*, 1731–2735.
39. Wolber, P.K.; Deininger, C.A.; Southworth, M.W.; Vandekerckhove, J.; Montagu, M.V.; Warren, G.J. Identification and Purification of a Bacterial Ice-Nucleation Protein. Proc. Natl Acad. Sci. U.S.A. **1986**, *83*, 7256–7260.
40. Fall, A.L.; Fall, R. High-Level Expression of Ice Nuclei in *Erwinia herbicola* Is Induced by Phosphate Starvation and Low Temperature. Curr. Microbiol. **1998**, *36*, 370–376.
41. Kawahara, H.; Mano, Y.; Obata, H. Purification and Characterization of Extracellular Ice-Nucleating Matter from *Erwinia uredovora* KUIN-3. Biosci. Biotechnol. Biochem. **1993**, *57*, 1429–1432.
42. Li, J. *Characterization of Bacterial Extracellular Ice Nucleators and Their Effects on the Freezing of Foods*; Rutgers, The State University of New Jersey: New Jersey, 1998; 1–142.
43. Li, J.; Lee, T.-C. Enhanced Production of Extracellular Ice Nucleators from *Erwinia herbicola*. J. Gen. Appl. Microbiol. **1998**, *44*, 405–413.
44. Franks, F. Nucleation: A Maligned and Misunderstood Concept. Cryo-Letters **1987**, *8*, 53–59.
45. Evans, T.W.; Margolis, G.; Sarofim, A.F. Mechanisms of Secondary Nucleation in Agitated Crystallizers. AICHE J. **1974**, *20*, 950–958.
46. Shirai, Y.; Nakanishi, K.; Matsuno, R.; Kamikubo, T. Effects of Polymers on Secondary Nucleation of Ice Crystals. J. Food Sci. **1985**, *50*, 401–406.
47. Estrin, J.; Wang, M.L.; Youngquist, G.R. Secondary Nucleation Due to Fluid Forces upon a Polycrystalline Mass of Ice. AICHE J. **1975**, *21*, 392–395.
48. Evans, T.W.; Sarofim, A.F.; Margolis, G. Models of Secondary Nucleation Attributable to Crystal–Crystallizer and Crystal–Crystal Collisions. AICHE J. **1974**, *20*, 959–966.
49. Kane, S.G.; Brian, P.L.T.; Sarofim, A.F. Determination of the Kinetics of Secondary Nucleation in Batch Crystallizers. AICHE J. **1974**, *20*, 855–862.

50. Hartel, R.W.; Chung, M.S. Contact Nucleation of Ice in Fluid Dairy Products. J. Food Eng. **1993**, *18*, 181–192.
51. Blond, G.; Colas, B. Freezing in Polymer–Water Systems. In *Food Freezing: Today and Tomorrow*; Bald, W.B., Ed.; Springer-Verlag London Limited: Great Britain, 1991; 27–44.
52. Franks, F.; Mathias, S.F.; Trafford, K. Nucleation of Ice in Undercooled Water and Aqueous Polymer Solutions. Colloids Surf. **1984**, *11*, 275–281.
53. Michelmore, R.W.; Franks, F. Nucleation Rates of Ice Undercooled Water and Aqueous Solutions of Polyethylene Glycol. Cryobiology **1982**, *19*, 163–175.
54. Franks, F. The Nucleation Kinetics in Model Food Emulsions. Cryo-Letters **1981**, *2*, 27–35.
55. Gavish, M.; Wang, J.L.; Eisenstein, M.; Lahav, M.; Leiserowitz, L. The Role of Crystal Polarity in α-Amino Acid Crystals for Induced Nucleation. Science **1992**, *256*, 515.
56. Mackenzie, A.P. Nonequilibrium Freezing Behavior of Aqueous Systems. Phil. Trans. R. Soc. Lond. B **1977**, *278*, 167–189.
57. Muhr, A.H.; Blanshard, J.M.V.; Sheard, J.S. Effects of Polysaccharide Stabilizers on the Nucleation of Ice. J. Food Technol. **1986**, *21*, 587–595.
58. Ozilgen, S.; Reid, D.S. The Use of DSC to Study the Effects of Solutes on Heterogeneous Ice Nucleation Kinetics in Model Food Emulsions. Lebensm. Wiss. Technol. **1993**, *26* (2), 116–120.
59. Franks, F.; Mathias, S.F.; Parsonage, P.; Tang, T.B. Differential Scanning Calorimetric Study on Ice Nucleation in Water and in Aqueous Solutions of Hydroxyethyl Starch. Thermochim. Acta **1983**, *61*, 195–202.
60. Franks, F.; Darlington, J.; Schenz, T.; Mathias, S.F.; Slade, L.; Levine, H. Antifreeze Activity of Antarctic Fish Glycoprotein and a Synthetic Polymer. Nature **1987**, *325*, 146–147.
61. Boonsupthip, W.; Lee, T.-C. Technical Poster Session: Refrigerated and Frozen Foods. In *Application of Antifreeze Proteins: Efficiency of Antifreeze Proteins in Actomyosin Gel-Forming Preservation*, IFT Annual Meeting Program & Food Expo Exhibit, Dallas, TX, June 10–14, 2000; Institute of Food Technologists: Chicago, IL, 2000; 78B-8.
62. Griffith, M.; Ewart, K.V. Antifreeze Proteins and Their Potential Use in Frozen Food. Biotechnol. Adv. **1995**, *13*, 375–402.
63. Feeney, R.E.; Yeh, Y. Antifreeze Proteins: Properties, Mechanism of Action, and Possible Applications. Food Technol. **1993**, *1*, 83–88, 90.
64. Hew, C.L.; Yang, D.S.C. Protein Interaction with Ice. J. Biochem. **1992**, *203*, 33–42.
65. Rasmussen, D.H.; MacKenzie, A.P. Effect of Solute on Ice–Solution Interfacial Free Energy; Calculation from Measured Homogenous Nucleation Temperatures. In *Water Structure at the Water–Polymer Interface*; Jellinek, H.H.G., Ed.; Plenum Press: New York, 1972; 126–182.
66. Hartel, R.W. Evaporation and Freeze Concentration. In *Handbook of Food Engineering*; Heldman, D.R., Lund, D.B., Eds.; Marcel Dekker, Inc.: New York, 1992; 341–392.
67. Hobbs, P.V. *Ice Physics*; Claredon Press: Oxford, 1974; 837.
68. Blanshard, J.M.V.; Muhr, A.H.; Gough, A. Crystallization from Concentrated Sucrose Solutions. In *Water Relationships in Foods*; Levine, H., Slade, L., Eds.; Plenum Press: New York, 1991; 639–762.
69. Muhr, A.H.; Blanshard, J.M.V. Effect of Polysaccharide Stabilizers on the Rate of Growth of Ice. J. Food Technol. **1986**, *21*, 683–691.
70. Fletcher, N.H. Crystal Growth. In *The Chemical Physics of Ice*; Fletcher, N.H., Ed.; Cambridge University Press: Great Britain, 1970; 104–146.
71. Fennema, O.R. Nature of Freezing Process. In *Low-Temperature Preservation of Foods and Living Matter*; Fennema, O.R., Powrie, W.D., Marth, E.H., Eds.; Marcel Dekker, Inc.: New York, 1973; 153–220.
72. Angell, C.A.; Choi, Y. Crystallization and Vitrification in Aqueous Systems. J. Microsc. **1986**, *141*, 251–263.
73. Langer, S. Dendritc Solidification of Dilute Solutions. Phys.-Chem. Hydrodyn. **1980**, *1*, 44–53.
74. MacFarlane, D.R.; Kadiyala, K.R.; Angell, C.A. Homogeneous Nucleation and Growth of Ice from Solutions. J. Chem. Phys. **1983**, *79*, 3921–3932.
75. Budiaman, E.R.; Fennema, O. Linear Rate of Water Crystallization as Influenced by Temperature of Hydrocolloid Suspensions. J. Dairy Sci. **1987**, *70*, 534–546.
76. Shukla, T.P. A New Development in Carbohydrate Research. Cereal Foods World **1991**, *36*, 251–252.
77. Slade, L.; Levine, H. Thermal Analysis of Starch and Gelatin. In *Proceedings 13th Annual Conference, North American Thermal Analysis Society*; McGhie, A.R., Ed.; NATAS: Philadelphia, PA, 1984.
78. Slade, L.; Levine, H. Polymer-Chemical Properties of Gelatin in Foods. Adv. Meat Res. **1987**, *4*, 251–266.
79. Wunderlich, B. *Macromolecular Physics: Crystal Nucleation, Growth, Annealing*; Academic Press: New York, 1976; 135.
80. Mansfield, M.L. An Overview of Theories of the Glass Transition. In *The Glassy State in Foods*; Blanshard, J.M.V., Lillford, P.J., Eds.; Nottingham University Press: Nottingham, 1993; 103–122.
81. Simatos, D.; Blond, G. Some Aspects of the Glass Transition in Frozen Food Systems. In *The Glassy State in Foods*; Blanshard, J.M.V., Lillford, P.J., Eds.; Nottingham University Press: Nottingham, 1993; 395–416.
82. Singh, R.P.; Heldman, D.R. Heat Transfer in Food Processing. In *Introduction to Food Engineering*; Singh, R.P., Heldman, D.R., Eds.; Academic Press, Inc.: San Diego, 1993; 129–224.
83. Ben-Yoseph, E.; Hartel, R.W. Computer Simulation of Ice Recrystallization in Ice Cream During Storage. J. Food Eng. **1998**, *38*, 309–329.
84. Donhowe, D.P.; Hartel, R.W. Recrystallization of Ice in Ice Cream During Controlled Accelerated Storage. Int. Dairy J. **1996**, *6*, 1191–1208.

85. Carpenter, J.F.; Hansen, T.N. Antifreeze Protein Modulates Cell Survival During Cryopreservation: Mediation Through Influence on Ice Crystal Growth. Proc. Natl Acad. Sci. U.S.A. **1992**, *89*, 8593–8597.
86. Clemmings, J.F.; Zoerb, H.F.; Rosenwald, D.R.; Huang, V.T. Method of Making Ice Cream. US Patent 5,620,732, April 15, 1997.
87. Fletcher, G.L.; Hew, C.L.; Joshi, S.B.; Wu, Y. Antifreeze Polypeptide-Expressing Microorganisms Useful in Fermentation and Freezing of Food. US Patent 5,676,985, October 14, 1997.
88. Knight, C.A.; Duman, J.G. Inhibition of Recrystallization of Ice by Insect Thermal Hysteresis Proteins: A Possible Cryoprotective Role. Cryobiology **1986**, 256–262.
89. Mueller, G.M.; McKown, R.L.; Corotto, L.V.; Hague, C.; Warren, G.J. Inhibition of Recrystallization of Ice by Chimeric Proteins Containing Antifreeze Domains. J. Biol. Chem. **1991**, *266*, 7339–7344.
90. Payne, S.R.; Sandford, D.; Harris, A.; Young, O.A. The Effects of Antifreeze Proteins on Chilled and Frozen Meat. Meat Sci. **1994**, *37*, 429–438.
91. Yeh, Y.; Feeney, R.E. Antifreeze Proteins: Structures and Mechanisms of Function. Chem. Rev. **1996**, *96*, 601–617.
92. Yeh, Y.; Feeney, R.E.; McKown, R.L.; Warren, G.J. Measurement of Grain Growth in the Recrystallization of Rapidly Frozen Solutions of Antifreeze Glycoproteins. Biopolymers **1994**, *34*, 1495–1504.

Image Analysis

Iyad Hatem
Jinglu Tan
University of Missouri–Columbia, Columbia, Missouri, U.S.A.

I

INTRODUCTION

Image analysis is the discipline in which images of objects are investigated in order to help perform a specific task. This task is usually one of simulated human vision. Examples include human inspection of food quality, diagnosis through ultrasound imaging by a trained physician, and monitoring cell growth in a cell culture by a biologist. These examples are just a few of many applications in the field of food quality and biological engineering. Recent and previous research has proven the potential of using image analysis to develop objective tools for such tasks.

This article introduces the basics of image analysis [1–4]. It starts with the main components of an imaging system and then describes the typical steps in image analysis. At the end of the article, a biological application is presented as a simple illustration of image segmentation.

IMAGE ACQUISITION AND REPRESENTATION

Images can be formed by sensing electromagnetic radiation reflected off objects. Certain regions of the electromagnetic spectrum are commonly used for light sources and detectors, which include visible, infrared, x-rays, and radio waves. Regardless of the sensing technology, an image is converted into a standard numerical form to make it amenable to computer analysis. Fig. 1 shows the main parts of a digital image analysis system. It includes a camera with lenses for image formation and a radiation sensor that converts irradiance at the image plane into an electrical signal. An example radiation sensor is the charge-coupled device (CCD). It is basically a light-sensitive crystalline silicon chip, which collects and counts the photoelectrons freed from a silicon layer by the incident photons reflected off a physical object. The number of electrons is transferred as an electrical signal to a frame grabber where it is digitized and stored in a device such as a computer. The computer provides a platform for digital analysis of the image using image processing software. The digitization consists of dividing an image into small rectangular elements, called pixels, and using integer values to represent the degree of brightness, called gray levels. This array of integer values are then either saved in a computer as a file with a special format that carries the spatial and gray-level information for later processing, or processed directly through algorithms implemented in hardware or software, the latter of which is called real-time processing.

The fidelity of a digital image in representing the original scene is related to the number of pixels determined by the spatial resolution and the range of gray values determined by the brightness resolution. The higher these resolution values are, the more information and details the image carries.

Many imaging systems acquire gray images and others produce multispectral images. In multispectral imaging, a number of gray images are taken at the same time for the same scene by different detectors sensitive in different regions of the electromagnetic spectrum. An example is a color imaging system in which three spectral bands are used, corresponding to red, green, and blue colors. The selection of these bands is based on the human perception of color. This perception results from the fact that there are three types of cone cells on the human retina that are tuned to these three bands in the visible portion of the electromagnetic spectrum.

IMAGE ENHANCEMENT AND RESTORATION

The main goal of image enhancement and restoration is to improve the quality of an image. Improvement is achieved by removing or reducing the degradations an image may have incurred. Degradations can occur because of imperfections in the lighting system, lens, photodetector, and digitizer. Geometric distortions can be caused by the geometry of the imaging system. For example, a spacecraft camera may produce an image with curvilinear sides. Image enhancement operations are classified into two categories: gray-level-based pixel operations and geometric operations.

Encyclopedia of Agricultural, Food, and Biological Engineering
DOI: 10.1081/E-EAFE 120007223
Copyright © 2003 by Marcel Dekker, Inc. All rights reserved.

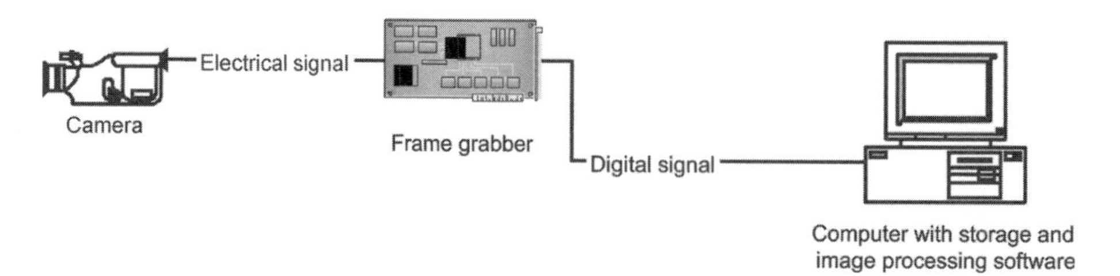

Fig. 1 Digital image analysis system.

Gray-Level-Based Pixel Operations

This type of operation is applied on the gray level of each pixel in an image. When such operations are applied as a pixel-to-pixel mapping between the input and output images, they are called *point operations*. When they operate on a group of pixels surrounding a center pixel, they are called *local operations*.

A good example of point operation is contrast enhancement by histogram stretching. An image histogram is a distribution graph of the gray level of pixels within the image. It is useful in determination of the image contrast. Fig. 2 shows a low-contrast cell image. The corresponding histogram of this image shows that the pixel values are tightly distributed. By enlarging the range of pixel values (stretching the histogram), the resulting image shows much better contrast (Fig. 3).

Point operations also work on multiple input images. For example, algebraic operations such as pixel-by-pixel sum, difference, product, or quotient of two input images can be used to produce an output image. One important application of such operations is motion or volume change detection. By subtracting two sequential images of a moving or growing object, the difference image shows the movement or growth of the object.

Local operations consider neighborhood pixels surrounding a center pixel. They are usually implemented through a process called spatial convolution. Spatial convolution multiplies each pixel in the range of a filter mask, also called kernel, with the corresponding weighting factors in the mask, adds up the products, and writes the final result to the position of the center pixel. In other words, convolving image F by kernel $G(M,N)$ gives the resultant image H as follows,

$$H(i,j) = \sum_{m}^{M} \sum_{n}^{N} F(m,n)G(i-m,j-n) \tag{1}$$

where i, j are the image pixel indices; m, n the kernel value indices; and M, N the sizes of kernel.

An important class of local operations consists of those used for spatial filtering, which include high-pass and low-pass filters. High-pass filters attenuate low-spatial-frequency components in an image to increase the contrast. On the other hand, low-pass filters attenuate the high-spatial-frequency components, which have a blurring effect. By subtracting a low-pass filtered image from its original, a sharpened image can be obtained. Two common 3×3 high-pass and low-pass masks are given in Fig. 4.

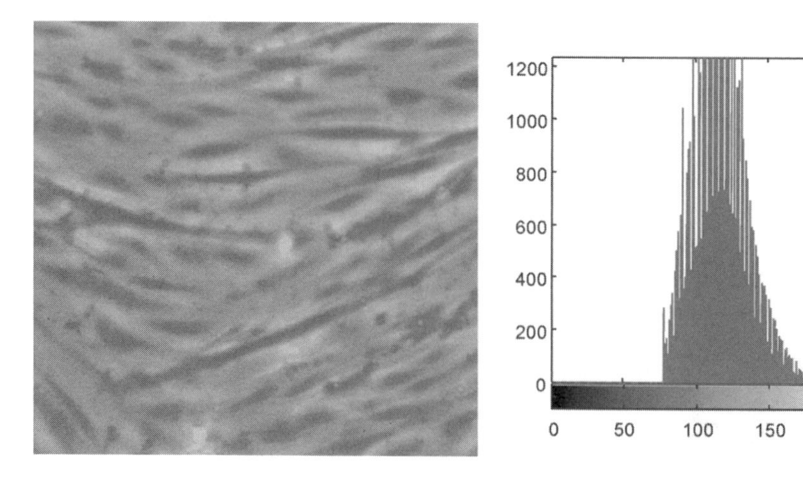

Fig. 2 A low-contrast cell image and its histogram.

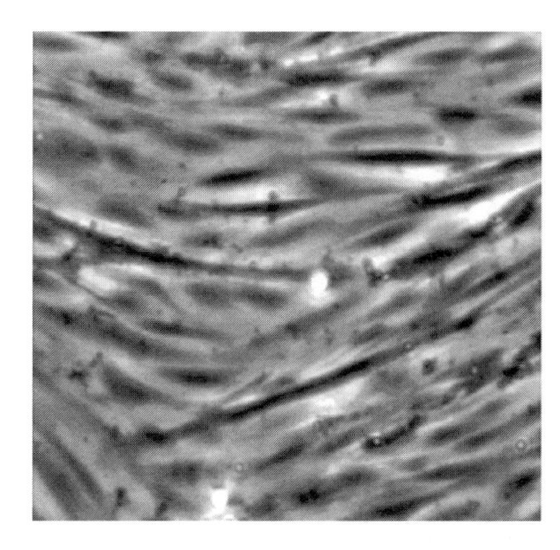

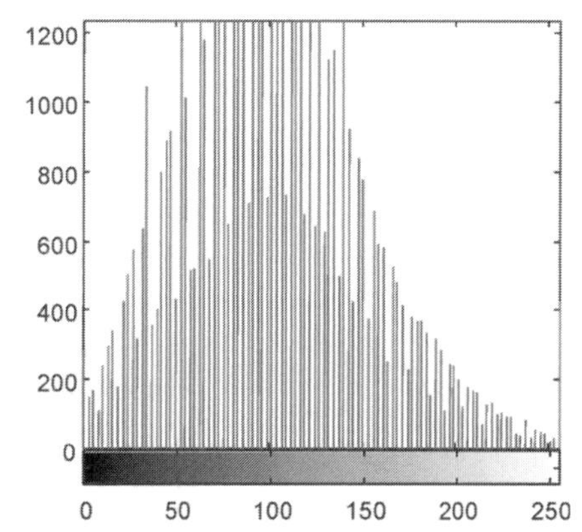

Fig. 3 The cell image in Fig. 2 after contrast enhancement (stretching its histogram).

Geometric Operations

Geometric operations adjust the spatial relationships among objects in an image in a constrained manner that preserves some resemblance of order. The adjustments usually require spatial transformation of pixels and gray-level interpolation to preserve the continuity of curvilinear features and the connectivity of objects within an image.

The basic spatial transformation operations are translation, rotation, and scaling. For an image located in the x–y plane, we can write the translation operation of pixel (x_i, y_i) to pixel (x'_i, y'_i) as

$$\begin{bmatrix} x'_i \\ y'_i \\ 1 \end{bmatrix} = \begin{bmatrix} 1 & 0 & T_x \\ 0 & 1 & T_y \\ 0 & 0 & 1 \end{bmatrix} \begin{bmatrix} x_i \\ y_i \\ 1 \end{bmatrix} \tag{2}$$

where T_x and T_y are the amounts of translation in the x and y directions, respectively. In a similar way, the scaling operation, which either magnifies or shrinks the image, depending on the scaling factors c and d, is

$$\begin{bmatrix} x'_i \\ y'_i \\ 1 \end{bmatrix} = \begin{bmatrix} 1/c & 0 & 0 \\ 0 & 1/d & 0 \\ 0 & 0 & 1 \end{bmatrix} \begin{bmatrix} x_i \\ y_i \\ 1 \end{bmatrix} \tag{3}$$

Low-pass:

1/9	1/9	1/9
1/9	1/9	1/9
1/9	1/9	1/9

High-pass:

-1	-1	-1
-1	9	-1
-1	-1	-1

Fig. 4 Two commonly used low-pass and high-pass masks.

and the rotation operation by an angle θ is

$$\begin{bmatrix} x' \\ y' \\ 1 \end{bmatrix} = \begin{bmatrix} \cos(\theta) & -\sin(\theta) & 0 \\ \sin(\theta) & \cos(\theta) & 0 \\ 0 & 0 & 1 \end{bmatrix} \begin{bmatrix} x_i \\ y_i \\ 1 \end{bmatrix} \tag{4}$$

IMAGE SEGMENTATION

Image segmentation is usually an imperative step in image analysis. It is the process of separating the object of interest in an image from the background and other entities. There are many techniques for image segmentation. Selection of techniques is dependent on the specific problem and the characteristics of the target objects. In addition, the application sequence of these techniques plays an essential role in achieving good segmentation results.

Three major approaches are used for image segmentation: the *region approach*, which assigns each pixel to a particular region; the *boundary approach* in which the boundaries between regions are determined; and the *edge approach* in which the edge pixels are identified and then linked together to form a boundary. The following sections confer examples of the classic techniques used for image segmentation.

Thresholding

Thresholding is a region approach. It is very useful for scenes containing solid objects resting upon a contrasting background. A simple threshold rule allocates all the pixels below or above a constant threshold gray value to the object. The remaining pixels are allocated to the background. In many cases, it may not be possible to use a single threshold to produce acceptable results. An example is when illumination is not homogeneous throughout the image. In such cases, it is necessary to use different threshold values for different image regions. This is called adaptive thresholding.

A major problem in image thresholding is the selection of the optimal threshold value. It is sometimes done manually by trial-and-error. There are many automatic threshold selection methods in the literature, such as histogram techniques and watershed algorithms. The simplest histogram technique is one applicable to images with a bimodal gray-level histogram. This technique finds two peaks in the histogram and then places the threshold at a position between the two peaks.

Gradient-Based Segmentation

This method of segmentation tries to find the boundaries between two objects by using the gradient magnitude. The gradient magnitude of a scalar function $f(x,y)$ is given by:

$$|\nabla f(x,y)| = \sqrt{\left(\frac{\partial f}{\partial x}\right)^2 + \left(\frac{\partial f}{\partial y}\right)^2} \quad (5)$$

It represents the steepness of the slope at every point of function f. In a pixel idiom, Eq. 5 is approximated by the following

$$|\nabla f(x,y)| = \max[|f(x,y) - f(x+1,y)|, \; |f(x,y) - f(x,y+1)|] \quad (6)$$

which is the maximum of the absolute vertical and horizontal neighboring pixel differences. This algorithm will trace out the maximum gradient boundaries given that the image is free of monotone spots and noise. A small amount of noise can send the tracking off the boundary. Noises can be reduced by a variety of techniques such as low-pass filtering.

Region Growing

This method of image segmentation is also a region-based approach. An image is divided into different regions depending on defined properties that distinguish the pixels inside these regions. The algorithm starts with initial regions of small neighborhoods and then examines the boundary strength in terms of the average properties between two adjacent regions. It will dissolve the weak boundaries and keep the strong ones. The region-merging process is iterated until no boundaries are weak enough to dissolve. One drawback of this method is its extensive computational needs. Its advantage lies in its ability to employ several image properties directly and simultaneously to determine the final boundaries.

Edge Detection and Linking

Edge-detection methods are based on the fact that boundary pixels on an object define a gray-level transition to the neighbors. By convolution with a set of directional derivative masks, the slope and the direction of this transition are quantified. There are several operators for edge detection, including Robert, Sobel, Prewitt, and Kirsch edge operators. For example, Sobel uses two convolution kernels shown in Fig. 5. Both kernels are applied to an image and the maximum value of the two convolutions is taken to create the edge-magnitude image.

In some cases, the resulting edge-magnitude image can have gaps, which may require filling. This is done with an edge-linking operator through a search process.

MORPHOLOGICAL IMAGE PROCESSING

These operations are commonly used on binary images and can be extended to gray-level images. They provide many useful tools for image analysis. Morphological operations generally follow an initial segmentation operation that clarifies the underlying structure of objects.

Binary morphological operations pass a structuring element over an image much like the mask used in spatial convolution. The structuring element is generally composed of an array of logical values of size 3 × 3 or greater. At each pixel, a specified logical operation (AND, OR, or NOT) is performed between the structuring element and the underlying binary image. Depending on the size of the structuring element and the logical operation, different morphological effects can be created. The following will briefly introduce the basic morphological operations commonly used in image processing applications.

Dilation and Erosion

Dilation and erosion are the most fundamental morphological operations. Dilation causes the expansion of shapes in the input image. It is useful for filling holes in

-1	-2	-1
0	0	0
1	2	1

-1	0	1
-2	0	2
-1	0	1

Fig. 5 Sobel edge operators.

segmented objects. In general, dilation can be defined as

$$\mathbf{E} = \mathbf{B} \oplus \mathbf{S} = \{x, y | \mathbf{S}_{xy} \cap \mathbf{B} \neq \mathbf{\Phi}\} \quad (7)$$

where **E** is the binary image that results from dilating image **B** with structuring element **S**, and **Φ** is the empty set. The dilated image is the set of points (x, y) such that **S** is translated so its origin is located at (x, y) and its intersection with **B** is not empty.

Erosion is the process of eliminating the boundary points from an object, shrinking the object by one or more pixels all around its perimeter. It is useful for removing noises segmented image objects. Its general formula is given by

$$\mathbf{E} = \mathbf{B} \otimes \mathbf{S} = \{x, y | \mathbf{S}_{xy} \subseteq \mathbf{B}\} \quad (8)$$

where the condition is that the translated **S** is contained within **B**.

Opening and Closing

Opening is a process of erosion followed by dilation. This operation can be used to separate objects with narrow connections and eliminate small objects in an image. It is defined as

$$\mathbf{B} \circ \mathbf{S} = (\mathbf{B} \otimes \mathbf{S}) \oplus \mathbf{S} \quad (9)$$

Closing is dilation followed by erosion. It is useful for filling small and thin holes in objects. It is defined as

$$\mathbf{B} \bullet \mathbf{S} = (\mathbf{B} \oplus \mathbf{S}) \otimes \mathbf{S} \quad (10)$$

Skeletonization

This operation is an application of erosion operations. It uses numerous different erosion masks oriented in various directions to eat away the objects in an image. At the end of this operation, each object becomes a thin skeleton. Skeletonization is very useful in revealing the underlying spatial relationships among the objects.

FEATURE EXTRACTION AND OBJECT CLASSIFICATION

Feature extraction is the process of calculating feasible measurements, which can describe the object of interest and discriminate it from other entities in an image. There are no generally applicable techniques for extracting features. The objective and nature of the image processing problem define what features may be useful. There are, however, commonly used object measurements like *size*, *shape*, and *texture*.

Size measurements include area, perimeter, length, and width. For these measurements, it is necessary to label the pixels of every object. A simple way to measure the area of an object is to count the number of pixels in an object. To measure the perimeter, one way is to apply an edge-detection operation and then count the boundary pixels of an object. The length and width of an object can be computed by locating and measuring the major and minor axes of the object.

Shape measurements are very useful in describing objects. An example is elongation, which distinguishes between slender objects from roughly square or circular objects. Elongation is often computed as the difference between the lengths of the major and minor axes of the best ellipse fit to an object, divided by the sum of the lengths. This measure is 0 for a circle and approaches 1.0 for long and narrow ellipse. This kind of measurement does not give detailed description of the shape of an object. A type of shape feature that gives more details is called shape descriptor. An example is the differential chain code. To compute this code, a boundary chain code is found that defines the direction of the boundary path of an object and then the derivative of this code is calculated to reflect the curvature of the boundary. This code can be further analyzed to obtain more detailed measurements of shape.

Texture is the attribute that measures the spatial arrangement or variation of the gray level of pixels in an object. A simple set of texture measurements is the statistical moments of histogram. For an image with K

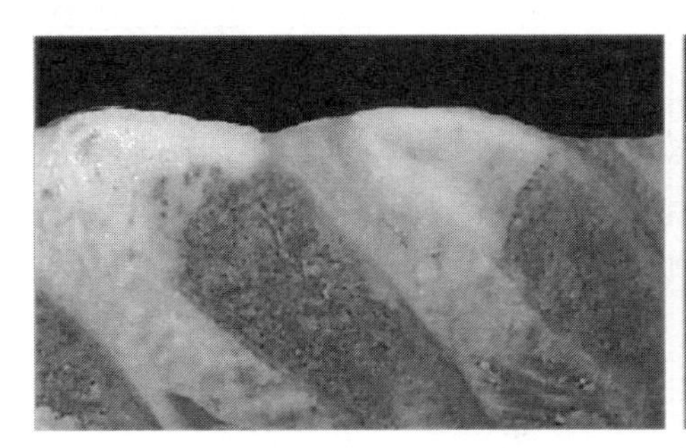

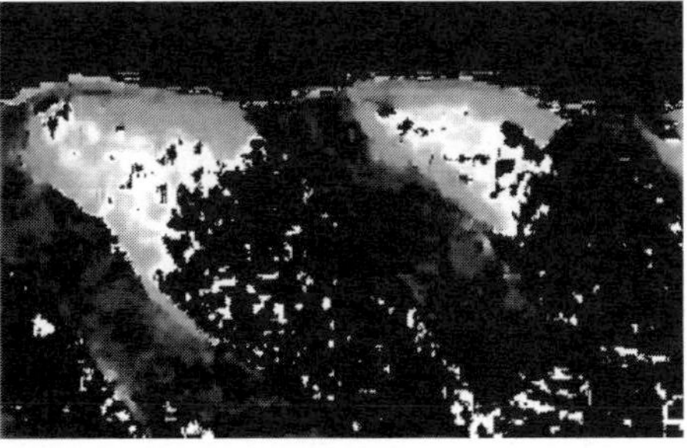

Fig. 6 An original vertebra image (left) and its hue image (right).

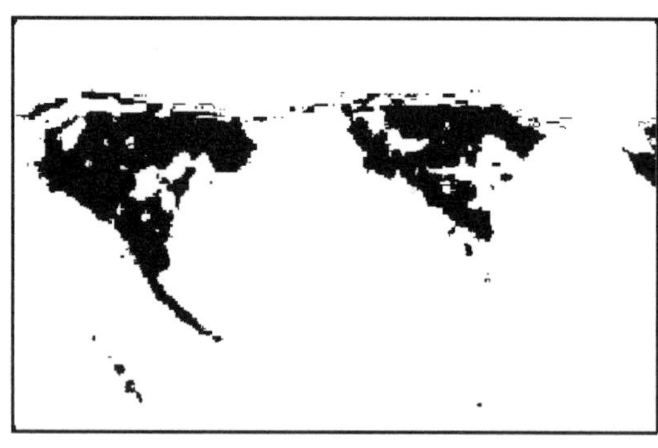
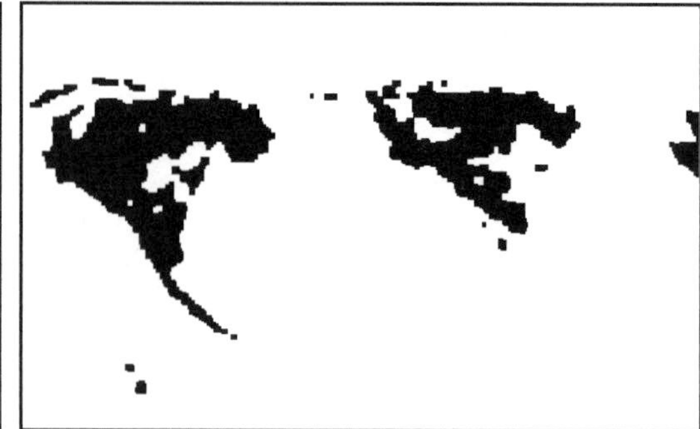

Fig. 7 Binary image of interest after a thresholding operation on the hue image (left) and the same image after opening and closing operations (right).

gray levels and an object within the image with mean gray value μ and gray-value histogram $h(k)$ with k ranging from 0 to $K-1$, the nth moment about the mean for this histogram is given by:

$$m_n = \frac{1}{K}\sum_{k=0}^{K-1}(k-\mu)^n h(k) \tag{11}$$

The second moment, m_2, or variance is a commonly used texture measure. It has good correlation with the human perception of roughness.

After collecting measurements on the characteristics of objects of interest, a feature selection procedure is usually applied. Different criteria can be used to select features, which include discrimination, reliability, independence, and size. An example method is the principal component analysis (PCA). PCA aims to restructure the data to reduce the test features to a smaller number of principal components, which account for most of the variations in the given data. It linearly transforms data from the original space into a new space, which minimizes the covariance among the transformed features (principal components) and maximizes the variance of each new feature. For more information about data reduction techniques, see Refs. 5,6.

After selecting the most useful features, objects or images can be classified. The classification process involves designing a classifier and testing its performance with known data.

Classifier design seeks decision boundaries that separate M object/pattern classes based on the computed features. Suppose there are P features $\mathbf{x} = (\mathbf{x}_1, \mathbf{x}_2, \ldots, \mathbf{x}_P)'$ extracted to discriminate M classes, the boundaries are defined by decision functions $d_1(\mathbf{x}), d_2(\mathbf{x}), \ldots, d_M(\mathbf{x})$. These functions can be linear or nonlinear. A pattern with feature values $\mathbf{x}$ is classified to class k if $d_k(\mathbf{x})$ has the largest value among the decision functions.

Different techniques are used to generate decision functions. If a priori knowledge about a pattern is available, the *Bayes rule* is used. When qualitative knowledge about the patterns is available, which is the common case, a classifier is designed by using a training or learning procedure. Different deterministic and statistical approaches have been used to design learning algorithms for classification. A simple way for training a classifier is to minimize the total number of classification errors based on a cost function. Decision functions are then located such that a minimum cost is generated by the classifier. For more details about classification, see Refs. 7–9.

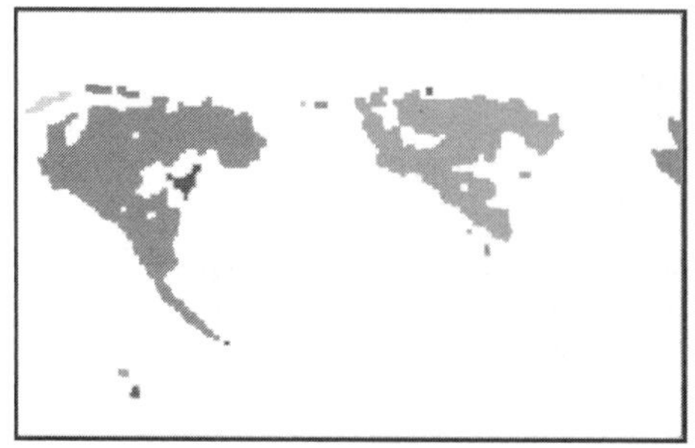
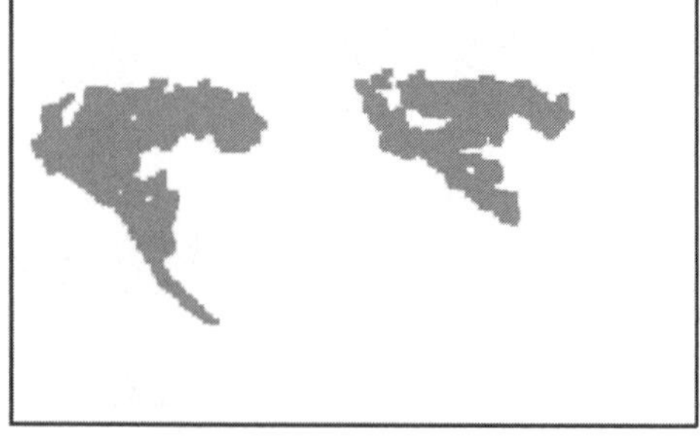

Fig. 8 Labeled objects in different gray levels (left) and the final objects after filtering by size (right).

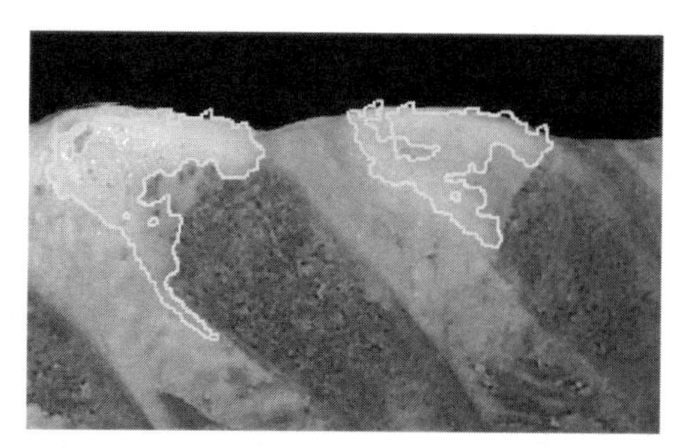

Fig. 9 The two final objects highlighted in the original image.

EXAMPLE APPLICATION

As an illustration, an image processing application in meat quality is presented here. Fig. 6 shows a digital image of a cut surface in the thoracic vertebrae of a beef carcass. This image contains information, a trained meat grader uses, to determine the maturity or physiological age of the animal. The white triangular-shaped areas in the image are cartilage. As an animal gets older, the cartilage ossifies and darkens. The color of cartilage is an important indicator of maturity. To measure the cartilage color, it is very necessary to isolate the cartilage areas from the surrounding bones, tissues, and fats, i.e., image segmentation is needed.

Analysis showed that the hue value of the HSI (hue, saturation, and intensity) color system is useful in discriminating cartilage from the surrounding tissues as shown in Fig. 6. The HSI values can be easily computed from the RGB (red, green, and blue) color system. The HSI system is useful in distinguishing one color from another by its chromaticity (hue and saturation) and brightness (intensity) values.

By applying a simple thresholding operation on the hue image, can be isolated the regions of interest (cartilage) as shown in Fig. 7 (left). An opening operation eliminates some extraneous small objects and a closing operation fills some of the holes in the big objects (Fig. 7, right).

An object-labeling algorithm is then applied to label the objects in the image. Fig. 8 (left) shows the labeled objects in different gray levels. Labeling allows computations of different features to be performed on individual objects. To eliminate the small objects, a filtering operation based on the size of objects is applied. This results in the two final objects shown in Fig. 8 (right). Fig. 9 shows the isolated cartilage objects in the original color image. Color and spatial features that characterize the cartilage are computed from this image. These features are used to develop a classifier for animal maturity evaluation.

REFERENCES

1. Baxes, G.A. *Digital Image Processing: Principles and Applications*; John Wiley & Sons, Inc.: New York, 1994.
2. Castleman, K.R. *Digital Image Processing*; Prentice-Hall, Inc.: New Jersey, 1996.
3. Costa, L.F.; Cesar, R.M., Jr. *Shape Analysis and Classification: Theory and Practice*; CRC Press LLC.: New York, 2001.
4. Jahne, B. *Image Processing for Scientific Applications*; CRC Press LLC.: New York, 1997.
5. Ehrenberg, A.S.C. *Data Reduction: Analyzing & Interpreting Statistical Data*; John Wiley & Sons, Inc.: New York, 1975.
6. Cooley, W.W.; Lohnes, P.R. *Multivariate Data Analysis*; John Wiley & Sons, Inc.: New York, 1971.
7. Duda, R.O.; Hart, P.E.; Stork, D.G. *Pattern Classification*; John Wiley & Sons, Inc.: New York, 2001.
8. Tou, J.T.; Gonzalez, R.C. *Pattern Recognition Principles*; Addison-Wesley Publishing Company, Inc.: Massachusetts, 1974.
9. Schalkoff, R. *Pattern Recognition: Statistical, Structural and Neural Approaches*; John Wiley & Sons, Inc.: New York, 1992.

Indirect Contact Freezing Systems

Elaine P. Scott
Virginia Polytechnic Institute and State University, Blacksburg, Virginia, U.S.A.

INTRODUCTION

As opposed to direct contact freezing systems, the product and refrigerant are separated in indirect contact freezing systems. Consequently, all indirect contact freezing systems involve an interface or barrier between the refrigerant or cooling medium and the food product. This interface could be either as simple as a metal plate, which separates the refrigerant and the food, or a product's packaging material, which prevents direct contact between the food and the cooling medium, such as a forced cold-air stream. Thus, the presence of the interface or barrier material, although practically necessary in many instances, adds an additional resistance to heat transfer in the process of removing sensible and latent heat from the product. As illustrated in Fig. 1, the barrier can be thought of as an added resistance in a thermal circuit, and the overall resistance from the food product to the surrounding air in this case is a result of: 1) the convection resistance from the air stream to the packaging layer (characterized by the inverse of the convective heat transfer coefficient); 2) conductive resistance of the packaging material itself; and 3) resistance at the package–food interface.

INDUSTRIAL SYSTEMS

Indirect contact freezing processes are used for both solid and liquid products. Typical indirect systems for solid products include plate freezers and air blast freezers while scraped surface heat exchanger systems are used for liquid products. Liquid immersion freezers are used for both solid and liquid products. These different industrial systems are discussed in the following sections.

Plate Freezers

Industrial plate freezers have been used extensively to freeze a wide variety of food products. The basic system consists of flat hollow plates; refrigeration coils run through these plates to cool the surface in contact with the food product. The food products are placed between stacked parallel plates and then pressure is applied to the overall stack to minimize the thermal contact resistance between the plates and the product, as shown in Fig. 2. Thus, the product is restricted to having a planar geometry; and therefore, unpackaged meat and fish products such as fish fillets, ground beef patties, and boned meat are well suited for this method. The method is also used to harden ice cream packaged in brick containers.[1] Other irregularly shaped products including vegetables, such as cauliflower, spinach, and broccoli, and shrimp, can be frozen using this method by packaging the product in brick shaped containers prior to freezing. In some instances, two levels of packaging are involved as servings or pieces are packaged individually in inner packs and then these inners are placed in an outer pack. In some cases, the inner packs are frozen prior to packing in the outer packs. In these cases, such as the freezing of offal, plate freezing has been shown to provide rapid freezing and flat packs, which can be easily packaged into the outer packs.[2] Overall, the primary attractions of plate freezing are the uniform high quality of the product, speed of freezing, and significant reductions in energy, packaging, and resources consumption.[3]

The plates in these freezing systems can be stacked horizontally or vertically. Horizontal plate freezers consist of a set of parallel, refrigerated plates within an insulated enclosure. The refrigeration system is typically in a bottom section beneath the enclosure. The plates are opened and closed hydraulically. These freezers are operated in both batch and continuous modes. In the batch mode, the spacing in the plates is expanded to allow the product to be loaded on large trays. Once the product is loaded, the plates are hydraulically closed and pressure is applied to assure good thermal contact between the refrigerated plate and the food product. Spacers (slightly thinner than the product thickness) are sometimes used for packaged foods to prevent the cartons from being crushed. Once the freezing process is complete, the plates are opened and the product is removed. Efforts have also been made to automatically load and unload the freezer to further facilitate the process and increase processing efficiencies.[4] This includes the use of automatic conveying systems,[5] which would also facilitate continuous operations. In the continuous mode, the plates are moved through the enclosed system as the product freezes and

Encyclopedia of Agricultural, Food, and Biological Engineering
DOI: 10.1081/E-EAFE 120007030
Copyright © 2003 by Marcel Dekker, Inc. All rights reserved.

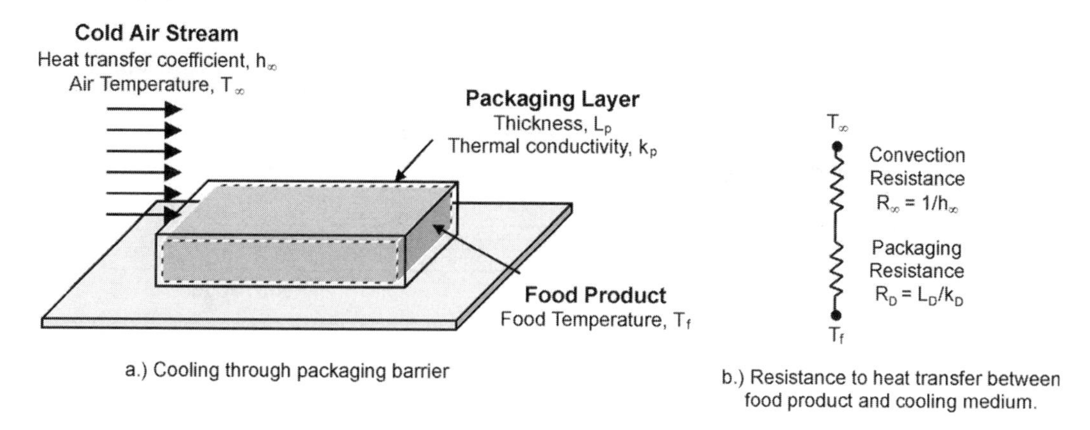

Fig. 1 Illustration of an indirect contact freezing system.

then the product between the plates is removed one at a time.

Vertical plate freezers are primarily used to freeze unpacked foods such as fish. This method is particularly suited to use on board freezer trawlers. Here, the food product is placed directly between the plates and then pressure is applied to again assure proper thermal contact. Once the process has been completed, the plates can be opened to remove the product. In some cases, a lifting device is included to facilitate the unloading operation.[6]

Air Blast Freezers

Air blast freezers are used for both direct and indirect freezing applications. In the latter, the geometry of the product might not be conducive to plate freezing, and other factors such as excessive moisture loss might prevent use of direct freezing methods. As a result, many products are often packaged with a protective film and then frozen using an air blast system.

In a basic air blast freezer, the product is exposed to a low temperature, high velocity air stream. The process can be continuous or batch; however, most systems are continuous. In these systems, a refrigeration system first cools an air stream to a minimum of approximately −40°C. The air flows (via fans) through an enclosed system in which the food product is moved through the system on a conveyor belt, as illustrated in Fig. 3. As shown in Fig. 1, the resistance to heat transfer between the air stream and the packaging layer is characterized by the inverse of the convective heat transfer coefficient, which is a function of the temperature and the velocity of the air. Thus, as the velocity increases, the convective heat transfer coefficient increases and the resistance to heat transfer decreases. As the velocity increases, however, the heat contribution from the fans also increases, thereby adding to the overall heating load of the system. Therefore,

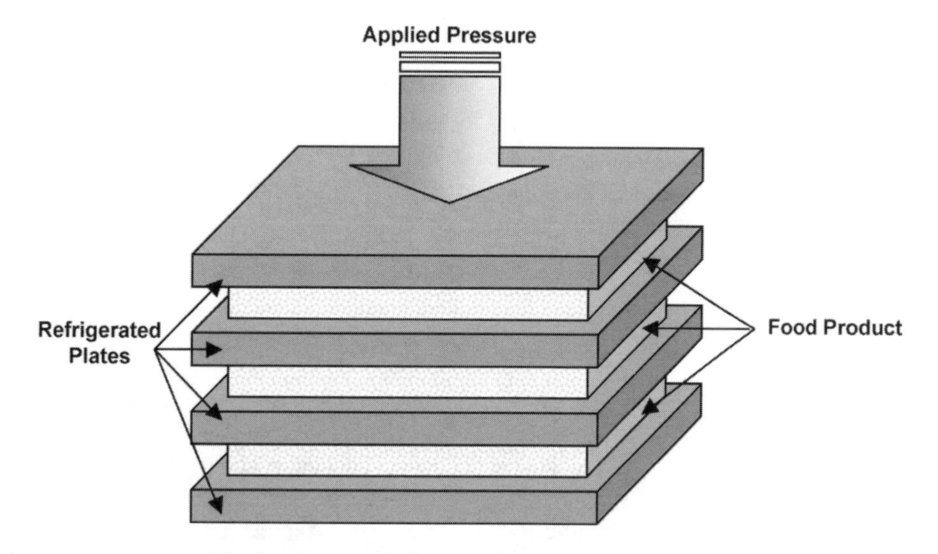

Fig. 2 Schematic of horizontal plate freezing system.

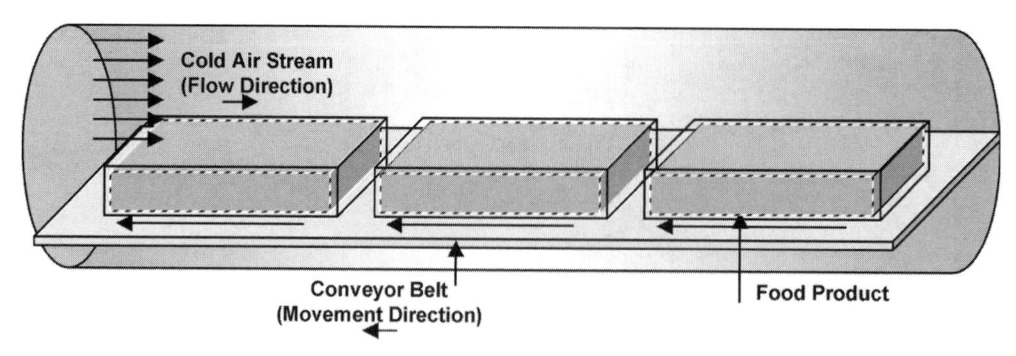

Fig. 3 Illustration of tunnel blast air freezing system.

there are practical limits to the magnitude of the velocity. The length and speed of the conveyor belt is dictated by the required residence time for the particular product to ensure proper freezing.

The configuration of the conveyor belt system can vary significantly. In the simplest case, the product is conveyed along a straight belt through a tunnel freezer. In more complex cases, multitiered and spiral designs are often used to minimize the space requirements of the system. It is important in such designs to make certain that the air flow is uniform over the product to ensure uniform freezing. The direction of the air flow is also an important factor. The basic possibilities are parallel flow (i.e., the air flows parallel to the product in the same direction), counter-flow (i.e., air flows in the opposite direction), and cross-flow (i.e., air flows perpendicular to the product). Parallel flow systems are not as efficient as counter-flow or cross-flow, and cross-flow systems are advantageous over counter-flow systems as they provide an even pressure balance between the product inlet and outlet openings, minimizing heat losses due to exchange with outside air at these locations. Current systems also provide for precise temperature and humidity control of the inlet air stream to prevent frost or ice accumulation.

Batch blast freezers usually consist of a well-insulated container with an air cooler and fans. The product is loaded on a movable cart with stacked trays and the cart is moved into the freezing container. A suitable duct system is also required to ensure uniform air flow over all the trays for proper freezing. Maintaining uniform air flow is very important. If partially filled or empty trays are present or if the container is poorly designed, air can bypass the filled trays to the path of least resistance, over the empty or partially filled trays or gaps,[7,8] and jeopardize the quality of the product in the filled trays.

Liquid Food Freezers

Another type of indirect freezing system is specifically designed for liquid foods. In many cases, the product is frozen in a two-step process. The liquid product is first chilled to form frozen slurry prior to packaging and then the freezing process is completed by placing the packaged product in a refrigerated compartment. An example of a product frozen in this manner is ice cream. Here, air is forced into the slurry to improve texture, and the product is then packaged for the final freezing step. Refrigerated scraped surface heat exchangers are typically used in the initial phase. Here, the product is cooled to a temperature several degrees below the initial freezing point such that 60%–80% of the total latent heat is removed.[9] In this process, a rotor equipped with scraper blades mixes the liquid to enhance heat transfer to the refrigerated shell, ensures a uniform temperature throughout the product, and prevents the product sticking to the shell surface. This process can be accomplished in either a batch or continuous mode.

In the final step, the packaged food product can simply be placed in a refrigerated room or processed using other methods such as plate freezer. This method is used, e.g., in the hardening of ice cream.[7]

Liquid Immersion Freezers

Another technique to freeze food products is by immersion in a cooled liquid such as eutectic NaCl brine.[10] This method offers high heat transfer coefficients and the ability to handle irregularly shaped geometries. Although this method is used for a number of direct freezing applications, it is also used to freeze citrus concentrate and to initially freeze the surface of poultry packaged in plastic film.[7] In the latter case, the process is usually completed in an air blast freezer. One drawback to this method is the possible accumulation of the salt brine (or other medium) on the surface of the product. Although the packaging material serves to protect the product from direct contact with the brine, it is often necessary to remove it by rinsing the product with water. Note that care must be taken to prevent the surface from warming significantly during the rinsing process.

REFERENCES

1. Mitten, H.L. Proper Hardening Ensures Quality. Dairy Field **1983**, *166* (6), 58–61.
2. Downey, J. *Comparison of Offal Chilling and Freezing Methods*; Meat Industry Research Institute of New Zealand: New Zealand, 1988; Vol. 853, 32.
3. Lassen, O. New Plate Freezer Applications. Scan Ref, *24* (5), 10–13.
4. Visser, K. A Large Scale Automatic Plate Freezer for Cartons of Meat. Inst. Chem. Eng. Symp. Ser. **1984**, (84), 219–229.
5. Marriott, D.; Kothuis, B.J.A. Fast and Efficient Freezing of Foods. Keoltech. Kilimaatregeling **1981**, *74* (5), 100–105.
6. Fenner, F.R. Vertical Top Discharge Plate Freezer. UK Patent Application, 1985, Jackstone Forster Ltd.
7. Brennan, J.G.; Butters, J.R.; Cowell, N.D.; Lilly, A.E.V. Freezing. *Food Engineering Operations*; Applied Science Publishers Limited: London, 1976; Chap. 14, 370–380.
8. Oliver, D.W. *Air Flows in Automatic Air Blast Carton Freezers*; Meat Industry Research Institute of New Zealand; MIRINZ 844; 1–16.
9. Singh, R.P.; Heldman, D.R. Food Freezing. *Introduction to Food Engineering*; 1984; Chap. 6, 191–212.
10. Jac, A. Freezing and Thawing in the Food Industry. Ind.-Aliment.-Agric., *93* (9/10), 1081–1091.

Intelligent Packaging

Evangelina T. Rodrigues
J. H. Han
The University of Manitoba, Winnipeg, Manitoba, Canada

INTRODUCTION

As consumers become more health conscious and demand better quality,[1] the food industry needs to improve the quality and safety of its products[2]; improved quality and safety demands that processors, distributors, storage conditions, and retailers be more precise and stringent in their operational procedures which may be achieved using more intelligent packaging.[1,2] Traditionally, packages were produced to contain and protect the product, provide convenience as well as some product information.[3] Initially, packaging systems were designed to minimize the interaction between the package itself and its content.[4] New packaging systems today are produced to act as "intelligent messengers,"[3] to increase the quality of foods,[2] and to provide more product information for processors and consumers.[1,2] Furthermore, technological trends now produce packaging systems that have increased the interaction between the package itself and the product.[4] Developing new packaging systems requires that the process be functionally effective and cost effective, so as to increase product safety and shelf life, as well as being appealing to the consumer.[5] Intelligent packaging and storage containing "sensing" and "interactive" devices are increasingly being developed for these purposes.[5]

NEW PACKAGING SYSTEMS

To date, many terms have been used to describe new packaging systems including terms such as "active," "smart," "clever," "interactive," and "intelligent" packaging.[4] The term "smart" packaging was first introduced in the mid-1980s to describe packages that were able to sense changes in the internal or surrounding environment and to alter some properties in response.[6] The term was then changed to "interactive" and then shortened to "active."[6] However, these terms are still being used inconsistently. In addition, many packages that are considered "active" are neither "smart" nor "intelligent" and simply act in a less passive way.[6] Packaging systems can be passive, active, or intelligent. "Passive" packaging is a conventional packaging system in which the package simply acts as a physical barrier to protect food commodities from the environmental conditions.[6] "Active" packaging possesses attributes beyond the basic physical barrier property[6] and can be classified into two types of systems. The first is simple active packaging, where the packaging system contains no active ingredient and/or no actively functional polymer. The second system, known as the true active packaging or advanced active packaging contains an active ingredient and/or an actively functional polymer. However, neither passive nor active packaging systems are considered intelligent as they do not sense changes in the environment and respond accordingly.[6]

Intelligent Packaging

Although many individuals have attempted to define intelligent packaging, some of their descriptions are incomplete and/or too simplified (Table 1). Therefore, we have attempted to amalgamate these definitions and create a more complete, clear, and precise definition of intelligent packaging that excludes active packaging.

"Intelligent," "smart," or "clever" packaging is an advanced active packaging system in that it contains attributes beyond the basic barrier properties, but also contains additional functions. Similar to active packaging, it can be divided into two types of systems. The first type is the "simple" intelligent packaging including sensor-incorporated packaging, environmentally reactive (not responsive) packaging, and computer communicable device equipped packaging. The packaging material contains a sensor, indicator, or an integrator that senses changes in the environment and consequently will signal those changes. The second type is the "interactive" or "responsive" intelligent packaging. It encompasses the characteristics of the first type and in addition, contains a response mechanism that responds to the signal (sensor, indicator, or integrator) mechanism. Interactive or responsive intelligent packaging has sensors incorporated into the package that start to neutralize or undo the negative changes occurring in the food product.[5]

To date, four types of applications of intelligent packaging systems are used to: 1) improve product quality and product value (quality indicators, temperature and time–temperature indicators, and gas concentration indicators); 2) provide more convenience (quality, distribution, and preparation/cooking methods); 3) change gas permeability properties; and 4) provide protection against theft,

Encyclopedia of Agricultural, Food, and Biological Engineering
DOI: 10.1081/E-EAFE 120007227
Copyright © 2003 by Marcel Dekker, Inc. All rights reserved.

Table 1 Collection of definitions of intelligent packaging

Name	Definition	Reference
CEST (Center for Exploitation of Science and Technology)	An integral component or inherent property of a pack, product or pack/product configuration which confers intelligence appropriate to the function and use of the product itself	1
Pault, Howard (Trigon Smartpak)	Has two important attributes: 1) it must be able to monitor the conditions of the package and the product and communicate the conditions to the package user and 2) it is made to control to the maximum extent the conditions of the product inside the package	2
Yam, Kit L.	Acts as an intelligent messenger or as an information link	3
Krumhar, K.C. and Karel, M.	Packaging technique containing an internal or external indicator for the active product history and quality determination	17
Brody, Aaron L.	Senses changes in the internal or surrounding environ ment and alters some of their relevant properties in response	6
Brody, Aaron L.	Measures a component and signals the result	7
Daniel, Carol David (CEST)	It is one that senses the environment and/or is able to convey information to the user	8
Dainelli, Dario (Cryovac Sealed Air Corporation)	Packaging material able to monitor the conditions to which food is packaged, thus providing information on its quality	15
de Kruijf, Nico	Concepts that monitor the condition of packed foods to give information about the quality of the packaged food during transport and storage	16

counterfeiting, and tampering. Within these applications, several other examples of intelligent packaging exist and will be discussed in the next few paragraphs. Nevertheless, all types of intelligent packaging systems are fabricated to be very user friendly and cost effective[2] and to benefit the processor, distributor, and consumer.[2] Table 2 summarizes the applications of passive, active, and intelligent packaging.

INTELLIGENT PACKAGING APPLICATIONS AND TECHNOLOGIES

Intelligent Packaging for Improving Product Quality and Product Value

Quality indicators

In this application of intelligent packaging, quality or freshness indicators are used to indicate if the product's quality is becoming impaired when it has been exposed to unfavorable conditions in storage, when being transported, on retail shelves and/or in consumers' homes.[1] These are internal or external indicators/sensors used to indicate elapsed time, temperature, humidity, time–temperature, shock abuse, as well as gas concentration changes.[1,4,8] Their main function is to show if the packaged products have deteriorated in quality[4] and to show a partial or complete history of the product.[1] This is usually indicated by a color change that may be reversible. However, intelligent indicators display a clear color change that remains permanent and is easy to read.[4] Here are some examples of indicators used to detect quality or freshness changes.

Color indicating film for the packaging of Kimchi (fermented Korean cabbage): the color indicating film is made of a resin (polypropylene), a carbon dioxide absorber (calcium hydroxide), and a chemical dye (bromocresol purple or methyl red) which indicate the freshness of Kimchi.[9]

Physical shock indicators: an example of this type of indicator is Shockwatch™ indicators produced by 3M Co. This indicator is made of a closed glass capillary tube. Inside, at one end of the tube, there is a red liquid and at the other end a dispersive material.[1] When the tube is subjected to a shock, collision, or acceleration, the contents

Table 2 Comparison of intelligent packaging applications with conventional/passive and active packaging

Packaging systems		Applications
Intelligent packaging	Sensing	Computer communicable device equipped packaging, quality indicators, time–temperature indicators, internal gas control indicators, etc.
	Responsive	Internal gas concentration controlling, moisture level controlling, etc.
Conventional/passive packaging		Jar, can, box, carton, pouch, bottle, cushioning foam, and other containers
Active packaging	Simple	Child proof, tampering proof, modified atmosphere packaging, edible films, biodegradable films, gas fluxing, and other advanced passive packaging
	Advanced	Oxygen scavenging, antimicrobial films, ethylene scavenging, moisture absorbing, carbon dioxide emitting/absorbing, ethanol releasing, etc.

of the two ends mix and the entire tube becomes red. A variety of indicators exist by simply changing the properties of the tube.[1]

Microbial growth and/or spoilage indicators: these are internal indicators that interact with components given off from microbial-contaminated foods.[4] The Lawrence Berkeley National Laboratory, among others, has developed an indicator to detect the presence of *Escherichia coli* O157 enterotoxin.[10] The indicator is incorporated into the packaging material and interacts with the toxin. The toxin binds to the film causing the film's color to change permanently (from blue to red).[10]

Temperature and time–temperature indicators

Temperature indicating labels are ideal for food products that need to be packaged in a temperature-controlled environment.[2] These external indicators are placed on the outside of the package and indicate the temperature that the package has been subjected to.[4] Temperature indicators can be classified into two categories.[2,4] The first category is the time–temperature integrators. They monitor the entire temperature history of the product and provide continuous information on the times and the temperatures that the product has been subjected to.[2] The second category is the temperature indicators. They show if the product has been exposed to increases or decreases in temperatures above or below the critical temperature.[2] In the United States, these indicators are used in chilled ready-made meat and dairy foods such as ice cream.[2,4] Newer intelligent indicators can now be kept at room temperature and will self-activate when they go above or below the critical temperature without any manual activation (unlike old temperature indicators which needed to be kept at cold temperatures and required physical activation).[2,4] However, these indicators have one disadvantage. They only indicate package surface temperatures and not the actual temperature of the product.[4] Therefore, the temperature of the environment in which the packaged product is contained in, mostly determines the color change and not necessarily the temperature of the product itself.[4]

I Point™ labels made by I Point AB Technology (Sweden) function as a complete temperature history indicator.[1,4] It is made of a capsule that consists of two parts. One part contains an enzyme solution and the other part contains a lipid substrate and indicator.[1] The indicator is activated when the portion between the two parts is broken and the contents of the two parts mix. The lipid substrate becomes hydrolyzed and the consequent change in pH causes a permanent color change.[1] This color is compared to a reference color scale on the label. In addition, the faster the temperature rises the faster the reaction proceeds.[1]

Fresh-Check™ indicator labels made by Lifelines Co. (U.S.) are based on a color change due to the polymerization of diacetylenic monomers.[1] As the label becomes exposed to temperature increases, the polymer changes from a lighter to a deeper color. The shelf life is over once the color of the polymer matches the reference color.[1] Similar to the previous indicator, the faster the temperature increases, the faster the polymer changes color. Every indicator is made to react according to the specific product and its specific shelf life.[1]

Gas concentration indicators

Internal gas level indicator labels are put into the package to control the inside atmosphere.[2,4] The labels change color at the correct gas level primarily due to enzymatic and chemical reactions.[4] They are present in the forms of color labels or tablets.[4] The advantage of this system is that it allows for a fast visual check without package destruction[2,4] and every individual package is controlled. In addition to controlling the food product, it also allows for package damage control. Furthermore, consumers will also be able to monitor product quality themselves.[2] The following are gas concentration indicators.

Oxygen (O_2) indicators interact with the O_2 entering the package via leaks. In addition, they are meant to be used with O_2 absorbers, which absorb any O_2 left inside the headspace, while the indicator verifies that no O_2 is present in the package.[4] One example is Ageless Eye™, introduced by the Mitsubishi Gas Chemical Company Inc. (Japan), which is used for the master batch package of O_2 scavenging sachets (Ageless™). When the environment is void of O_2 ($\leq 0.1\%$) the indicator is pink, and when O_2 is present ($\geq 0.5\%$) it is blue.[4] O_2 indicators and absorbers are commonly used together, especially in Japan for chilled and shelf-stable ready-made packaged foods.[4] Carbon dioxide (CO_2) indicators also exist and are used in packages in which high CO_2 levels are desired. They are used to indicate that the correct concentrations are present.[4] Although these O_2 and CO_2 indicators have many advantages, they also have some disadvantages. For example, some O_2 indicators are very sensitive while others have reversible color change reactions.[4] Reversible color change reactions in some CO_2 indicators also pose problems and it is therefore possible to obtain false readings. If micro-organisms contaminate a food product, they will cause a CO_2 increase and will cause the CO_2 levels to remain high in the package and although it is desired to have high CO_2 levels, the food product is no longer safe to consume[4] because the increase in CO_2 is due to microbial contamination and not to optimal package performance. Therefore, recent patented indicators were developed in which color changes remain permanent.[4]

Gas sensitive dyes, designed by Moonstone Co., are placed onto the label, which is then put into the package. The dyes are designed to produce different colors at different gas concentrations.[1] In the case of modified atmosphere package (containing CO_2 inside), when the package is opened and all the CO_2 has been liberated the dye changes from dark blue to a permanent yellow color. The dyes can also help indicate package integrity such as gas leaking through seal defects. A variation of this system could indicate CO_2 increases due to microbial growth.[1]

Table 3 shows examples of intelligent indicator systems used to improve product quality and product value.

Intelligent Packaging to Provide More Convenience

Convenience has always been an essential element in the food industry. New innovative ideas are introduced every day to make consumers happier by providing them with easier and quicker access to better quality foods. Convenience at the distribution level is also a benefit, which can be increased by using intelligent packaging.

Convenience indicators for quality

Temperature sensitive inks: "Smart Can" developed at the Packaging Steels Development Center (PACS) of the British Steel Tinplate (South Wales, UK) uses temperature-sensitive thermochromic inks. The ink is printed onto shrink sleeves before it is added onto the steel beverage cans.[11] As the temperature of the cans decrease, the thermosensitive ink changes from a white color to a blue color and exposes the words "READY TO SERVE."[11] This change occurs between 5 and 8°C. This allows the consumer to see when a product is cold enough to be served.[11]

Convenience indicators for distribution

Supply chain management and traceability uses automatic data capture, distribution, and location finding and can be connected to the Internet.[8] With this technology, it is easier to trace products using various data tag technologies such as bar codes and radio data tags.[8] Two examples of this technology are already in place. The first system is the Farm Advisory Services Team (FAST). This system is designed to trace fruits from the field farms to the customer. Information such as weight, time, and location is obtained from each item and placed into a database, which then gives an identification number for each fruit.[8] The second system is called the Scottish Courage (UK). This system includes beer keg track and trace programs. The technology uses radio frequency data tags to obtain information on beer kegs, such as its content, weight, and when and where it was filled.[8]

Improved convenience for preparation and cooking methods

K. Yam and his colleague suggested an intelligent cooking appliance system.[3] This system is an intelligent messenger system that carries essential information about the food product and the package through the printed extra bar code. Smart pantries, kitchens, and cleaning systems will eventually be introduced where encoded information in bar codes on packaged foods will allow automatic instructions to be read by microwave ovens and other cooking appliances.[5] Intelligent pantries and refrigerated units will eventually exist and will be able to take inventory of food products and expiry dates.[5] Smart cleaning and garbage disposal systems have a potential to decrease pollution, increase safety, and conserve water.[5] In addition, a time–temperature indicator can also be added into the bar code. The system will, therefore, have an added function of examining the quality and the safety of the food package.[3]

Table 3 Examples of indicator systems for food packages[a]

Manufacturer (trade name)	Compound detected	Mechanism of the indicator	References
Cox Technologies (Fresh Tag®)	Volatile amines	Color change of a food dye	18
Sira Technologies (Food Sentinel™ System)	Various microbial toxins	Bar code detector comprised of an immunochemical reaction; when a toxin is present the bar code is deteriorated	19
Toxin Alert (Toxin Guard™)	Various pathogens	Formation of a colored pattern when a target analyte is first bound to a specific labeled antibody	20
N/P	CO_2	Color change	21
N/P	CO_2, SO_2, NH_4	Indicator (attached to the packaging material) changes color	22
N/P	CO_2, H_2, NH_4	Color change of liquid crystal/liquid crystal + indicator	23
AVL Medical Instruments	CO_2, NH_4, amines, H_2S	Color change of CO_2, NH_4, and amine-sensitive dyes, formation of color of heavy-metal sulfides (H_2S)	24
VTT Biotechnology	H_2S	Color change of myoglobin	25
Aromascan	Volatile compounds	Volatile compound associated with spoilage affects a miniaturized electronic component possessing electrical properties	26
N/P	Diacetyl	Detection of optical changes in aromatic orthodiamine	27
Biodetect Corporation	e.g., acetic acid, lactic acid, acetaldehyde, ammonia, amines	Visually detectable color change of a pH-dye	28
N/P	Various (e.g., botulinum toxin)	Analyte causes an ion current across a gate membrane and it is electrochemically measured	29
N/P	Microbial enzymes	Color change of chromogenic substrates of the microbial enzymes	30
N/P	Micro-organisms	Colored compound diffuses due to microbial degradation of the lipid membrane	31
Lawrence Berkley National Laboratory	*E. coli* 0157 enterotoxin	Color change of polydiacetylene-based polymer	32
N/P	CO_2	Color change of bromothymol blue	33
N/P	Not specified	Color change of methylene blue or 2,6 dichlorophenol indophenol	34, 35
N/P	Ethanol	Alcohol oxidase–peroxidase–chromogenic substrate system	36
Lifelines Technology Inc. (Fresh-Check™)	Time–temperature	Color change due to chemical polymerization	1, 4
Trigon Smartpak Ltd. (Smartpak)	Time–temperature		4
3M Packaging Systems Division (MonitorMark®)	Time–temperature	Physical diffusion of a chemical solute causing a color change	4
I Point AB (I Point)	Time–temperature	A lipid substrate indicator becomes hydrolyzed, causes a pH change and consequently a permanent color change	1, 4

(*Continued*)

Table 3 Examples of indicator systems for food packages[a] (*Continued*)

Manufacturer (trade name)	Compound detected	Mechanism of the indicator	References
Mitsubishi Gas Chemical Co., Ltd. (Ageless Eye)	O_2	Color change	4
Toppan Printing Co., Ltd. (Freshilizer)	O_2	Color change	4
Toagosei Chemical Industry Co., Ltd. (Vitalon)	O_2	Color change	4
Finetec Co., Ltd. (Sanso-Cut)	O_2	Color change	4
Sealed Air Ltd. (Tufflex GS)	CO_2	Color change	4

[a] (Adapted from Ref. [10].)
N/P: trade name not provided.

Intelligent Packaging to Change Gas Permeability Properties

In this type of application of intelligent packaging, the packaging system can change its permeability properties in order to accommodate the freshness and quality of the packaged food product when the commodity is being transported, on retail shelves and/or in consumer homes. New and improved intelligent breathable films change their permeability with different fresh produce and as the temperature of the produce changes.[2] The films were developed given that respiration rates change with temperature.

Intellipac™ polymeric package materials produced by Landec Corp. (Menlo Park, California) are made by side-chain-crystallizable (SCC) polymers that are able to adequately and efficiently melt when the temperature increases[12,13] and therefore, permitting increased gas transmission through them.[12] These side chains vary in length and thus, their melting points can be changed. In addition, the reaction is reversible so that when the temperature decreases, the polymer changes its permeation properties and the gas transmission rate decreases.[13] This packaging system is ideal for high-respiratory-rate fresh and fresh-cut produce such as cut broccoli and cauliflower and has been commercialized since the 1990s.[6,12]

Intelligent Packaging to Provide Protection Against Theft, Counterfeiting, and Tampering

Although theft and counterfeiting are not too common in the food industry, it still poses a huge economical burden on society. On the other hand, tampering is more evident in these sectors and in the pharmaceutical industries. It is therefore necessary to introduce more intelligent and sophisticated devices to control and minimize these problems. These applications are different from the tamper-proof sealed foils or child-proof medicine bottles because intelligent tamper-proof packages function responsively with respect to tampering.

More intelligent tamper evidence, counterfeiting, and theft protection technologies (for example, tamper evidence labels or tapes) are being developed. These labels or tapes are invisible before tampering but change their color permanently and leave behind a "STOP" message when the package is opened.[2] Thermochromic inks, tear labels and tapes, void labels, holograms, and foil seals with some type of dye are some of the technologies being used today.[8] In addition, new devices are being introduced to improve problems of counterfeiting and theft. Holograms, special inks and dyes, laser labels, bar codes, and electronic data tags are all new methods attempting to control this problem.[8]

RECOMMENDATIONS AND CONCLUSIONS

Recommendations

In the future, advanced intelligent packaging systems will likely change properties in response to some type of environmental change. This can be achieved using material science or electronic engineering technology. Various polymers such as phase transition polymers, shape memory polymers (which change shape when the temperature changes), liquid crystal polymers, and hydrogels will be more consistently used to create intelligent packages.

Various suggestions exist as to what will become of intelligent packaging technology in the future. Some predict that packages will be able to talk using similar technology used in birthday cards.[11] This will be used for product advertising, which will benefit retailers and provide consumers with cooking instructions and expiry date information.[14] Thermochromic materials will be used to produce containers that darken in the sunlight to protect the product and return to a clear colored package under normal light making the product visible such as soft drink containers.[11] Glow in the dark packages will be introduced to describe serving temperatures using thermosensitive inks.[11] Holographics can be used on a variety of food commodity packages such as frozen food cartons, cereal boxes, and baked goods containers.[11,14] Intelligent temperature sensors coated over paperboard cartons will be able to detect when a meal is cooked by a color change or by a display saying "DONE."[14]

Conclusions

The role of packaging technologists is to produce packaging systems that provide product safety and maximize product quality[6] in a convenient and cost-effective way. The introduction of quality and freshness indicators (temperature and time–temperature indicators, gas level controls), increased convenience (quality, distribution, kitchens), smart permeability films, and tamper, theft, and counterfeiting evidence systems permit this. Furthermore, these systems allow each individual package to be controlled and monitored in a nondestructive manner. Food products will therefore be maintained in higher quality and value.

DISCLAIMER

Commercial products and manufacturer's names listed in this article are not the only ones which have been commercialized or developed in the world. Authors are not related to any of the commercial products/manufacturers that were referenced in this article.

REFERENCES

1. Summers, L. Intelligent Packaging for Quality. Soft Drinks Manag. Int. **1992**, *May*, 32–33, 36.
2. Pault, H. Brain Boxes or Simply Packed? Food Process.—UK **1995**, *64* (7), 23–24, 26.
3. Yam, K.L. Intelligent Packaging for the Future Smart Kitchen. Packaging Technol. Sci. **2000**, *13* (2), 83–85.
4. Ahvenainen, R.; Hurme, E. Active and Smart Packaging for Meeting Consumer Demands for Quality and Safety. Food Addit. Contam. **1997**, *14* (6/7), 753–763.
5. Karel, M. Tasks of Food Technology in the 21st Century. Food Technol. **2000**, *54* (6), 56–58, 60, 62, 64.
6. Brody, A.L. Smart Packaging Becomes Intellipac Registered. Food Technol. **2000**, *54* (6), 104–106.
7. Brody, A.L. What 's the Hottest Food Packaging Technology Today? Food Technol. **2001**, *55* (1), 82–84.
8. Daniel, C.D. Intelligent Packaging—An Overview. In *International Conference on Active and Intelligent Packaging*, Conference Proceedings, United Kingdom, Sept 7–8, 2000; Campden & Chorleywood Food Research Association Group: Chipping Campden GL55 6LD, UK, 1–16.
9. Hong, S.; Park, W. Development of Color Indicators for Kimchi Packaging. J. Food Sci. **1999**, *64* (2), 255–257.
10. Smolander, M. Freshness Indicators for Direct Quality Evaluation of Packaged Foods. In *International Conference on Active and Intelligent Packaging*, Conference Proceedings, United Kingdom, Sept 7–8, 2000: Campden & Chorleywood Food Research Association Group: Chipping Campden GL55 6LD, UK, 1–16.
11. Berragan, G. Innovation Accelerates. Soft Drinks Manag. Int. **1996**, *October*, 26, 29.
12. Hoofman, A.S. "Intelligent" Polymers. In *Controlled Drug Delivery Challenges and Strategies*; Park, K., Ed.; American Chemical Society: Washington, D.C., 1997; 485–497.
13. Brody, A.L. Innovations in Active Packaging—A United States Perspective. In *International Conference on Active and Intelligent Packaging*, Conference Proceedings, United Kingdom, Sept 7–8, 2000; Campden & Chorleywood Food Research Association Group: Chipping Campden GL55 6LD, UK, 1–16.
14. Anonymous. Talking Boxes, Food Temperature Sensors and Other 'Smart' Packaging Is Not Far Off. Quick Frozen Foods Int. **1994**, *36* (2), 114.
15. Dainelli, D. Regulatory Aspects of Active and Intelligent Packaging. In *International Conference on Active and Intelligent Packaging, Conference Proceedings, United Kingdom*, Sept 7–8, 2000; Campden & Chorleywood Food Research Association Group: Chipping Campden GL55 6LD, UK, 1–16.
16. de Kruijf, N. Objectives, Tasks and Results from the EC FAIR "Actipak" Research Project. In *International Conference on Active and Intelligent Packaging*, Conference Proceedings, United Kingdom, Sept 7–8, 2000; Campden & Chorleywood Food Research Association Group: Chipping Campden GL55 6LD, UK, 1–16.
17. Krumhar, K.C.; Karel, M. Visual Indicator Systems. US Patent 5,096,813, 1992.
18. Miller, D.W.; Wilkes, J.G.; Conte, E.D. Food Quality Indicator Device. PCT International Patent Application WO 99/0456, 1999.
19. Goldsmith, R.M. Detection of Contaminants in Food. US Patent 5,306,466, 1994.
20. Bodenhamer, W.T. Method and Apparatus for Selective Biological Material Detection. US Patent 6,051,388, 2000. (Toxin Alert, Inc., Canada).
21. Holte, B. An Apparatus for Indicating the Presence of CO_2 and a Method of Measuring and Indicating Bacterial Activity Within a Container or Bag. PCT International Patent Application WO 93/15402, 1993.
22. Horan, T.J. Method for Determining Bacteria Contamination in Food Package. US Patent 5,753,285, 1998.

23. Neary, M.P. Food Spoilage Indicator. US Patent 4,285,697, 1981.
24. Wolfbeis, O.S.; List, H. Method for Quality Control of Packaged Organic Substances and Packaging Material for Use with this Method. US Patent 5,407,829, 1995. (AVL Medical Instruments, AG, Schaffhausen, Switzerland).
25. Ahvenainen, R.; Pullinen, T.; Hurme, E.; Smolander, M.; Siika-aho, M. Package for Decayable Foodstuffs. PCT International Patent Application WO 98/21120, 1997. (VTT Biotechnology and Food Research, Espoo, Finland).
26. Payne, P.A.; Persaud, K.C. Condition Indicator. PCT International Patent Application WO 95/33991, 1995. (Aromascan PLC).
27. Honeybourne, C.L. Food Spoilage Detection Method. PCT International Patent Application WO 93/15403, 1993. (British Technology Group Ltd., London, UK).
28. Wallach, D.F.H.; Novikov, A. Methods and Devices for Detecting Spoilage in Food Products. PCT International Patent Application WO 98/20337, 1998. (Biodetect Corporation).
29. Case, G.D.; Worley, J.F. Thin Membrane Sensor with Biochemical Switch. PCT International Patent Application WO 93/10212, 1993.
30. DeCicco, B.T.; Keeven, J.K. Detection System for Microbial Contamination in Health-Care Products. US Patent 5,443,987, 1995.
31. Namiki, H. Food Freshness Indicators Containing Filter Paper, Lipid Membranes and Pigments. JP 08015251, 1996.
32. Quan, C.; Stevens, R. Protein Coupled Colorimetric Analyte Detectors. PCT International Patent Application WO 98/36263, 1998.
33. Mattila, T.; Tawast, J.; Ahvenainen, R. New Possibilities for Quality Control of Aseptic Packages: Microbiological Spoilage and Seal Defect Detection Using Head-Space Indicators. Lebensm.-Wiss. Technol. **1990**, *23*, 246–251.
34. Mattila, T.; Auvinen, M. Headspace Indicators Monitoring the Growth of *B. cereus* and *Cl. perfringens* in Aseptically Packed Meat Soup (Part I). Lebensm.-Wiss. Technol. **1990**, *23*, 7–13.
35. Mattila, T.; Auvinen, M. Indication of the Growth of *Cl. perfringens* in Aseptically Packed Sausage and Meat Ball Gravy by Headspace Indicators (Part II). Lebensm.-Wiss. Technol. **1990**, *23*, 14–19.
36. Cameron, A.C. Talasila. Modified Atmosphere Packaging of Fresh Fruits and Vegetables. In *IFT Annual Meeting/Book of Abstracts*, 1995; 254.

Ionizing Irradiation, Treatment of Food

Donald W. Thayer (Retired)
United States Department of Agriculture (USDA), Wyndmoor, Pennsylvania, U.S.A.

INTRODUCTION

Food irradiation is the treatment of a food with radiant energy to obtain a beneficial effect. These effects may include: 1) disinfestation of grains, fruits, and vegetables; 2) improvement of the shelf life of fruits and vegetables by inhibiting sprouting or by altering their rate of maturation and senescence; 3) improvement of shelf life of foods by the inactivation of spoilage organisms; and 4) improvement of the safety of foods by inactivating food-borne pathogens. The following forms of ionizing radiation are approved for use in the United States: the isotopic sources of gamma radiation cobalt-60 and cesium-137, accelerated electrons with energies of less than 10 MeV, and Bremsstrahlung (x-rays) with energies of less than 5 MeV (million electron volts).

PURPOSE

Food can be treated effectively with modest doses of ionizing radiation to eliminate food-borne pathogens. As these products are not sterile, the process can be described as being analogous to pasteurization. But unlike thermal pasteurization, there is very little increase in the temperature of the product and the process is sometimes called "cold pasteurization." Vegetative bacterial food-borne pathogens, such as *Campylobacter jejuni, E. coli* O157:H7, *Listeria monocytogenes, Salmonella* spp., and *Staphylococcus aureus*; protozoans, such as *Cyclospora cayetanensis* and *Toxoplasma gondii*; and nematodes, such as *Trichinella spiralis* can be killed with ionizing irradiation. In general the resistance to ionizing radiation increases in the following order: protozoans < vegetative bacteria < bacterial spores < viruses. The radiation dose required to inactivate 90% of the population of an organism is described as its *D*-value. The radiation D_{10} values for the inactivation of *E. coli* O157:H7, *L. monocytogenes, Salmonella* spp., *Staphylococcus aureus*, and the spore of *Bacillus cereus* when irradiated on beef at 5°C are: 0.30 kGy ± 0.02 kGy, 0.46 kGy ± 0.003 kGy, 0.70 kGy ± 0.04 kGy, 0.46 kGy ± 0.02 kGy, and 2.78 kGy ± 0.17 kGy, respectively. (Absorbed dose is the quantity of ionizing radiation energy absorbed per unit mass. The SI unit of absorbed dose is the gray (Gy) and is equivalent to the absorption of $1\,J\,kg^{-1}$. The gray has replaced the older term the rad, which equals an absorbed dose of $0.01\,J\,kg^{-1}$.) Inactivation of pathogens or spoilage organisms is influenced by the temperature and atmosphere during irradiation and by the type of product upon which the organisms are present. Irradiated foods have been demonstrated to be wholesome and nutritious.

REGULATION

Permitted radiation sources for the treatment of foods are described in the U.S. Code of Federal Regulations, Irradiation in the Production, Processing and Handling of Food.[1] The approved sources are the following: 1) gamma rays from sealed encapsulations of the radionuclides of cobalt-60 or cesium-137; 2) electrons generated from machine sources at energies not to exceed 10 MeV; and 3) x-rays generated from machine sources at energies not to exceed 5 MeV.

The purposes for which food irradiation may be used are defined in the Code of Federal Regulations (Table 1).[1] The USDA, Food Safety and Inspection Service regulations for the irradiation of poultry, meat, and meat products are found primarily in 9 CFR Parts 381.145[2] and 424.22.[3] The Animal and Plant Health Inspection Service regulation for disinfestation is 7CFR318.13-4f.[4] The packaging materials approved for irradiated in contact with foods and in general are pure monomer films and are described in 21 CFR 179.45.[5]

FDA regulations require that the label and labeling of retail packages or displays of foods treated with ionizing radiation include both the radura logo (see Fig. 1) and a disclosure statement "Treated with radiation" or "Treated by irradiation."[1] The radiation disclosure statement is not required to be any more prominent than the declaration of ingredients. Other Federal, State, and Local regulations may be applicable; e.g., United States food processing regulations require that the temperature of meat and poultry products must not exceed 12.7°C (55°F) during processing.

Encyclopedia of Agricultural, Food, and Biological Engineering
DOI: 10.1081/E-EAFE 120007225
Published 2003 by Marcel Dekker, Inc. All rights reserved.

Table 1 Permitted uses for ionizing radiation treatment of foods (21CFR179.26)

Use	Dose limitations
1. Control of *Trichinella spiralis* in pork carcasses or fresh cuts of pork carcasses	Minimum dose 0.3 kGy; maximum dose not to exceed 1 kGy
2. Growth and maturation inhibition of fresh foods	Not to exceed 1 kGy
3. Disinfestation of arthropod (insects, mites, and spiders) pests in food	Not to exceed 1 kGy
4. Control of food-borne pathogens in fresh or frozen, uncooked poultry products	Not to exceed 3 kGy; any packaging used shall not exclude oxygen
5. Control of *Salmonella* in fresh shell eggs	Not to exceed 3.0 kGy
6. Control of food-borne pathogens in and extension of shelf life of refrigerated or frozen, uncooked meat, ground meat, meat byproducts, or both meat and meat byproducts	Not to exceed 4.5 kGy maximum for refrigerated products; not to exceed 7.0 kGy maximum for frozen products
7. Control of microbial pathogens on seeds for sprouting	Not to exceed 8.0 kGy
8. Microbial decontamination of dry or dehydrated enzyme preparations (including immobilized enzymes)	Not to exceed 10 kGy
9. Microbial decontamination of dry or dehydrated culinary herbs, seeds, spices, and vegetable seasonings	Not to exceed 30 kGy
10. Sterilization of frozen packaged meats used solely in the National Aeronautics and Space Administration space flight programs	Minimum dose 44 kGy

GAMMA RADIATION SOURCE CHARACTERISTICS

Gamma rays are electromagnetic radiations or photons emitted by a nucleus in an excited state, which permits the nucleus to go to its lowest energy or ground state. The radionuclides Cobalt-60 (^{60}Co) (half-life 5.3 yr, 1.17 MeV and 1.33 MeV γ) and Cesium-137 (^{137}Cs) (half-life 30.2 yr, 0.66 MeV γ) are approved for the irradiation of foods. The radionuclides are doubly encapsulated within stainless steel tubes and arranged into a rectangular or cylindrical array. Most commercial cobalt-60 facilities will be charged with approximately 1 million curies (Ci) (37 petabecquerels) or greater, though there are many facilities with far less than this. A curie was originally defined as the number (3.7×10^{10}) of disintegrations per second occurring in 1 g of radium. The SI unit of radioactivity, the becquerel (Bq), replaces the older unit and is equal to one nuclear transformation per second. Cesium-137 is suitable only for dry storage designs because of its solubility in water, should the encapsulation fail.

Fig. 1 The radura symbol contains simple petals (representing the food) in a broken circle (representing the rays from the energy source).[3] It is usually printed in green on a white background. This figure is available for download at www.fsis.usda.gov/images/rdura.gif.

The shielding requirements for ^{60}Co are water 7 m and concrete 1.5 m. The tenth-value shielding thickness(es) is the distance the radiation must travel to be reduced to 10% of its initial value. The tenth-value shielding thicknesses for ^{60}Co are 20.6 cm, 6.9 cm, and 4.0 cm and for ^{137}Cs are 15.7 cm, 5.3 cm, and 2.1 cm for concrete, steel, and lead, respectively. Most commercial irradiators using ^{60}Co as the radiation source will store it under water and use a maze of concrete or steel to protect operators during transport of product into the irradiator. Four types of product transport systems are in common use, the tote box, carrier, pallet carrier, and pallet conveyor systems. Gamma sources can be used to irradiate pallet loads of food. Because the product absorbs radiation, the absorbed dose decreases in an exponential manner when exposed on only one side. So methods are used that expose at least two sides of the product to the radiation source. Even with these precautions the dose will tend to be lowest in the center of the product and the minimum dose somewhere on the top or bottom centerline of the stack. The ratio of the maximum to the minimum absorbed dose (max/min or overdose ratio), in practice, will be determined by the dose required to ensure that every part of the product receives the specified minimum dose required to obtain the desired effect and that no part of the product exceeds the defined maximum dose. The half-life of cobalt-60 dictates that approximately 12% of the source must be replenished each year to maintain throughput.

ELECTRON BEAM SYSTEMS RADIATION SOURCE CHARACTERISTICS

The maximum energy of accelerated electrons used for the irradiation of food is restricted by regulation to a maximum energy of 10 MeV. The intensity in amperes for a pulsed beam system is described as the time averaged intensity.

$$(I_{av}) = I_P \tau R$$

where I_P is the average peak amplitude of the pulse for the time period τ in sec; R is the number of pulses per second. The energy of the system is described in terms of electron volts (eV). The power of the system is the product of intensity and energy, i.e.,

$$1\,\text{MeV} \times 100\,\text{mA} = 100\,\text{kW}\,(100 \times 10^{-3} \times 1 \times 10^{6}).$$

The energy deposition is

$$D_x = K(I/A)(\mathrm{d}E/\mathrm{d}x)t$$

where D_x is the deposition of energy at depth x; I/A is the intensity density; $\mathrm{d}E/\mathrm{d}x$ is the incremental energy loss; t is the duration time of the irradiation; and K is the constant relating to the units of computation.

The actual dose will be influenced by the angle of the beam, scattering effects, gas gap between window and product, back scatter, product density, and product homogeneity. The penetration of electron beams of energy of 4 MeV, 6 MeV, 8 MeV, and 10 MeV is approximately 1.75 cm, 2.75 cm, 3.4 cm, and 4.5 cm in water, respectively. Most commercial electron beam systems use some form of conveyor belt to transport the product through the beam at a uniform rate.

BREMSSTRAHLUNG (X-RAY) RADIATION SOURCE CHARACTERISTICS

Bremsstrahlung radiation is produced when an electron beam strikes a converter such as tantalum, tungsten, or stainless steel. The photons produced cover a continuous range of energies up to the energy of the incident electrons. The common name x-ray includes both bremsstrahlung (braking) and characteristic monoenergetic radiation emitted when atomic electrons make transitions to more tightly bound states. X-ray sources used to irradiate food are restricted to maximum energies of 5 MeV or less. X-rays are generated by machine sources when accelerated electrons strike a target such as tantalum, tungsten, or stainless steel. Bremsstrahlung is broad-spectrum electromagnetic radiation emitted when an energetic electron is influenced by a strong electric field such as that in the vicinity of an atomic nucleus. It is produced when an energetic electron strikes a converter material, which is usually a high Z metal such as tantalum. The spectrum of energies that will be produced depends on the converter material and its thickness; it will contain energies up to the maximum kinetic energy of the electron. Pallets of food may be treated. The shielding requirements are related to the energy of the electron. The approximate half thickness value for 5 MeV photons are 22.1 cm, 9.6 cm, 2.88 cm, and 1.42 cm in water, concrete, iron, and lead, respectively.[6] Power requirements are high because of the low conversion efficiency (6%–12%) and the high cooling requirements, because 88%–94% of the energy is converted to heat in the target. Conversion efficiency is dependent on angle of impingement of electrons striking the target, the target metal, target thickness, and the acceleration energy of the electrons.

DOSIMETRY, AND DOSIMETRY SYSTEMS

Appropriate dosimetry systems must be used to establish the absorbed dose. A dosimetry system consists of dosimeters, measurement instruments, their reference standards, and the procedures for its use. A dosimeter is a device or material that when irradiated, exhibits a quantifiable change in some property that can be related to absorbed dose in a given material using appropriate analytical instrumentation and techniques. Appropriate references to dosimeters, dosimetry systems, and the determination of absorbed dose throughout the product are published by the American Society for Testing and Materials.[7,8,9]

PERSONNEL PROTECTION

All irradiators prevent exposure of personnel to radiation during operation of the unit. This is accomplished through the use of appropriate shielding, the construction of a maze for the transport of product in and out of the radiation field, and the installation of interlocks to prevent the accidental entry of personnel. These safety measures are mandated by regulation. Ozone is generated by ionizing radiation, and both ozone and nitrogen oxides are generated by high energy electrons in the presence of air; and adequate provisions must be made to eliminate it from the cell before personnel may enter. The rate of production of ozone by an electron beam can be calculated from the following equation: $O_3 = (600 \times G \times I \times d)/V$; where O_3 is the production rate of ozone in $cm^3 \, sec^{-1}$ per m^3 of air, G is the number of ozone molecules formed per 100 eV of radiation energy absorbed, I is the electron beam current in A, d is the path length in m of the electron beam in air, and V is the cell volume in m^3. The G values and the decomposition rates for ozone are dependent upon the temperature, the dose rate, the composition of the radiation cell walls, and the surface area of the cell. A typical G value for ozone production is 6.[10]

The references cited here can provide in-depth details about the chemistry, food science, microbiology, and physics of food irradiation.[11–16]

REFERENCES

1. Code of Federal Regulations. 21, Chapter 1 (4-1-2001 ed), Part 179—Irradiation in the Production, Processing and Handling of Food, Sec. 179.26 Ionizing Radiation for the Treatment of Food, U.S. Government Printing Office: Washington, D.C.
2. Code of Federal Regulations. 9 Chapter 3 (1-1-01 ed), Part 381—Poultry Products Inspection Regulations, Sec. 381.145 Poultry Products and Other Articles Entering or at Official Establishments; Examination and Other Requirements, U.S. Government Printing Office: Washington, D.C.
3. Code of Federal Regulations. 9 Chapter 3 (1-1-01 ed), Part 424—Preparation and Processing Operations, Sec. 424.22 Certain Other Permitted Uses, U.S. Government Printing Office: Washington, D.C.
4. Code of Federal Regulations. 7 Chapter 3 (1-1-01 ed), Part 318.13-4f—Administrative Instructions Prescribing Methods for Irradiation Treatment of Certain Fruits and Vegetables from Hawaii, U.S. Government Printing Office: Washington, D.C.
5. Code of Federal Regulations. 21, Chapter 1 (4-1-2001 ed), Part 179—Irradiation in the Production, Processing and Handling of Food, Sec. 179.45 Packaging Materials for Use During the Irradiation of Prepackaged Food, U.S. Government Printing Office: Washington, D.C.
6. NCRP. Structural Shielding Design and Evaluation for Medical Use of X Rays and Gamma Rays of Energies up to 10 MeV, Report 49; National Council on Radiation Protection and Measurements: Washington, D.C., 1976.
7. ISO/ASTM 51204 Practice for Dosimetry in Gamma Irradiation Facilities for Food Processing. *ASTM Standards on Dosimetry for Radiation Processing*; ASTM International: West Conshohocken, PA, 2002; 1–9.
8. ISO/ASTM 51608 Practice for Dosimetry in an X-Ray (Bremsstrahlung) Facility for Radiation Processing. *ASTM Standards on Dosimetry for Radiation Processing*; ASTM International: West Conshohocken, PA, 2002; 110–120.
9. ISO/ASTM 51649 Practice for Dosimetry in an Electron Beam Facility for Radiation Processing at Energies Between 300 keV and 25 MeV. *ASTM Standards on Dosimetry for Radiation Processing*; ASTM International: West Conshohocken, PA, 2002; 130–149.
10. NCRP. *Radiation Protection Design Guidelines for 0.1–100 MeV Particle Accelerator Facilities*, Report 51; National Council on Radiation Protection and Measurements: Bethesda, MD, 1977.
11. Diehl, J.F. *Safety of Irradiated Foods*, 2nd Ed.; Marcel Dekker Inc.: New York, 1995.
12. Fairand, B.P. *Radiation Sterilization for Health Care Products X-Ray, Gamma, and Electron Beam*; CRC Press: Boca Raton, FL, 2001.
13. Molins, R.A., Ed. *Food Irradiation: Principles and Applications*; Wiley-Interscience: New York, 2001.
14. Pauli, G.H. *U.S. Regulatory Requirements for Irradiating Foods*, Office of Premarket Approval, U.S. Food and Drug Administration, Center for Food Safety & Applied Nutrition, 1999; http://vm.cfsan.fda.gov/~dms/opa-rdtk.html.
15. Shultis, J.K.; Faw, R.E. *Radiation Shielding*; American Nuclear Society, Inc.: La Grange Park, IL, 2000.
16. Wilkinson, V.M.; Gould, G.W. *Food Irradiation*; Butterworth-Heinemann: Oxford, 1996.

Irrigation System Components

Robert G. Evans
United States Department of Agriculture (USDA), Sidney, Montana, U.S.A.

INTRODUCTION

The purpose of an irrigation system is to supplement the amount and timing of natural precipitation in order to fully or partially satisfy crop water requirements. Irrigation systems can serve tens of thousands of hectares with many hundreds of farms or be as small as a watering system for a home garden or even a single potted plant. The common components of an irrigation system, whether large or small, are defined not by size, but by function. Functional components of every irrigation system include diversion, delivery, distribution, and drainage subsystems.

The primary purpose of an irrigation system is to effectively and efficiently: 1) divert and deliver water from a source; 2) convey and deliver water to the cropped areas; and 3) uniformly and adequately distribute the water over the land being irrigated. It is essential that each subsystem contain elements capable of measuring and controlling flows for efficient management. Provisions to remove excess water in the delivery channels and fields, referred to as the drainage system, are equally important to the sustainable operation of any irrigation system, and therefore must also be considered a fundamental functional component.

In general, irrigation systems can be defined on at least three different levels (Table 1) and each level will have the same basic set of components regardless of scale. For example, a large irrigation "project" will have a network of reservoirs, diversion dams, and large canals to deliver water to a smaller network of canals and laterals that, in turn, distribute water to the farms. The "farm" system can also have diversion and delivery subsystems. Likewise, the individual "field" irrigation system will have analogous subsystems. Pressurized sprinkler and microirrigation systems will have pumps and mainlines to divert water to the submains and lateral lines that distribute and deliver water to the crop using "individual application (emission) devices." Surfacewater application systems divert water onto the soil surface where it is distributed and delivered to the point of infiltration.

There are numerous alternatives by which a component can be developed and each will have specific hardware and physical features depending on scale and where it is located in the system. Individual component subsystems may range from simple to extremely complex depending on location-specific water supply, engineering and cropping patterns, as well as environmental, safety, legal, social, and economic criteria.

DIVERSION SUBSYSTEM

The diversion subsystem diverts water from a water supply source into a delivery subsystem. At the project level, special structures (e.g., diversion dams, intakes, weirs, barrages) divert and control water from reservoirs, rivers or streams, and from groundwater (e.g., wells) sources.[1,2] Wide fluctuations in the temporal distributions of water supply quantity and quality during the growing season may greatly affect the system design and management. Fish passage and screening measures are usually part of surfacewater diversion systems. Recreational and fishery concerns regarding adequate in-stream flows may have large impacts on the system design as well as on diversion amounts and timing.

DELIVERY SUBSYSTEM

The delivery subsystem at the project level, often referred to as the water conveyance subsystem, delivers water from the diversion to the farm or from the farm inlet to each field. These flows are conveyed either solely by a combination of gravity flow through open channels or a closed conduit (pipe) systems, or a combination of both.

Open-channel canals and ditches are earthen water conveyance structures with the excavated spoil being used as bank material. These gravity-flow channels are usually constructed with a slight downward slope along general topographic contours. Often elevated sections are built from steel and concrete to convey water along steep hillsides or over streams. Sometimes large inverted siphons are constructed underneath natural drainage ways and streams in place of elevated sections for environmental, hydrologic, aesthetic, and economic reasons. In addition, channels are often lined with concrete or plastic membranes in areas with high seepage losses.

Encyclopedia of Agricultural, Food, and Biological Engineering
DOI: 10.1081/E-EAFE 120006926
Copyright © 2003 by Marcel Dekker, Inc. All rights reserved.

Table 1 Summary of irrigation system components at the project, farm, and field levels

Level	System component: Diversion	Delivery	Distribution	Drainage
Project	Rivers, streams, reservoirs	Canals, pipelines	Smaller open channels, pipelines	Open channels or buried pipelines
Farm	Project water, wells, ponds	Ditches, pipelines	Farm ditches and pipelines	Open channels or subsurface pipes
Field	Farm delivery, wells, ponds	Field ditches, pipelines, gated pipe, etc.	*Surface*: Furrows, borders, basins, contour ditches, etc. *Pressurized*: Sprinkler or microirrigation methods—pipelines and water emission devices	Open channels or subsurface perforated pipe, and required disposal technique

Some systems include relatively small internal regulating reservoirs to maximize management flexibility. Piped systems may be constructed from either concrete or plastic and operate under either pressurized or gravity flow conditions.

Special structures for flow measurement and water elevation control (e.g., drops, gates, chutes, flumes) are normally required throughout the delivery subsystem.[3–6] Water-level control structures for open canals can be self-regulating, manually operated or adjusted via complex telemetry systems using water level sensor and flow measurement data.[7]

Farm deliveries typically range from 0.8 $L\,sec^{-1}\,ha^{-1}$ to 1.2 $L\,sec^{-1}\,ha^{-1}$ for general irrigation on a whole farm basis. However, a farm may have several fields that are irrigated at different times and/or by different methods. Some types of irrigation systems require larger instantaneous flows (e.g., borders and level basins) for short time periods whereas others such as microirrigation systems generally require smaller, more constant deliveries over longer time periods.

A water delivery system may also include on-farm water storage facilities (e.g., small ponds) that collect water deliveries from the canal system or flows from on-farm wells to supplement project or farm water supplies. Ponds are used when water deliveries are too infrequent, sporadic, or unpredictable in both amount and timing to accumulate water for microirrigation or other needs requiring a constant water supply. These small storages may also be used to increase the amount of water available for special activities such as frost protection and crop cooling which utilize large amounts (e.g., 6 $L\,sec^{-1}\,ha^{-1}$–12 $L\,sec^{-1}\,ha^{-1}$) of water for relatively short periods (i.e., 4 $hr\,day^{-1}$–10 $hr\,day^{-1}$ several times a week) over large areas. On-farm or project storage facilities may also be used to collect runoff water from higher lands for reuse.

DISTRIBUTION SUBSYSTEM

On-farm distribution or the water application subsystem is divided into two basic categories according to the method used: gravity flow or pressurized.[8] Gravity flow systems, also called surface irrigation systems, apply water directly to the highest elevation areas of a field and then utilize gravity to redistribute the water across the field surface where it infiltrates into the soil. In pressurized systems, including various sprinkler and microirrigation systems, water is delivered through closed pipe systems pressurized by pumps or elevation differences (gravity) to the point of application. There are many variations of both surface and pressurized systems that have been developed over many years and in numerous locations for specific situations.

In selecting an irrigation system, the irrigator must consider crop and cropwater requirements, water supply, soil characteristics, expected efficiencies, field topography, field size and shape, area climate, and a number of economic factors such as labor requirements, available capital, and resource costs. Cultivation requirements or other farm practices may preclude the use of certain methods. Social and legal issues can also be important. Many of these factors are interdependent, and while one may or may not indicate a definite need for a particular irrigation method or practice (or even the need for irrigation), assessing these relationships is critical for the final selection of a specific application technology and design of the entire system.

DRAINAGE SUBSYSTEM

In almost all cases, some of the water applied to a field will not be directly usable by the crop, and this unneeded water must be removed by the drainage subsystem. At the project and farm levels, it is sometimes necessary to temporarily spill excess surface water into various types of drainage systems to protect project or farm facilities. Excessive surface water may occur due to administrative factors, equipment or structural failures, electrical outages, field runoff, runoff from other farms, and many other reasons. Some of the applied water may pass below the crop root zone (subsurface deep percolation) and need to be removed. Water may be required to leach excessive salts from the root zone. Other water may leave the field as surface runoff. Thus, it is usually necessary to develop structural systems to control and drain (remove) excess surface and subsurface water to prevent localized flooding and saturated soils that limit crop production. The surface runoff and subsurface drainage water must be collected and conveyed to a point where it can be pumped, reused or released to another irrigation system, lower lying natural drainage ways, lakes, ponds, rivers, streams, or wildlife habitat areas. There are many variations of surface and/or subsurface drainage systems and all require careful design, installation, and integration with the other components.[9]

Irrigation activities over many years have caused groundwater levels to rise in many areas around the world. In some cases, the groundwater is sufficiently near the surface to cause saturated soils in the crop root zone. Soils under this condition are referred to as waterlogged. In addition to poor aeration which limits productivity, these soils also often develop high levels of salts which can be detrimental to crop growth. Subsurface drainage networks may have to be installed to lower groundwater to levels that permit good aeration and favorable salt balances in the crop root zone. However, subsurface drainage waters may become seriously degraded due to high general salt levels, specific salts (e.g., selenium), pesticides, or fertilizers that may require special treatment measures and considerations before, or if, they are discharged into open waterways. Specially designed and managed evaporation ponds may be an acceptable alternative for drainwater disposal in some areas.

REFERENCES

1. Foy, T.; Green, H.S. Barrages and Dams on Permeable Foundations. In *Section 17, Handbook of Applied Hydraulics*, 3rd Ed.; Davis, C.V., Sorensen, K.E., Eds.; McGraw-Hill Book Company: New York, NY, 1969; 17-1–17-24.
2. Hernandez, N.M. Irrigation Structures. In *Section 34, Handbook of Applied Hydraulics*, 3rd Ed.; Davis, C.V., Sorensen, K.E., Eds.; McGraw-Hill Book Company: New York, NY, 1969; 34-1–34-53.
3. Bos, M.G., Ed. *Discharge Measurement Structures*, 3rd Ed.; International Institute for Land Reclamation and Improvement Publication 20: Wageningen, The Netherlands, 1989; 401 pp.
4. Kraatz, D.B.; Mahajan, I.K. Small Hydraulic Structures. Food and Agriculture Organization of the United Nations, Irrigation and Drainage Papers 26/1 and 26/2, Rome, Italy, 407 pp. and 293 pp., respectively.
5. Replogle, J.A.; Merriam, J.L.; Swarner, L.R.; Phelan, J.T. Farm Water Delivery Systems. In *Design and Operation of Farm Irrigation Systems*, Jensen, M.E., Ed.; Monograph No. 3; American Society of Agricultural Engineers: St. Joseph, MI, 1980; 317–343.
6. US Bureau of Reclamation, *Water Measurement Manual*, 3rd Ed.; US Government Printing Office: Washington, D.C., 1997; 396.
7. Duke, H.R.; Stetson, L.E.; Ciancaglini, N.C. Irrigation System Controls. In *Managment of Farm Irrigation Systems*, Monograph, Hoffman, G.J., Howell, T.A., Soloman, K.H., Eds.; American Society of Agricultural Engineers: St. Joseph, MI, 1990; 265–312.
8. Kruse, E.G.; Humphrys, A.S.; Popa, E.J. Farm Water Distribution Systems. In *Design and Operation of Farm Irrigation Systems*; Jensen, M.E., Ed.; Monograph No. 3; American Society of Agricultural Engineers: St. Joseph, MI, 1981; 395–446.
9. Van Schifgaarde, J., Ed. *Drainage for Agriculture*, American Society of Agronomy: Madison, WI, 1974; 700 pp.

Irrigation System Efficiency

R. E. Yoder
The University of Tennessee, Knoxville, Tennessee, U.S.A.

I

INTRODUCTION

Irrigation system efficiency is a measure of how effectively water extracted from a water supply source is used to produce the target crop to which the water is applied. When defining irrigation system efficiency, it is important to note that only applied irrigation water is considered; water from other sources, e.g., rainfall or a shallow water table, is not considered.

EFFICIENCY TERMS USED IN IRRIGATION

Numerous definitions of irrigation efficiencies are calculated and used to design and manage irrigation systems. Each efficiency is determined by partitioning water for a specific part of an irrigation system. Total *irrigation system efficiency* is the fraction of water extracted from a source, e.g., a reservoir or a well, that is beneficially used (leaching, climate control, seedbed preparation, germination of seeds, softening of soil crust for seedling emergence, and ET of other plants beneficial to the crop, e.g., wind breaks and cover crops for orchards) to produce the crop to which the water is applied. Other irrigation efficiencies are calculated based on the fraction of water that leaves a given point in an irrigation system and is delivered to a second point in the system. Efficiencies that have been used traditionally in irrigation system design and management are *conveyance* efficiency, *distribution* efficiency, and *field application* efficiency; taken together these measures of efficiency were used to define the overall irrigation system efficiency.[1] More recently standardized definitions and approaches for a comprehensive approach for determining irrigation performance measures were offered by Burt et al.[2]

The concept of efficiency of water use in irrigation is important, particularly in regions where water resources are scarce. While it is seldom possible to deliver every drop of water extracted from a source to the desired delivery point, that is a worthy goal for most irrigation systems. However, there are scenarios for which achieving complete irrigation efficiency is not desirable, e.g., when in-stream flows must be maintained for recreation or maintenance of aquatic populations, and those flows are dependent on return flows from irrigation projects.

The basic definition of any type of efficiency is the amount of output obtained from a given input:

$$\text{Efficiency} = \frac{\text{Output}}{\text{Input}} \times 100\%$$

The general relationship used to evaluate efficiencies between any two points within an irrigation system is:

$$[\text{calculated}]\ \text{Efficiency} = \frac{\text{Water Delivered to the Final Point}}{\text{Water Delivered to the Initial Point}} \times 100\%$$

To evaluate irrigation efficiency for any part of an irrigation system, the general relationship is:

$$\text{Irrigation Efficiency} = \frac{\text{Vol. Irrigation Water Beneficially Used by Target Crop}}{\text{Vol. Irrigation Water Applied–Change in Storage of Irrigation Water}} \times 100\%$$

While it is usually desirable to strive for high irrigation efficiency, that alone is not always indicative of a well-designed irrigation system or of a system that provides the best water application to a crop. It is possible that when all the water extracted from a source is beneficially used in the crop production—a perfect irrigation system efficiency—some or all of the crop may not receive an adequate amount of water for optimum growth and yield. Two concepts that must also be considered, which cannot be discussed in detail here, are irrigation distribution uniformity and irrigation adequacy. At a given point in a field, adequacy is a measure of whether the water requirement of the crop at that point in the field is satisfied; distribution uniformity is a measure of variation, or lack thereof, of the applied depth of water across an area, usually a field. If the water requirements of all plants in a field are met, the objective of the irrigation is usually met and the adequacy is high; however, a low irrigation efficiency may occur if the water application exceeds the crop production needs in parts of the field. If the water is applied nonuniformly, part of the crop in the field may be stressed and yields reduced, and the overall adequacy may be low. Conversely, if the water needs are met for every plant in a field, the irrigation efficiency will

Encyclopedia of Agricultural, Food, and Biological Engineering
DOI: 10.1081/E-EAFE 120006928
Copyright © 2003 by Marcel Dekker, Inc. All rights reserved.

likely be low because some portion of the field will receive an excess application of water. While both uniformity and adequacy are important considerations in irrigation system design and operation, because of natural variations in soil type, topography, field shape, etc. it is generally not cost-effective to design an irrigation system to achieve maximum values of both uniformity and adequacy. It is nevertheless the objective of most irrigation system designers and managers to optimize uniformity and adequacy.

Numerous factors can remove water from the irrigation system between the extraction point and the target area for delivery, thus reducing the efficiency of the system. Evaporation of water from open surfaces during delivery, transpiration by noncrop plants along ditch banks and field borders, infiltration from unlined delivery ditches, deep percolation beyond the root zone of the target crop, and surface runoff will all remove water from the system, and will reduce irrigation system efficiency. The water that is removed by any of these factors is not available for production of the target crop.

APPLICATION EFFICIENCY

The fraction of the total water volume delivered to the field border that is beneficially used to produce the target crop is usually termed the application efficiency (AE). AE is used to define irrigation system performance for a single irrigation event, and is defined as:

$$\mathrm{AE} = \frac{\text{Avg. Depth of Irrigation Water Contributing to Target}}{\text{Avg. Depth of Irrigation Water Applied}} \times 100\%$$

Irrigation water contributing to target is defined as the desired target depth of irrigation water to be applied. AE varies widely with the method of application of the irrigation water. Many surface irrigation systems, e.g., graded furrows and borders, will typically have AEs efficiencies that are lower than most sprinkler irrigation systems, while sprinkler irrigation systems will in turn typically have lower AEs efficiencies than most micro-irrigation (drip, trickle, microsprinkler, bubbler, etc.) systems. There are exceptions to these generalities, e.g., a well-designed and operated level basin surface system can have AEs higher than many sprinkler irrigation systems and may have efficiencies approaching those for micro-irrigation systems.

It is inherently impossible to achieve high field AEs in all graded surface irrigation systems. This is because the surface of the soil is used to convey the water from the upper end of the field to the lower end of the field; the surface of the soil thus serves both as a conveyance surface and an infiltration surface. This dual function of the soil surface causes excess water to be infiltrated at the upper end of the field so that an adequate volume of water for crop growth can be infiltrated at the lower end of the field. Additionally, to infiltrate an adequate volume of water at the lower end of the field requires that surface runoff will occur because of the infiltration opportunity time required for a full irrigation. Both of these factors serve to reduce greatly the field AE of graded surface irrigation systems.

Application of irrigation water through pressurized pipes to sprinkler irrigation systems eliminates many of the losses of water inherent in surface irrigation systems. However, there are some characteristics of sprinkler irrigation systems that generally prevent nearly perfect field AEs. In typical sprinkler irrigation systems, water is forced through an orifice to disperse the water (often at heights of several meters) and distribute it to the soil surface. The combination of the pressure that produces small droplets to disperse the water and the discharge of water at some height above the surface of the soil greatly increases the losses of water by evaporation and wind drift; evaporation losses are particularly high in arid and windy climates, and increase with irrigation frequency.

Because microirrigation systems typically apply water on, or very near, the soil surface, high field AEs can be achieved with these systems. While water losses by wind drift are generally eliminated, some evaporation will occur from the soil surface; in many drip and trickle irrigation systems, evaporation losses are nearly eliminated by placing the irrigation emitters beneath plastic film, which also serves as a mulch for the crop.

IRRIGATION SYSTEM EFFICIENCY

Irrigation system efficiency is the extension to the entire irrigation system of the concept of irrigation efficiency defined previously for a single field or area. From this concept, irrigation system efficiency can be defined as the ratio of the volume of irrigation water beneficially used to produce the target crop to the volume of water entering the irrigation system from the water source, with an adjustment for the volume of irrigation water that is stored in the system.

$$\text{Irrigation System Efficiency} = \frac{\text{Vol. Irrigation Water Beneficially Used by Target Crop}}{\text{Vol. Irrigation Water Entering System} - \text{Change in Storage of Irrigation Water}} \times 100\%$$

This definition is quite broad and when it is applied in practice it is understood that each component of the irrigation system must be considered. It is also essential to understand and consider beneficial and nonbeneficial uses of water in the production of crops, and to accurately partition the volume of water for each case being considered.

Water leaving a defined system area used to determine irrigation system efficiency will reduce the calculated irrigation system efficiency, but that same water volume may provide benefits to society or to the grower. The term *irrigation sagacity* was introduced by Burt et al.[2] to include consideration of uses of irrigation water for which benefits accrue to society, but that are not beneficially used to produce the target crop; these uses are defined as reasonable uses. Irrigation sagacity is used to consider all beneficial and reasonable uses of irrigation water within a defined irrigation system area, whereas, irrigation system efficiency is used to consider only the beneficial use of irrigation water to produce the target crop.

REFERENCES

1. Bos, M.G. Standards for Irrigation Efficiencies of ICID. J. Irrig. Drainage **1979**, *105* (1), 37–43.
2. Burt, C.M.; Clemmens, A.J.; Strelkof, T.S.; Solomon, K.H.; Bliesner, R.D.; Hardy, L.A.; Howell, T.A.; Eisenhauer, D.E. Irrigation Performance Measures: Efficiency and Uniformity. J. Irrig. Drainage **1997**, *123* (6), 423–442.

FURTHER READING

Clemmens, A.J.; Burt, CM. Accuracy of Irrigation Efficiency Estimates. J. Irrig. Drainage **1997**, *123* (6), 435–443.

Heermann, D.F.; Wallender, W.W.; Bos, M.G. Irrigation Efficiency and Uniformity. In *Management of Farm Irrigation Systems*; ASAE, 1990; Chap. 6, 125–149.

Merriam, J.L. Irrigation Performance Measures: Efficiency and Uniformity. J. Irrig. Drainage **1999**, *125* (2), 97–100.

Solomon, K.H.; Burt, C.M. Irrigation Sagacity: A Measure of Prudent Water Use. Irrig. Sci. **1999**, *18* (3), 135–140.

Irrigation System Operations

James E. Ayars
United States Department of Agriculture (USDA), Parlier, California, U.S.A.

INTRODUCTION

Irrigation is the deliberate application of water to soil; to increase water storage in the soil profile, to meet crop water requirements, and to increase the growth potential above that possible only with rainfall. Irrigation systems have been classified as surface, sprinkler, microirrigation, and subsurface based on their physical and operational characteristics. System selection and operation depends on topography, water quality and quantity, soil type and quality, cropping pattern, and crop value. This article will discuss operational objectives and characteristics of the major irrigation systems. Other articles in this encyclopedia discuss individual aspects of irrigation system management, design, and operation in greater detail.

OPERATIONAL OBJECTIVES

Irrigation systems should be designed and operated to meet the following objectives; to apply the amount of water required to grow a healthy crop, to apply the water in a timely manner based on crop demand, to create optimum soil water content, to apply water uniformly across the field, to maintain soil quality, and to be sustainable. Many of these objectives are subjective and depend on the production goal established by the manager, i.e., achieving maximum production might require a different management strategy than achieving maximum water use efficiency. The selected goal might affect the timing and depth of application but the need to apply water uniformly is constant regardless of the production goal.

It is necessary to characterize irrigation performance to determine whether the objectives are being met. Parameters developed for this purpose include irrigation efficiency, application efficiency, distribution uniformity (DU), and Christiansen's uniformity coefficient (CUC).

Irrigation efficiency is defined as the ratio of water consumed by the crop to water extracted for this purpose. This efficiency includes all water lost from the diversion of water from a river, stream, lake, or well until it is delivered to the soil surface. Depending on the delivery system the source of water and the irrigation system used, the irrigation efficiency can range from 28% to 84%. The average irrigation efficiency is in the range of 50%–60%.[1]

The application efficiency is the ratio of average depth of water infiltrated and stored in the root zone to the average applied expressed as a percentage. An application efficiency of one indicates that all the water infiltrated is stored in the soil and there is no deep percolation. If the amount applied is less than needed to replenish the soil water, the value might still be one but the crop is under irrigated. It is also possible that areas of the field area are under irrigated and others are fully irrigated. The application efficiency is only part of the analysis and a second parameter is also needed.

The DU is also used to characterize the quality of the irrigation management and system operation. This is the ratio of the average of the low quarter of the water infiltrated to the average water infiltrated for the entire field and is determined through extensive field sampling and statistical analysis. The low quarter is computed using all the samples in the lowest measured quartile. A DU of 100% indicates that there is perfect distribution of water across the field, a value which is never achieved.

The CUC was developed for use with sprinkler irrigation systems and was based on the water distribution patterns of sprinklers. The equation is:

$$\text{CUC} = 100(1.0 - \Sigma x/mn)$$

where x is the absolute deviation from the mean application (m); and n is the number of observations. The maximum value is 100% but this is never met and values greater than 85% are considered very good.

The volume of surface water runoff and deep percolation (water moving down past the plant root zone which is lost) resulting from irrigation depend on the system and its management. Surface runoff is a major consideration in the proper operation of an irrigation system because it carries sediment, fertilizer, salt, and pesticides. Options for dealing with runoff include collection and/or reuse on the same field, on another field, discharge to a drain, or return to a river for use by another irrigator. How runoff is handled is a major factor in characterizing the irrigation efficiency.

Deep percolation is needed to sustain irrigated agriculture but excessive loss can cause waterlogging,

Encyclopedia of Agricultural, Food, and Biological Engineering
DOI: 10.1081/E-EAFE 120006930
Published 2003 by Marcel Dekker, Inc. All rights reserved.

high water tables, long-term soil salinization, and transport of pollutants to groundwater. The objective in irrigation is to refill the soil water profile plus an additional amount equal to the leaching requirement, that amount of water needed to maintain the salt balance in the root zone. To control salt levels in the root zone, the leaching requirement is increased as the irrigation water salinity increases. Systems with poor DU have excessive losses in parts of the field compared to the remainder of the field.

The energy and labor requirements needed to operate a system have significant impacts on the initial and operational costs and the potential efficiency of a system. The energy requirement relates to the delivery and distribution of water to and onto the field. The options are gravity flow or a pressurized system. Gravity systems distribute water by overland flow on the soil surface. Pressurized systems distribute water using pipes and pumps to apply under operating pressures ranging from 70 kPa to 500 kPa. Labor requirements will vary from several persons moving siphons in a surface system and sprinkler pipe in a hand move sprinkler system to almost nil in a drip or microspray system.

Irrigation system selection and operation are determined by crop value, water quality, plant characteristics, and system efficiency. A crop such as pasture in arid areas requires irrigation but would not necessarily require a sophisticated system, while a perennial high value crop such as grapes would warrant the selection of a well designed and operated system.

Crops have different salt tolerances and when the salinity of the irrigation water approaches the crop salt tolerance care has to be exercised to avoid plant damage and salt accumulation in the soil profile. This might require that water is not sprayed on the leaves and that the frequency of irrigation be increased. Increasing the irrigation frequency keeps the soil water content high and minimizes the resident concentration of salt in the soil water.

Irrigation scheduling based on the detailed knowledge of crop water requirements is required for proper operation of an irrigation system. Irrigation scheduling requires knowledge of both how much to apply and when to apply it. The minimum depth of application used in scheduling is determined by the operational characteristics of the system. Surface systems generally cannot apply less than 50 mm depth with good uniformity. Sprinkler and drip systems can apply very small amounts effectively.

SURFACE IRRIGATION

These systems rely on the soil surface to distribute the water across the field. Water is applied at the head end of the field and moves across the field by gravity. There is a significant difference in the time that water is present at the head end of the field compared to the bottom end of the field. This time difference affects the total water infiltration across the field and thus the uniformity. The DU will be a function of the care taken in shaping the land surface, the soil type, and the rate of application of water to the field. Surface irrigation systems include; water spreading, contour ditch, contour levee, level basin, graded border, furrow and corrugation, surge, and cablegation. Surface systems generally have the lowest irrigation efficiencies compared to other major irrigation systems. These systems generally have large labor requirements and the lowest initial cost. These systems are best suited for areas having gentle slopes and "heavy" soils, i.e., clay loam, silty clay loam, and clay.

SPRINKLER IRRIGATION

Sprinkler systems eliminate many of the limitations of surface systems by distributing the water over the field surface with pipes and then spraying the water in the air to fall on the soil surface similar to rain. If the system has been properly designed the application of water will be less than the infiltration rate of the soil and there will be little or no runoff. These systems can be automated and controlled to apply the exact amount of water required to replenish the soil profile plus the leaching requirement. The uniformity of application is controlled by the overlapping of the sprinkler spray patterns. The CUC is generally used to characterize the uniformity and it should be in the range of 80%–90%. High wind will reduce the overall uniformity as will improper spacing between the sprinklers and the laterals and poor pressure uniformity. Proper operation of these systems depends on proper pressure distributions throughout the system. Sprinkler types include; hand move, solid set, center pivot, linear move, big gun, and side roll. These systems are adaptable to most soil types and topographies.

Sprinkler irrigation systems generally have higher costs than surface systems. Labor requirements are quite variable. Hand move systems have large labor costs while the self-propelled systems, center pivots, and lateral moves require more management time and less field labor.

MICROIRRIGATION

Microirrigation, the newest innovation in irrigation, has developed in the past 30 yr with the advent of PVC and plastics. These systems were initially designed for application in perennial crops and were permanent installations. Recent applications have been on annual crops with systems that are installed and removed annually and, in the Southwestern United States, Israel, and other locations where freezing is not a problem, with subsurface (buried) systems that are left in place permanently allowing farming operations to be carried out without annual removal and replacement. Microirrigation systems are designed to be point source applications of water and as a result, part of the soil surface remains dry. This reduces evaporation losses and also deep percolation losses when operated properly. This differs significantly from the other systems that completely wet the soil surface.

Microirrigation systems are pressurized and are operated at pressures and flow rates significantly lower than sprinkler systems and have lower energy costs. These systems are generally automated and operated on a much higher frequency than sprinkler systems and replace crop water use on a daily basis. The application rate is very low and infiltration is not normally a problem. Deep percolation is generally not a problem with high frequency irrigation that only replaces the previous day water use. The traditional measures for uniformity do not apply since the total surface area is not covered with water and total water application goals are different than other systems. Uniformity is characterized using the pressure distribution in the delivery lines and the manufacturing characteristics of the emitters. Microirrigation systems are labeled as surface drip, subsurface drip, microsprays, bubblers, and minisprinklers.

Microirrigation systems are suitable for use on most soils and topographies. These are generally the most costly system. After installation, the labor requirements are very low and the management requirements are high.

SUBSURFACE IRRIGATION

These irrigation systems have a very limited application. The method of irrigation is to raise the water table into the root zone quickly and then drain it back down. These systems are found on both organic and inorganic soils. Either closely spaced ditches or shallow subsurface drains are used and water is pumped into the drains increasing the water level and then the process is reversed. Typical measures used to characterize irrigation are generally not applied to these systems.

FURTHER READING

Jensen, M.E., Ed. *Design and Operation of Farm Irrigation Systems*; American Society of Agricultural Engineers: St. Joseph, MI, 1980, 829 p.

Hoffman, G., Howell, T.A., Solomon, K., Eds. *Management of Farm Irrigation System*; American Society of Agricultural Engineers: St. Joseph, MI, 1990, 1040 p.

WEB SITES

www.wcc.nrcs.usda.gov/nrcsirrig (accessed Nov 2000).

www.wateright.org/wateright2/index.html (accessed Nov 2000).

REFERENCE

1. Smith, M. Optimising Crop Production and Crop Water Management Under Reduced Water Supply. Proceedings of the 6th International Micro-Irrigation Congress (Micro2000), 2000, CD-Rom.

Irrigation Water Requirements

Edward M. Barnes
Douglas J. Hunsaker
United States Department of Agriculture (USDA), Phoenix, Arizona, U.S.A.

INTRODUCTION

Irrigation water requirements can be defined as the amount of water needed to supplement natural precipitation so crops do not experience water stress that would result in yield loss or crop-quality degradation. Additionally, irrigation water can be used to insure adequate seed germination, prevent heat stress, protect against damage from freezing, apply chemicals, suppress dust, and prevent excessive salt accumulation in the root zone. The total amount of irrigation required includes the water needed to meet the management objective and account for any inefficiencies or losses in the water delivery system. Consideration must also be given to the boundaries for which the irrigation requirements are defined. For example, irrigation requirements can be established for the growing season over an entire watershed, or defined for a particular field and single irrigation event.

WATER REQUIREMENTS

Crop-Water Requirements

Crop evapotranspiration (ET) demand is the dominant water requirement for most irrigation systems. Daily crop ET rates vary with meteorological conditions, crop type, and crop developmental stage. Management and soil environmental conditions, such as limited soil moisture, high soil salinity, and nutrient deficiencies, can also influence ET. Although a number of ET estimation techniques are available, the crop coefficient (K_c) approach has emerged as the most widely used method for determining crop-water requirements.[1] The K_c is an empirical ratio of crop ET to reference crop ET. A K_c curve, when constructed for an entire crop growing period, attempts to relate the daily water use characteristics of the specific crop to that of a specific reference crop. Simplified forms of the Penman-Monteith equation[2] that only require standard weather data for input have been developed for reference ET based on either a short reference crop (similar to a clipped grass kept at height of 0.12 m) or a tall reference crop (similar to 0.50-m tall alfalfa). These formulae are currently being recommended as standardized reference equations[2] to establish a universally consistent methodology for the crop coefficient approach. Crop coefficients are based on the assumptions that under conditions of equal evaporative demand, the ET from one crop will be related to another by a factor that is a function of growth stage, and that both crops are under no water stress. An example of the variation in crop coefficients with time for early-maturity cotton based on data collected in Arizona[3] is depicted in Fig. 1. The coefficients are typically derived using concurrent lysimeter measurements of a reference crop and the crop of interest. Crop coefficients should be applied with the specific reference crop for which they were developed.[4] Allen et al.[5] provide crop coefficients for a variety of crops and include procedures to account for regional climatic conditions. Under conditions of a wet soil surface, such as after irrigation or rainfall, additional water will be lost to the atmosphere through evaporation; therefore, the crop coefficient should be adjusted to reflect this additional moisture loss.[5,6]

Other Water Requirements

When determining irrigation requirements, it is important to consider uses of water for applications other than plant uptake. Many soils contain soluble salts that can reduce crop production when they are present in high concentrations. As these salts are left behind when water evaporates or is used by crops, many areas require additional irrigation water to leach the salts from the root zone. The actual concentration that can be tolerated before yield loss occurs varies among crops. For example, cotton and barley are fairly tolerant of high salinity levels, while corn and beans are much more sensitive.[7] The water requirements for leaching increase as the salinity of irrigation water increases. For a more detailed review of salinity management issues, see Hoffman et al.[8]

There are several other possible uses of irrigation water, including applications to insure complete germination of newly planted seeds and reduce soil surface crusting during emergence; climate modification to minimize the impact of frost or for freeze protection[9] or reduce crop heat stress; and to apply fertilizers or pesticides. All of these uses should be considered when determining the irrigation requirements.

Encyclopedia of Agricultural, Food, and Biological Engineering
DOI: 10.1081/E-EAFE 120006932
Published 2003 by Marcel Dekker, Inc. All rights reserved.

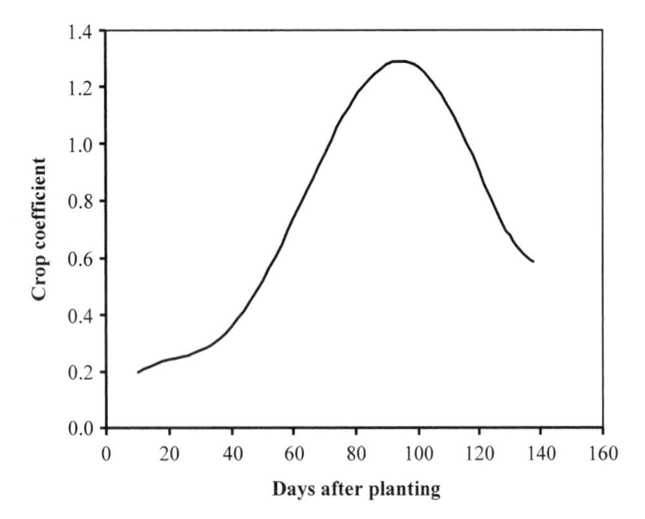

Fig. 1 Crop coefficient curve for early-maturity cotton grown in central Arizona based on data from Hunsaker.[3]

NATURAL WATER SOURCES

In arid regions, crop production relies almost entirely on irrigation water; however, in more humid regions, it is common to irrigate only at water-sensitive growth stages or when rainfall has become infrequent. Therefore, when determining irrigation water requirements for an entire season, some estimate of the expected contribution from precipitation must be considered. All of the rainfall recorded by a rain gage does not necessarily contribute to crops in the area where it falls. Especially during intense storms and in areas with variable topography, a portion of the rainfall can leave the field through runoff. Also, the soil root zone can become saturated; in which case any additional rainfall will travel below the root zone and lost to fulfill crop-water needs. The deep percolating soil water, however, does recharge the ground water table, particularly in the winter months when ET is low. Therefore, it is important that an *effective* precipitation amount be used in determining how much rainfall will lessen irrigation requirements. Seasonal rainfall estimates can be determined by using historical records for the area of interest, as described in Ref. [6].

Some irrigated lands will have relatively high water tables during part or all of the growing season that may contribute to the water requirements of the crop. The amount of contribution will depend on the depth of the rooting zone, water table depth, and hydraulic properties of the soil. Models are available for predicting the contribution of ground water to crops, and Allen et al.[5] provide some estimate of this contribution by a simple water balance equation. In areas where the water table does not contribute water to the root zone, consideration should also be given to moisture stored in the soil at the time of planting. Rainfall, or water left over from irrigation of a previous crop, could be stored in the soil and available for use by the newly planted crop. For winter grains, water from snowmelt can also contribute to crop-water requirements.

SYSTEM LOSSES AND INEFFICIENCIES

However excellent the design and operation of a given irrigation system, not all the irrigation water will be used as intended. Burt et al.[10] discuss a number of factors that can contribute to system losses. When considering irrigation requirements at a watershed level, it is important to include evaporation and seepage losses of water from reservoirs and conveyance ditches. At the field scale, losses include evaporation following emission from sprinklers and before infiltration, as well as runoff and deep percolation (i.e., water that travels below the root zone and is no longer available to the crop).

For a single irrigation event, the net irrigation requirement is the average volume of irrigation water needed to satisfy crop ET and/or other miscellaneous requirements. The total irrigation requirement is the average volume of water applied to the field to meet the net requirement plus losses due to application inefficiency. The application efficiency (AE) of the irrigation system is the average volume of water contributing to the net requirement divided by the average volume of water applied. Therefore, an estimate of AE is needed to determine the total irrigation requirement. For a given irrigation system, the AE will usually vary among irrigation events and will be affected by management practices (see previous section on irrigation system efficiency). The total irrigation requirement also depends on the ability of the irrigation system to apply water uniformly over the field, as each area of the field rarely receives exactly the same amount of irrigation. Excessive irrigation in some portions of the field and insufficient irrigation in others indicate poor uniformity. The irrigation uniformity can be characterized by the distribution uniformity (DU) term, defined as the minimum infiltrated depth in the field divided by the average depth,[11] or by the low-quarter DU term, defined as the average of the infiltrated depths occurring in quarter of the field with the smallest applications divided by the average depth of application.[6] Evaluation of the DU for the irrigation system will give an indication of the total irrigation requirement needed to insure sufficient uniformity across the field. Use of DU in conjunction with AE allows irrigators to make appropriate decisions on the irrigation requirements needed to achieve their goals. For example, depending on the value of the crop and cost of water, it may be acceptable to reduce the irrigation application amount somewhat, resulting in lower uniformity and under irrigation to a portion of the field. This will generally increase AE without a significant loss in yield. However, pronounced under irrigation to increase AE is typically uneconomical.

REFERENCES

1. Jensen, M.E.; Allen, R.G. Evolution of Practical ET Estimating Methods. In Proceedings of the 4th Decennial National Irrigation Symposium, Phoenix, AZ, Nov 14–16, 2000; Evans, R.G., Benham, B.L., Trooien, T.P., Eds.; ASAE: St. Joseph, MI, 2000; 52–65.
2. Walter, I.A.; Allen, R.G.; Elliott, R.; Jensen, M.E.; Itenfisu, D.; Mecham, B.; Howell, T.A.; Synder, R.; Brown, P.; Echings, S.; Spofford, T.; Hattendorf, M.; Cuenca, R.H.; Wright, J.L.; Martin, D. ASCE's Standardized Reference Evapotranspiration Equation. In Proceedings of the 4th Decennial National Irrigation Symposium, Phoenix, AZ, Nov 14–16, 2000; Evans, R.G., Benham, B.L., Trooien, T.P., Eds.; ASAE: St. Joseph, MI, 2000; 209–215.
3. Hunsaker, D.J. Basal Crop Coefficients and Water Use for Early Maturity Cotton. Trans. ASAE **1999**, *42* (4), 927–936.
4. Hargreaves, G.H. Defining and Using Reference Evapotranspiration. J. Irrig. Drain. Eng. **1994**, *120* (6), 1132–1139.
5. Allen, R.A.; Pereira, L.S.; Raes, D.; Smith, M. *Crop Evapotranspiration. (Guidelines for Computing Crop Water Requirements)*, FAO Irrigation and Drainage Paper No. 56; Food and Agriculture Organization of the United Nations: Rome, Italy, 1998; 300.
6. Martin, D.L.; Gilley, J.R. Irrigation Water Requirements. In *National Engineering Handbook, Part 623*; USDA, Soil Conservation Service: Washington, D.C., 1993; 284.
7. Maas, E.V.; Hoffman, G.J. Crop Salt Tolerance—Current Assessment. J. Irrig. Drain. Eng. **1977**, *103* (2), 115–134.
8. Hoffman, G.L.; Rhoades, J.L.; Letey, J.; Sheng, F. Salinity Management. In *Management of Farm Irrigation Systems*; ASAE Monograph No. 9; Hoffman, G.J., Howell, T.A., Solomon, K.H., Eds.; American Society of Agricultural Engineers: St. Joseph, MI, 1990; 665–715.
9. Rieger, M. Freeze Protection of Horticultural Crops. Hortic. Rev. **1989**, *11*, 45–109.
10. Burt, C.M.; Clemmens, A.J.; Strelkoff, T.S.; Solomon, K.H.; Bliesner, R.D.; Hardy, K.A.; Howell, T.A.; Eisenhauer, D.E. Irrigation Performance Measures—Efficiency and Uniformity. J. Irrig. Drain. Eng. **1997**, *123* (6), 423–442.
11. Clemmens, A.J.; Dedrick, A.R. Limits for Practical Level-Basin Design. J. Irrig. Drain. Eng. **1982**, *108* (2), 127–141.

Kinetics of Soil Removal During Cleaning

Kazuhiro Nakanishi
Takaharu Sakiyama
Koreyoshi Imamura
Okayama University, Okayama, Japan

INTRODUCTION

In food manufacturing processes, "soil" is defined as food residues which adhere on the surface of the equipment. Soils that adhere on the equipment wall surface provide surfaces for micro-organism growth and thus are unsanitary unless they are properly removed after processing. In small pipes, the adherence of soils on walls increases hydraulic and heat transfer resistances, which is undesirable. Thus, cleaning of equipment is conducted frequently during food manufacturing. After cleaning, the equipment is sterilized for the next batch. Cleaning is necessary to reduce the amount of disinfectants needed and improve sterilization efficiency.

In this article, first, characteristics of soil deposits are briefly summarized, particularly with respect to the difficulties of cleaning and then, cleaning kinetics, i.e., removal kinetics of the soil deposits from the surface, are reviewed. Finally, conventional and newly developing cleaning methods are introduced with their characteristics and cleaning behaviors.

CHARACTERISTICS OF SOIL DEPOSITS

Soil deposits generated during the food manufacturing processes arise from food itself and are composed of proteins, carbohydrates, fats, and inorganic substances (mineral compounds). However, the composition of the soil deposits will usually differ from the food being processed. For instance, the major components of soil deposits formed on the surface of a heat exchanger in the pasteurization (70–90°C) of skimmed milk were 44.4% protein and 45.0% mineral compounds in contrast with the composition in skimmed milk, i.e., 37.2% protein, 7.4% mineral compounds, and 52.6% lactose.[1] The major protein in the soil deposits was β-lactoglobulin (β-Lg) while the major protein in skimmed milk is casein.

Protein is typically the major component of soil deposits. At ordinary temperatures, proteins adhere to solid surfaces in a monomolecular layer.[2] Thus, the amount of soil deposits is small and can be removed by cleaning with dilute caustic solution. However, with increasing temperatures, the adsorbed amount of proteins, particularly those containing free cysteine residues and disulfide bonds such as β-Lg, bovine serum albumin, and ovalbumin, increases by forming multilayer aggregates through disulfide bond formation and/or disulfide exchange reaction.[3,4]

When the soils contain ingredients in addition to protein, the adsorbed amount is sometimes much higher than that observed when there is a single protein component. For example, the amount of milk deposit formed at the wall surface of a heat sterilizer is much higher than that obtained with β-Lg alone.

Information on the distribution of soil components in the deposited layer formed on the equipment surfaces is important not only to clarify the mechanism for soil formation but also to determine the cleaning conditions and detergents needed. In Fig. 1, three typical configu rations of a soil deposit composed from two components are depicted schematically. In Fig. 1a, the soil component B is embedded in the soil component A, and no specific interaction exists between the two components. In this case, the component B would be removed concomitantly when removing the component A. In Fig. 1b, the soil component B forms deposit near the surface, over which the component A accumulates. Here, removal of the component B would occur after that of component A. In Fig. 1c, the components A and B are interacting strongly with each other and have formed a chemical bond in the deposit. In this case, detergents that would chemically decompose the bond should be used. When coffee drinks (coffee with milk), which is a popular soft drink commercially produced in Japan, is heated to over 80°C using a heat exchanger, the soil deposits which form on the equipment wall are difficult to remove.[5] It is thought that tannin chemically interacts with milk protein (β-Lg), particularly at high temperatures. In the pasteurization of milk at 70–90°C, soil composed of proteins, fat globules, and mineral compounds deposit on the wall surface of the heat exchanger. The mineral compounds, mainly calcium phosphate, adhere on the wall surface, over which proteinaceous soil deposits accumulate.[6] Fat globules

Encyclopedia of Agricultural, Food, and Biological Engineering
DOI: 10.1081/E-EAFE 120007145
Copyright © 2003 by Marcel Dekker, Inc. All rights reserved.

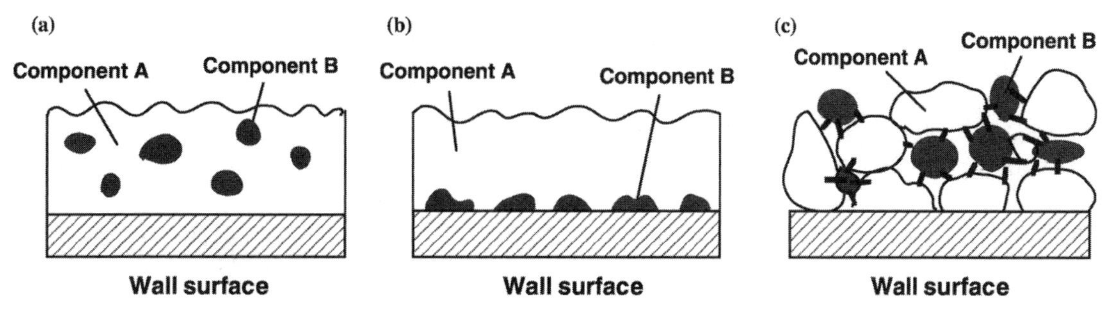

Fig. 1 Three typical deposit configurations when two compounds contribute to the surface soil.

are enclosed in the proteinaceous layer. Thus, this case corresponds to the combined case for cases (a) and (b) shown in Fig. 1. Furthermore, the distribution of the composition in the deposits sometimes changes with processing time. Such information on the change in the soil components is important to perform effective cleaning.

CLEANING STEPS

Cleaning Steps in Food Manufacturing

Cleaning of food equipment starts with rinsing to push out the remaining products of the manufacturing process. Then, cleaning with caustic or acid detergent follows to remove the soil deposits at the wall surface. The sequence of caustic and acid cleanings is determined based on the soil component distribution as described previously. For instance, in the case of cleaning of the heat exchanger used for milk pasteurization at low temperatures (70–90°C), caustic cleaning is first carried out, as the proteinaceous layer is formed near the outer surface of the deposit layer. After removal of the protein layer, the mineral compounds that exist near the wall surface of the heat exchanger should be removed using acid cleaning agents. Between caustic and acid cleanings, the system is rinsed to expel the detergent. Finally, rinsing with sterilized water is done before the start of the next batch.

Kinetics of Rinsing with Water

Fig. 2 shows a typical rinsing curve according to Plett and Loncin,[7] which indicates the relationship between the logarithms of the concentration of product (for prerinsing) or detergent (for intermediate rinsing) at the outlet of the equipment vs. time. The rinsing curve is typically divided into four major regions shown in Fig. 2. The rinsing curves for regions 1–3 reflect the residence time distribution of the product or detergent in the equipment in each region. Namely, region 1 is the period before the bulk of product or detergent starts to disappear at the outlet of the equipment. Region 2 indicates fluid mixing and dispersion. Region 3 results from dead spaces in the equipment. Region 4 occurs because solutes adhering on the surface gradually desorb. Usually, rinsing is stopped around the end of region 3. Optimizing the rinsing conditions is important to save energy and water.

The rinsing kinetics have been described mainly by the following two models. In one model, first-order kinetics is assumed for regions 2 and 3. Thus, the relationship between the logarithms of the solute concentration at the equipment outlet vs. time is linear, expressed by an equation, i.e., $\ln(C/C_0) = -k_i t$, where C_0 is the solute concentration in the bulk solution of product and k_i is the first-order rate constant for region i. The effects of various factors on the k_i values have been investigated. Plett and Loncin[7] studied the effects of the Reynolds number (Re) on the k_2 and k_3 values. With increasing Re number, the solute concentration at the boundary between regions 2 and 3 decreases and the k_2-value tends to increase. The k_3-value increases with increasing Re number when the ratio of length/diameter (L/D) of pipe in the dead-space is smaller than three. However, k_3 is nearly constant, irrespective of the Re number, when the L/D ratio is around four or higher.

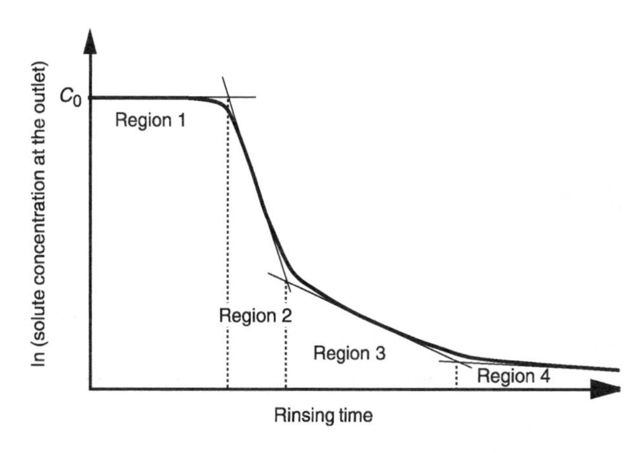

Fig. 2 A typical rinsing curve. (Drawn based on Ref. 8.)

In another model, the residence time distribution of soil in the equipment is analyzed on the basis of a fluid dispersion model. Kessler[8] showed that a dispersion model, which considered only the longitudinal dispersion of fluid, could not explain the residence time distribution in a long pipe with valves. He explained the residence time distribution by a model considering radial dispersion in addition to longitudinal dispersion.

Kinetics of Cleaning with Detergents

A typical cleaning curve is shown in Fig. 3, which indicates the relationship between the logarithms of the average soil concentration on the equipment surface vs. time. The cleaning curve does not simply reflect the residence time distribution of fluid like rinsing essentially did. The cleaning curve is divided into four major regions. Region I occurs when some fraction of the soil deposits are quickly removed. In region II, the soil deposit layer on the surface swells due to penetration of the detergent into the layer.[8,9] In particular, proteinaceous soil deposits are apt to swell when contacted with caustic detergents. Cracks then propagate in the deposits, which allow the detergent to penetrate further into the deposit and further accelerate the formation of cracking. Finally in the next region, the soil deposits layer is removed in lumps by fluid shear stress.[9] Thus, in region II, removal of soil deposits is negligible. The length of region II is not affected by fluid shear but strongly depends on the temperature. With increasing temperature, the length of region II decreases. In region III, the soil concentration on the surface decreases according to first-order kinetics, and removal rates depend on fluid shear. In region IV, the soil component adsorbed directly on the surface gradually desorbs. Usually, cleaning is finished somewhere in region IV. However, it is unclear how much surface soil should be removed to prevent growth of micro-organisms.

In the treatments described earlier, the spatial variation of the soil concentration was ignored. In an actual case, however, removal of soils from wall surface in dead spaces is much more difficult and sometimes, disassembling the equipment is necessary to remove the soils.

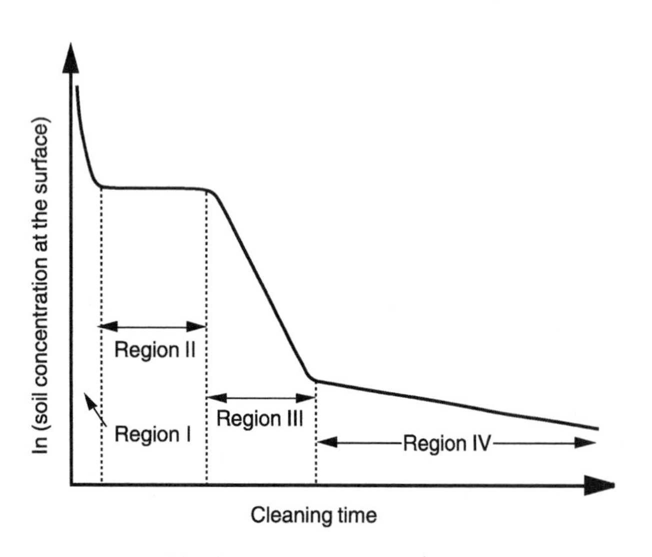

Fig. 3 A typical cleaning curve.

KINETICS IN CONVENTIONAL AND PROMISING TECHNOLOGIES

Caustic and Enzymatic Cleaning

Enzymes have been used for cleaning membranes used for concentrating liquid foods such as milk and fruit juice, but enzymes have not typically been used for cleaning equipment and pipelines. The use of enzymes may reduce the amount of detergent and energy required for cleaning because enzymes are generally effective under mild conditions. Therefore, the use of enzymes for the cleaning of the equipment is a promising technology.

Removal rates of proteinaceous soil in caustic and enzymatic cleanings are compared in Fig. 4. The ordinate shows the first-order rate constant at the early stage of desorption (referred to as initial desorption rate constant hereafter) of β-Lg evaluated at 50°C and at various pHs.[10] For caustic cleaning, the initial desorption rate constant increased prominently with pH in the range of pH above 12 (NaOH concentrations above 0.01 *N*), though it was very low at pHs below 12. For enzymatic cleaning, the initial desorption rate constant was maximal

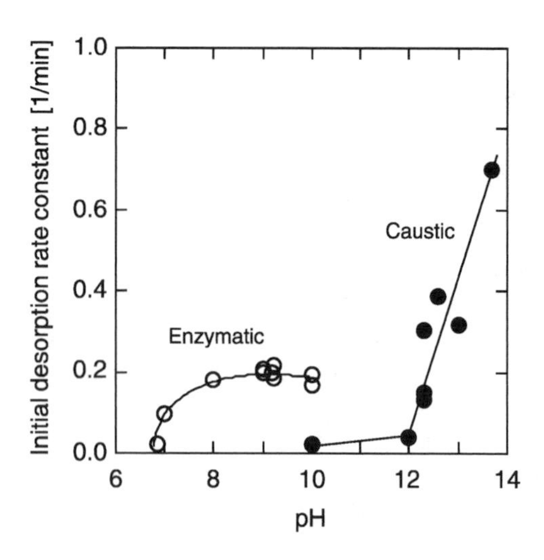

Fig. 4 Comparison of the initial desorption rate constants of β-Lg in caustic and enzymatic cleanings at 50°C.

at around the optimum pH for proteolytic reaction catalyzed by the protease (pH 9–10). The highest value of the initial desorption rate constant was comparable to the value obtained for caustic cleaning with 0.02 *N* NaOH (pH 12.3), while the final amount of residual β-Lg was less after the enzymatic cleaning at pH 9 than after the caustic cleaning with 0.02 *N* NaOH (data not shown).

Characteristics of caustic and enzymatic cleanings are summarized and compared in Table 1. Temperature dependence of the initial desorption rate constant is much stronger in caustic cleaning than in enzymatic cleaning. This suggests a difference of the rate-limiting step of desorption between caustic and enzymatic cleanings. In enzymatic cleaning, the most probable rate-determining step is the hydrolysis of the protein adsorbed on the surface. On the other hand, caustic cleaning does not degrade protein molecules into fragments. It should be noted that the initial desorption rate constant in enzymatic cleaning is also affected by the type and concentration of the enzyme used. The choice of enzyme with a high affinity and a high proteolytic activity for target components is important to decrease the amount of cleaning agents without lowering the efficiency of removal. Moreover, thermal stability of the enzyme affects the efficiency.

In connection with the structure of deposits shown in Fig. 1, proteolytic as well as caustic cleaning could be applied to cases (a) and (b) assuming the component A to be a proteinaceous layer. However, it might be difficult to apply caustic cleaning, in particular to case (c). As a matter of fact, caustic cleaning at 50°C scarcely removed the deposit from the stainless particles fouled with β-Lg and tannic acid at 90°C. Cleaning with a chlorinated agent was found to be the most effective because of its oxidative decomposing ability. In fact, chlorinated agent is said to be used in cleaning of the commercial plant for coffee drinks (coffee with milk) in Japan. Proteolytic cleaning was also effective at higher enzyme concentrations although its efficiency was somewhat lower than the cleaning with chlorinated agent.

Cleaning Methods Using Hydroxyl Radicals

Hydroxyl radical (•OH) has a high oxidation–reduction potential (2.80 V)[11] and thus in principle decomposes organic matter with an extremely high rate constant.[12] Therefore, •OH seems useful for cleaning. A cleaning system using •OH would minimize generation of waste water containing nonbiodegradable substances such as synthetic detergents and may reduce the time, labor, and energy needed for the cleaning in addition to the amount of waste water. Here, characteristics of two cleaning techniques using •OH generated by different reactions are shown.

One method to generate •OH is a well-known photolysis of hydrogen peroxide (H_2O_2) by irradiation of UV rays (< 254 nm).

$$H_2O_2 \rightarrow 2\,{\bullet}OH \qquad (1)$$

In this cleaning system, H_2O_2 solution is in contact with an equipment wall surface fouled with organic soils or it is made to flow on the surface and UV rays are irradiated over the liquid. The residual amount of soil generally decreases linearly with time, suggesting that the removal of adsorbed soil during the UV-H_2O_2 treatment follows a zero-order reaction.[13] It is noticeable that the UV-H_2O_2 cleaning does not require any lag time. The removal rate in the UV-H_2O_2 cleaning is increased by increasing the UV illuminance and by decreasing the thickness of the H_2O_2 flow, while there is an optimum H_2O_2 concentration for the removal rate.[13] It should be noted that this cleaning method is only applicable to remove soil

Table 1 Comparison of caustic and enzymatic cleanings against proteinaceous deposits

	Caustic cleaning	Enzymatic cleaning
pH	> 12	Optimum pH for the enzymatic reaction (pH 9–10)
Temperature	50–90°C	Optimum temperature for the enzymatic reaction (50°C)
Detergent concentration	High	Very low
Period for swelling of deposits	Necessary before removal occurs	Not necessary
Residual amount of deposit	Relatively high	Low
Removal rate	High	Relatively low
Dependence of removal rate on temperature	High	Low
Dependence of removal rate on flow rate	High	Low

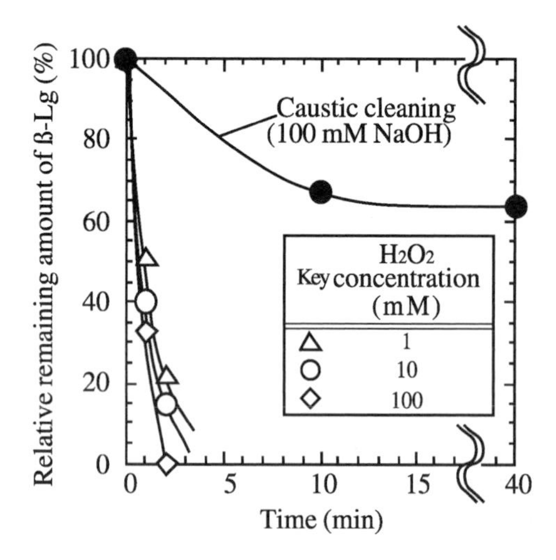

Fig. 5 Courses of the relative remaining amount of β-Lg adsorbed on the stainless steel plate during H_2O_2-electrolysis and alkaline cleaning using 100 mM NaOH at 22°C. The KCl concentration was 10 mM, and the applied potential was − 0.6 V vs. Ag/AgCl.

deposits from the surface of an apparatus such as a tank or belt conveyer on which UV rays can be irradiated.

Another method to generate •OH, based on the electrolysis of hydrogen peroxide has recently been developed by the authors.[14] In this method, •OH is generated according to Eq. 2:

$$H_2O_2 + e^- \rightarrow \bullet OH + OH^- \quad (2)$$

In this cleaning system, a metal surface fouled with organic soils is contacted with an aqueous solution containing hydrogen peroxide and supporting electrolyte. A slightly negative potential (above − 1 V vs. Ag/AgCl) is then applied to the metal. The •OHs that are generated by the electrolysis of hydrogen peroxide on the metal surface attack the organic soil deposits. Fig. 5 shows the experimental results in which the stainless steel surface fouled with β-Lg is cleaned up in a few minutes even at room temperature,[14] while most of β-Lg cannot be removed by caustic cleaning. The removal rate increases with an increase in the concentrations of H_2O_2 and supporting electrolyte and with a lowering in the applied potential. The H_2O_2-electrolysis cleaning does not require a large amount of water, energy, or chemicals and in principle could be applied to cleaning any metal surfaces fouled with organic soils.

REFERENCES

1. Jeurnink, Th.J.M. Milk Fouling in Heat Exchangers. Ph. D. thesis, Wageningen Agricultural University, The Netherlands, **1996**.
2. Wahlgren, M.; Arnebrant, T. Protein Adsorption to Solid Surfaces. Trends Biotechnol. **1991**, *9*, 201–208.
3. Arnebrant, T.; Barton, K.; Nylander, T. Adsorption of α-lactoalbumin and β-lactoglobulin on Metal Surfaces Versus Temperature. J. Colloid Interface Sci. **1996**, *119*, 383–390.
4. Itoh, H.; Nagata, A.; Toyomasu, T.; Sakiyama, T.; Nagai, T.; Saeki, T.; Nakanishi, K. Adsorption of β-Lactoglobulin onto the Surface of Stainless Steel Particles. Biosci. Biotechnol. Biochem. **1995**, *59*, 1648–1651.
5. Takahashi, T.; Nagai, T.; Sakiyama, T.; Nakanishi, K. Formation of Fouling Deposits from Several Soft Drinks on Stainless Steel Surfaces. Food Sci. Technol. Int. **1996**, *2*, 116–119.
6. Jeurnink, Th.J.M.; Brinkman, D.W. The Cleaning of Heat Exchangers and Evaporators after Processing Milk of Whey. Int. Dairy J. **1994**, *4*, 347–368.
7. Plett, E.A.; Loncin, M. Removal of Cleaning Agents and Disinfectants from Tubular and Plate Heat Exchangers. Chem.-Ing.-Tech. **1984**, *56*, 306–308.
8. Kessler, H.G. *Food Engineering and Dairy Technology*; Verlag A. Kessler, Freising, 1981.
9. Grasshof, A. Modellversuche zur Ablosoeung Testverkrusteter Milchbelaege von Erhizerplatten im Zirkulations-Reinigungsverfahren. Kiel. Milchwirtsch. Ber. **1983**, *35*, 493–502.
10. Nagata, A.; Sakiyama, T.; Itoh, H.; Toyomasu, T.; Enomoto, E.; Nagai, T.; Saeki, T.; Nakanishi, K. Comparative Study on Caustic and Enzymatic Cleanings of Stainless Steel Surface Fouled with β-Lactoglobulin. Biosci. Biotechnol. Biochem. **1995**, *59*, 2277–2281.
11. Schwarz, H.A.; Dodson, R.W. Equilibrium Between Hydroxyl Radicals and Thallium(II) and the Oxidation Potential of OH(aq). J. Phys. Chem. **1984**, *88*, 3643–3647.
12. Buxton, G.V.; Greenstock, C.L.; Helman, W.P.; Ross, A.B. Critical Review of Rate Constants for Reactions of Hydrated Electrons, Hydrogen Atoms, and Hydroxyl Radicals (•OH/•O^-) in Aqueous Solutions. J. Phys. Chem. Ref. Data **1988**, *17*, 513–886.
13. Imamura, K.; Tada, Y.; Tanaka, H.; Sakiyama, T.; Tanaka, A.; Yamada, Y.; Nakanishi, K. Cleaning of a Stainless Steel Surface Fouled with Protein Using a UV-H_2O_2 Technique. J. Chem. Eng. Jpn **2001**, *34*, 869–877.
14. Imamura, K.; Tada, Y.; Tanaka, H.; Sakiyama, T.; Nakanishi, K. Removal of Proteinaceous Soils Using Hydroxyl Radicals Generated by the Electrolysis of Hydrogen Peroxide. J. Colloid Interface Sci. **2002**, *250*, 409–414.

Laminar Flow

K. Belkacemi
J. Arul
Université Laval, Sainte-Foy, Quebec, Canada

INTRODUCTION

Many food and chemical processing operations involve fluid transport and flow of fluids through pipes, conduits, and processing equipments. Laminar flow characterizes the flow at low fluid velocities where inertial forces are relatively small compared to viscous forces. It occurs in narrow pipes, descending films, flow through porous media, flow around immersed bodies, and very viscous fluids. These flow situations are often encountered in food processing operations such as fluid transport, mixing, heating and cooling, extrusion, and various separation processes such as crystallization, falling and wiped-film evaporation, sedimentation, extraction, filtration, and drying. This entry presents a brief description of laminar flow of incompressible fluids in various geometries; flow through porous media, around immersed bodies, and in mixing; blood flow as well as heat and mass transfer in laminar flow regime. A deeper understanding of flow properties as well as other transport phenomena under laminar flow conditions is required in the design of various processes and equipments. Food materials being complex, it may not always be possible to describe fluid flows by fluid mechanical models. Dimensional analysis of specific flow problems, correlation between dimensionless groups, and extraction of analogies between momentum, heat and mass transfer phenomena can be useful.

BACKGROUND

The flow of fluids, a major aspect of study in fluid mechanics, is encountered in many areas of engineering practice, including food engineering. There are many situations where the details of fluid flow are important. The knowledge of how the velocity varies over the cross-section of a pipe and how the pressure and shear stress influence fluid flow are important in many process situations.

Gases and liquids exhibit viscosity, unlike solids while they lack elasticity that characterizes solids. While solids resist deformation under shear forces, fluids do not; but they respond to shear differently that their rate of strain or distortion depends on the force, i.e., the resistance of the fluids depends on the rate of strain. This resistance of fluids to shear rate is called *viscosity*. Thus, viscosity is a material property of a fluid, which leads to viscous forces that resist the movement of adjacent fluid layers. These forces arise from intermolecular attraction existing between molecules in a fluid.

When fluids flow through a pipe or open channel, two types of flow or fluid motion can occur, depending on their velocity. At low velocities, the flow is characterized by smooth flow in which fluid layers do not mix. This type of flow is called *laminar flow*. At higher velocities, eddies and swirls develop that tend to mix layers of fluids together, a characteristic of *turbulent flow*. Osborne Reynolds (1842–1912) was the first to observe that this instability in flow is determined by the relative importance of kinetic or inertial forces to the viscous forces in the fluid medium. The inertial forces tend to maintain the flow; whereas viscous forces tend to retard it and dampen incipient eddies. High inertial forces favor turbulence, and high viscous forces prevent turbulence by eliminating discontinuities in the flow. Thus, the laminar flow occurs when the inertial forces are relatively smaller compared with viscous forces. Laminar flow occurs in narrow pipes, descending films, flow through porous media, and very viscous fluids. Generally, laminar flow problems are solved by constructing relations between flow velocity, density, and pressure distributions from equations of momentum transfer or motion, the *Navier–Stokes* equations:[1,2]

$$\rho \frac{\partial \vec{\mathbf{U}}}{\partial t} = -\nabla p + \rho g + \mu \nabla^2 \vec{\mathbf{U}} \tag{1}$$

where ρ is the fluid density, $\vec{\mathbf{U}}$ the fluid velocity vector, p the fluid pressure, g the gravity force per unit mass, μ the fluid viscosity, and t the time. The equation of motion is truly the equation for conservation of momentum:

$$\begin{aligned} &(\textit{rate of momentum in}) \\ &\quad - (\textit{rate of momentum out}) \\ &\quad + (\textit{forces acting on the system}) \\ &\quad = (\textit{rate of momentum accumulation}) \end{aligned} \tag{2}$$

Fluids can be grouped into two classes according to their

Encyclopedia of Agricultural, Food, and Biological Engineering
DOI: 10.1081/E-EAFE 120006955
Copyright © 2003 by Marcel Dekker, Inc. All rights reserved.

response to: 1) externally applied pressure and 2) shear stress. In the first class of fluids, two types can be distinguished, i.e., those whose volume is dependent on the applied external pressure, the compressible fluids, and those which are not influenced by the applied pressure, the incompressible fluids. Based on their flow behavior to shearing forces, incompressible fluids are grouped into Newtonian and non-Newtonian fluids (Fig. 1). All Newtonian fluids are time independent. In contrast, non-Newtonian fluids are subdivided into: 1) time-independent fluids including Bingham bodies, Herschel-Bulkley bodies, pseudoplastic and dilatant fluids and 2) time-dependent fluids exhibiting thixotropy and rheopexy or negative thixotropy behavior. Details on fluid classification can be found elsewhere.[3,4]

Fluids encountered in food technology are both Newtonian and non-Newtonian. In addition, many food materials, called visco-elastic materials, exhibit both fluid and solid-like properties. Often the deformation and flow of these materials are of importance in process design. Most low-molecular weight liquids and solutions, all gases, and certain homogeneous dispersions display Newtonian behavior. Pseudoplastic and structured fluids, such as biopolymer solutions, and melts exhibit shear thinning. Dilatancy describes shear-thickening behavior of fluids, which occurs in concentrated suspensions of irregular particles in liquids. Some fluid foods exhibit plastic or Bingham behavior such as ketchup, vegetable fats, and margarine; some exhibit time-dependent thixotropic behavior such as mayonnaise and apple sauce. Some foods such as melted cheese and bread dough are visco-elastic. What follows is a description of laminar flow characteristics of incompressible fluids in different geometries relevant to applications in food and biological engineering. Heat and mass transfer under laminar flow conditions will also be presented. Although laminar flow is encountered in various areas of biology, this aspect has been intentionally excluded here, except for blood circulation.

REYNOLDS NUMBER

Dimensional Analysis

Dimensional analysis or *mathematics of the dimensions and quantities* is a very helpful tool in the analysis of fluid mechanics. It is built on Fourier's principle of dimensional homogeneity, which states that an equation expressing a physical relationship between quantities must have dimensional homogeneity so that each term in an equation have the same units. Thus, the ratio of one term to another in the equation is dimensionless. Furthermore, dimensional analysis provides a means of ascertaining the forms of physical equations from the knowledge of relevant variables and their dimensions. Although dimensional analysis cannot be used to produce analytical solutions to physical problems, it is a powerful tool in formulating them. For details on dimensional analysis, the reader is referred to Ipsen.[5] From the knowledge of the physical meaning of each term in the equation, we are able to interpret each of the dimensionless parameters or numbers

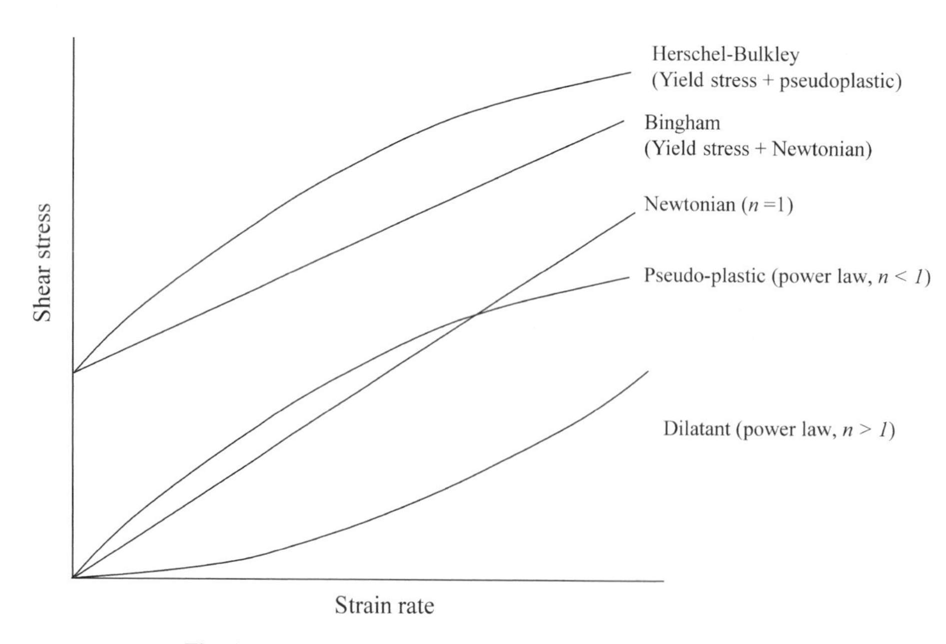

Fig. 1 Flow behavior of inelastic time-independent materials.

formed. Dimensionless numbers such as *Reynolds number*, *Froude number*, and *Euler number* are useful in correlating or predicting transport phenomena in laminar and turbulent flows. Reynolds number is the ratio of inertial or kinetic forces to viscous or friction forces. At very high Reynolds number, the inertial forces dominate, but at low values, the friction forces dominate. Its usefulness as a criterion to distinguish between laminar flow and turbulent flow has been demonstrated. Froude number is the ratio of inertial forces to gravity force and is useful in characterizing the fluid velocity. Euler number is the ratio of pressure force to inertial forces and characterizes pressure difference, Δp. For fluid flow in pipes, $\Delta p/L$ represents the pressure or head loss per unit of pipe length, and relates to head loss due to friction. The definition and physical significance of these numbers can be found elsewhere.[3,6]

Dimensional analysis in fluid mechanics permits the establishment of useful correlations of experimental data for a given situation. Experimental data obtained from small model systems can be used to scale up to large prototypes, provided dynamic similarity is maintained between the two systems, in addition to geometric similarity. Dynamic similarity is maintained when Reynolds, Froude, or Euler numbers are equal between the two systems.

Reynolds Number Criterion for Laminar Flow

For all fluids, the nature of the flow is governed by the relative importance of the viscous and the inertial forces. For Newtonian fluids, the balance between these forces can be represented by the value of Reynolds number, *Re*, as:

$$Re = \frac{\rho D U}{\mu} \tag{3}$$

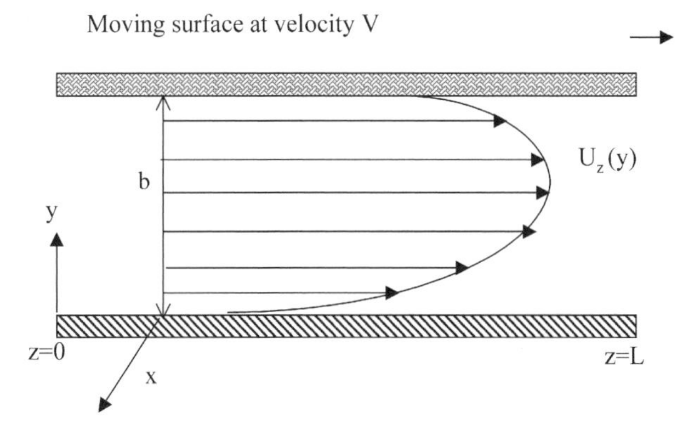

Fig. 2 Couette flow between parallel plates.

where D is characteristic dimension, U the fluid velocity, ρ the fluid density, and μ the fluid viscosity.

From experimental measurements on fluid flow in pipes, it has been found that the flow remains stable, laminar or streamline for values of *Re* up to about 2100. For values above 4000, the flow becomes turbulent. Between 2100 and about 4000, the flow may be either laminar or turbulent, depending on the condition at the entrance region of the tube. In fact, under these flow conditions the fluid is in a transition flow regime. However, laminar flow can persist up to *Re* of several thousands under very special conditions of well-rounded tube entrance.

For time-independent non-Newtonian fluids, the critical value of *Re* depends on the type and the degree of non-Newtonian behavior. Since the viscosity of non-Newtonian fluids is not constant, the Reynolds number as defined in Eq. 3 cannot be used. Several arbitrary definitions of Reynolds number for these fluids have been proposed. A widely used definition of Reynolds number for power-law fluids, Re, n is expressed as:[7]

$$Re, n = \frac{D^n \rho U^{2-n}}{K' \, 8^{n-1}} \tag{4}$$

with $K' = K[(1+3n)/4n]^n$, where K' is consistency index and n is flow or power-law index. For Newtonian fluids, $K' = \mu$ and $n = 1$.

The critical *Re* value increases with decreasing value of the power-law index, n, of the non-Newtonian fluid reaching a maximum of 2400 at $n = 0.4$ and dropping to 1600 at $n = 0.1$. Despite the complex dependence of the critical Reynolds number on the flow behavior of the non-Newtonian fluids, it is acceptable to assume that the laminar flow conditions cease to prevail at Reynolds number above 2000. For the purpose of process calculations, the widely accepted value of 2100 can also be used for time-independent non-Newtonian fluids as a good approximation.

LAMINAR FLOW OF INCOMPRESSIBLE FLUIDS

Couette Flow

The simplest case of Newtonian fluid flow is *Couette flow*, the low-velocity steady motion of a viscous fluid between two infinite plates moving parallel to each other (Fig. 2). The movement of the upper plate at velocity V drives the flow of fluid and induces a pressure gradient $\Delta p/L$ in the direction of flow. The direction of this flow depends on the magnitude and sign of Δp. In dimensionless form, the velocity of the fluid U_z/V is:

$$\frac{U_z}{V} = \tilde{K}\frac{y}{b}\left(1 - \frac{y}{b}\right) + \frac{y}{b} \tag{5}$$

with

$$\tilde{K} = \frac{b^2 \Delta p}{2\mu VL} \tag{6}$$

where $\tilde{K}$ is a constant, L the characteristic length, and b the width between the parallel plates. Depending on the magnitude of $\tilde{K}$, the velocity profile takes several forms as seen in Fig. 3.

When $\tilde{K} = 0$, the velocity profile is linear and corresponds to simple Couette flow where the velocity U_z is given by:

$$U_z = V\frac{y}{b} \tag{7}$$

The volumetric flow rate Q is given by:

$$Q = \int_0^b U_z w \, dy = \frac{b^3 w \Delta p}{12\mu L} + \frac{1}{2} bwV \tag{8}$$

When $V = 0$ in the Couette flow, the fluid flow becomes a pressure-driven *Hagen-Poiseuille flow*. Under such conditions, the velocity profile is parabolic (Fig. 4) with a symmetry axis at $y = b/2$ and the flow velocity is given by Eq. 9:

$$U_z = \frac{b^2 \Delta p}{8\mu L}\left[1 - 4\left(\frac{y - b/2}{b}\right)^2\right] \tag{9}$$

The volumetric flow rate is proportional to $\Delta p/L$ as given by:

$$Q = U_{\text{av}} bw = \frac{b^3 w \Delta p}{12\mu L} \tag{10}$$

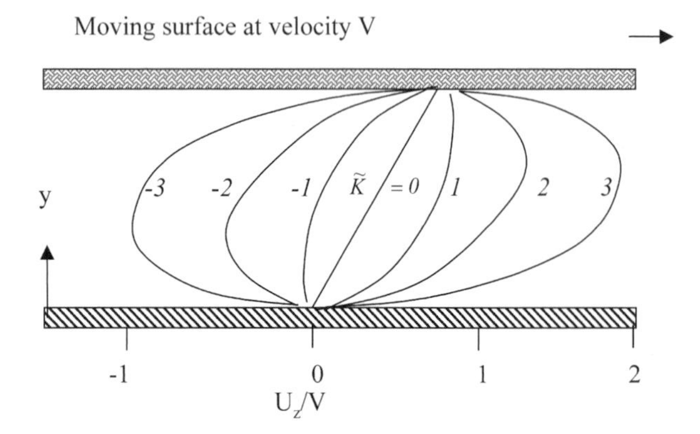

Fig. 3 Dimensionless velocity profile for generalized Couette flow.

Laminar Flow in Pipes

For laminar flow of Newtonian fluids, shearing stress exists at the boundary layer of flowing fluid within the tubes. *Poiseuille's law* describes laminar flow through a tube, and the resistance R to laminar flow of an incompressible fluid having viscosity μ through a horizontal tube of uniform diameter D and length L is given by:

$$R = \frac{128\mu L}{\pi D^4} \tag{11}$$

and the flow rate Q is given by:

$$Q = \frac{\Delta p}{R} \tag{12}$$

or

$$\frac{Q}{D^3} = \frac{\pi}{128\mu}\frac{\Delta p D}{L} \tag{13}$$

For laminar flow in pipes, the velocity profile is parabolic as in Couette flow.

The pressure loss or flow resistance is proportional to fluid velocity (Eq. 9) and it results mainly due to molecular friction. On the other hand, the pressure loss is practically proportional to the square of velocity for turbulent flow. In the latter, molecular friction contributes to pressure loss to a small degree, but turbulence plays a major part because the fluid elements at higher velocity continually enter lower velocity zones and they lose kinetic energy (proportional to square of velocity) by collision. Thus, flow resistance arises from the friction between fluid layers in laminar flow while in turbulent flow the shear stress at the wall is dominant. The (Fanning) *friction factor*, f can be defined as the ratio of shear stress to the velocity pressure; the former hinders the flow and the latter promotes it. (Some sources use 1 as a numerical constant for the ratio of shear stress to velocity pressure but others use 4. The constant 4 is associated with Fanning friction factor. Then $f = 4f'$, where f' is simply called the friction factor.)

$$f = \frac{\Delta p D}{L} / \frac{\rho U^2}{2} = \frac{2\Delta p D}{L\rho U^2} \tag{14}$$

where f is a dimensionless number, called the friction factor.

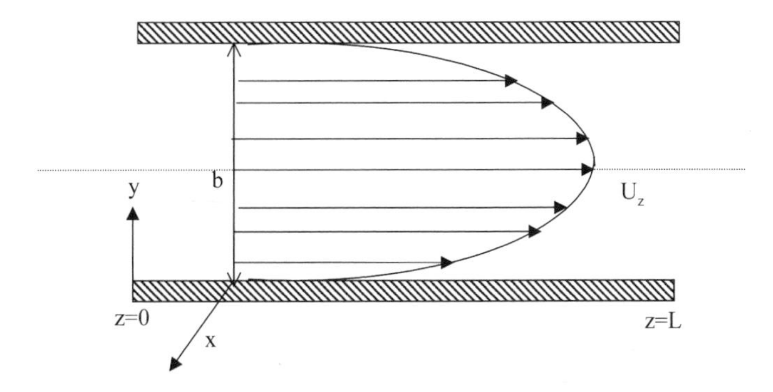

Fig. 4 Parabolic velocity profile of Newtonian fluids in laminar flow in a circular pipe.

Friction factor is a function of *Re* and under laminar flow regime:

$$f = \frac{\text{Constant}}{Re} = \frac{64}{Re} \tag{15}$$

and

$$f' = \frac{16}{Re} \tag{16}$$

The constant takes different values depending on the shape of the conduits. Setting $f = 64/Re$ in Eq. 14, one obtains Poiseuille's law for tube flow:

$$\frac{\Delta p}{L} = \frac{128\mu U}{\pi D} \tag{17}$$

Thus the pressure loss in laminar flow is proportional to fluid viscosity and velocity. In turbulent flow, the pressure drop is nearly proportional to the fluid density and to the square of the velocity, and the flow resistance is not linear:

$$\frac{\Delta p}{L} = f\frac{\rho U^2}{2D} \tag{18}$$

Fig. 5 depicts the relationship between friction factor *f* and *Re*. In laminar flow, *f* is simply $64/Re$ but it is little more complicated in turbulent flow. The transition to turbulence involves an abrupt rise in the flow resistance. As turbulence increases, the thickness of the laminar boundary layer vanishes, viscosity has less and less impact on Δp, and inertial forces dominate. For very high *Re*, friction factor becomes constant and Δp is practically proportional to U^2. Since at high *Re*, the fluid velocity dominates, the shear stress at the wall becomes important with increasing surface roughness of the pipe, ξ. The rougher the pipe and smaller the pipe diameter *D*,

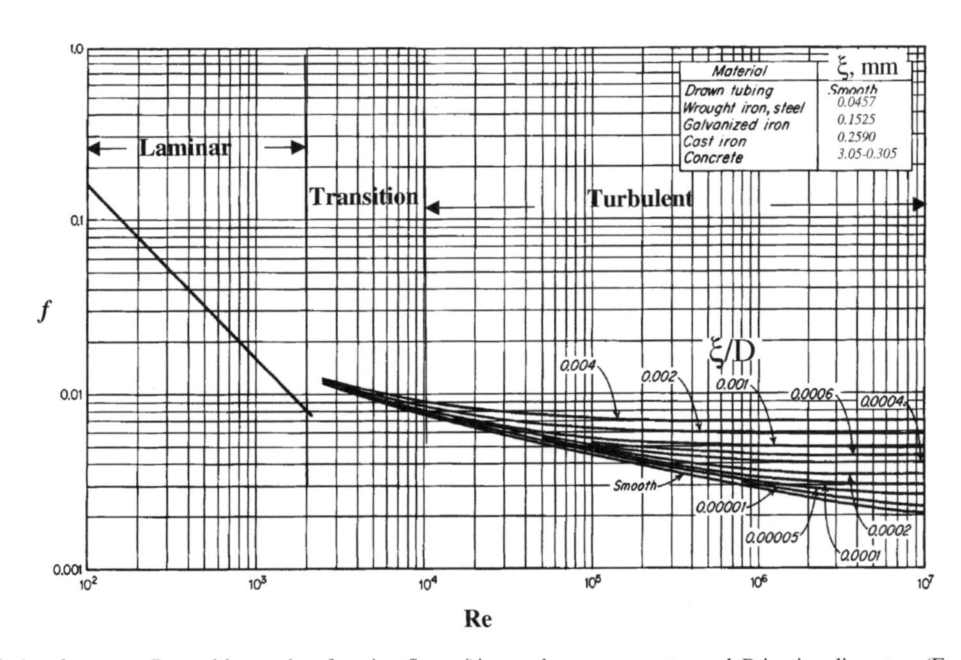

Fig. 5 Friction factor vs. Reynolds number for pipe flow. ξ is roughness parameter and *D* is pipe diameter. (From Ref. [8].)

the earlier the transition from a diminishing friction factor to a constant value, i.e., friction factor increases with ξ/D.

A flow situation can also be characterized by the friction factor, instead of Reynolds number. For cylindrical tubes, critical $Re = 2100$ and this corresponds to $f = 0.03$. Substitution for $f > 0.03$ in Eq. 18 gives:

$$\frac{\rho Q^2 L}{\Delta p D^5} < 48 \tag{19}$$

Since the viscosity term is absent in Eq. 19, this criterion can be applied to non-Newtonian fluids.

In laminar flow of time-independent non-Newtonians or power-law fluids, the flow can take place in several geometries such as tubes, annuli, falling films, and parallel plates. Fig. 6 shows the typical velocity and shear-stress profiles for a given time-independent fluid flowing in a tube.

The equation of motion in the direction z for a power-law fluid with the nonzero component of shear stress τ_{rz} for flow in cylinder with radius r gives:

$$\frac{1}{r}\frac{\mathrm{d}}{\mathrm{d}z}(r\tau_{rz}) = \frac{\mathrm{d}p}{\mathrm{d}z} \tag{20}$$

and the constitutive equation for a power-law fluid is:

$$\tau_{rz} = K\left(-\frac{\mathrm{d}U_z}{\mathrm{d}r}\right)^n \tag{21}$$

Applying the limiting condition that $\tau_{rz} = 0$ at $r = 0$, and rearranging Eqs. 20 and 21, the expression of the velocity of the fluid flow in the direction z is given by:

$$U_z = \left(\frac{\Delta p}{2KL}\right)^{1/n}\frac{nR^{(n+1)/n}}{n+1}\left[1 - \left(\frac{r}{R}\right)^{(n+1)/n}\right] \tag{22}$$

The velocity profile varies with the flow index n of the fluid, and it is shown for several fluids in Fig. 7. For pseudoplastic fluids ($n < 1$), the profiles are flatter than Newtonian fluids ($n = 1$). When n approaches zero, the fluid flow tends to plug flow. For dilatant fluids ($n > 1$), the profiles are less flat. When n becomes large, a triangular profile is approached.

The expression of friction factor f for power-law fluids is identical to the friction factor of Newtonian fluids. The friction factor f for non-Newtonian fluids varies with the flow index n.

$$f = \frac{64}{Re, n} \tag{23}$$

More information on the velocity distributions and friction factors for Newtonian and other non-Newtonian (Bingham and pseudoplastic) fluids in circular tubes are found in classical treatises.[4,9]

The flow resistance of various fittings in pipelines such as expansions, valves, orifices, and bends can be expressed by a general equation:

$$\Delta p = K\frac{\rho U^2}{2} \tag{24}$$

where K is the resistance coefficient dependent on Re. Information on K can be found in Cheremisinoff[9] and Tilton.[7]

Laminar Flow Between Parallel Planes

The consideration of fluid flow between two parallel planes constitutes a simplified approach of fluids flowing in ducts of various cross-sections. Fig. 8 illustrates the flow between two plates of length L separated by a distance b.

For Newtonian fluids, the velocity distribution is given by:

$$U_z(y) = \frac{\tau_w b}{4\mu}\left[1 - \left(\frac{2y}{b}\right)^2\right] \tag{25}$$

where τ_w is the shear stress at the wall. The velocity distribution in this geometry is close to that of Newtonian fluid in tubes. For non-Newtonian fluids the expressions

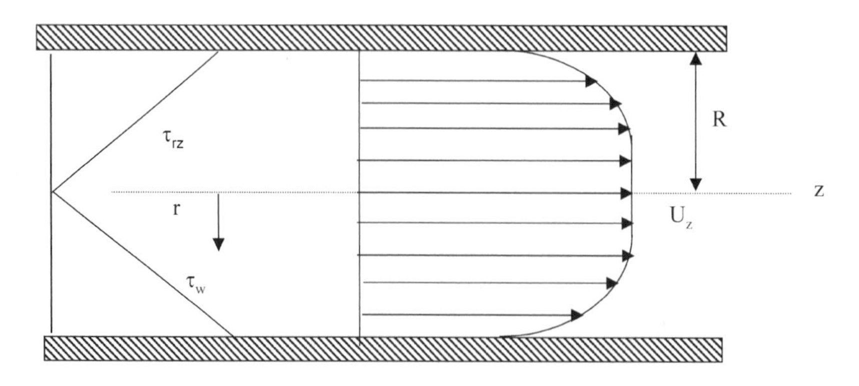

Fig. 6 Velocity and shear-stress profiles for a non-Newtonian fluid in laminar regime.

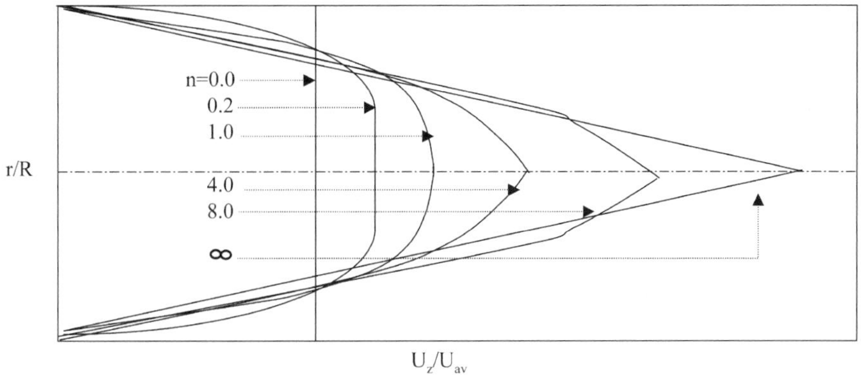

Fig. 7 Velocity profiles for flow tube of different power-law fluids.

are complex and vary with the flow index n. For instance, the expression of velocity distribution for a power-law fluid has identical expression as Eq. 22. The determination of friction factor f is analogous to the determination of f in the case of fluid flow in tubes. For calculation of Re, the hydraulic diameter D_H is used, instead of diameter D.[7,10,11] For tubes, D_H is 4 times the so-called hydraulic radius (area of cross-section A divided by the wetted perimeter P_w)

$$D_H = \frac{4A}{P_w} \tag{26}$$

and f is given by:

$$f = \frac{24}{Re} \tag{27}$$

Laminar Flow in a Concentric Annulus

The flow of Newtonian and non-Newtonian fluids through concentric annuli is an idealization of flows encountered in several industrially important processes, such as extrusion, and in heat exchangers. The flow pattern of this type of flow is shown in Fig. 9.

The evaluation of the velocity distribution in this case for a fluid flowing through an annulus of outer radius R and inner radius R_i, where $R_i/R < 1$, is a complex task. The mathematical description is more complex than that for flow in a simple pipe or between two parallel planes, even for the simple case of Newtonian fluids. The reader is referred to Chhabra and Richardson[4] for more details. For the simple case of Newtonian fluid, the friction factor for laminar flow in annular cross-section is given by Eq. 15, where the friction factor is corrected by a factor Φ, a function of R_i/R. The value of Φ is a constant whose value is 1.0 for circular sections and 1.5 for parallel planes. The hydraulic diameter D_H is given by $(D_o - D_i)$, where D_o and D_i are the outer and inner diameters, respectively. Equations for estimating friction factor for other cross-sections in laminar regime can be found elsewhere.[12]

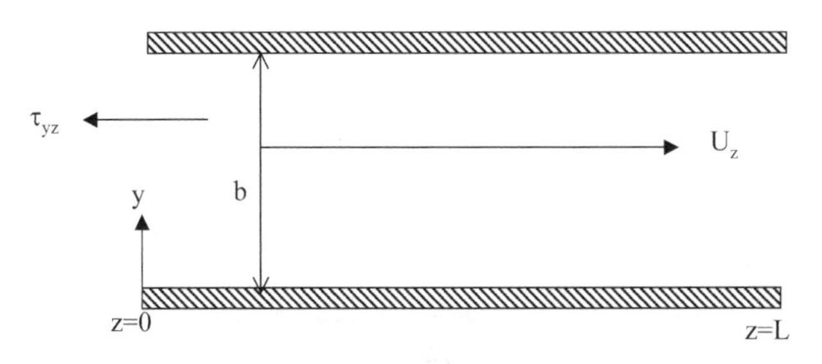

Fig. 8 Laminar flow between two parallel planes.

Laminar Flow in Other Cross-Sections

A valid approximation is that for $Re < 2100$, the flow is laminar and stable. Re is expressed in a general form:

$$Re = \frac{\rho D_H U}{\mu} \tag{28}$$

where D_H is the hydraulic diameter. This diameter for film on vertical wall is given by 4δ, where δ is the film thickness. For film on vertical tube wall, with the tube diameter D, the hydraulic diameter is given by $4[\delta - (\delta^2/D)]$. Thus, Reynolds number for a descending film, Re_{film}, on a vertical wall is given by:

$$Re_{film} = \frac{\rho 4\delta U}{\mu} \tag{29}$$

Flow Through Porous Media

The flow of fluids through porous particulate materials is of interest to agricultural, chemical, and water engineers. This flow takes place through interstices between particulates. At low Reynolds numbers, the pressure loss is approximately proportional to the fluid velocity and

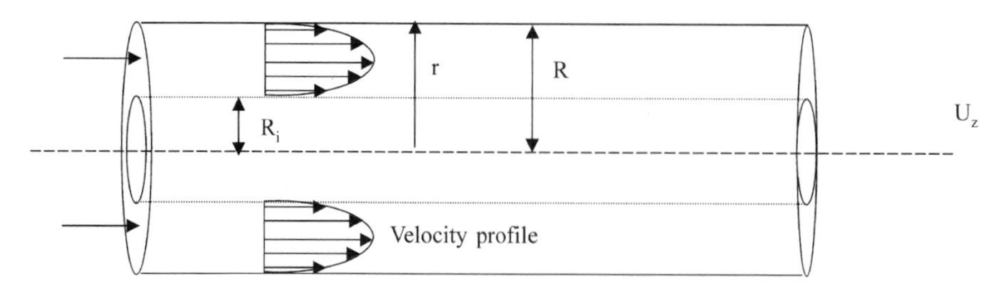

Fig. 9 Laminar flow in the concentric annulus.

viscosity. For high Reynolds numbers, as in the case of pressure filtration, losses in kinetic energy occur; and the pressure loss will be nearly proportional to the square of fluid velocity. Since the interstitial channels may have different diameters, laminar flow in narrow channels with turbulent flow in wider channels can occur together.

For flow through porous beds, a modified Reynolds number Re' is used:

$$Re' = \frac{D_{\mathrm{H}} \rho \overline{U}}{\mu} \tag{30}$$

where $\overline{U}$ is the average velocity, related to superficial velocity U, which would exist with no porous material present,

$$\overline{U} = \frac{U}{\varepsilon} \tag{31}$$

with ε, the porosity of the bed. The hydraulic diameter D_{H} is given by:

$$D_{\mathrm{H}} = \frac{4\varepsilon}{S} \tag{32}$$

where S is the specific surface of the porous material. For laminar flow, $Re' < 1$. Alternately, the flow through porous bed can be analyzed using Darcy's law, which is a low Reynolds number relationship.[13] Darcy's law states that volume flow $\tilde{Q}$ is directly proportional to the applied pressure and cross-sectional area of the bed A and inversely proportional to the depth of the bed L; and K, the proportionality constant (permeability or hydraulic conductivity), contains a factor inversely proportional to the fluid viscosity.

$$\tilde{Q} = \frac{K}{\mu} \frac{\Delta p A}{L} \tag{33}$$

A modified form of Darcy's equation is useful in describing fluid flow in filtration.

Laminar Flow in Stirred Vessels

The presence of laminar fluid motion or turbulence in an impeller-stirred vessel can be determined with impeller Reynolds number, N_{Re} defined as:[14]

$$N_{\mathrm{Re}} = \frac{D_m^2 N \rho}{\mu} \tag{34}$$

where D_{m} is the impeller diameter and N is the rotational speed. When $N_{\mathrm{Re}} < 10$, the flow is laminar, and when $> 10{,}000$, it is turbulent. Between the value of 10 and 10,000, the flow may be turbulent at the impeller and laminar at remote parts. Under laminar flow conditions, the impeller drags the fluid with it, resulting in a circular motion. With increase in fluid viscosity, the circular motion, imposed by the rotating impeller, will change to radial direction by the centrifugal force on the fluid layer. With non-Newtonian fluids such as suspensions of fine particles, motion in remote parts of the vessel may be negligible due to low shear rates. In such cases, the fluid circulation in laminar flow can be increased by increasing impeller diameter and reducing the rotational speed for a given power consumption and viscosity. In laminar flow, the power number N_{P} is inversely proportional to impeller Reynolds number and is given by:

$$N_{\mathrm{P}} = \frac{P}{\rho N^3 D_m^5} \tag{35}$$

where P is the power.

FLOW AROUND IMMERSED BODIES

Drag Coefficient, Newton, and Stokes Flow

This type of flow where solid particles are immersed and surrounded by fluids (Fig. 10) is often encountered in many food processes, particularly in settling operation, homogeneous fluidization (solid/liquid), heterogeneous fluidization (solid/gas), filtration, absorption, drying, ion-exchange/adsorption columns, and fixed-bed catalytic reactions. These operations may also involve more than one fluid phase.

An immersed body causes the removal of momentum from a moving fluid at a rate depending on the velocity pressure of the fluid. The rate of removal of momentum is called viscous drag as it is the consequence of fluid viscosity. The viscous drag or drag force, F_D, can then be defined as the force exerted by a moving fluid on the solid particle in the direction of flow and is given by:[15]

$$F_D = \frac{1}{2}\rho A_p U^2 (R_{e,p})^a \tag{36}$$

where A_p is frontal area of the particle perpendicular to the flow and $R_{e,p}$ is Reynolds number for a particle, which is given by:

$$R_{e,p} = \frac{\rho D_p U}{\mu} \tag{37}$$

where D_p is the particle diameter for spheres or characteristic dimension for other shapes. A factor analogous to friction factor f (which describes the resistance in fluid flow through pipes) is *drag coefficient*, C_D, for flow around immersed bodies:

$$C_D = (R_{e,p})^a = f(R_{e,p}) \tag{38}$$

Combining Eqs. 36 and 38, we obtain:

$$F_D = \frac{1}{2} C_D \rho A_p U^2 \tag{39}$$

or

$$C_D = \frac{F_D/A_p}{\rho U^2/2} \tag{40}$$

For $R_{e,p} < 1$, the flow around the object is laminar; and for $R_{e,p}$ between 1 and 10^3, transition from laminar to turbulent flow may occur depending on the surface roughness of the particle. For $R_{e,p}$ between 10^3 and 10^5, the flow becomes turbulent. For $R_{e,p} > 10^5$, the flow is entirely turbulent. Laminar flow occurs when the particles in the fluid are small. The flow patterns and drag forces for flow around spheres and other regular shapes at low flow velocities can be determined by numerical computations.[16]

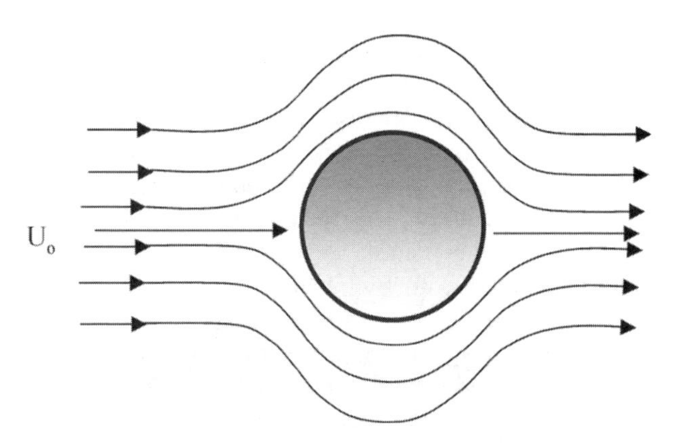

Fig. 10 Flow around immersed sphere.

Fig. 11 shows the relationship between the drag coefficient C_D and $R_{e,p}$ for spheres, disks, and cylinders. C_D is inversely proportional to $R_{e,p}$ for values of $R_{e,p}$ up to 1. This region is referred to as *Stokes' region* and C_D is $24/R_{e,p}$. C_D is constant ($C_D = 0.43$) for $R_{e,p}$ numbers $10^3 < R_{e,p} < 10^5$, and this region is referred to as *Newton's region*. The region between $R_{e,p}$ of 1 and 10^3 is intermediate range. For $R_{e,p} > 10^5$, C_D decreases further and the flow becomes turbulent everywhere on the surface of the moving object and behind the object. Eq. 39 indicates that viscous drag is proportional to the square of fluid velocity for $R_{e,p} > 1$. Substituting for $C_D = 24/R_{e,p}$ for laminar flow with spherical objects in Eq. 39, we obtain:

$$F_D = 9.42\mu D_p U \tag{41}$$

Thus, for laminar flow around a sphere, F_D is proportional to viscosity, characteristic dimension of the object, and fluid velocity.

Terminal Velocity and Settling

A consequence of increase in F_D with velocity is that an object falling through a fluid will not accelerate indefinitely because the viscous drag opposes the acceleration until a critical velocity is reached. After this point, the object continues to fall at a constant velocity, called the *terminal velocity*. The equilibrium between the drag force and the gravitational force can be written as:[15]

$$C_D = \frac{\rho_f U_t^2}{2}\frac{\pi D_p^2}{4} = \frac{\pi D_p^3 g(\rho_p - \rho_f)}{6} \tag{42}$$

where U_t is terminal velocity, ρ_p and ρ_f the particle and fluid densities, respectively, and g the acceleration due to gravity.

$$U_t^2 = \frac{4D_p g(\rho_p - \rho_f)}{3C_D \rho_f} \tag{43}$$

With $C_D = 0.43$ in the Newton law region,

$$U_t = 1.73\left(\frac{gD_p(\rho_p - \rho_f)}{\rho_f}\right)^{1/2} \tag{44}$$

For particle Reynolds numbers < 1 (termed creeping flow or Stokes flow region), Stokes derived the relation for the drag force on a sphere:

$$F_D = 3\pi\mu D_p U \tag{45}$$

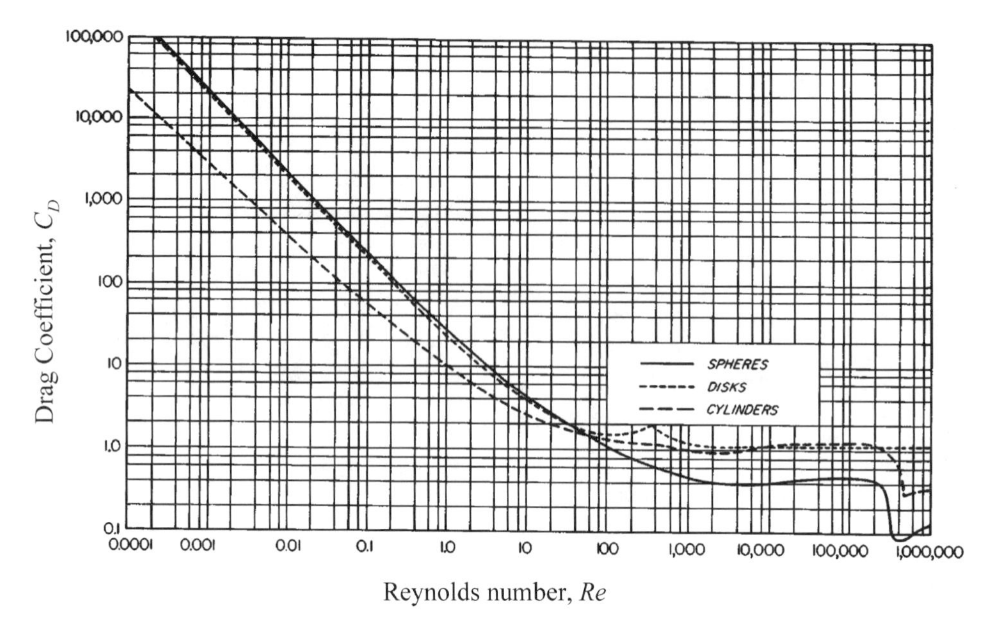

Fig. 11 Drag coefficients for spheres, disks, and cylinders. (From Ref. [17].)

Balancing of drag and gravity forces yields:

$$U_t = \frac{gD_p^2(\rho_p - \rho_f)}{18\mu} \tag{46}$$

Knowledge of terminal velocity is useful in settling and sedimentation problems, pneumatic transport, packed-bed and fluid-bed processes, and emulsion stability. The reader is referred to Happel and Brenner[18] and White[19] for more details on the hydrodynamics of flow at very low Reynolds numbers and Chhabra[20] for numerical solutions of the flow equations governing slow flows.

Stokes' law can be applied for particle diameters up to about 75 μm settling in air and 100 μm–200 μm particles settling in water. Newton's law applies when $D_p > 2$ mm in air and 4 mm–5 mm in water. Stokes' relation is not valid when the settling particles are very small as particle collisions with fluid molecules interfere with settling. Second, when the particle concentration is high, the settling velocity of particles diminishes due to hydrodynamic interaction between particles (hindered settling). Finally, the influence of shape on the flow resistance is small in the Stokes range but more significant in the Newton range.

Sedimentation of particles in a fluid may be carried out in a centrifugal field, instead of gravity field. The above principles are still valid as long as acceleration due to gravity is replaced by centrifugal acceleration, which is many times larger than when a sedimenting particle under gravity that falls under Stokes range may move to Newton's range in centrifugal field.

BLOOD FLOW

William Harvey (1578–1657) was the first to describe the functioning of heart, blood vessels, and circulation in 1628. The circulation consists of a double pump system, the right and left ventricles, and a distribution system. The latter consists of two major branches of elastic tubes. Typical branching consists of arteries leading to smaller vessels (the smallest are the capillaries), rejoining of small veins into larger ones, and back to the heart, thus enclosing a hydrodynamic loop.[21–23]

Blood is a visco-elastic, nonhomogeneous non-Newtonian fluid. The apparent viscosity of blood depends on: 1) the viscosity of plasma; 2) the ratio of erythrocyte volume to the total volume of blood (hematocrit); and 3) the flexibility and aggregation of erythrocytes. Apparent viscosity can reach the values of plasma viscosity in small capillaries. Since very small capillaries have a smaller diameter than erythrocytes, blood flow in such vessels can occur only by distortion of erythrocytes. Branching of larger vessels into smaller vessels increases the total cross-sectional area of the tubes. Thus, the average velocity of the blood in the smaller vessels and consequently, the flow resistance are reduced. This reduced velocity allows the blood to exchange substances with the cells in the capillaries and alveoli. Blood flow is also regulated by changes in vessel size and blood pressure. During exercise, blood vessels are selectively dilated to muscles and organs and blood pressure increases, creating greater blood flow. Decrease in vessel radius, due to plaque formation in the arteries, can greatly reduce the blood flow.

In small arteries and capillaries, the Reynolds number Re is below 2000 indicating blood flow is *laminar* in these vessels. Even in larger vessels, the flow is rarely turbulent under normal circumstances. Blood flow resistance R is defined as the ratio of arterio-venous pressure gradient $(p_a - p_v)$ and blood flow rate Q is expressed as:

$$R = \frac{(p_a - p_v)}{Q} \tag{47}$$

where p_a is arterial pressure, p_v venous pressure, and R peripheral resistance. Poiseuille equation describes blood flow reasonably well:

$$Q = \frac{(p_a - p_v)\pi r^4}{8\mu L} \tag{48}$$

It is evident from the equation that the radius of the vessel exerts a greater influence than either viscosity or length. Thus narrowing of an artery can cause turbulence, and the flow is characterized by sudden compression of the stream lines and formation of jets.

HEAT TRANSFER IN LAMINAR FLOW REGIME

Many situations are encountered in food processing where materials have to be heated or cooled in laminar flow regime because it is difficult to induce turbulence in them. Such materials include viscous liquids and dispersions with high solid content. Heat transfer may take place without phase change; and in some instances, heating and cooling may be accompanied by a phase change: freezing and thawing of food products, and melting and solidification of fats as in butter and chocolate processing. The heating or cooling medium is separated from the stream of fluid being heated or cooled by a wall, which may be a circular tube, parallel plate surfaces (plate heat exchanger), or rectangular duct. In laminar flow, due to the absence of eddies, heat transfer occurs mainly by conduction. Hence, knowledge of unsteady-state heat transfer by conduction is important. Since the flow influences the temperature gradients in the fluid, the transient heat conduction depends on the convection boundary condition at the surface of the solid. Predicting the heat transfer rate from a heat surface to a flowing fluid requires an energy balance along with an analysis of fluid dynamics and boundary layer analysis.

Ludwig Prandtl[24] recognized that viscosity could not be ignored in the gradient region near surfaces. The region near the surface, although small at high Reynolds numbers, is where viscosity is significant and shear rates are high. He called that region *hydrodynamic boundary layer*. The kinematic viscosity, ν, of the fluid carries information about the rate at which momentum may transfer through the fluid due to molecular motion. Just as the hydrodynamic boundary layer is that region where viscous forces are felt, a *thermal boundary layer* exists and it is the region where temperature gradients are present in the flow, contributing to natural convection in this region. Like kinematic viscosity, the thermal diffusivity, α, provides the information about the rate at which temperature differences within the fluid are dissipated by conduction. Thus, knowing the temperature gradient at the wall and thermal boundary layer thickness will enable us to evaluate the surface heat transfer coefficient. An energy balance can be imposed:

$$\begin{aligned}&(\textit{Energy diffused in})\\ &\quad + (\textit{viscous work within the fluid})\\ &\quad + (\textit{heat transfer at the surface})\\ &\quad = (\textit{energy diffused out})\end{aligned} \tag{49}$$

Prandtl number, Pr, has been found to relate the relative thicknesses of the hydrodynamic and thermal boundary layers:

$$Pr = \frac{\nu}{\alpha} = \frac{\mu/c_p}{k/\rho c_p} = \frac{\mu c_p}{k} \tag{50}$$

where k is the thermal conductivity and c_p the specific heat of the fluid. The hydrodynamic and thermal diffusion rates are the quantities that determine the thickness of the boundary layers in a given flow field; larger diffusivities signify that the viscous or temperature influence is felt farther. Thus, Pr describes the link between velocity distribution (flow) and temperature distribution.

A dimensionless measure of heat transfer is the *Nusselt number*, Nu, and is given by:

$$Nu = \frac{qL}{kA\Delta T} = \frac{hL}{k} \tag{51}$$

where q is the rate of heat transfer by convection, A the area of the heating wall, and h the heat transfer coefficient. Physically Nu denotes the ratio of the temperature gradient in the fluid at the interface to the total temperature difference between the wall and fluid stream. It can be regarded as the ratio of the length of the heating wall to the thickness of the thermal boundary layer.[25] Should Nu be constant, the heat transfer is proportional to the temperature difference (Newton's law of cooling). Experimental data can be correlated by using the general Eq. 52 for various situations such as heat transfer in the tube flow.

$$Nu = KPr^{m}Re^{n} \quad (52)$$

where K, m, and n vary with heat transfer conditions.

MASS TRANSFER IN LAMINAR FLOW REGIME

Diffusion, evaporation, and other molecular transport processes are facilitated by convective mass transfer when fluid motion is present. Convective mass transport and convective heat transport are closely analogous where heat is replaced by mass, temperature gradients by concentration gradients, and thermal conductivity by diffusion coefficient. However, unlike heat transfer, mass transfer may involve several components diffusing simultaneously.

Operations such as dehydration, hydration, crystallization, dissolution, evaporation, distillation, adsorption, membrane separation, and extraction may involve simultaneous heat and mass transfer. Mass transfer takes place within the material or at interfaces by diffusion comparable to heat conduction. Mass transfer occurs between solids and vapors/gases (drying, absorption, and adsorption); solids and liquids (leaching, crystallization, and dissolution); liquids and gases (evaporation and condensation); solids, liquids, and gases as distinct phases (fermentation, vegetable oil hydrogenation); and immiscible liquids as in liquid–liquid extraction. Under turbulent conditions, mass transfer occurs by diffusion through a laminar boundary layer followed by convective mass transfer in the bulk. In mass transfer, the usual driving force for diffusion is the concentration gradient. Nevertheless, diffusion also occurs under applied fields such as pressure gradient in reverse osmosis, temperature gradient in evaporation, by the application of centrifugal force in centrifugation or electric field in electrophoresis.

Although it is possible to describe fluid motion and diffusion (*Fick's laws*) under laminar regime, it is often difficult to do it for certain situations such as flow past objects as in packed beds or mass transfer across interfaces as in gas absorption and extraction. In addition, mass transfer may involve other mechanisms than simple diffusion. Hence mass transfer coefficients are often obtained experimentally and correlated in terms of dimensionless numbers (Table 1).

In mass transfer in fluids, nonsteady conditions exist. In operations such as drying, diffusion takes place from inside the solid to the surface and mass transfer from the surface to the flowing fluid by convective transport. For diffusion within the solid, the diffusion coefficient D describes the diffusion with the solid; and κ, the mass transfer coefficient describes the mass transfer from the surface. Mass transfer is more complicated in liquid/liquid and gas/liquid systems. They involve partition between phases. In such situations, it is preferable to use overall mass transfer coefficient, analogous to the overall heat coefficient. In certain situations like packaging, diffusion is rapid within the package but at the surface (packaging film) there is higher resistance to transport, and steady-state condition applies.

There is much similarity between heat transfer and mass transfer properties: h, heat transfer coefficient vs. κ the mass transfer coefficient and k, thermal conductivity vs. D, diffusion coefficient. In heat transfer, the relation between Nu, Re, and Pr is:

$$Nu = f(Re, Pr) \quad (53)$$

Table 1 Dimensionless groups for mass transfer and their significance

Name (symbol)	Formula	Physical significance	Comments
Sherwood (*Sh*)	$\frac{\kappa L}{D}$	Mass diffusivity/molecular diffusivity	Used in mass transfer
Schmidt (*Sc*)	$\frac{\mu}{\rho D}$	Viscosity force/molecular diffusivity	Used in mass transfer and diffusion in flowing systems
Stanton (*St*)	$\frac{h}{C_p \rho U}$	Heat transfer/thermal capacity of fluid	Used in heat transfer and forced convection
Reynolds (*Re*)	$\frac{\rho UL}{\mu}$	Inertial force/viscous force	Used in heat, mass, and momentum transfer to account for dynamic similarity
Peclet (*Pe*)	$\frac{LU}{\alpha}$	Convective transport/diffusive transport	Used in heat transfer and forced convection
Lewis number (*Le*)	$\frac{\alpha}{D}$	Thermal diffusivity/molecular diffusivity	Used in combined heat and mass transfer
Weber number (*We*)	$\frac{LU^2\rho}{\sigma}$	Inertial force/surface tension force	Used in momentum transfer, gas bubble, liquid droplet formation, and breakage of liquid jets

By analogy, the relation for mass transfer is:

$$Sh = f(Re, Sc) \tag{54}$$

where Sh is the *Sherwood number* ($\kappa L/D$) and Sc is the *Schmidt number* (k/D).

Equation for mass transfer is:

$$Sh = K\,Sc^m\,Re^n \tag{55}$$

Various mass transfer correlations exist in the literature for laminar flow regime for various geometries.[3,14,26–29] Most of them have a general expression of Eq. 55. A useful correlation for a laminar flow along flat plate[29] is widely utilized:

$$Sh = 0.323 Re^{1/2} Sc^{1/3} \tag{56}$$

Similar correlations were developed for other situations where laminar flow occurs as for liquid motion on spinning discs:[28]

$$Sh = 0.62 R_{\omega}^{1/2} Sc^{1/3} \tag{57}$$

where Re_{ω} is Reynolds number based on the rotation speed, ω, of the spinning disc of diameter D defined as:

$$Re_{\omega} = \frac{\rho D \omega}{\mu} \tag{58}$$

This correlation is valid for Re_{ω} between 100 and 20,000. For laminar flow through circular pipes, Schlichting[27] developed a correlation which depends only on Reynolds number Re:

$$Sh = 1.86 Re^{1/3} \tag{59}$$

This correlation is not valid for $Re < 10$ because of free convection. A similar correlation was developed for falling films:[29]

$$Sh = 0.69 Re^{1/2} \tag{60}$$

Other correlations, which slightly differ from Eq. 55, take the following general expression:

$$Sh = K' + K\,Sc^m Re^n \tag{61}$$

where K' is a constant generally equal to 2.0. Among the reported correlations, it is worth mentioning the correlation developed for the fluid motion in forced convection around a solid sphere:[26]

$$Sh = 2.0 + 0.6 Sc^{1/3} Re^{1/2} \tag{62}$$

Stoodley et al.[30] reported a similar correlation for the mass transfer in heterogeneous biofilms using microelectrodes and confocal microscopy. The values of the constants ranged from 1.45 to 2.0 for K', 0.22 to 0.28 for m, and 0.21 to 0.60 for n. Similar form of Eq. 61 has been applied to flow through packed beds.[31] This mass transfer equation is particularly useful at low Reynolds numbers, when the K' term becomes significant and accounts for the limiting Sh that occurs when there is no flow.

CONCLUSION

Many food processing operations involve laminar flow, particularly those dealing with high viscosity Newtonian and non-Newtonian fluids or suspensions. A large number of foods fall in this category: processed cheese, ice cream, plastic fats, butter, fruit juice concentrates, fruit preparations, syrups, sauces, chocolate products, etc. Laminar flow occurs in general operations such as fluid transport (pumping), mixing, heating and cooling during pasteurization, sterilization, and freezing. For heating and cooling of viscous products, scraped-surface heat exchangers are employed, involving Poiseuille flow in concentric annular or Couette geometries. Extrusion and extrusion cooking of starchy and proteinacious materials and cereals also involve laminar flow, often in Couette geometry. Laminar flow occurs in a variety of separation processes such as crystallization, falling-film and wiped-film evaporation, steam distillation, settling and sedimentation, extraction, and cross-flow filtration. Laminar flow past particulates occurs in drying. In many processes, heat transfer or mass transfer or both occur, in addition to fluid flow (momentum transfer). Ability to predict the effects of various variables in these processes and design equipments requires a deeper understanding of flow properties as well as other transport phenomena under laminar flow conditions. Given the complexity of food materials, it may not be possible always to describe fluid flows by fluid mechanical models. Dimensional analysis of specific flow problems, correlation between dimensionless groups, and extraction of analogies between momentum, heat, and mass transfer phenomena can be useful.

REFERENCES

1. Bird, R.B.; Stewart, W.E.; Lightfoot, N. *Transport Phenomena*; John Wiley: New York, NY, 1960.
2. Batchelor, G.K. *An Introduction to Fluid Dynamics*; Cambridge University Press: Cambridge, U.K., 1967.
3. Gekas, V. *Transport Phenomena of Foods and Biological Materials*; CRC Press: Boca Raton, FL, 1992.
4. Chhabra, R.P.; Richardson, J.F. Non-Newtonian Fluid Behaviour. *Non-Newtonian Flow in the Process Industries, Fundamentals and Engineering Applications*; Butterwoth-Heinmann, Jordan Hill: Oxford, U.K., 1999; 1–36.

5. Ipsen, D.C. *Units, Dimensions, and Dimensionless Numbers*; McGraw-Hill: New York, NY, 1960.
6. Munson, B.R.; Young, D.F.; Okiishi, T.H. Similitude, Dimensional Analysis and Modeling. In *Fundamentals of Fluid Mechanics*; Wiley: New York, NY, 1994; 401–463.
7. Tilton, J.N. Fluid and Particle Dynamics. In *Perry's Chemical Engineers' Handbook*, 7th Ed.; Perry, R.H., Green, D.W., Eds.; McGraw-Hill: New York, NY, 1997; 6.1–6.54.
8. Moody, L.; Princeton, N.J. Friction Factors for Pipe Flow. Trans. Am. Soc. Mech. Eng. **1994**, *66*, 671–684.
9. Cheremisinoff, N.P. Properties and Concepts of Single Fluid Flows. In *Encyclopedia of Fluid Mechanics*; Cheremisinoff, N.P., Ed.; Gulf Publishing Company: Houston, TX, 1986; Vol. 1, 277–352.
10. Lamb, H. *Hydrodynamics*; 6th Reprint, Dover Publications: New York, NY, 1986.
11. Dury, G.H. Hydraulic Geometry. In *Introduction to Fluvial Processes*; Chorley, R.J., Ed.; Barnes and Noble: New York, NY, 1969; 146–156.
12. Knudsen, J.G.; Katz, D.L. *Fluid Dynamics and Heat Transfer*; McGraw-Hill: New York, NY, 1958.
13. Leyton, L. *Fluid Behaviour in Biological Systems*; Clarendon Press: Oxford, U.K., 1975.
14. Uhl, V.W. *Mixing: Theory and Practice*; Academic Press: New York, NY, 1966.
15. Vennard, J.K. *Fluid Mechanics*, 4th Ed.; John Wiley: New York, NY, 1961.
16. Masliyah, J.H.; Epstein, N. Numerical Study of Steady Flow Past Spheroids. J. Fluid Mech. **1970**, *44*, 493–512.
17. Lapple, C.E.; Shepherd, C.B. Calculation of Particle Trajectories. Ind. Eng. Chem. **1940**, *32*, 605–616.
18. Happel, J.; Brenner, H. *Low Reynolds Number Hydrodynamics with Special Applications to Particulate Media*; Prentice-Hall, Inc.: Englewood Cliffs, NJ, 1965.
19. White, F.M. *Viscous Fluid Flow*; McGraw-Hill: New York, NY, 1974.
20. Chhabra, R.P. Steady Non-Newtonian Flow About a Rigid Sphere. In *Encyclopedia of Fluid Mechanics*; Cheremisinoff, N.P., Ed.; Gulf Publishing Company: Houston, TX, 1986; Vol. 1, 983–1033.
21. Burton, A.C. *Physiology and Biophysics of the Circulation*; Year Book Medicinal Publishers: Chicago, IL, 1972.
22. Caro, C.G.; Pedley, T.J.; Schroter, R.C.; Seed, W.A. *The Mechanics of the Circulation*; Oxford University Press: New York, NY, 1978.
23. Kenner, T.; Physiology of Circulation. In *Cardiology*; McGraw-Hill Clinical Medicine Series, Dalla-Volta, S. and Braunwald, E. Eds.; McGraw-Hill: New York, NY, 1999; 15–25.
24. Prandtl, L.; Tietjens, O.G. *Applied Hydro- and Aeromechanics*; Dover Publications: New York, NY, 1957.
25. Holman, J.P. *Heat Transfer*, 2nd Ed.; McGraw-Hill: New York, NY, 1968.
26. Sherwood, T.K.; Pigfort, R.L.; Wike, C.R. *Mass Transfer*; McGraw-Hill: New York, NY, 1975.
27. Schlichting, H. *Boundary Layer Theory*, 7th Ed.; McGraw-Hill: New York, NY, 1979.
28. Treybal, R.E. *Mass Transfer Operations*, 3rd Ed.; McGraw-Hill: New York, NY, 1980.
29. Knudsen, J.G.; Hottel, H.C.; Sarofim, A.F.; Wankat, P.C.; Knaebel, K.S. Heat and Mass Transfer. In *Perry's Chemical Engineers' Handbook*, 7th Ed.; Perry, R.H., Green, D.W., Eds.; McGraw-Hill: New York, NY, 1997; 5.1–5.79.
30. Stoodley, P.; Yang, S.; Lappin-Scott, H.; Lewandowski, Z. Relationship Between Mass Transfer Coefficient and Liquid Flow Velocity in Heterogeneous Biofilm Using Microelectrodes and Confocal Microscopy. Biotechnol. Bioeng. **1997**, *56*, 681–688.
31. Kennedy, C.A.; Lennox, W.C. A Pore-Scale Investigation of Mass Transport from Dissolving DNAPL Droplets. J. Contam. Hydrol. **1997**, *24*, 221–246.

Liquid Food Transport

K. P. Sandeep
North Carolina State University, Raleigh, North Carolina, U.S.A.

L

INTRODUCTION

Liquid transport systems for food processing applications generally consist of a pump, connecting pipes, valves, and fittings (elbows, reducing/expanding joint, tees, couplings, and unions). Depending on the product characteristics (such as viscosity and density), process parameters (such as temperature and flow rate), and system components (length and diameter of pipes, types and number of fittings), the flow characteristics (laminar, transition, or turbulent) of the product in various sections of the system will change. This in turn affects the energy required (calculated using the Bernoulli equation) to transport the food product through the system.

TYPES OF PUMPS

Pumps are classified into various categories based on their principle of operation, the types of products they can handle, the regulatory requirements they meet, and other criteria. However, the most common classification of pumps is based on its principle of operation. Based on this criterion, pumps are classified into two broad categories—dynamic (Table 1) and displacement (Table 2). In dynamic pumps, energy is constantly supplied to the product and the speed of the product within the pump reaches a value much higher than that at the discharge port so that a reduction in speed at or beyond the discharge port causes an increase in pressure. In a displacement pump, energy is intermittently supplied to the product by applying force to one or moving parts of the pump that directly increases the pressure of the product to the value required to move the product to the discharge port. Karassik et al.[1] and Perry et al.[2] have described various types of pumps in detail.

Dynamic Pumps

Dynamic pumps generally can be used to provide very high flow rates and at a steady rate with minimal increase in pressure. They are, however, quite ineffective in handling high-viscosity products and also require priming (venting of trapped gases and replacement with product—this can be done manually or automatically) in many applications. Dynamic pumps are further divided into two broad categories—centrifugal pumps and special effect pumps.

Centrifugal Pumps

A basic centrifugal pump consists of an impeller (consists of a number of blades) rotating within a casing (circular, volute, or diffuser type) and the product being delivered by the action of centrifugal force. Centrifugal pumps can be divided into three categories—axial flow, radial flow, and peripheral flow. Depending on the shape of the impeller, flow takes pace in the axial direction, radial direction, or a mixed direction (both axial and radial directions). The blades of the impeller are either open or closed and are mounted on a shaft that projects outside the casing. Closed impellers are more efficient than open or semiopen impellers. Open impellers (fixed or variable pitch) are used to handle viscous or particulate products. Impellers are called single suction impellers if the product enters from only one side and double suction if the product enters from both sides. Single stage pumps have only one impeller while multistage impellers have several impellers in series and are used to handle relatively high pressures. Centrifugal pumps are also classified as horizontal or vertical types depending on the orientation. A centrifugal pump is simple in construction, relatively light, has low wear and tear, requires low maintenance, but needs priming and has a lower efficiency than a reciprocating pump.

Special Pumps

Jet pumps make use of the momentum of one fluid to move another. If a liquid is used as the motive fluid, it is called an eductor jet pump and if a condensable gas is used, it is called an injector jet pump. An air-lift (or gas-lift) pump consists of a vertical pipe submerged in a product, with air being pumped into the pipe from the bottom, resulting in a pumping action that discharges an air-product stream. In an electromagnetic pump (conduction, induction, or linear induction), a magnetic field is applied to the product, which results in a force (or pressure) that pumps the product. A reversible

DOI: 10.1081/E-EAFE 120006953
Copyright © 2003 by Marcel Dekker, Inc. All rights reserved.

Table 1 Categories of dynamic (or momentum change) pumps

- Centrifugal
  - Axial flow (propeller)
    - Single and multistage
      - Closed impeller
      - Open impeller
        - Fixed and variable pitch
  - Radial flow, mixed flow
    - Single and double suction
      - Self priming and nonpriming
        - Open, semiopen, and closed impeller
      - Single and multistage
        - Open, semiopen, and closed impeller
  - Peripheral flow
    - Single stage and multistage
      - Self priming and nonpriming
- Special
  - Jet
    - Ejector (or siphon or exhauster or eductor)
    - Injector
  - Gas lift (or air lift)
  - Hydraulic ram
  - Electromagnetic
    - Conduction (AC or DC)
    - Induction (AC only)
    - Linear induction
  - Reversible centrifugal
  - Hermetically sealed magnetic-drive centrifugal
  - Vortex (shear-lift)
  - Laminated rotor
  - Inclined rotor
  - Regenerative turbine
  - Rotating casing (pitot tube)

Table 2 Categories of displacement pumps

- Reciprocating
  - Piston
    - Direct-acting steam—double acting
      - Simplex, duplex
    - Power (Crank and flywheel)—single acting and double acting
      - Simplex, duplex, triplex, multiplex
  - Plunger—single acting
    - Simplex, duplex, triplex, multiplex
  - Diaphragm
    - Simplex and multiplex
      - Fluid (or pneumatically) operated
      - Mechanically operated
- Rotary
  - Single rotor
    - Vane (or sliding vane)
      - Internal
      - External
    - Piston
      - Axial
      - Radial
    - Flexible member
      - Tube (or peristaltic or wave contraction)
      - Liner
      - Vane
    - Screw and wheel
  - Multiple rotor
    - Gear
      - External
      - Internal with fixed crescent
    - Lobe
      - Single
      - Multiple
    - Circumferential piston
      - Internal
      - External
    - Screw

centrifugal pump is similar to a centrifugal pump, with the difference that the pump delivers the same amount of product for both directions of shaft rotation. A hermetically sealed magnetic drive centrifugal pump is completely closed except at the suction and discharge end. The impeller and motor drives have a circular magnet attached to them and the impeller is driven by the magnetic interaction of the two magnets. A vortex (or shear-lift) pump consists of a concentric casing with axial suction and tangential discharge. The rotating impeller creates a vortex in the casing and moves the product to the exit. A laminated rotor pump consists of flat discs with one or more holes at the center, stacked axially, and separated by small spacers. The product enters through the holes, experiences centrifugal force and viscous drag against the disc, and exits the rotor at the periphery. An inclined rotor pump consists of a flat, elliptical plate (with teeth at periphery) mounted at an angle ($\sim 45^\circ$) to a rotating shaft which is in a tubular casing with its nozzle axial and in-line with the shaft. The product experiences centrifugal force due to the rotating plate and exits through the nozzle. A regenerative turbine pump consists of an impeller with several vanes at the periphery in a radially split casing with radial suction and discharge. The product experiences centrifugal force, strikes the casing, rebounds, and enters the impeller at the next vane, and this continues till the product reaches the discharge port. A rotating casing (pitot tube) pump consists of a rotating casing containing a stationary pitot tube at its center. The product enters at the center of the pump and the pitot tube converts kinetic energy of the product into static pressure and the product reaches the discharge port.

Displacement Pumps

Displacement pumps generally can be used to provide relatively low-flow rates and at rates not as steady as a dynamic pump. It can, however, operate at high pressures. They can be very effectively used for handling high-viscosity products and are generally self-priming for most applications. Displacement pumps are divided into two broad categories—reciprocating and rotary.

Reciprocating pumps

In a reciprocating pump, a piston, plunger, or a diaphragm either passes or flexes through a chamber back and forth with valves or other types of mechanisms controlling the flow product at the inlet and outlet. The three main categories of reciprocating pumps are piston, plunger, and diaphragm. Piston type pumps can be single acting (pumping is done on one side of piston only) or double-acting (pumping is done on both sides of piston). Piston type pumps can be direct-acting (directly connected to a steam cylinder) or power driven (by a crank and flywheel from the cross head of a steam engine). A plunger pump (always single-acting) has one or more constant diameter plungers reciprocating through one or more packing glands and displacing the product from cylinders in which there is considerable clearance. A diaphragm pump operates in a similar manner to piston and plunger pumps with a flexible diaphragm (metal, rubber, or plastic) instead or a piston or a plunger, with the diaphragms being operated pneumatically or mechanically. Reciprocating pumps are also classified as simplex (single cylinder), duplex (two cylinders), or multiplex (many cylinders).

Rotary pumps

In a rotary pump, the product is transferred by rotation of one or more moving parts (rotor) within a stationary housing (stator). The rotor rotates in one direction in a continuous manner and maintains a seal (no clearance if made of a resilient material or small clearance if made of rigid material) against the stator, thereby eliminating the need for valves. The main categories of rotary pumps are the single rotor and multiple rotor.

Rotary pumps with a single rotor are further divided into four categories—vane, piston, flexible member, and screw and wheel. In a vane pump, blades or vanes are moved radially (generally inward and outward) by cam surfaces and this action transfers the product from the inlet to the outlet. Vane pumps can be internal (cam surface is inside the stator and vanes are mounted in or on the rotor) or external (cam surface is the external radial surface of the rotor and vanes are mounted in the stator). In a rotary piston pump, pistons operating off cam surfaces, reciprocate within bores or the rotor and require no inlet or outlet valves. Rotary piston pumps are further classified as axial (piston elements move axially) or radial (piston elements have a radially reciprocal motion caused by an eccentric cam surface). In a flexible member pump, transfer of the product from the inlet to outlet is accomplished by making use of the elasticity of parts of the pump. In a flexible tube pump, all three chamber volumes are bounded by the inner surface of the flexible tube and the product volume is contained between two contact points of the rotor and the inner surface of the tube. In a flexible linear pump, pumping is accomplished in a manner similar to the external vane pump with all three chamber volumes being defined by the inner surface of the body, the outer surface of the liner, and the liquid seal contact between the liner and the body bore. A flexible vane pump is similar to an internal vane pump, with the vanes being flexible. A screw and wheel pump consists of a helical driving gear (screw) and a special form of spur gear (wheel), which is the driven gear.

Rotary pumps with multiple rotors are further divided into four categories—gear, lobe, circumferential piston, and screw. In a gear pump, two (or more) gears (or toothed wheels) mesh together, with each gear being mounted on a separate rotor. The gears could be internal (one gear inside another) or external (the gears mesh at their periphery). The rotation of the gears transfers the product from the inlet to the outlet. The lobe pump is similar to the gear pump, but has a rounded surface at the radial edge of the rotor surface (or wheel). The wheels have single or multiple lobes, which trap the product and convey it from the inlet to the outlet. The circumferential piston pump (internal, or external has rotor elements (which do not mesh or contact one another) shaped in the form of pistons which rotate and are supported by cylindrical hubs inset into the end plate of the pump. The screw pump is a variation of the gear pump with the threads of the screw forming cavities with another screw, and the product progressing axially within the cavities formed.

TYPES OF PIPES, VALVES, AND FITTINGS

There are several types of pipes, valves, and fittings used in a liquid transport system based on need. Some of the types of pipes used in liquid transport systems include plastic, rubber, and metal pipes. For sanitary applications, stainless steel pipes are used. These pipes are available in different grades and gauges. Some of the types of valves used in liquid transport systems are the aseptic, routing, piston, plug, compression type rising stem, mix-proof, gate, diaphragm, globe, angle, plug cock, butterfly, check, foot, ball, caged ball, disc, relief, needle, spool, and back pressure valve. Some of the types of fittings commonly used in liquid transport systems are the elbow (45, 90, and 180 degree bends of standard, square, long, or close return radius), tee, coupling, union, and reducer.

REFERENCES

1. Karassik, I.J.; Krutzsch, W.C.; Fraser, W.H.; Messina, J.P. *Pump Handbook*; McGraw-Hill Book Company: New York, 1976.
2. Perry, R.H.; Green, D.W.; Maloney, J.O. *Perry's Chemical Engineer's Handbook*, 6th Ed.; McGraw-Hill Book Company: Singapore, 1984.

Machine Vision in Precision Agriculture

Monte Dickson
Deere and Company, Des Moine, Iowa, U.S.A.

John F. Reid
Deere and Company, Moline, Illinois, U.S.A.

INTRODUCTION

Machine vision is the use of digital imaging for measurement, inspection, and control. Machine vision applications use the optical properties of a material to provide information that can be used to define material and feature characteristics. For agriculture, the optical properties may be related to physical characteristics of the product or system (e.g., size, shape, projected area, etc.) and the electromagnetic spectral characteristics.

The elements of a machine vision sensing system are represented in Fig. 1. The primary elements are the scene, the illumination system, the image of the scene, a description of the scene formed from the image, and an application response to the description formed from the image. These elements form a model by which a machine vision application can be subdivided into smaller tasks. Based on the agricultural application, this model must be configured with the proper components and algorithms.

Humans with eyesight make a majority of their decisions and act based on sight. In agriculture today, human vision is used to operate and control equipment, analyze crop and soil conditions, detect weeds and insects, and more. Machine vision provides a means to increase productivity and improve precision for many agriculture applications.

ILLUMINATION

Machine vision is a unique sensing method because excitation energy is often regulated and even structured in a pattern to get a response from the biological material in the scene. The illumination system is designed to consider the spectral quality of the illumination and the spatial distribution of illumination across the scene. However, in most agricultural applications where natural lighting is dominant, it can be difficult to control the spectral quality and spatial distribution of illumination.

The spectral properties of a biological material can be used to identify regions where the material exhibits unique electromagnetic responses. The illumination system provides electromagnetic radiation to excite the image sensor to form an image. Configuring the elements of a machine vision application requires an understanding of the electromagnetic response of the biological material, spectral characteristics of illumination sources, and spectral sensitivities of machine vision sensors. Sometimes the engineer or scientist needs to match these elements so that the image measures the desired biological material characteristics. Fig. 2 illustrates the importance of understanding the responses of the individual elements in an application.

Ambient solar illumination is the source for outdoor applications. The consideration of the effects of solar zenith and the influence of atmospheric conditions are important for these outdoor ambient applications. Understanding the interactions between the illumination source and the biological material is essential for developing a machine vision system for an agricultural application. As solar atmospheric conditions change, different levels of diffuse or specular incident radiation is emitted on the scene.

In applications where reflected light is being quantified, such as measuring reflectance off of crop canopies, a robust system will adjust to the changes and account for them in the measurement.

SCENE

The scene of a machine vision application is defined by the field-of-view of the image sensor. The complexity of a scene can be classified as a controlled scene or an uncontrolled scene. In a controlled scene, there is control over many of the environmental variables like illumination, image contrast, position and orientation of objects, or knowledge about the objects being imaged that can simplify the image processing. An uncontrolled scene is where there is less precise knowledge of the characteristics of the scene (e.g., unregulated illumination or uncertainty about the number or location of objects in the scene,

Encyclopedia of Agricultural, Food, and Biological Engineering
DOI: 10.1081/E-EAFE 120007237
Copyright © 2003 by Marcel Dekker, Inc. All rights reserved.

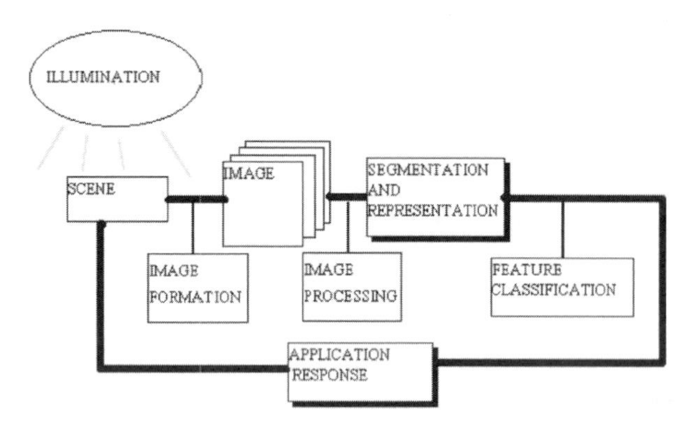

Fig. 1 A model of machine vision applications.

unknown scene boundaries). Most applications in this latter category are the challenging applications for machine vision and are what exist in agriculture. Fig. 3 illustrates an image from a machine vision sensor developed to sense information for selective harvest sensor for green asparagus in the field. The machine vision sensor is part of a system to determine if spears are harvestable and to determine the precise location of spears in the image. This application represents an uncontrolled scene environment since the types of objects that could show up in the image are unpredictable.

Optical characteristics of the biological material are important in the selection of both the illumination source and the image sensor. An understanding of these scene properties (like response to illumination) can be used to simplify the image processing requirements of an application. When the incident radiation is primarily diffused, the light across the scene is scattered and shadows are minimized. Inversely, such as on clear, sunlit days, much of the incident radiation is specular, which can result in shadows and bright spots in the image.

THE IMAGE, IMAGE FORMATION, AND IMAGE SENSORS

An image sensor measures a response from the scene in the form of a 2-D digital image. A signal transmission technique sends the analog image to a digitizer to convert the signal in a form that can be processed by a computer. The array of image pixels form a signal that can be used to measure photometric and morphometric characteristics of objects within the image. Photometric characteristics are properties of image pixels that are related to the response to the illumination. Morphometric characteristics are size and shape characteristics of objects represented by collections of pixels in the image.

Two common sensors used in machine vision applications are monochrome cameras and color cameras. Monochrome cameras are used to measure an image intensity response most often related to the reflected or transmitted light from a scene. Solid-state image sensors are more practical for agricultural applications because of their ruggedness and durability. Solid-state cameras typically are sensitive to light beyond the visible spectrum into the near infrared (NIR). Optical filters may be used with cameras to isolate portions of the spectrum where a scene has a response that simplifies the machine vision application. The asparagus image in Fig. 3 was captured

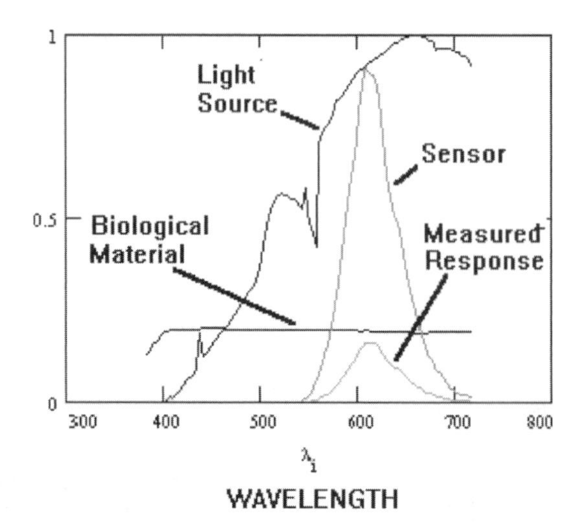

Fig. 2 The sensor response depends on characteristics of the source, receiver, and the biological material.

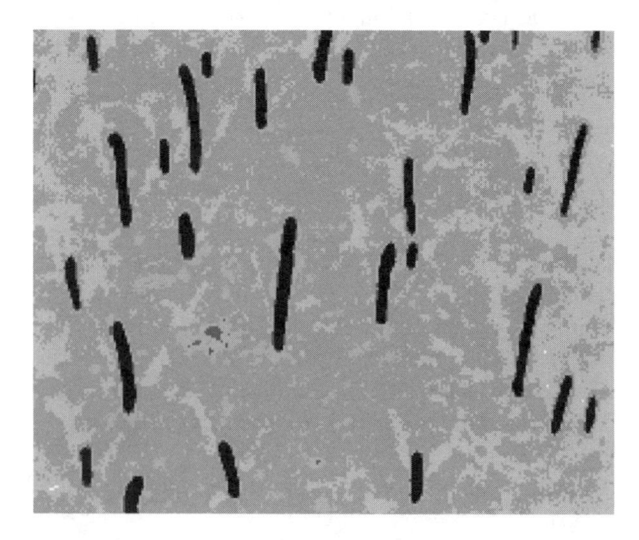

Fig. 3 An image of field grown asparagus (black) from a vision sensor mounted on an automatic harvester.

using a camera setup to measure NIR radiation at a narrow bandwidth around 800 nm.

Lenses and filters are important elements for the imaging system. Lenses control the portion of the scene, or field-of-view, that is projected onto the sensing elements of the camera. Filters control the spectral characteristics of the illumination sensed by the camera. For many applications, optical filters are important for enhancing features of interest in the image scene before it strikes the photosensitive digital array. It is very common in many agricultural applications to filter (pass through only) NIR light. Plant biological material reflects a large amount of NIR, while soils will absorb more NIR.

Most applications require some knowledge of the geometrical relationship between the image sensor and the scene. Spatial or geometric calibration permits the measurement of characteristics of the scene from the image. Intensity or color response calibration is used to characterize the response of individual pixels to known response levels.

After the signal is sensed at the camera, it is transmitted to an image digitizer to form an array of numbers representing the scene. An A/D converter processes the transmitted signal from the camera and converts it into an array of picture elements (pixels) representing the 2-D projection of the scene. The primary considerations of the selection of the image digitization system are the spatial resolution and the intensity resolution (Fig. 4). The spatial resolution is the number of pixels used to represent the 2-D image relative to a given scene dimension. The intensity resolution is the number of gray levels used to represent the transmitted image signal.

Commercial vision systems have spatial resolutions up to 1024 by 1024 pixels in an image or higher in some specialized sensors. Many vision systems have 256 gray levels for each pixel. A color vision system would have three values per pixel to represent the red, green, and blue responses of the visible spectrum. A single 1024×1024 color image could occupy over 3 Mb of memory. Increased spatial (256×240 vs. 1024×768) and intensity resolution (monochrome vs. color) generally increase the image storage requirements and the image processing time.

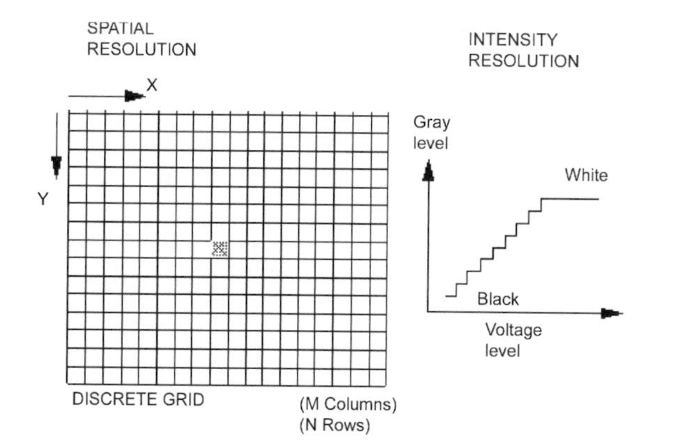

Fig. 4 An image is made up of an array of pixels representing a sensor excitation level.

IMAGE PROCESSING AND REPRESENTATION

The goal of image processing in a machine vision application is to extract information from the image that can be used in the application response or understanding of the scene. The operations of image analysis can be performed in hardware or software.

Image processing simplifies the image so that features related to the desired signal are enhanced. Typically, the data goes through a reduction process where the enhanced image is represented in a simpler form. One method of representation is the calculation of properties of connected pixels. Represented features are used to make decisions from the image. Features that represent useful properties of the scene have to be classified based on the arrangement of pixels and pixel values in the image. A representation process in image processing transforms the image from a connected set of pixels into higher-level descriptions of size and shape (area, length, location, elongation, shape factors, etc.).

Image classification techniques can be very simple rules based on observed image properties and statistics or they can be more sophisticated relationships derived from artificial intelligence analysis of image features. Artificial intelligence methods include techniques for pattern recognition, artificial neural networks, and fuzzy logic, among others. An engineer has to be able to select the key features that will make the system operate successfully.

APPLICATION RESPONSE

The machine vision system utilizes the information extracted from images for controlling machine functions or for providing data to an information management system. The application response in agriculture may be automatic steering control for a tractor or a field map of crop nutrient stress.

In measurement, information can be collected on the photometric or morphometric characteristics of the image. This information is often an essential element to learn the characteristics that are useful for an automated system. In an application where a machine vision system

Fig. 5 Image from a machine vision guidance system for automatic steering in row crops. The red lines are the detected row centers after processing the image using morphometric characteristics.

is used to automatically steer a tractor in a row crop field, morphometric characteristics are used to determine the locations of the crop row centers (Fig. 5). The location information is provided to the steering control system that actuates the turning of the tractor. In another application where a machine vision system is used to measure crop nutrient stress based on reflected light, photometric characteristics are used (Fig. 6). Chlorophyll can be a good indicator of the availability of nitrogen fertilizer to plants. Green is the primary pigment of chlorophyll. Therefore, measuring the amount of reflected green light provides a measurement of chlorophyll.

This information can be provided to a variable rate control system that applies nitrogen fertilizer or to a data recording system that integrates a Global Positioning System (GPS) for making maps. Engineering in these areas requires an understanding of controls, actuators, computer interfacing, information management, and interactions among biological materials.

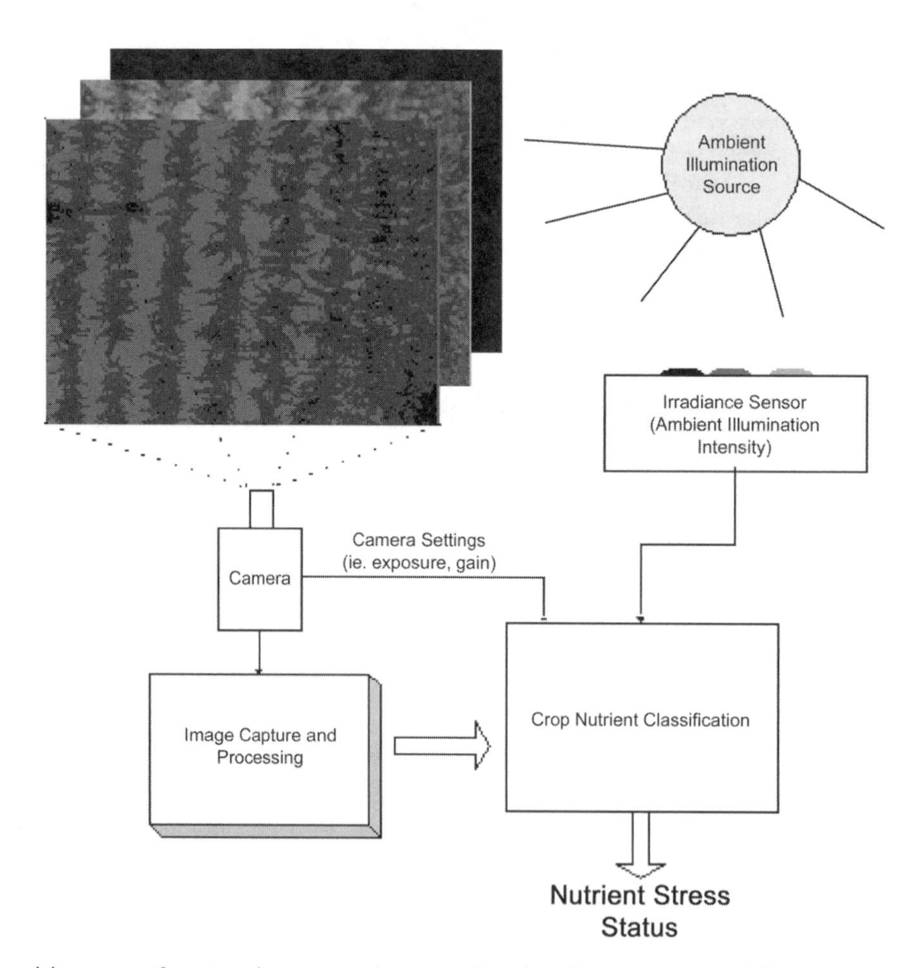

Fig. 6 A machine vision system for measuring crop nutrient status based on the measurement of light reflected off of crop canopies.

CONCLUSION

Applied machine vision for agricultural and biological systems can provide new and innovative sensing for inspection, measurement, and control. Machine vision technology integrates many fields of engineering together. Some useful areas of study for machine vision practitioners include computer programming, artificial intelligence, image processing, optical sensing, controls, computer interfacing, and mechanical systems design. Additional requirements for agricultural applications include an understanding of the physical properties of biological materials, agronomic principles, farming practices, and agricultural equipment.

FURTHER READING

Chung, S.O.; Sudduth, K.A.; Drummond, S.T. Determining Yield Monitoring System Delay Time with Geostatistical and Data Segmentation Approaches. Trans. ASAE **2002**, *45* (4), 915–926.

Kacira, M.; Ling, P.P.; Short, T.H. Machine Vision Extracted Plant Movement for Early Detection of Plant Water Stress. Trans. ASAE **2002**, *45* (4), 1147–1153.

Steward, B.L.; Tian, L.F.; Tang, L. Distance-Based Control System for Machine Vision-Based Selective Spraying. Trans. ASAE **2002**, *45* (5), 1255–1262.

Yang, C.; Everitt, J.H.; Bradford, J.M. Optimum Time Lag Determination for Yield Monitoring with Remotely Senses Imagery. Trans. ASAE, **2002**, *45* (6), 1737–1745.

Magnetic Field Applications to Foods

Fernanda San Martín
Federico Harte
Gustavo V. Barbosa-Cánovas
Barry G. Swanson
Washington State University, Pullman, Washington, U.S.A.

INTRODUCTION

Traditional methods for food preservation use heat for enzyme and microbial inactivation. Thermal treatments required for rendering a product safe often result in degraded sensory and nutritional attributes. Hence, processing methods that use little or no heat are currently under research. The use of magnetic fields is one of these novel processing methods known as nonthermal technologies. The advantages of using magnetic fields for food processing are the preservation of the fresh-like attributes as well as thermolabile nutrients, reduced energy requirements for processing, and, in some cases, the potential treatment of foods inside flexible packages.

TYPES OF MAGNETIC FIELDS

A magnetic field is the region of space in which a magnetic body is capable of magnetizing the surrounding bodies.[1] The density (H) of a magnetic field is measured in Oersteds. One Oersted is defined as one line of strength per cm^2. Magnetic field intensity (B) is measured in gauss or Tesla ($1\,T = 10{,}000\,G$). The relationship between magnetic field intensity and density is given by $B = \mu H$, where μ is the magnetic permeability of the media, which for a vacuum is equal to 1 and for air is 1.000024—for practical purposes approximated to 1. Therefore, the terms magnetic field density and intensity or strength are sometimes used interchangeably.[2]

Magnetic fields can be classified as homogeneous or heterogeneous, or as static and pulsed. Homogeneous magnetic fields are magnetic fields with a constant intensity over space, whereas heterogeneous fields exhibit an intensity gradient over space depending on the nature of the magnet. Static fields exhibit a constant strength over time and are generated by permanent magnets or direct current electromagnets. The intensity of pulsed magnetic fields increases quickly, reaches a peak, and then decreases in a very short period of time. Pulsed magnetic fields are generated with electromagnets as well. Within pulsed fields, oscillating magnetic fields are generated by alternate current electromagnets and the magnetic field intensity varies periodically depending on the frequency and type of wave resulting from the electric current in the magnet.[3] According to their relative intensity, magnetic fields are classified as low or high intensity. Low intensity magnetic fields exhibit strength on the order of tenths of gauss, whereas high intensity fields exhibit strength on the order of thousands of gauss and greater.[4]

MAGNETIC PROPERTIES OF MATTER

Materials are classified as diamagnetic, paramagnetic, or ferromagnetic according to their response to external magnetic fields. All materials exhibit diamagnetism as a result of the interference between the external magnetic field and the electrons orbiting atoms in molecules, elements, and compounds. The magnetic force of an external field acts upon the moving electrons in a way such that some electrons are accelerated whereas the velocity of others is decreased. The movement of the electrons may oppose the field, causing a slight repulsion of the material by the magnetic field. Diamagnetism is a property of most organic and inorganic compounds having paired electrons. On the other hand, materials that exhibit paramagnetism experience a magnetization proportional to the applied field. Paramagnetism usually occurs in elements or compounds with unpaired electrons such as free radicals and compounds of transition elements. Such unpaired electrons generate a small magnetic field that, when exposed to the external magnetic field, align in the magnetic field causing a slight attraction of the atom to the magnetic field. Ferromagnetic materials such as iron, nickel, cobalt, and some rare earths exhibit a long-range ordering phenomenon at the atomic level, which causes the unpaired electron spins to align parallel with each other in a region called domain. The magnetic field within domains is strong, however, in the absence of an external magnetic field the bulk material is not magnetized due to random orientation of the domains. Under external magnetic fields, domains align with each other, increasing the magnetic field by a large factor. Hence, ferromagnetic

Encyclopedia of Agricultural, Food, and Biological Engineering
DOI: 10.1081/E-EAFE 120007118
Copyright © 2003 by Marcel Dekker, Inc. All rights reserved.

effects are more noticeable than diamagnetic or paramagnetic effects.[5]

The magnetization of a material is expressed in terms of density of net magnetic dipole moments (μ) in the material as $\mu = K_m \mu_0$ where μ_0 is the magnetic permeability of space and K_m is the relative permeability. If under the influence of an external magnetic field the material does not show any magnetization then $K_m = 1$. For paramagnetic and diamagnetic materials, the relative permeability is very close to 1, whereas for ferromagnetic materials the relative permeability is very large. Another important concept is the magnetic susceptibility (χ) used to quantify the variability of relative permeability from one: $\chi = K_m - 1$. When magnetic susceptibility is equal in the three orthogonal axes x, y, and z, the material exhibits isotropic susceptibility. If the magnetic susceptibility is not equal along the x, y, and z axes, the material exhibits anisotropic susceptibility.[3,5] The type of susceptibility will determine the alignment of a molecule exposed to a magnetic field.

INTERACTION OF MAGNETIC FIELDS AND BIOLOGICAL MEMBRANES

The cellular membrane is hypothetically the primary site of interaction between cells and magnetic fields.[6–8] Biological membranes possess a fundamental characteristic structure described as the fluid mosaic model proposed by Singer and Nicolson in 1972.[9] The membrane is composed of a bilayer of phospholipids with embedded functional proteins that play an important role in transporting ions and other substances across the membrane. Biological membranes exhibit an intrinsic anisotropic structure and strong orientation in a magnetic field. In homogeneous fields, orientation of the cell membranes—parallel or perpendicular to the magnetic field—depends on constituent proteins, and the degree of orientation varies whenever phase transitions occur in the membrane. In heterogeneous fields, a translational force acts on anisotropic as well as on isotropic particles, depending on the variation of the intensity across particles, on the excess magnetic susceptibility of particles as compared to the magnetic susceptibility of the surrounding medium, and on particle volume.[10] The generation of metastable pores in cell membranes is attributed to the presence of magnetic particles in the membrane, such as biologically synthesized magnetite or contaminant particles. The basic premise is that the rotational motion of membrane-bound particles can transfer enough energy to the membrane to open aqueous pathways or form pores. Pore opening will result in significant influx of foreign molecules into the cytoplasm, alter cell biochemistry, and lead to the demise of the cell.[11]

BIOLOGICAL EFFECTS OF MAGNETIC FIELDS

The effects of magnetic fields on living organisms were observed as early as 1938. Kimball[12] observed that the rate of protoplasmic streaming in algae could be accelerated or retarded depending on the direction of the applied magnetic fields. Certain organisms are sensitive to very low intensity magnetic fields such as the magnetic field surrounding the earth. Aquatic magnetotactic bacteria move along the magnetic field lines of the earth.[13] Many animals also sense the magnetic field lines of the earth and are hypothesized to possess an internal magnetic compass.[14] However, no general consensus is yet established on whether exposure of living organisms to magnetic fields of high intensity is harmful, beneficial, or innocuous.

Calcium Transport Across Cellular Membranes

Alteration of ion transport across membranes as a consequence of magnetic field exposure is widely studied. Intracellular calcium concentration in lymphocytes exposed to combined a.c. and d.c. magnetic fields at 16 Hz (20.9 μT) and 50 Hz (65.3 μT) remained unchanged for exposure up to 60 min.[15] A later study reported increased ^{45}Ca uptake by human osteosarcoma cells exposed to simultaneous 40 μT a.c. and 20 μT d.c. magnetic fields. However, increased calcium uptake was only observed when cultured cells were exposed to the magnetic field, and returned to control levels once the magnetic field was removed. The ion uptake was frequency dependent, and exhibited a peak at 16.3 Hz. The existence of "windows," i.e., a range of frequencies at which increased calcium uptake due to magnetic field exposure are noticeable is hypothesized by Fitzsimmons et al.[16]

Magnetic Field Effects on DNA and Gene Transcription

Even though cellular membranes are theorized as the main interaction site between living organisms and magnetic fields, altered synthesis of proteins of eukaryotic and prokaryotic cell cultures exposed to magnetic fields has been reported.[17,18] Cells of *Escherichia coli* in the log growth phase exposed to a magnetic field of 3 mT and 60 Hz for 2 hr at 30°C did not exhibit any change in protein synthesis, but if the temperature was increased to 33°C for only 70 min, protein synthesis was altered.

Table 1 Effect of magnetic fields on different micro-organisms

Micro-organism	Type of MF	Field strength (T)	Effect	References
Wine yeast cells	Homogeneous SMF	1.1	No effect for 5 min, 10 min, 20 min, 40 min, or 80 min exposure	12
Serratia marcescens	Heterogeneous SMF	1.5	Growth rate equivalent to controls up to 6 hr; decreases between 6 hr and 7 hr; increases between 8 hr and 10 hr. At 10 hr, cell population equivalent to controls	19
Staphylococcus aureus	Heterogeneous SMF	1.5	Growth rate increases between 3 hr and 6 hr; then decreases between 6 hr and 7 hr. Cell population at 7 hr is same as in controls	19
Staphylococcus aureus	Homogeneous SMF	1.4	Inhibition of growth after 16 hr. No effect observed when exposure was interrupted hourly for 3 sec	20
Saccharomyces cerevisiae	Homogeneous SMF	1.49	No statistical differences from controls	21
Saccharomyces cerevisiae	Homogeneous OMF	0.0005	Cell concentration was 20%–30% higher than unexposed culture after 9 hr	22
Saccharomyces ellipsoide	Homogeneous SMF	0.1–1.2	No observable differences	23
Claviceps purpurea	Homogeneous SMF	1.1–1.4	Possible slight influence of MF on germination and growth direction of germ tubes	23
E. coli	Homogeneous SMF	11.7	Neither mutagenic nor lethal effect of the MF was observed	24
E. coli	Homogeneous SMF	11.7	Super high MF accelerates or reduces the growth rate depending on the media. When temperature was increased from 30 to 40°C, the growth rate was decreased in both media	25
E. coli	Homogeneous SMF	7	Comparing the ratio of number of cells under HMF/number of cells control: before 6 hr of treatment ratio < 1, after 24 hr of treatment ratio > 1	26
E. coli	Heterogeneous SMF	5.2–6.1		
E. coli	Heterogeneous SMF	5.2–6.1	The number of cells in stationary phase under the high magnetic field was 100,000 times higher than control	27
Bacillus subtilis	Homogeneous SMF	7	No difference in cell number between treated and untreated samples	28
Bacillus subtilis	Heterogeneous SMF	5.2–6.1	Twofold cell number with respect to the control due to a slower death rate of the vegetative cells	28
Pseudomonas stutzeri	PMF (10 μsec pulse width)	0.0013	After 7 hr, biomass was 25%–30% higher than in controls	29

SMF is the static magnetic field, OMF is the oscillating magnetic field, and PMF is the pulsed magnetic field.

Table 2 Inactivation of food spoilage micro-organisms

Micro-organism	Product	Field intensity (T)	Number of pulses	Frequency (kHz)	Temperature		Micro-organism per cm^3		Log cycles reduced
					Initial	Final	Before	After	
Streptococcus thermophilus	Milk	12	1	6	23	24	25,000	970	1.4
Saccaromyces cerevisiae	Yogurt	40	10	416	4	6	3,500	25	2.1
Saccaromyces cerevisiae	Orange juice	40	1	416	20	21	25,000	6	3.6
Mold spores	Dough	7.5	1	8.5	—	—	3,000	1	3.5

(From Ref. [30].)

Effect of Magnetic Fields on Micro-organisms

Most reports on the effect of magnetic fields on micro-organisms selected extremely low frequency (15 Hz–60 Hz) and very low intensities (micro Tesla to mili Tesla) to study effects on growth rate. The use of extremely low frequency and low intensity magnetic fields to investigate alteration of cellular growth by exposure to magnetic fields resulted from the concern of whether the magnetic fields generated from household appliances are harmful for people. Results are varied and the effects of selected magnetic fields range from inhibition to stimulation, including also the absence of any effect.

In some experiments, the observed result was dependent on the specific combinations of frequency and intensity of the magnetic field used. Results with higher intensity magnetic fields are variable as well. Table 1 summarizes published results on selected micro-organisms, type of magnetic field, and observed effects of several research groups.

MAGNETIC FIELDS AND FOOD PROCESSING

The use of magnetic fields for food preservation was first proposed by Hofmann[30] in a 1985 patent, which describes inactivation of selected micro-organisms by oscillating magnetic fields with intensities greater than 2 T. Table 2 presents results obtained under various treatment conditions. Some restrictions for microbial inactivation were noted, and the most important requirement for achieving microbial inactivation is the high electrical resistivity, i.e., greater than 10 ohm cm–25 ohm cm, in the food product. Metallic packaging materials cannot be used for magnetic field treatment, however, the microbial inactivation was successful when foods hermetically sealed in flexible film packages were used. So far, no other studies on the use of magnetic fields for food preservation other than Hofmann's Patent are known.

FINAL REMARKS

The use of magnetic fields as a technology for consistent food preservation is not yet possible. There is little evidence supporting the usefulness of magnetic fields as a food processing technology, but little research substantiates or contradicts potential uniform inactivation of micro-organisms in food. The mechanism of interaction between magnetic fields and living organisms is very complex and not completely understood at this time. Moreover, the lack of consistent experimental results makes elucidating interaction mechanisms an even more difficult task. However, the usefulness of magnetic field technology by the food industry should not be neglected. Although microbial inactivation is the most desirable attribute for using magnetic fields in the food industry, other applications should not be overlooked. The use of low intensity static magnetic fields as biomass growth promoters in fermentation and biotechnology processes either in the food or pharmaceutical industries may have potential use. The structural alteration of food constituents such as proteins or fats may, in theory, be achieved by exposure to magnetic fields, and although no research exists yet, products with specific properties could be obtained.

REFERENCES

1. Mulay, L.N. Basic Concepts Related to Magnetic Fields and Magnetic Susceptibility. In *Biological Effects of Magnetic Fields*; Barnothy, M.F., Ed.; Plenum Press: New York, 1964; Vol. 1, 33–55.
2. Abler, R. Magnets in Biological Research. In *Biological Effects of Magnetic Fields*; Barnothy, M.F., Ed.; Plenum Press: New York, 1969; Vol. 2, 1–27.
3. Pothakamury, U.R.; Barbosa-Cánovas, G.V.; Swanson, B.G. Magnetic-Field Inactivation of Microorganisms and Generation of Biological Changes. Food Technol. **1993**, *47* (12), 85–93.
4. Barbosa-Cánovas, G.V.; San Martín, M.F.; Harte, F.; Swanson, B.G. Magnetic Fields as a Potential Nonthermal

Technology for the Inactivation of Microorganisms. In *Control of Foodborne Microorganisms*; Juneja, V.K., Sofos, J.N., Eds.; Marcel Dekker, Inc.: New York, 2001; 399–418.
5. Serway, R.A. *Physics: For Scientists and Engineers*; CBS College Publishing: New York, 1982; 624–640.
6. Liboff, A.R. Cyclotron Resonance in Membrane Transport. In *Interactions Between Electromagnetic Fields and Cells*; Chiabrera, A., Nicolini, C., Schwan, H.P., Eds.; Plenum Press: New York, 1985; 281–296.
7. Phillips, J.L.; Haggren, W.; Thomas, W.J.; Ishida-Jones, T.; Adey, W.R. Effect of 72 Hz Pulsed Magnetic Field Exposure in RAS p21 Expression in CCRF-CEM Cells. Cancer Biochem. Biophys. **1993**, *13*, 187–193.
8. Tuinstra, R.; Greenebaum, B.; Goodman, E.M. Effects of Magnetic Fields in Cell-Free Transcription in *E. coli* and HeLa extracts. Bioelectromagnetics **1997**, *43*, 7–12.
9. Singer, S.J.; Nicholson, G.L. The Fluid Mosaic Model of the Structure of Cell Membranes. Science **1972**, *175* (23), 720–731.
10. Maret, G.; Dransfeld, K. Biomolecules and Polymers in High Steady Magnetic Fields. In *Strong and Ultrastrong Magnetic Fields and Their Applications*; Herlach, F., Ed.; Springer-Verlag: Berlin, Germany, 1985; 143–204.
11. Vaughan, T.E.; Weaver, J.C. Molecular Change Due to Biomagnetic Stimulation and Transient Magnetic Fields: Mechanical Interference Constrains on Possible Effects by Cell Membrane Pore Creation via Magnetic Particles. Bioelectrochem. Bioenerg. **1998**, *46*, 121–128.
12. Kimball, G.C. The Growth of Yeast in a Magnetic Field. J. Bacteriol. **1938**, *35*, 109–122.
13. Blakemore, R.P.; Frankel, R.B. Magnetic Navigation in Bacteria. Sci. Am. **1981**, *13*, 58–65.
14. Edmonds, D.T. *Electricity and Magnetism in Biological Systems*; Oxford University Press: Great Britain, 2001; 208–222.
15. Coulton, L.A.; Barker, A.T. Magnetic Fields and Intracellular Calcium: Effects on Lymphocytes Exposed to Conditions for 'Cyclotron Resonance.' Phys. Med. Biol. **1993**, *38*, 347–360.
16. Fitzsimmons, R.J.; Ryaby, J.T.; Magee, F.P.; Baylink, D.J. Combined Magnetic Fields Increased Net Calcium Flux in Bone Cells. Calcif. Tissue Int. **1994**, *55*, 376–380.
17. Mittenzwey, R.; Sübmuth, R.; Mei, W. Effects of Extremely Low-Frequency Electromagnetic Fields in Bacteria—The Question of a Co-stressing Factor. Bioelectromagnetics **1996**, *40*, 21–27.
18. Blank, M. Biological Effects of Electromagnetic Fields. Bioelectrochem. Bioenerg. **1993**, *32*, 203–210.
19. Gerencser, V.F.; Barnothy, M.F.; Barnothy, J.M. Inhibition of Bacterial Growth by Magnetic Fields. Nature **1962**, *196*, 539–541.
20. Hedrick, H.G. Inhibition of Bacterial Growth in Homogenous Fields. In *Biological Effects of Magnetic Fields*; Barhothy, M.F., Ed.; Plenum Press: New York, 1964; Vol. 1, 240–245.
21. Malko, J.A.; Constantinidis, I.; Dillejay, D.; Fajman, W.A. Search for Influence of 1.5 Tesla Magnetic Field in Growth of Yeast Cells. Bioelectromagnetics **1994**, *15*, 495–501.
22. Mehedintu, M.; Berg, H. Proliferation Response of Yeast *Saccharomyces cerevisiae* on Electromagnetic Field Parameters. Bioelectrochem. Bioenerg. **1997**, *43*, 67–70.
23. Montgomery, D.J.; Smith, A.E. A Search for Biological Effects of Magnetic Fields. Biomed. Sci. Instrum. **1963**, *1*, 123–125.
24. Okuno, K.; Ano, T.; Shoda, M. Effect of Super High Magnetic Field on the Growth of *Escherichia coli*. Biotechnol. Lett. **1991**, *13* (10), 745–750.
25. Okuno, K.; Tsuchiya, K.; Ano, T.; Shoda, M. Effect of Super High Magnetic Field on the Growth of *Escherichia coli* Under Various Medium Compositions and Temperatures. J. Ferment. Bioeng. **1993**, *75* (2), 103–106.
26. Tsuchiya, K.; Nakamura, K.; Okuno, K.; Ano, T.; Shoda, M. Effect of Homogeneous and Inhomogeneous High Magnetic Fields on the Growth of *Escherichia coli*. J. Ferment. Bioeng. **1996**, *81* (4), 343–346.
27. Horiuchi, S.; Ishizaki, Y.; Okuno, K.; Ano, T.; Shoda, M. Drastic High Magnetic Field Effect on Suppression of *Escherichia coli* Death. Bioelectrochemistry **2001**, *53* (2), 149–153.
28. Nakamura, K.; Okuno, K.; Ano, T.; Shoda, M. Effect of High Magnetic Field on the Growth of *Bacillus subtilis* Measured in a Newly Developed Superconducting Magnet Biosystem. Bioelectrochem. Bioenerg. **1997**, *43*, 123–128.
29. Hönes, I.; Pospischil, A.; Berg, H. Electrostimulation of Proliferation of the Denitrifying Bacterium *Pseudomonas stutzeri*. Bioelectrochem. Bioenerg. **1998**, *44*, 275–277.
30. Hofmann, G.A. Deactivation of Microorganisms by an Oscillating Magnetic Field. US Patent 4,524,079, June 18, 1985.

Mass Transfer in Biological Membranes

Arthur T. Johnson
University of Maryland, College Park, Maryland, U.S.A.

INTRODUCTION

Consider trying to separate small food flavor molecules from proteins and polysaccharides in solution in fruit juice. Or consider implanting porcine pancreas cells into a diabetic human to produce insulin when needed without allowing the human antibodies to contact directly with the foreign swine cells. Or consider removing pollutants and impurities from wastewater before recycling. If any of these were needed, you probably would use a membrane to achieve your purposes.

Membranes and films are thin layers of materials that can be used to control access of gases, liquids, or solids in solution on one side to the space on the other side. Biological engineers with many different interests use membranes and films to produce different effects. Companies often find membrane processing to be a less expensive alternative to other energy-intensive techniques, such as evaporation, distillation, or chemical separations. The energy cost of a membrane separation is only 5%–10% of that for evaporation of water.

There are two general classes of membranes: porous and nonporous. The former act as molecular sieves, allowing particles smaller than the pore diameters to pass and retaining the remainder of the particles. Diffusion occurs through these membranes just as it would through any other porous solid. Some membranes are nonporous, where the solute dissolves in the membrane material on its way through.

NONPOROUS MEMBRANES

Diffusion through nonporous membranes requires two steps. First, the molecular species passing through the membrane must dissolve in the membrane material. At this point it usually forms a solid-within-solid solution, although it is possible that the membrane behaves as a liquid. Second, the solute must move by molecular diffusion through the membrane material. Because of the first step, it is often possible that the limitation to movement of the diffusing species is not diffusion, but is instead due to limited solubility.

As long as the membrane can be considered to be homogenous (no pores, solute dissolved in membrane) diffusion occurs according to the steady-state Fick equation

$$\dot{m} = D_{\mathrm{ms}} A \frac{\Delta c_{\mathrm{m}}}{L} \tag{1}$$

where $\dot{m}$ is the mass flow rate (kg sec^{-1}); D_{ms}, the diffusion coefficient of the solute in the membrane (m^2 sec^{-1}); Δc_{m}, the concentration difference within the membrane (kg m^{-3}); L, the membrane thickness (m); and A, the membrane area (m^3).

POROUS MEMBRANES

There are other membranes that are porous, but the pores are so small that they effectively block some molecules from moving through the membrane while allowing smaller molecules to pass relatively freely. When considering food and biological materials, these membranes are used to distinguish between water (as the solvent) and solute molecules or ions. Such membranes are called semipermeable.

For a membrane that is partially solute permeable, solute particles flow through the membrane because of a concentration gradient across the membrane. Fick's equation can be used to describe this flow. Membrane pores are large enough, however, that solvent particles flow because of the pressure difference from one side of the membrane to the other. They flow as a streamlined fluid through a porous medium and obey Darcy's law

$$\dot{V} = \frac{-kA}{\mu} \frac{\partial p}{\partial L} \tag{2}$$

where $\dot{V}$ is the volume flow rate (m^3 sec^{-1}); k, the permeability (m^2); A, cross-sectional area (m^2); μ, fluid viscosity (N sec m^{-2}); and $\partial p/\partial L$ is the pressure gradient along the flow path (N m^{-3}).

The terms microfiltration, ultrafiltration, and nanofiltration are used to designate membrane pore sizes. In general, microfiltration is used to clarify slurries or remove suspended particulate matter; ultrafiltration is used to separate simple molecules, such as salts and sugars, from macromolecules; nanofiltration separates the smaller salt molecules from larger sugars and organic compounds; reverse osmosis separates water (or other solvent) from solutes. (Fig. 1)

Encyclopedia of Agricultural, Food, and Biological Engineering
DOI: 10.1081/E-EAFE 120007191
Copyright © 2003 by Marcel Dekker, Inc. All rights reserved.

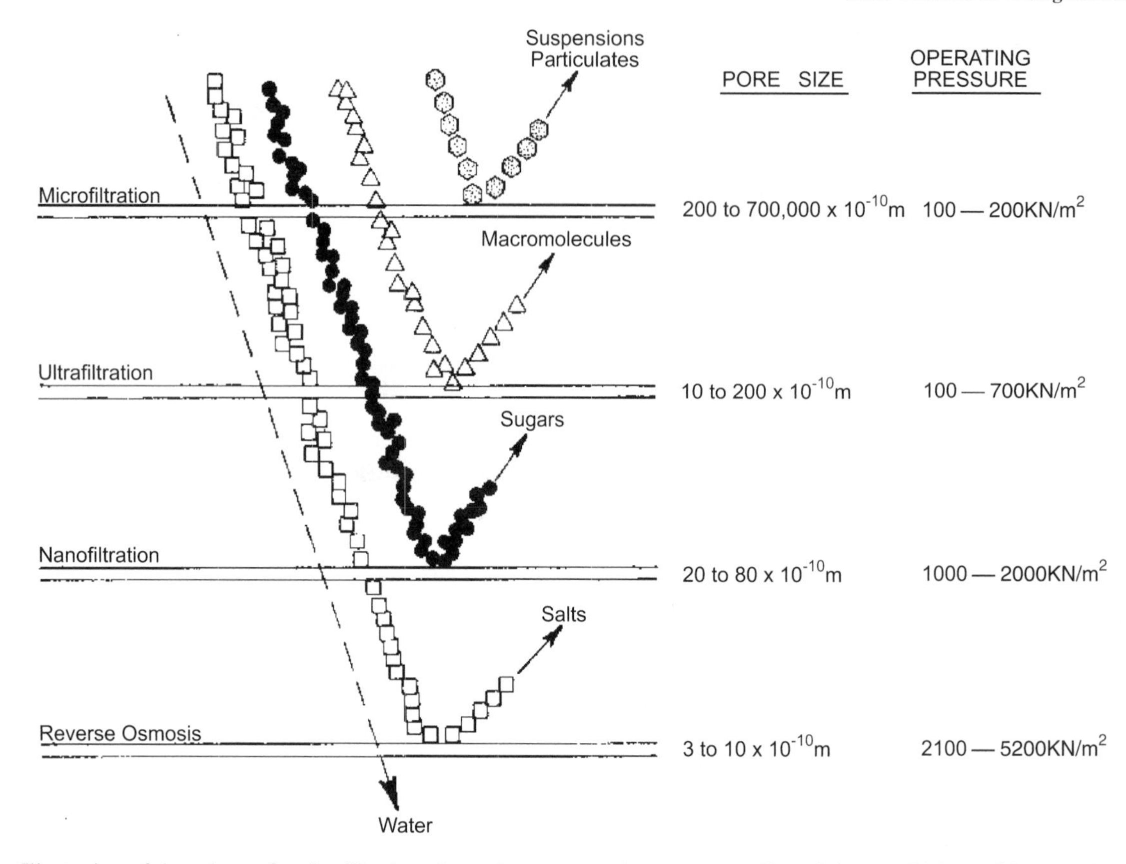

Fig. 1 Illustration of the scheme for classification of membrane separation processes. From Johnson, *Biological Process Engineering* (1999). Reprinted by permission of John Wiley & Sons, Inc.

The unit of measurement of the pore size of an ultrafiltration membrane is the Dalton, which is defined as the number of grams per gram mole (or kg (kg mol)$^{-1}$) of a molecule, equivalent to molecular mass. The minimum size of blood antibodies is usually taken to be 100 kDa, so an ultrafiltration membrane that excludes compounds this large and larger protects foreign tissues contained inside the membrane from attack and destruction by the new host into which a bioartificial organ is implanted.

Removal of solvent ions from the high concentration side of the membrane raises the local solute concentration near the membrane face, raising the driving osmotic pressure as well. This tends to reduce solvent flow through the membrane. This action has been termed "concentration polarization," and is responsible for slowing mass transfer from one side of the membrane to the other. When concentration polarization occurs, solvent flow becomes nearly independent of applied pressure. Concentration polarization is corrected largely by allowing the solution to flow parallel to the membrane face. As the rate of solvent flow through the membrane is often determined more by the convective mass transfer of concentrated solution away from the membrane than by the membrane itself, turbulent flow is often used to promote mixing.

MEMBRANE MATERIALS

Membranes are made from many types of materials and can be fabricated according to required specifications. Specialized films include biodegradable ones often produced from starch compounds, soluble films that dissolve when placed in water, and edible films for packaging food materials and that need not be removed before eating.

One material does not often possess all desired characteristics of strength, clarity, impermeability, stability, and others. An approach to impart all desired characteristics to one material is to laminate several materials together to form a composite. Reverse osmosis membranes are very thin (0.1 μm–0.25 μm) and require a porous support layer to resist the large, imposed physical pressure differences used. Since resistances to diffusion in the composite material are in series (Fig. 2), the overall resistance is the sum of the individual resistances.

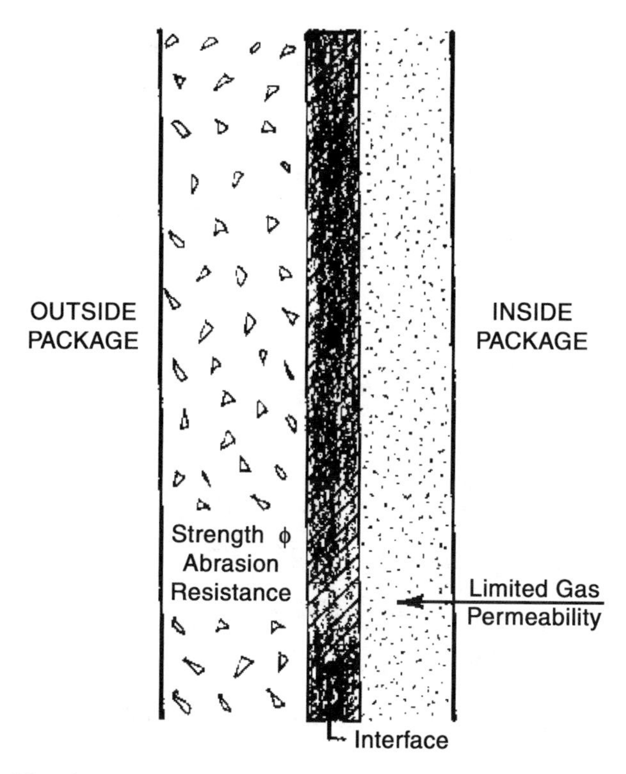

Fig. 2 Schematic of a composite film. From Johnson, *Biological Process Engineering* (1999). Reprinted by permission of John Wiley & Sons, Inc.

One means to increase flow rate is to increase surface area. Membranes for all levels of filtration are manufactured with large surface areas included in very small volumes. Thus, there are tubular membranes, membrane plates, spiral-wound membranes, and pleated sheets. One popular configuration is the hollow fiber type (Fig. 3). Small diameter (200 μm inside diameter, 215 μm–260 μm outside diameter, is typical for hemodialysis) hollow fibers form a mass inside a protective shell. The high pressure permeate is usually forced through the hollow fibers while the low-pressure solvent flows through the shell. This design is easy to troubleshoot, and the hollow fiber bundle is easy to change.

OTHER CONSIDERATIONS

In addition, other factors are important:

1. Cleaning. Membranes easily foul with retentate particles and must be cleaned with solutions that do not damage the membranes.
2. Temperature. Membranes may decompose or lose their physical strength at high temperatures. Permeate resistance decreases about 3% per °C rise.
3. Chemical and biological resistance. Some membrane types are much more susceptible to chemical or biological activity than others.
4. Aging. There is an average increase of 4%–6% in membrane permeate resistance for each year in service. Typical membranes have useful lives of 3 yr.
5. Cost. Cellulose acetate membranes are particularly inexpensive.
6. Pressures. Two pressures are important: 1) the pressure drop along the flow path, which influences the size of the pump required and 2) the pressure drop across the membrane, determined by membrane strength, and which must be resisted by the membrane.

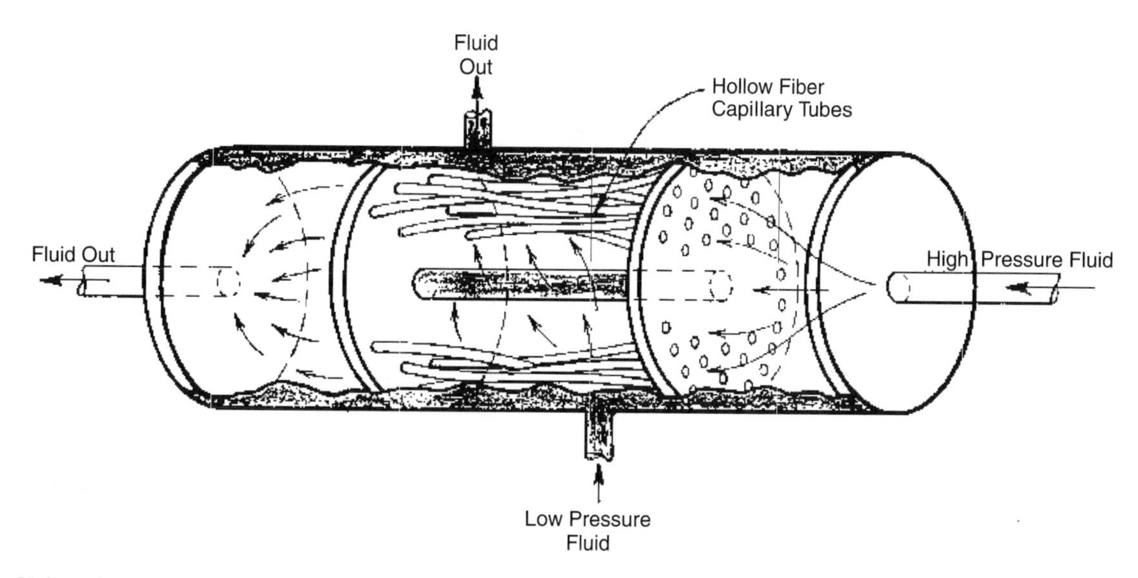

Fig. 3 Hollow fiber membrane device. From Johnson, *Biological Process Engineering* (1999). Reprinted by permission of John Wiley & Sons, Inc.

BIOLOGICAL MEMBRANES

Cell membranes are formed from a lipoprotein bilayer with lipid portions facing both the inside of the cell and the outside of the cell. This should make the membrane hydrophobic.

The traditional view of the cell membrane is that it acts as a barrier to substances that would otherwise pass freely between the cellular cytoplasm and the surrounding interstitial fluid. Thus, water, proteins, fatty acids, ions, and other materials are locked in the cell, whereas many external macromolecules and certain ions are excluded. Of particular importance are sodium and potassium ions, which play an important part in the maintenance of the cellular resting potential and in the propagation of neuromuscular action potentials. It is thought that both sodium and potassium ions migrate through the cell membrane by passive diffusion (according to Fick's law), but that there is an additional metabolically powered sodium–potassium pump in the membrane that maintains sodium concentration low and potassium concentration high inside the cell.

The difficulties with this view are that the cell membrane is full of holes, and that integrity of the membrane is not required for proper functioning of the cell.[1] A sodium–potassium pump, if it exists, would be overwhelmed by passive diffusion of each of these ions. Properties, traditionally attributed to the cell membrane must, therefore, be due to the cytoplasmic gel inside the cell.

REFERENCE

1. Pollack, G.H. *Cells, Gels, and the Engines of Life*; Ebner and Sons: Seattle, 2001.

FURTHER READING

Johnson, A.T. *Biological Process Engineering*; John Wiley & Sons: New York, 1999.

Maralidhara, H.S. Using Membranes for Separation Processing. Genet. Eng. News **2000**, *20* (14), 46, 77.

Mechanical Energy Balance

Yanyun Zhao
Oregon State University, Corvallis, Oregon, U.S.A.

M

INTRODUCTION

The purpose of this section is to provide the practice information necessary to predict pressure drop for non-time-dependent, homogeneous, Newtonian, and non-Newtonian fluids in fluid handling system. The intended application of this material is pipeline design and pump selection. A comprehensive discussion of pipeline design calculation can be found in Refs. 1, 2.

ENGINEERING BERNOULLI EQUATION

The equation used to describe the mechanical energy balance for fluids is commonly called the "engineering Bernoulli equation."[3,4] Numerous assumptions are made in developing the equation, including constant fluid density, absence of thermal energy effects, single phase, uniform material properties, and uniform equivalent pressure. The mechanical energy balance for pumping an incompressible fluid at steady-state conditions through a pipeline (Fig. 1) can be expressed as[2]:

$$\frac{u_2^2 - u_1^2}{\alpha} + g(Z_2 - Z_1) + \frac{P_2 - P_1}{\rho} + E_f + W = 0 \qquad (1)$$

where Z is the height above a reference point (m), P, the pressure (Pa), u, the fluid velocity (m sec^{-1}), W, the work output per unit mass (J kg^{-1}), E_f, the friction loss per unit mass (J kg^{-1}), α, the kinetic energy correction factor (dimensionless), ρ, the density (kg m^{-3}), g, acceleration due to gravity (9.81 m sec^{-2}), and subscripts 1 and 2 refer to two locations in the pipe system as shown in Fig. 1. The friction losses E_f include those from pipes of different diameters and from individual valves and fittings, and can be calculated as:

$$E_f = \sum_1^a \frac{2fu^2L}{D} + \sum_1^b \frac{k_f u^2}{2} \qquad (2)$$

where f is the friction factor (dimensionless), L, the length of straight pipe of diameter D (m), k_f, the friction coefficient for a fitting (dimensionless), and a and b are the numbers of straight pipe and valves or fittings (dimensionless).

KINETIC ENERGY CALCULATION

The kinetic energy term in the mechanical balance equation can be calculated by knowing the kinetic energy correction factor, α. In turbulent flow of any fluid, $\alpha = 2$. For various fluids in laminar flow, equations for calculating α values are given in Table 1.

FRICTIONAL LOSSES IN PIPES

As many fluid foods are non-Newtonian in nature, the friction losses in straight pipes and tubes can be estimated from the Fanning friction factor, f, which is defined as the ratio of the wall shear stress in a pipe to the kinetic energy per unit volume[2]:

$$f = \frac{2\sigma_w}{\rho u^2} \qquad (3)$$

f can be considered in terms of pressure drop by substituting the definition of the shear stress at the wall[2]:

$$f = \frac{(\Delta P)R}{\rho L u^2} = \frac{(\Delta P)D}{2\rho L u^2} \qquad (4)$$

where $\Delta P = P_2 - P_1$ (Pa). Simplification yields the energy loss per unit mass required in the mechanical energy balance

$$\frac{\Delta P}{\rho} = \frac{f 2 L u^2}{D} \qquad (5)$$

Steffe and Steffe and Singh[1,2] described the calculation of the Fanning friction factor f in great detail. This article provides a brief summary.

In laminar flow, f values can be determined from the equations describing the relationship between pressure drop and flow rate for a particular fluid.[1] Following are

Encyclopedia of Agricultural, Food, and Biological Engineering
DOI: 10.1081/E-EAFE 120006959
Copyright © 2003 by Marcel Dekker, Inc. All rights reserved.

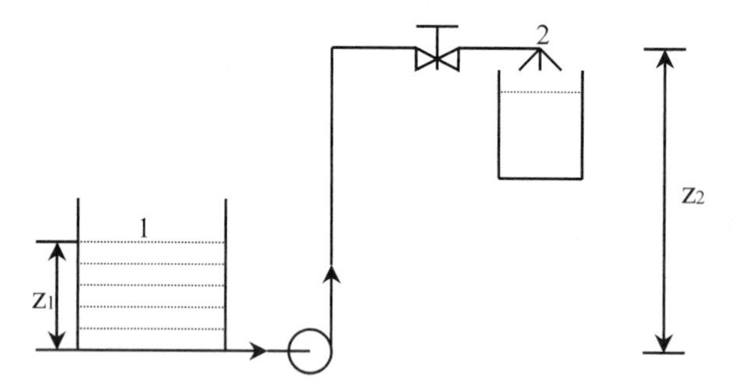

Fig. 1 Schematic illustration of pipelines for liquid food.

several examples:

For a Newtonian fluid when $N_{Re} < 2100$,

$$f = \frac{16}{N_{Re}} \tag{6}$$

For power law fluid,

$$f = \frac{16}{N_{Re,PL}} \tag{7}$$

For Bingham plastic fluid,

$$f = \frac{16(6N_{Re,B} + N_{He})}{6N_{Re,B}^2} \tag{8}$$

where N_{Re} is the Newtonian fluid Reynolds number, equal to $Du\rho/\mu$ (dimensionless), $N_{Re,PL}$, the power law fluid Reynolds number, $N_{Re,B}$, the Bingham plastic fluid Reynolds number, $Du\rho/\mu_{pl}$, and N_{He} is the Hedstrom number, $D^2\sigma_0\rho/\mu_{pl}^2$.

Table 1 Kinetic energy correction factors for laminar flow in tubes

Fluid	α, dimensionless
Newtonian $\sigma = \mu\gamma$	1.0
Power law $\sigma = K(\gamma)^n$	$\alpha = \frac{2(2n+1)(5n+3)}{3(3n+1)^2}$
[a]Bingham Plastic $\sigma = \mu_{pl}\gamma + \sigma_0$	$\alpha = \frac{2}{2-c}$
[a]Herschel–Bulkley $\sigma = K(\gamma)^n + \sigma_0$	$\alpha = \exp(0.168c - 1.062nc - 0.954n^5 - 0.115c^5 + 0.831)$ for $0.06 \leq n \leq 0.38$ and $\alpha = \exp(0.849c - 0.296nc - 0.600n^5 - 0.602c^5 + 0.733)$ for $0.38 < n \leq 1.60$

$c = \sigma_0/\sigma_w$ (dimensionless), σ_0, yield stress (Pa), σ_w, shear stress at the wall of the tube or slit (Pa), n, flow behavior index (dimensionless), γ, shear rate (1/sec), and μ_{pl} is plastic viscosity of Bingham fluid (Pa sec).
[a]Solution for the Bingham plastic materials is within 2.5% of the true solution.[9] Errors in using the Herschel–Bulkley solution are less than 3% for $0.1 \leq c \leq 1.0$ but as high as 14.2% for $0.0 \leq c \leq 0.1$.[10]
(From Ref. 1.)

Table 2 Fanning friction factor equations for turbulent flow in smooth tubes

Fluid	Fanning friction factor
Newtonian $\sigma = \mu\gamma$	$\frac{1}{\sqrt{f}} = 4.0\log_{10}(N_{Re}\sqrt{f}) - 0.4$ where: $N_{Re} = \frac{\rho D u}{\mu}$
Power law $\sigma = K(\gamma)^n$	$\frac{1}{\sqrt{f}} = \left(\frac{4}{n^{0.75}}\right)\log_{10}[(N_{Re,PL})f^{(1-(n/2))}] - \left(\frac{0.4}{n^{1.2}}\right)$ where: $N_{Re,PL} = \left(\frac{D^n(u)^{2-n}\rho}{8^{n-1}K}\right)\left(\frac{4n}{3n+1}\right)^n$
Bingham plastic $\sigma = \mu_{pl}\gamma + \sigma_0$	$\frac{1}{\sqrt{f}} = 4.53\log_{10}(1-c) + 4.53\log_{10}((N_{Re,B})\sqrt{f}) - 2.3$ where: $N_{Re,B} = \frac{Du\rho}{\mu_{pl}}$ and $c = \frac{\sigma_o}{\sigma_w} = \frac{2\sigma_0}{f\rho(u)^2}$

(From Ref. 1 with original data from Ref. 11.)

For turbulent flow, friction factors may be determined from empirical equations as shown in Table 2. The equations are only applicable to smooth pipes that include sanitary piping systems for foods.

FRICTION LOSS COEFFICIENT FOR VALVES, FITTINGS, AND SIMILAR PARTS

The friction loss coefficient, k_f must be determined from experimental data, summarized in various food engineering handbooks.[1,2,5,6] Friction loss coefficients for many valves and fittings for both laminar flow and turbulent flow are summarized in Tables 3 and 4. Laminar flow data are much more limited. In addition, Steffe, Mohamed, and Ford[7] determined magnitudes of the coefficient for a fully open plug valve, a tee with flow from line to branch, and a 90° short elbow as a function of *GRe* using applesauce as the test fluid. The regression equations were:

$$\text{Three-way plug valve}: \quad k_f = 30.3GRe^{-0.492} \tag{9}$$

$$\text{Tee}: \quad k_f = 29.4GRe^{-0.504} \tag{10}$$

Table 3 Friction loss coefficients (k_f) for the laminar flow of Newtonian fluids

Types of fitting or valve	N_{Re} = 1000	500	100
90° elbow, short radius	0.9	1.0	7.5
Tee, standard, along run	0.4	0.5	2.5
Branch to line	1.5	1.8	4.9
Gate valve	1.2	1.7	9.9
Globe valve, composition disk	11	12	20
Plug	12	14	19
Angle valve	8	8.5	11
Check valve, swing	4	4.5	17

(From Ref. 2 with original data from Ref. 5.)

Table 4 Friction loss coefficients k_f for the turbulent flow of Newtonian fluids through valves and fittings

Type of fitting or valve	k_f
45° elbow standard	0.35
Long radius	0.2
90° elbow standard	0.75
Long radius	0.45
Square or miter	1.3
180° bend close return	1.5
Tee, standard, along run, branch blanked off	0.4
Used as elbow, entering run	1.0
Used as elbow, entering branch	1.0
Branching flow	1.0
Coupling	0.04
Union	0.04
Gate valve, open	0.17
3/4 open	0.9
1/2 open	4.5
1/4 open	24.0
Diaphragm valve, open	2.3
3/4 open	2.6
1/2 open	4.3
1/4 open	21.0
Globe valve, bevel seat, open	6.0
1/2 open	9.5
Composition seat, open	6.0
1/2 open	8.5
Plug disk, open	9.0
3/4 open	13.0
1/2 open	36.0
1/4 open	112.0
Angle valve, open	2.0
Y or blowoff valve, open	3.0
Plug cock $\theta = 0°$ (fully open)	0.0
$\theta = 5°$	0.05
$\theta = 10°$	0.29
$\theta = 20°$	1.56
$\theta = 40°$	17.3
$\theta = 60°$	206.0
Butterfly valve $\theta = 0°$ (fully open)	0.0
$\theta = 5°$	0.24
$\theta = 10°$	0.52
$\theta = 20°$	1.54
$\theta = 40°$	10.8
$\theta = 60°$	118.0
Check valve, swing	2.0
Disk	10.0
Ball	70.0
Foot valve	15.0
Water meter, disk	7.0
Piston	15.0
Rotary (star-shaped disk)	10.0
Turbine-wheel	6.0

(From Ref. 2 with original data from Ref. 5.)

$$\text{Elbow}: \quad k_f = 191.0 GRe^{-0.896} \tag{11}$$

where GRe is a generalized Reynolds number, equal to $4n(D^n u^{2-n}\rho)/(8^{n-1}K)(3n+1)$ (dimensionless) and K is the consistency index (dimensionless). The k_f values for the sudden contraction or expansion of a Newtonian fluid in turbulent flow[8] may be calculated as:

$$\text{Contraction}: \quad k_f = 0.5\left(1 - \left(\frac{D_{small}}{D_{large}}\right)^2\right) \tag{12}$$

$$\text{Enlargement}: \quad k_f = 0.5\left(1 - \left(\frac{D_{small}}{D_{large}}\right)^2\right)^2 \tag{13}$$

where D_{large} and D_{small} are the diameters of the two pipes (m). The greatest velocity, which is the mean velocity in the smallest diameter pipe, should be used for both contraction and expansion in calculating the friction loss term ($k_f u^2/2$).

After evaluating the available data for friction loss coefficients in laminar and turbulent flow, Steffe and Steffe and Singh[1,2] provided the following "rule-of-thumb" guidelines, conservative for shear-thinning fluids, for estimating k_f values:

1. For Newtonian fluids in turbulent or laminar flow use the data listed in Tables 3 and 4.
2. For non-Newtonian fluids above a Reynolds number ($N_{Re,PL}$ or $N_{Re,B}$) of 500, use data for Newtonian fluids in turbulent flow (Table 4).
3. For non-Newtonian fluids in the Reynolds number range of $20 \leq N \leq 500$ use the following equations:

$$k_f = \frac{\beta}{N} \tag{14}$$

where N is N_{Re}, $N_{Re,PL}$, or $N_{Re,B}$ depending on the type of fluid in question. The constant, β, is found for a particular valve or fitting (or any related item such as contraction) and calculated as:

$$\beta = (k_f)_{turbulent}(500) \tag{15}$$

Values of β may be calculated from the k_f values provided in Table 3. The value of 20 was arbitrarily set as a lower limit of N in Eq. 14 because very low values of the Reynolds number in that equation will generate unreasonably high values of the friction loss coefficient.

According to Steffe and Steffe and Singh,[1,2] the above guidelines should only be used in the absence of actual experimental data. Many factors, such as a high extensional viscosity, may significantly influence k_f values.

COMPUTATION OF ENERGY REQUIREMENTS FOR PUMPING

The total power requirements for pumping are calculated from Eq. 1 written for the work input, $-W$ ($\mathrm{J\,kg^{-1}}$).

$$-W = g(Z_2 - Z_1) + \frac{(P_2 - P_1)}{\rho} + \frac{(u_2^2 - u_1^2)}{\alpha} + E_f \tag{16}$$

The actual power requirement is computed from the mass flow rate and the pumping energy given by

$$\text{Power} = m(W) \tag{17}$$

where m is mass flow rate of the fluids ($\mathrm{kg\,sec^{-1}}$). To calculate pump sizes, the accurate sizes of pipes being utilized must be incorporated into the computations. Sample calculations for power requirements in different types of fluids can be found in Refs. 2, 6.

REFERENCES

1. Steffe, J.F. Pipeline Design Calculations. In *Rheological Methods in Food Process Engineering*, 2nd Ed.; Steffe, J.F., Ed.; Freeman Press: Michigan, 1996; 128–138.
2. Steffe, J.F.; Singh, R.P. Pipeline Design Calculations for Newtonian and Non-Newtonian Fluids. In *Handbook of Food Engineering Practice*; Valentas, K.J., Rotstein, E., Singh, R.P., Eds.; CRC Press: Boca Raton, NY, 1997; 1–35.
3. Denn, M.M. *Process Fluid Mechanics*; Prentice-Hall: Englewood Cliffs, NJ, 1980.
4. Brodkey, R.S.; Hershey, H.C. *Transport Phenomena*; McGraw-Hill: New York, 1988.
5. Sakiadis, B.C. Fluid and Particle Mechanics. In *Perry's Chemical Engineers' Handbook*, 6th Ed.; Perry, R.H., Green, D.W., Maloney, J.O., Eds.; McGraw-Hill: New York, 1984; 5-4–5-68.
6. Rao, M.A. Transportation of Fluid Foods. In *Handbook of Food Engineering*; Heldman, D.R., Lund, D.B., Eds.; Marcel Dekker, Inc.: New York, 1992; 204–211.
7. Steffe, J.F.; Mohamed, I.O.; Ford, E.W. Pressure Drop Across Valves and Fittings for Pseudoplastic Fluids in Laminar Flow. Trans. ASAE **1984**, *27* (2), 616–619.
8. Crane, Co. *Flow of Fluids Through Valves, Fittings, and Pipe*; Technical Paper No. 401M; Twenty-first printing, Crane Co.: New York, 1982.
9. Metzner, A.B. Non-Newtonian Technology: Fluid Mechanics, Mixing, Heat Transfer. In *Advances in Chemical Engineering*; Drew, T.B., Hoopes, J.W., Eds.; Academic Press: New York, 1956; Vol. 1.
10. Briggs, J.L.; Steffe, J.F. Kinetic Energy Correction Factor of a Herschel–Bulkley Fluid. J. Food Proc. Eng. **1995**, *18*, 115–118.
11. Grovier, G.W.; Aziz, K. *The Flow of Complex Mixtures in Pipes*; R.E. Krieger Publishing Co.: Malabar, FL, 1972.

Mechanical Recompression

William Kerr
The University of Georgia, Athens, Georgia, U.S.A.

M

INTRODUCTION

Methods for removing water in order to concentrate solids are important for producing foods such as concentrated juices, sugar syrups, whey concentrate, and milk. In addition, it is often advantageous to concentrate wastewater streams prior to disposal. However, removal of water constitutes a major use of energy in the food industry, and methods to improve the efficiency of water removal are desirable. Among the methods available for concentrating liquid foods are freeze concentration, reverse osmosis, and evaporation. Mechanical vapor recompression (MVR) is one method for increasing the efficiency of evaporators.

EVAPORATION

Evaporation is the removal of water as vapor in order to form a concentrated liquid product. The boiling temperature is that temperature at which the vapor pressure of water in the product is equal to the surrounding total pressure. At this temperature and pressure, liquid water in solution exists in equilibrium with pure water vapor. As the saturation temperature decreases with pressure, it is often advantageous to conduct evaporation under vacuum, particularly for those products that would suffer quality loss at high temperatures. The presence of dissolved solutes, such as sugars or salts, causes the boiling temperature to be higher than that for pure water. For an ideal solution, the boiling point elevation is $\Delta T_B = 0.51m$, where m is the solute concentration in moles per kg of water. As evaporation continues, the remaining liquor becomes increasingly concentrated, and thus the boiling temperature continues to increase.

Approximately 2350 kJ of energy must be supplied to evaporate each kilogram of water. Typically, only liquid products, such as juices or milk, are evaporated. Due to the increasing temperature and viscosity of the evaporating liquid, solute concentrations greater than 40% are not easily attained. Industrial evaporators have a vessel for containing product, and a means of heating it. For systems under vacuum, a vapor chamber and means of condensing or compressing water vapor for disposal are required. Product may be heated in batches, or pumped through continuous falling or rising film heat exchangers.

The vapor leaving the product still has substantial enthalpy content. If condensed and dumped, this heat is lost. However, the vapor could be used to heat product in a subsequent evaporator, provided the product in that chamber boils at a lower temperature than that of the vapor. Thus, to use this method of heat capture subsequent chambers need to be operated at lower pressure. This is the principle of the multieffect evaporator, and several effects may be used in series to enhance energy efficiency. However, the number of effects is practically limited by capital costs and requirements for high vacuum.

MECHANICAL VAPOR RECOMPRESSION

Another approach to recover energy from escaping water vapor is to direct it back to the steam jacket of the same effect, thus providing heat for further evaporation of water from the product. The temperature of the vapor must first be increased and this is accomplished by passing it through a compressor. For example, consider a system in which juice evaporates at 50°C, and which is heated by steam at 70°C. From a steam table, it could be found that steam at 50°C has a vapor pressure of 12.35 kPA, and at 70°C has a vapor pressure of 31.19 kPa. Thus, the compressor must increase the pressure of exiting vapor by 18.84 kPA before it can be passed to the steam chest.

The compressor may be either a positive-displacement, centrifugal, or axial-flow type. The compressor operates by an electrically driven motor or by steam propelled turbine. Vapor exiting the compressor is often superheated, and condensate may be sprayed into the vapor to reduce the superheat. In some cases, additional make-up steam may be required to keep the system going.

ENERGY EFFICIENCY

Conversion of water to steam takes approximately 2350 kJ kg^{-1} of water. A compressor operating at 70%–75% efficiency, and motor efficiency of 90%, requires approximately 70 kJ kg^{-1}–140 kJ kg^{-1} of compressed vapor.[1,2] In theory, once the vaporization reaches steady

Encyclopedia of Agricultural, Food, and Biological Engineering
DOI: 10.1081/E-EAFE 120007064
Copyright © 2003 by Marcel Dekker, Inc. All rights reserved.

state, no additional steam is needed for heating. Thus, the $70\,kJ\,kg^{-1}-140\,kJ\,kg^{-1}$ represents the sustained energy costs, and is normally provided by electricity running the compressor motor. Calculations indicate that the energy efficiency is equivalent to that attained with a 10–20 effect evaporator. Estimates show that while reverse osmosis is more cost effective at water removal rates less than $5\,kg\,hr^{-1}-10\,kg\,hr^{-1}$, the costs of equipment and annual operating costs favor MVR when rates are higher than $10\,kg\,hr^{-1}$.[2] In addition to recovering energy from product vapors, MVR systems do not require a steady flow of water to a condenser.

PRODUCTS

A variety of food products are subject to evaporation with MVR including citrus juice,[1] sugar solutions,[2] milk, coffee, corn syrup, and cheese whey.[3] In other areas, MVR has been used to concentrate acid hydrolysates from wood processing,[4] waste water from fish processing,[5] and to produce soda ash and other chemicals.

ADVANTAGES AND DISADVANTAGES

MVR offers several advantages to single or multieffect evaporators. Energy efficiency is equivalent to that of a 10–20 effect evaporator, but with less initial cost and space requirements. In addition, compressors can be run on electricity, which in some cases offers cost savings when compared to oil or natural gas. In a single effect MVR, the product temperature can be kept well below the maximum temperature found in multieffect units. MVR is not limited to a single type, or even, a single effect, and a multieffect evaporator may incorporate one or more effects with mechanical recompression. MVR seems to be most practical for evaporators with large heat transfer areas, and products with low boiling points.

REFERENCES

1. Worral, G.P. Mechanical Vapor Recompression Conserves Energy in Citrus Juice Concentration. Food Technol. **1982**, *36* (5), 234–238.
2. Teixeira, A.A. Reverse Osmosis and Mechanical Vapor Recompression Evaporation for Liquid Food Concentration. In *Proceedings of the Special Food Engineering Symposium Held in Conjunction with Food and Dairy Expo '83*; ASAE Press, 1983.
3. Kearney, O. *The Use of Mechanical Vapour Recompression Evaporator to Concentrate Acid Casein Whey*; Commission of the European Communities: Luxembourg, 1989.
4. Blomgren, Y.; Jonsson, A.S.; Wimmerstedt, R. Concentration of Acid Hydrolyzate by Reverse Osmosis and Mechanical Vapor Recompression Evaporation. J. Wood Chem. Technol. **1991**, *11* (1), 117–135.
5. O'Neill, G. *Concentration of Fish Stickwaer by Mechanical Vapour Recompression*; Commission of the European Communities: Luxembourg, 1989.

Mechanical Separation Systems Design

M

J. Peter Clark III
Consultant, Oak Park, Illinois, U.S.A.

INTRODUCTION

Mechanical separation systems include sedimentation, centrifugation, and filtration. These have in common the fact that they are usually applied to the removal of solids from liquids, though they can be applied to the separation of liquids from each other and the separation of solids from gases, and the fact that design of such systems relies heavily on empirical relationships and experimental measurements rather than on predictive theory. Theory is used to suggest the form of useful correlations, but of necessity is highly idealized. This article describes suggested approaches for each of the systems as most often applied in the food industry. Extensive references are provided for further detail. Some of these references are relatively old, but are still accurate and relevant.

Some specific examples for foods include: clarification of juices, wine and beer; removal of excess pulp from fruit juice; separation of solvents from oil seed meal; separation of water from oils; removal of dust from air; removal of sugar crystals from molasses; and removal of decolorizing carbon from sugar solutions.

SEDIMENTATION

Sedimentation relies on gravitational force to effect separation of particles from a fluid because of a difference in density. The gravitational force is balanced by a drag force applied by the fluid. A constant velocity, the terminal velocity, results and for an isolated sphere in laminar flow is expressed by Stoke's Law:[1]

$$v_t = gD^2(\rho_p - \rho_l)/18\mu \tag{1}$$

where, v_t is the terminal velocity, m sec^{-1}; g the gravitational acceleration, 9.80665 m sec^{-2}; D the particle diameter, m; ρ_p the particle density, kg m^{-3}; ρ_l the fluid density, kg m^{-3}; and μ the fluid viscosity, kg m^{-1} sec^{-1}.

This equation can be used to measure viscosity when all other parameters are known, or to define an equivalent diameter of irregular particles. In most realistic suspensions of particles, hindered settling occurs in which the observed terminal velocity is reduced from the Stokes value because of interaction among particles and increased fluid flow as the fluid is displaced by falling particles.

Separation of particles from fluids can be intended to remove turbidity from the fluid, in which case it is called clarification, and is common in treating water, wine, and beer. If the focus is on concentrating the solids, the process is called thickening. In clarification, it is common to add chemicals that promote aggregation of particles and thus increases settling rates. Since such chemicals may be expensive, it is common to conduct settling rate tests to establish minimum effective addition rates. Some additives used in foods include egg white, gelatin, and approved polymers. Observation of the height of the interface between settled solids and clear liquid at various times gives a settling curve from which an overflow rate is determined[2] for continuous thickening or clarification. In batch processes, a corresponding time is determined.

There is considerable experience in water and sewage treatment using sedimentation to remove solids. Typical values of loading rates are 600 gpd ft^{-2}–1200 gpd ft^{-2} for untreated waste water and 360 gpd ft^{-2}–1200 gpd ft^{-2} for flocs treated with alum, iron, or lime.[3](1gpd ft^{-2} = 4.717×10^{-7} m sec^{-1}).

CENTRIFUGATION

Centrifugation increases the available force for separation by rapidly rotating a cylinder in which a suspension or slurry is placed, continuously or batch-wise. Examples include recovering sugar crystals from concentrated syrup, recovering yeast from fermentation broths, and clarifying fruit juices. The design parameters for centrifugal separation include: length and diameter of bowl, rotation speed, flow rate of suspension, and properties of suspension. There are highly automated batch centrifuges in which collected solids are automatically discharged, sometimes after washing and partial drying. The walls of the centrifuge may be porous, in which case the mother liquor and wash liquid are removed through the bed of solids. In a solid bowl, the liquids discharge from the center.

In a continuous centrifuge, the solids may be removed by a counter-rotating screw conveyor. No matter the design,

DOI: 10.1081/E-EAFE 120007079
Copyright © 2003 by Marcel Dekker, Inc. All rights reserved.

experiments are usually necessary to establish operating conditions that provide desired performance. A scale-up procedure[1] computes the following parameter:

$$\Sigma = (\omega^2(\pi b(r_2^2 - r_1^2)))/(2g\ln(2r_2/(r_1 + r_2))) \quad (2)$$

where ω is the rotational velocity, radian $\sec^{-1}$ ($= 2\pi N/60$, $N = \text{rev min}^{-1}$); b the length, m; r_2 the radial distance to interface of heavy phase with light phase (often set by a dam or valve) or distance to wall, for solids removal, m; r_1 the radial distance to surface of light phase, m.

To scale up from a flow rate, q_1 and Σ to a new flow rate, q_2, compute $\Sigma_2 = \Sigma_1\ (q_2/q_1)$ (Fig. 1).

This implies that flow rate is proportional to speed squared, but experience suggests[4] that the true exponent is about 1.5. Experimentation with actual suspensions and machines is usually necessary to define the range of operating conditions.

FILTRATION

Filtration uses a porous medium, such as woven cloth, wire mesh, or a bed of solids to remove fine particles from a fluid stream. The exact mechanism depends on the particles and medium, and the driving force can be imposed pressure, centrifugal acceleration, vacuum, or gravity. It is common and convenient to operate under constant pressure drop, in which case the process is described by the following equation:[1,5]

$$t = (\mu/g_c\Delta p)((c\alpha/2)(V/A)^2 + (R_m V/A)) \quad (3)$$

where t is the time taken to collect a volume, V of filtrate, sec; μ the fluid viscosity, kg m $\sec^{-1}$; g_c the conversion constant, 1 in SI units [32.17 ft lb (lbf $\sec^2)^{-1}$ in English units]; Δp the pressure difference, N m^{-2}; c the mass fraction solids in suspension; V the filtrate volume, m^3; A the filter area, m^2; α the specific cake resistance, m kg^{-1} and R_m the media resistance (including system losses), m^{-1}.

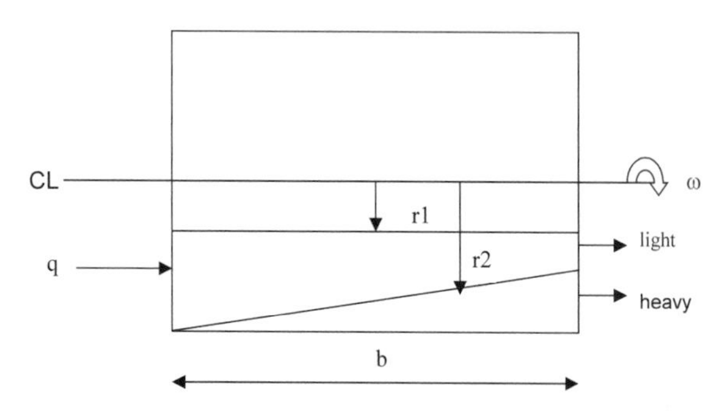

Fig. 1 Schematic of tubular centrifuge.

Data on a suspension of interest are collected using the medium of interest at various pressure drops. When plotted as t/V vs. V, a straight line should result for each Δp. From the slope and intercept, the various quantities in Eq. 3 are established and can be used to calculate required filter area or time for a given flow rate. Specific cake resistance may be a function of pressure drop, often taking the form[1,5]:

$$\alpha = \alpha_0(\Delta p)^s \quad (4)$$

If $s = 0$, the cake is incompressible, but it often falls between 0.2 and 0.8. In batch filtration, there is a trade-off between run length and capital cost. A large filter will hold more cake, but also will have more unproductive time in washing, drying, removing the cake, and reassembling the filter. Filtration time should be at least equal to the sum of all other times in the total cycle.[6]

The rotary vacuum filter is a common solution to obtaining continuous filtration[7–10] and is found in many food applications. Equations describing the performance of a rotary filter subject to the assumptions of negligible filter resistance and incompressible cake are:[7]

$$V_r N_r = A_d((2\Psi_f N_r \Delta p)/(\alpha w \mu))^{.5} \quad (5)$$

where V_r is the volume of filtrate per revolution, m^3; N_r the revolutions per time, $\sec^{-1}$; A_d the total area of drum, m^2; ψ the fraction of drum submerged in slurry tank (often 0.3–0.4); Δp the pressure drop, N m^{-2}; α the specific cake resistance, m kg^{-1}; w the mass of solids per volume of filtrate, kg m^{-3}; μ the viscosity of fluid, kg m^{-1} $\sec^{-1}$.

As a rule of thumb or sanity check, rotary vacuum filters can produce about 1500 lb ft^{-2} day^{-1} (0.085 kg m^{-2} $\sec^{-1}$) of finely ground wet solids or soft precipitates and about four times this rate of coarser solids and crystals. A rotation rate of 0.33 rpm and vacuum of 18 in.–25 in. of mercury is typical for the fine solids and a lower vacuum of 2 in.–6 in. of mercury for coarse solids.[11]

CONCLUSION

Useful descriptions of equipment for all three systems discussed can be found in other texts.[12,13] For food applications, there are additional design considerations, such as cleanability, ease of inspection, and the absence of

corrosion. Filter aids (substances added to suspensions to assist in filtration) must be approved for food contact. Some examples are rice hulls, cellulose fibers, activated carbon, and diatomaceous earth.

REFERENCES

1. Geankoplis, C.J. *Transport Processes and Unit Operations*, 3rd Ed.; Prentice Hall: Englewood Cliffs, NJ, 1993; 800–838.
2. Scott, K.J. Sedimentation. In *Handbook of Powder Science and Technology*; Fayed, M.E., Otten, L., Eds.; Van Nostrand Reinhold: New York, 1984; 607–686.
3. Metcalf & Eddy, Inc., *Wastewater Engineering*; McGraw-Hill: New York, 1972; 448.
4. Schnittger, J.R. Integrated Theory of Separation for Bulk Centrifuges. Ind. Eng. Chem. Process Des. Develop. **1970**, *9* (3), 407–413.
5. McCabe, W.L.; Smith, J.C. *Unit Operations of Chemical Engineering*, 3rd Ed.; McGraw-Hill: New York, 1976; 922–971.
6. Svarovsky, L. Filtration. *Kirk–Othmer Concise Encyclopedia of Chemical Technology*, 4th Ed.; Wiley-Interscience: New York, 1999; 847–849.
7. Peters, M.S.; Timmerhaus, K.D. *Plant Design and Economics for Chemical Engineers*, 2nd Ed.; McGraw-Hill: New York, 1968; 478–492.
8. Perry, J.H., Ed. *Chemical Engineer's Handbook*, 3rd Ed.; McGraw-Hill: New York, 1950; 937–955 (sedimentation), 964–992 (filtration), 992–1013 (centrifugation).
9. Badger, W.L.; Banchero, J.T. *Introduction to Chemical Engineering*; McGraw-Hill: New York, 1955; 553–601.
10. Considine, D.M., Ed. *Chemical and Process Technology Encyclopedia*; McGraw-Hill: New York, 1974; 240–245 (centrifugation), 487–503 (filtration), 1004–1013 (sedimentation).
11. Happel, J.; Jordan, D.G. *Chemical Process Economics*, 2nd Ed.; Marcel Dekker: New York, 1975; 469–470.
12. Charm, S.E. *The Fundamentals of Food Engineering*, 2nd Ed.; AVI: Westport, CT, 1971; 101–106 (sedimentation), 504–535 (filtration and centrifugation).
13. Loncin, M.; Merson, R.L. *Food Engineering Principles and Selected Applications*; Academic Press: New York, 1979; 98–101 (settling), 102–111 (centrifugation), 116–120 (filtration).

Membrane Concentration Systems

Munir Cheryan
University of Illinois, Urbana, Illinois, U.S.A.

INTRODUCTION

Membrane separation is based on the ability of semipermeable membranes to selectively separate molecules on the basis of size, shape, and chemical composition. It was originally developed in the early 1960s for the production of potable water from sea and brackish water. Since then, it has led to a vast array of applications unmatched by any other processing technique in its variety and versatility, especially with the development and maturation of sister processes, ultrafiltration (UF), microfiltration (MF), and nanofiltration (NF).

PRINCIPLES

A membrane's role is to act as a selective barrier, enriching certain components in a feed stream, and depleting it of others. Depending on the chemical nature and physical properties of the membrane, liquid streams can be concentrated or dewatered by reverse osmosis (RO), components in solution fractionated by UF or NF, slurries clarified or suspended matter removed by MF, as shown in Fig. 1.

In all four cases, hydraulic pressure (through the pump) is used to provide the driving force for permeation. In the case of RO and NF, it is primarily to overcome the chemical potential difference between the concentrate and the permeate which is expressed in terms of osmotic pressure.[1] Due to the high osmotic pressures of small soluble solutes, pressures in RO are frequently of the order of 20 bar–50 bar (300 psi–750 psi). Since UF and MF are designed to retain macromolecules and submicron particles, respectively, which exert little osmotic pressure, pressures required are much lower (1 bar–7 bar, 15 psi–100 psi), primarily to overcome hydraulic resistance of the polarized macromolecular layer on the membrane surface, a phenomena known as "concentration polarization." High cross flow velocities are required to minimize this mass transfer limitation.

The most appealing feature of membrane technology is its simplicity. It involves only the bulk movement (i.e., pumping) of fluids using mechanical energy (Fig. 2). Since they are continuous molecular separation processes that do not involve a phase change or interphase mass transfer, membrane processes are low in energy consumption, as shown in Table 1, and can be operated at ambient temperatures if necessary. This avoids product degradation problems associated with thermal processes, thus resulting in products with better functional and nutritional properties. In addition, no complicated heat transfer or heat generating equipment is needed, and the membrane operation requires only electrical energy to drive the pump motor.

There are some limitations with membranes as a liquid concentration process. It cannot take a liquid stream to very high solids concentrations, especially compared to an evaporator. In the case of RO and NF, it is the osmotic pressure of the concentrated solutes that limits the upper concentration that can be comfortably handled. For example, milk and whey exert an osmotic pressure of about 7 bar (100 psi) at room temperature. Since RO is commonly conducted at pressures of 40 bar (600 psi), it means a practical 4× concentration of milk and whey. In UF, it is rarely the osmotic pressure, but rather the low mass transfer rates and high viscosity of the concentrate that limits the process. Substantial improvements in membrane chemistry, systems engineering, and fouling/cleaning occurred in the 1980s and 1990s, and thus good membranes in well-engineered systems that can meet the rigorous sanitation and safety demands of the food industry are available at reasonable cost.

MEMBRANES

Membranes have been manufactured from over 150 materials. They can be classified into two general categories: polymeric and inorganic. The RO and NF membranes are made almost exclusively from polymers, the most common being cellulose acetate and polyamide. The most important properties are its permeability (measure of the rate at which a given molecule or the solvent permeates through the membrane) and its permselectivity (measure of the rate of permeation of one molecule relative to another). These characteristics are more commonly termed as "flux" and "rejection." Since, these properties can change with time and environmental conditions, secondary properties such as resistance to compaction, temperature and chemical stability, resistance

Encyclopedia of Agricultural, Food, and Biological Engineering
DOI: 10.1081/E-EAFE 120007067
Copyright © 2003 by Marcel Dekker, Inc. All rights reserved.

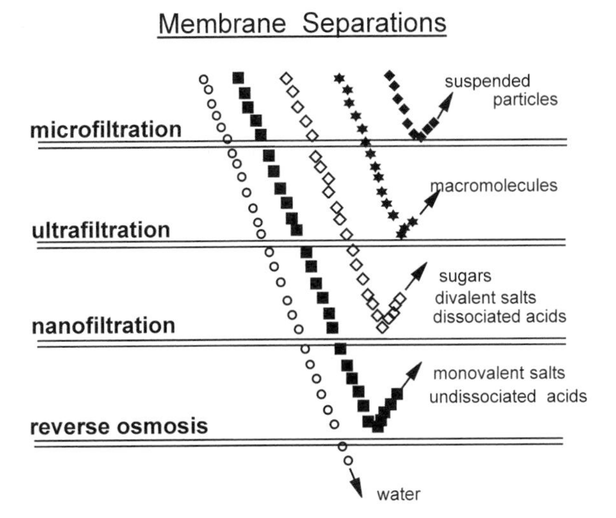

Fig. 1 Pressure-driven membrane processes.

Table 1 Comparison of energy consumption for concentration of whole milk from 15% to 31% total solids at a feed rate of 1000 kg hr^{-1}

Process	Energy (kcal kg^{-1} milk)
Thermal concentration	
Open-pan boiling	455
Double-effect evaporator	209
MVR evaporator	136
Membrane process	
Batch, single pump	80
Batch, dual pump	7
Continuous, one-stage	16
Continuous, three-stage	7

(From Ref. [2].)

to microbial attack, tolerance to cleaning and disinfecting solutions, and lack of toxicity of the contact materials, are also important.

The exact mechanism of transport through RO membranes is still a matter of controversy. Transport could be due to diffusion and/or convective flow through very fine pores in the "skin" of the asymmetric membrane. The sieve mechanism assumes that the membrane has pores intermediate in size between the solvent and solute molecules. Separation occurs because solute molecules are blocked out of the pores while smaller solvent molecules are able to enter the pores. The hydrogen bonding mechanism assumes permeation occurs in the noncrystalline portions of the membrane and is much faster for molecules that can form hydrogen bonds with the membrane material. The solution-diffusion mechanism attributes solute rejection to large differences in the diffusivity of solvent and solute in the membrane and/or to differences in their solubility in the membrane material. Both solvent and solute dissolve in the homogeneous nonporous surface layer of the membrane and then diffuse through the membrane in an uncoupled manner.[2]

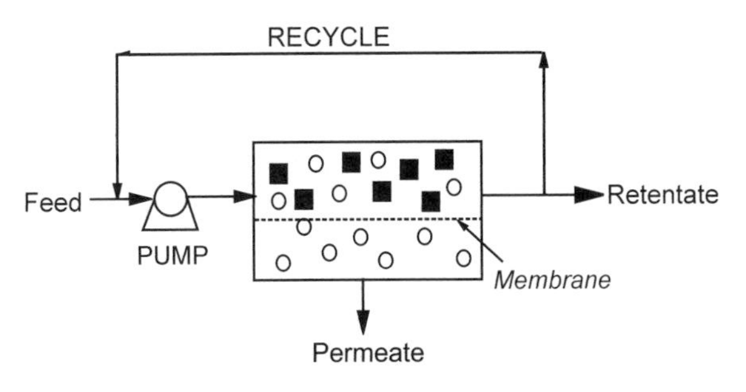

Fig. 2 Flow diagram of a membrane process.

The preferential sorption-capillary flow mechanism considers RO to be governed by two distinct factors: 1) an equilibrium effect involving preferential sorption at the membrane surface, which is governed by repulsive or attractive potential force gradients at the membrane surface, and 2) a kinetic effect which is concerned with the movement of solute and solvent molecules through the membrane pores. The kinetic effect is governed by potential force gradients and steric effects associated with the structure and size of the solute and solvent molecules relative to the membrane pores. Consequently, the chemical nature of the membrane surface and the size, number, and distribution of its pores will determine the success of the RO process. This mechanism could explain all the variations in RO separation that have been experimentally observed. Since RO membranes are made from several different materials and the physicochemical parameters are unique to each solution-membrane system, it is reasonable to assume that various mechanisms are possible.[2,3]

EQUIPMENT

There are four different types of equipment commonly used for RO and NF: tubular, hollow fine fiber (external feed), plate and spiral-wound (Fig. 3). Each design has its own special applications, advantages, and disadvantages.[1,2] In the tubular configuration, the membrane is cast inside a porous support tube, several of which are housed within one pressure vessel in a shell-and-tube arrangement. The flow is through the bore of the tubes that are usually connected in series with the appropriate end-cap. The permeate collects in the shell side from where it is pumped away. Most manufacturers use 18–19 tubes per module, each of 12.5 mm (0.5 in.) internal diameter and

TUBULAR

Feed

Retentate

Membrane tubes

Permeate

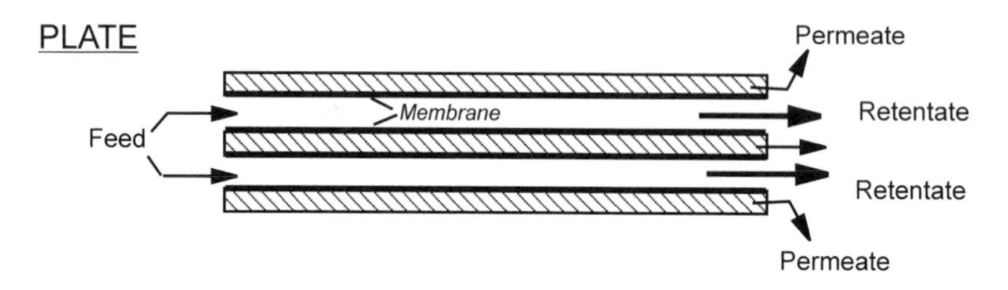

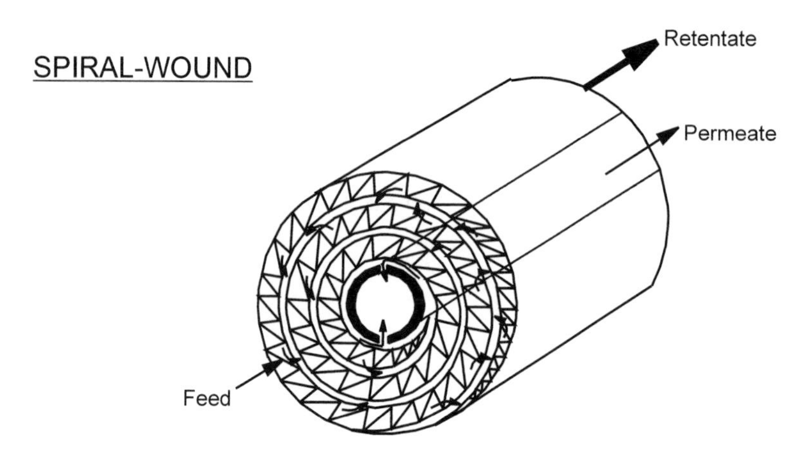

Fig. 3 Schematic of membrane equipment.

length of 4 ft–12 ft (1 m–3 m). Advantages of tubular systems include high turbulence, which generally leads to higher flux with polarizing feed streams; ability to handle suspended matter with particle sizes as large as 1.5 mm; and they are easy to clean. The major limitations are their low surface area to volume ratio (i.e., low "packing density"), which implies that a large floor area is required and there will be high hold-up in the modules, and high energy costs for pumping.

The hollow (fine) fibers typically have self-supporting polymeric tubes with internal diameters of 0.040 mm–0.050 mm. The feed is pumped from the shell side and permeate is withdrawn from the inside of the fibers. The hollow fibers used for RO are different from those used for UF and MF. The latter are larger in diameter (0.2 mm–3 mm) and the feed is pumped through the tube side, with the permeate withdrawn from the shell side. Some advantages of hollow fine fibers are that they are compact with high packing density, resulting in low holdup, and they have a high resistance to compression and thus very high pressures can be used. This is a particular advantage in applications where osmotic pressure is the main limiting factor, such as sugar and salt solutions. The main disadvantage is that it is easily fouled by suspended matter, and thus the feed requires good prefiltration.

Plate and spiral modules utilize flat-sheet membranes configured either vertically or horizontally, or in a spiral arrangement (Fig. 3). Spiral wound modules have a relatively high packing density, depending primarily on the feed channel spacer thickness; low cost per unit

Table 2 Selected uses of RO and NF in the food industry

Dairy
RO: Preconcentration of milk and whey prior to evaporation; bulk transport; specialty fluid milk products
NF: Partial demineralization and concentration of whey
Fruits and vegetables
RO: Juices (apple, citrus, grape, peach, and tomato); pigments (anthocyanins and betanins); wastewater (apple, pineapple, and potato)
NF: Beta-carotene from palm oil
Animal products
Eggs: concentration
Beverages
Low-alcohol beer
Sugar refining
Beet and cane extracts, maple syrup, candy wastewaters, and preconcentration
Oilseeds, cereals, and legumes
Soybean processing: oil degumming and refining (NF); recovery of soy whey proteins (RO); wastewater treatment (NF, RO)
Corn refining: steepwater concentration (RO); light-middlings treatment for water recycle (RO); concentration of dextrose (NF, RO); downstream processing of fermentation broths (NF, RO); wastewater treatment (NF, RO)

(From Ref. 4.)

membrane area; relatively easy replacement of modules from the pressure vessels; and low energy consumption per unit energy. One disadvantage is that with feeds containing large amount of suspended matter, good prefiltration is needed, to 1/20–1/50 the feed channel height.

APPLICATIONS

Table 2 lists selected applications of RO and NF in the food industry. Reverse osmosis has been used successfully to concentrate egg whites prior to spray drying, maple sap, citrus juices, tomato juice, coffee and tea extracts, sugar solutions, color pigments, and low-alcohol beer.[4] In the dairy industry the major use of RO is to concentrate whey, a byproduct of the cheese-making process, and concentrating milk prior to evaporation or manufacturing selected dairy products using less energy. Reverse osmosis has been used to recover valuable foodstuffs from and/or to decrease pollution by food wastes. These include effluents from potato starch processing, sucrose from candy-making waste water, protein from soy whey solutions, and anthocyanins from cranberry pulp wastes.

REFERENCES

1. Cheryan, M. *Ultrafiltration and Microfiltration Handbook*; CRC Press: Boca Raton, FL, 1998.
2. Cheryan, M. Concentration of Liquid Foods by Reverse Osmosis. In *Handbook of Food Engineering*; Lund, D.B., Heldman, D.R., Eds.; Marcel Dekker, Inc.: New York, 1992; 393–436.
3. Cheryan, M.; Nichols, D.J. Modelling of Membrane Processes. In *Mathematical Modelling of Food Processes*; Thorne, S., Ed.; Elsevier: London, 1992; 49–98.
4. Cheryan, M. Food Technology. Membrane Separations. In *Encyclopedia of Separation Science*; Wilson, I.D., Adlard, E.R., Cooke, M., Poole, C.F., Eds.; Academic Press: New York, 2000; 2849–2855.

Membrane Separation Mass Transfer

D. Vidal-Brotóns
P. Fito
M. Gras
Universidad Politécnica de Valencia (UPV), Valencia, Spain

INTRODUCTION

A membrane is a selective barrier between two bulk phases, through which mass transfer may occur under a variety of driving forces (pressure, concentration, temperature, electrical potential). Thus, a membrane process allows the selective transfer of some species of a mixture from one bulk phase to the other.

At present, membrane technology is being used[1–3] in all industrial areas such as food and beverages, biotechnology, pharmaceuticals, metallurgy, pulp and paper, textile, automotive, chemical industry. It is involved in water treatment for domestic and industrial water supply, and has also environmental applications.

The industrial application of a first group of membrane processes has undergone an impressive development since the early 1960s. It is the case for microfiltration (MF), ultrafiltration (UF), nanofiltration (NF), reverse osmosis (RO), and electrodialysis (ED). Dialysis (D) is not used industrially but is used on a large scale in the field of medicine: in terms of membrane area used and economical value of the membrane produced, the artificial kidney is the single largest application of membranes.[1] Other membrane processes applications are rapidly expanding: gas separation (GS) and pervaporation (PV). Another group is still at a developing stage: membrane contactors (MC), membrane distillation (MD), carrier-mediated processes (CMP). The combination of a membrane process and a chemical or biochemical reaction has given rise to the rather new concept of membrane reactor and membrane bioreactor.

Some membrane processes are particularly well suited for the separation and purification of biological molecules[4] since they operate at relatively low temperatures and pressures and involve no phase changes or chemical additives. Thus, these processes cause minimal denaturation, deactivation, and/or degradation of highly labile biological cells or macromolecules.

The aim of this contribution is to present general information on membrane processes, in the context of their biological, agricultural, and food applications, with major reference to developed industrial processes.

MEMBRANE PROCESSES

Membrane processes can be classified according to different view points (Table 1): nature and range of the driving force, range size of particles to be retained, molecular weight of compounds to be separated, etc.

Microfiltration and ultrafiltration are basically similar in that the mode of separation is molecular sieving through increasingly fine pores. Microfiltration membranes filter colloidal particles and bacteria, from 0.1 μm to 10 μm in diameter. The hydrodynamic resistance of such membranes is small, so low hydrostatic pressures are sufficient to obtain high fluxes. Ultrafiltration membranes can be used to filter dissolved macromolecules from solutions, such as proteins, with molecular weights ranging from about 10^4 Da to more than 10^6 Da. The membrane structure is denser, the hydrodynamic resistance increases, and the applied pressure must be greater than in MF.

In NF and RO, the membrane pores are within the range of thermal motion of the polymer chains that form the membrane. According to the so-called solution-diffusion model (SDM), solutes permeate the membrane by dissolving in the membrane material and diffusing down a concentration gradient. Separation occurs because of the difference in solubility and mobility of different solutes in the membrane.

Electrodialysis uses charged membranes to separate ions from aqueous solutions under the action of an electrical potential difference. The principal application of ED is the desalting of brackish groundwater, and, in the food industry, the deionization of cheese whey.

In GS, a gas mixture at an elevated pressure is passed across the surface of a membrane that is selectively permeable to one component of the feed mixture. Gas separation is used in the production of sterile oxygen-enriched air for aerobic fermentation processes, in the production of nitrogen-enriched air for the storage of food and agricultural products, and in the dehydration of air (compressed air and air conditioning).

In PV, a liquid mixture at atmospheric pressure contacts one side of a membrane, and permeate is removed as a vapor from the other side because of the low partial vapor

Encyclopedia of Agricultural, Food, and Biological Engineering
DOI: 10.1081/E-EAFE 120007070
Copyright © 2003 by Marcel Dekker, Inc. All rights reserved.

Table 1 Membrane processes and driving forces

Membrane process	Feed phase	Permeate phase	Driving force	Membrane pore sizes	Separated particles
Microfiltration	L	L	Pressure (< 2 bar)	0.05 μm–10 μm	> 0.1 μm, cells
Ultrafiltration	L	L	Pressure (1–10 bar)	1 nm–100 nm	Macromolecules, bacteria, yeasts
Nanofiltration	L	L	Pressure (10–25 bar)	< 2 nm	Low MW solutes: salts, glucose, lactose,
Reverse osmosis	L	L	Pressure (15–80 bar)	< 2 nm	Micropollutants
Peizodialysis	L	L	Pressure (up to 100 bar)	Nonporous	Ionic solutes
Gas separation	G	G	Concentration (activity) (up to 100 bar upstream, or vacuum downstream)	< 1 μm or nonporous	H_2, He, CH_4, CO_2, N_2, H_2S, H_2O, SO_2
Vapor permeation	G	G	Concentration (activity)		
Pervaporation	L	G	Concentration (activity), Partial vapor pressure	Nonporous	Alcohols, aromatics, chlorinated hydrocarbons
Electrodialysis	L	L	Electrical potential	Ion-exchange nonporous	Ionic solutes
Membrane electrolysis	L	L	Electrical potential	Cation exchange nonporous	Heavy metals, hydroxyl ions
Dialysis	L	L	Concentration (activity)	Nonporous	Urea, creatinine, phosphates, uric acid, caustic soda, alcohol, salts
Diffusion dialysis	L	L	Concentration (activity)	Ionic nonporous	Ionic solutes (acid, alkali)
Membrane contactors	L	L	Concentration (activity) Vapor pressure difference	0.05 μm–1 μm	Heavy metals; citric, acetic, lactic acids, penicillin, phenolics
	G	L	Concentration (activity)	or nonporous	SO_2, CO_2, CO, NO_x, H_2S, alcohols,
	L	G	Concentration (activity)		aroma compounds, O_2
Thermo-osmosis	L	L	Temperature		
Membrane distillation	L	L	Temperature, Vapor pressure difference	0.2 μm–1 μm	Water, benzene, TCE, ethanol, butanol aroma compounds

(From Refs. [1–3].)

pressure generated on the permeate side of the membrane by cooling and condensing the permeate vapor. The main industrial application of PV is in the dehydration of organic solvents; PV membranes can produce more than 99.9% ethanol from a 90% ethanol feed solution. Pervaporation processes are being developed to concentrate heat-sensitive products or remove and concentrate aroma compounds, and for the removal of dissolved volatile organic contaminants from wastewaters.

Carrier-mediated processes often employs liquid membranes containing a complexing or carrier agent, which reacts with one component of a mixture on the feed side of the membrane and then diffuses across the membrane to release the permeant on the product side of the membrane. The reformed carrier agent then diffuses back to the feed side of the membrane.

Membrane contactors (also called pertaction, perstraction, gas absorption, membrane-based solvent extraction, liquid–liquid extraction, membrane-based gas absorption and stripping, hollow-fiber contained liquid-membrane): the membrane acts only as an interface between two phases, but does not control the rate of passage of permeants across the membrane. The function of the membrane is to provide a large surface area for contact between the phases. Some examples of gas–liquid MC are blood oxygenation, removal of acid gases (CO_2, SH_2, CO, SO_2, NO_x) from flue gas, biogas, and natural gas, O_2 transfer (aerobic fermentation), CO_2 transfer (beverages),

NH_3 from air (intensive farming). In the case of liquid–gas MC: recovery of volatile bioproducts (alcohols, aroma compounds), and O_2 removal from water. Membrane contacters can also be used to separate two immiscible liquids (liquid–liquid contractors) or two miscible liquids (usually called MD). Some examples of MD applications are: production of pure water (boiler feed water for power plants, desalination of seawater), concentration of solutions (wastewater treatment, concentration of salts, acids, etc.), and removal of volatile bioproducts.

MEMBRANE TYPES

Membranes can be classified based on various criteria[5]: their geometry, bulk structure, composition, production method, separation regime, and application.

Most commonly, membrane are produced in flat-sheet or tubular (hollow-fiber) geometry.

Membranes either have a symmetric (isotropic) or an asymmetric (anisotropic) structure. The structure of a symmetric membrane (porous or dense) is uniform throughout its entire thickness, and the entire membrane thickness contributes a resistance to mass transfer, so their fluxes are typically relatively low. Asymmetric membranes consist of two structural elements, that is, a thin (0.1 μm–0.5 μm thick) microporous or dense selective layer, and a highly porous substructure, which provides only mechanical strength. It is also possible to employ thin-film composite membranes, which are porous asymmetric membranes skinned with an ultra thin (0.05 μm thick) dense layer.

Synthetic membranes are manufactured with a wide variety of materials (Table 2), which can be classified into organic (polymeric or liquid) and inorganic (ceramic, metal, carbon, etc.) materials. Most commercial membranes are made from polymers.

A number of different techniques are available to prepare synthetic membranes. Some of these techniques can be used to prepare polymeric as well as inorganic membranes. The most important techniques are[1,3,5]: sintering, stretching, tract-etching, template leaching, phase inversion, sol–gel process, vapor deposition, and solution coating.

It is possible to modify preformed membranes to enhance its overall performance, increasing flux and/or selectivity and increasing chemical resistance (solvent resistance, swelling, or fouling resistance). Heat-treatment (with hot water or steam treatment), drying, surface coating, chemical surface modification by fluorine, chlorine, bromine, or ozone treatments, are some of the methods used.

Table 2 Materials used for production of commercial membranes

Membrane material	Applications
Polymeric membranes:	
Cellulose acetate	GS, RO, D, UF, MF
Cellulose nitrate	MF
Cellulose regenerated	D, UF, MF
Polyamide	RO, NF, D, UF, MF
Polysulfone	G, UF, MF
Polyethersulfone	UF, MF
Polycarbonate	GS, D, UF, MF
Polyetherimide	UF, MF
Poly-2,6-dimethyl-1, 4-phenylene oxide	GS
Polyimide	GS
Polyvinylidene fluoride (PVDF)	UF, MF
Polytetrafluoroethylene (PTFE)	MF
Polypropylene	MF
Polyacrylonitrile	D, UF, MF
Polymethylmethacrylate	D, UF
Polyvinyl alcohol	PV
Polydimethylsiloxane	PV, GS
Ceramic membranes:	UF, MF
Alumina, titania, zirconia	
Metal membranes:	GS, MF
Palladium, palladium–silver alloy, palladium–gold alloy, stainless steel	
Anodic membranes:	
Alumina	
Carbon membranes:	GS, MF
Polyacrylonitrile and polyimide, polyvinylidene chloride acrylate	
Glass membranes:	MF
$SiO_2 + B_2O_3 + Na_2O$	

(From Refs. [1,3,5].)

Porous membranes induce separation by discriminating between particle sizes. Such membranes are used in MF and UF. Nonporous membranes can separate molecules of approximately the same size from each other. Separation takes place through differences in solubility and/or diffusivity. Such membranes are used in PV, VP, GS, and D. Reverse osmosis membranes can be considered as being intermediate between porous and nonporous membranes.

Originally, the main goal in characterization of porous membranes was to determine the pore-size distribution. However, membrane surface properties, such as hydrophobicity, zeta potential, and surface roughness, play an important role in fouling and retention properties of membrane processes. Characterization is therefore nowadays performed by a number of techniques.[1,6]

Ion-exchange membranes, used in ED and other electrically driven membrane processes, are described in the article *Electrodialysis*.

Liquid membranes [7] have potential applications in a number of industrial areas, in particular in biotechnology (pertraction).

MODULES

In order to apply membranes on a technical scale, large membrane areas are normally required. The smallest unit into which the membrane area is packed is called a module. The special case of the stacks of cell pairs is described in the article *Electrodialysis*.

Flat-sheet membranes are either packaged in plate-and-frame or spiral-wound modules, whereas tubular membranes are packaged in tubular, capillary, or hollow-fiber modules.

In plate-and-frame modules,[1,3] membrane, feed spacers, and permeate spacers are layered together between two end plates. The feed mixture is forced across the surface of the membrane. A portion passes through the membrane, enters the permeate channel, and makes its way to a central permeate collection manifold. The packing density of such modules is about $100\,m^2\,m^{-3}$–$400\,m^2\,m^{-3}$. Plate-and-frame modules are now only used in ED and PV systems and in a limited number of RO and UF applications with highly fouling feeds.

The spiral-wound module[1,3] is in fact a plate-and-frame system wrapped around a central collection pipe. Membrane and permeate-side spacer material are then glued along three edges to build a membrane envelope. The feed-side spacer separating the top layer of the two flat membranes also acts as a turbulence promoter. The module is placed inside a tubular pressure vessel. The feed flows axial through the cylindrical module parallel along the central pipe whereas the permeate flows radially toward the central pipe. The packing density of this module ($300\,m^2\,m^{-3}$–$1000\,m^2\,m^{-3}$) depends very much on the channel height, determined by permeate and feed-side spacer material. The standard industrial spiral-wound module is 203 mm in diameter and 1.016-m long.

Tubular membranes,[1,3] which are not self-supporting, are placed inside a porous paper, fiber glass, stainless steel, ceramic or plastic tube, with the diameter of the tube being, in general, more than 10 mm. The number of tubes put together in the module may vary from 4 to 18, but is not limited to this number. The feed solution always flows through the center of the tubes while the permeate flows through the porous supporting tube into the module housing. The packing density of the tubular module is rather low (less than $300\,m^2\,m^{-3}$). Ceramic membranes are mostly assembled in such tubular module configuration. The monolithic module is a special type of ceramic module, in which a number of tubes have been introduced in a porous ceramic block and the inner surfaces of the tubes are then covered by a thin top layer of aluminum or zirconium oxides. Tubular modules are now generally limited to UF applications, for which the benefit of resistance to membrane fouling due to good fluid hydrodynamics outweighs the high cost.

Hollow-fiber membrane modules[1,3] are formed in two basic geometries. In the first, the shell-side feed design, a loop, or a closed bundle of fibers is contained in a pressure vessel; the system is pressurized from the shell side; permeate passes through the fiber wall and exits the pressure vessel through the open fiber ends, which are potted with epoxy resins, polyurethanes, or silicone rubber. In the second, the bore-side feed type, the fibers are open at both ends, and the feed fluid is circulated through the bore of the fibers. Depending on the geometry chosen, asymmetric hollow-fibers are used with their skin on the inside or on the outside. The hollow-fiber module has the highest packing density, which can attain $30{,}000\,m^2\,m^{-3}$. It is used when the feed stream is relatively clean, as in GS (shell-side feed) and PV (bore-side feed).

The difference between the capillary module and the hollow-fiber module is simply a matter of dimensions since the module concept is the same. The capillary module has a packing density of about $600\,m^2\,m^{-3}$–$1200\,m^2\,m^{-3}$.

CONCENTRATION POLARIZATION

The feed mixture components permeate at different rates, so concentration gradients form in the solutions on either side of the membrane, unless the solutions are extremely well stirred. The layer of solution immediately adjacent to the membrane surface becomes depleted in the permeating solutes on the feed side of the membrane and enriched in these components on the permeate side. The phenomenon is called concentration polarization (CP). An exhaustive of CP is given in Baker's excellent monograph (see Ref. [1]) and also in Ref. [8].

Concentration polarization is generated by the separation performed by the membrane and, as such, cannot be avoided. As a result, the difference of concentration of a permeating component at both sides of the membrane is lower than the difference of concentration of this component in the bulk feed solution and in the bulk permeate solution. Concentration polarization lowers the flux of the permeating components and the membrane selectivity, and increases the potential for membrane fouling. Therefore, minimizing CP is one of the most important objectives in designing and engineering membrane separation systems. The importance of CP depends on the membrane separation process. It can significantly affect membrane performance in RO, but it is usually well controlled in industrial systems. On the other hand, UF, ED, and some PV processes, are seriously affected by CP. The CP phenomenon in ED is described in the article *Electrodialysis*.

The boundary layer film model[1] assumes that the concentration gradients, which control CP form in the laminar flow boundary layers that exist between the two membrane surfaces and each bulk solution (feed and permeate). The increase or decrease in the permeate concentration at the membrane surface c_{i_0} (kg m^{-3}), compared to the bulk feed solution concentration c_{i_b} (kg m^{-3}), determines the extent of CP. The ratio of the two concentrations is called the concentration polarization modulus (CPM):

$$\mathrm{CPM} = c_{i_0}/c_{i_b}$$

When CPM is 1, no CP occurs, but as CPM deviates significantly from 1, the effect of CP on membrane selectivity and flux becomes increasingly important.

The basic equation for CP (Eq. 1) can be easily derived[1]:

$$\mathrm{CPM} = [E_0 + (1 - E_0)\exp(-J_v\delta/D_{i_{BL}})]^{-1} \tag{1}$$

E_0 is the membrane's intrinsic enrichment (enrichment factor in the absence of a boundary layer), defined as

$$E_0 = c_{i_p}/c_{i_0}$$

where c_{i_p} (kg m^{-3}) is permeate salt concentration; J_v (m^3 m^{-2} sec^{-1}) is the permeate volume flux; δ (m) is the thickness of the boundary layer; and $D_{i_{BL}}$ (m^2 sec^{-1}) is the diffusion coefficient of the salt in the boundary layer.

Of the four factors that affect CP, the one most easily changed is the boundary layer thickness, δ, which is inversely proportional to the turbulence of the fluid flow. The most direct technique to enhance turbulence is to increase the fluid flow velocity past the membrane surface. Therefore, most membrane modules operate at relatively high feed fluid velocities. Membrane spacers are also widely used to promote turbulence. Pulsing the feed fluid flow through the membrane is another technique. However, the energy consumption of the pumps required and the pressure drops produced limit the turbulence that can be obtained in a membrane module.

The diffusivity of a solute can only be increased by changing the temperature. The diffusion coefficients of macromolecules or suspended colloidal particles are about 100 times smaller than those of salts. In addition, the fluxes in MF and UF are large relative to those in PV and GS. Hence, the consequences of CP are very severe in the case of MF and UF.

MEMBRANE TRANSPORT THEORY

Mathematical models of membrane processes are rarely used to predict permeate and solute fluxes, due to unavailability of parameters required by the models.[9] However, the models provide extremely useful insight of the processes, through the prediction of parametric dependence of processes performance, namely permeate flux and rejection coefficients, which allow evaluation of the ability of a membrane to control the rate of permeation of different species.

Three groups of these models can be distinguished: phenomenological models, models for dense membranes, and models for porous membranes.

Phenomenological transport models are black box equations that tell us nothing about the chemical and physical nature of the membrane or how mass transfer is related to the membrane structure.

During the last 30 yr, the mechanism of permeation through membranes has become much clear. This is particularly true for nonporous membranes, used in RO, GS, and PV, for which the SDM is now almost universally accepted and well supported by a body of experimental evidence.[1]

The theory of permeation through microporous membrane, in MF and UF, is much less developed, and current theories cannot predict the permeation properties of these membranes. This is due to the extremely heterogeneous nature of microporous membranes.

A very clear and didactic presentation of the SDM can be found in Baker's "Membrane Technology and Applications" (see Ref. [1]). The second of its 13 chapters is entirely dedicated to Membrane Transport Theory, with special reference to the SDM, giving its basic concepts and applying it to D, RO, GS, and PV. With some minor changes, a part of this recommendable reference book is next given. Mannapperuma's review "Design and Performance Evaluation of Membrane Systems" (see Ref. [9]) has been selected in the case of transport in microporous UF and MF membranes. Some general considerations proceed from Mulder's "Basic Principles of Membrane Technology" (see Ref. [3]).

Basic Concepts

A membrane may be defined as a permselective barrier between two homogeneous phases. A molecule or a particle is transported across a membrane from one phase to another because a force acts on that molecule or particle. The extent of this force is determined by the gradient in

potential across the membrane. Two main potentials are important in membrane processes, the chemical potential (μ) and the electrical potential (F). The electrochemical potential is the sum of the chemical potential and the electrical potential. The potential gradient arises as a result of differences in either pressure, concentration, temperature, or electrical potential. Membrane processes involving an electrical potential occur in ED and other related processes. The nature of these processes differs from that of other processes involving a pressure or concentration difference as the driving force since only charged molecules or ions are affected by the electrical field. Other possible forces such as magnetic fields, centrifugal fields, and gravity are generally not considered.

Most transport processes take place because of a gradient in chemical potential. The mass flux density of a component i, J_i (kg m^{-2} sec^{-1}), is described by Eq. 2:

$$J_i = -L_i \frac{d\mu_i}{dx} \tag{2}$$

where $\frac{d\mu_i}{dx}$ is the chemical potential gradient of component i with respect to x- direction, and L_i is a parameter linking this chemical potential driving force to the flux. Driving forces, such as gradients in concentration, pressure, temperature, and electrical potential, can be expressed as chemical potential gradients, and their effects on flux can be expressed by Eq. 2.

Many processes involve more than one driving force, e.g., both pressure and concentration in RO. Under isothermal conditions, pressure and concentration contribute to the chemical potential of component i according to

$$d\mu_i = RT\, d[\ln a_i] + v_i\, dp \tag{3}$$

where R is gas constant (8.314 J kmol^{-1} K^{-1}), T is absolute temperature (K), v_i is the molar volume of component i (m^3 kmol^{-1}), and p is the pressure (Pa). The concentration or composition is given in terms of activities a_i, in order to express nonideality.

$$a_i = \gamma_i x_i$$

where γ_i is the activity coefficient and x_i, the mole fraction. For ideal solutions, the activity coefficient $\gamma_i \Rightarrow 1$, and the activity becomes equal to the mole fraction.

In incompressible phases, such as a liquid or a solid membrane, volume does not change with pressure. In this case, integrating Eq. 3 with respect to concentration and pressure gives

$$\mu_i = \mu_i^o + RT \ln(\gamma_i x_i) + v_i(p - p_i^o) \tag{4}$$

The first term on the right hand side (μ_i^o) is a constant. It is the chemical potential of pure i at saturation vapor pressure of i, p_i^o.

In compressible phases (gases), the molar volume changes with pressure. Using the ideal gas laws in integrating Eq. 3 gives

$$\mu_i = \mu_i^o + RT \ln(\gamma_i x_i) + RT \ln \frac{p}{p_i^o} \tag{5}$$

Transport Through Dense Membranes: The SDM

In the SDM, permeants dissolve in the membrane material and then diffuse through the membrane, down a concentration gradient. The permeants are separated because of the differences in their solubility in the membrane and the differences in the rates at which they diffuse through the membrane.

Several assumptions must be made. Usually, the first assumption is that the fluids on either side of the membrane are in equilibrium with the membrane material at the interface. This assumption means that the gradient in chemical potential from one side of the membrane to the other is continuous. Implicit in this assumption is that the rates of mass transfer at the membrane interface are much higher than the rate of diffusion through the membrane. This appears to be the case in almost all membrane processes, but may fail in transport processes involving chemical reactions, such as facilitated transport, or in diffusion of gases through metals, where interfacial absorption can be slow.

The second assumption of SDM is that the pressure within a membrane is uniform, and that the chemical potential gradient across the membrane is expressed only as a concentration gradient. The difference in pressure across the membrane, $p_0 - p_1$, produces a gradient in chemical potential according to Eq. 3. The SDM assumes that when pressure is applied across a dense membrane, the pressure throughout the membrane is constant at the highest value. Consequently, the chemical potential difference across the membrane is expressed as a concentration gradient within the membrane, that is to say, a smooth gradient in solvent activity $\gamma_i x_i$. The flow that occurs down this gradient is again expressed by Eq. 2, but, since no pressure gradient exists within the membrane, Eq. 6 can be written by combining Eqs. 2 and 3 and assuming that γ_i is constant:

$$J_i = -\frac{RTL_i}{x_i}\frac{dx_i}{dx} \tag{6}$$

Using the concentration c_i (kg m^{-3}), defined as

$$c_i = M_i \rho x_i \tag{7}$$

where M_i is the molecular weight of i (kg kmol^{-1}) and ρ, the molar density (kmol m^{-3}), Eq. 8 is obtained

$$J_i = -\frac{RTL_i}{c_i}\frac{dc_i}{dx} \tag{8}$$

This has the same form as Fick's law, in which the term (RTL_i/c_i) can be replaced by the diffusion coefficient D_i. Thus,

$$J_i = -D_i \frac{dc_i}{dx}$$

Integrating over the thickness, d, of the membrane then gives

$$J_i = \frac{D_i(c_{i_{o(m)}} - c_{i_{l(m)}})}{d} \tag{9}$$

where $c_{i_{o(m)}}$ is the concentration of component i in the membrane at the feed interface (point o) and $c_{i_{l(m)}}$ at the permeate interface (point l).

Fig. 1a shows a semipermeable membrane separating a salt solution from the pure solvent. The membrane is assumed to be very selective, so the concentration of salt within the membrane is small. The pressure is the same on both sides of the membrane. The difference in concentration across the membrane results in a continuous, smooth gradient in the chemical potential of the water (component i) across the membrane, from μ_{i_l} on the water side to μ_{i_o} on the salt side. The pressure within and across the membrane is constant (that is, $p_o = p_m = p_l$), and the solvent activity gradient $\gamma_{i_{(m)}} x_{i_{(m)}}$ falls continuously from the pure water (solvent) side to the saline (solution) side of the membrane. Consequently, water passes across the membrane from right to left.

Fig. 1b shows the situation at the point of osmotic equilibrium, when sufficient pressure has been applied to the saline side of the membrane to bring the flow across the membrane to zero. The pressure within the membrane is assumed to be constant at the high-pressure value p_o. There is a discontinuity in pressure at the permeate side of the membrane, where the pressure falls abruptly from p_o to p_l, the pressure on the solvent side of the membrane. This pressure difference $p_o - p_l$ is equal to the osmotic pressure difference $\Delta\pi$. Equating the chemical potential on either side of the permeate interface, from Eq. 4, at osmotic equilibrium

$$RT \ln(\gamma_{i_{l(m)}} x_{i_{l(m)}}) - RT \ln(\gamma_{i_l} x_{i_l}) = -v_i(p_o - p_l) \tag{10}$$

At osmotic equilibrium $\Delta(\gamma_i x_i)$ can also be defined by

$$\Delta(\gamma_i x_i) = \gamma_{i_l} x_{i_l} - \gamma_{i_{l(m)}} x_{i_{l(m)}} \tag{11}$$

and since

$$\gamma_{i_l} x_{i_l} \approx 1$$

it follows, on substituting Eq. 11 into Eq. 10 that

$$RT \ln[1 - \Delta(\gamma_i x_i)] = -v_i(p_o - p_l) \tag{12}$$

Since $\Delta(\gamma_i x_i)$ is small

$$\ln[1 - \Delta(\gamma_i x_i)] \approx \Delta(\gamma_i x_i)$$

and Eq. 12 reduces to

$$\Delta(\gamma_i x_i) = \frac{-v_i(p_o - p_l)}{RT} = \frac{-v_i \Delta\pi}{RT}$$

Thus, the pressure difference across the membrane, $p_o - p_l = \Delta\pi$, balances the solvent activity difference $\Delta(\gamma_i x_i)$ across the membrane, and the flow is zero.

If a pressure higher than the osmotic pressure is applied to the feed side of the membrane, as shown in Fig. 1c, then the solvent activity difference across the membrane increases further, resulting in a flow from left to right. This is the process of RO.

The important conclusion illustrated by Fig. 1 is that although the fluids on either side of a membrane may be at different pressures and concentrations, within a perfect solution-diffusion membrane there is no pressure gradient, only a concentration gradient. Flow through this type of membrane is expressed by Fick's law, Eq. 9.

Application of the SDM to Specific Processes

Dialysis (direct osmosis)

Dialysis is the simplest application of the SDM because only concentration gradients are involved. In D, a membrane separates two solutions of different composition. The concentration gradient across the membrane causes a flow of solute and solvent from on side of the membrane to the other.

Equating the chemical potentials in the solution and membrane phases at the feed-side of the membrane and substituting the expression for the chemical potential of incompressible fluids from Eq. 4 gives

$$\ln(\gamma_{i_o} x_{i_o}) = \ln(\gamma_{i_o(m)} x_{i_o(m)})$$

and thus

$$x_{i_o(m)} = \frac{\gamma_{i_o}}{\gamma_{i_o(m)}} x_{i_o}$$

or, from Eq. 7

$$c_{i_o(m)} = \frac{\gamma_{i_o} \rho_m}{\gamma_{i_o(m)} \rho_o} c_{i_o} \tag{13}$$

Hence, defining a sorption coefficient K_{i_o} as

$$K_{i_o} = \frac{\gamma_{i_o} \rho_{o(m)}}{\gamma_{i_o(m)} \rho_o} \tag{14}$$

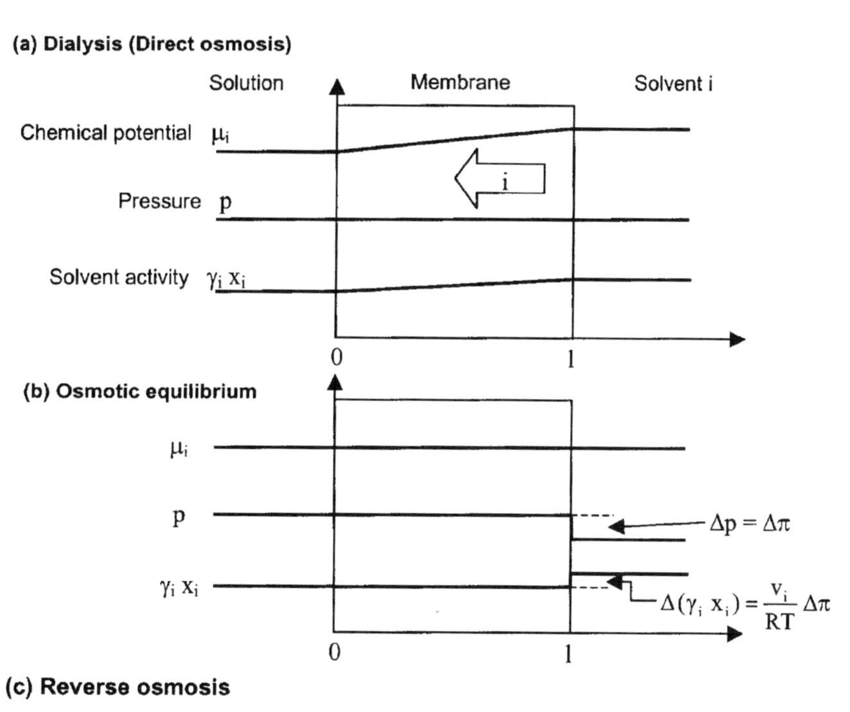

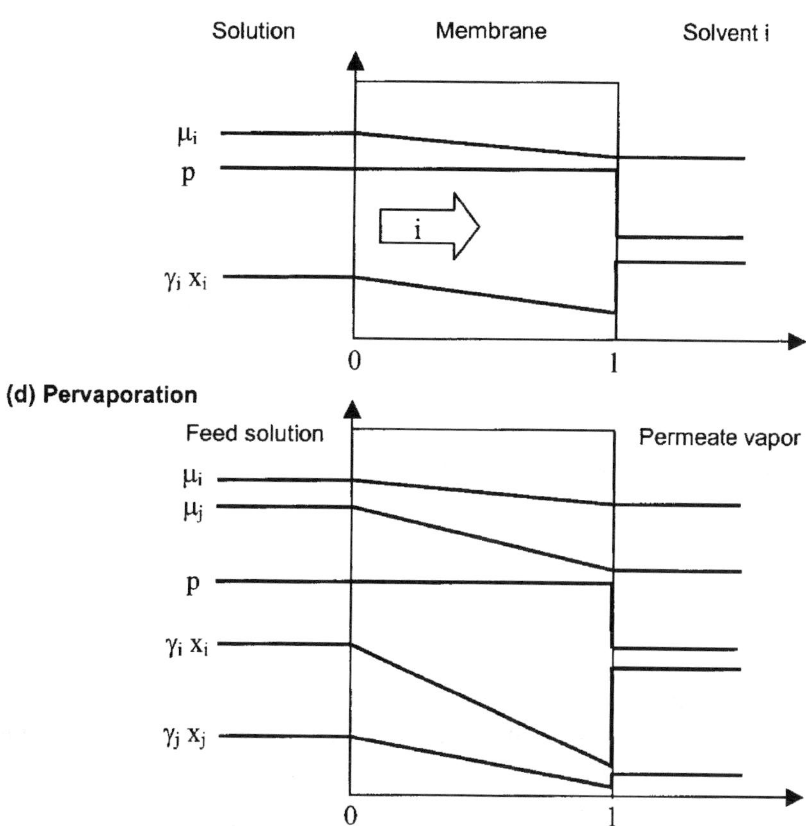

Fig. 1 Chemical potential, pressure, and solvent activity profiles through an osmotic membrane following the SDM. (From Ref. [1].)

Eq. 13 becomes

$$c_{i_o(m)} = K_{i_o} c_{i_o} \tag{15}$$

On the permeate side of the membrane, the same procedure can be followed, leading to an equivalent expression

$$c_{i_l(m)} = K_{i_l} c_{i_l} \tag{16}$$

where

$$K_{i_l} = \frac{\gamma_{i_l} \rho_{l(m)}}{\gamma_{i_l(m)} \rho_l} \tag{17}$$

The concentrations of permeant within the membrane phase at the two interphases can then be substituted from Eqs. 15 and 16 into Fick's law expression, Eq. 9, to give the familiar expression describing permeation through D membranes:

$$J_i = \frac{D_i(K_{i_o} c_{i_o} - K_{i_l} c_{i_l})}{d} \approx \frac{D_i K_i^L}{d}(c_{i_o} - c_{i_l}) = \frac{P_i^L}{d}(c_{i_o} - c_{i_l}) \tag{18}$$

The product $D_i K_i^L$ is normally referred to as the permeability coefficient P_i^L. For many systems, D_i, K_i^L, and thus P_i^L, are concentration dependent. Thus, Eq. 18 implies the use for these parameters of values that are averaged over the membrane thickness.

The permeability coefficient P_i^L is often treated as a pure materials constant, depending only on the permeant and the membrane material, but, in fact, the nature of the solvent used in the liquid phase is also important. From Eqs. 14, 17, and 18, P_i^L can be written as

$$P_i^L = D_i \frac{\gamma_i}{\gamma_{i(m)}} \frac{\rho_m}{\rho_o}$$

The presence of the term γ_i makes the permeability coefficient a function of the solvent used as a liquid phase.

Reverse osmosis

Reverse osmosis usually involves two components, water (i) and salt (j). Following the general procedure, the chemical potentials at both sides of the membrane are first equated. At the feed interface, the pressures in the feed solution and within the membrane are identical (as shown in Fig. 1c). Equating the chemical potentials at this interface gives the same expression as in D

$$c_{i_o(m)} = K_{i_o} c_{j_o} \tag{19}$$

A pressure difference exists at the permeate interface from p_o within the membrane to p_l in the permeate solution. Equating the chemical potentials across this interface, and substituting the appropriate expression for the chemical potential of an incompressible fluid to the liquid and membrane phases leads to

$$\ln(\gamma_{i_l} x_{i_l}) = \ln(\gamma_{i_l} x_{i_{l(m)}}) + \frac{v_i(p_o - p_l)}{RT}$$

Rearranging and substituting for the sorption coefficient K_i^L give the expression

$$c_{i_l(m)} = K_i^L c_{i_l} \exp\frac{-v_i(p_o - p_l)}{RT} \tag{20}$$

The expressions for the concentrations within the membrane at the interface in Eqs. 19 and 20 can now be substituted into Fick's law expression, Eq. 9, to yield

$$J_i = \frac{D_i K_i^L}{d}\left[c_{i_o} - c_{i_l}\exp\frac{-v_i(p_o - p_l)}{RT}\right] \tag{21}$$

Eq. 21, and the equivalent expression for component j, gives the water flux and the salt flux across the RO membrane in terms of the pressure and concentration difference across the membrane. However, Eq. 21 can be simplified further. Consider the water flux first. At the point at which the applied hydrostatic pressure balances the water activity gradient, the flux of water across the membrane is zero. Thus, Eq. 21 leads to

$$c_{i_l} = c_{i_o} \exp\frac{v_i \Delta\pi}{RT} \tag{22}$$

At hydrostatic pressures higher than $\Delta\pi$, Eqs. 21 and 22 can be combined to yield

$$J_i = \frac{D_i K_i^L c_{i_o}}{d}\left[1 - \exp\frac{-v_i[(p_o - p_l) - \Delta\pi]}{RT}\right]$$

or

$$J_i = \frac{D_i K_i^L c_{i_o}}{d}\left[1 - \exp\frac{-v_i(\Delta p - \Delta\pi)}{RT}\right] \tag{23}$$

where

$$\Delta p = p_o - p_l$$

is the difference in hydrostatic pressure across the membrane.

Under the normal conditions of RO the term $[-v_i(\Delta p - \Delta\pi)/RT]$ is small: for e.g., when $\Delta p = 100$ atm, $\Delta\pi = 10$ atm, and $v_i = 0.018\,\mathrm{m^3\,kmol^{-1}}$, the term $[v_i(\Delta p - \Delta\pi)/RT]$ is about 0.06. Under these conditions, the simplification

$$\lim_{y\to 0}(1 - e^y) = y$$

can be used, and Eq. 23 can be written to a very good

approximation as

$$J_i = \frac{D_i K_i^L c_{i_o} v_i (\Delta p - \Delta\pi)}{RT\, d}$$

This equation can be simplified to

$$J_i = A(\Delta p - \Delta\pi) \tag{24}$$

where

$$A = \frac{D_i K_i^L c_{i_o} v_i}{RT\, d}$$

In the RO literature, A is usually called the water permeability constant.

Similarly, a simplified expression for the salt flux J_j through the membrane can be derived, starting with the equivalent to Eq. 21

$$J_j = \frac{D_j K_j^L}{d}\left[c_{j_o} - c_{j_l} \exp\frac{-v_j(p_o - p_l)}{RT}\right] \tag{25}$$

Because the term $[-v_j(p_o - p_l)/RT]$ is small, the exponential term in Eq. 25 is close to 1, and Eq. 25 can then be written as

$$J_j = \frac{D_j K_j^L}{d}[c_{j_o} - c_{j_l}]$$

or

$$J_j = B(c_{j_o} - c_{j_l}) \tag{26}$$

where B is usually called the salt permeability constant and has the value

$$B = \frac{D_j K_j^L}{d}$$

Predictions of salt and water transport can be made from this application of the SDM to RO (first derived by Merten and coworkers). According to Eq. 24, the water flux through an RO membrane remains small up to osmotic pressure of the salt solution and then increases with applied pressure, whereas, according to Eq. 26, the salt flux is essentially independent of pressure.

A measure of the ability of the membrane to separate salt from the feed solution is the term called intrinsic rejection coefficient R, which is defined as

$$R(\%) = \left(1 - \frac{c_{j_l}}{c_{j_o}}\right) \times 100$$

For a perfectly selective membrane, the permeate salt concentration

$$c_{j_l} = 0$$

and

$$R = 100\%$$

For a completely unselective membrane, the permeate salt concentration is the same as the feed salt concentration

$$c_{j_l} = c_{j_o}$$

and

$$R = 0\%$$

The rejection coefficient increases with applied pressure because the water flux increases with pressure but the salt flux does not, so c_{j_l} decreases.

Transport through porous membranes

No unified theory, equivalent to the SDM, has been developed to describe transport in microporous membranes. These membranes consist of a polymeric matrix in which a large variety of pores is present, within the range 2 nm to 10 μm. In MF membrane, the pores exist over the whole membrane thickness, whereas UF membranes generally have asymmetric structure and the porous top-layer mainly determines the resistance to transport. The existence of different pore geometries implies that different models have been developed to describe transport adequately.

Microfiltration

Microfiltration is a sieving process where particles, from about 0.1 μm–10 μm, are physically retained by the membrane. Screen filters used in crossflow MF systems have pores smaller than the particles. The osmotic pressure becomes insignificant, since the particles are relatively large and their number density is small. The leakage of particles through the membrane is insignificant. Rejected particles accumulate at the membrane surface forming a cake layer due to which hydraulic resistance for convective flow of the liquid becomes significant.

The transport model for MF involves an equation for the flow of liquid through the membrane:

$$J_v = \frac{1}{r_m}(P_2 - P_3) \tag{27}$$

where J_v is the solvent flux ($m^3\,m^{-2}\,sec^{-1}$), P_2, and P_3 (Pa) are the pressure values at both sides of the membrane surface, at the feed/retentate side and at the permeate side, respectively.

Membrane resistance, r_m ($Pa\,sec\,m^{-1}$), can be determined easily be conducting an experiment using pure solvent and calculating the slop of the flux–pressure relation. It can also be predicted, if the pore-size distribution is known, using Hagen–Poiseuille equation for capillary flow (Eq. 28):

$$r_m = \frac{128 \times \mu \times d_m}{\pi \times \sum n_p \phi_p^4} \tag{28}$$

where μ is solvent dynamic viscosity (Pa sec), d_m is membrane thickness (m), and n_p (pores m^{-2}) is the number of pores of diameter ϕ_p (m) per unit area of membrane surface.

The transport model also involves an equation for the flow of liquid through the cake layer (Eq. 29)

$$J_v = \frac{1}{r_c}(P_1 - P_2) \tag{29}$$

where r_c (Pa sec m^{-1}) is the hydraulic resistance of the cake. When the cake is incompressible, r_c can be estimated by the Carman–Kozeny relationship (Eq. 30)

$$r_c = \frac{K \times (1 - \varepsilon_c)^2 \times S_c^2 \times \mu \times d_c}{\varepsilon_c^3} \tag{30}$$

The void fraction (ε_c) of a randomly packed cake is about 0.4, while the specific surface area S_c (m^{-1}) is $6/\phi_c$ for rigid spheres with diameter ϕ_c. The coefficient K has a value of about 5. The dynamic viscosity of the permeant is μ (Pa sec), and d_c (m) is cake layer thickness.

It is customary to add Eqs. 27 and 29 to eliminate P_2:

$$J_v = \frac{1}{r_c + r_m}(P_1 - P_3) \tag{31}$$

The transport model involves also Fick's equation for the diffusion of the solute through the boundary layer (Eq. 32):

$$J_i = -D_{i_{BL}} \frac{dc_i}{dx} \tag{32}$$

The diffusive flow of solute across any plane in the boundary layer, given by Eq. 32, is equated to the solute flow due to bulk flow, $J_v c_i$, and integrated over the thickness of the boundary layer, δ (m), to arrive at Eq. 33.

$$J_v = \frac{D_{i_{BL}}}{\delta} \ln \frac{c_1}{c_2} \tag{33}$$

Since the boundary layer thickness is not easily defined or determined, the practice is to replace ($D_{i_{BL}}/\delta$) by h_b, which is the mass transfer coefficient of the boundary layer:

$$J_v = h_b \ln \frac{c_2}{c_1} \tag{34}$$

In conditions encountered in MF, the concentration of rejected particles at the membrane surface reaches saturation level very early. When this condition is met, c_2 is replaced by a constant c_c, which is the cake layer concentration:

$$J_v = h_b \ln \frac{c_c}{c_1} \tag{35}$$

Cake layer concentration c_c has a theoretical maximum value of 0.74 (v/v) for hexagonally packed rigid spheres. Values as high as 0.9 have been observed for flexible particles such as red blood cells. Mixtures of nonuniform size particles also can produce higher values.[10] Macromolecular solutions such as proteins and starches have lower values in the range 0.2–0.4.

The mass transfer coefficient h_b (m sec^{-1}) can be estimated using equations like Eq. 36,[10] derived for analogous heat transfer problem in developing laminar flow in tubes, where L (m) is tube length, and γ_w (sec^{-1}) is shear rate at the wall:

$$h_b = 0.807\left(\frac{\gamma_w D_{i_{BL}}^2}{L}\right)^{1/3} \tag{36}$$

Diffusivity of the solute across the boundary layer, $D_{i_{BL}}$, which should be used in Eq. 36, may be calculated[10] as:

$$D_{i_{BL}} = 0.0075\phi_p^2 \gamma_w$$

so the mass transfer coefficient h_b is expressed by

$$h_b = 0.078\left(\frac{\phi_p^4 \gamma_w^3}{16L}\right)^{1/3}$$

Eq. 34 is valid before cake layer formation, while Eq. 35 is valid after cake layer formation. Eqs. 31 and 34 or 35 constitute a transport model of MF process. This model contains only one property of the membrane but three properties of the boundary/cake layer (r_c, c_c, h_b). The only membrane property (r_m) also becomes insignificant compared to r_c in practical situations. Therefore, MF process is controlled almost entirely by the boundary/cake layer.

Ultrafiltration

The transport model for UF consists of Eqs. 31, 34, and 35, derived for MF process. The pressure and concentration profiles and the mathematical model of UF are similar to those of MF. However, the osmotic pressure becomes somewhat more significant since the particles are relatively smaller and their number density is higher. Therefore, the pressure dependence of the permeate flux in UF can be represented by two alternative models, known as the resistance model and the osmotic pressure model.

The resistance model[11] assumes that the hydraulic resistance of the boundary layer, r_b, increases proportionately with pressure:

$$r_b = \Psi(P_1 - P_3) \tag{37}$$

Substituting Eq. 37 in Eq. 31 and rearranging gives Eq. 38:

$$J_v = \frac{(P_1 - P_3)}{r_m + \Psi(P_1 - P_3)} \tag{38}$$

The osmotic pressure model[9] does not take into account the hydraulic resistance of the boundary layer, and hypothesizes that the increase in pressure increases the concentration of solute at the membrane surface, hence the osmotic pressure.

Both the resistance model and the osmotic pressure model correctly emulate the experimentally observed flux pressure behavior.

Transport in electrically driven membrane processes

Mass transfer in electrolyte solutions is determined by the driving forces acting on the individual ions of the solution and by the friction of the ions with other components in the solution. The driving forces can be expressed by gradients in the electrochemical potential of individual components, and the friction that has to be overcome by the driving force can be expressed by the ion mobility or diffusivity. A very good overview of mathematical relationships, which can be applied to study energy and mass transfer phenomena that occur in ED processes, is provided in Ref. [12].

ACKNOWLEDGMENTS

Membrane technology is increasingly expanding, and the number of people dealing with membranes is growing rapidly. For those of us who are involved in helping students to learn the basics and some applications of membrane technology, the invaluable help of books like Richard W. Baker's "Membrane Technology and Applications" (McGraw-Hill) (see Ref. [1]), Ian D. Wilson's "Encyclopedia of Separation Science" (Academic Press) (see Ref. [2]), and Marcel Mulder's "Basic Principles of Membrane Technology" (Kluwer Academic Publishers) (see Ref. [3]), must be deeply acknowledged.

REFERENCES

1. Baker, R.W. *Membrane Technology and Applications*; McGraw-Hill: New York, 2000.
2. Wilson, I.D. In *Encyclopedia of Separation Science*; Wilson, I.D., Ed.; Academic Press: London, 2000.
3. Mulder, M. *Basic Principles of Membrane Technology*, 2nd Ed.; Kluwer Academic Publishers: Dordrecht, 1998.
4. Zydney, A.L. Membrane Bioseparations. In *Encyclopedia of Separation Science*; Wilson I.D., Ed.; Academic Press: London, 2000; 1748–1755.
5. Pinnau, I. Membrane Preparation. In *Encyclopedia of Separation Science*; Wilson, I.D., Ed.; Academic Press: London, 2000; 1755–1764.
6. Huisman, I.H. Microfiltration. In *Encyclopedia of Separation Science*; Wilson, I.D., Ed.; Academic Press: London, 2000; 1764–1777.
7. Boyadzhiev, L. Liquid Membranes. In *Encyclopedia of Separation Science*; Wilson, I.D., Ed.; Academic Press: London, 2000; 1739–1748.
8. Wijmans, H. Concentration Polarization. In *Encyclopedia of Separation Science*; Wilson, I.D., Ed.; Academic Press: London, 2000; 1682–1687.
9. Mannapperuma, J.D. Design and Performance Evaluation of Membrane Systems. In *Handbook of Food Engineering Practice*; Valentas, K.J., Rotstein, E., Singh, R.P., Eds.; CRC Press LLC: Boca Raton, FL, 1997; 167–209.
10. Zydney, A.L.; Colton, C.K. A Concentration Polarization Model for the Filtrate Flux in Cross Flow Microfiltration of Particulate Suspensions. Chem. Eng. Commun. **1986**, *47*, 21.
11. Cheryan, M. *Ultrafiltration Handbook*; Technomic Publishing: Lancaster, 1986.
12. Strathman, H. Electrodialysis. In *Encyclopedia of Separation Science*; Wilson, I.D., Ed.; Academic Press: London, 2000; 1707–1717.

Membrane Separation System Design

Sean X. Liu
Rutgers, The State University of New Jersey, New Brunswick, New Jersey, U.S.A.

INTRODUCTION

Membrane processes in the food industry are primarily used to concentrate or fractionate a liquid to generate two liquids or one liquid and one solid that differ in their compositions. The concentration or fractionation is achieved by diffusion of some components of the liquid feed across the thin membranes that sometimes produce chemical and physical separations at lower costs. The advantages of membrane technology are:

- Most systems are simple, modular in nature, and can be retrofit into existing processes.
- Membrane processes are nondestructive for thermally labile foods and flavors, and able to recover valuable products from waste streams and by-products.
- Most of membrane processes do not involve a phase change and therefore energy efficient, and for those membrane processes that do involve phase changes, the energy requirement is still far less than that of a typical conventional separation technology.
- Membrane systems can be operated either continuously (single or multiple staged) or batchwise.
- Membranes can be used to improve food product quality or achieve separations that were previously impossible or uneconomical.

Potential disadvantages are:

- Concentration polarization and fouling result in various degrees of performance reduction.[1–3]
- Most polymeric materials cannot maintain mechanical stability under conditions of high temperature, high pH, chlorine, and organic solvents.
- In some cases, sufficient separations cannot always be achievable.

It should be noted that membrane separation is not an ideal new technology, but just another type of unit operation, whose attractiveness must be weighed against other separation technologies when evaluating the advantages and disadvantages of membrane applications in a particular food processing operation.

The most important qualities of a membrane are high selectivity, high permeability, mechanical stability, temperature stability, and chemical resistance.[1] Selectivity is listed as the number one criterion in considering using membrane processes to perform a certain task. This is because using a multistage membrane process or increasing membrane surface can sometimes overcome the shortfall of low permeability (flux), however, in doing so, one must consider the possibility of a better and more established separation technology that may exist and provide better economical and technological advantages. Although suitable membranes are at the heart of a successful application of membrane technology, other aspects, especially the hydro-dynamically and economically optimal membrane arrangement (module design) and the module arrangement (plant design) are of at a minimum equal importance.

BACKGROUND

Membrane technology is an emerging separation technology, and because its multidisciplinary characters, it can be used to perform a large number of separations in food processing. The membrane processes that are commonly found in food processing plants or research laboratories include microfiltration (MF), reverse osmosis (RO), ultrafiltration (UF), nanofiltration (NF), electrodialysis (ED), membrane distillation (MD), and pervaporation (PV). Membrane processes are based upon different separation principles or mechanisms and their applications in food processing range from concentration of food fluids to aromatic flavor recovery. Despite these differences, all membrane processes have one thing in common—they all have a membrane that acts as a permselective barrier segregating permeate from feed. The membrane is at the center of every membrane process. This is because the membrane not only functions as a gatekeeper to the retained species, and at the same time allows one or more components to transport across it in a liquid feed stream, but also provides a large contacting area in which mass transfer can take place.[2] However, membrane separation can only be achieved when a driving force is applied to the underlying membrane process. A schematic diagram

Encyclopedia of Agricultural, Food, and Biological Engineering
DOI: 10.1081/E-EAFE 120007075
Copyright © 2003 by Marcel Dekker, Inc. All rights reserved.

of a two-phase conceptual system is shown in Fig. 1. It should be also reminded that there is no perfect man-made membrane ever existed. This situation will be with us in the foreseeable future until perhaps we fully understand the mechanisms that regulate the mass transfer in the membrane, and we are able to tailor the membrane structures to the need of separation of molecules of interest by using the latest advancement in nanotechnology. In assessing membrane systems two experimental parameters that determine the overall performance of membrane processes should be the main focus of designers' attention. The first one is selectivity and the other is permeation flux.

The selectivity of a membrane towards a mixture, which characterizes the extent of separation, is customarily expressed by one of the two quantities: the retention, R; and the separation factor, α. The retention, R, is more suitable for the membrane separation of a dilute binary system and given by

$$R = \frac{C_f - C_p}{C_f} \tag{1}$$

where C_f is the solute concentration in the feed stream and C_p the solute in the permeate. The value of R varies between 100% (complete rejection or retention) and 0% (complete permeation). For most mixtures, however, separation factor is more adequate:

$$\alpha_{ij} = \frac{(C_i/C_j)^p}{(C_i/C_j)^f} \tag{2}$$

where C_i and C_j are the concentrations of components i and j in the permeate and in the feed. The value of α is greater than one, if the component i is more readily transported across the membrane than component j and if the separation occurs.

The other parameter, permeation flux, defined as mass (or volume) of the permeate across a unit area of the membrane per unit time (common SI unit: $\mathrm{kg\,s^{-1}\,m^2}$), takes many forms depending upon the underlying membrane processes. It is normally expressed as

$$J_i = -K\frac{dg}{dz} \tag{3}$$

where K is phenomenological coefficient and dg/dz is the driving force expressed as the gradient of g (concentration, temperature, pressure) in the z direction toward the membrane. The phenomenological coefficient, K is strongly related to the driving force, module configuration, and operating conditions.

Membrane processes can be classified according to the nature of their driving forces and pore size of the membrane.[1,2] Although all membrane processes are driven by electrochemical potential gradient, one particular driving force is usually dominated in a membrane process. Three types of membrane separation processes relevant to the food industry can be considered: those that are driven by hydrostatic pressure difference, by partial vapor pressure gradient, or by electrical potential differences. Table 1 lists common membrane processes that find the use in food processing and the classification of these processes according to their driving forces.

DESIGN CONSIDERATIONS

In many cases, it is still true to say that the design of a membrane process or/and the selection of a membrane

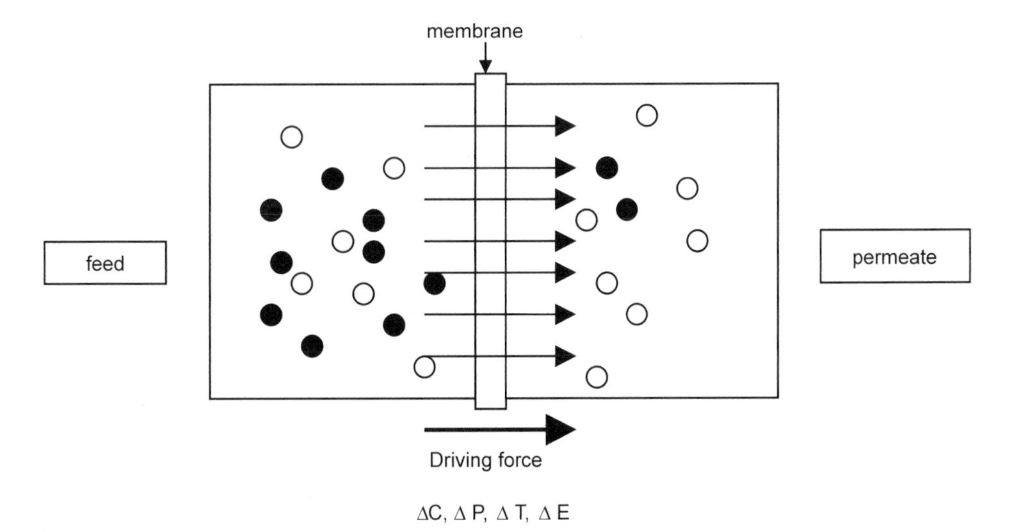

Fig. 1 Schematic diagram of a conceptual two-phase membrane system.

Table 1 Classification of membrane processes relevant to the food industry

Membrane processes	Morphology and type	Driving force	Typical food applications
Microfiltration	Symmetric and porous	Pressure	Sterile filtration and colloid removal
Ultrafiltration	Asymmetric and porous	Pressure	Protein concentration and separation of macromolecules
Reverse osmosis	Asymmetric/composite and nonporous	Pressure	Desalination and concentration of fruit juices
Nanofiltration	Asymmetric/composite and porous	Pressure	Separation of small organics and some salts
Electrodialysis	Ion-exchange and porous or nonporous	Electrical potential	Separation of salts from water and nonionic species
Membrane distillation	Hydrophobic, symmetric, and porous	Temperature	High-purity water and concentration of food fluids
Pervaporation	Asymmetric/composite and nonporous	Partial vapor pressure	Dehydration of alcohols and separation of volatile flavors and aromas

module/material for desired separations remains predominantly an art, in which knowledge, experience, and science play important roles. More often than not, there is no "right answer" in the absolute sense, as more than one solution is both technically and economically viable. However, a careful evaluation at the outset of as many as possible of the factors influencing the choices will help narrow down the items on the list.

When contemplating the use of any particular membrane process for the separation of components in a liquid food stream after the initial assessment, several process issues must be evaluated. The first step in doing so is to draw up the detailed requirements for the process. Accurate, qualitative, and, where possible, quantitative information on the following aspects should therefore be specified:

- The components and range of concentrations in the feed.
- The intended use or fate of the treated feed liquids (i.e., final products, further processing, etc).
- The intended use of or fate of the permeate (i.e., disposal, reuse, further processing, etc).
- Permeation flux.
- The minimum properties of the treated food fluids and permeate that will make the intended use or fate possible.
- Membrane transport mechanism.
- Cost-effective and environmentally friendly alternative solution.

Any one or more of the above factors may, depending upon circumstances, influence the design of a membrane process. In the case of foods and other kindred products, it is normal for quality considerations to override economical factors. This is particularly true for aroma compound recovery in premium orange juice production using PV, in which the preservation of the permeate is the key to exerting a major influence on the ultimate quality of the product. In other food processing applications, a membrane process can be an important intermediate operation that is vital to subsequent processing operations.

The next issue to be addressed is whether the membrane processes are actually capable of separating the components from liquid foods. The answer to this question for the pressure-driven membrane processes such as MF, UF, NF, and RO is generally affirmative, provided that appropriate membranes (pore size, and for NF and RO membrane properties such as charge, hydrophilic tendency) are used. For PV, the answer is more complicated and conditional. It is well documented that PV works well when the compound to be removed has a high vapor pressure relative to the background material and a low solubility in the background material. In dilute aqueous solutions such as aromas in orange juice, it is generally the Henry's law constant that determines whether an aroma compound can effectively separated by PV. The Henry's law constant represents the vapor–liquid partitioning of organic compounds in an aqueous system. The general rule of thumb is: the more dissimilar the components, the easier it will be to separate them.

Once the process designer has determined that a particular membrane process will, theoretically, work. The subsequent questions to be answered are:

- Does a membrane material exist which will do the job?
- Is this membrane material available in a membrane module?

The answer to the first question is usually positive. A great deal of membrane research has been performed on many membrane materials and feed mixtures.

In addition, a wide array of membrane materials is available which may achieve the desired separation, but which have not been tested in a membrane mode of interest.

Another variable in the selection of membrane materials is whether a single layer membrane or a multilayer membrane is to be used. Membranes used in an MF are normally single layer, isotropic, while membranes in other pressure-driven membrane filtrations and PV are composed of composite or multiplayer, nonhomogenous materials. This is because a membrane with desired selectivity may require a significant thickness to deliver the desired physical properties such as burst pressure, however, improving membrane mechanical stability by increasing membrane thickness would inadvertently reduce permeation flux. To get around this problem, a composite or inhomogeneous membrane is employed where a thick layer of polymer material with large pore size supports a top thin layer of the active membrane.

The second question is about the issue of commercial availability of membrane configurations or membrane modules for particular membrane materials. A module is the smallest unit into which the membrane material is packed. The reason for using modules is because although polymer membranes are made in two basic physical forms: flat sheet and tubular, many practical membrane systems that need large membrane areas can only be accommodated in membrane modules. For pressure-driven membrane processes, MD, and PV there are four primary configurations (modules), each with inherent advantages and weaknesses. These four are: spiral wound; hollow fiber; plate and frame; and tubular. Two special modules, rotary and vibrating, have been used to reduce concentration polarization in MF, UF and PV.[3,4]

Membrane Modules

Spiral wound

A spiral wound module is a logical step from a flat sheet membrane. In spiral wound modules, a flat membrane envelope or set of envelopes is rolled into a cylinder as shown in Fig. 2a. The envelope is constructed from two sheets of membrane, sealed on three edges. The inside of the envelop is the permeate side of the membrane. A thin porous spacer inside the envelope keeps the two sheets separated. The open end of envelope is sealed to a perforated tube (the permeate tube) with a proper adhesive so that the permeate can pass through the perforations and, for PV, it is also the place to which the vacuum or sweep gas is applied. Another spacer is laid on top of the envelope before it is rolled, creating the flow path for the feed liquid. This feed spacer generates turbulence due to the undulating flow path that disrupts the liquid boundary layer, thereby enhancing the feed side mass transfer rate. The fact that the envelopes and spacers are wrapped around the permeate tube gives the module its name, spiral wound module. The spiral wrapped envelopes and spacers are then wrapped again with tape, glass, or net-like sieve before fitting into a pressure vessel. In this way, a reasonable membrane area can be housed in a convenient module, resulting in a very high surface area to volume ratio. One noticeable drawback lies in the permeate path length. A permeating component that enters the permeate envelope farthest from the permeate tube must spiral inward several feet. Depending upon the path length, permeate spacer design, gel layer, and permeate flux, significant permeate side pressure drops can be encountered. The other disadvantage of this module is that it is a poor choice for treating fluids containing particulate matters.

Hollow fiber

In a hollow fiber configuration, small diameter polymer tubes are bundled together to form a hollow fiber module like a shell and tube heat exchanger (Fig. 2b). These modules can be configured for liquid flow on either the tube side, or the lumen side (inside the hollow fibers). These tubes have diameters on the order of 100 μm. As a result, they have a very high surface area to module volume ratio. This makes it possible to construct compact modules with high surface areas. The drawback is that the liquid flow inside the hollow fibers is normally within the range of laminar flow regime due to its low hydraulic diameter. The consequence of prevalent laminar flows is high mass transfer resistance on the liquid feed side. However, because of laminar flow regime, the modeling of mass transfer in a hollow fiber module is relatively easy and the scale-up behavior is more predictable than those in other modules. One noticeable problem with a hollow fiber module is that a whole unit has to be replaced if failure occurs.

Plate and frame

Plate and frame configuration is a migration from filtration technology, and is formed by the layering of flat sheets of membrane between spacers. The feed and permeate channels are isolated from one another using flat membranes and rigid frames (Fig. 2c). This configuration was an early favorite it is a natural scale-up from bench-scale laboratory membrane cells that have one feed chamber and one permeate chamber separated by a flat sheet of membrane. A single plate and frame unit can be used to test different membranes by swapping out the flat sheets of membrane. Further, it allows for the use for membrane materials that cannot be conveniently produced as hollow fibers or spiral wound elements.

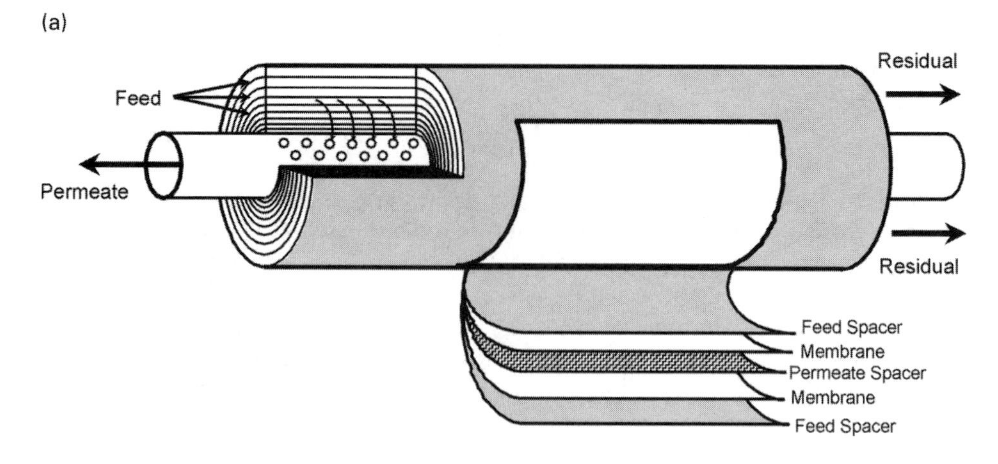

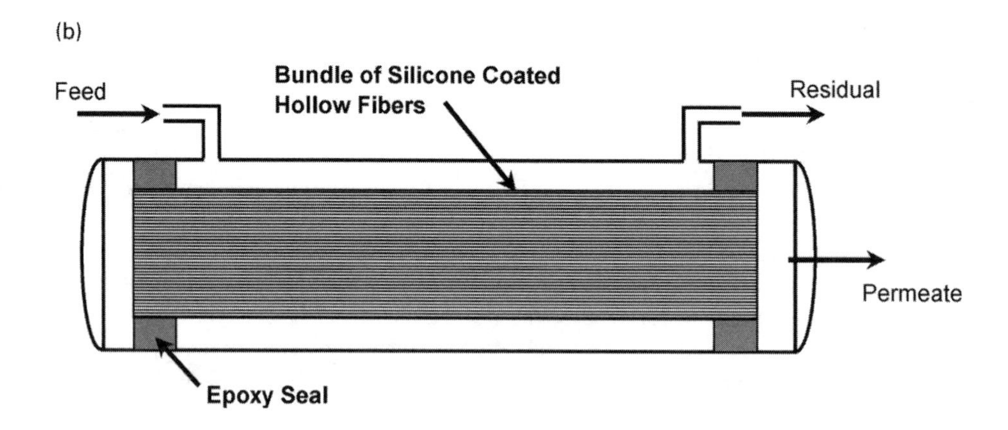

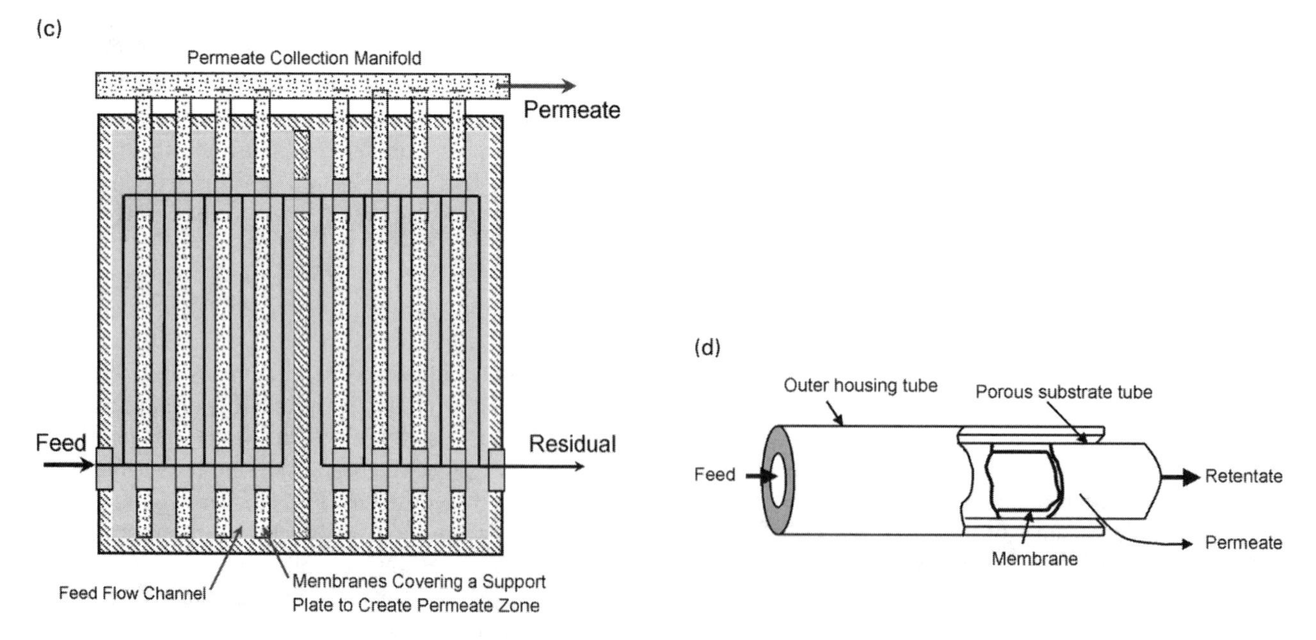

Fig. 2 a) A schematic illustration of a spiral wound module (Courtesy of Dr. Leland Vane of USEPA). b) A schematic illustration of a hollow fiber module (Courtesy of Dr. Leland Vane of USEPA). c) A schematic illustration of a plate and frame module (Courtesy of Dr. Leland Vane of USEPA). d) A schematic illustration of a tubular module.

The disadvantages are that the ratio of membrane area to module volume is low compared to spiral wound or hollow fiber modules, dismounting is time-consuming and labor-intensive, and higher capital costs associated with the frame structures.

Tubular

Polymeric tubular membranes are usually made by casting a membrane onto the inside of a preformed tube, which is referred to as the substrate tube. The tube is generally made from one or two piles of nonwoven fabric such as polyester or polypropylene. The diameters of tubes range from 5 mm to 25 mm (Fig. 2d). A popular method of construction of these tubes is a helically wound tape that is welded at the edges. The advantage of the tubular membrane is its mechanical strength if the membrane is supported by porous stainless steel or plastic tubes. Tubular arrangements often provide good control of flow to the operators and are easy to clean. Additionally, it is the only membrane format for inorganic membranes, particularly ceramics. The disadvantage of this type of modules is mainly higher costs in investment and operation. The arrangement of tubular membranes in a housing vessel is similar to that of hollow fiber element. Tubular membranes sometimes are arranged helically to enhance mass transfer by creating a second flow (Dean vortex) inside the substrate tube.[5]

Although, the specification for the process is the most critical issue in process design of membrane systems, certain auxiliary steps must also be considered in the planning of a membrane system. For example, temperature, pH limits and tolerance to certain chemicals, particularly cleaning agents such as alkalis and detergents should be considered before a process is put online. These cleaning chemicals as well as seals and adhesives used in the membrane modules have to be approved by FDA or other regulatory agencies for used in food processing—an aspect of process deign that is often neglected by some designers.

Ultimately, it is the outcome of an economic evaluation of the conceptual membrane process that often foretells the fortune of the process implementation. After initial selection of the operational mode (continuous or batchwise), membrane materials/modules, auxiliary materials/procedures, and the total membrane area (the product of permeation flux and membrane area is the process throughput) of a single staged or multiple staged process, an overall cost structure should be evaluated according to the following factors:

- Membrane material and its availability
- Module housing
- Pretreatment
- Instrumentation, piping, and pumping
- Labor
- Washing chemicals, water, and down-time
- Polishing step
- Disposal or treatment of waste streams and wastewater
- Replacement frequency of membranes or membrane modules

Like implementation of any new technologies, the financial representation associated with the use of a membrane-based technology should take into account of additional capital investments and/or operational costs incurred.

FOUL PLAY: MEMBRANE FOULING

As described in the previous sections, concentration polarization phenomena of membrane processes cause noticeable decline of membrane performance. In membrane filtration processes such as MF and UF concentration polarization phenomena always accompany by the formation of a gel layer that is either irreversible or reversible. The cause of gel layer formation is thought to be the result of the rapid accumulation of retained solutes near the membrane surface to the point that the concentration of macromolecule solute reaches the gel forming concentration. High retention of solutes near the membrane surface inevitably also leads to concentration polarization and as a result of that the performances of membrane filtration processes (pressure-driven processes) suffer. The version of concentration polarization in PV is slightly different from its kindred in membrane filtration as stated in the other entries. Concentration polarization also negatively affects the performance of an ED process. For MD temperature polarization is the main culprit for the decline in the process performance.

Membrane fouling is suspected if the membrane flux is continuously declining after a period of time of operation. This is usually an irreversible, partially concentration dependent, and time-dependent phenomenon, which distinguishes it from concentration polarization. The identification of membrane fouling is imprecise and often based upon operator's experience, performing fouling tests with membrane filtration index apparatus, and membrane vendor's recommendations. Membrane fouling is intimately related to concentration polarization but the two are not exactly interchangeable in our description of membrane performance deterioration. Moreover, we now know that all membrane filtration processes experience some degree of concentration polarization but fouling occurs mainly in MF and UF. Relatively large pores in these membranes are implicitly vulnerable to fouling agents such as organic and inorganic precipitates, and fine particulate matters that could lodge in these pores or deposit irreversibly on the membrane surface. The exact cause of membrane fouling is very complex and therefore difficult to depict in full confidence with available theoretical understandings. Even for a known solution, fouling is influenced by a number of chemical and physical parameters such as concentration, temperature, pH, ionic strength, and specific interactions (hydrogen bonding, dipole–dipole interactions).[2,3]

Membrane fouling in membrane filtration processes like concentration polarization is unavoidable—this is particularly true for protein concentration or fractionation. However, certain steps that will greatly reduce the severity of membrane fouling can still be achieved. One effective way of reducing membrane fouling is to provide pretreatment to the feed liquids. Some simple adjustments such as varying pH values and using hydrophilic membrane materials can also do wonders in protein concentration operations. There are persistent interests in modifying membrane properties to minimize the membrane-fouling tendency around the world. Since membrane fouling is intimately associated with concentration polarization phenomenon, any action taken to minimize concentration polarization will also benefit the fight against membrane fouling. Unfortunately, no matter how much effort is put forward, fighting against membrane fouling, it will eventually occur. The only solution then is employing cleaning regimen. The frequency of cleaning depends upon many factors and should be considered as a part of process optimization exercise. There are three basic types of cleaning methods currently used: hydraulic flushing (back-flushing); mechanical cleaning (only in tubular systems) with sponge balls; and chemical washing. When using chemicals to perform de-fouling, caution must be observed since many polymeric membrane materials are susceptible to chlorine, high pH solutions, organic solvents, and a host of other chemicals.

REFERENCES

1. Rautenbach, R.; Albreht, R. *Membrane Processes*, John Wiley & Sons: Chichester, England, 1989.
2. Mulder, M. *Basic Principles of Membrane Technology*, Kluwer Academic Publishers: Dordrecht, The Netherlands, 1991.
3. Cheryan, M. *Ultrafiltration and Microfiltration Handbook*, Technomic Publishing Co.: Lancaster, PA, 1998.
4. Vane, L.M.; Alvarez, F.R.; Giroux, E.L. Reduction of Concentration Polarization in Pervaporation Using Vibrating Membrane Module. J. Membr. Sci. **1999**, *153*, 233.
5. Moulin, P.; Manno, P.; Rouch, J.C.; Serra, C.; Clifton, M.J.; Aptel, P. Flux Improvement by Dean Vortices: Ultrafiltration of Colloidal Suspensions and Macromolecular Solutions. J. Membr. Sci. **1999**, *156*, 109.

Membrane Structure

Gregory R. Ziegler
The Pennsylvania State University, University Park, Pennsylvania, U.S.A.

M

INTRODUCTION

Pressure-driven membrane separation processes employ membranes that act as selective filters to separate colloidal particles or dissolved species based largely on differences in size. Regardless of the material of construction, several general features of membrane structure determine performance. These membranes act as *screen filters*, like sieves, that retain particles on their surface, vis-à-vis, *depth filters* typically used in dead-end filtration that trap particles within a tortuous matrix of randomly-oriented fibers or granules. Consequently, membranes can be assigned a quantitative rating related to the pore size distribution on the separation surface. The filtration spectrum is divided into the processes known as *microfiltration* (MF), *ultrafiltration* (UF), *nanofiltration* (NF), and *reverse osmosis* (RO) based principally on the size of the solute or particle retained by the membrane. MF, UF, and NF are membrane-based separations processes capable of rejecting particles and dissolved molecules larger than 0.1 μm, 2 nm, and < 2 nm, respectively. RO is a membrane-based process in which an applied transmembrane pressure causes selective transport of solvent against its osmotic pressure gradient. The properties of a membrane that affect its retention and flux characteristics include: chemical composition, pore size, and the distribution of pore diameters, pore density (number of pores per unit of membrane surface area), porosity or void volume (fraction of the membrane not occupied by the membrane material), and the structural relationship between the active surface and the support layers. Membranes are formed as either flat sheets, hollow fibers, or coated tubes.

MICROSTRUCTURE

In the context of separations processes, a membrane is a structure having lateral dimensions much greater than its thickness (ca. 100 μm), through which mass transfer may occur under a variety of driving forces. Membranes may be either *homogeneous* (Fig. 1), with essentially the same structural and transport properties throughout their thickness, or *asymmetric* (Fig. 2), comprising two or more structural planes of nonidentical morphologies. Homogeneous membranes are often referred to as isotropic microporous membranes, and asymmetric membranes are sometimes called skinned membranes. The selective, or semipermeable, *membrane skin* is the thin (0.1 μm–0.2 μm) distinguishable layer on the upstream face of an asymmetric membrane that is primarily responsible for determining the permeability characteristics. If the skin and support layers are formed in a single operation from a single material the membrane is said to have an *integrally-skinned* structure. If the skin is a coating on the support layer then the membrane is a composite structure. *Composite membranes* have chemically or structurally distinct layers. Homogeneous membranes retain particles above a characteristic pore size, e.g., 0.2 μm, and so find application in MF processes, while asymmetric membranes are generally rated in terms of the nominal molecular weight cutoff, and are employed in UF processes. *Dense* or nonporous membranes have no detectable pores. The "pores" of RO membranes are below typical limits of resolution, and are generally accepted to be in the 4 Å–8 Å range.

Pressure-driven membrane separations are nonequilibrium processes. Separation occurs because one species permeates the membrane at a greater rate than another. The permeation rate is, to a first approximation, based on molecular size or weight. If the membrane contains pores large enough to allow for convective flow, no separation occurs (Fig. 3, A-1). Knudsen diffusion occurs in pores with dimensions smaller than the mean free path of the molecules (Fig. 3, A-2). Smaller species migrate more readily than larger ones. When the pores are small enough to physically exclude the larger species separation occurs by *molecular sieving* (Fig. 3, A-3). If the mechanism of separation involves *solution–diffusion* (Fig. 3, B), then both the molecular size and its solubility in the membrane material are important.

MATERIALS OF CONSTRUCTION

The majority of membranes in service are polymeric. While numerous polymers have been investigated for their potential as separation membranes, relatively few have

Encyclopedia of Agricultural, Food, and Biological Engineering
DOI: 10.1081/E-EAFE 120007069
Copyright © 2003 by Marcel Dekker, Inc. All rights reserved.

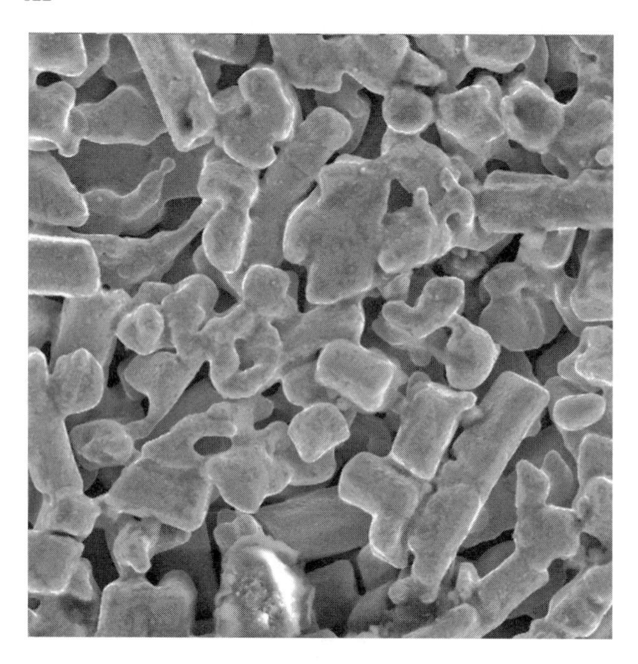

Fig. 1 Microstructure of a sintered stainless steel UF membrane, 1000× magnification (courtesy of Millipore).

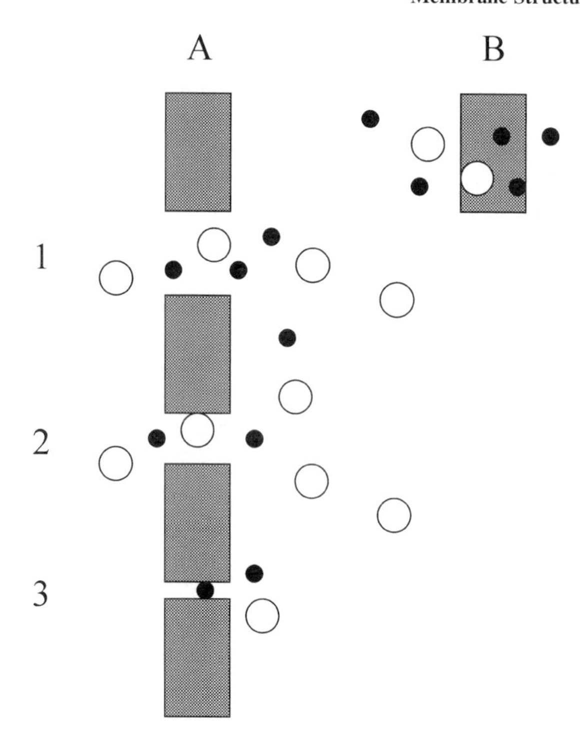

Fig. 3 Mechanisms for molecular separation with membranes (adapted from Ref. 1).

been commercialized. Important properties of polymers for use as membranes include high water affinity, and thermal, physical, and chemical stability. RO membranes, requiring both small pores and significant water sorption, are manufactured mainly from cellulose acetate and polyamide polymers. Cellulose acetate, polyvinylidene fluoride, polyacrylonitrile and, most commonly, polysulfone are used to manufacture UF membranes. Materials for MF applications include those previously mentioned for RO and UF as well as polycarbonate, polypropylene, polyethylene, and PTFE (Teflon®). Selectivity of a membrane results from steric exclusion and surface force interactions. The base polymer surface chemistry can be modified to alter hydrophilicity. Charge repulsion between anions and a hydrophilic (anionic) membrane surface complements separation due to size exclusion, and increases flux and reduces fouling in most aqueous systems.

Fig. 2 Microstructure of a asymmetric polyethersulfone (PES) UF membrane, 500× (courtesy of Millipore).

The productivity of a membrane system will normally decrease with time as the membrane densifies (compacts) under the applied transmembrane pressure. Collapse or *compaction* of the porous support layer adjacent to the membrane skin increases the thickness of the selective skin. Solvents that can plasticize the polymer tend to accelerate membrane compaction and shorten membrane life, as does higher temperatures. The temperature limits for membranes are related to the glass transition temperature of the polymer. Rigid polymers with high glass transition temperatures are preferred materials.

Membranes can be manufactured from nonpolymeric materials such as metals (stainless steel), metal oxides (alumina, zirconia), carbon, and glass. These *ceramic* and *metallic* membranes are resistant to chemical and mechanical degradation and perhaps most important for food applications, can be heat sterilized. Ceramic membranes have been manufactured with pores of 20 nm. Pleated stainless steel membranes are available for gas and steam filtration.

FORMING TECHNIQUES

Polymeric membranes are formed using a variety of techniques that determine if the surface is either porous or dense. Membranes are fabricated by casting films from polymer solutions close to the point of phase separation (20%–40%). Precipitation of the polymer results in polymer-rich and polymer-poor regions that become the walls and pores of the membrane, respectively. In the *dry-phase separation process*, dissolved polymer is precipitated by evaporation of a sufficient amount of solvent to form a membrane structure. Mixtures of additives are present in the solution to alter its precipitation properties during solvent evaporation. The selective skin layer forms rapidly at the solvent–air interface as solvent is evaporated, impeding further evaporation, and allowing phase separation to proceed to a greater extent within the depth of the membrane. This creates asymmetry in the membrane structure. In the *wet-phase separation process*, the dissolved polymer is precipitated by immersion in a nonsolvent bath. Dry- and wet-phase process may be combined into a dry–wet-phase process. Temperature change may be used to precipitate the polymer in *thermally-induced separation*, or to modify the rate of precipitation in the dry- or wet-phase processes. Thermal annealing treatments may be applied after membrane formation. The dry process produces membranes that are more robust, but flux and rejection properties may be inferior to wet-process membranes.

Track-etch membranes with well-defined pore sizes are produced by exposing a dense, thin (15 μm) polymeric film to ion bombardment followed by etching of the damaged region in a chemical bath (usually alkali). This process produces a narrow pore size distribution with low surface pore density (10%–15%). However, since the overall membrane-thickness is much less and the flow path less tortuous, flux characteristics are similar to other microfiltration membranes with greater pore density. Polycarbonate and polyester are commonly used for track-etch membranes.

Composite, asymmetric membranes are manufactured of different layers of material of decreasing pore size applied in stages. *Langmuir-Blodgett membranes* are synthetic composite membranes produced by sequential deposition of one or more monolayers of surface-active component onto a porous or dense support. Composite membranes may be formed by separately casting the selective and support layers followed by lamination, dip coating the support, vapor phase deposition of the selective layer, or interfacial polymerization of reactive monomers onto the surface of the support film. Currently, many RO membranes are thin-film composites.

Surface treatments, e.g., surface charge, add unique separation characteristics to existing membranes. This can be accomplished by reacting the base polymer with hydrophilic groups prior to precipitation, surface grafting hydrophilic groups to previously fabricated membrane, or employing polymer blends. Self-cleaning membranes have been produced by attaching enzymes to the membrane surface.

Sol–gel membrane formation is a multistep process for making membranes by reaction between two chemically multifunctional materials dissolved in a solvent that results in a network structure with solvent retained in the network. This is followed by heat treatment to achieve a desired pore structure. The sol–gel process can produce NF membranes from ceramic materials. *Slip cast* alumina membranes have been manufactured by applying successive layers of uniform particles to a porous support of α-Al_2O_3. These membranes have pore sizes in the UF range (40 Å–1000 Å). Membranes formed via sintering techniques, whether metallic, ceramic, or polymeric, typically have homogeneous microstructure with pore diameters of 0.1 μm or larger and are most appropriate for MF applications.

The selective layer of *dynamically formed membranes* is fabricated by the deposition of substances contained in the fluid being processed onto a porous support layer. Dynamically formed zirconia membranes have been commercialized. Supports for these membranes include stainless steel and alumina.

TECHNIQUES FOR STRUCTURAL CHARACTERIZATION

Techniques used to characterize the porosity and morphology of membranes include: bubble-pressure breakthrough, mercury porosimetry, solute retention challenge, electron microscopy, adsorption-based methods, and nuclear magnetic resonance.[2] The first three techniques work best for pore sizes greater than 10 nm.

The *bubble-pressure breakthrough* test, based on Eq. 1, was the method for determining membrane pore size until ASTM F316 was abandoned in 1995.

$$d_p = \frac{4\gamma\cos A}{P} \tag{1}$$

where d_p is the pore diameter, γ is the surface tension at the solvent–air interface, A is the liquid–solid contact angle, and P is the applied pressure. The method measures the pressure required to force one immiscible fluid (usually air) through the pores of a membrane previously filled with a second immiscible fluid (often water). Practically, artifacts may be introduced by the compaction of the membrane due to the high pressures required, especially for UF membranes. Furthermore, the bubble point depends on parameters such as the rate of pressure increase, polymer material, pore length, and temperature (through its affect on viscosity and surface tension). However, a variation of the test is still useful for evaluating membrane integrity.

Although standard procedures have not been defined, the most common technique for characterizing UF membranes is the *solute rejection challenge*. The rejection factor, R, is determined for a set of solutes of varying molecular weight or hydrodynamic radius. A plot of molecular weight vs. R is constructed. The molecular weight cutoff (M.W.C.O.) is defined as the molecular weight of a solute with a rejection factor of 0.9. The test should be conducted at low solute concentrations (0.1%), high agitation rates, and low pressure to avoid concentration polarization. Solutes may be tested individually or as a mixture. However, the presence of cosolutes, especially high molecular weight species, may alter the rejection properties through the formation of a "dynamic" membrane, and low molecular weight solutes, particularly ions, may alter the conformation of polymers. RO membranes are generally rated on the basis of % NaCl rejected.

Scanning or transmission electron microscopy can be used to directly measure pore sizes greater than about 5 nm. However, samples must be dry, so shrinkage of wet-phase process membranes may distort the morphology. Furthermore, the measurement of a statistically relevant number of pores to obtain an accurate size distribution is extremely tedious.

A measure of membrane-solute *surface force interactions* can be obtained by injecting solutes into a liquid chromatography column filled with small particles of the membrane material. A relative scale of adsorption affinity can be constructed by comparing the partition coefficients obtained from retention volumes.

REFERENCES

1. Zolandz, R.R.; Fleming, G.K. Definitions. In *Membrane Handbook*; Ho, W.S.W., Sirkar, K.K., Eds.; Van Nostrand Reinhold, Inc.: New York, 1992; 19–24.
2. Kulkarni, S.S.; Funk, E.W.; Li, N.N. Membranes. In *Membrane Handbook*; Ho, W.S.W., Sirkar, K.K., Eds.; Van Nostrand Reinhold, Inc.: New York, 1992; 408–431.

Membrane System Operation

D. Vidal-Brotóns
P. Fito
M. Gras
Universidad Politécnica de Valencia (UPV), Valencia, Spain

M

INTRODUCTION

In membrane processes, the feed stream is divided into two streams: the retentate stream and the permeate stream. If the aim is concentration, the retentate will usually be the product stream. In the case of purification/fractionation, both retentate or permeate can hold/be the desired product.

The performance of a membrane system is measured in terms of its ability to produce large volumes of permeate in a short period of time and the degree of purity of permeate with respect to the solute concentration. Permeate flux and solute rejection are the two parameters universally used for this purpose.

PERFORMANCE PARAMETERS

The permeate flux, J_v [$m^3 m^{-2} sec^{-1}$], is defined as the volume of permeate that flows through a unit area of membrane in a unit time period.

If c_{if} ($kg\,m^{-3}$) is the concentration of the solute i in the bulk feed solution and c_{ip} ($kg\,m^{-3}$) is its concentration in the bulk permeate solution, then the apparent solute rejection coefficient (or retention coefficient), R (%), is defined as:

$$R = \frac{c_{if} - c_{ip}}{c_{if}} \times 100 = \left(1 - \frac{c_{ip}}{c_{if}}\right) \times 100 \qquad (1)$$

If c_{ir} ($kg\,m^{-3}$) is the concentration of the solute i in the bulk retentate solution, the concentration factor (CF) is given by:

$$CF = \frac{c_{ir}}{c_{if}} \qquad (2)$$

Other parameters are also used. The ratio of the permeate volume (or flow rate) (V_p) to the feed volume (or flow rate) (V_f) is called the recovery rate (RR), whereas the ratio of the feed volume to the retentate volume (V_r) is the volume concentration ratio (VCR):

$$RR = \frac{V_p}{V_f} \qquad (3)$$

$$VCR = \frac{V_f}{V_r} = \frac{V_f}{V_f - V_p} = \frac{1}{1 - RR} \qquad (4)$$

Assuming all the ions remain in the retentate solution ($c_{ip} \approx 0$), CF $\approx$ VCR.

MEMBRANE SYSTEMS

Batch Systems

Batch processing systems (Fig. 1)[1] are the simplest in design. A limited volume of feed solution is recirculated through a module at a high flow rate. The process continues in closed-circuit until the required separation is achieved, after which the concentrate solution is drained from the feed tank. Batch processes are particularly suited to the small-scale operations common in the biotechnology and pharmaceutical industries. They require the least membrane area to achieve a given concentration in a given time. Batch system with total recycle is the optimal configuration for low-pressure operation where operating pressure is in the same range as the pressure drop along the membrane module. This situation occurs in microfiltration (MF) systems. In ultrafiltration (UF) applications, where the system pressure is significantly higher than the pressure drop along the module, it is more economical to use a separate recirculation pump. This is the batch system with partial recycle.

Feed and Bleed Systems

In the feed and bleed systems (Fig. 2), a large flow of solution is circulated continuously through a bank of membrane modules. The feed gets concentrated in the recirculation loop, and a part of it is bled off

Encyclopedia of Agricultural, Food, and Biological Engineering
DOI: 10.1081/E-EAFE 120016147
Copyright © 2003 by Marcel Dekker, Inc. All rights reserved.

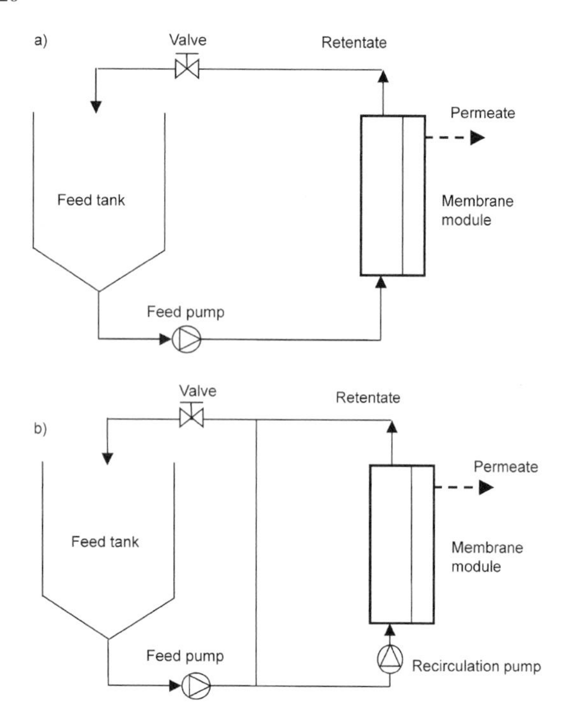

Fig. 1 Batch processing systems: (a) total recycle and (b) partial recycle. (Modified from Ref. [1].)

continuously. A small amount of fresh feed is pumped into the loop, just before the recirculation pump, to balance the retentate bleed and permeation rate. Feed and bleed systems are continuous processing systems. In this configuration, the membrane always encounters the highest concentration, which generally corresponds to the lowest flux. Therefore, it has the highest possible membrane area. To overcome the inefficiency of feed and bleed designs, industrial systems normally use the multistage feed and bleed configuration, which consists of several feed and bleed loops in series. This system reduces the membrane area drastically. Three or four loop feed and bleed design has a membrane area very close to batch design. It is invariably the most expensive of all the design configurations.

Single Pass Systems

In single pass design (Fig. 3), the feed enters from one end and exits from the other end having achieved the desired concentration ratio or the permeate recovery. This is the most economical of all membrane system configurations, due to low membrane area, lack of recirculation pumps,

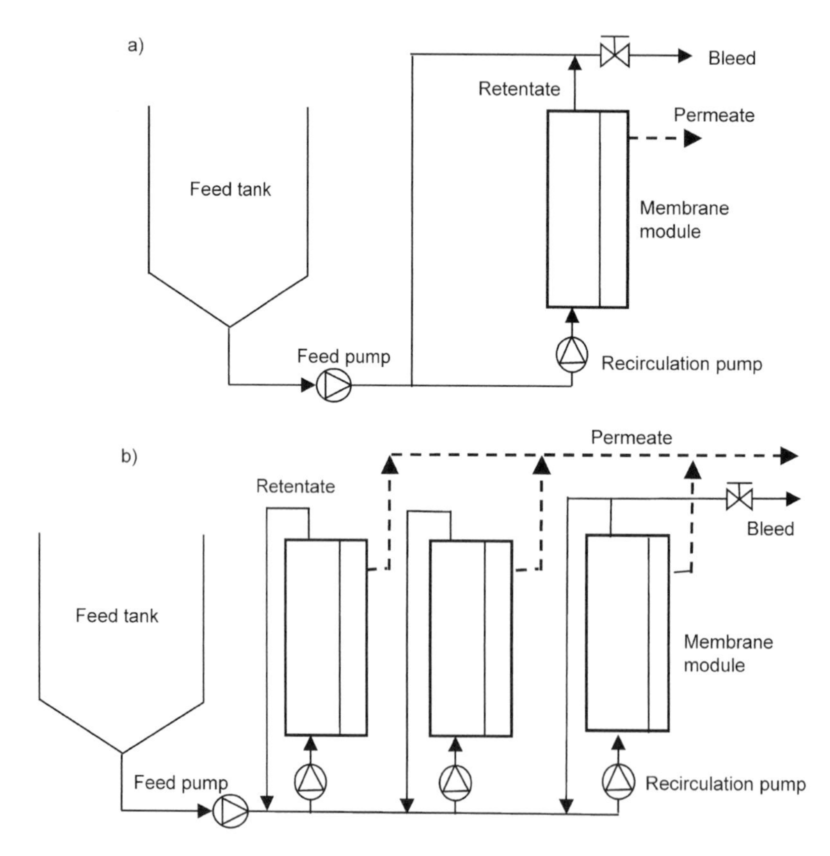

Fig. 2 Feed and bleed systems: (a) single-loop system and (b) multiloop system. (Modified from Ref. [1].)

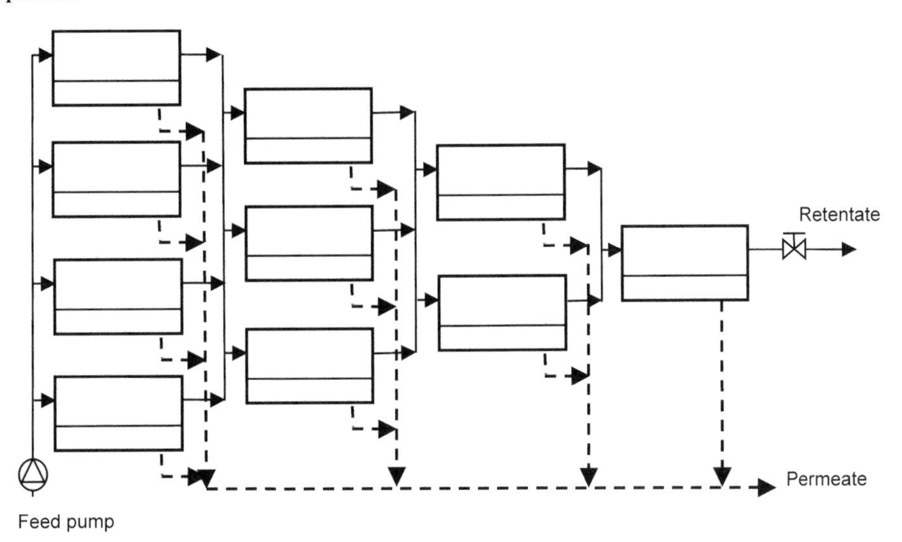

Fig. 3 Single pass system. (Modified from Ref. [1].)

and simplicity of design. This design is restricted to situations with low concentration polarization that does not require high cross-flow velocities. Reverse osmosis (RO) of brackish water and high purity water production are common applications. Pressure vessels containing several spiral membrane modules are the norm in single pass systems. The cross-flow velocity decreases as feed passes through the membranes, due to feed volume reduction through permeation. This is compensated by the tapered configuration where several stages, with progressively decreasing number of pressure vessels in parallel, are arranged in series. Single pass design has the lowest retention time, an advantage in food processing, since many food components deteriorate with time. However, the only known food process application using single pass design is the RO treatment of evaporator condensate to produce boiler feed water.

Diafiltration Systems

Diafiltration involves the addition of solvent (diluent) to the retentate, and removing it as permeate, together with a higher level of purity (Fig. 4). Diafiltration can be done sequentially, by alternating concentration and dilution several times, or continuously, by adding diluent to makeup for the permeate volume that is being removed. Diafiltration can be done in any filtration range, from MF to RO, in batch systems as well as in continuous systems.

Cocurrent Permeate Flow System

Cocurrent permeate flow system (Fig. 5) incorporates a permeate pumping loop parallel to the retentate pumping loop, so that the pressure profile in the permeate closely simulates the profile in the retentate loop. This results in a virtually uniform pressure drop along the length of the membrane module. Ceramic MF systems incorporating this design have reported permeate fluxes several folds higher compared to alternative designs.

FOULING CONTROL

Membrane fouling is the main cause of permeate flux decline and loss of product quality in membrane systems, so fouling control dominates membrane system design and operation. The cause and prevention of fouling depend greatly on the feedwater being treated, and appropriate control procedures must be devised for each plant.

Membrane fouling may be defined[2] as the reversible or irreversible deposition of retained particles, colloids, emulsions, suspensions, macromolecules, salts, etc. on or in the membrane. This includes adsorption, pore blocking, precipitation, and cake formation. Fouling occurs mainly in MF and UF, where porous membranes, which are implicitly susceptible to fouling, are used. In pervaporation (PV) and gas separation (GS), with dense membranes, fouling is virtually absent. Sources of fouling can be divided into four principal categories[3]: scale, silt, bacteria, and organic compounds.

Scale

Scale is caused by precipitation of dissolved metal salts in the feedwater on the membrane surface. As salt-free water is removed in the permeate, the concentration of ions in the feed increases until at some point the solubility limit is exceeded. The salt then precipitates on the membrane

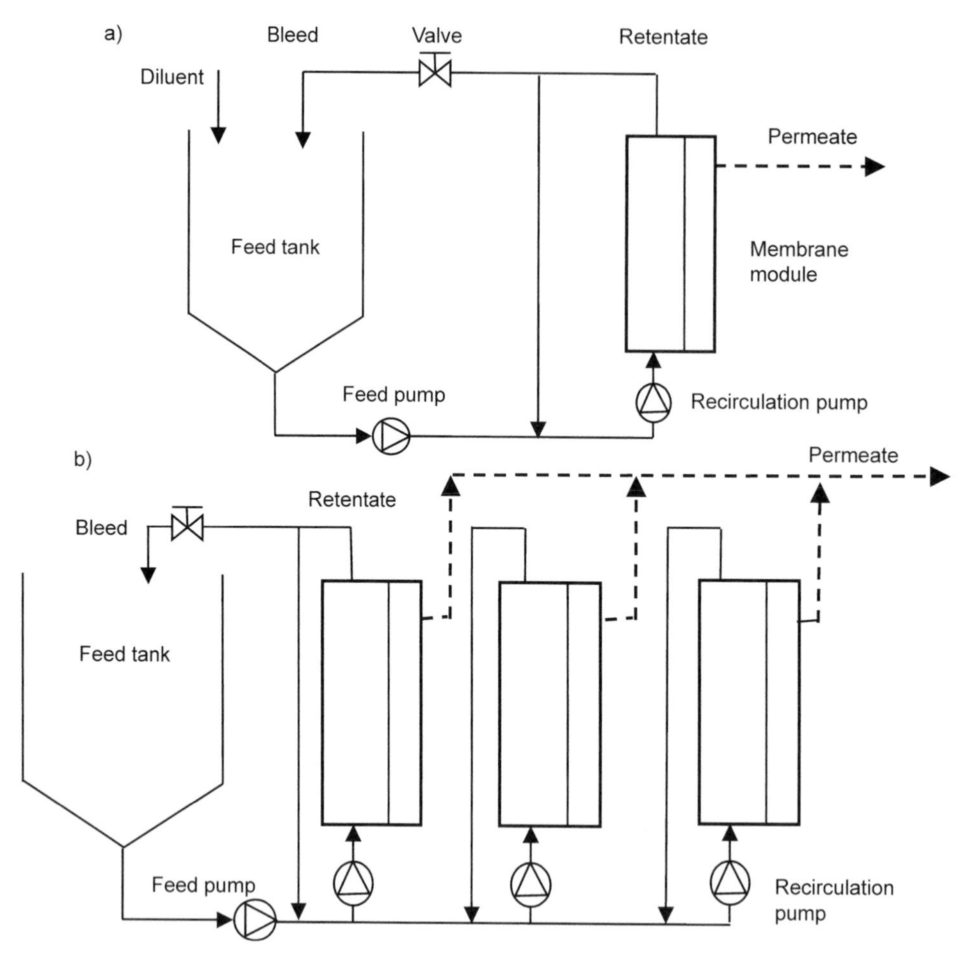

Fig. 4 Diafiltration systems: (a) single-loop system and (b) multiloop system. (Modified from Ref. [1].)

surface as scale. Many brackish water RO plants operate at RR of 80% or 90% (CF of 5 or 10, respectively). Salt concentrations on the brine side of the membrane may then be far above the solubility limit. pH adjustment by addition of acids is used to prevent carbonate scaling while sequestrants are used to prevent sulfate scaling.

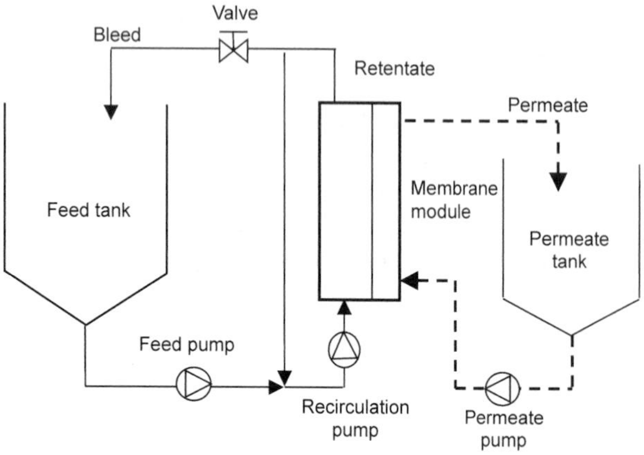

Fig. 5 Cocurrent permeate flow system. (Modified from Ref. [1].)

Silt

Silt is formed by suspended particulates of all types that accumulate on the membrane surface: organic colloids, iron corrosion products, precipitated iron hydroxide, algae, and fine particulate matter. To avoid fouling by suspended solids, some form of feedwater filtration is required. All RO units are fitted with a 0.45-μm cartridge filter in front of the high-pressure pump, and a sand filter, sometimes supplemented by addition of a flocculating chemical, may be required. Centrifuging is also used to separate fine cheese particles left over in whey before UF. Flocculation of colloidal suspensions by treating with polyelectrolytes is also common in RO. Pectin is a major foulant in fruit juice clarification and concentration. Depectinization using enzymes helps alleviate membrane fouling and also increases juice yield. High temperature helps reduce fouling in applications involving hydrocolloids like

gelatinized starch and gelatin. High temperature also increases permeate fluxes by reducing the viscosity.

Biofouling

Biofouling is the growth of bacteria on the membrane surface. The susceptibility of membranes to biological fouling is a strong function of the membrane composition. Feedwater to cellulose acetate (CA) membranes must always be sterilized. Because CA can tolerate up to 1 ppm chlorine, sufficient chlorination is used to maintain 0.2 ppm free chlorine. Periodic treatment of polyamide hollow fibers and thin-film composite membranes with a bactericide usually controls biological fouling. Chlorination can also be used to sterilize the feedwater to these membranes, but must then be followed by effective dechlorination, generally achieved by adding sodium metabisulfate.

Organic Fouling

Organic fouling is the attachment of materials such as oil or grease to the membrane surface. Removal of the organic material from the feedwater by filtration or carbon adsorption is required.

Membrane Cleaning

Acids (nitric, phosphoric, citric) are used to clean inorganic fouling (calcium phosphate in dairy applications, calcium carbonate in desalination applications). Caustic cleaning is the most effective method when fouling is organic in nature. Proteins and fats that cause fouling in dairy applications are cleaned using caustics or chlorine. Chemical cleaning of membranes requires close attention to chemical compatibility of the membranes.

MEMBRANE SYSTEM CONTROL

The objective of membrane system control[1] is to maintain required flow rate and quality of a process stream, responding to variations in the feed stream and in the processing system. The primary control variable in a pressure driven membrane system is the pressure. By means of a pneumatically actuated back pressure valve, the control system adjusts the pressure as permeate flux changes, due to fouling or to variation of feed stream concentration. The quality of the output streams is controlled by maintaining a given ratio between flow rates of feed stream and output streams, actuating upon the speed of the volumetric pumps motor, or upon control valves in the case of centrifugal pumps. Some membrane systems are designed to introduce additional membrane banks as the demand increases or the fluxes decrease.

ACKNOWLEDGMENTS

Figs. 1–5 are modified from Ref. 1, *Handbook of Food Engineering Practice* (Valentas, K.J., Rotstein, E., Singh, R.P., Eds.), Mannapperuma, J.D., Design and Performance Evaluation of Membrane Systems, p. 190–194, copyright (1997), with permission from CRC Press LLC.

REFERENCES

1. Mannapperuma, J.D. Design and Performance Evaluation of Membrane Systems. In *Handbook of Food Engineering Practice*; Valentas, K.J., Rotstein, E., Singh, R.P., Eds.; CRC Press LLC: Boca Raton, FL, 1997; 167–209.
2. Mulder, M. *Basic Principles of Membrane Technology*, 2nd Ed.; Kluwer Academic Publishers: Dordrecht, The Netherlands, 1996.
3. Baker, R.W. *Membrane Technology and Applications*; McGraw-Hill: New York, 2000.

Membrane Transport Models

Takeshi Matsuura
University of Ottawa, Ottawa, Ontario, Canada

INTRODUCTION

Food processing is one of the major applications of membrane separation processes. The advantages of membrane separation over other separation processes such as distillation are: 1) volatile flavor components do not escape in the process of concentrating liquid food; 2) the membrane process is intrinsically an energy-saving process without involving any phase change[1] with an exception of pervaporation process; and 3) the gentleness of the process since many food ingredients are thermally labile. Though there are a number of writings on the experimental testing of membrane food processing, very little work has been done to elucidate the fundamental principles involved therein.

BASIC TRANSPORT EQUATIONS

A complete reverse osmosis (RO) experiment involves obtaining data on pure water permeation rate (PWP, $\mathrm{kg\,hr^{-1}}$), product permeation rate (PR, $\mathrm{kg\,hr^{-1}}$) in the presence of solute with respect to a given area of membrane surface, and solute separation (f) under the specified operating condition of temperature, pressure, solute concentration in the feed solution, and feed flow rate. A large number of data for PWP, PR, and f are available in Ref.[1]

At any given operating temperature and pressure, each set of RO data can be analyzed on the basis of: 1) PWP is directly proportional to the operating gauge pressure P; 2) the solvent flux N_B through the membrane is proportional to the effective driving pressure for fluid flow through the membrane (assumed to be practically the same as $P - \Delta\pi$); 3) the solute flux N_A through the membrane is due to the pore diffusion and hence proportional to the concentration difference across the membrane; 4) the mass transfer coefficient k on the high-pressure side of the membrane is given by the film theory.[2] This analysis gives rise to the following basic transport equations for RO[3]:

$$A = \frac{\mathrm{PWP}}{M_B \times S \times 3600 \times P} \tag{1}$$

$$N_B = A[P - \pi(X_{A2}) + \pi(X_{A3})] \tag{2}$$

$$N_B = \left(\frac{D_{AM}}{K\delta}\right)\left(\frac{1 - X_{A3}}{X_{A3}}\right)(c_2 X_{A2} - c_3 X_{A3}) \tag{3}$$

$$N_B = c_1 k (1 - X_{A3}) \ln\left(\frac{X_{A2} - X_{A3}}{X_{A1} - X_{A3}}\right) \tag{4}$$

All the symbols are defined in the list of symbols at the end of this article.

Eq. 1 defines the pure water permeability constant A for the membrane, which is a measure of its overall porosity; Eq. 2 defines the solute transport parameter $D_{AM}/K\delta$ of the solute for the membrane, which is also a measure of the average pore size of the membrane surface. Under steady-state operating conditions, a single set of experimental PWP, PR, and f data enables one to calculate the quantities A, X_{A2}, $D_{AM}/K\delta$, and k at any point (position or time) in a RO system via Eqs. 1–4. Conversely, PWP, PR, and f can be calculated from a given set of A, $D_{AM}/K\delta$, and k data under a given operating condition of the feed solution and operating pressure.[3]

RELATIONSHIP BETWEEN $(D_{AM}/K\delta)_{NaCl}$ AND $(D_{AM}/K\delta)$ FOR OTHER SOLUTES

For completely ionized electrolyte solutes

$$\left(\frac{D_{AM}}{K\delta}\right)_{\mathrm{solute}} = \mathrm{Const.} \times \exp\left[n_c\left(-\frac{\Delta\Delta G}{RT}\right)_{\mathrm{cation}} + n_a\left(-\frac{\Delta\Delta G}{RT}\right)_{\mathrm{anion}}\right] \tag{5}$$

where n_c and n_a represent the number of moles of cations and anions, respectively, in 1 mol of ionized solute. Applying Eq. 5 to $(D_{AM}/K\delta)_{NaCl}$,

$$\ln\left(\frac{D_{AM}}{K\delta}\right)_{\mathrm{NaCl}} = \ln C^*_{\mathrm{NaCl}} + \left[\left(-\frac{\Delta\Delta G}{RT}\right)_{\mathrm{Na^+}} + \left(-\frac{\Delta\Delta G}{RT}\right)_{\mathrm{Cl^-}}\right] \tag{6}$$

where $\ln C^*_{\mathrm{NaCl}}$ is a constant representing the porous structure of the membrane surface in terms of

Encyclopedia of Agricultural, Food, and Biological Engineering
DOI: 10.1081/E-EAFE 120007071
Copyright © 2003 by Marcel Dekker, Inc. All rights reserved.

$(D_{AM}/K\delta)_{NaCl}$. By using the data on $-\Delta\Delta G/RT$ for Na^+ and Cl^- ions for the membrane material–solution system involved, the value of $\ln C^*_{NaCl}$ for the particular membrane employed can be calculated. Then, $D_{AM}/K\delta$ values for any other electrolyte solutes can be calculated from the relation,

$$\ln\left(\frac{D_{AM}}{K\delta}\right)_{solute} = \ln C^*_{NaCl} + \left[n_c\left(-\frac{\Delta\Delta G}{RT}\right)_{cation} + n_a\left(-\frac{\Delta\Delta G}{RT}\right)_{anion}\right] \quad (7)$$

Available data on $-\Delta\Delta G/RT$ for different ions appliable for cellulose acetate (acetyl content 39.8%) membrane–aqueous solution systems are listed in Table 1.

Table 1 Data on free energy parameter $(-\Delta\Delta G/RT)_i$ for some inorganic ions at 25°C, applicable for interfaces involving aqueous solutions and cellulose acetate (CA-398) membranes in RO/UF transport

Inorganic cations		Inorganic anions	
Species	$(-\Delta\Delta G/RT)_i$	**Species**	$(-\Delta\Delta G/RT)_i$
H^+	6.34	OH^-	− 6.18
Li^+	5.77	F^-	− 4.91
Na^+	5.79	Cl^-	− 4.42
K^+	5.91	Br^-	− 4.25
Rb^+	5.86	I^-	− 3.98
Cs^+	5.72	IO_3^-	− 5.69
NH_4^+	5.97	$H_2PO_4^-$	− 6.16
Mg^{2+}	8.72	BrO_3^-	− 4.89
Ca^{2+}	8.88	NO_2^-	− 3.85
Sr^{2+}	8.76	NO_3^-	− 3.66
Ba^{2+}	8.50	ClO_3^-	− 4.10
Mn^{2+}	8.58	ClO_4^-	− 3.60
Co^{2+}	8.76	HCO_3^-	− 5.32
Ni^{2+}	8.47	HSO_4^-	− 6.21
Cu^{2+}	8.41	SO_4^{2-}	− 13.20
Zn^{2+}	8.76	$S_2O_3^{2-}$	− 14.03
Cd^{2+}	8.71	SO_3^{2-}	− 13.12
Pb^{2+}	8.40	CrO_4^{2-}	− 13.69
Fe^{2+}	9.33	$Cr_2O_7^{2-}$	− 11.16
Fe^{3+}	9.82	CO_3^{2-}	− 13.22
Al^{3+}	10.41	$Fe(CN)_6^{3-}$	− 20.87
Ce^{3+}	10.62	$Fe(CN)_6^{4-}$	− 26.83
Cr^{3+}	11.28		
La^{3+}	12.89		
Th^{4+}	12.42		

Source: (From Ref. [9].)

For completely nonpolar organic solutes,

$$\ln\left(\frac{D_{AM}}{K\delta}\right)_{solute} = \ln C^*_{NaCl} + \ln\Delta^* + \left(-\frac{\Delta\Delta G}{RT}\right) + \delta^*\Sigma E_s + \omega^*\Sigma s^* \quad (8)$$

where $\ln\Delta^*$ sets a scale for $\ln(D_{AM}/K\delta)_{solute}$ in terms of $\ln C^*_{NaCl}$, $-\Delta\Delta G/RT$, $\delta^*\Sigma E_s$, and $\omega^*\Sigma s^*$, represent the difference in free energy of hydration in the bulk solution phase and the membrane/solution interface, the steric effect and the nonpolar effect, respectively, on the solute transport parameter.[4–7] Some transport parameters pertinent to organic solutes involved in liquid food are listed in Table 2. The mass transfer coefficient, k, for a solute can be calculated by

$$k_{solute} = k_{NaCl}\left[\frac{(D_{AB})_{solute}}{(D_{AB})_{NaCl}}\right]^{2/3} \quad (9)$$

Osmotic pressure data relevant to some food processing are given in Tables 3 and 4.

APPLICATION OF TRANSPORT EQUATIONS TO REAL FRUIT JUICE CONCENTRATION

The basic transport equations given as Eqs. 1–4 can be used by rewriting the equations on weight basis in the following way:

$$A_{(wt)} = \frac{PWP}{S \times 3600 \times P} \quad (10)$$

$$N_{B(wt)} = A_{(wt)}[P - \pi(X_{C2}) + \pi(X_{C3})] \quad (11)$$

$$N_{B(wt)} = \left(\frac{D_{AM}}{K\delta}\right)\left(\frac{1 - X_{C3}}{X_{C3}}\right)[c_{(wt)2}X_{C2} - c_{(wt)3}X_{C3}] \quad (12)$$

$$N_{B(wt)} = kc_{(wt)}(1 - X_{C3})\ln\left(\frac{X_{C2} - X_{C3}}{X_{C1} - X_{C3}}\right) \quad (13)$$

where $A_{(wt)}$, $N_{B(wt)}$, and $c_{(wt)}$ are in units of $kg\,m^{-2}\,sec^{-1}\,kPa^{-1}$, $kg\,m^{-2}\,sec^{-1}$, and $kg\,m^{-3}$, respectively, and X_C represents the carbon weight fraction in solution.[8] Some $D_{AM}/K\delta$ values applicable to juice concentration are given in Table 5. Data for k are available in the literature.[8] In order to apply the above equations, osmotic pressure data of fruit juices as a function of carbon weight fractions are

Table 2 Physicochemical and transport parameter $D_{AM}/K\delta$[a] for some organic solutes

Solute	ln ($D_{AM}/K\delta$)		Solute	ln ($D_{AM}/K\delta$)
	CA	PA		PA
Alcohols				
2-Pentanol	−7.96	−11.23	*n*-Butyl butyrate	−13.08
s-Butyl alcohol	−7.84	−10.88	Ethyl heptanoate	−11.58
1-Propyl alcohol	−7.72	−10.31	Ethyl octanoate	−11.26
n-Hexyl alcohol	−6.26	−9.97	Ethyl decanoate	−10.62
n-Octyl alcohol	−5.93	−9.28	Ethyl caproate	−11.90
n-Nonyl alcohol		−9.01	*n*-Butyl propionate	−12.20
n-Decyl alcohol		−8.69	Ethyl pentanoate	−12.20
n-Pentanol	−6.41	−10.29	Ethyl 3-methyl butyrate	−13.08
n-Butyl alcohol	−6.57	−10.61	Ethyl butyrate	−11.85
i-Butyl alcohol	−7.84	−10.74	Methyl 2-methyl butyrate	−13.06
n-Propyl alcohol	−6.94	−10.29	Ethyl propionate	−10.97
Ethyl alcohol	−6.24	−9.79	2-Propyl acetate	−11.89
2-Methylbutan-1-ol	−7.96	−11.27	Methyl 2-methyl propionate	−13.06
3-Methylbutan-1-ol	−7.96	−10.63	Ethyl propionate	−10.97
Methyl alcohol	−5.40	−9.44	2-Propyl acetate	−11.89
Aldehydes			Methyl 2-methyl propionate	−11.89
1-Octanal	−6.07	−10.65	Methyl heptanoate	−11.15
1-Nonanal	−5.91	−10.16	Methyl octanoate	−10.83
1-Decanal		−10.01	Methyl decanoate	−10.19
1-Undeanal		−9.69	*n*-Hexyl acetate	−11.15
1-Hexanal	−6.39	−11.29	Methyl caproate	−11.47
Acetaldehyde	−7.50	−9.75	*n*-Butyl acetate	−11.77
			2-Methyl-1-propyl acetate	−12.65
Ketones			Methyl 3-methyl butyrate	−12.65
3-Pentanone	−7.12	−10.61	1-Propyl acetate	−11.42
2-Pentanone	−7.23	−11.06	Methyl butyrate	−11.42
Acetone	−6.83	−10.03	Ethyl acetate	−10.54
			Methyl propionate	−10.54
Esters			2-Methyl 1-butyl acetate	−13.60
2-Propyl 2-methyl propionate		−13.68	3-Methyl 1-butyl acetate	−11.75
2-Propyl caproate		−13.26	Methyl 4-methyl pentanoate	−11.75
Ethyl 2-methyl butylate		−13.49	Methyl acetate	−10.10
2-Propyl butylate		−13.31		
2-Propyl propionate		−12.33		
Ethyl 2-methyl propionate		−12.33		
Pentyl hexanoate		−12.83		
n-Butyl caproate		−13.13		

Solute	ln ($D_{AM}/K\delta$) CA
Acids	
Benzoic acid	−11.82[b]
	−6.58[c]
Acetic acid	−12.11[b]
	−6.81[c]
Propionic acid	−12.30[b]
	−6.93[c]
Butyric acid	−12.22[b]
	−7.05[c]
Valeric acid	−12.27[b]
	−7.16[c]

(*Continued*)

Table 2 Physicochemical and transport parameter $D_{AM}/K\delta^a$ for some organic solutes (*Continued*)

Solute	ln ($D_{AM}/K\delta$) CA	ln ($D_{AM}/K\delta$) PA	Solute	ln ($D_{AM}/K\delta$) PA
Lactic acid	− 12.46[b]			
	− 8.73[c]			
Malic acid	− 12.13[b]			
	− 9.66[c]			
Tartaric acid	− 12.56[b]			
	− 10.73[c]			
Citric acid	− 12.40[b]			
	− 11.88[c]			
Sugars				
D-Glucose	− 12.97			
D-Fructose	− 12.97			
Sucrose	− 14.71			
Maltose	− 14.71			
Lactose	− 14.71			

[a] Data based on $\ln C^*_{NaCl} = -12.5$ (for CA membrane sodium chloride separation of 97.9% and for PA membrane sodium chloride separation of 99.2%).
[b] For solute in ionized form.
[c] For solute in nonionized form.
Source: (From Ref. [10].)

needed. It was found that the relation could be expressed by

$$\frac{\pi}{X_C} = a\pi + b$$

The constants a and b are given in Table 6.

FRACTIONATION OF PROTEIN–PROTEIN MIXTURES IN AQUEOUS SOLUTIONS

Fractionation of protein–protein is of interest in the membrane separation process of dairy products. Unfortunately, the latter problem has not yet been studied in the framework of transport theory. The data given in

Table 3 Osmotic pressure (kPa) data for sodium chloride and some food sugars at 25°C

Molality	NaCl	Glucose	Fructose	Sucrose	Maltose	Lactose
0	0	0	0	0	0	0
0.1	462	259	253	248	214	214
0.2	917	517	496	503	455	455
0.3	1,372	776	790	758	724	724
0.4	1,820	1,034	1,013	1,020	933	
0.5	2,282	1,293	1,307	1,282		
0.6	2,744	1,517	1,611	1,551		
0.7	3,213	1,744	1,824	1,827		
0.8	3,682		2,067	2,103		
0.9	4,158		2,310	2,379		
1.0	4,640		2,564	2,668		
1.2	5,612		3,101	3,241		
1.4	6,612		3,587	3,840		
1.6	7,646			4,447		
1.8	8,701			5,061		
2.0	9,784			5,695		
3.0	15,651			9,128		
4.0	22,326			12,866		

Source: (From Ref. [1].)

Table 4 Osmotic pressure (kPa) data for some proteins at 25°C

Molality × 10^3	Bovine serum albumin	γ-Casein
0	0	0
0.2	0.5	0.5
0.4	1.1	1.0
0.6	1.7	1.6
0.8	2.6	2.2
1.0	3.4	2.9
1.5	6.0	4.0
2.0	—	6.0
6.0	48	—

Source: (From Ref. [1].)

Table 7 shows the difficulty to predict the fractionation of albumin/γ-globulin mixture on the basis of separation data for individual proteins.

The first experiment shows that both albumin and NaCl permeated through the membrane freely. The second experiment shows that γ-globulin was retained completely by the membrane, accompanied by severe reduction in permeation rate. These experiments indicate the possibility to fractionate the two proteins. The third experiment shows, however, that a significant amount of albumin was retained on the feed side of the membrane, rendering the fractionation impossible. Severer reduction in permeation rate was observed.

Table 6 Data on constants *a* and *b* for osmotic pressure calculation at 25°C

Juice	*a*	*b*
Lime	3.31	3,997
Lemon	2.59	4,442
Prune	3.31	4,217
Carrot	4.93	3,088
Tomato	8.95	4,187
Other	3.94	3,560

Source: (From Refs. [8], [11].)

LIST OF SYMBOLS

A	Pure water permeability constant, kmol $H_2O/(m^{-2}\,sec^{-1}\,Pa^{-1})$
$\ln C^*_{NaCl}$	Constant defined by Eq. 6
c	Total (including solute and solvent) molar concentration of solution, gmol m^{-3}
D_{AB}	Diffusivity of solute in water, $m^2\,sec^{-1}$
$D_{AM}/K\delta$	Solute transport parameter, m sec^{-1}

Table 5 Effect of feed concentration on $D_{AM}/K\delta$ for fruit juices at 4137 kPa g (600 psig) and 25°C

Juice	Carbon content in feed solution (ppm)	$(D_{AM}/K\delta) \times 10^5$ (cm sec^{-1})
Apple juice	29,900	0.81
	43,800	0.84
	61,900	0.66
	84,800	0.36
Pineapple juice	29,800	0.64
	47,300	0.43
	62,200	0.24
	80,400	0.35
Orange juice	30,800	1.32
	45,000	0.97
	80,200	1.18
Grapefruit juice	31,700	0.66
	45,900	0.35
	58,500	0.77
	86,900	0.43
Grape juice	33,300	1.12
	48,100	0.63
	62,700	0.39
	81,500	0.69

Source: (From Ref. [8].)

Table 7 Experimental data for the separation of albumin, γ-globulin, and sodium chloride

Concentration in feed (wt%)					Solute separation (%)		
Albumin	**γ-globulin**	**NaCl**	**PWP × 10^3 (kg hr^{-1})**	**PR × 10^3, (kg hr^{-1})**	**Albumin**	**γ-globulin**	**NaCl**
1.85	0.0	0.788	1,227	1,066	≈ 0	—	≈ 0
0.0	0.936	0.796	1,214	12.0	—	≈ 100	—
2.37	0.790	0.672	1,360	5.7	42.0	≈ 100	≈ 0

Source: (From Ref. [1].)

f	Solute separation = (solute concentration in feed − solute concentration in permeate)/(solute concentration in feed)
$\Delta\Delta G/RT$	Free energy parameter
k	Mass transfer coefficient, m sec^{-1}
M_B	Molecular weight of solvent
N_B	Solvent water flux through membrane, gmol m^{-2} sec
P	Operating pressure, Pa
S	Effective membrane area, m^2
X	Mole fraction
A	Solute
1, 2, 3	Bulk feed, concentrated boundary layer, and permeate

REFERENCES

1. Matsuura, T.; Sourirajan, S. Membrane Separation Processes. In *Engineering Properties of Foods*, 2nd Ed.; Rao, M.A., Rizvi, S.S.H., Eds.; Marcel Dekker: New York, NY, 1995; 311–388.
2. Sherwood, T.K. *Mass Transfer Between Phases*; 33rd Annual Priestley Lectures; Pennsylvania State University, 1959; 38.
3. Sourirajan, S. *Reverse Osmosis*; Academic: New York, NY, 1970.
4. Matsuura, T.; Bednas, M.E.; Dickson, J.M.; Sourirajan, S. Polar and Steric Effects in Reverse Osmosis. J. Appl. Polym. Sci. **1974**, *18*, 2829–2846.
5. Matsuura, T.; Dickson, J.M.; Sourirajan, S. Free Energy Parameters for Reverse Osmosis Separations of Undissociated Polar Organic Solutes in Dilute Aqueous Solutions. Ind. Eng. Chem. Process Des. Dev. **1976**, *15*, 149–161.
6. Pereira, E.N.; Matsuura, T.; Sourirajan, S. Reverse Osmosis Separations and Concentrations of Food Sugars. J. Food. Sci. **1976**, *41*, 672–680.
7. Matsuura, T.; Baxter, A.G.; Sourirajan, S. Predictability of Reverse Osmosis Separations of Higher Alcohols in Dilute Aqueous Solutions Using Porous Cellulose Acetate Membranes. Ind. Eng. Chem. Process Des. Dev. **1977**, *16*, 82–89.
8. Matsuura, T.; Baxter, A.G.; Sourirajan, S. Concentration of Fruit Juices by Reverse Osmosis Using Porous Cellulose Acetate Membranes. Acta Aliment. **1973**, *2*, 109–150.
9. Sourirajan, S. *Lectures on Reverse Osmosis*; Division of Chemistry, National Research Council: Ottawa, Canada, 1983.
10. Matsuura, T.; Sourirajan, S. A Fundamental Approach to Application of Reverse Osmosis for Food Processing. AIChE Symp. Ser. **1978**, *74*, 196–208.
11. Matsuura, T.; Baxter, A.G.; Sourirajan, S. Studies on Reverse Osmosis for Concentration of Fruit Juices. J. Food Sci. **1974**, *39*, 704–711.

Metal Containers

Felix H. Barron
Joel D. Burcham
Clemson University, Clemson, South Carolina, U.S.A.

INTRODUCTION

Rigid packages such as metal cans and glass jars are containers whose shape or contour is neither affected by its contents nor deformed by external mechanical pressure of up to 10 lb per square inch gauge (68,947.3 Pa), which is the normal firm finger pressure.[1] These containers are especially appropriate to thermal processing where the risk of potential changes in shape or contour exists causing seal failure due to high internal or external pressures. Their impermeability and low chemical reactivity make rigid containers a good choice to package foods and beverages where an extended shelf life is desired.

HERMETICALLY SEALED RIGID CONTAINERS

The preservation of food safety and quality in packaged foods depends on hermetically sealed packages. A hermetically sealed package can be considered as an air or liquid tight container that is designed and intended to protect its contents against the entry of micro-organisms during and after processing.[1]

The purpose of food preservation is to maintain the quality and safety of the food product through its shelf life cycle including processing, distribution, and storage. Preservation through thermal processing can be achieved by several means including sterilization and pasteurization. In general, all foods in glass containers are processed in water heated to a desired temperature for a known time using pressurized steam in a retort, which has air under pressure added to the vessel to maintain the hermetic seal in the glass container.

Commercial sterility of thermally processed foods is the condition achieved by the application of heat, which renders the food free of micro-organisms capable of reproducing in the food under normal nonrefrigerated conditions of storage and distribution or free of viable micro-organisms (including spores) of public health significance.[1]

PROFILE OF A HERMETIC SEAL

The integrity of any rigid container in food packaging is commonly achieved by a hermetic seal, which ensures the safety of the food product. Because of this critical importance, U.S. federal agencies such as FDA and USDA[2] have established basic can seam or seal measurements requirements. The profile of a double seam for metal cans is shown in Fig. 1. Basic required measurements will depend on the tool used as follows:

- Using a micrometer: Cover hook, body hook and width (length, height), tightness (observation for wrinkles), and thickness. Optional parameters are overlap and countersink.
- Using a scope or projector: Body hook, overlap, tightness, and thickness (micrometer). Optional parameters are width, cover hook, and countersink.

Two measurements at different locations, excluding the side seam, shall be made for each double seam characteristic if a seam scope or seam projector is used. When a micrometer is used, three measurements shall be made at points approximately 120° apart, excluding the side seam.

The theoretical overlap length can be calculated by the following formula:

$$\text{Theoretical overlap length} = \text{CH} + \text{BH} + T - W$$

where CH is cover hook, BH is body hook, T is cover thickness, and W is seam width (height, length).

The following is a basic terminology related to double seam measurements (Fig. 1)[1]:

- "Crossover": The portion of a double seam at the lap.
- "Cutover": A fracture, sharp bend, or break in the metal at the top of the inside portion of the double seam.
- "Deadhead": A seam, which is incomplete due to chuck spinning in the countersink.
- "Droop": Smooth projection of the double seam below the bottom of the normal seam.
- "False seam": A small seam breakdown where the cover hook and the body hook are not overlapped.
- "Lap": Two thicknesses of material bonded together.

Encyclopedia of Agricultural, Food, and Biological Engineering
DOI: 10.1081/E-EAFE 120014716
Copyright © 2003 by Marcel Dekker, Inc. All rights reserved.

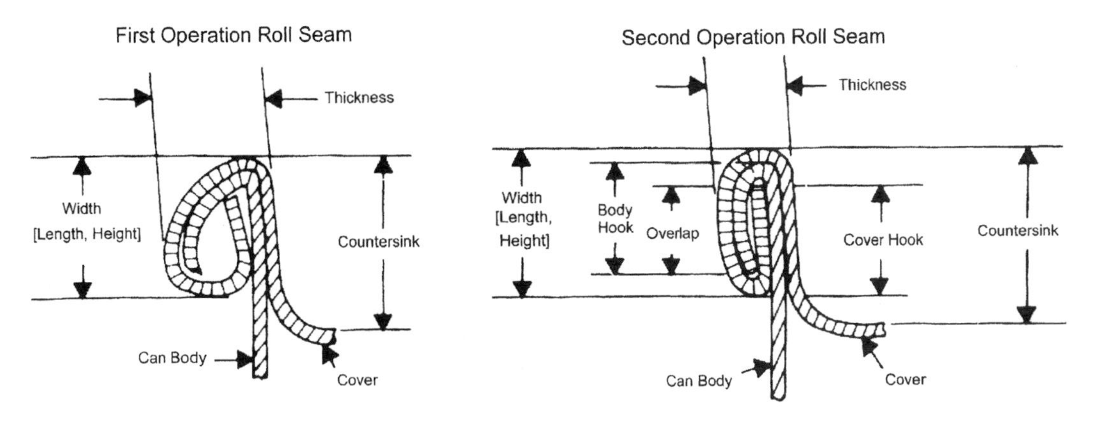

Fig. 1 Profile of a metal double seam.

METAL CONTAINERS

Practically all metal cans are made of aluminum or steel. Their bodies and ends may also be made from different materials or alloys. The materials available[3] for steel cans include the following:

- Black plate or bare steel: Steel sheets coated to prevent corrosion and contamination. The coatings may be epoxy or vinyl. The exterior of the cans may also be coated with organic compounds such as lacquers, varnishes, or enamels for protection and to avoid food–container interactions.
- Tinplate: Steel sheets coated with a layer of commercially pure tin to protect against corrosion and to provide a bright appearance.
- Tin-free steel (TFS): Steel sheets coated with a very thin layer of chromium to replace tin.

Can Sizes and Dimensions

Metal cans can be of many sizes and shapes; they can be three-piece or two-piece cans, and they can be soldered, cemented, or welded. A comparison of reported can sizes between 1950 and 2001[4–6] reflects no significant changes as indicated in Fig. 2 and Table 1.[4]

Basic Manufacturing Process for Metal Cans

The manufacturing process will depend on the type of can being produced. The basic steps[3] to produce metal cans are briefly discussed as follows:

Three-piece cans: The name derives from the fact that these cans have three parts: body, bottom, and top. They can have different shapes, the most common is the cylindrical, but square and rectangular shapes are also available. The bottom end of the can is joined during

Table 1 Typical can sizes

Can name	Diameter (in.)	Height (in.)	Canner's designation	Approximate net weight	Net contents liquid products
6Z	2-1/8	3-1/2	202 × 308	5-3/4 oz	
8Z tall	2-11/16	3-1/4	211 × 304	8-1/2 oz	7-3/4 fl. oz
No. 1 picnic	2-11/16	4	211 × 400	10-1/2 oz	9-1/2 fl. oz
No. 211 cylinder	2-11/16	4-7/8	211 × 414		12 fl. oz
No. 300	3	4-7/16	300 × 407	14-1/2 oz	13-1/2 fl. oz
No. 1 tall	3-1/16	4-11/16	301 × 411	1 lb	15 fl. oz
No. 303	3-1/16	4-3/8	303 × 406	1 lb	15 fl. oz
No. 303 cylinder	3-3/16	5-9/16	303 × 509	1 lb 5 oz	1 pt 3 fl. oz
No. 2 vacuum	3-7/16	3-3/8	307 × 306	12 oz	14 fl. oz
No. 2	3-7/16	4-9/16	307 × 409	1 lb 4 oz	1 pt 2 fl. oz
No. 2 cylinder	3-7/16	5-3/4	307 × 512	1 lb 9 oz	1 pt 7 fl. oz
No. 2-1/2	4-1/16	4-11/16	401 × 411	1 lb 13 oz	1 pt 10 fl. oz
No. 3 cylinder	4-1/4	7	404 × 700		1 qt 14 fl. oz
No. 5	5-1/8	5-5/8	502 × 510	3 lb 9 oz	1 qt 1 pt 4 fl. oz
No. 10	6-3/16	7	603 × 700	6 lb 10 oz	3 qt

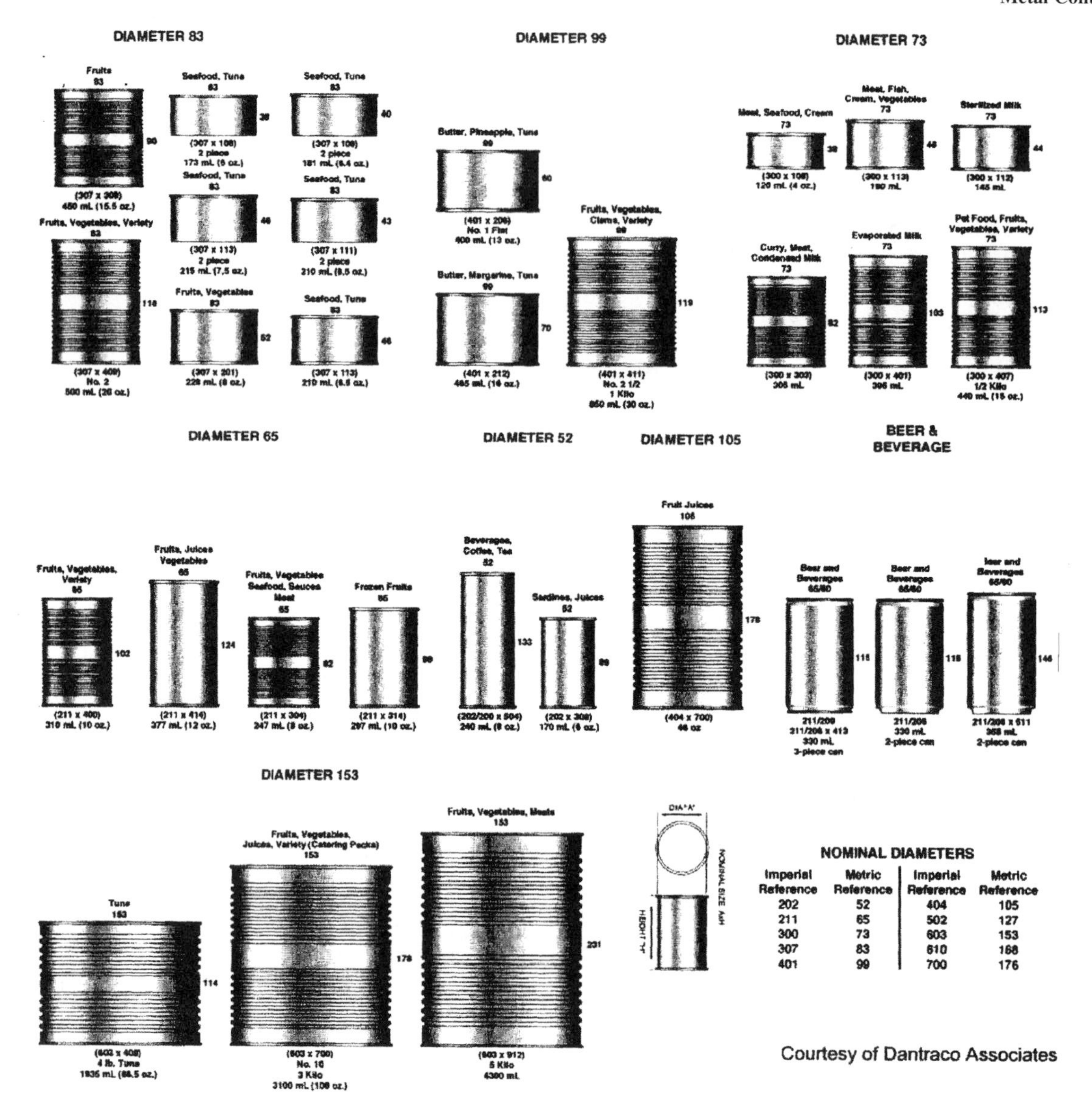

Imperial Reference	Metric Reference	Imperial Reference	Metric Reference
202	52	404	105
211	65	502	127
300	73	603	153
307	83	610	168
401	99	700	176

Fig. 2 Guide to principal can sizes.

the can-making process. Side seaming can be achieved through soldering, cementing (epoxy or vinyl), or welding. The top end is joined at the cannery. These cans are also known as "Tin Cans" because they were originally made from tin-plated iron sheets. Three-piece metal cans were recognized as "Sanitary Cans" when solder was applied only to the exterior of the container to prevent contamination of its contents.

Two-piece cans—Drawn and Ironed (D&I): This type of cans have the body and the bottom integrated into one piece. The top is joined at the processor's facility. During the can-making process, the sidewall is thinned and stretched while the thickness of the bottom is maintained. These cans are most widely used for beer and beverages, and they are capable of withstanding internal pressures. Continuous improvements have been made in the speed to manufacture cans, which in 1980 was typically in the order of 1200 cans per minute.

Two-piece cans—Drawn and Redrawn (D&R): These cans also have the body and the bottom integrated into one piece creating a "sanitary" seamless container (except for the top closure); both parts have essentially the same thickness. The top end can be solid or with an easy-opening feature. These cans are also capable of resisting internal (vacuum) or external pressures.

Food canners have been switching from three-piece to two-piece cans in their operations.

PROTECTIVE CAN COATINGS

Can enamels are organic coatings applied to the steel base to prevent corrosion and chemical interactions between the food product and the container. Each type of food requires a specific enamel depending on the nature of the product: acidic, alkaline, highly colored, and fatty foods among others. The use of universal coatings (good for general use) by manufacturers of aluminum[7] and steel cans is one of the latest research developments. Vegetable canners[8] are now replacing specific enamels by universal coatings. A similar situation is being implemented in the beverage industry.

An important advantage of the use of universal coatings is reflected in transportation and warehousing costs.

The use of specific enamels or universal coatings should be in accordance with food laws and regulations and with good business practices. Basic requirements include: 1) nontoxic; 2) not reactive with the contents; 3) not to affect the content's quality; 4) easily applicable to the steel plate; 5) capable to withstand thermal processing and storage conditions; 6) be mechanically resistant; and 7) economic.

The following are common specific enamels[6] used in the food industry:

- Oleoresinous: Such as the R and C types. The R enamels are commonly used for highly colored fruits; the C enamels are used for corn, peas, poultry, and seafood or foods with high content of sulfur compounds.
- Phenolic: More chemical resistant than oleoresinous enamels, generally used for seafood packaging.
- Epoxy: It may be combined with phenolic and oleoresinous enamels to be used for fruits and foods

Table 2 Corrosion of metal cans and common causes

Corrosion in food cans	Common causes due to canning practices
Internal (mostly food product related):	Internal:
Pitting (pinholes)	Food product: pH, nature of acid, presence of corrosion accelerators such as copper, nitrates, and phosphates
Swelling (bulging of cans)	Filling and vacuum: excessive air in headspace
Detinning (dissolution of coated tin)	Thermal exhausting: air presence if below 250°F
Rusting (rust formation)	
Enamel peeling (coating detachment)	
Staining (dark sulfide deposits)	
Discoloration (often black deposits)	
External (mostly atmospheric conditions):	External:
Detinning (due to alkaline retort waters)	Code marking: sharp imprints
Rusting (due to corrosive water or poor storage)	Closing: faulty seams, leaking
Staining (surface changes affecting brightness)	Washing sealed cans: salty or acidic solutions
	Steam retort operations: inadequate air removal, long come-up time, low-pressure steam supply
	Water cooling: water on can may not evaporate if cans are cooled below 95°F; corrosive (alkaline) water
	Scratches and abrasions: expose the steel base plate
	High storage temperatures: above 75°F may result in swollen cans and perforations
	Sweating: if the temperature of cans is less than the atmosphere. High relative humidity

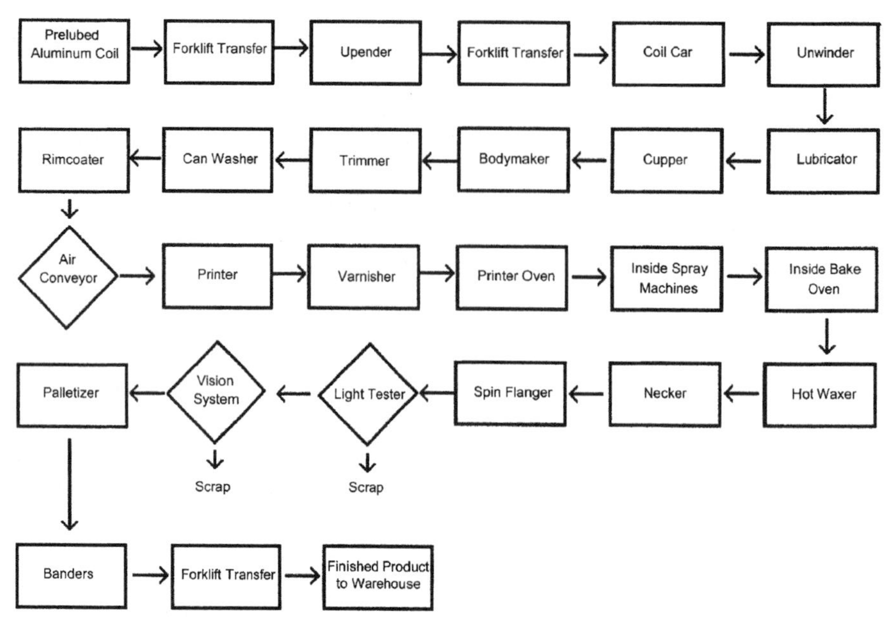

Fig. 3 D&I flow diagram.

with a high fat content. They are stable at high temperatures.

- Vinyl: It is usually combined with phenolic and oleoresinous enamels to be used in foods to be thermally processed at 200°F (93.3°C) or less.

CORROSION IN FOOD CANS

The gradual deterioration of metal cans due to chemical reactions is undesirable to food processors or can manufacturers mostly because of economic reasons. The unpleasant dark colored residues (rust) resulting from corrosion do not usually represent a health hazard; however, resulting product recalls end up in lawsuits and economic losses.

Presently, the corrosion of canned foods represents isolated cases. The technology to prevent it has been very effectively applied during the can-manufacturing process.

Table 2 shows common causes of corrosion in metal cans used for food.

Fig. 4 Unwinder in a D&I process.

Fig. 5 Necker in a D&I process.

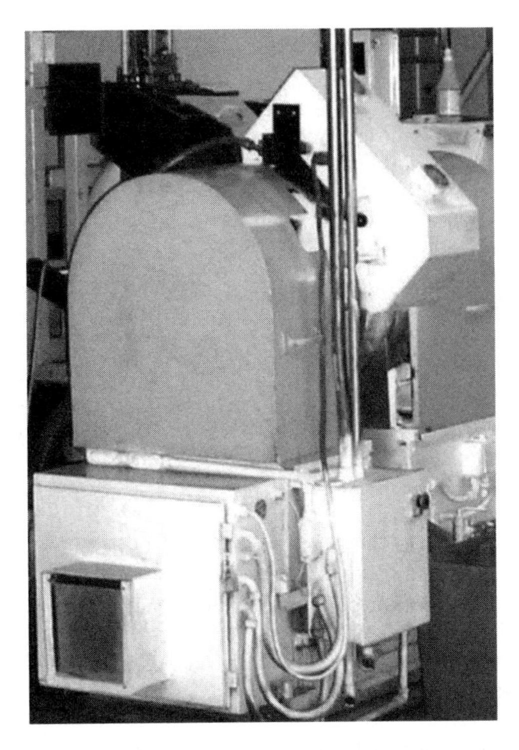

Fig. 6 Light tester in a D&I process.

ALUMINUM CANS

The body and ends of these cans may be made from different aluminum alloys. A hard material such as H-19 can be used for the body and a soft material such as H-32 for the end, especially with easy-opening features. The D&I or the D&R method is commonly used to make beverage aluminum cans. Types of aluminum alloys generally used are the 3000 and 5000 series depending on the amount of manganese used.[9]

In co-operation with Crown Cork and Seal Company, Figs. 3–8 were prepared to show a typical D&I process for aluminum cans, some machines used in the operations, and some of the intermediate can products from the forming steps.

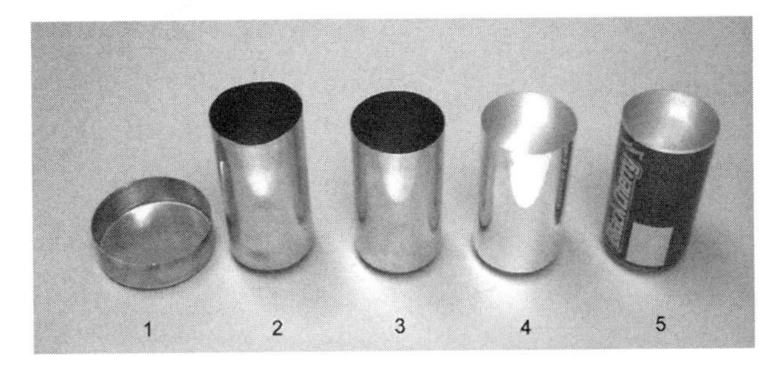

Fig. 7 Forming steps (I) of a D&I aluminum can: (1) cupper; (2) bodymaker; (3) trimmer; (4) can washer; and (5) printer.

Fig. 8 Forming steps (II) of a D&I aluminum can: (6) hot waxer; (7) necker, first stage; (8) necker, second stage; and (9) spin flanger.

Typical product (Fig. 9) control parameters in a D&I process are: dome depth, wall thickness, trim height, coating amount and distribution, physical defects, metal weight, color labeling, and dome inversion.

Some advantages[3] of aluminum cans include the following: they do not impart taste or flavor to the contents, do not cause staining due to sulfur compounds in foods, are corrosion resistant, have a light weight, and are recyclable.

A major disadvantage is their weak resistance to acid and chlorine solutions; however, proper coatings on the can body and can ends solve this problem.

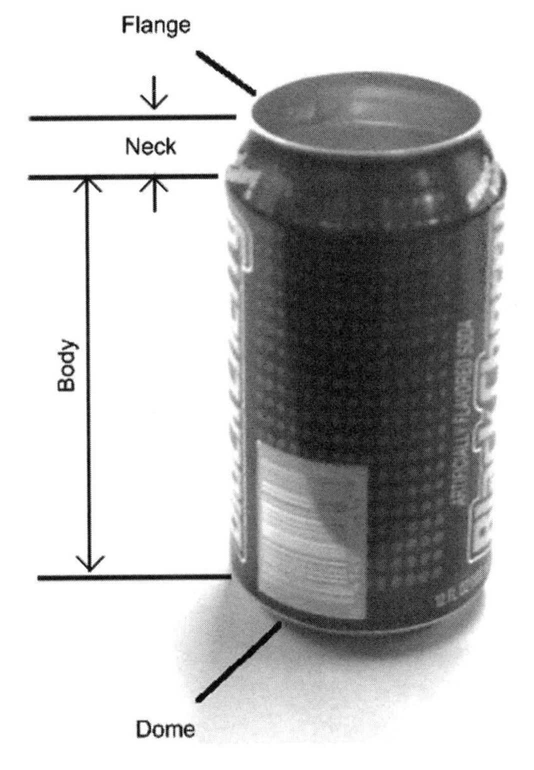

Fig. 9 Profile of an aluminum can.

Beverage processors using aluminum cans do not face major problems with these containers.[10] The hermetic seals and high chemical resistance allow for the preservation of a desired shelf life of soft drinks, which are typically between 17 and 39 weeks depending on the type of product.

Although deformed cans represent a critical control point at receiving, its occurrence is considered isolated. Other potential problems include the presence of broken pull-tabs or can leaking causing corrosion in other adjacent cans.

Some control parameters of importance during beverage processing with aluminum cans are related to the seaming operation. Manufacturers of can ends provide specifications for important seam measurements that need to be met in order to ensure package integrity. These include: counter sink, width, thickness, cover hook, body hook, "wrinkle," and overlap (Fig. 1).

REFERENCES

1. Code of Federal Regulations. *Title* 9, *Part* 318, Office of the Federal Register National Archives and Records Administration; U.S. Government Printing Office: Washington, 1992.
2. Code of Federal Regulations. *Title* 21, *Part* 113, Office of the Federal Register National Archives and Records Administration; U.S. Government Printing Office: Washington, 1992; 133–136.
3. Winship, J.T. How Metal Containers Are Made. Am. Mach. **1980**, *April*, 155–166.
4. Campbell, C.H. *Campbell's Book. A Manual on Canning, Pickling and Preserving*, 3rd Ed.; Vance Publishing Corporation: Chicago, 1950; 4.
5. Dantraco Associates. http://www.canmachines.com (accessed March 2001).
6. Lopez, A. Containers for Canned Foods. In *A Complete Course in Canning, Book 1—Basic Information on Canning*, 11th Ed.; The Canning Trade, Inc.: Baltimore, Maryland, 1981; 168.
7. King, L. Personal communication. Crown Cork and Seal Company, Cheraw, SC, February 2001.
8. Swink, M. Personal communication. McCall Farms, Effingham, SC, February 2001.
9. Matsubayashi, H. Metal Containers. In *Food Packaging*; Kadoya, T., Ed.; Academic Press, Inc.: New York, 1990; 90–91.
10. Sellers, D.F. Personal communication. Carolina Canners, Inc., Cheraw, SC, February 2001.

Microbial Genetics

Benjamin C. Stark
Illinois Institute of Technology, Chicago, Illinois, U.S.A.

M

INTRODUCTION

Bacteria have features of their genetics that are common to all organisms, as well as those more characteristic of prokaryotes. Their genomes vary over a more than a factor of 10 in size and gene number, and can consist of both main chromosome and plasmids. Expression of these genes is controlled mainly at the level of transcription by both positive and negative means. The bacterial genetic code conforms nearly completely to the universal genetic code. Although bacteria do not have sexual reproduction, they can exchange genes by a number of mechanisms, which are thus important in evolution.

BACTERIAL GENOMES: GENERAL PROPERTIES

Bacterial genomes are generally single, circular, double-stranded DNA molecules (chromosomes). There are rare cases of linear chromosomes or a genome comprised of several circular chromosomes.[1] As of May 2001, the complete sequences of 50 prokaryotic genomes were available in the database of the National Center for Biotechnology Information (NCBI); 10 were from members of the archaebacteria and 40 were from eubacterial species.[2] Among the eubacteria the genome sizes ranged from 0.58 million base pairs (484 genes) for *Mycoplasma genitalium* to 7.04 million base pairs (at least 6752 genes) for *Mesorhizobium loti*. The *Escherichia coli* K12 genome has 4.64 million base pairs and 4404 genes, 4289 of which encode proteins and 115 of which encode structural RNAs (mostly transfer RNAs and ribosomal RNAs). The very small number of genes in *M. genitalium* is likely a result of its being a parasite, having evolved so as to lose genes for functions that can be supplied by the host; free living bacteria require many more functions to be encoded by their own genomes.

The average size of the *E. coli* protein-encoding genes is about 950 base pairs, while the intergene distance averages 118 base pairs.[3] These figures are probably typical for a wide range of bacteria. Introns are rare in the archaebacteria and extremely rare in the eubacteria.

In addition to genomic DNA, eubacteria may contain plasmids, double-stranded, almost always circular DNA molecules that can exist independently of the genome; certain plasmids called "episomes" can also exist integrated into the chromosome. The sizes of plasmids vary from slightly more than one thousand to hundreds of thousands of base pairs; the number of copies per cell varies from one to several dozen (plasmids engineered for recombinant DNA work in *E. coli* can exist at hundreds of copies per cell). A single cell may contain up to 12 different plasmids,[4] each existing at its characteristic number per cell. *M. loti*, for example, in addition to its chromosome, has two large plasmids of 0.35 and 0.21 million base pairs, the larger of which encodes 320 proteins.[2] Plasmids may carry genes encoding toxins as well as functions such as resistance to heavy metals and antibiotics, nitrogen fixation, degradation of hydrocarbons (especially in the Pseudomonads), and induction of tumors in plants (for example the Ti-plasmid of *Agrobacterium tumefaciens*). "Lysogenic" bacteria carry bacteriophage genomes, the expression of which is repressed, integrated into the host chromosome.

Bacterial chromosomal DNA is replicated bidirectionally from a single origin in a semiconservative manner. At each of the two resulting replication forks, one newly synthesized single strand is polymerized as a long, continuous piece and the other in small pieces that are then ligated together. Thus, replication is also said to be semidiscontinuous. Both single strands at each replication fork are synthesized by a single replication apparatus, which contains two catalytic centers, one for each strand. Plasmids are also replicated by semiconservative mechanisms from single origins, although with details that may differ from those of chromosomal replication.

TRANSCRIPTION (RNA SYNTHESIS) AND CONTROL OF GENE EXPRESSION

Transcription, or copying of a gene sequence into RNA, is the initial step in gene expression in all organisms. Eubacteria use a single RNA polymerase to transcribe all their genes. A particular subunit of the polymerase called "sigma factor" is responsible for binding of the enzyme to each gene to initiate transcription. This occurs at sites called "promoters." In *E. coli* most promoters are recognized by the sigma factor, σ^{70}; these promoters consist of two 6 bp

Encyclopedia of Agricultural, Food, and Biological Engineering
DOI: 10.1081/E-EAFE 120007195
Copyright © 2003 by Marcel Dekker, Inc. All rights reserved.

regions located in the region 1–45 bp upstream of the transcriptional start site (" − 1 to − 45").[5] *Escherichia coli* also has subsets of genes with alternative promoter sequences, which can be expressed only in the presence of corresponding, alternative sigma factors. Other bacterial species may have a subset of promoters similar to those recognized by *E. coli* σ^{70} as well as a variety of other promoter types[6,7]; regardless, the promoters are comprised of one or more short sequences in the −1 to −45 region.

The most important level of control of gene expression is initiation of transcription. This allows bacteria to express genes at various levels, appropriate to the level of each gene product needed; it also allows cells to have genes, the expression of which is needed only under certain circumstances, and expend the energy to express these genes only when their products are needed. The transcription level is controlled to a certain extent by the actual sequence or "strength" of each promoter (i.e., its inherent ability to bind RNA polymerase). It may also be controlled by the binding of regulatory proteins to specific DNA sequences near to or within the promoter. Some of these regulatory proteins ("repressors") can act negatively, that is when bound to their sites on the DNA they inhibit initiation of transcription by RNA polymerase; positive acting regulatory proteins enhance transcriptional initiation when they bind to their sites on DNA. Some promoters control expression of a single gene; others control expression of a cluster of two or more genes. A unit containing a promoter, binding site(s) for regulatory proteins, and one or more genes is known as an "operon." A group of operons subject to the control of a common regulatory protein is called a "regulon."

Transcription can also be controlled by modulation of termination of RNA synthesis. In antitermination, normal termination of transcription is inhibited; this allows transcription to continue into a following gene so that it is also expressed. Attenuation is a control mechanism seen in several operons encoding enzymes of amino acid biosynthesis. Transcription of such operons can be down regulated by formation of a transcriptional terminator just following the transcriptional start. Post-transcriptional processing of eubacterial mRNAs is very rare (in large part because there are almost no introns in eubacteria). Mature transfer and ribosomal RNAs, as well as other RNAs, however, are produced by specific post-transcriptional cleavages of the initial transcripts ("RNA precursors"). In the case of tRNAs, post-transcriptional maturation also includes addition of the 3′ CCA end to which the amino acid is attached.

The expression of all genes of a bacterium can be measured simultaneously using "microarray" technology.[8] This technique is based on complementary base-pairing between mRNA preparations and a collection of DNA probes which are fixed to a solid support; each of the probes is complementary to a specific mRNA. Variations in levels of all mRNAs in response to environmental changes or changes in the growth phase, for example, can be easily measured in this way.

TRANSLATION (PROTEIN SYNTHESIS)

Bacterial ribosomes are somewhat smaller than those of eukaryotes. The complete 70S ribosome (2.5 million Daltons) is made up of 30S and 50S subunits. Initiation of translation requires base pairing between a six base sequence in the mRNA (the "Shine-Dalgarno sequence") located a few bases upstream of the start codon and a complementary sequence in the 16S rRNA of the 30S subunit. Energy for translation comes from both ATP (required to charge tRNAs with amino acids) and GTP (to run the actual steps of translation). Messenger RNAs can contain sequences corresponding to single genes or, if transcribed from operons containing more than one gene, multiple coding sequences (these are known as "polycistronic mRNAs"). In either case, multiple ribosomes translate each individual mRNA molecule (forming "polysomes"). Bacteria use the universal genetic code; thus, they use 61 codons to represent the 20 amino acids. The resulting "degeneracy" ranges from 1 to 6 codons per amino acid. There are three chain-termination codons, UGA, UAA, and UAG. AUG is used most often as the initiation codon, although GUG and UUG can also be used; in all cases, however, the first amino acid translated in each polypeptide is methionine.

GENE TRANSFER

In place of sexual reproduction and genetic recombination during gamete formation as mechanisms of introducing genetic variation, eubacteria have several methods of transferring genes between cells. Typically, all methods would not exist for every species. In transformation, naked DNA (produced, for example, by cell lysis) can be taken up and incorporated into the host cell chromosome by homologous recombination; about 20,000 bp would be a typical amount to be transferred in a single event. In transduction, bacterial DNA is transferred to a host cell in a bacteriophage particle as a byproduct of a bacteriophage infection; up to about 50,000 bp of DNA can be transferred in this way and the transferred DNA integrated into the host chromosome by subsequent recombination.

Larger amounts of DNA can be transferred by the mechanism of conjugation. This is active transfer of DNA from a donor to a recipient cell mediated by functions encoded on conjugative plasmids, some of which can

transfer DNA only within a narrow range of species and some of which have a broad host range. Conjugative plasmids, if not integrated into the host chromosome, can transfer themselves from cell to cell; those that are episomes may also integrate into the host chromosome, and in this way transfer chromosomal DNA from donor to recipient cells. Smaller amounts of host DNA are transferred more readily than large amounts, but it is possible to transfer an entire chromosome (many millions of base pairs). Subsequent homologous recombination between host and transferred DNA will replace host with donor sequences.

Transposable elements (transposons) are mobile, being able to move, for example, from one site to another within a chromosome, or from a chromosome to a plasmid. Some transposons simply move from one location to another; others "move" by remaining in the original location while synthesizing a copy that is inserted elsewhere. Transposons typically choose their target sites in the host DNA at random, although some may show a preference for certain sequences.[3]

EVOLUTION

Evolution of bacteria occurs by slow accrual of mutations, but apparently much more rapidly by so-called "horizontal gene transfer" between species. A striking example is shown by a comparison of the genomes of *E. coli* K12 and the pathogenic *E. coli* 0157:H7.[7] The two strains share a common "backbone" sequence of 4.1 million base pairs, comparison of which indicates that they split from each other 4.5 million years ago. In this time, hundreds of segments of DNA that are not related to the backbone sequence have accumulated individually in the genome of each strain. The most likely explanation is that these DNA fragments have been acquired by horizontal transfer from different species. In each strain there are more than 175 such fragments, totaling 1.34 and 0.53 million base pairs for the 0157:H7 and K12 strains, respectively. In the case of the 0157:H7 strain, 26% of the bacterium's total genes are within the (presumably) horizontally transferred DNA. There is also evidence that horizontal transfer can occur from eukaryotes to bacteria.[9,10]

Transposable elements often carry genes that confer resistance to antibiotics. Several of these can accumulate in conjugative plasmids (called R plasmids), which are able to move via conjugation from cell to cell within and between species. In this way, resistance to multiple antibiotics can be spread throughout bacterial populations.

REFERENCES

1. Lengeler, J.W.; Drews, G.; Schlegel, H.G. *Biology of the Prokaryotes*; Blackwell Science: Thieme, 1999; 955.
2. http://www.ncbi.nlm.nih.gov (accessed May 2001).
3. Lewin, B. *Genes VII*; Oxford University Press: New York, 2000; 990.
4. Snustad, D.P.; Simmons, M.J. *Principles of Genetics*; John Wiley & Sons: New York, 2003; 840.
5. Lisser, S.; Margalit, H. Compilation of *E. coli* mRNA Promoter Sequences. Nucleic Acids Res. **1993**, *21*, 1507–1516.
6. Strohl, W.R. Compilation and Analysis of DNA Sequences Associated with Apparent Streptomycete Promoters. Nucleic Acids Res. **1992**, *20*, 961–974.
7. Perna, N.T., et al. Genome Sequence of Enterohaemorrhagic *Escherichia coli* 0157:H7. Nature **2001**, *409*, 529–533.
8. Gmuender, H. Perspectives and Challenges for DNA Microarrays in Drug Discovery and Development. Biotechniques **2002**, *32*, 152–158.
9. Moens, L.; Vanfleteren, J.; Van de Peer, Y.; Peeters, K.; Kapp, O.; Czeluzniak, J.; Goodman, M.; Blaxter, M.; Vinogradov, S. Globins in Nonvertebrate Species: Dispersal by Horizontal Gene Transfer and Evolution of the Structure–Function Relationships. Mol. Biol. Evol. **1996**, *13*, 324–333.
10. Lander, E.S., et al. Initial Sequencing and Analysis of the Human Genome. Nature **2001**, *409*, 860–921.

Microbial Metabolism

Glen H. Smerage
Arthur A. Teixeira
University of Florida, Gainesville, Florida, U.S.A.

INTRODUCTION

Microbial metabolism is metabolism of independent, individual cells—the totality of physical and chemical processes by which a cell executes activities of its life cycle, including the ultimate, reproduction. Two major facets of metabolism are biosynthesis within a cell of the new matter of growth and development and procurement of energy necessary to drive biosynthesis and other cellular work. A cell obtains all the materials (nutrients) for biosynthesis and energy from its environment; waste products of metabolism are released by the cell into its environment. Many of the possible wastes are of great industrial interest. Functionally, the internal structure of a cell, its production of cell matter (product) and wastes (byproducts), and the input of raw materials and output of wastes are analogous to industrial factories.

CELL ARCHITECTURE

The cell is the canonical structural and functional unit of all living organisms. Microbes exist as individual or clustered, independent cells. Every cell has a cytoplasmic membrane defining its boundary, separating its interior and environment, and regulating material transfers between the two. Membranes of many microorganisms are surrounded by walls for strength and protection. Membrane and wall correspond to walls of an industrial factory.

Internal structures and organizations of microbial cells are prokaryotic (all bacteria and archaea) or eukaryotic (algae, fungi, and protozoa). All macroorganisms (plants and animals) are eukarya. Prokarya and eukarya conduct the same overall life processes but differ in enabling structural organizations. A prokaryotic cell corresponds to a large, one room factory with production machinery and management dispersed openly throughout; a eukaryotic factory is highly partitioned by membranes (walls) into several, distinct functional units.

Cytoplasmic membranes of both cell types are phospholipid bilayers. Within a membrane is a colloidal fluid, called cytoplasm, consisting of an aqueous solution (cytosol) of a multitude of organic compounds and inorganic ions and suspended bodies and particles. Primary suspended bodies in prokarya are a single chromosome, many ribosomes, and granules storing carbon, nitrogen, sulfur, or phosphorus compounds. The prokaryotic chromosome is a single, circular deoxyribonucleic acid (DNA) molecule highly twisted to fit in the cell; referred to as the nuclear region or nucleoid, it is not enclosed in a membrane. Ribosomes are small particles composed of ribonucleic acid (RNA) and protein where structural and enzymatic protein syntheses occur.

Primary suspended bodies in eukarya are membrane bound organelles—the nucleus, mitochondria, chloroplasts, endoplasmic reticulum, Golgi complex, vacuoles, lysosomes—in addition to ribosomes and granules. Organelles are major functional units of eukaryotic metabolism; mitochondria or chloroplasts produce cellular energy from nutrients or light, respectively; the endoplasmic reticulum and Golgi complex are extensive, coordinated membrane structures for material storage, processing, and transport within the cell; the nucleus contains genetic machinery of the cell.

The prokaryotic nucleoid and eukaryotic nucleus are repositories and exporters of genetic information governing the structure, functions, and life cycle of the cell. They correspond to blueprints, processing steps, and management of a factory.

CELLULAR ENERGY PROCUREMENT

Microorganisms of both cell types procure cellular energy to drive the processes of life from the Gibbs free energy of chemical compounds released by breaking them into simpler constituents (catabolism) or, less commonly, from light. Microorganisms employing catabolism are known as chemotrophs; phototrophs employ light. The source energy in each case is transformed and stored within a cell in carriers that are coenzyme molecules: adenosine triphosphate (ATP), reduced nicotinamide-adenine dinucleotide (NADH), and reduced flavin adenine dinucleotide ($FADH_2$) for subsequent use in biosynthesis and other microbial work. Two different processes, substrate level phosphorylation and oxidative phosphorylation are used to form ATP, the primary energy carrier.

Encyclopedia of Agricultural, Food, and Biological Engineering
DOI: 10.1081/E-EAFE 120007193
Copyright © 2003 by Marcel Dekker, Inc. All rights reserved.

Catabolism consists of oxidation–reduction reactions. Most microorganisms are chemoorganotrophic, catabolizing organic compounds—carbohydrates, lipids, proteins, and nucleic acids. Chemolithotrophs catabolize inorganic compounds. Two types of organic catabolism, fermentation and respiration, are used by microbes to generate ATP. Fermentation, the simpler of the two, is a sequence of anaerobic oxidation–reductions many bacteria and microbial eukarya employ in anaerobic conditions. Specific reactions depend on genus and species of microorganism but have a basic structure: glycolysis followed by final product formation. A sequence of enzyme catalyzed metabolic reactions is a metabolic pathway.

Glycolysis is a 10 step catabolic path occurring in the cytosols of both prokarya and eukarya catalyzed by 10 cytosolic enzymes that converts (breaks down) one molecule of glucose into two molecules of pyruvate (pyruvic acid) and two of ATP. The sequence of glycolytic reactions involves glucose, adenosine diphosphate (ADP), inorganic phosphorous (P_i), and coenzyme nicotinamide-adenine dinucleotide (NAD) in oxidized (NAD^+) and reduced (NADH) forms. Overall, it is equivalent to conversion of glucose into pyruvate with formation of ATP from ADP and P_i. This generation of ATP is substrate level phosphorylation.

$$\text{glucose} + 2NAD^+ + 2ADP + 2P_i \rightarrow 2\,\text{pyruvate} + 2NADH + 2H^+ + 2ATP + 2H_2O$$

Final product formation reduces pyruvate to lactate (lactic acid) or ethanol and CO_2 or another organic compound while reoxidizing NADH to restore NAD^+ consumed early in glycolysis, and no ATP is produced. The final product is to the microorganism a waste in its central task of fermentation—production of ATP, but many of those wastes are commercially important to humans. The final product in any fermentation attests to incomplete oxidation of carbon atoms in the initial organic substrate (glucose) and partial harvest of its free energy. The substrate is the acceptor and the final product is the donor of electrons in the overall catabolic path.

The equation for reduction of pyruvate to lactate by lactic acid bacteria is

$$\text{pyruvate} + NADH + H^+ \rightarrow \text{lactate} + NAD^+$$

and the overall equation for fermentation of glucose to lactic acid is

$$\text{glucose} + 2ADP + 2P_i \rightarrow 2\,\text{lactate} + 2H_2O + 2ATP$$

Similarly, the overall equation for fermentation of glucose to ethanol by anaerobic yeast is

$$\text{glucose} + 2ADP + 2P_i \rightarrow 2\,\text{ethanol} + 2CO_2 + 2H_2O + 2ATP$$

An alternative organic catabolism is aerobic cellular respiration. In contrast to fermentation, it accomplishes complete oxidation of glucose to CO_2 and harvest of its considerable free energy by using oxygen as the terminal electron acceptor. Aerobic respiration begins with glycolysis to produce pyruvate, which is then oxidized into acetyl CoA plus CO_2 in a sequence of reactions involving coenzyme A (CoA), NAD^+, and NADH summarized by

$$\text{pyruvate} + CoA + NAD^+ \rightarrow \text{acetyl CoA} + NADH + CO_2$$

Acetyl CoA is then enzymatically oxidized completely to CO_2 with reduction of coenzymes FAD and NAD^+ by the tricarboxylic acid (TCA or citric acid or Krebs) cycle of eight steps, summarized by

$$\text{acetyl CoA} + 3NAD^+ + FAD + ADP + P_i + 2H_2O \rightarrow 2CO_2 + COA + 3NADH + 2H^+ + FADH_2 + ATP$$

where FAD is flavin adenine dinucleotide, and $FADH_2$ is reduced FAD. Oxidations of pyruvate and acetyl CoA occur within the cytosols of prokarya and within the mitochondrial matrices of eukarya.

As in fermentation, NAD^+ and FAD reduced in respiratory reactions must be restored by oxidation of the NADH and $FADH_2$ produced. This is accomplished via the electron transport (respiratory) chain located in the cytoplasmic membrane of prokarya and the inner mitochondrial membrane of eukarya. Energy released by these oxidations in the transport chain is transferred to ATP molecules synthesized from ADP via oxidative phosphorylation coupled with the electron transport.

Aerobic respiration yields considerably more of the free energy of glucose than fermentation. Glycolysis produces two molecules of pyruvate, two of NADH, and a net of two ATP from each molecule of glucose. Reduction of one molecule of pyruvate to lactate or ethanol yields one molecule of product, one of NAD^+, and no ATP. Thus, either fermentation yields just two ATP molecules for biological work from each glucose molecule fermented.

Oxidation of one molecule of pyruvate yields one molecule of acetyl CoA and one of NADH; TCA cycle oxidation of one molecule of acetyl CoA yields three of NADH, one $FADH_2$, and one ATP; aerobic respiration oxidizes two additional NADH molecules created in glycolysis. The conversion of one glucose to two pyruvate yields 10 NADH and two $FADH_2$ molecules for reoxidation via the electron transport system; that, in turn, yields up to three and two ATP molecules, respectively, per molecule of NADH and $FADH_2$. Thus, aerobic respiration may produce 38 ATP molecules from one glucose molecule while fermentation produces only two. The free energy of respiratory ATP, about 1200 kJ/mol, is about 40% of the free energy of glucose available by complete oxidation by oxygen; the remainder is lost as heat.

Glucose is not the only organic substrate amenable to fermentation and aerobic respiration. Many microorganisms use other carbohydrates, lipids, or proteins, breaking them via preliminary catabolic paths into intermediate metabolites of fermentation or respiratory paths that subsequently enter and complete the remainders of those catabolic paths. Carbohydrates enter various points of glycolysis; proteins typically enter as pyruvate or acetyl CoA; lipids enter near the end of glycolysis or as acetyl CoA.

CELLULAR BIOSYNTHESIS

Cytosolic water comprises about 90% of the mass of a microbial cell; the remaining mass pertains mainly to myriad organic compounds, predominantly proteins, comprising internal functional units and the cytoplasmic membrane. Elemental components, in diminishing percentage by weight, are carbon, oxygen, nitrogen, hydrogen, phosphorus, and sulfur; small amounts of several metals are essential to biosynthesis. A common, empirical formula for a microbial cell, $C_5H_7NO_2$, expresses the aggregate essence of many distinct constituent compounds, all of which contain carbon. Those compounds are organic macromolecules: carbohydrates (cell walls), lipids (membranes), proteins (enzymes, transmembrane carriers, structural proteins, and signal receptors), and nucleic acids (store and express genetic information).

Constituent macromolecules of cells are synthesized (anabolism) by heterotrophs from organic substances (nutrients) available in their environments as sources of carbon; the same is accomplished by autotrophs using inorganic carbon. Anabolic pathways generally are reductive rather than oxidative, produce macromolecules from small, simple precursors, and employ energy from ATP and NADH produced during catabolism. Less energy is required for cell synthesis from organic carbon sources than inorganic sources.

Focusing on heterotrophs, nutrients may be simple, ideally glucose for carbon and ammonia (NH_3) or nitrate (NO_3^-) for nitrogen, or complex molecules that must first be catabolized to simpler forms to obtain carbon and nitrogen for syntheses of macromolecules. Many precursors of macromolecules are available as metabolic intermediates of catabolism for energy procurement. Furthermore, anabolic pathways are substantially reversals of catabolic pathways between common small and large molecule endpoints, with several reversible reactions between intermediate metabolites shared by paired paths. However, each pathway of a pair has at least one intermediate reaction that is essentially irreversible and mediated by different enzymes. Paired anabolic and catabolic pathways are, thereby, rendered irreversible, independent, and reciprocally controlled.

For example, gluconeogenesis is the anabolic path paired with glycolysis; both work between glucose and pyruvate. Three of the 10 glycolysis reactions are essentially irreversible and must be bypassed in gluconeogenesis by different enzymatically catalyzed reactions that are essentially irreversible in the direction of glucose synthesis. The other seven reactions of glycolysis are reversible and shared with gluconeogenesis. Overall, glycolysis and gluconeogenesis are irreversible and independently regulated.

CONCLUSION

This discussion introduced two facets of microbial metabolism, energy procurement and biosynthesis, with emphasis on the former. Other important and interesting facets are: mechanisms and dynamics of molecular transport through the cytosol and various membranes; roles of genes in the nucleus or nucleoid in specifying cell structure and controlling dynamics of energy procurement, biosynthesis, growth, and reproduction; chemolithotrophic and photolithotrophic metabolisms; and effects of temperature and other environmental factors on metabolism. Several mechanisms of membrane transport are particularly interesting. Many engineers will be interested in exploring mathematical descriptions of processes and system models and analyses of behaviors of models of microbial metabolism.

BIBLIOGRAPHY

Atlas, R.M. *Principles of Microbiology*, 2nd Ed.; Wm. C. Brown Publishers: Dubuque, IA, 1997.

Brock, T.D.; Madigan, M.T.; Martinko, J.M.; Parker, J. *Biology of Microorganisms*, 7th Ed.; Prentice Hall: Englewood Cliffs, NJ, 1994.

Shuler, M.L.; Kargi, F. *Bioprocess Engineering: Basic Concepts*; Prentice Hall: Englewood Cliffs, NJ, 1992.

Blanch, H.W.; Clark, D.S. *Biochemical Engineering*; Marcel Dekker: New York, 1996.

Stoker, H.S. *General, Organic, and Biological Chemistry*; Houghton Mifflin: Boston, 2001.

Microbial Population Dynamics

Arthur A. Teixeira
Glen H. Smerage
University of Florida, Gainesville, Florida, U.S.A.

INTRODUCTION

The manner in which populations of micro-organisms increase or decrease in response to controlled environmental stimuli is fundamental to the engineering design of bioconversion processes (fermentations) and thermal inactivation processes (pasteurization and sterilization) important in the food and bioprocess industries. In order to determine optimum process conditions and controls to achieve desired results, the effect of process conditions on rates (kinetics) of population increase or decrease need to be characterized and modeled mathematically. A brief overview of microbial growth kinetics modeling important in bioprocess fermentations is given, followed by a similar overview of modeling thermal inactivation kinetics of bacterial spores important to sterilization treatments in food, pharmaceutical, and bioprocess industries.

MICROBIAL GROWTH KINETICS

Industrial fermentations may be carried out as batch, continuous, or fed-batch processes. Only the batch process is considered here as an example of a closed culture system containing an initial limited amount of nutrient. The inoculated culture will experience a number of phases, as illustrated in Fig. 1. After inoculation, there is a period during which no growth appears, referred to as the lag phase. It is influenced by history of the inoculum development. This lag period can be reduced substantially by use of a suitable inoculum.[1] As growth in population begins to appear, the growth rate gradually increases reaching a constant maximum rate causing the number of cells in the population to increase exponentially with time as indicated by the inclined straight line on the semilog plot in Fig. 1. This period is known as the logarithmic or exponential phase. The population increase with time can be described by the equation:

$$\frac{\mathrm{d}x}{\mathrm{d}t} = kx \tag{1}$$

where x is the number of microbial cells (or concentration of microbial biomass), t is time in hr, and k is the specific growth rate constant in logarithm cycles per hour.

On integration, Eq. 1 gives:

$$X_t = X_0 e^{kt} \tag{2}$$

where X_t is the number of microbial cells at time t, X_0 is the initial number of cells at beginning time t_0, and e is the base of the natural logarithm.

On taking natural logarithms, Eq. 2 becomes:

$$\ln\left(\frac{X_t}{X_0}\right) = k(t - t_0) \tag{3}$$

Thus, a plot of natural logarithm of population size against time should yield a straight line, the slope of which should equal k.

Eqs. 2 and 3 predict that growth will continue indefinitely. However, growth results in the consumption of nutrients and excretion of products from microbial metabolism. Thus, after a certain time in a batch reactor (fermentor), the growth rate decreases until population growth ceases altogether. The growth curve becomes flat, indicating a stationary phase that is often followed by a decline indicating a "death" phase (not shown). This may be due to depletion of some essential nutrient in the medium (substrate limitation) and/or accumulation of some autotoxic metabolic product in the medium (toxin limitation). It is common to relate substrate depletion to population growth through a stoichiometric relationship or yield constant:

$$\frac{\mathrm{d}s}{\mathrm{d}t} = -Y_{s/x}\frac{\mathrm{d}x}{\mathrm{d}t} \tag{4}$$

where s is the substrate nutrient concentration, x is the cell population (concentration), $Y_{s/x}$ is the yield constant for substrate, and t is the time during process.

For metabolic products that are growth associated, the formation of product can be related to population growth by the expression:

$$\frac{\mathrm{d}P}{\mathrm{d}t} = Y_{P/x}\frac{\mathrm{d}x}{\mathrm{d}t} \tag{5}$$

where P is the product formed at any time, x is the cell population (concentration), $Y_{P/x}$ is the yield constant for the product, and t is time during process.

Encyclopedia of Agricultural, Food, and Biological Engineering
DOI: 10.1081/E-EAFE 120007194
Copyright © 2003 by Marcel Dekker, Inc. All rights reserved.

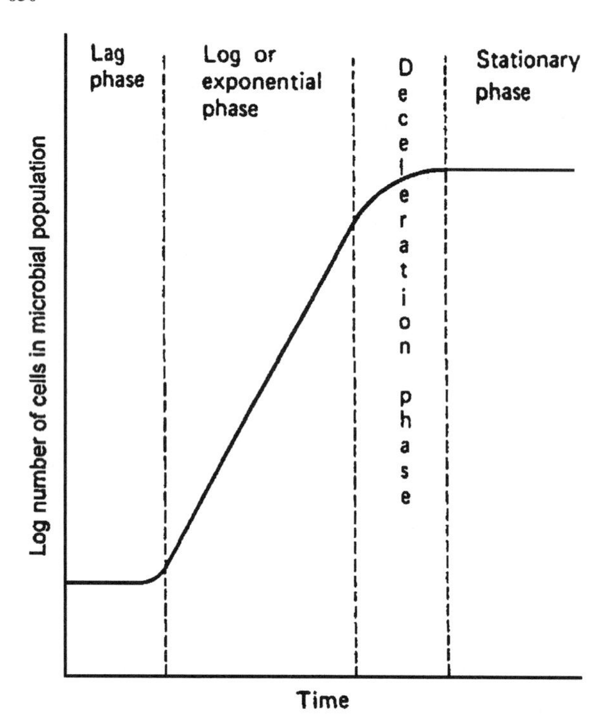

Fig. 1 Population growth of a typical microbial culture in batch conditions.[1]

These on-going simultaneous dynamics of growth in cell population (concentration), depletion of substrate, and formation of product are shown in Fig. 2 for the case of fermenting grape juice into wine.[2] In this case, population growth of yeast (*Saccharomyces cerevisiae*) is related to depletion of sugar concentration in the juice (substrate) and increasing ethanol concentration (product).

THERMAL INACTIVATION KINETICS OF BACTERIAL SPORES

The design of thermal sterilization processes has been traditionally based on the assumption that thermal inactivation of bacterial spores can be modeled by a single first-order reaction.[3] This can be described as a straight-line survivor curve when the logarithm of the number of surviving spores is plotted against time of exposure to a lethal temperature, as shown in Fig. 3 for survivor curves obtained at three different lethal temperatures. The decline in population of viable spores can be described by the following exponential or logarithmic equations:

$$C = C_0 e^{-kt} \tag{6}$$

$$\ln\left(\frac{C}{C_0}\right) = -kt \tag{7}$$

where C is the concentration of viable spores at any time t, C_0 is the initial concentration of viable spores, and k is the first-order rate constant and is temperature dependent. This temperature dependency is illustrated in Fig. 3, in which T_1, T_2, and T_3 represent increasing lethal temperatures with corresponding rate constants, $-k_1$, $-k_2$, and $-k_3$.

The temperature dependency of the rate constant is also an exponential function that can be described by a straight line on a semilog plot when the natural log of the rate constant is plotted against the reciprocal of absolute temperature. The equation describing this straight line is known as the Arrhenius equation:

$$\ln\frac{k}{k_0} = -\frac{E_a}{R}\left[\frac{T_0 - T}{T_0 T}\right] \tag{8}$$

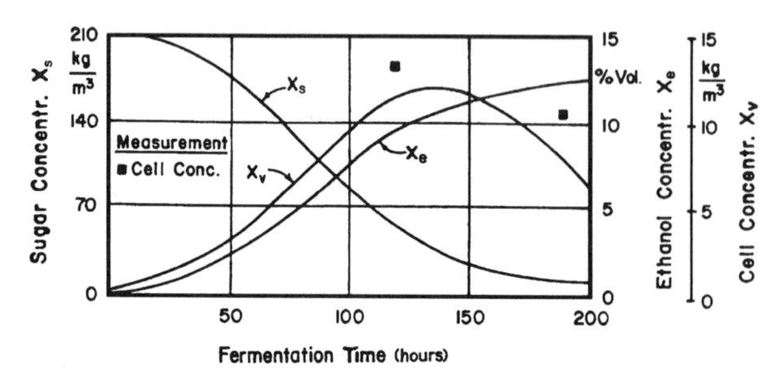

Fig. 2 Batch fermentation of grape juice (must) with *Saccharomyces cerevisiae*,[2] comparing model-predicted cell concentration over time with measured data points (dark squares).

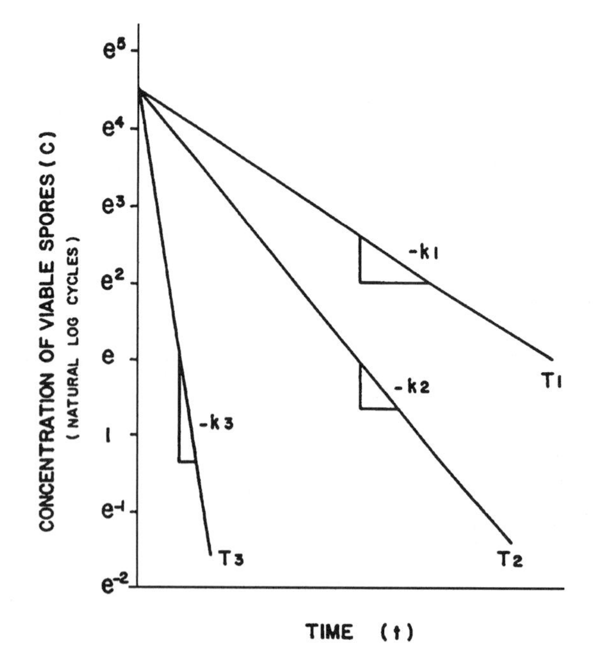

Fig. 3 Family of spore survivor curves on semilog plot showing viable spore concentration vs. time at different lethal temperatures. (From Ref. [4], p. 567.)

where k is the rate constant at any temperature T and k_0 is the reference rate constant at a reference temperature T_0. The slope of the line produces the term E_a/R, in which E_a is the activation energy and R is the universal gas constant. Thus, once the activation energy is obtained in this way, Eq. 8 can be used to predict the rate constant at any temperature. Once the rate constant is known for a specified temperature, Eq. 6 or 7 can be used to determine the time required for exposure at that temperature to reduce the initial concentration of viable bacterial spores by any number of log cycles.[4] The objective in most commercial sterilization processes is to reduce the initial spore population by a sufficient number of log cycles so that the final number of surviving spores is one millionth of a spore, interpreted as probability for survival of one in a million in the case of spoilage-causing bacteria and one in a trillion in the case of pathogenic bacteria such as *Clostridium botulinum*.

Actual survivor curves plotted from laboratory data often deviate from a straight line, particularly during early periods of exposure. These deviations can be explained by the presence of competing reactions taking place simultaneously, such as heat activation causing "shoulders" and early rapid inactivation of less heat-resistant spore fractions causing biphasic curves with tails, as shown in Fig. 4. Activation is a transformation of viable, dormant spores enabling them to germinate and grow in a substrate medium. This transformation is part of the life cycle of spore-forming bacteria as shown in Fig. 5. Under hostile environments, vegetative cells undergo sporulation

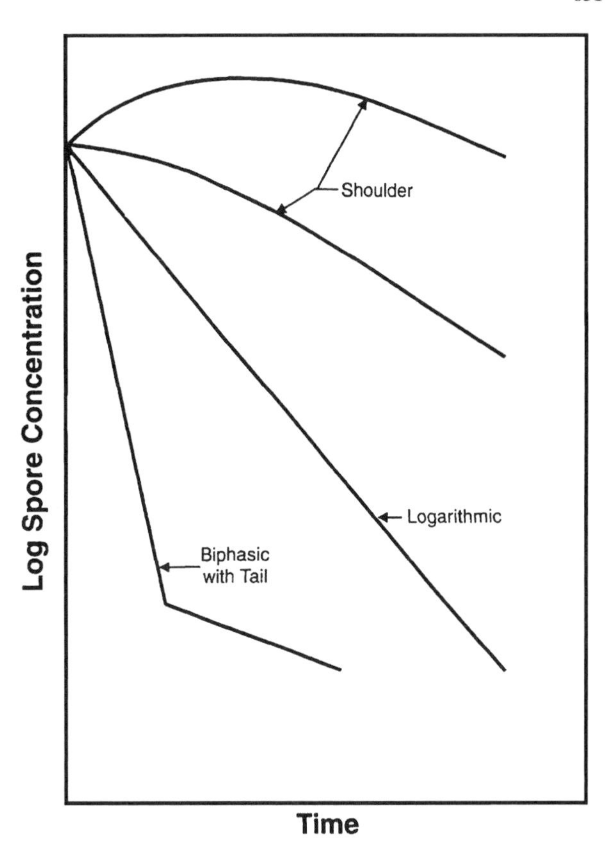

Fig. 4 Schematic nonlinear survivor curves for bacterial spores illustrating "shoulders" and "biphasic" population responses to exposure at constant lethal temperatures.

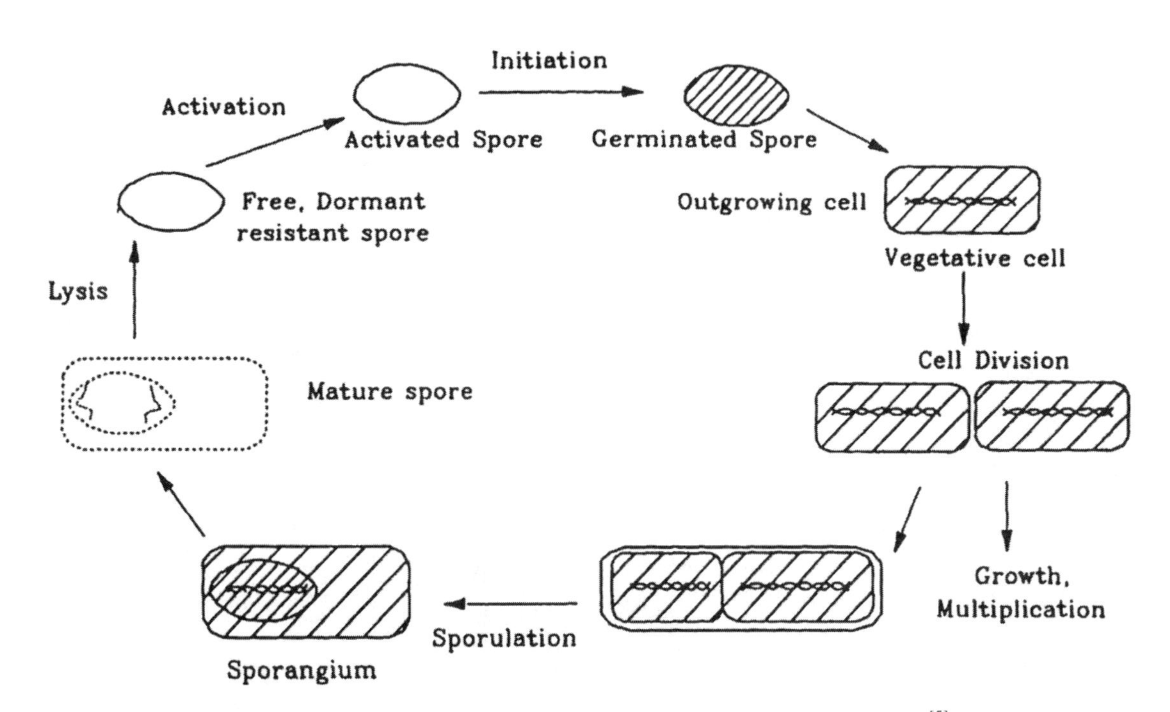

Fig. 5 Cycle of spore formation, activation, germination, and outgrowth.[5]

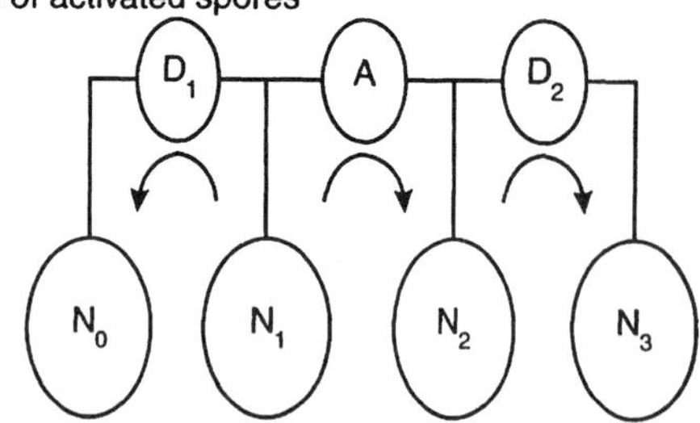

Fig. 6 Process diagram of the Sapru model of bacterial spore populations during heat treatment.[9]

to enter a state of dormancy as highly heat-resistant spores. When subjected to heat in the presence of moisture and nutrients, they must undergo the activation transformation in order to germinate once again into vegetative cells and produce colonies on substrate media to make their presence known.[5–7]

In recent years, several workers have approached the mechanistic modeling of shoulders in survivor curves of bacterial spores subjected to lethal heat.[8–12] The most recent model of Sapru et al. is based on the system diagram presented in Fig. 6.[8,9] The total population at any time is distributed among storehouses (stores) containing the fraction of population in a given stage of the life cycle. Dormant, viable spores, potentially capable of producing colonies in a growth medium after activation, comprise population N_1. Activated spores, capable of forming colonies in a growth medium, comprise population N_2. Inactivated spores, incapable of germination and growth,

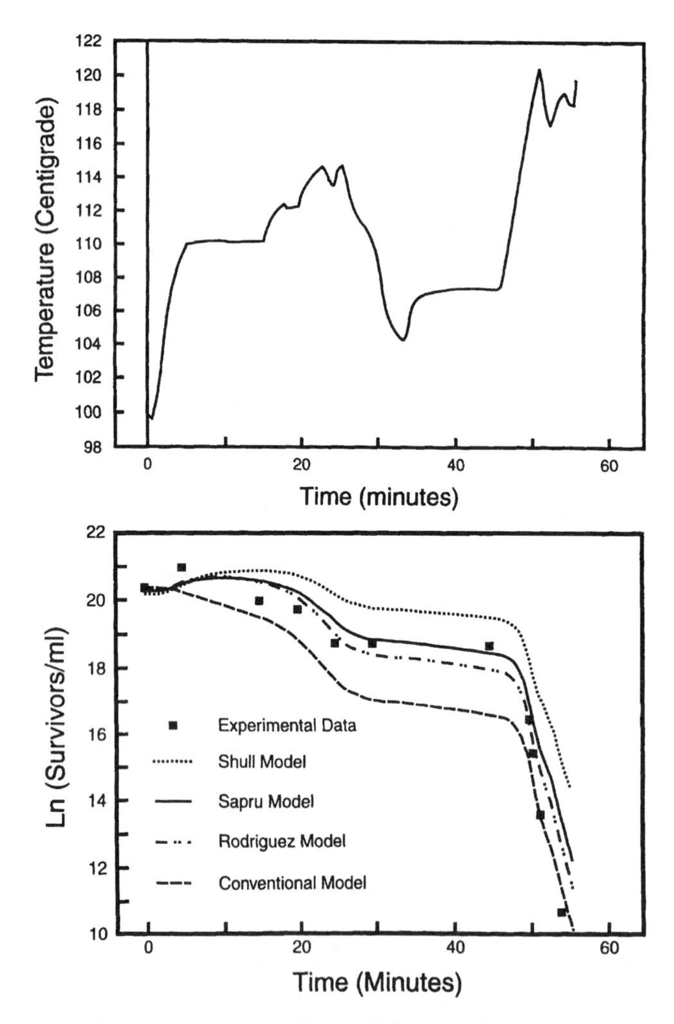

Fig. 7 Top: Temperature history experienced by spore population during experimental heat treatment. Bottom: Experimental data points and model-predicted survivor curves in response to temperature history above for the Sapru, Rodriguez, Shull, and conventional models.[9]

Table 1 Comparison of mathematical models of Sapru, Rodriguez, Shull, and conventional models. Dormant population is N_1, activated population is N_2, and survivors $N = N_2$

Model	Mathematical model
Sapru[8,9]	$\frac{dN_1}{dt} = -(K_{d1} + K_a)N_1$
	$\frac{dN_2}{dt} = K_aN_1 - K_{d2}N_2$
Rodriguez[11,12]	$\frac{d}{dt}N_1 = -(K_d + K_a)N_1$
	$\frac{d}{dt}N_2 = K_aN_1 - K_dN_2$
Shull[10]	$\frac{d}{dt}N_1 = -K_aN_1$
	$\frac{d}{dt}N_2 = K_aN_1 - K_dN_2$
Conventional[4]	$\frac{dN}{dt} = -K_dN$

(From Ref. [9].)

comprise populations N_3 and N_0. Activation, A, transforms many members of N_1 to N_2 while inactivation, D_1, transforms other members of N_1 to N_0. Inactivation, D_2, transforms activated spores from N_2 to N_3. It is assumed that A, D_1, and D_2 are independent, concomitant, and first-order with respective rate constants K_a, K_{d1}, and K_{d2}. The Rodriguez model[11,12] is obtained from the Sapru model by assuming that inactivation transformations D_1 and D_2 have identical rate constants, i.e., $K_{d1} = K_{d2} = K_d$. The Shull model[10] is obtained from the Sapru model by deleting D_1 (K_{d1}=0) and N_0, and setting $K_{d2} = K_d$. Finally, the conventional model is obtained by deleting N_1, N_0, D_1, and A ($K_{d1} = K_a$), so that only $N = N_2$, $D = D_2$, and N_3 remain, and setting $K_d = K_{d2}$ (see Table 1).

These models were tested by comparing model-predicted with experimental survivor curves in response to constant and dynamic lethal temperature histories. For dynamic temperatures, the curves were predicted by simulation, using Arrhenius equations to vary rate constants with temperature. In separate experiments with dynamic temperature, sealed capillary tubes containing bacterial spores suspended in phosphate buffer solution were placed in an oil bath, and the temperature was varied manually. Tubes were removed at times spaced across the test interval and analyzed for survivor counts. Temperature histories of the oil bath were recorded.[9] The dynamic, lethal temperature for one of these experiments is shown at the top of Fig. 7. Corresponding survivor data and survivor curves predicted by the various models in response to this dynamic temperature are given at the bottom of Fig. 7. Note that activation of the significant subpopulation of dormant spores was predicted well by the early increase in population response shown by the Rodriguez and Sapru models. Both models performed much better overall than the other two. The conventional model, being limited to a single first-order death reaction, was incapable of predicting any increase in population.

REFERENCES

1. Stanbury, P.F.; Whitaker, A. Microbial Growth Kinetics. *Principles of Fermentation Technology*, 1st Ed.; Pergamon Press: New York, 1993; Chap. 2, 11–12.
2. Teixeira, A.A.; Shoemaker, C.F. On-Line Control of Unit Operations. *Computerized Food Processing Operations*; Van Nostrand Reinhold: New York, 1989; Chap. 4. 123–128.
3. Stumbo, C.R. *Thermobacteriology in Food Processing*, 2nd Ed.; Academic Press: New York, 1973.
4. Teixeira, A.A. Thermal Process Calculations. In *Handbook of Food Engineering*; Heldman, D.R., Lund, D.B., Eds.; Marcel Dekker, Inc.: New York, 1992; Chap. 11. 565–567.
5. Gould, G.W. Injury and Repair Mechanisms in Bacterial Spores. In *The Revival of Injured Microbes*; Andrew, M.H.E., Russel, A.D., Eds.; Academic Press: New York, 1984; 199–218.
6. Lewis, J.C.; Snell, N.S.; Alderton, G. Dormancy and Activation of Bacterial Spores. In *Spores III*; Campbell, L.L., Halvorson, H.O., Eds.; American Society for Microbiology: Washington, D.C., 1965; 47–62.
7. Keynan, A.; Halvorson, H. Transformation of a Dormant Spore into a Vegetative Cell. In *Spores III*; Campbell, L.L., Halvorson, H.O., Eds.; American Society of Microbiology: Washington, D.C., 1965; 174–189.
8. Sapru, V.; Teixeira, A.A.; Smerage, G.H.; Lindsay, J.A. Predicting Thermophilic Spore Population Dynamics for UHT Sterilization Processes. J. Food Sci. **1992**, *57* (5), 1248–1252, 1257.
9. Sapru, V.; Smerage, G.H.; Teixeira, A.A.; Lindsay, J.A. Comparison of Predictive Models for Bacteria Spore Population Response to Sterilization Temperatures. J. Food Sci. **1993**, *58* (1), 223–228.
10. Shull, J.J.; Cargo, G.T.; Ernst, R.R. Kinetics of Heat Activation and Thermal Death of Bacterial Spores. Appl. Microbiol. **1963**, *11*, 485–487.
11. Rodriguez, A.C.; Smerage, G.H.; Teixeira, A.A.; Busta, F.F. Kinetic Effects of Lethal Temperatures on Population Dynamics of Bacterial Spores. Trans. ASAE **1988**, *31* (5), 1594–1606.
12. Rodriguez, A.C.; Smerage, G.H.; Teixeira, A.A.; Lindsay, J.A.; Busta, F.F. Population Model of Bacterial Spores for Validation of Dynamic Thermal Processes. J. Food Proc. Eng. **1992**, *15*, 1–30.

Microfiltration

D. Vidal-Brotóns
P. Fito
M. Gras
Universidad Politécnica de Valencia (UPV), Valencia, Spain

INTRODUCTION

Some membrane processes are particularly well suited for the separation and purification of biological molecules[1] since they operate at relatively low temperatures and pressures and involve no phase changes or chemical additives. Thus, these processes cause minimal denaturation, deactivation, and/or degradation of highly labile biological cells or macromolecules. Microfiltration (MF) is one of these processes, and this explains the impressive development of its industrial applications. Microfiltration is the largest industrial market within the membrane field,[2,3] responsible for about 40% of total sales, both in Europe and in the United States.

Microfiltration is an operation that uses porous membranes to remove suspended particles from liquids or gases. Microfiltration and ultrafiltration (UF) are primarily similar in that the mode of separation is molecular sieving through increasingly fine pores. The MF membranes filter colloidal particles and bacteria, from 0.1 μm to 10 μm in diameter. The hydrodynamic resistance of such membranes is small, so low hydrostatic pressures (<2 bar) are sufficient to obtain high fluxes.

MF OPERATION MODES

Microfiltration can be carried out in two different operation modes.[3] The most widely used process design is dead-end (or in-line) filtration, in which the entire fluid flow is forced through the membrane under pressure. The suspended particles accumulate continuously on the membrane surface or in its interior, so the pressure required to maintain a given flow increases, or the flow produced by a given applied transmembrane pressure decreases, until at some point the membrane must be replaced.

To reduce this deposition process, MF is often carried out in the crossflow (tangential flow) mode, in which the feed solution is circulated across the surface of the filter, in a tangential direction, scouring away particles from the membrane surface, and producing two streams: a clean, particle-free permeate, and a concentrated retentate containing the particles. The application of crossflow filtration has increased, particularly for solutions with high particle concentrations.

MF MEMBRANES

The two principal types of MF membrane filters in use are screen filters and depth filters. Screen filters collect the retained particles on the surface of the membrane, primarily by a sieving mechanism. Because they become rapidly plugged by the accumulation of retained particles at the top surface, they are preferred for the crossflow MF systems, where the circulating fluid helps to keep the filter clean.

Depth filters have relatively large pores on the top surface and contain a random, tortuous porous structure, into which particles are retained through adsorption onto the pore walls or by mechanical entrapment at constrictions in the membrane pores. Depth filters can retain a high particle load before fouling, so they are preferred for dead-end filtration.

The MF membranes are available in a wide variety of materials (Table 1). Advantages of inorganic materials include higher stability towards extreme process conditions, such as high temperature, extreme pH values, and solvents other than water. Some novel membranes are prepared by lithographic techniques.[3]

MF MODULES AND PROCESSES

The in-line plate-and-frame module, the first available MF module, is still widely used to process small volumes of solution.

More recently, a variety of cheap and reliable in-line cartridges have been produced that allow a much larger area of membrane to be incorporated into a disposable unit. A typical disposable pleated cartridge is 254-mm long and 50 mm–64 mm in diameter and contains about 0.28 m^2 of membrane. The membrane pleated and then folded around the permeate core consists of several layers: an outer prefilter facing the solution to be filtered, followed

Encyclopedia of Agricultural, Food, and Biological Engineering
DOI: 10.1081/E-EAFE 120016146
Copyright © 2003 by Marcel Dekker, Inc. All rights reserved.

Table 1 Materials used for production of commercial microfiltration membranes

Polymeric membranes:	*Ceramic membranes*:
Cellulose acetate	Alumina, titania, zirconia
Cellulose nitrate	*Metal membranes*:
Cellulose regenerated	Palladium, palladium–silver alloy, palladium–gold alloy, stainless steel
Polyamide	
Polysulfone	*Anodic membranes*:
Polyethersulfone	Alumina
Polycarbonate	*Carbon membranes*:
Polyetherimide	Polyacrylonitrile and polyimide, polyvinylidene chloride acrylate
Polyvinylidene fluoride (PVDF)	
Polytetrafluoroethylene (PTFE)	*Glass membranes*:
Polypropylene	$SiO_2 + B_2O_3 + Na_2O$
Polyacrylonitrile	

From Refs. 4, 6.

by a finer polishing membrane filter. The cartridge fits inside a specially designed housing into which the feed solution enters at a pressure of 0.7 bar–8.3 bar. Initially, the pressure difference across the filter is small but as retained particles block the filter, the pressure difference increases until a predetermined limiting pressure is reached, and the cartridge is changed. The use of a prefilter extends the life of the MF cartridge significantly. The correct combination of prefilter and final membrane must be determined for each application. The volume of solution that can be treated by a MF membrane is directly proportional to the particle level in the feed solution. As a rough rule of thumb,[2] the particle-holding capacity of a cartridge filter for a noncritical use is between $100\,g\,m^{-2}$ and $300\,g\,m^{-2}$ of membrane area, so MF membrane lifetimes are often measured in hours.

Recent innovation in MF has mainly concerned the development of crossflow filtration technology and membranes, for highly concentrated feed streams. Membranes must retain particles at the membrane surface, to reduce internal fouling, so only screen filters or asymmetric membranes (refer to the article *Membrane Separation Mass Transfer*) can be used. The design of these processing systems closely follows that of UF.

In many crossflow MF systems and in some dead-end systems, backflushing is applied to remove the fouling layer from the membrane, forcing periodically the permeate back through the membranes by using high counterpressures (about 0.5 bar) for several seconds, every few minutes. Each backflush step promotes a new flux increase, without changing the membrane. This procedure cannot be used with thin film membranes, due to tendency for delamination, but it is feasible with inorganic (ceramic) membranes.

Fouling is reduced by high crossflow velocities and low transmembrane pressures. The MF processes have been developed, which facilitate a crossflow both on the feed side and on the permeate side, guaranteeing a uniform transmembrane pressure. Batch system with total recycle and cocurrent permeate flow system (refer to the article *Membrane System Operation*) are particularly well suited for MF processes. Ceramic MF systems incorporating the cocurrent permeate flow design have reported permeate fluxes several folds higher compared to alternative designs. Other process techniques to reduce fouling are the use of pulsed flow, gas scattering, and electric or acoustic fields, the use of flow geometries that create high shear rates, and the use of two-phase flow, generated by introducing air bubbles into the feed solutions.[5]

MF APPLICATIONS

One of the main industrial applications of MF[2,3,6] is the sterilization and clarification of all kinds of beverages and pharmaceuticals in the food and pharmaceutical industries. Cold sterilization of beer using MF was introduced on a commercial scale in 1963. The process was not generally accepted at that time, but has recently become more common. Filters (1 μm) can remove essentially all yeasts as well as provide 10^6 reduction in the common bacteria found in beer and wine. The filtration system typically involves one or more prefilters to extend the life of the final polishing filter. A great number of trials are being done at an industrial scale to round-off new applications: MF of corn steep water as an alternative to heat pasteurization;[7] clarification of pulpy fruit juices;[8] skim milk MF with in-process pH adjustment to produce highly concentrated retentate reduced in Ca and whey protein content with good potential for cheese manufacture.[9] Current developments of MF in the dairy industry have been recently reviewed.[10] As a general rule, the more difficult (and specific) steps in the design of new processes are fouling minimization and cleaning procedures.

Microfiltration is also industrially used to remove particles during the processing of ultrapure water in the semiconductor industry. Some new applications in the area of drinking water production are being studied.[11]

In biotechnology, MF is especially suitable in cell harvesting, and as a part of a membrane bioreactor, involving a combination of biological conversion and separation.

In the biomedical field, plasmapheresis (which involves the separation of plasma with its value products from blood cells) appears to have an enormous potential.

CONCLUSIONS

In the last 20 yr, new applications of MF have become possible, due to important improvements in membrane fabrication (for e.g., ceramics) and in system operation (for e.g., the use of back pulsing and of low and uniform transmembrane pressure). Membrane fouling is still frequently a major problem (for e.g., in the processing of beverages, such as fruit juices, milk, and beer). A lot of work is being done to combine MF with good pre- and post-treatments and with other separation processes, in order to achieve better and more economic separations.[3]

ACKNOWLEDGMENTS

Data in Table 1 are compilated from Ref. 4—Pinnau, I. Membrane Preparation, In *Encyclopedia of Separation Science*; Wilson, I.D., Ed.; p. 1758, copyright (2000), with permission from Elsevier Science, and from Ref. 6—Mulder, M. *Basic Principles of Membrane Technology*, 2nd Ed., p. 288–292, copyright (1996), with kind permission of Kluwer Academic Publishers.

REFERENCES

1. Zydney, A.L. Membrane Bioseparations. In *Encyclopedia of Separation Science*; Wilson, I.D., Ed.; Academic Press: London, 2000; 1748–1755.
2. Baker, R.W. *Membrane Technology and Applications*; McGraw-Hill: New York, 2000.
3. Huisman, I.H. Microfiltration. In *Encyclopedia of Separation Science*; Wilson, I.D., Ed.; Academic Press: London, 2000; 1764–1777.
4. Pinnau, I. Membrane Preparation. In *Encyclopedia of Separation Science*; Wilson, I.D., Ed.; Academic Press: London, 2000; 1755–1764.
5. Liao, W.C.; Martínez-Hermosilla, A.; Hulbert, G.J. Flux Enhancements in Cross-Flow Microfiltration of Cheese Whey Solutions. Trans. ASAE **1999**, *42* (3), 743–748.
6. Mulder, M. *Basic Principles of Membrane Technology*, 2nd Ed.; Kluwer Academic Publishers: Dordrecht, The Netherlands, 1996.
7. Rane, K.D.; Cheryan, M. Membrane Filtration of Corn Steep Water. Cereal Chem. **2001**, *78* (4), 400–404.
8. Vaillant, F.; Millan, A.; Dornier, M.; Decloux, M.; Reynes, M. Strategy for Economical Optimisation of the Clarification of Pulpy Fruit Juices Using Crossflow Microfiltration. J. Food Eng. **2001**, *48* (1), 83–90.
9. Brandsma, R.L.; Rizvi, S.S.H. Depletion of Whey Proteins and Calcium by Microfiltration of Acidified Skim Milk Prior to Cheese Making. J. Dairy Sci. **1999**, *82* (10), 2063–2069.
10. Saboya, L.V.; Maubois, J.L. Current Developments of Microfiltration Technology in the Dairy Industry. Lait **2000**, *80* (6), 541–553.
11. Zheng-Teng; Jian-Yuan-Huang; Fujita, K.; Takizawa, S. Manganese Removal by Hollow Fiber-Filter. Membrane Separation for Drinking Water. Desalination **2001**, *139* (1/3), 411–418.

Microwave Food Preservation

Ashim K. Datta
Cornell University, Ithaca, New York, U.S.A.

M

INTRODUCTION

Microwave and radio frequency heating refer to the use of electromagnetic waves of certain frequencies to generate heat in a material. Microwave and radio frequency heating for pasteurization and sterilization are preferred for the primary reason that they are *rapid* (as they do not depend as much on the diffusional limitations) and therefore require less come-up time. They can approach the benefits of high temperature–short time processes—thermal degradation of the desired components is reduced (see Fig. 1). The other significant advantage of these processes is that they are relatively *more uniform.* The heating systems can also be turned on or off instantly, which is an advantage. In addition, product can be pasteurized after being packaged, which is a significant advantage. Microwave processing systems can also be more energy efficient.

The temperature rise caused by the energy absorption from microwaves and radio frequency can be used for destruction of micro-organisms. This is the basis of pasteurization and sterilization using these two types of electromagnetic waves. Thus, the mechanism of destruction is heat, exactly as in conventional thermal processing.[1] There has been controversy, however, since the beginning of microwave processing, about the possible nonthermal effects of microwave processing—these are effects unrelated to the lethality caused by the heat. It is generally accepted that the nonthermal effects, even if they exist, are relatively insignificant.

HOW MICROWAVES HEAT FOOD

Microwave and radio frequency heating refer to the use of electromagnetic waves of certain frequencies to generate heat in a material. Typically, food processes involve the two frequencies of 2450 MHz and 915 MHz.[2–4] Of these two, the 2450 MHz frequency is the one most commonly used. Heating in microwave and radio frequencies involve primarily two mechanisms—dielectric and ionic. Water in food is often the primary component responsible for the dielectric heating. Due to their dipolar nature, water molecules try to follow the electric fields as they alternate at the very high frequencies. Such vibrations of the water molecules produce heat. Ions, such as those present in a salty food, migrate under the influence of electric field, generating heat. This is the second major mechanism of heating in microwave and radio frequencies.

Electromagnetic waves consist of electric and magnetic fields. It is the electric field that reacts with dielectric and ionic components of the food. Depending on the design of the heating system, a wave pattern is setup. From an engineering standpoint, the wave pattern in this system can be computed from the Maxwell's equations of electromagnetics, which will provide the electric field E. This electric field E depends on the dielectric properties of the material called the dielectric constant and the dielectric loss. The rate of heat generation per unit volume, Q, at a location inside the food during microwave and radio frequency heating can be characterized by

$$Q = 2\pi f \varepsilon_0 \varepsilon'' E^2 \tag{1}$$

where E is the strength of electric field of the wave at that location, f the frequency of the microwaves or the radio frequency waves, ε_0 the permittivity of free space (a physical constant), and ε'' the dielectric loss representing the material's ability to absorb the wave. The dielectric properties of the food vary with its composition (or formulation), moisture and salt being the two primary determinants. The heat generated in the food, Q, raises its temperature, the exact magnitude of which depends on the duration of heating, the location (e.g., the cold point) in the food, convective heat transfer at the surface, and the extent of evaporation of water inside the food and at its surface. Conceptually, the temperature during the heating process can be obtained from the energy equation

$$\rho c_p \frac{\partial T}{\partial t} = \underbrace{k\nabla^2 T}_{\text{diffusion}} + \underbrace{Q}_{\text{heat generation}} - \underbrace{\lambda I}_{\text{evaporation}} \tag{2}$$

Here, ρ, c_p, and k are the density, specific heat, and thermal conductivity, respectively, of the food being heated. The rate of evaporation at any location in the food is given by I, and λ is the latent heat of water. The difficulty in using Eq. 2 arises due to several factors including: 1) the need to solve Maxwell's equations to obtain the heating

Encyclopedia of Agricultural, Food, and Biological Engineering
DOI: 10.1081/E-EAFE 120007113
Copyright © 2003 by Marcel Dekker, Inc. All rights reserved.

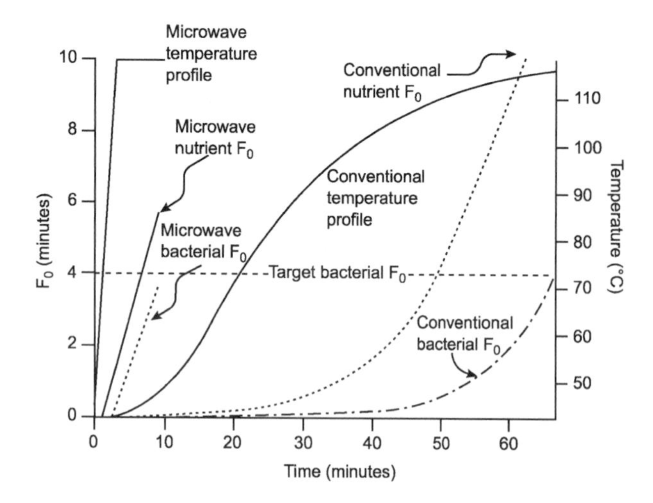

Fig. 1 Quality parameters (nutrient F_0 or cook values) for same target bacterial sterilization (F_0 value of 4) in microwave and conventional heating showing a much greater destruction of nutrients in conventional heating (represented by a substantially higher nutrient F_0). The temperature profiles used are typical of microwave and conventional heating.

rate Q; 2) the difficulty in handling the internal evaporation rate I; 3) complex boundary conditions that include moisture transport at the surface; and 4) coupling of the energy equation with Maxwell's equations as heating changes the dielectric properties. When heating is rapid, sometimes a straight-line temperature profile is obtained, as shown in Fig. 1. Perhaps in such a situation, the diffusion and evaporation are not so significant, i.e., the temperature profile is given by

$$\rho c_{\mathrm{p}} \frac{\mathrm{d}T}{\mathrm{d}t} = Q \tag{3}$$

which is linear with time. If Q is unavailable from computations, it is possible to measure the temperature T as a function of time, t, using fiber-optic temperature sensors readily available (although somewhat expensive). Using the measured temperature, the duration of heating can be established for a desired level of sterilization, as discussed later. Detailed engineering analysis of heat transfer (and associated moisture transfer) in microwave heating of foods is presented in Refs. 5 and 6.

ADVANTAGES OF MICROWAVE AND RADIO FREQUENCY PROCESSING

Microwave and radio frequency heating for pasteurization and sterilization are preferred for the primary reason that they are rapid (as they do not depend as much on the diffusional limitations) and therefore require less come-up time. They can approach the benefits of high temperature–short time processes—thermal degradation of the desired components is reduced (see Fig. 1). The other significant advantage of these processes is that they are relatively more uniform. The heating systems can also be turned on or off instantly, which is an advantage. In addition, product can be pasteurized after being packaged, which is a significant advantage. Microwave processing systems can also be more energy efficient.

PRIMARILY A THERMAL EFFECT

The temperature rise caused by the energy absorption from microwaves and radio frequency can be used for destruction of micro-organisms. This is the basis of pasteurization and sterilization using these two types of electromagnetic waves. Thus, the mechanism of destruction is heat, exactly as in conventional thermal processing.[1] There has been controversy, however, since the beginning of microwave processing, about the possible nonthermal effects of microwave processing—these are effects unrelated to the lethality caused by the heat. It is generally accepted that the nonthermal effects, even if they exist, are relatively insignificant.

CALCULATION OF STERILIZATION

As thermal effect is the sole lethal mechanism assumed in this processing technology, temperature–time history at the coldest location determines the microbiological safety of the process. Once temperature is known at the coldest point as a function of time, accumulated lethality can be calculated following the well-known equation

$$F_0 = \int_0^{t_{\mathrm{p}}} 10^{(T-121.11)/Z}\, \mathrm{d}t \tag{4}$$

where T is the cold point temperature at any time t, Z the Z-value in °C, and t_{p} the total duration of heating. For a desired process lethality, F_0, the process time t_{p} can be found from Eq. 4 using measured or computed temperature T as a function of time. The desired process lethality is based on the same guidelines as for conventional thermal processing, set by the Food and Drug Administration in the United States. Both the magnitude of temperature–time history and the location of the cold point are functions of composition (ionic content, moisture, density, and specific heat), shape and size, microwave frequency, and the applicator (oven) design. There are major differences between conventional and microwave heating in terms of the location of the cold point and how temperature–time history there is affected by a number of critical process factors, which are now discussed.

NATURE OF MICROWAVE HEATING CONTRASTED WITH CONVENTIONAL HEATING

Temperature–time history at the coldest point for a conventional thermal process is quite predictable. For example, for a conduction-heated (solid) food, it is usually the geometric center. In microwave heating, the coldest point is difficult to predict and it can even change during the heating process (see Fig. 2), depending on a number of food and oven factors. As the coldest point location and its temperature–time history are not easily predictable, it is quite difficult (if not impossible) to develop simple procedure such as the Ball formula for process calculations.

Initially, at lower temperature, microwave absorption is lower. Thus, the waves are able to penetrate further into the material. As the material heats up, it absorbs microwaves more readily and the waves are not able to penetrate as far. Especially in foods with significant ions, the surface at higher temperatures can act as a shield.

As heat is constantly generated everywhere in the food, but at different rates, the difference between the temperature at the coldest and the warmest point in the food, i.e., the nonuniformity of temperatures in the food, keeps increasing with time. This is unlike conventional heating where the coldest point approaches the warmest temperature of the system (typically the heating medium temperature) with time. Eventually, however, temperatures in microwaves tend to approach

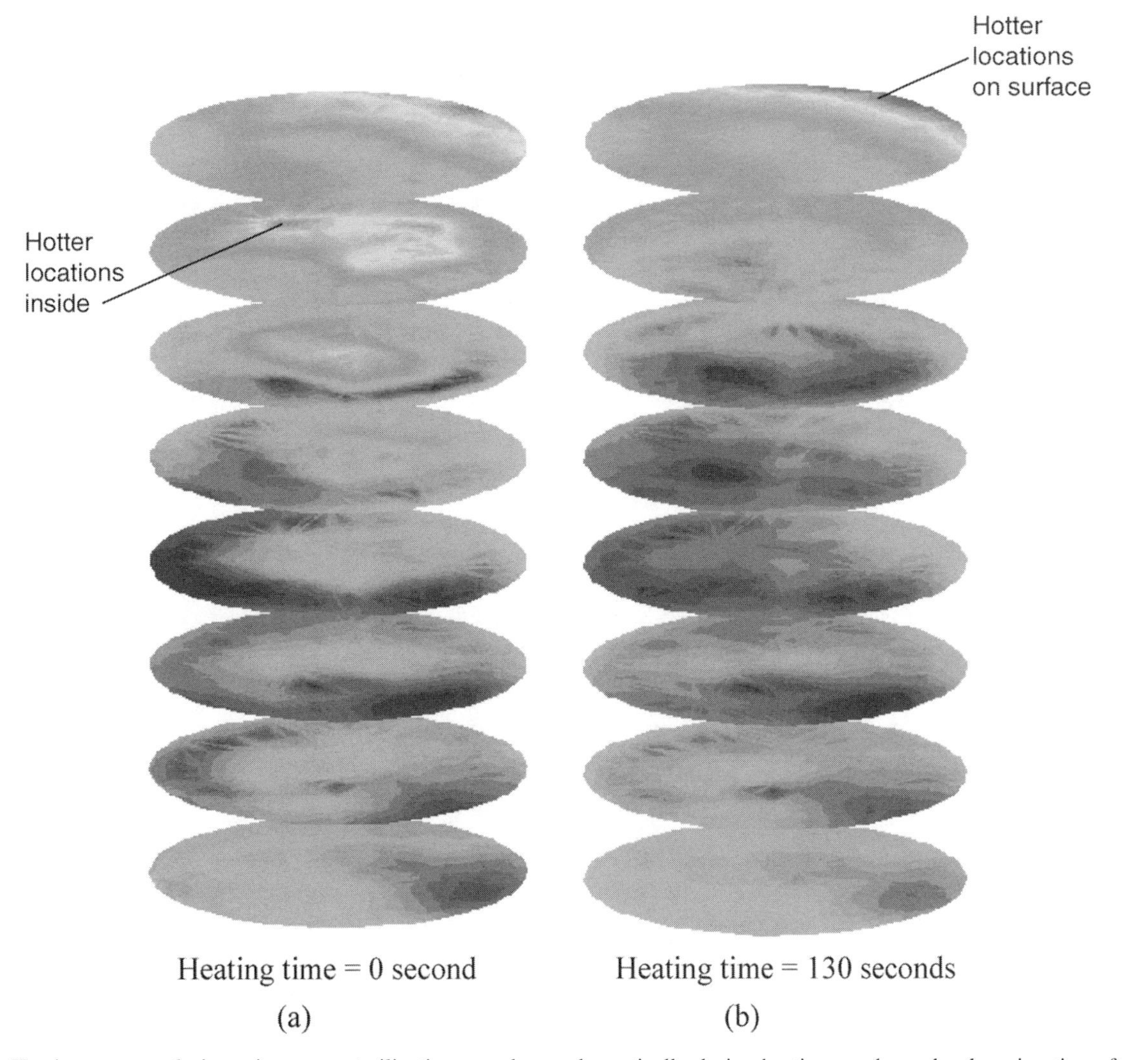

Fig. 2 Heating patterns during microwave sterilization can change dramatically during heating, as shown by the migration of hottest locations (in white) from (a) interior to (b) surface. Shown are the computed results for a ham cylinder (0.7% salt) heated in a microwave oven similar to domestic microwave ovens. (From Ref. 22.)

the boiling temperature and therefore approach uniformity.

As food is heated more uniformly, its sensory quality increases as overprocessing can be decreased. In microwave heating, the two most common ways to improve uniformity of heating are to use lower power (or power cycling) and provide for oscillatory physical movement such as that provided by the turntable in the home microwave oven.

In conventional heating, the surface is at the highest temperature. In microwave heating, typically the food heats up while the surrounding air stays cold. The cold air keeps the surface temperature lower than locations near the surface of food. Surface evaporation, especially when heating an unpackaged food, can also decrease the surface temperature significantly.

Come-up time in microwave heating is short. Shorter come-up time can retain more of the sensory qualities as the total time of heating is reduced. This is often the basis for preferring microwaves to conventional heating. In calculating process time, the come-up time cannot be given nearly so much importance in microwave heating as in conventional heating.

PRODUCT AND PACKAGE FACTORS

Food shape, volume, surface area, and composition are critical factors in microwave heating.[7–9] These factors can affect the total energy absorption as well as its spatial distribution (i.e., location of cold point), leading to effects such as corner and edge overheating, focusing, and resonance. Composition, in particular, moisture and salt have a much greater influence on microwave processing than on conventional processing, due to its influence on dielectric properties.

Multicomponent foods such as multicompartment frozen dinners make different food components heat differently.[7,10] Packaging material is also a critical process factor. Packaging materials for in-package pasteurization or sterilization need to be microwave transparent and have a high melting point. Metal reflects microwaves. Packages with some metal component such as aluminum foil and susceptor can change the food temperatures considerably.

PROCESS AND EQUIPMENT FACTORS

Several process and equipment factors are critical in microwave heating.[7] Design (size, geometry, etc.) of the microwave oven can completely change the magnitude and spatial variation of power absorption in the food. Presence or absence of devices that are added to improve uniformity, such as mode stirrers or turntables, are major factors. Placement of the food inside the oven can also have significant influence on the magnitude and uniformity of power absorption.

Heating of containerized liquid in a microwave without agitation causes flow of the liquid and its thermal stratification inside the container. Warmer liquid moves to the top, as in conventional heating. Combined with product (such as viscosity), package (such as shielding with aluminum foil), and equipment factors, temperature in heating of liquids can be quite complex and cold points need to be determined experimentally for each situation.[11–13]

Frequency of the microwaves can dramatically change the heating rates and their spatial distribution. Taking a simplified view, a lower frequency of 915 MHz has a higher depth of penetration than the 2450 MHz that is used in home microwaves.

USE OF MATHEMATICAL SIMULATION IN STUDYING THE COMPLEX HEATING PROCESS

Due to the complexity of the system where heating pattern depends on such a large number of factors, simulation-based design can save significant time and resources in developing microbiologically safe processes. State-of-the-art commercial software simulating the electromagnetics and heat transfer has been used for microwave food process design.[14,15] Such software can provide a comprehensive insight into the heating process by showing interior power depositions (heating rates) in a 3-D object, as shown in Fig. 2. Details of heating, as shown in Fig. 2, are not possible using experimentation. Simulation-based design can allow the process and equipment designers to avoid obviously wrong combinations of food and process parameters. Location of the cold point (the critical process factor) and its time–temperature history can be designed this way.

MICROBIOLOGICAL ISSUES

Several microbiological issues associated with microwave pasteurization and sterilization are discussed in Refs. 1 and 16. Evaluation of microwave sterilization, including alternatives to microbiological studies,[17] are also discussed in Ref. 1.

INDUSTRIAL PASTEURIZATION AND STERILIZATION SYSTEMS

Industrial microwave (and radio frequency) pasteurization and sterilization systems have been reported off and on for over 30 yr (see, e.g., Ref. 18). Implementation of a microwave sterilization process can vary significantly with various manufacturers. Unlike conventional heating, the design of the equipment can have more dramatic influence on the critical process parameter—temperature at the coldest point and its location. At the time of writing, there are two companies producing pasteurized and sterilized food using microwave heating (see, e.g., Ref. 19). A major industry–university collaboration to further commercialize this heating technology is also underway at the Washington State University (see Ref. 20).

ACKNOWLEDGMENTS

Much of the engineering guidelines presented here are abridged from the work of Datta and Davidson[1] with the permission of the Institute of Food Technologists, Chicago.

REFERENCES

1. Datta, A.K.; Davidson, M. Kinetics of Microbial Inactivation for Alternative Processing Technologies: Microwave and Radio-Frequency Processing. J. Food Sci. **2000**, *65* (8), 32S–41S.
2. Mudgett, R.E.; Schwartzberg, H.A. Microwave Food Processing: Pasteurization and Sterilization. A Review. AIChE Sym. Ser. **1982**, *218* (78), 1–11.
3. Decareau, R.V. Pasteurization and Sterilization. *Microwaves in the Food Processing Industry*; Academic Press, 1985; 182–202.
4. Buffler, C.R. *Microwave Cooking and Processing: Engineering Fundamentals for the Food Scientist*; Van Nostrand Reinhold: New York, 1993.
5. Datta, A.K. Fundamentals of Heat and Moisture Transport for Microwave Processing of Foods. In *Handbook of Microwave Technology for Food Applications*; Datta, A.K., Anantheswaran, R.C., Eds.; Marcel Dekker, Inc.: New York, NY, 2001.
6. Datta, A.K. Analysis of Microwave Heating of Foods. In *Food Processing Operations Modeling: Design and Analysis*; Irudayaraj, J., Ed.; Marcel Dekker, Inc.: New York, NY, 2001.
7. Zhang, H.; Datta, A.K. Electromagnetics of Microwave Heating: Magnitude and Uniformity of Energy Absorption in an Oven. In *Handbook of Microwave Technology for Food Applications*; Datta, A.K., Anantheswaran, R.C., Eds.; Marcel Dekker, Inc.: New York, NY, 2001.
8. Fleischman, G.J. Predicting Temperature Range in Food Slabs Undergoing Long-Term Low Power Microwave Heating. J. Food Eng. **1996**, *27* (4), 337–351.
9. Fakhouri, M.O.; Ramaswamy, H.S. Temperature Uniformity of Microwave Heated Foods as Influenced by Product Type and Composition. Food Res. Int. **1993**, *26*, 89–95.
10. Ryynanen, S.; Ohlsson, T. Microwave Heating Uniformity of Ready Meals as Affected by Placement, Composition and Geometry. J. Food Sci. **1996**, *61* (3), 620–624.
11. Prosetya, H.; Datta, A.K. Batch Microwave Heating of Liquids: An Experimental Study. J. Microw. Power Electromagn. Energy **1991**, *26* (3), 215–226.
12. Anantheswaran, R.C.; Liu, L.Z. Effect of Viscosity and Salt Concentration on Microwave Heating of Model Non-Newtonian Liquid Foods in Cylindrical Containers. J. Microw. Power Electromagn. Energy **1994**, *29*, 127.
13. Knutson, K.M.; Marth, E.H.; Wagner, M.K. Use of Microwave Ovens to Pasteurize Milk. J. Food Protect. **1988**, *51* (9), 715–719.
14. Dibben, D. Electromagnetics: Fundamentals and Numerical Modeling. In *Handbook of Microwave Technology for Food Applications*; Datta, A.K., Anantheswaran, R.C., Eds.; Marcel Dekker, Inc.: New York, NY, 2000.
15. Burfoot, D.; Railton, C.J.; Foster, A.M.; Reavell, R. Modeling the Pasteurization of Prepared Meal with Microwaves at 896 MHz. J. Food Eng. **1996**, *30*, 117–133.
16. Heddleson, R.; Doores, S. Factors Affecting Microwave Heating of Foods and Microwave Induced Destruction of Foodborne Pathogens—A Review. J. Food Protect. **1994**, *57* (11), 1023–1025.
17. Prakash, A.; Kim, H.J.; Taub, I.A. Assessment of Microwave Sterilization of Foods Using Intrinsic Chemical Markers. J. Microw. Power Electromagn. Energy **1997**, *32* (1), 50–57.
18. Schlegel, W. Commercial Pasteurization and Sterilization of Food Products Using Microwave Technology. Food Technol. **1992**, *46* (12), 62–63.
19. Tops, R. Personal communication. Tops Foods, Olen, Belgium, 2000.
20. Lau, M.H.; Tang, J. Pasteurization of Pickled Asparagus Using 915 MHz Microwaves. J. Food Eng. **2002**, *51*, 283–290.
21. Datta, A.K.; Hu, W. Quality Optimization of Dielectric Heating Processes. Food Technol. **1992**, *46* (12), 53–56.
22. Zhang, H.; Datta, A.K.; Taub, I.A.; Doona, C. Electromagnetics, Heat Transfer, and Thermokinetics in Microwave Sterilization. Am. Inst. Chem. Eng. J. **2001**, *47* (9), 1957–1968.

Mixing of Air

Yanbin Li
University of Arkansas, Fayetteville, Arkansas, U.S.A.

INTRODUCTION

Mixing of air is practiced in air conditioning, grain drying, food processing, and bioprocessing. In many cases, a hot/wet air stream is partially reused and mixed with fresh air.[6] For example, the use of recycled air during a drying process results in a less expensive operation. Air is used as a heat-transfer medium, a source or sink for water vapor, a source of oxygen for micro-organisms, and a vehicle for vapors to be removed or used as media during processing.[4] The relationships of pressure, volume, and temperature in air mixtures are defined by the ideal gas law and partial pressure and partial volume laws. Based on the concepts of mass and heat transfer balances, a psychrometric chart is available and commonly used to solve the problems in mixing of air–water vapor. Mixing of two air streams is discussed for theories, calculation, and outcomes of adiabatic and nonadiabatic air mixing.

AIR MIXTURES

Composition of Dry Air

Dry air is defined as a mixture having the composition by volume: nitrogen, 78.09%; oxygen, 20.95%; argon, 0.93%, carbon dioxide, 0.03%, and other minor components (Table 1). Usually, air is treated as a mixture of many gases, principally a mixture of nitrogen and oxygen with some water vapor. Very often, air is also treated as a single gas or substance with a molecular weight of 28.97 in air–water vapor mixtures.

The Ideal Gas Law

The ideal gas law is used to mathematically predict the behavior of air mixtures in terms of pressure, volume, mass, and thermal properties. The pressure–volume–temperature relationship of air mixtures can be expressed as:

$$pV = nRT \tag{1}$$

and

$$p_1V_1/T_1 = p_2V_2/T_2 = p_3V_3/T_3 = \cdots p_nV_n/T_n \tag{2}$$

where p is the pressure (N m^{-2}), V the gas volume (m^3), T the absolute temperature (K), n the number of moles (kg mol), R the universal gas constant [8314.34 N m (kg mol K)$^{-1}$], and subscripts 1,2,3…n refer to each individual component in the air mixture.

In Eq. 1, the gas constant is dependent upon the term dimensions, and in Eq. 2, any consistent set of terms can be applied as it is based on a ration. The ideal gas law does not hold perfectly for extreme conditions. However, it is entirely suitable for most conditions in food and biological processing.[3]

Pressure of a Mixture

If the gas mixture is homogeneous and the component gas molecules are in constant random motion, Dalton's law of partial pressures is stated as: 1) each gas behaves as if it alone occupies the volume; 2) each gas behaves as if it exists at the temperature of the mixture and fills the entire volume by itself; and 3) the pressure of the gas mixture is the sum of the pressures of each of its components when each behaves as described in statements 1 and 2. Each component in a mixture of gases exerts the same pressure it would exert if it alone occupies the same volume at the same temperature, expressed as follows:

$$p_t = p_1 + p_2 + p_3 + \cdots p_n \tag{3}$$

where p_t refers to the total pressure of a mixture of gases.

Volume of a Mixture

The Amagat's law of partial volumes states that the volume of a mixture of gases at a certain temperature and pressure is equal to the sum of the volumes of the individual gases at the same conditions as follows:

$$V_t = V_1 + V_2 + V_3 + \cdots V_n \tag{4}$$

where V_t is the total volume of the mixture after mixing. Using the ideal gas law, the general expression for V is:

$$V = (mRT)/(pM) \tag{5}$$

where m is the mass of gas (kg of dry air) and M the molecular weight of gas (kg kg^{-1} mol^{-1}).

Encyclopedia of Agricultural, Food, and Biological Engineering
DOI: 10.1081/E-EAFE 120007056
Copyright © 2003 by Marcel Dekker, Inc. All rights reserved.

Table 1 The composition of dry air

Constituent	Chemical formula	Molecular mass ($kg\,kg^{-1}\,mol^{-1}$)	Volume fraction in air ($m^3\,m^{-3}$)
Air		28.97	1.0000
Ammonia	NH_3	17.02	0.0000
Argon	Ar	39.90	0.0093
Carbon dioxide	CO_2	44.00	0.0003
Carbon monoxide	CO	28.00	0.0000
Helium	He	4.00	0.0000
Hydrogen	H_2	2.02	0.0000
Nitrogen	N_2	28.02	0.7809
Oxygen	O_2	32.00	0.2095
Water vapor	H_2O	18.01	—

AIR–WATER VAPOR MIXTURES

Psychrometric Properties

The psychrometric properties of the air–water vapor mixtures include dry bulb temperature, wet bulb temperature, and dew point temperature, relative humidity, humidity ratio, enthalpy, density or specific volume, and atmospheric pressure. A psychrometric chart is a graphical representation of these properties calculated using basic thermodynamic equations. The tables of psychrometric properties and psychrometric chart are available from several sources including *ASHRAE Handbook of Fundamentals*.[1]

A state point, any point on the psychrometric chart, can be defined and located by any two properties for a mixture of air–water vapor. Once the state point is determined, the values of all other properties can be obtained.[1] Therefore, problems in air–vapor mixtures, including heating, cooling, humidification, dehumidification, and mixing can be solved using the psychrometric chart.

Adiabatic Mixing of Two Air Streams

The mixing of two air streams at different initial conditions is commonly used in food and biological processing, as schematically shown in Fig. 1. In an adiabatic mixing process, the heat, q, is not added to or removed from the mixing system. The mass balance for dry air is:

$$m_{a1} + m_{a2} = m_{a3} \tag{6}$$

where m_a is the mass of dry air (kg of dry air) and 1, 2, and 3 refer to original (unmixed) air with properties at states 1 and 2, and to air properties of the mixed mass at state 3. The mass balance for water vapor is:

$$m_{a1}W_1 + m_{a2}W_2 = (m_{a1} + m_{a2})W_3 \tag{7}$$

where W is the humidity ratio (kg H_2O/kg^{-1} of dry air). If velocity and elevation terms are negligible and no work enters or leaves the system, the energy balance is:

$$m_{a1}h_{a1} + m_{a2}h_{a2} = (m_{a1} + m_{a2})h_{a3} \tag{8}$$

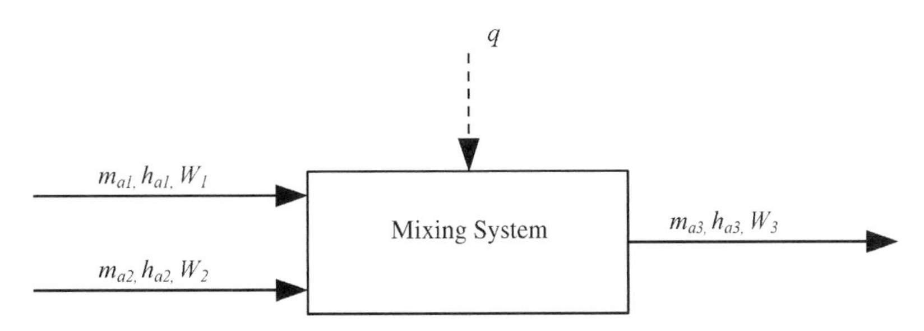

Fig. 1 Mixing of two air streams at two different states 1 and 2 to a final mixture at state 3. For an adiabatic mixing process, the heat q is not added to or removed from the system.

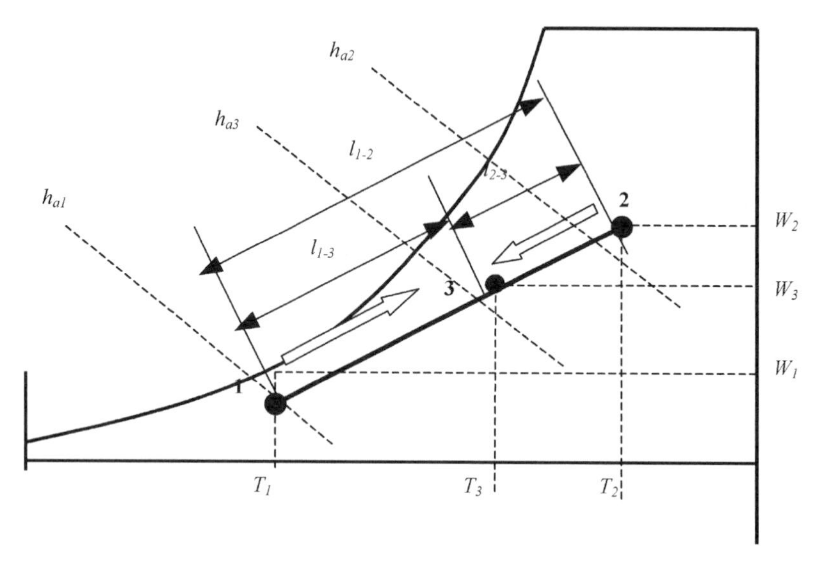

Fig. 2 Mixing of two air streams at two different original states 1 and 2 to a final mixture at state 3.

where m_a is the mass of dry air (kg of dry air). The enthalpy differences between the mixture and each of the sources of air are proportional to the flow rate in a continuous-flow process.[2] A combination of Eqs. 6–8 results in:

$$(h_{a2} - h_{a3})/(h_{a3} - h_{a1}) = (W_2 - W_3)/(W_3 - W_1)$$
$$= m_{a1}/m_{a2} = l_{2-3}/l_{1-3} \quad (9)$$

where h_a is the specific enthalpy (N m kg^{-1} of dry air), and l_{2-3} and l_{1-3} are the corresponding lines shown in Fig. 2. When a continuous-flow mixing process is considered, the mass flow rates, $\dot{m}_{a1}$ and $\dot{m}_{a2}$, may be used to substitute the masses, m_{a1} and m_{a2}, in Eqs. 6–9.

The state point 3 of an air–vapor mixture resulting from mixing airs of different state points 1 and 2 falls very nearly on a straight line connecting the two initial states as shown in Fig. 2. In other words, when mixing two air streams, each at different temperatures and pressures, the resultant mixture should have the temperature and moisture properties that locate somewhere between the temperatures and humidities of the two unmixed streams. As shown in Fig. 2, the state point of the resultant mixture, point 3, is on the straight line connecting the state points of the two air streams, points 1 and 2.

Quantitatively, the mass properties of mixed air are closer to the properties of the unmixed air stream that contributes more mass to the mixture. On the psychrometric chart (Fig. 2), if the length of the line connecting the two state points 1 and 2 is l_{1-2}, the state point 3 of the mixed air can be located at a length of $(m_{a2}/(m_{a1} + m_{a2}))l_{1-2}$ from state point 1 according to Eq. 9.

Outcomes of Air Mixing Process

Three possible outcomes can be expected from the mixing process.[5] The first possible outcome is that the resultant mixture is not saturated when the line connecting state points 1 and 2 of the unmixed air streams, l_{1-2}, is located completely in the psychrometric chart (below the saturation curve). The second possible outcome is that when the line, l_{1-2}, just touches the saturation curve (100% relative humidity) of the psychrometric chart, the mixture may leave the process at the saturation and with condensation imminent. The third possible outcome is from the case when part of the connecting line, l_{1-2}, is located beyond the saturation curve, which is due to mixing of very warm air with very cold air, and both with high moisture contents. When the state point of the mixed air is predicted to be above the saturation curve, the mixture becomes thermodynamically unstable, resulting in condensation.

In the mixing of three or more air streams, determination of the state point for the mixture requires that any two state points be used first, with the resultant mixture combined with the third state point, etc.

Nonadiabatic Mixing

Problems in nonadiabatic mixing of two air streams can be solved using the method similar to that for adiabatic mixing. All the equations used for the adiabatic mixing remain the same except for the energy balance[7] as shown in Fig. 1 and the following equation:

$$q + m_{a1}h_{a1} + m_{a2}h_{a2} = (m_{a1} + m_{a2})h_{a3} \quad (10)$$

where q is the heat added or removed per unit mole leaving. The rates of masses and heat, $\dot{m}_{a1}$, $\dot{m}_{a2}$ and $\dot{q}$, may be used in Eq. 10 for a continuous mixing process.

REFERENCES

1. ASHRAE. *ASHRAE Handbook 1997 Fundamentals*; The American Society of Heating, Refrigerating and Air-Conditioning Engineers, Inc.: Atlanta, GA, 1997.
2. Cook, E.M.; DuMont, H.D. *Process Drying Practice*; McGraw-Hill: New York, NY, 1991.
3. Heldman, D.R., Lund, D.B., Eds. *Handbook of Food Engineering*; Marcel Dekker: New York, NY, 1992.
4. Henderson, S.M.; Perry, R.L. *Agricultural Process Engineering*; The AVI Publishing Company: Westport, CN, 1976.
5. Johnson, A.T. *Biological Process Engineering*; John Wiley & Sons: New York, NY, 1999.
6. Toledo, R.T. *Fundamentals of Food Process Engineering*; An Aspen Publication: Gaithersburg, MD, 1999.
7. Wark, K. *Advanced Thermodynamics for Engineers*; McGraw-Hill: New York, NY, 1995.

Modified Atmosphere Packaging

Aaron L. Brody
Packaging/Brody, Inc., Duluth, Georgia, and
The University of Georgia, Athens, Georgia, U.S.A.

INTRODUCTION

Modified atmosphere packaging is a relatively new term referring to a series of technologies intended to prolong the quality retention of foods, beverages, and other contained products. Coined about 20 years ago, the term derived from the original "controlled atmosphere" preservation to differentiate it from methods developed to extend shelf life of fresh produce in distribution environments as the latter employed mechanical and chemical procedures to literally control the temperature and gaseous environment of the product. Because gas-permeable plastic package materials are always used to contain respiring food or food containing respiring micro-organisms in modified atmosphere packaging, total control was lost and thus the descriptor, "modified," to convey the notion of dynamic change. Some professionals included vacuum packaging of respiring foods or foods containing viable micro-organisms such as primal cuts of fresh red meat within the modified atmosphere term because, despite original reduced oxygen, biological action by the food continued to alter the internal package environment. One area generally not encompassed by modified atmosphere packaging is reduced oxygen in foods and other products that are biologically (but not biochemically) inert such as instant coffee or canned foods.

Many other terms such as microenvironmental control and Tectrol® have been applied by various people and organizations, and modified atmosphere has been used across a far broader field, but most professionals working within the discipline accept the definition of altering the internal gaseous environment of packages under controlled temperature conditions to beneficially retain initial microbiological, enzymatic, and/or biochemical quality. Although some would argue the concept might be better described by "reduced oxygen," too many recent developments such as with fresh red meat color retention and fresh produce quality retention have clearly indicated that elevated oxygen may also impart beneficial shelf life effects.

HISTORY

Classicists usually date the beginnings of modified atmosphere packaging to the publications of Kidd and West for meats and Smock for apples, but, however important these peer review efforts were, these represent post facto reflections of commercial actions arising from empirical observations. Carcass meat shelf life in shipboard distribution from the Antipodes to England during the early 20th century was extended by the presence of carbon dioxide. Closed warehouses in upstate New York were employed for apple storage employing normal fruit respiration to reduce oxygen and increase carbon dioxide.

The real commercial beginnings, however, date from research conducted on fresh meats and produce by professionals at Whirlpool Corporation during the 1950s to enhance refrigeration as a food preservation technology. The resulting patents and information under the Tectrol name were employed to construct warehouses and later truck transports for preservation of apples, fresh head lettuce, and a host of other fresh foods. During the 1960s, Whirlpool Corporation spun off the technology to a joint venture company called Transfresh, a major supplier of transport control for fresh food and the predecessor to today's Fresh Express, the world's leading producer of fresh cut vegetables. Fresh Express also holds the basic patents describing the interactions of package structures and respiring fresh food contents. Thus, modified atmosphere packaging of fresh food in the 21st century traces its lineage directly to Whirlpool Corporation's original 1950s research.

Paralleling the Whirlpool Corporation efforts (which culminated in a 1964 Institute of Food Technologists' Industrial Achievement Award) was work by the ancestor company, what is now Cryovac Sealed Air, on vacuum packaging of poultry and later primal cuts of fresh beef and pork to extend refrigerated shelf life. The now mainstream "Barrier Bag" or "boxed beef" concept represented the first significant change in fresh meat preservation since the invention and application of mechanical refrigeration, and truly, totally revolutionized the world meat industry by providing means to deliver safe and high quality fresh meat to retailers and hotel/restaurant/institutional outlets for reduction into consumer cuts. The era of handling open-carcass beef and pork ended only gradually because of resistance to change but, for practical purposes, Cryovac technology of total systems of controlled temperature, gas barrier plastic packaging, and vacuum packaging equipment has become the mainstream method of handling beef, pork, and other fresh meats. These technologies, in turn, spawned reduced oxygen packaging for retail cuts of fresh meat and

Encyclopedia of Agricultural, Food, and Biological Engineering
DOI: 10.1081/E-EAFE 120007139
Copyright © 2003 by Marcel Dekker, Inc. All rights reserved.

prepared foods, and high oxygen packaging of fresh red meats.

During the 1960s, research in the United Kingdom Chorleywood Research on preservation of soft bakery goods resulted in the application of elevated carbon dioxide to retard mold growth on the products. This effort was applied in Europe to avoid the introduction of preservatives whose presence had to be declared on bread, pastry, and cake labels. In a social environment that rejected food additives, the alternative that delivered better results without the undesirable stigma of "preservatives" was welcomed by bakers who rapidly implemented the technology.

Advances in altered gaseous environment technologies to extend shelf life of fresh and minimally processed foods have increased during the last years of the 20th century with concepts such as hot fill, pasteurization, hurdle technologies, and the introduction of other gases. A cadre of organizations led by Transfresh and Cryovac have been instrumental in sparking and performing the research and development that has caused modified atmosphere packaging to quietly leap to the leading position among food preservation processes in this 21st century: CVP, M-Tek, Multivac, Tiromat, Harpak, Ross Reiser, MAP Systems, Campden and Chorleywood Food Research, Landec and University of California Davis.

Surprising to many is that modified atmosphere packaging in its various manifestations is now well ahead of the more widely publicized canning, freezing, aseptic packaging, and retort pouch and tray packaging in terms of volume of food preserved. Vacuum barrier bag packaging of fresh red meat primal cuts for distribution packaging is employed for virtually all meat in industrialized societies. Much of our poultry and all the growing ground poultry is packaged under some form of altered gas. Fresh cut produce has completely changed the complexion of the fresh produce industry and is one major reason for the major increase in per capita consumption of fresh vegetables and fruit. Minimally processed foods represent the newest entry for modified atmosphere packaging in foods such as moist pasta, entrees, and lunch kits, and is partially responsible for the remarkable, almost exponential increase in "home replacement meals" in the industrialized dietary. And modified atmosphere packaging is the basis for all of the many alternative technologies offered for case ready fresh meat.

PRINCIPLES

All foods deteriorate by microbiological, enzymatic, biochemical, and/or physical means, losing quality, spoiling, or even becoming potentially hazardous for consumption. Virtually all the deteriorative vectors are accelerated by temperature following Arrenhius equations, i.e., increasing exponentially with linear increases in temperature. Further, most deteriorative reactions are fundamentally oxidative, meaning they generally require oxygen to occur. Many deteriorative reactions involve aerobic respiration in which the food or micro-organism consumes oxygen and produces carbon dioxide and water. By reducing oxygen, or significantly changing its concentration away from optimum of atmospheric 20.9% in the tissue, aerobic respiration can be slowed, with the rate of decrease increasing with reduction of temperature.

Either simultaneously or independently, increasing carbon dioxide leads to mass action reaction rate reduction, coupled with direct alteration of pH due to dissolution of the gas in tissues. The rate of dissolution of gas increases with reduced temperature explaining the benefit of maintaining reduced temperatures for modified atmosphere packaging. Thus, one general premise of modified atmosphere food packaging is that the lowest temperature above freezing should be maintained for the gas alteration to provide any beneficial effect; modified atmosphere packaging requires initial quality and maintenance of reduced temperature. In fact, many benefits of modified atmosphere packaging are primarily due to the application of effective temperature control simultaneously with the gas atmosphere changes.

Similarly, because water enters into respiration reactions, the presence of high water vapor both slows aerobic respiration by mass action laws and retards moisture evaporation that, of course, leads to quality loss.

Unfortunately, not all fresh and minimally processed food micro-organisms and enzyme systems respond equally to altered oxygen and carbon dioxide, and some do not respond measurably within the parameters of commercial need. For example, produce with origins in tropical regions usually requires temperatures of about 10°C to avoid chill damage, and so the rule of low temperature does not apply. Although each different fruit and vegetable exhibits oxygen/carbon dioxide concentrations optimum to preservation, measurable benefits can be derived from reducing the oxygen and increasing the carbon dioxide concentrations, provided, of course, temperature is reduced to optimum levels. Although optimum shelf life extensions are not achieved, some advantage occurs over just temperature control. Modified atmosphere packaging is capable of multiples of shelf life of 2–10 times over equivalent commercial practice for fresh produce.

Fresh and minimally processed produce is subject to undesirable anaerobic respiration if the oxygen concentration is reduced to near zero either by intentional alteration or by permitting the natural aerobic respiratory processes to proceed. Although the deteriorative reactions are slowed by reduced oxygen (and accompanying elevated carbon dioxide) the secondary respiratory anaerobiosis becomes dominant and so should be obviated. This problem has tended to dominate modified atmosphere packaging research and development. Solutions, such as high gas permeability package materials (e.g., ethylene vinyl acetate), micropores, doping with mineral fill, actually puncturing the material and side chain crystallizable polymers that melt at specific temperatures to increase gas and therefore oxygen permeability, have been introduced. Most have demonstrated to be somewhat effective, but only the Landec side chain crystallizable polymers have proven to provide reversible and safe mechanisms for high respiration rate fresh and minimally processed produce.

Related in principle is the hazard arising from the potential growth of pathogenic anaerobic micro-organisms in low acid, high water activity foods such as minimally processed entrees for home meal replacement, and especially those receiving mild heat "pasteurization" heat treatments to reduce aerobic microbial spoilage. Application of modified atmosphere packaging to prolong quality retention often leads to anoxic conditions within the product particularly when oxidation reactions consume residual oxygen that is not replenished. With the paucity of aerobic micro-organisms due to the heat treatment or modified atmosphere conditions, little competition exists for anaerobic pathogens whose growth is fostered by the reduced oxygen conditions. If, as too often occurs, the temperature rises above 3°C, some anaerobic pathogens can grow and produce toxin and not signal their hazardous condition by odor or appearance. This phenomenon, although rare in commerce, has led to hesitation by processor/packagers and concern by regulatory officials to encourage the application of reduced oxygen modified atmosphere packaging for home meal replacement foods. The judicious introduction of hurdle technologies, coupling reduced pH and water activity with clean room processing plus good temperature and distribution control, has helped to significantly reduce the potential hazard.

COMMERCIAL APPLICATIONS

Red Meat

Fresh red meat contains aerobic spoilage micro-organisms whose growth is significantly slowed by reduced temperature, reduced oxygen from vacuum in gas barrier packaging, and elevated carbon dioxide resulting from normal respiratory processes. The presence of a natural aerobic microbiological load is regarded as sufficient to obviate anaerobic pathogenic microbial growth. About 40% of fresh beef is in the ground form with heavy microbiological counts throughout the meat. Ground beef is pressure stuffed into "keeper casings" to force out most of the air for reduced oxygen and eventually elevated carbon dioxide distribution. At retail level, the gas barrier flexible packages are opened to expose the meat to air and to thus oxygenate the pigment to the desired bright "cherry" red color desired by most consumers.

Centrally packaged meat has been under development for more than 40 yr with relatively little market penetration to date apparently due to market resistance. Whether or not case ready beef will become a major product in the coming years continues to be a subject of considerable debate. The North American conversion of pork to case ready under a version of modified atmosphere packaging is well underway. Perhaps 30 different technologies to deliver safe red retail cuts to stores for immediate consumer display, all variations of modified atmosphere packaging, have been offered. Color is a major challenge since the desirable oxymyoglobin red color is elusive and vulnerable to oxidation to undesirable brown metmyoglobin. Microbiological growth must be retarded to be able to achieve one or more weeks shelf life required for distribution from central locations to labor-intensive retail store backroom cutting and packaging.

In Europe where centralized red meat packaging is well established, the most widely used technology is high oxygen/high carbon dioxide. The 70%–80% oxygen level retains the red color while the 20+% carbon dioxide suppresses microbiological growth under refrigerated distribution. Elevated oxygen leads to fat oxidation and thus limits the time. Although used in North America, high oxygen packaging is hardly the technology of choice since the time is perceived to be too short for the long distances prevalent in North America.

To attempt to achieve the two or more weeks desired, reduced oxygen packaging in either master packs containing several retail primary packages or barrier primary packages are employed. The retail versions may be under vacuum or under elevated carbon dioxide. In the retail store, the package gas barrier is compromised thus exposing the red meat to air (or even oxygen) and subsequently rejuvenating the desired bright red color.

Poultry

Case ready poultry has been a commercial reality in North America since the 1980s and is often credited with being a major contributory factor to the increase in poultry consumption to rival that of red meat. Some primary packaged intact pieces of poultry are master packed under elevated carbon dioxide to prolong controlled refrigeration shelf life to more than two weeks. All the increasingly popular ground poultry is packaged under elevated carbon dioxide in gas barrier primary packaging such as barrier film laminated trays or expanded polystyrene trays overwrapped with barrier shrink film.

Seafood

Relatively little seafood in the United States is packaged under modified atmosphere because of a regulatory fear that pathogenic anaerobic micro-organisms indigenous to fish could grow and produce toxin without overt spoilage to signal consumers that the product is hazardous. The application of modified atmosphere packaging for seafood is much more prevalent in Europe signifying that the technology is sound under controlled distribution conditions.

Produce

Perhaps the most dramatic marketing success of the 1990s in industrialized nations was the growth of fresh cut vegetables, a visible reminder of the application of modified atmosphere packaging systems. Begun as a technology for intact vegetables and fruits in distribution channels, altered gaseous environment preservation with initial gas mixtures specific for each type of produce moved into vacuum packaging for cut salad vegetables for quick service restaurants during the 1970s. Initial vacuum or reduced oxygen permits aerobic respiration of both produce and contained micro-organisms, with consequent production of carbon dioxide that is effective in suppressing aerobic microbial growth. This technology was succeeded by injection of either a carbon dioxide gas mix or nitrogen into the package to more rapidly permit the internal gas to reach an equilibrium nearly optimum for the shelf life extension.

Obviation of respiratory anaerobiosis was first accomplished by actually puncturing the plastic film packages and later by very significantly increasing the surface to volume ratio of the film to increase the total entry of air. This method was succeeded by employing film materials engineered for high gas permeation rates that would permit reaching equilibrium internal gas ratios within a day of packaging. Although not ideal, these package structures, coupled with good quality raw material and the generous use of chilled water to clean and reduce temperature, deliver modified atmosphere packaging capable of shelf life of two weeks under good commercial distribution temperatures.

Prepared Foods

However complex is the extension of shelf life of case ready red meat, prepared foods probably present a more difficult challenge. Home meal replacement foods are a mix of many different ingredients and components and are intended to be ready-to-eat or ready-heat-and-eat, that is, with no added cooking, dressing, or flavors by the consumer. The diversity of components immediately suggests a microbiological mix with some not necessarily susceptible to a gas blend to which others might respond. Further, most minimally processed prepared foods are generally low acid and so are potentially susceptible to anaerobic pathogenic microbiological growth if packaged under reduced oxygen. From a positive perspective, however, reduced oxygen plus carbon dioxide in the 25%–35% range suppresses growth of aerobic micro-organisms, retards biochemical oxidation, and generally offers the opportunity of delivery of relatively high-quality prepared foods. Some processors have been adding a mild heating hurdle that further enhances the shelf life by eliminating almost all aerobic spoilage micro-organisms.

Among the prepared food products that are being processed and packaged under modified atmosphere packaging are moist pasta, egg rolls, sausage and biscuit and related sandwiches, pates, soups and soup concentrates, entrees, beef stews, ethnic dishes, mashed potatoes, and pasta sauces. All are, of course, distributed under refrigeration as an indispensable element of the modified atmosphere packaging system.

Soft Bakery Goods

The only product category to date that does not benefit from chilling is soft bakery goods such as bread, pastries, and cakes. In this situation, modified atmosphere packaging and, in particular, elevated carbon dioxide functions very effectively to retard mold growth. The problem is that as temperature is reduced, the rate of staling is increased, and so chilling actually increases the rate of product quality deterioration. Several European bakers have developed an interesting product category from this anomaly: par-baked soft bakery goods. Modified atmosphere packaging retards mold growth and moisture loss at ambient temperatures. Staling reactions are slowed by the elevated temperature conditions. Consumers finish bake the product, an action that compensates for much of the mouth feel deterioration of crumb staling by driving moisture back into the starch matrix. To date, very little modified atmosphere packaging of soft bakery goods is practiced in the United States.

CONCLUSION

Modified atmosphere packaging for chilled foods can only grow in the future because it is the technology that causes the least alteration in the product itself while enhancing preservation. As attention is directed to the preservation aspects of the technology as contrasted to the issues of avoidance of anaerobiosis, as newer active packaging structures are developed, as the comprehension of hurdle technologies becomes the foundation, and as antimicrobial gases enter the picture, growth will accelerate to capture entirely new food groups and to grow in existing categories.

FURTHER READING

Blakistone, B. *Principles and Applications of Modified Atmosphere Packaging of Foods*, 2nd Ed.; Blackie Academic and Professional: London, 1998.

Brody, A.L. *Controlled/Modified Atmosphere/Vacuum Packaging of Foods*; Food and Nutrition Press: Trumbull, Connecticut, 1989.

Brody, A.L. *Modified Atmosphere Food Packaging*; Institute of Packaging Professionals: Herndon, Virginia, 1994.

Day, B.P.F. *Proceedings of Modified Atmosphere Packaging (MAP) and Related Technologies*; Campden and Chorleywood Food Research Association: Campden, UK, 1995.

Moisture Content Measurement

M

Carl J. Bern
Thomas Brumm
Iowa State University, Ames, Iowa, U.S.A.

INTRODUCTION

Water is more abundant than any other substance on earth and is one of the most important constituents of biological materials. In grains and seeds, moisture is the most important factor influencing suitability for harvest, storage, handling, and other processes. In the grain trade, moisture directly influences price.

Farmers recognized the importance of crop moisture to storability long before quantitative measurement techniques were available. Through experience, they related moisture to the sensations felt as they bit a kernel or thrust an arm into a grain mass. This article will discuss moisture in biological materials and the principles and use of some moisture measurement techniques available today.

HYGROSCOPICITY

Moisture is held in cells of biological materials in three forms:

- A layer of adsorbed "bound" moisture molecules on the outer surface of cells.
- Adsorbed moisture beyond the bound moisture layer.
- Moisture absorbed into the cell.

When a biological material is placed in air at some constant defined temperature and relative humidity, there is a net exchange of moisture between the material and the air until the material reaches a constant moisture level, known as its equilibrium moisture content for that air condition. Equilibrium moisture properties of different materials can be defined by isotherms (Fig. 1). A point on an isotherm defines the equilibrium moisture content the material will come to in an atmosphere defined by air temperature and air relative humidity. The point will be on a desorption isotherm if the material is losing moisture, and on an adsorption isotherm if moisture is being gained. The desorption line predicts, e.g., that if a wet sample of yellow dent corn is placed in an atmosphere at 40% relative humidity and 22°C, it will eventually come to a constant moisture of about 10.8% w.b. The separation of desorption and adsorption isotherms illustrates a hysterisis effect. Empirical-derived based isotherms have been defined for many biological materials.[2]

MOISTURE MEASUREMENT METHODS

Several methods of moisture content measurement are classified in Fig. 2. Direct methods calculate moisture content based on a mass change or chemical reaction when water is removed, assuming the material consists only of dry matter and water, while indirect methods measure a property known to be related to moisture content.

Air Oven

Most prominent of the direct methods is the air-oven method in which the mass change of a sample held in an oven set at a prescribed temperature for a prescribed time is used to compute moisture content. At the end of the time in the oven, all water is assumed to be gone and all dry matter is assumed to remain. Many organizations have adopted standards for air-oven moisture measurement methods.[3–7] Table 1 shows heating times and oven temperatures recommended by ASAE for some grains and seeds.

Distillation

Distillation methods involve removal of moisture by heating sample material in liquids such as oil, which has a higher boiling point than water. The Brown–Duvel method[8] involves heating sample material in a hydrocarbon oil until a temperature of 190°C is reached. Volatilized moisture is condensed and measured in a graduated cylinder. The procedure can be completed in 25 min or less. A variation of this method was developed for use by farmers for grain or hay.[9] A 100-g sample is ground in a food chopper and then submerged in vegetable oil and heated to 375°F. This takes about 20 min. Moisture content is calculated from the change in weight of the oil plus sample.

Encyclopedia of Agricultural, Food, and Biological Engineering
DOI: 10.1081/E-EAFE 120006889
Copyright © 2003 by Marcel Dekker, Inc. All rights reserved.

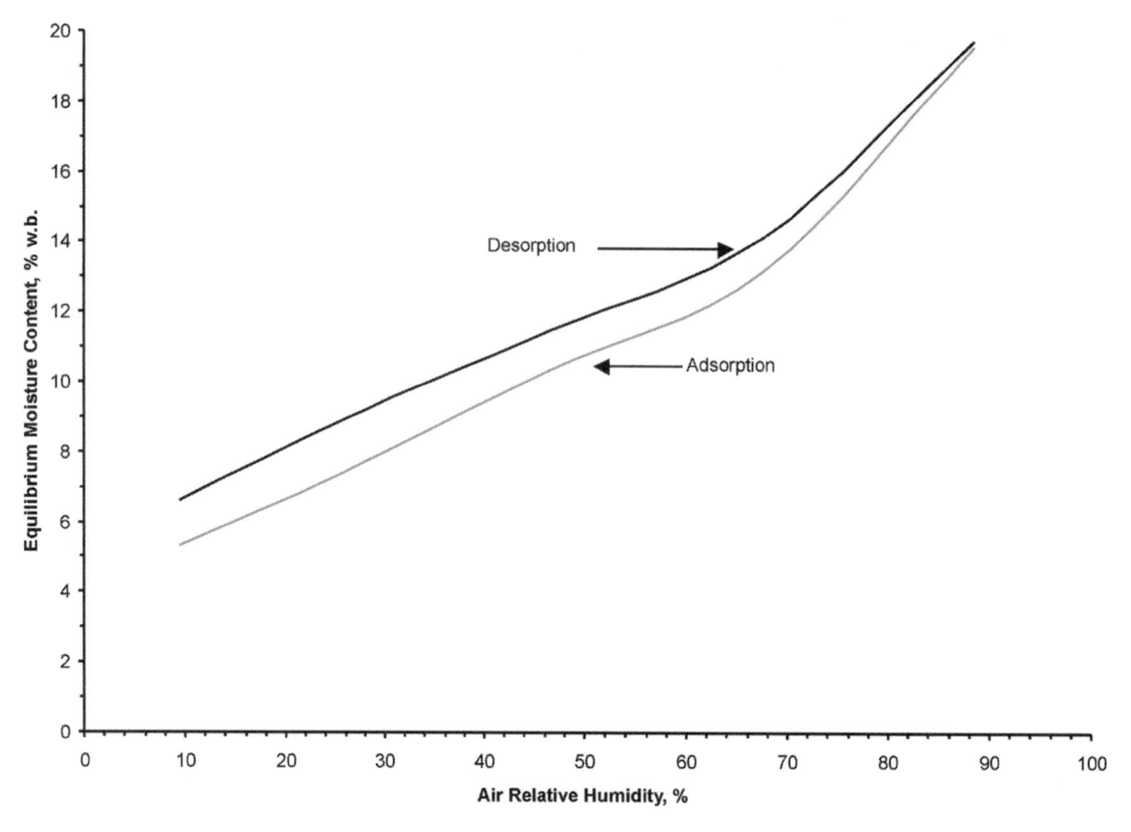

Fig. 1 Equilibrium moisture isotherms for yellow dent corn at 22°C. (From Ref. 1.)

Chemical Reaction

Chemical reaction methods rely on chemical reactions with moisture in the material and moisture content is calculated from chemical reaction equations. The Karl Fischer method uses the reaction of iodine with water in the presence of sulfur dioxide and pyridine to form hydriodic acid and sulfuric acid. Water is first extracted from finely ground material with anhydrous methanol.[10] The Karl Fischer method is among the most accurate methods, and agreement with it was used as a criterion for establishing air-oven standard procedures.[11] The complexity of the procedure limits its use.

Electrical Resistance and Capacitance

Researchers discovered nearly 100 yr ago that some electrical properties of grains are closely related to

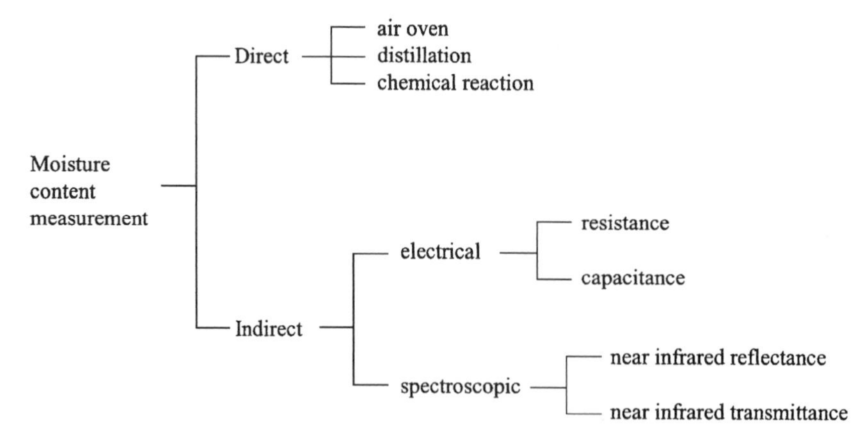

Fig. 2 Classification of some common moisture content measurement methods.

M

Table 1 Oven temperature and heating time for moisture content determination of some unground seeds and grains

Seed	Oven temperature ∀1°C	Heating time hr	Heating time min	Sample size (g)
Alfalfa	130	2	30	10
Barley	130	20	0	10
Beans, edible	103	72	0	15
Corn	103	72	0	15 or 100[a]
Flax	103	4	0	5.7
Kale	130	4	0	10
Oats	130	22	0	10
Rape (canola)	130	4	0	10
Rye	130	16	0	10
Safflower	130	1	0	10
Sorghum	130	18	0	10
Soybeans	103	72	0	15
Sunflower	130	3	0	10
Wheat	130	19	0	10

[a] Use 100 g if moisture exceeds 25%.
(From Ref. 4.)

moisture content. Development over the years has led to modern meters, which relate moisture to the resistance of materials placed between and in contact with two electrodes, or to the capacitance of a capacitor having material packed between its plates. A different calibration is derived for each type of grain or seed. (Calibrations are empirically derived equations that relate the measured property to the true moisture content as determined by a direct method.) Besides moisture, these electrical properties also depend on bulk density and temperature, and these factors must be held constant or compensated for. Furthermore, both resistance and capacitance of materials are affected differently by bound water and free water.

The Weston Moisture Meter, popular through the 1950s, operated by measuring electrical resistance of grain crushed between counter-rotating corrugated steel rolls, which served as electrodes. Modern single-kernel testers use a similar procedure. Wood moisture sensors relate moisture content to resistance between two pins inserted into wood at a precise distance from each other.

Most modern moisture meter models utilize the capacitance principle. Compared to resistance meters, they are less subject to errors from uneven moisture distribution within a sample. Furthermore, they can be operated over a greater moisture range and are less sensitive to the degree of physical contact with material.[11] Modern capacitance meters achieve relatively uniform bulk density during loading and automatically compensate for temperature effects. Fig. 3 shows a Dickey-John hand-held capacitance tester and also a Dickey-John GAC2100 capacitance meter. The GAC2100 has been the U.S. Federal Standard tester for USDA-licensed grain graders since 1997. Capacitance meters use different calibrations for each grain being tested.

Near-Infrared Spectrometry (NIRS)

The moisture content of biological materials is related to their optical properties, i.e., the ability to absorb selective wavelengths of light. The amount of absorbance of certain wavelengths of near-infrared light (750 nm–2600 nm) is proportional to the levels of moisture and other chemical constituents. Williams and Norris[12] provide an excellent overview of the use of near-infrared technology for moisture measurement in agricultural and food industries.

NIRS methods have the advantage of being fast (often less than 1 min), relatively accurate and precise,

Dickey-john hand-held

Dickey-john GAC 2100

Fig. 3 Capacitance type moisture testers.

cost-effective over many samples, nondestructive, and capable of determining multiple constituents in one measurement. Operation of NIRS devices requires minimal training. However, such devices are often expensive relative to other indirect moisture determination methods. NIRS devices must be carefully calibrated to a direct method, often using complex statistical procedures.

Various configurations of NIRS devices are in use today. Table-top units include reflectance-types, which may require sample grinding, and transmittance-type units that require little or no sample preparation. On-line devices, both reflectance and transmittance types, are used to measure moisture in moving material in manufacturing or processing situations.

SAMPLING

Material is routinely sampled for moisture analysis because it is usually impractical to analyze an entire lot. The accuracy of any moisture determination is first dependent upon having a representative sample, i.e., one that possesses the same properties as the lot from which it is drawn.

"Hand-grab" or "scoop" samples are rarely representative, although samples taken by hand can be representative if a good sampling method is used. Three things should be considered to ensure that any type of sample is representative[13]: 1) appropriate equipment; 2) multiple samples; and 3) careful sample handling.

Hand or mechanical probes are used to draw a sample "core" from material at rest. Hand or automatic (mechanical) diverters are used to collect samples at intervals from flowing material. Subdividing samples should be done with a Boerner divider or other similar mechanical device so that the subsamples are also representative.

A single drawn sample cannot fairly represent an entire lot, especially if there is moisture variability throughout the lot. Combining multiple samples helps overcome and average this variability. Random sampling at different locations and intervals helps eliminate systematic biases.

Sample handling procedures should deliver the sample unchanged for analysis. Clean and appropriate containers should be used. Rough handling can cause sample degradation or loss. Storage conditions should minimize changes in moisture due to microbial respiration or exchange of moisture with the surrounding environment.

PRECISION AND ACCURACY

Accuracy is a measure of the nearness of a value to the correct or true value. Precision refers to the repeatability of a measure. Both are important in evaluating the results of moisture determinations.

Accuracy is often represented by the "bias," i.e., the mean ($\overline{D}$) of the differences (D) between the measured values and the true value. Bias is often interpreted to represent a constant systematic error that is not expected to change.

Precision is measured by the standard deviation of repeated measurements with an "ideal" of zero. The standard deviation of differences ($\text{SDD} = \sqrt{[\Sigma(D - \overline{D})^2/(n-1)]}$) is statistically more accurate in describing the precision of indirect moisture determinations that are calibrated to a direct method than the standard deviation of the measurements themselves.

For indirect moisture analysis methods, if samples used to determine the bias and SDD are representative of the population to be measured, then 95% of future measurements will be within $2 \times \text{SDD}$ of the true value plus bias, and 99% will be within $3 \times \text{SDD}$ of the true value plus bias.[14]

It is unreasonable to assume that the SDD of measurements made with an indirect method can be smaller than the standard deviation of repeated determinations of the direct method to which it is calibrated.

REFERENCES

1. Chung, D.S.; Fost, H.B.P. Adsorption and Desorption of Water Vapor by Cereal Grains and Their Products. Trans. ASAE **1967**, *10* (2), 552–575.
2. ASAE. D2445.5 Moisture Relationships of Plant-Based Agricultural Products. *ASAE Standards 2000*; ASAE: St. Joseph, MI, 2000; www.asae.org.
3. AACC. *Approved Methods of the American Association of Cereal Chemists*, 10th Ed.; American Association of Cereal Chemists: St. Paul, MN, 2000; www.scisoc.org/aacc.
4. ASAE. S352.2 Moisture Measurement—Unground Grain and Seeds; S353 Moisture Measurement—Meat and Meat Products; S358.2 Moisture Measurement—Forages; S410.1 Moisture Measurement—Peanuts; S487 Moisture Measurement—Tobacco. *ASAE Standards 2000*; ASAE: St. Joseph, MI, 2001; www.asae.org.
5. AOCS. *Official Methods and Recommended Practices of the AOCS*; American Oil Chemists' Society: Champaign, IL, 2001; www.aocs.org.
6. AOAC. *Official Methods of Analysis of AOAC International*, 17th Ed.; Association of Official Analytical Chemists International: Gaithersburg, MD, 2000; www.aoac.org.
7. USDA. *Moisture Handbook*; U.S. Department of Agriculture, Grain Inspection, Packers and Stockyard Administration, Federal Grain Inspection Service: Washington, D.C., 1986; www.usda.gov/gipsa.

8. Brown, E.; Duvel, J.W.T. *A Quick Method for the Determination of Moisture in Grain*; Bulletin 99, USDA Bureau of Plant Industry: Washington, D.C., 1907.
9. Hull, D.O.; Van Fossen, L.D. *A Simple Way to Test Your Grain and Hay for Moisture*; Pamphlet 275, Agricultural Engineering Information Series 2, Cooperative Extension Service, Iowa State University: Ames, IA, 1967.
10. Hart, J.R.; Neustadt, M.H. Application of the Karl Fischer Method to Grain Moisture Determination. Cereal Chem. **1957**, *34* (1), 27–37.
11. Young, J.H. Moisture. In *Instrumentation and Measurement for Environmental Sciences*; Henry, Z.A., Zoerb, G.C., Birth, G.S., Eds.; ASAE: St. Joseph, MI, 1991.
12. Williams, P.C., Norris, K.H., Eds. *Near Infrared Technology in the Agricultural and Food Industries*; American Association of Cereal Chemists: St. Paul, MN, 1987; 38.
13. USDA. *Practical Procedures for Grain Handlers*; U.S. Department of Agriculture, Grain Inspection, Packers and Stockyard Administration, Federal Grain Inspection Service: Washington, D.C., 1999; www.usda.gov/-gipsa.
14. Hruschka, W.R. Data Analysis: Wavelength Selection Methods. In *Near Infrared Technology in the Agricultural and Food Industries*; Williams, P.C., Norris, K.H., Eds.; American Association of Cereal Chemists: St. Paul, MN, 1987; 38.

Moisture Sorption Isotherms

José Miguel Aguilera
Pontificia Universidad Católica de Chile, Santiago, Chile

INTRODUCTION

At constant temperature and after long-term equilibration conditions a unique relationship sets in between the total moisture content of a food (W) and a variable defined alternatively as the equilibrium relative humidity (ERH) of the surrounding space, the relative vapor pressure of water (RVP), or the measured water activity of the food (a_w). This relation is called the *moisture sorption isotherm* (MSI) and it is depicted graphically in the literature as a curve of moisture content W vs. ERH, RVP, or a_w. MSI represents equilibration data (although not necessarily thermodynamic equilibrium) obtained at constant temperature either by *adsorption* (moisture gain) or *desorption* (moisture loss). MSIs of food products are generally sigmoidal in shape, a result of several basic interacting mechanisms of water binding. A typical MSI that would apply to many foods as well as MSIs of food components or additives are shown in Fig. 1. As temperature increases the amount of water sorbed by a food decreases and the isotherms shift downwards. Experimental data can be fitted into models or equations that describe the MSI in different ranges of a_w.

Water sorption by foods depends, among others, on: 1) the microstructure of the product; 2) the physical–chemical state of food components (e.g., amorphous and crystalline sucrose, Fig. 1); and 3) the chemical composition (e.g., protein, starch, and oil, Fig. 1). Hence, it is impossible to predict a priori the water vapor pressure exerted by a complex food system and MSIs have to be determined experimentally. Published data for sorption isotherms of many foods are available[1] as well as a comprehensive bibliography on the subject.[2] The practical use of sorption isotherms in food applications can be found in Ref. 3.

INTERPRETATION OF SORPTION ISOTHERMS

It is useful to analyze some basic mechanisms responsible for depressing the vapor pressure of water in foods and their bearing on the shape of MSI. Such hypothetical analysis distinguishes three regions in the MSI depending on whether the prevailing effect on a_w is due to pure sorption, capillary condensation, or solute effect (Fig. 2). In region I minimal water is contained in the product and water molecules present are bound to active sites such as polar groups in molecules, mainly by hydrogen bonding. The moisture content (W_0) theoretically representing the adsorption of the first layer of water molecules (point A in the isotherm) is usually associated with a *monolayer value*, as it is determined from the BET isotherm equation. It is generally found for a_w of 0.2–0.4 and below a moisture content of 0.1 g/g solids.

Region II represents water possibly filling micro- and macropores in the system and subject to capillary effects. In this region, chemical and biochemical (e.g., enzymatic) reactions, requiring solvent water, start to take place due to increased mobility of solutes. In region III, excess water is present as part of the fluid phase in high-moisture materials, exhibiting nearly all the properties of bulk water. The presence of solutes is responsible for the depression in water vapor pressure. Microbial growth becomes a major deteriorative reaction in this region and a_w gives a rough indication of the lower limit for growth of salt-tolerant micro-organisms ($a_w = 0.60$), most molds ($a_w = 0.80$), yeasts (about $a_w = 0.87$) and pathogenic bacteria ($a_w = 0.91$). A schematic representation of the three main water activity-depressing mechanisms that may act in a food as well as the basic equations is presented in Fig. 3.[4]

HYSTERESIS

In practical sorption experiments the desorption isotherm usually lies above the adsorption branch, a phenomenon called *hysteresis* (Fig. 2). This means that a lower vapor pressure is needed to achieve a given moisture content by desorption than by adsorption. Some explanations for the presence of hysteresis involve factors such as supersaturation of solutions during desorption, presence of metastable phases (e.g., amorphous domains), contact angle phenomena, phase transitions, and irreversible structural changes caused by sorption–desorption cycles. Some authors establish that hysteresis proves that foods are generally not in thermodynamic equilibrium.

At this point it should be clear that MSIs are highly product specific and depend on how products are prepared,

Encyclopedia of Agricultural, Food, and Biological Engineering
DOI: 10.1081/E-EAFE 120007224
Copyright © 2003 by Marcel Dekker, Inc. All rights reserved.

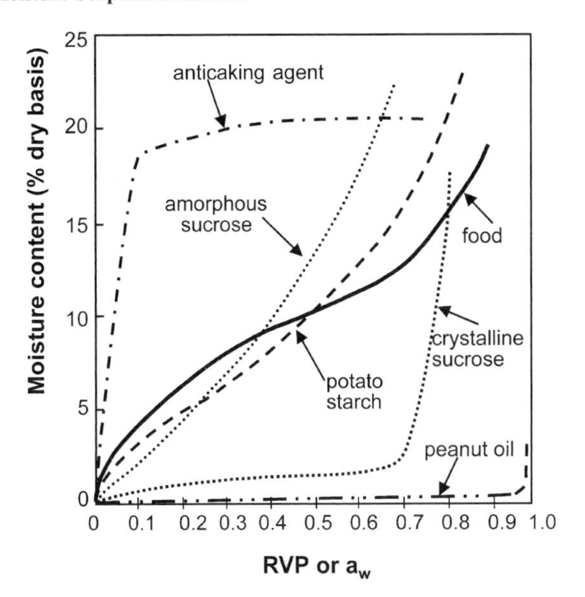

Fig. 1 Typical moisture isotherms for foods.

protected from the environment, and handled during storage, as moisture may be continuously gained or lost.

DETERMINATION OF MSI[5,6]

Among the several methods available for constructing a water sorption isotherm, gravimetric methods are the simplest to implement in a laboratory. Basically, dried (for adsorption) or moist samples (for desorption) are placed

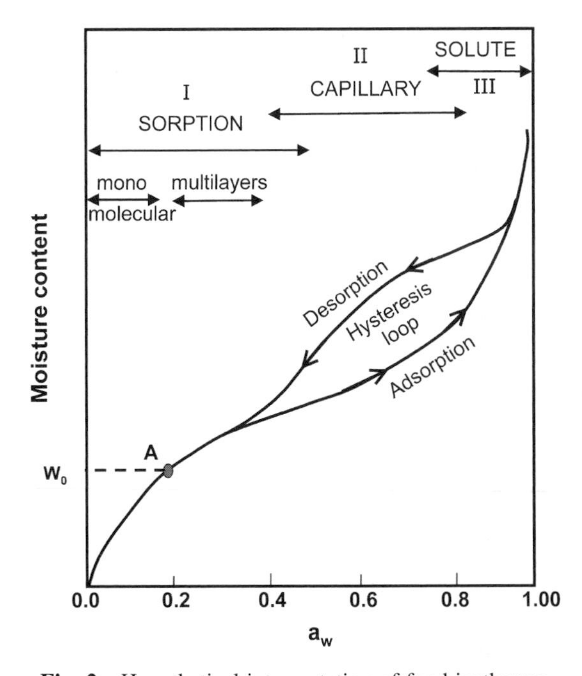

Fig. 2 Hypothetical interpretation of food isotherms.

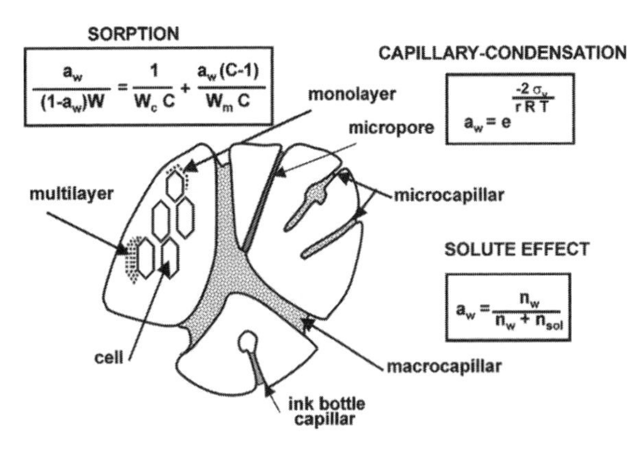

Fig. 3 Main water activity—depressing mechanisms in sorption isotherms.

in Petri dishes, aluminum pans, or weighing bottles and exposed to the humid atmosphere in a container, under constant temperature conditions. Care has to be taken in sample preparation because drying may induce structural changes that affect the results of the isotherm.

The sorption containers are usually glass desiccators or preserving jars provided with a perforated support to accommodate the sample holders. The bottom of the container is filled with saturated salt solutions (slurries), which provide different relative humidity in the headspace (see Table 1) or solutions of sulfuric acid of known concentration, which also give constant RH. Temperature control is achieved by placing the containers in a thermostated cabinet. Moisture content may be determined by any official method (e.g., drying for 3 hr at 100°C) or by vacuum drying at 80°C. Equilibration is supposed to have been reached after several days or when the difference in weight of two successive measurements is less than 1 mg/g of solids. A final drying step over P_2O_5 is recommended to report the final moisture.

In order to judge the reliability of the acquired sorption data, construction of the MSI of a standard reference material is recommended. The reference material most widely utilized is microcrystalline cellulose Avicel PH 101, manufactured by the FMC Company, whose MSI can be found in Ref. 6.

USE OF MSI

Although it is questionable whether true thermodynamic equilibrium is achieved during actual determination of MSIs their use in food technology and in engineering applications is extensive. Knowledge of the MSI of a food permits rapid translation of moisture data, which are easy to obtain, to the respective ERH or a_w.

Table 1 ERH of the headspace over saturated salt solutions

	% RH							
Temperature (°C)	5	10	15	20	25	30	35	40
$LiCl{\cdot}H_2O$	16	14	13	12	11	11	11	11
CH_3COOK	25	24	24	23	23	23	23	23
$MgBr_2{\cdot}6H_2O$	32	31	31	31	31	30	30	30
$MgCl_2{\cdot}6H_2O$	33	33	33	33	33	32	32	31
$K_2CO_3{\cdot}2H_2O$	—	47	45	44	43	42	41	40
$Mg(NO_3)_2{\cdot}6H_2O$	54	53	53	52	52	52	51	51
NaBr	59	58	58	57	57	57	57	57
$CuCl_2$	65	68	68	68	67	67	67	67
$C_2H_3LiO_2{\cdot}2H_2O$	72	72	71	70	68	66	65	64
$SrCl_2{\cdot}6H_2O$	77	77	75	73	71	69	68	68
NaCl	76	75	75	75	75	75	75	75
$(NH_4)_2SO_4$	81	80	79	79	79	79	79	79
$CdCl_22.5H_2O$	83	83	83	82	82	82	79	75
KBr	—	86	85	84	83	82	81	80
$Li_2SO_4{\cdot}H_2O$	84	84	84	85	85	85	85	81
KCl	88	87	87	86	86	84	84	83
K_2CrO_4	89	89	88	88	87	86	84	82
$C_7H_5NO_2$	88	88	88	88	88	88	86	83
$BaCl_2{\cdot}2H_2O$	93	93	92	91	90	89	88	87
KNO_3	96	95	95	94	93	92	91	89
K_2SO_4	98	97	97	97	97	97	96	96
$Na_2H_2P_2O_6.6H_2O$	98	98	98	98	97	96	93	91
$Pb(NO_3)_2$	99	99	98	98	97	96	96	95
$NaNO_2$	—	—	—	—	64	—	—	—

In air drying the difference between the relative humidity of the surrounding air and ERH corresponding to the average moisture content of the solid (which can be determined at any time) is fundamental as it defines a driving force for mass transfer. Also, through the MSI it is possible to predict the theoretical final moisture of the product (or *equilibrium moisture content*) as it has to be in equilibrium with the surrounding air. Large moisture gradients inside large food pieces make these predictions somewhat more complicated.

During storage of unpackaged dry foods, the adsorption branch of the MSI is important since it defines the path (at constant temperature) for moisture changes when exposed to a moist atmosphere. As moisture increases by moisture pick-up, critical moisture content may be attained at which the food becomes unacceptable (W_{crit}) from a quality viewpoint. Values of W_{crit} may be as low as those in equilibrium with air at 10%–20% RH for potato flakes or orange powder, and 35%–50% RH for most dry materials prone to structural changes (e.g., stickiness or caking).

Engineering calculations of permeable packaging material, often the economic solution to retard moisture pickup, make use of the point in the isotherm corresponding to the relative humidity of the ambient atmosphere to define a driving force for moisture transfer through the packaging film. Similarly, the MSI can be used to predict the change in weight from the legal net weight and to locate the critical moisture or a_w above which the product should not be sold.

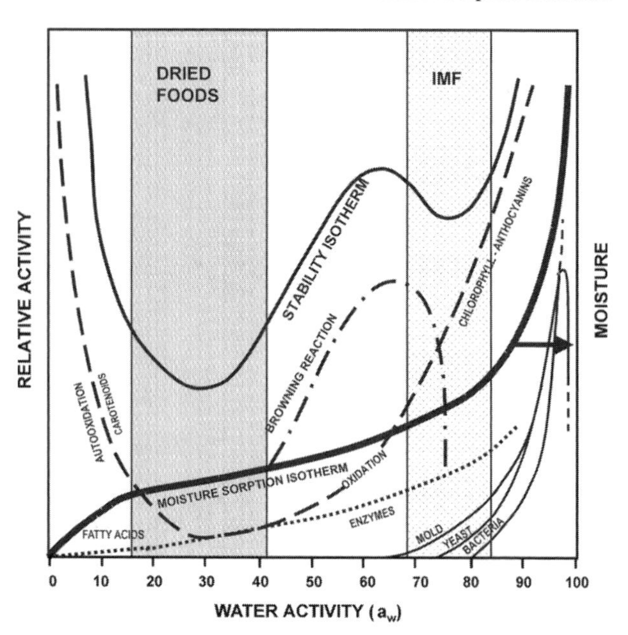

Fig. 4 Water activity—stability diagram for foods.

MSI data have been crucial in understanding the stability of dried foods and in the development of intermediate moisture foods (IMF). The so-called stability diagram (Fig. 4) summarizes the typical qualitative relationships between the most important deleterious reactions in foods (i.e., physical, chemical, and microbiological) and water activity.[7] Analysis of such a graph shows two minima of the stability isotherm: one at low moisture contents corresponding to dried foods, and another one at higher moisture contents, typical of IMF.

IMF have a_w in the range 0.65–0.95, which usually corresponds to the relative humidity of the ambient air. IMF are stable without refrigeration despite their relatively high moisture content because they are usually further stabilized against microbial and chemical changes by adjustment of the pH, mild preheating, addition of preservatives, or control of the redox potential.

REFERENCES

1. Iglesias, H.A.; Chirife, J. *Handbook of Food Isotherms: Water Sorption Parameters for Food and Food Components*; Academic Press: New York, 1982.
2. Wolf, W.; Spiess, W.E.L.; Jung, G. *Sorption Isotherms and Water Activity of Food Materials*; Science and Technology Publishers Ltd.: London, 1985.

3. Labuza, T.P. *Moisture Sorption: Practical Aspects of Isotherm Measurement and Use*; American Association of Cereal Chemists: St. Paul, MN, 1984.
4. Aguilera, J.M.; Stanley, D.W. *Microstructural Principles of Food Processing and Engineering*; Aspen Publishers, Inc.: Gaithersburg, PA, 1999.
5. Gal, S. Recent Developments in Techniques for Obtaining Complete Sorption Isotherms. In *Water Activity: Influences in Food Quality*; Rockland, L.B., Stewart, G.E., Eds.; Academic Press: New York, 1981; 89–110.
6. Spiess, W.E.L.; Wolf, W. Critical Evaluation of Methods to Determine Moisture Sorption Isotherms. In *Water Activity: Theory and Applications to Food*; Rockland, L.B., Beauchat, L.R., Eds.; Marcel Dekker, Inc.: New York, 1987; 215–233.
7. Rockland, L.B.; Beauchat, L.R. *Water Activity: Theory and Applications to Food*; Marcel Dekker, Inc.: New York, 1987; vii.

Multiple-Effect Evaporators

Amarjit S. Bakshi
ConAgra Foods, Irvine, California, U.S.A.

INTRODUCTION

Water is the predominant ingredient in most foods and exceeds 85% in milk, fruit, and vegetable juices. This water content can frequently be reduced to lower container, storage, and shipping costs or to achieve desirable sensory attributes in the food. The oldest method of removing water is sun evaporation. Even today over 1,000,000 tn of salt is produced each year in San Francisco Bay by this method. However, most of the liquid foods are concentrated by heating with steam in an evaporator. Evaporative concentrators for liquid foods evaporate the water and separate the vapors from the residual liquid.[2]

OVERVIEW

There are two types of basic evaporation systems:

- Single effect.
- Multiple effect.

When vapors from concentrated liquid are not used to heat additional product, the evaporator system is known as single effect. A single-effect evaporator may be modified to furnish vapors to a second evaporator operating at lower pressure or to use vapors coming from another evaporator operating at higher pressure. A major consideration in the design of an evaporator is maximization of the steam economy, i.e., maximizing evaporation of product per unit of steam supplied. The energy contained in the vapors discharged from an evaporator can be partially recovered using the vapors as a heat source in another evaporator called "effect" operating at lower pressure (lower boiling point). If the feed to the first effect is at a temperature near the boiling point corresponding to the first effect, 1 kg of steam will evaporate almost 1 kg of water. The first effect operates at a boiling temperature high enough so that the evaporated water vapors can serve as the heating medium of the second effect. Here, another 0.5–1-kg water is evaporated, which may go to the condenser if the evaporator is double effect or may be used as a heating medium of the third effect. This may be repeated for any number of effects. Large evaporators having 6 or 7 effects are common in the chemical industry and evaporators having up to 10 or more effects have been built.[5] Increased steam economy of a multiple-effect evaporator is gained at the expense of evaporator cost. The total heat-transfer area will increase substantially in proportion to the number of effects in the evaporator. The following table provides estimated steam economy for different effect evaporators for fluid milk[3]:

Number of effects	Water evaporated (kg)	Steam to the primary effect (kg)
1	1.0	0.95
2	2.0	0.85
3	3.0	0.8

Different arrangements for multiple evaporators are possible (Fig. 1): forward feed; backward feed; mixed feed; and parallel feed. In *forward feed*, raw food product is introduced in the first effect and passed from effect to effect parallel to the steam flow. Concentrated product is withdrawn from the last effect. This is the preferred arrangement when the feed is hot or when the concentrated product would be damaged or would deposit scale at the high temperature as in case of milk pudding. Forward feed simplifies the operation when liquid can be transferred by pressure difference alone, thus eliminating the need for intermediate pumps. In *backward feed*, raw liquid food product enters the last (coldest) effect, the discharge from this effect becomes the feed to the next to last effect, and so on until concentrated product is discharged from the first effect. This method of operation is normally used when the feed is cold. This arrangement is also used when the product is viscous and high temperature is needed to keep the viscosity low enough to give reasonable heat-transfer coefficients. Normally this type of arrangement is used for nonfood products. These two types of feed flows, however, are different with respect to their effects on the chemical changes in food quality; e.g., nonenzymatic browning proceeds most rapidly at higher temperature concentration. Thus, a backward feed arrangement may be unacceptable for

Encyclopedia of Agricultural, Food, and Biological Engineering
DOI: 10.1081/E-EAFE 120007061
Copyright © 2003 by Marcel Dekker, Inc. All rights reserved.

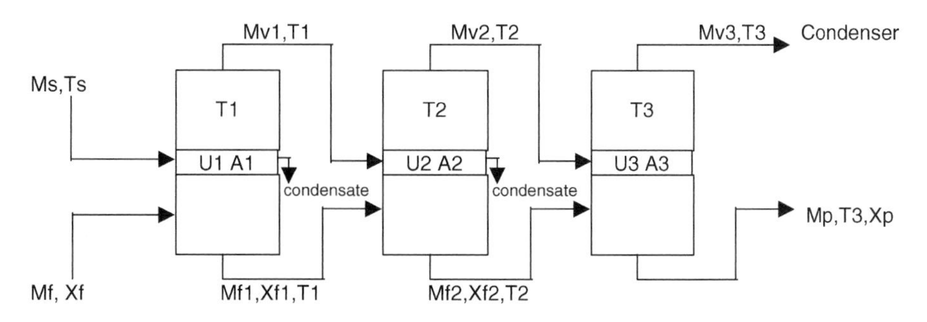

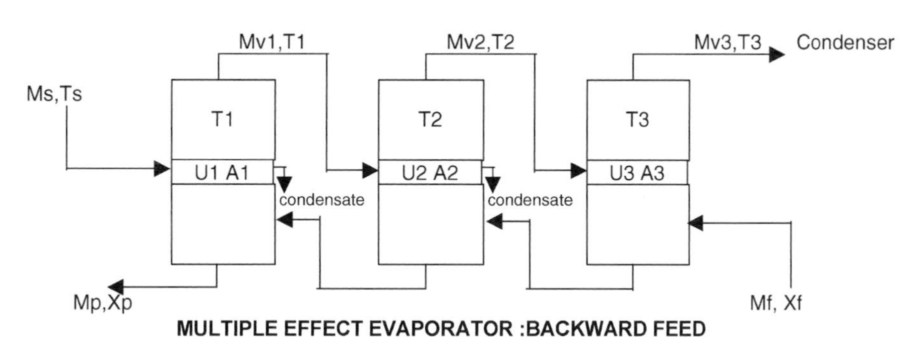

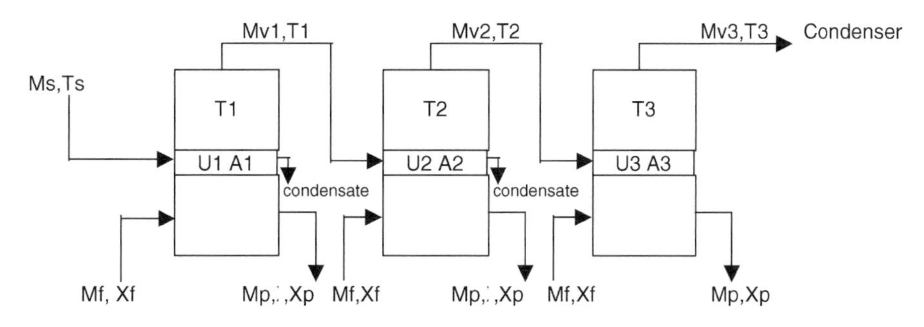

Fig. 1 Schematics of multiple-effect evaporators.

temperature sensitive food products as concentrated product is exposed to higher temperature in this arrangement. In addition, a high solid content food product at the high temperature may result in denaturation of the protein and thus formation of scale in the backward flow arrangement. The *parallel feed* involves the introduction of raw feed and withdrawal of concentrated product at each effect. It is primary used when the feed is substantially saturated.

DESIGN CONSIDERATIONS

A number of approximate methods have been published for estimating multiple-effect evaporator performance and heat-surface requirements.[5] However, because of the wide variety of methods of feeding and the added complication of feed heater and condensate flash system, the only certain way of determining performance is by detailed heat and material balance calculations.

In designing a multiple-effect evaporator, the needed parameters are the amount of steam consumed, the area of the heating surface, the approximate temperature in each effect, and the amount of vapors leaving each effect. In a single-effect evaporator, these quantities are calculated from the material and enthalpy balances. In a multiple-effect evaporator, however, a trial-and-error method is used in place of a direct algebraic solution. Consider, e.g., a forward feed triple-effect evaporator as shown in Fig. 1. Liquid food is introduced into the evaporator chamber of the first effect. Steam enters the heat exchanger in the first effect and condenses to impart its heat to the incoming product. The condensate is discarded. The vapors produced from the first effect are used as the heating medium in the second effect and the feed for the second effect is the partially concentrated product from the first effect. The vapors and the concentrated product from the second effect are used in the third effect. The vapors produced in the final effect are conveyed to a condenser and a vacuum system. The following equations may be written: an enthalpy and mass balance for each effect; heat-transfer equations for each effect; and known total evaporation. If the heating surface is assumed to be same in each effect, the unknowns in these equations are the rate of steam flow from each effect and the boiling point in each effect. It is possible to solve these simultaneous equations using computers. Most of the equipment suppliers do have software for this purpose. Another method for calculation is:

1. Assume boiling point in each effect.
2. From enthalpy balance find the rate of steam and liquid flow from each effect.
3. Calculate heating surface in each effect.
4. If the heating surface areas are not approximately equal, estimate new values for the boiling temperature in each effect and repeat the above calculations.

Material Balance

The total product mass balance is:

$$M_f = M_{v1} + M_{v2} + M_{v3} + M_p$$

where M_f is the feed rate to the first effect; M_{v1}, M_{v2}, and M_{v3} are the vapor flow rates from the 1st, 2nd, and 3rd effects; and M_p is the final concentrated product.

The solid mass balance is:

$$M_f X_f = M_p X_p$$

X_f and X_p are the solid fractions of the feed and concentrated product, respectively.

Enthalpy in the three effects can be written as:

$$M_f H_f(T_f, X_f) + M_s H_s(T_s) = M_{v1} H_v(T_1) + M_{f1} H_{f1}(T_1, X_{f1}) + M_s H_c(T_s)$$

$$M_{f1} H_{f1}(T_1, X_{f1}) + M_{v1} H_v(T_1) = M_{v2} H_v(T_2) + M_{f2} H_{f2}(T_2, X_{f2}) + M_{v1} H_c(T_1)$$

$$M_{f2} H_{f2}(T_2, X_{f2}) + M_{v2} H_v(T_2) = M_{v3} H_v(T_3) + M_p H_p(T_3, X_p) + M_{v2} H_c(T_2)$$

H refers to the enthalpy, $\mathrm{kJ\,kg^{-1}}$; f is the feed; s is the steam; p is the concentrated product; v is the vapor from 1st, 2nd, and 3rd effects; and c is the condensate.

The enthalpy of the fluids entering or leaving each effect can be estimated by:

$$H(T, X) = C_p(T)$$

C_p is the specific heat and can be calculated by the relationship given below. Specific heat for plant concentrate from fruit and vegetable can be estimated by the following relationship[6]:

$$C_p = 33.49M + 837.36$$

C_p is the specific heat $\mathrm{J\,kg^{-1}\,K^{-1}}$. M is the moisture content in percentage. The specific heat can also be calculated by its composition.[4]

Heat transfer across heat exchanger in each effect can be expressed as:

$$Q_1 = U_1 A_1 (T_s - T_1) = M_s H_v(T_s) - M_s H_c(T_s)$$

$$Q_2 = U_2 A_2 (T_1 - T_2) = M_{v1} H_v(T_1) - M_{v1} H_c(T_1)$$

$$Q_3 = U_3 A_3 (T_2 - T_3) = M_{v2} H_v(T_2) - M_{v2} H_c(T_2)$$

A is the area of heat exchanger in each effect and U the overall heat-transfer coefficient

Overall heat-transfer coefficients can be calculated from relationships given in such books as the *Chemical Engineering Handbook*.[5] The following table provides the order of magnitude for overall heat-transfer coefficients associated with falling film evaporators[3]:

Boiling Point (T_b)

The factors effecting the boiling point during the concentration of liquids containing the dissolved solids are:

- Hydrostatic pressure exerted by column of liquid.
- Vacuum in each effect.
- Concentration in each effect.

	Skim milk ($W m^{-2} K^{-1}$)	Whole milk ($W m^{-2} K^{-1}$)	Typical operating temperature (°C)
1st effect	2,300–2,600	2,000–2,200	60–70
2nd effect	1,900–2,200	1,700–1,900	50–60
3rd effect	1,000–1,200	900–1,100	30–40

The pressure exerted by a column of liquid of height h and density ρ is:

$$P = \rho h(g/g_c)$$

Steam tables can be used to estimate the boiling point of water at a given pressure in each effect. Boiling point elevation due to solid concentration in each effect can be estimated by the following equation[6]:

$$\Delta T_b = 0.51m$$

m is the molality = (moles of solvent)/(1000 g solvent)

Using the above relationship, Fig. 2 shows the boiling point elevation of a sugar solution at different concentrations.

Similar equations can be written for backward and parallel feed multiple-effect evaporator arrangements.

CLEANING OF MULTIPLE-EFFECT EVAPORATORS

Because of the close design and large thermal capacity of multiple-effect evaporators, the use of steam or hot water is ineffective for sterilization. The following cleaning sequence is recommended by [1]:

1. Warm water rinse (45°C).
2. After the rinse water has run clear, clean with 1%–4% alkaline solution (80°C) by recirculating for 45 min–50 min.
3. Discharge the alkaline solution and rinse with warm water (60°C).
4. Recirculate a 0.3%–0.5% acid solution (70°C) for 20 min–30 min.

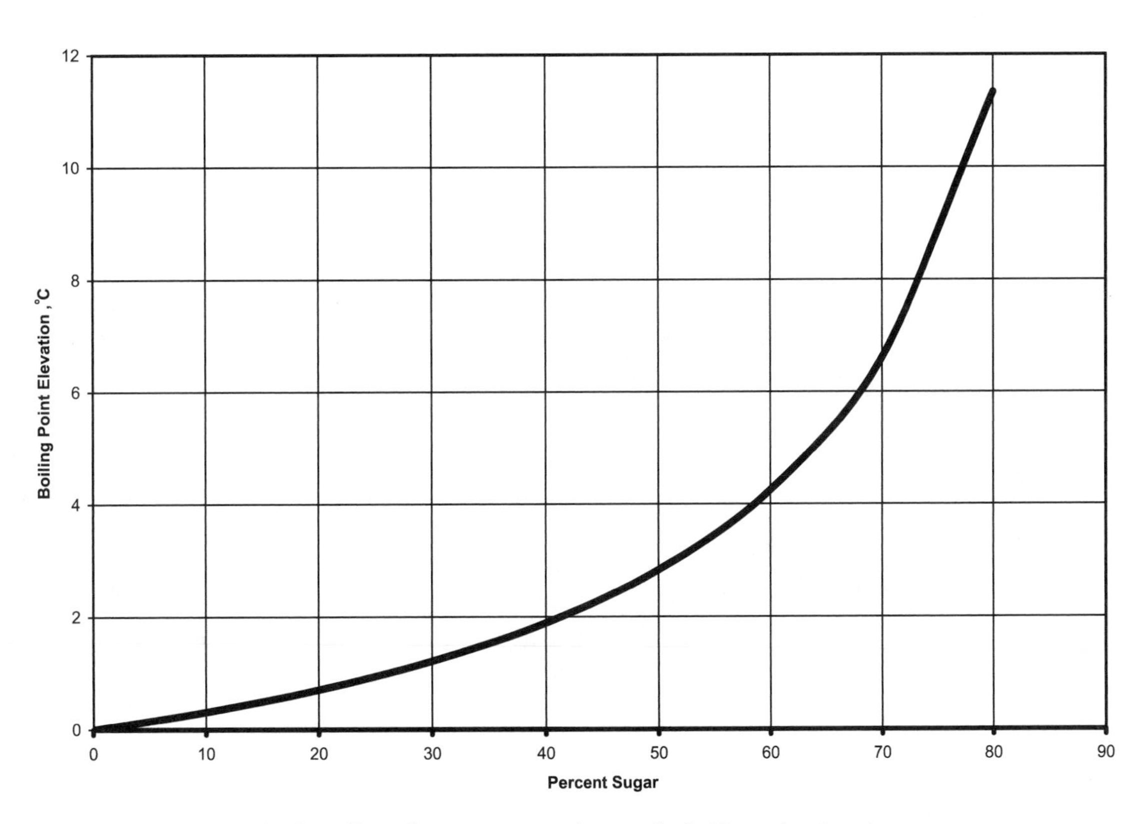

Fig. 2 Effect of sugar concentrations on the boiling point elevation.

5. Discharge the acid solution and follow with warm rinse (60°C).
6. Immediately before start up, sterilize the evaporator with water containing 75 ppm active chlorine for 5 min and followed with a warm water rinse.

REFERENCES

1. Hall, C.W.; Hedrick, T.I. *Drying Milk and Milk Products*; The AVI Publishing Co.: Westport, CT, 1966.
2. Heid, J.L.; Maynard, A.J. *Fundamental of Food Processing Operations: Ingredients, Methods and Packaging*; AVI Publishing Co.: Westport, CT, 1976.
3. Kessler, H.G. *Lebensmittel und Bioverfahrenstechnick*; Technische Universital Munchen-Weihenstephan: Germany, 1996.
4. McCabe, W.L.; Smith, J.C. *Unit Operations of Chemical Engineering*, 3rd Ed.; McGraw-Hill Book Company: New York, NY, 1976.
5. Perry, R.H.; Chilton, C.H. *Chemical Engineering Handbook*, 5th Ed.; McGraw-Hill Book Company: New York, NY, 1973.
6. Toledo, R.T. *Fundamentals of Food Process Engineering*; AVI Publishing Co.: Westport, CT, 1980.

Natural Convection

John J. Hahn
William A. Jacoby
University of Missouri–Columbia, Columbia, Missouri, U.S.A.

N

INTRODUCTION

Natural convection refers to flow induced by the heating of a fluid that would otherwise be motionless in the absence of heat transfer. This article will introduce the basic principles of heat and momentum transport. These concepts will be developed into governing equations used to solve basic natural convection problems. This leads to a discussion of dimensionless numbers and their usefulness in solving problems that are more complex. Finally some biological applications will be introduced.

BACKGROUND

Natural or free convection occurs when a fluid is set in motion by coming into contact with a surface of different temperature. The motion is a result of the imbalance in gravity and pressure forces on a fluid particle arising from differences of density associated with the transfer of heat. The fluid motion increases the rate of heat transfer above the value that would be experienced in the absence of motion. There is a close connection between how fluid moves and how the heat is transferred within the moving fluid. In natural convection, heat transfer drives fluid flow. Forced convection, however, relies on fluid flow to enable heat transfer.

The density of most fluids decreases as their temperature increases. When heating a pot of water on a stove, the heat transfer to the water forms a layer of warm, less dense fluid immediately adjacent to the surface of the bottom of the pot. This warmed fluid rises toward the upper surface of the water inducing the flow of cooler water toward the bottom, where it gets heated and renews the cycle of motion.

Differences in density lead to stratification in natural system when fluids such as air and water are heated by sunlight from above. Colder, more dense fluids are "trapped" beneath warmer, less dense fluids. Conversely, natural convection occurs when, on cool summer evenings following hot days, the ground radiates heat and warms air near its surface. Therefore, air quality indices are typically higher in summer than in winter.

Natural convection is vital in combustion processes. The heat released induces motion in the surrounding air that provides the necessary supply of oxygen. Natural convection also enables heat transfer in the cooling fins of man-made objects such as electronic components, as well as flora and fauna that exhibit appendages adapted for this purpose. Further, natural convection is vital in temperature regulation in all warm-blooded species.

Natural convection plays an important role in biological systems and processes. Certain plants and crops require specific temperature control to thrive. In a controlled environment such as a greenhouse, heat can be evenly distributed using natural convection. Heat loss by animals and livestock can be calculated based on convective heat transfer, which may have significant economic impact. Certain sterilization techniques used in food processing also use natural convection to transport heat where other heat transport techniques may not be possible or desirable. In an environmental sense, wind energy can be extracted from the heat engine generated by natural convection currents. We will discuss the fundamental principles behind natural convection and some of its applications in biological systems.

BASIC HEAT TRANSPORT PHENOMENON

The first step in understanding natural convection is developing an appropriate form of the thermal energy equation. This derivation is in the context of a Lagrangian or closed system. Similar derivations using a Eulerian or open, steady-state system are found in the literature.[1–5] Both derivations lead to the same result.

The first law of thermodynamics states that the change in energy is equal to the work done on the system, the total heat transferred to the system, and any heat sources and sinks.[1] This concept is stated mathematically in Eq. 1, which is an energy balance around a material element, V. Vector tensor notation, symbols, and operators are defined as in Ref. [1]. In this section, vectors will be denoted in

Encyclopedia of Agricultural, Food, and Biological Engineering
DOI: 10.1081/E-EAFE 120007186
Copyright © 2003 by Marcel Dekker, Inc. All rights reserved.

bold type and variables in italics.

$$\frac{D}{Dt}\int_V \rho\left(\frac{1}{2}v^2+\hat{U}\right)\mathrm{d}V=\int_A \mathbf{v}\cdot\mathbf{t}\,\mathrm{d}A+\int_V \mathbf{v}\cdot\rho\mathbf{F}\,\mathrm{d}V-\int_A \mathbf{q}\cdot\mathbf{n}\,\mathrm{d}A+\int_V \rho Q\,\mathrm{d}V \quad (1)$$

The operator D/Dt represents the substantial time derivative also known as the Lagrangian derivative. The term on the left hand side of the equals sign is the change in energy of the system as a function of time where ρ is the fluid density, $\mathbf{v}$ the velocity of the fluid, U the internal energy, V the control volume, $\mathbf{F}$ the body or gravity forces acting on the control volume, $\mathbf{q}$ the heat flux, A the surface area of the control volume, and Q the heat generated. The first term on the right hand side of the equals sign represents the work done by surface forces where $\mathbf{t}$ is the stress vector. The second term is work done by body forces. The third term is heat conduction. The last term represents a heat source or sink.

Applying Reynolds Transport Theorem to Eq. 1 results in:

$$\int_V \rho\frac{\mathrm{d}}{\mathrm{d}t}\left(\frac{1}{2}v^2+\hat{U}\right)\mathrm{d}V=\int_V \nabla\cdot(\mathbf{T}\cdot\mathbf{v})\mathrm{d}V+\int_V \mathbf{v}\cdot\rho\mathbf{F}\,\mathrm{d}V-\int_V \nabla\cdot\mathbf{q}\,\mathrm{d}V+\int_V \rho Q\,\mathrm{d}V \quad (2)$$

where $\mathbf{T}$ is the second order stress tensor. For a differential balance, the integrals can be removed and this equation can then be written as:

$$\rho\frac{D}{Dt}\left(\frac{1}{2}v^2+\hat{U}\right)=\nabla\cdot(\mathbf{T}\cdot\mathbf{v})+\mathbf{v}\cdot\rho\mathbf{F}-\nabla\cdot\mathbf{q}+\rho Q \quad (3)$$

Introduction of the mechanical energy balance can lead to further simplification of Eq. 3. Starting with Cauchy's equation of motion

$$\rho\frac{D\mathbf{v}}{Dt}=\rho\mathbf{F}+\nabla\cdot\mathbf{T} \quad (4)$$

and forming the scalar product of Eq. 4 and the velocity vector results in Eq. 5.

$$\rho\frac{1}{2}\frac{D}{Dt}(v^2)=\rho\mathbf{v}\cdot\mathbf{F}+\mathbf{v}\cdot(\nabla\cdot\mathbf{T}) \quad (5)$$

Subtracting Eq. 5 from Eq. 3 yields Eq. 6.

$$\rho\frac{D\hat{U}}{Dt}=\mathbf{T}^t:\nabla\mathbf{v}-\nabla\mathbf{q}+\rho Q \quad (6)$$

Limiting the discussion to Newtonian fluids allows Eq. 7

$$\mathbf{T}=-p\mathbf{I}-\boldsymbol{\tau} \quad (7)$$

where $\mathbf{I}$ is the identity matrix and $\boldsymbol{\tau}$ the second order sheer stress tensor. τ_{ii} represents normal stress and τ_{ij} where ij represents tangential or sheer stress. Substituting the transpose of $\mathbf{T}$ into Eq. 6 results in Eq. 8.

$$\rho\frac{D\hat{U}}{Dt}=-p\nabla\cdot\mathbf{v}-\tau:\nabla\mathbf{v}-\nabla\mathbf{q}+\rho Q \quad (8)$$

The expression on the left side of the equation is the time rate of change of the internal energy, measured at a point moving with the fluid. The first term on the right hand side is the rate of reversible increase in internal energy by compression. The second term is the rate of irreversible increase in internal energy by viscous dissipation, which is always positive. The third term represents the rate of energy input by conduction. The fourth term represents the heat source or sink.

Eq. 8 represents the equation of thermal energy in its general form. However, this equation involves the substantial derivative of the internal energy. Since internal energy is not readily measured, it would be easier to work with some other variables. We start by defining the internal energy to be a function of specific volume and temperature:

$$d\hat{U}=\left(\frac{\partial\hat{U}}{\partial T}\right)_{\hat{V}}\mathrm{d}T+\left(\frac{\partial\hat{U}}{\partial\hat{V}}\right)_T\mathrm{d}\hat{V} \quad (9)$$

$$d\hat{U}=C_V\mathrm{d}T+\left(-p+T\left(\frac{\partial p}{\partial T}\right)_{\hat{V}}\right)\mathrm{d}\hat{V} \quad (10)$$

Substituting Eq. 10 into the left side of Eq. 8

$$\rho\frac{D\hat{U}}{Dt}=\rho\left[C_V\frac{DT}{Dt}+\left[-p+\left(\frac{\partial p}{\partial T}\right)_{\hat{V}}\right]\frac{D\hat{V}}{Dt}\right] \quad (11)$$

Recalling that the specific volume is inversely proportional to the density and utilizing the equation of continuity, Eqs. 8 and 11 combine to form

$$\rho C_V\frac{DT}{Dt}=-T\left(\frac{\partial p}{\partial T}\right)_{\hat{V}}(\nabla\cdot\mathbf{v})-\tau:\nabla\mathbf{v}-\nabla\mathbf{q}+\rho Q \quad (12)$$

Similarly in terms of enthalpy, H,

$$\mathrm{d}\hat{H}=C_P\mathrm{d}T+\left[\frac{1}{\rho}-T\frac{\partial}{\partial T}\left(\frac{1}{\rho}\right)_P\right]\mathrm{d}p \quad (13)$$

Combining terms gives us

$$\rho C_P\frac{DT}{Dt}=\left(\frac{\partial\ln\frac{1}{\rho}}{\partial\ln T}\right)_P\frac{Dp}{Dt}-\tau:\nabla\mathbf{v}-\nabla\mathbf{q}+\rho Q \quad (14)$$

Eq. 14 can be simplified even further if we make the following assumptions. Many of the terms on the right side of the equation are negligible or zero. Most fluids tend to be incompressible making the first term zero. Because we will be dealing with natural convection, we will assume that the fluid is not highly viscous and therefore the viscous dissipation term can also be assumed to be negligible. If there is no heat generation, then Q is also zero, which only leaves us the conduction term. Using Fourier's Law

$$\mathbf{q} = -k\nabla T \tag{15}$$

If k is not a function of T, then Eq. 14 can be rewritten as

$$\rho C_P \frac{DT}{Dt} = k\nabla^2 T \tag{16}$$

Eq. 16 is the energy balance equation used in many heat transfer applications including natural convection. Depending on the application, this equation can be written in any appropriate coordinate system.

BASIC MOMENTUM TRANSPORT THEORY

Before exploring natural convection applications, we must first develop an equation of motion to describe the velocity profiles of the convection. We will approach this derivation with a Lagrangian description similar to the equation of thermal energy as follows

$$\frac{D}{Dt}\int_V \rho\mathbf{v}\,\mathrm{d}V = \int_V \rho\mathbf{F}\,\mathrm{d}V + \int_A \mathbf{t}\,\mathrm{d}A \tag{17}$$

This equation represents the total force acting on the control volume. The first term on the right hand side is the total body force and the second term represents the total surface forces acting on the volume. Applying the Reynolds transport theorem to the left side and the Gauss divergence theorem to the surface integral to convert it to a volume integral, Eq. 17 can be rewritten as

$$\rho\frac{D\mathbf{v}}{Dt} = \rho\mathbf{F} + \nabla\cdot\mathbf{T} \tag{18}$$

In natural convection, the only body force, **F**, will be gravity. Assuming the fluid is incompressible and the viscosity is constant, the second order stress tensor can be expanded so Eq. 18 can be rewritten as

$$\rho\frac{D\mathbf{v}}{Dt} = \rho\mathbf{g} - \nabla p + \mu\nabla^2\mathbf{v} \tag{19}$$

Eq. 19 is also known as the Navier-Stokes equation. $\mathbf{v}$ is the velocity of the fluid, $\mathbf{g}$ the acceleration due to gravity, p the pressure, and μ the viscosity.

In natural convection systems we will assume the pressure gradient is only due to the weight of the fluid so that Eq. 19 is written as

$$\rho\frac{D\mathbf{v}}{Dt} = \mathbf{g}(\rho_b - \rho) + \mu\nabla^2\mathbf{v} \tag{20}$$

where ρ_b and ρ are the density of the bulk fluid and fluid near the surface, respectively. Expanding density in a Taylor series with respect to temperature and eliminating the higher order terms, the change in buoyancy will be approximated as

$$\beta = \frac{\rho_b - \rho}{\rho(T - T_b)} \tag{21}$$

This is known as the Boussinesq approximation. Substituting this into Eq. 20 gives

$$\rho\frac{D\mathbf{v}}{Dt} = \mathbf{g}\beta\rho(T - T_b) + \mu\nabla^2\mathbf{v} \tag{22}$$

Eq. 22 will be the governing equation for many typical natural convection problems.

NATURAL CONVECTION BETWEEN PARALLEL PLATES

We can now study the simple natural convection problem of a fluid between two parallel plates, with one plate being kept at a constant temperature T_2 and the other kept at temperature T_1, as seen in Fig. 1. The temperature distribution is first determined using Eq. 16. Because the plates are assumed to be infinitely wide in the x and z direction, the temperature only varies in the y direction. For steady state, the time dependent term can be eliminated so that Eq. 16 simplifies to

$$k\frac{\mathrm{d}^2T}{\mathrm{d}y^2} = 0 \tag{23}$$

The boundary conditions are

$$T = T_2 \text{ at } y = -b \tag{24a}$$

$$T = T_1 \text{ at } y = +b \tag{24b}$$

The solution to this differential equation is

$$T = \frac{T_1 + T_2}{2} - (T_2 - T_1)\left(\frac{y}{2b}\right) \tag{25}$$

Because of the way this problem is set up, the temperature profile is a linear relationship between the temperature and distance.

Examining the momentum balance, we see that the z-component of the velocity only varies in the y

N

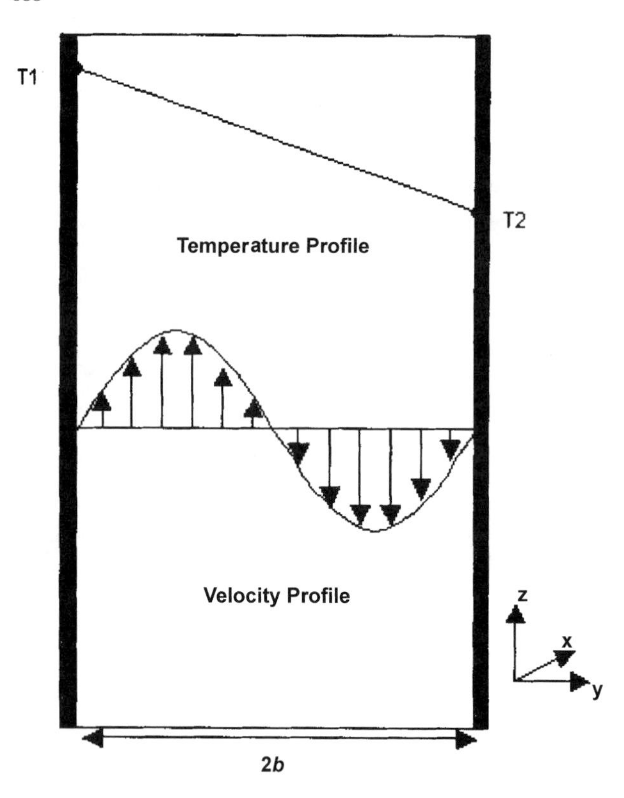

Fig. 1 Temperature and velocity profile between two infinite parallel plates.

direction and Eq. 22 simplifies to

$$\mu \frac{d^2 v_z}{dy^2} = -\mathbf{g}\beta\rho(T - T_b) \tag{26}$$

The boundary conditions for this equation are

$$v_z = 0 \text{ at } y = -b \tag{27a}$$

$$v_z = 0 \text{ at } y = +b \tag{27b}$$

Substituting in Eq. 25 for T and solving the differential equation gives us

$$v_z = \frac{1}{12}\frac{\rho\beta\mathbf{g}b^2\Delta T}{\mu}(\eta^3 - \eta) \tag{28}$$

Using the following dimensionless terms

$$\phi = \frac{bv_z\rho}{\mu} \tag{29a}$$

$$\eta = \frac{y}{b} \tag{29b}$$

Eq. 28 is rewritten as

$$\phi = \frac{1}{12}\frac{\rho^2\beta\mathbf{g}b^3\Delta T}{\mu^2}(\eta^3 - \eta) = \frac{1}{12}Gr(\eta^3 - \eta) \tag{30}$$

Gr is known as the Grashof number, a dimensionless number which represents the ratio of the buoyancy forces to the viscous forces. This dimensionless number will be important in solving most natural convection problems.

Fig. 1 shows the graphical solution to both the temperature profile and velocity profile. On the hot side, the velocity profile shows the fluid moving up as its density decreases because of heating. The opposite effect is seen on the cold side. In all these natural convection problems, the heat and the momentum equations (Eqs. 16 and 22) are always coupled together since the velocity is a function of buoyancy, which is dependent upon the temperature.

SOLVING MORE COMPLICATED NATURAL CONVECTION PROBLEMS

Solutions to other simple geometries are well documented in a number of sources. However, when dealing with more complicated systems, it is often easier to use empirical formulas based on key dimensionless groups such as the Grashof number.

One of the most significant dimensionless numbers used in convection problems is the Nusselt number. The Nusselt number is the ratio of the heat transfer due to convection represented by the heat transfer coefficient, h, to the heat transfer due to thermal conductivity of the fluid alone.

$$Nu = \frac{hL}{k} \tag{31}$$

The term, L, is the characteristic length and is dependent on the geometry. In the case of fluid passing over a plate, the characteristic length would simply be the length of the plate traveled by the fluid. In the case of a pipe, the characteristic length would be the length of the pipe if the fluid is traveling in the direction of the pipe or the diameter of the pipe if the fluid is traversing the pipe.

The Nusselt number can be expressed as a function of the Grashof number and the Prandtl number. The Prandtl number is the ratio of the kinematic viscosity (momentum diffusivity), μ/ρ, to the thermal diffusivity, $k/\rho c_p$.

$$Pr = \frac{\mu/\rho}{k/\rho c_p} = \frac{c_p\mu}{k} \tag{32}$$

μ represents the viscosity of the fluid, ρ the density, k the heat transfer coefficient, and c_p the heat capacity of the fluid. For many problems, the Nusselt number can then be expressed as a function of the Grashof number and the Prandtl number as follows:

$$Nu = f(Gr, Pr) = c(Gr\,Pr)^n = c(Ra)^n$$

$$= c\left(\frac{\rho^2 \beta g L^3 \Delta T}{\mu^2}\frac{c_p \mu}{k}\right)^n \quad (33)$$

where c and n are empirically derived constants based on the shape factor of the transport problem. The product of the Grashof number and the Prandtl number is also referred to as the Rayleigh number. Table 1 shows some constants used with natural convection problems.

Example—Convection from a Cow

A 600 kg cow is standing in still air at 15°C. For simplicity, we assume the cow to be a horizontal cylinder with a length of 1.64 m and diameter of 40 cm. We will assume ideal gas and start with an initial guess of 19°C. Using obtained values for air density and viscosity,[2,6] we can calculate the following:

$$\rho = 1.22 \text{kg m}^{-3} \text{ at } 290\text{K} \quad (34a)$$

$$\mu = 1.81 \times 10^{-5} \text{kg m}^{-1}\text{sec}^{-1} \text{ at } 290\text{K} \quad (34b)$$

$$k = 0.0252\,\text{N m sec}^{-1}\,\text{m}^{-1}\,\text{K}^{-1} \quad (34c)$$

$$Pr = 0.7 \quad (34d)$$

$$Gr = \frac{\rho^2 \beta g b^3 \Delta T}{\mu^2} = \frac{(1.22\,\text{kg m}^{-3})^2(9.81\,\text{m sec}^{-2})(3.42 \times 10^{-3}\text{K}^{-1})(0.8\,\text{m})^3(292 - 288\text{K})}{(1.81 \times 10^{-5}\,\text{kg m}^{-1}\,\text{sec}^{-1})^2} = 3.12 \times 10^8 \quad (34e)$$

Using Table 1, we calculate:

$$Nu = 0.53(Gr\,Pr)^{0.25} = 0.53((3.12 \times 10^8)(0.7))^{0.25}$$

$$= 64.4 \quad (34f)$$

Rearranging Eq. 31

$$h = \frac{Nu{\cdot}k}{L} = \frac{(64.4)(0.0252)}{0.8} = 2.02\,\text{W m}^{-2}\,\text{K}^{-1} \quad (34g)$$

If the cylindrical shape is still considered, the total heat loss can be estimated as:

$$q = hA(T - T_b) = (2.02)(5.12)(17) = 176\,\text{W} \quad (34h)$$

Other biological systems can be approximated using similar techniques as shown above. More complicated systems will have specific empirical correlations, which extend beyond the scope of this article.[7–11]

Example—Convective Heating in a Greenhouse

In order to provide heat in the wintertime for optimal plant growth, some greenhouses employ the use of natural convection to distribute warm air. A network of heating

Table 1 Constants used for Eq. 33 for various shapes and Rayleigh numbers

Geometry	$Gr\,Pr$	c	n
Vertical planes and cylinders	$<10^4$	1.36	.2
	10^4–10^9	.59	.25
	$>10^9$	.13	.33
Horizontal cylinders	$<10^{-5}$	.49	0
	10^{-5}–10^{-3}	.71	1/25
	10^{-3}–1	1.09	0.1
	1–10^4	.53	.2
	10^4–10^9	.53	.25
	$>10^9$	.13	.33
Horizontal plates			
—Upper surface heated	10^5–2×10^7	.54	.25
	2×10^7–3×10^{10}	.14	.33
—Lower surface heated	10^5–10^{11}	.58	.2

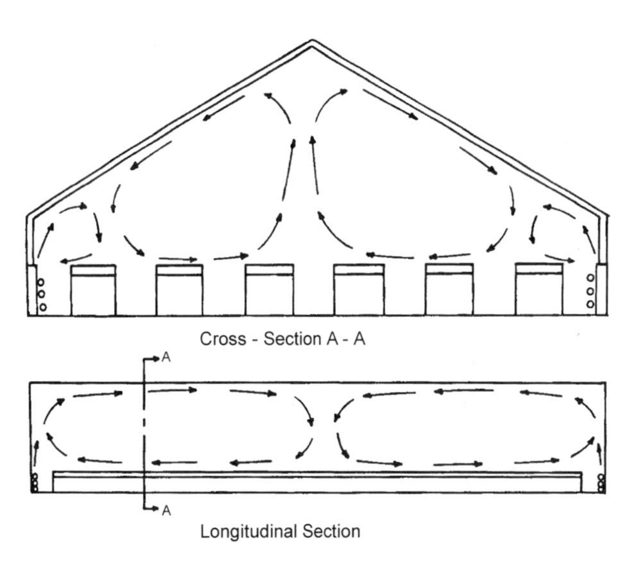

Fig. 2 Airflow pattern generated by heating pipes in a greenhouse. (From Ref. [12].)

pipes, set up along the perimeter of the greenhouse, can generate heat by passing hot water or steam through them to act as a radiator. The heated air would rise due to the change in buoyancy and circulate within the greenhouse. The location of the pipe network is critical for insuring proper airflow. Fig. 2 shows an example of undesirable airflow.[12] The warmer air rising from the pipes is met by falling air that has been cooled from the glass roof. This airflow pattern creates cold spots within the greenhouse, which is detrimental to some plant development. Heating pipes can be placed overhead to minimize the cold down drafts. Another technique is to employ a combination forced and natural convection system by installing fans in strategic places to distribute the warmer air. Other techniques of heat distribution are used in greenhouse design, and choosing what techniques to use will depend on the climate, region, and plant and soil conditions.[12,13]

ACKNOWLEDGMENT

We would like to thank Paul Chan for all his help.

REFERENCES

1. Bird, R.B.; Stewart, W.E.; Lightfoot, E.N. *Transport Phenomena*; John Wiley & Sons: New York, 1960.
2. Geankoplis, C.J. *Transport and Unit Operations*, 3rd Ed.; Prentice Hall PTR: Englewood Cliffs, New Jersey, 1993.
3. Cengel, Y.A. *Heat Transfer: A Practical Approach*; WCB McGraw-Hill: Boston, 1998.
4. Leal, L.G. *Laminar Flow and Convective Transport Processes: Scaling Principles and Asymptotic Analysis*; Butterworth-Heinemann: Boston, 1992.
5. Arpaci, V.S.; Larson, P.S. *Convection Heat Transfer*; Prentice-Hall, Inc.: Englewood Cliffs, New Jersey, 1984.
6. Johnson, A.T. *Biological Process Engineering*; John Wiley & Sons, Inc.: New York, 1999.
7. Heldman, D.R.; Singh, R.P. *Food Process Engineering*, 2nd Ed.; AVI Publishing Company, Inc.: Westport, Connecticut, 1981.
8. Bailey, J.E.; Ollis, D.F. *Biochemical Engineering Fundamentals*, 2nd Ed.; McGraw-Hill, Inc.: New York, 1986.
9. Kakac, S.; Aung, W.; Viskanta, R. *Natural Convection Fundamentals and Applications*; Hemisphere Publishing Corporation: Washington, 1985.
10. Jaluria, Y. Natural Convection Heat and Mass Transfer. In *The Science and Application of Heat and Mass Transfer*; Spalding, D.B., Ed.; Pergamon Press: Oxford, 1980.
11. Goldman, C.R.; Horne, A.J. *Limnology*; McGraw-Hill Book Company: New York, 1983.
12. Grey, H.E. *Greenhouse Heating and Construction*; Florists' Publishing Company: Chicago, 1956.
13. Nelson, P.V. *Greenhouse Operation and Management*; Reston Publishing Company: Reston, VA, 1981.

Neural Networks

K. Chao
United States Department of Agriculture (USDA), Beltsville, Maryland, U.S.A.

K. C. Ting
The Ohio State University, Columbus, Ohio, U.S.A.

N

INTRODUCTION

Quality is increasingly important in food and agricultural production as well as other manufacturing processes. In continuous processes, the goal in most cases is to keep the process composition steady and close to the optimum conditions. Uniform quality is also a required aspect of the process. Furthermore, there are frequently legal obligations that have to be fulfilled by product composition, and in many cases the most economical product is the one closest to the legal limit. There are requirements for environmental protection, production, and plant safety. All of these require that the composition of various products be kept stable. These principles have been considered for many years in the development of control theory, controllers, and actuators in parallel with the growth of manufacturing industries.

Because the behavior of most bioproduction processes are usually characterized by the interactions of many components, large number of influences, and nonlinearity, the modeling approach based on the stable composition principle is insufficient to represent the behavior of the process. On the other hand, measurements are usually available for each process cycle, which can be used to monitor the process. Neural networks have been shown to be good predictors of relations between process data and can be effectively used for modeling plant processes. They are thus an appropriate candidate for inclusion in process control designs.

In the following sections, process control designs are discussed with regard to why neural networks are appropriate to be included in process control, the backpropagation learning algorithm and a design example of neural networks based process control in a continuous food frying process are presented.

PROCESS CONTROL DESIGN

There are three levels of monitoring and controlling of a product in a processing plant. The simplest monitoring is achieved by randomly pulling samples from a production line and analyzing these samples offline in a quality control laboratory. These results are then studied using statistical processing control[1] to ensure the consistency of the plant output. The next level is a continuous or semicontinuous auditing function done by an automatic analyzer. A continuous monitoring of the finished products can alleviate the product liability problem but cannot prevent the large-scale massive production of faulty finished products. Therefore, there is a need in manufacturing processes to monitor and control the intermediate products, as well as the finished products, and make process control decisions based on the results of on-line measurements. This type of on-line process control can be visualized as a single-loop-feedback control system,[2] as shown in Fig. 1. The goal of the feedback control is to design the controller transfer function, G_c, to maintain the output, y, at a desired set point value, r, in spite of the disturbance, d, that affect the process.

Most classical feedback control concepts have been developed based on the study of second-order linear systems. A general second-order linear system is given by the linear differential equation,

$$\frac{d^2y(t)}{dt^2} + 2\xi\omega_0\frac{dy(t)}{dt} + \omega_0^2 y(t) = \omega_0^2 u(t) \tag{1}$$

By using operator $s = d/dt$, the normalized transfer function, $H(s)$, of a second-order system is obtained:

$$H(s) = \frac{\omega_0^2}{s^2 + 2\xi\omega_0 s + \omega_0^2} \tag{2}$$

where ω_0 is natural frequency in rad sec^{-1} and ξ is the damping factor. The root of the characterization equation (the denominator of equation), s_1 and s_2 can be real or complex, depending on the value of the damping factor, ξ. Fig. 2 shows the effect of different values of the damping constant. When $\xi < 1$, the roots s_1 and s_2 are complex so that the system response is underdamped and oscillatory. When $\xi > 1$, the roots are real so that the system is

Encyclopedia of Agricultural, Food, and Biological Engineering
DOI: 10.1081/E-EAFE 120007216
Copyright © 2003 by Marcel Dekker, Inc. All rights reserved.

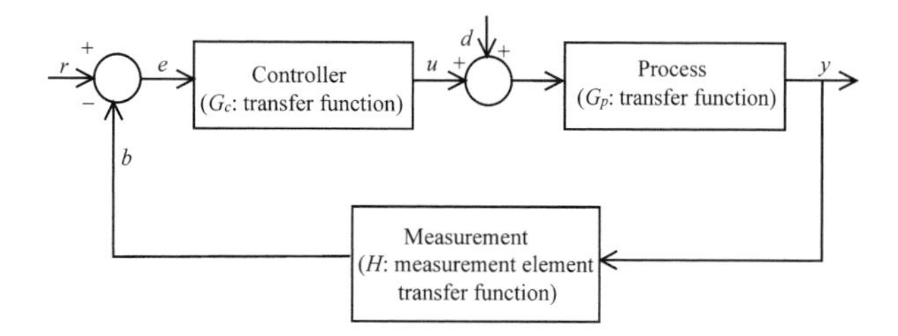

Fig. 1 Single-loop feedback control.

overdamped. When $\xi = 1$, the system response is critically damped.

Control system engineers often attempt to make the set point step response approximate the step response of an underdamped second-order system. A number of design procedures[3,4] have been developed to tune proportional-integral-derivative (PID) controllers for single-loop-feedback systems. The PID controller is given by

$$u = -K_P\left[e + K_I\int e\,dt + K_D\frac{de}{dt}\right] \tag{3}$$

where K_P is the proportional gain constant, K_I is the integral gain constant, and K_D is the derivative gain constant. These single-loop PID controllers have been widely used in the process industry. However, the inability of this type of controller to handle multivariable systems and/or interactions between state and control variables has led to the development of new types of process control designs. There are modern engineering and mathematical approaches that can overcome these shortcomings of the single-loop design procedure. Multivariable models can be formulated for process problems. Dynamic models can be written in terms of the state variables that describe the actual dynamic response of the system.

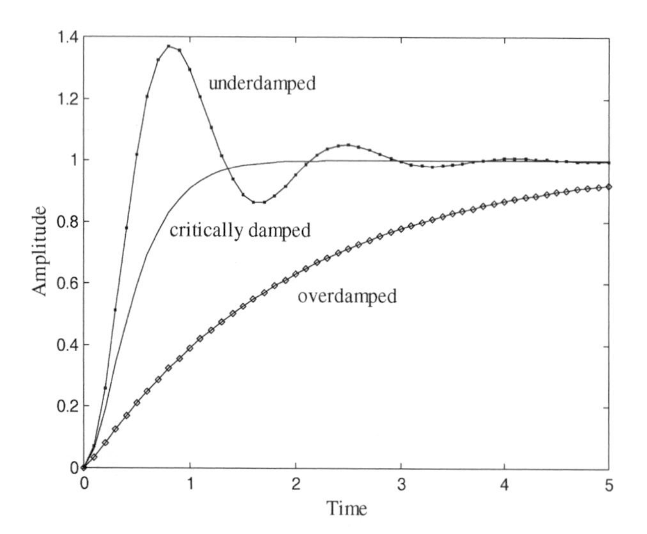

Fig. 2 Effect of the damping constant on second-order systems.

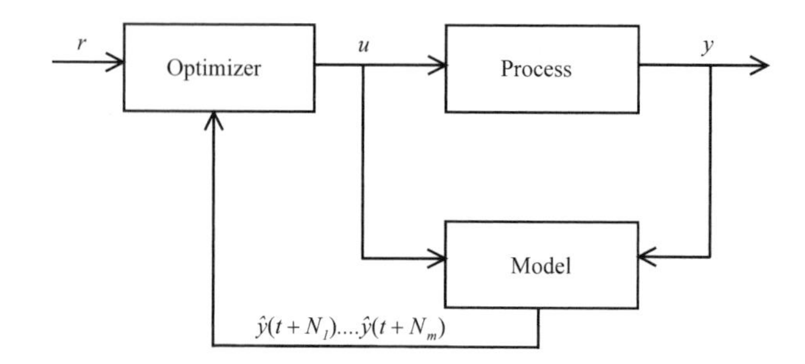

Fig. 3 Model-based predictive control.

One of the advances in control theory in recent years has been the inclusion of model predictions in the design of process control systems. This idea has been generalized by Clarke, Mothadi, and Tuffs,[5] to deal with complex dynamics (e.g., unstable inverse systems, time-varying, time delay, etc.). A model-based predictive control is composed of three main components: the system, its model, and a function optimizer (Fig. 3). The model is used to predict future plant behavior. According to the predicted behavior of the plant, the optimizer defines the required sequence of action u to make the system behave as desired. This optimizer takes the form of a quadratic cost function.

This control scheme can be considered an open loop since the plant output is not required. However, the plant output can be used in later adjustments to the model. This is important because it must be assumed that after a while, due to discrepancies in the model, the plant behavior is going to diverge from that of the model. An important characteristic of this scheme is that no controller design is required, only the feedforward model of the plant is going to affect the efficiency of the model-based predictive control. Hence, it is straightforward to apply neural networks in this context. Actually the problem of lack of transparency of the model has a reduced impact on this scheme because what is most important here is the accuracy and the robustness of the model.

FEEDFORWARD-BACKPROPAGATION (FFBP) NEURAL NETWORKS

Neural networks are mathematical black boxes that map inputs to outputs.[6] The mapping in the neural networks is achieved by a network of nodes with activation functions, and interconnected with links with various weights. As the neural network is modeled after the brain, it needs to undergo the same process as the brain in order to work, i.e.,

1. Accumulation of experience, i.e., input–output data via sensors.
2. Learning the mapping between the input and output.

Thus to model a process, a neural network must be trained with data sets representing input and output of interest that describe the process to be modeled and controlled.

A typical FFBP neural network (Fig. 4) consists of a number of layers (the input, hidden, and output layers) each consisting of a number of nodes or processing units. Each node in the input layer brings into the network the value of one independent variable. The nodes in the hidden layer do most of the work. Each of the output nodes computes a dependent variable. The network is fully connected in that there are links between all the nodes in adjacent layers. There is a separate link from each input node to each hidden node and from each hidden node to each output node. Each link has a weight, which is stored in and maintained by the node on the receiving end of the link. The network operates in two modes: mapping and learning modes. In mapping mode, the network processes one set of inputs at a time, producing an estimate of the values of output variables. The mapping carried out by the network depends on the values of the weights. Backpropagation is the method of finding the optimum values for these weights. It involves training the network based on input–output data. The basic neural networks applied for the backpropagation learning algorithm is illustrated in Fig. 4.

The notation conventions are shown in the figure: output layer is denoted by O_i, hidden layer by V_j, and input terminals by ξ_k. There are connections (weights and biases) ω_{jk} from the inputs to the hidden layer, and W_{ij} from the hidden layer to the output layer. Note that the index i refers to an output unit, j to a hidden unit, and k to an input terminal. The backpropagation algorithm[7] can be summarized as follows:

- Step 1: Initialize weights from N inputs to M output nodes to small random values.

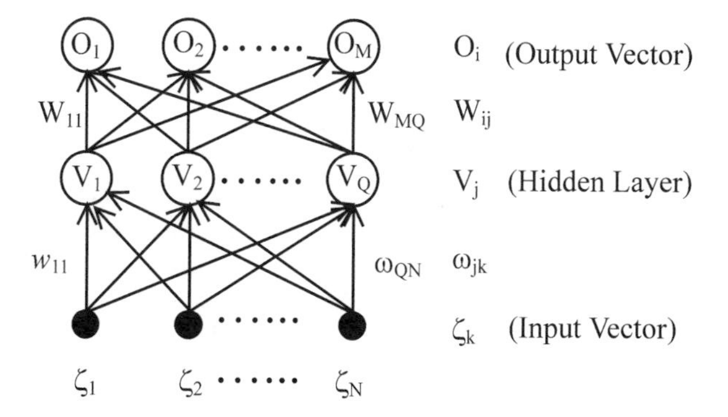

Fig. 4 Structure of a multilayer feedforward network.

- Step 2: Choose an input pattern ζ_k^p, where p is the input number (e.g., for r input vectors, $p = 1,\ldots,$r), $k = 1,\ldots,N$, and apply it to the input layer.
- Step 3: Feed the signal forward through the network using the relation

$$O_i^p = f(h_i^p) = f\left(\sum_j W_{ij} V_j^p\right)$$
$$= f\left(\sum_j W_{ij} f\left(\sum_k \omega_{jk} \zeta_k^p\right)\right),$$

for each i, j, and k until the final outputs O_i^p have been calculated. Note that $f(\cdot)$ is called the activation function, which must be continuous and differentiable. The notation (h_i^p) indicates the value transmitted to output node p for sample i.
- Step 4: Compute the deltas for the output layer, using $\delta_i^p = f'(h_i^p)\left[\Omega_i^p - O_i^p\right]$ by comparing the desired outputs Ω_i^p with the calculated outputs O_i^p for the pattern p being considered.
- Step 5: Compute the deltas for the preceding (hidden-to-input) layer by propagating the error backwards, using

$$\delta_j^p = f'(h_i^p)\sum_i W_{ij}\delta_i^p.$$

- Step 6: Update weights associated with each layer by the learning rule according to the relation

$$\varpi_{pq}^{new} = \varpi_{pq}^{old} + \Delta\varpi_{pq},$$

where output and input refer to the two ends p and q of the connection connected; the learning rule always has the form:

$$\Delta\varpi_{pq} = \eta\sum_{pattern}\delta_{output}\Phi_{input}$$

where Φ stands for the appropriate input-end activation from a hidden unit or a real input. The gain term η decreases in time provided that $0 \leq \eta \leq 1$.

NEURAL NETWORK MODEL-BASED PROCESS CONTROL

Artificial neural networks (ANN) have been used successfully to solve real-world process control problems.[8–12] These applications include temperature control, chemical process system control, and theoretical development.

In ANN-based process control, dynamic models that include past values of the system input and outputs are typically used. A single-input–single-output dynamic model can be represented as follows:

$$\hat{y}(t) = f\big(y(t-1),\ldots,y(t-n);\, u(t-k),\ldots,u(t-k-n) + e(t)\big) \tag{4}$$

where $y(t)$ and $u(t)$ are the sampled process output and input at time t, k is an input–output time delay, n is the number of past inputs and outputs used in the model, f is a nonlinear function describing the system behavior, and $e(t)$ is the approximation error.

To illustrate the ANN modeling, we present a detailed analysis of a continuous, snack food frying process.[13] The product quality of snack food frying process is affected by two factors: the inlet temperature of the frying oil and the residence time of the product in the fryer. Color and moisture content are two quality indicators for the process. Sensors are placed at the end of the production line to directly measure the quality attributes, so there are time lags between the inputs and outputs of the process. This process can be modeled by a FFBP neural network:

$$\hat{\mathbf{y}}(t) = \hat{f}\big[\mathbf{y}(t-1), \mathbf{y}(t-2),\ldots,\mathbf{y}(t-p), \mathbf{u}(t-d-1), \times \mathbf{u}(t-d-2),\ldots,\mathbf{u}(t-d-q), \mathbf{w}, \mathbf{e}(t)\big] \tag{5}$$

where $\mathbf{y}(t) = [y_1(t), y_2(t)]^T$ is the process output vector for color and moisture content (%) at time t; $\mathbf{u}(t) = [u_1(t), u_2(t)]^T$ is the process input vector for inlet temperature (°C) and residence time (sec) at time t; $\mathbf{w} = [w_1(t), w_2(t)]^T$ is the set of weights and bias terms for the network model; p represents the order of the past outputs; q represents the order of the past inputs; and d is the time lag from the process input to output. The training of the network (Eq. 5) for the snack food frying process is an optimization problem. The weights and bias terms can be adapted to minimize the square error of the network output as follows:

$$J = \frac{1}{2}\sum_{t=1}^{N} \big[\mathbf{y}(t) - \hat{\mathbf{y}}(t)\big]^T \big[\mathbf{y}(t) - \hat{\mathbf{y}}(t)\big] \tag{6}$$

Table 1 shows that optimal number of hidden nodes[3] can be determined based on the minimum mean square error (MSE). Table 2 shows that the validating MSE reaches minimum values at the model order (p_1, p_2, q_1, q_2) of (2,2,2,2). For the meaning of p and q, refer to equation Eq. 5. The subscripts of p and q refer to inputs 1 and 2, as well as outputs 1 and 2. The resulting smallest structure of the network for the neural network process model is $8 \times 3 \times 2$, i.e., 8 (the result of $2+2+2+2$) inputs by 3 hidden nodes by 2 outputs. Each of the predictions appears to be good (13: color MSE = 0.039, moisture MSE = 0.0117). The neural network prediction model is therefore ready for the design of the neural network based process controller.

Table 1 Results of determination of number of hidden nodes

Number of hidden nodes	Training MSE[a]	Validating MSE
1	0.107641	0.132214
2	0.031462	0.051293
3	0.029928	0.050594
4	0.029437	0.050692
5	0.028460	0.052278

[a] Mean square error.
(Adapted from Ref. 13.)

Once a trained FFBP network model is available, it can be used in a straightforward manner for process control. In most applications, the networks were trained to become a system inverse. To illustrate the neural model predictive control, a continuous, snack food frying process control[14] is discussed. Eq. 5 can be further rewritten in the form of one-step-ahead prediction:

$$\hat{\mathbf{y}}(t+d+1) = \hat{f}\,[\,\mathbf{y}(t+d), \mathbf{y}(t+d-1),\ldots,\mathbf{y}(t+d-p+1), \mathbf{u}(t), \mathbf{u}(t-1),\ldots,\mathbf{u}(t-q-1), \mathbf{w}] \tag{7}$$

This equation contains the input vector, $\mathbf{u}(t)$, so it can be used to compute the control action. A function can be defined to compute the control action.

$$\mathbf{F}[\mathbf{u}(t)] = \mathbf{r}(t) - \hat{f}\big[\mathbf{y}(t+d), \mathbf{y}(t+d-1),\ldots,\mathbf{y}(t+d-p+1), \mathbf{u}(t), \mathbf{u}(t-1),\ldots,\mathbf{u}(t-q-1), \mathbf{w}\big] \tag{8}$$

where $\mathbf{r}(t)$ is the tracking signal of the network model predicted output $\hat{\mathbf{y}}(t+d+1)$. The control command can be computed with the inverse of the function $\mathbf{F}[\mathbf{u}(t)]$. However, the future process outputs, i.e., $\mathbf{y}(t+d-p+1),\ldots,\mathbf{y}(t+d)$ need to be estimated in order to realize the computation. The control law can be calculated by

Table 2 Results of order determination of neural network process model

Model order	Training MSE[a]	Validating MSE
(1,1,1,1)	0.032226	0.055117
(2,2,2,2)	0.029928	0.050594
(3,3,3,3)	0.031292	0.052863
(4,4,4,4)	0.034187	0.054443

[a] Mean square error.
(Adapted from Ref. 13.)

Table 3 Three integral error values for tuning neural network model-based process controller

γ	ISE	IAE	ITAE
0.001	0.026086	0.113393	8.855017
0.002	0.015497	0.077924	6.807420
0.003	0.012147	0.066104	6.449880
0.004	0.010648	0.060118	6.316867
0.005	0.009940	0.058497	6.308353
0.006	0.009683	0.059158	6.326198
0.007	0.009762	0.060914	6.372906
0.008	0.010158	0.063477	6.436001
0.009	0.010939	0.067334	6.529615
0.010	0.012320	0.073170	6.734011

(Adapted from Ref. 14.)

solving the nonlinear equation, the process prediction model, at each time instant interactively in a numerical fashion.

$$\mathbf{u}^k(t) = \mathbf{u}^{k-1}(t) + \Delta^{k-1}\mathbf{u}(t) \quad (9)$$

where $\mathbf{u}^k(t)$ is the computed $\mathbf{u}(t)$ at the kth iteration, and $\Delta^{k-1}\mathbf{u}(t)$ is the updating increment of $\mathbf{u}(t)$ at the k-1 step iteration, which can be determined by a numerical method. The gradient descent method can be applied to optimize the appropriate objective function. The control action can then be updated and expressed using a Jacobian matrix. Three error integral objective functions are minimized. They are the integral of square error (ISE), the integral of absolute error (IAE), and the integral of absolute error multiplied by time (ITAE). The minimization is achieved by tuning a single parameter (γ) in the Jacobian matrix, described in detail in Huang, Whittaker, and Lacey.[14] The three objective functions test and verify each other. The controller tuning results are shown in Table 3. With the optimal value γ of 0.005, the set point tracking and step disturbance rejection response of the controller are shown in Figs. 5 and 6. The set point tracking responses show how the process output tracks the changes of the set point values. The step disturbance rejection responses show how the process output sustains the step disturbances during the controller operations.

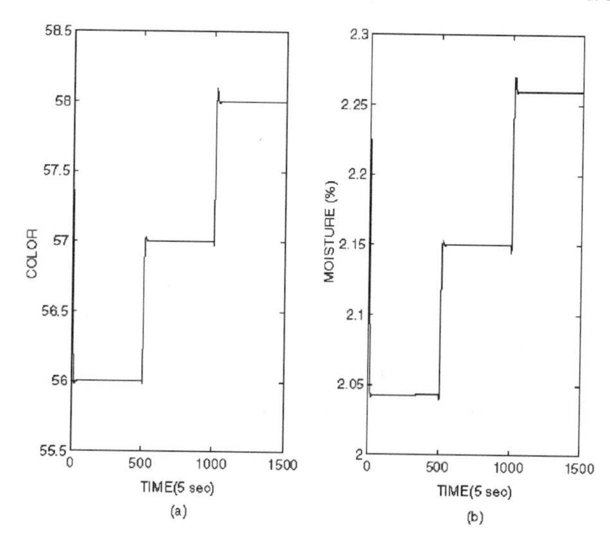

Fig. 5 Neural network model-based control in continuous food frying process[14]: (a) setpoint tracking of color response (setpoint = 57), (b) setpoint tracking of moisture content response (setpoint = 2.15%).

SUMMARY

Model-based control strategies are needed for predictive control of a system. The model used in the strategy correlates the input to and output from the system. This predictive power enables the control strategy to determine the change of input in anticipation of future output. The advantage is most obvious when time delay between the input and output of a system is a concern. ANN are a good modeling tool for establishing the relationships between system input and output based on existing data.

This article outlines the basic mathematics used in establishing and training neural network based control systems. They are illustrated with two systems studied in the fast food industry. Other results were mentioned briefly. A search of the literature showed hundreds of such studies, and the method has already been incorporated into successful commercial applications. A rapid increase in both the applications and the theoretical development can be reasonably expected in the near future.

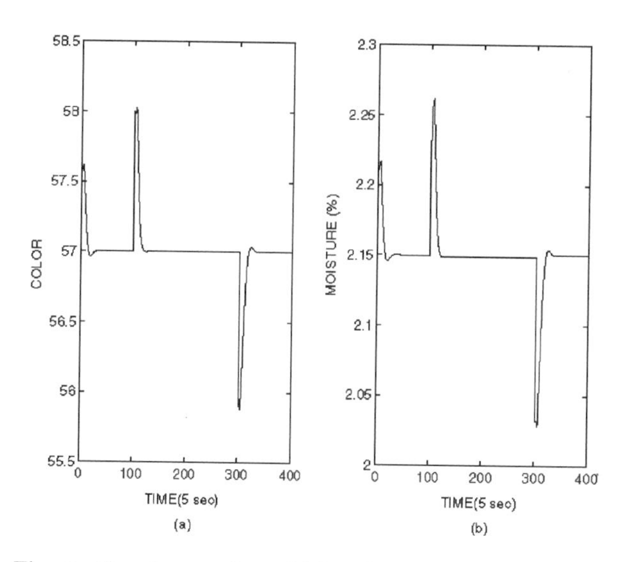

Fig. 6 Neural network model-based control step disturbance rejection in continuous food frying process[14]: (a) color response, (b) moisture content response.

REFERENCES

1. Himmelblau, D.M. *Process Analysis by Statistical Method*; John Wiley and Sons: New York, NY, 1970.
2. Kailath, T. *Linear Sysems*; Prentice-Hall: Englewood Cliffs, NJ, 1980.
3. Coughanowr, D.R.; Koppel, L.B. *Process Systems Analysis and Control*; McGraw-Hill, Inc.: New York, NY, 1965.
4. Seborg, D.E.; Edgar, T.F.; Mellichamp, D.A. *Process Dynamics and Control*; John Wiley and Sons: New York, NY, 1989.
5. Clarke, D.W.; Mothadi, C.; Tuffs, P.S. Generalized Predictive Control. Automatica **1987**, *23*, 137–148.
6. Kosko, B. *Neural Networks and Fuzzy Systems: A Dynamical Systems Approach to Machine Intelligence*; Prentice-Hall: Englewood Cliffs, NJ, 1992.
7. Rumulhart, D.E.; McClelland, J.L. *Parallel Distributed Processing*; MIT press: Cambridge, MA, 1986; Vol. 1.
8. Bhat, M.; McAvoy, T.J. Use of Neural Nets for Dynamic Modeling and Control of Chemical Process Systems. Comput. Chem. Eng. **1990**, *14* (5), 573–583.
9. Khalid, M.; Omatu, S. A Neural Network Controller for a Temperature Control System. IEEE Control Syst. **1992**, *12* (3), 58–64.
10. Hoskins, J.C.; Himmelblau, D.M. Process Control Via Artificial Neural Networks and Reinforced Learning. Comput. Chem. Eng. **1992**, *16* (4), 241–251.
11. Steck, J.E.; Rokhsaz, K.; Shue, S.P. Linear and Neural Network Feedback for Flight Control Decoupling. IEEE Control Syst. **1996**, *16* (4), 22–30.
12. Ortega, J.G.; Camacho, E.F. Mobile Robot Navigation in a Partially Structured Static Environment, Using Neural Predictive Control. Control Eng. Pract. **1996**, *4* (12), 1669–1679.
13. Huang, Y.; Whittaker, A.D.; Lacey, R.E. Neural Network Prediction Modeling for a Continuous, Snack Food Frying Process. Trans. ASAE **1998**, *41* (5), 1511–1517.
14. Huang, Y.; Whittaker, A.D.; Lacey, R.E. Internal Model Control for a Continuous, Snack Food Frying Process Using Neural Networks. Trans. ASAE **1998**, *41* (5), 1519–1525.

Newtonian Models

Ellen K. Chamberlain
Kraft Foods, Glenview, Illinois, U.S.A.

N

INTRODUCTION

In 1687, Sir Isaac Newton wrote his famous *Philosophiae Naturalis Principia Mathematica* in which he expressed his idea for an ideal fluid: "The resistance (*sic* stress) which arises from the lack of slipperiness (*sic* viscosity) originating in a fluid, other things being equal, is proportional to the velocity by which the parts of the fluid are being separated from each other (*sic* velocity gradient)."[1] Newton seemed to think that repulsive forces were due to the action of atomic springs in contact with each other. Although Newton was correct in his physical theory, there was no experimental evidence to prove or disprove Newton's position for over a century.[2] In 1845, Stokes wrote out Newton's concept in 3-D mathematical form.[1] Then in 1856 Poiseuille's capillary flow data were analyzed to prove Newton's relation experimentally.[1] Couette used a concentric cylinder apparatus to test the relation and found that his results agreed with the viscosities he measured in capillary flow experiments.[1]

DEFINITION

The characteristic property of an ideal fluid is that the rate at which the material deforms ($\dot{\gamma}$) is proportional to the applied force (σ) and the constant of proportionality is known as the Newtonian viscosity (η), dynamic viscosity or, simply, the viscosity. In other words, for a Newtonian fluid the shear stress is always proportional to the shear rate.[3] Mathematically this is written as:

$$\sigma = \eta\dot{\gamma}$$

where σ is the shear stress, η is the viscosity, and $\dot{\gamma}$ is the shear rate. Elements of ideal fluids or Newtonian fluids are in a continuous state of thermal agitation or Brownian movement. Upon applying an external force, there is no static equilibrium in the ideal fluid to be disturbed, so the elements move to new positions and remain randomly distributed due to Brownian motion.[4] This repositioning without change in structure will continue as long as the external force is being applied since there is no counteraction to limit it.[4]

Shear stress, denoted here as σ, is sometimes written as τ. The two symbols have the same meaning, however, the symbol σ is frequently used in current literature.[5]

Newtonian fluids may also be defined in terms of their kinematic viscosity (ν) which is equal to the dynamic viscosity divided by density (η/ρ). Kinematic viscosity is typically measured using an Ubbelohde or Cannon Fenske capillary viscometer.[6] The force of gravity acts as the force driving the liquid sample through the capillary. The viscometer is filled by inverting the capillary tube into a sample and suctioning fluid into the fixed sample bulb. The viscometer is turned upright and placed into a controlled temperature bath. After the fluid has reached thermal equilibrium, the fluid is allowed to flow down through the capillary. The fluid travel time between two etched lines on the glass viscometer is measured. The resulting time is considered the efflux time and fluid viscosity is calculated from this value.[6] As with any fluid, it is important to indicate (and, for that matter, control) the temperature at which a viscosity is measured since viscosity is strongly affected by temperature.

GRAPHICAL REPRESENTATION

When the correlation between shear stress and shear rate is graphically displayed with shear stress on the ordinate and shear rate on the abscissa, the diagram is called the flow curve. Newtonian fluids have a flow curve with a linear relationship between the shear stress and the shear rate with zero intercept (Fig. 1). The slope of the Newtonian line increases with increasing viscosity. Another common diagram used to describe materials is the viscosity curve with viscosity plotted vs. shear rate. Newtonian fluids have a constant viscosity even with changing shear rate, therefore, the viscosity curve is a straight line parallel to the abscissa (Fig. 2).

EXAMPLES

Materials that are made up of a single liquid phase and contain only low molecular weight, mutually soluble components, are Newtonian.[3] Most products that food rheologists encounter are non-Newtonian, with a few

Encyclopedia of Agricultural, Food, and Biological Engineering
DOI: 10.1081/E-EAFE 120006949
Copyright © 2003 by Marcel Dekker, Inc. All rights reserved.

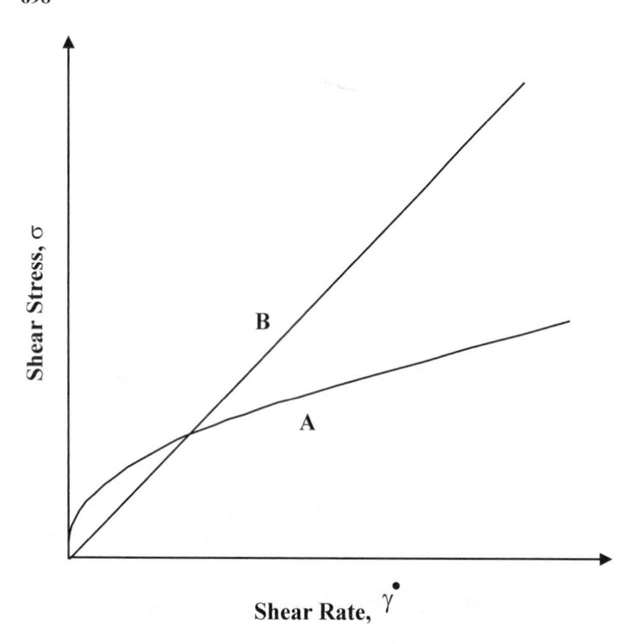

Fig. 1 Flow curves for fluids. Curve A is a typical shear thinning liquid which is a common flow characteristic in fluid foods. Curve B is a Newtonian fluid.

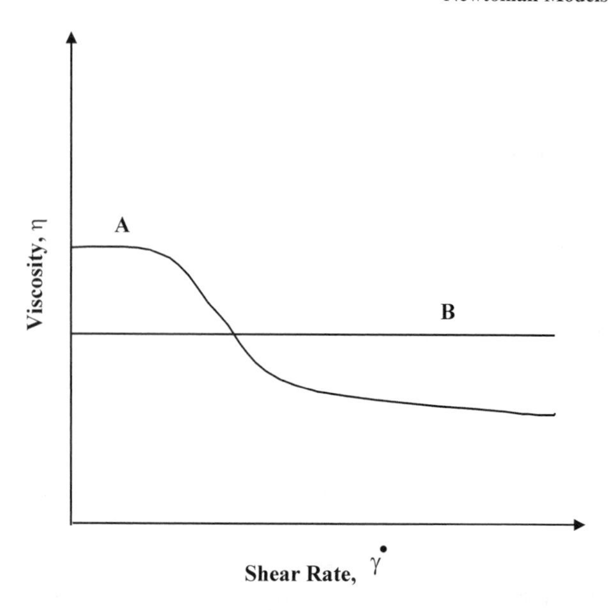

Fig. 2 Viscosity curves for fluids. Curve A is a typical shear thinning liquid which is a common flow characteristic in fluid foods. Curve B is a Newtonian fluid.

exceptions. This is due to the fact that foods are complex materials consisting of varying particle sizes, different phases, and multiple components. Some Newtonian materials found in the food and agriculture industry are: water, oils, gasoline, glycerine, ethylene glycol (antifreeze), ethanol, filtered juices, most honeys, and syrups.

REFERENCES

1. Macosko, C. Viscous Liquid. *Rheology: Principles, Measurements, and Applications*; VCH Publishers, Inc.: New York, NY, 1994; 65–106.
2. Rohn, C.L. History of Rheology and Macromolecular Science. *Analytical Polymer Rheology*; Hanser Publishers: New York, NY, 1995; 1–11.
3. Dealy, J.M.; Broadhead, T.O. Rheometry for Process Control. In *Techniques in Rheological Measurement*; Collyer, A.A., Ed.; Chapman & Hall: New York, NY, 1993; 286–318.
4. Prentice, J.H. A Few Basic Ideas. *Measurements in the Rheology of Foodstuffs*; Elsevier Applied Science Publishers: New York, NY, 1984; 7–13.
5. Dealy, J.M. Official Nomenclature for Material Functions Describing the Response of a Viscoelastic Fluid to Various Shearing and Extensional Deformations. J. Rheol. **1993**, *37*, 136–148.
6. Steffe, J.F. Introduction to Rheology. *Rheological Methods in Food Process Engineering*, 2nd Ed.; Freeman Press: East Lansing, MI, 1996; 1–93.

Non-Newtonian Models

Maria Elena Castell-Perez
Texas A&M University, College Station, Texas, U.S.A.

N

INTRODUCTION

Many fluid materials of agricultural origin such as creamy salad dressings, ketchup, and yogurt, usually show interesting flow behavior. This simply means that they do not behave in the well-ordered fashion of Newtonian fluids, such as water, honey, milk, and vegetable oil, when pumped, agitated, or subjected to any processing application. These fluids are known as *non-Newtonian fluids*. Most of these fluids are transported by pumping at some stage during processing or packaging and therefore a knowledge of their flow behavior is important for determining the power requirements for pumping and for sizing pipes. Furthermore, the flow behavior may be related to sensory characteristics such as texture. Thus, once the type of flow behavior has been identified, more can be understood about the way the fluid responds under specific system characteristics. Models have been developed which quantitatively describe this behavior. Such mathematical models range from the very simple to the very complex. This article summarizes the most common models used to describe non-Newtonian fluid behavior, with specific examples from the food and agricultural fields.

BACKGROUND

Not all fluids and semifluids of food importance begin to flow when a shear stress is applied or have constant viscosity with increasing shear stress. For instance, pastes will not spread easily on a cracker or bread until a certain additional force is applied in the form of spreading. The flow behavior of pastes, suspensions, dispersions, and some syrups fall under the category of non-Newtonian rheology. Examples are whipped cream, mayonnaise, ketchup, fruit juice concentrate, vegetable purees, and corn flour mixed with a little water. Thus, the study of viscosity and flow behavior of fluids and semifluids may be extremely complex. They display increasing or decreasing viscosity with increasing shear, a yield stress, time-dependency, or viscoelastic effects.

The flow behavior of inelastic fluid materials is usually visualized as a plot of shear stress vs. shear rate. A Newtonian fluid is broadly defined as one for which the relationship between shear stress and shear rate, $\sigma/\dot{\gamma}$, is linear with a zero intercept and the slope, or viscosity μ, is a constant. The main characteristic of a non-Newtonian fluid is that viscosity is not a constant and it depends on: 1) the velocity gradient (i.e., shear rate) and 2) the previous history of the material, in addition to temperature. Thus, the viscosity of these materials may be highly dependent on moisture content and concentration or composition. In some cases it is also affected by prior treatment such as freezing followed by thawing or heating followed by cooling.[1]

NON-NEWTONIAN MODELS

A non-Newtonian model is a mathematical expression used to describe the behavior of a unique type of fluids that deviate from the Newtonian model. These models have been developed from empirical correlations to more fundamental equations. They are necessary to estimate changes in the fluid's viscosity as a function of the shear rate.

The Newtonian model

$$\sigma = \mu\dot{\gamma} \tag{1}$$

describes the simplest type of fluid and it can be used for a wide variety of fluids of importance to engineers and scientists. This model may be used to describe many agricultural materials and food products. However, for many such materials (so called non-Newtonian fluids), more complicated models are required. This is because when the shear rate is varied, the shear stress does not vary in the same proportion (or even necessarily in the same direction) and the relationship $\sigma/\dot{\gamma}$ is not linear [Fig. 1(a)]. Some of these materials also have a yield stress, σ_0, which must be attained before flow begins. The viscosity of such fluids will therefore change as the shear rate is varied, i.e., the velocity affects the viscosity resulting in a much higher (or in some cases lower) rate of shear than for a Newtonian fluid.[2]

If a fluid were known to be Newtonian, its viscosity μ could be determined by applying a single shear rate, measuring the corresponding shear stress σ, and calculating μ as the ratio of σ to the shear rate $\dot{\gamma}$ (Eq. 1). If the fluid is non-Newtonian, the ratio of shear stress to shear rate will

Encyclopedia of Agricultural, Food, and Biological Engineering
DOI: 10.1081/E-EAFE 120006950
Copyright © 2003 by Marcel Dekker, Inc. All rights reserved.

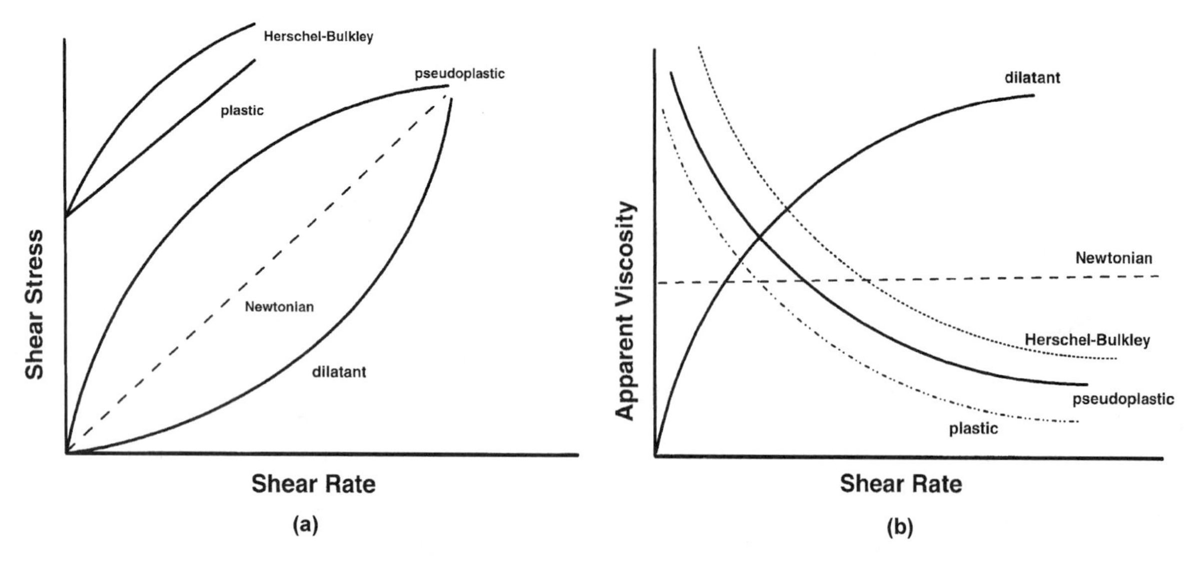

Fig. 1 Time-independent non-Newtonian models: (a) rheogram; (b) viscosity dependence on shear rate.

change with shear rate. The ratio for any given shear rate is therefore called the *apparent viscosity* at that shear rate [Fig. 1(b)]. The most commonly encountered type of inelastic, time-independent non-Newtonian behavior is *pseudoplastic*, when the flow begins as soon as the stress is applied, but the viscosity of the fluid decreases with increase in shear rate. This type of behavior is known as *shear-thinning*. Examples are mayonnaise, mustard, baby foods, and vegetable purees. Many water-soluble proteins and polymer solutions (e.g., albumin, gelatin) also exhibit pseudoplastic flow.[2]

Dilatant or *shear-thickening* behavior in foods is rarely encountered. The apparent viscosity increases with an increase in shear rate. A stiff paste slurry of maize or corn flour in water can appear to be quite liquid when swirled around in a cup. However, on pouring some out onto a hard surface and applying extreme shear forces can cause a sudden increase in viscosity due to its dilatancy. The viscosity can become so high as to make it appear solid. The "liquid" then becomes very stiff for an instant and can shatter just like a solid material.

The shear stress, σ, vs. shear rate, $\dot{\gamma}$ or velocity gradient, dv/dy, curves for pseudoplastic and dilatant materials can be described using the power-law model[1]:

$$\sigma = K\left[\frac{dv}{dy}\right]^n = K\dot{\gamma}^n \tag{2}$$

where K is the consistency coefficient (Pa sec^n) and n is the flow behavior index (dimensionless). For a pseudoplastic fluid, $0 < n < 1$ and for a dilatant fluid $1 < n < \infty$. Newtonian fluids can be considered a special case of this model in which $n = 1$ and K is the dynamic viscosity, μ. If values for shear stress at various shear rates are known, a plot of the logarithm of shear stress vs. the logarithm of shear rate will give a straight line with slope n and an intercept equal to K. The apparent viscosity, η_a, of a power-law fluid is then expressed as:

$$\eta_a = K\dot{\gamma}^{n-1} \tag{3}$$

For a *plastic* material the shear stress vs. shear rate curve is a straight line, but there is a yield stress which is the intercept of the straight line with the shear stress axis. Also, the apparent viscosity decreases with an increase in shear rate after the initial stress has been overcome. Examples are toothpaste, and purees. If the fluid has a yield stress and the shear stress vs. shear rate curve is convex toward the shear stress axis, then the fluid is called *Herschel-Bulkley* (ketchup) or *Casson-type* plastic (chocolate). Again, the apparent viscosity decreases with an increase in shear rate after the initial stress has been overcome.[2]

An equation similar to Eq. 2 can also be used to describe plastic and Herschel-Bulkley behavior. In fact, Eq. 2 can be considered a special case of the more general Herschel-Bulkley model:

$$\sigma = K\left[\frac{dv}{dy}\right]^n + \sigma_0 \tag{4}$$

with σ_0 as the yield value and the model describing plastic flow is

$$\sigma = \eta_{pl}\left[\frac{dv}{dy}\right]^n + \sigma_0 \tag{5}$$

The respective expressions for the apparent viscosity are:

$$\eta_a = K\dot{\gamma}^{n-1} + \sigma_0\dot{\gamma}^{-1} \tag{6}$$

and

$$\eta_a = \eta_{pl} + \sigma_0 \dot{\gamma}^{-1} \quad (7)$$

where η_{pl} is the plastic viscosity.

In some cases the shear stress vs. shear rate curve can be transformed into a straight line by plotting the square of the shear stress vs. the square root of the shear rate. This special case is called a Casson fluid.[3]

$$\sigma^{1/2} = K\left[\frac{dv}{dy}\right]^{1/2} + \sigma_0^{1/2} \quad (8)$$

Chocolate is one of the most notable examples of this type of fluid.

TIME-DEPENDENT NON-NEWTONIAN MODELS

In some cases, the apparent viscosity of a fluid changes with time as the fluid is continuously sheared at constant temperature (Fig. 2). If the apparent viscosity decreases with time, the fluid is called *thixotropic* and if it increases with time, it is called *rheopectic*. Similarly, if the shear stress is measured as a function of shear rate as the shear rate is first increased and then decreased, there will be *hysteresis* in the shear stress vs. shear rate plot. Thixotropic behavior is probably the result of a breakdown in the structure of the material as shearing continues. A starch-based food material will have an initial weak structure prone to breakdown with time or rapid shearing. This behavior is usually reversible for many foods. This property is extremely important in industrial products, e.g., to prevent settling of dispersed solids on storage.

In the case of rheopectic fluids, the structure builds as shearing continues. This type of behavior is rare but can occur in highly concentrated starch solutions over long periods of time.

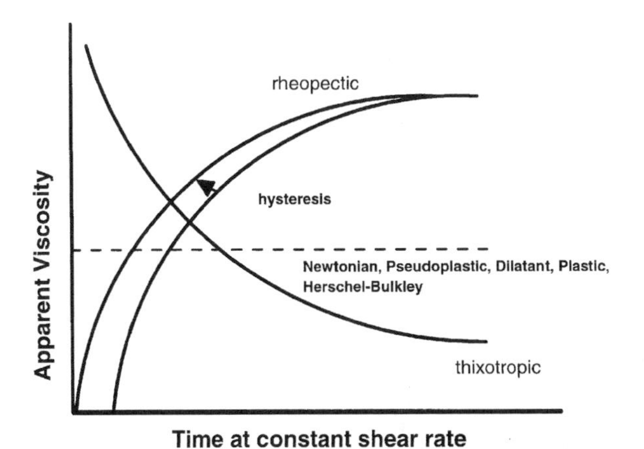

Fig. 2 Time-dependent non-Newtonian models.

Time-dependent behavior is more difficult to model. The most commonly used model to describe thixotropic behavior uses a structural decay parameter λ in the Herschel-Bulkley model to account for structural breakdown as[4]:

$$\sigma = f(\lambda, \dot{\gamma}) = \lambda\left(\sigma_0 + K(\dot{\gamma})^n\right) \quad (9)$$

where λ is a function of time and $\lambda = 1$ at the onset of shearing and $\lambda = \lambda_{equilibrium}$ after complete breakdown.[2]

VISCOELASTICITY

Some non-Newtonian fluids may appear liquid-like and capable of indefinite deformation but on release of the deforming stress show some recovery of shape. Such materials are known as viscoelastic fluids. All viscoelastic fluids are non-Newtonian, but not all non-Newtonian fluids are viscoelastic. Dairy cream, ketchup, pastes, soft cheese, some food suspensions, and particulate systems (dispersions) are examples of viscoelastic fluids. For these materials, the viscosity is dependent on the total strain (deformation) imposed as well as the strain rate (shear rate/time). This is often described by saying the fluid has a memory of all past strains. When shearing a viscoelastic fluid, so called normal stresses will appear. These normal stresses can result in flow behavior quite different from that of Newtonian fluids.[5]

APPLICATIONS

From an application point of view, the models of non-Newtonian flow are required to solve food industry problems in numerous areas. Understanding the flow behavior of a fluid food material allows the scientist and engineer to predict its performance in a mixing, piping, dipping, or coating operation, and the general way in which it will handle during processing. The knowledge of the flow behavior of a non-Newtonian fluid is indispensable for design of improved processing methods, or products that can perform well in use under various conditions of stress, strain, temperature, humidity, or other environmental variables.

REFERENCES

1. Kokini, J.L. Rheological Properties of Foods. In *Handbook of Food Engineering*; Heldman, D.R., Lund, D.B., Eds.; Marcel Dekker, Inc.: NY, 1992; 1–38.

2. Steffe, J.F. *Rheological Methods in Food Process Engineering*, 2nd Ed.; Freeman Press: East Lansing, Michigan, 1996; 13–21.
3. Casson, N. A Flow Equation for Pigmented-Oil Suspension of the Printing Ink Type. In *Rheology of Dispersed Systems*; Mill, C.C., Ed.; Pergamon Press: New York, 1959; 84–104.
4. Tiu, C.; Boger, D.V. Complete Rheological Characterization of Time-Dependent Food Products. J. Texture Stud. **1988**, *5*, 329–338.
5. Boger, D.V.; Walters, K. *Rheological Phenomena in Focus*; Rheology Series; Elsevier Science Publishers B.V.: Amsterdam, The Netherlands, 1993; Vol. 4.

Nusselt Number

Kevin Cronin
Diarmuid MacCarthy
University College, Cork, Cork, Ireland

N

INTRODUCTION

Convection is one of the basic mechanisms of heat transfer between a solid body and a fluid and can be quantified by the Convective Heat Transfer Coefficient. This coefficient is in turn determined by a quantity known as the Nusselt number. The Nusselt number is dependent on the geometry of the heat transfer system and on physical properties of the fluid. There are a large number of correlation formulae available to determine the Nusselt number for different heat transfer situations. The workings of these correlations are illustrated by sample calculations.

CONVECTION

There are three basic mechanisms of heat transfer: conduction, convection, and radiation. Heat transfer through a solid is by conduction only as the molecules occupy fixed positions. Heat transfer through a liquid or gas (collectively known as fluids) can be by conduction or convection depending on whether bulk motion of the fluid takes place. If there is bulk fluid motion, the heat transfer will be by convection; without bulk fluid motion, heat transfer is by conduction. Fluid motion enhances the rate of heat transfer and so, convective heat transfer through a fluid is always greater than conductive heat transfer through a fluid.

Heat transfer by convection can be classified as being either forced convection or natural (also known as free) convection. Forced convection occurs when fluid movement is artificially produced by some device, such as a fan or pump. Free convection occurs when fluid motion is the result of natural buoyancy effects, i.e., the tendency for a warm part of the fluid to rise (move) relative to a colder portion. In addition, convection can also be categorized as to whether it is external or internal. External convection is when the fluid flows on the outside of the solid surface of interest, for instance over a pipe or food product. Internal convection is the case where the fluid flows inside the surface and is entirely enclosed by the surface, for instance the flow inside a pipe or duct. In many items of plant, both occur simultaneously. Fig. 1 illustrates external forced convection.

CONVECTIVE HEAT TRANSFER COEFFICIENT

The heat transfer between a solid surface and a contacting fluid is also by convection. The rate of heat transfer is a function of the temperature difference between the solid and the fluid, the surface area available for heat transfer, and a constant known as the convective heat transfer coefficient. Mathematically

$$Q = hA(T_s - T_\infty) \tag{1}$$

where Q is the rate of convective heat transfer (W), h the convective heat transfer coefficient (W m^{-2} °C^{-1}), A the heat transfer surface area (m^2), T_s the temperature at the surface of the solid (°C), and T_∞ the bulk temperature of the fluid (°C).

This is the appropriate formula when the solid is hotter than the fluid and is being cooled by losing heat to the fluid. For the case of heating of the solid by heat transfer from the fluid, the temperature symbols are reversed.

The convective heat transfer coefficient is a complex quantity and in general varies from point to point over the heat transfer surface. It depends on the fluid properties of dynamic viscosity, thermal conductivity, density, specific heat, and fluid velocity. It is sensitive to whether the fluid flow is laminar or turbulent. It is also a function of dimensions and geometry of the solid, e.g., whether the solid is a flat slab or a sphere, etc.

NUSSELT NUMBER

The convective heat transfer coefficient is related to the many physical variables that influence it through the concept of Nusselt number, Nu. In a sense then, the Nusselt number is an attempt to capture in one quantity all the important factors that determine convective heat transfer rates and thus reduce the total number of variables that are needed to define the convective heat transfer coefficient. The relationship between both quantities is

$$h = \frac{Nu\,k}{\delta} \tag{2}$$

Encyclopedia of Agricultural, Food, and Biological Engineering
DOI: 10.1081/E-EAFE 120006998
Copyright © 2003 by Marcel Dekker, Inc. All rights reserved.

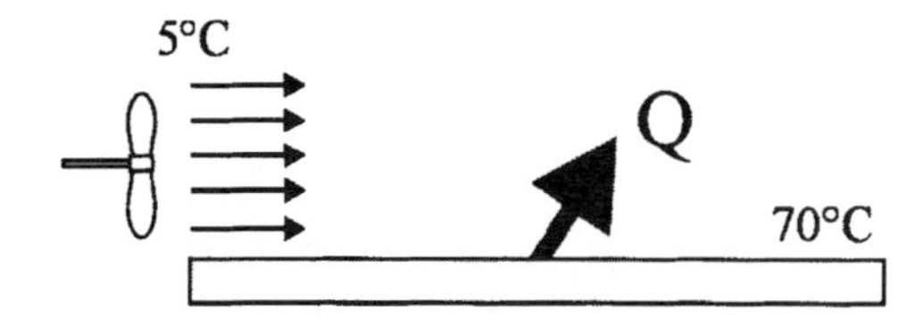

Fig. 1 External forced convection.

which can be used to define the Nusselt number as

$$Nu = \frac{h\delta}{k} \tag{3}$$

where k is the thermal conductivity of the fluid ($\mathrm{W\,m^{-1}\,°C^{-1}}$) and δ the characteristic length (m).

The characteristic length is some physical dimension that defines the geometry of the heat transfer situation, for instance the length of the flat slab or diameter of the sphere that is being heated or cooled. The Nusselt number is in effect a dimensionless version (meaning it has no units of measurement) of the convective heat transfer coefficient. The number takes its name from Wilhelm Nusselt, a German engineer who worked in the field of heat transfer in the early decades of the 20th century.

The Nusselt number is equal to the ratio of heat transfer by convection through the fluid to heat transfer by conduction. It is a measure of the increase in heat transfer through a fluid layer as a result of convection occurring compared to conduction across the same layer of fluid. A Nusselt number equal to one means that heat transfer through the fluid is by conduction only and the rate of heat transfer will be low. This would occur if the fluid is totally immobile. The larger is the Nusselt number, the more significant is the convection and the greater is the heat transfer.

The Nusselt number is used to determine the rate of heat transfer to or from a solid surface to an adjacent fluid. If the Nusselt number can be calculated for any given condition, then the convective heat transfer coefficient and hence the convective heat flux can be calculated using Eqs. 2 and 1, respectively. Various standard formulae (known as correlations) that have been derived from theoretical considerations and experimental work are available for the Nusselt number. Each particular heat transfer case (for instance, whether fluid flow is laminar or turbulent, whether convection is free or forced, whether the solid surface is flat or curved, etc.) will have its own defining correlation.

FORCED CONVECTION CORRELATIONS

For forced convection, the Nusselt number correlations are always expressed in terms of two other dimensionless numbers: the Reynolds number, Re and the Prandtl number, Pr. For most cases of forced convection, the general form of the correlation is

$$Nu = a\,Re^{b}\,\mathrm{Pr}^{c} \tag{4}$$

where a, b, and c are numerical constants that reflect the particular conditions that prevail. The Reynolds number and Prandtl number quantify the behavior of the fluid close to the solid surface of interest. They can be either calculated from fluid properties or read from tables. Fluid properties are usually evaluated at a temperature known as the film temperature, which is equal to the average of the solid and fluid temperatures. Knowing their magnitudes and the appropriate values of the three constants, the Nusselt number can be determined.

Each fluid, arising from its intrinsic transport properties, will have its own characteristic Prandtl number although these will vary with the temperature of the fluid. Gases and dry vapors (including air, steam, carbon dioxide, etc.) tend to have Prandtl number between 0.65 and 1.0. Free flowing liquids (such as water, common refrigerants, etc.) will have Prandtl number lying between 1 and 10 while thick (i.e., highly viscous) oils can have Prandtl number reaching up to 100,000.

The Reynolds number is more wide ranging as it depends on the velocity of the fluid and the geometry of the problem in addition to the properties of the fluid. It is used to define whether the fluid flow is laminar or turbulent. Laminar flow takes place when the fluid particles travel in smooth, orderly, parallel streamlines. In turbulent flow, the paths of the particles of fluid are no longer straight but have an irregular motion superimposed on the flow. The streamlines cross each other in a random fashion producing much greater mixing of the fluid and hence increasing the rate of heat transfer. The Reynolds number is defined as

$$Re = \frac{\rho u \delta}{\mu} \tag{5}$$

where ρ is the density of fluid ($\mathrm{kg\,m^{-3}}$), u the bulk fluid velocity ($\mathrm{m\,s^{-1}}$), δ the characteristic length (m), and μ the dynamic viscosity of fluid ($\mathrm{kg\,m^{-1}\,s^{-1}}$).

Note that the characteristic length is the same physical dimension as is used in the Nusselt number equation. Fluid

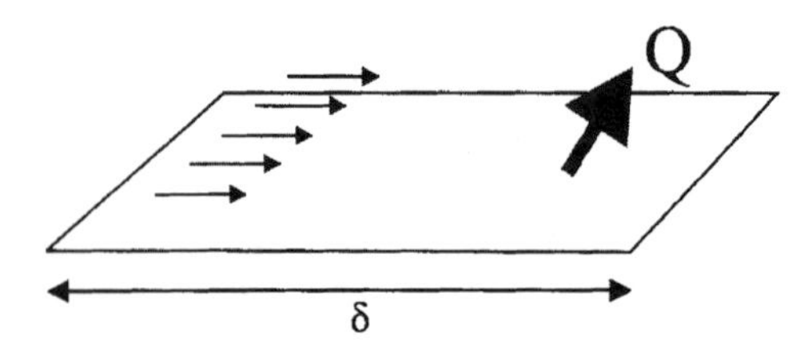

Fig. 2 Flow over a flat surface.

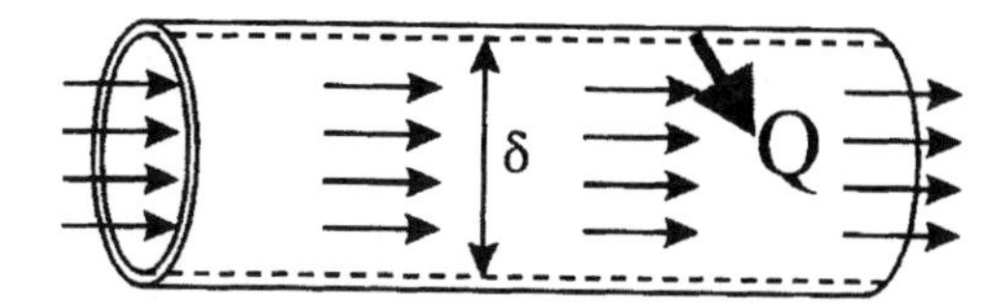

Fig. 3 Flow inside a tube.

flow parallel to a flat surface will be laminar in nature when the Reynolds number is less than 500,000 and turbulent if the Reynolds number is greater. For fluid flow inside a tube, this critical Reynolds number that separates laminar from turbulent flow is in the region of 2300.

To illustrate the formulation and calculation of these Nusselt number correlations, two specific examples will be given:

1. For laminar fluid flow over (i.e., along or parallel to) a flat surface, as shown in Fig. 2, a correlation is

$$Nu = 0.664\,Re^{1/2}Pr^{1/3} \tag{6}$$

   Note that this gives the average Nusselt number over the surface in the case where the Reynolds number is less than 500,000 and the Prandtl number is greater than 0.6.
2. For fully developed turbulent fluid flow inside a smooth pipe, as depicted in Fig. 3, a popular correlation is

$$Nu = 0.023\,Re^{0.8}Pr^{0.4} \tag{7}$$

   and is valid where the Reynolds number is above 10,000 and the Prandtl number is bounded within 0.7 and 160. Note that the index of 0.4 on the Prandtl number is applicable for the case of heat flow from the pipe to the fluid. If the opposite occurs then this index will have a magnitude of 0.3.

It can be seen that the first correlation is for external convection while the second is a case of internal convection.

FREE CONVECTION CORRELATIONS

For free convection, the Nusselt number correlations have a different formulation. They are given in terms of the Grashof number, *Gr* and the Prandtl number, *Pr*. The Grashof number in free convection corresponds to the Reynolds number in forced convection. By convention in the study of free convection, the Grashof number and Prandtl number are amalgamated together and known as the Raleigh number, *Ra*. The general form of the Nusselt number correlation is

$$Nu = a(Gr\,Pr)^b = a\,Ra^b \tag{8}$$

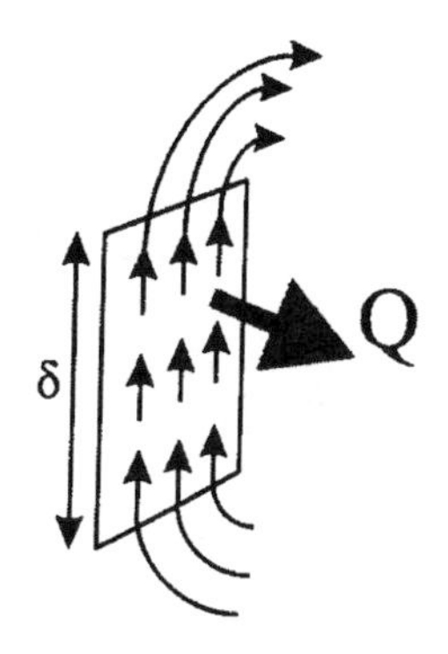

Fig. 4 Free convection over a vertical wall.

where a and b are again constants (not the same as previously used) that reflect the particular conditions of geometry and fluid flow regime.

As an illustration, a correlation for turbulent flow parallel to a vertical surface (external convection), illustrated in Fig. 4, is

$$Nu = \left(0.825 + \frac{0.387\,Ra^{1/6}}{\left[1 + \left(\frac{0.492}{Pr}\right)^{9/16}\right]^{8/27}}\right)^2 \tag{9}$$

and this is valid for all ranges of the Raleigh number.

COMBINED FORCED AND FREE CONVECTION SYSTEMS

In many cases, where heat transfer by convection occurs, the mechanism will be either forced or free and the appropriate correlation for the Nusselt number can be adopted. In certain instances though, for instance in a fan assisted oven where the fan is on a low setting, both mechanisms will be present and it is necessary to determine an overall Nusselt number that takes into account the respective contributions to heat flow of both phenomena. One approach is to determine the Nusselt number for forced and free convection separately and then combine them using

$$Nu_{co} = \left(Nu_{fo}^n \pm Nu_{fr}^n\right)^{1/n} \tag{10}$$

where Nu_{co} is the combined Nusselt number, Nu_{fo} the forced convection Nusselt number, and Nu_{fr} the free convection Nusselt number.

The addition sign in the equation is correct when free convection assists forced convection and the subtraction sign should be used when the mechanisms oppose each other. The term n is a constant that depends on the system geometry and is usually in the region of 3.

SUMMARY

To complete this article, a sample calculation to determine the Nusselt number for the case of forced external convection over a flat plate is performed.

Water at 50°C flows over a 2 m long plate with a velocity of 0.1 m s^{-1}. The plate has a dimension of 1.5 m perpendicular to the flow and is maintained at a temperature of 20°C. Calculate the Reynolds number, the Nusselt number, the convective heat transfer coefficient, and the heat transfer from water to plate.

$$\text{Film temperature} = \frac{20 + 50}{2} = 35°\text{C}$$

From thermodynamic tables, the properties of water at this temperature are: Prandtl number = 4.8; density = 994 kg m^{-3}; dynamic viscosity = 718×10^{-6} kg m^{-1} s^{-1}; thermal conductivity = 0.625 W m^{-1} °C^{-1}.

Eq. 5 gives the Reynolds number as

$$Re = \frac{994 \times 0.1 \times 2}{718 \times 10^{-6}} = 276,880$$

This is less than the critical Reynolds number for this flow geometry of 500,000, so the flow is laminar. The appropriate Nusselt number correlation is given by Eq. 6.

$$Nu = 0.664 \times 276,880^{1/2} \times 4.8^{1/3} = 589.4$$

The convective heat transfer coefficient can be evaluated using Eq. 2

$$h = \frac{0.625 \times 589.4}{2} = 184.2\,\text{W m}^{-2}\,°\text{C}^{-1}$$

The heat flow can then be calculated with Eq. 1

$$Q = 184.2 \times 2 \times 1.5 \times (50 - 20) = 16,578\,\text{W}$$

The result predicts that the heat flow from the water to the plate will be about 16.5 kW.

CONCLUSION

On a final point it is important to remember that analysis of heat transfer can be an inexact science. Results based on Nusselt number correlations can deviate by up to 20% or 30% from the actual case.

FURTHER READING

Cengel, Y.A. *Introduction to Thermodynamics and Heat Transfer*; Irwin/McGraw-Hill: Boston, 1997.

Jenna, W.S. *Engineering Heat Transfer*; PWS Publishers: Boston, 1986.

Welty, J.R. *Engineering Heat Transfer*; John Wiley and Sons: New York, 1974.

Hallstrom, B.; Skjoldebrand, C.; Tragardh, C. *Heat Transfer and Food Products*; Elsevier Applied Science: London, 1988.

Ohmic Heating

Sudhir K. Sastry
The Ohio State University, Columbus, Ohio, U.S.A.

O

INTRODUCTION

In 1827, George Ohm first outlined what is now known as Ohm's Law. However, it was James Prescott Joule, in 1840, who first recognized the thermal effects of electricity within a conductor. The technology of heating of materials by passing current through them has been termed as Ohmic or Joule heating, in their honor. In the latter part of the 19th century, a number of patents were issued for the heating of flowable materials. The technology has since been revived periodically—having seen industrial application for milk pasteurization in the 1930s, before being discontinued. In the 1980s, the technology was once again revived and has achieved some industrial applications, including the pasteurization of liquid eggs and the processing of fruit products.

BASIC PRINCIPLES

The basic principle of ohmic heating is the dissipation of electrical energy into heat, resulting in internal energy generation, which is proportional to the square of the electric field strength and the electrical conductivity.[1,2]

$$\dot{u} = |\nabla V|^2 \sigma \tag{1}$$

where the electrical conductivity σ is a function of temperature, material, and the method of heating. For cellular materials, the electrical conductivity undergoes a significant increase in the temperature range of 70°C, due to denaturation of cell-wall constituents. However, when an electric field is applied, cell-wall breakdown occurs at lower temperatures, thus the increase of electrical conductivity occurs over a wider range of temperatures[3] (Fig. 1).

Above a certain electric field strength or if the material has been previously thermally treated,[4] the electrical conductivity–temperature curve often becomes linear.

As the electrical conductivity increases with temperature, ohmic heating becomes more effective at higher temperatures. The electrical conductivity of liquid foods tends to follow a linear trend, regardless of the mode of heating. As no cellular structure exists, the properties remain essentially the same[5] (Fig. 2).

As the rate of heating may be affected by varying either the electric-field strength or product electrical conductivity, the technology represents many opportunities for creativity on the part of the process engineer or product developer. It is possible to design heaters even for materials of relatively low electrical conductivity if the electric field strength is made sufficiently large. It is also possible to heat materials at extremely rapid rates. Further, for materials of uniform electrical conductivity, the energy generation is more uniform than microwave heating. The basic principles have been addressed in a number of publications including Refs. 3,6–8.

MICROBIAL DEATH KINETICS

The question of whether or not ohmic heating results in a nonthermal contribution to microbial lethality has been addressed in a number of studies in the literature.

Early literature on this topic was inconclusive,[9] as most studies either did not specify sample temperatures or failed to eliminate it as a variable. It is essential that studies comparing conventional and ohmic heating be conducted under temperature histories that are as identical as possible. Palaniappan, Sastry, and Richter[10] found no difference between ohmic and conventional heat treatments under identical thermal histories on the death kinetics of yeast cells (*zygo Saccharomyces baillii*) (Table 1). They found, however, that a mild electrical pretreatment of *Escherichia coli* decreased the subsequent inactivation requirement in certain cases.

More recent studies suggest that mild electroporation might occur during ohmic heating. The presence of pore-forming mechanisms on cellular tissue has been confirmed by recent work.[11–13] Another recent study,[14] conducted under near-identical temperature conditions, indicates that the kinetics of inactivation of *Bacillus subtilis* spores can be accelerated by an ohmic treatment (Table 2). A two-stage ohmic treatment (ohmic treatment, followed by a holding time prior to a second heat treatment) was found to further accelerate death rates. Lee and Yoon[15] have indicated that leakage of intracellular constituents of *S. cerevisiae* was found to be enhanced under ohmic heating as compared to conventional heating in boiling water.

Encyclopedia of Agricultural, Food, and Biological Engineering
DOI: 10.1081/E-EAFE 120007114
Copyright © 2003 by Marcel Dekker, Inc. All rights reserved.

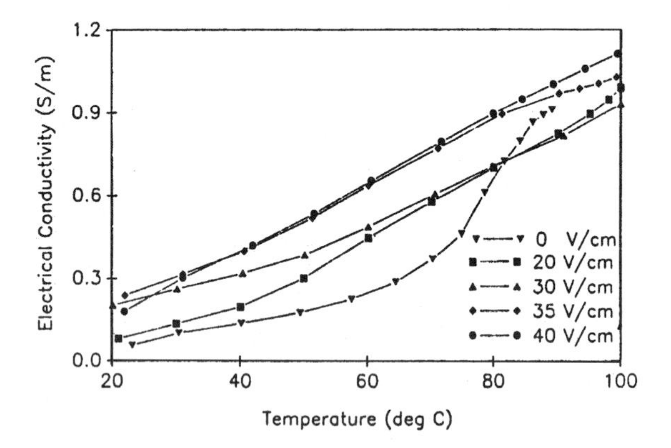

Fig. 1 Electrical conductivity of carrot subjected to various electric field strengths. (From Ref. [3].)

The additional effect of ohmic treatment may be due to the low frequency (50 Hz–60 Hz), which allows cell walls to build up charges and form pores. This effect is less apparent in high-frequency methods such as radio-frequency or microwave heating, where the electric field is reversed before sufficient charge buildup.

ELECTROLYTIC EFFECTS

When a current flows through an electrolyte, electrolytic reactions may occur at the electrode–solution interface. When direct current is used, different reactions occur at the cathode and anode; however, under alternating currents, the cathode and anode are periodically reversed and both cathodic and anodic products may occur at either electrode. If the potential drop at the electrode–solution interface can be kept below the critical electrode potential for the system, Faradaic electrolytic processes

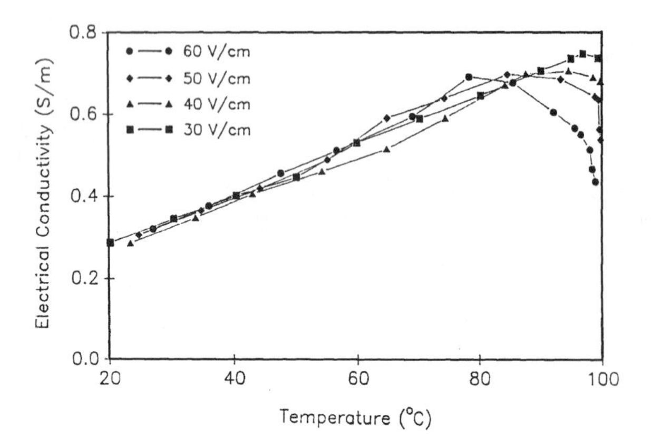

Fig. 2 Electrical conductivity of orange juice subjected to various electric field strengths. The decrease at high temperatures is due to evaporation and boiling within an unpressurized heater. (From Ref. [5].)

can be prevented. This implies that either the frequency is increased sufficiently or the electrode capacitance is increased for such minimization to occur. A detailed analysis of these phenomena has been presented by Amatore, Berthou, and Hebert.[16] An important consideration is to address such issues economically. Electrodes will also undergo wear over time, and the use of wear-resistant, inert materials is desirable.

APPLICATIONS

While a variety of potential applications exists for ohmic heating, most of them are yet to be commercially exploited. These include sterilization, pasteurization, processing of fouling-sensitive material, blanching, thawing, on-line detection of starch gelatinization, and as a pretreatment for drying and extraction. In addition, the presence of electric fields has various interesting effects on bacterial cells.

Sterilization

The above advantages have resulted in great expectations for the technology for sterilization of particulate foods. Typical systems for this purpose involve a material flowing through a suitably designed set of electrodes. One common commercial embodiment has been the APV ohmic heater, where the field is aligned with the flow [Fig. 3(a)]. A coaxial design is shown in Fig. 3(b).

Commercial sterilization requires that all parts of the flowing material be exposed to a sufficient time and temperature to destroy pathogenic spore-forming microorganisms. When the material contains solid material of significant size (about 1 cm–2 cm), the problem is considerably more difficult as it is not possible to noninvasively measure temperatures within the flowing solids. A mathematical model is necessary to characterize the worst-case situation.

The approaches for modeling have involved the static system models of De Alwis and Fryer,[6] for a limited number of particles and Zhang and Fryer,[17] for a set of static particles in a lattice. Sastry and Palaniappan[18] have analyzed cases involving high solid concentration in a mixed fluid. For a flowing system, models have been developed by Sastry[8] for a plug-flow system and Orangi, Sastry, and Li[19] for a system involving velocity profile.

Models have yielded results on heating of solid–liquid mixtures, which have been subsequently verified. For example, it has been shown[18] that if a single solid particle is of a lower electrical conductivity than the surroundings, its heating rate tends to lag the fluid. However, as

O

Table 1 Kinetic reaction rate constants (k) for *zygo S. baillii* under conventional and ohmic heating

Temperature (°C)	D-values for conventional heating (min^{-1})	k for conventional heating (sec^{-1})	D-values for ohmic heating (min^{-1})	k for ohmic heating (sec^{-1})
49.8	294.6	0.008	274.0	0.009
52.3	149.7	0.016	113.0	0.021
55.8	47.21	0.049	43.11	0.054
58.8	16.88	0.137	17.84	0.130
Z values (°C) or activation energy (E_a) (Cal mol^{-1})	7.19[a]	29.63[b]	7.68[a]	27.77[b]

[a] Z value.
[b] Activation energy.
(From Ref. [10].)

the concentration of low electrical conductivity of the solid increases, its heating rate bypasses that of the fluid.

For sterilization, the crucial items of interest are the worst case heating scenario and the location of cold zones. If the solid and liquid phases are of equal electrical conductivity, the heating is relatively uniform. A problem may arise when individual inclusion particles of electrical conductivity, significantly different from its surroundings, enter the system. Under such conditions, one of the phases will lag the other.

The question of which phase lags which one is dependent on the electrical conductivity of the respective phases and the extent of fluid motion (fluid–solid convective heat transfer coefficient). A number of other studies have attempted to address the use of chemical markers[20] or temperature measurement within ohmic heaters.[21] Clearly, model improvement and temperature measurements will be crucial in the future development of ohmic heating for sterilization.

Fouling by Proteinaceous Materials

One advantage claimed for ohmic heating has been its ability to heat proteinaceous materials, which represent a challenge to conventional processing. However, protein deposits can adhere to electrode surfaces and create a high electrical resistance film at the interface. This may result in arcing, as discussed by Reznik[22] in one of his patents. Studies by Wongsa-Ngasri[23] have shown that arcing depends on the protein content, initial temperature, and flow rate. Wongsa-Ngasri has also presented arcing diagrams, which show the range of conditions under which arcing may occur.

Seafood Processing

Research at Oregon State University[24] and in Japan indicate that ohmic heating is a useful method for protease inactivation in restructured seafood products, permitting

Table 2 D-values and kinetic reaction rate constants (k) for *B. subtilis* spores under conventional and ohmic heating

Temperature (°C)	D-values for conventional heating (min^{-1})	k for conventional heating (sec^{-1})	D-values for ohmic heating (min^{-1})	k for ohmic heating (sec^{-1})
88	32.8	0.00117	30.2	0.001271
90 (stage 1 of two-stage heating)	17.1	0.002245	14.2	0.002703
90 (stage 2 of two-stage heating)	9.2	0.004172	8.5	0.004516
92.3	9.87	0.003889	8.55	0.004489
95	5.06	0.007586		
95.5			4.38	0.008763
97	3.05	0.012585		
99.1			1.76	0.021809
Z value (°C) or activation energy (E_a) (Cal mol^{-1})	8.74[a]	70.0[b]	9.16[a]	67.5[b]

[a] Z value.
[b] Activation energy.
(From Ref. [14].)

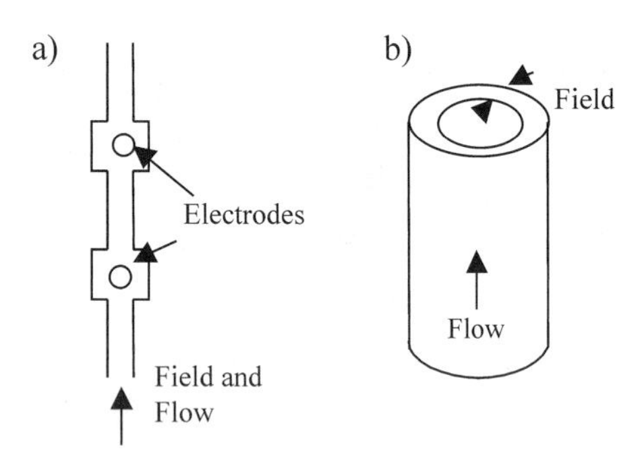

Fig. 3 Designs of ohmic heaters: (a) in-line field and (b) coaxial.

rapid heating to inactivation temperatures. Balaban and coworkers, at the University of Florida, have thawed large blocks of shrimp using a specially designed ohmic heating process.

Pretreatments for Water Removal

Wang[12] showed that a short pretreatment of vegetable tissue with ohmic heating up to 80°C resulted in a significant acceleration of the drying process when compared to control (untreated), conventionally heat pretreated, and microwave pretreated samples. Wang also found that apple juice expression could be made more efficient by such a permeabilization process (similar to an electroplasmolysis process previously developed in the former Soviet Union[25]). Further, the mechanical energy required for juice expression was significantly reduced. Lima and Sastry,[26] following up on Wang's work, showed that the waveform and frequency also had significant effects.

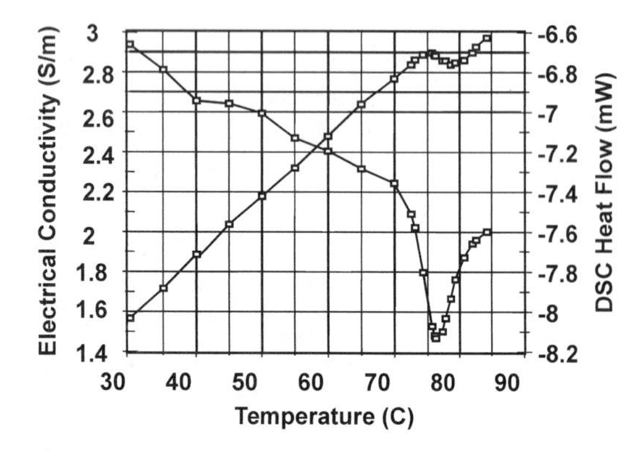

Fig. 4 Superimposed DSC curve and electrical conductivity–temperature relationship for corn-starch suspension. (From Ref. [12].)

Ohmic Heating for Detection of Starch Gelatinization

Wang[12] and Wang and Sastry[27] showed that the electrical conductivity–temperature curve of starch solutions showed negative peaks at temperatures corresponding to similar peaks of a differential scanning calorimetric thermograph (Fig. 4). This led to the possibility of monitoring gelatinization from electrical conductivity measurements, and suggests that on-line techniques could be developed for this purpose.

Extraction Enhancement

Bazhal and Guly,[28,29] among others, have shown the effectiveness of ohmic or moderate electric field (MEF) pretreatment in improvement of sugarbeet extraction. Work in Korea[30] has indicated that soymilk extraction can be improved by ohmic/MEF treatment. Schreier, Reid, and Fryer[31] reported the increase of diffusion by ohmic heating. Recent work by Kulshrestha and Sastry[13] shows that mild treatment of beet tissue (temperature rise of 1–2°C) results in leaching of dyes from cells.

CONCLUSION

The effects of ohmic heating appear to have application for a variety of processes. However, the exploitation of positive effects, while minimizing and eliminating negative aspects, represents interesting engineering challenges for the future.

NOMENCLATURE

D	Time in minutes required for a log-cycle reduction in bacterial population
$\dot{u}$	Internal energy generation rate
V	Voltage

Greek Letters and other Symbols

σ	Electrical conductivity
∇	Gradient

REFERENCES

1. Sastry, S.K.; Palaniappan, S. Ohmic Heating of Liquid-Particle Mixtures. Food Technol. **1992**, *46* (12), 64–67.
2. Sastry, S.K.; Li, Q. Modeling the Ohmic Heating of Foods. Food Technol. **1996**, *50* (5), 246–248.
3. Palaniappan, S.; Sastry, S.K. Electrical Conductivity of Selected Solid Foods During Ohmic Heating. J. Food Process Eng. **1991**, *14*, 221–236.

4. Wang, W.-C.; Sastry, S.K. Changes in Electrical Conductivity of Selected Vegetables During Multiple Thermal Treatments. J. Food Process Eng. **1997**, *20*, 499–516.
5. Palaniappan, S.; Sastry, S.K. Electrical Conductivity of Selected Juices: Influences of Temperature, Solids Content, Applied Voltage and Particle Size. J. Food Process Eng. **1991**, *14*, 247–260.
6. De Alwis, A.A.P.; Fryer, P.J. A Finite Element Analysis of Heat Generation and Transfer During Ohmic Heating of Food. Chem. Eng. Sci. **1990**, *45* (6), 1547–1559.
7. Halden, K.; De Alwis, A.A.P.; Fryer, P.J. Changes in the Electrical Conductivity of Foods During Ohmic Heating. Int. J. Food Sci. Technol. **1990**, *25*, 9–25.
8. Sastry, S.K. A Model for Heating of Liquid-Particle Mixtures in a Continuous Flow Ohmic Heater. J. Food Process Eng. **1992**, *15*, 263–278.
9. Palaniappan, S.; Sastry, S.K.; Richter, E.R. Effects of Electricity on Microorganisms: A Review. J. Food Process Preserv. **1990**, *14*, 393–414.
10. Palaniappan, S.; Sastry, S.K.; Richter, E.R. Effects of Electroconductive Heat Treatment and Electrical Pretreatment on Thermal Death Kinetics of Selected Microorganisms. Biotech. Bioeng. **1992**, *39*, 225–232.
11. Imai, T.; Uemura, K.; Ishida, N.; Yoshizaki, S.; Noguchi, A. Ohmic Heating of Japanese White Radish *Rhaphanus sativus* L. Int. J. Food Sci. Technol. **1995**, *30*, 461–472.
12. Wang, W.-C. Ohmic Heating of Foods: Physical Properties and Applications Ph.D. Dissertation, The Ohio State University, Columbus, OH, 1995.
13. Kulshrestha, S.A.; Sastry, S.K. Low-Frequency Dielectric Changes in Vegetable Tissue from Ohmic Heating. In The 1999 Annual IFT Meeting, Chicago, IL, July 24–28, 1999; Abstract No. 79 B-3.
14. Cho, H.-Y.; Yousef, A.E.; Sastry, S.K. Kinetics of Inactivation of *Bacillus subtilis* Spores by Continuous or Intermittent Ohmic and Conventional Heating, 1999.
15. Lee, C.H.; Yoon, S.W. Effect of Ohmic Heating on the Structure and Permeability of the Cell Membrane of *Saccharomyces cerevisiae*. In The 1999 Annual IFT Meeting, Chicago, IL, July 24–28, 1999; Abstract No. 79 B-6.
16. Amatore, C.; Berthou, M.; Hebert, S. Fundamental Principles of Electrochemical Ohmic Heating of Solutions. J. Electroanal. Chem. **1998**, *457*, 191–203.
17. Zhang, L.; Fryer, P.J. Models for the Electrical Heating of Solid–Liquid Food Mixtures. Chem. Eng. Sci. **1993**, *48*, 633–643.
18. Sastry, S.K.; Palaniappan, S. Mathematical Modeling and Experimental Studies on Ohmic Heating of Liquid-Particle Mixtures in a Static Heater. J. Food Process Eng. **1992**, *15*, 241–261.
19. Orangi, S.; Sastry, S.K.; Li, Q. A Numerical Investigation of Electroconductive Heating of Solid–Liquid Mixtures. Int. J. Heat Mass Transfer **1998**, *41* (14), 2211–2220.
20. Kim, H.-J.; Choi, Y.-M.; Yang, T.C.S.; Taub, I.A.; Tempest, P.; Skudder, P.; Tucker, G.; Parrott, D.L. Validation of Ohmic Heating for Quality Enhancement of Food Products. Food Technol. **1996**, *50* (5), 253–261.
21. Ruan, R.; Chen, P.; Chang, K.; Kim, H.-J.; Taub, I.A. Rapid Food Particle Temperature Mapping During Ohmic Heating Using FLASH MRI. J. Food Sci. **1999**, *64* (6), 1024–1026.
22. Reznik, D. Electroheating Apparatus and Methods. US Patent No. 5,583,960, 1996.
23. Wongsa-Ngasri, P. Effects of Inlet Temperature, Electric Field Strength and Composition on the Occurrence of Arcing During Ohmic Heating. M.S. Thesis, The Ohio State University, Columbus, OH, 1999.
24. Yongsawatdigul, J.; Park, J.W.; Kolbe, E. Electrical Conductivity of Pacific Whiting Surimi Paste During Ohmic Heating. J. Food Sci. **1995**, *60* (5), 922–925, 935.
25. Grishko, A.A.; Kozin, M.; Chebanu, V.G. Electroplasmolyzer for Processing Raw Plant Material. US Patent No. 5,031,521, 1991.
26. Lima, M.; Sastry, S.K. The Effects of Ohmic Heating Frequency on Hot-Air Drying Rate and Juice Yield. J. Food Eng. **1999**, *41*, 115–119.
27. Wang, W.-C.; Sastry, S.K. Starch Gelatinization in Ohmic Heating. J. Food Eng. **1997**, *20* (6), 499–516.
28. Bazhal, I.G.; Guly, I.S. Effect of Electric Field Voltage on the Diffusion Process. Pishch. Promst. **1983**, *1*, 29–30.
29. Bazhal, I.G.; Guly, I.S. Extraction of Sugar from Sugarbeet in a Direct Current Electric Field. Pishch. Tekhnol. **1983**, *5*, 49–51.
30. Kim, J.-S.; Pyun, Y.-R. Extraction of Soymilk Using Ohmic Heating. In 9th World Congress of Food Science and Technology, Budapest, Hungary, July 30–August 4, 1995; Abstract No. P125.
31. Schreier, P.J.R.; Reid, D.G.; Fryer, P.J. Enhanced Diffusion During the Electrical Heating of Foods. Int. J Food Sci. Technol. **1993**, *28*, 249–260.

Osmotic Pressure

Ernest W. Tollner
The University of Georgia, Athens, Georgia, U.S.A.

INTRODUCTION

Osmosis is fundamental to life processes. Osmotic phenomena also lie at the core of agricultural production and feed, food, and fiber processing operations. Phenomena related to osmosis impact the shelf life and consumer acceptance of food products. The processing of organic and inorganic wastes and byproducts frequently involves osmotic phenomena. This discussion is confined to aqueous solutions usually found in food and agriculture operations.

THERMODYNAMICS OF OSMOSIS

One may easily derive an expression for osmotic pressure starting from equilibrium thermodynamics. Suppose there are two vessels connected by a semipermeable membrane as shown in Fig. 1. One vessel in Fig. 1A–C contains pure water while the other vessel contains a dissolved solid in water. In Fig. 1A, pure water moves toward the water with solute in response to an osmotic pressure gradient, which is maintained by the semipermeable membrane. Fig. 1B shows the same system in equilibrium, with an osmotic pressure denoted by π. Fig. 1C portrays a reverse osmosis system where an opposing pressure P in excess of π has been applied. Reverse osmosis systems are useful for removal of solutes from water. Both vessels and contents are at the same temperature. At thermal and chemical equilibrium, the total electrochemical potential of the pure and solute bearing water is the same as shown in Eq. 1.

$$f_{\mathrm{W}}^{\mathrm{A}}(T, P^{\mathrm{A}}) = f_{\mathrm{W}}^{\mathrm{B}}(T, P^{\mathrm{B}}) \tag{1}$$

In Eq. 1, f denotes the total electropotential of pure water (left side, superscript A) and water with dissolved substrates (right side, superscript B) and T and P denote temperature and pressure, respectively.[1] One may transform the total potential of pure water to that for water with dissolved solutes as shown in Eq. 2.

$$\gamma_{\mathrm{W}}^{\mathrm{A}} x_{\mathrm{W}}^{\mathrm{A}} f_{\mathrm{W}}^{\mathrm{A}}(T, P^{\mathrm{B}}) \exp[V_{\mathrm{W}}^{L}(P^{\mathrm{A}} - P^{\mathrm{B}})/RT] = f_{\mathrm{W}}^{\mathrm{B}}(T, P^{\mathrm{B}}) \tag{2}$$

In Eq. 2, $x_{\mathrm{W}}^{\mathrm{A}}$ is the mole fraction of water and $\gamma_{\mathrm{W}}^{\mathrm{A}}$ the chemical activity of the solute. The notion of chemical activity arises from the fact that actual solute concentrations do not behave as ideal concentrations. The chemical activity coefficient in effect adjusts concentration parameters in chemical reaction equations to better agree with observed behavior. For dilute solutions, the chemical activity approaches unity. As we are addressing water, the chemical activity and water activity are synonymous. The water activity coefficient, further discussed in works such as Toledo[2] and Stumm and Morgan,[3] accounts for departures from ideality and is important with respect to food packaging, discussed later. The term within the exponential in Eq. 2 is referred to as the *Poynting factor*. One may derive the Poynting factor in a straightforward manner starting from Gibbs relationships for nonreacting materials (e.g., see Ref. 4). Simplifying Eq. 2 as shown below results in the fundamental relationship for osmotic pressure.

$$\pi = (P^{\mathrm{A}} - P^{\mathrm{B}}) = \frac{-RT}{V_{\mathrm{W}}^{L}} \ln(\gamma_{\mathrm{W}}^{\mathrm{A}} x_{\mathrm{W}}^{\mathrm{A}}) \tag{3}$$

In many cases, the water activity $\gamma_{\mathrm{W}}^{\mathrm{A}}$ is close to one. Water activity may be defined as the ratio of the solute bearing vapor pressure to the vapor pressure of pure water.[2] Note that the sum of the mole fraction of water $x_{\mathrm{W}}^{\mathrm{A}}$ (close to one) and the mole fraction of the solute $x_{\mathrm{S}}^{\mathrm{A}}$ (close to zero) is unity. The negative of the natural logarithm of $x_{\mathrm{W}}^{\mathrm{A}}$ is approximated by $x_{\mathrm{S}}^{\mathrm{A}}$ to within 5% when $x_{\mathrm{S}}^{\mathrm{A}}$ is less than 0.1. Further, one may write $x_{\mathrm{S}}^{\mathrm{A}}$ as $V_{\mathrm{W}}^{L} \times C_{\mathrm{S}}$, where C_{S} is the solute concentration in $\mathrm{g\,mol\,L^{-1}}$. These substitutions lead to Eq. 4, the *Van't Hoff* equation.

$$\pi = \frac{RT}{V_{\mathrm{W}}^{L}} x_{\mathrm{S}}^{\mathrm{A}} = RTC_{\mathrm{S}} \tag{4}$$

One may extend Eq. 4 to the case of multiple solutes as shown in Eq. 5, assuming the total solute concentration is small.

$$\pi = RT \sum_{i=1}^{n} C_{\mathrm{S}i} \tag{5}$$

Encyclopedia of Agricultural, Food, and Biological Engineering
DOI: 10.1081/E-EAFE 120007177
Copyright © 2003 by Marcel Dekker, Inc. All rights reserved.

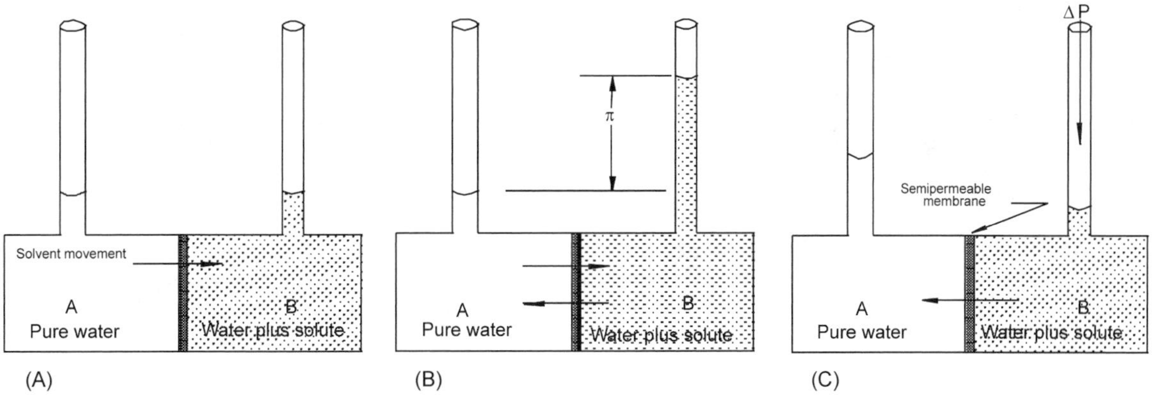

Fig. 1 Osmotic scenarios showing (A) osmosis; (B) osmotic equilibrium; and (C) reverse osmosis.

The semipermeable membrane in Fig. 1 allows solvent to pass while blocking the passage of solutes. The blockage mechanism is rooted in inorganic and/or organic chemistry relations. The chemistry of the semipermeable membrane to be used in a separation process thus depends on the nature of the solute to be blocked.

Flow (kg per m^2 of membrane area) across the membrane depends on the membrane permeability k_w ($kg\,m\,N^{-1}$), pressure across the membrane (ΔP, $N\,m^{-2}$), osmotic pressure ($\Delta\pi$, $N\,m^{-2}$), and thickness of the membrane L (m) as shown in Eq. 6.

$$N_W = k_W \frac{\Delta P - \Delta\pi}{L} \tag{6}$$

Summarizing, Eq. 3 is the basic equation for osmotic pressure, which comes from basic thermodynamic considerations. The semipermeable membrane is required for the expression of osmotic pressures. The Van't Hoff relationship (Eqs. 4 and 5) assumes a water activity of unity and one or more small solute concentrations. Eq. 6 shows how flux across a semipermeable membrane may be computed with knowledge of external pressure, osmotic pressure, membrane thickness, and membrane permeability. The remainder of this article explores some of the many ramifications of osmosis and osmotic pressure.

OSMOTIC PRESSURE AND LIFE PROCESSES

The cell is a fundamental structure in the organization of biological entities from the bacterium to the human being. The cell wall functions as a controllable semipermeable membrane. The function of the particular cell determines the nature of the materials to be blocked, thus cell wall composition depends on the cell function. Many cells have lipids (phospholipids, cholesterol, and glycolipids), forming a bimolecular structure. The lipids are hydrophobic on one end and hydrophilic on the other. The hydrophilic ends meet between the bilayer, leaving the hydrophobic ends facing away from the cell and into the cell (see Fig. 2). Special carrier proteins are interspersed, which allow for transport of specific molecules across the cell membrane. Other proteins (enzymes) have catalytic activity and mediate chemical reactions within the membrane. Yet other proteins provide structural support for the membrane. Membranes tend to be permeable to oxygen, nitrogen, carbon dioxide, water, and other smaller constituents. They tend to be impermeable to large polar molecules such as glucose and to charged ions. Transport of essential elements and ions across the membrane is mediated by the carrier proteins. Thus, the cell membrane is selectively semipermeable. The cell can actively maintain an osmotic pressure by regulating the passage of specific ions through the carrier proteins. The degree of semipermeablity with respect to each element involved in the bioprocesses of the cell is mediated by the needs of

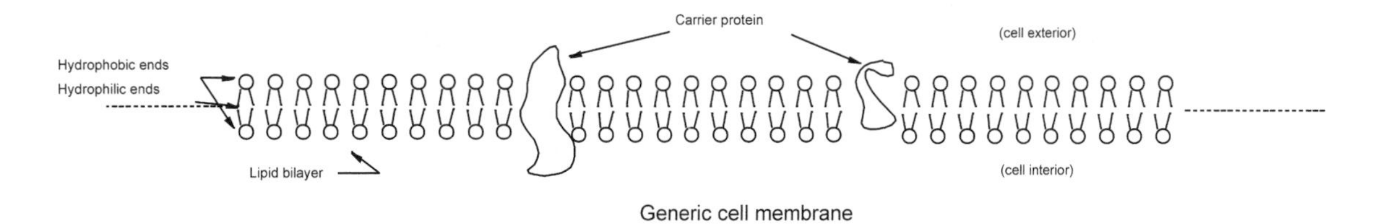

Fig. 2 Hypothetical cell membrane.

the cell and the availability of particular solutes. The ability of a cell to regulate the degree of semipermeability (e.g., P_W in Eq. 6) is known as *osmoregulation*. Works such as that of Krogh[5] gives an exhaustive treatment of osmotic regulation in 16 Phyla of the animal kingdom.

OSMOSIS IN SPECIFIC PLANT PROCESSES

Saline Water Use

The ability to influence the selective semipermeability provides points of influence on plant productivity for food, chemicals, and energy. For example, modifying cells to remove water from a saline source has tremendous possibilities for increasing crop production in areas where fresh water is scarce. Rains, Valentine, and Hollaender[6] describe efforts to modify cells to adapt to saline environments using genetic engineering techniques. These studies build on fundamental investigations into osmosis by Pfeffer,[7] who essentially defined the role of the cell membrane as a semipermeable osmotic membrane.

Drought Tolerance

Drought stress is related to cell function at the osmotic level in an opposite but complimentary manner. Instead of adding or subtracting various solutes from the system, drought stresses look at the effect of removing water from the system, with the roots and other plant membranes functioning as osmotic membranes. Water movement in plants is regulated by the diffusion pressure deficit, which is directly related to osmotic pressure in the xylem cells. Most plants transpire at optimum levels near the 0.3 bar–1.0 bar range. Referring to Eq. 6, the osmotic pressure π is increased above the water tension P to keep flow going in a positive direction. As the soil dries out, the soil water potential moves above 1 bar and the flow of water into the roots and the transpiring leaves decreases due to inability of most plants to achieve the intracellular solute concentrations required to maintain a suitable osmotic gradient. Thus, transpiration decreases and leaves show curling, wilting, or other signs of water stress. The stomata, structures that regulate water movement from leaves to the atmosphere, then begin to close. As the soil dries to 15 bar, the wilting point, water movement ceases and the plant dies (e.g., see Ref. 8).

Freeze Hardiness

Freeze hardiness is related to osmotic phenomenon. Just as antifreeze depresses the freezing point of water in an automobile, solute concentrations developed across plant osmotic membranes depress the freezing point of water in plants. Plants, which are actively growing, are much more vulnerable to freeze damage than plants, which are not growing. Fruit and ornamental producers are concerned that seasonal freezing be past before plants are induced to enter rapidly growing blooming phases. Researchers have been (and currently are) evaluating genetic engineering approaches for modifying osmoregulation in the cell for increasing drought hardiness and freeze hardiness for the past 25 yr.[6]

OSMOTIC PROCESSES IN FOOD PROCESSING

Food dehydration causes significant departures from the assumption that the mole fraction for water is near 1 and the mole fractions of solids is near zero. Indeed, the opposite is true. Thus, one cannot assume the water activity to be near unity. As discussed above, water activity is a ratio of the vapor pressure of the solute bearing material to that of pure water. Toledo[2] provides a derivation of the water activity coefficient that is very similar to that of the Poynting factor discussed in Eq. 2. Knowledge of water activity enables prediction of degrees of lipid oxidation, nonenzymatic browning, mold, yeast, and bacterial activity. Bacterial, yeast, and mold activity generally stops when water activity is below 0.7. Lipid and other browning reactions are generally highest at a water activity of around 0.7. Water activity levels of around 0.3 minimize lipid oxidation and stop nonenzymatic browning reactions. At levels below 0.3, lipid oxidation may increase. A search of the web will reveal several commercially available instruments for measuring water activity.

Numerous approaches for relating the water activity coefficient to the mole fraction of solutes are summarized by Toledo.[2] Approaches exist for calculating the water activity in food products based on knowledge of multiple dissolved constituents.

Suppose a dry cereal has multiple ingredients (e.g., raisins, dates, various nuts, grains, etc.). The atmosphere inside the package determines the basic water potential. Each dry constituent will equilibrate with the package atmosphere. Some materials may maintain a high affinity for moisture in poorly sealed package and thereby provide locations where water activity may increase to levels high enough for microbial action to begin.

One application of reverse osmosis application in the food industry is in maple syrup production given by Koelling and Heiligmann.[9] Reverse osmosis has been used commercially in the maple syrup production since the mid-1970s, though less efficient technology dates back to the late 1950s. The objective of using reverse osmosis is

to remove a substantial portion (approximately 75%) of the water from the sap, concentrating it to between 7°Brix and 10°Brix before it enters the evaporator, thereby reducing evaporator fuel costs and boiling time. Other advantages frequently cited for reverse osmosis include shortening the holding time for unprocessed sap (less spoilage) and shortening the time sap is processed at high temperatures.

OSMOTIC PROCESSES IN WATER TREATMENT AND WASTEWATER RECLAMATION

Reverse osmosis (Fig. 1C) is widely used in water and wastewater reclamation. It is used for removing specific solutes. Reverse osmosis is a straightforward application of Eq. 6, where P exceeds the osmotic pressure to cause solvent (water) to leave the solute. Incoming solute bearing water is pressurized to a level ranging from 200 psi to 400 psi.[10] Most of the water passes in reverse through the membrane, leaving a small quantity of concentrated solute, which is properly disposed. A wide range of inorganic and organic solutes may be removed using reverse osmosis. Reverse osmosis is used after other physical and chemical processes are employed for removing most of the solute. Pretreatment is necessary to preserve the permeability, thus the life of the membrane. A web search using "reverse osmosis" will reveal several excellent pages relating to reverse osmosis.

REFERENCES

1. Fournier, R.L. *Basic Transport Phenomena in Biomedical Engineering*; Hemisphere Publishing Corp.: Philadelphia, PA, 1998.
2. Toledo, R.T. *Fundamentals of Food Process Engineering*, 2nd Ed.; Van Nostrand Reinhold: New York, NY, 1991.
3. Stumm, W.; Morgan, J.J. *Aquatic Chemistry: An Introduction Emphasizing Chemical Equilibria in Natural Waters*; John Wiley & Sons: New York, NY, 1981.
4. Reynolds, W.C. *Thermodynamics*, 2nd Ed.; McGraw-Hill: New York, NY, 1965.
5. Krogh, A. *Osmotic Regulation in Aquatic Animals*; Dover Publications: New York, NY, 1965.
6. Rains, D.W.; Valentine, R.C.; Hollaender, A. *Genetic Engineering of Osmoregulation: Impact on Plant Productivity for Food, Chemicals and Energy*; Plenum Press: New York, NY, 1980.
7. Pfeffer, W. *Osmotic Investigations, Translated by G. R. Kepner and J. Stadelmann*; Van Nostrand Reinhold Co.: New York, NY, 1985; Translated from a Work Published in 1877.
8. House, C.R. *Water Transport in Cells and Tissues*; Edward Arnold Ltd: London, 1974.
9. Koelling, M.R.; Heiligmann, R.B. *North American Maple Syrup Producers Manual*; Bulletin 856; Cooperative Extension Service, Ohio State University: Columbus, OH, 1996.
10. Tchobanoglous, G.; Burton, F.L. *Wastewater Engineering: Treatment, Disposal and Reuse*, 3rd Ed.; McGraw-Hill: New York, NY, 1991.

Package Functions

Matthew D. Steven
Joseph H. Hotchkiss
Cornell University, Ithaca, New York, U.S.A.

INTRODUCTION

Packaging can be viewed as both necessary and wasteful. Consumers and industry have expectations of packaging and criticisms when it fails to meet these expectations. These expectations relate to the functions of packaging, but what are these functions? This is the question we attempt to answer here.

Packaging has been defined as a bundle of something packed, wrapped or boxed for distribution. However, such definitions do not cover the full range of functions and expectations. A functional definition is favored: Packaging is a multifunctional technical system used to contain, distribute, and market products. It interacts with the converter, packer, distributor, retailer and consumer, as well as with the product and the environment.

FUNCTIONS

The functions of packaging are the services that packaging is expected to deliver, be that during shipment or during the useable life of the product. There are seven major functions or considerations of food packaging: containment, protection, communication, functionality, production efficiency, environmental impact, and consumer safety/regulation.

Containment

Containment is probably its earliest and most commonly noted function. It would be impossible to buy foods if they were not first contained in some form of package. Without a container, most products would not be readily available. In fact, it is hard to find a food, including fresh fruits and vegetables, which is not packaged in some form during distribution.

Important factors in containment depend on the product and its distribution. Paper-based packages are fine for containing powdered and granular products, but require an adequate sealing system and are useless for liquids unless coated to provide moisture resistance. The material of package construction is therefore an important factor in containment. Another example can be found in carbonated beverages—high gas barrier packages are required to contain carbonation between the point of production and the point of consumption.

The sealing method is also a key factor in containment.[1] Choosing the best material will be of little value if the closure and sealing are not effective. Glass is essentially impervious to gases yet glass packages, e.g., beer bottles, would not retain CO_2 if inadequately sealed. One of the major challenges in packaging research is the nondestructive inspection of packages for leaks.[2]

Package durability plays an important role in containment during distribution. Resistance to dynamic forces (shock and vibration) and static loads (crush tolerance) encountered during distribution is required.[3–5] Dynamic loads are often found during handling (and mishandling) of products, e.g., transportation, loading, and unloading, whereas static loads are frequently encountered during storage when cases of product are stacked. There has been considerable research conducted on the effect of package design on the ability to tolerate distribution.[3]

Milk provides an example of the range of containment systems that can be used for a single product: metal cans, plastic pouches, coated paperboard cartons, plastic bottles, glass bottles, foil-laminated paperboard cartons. Milk is sold in all of these. The security of containment is different for each package: cans are very strong while the others may be more prone to leakage and damage in transit.

Protection

Food packaging must protect against several factors that will make the food unsuitable for consumption. Deteriorative mechanisms can be divided into four types: biological agents, mechanical damage, chemical degradation, and physical damage. Biological agents include micro-organisms, insects, and rodents. Protection from these takes several forms—the primary one being a physical barrier. For rodents and insects, both of which can penetrate all but glass and metal, barrier is critical. Protection from micro-organisms is more complex. Packaging must not directly contribute micro-organisms and must provide a physical barrier. Packaging is often designed to actively inhibit microbial growth in food. Technologies employed include gas and moisture barriers,

Encyclopedia of Agricultural, Food, and Biological Engineering
DOI: 10.1081/E-EAFE 120007126
Copyright © 2003 by Marcel Dekker, Inc. All rights reserved.

modified atmospheres, and antimicrobial polymers. Barrier packaging is designed to prevent the ingress of vital oxygen to aerobic bacteria. This concept has been extended in modified atmosphere packaging (MAP) by tailoring the internal atmosphere of the package to inhibit microbial growth.

The product can be protected from mechanical damage by appropriately designed packaging. Corrugated board cases are typically used to help protect packaged foods from mechanical damage. Specially designed trays are used to protect raw produce by isolating individual items within the shipping container and ensuring that the shipping container, not the produce, supports the compressive loads developed during stacking. Compressive forces can become very large as the height of product stacked increases. Vibration can cause abrasion, resulting in damage to the surface designs of packaged goods and in deterioration (bruising, surface damage) of produce.

Chemical degradation is inherent in foods. However, environmental factors can alter the rate of these reactions. Many reactions require either water or oxygen to progress—barrier materials prevent the ingress of required reaction components and are a common approach in preservation. Chemical degradation due to the loss or gain of various chemical compounds can affect sensory characteristics. The most obvious examples are water and flavor compounds. Loss or gain of water from a product can result in significant changes in product texture and acceptability, as well as chemical and microbiological deterioration. Again, the use of barrier materials can provide protection. Permeation of gases and vapors typically follows Fickian diffusion.[6,7] The loss of flavor compounds can result in perceptible changes in sensory characteristics. Suitable barrier packaging can be used, but care needs to be taken as some packaging materials can be a major cause of such loss. This is referred to as flavor "scalping" where the packaging material absorbs flavor compounds from the product.[8] Similarly food products can absorb flavor compounds from the packaging—leading to off-flavors in the product.

UV or visible light can result in product deterioration, e.g., loss of riboflavin in milk.[9] Packaging can protect the product from these factors by acting as a physical barrier, through opacity or by including UV blockers in transparent materials.

Communication

The primary communication function of packaging is product identification and use. This identification is not solely through text or graphics, but can be incorporated into container shape and color. A prime example is the shape of the classic Coca-Cola bottle, which is synonymous with the brand name and identifies the product without labels. Product categories are often associated with package shape, e.g., the generic potato chip bag and the returnable glass milk bottle.

Graphics and text combine to convey information and persuade the consumer to purchase a product. Graphical cues help identify brand and special characteristics of the product—kosher designations for example. The textual component fulfills many legal requirements such as declarations of package size, nutritional information, and manufacturer.[10–12]

It is also important for secondary cartons to communicate, particularly in automated distribution systems. The Universal Product Code (UPC) identifies secondary cases. These are required by the distribution system and carry information about identity, pack size, batch number, sell by date, and more. UPC codes can be used to trace a product back to the plant of origin should the need arise.

Functionality

Functionality concerns those features designed to facilitate the product's use by both consumers and manufacturers. Consumer functionality relates to ease of use of the package, special features that may aid in the use of the product, and in cases where more product is packaged than is used at a single occasion, reclosure. Packaging is often used to unitize multiple units into a single unit to make them easier to carry and more convenient to purchase, e.g., a box of hot chocolate containing several single-serving sachets.

Microwaveability is an example of a package-convenience feature and requires consideration of material tolerances to the microwave heating environment.[13] Packaging for use during product preparation in conventional ovens leads to similar consideration of material characteristics. Other examples of special features include boil-in-the-bag type systems, aeration systems for draught beers and whipped toppings, and self heating or cooling packages.

Ease of use is concerned with features such as opening and dispensing. While rugged sealing/closure is critical, the product must be easily accessible to the consumer. Maintaining package integrity during distribution while allowing customers easy access has precipitated many novel solutions. Pouches with an internal zipper seal, notches in sachets to provide tear initiation, and pull-tabs are all examples of systems designed to allow ready access, yet maintain sealed integrity.

Tamper resistance and dispensing are further examples of product use features. For example, plastic bottles can readily dispense viscous products by applying pressure to cause product flow. Pour spouts, integral scoops and spoons, wide mouth jars, and flip top caps are all features designed to enhance dispensing.

Functionality for the manufacturer relates to production and handling efficiency, e.g., suitability for high-speed automated filling, closing, and handling.

Environmental Considerations

Packaging needs to have minimal environmental impact.[14] Three R's are often quoted: Reduce, Reuse, and Recycle. Reduction is the best way to reduce packaging waste, but it is important that the packaging fulfills the functions described above—too much reduction may result in package failure. Reuse (refilling) is the second stage in considering environmental impact. When evaluating reuse, it is important to remember the environmental costs in shipping and cleaning returned containers, and that containers cannot be reused indefinitely without failing. Recycling is the retrieval of package materials and remanufacture into a different form. For example, paper is reground into pulp and reformed into new products; PET bottles are recycled into fabrics.

Litter is a related but different issue. It is important to consider the ease of disposal and the contribution of the package to litter if improperly discarded. The use of readily degradable plastics can reduce the consequences of improper disposal.

Package Safety and Regulation

Food packaging is highly regulated for safety and honesty throughout the world.[6,11] The types of hazards vary with material and package type, but the most significant issue is the migration of small molecules into food. This is mainly a concern with plastics, but can also be an issue with metals. Approval for food contact for a specific material is granted by regulatory bodies within each country, e.g., the FDA in the United States. The criteria for approval are normally concerned with levels of migration of various small molecular weight compounds, e.g., monomer units or material additives, from the material into food stimulants placed in direct contact with the material. Examples of compounds of concern include styrene from poly(styrene) and dioxin from paperboard.[15]

NEWER TECHNOLOGIES

Combining Materials

Knowledge of specific product requirements and changing preservation technology has made it necessary to use combinations of materials.[16] Often, no one material is able to provide all the required functions. The technology has become sophisticated as evidenced by the widespread use of multilayer films. These multilayer materials can be produced by lamination, in the case of dissimilar materials such as paper, foil, and polymers or by coextrusion where two or more different polymers are extruded simultaneously to form a film with a layer of each. Some polymers are not suitable for coextrusion and must be laminated. An example of multilayer materials can be found in applications where heat sealing, easy printing, and high gas and water barrier are required. Paper products provide the printing surface and strength, foil provides the barrier, but neither can be readily heat-sealed. The resulting package in this case would be an outer paper layer, for printing, laminated to a center foil barrier layer laminated to an inner polymer layer for heat sealing. An example of this, somewhat modified, can be found in single-serve aseptic "boxes" for beverages.[17]

Modified Atmosphere Packaging

MAP involves a change in the internal atmosphere to enhance shelf life.[18,19] The most common application is in refrigerated perishable foods and minimally processed fruits and vegetables. The simplest MAP is gas flushing prior to sealing. Many products are flushed with nitrogen to remove oxygen and prevent oxidation during storage. A more sophisticated version involves flushing the package with a gas mixture. The respiration of the food product, micro-organisms, or various chemical reactions can change the composition of the atmosphere inside a package. More advanced forms of MAP involve the rational design of barrier properties in combination with gas flushing to make use of respiration to achieve a desired steady state atmosphere composition. The design of such systems can be complex due to the number of variables involved but can also result in significant improvement in the shelf life.

Active Packaging

The next generation of packaging may have additional functions to those discussed above. One concept under widespread development is termed "active packaging." Rather than passively separating a product from its environment, active packaging fosters desirable interactions, either between the packaging and the environment, the food and the environment, or the packaging and the food.[20] This is in contrast to traditional packaging interactions, which involve migration, sorption, and permeation.[21] Active packaging can take many forms including antimicrobial materials, incorporation of enzymes, improvement of flavor, indication of contamination, and others.[22] Current commercial applications are limited to antimicrobial films and oxygen, moisture, and ethylene absorbers and ethanol emitters.

P

CONCLUSION

A frequent consumer complaint with packaging is that there is too much of it, but is that really the case? The above outline of the seven major functions and considerations of food packaging—containment, protection, communication, functionality, production efficiency, environmental impact, and consumer safety/regulation—apply regardless of the form of the packaging. In evaluating the adequacy or excess of any packaging system, it is important to reflect upon all these areas. Failure to do so could result in failure of the package and loss or failure of the product contained.

REFERENCES

1. de Oliviera, L.M. Flexible Packaging Heat Sealing. Coletanea Inst. Tecnol. Aliment. **1996**, *26* (2), 111–117.
2. Safvi, A.A.; Meerbaum, H.J.; Morris, S.A.; Harper, C.L.; OBrien, W. Acoustic Imaging of Defects in Flexible Food Packages. J. Food Prot. **1997**, *60* (3), 309–314.
3. Brandenburg, R.K.; Lee, J.J.-L. *Fundamentals of Packaging Dynamics*; MTS Systems Corporation: Minneapolis, MN, 1985.
4. Griffin, R.C., Jr.; Sacharow, S.; Brody, A.L. *Principles of Package Development*; Van Nostrand Reinhold Co. Inc.: New York, 1985.
5. Hanlon, J.F. *Handbook of Package Engineering*; Technomic: Lancaster, PA, 1992.
6. Piringer, O.G., Baner, A.L., Eds. *Plastic Packaging Materials for Food: Barrier Function, Mass Transport, Quality Assurance and Legislation*; Wiley-VCH: Weinheim, Germany, 2000.
7. Robertson, G.L. *Food Packaging: Principles and Practise*; Marcel Dekker: New York, 1993.
8. Hotchkiss, J.H., Ed. *Food and Packaging Interactions*; American Chemical Society: Washington, D.C., 1988.
9. Fennema, O.R., Ed. *Food Chemistry*, 3rd Ed.; Marcel Dekker: New York, 1996.
10. Anon. *A Consolidation of the Food Regulations 1984 Incorporating Amendments 1 to 10*; N.Z. Ministry of Health: Wellington, New Zealand, 1995.
11. Joint FAO/WHO Food Standards Programme: Codex Alimentarius Commission, *Codex Alimentarius: Food Labelling Complete Texts*; Food and Agriculture Organization of the United Nations, World Health Organization: Rome, 1998.
12. Anon. *A Food Labeling Guide*; U.S. Department of Health and Human Services: Washington, D.C., 1999.
13. Decareau, R.V. *Microwave Foods: New Product Development*; Food and Nutrition Press: Trumbull, CT, 1992.
14. Stilwell, E.J.; Canty, R.C.; Kopf, P.W.; Montrone, A.M. *Packaging for the Environment: A Partnership in Progress*; American Management Association: New York, 1991.
15. Eun-Ju-Seog; Jun-Ho-Lee; Singh, R.K. Migration of Styrene in Relation to Food Packaging Materials. J. Food Sci. Nutr. **1999**, *4* (2), 152–158.
16. Brown, W.E. *Plastics in Food Packaging: Properties, Design and Fabrication*; Marcel Dekker: New York, 1992.
17. Chambers, J.V., Nelson, P.E., Eds. *Principles of Aseptic Processing and Packaging*; Food Processors Institute: Washington, D.C., 1993; 126–127.
18. Farber, J.M., Dodds, K.L., Eds. *Principles of Modified-Atmosphere and Sous Vide Product Packaging*; Technomic: Lancaster, PA, 1995.
19. Parry, R.T., Ed. *Principles and Applications of Modified Atmosphere Packaging of Foods*; Blackie Academic and Professional: Glasgow, 1993.
20. Labuza, T.P.; Breene, W.M. Applications of Active Packaging for Improvement of Shelf Life and Nutritional Quality of Fresh and Extended Shelf Life Foods. J. Food Process. Preserv. **1989**, *13* (1), 1–70.
21. Appendini, P.; Hotchkiss, J.H. Immobilisation of Lysozyme on Food Contact Polymers as Potential Antimicrobial Films. Packag. Technol. Sci. **1997**, *10* (5), 271–279.
22. Rooney, M.L. Overview of Active Food Packaging. In *Active Food Packaging*; Rooney, M.L., Ed.; Blackie Academic and Professional: Glasgow, 1995; 1–37.

Package Permeability

John M. Krochta
University of California, Davis, California, U.S.A.

INTRODUCTION

Permeability is a process that involves adsorption (dissolution) of a low-molecular-weight gas (e.g., oxygen or carbon dioxide) or vapor (e.g., water vapor or food aroma) onto one side of a plastic polymer (e.g., polyethylene) film, diffusion through the film, and then desorption (evaporation) from the other side of the film. The net transfer of the gas or vapor (permeant) is from the film side where it is at higher partial pressure to the film side where it is at lower partial pressure.

Permeability is a process unique to polymers, since such adsorption–diffusion–desorption does not occur with metal or glass materials. Permeability has no relevance either for paper materials, where transport through pores is the mechanism for transfer of gas or vapor from high to low concentration sides. Transport of gas or vapor through small pores and cracks in polymeric materials is also possible. However, transport through such pores and cracks occurs by mechanisms different from permeability.[1–4]

Analysis of polymer permeability has become increasingly important to Agricultural and Food Engineering. Polymeric materials have found many uses, ranging from controlled release of drugs or agricultural chemicals,[5] to plastic alternatives to metal and glass in food packaging.[6] In the former, selection of a polymer with the desirable permeability allows release at the optimum rate. In the latter, the plastic polymer packaging permeability must be taken into account to determine the shelf life of the food product. Polymer films with selected permeabilities to oxygen and carbon dioxide have also allowed development of the whole area of modified atmosphere packaging (MAP) of fresh-whole and fresh-cut fruits and vegetables.

PERMEABILITY MECHANISM

In 1855, Fick published his first law of diffusion in which permeant mass flux through a section of isotropic material was related to concentration gradient measured normal to the section by a diffusion coefficient[7]:

$$F = -D\frac{\partial C}{\partial x} \qquad (1)$$

Here, F is permeant flux or transmission rate (quantity of permeant per unit time per unit area), C is permeant concentration, $\partial C/\partial x$ is permeant concentration gradient in the x direction over thickness ∂x, and D is the diffusion coefficient. Fick's First Law is thus analogous to Ohm's Law for electrical conduction, Fourier's Law for heat conduction, and Newton's law for momentum transfer. At steady state with D constant through a sheet of thickness L:

$$F = D(C_1 - C_2)/L \qquad (2)$$

where C_1 and C_2 are the concentrations at surfaces 1 and 2 of the sheet, respectively.

In 1866, Graham proposed that the permeability through a polymer membrane included adsorption of a permeant in the polymer matrix at the membrane surface, diffusion through the membrane because of a permeant concentration gradient, and desorption from the low-concentration side of the membrane.[6] In 1879, Von Wroblewski built on this concept by showing that sorption of gas in rubber followed Henry's Law (linear isotherm)[8]:

$$C = Sp \qquad (3)$$

Here, p is partial pressure of the permeant in equilibrium with the polymer containing permeant with concentration C. The factor relating p and C is the solubility coefficient, S. If the solubility coefficient is constant, Eq. 2 can be combined with Eq. 1 to obtain:

$$F = DS(p_1 - p_2)/L \qquad (4)$$

The product DS is defined as the permeability coefficient (or permeability), P:

$$F = P(p_1 - p_2)/L = P\,\Delta p/L \qquad (5)$$

Using this expression is convenient when C_1 and C_2 are not known, but the partial pressures of the permeant in equilibrium with the sheet surface on both sides of the sheet, p_1 and p_2, are known. However, $P = DS$ only when D and S are both constant over the concentration range C_1–C_2 and partial pressure range p_1–p_2, respectively. Otherwise, P is an average or effective permeability coefficient that applies only over the partial pressure range p_1–p_2 in relating F to $(p_1 - p_2)/L$. Thus, depending on the permeant–polymer pair, P may be constant or not. Generally, D, S, and P are relatively constant when little

Encyclopedia of Agricultural, Food, and Biological Engineering
DOI: 10.1081/E-EAFE 120007129
Copyright © 2003 by Marcel Dekker, Inc. All rights reserved.

interaction occurs between the permeant (e.g., water vapor, which is polar) and polymer (e.g., low density polyethylene, which is nonpolar). In these cases, the permeant does not serve to plasticize the polymer. For cases where the permeant (e.g., water vapor) plasticizes the polymer (e.g., ethylene-vinyl alcohol copolymer, which is polar), the result is that permeant concentration has an effect on *D*, *S*, and *P*, thus they are not constant.[9]

Transmission Rate Vs. Permeability Coefficient Vs. Permeance

There are several terms in the literature, which refer to the permeation process described in Eq. 5, that are important to distinguish.[3,10,11] *F*, as mentioned previously, is the permeant flux or transmission rate (quantity/area-time). To determine *F*, one would measure the quantity of permeant (q) that permeates a film of a given area (A) over a specified time (t) at steady state:

$$F = q/At \tag{6}$$

As calculation of the permeant flux or transmission rate does not include the partial pressure driving force, Δp, or the film thickness, L, these must be specified when reporting permeant flux (transmission rate) data. Frequently, the transmission rate is multiplied by the film thickness (L) to obtain a thickness-normalized transmission rate, N (quantity-thickness/area-time) in which case only the driving force must then be specified.

$$N = FL = qL/At \tag{7}$$

In either case, if the transmission rate is a function of permeant concentration, the actual p_1 and p_2 used for its determination must be indicated.

As can be seen in Eq. 5, if one multiples the flux (transmission rate) times the film thickness (L) and divides by the driving force (Δp), the result is the permeability coefficient or permeability, P (quantity-thickness/area-time-Δp):

$$P = FL/\Delta p = qL/At\Delta p \tag{8}$$

As calculation of the permeability coefficient includes transmission rate (F), partial pressure driving force (Δp), and film thickness, nothing else must be specified when reporting P values. The only exception is if P (i.e., D and/or S) is a function of permeant concentration, the actual p_1 and p_2 used for its determination must be indicated.

In Eq. 5, P/L is sometimes combined into the term Permeance (quantity/area-time-Δp):

$$R = F/\Delta p = q/At\Delta p \tag{9}$$

As calculation of R does not include the film thickness, L, it must be specified when reporting R data. If the transmission rate is a function of permeant concentration, the actual p_1 and p_2 used for its determination must be indicated.

Careful attention to the definitions and units of these terms is critical for correct use. Sometimes, the terms transmission rate, thickness-normalized transmission rate, permeability coefficient (or permeability), and permeance are used interchangeably in which case careful attention must be made to the actual units used.

Permeability Vs. Migration and Scalping

Permeability is to be distinguished from migration and sorption (also called scalping).[11–17] In migration, a component of a plastic package (e.g., residual monomer, plasticizer, antioxidant) transfers from the package material to the food product and possibly to the external environment.[18] In sorption (scalping), a component of food product is sorbed by a plastic packaging material without transfer to the surrounding atmosphere.[10] Besides producing loss in food quality, scalping can compromise the package integrity and/or barrier properties.[19]

MEASUREMENT OF PERMEABILITY COEFFICIENTS

Many methods have been developed for measurement of permeability coefficients.[20] Table 1 lists the methods that have been accepted as standard methods and/or developed with modern instruments, for measuring the permeabilities of both films and formed packages. The permeability coefficients of most common interest in Agricultural and Food Engineering are the water vapor permeability (WVP), oxygen permeability (O2P), carbon dioxide permeability (CO2P), and organic compound (e.g., aroma) permeability. For measuring film permeability, the film is sealed in a permeability test cell at a controlled constant temperature and pressure, and each side of the film is exposed to controlled water vapor, oxygen, carbon dioxide, or organic compound partial pressure to create a defined Δp. In the case of O2P, CO2P, and organic compound permeability, the relative humidity must also be defined and controlled if the polymer is plasticized by absorption of water vapor. After the film has reached equilibrium with the conditions of the test and steady-state achieved, the quantity of permeant transferring through the film is measured by some method and then converted into a permeability coefficient according to Eq. 8.

Table 1 Permeability coefficient measurement methods

Permeability coefficient	Measurement method
WVP	ASTM E 96[21]
	ASTM F 1249[22]
	McHugh, Avena-Bustillos, and Krochta[23]a
	Gennadios, Weller, and Gooding[24]a
	ASTM D 895[25]b
	ASTM D 1251[26]b
	ASTM D 3079[27]c
	ASTM D 2684[28]d
O2P	ASTM D 3985[29]
	Gilbert and Pegaz[30]
	ASTM F 1307[31]e
CO2P	Johnson and Demorest[4]
	Gilbert and Pegaz[30]
Aroma permeability	Johnson and Demorest[4]
	Hernandez, Giacin, and Baner[32,33]
	DeLassus, Standburg, and Howell[34]
	Hatzidimitriu, Gilbert, and Loukakis[35]
	Miller and Krochta[36]
	Seeley[37]

[a] Hydrophilic films.
[b] Package "WVP" reported as g/30d-package.
[c] Package "WVP" reported as g/30d-package.
[d] Package "permeability" reported as g-cm/d-m^2.
[e] Package "O2P" reported as cc-thickness/d-package-Δp.

PERMEABILITY COEFFICIENT VALUES

Polymer permeability coefficients can be found in a number of references.[3,4,8,10,38,39] Table 2 lists O2P, CO2P, and WVP values for a number of common polymers. Many combinations of units are used in the polymer literature.

Tables are available for conversion from one set of units to another.[3,8,38] Nonetheless, one must be careful in converting from one set of units to another. Furthermore, literature values of permeability should be used only for comparisons of different polymers and rough design estimates, and only for the conditions of permeability measurement. Considerable variation in

Table 2 Permeabilities of plastic films used in packaging

Polymer	Oxygen[a]	Carbon dioxide[a]	Water vapor[b]
Vinylidene chloride copolymers	0.039–0.58	0.19–2.9	0.001–0.01
Ethylene-vinyl alcohol copolymers	0.027–0.18[c]		
Ethylene-vinyl alcohol copolymer	4.3–2.1[d]		0.091–0.24
Nylon-6	7.8–11.6	39–47	0.70
Poly(ethylene terephthalate) (PET)	12–16	58–97	0.12
Poly(vinyl chloride) (PVC)	19–78	78–190	0.14
High density polyethylene (HDPE)	390–780	2300–2700	0.025
Polypropylene (PP)	580–970	1900–2700	0.041
Low density polyethylene (LDPE)	970–1400	3900–7800	0.091
Polystyrene (PS)	970–1600	2700–5800	0.047

[a] cc-μm/m^2-d-kPa at 20°C and 75% RH, unless otherwise noted.
[b] g-mm/m^2-d-kPa at 38°C and 90%–0% RH driving force.
[c] 0% RH.
[d] 100% RH.
(Adapted from Ref. [38].)

the permeability of a given material can result from differences in polymer molecular structure and weight, additives, and polymer-product- (e.g., film-) formation conditions.[18] Permeability for a selected commercial polymer and polymer product should be obtained from the supplier and/or measured by the user.

FACTORS AFFECTING PERMEABILITY

The permeability coefficient is influenced by both diffusion coefficient (reflecting the *speed* at which the permeant molecules diffuse through the polymer) and solubility coefficient (reflecting the *amount* of permeant in the polymer).[38] The factors which affect permeability can be grouped into compositional factors and environmental factors.[8–11,38–43] Compositional factors include permeant size and shape, polymer morphology, polymer additives, and permeant–polymer interaction.[3,6,8,9,18,38,39,44]

Temperature always has an effect on the permeability coefficient, with the effect reflecting the temperature effect on the solubility coefficient and the diffusion coefficient. An Arrhenius type equation describes the relationship between the permeability coefficient and temperature[8]:

$$P = DS = D_0 S_0 \exp\{-(E_d + \Delta H_s)/RT\}$$
$$= P_0 \exp(-E_p/RT) \quad (10)$$

ΔH_s is the heat of solution of the permeant in the polymer. ΔH_s is small and positive for permanent gases like oxygen, with the result that S increases slowly with temperature. However, ΔH_s is negative and larger for condensable vapors like water, with the result that S decreases with temperature. E_d is the activation energy (necessary for a hole to appear for diffusion jump) for diffusion of the permeant in the polymer. Thus, E_d is always positive and D always increases with temperature. E_p is an apparent activation energy for permeability of the permeant in the polymer. Thus, depending on whether ΔH_s is positive or negative, and on the relative size of ΔH_s and E_d, the permeability coefficient theoretically may increase or decrease with temperature. However, for all known permeant–polymer pairs, the permeability coefficient increases with temperature.[38]

Depending on the polymer material, the relative humidity can also affect permeability. Humidity has no effect on the polyolefins (nonpolar polymers), vinylidene chloride copolymers, and acrylonitrile copolymers (non-hydrophilic, polar polymers). However, the permeability of hydrophilic polymers such as ethylene-vinyl alcohol copolymer and most polyamides increases with the amount of moisture absorbed as water acts as a plasticizer for these polymers. Some polymers display a small decrease in permeability with absorption of moisture, including polyethylene terephthalate and amorphous nylons.[38]

MULTILAYER FILMS

Multilayer films are often fabricated to combine the unique properties of each layer into one structure. The relationship between the permeability coefficient for the total multilayer film, P_T, of total thickness, L_T, and the permeability coefficients and thicknesses of the layers, $1 \ldots n$, can be shown[3,8,39] to be

$$L_T/P_T = (L_1/P_1) + (L_2/P_2) + \cdots (L_n/P_n) \quad (11)$$

The above relationship is easy to use if the permeability coefficients of the individual layers are independent of the permeant partial pressure and of the water vapor partial pressure (if a gradient of the latter exists across the film). Otherwise, the individual permeability coefficients will depend on the thickness and positioning of the layers.

PREDICTION OF SHELF LIFE FROM PERMEABILITY COEFFICIENTS

A number of mathematical models exist, based on the assumption that the product shelf life depends on the permeability of the container to moisture and/or oxygen.[45] The models also require knowledge of the amount of moisture or oxygen transfer that a preserved product can endure.

A simple steady-state model for predicting the shelf life of oxygen-sensitive products uses Eq. 8 derived above.[3,18,46] More complicated, unsteady-state models are available, which take into account the kinetics of oxygen reacting with the product.[11,47]

An unsteady-state model has been developed that allows prediction of the shelf life of moisture-sensitive products.[3,11,18,46,48]

Additional models are available for MAP of fresh fruits and vegetables. Because of the complexity of MAP of fruits and vegetables, quite a number of different models with different assumptions exist.[49–59]

CONCLUSION

Permeability involves the transfer of a particular gas or vapor from one side of a particular polymeric material to the other. Plastic polymer packaging permeability must be taken into account to determine the shelf life of the food product being protected. Research continues to develop polymers with improved properties and improved fabrication methods.[38] Environmental goals such as reducing the amount of packaging required (source reduction), recycling, and degradability have added to the challenge of this area.[18,38,60–64] Utilization of moisture scavengers, oxygen scavengers, and other active packaging concepts can enhance the protective functions of polymer-based packaging.[65–68]

REFERENCES

1. Geankoplis, C.J. *Transport Processes and Unit Operations*, 2nd Ed.; Prentice-Hall, Inc.: Englewood Cliffs, NJ, 1983; 451–455.
2. Vieth, W.R. *Diffusion in and Through Polymers*; Hanser Publishers: New York, 1991; 322.
3. Hernandez, R.J. Food Packaging Materials, Barrier Properties and Selection. In *Handbook of Food Engineering Practice*; Valentas, K.J., Rotstein, E., Singh, R.P., Eds.; CRC Press: New York, 1997; 291–360.
4. Johnson, B.; Demorest, R. Testing, Permeation and Leakage. In *The Wiley Encyclopedia of Packaging Technology*, 2nd Ed.; Brody, A.L., Marsh, K.S., Eds.; John Wiley & Sons, Inc.: New York, 1997; 895–900.
5. Richards, J.H. The Role of Polymer Permeability in the Control of Drug Release. In *Polymer Permeability*; Comyn, J., Ed.; Elsevier Applied Science: New York, 1985; 217–268.
6. Ashley, R.J. Permeability and Plastics Packaging. In *Polymer Permeability*; Comyn, J., Ed.; Elsevier Applied Science: New York, 1985; 269–308.
7. Crank, J. *The Mathematics of Diffusion*; Oxford University Press: New York, 1975; 414.
8. Robertson, G.L. *Food Packaging Principles and Practice*; Marcel Dekker, Inc.: New York, 1993; 73–110.
9. Rogers, C.E. Permeation of Gases and Vapours in Polymers. In *Polymer Permeability*; Comyn, J., Ed.; Elsevier Applied Science: New York, 1985; 11–74.
10. Giacin, J.R.; Hernandez, R.J. Permeability of Aromas and Solvents in Polymeric Packaging Materials. In *The Wiley Encyclopedia of Packaging Technology*, 2nd Ed.; Brody, A.L., Marsh, K.S., Eds.; John Wiley & Sons, Inc.: New York, 1997; 724–733.
11. Hernandez, R.J.; Giacin, J.R. Factors Affecting Permeation, Sorption, and Migration Processes in Package-Product Systems. In *Food Storage Stability*; Taub, I.A., Singh, R.P., Eds.; CRC Press: New York, 1998; 269–330.
12. Gray, J.I., Harte, B.R., Miltz, J., Eds. *Food Product-Package Compatibility*; Technomic Publishing Co, Inc.: Lancaster, PA, 1987; 286.
13. Hotchkiss, J.H. Overview on Chemical Interactions Between Food and Packaging Materials. In *Foods and Packaging Materials—Chemical Interactions*; Ackermann, P., Jagerstad, M., Ohlsson, T., Eds.; The Royal Society of Chemistry: Cambridge, England, 1995; 3–11.
14. Risch, S.J., Hotchkiss, J.H., Eds. *Food and Packaging Interactions II*; ACS Symposium Series 473; American Chemical Society: Washington, D.C., 1991; 262.
15. Hotchkiss, J.H., Ed. *Food and Packaging Interactions*; ACS Symposium Series 365; American Chemical Society: Washington, D.C., 1988; 305.
16. Giacin, J.R. Factors Affecting Permeation, Sorption and Migration Processes in Package-Product Systems. In *Foods and Packaging Materials—Chemical Interactions*; Ackermann, P., Jagerstad, M., Ohlsson, T., Eds.; The Royal Society of Chemistry: Cambridge, England, 1995; 12–22.
17. Linssen, J.P.H.; Roozen, J.P. Food Flavour and Packaging Interactions. In *Food Packaging and Preservation*; Mathlouthi, M., Ed.; Blackie Academic & Professional: New York, 1994; 48–61.
18. Selke, S.E.M. *Plastics Packaging Technology*; Hanser/-Gardner Publications, Inc.: Cincinnati, 1997; 206.
19. Harte, B.R.; Gray, J.I. The Influence of Packaging on Product Quality. In *Food Product-Package Compatibility*; Gray, J.I., Harte, B.R., Miltz, J., Eds.; Technomic Publishing Co., Inc.: Lancaster, PA, 1987; 17–29.
20. Felder, R.M.; Huvard, G.S. Permeation, Diffusion, and Sorption of Gases and Vapors. In *Methods of Experimental Physics*; Fava, R.A., Ed.; Academic Press: New York, 1980; 315–377.
21. ASTM. Designation E 96-95: Standard Test Methods for Water Vapor Transmission of Materials. In *Annual Book of ASTM Standards*; American Society for Testing and Materials: Philadelphia, PA, 1995; 785–792.
22. ASTM. Designation F 1249-90: Standard Test Method for Water Vapor Transmission Rate Through Plastic Film and Sheeting Using a Modulated Infrared Sensor. In *Annual Book of ASTM Standards*; American Society for Testing and Materials: Philadelphia, PA, 1995; 1131–1135.
23. McHugh, T.H.; Avena-Bustillos, R.; Krochta, J.M. Hydrophilic Edible Films: Modified Procedure for Water Vapor Permeability and Explanation of Thickness Effects. J. Food Sci. **1993**, *58* (4), 899–903.
24. Gennadios, A.; Weller, C.L.; Gooding, C.H. Measurement Errors in Water Vapor Permeability of Highly Permeable, Hydrophilic Edible Films. J. Food Eng. **1994**, *21*, 395–409.
25. ASTM. Designation D 895-94: Standard Test Method for Water Vapor Permeability of Packages. In *Annual Book of ASTM Standards*; American Society for Testing and Materials: Philadelphia, PA, 1995; 118–119.
26. ASTM. Designation D 1251-94: Standard Test Method for Water Vapor Permeability of Packages by Cycle Method. In *Annual Book of ASTM Standards*; American Society for Testing and Materials: Philadelphia, PA, 1995; 202–203.
27. ASTM. Designation D 3079-94: Standard Test Method for Water Vapor Transmission of Flexible Heat-Sealed Packages for Dry Products. In *Annual Book of ASTM Standards*; American Society for Testing and Materials: Philadelphia, PA, 1995; 328–329.
28. ASTM. Designation D 2684-95: Standard Test Method for Permeability of Thermoplastic Containers to Packaged Reagents or Proprietary Products. In *Annual Book of ASTM*

Standards; American Society for Testing and Materials: Philadelphia, PA, 1995; 93–96.

29. ASTM. Designation D 3985-95: Standard Test Method for Oxygen Gas Transmission Rate Through Plastic Films and Sheeting Using a Coulometric Sensor. In *Annual Book of ASTM Standards*; American Society for Testing and Materials: Philadelphia, PA, 1995; 491–496.
30. Gilbert, S.; Pegaz, D. Find New Way to Measure Gas Permeability. Packaging Eng. **1969**, *January*, 66–69.
31. ASTM. Designation F 1307-90: Standard Test Method for Oxygen Transmission Rate Through Dry Packages Using a Coulometric Sensor. In *Annual Book of ASTM Standards*; American Society for Testing and Materials: Philadelphia, PA, 1995; 1150–1155.
32. Hernandez, R.J.; Giacin, J.R.; Baner, A.L. The Evaluation of the Aroma Barrier Properties of Polymer Films. J. Plast. Film Sheeting **1986**, *2* (3), 187–211.
33. Hernandez, R.J.; Giacin, J.R.; Baner, A.L. The Evaluation of the Aroma Barrier Properties of Polymer Films. In *Plastic Film Technology*; Technomic Publishing Co., Inc.: Lancaster, PA, 1989; 107–131.
34. DeLassus, P.T.; Standburg, G.; Howell, B.A. Flavor and Aroma Permeation in Barrier Film: the Effects of High Temperature and High Humidity. Tappi J. **1988**, *11*, 177.
35. Hatzidimitriu, E.; Gilbert, S.G.; Loukakis, G. Odor Barrier Properties of Multi-layer Packaging Films at Different Relative Humidities. J. Food Sci. **1987**, *52*, 472.
36. Miller, K.S.; Krochta, J.M. Measuring Aroma Transport in Polymer Films. Trans. ASAE **1998**, *41* (2), 427–433.
37. Seeley, D. Aroma Barrier Testing. In *The Wiley Encyclopedia of Packaging Technology*, 2nd Ed.; Brody, A.L., Marsh, K.S., Eds.; John Wiley & Sons, Inc.: New York, 1997; 39–41.
38. Delassus, P. Barrier Polymers. In *The Wiley Encyclopedia of Packaging Technology*, 2nd Ed.; Brody, A.L., Marsh, K.S., Eds.; John Wiley & Sons, Inc.: New York, 1997; 71–77.
39. Paine, F.A.; Paine, H.Y. *A Handbook of Food Packaging*; Blackie Academic & Professional: New York, 1992; 390–425.
40. Hernandez, R.J. Polymer Properties. In *The Wiley Encyclopedia of Packaging Technology*, 2nd Ed.; Brody, A.L., Marsh, K.S., Eds.; John Wiley & Sons, Inc.: New York, 1997; 758–765.
41. Pascat, B. Study of Some Factors Affecting Permeability. In *Food Packaging and Preservation—Theory and Practice*; Mathlouthi, M., Ed.; Elsevier Applied Science Publishers: New York, 1986; 7–24.
42. Jasse, B.; Seuvre, A.M.; Mathlouthi, M. Permeability and Structure in Polymeric Packaging Materials. In *Food Packaging and Preservation*; Mathlouthi, M., Ed.; Blackie Academic & Professional: New York, 1994; 1–22.
43. Halek, G.W. Relationship Between Polymer Structure and Performance in Food Packaging Applications. In *Food and Packaging Interactions*; Hotchkiss, J.H., Ed.; American Chemical Society: Washington, D.C., 1988; 195–202.
44. Moisan, J.Y. Effects of Oxygen Permeation and Stabiliser Migration on Polymer Degradation. In *Polymer Permeability*; Comyn, J., Ed.; Elsevier Applied Science: New York, 1985; 119–176.
45. Chao, R.R.; Rizvi, S.S.H. Oxygen and Water Vapor Transport Through Polymeric Film. In *Food and Packaging Interactions*; Hotchkiss, J.H., Ed.; American Chemical Society: Washington, D.C., 1988; 217–242.
46. Robertson, G.L. *Food Packaging Principles and Practice*; Marcel Dekker, Inc.: New York, 1993; 288–297, 354–363.
47. Khanna, R.; Peppas, N.A. Mathematical Analysis of Transport Properties of Polymer Films for Food Packaging: III. Moisture and Oxygen Diffusion. AIChE Symp. Ser. 218 **1982**, *78*, 185–191.
48. Taoukis, P.S.; El Meskine, A.; Labuza, T.P. Moisture Transfer and Shelf Life of Packaged Foods. In *Food and Packaging Interactions*; Hotchkiss, J.H., Ed.; American Chemical Society: Washington, D.C., 1988; 243–261.
49. Zagory, D.; Kader, A.A. Modified Atmosphere Packaging of Fresh Produce. Food Technol. **1988**, *42* (9), 70–77.
50. Kader, A.A.; Zagory, D.; Kerbel, E.L. Modified Atmosphere Packaging of Fruits and Vegetables. CRC Crit. Rev. Food Sci. Nutr. **1989**, *28* (1), 1–30.
51. Kader, A.A.; Singh, R.P.; Mannapperuma, J.D. Technologies to Extend the Refrigerated Shelf Life of Fresh Fruits and Vegetables. In *Food Storage Stability*; Taub, I.A., Singh, R.P., Eds.; CRC Press: New York, 1998; 419–434.
52. Robertson, G.L. *Food Packaging Principles and Practice*; Marcel Dekker, Inc.: New York, 1993; 484–499.
53. Talasila, P.C.; Cameron, A.C. Free-Volume Changes in Flexible, Hermetic Packages Containing Respiring Produce. J. Food Sci. **1997**, *62* (4), 659–664.
54. Talasila, P.C.; Cameron, A.C. Prediction Equations for Gases in Flexible Modified-Atmosphere Packages of Respiring Produce Are Different than Those for Rigid Packages. J. Food Sci. **1997**, *62* (5), 926–930.
55. Segall, K.I.; Scanlon, M.G. Design and Analysis of a Modified-Atmosphere Package for Minimally Processed Romaine Lettuce. J. Am. Soc. Hort. Sci. **1996**, *121* (4), 722–729.
56. Zagory, D. Modified Atmosphere Packaging. In *The Wiley Encyclopedia of Packaging Technology*, 2nd Ed.; Brody, A.L., Marsh, K.S., Eds.; John Wiley & Sons, Inc.: New York, 1997; 650–656.
57. Garrett, E.H. Fresh-Cut Produce. In *Principles and Applications of Modified Atmosphere Packaging of Foods*; Blakistone, B.A., Ed.; Aspen Publishers, Inc.: Gaithersburg, MD, 1999; 125–134.
58. Powrie, W.D.; Skura, B.J. Modified Atmosphere Packaging of Fruits and Vegetables. In *Modified Atmosphere Packaging of Food*; Ooraikul, B., Stiles, M.E., Eds.; Ellis Horwood Ltd: New York, 1991; 169–245.
59. Mannapperuma, J.D.; Zagory, D.; Singh, R.P.; Kader, A.A. Design of Polymeric Packages for Modified Atmosphere Storage of Fresh Produce. In *5th Controlled Atmosphere Research Conference*, Wenatchee, WA, 1989.
60. Selke, S.E. Environment. In *The Wiley Encyclopedia of Packaging Technology*, 2nd Ed.; Brody, A.L., Marsh, K.S., Eds.; John Wiley & Sons, Inc.: New York, 1997; 343–348.

61. Krochta, J.M. Film, Edible. In *The Wiley Encyclopedia of Packaging Technology*, 2nd Ed.; Brody, A.L., Marsh, K.S., Eds.; John Wiley & Sons, Inc.: New York, 1997; 397–401.
62. Krochta, J.M.; De Mulder-Johnston, C. Edible and Biodegradable Polymer Films: Challenges and Opportunities. Food Technol. **1997**, *51* (2), 61–74.
63. Bastioli, C. Biodegradable Materials. In *The Wiley Encyclopedia of Packaging Technology*, 2nd Ed.; Brody, A.L., Marsh, K.S., Eds.; John Wiley & Sons, Inc.: New York, 1997; 77–82.
64. Borchardt, J.K. Recycling. In *The Wiley Encyclopedia of Packaging Technology*, 2nd Ed.; Brody, A.L., Marsh, K.S., Eds.; John Wiley & Sons, Inc.: New York, 1997; 799–805.
65. Rooney, M.L., Ed. *Active Food Packaging*; Blackie Academic and Professional: New York, 1995; 260.
66. Rooney, M.L. Active Packaging. In *The Wiley Encyclopedia of Packaging Technology*, 2nd Ed.; Brody, A.L., Marsh, K.S., Eds.; John Wiley & Sons, Inc.: New York, 1997; 2–8.
67. Brody, A.L.; Strupinsky, E.R.; Kline, L.R., Eds. *Active Packaging for Food Applications*; Technomic Publishing Co., Inc.: Lancaster, PA, 2001; 280.
68. Miltz, J.; Passy, N.; Mannheim, C.H. Trends and Applications of Active Packaging Systems. In *Foods and Packaging Materials—Chemical Interactions*; Ackermann, P., Jagerstad, M., Ohlsson, T., Eds.; The Royal Society of Chemistry: Cambridge, England, 1995; 201–210.

Package Properties

Felix H. Barron
Joel D. Burcham
Clemson University, Clemson, South Carolina, U.S.A.

P

INTRODUCTION

The knowledge of properties of food packaging materials and packages is critically important to manufacturers and food processors. Only the optimum properties will effectively perform the packaging functions[1]: protection, containment, information, and convenience or utility of use. For example: a canned soup is very well protected, fresh oranges in a holding net are well contained, packaged foods with nutritional labels provide useful information, and fruit juices in easy open pouches are convenient.

The importance of properties becomes obvious when it is realized that packaging materials are exposed to various environmental or processing conditions such as cold and hot temperatures, humid or dry environments, high and low altitudes, dark or lighted areas, corrosive or oxidizing agents such as acids and oxygen.

Using the knowledge of properties of packaging materials, processors can meet requirements for consistency, compliance to laws, customer specifications, and make choices based on cost and attributes not related to performance (such as corporate relationships). Package materials have many properties, but only few need to be tested. Converters, suppliers, or distributors of packaging materials often have specification sheets available describing the important material properties. It is necessary to know how materials will perform in any situation, whether manufacturing, distribution, processing, or storage, to ultimately achieve the optimum performance of packaging functions.

PACKAGING MATERIALS AND IMPORTANT PROPERTIES

From a processor's point of view, food safety is most important; therefore, the assurance of the integrity of packaged foods is critical to the acceptability of closures and containers (packages). The integrity of a package prevents the biological, chemical, or physical contamination of food from or to the environment. For this reason, there are some fundamental evaluations or tests required by federal agencies, including the FDA and USDA. Some examples are shown in Table 1.

The mechanical properties, especially strength of packaging materials, are also important for protection during distribution and processing of packaged foods. Acceptable barrier properties, on the other hand, would be important in preventing chemical or physical degradation of food including gain or loss of flavors and aromas and changes in color and texture.

The performance of packages depends on the packaging material properties (after conversion) and the converting processes used to manufacture the package. Many materials can be modified to improve package performance. Generally, packaging materials can be grouped in four major categories: polymers, metals, paper and paperboard, and glass.

The selection of packages and packaging materials should be based primarily on considerations of food safety, followed by quality, cost, legal and international issues.

Important properties of common packaging materials generally are mechanical, optical, and thermal properties. Table 2 is a list of common properties identified in the literature.

Common package properties are:

1. Bursting strength—the resistance of a packaging material to a sudden rupture especially due to internal pressure.
2. Coefficient of friction—a measure of the force opposing an applied force parallel to a surface; it is dependent upon the perpendicular force between the material and another material surface.
3. Density—the mass of a material per unit volume.
4. Elongation—the change in length of a material resulting from tensile stress.
5. Folding endurance—a measure of the number of double folds required to break a strip of a material, written as a common logarithm.
6. Gauge—a unit length, 1×10^{-4} in.
7. Gloss—the ratio of light flux specularly reflected from a surface to the total reflected flux.
8. Haze—material opacity due to internal and surface reflections of incident light.
9. Light transmission—the light flux through a material over an interval of time.
10. Modulus of elasticity—a general ratio of a specific form of stress to a specific form of strain; specific

DOI: 10.1081/E-EAFE 120007128
Copyright © 2003 by Marcel Dekker, Inc. All rights reserved.

Table 1 Examples of selected package properties related to package integrity required by USDA and FDA (9CFR, 318 (381) and 21 CFR, 113)

Polymeric, paper and paper board containers	Metal containers	Glass containers
Testing: Burst strength, tensile seal strength	Testing: *Double seam measurements using a micrometer*: Cover hook, body hook, width, tightness, thickness, and side seam juncture	Testing: *Capper efficiency*: Cap tilt, vacuum, security, gasket impression, removal torque, and button position
Property related: Burst strength, tensile strength	Property related: Compatibility of the sealing compound, base weight (thick ness), and hardness (temper)	Property related: Compatibility of the sealing compound; glass finish

examples are Young's modulus, bulk modulus, and shear modulus.

11. Opacity—the ratio of the amount of incident light reflected from a material to the amount of incident light transmitted through the material.
12. Oxygen transmission rate (OTR)—the amount of oxygen passing through a material under specified conditions of time, temperature, pressure, and relative humidity.
13. Rockwell hardness test—the relative hardness of a material using a specified machine that indents the surface of the material under controlled conditions.
14. Stress—force absorbed by a body as a result of external forces applied to it.

Table 2 Selected properties of packaging materials

Properties	Polymer	Paper and paperboard	Metal	Glass
Tensile strength	✓	✓	✓	✓
Tear initiation	✓	✓		
Modulus of elasticity (stiffness)	✓	✓		✓
WVTR	✓			
OTR	✓	✓		
Burst	✓	✓		
Coefficient of friction	✓			✓
Permeability to gases	✓	✓		
CO_2 transmission rate	✓			
Resistance to fats, oils, and greases	✓	✓		
Density	✓			
Optical properties	✓	✓	✓	✓
Unrestrained shrink	✓			
Shrink tension	✓			
Shrink temperature	✓			
Peak load (instrumented impact)	✓			
Energy to break (J)	✓			
Thermal strength				✓
Thermal shock resistance				✓
Rockwell 30 T hardness (HR30T)			✓	
Folding strength		✓		
Scuffing resistance		✓		
Smoothness		✓		
Ink receptivity		✓		
Firmness of the surface		✓		

15. Tear strength—the force required to tear a material; tearing initiation and tearing propagation are commonly measured.
16. Tensile strength—the maximum force that a material specimen can resist under tension.
17. Tensile stress—a stretching force along an axis.
18. Thermal endurance—the amount of thermal shock a material can withstand.
19. Thermal shock—the application of a temperature change to a material over a short time period.
20. Water vapor transmission rate (WVTR)—the amount of water vapor that passes through a specimen in a set time period under controlled conditions of time, temperature, and relative humidity.

Other important food packaging material properties include: appearance, machinability and the material forms available (films, bags, jars, and bottles), width, length, thickness (gauge), and compatibility with conventional converting equipment and sealing technique.

PACKAGE PROPERTIES AND PACKAGING FUNCTIONS

Packaging manufacturers and food processors are interested in knowing how package properties relate to package functions in order to optimally fulfill food requirements. Selected examples of these relationships are shown in Table 3.

SPECIFICATIONS OF PACKAGE PROPERTIES

The following issues should be considered for packaging specifications[1]:

1. The purpose of the package or packaging material and a general description of the package.
2. Construction materials and dimensions.
3. Protection requirements, related to the expected surrounding conditions, such as temperature, relative humidity, light, and oxygen; e.g., 85°F and 90% relative humidity. Physical protection during handling and transportation should also be included.
4. Description of the functionality of the package, such as ease of opening, labels that can withstand freezing temperatures, easy disposal or recycling.
5. Graphics and other information requirements.
6. Names and addresses of suppliers or contractors.

Table 3 Selected package properties and their relationship to packaging functions

Tensile strength	Protection and containment; package, seal, or seam integrity; prevents external contamination
WVTR	Protection against food degradation by water
OTR	Protection against food oxidation; prevents growth of micro-organisms
Burst strength	Protection and containment; seal and package integrity; prevents external contamination
Tear strength	Protection and containment; package integrity; packaging operations, distribution

Processors and converters of plastics (polymers) are recognized for their efforts in developing good product specifications to facilitate package selection. Table 4 shows typical information from specifications of polymer packages and their applications in the food industry.

MODIFICATION OF PACKAGE PROPERTIES

Package properties or the functionalities of package materials are often improved during their processing or converting operations by the addition of agents or substances. In other cases, several layers of various packaging materials are bound together to improve the permeability to water vapor or gases.

Metal containers are commonly coated with lacquers to prevent food deterioration; the external surfaces of glass containers are treated to reduce abrasion and breakage.

Other examples related to protecting packages and/or their contents from degradative agents or processes such as oxygen in air, UV exposure, reactions between cans and acidic contents or micro-organisms are presented in Table 5.

The following is a general list of other packaging-material modifying agents used in the food or nonfood industry: adhesion promoters; antistatic agents; blowing agents; brighteners; catalysts and promoters; chelating agents; clarifiers; cling agents; colorants; compatibilizers; degradable agents; fillers; fragrance enhancers; heat stabilizers; hydrophilic modifiers; hydrophobic agents; impact modifiers; low-profile agents; lubricants; plasticizers; reinforcement agents; surfactants; and viscosity depressants.

PACKAGE PROPERTIES AND TESTING

The testing of package properties is important to both the converter and the food processor. As mentioned earlier, the main concern for food processors is food safety and

Table 4 Values of selected properties of polymer packages

Packaging material	WVTR ($g/m^2/24\,hr$) [$g/100\,in.^2/24\,hr$]	OTR ($cc/m^2/24\,hr$ @ 73°F, 0% RH, 1 atm) [$cc/m^2/24\,hr$ @ 73°F, 0% RH, 1 atm]	Tear (g)	Tensile strength (psi @ 73°F)	Ball burst strength (cm-kg)	Haze (%)	Gloss (%)	Temperature usage range (°F)	Applications
Films[a]									
HS-3000 (2 mils)	0.8–1.0 (@ 100°F, 100% RH)	< 10 (high barrier)	(Initiation) very difficult	Longitudinal 6,500–7,500 Transverse 5,500–6,200	22	Appearance: white, opaque, or clear	Appearance: white, opaque, or clear	0–150	Chub packed fresh packed ground beef
BDF-2050 (75–100 gauge, antifog)	[1.2–1.0]	26.0–19.4	15.0–20.0	10,000	16.0–18.5	4.0–4.7	80	− 60–90	Fresh meat and poultry
BDF-2001	[1.0–0.75]	5.0–4.0	8.0–14.0	10,000	8.0–12.0	3.0	85	− 40–90	Bakery goods, fresh pizza, fruit juices
PD-960 (125 gauge)	[0.9–1.10] (@ 73°F, 100% RH)	6,000–8,000	30	12,000	26	6.5	84	0–90	Fresh produce
FS 7100 series (3.5–5.5 mils)	[0.28–0.20] (@ 73°F, 100% RH)	0.5–0.2	270–350	4,700–6,000			Clear	40–205	Extended shelf life of cold, hot, and pumpable foods, syrups, toppings, fillings
FS 7000 (3.5–5.5 mils)	[0.50–0.20] (@ 73°F, 100% RH)	< 4–2	250–425	6,200–8,000			Clear	40–205	Extended shelf life of cold, hot, and pumpable foods, syrups, toppings, fillings

BDF-4000 (100 gauge)	[0.75] (@ 73°F, 100% RH)	4	14	10,000	12.0	3.0	85	− 60–90	Cheese cuts
B-series bag	[0.5–0.6] (@ 100°F, 100% RH)	[3–6]	~ 20–30	6,500–9,000		Transparent	High gloss		Poultry, pork, corned meats, smoked and processed meats
E-series bag	[0.65] (@ 100°F, 100% RH)	[4,000]	15–20	9,722–8,450		Clear	High	(minimum) − 20	Fresh poultry and seafood
CN-530/CN-590 cook-in film		20	(initiation) very difficult	6,000			Clear or printed	(maximum, cooking) 200 +	Cook and ship, cook and strip applications, microwave-ready foods
C-5045/C-5030 cook chill casings (4.5/3.0 mils	[0.4/0.57] (@ 100°F, 100% RH)	46/73	1,860/1,400			Translucent			Kettle cooked foods into casings
Edible films[2]									
Polyol-plasticized at various levels	0.25×10^{-11}–24.6×10^{-11} (g/m/sec/Pa @ 22°C)	4.1×10^{-16}–7.3×10^{-12} (cm^2/sec/Pa @ 22°C)							Packaging or coatings

[a] From Cryovac/Sealed Air Corporation, P.O. Box 464, Duncan, SC 29334.

Table 5 Selected agents to modify package properties or functionality

Modifying agent	Function	Compound	Application
Antiblocking agent[3]	It reduces the sticking between film layers	Diatomaceous earth, talc, calcium carbonate, etc.	Plastic films such as LDPE and PP
Antifogging agent[4]	Lowers surface tension of water droplets to maintain clarity	Alkylphenol ethoxylates, polyoxyethylene esters of oleic acid	Plastic films such as LDPE, PVC, EVA, OPS, polyester
Antimicrobial (Biocides)[5]	Destroys or inhibits growth of micro-organisms on material surfaces	2-*n*-Octyl-4-isothiazolin-3-one	Toys, kitchen utensils, door handles
Antioxidants[6]	Inhibits oxidative degradation of polymers	Anox 29, BHA, CAO-3, BHT, etc.	Polymers (polyolefins) (FDA 178.2010)
UV stabilizers[7]	Inhibits or slows polymer degradation from UV exposure	Benzophenones and hindered amine	Polymers (polyolefins)
Can enamels (lacquers)[8]	Protection against food degradation	Epoxi-phenolic	Metal cans for fruits and fatty foods
	Good for sterilization	Oleoresins, phenolics	Acid fruits, seafood, and meats
Surface strengtheners and lubricators[9]	Strengthens glass surface	Tin or titanium tetrachloride	Glass containers
	Reduces coefficient of friction	Water emulsions of waxes and PE	

Table 6 Common standardized testing of packaging materials

Property	Polymer	Paper and paperboard	Metals	Glass
Tensile strength	D638M-90, D882-97, D1623-95, D5026-95a, R527-1996 (I), 1184-1983 (I), 1926-1979 (I), 6239-1986 (I)	D828-93, D829-95, T404, T494, T541	E8M-98, E345-93, E646-63	
Tear	D1004-94a, D1922-94a, D1938-94, 6383/1-1983 (I)	D689-96a, T414	E436-91, E604-94	
WVTR	F372-94, F1249-95, T557, 1663-1981 (I)	T464, T523		
Burst strength	D3420-95	D774M-96a, T403		
OTR	D3985-95			
Gas transmission rate	D1434-92			
Resistance to fats and oils	F119-92	D722-97, T454		
Blocking resistance	D3354-96, 11502 (I)	D918-93		
Transparency	D542-95, D1746-97, 489-1983 (I)			
Heat seal seam	F88-94			
Brightness		D985-93, T452, T560, T562		
Thickness, density	D792-91, D1505-96, D1622-93, D1895-96, 61-1976 (I), 845-1988 (I), 1183-1987 (I), 4593-1979 (I)	D645M-96, T411, T551		C693-93, C729-95
Stiffness	D747-95, D1043-92, 458/1-1985 (I)	T451, T489, T495, T535, T543, T556, T566		
Hardness (Rockwell)	D785-93, 2039/2-1987 (I)		E18-97a, E1842-96	C730-85 (Knoop)
Thermal shock resistance				C149-86
Internal pressure strength				C147-86
Modulus of elasticity	D882-97, D5418-95a	T451, T489, T535	E111-97, E1876-97, E1875-97	C623-92

C–F listings pertain to ASTM, (I) to ISO standards, and T to TAPPI.

package integrity; secondary concerns, such as appearance, still need to be satisfied. It is the responsibility of the converter to ensure the packaging material has been tested to meet the required specifications. Commonly, packaging manufacturers use standardized testing (Table 6) to achieve consistency throughout the industry. The food industry relies on testing of package material properties and packaged samples to ensure that food products are preserved at least as long as their specified shelf life.

REFERENCES

1. Barron, F.H. *Food Packaging and Shelf Life: Practical Guidelines for Food Processors*; Extension Publication EC 686, Cooperative Extension Service, Clemson University: Clemson, SC, January 1995.
2. Arvanitoyannis, I.; Biliardis, C.G. Physical Properties of Polyol-Plasticized Edible Films Made from Sodium Caseinate and Soluble Starch Blends. Food Chem. **1998**, *62* (3), 333–342.
3. Fazzari, A.M. Antiblocking Agent. In *Modern Plastics World Encyclopedia 2000*; Chemical Week Associates: Danvers, MA, 1999; Vol. 76, No. 12, B-3.
4. Unknown, Antifogging Agent. In *Modern Plastics World Encyclopedia 2000*; Chemical Week Associates: Danvers, MA, 1999; Vol. 76, No. 12, B-4.
5. Ice-Pettegrew, M.L. Antimicrobial. In *Modern Plastics World Encyclopedia 2000*; Chemical Week Associates: Danvers, MA, 1999; Vol. 76, No. 12, B-6.
6. Hendrix, R. Antioxidant. In *Modern Plastics World Encyclopedia 2000*; Chemical Week Associates: Danvers, MA, 1999; Vol. 76, No. 12, B-7.
7. Horsey, D. Light Stabilizer. In *Modern Plastics World Encyclopedia 2000*; Chemical Week Associates: Danvers, MA, 1999; Vol. 76, No. 12, B-70.
8. Lopez, A. Containers for Canned Foods. In *A Complete Course in Canning, Book 1—Basic Information on Canning*, 11th Ed.; The Canning Trade, Inc.: Baltimore, MD, 1981; 159–160.
9. Cavanagh, J. Glass Container Manufacturing. In *The Wiley Encyclopedia of Packaging Technology*, 2nd Ed.; Brody, A.L., Marsh, K.S., Eds.; John Wiley and Sons, Inc.: New York, 1997; 483.

Pasteurization Systems

Thomas M. Gilmore
International Association of Food Industry Suppliers, McLean, Virginia, U.S.A.

INTRODUCTION

This entry provides an overview of the history, present, and a peek at emerging nonthermal pasteurization technique. Historically, "pasteurization" was applied to wine and preserving foods. (Today, it is correctly called "canning.") Pasteurization of milk and milk products took center-stage in the first half of the 21st century and remains the primary public health measure used by the dairy industry, and most recently, the juice industry. Thermal destruction of pathogens and, in certain products, nonpathogens meeting prescribed legal requirements is the only acceptable method for pasteurization and high-temperature processing of milk, milk products, and other comestibles.

The entry covers the most basic equipment for pasteurization and the minimum legal requirements.

OVERVIEW

In the Beginning There Was Louis Pasteur

In 1856 Pasteur discovered that heating wine to 50–60°C (122–140°F) prevented spoilage. Thus the most significant thermal process to protect against food-borne disease outbreaks was invented.

In the early 20th century, 25%–35% of all food-borne disease outbreaks were attributed to milk or milk products—eight times more than water-borne outbreaks. Milk cannot be harvested in a sterile form and is nearly always contaminated during collection. Milk is a first or early food for the very young and is often a staple for the elderly. Both age groups are frequently immune compromised. Actually, all pathogens thrive or at least survive in milk. Pasteurization is the only public health measure, if properly applied, that will protect against all infectious milk-borne disease organisms which are in the raw milk supply.

Pasteurization as Defined in the United States

The application of a heat process to good quality milk for the purpose of rendering it a safe and nutritious food product, which will survive on the shelf for a 10-day to 20-day period under refrigerated conditions has been the industry standard for over five decades. Pathogens are destroyed, industry and the consumer are happy and healthy, and the nation's milk supply is safe and wholesome.

Pasteurization Conditions

Pasteurization time/temperature requirements are shown in Table 1.

Vat (Batch) Pasteurization

The product is heated in a jacketed stainless steel vat by hot water and/or steam in the jacket liner and fitted with thermometers to monitor and record product temperatures, and some means of agitation to assure uniformity in temperature distribution. Other requirements include properly designed valves, time/temperature recording requirements, and methods of operation.

Generally, all vat- or batch-type pasteurizers should conform to the "3-A Sanitary Standards for Non-Coil Type Batch Pasteurizers For Milk and Milk Products," Number 24-. This standard provides guidelines for the installation, approved materials, finish, and fabrication of vat pasteurizers.

HTST Systems

Vat pasteurization (Fig. 1) has limited use because of the mostly large-volume dairies found in the United States. After the introduction of plate heat exchangers into the United States in 1928, continuous flow pasteurization of milk was possible. In 1931, Pennsylvania conducted thermal destruction of pathogens at 72°C (161°F) for 15-sec hold. Studies at the University of California—Davis and the University of Maryland established the currently recognized minimum time/temperature requirements for milk pasteurization. By 1938, HTST systems (Fig. 2) with controls and diversion valves were used—they are now required.

In the most basic systems, cold raw milk at about 4.4°C (40°F) enters the constant-level (balance) tank and is drawn into the regenerator section of the heating/cooling unit (a plate and frame unit or a tubular unit) using reduced

Encyclopedia of Agricultural, Food, and Biological Engineering
DOI: 10.1081/E-EAFE 120007104
Copyright © 2003 by Marcel Dekker, Inc. All rights reserved.

Table 1 Pasteurization time/temperature requirements

Product	Time–Temp.		
	VAT	HTST	HHST
Whole milk, low fat, skim	30 min 63°C (145°F)	15 sec 72°C (161°F)	1.0 sec 89°C (191°F), 0.5 sec 90°C (194°F), 0.1 sec 94°C (201°F), 0.05 sec 96°C (204°F), 0.01 sec 100°C (212°F)
Milk products—with increased viscosity, added sweetener, or fat content 10% or more	30 min 66°C (150°F)	15 sec 75°C (166°F)	Same as above
Egg nog, frozen dessert mixes	30 min 69°C (155°F)	25 sec 80°C (175°F), 15 sec 83°C (180°F)	Same as above

Note: Those pasteurized milk products that are further heated in an acceptable system to a minimum of 138°C (280°F) for a minimum of 2.0 sec are to be labeled as "Ultra Pasteurized" (U.S.A. only).

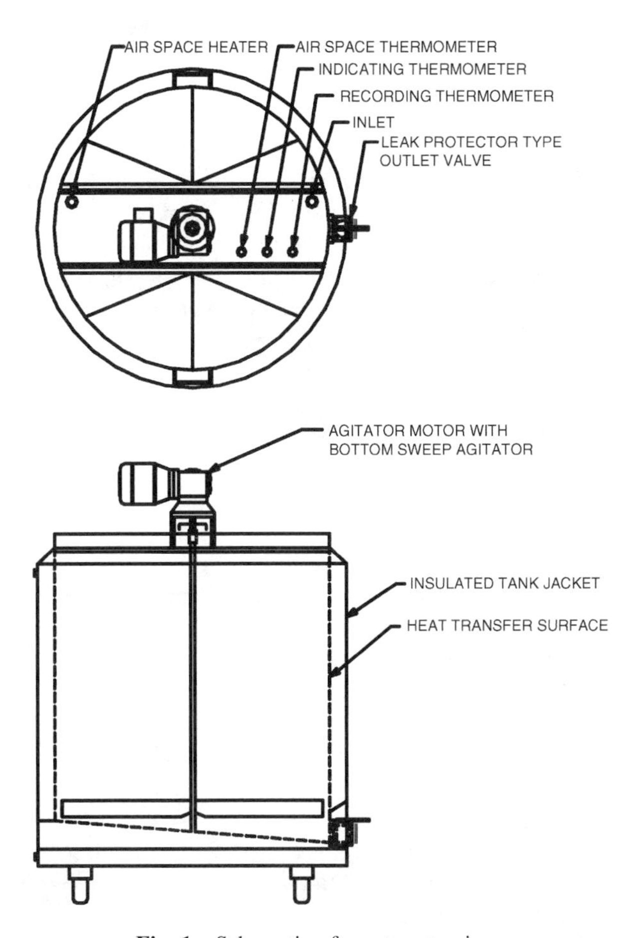

Fig. 1 Schematic of a vat pasteurizer.

pressure. The regenerator section is to prewarm the incoming cold raw milk or raw milk product with countercurrent flow of the hot pasteurized milk or milk product. Typical systems reclaim 85%–90% of the latent heat, which represents a considerable energy and cost savings. Since there is raw milk and pasteurized product on opposite sides of thin plates or thin-walled tubing, a pressure differential between pasteurized side of the plate or tube and the raw side is required. The Grade "A" Pasteurized Milk Ordinance (PMO) Item 16p (D)[2] requires a pressure differential of 7.0 kPa (1 psi)—the higher pressure on the pasteurization side. The 3-A HTST/higher-heat, shorter-time (HHST) Practice requires a 14.0 kPa (2 psi) differential.[3]

From the regenerator, the warmed milk is drawn through a positive displacement timing pump, which delivers it under positive pressure to the heating and cooling sections of the heating/cooling unit. In the heating section, the warmed milk and steam-heated hot water flow

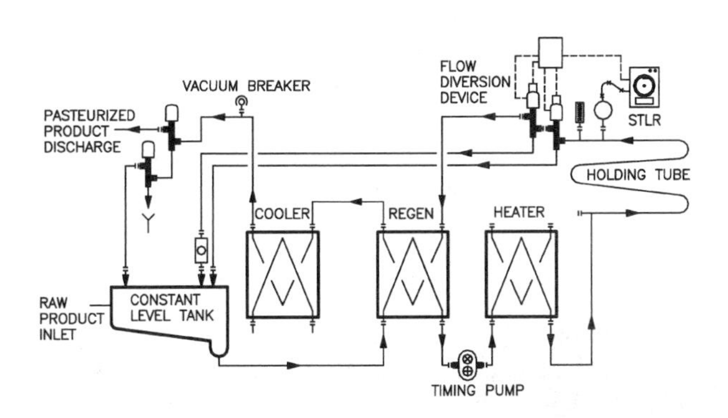

Fig. 2 HTST pasteurizer with positive displacement rotary timing pump.

counter-current on opposite sides of a plate or tube and the milk continues to heat equal to or, usually, above the legal minimum pasteurization temperature in Table 2. No pressure differential is required in the final heating section.

The milk, now at or above the legal pasteurization temperature, flows under pressure through the holding tube where its resident time must be at least 15 sec and the temperature maintained at greater or equal to 72°C (161°F), or for certain milk products, at higher temperatures and longer times found in Table 2. The rate of flow is governed by the pumping speed of the timing pump. The resident time is then a function of the pumping rate of the pump, the length of the holding tube, and the surface friction of the particular milk product. A detailed discussion of the holding tube is found in a section elsewhere in this encyclopedia. The timing device is designed and sealed to insure the minimum holding time.

The next point of contact for the milk is the sensors of the indicating thermometer and the recording thermometer. At this point, the milk will continue movement through the flow divert valve to the regenerator section if at or above the appropriate legal minimum temperatures found in Table 1 or other temperatures acceptable to the regulatory control agency. If the milk or product is not at or above the required minimum temperature, it will be returned (diverted) to the raw milk constant level tank via action of the flow-diversion valve and through the divert line. This is the most important critical control point in the pasteurization of milk, eggs, or other pumpable foods that are pasteurized. However, flow divert valves and controls are only required on systems for milk pasteurization.

In diverted flow, temperature adjustments or other corrective actions may be made to achieve the minimum time/temperature relationship followed by re-pasteurizing the milk. In forward flow, the safety thermal limit recorder signals the flow-diversion valve to be in the forward flow position, which allows the milk to continue its flow as a legally pasteurized product.

The hot pasteurized milk or product then passes through the product-to-product regenerator as described above and is partially cooled, typically to 9 to 12°C (48 to 54°F). Then the partially cooled product passes through the cooling section of the heat transfer system. The removal of the remaining heat is effected by water or a propylene glycol/water solution flowing counter-current on the opposite side of a plate or tube. It is to be noted that ethylene glycol cannot be used due to its toxicity. The product exiting the cooler section legally must be at or below 7.2°C (45°F), preferably at 1.1–4.4°C (34–40°F). The current 3-A Accepted Practices for HTST/HHST Systems[3] require a minimum pressure differential of 14.0 kPa (2 psi) in cooler sections with the higher pressure on the product side. This is to protect against propylene glycol or cooling water movement through any pinholes into the product. The current PMO requires this pressure differential a requirement in HHST systems.

The cold pasteurized product exits the cooler and immediately rises at least 305 mm (12 in.) above any raw milk in the HTST system and is open to the atmosphere through a sanitary vacuum breaker at the minimum height or higher.

From this point, the pasteurized milk or product may be pipeline transferred to either a pasteurized product storage tank, a filler tank, or a surge tank for subsequent packaging.

HTST Auxiliary Equipment

Various product treatments may be applied in addition to the basic HTST pasteurization and are often incorporated into the system. The most common examples are homogenizers, separators and clarifiers, auxiliary raw product pumps, and flavor control equipment.

When any of this equipment is added to an HTST system, it must be designed, installed, and operated so that it will not: reduce the holding time below the legal minimum; interfere with the required pressure relationships within the product-to-product regeneration section and the product-to-media in the cooling section; and adversely affect the minimum required product temperature, proper operation of the flow-diversion valve, or the functions of the recorder controller.

Homogenization is the process of reducing the fat globule size to such an extent that after 48 hr of storage no visible separation will occur. The fat content also must not differ by more than 10% throughout the product. This process is accomplished by forcing whole milk through small openings at extremely high pressures.

The homogenizer is a high-pressure piston pump containing three to five pistons driven by a crankshaft. The fat globule size is reduced by forcing the milk through a small orifice known as the homogenizing valve. This boosts pressure from 300 kPa (44 psi) at the inlet to a homogenization pressure of 10 Mpa–20 Mpa (1500 psi–3000 psi), although high pressures of homogenization up to 70 MPa (10,000 psi) is used for some applications. The product temperature must be high enough to have all or most of the fat as a liquid for efficient homogenization.

Since homogenizers are positive displacement pumps, they are flow-promoting devices, and their placement in the HTST system is important. Unless used as a timing pump, homogenizers must be installed and operated as a nonflow promoting device. This requires a nonrestricted re-circulation line. A common placement is between the heating section outlet and the flow-divert valve. If the homogenizer is of a smaller capacity than the timing pump, a pressure-release valve is also necessary to allow product to return to the balance tank if the flow of product exceeds the homogenizer's capacity.

If the homogenizer is used as the timing pump along with a conventional timing pump, a bypass line equipped with restricted manual valve or a positive, fail-safe, air-operated shut-off valve is necessary. This is to prevent product from flowing too fast and thus being considered sublegal with respect to the lack of holding time. Both the homogenizer and the timing pump must be sealed at their fastest speeds to maintain legal holding time.

Separators and clarifiers

Separation of whole milk is used for the partial or complete removal of milk fat particles (cream) from the product. This process is accomplished by exposing whole milk to a centrifugal force through a series of high-speed rotating discs or plates. Separation of whole milk is done to adjust milk fat content to a desired or legally required level less than that of the incoming milk or to standardize the fat level content.

Small capacity units of early design require manual cleaning while modern, high-capacity units are CIP cleaned.

Clarifiers are often used for separation of solid particles from raw milk as it is received to improve the keeping quality of the raw milk or in the process to recover cheese fines. The principal difference between a centrifugal clarifier and a separator is the design of the disc stack. Flow of product into a clarifier will be substantially the same at the discharge less any unwanted solids. Alternatively, separators function not only to clarify product but also to create heavy and light phase discharge streams, i.e., skim milk and cream.

The regulatory concerns for both are that they are powerful flow promoters, and separators alter the flow of product by removing fat. Separators located on the raw side of the HTST system must be installed before the timing pump and must be automatically valved out of the system during shutdown, power interruption, or manual changing of the flow-diversion valve switch to the "inspect" position. Separators located between two milk-to-milk regenerators must be interwired to the flow-diversion valve, timing pump, and the pressure differential control device. The reason is to prevent the pumping of raw milk into pasteurized milk during diverted flow or shutdown.

Separators may also be located on the pasteurized side of the system and must also be automatically valved out of the system during periods of diverted flow, loss of power or shutdown, and when the flow-diversion valve is manually manipulated. This is to assure proper pressure relationships in the regenerator section and prevent negative pressure on the forward-flow port of the flow-diversion device.

Auxiliary pumps

Sanitary centrifugal pumps and sanitary positive rotary pumps are used as timing pumps, booster pumps, and feed pumps in HTST systems. They are usually coupled directly to a motor with a shaft or indirectly through a gearbox.

A booster pump may be installed in continuous pasteurizer systems under certain closely controlled conditions. Booster pumps must always be of the centrifugal design and they serve several functions in modern HTST systems including: 1) providing pressure to the timing pump inlet by moving raw milk from the balance tank to the raw regenerator; 2) providing pressure to the homogenizer inlet when the homogenizer serves as the timing pump; 3) increasing regenerator efficiency; and 4) reducing excessive vacuum and the associated "flashing" of raw milk in the regenerator.

Booster pumps are usually installed between the constant level tank and the raw regenerator section. A booster pump may operate only when the timing pump is running, the system is in forward flow, and the required regenerator differential pressure is greater than or equal to 7.0 kPa (1 psi)[2] or greater than or equal to 14.0 kPa (2 psi).[4]

A stuffing pump is a centrifugal pump placed in the HTST system to feed ("stuff") or remove product from auxiliary pumps. They are commonly used to feed product to separators, vacuum chambers, and homogenizers. In meter-based timing systems (MBTS), centrifugal feed pumps are used as the timing pump. When the homogenizer is a timing pump, product feed pumps are used, which must be interwired so that it cannot run unless the homogenizer is operating. Product removal pumps are used to remove product from vacuum flavor enhancing equipment and must meet the same flow-promoting limitation as above.

Flavor control equipment

This equipment includes vacuum and steam-assisted vacuum systems whose function is to remove volatile, unwanted flavors or to expel dissolved air. When installed in an HTST system, it must meet the same restrictions as for all other auxiliary equipment outlined in "HTST Auxiliary Equipment" and must not adulterate the finished product with unwanted contaminants including water.

If the vacuum equipment is installed downstream from the flow-diversion valve, a means of preventing negative pressure is required between the forward-flow port of the flow-diversion valve and the inlet to the vacuum chamber. A vacuum breaker and a fail-safe, power-activated shut-off valve located at the entrance to the vacuum chamber are necessary. If steam is used, it shall be of culinary quality, meeting 3-A Accepted Practices for Culinary Steam[5] and only contain boiler compounds that comply with 21 CFR Part 173.310.[6]

MBTS

Magnetic flow meters are frequently used to replace timing pumps in HTST and aseptic systems. In conventional pump-timed systems, only the temperature and pressure differential is monitored to control the flow-diversion device. The product flow rate is maintained mechanically at a set speed by the time pump, which is sealed to prevent operator tampering. In meter-based systems, the product flow rate through the holding tube is constantly monitored and controlled. Forward flow is only possible when a preset acceptable temperature (minimum legal) is reached and maintained.

There are two conditions that will result in diverted flow: one is if the product exceeds the preset flow rate; the second being when the flow rate drops below what can be accurately measured.

Meter-based timing systems by U.S. regulations must be installed as complete systems as submitted to and reviewed by the FDA.

Advanced Thermal Systems

With the trend toward consolidation of dairy plants and the resultant increase in distances required for distributing the pasteurized products, processors are opting for higher temperature pasteurization systems, which in most cases greatly enhance keeping quality and shelf-life of processed milk products.

The industry has expressed interest in processes that use higher temperatures 89°C (191°F) and above for shortened times (1 sec and less) for the processing/pasteurization of Grade A milk products. This process is appropriately called HHST or extended shelf life (ESL) systems (see Table 1 for time and temperature). Most of these systems are being used to process ultra pasteurized (UP) products and require normal legal flow-diversion devices. Also most of them operate at temperatures in the 132–149°C (270–300°F) range.

These higher processing temperatures with shorter times, produce an increase in product shelf-life without significantly affecting the desirable flavor of the milk product. The minimum required times and temperatures are based on the ice cream thermal death curve and the computations assume full laminar flow. The calculated holding times are not required to be adjusted for higher product viscosities.

HHST, UP, or ultra high temperature (UHT) systems

Indirect heated systems use the same heat exchange equipment as for HTST. Because of the high temperatures and relatively short hold times, the flow-diversion valve is located at the end of the cooler or final regenerator section. Holding times are determined from the product pumping rate rather than by laminar flow.

Direct-heated HHST systems may be by steam injection or steam infusion. Steam injection is introduced into the milk by a sanitary steam nozzle or the milk is infused into an atmosphere of steam. Both direct types must have steam-rated vessels and a water removal system.

UHT

The UHT process pumps raw milk or product through a closed system. The product is preheated, highly heat-treated, homogenized, cooled, and usually aseptically packaged. Low acid (pH above 4.5 and for milk above pH 6.5) products are heated to 135–160°C (275–320°F) and held for a few seconds. Heat transfer is either direct or indirect. High acid (pH less than 4.5) products such as juice are normally heated at 90–94°C (194–201°F) for 15 sec–20 sec. All processing and equipment surfaces downstream from those used for heat treatment, including packaging materials, must be maintained aseptic. The brief but high heat treatment destroys all vegetative micro-organisms and their inactive spores. Low-acid, aseptically processed foods in the United States must meet 21 CFR Part 113[7] in addition to other applicable regulations such as the Grade A Pasteurized Milk Ordinance for Milk and Milk Products.[2]

The UHT systems use much of the same equipment and controls as HTST/HHST systems plus that necessary to insure continuous aseptic conditions or a fail-safe shutdown. Aseptic surge tanks and packaging equipment are required.

UP and ESL

These processes and terminology are currently used in the dairy industry within the U.S.A. Until recently in the United States, UP/ESL products were limited to low-volume and seasonal products such as coffee cream, heavy cream products, and eggnog. However, with the consolidation of the dairy industry, longer shelf life is not only desirable, but a necessity.

The UP heat treats milk or dairy products to UHT conditions to achieve the same micro-organism destruction in much of the same equipment heretofore described. Aseptic conditions are required for the filler. The packaging materials or the packaging equipment is not

aseptic, but must be "ultra clean." As of now, there is no legal U.S. definition for ultra clean. The packaging equipment must protect from the product recontamination of the UP product.

Beyond Pasteur

As of June 1, 2001, there were 15,000 references on worldwide websites for pasteurization: 498 for HTST pasteurization, the remainder for nonthermal processing.

Of the many nonthermal processes available, membrane filtration has been used successfully to separate raw milk to substantially reduce micro-organism populations, the cream and milk recombined, and then HTST pasteurized at minimum or near minimum time/temperature conditions. The advantage is a better tasting product while maintaining the same shelf life as that for longer times and higher temperatures normally used in HTST processing to achieve a 15-day to 20-day shelf-life.

High-pressure processing (HPP) is another method being used in minimal or nonthermal processing. It has been applied to a variety of low-acid, ready-to-eat foods. Pressures range from 100 Mpa to 800 Mpa (14,500 psi–116,000 psi).

Pulsed electric field (PEF) is another emerging method of nonthermal processing, which is being investigated for juice and drinks, and solids and semisolids such as orange juice to inactivate pectin methylesterase, tomato-based products, rice pudding, and skim milk. In some cases, PEF is used with other barrier technologies.

The case of low frequency, ultrasound (less than or equal to 20 kHz) has been long known for its destructive effects on buildings. It is now being applied to microbial destruction. The lethality is associated with cavitation of nuclei. The inactivation of pectin methylesterase and yeast destruction in orange juice are examples of PEF applications.

Ozonation is in commercial use for water purification and is now being used for equipment disinfection. Ozone has the ability to deactivate viruses, bacteria, and amoebocytes. Ozone is a strong oxidant and may be limited to surface decontamination of whole raw fruits and vegetables.

Ohmic heating is being studied as a cooking and sterilizing method for starch and protein gel. At this time ohmic heating has the most potential for cooking.

These and other nonthermal processes such as magnetic resonance and microwave–infrared technology used alone or in combinations will supplement traditional thermal pasteurization.

REFERENCES

1. Dubos, R. *Pasteur and Modern Science*; Doubleday and Co., Inc.: New York, 1960.
2. Anon. Grade "A" Pasteurized Milk Ordinance (PMO), Publication No. 229, Public Health Service/Food and Drug Administration, Department of Health and Human Services, Public Health Services, Food and Drug Administration (HFS-626), 22 C St., SW, WDC 20204, USA, 1999 rev., 4.
3. Anon. 3-A Sanitary Standards for Non-coil Type Batch Pasteurizers for Milk and Milk Products, Number 24-02.
4. Anon. 3-A Accepted Practices for the Sanitary Constructions, Installation, Testing and Operation of High-Temperature, Short-Time and Higher-Heat, Shorter-Time Pasteurization Systems, Number 603-07.
5. Anon. 3-A Accepted Practices for a Method of Producing Steam of Culinary Quality, Number 609-02.
6. Anon. Title 21 of the Code of Federal Regulations, Part 173.310, Food and Drug Administration, 5600 Fishers Lane, HC-61 Rockville, MD 20857, USA.
7. Anon. Ibid, Part 113.

ADDITIONAL SOURCES

www.3-A.org
www.fda.gov
www.usda.gov
Anon. Milk Pasteurization Controls and Tests, 8th Edition, 2001, Food and Drug Administration, State Training Team, 5600 Fishers Lane, HC-61, Rockville, MD 20857, USA.
Anon. Milk and Milk Product Equipment—A Guideline for Evaluating Construction, Aug. 2000 revision, Public Health Service/Food and Drug Administration; Department of Health and Human Services, Public Health Service, Food and Drug Administration (HFS-626), 200 C St., SW, WDC 20204, USA.

Pesticide Application

Andrew J. Hewitt
Stewart Agricultural Research Services, Inc., Macon, Missouri, U.S.A.

INTRODUCTION

The effective management of pests, diseases, and weeds requires the optimization of control strategies with respect to timing, delivery systems, and minimization of adverse effects, for example through environmental protection. Different targets have different optimal control methods, so there is not a "one size fits all" scenario. Control strategies can include chemical and nonchemical systems. Chemical systems using pesticides and agricultural chemicals usually provide the most rapid, economical method of pest control in large areas for multiple pest types. Biological control systems and integrated pest management programs have great value for control of some pests in specific areas. Given the wider use of chemical methods, the present article focuses on such methods, discussing current characteristics of spray application methods, drift management, and modeling.

SPRAY APPLICATION METHODS

Pesticides are usually applied as liquid sprays or solids (e.g., dusts and granules). Vehicle-mounted application platforms usually involve spraying using aircraft (aerial application), ground rigs, orchard airblast, or chemigation. Hand-held equipment includes lever- or engine-operated knapsack sprayers, and a variety of other sprayers such as spinning disc and electrostatic sprayers. These tend to be more common in developing countries where labor is often cheaper than aircraft or ground equipment. Information on hand-held applications can be found elsewhere (see Ref. [1]).

The difference in spray characteristics from different application platforms causes differences in spray movements and deposition. In aerial and ground rig applications, spraying is usually conducted from above the canopy, in the downward direction. In orchard airblast applications, spraying may be from above if using towers, or from the ground upwards. A third type of application involves spraying from the top and sides of the canopy using "wrap-around" sprayers for crops such as grapes.

DROPLET SIZE

The main factor affecting the performance of a spraying operation with respect to collection of particles at the target surface, minimization of losses by drift and loss to the ground, and optimization of dose-transfer and efficacy, is the droplet size spectrum of the applied product. Applicators have a wide range of nozzle and atomizer types for use in spraying. In agricultural and biological applications, these may span a range of droplet size spectra from aerosol for vector control to Very Fine or Fine for some insecticide products, to Medium for typical arable applications, to Coarse or Very Coarse for most herbicide applications, and Extra Coarse in chemigation and irrigation. These droplet size categories have been defined in reference categories that facilitate standardization across measurement systems. The original concept was developed in Europe in the 1980s by the British Crop Protection Council (BCPC),[2] and adopted as a standard in the United States for labeling of pesticides early in the 20th century.[3] Fig. 1 shows an example of different droplet size data being measured for a given reference spray by two different laser diffraction droplet size measurement instruments in two different wind tunnels. Although, system 1 was always slightly coarser than system 2, the reference category separations were similar, indicating that spray classifications according to these reference categories would be similar for both systems. An extensive review of the importance of droplet size in agricultural spraying was conducted by Hewitt.[4] Information on droplet size performance for different types of nozzle and atomizer can be found in many nozzle catalogs, as well as in databases (see Ref. [5]) and models (see Refs. [6,7]). In general, sprays become finer with the following factors:

- Smaller orifice size.
- Wider fan angle.
- Higher pressure (however, with solid stream nozzles, lower pressures usually produce finer sprays).
- Greater air shear, e.g., faster flight speed in aerial spraying, greater nozzle angle on the boom, or greater fan speed in airblast applications.

In addition, some nozzle types produce finer sprays than other types, due to the different mode of atomization (sheet breakup, ligament disintegration or direct droplet formation; air-shear atomization, electrohydrodynamic or sonic atomization) or energy input.

Although droplet size is an important factor in spray performance, other factors may also be important in many

Encyclopedia of Agricultural, Food, and Biological Engineering
DOI: 10.1081/E-EAFE 120006944
Copyright © 2003 by Marcel Dekker, Inc. All rights reserved.

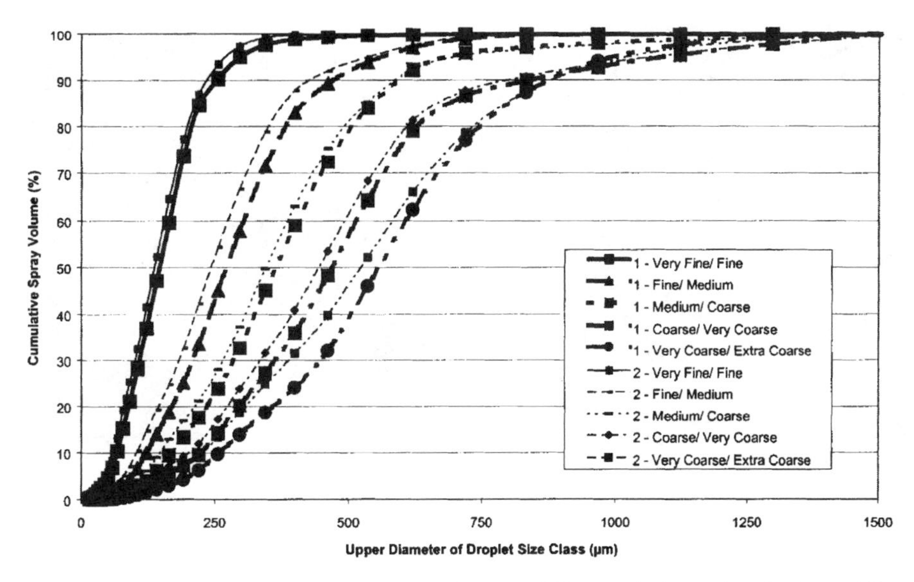

Fig. 1 Droplet size classification for reference sprays measured at two different wind tunnels using laser diffraction instruments.

types of application. Droplet velocities, trajectories, evaporation rates, densities, electrostatic charge, and other characteristics can affect movements and deposition independently of droplet size. Recognizing these factors can affect spray performance; the international BCPC spray classification system is considering additional variables to droplet size, such as drift potential.[8]

TANK MIX

The tank mix that is applied for pest control can have an effect on spray performance. The tank mix usually includes one or more active ingredient products, a carrier (water and/or oil), and any adjuvants that are included for spray retention, spreading/sticking, uptake, drift control, fertilizing, or other improvements. Although, the selection of nozzle and use parameters has the most significant effect on atomization, the tank mix is also important. Fig. 2 shows atomization of a herbicide from flat fan and disc-core (swirl) nozzles. Atomization is primarily by sheet breakup. If a polymeric adjuvant is included in the spray mixture causing an increase in extensional viscosity, the sheet breakup changes are as shown in the corresponding Fig. 3. The effect of the polymer was more pronounced on the swirl nozzle, where the air core in the center of the spray was reduced. With both nozzles, there was a decrease in spray angle and an increase in breakup length and average droplet size. However, due to the change in the breakup of the liquid sheet, there was also an increase in the range of droplet sizes in the spray, or relative span. In thousands of atomization tests with a range of tank

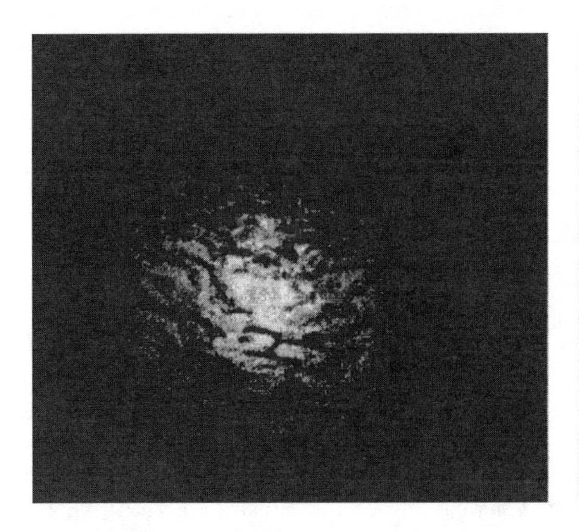

Fig. 2 Atomization of herbicide through flat fan (left) and disc-core (right) nozzles by sheet breakup.

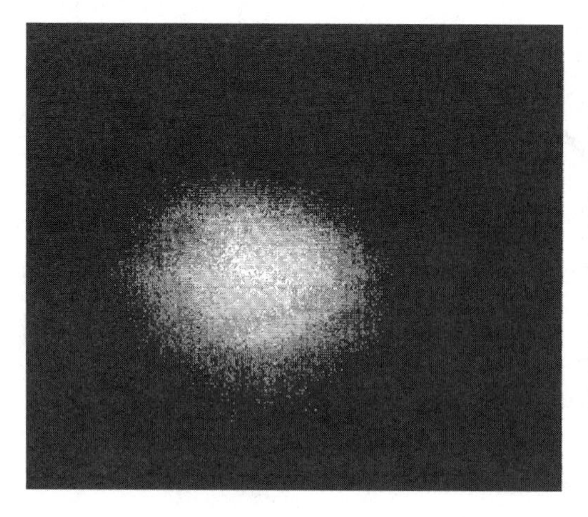

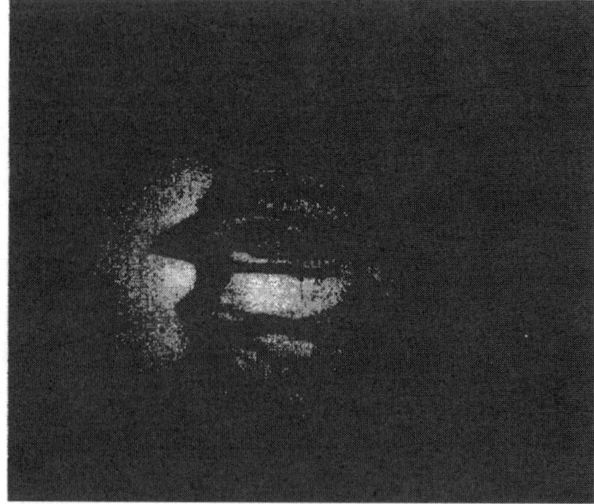

Fig. 3 Atomization of herbicide with polymer adjuvant through flat fan (left) and disc-core (right) nozzles.

mixes spanning the physical properties of most commercial tank mixes used in the United States, and including most of the major nozzle types used in conventional spray applications in the United States, the spray drift task force (SDTF), a consortium of pesticide registrants, showed that several major physical properties have the greatest effect on atomization. These include the dynamic surface tension at surface lifetime ages close to those of the atomization process (< 20 msec), shear viscosity and extensional viscosity.[9] Subsequent research into the effects of adjuvants on sprays has shown that additional physical property effects may also be important, with emulsion and solution adjuvants exhibiting different behaviors. A review of these effects is given elsewhere.[10]

SPRAY DRIFT MANAGEMENT

The issue of spray drift has received considerable attention in most countries where agricultural and biological sprays are applied. Extensive studies have been conducted (see Ref. 5, and the literature review in Ref. 11) to determine the main factors affecting spray drift. These have been shown to include the emission droplet size spectrum, spray release position, meteorological conditions, and canopy. Drift tends to increase as droplet size decreases (unless a shield or effective air-assistance is used to help prevent lateral movements of droplets), wind speed increases, release height increases, and canopy or downwind vegetation decrease. Summaries of drift from the extensive SDTF studies can be found in the following publications. Drift from aerial applications depended mainly on the droplet size spectrum, aircraft height, boom length, and wind speed.[5] Drift in ground spraying using hydraulic sprayers was related mainly to the boom height and emission droplet size spectrum.[12] Drift in orchard airblast applications was related to canopy characteristics (leaf area index, tree height, row spacing, foliage density), sprayer setup (sprayer type, nozzle types, directions, air speed), and in dormant canopies, wind speed.[13] The SDTF atomization and physical property studies have been described in Ref. [14] Other drift studies from Germany, the United Kingdom, United States, and other countries were described by Hewitt[15] and van de Zande.[16] One of the largest sets of drift studies for regulatory use in ground and orchard spraying in Europe and Canada was reported in Ref. [17].

Many regulatory agencies have set restrictions/requirements on application specifications, meteorological conditions, or "no application" (or "buffer") zones. The impact of such proposals in the United States was discussed in Ref. [18]. Labeling approaches in the United States include consideration of droplet size spectra, release height, and wind speed. In the United Kingdom, risk assessment can allow reductions in generic buffer zones if low drift sprayers, reduced spray volumes, larger water body sizes, or other drift mitigation approaches apply.[20] Similar approaches are being implemented and developed in other countries. In early 2002, the U.S. environmental protection agency was finalizing its pesticide regulation (PR) notice on spray and dust drift. The latest version of this document can be seen at the web site *www.epa.gov/pesticides*.

SPRAY MODELING

Models have been developed to predict the fate of pesticides in the environment. The main model that is used in the United States for predicting drift exposure risk is AgDRIFT®,[19,21,22] which is based on a previous model, AGDISP developed by the U.S. Forest Service, National American Space Administration (NASA), and the U.S. Army.[23] These models use Lagrangian equations to track the movement of droplets released from aircraft, including considerations of the wake effect of the sprayer that are not included in many other gaussian dispersion models. AgDRIFT also includes improved evaporation algorithms and extensive databases of nozzles, tank mixes, and aircraft. The use of AgDRIFT for risk assessment requires additional information on the toxicity of pesticides to specific organisms requiring protection, since the model only provides a worst-case estimate of drift exposure, not toxicity.

Other models that have been used for modeling agricultural spray applications include regression models developed in the United States for aerial spray dispersal,[24] random walk models developed in the United Kingdom for ground spraying,[25] empirical models for ground spraying in the United States,[26] and gaussian plume models for aerial application in Australia.[27]

FUTURE ACTIVITIES

Recent activities are utilizing geographic information systems (GIS) frameworks for integrating spatial information on areas being sprayed and sensitive and other adjacent areas with real-time meteorological monitoring, drift modeling, and/or efficacy modeling to optimize spray applications.[28]

REFERENCES

1. Matthews, G.A. *Pesticide Application Methods*; Longman Scientific and Technical: New York, NY, 1992; 405.
2. Doble, S.J.; Matthews, G.A.; Rutherford, I.; Southcombe, E.S.E. A System for Classifying Hydraulic Nozzles and Other Atomizers into Categories of Spray Quality. Proc. 1985 Brit. Crop Prot. Conf.—weeds vol. 9A-6., 1985; 1125–1133.
3. ASAE. American Society for Agricultural Engineers Standard S-572, Spray Nozzle Classification by Droplet Spectra. American Society of Agricultural Engineers, St Joseph, Michigan, USA. 1999.
4. Hewitt, A.J. The importance of droplet size in agricultural spraying. Atomization Sprays **1997**, *7* (3), 235–244.
5. Hewitt, A.J.; Johnson, D.R.; Fish, J.D.; Hermansky, C.G.; Valcore, D.L. Development of the Spray Drift Task Force Database for Aerial Applications. J. Environ. Toxicol. Chem. **2002**.
6. Hewitt, A.J.; Hermansky, C.; Valcore, D.L.; Bryant, J.E. Modeling Atomization and deposition of agricultural sprays. Proc. ILASS-Americas '97, 178–182, Ottawa, Canada, Institute for Liquid Atomization and Spray Systems Americas (ILASS Americas), c/o Norman Chigier, Carnegie-Mellon University, Pittsburgh, PA, USA, 1997.
7. Kirk, I.W. *Application Parameters for CP Nozzles*, Paper No. AA97-006; ASAE/NAAA Joint Technical Session: Las Vegas, NV, 1997.
8. Southcombe, E.S.E.; Miller, P.C.H.; Ganzelmeier, H.; Van de Zande, J.C.; Miralles, A.; Hewitt, A.J. The International (BCPC) Spray Classification System Including a Drift Potential Factor. Proc. Br. Crop Prot. Conf. 1997; Vol. 5A-1, 371–380.
9. Hewitt, A.J.; Hermansky, C. Effect of Liquid Properties on Spray Performance. Proc. Chem Show, New York, NY, 1997.
10. Hewitt, A.J; Miller, P.C.H., Dexter, R.W.; Bagley, W.E. The influence of Tank Mix Adjuvants on the Formation, Characteristics and Drift Potential of Agricultural Sprays. Proceedings International Symposium on Adjuvants for Agrichemicals, Amsterdam, Netherlands, 2001.
11. Bird, S.L.; Esterly, D.M.; Perry, S.G. Off-Target Deposition of Pesticides from Agricultural Aerial Spray Applications. J. Environ. Qual. **1996**, *25*, 1095.
12. Hewitt, A.J.; Valcore, D.L.; Barry, T. Analyses of Equipment, Weather and Other Factors Affecting Drift from Applications of Sprays by Ground Platforms. 20th Symposium on Pesticide Formulations and Application Systems, American Society for Testing and Materials, New Orleans, 1999.
13. Hewitt, A.J. Spray Drift Studies with Orchard Airblast Sprayers. Phytoparasitica **2002**.
14. Hewitt, A.J.; Valcore, D.L. *The Measurement, Prediction and Classification of Agricultural Sprays*; ASAE Paper No. 981003, 1998.
15. Hewitt, A.J. Guest Editorial. *Phytoparasitica*; Developments in International Harmonization of Pesticide Drift Management; 2001; Vol. 29, No. 2.
16. Van de Zande, J.J.C. Environmental Risk Control. International Advances in Pesticide Application. Aspects of Applied Biology, 2002; Vol. 66.
17. Ganzlemeier, H.; Rautmann, D.; Spangenberg, R.; Streloke, M.; Herrmann, M.; Wenzelburger, H.; Walter, H. *Studies on the Spray Drift of Plant Protection Products*; Blackwell Wissenschafts-Verlag GmbH: Berlin/Wien, 1995.
18. Hewitt, A.J. Spray Drift: Impact of Requirements to Protect the Environment. Crop Protection, 2000; Vol. 19, 623–627.
19. Hewitt, A.J. Spray Drift Modeling, Management and Labeling in the U.S. Aspects of Applied Biology, 2000; 57, 11–20.
20. Gilbert, A.J. Local Environmental Risk Assessment for Pesticides (LERAP) in the UK. Aspects Applied Biology, 2000; 57, 83–90.
21. Teske, M.E.; Bird, S.L.; Esterly, D.M.; Curbishley, T.B.; Ray, S.L.; Perry, S.G. AgDRIFT: An Update of the Aerial Spray Model AGDISP. J. Environ. Toxicol. Chem. **2002**.
22. Bird, S.L.; Perry, S.G.; Ray, S.; Teske, M.E. An Evaluation of AgDRIFT 1.0 for Use in Aerial Applications. J. Environ. Toxicol. Chem. **2002**.
23. Bilanin, A.J.; Teske, M.E.; Barry, J.W.; Ekblad, R.B. AGDISP: The Aircraft Spray Dispersion Model, Code

P

Development and Experimental Validation. Trans. ASAE **1989**, *32*, 327–334.
24. Akesson, N.B.; Yates, W.E.; Smith, N.; Cowden, R.E. *Rationalization of pesticide drift-loss accountancy by regression models*; Paper No. 81-1006; American Society of Agricultural Engineers: St. Joseph, MI, 1981.
25. Walklate, P.J. A Random-Walk Model for Dispersion of Heavy Particles in Turbulent Air Flow. Boundary-Layer Meteorol. **1987**, *39*, 175–190.
26. Smith, D.B.; Bode, L.E.; Gerard, P.D. Predicting Ground Boom Spray Drift. TRANS. ASAE **2000**, *43* (3), 547–553.
27. Woods, N.; Craig, I.P.; Dorr, G. Measurement of Spray Drift of Pesticides Arising from Aerial Application in Cotton. J. Environ. Qual. **2001**, *30* (3), 697–701.
28. Hewitt, A.J.; Maber, J.; Praat, J.P. Drift Management Using Modeling and GIS Systems, Proc. World Congress of Computers in Agriculture, Iguassu, Brazil, 2002.

Phase Diagrams

Gönül Kaletunç
The Ohio State University, Columbus, Ohio, U.S.A.

P

INTRODUCTION

Physical characterization of food components has received growing attention because the nature of the physical states can be related to the quality and shelf-life of the food products. Development of a fundamental understanding of the influence of processing and storage on quality parameters of food products requires identification and investigation of measurable physical properties as a function of conditions relevant to processing and storage.

Thermodynamic phase diagrams have been used to evaluate the equilibrium properties of a material as a function of composition and environmental parameters. Phase diagrams in the form of temperature–composition plots were generated for aqueous model systems of carbohydrates and polymers in order to interpret the freezing behavior of cells and tissues.[1] These plots showed the boundaries of solid and liquid phases, which are in thermodynamic equilibrium. In food systems, kinetically controlled metastable states also exist due to rapid heating or cooling.[1–3] MacKenzie[1] included properties of nonequilibrium behavior on phase diagrams referring to the total display as "supplemented phase diagrams." These diagrams are also called as extended phase diagrams, dynamic phase diagrams, or more commonly state diagrams.[4,5]

State diagrams have been proposed for food products and processing.[6–11] White and Cakebread[12] characterized the physical state(s) of sugars in food products to interpret some of the defects associated with changes in storage conditions. MacKenzie[1] and Franks et al.,[6] developed state diagrams for hydrophilic polymers, in connection with their use as cryoprotectants. Levine and Slade[7] applied an idealized state diagram for a hypothetical small carbohydrate to explain and/or predict the functional behavior of such carbohydrates in frozen foods. Roos and Karel[8] discussed the use of a "generic," simplified state diagram for a water-soluble food component to show the effect of temperature and moisture content on food stability. State diagrams were also applied to map the path of various food processes.[9,11]

DEVELOPMENT OF STATE DIAGRAMS

The *state of the system* is defined by the condition of the system at a given time. The values of the variables including temperature, pressure, and composition specify the state of the system. Processes can change the state of the system because the variables specifying the state are changed. If the two states are in thermodynamic equilibrium, in which the chemical potential of any substance in both states has the same value, they are referred to as phases rather than states. In kinetically controlled metastable states, which frequently occur in biological systems, the physical properties of a given material can be time, temperature, and composition dependent.[1,3,6] A complete understanding of the behavior of a given system requires the study of phases in equilibrium as well as nonequilibrium, time-dependent phenomena. Kinetically controlled states can be investigated under conditions relevant to processing and storage.

Techniques for Determining the Physical State of a Material

Thermal, rheological, and spectroscopic techniques are employed to characterize the physical state of materials. Thermal analysis techniques, specifically differential scanning calorimetry (DSC) is commonly used to determine the transformations of both the equilibrium and nonequilibrium states. Thermally induced transitions can be determined from specific heat capacity vs. temperature curves. Transformations between the states appear as apparent anomalies in the temperature dependence of the specific heat capacity of the material. The transition temperature defines the point at which the two states coexist at a given moisture content. The DSC detects both first-order and second-order transitions. First-order transitions include starch melting, starch gelatinization, ice melting, crystallization, protein denaturation, protein aggregation, and lipid melting. While melting is a true phase transition, crystallization[1] and aggregation[13] are kinetically controlled.

The glass transition observed in food systems is a second-order transition, which appears as a discontinuity

Encyclopedia of Agricultural, Food, and Biological Engineering
DOI: 10.1081/E-EAFE 120006986
Copyright © 2003 by Marcel Dekker, Inc. All rights reserved.

in the specific heat in a DSC thermogram. Rasmussen and MacKenzie[14] demonstrated that the glass transition of pure water was affected by the heating rate because lower heating rates allow longer times for the material to relax resulting in the detection of the glass transition at lower temperatures. Therefore, it is essential that nonequilibrium transitions such as the glass transition be reported with the experimental time-scale. In addition to the discontinuity in specific heat, the glass transition can be determined as a change in the slope of the volume expansion, or a discontinuity in the thermal expansion coefficient or a reduction of about three orders of magnitude in viscosity.[15] Young's modulus and dynamic viscoelastic properties also display changes during glass transition.[16] Nuclear magnetic resonance (NMR) has been used to describe the glass transition of cereal systems by measuring relaxation times after application of a radio frequency pulse.[17]

The value of glass transition temperature has been reported to depend on the technique used.[17] The measured glass transition temperature of amylopectin increased with the technique used in order from NMR, to Instron universal texturometer, to DSC, to dynamic mechanical thermal analysis. The differences may be due to the sensitivity of a specific observable to the degree of mobility at the level of measurement. While NMR can detect a local mobility transition of a side chain, other techniques measure the glass transition temperature averaged over the entire sample. Because the glass transition is kinetically controlled, differences in the frequency of the measurements and the rate of heating are expected to influence the measured glass transition temperature. Furthermore, transitions may or may not be observable depending on the compatibility of the relaxation time of measured transition and the experimental time domain and the sensitivity of the measurement technique.

Construction of State Diagram

Thermal and rheological data collected as a function of processing and storage conditions can be used to construct state diagrams. Similar state diagrams can be constructed for the corresponding products to assess the physicochemical changes during processing and product storage. Given the number of parameters associated with processing and storage, multidimensional state diagrams, which include temperature, pressure, moisture content, concentrations of additives, and a variety of processing parameters as dimensions are needed. Diagrams of more than 3-D cannot be represented in space, thereby making interpretation and application of the complete diagram difficult. Therefore, simplified 2- or 3-D subdiagrams of the parent state diagrams are constructed.

The Gibbs phase rule specifies the minimum number of variables that are sufficient to describe the system.[18] The phase diagram describing a system will have the same number of coordinate axes as the number of variables prescribed by the phase rule. Pressure and temperature typically are selected and are particularly relevant variables for food applications. For a binary system, the relationship between composition and the phase transition temperatures can be displayed as a 2-D phase diagram at constant pressure.

MacKenzie[1] constructed state diagrams of sucrose–water system by incorporating equilibrium and nonequilibrium data observed either during cooling or subsequent heating scans. Except for melting, all of the transitions in the state diagram are kinetically controlled and cannot be described as true phase transitions. The curve defined by the calorimetrically measured melting points is called the liquidus curve. The state diagram for the sucrose–water system was expanded by Levine and Slade[9] to include solidus and vaporus curves (Fig. 1). The intersection of the extrapolated equilibrium melting curve and the glass curve corresponds to a maximally freeze concentrated solution. The maximally freeze concentrated state is reached as a result of gradual concentration due to ice crystallization and separation during slow cooling. The conditions defining this state have been shown to be important for designing processes for frozen food technology as well as for maintaining

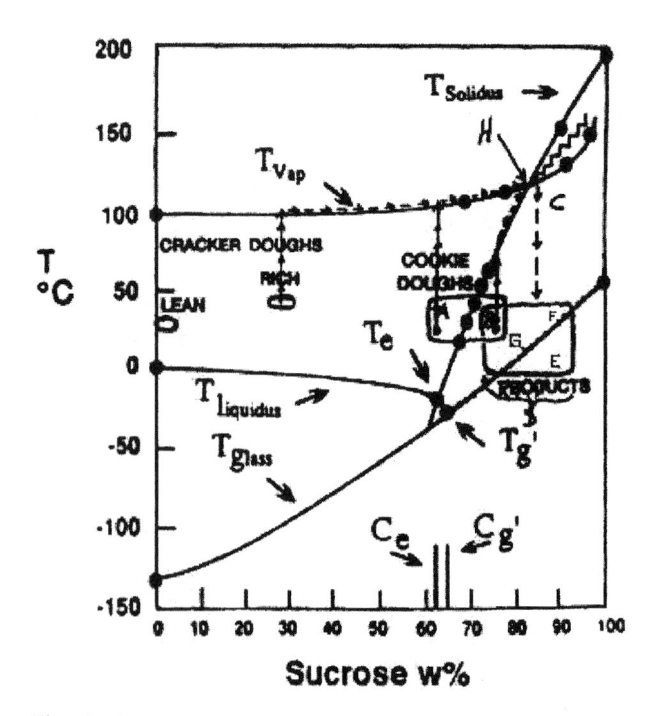

Fig. 1 Sucrose–water state diagram on which the paths of cookie and cracker baking process are superimposed. (From Ref. 23.)

the physical stability of food products during frozen storage.[5,7]

A number of state diagrams have been proposed for oligomeric carbohydrates[19,20] and proteins[21,22] due to the significance of their structure–property relationships in food processing. Kaletunç and Breslauer[11] cautioned that the use of thermal data on the individual components to interpret complex systems ignores interactions between the components and the potential alteration of component thermal properties, which might occur as a result of the separation process. These investigators stated that the complex system itself should provide the appropriate thermodynamic reference state for processing-induced alterations in the thermal properties of the post-processed material. Kaletunç and Breslauer[11] developed a state diagram for wheat flour, with the goal of characterizing the physical state of the wheat flour prior to, during, and after extrusion processing (Fig. 2). Curves on the state diagram, which define the moisture-content dependence of the transition temperatures, form the boundaries between regions which correspond to particular states of the wheat flour. Fig. 2 shows the transition temperatures corresponding to melting, gelatinization, and glass curves for wheat flour and a melting curve for the freezable water.

The glass transition occurs over a temperature range rather than at a well-defined temperature for a given water content. This becomes more apparent for high molecular weight polymers with a high polydispersity ratio (the ratio of weight average molecular weight to number average molecular weight). In extruded materials, in addition to the heterogeneous nature of pre-extruded material, the breadth of the transition may originate from polymer fragmentation during extrusion. In the case of a broad molecular weight distribution, the glass transition may occur as a continuum over a broad temperature range. One should, therefore, recognize when constructing and using state diagrams, that the glass transition occurs over a range of temperature and that regions of the diagram near T_g-defined boundaries may actually represent states, which are mixtures of glassy and rubbery material.

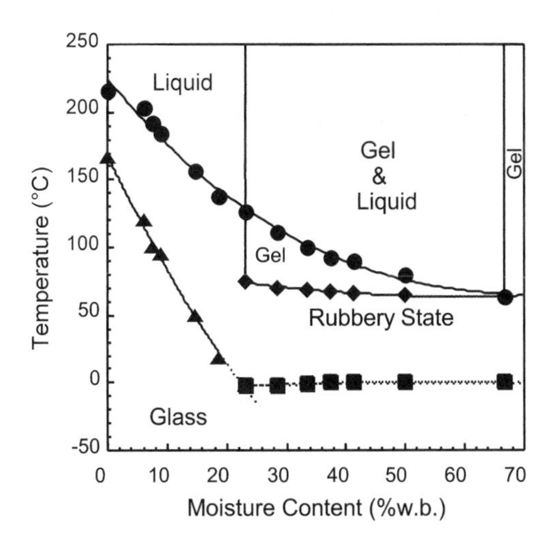

Fig. 2 Wheat flour–water state diagram at 30 atm. (From Ref. 11.)

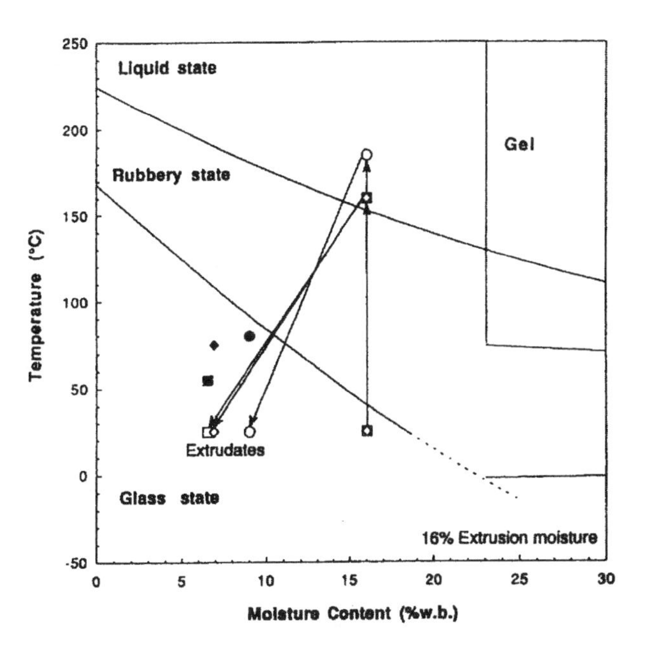

Fig. 3 Wheat flour–water state diagram on which the paths of the extrusion process at various SME values are superimposed. T_g values of the extrudates are indicated as the corresponding filled symbols. (From Ref. 11.)

APPLICATION OF PHASE DIAGRAMS

Process Analysis and Design

Slade and Levine[23] demonstrated that a sucrose–water state diagram can be used to understand the cookie and cracker baking process. These investigators mapped the various steps of cookie and cracker manufacturing, including dough mixing, lay time, machining, and baking on the sucrose–water state diagram (Fig. 1). For baked product, while point E represents a product with optimum initial quality and storage stability, a product described by points F or G is clearly in an unstable rubbery state and is expected to have a shorter shelf-life. Depending on the temperature and relative humidity conditions during distribution and storage, the amount of water removed during baking must be adjusted to attain the desired glass transition temperature.

The temperature and moisture content ranges covered by the wheat-flour state diagram in Fig. 2 correspond to those applied to cereal flours as part of many food-processing operations, such as baking, pasta extrusion, and high-temperature extrusion cooking. Kaletunç and Breslauer[11] mapped on this state diagram the path of a high-temperature extrusion cooking process. They also placed the T_g values of wheat flour extrudates on the state diagram to assess the impact of shear, in terms of specific mechanical energy (SME), on the wheat flour. Fig. 3 shows the path of the extrusion process (open symbols) on the wheat flour state diagram, at the various SME conditions and the T_g values of the corresponding extrudates (the corresponding filled symbols). Different SME values lead to different extents of fragmentation of the wheat flour, as reflected in the difference between the T_g values of the extrudate and the corresponding temperature on the glass curve of wheat flour at the same moisture content. The difference between the ambient temperature and the T_g values of the extrudates can be used to assess the storage stability of the extrudate. For highly fragmented extrudates, which display low T_g values, elevated storage temperature may induce the rubbery state.

Kaletunç and Breslauer[11] noted that, based on the wheat-flour state diagram and the SME dependence of T_g, processing conditions can be adjusted to yield extrudates with desired T_g values. To make these correlations of practical value, a criterion which relates the target T_g to the end-product attributes of significance to the consumer should be defined. Kaletunç and Breslauer[24] demonstrated for corn flour extrudates that an increase in T_g is related to an increase in the sensory textural attribute of crispness, which is one of the major criteria by which a consumer judges the quality of an extruded product.

The state diagram displayed in Fig. 2 also can be utilized to map the path of pasta-extrusion and drying processes (Fig. 4).[25] Pasta is in the rubbery state during the extrusion process, drying, and part of the cooling process. As the moisture content of the pasta decreases, the difference between the temperature of the pasta and its glass transition temperature at given moisture content decreases. A greater difference between the drying temperature and the glass transition of the pasta, will result in a faster rate of dehydration due to the increased water mobility and diffusion rates in the rubbery state. Zweifel et al.[26] used the state diagram of starch to map the pasta making process in order to evaluate the thermal modifications of starch during high-temperature drying of pasta.

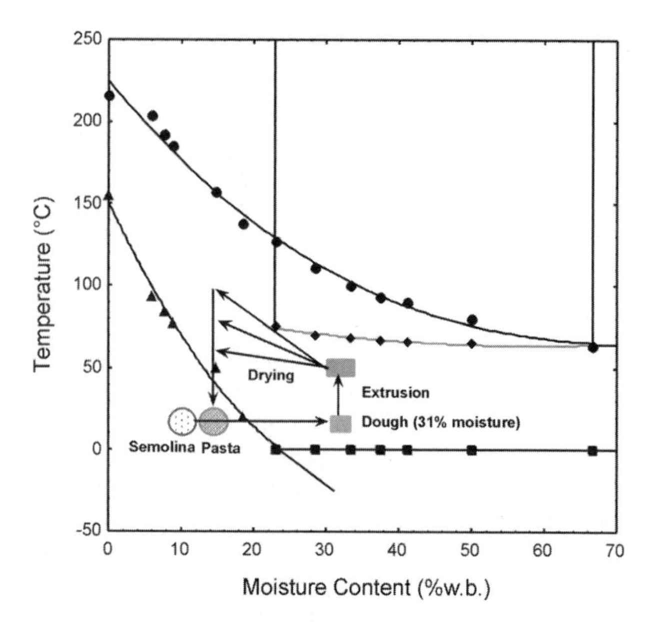

Fig. 4 Wheat flour–water state diagram on which the path of the pasta extrusion process with three drying schemes are superimposed.

Product Development

State diagrams may be utilized in product development by evaluating the effect of additives on the physical state of the material during processing. The curves which define the boundaries of the state diagram may shift in the presence of additives and as a result processing conditions may need to be modified. Barrett et al.[27] reported the plasticization effect of sucrose on corn extrudates in the presence of moisture as well as its effect on reduction of shear (lowering of the specific mechanical energy) in the extruder. These investigators stated that sucrose plasticizes corn extrudates causing a decrease in T_g of extrudates and a loss of crispness. Therefore, addition of sucrose requires modification of the operating conditions (e.g., by decreasing SME) to compensate for the T_g-depressing influence of sucrose in order to produce extrudates with desirable attributes.

First and Kaletunç[28] investigated the efficacy of using semolina pasta processing conditions for composite flour (semolina and soy) pasta processing utilizing a state diagram. The results showed that the addition of soy decreased the T_g of the semolina dough. These investigators concluded that the redness developed during drying of soy-added pasta can be avoided by drying at lower temperature, which is possible due to the lower T_g of the composite flour dough.

CONCLUSION

Improvement of the quality of existing products or processes, as well as development of new products and design of novel processes requires an understanding of the impact of processing and storage conditions on the physical properties and the structures of pre- and post-processed materials. To achieve this fundamental understanding one needs to study the physical properties of

such materials under conditions simulating processing and storage. The database generated from such studies can be used to construct state diagrams. Quantitatively accurate state diagrams, and kinetic data for the physical and chemical phenomena that occur in the rubbery state, are essential in order to assess the impact of processing on the properties and storage stability of products.

By superimposing a processing path on a temperature vs. moisture state diagram, one can determine the physical state of the material by locating the intersection of the temperature and moisture content values corresponding to a given stage of the process. For any raw material, such a diagram could be used as a predictive tool for evaluating the performance of that material during processing. Since they can provide a rational basis for designing processing conditions and/or raw material formulations, state diagrams have great potential value in analysis of food manufacturing processes, product development, and in the design of processes and storage conditions.

REFERENCES

1. MacKenzie, A.P. Non-equilibrium Behavior of Aqueous Systems. Phil. Trans. R. Soc. Lond. B **1977**, *278*, 167–189.
2. Levine, H.; Slade, L. Non-equilibrium Behavior of Small Carbohydrate–Water Systems. Pure Appl. Chem. **1988**, *60*, 1841–1864.
3. Franks, F.; Mathias, S.F.; Hatley, R.H.M. Water, Temperature and Life. Phil. Trans. R. Soc. Lond. B **1990**, *326*, 517–533.
4. MacKenzie, A.P. The Physico-chemical Basis for the Freeze Drying Process. Dev. Biol. Stand. **1977**, *36*, 51–67.
5. Franks, F. Freeze Drying. From Empiricism to Predictability. CryoLetters **1990**, *11*, 93–110.
6. Franks, F.; Asquith, M.H.; Hammond, C.C.; Skaer, H.B.; Echlin, P. Polymeric Cryoprotectants in the Preservation of Biological Ultrastructure. J. Microsc. **1977**, *110*, 233–238.
7. Levine, H.; Slade, L. Principles of "Cryostabilization" Technology from Structure/Property Relationships of Carbohydrate/Water Systems. CryoLetters **1988**, *9*, 21–63.
8. Roos, Y.; Karel, M. Applying State Diagrams to Food Processing and Development. Food Technol. **1991**, *45*, 66–71.
9. Slade, L.; Levine, H. The Glassy State Phenomenon in Food Molecules. In *The Glassy State in Foods*; Blanshard, J.M.V., Lillford, P.J., Eds.; Nottingham University Press: Loughborough, 1993; 35–101.
10. Roos, Y.H. *Phase Transitions in Foods*. Academic Press: San Diego, 1995.
11. Kaletunç, G.; Breslauer, K.J. Construction of a Wheat-Flour State Diagram: Application to Extrusion Processing. J. Therm. Anal. **1996**, *47*, 1267–1288.
12. White, G.W.; Cakebread, S.H. The Glassy State in Certain Sugar-Containing Food Products. J. Food Technol. **1966**, *1*, 73–82.
13. Kaletunç, G.; Breslauer, K.J. *Heat Denaturation Studies of Zein by Optical and Calorimetric Methods*, IFT Annual Meeting, New Orleans, Louisiana, June 20–24, 1992; abs. 289.
14. Rasmussen, D.H.; MacKenzie, A.P. The Glass Transition in Amorphous Water. Application of the Measurements to Problems Arising in Cryobiology. J. Phys. Chem. **1971**, *75*, 967–973.
15. Eisenberg, A. The Glassy State and the Glass Transition. In *Physical Properties of Polymers*; Mark, J.E., Eisenberg, A., Graessley, W.W., Mandelkern, L., Koenig, J.L., Eds.; Am. Chem. Soc.: Washington, DC, 1993; 61–97.
16. Blanshard, J.M.V. The Glass Transition, Its Nature and Significance in Food Processing. In *Physico-chemical Aspects of Food Processing*; Beckett, S.T., Ed.; Blackie Academic & Professional: London, 1995; 17–49.
17. Kalichevsky, M.T.; Jaroszkiewicz, E.M.; Ablett, S.; Blanshard, J.M.V.; Lillford, P.J. The Glass Transition of Amylopectin Measured by DSC, DMTA, and NMR. Carbohydr. Polym. **1992**, *18*, 77–88.
18. Mortimer, R.G. *Physical Chemistry*; The Benjamin/Cummings Publishing Co, Inc: California, 1993.
19. Ablett, S.; Darke, A.H.; Izzard, M.J.; Lillford, P.J. Studies of the Glass Transition in Malto-oligomers. *The Glassy State in Foods*; Nottingham University Press: Loughborough, 1993; 189–206.
20. Kajiwara, K.; Franks, F. Crystalline and Amorphous Phases in the Binary System Water-Raffinose. J. Chem. Soc. Faraday Trans. **1997**, *93*, 1779–1783.
21. Madeka, H.; Kokini, J.L. Effect of Glass Transition and Crosslinking on Rheological Properties of Zein: Development of a Preliminary State Diagram. Cereal Chem. **1996**, *73*, 433–438.
22. Kokini, J.L.; Cocero, A.M.; Madeka, H. State Diagrams Help Predict Rheology of Cereal Proteins. Food Technol. **1995**, *49* (3), 74–82.
23. Slade, L.; Levine, H. Water and the Glass Transition. Dependence of the Glass Transition on Composition and Chemical Structure: Special Implications for Flour Functionality in Cookie Baking. J. Food Eng. **1995**, *24*, 431–509.
24. Kaletunç, G.; Breslauer, K.J. Glass Transitions of Extrudates: Relationship with Processing-Induced Fragmentation and End Product Attributes. Cereal Chem. **1993**, *70*, 548–552.
25. Kaletunç, G. Construction of State Diagrams for Cereal Processing. In *Characterization of Cereals and Flours: Properties, Analysis, and Applications*; Kaletunç, G., Breslauer, K., Eds.; Marcel Dekker: New York, 2003; 151–173.

26. Zweifel, C.; Conde-Petit, B.; Escher, F. Thermal Modifications of Starch During High-Temperature Drying of Pasta. Cereal Chem. **2000**, *77*, 645–651.
27. Barrett, A.; Kaletunç, G.; Rosenberg, S.; Breslauer, K. Effect of Sucrose and Moisture on the Mechanical, Thermal, and Structural Properties of Corn Extrudates. Carbohydr. Polym. **1995**, *26*, 261–269.
28. First, L.; Kaletunç, G. *Application of State Diagram to Pasta Processing*, IFT Annual Meeting, New Orleans, Louisiana, June 23–27, 2001; abs. 88C-5.

Physical Properties of Agricultural Products

Yubin Lan
Fort Valley State University, Fort Valley, Georgia, U.S.A.

Qi Fang
Banner Pharmacaps, Inc., High Point, North Carolina, U.S.A.

INTRODUCTION

Knowledge of the physical properties of agricultural products is being used in the planting, harvesting, drying, storing, and processing of the agricultural products. Procedures have been developed that take the physical characteristics of the agricultural products into account. Producers, processors, and consumers have developed new equipment and techniques or modified existing ones so that agricultural products can be handled effectively. Therefore, proper design of machines and processes to harvest, handle, and store agricultural materials and to convert these materials into food and feed requires an understanding of their physical properties. These properties include size, shape, and density; deformation in response to applied static and dynamic forces; moisture adsorption and desorption hydrodynamic properties; and response to electromagnetic radiation.[1]

This chapter presents an introduction to the field of physical properties of agricultural products. This topic has been comprehensively covered by a few books available in market,[2–15] and not published books.[1,6] The authors encourage readers to check with these books for tables, figures, and comprehensive understanding related to physical properties of agricultural materials. When physical properties of agricultural products, such as grains, seeds, fruits and vegetables, eggs, forage, and fibers, are studied by considering either bulk or individual units of the material, it is important to have an accurate estimate of shape, size, volume, specific gravity, surface area, and other physical characteristics, which may be considered as engineering parameters for those products. The general characteristics of agricultural products, definitions, and methods for determination of the physical properties of agricultural products are discussed.

PHYSICAL CHARACTERISTICS

Shape, size, volume, surface area, density, porosity, color, and appearance are some of the physical characteristics which are important in many problems associated with design of a specific machine or analysis of the behavior of the product in handling of the material.[10]

Stroshine and Hamann[1] explained clearly for the characteristics of agricultural products as follows. Agricultural materials and food products have several unique characteristics that set them apart from engineering materials. This is a brief introduction to these characteristics.

SHAPE AND SIZE

Shape and size are inseparable in a physical object, and both are generally necessary if the object is to be satisfactorily described. Further, in defining the shape some dimensional parameters of the object must be measured.

Agricultural materials are heterogeneous by nature. Their basic structure is the cell and these cells are organized into tissues. Fruits, vegetables, grain kernels, and forage leaves are each composed of several types of tissue with widely varying characteristics. Stroshine and Hamann[1] stated that one commonly used technique for qualifying differences in shape of fruits, vegetables, grains, and seeds are to calculate sphericity. Sphericity is defined as the ratio of this volume to the volume of a sphere which circumscribes the object: Sphericity = [(Volume of ellipsoid with equivalent diameters)/(Volume of circumscribed sphere)]$^{1/3}$.

Food products are usually more homogeneous than the agricultural materials from which they are made. However, variations in composition and physical properties of the agricultural materials can cause significant changes in the food properties. In some cases, continuous monitoring of these properties will allow processors to make adjustments in processes and equipments so that the quality of the final product is consistent.

The moisture contents of agricultural materials can drastically affect their properties. For example, the cohesiveness of forages compacted into wafers or pellets vanes greatly with the moisture of the forage. Cereal grains, high in moisture (18%–20% water by weight), may

Encyclopedia of Agricultural, Food, and Biological Engineering
DOI: 10.1081/E-EAFE 120006901
Copyright © 2003 by Marcel Dekker, Inc. All rights reserved.

be more difficult to grind into feed than grains at lower moisture (12%–14% water by weight). Such moisture differences will also cause significant changes in kernel density, thermal conductivity, specific heat, deformation in response to a compressive load, and dielectric properties.

Most agricultural materials and food products are susceptible to changes due to chemical reactions involving enzymatic activity, respiration, or attack by microbes or insects. The moisture and temperature of these products must be controlled in order to prevent or retard deterioration. An example of a chemical reaction is the change in texture of starchy baked goods after baking. Although loss of moisture can make bread hard, staling of bread is not caused by moisture loss. Cereal scientists have found that it is at least partially caused by recrystallization of starch, a process called retrogradation. Heating at temperatures between 60 and 100°C can partially reverse this recrystallization. That is why toasting freshens bread. Furthermore, retrogradation occurs faster at temperatures slightly above freezing than at room temperature.

Many agricultural products contain living cells, which are still respiring. The heat produced by respiration of fruits and vegetables must be taken into account when designing cold storage facilities. Respiration can also consume dry matter and cause changes in texture, which can be measured with force-deformation tests. However, a low level of respiratory activity is needed to preserve quality. When bacteria and mold grow on foods and agricultural materials, they produce substances that change flavor and may also be toxic. The resulting changes in optical and mechanical properties can be used to monitor and detect this deterioration. When fruits and vegetables are bruised or sliced, enzymatic reactions cause discoloration of the damaged tissue. Considerable research effort has focused on developing methods of detecting bruises and cuts by scanning for differences in optical or mechanical properties.

The mechanical properties of most agricultural materials and food products are also unique. A plot of the deformation of a block of steel as a function of applied force will give a straight line over a wide range of applied force. A similar plot for a cube cut from a fruit, vegetable, cereal grain, or food product will be linear for a very small range in applied loads and then it will become concave toward the force axis. Furthermore, most agricultural materials exhibit what is known as viscoelastic behavior—the amount of force needed to give a certain deformation is dependent on loading rate. When the force is applied rapidly, more force is needed to achieve a given amount of deformation. If a constant force is applied for several minutes or hours, the material will exhibit what is known as creep; it will continue to deform with time. If the deformation is held constant, the amount of force required to maintain that deformation decreases with time; this is called stress relaxation.[1]

Rahman[13] explained clearly different forms of density, porosity, and shrinkage used in process calculations and these definitions are as follows.

DENSITY

Density is one of the most important transport properties and so is widely used in process calculations. It is the unit mass per unit volume and SI unit of density is $\mathrm{kg\,m^{-3}}$

1. True density is the density of a pure substance or a material calculated from its components' densities considering conservation of mass and volume.
2. Substance density is the density measured when a substance has been thoroughly broken into pieces small enough to guarantee that no pores remain.
3. Particle density is the density of a sample that has not been structurally modified, so will include the volume of all closed pores but not the externally connected pores.
4. Apparent density is the density of a substance including all pores remaining in the material.
5. Bulk density is the density of a material when packed or stacked in bulk.

POROSITY

Porosity indicates the volume fraction of void space or air. Porosity can also be defined in various ways:

1. Apparent porosity is the ratio of total enclosed air space or void volume to the total volume of a material.
2. Open pore porosity is the ratio of the volume of pores connected to the outside to the total volume.
3. Closed pore porosity can be defined as the result of apparent porosity minus open pore porosity.
4. Bulk porosity includes the air or void volume outside the individual material when packed or stacked as bulk. Porosity is usually measured directly by measuring the volume fraction of air or it is derived from density and shrinkage data.
5. Total porosity is the total volume fraction of air or void space (i.e., inside and outside of the materials) when material packed or stacked as bulk.

SHRINKAGE

Shrinkage is the change of volume during processing such as due to moisture loss during drying, ice formation during freezing, and formation of pore by puffing, and can be defined as follows:

1. Apparent shrinkage during processing can be defined as the ratio of the apparent volume at given moisture content and initial apparent volume of the materials before processing.
2. Isotropic shrinkage can be described as the uniform shrinkage in all dimensions of the materials.
3. Anisotropic shrinkage can be described as the nonuniform shrinkage in different dimensions.

APPLICATIONS

Rahman[12] summarized the physical property applications in process design and food products quality determination.

For food products quality determination, one of the application examples is that the solubility of instant food powder depends on the porosity of the granules. The puffiness of foods also depends on the porosity. Specific processing can be applied to get specific porosity for better texture and moth feel of foods.

For process design, size, shape, volume, surface area, density, and porosity are the physical characteristics important in many food materials.

Fissures in Rice Kernels During Moisture Adsorption

Rice grains are hygroscopic and respond dynamically and physically to moisture and temperature changes in the environment. Rapid moisture adsorption by low-moisture rice grains may cause the grain to fissure. Moisture adsorption initially causes a rice kernel to expand or swell at its surface. Expansion during moisture adsorption induces complicated stress patterns in rice kernels when the stresses exceed the failure strength of the grain.[16,17] The Combination of physical properties such as stress and strain has a continuous influence on the rice grain. Rice grain with stress fissures break more readily than sound kernels during harvesting, handling, milling, and transporting, and thereby reduce the quality and market value of the rice grain.

REFERENCES

1. Stroshine, R.; Hamann, D. *Physical Properties of Agricultural Materials and Food Products*; Department of Agricultural and Biological Engineering, Purdue University: West Lafayette, IN, 1993.
2. ASAE. D241.1. Density, Specific Gravity, and Mass–Moisture Relationships of Grains for Storage. In *Standards Engineering Practice and Data Adopted by the American Society of Agricultural Engineers*; American Society of Agricultural Engineers: St. Joseph, MI, 2000; 504–506.
3. ASAE. D243.3 Thermal Properties of Grain and Grain Products. In *Standards Engineering Practice and Data Adopted by the American Society of Agricultural Engineers*; American Society of Agricultural Engineers: St. Joseph, MI, 2000.
4. ASAE. D245.5. Moisture Relationships of Plant-Based Agricultural Products. In *Standards Engineering Practice and Data Adopted by the American Society of Agricultural Engineers*; American Society of Agricultural Engineers: St. Joseph, MI, 2000; 508–524.
5. ASAE. D239.2. Dielectric Properties of Grain and Seed. In *Standards Engineering Practice and Data Adopted by the American Society of Agricultural Engineers*; American Society of Agricultural Engineers: St. Joseph, MI, 2000; 549–558.
6. Garrett, R.E. *Properties of Materials in Biological Systems*; University of California: Davis, 1996.
7. Heldman, D.R.; Lund, D.B. *Handbook of Food Engineering*; Marcel Dekker: New York, 1992.
8. Mohsenin, N.N. *Thermal Properties of Foods and Agricultural Materials*; Gordon and Breach Science Publishers: New York, 1980.
9. Mohsenin, N.N. *Electromagnetic Radiation Properties of Foods and Agricultural Products*; Gordon and Breach Science Publishers: New York, 1984.
10. Mohsenin, N.N. *Physical Properties of Plant and Animal Materials*; Gordon and Breach Science Publishers: New York, 1986.
11. Nelson, S.O. Electrical Properties of Agricultural Products—A Critical Review. Trans. ASAE **1973**, *16* (2), 384–400.
12. Okos, M.R. *Physical and Chemical Properties of Food*; American Society of Agricultural Engineers: St. Joseph, MI, 1986.
13. Rahman, S. Physical Properties of Foods. In *Food Properties Handbook*; Rahman, S., Ed.; CRC Press: Boca Raton, FL, 1995; 179–224.
14. Rao, M.A.; Rizvi, S.S.H. *Engineering Properties of Foods*; Marcel Dekker: New York, 1986.
15. Sitkei, G. *Mechanics of Agricultural Materials*; Translated by S. Bars; Elsevier Science Publishers: New York, 1986.
16. Lan, Y.; Kunze, O.R. Fissure Characteristics Related to Moisture Adsorption Stresses in Rice. Trans. ASAE **1996**, *39* (6), 2169–2174.
17. Lan, Y.; Kunze, O.R.; Lague, C.; Kocher, M.F. Mathematical Model of the Distribution of Stress Within a Rice Kernel from Moisture Adsorption. J. Agric. Eng. Res. **1999**, *72*, 247–257.

Physical States

Pavinee Chinachoti
University of Massachusetts, Amherst, Massachusetts, U.S.A.

INTRODUCTION

Physical states of matter, solid, liquid, and gas, exist because of intermolecular forces that determine the distance of separation between the molecules. While molecules in solid are close together in an orderly fashion, those in liquid are not held so rigidly in position and hence they can move around. In gas, the molecules are separated by a large distance, and they move independent of one another because there is no appreciable intermolecular interaction among them. Gas is a substance that is in the gaseous state at ordinary temperature and pressure whereas a vapor is the gaseous state of any substance that is either a liquid or a solid in ordinary conditions. For example, water at 25°C and 1 atm pressure is a liquid and therefore, its gaseous form is a vapor, and oxygen at same conditions is a gas. Water is in the solid state as ice, in the liquid state as water, and in the gaseous state as steam or water vapor. These physical states exhibit different properties.[1]

Most substances can exist in one of these three states and their physical properties are highly dependent on the physical state.[2] In a gas, molecules are far apart (compared to their diameters) making it very compressible, low in density, assuming the volume and shape of its container, and very free in molecular motion. In a liquid, the molecules are held close together by one or more types of attractive forces with very little empty space. This makes liquid only slightly compressible, high in density, and free in molecular motion when molecules slide past one another freely and molecules do not break away from attractive forces so liquid can flow and has definite volume but assumes the shape of its container. In a solid, because molecules are held rigidly in position with virtually no freedom of motion (molecules vibrate about fixed positions). A number of solids are characterized by long-range order, i.e., molecular arrangements are in regular configurations in three dimensions.[2] With very little intermolecular space, a solid is virtually incompressible, high in density, and has a definite volume and shape. Water is a rare exception, where the density of its solid state (ice) is lower than the liquid state (4°C) creating interesting natural events, such as ice floating in the ocean, creating unique habitats for living animals both above and under the water.

A SIMPLE PHASE/STATE DIAGRAM FOR SINGLE COMPONENT SYSTEMS

Physical state of a matter can change from one to another when a system changes in properties according to variable of states (such as temperature or pressure). During two different times, if one or two properties of a system are found to be different, then in this time interval a process has taken place and a change of state has occurred. Fig. 1 shows a simple relationship among temperature and pressure and the physical states of water. This diagram is commonly named a Phase Diagram. However, in fact, the word *phase* here means a system or a portion of a system that is spatially homogeneous and has a boundary. A phase boundary is a surface that separates substances that are dissimilar in all of their structural elements (including composition, crystal structure, bond lengths, mass density, and the more obvious discontinuities between differing physical states of matter). Fig. 1 shows three states of water.

The physical conditions where there is a phase transition are those where the solid lines are drawn in Fig. 1. For the case of water, the diagram is divided into three regions and the line separating any two regions indicates conditions under which these two phases can coexist in equilibrium. The point at which all three curves meet is called the triple point. For water, this point is at 0.01°C and 0.006 atm, the condition under which all three phases can be in equilibrium with one another. The line where solid and liquid coexists is called the melting curve or the fusion curve. The curve where liquid and vapor coexists is called the vaporization or the condensation curve.

For carbon dioxide (Fig. 2), the triple point is at − 57°C and 5.2 atm. Unlike water, which can have stable solid, liquid, and vapor at 1 atm pressure, carbon dioxide is not stable in the liquid phase at 1 atm, which is far below 5.2 atm, and thus only solid and vapor phases can exist under atmospheric conditions. The phase change from solid directly to vapor (e.g., solid carbon dioxide or dry ice) is called sublimation. Sublimation for dry ice can occur at an atmospheric pressure whereas sublimation for water normally occurs under a vacuum at < 0.006 atm (Fig. 1).

Along the evaporation line, at temperature below the critical point temperature, T_c, a meaningful distinction can

Encyclopedia of Agricultural, Food, and Biological Engineering
DOI: 10.1081/E-EAFE 120006984
Copyright © 2003 by Marcel Dekker, Inc. All rights reserved.

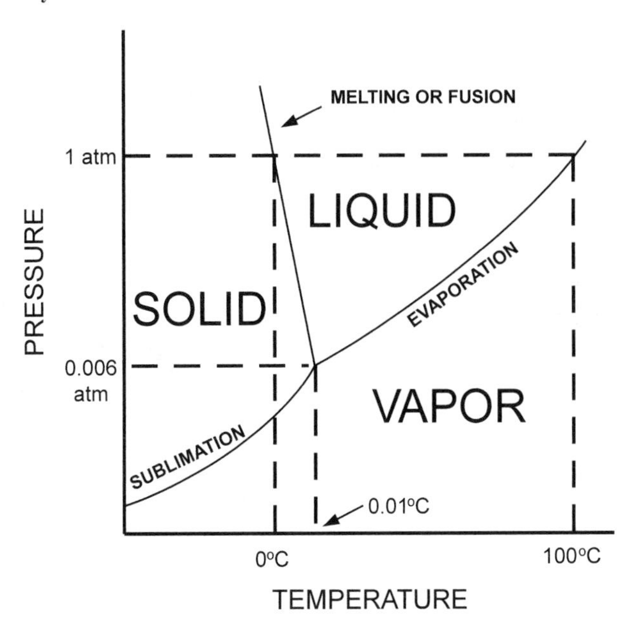

Fig. 1 The phase diagram of water. (From Ref. [2].)

be made between liquid and vapor phase. Going from the Triple Point on this line, the liquid density decreases with increase in temperature despite an increase in pressure. Simultaneously, the vapor density increases due to the compression as the pressure increases. When the critical point is reached, the two densities converge into a single value, 400 kg m^{-3}. At above T_c, water exists as a "fluid" that has the volume-filling property of a gas but cannot form droplets. Therefore, at below T_c, the liquid and vapor coexist as two distinct phases different in density and, at above T_c, only one "fluid" phase exists. More detailed description of phase equilibrium is given elsewhere.[3–5]

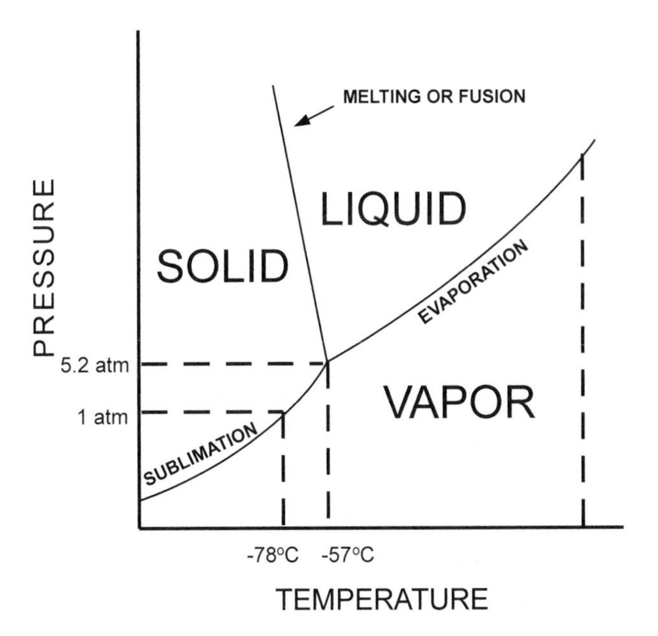

Fig. 2 The phase diagram of carbon dioxide. (From Ref. [2].)

PHASE CHANGE

At the melting point of ice (273.15°K at 1 atm), or the boiling point of water (373.12°K at 1 atm), or the point of sublimation, transition between corresponding phases are accompanied by volume and entropy discontinuities. They are first-order phase transitions because entropy, *S*, and volume, *V*, are first-order derivatives of the Gibbs free energy.[6] The entropy discontinuity upon a phase transition has measurable enthalpy effects, such as enthalpy of freezing or evaporation. Freezing and boiling can be partial or total, depending on the external constraints such as constant pressure or constant volume.

Metastability is a situation immediately prior to a phase transition when homogeneous liquid or vapor still exists inside a coexistence region. Since a metastable state is in a nonequilibrium situation, its free energy is greater than the equilibrium free energy. Superheated and supercooled liquids are metastable.[7] A superheated liquid is obtained when a liquid (in a liquid–gas coexistence region) is exposed to a pressure lower than its vapor pressure at a given temperature or to a temperature higher than the boiling temperature at a given pressure. In food processing, applying vacuum can create a superheated liquid situation and this liquid will boil or evaporate leading to evaporative cooling and increased solid concentration. Likewise, a liquid entering into a liquid–solid coexistence region without solidification is called a supercooled liquid. Supercooled water is of a particular importance in food freezing as it is related to the ice crystallization kinetics that can impact the quality of frozen foods. Here, thermophysical properties of multiple components systems and phase change may impact properties of bioproducts.[8,9]

Although a phase transition is described in thermodynamic terms, phase changes are observed experimentally and expressed as a kinetic process, such as sugar and fat crystallization. They are either endothermic or exothermic with energy called latent heat.[5] For evaporation, in general, the amount of energy that must be supplied for the liquid-to-vapor phase change is called the latent heat of evaporation. This is the point when a liquid is heated to a temperature when its vapor pressure equals to the surrounding pressure. For melting, the amount of energy to be supplied for the solid-to-liquid transition is called the latent heat of melting. Latent heats of condensation and fusion are the exothermic energy released upon condensation of vapor and crystallization, respectively. Molar energy of sublimation equals molar energy of melting plus molar energy of evaporation.

In the case of water,[6] the fundamental characteristics of water is its ability to form a number of strong hydrogen bonds that are directional (tetrahedral) arrangement. Hexagonal ice forms a 3-D network held together by hydrogen bonds. When it melts at 1 atm, it loses its long-ranged order which is accompanied by a 9% increase in density. But melting of ice does not necessarily involve breaking of many great hydrogen bonds. The latent heat of melting being only 13% less than the latent heat of sublimation indicates that the majority of hydrogen bonds are unbroken upon melting of ice. The enthalpy of sublimation is required to break bonds when tightly held molecules of solids become vapor. Thus, liquid water at near zero and supercooled water is described to exhibit local tetrahedral arrangement although this order is short-ranged and short-lived or transient.[6] In a normal circumstance, heating leads to an increase in temperature. But under certain circumstances, the system temperature remains unchanged as the energy provided is used in the phase change process. The effect of supplying energy remained "latent" (with respect to temperature). Latent heat of evaporation (40.79 kJ mol^{-1}) is many times higher than that for fusion (6.01 kJ mol^{-1}). For the case of water, the energy costs for evaporation and dehydration are high in comparison with processes involving no evaporation. Table 1 demonstrates some examples of latent heat of fusion for some food materials. Steam is also an effective heating medium since it gives out large amount of energy upon condensing. Fig. 3 shows a typical heating curve from the solid phase through the liquid phase to the gas phase of a substance. The steepness of the solid, liquid, and vapor heating lines is dependent on the specific heat of the substance in each state. More examples of phase changes in food (such as lipid crystallization, crystallization, and food freezing) are given elsewhere.[13–15]

Table 1 Latent heat of fusion of some food and food components

	Latent heat of fusion		
Saturated monoacid triglycerides (KJ mol^{-1})	α [10]	β'[10]	β [11]
Fatty acid chain length			
11	55.3	71.6	84.6
20	122.2	160.4	220.6
30	96.3	150.3	n/a
Water[12] (KJ kg^{-1})	335		
Apples[12] (KJ kg^{-1})	280		
Corn[12] (KJ kg^{-1})	251		
Fish[12] (KJ kg^{-1})	276		
Beef[12] (KJ kg^{-1})	255		
Bread[12] (KJ kg^{-1})	109–121		
Ice cream[12] (KJ kg^{-1})	222		

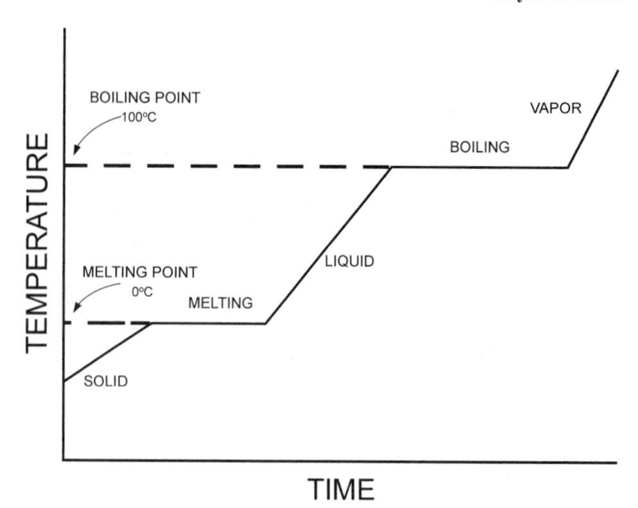

Fig. 3 A typical heating curve of water.

POLYPHASE SYSTEMS

Single component systems may exist in multiple phases under a given P-T-V condition. Pure water has been reported to exist in as many as ten ice crystalline structures. Some of these may coexist and some predominates depending on the surrounding conditions. Multiple component systems, depending on molecular miscibility and thermodynamic and kinetic factors, may exhibit a significant degree of heterogeneity in phases or domains. As the number of components increases, the number of phases increase. Phase change as a function of P-V-T and component concentration, can no longer be represented in a simple phase diagram.

Additionally, biological systems such as food are usually far from a true equilibrated state and at a given time they are not at their most thermodynamically favorable state. Food physical states and phases often times are controlled by kinetic parameters. Manifestation of selective distribution of the surrounding solvent (i.e., water) and phase-separated entities are some of the challenges. The current state of knowledge of a glass transition and the nature of amorphous glassy structure is not yet fully understood.

CRYSTALLIZATION VS. VITRIFICATION

When a liquid is cooled isobarically passing its theoretical melting temperature (T_m), it contracts accordingly to its positive thermal expansion coefficient. If crystalline solids are formed (such as in some cases when the cooling rate is adequately slow or the samples is seeded with crystals), its volume will decrease discontinuously as it crystallizes at (or near) T_m (dashed vertical line in Fig. 4). If not

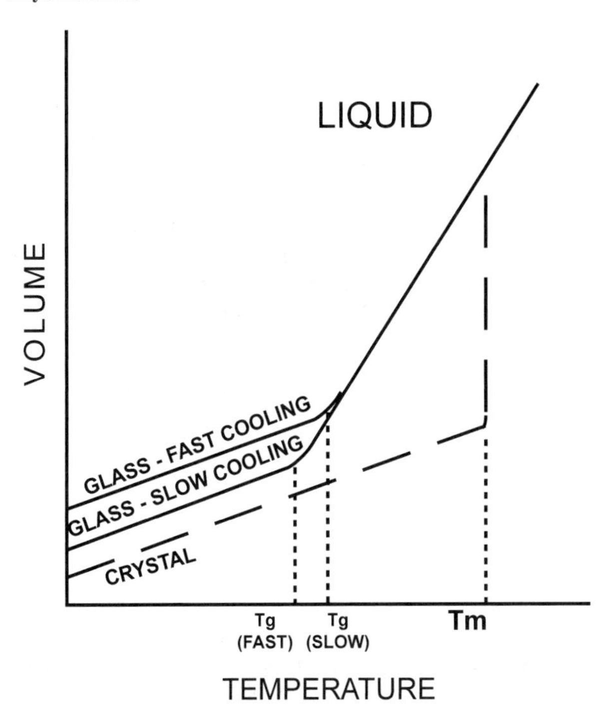

Fig. 4 Volumetric change upon cooling of a liquid at different cooling rates. (Modified from Ref. [6].)

crystallized, the increase in molecular tardiness in the reorganization sometimes manifested by an increase in macroscopic viscosity with decreasing temperature. This is called vitrification. Not all molecules have properties that favor vitrification under experimental conditions. Conditions that favor vitrification are in general those that inhibit crystallization (particularly for molecules that have relatively large surface tension and large entropy change upon fusion). The crystallization rate and the extent of supercooling has been described as the result of the competition between thermodynamic driving force for nucleation and the kinetics of growth.[6] Vitrification and formation of a glass can only be described meaningfully not only from temperature but also from the specific time frame involved in cooling or heating. Macroscopic properties, such as viscosity and storage modulus, are measured with methods that take relatively longer time in comparison to molecular movement that is measured in only a microsecond or millisecond range of time scale. A single-component liquid may show a significant decrease in molecular mobility (an increase in relaxation time) as the temperature is decreased below the freezing point. If the liquid is prevented from crystallizing within the time scale comparable to that required for the necessary molecular rearrangement, molecular structure is "frozen" in time and said to have been vitrified. This is called a glass.

Since time required to crystallize and the tendency to form a glass is, in theory, inversely proportional to the molecular relaxation time, it has been emphasized in literature that parameters like glass transition temperature (T_g) is an indicator of food stability (e.g., see Ref. 16). This may be the case of truly miscible systems when molecular relaxation time correlates with structural relaxation on a macroscopic level, e.g., shear or storage moduli. However, this is only representing some narrow classification of foods.

GLASS TRANSITIONS OF BIOPOLYMERS

For more complex biopolymer systems, the vitrified or glassy state contains conformationally disordered arrangement. Lacking of cooperative chain motion, mobility (that may exist) is one of the side chains and surrounding solvent.

Glass transition theories are based on a number of hypotheses (as described in Ref. 17). The free volume theory introduces the viscosity as an activation energy for diffusion. Cavities or segment voids can be described macroscopically as coefficients of expansion below and above the glass transition temperature, T_g.[17,18] The Williams–Landel–Ferry (WLF) equation provides some form of an analytical relationship between polymer melt viscosity and free volume.[19] It is a highly generalized assumption only for a linear amorphous polymer at above T_g. This is a kinetic theory and T_g described and measured is meaningless without disclosing the kinetic parameters related to the observed events, such as heating and cooling rates, sample history, frequency or time frame of the experiments, presence and fate of compositional change, and so on. There are thermodynamic theories describing a true second-order transition as an equilibrium property which is difficult to achieve.[20]

T_g value is a kinetic term and faces a real challenge in application as a reference value for food stability determination. Additionally, T_g is not a single precise term as shown in literature, it may spread over a large temperature range (up to 50–100°C in some cases) much to do with heterogeneity among various domains and phase separated regions. If ever used, T_g is better described as a distribution function with midpoint and breadth of the transition temperature.[21]

Although a glass macroscopically appears rigid, it may exhibit molecular motions of a liquid like property. This is possible if the motions are not molecularly miscible with and not coupled with the stiff backbone motion of glass. Characterization of glassy and rubbery states of a polymer is therefore highly dependent on the relaxation phenomena and associated time scale being observed. This means it is dangerous to extrapolate properties beyond the observed time scale (e.g., using rheological or thermal analysis relaxation data to mean all molecular order–disorder mobility). It is also dangerous to assume that a system that exhibits a single glass transition observed in a longer time scale behaves homogeneously on a molecular level. For example, water in a glassy and rigid biopolymer matrix can have a liquid like mobility, such as a case of unfreezable water in a frozen starch system.[22]

REFERENCES

1. Chang, R. *Physical Chemistry with Applications to Biological Systems*; Macmillan Publishing Co., Inc.: New York, 1981; 501.
2. Chang, R. *Intermolecular Forces.* Chemistry; Random House, Inc.: New York, 1988; 395–443.
3. Debenedetti, P.G. *Metastable Liquids*; Princeton University Press: Princeton, NJ, 1996; 411.
4. Klotz, I.M.; Rosenberg, R.M. *The Phase Rule.* Chemical Thermodynamics: Basic Theory and Methods; John Wiley & Sons: New York, 1994; 293–306.
5. Borg, R.J.; Dienes, G.J. Phase Diagrams for Single Component Systems. *The Physical Chemistry of Solids*; Academic Press: San Diego, CA, 1992; 340–345.
6. Baierlein, R. Phase Equilibrium. *Thermal Physics*; Cambridge University Press: New York, NY, 1999; 270–305.
7. Angell, C.A. Supercooled Water. Annu. Rev. Phys. Chem. **1983**, *34*, 593.
8. Franks, F. Protein Destabilization at Low Temperatures. Adv. Prot. Chem. **1995**, *46*, 105–139.
9. Franks, F. Phase Changes and Chemical Reactions in Solid Aqueous Solutions: Science and Technology. Pure Appl. Chem. **1997**, *69*, 915–920.
10. Ollivon, M.; Perron, R. Measurements of Enthalpies and Entropies of Unstable Crystalline Forms of Saturated Even Monoacid Triglycerides. Thermochim. Acta **1982**, *53*, 183–194.
11. Timms, R.E. Heats of Fusion of Glycerides. Chem. Phys. Lipids **1978**, *21*, 113–129.
12. Earle, R.L. *Unit Operations in Food Processing*; Pergamon Press: New York, 1983; 187.
13. Lewis, M.J. Sensible and Latent Heat Change. *Physical Properties of Foods and Food Processing Systems*; Ellis Horwood Ltd.: Chichester, UK, 1987; 220–245.
14. Rao, M.A.; Hartel, R.W. *Phase/State Transitions in Foods*; Marcel Dekker: New York, NY, 1998.
15. Roos, Y.H. Phase Transitions and Transformations in Food Systems. In *Handbook of Food Engineering*; Heldman, D.R., Lund, D.B., Eds.; Marcel Dekker: New York, NY, 1992; 145–197.
16. Slade, L.; Levine, H. Beyond Water Activity: Recent Advances Based on an Alternative Approach to the Assessment of Food Quality and Safety. Crit. Rev. Food Sci. Nutr. **1991**, *30*, 115–360.
17. Eyring, H.; John, M.S. *Significant Liquid Structure*; John Wiley & Sons: New York, 1969.
18. Doolittle, A.K. Studies in Newtonian Flow. II. The Dependence of the Viscosity of Liquids on Free Space. J. Appl. Phys. **1951**, 1471.
19. William, M.L.; Landel, R.F.; Ferry, J.D. The Temperature Dependence of Relaxation Mechanisms in Amorphous Polymers and Other Glass-Forming Liquids. J. Am. Chem. Soc. **1955**, *77*, 3701–3707.
20. Cesaro, A.; Sussich, F. Plasticization: The Softening of Materials. In *Bread Staling*; Chinachoti, P., Vodovotz, Y., Eds.; CRC Press: Boca Raton, FL, 2000; 19–60.
21. Peleg, M. On Modeling Changes in Food and Biosolids at and Around Their Glass Transition Temperature Range. CRC Crit. Rev. Food Sci. Nutr. **1996**, *36*, 49–67.
22. Li, S.; Dickinson, L.C.; Chinachoti, P. Water Mobility in Waxy Corn Starch by ^{2}H and ^{1}H NMR. J. Agric. Food Chem. **1998**, *46*, 62–71.

Plasticization

Paul Cornillon
Danone Vitapole, Palaiseau, France

P

INTRODUCTION

The application of polymer science concepts to food science problems has helped scientists and engineers redefine the physics, chemistry, and engineering of food products. Glass transition plays a major role in the evaluation of the physical and chemical stability of food products. As water is the major plasticizer in food and biological products, it is of primary importance to understand its effect on the physical and chemical stability of food. Thus, characterizing the plasticization effect and function for food systems has been a major thrust over the last 10 yr–15 yr. This article is focused on presenting the major effects of and equations that define plasticization.

BACKGROUND

Over the last 10 yr–15 yr, food scientists have changed the nature of food. Indeed, they changed the quality and consumer acceptance of food products by applying and using basic scientific concepts. Chemistry, physics, biology, microbiology, and engineering are among the most famous fields used when studying foods. Of primary importance in the search for better and safer food products, preservation techniques have evolved to unit operations more or less specific to each product.

During storage and processing, phase transitions of particular components can occur in foods.[1] Glass transition is among the most studied transitions. Its importance has been highlighted many times, especially by Levine and Slade.[2–5]

The glass transition temperature of food biopolymers influences the physical and chemical stability of final products as well as their mechanical properties. Glass transition is a characteristic transition in food products that occurs over a range of temperature where motion of segments of polymer becomes more and more enhanced and activated. The length of each segment will influence its flexibility and T_g.[6] The mechanical, physical, and chemical properties depend upon weight or volume fraction of water inducing plasticization of polymer chains. The depression of the glass transition temperature is sometimes called "plasticization function" and is very important when studying basic processing operations like drying, freezing, freeze-drying, extrusion, and mixing.

APPLICATIONS

A food product in the glassy state presents a decrease in molecular mobility, as compared to that in the rubbery state, which affects the rates of reactions and relaxation processes. Managing this decrease would help protect foods: during processing and storage, variations of water content and/or temperature will dramatically affect the stability of food glasses.

Such plasticization will influence the enzymatic reactions like sucrose hydrolysis. It was found that sucrose hydrolysis in a lactose/sucrose system did not occur below the glass transition.[7] Stability maps, based on either the water activity or the glass transition temperature, were obtained to characterize the processing and storage conditions of foods.[8] Moreover, such approach was applied to the determination of mechanical properties of cereal products around the glass transition.[9–11] In another report, the stability of food powders in relation to caking was investigated in relation to moisture and temperature variations that would enhance aggregation of particles.[12] Similar theory was applied to evaluate the thermal inactivation of *Bacillus Stearothermophilus* spores in the glassy state of corn embryos.[13] Data showed that above the glass transition temperature the rate of thermal inactivation depended on the free volume. Similarly, mold growth could be observed in bread and maltodextrin samples below the glass transition temperature.[14] Finally, relationships between molecular mobility and reactivity were found in frozen foods[15] and the hydration mechanism of proteins was analyzed near T_g as it related to the plasticizing effect of water.[16] A more comprehensive review of these and other applications can be found elsewhere.[1]

Encyclopedia of Agricultural, Food, and Biological Engineering
DOI: 10.1081/E-EAFE 120006987
Copyright © 2003 by Marcel Dekker, Inc. All rights reserved.

MODELING

The Gordon–Taylor Model

The Gordon–Taylor model[17] has been applied to food and biological constituents to predict the depression of their glass transition temperature from addition of water. It assumes an additive law of volume of the repeating monomer units in the polymer chains and has been used to predict water plasticization in carbohydrate and protein systems.

For a binary polymer/water system, the glass transition temperature can be obtained by applying Eq. 1:

$$T_{\mathrm{g\,system}} = \frac{w_1 T_{\mathrm{g1}} + k w_2 T_{\mathrm{g2}}}{w_1 + k w_2} \quad (1)$$

where w_i and T_{gi} are the weight fraction and glass transition temperature of component i, respectively. k is the Gordon–Taylor parameter that is defined by Eq. 2:

$$k = \frac{\rho_1 \Delta\alpha_2}{\rho_2 \Delta\alpha_1} \quad (2)$$

where $\Delta\alpha_i$ and ρ_i are the variations of the expansion coefficient and the density of constituent i, respectively. Note that this model cannot be applied to systems other than binary mixtures.

The Couchman–Karasz Model

This model is equivalent to the Gordon–Taylor equation where k is replaced by:

$$k = \frac{\Delta C_{\mathrm{p2}}}{\Delta C_{\mathrm{p1}}} \quad (3)$$

where $\Delta C_{\mathrm{p}i}$ is the change in heat capacity at T_{gi} for constituent i. Hence, Eq. 1 becomes:

$$T_{\mathrm{g\,system}} = \frac{w_1 \Delta C_{\mathrm{p1}} T_{\mathrm{g1}} + w_2 \Delta C_{\mathrm{p2}} T_{\mathrm{g2}}}{w_1 \Delta C_{\mathrm{p1}} + w_2 \Delta C_{\mathrm{p2}}} \quad (4)$$

This model[18] is based on the continuity of entropy and volume and assumes that the constituents of the mixture are compatible (i.e., miscible).

For a system with n components, the Couchman–Karasz model is as follows:

$$T_{\mathrm{g\,system}} = \frac{\sum_{i=1}^{n} w_i \Delta C_{\mathrm{p}i} T_{\mathrm{g}i}}{\sum_{i=1}^{n} w_i \Delta C_{\mathrm{p}i}} \quad (5)$$

Water is the most common plasticizer in food and biological products. Its ΔC_{p} and T_{g} are $1.94\,\mathrm{J\,g^{-1}\,K^{-1}}$ and $-135°\mathrm{C}$, respectively.[5,19,20]

Even though the Gordon–Taylor and the Couchman–Karasz equations are often used, they are not appropriate for many systems like complex mixtures of three or more components and mixtures that present very broad glass transitions (e.g., gluten-based materials). Such disagreement is often due to phase separation in the system, and in such cases, other models are needed.[6]

The Group Contribution Approach

Group contribution theories can predict compound properties from their molecular structures and interactions of molecules in the mixture. The method is quite fast and does not require the use of complicated computer algorithms.[21–26]

Group contribution theory was first developed for simple organic compounds. However, with the development of the well-known Lattice Fluid (LF) theory,[27] prediction of many temperature and pressure dependent physical properties has been made for simple organic compounds and synthetic polymers. The theory requires the determination of parameters, called "scaling constants," for temperature, pressure, and density.

LF theory has been applied successfully to the determination of thermodynamic properties of nonwater soluble and synthetic polymers and polymer mixtures[28] and their glass transition temperatures[29] using the Gibbs–DiMarzio approach.[30,31] This approach of zero configurational entropy at glass transition has been used previously for the prediction of the glass transition temperatures of pure polymers and polymer mixtures.[32–34]

With another approach,[6] a group contribution theory was proposed and applied to complex biological mixtures. It was based on the analysis of the plasticizing effect of water on the weakening hydrogen bonds and the dipole–dipole intra- and inter- macromolecular interactions from shielding effects of attractive forces by water molecules.

Results from these studies are very encouraging for the application of the group contribution approach to the modeling and prediction of the plasticization effect of water in food and biological systems.

FUTURE NEEDS

As indicated above, plasticization plays an important role in the design and development of new food products. Hence, a very precise and complete understanding of this phenomenon is critical. This can be achieved by improving calculation methods and models, by developing new analytical methods of analysis, and by standardizing the commonly used ones like differential scanning calorimetry (DSC), dielectric thermal analysis (DEA), dynamic mechanical analysis (DMA), electron spin resonance (ESR), and nuclear magnetic resonance (NMR).

REFERENCES

1. Roos, Y.H. *Phase Transitions in Foods*; Academic Press: San Diego, CA, 1995.
2. Levine, H.; Slade, L. A Polymer Physico-chemical Approach to the Study of Commercial Starch Hydrolysis Products (SHPs). Carbohydr. Polym. **1986**, *6*, 213–244.
3. Levine, H.; Slade, L. Principles of Cryostabilization Technology from Structure/Property Relationships of Carbohydrate/Water Systems. Cryo-Lett. **1988**, *9*, 21–63.
4. Slade, L.; Levine, H. A Food Polymer Science Approach to Selected Aspects of Starch Gelatinization and Retrogradation. In *Frontiers in Carbohydrate Research-1: Food Applications*; Millane, R.P., BeMiller, J.N., Chandrasekaran, R., Eds.; Elsevier Science: London, England, 1989; 215–270.
5. Slade, L.; Levine, H. Beyond Water Activity: Recent Advances Based on an Alternative Approach to the Assessment of Food Quality and Safety. Crit. Rev. Food Sci. Nutr. **1991**, *30*, 115–360.
6. Matveev, Y.I.; Grinberg, V.Y.; Tolstoguzov, V.B. The Plasticizing Effect of Water on Proteins, Polysaccharides and Their Mixtures. Glassy State of Biopolymers, Food and Seeds. Food Hydrocoll. **2000**, *14*, 425–437.
7. Kouassi, K.; Roos, Y.H. Glass Transition and Water Effects on Sucrose Inversion by Invertase in a Lactose–Sucrose System. J. Agric. Food Chem. **2000**, *48*, 2461–2466.
8. Karel, M.; Buera, M.P.; Roos, Y.H. Effects of Glass Transitions on Processing and Storage. In *The Glassy State in Foods*; Blanshard, J.M.V., Lillford, P.J., Eds.; Nottigham University Press: Loughborough, England, 1993; 13–34.
9. Attenburrow, G.; Davies, A.P. The Mechanical Properties of Cereal Based Foods in and Around the Glassy State. In *The Glassy State in Foods*; Blanshard, J.M.V., Lillford, P.J., Eds.; Nottigham University Press: Loughborough, England, 1993; 317–332.
10. Nelson, K.A.; Labuza, T.P. Glass Transition Theory and the Texture of Cereal Foods. In *The Glassy State in Foods*; Blanshard, J.M.V., Lillford, P.J., Eds.; Nottigham University Press: Loughborough, England, 1993; 513–518.
11. Peleg, M. Mechanical Properties of Dry Brittle Cereal Products. In *The Properties of Water in Foods: ISOPOW 6*; Reid, D.S., Ed.; Blackie Academic & Professional: London, England, 1998; 233–254.
12. Peleg, M. Glass Transitions and the Physical Stability of Food Powders. In *The Glassy State in Foods*; Blanshard, J.M.V., Lillford, P.J., Eds.; Nottigham University Press: Loughborough, England, 1993; 435–454.
13. Sapru, V.; Labuza, T.P. Temperature Dependence of Thermal Inactivation Rate Constants of *Bacillus Stearothermophilus* Spores in a Glassy State. In *The Glassy State in Foods*; Blanshard, J.M.V., Lillford, P.J., Eds.; Nottigham University Press: Loughborough, England, 1993; 499–506.
14. Chirife, J.; Buera, M.P.; Gonzalez, H.H.L. The Mobility and Mold Growth in Glassy/Rubbery Substances. In *Water Management in the Design and Distribution of Quality Foods: ISOPOW 7*; Roos, Y.H., Leslie, R.B., Lillford, P.J., Eds.; Technomic Publishing Co.: Lancaster, PA, 1999; 285–298.
15. Le Meste, M.; Champion, D.; Roudaut, G.; Contreras-Lopez, E.; Blond, G.; Simatos, D. Mobility and Reactivity in Low Moisture and Frozen Foods. In *Water Management in the Design and Distribution of Quality Foods: ISOPOW 7*; Roos, Y.H., Leslie, R.B., Lillford, P.J., Eds.; Technomic Publishing Co.: Lancaster, PA, 1999; 267–284.
16. Gregory, R.B. Protein Hydration and Glass Transitions. In *The Properties of Water in Foods: ISOPOW 6*; Reid, D.S., Ed.; Blackie Academic & Professional: London, England, 1998; 57–100.
17. Gordon, M.; Taylor, J.S. Ideal Copolymers and the Second-Order Transitions of Synthetic Rubbers. I. Non-crystalline Copolymers. J. Appl. Chem. **1952**, *2*, 493–500.
18. Couchman, P.R.; Karasz, F.E. A Classical Thermodynamic Discussion of the Effect of Composition on Glass Transition Temperatures. Macromolecules **1978**, *11*, 117–119.
19. Sugisaki, M.; Suga, H.; Seki, S. Calorimetric Study of the Glassy State. IV. Heat Capacities of Glassy Water and Cubic Ice. Bull. Chem. Soc. Jpn. **1968**, *41*, 2591–2599.
20. Roos, Y.H.; Karel, M.; Kokini, J.L. Glass Transitions in Low Moisture and Frozen Foods: Effect on Shelf-life and Quality. Food Technol. **1996**, *11*, 95–108.
21. Derr, E.L.; Deal, C.H., Jr. Analytical Solutions of Groups. Correlation of Activity Coefficients Through Structural Group Parameters. In *Proceedings*, International Symposium on Distillation, London, England, 1969; Vol. 3, 40–51.
22. Fredenslund, A.A.; Gmehling, J.; Rasmussen, P. *Vapor–Liquid Equilibria Using UNIFAC*; Elsevier Science: Amsterdam, The Netherlands, 1977.
23. Horvath, A.L. Molecular Design, Chemical Structure Generation from the Properties of Pure Organic Compounds. In *Studies in Physical and Theoretical Chemistry*, 1st Ed.; Elsevier Science: Amsterdam, The Netherlands, 1992; 75.
24. Bicerano, J. *Prediction of Polymer Properties*; Marcel Dekker: New York, NY, 1993.
25. Constantinou, L.; Jaksland, C.; Bagherpour, K.; Gani, R.; Bogle, I.D.L. Application of the Group Contribution Approach to Tackle Environmentally Related Problems. AIChE Symp. Ser. **1994**, *303*, 105–116.
26. Constantinou, L.; Gani, R. New Group-Contribution Method for Estimating Properties of Pure Compounds. AIChE J. **1994**, *40* (10), 1697–1710.
27. Sanchez, I.C.; Lacombe, R.H. Statistical Thermodynamics of Polymer Solutions. Macromolecules **1978**, *11*, 1145–1156.
28. Boudouris, D.; Constantinou, L.; Panayiotou, C. Group Contribution Estimation of the Thermodynamic Properties of Polymers. Ind. Eng. Chem. Res. **1997**, *36* (9), 3968–3973.
29. Boudouris, D.; Constantinou, L.; Panayiotou, C. Prediction of Volumetric Behavior and Glass Transition Temperature

P

of Polymers: A Group Contribution Approach. Fluid Phase Equi. **2000**, *167*, 1–19.
30. Gibbs, J.H.; DiMarzio, E.A. Nature of the Glass Transition and the Glassy State. J. Chem. Phys. **1958**, *28*, 373–383.
31. DiMarzio, E.A.; Gibbs, J.H. Molecular Interpretation of Glass Temperature Depression by Plasticizers. J. Polym. Sci. **1963**, *A-1*, 1417–1428.
32. Panayiotou, C.; Vera, J.H. On the Fluid Lattice and Gibbs–DiMarzio Theories. J. Polym. Sci. Pol. Lett. **1984**, *22* (11), 601–606.
33. Panayiotou, C.G. Glass Transition Temperature in Polymer Mixtures. Polym. J. **1986**, *18* (12), 895–902.
34. Prinos, J.; Panayiotou, C. Glass-Transition Temperature in Hydrogen-Bonded Polymer Mixtures. Polymer **1995**, *36* (6), 1223–1227.

Plate Heat Exchangers

R. Simpson
Universidad Técnica Federico Santa María, Valparaíso, Chile

S. Almonacid
Oregon State University, Astoria, Oregon, U.S.A.

P

INTRODUCTION

In the modern industrialized food industry, it is common to find unit operations such as refrigeration, freezing, thermal sterilization, drying, and evaporation.[17] The referred unit operations consider heat transfer from a heating or cooling medium to a food product. The commercially successful Plate Heat Exchangers (PHEs) were developed in the early 1920s by Dr. Richard Seligman, founder of the APV Company. These equipments were first introduced to meet hygienic demands of the dairy industry; however, with time, these PHEs have undergone continuous refinements and innovations, leading to today's range of thin gage-plate machines capable of operating at pressures up to 2.5 MPa (360 $lb_f\,in^{-2}$) and temperatures up to 260°C (500°F) with a heat transfer area of 650 m^2. In some specific cases though, the transfer area can reach values[1] up to 2500 m^2. The PHEs are widely used in the food and chemical processing industries, mainly due to their various advantages displayed in specific cases. A PHE consists of a number of corrugated metal plates (the heat transfer area) that include ports for fluid entry and fluid discharge at each corner, which are all bound (or clamped) together in a rigid frame to give them support.[9] The construction principle is similar to a filter press, i.e., the pressure drop of the fluid passing through the PHE is normally small, with a well-controlled leakage.

Fig. 1 shows a schematic diagram of a PHE. The stationary part of the unit consists of a frame to which a fixed end-cover is attached (c and b), together with a handlebar that allows transportation (d). The thin rectangular metal plates are sealed around the edges by gaskets and kept in position with a movable end-cover (a), which is adjusted with a compression bolt.

Standard plates are made of stainless steel, titanium, nickel, metal monel, hastelloy C, phosphorous bronze, cupronickel, etc. Standard plates constitute a compressed single piece ranging from 0.4 mm to 0.5 mm of thickness with corrugated shape. The corrugated design of plates assures rigidity; at the same time, it enhances the turbulence performance and maximizes the flow distribution. In this design, the heat transfer area is easily adjusted by adding or subtracting plates, making this a very versatile equipment, which can be used for multiple services and several fluids. These fluids can flow through several different parts of the heat exchanger allowing different thermal effects in the outlet stream. High-viscosity fluids (up to 30,000 cP) traveling through these equipments, produce a relatively high heat transfer coefficient, mainly because PHEs assure turbulent flows with Reynolds numbers as low as 150. One of the design limitations of the PHE is the fact that liquids should not have more than 5% solids and should have particle sizes up to 1 mm in diameter.[18]

DESIGN AND CONSTRUCTION OF PHEs

Pressure Drop

In a PHE, there are three components that contribute to the total pressure drop:

1. Pressure drop generated at the fluid entrance and exit points.
2. Pressure drop due to the corrugated plate design.
3. Pressure drop due to the fluid height change.

The inlet and outlet plate passage pressure drop effect can be assessed as 1.5 times the inlet velocity head per mass value.[19] Another way of taking into account this effect is to include it in the friction factor expression obtained.

The equivalent diameter is 2.0 times the gap between the plates. The corrugated shape increases the friction factor value up to 60 times as compared to the factor recorded in a straight pipe of similar size and same Reynolds number.

The friction factor can be described by the following expressions:

$$f = a + \frac{b}{Re} \tag{1}$$

$$f = \frac{c}{Re^n} \tag{2}$$

DOI: 10.1081/E-EAFE 120007005
Copyright © 2003 by Marcel Dekker, Inc. All rights reserved.

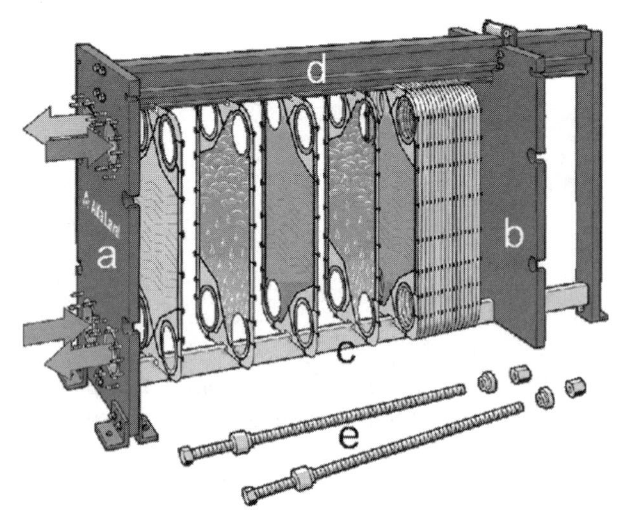

Fig. 1 A PHE: a) movable end-cover; b) fixed end-cover; c) plate pack; d) carrying bar; and e) compression bolt. (Courtesy of Alfa Laval AB, Chile.)

where the constants a, b, c, and n must be experimentally determined.

Eq. 1 is suitable for low Reynolds numbers (Re from 100 to 3000), whereas for turbulent flow ($Re > 3000$), Eq. 2 should be used.

Typical values

a	b	c	n	Remarks
0.093	57.5			Chevron (with $\beta = 30°$); for $260 < Re < 3000$[1]
		0.898	0.16	Idem; for $3000 < Re < 50000$[1]
0.52	34.80			Chevron (47.2 cm × 5.3 cm), criss–cross arrangement $100 < Re < 600$[13]
		3.067	0.155	Idem; $600 < Re < 3000$[13]
		0.325	0.181	Idem, criss distribution; for $100 < Re < 7800$[13]
		2.5	0.3	For commercially used PHEs[3]

Due to the design variety and various flow arrangements possible for PHEs, the values of these parameters fluctuate in a wide range.

Some specific correlations for PHE condensation (two-phase flow) have been obtained, but these are valid only for particular systems.

Heat Transfer

The heat transfer in a PHE is defined in the usual form by:

$$Q_T = UA_T(\Delta T)_{HT} \tag{3}$$

The heat transfer area is calculated by the following expression:

$$A_T = (2n - 1)a_p \tag{4}$$

where a_p is the plate area and n the number of plates.

The term $(\Delta T)_{HT}$ is the logarithmic mean temperature difference (LMTD).

In general, the heat transfer coefficients for turbulent flow or transition flow can be correlated by the following expression:

$$Nu = a\,Re^b Pr^c \left(\frac{\mu_m}{\mu_w}\right)^n \tag{5}$$

where the constant values will depend on the type, plate arrangement, and flow regime in the heat exchanger. Usual values for turbulent or transition flows are as follows[9]: $a = 0.15–0.40$; $b = 0.65–0.85$; $c = 0.30–0.45$; and $n = 0.05–0.20$.

Similar correlations are cited in the literature, but $n = 0$ is considered. The $n = 0$ criterion is valid for fluids with a high Prandtl number ($Pr > 1$, as it is the case for most liquids). One of the proposed expressions found in the literature is[3]:

$$Nu = 0.44Re^{0.672}Pr^{0.4} \tag{6}$$

Models based on Eq. 5 or on the corrected Nusselt Model have been developed for condensation (phase change).

Design Conclusion

When estimating the pressure drop in heat exchangers, the contribution made by the plate's inlets and outlets should not be neglected. The scarcity/paucity of data should not overcome the benefits exhibited by the PHEs from a thermal design point of view. Given the high heat transfer coefficients achieved, it is possible to operate with temperature differences as low as 1°C between the heat exchanger inlet and outlet.

FOULING AND MILK FOULING

Fouling of heat transfer equipment by food products is a serious problem in the food industry.[5] This deposit formation has detrimental effects; it causes a significant decrease of the heat transfer performance, promoting at the same time an increase in the microbiological population, causing quality problems.[5] Fouling, and particularly milk fouling, has been studied by several researchers.[2,4–7,11] The deposit composition, in milk fouling, is well known and so are the chemical changes that milk undergoes when it is heated.[8] As reported by Georgiadis, Rotstein, and Macchietto,[8] the two major contributions in this field of study are the articles of Fryer et al.[6] and the researches performed at the process-engineering laboratory in France.[10]

As reported by Sandu and Singh,[16] the marginal costs associated with fouling are: 1) increased capital cost; 2) increased energy costs in operation; 3) higher maintenance costs; 4) production losses during plant downtime for cleaning; and 5) energy losses.

ADVANTAGES, LIMITATIONS, AND COMPARISONS OF PHEs VS. SHELL AND TUBE HEAT EXCHANGERS

In general, PHEs feature the following advantages:

- The product volume held up in a PHE equipment is small, therefore, the transient response of control parameters such as temperature and flow feed occurs very quickly.
- PHEs are very flexible as the heat transfer area can be easily adjusted by adding or removing plates.
- Maintenance of PHEs is simple. They can easily be dismantled for product surface inspection, plus the cleaning can be done in a short period of time.
- A PHE usually uses less floor space and also weighs less than other types of heat exchangers.
- PHEs contribute to energy conservation as they allow regeneration.
- The high turbulence generated by the corrugated design of the plates allows the fouling build up to be retarded.

Limitations

Some characteristic and intrinsic limitations of PHEs are:

- Maximum operation pressures up to about 2.5 MPa (360 $lb_f\,in^{-2}$).
- Maximum operation temperatures, which can reach values up to 260°C (500°F) but normally lower than 160°C.
- Usually fluids with suspended particles cannot be easily handled, so, particle sizes that do not exceed one third of the channel gap (about 1 mm) are recommended.
- PHEs do not allow economic large-scale handling of low-density fluids (i.e., steam).

APPLICATIONS

PHEs are commonly used in the dairy industry for pasteurization and sterilization processes.[12] In general, they are widely used for fluid food pasteurization, also being recently applied to sterilization processes under pressure.[14] The beer industry has made extensive use of this type of heat exchanger as well.

A nontraditional application was developed to automatically process ricotta cheese. The advantages of this system include product homogeneity, improved shelf life, high recovery degree of protein and milk fats ($>$ 95%), increased yields, and reduced labor costs.

Recently, a beet sugar factory in southern Poland launched an ambitious energy saving program to reduce fuel consumption. Factory managers struck upon the idea of utilizing waste heat from the vacuum pans to heat raw juice in their double-sided Alfa Laval Wide-gap PHE, generating considerable savings.

PHEs have also been used for *Salmonella enteriditis* elimination from liquid egg products. For this task, capillary tubes and a plate exchanger are used.[15]

Among other nontraditional applications, fatty materials can be hydrogenated using PHE at a pressure of 150 psig.

ACKNOWLEDGMENTS

The authors wish to acknowledge the support from Mrs. Viviana Miranda (Ph.D. Student at Universidad Católica de Chile) and Mr. Alonso Jaques (Chemical Engineering Student at Universidad Técnica Federico Santa María).

REFERENCES

1. Bond, M.P. Plate Heat Exchangers for Effective Heat Transfer. Chem. Eng. **1981**, *367*, 162–167.
2. Burton, H. Deposit of Whole Milk in Treatment Plants—A Review—and Discussion. J. Dairy Res. **1968**, *34*, 317–330.
3. Cooper, A. Recover More Heat with Plate Heat Exchanger. Chem. Eng. (Lond.) **1977**, *285*, 280–289.
4. De Jong, P.; Bowman, S.; Van Der Linder, H.J.L.J. Fouling of Heat Treatment Equipment in Relation to Denaturation of β-Lactoglobulin. J. Soc. Dairy Technol. **1992**, *45* (1), 3–8.
5. Delplace, F.M.; Leuliet, J.C.; Tissier, J.P. Fouling Experiments of a Plate Heat Exchanger by Whey Protein Solutions. Trans. IchemE **1994**, *72*, 163–169.
6. Fryer, P.J.; Robins, P.T.; Green, C.; Schreier, A.; Pritchard, M.; Hasting, D.; Royston, G.; Richardson, J.F. A Statistical Model for Fouling of a Plate Heat Exchanger

by Whey Protein Solution at UHT Conditions. Trans. I chem E Part C **1996**, *74*, 189–199.

7. Georgiadis, M.; Rotstein, G.; Macchietto, S. Modeling and Simulation of Shell and Tube Heat Exchangers Under Milk Fouling. AICHE J. **1998**, *44*, 959–970.
8. Georgiadis, M.; Rotstein, G.; Macchietto, S. Optimal Design and Operation of Heat Exchangers Under Milk Fouling. AICHE J. **1998**, *44*, 2099–2111.
9. Holdsworth, S.D. *Aseptic Processing and Packaging of Food Products*; Elsevier Applied Science: England, 1996.
10. Lalande, M.; René, F.; Tissier, J.P. Fouling and Its Control in Heat Exchangers in the Dairy Industry. Biofouling **1989**, *1*, 233–250.
11. Lalande, M.; Tissier, J.P.; Corrieu, G. Fouling of Heat Transfer Surfaces Related to β-Lactoglobulin Denaturation During Heat Processing of Milk. Biotechnol. Prog. **1985**, *1* (2), 131–139.
12. Lalande, M.; Tissier, J.P.; Corrieu, G. Fouling of a Plate Heat Exchanger Used in UHT Sterilization of Milk. J. Dairy Res. **1984**, *51*, 123–142.
13. Lopéz, J. Pressure Drop in PHE's. Degree Thesis Chemical Engineering, Universidad Técnica Federico Santa María, Valparaíso, 1981.
14. Louka, M. Simulation of Dynamic Properties of Batch Sterilizer with External Heating. J. Food Eng. **1997**, *20*, 91–106.
15. Michalski, C.B. Use of Capillary Tubes and Plate Heat Exchanger to Validate U.S. Department of Agriculture Pasteurization Protocols for Elimination of *Listeria monocytogenes* in Liquid Egg Products, **2000**, *63* (7), 921–925.
16. Sandu, C.; Singh, R.K. Energy Increases in Operation and Cleaning Due to Heat Exchanger Fouling in Milk Pasteurization. Food Technol. **1991**, *45*, 84–91.
17. Singh, R.P.; Heldman, D.R. *Introduction to Food Engineering*; Academic Press: New York, 1992.
18. Thompson, D. Energy Analysis in Heating and Cooling Processes. Food Technol. **1977**, *31*, 51–56.
19. *Ullman's Encyclopedia of Industrial Chemistry*; VCH Publisher: Federal Republic of Germany, 1988; Vol. B3, 2-10, 2-11, 2-56, 2-66, 2-67.

Poultry Production Systems

Joseph M. Zulovich
University of Missouri, Columbia, Missouri, U.S.A.

P

INTRODUCTION

Poultry production systems house laying hens, broiler, and turkeys to produce the following common food products: eggs, chicken products, and turkey products. Gamebirds and other poultry include the raising of any of the following: ducks, geese, quail, and guinea fowl. This article focuses on egg, chicken, and turkey production. The various stages of egg and poultry meat production are introduced, and brief descriptions of the different types of facilities are presented.

EGG PRODUCTION

The production of eggs incorporates several stages of production. The most common stage includes the layer operations that actually produce the eggs. The other egg production stages develop and grow the laying hens used for egg production.

Layer Operations

Layer operations house laying hens for egg production. Housing systems used for layer hens have evolved over the years into the multiple types of housing systems used today. All housing systems for layer hens provide feed and water access for the birds while also providing for egg collection and environmental temperature control. The specific design and operation of a housing system significantly depends upon the corresponding marketing emphasis of the eggs produced by the system. The housing systems used by the various current marketing themes are discussed below.

Commodity egg production

Commodity egg production systems produce a significant portion of the eggs marketed as fresh eggs and most of the eggs used in various food products. Eggs from commodity egg layer operations are shipped either to grocery chains or to egg processing plants that will process the eggs into other forms or products. The commodity egg production system implements a housing system that minimizes labor requirements of the caretakers for care of the laying hens and collection of eggs and maximizes the use of equipment used by caretakers to care for the birds. Minimizing labor input while maximizing equipment use creates a least cost system to produce eggs.

A commodity egg production housing system typically uses a cage arrangement to house the laying hens. The cages are arranged in rows and in a stair step fashion such that no cages are directly above another. The rows of cages are located on the first floor of a cage layer building and the manure from the hens is collected and stored in the basement of the building. Fresh air for ventilation enters the building through inlets located in the ceiling of the first floor and leaves the building through fans located in the basement walls. The inlets are arranged in the ceiling to provide fresh air for all hens in the cages. The fans pull the fresh air down through the cages and out of the buildings. The air movement created by this orientation of the ventilation system components ensures that fresh air is accessible to all birds and no stale air is forced into the hen space of the cages. Temperature control is provided by the ventilation system by controlling the quantity of air moved through the building system. Feed and water is provided for each individual cage by automatic feed and watering systems. Eggs are collected from the cages and transported to a collection room by an automatic conveyer system. The caretaker for this type of housing system monitors the hens to ensure the hens are provided adequate feed and water, and are receiving fresh air. If an equipment problem is identified the caretaker can either contact the repair service or make minor repairs and adjustments.

Cage-free egg production

Cage-free egg production systems produce eggs that are primarily marketed as fresh eggs in stores. As indicated by the name, cage-free egg production does not use cages. A cage-free system was used as the commodity egg production system prior to the development of the cage housing system. However, current cage-free systems typically have incorporated improvements developed for cage systems to aid in the care of the housed laying hens. Ventilation systems providing fresh air to the hens have made major improvements since cage-free laying hen facilities were the primary system for egg production. Automatic feed delivery and egg collection systems may

DOI: 10.1081/E-EAFE 120006916
Copyright © 2003 by Marcel Dekker, Inc. All rights reserved.

or may not be incorporated in a given cage-free housing system, depending upon the specific marketing emphasis coordinated with the housing system.

Cage-free egg production housing systems will have the following common characteristics. Laying hens typically have access to two different areas of the building. First, hens can access feed and water which is provided on the floor of the building. The floor is usually covered with a bedding material like wood shavings to incorporate the manure produced from the hens. Second, the hens have access to nesting boxes for laying eggs. Most eggs are laid in the enclosed nesting boxes, but some eggs may be laid on the floor of the building. Eggs laid on the floor can be dirty and may be lost depending upon egg collection system. Some cage-free systems have a third area called a roosting area that the hens use for sleeping. A ventilation system is incorporated into the building to provide fresh air and temperature control for the hens housed inside the building. The caretaker monitors the hens to ensure adequate feed and water and functioning of ventilation system. Depending upon the exact configuration of the cage-free system, the caretaker may also be responsible for collecting all eggs and possibly delivering feed to feeders. Increased responsibilities of the caretaker typically results in a system that costs more to produce an egg than the commodity cage system described above.

Small flock egg production

Small flock egg production typically produces eggs for use by the caretaker and a potentially relatively small additional market. Therefore, small flock egg production is not expected to produce an income of sufficient size to solely support the financial needs of the caretaker. Small flock egg production uses a simple hen house design to provide for the housing of the laying hens. The simple hen house will provide a few nesting boxes, a roosting area and feed and water access located on the floor. Opening a couple of windows or ventilation doors provides ventilation. Temperature control is typically not incorporated into the simple hen house. The size of a simple hen house can vary to house 25 hens up to a couple hundred hens. The caretaker typically provides labor for feed and water delivery and egg collection as well as monitoring the hens. No automatic systems are incorporated into simple hen house designs.

Supporting Operations

Supporting operations for egg production produce and raise laying hens that are used for egg production. Commodity egg production systems and cage-free systems can house thousands of birds per building. When a laying hen building is stocked with new hens, the entire building is filled at one time so a significant number of new pullets (young birds about to start laying eggs) are required as a large flock to fill the larger capacity buildings.

Breeder and hatching operations

Fertile eggs are required to produce new chicks that will be grown into new laying hens. Fertile eggs are produced in cage-free egg production systems described above. However, roosters are also housed with the hens to produce fertile eggs. The fertile eggs are collected and transported to a hatchery. The eggs are placed in incubators that are located in the hatchery. The incubators keep the eggs warm so that chicks can develop from the embryo located inside the fertile egg. After the chicks hatch from the eggs, they are separated by sex, vaccinated and placed into transport boxes. The female chicks are transported to pullet rearing operations to be raised into replacement laying hens. The male birds are transported to broiler operations that can raise the male chicks for meat production. Some hatcheries sell chicks in various quantities that are all vaccinated but may or may not be separated by sex.

Pullet rearing operations

Pullet rearing operations raise female baby chicks into replacement laying hens. The birds are raised inside enclosed single story buildings. The floor is covered with a bedding material like wood shavings to incorporate the manure produced from the growing birds. Feed and water are provided in feeders and waterers using automatic delivery systems. A ventilation system provides fresh air for the birds and controls the inside temperature. The building itself and all incorporated systems are designed such that no outside light can enter the building. The light is controlled during the entire growing period of about 20 to 24 weeks. When the birds mature, they are ready to lay eggs and they believe that it is spring regardless when during the year birds actually mature. Replacement laying hens that believe it is spring will typically begin laying eggs once they reach maturity. If a bird matures and she thinks it is fall or winter, she will not begin laying eggs until she believes it is spring.

BROILER PRODUCTION

Broiler production produces most of the chicken marketed as meat and meat products. The breed of bird used for broiler production is typically different from breeds used for egg production. Birds used for broiler production typically are more muscular or meatier and will grow quickly. Baby chicks raised as broilers are from hatcheries.

The hatcheries incubate fertile eggs produced from breeder operations. Breeder operations use cage-free housing systems similar to ones described above to produce fertile eggs for incubating broiler chicks. Some hatcheries producing broiler chicks have the baby chicks available for sale and shipping. Most baby broiler chicks are transported to broiler growers who raise the chicks for the company that produced the chick.

Contract Broiler Production

A broiler building used by growers raising broilers under a contract for a company has the ability to provide the necessary conditions inside for a broiler to grow regardless of outside conditions. The broilers are raised in single story buildings. The floor is covered with a bedding material like wood shavings to incorporate the manure produced from the growing broilers. Feed is provided in feeders connected to automatic delivery systems. Water is provided with a nipple type watering system designed to minimize spillage of water. The ventilation system is designed to provide fresh air and control inside temperature. Heaters are used to heat the building in the winter and when the birds are small. Circulating fans and a misting system are often incorporated to provide cooling to the broilers during hot weather. The building capacity typically ranges from 20,000 to 40,000 birds per building. The grower caretaker monitors the broilers to ensure feed and water are provided and the inside conditions are adequate.

Free Range/Pasture and Small Flock Production

Some broilers are raised in outside free range or pasture systems. These outside systems limit raising broilers to the warm season of the year. Feed and water is provided, but no heating or supplemental cooling potential is typically provided. Free range or pasture-raised broilers usually have to develop their own market access. Small flock broiler production typically uses simple house designs. The floor is covered with bedding material. Feed and water is provided in feeders and waterers and are refilled by the caretaker. Windows and ventilation doors that are manually opened or closed usually provide ventilation. A brooder heater can provide supplemental heat when the broilers are small.

TURKEY PRODUCTION

Turkey production produces turkeys marketed as turkey meat and meat products. Turkeys are raised in housing systems and production arrangements very similar to broiler production. A company produces turkey poults that are raised by contract turkey growers for the company. The only difference is that the baby turkeys are typically raised initially in a brooder building that is similar to a broiler building. Then the partially grown turkeys as transferred to a larger, simpler building for the remainder of the growout period. This larger building provides more space for the larger growing turkeys and uses essentially the same functional design as the brooder building or a broiler building.

In summary, poultry production systems, that minimize unit production costs, utilize extensive building systems to minimize labor required by caretakers to produce eggs and meat. The specific design depends upon the stage of production and any specific marketing effort related to the housing system. Small flock production typically uses less technology and more labor to produce a given unit of production.

FURTHER READING

NRAES. Animal Behavior and the Design of Livestock and Poultry Housing Systems; Cornell University: 152 Riley-Robb Hall, Ithaca, New York, 1995; 14853–5701.

United Egg Producers. Animal Husbandry Guidelines-Overview of Best Management Practices for United States Egg Laying Flocks. 1720 Windward Concourse, Suite 230. Alpharette, GA; 2003; 30005.

SAN. Profitable Poultry: Raising Birds on Pasture. Sustainable Agriculture Network. 2002.

Poultry Housing Websites January 21, 2002

http://www.ext.vt.edu/pubs/poultry/factsheets/10.html. Small-scale poultry coops seem to be built in almost every possible shape and size.

http://www.ext.vt.edu/pubs/poultry/factsheets/designs.html. Designs for small-scale poultry housing.

http://www.aces.edu/poultryventilation/. Sites on ventilation and preparation of housing for weather are shown.

http://attra.ncat.org/attra-pub/poulthous.html. Examines different options for housing. Poultry housing design will depend on the type of production system used.

http://www.ae.iastate.edu/aen159.htm. Air quality within livestock and poultry confinement housing has long been known to affect animal performance.

http://www.dpi.qld.gov.au/poultry/5125.html. Pens for show birds, regulation and environmental needs.

http://www.mywebpage.net/domestic-fowl-trust/poultry_houses.html. The Domestic Fowl Trust has plans for housing. You have to order but there are pictures of housing options.

http://www.poultryclub.org/ACHousing.htm. Gives advice on poultry housing.

http://www.backyardchickens.com/. Gives Coop designs for roosts and pens for poultry.

http://www.geocities.com/Heartland/Plains/4175/henhouse.html. Plan for a small hen house.

http://www.aces.edu/department/poultryventilation/InsulationPVP.pdf. Need for insulation in warm climate Poultry housing.

http://www.dashlink.com/~rockingt/link-housing.html. Free plans for housing-boxes, cages, etc.

http://www.jaquesint.com/poultry_housing.htm. larger scale poultry housing descriptions.

http://www.pakissan.com/allabout/livestock/poultry/housing.systems.shtml. Different types of housing and advantages and disadvantages of each.

http://www.freerange.org/nfp/poultry/list%20of%20pages%20domestic.htm. Plans for poultry housing of different scales.

http://www.melrosechem.com/english/publicat/farms/poultry/biosecur.pdf. Bio security in poultry housing.

http://www.cps.gov.on.ca/english/plans/E5000/5000/M-5000L.pdf. Article on poultry housing.

http://www.melodious.com/gate/howto/coop/. plans for coops and other poultry equipment.

http://www.agr.gov.sk.ca/DOCS/Econ_Farm_Man/Planning/CanadaPlan/5000/5602.pdf. Lighting for poultry housing.

http://home.wanadoo.nl/gjosinga/page1.htm. Plans for care and accommodations for poultry.

Power Measurement

Lon R. Shell
Southwest Texas State University, San Marcos, Texas, U.S.A.

P

INTRODUCTION

Power is defined as the rate of doing *work* and is measured in English customary units of ft lb min^{-1}. Work is considered to be a force acting through a distance and is measured in ft lb. For example, if a force is used to move a 10-lb load (mass) through a distance of 100 ft, 1000 ft lb of work is accomplished. In this example, 1000 ft lb of work can be accomplished by moving a 100-lb load 10 ft or any combination of distance and load that has a product of 1000 ft lb. The time, seconds or minutes, expended when a force moves a load through a given distance is considered when determining power. Work is accomplished only if the mass is moved. *Energy*, the capacity for doing work, has the same units of measure as work.[1]

Torque is a load applied to a lever arm to produce or tend to produce rotary force. Torque is similar to work in that its English customary unit of measure is ft lb. Torque is sometimes defined simply as a turning effort.[2] Application of torque does not ensure moving the lever arm. For example, if a 100-lb load is applied to a 1-ft lever arm, 100 ft lb of torque is produced regardless of any rotation. It will produce rotation only if it is not balanced by an equivalent torque.[3] To differentiate torque from work, some texts reverse the units using lb ft for torque and ft lb for work. If torque produces rotation, work is accomplished. Torque is an important concept when defining and measuring rotary power.

HISTORY OF POWER MEASUREMENT

The problem of defining power came about when engines, especially steam engines, were introduced. James Watt developed one of the first successful steam engines in England in the latter part of the 18th century. Since most of the power at that time was produced by men and horses, he decided to compare the power of his steam engines to that of horses. He conducted a series of tests with draft horses and determined the average amount of work a draft horse could do in a certain length of time.[4] Watt observed that a horse could lift 366 lb of coal from a mineshaft at the rate of 1 ft sec^{-1} or 60 ft min^{-1}. In other words, the work accomplished was equal to 21,960 ft lb (366 lb × 60 ft). He rounded the 21,960 to 22,000 and increased this value by 50% (11,000 lb ft) to 33,000 ft lb min^{-1}, deliberately underrating his steam engines. One *horsepower* (Hp) is defined as 33,000 ft lb min^{-1} or 550 ft lb sec^{-1}. This basic unit is still used today; but because Hp was developed in England using English customary units, the term power, a nonunit-specific term, has replaced it. The SI unit of power is named W (Watt) for James Watt.

POWER MEASUREMENT—SI UNITS

Work as explained in "Introduction" is a force acting through a distance. *Force* in SI units is measured in N (Newtons) named after scientist Sir Isaac Newton. A Newton is the unit of force required to accelerate one kilogram of mass one meter per second per second. The derived unit for work in SI units is N m (Newton meter), and this combination has been given the name J (joule). As 1 J is so small the kJ (kilojoule) or MJ (megajoule) is often used as measures of work or energy.[1,3] To convert ft lb to J, multiply by 1.355818.[5]

Power is identified as W to honor James Watt and is equal to a Newton meter per second (N m sec^{-1}). One customary Hp is the equivalent of 745.7 W and 1 kW (kilowatt) is the equivalent of 1.341 Hp.[3]

Applications that involve fluid power, hydraulic power, heat exchange capacity, water power, and electrical power are also measured. These applications may use the customary power measurement units Hp, Btu min^{-1}, or Btu hr^{-1}. They may also be measured in SI units, J sec^{-1}, W, or kW. Electrical power is measured in W or kW.[5] This article will focus on mechanical power, rotary and linear measurements.

POWER MEASUREMENT TERMS

It is important to understand that the amount of power varies depending on how and where it is measured.[6] To clearly understand power measurement, additional terms must be defined.

Dynamometer—an instrument used to measure linear or rotary power. Both types determine power by measuring force, time, and the distance through which the force is moved. *Rotary* dynamometers may be

Encyclopedia of Agricultural, Food, and Biological Engineering
DOI: 10.1081/E-EAFE 120006849
Copyright © 2003 by Marcel Dekker, Inc. All rights reserved.

classified as brake or torsion. *Linear* dynamometers are called drawbar (Db) dynamometers.

- Linear—Db dynamometers measure pull and the velocity of the vehicle making the pull.
- Rotary dynamometers measure power at the flywheel, power take off (PTO) shaft, or drive wheel. Dynamometers may employ electric generators or eddy currents, air, water, or hydraulics to apply the load. All rotary absorption type dynamometers generate heat when the load is applied and this heat may be dissipated by circulating water through the dynamometer.[7]

Brake Power—power output of the crankshaft and measured at the flywheel. Brake power is also called flywheel or engine power. (The term brake power came about because the first devices for measuring power were Prony Brakes.) Brake power may be measured with the engine stripped of accessories such as alternator, fan, water pump, exhaust system, air conditioner, and hydraulic and pneumatic systems. Engines are tested on stands in the laboratory provided with systems needed by the engine to start and run.

Maximum Brake Power—the maximum power an engine will develop with the throttle fully open at a specific speed. Automobile engines are usually rated with throttles fully open at the speed (rpm) their power curve peaks. Automobile engines can maintain maximum power for only a short duration. Tractor engines are designed with governors to maintain consistent operating speeds in a range that will allow the engine to be operated at demand loads over extended time periods.

Observed Power—power without any correction for atmospheric temperature pressure (barometric) and vapor pressure.

Corrected Power—power has been adjusted (corrected) for atmospheric conditions. Engines perform better at cooler temperatures and higher barometric pressures because more oxygen can enter the combustion chamber and more fuel can be burned.

Hydraulic Power—power developed by a hydraulic pump. Pump power is calculated using flow rate and pressure.

Torque Reserve—also known as Torque Rise, the additional torque that is available beyond torque at maximum power. Torque reserve is very important in tractors in that it provides what is called lugging ability so the tractor can continue through high pulls or loads without killing the engine. It is not uncommon for tractor engines today to exhibit a 30%–40% torque rise.

Indicated Power—power generated in the engine on top of the pistons. It takes into account mean effective pressure on each cylinder, length of stroke, area on top of piston, and speed.

Friction Power—the power required to run the engine at any given speed without production of useful work. Friction power of an engine can be found by subtracting brake power from indicated power.

Fuel Equivalent Power—computed from the product of consumption rate and heating value of the fuel.

There are many other power terms used in advertising and service literature, e.g., *advertised*, *certified*, *effective*, *guaranteed*, *net engine*, and *rated*, to name a few. These terms can be confusing and are sometimes used in place of terms above. Because of this, they should be used judiciously and only if they are defined.

If the objective of measuring power is to determine internal combustion engine power efficiencies then fuel equivalent power, indicated power, friction power, and brake power, all examples of engine power, can be measured or calculated. These powers in turn can be used to ascertain engine power efficiencies such as *indicated thermal efficiency*, *brake thermal efficiency*, and *mechanical efficiency*.[8] It is beyond the scope of this article to describe these and other engine power efficiency terms.

Transmitted Power—engine power transmitted from flywheel to a point on a machine or vehicle. Axle, PTO, drawbar, and chassis are examples of transmitted power. Transmitted power may also be called *Output Power*. Transmitted powers will always be less than flywheel power because of the losses in the drive train. Of the transmitted powers identified, namely, axle, wheel, PTO, and drawbar, drawbar power will be the lowest and most variable because of losses due to the power train, tractive elements of the vehicle, and the surface it is operating on. PTO power is approximately 96% of flywheel power and Db power is approximately 86% of PTO power.[9] *ASAE D497.4 Agricultural Machinery Management Data* gives more specific ratios that take into account tractor type, 2WD, MFWD, 4WD, track (belted), and tractive conditions, concrete or firm, tilled, and soft soils.[10]

MEASURING LINEAR AND ROTARY POWER

Mechanical power may be linear or rotary. Linear power occurs when a force is exerted with a linear velocity. A tractor pulling a plow (power transmitted to the drawbar) is an example of linear power and in English units is usually defined as DbHp (drawbar horsepower) in that the force or load component of power is measured at the tractor's drawbar with a pull meter in pounds. Power transmitted to the PTO shaft is rotary power and is called PTO power. Older tractors produced a similar rotary power called *belt* (brake) horsepower. On these older tractors, the power was measured from the belt pulley because they were not equipped with a PTO shaft. Today's larger tractors may not be equipped with PTO shafts because they are used for pulling large loads such as primary tillage tools. Their power is measured at the engine flywheel and is published as flywheel power. Chassis dynamometers also measure rotary power and are most commonly used for measuring truck or automobile power at the drive wheels.

Linear Power

The following formula is convenient for measuring linear power using either SI or English units DbHp.[11,12] Because W is so small (745.7 W Hp^{-1}), kW is normally used to identify power in SI units. The constants shown in formulas below assume these units.

DbP = drawbar power expressed in kW [Hp]

DbP = FS/K

F = force measured in kN [lb]

S = forward speed, km hr^{-1} [mph]

K = units constant, 3.6 [375]

Example problem

A tractor is pulling a tillage tool that exhibits 53.3 kN [12000 lb] load at 8.05 km hr^{-1} [5 mph] speed.

Solving problem using SI units:

$$119.3\,\text{kW} = \frac{53.38\,\text{kN} \times 8.05\,\text{km hr}^{-1}}{3.6}$$

Working the same problem using customary English units:

$$160\,\text{DbHp} = \frac{12,000\,\text{lb(Pull)} \times 5\,\text{mph}}{375}$$

Solving the problem in customary units using unit factoring without the 375 conversion factor:

$$160\,\text{DbHp} = \frac{12{,}000\,\text{lb}}{\text{Pull}}\,\frac{5\,\text{mi}}{\text{hr}}\,\frac{\text{hr}}{60\,\text{min}}\,\frac{5280\,\text{ft}}{\text{mi}}\,\frac{\text{Hp}}{33{,}000\,\text{ft lb min}^{-1}}$$

Rotary Power

Rotary power is the product of work/revolution and rotary speed. Torque, an important concept when measuring rotary power, is the product of lever arm length and a force acting perpendicular to it. Rotary power may be measured with a Prony brake, a simple form of an absorption dynamometer. A Prony brake consists of an adjustable brake band or friction device that contacts a rotating member of the engine usually the flywheel. The lever arm is part of the friction device. As the adjustable friction device is tightened the load on the engine is increased. This load is sensed by a scale or load cell attached to the lever arm preventing it from rotating with the flywheel. A tachometer measures the rotary speed of the flywheel in rpm.

To measure brake or rotary power the following information is needed.

- *Rotary speed* of the flywheel or rotating member. Rotary speed is the amount of angular rotation per unit of time. The most common unit for rotary speed is rpm.
- Length of lever arm.
- Load on lever arm.

Example problem

Find the brake power (BP) in SI units (kW) and English customary units (Hp) of a small engine with a Prony brake attached.

Given:

Rotary speed—engine is turning 3000 rpm after the brake band has been adjusted to obtain force on lever arm.

Length of lever arm—lever arm is 0.3048 m [1 ft] long.

Load—force (load) 44.48222 N [10 lb] is measured by scale attached to end of lever arm.

A convenient formula to find the BP of small engine identified above using SI or English units is:

BP = brake power expressed in kW [BHp]

BP = $2\pi TN/K$

T = engine torque N m [lb-ft]

N = number of revolutions per minute, rpm

K = units constant = 60,000 [33,000]

Solving problem in SI units:

$$4.26\,\text{kW} = \frac{(2 \times 3.14) \times 44.48222\,\text{N} \times 0.3048\,\text{m} \times 3000\,\text{rpm}}{60,000}$$

Solving problem using English units:

$$5.71\,\text{BHp} = \frac{(2 \times 3.14) \times 10\,\text{lb} \times 3000\,\text{rpm}}{33,000}$$

TRACTOR TEST CODES—STANDARDS

Procedures and codes for testing the power and performance of tractors have been developed by a number of organizations including the Society of Automotive Engineers (SAE), The Society for Engineering in Agricultural, Food, and Biological Systems (ASAE), Nebraska Tractor Test Center (NTTC), and the Organization for Economic Co-operation and Development (OECD).[12–14] All these organizations test both linear and rotary powers of tractors and their codes identify standards for PTO (rotary) and drawbar (linear) power in addition to three-point hydraulic lift capacities and sound level measurements. They also record, in addition to the environmental conditions, fuel type, fuel temperature, fuel and lubricants used and consumed, and other variables that impact the performance of the tractor. When measuring Db power, all factors considered in the PTO tests are accounted for plus those that impact traction, which include, gear, tractor ballast, tires, and surface. Drawbar performance is measured on a concrete track for consistency.

Because tractor drawbar performances are determined on concrete, Frank Zoz, retired John Deere Product Engineer has developed *Predicting Field Performance*, an "Excel" template to predict tractive performance and efficiency of tractors on dirt using the primary inputs, PTO power, ground speed, ballast, tractor dimensions, implement hitch, tire size, and pressure configuration that are used in the SAE, NTTC & ASAE official tests. Tractor performance including pull, drive wheel slip, and forward speed are predicted on soils with different penetrometer readings. The spreadsheet predicts performance at given weights or can be used to calculate required tractor ballasting, front and rear.[15]

TRACTOR POWER EFFICIENCIES

Tractor power efficiencies consider drive train and traction elements. The efficiency of transmitting engine, PTO or axle (input) power to the tractor's drawbar (output) power may be identified as Power Delivery Efficiency (PDE) or Tractive Efficiency (TE). PDE is the ratio of Db to flywheel or Db/PTO. TE is the ratio of drawbar to axle power, Db/Ax.[15,16] (Since PTO is approximately 96% of flywheel it can be used for flywheel in determining PDE.) PDE is more meaningful than TE when comparing the power efficiencies of tractors because the entire drive train is taken into account. For example, when comparing the power efficiency of a wheel tractor to a belted (track) tractor, the effects of the drive train, clutches, transmission, as well as the tractive elements are included in PDE but not TE.

REFERENCES

1. Goering, C.E. Basic Thermodynamics of Engines. *Engines & Tractor Power*, 3rd Ed.; ASAE Textbook Number 3, American Society of Agricultural Engineers: St. Joseph, MI, 1992; 22–23.
2. Parady, W.H. Understanding Power. *Understanding & Measuring Power, Motors, Engines, Automobiles, Trucks, Tractors, SI (Metric) Terms Included*, 2nd Ed.; American Association for Vocational Instructional Materials (AAVIM): Athens, GA, 1978; 11–12.
3. Hunt, D. Power Performance. *Farm Power and Machinery Management*, 9th Ed.; Iowa State University Press: Ames, IA, 1995; 26–30.
4. Parady, W.H. Understanding Power. *Understanding & Measuring Power, Motors, Engines, Automobiles, Trucks, Tractors, SI (Metric) Terms Included*, 2nd Ed.; American Association for Vocational Instructional Materials (AAVIM): Athens, GA, 1978; 14–15.
5. ASAE Engineering Practice ASAE EP285.7. Use of SI (Metric) Units. *ASAE Standards*, 35th Ed.; ASAE—The Society for Engineering in Agriculture, Food and Biological Systems: St. Joseph, MI, 1988; 12–19.
6. Goering, C.E. Power Efficiencies and Measurements. *Engines & Tractor Power*, 3rd Ed.; ASAE Textbook Number 3; American Society of Agricultural Engineers: St. Joseph, MI, 1992; 92.
7. Liljedahl, J.B.; Turnquist, P.K.; Smith, D.W.; Hoki, M. Tractor Tests and Performance. *Tractors and Their Power Units*, 4th Ed.; AVI, Van Nostrand Reinhold: New York, NY, 1989; 404–415.
8. Goering, C.E. Power Efficiencies and Measurement. *Engines & Tractor Power*, 3rd Ed.; ASAE Textbook Number 3; The Society for Engineering in Agriculture, Food and Biological Systems: St. Joseph, MI, 1992; 99–104.
9. Roth, L.O.; Field, H. Tractors and Power Units. *Introduction to Agricultural Engineering*, 2nd Ed.; Chapman & Hall: New York, 1991; 63–74.
10. http://www.asae.frymulti.com/standards.asp (accessed January 2001).
11. Goering, C.E. Power Efficiencies and Measurement. *Engines & Tractor Power*, 3rd Ed.; ASAE Textbook Number 3; The Society for Engineering in Agriculture, Food and Biological Systems: St. Joseph, MI, 1992; 91–96.
12. www.ianr.unl.edu/ianr/bse/ttl/ (accessed January 2001).
13. Roth, L.O.; Field, H. Tractors and Power Units. *Introduction to Agricultural Engineering*, 2nd Ed.; Chapman & Hall: New York, 1991; 63–76.
14. Parady, W.H. Determining the Size of Power Unit to Use. *Understanding & Measuring Power, Motors, Engines, Automobiles, Trucks, Tractors, SI (Metric) Terms Included*, 2nd Ed.; American Association for Vocational Instructional Materials (AAVIM): Athens, GA, 1978; 22–24.
15. Zoz, F.M. Predicting Tractor Field Performance. Computer Spreadsheet Template. Waterloo, IA, 1999.
16. Shell, L.R.; Zoz, F.M.; Turner, R. Field Performance of Rubber Belt and MFWD Tractors in Texas Soils. *Belt and Tire Traction in Agricultural Vehicles, SAE SP-1291*; S.A.E. Paper 972729, Society of Automotive Engineers, Inc.: Warrendale, PA, 1997; 65–73.

Prandtl Number

H. S. Ramaswamy
McGill University, Ste Anne de Bellevue, Quebec, Canada

G. B. Awuah
National Food Processors Association, Washington, District of Columbia, U.S.A.

C. R. Chen
McGill University, Ste Anne de Bellevue, Quebec, Canada

P

INTRODUCTION

A perfect fluid by definition is one that is inviscid and whose density is constant. Such fluids are ideal and indicate the absence of shear stresses between the fluid layers and therefore, cannot be found in nature.[1] Therefore, two adjacent layers of an ideal fluid can move at different velocities (slip flow) without any internal frictional forces. Under certain conditions, however, such fluids provide valuable information when analyzing situations involving real fluids. Real fluids move with a finite velocity difference over adjacent layers or solid boundaries. The viscosity of the fluid is responsible for the gradual velocity profile across the layers. In real fluids, the effect of viscosity starts at the leading edge of the solid body and increases downstream. In the proximity of a stationary boundary, the velocity of a real fluid must start at zero at the boundary to a finite stream velocity in the fluid layer. Ludwig Prandtl, a German teacher and scientist, in 1904, conceived the idea of a boundary layer that adjoins the surface of a body moving through a fluid. The treatment of the boundary layer near the solid boundary as a viscous fluid and the flow outside the boundary layer as a frictionless fluid was one of Prandtl's significant contribution and perhaps the greatest single discovery in the history of fluid mechanics.[1,2] Prandtl's hypothesis on fluid flow around an object reconciled two contradictory facts.[1] He supported the idea that the effect of viscosity is negligible in most flow fields if the momentum diffusivity or kinematic viscosity (μ/ρ) is small. At the same time, he accounted for drag by insisting that the no-slip condition must be satisfied at the wall, no matter how small the velocity is.

Heat transfer problems involving fluid flow require simultaneous solution of the Navier–Stokes equation of motion and continuity, and the energy equations for four dependent variables including 2-D velocity components, pressure, and temperature. It is sufficiently complex to integrate the governing differential equation describing a flow situation. However, one can use such information to find out which dimensionless numbers can be used to correlate experimental data. Again, Prandtl demonstrated how the equations of motion within the boundary layer could be simplified. In recognition of his contributions to fluid dynamics, a dimensionless parameter, which describes the effect of fluid thermophysical properties on heat transfer, and which traditionally is combined with other dimensionless parameters for evaluating fluid-to-surface interfacial heat transfer coefficient, was named after Ludwig Prandtl and was acronymed *Pr*. In this article, the role of Prandtl number in heat transfer applications and some of the associated correlations are discussed.

PRANDTL NUMBER (*Pr*)

Prandtl number (*Pr*) represents the ratio of the shear component of diffusivity for momentum (μ/ρ) to the diffusivity for heat ($k/\rho C_p$). In simple terms, *Pr* represents the ratio of kinematic viscosity (ν) to thermal diffusivity (α) of a fluid. Mathematically, the Prandtl number may be expressed as follows for Newtonian fluids:

$$Pr = \frac{\mu C_p}{k} = \frac{\mu/\rho}{k/\rho C_p} = \frac{\nu}{\alpha} \tag{1}$$

For non-Newtonian fluids flowing through circular tubes, five different forms of the Prandtl number corresponding to five definitions of the Reynolds number have been used by researchers[3]: 1) generalized Reynolds number (*GRe*); 2) Reynolds number based on the apparent viscosity at the wall, Re_a; 3) Reynolds number (Re_{gen}), derived from the nondimensional momentum equation; 4) Reynolds number based on the solvent viscosity (Re_s); and 5) Reynolds number (Re_{eff}) based on the effective viscosity. For experimental or analytical studies on drag and heat transfer for non-Newtonian fluids under laminar flow

DOI: 10.1081/E-EAFE 120006999
Copyright © 2003 by Marcel Dekker, Inc. All rights reserved.

conditions, GRe and Re_a with their corresponding Prandtl numbers are recommended, while the Re_a, Pr_a combination is more practical for turbulent pipe flow, as it allows comparison of experimental data with analytical predictions.[3] The generalized Prandtl number (GPr) can be expressed as follows to conform with Eq. 1:

$$GPr = \left[\frac{C_p m \left[(3n+1)/n \right]^n 2^{n-3} (D/V)^{1-n}}{k_f} \right] \tag{2}$$

where n, m(Pa sec^n), D (m), and V (m sec^{-1}) are the fluid behavior index, consistency coefficient, tube diameter, and average velocity, respectively. The average velocity $V = 4Q/(\pi D^2)$, while Q (m^3 sec^{-1}) represents the volumetric through flow rate. For conventional canning of foods in rotary autoclaves, the apparent viscosity has been calculated at a shear rate commensurate with the rotation speed of the autoclave,[4,5] with the diameter of rotation representing the characteristic dimension.

Physically, Prandtl number relates the relative thickness of the hydrodynamic layer to the thermal boundary layer and represents a combination of fluid thermophysical properties. Therefore, Prandtl number may be thought of as a property.[6] It quantifies the influence of fluid physical properties during the transfer of heat involving fluids, and plays an important role by way of contributing to mathematical formulations that describe the transfer of heat involving fluid flow in industrial unit operations. For instance, large Pr values (i.e., for small thermal conductivity) associated with unit operations imply a relatively small rate of heat transfer. For liquid metals, Pr is less than one, and greater than one for oils. For air, Pr is approximately 0.7. Values for other substances can be found in Refs. 6 and 7.

THE BOUNDARY LAYER CONCEPT AND PRANDTL NUMBER

The effect of fluid friction at high Reynolds number, as hypothesized by Prandtl, is limited to a thin layer near the boundary of a body. This idea gave birth to the term boundary layer. In addition, there is no significant pressure change across the boundary layer. This means that pressure across the boundary is the same for inviscid flow outside the boundary layer.[6] These findings simplified analytical solutions for viscous flows, as the pressure can be determined from experiments or the inviscid flow theory.

Analytical estimates and experiments have indicated that adequate representation of physical reality is contained in the assumptions that the fluid at the surface loses all its momentum relative to the surface, resulting in the "no-slip" approximation at the fluid/surface interface. Therefore, the velocity and temperature profiles resemble those depicted in Fig. 1 where u and y are the local velocity and distance normal to the surface, respectively. The no-slip (zero surface velocity) condition led to the notion that a surface is a sink for fluid momentum.[8] The thickness of the boundary layer, δ, is arbitrarily taken to represent the distance away from the surface where the velocity reaches 99% of the free-stream velocity (U_e). The velocity (u) changes smoothly from 0 to U_e due to the presence of fluid friction in sublayers just below and above a given point of interest. Similarly, the temperature profile will change from T_w at the wall to $T = T_e$ at the edge of the so-called thermal boundary layer (δ_T). The thermal boundary layer is defined as the point where $T - T_w = 0.99(T_e - T_w)$. It is imperative to note that thermal and velocity boundary layers are not necessarily equal. Therefore, the thermal boundary layer thickness δ_T may be thicker or thinner than the velocity boundary layer (δ), depending on the Prandtl number. Pohlhausen[7] demonstrated that the relationship between the hydrodynamic and thermal boundary layers for fluid with $Pr > 0.6$ gives approximately $\delta = \delta_T Pr^{1/3}$ for laminar flow past a flat plate. As a result, the average heat transfer coefficient (h) over a flat surface could be estimated from the relationship[7]:

$$h = 0.644 \frac{k}{L} Re^{1/2} Pr^{1/3} \tag{3}$$

where L is the total length of the surface, with properties estimated at film temperature for Re of up to 5×10^5. Surprisingly, most common gases have a Prandtl number of approximately 0.7. This means that heat will diffuse faster than momentum for gases. Therefore, all other things being equal, the thermal boundary layer in gas flow is more likely to be thicker than the velocity boundary

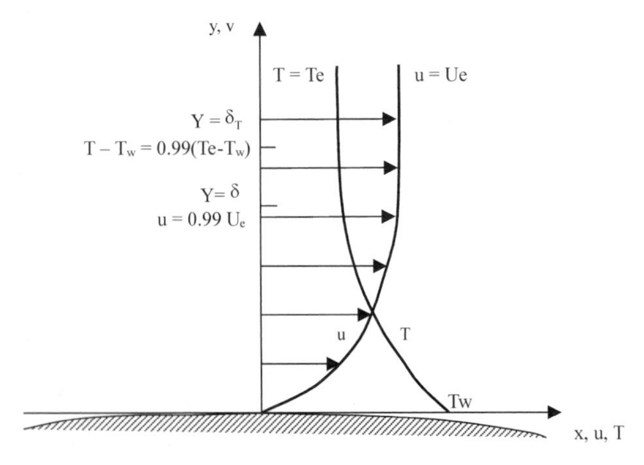

Fig. 1 Typical velocity and temperature profiles in a boundary layer showing the definition of velocity and temperature boundary layer thickness.

layer in a ratio of approximately 1: 0.7.[8] For water and oil, Pr is of the order 10.

Different fluids exhibit different properties, and the Prandtl number somewhat plays an important role in heat transfer applications in the sense that correlations developed using a particular fluid may not be applicable for other applications if the Prandtl number falls outside the applicable range for the reference fluid. As Pr for gases (~ 0.7) are very close to unity, a special case of $Pr = 1$ has been used for mathematical modeling of low and high-speed flow to understand the effect of fluid velocity on heat transfer coefficient under zero pressure gradient.[8] Such special cases are not generally found in the real world. However, for low-speed flow with $Pr = 1$, the nondimensional profiles for velocity and temperature are identical. This means that the transfer of momentum and heat are directly analogous, and the boundary layer thickness for velocity and thermal profiles are equal.

MATHEMATICAL MODEL FOR HEAT TRANSFER AND PRANDTL NUMBER DEPENDENCE

The governing conservation equations for modeling heat transfer and Prandtl number dependence include[9,10]:

1. The classical incompressible Navier–Stokes equation:

$$\frac{\partial(\rho U)}{\partial t} + \nabla\cdot(\rho U \otimes U) = \rho g + \nabla\cdot(-P\delta + \mu \nabla U) \qquad (4)$$

2. The energy equation:

$$\frac{\partial(\rho h)}{\partial t} + \nabla\cdot(\rho U h)) - \nabla\cdot(k\nabla T) = \frac{\partial P}{\partial t} + q_v \qquad (5)$$

3. The continuity equation:

$$\frac{\partial \rho}{\partial t} + \nabla\cdot(\rho U) = 0 \qquad (6)$$

To investigate the effect of Pr on heat transfer in internally heated liquid pools with Ra up to 10^{12}, solutions for time-independent flow, temperature fields, and heat transfer characteristics were averaged over time to produce quasi-steady values.[9] By introducing a dimensionless temperature, time, pressure, velocity, and coordinates (x,y,z), the conservation equations were simplified and written in dimensionless forms as follows:

$$\nabla\cdot(\Psi) = 0 \qquad (7)$$

$$\frac{\partial \Psi}{\partial \mathrm{Fo}} + \nabla\cdot(\Psi \otimes \Psi) = \frac{Ra \cos(\Psi, g)\cdot\theta}{Pr} + \nabla\cdot(-\xi\delta + \nabla\Psi) \qquad (8)$$

$$\frac{\partial \theta}{\partial \mathrm{Fo}} + \nabla\cdot(\Psi\cdot\theta)\frac{\nabla\cdot(\nabla\theta)}{Pr} = \frac{1}{Pr} \qquad (9)$$

with boundary conditions as follows for semicircular and hemispherical geometries[9]:

$$\Psi|_{\text{all walls}} = 0; \qquad \theta|_{\text{top, bottom side: walls}} = 0; \qquad \frac{\partial\theta}{\partial Z}\Big|_{\text{face, back walls}} = 0 \qquad (10)$$

With the appropriate boundary conditions and numerical approximations, solutions to Eqs. 4–9 can provide useful information regarding the effect of the Prandtl number on fluid flow dynamics and heat transfer for simple and complex geometries under practical conditions.

Most heat transfer data for fluid flow through conduits have been correlated empirically with all property values evaluated at bulk fluid temperature. To correlate variables and their influence on h, the Reynolds (Re), Grashof (Gr), and Prandtl (Pr) numbers are grouped and related to the Nusselt number (Nu) as follows:

$$Nu = \frac{hD}{k} = f(Re, Pr, Gr) \qquad (11)$$

In reality however, fluid temperature varies across the cross-section of conduits, and there is often the need to refine Eq. 11 by way of additional parameters to improve correlations. In addition, fluid velocity distribution for laminar flow will likely deviate from the parabolic profile.[11] The most commonly used factor is the dimensionless viscosity ratio (μ_b/μ_w), where μ_b and μ_w represents the viscosity at fluid bulk and wall temperatures, respectively. Another option is to evaluate fluid properties at the film temperature, which is the arithmetic mean between the bulk and wall temperatures. For sharp changes of fluid physical properties in the boundary layer, Mikheev[12] suggested the use of the ratio (Pr/Pr_w) as a correlating parameter. For viscous fluids, the ratio Pr/Pr_w is the same as the viscosity ratio (μ_b/μ_w) as the property showing significant variation is the viscosity.[13] According to the authors, the ratio Pr/Pr_w is more reasonable because heat transfer is governed by temperature fields, which are described by the Prandtl number.

Table 1 Dimensionless correlations involving the Prandtl number

Application	Product/shape	Correlation	*Pr*	*Re*	Reference
Retorting axial rotation	Lead sphere in water/sucrose	$Nu = -33 + 53Re^{0.28}Pr^{0.14}D_p/R_r(1-\varepsilon)^{0.46}$	—	—	14
Retorting still	Whole mush room	$Nu = 0.01561(Gr\,Pr)^{0.529}$	—	—	15
Retorting end-over-end	Single sphere, cube, or cylinder in water/oil	$Nu = 0.93Re^{0.51}Pr^{0.36}(h_s/H)^{0.21}$	2.6–90.7	520–5.4 × 10^5	16
Retorting end-over-end		$Nu = (17 \times 10^5)Re^{1.449}Pr^{1.19}We^{-0.551} \times (D_c/2H)^{0.932}(V_h/V_c)^{0.628}$	—	—	17
Retorting end-over-end	Non-Newtonian Guar gum	$Nu = 1.41GRe^{0.482}GPr^{0.355}$	48–57,000	0.1–12,000	18
Retorting axial rotation		$Nu = 2.7 \times 10^4 Re^{0.294}Pr^{0.33}\phi^{6.98}$	—	—	19
Retorting axial rotation	Silicone oil	$Nu = 0.434Re^{0.571}Pr^{0.278}(L/D)^{0.3565}(\mu_b/\mu_w)^{0.154}$	2.2–2300	12–44,000	20
Oscillatory motion	Nylon sphere in water/CMC	$Nu = 2 + 4.45E^{-2}GRe^{0.555}GPr^{-0.294}GGr^{-0.352} (d/D)^{1628}(V_p/V_f)^{-1.748}(V_p/V_s)^{-0.533}(\alpha_f/\alpha_p)^{0.227}$	2.5–3077	2.9–7660	21
Continuous tube flow	Mushroom-shaped aluminum casting in non-Newtonian fluid	$Nu = 2 + 7.9464Re^{0.208}Pr^{0.144}$	(1.048–8.364) × 10^4	0.111–2.001	22
		$Nu = 22.7 + 0.036Re^{0.45}Pr^{0.53}$			
Continuous tube flow	Teflon spheres in CMC solutions	$Nu = 2 + 0.07GRe^{0.62}GPr^{0.53}(d/D)^{0.16}$	1.8–332	7.7–2693	23
Continuous tube flow	Silicone cube in starch	$Nu = 2 + 2.8 \times 10^{-2}Re^{1.6}Pr^{0.89}$	9.47–376.18	1.23–27.38	24

PRANDTL NUMBER IN CORRELATIONS FOR ESTIMATING HEAT TRANSFER

Although an exhaustive review of dimensionless correlations involving Prandtl number will not be presented here, Table 1 attempts to present a spectrum of correlations that have been developed to address specific problems related to the food processing industry. It is evident from Table 1 that *Pr* range depends on the type of application and most importantly, the type of fluid used to model and/or reflect the problem under consideration. Unfortunately, some researchers have presented correlations without explicitly quoting the applicable range for the Prandtl number. In addition, most correlations place more emphasis on the Reynolds number, which obviously is very critical in establishing fluid flow regimes, with little or no recognition of the effect of the Prandtl number in developed models. However, the Prandtl number plays an equally important role by defining the relationship between diffusivities for momentum and that for heat in energy transfer applications. It is prudent to emphasize among other things, the applicable range for all dimensionless parameters used in developed expressions that establish the dimensionless heat transfer coefficient (Nusselt number).

NOMENCLATURE

A_p	Surface area of particle (m^2)
A_{sph}	Surface area of an equivalent sphere (m^2)
C_p	Heat capacity ($kJ\,kg^{-1}\,C^{-1}$)
D_c	Diameter of can (m)
d, D	Diameter of particle; inside tube diameter (m)
Fo	Dimensionless time

GGr	Generalized Grashof number $[g\beta\rho^2 \Delta T D^3]/[m\{(3n+1)/n\}^n 2^{n-3}(D/V)^{1-n}\}]^2$
GPr	Generalized Prandtl number $[C_p\{m[(3n+1)/n]^n 2^{n-3}(D/V)^{1-n}\}/k]$
Gr	Grashof number $(L^3\rho^2 g\beta\Delta T/\mu^2)$.
Gre	Generalized Reynold number $[\rho VD/\{m[(3n+1)/n]^n 2^{n-3}(D/V)^{1-n}\}]$
g	Gravitational acceleration $(m^2 sec^{-1})$.
H	Height of can (m)
h	Heat transfer coefficient $(W m^{-2} C^{-1})$; total enthalpy $(J kg^{-1})$ in Eq. 5
h_s	Can headspace (m)
K'	Constant in power law model defined as: $m[(3n+1)/4n]^n$
k	Fluid thermal conductivity $(W m^{-1} C^{-1})$
L	Length (m)
m	Consistency coefficient $(Pa{\cdot}s^n)$
Nu	Nusselt number (hD/k)
n	Flow behaviour index
P	Pressure (Pa)
Pr	Prandtl number $(\mu C_p/k)$
Q	Volumetric flow rate $(m^3 s^{-1})$
q_v	Volumetric heat generation rate $(W m^{-3})$
Re	Reynold number $(\rho VD/\mu)$
Ra	Rayleigh number $(Gr.Pr)$
R_r	Radius of the agitated retort (m)
T	Temperature (°C)
t	Time (s)
U	Velocity vector $(m s^{-1})$
V	Mean velocity $(m s^{-1})$; volume (m^3)
V_c	Volume of can (m^3)
V_f	Fluid velocity $(m sec^{-1})$
V_h	Volume of headspace (m^3)
V_p	Particle velocity $(m sec^{-1})$
V_s	Slip velocity $(m sec^{-1})$
We	Weber number $(\omega^2 H^2 \pi D_c/\sigma)$
Z	Dimensionless length

Greek letters

β	Volumetric coefficient of expansion (1/K)
ρ	Density $(kg m^{-3})$
μ	Dynamic viscosity (Pa·sec)
ν	Kinematic viscosity $(m^2 sec^{-1})$
δ	Kroenecker's delta
θ	Dimensionless temperature
σ	Surface tension $(kg sec^{-2})$
α_p	Thermal diffusivity of particle $(m^2 sec^{-1})$
α_f	Thermal diffusivity of fluid $(m^2 sec^{-1})$
ε	Particle concentration (%) (volume of particle(s)/total volume of can)
Ψ	Sphericity of particle (A_{sph}/A_p); Dimensionless velocity
$\otimes$	Tensor product, defined as $(A\otimes B)_{ij} = (A_i B_j)$
ξ	Dimensionless pressure
ω	Can angular velocity $(2\pi N/60)$, (sec^{-1})

Subscript

b	bulk fluid
f	Fluid or subscript representing constant
p	Particle
w	Wall

REFERENCES

1. Eskinazi, S. *Principles of Fluid Mechanics*; Allyn and Bacon, Inc.: Boston, 1962.
2. Kundu, P.K. *Fluid Mechanics*; Academic Press Inc.: New York, 1990.
3. Cho, Y.I.; Hartnett, J.P. Non-Newtonian Fluids. In *Handbook of Heat Transfer Applications*; Rohsenow, W.M., Hartnett, J.P., Ganic, E.N., Eds.; McGraw-Hill Book Co.: New York, 1985.
4. Rao, M.A.; Anantheswaran, R.C. Convective Heat Transfer in Fluid Foods in Cans. Adv. Food Res. **1988**, *32*, 39–84.
5. Ramaswamy, H.S.; Abbartemarco, C.; Sablani, S.S. Heat Transfer Rates in a Canned Food Model as Influenced by Processing in an End-Over-End Rotary Steam/Air Retort. J. Food Process Preserv. **1993**, *17*, 269–286.
6. Welty, J.R.; Wicks, C.E.; Wilson, R.E. *Fundamentals of Momentum, Heat and Mass Transfer*; John Wiley & Sons: New York, 1969.
7. Geankoplis, C.J. *Transport Processes and Unit Operations*, 3rd Ed.; Prentice Hall Inc.: New Jersey, 1993.
8. Schetz, J.A. *Foundations of Boundary Layer Theory: For Momentum, Heat and Mass Transfer*; Prentice-Hall, Inc.: Englewood Cliffs, NJ, 1984.
9. Nourgaliev, R.R.; Dinh, T.N.; Sehgal, B.R. Effect of Fluid Prandtl Number on Heat Transfer Characteristic in Internally Heated Liquid Pools with Rayleigh Numbers up to 10^{12}. Nucl. Eng. Design **1997**, *169*, 165–184.
10. Kerr, R.M.; Herring, J.R. Prandtl Number Dependence of Nusselt Number in Direct Numerical Simulations. J. Fluid Mech. **2000**, *419*, 325–344.
11. Kay, J.M.; Nedderman, R.M. *An Introduction to Fluid Mechanics and Heat Transfer: With Applications in Chemical & Mechanical Process Engineering*; Cambridge University Press: Cambridge, 1974.
12. Mikheev, M.A. Teplootdacha pri Turbulentnom Dvizhenii Zhidkosti v Trubkakh (Heat Transfer in Turbulent Pipe Flows). Izv. Akad. Nauk SSR (Otdel Tekhnicheskikh Nauk) **1952**, *10*, 128–139.
13. Zukauskas, A.; Ziugzda, J. *Heat Transfer of a Cylinder in Crossflow*; Hemisphere Pub. Co.: New York, 1985.
14. Lenz, M.K.; Lund, D.B. The Lethality-Fourier Number Method. Heating Rate Variations and Lethality Confidence Intervals for Forced-Convection Heated Foods in Containers. J. Food Process Eng. **1978**, *2*, 227–271.

15. Sastry, S.K. *Convective Heat Transfer Coefficient for Canned Mushrooms Processing in Still Retorts*; ASAE Paper 84-6517, ASA: St. Joseph, MI, 1984.
16. Sablani, S.S.; Ramaswamy, H.S.; Mujumdar, A.S. Dimensionless Correlations for Convective Heat Transfer Coefficients in Cans with End-Over-End Rotation. J. Food Eng. **1997**, *34*, 453–472.
17. Duquenoy, A. Heat Transfer to Canned Liquids. In *Food Process Engineering, Vol 1: Food Processing Systems*; Linko, P., et al., Eds.; Applied Science Publishers Ltd: London, 1980; vol 1, 483–489.
18. Anantheswaran, R.C.; Rao, M.A. Heat Transfer to Model Non-Newtonian Liquid Foods in Cans During End-Over-End Rotation. J. Food Eng. **1985**, *4*, 21–35.
19. Fernandez, C.L.; Rao, M.A.; Rajavasireddi, S.P.; Sastry, S.K. Particulate Heat Transfer to Canned Snap Beans in Steritort. J. Food Process Eng. **1988**, *10*, 183–198.
20. Soule, C.L.; Merson, R.L. Heat Transfer Coefficient to Newtonian Liquids in Axially Rotated Cans. J. Food Process Eng. **1985**, *8*, 33–46.
21. Ramaswamy, H.S.; Zareifard, M.R. Dimensionless Correlations for Forced Convection Heat Transfer to Spherical Particles Under Tube-Flow Heating Conditions. In *Transport Phenomena in Food Processing*; Welti-Chanes, J., Velez-Ruiz, J.F., Barbosa-Canova, G., Eds.; Technomic Publishing Co.: Lancaster, PA, 2000.
22. Zuritz, C.A.; McCoy, S.; Sastry, S.K. Convective Heat Transfer Coefficients for Irregular Particles Immersed in Non-Newtonian Fluid During Tube Flow. J. Food Eng. **1990**, *11*, 159–174.
23. Awuah, G.B.; Ramaswamy, H.S. Dimensionless Correlations for Mixed and Forced Convection Heat Transfer to Spherical and Finite Cylindrical Particles in an Aseptic Processing Holding Tube Simulator. J. Food Process Eng. **1996**, *19*, 241–267.
24. Chandarana, D.I.; Gavin, A., III.; Wheaton, F.W. Particle/Fluid Interface Heat Transfer Under UHT Conditions at Low Particle/Fluid Relative Velocities. J. Food Process Eng. **1990**, *13*, 191–206.

Primary Packaging

Manjeet S. Chinnan
Dong S. Cha
The University of Georgia, Griffin, Georgia, U.S.A.

P

INTRODUCTION

A food package is a structure designed to contain a product for easier and safer transport, protect the product against contamination and damage, and provide a convenient means for dispensation. An understanding of the product to be packaged is vital for finding the perfect match among the wide array of materials being employed these days. Steel and aluminum cans are used to pack carbonated drinks while most noncarbonated drinks are packed in polyvinyl chloride (PVC) bottles. Wines and most carbonated beverages are packed in glass bottles. All dairy products, except milk, are packed in either polystyrene (PS) or polypropylene (PP) packaging.

Details of various packaging materials and packages are given below, and illustrations of selected packages and package concepts are presented in Fig. 1.

GLASS

Glass is the oldest and the least expensive of all packaging materials. Furthermore, glass has excellent vertical compressive strength and is a perfect barrier material against gas, water vapor, micro-organisms, odors, etc.[1] Marketers and consumers often regard the transparency of glass as a desirable property. However, technologists may view the transparency as less than desirable because visible and UV radiations accelerate biochemical reactions. Plastic materials are displacing glass in industrialized societies because glass is energy intensive to produce and heavy and vulnerable to impact and vibration.

Glass Bottles and Jars

The glass bottles used in packaging vary greatly in volume, weight, and shape. This variety is the result of marketing considerations and strength requirements. We can classify bottles into various types based on the specific product application. The type of bottle used to pack soft drinks and milk is somewhat different from the type often used to pack beer. Yet another type of bottle is used in packaging nonliquids such as jelly and vegetables.

PAPER

Paper is the most widely used packaging material in the world. It is generally termed board when its substance exceeds 250 μ. Technically speaking the protective properties of paper are almost nonexistent and its usefulness is almost solely as decoration and dust cover. Paperboard can be an effective structural material to safeguard contents against impact, compression, and vibration. When coated with plastic, paper or paperboard can provide protection against other environmental variables such as moisture. Despite their long history as packaging materials, paper is only infrequently used as protective packaging against moisture, gas, odors, or micro-organisms.

Boxes

Boxes may be categorized into three kinds based on the wide variety of food or nonfood products being packed. The first kind is solely made out of cardboard and represents the average cardboard package used for many food and nonfood products. The second kind has the same characteristics but contains an inner PE bag. Inner bags or liners are often used to keep foodstuffs fresh. The third kind represents blister packaging, which is often used to pack small nonfood products and provides protection against mechanical damage from vibrations during transportation and handling.

Liquid Board Package

Cardboard as a packaging material for liquids has been used for several decades. The "Tetra Classic" was introduced as early as 1952.[2] The most important markets for liquid carton board are the milk and juice industries and to a lesser extent wine, water, and soup.[2] In order to hold liquids, liquid board is laminated with other materials such as PE and aluminum. For example, Tetra Pak's Tetra-Briks used for juice packaging contain 75% cardboard, 20% PE, and 5% aluminum.[3] Cardboard is used as the middle layer with PE, aluminum layer on the inside and PE alone on the outside.

Encyclopedia of Agricultural, Food, and Biological Engineering
DOI: 10.1081/E-EAFE 120007134
Copyright © 2003 by Marcel Dekker, Inc. All rights reserved.

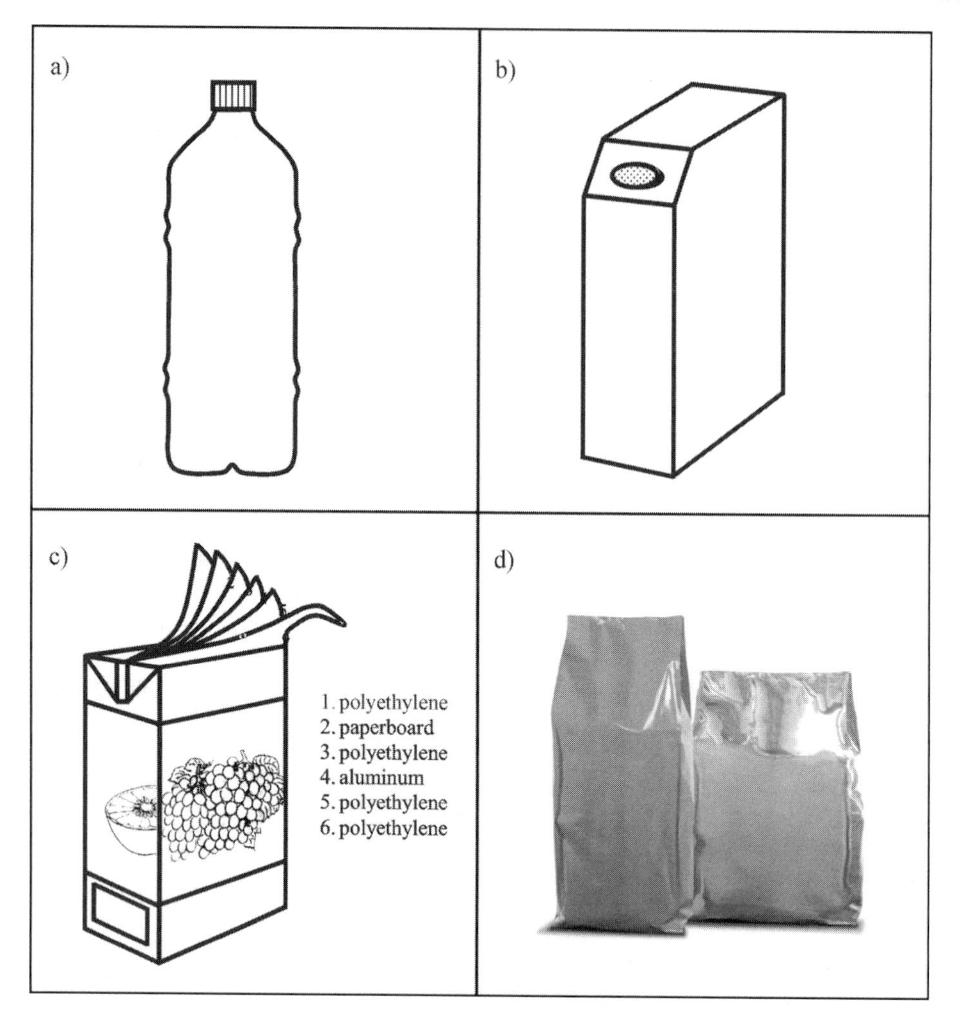

Fig. 1 Illustration of various packages: a)PET plastic bottle for beverages; b) paper board carton with plastic insert [a) and b) from Resurreccion, A.V.A.; Chinnan, M.S.; Erickson, M.C.; Hashim, I.B.; Balasubramaniam, V.; Mallikarjunan, P.; Liao, J.-Y. Consumer-Based Approach in Developing Alternative Packaging Systems for Fluid Milk for the Elderly. In *Packaging Year Book, 1996*; Blakistone, B., Ed.; National Food Processors Association: Washington, D.C., 1997; 16–37]; c) multiple layer package for aseptic packaging of beverages [http://www.aseptic.org/Award%20Main.htm (accessed October 2002)]; and d) stand-up pouches [http://www.packaging-technology.com/contractors/containers/elagverpac/elagverpac4.html (accessed October 2002)]. (From Chinnan and Cha.)

METALS

Steel and Aluminum Cans

Steel, tin, aluminum, and chromium are commonly used packaging materials for foods. Steel is rigid, which makes it the perfect microbial, gas, and water vapor barrier that resists every temperature to which a food may be subjected. It features traditionally in cans and glass bottle closures, but is subject to corrosion in the presence of air and moisture and so is almost always protected by other materials. Until the 1980s, the most widely used steel protection was tin, which also acted as a base for lead soldering of the side seams of tin cans. When lead was declared toxic and removed from cans during the 1980s in the United States, tin was also found to be superfluous and its use as steel can liner declined.[4] In almost every instance, organic coatings on steel such as vinyl and epoxies provide the principal protection. Aluminum is lighter in weight than steel and easier to fabricate; it has therefore become the metal of choice for beverage containers. As with steel, aluminum must be coated with plastic to safeguard against corrosion. Aluminum cans must have internal pressure from CO_2 or N_2 to maintain their structure, and so it is not widely used for food canning applications in which internal vacuum and pressure fluctuate as a result of retorting.

There is a strong competition between steel and aluminum for beverage cans. Almost all lids of beverage cans are made out of aluminum while 50% of the body structure of all cans is made out of steel and the rest out of aluminum. For food cans, the current situation calls for improvement. The way to improve the cans is to make them lighter. Many developments, currently underway, aim at reducing the weight of steel beverage cans to save material costs. In the last decade, the weight of steel beverage cans has been reduced by 20%. The lighter version of aluminum and steel cans is expected to replace the current models completely. Furthermore, there are some emerging technological developments that will influence the weight of food cans.

PLASTICS

The term "plastics" describes a number of families of polymeric materials, each with different properties. Most plastics are not suitable as packaging materials because they are either too expensive or considered inappropriate for toxic food. Various kinds of plastics may be combined or used with other materials to deliver the desired properties.

Flexible Packaging

Type of material used in flexible packaging (films and bags) depends largely on the products that are packed. For many food products, barrier properties for moisture and gasses, especially oxygen and carbon dioxide, play a crucial role. Other products may not need high barrier properties. High barrier films are typically multilayer films, either coextruded laminates or coated films. Typical laminates consist of a supporting/carrying layer made from PP and a super thin barrier layer made from polyvinylidene chloride (PVDC) to provide excellent barrier properties. The thickness of the supporting layer may be reduced using PET. The use of PP metallocene may also provide a thinner carrying layer. Generally, coated films consist of a carrying layer made from PP and PET and a coating of aluminum or silicon oxide. Coated films are an improvement compared to laminates, as the barrier layer is extremely thin, leading to a low weight package. These super thin films can be further improved by the substitution of PP with PET and by the use of PP metallocene.

Polyethylene Terephthalate (PET) Bottles

PET bottles were introduced in the soft drinks sector to replace glass bottles. PET bottles are especially suited to pack carbonated soft drinks. PET bottles have also replaced PVC bottles that are often used in the packaging of mineral water. In Europe, 50% of the PET packaging is used for soft drinks, 27% for pack mineral water, and 5% for other beverages. The rest (18%) is used for other purposes like food and nonfood packaging.[5] Most PET bottles used are one-way (nonreturnable) containers. An improvement in PET bottles has led to the designing of refillable bottles. This development was made possible by the emergence of new types of PET bottles that can be cleaned at temperatures up to 75°C. The refillable PET bottles are designed to last 25 trips during a lifespan of 4 yr. Many bottles, however, make fewer trips because of damage during the refill process (scuffing). PET bottles are normally made out of virgin PET.[5]

Polystyrene and Polypropylene Cups

PS and PP cups, made from thermoformed sheets are used in the liquid food market to pack yogurt and butter. PP cups are lighter than PS cups.

Pouch and Polycarbonate (PC) Bottles

Besides improvements in traditional packaging, some new packages have been developed for the beverage sector. Both Tetra Pak and Elopak introduced the plastic pouch for the packaging of milk and juice. Tetra Pak uses linear low-density polyethylene (LLDPE) while Elopak uses multiple-layer PP laminates. The main advantage of using pouches for liquid packaging is that they are extremely light. However, the pouches are harder to handle than nonflexible packaging; after opening they need to be emptied into another container. Although the pouches have a very small cost price, the unique handling characteristics may prevent the pouch from gaining a large market share.

The PC bottle was introduced in the Dutch market in 1996 for the packaging of milk. The advantages of the PC bottle are that it is lightweight, refillable, and has a trip number of 30.[3] Moreover, the square shape of the bottle leads to saving of shelf space.

Stand-up Pouches

These are being used for a variety of products, including cereal, snacks, biscuit, candy, dried fruit, fruit juice, and side dishes. Constructed of a durable multi-ply laminate, stand-up pouches have a wide face perfect for an elaborate label or print design or for simply displaying the product through a clear faced bag. This new form of packaging gives creative retailers new opportunities to display their products in customized racks or innovative pallet displays. They are now found in many everyday applications and are in fact replacing rigid containers altogether in some product lines.

REFERENCES

1. Brody, A.L.; Marsh, K.S. *Wiley Encyclopedia of Packaging Technology*, 2nd Ed.; John Wiley: New York, NY, 1997.
2. Freeman, M. *PPI International Fact and Price Book*; Pulp Paper Int.: San Francisco, 1997.
3. Buelens, M. Tetra Pak: Verpakking en Milieu. Energy Milieu **1997**, 131–135.
4. Robertson, G.L. *Food Packaging; Principle and Practice*; Marcel Dekker: New York, NY, 1993.
5. Hekkert, M.P.; Joosten, L.A.J.; Worrell, E.; Turkenburg, W.C. Reduction of CO_2 Emissions by Improved Management of Material and Product Use: The Case of Primary Packaging. Resour. Conserv. Recycl. **2000**, 33–64.

Process Control Systems

Yanbo Huang
Texas A&M University, College Station, Texas, U.S.A.

Yubin Lan
Fort Valley State University, Fort Valley, Georgia, U.S.A.

INTRODUCTION

Modern process control systems are the systems that use computers to realize automatic control of production processes. A process control system consists of a computer and industrial production equipment. The computers in process control systems are usually called industrial control computers. In each process control system one or several computers may be needed depending on the characteristics of the process and control strategy. Input and output channels are important for communication between the computer and production equipment. In food and agricultural engineering, various parameters in processes are desired to be controlled. When human operators are involved in controlling the parameters, they may be subject to overcorrection and overreaction to normal process variability. In order to ensure consistency in product quality, automatic control can be configured on the process with a computerized system.

PROCESS CONTROL COMPUTERS

Figure 1 shows the relationship between the computer and the production equipment in the process. Computers are the key to realize automatic process control. A computer system for process control is similar to the computer systems for data processing. It includes hardware and software to meet the needs of process control.

Process Control Computer Hardware

The hardware includes one or several computers, peripherals, instrumentation, and input–output equipment, which is the basis of a computerized process control system.

Computer

The computer is the center of the automatic control system. It mainly consists of Central Processing Unit (CPU) and memory. The task of the computer is, based on the data measurement and information feedback or operator's control signals from the input equipment, to perform data processing by numerical computing in terms of designated control algorithms, and then send control commands out to the output equipment.

Peripherals

In process control, peripherals include disks, monitor, mouse, keyboard, printers, and other man-machine communication equipment. With them, human operators are able to interfere with the operation of the process, such as start/stop the process, query and output information about the process, modify computer programs, etc.

Instrumentation

Input and output equipment should be connected with the process through instrumentation. Instrumentation includes measurement instrumentation, display instrumentation, and actuators. The reliability and control quality of a process control system are closely related with the quality and precision of these instrumentations.

Input–output equipment

In general, in a process control system, the computer controls the production process through input–output channels. Input–output equipment is the "bridge" between the computer and the process. The functions of the equipment are:

1. Convert parameters in the production line (analog quantity, binary quantity, pulse quantity, and digital quantity) into the codes the computer can recognize, and then transmit them to the main computer for later processing.
2. Convert the computing results from control algorithms on the main computer into the control signals of the controlled variables in the production process. These control signals are for the operations of actuators.

Encyclopedia of Agricultural, Food, and Biological Engineering
DOI: 10.1081/E-EAFE 120007215
Copyright © 2003 by Marcel Dekker, Inc. All rights reserved.

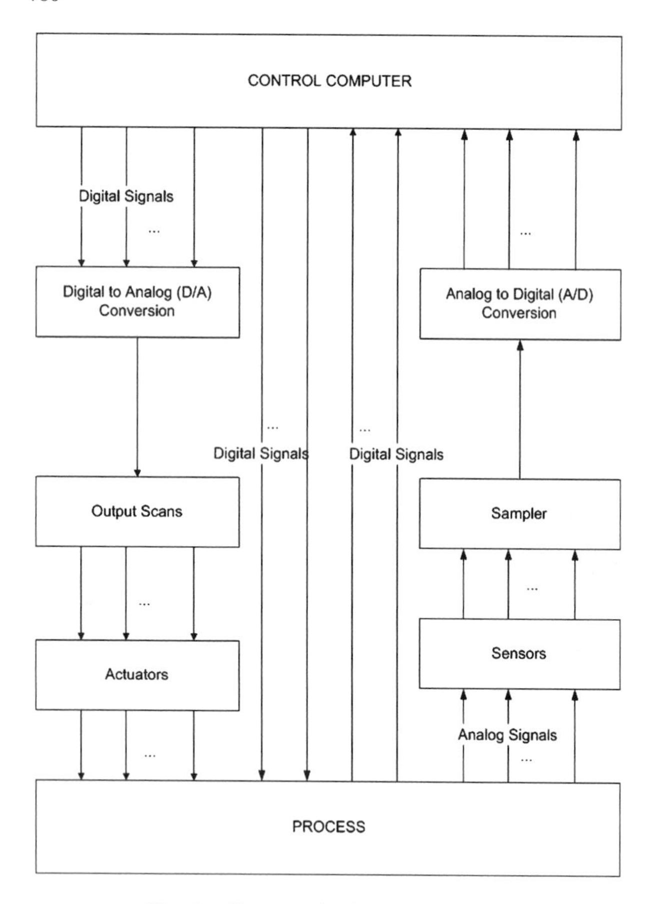

Fig. 1 Computerized process control.

Process Control Computer Software

The software can be classified into two categories: system software and application software.

System software

It is provided by computer manufacturers. It includes language (assembler and compiler), management (operating system), and service (diagnosis, fault-tolerance, and editing) programs.

Application software

It is developed by technical personnel in the food and agricultural manufacturers and process control scientists and engineers. It includes data processing and analysis, process modeling, and process control programs. These programs describe the process and control laws and realize control actions.

PROCESS CONTROL SYSTEM STRUCTURE

Process control systems can be basically divided into the following two types: open-loop and closed-loop control systems.

Open-Loop Systems

The structure of open-loop control systems is shown in Fig. 2. In these systems, the computer measures the process parameters and releases warnings for abnormal process conditions through instrumentation, signal transform, and analog to digital (A/D) converters, and then it compares the measurement with the specified values (setpoints) and outputs the results (printing and display). Human operators can manage the process in terms of the output results. The computer output does not act on the process directly.

Closed-Loop Systems

The structure of closed-loop control systems is shown in Fig. 3. In these systems, besides printing and displaying the computing results, the computer sends commands,

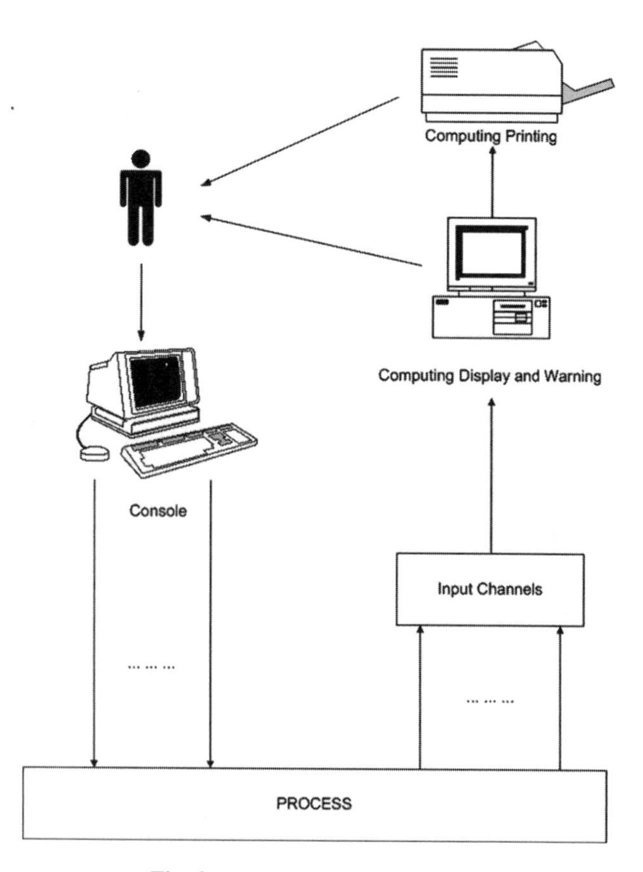

Fig. 2 Open-loop process control.

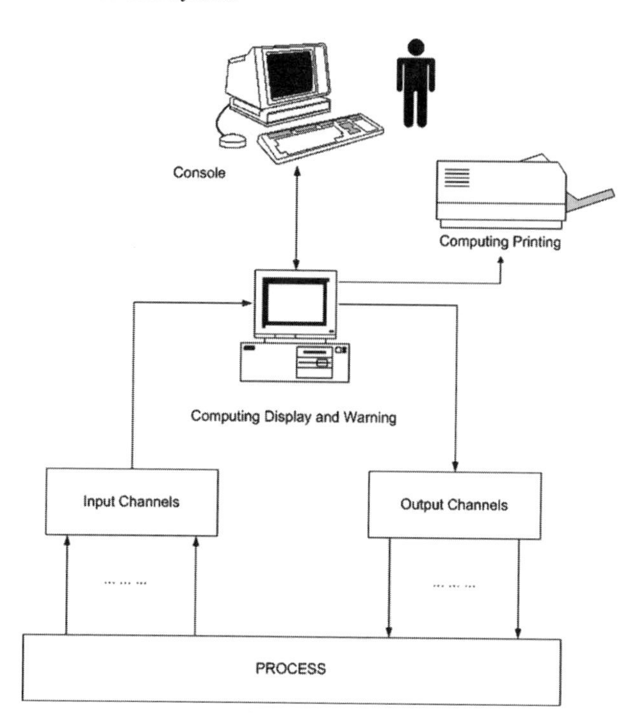

Fig. 3 Closed-loop process control.

based on the computing results directly, to the instrumentation and actuators in order to control the process.

AN EXAMPLE

In the snack food industry, continuous frying is an important process for snack production. In the frying process a continuous input of snack food material is put in one end of a fryer, pushed through by a submerger and oil flow, and then extracted at the other end. The raw material exists in an extruder in the form of a ribbon. Along the inlet belt, a cutter slices the ribbon into set lengths. The cut ribbon is dropped into the oil of the fryer where it encounters a short free-float zone. A submerger that covers most of the length of the fryer pushes the chips below the surface of the oil and carries them to the other end. After the submerger, the chips are taken out by a conveyor. Then, the chips are placed on a separate cooling conveyor where ambient airflow cools them. Finally the chips are transferred to a weight-belt where sensors measure the brightness (or color intensity) and moisture content, indicators of the final product quality. The quality control of the snack products is maintained by monitoring and stabilizing the levels of these indicators in terms of setpoints of the process operation.

A computerized closed-loop control system was configured to implement the controlling functions described above.[1] The process parameters, submerger speed, material residence time and inlet oil temperature, and output measurements, color, moisture content and oil content, went to the computer through the input channels by sequential sampling and A/D conversions. The computer computed the results of data processing, data correlation analysis, modeling, prediction, and control with the algorithms coded in computer programs. The computing results were printed and displayed on the console computer as the reference for human operators. At the same time, the computing results of control variables were sequentially sent to the corresponding speed and temperature actuators through D/A conversions to realize the closed-loop control of the final snack product quality.

A PC platform was used for the control system. The computer system was equipped with the system software, such as C and C + + compilers, DOS operating system, and Microsoft Windows system. The application software were developed for data preprocessing, data correlation analysis, modeling, prediction (one-step-ahead and multi-step-ahead), control (internal model control and predictive control) in the linear and nonlinear (artificial neural networks) modes.

REFERENCE

1. Huang, Y.; Whittaker, A.D.; Lacey, R.E. *Automation for Food Engineering: Food Quality Quantization and Process Control*; CRC Press LLC: Boca Raton, Florida, 2001.

Processing Waste Land Application

Christina A. Mireles DeWitt
Oklahoma State University, Stillwater, Oklahoma, U.S.A.

Michael T. Morrissey
Oregon State University, Astoria, Oregon, U.S.A.

INTRODUCTION

Food processors are being increasingly faced with more stringent regulations concerning the disposal of their discharges. Food processing residuals are by-products that are usually more suitable for land application than many other industrial and municipal wastewaters. As a result, recycling of food processing wastewaters and residuals through land application is becoming a more viable and economic alternative to the discharge of waste to municipal treatment systems.[1] Currently, land application is being utilized by many different kinds of food processing companies. However, it is usually the larger food processors that take advantage of this type of effluent treatment.

Land application of waste is a process that has been practiced in various locations throughout the United States and the world.[2,3] Land application describes the spreading of wastes on land intended for the production of agricultural and nonagricultural (e.g., forest application) crops or the use as a soil amendment/fertilizer to reclaim areas that have been disturbed by mining or other activities.[3] In many instances, the waste utilized for land application is often referred to as "biosolids." Biosolids is a term that is commonly used to describe the organic product from the treatment of municipal wastewater that can be beneficially used for soil enhancement and fertilization. In this article, biosolids will refer to the organic product from either food processing liquid waste that has been treated to reduce oxygen consuming organic pollutants or solid waste that has been lime stabilized or composted.

TREATMENT

The oxygen consuming potential of organic material in liquid waste is typically measured in terms of biological oxygen demand (BOD) or chemical oxygen demand (COD). There are many different ways liquid waste can be treated in order to reduce its organic load and thus achieve a reduction in BODs and CODs. Pretreatment of liquid waste usually involves screening to remove large solids. Primary treatments can include dissolved air flotation and/or gravity sedimentation to remove suspended solids. In some cases, screening is also referred to as a primary treatment. Screening is normally achieved with either a flat-bed vibrating type screen, a gyrating screen, an internally or externally fed rotary drum screen, or a static screen usually providing at least 20-mesh (0.85-mm) separation of the solids from the wastewater flow.[4,5] Dissolved air flotation[6,7] is a process that removes suspended solids by flotation using micrometer-sized air bubbles. Wastewater or a recycle stream is saturated with air and pressurized at 3 atm–5 atm. When the pressurized dissolved air waste stream enters the bulk tank, there is a decrease in pressure, which results in release of tiny air bubbles. Suspended solids adhere to the bubbles through hydrophobic interactions, float to the surface, and are recovered with scrapers.[5] Primary treatment can remove about 40% of the organic load.[8] Depending on the type of food processing waste, secondary or tertiary treatments may also be warranted to further reduce organic loads and/or provide a pathogen reduction step. These treatments may include chemical precipitation, trickling filtration (also referred to as attached-growth systems), dissolved air flotation, activated sludge, aerobic lagoons, anaerobic lagoons, and anaerobic filters. Reported average reductions in food processing liquid waste organic loads as measured by the 5-day biological oxygen demand (BOD_5) are 0%–10% for screening and 10%–30% for sedimentation.[4] The BOD_5 can be reduced by 35%–97% using secondary and tertiary treatments (Table 1).

APPLICATION

Biosolids can be land applied in either solid or liquid form. Liquid waste can be applied through direct injection into ground, irrigation, or spray runoff (also called overland flow). Direct injection of wastewater into the soil, or immediately covering it with dirt, can help reduce odor and runoff problems. An advantage of the liquid method is that waste can be stored and the time of spreading can be controlled to minimize pollution. Transportation problems are also simpler than with solid materials.[3] In the food

Encyclopedia of Agricultural, Food, and Biological Engineering
DOI: 10.1081/E-EAFE 120007156
Copyright © 2003 by Marcel Dekker, Inc. All rights reserved.

Table 1 Percent reduction of BOD_5 from treatments of wastewater using different unit operations

Treatment method	% BOD_5 reduction
Screening	0–10
Sedimentation	10–30
Chemical precipitation	35–80
Trickling filtration	40–96
Activated sludge	50–97
Aerobic lagoon	40–85
Anaerobic lagoon	40–90
Anaerobic filter	50–80

(Adapted from Ref. 4.)

processing industry, the spray runoff method is commonly practiced. This involves spraying wastewater on upper reaches of a slope and then allowing it to flow down through vegetation on the sloped surface. Soils best suited to overland flow methods are clays and clay loams with limited drainability. Slope should be between 2% and 6%.[2] Solid waste obtained either directly from processing (e.g., cereals, fruits, vegetables) or after secondary or tertiary treatment (in the form of sludge or compost) can be spread over fields to provide nutrients for plants.

BENEFITS

Benefits of land application of food waste include addition of nutrients important to plant growth. Nitrogen, phosphorous, and water-soluble potassium (or potash) are three major plant nutrients that can be supplied from food processing waste. In addition, land application supplements irrigation water, increases groundwater supply, and is more economical than conventional treatment.[9] Land application requirements include enough suitable land and a balance of wastewater application with nutrient uptake by plants. Pollutant removal is achieved with land application through plant uptake and by physical, chemical, and microbial processes in the soil. The soil, plants, bacteria, and air help further purify the waste and serve, basically, as a complex biological filter. However, there is a limit to the ability of this biological filter to absorb nutrients from waste. An excess nutrient application can result in runoff into the environment.[10]

NUTRIENT CONCERNS

Problems that can result from land application of food wastes include groundwater contamination, nuisance odors, aerosol drift, surface run-off into waterways, and degradation of soil structure and quality. When evaluating a site for possible application of wastewater, climate, topography, soil characteristics, geological formations, groundwater, and receiving waters are all site characteristics that should be taken into account.[11,12]

Nitrogen—Some of the nutrient problems that have to be considered when land-applying food wastes are nitrogen and phosphorous loading. Application rate is a primary consideration that must be taken into account. Nitrogen is the primary nutrient that contributes to plant growth. In biosolids, it exists as organic and inorganic compounds. Organic nitrogen is not available to plants; it must be converted by bacteria to ammonia and then oxidized to nitrate to become biologically available. Heavy application of nitrogen-containing biosolids can result in unused nitrate migrating to surface or ground water. Nitrates entering the human food chain can result in adverse health effects and other environmental effects.[13] Nitrate, which is highly mobile in soils, can contaminate ground water. It is formed in soil from oxidation of organic nitrogen and ammonia present in wastewater or it is already present in aerobically treated wastewater. If it is not taken up by plants, it leaches into groundwater or is converted to nitrogen gas by denitrification[14,15] and returned to the atmosphere. When processing wastewater is applied to grazed pasture, the grazing animals recycle about 90% of their nitrogen intake back into the pasture. Only 10% of the nitrogen taken up by the plants is actually removed from the site in the form of animal products. Therefore, better nitrogen removal can be achieved if pasture crop is harvested (hay or silage). Harvesting also allows higher wastewater nitrogen application rates.[5]

Phosphorous—Phosphorous follows nitrogen in regards to its importance in plant nutrition. Decomposition of organic phosphorous by bacteria or hydrolysis of polyphosphates in water forms a water soluble, stable orthophosphate (PO_4). It is in the orthophosphate form that phosphorous is predominantly absorbed by plants. However, phosphorous can build up in soil and become excessive. Repeated application of phosphates can overload the soil and result in leaching.[13] Leaching of phosphorous is a concern because it can lead to over fertilization and excessive plant growth which results in an acceleration of the eutrophication process in fresh receiving waters. Eutrophication is the natural aging of lakes and streams because of nutrient enrichment. However, human activities can result in increased nutrient loading rates and thus an acceleration of this natural process. Accelerated eutrophication occurs when the algae in an aquatic system rapidly multiply because of fertilization. The algae create a dense upper layer and make it difficult for larger submerged aquatic vegetation to get enough light. Dieback of aquatic vegetation reduces available aquatic habitat. In addition, when alga dies it decomposes, removing dissolved oxygen from the water. This in turn, can lead to suffocation of the aquatic life.[16,17]

CONCLUSION

Effluent treatment using land application can be successfully applied to food processing waste provided companies properly identify the nutrient concerns of the waste generated at their facility. For example, the major wastewater nutrient of concern is nitrogen.[18–20] After primary treatment (sedimentation), wastewater from these industries can be spray irrigated provided the loading rate of nitrogen does not exceed agronomic rates (300 kg N/ha-yr–600 kg N/ha-yr). Lime-stabilized meat packing sludges can also be land applied.[20,21] Land application from nonmeat industries has also been demonstrated in corn- and potato-based industries,[22,23] dairy processing,[24] and many other food processing industries.[25]

REFERENCES

1. Ritter, W.F. Innovative Land Application Systems for Food Processing Wastes. In *Utilization of Food Processing Residuals: Selected Papers Representing University, Industry, and Regulatory Applications*, NRAES-69; Robillard, P.B., Martin, K.S., Eds.; Northeast Regional Agriculture Engineering Service: Ithaca, NY, 1993; 81–85.
2. Kerns, W.R. *Land Application of Wastewater: Systems and Technologies*; Series 3 of 3; Virginia Polytechnic Institute and State University: Blacksburg, VA, 1976; 1.
3. Forste, J.B. Land Application. In *Biosolids Treatment and Management*; Girovich, M.J., Ed.; Marcel Dekker, Inc.: New York, 1996; 389–448.
4. Russell, P. Food Processing Wastewater Treatment. In *Agriculture Wastes: Principles and Guidelines for Practical Solutions*, Cornell University Conference on Agriculture Waste Management, Syracuse, NY, Feb 10–12; Ludington, D.C., Ed.; New York State College of Agriculture and Life Sciences: Ithaca, NY, 1971; 167–172.
5. van Oostrom, A.J. Waste Management. In *Meat Science and Applications*; Hui, Y.H., Nip, W.-K., Rogers, R.W., Young, O.A., Eds.; Marcel Dekker, Inc.: New York, 2001; Chap. 27, 635–671.
6. ConSep. Dissolved Air Flotation. http://www.consep.com.au/Prod_westech_daf.html (accessed August 2002).
7. Tramfloc, Inc. Emulsion Breaking in DAF and API Equipped Systems. http://www.tramfloc.com/tf15.html (accessed August 2002).
8. Gillies, M.T. Utilization Versus Pollution. In *Whey Processing and Utilization: Economic and Technical Aspects*; Noyes Data Corporation: Park Ridge, NJ, 1974; 3–23.
9. Kerns, W.R. *Land Application of Wastewater: An Overview*; Series 1 of 3; Virginia Polytechnic Institute and State University: Blacksburg, VA, 1976; 1–2.
10. Russell, J.M.; Cooper, R.N.; Lindsey, S.B. Reuse of Wastewater from Meat Processing Plants for Agricultural and Forestry Irrigation. Water Sci. Technol. **1991**, *24* (9), 277–286.
11. U.S. Environmental Protection Agency. *Evaluation of Land Application Systems*, EPA-430/9-75-001; Environmental Protection Agency, 1975; 21.
12. U.S. Environmental Protection Agency. Land Application of Sewage Sludge: A Guide for Land Appliers on the Requirements of the Federal Standards for the Use or Disposal of Sewage Sludge, 40 CFR Part 503, EPA/831-B-93-002b; December 1994.
13. Girovich, M.J. Biosolids Characterization, Treatment and Use. In *Biosolids Treatment and Management*; Girovich, M.J., Ed.; Marcel Dekker, Inc.: New York, 1996; 1–45.
14. Pidwirny, M.J. Fundamentals of Physical Geology: The Nitrogen Cycle. http://www.geog.ouc.bc.ca/physgeog/contents/9s.html (accessed March 2002).
15. Kimball, J.W. The Nitrogen Cycle. http://www.ultranet.com/~jkimball/BiologyPages/N/NitrogenCycle.html (accessed March 2002).
16. The Impact of Phosphorous on Aquatic Life: Eutrophication. http://www.agnr.umd.edu/users/agron/nutrient/Factshee/Phosphorus/Eutrop.html (accessed March 2002).
17. Sharpley, A.N.; Daniel, T.; Sims, T.; Lemunyon, J.; Stevens, R.; Parry, R. *Agricultural Phosphorus and Eutrophication, ARS*-149; USDA Agriculture Research Service: Springfield, VA, 1999. http://www.ars.usda.gov/is/np/Phos&Eutro/phos%26eutro.pdf (accessed March 2002).
18. Cooper, R.N.; Russell, J.M. Meat Industry Processing Wastes: Characteristics and Treatment. In *Encyclopedia of Food Science and Technology*; John Wiley and Sons, Inc.: New York, 1991; 1683–1691.
19. Russell, J.M.; Cooper, R.N.; Lindsey, S.B. Soil Denitrification Rates at Wastewater Irrigation Sites Receiving Primary-Treated and Anaerobically Treated Meat-Processing Effluent. Bioresour. Technol. **1993**, *43*, 41–46.
20. Hepner, L.; Johnson, R.; Weber, C. Land Application of Meat Packing Plant Waste. In *Utilization of Food Processing Residuals: Selected Papers Representing University, Industry, and Regulatory Applications*, NRAES-69; Robillard, P.B., Martin, K.S., Eds.; Northeast Regional Agriculture Engineering Service: Ithaca, NY, 1993; 62–66.
21. Witmayer, G. Nutrient Management: The Key to Food Processing Waste Disposal. In *Utilization of Food Processing Residuals: Selected Papers Representing University, Industry, and Regulatory Applications*, NRAES-69; Robillard, P.B., Martin, K.S., Eds.; Northeast Regional Agriculture Engineering Service: Ithaca, NY, 1993; 76–80.
22. Brandt, R.C. Wise Foods Cooperative Land Application Recycling Program—A Case Study. In *Utilization of Food Processing Residuals: Selected Papers Representing University, Industry, and Regulatory Applications*,

NRAES-69; Robillard, P.B., Martin, K.S., Eds.; Northeast Regional Agriculture Engineering Service: Ithaca, NY, 1993; 67–75.

23. de Haan, F.A.M.; Zwerman, P.J. Land Disposal of Potato Starch Processing Wastewater in the Netherlands. In *Food Processing Waste Management*, Proceedings of the 1973 Cornell Agriculture Waste Management Conference, Syracuse, NY, 1973; 222–228.
24. Gillies, M.T. Utilization Versus Pollution. In *Whey Disposal and Utilization: Economic and Technical Aspects*; Noyes Data Corporation: Park Ridge, NJ, 1974; 15.
25. Chawla, V.K; P.J. Treatment of Fish and Vegetable Processing Waste-Lagoon Effluent by Soil Bio-filtration. In *Food Processing Waste Management*, Proceedings of the 1973 Cornell Agriculture Waste Management Conference, Syracuse, NY, 1973; 74–85.

Processing Waste Pretreatment

Sang Hun Kim
Kangwon National University, Kangwon-do, South Korea

INTRODUCTION

The conventional treatment systems are designed to biodegrade and stabilize waste solids for disposal. The food processor has the opportunity to use the same techniques and others, to recover waste solids as by-products.

Despite efforts to reduce the pollution load by in-plant and process modifications, the food processor may still find it necessary to treat the plant's wastewater effluent. Alternatives lie between discharge to the publicly owned treatment works and treatment by the processor and discharge to receiving waters. Wastewater treatments are usually categorized into pretreatment, primary, secondary, and tertiary referring to the order of treatment. In industrial waste treatment systems the conventional order of unit processes may be altered or even reversed. This is because the characteristics of the industrial waste differ from municipal wastes.

The primary purpose of pretreatment is to condition the waste effluent and to remove materials that may interfere with subsequent treatment process and equipment.

Food processors are in a more favorable situation. Screen can be located near the source of solids discharge, and solids can be removed before they leach excessive soluble or colloidal particles into the wastewater stream. Screening may, in fact, be considered part of in-plant waste management rather than a part of the treatment system. A variety of screens have been developed for use in pretreatment of the food processing waste; they include static, vibrating, rotating, and centrifugal screens.[1]

CHARACTERISTICS OF THE SCREENING DEVICES

The screening element may consist of parallel bars, rods or wires, grating, wire mesh, or perforated plate, and the openings may be of any shape but generally are circular or rectangular slots. A screen composed of parallel bars or rods is called a bar rack (or a bar screen). The term "screen" is used for screening devices consisting of perforated plates, wedge wire elements, and wire cloth. The materials removed by these devices are known as screenings. According to the method used to clean them, bar racks and screens are designed as hand-cleaned or mechanically cleaned. Typically, bar racks have clear openings (spaces between bars) of 5/8 in. (15 mm) or more. Screens have openings of less than 5/8 in. (15 mm). The screens with openings greater than 1/4 in. (6.3 mm) are considered coarse. The principal types of screening devices now in use are described in Table 1.

BAR RACKS

In pretreatment, bar racks are used to protect pumps, valves, pipelines, and other appurtenances from damage or clogging by rags and large objects. For the treatment of food processing waste may or may not need them, depending on the character of the wastes. A typical bar rack used for wastewater treatment is shown in Fig. 1.

STATIC SCREENS

Static screen have no moving parts except for possible backwashing mechanisms. An ordinary kitchen strainer is a static screen. The hydraulic head or pumping pressure is used to move wastewater over the surface and through the screen. Screens have been developed to increase hydraulic throughput, to selectively separate out specific particulate matter, and to be trouble-free.

The static wedge wire screen characterized by its cross sectional slope is one such fast, nearly trouble-free screen. An occasional use of a spray backwash may be necessary. Wastewater flows down over the screen and coarse solids and denser material follow the tangential path while water with dissolved and finer solids pass through the screen. Fig. 2 shows wastewater screening for carrot processor in a farm of Santa Fe Spring, California.

Hydraulic loading rate depend upon the width of the screen, surface area, and screen size. Solid removal efficiencies depend upon the coarseness of the wastewater solids and screen size. Most screens used in pretreatment

Encyclopedia of Agricultural, Food, and Biological Engineering
DOI: 10.1081/E-EAFE 120007153
Copyright © 2003 by Marcel Dekker, Inc. All rights reserved.

Table 1 Description of screening devices used in pretreatment of wastewater

Screening devices	Screen size classification	Size range (in.)	Screen material
Bar racks	Coarse	0.6–3.0	Bars
Fine bar	Fine coarse	0.125–0.50	Thin bars
Screens			
Static parabolic	Fine	0.01–0.125	Stainless steel wedge wire
Static perforated	Fine coarse	0.125–0.375	Perforated plate
Vibrating separator	Very fine	0.002–0.02	Stainless steel cloth
Rotating drum	Fine	0.01–0.125	Stainless steel wedge wire
Rotating disk	Very fine	0.006–0.015	Stainless steel woven wire
Centrifugal	Very fine	0.002–0.02	Stainless steel, polyester, and fabric screen cloth

are not intended to remove a significant amount of suspended solids, but only coarse solids.

VIBRATING SCREENS

These are separator devices with vertical and tangential motions, which cause the retained material to move across the screen to the periphery where it is collected and discharged. Screens of decreasing sieve size can be stacked to continually separate out smaller sized solids. Fig. 3 illustrates a single vibrating screen unit.

Basically, the vibrating separator is a screening device that vibrates about its center of mass. Vibration is accomplished by eccentric weights on the upper and lower ends of the motion-generator shaft. Rotation of the top weight creates vibration in the horizontal plane, which causes material to move across the screen cloth to the periphery. The lower weight acts to tilt the machine, causing vibration in the vertical and tangential planes. The angle of lead, given the lower weight with relation to the upper weight, provides variable control of the spiral-screening pattern. Speed and spiral pattern of material travel over the screen cloth can be set by the operator for maximum throughput and screening efficiency of any screenable product.

The advantages of vibrating screens are fewer tendencies to blind and more throughputs for given screen size and area. The disadvantages are accidental rupture of a screen, occasional mechanical failure, and required maintenance.

ROTATING SCREEN

These are made in a variety of model designs. Some are center fed and self cleaning (similar in concept to micro strainers except that the screen size is larger). One rotating screen system loads the wastewater on the outside of the screen drum, removes the collected solids by a doctor blade and effluent wastewater cleans the screen as it leaves the bottom of the drum. Fig. 4 shows in-operation on food processing waste and Fig. 5 illustrates this concept. Rotary speed and hydraulic loading can be controlled.

The advantages are continual operation with little or no blinding (self cleaning), minimal operator attendance, and

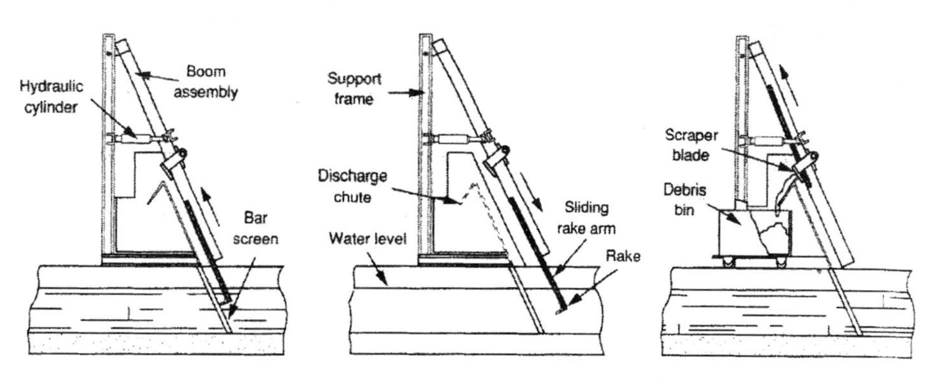

Fig. 1 Mechanically cleaned bar rack.[2]

Fig. 2 Static wedge wire screen.

Fig. 4 Externally fed wedge wire rotating screen.

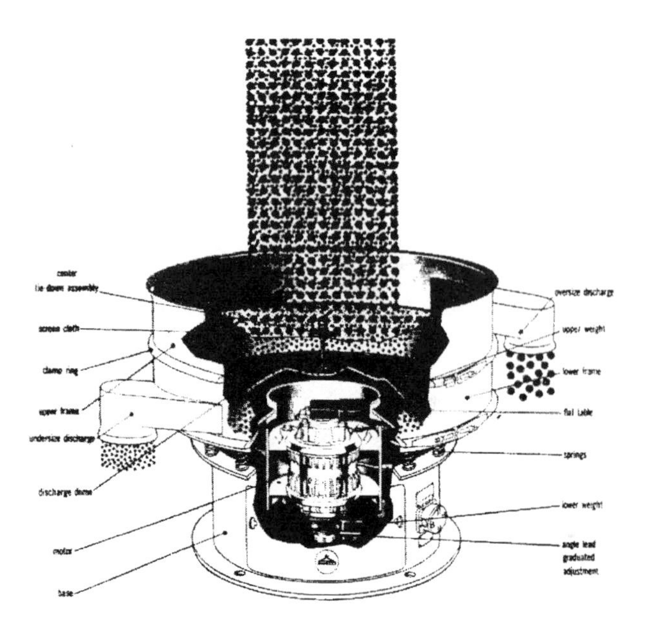

Fig. 3 Vibrating screen.

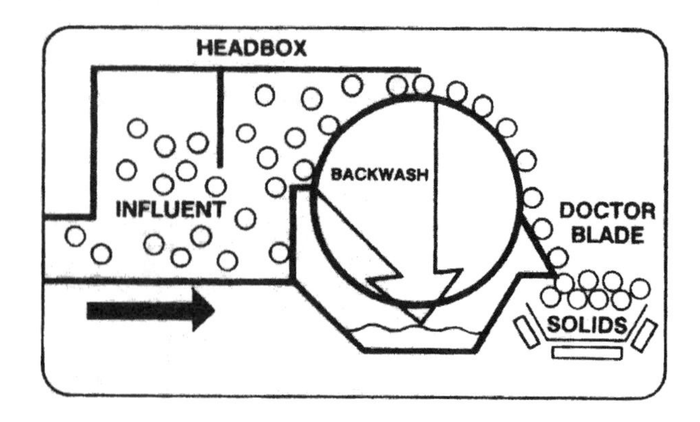

Fig. 5 Schematic diagram of self-cleaning, rotating screen.

Fig. 6 Rotating screen used as a replacement for primary sedimentation.[3]

availability of different screen size for different functions, and easy handle of greasy, oily, and stringy solid. Hydraulic loading rates vary with the screen size and surface area.

Disadvantages would be the potential for mechanical failures and required maintenance assorted with mechanized equipment.

ROTATING DISK SCREEN

In recent times very fine screens, typically with openings of about 0.01 in., have been used as replacement for primary sedimentation facilities. A rotary disk screen has been used in the City of San Diego aquaculture facility for more than 10 yr (Fig. 6).

CENTRIFUGAL SCREENS

Most screens described above rely on hydraulic head to move wastewater through the screening mechanism. Several screens are produced which force wastewater through the screen by centrifugal force. The general advantage of centrifugal screens is higher hydraulic loading rates.

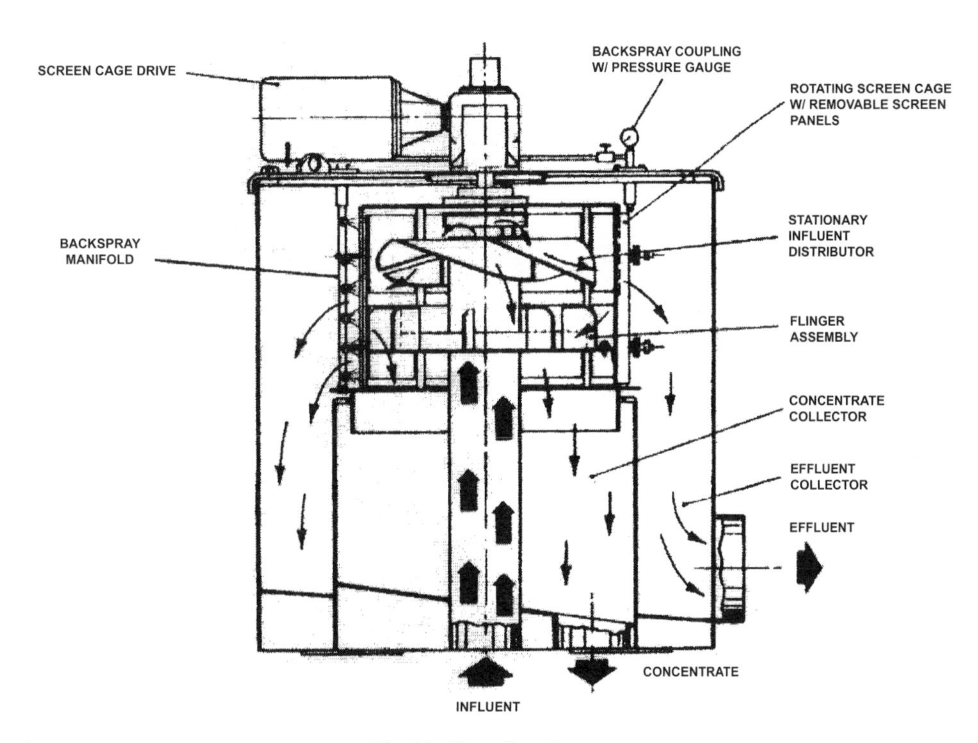

Fig. 7 Centrifugal screen.

A centrifugal screen[1] is illustrated in Fig. 7. Influent coming up the center is distributed onto the screen by a stationary baffle. The rotating screen forces water through but does not have enough speed to establish a strong centrifugal force. Thus, retained solids are not strongly impacted against the screen surface and a continuous backwash spray mechanism removes them to a collector. These screens have been successfully used in fruit, vegetable, meat, poultry, and fish processing operations. The units are expensive and subject to routine mechanical failures.

REFERENCES

1. Green, J.H.; Kramer, A. Pre- and Primary Treatment. In *Food Processing Waste Management*; Avi: Westport, Connecticut, 1979; 299–306.
2. Tchobanoglous, G.; Burton, F.L. Physical Unit Operations. *Waste Water Engineering*, 3rd Ed.; Irwin/McGraw-Hill: Boston, 1991; 200–203.
3. Crites, R.W.; Tchobanoglous, G. Wastewater Pretreatment Operations and Processes. In *Small and Decentralized Wastewater management Systems*; WCB/McGraw-Hill: Boston, 1998.

Processing Waste Primary Treatment

Christopher C. Miller
Rose Acre Farms, Social Circle, Georgia, U.S.A.

INTRODUCTION

Primary treatment systems are the main systems in removing solids in wastewater. This is done in one of the three ways: settlement; flotation (fat traps, air flotation, and inclined plate separators); and chemical treatment (pH neutralization, coagulation, flocculation, oxidation, and nutrients). The three systems listed are used singly or combined depending upon the food plant's commodity (dairy, meat, poultry, fruits, vegetables, etc.), location (rural, urban), and environmental permit limits (local, state, federal EPA).

SETTLEMENT

Theory

Settlement is the oldest and most common method in Primary treatment systems, removing solids larger than 10 mm (0.39 in.).[2] This method removes 35% of Biological Oxygen Demand (BOD) and 60% of Total Suspended Solids (TSS) in meat and vegetable processing wastes.[2] Solids smaller than 1 mm (0.04 in.) do not settle by gravity and must use chemical treatments to settle.[2] Soluble and colloidal foods like soft drinks and salad dressing waste do not settle out well either.[2] Stokes Law is the basis for settlement theory, calling on the density differences between the particles and water, flow velocity, and temperatures.[1,2] The density differences are one of four types of settlement behavior noticed.[1,2] The second is where settlement particles have greater velocity than theory states where the flocculent settles faster than the traditional settlement behavior is formed.[1,2] The remaining two settlement behaviors, zone settlement and compressive settlement, occur when activated sludge is concentrated in the system.[1,2]

Types

The two types of settlement tanks used are circular and rectangular. Most tanks are circular because they are cheaper and have more reliable scrapers, though rectangular tanks are common to reduce the required operational land area.[2] These tanks are designed for low velocity and uniform flow, with sidewall depth of 2.4 m–4.5 m (7 ft–15 ft) with slope to bottom of 5°–10°.[1,2] The tank is traditionally made of reinforced concrete though prefabricated tanks are made either of plastic or of steel depending upon tank volume.[2] The scrapers usually are made of galvanized steel, and the weirs used for uniform flow within the tank are made of plastic.[1,2] A hydrostatic head periodically removes any accumulated sludge with decanting through a screw-threaded bell valve.[2] Extra removal of fatty wastes uses a screen baffle and upper scraper to combine the sludge.[1,2]

Loading Rates

The common loading rates for settlement tanks are shown in Table 1.[2]

FLOTATION

Introduction

A large variety of food processing wastewaters contain fat, oil, and grease (FOG).[1,2] Most FOG is biodegradable if soluble or dispersed to even concentrations for bacteria usage.[2] The rest separates from the main wastewater, creating problems for bioreactors, process instruments, and aeration gas transfer.[2] For aerobic treatments, the maximum concentration is 150 mg L^{-1}.[2] To reach this standard, flow is reduced in a single chamber and oil is trapped in underwater weirs, causing it to float to the surface.[2] If the wastewater has detergents or surface-active chemicals in them FOG emulsifies and separation efficiency is reduced.[2] Flotation overcomes this in three ways: fat traps; dissolved air flotation (DAF); and inclined plate separators.[2]

Fat Traps

Fat traps are designed for even flow through a tank, allowing density differences to raise fat to the surface without disturbing any sludge and scum already settled at the bottom.[2] A drain valve removes the built-up sludge and scum at the tank bottom. Fat traps are designed similar to settling principles.[2] Their typical length/width ratio is 2:1. Table 2 shows the fat trap load rates.[2]

DOI: 10.1081/E-EAFE 120007154
Copyright © 2003 by Marcel Dekker, Inc. All rights reserved.

Table 1 Loading rate settlement tank design

Type of waste	Retention time (hr)	Surface load [$m^3 m^{-2} day^{-1}$ ($ft^3 ft^{-2} day^{-1}$)]
Meat processing	2–3	45 (147.6)
Vegetable	2.5	25 (82)
Cannery	4	40 (131.2)
Typical	4	40 (131.2)

(From Ref. [2], p. 162.)

Dissolved Air Flotation

Dissolved air flotation, like sedimentation, is particle size dependent and also depends upon gas transfer efficiency and bubble attachment to the particle.[1,2] This lowers the required tank size and avoids any potential anaerobic conditions.[2] Compared to gravity separation, DAF is more complex and less reliable.[2] DAF uses pressure release from a supersaturated solution of air.[1,2] The most common DAF is where pressure supersaturates the wastewater and is released to remoisten the dissolved air.[2] Design is based on flow velocity and retention, but also on a minimum air/solids ratio. (The last is difficult to set by theory and is optimized prior to commissioning.)[2] Many DAF units are purpose built for great flexibility, requiring periodic adjustment.[1,2] Typical DAF operating characteristics are $30\,m^3 m^{-2} day^{-1}$–$60\,m^3 m^{-2} day^{-1}$ ($98.4\,ft^3 ft^{-2} day^{-1}$–$196.8\,ft^3 ft^{-2} day^{-1}$) surface load, 20 min–30 min retention time, 25%–100% recycle rate, and 3 atm–6 atm (43.1 psig–86.2 psig).[2] Problems with DAF include poor reliability and higher operational costs than gravity separation, according to operational surveys.[2] DAF is also not suitable to the constant changes in wastewater characteristics.[2] Table 3 shows the ideal DAF performance compared to gravity separation.[2]

Incline Plate Separators

Incline plate separators are settlement tanks with a series of incline parallel plates inside a tank.[2] The flotation is new to treating food-processing wastewaters, so limited research has been done on this.[2] Startup costs for inclined plate separators are less than conventional settling tanks though maintenance and sanitation costs are higher.[2] Inclined plate separators use the hydrophobic interface of surfaces and particles to allow them to coalesce on the separator's surface.[2] The result is a smaller tank size, with retention time of 30 min and increased surface loads of $2.5\,m^3 m^{-2} hr^{-1}$–$3.0\,m^3 m^{-2} hr^{-1}$($8.10\,ft^3 ft^{-2} hr^{-1}$–$9.84\,ft^3 ft^{-2} hr^{-1}$).[2]

CHEMICAL TREATMENT

Introduction

Chemical treatment in primary treatment systems are used to correct pH and improve settling rates with increased particle density.[2] It is used when an upset condition or seasonal changes occur.[2] Chemical treatment's big disadvantages are chemical costs, good control to optimize performance, and higher operational and maintenance costs.[2] The four most common methods are: pH neutralizing; coagulation/flocculation; oxidation; and nutrients.[2]

pH Neutralization

pH neutralization's cheapest and most common chemicals are sulfuric acid and sodium hydroxide, each sold at a maximum 50% v/v solution.[2] Other products considered may be lime (cheaper base, but poor solubility and more difficult to prepare), commercial slurries, and phosphoric acid and carbonic acid for anaerobic treatments in lieu of sulfuric acid.[2]

Table 2 Loading rate fat trap design

Product	Retention Time (min)	Surface load [$m^3 m^{-2} hr^{-1}$ ($ft^3 ft^{-2} hr^{-1}$)]
Meat processing	20	2 (6.56)
Margarine processing	20–40	3 (9.84)
Milk, butter, and cheese	30	1 (3.28)
Milk processing	20	0.4 (0.131)

(From Ref. [2], p. 164.)

P

Table 3 Ideal DAF performance

	Infeed ($mg\,L^{-1}$)			Discharge ($mg\,L^{-1}$)			% Removed ($mg\,L^{-1}$)		
Wastewater type	**TSS**	**FOG**	**BOD**	**TSS**	**FOG**	**BOD**	**TSS**	**FOG**	**BOD**
Cooking oil	230	460	2900	20	25	94	91.3	94.6	96.9
Margarine	5000	3900	NDA	200	40	NDA	96	99	NDA
Meat slaughter houses	7428	3110	NDA	712	97	NDA	90.4	96.9	NDA
Poultry waste	1690	331	1075	275	74	86	83.7	77.6	92
Gelatine waste	2680	2825	NDA	458	315	NDA	82.9	88.9	NDA
Jams/pickles	1350	NDA	790	270	NDA	315	80	NDA	60.1

NDA-No data available.
(From Ref. [2], p. 165.)

Coagulation/Flocculation

Coagulation is used for very small, positively charged particles less than 5 mm (0.2 in.) to absorb negatively charged ions in the mixing solution.[1,2] Flocculation bridges these small particles together into larger particles to create a sludge blanket by straining and enmeshment.[1,2] Polymers used for flocculation include polyacrimide, aluminum hydroxide, and ferric hydroxide, though they may not be needed because of the great mixing between the chemicals and the particles.[2] Fig. 1 shows a typical flocculation system.[2] A typical flocculation system requires a rapid mix tank to completely mix the waste and the reagents.[1,2] This forms the flocculant that is transferred to a flocculation tank for gentler mixing to ensure good enmeshment.[2] This gentle mixing is done in a second, slow-stirring flocculation tank (mixer speed of 1 rpm–2 rpm) before the flocculant's final settlement or flotation.[2] To ensure correct flocculant dosage and conditions, lab and pilot plant testing is required.[1,2] Cost-benefit analysis of chemicals may be required for sludge consolidation.[2] Sufficient polymer concentrations are $1\,mg\,L^{-1}$–$5\,mg\,L^{-1}$ organic and/or $10\,mg\,L^{-1}$–$100\,mg\,L^{-1}$ of iron or aluminum-based compounds.[2] If there are recycled solids involved, food-grade flocculants like cellulose, starch, lignosulfonic acid, and calcium polyphosphate would be used.[2]

Oxidation

Oxidation is used during a very short operating season or when a short-term problem occurs, like toxic wastes or

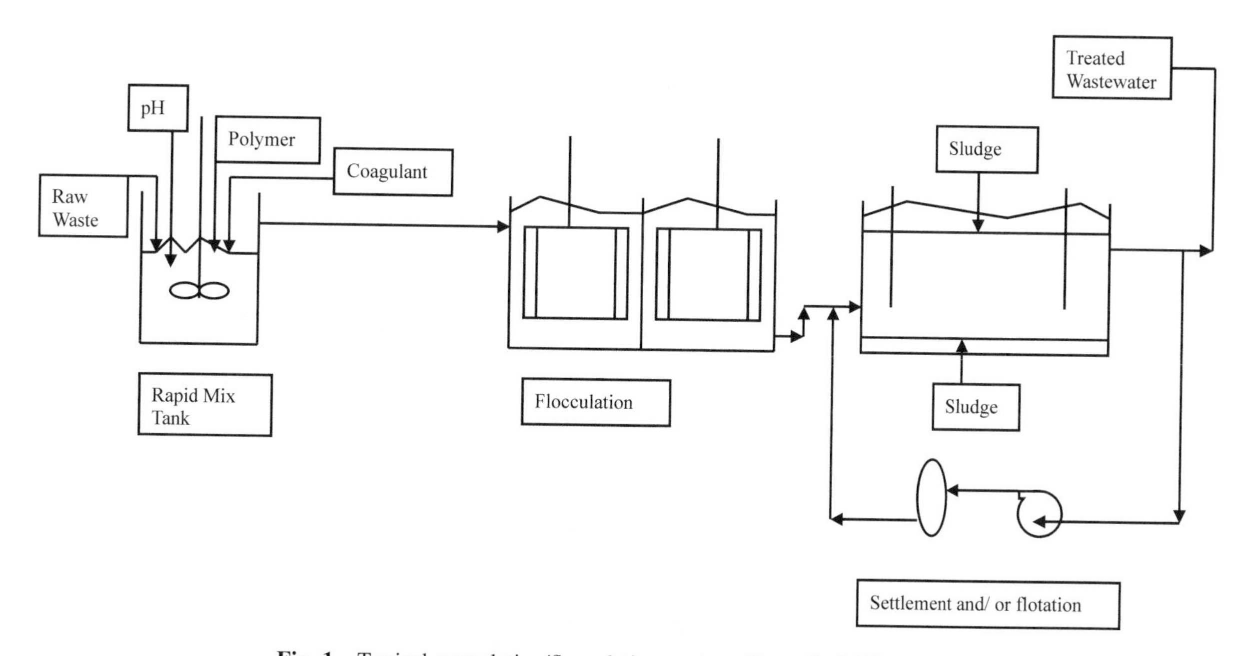

Fig. 1 Typical coagulation/flocculation system. (From Ref. [2], p. 167.)

a biological malfunction.[1,2] It is also useful when there is not enough aeration.[2] Pure oxygen, ozone, chlorine, and permanganate are usually used during oxidation.[2] If wastewater is being recycled, then chlorine, hypochlorite, and ozone are used as wastewater sterilizer.[2]

Nutrients

To improve the performance of some wastes, supplements of nitrogen, phosphorous, trace metals, and vitamins are added.[2] For successful aerobic treatment, a carbon (BOD), nitrogen (ammonia-based), and phosphorous ratio of 100:5:1 is established.[1,2] If organic nitrogen is in this as a protein, the microbes in the wastewater can use it as a nitrogen source.[2] Urea is added only if nitrogen is required. If both nitrogen and phosphorous are needed, inorganic fertilizer is added.[2] When biomass requirements are met, the nutritional need is lowered because the recycled nutrients are controlled by ensuring a $1\,mg\,L^{-1}$–$2\,mg\,L^{-1}$ residue of nutrients in the treated wastewater.[2] For anaerobic bacteria, the carbon–nitrogen–phosphorous ratio is 300:5:1 because of the lower growth rate and nutritional requirements.[2] Anaerobic bacteria are vulnerable to shortages of iron, cobalt, nickel, and manganese though all are commercially available and commonly added to industrial digesters.[2] Extra nutrients also assist aerobic treatment with complex nutrients and in addition are an effective method in countering any mild sulfur toxicity in the system.[2]

CONCLUSION

Primary treatment systems use settling, flotation, and chemical treatments to reduce BOD, TSS, and FOG in food processing wastes in order to keep a facility in compliance with local, state, and federal regulations. As environmental regulations tighten worldwide, the primary treatment systems must adapt to meet the challenges ahead.

REFERENCES

1. Corbitt, R.A. Wastewater Disposal. In *Standard Handbook of Environmental Engineering*, 2nd Ed.; Corbitt, R.A., Ed.; McGraw-Hill: New York, 1999; 6.75–6.90.
2. Wheatley, A.D. Food and Wastewater. In *Food Industry and the Environment in the European Union: Practical Costs and Implications*, 2nd Ed.; Dalzell, J.M., Ed.; Aspen Publications: Gaithersburg, MA, 2000; 160–169.

Processing Wastewater Recovery

D. Raj Raman
The University of Tennessee, Knoxville, Tennessee, U.S.A.

P

INTRODUCTION

Wastewater recovery is fundamentally a separation process in which undesirable compounds (pollutants) are removed from a valuable commodity (water), and/or in which pollutants are concentrated or modified through physicochemical or biological means to produce valuable commodities. Several benefits may be realized by incorporating wastewater recovery into a food processing operation, including reduced freshwater consumption, reduced wastewater treatment volumes, reduced chemical requirements, production of useful byproduct streams, and reduced energy requirements.[1–5] Conversely, both capital and operating costs are associated with wastewater recovery technologies, so that the optimal degree of recovery, as well as the specific recovery methods employed, will depend on the local costs of labor, energy, raw materials, and waste disposal.

Wastewater recovery is not unique to the food processing industry; entities ranging from municipal wastewater treatment plants to semiconductor manufacturing plants employ wastewater recovery techniques. However, the potential for deploying wastewater recovery techniques is great in the food processing sector because of the large volumes of water involved in most food processing operations, and because the contaminants in food processing wastewater are typically foodstuffs themselves, and are thus not particularly toxic compounds[6]—pathogen-laden wastes arising from certain processing operations are one exception to this generalization.[7,8]

Implementing a successful wastewater recovery system at a food processing plant requires more than the selection of a series of unit operations for water purification; significant recovery may be realized by clever rearrangement of process water flow, e.g., by using relatively clean process water for an initial washdown of work areas.[6] Therefore, consultation with plant managers and workers, and implementation of a systems-level approach to the problem, may contribute greatly to the design and implementation of successful wastewater recovery systems.[9]

UNIT OPERATIONS

Conventional wastewater treatment, as applied to both domestic and industrial wastewaters, is broken into three categories: primary treatment, involving gross physical treatment of a wastewater, such as by screening or settling; secondary treatment, involving chemical treatment of the wastewater, such as flocculation or chlorination; and tertiary treatment, involving biological treatment of the wastewater, e.g., through an activated sludge process (see "Waste Reduction Systems," "Pre-treatment Systems," "Primary Treatment Systems," "Biological Oxidation," for details on these processes). While any combination of primary, secondary, and tertiary treatment can be used as part of the recycling stream, this section focuses on unit operations that are most typically employed to deliver reuse-quality water to a food processing plant. Furthermore, this section does not focus on agricultural reuse of food-processing wastewaters, such as that realized in waste irrigation schemes (see "Land Applications" for details on land application of wastewaters).

Dissolved Air Flotation (DAF)

DAF systems are primarily used for the removal of fats, oils, and grease (FOG), and other suspended solids (TS) from food processing wastewaters.[10–13] To accomplish this task, DAF systems harness the buoyancy of FOG, augmented by microscopic air bubbles, to bring the contaminants to the top of a tank, where they can be skimmed off (Fig. 1). Air is introduced into the wastewater stream in a pressure vessel, operating at several atmospheres, and with air to solids mass ratios of 1%–4%.[14,15] When the aerated liquid enters the main DAF clarifier, which is at atmospheric pressure, it degasses and the air bubbles rise through the liquid, entrapping suspended solids and FOG on the way to the reactor surface. Chemicals such as coagulants and or surfactants may be introduced as part of the DAF unit operation to assist in solids removal. The remaining aqueous phase (subnatant) effluent typically contains only

Encyclopedia of Agricultural, Food, and Biological Engineering
DOI: 10.1081/E-EAFE 120007157
Copyright © 2003 by Marcel Dekker, Inc. All rights reserved.

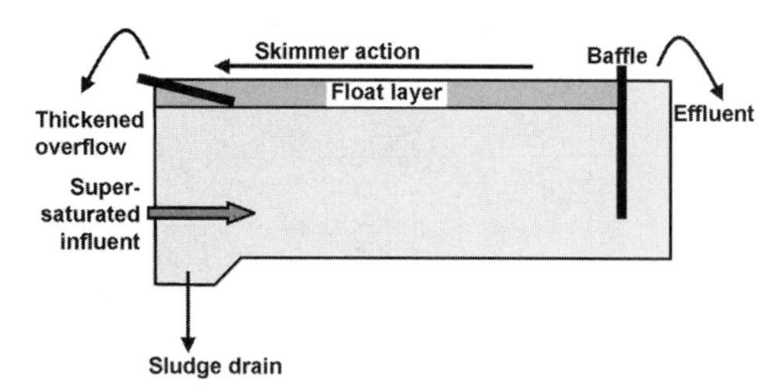

Fig. 1 Schematic of DAF operation. (Adapted from Ref. 10.)

a small fraction of the original FOG and TS; pathogenic organism counts may also be significantly decreased by DAF treatment.[16] However, DAF systems neither remove pathogenic organisms sufficiently for final reuse, nor do they effectively remove small non-FOG molecules, including salts, sugars, and starches. Despite this rather incomplete treatment, DAF systems may be an important front-end of a wastewater recycling system in which membrane processes (described below) are used downstream of the DAF to achieve further water purification. DAF units typically achieve solids removal with considerably smaller footprints than comparable gravity settling clarifiers—an important advantage at space limited facilities.[12,13]

DAF systems are typically sized empirically, based on pilot studies to determine the ratio of air to solids necessary for a desired level of contaminant removal.[10] Approximate DAF sizing may also be done based on prior experience, e.g., typical hydraulic loading rates are $0.7\,L\,min^{-1}\,m^{-2}$–$4\,L\,min^{-1}\,m^{-2}$[12]; typical solids loading rates are $0.04\,kg\,min^{-1}\,m^{-2}$–$0.28\,kg\,min^{-1}\,m^{-2}$.[14] DAF is a well-established technology, having been successfully employed in a wide variety of food processing operations.[11–13]

Membrane Filtration

Membrane filtration systems rely upon a physical barrier (the membrane) to separate contaminants from water (see "Mass Transport in Membranes," "Membrane Transport Models," "Reverse Osmosis," "Electrodialysis," "Ultrafiltration," "Design of Membrane Separation Systems" for details on membrane separation processes). Crossflow is typically employed to clear the concentrated contaminants from the influent side of the membrane. In the context of food processing water reclamation, membrane systems appear to be generally superior to evaporation and distillation separation processes, and the implementation of membrane separation technologies for water reuse has proved economically sound in a number of applications.[6] Membrane filtration processes are classified on the basis of the particle sizes removed: ultrafiltration removes particles down to 5 nm–100 nm, nanofiltration in the range 0.5 nm–5 nm, and reverse osmosis (RO) removing particles down to the 0.1 nm range.[6] Tradeoffs exist between membrane filter efficiency and cost. As filter pore size decreases (i.e., as one proceeds from ultrafiltration to RO systems), filter operating pressures, fouling rates, and capital costs all increase. More importantly, installing membrane filtration systems without regard to the influent stream composition may cause extremely poor performance and reliability. Therefore, successful deployments of ultrafiltration, nanofiltration, or RO in food processing plants have typically relied on pretreatment to reduce the contaminant loading to the filters.[5,14,15] Critical design parameters for filter systems include transmembrane pressures, number of stages, crossflow rates, membrane flux rates, permeate recovery rates, and power consumption.[16]

Ozonation

Ozone (O_3) is a highly unstable and strong oxidant that can be employed to remove pathogens and to oxidize organic carbon in wastewater. Upon dissolving in water, ozone reacts with water to form several intermediate species, including the free radicals HO_2 and HO.[10] Unlike chlorine disinfection, ozone does not raise concerns about residual organochloride compounds with potentially serious health risks,[6,20] and unlike ultraviolet (UV) sterilizations, ozone will work in turbid wastewaters where UV radiation does not penetrate effectively.[6] However, in waters containing bromide, ozone treatment causes the formation of carcinogenic bromate ions, the importance of which depends on the intended reuse of the water. Control measures for bromate ion include filtration, chemical additions, reductions in pH, addition of ammonia, and reductions in O_3 concentrations,[21–24] but a definitive strategy that is appropriate to food processing wastewater reuse applications does not appear to have emerged at the time of this writing. Critical design parameters for ozonoation systems include ozone delivery rate (mass/time), contact-chamber sizing, and contact time. Process variables, such as water temperature, pollutant concentration, and pH will significantly affect these parameters.

UV Sterilization

UV sterilization is a method for pathogen removal from certain food processing wastewaters.[25] UV radiation disrupts nucleic acids, thus killing or sterilizing bacteria and viruses.[2,10] UV radiation is generated by special high-intensity UV lamps, and is typically applied to a thin layer of flowing water. Water column thickness is critical,

Table 1 Typical removal efficiencies for wastewater recycling unit operations

Technology	FOG removal	TS removal	Salt removal	COD removal	Pathogenic bacteria	Pathogenic viruses
DAF	E	E	P	F	P	P
Ultrafiltration	N/A	F	P	P	F	P
Nanofiltration	N/A	G	F	F	E	F
R/O	N/A	E	E	E	E	E
Ozonation	P	P	P	F	E	E
UV	P	P	P	P	E	G

Key: E—excellent; G—good; F—fair; P—poor; N/A—not typically applicable.

because UV radiation intensity decreases with penetration depth into the liquid, as per Beers Law. For this same reason, UV sterilization is most appropriate for waters that transmit UV radiation effectively (i.e., are not opaque in UV wavelengths around 254 nm); otherwise the depth of penetration and sterilization effectiveness will be minimal.[10] Critical design parameters include UV energy delivered per volume of water, lamp type, lamp life expectancy, and lamp performance vs. time.[26] Process variables such as target organisms (and required irradiance to destroy them) and water turbidity will determine these parameters.

CONCLUSION

Table 1 provides a brief overview of the technologies described in this section along with estimates of their performance at removing certain pollutants from wastewaters. However, Table 1 should be interpreted with the caveats noted in the text.

The importance of performing a systems-level analysis of the water flow requirements in a facility, and of incorporating realistic economic analyses in the design process, cannot be overemphasized. Most importantly, careful engineering design must be conducted to ensure that any water recycling technologies protect the health of the public and of the plant employees, especially in light of threats from emerging organisms and bioterrorism.

ACKNOWLEDGMENTS

The thoughtful comments of Dr. S. Edward Law and the financial support of the University of Tennessee are gratefully acknowledged by the author.

REFERENCES

1. Mavrov, V.; Belieres, E. Reduction of Water Consumption and Wastewater Quantities in the Food Industry by Water Recycling Using Membrane Processes. Desalination **2000**, *131* (1–3), 75–86.
2. Palumbo, S.A.; Rajkowski, K.T.; Miller, A.J. Current Approaches for Reconditioning Process Water and Its Use in Food Manufacturing Operations. Trends Food Sci. Technol. **1997**, *8* (3), 69–74.
3. Chmiel, H. The EU Eco Audit Directive and Production-Integrated Environmental Protection as Illustrated by the Food Industry. Chem. Ing. Tech. **1995**, *67* (12), 1595–1602.
4. Moore, T. A Separable Feast—Membrane Applications in Food-Processing. EPRI J. **1994**, *19* (6), 17–23.
5. Miyaki, H.; Adachi, S.; Suda, K.; Kojima, Y. Water Recycling by Floating Media Filtration and Nanofiltration at a Soft Drink Factory. Desalination **2000**, *131* (1–3), 47–53.
6. Food Manufacturing Coalition for Innovation and Technology Transfer. Wastewater Reduction and Recycling in Food Processing Operations; 1997. http://www.fpc.unl.edu/fmc/7wastewater.html (accessed Aug 2001).
7. Zhang, S.Q.; Kutowy, O.; Kumar, A.; Malcolm, I.A. Laboratory Study of Poultry Abattoir Wastewater Treatment by Membrane Technology. Can. Agric. Eng. **1997**, *39* (2), 99–105.
8. Rajkowski, K.T.; Rice, E.W. Recovery and Survival of *Escherichia coli* O157: H7 in Reconditioned Pork-Processing Wastewater. J. Food Prot. **1999**, *62* (7), 731–734.
9. Almato, M.; Sanmarti, E.; Espuna, A.; Puigjaner, L. Rationalizing the Water Use in the Batch Process Industry. Comput. Chem. Eng. **1997**, *21*, 971–976.
10. Tchobanoglous, G. *Wastewater Engineering: Treatment, Disposal, and Reuse*, 3rd Ed; Metcalf & Eddy, Inc., Revised by George Tchobanoglous, Frank Burton; McGraw-Hill, Inc.: New York, 1991; Sec. 6.7, 242–248.
11. Carawan, R.E.; Valentine, E.G. Dissolved Air Flotation Systems (DAFs) for Bakeries, North Carolina Cooperative Extension Service Publication Number CD-43; Revised March 1996. http://www.bae.ncsu.edu/programs/extension/publicat/wqwm/cd43/cd43.html (accessed Feb 2002).
12. Krofta Technologies Corp. Wastewater Treatment Systems, Industrial Municipal—Dissolved Air Flotation DAF. Advantages of Dissolved Air Flotation Clarifiers, 2001. http://www.wastewater-treatment-equipment.com/wastewater-treatment-equipment/index.html (accessed Feb 2002).

13. Hi-Tech Environmental, Inc. Dissolved Air Flotation, 1998. http://www.hi-techenv.com/DAF.htm (accessed Feb 2002).
14. ConSep Pty Limited. Dissolved Air Flotation—D.A.F Process Design Basics. http://www.consep.com.au/Prod_westech_daf.html (accessed Oct 2002).
15. Matsui, Y.; Fukushi, K.; Tambo, N. Modeling, Simulation and Operational Parameters of Dissolved Air Flotation. J. Water Serv. Res. Technol.-Aqua **1998**, *47* (1), 9–20.
16. Edzwald, J.K.; Tobiason, J.E.; Parento, L.M.; Kelley, M.B.; Kaminski, G.S.; Dunn, H.J.; Galant, P.B. Giardia and Cryptosporidium Removals by Clarification and Filtration Under Challenge Conditions. J. Am. Water Works Assoc. **2000**, *92* (12), 70–86.
17. Chmiel, H.; Mavrov, V.; Belieres, E. Reuse of Vapour Condensate from Milk Processing Using Nanofiltration. Filtration Sep. **2000**, *37* (3), 24–27.
18. Mavrov, V.; Fahnrich, A.; Chmiel, H. Treatment of Low-Contaminated Waste Water from the Food Industry to Produce Water of Drinking Quality for Reuse. Desalination **1997**, *113* (2–3), 97–203.
19. Cheryan, M. *Ultrafiltration and Microfiltration Handbook*; Technomic Pub. Co.: Lancaster, PA, 1998.
20. Rice, R.G. Ozone in the United States of America—State-of-the-Art. Ozone-Sci. Eng. **1999**, *21* (2), 99–118.
21. Gordon, G.; Gauw, R.D.; Emmert, G.L.; Walters, B.D.; Bubnis, B. Chemical Reduction Methods for Bromate Ion Removal. J. Am. Water Works Assoc. **2002**, *94* (2), 91–98.
22. Kirisits, M.J.; Snoeyink, V.L.; Chee-Sanford, J.C.; Daugherty, B.J.; Brown, J.C.; Raskin, L. Effect of Operating Conditions on Bromate Removal Efficiency in BAC Filters. J. Am. Water Works Assoc. **2002**, *94* (4), 182–193.
23. Galey, C.; Mary-Dile, V.; Gatel, D.; Amy, G.; Cavard, J. Controlling Bromate Formation. J. Am. Water Works Assoc. **2001**, *93* (8), 105–115.
24. Pinkernell, U.; von Gunten, U. Bromate Minimization During Ozonation: Mechanistic Considerations. Environ. Sci. Technol. **2001**, *35* (12), 2525–2531.
25. Lazarova, V.; Savoye, P.; Janex, M.L.; Blatchley, E.R.; Pommepuy, M. Advanced Wastewater Disinfection Technologies: State of the Art and Perspectives. Water Sci. Technol. **1999**, *40* (4–5), 203–213.
26. Unit Liner Co. Factors Affecting UV Sterilization: Technical Advice from Emperor Aquatics, 2000. http://www.pondliner.com/UVFactors.htm (accessed Feb 2002).

Properties of Concentrated Foods

Roberto A. Buffo
University of Minnesota, St. Paul, Minnesota, U.S.A.

P

INTRODUCTION

Concentrated foods are obtained by removing water by evaporation. Evaporative concentration is used to produce a concentrate containing the desired solids in solution. Its primary objectives are to reduce the weight and volume of products, thereby lowering packaging, transportation, and storage costs, reduce energy consumption if subsequently dried and reduce water activity to enhance storage stability.[1] A discussion of their thermal, rheological, and electrical properties follows.

THERMAL PROPERTIES

There are four basic properties linked to temperature: 1) density; 2) specific heat; 3) thermal conductivity; and 4) thermal diffusivity. Knowledge of these properties is essential to researchers and engineers for the optimum design of heat transfer processes.[2]

Density

The density of a substance is the ratio between its mass and the volume it occupies. Units are $g\,cm^{-3}$, $kg\,m^{-3}$, $lb\,ft^{-3}$. Density changes inversely with temperature: it decreases as temperature increases (above 4°C). Density differences in heated fluids provide the driving force for natural convection.[3]

Water has its maximum density of $1\,g\,cm^{-3}$ at 4°C. The addition of solids to water, with the exception of fat, will increase its density. The specific gravity (SG) of a liquid is defined as the dimensionless ratio between a certain mass of the liquid and an equal volume of water. Specific gravity can also be expressed in terms of density values:

$$SG = \rho_L/\rho_w \tag{1}$$

where ρ_L is the density of the unknown and ρ_w is the density of water. Evidently, any SG value applies only at a specific temperature.[3]

The volumetric pycnometer method is the most suitable SG measurement technique for liquid foods. Density of water and liquids of relatively low concentration (below 40% total solids) is measured using 28 mL glass pycnometers within the 0–100°C temperature range. For highly concentrated liquid foods (above 40% total solids), 12 mL pycnometers made of an aluminum alloy are used within the same temperature range.[2] The following readings are taken: 1) the weight of the empty bottle w_1; 2) the weight of the bottle full of water w_2; and 3) the weight of the bottle full of liquid w_3. Then, SG of the liquid is:

$$SG = (w_3 - w_1)/(w_2 - w_1) \tag{2}$$

Hydrometers work on the principle that a floating body displaces its own weight of fluid. The instrument is placed in the fluid and the density is directly read from the scale on the stem. Hydrometers are easy to use and available in a range of sizes for different applications. There are also special hydrometers and scales: the Brix saccharometer shows directly the percentage sucrose by weight in the solution at the temperature indicated on the instrument; lactometers measure the density of milk and alcohol-ometers the density of alcohol solutions (alcohol 0%–100% in volume); the Baumé scale for fluids heavier than water goes from 0° (SG = 1) to 66° (SG = 1.842).[3]

The literature reports numerous empirical equations to estimate density of concentrated products; two examples are provided:

Phipps[2]—density of cream between 40 and 80°C and fat content up to 40%:

$$\rho = 1038.2 - 0.17T - 0.003T^2 - [(133.7 - 475.5T^{-1})/X_F] \tag{3}$$

where T is temperature and X_F fractional content of fat.

Choi[2]—density of tomato juice between 30 and 80°C and water content range from 20% to 95.2%:

$$\rho = \rho_w X_w + \rho_s X_s \tag{4}$$

where X_w and X_s are fractional content of water and solids, and ρ_w and ρ_s are density of water and solids, which depend on temperature as follows:

$$\rho_w = 999.89 - 0.060334T - 0.003671T^2 \tag{5}$$

$$\rho_s = 1469.3 + 0.54667T - 0.006965T^2 \tag{6}$$

Encyclopedia of Agricultural, Food, and Biological Engineering
DOI: 10.1081/E-EAFE 120006962
Copyright © 2003 by Marcel Dekker, Inc. All rights reserved.

Specific Heat

Specific heat denotes the amount of heat required to change 1° of temperature of one unit mass of the substance. Most common units are cal g^{-1}°C but also J kg^{-1} K and BTU lb^{-1}°F.[4]

Since, water has a much higher specific heat than most other food constituents, the specific heat of a food is significantly affected by the amount of water present and the physical state of the water.[3] Specific heat is temperature dependent, although it may be considered a constant property over relatively small temperature intervals.[5] It is possible to predict the specific heat of a food from its composition:

$$C_p = m_w C_{pw} + m_c C_{pc} + m_p C_{ppr} + m_f C_{pf} + m_a C_{pa} \quad (7)$$

where w is water, c, carbohydrate, pr, protein, f, fat, a, ash, and m and C_p are mass fraction and specific heat at constant pressure of the respective components.[3]

The most common way of measuring specific heat of concentrated foods is by the method of mixtures, in which the specimen of a known mass and temperature is dropped into a calorimeter of known specific heat containing water of a known mass and temperature. The specific heat is then computed from a heat balance between the heat gained or lost by the water and calorimeter and that lost or gained by the specimen:

$$C_{pcl} m_c (T_i - T_e) + C_{ps} m_{sm} (T_i - T_e) = C_{pw} m_w (T_e - T_w) \quad (8)$$

where cl, sm, and w refer to calorimeter bucket, sample, and water, respectively, T_i is initial temperature of sample and bucket, T_w initial water temperature, and T_e equilibrium temperature.[2]

Thermal Conductivity

If the temperature gradient ΔT between the two surfaces through which heat is flowing is unity, the quantity of heat q that will flow in unit time t across unit area A is called thermal conductivity k. Units are W m^{-1} K, Btu hr^{-1} ft °F.[5]

Thermal conductivity of concentrated foods depends on their density, structure, and composition, primarily fat and water. Since, the thermal conductivity of fat is lower than that of water, high levels of fat will decrease thermal conductivity. This has an important impact on the rates of heating and cooling and, thus, on process efficiency and potential flavor deterioration due to overheating. Since, the thermal conductivity of ice is greater than that of water, frozen foods will exhibit greater conductivity than their unfrozen counterparts. On either side of the freezing point, thermal conductivity depends on temperature. It has also been found in the case of liquids to increase with pressure and decrease with increasing concentration of solutes.[4]

Methods of measurement of thermal conductivity can be divided into two broad categories: steady-state and unsteady-state heat transfer. Although the former are simpler, they are also limited by the fact that a steady-state condition may take several hours to attain. In a common steady-state procedure, the sample is confined in a cylinder that is assumed to be infinite (i.e., negligible end effects). A central heat source provides heat to be transferred through the material with a thermal conductivity computed as:

$$k = [P \ln(r_2/r_1)]/[2\pi L(T_1 - T_2)] \quad (9)$$

where P is the power used by the central heater, L, the length of the cylinder, and T_1 and T_2 are the temperatures of the specimen at radius r_1 and r_2, respectively.[2]

One of the most common unsteady-state procedures is the Fitch method.[2] A sample is placed between two copper blocks fitted with thermocouples. The top block serves as the base of a well-insulated isothermic vessel filled with the specimen of interest. The lower block is of known mass and embedded in insulation. The heat input necessary to maintain the vessel at the particular temperature is measured by the temperature rise with time of the lower copper block:

$$k = 2.303(m\, C_p\, L/A)[\log(T_1/T_2)/t] \quad (10)$$

where m and C_p are mass and specific heat of the heat sink, T_1 and T_2 temperature differences at the beginning and end of the experiment, t, time, and A and L are area and thickness of the test specimen, respectively.

Thermal Diffusivity

Thermal diffusivity α is the ratio of the thermal conductivity to the specific heat of the product multiplied by its density, as follows:

$$\alpha = k/(\rho C_p) \quad (11)$$

Units are $m^2\,sec^{-1}$, $ft^2\,sec^{-1}$. In physical terms, thermal diffusivity gives a measure of how quickly the temperature of a food product will change when it is heated or cooled. Materials with high α will heat or cool quickly; conversely, substances with low α will heat or cool slowly. Thus, thermal diffusivity is an important property when considering unsteady-state heat transfer situations. It is most commonly determined mathematically based on k, C_p, and ρ values.[3]

RHEOLOGICAL PROPERTIES

Fluid and semisolid foods exhibit a wide variety of rheological behavior ranging from Newtonian to time dependent and viscoelastic. Fluid foods containing relatively large amounts of dissolved low molecular weight compounds (e.g., sugars) and no significant amounts of polymers (e.g., proteins, pectins, and starches) or insoluble solids can be expected to exhibit Newtonian behavior (water, milk, sugar syrups, honey, edible oils, and filtered juices). A small amount ($\cong 1\%$) of a dissolved polymer can substantially increase viscosity and alter flow characteristics to a non-Newtonian behavior: shear thinning (salad dressings, concentrated fruit juices) or shear thickening (partially gelatinized starch dispersions). It is important to note that whereas rheological properties are changed, magnitudes of thermal properties remain relatively close to those of water. Rheological assessment of concentrated foods is important in quality control, texture, and processing.[6]

ELECTRICAL PROPERTIES

Electrical conductivity (ε) depends on food composition (minerals increase ε whereas fat decreases it), density (lower at higher densities due to lower molecular mobility), and temperature (lower at lower temperatures with a sharp drop at the freezing point).[4] Conductivity measurements are used to monitor the concentration of sugar liquor during the concentration process that proceeds to crystallization. There is an inverse relationship between conductivity and the degree of super saturation.[3] The oil phase of an emulsion can be determined by measuring electrical conductivity as there will be a decrease in ε as the oil content increases.[7]

Dielectric properties are important regarding the application of microwave energy. It is related to the oscillation of asymmetric molecules such as water, which causes intermolecular friction that is dissipated as heat.[4]

REFERENCES

1. Chen, C.S.; Hernández, E. Design and Performance Evaluation of Evaporation. In *Handbook of Food Engineering Practice*; Valentas, K.J., Rotstein, E., Singh, R.P., Eds.; CRC Press: New York, 1997.
2. Choi, Y.; Okos, M.R. Thermal Properties of Liquid Foods: Review. In *Physical and Chemical Properties of Food*; Okos, M.R., Ed.; American Society of Agricultural Engineers: St. Joseph, MO, 1986; 35–77.
3. Lewis, M.J. *Physical Properties of Foods and Food Processing Systems*; Ellis Horwood, Ltd.: Chichester, England, 1987.
4. Szczesniak, A.S. Physical Properties of Foods. In *Physical Properties of Foods*; Peleg, M., Bagley, E.B., Eds.; AVI Publishing Company, Inc.: Wesport, CT, 1983; 1–42.
5. Mohsenin, N.N. *Thermal Properties of Foods and Agricultural Materials*; Gordon and Breach, Inc.: London, 1980.
6. Rao, M.A. Introduction. In *Rheology of Fluid and Semisolid Foods*; Rao, M.A., Ed.; Aspen Publishers, Inc.: Gaithesburg, MD, 1999; 1–24.
7. McClements, D.J. *Food Emulsions: Principles, Practices and Techniques*; CRC Press: New York, 1999.

Properties of Food Powders

Hiromichi Hayashi
Tokyo University of Agriculture, Tokyo, Japan

INTRODUCTION

Human beings, nowadays, are able to enrich their eating habits by powdered food. For example, flour is changed into various processed food, such as bread, noodles, pasta, and confectionery by adding moisture, fabricating, and heating. Powdered food is made from various raw foods such as grain, milk, fruits, vegetables, marine products, and seasoned by spray drying or grinding.

It is difficult to obtain small and uniform particle sizes at the grinding and classifying stage while making powdered food from solid food. In order to obtain the required particle size, it is necessary to select suitable equipments for raw food and pulverize and classify them repeatedly. The characteristics of fine powder are its dust forming ability, cohesiveness, and easy absorption of moisture from the atmosphere; these characteristics can be improved if the fine powder is agglomerated to larger particle size by fluidized bed with spray water or steam. Powdered food are indispensable processed food.

Classification criteria of powdered food may vary based on convenience or applications. On the basis of processes, they are classified as follows:

1. Ground powder: powder sugar, spice.
2. Spray-dried powder: milk powder, egg powder.
3. Drum-dried powder: mashed potato, casein, lactose.
4. Agglomerated powder: instant coffee, instant milk.
5. Precipitated powder: protein isolate.
6. Crystalline powder: salt and sugar.
7. Mixture of powder: dry fruits drink.

Powdered-dried food are always compared with fresh foods for flavor, color, taste, and texture because the quality of products change relatively during unstable preservation.

PHYSICAL PROPERTIES

Particle Size and Shape

Processing methods such as grinding, drying, and sieving influence properties of a particle. The particle size of powdered food vary from 1 to 1000 μm, i.e., 1–99 μm in smaller-sized particles such as starch, flour obtained by grinding, 100–1000 μm in bigger-sized particles such as instant coffee and instant soup obtained by freeze drying.

Powdered food vary in size and shape based on the processing method used: irregular shape to flour and cocoa by grinding, globular shape in starch and dried milk by flash and spray drying, and grain shape to sugar and salt by crystallization.

Density

Density is of three types: loose, tapped, and particle density, and is expressed in kg m^{-3} in SI unit. Table 1 shows the particle size, bulk, and particle density of typical powdered foods.

Loose bulk density

Generally, the density is 300–800 kg m^{-3} in powdered food and the powder pours into a container without tapping. This density particles have considerable void outside the particle in a container. The void depends upon the particle size, size distribution, and shape. The particles with fine size or high fat contents show cohesive property. There is comparatively an open-bed structure in the container by the attraction of cohesive particles. Particles with loose-bulk density are affected by these properties.

Tapped bulk density

Powder is compressed by two methods: mechanical and natural compression, the first one vibrates with tapping, while the other occurs naturally when it is stored up in a silo tank and vibrated during conveying. The compressed stress of the powder is not very large during storage in the silo tank. The cohesive powder, however, encounters trouble during discharge at the outlet of a silo tank because of bridge formation.

Particle density

The particle solid density of powdered food is generally 1,400 kg m^{-3}–2,040 kg m^{-3} although it depends on their

Encyclopedia of Agricultural, Food, and Biological Engineering
DOI: 10.1081/E-EAFE 120006966
Copyright © 2003 by Marcel Dekker, Inc. All rights reserved.

P

Table 1 Physical property of powdered foods[1]

	D_a (μm)	B_l (kg m^{-3})	B_t (kg m^{-3})	P_d (kg m^{-3})	P_t (kg m^{-3})
Instant coffee (FD)	2100	238	262	1270	1600
Instant coffee (SD)	461	680	850	1300	1600
Whole dried milk (SD)	98	390	610	1250	1450
Nonfat dried milk (SD)[a]	218	594	730	1365	1625
Sugar	426	841	847	2039	2039
Wheat	23	422	858	1380	1470
Starch	30	690	1020	1620	1620
Salt	1180	1300	1370	2020	2020

Nomenclature: D_a, arithmetic mean particle size; B_l, loose bulk density; B_t, tapped bulk density; P_d, particle density; P_t, true particle density; FD, freeze drying; SD, spray drying.
[a] Agglomerated.

Table 2 Compressibility, void, and porosity of powdered foods[1]

	Instant coffee (FD)	Instant coffee (SD)	Whole dried milk	Nonfat dried milk	Sugar	Flour	Starch	Salt
Cp	9	20	36	19	0.7	51	32	5
V	81	86	60	59	62	69	57	41
Po	21	19	14	16	7	0	0	0

Nomenclature: Cp: Compressibility (%), V: Void (%), Po: Porosity (%).
Source: From Ref. 2.

composition. Food generally contains fat, protein, carbohydrate, water, and mineral. The densities of fat, protein, carbohydrate, water, and mineral is 930, 1350, 1670, 1000, and 4120 in kg m^{-3}, respectively.

Spray-dried food usually has air pores inside the particle as a result of puffing, because a liquid drop instantly is dried at high temperature. The pore volume depends on inlet air temperature and atomizing methods such as centrifugal or pressure. The pore volume by the centrifugal method is higher than that by pressure method.

Compressibility

This is calculated by the following equation:

$$C = (B_t - B_l)/B_t$$

where C is compressibility (%), B_t the tapped bulk density (kg m^{-3}), and B_l is the loose bulk density (kg m^{-3}).

The flow ability of powdered food reveals bad shape when the compressibility exceeds 20%. Table 2 shows compressibility, void, and porosity of powdered food.

Angle of Repose

The angle of repose is an indispensable parameter while designing the processing, storage, and conveying for

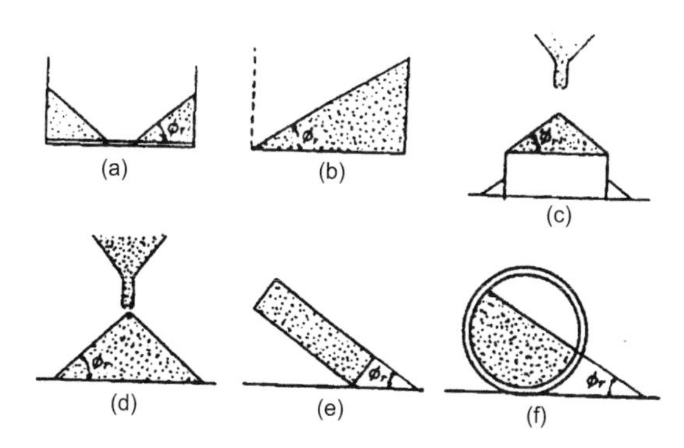

a) **Discharge method**
b) **Bed rapture method**
c) **Fall from fix funnel method**
d) **Tilting box method**
e) **Rotating drum method**

Fig. 1 Measuring methods for angle of repose.

Table 3 Angle of repose for powdered foods

Name of powder	Whey powder	Dried whole milk	Dried nonfat milk	Wheat flour	Wheat grain	Glutinous starch powder	Starch	Soy bean grain
Angle or repose	52	56	49	56	27	41	53	23

powdered food. This is a standard test for determining the free flow in powdered food.

Various methods in the test are as follows (shown in Fig. 1.)

1. Fall from fix funnel method
2. Bed ruptures method
3. Rotating drum method
4. Tilting box method

The angle of repose is an angle when powder accumulates in a cone shape on a level surface and may vary depending on the physical properties (particle size, density, shape etc.), composition of powder, and humidity in the atmosphere. Powder with high cohesiveness is difficult for determination of angle of repose because the angle is irregular and does not show reproduction. It corresponds generally to angle of repose and the internal friction in powder, but this relationship is not applicable in the case of wet and cohesive powder. Table 3 shows the angle of repose for typical powdered foods.

Carr[3] has distinguished the flow ability of powder based on angle of repose as follows:

1. Less than 35°: free flow ability.
2. 36–45°: a little cohesiveness.
3. 46–55°: cohesiveness.
4. More than 56°: high cohesiveness.

REFERENCES

1. Micha, P.; Mannheim, C.H.; Passy, N. Flow Properties of Some Food Powders. J. Food Sci. **1973**, *38*, 959–964.
2. Hayeshi, H. Physical Properties of Powdered Foods, Internal research paper in Food Science Dept., Tokyo University of Agriculture, 1998.
3. Carr, R.L. Chem. Eng. J. **1965**, *18*, 163.

Psychrometrics of Animal Environment

Steven J. Hoff
Iowa State University, Ames, Iowa, U.S.A.

P

INTRODUCTION

The production efficiency, comfort, and health status of animals reared for production agriculture are intimately tied with the total environment surrounding the animal. The environment includes the physical structure providing protection from the elements and the quality of air in which animals live. The quality of air is in turn associated with various gas concentrations, dust levels, and the mix of air and water vapor surrounding the animal. The science of *psychrometrics* refers to this latter component of the animal's environment; i.e., the relation between air, or more precisely "dry air (da)" and moisture, or more precisely "water vapor (wv)."

Psychrometrics is a branch of thermodynamics dealing specifically with the relation between dry air and water vapor mixtures and is a large component of heating, ventilating, and air-conditioning fields of engineering. Many aspects of an animal's production efficiency and comfort are directly tied to the relative mixtures of dry air and water vapor. This mixture, which we often take for granted, can be the difference between comfortable and efficiently producing animals and those that are stressed thermally by their environment. It is important to understand some underlying principles that define this mixture of dry air and water vapor and ultimately how these relative mixtures affect modern-day animal production systems.

WHAT CONSTITUTES AN OPTIMUM THERMAL ENVIRONMENT FOR ANIMALS?

An optimum thermal environment is an environment that does not require an extraordinary response by the animal to maintain a constant core temperature. This is a complicated way of saying that the animal, being warm blooded, is always in a mode of sustaining life. Life is sustained in one major way by keeping its core temperature as constant as possible. If the thermal environment is "cold," the animal responds by huddling, shivering, redirecting blood flow to the core, or excessively eating because too much heat energy is lost to the environment, and to make up this difference, more energy is needed or needs to be conserved. Likewise, a thermal environment that is "hot" invokes heat loss increasing mechanisms such as panting, lying spread out on cool floors, wallowing in mud, and suppressed eating, to name a few. These responses to "cold" and "hot" are all extraordinary measures required by the animal to maintain a constant core temperature, and all imply that an optimum thermal environment does not exist.

PSYCHROMETRIC FUNDAMENTALS

An animal will respond in adaptive ways to cold and hot environments. The definition of cold and hot environments varies considerably amongst animal species, animal age, etc. but ultimately, the dry air and water vapor mixture surrounding an animal needs to be assessed and understood before deciding upon the suitability of the environment. Anyone who has ever watched a weather report has already become familiar with many of the terms used to assess dry air and water vapor mixtures. What follows is a brief description of dry air and water vapor mixtures and the methods used to assess the suitability of this mixture for animal production systems.

The earth's atmosphere consists of a mixture of gases (i.e., the dry air part) plus a variable quantity of water vapor. The composition of dry air (volume basis) consists of nitrogen (78%) and oxygen (21%) with trace fractions of argon, carbon dioxide, and other less notable components. In addition to these "dry air" components, water vapor to varying percentages is present that combine to yield a mixture that defines, in part, the quality of the thermal environment for animals. This mixture is referred to moist air.

A thorough understanding of the physical and thermodynamic properties of dry air/water vapor mixtures, i.e., psychrometrics, is a fundamental requirement when designing and assessing an animal's environment. The energy exchange that an animal experiences with its environment is to a large degree a function of the dry air/water vapor mixture presented to the animal.

Encyclopedia of Agricultural, Food, and Biological Engineering
DOI: 10.1081/E-EAFE 120006911
Copyright © 2003 by Marcel Dekker, Inc. All rights reserved.

Properties of Dry Air and Water Vapor Mixtures

The weight of water vapor that can be contained in the air varies with the air temperature and barometric pressure. Given the same barometric pressure, warm air is capable of holding more water vapor than cold air. The water vapor content ranges from zero for pure dry air to a saturated state (i.e., 100% relative humidity) for a given level of temperature and barometric pressure. The barometric pressure, volume, weight, and thermal properties of dry air/water vapor mixtures can be related by a series of physical laws that, though developed for a "perfect" gas, are applicable to processes involving dry air/water vapor mixtures under normal conditions with sufficient accuracy.

Ideal Gas Law: The ideal gas law states that the thermodynamic property of moist air, treated as a perfect gas, can be determined by the barometric pressure and two other independent properties. The mathematical relationship among the properties of a perfect gas is[1]:

$$PV = MRT \tag{1}$$

where P is the barometric pressure, lb_f/ft^2; V is the volume, ft^3; M is the mass, lb_m; R is the gas constant, ft-lb_f (lb_m-°R); T is the absolute temperature, °R (°F+460)

A dry air/water vapor mixture behaves sufficient enough to a perfect gas that the ideal gas law can be used to analyze processes involving this mixture for assessing an animal's thermal environment. The gas constant for dry air, $R_a = 53.35$ ft-lb_f (lb_m-°R), and the gas constant of water vapor, $R_w = 85.78$ ft-lb_f (lb_m-°R), can be assumed as functionally accurate for the temperature and pressure ranges typical of animal production systems.[1]

Dalton's Law of Partial Pressures: Dalton's Law states that the total barometric pressure is equal to the added pressures from each component. In a mixture of water vapor with dry air, according to Dalton's Law of Partial Pressures, each contributes a partial pressure with the sum equal to the barometric pressure

$$P = P_a + P_w = M_a R_a T_a / V_a + M_w R_w T_w / V_w \tag{2}$$

where P_a is the partial pressure exerted by the dry air component of a mixture, lb_f/ft^2; P_w is the partial pressure exerted by the water vapor component of a mixture, lb_f/ft^2.

Eq. 2 can be rearranged by making the assumption that the dry air and water vapor components are uniformly mixed, thus both will experience the same temperature for a similar volume

$$P = T/V(M_a R_a + M_w R_w) \tag{3}$$

In addition, as the volume and temperature of the mixture are equal, from Eq. 1 the following can be written

$$P_w/P_a = (M_w R_w)/(M_a R_a) \tag{4}$$

Therefore, if the barometric pressure and the weight of water vapor are known, the partial pressures exerted by the dry air and water vapor components can be determined.

All psychrometric properties used to assess the thermal environment for an animal use these basic physical laws as a starting point for analysis. There exist seven basic properties used to define the "state" of dry air/water vapor mixtures. These are in turn grouped into "humidity" terms, "temperature" terms, and "auxiliary" terms. In addition to these seven basic properties, knowledge of the barometric pressure is required for a total of eight basic properties used to define the state of dry air/water vapor mixtures. Each grouping of terms is discussed in more detail below.

Humidity Psychrometric Properties

Two basic properties are used to define the water vapor content mixed with dry air. These are called relative humidity and absolute humidity.

Relative Humidity: The relative humidity (φ) is defined as the ratio of the actual water vapor pressure in the air (P_w) to the water vapor pressure if the air was saturated with moisture at the same temperature (P_{ws}). The relationship for relative humidity expressed as a percent is given by:

$$\varphi(\%) = 100(P_w/P_{ws}) \tag{5}$$

The water vapor pressures for a space saturated with water vapor (P_{ws}) can be found directly from any standard steam table. The relative humidity is commonly referred to in all weather forecasts. It is important to note that relative humidity is in fact relative. Warm air at 100% relative humidity will contain an absolute amount of water vapor that is higher than cold air at 100% relative humidity, assuming both are at the same barometric pressure.

Absolute Humidity: The absolute humidity (W) is the actual weight of water vapor, in pounds of water vapor per pound of dry air. The base of one pound of dry air is used since it is a constant for any change of condition. From Eq. 4, the absolute humidity for a dry air/water vapor mixture can be written as

$$W = M_w/M_a = (R_a P_w)/(R_w P_a)$$
$$= (R_a/R_w)\{P_w/(P - P_w)\} \tag{6}$$

Substituting the numerical values for the dry air and water vapor gas constants R_a (53.35 ft-lb/(lb-R)) and R_w

(85.78 ft-lb/(lb-R)), the absolute humidity becomes

$$W = 0.622\{P_w/(P - P_w)\} \tag{7}$$

where M_w is the mass of actual water vapor in mixture, lb_m, wv; M_a is the mass of dry air, lb_m, da.

Temperature Psychrometric Properties

Three basic properties are used to define the "temperature" of dry air/water vapor mixtures. These are called dry bulb, dew point, and wet bulb temperatures.

Dry Bulb Temperature (T_{db}): The dry bulb temperature is the temperature most of us are familiar with and can be measured directly with a common mercury thermometer (or equivalent). When weather forecasts are given, the temperature mentioned is technically the dry bulb temperature.

Dew Point Temperature (T_{dp}): The dew point temperature is the temperature of a dry air/water vapor mixture at which moisture will start to condense out of the air as the air is cooled *at constant barometric pressure and absolute humidity*. Experimentally the dew point temperature of a dry air/water vapor mixture can be determined by bringing the mixture into contact with a polished metal surface whose temperature can be both controlled and measured. If the metal surface is slowly cooled, the portion of mixture in contact with the surface will also be cooled. When the mixture reaches its dew point temperature, the metal surface will become fogged as the water vapor starts to condense from the air. Therefore, by measuring the temperature of the surface when the moisture first appears the dew point temperature of the mixture is determined. The polished surface makes the appearance of fog readily visible. The dew point temperature is commonly given on weather forecasts.

Wet Bulb Temperature (T_{wb}): The wet bulb temperature is measured with a common mercury thermometer (or equivalent) with the sensing bulb covered with a water-moistened cloth or wick. Surrounding air is allowed to move past this wetted wick allowing liquid water to evaporate from the wick. The evaporation of the water from the wick into the surrounding air attains a steady state temperature in which sensible heat is transferred just rapidly enough from the surroundings to provide energy for evaporation. The wetted bulb cools by evaporation of the water from the bulb, thus lowering the temperature indicated below the dry bulb temperature of the air. The drier the surrounding air, the greater the rate of evaporation and lower the wet bulb temperature relative to the dry bulb temperature. The wet bulb temperature was "invented" because, unlike dew point temperature, it is a relatively easy measurement to make in the field.

Auxiliary Psychrometric Properties

Two auxiliary properties are used in addition to the humidity and temperature properties described previously. These are called specific volume and enthalpy.

Specific Volume: The specific volume defines the volume of a dry air/water vapor mixture attained for each pound of dry air at a constant barometric pressure. The base of one pound of dry air is again used because the pounds of dry air in psychrometric processes remain constant after steady-flow conditions are established.

Standard air at a barometric pressure of 14.7 psia, 70°F dry bulb temperature, and zero water vapor content has a specific volume of 13.34 ft^3 per pound of dry air. However, as the temperature is increased the volume of the air will increase since the addition of heat causes it to expand at constant pressure. In addition, as water vapor is added to the air the volume will increase if the barometric pressure is to remain constant. The specific volume (V) of moist air (ft^3/lb_m, da) can be calculated from

$$V = V_a + \mu V_{as} \tag{8}$$

where V_a is the specific volume of dry air, ft^3/lb_m, da; V_{as} is the difference between the volume of moisture at saturation and the specific volume of dry air itself ($V_s - V_a$); V_s is the volume of moist air at saturation per pound of dry air, ft^3/lb_m, da; μ is the degree of saturation, the ratio of the specific humidity of moist air to the specific humidity of saturated air at the same dry bulb temperature and pressure.

Enthalpy: Enthalpy is a thermodynamic term used to define the amount of energy contained in a dry air/water vapor mixture. The energy contained in the mixture can be present both as sensible heat (indicated by dry bulb temperature) and latent heat of vaporization (energy content of the water vapor). For convenience, the enthalpy is expressed as Btu per pound of dry air. The values of heat content are neither necessarily absolute nor do they need to indicate the total heat contained. It is often more convenient to have the various expressions for heat content to represent the difference in heat energy between a fixed reference condition or datum which has an assigned enthalpy value of zero and the condition under consideration. Thus, commonly used datum for dry air is 0°F and for the water vapor it is 32°F. The enthalpy (H) of a dry air/water vapor mixture (Btu per pound of dry air) can be calculated from

$$H = H_a + H_w \tag{9}$$

where H_a is the energy content of the dry air component, Btu/lb_m, da; H_w is the energy content of the water vapor component, Btu/lb_m, wv.

In the form of Eq. 9, the units do not match and thus to convert H_w to a per unit mass of dry air, the absolute humidity is used as follows

$$H = H_a + WH_w \tag{10}$$

where W is the absolute humidity, lb_m, wv/lb_m, da.

The enthalpy of the dry air part of the mixture is determined by

$$H_a = C_{p,da}(T_{db}) \tag{11}$$

where $C_{p,da}$ is the specific heat of dry air, 0.240 Btu/lb_m, da-°F and the enthalpy of the water vapor part is determined by

$$H_w = H_{fg} + C_{p,wv}(T_{db}) \tag{12}$$

where H_{fg} is the latent heat of vaporization at 0°F (= 1061 Btu/lb_m, wv), $C_{p,wv} = 0.444$ Btu/lb_m,wv-°F.

Substituting Eqs. 11 and 12 into Eq. 10 results in a description of the energy content of a dry air/water vapor mixture

$$H = 0.240T_{db} + W(1061 + 0.444T_{db}) \tag{13}$$

The energy content of moist air (H) consists of the energy contained in the vapor as a result of vaporization (H_{fg}, assumed to occur at 0°F) and the sensible energy required to heat this water vapor to the current dry bulb temperature ($C_{p,wv}T_{db}$). In other words, the enthalpy "tracks" the energy gained by each mass of water vapor contained in a dry air/water vapor mixture. At some point in time, the water vapor in the air was liquid water. In psychrometrics, it is assumed that this liquid water was vaporized at 0°F, thus absorbing 1061 Btu of energy for every lb_m of water vapor vaporized. Once vaporized, this water vapor had to at some point gain sensible energy to achieve the current dry bulb temperature.

THE PSYCHROMETRIC CHART

The two humidity, three temperature, and two auxiliary properties are used, along with the barometric pressure, to define the state of dry air/water vapor mixtures. It turns out that if the barometric pressure is known with any two other properties (except for combinations of wet bulb temperature and enthalpy), the remaining five properties can be determined. A very useful tool for determining all properties associated with dry air/water vapor mixtures is a psychrometeric chart.[1] A generalized schematic of a psychrometric chart is given in Fig. 1. The coordinates have values of dry bulb temperature (T_{db}) along the x-axis and the absolute humidity (W) on the y-axis. The additional properties of moist air that can be obtained from the psychrometric chart are shown in Fig. 1. The other properties given are relative humidity, dew point temperature, wet bulb temperature, enthalpy, and specific volume. The intersection of any two property lines establishes a given state from which all other properties can be found.

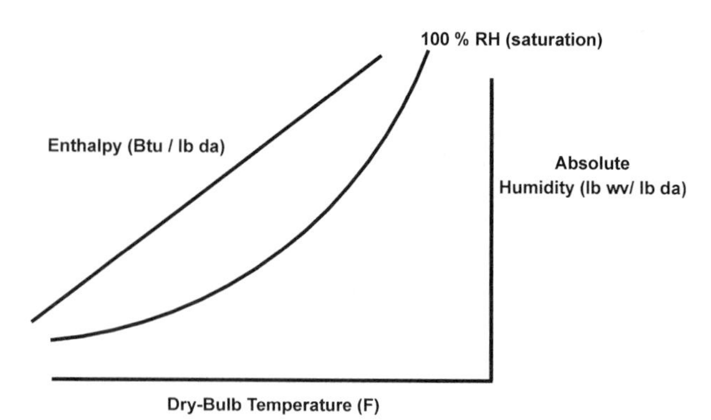

Fig. 1 Properties of dry air/water vapor mixtures given on a psychrometric chart.

USING THE PSYCHROMETRIC CHART TO ASSESS AN ANIMAL'S ENVIRONMENT

The psychrometric chart can be used to assess the thermal quality that an animal is subjected to. One of the primary uses is to assess the combined influence of dry bulb temperature (T_{db}) and water vapor content (W) on heat stress conditions for the animal. Several proposals have been made with the most common approach using the temperature–humidity index (THI). The THI combines dry and wet bulb temperatures (T_{db}, T_{wb}) into an equivalent temperature that defines how an animal feels when subjected to hot environments. An example THI proposal for dairy cows (°C) is given below.[2]

$$THI_{dairy\ cows} = 0.40(T_{db} + T_{wb}) + 4.8 \tag{14}$$

where T_{db} is the dry bulb temperature (°C); T_{wb} is the wet bulb temperature (°C).

With this proposal, it has been suggested that dairy cows will not be heat stressed if the THI is less than 23.5°C (74°F), will be moderately stressed if the THI is at about 25°C (77°F), and will be severely heat stressed if the THI is above 26°C (79°F). For example, if a barn's environment is at $T_{db} = 30$°C and the room's relative humidity is at 50%, the resulting wet bulb temperature, determined using a psychrometric chart, is $T_{wb} = 22$°C. Using these psychrometric properties, the resulting THI for a dairy cow under

these conditions is

$$\begin{aligned}\mathrm{THI}_{\text{dairy cows}} &= 0.40(T_{\mathrm{db}} + T_{\mathrm{wb}}) + 4.8 \\ &= 0.40(30 + 22) + 4.8 \\ &= 25.6°\mathrm{C}(78°\mathrm{F})\ (\text{moderate-to-severely heat stressed})\end{aligned} \quad (15)$$

Similar indices exist for poultry, pigs, and humans. Some modern-day control systems for animal housing applications incorporate measurements for the water vapor content in the air so that control decisions can be made based on THI levels in the barn. For the dairy barn conditions given above, a control system might activate cooling fans or water sprinkling systems to provide cooling for the animal.

CONCLUSION

Several factors need to be considered for housing animals. Production efficiency and animal comfort are a function of many environmental factors, both physical and thermal. The animal will respond to its environment by adjusting the transfer of heat energy between its body and the surroundings. The ability to make heat transfer adjustments is in turn a function of the thermal conditions surrounding an animal. The dry air/water vapor mixture affects the animal's ability to transfer heat and thus an understanding of psychrometrics is extremely important.

Many rather sophisticated housing designs have been developed to ensure that animals are given the chance to be comfortable and thus to produce at the highest level of efficiency. Unlike human housing design, animal housing effectiveness is difficult to assess directly. The astute producer will rely quite heavily on animal behavior and other trends to alert the producer to a comfortable or uncomfortable environment for the animal.

FURTHER READING

For those interested in more information on psychrometric properties and processes, the book by Albright[1] and the ASHRAE Handbook of Fundamentals[3] is recommended. Psychrometric properties and their relation to animal comfort and productivity is discussed in great detail in the book by Curtis.[2]

REFERENCES

1. Albright, L.D. *Environment Control for Animals and Plants*; The American Society of Agricultural Engineers: St. Joseph, MI, 1990.
2. Curtis, S.E. *Environmental Management in Animal Agriculture*; The Iowa State University Press: Ames, IA, 1983.
3. ASHRAE Handbook of Fundamentals; American Society of Heating, Refrigerating, and Air-Conditioning Engineers, Inc.: Atlanta, GA, 2001.

Pulsed Electric Fields Food Preservation

David R. Sepúlveda
Gustavo V. Barbosa-Cánovas
Barry G. Swanson
Washington State University, Pullman, Washington, U.S.A.

INTRODUCTION

The main objective of food preservation processes is to eliminate all pathogenic bacteria and to reduce and control the presence of spoilage bacteria. Quality attributes such as flavor, color, texture, and nutritive properties, however, need to be preserved as well in order to maintain the overall quality of the product. Some of the traditional food preservation processes like pasteurization and canning involve the use of heat to reduce the amount of bacteria present in raw products. The application of heat, although proven effective in controlling pathogenic and spoilage bacteria, is also responsible for deterioration of other quality attributes such as color and nutritional value. The constant search for better and more efficient technologies to process and preserve foods has led to the development of alternative preservation technologies. The idea behind their development is to obtain safer products with longer shelf life and improved sensory and nutritive quality characteristics through processes of higher efficiency.

Several attempts have been made to accomplish this task but only some have partially fulfilled the stated requirements. Preservation of food by Pulsed Electric Fields (PEF) is one alternative preservation process that has shown excellent potential for adoption as a method to preserve liquid and semiliquid food products. As in other novel preservation technologies, the use of an alternative source of energy that selectively targets micro-organisms and causes limited detrimental effects (or none at all) on the rest of the quality attributes is a principle of paramount importance in PEF processing.

HISTORICAL BACKGROUND

The use of electricity as an energy source to inactivate micro-organisms has been explored since the early 1900s. Around 1914, milk was processed by allowing it to run through a glass tube where a rapidly alternating current of 3600 V–4200 V was applied at a temperature of about 62°C in order to eradicate *Bacillus tuberculosis* from milk for infant feeding. A large-scale processing plant located in Liverpool was capable of processing about 30 gal hr^{-1}, which represented a considerable capacity at that time.[1] Posterior efforts in the United States can be found around 1919 in a study conducted at the request of the Surgeon General's office where an electrical "purification" process known as the Electro-Pure process was studied. In this process, milk was delivered by gravity to a series of porcelain cups where an electrical discharge of 2300 V was applied at a frequency of 25 Hz, producing an increase in temperature from 40 to 70°C. The Electro-Pure process was able to handle around 5000 lb of milk per hour at a reasonable price and it was believed that the heat produced by the electric current was directly responsible for the "purification" effect.[2]

However, although the purpose of these processes was to thermally inactivate bacteria, some researchers suggested that in addition to the thermal effect, the electric current itself was an important destroying agent, as bacteria could be destroyed at a temperature much below their thermal death point when electric currents were applied.[3] Subsequent studies conducted around the 1960s by Sale and Hamilton[4] renewed the interest in using PEF as a method to inactivate micro-organisms in foods. This research group gathered all the available information on PEF generated by then and conducted their own experiments, concluding that high electric fields up to 30 kV cm^{-1} had a lethal effect on a number of species of vegetative bacteria and yeasts suspended in liquids. They also determined a strong correlation between the degree of inactivation and the pulse length, number of pulses, and field strength in the suspension; higher electric fields and longer treatment times (pulse width times the number of pulses) produced a higher degree of inactivation.

Since then, a growing interest in the study and development of this technology has been observed and currently, there are more than 30 research groups around the world studying the viability and potential use of this technology as a food preservation method.[5]

TECHNOLOGICAL ASPECTS OF PEF

All PEF systems developed up to date are comprised of five major sections: control system, data gathering system, high voltage source, pulse generator, and treatment chamber (see Fig. 1).

Encyclopedia of Agricultural, Food, and Biological Engineering
DOI: 10.1081/E-EAFE 120007112
Copyright © 2003 by Marcel Dekker, Inc. All rights reserved.

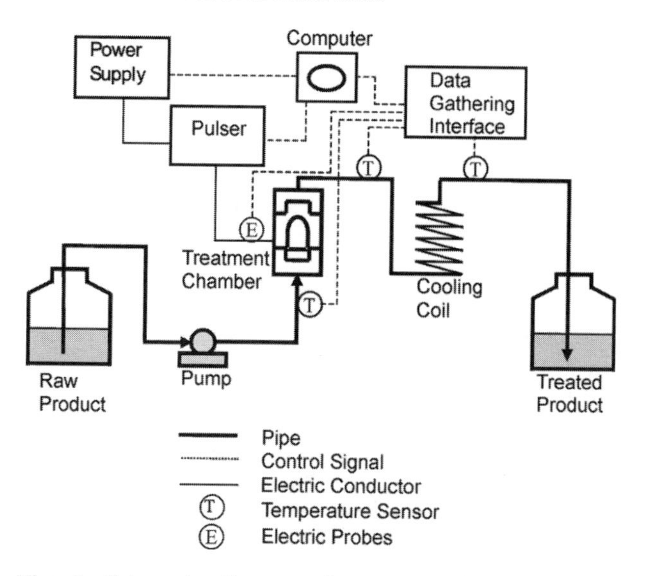

Fig. 1 Schematic diagram of a PEF unit and its main components.

PEF treatment basically involves placing the liquid food inside an electrically insulated container where high voltage pulses in the order of 20 kV–40 kV (usually a couple of microseconds) are applied. The applied high voltage results in the creation of an electric field that causes bacterial inactivation; the intensity of the electric field depends on the intensity of the applied voltage and on the distance between the electrodes. Electric fields in the range of $15\,kV\,cm^{-1}$–$90\,kV\,cm^{-1}$ are commonly used in this technology.[6] PEF treatment is accomplished via the following steps. After a high voltage has been generated by a high voltage source up to a level defined by the control system, the pulse generator produces a sudden discharge of this voltage into the treatment chamber. The pulse shape of the discharge is defined by the control system along with the pulse generator. Several different pulse shapes have been used yielding different results. Some of the most commonly used pulse shapes are the exponential decaying pulse, the square pulse, and the bipolar pulse.[7] It is generally accepted that square pulses have a higher inactivation efficacy than exponential decaying pulses; bipolar pulses are known to have an enhanced effectiveness, although direct comparison is not possible. Once in the treatment chamber, the high voltage pulse generates an electric field pulse. The pulse shape, as well as the treatment chamber design, defines the characteristics of the electric field. Continuous and batch chambers, parallel and concentric chambers, among others, are examples of the different models of treatment chambers available. Data gathering systems are used to monitor all relevant parameters such as voltage, temperature, and pulse shape during processing.[8]

MECHANISM OF MICROBIAL INACTIVATION BY PEF

Several mechanisms have been proposed to explain the action of PEF on micro-organisms.[9] Although some of these mechanisms differ in fundamental aspects, most indicate that the principal effect of PEF is on the disruption of the micro-organism cell membrane. Permeabilization of cell membranes by PEF is known as "electroporation" or "electropermeabilization" (see Fig. 2).

The cell membrane is a semipermeable barrier that isolates the cytoplasm and cell organelles from the surroundings. Membrane semipermeability allows the cell to interact with the environment by taking nutrients from it and excreting waste materials. Ions present in the cytoplasm as well as in the surrounding liquid environment are attracted to electric fields, causing a polarization of the membrane, since positive ions inside the cell tend to migrate to one end of the cell and negative ions to the opposite end. As the intensity of the electric field increases, the trans-membrane potential also increases, until the electrical breakdown of the membrane is reached. At this point, the force exerted by the ions over the membrane surpasses its strength. The membrane can no longer contain the ions and is punctured by the migration of ions out of the cell due to the action of the electric field.[11] Low-field intensities or short-pulse duration will cause reversible breakdown where small perforations can be quickly "repaired" by microbial cells. However, if the electric field intensity is high enough, and/or the pulse duration long enough, larger pores will form and irreversible breakdown will result. Formation of irreversible pores will lead to leaking of intracellular material, osmotic imbalance, swelling, and eventually to death. Cell size and shape as well as mechanical characteristics define membrane strength; hence, different external electric fields are needed to induce the critical trans-membrane potential in different types of bacteria. In general, bacteria have proven to be

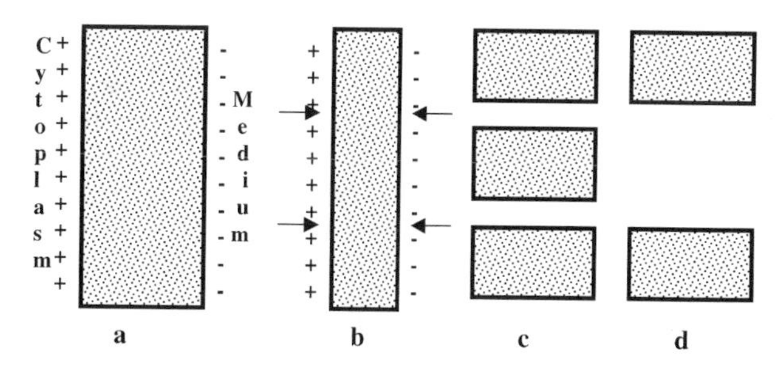

Fig. 2 Mechanism of electroporation in the cell membrane showing (a) the initial condition of the membrane, (b) the compression caused by the presence of an electric field, (c) the formation of small pores (reversible breakdown), and finally (d) the irreversible breakdown. (Adapted from Ref. 10.)

more resistant to electric fields than yeasts while spores are more resistant than vegetative cells. The effect of PEF on cells other than micro-organisms is being studied in order to apply the same principle of electroporation to promote the extraction of intracellular products and to improve the drying efficiency of vegetable tissues.[12]

CONCLUSION

Better understanding of PEF technology is still needed to take full advantage of all the desirable features this technology offers. The development of industrial-scale processing equipment and the creation of specific regulations for this emergent technology are important issues for future growth of PEF technology. Important developments applying PEF to the treatment of fresh fruit juices with low pH have been accomplished representing the most advanced area of industrial-scale use. Still ahead, several important challenges need to be addressed. Reliable generation of electric fields, a better understanding of inactivation kinetics, and potential presence of detrimental effects are some of the technical issues waiting to be clarified. Much of the current research in this area is focused on the preservation of specific food products instead of looking for generalizations. Dealing with one specific set of conditions at a time might be the best way to develop further knowledge in this area. The careful study of this technology will most likely lead to the next level where efficient technology in harmony with the environment will produce safer and more nutritious foods with increased shelf life and better quality characteristics.

REFERENCES

1. Beattie, J.M.; Lewis, F.C. Electric Treatment of Milk for Infant Feeding and Destruction of *Bacillus tuberculosis*. J. State Med. **1916**, (24), 174–177.
2. Anderson, A.K.; Finkelstein, R. A Study of the Electro-Pure Process of Treating Milk. J. Dairy Sci. **1919**, (2), 374–406.
3. Beattie, J.M.; Lewis, F.C. The Electric Current (Apart from the Heat Generated), a Bacteriological Agent in the Sterilization of Milk and Other Fluids. J. Hyg. **1925**, (24), 123–137.
4. Sale, A.; Hamilton, W. Effects of High Electric Fields on Microorganisms. I. Killing of Bacteria and Yeasts. Biochim. Biophys. Acta **1967**, (148), 781–788.
5. Barbosa-Cánovas, G.V.; Góngora-Nieto, M.M.; Pothakamury, U.R.; Swanson, B.G. *Preservation of Foods with Pulsed Electric Fields*; Academic Press: New York, 1999.
6. Barbosa-Cánovas, G.V.; Pierson, M.D.; Zhang, Q.H.; Schaffner, D.W. Pulsed Electric Fields. In *Kinetics of Microbial Inactivation for Alternative Food Processing Technologies*; Special Supplement. J. Food Sci. **2000**, 65–81.
7. Jeyamkondan, S.; Jayas, D.S.; Holley, R.A. Pulsed Electric Field Processing of Foods: A Review. J. Food Prot. **1999**, *62* (9), 1088–1096.
8. Góngora-Nieto, M.M.; Sepúlveda, D.R.; Pedrow, P.; Barbosa-Cánovas, G.V.; Swanson, B.G. Food processing by pulsed electric fields: treatment delivery, inactivation level, and regulatory aspects. Lebensm.-Wiss. Technol. **2002**, *35* (5), 375–388.
9. Qin, B.L.; Pothakamury, U.R.; Barbosa-Cánovas, G.V.; Swanson, B.G. Nonthermal Pasteurization of Liquid Foods Using High-Intensity Pulsed Electric Fields. Crit. Rev. Food Sci. Nutr. **1996**, *36* (6), 603–627.
10. Zimmermann, U. Electrical Breakdown, Electropermeabilization and Electrofusion. Rev. Phys. Biochem. Pharmacol. **1986**, (105), 176–256.
11. Zimmermann, U.; Pilwat, G.; Beckers, F.; Riemann, F. Effects of External Electric Fields on Cell Membranes. Bioelectrochem. Bioenerg. **1976**, *3*, 58–83.
12. Knorr, D.; Geulen, M.; Grahl, T.; Sitzmann, W. Food Application of High Electric Field Pulses. Trends Food Sci. Technol. **1994, March** (5), 71–75.

Pulsed X-Ray Treatments of Foods

William Kerr
The University of Georgia, Athens, Georgia, U.S.A.

P

INTRODUCTION

Since the discovery of x-rays in 1895 by W. C. Roentgen, many applications of x-rays have developed in chemistry, physics, medicine, and weapons research. In recent years, exposing foods to x-rays is one type of irradiation process that has been considered for assuring the safety and quality of foods.

WHAT ARE X-RAYS?

X-rays are part of the same electromagnetic spectrum that includes radio, microwave, infrared, visible light, ultraviolet, and gamma radiation. As such, x-rays are composed of oscillating and perpendicular electric and magnetic fields that propagate in a direction normal to both fields. X-ray wavelengths vary from approximately 10 nm (10^{-8} m) to 10^{-4} nm (10^{-13} m), with a corresponding frequency range of 3×10^{16} Hz–3×10^{21} Hz.

Wilhelm Roentgen discovered that x-rays were generated when a beam of high-speed electrons bombarded a target. In his original experiment, the glass walls of the cathode ray tube stopped the beam of electrons, but modern x-ray generating devices use a metal target such as copper, tungsten, or molybdenum. In general, high-atomic metals give the best conversion to x-rays. In high-vacuum x-ray tubes, the glass tube is evacuated to prevent collision of electrons with gas molecules. An electric current is used to heat a metal filament cathode to a high temperature, causing electrons to be emitted. These are focused and directed towards a positively charged anode, and accelerated by the high voltage (30 kV–100 kV) existing between the cathode and anode. The electrons collide with a metal target located in front of the anode, producing *Brehmsstralung* x-rays in all directions. An angular range of beams is allowed to pass through a beryllium window, a low molecular weight metal with few electrons to impede the x-rays. About 99% of the electron energy is converted to heat, so that the anode must be actively cooled (Fig. 1).

To produce more x-ray photons, a higher effective anode–cathode voltage must be applied. However, field currents limit the maximum voltage drop across a given electrode. Placing several anode–cathode pairs in sequence, each with a moderate voltage difference, can generate greater overall voltage potentials. The energy imparted to electrons is usually described in electron volts (eV). One eV is the kinetic energy acquired by an electron accelerating through a difference of 1 V. With multisection linear accelerators (linacs), energies of several MeV can be attained. The energy attained in high-voltage DC accelerators is limited by the maximum voltage difference that can be maintained.

Higher power can be attained by using linacs in which successive stages are powered by time varying power supplies. Although the voltage drop across any stage is alternately negative and positive, the gaps and voltage frequency are adjusted so that electrons experience only accelerating fields when they cross the gap. Many linacs are operated in pulsed mode, in which pulses of electrons are sent through the accelerator. Thus, power is supplied only in short periods of time, followed by a low power mode. Thus, very high-energy x-rays can be delivered in short bursts, but with less overall power consumption and heat build-up (Fig. 2).

Beam power up to several GeV can be attained using magnetic fields to direct electrons in a circular path. The electrons can make multiple passes through the accelerator before finally being directed at the target. In sychrotorons, high-energy electrons are delivered by a linac, then accelerated through one or more rf cavities arranged in a large circular accelerator. As the particles are accelerated through curved trajectories, they lose some energy in the form of radiation. This radiation is useful, as it is the source of high-energy x-rays. However, excessive costs and beam energy preclude the use of synhcrotrons in food processing.

Development of relatively high power, yet cost effective, x-ray sources for processing has been a major research thrust. Thompson and Cleland[1] describe a high-power dynamitron accelerator for x-ray processing. This is a DC linear accelerator that can produce beam powers up to 5.0 MeV. X-ray photon energy is limited to 5 MeV in several countries, due to concerns with induced radioactivity though neutron activation. However, evidence suggests that energies up to 7.5 MeV would be safe.[2] Mckeown et al.[3] have designed new x-ray converters to increase beam power up to 5 MeV–7 MeV. The development of pulsed x-rays is another important area for increasing beam power. In induction linacs, a pulse-forming network is used to convert AC power into

Encyclopedia of Agricultural, Food, and Biological Engineering
DOI: 10.1081/E-EAFE 120007116
Copyright © 2003 by Marcel Dekker, Inc. All rights reserved.

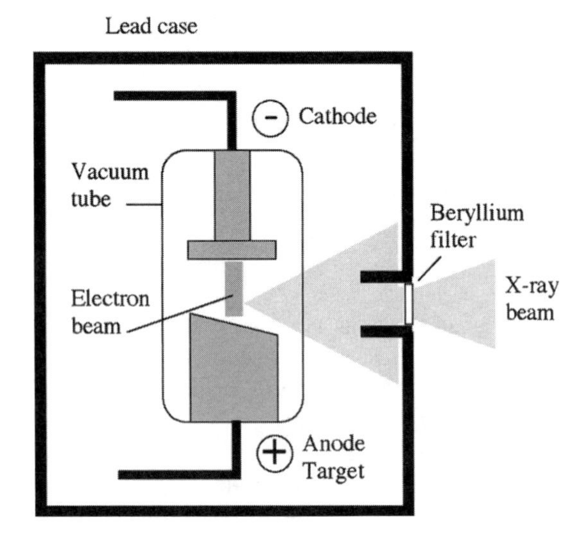

Fig. 1 X-rays, generated as electrons, are accelerated by a high voltage and bombard a metal target.

short, high-voltage pulses. These induce short voltage pulses ($>10\,\mu$sec at $100\,\text{sec}^{-1}$) in a secondary coil, causing clusters of electrons to be accelerated. Several of these modules are placed in line to accelerate electrons, with the pulses timed in each module so as to increase the total kinetic energy of the beam. With this configuration, beam energies up to 10 MeV have been attained.[4] Scientists at Sandia National Laboratories have developed a repetitive high-energy pulsed power (RHEPP) accelerator. This system uses a magnetically switched pulse forming line to produce >50 nsec pulses at frequencies up to $400\,\text{sec}^{-1}$. Peak dose rates measured by the gray (Gy), of up to $10^{11}\,\text{Gy sec}^{-1}$ have been attained, as compared to $10^{7}\,\text{Gy sec}^{-1}$ for RF linacs, and $10^{6}\,\text{Gy sec}^{-1}$ for continuos DC accelerators.[5]

X-RAY PROCESSING

X-ray technologies have been of tremendous value in chemistry, physics, astronomy, and medicine. In food processing, x-rays have been considered along with γ-rays and electron beams as a source for direct irradiation of foods, and as a means to pasteurize or sterilize select foods. In contrast to thermal processing, in which micro-organisms are exposed to high temperature for a period of time, x-ray photons transfer high amounts of energy to the food through collisions with atomic electrons. These collisions produce short-lived reactive chemicals deleterious to micro-organisms.

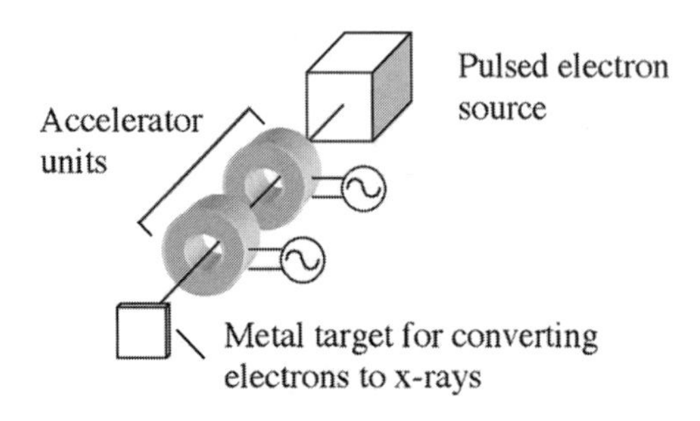

Fig. 2 Higher energy electrons are produced by a linac. Continuous or pulsed electrons are accelerated through several stages of alternating electric fields.

X-rays have been compared to electron beam and γ-rays for their suitability in food irradiation. X-rays are produced less efficiently than electron beams; indeed, electron beams are required to produce x-rays. Electron beams can also be aimed directly at the product, while x-rays and γ-rays scatter in all directions. However, electron beams do not penetrate well into the food. X-rays and γ-rays are similar in their properties for food irradiation, with photons that lose their energy exponentially with product depth. In water, 50% of photon energy is left at a depth of 10.9 cm, while 10% of photon energy is left at 36.3 cm. Typically, foods are irradiated on both sides to assure penetration of thick pieces, boxes, or pallets of product. While γ-ray photons are more energetic than x-ray photons, γ-rays are produced by radioactive decay of ^{60}Co. In addition to the extensive safety concerns in dealing with radioactive isotopes, some researchers believe that there is insufficient ^{60}Co for sustained operation of large-scale irradiation facilities. As mentioned previously, the conversion of electrons to x-rays is not efficient or highly directed. Thus, current research is directed toward producing high-power accelerators that give better conversion, but with facilities that are cost-effective.

EFFECTS ON FOOD

In general, x-ray, γ-ray, and electron beam irradiation have been used to kill micro-organisms including bacteria cells, bacterial spores, molds, yeasts, and viruses. Irradiation has also been used to eliminate parasites and insect infestations. Interestingly, complex organisms often have greater sensitivities to irradiation than simple ones. Whereas doses of up to 56 kGy are necessary to ensure food safety, 0.01 kGy may be lethal to humans. The exact lethal dose depends upon the type of organism and food. For example, the D value is 0.25 kGy for *Yersinia enterocolitica*[6] in ground beef at 25°C, and 0.77 kGy for *Listeria monocytogenes* in chicken.[7,8] Microbial inactivation occurs as ionizing radiation ejects electrons from the DNA of the organism, causing short-lived formation of ions, and disrupting DNA replication.[7,8]

Ionizing radiation is not selective and there have been some concerns that ionization of food constituents could be deleterious. Ionization of water may occur to form $H_2O\cdot^+$ radicals, which then react with other molecules. However, few such radicals are formed at reasonable radiation doses.[9] Irradiation of carbohydrates may cause some hydrolysis, polysaccharide cleavage, or formation of organic acid and ketones. At moderate doses, there seems to be little effect on nutritional quality, but minor effects on food texture have been noted.[10] Some evidence of protein denaturation, hydrolysis, and cleavage exists, but no effects on biological availability have been found.[11] At modest doses (1 kGy–10 kGy), there seems to be no more effect on carbohydrates and proteins than might be found with thermal processing. Lipids are perhaps more susceptible to degradation by irradiation, as ionization causes formation of lipid hydroperoxides. High-irradiation doses can lead to rancidity and lipid polymerization. Removal of oxygen and use of antioxidants during irradiation seems to help this problem. Vitamins are also degraded by irradiation, but low treatment doses, cold temperatures, and elimination of light and oxygen can minimize effects. The general consensus is that irradiated foods are wholesome and safe, particularly when reasonable doses are used. Any chemical or physical changes that may occur are generally less than, and certainly no worse than, those found in other processed foods.[12]

DOSES

As with most unit operations, sufficient treatment is needed to attain the desired effect, but excessive treatment may be deleterious. Optimum radiation depends on the commodity, temperature, effect to be attained, and desired shelf life. The Food and Drug Administration has issued approved irradiation doses through the years, and some of these are shown in Table 1. In general, low doses (< 1 kGy) are designed to control insects, sprouting, or trichinae. Medium doses (1 kGy–10 kGy) are used to control *Salmonella*, *Shigella*, and similar micro-organisms in meat, and to control mold growth on fruits. High doses (10 kGy–30 kGy) are used to kill micro-organisms on spices, or in limited cases, to sterilize foods.

FDA requires that irradiated foods be labeled as such. In addition, for a new food group to be legally irradiated, a petition must be submitted under the "food additive" category.

Table 1 FDA approved irradiation doses for select products

Food	Purpose	Dose (kGy)
Wheat and flour (1963)	Insect disinfestation	0.2–0.5
Potatoes (1964)	Sprout inhibition	0.05–0.15
Spices (1983)	Microbial inactivation	30
Spices (1984)	Disinfestation	
Dried enzymes (1985)	Microbial inactivation	10
Pork (1985)	Inactivate trichinosis	0.3–1.0
Poultry (1990)	Microbial inactivation	1.5–3.0
Meat (1997)	Pathogen reduction	4.5

REFERENCES

1. Thompson, C.C.; Cleland, M.R. High-Power Dynamitron Accelerators for X-Ray Processing. Nucl. Instrum. Methods Phys. Res. **1989**, B*40/41*, 1137–1141.
2. FAO/IAEA. Consultants Meeting on the Development of X-Ray Machines for Food Irradiation, Vienna, 1995.
3. Mckeown, J.; Armstrong, L.; Cleland, M.R.; Drewell, N.H.; Dubeau, J.; Lawrence, C.B.; Smyth, D. Photon Energy Limits for Food Irradiation: A Feasibility Study. Radiat. Phys. Chem. **1998**, *53*, 55–61.
4. Lagunas-Solar, M.C. Induction-Linear Accelerators for Food Processing with Ionizing Radiation. Nucl. Instrum. Methods Phys. Res. **1985**, B*10/11*, 987–993.
5. Schneider, L.X.; Reed, K.W.; Kaye, R.J. Repetitive High Energy Pulsed Power Technology Development for Industrial Applications. *Proceedings of the 14th International Conference on the Application of Accelerators in Research and Industry*, Denton, TX, 1996, 1085–1088.
6. El-Zawahry, Y.A.; Rowley, D.B. Radiation Resistance and Injury of *Yersinia enterocolitica*. Appl. Environ. Microbiol. **1979**, *37*, 50–54.
7. Huhtanen, H.; Jenkins, R.K.; Thayer, D.W. Gamma Radiation Sensitivity of *Listeria monocytogenes*. J. Food Prot. **1989**, *52*, 610–613.
8. CAST. Ionizing Energy in Food Processing and Pest Control: II. Applications Report No. 115; Council for Agricultural Science and Technology: Ames, Iowa, 1989.
9. Swallow, J. Wholesomeness and Safety of Irradiated Foods. In *Nutritional and Toxicological Consequences of Food Processing*; Friedman, M., Ed.; Plenum Press: New York, NY, 1991; 11–31.
10. Bhatty, R.; MacGregor, A. Gamma Irradiation of Hulless Barley: Effect of Grain Composition, β-Glucans and Starch. Cereal Chem. **1988**, *65* (6), 463–470.
11. Taub, I. Chemistry of Hydrated Muscle Proteins Irradiated at − 40°C. Proceedings of the Army Science Conference III, West Point, PA, 1976; 289.
12. Institute of Food Technologists. Perspective on Food Irradiation. Food Technol. **1987**, *41* (2), 100–101.

Q_{10}

Petros S. Taoukis
National Technical University of Athens, Athens, Greece

INTRODUCTION

Q_{10} expresses the effect of temperature on the rates of quality-related reactions occurring in food systems during processing and subsequent distribution and storage. It is a dimensionless number defined as the ratio of the reaction rate constants at temperatures differing by 10°C.

In this article, reaction kinetics in foods and kinetic modeling of food quality are addressed. Effective and quantitative knowledge of the modes and rate of food deterioration during and postprocessing is a valuable tool for food process design optimization and determination and modeling of the shelf life or keeping quality of the food products.

OVERVIEW

Food is a physicochemical system of high complexity involving numerous physical and chemical variables. Food quality change, in general, may be expressed as a function of composition and environmental factors:

$$\frac{dQ}{dt} = F(C_i, E_j) \tag{1}$$

where C_i are composition factors, such as concentration of reactive compounds, inorganic catalysts, enzymes, reaction inhibitors, pH, water activity, as well as microbial populations and E_j are environmental factors, such as temperature, relative humidity, total pressure and partial pressure of different gases, light and mechanical stresses. Even if this system could be explicitly expressed in terms of measurable parameters, no analytical solution is attainable and possible numerical solutions are often too elaborate for any practical purpose. The established methodology has been elaborated elsewhere in this encyclopedia—in the articles *Decimal Reduction Times*; *Activation Energy in Thermal Process Calculations*; and *Thermal Resistant Constant*—and consists of first identifying the chemical and biological reactions that influence the quality and the safety of the food. Then, through a careful study of the food components and the process, the reactions judged to have the most critical impact on the deterioration rate are considered.[1] Based on this analysis and without underestimating the underlying complexity of food systems, food degradation and shelf life loss are in practice represented by the loss of desirable quality factors A (such as nutrients, characteristic flavors) or the formation of undesirable factors B (such as off flavors, microbial load, and discoloration). The rate of loss of A (correspondingly formation of B) is expressed as:

$$r_A = \frac{-d[A]}{dt} = k[A]^m \tag{2}$$

The quality factors $[A]$ are usually quantifiable chemical, physical, microbiological, or sensory parameters characteristic of the particular food system, k the apparent reaction rate constants, and m the apparent reaction order. Methodology for determination of the apparent reaction order and reaction rate constant is described in the relevant sections. Regardless of the value of m (the order of the reaction), Eq. 2 can be expressed in the form:

$$f_q(A) = kt \tag{3}$$

where the expression $f_q(A)$ can be termed the *quality function* of the food. For a given extent of deterioration, translated to a value of the quality function, $f_q(A_t)$, the rate constant is inversely proportional to the time to reach that degree of quality loss. This holds also for t_s, the time for the quality to reach an unacceptable level, i.e., the shelf life.

The form of the quality function of the food for an apparent zero, 1st, 2nd, and mth order reaction is shown in Table 1 along with the half-life time of the reaction, i.e., the time for the concentration of the quality index A to reduce to half its initial value.[2]

In order to include in the quality function the effect of the environmental factors, the commonly used approach is to model it into the apparent reaction rate constant, i.e., expressing k of Eq. 3 as a function of E_j: $k = k(E_j)$. The factor most often considered and studied is temperature, T. This is justifiable because temperature strongly affects reaction rates during processing and subsequent distribution and storage. Additionally, during the postprocessing phase, it is imposed to the food externally (direct effect of the environment), the other factors being at least to some extent controlled by the food packaging. Several equations have been proposed and applied to express the $k = k(T)$ function. Some have theoretical foundations but most are applied as

Encyclopedia of Agricultural, Food, and Biological Engineering
DOI: 10.1081/E-EAFE 120006981
Copyright © 2003 by Marcel Dekker, Inc. All rights reserved.

Table 1 Quality function and half-life time for deterioration of quality index A

Apparent reaction order	Quality function $f_q(A_t)$	Half-life $t_{1/2}$
0	$A_0 - A_t$	$A_0/(2k_0)$
1	$\ln(A_0 - A_t)$	$\ln 2/k_1$
2	$1/A_0 - 1/A_t$	$1/(k_2 A_0)$
m ($m \neq 1$)	$\frac{1}{m-1}\left(A_t^{1-m} - A_0^{1-m}\right)$	$\frac{2^{m-1}-1}{k_m(m-1)} A_0^{1-m}$

semiempirical or empirical formulas that adequately express the temperature dependence of the rate constant k, within a stated temperature range of practical interest. The Arrhenius relation, developed theoretically for reversible molecular chemical reactions, has been applied widely for food quality loss reactions (see the article *Activation Energy in Thermal Process Calculations*).

Q_{10} has been traditionally used to express temperature dependence by the food industry and in the earlier food science and biochemistry literature. Q_{10} is a dimensionless number defined as the ratio of the reaction rate constants at temperatures differing by 10°C. Equivalently Q_{10} equals the reduction of shelf life t_s when the food is stored at a temperature higher by 10°C than the reference temperature. Namely:

$$Q_{10} = \frac{k(T+10)}{k(T)} = \frac{t_s(T)}{t_s(T+10)} \tag{4}$$

Similar to Q_{10} the term Q_A is sometimes used. The definition of Q_A is the same as Q_{10} with 10°C replaced by A°C.

The Q_{10} approach in essence introduces a temperature dependence equation of the form

$$k(T) = k_0\, e^{bT} \quad \text{or} \quad \ln k = \ln k_0 + bT \tag{5}$$

which implies that if $\ln k$ is plotted vs. temperature (instead of $1/T$ of the Arrhenius equation), a straight line is obtained; where k is the reaction rate constant at absolute temperature T, constant k_0 is a reference rate constant at $T = 0\,\text{K}$, and b the slope of the $k(T)$ vs. T plot. Equivalently, shelf life can be plotted vs. temperature:

$$t_s(T) = t_{s0}\, e^{-bT} \quad \text{or} \quad \ln t_s = \ln t_{s0} - bT \tag{6}$$

Such plots are often called *shelf life plots*, where b is the slope of the shelf life plot and t_{s0} is the intercept. It should be noted that Eqs. 5 and 6 also hold using temperature in °C (or F), which is often more convenient for conceptual and presentation purposes. In these cases, care should be taken to use the appropriate value and unit conversions (for °C, slope b same as in Eq. 5 and k_0, the rate constant value, for 0°C; for °F, slope equal to $b/1.8$ and k_0, the rate constant value, for 0°F).

The shelf life plots are true straight lines only for narrow temperature ranges of 10–20°C.[3] For such a narrow interval, data from an Arrhenius plot will give a relatively straight line in a shelf life plot. This implies that Q_{10} and b are actually functions of temperature:

$$\ln Q_{10} = 10b = \frac{E_A}{R} \times \frac{10}{T(T+10)} \tag{7}$$

where E_A (in J mol^{-1} or cal mol^{-1}) is the activation energy parameter of the Arrhenius equation of the reaction that controls quality loss and R the universal gas constant in $\text{mol J}^{-1}\text{K}^{-1}$ or $\text{mol cal}^{-1}\text{K}^{-1}$, respectively. This dependence of the Q_{10} value is more pronounced when the temperature "sensitivity" of the reaction is greater, i.e., the larger the activation energy. Table 2 summarizes the dependence of Q_{10} on temperature and E_A and important types of food reactions that fall in the respective range of values of Q_{10} and E_A.

The value of the quality function, $f_q(A)_t$, at time t, after exposure of the food at a known variable temperature exposure, $T(t)$, can be found based on Eq. 5 by calculating the integral of $k[T(t)]\,dt$, from 0 to time t.

$$f_q(A)_t = k_0 \int_0^t e^{bT}\, dt \tag{8}$$

The integral can be calculated analytically for simple $T(t)$ functions or numerically for more complex ones.[2] Controlled temperature functions like square, sine,

Table 2 Q_{10} dependence on E_A and temperature

E_A Cal mol^{-1} (kJ mol^{-1})	Q_{10} at 5°C	Q_{10} at 20°C	Q_{10} at 40°C	Typical food reactions
10 (41.8)	1.87	1.76	1.64	Diffusion controlled, enzymic, hydrolytic
20 (83.7)	3.51	3.10	2.70	Lipid oxidation, nutrient loss
30 (125.5)	6.58	5.47	4.45	Nutrient loss, nonenzymic browning
70 (293)	21.8[a]	16.9[a]	12.6[a]	Vegetative cell thermal destruction[a]

[a] Q_{10} values for 60, 75, and 95°C, respectively, relevant to pasteurization range.

Table 3 Analytical expressions for calculation of Γ for different temperature functions

Temperature function	Expression for Γ^{a}
Sine wave	$I_0(a_0 b)$
Square wave	$\frac{1}{2}(e^{a_0 b} + e^{-a_0 b})$
Spike wave	$(e^{a_0 b} + e^{-a_0 b})/2a_0 b$

$^{a}b = \ln Q_{10}/10$, a_0 the amplitude of the sine, square, and spike periodic functions, and $I_0(x)$ a modified Bessel function of zero order. (From Ref. 15.)

and linear (spike) wave temperature fluctuations can be applied to validate the temperature dependence model and Q_{10} value, obtained from several constant temperature shelf life experiments. To systematically approach the effect of variable temperature conditions, the concept of effective temperature, T_{eff}, can be introduced. T_{eff} is a constant temperature that results in the same quality change as the variable temperature distribution over the same period of time. T_{eff} is characteristic of the temperature distribution and the kinetic temperature dependence of the system. The rate constant at T_{eff} is analogously termed effective rate constant, and $f_q(A)_t$ of Eq. 8 is equal to $k_{\text{eff}}t$. If T_{m} and k_{m} are the mean of the temperature distribution and the corresponding rate constant, respectively, the ratio $\Gamma = k_{\text{eff}}/k_{\text{m}}$ is also characteristic of the temperature distribution and the specific system. For some known characteristic temperature functions, analytical expressions for the Q_{10} approach are tabulated in Table 3.

The estimation of the true Q_{10} of the quality loss reaction or the shelf life of a particular food system should be carefully and systematically approached. According to compiled databases,[3–5] canned products have Q_{10} values ranging from 1.1 to 4, dehydrated foods from 1.5 to 10, and frozen foods from about 3 to as much as 40. Thus, use of a "representative" mean Q_{10} for any of these categories, as is often applied in the food industry for practical estimations, can result in highly inaccurate predictions. If data for the specific food system in question is unavailable, in order to obtain reliable results, Q_{10} should be experimentally determined by conducting kinetic measurements or shelf life studies at least at two temperatures differing by 10°C. In practice, since there is experimental error involved in the determination of the values of k and t_s, calculations of Q_{10} from only two points may give a substantial error. It is recommended that the reaction rate or shelf life is measured at three or more temperatures. The measured k or t_s are plotted vs. T in a semilog graph and a least square linear regression fit to Eqs. 5 or 6 is employed (see Fig. 1). When applying regression techniques, statistical analysis can be used to determine the 95% confidence limits of the estimated Q_{10} value. An optimization scheme to estimate the number of experiments to obtain the maximum accuracy for the least possible amount of work showed that five experimental temperatures is the practical optimum.[6] However in most practical applications, measurements at three temperatures are sufficient and techniques to get narrower confidence limits can be employed.[7]

The practical use of the shelf life plot is shown in Fig. 2. If the targeted shelf life of a low moisture product under development is set at 20 mo at 22°C by using an accelerated test at 40°C, it can be evaluated whether this can be achieved. By placing a point on the plot at 20 mo and 22°C and drawing a straight line to the 40°C vertical line, using slopes dictated by the Q_{10} values, a corresponding shelf life of 33 day, 50 day, 83 day, or 172 day is obtained for Q_{10} of 5, 4, 3, or 2, respectively. This would be the minimum respective duration of the accelerated test needed to confirm the shelf life hypothesis.

Even when proper shelf life testing is conducted, caution should be drawn to the effect of the uncertainty in

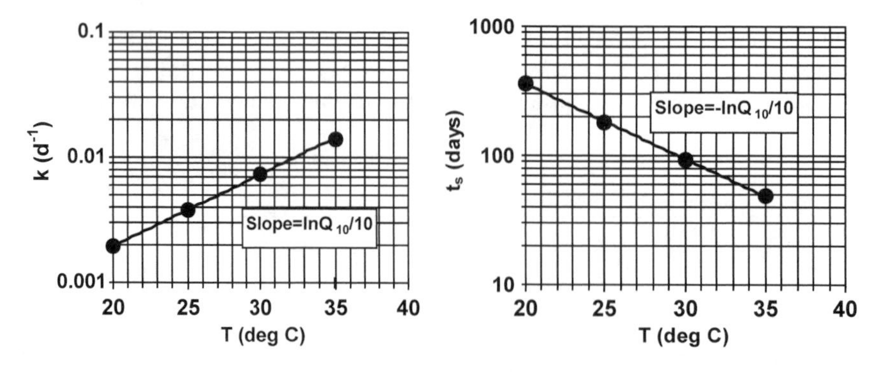

Fig. 1 Plots of measured reaction rate constant k vs. T and t_s vs. T (shelf life plot) of a hypothetical food system. (A UHT orange juice drink is assumed. Quality and shelf life is based on vitamin C loss, a first order reaction. Shelf life at 20°C is 1 yr based on 50% vitamin loss. Q_{10} from these plots by linear regression is calculated at 3.8.)

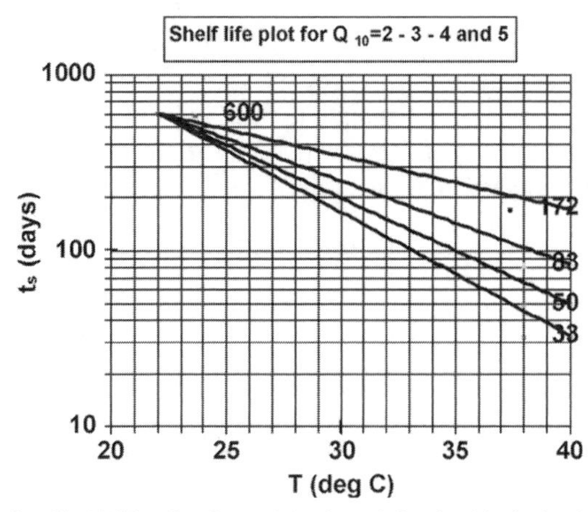

Fig. 2 Shelf life plot for a dehydrated food with desired shelf life of 20 mo at 22°C and equivalent accelerated shelf life test times at 40°C.

the determination of the Q_{10} value in shelf life prediction. As shown in Table 4, if a product has a Q_{10} of 2.5 and a measured shelf life of 20 day at 50°C the predicted shelf life at 20°C is 312 day compared to 540 days if Q_{10} is 3. Thus, substantial deviations in the predicted shelf life at low temperatures can arise if approximate Q_{10} values are used to make the extrapolation. The inaccuracy of the prediction is smaller when extrapolating from lower temperatures. In practice, one usually compromises between time, cost, and accuracy. Additionally, accelerated tests at temperatures of >40°C are not recommended because of the possibility that the reactions of critical importance may change between the high temperature and the normal storage temperature of the product. Application and limitations of Accelerated Shelf Life Testing (ASLT) based on Q_{10} has been reviewed.[1,8]

Another term used for temperature dependence of microbial inactivation kinetics in canning and sometimes of food quality loss[9] is the z-value. The value of z is the temperature change that causes a tenfold change in the thermal death time or the reaction rate constant. As in the case of Q_{10}, z depends on the reference temperature. Q_{10}, z, and E_A are related by the following equation:

$$z = \frac{23}{\ln Q_{10}} = \frac{\ln 10 RT^2}{E_A} \tag{9}$$

It should be noted that there are factors relevant to food and food quality loss reactions that can cause significant deviations from temperature dependent behavior that can be described by Q_{10} or the Arrhenius equation.[10] Phase changes are often involved. Fats may change to the liquid state contributing to the mobilization of organic reactants or vice versa.[11] In frozen foods, the effect of phase change of the water of the food is very pronounced in the immediate subfreezing temperature range. Generally, frozen systems follow a common pattern: 1) just below the initial freezing point the rate increases (in an almost discontinuous fashion) to values well above those obtained in the supercooled state at the same temperature; 2) passes through a maximum; and 3) finally declines at lower temperatures.[12] The rate increase is especially notable for reactants of low initial concentration. The rate enhancement induced by freezing is basically related to the freeze-concentration effect. This enhancement is prominent in the temperature zone of maximum ice formation. The width of this zone will depend on the type of food but generally will be in the range of -1 to -10°C. Other phase change phenomena are also important. Carbohydrates in the amorphous state may crystallize at lower temperatures, creating more free water for other reactions but reducing the amount of available sugars for reaction. Glass transition phenomena are also implicated in systems that, at certain temperature ranges, deviate significantly from Q_{10} or Arrhenius behavior. Certain processing conditions or drastic changes in storage conditions, such as rapid cooling and solvent removal, result in formation of metastable glasses, especially in carbohydrate containing foods.[13,14] In such systems, the WLF equation may more effectively describe temperature dependence (see the article *WLF Equation* elsewhere in this encyclopedia).

Table 4 Effect of Q_{10} on shelf life

	Shelf life (day)			
Temperature (°C)	$Q_{10} = 2$	$Q_{10} = 2.5$	$Q_{10} = 3$	$Q_{10} = 3.5$
50	20	20	20	20
40	40	50	60	70
30	80	125	180	245
20	160	313	540	858

REFERENCES

1. Taoukis, P.; Labuza, T.P. Summary: Integrative Concepts (Shelf-life Testing and Modelling). In *Food Chemistry*, 3rd Ed.; Fennema, O., Ed.; Marcel Dekker: New York, 1996; Chap. 17, 1013–1042.
2. Taoukis, P.; Labuza, T.P.; Saguy, I. Kinetics of Food Deterioration and Shelf-life Prediction. In *The Handbook of Food Engineering Practice*; Valentas, K.J., Rotstein, E., Singh, R.P., Eds.; CRC Press: Boca Raton, FL, 1997; Chap. 10, 361–403.

3. Labuza, T.P. *Shelf Life Dating of Foods*; Food and Nutrition Press: Westport, CT, 1982; 500.
4. Okos, M.R. *Physical and Chemical Properties of Food*; American Society of Agricultural Engineers: St. Joseph, MI, 1986; 407.
5. Man, C.M.D.; Jones, A.A. *Shelf Life Evaluation of Foods*; Chapman and Hall: New York, NY, 1994; 321.
6. Lenz, M.K.; Lund, D.B. Experimental Procedures for Determining Destruction Kinetics of Food Components. Food Technol. **1980**, *34* (2), 51–55.
7. Labuza, T.P.; Kamman, J. Reaction Kinetics and Accelerated Tests Simulation as a Function of Temperature. In *Applications of Computers in Food Research*; Saguy, I., Ed.; Marcel Dekker: New York, 1983; Chap. 4, 71–151.
8. Labuza, T.P.; Schmidl, M.K. Accelerated Shelf-life Testing of Foods. Food Technol. **1985**, *39* (9), 57–62, 64.
9. Hayakawa, K. New Procedure for Calculating Parametric Values for Evaluating Cooling Treatments Applied to Fresh Foods. Can. Inst. Food Sci. Technol. **1973**, *6* (3), 197–200.
10. Labuza, T.P.; Riboh, D. Theory and Application of Arrhenius Kinetics to the Prediction of Nutrient Losses in Food. Food Technol. **1982**, *36*, 66–74.
11. Templeman, G.; Sholl, J.J.; Labuza, T.P. Evaluation of Several Pulsed NMR Techniques for Solids in fat Determination in Commercial Fats. J. Food Sci. **1977**, *42*, 432–436.
12. Fennema, O.; Powrie, W.D.; Marth, E.H. *Low-Temperature Preservation of Foods and Living Matter*; Marcel-Dekker: New York, 1973.
13. Roos, Y.H.; Karel, M. Differential Scanning Calorimetry Study of Phase Transitions Affecting Quality of Dehydrated Materials. Biotechnol. Prog. **1990**, *6*, 159–163.
14. Levine, H.; Slade, L. "Collapse" Phenomena—A Unifying Concept for Interpreting the Behavior of Low Moisture Foods. In *Food Structure—Its Creation and Evaluation*; Blanshard, J.M.V., Mitchell, J.R., Eds.; Butterworths: London, 1988; 149–180.
15. Tuma, J.J. *Engineering Mathematics Handbook*, 3rd Ed.; McGraw Hill: New York, 1987; 498.

Refrigerants

Vikram Ghosh
R. C. Anantheswaran
John D. Floros
The Pennsylvania State University, University Park, Pennsylvania, U.S.A.

R

INTRODUCTION

Refrigerants are the working fluids that circulate around a refrigeration system, absorbing heat from areas to be cooled and transferring it to where the heat can be dissipated. A variety of refrigerants are available for different cooling/freezing applications. In the food industry, the most commonly used refrigerants are ammonia and freons, but in the recent years, the trend has been to move away from freons and to look for more environmental friendly compounds.

The first refrigerant used was ether in a hand operated vapor-compression machine. After many changes, ammonia took over as the main refrigerant in the later part of the 19th century. Ammonia has good refrigeration properties but is very toxic and so the search for a safer refrigerant continued. In the early 1930s, a breakthrough came when E. I. du Pont de Nemours developed Freons. Freons are one- and two-carbon compounds containing chlorine and fluorine atoms generally known as chlorofluorocarbons (CFCs). In the late 1980s and early 1990s, 75% of the food supply channels depended on CFC refrigerants for cooling during processing, storage, and distribution.[1]

In 1974, Rowland and Molina[2] proposed that CFCs moving into the atmosphere would remain there until transported to the stratosphere where they would be photolyzed, releasing chlorine atoms. Then, through a series of catalytic reactions, the free chlorine atoms might react with ozone converting it to oxygen. Their hypothesis was later confirmed and in recent years, environmental concerns led to political restrictions of the use of CFCs as the working fluids in refrigeration and air conditioning plants. This led to the development of alternative refrigerants. The majority of these new refrigerants fall into two catergories, hydrofluorocarbons (HDCs), which contain no chlorine and have zero ozone depletion potential and hydrochlorofluorocarbons (HCFCs), which contain chlorine, but the addition of hydrogen to the CFC structure allows them to be dispersed in significant proportions in the lower atmosphere before they can reach the ozone layer. Hydrochlorofluorocarbons therefore have much lower ozone depletion potential, ranging from 2% to 10% that of CFCs. Many nations have signed the Vienna Convention, intended to control the production of substances known to deplete the ozone layer. Research is being carried out to determine the properties of new ozone-friendly refrigerant fluids.[3]

DESIGNATION OF REFRIGERANTS

Refrigerants are designated as R followed by a numeral based on their chemical formula. For a CFC derived from a saturated hydrocarbon having a chemical formula $C_mH_nF_pCl_q$, in which $(n+p+q) = 2m+2$, the complete designation is $R(m-1)(n+1)(p)$. For brominated versions of a refrigerant, the letter B and a number follow the above representation, where the number denotes the number of chlorine atoms replaced by the bromine atoms. In the case of common inorganic refrigerants, a number indicating it's molecular weight is added to 700 after R. Azeotropic mixtures of refrigerants that behave like pure substances, are given arbitrary designations. For example R502 is a mixture of 48.8% R22 and 51.2% R115. Unsaturated freons for which $(n+p+q) = 2m$, are named with 1 before $(m-1)$. Thus ethylene is R1150.[4]

SELECTION OF A REFRIGERANT

There is no single refrigerant that is suitable for all types of refrigeration systems. The type of refrigerant that can be used depends on the system's requirements. The particular refrigerant chosen needs to satisfy the thermodynamic, chemical, and physical requirements as discussed in the sections below. The physical properties for the common refrigerants are given in Table 1.

PERFORMANCE REQUIREMENTS

1. *Boiling and condensing temperatures and pressures*: The evaporating and condensing temperatures of a refrigerant must be below the uncooled food, and above the ambient, heat discharge temperatures, respectively. As the uncooled food temperature is usually close to ambient,

Encyclopedia of Agricultural, Food, and Biological Engineering
DOI: 10.1081/E-EAFE 120007018
Copyright © 2003 by Marcel Dekker, Inc. All rights reserved.

Table 1 Physical properties of refrigerants[a]

	R-12	R-134a	R-22	R-123	R-717	R-502
Chemical formula	CCl_2F_2	CF_3CH_2F	$CHClF_2$	$CHCl_2CF_3$	NH_3	—
Molecular mass	120.93	102.3	86.48	152.93	17.03	111.63
Boiling point (°F)	− 21.62	− 15.8	− 41.36	82.17	− 28.0	− 49.8
Freezing point (°F)	− 252	− 141.9	− 256	− 160.87	− 107.9	—
Critical temperature (°F)	233.6	214.0	204.8	362.82	271.4	179.9
Critical pressure (psia)	596.9	589.8	721.9	532.87	1657	591.0
Crtical volume ($ft^3\,lb^{-1}$)	0.0287	0.029	0.0305	—	0.068	0.0286
Latent heat (Btu ($lb^{-1}\,mol^{-1}$)	8591	9531	8687	11,215	10036	8280

[a](From Ref. 5.)

there must be an operating pressure at which the temperatures are accessible to the refrigerant if it is to be a feasible choice. The operating pressure should be above the atmospheric pressure to avoid an inward leak, which is very difficult to detect. An inward leak will introduce air into the refrigeration circuit that will not condense and will interfere with the refrigerant's heat transfer capability.[6]

2. *Freezing temperature*: The refrigerant selected should have a freezing point much lower than the operating temperature. This will avoid any problems with the solidification of the refrigerant.

3. *Latent heat of vaporization*: The latent heat of vaporization is the amount of heat absorbed by the refrigerant during evaporation. Since heat is transferred through the refrigeration system by the evaporation/condensation cycle, a higher latent heat refrigerant will require less cycling than a refrigerant with a lower latent heat, and thereby will be more efficient.

4. *Specific volume*: The refrigerant mass flow rate around the cycle is directly related to the heat transferred. The product of vapor specific volume and flow rate relates directly to the compressor size. Reciprocating compressors normally use R-12, R-22, R-500, R-13, and R-717. Centrifugal compressors are adaptable to R-11, R-12, R-114, R-113, and for very large tonnage systems, R-22 is the preferable choice.[7]

5. *Compression ratio*: It is the ratio of the (absolute) discharge pressure to the suction pressure. A low compression ratio usually indicates efficient compressor operation. Compression ratios of six and below are in the low range. Higher compression ratios can result in decreased compressor efficiency.

INCIDENTAL REQUIREMENTS

The constraints on refrigerants incidental to their heat transfer properties pertain to flammability, toxicity, water solubility, partial miscibility with oil, and environmental safety.

1. *Inflammability*: Flammability or explosion hazard restricts the use of some refrigerant candidates that have excellent heat transfer capability. Hydrocarbons, such as methane, ethane, propane, and butane are highly explosive and flammable. Ammonia is explosive in an air mixture with concentrations of 16%–25% by volume ammonia. None of the fluorocarbons is explosive or flammable.

2. *Toxicity*: Based on hazard to life as gases or vapors, compounds have been divided into six groups by Underwriters Laboratories. Group six contains compounds with low toxicity. It includes R12, R114, R13B1, etc.[4]

3. *Action of refrigerant with water*: The presence of moisture is very critical in refrigeration systems operating below 0°C. If more water is present than can be dissolved by the refrigerant, then there is a danger of ice formation and consequent choking in the expansion valve or capillary tube used for throttling in the system. This is called moisture choking. In R12 and R114 systems, in which water has low solubility, moisture choking is a frequent problem.[4]

4. *Action with oil*: In compressors, some oil is carried by the high temperature refrigerant vapor to the condenser and ultimately to the expansion valve and evaporator. In the evaporator, as the refrigerant evaporates, a distillation process occurs and the oil separates from the refrigerant. A build-up of oil in the evaporator will result in a reduced heat transfer coefficient, *oil-choking* in the evaporator due to restricted refrigerant flow, even blockage and ultimately to oil starvation in the compressor.

The refrigerants, which are completely miscible, such as R11, R12, R113, etc., and those that are not miscible with oil, such as ammonia do not present such problems. In the case of the latter, an oil separator is installed close to the compressor in the discharge line and the separated oil is continuously returned to the crankcase of the compressor. In systems with refrigerants only partially miscible with oil, the return of oil to the compressor creates problems.[4]

PHYSICAL REQUIREMENTS

Generally recognized physical properties, such as dielectric strength, thermal conductivity, viscosity, and special properties that characterize leak tendency and detection are important, though not critical, to a compound's value as a refrigerant.

1. *Dielectric strength*: The dielectric strength of a refrigerant is primarily important in hermetically sealed units in which the electric motor is exposed to the refrigerant. Fluorocarbons are good electrical insulators as compared to nitrogen.

2. *Thermal conductivity*: High refrigerant thermal conductivity contributes to high heat transfer.

3. *Viscosity*: A lower viscosity refrigerant will have lower pumping losses and a higher heat transfer coefficient for transfer between the refrigerant (liquid) and the evaporation section tubing.

4. *Leaks*: Leak detection should be easy. The greatest drawback of the fluorocarbons in this regard is the fact that they are odorless. At times this results in a substantial loss of costly refrigerant due to leaks. An ammonia leak can be very easily detected by its pungent odor. Leaks in ammonia plants are common due to corrosion of susceptible components of the piping connections.

REFRIGERANTS COMMONLY USED IN THE FOOD INDUSTRY

Mechanical refrigeration systems provide by far the bulk of the industrial cooling capacity, although cryogenics (e.g., liquid nitrogen, and more often, solid CO_2) are used in some applications for food freezing. The commonly used refrigerants are Refrigerant 717 (R717–ammonia), Refrigerant 12 (R12), Refrigerant 22 (R22), and Refrigerant 502 (R502). The following descriptions are taken from Hallowell.[6]

1. *Ammonia–Refrigerant 717–NH_3*: During the early years of refrigeration, ammonia was the most widely used refrigerant and is still in widespread use in commercial and industrial systems. Ammonia has excellent refrigerant qualities. At atmospheric pressure, it evaporates at −28°F (−33°C). Condensing pressure at 95°F (35°C) is 181 psig (12.7 kg cm^{-2}). This is a convenient pressure and causes no difficulty in compressor construction or cost. Ammonia systems provide completely automated, reliable service. Most ammonia equipment is for industrial duty and has a long life expectancy.

Due to environmental concerns with the CFCs, ammonia is being reconsidered for many applications formerly filled by CFCs. It is irritating to the mucous membranes and eyes and still cannot be used in domestic refrigeration and home air conditioning. The use of ammonia is limited to relatively large installations because of the cost of the pressure containment. Ammonia works well with secondary refrigerants where the entire ammonia plant is confined to a "mechanical" room with chilled brine circulating from this room to separate heat exchangers in food chilling locations.

2. *Dichlorodiflouromethane–Refrigerant 12–CCl_2F_2*: R-12 was one of the first halocarbon refrigerants to be used extensively. It has a boiling point of −21.6°F (−30°C) at atmospheric pressure. Condensing pressure at 95°F (35°C) is 108.3 psig (7.6 kg cm^{-2}). It is useful from freezing to air conditioning temperatures. For many years, R12 was the principal halocarbon refrigerant used. Oil separation can cause problems if not handled properly, since oil is miscible in R-12. Line sizing is critical to insure that upward flowing suction lines maintain sufficient velocity to carry oil from the evaporators to the compressor. Compressors can be operated in parallel, but care must be taken to assure that oil flows back to all compressors. Line sizing and layout must also be treated carefully for parallel operation. Oil return, as in most halocarbon systems, is dependent on proper velocities in upflow risers and on proper slope and design of the suction lines.

R12 is odorless under normal conditions of use and presents no hazard, even in heavily occupied areas. In the presence of an open flame, it will break down into unpleasant components that can be toxic. It is nonflammable and is considered a safe refrigerant. It does not contaminate products stored near the refrigeration system even if a leak occurs. Minimum pressure drop and reduced pumping costs for economical operation require large pipe sizes and careful installation. Hence, cost efficient plant designs usually locate compressors near their evaporators. In larger plants, unless a brine system is employed, several compressor locations are preferable to a single large machine room.

3. *Chlorodifluoromethane–Refrigerant 22–$CHClF_2$*: R-22 is primarily used in freezer applications since it has a boiling point of −41.4°F (−41°C) at atmospheric pressure. This low boiling point allows R-22 to be used in low temperature applications even at ambient evaporation pressure. Discharge temperatures must be higher in low temperature operations with air-cooled condensers than when water cooling is used to keep the cost of the condensers down. This usually limits the usage of R-22 at low suction pressure to water-cooled condensers. Condensing pressure at 95°F (35°C) is 181.8 psig (12.8 kg cm^{-2}). In air-cooled condenser applications, R-22 systems can be operated at about 25°F (−4°C) and above without much difficulty, if condensers are sized generously, particularly at lower temperatures. Line sizing must be handled with the same care as in

Table 2 Areas of application of some less used refrigerants[a]

Refrigerant	Application
Water (H_2O)	Used only in the steam-ejector systems for air conditioning
Carbon dioxide	Used as solid carbon dioxide or *dry ice* in transport refrigeration only
Refrigerant 11 (CCl_3F)	Used in centrifugal compressors with large capacity central air conditioning plants and for cooling water and brine
Refrigerant 13 ($CClF_2$)	For low temperature refrigeration
Refrigerant 14 (CF_4)	Lowest boiling refrigerant. Used for temperatures as low as −125°C with R22 and R502 in the first stage and R13 and R503 in the second
Refrigerant 13B1 (CF_3Br)	For moderately low temperature (−45 to −60°C)
Refrigerant 115 (C_2ClF_5)	Similar applications as Refrigerant 22
R134a (CF_3CH_2F)	This refrigerant has shown the potential for replacing R12 as a ozone safe refrigerant
R123 (CF_3CHCl_2)	Suggested replacement for R22, though there is still controversy about the viability of replacing R-22 with R123[8]

[a](From Ref. 4.)

the case of R-12 to achieve good results. R-22 lines can be smaller than R-12 for the same pressure drops, but are still quite large. R-22 requires less compressor displacement than R-12 for the same amount of refrigeration, but the horsepower per ton is about the same.

4. *Refrigerant 22/115–Refrigerant 502–$CHClF_2$/$CClF_2CF_3$*: R-502 was developed specifically for use in freezer applications and it is used in large supermarket freezers. It has a boiling point of −49.8°F (−45°C) at atmospheric pressures and a condensing pressure of 199.7 psig ($14\,g\,cm^{-2}$) at 95°F (35°C). Even with air-cooled condensers, the temperature of the discharge gas is not excessively high. R-502 is an azeotrope and is composed of a mixture of R-22 and R-115. The same restrictions apply to this refrigerant as to other halocarbon refrigerants.

5. *Other refrigerants*: A number of other refrigerants have been listed in the ASHRAE handbook,[5] but are not used to any extent in food applications. A list of some other refrigerants and their applications is given in Table 2.

There is a group of refrigerants with very low boiling temperatures, which are called cryogens. They include carbon dioxide, nitrogen, nitrous oxide, and others. They are used in specialized very low temperature freezing applications, metallurgy, medicine, etc.

6. *Secondary refrigerants*: Secondary refrigerants are normally fluid solutions that are cooled by a refrigerant and circulated as a fluid to pipe coils or other types of coolers. Brine solutions are normally used as secondary refrigerants in most food applications.

REFERENCES

1. McFarland, M. Chlorofluorocarbons and Ozone. Environ. Sci. Technol. **1989**, *223* (10), 1203–1207.
2. Rowland, F.S.; Molina, M. Chlorofluoromethanes in the Environment. Rev. Geophy. Space Phy. **1975**, *13* (1), 1–35.
3. Hewitt, G.F.; Shires, G.L.; Polezhaev, Y.V. Refrigerants. *International Encyclopedia of Heat and Mass Transfer*; CRC Press: Boca Raton, FL, 1997; 942.
4. Arora, C.P. Refrigerants. *Refrigeration and Air Conditioning*; Tata McGraw-Hill Publishing Co. Ltd.: New Delhi, 1994; 106–154.
5. ASHRAE. *Handbook of Fundamentals*, SI Ed.; ASHRAE: Atlanta, GA, 1993.
6. Hallowell, E.R. Refrigerants and Refrigerant Characteristics. *Cold and Freezer Storage Manual*; AVI Publishing Co.: Westport, CT, 1980; 20–41.
7. Perry, R.; Chilton, C.H. Refrigeration. *Chemical Engineers Handbook*; McGraw-Hill Book Company: New York, NY, 1973; 12-29–12-31.
8. Srinivasan, K. CFC Alternatives—A Fresh Look. Environ. Conserv. **1994**, *19* (4), 339–341.

Refrigeration System Components

Q. T. Pham
University of New South Wales, Sydney, New South Wales, Australia

R

INTRODUCTION

During the refrigeration of foodstuffs, heat is first transferred from food to refrigerant, then it is removed from the refrigerant by a mechanical system. The first process takes place in freezers and chillers, while the second is carried out by compressors, condensers, and refrigerant pumps. A simple refrigeration system is shown in Fig. 1. Thus, food refrigeration equipment can be classified into refrigeration providers, which perform the second function, and refrigeration applications, which perform the first function. Standard mechanical components such as pumps, motors, and piping will not be considered in this article.

CHILLERS AND FREEZERS

Chillers and freezers can be classified into air-cooled, immersion, spray, cryogenic, and surface-contact chillers.

In *air chillers/freezers* (Fig. 2), cold air is blown around the product. Heat is transferred from food to air, then carried to cooling coils. Air is a rather inefficient cooling agent, but the equipment is relatively inexpensive and the risk of contamination of the food is low. Because heat transfer at the surface of the food is usually the limiting factor, a great deal of effort has gone into designing the air distribution system. Air has a tendency to form channels along "paths of least resistance" and to bypass the rest of the product. Stagnant and recirculation zones can also occur (Fig. 2). Efforts to make the chiller more compact often result in difficulties in distributing the airflow by forcing it through tight turns and sudden expansions. To overcome this, various air distribution systems have been devised involving diffusers, baffles, ducts, and meshes. Product may be not stacked uniformly, causing air channeling.

When air temperature varies with time or location, warm moist air may come into contact with cold surfaces (walls, ceilings, etc.) and moisture will condense on these surfaces and may drip on the product. Mold and micro-organisms may grow and get blown around the chamber. Condensation can be prevented by ensuring uniform air distribution to eliminate warm spots, insulation of cold surfaces such as cooling coil sides and bottom, and protection against ingress of warm air.

Food can be cooled efficiently by contacting the food with a cold liquid, using, for e.g., *immersion chillers* or *spray chillers*. Some commonly used coolants are water, brine, propylene glycol, and ice slurry. Water and ice slurry cannot cool food below 0°C. The liquid must be certified as safe in contact with food. Immersion chilling is faster than air chilling, especially for smaller products, but may have several problems: 1) absorption of liquid or solutes by the product, leading to undesirable appearance, discoloration, or other quality losses; 2) cross-contamination between products; 3) leaching of food components such as fat; and 4) effluent disposal.

In *spray chillers*, cold liquid is sprayed continuously or intermittently over the product. Intermittent spraying combined with air cooling is used for meat cooling, where evaporative cooling during the intervals between sprays increases the refrigeration rate. Other advantages of intermittent spraying are less liquid effluent to dispose of, and less absorption of water by the product.

Surface contact chillers are those in which the product is cooled by contact with a cold solid surface. They include plate chillers/freezers, mold freezers, belt chillers, and scraped surface freezers. Contact with a solid surface ensures a high heat transfer rate (similar to immersion freezers) but absorption of refrigerating liquid is not a problem and there is no liquid effluent. However, plate freezers can only be used on products with flat surfaces, such as cartons, while belt freezers are best used on thin or small products such as fish and peas. Plate and mold freezing can also be labor intensive or require sophisticated automation.

COMPRESSORS

Mechanical refrigeration works by evaporating a refrigerant liquid at low pressure and temperature, compressing the vapor, cooling and re-condensing it and throttling it through a valve to reduce the pressure. Cooling is obtained when heat is absorbed from the product to evaporate the liquid. Compressors are literally the heart of the refrigeration supply system, recirculating refrigerant vapor so that it can be condensed and used again.

The energy efficiency of a compressor is commonly measured by a quantity called the coefficient of performance, or COP. It is the ratio of refrigeration effect (i.e., heat removed) to the mechanical work input (or in some cases, to the electrical work input). A theoretical

Encyclopedia of Agricultural, Food, and Biological Engineering
DOI: 10.1081/E-EAFE 120007019
Copyright © 2003 by Marcel Dekker, Inc. All rights reserved.

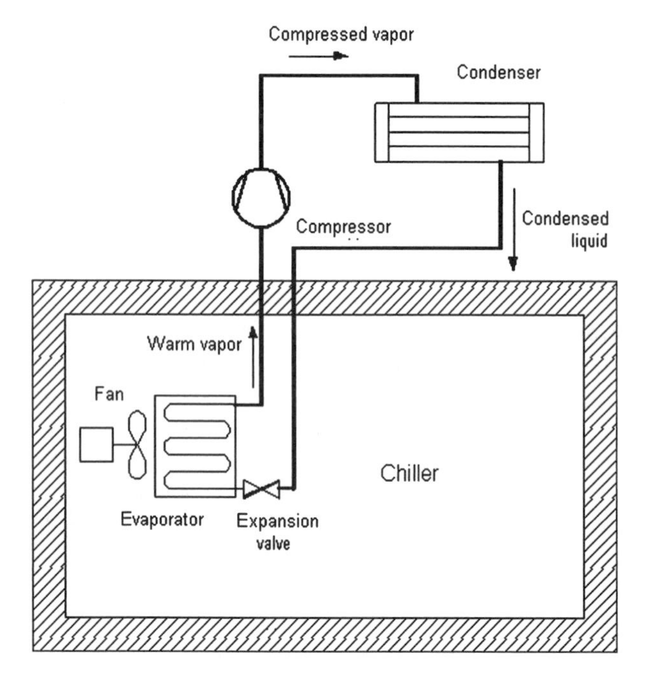

Fig. 1 Components of a typical refrigeration system.

COP can be calculated based on condensing and evaporating temperature and refrigerant type. The actual COP will be much lower than this, mainly because of various nonidealities in the design and operation of the compressor (friction, refrigerant recirculation, etc.). The deviation from ideality is mostly due to the energy efficiency or *isentropic efficiency* of the compressor (the ratio of the work of an ideal isentropic, i.e., no-loss, compressor to that of a real compressor).

The isentropic efficiency of a given compressor depends on the operating conditions; it is highest when the compressor is working at maximum capacity, and rapidly decreases with loading (Fig. 3). In most applications, the cooling load is not constant but varies during the day and with weather. A single compressor system designed to cope with the peak load will run partly loaded and hence at low efficiency for most of the time. Large installations will usually have several compressors that are manually or automatically cycled on and off in such a way that high loading is maintained.

The maximum capacity of a compressor is the product of its displacement (volume times revolution speed) and its *volumetric efficiency* (ratio of volume of gas pumped to the swept volume of the compressor), which is always less than 1 due to leakage and re-expansion of gas trapped inside the cavity. However, these effects are taken into account in manufacturers' charts and the user normally need not be concerned about them. Capacity control can be obtained with any type of compressor by varying the speed, using gears, adjustable belt drives, variable speed motors, or an inverter to control the electrical frequency. Also, each type of machine will usually have some built-in capacity adjusting mechanism. When using speed control, it is important not to go below a minimum speed, to avoid loss of sealing in positive displacement machines or unstable operation in centrifugal machines.

Refrigeration compressors in common use can be divided into positive displacement compressors and centrifugal compressors. The former category includes reciprocating (or piston) compressors, rotary vane compressors, single-screw and twin-screw compressors, scroll compressors, and trochoidal compressors.

Reciprocating (piston) compressors are widely used, especially in small- and medium-sized applications. Single-stage compressors are used for temperatures down to about − 20°C. For lower temperatures, it is more

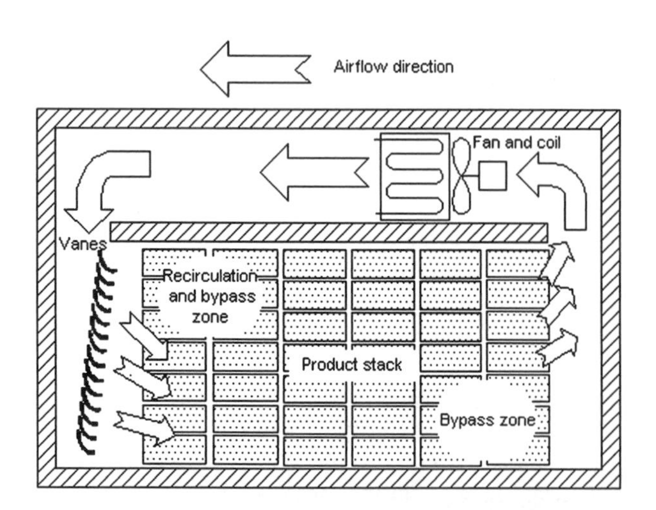

Fig. 2 Airflow pattern in a chiller.

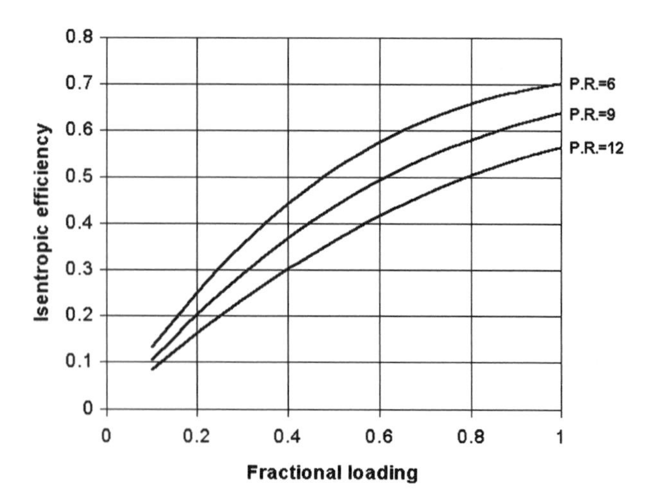

Fig. 3 Isentropic efficiency of a typical twin-screw compressor at various pressure ratios.

efficient to use integral two-stage compressors, in which some of the cylinders are used as boosters, delivering partly compressed vapor to the suction ports of the remaining cylinders. Capacity is controlled by bypassing some of the cylinders.

Rotary vane compressors have a rotor housed inside a cylindrical casing. The rotor axis is off center and has slots for sliding vanes, which divide the crescent-shaped space between rotor and casing into unequal-sized cells. The cells rotate with the rotor, changing its volume in the process. Low-pressure vapor is trapped in the cell at the point where the cell volume is largest, and discharged when the cell volume is smallest.

Single-screw compressors have a cylindrical rotor (screw) with spiral grooves and a pair of meshed star-shaped gate rotors. The gate rotors are positioned on opposite sides of the screw and rotate in a perpendicular direction to it. The vapor is trapped in the space between screw, casing wall, casing's discharge end, and a gate rotor tooth. As the screw rotates, the meshing gate rotor's tooth moves towards the casing's discharge end, reducing the volume of the trapped gas. Capacity is controlled by using a slide valve in the casing to bypass some partly compressed vapor back to the inlet.

Twin-screw compressors have two meshed screws, which rotate in opposite directions. The operating principle is similar to that of single-screw compressors, but the gas is trapped between the two meshing screws, the casing wall and the casing's discharge end. As the screws rotate, the point where the screw threads touch each other to enclose the vapor moves towards the casing's discharge end, causing a reduction in the volume of the trapped gas. Capacity is controlled by using a slide valve, as in single-screw compressors.

Both single- and twin-screw compressors are designed with a built-in *volume ratio* between suction and discharge gas. This is the ratio of the gas volume just after it is cut off from the inlet port to the volume just before it opens to the discharge port. Compressors work most efficiently if the inlet and outlet pressures allow the actual suction to discharge volume ratio of the gas to match the built-in volume ratio. Recently, compressor with variable volume ratio are also available.

In *scroll compressors*, two meshing scroll rotors trap the vapor in the space between two contact points. As the outer contact point moves from the outer edge towards the center, the volume of the trapped gas is progressively reduced, until it is discharged at the center.

Centrifugal compressors have rotating impellers that create a pressure difference between the center and the circumference. Vapor enters through a port near the axis and is discharged radially at the impeller tips. Flow is continuous and there are no contacting of moving solid surfaces as in positive-displacement compressors. Centrifugal compressors can be built with very large capacities and, by using high speeds and multistage arrangements (with as many as 10 stages), high pressure ratios can be produced. Capacity is controlled by adjusting the position of inlet vanes or suction dampers.

CONDENSERS

Condensers are heat exchangers in which compressed refrigerant vapor is cooled and condensed. Condensers must have enough capacity to remove the heat absorbed by the refrigeration applications plus the energy input at the compressors. Since the condenser's heat removal rate depends on the temperature difference between the condensing refrigerant and the cooling medium, insufficient condensing capacity will cause the condensing temperature and pressure to rise, increasing energy consumption by the compressors. When compressors cannot cope with the rising condensing temperature, the evaporation temperature will also rise and eventually the compressors may "trip out." On the other hand, if condensing temperature is too low, the condensing pressure will be insufficient to force condensed refrigerant through the circuit and into the evaporator, reducing capacity. Therefore, condenser capacity must always match refrigeration duty; it is controlled, say by turning the fans on and off or by isolating part of the condenser.

Condensers are classified by the cooling medium they use: air cooled, water cooled, or evaporative. Air-cooled condensers, with a fan blowing air past a finned coil, are used for small domestic and commercial systems. They require little maintenance apart from ensuring that the coils are clean and airflow is unobstructed. Water-cooled condensers are found in large installations; they are efficient but the recirculating water must be cooled in a cooling pond or cooling tower (unless a source of cool water is available, such as the sea). In evaporative condensers (Fig. 4), water is sprayed or cascaded over the refrigerant coils while air flows past, maintaining wet-bulb temperature on the outside. Since air dry-bulb temperature is usually several degrees higher than wet-bulb temperature and water conducts heat better than air, air-cooled condensers are less efficient than water-cooled condensers or evaporative condensers. Both evaporative and water-cooled condensers are liable to corrosion and fouling and must be inspected and cleaned regularly.

COOLING COILS

Cooling coils are the interface between the refrigerant and the food or its immediate surrounding environment. Most air-cooled facilities use forced draft cooling units, which

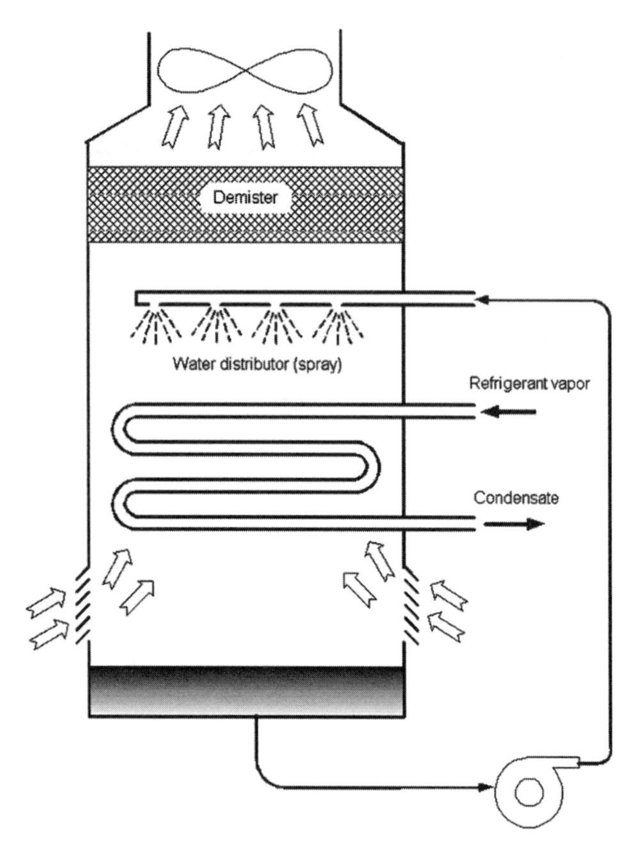

Fig. 4 Evaporative condenser.

consist of a finned coil and one or more fans. Direct drive propeller fans are the most popular type, although centrifugal fans may be used when there are large pressure losses (such as ducted air distributors or sock-type distributors). For given overall dimensions, cooling coil capacity can be increased by increasing fin density (and hence heat exchange area) and air velocity, but both these measures will also increase fan energy consumption, which will eventually be converted into heat and add to the refrigeration load. When the refrigerant temperature is below 0°C, frost will form on the surface and eventually block the airflow; if this is likely to happen, fin spacing must be balanced with the time between defrostings. When the refrigerant is above 0°C, water vapor will condense on the coil and must be collected in a tray and drained. The collecting tray itself will be colder than the air around it and condensation will form underneath then drip down. If this happens, the tray must be insulated or heated. The air velocity out of the coils should be kept at 2.5 m sec^{-1} or less to avoid blowing condensate droplets.

The cooling fluid in the coil may be water or glycol solution, which does not evaporate, or a refrigerant. In the latter case, the coil may operate as a dry expansion unit (refrigerant discharged as dry vapor), a liquid-recirculation unit (more refrigerant than needed is pumped through, so that the discharge from the coil is partly liquid), or a flooded unit (the coil is situated underneath a refrigerant tank and liquid fills the whole coil). The use of a nonrefrigerant cooling fluid reduces heat transfer efficiency and requires extra equipment such as pump, heat exchanger, and storage tank (Fig. 5), but may be justified on safety ground (for dangerous refrigerant such as ammonia), and also provides buffering capacity and better temperature control.

Cooling coils can be controlled in several ways. With dry expansion units, it is essential to maintain the correct refrigerant flow by means of an expansion valve; too much flow will allow liquid refrigerant to go back to the compressor and damage it. The air temperature can be controlled by a thermostat that turns the fans and refrigerant flow on and off. For increased efficiency, multi-speed fan control can be used in large systems.

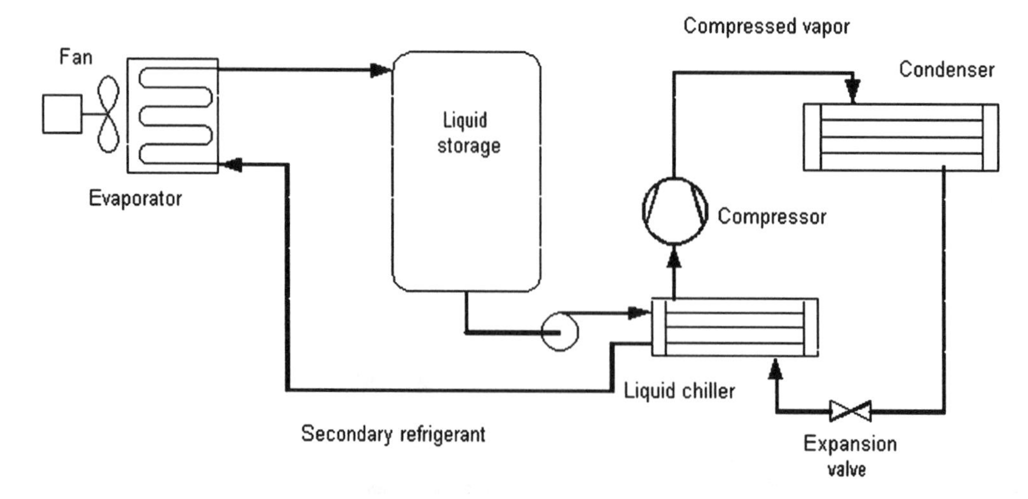

Fig. 5 Refrigeration system using secondary refrigerant.

The refrigerant pressure may also be controlled, usually with a pressure-regulating valve, to maintain a constant refrigerant temperature.

CONSIDERATIONS IN SELECTION AND OPERATION OF REFRIGERATION COMPONENTS

Criteria for selecting equipment will differ in priority depending on the size of the operation. Smaller operators will probably not value energy efficiency as much as larger ones, but for them low maintenance is a primary consideration. For some users, flexibility (the ability to handle different products or production rates) will be primary considerations while for others capital and operating costs are more important. However, in all food applications, the equipment must operate reliably and ensure that the food remains safe and wholesome according to specifications.

For freezers and chillers, the primary objective is to extract heat within a certain time from the product and other sources (lights, conduction, air infiltration, personnel, etc.), to do this uniformly for all the product, to avoid surface drying, contamination, microbial growth and other quality problems, and to avoid condensation. For compressors and condensers, the objectives are high-energy efficiency, ability to handle widely variable loads, and safe and environmentally acceptable operation with regards to leakage and noise. When putting a system together, it is important to ensure that the various components are balanced to give optimal performance for the given price. An undersized condenser will require oversize compressors, evaporators, and associated equipment.

FURTHER READING

American Society of Heating, Refrigerating and Air-Conditioning Engineers. *ASHRAE Handbook of HVAC Systems and Equipment*. ASHRAE, Atlanta, GA, 1996.

American Society of Heating, Refrigerating and Air-Conditioning Engineers. *ASHRAE Handbook of Refrigeration*. ASHRAE, Atlanta, GA, 1998.

Dossat, R.J. *Principles of Refrigeration*, 2nd Ed. Wiley: New York, 1978.

King, G.R. *Modern Refrigeration Practice*. McGraw-Hill: New York, 1971.

Perry, R.H.; Green, D.W. *Perry's Chemical Engineers' Handbook*, 7th Ed. McGraw-Hill: New York, 1997, pp. 11-77–11-118.

Trott, A.R.; Welch, T. *Refrigeration and Air-Conditioning*, 3rd Ed. Butterworth, Heineman: Oxford, 2000.

Remote Sensing in Agriculture

J. Alex Thomasson
Mississippi State University, Mississippi State, Mississippi, U.S.A.

INTRODUCTION

Remote sensing, which can be loosely defined as detection of ground-based things from air or space, is playing an increasingly important role in agricultural production. Sensor systems in satellites or aircraft collect image data in various portions of the electromagnetic (EM) spectrum, and these data contain much information about conditions of agricultural fields. The conditions of interest may include crop health or potential yield, soil conditions, weed infestations, etc. Image data can yield vital information about the variability of these conditions with respect to location, and this kind of information can be used by agricultural producers to optimize their management decisions with respect to location, thus maximizing profit and minimizing environmental effects.

SPACIAL VARIABILITY AGRICULTURE

In the modern agricultural age, farm producers are concerned with the spatial variability of croplands; i.e., the way characteristics of interest change in value and type from one location to another in a field. Examples of these characteristics include crop yield, soil fertility, density and variety of weeds, etc. Producers care about spatial variability in these characteristics because they would like to economically and environmentally optimize their use of crop inputs such as irrigation water, seed, fertilizer, etc. Knowledge of spatial variability requires measurement and record keeping of the characteristic of interest along with the location where the measurement was made. If such measurements can be made with reasonable accuracy from a distance, as from an airplane or satellite, then many measurements along with their locations can be collected instantaneously over a wide area in the form of an image.

ELECTROMAGNETIC ENERGY AS A SENSING TOOL

According to Lillesand and Kiefer,[1] remote sensing can be defined as "the science and art of obtaining information about an object, area, or phenomenon through the analysis of data acquired by a device that is not in contact with the object, area, or phenomenon under investigation." In current agricultural engineering investigations, the meaning of the term, remote sensing, is generally restricted to the detection of ground-based things from air or space. Energy in some form is the entity that is detected by any sensor. EM radiation is the most commonly detected form of energy in remote sensing. EM radiation includes ultraviolet, visible, and infrared energy, etc. The type of EM energy varies according to wavelength (or frequency, see Fig. 1). The sensitivity of most remote-sensing equipment (cameras, scanners, etc.) is described in terms of wavelength in fractions of meters like micrometers (1×10^{-6}m, μm) and nanometers (1×10^{-9}m, nm). Commonly used remote-sensing equipment is sensitive in the visible and infrared (IR) portions of the EM spectrum (Fig. 2).

Energy, such as from the sun, that illuminates a scene on the earth is modified by its interaction with the atmosphere in two ways, scattering and absorption. Atmospheric scattering is classified into the following categories: Rayleigh scattering, in which small particles such as air molecules affect short wavelengths, causing such phenomena as blue sky and haze; Mie scattering, in which the particles (water vapor, dust) are near the size of the longer wavelengths being scattered; and nonselective scattering, in which large particles such as water droplets scatter visible and IR equally, making fog and clouds appear white. Certain atmospheric constituents cause energy losses at specific wavebands: water vapor, CO_2, ozone, etc. The distribution of the sun's energy, the distribution at the earth's surface, and the sensitivity ranges of common remote-sensing equipment, are depicted in Fig. 3. Figure 3a is based on the concept that the sun's energy approximates that of a blackbody at 6000K, the distribution of which was modeled by Planck.

The incident energy reacts with each physical object on the ground in three ways: reflection, absorption, and transmission (Fig. 4). The magnitude of incident energy at any wavelength must equal the magnitude of the sum of energy at that wavelength reflected, absorbed, and transmitted. The ratio of reflected energy to incident energy is termed reflectance, and it typically varies by wavelength (i.e., spectrally). The spectral nature of an object's reflectance is generally its most important

Encyclopedia of Agricultural, Food, and Biological Engineering
DOI: 10.1081/E-EAFE 120006940
Copyright © 2003 by Marcel Dekker, Inc. All rights reserved.

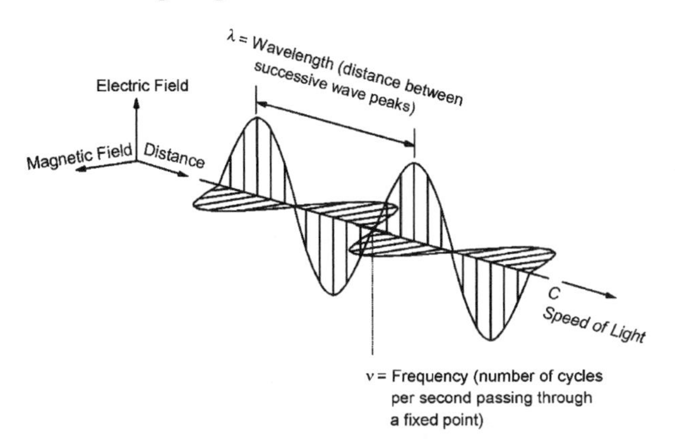

Fig. 1 Conceptual drawing of an EM wave. (From Ref. [1], p. 4.)

property in remote-sensing investigations. Reflectance varies not only by wavelength, but also by the angle of incident energy (Fig. 5). Objects that reflect energy at a mirror angle to that of the incident energy are called specular reflectors, and those that reflect energy at all angles regardless of the incident angle are called diffuse, or Lambertian, reflectors. The smoother an object's surface, the more it tends toward specular reflection, and the rougher an object's surface, the more it tends toward diffuse reflection. No natural objects are perfectly specular or perfectly diffuse reflectors, but they tend to approximate one or the other at certain combinations of incident angle and wavelength of energy. If one is careful to plan remote-sensing missions to be conducted at the proper time of day and year for a given location on earth, then one can usually assume that most objects on the ground will act in a way that approaches that of a diffuse reflector.

REFLECTANCE OF AGRICULTURAL MATERIALS

A commonly known fact that is very important to agricultural investigations is that plants and other materials vary in reflectance according to their type and condition. A good example of this is the large difference in infrared reflectance between green plant material and plant residues (Fig. 6). Differences in reflectance can also allow investigators to distinguish between different types of soils (Fig. 7), but they must consider the effects that soil moisture has on soil reflectance.

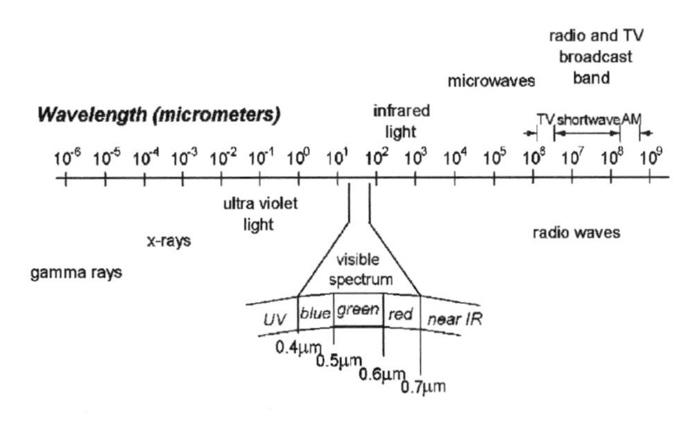

Fig. 2 The EM spectrum. (From Ref. [1], p. 5.)

SENSORS

Sensing devices used for remote sensing usually detect the variation in natural energy radiating from the ground toward the device. In most cases, this energy originates from the sun and is reflected by objects on the ground. Thus, when a camera suspended from an aircraft (or a sensor aboard a satellite) acquires an image of the ground below, the image is a representation of the reflectance of the various objects on the ground within the camera's field of view. Reflectance is a number that must be between zero and one, with the numbers nearer to one associated with high values of reflectance, which result in bright spots in an image. Thus, the bright areas in a remotely sensed image are associated with highly reflective objects on the ground, while the dark areas are associated with weakly reflective objects.

Two types of sensing devices are used in remote sensing, passive and active. Passive sensors are those that detect naturally occurring energy, such as that from the sun, or thermal energy emanating from an object because of its temperature. Passive sensors are the most commonly used and include film and digital cameras, and digital scanners. Active sensors are those that provide their own source of energy for illuminating a scene. A very simple example is a flash camera for taking pictures in a dark area. More complex examples include radio detection and ranging (radar) and light detection and ranging (lidar). Both of these include an emission source as well as a detector, and they measure the time lapse between the emission and the reflected return, and in some cases the strength of the reflected return.

IMAGERY

Remote sensing usually results in the creation of an image that represents the variation in reflectance (or radiance) over an area on the ground. Digital images are made up of picture elements, or pixels, that represent the smallest area on the ground that can be resolved by a particular sensor. Whereas reflectances are recorded numerically with digital sensors, each pixel in a digital image has a numeric value associated with it; the higher the value, the higher the reflectance of the represented ground location. Fig. 8 is

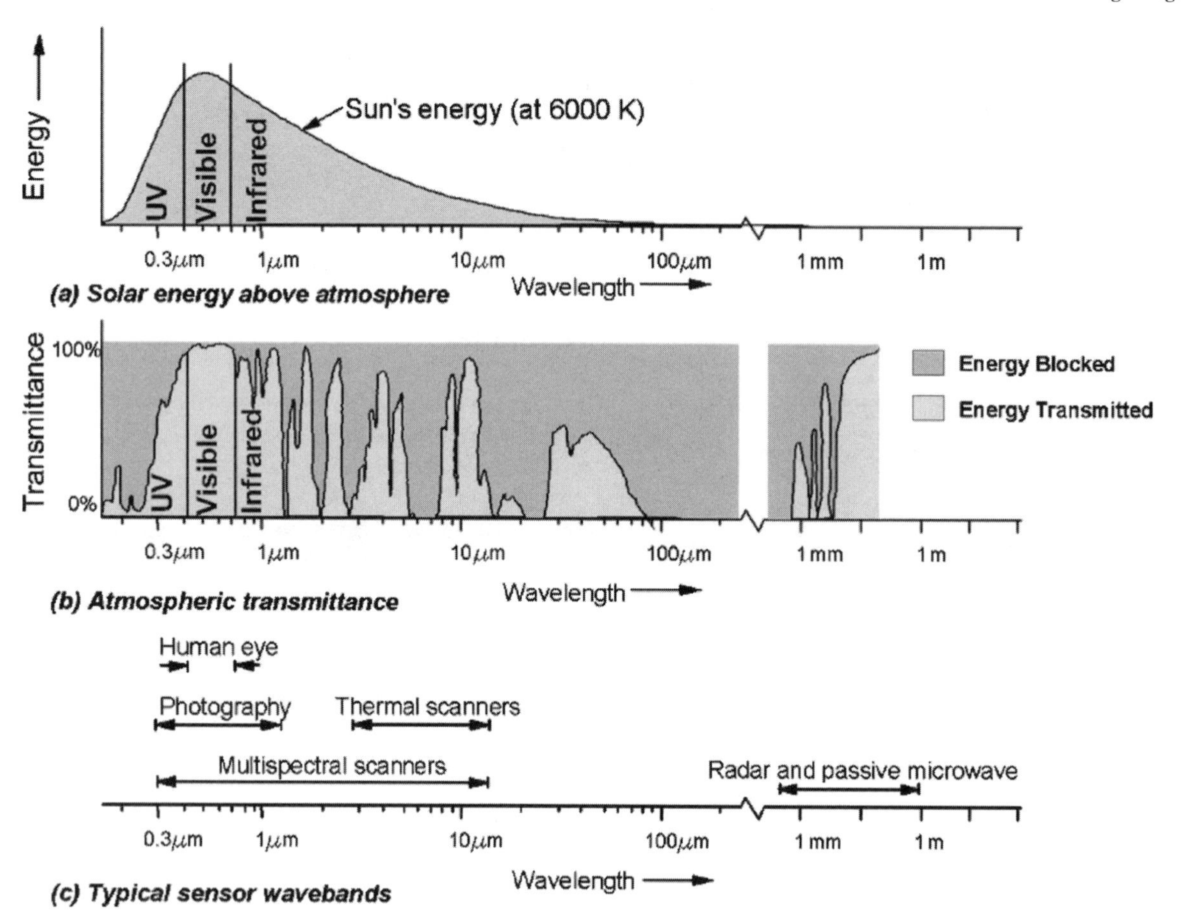

Fig. 3 Spectral distribution of (a) solar energy external to the earth's atmosphere, (b) transmittance of earth's atmosphere, and (c) sensitivity ranges for common remote sensors. (From Ref. [1], p. 11.)

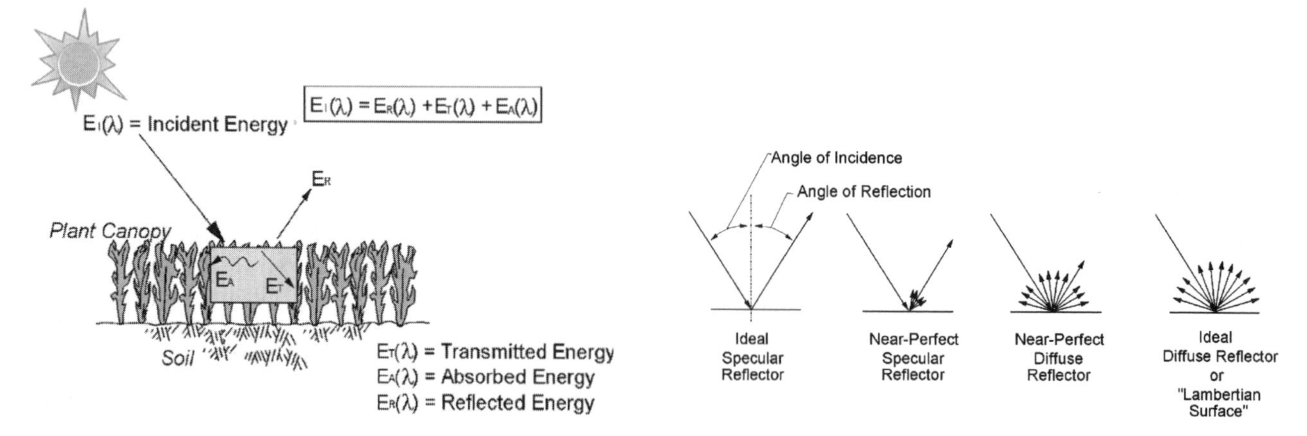

Fig. 4 Depiction of the interaction of a ray of EM energy with an object (a plant canopy in this case) on the earth's surface.

Fig. 5 Depiction of the angular variation of reflectance among objects. (From Ref. [1], p. 13.)

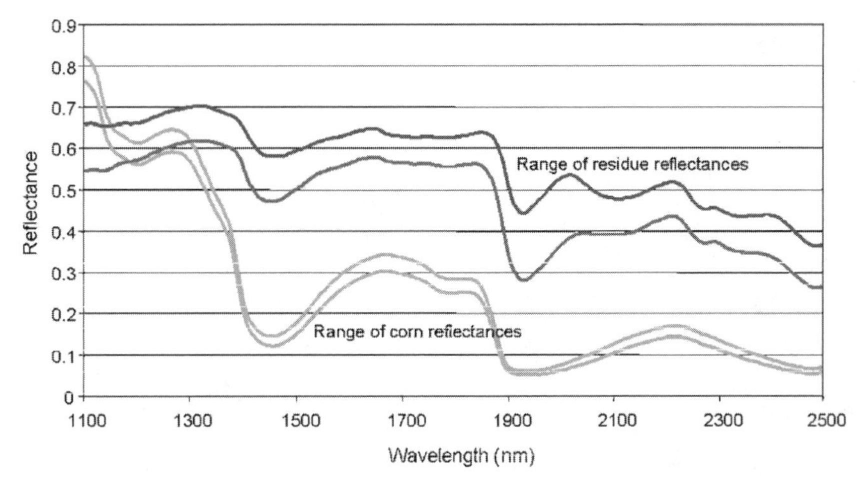

Fig. 6 Ranges of infrared reflectance vs. wavelength (1100 nm–2500 nm) for corn plant leaves and corn plant residues remaining in field. (Data from Shearer, S.A. Department of Biosystems and Agricultural Engineering, University of Kentucky.)

a representation of the numeric nature of image pixels. Film-based images, in a sense, contain pixels as well. However, their pixels are actually small metallic grains, unevenly spaced throughout the image, which retain the reflectance information in the scene by the degree of photochemical reaction that takes place when the grains are exposed to EM energy.

Remotely sensed images are often collected in what are called bands. A black-and-white image contains only one band of information, generally, representative of the total reflected energy in visible and/or infrared wavelengths. A color image contains three bands of information, with the first band representative of reflected energy in the blue wavelength range, the second band representative of reflected energy in the green wavelength range, and the third band representative of reflected energy in the red wavelength range. These bands of information are stored as separate layers of digital numbers (Fig. 9). While human vision can comprehend only three bands overlaid into a composite image, sensors are not restricted to any number, so images are often collected with many bands of information. Multispectral sensors are those that obtain image information in more than one band, up to several, say 20 or fewer, bands. Hyperspectral sensors are those that obtain image information in many, say 50 to several hundred, bands. The term, ultraspectral, has recently been coined to describe those sensors that obtain information in a very high number of bands, say several hundred or more. As human vision can comprehend only three image bands at a time, computer analysis methods are a practical requirement for images that contain more than three bands. Two satellites with sensors commonly used for agricultural study are Landsat and SPOT. Table 1 includes the temporal, spectral, and spatial capabilities of each.

A PRACTICAL EXAMPLE

A typical example of the use of remote sensing in agricultural research is in relating remotely sensed images to variations in crop yield. One way of doing this is by comparing, on a pixel-by-pixel basis, ground-based yield maps to remote-sensing-based normalized differential vegetative index (NDVI), which is calculated as follows:

$$\mathrm{NDVI} = \frac{\mathrm{NIR} - \mathrm{red}}{\mathrm{NIR} + \mathrm{red}}$$

where NIR is the value of a given pixel in the near-infrared band and red is the value of a given pixel in the red band.

This ratio is calculated for each pixel in an image, and the pixel values in the resulting image are subsequently

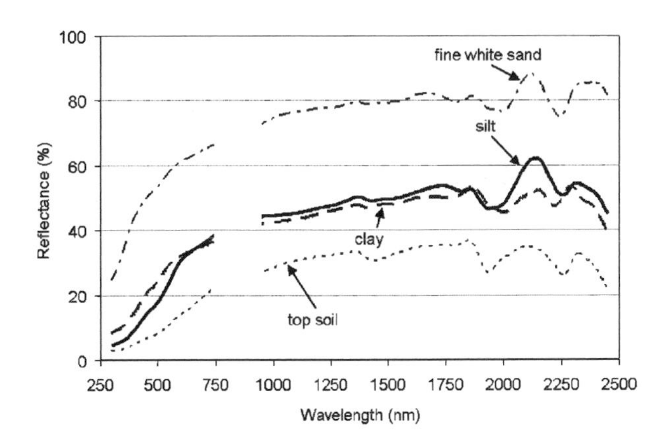

Fig. 7 Curves of reflectance vs. wavelength for commercially available reference soils. Gaps in curves reflect erroneous data produced by a detector change inherent in the spectrophotometer used to collect the data.

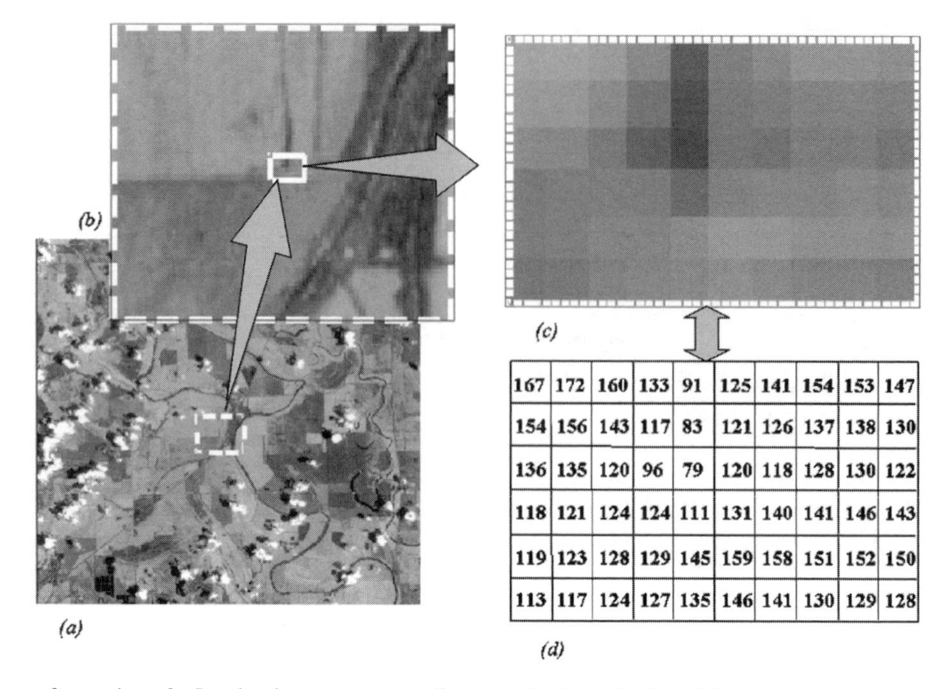

167	172	160	133	91	125	141	154	153	147
154	156	143	117	83	121	126	137	138	130
136	135	120	96	79	120	118	128	130	122
118	121	124	124	111	131	140	141	146	143
119	123	128	129	145	159	158	151	152	150
113	117	124	127	135	146	141	130	129	128

Fig. 8 Four views of a portion of a Landsat image corresponding to agricultural fields in Mississippi, depicting the numeric nature of pixels in images.

scaled to an appropriate level, such that they cover an appropriate range of possible pixel values. NDVI, generally, is strongly related to the amount of chlorophyll in plant leaves, which is a sign of how vigorously a plant is growing. Other indices have been derived, and sometimes individual bands are used as the primary source of information.

A satellite image of a cotton field at the full-canopy stage is shown in Fig. 10, and a strong linear relationship between average cotton yield in this field and pixel value in Landsat band 4 is shown in Fig. 11.

Shorterλ
UV
Blue
Green
Red
IR_1
IR_2
Longerλ

Fig. 9 Concept drawing of the nature of image bands.

Table 1 Basic specifications of remote-sensing systems aboard Landsat and SPOT satellites

Specification	Landsat Thematic Mapper	SPOT
Overflight frequency (day)	16	26
Spectral bands (nm): center wavelength–bandwidth	485–70	545–90
	560–80	645–70
	660–60	840–100
	830–140	1650–200
	1650–20	
	*11750–2100	
	2215–270	
Spatial resolution (m) *60	30	20

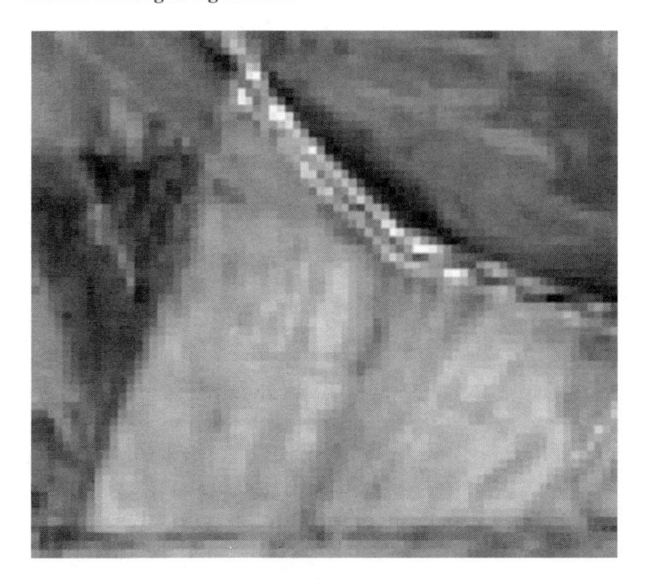

Fig. 10 Landsat band 4 (near-infrared) image of a cotton field during growing season.

CONCLUSIONS

It is clear that much important information about an agricultural field can be determined with remote sensing in a very rapid and efficient manner. As managers of agricultural production have become and are becoming more and more concerned with optimizing profits and environmental considerations with respect to spatial variability, remote sensing has and will continue to become more and more important as an agricultural production management tool.

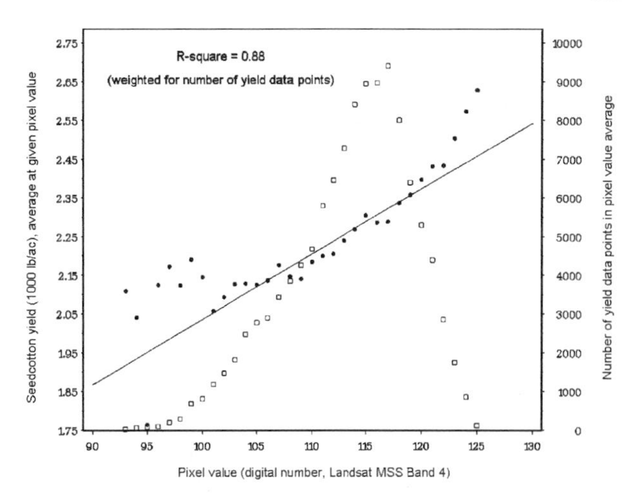

Fig. 11 Relationship between average cotton yield and remotely sensed crop reflectance (dots relate to left axis; squares relate to right axis).

ACKNOWLEDGMENTS

The author would like to express his appreciation to Mr. F. Paul Lee and Mr. James R. Wooten for their assistance in developing graphical materials for this article.

REFERENCE

1. Lillesand, T.M.; Kiefer, R.W. *Remote Sensing and Image Interpretation*, 2nd Ed.; John Wiley & Sons: New York, 1987.

Residence Time Distribution—Biomedical, Food, and Environmental Applications

Wei Wen Su
University of Hawaii at Manoa, Honolulu, Hawaii, U.S.A.

INTRODUCTION

In the article *Bioprocess Residence Time Distribution*,[1] a review on the applications of residence time distribution (RTD) analysis in industrial bioprocessing is presented. The scope of the present article is to review recent literature on the applications of RTD analysis in the biomedical, food process, and environmental fields. Particular attention is placed on RTD measurement techniques and flow models unique in these fields. RTD analysis offers a powerful tool to probe many biomedical problems, and to aid design and optimization of many food and environmental processes.

BIOMEDICAL APPLICATIONS

In the biomedical field, RTD has been widely applied in pharmacokinetic studies to examine the kinetics of drug absorption; it has also been employed to study transport physiology (e.g., assessment of changes in retinal blood flow and transport in lymphatic capillaries), and macromixing in bioartificial organs, just to name a few examples.

Study of Pharmacokinetics

The general application of RTD in pharmacokinetics has been reviewed by Weiss.[2,3] The RTD analysis generates information about the dynamics of the drug distribution process. Accurate assessment of the drug distribution is a prerequisite for the evaluation of drug elimination.[3] Mechanistic pharmacokinetic modeling is based on the circulatory multiorgan structure of the body and the underlying transport of drugs by blood flow to the various organs and tissues.[3] Kinetics of drug distribution at the organ level is mostly approximated as transfer between well-stirred compartments neglecting intravascular dispersion and diffusion within tissue parenchyma.[3] Mathematical models that adequately describe pharmacokinetic systems such as intact bodies or isolated perfused organs could, in principle, be identified by analyzing the system's response upon a drug dose. However, this is complicated by the complexity of the multiorgan structure of the body. Weiss[3] suggested two modeling strategies to simplify analysis of pharmacokinetics in organs and intact body based on residence time theory. One strategy calls for model decomposition into natural subsystems (e.g., isolated perfused organs), which can be identified separately; and the other relies on reduction of the complexity of the model by creating a simplified structural model such as the recirculatory model.[3] Building on the recirculatory model, Lansky[4] proposed a stochastic model for the RTD of a drug injected into the circulatory system. The properties of the residence time are derived from the assumptions made about the cycle time distribution and the rule for elimination (immediate vs. delayed elimination). This rule is given by the probability distribution of the number of cycles needed for elimination of a drug molecule.[4] Several discrete and continuous probability distributions were proposed by Lansky[4] to describe the cycle time distribution.

Assessment of Transport Physiology

Lee et al.[5] validated the use of RTD functions in human subjects to assess global changes in overall retinal flow, instead of analyzing just single vessel flow, by monitoring flow changes due to oxygen breathing. Changes in retinal blood flow could provide important information for clinical decision-making in patients with diabetic retinopathy, sickle cell disease, and *retinitis pigmentosa*.[5] Lee et al.[5] combined RTD functions for individual vessels into composite functions that potentially represent the overall retinal flow. These researchers applied scanning laser ophthalmoscopic and image processing techniques to acquire fluorescein tracer concentration at windows placed on the retinal vessels distal to intravenous injection. RTD functions can be used to obtain a value for mean residence time (t_m). The actual retinal vasculature consists of multiple arteries feeding into capillary beds with multiple veins returning flow to the heart. RTD functions from multiple arteries can be combined by weighting each artery according to its cross-sectional area. The composite arterial and venous curve data are then used to calculate a mean circulation time (MCT) based on all

Encyclopedia of Agricultural, Food, and Biological Engineering
DOI: 10.1081/E-EAFE 120016067
Copyright © 2003 by Marcel Dekker, Inc. All rights reserved.

vessels. Using this technique, Lee et al.[5] found that MCT increased for the group of subjects when breathing oxygen compared to normal air, representing a decrease in retinal blood flow. This method could be potentially useful for assessing hemodynamic changes in the retina associated with a wide range of eye diseases.[5]

Swartz, Berk, and Jain[6] presented a novel method for characterizing transport in the lymphatic capillaries in the tail of an anesthetized mouse. RTD measurements (using interstitially injected fluorescently labeled macromolecules) were used to determine net flow velocity in the lymphatic network as well as to provide a relative measure of lymphatic uptake of macromolecules from the interstitium. Using this method, Swartz, Berk, and Jain[6] examined the effects of particle size and injection pressure on the net velocity in the network and uptake rates. This method can be used to examine various aspects of transport physiology of the initial lymphatics.

Design of Bioartificial Organs

Understanding of hydrodynamic behavior is crucial for designing efficient artificial organs. A good example is the design of bioartificial liver or bioreactor-based hybrid liver support devices that are of great importance for liver assist.[7] In such systems, liver cells are typically cultured in 3D scaffolds. Catapano et al.[7] investigated the influence of bioreactor design and operating conditions on the hydrodynamics and reactor performance in two bioreactors proposed for liver assist. One of them was a clinical-scale bioreactor where liver cells are cultured around a 3D network of hollow fiber membranes and the other a laboratory-scale bioreactor with cells adherent on collagen-coated flat substrata. These reactors were characterized by RTD measurement and analyzed using the tank-in-series model.[8] Based on the estimated number of mixing tanks, Catapano et al.[7] showed that the two reactors exhibit different degrees of back-mixing. Switching from recycle to single-pass operation slightly reduced back-mixing in the clinical-scale bioreactor. Increasing feed flow rate significantly enhanced back-mixing in the laboratory-scale bioreactor.

FOOD PROCESSING APPLICATIONS

Food preservation by canning is widely practiced in the food industry. Common problems associated with heat sterilization during conventional canning include slow heat penetration to the slowest heating point in the container, the long processing times required to deliver the required lethality, destruction of the nutritional and sensory characteristics of the food, low productivity, and high-energy costs.[9] Aseptic processing provides an attractive alternative for food preservation via thermal processing, especially for high-quality food products. It is a continuous process that involves heating, holding, and cooling a food product followed by filling and sealing into presterilized containers under aseptic conditions.[9] Traditionally aseptic processing is applied to liquid food such as milk. There is also considerable interest in applying aseptic processing in low acid foods containing large particulates, such as pea soup, and this topic has been extensively reviewed by Ramaswamy et al.[9,10] As cited in Ref. 10, the most critical factors in analyzing aseptic processing of particulate food are particle size and shape, the fluid particle heat transfer coefficient, and the RTD in the heat exchangers and the holding tube determined with real food flowing in a commercial system. RTD is affected by various factors associated with the food material and the processing system. The factors associated with the food material may be further subdivided into the carrier fluid and food particles.[10] The fluid factors encompass type, concentration, density, flow rate, and rheological properties; while the particle factors include type, size, shape, density, and concentration with reference to the carrier fluid.[10] The system factors include: temperature, pumping system, mutator speed and configuration, heat exchanger orientation, holding tube length and diameter.[10] While the residence time of the fastest-moving particle is required for thermal process calculations, the residence time for particle-bulk will determine the extent of quality retention.[10]

Several research groups have investigated the RTD of carrier fluids as well as various simulated and real food particles in the heat exchanger and holding tube sections of the aseptic processing system. A detailed review of these studies was given by Ramaswamy et al.[10] In aseptic processing of food particulates, plug flow would lead to a product with adequate homogeneity, but heat transfer would be poor due to insufficient radial and axial mixing. Conversely, perfectly mixed flow is characterized by intensive mixing, which promotes heat transfer but suffered from increased dispersion. In reality, the flow conditions in typical aseptic processes are between plug flow and perfectly mixed flow. In general, the liquid phase has been modeled by axial dispersion or tank-in-series model, while RTD of food particles suspended in carrier fluids has been modeled using normal probability distribution, superposition of two or more normal distribution curves, or Gamma distributions.[10] In a typical aseptic processing system, the whole system has to be evaluated as an integral unit. The overall optimum processing condition can only be obtained by coupling heat transfer characteristics and RTD, with the destruction kinetics of enzymes/nutrients and heat-resistant micro-organisms.[10]

ENVIRONMENTAL APPLICATIONS

Knowledge of residence time and its distribution is of prime importance for the characterization and design of many environmental engineering processes. This is especially true in the field of wastewater treatment, where a homogeneous fluid distribution is often essential. Several reports exist in the literatures that deal with application of the RTD analysis to investigate the hydraulic characteristics of waste-treatment-related systems, including lagoons, anaerobic digesters,[11] constructed wetlands,[12] trickling filters,[13] sequencing batch biofilm reactors,[14] and novel ultrasonic reactors,[15] among others.

Newell et al.[16] used RTD data to develop wastewater process models that simulated the hydraulic characteristics of real biological nutrient removal pilot plants. The RTD tests provided useful information for evaluating mixing effectiveness, volume utilization, and for determining an appropriate hydraulic topology for the dynamic models of the pilot plants. The RTD analysis coupled with the process simulation also led to identification of short-circuiting and dead zones in the pilot plants. In the study of Seguret, Racault, and Sardin,[13] RTD was used to build hydrodynamic models for full-scale trickling filters. In order to explain the observed RTDs, a hydrodynamic model was developed assuming that the tracer is transported by an axially dispersed plug flow with molecular diffusion in and out of the biomass in the filter. Such model allowed identification of the volume of biomass and circulating liquid in the filter. Johnson et al.[17] determined the optimal theoretical RTD for chlorine contact tanks (CCT) that are commonly used for disinfections of potable water supplies. Through this analysis, Johnson et al. proposed strategies to enhance microbial inactivation within the CCTs by improving the hydraulic performance of the CCTs.

Lagoons are widely used to reduce organic pollution and bacteriological contamination.[18] Baleo, Humeau, and Le Cloirec[18] described two computational approaches for obtaining theoretical predictions of both the spatial distribution of the mean residence time and the temporal distribution of the exit residence times in lagoons. The first approach consists of solving a transport equation of the local mean age of the fluid. The result obtained is a spatial distribution of the local mean age of the fluid. The second approach involves injecting a virtual tracer and measuring the time elapsed between the injection and the termination of the tracer trajectory using a Lagrangian reference frame. The result obtained is an exit age distribution. The prediction enables the determination of the geometrical characteristics of the flow that affects the residence time dispersion. The approaches reported by Baleo, Humeau, and Le Cloirec[18] could be used to aid design of lagoon systems with desirable flow patterns to improve the performance of such systems.

The anaerobic fixed film reactors (AFFR) are reactors that contain a mixed population of bacteria immobilized on surfaces of solid support medium. They have been widely applied in the high-strength waste treatment. Chua and Fung[19] presented RTD studies to investigate the hydrodynamic characteristics in an AFFR equipped with an effluent recycle stream, which is used for diluting and distributing the organic constituents of the trade effluent. Tracer pulse-response technique and dispersion model were employed to probe the RTD characteristics in the AFFR. The results showed that significant dispersion could result from the effluent recycle, which in turn could negatively affect the AFFR performance. In addition to fix-beds, anaerobic fluidized-bed reactors (AFBR) are widely used as anaerobic digesters. One common problem in operating such reactors comes from the excessive biogas production and associated gas hold-up that may lead to bed contraction and reduced the contact time between the waste stream and the bioparticles.[11] Buffiere, Fonade, and Moletta[11] investigated the effect of biogas production on the performance of an AFBR in terms of mixing and phase hold-ups. RTD measurements in the AFBR were done using the tracer injection method. The tracer used was lithium chloride, the concentration of which was measured with a flame photometer. Dimensionless dispersion numbers were calculated from the RTD data.[8] The performance of the fluidized-bed reactor was shown to be greatly influenced by the presence of biogas since the biogas effervescence modifies the reactor hydrodynamics and the phase hold-ups. This is particularly problematic under high-organic loading rates, where a high level of biogas is produced.

Propagation of ultrasonic waves in water leads to H_2O sonolysis and production of radical species, especially the very strong oxidizing hydroxyl radicals.[15] As such, ultrasound has been proposed and proved to be an effective degradation method of organic pollutants.[15] This idea has led to the development of high-frequency ultrasonic reactors for wastewater treatment. Similar to other chemical reactors, the conversion yield (in this case the sonochemical degradation yield) of the ultrasonic reactor is related to reaction kinetics and reactor hydraulic behavior that determines contact time between reactants. In the work of Gondrexon et al.,[15] RTD analysis was used to investigate the hydrodynamic behavior of a high-frequency ultrasonic reactor for degradation of organic pollutants. RTD measurements were performed by means of a NaCl tracer, which was shown to be stable under sonication.[15] The high-frequency sonochemical reactor was found from the RTD analysis to behave like a completely mixed reactor as soon as ultrasonic irradiation was applied. Such efficient mixing, however, could not be explained by the single contribution of the acoustic streaming since the velocities remain relatively low.[15] While RTD analysis offers a simple means to characterize the apparent reactor hydrodynamics, more understanding on the underlying mixing mechanism is necessary in order to predict the hydrodynamics in sonochemical reactors, for effective design and operation of such reactors.

CONCLUSION

Expanding from its roots in chemical reactor analysis, over the years RTD analysis has also found its way to applications in many other fields. To this end, special techniques are required to collect the RTD data, especially in biomedical studies. Interpretation of the RTD data also calls for appropriate flow models. While the task of establishing these flow models may depend on the complexity of the systems in question, the selected RTD studies reviewed in this chapter provide a broad base of approaches to establish such models relevant to biomedical, food, and environmental applications. RTD analysis offers a powerful tool to better understand many biomedical problems, and to aid design and optimization of many food and environmental processes.

REFERENCES

1. Su, W. Bioprocess Residence Time Distribution. In *Encyclopedia of Agricultural, Food, and Biological Engineering*; Heldman, D., Ed.; Dekker: New York, 2003.
2. Weiss, M. The Relevance of Residence Time Theory to Pharmacokinetics. Eur. J. Clin. Pharmacol. **1992**, *43*, 571–579.
3. Weiss, M. Pharmacokinetics in Organs and the Intact Body: Model Validation and Reduction. Eur. J. Pharm. Sci. **1998**, *7*, 119–127.
4. Lansky, P. A Stochastic Model for Circulatory Transport in Pharmacokinetics. Math. Biosci. **1996**, *132* (2), 141–167.
5. Lee, E.T.; Rehkopf, P.G.; Warnicki, J.W.; Friberg, T.; Finegold, D.N.; Cape, E.G. New Method for Assessment of Changes in Retinal Blood Flow. Med. Eng. Phys. **1997**, *19* (2), 125–130.
6. Swartz, M.A.; Berk, D.A.; Jain, R.K. Transport in Lymphatic Capillaries. I. Macroscopic Measurements Using Residence Time Distribution Theory. Am. J. Physiol. **1996**, *270* (1 Pt 2), H324–H329.
7. Catapano, G.; Euler, M.; Gaylor, J.D.S.; Gerlach, J. Characterization of the Distribution of Matter in Hybrid Liver Support Devices Where Cells Are Cultured in a 3-D Membrane Network or on Flat Substrata. Int. J. Artif. Organs **2001**, *24* (2), 102–109.
8. Levenspiel, O. *Chemical Reaction Engineering*, 3rd Ed.; Wiley: New York, 1999.
9. Ramaswamy, H.S.; Awuah, G.B.; Simpson, B.K. Heat Transfer and Lethality Considerations in Aseptic Processing of Liquid/Particle Mixtures: A Review. Crit. Rev. Food Sci. Nutr. **1997**, *37* (3), 253–286.
10. Ramaswamy, H.S.; Abdelrahim, K.A.; Simpson, B.K.; Smith, J.P. Residence Time Distribution (RTD) in Aseptic Processing of Particulate Foods: A Review. Food Res. Int. **1995**, *28* (3), 291–310.
11. Buffiere, P.; Fonade, C.; Moletta, R. Mixing and Phase Hold-Ups Variations Due to Gas Production in Anaerobic Fluidized-Bed Digesters: Influence on Reactor Performance. Biotechnol. Bioeng. **1998**, *60* (1), 36–43.
12. Chendorain, M.; Yates, M.; Villegas, F. Fate and Transport of Viruses Through Surface Water Constructed Wetlands. J. Environ. Qual. **1998**, *27* (6), 1451–1458.
13. Seguret, F.; Racault, Y.; Sardin, M. Hydrodynamic Behaviour of Full Scale Trickling Filters. Water Res. **2000**, *34* (5), 1551–1558.
14. Morgenroth, E.; Wilderer, P.A. Controlled Biomass Removal—The Key Parameter to Achieve Enhanced Biological Phosphorus Removal in Biofilm Systems. Water Sci. Technol. **1999**, *39* (7), 33–40.
15. Gondrexon, N.; Renaudin, V.; Petrier, C.; Clement, M.; Boldo, P.; Gonthier, Y.; Bernis, A. Experimental Study of the Hydrodynamic Behaviour of a High Frequency Ultrasonic Reactor. Ultrason. Sonochem. **1998**, *5* (1), 1–6.
16. Newell, B.; Bailey, J.; Islam, A.; Hopkins, L.; Lant, P. Characterizing Bioreactor Mixing with Residence Time Distribution (RTD) Tests. Water Sci. Technol. **1998**, *37* (12), 43–47.
17. Johnson, P.; Graham, N.; Dawson, M.; Barker, J. Determining the Optimal Theoretical Residence Time Distribution for Chlorine Contact Tanks. AQUA (OXFORD) **1998**, *47* (5), 209–214.
18. Baleo, J.N.; Humeau, P.; Le Cloirec, P. Numerical and Experimental Hydrodynamic Studies of a Lagoon Pilot. Water Res. **2001**, *35* (9), 2268–2276.
19. Chua, H.; Fung, J.P.C. Hydrodynamics in the Packed Bed of the Anaerobic Fixed Film Reactor. Water Sci. Technol. **1996**, *33* (8), 1–6.

R

Retortable Pouches

Barbara Blakistone
National Food Processors Association, Washington, District of Columbia, U.S.A.

INTRODUCTION

Once the bastion of the military in the form of meals-ready-to-eat (MREs), retortable pouches, either sold refrigerated or as a shelf stable product, are becoming increasingly more popular and continue to impact the can market especially in hotel, restaurants, and institutions (HRI), markets in which the pouch is replacing the No. 10 can as a convenience to the end user. The pouches were developed in the 1950s by U.S. Army Natick Research, Development and Engineering (RD&E) Center but were not popular with HRI or consumers until recently. The packaging problem with pouches in the retail market was that high-impact graphics with extensive distribution of ink challenged the adhesives of the lamination procedure. Then came high-temperature urethane adhesives to prevent delamination, and the consumer was afforded maximum labeling information. Now, even the pharmaceutical industry is considering the advantages of pouches, and some companies are converting their packaging of enterals (nutrients/medicinals for direct delivery into the gastrointestinal tract) to pouches. The retail market offers consumers tuna fish for sandwich making and foods for their pets in stand-up pouches. Other retail products include chipped beef in gravy, a vegetable and rice dish, which is an extended shelf life product, sold refrigerated and is not shelf stable, Halal and Kosher foods, beef and broth, chili, hot dogs, and chicken patties, and new on the Canadian market, rice and pasta dishes.

PACKAGE CONSTRUCTIONS

Pouches may be either premade or formed from rollstock, the more attractive price alternative. Alternately, premades permit an increased line speed over that of rollstock, and mechanical issues of converting rollstock to pouches at the food plant disappear. In counter to that advantage are the problems higher line speed brings: contamination issues especially with possibilities of product sloshing into the seal area. Volume can be from 3 oz to 2 gal (88.7 mL – 7.57 L). Incoming materials should be checked for thickness, proper components, roll identification, and, if preformed, seal width and pouch size. Once sealed, the pouches should have a seal width of at least 1/16 in. on all sides for quality assurance.

Preformed pouches may be foil-based laminates in constructions such as polyester (PET)/nylon/foil/cast polypropylene (PP) or PET/foil/cast PP. Nonfoil constructions may be PET/nylon/ethylene vinyl alcohol (EVOH)/-cast PP or PET/high barrier Saran (PVDC)/cast PP. Rollstock may be coextruded without adhesives in a construction like polyethylene sealant/tie layer/nylon/E-VOH. PET is known for its strength and toughness, and its heat resistance (melt point of 264°C/507°F) makes it inert to food at elevated temperatures. The nylon (polyamide) provides strength and abuse resistance to piercing and flex-pinhole resistance. EVOH is the gas (i.e., oxygen) barrier. Foil is also a gas and moisture barrier, and it protects against light. PP is another heat-sealable film that resists retort temperature (up to 130°C/266°F), adds stiffness to the package, and resists staining. As mentioned, stand-up pouches have become popular and address the complaint that pouches cannot be displayed in the grocery store.

Stand-up construction is either 4 ply or 2 ply:

- 4 ply—Reverse printed PET/nylon/foil/cast PP. The foil is another enticement to the consumer who is looking for body feel and stiffness in a stand-alone package.
- 2 ply—Metallized or silicon oxide PET/cast PP.

PACKAGE DEFECTS

Critical defects which may compromise the integrity of the pouch were originally defined in the National Food Processors Association's (NFPA) Flexible Packaging Integrity Bulletin[1] and include the following:

1. Channel leaker
2. Cuts
3. Fracture
4. Leaker
5. Nonbonding
6. Notch leaker
7. Puncture

Encyclopedia of Agricultural, Food, and Biological Engineering
DOI: 10.1081/E-EAFE 120007133
Copyright © 2003 by Marcel Dekker, Inc. All rights reserved.

8. Swollen package
9. Wrinkle

Major defects are defined as those that result in a pouch not showing visible signs of loss of hermetic seal but are of such a magnitude that seal integrity may have been lost.[1] Such defects include:

1. Abrasion
2. Blister
3. Compressed seal
4. Contaminated seal
5. Delamination
6. Misaligned seal
7. Seal creep
8. Wrinkle

Some major defects may, in fact, be minor upon thorough examination of the pouch. A minor defect is defined as a defect that has no adverse effect on the hermetic seal.[1] The effects of minor defects are usually cosmetic. Defects, which may be major or minor, include abrasions, blisters, compressed seals, contaminated seals, misaligned seals, seal creeps, and wrinkles. Exclusively minor defects include convolutions/embossing (slight visual impressions in the seal indented on one side and raised on the other), crooked seals/short seals, flex cracks (small breaks in one or more layers of the package, due to flexing, not a leaker), hot folds, stringy seal, and waffling (embossing caused by racks during thermal processing that appears on the surface of the pouch).

The FDA Bacteriological Analytical Manual[2] lists some additional defects to those in the NFPA Bulletin,[1] but those additions will not be discussed in this review. However, Chapter 22 of the BAM is useful in its discussion of package integrity.

The food industry continues to ponder what is the critical defect rate for pouches and the other numerous types of plastic containers. The answer depends on the initiative a company wishes to expend on equipment and maintenance. One contract packager reported zero defects after spending time and money on making sure the packaging equipment was thoroughly optimized before production began and prompt maintenance of any tooling that was not performing properly during the day's run. Principal problems in the use of pouches that the food industry is currently experiencing are product in the seal area (termed contaminated seal in the NFPA document) and wrinkled seals. While collaborative studies between Virginia Polytechnic Institute and State University and NFPA[3–5] have established threshold sizes of microbial invasion into model packages with intentional defects, the next step is to consider these numbers in the context of risk. The theoretical studies need to be evaluated in terms of what happens in the processing plant.

Natick RD&E Center has reported a spoilage failure rate of 1.7/10,000 during a program to test reliability of the pouch.[6] In a separate comparative test between tin plate cans and the retort pouch, Natick found no significant difference between the two after distribution testing and biotesting the containers. The failure rate was 1.1/10,000, which included manufacturing defects as well as transportation-related failures. This fits with current pouch failure rates of which NFPA is aware and is a considerable improvement over the defect rate averaging 1.8% in MREs in the early 1990s.

PACKAGE INSPECTION AND REGULATION

Pouches are commonly inspected visually. Line speeds are slow enough to permit this (e.g., 60 packages/min). FDA regulations[7] have always required Good Manufacturing Practices and container closure inspections, which for pouches have not been specifically defined. The Agency says that appropriate tests must be conducted with sufficient frequency to ensure proper closing machine performance and consistently reliable hermetic seal production. USDA regulations[8] for the packaging of meat and poultry products require visual inspection of the seals and the entire container before and after retorting and physical testing after retorting, so postprocessing defects are very low. Incubation requirements for flexible containers have changed as the Agency moves away from the previous "command and control" policy. USDA Food Safety and Inspection Service's Office of Policy, Program Development and Evaluation permits: 1) the 10-day incubation as established in the regulations; 2) a modified incubation procedure (e.g., increase the sample percentage of a lot incubated for less than 10 days); or 3) release of the product without incubation if the company has a letter from a processing authority stating that safety and stability of the product has been provided for by a HACCP plan, QC program, or process schedule. Physical tests are required at least every 2 hr but may be defined by the plant to assess package integrity. U.S. military regulations require MRE pouches to be visually inspected, burst tested, incubated, dye tested, and tested for residual gas after retorting. National Food Processors Association (NFPA) guidelines[1] suggest squeeze testing of pouches, inspection of the head and side seals as well as the width of the seal area. This is to be done on all packages across the web or in each lane at start-up, every 30 min thereafter, and after splicing. Physical tests, described later, such as the seal tensile strength, squeeze test, or a burst test should be done at start-up and every 4 hr thereafter or more frequently if problems occur.

Squeeze testing is a manual kneading action, which forces product against the interior seal surface area. Burst testing,[9] a machine test, evaluates the peak or ultimate seal strength (rupture pressure) by exerting uniform pressure against all seals while filling the pouch with air. Inflation causes rupture. Tensile strength evaluates the resistance of the seal to separation using an Instron© (Canton, Massachusetts). However, tensile strength testing may not detect a weak area of seal unless multiple sections across the width of the seal are assessed. While the squeeze test is nondestructive on packages with good seals, the burst and tensile tests are, by their nature, destructive tests. Burst testing is becoming a very widely used food industry test. In a static variation on the burst test called the creep test, the internal pressure is increased by a needle delivering air into the package to a set pressure for a specified time. This type of testing examines package performance under stress conditions by measuring the resistance of the seals to shear load (pressure). Creep testing is considered static because it does not intentionally rupture the container. It may be recalled that creep may be visualized as the partial opening of the inner border of a seal by pressure. The creep test is typically performed for 30 sec at 80% of the burst value and is reported as pass/fail. Creep to failure is a dynamic test that determines time to fail as the variable when given a constant pressure that is set at slightly greater than the creep pressure. Pouches are then evaluated in terms of time to rupture against an average time. Decreasing time to rupture indicates the seals are becoming weaker. The test has been developed by TM Electronics Inc. (Worcester, Massachusetts) as a test suitable for statistical process control.

An improved version of the burst test uses a platen restraint allowing a maximum expansion of 10%. This test improves the distinction between good and bad seals. Pouches that fail to burst are examined for "creep" at the seals. Greater than 1/16-in. (1.6 mm) increase is cause for rejection. Semiporous platen restraining plates are available and may be used for indication of leakage in the package as a whole. A pressure decay rate is then established in this type of restraining test. Such testing indicates changes in seal strength compared to a standard set by the user.

Another seal test used is the traditional dye penetration test in which a dye solution is poured into the seal area and the package is observed for dye leaking through the seal. Dyes may be aqueous-based such as developed by the DuPont TYVEK Medical Packaging Group (0.5% Triton X-100 surfactant and 0.05% Toluidine blue)[10] or alcohol-based such as recommended by NFPA (0.5% methylene blue in isopropanol).

The drop test is a physical test in which the pouch is slid down a chute from a precalculated height based on gross container weight. The seals and the package are inspected after dropping. Residual gases in the headspace may be tested if the pouches are filled in a way that reduces the oxygen content.

PACKAGING MACHINERY

Machinery for filling is governed by the fact that there are preformed pouches or rollstock that are formed into pouches. Form/fill/seal (F/F/S) equipment may be either vertical or horizontal. While either style can be used for food applications, vertical F/F/S (VF/F/S) is used more for pumpable foods like stews while horizontal F/F/S (HF/F/S) is used for solid, placeable items like meat patties. Horizontal machinery is advantageous in that it can have a greater number of operating stations for multiple filling, headspace purging, etc.

In HF/F/S, a single roll of film material is fed over forming collars with heat seals setting off the appropriate sized pouches, which are then opened by air jet for filling at one or more stations. Fig. 1 depicts the HF/F/S operation. Pouches are characterized by a three-sided seal with the bottom being formed by the forming collar, so no sealing is necessary (one side remains open for filling). The formed pouch may be cut from the main web before or after sealing all three sides. The machinery will accommodate a gusseted construction to form stand-up pouches (Fig. 2).

VF/F/S machines offer the possibility of using two types of materials to form a top and bottom web (Fig. 3). Pouches made from a single web (Fig. 4) are characterized by a seal across the pouch top and bottom and a vertical seal through the center of the back. The vertical seal may be of the fin or lap type (Fig. 2). The fin brings together two inside surfaces with the net result of a small flap of material protruding along the vertical. More aesthetic is the lap seal, but the inside surface must be married with the outside surface. If the machinery is a multilane system, the pouches will have four seals.

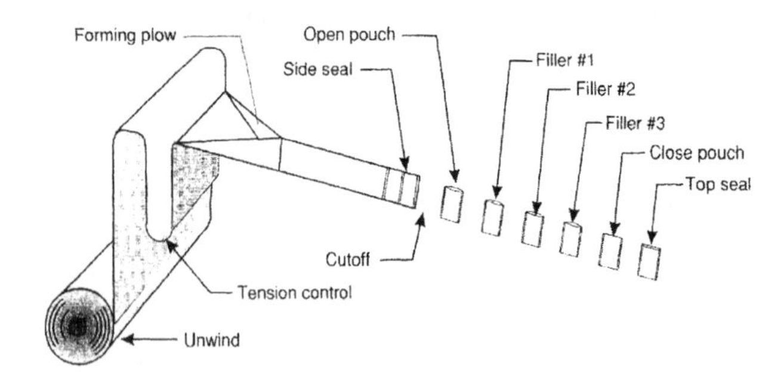

Fig. 1 Material and product flow in HF/F/S machine. (From Soroka, W. *Fundamentals of Packaging Technology*, 2nd Ed.; Institute of Packaging Professionals: Naperville, Illinois, 1999; 343–345.)

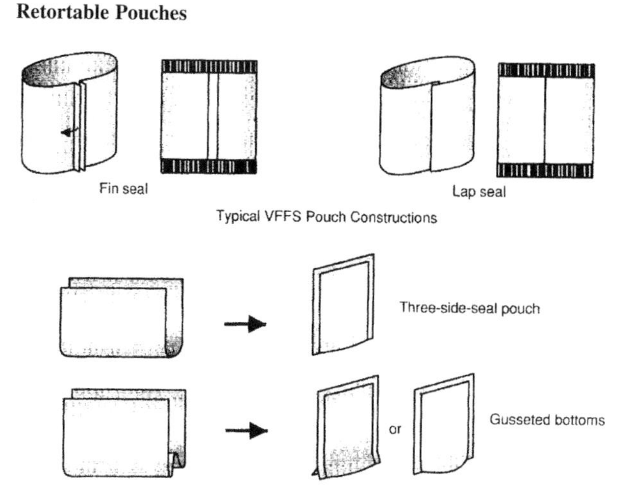

Fig. 2 VF/F/S and HF/F/S machines produce pouches with different seal geometries. (From Soroka, W. *Fundamentals of Packaging Technology*, 2nd Ed.; Institute of Packaging Professionals: Naperville, Illinois, 1999; 343–345.)

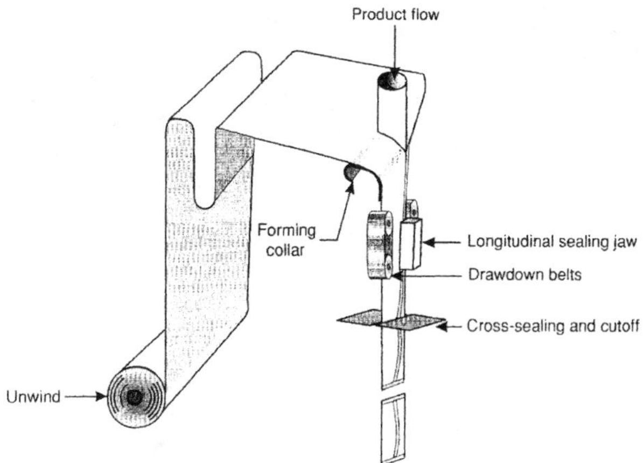

Fig. 4 Material and product flow in a VF/F/S machine using single rollstock. (From Soroka, W. *Fundamentals of Packaging Technology*, 2nd Ed.; Institute of Packaging Professionals: Naperville, Illinois, 1999; 343–345.)

To minimize headspace air to prevent expansion in the retort or ballooning of the pouch when placed in boiling water for heat and serve, several methods are used. The simplest is to squeeze the pouch with rollers, though the possibility of squeezing traces of product into the seal area must be considered. Vacuum may be drawn by a tube inserted inside the pouch, or the pouch may enter a vacuum chamber prior to sealing. Neither method is totally efficient. As with hot filling, saturated or superheated steam may be used to flush air, with condensation of water vapor upon cooling minimizing the amount of air. Superheated steam is less effective, but superheated steam causes less moisture condensation. Backflushing of the pouch with nitrogen is also done.

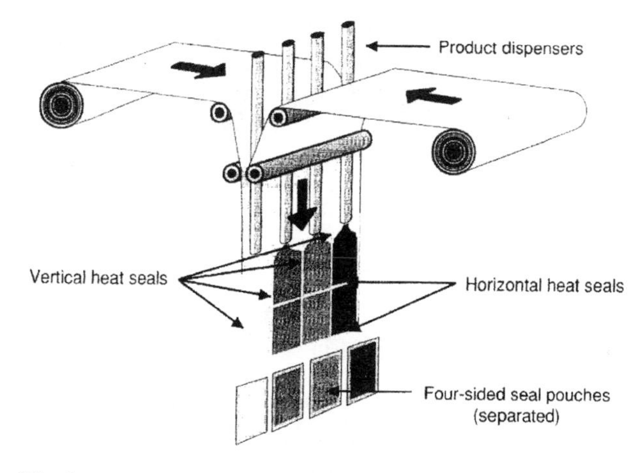

Fig. 3 A multilane VF/F/S machine accommodating two types of rollstock. (From Soroka, W. *Fundamentals of Packaging Technology*, 2nd Ed.; Institute of Packaging Professionals: Naperville, Illinois, 1999; 343–345.)

All pouches are heat sealed by either hot bar or impulse sealing. As with all heat sealing, the factors of time, temperature, and dwell time determine the quality of the seal. The quality, of course, must be the best. Retorting will easily stress weak seals to the point of failure.

PROCESSING

Retorts used in processing pouches can be batch or continuous, agitating or nonagitating, and they require air or steam overpressure (pressure supplied to a retort in excess of that exerted by the heating medium at a given process temperature) to control pouch integrity. There are three kinds of retorts advocated in processing pouches:

- Saturated, superheated steam/water spray (S/W), or cascading water (Fig. 5)
- Total immersion water with air or steam overpressure
- Steam/air (S/A)

Hydrostatic retorts that use cascading water with overpressure may also be used for retortable pouches, but their use is not common.

Both the S/W and S/A must have a system to circulate the heating media. With S/A systems, it can be difficult to maintain the steam:air ratio. The water circulation of the S/W retort solves that potential problem. The S/W system is particularly attractive in that there is no restriction on air pressure. The heat to cool transition is easy with no danger

Fig. 5 Example of a S/W spray retort. (Photo courtesy of FMC Technologies Inc., Madera, California.)

of a pressure drop that could affect package integrity. Agitating retorts may enhance pouch heat penetration and therefore processing efficiency, but pouches are not usually agitated so S/W systems are usually used in the nonagitating mode. Systems that control ramping of temperature and pressure during heating and cooling reduce the impact on the pouches and result in better control of lethality delivered during the cooling phase. Total immersion water systems pump in water at 220–270°F and then start the "come up" time, which can stress packages. Other stress conditions include poor overpressure control, which can lead to bad seals. While the retort process stresses even good pouch seals, when cooled back to room temperature, the seals regain 90% of their original strength.[6]

Design of racking systems for placement of pouches in the retort is critical for providing adequate circulation of the heating medium. Both vertical and horizontal designs are available, though horizontal racks are the most common. In spite of material improvements over the years, pouches still require careful handling. Avoid burrs and rough spots on transport holders and conveyor systems through proper maintenance.

Process filings are required by FDA on low acid and acidified foods and by USDA on low acid foods, and the filings must include factors critical to heat penetration and therefore commercial sterility of the product. Compliance with the defined critical factors during production ultimately determines whether process deviations occurred. These factors in retorting of pouches or any container include minimum headspace, product consistency, maximum filling or drained weight, initial temperature, processing temperature and time, temperature distribution, container orientation, residual gas in the headspace and in the food, processing and racking systems, processing media, product heating rates, and materials from which the pouch is constructed.

Flexible films offer little resistance to heat transfer as their geometry is usually that of a slab. The thin profile and flat shape of the pouch afford rapid heat penetration due to the high surface area to volume ratio. Container thickness is the most significant problem affecting product heating in flexibles.[11] Because thickness cannot be controlled by the container, it is desirable to physically define container thickness during processing by means of specially designed racks, trays, or cassettes.

When processing unconfined flexibles, thickness is dependent upon retort overpressure and factors mentioned above such as residual gas in the headspace, fill weight, container size and orientation, and product characteristics.[11] Excessive fill weights may significantly increase container thickness, which results in process deviations.

Physical and chemical characteristics of the product also dictate heating rates. Characteristics affecting heating rates include composition of the ingredients, particulate size, physical state (fresh, frozen, or cooked), homogeneity, specific gravity, soluble solids, occluded gases, viscosity, etc. In conduction heating foods, minor variations in the formula do not significantly impact the heating rate, but foods with sufficient free liquid to promote convection heating may be adversely affected by increases in percent starch or other water-binding ingredients with the result of overprocessing.

Air in the pouch not only can cause ballooning of the pouch during processing, affecting thickness, but also results in underprocessing as air is an excellent insulator against heat.

The attraction of pouches to the processor is minimal overcooking of the product at the peripheral areas of the container and overall minimizing of quality loss with enhanced retention of heat sensitive nutrients. Pouches can provide benefit when all phases of the production operation are closely controlled.

For further information on processing, the reader is referred to all of Chapter 5 in Ref. 6 and Chapters 10, 11, and 13 in Ref. 12.

CONCLUSION

Advances in package construction have permitted the retortable pouch, developed in the 1950s for the military, to penetrate the retail market as well as HRI. Machinery for filling the pouch is governed by the fact that there are preformed pouches or film rollstock. F/F/S equipment is either vertical or horizontal, the latter being used to prepare preforms that will likely contain placeable items like meat patties. Retorts for processing are either saturated, superheated S/W spray or cascading water, total immersion water with air or steam overpressure, or S/A. The thin profile and flat shape of the pouch afford rapid heat penetration due to the high surface to volume ratio. The attraction of pouches to the processor is minimal overcooking of the product at the peripheral areas of the container and overall minimizing of quality loss with enhanced retention of heat-sensitive nutrients.

REFERENCES

1. Flexible Package Integrity Committee. *Flexible Packaging Integrity Bulletin 41-L*; National Food Processors Association: Washington, D.C., 1989.
2. Lin, R.C.; King, P.H.; Johnston, M.R. Examination of Containers for Integrity. *Food and Drug Administration Bacteriological Analytical Manual*, 8th Ed.; Revision A, AOAC International: Gaithersburg, MD, 1998; Chap. 22, 40.
3. Keller, S.; Marcy, J.; Blakistone, B.; Hackney, C.; Carter, W.H.; Lacy, G. Application of Fluid Modeling to Determine the Threshold Leak Size for Liquid Foods. J. Food Prot. *in press*.
4. Keller, S.; Marcy, J.; Blakistone, B.; Hackney, C.; Carter, W.H.; Lacy, G. Application of Fluid and Statistical Modeling to Establish the Leak Size Critical to Package Sterility. J. Food Prot. *in review*.
5. Keller, S.; Marcy, J.; Blakistone, B.; Hackney, C.; Carter, W.H.; Lacy, G. Effect of Microorganism Characteristics on the Leak Size Critical to Predicting Package Sterility. J. Food Prot. *in press*.
6. Downing, D.L. Retortable Flexible Containers. *A Complete Course in Canning, Book II, Microbiology, Packaging. HACCP, and Ingredients*, 13th Ed.; CTI Publications, Inc.: Timonium, MD, 1996; Chap. 5, 220.
7. Code of Federal Regulations. Title 21, Food and Drugs. Part 113.5, Current Good Manufacturing Practice and 113.60 (a)(ii)(3), Containers; U.S. Government Printing Office: Washington, D.C., April 1, 1997.
8. Code of Federal Regulations. Title 9, Animals and Animal Products. Part 318.301, Containers and Closures; U.S. Government Printing Office: Washington, D.C., January 1, 2001.
9. Franks, M.; Franks, S. Testing Medical Device and Package Integrity. http://tmelectronics.com (accessed September 2001).
10. Hackett, E.T. Dye Penetration Effective for Detecting Package Seal Defects. Pkg. Technol. Eng. **1996**, *5* (8), 49–52.
11. *Guidelines for Thermal Process Development for Foods Packaged in Flexible Containers*; National Food Processors Association: Washington, D.C., 1985.
12. Gavin, A., Weddig, L.M., Eds. Chapter 10: Still Retorts—Processing with Overpressure and Chapter 11: Hydrostatic Retorts—Continuous Container Handling. *Canned Foods, Principles of Thermal Process Control, Acidification and Container Closure Evaluation*, 6th Ed.; The Food Processors Institute: Washington, DC, 1995.

Reverse Osmosis

Zeki Berk
Technion, Israel Institute of Technology, Haifa, Israel

INTRODUCTION

The French physicist Abbé Jean Antoine Nollet (1700–1770) was the first to observe that when an animal membrane is placed between pure water and alcohol, water passes through it into the alcohol, causing an increase of pressure, but alcohol does not pass through it into the water. Subsequently the term "osmosis" was coined to describe the spontaneous passage of water from a dilute into a more concentrated solution through certain membranes, later qualified as "semipermeable membranes." To stop the spontaneous passage of water into a solution through osmosis, a certain pressure, equal to the osmotic pressure of the solution, must be applied against the direction of flow. Applying a pressure in excess of the osmotic pressure will result in the reversal of the direction of flow, from the more to the less concentrated solution. This is "reverse osmosis" or RO (Fig. 1).

Reverse osmosis belongs to a group of processes known as "pressure-driven membrane separation processes." Other processes in this group are: microfiltration, ultrafiltration, and nanofiltration. These processes differ not only in the size range of the particles retained by the membrane (Table 1) but also in the types of membranes used and in the mechanism of mass transport through the membrane. Reverse osmosis membranes are able to retain most low molecular weight solutes and salt ions. Nanofiltration, sometimes called "loose RO" retains low molecular weight solutes but permits the selective passage of certain ions.

A simplified diagram of an RO operation is shown in Fig. 2. In the case of water desalination or purification (polishing), the permeate (pure water) is the product and the retentate (brine) is a waste. In contrast, when RO is used for concentrating a dilute stream (e.g., pre-concentration of tomato juice), the retentate (concentrate) is the product and the permeate (water) is the waste. In some applications both the retentate and the permeate may be useful products of different characteristics, or intermediate product streams.

Modern RO research began with experiments on membrane desalination of water in the late 1950s, by Reid and Breton at the University of Florida and by Loeb and Sourirajan at the University of California, Los Angeles.[1] Research on the concentration of fruit juices by RO was reported already in 1965.[2] Water desalination is still by far the most widespread RO application, followed by other water-related processes such as wastewater treatment, production of ultra-pure water for the semiconductor industry, and drinking water purification. With the development of more suitable membrane materials, applications in the food industry have been gaining in importance.[3]

REVERSE OSMOSIS MEMBRANE SYSTEMS

Membrane Materials

The principal characteristics of a good RO membrane for food application are:

- High permeate flux (flow rate per unit membrane area) at moderate pressure.
- High solute rejection (retention).
- Good mechanical strength.
- Resistance to chemical and microbial attack.
- Resistance to common cleaning agents and disinfectants.
- Tolerance to high temperatures.
- Compliance with food safety requirements.

Some of these requirements are mutually contradictory. Thus, high flux (more porous structure) often results in poor solute retention and vice versa. The selection of the best membrane for a given use is, therefore, a matter of optimization. Tolerance to high temperature is an advantage because of the tendency to operate, whenever possible, at relatively high feed temperature in order to increase the permeate flux and also because of the higher efficiency of the cleaning cycles at higher temperature.

Most RO membranes are anisotropic; i.e., they consist of layers with different structures. "Asymmetric membranes" consist of a dense "skin," 0.1 μm–0.5 μm in thickness, on top of a more porous, much thicker supporting layer, which itself rests on a very porous thick fabric backing. The mass transfer properties of the membrane reside almost entirely in the thin skin, while the support layer and fabric backing merely provide mechanical strength. Both the skin and the layer underneath consist of the same polymer, e.g., cellulose acetate. In contrast, in the case of membranes known as "thin film

Encyclopedia of Agricultural, Food, and Biological Engineering
DOI: 10.1081/E-EAFE 120007072
Copyright © 2003 by Marcel Dekker, Inc. All rights reserved.

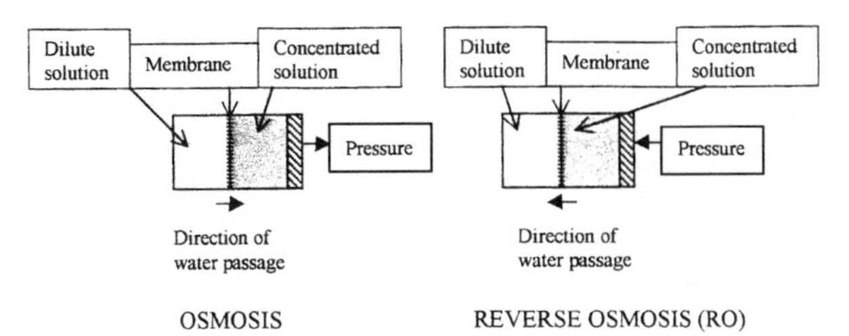

Fig. 1 Osmosis and reverse osmosis.

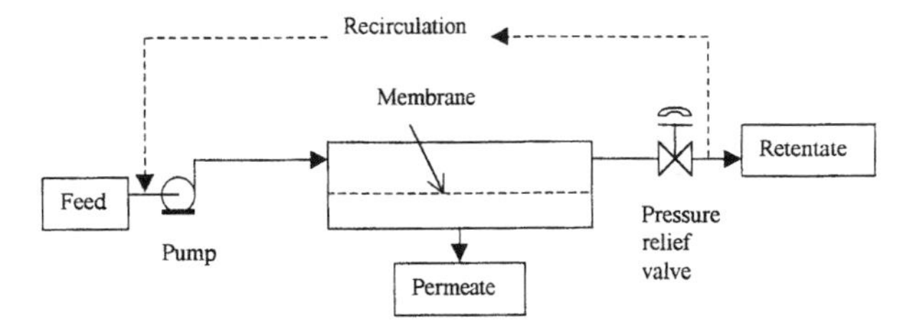

Fig. 2 Schematic presentation of a reverse osmosis module.

composite membranes," different polymers are used for the two layers. A common combination is a polyamide film on top of a microporous polysulfone support cast onto a nonwoven highly porous backing.

A completely different type of RO membranes, known as "dynamic membranes" are formed in situ as a gel (made of, e.g., hydrous zirconium oxide and polyacrylic acid) deposited on a tubular porous ceramic support.[4,5] The advantage of dynamic membranes is in their being renewable.

Membrane Configuration

The most common types of membrane configuration are: flat sheet, spiral-wound, tubular, and hollow fiber. With hollow fiber and spiral-wound configurations, it is possible to pack a large area of membrane into a compact module. Hollow fiber membranes are less convenient for food applications because of extensive fouling and difficult cleaning.

MASS TRANSFER IN RO

Mechanism of Solvent and Solute Transport

Unlike ultrafiltration and microfiltration membranes, RO membranes are, at least at their barrier skin layer, essentially nonporous. Thus, solvent flux and solute rejection at the barrier surface cannot be satisfactorily explained solely in terms of transport through pores and capillaries. Different theories have been proposed to explain and predict solute and solvent transport through RO membranes. The homogeneous solution–diffusion (HSD) model, proposed by Lonsdale, Merten, and Riley[6] assumes that both the solvent and the solute dissolve on the homogeneous (poreless) surface layer and then travel through the membrane by molecular diffusion. Separation is explained in the light of differences in the solubilities and diffusivities of the solvent and solute. Other variants of this model assume a combination of molecular diffusion and pore flow. The preferential sorption–capillary flow model (PSCF), proposed by Sourirajan[7] assumes that the membrane surface is essentially microporous. If the membrane material has preferential sorption for water and negative sorption (repulsion) for salt, then a monomolecular layer of pure water separates at the interface and is forced into the micropores. Another model, based on irreversible thermodynamics theory, was proposed by Kedem and Katchalsky.[8] An extensive review of theories relevant to mass transport in RO has been provided by Soltanieh and Gill.[9]

Permeate Flux

The rate of permeate flow per unit membrane area is the most important performance characteristic of an industrial RO operation. As explained above, RO is a pressure-driven process. The applied pressure difference (ΔP) must overcome the resistance of the membrane to flow, plus the osmotic pressure of the solution (π). In other words, the available "driving force" for the transmembrane

Table 1 Characteristics of pressure-driven membrane processes

	Size of retained particles		Typical applied pressure range	
Process	Nanometers	MW (Daltons)	bars	psig
Microfiltration	> 100		1–3	15–50
Ultrafiltration	2–100	> 5000	2–10	30–150
Nanofiltration	0.5–5	200–20,000	3–30	50–400
Reverse osmosis	< 1	< 200	30–70	300–1000

transport of water is the difference ($P - \pi$), also known as net applied pressure (NAP). The permeate flux (volumetric rate of permeate passage per unit membrane area), $J_w\,(m^3\,m^{-2}\,s^{-1})$ is proportional to NAP.

$$J_w = \frac{Q_w}{A} = K_w(\Delta P - \pi) \qquad (1)$$

where:

Q_w = permeate (water) flow rate

A = membrane area

Eq. 1 assumes total solute rejection.

If Q_f is the flow rate of the feed and assuming total solute rejection, the concentration ratio R achieved by RO is:

$$R = \frac{c_p}{c_f} = \frac{Q_f}{Q_f - Q_w} = \frac{Q_f}{Q_f - AJ_w} \qquad (2)$$

where c_p and c_f are the concentration of the solute in the product (retentate) and feed, respectively.

It does not follow from Eq. 1 that the flux can be increased indefinitely by increasing the applied pressure P. The actual permeate flux is lower than the theoretical value and the deviation from linearity increases with increasing applied pressure. The main reasons for less-than-theoretical flux are fouling, concentration polarization, gel polarization, and membrane compaction. Fouling is the irreversible accumulation of colloidal materials on the membrane surface. Concentration polarization results from the accumulation, near the membrane surface, of a layer of solution more concentrated than the bulk liquid. This results in an increase in the osmotic pressure π at the upstream face of the membrane, with a corresponding decrease in the NAP. Gel polarization occurs particularly with protein containing feeds such as milk and whey. A layer with gel-like properties accumulates near the membrane and constitutes an additional resistance to mass transfer. Membrane compaction is particularly serious in less rigid membranes submitted to very high applied pressure. Fouling, concentration polarization, and gel polarization may be somewhat counteracted by more intensive "sweeping" of the membrane surface, i.e., by increasing the velocity, hence the flow rate of the feed. This has the disadvantage of reducing the concentration ratio (see Eq. 2) and creating the need for recirculation.

In summary, permeate flux varies according to membrane type, feed composition, concentration ratio, applied pressure, and temperature. At constant operation conditions, the flux decreases with time as a result of membrane fouling. Some reported values of permeate flux are given in Table 2.

APPLICATION OF RO FOR THE CONCENTRATION OF LIQUID FOODS

General

An extensive review on the applications of RO in the food industry has been provided by Köseoğlu and Guzman.[10] The main advantages of concentration by

Table 2 Typical permeate flux values in different food RO applications

Membrane type	Feed	Concentration	Applied pressure (bar)	Temp. (°C)	Permeate flux ($1\,m^{-2}\,h$)	Ref.
CA	Whole milk	2-fold	34	30	4	[10]
CA	Whole milk	2.5-fold	48	30	13	[10]
CA	Skim milk	2.5-fold	28	25	10	[10]
CA	Sweet whey	2-fold	40	30	17	[10]
PA	Whole milk	2-fold	27	50	15	[10]
PA	Onion juice	From 10°Bx. to 18°Bx.	40	25	9.9	[57]
			40	35	11.1	
			46.7	25	13.2	
			46.7	35	16.9	
			53.3	25	15.6	
			53.3	35	19.2	
PA	Orange juice	1.25-fold	41.4	20	16	[27]
		2-fold	41.4	20	5	
		1.25-fold	62.1	20	20	
		2-fold	62.1	20	12	

CA = Cellulose acetate. PA = Polyamide.

RO in comparison with the conventional evaporation processes are:

- Low consumption of energy (5 kWh tn^{-1}–10 kWh tn^{-1} of water removed) and water.
- No thermal damage to the product.
- Better retention of volatile aroma components.
- Low capital cost and space requirement.

The principal shortcomings are:

- High maintenance cost due to the limited durability of membranes.
- Complexity of cleaning and sanitation procedures.
- Risk of rapid fouling, requiring a step of pretreatment where feasible (e.g., prefiltration in the case of pulpy juices, fat removal, filtration, and pH adjustment in the case of whey).

Concentration by RO is usually not suitable for the production of highly concentrated products because of the high osmotic pressure that must be overcome. Furthermore, flow and mass transfer rates are greatly impaired due to the extreme viscosity of highly concentrated food liquids. RO is therefore mainly used for the pre-concentration of dilute materials such as whey, maple sap, tomato juice, pulp-wash liquor etc. prior to final concentration by evaporation. High retentate concentration can be achieved by using "low retention" membranes.[11,12] The passage of small quantities of solutes through such membranes lowers the effective osmotic pressure across the membrane and results in acceptable permeation rates despite the high concentration of the retentate.

Dairy Applications

The largest user of RO in the food processing industry is the dairy sector. It is estimated that there is over 300,000 m^2 of RO membrane installed in that branch.[13] RO applications in the dairy industry include:

- Whole milk concentration to increase cheese yields.[14]
- Whole milk concentration for yogurt[15–17] and ice cream.[18]
- Whole milk, skim milk, and whey concentration to reduce transport cost or pre-concentration prior to conventional evaporation to reduce energy cost.[19–21]
- Concentration of whey ultrafiltration permeate for lactose recovery.[22]
- Evaporator condensate (cow water) polishing.[23]
- Waste effluent treatment for BOD/COD reduction and recovery of useful components.[24,25]

Vegetable and Fruit Juice Concentration

Morgan, Lowe, Merson, and Durkee[2] were among the first to investigate the concentration of fruit juices by RO. Citrus,[26,27] tomato,[28,29] apple,[30,31] pineapple,[32] and other juices[33,34] have been concentrated by RO. For reasons explained above, the upper concentration level practically attainable by RO is 20° Brix–40° Brix. However, the production of 60° Bx citrus concentrate with the help of novel RO membranes has also been reported.[35] In the case of pulpy or cloudy juices, such as citrus and tomato, pulp and suspended colloidal particles are usually removed by centrifugation or microfiltration before RO, to be added back to the concentrated retentate later. While most of the juice solutes are effectively retained by RO membranes, some of the water soluble low molecular weight aroma components are lost to the permeate. In the case of heat sensitive fruit juices, RO concentration is carried-out at low temperature to prevent browning and cooked taste. In contrast, vegetable juices and particularly tomato juice may be RO concentrated at high temperature (60–75°C) in order to achieve high permeation rates.

Beer, Wine, and Other Alcoholic Beverages

Reverse osmosis technology may be used to reduce the alcohol content of beer, wine, cider, etc. or, on the contrary, to produce a more concentrated drink. By virtue of its low molecular weight and its chemical characteristics, ethyl alcohol penetrates RO membranes to some extent. When an alcoholic beverage is submitted to RO through a membrane permeable to alcohol, both water and alcohol pass to the permeate. Pure water can now be added back to the concentrated retentate to produce a full-bodied beverage with a low alcohol content.[36–38] To produce a concentrated product with high alcohol content, alcohol is recovered from the permeate and added back to the retentate.[39] RO membranes with different relative permeabilities to water and alcohol are available. RO has been also applied for the concentration of wort (for beer)[40] and grape must (for wine)[41,42] and for improving feed-water quality in the brewing industry.[43]

Sugars and Syrups

In the manufacture of both cane and beet sugar, RO offers an attractive alternative to thermal evaporation for the initial concentration of thin juices.[44,45] The use of RO for the initial concentration of maple sap was one of the first applications of membrane technology in food processing.[46] RO concentration of birch sap has been reported.[47] In all these applications high operating temperatures (80°C and above) can be used with advantage, due to the thermal stability of the products. Solute retention is practically 100%.

Other Applications

Other applications of RO as a concentration process in the food industry include:

- Treatment of steep liquor and starch wash water in the wet-milling of corn, with the objective of reducing fresh water consumption and recovery of solutes.[48,49]
- Concentration of aqueous extracts of oilseeds in the production of isolated proteins.[50,51]
- Concentration of tea[52] and green tea[53,54] extracts.
- Concentration of egg white.[55]
- Concentration of aromas.[56–58]

A recent and promising application is the use of RO for the concentration of miscella in the vegetable oil industry instead of the conventional solvent distillation process.[59] The method is particularly suitable for the recovery of solvents when alcohols and not hexane are used for oil extraction.[60] In addition to energy savings, RO concentration of oil miscella offers the advantage of greater safety.

REFERENCES

1. Sourirajan, S. *Reverse Osmosis*; Logos Press Limited: London, 1970; 5–23.
2. Morgan, A.I., Jr.; Lowe, E.; Merson, R.L.; Durkee, E.L. Reverse Osmosis. Food Technol. **1965**, *19*, 1790.
3. Mohr, C.M.; Engelgau, D.E.; Leeper, S.A.; Charboneau, B.L. *Membrane Applications and Research in Food Processing*; Noyes Data Corporation: Park Ridge, NJ, 1989; 256–275.
4. Freedman, A.M.; Shaban, H.I. Protein, Lactose and Lactic Acid Separation from Cheese Whey Using Reverse Osmosis Dynamically Formed Membrane. J. Food Technol. **1971**, *6* (3), 309–315.
5. Watanabe, A.; Ohtani, T.; Kimura, S.; Kimura, S. Performance of Dynamically Formed Zr(IV)-PAA Membrane During Concentration of Tomato Juice. Rep. Natl Food Res. Inst. (Jpn) **1984**, *44*, 135–140.
6. Lonsdale, H.K.; Merten, U.; Riley, R.L. Transport Properties of Cellulose Acetate Osmotic Membrane. J. Appl. Polym. Sci. **1965**, *9*, 1341.
7. Sourirajan, S. The Science of Reverse Osmosis—Mechanisms, Membranes, Transport and Applications. Pure Appl. Chem. **1978**, *50* (7), 593–615.
8. Kedem, O.; Katchalsky, A. Thermodynamic Analysis of the Permeability of Biological Membranes to Non-electrolytes. Biochim. Biophys. Acta **1958**, *27*, 229.
9. Soltanieh, M.; Gill, W. Review of Reverse Osmosis Membranes and Transport Models. Chem. Eng. Commun. **1981**, *12* (1–3), 279–363.
10. Köseoğlu, S.S.; Guzman, G.J. Application of Reverse Osmosis Technology in the Food Industry. In *Reverse Osmosis*; Amjad, Z., Ed.; Chapman and Hall: New York, 1998; 300–333.
11. Gostoli, C.; Bandini, S.; di Francesca, R.; Zardi, G. Concentrating Fruit Juices by Reverse Osmosis—Low Retention–High Retention Method. Fruit Process. **1995**, *5* (6), 183–187.
12. Gostoli, C.; Bandini, S.; di Francesca, R.; Zardi, G. Analysis of a Reverse Osmosis Process for Concentrating Solutions of High Osmotic Pressure: The Low Retention Method. Food Bioprod. Process. **1996**, *74* (C2), 101–109.
13. Koch Membrane Systems Inc. Wilmington, MA. U.S.A. Private Communication.
14. Barbano, D.M.; Bynum, D.G. Whole Milk Reverse Osmosis Retentates for Cheddar Cheese Manufacture: Cheese Composition and Yield. J. Dairy Sci. **1984**, *67*, 2839–2849.
15. Jepsen, E. Membrane Filtration in the Manufacture of Cultured Milk Products—Yoghurt, Cottage Cheese. Cult. Dairy Prod. J. **1979**, *14* (1), 5–8.
16. Davis, F.L.; Shankar, P.A.; Underwood, H.M. Recent Developments in Yoghurt Starters: The Use of Milk Concentrated by Reverse Osmosis for the Manufacture of Yoghurt. J. Soc. Dairy Technol. **1977**, *30*, 23–28.
17. Guirguis, N.; Versteeg, K.; Hickey, M.W. The Manufacture of Yoghurt Using Reverse Osmosis Concentrated Skim Milk. Aust. J. Dairy Technol. **1987**, *42* (1–2), 7–10.
18. Bundgaard, A.G. Hyperfiltration of Skim Milk for Ice Cream Manufacture. Dairy Ind. **1974**, *39* (4), 119–122.
19. Fenton-May, R.I.; Hill, C.G., Jr.; Amundson, C.H.; Lopez, M.H.; Auchair, P.D. Concentration and Fractionation of Skim-Milk by Reverse Osmosis and Ultrafiltration. J. Dairy Sci. **1972**, *55* (11), 1561–1566.
20. Schmidt, D. Milk Concentration by Reverse Osmosis. Food Technol. Aust. **1987**, *39* (1), 24–26.
21. Nielsen, I.K.; Bundgaard, A.G.; Olsen, O.J.; Madsen, R.F. Reverse Osmosis for Milk and Whey. Process Biochem. **1972**, *7* (9), 17–20.
22. Morris, C.W. Plant of the Year: Golden Cheese Company of California. Food Eng. **1986**, *58* (3), 79–90.
23. Guengerich, C. Evaporator Condensate Processing Saves Money and Water. Bull. Int. Dairy Feder. **1996**, *311*, 15–16.
24. Wheatland, A.B. Treatment of Waste Waters from Dairies and Dairy Product Factories—Methods and Systems. J. Soc. Dairy Technol. **1974**, *27* (2), 71–79.
25. Re, G.; Giacomo, G.; Aloisio, L.; Terreri, M. RO Treatment of Waste Waters from Dairy Industry. Desalination **1998**, *119* (1–3), 267–271.
26. Anon; RO Membrane System Maintains Fruit Juice Taste and Quality. Food Eng. Int. **1989**, *14* (3), 54.
27. Medina, B.G.; Garcia, A. Concentration of Orange Juice by Reverse Osmosis. J. Food Process Eng. **1988**, *10*, 217–230.

28. Merlo, C.A.; Rose, W.W.; Pederson, L.D.; White, E.M. Hyperfiltration of Tomato Juice During Long Term High Temperature Testing. J. Food Sci. **1986**, *51* (2), 395–398.
29. Merlo, C.A.; Rose, W.W.; Pederson, L.D.; White, E.M.; Nicholson, J.A. Hyperfiltration of Tomato Juice: Pilot Plant Scale High Temperature Testing. J. Food Sci. **1986**, *51* (2), 403–407.
30. Moresi, M. Apple Juice Concentration By Reverse Osmosis and Falling Film Evaporation. In *Preconcentration and Drying of Food Materials*; Bruin, S., Ed.; Elsevier Science Publishers: Amsterdam, 1988; 61–76.
31. Sheu, M.J.; Wiley, R.C. Preconcentration of Apple Juice by Reverse Osmosis. J. Food Sci. **1983**, *48*, 422–429.
32. Bowden, R.P.; Isaacs, A.R. Concentration of Pineapple Juice by Reverse Osmosis. Food Aust. **1989**, *41* (7), 850–851.
33. Matsuura, T.; Baxter, A.G.; Sourirajan, S. Studies on Reverse Osmosis for Concentration of Fruit Juices. J. Food Sci. **1974**, *39* (4), 704–711.
34. Köseoğlu, S.S.; Lawhon, J.H.; Lusas, E.W. Vegetable Juices Produced with Membrane Technology. Food Technol. **1991**, *45* (1), 124, 126–128.
35. Cross, S. Membrane Concentration of Orange Juice. Proc. Fla State Hortic. Soc. **1989**, *102*, 146–152.
36. Gnekow, B.R. Simultaneous Double Reverse Osmosis Process for Production of Low and Non-alcoholic Beverages (Particularly Wines). US Patent 4,888,189, 1989.
37. Nielsen, C.E. Low Alcohol Beer by Hyperfiltration Route. Brew. Distill. Int. **1982**, *12* (8), 39–41.
38. Bui, K.; Dick, R.; Moulin, G.; Galzy, P. A Reverse Osmosis for the Production of Low Ethanol Content Wine. Am. J. Enol. Vitic. **1986**, *37* (4), 297–300.
39. Bui, K.; Dick, R.; Moulin, G.; Galzy, P. Partial Concentration of Red Wine by Reverse Osmosis. J. Food Sci. **1988**, *53* (2), 647–648.
40. Moffat, D.J. Improved Method and Apparatus for Processing a Preparation. PCT-International Patent Application, 1999.
41. Palmer-Benson, T. Improving Ontario Wine through Reverse Osmosis. Food Can. **1986**, *46* (8), 20–21.
42. Duitschaever, C.L.; Alba, J.; Buteau, C.; Allen, B. Riesling Wines Made from Must Concentrated by Reverse Osmosis. Am. J. Enol. Vitic. **1991**, *42* (1), 19–25.
43. Narziss, L. Das Brauwasser als bedeutener Rohstoff. Moeglichkeiten und Grenzen moderner Brauwasser Enthaertung. Brauwelt **1989**, *129* (5), 167–174.
44. Madsen, R.F. Application of Ultrafiltration and Reverse Osmosis to Cane Juice. Int. Sugar J. **1973**, *75*, 163–167.
45. Tragardh, G.; Gekas, V. Membrane Technology in the Sugar Industry. Desalination **1988**, *69* (1), 9–17.
46. Willits, C.O.; Underwood, J.C.; Merten, U. Concentration by Reverse Osmosis of Maple Sap. Food Technol. **1967**, *21* (1), 24–26.
47. Kallio, H.; Karppinen, T.; Holmbom, B. Concentration of Birch Sap by Reverse Osmosis. J. Food Sci. **1985**, *50* (5), 1330–1332.
48. Kollacks, W.A.; Rekers, C.J.N. Five Years of Experience with the Application of Reverse Osmosis on Light Middlings in a Corn Wet Milling Plant. Starch (Staerke) **1988**, *40* (3), 88–93.
49. Wu, Y.V. Reverse Osmosis and Ultrafiltration of Corn Light Steep-water Solubles. Cereal Chem. **1988**, *65* (2), 105–109.
50. Lawhon, J.T.; Lin, S.H.C.; Cater, C.M.; Mattil, K.F. Fractionation and Recovery of Cottonseed Whey Constituents by Ultrafiltration and Reverse Osmosis. Cereal Chem. **1975**, *52* (1), 34–43.
51. Lawhon, J.T.; Manak, L.J.; Lusas, E.W. Using Industrial Membrane Systems to Isolate Oilseed Protein without an Effluent Waste Stream. Abstr. Pap. Am. Chem. Soc.(Coll.) **1979**, *178* (1), 133.
52. Schreier, P.; Mick, W. Ueber das Aroma von Schwartzem Tee: Herstellung eines Teekoncentrates mittels Umgekehrosmose und dessen Analytische Characterisierung. Z. Lebensm.-Unters. Forsch. **1984**, *179*, 113–118.
53. Zhang, S.Q.; Fouda, A.E.; Matsuura, T. Reverse Osmosis Transport and Module Analysis for Green Tea Juice Concentration. J. Food Process Eng. **1992**, *16*, 1–20.
54. Zhang, S.Q.; Matsuura, T. Reverse Osmosis Concentration of Green Tea. J. Food Process Eng. **1991**, *14*, 85–105.
55. Conrad, K.M.; Mast, M.G.; Ball, H.R.; Froning, G.; MacNeil, J.H. Concentration of Liquid Egg White by Vacuum Evaporation and Reverse Osmosis. J. Food Sci. **1993**, *58* (5), 1017–1020.
56. Matsuura, T.; Baxter, A.G.; Sourirajan, S. Reverse Osmosis Recovery of Flavor Components from Apple Juice Waters. J. Food Sci. **1975**, *40* (5), 1039, 1046.
57. Nuss, J.S.; Guyer, D.E.; Gage, D.E. Concentration of Onion Juice Volatiles by Reverse Osmosis and its Effects on Supercritical CO_2 Extraction. J. Food Process Eng. **1997**, *20* (2), 125–139.
58. Braddock, R.J.; Sadler, G.D.; Chen, C.S. Reverse Osmosis Concentration of Aqueous-phase Citrus Juice Essence. J. Food Sci. **1991**, *56* (4), 1027–1029.
59. Köseoğlu, S.S.; Engelgau, D.E. Membrane Applications and Research in Edible Oil Industry: Assessment. J. Am. Oil Chem. Soc. **1990**, *67* (4), 239–245.
60. Köseoğlu, S.S.; Lawhon, J.H.; Lusas, E.W. Membrane Processing of Crude Vegetable Oils. II. Pilot Scale Solvent Removal from Oil Miscellas. J. Am. Oil Chem. Soc. **1990**, *67* (5), 281–287.

Reynolds Number

Christopher R. Daubert
North Carolina State University, Raleigh, North Carolina, U.S.A.

INTRODUCTION

Engineering effective food processing systems depend on the application of fluid dynamics to properly select pump and pipeline sizes. Many issues can confound the pipeline design problem, specifically material properties and quantity demands. Studying ideal fluids tends to simplify fluid flow, whereas real fluids exhibit more complex flow scenarios created by the presence of viscous behavior.[1] Viscosity is defined as the resistance to flow during exposure to shearing (or viscous) forces, resulting from molecule–molecule or molecule–boundary interactions. For fluid flow to occur, these resisting forces must be overcome, and in the process, two vastly different flow types, laminar or turbulent, may result (Fig. 1).

Laminar flow is synonymous with streamline or unidirectional flows, fluid layers of infinitesimal thickness essentially slide against adjacent layers. Turbulent flows, on the other hand, are characterized with multidirectional, irregular flows. The region between these flow types is known as the transitional flow period. Distinguishing between flow regimes has been summarized with a dimensionless parameter known as the Reynolds Number (*Re*). An understanding of the interplay between flow regimes and the Reynolds number enables the engineer to more efficiently design fluid processing systems.

HISTORY

A professor of engineering at Owens College, Manchester, England, Osborne Reynolds (1842–1912) made significant contributions to the field of fluid mechanics. His classical paper describing the law of resistance in parallel channels is credited with establishing the criterion for transition from laminar to turbulent flows.[2] In Reynolds' experiment, he injected colored dye into a rate-controlled stream of water, with the dye density matching that of water. The dye pattern in the flow field was monitored with changing flow rates. At low stream velocities, the dye traveled in a straight line. As flow rates increased, the dye no longer moved in a linear stream, but rather was diffused throughout the flow. A dimensionless, mathematical quantity, now called the Reynolds number (*Re*), exists to help predict laminar or turbulent flow conditions,

$$Re = \frac{\rho D \bar{v}}{\mu} \tag{1}$$

where ρ is the fluid mass density, D is the channel/pipe diameter, $\bar{v}$ is the average velocity, and μ is the Newtonian viscosity.[3] Eq. 1 represents the common form of *Re*, applicable to Newtonian, or time and shear independent fluids like water, honey, or skim milk whose behavior in laminar flow is characterized by Newton's Law:

$$\sigma = \mu \dot{\gamma} \tag{2}$$

the shear stress (σ) is proportional to the shear rate $(\dot{\gamma})$, and the Newtonian viscosity is the constant of proportionality.

RHEOLOGICAL IMPLICATIONS

Closer inspection of the dimensionless group reveals that *Re* is a ratio of two groups of forces[4]:

$$Re = \frac{\text{Inertial Forces}}{\text{Viscous Forces}} \tag{3}$$

The inertial force terms reflect the momentum associated with a moving mass, while the viscous component incorporates forces resisting fluid motion—as previously described. The denominator tends to complicate the expression, as the viscous forces are often influenced by extenuating circumstances, namely shear and temperature conditions. The vast majority of fluid foods behave as non-Newtonian liquids, in other words they display a dependence on the rate of shear. Discussion of Reynolds number for two common types of non-Newtonian flow responses, Bingham plastic and power law behavior, follows.

Bingham Plastic

Foods like dressings and sauces often incorporate a yield stress to minimize flow dispensation after application to a complementary food. A yield stress (σ_0) is a minimum threshold which must be overcome for a material to flow, and the Bingham plastic model is commonly used to

Encyclopedia of Agricultural, Food, and Biological Engineering
DOI: 10.1081/E-EAFE 120006957
Copyright © 2003 by Marcel Dekker, Inc. All rights reserved.

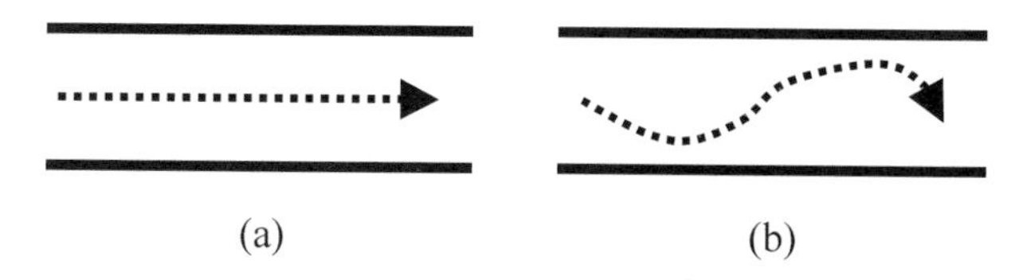

Fig. 1 Illustration of molecular pathways for (a) laminar and (b) turbulent flow regimes.

describe the rheological behavior,

$$\sigma = \sigma_0 + \mu_{pl}\dot{\gamma} \tag{4}$$

where the shear stress is related to the shear rate via yield stress and plastic viscosity (μ_{pl}) constants. The Reynolds number for Bingham-type flows (Re_{BP}) is similar to the Newtonian situation (Eq. 1) with the plastic viscosity replacing the Newtonian viscosity.

$$Re_{BP} = \left(\frac{\rho D\bar{v}}{\mu_{pl}}\right) \tag{5}$$

Power Law

When the fluid displays pseudoplastic (shear thinning) or dilatant (shear thickening) behavior without a yield stress, the apparent viscosity, η, is typically described by a power law function,

$$\eta = K\dot{\gamma}^{n-1} \tag{6}$$

where K and n are constants known as the consistency coefficient and flow behavior index, respectively. The power law or generalized Reynolds number (Re_{PL})

$$Re_{PL} = \left(\frac{\rho D^n(\bar{v})^{2-n}}{8^{n-1}K}\right)\left(\frac{4n}{3n+1}\right)^n \tag{7}$$

collapses to the Newtonian Re (Eq. 1) when the flow behavior index, n, equals 1.0. Steffe and Singh (1997) offer a detailed review of pipeline design applications for different rheological flows.[5]

Flow Regime Determination

A common Reynolds number misunderstanding is that Re of 2100 is a firm, critical criterion distinguishing between laminar and transitional flows, applying regardless of fluid flow behavior. Osborne Reynolds observed for a given system that a critical Reynolds number (Re_C) defined the upper and lower critical velocities for all fluids. The engineer may ascribe the upper limit of laminar flow to be defined by $2700 < Re < 4000$, and the lower limit for turbulent flows as set by $Re = 2100$.[1] Therefore, whenever Re is less than 2100, flow through straight tubes is laminar, and Re greater than 4000 should be turbulent flow. The region between these extremes is referred to as the transitional zone. In reality, the transition from laminar to turbulent flow regimes may occur in a range from $1000 < Re_C < 5000$.[6] A good rule of thumb: laminar conditions prevail when the following criterion is achieved:

$$Re \leq Re_C \tag{8}$$

Critical Reynolds numbers do in fact, fluctuate with material properties; Table 1 displays the equations for previously discussed rheological scenarios.

Particulate Complications

Two-phase flow prompts significant deviations from the flow profile of the carrier fluid. In essence, the introduction of particles adds barriers, impeding the development of streamline flow. Therefore, Reynolds number for particulate fluids (Re_p) is introduced to account for the two phases, solid particles (p) and continuum liquid (l).[8]

$$Re_p = \frac{\rho_l(\bar{v}_l)^{(2-n)}(D-D_p)^n}{K} \tag{9}$$

This expression collapses to the Newtonian Re for the scenario of zero particulates suspended in a Newtonian fluid. Other issues also complicate flow schemes, such as particulate loading or concentration. By definition,

Table 1 Critical Reynolds number calculations[7]

Flow behavior	Re	Re_C	Food
Newtonian	$\frac{\rho D\bar{u}}{\mu}$	2100	Water, milk, honey
Power law	$\left(\frac{\rho D^n(\bar{v})^{(2-n)}}{8^{n-1}K}\right)\left(\frac{4n}{3n+1}\right)^n$	$\frac{6464n(2+n)^{(2+n/(1+n))}}{(1+3n)^2}$ or $\frac{2100(4n+2)(5n+3)}{3(1+3n)^2}$	Salad dressing, apple sauce
Bingham plastic[a]	$\frac{\rho D\bar{v}}{\mu_{pl}}$	$\frac{He}{8C}(1-4C/3+C^4/3)$	Tomato paste, meat paste

[a] He (Hedstrom number) $= \frac{D^2\sigma_0\rho}{\mu_{pl}^2}$; $C = \frac{He(1-C)^3}{16,800}$.

Fig. 2 Parabolic velocity profiles for fully developed laminar flow conditions.

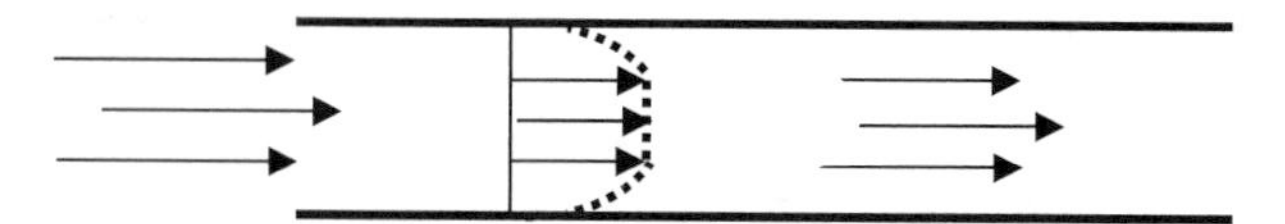

Fig. 3 Depiction of a plug flow velocity profile.

the volume fraction (ϕ) is dimensionless

$$\phi = \frac{\text{particle volume}}{\text{total volume}} \tag{10}$$

and is often included as a separate variable during analysis.

PROCESSING CONSIDERATIONS

Velocity Profile

Flow regimes and rheological properties dictate velocity profiles for fully developed flows, i.e., the flow is not influenced by effects from entrance, exit, or other fittings. These profiles predict fluid velocity as a function of channel geometry. Describing these profiles is crucial for understanding important issues facing the food engineer, such as heat and mass transfer calculations as well as residence time distribution.

Assumptions are required to establish a velocity profile or distribution during tube flow. First, the fluid must be considered incompressible: the material experiences no density change during processing. Another necessary condition requires "zero slip" to occur at the boundary between the tube and fluid. Essentially, a thin layer of fluid at the tube wall has no velocity in the axial direction ($v_x = 0$.) With these assumptions established, constitutive equations are used to predict velocity as a function of position for fully developed flow.

During laminar flow conditions ($Re < Re_C$), the fluid typically has a parabolic velocity profile with the maximum velocity found at the center of the tube. Fig. 2 depicts the general geometry and patterns associated with streamline, parabolic flows. Obeying the assumptions, velocity in the axial direction at the tube boundary $r = R$ is zero, and the maximum velocity is found at the geometric center of the tube $r = 0$.

For Newtonian liquids, the velocity profile is governed by a pressure drop per length $\left(\frac{\Delta P}{L}\right)$ and the viscosity (μ). Velocity in the axial (x) direction is determined as a function of radial position ($0 \leq r \leq R$).

$$v_x(r) = \left(\frac{\Delta P}{L}\right)\frac{(R^2 - r^2)}{4\mu} \tag{11}$$

$$v_{max} = \left(\frac{\Delta P}{L}\right)\frac{R^2}{4\mu} \tag{12}$$

The maximum velocity is twice the average velocity ($\bar{v}$), where the volumetric flow rate is the product of the cross sectional area and $\bar{v}$.

$$\frac{v_{max}}{\bar{v}} = 2 \tag{13}$$

Power law liquids incorporate flow behavior index and consistency coefficient into the velocity profile expression.

$$v_x(r) = \left(\frac{\Delta P}{L}\right)^{1/n}\frac{1}{(2K)^{1/n}}\left(\frac{n}{n+1}\right)\left[R^{(n+1)/n} - r^{(n+1)/n}\right] \tag{14}$$

$$v_{max} = \left(\frac{\Delta P}{L}\right)^{1/n}\frac{1}{(2K)^{1/n}}\left(\frac{n}{n+1}\right)\left[R^{(n+1)/n}\right] \tag{15}$$

Table 2 Fluid dynamics nomenclature

Symbol	Name	SI units
D	Channel dimension (tube diameter)	m
D_p	Particle dimension	m
Re	Reynolds number	—
Re_C	Critical Reynolds number	—
Re_{PL}	Power law Reynolds number	—
Re_{BP}	Bingham plastic Reynolds number	—
Re_p	Particle Reynolds number	—
σ	Shear stress	Pa
σ_0	Yield stress	Pa
$\dot{\gamma}$	Shear rate	sec^{-1}
μ	Newtonian viscosity	Pa sec
μ_{pl}	Plastic viscosity	Pa sec
η	Apparent viscosity	Pa sec
He	Hedstrom number	—
C	Critical parameter	—
R	Tube radius	m
r	Tube radial position	m
n	Flow behavior index	—
K	Consistency coefficient	Pa sec^n
v_x	Axial velocity	m sec^{-1}
v_{max}	Maximum velocity	m sec^{-1}
ρ	Density	kg m^{-3}
P	Pressure	Pa
L	Length	m
ϕ	Volume ratio	—

R

$$\frac{v_{max}}{\bar{v}} = \frac{3n+1}{n+1} \tag{16}$$

Notice how the equations collapse to the Newtonian profile (Eqs. 11–13) when $n = 1$ and $K = \mu$.

Materials incorporating a yield stress display plug flow: a flat, centralized region of the velocity profile. Plug flow simplifies problems associated with heat and mass transfer calculations, as the velocity distribution is more uniform (Fig. 3).

Complications arise, however, as the fluid flow behavior becomes more turbulent. Velocity profiles are no longer parabolic. Nevertheless, the fastest location still resides at the center of the tube. Steffe (1996) provides the derivation and additional details of velocity profiles during tube flow.[7]

Heat and Mass Transfer

As expected, turbulent flow enhances mechanisms for heat and mass transfer. Improved efficiency is achieved for mixing processes and thermal treatments when the flows are nonlaminar. For further information, Aravinth (2000) includes a physical representation of transport mechanisms encountered during turbulent flow through tubes[9] (Table 2).

REFERENCES

1. Vennard, J.K. Flow of a Real Fluid. In *Fluid Mechanics*, 4th Ed.; John Wiley & Sons, Inc.: New York, 1961, Chap. 7.
2. Reynolds, O. An Experimental Investigation of the Circumstances Which Determine Whether the Motion of Water Shall Be Direct or Sinuous and of the Law of Resistance in Parallel Channels. Philos. Trans. R. Soc. **1883**, *174* (III), 935–982.
3. Langhaar, H.L. Principles and Illustration of Dimensional Analysis. In *Dimensional Analysis and Theory of Models*; John Wiley & Sons, Inc.: New York, 1951, Chap. 2.
4. Azbel, D.S.; Cheremisinoff, N.P. Table 2.1 Various Dimensionless Groups and Their Physical Significance. In *Fluid Mechanics and Unit Operations*; Ann Arbor Science: Ann Arbor, MI, 1983.
5. Steffe, J.F.; Singh, R.P. Pipeline Design Calculations for Newtonian and Non-Newtonian Fluids. In *Handbook of Food Engineering Practice*; Valentas, K.J., Rotstein, E., Singh, R.P., Eds.; CRC Press: New York, 1997.
6. Daugherty, R.L.; Franzini, J.B.; Finnemore, E.J. *Fluid Mechanics with Engineering Application*, 8th Ed.; McGraw-Hill Book Company: New York, 1985; 203–205.
7. Steffe, J.F. Tube Viscometry. *Rheological Methods in Food Process Engineering*, 2nd Ed.; Freeman Press: East Lansing, MI, 1996, Chap. 2.
8. Sandeep, K.P.; Zuritz, C.A. Residence Times of Multiple Particles in Non-Newtonian Holding Tube Flow: Effect of Process Parameters and Development of Dimensionless Correlations. J. Food Eng. **1995**, *25*, 31–44.
9. Aravinth, S. Prediction of Heat and Mass Transfer for Fully Developed Turbulent Fluid Flow through Tubes. Int. J. Heat Mass Transfer **2000**, *43*, 1399–1408.

Rheometers

Sheryl Barringer
Puntarika Ratanatriwong
The Ohio State University, Columbus, Ohio, U.S.A.

INTRODUCTION

The rheometer, or viscometer, measures the rheological properties of fluids by either the resistance to flow under a known force or the force produced by a known amount of flow.[1-3] Selecting the right rheometer, which offers reliable and accurate data for the specific application, is the first step in solving rheological problems. Commercial rheometers can be divided into analytical and empirical rheometers. Analytical rheometers give the actual viscosity and consist of capillary tubes and rotational rheometers. Empirical rheometers are frequently the industry standard, and can in some cases be converted to viscosity. Many of the manufacturers are listed in Table 1.

RHEOMETERS FOR ANALYTICAL RESEARCH

There are many commercially available rheometers, which are based on the same principles. However, specific components may be added for special applications. All measuring geometries generate a specific deformation under three common requirements: laminar flow of fluid, isothermal operation, and no slip at solid–fluid interfaces. They typically require a small amount of sample. Corrections for errors and detailed calculations of viscosity can be found in Ref. 4.

Capillary Tube Rheometer

The time is measured for a standard volume of fluid to flow through a given length of capillary tube under a driving pressure generated by gravity, gas pressure, or a descending piston. Typical glass capillary tubes include Ostwald, Cannon–Fenske, Counter current, and Ubbelohde viscometers.

Ostwald is the simplest while Cannon–Fenske is modified for more accuracy and wider use by bending both arms to reduce error caused by bad alignment of the upper and lower bulbs. Capillary rheometers are mainly used for kinematic viscosity determination of low to medium viscosity Newtonian liquids, but may be used for non-Newtonian fluids by using the Rabinowitsch–Mooney equation. Since the viscosity of non-Newtonian fluids is a function of shear rate, it must be expressed as an apparent viscosity unless the pressure is varied to create a constant flow rate. A series of capillaries with a range of diameters is needed to measure a wide range of viscosities. The capillary tube offers a high degree of accuracy, easy operation, and low cost. It is not applicable for fluids with particulates unless they are filtered out prior to measurement.

Rotational Rheometers

There are many companies offering a variety of rotational rheometers with special optional features (Table 1). Continuous measurements under changing conditions or temperatures and measurement of time-dependent effects are possible. Oscillatory testing is also possible. Rotational rheometers typically have interchangeable measuring geometries: concentric cylinder (couette type), cone and plate, and parallel plate.

Concentric cylinder

The fluid sample is sheared in the gap between inner and outer cylinders where either the inner (Couette) or outer (Searle) cylinder rotates. The torque needed to prevent the fixed cylinder from moving is a measure of the shear stress while the speed of the moving cylinder is a measure of the shear rate. However, a narrow gap between the cylinders is required since a wide gap causes significant errors for highly shear-thinning food. Hence, it is not usable for particulate foods. Adding a well in the bottom of the inner cylinder or using a slight angle called a Mooney-couette cylinder helps minimize the end effects by trapping air bubbles, eliminating the torque on the bottom or creating a shear rate along the cone equal to that between the walls.[4,5]

Cone and plate

The sample is held by the surface tension between a small-angle rotating cone and a fixed plate. The torque caused by the drag of fluid on the cone is measured. The small angle between the cone and plate gives uniform shear stress throughout the material. The true shear rate is easily obtained, which is not true for capillary tube or parallel

Encyclopedia of Agricultural, Food, and Biological Engineering
DOI: 10.1081/E-EAFE 120006952
Copyright © 2003 by Marcel Dekker, Inc. All rights reserved.

Table 1 Manufacturers of various rheometers

Name and website	Phone	Type[a]	Special attachments
ATS Rheosystems (*www.atsrheosystems.com*)	(609) 298-2522	CC, CP, PP	Sealed cell, high pressure cell, extended temperature cell, heated couette cell, and electro rheology cell
Boekel Co. (*www.boekelsci.com*)	(800) 336-6929	Zahn cup	
Bohlin Instruments (*www.bohlinusa.com*)	(732) 254-7742	CC, CP, PP	Electro rheology and UV curing cells, and vane tools
Brookfield (*www.belgmbh.com*)	(800) 628-8139	CC, CP, spindles	
Cannon Instrument, Co.	(800) 676-6232	Capillary tube	
CSC Scientific Co., Inc. (*www.cscscientific.com*)	(800) 458-2558	Bostwick consistometer	
Haake (*www.haake-usa.com*)	(201) 265-7865	CC, CP, PP	
Paar Physica (*www.paarphysica.com*)	(847) 759-9657	CC, CP, falling ball	
Rheometric Scientific, Inc. (*www.rheosci.com*)	(732) 560-8550	CC, CP, PP spindles	Thermal cell
TA Instrument Ltd. (*www.tainst.com*)	(302) 427-4000	CC, CP, PP	Peltier plate and test chamber

[a] CC, CP, and PP denote concentric cylinder, cone and plate, and parallel plate, respectively.

plate. Thus, the cone and plate rheometer is particularly valuable for non-Newtonian fluids. Other advantages are that end effects are negligible and high shear rate measurement can be made without having to compensate for the heating effect since a thin layer of the fluid is in contact with a temperature-controlled metal plate.[6] However, this rheometer is not suitable for particulate foods due to the small gap between the cone and plate.

Parallel plate

Fluid is placed between two parallel disks or plates. One plate is rotating or oscillating while the other is fixed. The gap between the plates can be adjusted, thus it is applicable for highly viscous suspensions or particulate foods. However, shear stress is a function of radius, creating nonuniform shear stress, which makes calculation more complicated.

Special attachments: thermal and electro rheology cells

A special thermal cell enables rheological tests to be performed as a function of temperature. Thus, study of structural changes during phenomena such as crystallization, gelatinization, melting, or denaturation is possible. An electro rheology cell is used for magnetic fluid suspensions, which change their properties when an electric field is applied. Chocolate is one of the few food items that have been found to change viscosity under an electric field.

EMPIRICAL RHEOMETERS FOR QUALITY CONTROL

The quality control rheometer must be easy to use, fast, and durable for routine measurement. Precise data may not be needed; however, data should correlate to sensory results or processing characteristics.

Brookfield Viscometer

The Brookfield is the most commonly used rheometer in the food industry. It can be used on a wide range of liquid and semisolid foods. It is mainly used for quality control; however, the Brookfield DVIII can measure true viscosity. The Brookfield is a concentric cylinder type rheometer with various spindles. The correct spindle and speed must be chosen based on the viscosity of the food. Other accessories such as the Wells–Brookfield cone and plate or small sample adapter can be used for small samples. A heliopath and T-bar spindles are used to give a more accurate reading for gels.

Libby's Tube/Efflux Tubes

Libby's tube is typically used for the measurement of fruit juices especially tomato juice. It is a pipette with a stopcock on the orifice. The time for the juice to flow from the upper to the lower calibration mark is recorded. Libby's tube can be used for thicker samples than the capillary tube and allows for small particulates.[7]

Zahn Cup

This cup with a hole in the bottom is essentially a very short capillary rheometer. The time for a standard volume to flow through the Zahn cup is measured. Zahn cup (Fig. 1) is used as a rapid method in quality control of Newtonian or near-Newtonian fluids such as milk, syrup, and oil. It offers easy operation and low cost; however, extremely accurate data cannot be obtained.

Falling Ball Rheometer

This rheometer is designed based on Stokes' law. The time for a ball to fall through a vertical or tilted tube of the liquid under gravity is measured. It is necessary to select the proper size and density ball based on the viscosity. The falling ball rheometer is easy to use; however, only transparent Newtonian fluids can be used. Opaque or non-Newtonian fluids cannot be used because the ball is difficult to see or Stokes' law no longer applies.

Bostwick Consistometer

The Bostwick consists of two compartments. The first compartment is filled with approximately 100 ml of product and the other compartment is a trough 5 cm × 24 cm × 2.5 cm high (Fig. 2). The floor of the second compartment has a series of parallel lines 1 cm. apart from the gate to the other end. The consistency is the distance the product flows in 30 sec after releasing the gate. It is used for quality control of products such as fruit and vegetable puree, applesauce, tomato catsup, thick salad dressings, and baby foods. The data obtained from the Bostwick consistometer correlates well to sensory evaluation in terms of mouthfeel.

Adams Consistometer

A truncated cone is filled with 200 ml of product. The cone is placed on a table with a series of concentric circles so that when the cone is raised the product flows across the table. The extent of flow at four equidistant points is recorded after 30 sec and averaged. It is mainly used for pie fillings since it is one of the few rheometers that can handle large particulates.

Brabender Viscoamylograph

The Brabender visco/amylo/graph combining the Viscograph and Amylograph is the most common method to measure gelatinization properties of starch and starch-containing products. The pasting properties of the suspension are measured at a constant rate of shear with either changing or constant temperature. Seven suspended sensing pins are immersed in the sample bowl with counteracting pins for keeping the sample in suspension. The sample bowl rotates while being heated or cooled. The viscosity is continuously plotted by a strip chart recorder.[8]

Slit Rheometer

The slit rheometer is used for monitoring the rheological behavior of products during processing in extruders. Newtonian and non-Newtonian fluids can be measured.

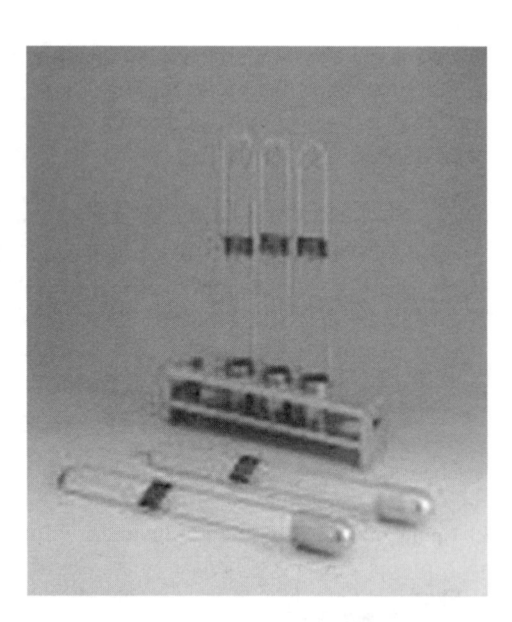

Fig. 1 Zahn cups (Boekel Co.).

Fig. 2 Bostwick consistometer (CSC Scientific Co., Inc.).

The slit rheometer is attached to the extruder allowing for online monitoring of extrudates prior to expansion and dehydration. Estimated shear rate and shear stress are obtained from measuring the flow rate and pressure drop over the entire slit die. The aspect ratio between the width and height of the slit should be greater than 10 to avoid edge effects.[4]

REFERENCES

1. Rao, M.A. Rheological Properties of Fluid Foods. In *Engineering Properties of Foods*; Rao, M.A., Rizvi, S.S.H., Eds.; Marcel Dekker, Inc.: New York, 1995; 1–54.
2. Urbicain, M.J.; Lozano, J.E. Definition, Measurement, and Prediction of Thermophysical and Rheological Properties. In *Handbook of Engineering Practice*; Valentas, K., Rotstein, E., Singh, R.P., Eds.; CRC Press: Boca Raton, FL, 1997; 427–488.
3. Whorlow, R.W. Deformation and Stress. *Rheological Techniques*; Ellis Horwood: New York, 1980; 18–58.
4. Steffe, J.F. *Rheological Methods in Food Process Engineering*; Freeman Press: East Lansing, MI, 1992; 1–228.
5. Ma, L.; Barbosa-Canovas, G.V. Review: Instrumentation for the Rheological Characterization of Foods. Food Sci. Technol. Int. **1995**, *1* (1), 3–17.
6. Bourne, M.C. *Food Texture and Viscosity*; Academic Press, Inc.: San Diego, CA, 1982; 1–325.
7. Sone, T. *Consistency of Food Stuffs*; D. Reidel Publishing Co.: Dordrecht, The Netherlands, 1972; 1–188.
8. Shuey, W.C.; Schmitz, A.O. Instrument Construction. In *Amylograph Handbook*; Shuey, W.C., Tipples, K.H., Eds.; American Association of Cereal Chemists: Minneapolis, MN, 1980; 1–37.

R

Schmidt Number

Bengt Hallström
Lund University, Lund, Sweden

INTRODUCTION

At the occasion of the 25th year celebration of the Technische Hochschule Danzig-Langfur, July 18–20, 1929, Professor Dr.-ing. Ernst Schmidt in his presentation, probably for the first time, introduced the dimensionless group *k/v* which later came to be the Schmidt number. His presentation was later published in the journal Gesundheits-Ingenieur.[1] At the beginning, this dimensionless group was, especially in Germany, referred to as the Prandtl number for mass transfer and was denominated Pr'. In Germany, this term was used for several years, and for instance as late as 1956 O. Krischer used the term in his book Die wissentschaftlichen Grundlagen der Trocknungstechnik. However, already in 1933 at the Round Table Conference of Chicago Meeting, American Institute of Chemical Engineers proposed the name Schmidt Number.

THE ORIGINAL VERSION BY SCHMIDT

In his lecture at the abovementioned occasion, Schmidt suggested the similarity between heat and mass transfer at forced as well as at free convection. By comparing the differential equations for heat flow and diffusion, he introduced the relationship *k/v* in the diffusion equation corresponding to *a/v* for heat flow. This relationship *k/v* later became the Schmidt number. In the paper mentioned above, he accordingly wrote these equations as follows, presented here both in the original way with symbols as Schmidt used and with symbols more common today.

Heat	$\alpha/C_p = a/l\,\Phi(wl/v, a/v)$		$h/c_p = \alpha/d\,F(ud/v, \alpha/v)$	
Mass	$\kappa = k/l\,\Phi(wl/v, k/v)$		$k_G = D/d\,F(ud/v, D/v)$	
	$a = \lambda/C_p$ m^{2}hr^{-1}	Temperaturleitzahl	$\alpha = k/\rho c_p$ m^{2}sec^{-1}	Thermal diffusivity
	C_p kcal m^{-3}C^{-1}	Spezifische Wärme	c_p J kg^{-1}K^{-1}	Specific heat
	k m^{2}hr^{-1}	Diffusionskonstante	D m^{2}sec^{-1}	Diffusivity
	l m	Kennzeichnende Abmessung	d m	Characteristic length
	w m hr^{-1}	Kennzeichnende Geschwindigkeit	u m sec^{-1}	Velocity
	α kcal m^{-2}hr^{-1}C^{-1}	Wärmeubergangszahl	h W m^{-2}K^{-1}	Heat transfer coefficient
	κ m hr^{-1}	Verdunstungsziffer	k_G m sec^{-1}	Mass transfer coefficient
	λ kcal m^{-1} hr^{-1}C^{-1}	Wärmeleitzahl	k W m^{-1}K^{-1}	Thermal conductivity
	v m^{2}hr^{-1}	Kinematische Zähigkeit	v m^{2}sec^{-1}	Kinematic viscosity
	ρ kg m^{-3}	Dichte	ρ kg m^{-3}	Density

The different groups in the equations are as follows:

	By Schmidt	Here
Nusselt (*Nu*)	$\alpha l/aC_p$	$hd/\alpha c_p$
Reynolds (*Re*)	wl/v	$ud\rho/\mu$
Prandtl (*Pr*)	a/v	$\alpha\rho/\mu$
Sherwood (*Sh*)	$\kappa l/k$	$k_G d/D$
Schmidt (*Sc*)	k/v	$D\rho/\mu$
Grashof (heat) (*Gr*)	$l^3g/v^2(T_w/T_o - 1)$	$gd^3\beta\Delta T/v^2$
Grashof (mass) (Gr')	$l^3g/v^3(m_o/m_w T_w/T_o - 1)$	$gd^3\zeta\Delta\rho/v^2$

β and ζ are the coefficients of volume expansion and density change, respectively.

Using these dimensionless groups, the above equations can be written as

$$Nu = F(Re, Pr) \quad \text{and} \quad Sh = F(Re, Sc)$$

For free convection, Schmidt also gave analogous relationships

$$Nu = F_1(Gr, Pr) \quad \text{and} \quad Sh = F_1(Gr', Sc)$$

The physical meaning of the Schmidt number is the ratio of momentum diffusivity to mass diffusivity.

Encyclopedia of Agricultural, Food, and Biological Engineering
DOI: 10.1081/E-EAFE 120007040
Copyright © 2003 by Marcel Dekker, Inc. All rights reserved.

S

DIMENSIONAL ANALYSIS

By means of dimensional analysis it is possible to identify dimensionless groups. This can for instance be done by means of the Buckingham theorem[2] according to which a functional relationship among q quantities or variables whose units may be given in terms of u fundamental units or dimensions may be written as $(q-u)$ dimensionless groups.

In the case of mass transfer, a fluid at forced convection flows through a cylindrical pipe and mass transfer takes place between the fluid and the wall of the pipe. The dimensions M for mass, L for length, and T for time will form the dimensional groups. The following variables can be expected to be involved in the system:

Mass transfer coefficient	k_G	L/T
Density	ρ	M/L^3
Viscosity	μ	M/LT
Velocity	u	L/T
Mass diffusivity	D	L^2/T
Pipe diameter	d	L

The number of variables is $q=6$ and the number of units is $u=3$. According to Buckingham,[2] there shall then be $6-3=3$ dimensionless groups π, which means

$$\pi_1 = F(\pi_2, \pi_3)$$

D, d, and ρ are such quantities that they alone cannot form a dimensionless group. In the following equations, h' is included with these three variables to form π_1, u to form π_2, and μ to form π_3.

$$\pi_1 = D^a d^b \rho^c k_G$$

$$\pi_2 = D^a d^b \rho^c u$$

$$\pi_3 = D^a d^b \rho^c \mu$$

The dimensional equation corresponding to the π_1-equation above is accordingly

$$M^0L^0T^0 = (L^2/T)^a (L)^b (M/L^3)^c (L/T)$$

Adding the exponents for each unit gives

$$M: c = 0$$

$$L: 2a + b - 3c + 1 = 0$$

$$T: -a - 1 = 0$$

which results in $a=-1$, $b=1$, $c=0$. Thus, the equation for π_1 will be

$$\pi_1 = D^{-1} d^1 h' = k_G d/D = Sh$$

In a similar way, the other groups are deducted

$$\pi_2 = ud/D \quad \text{and} \quad \pi_3 = \mu/\rho D = Sc$$

Dividing π_2 by π_3 gives

$$\pi_2/\pi_3 = ud\rho/\mu$$

which is identical with the Reynolds number and the final groups and their relationship are

$$Sh = F(Re, Sc)$$

Analogous to the Schmidt number for mass transfer is the Prandtl number relevant for heat transfer. This number is interpreted as the ratio of momentum diffusivity to heat diffusivity. The relation of heat diffusivity to mass diffusivity is called the Lewis number, Le. Accordingly

$$Le = \alpha/D$$

For an ideal fluid

$$v = \alpha = D, \quad \text{or} \quad Pr = Sc = Le = 1$$

Values of the Schmidt number for gases range from 0.5 to 2, and for liquids from about 100 to more than 10,000 for viscous liquids.[3] Some Sc-values are given in Ref. 4 and some examples are given below for gases diffusing in air:

Water vapor at 0°C, $Sc = 0.616$

100°C, $Sc = 0.608$

Ethanol 0°C, $Sc = 1.31$

Butanol 0°C, $Sc = 1.95$

HOW THE SCHMIDT NUMBER IS USED

Mass transfer may occur in several unit operations like evaporation, drying, distillation, condensation, crystallization, catalytic reactions, membrane technology, and so on. In such engineering problems, it is important to understand the mechanism of mass transfer and here the Schmidt number in combination with other dimensionless groups is important.

As demonstrated by Schmidt in his original presentation as shown above, the Schmidt number is one part of a dimensional equation which is used to understand mass transfer experiments and problems as well as to make such calculations. The formula results in the Sherwood number which in turn gives the mass transfer coefficient k_G. The numerical values of these mass transfer coefficients are thus related to several factors such as flow velocity, the velocity profile, viscosity, diffusivity, density as demonstrated by equations based on the Schmidt analogy. The equation describing the relationship between the Sherwood number and the Reynolds and Schmidt numbers is based on experiments and on the analogy with heat transfer problems and is valid only for such conditions (geometry, flow conditions) for which they are deducted. As the expressions for heat and mass transfer are identical, it is possible to use heat transfer correlations to derive mass transfer correlations simply by substituting Sh for Nu and Sc for Pr at forced convection, or Gr for Gr' at free convection.

The expressions for the heat and mass correlations generally have the form

$$Nu = C\,Re^n\,Pr^m \quad \text{and} \quad Sh = C\,Re^n\,Sc^m$$

As mentioned before, the constants C, n, and m are the same in both expressions when the geometry and the flow conditions are the same. For forced convection in channels, normally $C = 0.023$, $n = 0.8$, and $m = 0.33$ are used. For forced convection around a sphere, the expressions are[5]:

$$Nu = 2.0 + 0.60\,Re^{1/2}\,Pr^{0.33} \quad \text{and}$$

$$Sh = 2.0 + 0.60\,Re^{1/2}\,Sc^{0.33}$$

Several expressions for different geometries and different flow conditions are available in the literature (e.g., Ref. 3). It should be observed that in almost all cases the exponent for Sc is 0.33. According to Lydersen,[6] it may vary between 0.33 and 0.50, but even higher values are mentioned in the literature.[4]

REFERENCES

1. Schmidt, E. Verdunstung und Wärmeubertragung. Gesundheitswes-Ingenieur **1929**, *29. Heft*, 525–529.
2. *Perry's Chemical Engineers Handbook*; Mc Graw Hill Book Co., 1985.
3. Geankoplis, C.J. *Transport Processes and Unit Operations*, 3rd Ed.; Prentice-Hall International, Inc.: New Jersey, 1993.
4. Grigull, U. *Die Grundgesetze der Wärmeubertragung*; Springer-Verlag: Berlin, 1955.
5. Bird, R.B.; Stewart, W.E.; Lightfoot, E.N. *Transport Phenomena*; Wiley International Edition: Tokyo, 1960.
6. Lydersen, A.L. *Mass Transfer in Engineering Practice*; John Wiley & Sons Ltd.: Chichester, 1979.

Scraped Surface Heat Exchangers

Peter Fryer
Sandrine Rodriguez
University of Birmingham, Birmingham, United Kingdom

S

INTRODUCTION

Thermal processing is carried out in heat exchangers, chosen according to the product nature, the heat requirement, and the process operating conditions. Plate and tube exchangers are used for low viscosity products, which do not generate severe fouling during processing. High viscosity products and/or containing particles, which might give rise to fouling in practice, must be treated in scraped surface heat exchangers (SSHE). This type of exchanger is also commonly used in ice-cream manufacture, where freezing is carried out in a SSHE.

Many designs of SSHE exist. Common to them all, however, is that a SSHE consists of two coaxial cylinders with an internal rotating shaft fitted with blades at an angle β to the shaft, shown in cross section in Fig. 1. Fluid flows axially along the SSHE between the static and rotating cylinders. The blades improve the heat transfer between the heating or cooling medium by scraping the inner tube wall and, therefore, preventing fouling, burn-on, or freeze-on problems. This is achieved through a combined effect of turbulence, interfacial film removal, and mixing of the product. High heat transfer coefficients can be achieved as the boundary layer is continuously replaced by fresh material. As the product remains in contact with the heat transfer surface only for a few seconds, smaller heating areas are needed than for conventional heat transfer equipment and higher temperature driving forces between the cooling or heating medium and the product can be used.

Two main types of industrial SSHE exist: liquid full and thin film liquid heat exchangers. The liquid full SSHE is completely filled with the product and the blades scrape the inner tube wall. Vertical or horizontal units can be found. These exchangers are widely used in the food industry. In contrast, the thin (also called wiped) film heat exchanger is usually a vertical cylinder containing a rotating shaft fitted with wiping blades. Thin film heat exchangers are used in the food industry to evaporate (and therefore concentrate) milk products.

INDUSTRIAL APPLICATIONS OF SSHES

Scraped surface heat exchangers can process a wide variety of products such as margarine, peanut butter, salad dressing, jam, tomato paste, soups, baby food (cereal based food and fruit purées), processed cheese, meat products, and ice-cream.[1,2]

Their versatility relies on their ability to process viscous and solid materials and the ability to vary the ratio of tube wall to shaft diameter. The largest rotor diameters tend to be used with foods which do not contain particles, so that the annular gap between rotor and stator is small, giving a very high degree of turbulence.[3] On the other hand, a larger gap allows fluid containing particles to be treated, because turbulence and mixing is minimized so that the product is not excessively damaged through the process. The maximum particle size is 25 mm and the rotor size can be designed to suit a particulate food material.[3]

Typically, SSHEs are also used in ice-cream freezing, where the cooling medium is usually ammonia and the role of the blades is both to prevent freezing on the wall of the exchanger and to incorporate air in the mixture (up to approximately 50% v/v in ice cream). Rotors are designated in terms of cylinder volume. For example, an 80% displacement shaft (which occupies 80% of the barrel volume) is used in the production of low temperature ice-cream. A 15% displacement shaft produces a wetter ice-cream; this texture is more desirable when the product has to be fluid enough to be molded before the hardening stage. The foam mixture is frozen in the SSHE to an outlet temperature of between -4 and -9°C in order to get a large number of ice crystals. The size of the crystals, which is a function of the outlet temperature, governs the eating quality of the ice-cream.[4] Small crystals give a softer and smoother texture to the final product and it is generally agreed that ice crystals of size smaller than 25 μm are not perceived in the mouth when tasted as a food mixture. Lowering the outlet temperature of the SSHE has the advantage of lowering the thermal duty required in subsequent processes within ice-cream production (hardening and storage); however, when the outlet temperature is decreased, the viscosity of the product is also increased, necessitating a higher dasher (rotor) power. Lowering the outlet temperature thus reduces the heat transfer capacity of a given exchanger surface.[4]

The crystal size distribution of the product is also governed by its residence time distribution (RTD) in the freezer. For instance, an ice crystal that spends a long time in the freezer will be allowed to grow but will also

Encyclopedia of Agricultural, Food, and Biological Engineering
DOI: 10.1081/E-EAFE 120007006
Copyright © 2003 by Marcel Dekker, Inc. All rights reserved.

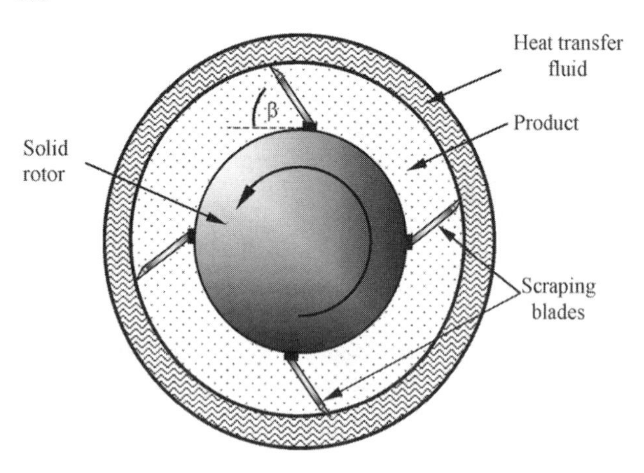

Fig. 1 Schematic of the cross section of a SSHE.

experience more shear from the scrapers. Research has also shown that average crystal size decreases with increasing shear rate. As a result of nonideal flows in a SSHE, volume elements spend different lengths of time in the equipment, and their thermal treatment is therefore affected. Understanding the flow behavior of a particular system and knowledge of its critical limits (i.e., the slowest volume element) is crucial to ensure efficient commercial production. As an example,[5,6] ice-cream quality which ideally requires ice crystals in the size range 45 μm–55 μm[7] is highly affected by the time the crystals spend in the barrel. A high residence time gives opportunity for the ice crystals to grow, but also ensures that the crystals are subject to shear for longer times, leading to greater physical damage and reduction of their size. Similarly, the air cell size and fat globules within the mixture are affected by the time they stay in the freezer, as fat tends to destabilize if sheared too long.

DESIGN FACTORS FOR SSHES

Flow Patterns

Flow patterns in the SSHE are complex. Couette flow between concentric cylinders can be characterized by the Taylor number: flow is laminar until a critical Taylor number, dependent on the gap width, is reached where vortices appear. Flow in an annulus with mixed axial flow is essentially a combination of Poiseuille and Couette flows. Härröd and Maingonnat[8] found that closest to plug flow was found when the rotational Reynolds number was slightly above critical, i.e., when neighboring vortices do not mix. Flow patterns in a SSHE are further complicated by[9]:

1. Developing velocity profiles;
2. The fitting and geometry of the blades;
3. Physical properties of the product. Temperature differences will affect product viscosity and thus the axial and tangential velocity profiles. Any shear-thinning behavior of the fluid will also affect flow patterns as rotational velocity and thus shear increases within the gap.[10]

Flows in an SSHE have been characterized in terms of

$$\text{Rotational Reynolds number}: \quad Re_{\mathrm{R}} = \frac{\rho N_{\mathrm{r}} D_{\mathrm{T}}^2}{\mu}$$

$$\text{Axial Reynolds number}: \quad Re_{\mathrm{A}} = \frac{\rho v_{\mathrm{bulk}}}{\mu}(D_{\mathrm{T}} - D_{\mathrm{S}})$$

Where μ is viscosity and ρ density of the material, D_{T} and D_{S} the diameter of the tube and shaft, respectively, v_{bulk} the mean axial flow, and N_{r} the rotational speed. Härröd[9] divided the flow patterns within an SSHE into several regimes.

$Re_{\mathrm{R}} < 250$:	laminar	$Re_{\mathrm{A}} < 1.5 \times 10^4$	laminar
$250 < Re_{\mathrm{R}} < 10^5$	laminar with vortex	$Re_{\mathrm{A}} > 1.5 \times 10^4$	turbulent
$Re_{\mathrm{R}} > 10^5$	turbulent		

Observations of laminar flow show an almost stationary film at the wall, only disturbed by the scraping action of the blades.[11] Radial mixing occurs at the tip of the blades and increases with rotor speed. Back-mixing takes place in the gap between the blades and increases with shaft speed. Stanzinger, Feigl, and Windhab[12] show that it is possible to simulate flow in an annulus with blades using computational methods; regions of high shear stress and energy dissipation are seem round the blades. Baccar and Abid[13] simulated the hydrodynamic and thermal behavior within a SSHE: stagnation zones were found both at the rear side of the blade, and in the vicinity of the shaft, leading to poorer heat transfer. Magnetic resonance imaging has been used by several groups[14–16] to show the flow profiles and identify mixing regions in the system.

Residence Times

The ideal flow in a SSHE would be no axial mixing with perfect radial mixing. Here all elements of the fluid remain in the exchanger for the same time and undergo the same temperature–time–shear process. Many authors have studied RTDs to represent experimental data. Russell, Burmester, and Winch[6] studied flows by following pulses of dye tracer, and concluded that a change in apparent viscosity (for similar flow behavior index) had no effect on the variance of the RTD curves. However, the shear-thinning behavior of the carbopol with a change in temperature would have a great effect. Lee and Singh[17–19] studied RTDs of potato cubes, up to particle concentration of 40% w/w. They found that an increase in shaft speed, viscosity, or particle size increased the mean residence time. The orientation of the SSHE had no significant influence on the minimum particle residence time, mean, and maximum residence times were significantly affected by orientation due to gravity. Rodruiguez[20] used positron emission particle tracking (PEPT) to study particle paths through the exchanger: the mean flow within the SSHE was found by averaging individual trajectories. Axial and radial trajectories are shown in Fig. 2. Results suggest that the flow in the barrel is close to plug flow, and that the high level of mixing identified in Ref.[6] result from inlet and outlet difficulties. These results suggest that if plug flow is required, inlet and outlet design should be addressed.

Heat Transfer and Power Consumption

A number of authors have discussed heat transfer in the SSHE. A three-step heat transfer mechanism was proposed by Trommelen, Beek, and Van De Westelaken[21] as:

1. Conduction heat transfer in a thin layer of material near the wall occurring between two scrapings.
2. Partial temperature equalization on the boundary layer that builds up on the blade, assuming the velocity of the liquid in that layer is relatively low compared to the tangential velocity of the shaft.
3. Convection heat transfer in the radial direction from the layer of material to the bulk of the fluid, enhanced by the presence of Taylor vortices.

Many correlations for heat transfer (such as[22–25]) have been given: most involve some combination of the axial and radial Reynolds numbers, together with the fluid Prandtl number[23] give:

$$Nu = 1.3\, Re_{\mathrm{R}}^{0.53} Pr^{0.33}$$

for flow of sodium alginate. The form of the equation varies with the geometry of the exchanger, such as the number of blades; increased numbers of blades increase the heat transfer and also decrease radial dispersion. This is critically related to power consumption: most power is consumed through the blade action. The relationship between power consumption and the radial Reynolds number has been studied (see Ref.[25,26]): power consumption scales with Re_{R}^{a}, where a is in the range -1 to -1.6, whilst $a = -1$ would be expected for laminar flow, the difference is due to the scraping effects.

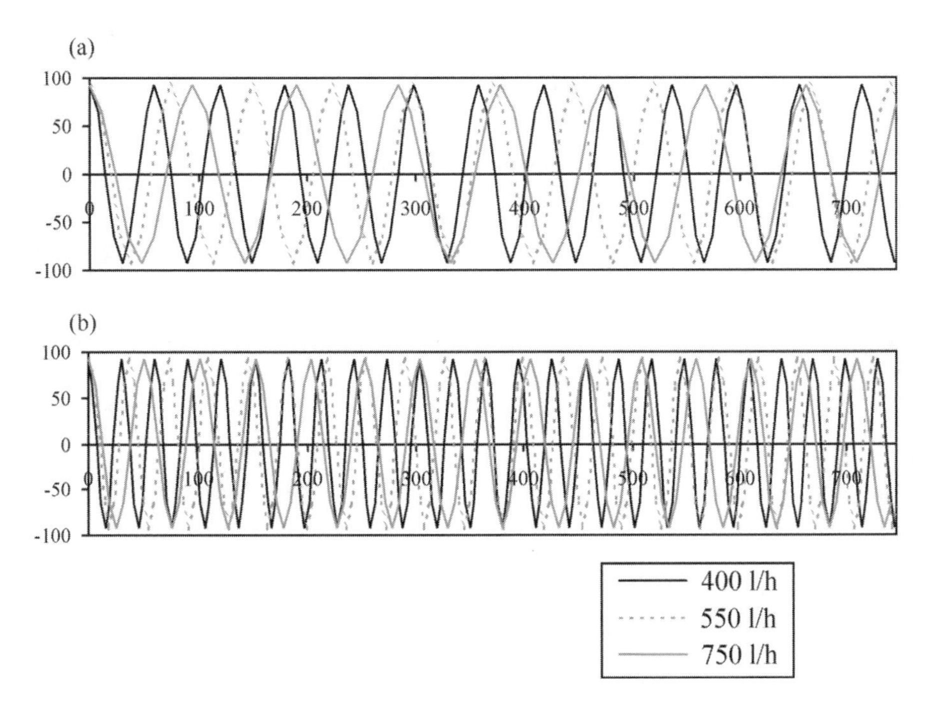

Fig. 2 Flow patterns of tracers inside an SSHE as a function of axial and rotational flow: (a) 30 rpm, (b) 60 rpm, from Ref.[22]; obtained using PEPT. Note that as throughput and rotational speed increase the number of spirals increases.

CONCLUSION

The design and use of SSHEs has been briefly reviewed. SSHEs are widely used to process fluids that foul heavily, such as ice cream. Extensive design information can be obtained from the literature; care must be taken to design for the correct residence time distribution and to determine the correct heat transfer behavior and power consumption.

REFERENCES

1. Darlington, R. Heat Transfer and Heat Exchangers. Food Manuf. **1972**, *47* (10), 31–34.
2. Day, R.H. Scraped Surface Heat Exchangers for High Temperature Short Time Processing of Foods. Food Trade Rev. **1970**, *40* (7), 33–38.
3. Hayes, J.B. Scraped Surface Heat Exchanger in the Food Industry. AICHE Symp. Ser. **1988**, *84*, 251–265.
4. Bald, W.B. *Food Freezing, Today and Tomorrow*; Springer Series in Applied Biology, New York, 1991; Chap. 11, 133–142.
5. Russell, A.B. Control of Ice Crystallisation in Scraped Surface Ice Cream Freezers. *Eng. and Food at ICEF 7*; B17-B20; University of Sheffield Press: Sheffield, UK, 1997.
6. Russell, A.B.; Burmester, S.S.H.; Winch, P.J. Characterisation of Shear Thinning Flow Within a Scraped Surface Heat Exchanger. Trans. Inst. Chem. Eng. **1997**, *75* (C3), 191–197.
7. Arbuckle, W.S. The Freezing Process. In *Ice Cream*, 4th Ed.; Marshall, R.T., Ed.; Chapman & Hall: London, 1991.
8. Härröd, M.; Maingonnat, J.F. Performance of Scraped Surface Heat Exchangers. In *Thermal Processing and Quality of Foods*; Zeuthen, P., et al. Eds.; Elsevier Applied Science: London, 1983; 318–323.
9. Härröd, M. Scraped Surface Heat Exchangers—A Literature Survey of Flow Patterns, Mixing Effects, Residence Time Distribution, Heat Transfer and Power Requirements. J. Food Process Eng. **1986**, *9*, 1–62.
10. Naimi, M.; Devienne, R.; Lebouche, M. Incidence du seuil d'écoulement sur les performances dynamiques et thermiques de l'échangeur de chaleur à surface raclée. Int. J. Heat Mass Transfer **1993**, *36* (5), 1413–1421.
11. Trommelen, A.M.; Beek, W.J. Flow Phenomena in Scraped Surface Heat Exchanger (Votator-Type). Chem. Eng. Sci. **1971**, *26*, 1933–1942.
12. Stanzinger, M.; Feigl, K.; Windhab, E. Non-Newtonian Flow Behaviour in Annular Gap Reactors. Chem. Eng. Sci. **2001**, *56*, 3347–3365.
13. Baccar, M.; Abid, M.S. Numerical Analysis of Three-Dimensional Flow and Thermal Behaviour in a Scraped Surface Heat Exchanger. Rev. Gén. Therm. Fr. **1997**, *36*, 782–790.
14. Corbett, A.M.; Phillips, R.J.; Kauten, R.J.; McCarthy, K.L. Magnetic Resonance Imaging of Concentration and Velocity Profiles of Pure Fluids and Solid Suspensions in Rotating Geometries. J. Rheol. **1995**, *39*, 907–924.
15. Wang, W.; Walton, J.H.; McCarthy, K.L. Flow Profiles of Power Law Fluids in Scraped Surface Heat Exchanger Geometry Using MRI. J. Food Process Eng. **1999**, *22*, 11–27.
16. Wang, W.; Walton, J.H.; McCarthy, K.L. Flow Profiles of Power Law Fluids in Scraped Surface Heat Exchanger Geometry Using MRI. J. Food Process Eng. **1999**, *22*, 11–27.
17. Lee, J.H.; Singh, K.H. Particle Residence Time Distributions in a Model Horizontal Scraped Surface Heat Exchanger. J. Food Process Eng. **1991**, *14* (2), 125–146.
18. Lee, J.H.; Singh, K.H. Process Parameters Effects on Particle Residence Time in a Vertical Scraped Surface Heat Exchanger. J. Food Sci. **1991**, *56* (3), 869–870.
19. Lee, J.H.; Singh, K.H. Scraped Surface Heat Exchanger Orientation on Particle Residence Time Distributions. J. Food Sci. **1991**, *56*, 1446–1447.
20. Rodruiguez, S.E.; PhD Thesis, University of Birmingham, Birmingham, UK, 2000.
21. Trommelen, A.M.; Beek, W.J.; Van De Westelaken, H.C. A Mechanism for Heat Transfer in a Votator-Type Scraped Surface Heat Exchanger. Chem. Eng. Sci. **1971**, *26*, 1987–2000.
22. De Goede, R.; De Jong, E.J. Heat Transfer Properties of a Scraped Surface Heat Exchanger in the Turbulent Flow Regime. Chem. Eng. Sci. **1993**, *48* (8), 1393–1404.
23. Maingonnat, J.F.; Corrieu, G. A Study of the Thermal Performance of a Scraped Surface Heat Exchanger. Part 1: Review of the Principal Models Describing Heat Transfer and Power Consumption. Int. Chem. Eng. **1986a**, *26* (1), 45–54.
24. Maingonnat, J.F.; Corrieu, G. A Study of the Thermal Performance of a Scraped Surface Heat Exchanger. Part 2: The Effect of the Axial Diffusion of Heat. Int. Chem. Eng. **1986b**, *26* (1), 55–68.
25. Abichandani, H.; Sarma, S.C. Heat Transfer and Power Requirements in Horizontal Thin Film Scraped Surface Heat Exchangers. Chem. Eng. Sci. **1988**, *43* (3), 871–881.
26. Cox, D.R.G.; Gerrard, A.J.; Wix, L. Power Consumption and Backmixing in Horizontal Scraped Surface Heat Exchangers. Trans. Inst. Chem. Eng. **1993**, *71* (C), 187–193.

Secondary Packaging

Raymond A. Bourque
Ocean Spray Cranberries, Inc., Lakeville-Middleboro, Massachusetts, U.S.A.

S

INTRODUCTION

Secondary packaging is that packaging used to contain one or a number of primary packages.

The role of the secondary package is generally to protect the primary package and thus the food product throughout distribution from the place and time of packaging the food in the primary package to the place and time it is offered for sale to the final consumer. In this role, the secondary package is the shipping container.

This article of the encyclopedia will focus on that role in which the secondary package is defined as the shipping container.

ROLE OF THE SECONDARY PACKAGE

The secondary package when performing as the shipping container must protect the primary packages and the food products throughout distribution. This is a critical and remarkable task considering the distances food products are shipped, the length of time from when they are manufactured until they get to the consumer, and the adverse and hostile conditions they encounter throughout distribution. In today's world of free trade and global commerce, the performance requirements of the shipping container is greater than ever before. Historically, the food distribution system was only a domestic system and now it is a global system. In the past, agricultural products were grown, processed, and packaged in the United States and distributed throughout the United States by rail or truck. A very demanding journey taking place over as much as a year before the food was delivered to the final consumer. Food was processed and packaged close to the growing areas in the South and West and shipped to the high population, industrial areas of the North. Crops were seasonal, thus had to be stored to last until the next harvest. Today, the above distribution system still exists, but in addition many food products are shipped by air and sea throughout the world. The food products are exposed to climatic, handling, and storage conditions that may be much more severe than encountered domestically. During all this time and over all this distance, the shipping container must withstand a variety of mechanical, chemical, and biological stresses. Not only must the shipping container withstand these stresses, but it must also do it in a way that assures that the primary packages and thus the food products are not damaged. The secondary package must assure that high quality and safe food is delivered to the consumer.

The Distribution Environment

To design and engineer an adequate secondary package, the stresses which will be encountered must be understood and provided for in the structural design. The following is an overview of a few of the typical types of stresses that a secondary package must endure.

Assembly and packaging

First the shipping container must withstand the mechanical abuse of the assembly and packing process. In the food industry, primary containers are received either in bulk or prepackaged in corrugated boxes, referred to as reshippers. Plastic and glass bottles, cans, and jars are commonly received on bulk pallets. They are automatically removed from the bulk pallets and conveyed to the filling, capping, sealing, and labeling operations and then on to a corrugated box or tray former and packer of a variety of different designs. The shipping container is formed automatically, packed with the primary packages and sealed at production rates of up to 100 shipping containers per minute. The mechanical abuse to the shipping container can be significant in these operations. Proper assembly, closing, and sealing of the shipping container is critical to ensure the integrity of the container and thus the physical strength of the finished shipping container. The secondary package is then conveyed to either a hand or automatic palletizing operation. Again the containers can be abused—either manually by being dropped on edges or corners or mechanically by poor adjustment of automatic palletizers.

When properly designing a secondary package, it is critical to size the package such that there is very little overhang or underhang on the pallet, usually less than 1 in. (25 mm) in either direction on a standard 40 in. × 48 in. (1000 mm × 1200 mm) grocery industry pallet. Too much overhang results in a loss of stack strength and

DOI: 10.1081/E-EAFE 120007135
Copyright © 2003 by Marcel Dekker, Inc. All rights reserved.

pallet-to-pallet impact damage; too much underhang allows loads to shift during shipment.

When bottles and jars are delivered to the packaging plant in reshippers (containers used to deliver empty packages as well as ship the filled packages), similar stresses on the packaging line are encountered. The primary packages are removed from the reshipper and the reshippers are transported considerable distances to re-meet the primary packages at the packer. Severe damage can occur in the packer if it is not properly adjusted.

Other primary packages such as paperboard cartons, trays, tubes, and pouches are handled differently. These types of packages are typically formed automatically on the packaging line, not received on bulk pallets or reshippers as are cans and bottles. These primary packages are packed into the secondary package on-line in a similar manner as for cans and bottles.

Warehousing

Palletized secondary packages are then transported, usually by forklift truck, into a warehouse where the pallets are stored either one-pallet high in racks or, in 2–4 pallet high stacks. Damage to the secondary packages can occur when the forklift assembles the stacks. Generally, finished goods warehouses are not temperature or humidity controlled, thus the secondary package can be exposed to a variety of temperature and humidity conditions from a few weeks to several months. Many primary packages such as plastic bottles and paperboard cartons cannot support weight without crushing, thus the secondary package must support the entire load without deforming during the warehousing period. Thus, the stacking strength of the shipping container, even under a variety of temperature and humidity conditions must be adequate to carry the weight without deformation. Defining the stacking weight requirements and the loss of structural strength of the corrugated board due to humidity and time are critical to engineering an adequate secondary package.

Additionally, with many food products, infestation (e.g., insects) can be a problem and thus the shipping container must protect the primary packages and products from any infestation potential that might occur during warehousing.

Transport

When the products are ready for shipment, the pallets are removed from the stacks or from the racks, usually by forklifts, transferred to a loading dock where they are staged for loading onto either rail cars or trucks for transport. The shipping containers must be capable of withstanding the impact and vibration stresses incurred in the forklift operations and the loading operations. The product is then loaded into the truck or railcar, 18–22 pallets per truck, 56–60 pallets per railcar. The pallets are stacked one-pallet high if the maximum weight of the truck—approximately 40,000 lb (18,140 kg) for a 40-foot trailer—is fully used. If the load is less than 40,000 lb (18,140 kg), pallets may be double-stacked creating even more opportunity for damage. Most damage in double stacking pallets occurs on the top layer of the bottom pallet. Occasionally, dunnage, material used to physically minimize the movement of pallets during transportation, is used in trucks on rail cars. Product is shipped anywhere from a few miles (km) to a few thousand miles (km). Coast-to-coast transport of finished food products is very common. A variety of stresses occur during shipping. Vibration is a major source of damage, particularly to fragile products. Pallet-to-pallet impacts occur when trucks stop quickly or railcars connect to one another. Many tests have been designed to measure these forces and their effect on packaging. These tests will be described later in this section.

Delivery to customer

Usually the shipping process does not consist of one shipment from the point of production to the point of consumption but instead involves several hand-offs at different transfer stations or warehouses along the way. This results in more handling of the pallets, more warehousing, and sometimes more multipallet stacking. Once delivered to the customers' warehouse, secondary packages are removed from the pallets in "picking operations" and placed on multiproduct pallets for final shipment to the retail facility. Secondary packages are placed on mixed pallets in a variety of different ways—on their side or even upside down. The purpose of "picking" is to load mixed pallets with a variety of products to meet the retail outlets' needs. Thus, there is little control by the manufacturer on how the secondary packages are placed on the mixed pallet. Heavier goods can even be placed on top of lighter more fragile goods. Mixed pallets are then shipped usually by truck to the final retail outlet where the product is off-loaded in a back room and transferred by grocery stockers to the retail shelf. Even at this point, the chance of damage still exists. Packages are usually opened by grocery stockers with case knives. If carelessly done, this can result in damage to the primary package if the secondary package is not properly designed and the cutting instructions are not communicated clearly.

Communication

In addition to protecting the product throughout distribution, the secondary package must also communicate critical information to many people along the way. The secondary package must identify the contents, the manufacturer's name and address, the code date, a universal product code (UPC), a box manufacturers certificate, and any other critical handling instructions such as "fragile" or "store at 40°F (4°C) or less." Instructions for opening the secondary package are also advisable. In some cases the secondary package will actually be merchandised and thus marketing copy will be included.

Protecting the Food

The role of secondary packages is not only to survive the above-described rigorous distribution system but also in so doing to protect the primary package so that the final consumer is provided with a high-quality, nutritious, safe food product. Proper design and engineering of the secondary package relies on the accurate identification of the requirements of the primary package and the hazards and stresses that will be encountered in the various aspects of the distribution process.

SECONDARY PACKAGING ALTERNATIVES

Historically, wooden boxes, wooden crates, burlap (woven cloth) bags, and canvas wraps were all used to protect the primary packages and food products throughout distribution. Today, most secondary packages are made of corrugated board or shrink films or a combination of both.

As corrugated board and shrink films are the materials in common use today, this section will focus on the design, construction, and performance of secondary packages made of these materials.

Corrugated Boxes

The corrugated box is the most prevalent secondary package for food products. It is made of a renewable, reasonably inexpensive, and recyclable material. Corrugated board and box design can be engineered in a variety of ways to provide the physical properties adequate to protect food products from the stresses and abuses that will be encountered in the distribution environment.

Materials

Corrugated board is manufactured from linerboard and corrugating medium.

Linerboard is made by the Kraft or sulfate paper making process. Softwoods are used as they have long fibers that produce stronger paperboard than short fibers. Linerboard is brown unless made of bleached virgin pulp in which case it is solid white. A mottled white linerboard is also made by applying a thin layer of bleached pulp on the outer surface of brown board made of nonbleached pulp. Linerboard is defined in pounds per $1000\,ft^2$ (lb/MSF^{-1}) or grams/sq. meter (g/m^{-2}). The weight is referred to as "Basis Weight." The most common linerboard used to make corrugated board is $42\,lb\,MSF^{-1}$ ($205\,g\,m^{-2}$). Other common linerboard grades are $69\,lb\,MSF^{-1}$ ($337\,g\,m^{-2}$), $33\,lb\,MSF^{-1}$ ($161\,g\,m^{-2}$), and $90\,lb\,MSF^{-1}$ ($439\,g\,m^{-2}$)—many other weights exist for more specialized uses.

Corrugating medium is made of shorter fiber pulp because it must be soft and pliable. Virgin hardwoods and recycled corrugated board are used primarily. Approximately, 25% of corrugated board is made of recycled material. Most of the recycled material is used in the medium because the recycling process reduces the fiber length and short fibers are best used in medium, while longer fibers are beneficial in the linerboard. Basis weights of $26\,lb\,MSF^{-1}$ ($127\,g\,m^{-2}$), and $33\,lb\,MSF^{-1}$ ($161\,g\,m^{-2}$), are the most common for corrugating medium.

Linerboard and corrugating medium are combined in a machine called a Corrugator. Single wall corrugated board consists of a layer of corrugating medium formed into fluted and sandwiched between two layers of linerboard. The basis weight of the linerboard layers and the fluted corrugating medium can differ and is defined by the properties required of the finished board. The basis weight of the most common corrugated boxboard is 42/26/42 (205/127/205), whereas the linerboards are $42\,lb\,MSF^{-1}$ ($205\,g\,m^{-2}$), and the corrugating medium is $26\,lb\,MSF^{-1}$ ($127\,g\,m^{-2}$) (Fig. 1).

Double-wall corrugated or triple-wall corrugated consists of two or three corrugating medium layers separated and faced on each side with layers of linerboard.

The Corrugator is fed with rolls of linerboard and medium. The Corrugator combines the layers of linerboard and the medium and forms the flutes or curves in the medium. The flutes are formed by softening the medium with steam and passing it through metal male/female rollers. Adhesive is applied to the flutes and the medium is combined with the inner and outer linerboard layers. The flutes create columns in the board resulting in greatly enhanced structural strength.

Flutes differ in size depending on the intended properties. The four standard sizes used are—A, B, C, and E. They differ in flute height and the number of flutes per foot (m). A is the largest with the least flutes per foot (m), C, the second largest, B smaller than A and C, and E is the smallest.

The terminology seems strange in that "C" flute is bigger than "B" flute. A and B were created first, C came later to meet the need for a flute size in-between A and B. The box designer will specify a flute size depending on

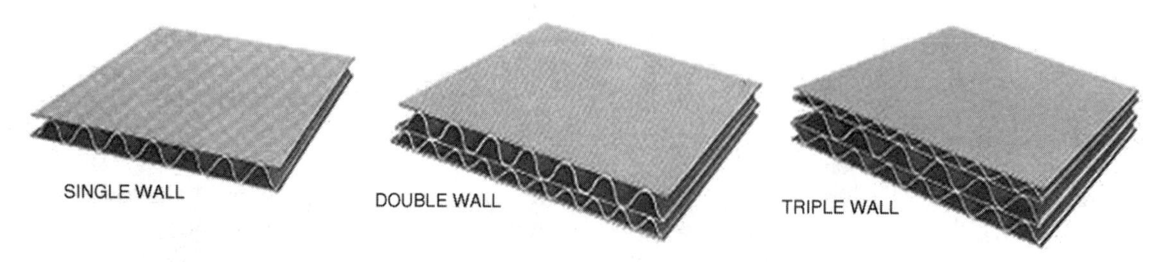

Fig. 1 Corrugated © 1992, Fibre Box Association.

the properties required. Larger flutes provide more cushioning and column strength. Smaller flutes crush less and can be folded more precisely.

The strength of corrugated board is a result of the linerboard and corrugating medium's basis weight, the number of layers (e.g., single wall, triple wall), and the flute size.

Historically, the most common method to measure and specify board strength has been the Mullen test. The Mullen test is a bursting test that measures the pounds (kg) of force required to puncture a specific board combination. Typical board grades as defined by the Mullen test are 125 lb (57 kg), 175 lb (79 kg), 200 lb (90.8 kg), 275 lb (125 kg), and 350 lb (159 kg). There are a number of other less common grades.

Edge Crush is a newer test method now becoming a common standard to measure and specify board strength. The Edge Crush test applies a compression or top load force on a section of board prepared in a precise manner with the flutes in the vertical position.

Today, corrugated box manufacturers may use either method to standardize and specify corrugated board grades and strength. The manufacturer's certificate, which appears on the bottom of the box, identifies the box manufacturer, the basis weight of the materials used, and the board strength (defined using either Mullen test or Edge Crush test terminology).

Corrugated box designs

There are a number of different box designs commonly used. The packaging engineer will select the proper design based on the product to be packaged, the type of equipment used to erect, load, and seal the box, the inner partitions provided for additional strength, the method to insert the partitions, and how the primary packages will be removed from the box (Fig. 2).

The regular slotted container (RSC) box is one of the most common designs. It is also the most efficient design as no material is wasted. The blank is rectangular and is slotted and scored such that all flaps are of the same length. Liners and partitions can be inserted into the RSC box to increase its strength. A bottom pad insert may also be used to fill the gap between the inner flaps. Partitions are commonly used in the RSC with glass jars or bottles to provide cushioning to prevent breakage during transit. Dividers are commonly used to increase the stacking strength for plastic bottles and paperboard cartons. The RSC blank is formed into a knock-down (KD) box by the manufacturer by folding the blank and joining the ends by gluing or stapling and thereby forming the manufacturers' joint. The KD is erected at the packaging plant just prior to being loaded with the primary packages (Fig. 3).

The center special slotted container (CSSC) is similar to the RSC except the inner flaps are cut to different

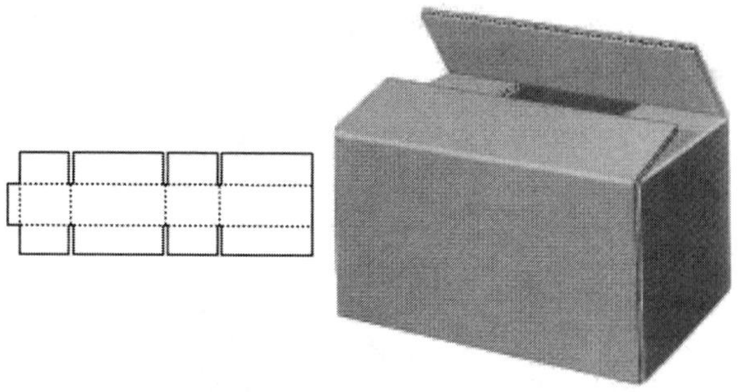

Fig. 2 Regular slotted container. © 1992, Fibre Box Association.

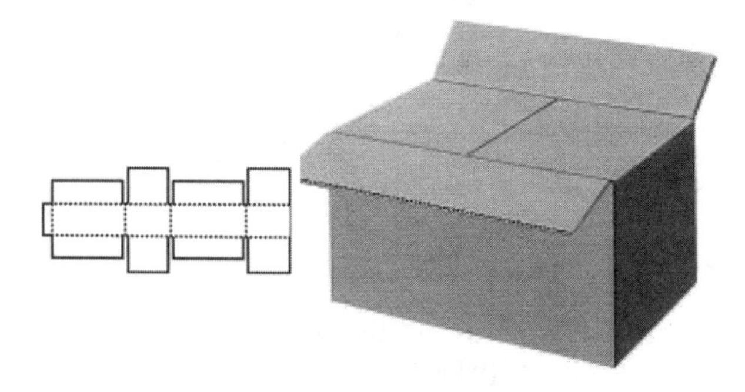

Fig. 3 Center special slotted container. © 1992, Fibre Box Association.

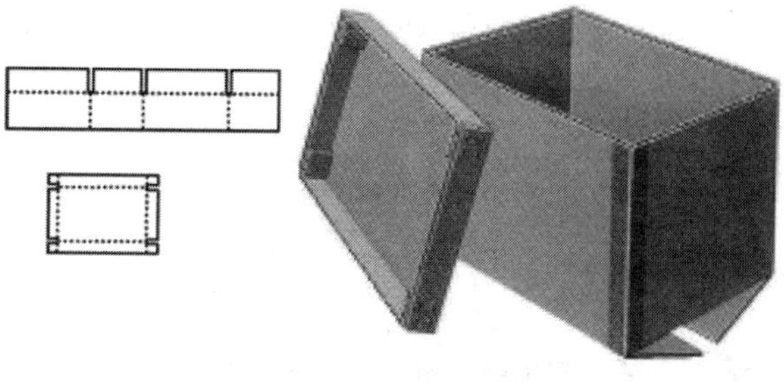

Fig. 4 Half-slotted container. © 1992, Fibre Box Association.

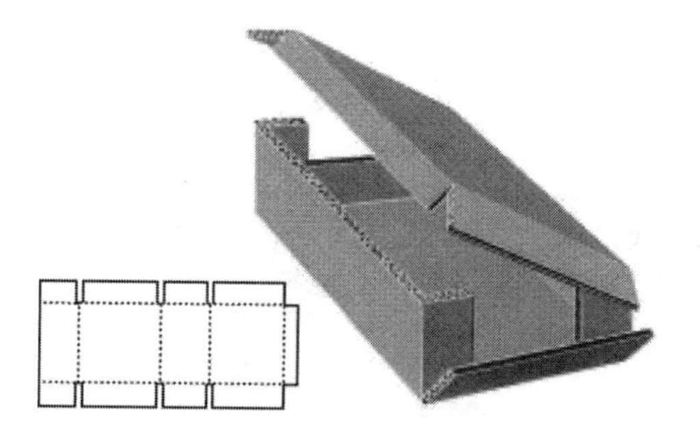

Fig. 6 Wraparound box. © 1992, Fibre Box Association.

dimensions than the outer flaps so that the inner flaps meet when the box is erected. This results in board being wasted when the blanks are cut. The CSSC is a very strong box because all flaps meet and there is no gap between the inner flaps. As with the RSC, liners and partitions can be added for extra strength or cushioning but pads are generally not used as the inner flaps meet. As with the RSC, the CSSC is received as a KD box and is formed, loaded, and sealed at the packaging plant.

There are a number of other slotted containers, which are modifications of the RSC and CSSC. Please refer to the references for more detail on these and other box designs (Fig. 4).

The half-slotted container (HSC) is half of an RSC blank. When the manufacturing joint is glued and the box erected, it is an RSC box without the top flaps. The HSC is usually closed by applying a telescope top over the HSC. The telescope top can extend over the entire height of the HSC, providing added stacking strength, or just enough to secure the telescope top onto the HSC as a cover. The HSC can also be closed with a dust cover that is glued to the length dimensions and simply covers the HSC (Fig. 5).

Unlike the slotted containers which are formed from one corrugated blank the Bliss box is made of two or three separate blanks. The blanks are scored such that the scored flaps can be glued to the body blank, forming very strong reinforced corners. The two or three blanks are assembled and glued on a "Bliss Machine." While the Bliss box uses more board than the RSC, it uses less than an RSC with partitions. The Bliss box is very cost effective for products that must be stacked 2–3 pallets high and contain primary packages that cannot carry topload without deforming. While the Bliss box was designed for heavy items such as frozen beef and poultry or fresh vegetables, recently it has become a popular alternative to the RSC with partitions for plastic bottles and jars. A Bliss box with a partition uses considerably less board and provides greater stacking strength than the RSC with partitions of equal topload strength (Fig. 6).

The wraparound box is assembled from a blank in the packaging plant. The wraparound is essentially an RSC, which is formed around the product and glued after the product is loaded. The major advantage of the wraparound is that a very tight (thus strong) container is produced because a loose fit is not necessary for loading the primary packages into the box after it is formed as with the RSC. Less board can also be used as the RSC flaps can be designed onto the box ends as opposed to the top and bottom (see Fig. 7).

The plastic shrink wrap, with or without a corrugated tray, is an alternative to the corrugated box. This is used generally with nonfragile, abuse-resistant primary packages. Primary packages packed in shrink wrap secondary packages must be strong enough to support the weight encountered in stacked pallets. The tray shrink

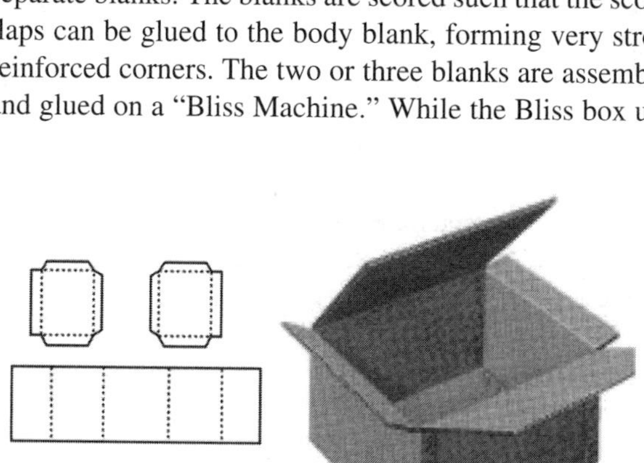

Fig. 5 Bliss box. © 1992, Fibre Box Association.

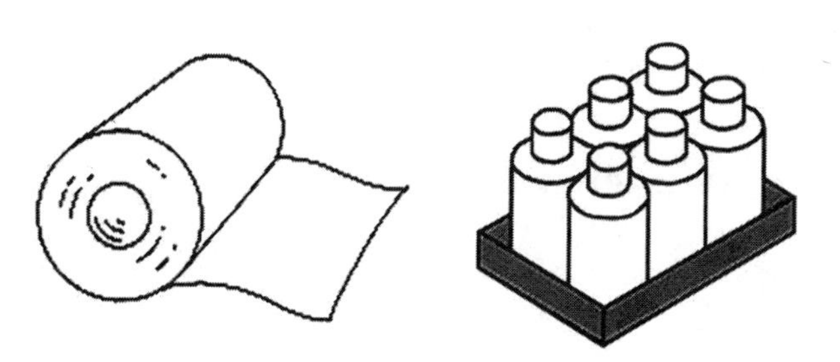

Fig. 7 Tray/shrink wrap.

wrap is commonly used on metal cans and some plastic bottles which can support the weight encountered in distribution.

Trays are usually 2 in.–3 in. (50 mm–75 mm) high and made of corrugated board. The tray material construction is similar to that already discussed.

The trays are packed with the primary packages and then the entire unit is wrapped with a shrink film. The wrapped tray and contents is then conveyed through a heat tunnel to shrink the film.

The film thickness will vary depending on the weight and size of the secondary package. Film thicknesses of 2 mil–3 mil (50 μm–75 μm) are common for food grocery products.

Historically, shrink films were polyvinyl chloride (PVC). Today, most shrink films are polyethylene. Polyethylene shrink films are made by a blowing process which produces a biaxially oriented film. By controlling the amount of stretch in each direction the amount of shrinkage of the film can be tailored to the needs of the secondary package.

Thus, the packaging engineer can specify the cross direction and machine direction shrinkage percentage depending on the nature of the primary packages and the tightness required of the secondary package.

TEST METHODS

Many tests have been developed to evaluate the performance of packages in general and specifically secondary packages. The American Society for Testing and Materials (ASTM, Philadelphia, PA) develops, standardizes, and publishes test methods, for evaluation and selection of packages and packaging materials. The testing Association of the Pulp and Paper Industries (TAPPI, Atlanta, GA) is another organization that develops and standardizes test methods. Common laboratory tests evaluate attributes such as compression strength, drop resistance, moisture effect, scuff resistance, impact resistance, vibration damage, etc. Field tests consist of pallet stacking tests and truck or rail shipping tests.

The following are a few tests commonly used to evaluate secondary packages. Please see the references for additional test methods.

LIST OF ABBREVIATIONS

ASTM D642-90	Test method for determining compressive resistance of shipping containers, components, and unit loads.
ASTM D775-80	Test method for drop test for loaded boxes.
ASTM D880-86	Test method for incline impact test for shipping containers.
ASTM D951-88	Test method for water resistance of shipping containers by spray method.
ASTM D999-86	Method of vibration testing of shipping containers.
ASTM D1596-91	Test method for shock absorbing characteristics of package cushioning materials.
TAPPI T803	Puncture test of containerboard.
TAPPI T810	Bursting strength of corrugated and solid fiberboard.
ASTM D882-90	Test method for tensile properties of thin plastic sheeting.
ASTM D4635-86	Specification for low-density polyethylene films for general use and packaging applications.

DESIGNING AND ENGINEERING A SECONDARY PACKAGE

The previous discussion of the distribution environment, the materials available, the secondary design options, and the common test procedures provide a basis for the packaging engineer to use in engineering a secondary package. The following is a summary of the steps involved in designing a proper secondary package:

1. Identify and quantify the physical properties of the primary packages. Quantify how much impact and stresses the primary packages can withstand without failure.
2. Identify and quantify the forces and stresses to be encountered in the distribution process.
3. Quantify the requirements of the secondary package. How much load or force must it withstand in distribution?
4. Identify the equipment constrains. The equipment used to assemble, load, and seal the secondary package will influence the design selected.
5. Identify alternative designs and material constructions based on the requirements and equipment constraints.
6. Establish a test protocol to evaluate the design alternatives against the performance requirements. Alternative designs can be tested to assure the design(s) properly protects the primary package throughout the expected distribution environment. Test protocols should include laboratory testing and field testing.
7. Identify the cost of the alternatives. Estimate total cost including materials usage, manufacturing requirements, and capital equipment costs.

8. Conduct tests to quantify performance of alternative designs. Use standard industry test methods whenever possible. Laboratory testing can be used to compare alternatives. Field testing should be used to confirm the performance of the final candidate(s).
9. Select design, establish specifications. Specifications must include material specifications, package drawings with dimensions, package forming and sealing specifications, and pallet patterns.

THE SECONDARY PACKAGE AS A MULTIPACK

Secondary packaging sometimes is also defined as a means to multipack a number of primary packages such that the unit of sale to the consumer comprises a number of primary packages. Shrink or stretch film wraps and paperboard sleeves, wraps, and cartons are examples of secondary packaging when the secondary package serves as a multipack. In this role, the secondary package is the display package and the unit for sale to the consumer. When used as a multipack the secondary package may also provide merchandising benefits—it may be printed and thus used to communicate to the consumer the brand, the contents, product benefits, use instructions, warning labels, and the manufacturer's name and address. In the role of a multipack, the secondary package also protects the primary package from environmental contaminants such as dust, moisture, and infestation.

FURTHER READING

Fibre Box Handbook; Fibre Box Association, Rolling Meadows, IL, 1992.

Foster, G. Boxes Corrugated. In *The Wiley Encyclopedia of Packaging Technology*, 2nd Ed.; Brody, A., Marsh, K., Eds.; John Wiley & Sons, Inc.: New York, 1997; 100–108.

Hanlon, J.F. *Handbook of Package Engineering*, 2nd Ed.; Technomic Publishing Company, Inc.: Lancaster, PA, 1992; 3-1–3-8; 14-1–14-8.

Selected ASTM Standards on Packaging, Third Edition; American Society for Testing and Materials: Philadelphia, PA, 1991.

Technical Association of the Pulp and Paper Industry; Atlanta, GA.

S

Shear Rheology of Liquid Foods

M. A. Rao
Cornell University, Geneva, New York, U.S.A.

INTRODUCTION

Rheological properties of foods are based on flow and deformation responses when subjected to stress. The study of viscosity is but one aspect of rheology, which by definition is the study of deformation and flow of matter. Most fluid foods contain water, and those containing relatively large amounts of dissolved low molecular weight compounds (e.g., sugars) and no significant amount of either polymers or insoluble solids can be expected to exhibit Newtonian behavior.

A small amount ($\sim 1\%$) of a dissolved polymer can substantially increase the viscosity of water and also alter the flow characteristics from Newtonian of water to non-Newtonian of the aqueous dispersion. Whereas the rheological properties of a fluid food are altered substantially, magnitudes of the thermal properties (e.g., density, heat capacity, and thermal conductivity) of the dispersion can be predicted from those of water.

Examples of application of shear rheological parameters, such as apparent viscosity and power law parameters, in food processing include[1]: estimation of friction loss in tubes, power requirement for pumping in flow and mixing systems, characterization of residence time distribution, and in correlation of heat transfer data and simulation of heat transfer phenomenon.

Shear flow properties of fluid foods are based on shear rate and shear stress data. Shear rate, denoted by the symbol $\dot{\gamma}$ is the velocity gradient established in a fluid as a result of a shear stress acting on it. It is expressed in units of reciprocal seconds, $\sec^{-1}$. Shear stress, denoted by the symbol σ, is the stress component applied tangentially; it is equal to the force vector (a vector has both magnitude and direction) divided by the area of application and is expressed in units of force per unit area, pascal (Pa).

Viscosity is the internal friction of a fluid or its tendency to resist flow. It is denoted by the symbol η for Newtonian fluids, whose viscosity does not depend on the shear rate, and for non-Newtonian fluids to indicate shear rate dependence by η_a, defined as:

$$\eta_a = \frac{\text{shear stress}}{\text{shear rate}} = \frac{\sigma}{\dot{\gamma}} \tag{1}$$

The preferred units of viscosity are Pa sec or mPa sec, but other units such as centipoise (cp) (1 cp = 1 m Pa sec) can be found in the literature.

NEWTONIAN AND NON-NEWTONIAN FOODS

Newtonian Foods

A plot of shear rate vs. shear stress can be used to classify flow behavior of foods (Fig. 1). With Newtonian fluids, the shear rate is directly proportional to the shear stress and the plot begins at the origin. Typical Newtonian foods are those containing only low-molecular weight compounds (e.g., water and sugar) and that do not contain large concentrations of either dissolved polymers (e.g., pectins, proteins, and starches) or insoluble solids (e.g., insoluble fiber): water, sugar syrups, most honeys, most carbonated beverages, edible oils, filtered juices, and pasteurized milk are Newtonian fluids.

All other types of fluid foods are non-Newtonian which means that either the shear stress–shear rate plot is not linear and/or the plot does not begin at the origin, or the material exhibits time-dependent rheological behavior as a result of structural changes. Flow behavior may depend only on shear rate and not on the duration of shear (time-independent) or may also depend on the duration of shear (time-dependent). Several types of time-independent flow behavior of foods have been encountered.

Shear-Thinning Foods

With shear-thinning fluids, popularly called pseudoplastic, the curve begins at the origin of the shear stress–shear rate plot but is concave upwards, i.e., an increasing shear rate gives a less than proportional increase in shear stress. Shear-thinning is due to the breakdown of structural units in a food due to the hydrodynamic forces generated during shear. Most non-Newtonian foods exhibit shear thinning behavior, including mayonnaises, salad dressings, concentrated fruit juices, and pureed fruits and vegetables.

Yield Stress

The flow of some materials may not commence until a threshold value of stress, called the yield stress (σ_0) is exceeded. Shear-thinning with yield stress behavior is exhibited by foods, such as tomato concentrates, tomato ketchup, mustard, and mayonnaise.

Encyclopedia of Agricultural, Food, and Biological Engineering
DOI: 10.1081/E-EAFE 120006948
Copyright © 2003 by Marcel Dekker, Inc. All rights reserved.

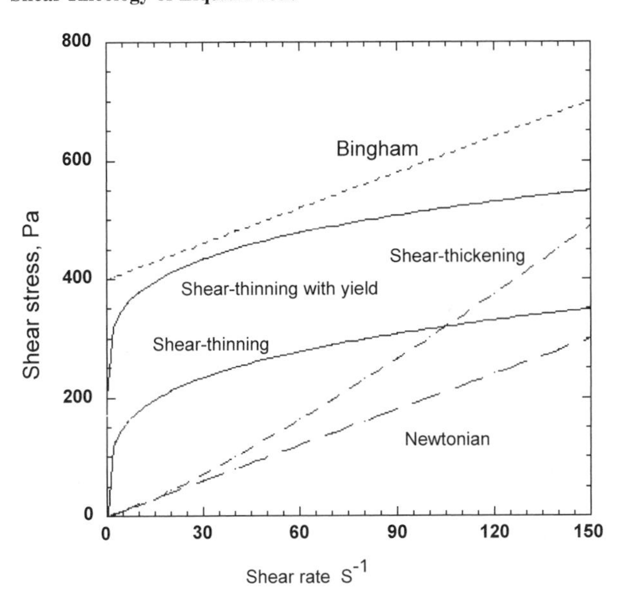

Fig. 1 Shear rate vs. shear stress data is used to describe flow behavior of foods. Most foods are shear-thinning in nature with or without a yield stress.

Shear-Thickening Foods

In shear-thickening behavior also, the curve begins at the origin of the shear stress–shear rate plot and is concave downwards, i.e., an increasing shear stress gives a less than proportional increase in shear rate. This type of flow has been encountered in partially gelatinized starch dispersions. The expression dilatant is popularly and incorrectly used to describe shear-thickening. Because dilatancy implies an increase in the volume of the sample during the test, it is incorrect to use it to describe shear-thickening rheological behavior.

MODELS FOR TIME-INDEPENDENT FLOW BEHAVIOR

A flow model may be considered to be a mathematical equation that can describe rheological data, such as shear rate–shear stress data in a basic shear diagram. In some instances, such as for the viscosity vs. temperature data during starch gelatinization, more than one equation may be necessary to describe the rheological data.

Newtonian Model

The model for a Newtonian fluid is described by the equation:

$$\sigma = \eta\dot{\gamma} \quad (2)$$

From the definition of a Newtonian fluid, the shear stress and the shear rate are proportional to each other, and a single parameter η characterizes the data. For a Bingham plastic fluid that exhibits a yield stress (σ_0), the model is:

$$\sigma - \sigma_0 = \eta_B\dot{\gamma} \quad (3)$$

In Eq. 3, η_B is called the Bingham plastic viscosity.

The Newtonian model and the Bingham plastic model can be illustrated by straight lines in terms of shear rate and shear stress (Fig. 1), and the former can be described by one parameter η and the latter by two parameters: η' and σ_0, respectively. However, the shear rate–shear stress data of shear-thinning and shear-thickening fluids are curves that require more than one parameter to describe their data.

Power Law Model

Shear stress–shear rate plots of many fluids become linear when plotted on double logarithmic coordinates and the power law model describes the data of shear-thinning and shear thickening fluids.

$$\sigma = K\dot{\gamma}^n \quad (4)$$

In Eq. 4, K is the consistency coefficient (Pa sec^n) and is the shear stress at a shear rate of 1.0 sec^{-1}, and the exponent n is the flow behavior index is dimensionless that also reflects the closeness to Newtonian flow. For the special case of a Newtonian fluid ($n = 1$), the consistency coefficient K is identically equal to the viscosity of the fluid. When the magnitude of $n < 1$ the fluid is shear-thinning and when $n > 1$ the fluid is shear-thickening in nature.

Many models have been developed to describe the shear rate vs. the shear stress data.[1] Because it contains only the two parameters K and n, the power law model has been used extensively to characterize fluid foods, and in studies on handling and heating/cooling of foods. However, it is an empirical relationship that may be applicable over one or two decades range of shear rates. Extensive compilations of the magnitudes of power law parameters can be found in Refs.[1,2] Rao[1] compiled the magnitudes of power law parameters of food commodities. In addition, the influence of temperature in quantitative terms of activation energies, and the effect of concentration of soluble and insoluble solids on the consistency coefficient are given.

Taking logarithms of both sides of Eq. 4:

$$\log \sigma = \log K + n \log \dot{\gamma} \quad (5)$$

The parameters K and n are determined from a plot of $\log \sigma$ vs. $\log \dot{\gamma}$, and the resulting straight line's intercept is $\log K$ and the slope is n.[1] Although the power law model is popular and useful, it is an empirical relationship. Another reason for its popularity is due to its applicability over the shear rate range: $10^1-10^4\,\text{sec}^{-1}$ that can be obtained in many commercial viscometers.

Herschel–Bulkley Model

When yield stress of a food is measurable, it can be included in the power law model and the model is known as the Herschel–Bulkley model:

$$\sigma - \sigma_0 = K_H \dot{\gamma}^{n_H} \tag{6}$$

In Eq. 6, $\dot{\gamma}$ is shear rate (sec^{-1}), σ is shear stress (Pa), n_H is the flow behavior index, K_H is the consistency coefficient, and σ_0 is yield stress. If the yield stress of a sample is known from an independent experiment, K_H and n_H can be determined from linear regression of $log(\sigma - \sigma_0)$ vs. log ($\dot{\gamma}$) as the intercept and slope, respectively. Values of yield stress estimated using a flow model often do not agree with those determined experimentally, e.g., using the vane method.

Casson Model

The Casson[3] model (Eq. 7) is a structure-based model that, although was developed for characterizing printing inks originally, has been used for a number of food dispersions.

$$\sigma^{0.5} = K_{0c} + K_c(\dot{\gamma})^{0.5} \tag{7}$$

For a food following the Casson model, a straight line results when the square root of shear rate, $(\dot{\gamma})^{0.5}$ is plotted against the square root of shear stress, $(\sigma)^{0.5}$ with slope K_c and intercept K_{0c}. The Casson yield stress is calculated as the square of the intercept, $\sigma_{0c} = (K_{0c})^2$ and the Casson plastic viscosity as the square of the slope, $\eta_{Ca} = (K_c)^2$. The International Office of Cocoa and Chocolate has adopted the Casson model as the official method for interpretation of flow data on chocolate samples.

One draw back of the power law model is that it does not describe the low-shear and high-shear rate constant-viscosity data of shear-thinning foods where limiting zero-shear (η_0) and infinite-shear viscosities (η_∞) are reached, respectively. The models of Cross and Carreau are capable of describing apparent viscosity vs. shear rate data over the entire ranges of data obtained on food polymer dispersions.[1]

TIME-DEPENDENT RHEOLOGICAL BEHAVIOR

Foods that exhibit time-dependent shear-thinning behavior are said to exhibit thixotropic flow behavior. Most of the foods that exhibit thixotropic behavior are heterogeneous systems containing small solid particles or droplets. When the fluid is at rest, the particles or molecules in the food are linked together by weak forces. When the hydrodynamic forces during shear are sufficiently high, the inter-particle linkages are broken resulting in reduction in the size of the structural units that in turn offer lower resistance to flow during shear. This type of behavior is common to foods such as salad dressings and soft cheeses where the structural adjustments take place in the food due to shear until an equilibrium is reached.

Time-dependent shear-thickening behavior is called antithixotropic behavior, formerly called rheopectic behavior. The primary difficulty in obtaining reliable thixotropic or antithixotropic data is that during loading of the test sample in to a measurement geometry structural changes occur that cannot be either controlled or expressed quantitatively.[1]

EFFECT OF CONCENTRATION ON VISCOSITY

In most fluid foods, often it is possible to identify the component(s), called key component(s) that play an important role in the rheological properties.[1] The effect of concentration (c) of soluble or insoluble solids on either apparent viscosity (η_a) or the consistency coefficient of the power law model (K) can be described by either exponential or power law relationships:

$$\eta_a \propto \exp(ac) \tag{8}$$

$$\eta_a \propto c^b \tag{9}$$

$$K \propto \exp(a'c) \tag{10}$$

$$K \propto c^{b'} \tag{11}$$

In Eqs. 8–11, a, a', b, and b' are constants to be determined from experimental data. For example, the effect of concentration (c) of soluble solids (°Brix) and insoluble solids (pulp) on either apparent viscosity or the consistency coefficient of the power law model of concentrated orange juice can be described by exponential relationships.[4]

EFFECT OF TEMPERATURE ON VISCOSITY

A wide range of temperatures is encountered during processing and storage of fluid foods, so that the effect

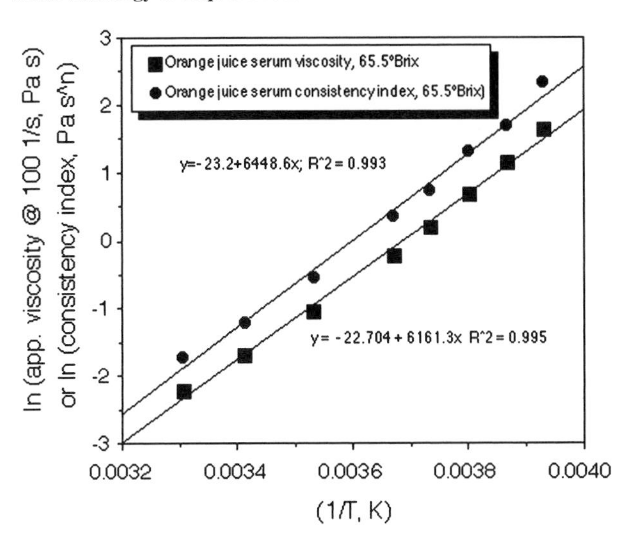

Fig. 2 An Arrhenius-type model can be used to describe the apparent viscosity and power law consistency coefficient data as a function of temperature data.

of temperature on rheological properties needs to be described quantitatively. The effect of temperature on the viscosity of Newtonian fluids, or on either the apparent viscosity at a specified shear rate (Eq. 12) or the consistency coefficient, K of the power law model (Eq. 13) of a fluid can be described often by an Arrhenius type relationship. For the apparent viscosity, the relationship is:

$$\eta_a = \eta_{\infty A} \exp(E_a/RT) \quad (12)$$

In Eq. 12, η_a is the apparent viscosity at a specific shear rate, $\eta_{\infty A}$ is the frequency factor, E_a is the activation energy ($\mathrm{J\,mol^{-1}}$), R is the gas constant ($\mathrm{J\,mol^{-1}\,K^{-1}}$), and T is temperature (K).

From the slope of a plot of $\ln \eta_a$ (ordinate) vs. $(1/T)$ (abscissa), E_a can be calculated as: *slope* $\times R$); we note that $\eta_{\infty A}$ is exponential of the intercept. The Arrhenius equation for the consistency coefficient is:

$$K = K_\infty \exp(E_{ak}/RT) \quad (13)$$

In Eq. 13, K_∞ is the frequency factor, E_{ak} is the activation energy ($\mathrm{J\,mol^{-1}}$), R is the gas constant, and T is temperature (K). A plot of $\ln K$ (ordinate) vs. $(1/T)$ (abscissa) results in a straight line, and $E_{ak} = (\text{slope} \times R)$, and K_∞ is exponential of the intercept. Applicability of the Arrhenius model to the apparent viscosity vs. temperature data on a concentrated orange juice serum sample[4] is shown in Fig. 2.

CONCLUSION

Most fluid foods are non-Newtonian in nature. Their rheological behavior depends on several factors, such as: rate of shear and temperature. A few rheological models were presented, but others are also being used. For any food, it would be desirable to understand the role of composition and structure on rheological behavior.

REFERENCES

1. Rao, M.A. *Rheology of Fluid and Semisolid Foods: Principles and Applications*; Kluwer Academic/Plenum Publishers: New York, 1999.
2. Holdsworth, S.D. Rheological Models Used for the Prediction of the Flow Properties of Food Products: A Literature Review. Trans. Inst. Chem. Eng. **1993**, *71* (Part C), 139–179.
3. Casson, N. A Flow Equation for Pigment-Oil Suspensions of the Printing Ink Type. In *Rheology of Disperse Systems*; Mill, C.C., Ed.; Pergamon Press: New York, 1959; 82–104.
4. Vitali, A.A.; Rao, M.A. Flow Properties of Low-Pulp Concentrated Orange Juice: Effect of Temperature and Concentration. J. Food Sci. **1984**, *49*, 882–888.

Shear Viscosity Measurement

M. A. Rao
Cornell University, Geneva, New York, U.S.A.

INTRODUCTION

A well-designed viscometer must be capable of providing readings that can be converted to shear rate ($\dot{\gamma}$, $\sec^{-1}$) and shear stress (σ, Pa), and allow for the recording of the readings so that time-dependent flow behavior can be studied. In the case of low-viscosity foods, the shear stresses will be low in magnitude so that instruments that minimize friction by the use of air bearings are preferable. The flow in the selected geometry should be steady, laminar and fully developed, and the temperature of the test fluid should be maintained uniform.[1]

VISCOMETER GEOMETRIES

For foods that exhibit Newtonian behavior, viscometers that operate at a single average shear rate, such as a glass capillary, are acceptable. A glass capillary dilution (Ubbelohde) viscometer is shown in Fig. 1. This viscometer is especially useful for determining the intrinsic viscosity of a food polymer:[1] 1) A measured volume is loaded through tube G into the reservoir J; 2) The viscometer is placed in a vertical orientation in a constant temperature bath; 3) Tube B is closed with a finger and suction is applied at A until the test fluid reaches the center of bulb C, suction at A is removed, and the finger from tube B is removed and immediately placed over tube A; 4) After removing finger from Tube A, the efflux time, t for the liquid meniscus to pass from D to mark F is measured to within 0.1 sec. To ensure reproducible results, steps 3 and 4 should be repeated.

The efflux time, t, for the liquid to pass from mark D to mark F is noted and used to calculate the viscosity using calibration constants provided by the manufacturer. Alternatively, the elapsed times for a test liquid (t) and a standard liquid (t_{st}) of the same density to flow between two etched lines are determined. Because the magnitude of viscosity is directly proportional to the time of flow, the viscosity of a fluid can be calculated as:

$$\eta = (t/t_{st})\eta_{st} \qquad (1)$$

Glass capillary viscometers are not suitable for liquids that deviate substantially from Newtonian flow or contain large particles or a high concentration of suspended solids.

For foods that exhibit non-Newtonian behavior, data should be obtained at several shear rates and the commonly used viscometric flow geometries in rheological studies on foods include: concentric cylinder (Fig. 2), plate and cone (cone–plate) (Fig. 3), parallel disc (also called parallel plate), capillary/tube/pipe, and slit flow. Derivation of the relationships for shear stress and shear rate can be found elsewhere.[1,2]

Modern commercial viscometers/rheometers are automated so that values of the shear rate and shear stress are computed by a computer. In a properly designed cone–plate geometry the shear rate depends only on the rotational speed and not on the geometrical characteristics. In all other flow geometries, the dimensions of the measuring geometry (capillary, concentric cylinder, parallel discs) play important roles.[1] Because the concentric cylinder and the plate–cone geometries are encountered commonly, the equations for shear stress and Newtonian shear rate are given.

Concentric Cylinder Viscometer

In a concentric cylinder geometry, the shear stress can be determined from the total torque (M) due to the test fluid:

$$\sigma = \frac{M}{2\pi r_i^2 h} \qquad (2)$$

In Eq. 2, r_i is the radius and h, the height of the rotating bob. The Newtonian shear rate in a concentric cylinder geometry can be calculated exactly from the expression:

$$\dot{\gamma}_N = \frac{2\Omega}{\left[1 - \left(\frac{r_i}{r_o}\right)^2\right]} \qquad (3)$$

In Eq. 3, Ω is the angular velocity of the rotating bob; r_i, the radius of the bob; and r_o, the radius of the cup. The viscosity of a Newtonian fluid from concentric

Encyclopedia of Agricultural, Food, and Biological Engineering
DOI: 10.1081/E-EAFE 120005736
Copyright © 2003 by Marcel Dekker, Inc. All rights reserved.

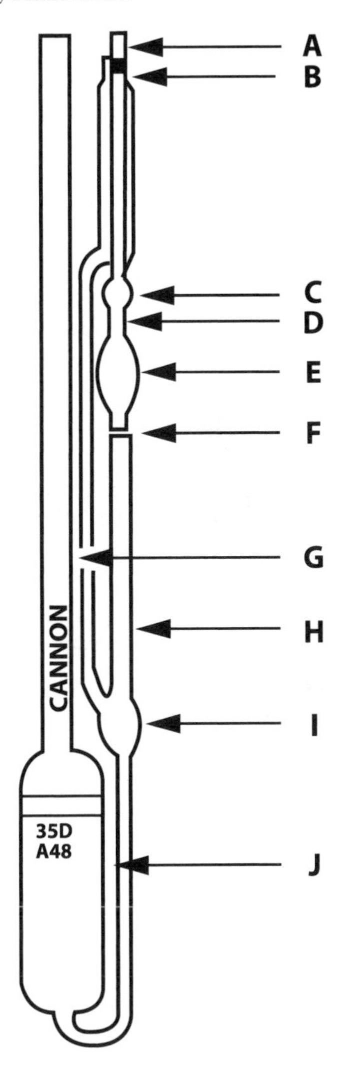

Fig. 1 Schematic diagram of a glass capillary dilution (Ubbelohde) viscometer; role of identified sections is given in the text.

cylinder flow data is given by the Margules equation:

$$\eta = \left(\frac{M}{4\pi h \Omega}\right)\left(\frac{1}{r_i^2} - \frac{1}{r_o^2}\right) \tag{4}$$

Cone–Plate Viscometer

The shear rate and shear stress are given by the following equations:

$$\text{Shear stress, } \sigma = \frac{3T_c}{D} \tag{5}$$

$$\text{Shear rate, } \dot{\gamma} = \frac{\Omega}{\theta_o} \tag{6}$$

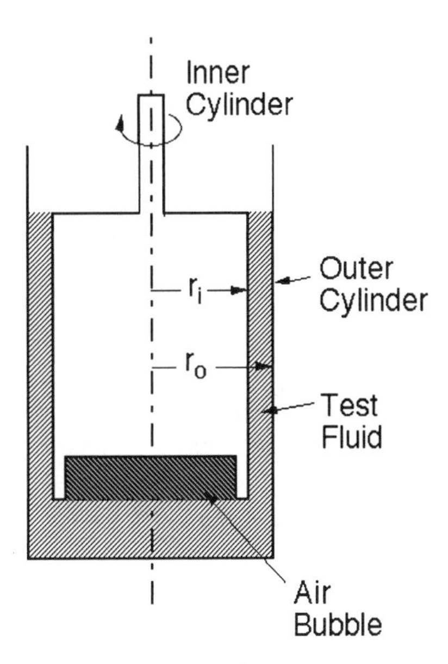

Fig. 2 Schematic diagram of a concentric cylinder viscometer; the bottom of the rotating cylinder is hollowed out to trap air and minimize friction.

In Eqs. 5 and 6, T_c is the torque per unit area (M area^{-1}); D, the diameter of the rotating cone or plate; Ω, the angular velocity; and θ_o, the cone angle in radians. In general, θ_o is usually quite small (2°–4°).

Mixer Viscometer

Some rotational viscometers employ a rotating disc, bar, paddle, or pin at a constant speed (or series of constant speeds). In such geometries, it is difficult to obtain true shear stress, and the shear rate usually varies from point to point in the rotating member. In particular, the velocity field of a rotating disc geometry can be considerably distorted in viscoelastic fluids. Nevertheless, because they are simple to operate and give results easily, and their cost

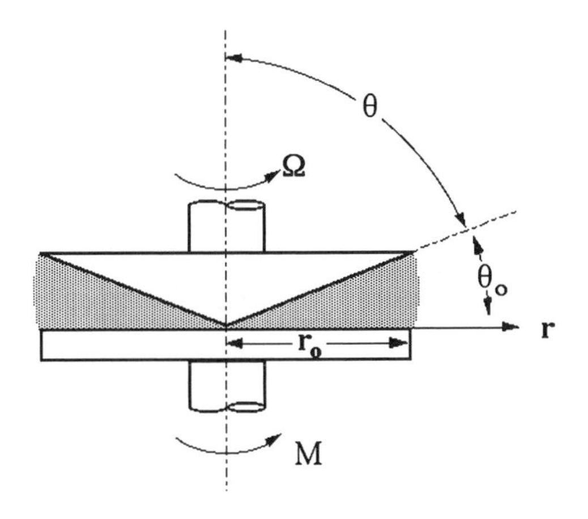

Fig. 3 Schematic diagram of a cone and plate geometry.

is low, they are widely used in the food industry. However, reliability of the data obtained with them should be verified by comparison with data obtained using well-defined geometries (capillary/tube, concentric cylinder, and cone–plate).

Many foods are suspensions of solid matter in a continuous medium. For these and other complex foods, approximate shear rate–shear stress data can be obtained by a technique using a rotating vane[1] that minimizes settling and separation of the product. The technique was developed for the determination of power consumption during mixing of fluids[3] and adapted to studying rheology of fermentation broths.[4] It is based on the assumption that the average shear rate around the paddle is directly proportional to the rotational speed (N).

MEASUREMENT OF YIELD STRESS OF FOODS WITH A VANE

Yield stress is an important rheological property of foods such as mustard, tomato ketchup, tomato sauce, and melted chocolate. A sample with undisturbed structure has a high yield stress, called static yield stress. In contrast, a sample whose structure has been disturbed by shear has a lower magnitude of yield stress, called dynamic yield stress. The shearing of a sample may take place during loading of the sample in to a measuring geometry and collection of rheological data.

A technique using a vane with at least four blades, such as that illustrated in Fig. 4, developed by Dzuy Nguyen and Boger,[5] was employed to examine yield stress of foods.[6,7] In this technique, the maximum torque reading is determined at a low rotational speed (e.g., 0.4 rpm) using a controlled-strain viscometer as illustrated in Fig. 5. The yield stress σ_{0v} is then calculated from the equation:

$$T_m = \frac{\pi D_v^3}{2}\left(\frac{H}{D_v} + \frac{1}{3}\right)\sigma_{0v} \qquad (7)$$

In Eq. 7, T_m is the maximum torque reading (N m), and D_v (m) is the diameter and H (m) the height of the paddle, respectively. Consistent results were obtained on yield stress of pureed fruits and vegetables.[1]

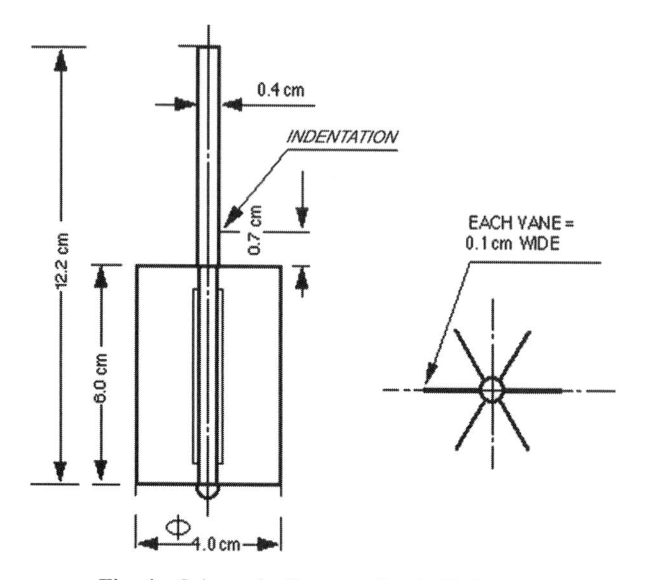

Fig. 4 Schematic diagram of a six-blade vane.

MEASUREMENT OF VISCOELASTIC BEHAVIOR OF FLUID FOODS

Viscoelastic behavior of many foods has been studied by means of dynamic shear, creep-compliance, and stress relaxation techniques. It is customary to employ different symbols for the various rheological parameters in different types of deformation: shear, bulk, or simple, and the symbols employed for the three types of deformation are given in Table 1.[8] The paradigm shift from emphasis on normal stress measurements to dynamic rheological technique took place gradually beginning in the late 1950s. Here, only the dynamic rheological technique will be discussed. Discussion of creep-compliance, and stress relaxation techniques can be found elsewhere.[1,2,8]

Oscillatory Shear Flow

Small amplitude oscillatory shear (SAOS), also called dynamic rheological experiment, can be used to determine viscoelastic properties of foods. The experiments can be conducted with the couette, plate and cone, and parallel plate geometries. In the SAOS tests, a food sample is subjected to a small sinusoidal strain or deformation $\gamma(t)$ at time t and the phase difference between the oscillating stress and strain as well as the amplitude ratio are measured. The information obtained should be equivalent to data from a transient experiment at time $t = \omega^{-1}$.

$$\gamma(t) = \gamma_0 \sin(\omega t) \qquad (8)$$

Expressions can be derived for the storage modulus, G' (Pa) and loss modulus, G'' (Pa), and the loss tangent, $\tan\delta = (G''/G')$. The storage modulus G' expresses the magnitude of the energy that is stored in the material or recoverable per cycle of deformation; G'' is a measure of the energy that is lost as viscous dissipation per cycle of deformation. Therefore, for a perfectly elastic solid, all the energy is stored, i.e., G'' is zero and the stress and the strain will be in phase. In contrast, for a liquid with no

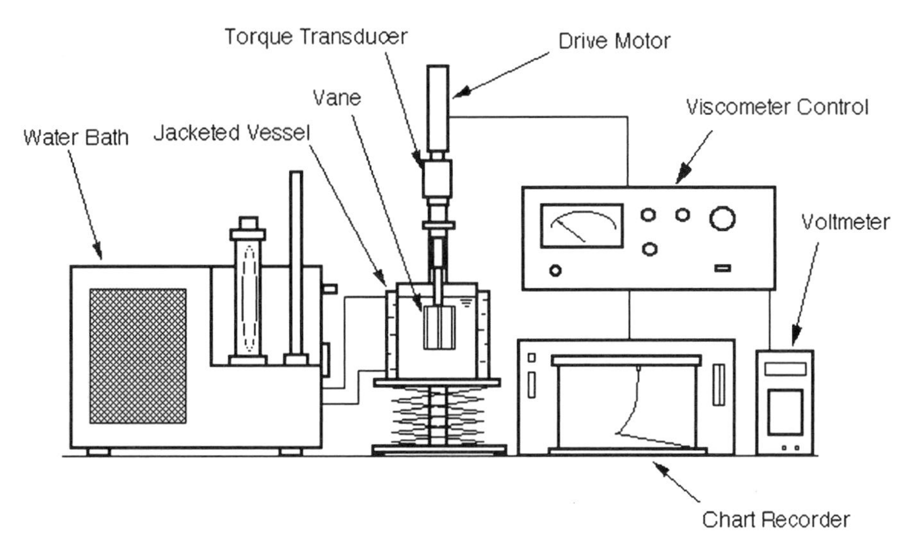

Fig. 5 Experimental set up used to determine yield stress of apple sauce. (From Ref. 6.)

elastic properties all the energy is dissipated as heat, i.e., G' is zero and the stress and the strain will be out of phase by 90° (Fig. 6).

For a specific food, magnitudes of G' and G'' are influenced by frequency, temperature, and strain. For strain values within the linear range of deformation, G' and G'' are independent of strain. The loss tangent is the ratio of the energy dissipated to that stored per cycle of deformation. These viscoelastic functions have been found to play important roles in the rheology of structured polysaccharides and proteins, and processed foods.

Types of Dynamic Rheological Tests

Dynamic rheological experiments provide suitable means for monitoring the gelation process of many biopolymers and for obtaining insight into gel/food structure because they satisfy several conditions[1]: they are nondestructive and do not interfere with either gel formation or softening of a structure, the time involved in the measurements is short relative to the characteristic times of the gelation and softening processes, and the results are expressible in fundamental terms so that they can be related to the structure of the network.[9] For the same reasons, they were used to follow the changes induced on potato cells by added cellulase[10] and α-amylase on wheat starch.[11]

A strain sweep in which the strain is varied over a range of values is an essential test to determine the linear viscoelastic range. With controlled stress rheometers, a stress sweep can be conducted. The limit of linearity can be detected when dynamic rheological properties (e.g., G' and G'') change rapidly from their almost constant values. Three other types of dynamic rheological tests can be conducted to obtain useful properties of viscoelastic foods, such as gels, and of gelation and melting[1]: Frequency sweep studies in which G' and G'' are determined as a function of frequency (ω) at a fixed temperature. Temperature sweep studies in which G' and G'' are determined as a function of temperature at fixed ω. This test is well suited for studying gel formation during cooling of a heated polymer (e.g., pectin, gelatin)

Table 1 Symbols for viscoelastic parameters from shear, simple extension, and bulk compression

Parameter	Shear	Simple extension	Bulk compression
Stress relaxation modulus	$G(t)$	$E(t)$	$K(t)$
Creep compliance	$J(t)$	$D(t)$	$B(t)$
Storage modulus	$G'(\omega)$	$E'(\omega)$	$K'(\omega)$
Loss modulus	$G''(\omega)$	$E''(\omega)$	$K''(\omega)$
Complex modulus	$G^*(\omega)$	$E^*(\omega)$	$K^*(\omega)$
Dynamic viscosity	$\eta'(\omega)$	$\eta'_e(\omega)$	$\eta'_v(\omega)$
Complex viscosity	$\eta^*(\omega)$	$\eta^*_e(\omega)$	$\eta^*_v(\omega)$

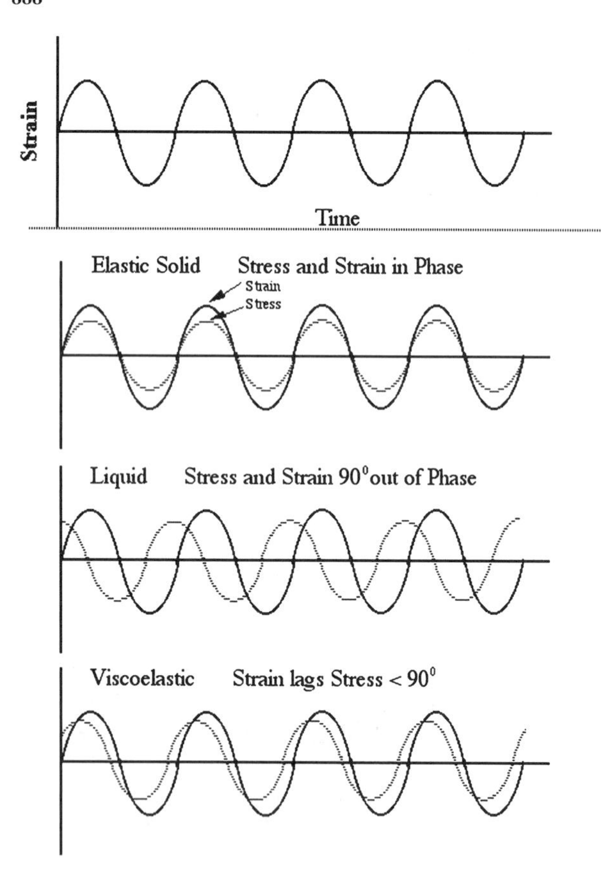

Fig. 6 Dynamic rheological test: applied sinusoidal strain and the stress responses of an ideal solid, a viscous fluid, and a viscoelastic fluid, respectively.

dispersion, and gelatinization of a starch dispersion during heating and gel formation of proteins. Time sweep study in which G' and G'' are determined as a function of time at fixed ω and temperature. This type of test, often called a gel cure experiment, is well suited for studying structure development in physical gels.

CONCLUSION

Rheological data on foods are best obtained using well-defined geometries, including capillary, concentric cylinder, and cone and plate. However, the presence of large solids in and the heterogeneous nature of some foods may preclude use of well-defined geometries necessitating approximate measurement methods. Other concerns include, wall slip in some foods and dehydration of samples during measurement. The vane method has been found to be well suited for measurement of yield stress.

REFERENCES

1. Rao, M.A. *Rheology of Fluid and Semisolid Foods: Principles and Applications*; Kluwer Academic/Plenum Publishers: New York, 1999.
2. Steffe, J.F. *Rheological Methods in Food Process Engineering*; Freeman Press: East Lansing, MI, 1996.
3. Metzner, A.B.; Otto, R.E. Agitation of Non-Newtonian Fluids. Am. Inst. Chem. Eng. J. **1957**, *3*, 3–10.
4. Bongenaar, J.J.T.; Kossen, N.W.F.; Metz, B.; Meijboom, F.W. A Method for Characterizing the Rheological Properties of Viscous Fermentation Broths. Biotech. Bioeng. **1973**, *15*, 201–206.
5. Dzuy, N.Q.; Boger, D.V. Direct Yield Stress Measurement with the Vane Method. J. Rheol. **1985**, *29*, 335–347.
6. Qiu, C.-G.; Rao, M.A. Role of Pulp Content and Particle Size in Yield Stress of Apple Sauce. J. Food Sci. **1988**, *53*, 1165–1170.
7. Yoo, B.; Rao, M.A.; Steffe, J.F. Yield Stress of Food Suspensions with the Vane Method at Controlled Shear Rate and Shear Stress. J. Texture Stud. **1995**, *26*, 1–10.
8. Ferry, J.D. *Viscoelastic Properties of Polymers*; John Wiley: New York, 1980.
9. Clark, A.H.; Ross-Murphy, S.B. Structural and Mechanical Properties of Biopolymer Gels. Adv. Polym. Sci. **1987**, *83*, 57–192.
10. Shomer, I.; Rao, M.A.; Bourne, M.C.; Levy, D. Rheological Behaviour of Potato Tuber Cell Suspensions During Temperature Fluctuations and Cellulase Treatments. J. Sci. Food Agric. **1993**, *63*, 245–250.
11. Champenois, Y.C.; Rao, M.A.; Walker, L.P. Influence of α-Amylase of the Viscoelastic Properties of Starch–Gluten Pastes and Gels. J. Sci. Food Agric. **1998**, *78*, 127–133.

Sheep Production Systems

Harvey J. Hirning
North Dakota State University, Fargo, North Dakota, U.S.A.

S

INTRODUCTION

Sheep and goats are produced throughout all of the major landmasses of the world. Because they are of similar size, most of the facilities for the two species are very similar.

Sheep may be raised for meat, milk, or wool while goats are typically raised for their meat or milk, although some breeds are raised for the hair or wool. Facilities may be very simple to very complex depending upon the environmental conditions encountered. In moderate climates where pastures are adequate year round, the only facilities used may be watering troughs or pails. In severe climates they may be raised in totally enclosed environmentally controlled buildings.

PASTURE PRODUCTION

In areas where there is adequate pasture land available, sheep may be raised in flocks tended by a shepherd who is responsible for the safety of the flock and to insure that they are not overgrazing any part of the range. Lambing is done in the pasture.

If pastureland is limited, the sheep may be raised in fenced pastures to limit the space required. Temporary fences may be used to insure that the sheep move to new grass on a regular basis. Fencing materials depend upon what materials are locally available. Producers may use stone, wood, barbed wire, woven wire, or combinations of materials. Fences must be high enough to keep adult sheep from jumping over the top (about 1.2 m) with openings small enough to keep lambs from escaping (about 17 cm by 17 cm). If the fence is to keep predators out, it must be high enough to keep the predators from jumping over the top (about 1.8 m for dogs, coyotes, and fox) with provisions to keep them from climbing the fence or digging under the fence. An electrified wire near ground level is often adequate to prevent digging under the fence, while a similar wire at the top will prevent them from climbing the fence.

An electrified fence can be used to confine sheep into pastures, however because of the high electrical resistance of the wool, some provision must be made to insure that the sheep will receive an adequate shock. This can be accomplished with specially designed wire that carries both an energized and a ground wire in the same strand or by alternating energized and ground wires in a 5 or 7 strand fence. Sheep should be given time to become acclimated to an electric fence before trying to limit them to a small space.

WATERING FACILITIES

Basic watering facilities include anything that will contain a small quantity of water such as a pail or trough. In areas where water supplies are scarce, a well with a storage tank may be needed, while in other areas, naturally occurring water such as streams or impoundments may be adequate.

Areas with significant time periods with below freezing temperatures may require that water be supplied in heated waterers. Commercial water fountains with heaters built into them are readily available. Proper wiring procedures must be followed so that the sheep do not receive an electrical shock when drinking.

LAMBING FACILITIES

If lambing occurs during mild weather (temperatures above 5°C) with little rain or wind, lambing may be successful under pasture conditions. Producers who desire to lamb during winter conditions or who are trying to get three batches of lambs in a two-year period may need to provide some environmental modification during the lambing period. For small flocks this is frequently a space in a building that is normally used for other purposes for most of the year. A space about 1.2 m by 1.2 m is adequate for a ewe and her lambs up to a week old.

When flocks are larger than 50 ewes, a special lambing shed is usually used. The lambing pens may be removed after lambing and the space is used for storage or shearing.

ELECTRIFIED FENCES

The high resistance of wool to the flow of electric current has limited the effectiveness of single strand electric fence. Sheep require a current flow of greater than 5 mA to elicit and aversion response.

Encyclopedia of Agricultural, Food, and Biological Engineering
DOI: 10.1081/E-EAFE 120006917
Copyright © 2003 by Marcel Dekker, Inc. All rights reserved.

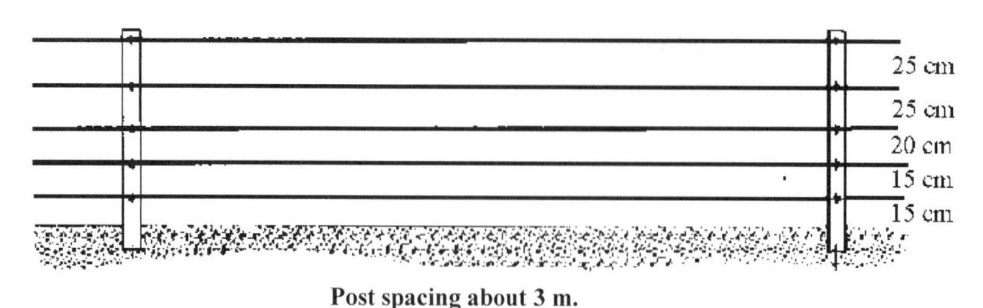

Fig. 1 A five wire electrified fence for sheep pastures. The top, bottom, and middle wires are energized. The intermediate wires are grounded. ©Eolss Publishers 2001 from Hirning, Harvey J., Encyclopedia of Life Support Systems, 2001, with permission from Eolss Publishers.

Fencing with 5 to 8 strands of high tensile wire is excellent for a sheep pasture. A 1.3 m high fence works well to keep sheep in and many predators out. The bottom wire should be charged and placed about 15 cm above the ground. Alternate electrified grounded wires spaced about 15 cm apart with the top wire at least 1.2 m above the earth (Fig. 1).

Wire should be at least 14 gauge. Heavier wire will conduct the voltage better and will also be stronger. Sheep must become accustomed to the electric fence. Freshly shorn sheep will have less wool for insulation and therefore more likely to experience the shock effect when exposed to the fence. It helps if the sheep are allowed to discover the electric fence on their own such as in a large enclosed area rather than being crowded against the fence.

A multiwire fence will teach the sheep to respect the fence more quickly than a single or two wire fence. Most animals can be trained in 48 hr. After the sheep have learned to respect an electric fence, they can be controlled in temporary enclosures that have only two or three wires. A two-wire system requires that the lower wire be about 25 cm above the ground and the upper wire 50 cm above the ground.

CORRAL FENCES

Corral fences are exposed to considerable pressure from the sheep. Post spacing is usually less than 3 m. The actual fencing material may be wood, prefabricated metal panels, or woven wire. Lambs will be able to escape if openings are larger than 15 cm. An adult with a lamb requires about 2 m^2 of space.

Wooden fences are typically made from 4 cm by 14 cm boards. Spacing between boards is 10 cm between the earth and lowest board, then a 15 cm opening. Other boards may be placed at increased spacing, but the upper board should be no more than 25 cm above the board below.

WATERING TROUGHS

Many types of containers can be used to supply water for the sheep. A simple trough is adequate in moderate and warm climates. In areas that frequently have temperatures colder than 0°C, heated waterers are needed. Many commercial units are available in area where heated waterers are needed. A concrete apron around the waterer will help prevent the development of a mud-hole around the waterer. Sloping the concrete 1 cm per 10 cm of width will help prevent the buildup of manure and ice on the concrete apron. Typical water requirements in a temperate climate will vary from 0.4 L per animal per day to 11 L per animal per day. Water intake is greatly influenced by season, with water intake in the summer as much as 12 times higher than that during the winter (Table 1).[1]

A water storage system can be accomplished with ground level tanks, elevated tanks, or earthen storage basins (dugouts). High quality water is particularly important to young lambs, however, all sheep will respond positively to high quality water. Brackish or water that is too hot or cold will cause sheep to reduce their daily water intake.

Several waterers are more desirable than one large waterer. The waterers should be located so that they are convenient to the sheep.

A pipe watering trough can be used to water several pens or large groups of sheep at one time. A 15 cm plastic

Table 1 Typical daily water requirements for sheep in L $animal^{-1}$ day^{-1}

Animal	L day^{-1}
Ram	7.5
Dry ewes	7.5
Ewes with lambs	11
2 kg–9 kg lambs	0.4–1.2
Feeder lambs	6

pipe with opening cut into the top works well and each opening is 10 cm–12 cm wide and about 20 cm–25 cm long. A slight slope to the pipe (about 1 cm in 6 m) will allow the operator to drain the pipe as needed.

Best results with nipple waterers occur with one nipple at ewe height and another at lamb height. Mount the waterers at 45 cm–50 cm above the floor for lambs and 55 cm–60 cm above the floor for ewes.

FEEDING EQUIPMENT

Pails and buckets are frequently used to feed small numbers of sheep. For large numbers some type of automation can greatly reduce the hand labor required. Feed mixing wagons and feed delivery wagons make fence line feed-bunks an attractive option. Self-feeders can be used to make feed available at all times and thereby reduce the amount of feeding space needed.

Creep Feeders

A creep feeder will allow lambs to access high energy, highly palatable feed without allowing the ewe to access this special feed. Openings from 15 cm to 25 cm wide and 35 cm to 50 cm high into the enclosure are large enough to allow the lamb to enter, but will prevent the adult sheep from entering.

Allow 5 cm of feed trough space per lamb (weighing less than 30 kg) in the creep feeder. If feed is continuously available, this may be reduced to 3 cm per lamb. The bottom of the feed trough can be at ground level; however the edge of the trough should not be more than 15 cm above the ground to allow the lambs to easily reach over the edge of the trough. Placing a raised board (20 cm–30 cm) centered down the length of the trough will help prevent lambs from standing up on top of the trough (Fig. 2).

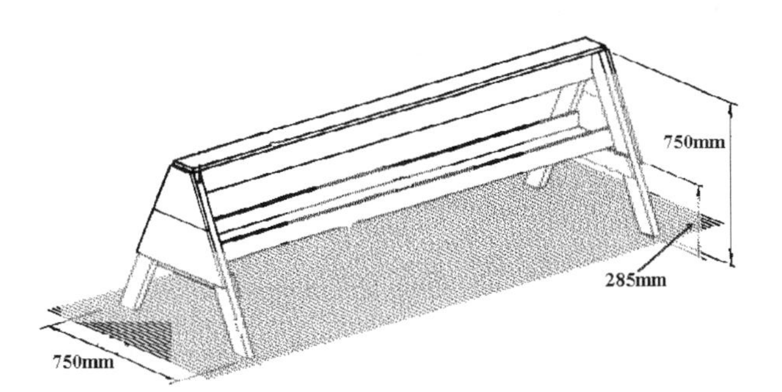

Fig. 2 A bunk for feeding grain to lambs. The center divider is used to discourage the lambs from standing in the bunk.

Bunk Feeders

Feed for finishing lambs and for confinement production operations can be offered in bunks placed within the sheep lots or along one of the fences. A fence line feed-bunk can easily be filled with a side discharge feed wagon. Bunks inside the pens are more difficult to access and are usually filled with an auger or belt type conveyor.

The floor of fence line feed bunks can be at ground level. Throat height of the bunk should not exceed 15 cm. Openings for the head can be 24 cm high, but small lambs may be able to climb into the bunk. Placing a stock panel with 10 cm by 20 cm opening over the top of the fenceline bunk will keep lambs in the pen.

Bunks need to be sturdy enough to withstand the pressure placed upon them by the sheep. While the pressure from one sheep may not seem like much, the pressure from 100 or more can destroy many fences. Concrete aprons adjacent to the feed bunk should slope away from the bunk about 1 cm vertically for each 12 cm horizontal. This will cause any manure and urine to be moved away from the bunk by hoof action and will lessen the amount being deposited into the bunk.

Allow 30 cm–35 cm of bunk space for rams and 40 cm–50 cm of space for ewes. Feeder lambs will need about 23 cm–30 cm of space, increasing as they grow. Self fed animals require only about 30%–50% of the space needed for animals that are limit fed.

Hay Feeders

Hay and forages can be fed in simple feeders made from stock panels or by placing the forage next to an opening in the fence. Opening in vertical panels needs to be about 20 cm wide to allow the sheep to get their head through the opening to feed and still restrain them so they cannot climb into the feeder.

Feeding large bales of hay will require that either the hay be moved closer to the sheep periodically or that the feeding panel be moveable so that the sheep can push it closer to the hay as they eat away from the edges.

LAMBING EQUIPMENT

In areas of the world where lambing can take place in the pasture, very little equipment is needed. However, in cold climates or during inclement weather some protection for the lambs is needed to get them dried off and suckling. An orphan adoption pen may be needed to get some ewes to accept their young or to adopt a lamb from another ewe.

Lambing pens (jugs) can be used to make it easier to tend the ewe during lambing and to improve the chances that a ewe will accept her lambs. A typical lambing pen will be about 1.25 m by 1.25 m and about 0.75 m high. Bedding can be used to provide a clean place for the ewe and lamb and to modify the temperature around the lamb. In cold climates, a small heater may be placed in one corner of the pen to dry the lamb more quickly after birth and provide increased comfort. Care must be exercised to prevent a fire from the heater unit. A better option is to remove the lamb from the ewe and dry it with a towel. If the lamb is suffering from hypothermia, it may be placed in a small heated box until normal body temperatures are restored and then returned to the ewe.

Nearly every flock of sheep will have some orphan lambs. Some adoption technique will be needed to get an ewe to accept a lamb that is not her own. This can sometimes be done with slime or wet grafting. Wet grafting is a process where some of the afterbirth is rubbed over a portion of the lamb so that the ewe will recognize the scent and accept the lamb.

If wet grafting does not work, an orphan adoption pen or crate can be used. In this pen, the ewe is placed with her head in a stanchion or headgate. The lamb is then given access to the rear of the ewe. A lamb adoption pen is about 80 cm wide and 1.25 m long. This allows the ewe to have access to water and feed. The stanchions allow the ewe to lie down or stand up as they desire, but they cannot see or smell the lambs. Modifying the smell of the lamb is not usually necessary and suckling training is seldom needed.

ARTIFICIAL REARING

In some cases there will be too many lambs for the ewes that are available. Lambs can be successfully reared on milk replacer diets. Lambs should be given dry feeds as early in their life as possible.

Feeding may be done with a bottle and nipple if there are only a few lambs. In large flocks, a large commercial feeder may be needed. One option is a 7 mm–8 mm diameter polyvinyl chloride (PVC) pipe fitted with nipples. A refrigerated bulk milk tank holds a one-day supply and a liquid level control regulates the flow into the pipe. Allow 3–5 lambs per nipple and limit the group size to about 25 lambs.

Mount the nipple just above the level of the milk in the reservoir to require the lambs to suck the milk up through the plastic tubes. When the lambs stop sucking, the milk flows out of the tube and into the reservoir.

CONFINEMENT PRODUCTION

Sheep raised in confinement are completely at the mercy of the shepherd for feed, water, manure management, environmental control, or any other needs. Feed and water can be provided in much the same manner as it is for sheep used in partial confinement systems.

Open front buildings or buildings with curtain sidewalls can be used to modify the environment to provide comfort for the sheep and shepherd. Totally enclosed and fan ventilated buildings require greater skill on the part of the operator to maintain a healthy environment. Optimum temperatures vary with the age of the sheep and length of the fleece. Freshly shorn sheep will require temperatures about 5–10°C warmer than sheep with full fleece. Adult sheep in full fleece can tolerate temperatures as low as −40°C if they are kept dry, but will perform much better if temperatures are kept above −15°C.

Newborn lambs need temperatures about 25°C until they are dry and then the temperature can be reduced to 12–20°C. Pens can be modified to provide the lambs with access to warmer temperatures than their mothers by placing baffles in the pens and adding supplemental heat in the lambs' quarters.

Some of the materials that have been used for slotted floors include: green undressed hardwood, native hardwoods, concrete, metal, and plastic. A high percentage of openings in the floor will result in cleaner fleece and floors. Flattened expanded metal X-plate, unflattened expanded metal, and expanded metal works well. Sheep are more sure footed and hooves wear more evenly on expanded metal than on the other materials. The material chosen will be a compromise between cost, desired life expectancy, labor availability, and traction.

Concrete slats about 8 cm wide need to be spaced about 1.25 cm apart. A smooth slightly crowned slat is more self-cleaning than a flat slat. However, animals will have less traction on the crowned slats.

X-plate with 11 cm long openings can damage the feet of young lambs and should only be used with larger sheep.

Unflattened 2 cm No. 9 expanded metal flooring works well, however during the expansion process some burrs are left on the material. Place older sheep on the flooring first to wear off these burrs before young lambs are placed on them.

If sheep are given a choice of a solid floor with straw bedding or wooden slats, they will spend more time on the straw than on the slats. The sheep will move around more in the straw bedded area than they will on the wooden slats (Table 2).

HANDLING FACILITIES

Hand wrestling techniques are often used to handle sheep in small flocks. However, in larger flocks this method is impractical. The use of handling facilities (corrals) greatly reduces the amount of labor, stress placed on the sheep during medical treatments and castration, and risk of injury during the handling operation.

Table 2 A summary of space requirements for sheep. ©Eolss Publishers 2001 from Hirning, Harvey J. Encyclopedia of Life Support Systems, 2001, with permission from Eolss Publishers

		Rams 80 kg–140 kg	Dry ewes, 70 kg–90 kg	Ewes + lambs	Prewean lambs, 2.5 kg–14 kg	Feeder lambs, 14 kg–50 kg
Building floor space ($m^2 hd^{-1}$)	Solid	1.8–2.7	1.1–1.5	1.4–1.8	0.15–0.18/lamb	0.7–0.9
	Slotted	1.3–1.8	0.7–0.9	0.9–1.1		0.37–0.45
Lot space ($m^2 hd^{-1}$)	Dirt	2.3–3.7	2.3–3.7	2.75–4.6	—	1.8–2.7
	Paved	1.5	1.5	1.8		0.9
Feeder space ($cm\,hd^{-1}$)	Limit-fed	30	40–50	40–50	5	22–30
	Self-fed	15	10–15	10–15		2.5–5
Water (head/bowl or nipple) ($head\,m^{-1}$)	Tank	10	40–50	40–50	Water available	50–75
		7	50–80	50–80		80–130

A good handling facility includes a gathering area, a working area, and a holding area. The gathering area is a place where the sheep are placed prior to entering the working area. Adult sheep need about $0.4\,m^2$–$0.5\,m^2$. Adult ewes with lambs require about $0.65\,m^2$.

The transition from the gathering pen to the working area needs to be designed to provide a funnel effect. A circular transition area limits the forward visibility of the sheep and helps them follow the sheep ahead of them.

The working area is merely a long narrow pen or chute. Many operators prefer a chute so narrow that sheep cannot turn around. Other operators will use a chute 1 m wide so that most sheep can turn around easily.

A footbath can be incorporated into the working chute or placed at the exit. A wooden shelf along the side of the working chute can be used to hold medications and treating equipment.

Avoid any sharp edges on all handling equipment. Sharp edges can cause injury to the sheep or the operator. It can also cause wool to be pulled from the sheep.

ASSOCIATED HANDLING EQUIPMENT

Handling pens near the shearing barn make handling and sorting quicker and easier. Small catch pens make it easier for the shearer to get the next sheep during shearing. A production line type arrangement reduces the amount of labor required for catching sheep during the shearing process. Use floors that are easy to keep clean. Plan for wool storage and bagging equipment in the shearing area.

A roof over the working area keeps the sun and rain off the operator and the sheep. Good lighting is essential to get the sheep to enter the area. A dark area appears to be a hole in front of the sheep, so they are reluctant to enter.

Nice to have items include an office and scale areas. A warming area for newborn lambs can easily be incorporated into this building.

COMPUTER IDENTIFICATION

The development of computer identification equipment has raised all kinds of interesting possibilities in the management of the sheep flock. Identification equipment may be in the form of an ear mounted computer chip or a bolus placed in the gastrointestinal tract.

These identification badges are then read by the computer and may be used to allow the sheep access to a particular pen or feeder. They may also be used to control the amount of feed supplied to each sheep, or to monitor the movements of the sheep.

REFERENCE

1. Committee on Animal Nutrition, National Research Council, Board of Agriculture, Subcommittee on Sheep Nutrition, *Nutrient Requirements of Domestic Animals: Nutrient Regulation of Sheep*, 6th Revised Ed.; National Academy Press: Washington D.C., 1985.

FURTHER READING

American Sheep Industry Association Inc. *Sheep Production Handbook*, 1997; 210–211, 717–719.

Berge, E. Housing of Sheep in Cold Climate. Livest. Prod. Sci. **1997**, *49* (2), 139–149.

Caja, G. *Electronic Identification of Sheep, Goats and Cattle Using Ruminal Bolus*. Proceedings of the 30th Biennial Session of ICAR,1996; 355–358.

Caja, G. *Comparison of Different Devices for Electronic Identification of Dairy Sheep*. Proceedings of the 30th Biennial Session of ICAR, 1996; 349–353.

Doane, T.H. Sheep Space Allotments. NebGuide **1979**, *G79-453-A*, 3.

Fitch, G. Electric Fencing for Sheep. OSU Extension Facts, 3855, 2.

Fraser, D.; Rushen, J. A Colostrums Feeder for Newborn Lambs. Appl. Anim. Behav. Sci. **1993**, *35* (3), 267–276.

Gordon, G.D.H.; Cockram, M.S. A Comparison of Wooden Slats and Straw Bedding on the Behaviour of Sheep. Anim. Welf. **1995**, *4* (2), 131–134.

Hargreaves, A.L.; Hutson, G.D. Handling Systems for Sheep. Livest. Prod. Sci. **1997**, *49* (2), 121–138.

Helberg, M. Plastic Livestock Housing—Further Development and Prospects—A Brief Account. KTBL-Arbeitspapier **1995**, *220*, 54–59.

Hirning, H.J., et al. *Sheep Housing and Equipment Handbook*, 1994; 90.

Hirning, H.J. ARTICLE 5.11.3.4 EQUIPMENT FOR SHEEP, Encyclopedia of Life Support Systems, 2001.

Kencove Farm Fences. Electric Fencing Manual. http://www.kencove.come/stafix/construct.htm.

Kott, R. Managing the Sheep Flock During the Lambing Season. *Montana Extension Factsheet*.

Martyn, E. Wall Ventilated Building. Agric. Eng. **1993**, *48* (4), 123.

McGregor, B.A. Observations on the Effectiveness of Prefabricated Wire Fences for Fibre Goats and Sheep. Proc. Aust. Soc. Anim. Prod. **1990**, *18*, 292–295.

Ong, R.M., et al. Behavioural and EEG Changes in Sheep in Response to Painful Acute Electrical Stimuli. Aust. Vet. J. **1997**, *75* (3), 189–193.

Osturk, T. Structural Features and Sufficiency of Sheep Barns in Samsun Region. Ondokuzmayis-Universitesi-Ziraat-Fakultesi-Dergisi **1996**, *11* (1), 117–126.

Sherwood Number

Fernanda A. R. Oliveira
Jorge C. Oliveira
University College Cork, Cork, Ireland

S

INTRODUCTION

The Sherwood number is a dimensionless group used in the analysis of mass transfer by convection. It is related to the resistance to mass transfer in a fluid phase, across a boundary layer next to the interface with another phase (solid or immiscible fluid) and expresses the ratio between the convective mass flux in the boundary layer and a pure diffusional flux. It is particularly useful for estimating the film mass transfer coefficient, specially when scaling up from laboratory to industrial scale.

DEFINITION

The Sherwood number (Sh) is defined by:

$$\mathrm{Sh} = \frac{KL}{D} \tag{1}$$

where D is the mass diffusivity of the component in the fluid, K is the convective mass transfer film coefficient, and L is a characteristic dimension that depends on the geometry of the system. It is important to note which dimension is used in the definition of a specific Sh number. Usually, for mass flux from/into immersed bodies the characteristic dimension of the body is used (e.g., diameter for spheres); for mass flux across a plane (e.g., a membrane) the length parallel to the hydrodynamic flow is used; and for flow inside pipes and ducts the hydraulic diameter is considered.

PHYSICAL MECHANISMS

When two immiscible phases that contain a component in concentrations that are not in equilibrium come into contact, a mass flux of the component is established across the interface (e.g., drying of food in hot air, leaching of water-soluble components in blanching, gas–liquid absorption). Due to the no-slip condition, there must be a stagnant layer at the interface, which will also impair the movement of other layers over it by viscous effects. In this region, mass flux in the direction normal to the interface is restricted (ultimately, there would be pure diffusional flux over a truly stagnant layer). Hence, there will be a concentration gradient across the boundary layer. Normally, there will be a mass flux and a concentration gradient in the other phase also, but not always (for instance, dissolution of a crystal in an aqueous solution). It is assumed that the interface itself does not present a resistance to mass transfer and that at this point the system is at equilibrium.

It must be noted that the thickness of the boundary layer may be significantly higher than that of the stagnant layer. Let us consider, for instance, that the fluid phase is moving over the interface. The layers next to the stagnant layer will move slowly and there must be a laminar flow region as well. In this region, there is no advection perpendicular to streamlines and there will be significant resistance to mass transfer towards/from the interface. Therefore, the flux across a mass transfer boundary layer is not pure diffusional flux and it is no longer considered correct to relate the convective mass transfer coefficient to pure diffusion over a (stagnant) boundary layer. The simplest definition is that the boundary layer thickness is that needed for the fractional concentration ($C*$) to reach 99%, as defined by:

$$C* = \frac{C - C_b}{C_i - C_b} = 0.99 \tag{2}$$

where C is the concentration at the edge of the boundary layer opposite the interface, C_b the concentration in the free fluid stream, and C_i the concentration at the interface, on the side of the fluid.

It is also important to note that the thickness of the boundary layer may vary substantially along the interface surface, as it is affected by the hydrodynamic conditions.

PHYSICAL SIGNIFICANCE OF THE SHERWOOD NUMBER

Dimensionless numbers are groups of system variables that combine in such a way that all dimensions cross out. Therefore, they can be interpreted as a ratio between two quantities that have the same units. In the case of

Encyclopedia of Agricultural, Food, and Biological Engineering
DOI: 10.1081/E-EAFE 120007039
Copyright © 2003 by Marcel Dekker, Inc. All rights reserved.

the Sherwood number, we can write Eq. 1 as:

$$\mathrm{Sh} = \frac{K}{D/L} \tag{3}$$

The numerator is related to the actual resistance to mass transfer between the fluid and the interface and the denominator to the resistance to diffusional mass transfer across a length L of stagnant fluid. Therefore, Sh indicates how the film resistance compares to resistance to diffusional mass transfer.

A more detailed theoretical analysis will yield a complementing perspective on the physical significance of Sh. For convective mass transfer in steady state, the mass flux (n) of the diffusing component is proportional to the concentration gradient across the boundary layer from C_i to C_b:

$$n = KA(C_i - C_b) \tag{4}$$

where A is the interface surface, perpendicular to the mass flux, and K the film coefficient.

Due to the no-slip condition, at the interface there is a stagnant layer on the side of the fluid, and therefore we can write that just next to the interface, the mass flux may also be given by:

$$n = -D{\cdot}A \left.\frac{\partial C}{\partial x}\right|_{x=0} \tag{5}$$

where x is the distance from the interface. Combining Eqs. 4 and 5:

$$-D\left.\frac{\partial C}{\partial x}\right|_{x=0} = K(C_i - C_b) \tag{6}$$

We can write Eq. 6 in dimensionless form by using dimensionless variables, normalising both C and x. We select a characteristic dimension of the system, L, to normalize the space co-ordinate. We do not wish to use the thickness of the boundary layer for that purpose, as it is difficult to determine experimentally, and because it may vary along the interface (for instance, at the entrance of a plane, around a small sphere). The dimensionless concentration will be the fractional difference from C_i to C_b. Mathematically, we have:

$$x^* = \frac{x}{L} \tag{7}$$

$$C^* = \frac{C - C_b}{C_i - C_b} \Rightarrow \partial C = (C_i - C_b)\partial C^* \tag{8}$$

where x^* and C^* are the dimensionless variables. Eq. 5 becomes:

$$\left.\frac{\partial C^*}{\partial x^*}\right|_{x^*=0} = \frac{KL}{D} = \mathrm{Sh} \tag{9}$$

and therefore, we can note that the Sherwood number is also the dimensionless concentration gradient at the interface.

APPLICATION

Determining mass transfer film coefficients is the major application of the Sherwood number. According to the principle of dynamic similarity, regardless of the individual values of specific system properties, relationships between dimensionless numbers are the same, and this is most useful for scale-up. The resistance to convective mass transfer across the boundary layer is related to the film coefficient K. Everything that affects the flow properties and streamlines of the fluid around the interface will affect the thickness of the boundary layer and hence the value of K: this includes not only the physical properties of the fluid (e.g., density, viscosity), but also size, shape, and smoothness of the surface. Furthermore, the thickness of the boundary layer is rarely constant along the whole interface. Therefore, K can only be determined in situ, or then estimated from dimensionless correlations.

DIMENSIONLESS CORRELATIONS

The Sherwood number is usually calculated as a function of other dimensionless numbers that depend on whether convection is forced or natural: in the former, the Reynolds (Re) and Schmidt (Sc) numbers are used while in the latter the Reynolds number is replaced by the Grashof number (Gr). Some theoretical results can be obtained for simplified situations, but in general, correlations will have been determined by fitting experimental data to a general expression. It is very important to note that dimensionless correlations are empirical and hence can only be used inside the range of values of the dimensionless numbers that were used in their determination. The situation in terms of geometries, types of materials (fluid, surfaces), and type of flow (cross-flow, longitudinal flow, internal flow in pipes, external flow around pipes, etc.) must also be the same. The most

Table 1a Dimensionless correlations for estimation of the Sherwood number.[1–3] Correlations of the form of equation 10

Situation	Characteristic dimension of Sh and Re	a	b	c	d	Range of validity
Laminar flow over a flat surface	Length parallel to flow	0	0.664	1/2	1/3	Sc > 0.6 Re < 2000
Turbulent flow over a flat surface	Length parallel to flow	0	0.037	4/5	1/3	0.6 < Sc < 3000 5×10^5 < Re < 10^8
Cross flow over a cylinder	Cylinder diameter	0	0.989 for 0.4 < Re < 4 0.911 for 4 < Re < 40 0.683 for 40 < Re < 4×10^3 0.193 for 4×10^3 < Re < 4×10^4 0.027 for 4×10^4 < Re < 4×10^5	0.989 for 0.4 < Re < 4 0.911 for 4 < Re < 40 0.683 for 40 < Re < 4×10^3 0.193 for 4×10^3 < Re < 4×10^4 0.027 for 4×10^4 < Re < 4×10^5	1/3	Sc > 0.6 0.4 < Re < 4×10^5
Cross flow over a cylinder	Cylinder diameter	0	0.75 for 1 < Re < 40 0.51 for 40 < Re < 1000 0.26 for 1000 < Re < 2×10^5 0.076 for 2×10^5 < Re < 10^6	0.4 for 1 < Re < 40 0.5 for 40 < Re < 1000 0.6 for 1000 < Re < 2×10^5 0.7 for 2×10^5 < Re < 10^6	0.37 for 0.7 < Sc < 10 0.36 for Sc > 10	Sc > 0.7 1 < Re < 10^6
Cross flow over a rectangular tube	Height of duct	0	0.246	0.588	1/3	5000 < Re < 10^5
Cross flow over a vertical plate	Height of plate	0	0.228	0.731	1/3	4000 < Re < 15000
Flow over a free falling sphere	Sphere diameter	2	0.6	1/2	1/3	0.71 < Sc < 380 3.5 < Re < 76000
Forced flow around a sphere	Sphere diameter	2	7.9464	0.20795	0.14413	1 < Sc < 8.4 0.1 < Re < 2
Forced flow around a sphere	Sphere diameter	2	$28.37*(D_s/D_t)^{1.787}$ D_s is the tube diameter, D_t is the sphere diameter	0.233	0.143	1.14 < Sc < 7.22 0.1 < Re < 2
Forced flow around a sphere	Sphere diameter	$2 + 1.3.Sc^{0.15}$ (modified Eq. 10)	0.66	0.50	0.31	0.71 < Sc < 380 0.42 < Re < 2100
Forced flow around small spheres	Sphere diameter	0	0.61	1/2	0	200 < Re < 3000
Forced flow around small spheres	Sphere diameter	0	0.085	0.78	0	3500 < Re < 15000
Continuous flow of spheres in a carrier fluid inside circular tubes	Sphere diameter	2	$8.4703(L/D)^{0.6272}$ $[(D - 2r_p)/D]^{-0.1142}$ L is the tube length, r_p is the radial position of the sphere	0.553	0.2716	185 < Sc < 1075 41 < Re < 478
Continuous flow of spheres in a carrier fluid inside circular tubes	Sphere diameter	2	$0.685(L/D)^{0.5314}$ L is the tube length	0.7023	0.3925	185 < Sc < 1075

(Continued)

Table 1a Dimensionless correlations for estimation of the Sherwood number.[1-3] Correlations of the form of equation 10 (*Continued*)

Situation	Characteristic dimension of Sh and Re	a	b	c	d	Range of validity
Continuous flow ofspheres in a carrier fluid inside circular tubes	Sphere diameter	0	0.1	0.58	0.33	41 < Re < 478 3.7 < Sc < 5.5
Continuous flow of spheres in a carrier fluid inside circular tubes	Sphere diameter	0	0.0336	0.8	0.32	563 < Re < 2000 3.7 < Sc < 5.5
Continuous flow of cubes in a carrier fluid inside a scrapped surface heat exchanger	Equivalent diameter of the cube	2	2.58.$(L/D)^{0.5314}$ L is the tube length, r_p is the radial position of the sphere	0.43	0.17	2000 < Re < 6000 185 < Sc < 1075
Spheres rotating in an otherwise static fluid	Sphere diameter	10	0.24	1/2 (Re for angular velocity)	1/3	41 < Re < 478 500 < Sc < 2000
Turbulent flow inside tubes, entrance region	Hydraulic diameter	0	1.86*D/L (D tube diameter L tube length)	1/3	1/2	2500 < Re < 85000 0.48 < Sc < 16700
Fully developed turbulent flow inside tubes	Hydraulic diameter	0	0.023	4/5	0.4 for heating O.3 for cooling	entrance region 0.7 < Sc < 160 Re > 10000 L/D > 10
Fully developed turbulent flow inside tubes	Hydraulic diameter	0	0.027	4/5	1/3	0.7 < Sc < 16700 Re > 10000 L/D > 10
Fully developed laminar flow inside a circular annulus, Sh at the surface of the inner cylinder	Diameter of inner cylinder	17.46 for $D_{in}/D_{out} = 0.05$ 11.56 for $D_{in}/D_{out} = 0.1$ 7.37 for $D_{in}/D_{out} = 0.25$ 5.74 for $D_{in}/D_{out} = 0.5$ 4.86 for $D_{in}/D_{out} = 1$	0	—	—	Re < 2000
Fully developed laminar flow inside a circular annulus, Sh at the surface of the outer cylinder	Diameter of outer cylinder	3.66 for $D_{in}/D_{out} = 0$ 4.06 for $D_{in}/D_{out} = 0.05$ 4.11 for $D_{in}/D_{out} = 0.1$ 4.23 for $D_{in}/D_{out} = 0.25$ 4.43 for $D_{in}/D_{out} = 0.5$ 4.86 for $D_{in}/D_{out} = 1$	0	—	—	Re < 2000

Table 1b Correlations of the form of Eq. 11

Situation	Characteristic dimension of Sh and Re	a	b	c	d	Range of validity
Natural convection over a heated horizontal flat plate or under a cooled horizontal flat plate	Plate width	0	0.54	1/4	1/4	$10^4 < GrSc < 10^7$
Natural convection over a heated horizontal flat plate or under a cooled horizontal flat plate	Plate width	0	0.15	1/3	1/3	$10^7 < GrSc < 10^{11}$
Natural convection over a cooled horizontal flat plate or under a heated horizontal flat plate	Plate width	0	0.27	1/4	1/4	$10^7 < GrSc < 10^{11}$
Natural convection over an immersed cylinder	Cylinder diameter	0	0.675 for $10^{-10} < GrSc < 10^{-2}$ 1.02 for $10^{-2} < GrSc < 10^2$ 0.850 for $10^2 < GrSc < 10^4$ 0.480 for $10^4 < GrSc < 10^7$ 0.125 for $10^7 < GrSc < 10^{12}$	0.058 for $10^{-10} < GrSc < 10^{-2}$ 0.148 for $10^{-2} < GrSc < 10^2$ 0.188 for $10^2 < GrSc < 10^4$ 0.250 for $10^4 < GrSc < 10^7$ 0.333 for $10^7 < GrSc < 10^{12}$	0.058 for $10^{-10} < GrSc < 10^{-10}$ 0.148 for $10^{-2} < GrSc < 10^2$ 0.188 for $10^2 < GrSc < 10^4$ 0.250 for $10^4 < GrSc < 10^7$ 0.333 for $10^7 < GrSc < 10^{12}$	$10^{-10} < GrSc < 10^{12}$
Natural convection over carrot cylinders	Cylinder diameter	0	2.45	0.108	0.108	Not known
Natural convection over potato cylinders	Cylinder diameter	0	2.02	0.113	0.113	Not known
Natural convection inside a rectangular container	Width of container	0	$0.42\ (w/H)^{0.3}$ H is height and w width	1/4	0.262	$1 < Sc < 20000$ $10^4 < GrSc. < 10^7$ $10 < H/w < 40$
Natural convection inside a rectangular container	Width of container	0	0.046	1/3	1/3	$1 < Sc < 20$ $10^6 < GrSc < 10^9$ $1 < H/w < 40$

common correlations have the form:

$$\mathrm{Sh} = a + b\mathrm{Re}^c \cdot \mathrm{Sc}^d \quad \text{(forced convection)} \tag{10}$$

$$\mathrm{Sh} = a + b\mathrm{Gr}^c \cdot \mathrm{Sc}^d \quad \text{(natural convection)} \tag{11}$$

where Re is the Reynolds number ($\mathrm{Re} = \rho v \phi / \mu$), Sc the Schmidt number ($\mathrm{Sc} = \mu / [\rho D]$), and Gr the Grashof number ($\mathrm{Gr} = \Delta C \beta g L^3 \rho^2 / \mu^2$). In these definitions, ρ and μ are the average fluid density and viscosity, respectively, v is the average velocity of the free fluid stream, g is the acceleration of gravity, $\Delta C = |C_i - C_b|$, $\beta = |\rho_b - \rho_i|/\rho_i$, D is the diffusivity of the component in the stagnant fluid, and ϕ is the characteristic dimension of the environment where the flow takes place. It must be noted that while the same characteristic dimension is generally used in the Reynolds number (ϕ) and in the Sherwood and Grashof numbers (L), they could also be different: for instance, for mass transfer to spheres immersed in a fluid that is circulating inside a pipe, the characteristic dimension for Sh might be the diameter of the spheres and for Re it could be the diameter of the pipe. This should be carefully verified.

The Sherwood number may have a lower limit (parameter a), for instance, the theoretical value for spheres immersed in a fluid is 2 (using the diameter for characteristic dimension). However, some authors have proposed a totally empirical fit and allowed the parameter a to be given by regression. For other geometries, a is often taken to be zero. The exponents of Gr, Re, and Sc may also have theoretical values for simplified situations, for instance, for constant properties and laminar Newtonian flow over a horizontal flat surface c would be 1/2 and d 1/3. Table 1(a–c)shows correlations that have been proposed for various situations, for isothermal conditions. Details can be found in comprehensive reviews that collected various correlations published in literature.[1–3]

It should be noted that some correlations were actually developed for heat transfer, for the similar expression to Eq. 8, which is:

$$\mathrm{Nu} = a + b\mathrm{Re}^c \cdot \mathrm{Pr}^d \tag{12}$$

where Nu is the Nusselt number and Pr the Prandtl number. The Nusselt number is in fact the same as

Table 1c Other correlations for Sh

Situation	Correlation	Range of validity
Cross flow over a cylinder	$0.3 + \frac{0.62Re^{1/2}Sc^{1/3}}{[1+(0.4/Sc)^{2/3}]^{1/4}}\left[1 + \left(\frac{Re}{282000}\right)^{5/8}\right]^{4/5}$	ReSc > 0.2
Flow over an immersed sphere	$2 + \left(0.4Re^{1/2} + 0.06Re^{2/3}\right)Sc^{0.4}$	0.71 < Sc < 380 3.5 < Re < 76000
Fully developed turbulent flow inside tubes	$\frac{(f/8)\mathrm{ReSc}}{1.07+12.7(f/8)^{1/2}(\mathrm{Sc}^{2/3}-1)}$ f is the friction factor, given by Moody's diagram	0.5 < Sc < 2000
Fully developed turbulent flow inside tubes	$\frac{(f/8)(\mathrm{Re}-1000)\mathrm{Sc}}{1+12.7(f/8)^{1/2}(\mathrm{Sc}^{2/3}-1)}$ f is the friction factor, given by Moody's diagram	$10^4 < \mathrm{Re} < 5 \times 10^6$ 0.5 < Sc < 2000
Natural convection over a vertical plate, laminar flow	$0.68 + \frac{0.670Gr^{1/4}Sc^{1/4}}{[1+(0.492/Sc)^{9/16}]^{4/9}}$	$3000 < \mathrm{Re} < 5 \times 10^6$ $\mathrm{GrSc} < 10^9$
Natural convection over a vertical plate	$\left\{0.825 + \frac{0.387Gr^{1/4}Sc^{1/4}}{[1+(0.492/Sc)^{9/16}]^{8/27}}\right\}^2$	$\mathrm{GrSc} < 10^{12}$
Natural convection over an immersed cylinder	$\left\{0.60 + \frac{0.387Gr^{1/6}Sc^{1/6}}{[1+(0.559/Sc)^{9/16}]^{8/27}}\right\}^2$	$\mathrm{GrSc} < 10^{12}$
Natural convection over an immersed sphere	$2 + \frac{0.589Gr^{1/4}Sc^{1/4}}{[1+(0.469/Sc)^{9/16}]^{4/9}}$	Sc > 0.7
Natural convection inside a rectangular container	$0.22\frac{Gr^{0.28}Sc^{0.56}}{(0.2+Sc)^{0.28}}(w/H)^{1/4}$ H is height and w width	$\mathrm{GrSc} < 10^{11}$ $\mathrm{Sc} < 10^5$
Natural convection inside a rectangular container	$0.18\frac{Gr^{0.29}Sc^{0.58}}{(0.2+Sc)^{0.29}}$	$10^3 < \mathrm{GrSc} < 10^{10}$ $2 < H/w < 10$ $10^{-3} < \mathrm{Sc} < 10^5$ $\mathrm{GrSc}^2/(0.2 + \mathrm{Sc}) < 10^3$ $1 < H/w < 2$ H is height and w width

Sherwood, but for heat transfer, and this can also be said for the Prandtl (heat transfer) and Schmidt (mass transfer) numbers.

Other correlations using related dimensionless numbers may also be found for forced convection. The Stanton number (St), sometimes referred to as a modified Sherwood number, may be used instead of Sh. By definition $\mathrm{St} = K/v$, and therefore:

$$\mathrm{St} = \frac{K}{v} = \frac{\mathrm{Sh}}{\mathrm{Re{\cdot}Sc}} \tag{13}$$

Correlations may also be found for the so-called Colburn factor for mass transfer (j_D) as a function of the Reynolds number, which allow us to estimate the film coefficient, as according to the Chilton–Colburn analogy:

$$j_D = \mathrm{St{\cdot}Sc}^{2/3} \tag{14}$$

Eq. 14 is valid in the range $0.6 < Sc < 3000$ and turbulent flow, and also in laminar flow if the pressure gradient in the flow direction is negligible.

In some situations, it may be necessary to consider both forced and natural convection. The general practice is to sum (or subtract) Eqs. 10 and 11, but this is an approximation prone to errors. More precise analysis can be found in literature.[4]

S

REFERENCES

1. Incropera, F.P.; DeWitt, D.P. *Fundamentals of Heat and Mass Transfer*, 4th Ed.; John Wiley & Sons: New York, 1996; 347–516.
2. Treybal, R.E. *Mass Transfer Operations*, 3rd Ed.; McGraw-Hill: New York, 1980; 50–200.
3. Baptista, P. Flow and Heat Transfer Analysis of Two-Phase Systems with Large Solid Particles Moving in Carrier Fluids. PhD dissertation Catholic University of Portugal, College of Biotechnology: Porto, Portugal, 1995; 76–90.
4. Kakac, S.; Shah, R.K.; Aung, W. *Handbook of Single-Phase Convective Heat Transfer*, 1st Ed.; John Wiley & Sons: New York, Chap. 14,15; 1987.

Soil Compaction Management

Randy L. Raper
United States Department of Agriculture (USDA), Auburn, Alabama, U.S.A.

INTRODUCTION

Soil compaction plagues American agriculture by reducing yields and increasing soil erosion. During temporary droughts that often limit agricultural production, soil compaction can hinder plant roots from reaching depths of soil where moisture is available. Measurements of soil compaction taken with a soil cone penetrometer indicate that compaction can vary greatly across fields due to past tillage, traffic, and natural conditions. Various site-specific technologies such as variable-depth tillage, seeding, or cover crops may offer valuable alternatives for management of soil compaction, which allow energy and soil to be conserved and crop yields to be maximized.

OVERVIEW

For optimal crop production, plants need an adequate rooting environment to provide sufficient water and nutrients. During temporary droughts that often limit agricultural production, soil compaction can hinder plant roots from reaching depths of soil where moisture is available. Photosynthesis is reduced, which therefore reduces crop yields.

Soil that is excessively compacted and lying near the soil surface is often responsible for inadequate rooting depth by plants. As soil particles are packed close together, less room exists for soil moisture and air that are essential for plant growth. Soil compaction also limits root growth potential by increasing mechanical strength of soil. Soil moisture is unavailable to plant roots if there is no loosened pathway from the soil surface through the layers of compacted soil.

Soil compaction reduces infiltration of water and air through the soil surface and can lead to greater erosion. As the sizes and volumes of pores are reduced, so are the pathways for water to infiltrate into soils. Rainfall ponds on the soil surface for longer periods of time, which often results in increased runoff and soil erosion.

Unfortunately, all regions of the United States are susceptible to soil compaction with extreme cases being reported in the Southeast, the upper Midwest, and the Pacific Northwest. In many cases, soil compaction occurs naturally, particularly in soils degraded by excessive erosion. Many Southeastern U.S. soils suffer from soil compaction, being very old, shallow, and severely eroded.

Excessive vehicle traffic is one of the main causes of soil compaction. Numerous passes, common in conventional tillage systems, compact the soil. Heavy equipment that does not adequately spread out the load on the soil surface can increase soil compaction. Trafficking the soil when it is in a compactable state, like when it is wet, can increase soil compaction. Some tillage practices, such as annual moldboard plowing and surface tillage can also severely compact layers of soil. Many soils of the United States have been conventionally farmed using these techniques for more than 100 yr. The result is a soil condition that has poor internal structure, poor aggregation, and is easily compacted by vehicle and animal traffic.

Prevention and management of soil compaction is possible using strategies to either manage the location of vehicle traffic or minimize the loads applied to the soil surface. Controlled traffic is a concept that has been utilized in some locations to segregate the zones of vehicle traffic from plant growth zones. Decreasing vehicle size or load has been shown to reduce soil compaction in wheel tracks. Reduced tire inflation pressure, increased tire size, or using tracks are methods of increasing the contact area between the tire/track and soil, which may translate into reduced soil compaction.

Soils in transition from conventional tillage systems to reduced tillage systems seem to be particularly susceptible to soil compaction. Strategies for management of soils in this transitory phase need to be developed and may consist of methods other than mechanical tillage. Cover crops and crop rotations may offer alternative methods of capturing additional rainfall into the rooting profile, effectively reducing soil strength and increasing root proliferation. Research suggests that increasing levels of organic matter through use of soil amendments may also increase soil moisture and plant rooting activity.

Soils compacted by traffic and/or natural forces can be loosened and appropriately managed by tillage, which usually consists of subsoiling or chiseling. However,

Encyclopedia of Agricultural, Food, and Biological Engineering
DOI: 10.1081/E-EAFE 120007239
Published 2003 by Marcel Dekker, Inc. All rights reserved.

the benefits of such tillage are mostly temporary and annual treatments are often necessary. Tillage should also be delivered to the appropriate depth of compaction. Tillage shallower than necessary shows little benefit while tillage deeper than necessary wastes energy and can do excessive soil disturbance and increase erosion. Site-specific management of tillage may prove to optimize energy inputs while maximizing crop yields and minimizing soil erosion.

MEASUREMENT OF SOIL COMPACTION

Soil compaction is commonly measured by means of a soil cone penetrometer (Fig. 1), which is defined by ASAE Standard S313.2 (ASAE, St. Joseph, Michigan). This device is composed of a cone attached to a shaft. A handle is attached to the shaft through a load-measuring device. The handles are used to push the soil cone penetrometer into the soil while the load-measuring device registers the force necessary for insertion. The force required for insertion is divided by the cross-sectional area of the top of the cone and is referred to as "cone index." Values of cone index that are typically used to indicate problematic rooting environments are 2 MPa (300 psi).

Automated methods of obtaining several measurements of cone index at the same time have been developed because a large amount of time and effort are required to obtain cone index measurements and significant variability of soil compaction is present in most fields. A multiple-probe soil cone penetrometer has been used to obtain site-specific soil compaction information (Fig. 2). An added benefit of the multiple-probe unit is that measurements of soil compaction are also measured across the row from the no trafficked row middle to the trafficked row middle. Frequently, soil compaction is found to vary more across a row from a no trafficked row middle to a trafficked row middle, due to the presence of traffic, than across the entire field.

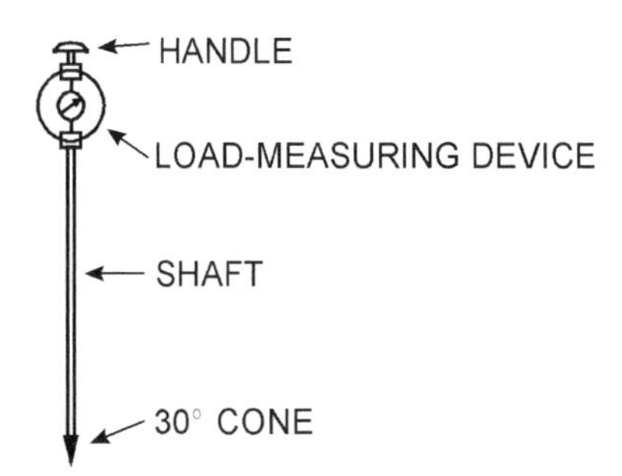

Fig. 1 Diagram of soil cone penetrometer used to measure the depth and degree of soil compaction.

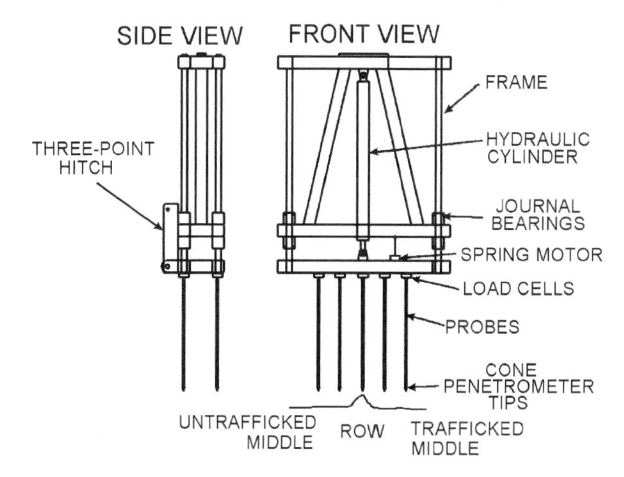

Fig. 2 Side and front view of multiple-probe soil cone penetrometer created to expedite the measurement of soil compaction across a row and throughout a field.

SITE-SPECIFIC VARIATION OF SOIL COMPACTION

Many reasons are hypothesized for the variation that we find in soil compaction across fields. Many fields have been site-specifically measured using the soil cone penetrometer and field maps created of the varying soil compaction. Many of these maps show shallow compacted soil, which limits root growth and crop yield in large areas of the field (Fig. 3). These data can be analyzed to determine the optimal sampling distance between cone index measurements. Spatial statistics can be used to calculate the "range," which is the approximate distance from one point to another within a field that can be assumed to be correlated. In the data shown in Fig. 3, the range of the data is approximately 13 m, which indicates the frequency at which soil compaction changes across this field. This relatively small value indicates a great amount of variation present in this field.

Determining the causes of site-specific soil compaction can be very difficult when the history of the field is not known. Past tillage and vehicle traffic practices may be responsible, but a significant amount of site-specific soil compaction may be due to the inherent variability of soils. Within many fields, several soil types are commonly found, which may contain a broad range of percentages of sand, silt, and clay, as well as many different types of clay. Due to past tillage systems, soil has probably been redistributed from higher elevations to low-lying areas because of erosion from wind, water, and tillage. Many subsoils are now being tilled and treated as topsoils because the topsoil has long since eroded.

Due to the different soils that may be present in fields, a great degree of variation in water holding capacity of soils

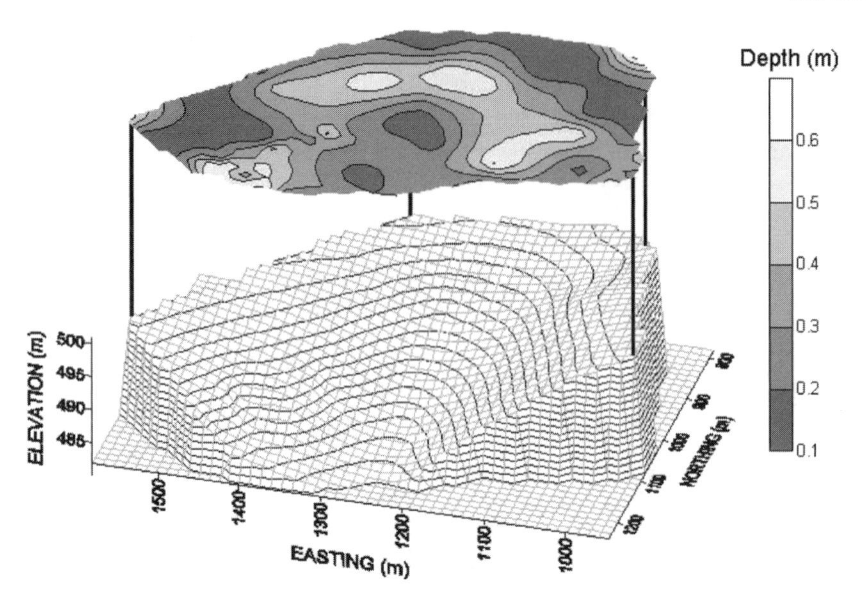

Fig. 3 Contour graph of depth of 2 MPa cone index value plotted over the elevation of the field.

may also be found. Variations in soil types, clay amounts, clay types, elevation, and infiltration rates may be responsible for nonuniform soil compaction. As traffic is applied to these various sites, soils will respond differently with some being compacted and some being unaffected.

Some areas of the field may also have been subject to different amounts of vehicle traffic. Not all row middles are equally trafficked, with some being heavily trafficked and others receiving no traffic at all. Headlands (areas near the end of the field where vehicles turn) are also some of the most compacted regions of fields because of the intensive traffic and also due to the turning action of vehicle tires.

MANAGEMENT OF SITE-SPECIFIC SOIL COMPACTION

It is now being recognized that soil compaction may be managed using site-specific technologies previously unavailable. Once an area of the field has been measured and found to exhibit excessive soil compaction, this area may need to receive special amelioration practices not used throughout the entire field. These can include tillage to the appropriate depth to provide root channels for restricted plant roots or cover crops to enhance infiltration of rainfall.

Site-specific tillage is a concept that is being investigated by many researchers to see if it is practical to routinely map problematic fields with a soil compaction-measuring device. This concept calls for the depth of tillage to be varied based on the needs of the soil. For example, if soil compaction was found to extend downward into the soil for 20 cm, it would be a waste of tillage energy to till above that depth and not provide loosened soil for root channels through this compacted region. In addition, tilling too deeply would again waste energy when tillage was only necessary to a 20-cm depth. Tillage forces increase dramatically when tillage depth increases, so it is important to till only as deep as necessary to prevent wasting energy.

Based on the estimated range of 13 m calculated from the data presented in Fig. 3, adjustments in tillage depth would be necessary at this same distance. Some fields with lesser amounts of variability would require fewer changes in tillage depth, perhaps spaced farther apart than 13 m, while increased amounts of variability would require more frequent adjustments in tillage depth.

In addition to controlling tillage depth, other aspects of tillage may prove useful to manage site-specific soil compaction. Many soils respond differently to tillage. This lack of a uniform response may be due to the shape of the tillage tool, which causes a wider zone of soil to be disturbed, larger clods to be brought to the soil surface, and more burial of surface residue. New developments in tillage tool design may allow these parameters to be changed on the go, so that site-specific changes can be made.

Other nontillage approaches may be used to manage site-specific soil compaction. These may include using cover crops in problematic areas to assist infiltration and reduce runoff. Although most fields would benefit from increased organic matter resulting from degradation of a cover crop, some cover crops have very aggressive rooting systems that may prove especially beneficial to subsequently planted cash crops. These cover crops may be used selectively in those areas with site-specific soil compaction where their aggressive roots can "till" the soil. However, costs may prohibit their use over the entire field.

SUMMARY

Soil compaction plagues American agriculture by reducing yields and increasing soil erosion. Measurements of soil compaction indicate that it can vary greatly across fields due to past tillage, traffic, and natural conditions. New technologies, including site-specific tillage, may offer valuable alternatives for management of soil compaction that allow energy to be conserved, yields to be maximized, and soil to be conserved.

FURTHER READING

ASAE. Procedures for Obtaining and Reporting Data with the Soil Cone Penetrometer EP542. *ASAE Standards*; ASAE: St. Joseph, MI, 1999; 964–966.

ASAE. Soil Cone Penetrometer S313.2. *ASAE Standards*; ASAE: St. Joseph, MI, 1999; 808–809.

Gill,W.R.; Vanden Berg, G.E. *Soil Dynamics in Tillage and Traction*; USDA: Auburn, AL, 1966.

Larson, W.E.; Eynard, A.; Hadas, A.; Lipiec, J. Control and Avoidance of Soil Compaction. In Soil Compaction in Crop Production, Soane, B.D., van Ouwerkerk, C., Eds.; Elsevier: Amsterdam, 1994; 597–625.

Perumpral, J.V. Cone Penetrometer Applications—A Review. Trans. ASAE **1987**, *30*, 939–944.

Schuler, R.T.; Casady, W.W.; Raper, R.L. Soil Compaction. In *Conservation Tillage Systems and Management*; Reeder, R.C., Ed.; Midwest Plan Service: Ames, IA, 2000.

Soane, B.D.; van Ouwerkerk, C. Soil Compaction Problems in World Agriculture. In *Soil Compaction in Crop Production*; Soane, B.D., van Ouwerkerk, C., Eds.; Elsevier: Amsterdam, 1994; 1–22.

Soil Dynamics

Ernest W. Tollner
The University of Georgia, Athens, Georgia, U.S.A.

INTRODUCTION

Soil dynamics is a soil science and mechanics concerned with soils in motion. Soil dynamics is defined as the relation between forces applied to the soil and the resultant soil reaction.[1] Forces applied to the soil may originate from the natural forces of wind, water, or other natural forces such as earthquakes. They may also originate from animals and tillage equipment, the primary concern of agricultural soil dynamics. Soil is a granular material that varies in composition from organic peat to gravel and may contain varying amounts of water. The dynamic properties of soil are properties made manifest by the action of forces creating movement of soil.

THE STATE OF THE DISCIPLINE

Comprehensive reviews of the soil dynamics discipline are contained in Refs. [1–3]. An updated monograph series, *Advances in Soil Dynamics*, is currently in production by the American Society of Agricultural Engineers.[4] Forces resulting in soil motion include those associated with tillage systems and those resulting from gravity, wind, and water movement. Tillage machine performance in the tilled zone and frequently resulting compaction in surrounding soil regions are the purview of soil dynamics. Compaction induced by tractors and other prime movers is central to soil dynamics. How to creatively separate desired soil tillage from soil compaction induced by the tillage implement and prime mover system has been an overarching goal of many soil dynamic investigations. For example, repeated use of the moldboard plow in the southeastern United States resulted in tillage pans at approximately 15-cm plow depth that persist long after the field may have been converted from conventionally tilled agriculture to other uses. The use of disk tillage likewise results in a tillage hardpan below the zone of disk operation. Increasing size of tillage implements and prime movers frequently results in high compaction levels under the prime mover travel path. Compaction zones near the surface in some fields in conservation tillage, which result from long-term wetting–drying forces, have also correlated with reduced crop yields, particularly in cases where mulch persistence and earthworm counts were low.

Deformation of soil depends on the dynamic parameters of shear, tension, compression, plastic flow, friction, adhesion, and perhaps other phenomena awaiting identification. Shear is a failure or yield condition wherein the soil responds by slipping along defined planes. The tensile strength of a soil is the force required to pull the soil apart at incipient failure. Compression is a condition associated with a volume change in soil. Shatter resistance is related to compression failure. Plastic flow is a failure condition often associated with moist, clayey soils in which the soil behaves as a non-Newtonian liquid. Shear, tension, compression, and plastic flow represent distinctly different response modes for soil. Soil texture, mineralogy, moisture content, loading patterns, profile position, and past history affect which mode may predominate. Soil–metal friction and adhesion play an important part in how a tillage implement imparts forces (e.g., loading pattern) to the soil.

Over the 20th century, researchers worked diligently to develop a complete, high-utility soil mechanics that includes static and dynamic soil properties. A complete soil mechanics is the description of soil movements coupled with forces via an independently measurable parameter set. The complete soil mechanics should bridge shear, tension, compression, and plastic flow failure mechanisms. A complete, high-utility soil mechanics provides force–deformation descriptions with many parameters being constant over a wide range of soil textures, mineralogy, moisture regimes, and loading combinations such that many parameters were approximately constant and thus not requiring measurement at each application.

A complete, high-utility soil mechanics valid over a range of moisture and texture regimes has proven extremely illusory. Likewise, the identification of soil dynamic properties is uncertain.[1] Thus, there has been an emphasis on simple mechanics describing particular situations.

HISTORICAL OVERVIEW

The development of the soil dynamics discipline began with efforts to describe the effect of the moldboard plow on soil motion. Doner and Nichols[5] developed a partial

Encyclopedia of Agricultural, Food, and Biological Engineering
DOI: 10.1081/E-EAFE 120007234
Copyright © 2003 by Marcel Dekker, Inc. All rights reserved.

mechanics of the moldboard plow based on the observed path of soils shown in Fig. 1. Kawamura[6] extended the work to include a soil compressibility factor, which was not readily measured. The limited mechanics was centered on soil–metal friction and normal soil pressures. This work does not represent a complete mechanics for the moldboard plow because of empirical constants, which vary widely with soil type and moisture regime and thus require measurement each time the mechanics is applied. However, the semiempirical relationship provided the basis for comparing numerous variations in moldboard shape. The moldboard plow loosened the tillage zone and covered surface residues. Repeated usage often caused development of a tillage pan.

An understanding of tension and compression processes has lead to the following tillage implement design principle: one should place soil in tension at the bottom of the tillage zone to minimize compaction pan formation. Compression failure should be directed from the bottom of the tillage zone upward. For example, the Paraplow® accomplishes general soil loosening without developing a tillage pan (e.g., Ref. [7]) in many soil textures and mineralogies if the moisture content is in an acceptable tillage regime.

A major step forward in the development of soil dynamic mechanics was to borrow the Mohr–Coulomb failure concept from civil engineering. Failure, defined by the state of stress at incipient volume change, is a concept borrowed from civil engineering soil mechanics. Shear, tensile, compression, and plastic flow represent failure modes. The predominant model for describing soil shear failure is the Mohr–Coulomb relationship given in Eq. 1:

$$\tau_{\max} = c + \sigma_n \tan \phi \qquad (1)$$

where $\tau_{\max}$ is the shear stress at failure, c the cohesion stress, σ_n the summation of normal forces in a given plane when the normal stress is zero, and φ is the coefficient of friction of soil grains sliding on soil grains.

Fig. 2 shows a schematic representation of the forces and deformation under shear flow. The parameters of Eq. 1 vary widely with soil type, moisture content, and past history, which reduces the utility of shear-based mechanics. Shear flow describes force–deformation relations at failure fairly well in most dry soils. Increasing empiricism is required to apply the shear mechanism to soils with increasing moisture and compressibility levels. Additional details are provided by Chancellor.[8] Tensile strength is related to the cohesion coefficient in Eq. 1; however, other approaches borrowed from beam theory are most often used to describe tensile strength mechanics. Shear failure has become fairly well understood because it

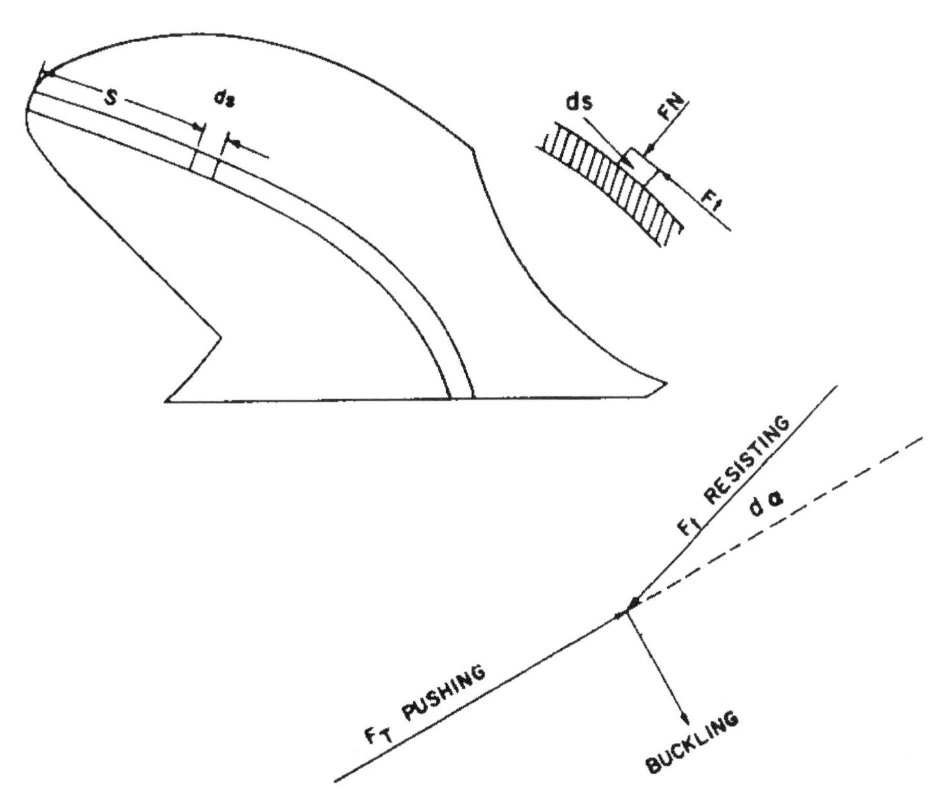

Fig. 1 Path of soil particles along a moldboard plow, including force balances. The forces change as distance traveled changes. (From Ref. [1].)

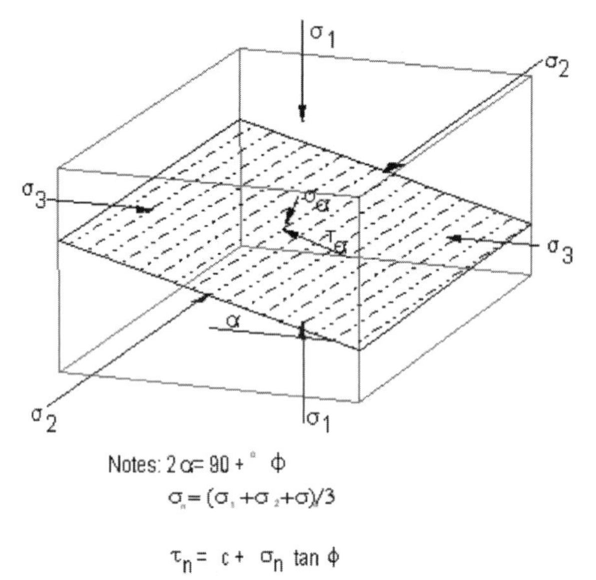

Fig. 2 Isometric cube showing shear failure plane and forces.

predominates in most civil engineering soil mechanics problems.

Compression is a failure condition associated with a volume change in soil. Approaches borrowed from beam theory may describe compression mechanics for dry, rigid soils. Empiricism increases when the soil is nonrigid (e.g., freshly tilled) and moist. Eq. 2 represents one such model describing force–compression relationships (see Ref. [9]):

$$\ln\left(1+\frac{\Delta V}{V_0}\right)=\ln\left(\frac{\rho_0}{\rho}\right)=(A+B\sigma_{\rm g})\left(1-{\rm e}^{C\sigma_{\rm h}}\right) \quad (2)$$

where ΔV is the volume change, V_0 the initial sample volume, ρ the resulting sample bulk density, ρ_0 the initial sample bulk density, $\sigma_{\rm h}$ the soil stress, and A, B, C constants.

Establishing compression failure criteria becomes problematic with tilled, moist soils.

Plastic flow is a failure condition often associated with moist, clayey soils in which the soil behaves as a non-Newtonian liquid. The Atterburg limits (liquid limit, plastic limit) are moisture contents that define plastic flow behavior for soils. Designation of failure criteria is difficult with both compression and plastic flow. The constants A, B, C are empirical and vary significantly with soil mineralogy and moisture content. Soil pressure–sinkage relations, soil cutting and soil traction relationships are replete with empirical coefficients especially when failure mechanisms other than soil shear dominate. Soil-implement frictional forces and adhesion further complicate the effort to develop workable soil mechanics. Compression and plastic flow mechanisms are similar for moist, low-density clay soils. Soil texture, moisture, and loading patterns determine the mode or combination of modes that predominate.

Soil cutting and traction mechanics with shear being the primary failure mechanism is fairly well understood. Reece[10] recognized that soil-cutting (or earthmoving) mechanics is similar to the bearing capacity of shallow foundation mechanics worked out by Terzaghi in 1943. The Reece[10] equation appears as shown in Eq. 3:

$$P=w(\gamma g d^2 N_\gamma + c d N_c + q d N_q) \quad (3)$$

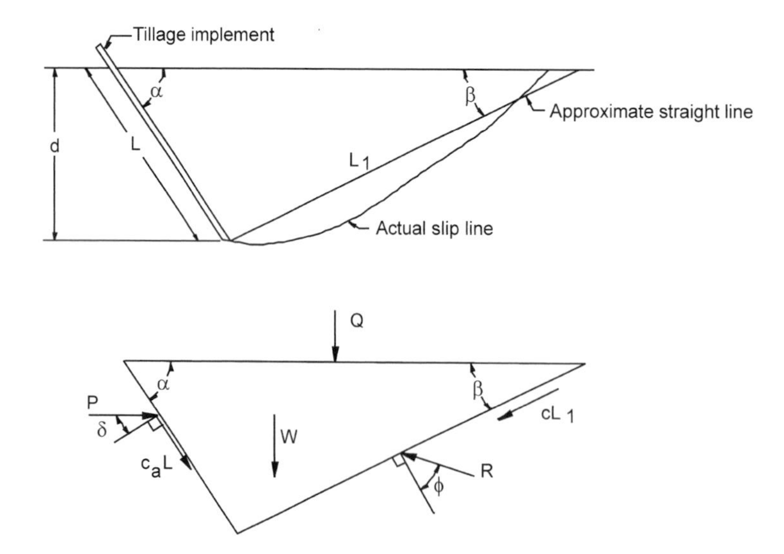

Fig. 3 Example of soil shear failure mechanics. The soil cohesion c and the soil friction angle φ are directly from the Mohr–Coulomb relationship in Eq. 1. Notes: c = soil cohesion; $c_{\rm a}$ = soil–metal adhesion; Q = weight of overburden soil (frequently zero); R = the soil resultant force; P = force exerted by the implement; δ = soil–metal friction angle; α = implement angle; ϕ = soil–soil friction angle; and β = soil wedge angle.

S

where P is the total tool force, d the tool working depth, c the soil cohesion, q the surcharge pressure along the soil surface, g the acceleration of gravity, w the tool width, and N_γ, N_c, N_q the factors depending on tool geometry, soil–metal friction, soil–metal adhesion, soil cohesion, and soil–soil friction.

Fig. 3 shows a shear failure wedge model that has provided useful approximations of tillage and earthmoving mechanics when soils fail by shear. Tool width, depth, and other geometry determine whether a 2-D or 3-D approach is appropriate for describing the cutting action of a tillage or earthmoving tool. Slight modifications of Eq. 3 may be employed to describe some tools. 2-D and 3-D implementations of Eq. 3 are discussed at length by McKyes.[11]

The predominantly shear-based tillage mechanics has provided a basis for several tillage implement design principles that are useful over a wide range of soils, which exhibit shear failure. Optimal soil–tool angles can be determined for soils as a function of cohesion and soil–soil friction. There exists a critical depth, reviewed by McKyes,[11] below which the wedge shown in Fig. 3 does not explain soil displacement. The critical depth to tool-width ratio may vary from 1 (soils exhibiting plastic flow) to 16 depending on tool geometry and soil conditions. Below critical depth, there is a zone of failure resulting in horizontal compression. The resulting compaction is undesirable for agriculture.

Traction mechanics began by studying shear-based approaches for predicting forces at incipient wheel slippage. Rolling resistance of wheels was added, building on the work of Bekker.[12] The Bekker mechanics accounts for maximum traction forces, rear wheel sinkage, rolling resistance of each machine, and wheel slip rates. McKyes[11] provides an example of how the Bekker mechanics can be used to match prime mover to tillage tool. Wismer and Luth[13] provided a convenient approach for evaluating traction capabilities of prime movers. Studies of traction mechanics have resulted in soil-engaging grousers and tire lug designs that are generalized over a wide variety of soils.

Prime mover tillage implement matches should accomplish the following objectives.[11]

1. The prime mover must have the traction capacity necessary to provide the draft required by the soil-cutting implement.
2. The prime mover should have sufficient reserve tractive capability such that slip is not excessive, and energy use and tire wear will not be excessive.
3. The prime mover should not be overly heavy or powerful such that it operates away from its optimal range.
4. The prime mover should not have more weight or ground pressure, lest compaction be excessive.

A complete soil mechanics must bridge from shear to other failure modes. The existence of multiple failure modes and the dependence of failure mode on soil moisture and texture are confounding. Ultimately, the 3-D stress–deformation constitutive relationship is pivotal for describing the relationship between the forces and resulting soil displacement. Once an adequate force–deformation relationship is known, then complex problems may be solved using sophisticated finite difference/finite element modeling techniques appropriate for nonlinear materials. Karafiath and Nowatzki[14] developed a finite difference model of the soil–vehicle interaction using plasticity theory, an extension of Mohr–Coulomb shear mechanics. Schafer[3] summarizes other results using similar approaches for modeling vehicle mobility. These approaches enable detailed evaluation of alternative tire lug and track grouser design.

The introduction of critical state soil mechanics (CSSM), cam-clay, and related approaches developed at Cambridge University over the latter 20th century, represented a step forward in the attempt to bridge across soil failure modes. CSSM successfully models many soil failure modes and is providing insights to more general force deformation models (e.g., Ref. [15]). A detailed presentation of CSSM is beyond the scope of this article. Atkinson and Bransby[16] introduced CSSM. Prospects for improvements in the force–deformation models bode well for soil dynamics progress. Volumes of the *Advances in Soil Dynamics* series that address CSSM and related approaches are currently in progress.

FUTURE DIRECTIONS

The soil dynamics discipline continues to affect, directly and indirectly, feed, food, and fiber production. Soil dynamics indirectly relates to animal grazing management via the compaction effects of animal hooves and resulting effect on forage plant growth. Plant response in the presence of hard pans is influenced by soil dynamics in that root grow in response to soil deformations. Soil dynamics is an integral part of understanding the reaction of soils to natural phenomena such as freezing, thawing, and shrinking–swelling. Barnes et al.[2] developed a monograph summarizing in an empirical manner how natural and man-induced soil forces and resulting compaction affects water, nutrients, heat, and air transport in soils. Soil transport phenomena and strength affects plant establishment, growth, and yield. The lack of a complete, high-utility mechanics explains the high level of empiricism. On the global scale, soil dynamics and compaction management are nearly identical in scope.

Recent trends toward greatly reduced mechanical tillage for crop production is shifting priorities in soil dynamics from mechanical tillage investigations to biologically based soil dynamics phenomena. The utility of selected plants for removing the effects of hardpans via root penetration is one example. The effect of soil hardness on earthworm and other biological entity activity is another example where contemporary knowledge of soil dynamics may contribute. Thus, the problem scale is shifting from farm-field scale to an individual biological unit scale. Continued advances in the development of a complete soil mechanics will no doubt foster understanding the relation of soil mechanics to feed, food, and fiber production.

REFERENCES

1. Gill, W.R.; VandenBerg, G.E. *Soil Dynamics in Tillage and Traction*; USDA-ARS Handbook No. 316; United States Department of Agriculture: Washington, D.C., 1968.
2. Barnes, K.K.; Carlton, W.M.; Taylor, H.M.; Throckmorton, R.J.; VandenBerg, G.E. *Compaction of Agricultural Soils*; ASAE Monograph No. 1; ASAE: St. Joseph, MI, 1971.
3. Schafer, R.L., Ed. Proceedings of the International Conference on Soil Dynamics, Sponsored by the USDA National Tillage Machines Laboratory and the Auburn University Agricultural Experiment Station, Auburn, AL, 1985; 5 Volumes.
4. Chancellor, W.J. *Advances in Soil Dynamics*; ASAE: St. Joseph, MI, 1994; Vol. 1.
5. Doner, R.D.; Nichols, M.L. The Dynamic Properties of Soil: V. Dynamics of Soil on Plow Moldboard Surfaces Relating to Scouring. Agric. Eng. **1934**, *15*, 9–13.
6. Kawamura, N. Study of the Plow Shape I: Analysis of the Sod Plow Considering Strength of Soil. Soc. Agric. Mach. J. (Jpn) **1952**, *14* (3), 65–71.
7. Clark, R.L.; Radcliffe, D.E.; Langdale, G.W.; Bruce, R.R. Soil Strength and Water Infiltration as Affected by Paratillage Frequency. Trans. ASAE **1993**, *36* (5), 1301–1305.
8. Chancellor, W.J. Soil Physical Properties. In *Advances in Soil Dynamics*; Chancellor, W.J., Ed.; ASAE: St. Joseph, MI, 1994; Vol. 1, 21–245.
9. Bailey, A.C.; Johnson, C.E.; Schafer, R.L. A Model for Agricultural Soil Compaction. J. Agric. Eng. Res. **1986**, *33* (4), 257–262.
10. Reece, A.R. The Fundamental Equation of Earthmoving Mechanics. *Symposium on Earthmoving Machinery*; Institute of Mechanical Engineers: London, 1965; Part 3F, 179.
11. McKyes, E. *Soil Cutting and Tillage*; Elsevier: Amsterdam, Netherlands, 1985.
12. Bekker, M.G. *Theory of Land Locomotion—The Mechanics of Vehicle Mobility*; Univ. of Michigan Press: Ann Arbor, MI, 1960.
13. Wismer, R.D.; Luth, H.J. Off-Road Traction Prediction for Wheeled Vehicles. Trans. ASAE **1974**, *17* (1), 8–12.
14. Karafiath, L.L.; Nowatzki, E.A. *Soil Mechanics for Off-Road Vehicle Engineering*; Trans Tech Publications: Aedermannsdorf, Switzerland, 1978.
15. Grisso, R.D.; Johnson, C.E.; Bailey, A.C. Soil Compaction by Continuous Deviatoric Stress. Trans. ASAE **1987**, *30* (5), 1293–1301.
16. Atkinson, J.H.; Bransby, P.L. *The Mechanics of Soils: An Introduction to Critical State Soil Mechanics*; McGraw-Hill: London, 1978.

Soil Properties

Richard Cooke
University of Illinois, Urbana, Illinois, U.S.A.

S

INTRODUCTION

Soil is a heterogeneous, multiphase, disperse, and porous system. It is the most common medium for plant growth. Plants extract nutrients from the soil, and the soil also provides rigid support for plant growth.

SOIL PHASES

- *Soil particles*: These are mainly alumino-silicates from weathered rocks and organic matter from the decay of vegetation. Organic matter is usually limited to the top layers of the soil.
- *Soil water*: Soil water contains dissolved minerals which plants need for growth and cell integrity. Almost all the nutrients used by the plants are obtained from the soil solution.
- *Soil air*: Soil air provides oxygen for root respiration and for microbial activity.

Irrigation and drainage are concerned with maintaining the optimum balance between soil water and soil air. If there is too much water, root growth tends to be shallow, roots may rot, and anaerobic reactions may produce toxic by-products that reduce growth. If there is too little water, the supply of nutrients to the plants may be limited and yields are reduced. If conditions are dry enough, wilting may occur and the plants may even die. Some plants have evolved mechanisms for acquiring oxygen under extremely wet conditions or for conserving water under extremely dry conditions.

SOIL TEXTURE

Soil particles are mainly weathered rock particles. These particles are classified according to their diameters (Table 1).

Soil texture can be determined in either a qualitative or a quantitative manner.[1] Qualitatively, soil texture can be determined by how the soil feels when it is rubbed between the fingers. In general, soil behavior can be inferred from such a qualitative classification. A coarse-grain sandy soil tends to be loose, well aerated, and easy to cultivate. Water moves through sands relatively quickly and they do not retain much water, since they have few small pores. A fine-textured clay soil tends to absorb much water and become plastic and sticky when wet, and tight, compact, and cohesive when dry. Water moves through clays more slowly, but because they have many small pores, they tend to retain more water.

Quantitatively, soil texture is determined by the proportions of sand, silt, and clay that make up the mineral portion of the soil (i.e., excluding the organic matter in the soil). Soil texture information is conveniently displayed in a textural triangle, one version of which is shown in Fig. 1. In this triangle, the axis for each component runs parallel to the baseline that is opposite the apex representing 100% of that component. In the example shown on the triangle, the soil contains 30% sand, 35% silt, and 35% clay and is thus classified as a clay loam.

Clay and Clay Mineralogy

Clay plays an important role in soil behavior. Soils may be classified based on proportions of sand, silt, and clay particles[3]:

- Below 30%–35% clay: Clay particles are dispersed in a matrix formed by the coarse particles. Soil properties are determined by the clay content.
- Above 30%–35% clay: Coarse particles are dispersed in a continuous clay matrix. Properties are determined by the clay mineralogy. These soils are known as *clayey soils*.

Clay minerals are differentiated on the basis of the arrangement of the aluminum and the silicon ions in the molecules that make up the particles. The main clay minerals are kaolinite, illite, and montmorillonite, and soils are classified according to the dominant mineral. Kaolinites undergo very little volume change when they are wetted while montmorillonites undergo dramatic volume changes. Illites are intermediate between the two extremes. Montmorillonites tend to form irreversible cracks from repeated wetting and drying. These cracks can influence water movement and water holding capacity in these soils. Pottery and other clayware are made mainly from kaolinitic soils.

Encyclopedia of Agricultural, Food, and Biological Engineering
DOI: 10.1081/E-EAFE 120006920
Copyright © 2003 by Marcel Dekker, Inc. All rights reserved.

Table 1 Classification of soil particles

	USDA classification (mm)	International classification (mm)
Clay	< 0.002	< 0.002
Silt	0.002–0.05	0.002–0.02
Very fine sand	0.05–0.10	
Fine sand	0.10–0.25	0.02–0.20
Medium sand	0.25–0.50	
Coarse sand	0.50–1.00	0.20–2.00
Very coarse sand	1.00–2.00	
Gravel	> 2.00	> 2.00

Water Related Soil Properties

Water is essential for plant growth. Plants extract mineral laden water from the soil. For effective irrigation and drainage it is important that the processes governing soil water storage and soil water movement be understood.

- In a *saturated* soil all the pore space is occupied by water. Under field conditions full saturation is hardly ever achieved, as there is almost always some air trapped in the soil. The approximate saturation achieved in practice is referred to as *field saturation*.
- As the soil drains, air occupies the biggest pores in the soil. As more and more water is removed from the soil, smaller and smaller pores are drained.
- *Field capacity* is defined as the soil water content when the gravity filled pores are drained. This occurs one day after field saturation in a sandy soil and 3 day after field saturation in a clay soil.
- Below field capacity, for all intents and purposes, no more gravity drainage occurs. Water is removed from the soil by evaporation at the soil surface or through uptake by plants.
- *Permanent wilting point* is defined as the soil water content below which plants cannot extract any water from the soil. Water is then held in very small pores or chemically bound to the soil particles. Permanent wilting point corresponds to a suction of 15 bar (Fig. 2).

In irrigation and drainage, the main objective is to maintain the soil water content between field capacity and permanent wilting point.

Several other important soil properties are defined below. These properties can be visualized with the help of the schematic diagram (Fig. 3). This diagram shows the soil as a three-phase system. For convenience the phases are separated.

Porosity (n)

$$\text{Porosity}\,(n) = \frac{\text{Volume of pores}}{\text{Volume of soil}} = \frac{V_p}{V}$$

In general, porosity ranges from 0.3 to 0.6.[1] Coarse-textured soils have larger pores than fine textured soils, but with smaller overall porosity.

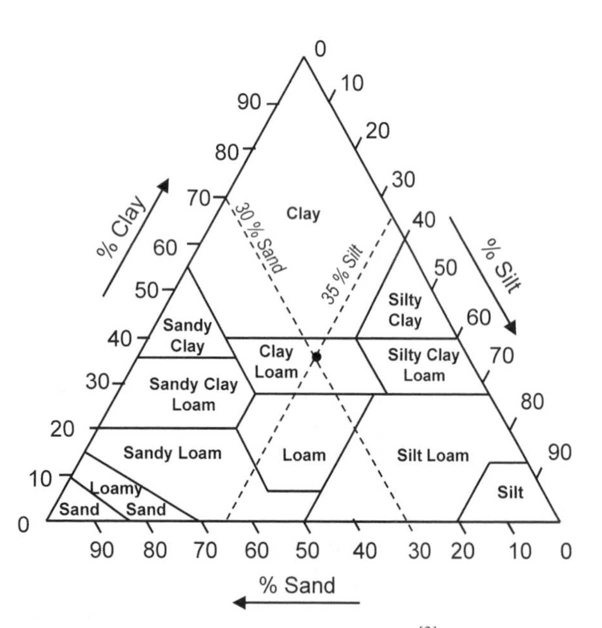

Fig. 1 Soil textural triangle[2].

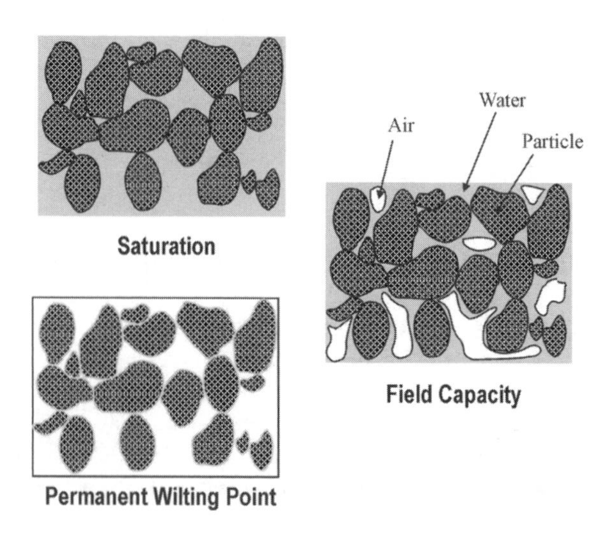

Fig. 2 Soilwater level under different field conditions.

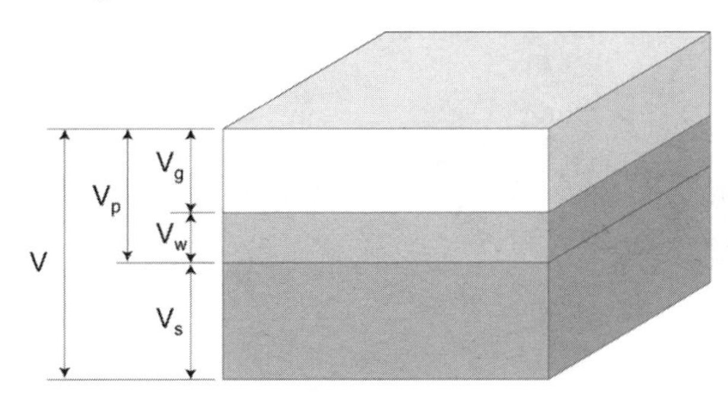

Fig. 3 Schematic diagram of soil phases.

Bulk density (ρ_b)

$$\text{Bulk density}\,(\rho_b) = \frac{\text{Mass of soil particles}}{\text{Volume of soil}} = \frac{M_s}{V}$$

Sandy soils have higher bulk density than clayey soils. The density of the alumino-silicate particles in the soil is 2.6 gm cm^{-3}. Thus, if a soil has a porosity of 0.5, the bulk density will be 1.3 gm cm^{-3}. In general,

$$\text{Porosity} = 1 - \frac{\text{Bulk density}}{2.6}$$

Volumetric water content (θ)

$$\text{Soil water content}\,(\theta) = \frac{\text{Volume of water}}{\text{Volume of soil}} = \frac{V_w}{V}$$

In nonswelling soils, the value of soil water content at saturation is equal to the porosity. In swelling soils, however, the volume of water at saturation may be much more than the porosity of the dry soil.

Saturated hydraulic conductivity

Saturated hydraulic conductivity is the rate at which water moves through saturated soils. It is one of the most

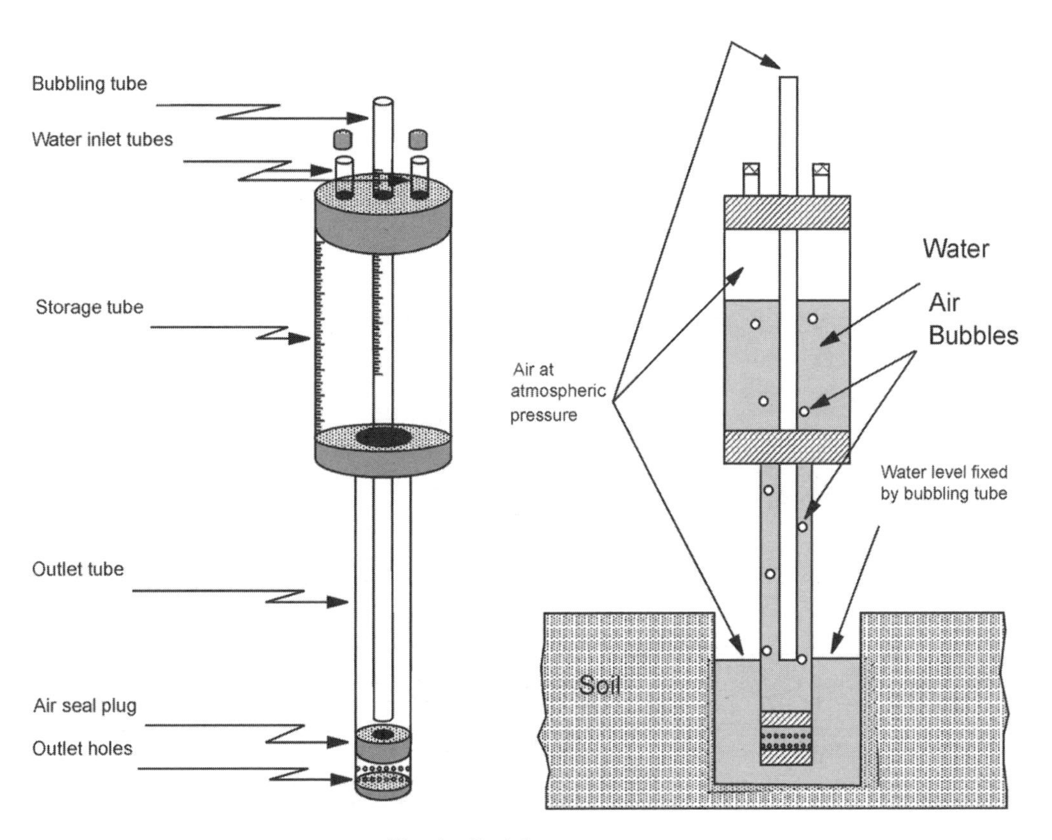

Fig. 4 Guelph permeameter.

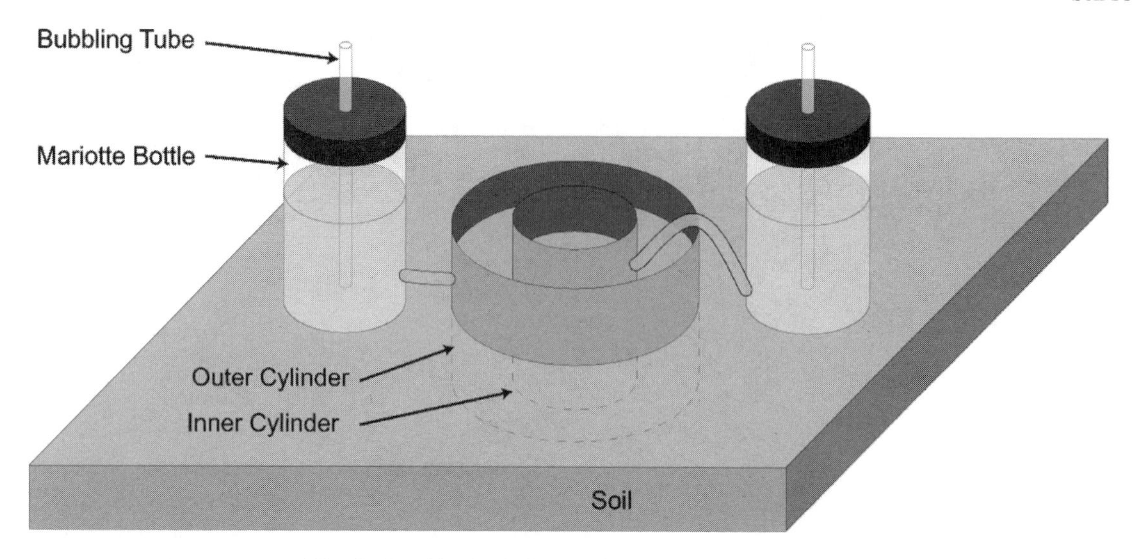

Fig. 5 Double ring infiltrometer with Mariotte bottles.

used soil properties in irrigation and drainage. It is dependent on the total porosity of the soil and also on the arrangement of the pores. It has values of about $1\,\mathrm{cm\,hr^{-1}}$–$10\,\mathrm{cm\,hr^{-1}}$ in sandy soils and $0.0001\,\mathrm{cm\,hr^{-1}}$–$0.1\,\mathrm{cm\,hr^{-1}}$ in clayey soils. As soils dry out, the cross-sectional area for water transport decreases. Thus, the rate at which water moves through the soil also decreases. The unsaturated hydraulic conductivity is normally orders of magnitude less than the saturated hydraulic conductivity.

While there are methods for measuring hydraulic conductivity in the laboratory, field methods produce more representative values. Two such methods are described later.

Guelph Permeameter. The first step in using the permeameter is to make a hole in the ground (Fig. 4). The bubbling tube is then adjusted to maintain the water level in the hole at a fixed level. Eventually, the rate at which the water level in the storage tube falls becomes constant. This constant rate, along with the geometry of the hole and the water depth in the hole, is used in the determination of hydraulic conductivity.

Cylinder Infiltrometers. Cylinder infiltrometers consist of metal cylinders, partially driven into the ground, filled with water, with the hydraulic conductivity being the rate at which the water level falls (Fig. 5). There is usually some lateral seepage below the bottom of the cylinder. This shortcoming is usually eliminated by using two concentric cylinders, with the outer cylinder acting as a buffer. Since, there is some difficulty in maintaining the same water level in both cylinders under falling head conditions, modern double cylinder (ring) infiltrometers use Mariotte bottles to maintain a constant level of water in both cylinders.

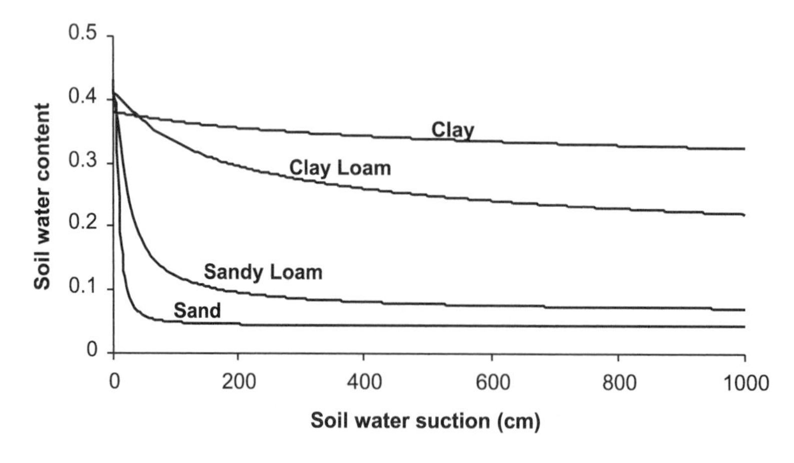

Fig. 6 Soil water retention curves.

SOIL WATER RETENTION

Soil Suction

Consider a saturated soil sample in contact with a porous ceramic membrane that allows the passage of the soil solution but not the soil particles. If the plate is placed under suction (application of a subatmospheric pressure) then water will flow out of the soil, with the volume removed being a function of the soil water suction. The curve showing the relationship between soil water suction and soil water content for a soil is called the *characteristic curve* or the *retention curve*. In actual fact, the shape of the curve depends on whether suction is being increased or decreased, that is, whether the soil is being dried or being wetted. This phenomenon, termed hysteresis, is due to contact angle effects, the occurrence of entrapped air, swelling and shrinking, and inkbottle effect in the soil.

Retention curves for four soil types are shown in Fig. 6. The sand has the greatest drop-off in water content at low suctions, a reflection of the relatively low suction required to empty large pores. The clay exhibits very little change in water content over the range of suctions shown. The loams are intermediate between the two extremes.

REFERENCES

1. Hillel, D. *Soil and Water. Physical Principles and Processes*; Academic Press: New York, 1971; 288 pp.
2. Hudson, N.W. *Field Engineering for Agricultural Development*; Clarendon Press: Oxford, 1975; 226 pp.
3. Warkentin, B.P. Clay Soil Structure Related to Soil Management. Trop. Agric. (Trinidad) **1982**, *59*, 82–91.

Solar Energy

John W. Bartok, Jr.
University of Connecticut, Storrs, Connecticut, U.S.A.

INTRODUCTION

The main applications for solar energy use in agriculture have been in crop and grain drying, greenhouse and nursery crop production, space heat for buildings, remote electricity supply, and water pumping. Although solar energy is classified as a renewable, nonpolluting source, its use has been limited by climate and location. In agriculture, collectors have been built utilizing low-cost plastic and sheet metal materials and many applications use existing structures. For most applications, it is most economical to reduce the need for energy first by installing insulation, purchasing efficient motors, making good use of space, and servicing environment control equipment.

OVERVIEW

Grain and Crop Drying

Most solar collectors designed for grain and crop drying are active air systems without heat storage. They consist of a single or double pass collector, fan or blower, and ducting. Plans are available for fixed and movable flat plate collectors, solar attics and walls, inflated tube collector (Figs. 1 and 2), and wrap-around solar bin collector[1] (Fig. 3). These provide low-temperature heat that is adequate for drying crops. Additional heat may be necessary from a fossil fuel heater or other source. If possible, the collector should be located so that it can provide heat for the home or work area when not in use for crop drying. Some portable collectors have been developed that can be moved to another location and application (Fig. 4).

Collectors can be sized based on maximum temperature rise (outlet minus inlet), average 24-hr temperature rise or peak or average heating load. Common cover materials are polycarbonate, fiberglass reinforced plastic (FRP), or polyethylene. Collectors are usually shaded during the summer when not in use to prevent excessive heat and extend the life of the cover.

Recommendations for hot air collectors are to maintain an air velocity of $2.54\,\mathrm{m\,sec^{-1}}$–$5.08\,\mathrm{m\,sec^{-1}}$ [500 feet/min–1000 feet/min (fpm)] past the absorber. Axial-flow and backward-curved centrifugal fans are the most common for air movement. Sheet metal, plywood or rigid foil-faced fiberglass are suitable for ducts. All ducts should be airtight, as short as possible, and insulated to reduce heat loss. They should be sized for at least $1\,\mathrm{m^2}/2.5\,\mathrm{m^3\,sec^{-1}}$–$3.0\,\mathrm{m^3\,sec^{-1}}$ [$1\,\mathrm{ft^2}$/500 cubic feet/min–600 cubic feet/min (cfm)] air flow. The air is not recirculated as it contains moisture and dust.

Passive and active solar dryers have been built to dehydrate fruit and vegetables. They contain screened trays or racks covered by a structure with a glazed south-facing roof. Air movement removes the moisture that evaporates from the crop.

Although tobacco has been traditionally dried with fossil fuels, there is an increasing interest in the use of solar energy. Ideal curing conditions are a temperature of 18–35°C (65–95°F) and relative humidity of 75%–80%. Conventional hot air solar collectors and greenhouse structures are being used to provide the solar heat.

A solar kiln can provide an environment where moisture extraction from the lumber can be controlled. These are active systems with a fan that draws air through the solar panels, which is part of the roof (Fig. 5). The fan circulates the hot, dry air (43–60°C) through the stacked lumber where it absorbs moisture. It is then expelled outside. Moisture content of the lumber is reduced to 6%–8%. No heat storage is required. Commercial kilns and plans for home built kilns are available.

Greenhouses

Greenhouses depend on solar radiation for light and heat during the day. They need fossil fuel heat during the night when about 80% of the heat load occurs. Research at Rutgers University[2] and other universities have evaluated low-cost external solar water systems (Fig. 6). Heated water is stored in the floor below the crop.

Systems for collecting the excess heat from the sun's radiation that is trapped within the greenhouse have been developed. These can be either passive (water barrels, rock collectors, or phase change salts), or active (EPDM water tubing or air ducts). Storage can be in water tanks or rock beds located below the floor. The high initial cost and maintenance have limited their use. Solar heated salt ponds covered by a greenhouse structure have been developed by the Ohio Agricultural and Research Development Center (OARDC) at Wooster[3] and other research stations (Fig. 7). Their advantages include

Encyclopedia of Agricultural, Food, and Biological Engineering
DOI: 10.1081/E-EAFE 120006844
Copyright © 2003 by Marcel Dekker, Inc. All rights reserved.

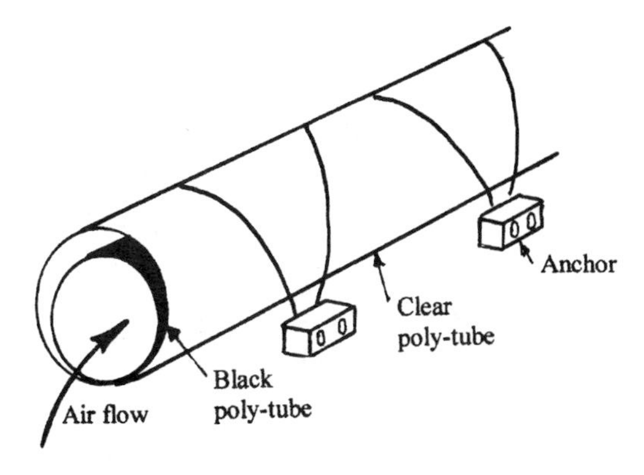

Fig. 1 Poly-tube air collector.

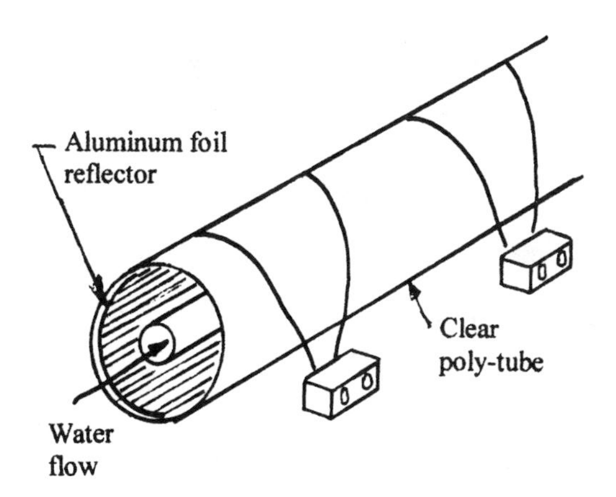

Fig. 2 Poly-tube water heater.

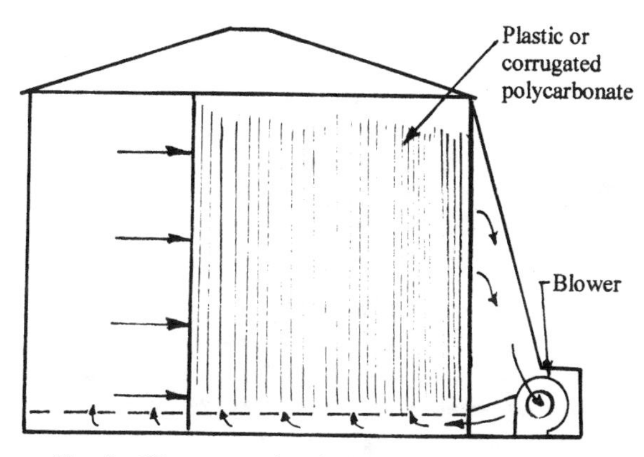

Fig. 3 Wrap-around grain bin air hot collector.

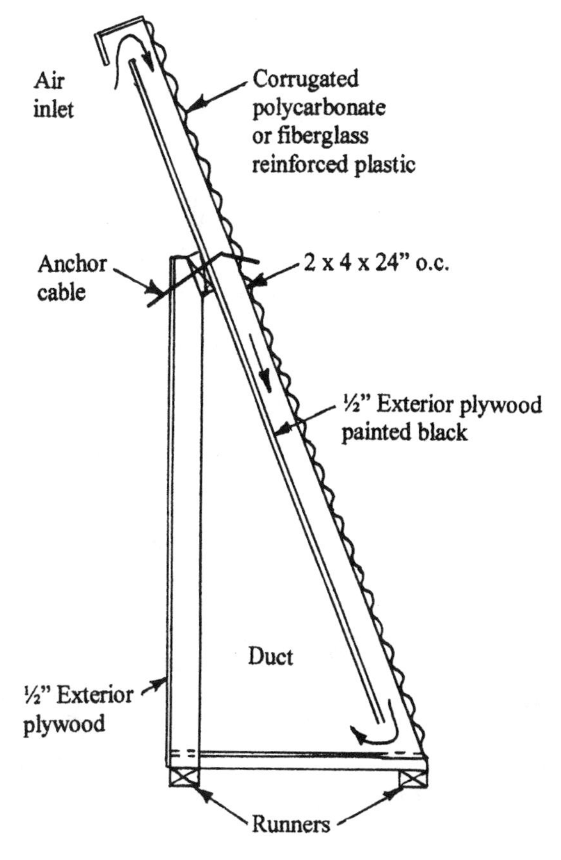

Fig. 4 Portable air collector.

relatively low cost, passive operation, and the ability to collect and store summer radiation for use during the winter. They have not been adopted by the commercial greenhouse industry due to space and management considerations.

The pond that is similar to an in-ground swimming pool is filled with water. Sodium chloride or other salt is dissolved in the water to form a uniform concentration in the lower half and decreasing concentration gradient from the pond middepth to the surface. The water, which is heated all summer reaches a temperature above 66°C, is drawn off when heat is needed during the cooler parts of

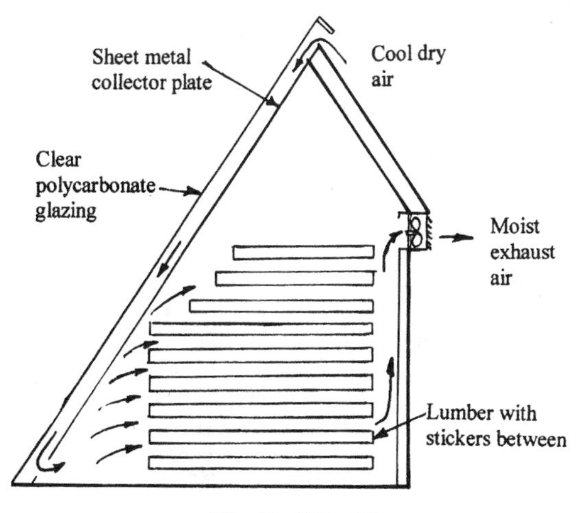

Fig. 5 Solar kiln.

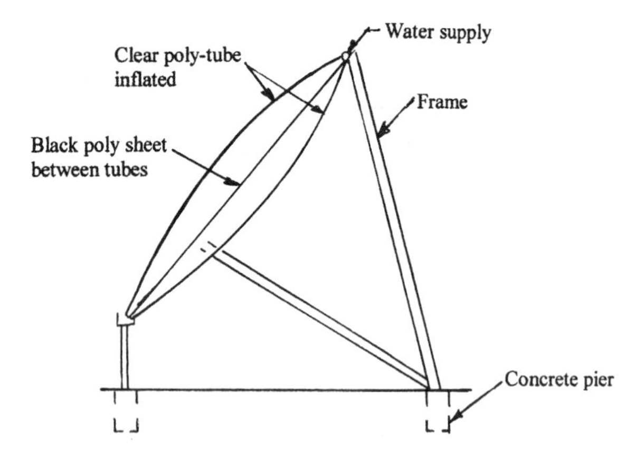

Fig. 6 Rutgers University water collector for greenhouse floor heat.

the year. Water is pumped from the pond to air heat exchangers in the greenhouse. An alternative method utilizes a heat pump and a direct hot water heating system.

Solar design greenhouses for hobby use are common.[4] They employ an insulated building with south facing glazing to capture the heat. Barrels or cans of water, bins filled with rock or tubes filled with Glauber salt ($Na_2SO_4 \cdot 10H_2O$) or other phase change salts, provide heat storage.

Space Heating

Indoor air in livestock buildings contains moisture, odors, toxic gases, and dust. This has to be removed and replaced with clean air that may have to be tempered. Solar energy has been utilized to preheat the make-up air and supplement a fossil fuel heating system. A limited number of commercial collector systems have been installed. More common systems are farm built as a solar roof or wall on the building or a separated collector system[5] (Fig. 8).

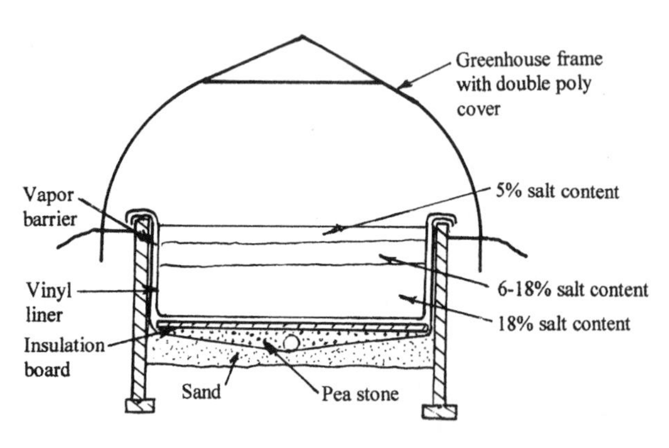

Fig. 7 Solar pond.

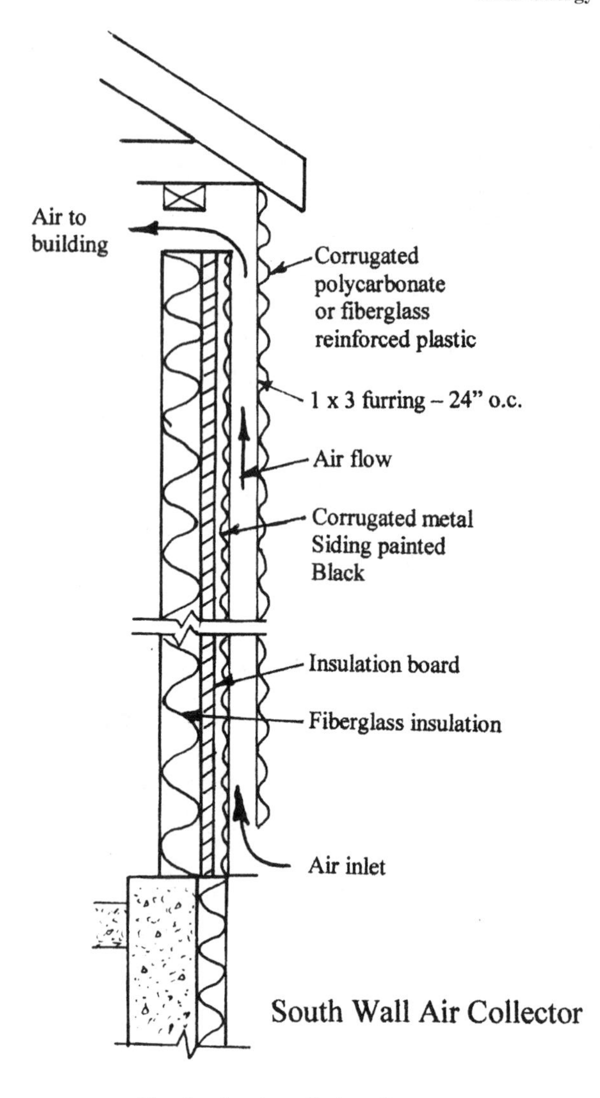

Fig. 8 South wall air collector.

Most are air systems that incorporate a rock storage to make the heat available at night.

Depending on location and design, these systems have efficiencies of 10%–60%. Research has shown that higher airflow increases transfer coefficients and efficiency but reduces peak temperature. An airflow of $0.01\ m^3\ sec^{-1}$–$0.015\ m^3\ sec^{-1}$ per m^2 ($2\ cfm\ ft^{-2}$–$3\ cfm\ ft^{-2}$) of collector has been found to be best.

Rock storages are normally sized at about 60 cu m of 50 mm–100 mm diameter rock for each cu m/sec (1 cu ft/cfm) of airflow required for the livestock ventilation system. The maximum storage temperature will be a few degrees less than the maximum air temperature. Fan capacities should be based on delivery against 6.35 mm static pressure. The storage should be well insulated to retain the heat.

Water Heating

Solar hot water heaters have been installed mainly to preheat water for animal preparation and cleanup in animal facilities and for tempering water in plant production facilities. Commercially available units mounted on a barn or shed roof work well and are easy to install.

Photovoltaics

Most farms require considerable electricity for environment control and materials handling. Photovoltaics, the direct conversion of solar energy to electricity, is starting to provide this need and has great potential for future application.[6] It is one of the best technologies that result in the least disruption to current energy use practices.

Photovoltaics can be installed as an independent system with battery storage or with an interface to the electric utility grid. Battery systems include photovoltaic modules, deep cycle batteries (6 V, 12 V, 24 V, or 48 V), a charge regulator, and a load center that monitors system operation. Addition of an electronic tracker increases power output as the modules stay fixed on the sun. The electricity can be used directly for lighting and direct current motors or can be converted by a 120/240 V synchronous power inverter to alternating current for existing equipment. The equipment has to be matched to the load requirements. In most states the excess electricity generated from alternative systems has to be purchased by the local utility. Some utilities purchase this power at net metering (retail price) rather than wholesale or avoided cost. This provides a substantial improvement in net returns.

Present uses include power for electric fences, lighting, irrigation, and remote equipment operation. Off the shelf systems for fencing and lighting are competitive with conventional electric systems.

Photovoltaic systems for irrigation, livestock water supply, and pond aeration may be the most cost effective option for locations where there is no existing power supply. The size and cost of a system depends on the water supply and peak demand.

Photovoltaic system costs have decreased from about $35/peak watt to a present cost of about $5/peak watt. Further reductions to less than $1/watt would greatly increase the use of photovoltaics. Conversion efficiency of sunlight ranges between 10% and 25%.

Economics

Solar energy development and use in agriculture has generally been tied to fossil fuel prices, government support for research or in some parts of the world, the availability of energy resources. Major research and development took place in the 1980s when fossil fuels were in short supply and prices increased significantly. Since the 1990s, research has been limited and the number of manufacturer's and suppliers has decreased.

Decisions in planning a solar heating system are based on the nature and size of the demand for heat, hot water, and electricity. The selection is also based on whether the transfer medium will be water, air, or if electricity is needed. A next step is to decide if storage is required. Alternative systems should be compared on cost and the amount of energy collected. The bottom line can be evaluated by looking at the economics.

Although the net present value or a levelized cost of energy technique for evaluating solar system economics gives more accurate costs, most of the research has used the simple payback method.[7] To improve the economics, most agriculture applications are designed to provide only part of the total load. Another factor that needs to be considered is the large amount of space needed in many installations. For example, in greenhouse heating, about 1 m^2 of collector is needed for each square meter of greenhouse floor area. The collector space has a return of about $10/$m^2$ whereas using it for additional greenhouse production area will return $50/$m^2$–$150/m^2.

REFERENCES

1. *Low Temperature & Solar Grain Drying*; MWPS-22 Midwest Plan Service, Iowa State University: Ames, IA, 1980; 86.
2. Mears, D.R.; Roberts, W.J.; Cipolletti, J.P. *Solar Heating of Commercial Greenhouses*; Dept. of Biological and Agricultural Engineering, Rutgers University: New Brunswick, NJ, 1981; 71.
3. Husseini, I.; Short, T.; Badger, P. *Radiation on a Solar Pond with Greenhouse Covers and Reflectors*; Transactions of the American Society of Agricultural Engineers, 1979; Vol. 22, No. 6, 1385–1389.
4. Yanda, B.; Fisher, R. *Solar Greenhouse—Design, Construction and Operation*; John Muir Publications, Inc.: Santa Fe, NM, 1980.
5. USDA. On-Farm Demonstration of Solar Heating of Livestock Shelters. Prepared for US Dept. of Energy, Division of Solar Thermal Energy Systems, Interagency Agreement No. DE-A101-78CS35149; 1982.
6. Garg, H. *Solar Energy*; Reidel Publishing Co., 1987; Vol. 3, 174.
7. Duffie, J.A.; Beckman, W.A. *Solar Energy and Thermal Processes*; John A. Wiley and Sons: New York, NY, 1974.

Solid Food Rheology

V. N. Mohan Rao
Ximena Quintero
Frito-Lay, Plano, Texas, U.S.A.

INTRODUCTION

In order to provide foods of higher quality a more extensive knowledge of their physical and mechanical properties is needed. The field of rheology encompasses the mechanical properties of solids, semisolids, and liquids. The mechanical properties of foods are extensively used in the design of food handling equipment and in the development of process conditions for unit operations in the food industry. An example is the use of modulus of elasticity of grains at various temperatures and relative humidities in predicting internal stresses during drying. The drying process is then optimized so that the grains can be dried minimizing any stress cracks due to thermal stresses. Additionally, some mechanical properties of finished food products (percent sag or creep of the food gel made from pectin or a similar source) are used in quality control to ensure consistent production. The mechanical properties of foods (solids and semisolids) will be discussed in this article.

STRESS AND STRAIN

Stress can be defined as the response or internal reaction of a material to applied forces. It is a force intensity reaction dependent on the area on which the forces are acting and is expressed as a force per unit area as area (A) approaches 0. Stress is a tensor quantity and is completely defined by a 3 × 3 matrix in a Cartesian coordinate system. Stress can be classified as normal stress (σ) or shear stress (**⓿**) and can be generally expressed as follows.

$$\sigma \quad \text{or} \; \textbf{⓿} = \lim_{\delta A \to 0} (F/\delta A)$$

Strain is the relative change in dimension or shape of a body subjected to stress. Strain is also a tensor quantity that is expressed as the ratio of the change in dimension to the original dimension as is therefore dimensionless.[1]

The normal strain component is defined as the fractional change in the original length of line and is designated by ε with subscripts to indicate the direction of the line for which the strain is measured. Thus, from Fig. 1, the value of ε_{11} in the case of plane strain at a point O is

$$\varepsilon_{11} = \lim_{\mathrm{OC} \to 0} \frac{\mathrm{O'C'} - \mathrm{OC}}{\mathrm{OC}}$$

Similarly, the other normal strain ε_{22} is also defined as

$$\varepsilon_{22} = \lim_{\mathrm{OE} \to 0} \frac{\mathrm{O'E'} - \mathrm{OE}}{\mathrm{OE}}$$

The normal strain is positive when the line elongates and negative when the line contracts. The shear strain component is specified with respect to two axes, which are perpendicular in the undeformed body, and is designated by γ_{ij}, where i and j are not the same number. For small shear strains it is adequate to define shear strain in terms of the change in angle (Θ, in radians) or as the tangent of this change in angle. Using this definition, from Fig. 1, the value of shear strain at point O is

$$\gamma_{ij} = \Theta = \lim_{\mathrm{OC} \to 0} (\angle \mathrm{COE} - \angle \mathrm{C'O'E'})$$

ELASTIC (HOOKEAN) SOLIDS

A Hookean solid has a magnitude of deformation proportional to the magnitude of the applied force. It is rheologically represented by a spring and characterized by a constant called the modulus or elastic modulus, which is defined as the ratio of the stress to the strain. Depending on the method of force application, three kinds of moduli may be computed for a Hookean solid.[2]

The modulus determined by applying a force perpendicular to the area defined by the stress (Fig. 2) is called the modulus of elasticity or Young's modulus (E) expressed as,

$$E = \sigma/\varepsilon$$

and

$$\varepsilon = \delta L/L$$

where σ is the stress, ε is the strain, L is the original length, and δL is the change in length. The modulus of elasticity is a measure of the stiffness of the material.

Encyclopedia of Agricultural, Food, and Biological Engineering
DOI: 10.1081/E-EAFE 120006971
Copyright © 2003 by Marcel Dekker, Inc. All rights reserved.

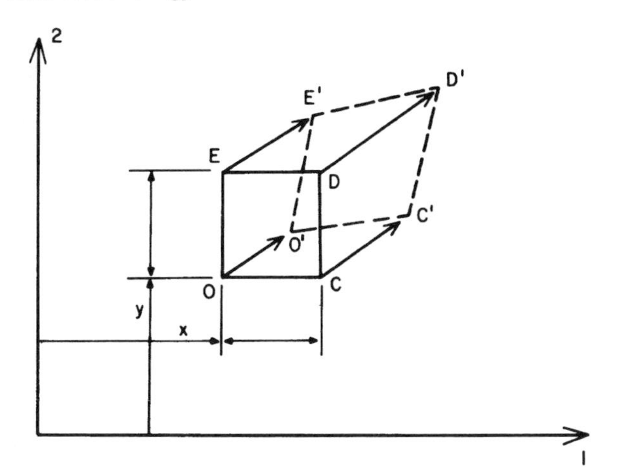

Fig. 1 Deformation of a small element (OCDE) along a plane surface.

The modulus computed by applying a force parallel to the area defined by the stress, or a shearing force, is called the shear modulus or the modulus of rigidity (G). G is a measure of material resistance to change in shape. It is defined as

$$G = \tau/\gamma$$

where τ is the shear stress and γ is the shear strain.

If the force is applied from all directions (isotropically) and the change in volume per original volume obtained, then one can compute the bulk modulus (K) as

$$K = \sigma/\varepsilon_v$$

where σ is the isotropic stress and ε_v is the volumetric strain (change in volume/original volume).

Another quantity that is commonly used is called the Poisson's ratio (ν) which can be defined from compression data (Fig. 1):

$$\nu = -\varepsilon_{22}/\varepsilon_{11}$$

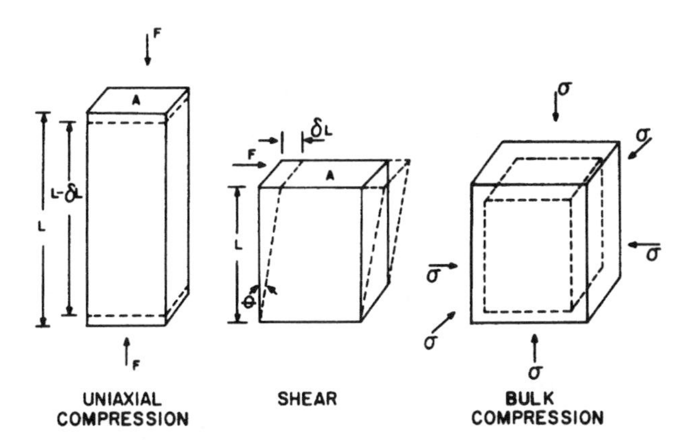

Fig. 2 Uniaxial compression, shear and bulk (isotropic) compression of an elastic solid.

Theoretically, Poisson's ratio may vary from -1 to 0.5, but negative values are very uncommon. Typically, ν varies from 0.0 to near 0.5 for materials like water (0.5 represents incompressible materials).

RHEOLOGICAL TESTING OF IDEAL SOLIDS

Uniaxial compression–tension tests are the simplest of all quasistatic tests. In these tests, a sample in the form of a cylinder or a cube is deformed at a constant deformation rate. If the magnitudes of force and deformation are small, then the body may be assumed to be elastic.[3] In addition, the behavior may be linear elastic if the stress and strain are linearly related and nonlinear if the relationship is not linear.

Another test that can be applied to foods, which are fairly brittle, (e.g., spaghetti) is the bending test (Fig. 3). The formula for the modulus of elasticity can be derived[4] yielding

$$E = \frac{FL^3}{48I\delta}$$

where E is the modulus of elasticity, F is the force, L is the length of the sample, I is the moment of inertia, and δ is the maximum deflection.

For materials that cannot be modified to yield a sample possessing convenient geometry the application of Hertz's equations is appropriate.[5] The necessary equations for a spherical and convex body (Fig. 3) are as follows.

For axial loading of a spherical sample between flat plates

$$E = \frac{0.531\,F(1-\nu^2)}{D^{1.5}}[4/d]^{0.5}$$

For a plate on a convex body

$$E = \frac{0.531\,F(1-\nu^2)}{D^{1.5}}[1/R_1 + 1/R'_1]^{0.5}$$

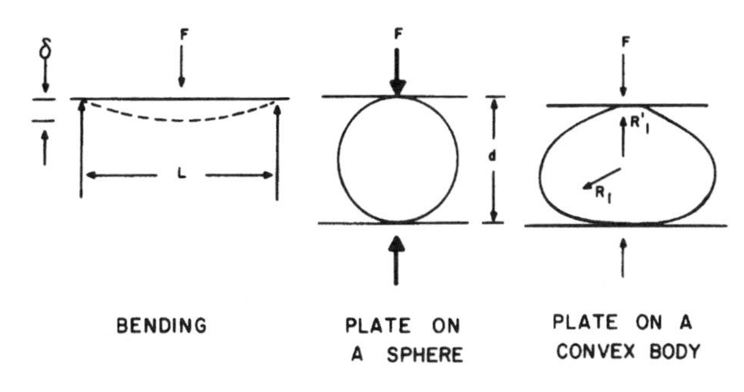

Fig. 3 Bending and application of Hertz's equation to convex bodies to measure the modulus of elasticity.

where F is the force corresponding to deformation D, d is the diameter of the spheres, R_1 and R'_1 are radii of curvature, and ν is the Poisson's ratio.

VISCOELASTIC MATERIALS

Viscoelastic materials exhibit both elastic and viscous properties at the same time. Most biological materials are viscoelastic (e.g., dough). A viscoelastic material may be one of two general types, either linear or nonlinear. A linear viscoelastic material has properties which are dependent upon time alone and not the magnitude of stress that is applied to the material. The other class of viscoelastic materials are nonlinear viscoelastic. These materials exhibit mechanical properties that are a function of time and the magnitude of stress applied. They may possess no elastic zone and will in general be represented by nonlinear relationships between stress, strain, and their derivatives. However, in some definitions the classification of linear and nonlinear viscoelastic behavior may depend on the nature of the relationship between stress, strain, and time.

SOME VISCOELASTICITY TESTS

Creep

In the creep test, the sample is exposed to an initial constant load or stress and the resultant sample deformation is monitored as a function of time.

Stress Relaxation

Stress relaxation involves applying an instantaneous deformation to a body and keeping this deformation or strain constant throughout the test. The way that the resultant stress is alleviated by the material is monitored as a function of time. Stress relaxation is commonly achieved by allowing the crosshead of a universal testing machine to compress the sample to a particular deformation and then stopping it. The constant strain is therefore maintained and the chart will record the manner by which the sample attempts to relieve the imposed stress. This test may be performed under conditions of compression or tension.

Stress relaxation can be easily performed on a material whose behavior is represented by a single Maxwell body. This body possesses the ability to undergo an initial instantaneous deformation either in tension or compression.

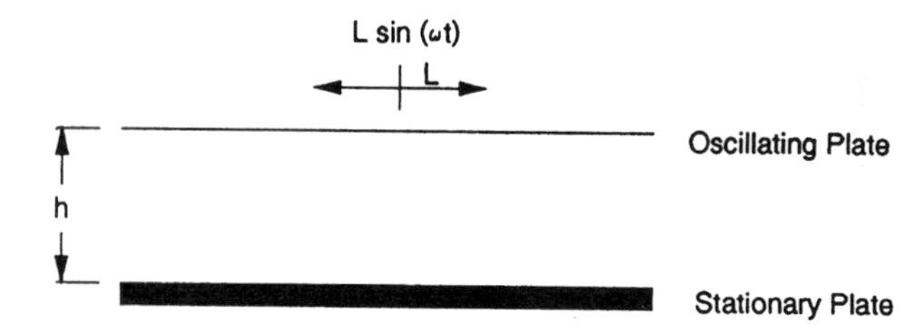

Fig. 4 Oscillatory strain between parallel plates.[6]

Dynamic Testing

Dynamic testing is used to study the viscoelastic behavior of food by harmonically deforming the materials with time.[6] The most typical dynamic test is done by applying sinusoidal simple shear. Consider a parallel plate system where the lower plate is fixed and the upper plate moves back and forth (Fig. 4). By supposing the strain in the material between plates is a function of time defined as

$$\gamma = \gamma_0 \sin(\omega t)$$

where γ_0 is the amplitude of the strain equal to L/h when the motion of the upper plate is $L\sin(\omega t)$. ω is the frequency in rad sec^{-1} which is equivalent to $\omega/(2\pi)$ cycles sec^{-1}. If the two plates were separated by a distance of 1.5 mm and the upper plate moves 0.03 mm from the center line, then the strain amplitude may be calculated as 2%: $\gamma_0 = L/h = 0.03/1.5 = 0.02$. A 1% strain could be achieved by maintaining $h = 1.5$ mm and moving the plate 0.015 mm.[6]

Using a sine wave for strain input results in a periodic shear rate:

$$\frac{d\gamma}{dt} = \dot{\gamma} = \frac{d(\gamma_0 \sin(\omega t))}{dt}$$

which can be evaluated as

$$\dot{\gamma} = \gamma_0 \omega \cos(\omega t)$$

with a small strain amplitude (so the material will behave in a linear viscoelastic manner), the following shear stress is produced by the strain input:

$$\sigma = \sigma_0 \sin(\omega t - \delta)$$

where σ_0 is the amplitude of the shear stress and δ is the phase shift relative to the strain.[6]

The results of the amplitude oscillatory tests can be described by plots of the amplitude ratio (σ_0/γ_0) and the phase shift (δ) as frequency dependent functions (Fig. 5). The following material functions, however, are used to describe the results.

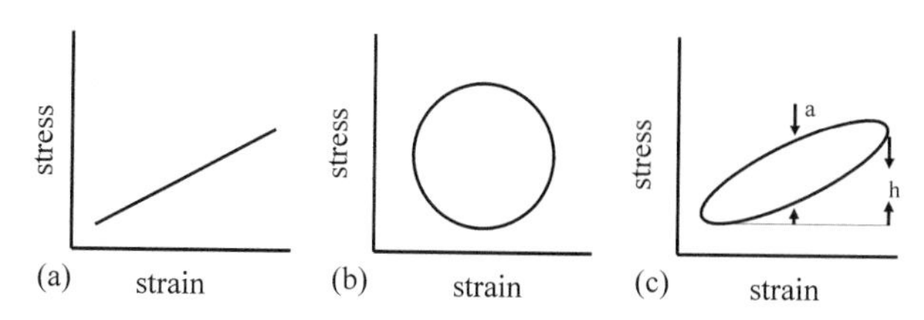

Fig. 5 Relation between stress and strain in dynamic tests for elastic and viscoelastic materials: (a) elastic, (b) viscous, and (c) viscoelastic sin $\eta = a/h$.

The shear stress equation may be written as

$$\sigma = G'\gamma + (G''/\omega)\gamma$$

G' is called storage modulus and G'' is called the loss modulus. They are both functions of frequency, which can be expressed in terms of amplitude ratio and phase shift:

$$G' = (\sigma_0/\gamma_0)\cos(\delta)$$

$$G'' = (\sigma_0/\gamma_0)\sin(\delta)$$

Additional frequency dependent material functions include the complex modulus (G^*), complex viscosity (η^*), the dynamic viscosity (η'), and the out of phase component of the complex viscosity (η'')[6]:

$$G^* = \sigma_0/\gamma_0 = \sqrt{(G')^2 + (G'')^2}$$

$$\eta^* = G^*/\omega = \sqrt{(\eta')^2 + (\eta'')^2}$$

$$\eta' = G''/\omega$$

$$\eta'' = G'/\omega$$

Another material function used to describe viscoelasticity is

$$\tan(\delta) = G''/G' = \eta'/\eta''$$

Tan delta is very high for dilute solutions, 0.2–0.3 for amorphous polymers, low (near 0.01) for gels and glassy crystalline polymers.[6]

If a material is a Hookean solid, the stress and strain are in phase and $\delta = 0$. G'' and η' are also equal to 0 because there is no viscous dissipation of energy. G' is a constant equal to the shear modulus (G).

If a material behaves as an ideal Newtonian material, the stress and strain are 90° out of phase and $\delta = \pi/2$. G' and η'' are zero because the material does not store energy. η' is constant and equal to the Newtonian viscosity (μ). Non-Newtonian fluids show similar behavior as the frequency approaches zero.[6] Dynamic properties of selected foods measured by direct stress–strain tests are found in Rao and Rizvi's Engineering Properties of Foods.[7]

REFERENCES

1. Timoshenko, S.P.; Goodier, J.N. Analysis of Stress and Strain in Three Dimensions. In *Theory of Elasticity*, 3rd Ed.; McGraw-Hill, Inc.: New York, 1970; 219–234.
2. Wang, C.-T. Stress–Strain Relations and the General Equations of Elasticity. In *Applied Elasticity*; McGraw-Hill, Inc.: New York, 1953; 24–41.
3. Rao, V.N.M. Properties of Solid Foods. In *Viscoelastic Properties of Foods*; Rao, M.A., Steffe, J.F., Eds.; Elsevier Applied Science: London, 1992; 3–47.
4. Skinner, G.E.; Rao, V.N.M. Linear Viscoelastic Behavior of Frankfurters. J. Texture Stud. **1986**, *17*, 421–432.
5. Mohsenin, N.N. *Physical Properties of Plant and Animal Materials*; Gordon & Breach: New York, 1986.
6. Steffe, J.F. Viscoelasticity. In *Rheological Methods in Food Process Engineering*; Freeman Press: East Lansing, Michigan, 1992; 176–185.
7. Rao, V.N.M.; Delaney, R.A.M. Rheological Properties of Solid Foods. In *Engineering Properties of Foods*, 2nd Ed.; Rao, M.A., Rizvi, S.S.H., Eds.; Marcel Dekker, Inc.: New York, USA, 1995; 76–81.

Solid Food Transport

Jenni L. Briggs
Iowa State University, Ames, Iowa, U.S.A.

INTRODUCTION

In food processing operations, solid food transport is the movement of raw ingredients, materials in process, and finished product. Transport systems require efficiency in utilized plant space, in timeliness of material delivery, and in meeting capacity demands of the plant. In addition, the physical characteristics of the material and safety concerns associated with the product and process must be considered when designing transport systems. Fragility, stickiness, and moisture content are some physical properties of the transported material that must be taken into account. Dust generation due to the transportation of materials and the likelihood of food contamination are safety aspects to consider prior to choosing a transport system.

Material handling equipment may be classified into the following categories: 1) conveyors; 2) elevators; 3) pneumatic conveying; 4) cranes and hoists; and 5) trucks/forklifts.

CONVEYORS

Belt Conveyors

Generally, belt conveyors are used to transport materials horizontally or at inclines of less than 15° using plain belts, and up to 45° using cleats or crossbars. The maximum incline angle is dependent on material size and center of gravity of a package, and particle size, particle shape, moisture content, angle of repose, and flowability in the case of bulk materials.[1] Belt conveyor systems consist of an endless belt placed on top of rollers (idlers). Rollers may be flat or arranged in a trough configuration, and are used to guide and support the belt using one or more pulleys connected to a drive unit. Proper tensioning and correct tracking are essential in belt conveying systems.[2]

Belt conveyors are able to transport a variety of materials including wet or dry materials, or packaged goods. In addition to moving materials for transport purposes, conveyors may also be used in transit processing such as baking, blanching, sorting, sieving, and coating.[2] The proper belt material may be determined after considering the application and properties of the transported material. Consideration must also be given to the cleanability of the belt surface as well as the drive and rollers particularly when transporting in-process food materials. Construction materials for belts are canvass, metal, and plastic. Flexible metal band and wire mesh belts are generally made from steel, or stainless steel when in direct contact with food. The heat resistance and tensile strength of these belts make them desirable for a wide range of in-process transport. Woven cotton fiber remains in use for many dough operations. Problems that may occur using this material include bacterial growth in the fiber, which may transfer to the food product, and frayed edging of the belt. The latter difficulty may be overcome by using belts that have had the edges sealed by a thermoplastic.[3] The most common plastic materials are acetal, polypropylene, and polyethylene. As an additional food safety precaution, antimicrobial belts have been developed, which incorporate a biocide into the plastic.[4]

Vibratory Conveyors

Vibratory conveyors are designed to toss material upwards and forward by oscillating a trough. The capacity of this system, which can range from tons to grams, is dependent on the magnitude of trough displacement, frequency of displacement, slope of the trough, angle of throw, and the material being conveyed. Material should exhibit a high friction factor against the trough surface and a high internal friction factor, especially when conveying granular materials.[5] The conveyed material should also be dense enough to minimize air resistance. Foods and ingredients that are conveyed by this means include granular materials and fragile foods. Conveying is gentle on the materials; therefore, vibratory conveyors are used in transporting delicate food products such as potato chips.

Magnetic Conveyors

Magnetic conveyors are used to transport ferromagnetic containers such as canned goods. Cans may be moved vertically, horizontally, or rotated for transport or for orienting the product. Compared to other conveying methods, this system is virtually noiseless.[6]

Encyclopedia of Agricultural, Food, and Biological Engineering
DOI: 10.1081/E-EAFE 120006973
Copyright © 2003 by Marcel Dekker, Inc. All rights reserved.

Gravity Conveyors

Gravity conveyors rely on downward inclines of 2° or greater to transport packages, cases, trays, barrels, and drums. The conveyor consists of 2–3 rollers or skate wheels mounted on an axis, which are connected to a supporting frame. This construction of gravity conveyors also allows the system to be used as a manual conveyor for material on a horizontal track.

ELEVATORS

Elevators are used to move material vertically. They consist of buckets made from plastic or metal running on an endless chain. Generally, the elevator is enclosed. Material is scooped or loaded into the bucket at the lower level, also referred to as the boot, transported up the vertical elevation, and discharged. Bucket elevators are categorized based on their discharge mechanism:

1. *Gravity discharge*: These discharge buckets tend to be high capacity, and the system operates at relatively low velocities ($0.25\,\mathrm{m\,sec^{-1}}$). At the point of discharge the buckets are mechanically tipped. Gravity discharge is used less often compared to other discharging methods because it is slow and expensive.
2. *Positive displacement discharge*: Positive displacement buckets hold small volumes compared to other types of discharge buckets. The bucket is tipped by changing direction of the bucket path using sprockets on the down side of the conveyor chain. This discharging method should be considered when transporting sticky or fluffy material.[1]
3. *Centrifugal discharge*: Centrifugal discharge buckets are high capacity, and discharge by centrifugal force into an outlet shoot. Operating speed is about $1.5\,\mathrm{m\,sec^{-1}}$. Due to the relatively high speeds, wear may become problematic on centrifugal discharge elevators. The percent and size of lumps in the material should be kept at a minimum when using this system. Furthermore, the generation of dust may be hazardous when transporting material that may result in a dust explosion.
4. *Continuous displacement discharge*: Continuous displacement buckets operate near a speed of $0.5\,\mathrm{m\,sec^{-1}}$. Buckets are closely packed, and the material is loaded into the bucket as opposed to being scooped. At the discharge point, the material flows across the back of the previous bucket. The flow of material acts as a continuous stream as consecutive buckets are emptied.

PNEUMATIC CONVEYORS

In pneumatic conveying, material is moved in a gas stream. Air is generally used as the gas; however, nitrogen and other gases may be used when transporting oxygen sensitive material. The four modes of pneumatic conveying are summarized as follows:[7]

1. *Dilute phase*: Material is suspended in the gas stream, and air velocities are high.
2. *Continuous dense phase*: Material flows as a continuous moving bed, and velocities are lower than for the dilute phase.
3. *Discontinuous dense phase*: Material flows as a plug, and air velocities are lower than for the continuous dense phase.
4. *Solid dense phase*: Material occupies 80%–100% of the pipe flowing as a plug, and the air velocity is very low.

The terminology dilute and dense phase depends on the mass flow ratio, which is the ratio of the mass of solids to the mass of conveying gas. Dilute phase refers to mass flow ratio from 0 to 15, and dense phase refers to mass flow ratio greater than 15.[8]

Almost all powders and granular materials may be pneumatically conveyed. Among some of the foods that have been transported using this mode are sugar, salt, flour, starch, wheat, corn, rice, and coffee. The type of flow regime chosen will generally depend on material fragility. Dilute phase conveying is not appropriate for materials subject to degradation due to abrasive action during transport. The system most appropriate for fragile materials is the solid dense phase because of its low air velocity. This mode has been used to transport nuts (peanuts, almonds, and hazelnuts), rice, cereals, and pet foods.[7]

CRANES AND HOISTS

Cranes and hoists are used to move a range of materials, and differentiated from each other by their lifting mechanism. The power mechanism of a crane is separate from the lifting hook, and with a hoist, the power mechanism is with the lifting hook device. A crane has a boom, which is also referred to as a jib, and generally pivots. Furthermore, the jib may elevate as found with derrick cranes, or be fixed in elevation as found with pillar cranes.[6] These have found use in loading and unloading of railcars, and ships. Hoists, on the other hand, are used to move materials vertically and horizontally. Their motion is guided by suspended railings covering the area of transport.[6]

FORKLIFTS

Forklifts may be considered as trucks with vertical lift capabilities, and are mostly used for stacking and moving pallets and boxes. A common lifting height is 3 m, but lifting heights of 8 m–10 m are available as well as low-lift trucks, which provide lifts between 0.05 m–0.10 m. Their tilting capacities generally range from 10 to 15° backwards and 3–5° forward.[6] Attachments for forks may be obtained, which will allow for the handling of alternative materials. For example, scoops may be used for bulk material handling.

REFERENCES

1. Fruchtbaum, J. *Bulk Materials Handling Handbook*; Van Nostrand Reinhold Co.: New York, 1988.
2. Brennan, J.G.; Butters, J.R.; Cowell, N.D.; Lilley, A.E.V. *Food Engineering Operations*; Elsevier Applied Science: New York, 1990.
3. Cowey, P. Watch Your Belts. Food Manuf. **1999**, *74* (2), 41–42.
4. Cowey, P. Anti-microbial Conveyor Belts. Food Process. **1997**, *66* (8), 10–11.
5. Raymus, G.J. Handling of Bulk Solids and Packaging of Solids and Liquids. In *Perry's Chemical Engineers' Handbook*, 7th Ed.; Perry, R.H., Green, D.W., Eds.; McGraw-Hill: New York, 1997.
6. Cowell, N.D. Storage, Handling, and Packaging. In *Food Industries Manual*; Ranken, M.D., Kill, R.C., Eds.; Blackie Academic and Professional: New York, 1993.
7. Bell, J. Pneumatic Conveying of Fragile Foods. Food Process. **1997**, *66* (8), 14–15.
8. Klinzing, G.E.; Marcus, R.D.; Rizk, F.; Leung, L.S. *Pneumatic Conveying of Solids*, 2nd Ed.; Chapman & Hall: New York, 1997.

Specific Heat Capacity Measurement

Chang Hwan Hwang
Sundaram Gunasekaran
University of Wisconsin–Madison, Madison, Wisconsin, U.S.A.

S

INTRODUCTION

Specific heat capacity is defined as the amount of heat energy required by a unit mass of a material to raise its temperature by a unit degree. This is described by Eq. 1.

$$Q = \int_0^T mC_p \, dT \tag{1}$$

When C_p is constant within the temperature range of interest, the equation can be further simplified.

$$Q = mC_p \Delta T \tag{2}$$

As food and biological materials are processed mostly under constant pressure during heat transfer, C_p is normally used instead of C_v, the specific heat capacity at constant volume, which is usually used for gases. The SI and English units for specific heat capacity are $\mathrm{kJ\,kg^{-1}\,K^{-1}}$ and $\mathrm{Btu\,lb^{-1}\,{}^\circ F^{-1}}$, respectively.

In the literature, the term "heat capacity" and "specific heat" are often used when specific heat capacity is meant. Heat capacity ($\mathrm{kJ\,K^{-1}}$) is the amount of energy supplied to corresponding temperature rise of a material regardless of its mass, i.e., it is an extensive property. Specific heat is the ratio of heat capacity of a material to the heat capacity of water at a reference temperature when mass of 1 g and temperature rise of 1°C is used. Therefore, specific heat is a dimensionless number.[1]

MEASUREMENT OF SPECIFIC HEAT CAPACITY

Method of Mixtures

It is the most commonly used method. In this, m_{ex} grams of a heat exchange liquid (usually water or oil) of known specific heat capacity $C_{p,ex}$ and temperature $T_{i,ex}$ are mixed with m_s grams of the sample at a different temperature $T_{i,s}$ in a vacuum-jacketed container (Fig. 1). The mixture is allowed to reach an equilibrium temperature T_{eq}. As the container also participates in temperature equilibration, mass (m_c), specific heat capacity (C_c), and initial temperature ($T_{i,c}$) of the container should be known. Assuming adiabatic conditions (i.e., no energy flow in or out of system), C_p can be calculated based on energy balance as follows:

$$C_p = \frac{C_{p,ex} m_{ex}(T_{i,ex} - T_{eq}) - C_c m_c (T_{eq} - T_{i,c})}{m_s (T_{eq} - T_{i,s})} \tag{3}$$

Water is usually used as the heat exchange medium. When some ingredients of the sample are water-soluble, oil is recommended as the heat exchange medium to avoid heats of solution.[2,3] Thermal leakage of the container is a possible source of error and can be corrected by determining instantaneous temperature increase. It can be graphically performed by extrapolating temperature curves immediately before and after mixing according to Stitt and Kennedy.[4]

Method of mixtures is simple to use and can be used for testing large amounts of samples necessary to assure sufficient accuracy when testing inhomogeneous samples like foods and biomaterials. The measurement is more accurate when the system reaches steady state heat transfer condition and when phase transitions occur slowly over time. However, the measurement time is too long to prevent food spoilage and not good over a large temperature range for samples whose C_p is highly temperature dependent.[5]

Method of Guarded Plates

The sample is loaded on thermally shielded guard plates and heated by electric heaters (Fig. 2). The temperature increase from T_1 to T_2 is measured over time, t. The electrical energy input is determined from voltage (V_t) and current (I) values. The energy balance calculation provides C_p as follows:

$$C_p = \frac{V_t \times I \times t}{m_s (T_2 - T_1) \times 1000} \tag{4}$$

Differential Scanning Calorimetry (DSC)

In a DSC, the difference in energy supplied between a sample cell and a reference cell to maintain two cells at the same temperature is measured. The cells are heated by

DOI: 10.1081/E-EAFE 120006989
Copyright © 2003 by Marcel Dekker, Inc. All rights reserved.

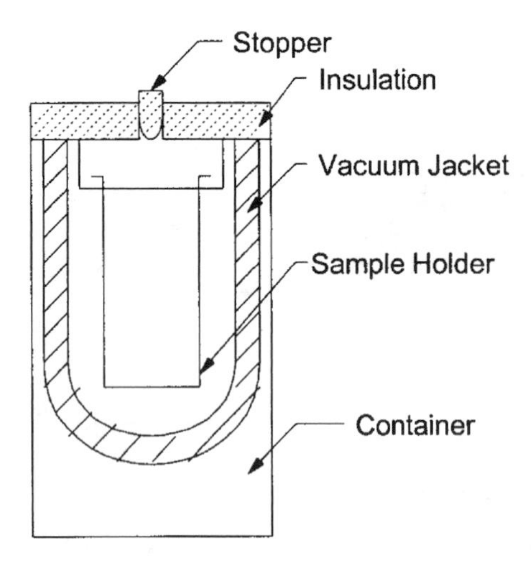

Fig. 1 A vacuum-jacketed calorimeter typically used for specific heat capacity measurement.

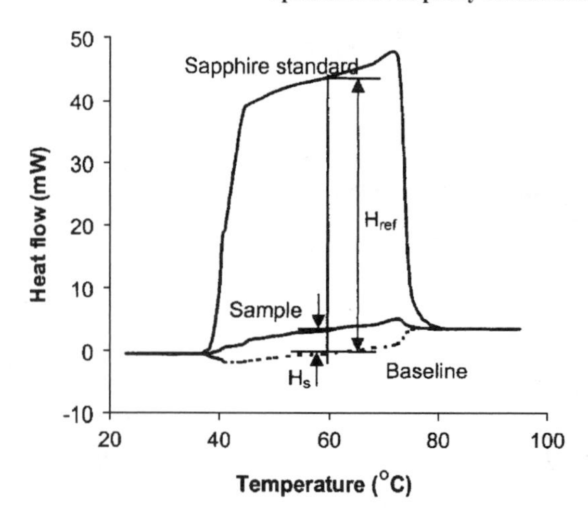

Fig. 3 Differential scanning calorimeter thermograms for specific heat capacity measurement.

heating elements underneath the cells. The difference in energy input to the cells is directly proportional to specific heat capacity of the sample. The measurements include determinations of energy flow curves for baseline, sample, and the reference material (Fig. 3). The magnitude of energy input for sample with respect to baseline is compared with that of the reference. Sapphire is commonly used as the reference due to its linear response over a wide temperature range and high purity. The specific heat capacity of sapphire is $0.906 + 0.000368T_k - 21440.884/T_k^2$, $273\,K < T_k < 1973\,K$.[6] The specific heat capacity of sample is calculated as:

$$C_p = \frac{m_{ref} H_s}{m_s H_{ref}} C_{p,ref} \tag{5}$$

This test is performed quickly with small sample size and can easily determine over a wide range of temperatures. It is frequently used to determine the temperature dependence of specific heat capacity. Despite of equipment cost, many researchers currently use this method for specific heat capacity determination. However, the sample must be well homogenized.[2,7]

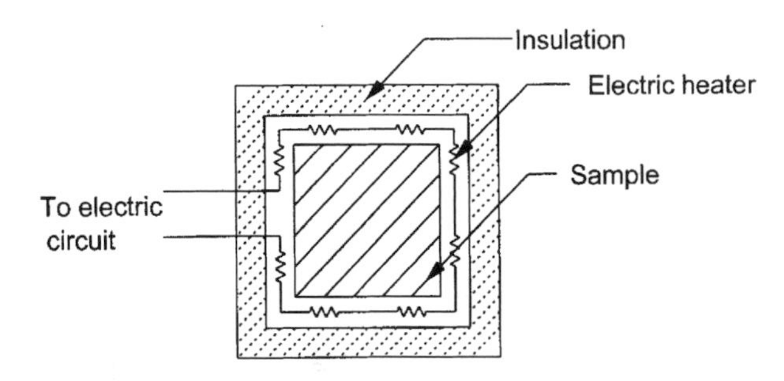

Fig. 2 Diagram of guarded-plate for specific heat capacity measurement.

Other Methods

Comparison calorimeter and adiabatic calorimeter[8] are also used to determine specific heat capacity. Indirect calculation is useful if density (ρ), thermal conductivity (k), and thermal diffusivity (α) are available for the samples as $C_p = k/\rho\alpha$.[2] Differential thermal analysis (DTA) is used since it generates the same degree of energy transfer to both sample and reference with temperature monitoring.[9,10] Nuclear magnetic resonance (NMR) is used to model specific heat capacity calculation due to its ability to determine water status in foods.[11–13]

ESTIMATION OF SPECIFIC HEAT CAPACITY

Specific heat capacity of composite foods can be estimated if individual component's mass fraction (x_i) and corresponding specific heat capacity (C_{pi}) are known (Eq. 6):

$$C_p = \sum x_i C_{pi} \tag{6}$$

As water is the major constituent in most foods, Eq. 6 can be simplified by considering foods as a two-component system—water and dry solids. Because the specific heat capacity of water is $4.1861\,kJ\,kg^{-1}\,K^{-1}$, the specific heat capacity of high moisture foods can be estimated as:

$$C_p = 4.1868x_w + C_{p,s}x_s \tag{7}$$

$C_{p,s}$ can then be easily obtained by extrapolating from a previously reported single value[14] to the moisture-free value. This estimation is useful if there is no strong temperature dependence.

Simple Equations for High Moisture Foods

Siebel[15] estimated C_p of dry solids of fat-free fruits and vegetables, purees, and plant concentrates to be $0.8374\,\text{kJ}\,\text{kg}^{-1}\text{K}^{-1}$. The Siebel's equation for above freezing ($T_k > 274\,\text{K}$) is:

$$C_p = 4.1868x_w + 0.8374(1 - x_w) = 3.3494x_w + 0.8374 \quad (8)$$

Similarly, using specific heat capacity of ice = 2.0934 $\text{kJ}\,\text{kg}^{-1}\text{K}^{-1}$, Siebel[15] estimated C_p of food below freezing ($T_k > 274\,\text{K}$):

$$C_p = 1.2560x_w + 0.8374 \quad (9)$$

When fat is present, the specific heat capacity of foods may be estimated by accounting for mass fraction of fat (x_f), solids nonfat (x_{snf}), and moisture (x_w) (Eqs. 10 and 11).

$$C_p = 1.6747x_f + 0.8374x_{snf} + 4.1868x_w \quad \text{for } T_k > 273\,\text{K} \quad (10)$$

$$C_p = 1.6747x_f + 0.8374x_{snf} + 2.0934x_w \quad \text{for } T_k < 273\,\text{K} \quad (11)$$

The Siebel's equation is adopted by the American Society for Heating, Refrigeration, and Air Condition Engineers[16] for estimating specific heat capacity of foodstuffs. It is oversimplified and the assumption that all types of nonfat solids have the same specific heat capacity may not be correct. Also, all the water may not be frozen below the initial freezing point. However, estimates based on Siebel's equations match well with actual specific heat capacity when moisture content is more than 0.7 and no fat is present.[17]

Generalized Predictive Equations

Many generalized predictive models for specific heat capacity of foods based on composition have been proposed.[15,18–21] The model proposed by Choi and Okos[21] is presented in Table 1 along with equations for individual component specific heat capacity that are temperature dependent. For frozen foods, the unfrozen water fraction is predicted by the projected freezing point depression as well as ice fraction. The predictable temperature range is −40 to 150°C and the prediction error is within ±4.7%. Specific heat capacity values calculated based on Choi and Okos model[21] are generally higher than those calculated by Siebel equations for high moisture foods.[17]

Riedel[8] reported that simple linear models such as Siebel model[15] are good only for foods containing a small range of moisture contents. He proposed a model that accurately estimates the specific heat capacity of foods over wide range of moisture contents (Eq. 12). Miles, van Beek, and Veerkamp[22] determined the C_p values, predicted by Eq. 12, that deviates less than 10% compared to measured C_p values over a moisture range of 0%–100% wet basis.

$$C_p = C_{p,w}x_w + 4.19(a + 0.001T)(1 - x_w) - b\exp(-43x_w^{2.3}) \quad (12)$$

where a and b are food-specific constants listed in Table 2.

Specific Heat Capacity Model for Different Food Materials

Predictive specific heat capacity equations available in the literature for various foods are summarized in Table 3. For

Table 1 Generalized prediction equation for specific heat capacity of a composite food material by Choi and Okos[21]

Component	Specific heat capacity equations	Condition
Protein	$C_{p,p} = 2.0082 + 1.2089 \times 10^{-3}T - 1.3129 \times 10^{-6}T^2$	
Fat	$C_{p,fat} = 1.9842 + 1.4733 \times 10^{-3}T - 4.8008 \times 10^{-6}T^2$	
Carbohydrate	$C_{p,c} = 1.5488 + 1.9625 \times 10^{-3}T - 5.9399 \times 10^{-6}T^2$	
Fiber	$C_{p,fi} = 1.8459 + 1.8306 \times 10^{-3}T - 4.6509 \times 10^{-6}T^2$	
Ash	$C_{p,a} = 1.0926 + 1.8896 \times 10^{-3}T - 3.6817 \times 10^{-6}T^2$	
Water	$C_{p,waf} = 4.1762 - 9.0862 \times 10^{-5}T + 5.4731 \times 10^{-6}T^2$	$T > 273\,\text{K}$
	$C_{p,wbf} = 4.0817 - 5.3062 \times 10^{-3}T + 9.9516 \times 10^{-4}T^2$	$T < 273\,\text{K}$
Ice	$C_{p,i} = 2.0623 + 6.0769 \times 10^{-3}T$	
Composite food	$C_p = PC_{p,p} + FC_{p,fat} + CC_{p,c} + FiC_{p,fi} + AC_{p,a} + MC_{p,waf}$	$T > 273\,\text{K}$
	$C_p = PC_{p,p} + FC_{p,fat} + CC_{p,c} + FiC_{p,fi} + AC_{p,a} + MC_{p,wbf} + IC_{p,i}$	$T < 273\,\text{K}$

Table 2 Food-specific constants in Riedel's formula (Eq. 12) for specific heat capacity calculation over a wide range of moisture contents[8]

Food	*a*	*b*
Beef	0.385	0.08
Saltwater fish	0.410	0.12
Egg white	0.330	0.06
Low-fat cheese	0.390	0.10
Baker's yeast	0.395	0.10
White bread	0.350	0.09
Potato starch	0.340	0.06
Coffee extract	0.390	0.13

some foods, several equations are available. These equations should be used only in the temperature and moisture content range specified. If one value is necessary for calculation, thermal properties complied by Polley, Snyder, and Kotnour[14] may be referred.

ADDITIONAL FACTORS TO INFLUENCE C_p

Temperature

The specific heat capacity of high-moisture, low-fat foods is nearly independent of temperature because C_p of water varies only 1% between 0 and 100°C. The specific heat capacity of dry solids also varies less than 1% except when phase transitions occur.[3] For most foods, the predictive equations are function of temperature and valid only in the specific temperature range because phase and state transitions of the ingredients cause deviations in specific heat capacity measurements.

Phase and State Transitions

Specific heat capacity of foods often changes due to first- and second-order phase transitions.[23] The change in specific heat capacity due to a second-order transition can be determined using a DSC which shows a sudden change at the transition temperature.[23] The specific heat capacity of an amorphous material may be higher than that in its crystalline state. Hwang et al.[24] reported that specific heat capacity of gelatinized starch is higher than that of granular starch. AbuDagga[25] observed an increase in specific heat capacity due to fish muscle protein denaturation.

When phase transitions occur, energy potential of the materials change. Compared to small energy changes due to crystallization, starch gelatinization, and protein denaturation, etc. energy changes during ice and fat melting are large. Freezing and thawing of foods and biomaterials occur over a wide temperature range. This causes the specific heat capacity to deviate from its normal value during freezing and thawing. Therefore, an apparent specific heat capacity has to be determined.[26] van Beek[27] developed an equation for apparent specific heat capacity, $C_{p,app}$ at temperatures below the initial freezing point (Eq. 13).

$$C_{p,app} = C_{p,s}(1 - x_w) + C_{p,w}x_w\frac{T_f}{T} + C_{p,i}\left(1 - \frac{T_f}{T}\right)x_w - \Delta H_f x_w \frac{T_f}{T^2} \tag{13}$$

Schwartzberg[28] accounted for unfreezable water as bound water and treated it as a part of the solid fraction.

$$C_{p,app} = C_{p,f} + (x_w - m_u x_s)\Delta H_f(T_0 - T_f)(T_0 - T)^2 \tag{14}$$

The unfreezable water could roughly be estimated as $x_b = 0.5x_p + 0.3x_c$, where x_p and x_c are mass fraction of proteins and carbohydrates, respectively.[5]

Latyshev and Ozerova (cited by Miles, van Beek, and Veerkamp[22]) estimated specific heat capacity of animal fats as:

$$C_{p,fat} = A + BT + \sum_{i=1}^{2} \frac{A_i}{1 + \frac{B_i}{A_i}(T - C_i)^2} \tag{15}$$

where A, B, A_i, B_i, and C_i are the fat-specific constants listed in Table 4. As above equation is merely empirical based on specific heat capacity traces, a new set of constants should be determined in case of severe thermal history changes and composition difference depending on animal species.

Pressure

Specific heat capacity of food materials is generally assumed constant regardless of the pressure. However, the assumption may not be true always since unusual process conditions are increasingly used. Reddy et al.[29] performed high-pressure sterilization of bacterial spores at 414 MPa–827 MPa and Knorr, Schlueter, and Heinz[30] proposed high pressure freezing and thawing at 70 MPa–350 MPa compared to normal food processing conditions such as autoclaving at 0.2 MPa. Denys et al.[31] developed a heat transfer model of high-pressure-shift freezing and thawing on thylose gel. They approximated apparent specific heat capacity of the gel by shifting the known value at atmospheric pressure on the temperature scale according to the pressure applied without considering intrinsic effect of pressure. Chourot et al.[32] performed numerical approximation with shifted specific heat capacity to model high-pressure thawing on a model

S

Table 3 Predictive models for specific heat capacity (C_p, kJ kg^{-1} K^{-1}) of various foods

Food type	Equation and condition	Source
Apple	$C_p = 1.51 + 2.32x_w$, $0.50 < x_w < 0.88$	34
Apple, granny smith	$C_p = 3.400 + 0.0049T$, $-1 < T < 60°C$	35
Apple, golden delicious	$C_p = 3.3600 + 0.0075T$, $-1 < T < 60°C$	35
Apple, juice	$C_p = 3.3846 - 1.8177 \times 10^{-2}B_x + 2.3472 \times 10^{-3}T$, $6 < x_w < 75°$Brix, $30 < T < 90°C$	36
Apples	$C_p = 1.18134 + 0.000414T + 2.3900x_w$, $0.5 < x_w < 0.9$, $0 < T < 90°C$	3
Avocado	$C_p = 0.92 + 3.32x_w$, $0.44 < x_w < 0.74$	34
Babaco	$C_p = -0.53 + 4.63x_w$, $0.72 < x_w < 0.93$	34
Banana	$C_p = 1.34 + 2.62x_w$, $0.45 < x_w < 0.76$	34
Beef	$C_p = 4.187(x_w + (0.385 + 0.001T)(1 - x_w) - 0.08\exp(-43x_w^{2.3}))$, $0 < x_w < 1, T > 0°C$	8
Beef, raw	$C_p = 1.672 + 2.508x_w$	22
Bread, white	$C_p = 4.187(x_w + (0.35 + 0.001T)(1 - x_w) - 0.09\exp(-43x_w^{2.3}))$, $0 < x_w < 1, T > 0°C$	8
Cabbages	$C_p = -1.2880 - 0.001088T + 5.7020x_w$, $0.8 < x_w < 0.9$, $0 < T < 90°C$	3
Candletree	$C_p = 0.55 + 3.32x_w$, $0.55 < x_w < 0.83$	34
Cassava	$C_p = 1.577 + 2.5x_w$, $0.1 < x_w < 0.68$, $36 < T < 51°C$	37
Cheese, cheddar	$C_p = 14.040 - 0.128x_w - 0.187x_f - 0.388x_p + 0.00097x_w{*}x_f + 0.00476x_w{*}x_p + 0.00425x_f{*}x_p$	38
Cheese, low fat	$C_p = 4.187(x_w + (0.39 + 0.001T)(1 - x_w) - 0.10\exp(-43x_w^{2.3}))$, $0 < x_w < 1$, $T > 0°C$	8
Cherry, black	$C_p = 1.47 + 2.37x_w$, $0.41 < x_w < 0.78$	34
Coffee, Columbian, powdered	$C_p = 1.2729 + 0.006459T$, $45 < T < 150°C$	39
Coffee, Mexican, powdered	$C_p = 0.9210 + 0.007554T$, $45 < T < 150°C$	39
Corn, yellow dent	$C_p = 1.461 + 3.555x_w$, $0.1 < x_w < 0.30$, $12 < T < 29°C$	40
Cumin, seed	$C_p = 1.574 + 0.00351T + 0.06459x_w/(1 - x_w)$, $0.017 < x_w < 0.17$, $-70 < T < 50°C$	41
Egg, white	$C_p = 4.187(x_w + (0.33 + 0.001T)(1 - x_w) - 0.06\exp(-43x_w^{2.3}))$, $0 < x_w < 1$, $T > 0°C$	8
Fat, animal	$C_p = 1.884 + 0.007008T + 2.24/(1 + 0.0089(T - 13.8)^2) + 4.280/(1 + 0.0234(T - 47.85)^2)$	22
Fish, saltwater	$C_p = 4.187(x_w + (0.41 + 0.001T)(1 - x_w) - 0.12\exp(-43x_w^{2.3}))$, $0 < x_w < 1$, $T > 0°C$	8
Fruit, juice	$C_p = 1.6747 + 2.5120x_w$, $x_w > 0.5$	18
Grape	$C_p = 0.68 + 3.46x_w$, $0.48 < x_w < 0.81$	34
Grapefruit	$C_p = 1.41 + 2.29x_w$, $0.01 < x_w < 0.89$	34
Green beans	$C_p = 0.9634 + 0.0138T + 1.8649x_w$, $0.1 < x_w < 0.9$, $-40 < T < -10°C$	3
Guava	$C_p = 0.97 + 2.94x_w$, $0.51 < x_w < 0.87$	34
Lemon	$C_p = 1.51 + 1.79x_w$, $0.01 < x_w < 0.11$	34
Lime	$C_p = 0.66 + 3.34x_w$, $047 < x_w < 0.90$	34
Meat	$C_p = 0.979 + 3.1754x_w$, $T > 0°C$	42
Milk	$C_p = 4.187x_w + (1.373 + 0.0113T)(1 - x_w)$, $0.6 < x_w, < 0.92$	43
Mulberry	$C_p = 0.55 + 3.53x_w$, $0.59 < x_w < 0.87$	34
Mushrooms	$C_p = 1.5400 + 0.000203T + 2.6270x_w$, $0.3 < x_w < 0.95$, $0 < T < 90°C$	3
Mushrooms	$C_p = 1.0217 + 0.0092T + 2.47x_w$, $0.1 < x_w < 0.90$, $40 < T < 70°C$	44
Muskmelon	$C_p = -0.52 + 4.47x_w$, $0.77 < x_w < 0.94$	34
Naranjilla	$C_p = 0.85 + 3.03x_w$, $0.63 < x_w < 0.92$	34
Nectarine	$C_p = 1.56 + 2.32x_w$, $0.44 < x_w < 0.87$	34
Oat	$C_p = 1.2770 + 3.266x_w$, $0.1 < x_w < 0.18$, $20 < T < 24°C$	45
Onion, White	$C_p = 1.8400 + 2.3400x_w$, $0.0 < x_w < 0.692$, $T = 20°C$	46
Orange	$C_p = 1.52 + 2.19x_w$, $0.01 < x_w < 0.83$	34
Papaya	$C_p = 1.23 + 2.48x_w$, $0.56 < x_w < 0.90$	34
Peach	$C_p = 1.51 + 2.18x_w$, $0.01 < x_w < 0.88$	34
Peanut oil, hydrogenated	$C_p = 1.970 + 0.00489T$, $46.84 < T < 76.84°C$	47
Peanut oil, unhydrogenated	$C_p = 2.057 + 0.00167T$, $26.84 < T < 56.84°C$	47
Pear	$C_p = 1.70 + 2.30x_w$, $0.33 < x_w < 0.81$	34
Pineapple	$C_p = 0.98 + 2.86x_w$, $0.46 < x_w < 0.85$	34
Pistachio	$C_p = 1.074 + 0.2779x_w$, $0.05 < x_w < 0.4$	48
Plantain	$C_p = 1.676 + 2.2x_w$, $0.1 < x_w < 0.68$, $36 < T < 51°C$	37
Plum	$C_p = 0.11 + 4.01x_w$, $0.68 < x_w < 0.89$	34

(Continued)

Table 3 Predictive models for specific heat capacity (C_p, kJ kg^{-1}K^{-1}) of various foods (*Continued*)

Food type	Equation and condition	Source
Potato	$C_p = 1.6998 + 0.006113T + 0.8499x_w/(1-x_w) - 0.1042x_w^2/(1-x_w)^2$, $-40 < T < 70°C$	49
Potato	$C_p = 0.9043 + 3.266x_w$	50
Potato mix, fried	$C_p = 3.0363 + 0.005951T$, $50 < T < 100°C$ $C_p = 2.4036 + 0.00506T$, $T > 100°C$	51
Rasins	$C_p = 1.4400 + 2.782x_w$, $T = 20°C$	52
Rice, finished	$C_p = 1.180 + 3.768x_w$, $0.1 < x_w < 0.18$, $20 < T < 24°C$	45
Rice, hulled	$C_p = 1.202 + 3.810x_w$, $0.1 < x_w < 0.18$, $20 < T < 24°C$	45
Sorghum, flour	$C_p = -5.799 + 3.247 \times 10^{-2}T_k - 3.370 \times 10^{-5}T_k^2 + 6.443x_w - 9.039 \times 10^{-3}T_k x_w$, $0.02 < x_w < 0.29$, $T = 24°C$	53
Sorghum, grain	$C_p = 0.6979 + 0.0039T + 0.0592x_w$, $0.086 < x < 0.163$, $10 < T < 65°C$	54
Soybean	$C_p = 1.444 + 0.0536x_w/(1-x_w)$, $0.08 < x_w < 0.25$, $T = 42°C$	55
Soybean flour, defatted	$C_p = 1.650 + 3.211x_w$, $0.136 < x_w < 0.362$, $57 < T < 77°C$	56
Starch, dried corn	$C_p = 1.0185 + 0.0077893T$, $30 < T < 90°C$, $x_w = 0$	57
Starch, potato	$C_p = 4.187(x_w + (0.34 + 0.001T)(1-x_w) - 0.06\exp(-43x_w^{2.3}))$, $0 < x_w < 1$, $T > 0°C$	8
Strawberry	$C_p = 0.68 + 3.29x_w$, $0.67 < x_w < 0.92$	34
Surimi from Pacific whiting	$C_p = 2.33 + 0.006T + 1.49x_w$, $0.74 < x_w < 0.84$, $25 < T < 90°C$	25
Tangerine	$C_p = 1.52 + 2.21x_w$, $0.01 < x_w < 0.87$	34
Tomato	$C_p = 0.71 + 3.39x_w$, $0.64 < x_w < 0.95$	34
Watermelon	$C_p = 0.33 + 3.85x_w$, $0.67 < x_w < 0.92$	34
Wheat, hard red spring	$C_p = 1.394 + 3.220x_w$, $0.01 < x_w < 0.19$, $0.6 < T < 21°C$	58
Wheat, soft white	$C_p = 1.394 + 4.082x_w$, $0.07 < x_w < 0.20$, $11 < T < 32°C$	40
Yam	$C_p = 1.606 + 2.4x_w$, $0.1 < x_w < 0.68$, $36 < T < 51°C$	37
Yeast, baker's	$C_p = 4.187(x_w + (0.395 + 0.001T)(1-x_w) - 0.10\exp(-43x_w^{2.3}))$, $0 < x_w < 1$, $T > 0°C$	8
Yogurt, plain	$C_p = 1.4694 + 2.5984x_w$, $0.05 < x_w < 0.7$	59

food and concluded with inaccurate temperature distribution profiles due to the lack of thermophysical properties. Takagi and Teranishi[33] determined heat capacities of pure water at 25°C as 4.18 kJ kg^{-1}K^{-1}, 4.05 kJ kg^{-1}K^{-1}, and 3.95 kJ kg^{-1}K^{-1} at 0.1 MPa, 50 MPa, and 100 MPa, respectively, which is the major ingredient for most foods and changed about 5.5% between 0.1 MPa and 100 MPa. Miles[22] determined the relationship between C_p and C_v, which shows the influence of adiabatic compressibility, K_T and cubical expansion coefficient, θ on specific heat capacity (Eq. 16). This would yield a specific heat capacity modification based on structural changes by pressure and temperature for applications such as high-pressure processing.

$$C_v = C_p - \frac{T\theta^2}{\rho K_T},$$
$$\text{where } \theta = \frac{1}{V}\left(\frac{\partial V}{\partial T}\right)_P, \quad K_T = -\frac{1}{V}\left(\frac{\partial V}{\partial P}\right)_T \tag{16}$$

Table 4 Constants in Latyshev and Ozerova's formula for specific heat capacity of fats[22]

Constant	Beef fat	Pork fat
A (kJ kg^{-1}°C^{-1})	1.884	1.420
B (kJ kg^{-1}°C^{-2})	0.007008	0.00367
A_1 (kJ kg^{-1}°C^{-1})	2.24	3.73
B_1 (kJ kg^{-1}°C^{-3})	0.02	0.05
C_1 (°C)	13.8	1.18
A_2 (kJ kg^{-1}°C^{-1})	4.28	4.5
B_2 (kJ kg^{-1}°C^{-3})	0.1	0.15
C_2 (°C)	47.85	26.85

CONCLUSIONS

Determination of specific heat capacity of foods is fairly simple using conventional methods such as the method of mixtures or the DSC. Nonetheless, much work has been focused on developing a generalized prediction model for all foods based on composition. This has resulted in several such models accounting for various food constituents. Among these, the model generated by Choi and Okos[21] appears to be the most comprehensive. If such a model were to account for phase and state changes and pressure, it would prove to be very valuable. Such predictions can be extended to manufactured foods and food mixtures as long as their composition is known.

NOTATIONS

A, A_i, B, B_i, C_i	Fat-specific constants in Latyshev and Ozerova's equation for specific heat capacity of fats
a, b	Food-specific constants in Riedel's equation
B_x	Concentration, °Brix
C_c	Specific heat capacity of container, $kJ\,kg^{-1}\,K^{-1}$
C_p	Specific heat capacity under constant pressure, $kJ\,kg^{-1}\,K^{-1}$
$C_{p,a}$	Specific heat capacity of ash, $kJ\,kg^{-1}\,K^{-1}$
$C_{p,app}$	Apparent specific heat capacity of food, $kJ\,kg^{-1}\,K^{-1}$
$C_{p,c}$	Specific heat capacity of carbohydrates, $kJ\,kg^{-1}\,K^{-1}$
$C_{p,ex}$	Specific heat capacity of heat exchange liquid, $kJ\,kg^{-1}\,K^{-1}$
$C_{p,f}$	Specific heat capacity of fully frozen food, $kJ\,kg^{-1}\,K^{-1}$
$C_{p,fat}$	Specific heat capacity of fats, $kJ\,kg^{-1}\,K^{-1}$
$C_{p,fi}$	Specific heat capacity of fibers, $kJ\,kg^{-1}\,K^{-1}$
$C_{p,i}$	Specific heat capacity of ice, $kJ\,kg^{-1}\,K^{-1}$
$C_{p,p}$	Specific heat capacity of proteins, $kJ\,kg^{-1}\,K^{-1}$
$C_{p,ref}$	Specific heat capacity of reference, $kJ\,kg^{-1}\,K^{-1}$
$C_{p,s}$	Specific heat capacity of dry solid content in food, $kJ\,kg^{-1}\,K^{-1}$
$C_{p,waf}$	Specific heat capacity of water above freezing, $kJ\,kg^{-1}\,K^{-1}$
$C_{p,wbf}$	Specific heat capacity of water below freezing, $kJ\,kg^{-1}\,K^{-1}$
C_{pi}	Specific heat capacity of component i, $kJ\,kg^{-1}\,K^{-1}$
C_v	Specific heat capacity under constant volume, $kJ\,kg^{-1}\,K^{-1}$
H_{ref}	Magnitude difference from the base line for the reference
H_s	Magnitude difference from the base line for the sample
I	Average current, A
k	Thermal conductivity, $W\,m^{-1}\,K^{-1}$
K_T	Adiabatic compressibility, Pa^{-1}
m	Mass of the material
m_c	Mass of the container, kg
m_{ex}	Mass of the heat exchange liquid, kg
m_{ref}	Mass of the reference, kg
m_s	Mass of the sample, kg
m_u	Mass of unfreezable water, kg
P	Pressure, Pa
Q	Heat energy, kJ
T	Temperature, °C
t	Time, sec
T_{eq}	Equilibrium temperature of the mixture, °C
T_f	Initial freezing temperature, °C
$T_{i,c}$	Initial temperature of the container, °C
$T_{i,ex}$	Initial temperature of the heat exchange liquid, °C
$T_{i,s}$	Initial temperature of the sample, °C
T_k	Absolute temperature, K
T_0	Melting temperature of pure water, °C
V	Volume, m^3
V_t	Average voltage, V
x_b	Mass fraction of water bound in an unfreezable form per unit mass of solid foods
x_c	Mass fraction of carbohydrates
x_f	Mass fraction of fat
x_i	Mass fraction of component i
x_p	Mass fraction of proteins
x_s	Mass fraction of solid content in the food
x_{snf}	Mass fraction of solids nonfat
x_w	Mass fraction of water
θ	Cubical expansion coefficient, K^{-1}
ρ	Density, $kg\,m^{-3}$
α	Thermal diffusivity, $m^2\,sec^{-1}$
ΔH_f	Latent heat of ice melting, $kJ\,kg^{-1}$
ΔT	Temperature difference, °C

REFERENCES

1. Pohl, R.O. Specific Heat. *McGraw-Hill Encyclopedia of Science and Technology*, 7th Ed.; McGraw-Hill, Inc.: New York, 1992; Vol. 17, 208–209.
2. Mohsenin, N.N. *Thermal Properties of Foods and Agricultural Materials*; Gordon and Breach: New York, 1980; 25–82.
3. Vagenas, G.K.; Drouzas, A.E.; Marinos-Kouris, D. Predictive Equation for Thermophysical Properties of Plant Foods. In *Engineering and Food.-1 Food Technology*; Spiess, W.E.L., Schubert, H., Eds.; Elsevier Science Pubs., Ltd.: Barking, Essex, England, 1990.
4. Stitt, F.; Kennedy, E.K. Specific Heats of Dehydrated Vegetables and Egg Powder. Food Res. **1945**, *10*, 426–436.
5. Lind, I. The Measurement and Prediction of Thermal Properties of Food During Freezing and Thawing—A Review with Particular Reference to Meat and Dough. J. Food Eng. **1991**, *13*, 285–319.

6. Perry, R.H.; Green, D.W.; Maloney, J.O. *Perry's Chemical Engineer's Handbook*; McGraw-Hill Book Company: New York, 1987; 3–129.
7. Peralta Rodriguez, R.D.; Rodrigo, M.; Kelly, P. A Calorimetric Method to Determine Specific Heats of Prepared Foods. J. Food Eng. **1995**, *26*, 81–96.
8. Riedel, L. Eine formel zur Berechnung der Enthapie Fettarmer Lebensmitteln in Abhängigkeit von Wassergehalt and Tempertur. Chem. Mikrobiol. Technol. Lebensm. **1978**, *5*, 129–133.
9. Wright, D.J. Thermoanalytical Methods in Food Research. Crit. Rep. Appl. Chem. **1984**, *5*, 1–36.
10. Yoncoskie, R.A. The Determination of Heat capacities of Milk Fat by Differential Thermal Analysis. J. Am. Oil Chem. Soc. **1969**, *46* (1), 49–51.
11. Haly, A.R.; Snaith, J.W. Calorimetry of Rat Collagen Before and After Denaturation: The Effect of Fusion of the Absorbed Water. Biopolymer **1971**, *10*, 1681–1699.
12. Kerr, W.L.; Kauten, R.J.; Ozilgen, M.; McCarthy, M.J.; Reid, D.S. NMR Imaging, Calorimetric, and Mathematical Modeling Studies of Food Freezing. J. Food Process. Eng. **1996**, *19*, 363–384.
13. Cornillion, P.; Andrieu, J.; Duplan, J.C.; Laurent, M. Use of Nuclear Magnetic Resonance to Model Thermophysical Properties of Frozen and Unfrozen Model Food Gels. J. Food Eng. **1995**, *25* (1), 1–19.
14. Polley, S.L.; Snyder, O.P.; Kotnour, P. A Compilation of Thermal Properties of Foods. Food Technol. **1980**, *34*, 76–80, 82–84, 86–88, 90–92, 94.
15. Siebel, J.E. Specific Heat of Various Products. Ice Refrigeration **1892**, *2*, 256–257.
16. ASHARE. *ASHARE Handbook, Fundamentals*; American Society of Heating, Refrigerating and Air Conditioning Engineers: Atlanta, GA, 1981.
17. Toledo, R.T. *Fundamentals of Food Process Engineering*; Van Nostrand Reinhold: New York, 1991; 132–143.
18. Dickerson, R.W. Thermal Properties of Foods. In *The Freezing Preservation of Foods, Factors Affecting Quality in Frozen Foods*, 4th Ed.; Tressler, D.K., van Arsdel, W.B., Copley, M.J., Eds.; AVI Publishing Company, Inc.: Westport, CT, 1968; Vol. 2.
19. Charm, S.E. *The Fundamentals of Food Engineering*, 2nd Ed.; AVI Publishing Company, Inc.: Westport, CT, 1971.
20. Lamb, J. Influence of Water on the Thermal Properties of Foods. Chem. Ind. **1976**, *24*, 1046–1048.
21. Choi, Y.; Okos, M.R. Effects of Temperature and Composition on Thermal Properties of Foods. In *Food Engineering and Process Applications*; Le Maguer, M., Jelen, P., Eds.; Elsevier Applied Science Publishers: New York, 1986; Vol. 1, 93–101.
22. Miles, C.A.; van Beek, G.; Veerkamp, C.H. Calculation of Thermophysical Properties of Foods. In *Physical Properties of Foods*; Jowitt, R., Escher, F., Hallstrom, B., Meffert, H.F.Th., Spiess, W.E.L., Vos, G., Eds.; Applied Science Publishers: New York, 1983; 269–312.
23. Ross, Y.; Karel, M. Crystallization of Amorphous Lactose. J. Food Sci. **1992**, *57* (3), 775–777.
24. Hwang, C.H.; Heldman, D.R.; Chao, R.R.; Taylor, T.A. Changes in Specific Heat of Corn Starch due to Gelatinization. J. Food Sci. **1999**, *64* (1), 141–144.
25. AbuDagga, Y.; Kolbe, E. Thermophysical Properties of Surimi Paste at Cooking Temperature. J. Food Eng. **1997**, *32*, 325–337.
26. Pham, Q.T. Prediction of Calorimetric Properties and Freezing Time of Foods from Composition Data. J. Food Eng. **1996**, *30*, 95–107.
27. van Beek, G. Berekening van Thermofysische Eigenschappen van Tuinbouwprodukten uit de Samenstelling en Toepassing Daarvan bij de Berekening van de Veldwarmte. Koeltechniek **1979**, *71* (1), 3–9.
28. Schwartzberg, H. Mathematical Analysis of the Freezing and Thawing of Foods, AIChE Summer Meeting, Detroit, Michigan, 1981.
29. Reddy, N.R.; Solomon, H.M.; Fingerhut, G.A.; Rhodehamel, E.J.; Balasubramaniam, V.M.; Palaniappan, S. Inactivation of Clostridium *botulinum* Type E Spores by High Pressure Processing. J. Food Safety **1999**, *19* (4), 277–288.
30. Knorr, D.; Schlueter, O.; Heinz, V. Impact of High Hydrostatic Pressure on Phase Transitions of Foods. Food Technol. **1998**, *52* (9), 42–45.
31. Denys, S.; van Loey, A.M.; Hendrickx, M.E.; Tobback, P.P. Modeling Heat Transfer During High-Pressure Freezing and Thawing. Biotechnol. Prog. **1997**, *13*, 416–423.
32. Chourot, J.M.; Lemaire, R.; Cornire, G.; le Bail, A. Modeling of High Pressure Thawing. In *High Pressure Bioscience and Biotechnology*; Hayashi, R., Balny, C., Eds.; Elsevier: Amsterdam, Netherlands, 1996; 439–444.
33. Takagi, T.; Teranishi, H. Measurements of Pressure Effects on Excess Molar Enthalpies for Binary Mixtures of Benzene and *n*-Hexane with Cyclohexane Using a New Calorimeter. Fluid Phase Equilibria **1991**, *61*, 299–307.
34. Alvarado, J. de D. Specific Heat of Dehydrate Pulps of Fruits. J. Food Process. Eng. **1991**, *14*, 189–195.
35. Ramaswamy, H.S.; Tung, M.A. Thermophysical Properties of Apple in Relation to Freezing. J. Food Sci. **1981**, *46*, 724–728.
36. Constenla, D.T.; Lozano, J.E.; Crapiste, G.H. Thermophysical Properties of Clarified Apple Juice as a Function of Concentration and Temperature. J. Food Sci. **1989**, *54* (3), 663–668.
37. Njie, D.N.; Rumsey, T.R.; Singh, R.P. Thermal Properties of Cassava, Yam and Plantain. J. Food Eng. **1998**, *37*, 63–76.
38. Marschoun, L.T.; Muthukumarappan, K.; Gunasekaran, S. Thermal Properties of Cheddar Cheese: Experimental and Modeling. Int. J. Food Properties **2001**, *in press*.
39. Singh, P.C.; Singh, R.K.; Bhamidipati, S.; Singh, S.N.; Barone, P. Thermophysical Properties of Fresh and Roasted Coffee Powders. J. Food Process. Eng. **1996**, *20*, 31–50.
40. Kazarian, E.A.; Hall, C.W. Thermal Properties of Grains. Trans. ASAE **1965**, *8* (1), 33–48.
41. Singh, K.K.; Goswami, T.K. Thermal Properties of Cumin Seed. J. Food Eng. **2000**, *45*, 181–187.

42. Levy, F.L. Enthalpy and Specific Heat of Meat and Fish in the Freezing Range. J. Food Technol. **1979**, *14* (6), 549–560.
43. Fernandez-Martin, F.; Montes, F. Influence of Temperature and Composition on Some Physical Properties of Milk and Milk Concentrates. I. Heat Capacity. J. Dairy Res. **1972**, *39*, 65–73.
44. Shrivastava, M.; Datta, A.K. Determination of Specific Heat and Thermal Conductivity of Mushrooms (*Pleurotus florida*). J. Food Eng. **1999**, *39*, 255–260.
45. Haswell, Q.A. A Note on the Specific Heat of Rice, Oats, and Their Products. Cereal Chem. **1954**, *31* (4), 431–432.
46. Rapusas, R.S.; Driscoll, R.H. Thermophysical Properties of Fresh and Dried White Onion Slices. J. Food Eng. **1995**, *24*, 149–164.
47. Ward, T.L.; Singleton, W.S. Thermal Properties of Fats and Oils. VII. Hydrogenated and Unhydrogenated Oils. J. Am. Oil. Chem. Soc. **1950**, *27*, 423.
48. Hsu, M.H.; Mannapperuma, J.D.; Singh, R.P. Physical and Thermal Properties of Pistachios. J. Agric. Eng. Res. **1991**, *49* (4), 311–321.
49. Wang, N.; Brennan, J.G. The Influence of Moisture Content and Temperature on the Specific Heat of Potato Measured by Differential Scanning Calorimetry. J. Food Eng. **1993**, *19*, 303–310.
50. Yamada, T. Thermal Properties of Potato. J. Agric. Chem. Soc. Jpn **1970**, *44* (12), 587–590.
51. Buhri, A.B.; Singh, R.P. Thermal Property Measurements of Fried Foods Using Differential Scanning Calorimeter. In *Developments in Food Engineering: Proceedings of the 6th International Congress on Engineering and Food*; Yano, T., Matsuno, N., Eds.; Blackie Academic and Professional: New York, 1994; 283–285.
52. Vagenas, G.K.; Marino-Kouris, D.; Saravacos, G.D. Thermal Properties of Raisins. J. Food Eng. **1990**, *11*, 147–158.
53. Palacios, L.G. Measurement of Modeling of Thermal Properties of Sorghum and Soy Flour. Master's Thesis, Texas A&M University, College Station, TX, 1981.
54. Gennadios, A.; Bhatnagar, S.; Weller, C.L.; Hanna, M.A. Specific Heat of Sorghum Grain by Differential Scanning Calorimetry. ASAE Paper, International ASAE Summer Meeting, Kansas City, MO, June 19–22, 1994; 94–6036.
55. Deshpande, S.D.; Bal, S. Specific Heat of Soybean. J. Food Process. Eng. **1999**, *72*, 469–477.
56. Arce, J.A.; Sweat, V.E.; Wallapapan, K. Thermal Diffusivity and Conductivity of Soy Flour. ASAE Paper, Annual Meeting of the American Society of Agricultural Engineers, Orlando, FL, June 1981; 81-6528.
57. Drouzas, A.E.; Maroulis, Z.B.; Karathanos, G.D. Direct and Indirect Determination of the Effective Thermal Diffusivity of Granular Starch. J. Food Eng. **1991**, *13*, 91–101.
58. Muir, W.E.; Viravanichai, S. Specific Heat of Wheat. J. Agric. Eng. Res. **1972**, *17* (4), 338–342.
59. Kim, S.S.; Bhowmik, S.R. Thermophysical Properties of Plain Yogurt as Functions of Moisture Content. J. Food Eng. **1997**, *32*, 109–124.

Spray Drying

Hidefumi Yoshii
Takeshi Furuta
Apinan Soottitantawat
Tottori University, Tottori, Japan

Pekka Linko
Helsinki University of Technology, Espoo, Finland

INTRODUCTION

Spray drying is by definition the transformation of a feed of liquid or paste material (solution, dispersion, or paste) into a dried particulate powder by spraying the feed into a hot drying medium.[1,2] Spray drying can form a powdered spherical product directly from a solution or dispersion. The main advantages of spray drying over other drying methods are rapid drying and minimal temperature increase of the material. Every spray dryer consists of a feed pump, atomizer, air heater, air dispenser, drying chamber, and systems for exhaust air cleaning and powder recovery, as shown in Fig. 1. There are large variations in size, in detailed designs, including combinations with other types of dryers, and in the modes of operation depending on the large variety of materials to be processed and product properties desired. Further, modern spray drying systems are highly automated.

OVERVIEW

Spray drying involves the atomization of a liquid or a fine dispersion of preconcentrated food material into a spray of tiny droplets that are contacted with a current of hot air typically of 150–300°C in a large drying chamber. Pressure spray nozzle or centrifugal disc type atomizers are most commonly employed in producing sprayed droplets. Atomization is the key function in spray drying that determines the droplet size distribution. The feed liquid is atomized to the greatest possible degree in order to increase its surface area. For instance, to process 1 cm^3 of a solution into fine droplets of 100 μm or 1 μm diameter, the surface area increases 100 or 10,000 fold, respectively. Usually, the atomized spray has a droplet diameter of 10 μm–200 μm, and a typical drying speed of 5 sec–30 sec.[3] Small droplets with large surface area allow rapid heat transfer and drying. Both rapid drying and the maintaining of the product at the wet bulb temperature minimize thermal damage to the product. In spray drying of food materials the feed rate is usually controlled to reach an outlet air temperature of less than 100°C and a product (or wet-bulb) temperature of around 50°C or less. The conditions of typical spray drying are given in Table 1.

The initial contact between spray droplets and drying air controls the evaporation rate and product temperature in the dryer. There are three modes of contact: 1) cocurrent (drying air and particles move through the drying chamber in the same direction); 2) counter-current (drying air and particles move through the drying chamber in opposite directions); and 3) mixed flow (particle movement through the drying chamber experiences both cocurrent and counter-current phases). The dried particle in which the core material is held in a micro dispersion, fall through the gaseous medium to the bottom of the dryer and is collected.

Atomization—Most foods are sensitive to heat. Therefore, it is important to select a proper drying system based on the physical and chemical characteristics. When the sprayed droplet is in the constant drying rate period, the temperature of the droplet is said to be equal to the wet-bulb temperature of air, which is substantially lower than the inlet temperature. Because the product temperature can be controlled to drop substantially at the exit, spray drying is one of the most suitable methods for drying of heat sensitive food materials. The quality of spray-dried foods is quite dependent on the atomization characteristics and heat and mass transfer of droplets inside the spray dryer. The atomization depends on the pressure of high-pressure pumps, viscosity, density, and surface tension of the solution. The experimental formulas were obtained as Eq. 1 by Turner and Moulton[4] and as Eq. 2 by Ishikawa,[5] respectively, for the mean diameter, $\bar{D}_p$ of droplets atomized by the pressure nozzle:

$$\bar{D}_p = 16.56 d_e^{1.54} F^{-0.44} \sigma^{0.7} \mu_l^{0.16} \quad (1)$$

where $\bar{D}_p$ is the drop mean diameter (μm), d_e, orifice diameter (0.723 μm–1.076 μm), F, mass feed rate (7403 g sec^{-1}–28,476 g sec^{-1}), μ_l, viscosity of liquid

Encyclopedia of Agricultural, Food, and Biological Engineering
DOI: 10.1081/E-EAFE 120007088
Copyright © 2003 by Marcel Dekker, Inc. All rights reserved.

Fig. 1 Spray dryer.

(0.84 c.p.–1.98 c.p.), and σ is the surface tension (27.3 dyn cm^{-1}–36.9 dyn cm^{-1}).

$$\bar{D}_p = 98.4\, d_e^{1.35} S_i^{1.0} Q^{-1.0} \sigma^{0.62} \mu_l^{0.26} \quad (2)$$

where $\bar{D}_p$ is the drop mean diameter (μm), d_e, orifice diameter (0.5 mm–1.0 mm), Q, volume feed rate (30 l hr^{-1}–140 l hr^{-1}), S_i, inlet section area of insert core (0.870 mm^2–1.988 mm^2), μ_l, viscosity of liquid (44 c.p.–250 c.p.), and σ is the surface tension (39 dyn cm^{-1}–55 dyn cm^{-1}). Friedman, Gluckert, and Marshall[6] presented the following dimensionless equation for the rotary disc atomizer with a multiblade disc.

$$\frac{\bar{D}_p}{D_N} = 0.4\left(\frac{\Gamma}{\rho_l N D_N^2}\right)^{0.6}\left(\frac{\mu_l}{\Gamma}\right)^{0.2}\left(\frac{\sigma \rho_l L}{\Gamma^2}\right)^{0.01} \quad (3)$$

where $\bar{D}_p$ is the drop mean diameter (m), D_N, blade diameter (0.02 m–0.20 m), L, length of wet peripheral of the blade (0.08 m–0.50 m), N, rotation speed (14.3 l sec^{-1}–3000 l sec^{-1}), Γ, feed rate per length of wet peripheral of the blade (0.036 kg m^{-1} sec^{-1}–2.28 kg m^{-1} sec^{-1}), ρ_l, density of liquid (1000 kg m^{-3}–1400 kg m^{-3}), μ_l, viscosity of liquid (0.001 Pa sec–9 Pa sec), and σ is the surface tension (74 dyn cm^{-1}–100 dyn cm^{-1}).

Table 1 Conditions of spray drying for skim milk powder

Material	**Defatted milk**
Material flow rate	45,000 kg-material hr^{-1}
Product flow rate	3,700 kg-product hr^{-1}
Material temperature	40°C
Feed water content (dry-base)	1.08
Product water content (dry-base)	0.03
Product average diameter	85 μm
Product apparent density	650 kg m^{-3}
Flow rate of hot air	2,500 m^3 min^{-1}
Inlet air temperature	185°C
Outlet air temperature	95°C
Pressure	250 kg cm^{-2}
Orifice diameter	10 mm
No. of nozzle	1
Flow type	Cocurrent
Tower diameter	8 m
Length of cylindrical part	10 m
Length of conical part	6 m
Viscosity of material	100 mPa sec

A solution can be atomized into 10 μm–200 μm diameter droplets by the spray pressure (100 kg cm^{-2}–300 kg cm^{-2}) nozzle or a centrifugal disc type atomizer (10,000 rpm–30,000 rpm). The processing capacity of a commercial spray dryer has gradually increased to an evaporation rate of around 7000 kg hr^{-1} (with multiple disc systems feed rates of up to 30,000 kg hr^{-1}).

Microencapsulation of food flavors by spray drying—Microencapsulation of volatile materials by spray drying presents the challenge of removing water by vaporization, while retaining substances that are much more volatile than water. Since, most flavors are volatile and labile compounds prone to evaporation and degradation, microencapsulation is important in providing protection against degradative reactions and loss of flavors during food processing. During the drying process, at least 90% of water is evaporated yet the more volatile flavor constituents are retained when appropriate optimum drying conditions are followed. The accepted explanation for this phenomenon relates to the fact that as an atomized droplet of the feed material makes contacts with the hot dry air, it quickly starts to dry from the outside. The surface moisture content of the drying droplets decreases very rapidly. When the surface reaches a moisture content of less than 10%, it is no longer permeable to most flavor compounds but remains quite permeable to water molecules. The concept is based on the fact that the diffusivities of water and the flavor differ and depend strongly on the water content. The diffusivity of the flavor is substantially lower than that of water at low moisture contents.[7,8] It has been observed that the droplet surface is covered by a film of low moisture content (case hardening) as the drying proceeds. This theory is known as "the selective-diffusion theory," which was first proposed by Thijssen and Rulkens.[9] Fig. 2 shows a schematic illustration of selective diffusion model for drying of liquid foods. The theory was successfully applied to estimate the retention of hydrophilic (water soluble) flavor-like ethanol and acetone during a slab drying and a single droplet drying.[10]

Coumans et al.[8] and Ré[11] have recently given extensive overviews on theoretical and practical aspects of flavor retention during spray drying. King[12,13] has also published critical reviews on the factors and mechanisms determining the loss of flavor substances during spray drying. Recently, a symposium proceeding has also been published concerning the flavor encapsulation and the release of encapsulated flavors.[14]

The main carriers used to encapsulate flavors have been maltodextrins, gum arabic, and modified starch. The retention of aroma compounds by these carbohydrates has been discussed from the point of view of their physicochemical interactions by Goubet, Quere, and Violey.[15] Hydrophobic flavors, on the other hand, should be first solubilized in a carrier solution. Typically, the flavor compound to be encapsulated is added to the carrier solution and homogenized with a mixer (homogenizer) to create small flavor emulsion droplets within the carrier solution (O/W emulsion). Shue and Rosenberg[16] investigated the retention of a few esters during spray drying, indicating that the stability of an emulsion droplet is an important factor in flavor retention. Bhandari et al.[17] found that the proportion of maltodextrin and gum arabic markedly influences flavor retention. Rish and Reineccius[18] and

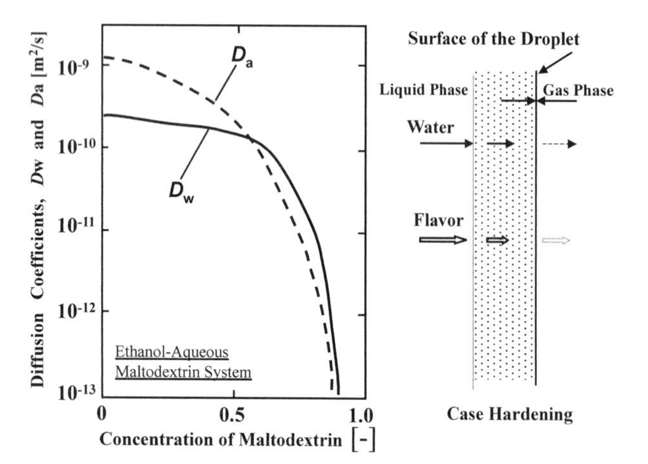

Fig. 2 Schematic model of selective diffusion model for drying liquid food. (D_a: diffusion coefficient of ethanol, D_w: diffusion coefficient of water.)

Ré and Liu[19] examined the effect of emulsion size on the retention and shelf life of spray dried orange oil. They observed that flavor retention increases with decreasing emulsion size. Rosenberg, Kopelman, and Talmon[20] observed that the release of the encapsulated flavor during storage increases with increasing relative humidity. Whorton and Reineccius[21] have correlated the flavor release rate with the glass transition temperature and the collapse temperature of wall materials. Gibbs, Kermasha, and Mulligan[22] reviewed several techniques of encapsulation used in the food industry. The process variables which have been said to influence flavor retention are the solid content of the feed material, type and molar mass of the carrier, concentration of the flavor (flavor load), dryer inlet and exit temperatures, relative humidity of the inlet air, and the particle size of the atomized droplet.

REFERENCES

1. Masters, K. *Spray Drying Handbook*, 5th Ed.; Longman Group: UK, 1991; p. 51.
2. http://www.niro.com/html/drying/fdspraychem.html.
3. Furuta, T.; Hayashi, H.; Ohashi, T. Some Criteria of Spray Dryer Design for Food Liquid. Drying Technol. **1994**, *12*, 157–177.
4. Turner, G.M.; Moulton, R.W. Drop-size Distributions from Spray Nozzles. Chem. Eng. Prog. **1953**, *49*, 185–196.
5. Ishioka, Y. Effect of Spraying Conditions on Drop-sizes of Concentrated Milk. Kagaku-Kougaku (in Japanese) **1964**, *28*, 52–58.
6. Friedman, S.J.; Gluckert, F.A.; Marshall, W.R., Jr. Centrifugal Disk Atomization. Chem. Eng. Prog. **1952**, *48*, 181–191.
7. Furuta, T.; Tsujimoto, S.; Makino, H.; Okazaki, M.; Toei, R. Measurement of Diffusion Coefficients of Water and Ethanol in Aqueous Maltodextrin Solution. J. Food Eng. **1984**, *3*, 169–186.
8. Coumans, W.J.; Kerkhof, Piet J.A.M.; Bruin, S. Theoretical and Practical Aspects of Aroma Retention in Spray Drying and Freeze Drying. Drying Technol. **1994**, *12*, 99–149.
9. Thijssen, H.A.C.; Rulkens, W.H. Retention of Aromas in Drying Food Liquid. De Ingenieur **1968**, *80*, 45–56.
10. Furuta, T.; Tsujimoto, S.; Okazaki, M.; Toei, R. Effect of Drying on Retention of Ethanol in Maltodextrin Solution during Drying of a Single Droplet. Drying Technol. **1983**, *2*, 311–327.
11. Ré, M.I. Microencapsulation by Spray Drying. Drying Technol. **1996**, *16*, 1195–1236.
12. King, C.J. Spray Drying Food Liquids and the Retention of Volatiles. Drying Technol. **1995**, *13*, 1221–1240.
13. King, C.J. Spray Drying Food Liquids and the Retention of Volatiles. Chem. Eng. Prog. **1990**, *86*, 33–39.

monitoring of temperature, pressure, and flow are critical. A means to recycle or remove product that does not meet process requirements should be included.

CONCLUSION

Steam infusion heating is probably the most gentle, direct-contact method available for heating homogeneous, liquid food products. It is efficient, cost-effective, and a time-proven technology, especially for use in pasteurization processes.

REFERENCES

1. Waukesha Cherry-Burrell, A United Dominion Company. Direct Steam Incorporation Systems 1999; Brochure PE-2516, TA-3M-899JA.
2. Sanders, M.J. Infusion Heating, the Thermal Technology for Premium Dairy Products. DFI News for the Dairy/Food Industry, Vol. 10, No. 4.
3. Nahra, J.E.; Woods, W. Method and Apparatus for Treating Liquid Materials. US Patent 4,591,463, May 27, 1986.
4. Nahra, J.E.; Zimmer, A.G. Apparatus for Treating Fluent Material. US Patent 5,544,571, August 13, 1966.
5. Kj.ae butted.rulff, G.; Poulsen, O. Plant for Treating Foodstuffs. US Patent 5,881,638, March 16, 1999.
6. Nahra, J.E.; Woods, W. Apparatus for Treating Fluent Materials. US Patent 4,310,476, January 12, 1982.
7. Bronnert, H.X. Steam Infusion Float Control. US Patent 5,092,230, March 3, 1992.
8. Catelli, C. Contrivance for Heating, Pasteurizing and Sterilizing Fluid Foodstuffs. US Patent 4,432,276, February 21, 1984.
9. Toledo, R.T. *Fundamentals of Food Process Engineering*, 2nd Ed.; Aspen Publishers: Maryland, 1999; 66–159.
10. Invensys APV. Packaged Process Systems 2001; Sales Brochure PkgSys01.0.

S

Steam Injection Heating

Timothy J. Bowser
Oklahoma State University, Stillwater, Oklahoma, U.S.A.

INTRODUCTION

Definition

Steam injection heating is a direct-contact process in which steam is mixed with a pumpable food product. Heating occurs when steam transfers energy directly to the food. Steam gives up its latent heat of vaporization and, depending upon system pressure, some sensible heat. Injection implies that steam is forced into the food product under pressure. As steam directly contacts the food, and the condensate becomes incorporated into it, the steam source must be sufficiently "clean" for human consumption. Steam injection heating for food products requires:

- Culinary steam source of sufficient pressure and volume for injection into the pumped product.
- Pumpable food product.
- Mechanical device(s) to facilitate steam injection.

Benefits

Many benefits are cited for steam injection heating, including an increase in energy efficiency and improved product quality with minimum loss of nutrients. Other potential benefits include reduced equipment size, mass, and expense; improved control of product temperature; simplified operation; reduced maintenance costs; less product burn-on; rapid heating time; and increased process flexibility.[1–3]

Purpose

The purpose of this article is to review the use and application of steam injection heating in the food processing industry.

REQUIREMENTS

Steam

The first requirement for steam injection heating of food products is a stable, culinary steam source. Piping systems play an important role in the delivery of steam to injection equipment. Corrosion resistant piping materials are desired for improved product quality and shelf life, but may not be required, depending on the physical properties of the product. Dirt legs and filters may be used to remove most rust and scale particles (common with steel piping) if performance is acceptable.

Several aspects of steam usage should be considered for direct injection applications.

- Flow and pressure control.
- Air removal.
- Water (condensate) removal.

Steam flow is controlled using automated and manual valves. Automated valves are recommended for variable flow conditions. Constant, upstream steam pressure permits automated valves to function more effectively. A pressure control valve (PCV) is used to maintain constant, downstream pressure. Steam injection heating systems should include a PCV when accurate product temperature control is desired. Pressure upstream of a PCV must always be greater than the pressure setting of the PCV. When multiple demands are placed on a common steam system, direct measurement of upstream line pressure (over time) is needed to identify the lowest pressure. This lowest pressure then, along with the PCV specifications, is used to determine the highest, constant steam pressure for operation of the steam injection system.

Air is present in equipment and pipelines before steam is added and can continue to enter the system with steam. For some food products, air is objectionable and must be excluded from the piping system and removed by air vents. Stagnant air–steam mixtures tend to settle out over time, with air dropping to the bottom of an enclosure. Air vents must be installed at the end, or remote point of steam lines, before the line connects to equipment. This allows steam to push air toward the vent for removal. Equipment air vents should be located at a point farthest from the steam entry, at a low elevation. Product or an inert gas purge can be used to remove air at startup.

Water or condensate should be removed from steam prior to its passing through a flow control valve. Water carries far less internal energy than steam (at a given pressure) and reduces the capacity of the heater. Slugs of

Encyclopedia of Agricultural, Food, and Biological Engineering
DOI: 10.1081/E-EAFE 120007007
Copyright © 2003 by Marcel Dekker, Inc. All rights reserved.

water result in poor process control and dilute the food product.

Food Products

Food products selected for steam injection should be pumpable with respect to the entire steam injection system. Particles must flow freely through piping and equipment (which must be designed and installed for this purpose). Some experimentation may be needed to determine the effects of steam injection on product consistency and texture. High shearing forces caused by steam injection may change particle size and shape, or break down emulsions.

Dilution is often a major consideration when using steam injection heating for foods. Dilution can be calculated using the energy and mass relationships given in the following section. A flash chamber or deaerator can be used to cool product and remove moisture added by steam.

Product temperatures above the atmospheric boiling point in a steam injection heating system are maintained by internal pressure. Pressure requirements are often achieved by using a throttling or backpressure valve.[4] Forcing a product to accelerate through the narrow passage of a throttling valve may be detrimental to its quality. Other means of maintaining system pressure include synchronized pumps[4] and pressurized tanks.[5] The last two methods may function to maintain superior product consistency and quality, but are more expensive and difficult to control. Abrupt changes in product temperature and pressure should be controlled to prevent cavitation and product quality losses, which are especially important for food products containing thermally sensitive nutrients. Cavitation can damage piping and equipment[6] and volatile components (e.g., flavor) of the product may be lost or require steps for recovery.

Equipment

Equipment for steam injection heating may be classified into four basic types: spargers, mixing tees, venturis, and modulating systems. The first three types perform poorly under conditions of variable product flow, changes in product temperature and pressure, and changes in steam pressure. Hammering is often observed in these systems.

Modern steam injection heaters include some means of steam flow modulation. One design includes a spring-loaded piston, which maintains a pressure differential between steam and water to help prevent hammering. This system operates automatically for changes in steam and product pressure and flow, allowing a wide range of turndown and accurate temperature control.[7]

Custom-made steam injection heating devices have been built for years. Steam injection heaters, in their simplest form, are economical and simple to fabricate. Commercial units have the advantage of immediate availability, a history of applications and possibly 3A or other approval. Fig. 1 shows a pilot-sized, skid-mounted steam injection heater (model SCR, Pick Heaters, Inc., West Bend, WI); inset shows the disassembled heater. Fig. 2 is a flow diagram of a typical direct steam heating process.

Control

Control of steam injection heaters is normally configured as a feedback loop. Temperature of the heated product is an input to a P&ID or other controller. Controller settings are optimized to achieve desired steam valve control. This arrangement is simple, inexpensive, and functional in the absence of wide product flow and temperature fluctuations. If product flow rate and temperature vary widely, a feed-forward control loop[8] can be included.

When steam injection heating is part of a pasteurization process, the product temperature of interest is at the end of a hold tube. This introduces a significant lag time into

Fig. 1 Steam injection heater. (Photograph courtesy of the Oklahoma Food and Agriculture Products Research and Technology Center, Oklahoma State University.)

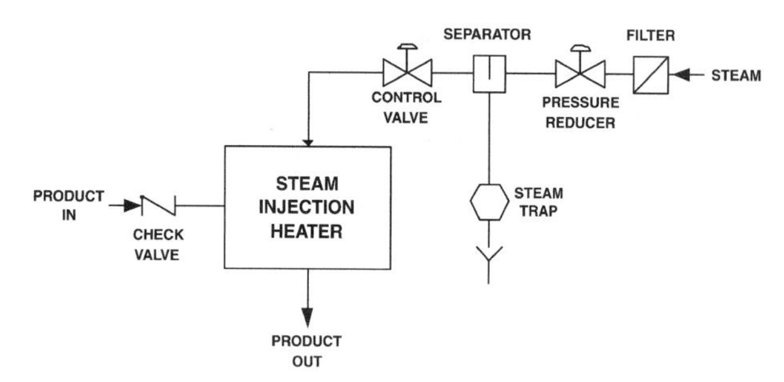

Fig. 2 Flow diagram of typical steam injection heating process.

the control loop. Lag time can be compensated for by insulating the hold tube, avoiding over-design (unnecessary safety factors), and adding feed forward control.

MASS AND ENERGY BALANCE

Using the method described by Toledo,[9] the mass balance of a system is given by: *Inflow = outflow + accumulation.* For the case of steam injection heating of a pumpable food product, the mass balance is written as Eq. 1.

$$m_s + m_f = m_p \quad (1)$$

where m_s is the mass flow rate of steam injected into the food, $kg\,sec^{-1}$, m_f the mass flow rate of feed product introduced into the system, $kg\,sec^{-1}$, m_p the mass flow rate of product exiting the system, $kg\,sec^{-1}$.

A method of determining an energy balance is also described in Ref. 9. The energy balance of a system is given by: *Energy in = energy out + accumulation.* For steam injection heating, the energy balance is given by Eq. 2

$$q_s + q_f = q_p \quad (2)$$

where q_s is the heat transfer rate of steam injected into the system, W; q_f the heat transfer rate of feed product entering the system, W; q_p the heat transfer rate of product leaving the system, W.

The heat transfer rate of each component of the energy balance must be determined. For convenience, system losses are ignored, the sensible heat of steam is neglected and saturated steam is assumed.

$$q_s = m_s H_v$$

$$q_f = m_f c_{pf}(T_f - T_{ref})$$

$$q_p = m_f c_{pp}(T_p - T_{ref})$$

where H_v is the latent heat of vaporization of steam, $kJ\,kg^{-1}$; c_{pf} the specific heat of feed product, $kJ\,kg^{-1}\,K^{-1}$; c_{pp} the specific heat of product, $kJ\,kg^{-1}\,K^{-1}$; T_f the temperature of feed product, K; T_p the temperature of product, K; T_{ref} the temperature of reference state (water at 273), K.

Substituting the heat transfer rate of each component into Eq. 2 yields the overall energy balance for steam injection heating shown by Eq. 3.

$$m_s H_v + m_f c_{pf}(T_f - T_{ref}) = m_f c_{pp}(T_p - T_{ref}) \quad (3)$$

For product formulation purposes, the mass of steam (condensate) added to the product can be obtained by solving Eq. 3 for m_s.

APPLICATION

Food Products

Steam injection heating has been used to process many food products. The list includes animal feeds,[10] baby foods, cereal,[11] cheese sauce,[12] coffee,[13] milk and cereal based slurries,[14] peanuts,[15] texturized protein,[16] sauces, relishes, jams, jellies and salad dressing,[17] soy products,[18] vegetables and meat,[19] and whey-derived fat substitute.[20]

Selection

Selection of a steam injection heater involves defining the process requirements, and conducting a mass and energy balance. Process requirements include initial and final product temperature, minimum, maximum, and design product flow rates, velocities and pressures, product properties such as specific heat, density, viscosity, corrosiveness, and solids content and friability.

Stability

Stability of steam injection systems has historically been an issue, which remains unsolved. Mechanical or sonic vibrations are common, resulting in pressure fluctuations and cavitation. Mixing of steam and viscous fluids also contribute to system instability.[17] Pick[21] claimed that the reduction of bubble size eliminated water hammer noise. Charland[22] noted that water hammer resulted from pressure waves created during the mixing of steam and water. Backflow of product into the steam supply line also contributes to instability problems.[17] Bowser, Weckler, and Jayasekara[23] recommended a correlation between two dimensionless values, the Peclet number (heat transfer coefficient) and the thermodynamic ratio (heat energy required to boil the product divided by the energy of the injected steam) to design stable, steam injection systems. While this method is new and requires further development, it represents an important step in refining the selection process and operation of steam injection equipment.

Safety

Steam control valves should be specified as "fail-close." One or more check valves are recommended for the steam supply line to prevent product entry. Also, a check valve is necessary in the product line to prevent steam entry. Steam flow in product lines (especially stagnant or empty lines) can be dangerous to operators. Steam flow can cause positive displacement pumps and their drive systems to rotate and potentially explode. The outside surface of a sanitary steam heater may not be insulated to accommodate cleaning, presenting a burn hazard to operators. Product safety is an issue when steam injection heating is part of a pasteurization process. For this case, control and monitoring of temperature, pressure, and flow are critical. A means to recycle or remove product that does not meet process requirements should be included.

CONCLUSION

Steam injection heating is a simple, direct-contact method for heating pumpable food products. It is efficient, cost-effective, and a time-proven technology, especially for products requiring or capable of withstanding rapid heating and the shearing forces of steam injection.

REFERENCES

1. Direct Steam Injection Cuts CIP Costs at Waterford Foods. Food Eng. **1997**, *1*, 12–13.
2. Pick, A.E. Producing Hot Water by Direct Steam Injection. Specifying Eng. **1984**, *1*, 80–84.
3. Schroyer, J.A. Understanding the Basics of Steam Injection Heating. Chem. Eng. Prog. **1997**, *5*, 52–55.
4. Hildebolt, W.M.; Hundt, M.T.; Small, R.E. Protein Texturization by Steam Injection. US Patent 4,200,041, April 29, 1980.
5. Long, M. Method and Apparatus for Sterilization with Incremental Pressure Reduction. US Patent 5,344,609, September 6, 1994.
6. Fraser, W.H. Centrifugal Pump Hydraulic Performance and Diagnostics. In *Pump Handbook*, 2nd Ed.; Karassik, I.J., Krutzsch, W.C., Fraser, W.H., Messina, J.P., Eds.; McGraw Hill Book Co.: New York, 1986; 2.266–2.273.
7. King, L.T. Steam Injection and Mixing Apparatus. US Patent 5,066,137, November 19, 1991.
8. Burnham, G.; Khars, J.; Posluszny, A.T. Steam Injection Water Heater. US Patent 4,732,712, March 22, 1988.
9. Toledo, R.T. *Fundamentals of Food Process Engineering*, 2nd Ed.; Aspen Publishers: Maryland, 1999; 66–159.
10. Deyoe, C.W.; Bartley, E.E. Liquid Starch–Urea Ruminant Feed and Method of Producing Same. US Patent 3,988,483, October 26, 1976.
11. Badertscher, E. Hydrolyzing Cereal with Fruit or Honey Present and Apparatus Therefore. US Patent 6,017,569, January 25, 2000.
12. Collyer, S.G.; Hersom, A.C. Steam Injection Process. US Patent 4,752,487, June 21, 1988.
13. Siccardi, A. Steam Injection Nozzle for Beverages. US Patent 5,233,915, August 10, 1993.
14. Badertscher, E.; Poget, P. Tubular T-Shaped Nozzle Assembly for Treating Fluids. US Patent 5,395,569, March 7, 1995.
15. Harris, H. Method and Apparatus for the Continuous Production of Thermally Processed Food Slurries. US Patent 4,302,111, November 24, 1981.
16. Hildebolt, W.M.; Hundt, M.T.; Small, R.E. Protein Texturization by Steam Injection. US Patent 4,200,041, April 29, 1980.
17. White, D.R.; Hobgood, D.F.; Swim, L.E.; Staley, L.D. Methods and Apparatus for Sanitary Steam Injection. US Patent 4,614,661, September 30, 1986.
18. Melcer, I.; Sair, L. Steam Injection and Flash Heat Treatment of Isoelectric Soy Slurries. US Patent 4,054,679, October 18, 1977.
19. Long, M. Method and Apparatus for Sterilization with Incremental Pressure Reduction. US Patent 5,344,609, September 6, 1994.
20. Rhodes, K.H. Process for Making Whey-Derived Fat Substitute Product and Products Thereof. US Patent 5,413,804, May 9, 1995.
21. Pick, A.E. Method and Apparatus for Preventing Water Hammer in High Pressure Steam Injection Water Heaters. US Patent 3,984,504, October 5, 1976.
22. Charland, L. Fluid Mixing Valve. US Patent 4,311,160, January 19, 1982.
23. Bowser, T.J.; Weckler, P.R.; Jayasekara, R. Design Parameters for Operation of a Steam Injection Heater Without Water Hammer When Processing Viscous Food and Agricultural Products. Paper No. 026023, 2002 ASAE International Meeting, Chicago, IL, July 28–31, 2002; The Society for Engineering in Agriculture, Food and Biological Systems: St. Joseph, MI, 2002.

Storage Vessels for Suspensions and Concentrates

Jorge E. Lozano
PLAPIQUI (UNS-CONICET), Bahía Blanca, Argentina

INTRODUCTION

Food engineering involves the conversion of foods by physical, chemical, or biochemical means. These processes require the handling and storage of large quantities of raw and processed foods in vessels of different characteristics, depending upon the physical state (liquid, solid, or semisolid). Liquid foods present no unusual storage problems since the product does not freeze at ambient temperature and has practically the vapor pressure of water, which makes handling relatively easy. However, many liquid foods become highly viscous at low temperatures due to the presence of solids in suspension, and the storage system may need to be designed for this event. In order to determine internal pressure in vessels and/or mixing or agitation condition of highly viscous liquid food, some physical and rheological properties of the stored product are required.

OVERVIEW

Definition and Classification of Storage Vessels

Vessels are in general installations that serve in the reception of the foodstuffs (raw food, intermediate, or finished product). Materials in contact with foods must be nontoxic and should be inert to foods or should not influence foods negatively. Vessels for food storage are made in stainless steel, aluminum, wood, and fiberglass reinforced thermoplastics. Concrete and carbon steel tank, lined with a food grade coating are also very satisfactory. Stainless steel tanks are easy to clean. Most foodstuffs require well ground or electrolytically polished finishing. Materials not in contact with foods have to comply with special requirements (i.e., should be solids and corrosion proof) and must in general have a uniform appearance and some treatment (e.g., lacquered).

According to the type and utilization, storage vessels can be classified as open (storage silo) and closed (tanks), respectively.[1] The capacity of food vessels ranges from few liters to hundred of tons.

Vessels or containers for liquid foods can also be classified by form or function. Containers by form are beaker, bottle, cup, and jar. Containers by function may be drinking vessel (cup), food serving container (plate and tray), food gathering and preparation container (cooking vessel and mortar), and scientific and industrial container (alembic and retort). Typical food liquid storage vessels are defined in Table 1. Tanks are by far the more important vessels dedicated to liquid food storage. Principal storage tank dimensions are given in Fig. 1.

Internal Pressure in Vessels

A tank classification commonly used is based on the internal pressure.[2] Pressure was defined as force per unit area. A standard value of 101.3 kPa is the accepted atmospheric pressure at sea level. Tanks built to pressures exceeding this value are called pressure vessels. Values below atmospheric pressure is considered vacuum. For tank work, pressure or vacuum is measured in the vapor space. On the other hand, the pressure measured at the bottom of a liquid filled tank is proportional to both the height (head) and density of the liquid. Pressure can be determined from liquid density (p), gravitational force (g), and liquid column height (h) with the formula:

$$P = \rho g h \qquad (1)$$

Fermentation may also be retarded by maintaining a headspace of carbon dioxide (CO_2) under pressure. In such a case, vessels must be of heavy construction, since pressures of 827 kPa gauge are necessary to maintain the necessary CO_2 concentration (at 15°C).

When the pressure outside the vessel is greater than the pressure inside the vessel, it may result in serious damage because of the large area involved. In some cases, rapid cooling of concentrated fruit juice tanks have produced vacuum enough to collapse the dome.

Food Properties Associated with Tank Design and Operation

In the case of liquid food suspensions and concentrates, some considerations must be contemplated in the designing of a storage tank. Suspensions may be defined as preparations containing finely divided particles (between 1 μm and 50 μm in diameter) distributed uniformly throughout a fluid. These particles may be large enough to sediment under the influence of gravity.

Encyclopedia of Agricultural, Food, and Biological Engineering
DOI: 10.1081/E-EAFE 120006965
Copyright © 2003 by Marcel Dekker, Inc. All rights reserved.

Table 1 Vessels used to store liquid foods

Name	Definition
Barrel	A cylindrical container, usually bulging outward in the middle and usually held together by metal hoops
Tank	A large container or reservoir for liquids or gases
Tub	A low, wide, open container
Vat	A large container for holding or storing liquids

The velocity that an isolated particle suspended in a Newtonian liquid moves due to gravity is given by Stokes' equation (Eq. 1). Stokes' equation states that the rate of sedimentation of particles in a fluid (v) is a function of the size of the particle (d), the force of gravity (g), the thickness or viscosity (μ) of the fluid, and the density difference between the particle (ρ_p) and fluid (ρ_f).

$$v = 2d^2(\rho_p - \rho_f)g/9\mu \tag{2}$$

When particles of a suspension come close together, they can form aggregates called flocculates that will settle more rapidly. Storage vessels for food suspensions or concentrated liquid foods must be designed by following careful studies of the physical and rheological behavior of each particular product.

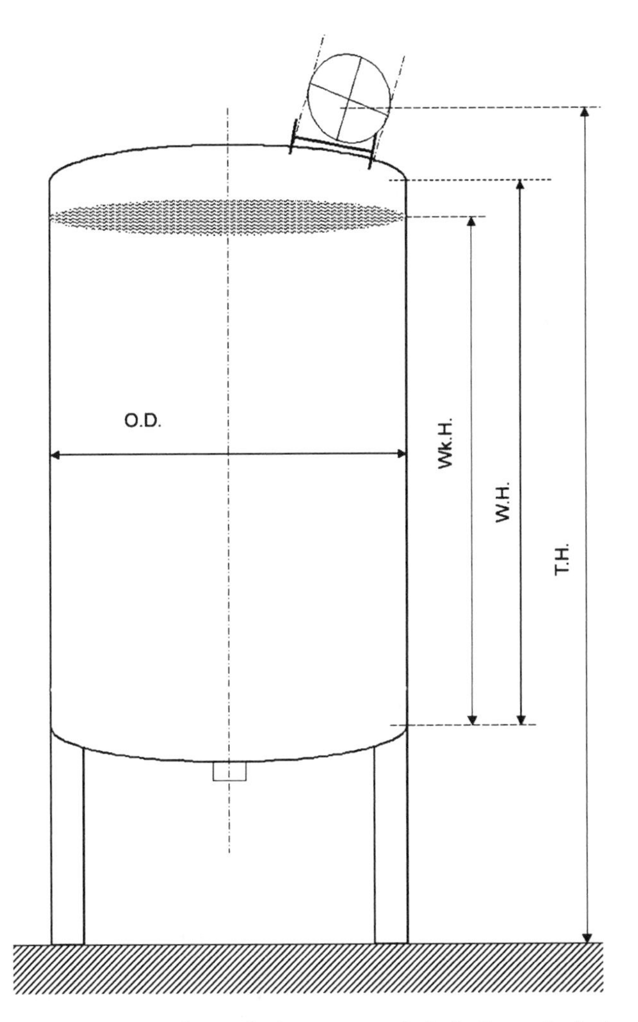

Fig. 1 Sketch of a typical process tank including principal dimensions: diameter (OD), wall height (WH), working height (WkH), and total height (TH).

Density

Care must be exercised when a liquid food is going to be stored in a vessel designed for another one. If there is a significant increase in density of the new liquid, the hydrostatic pressure acting on the tank walls is greater and the design liquid level must be reduced. There is a very strong dependence of density with concentration (Fig. 2) and a noticeable decrease as temperature is increased. Density (ρ) is the unit mass per unit volume (kg m^{-3}). Choi and Okos[4] correlated experimental data of density using a model based on the mass fraction of major food components:

$$\rho = 1/\sum \frac{w_i}{\rho_i} \tag{3}$$

where w_i is the mass fraction and ρ_i (kg m^{-3}) is the individual component density (Table 2).

Viscosity

Viscosity is a measure of a fluid's resistance to flow. It describes the internal friction of a moving fluid. Viscosity of a fluid is defined by the Newton's law of

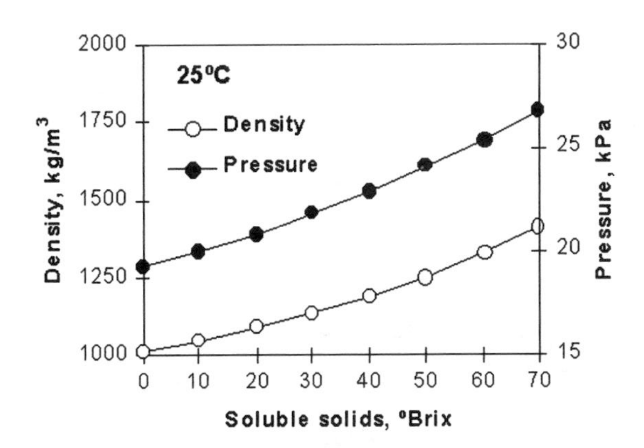

Fig. 2 Increase in density and pressure at the bottom in a 2 m high apple juice tank, with soluble solids. (From Ref. [3].)

Table 2 Parameters of equation $\rho_i\,(\mathrm{kg\,m^{-3}}) = a + bT(°\mathrm{C}) + cT(°\mathrm{C})^2$

Component	a	b	c
Protein	1.3299×10^{-3}	0.5184	—
Carbohydrate	1.5991×10^{-3}	0.3105	—
Fiber	1.3115×10^{-3}	0.3659	—
Ash	2.4238×10^{-3}	0.2806	—
Fat	9.2559×10^{-2}	0.4176	—
Water	997.18	3.1439×10^{-3}	3.7574×10^{-3}

friction as the ratio of the shear stress (τ) to strain rate.

$$\mu = \tau/(\delta u/\delta y) \tag{4}$$

The SI units of μ are $\mathrm{Nsec\,m^{-2}}$. Viscosity is a function of temperature (decreases with increasing temperature). Almost all fluid motions of practical interest involve interaction between inertial and viscous forces. For such flows, it is convenient to consider the ratio of the viscosity (μ) to the density (ρ) of the fluid. This ratio is called *kinematic viscosity* and is denoted by ν.

Most liquid foods are polydisperse systems. Particularly, cloudy fruit juice has solids of various dimensions distributed in a serum, mainly sugars and organic acids. One of the main problems with cloudy juice production is the assurance of cloud stability.[5] Particle size, shape, and volume fraction of particles (ϕ), serum viscosity (η_f), pH, and electrolyte concentration as well as electro-viscous effects, modify juice viscosity compromising the colloidal system stability.[6] Fig. 3 compares viscosity with soluble solids for both a clarified and a cloudy apple juice at 25°C. When size of particles is considered below about 0.5 μm diameter, a higher relative viscosity is always to be expected.

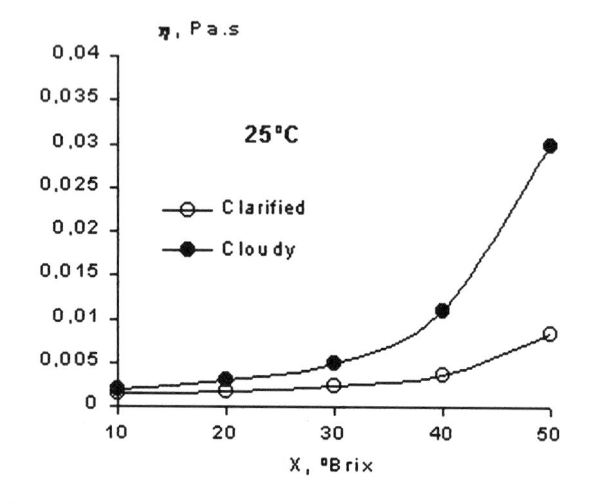

Fig. 3 Viscosity of cloudy and clarified apple juice as a function of soluble solids at 25°C. (From Ref. [6]; with permission.)

Agitation and Removal of Liquid Food Suspensions from Storage

Agitation and mixing of high-viscosity fluids occurs in food industry. Poor storage equipment design or use of improper equipment in the agitation process of liquid suspensions may result in build-up of nonpumpable solids within storage. As a result, storage capacity is reduced and an increasing difficulty to remove the solids due to "cementing" may occur. Thorough agitation of the entire storage prior to pumping is the best way to reduce the accumulation of nonpumpable solids. Agitation is accomplished by using high-horsepower, propeller-type agitators, or recirculation with high-capacity pumps. Although, a wide variety of mixer impellers are available, turbine and anchor type are most commonly used in the food industry (Fig. 4).

Mixing used to be mainly a empirical field of study. During the last years, more fundamental studies were performed.[7] For a tank with smooth walls without baffles and the impeller completely immersed, the power needed for agitation was estimated as the Newton number of power (N_p) as a function of the Reynolds number (Re)[8]:

$$N_p = P/w^3hD^4\rho \tag{5}$$

$$\mathrm{Re} = wD^2/\upsilon \tag{6}$$

where P is the power (W), w, the angular velocity ($\mathrm{rad\,sec^{-1}}$), h, the height (m), D, the diameter (m), ρ, the density ($\mathrm{kg\,m^{-3}}$), and υ is the kinematic viscosity ($\mathrm{m^2\,sec^{-1}}$). Relation between N_p and Re numbers for a single blade, in the range ($10 < \mathrm{Re} < 10^7$), was approximately given by[8]:

$$\mathrm{Log}\,N_p = 0.0642\,(\log \mathrm{Re})^2 - 0.8167\,(\log \mathrm{Re}) + 0.3621 (r^2 = 0.9989) \tag{7}$$

Calculations, modeling, and economics in design and selection of mixers for high viscous and suspensions was recently published.[7,9] In summary, to properly store, mix, or pump liquid food concentrates and suspensions, a knowledge of the characteristics and properties of the stored product such as density, viscosity, vapor pressure,

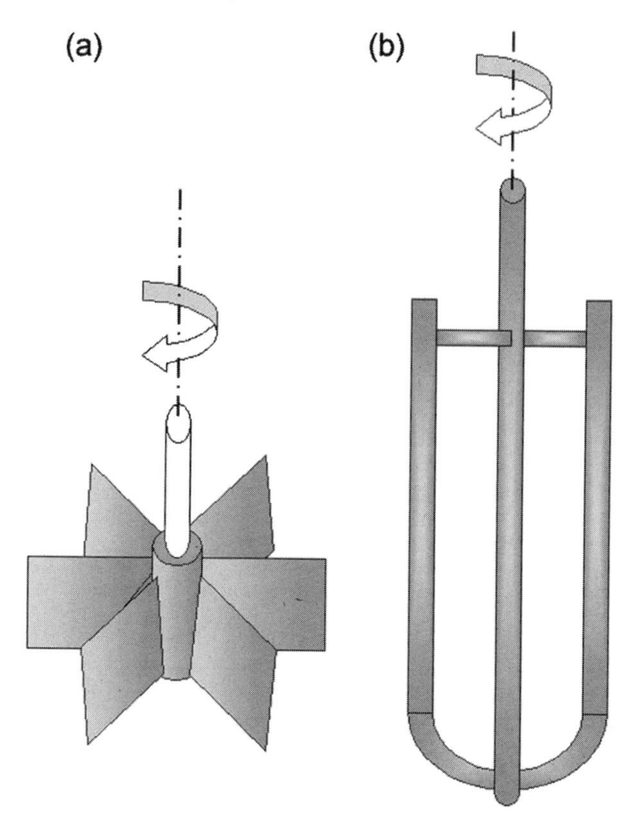

Fig. 4 Two types of impellers used for mixing or agitation of fluid foods: (a) turbine-type impeller and (b) anchor-type impeller.

and tendency to sediment are vital for both the tank designer and the process operator.

A Concluding Example

Consider, for instance, the case of a tank 1 m high stirred with a turbine-type impeller formed by a single blade, 0.1 m high and 0.25 m across, turning at a rate of 50 rad sec^{-1}. Pressure at the bottom and agitation power requirement must be determined when the tank contains: 1) water; 2) a 50°Brix clarified apple juice; and 3) a 50°Brix cloudy apple at the same temperature (25°C). Density and viscosity of juices are given in Figs. 2 and 3. Water properties are given elsewhere.[8] Calculated rotary Re (Eq. 6) resulted 3.1×10^6; 4.9×10^5; and 1.2×10^5 for water, clarified, and cloudy juice, respectively. With Re values in Eq. 7 N_p number was estimated. Finally, agitation power P (Eq. 5) resulted: 283 W for water, 378 W for clarified juice, and 451 W in the case of cloudy juice. Calculated pressure (Eq. 1) was 9.8×10^3 Pa for the tank with water and 15.9×10^3 Pa in the case of a 50°Brix juice.

REFERENCES

1. Spreer, E. Dairy Equipment. In *Milk and Dairy Product Technology*; Marcel Dekker, Inc.: New York, 1998; 133–138.
2. Myers, P. Tanks and Pressure Vessels. In *Encyclopedia of Chemical Technology*; Kirk-Othmer, Ed.; John Wiley & Sons: New York, 1994; Vol. 9, 623–657.
3. Constenla, D.T.; Crapiste, G.H.; Lozano, J.E. Thermophysical Properties of Clarified Apple Juice as a Function of Concentration and Temperature. J. Food Sci. **1989**, *54* (3), 663–669.
4. Choi, Y.; Okos, M.R. Effects of Temperature and Composition on the Thermal Properties of Foods. In *Food Engineering and Process Applications*; Le Maguer, M., Jensen, J., Eds.; Elsevier: London, 1986; Vol. 1, 93–101.
5. Genovese, D.B.; Elustondo, M.P.; Lozano, J.E. Color and Cloud Stabilization in Cloudy Apple Juice by Steam Heating During Crushing. J. Food Sci. **1997**, *62* (6), 1171–1175.
6. Genovese, D.B.; Lozano, J.E. Effect of Cloud Particle Characteristics on the Viscosity of Cloudy Apple Juice. J. Food Sci. **2000**, *64* (4), 641–645.
7. Bakker, A.; Gattes, L.E. Properly Choose Mechanical Agitators for Viscous Liquids. Chem. Eng. Prog. **1995**, *91* (12), 25–34.
8. Loncin, M.; Merson, R.L. Momentum Transfer in Fluids. In *Food Engineering*; Academic Press Inc.: London, 1979; 30–38.
9. vonEssen, J.A.; Ricks, B. Design Agitated Slurry Storage Tanks to Minimize costs. Chem. Eng. Prog. **1999**, *95* (11), 51–55.

Substrate Kinetics

C. K. Bower
J. McGuire
M. K. Bothwell
Oregon State University, Corvallis, Oregon, U.S.A.

INTRODUCTION

Enzymes combine with substrates to lower the activation energy of a chemical reaction, thereby increasing the rate at which the reaction will proceed. The substrate kinetics characterizing such reactions are commonly quantified by recording the increase in amount of product, decrease in amount of substrate, or appearance of a chromogenic compound from a coupled secondary reaction. The resulting data can be used to estimate the kinetic parameters governing the reaction.

MODELING HOMOGENEOUS SUBSTRATE KINETICS

Enzyme-catalyzed reactions can be classified as homogeneous and heterogeneous depending on whether or not they take place within a single phase (such as in solution) or involve more than one phase (such as systems involving insoluble substrates or enzymes immobilized on a solid support). When these reactions occur entirely in solution, in general, evaluation of the relevant dynamics can be fairly straightforward.

The classic mechanism describing enzyme-catalyzed reactions states that free enzyme (E) reversibly forms a complex with substrate (S), and this complex then breaks down to produce a product (P), with k_1, k_{-1}, and k_2 representing rate constants for each step of the reaction:

$$E + S \underset{k_1}{\overset{k-1}{\leftrightarrows}} ES \overset{k_2}{\longrightarrow} E + P$$

As fully developed in standard texts of biochemistry and biochemical engineering,[1,2] two major approaches are used in developing a rate expression for the enzyme-catalyzed reaction: 1) a rapid equilibrium approach and 2) a quasi-steady-state approach. With the rapid equilibrium approach we assume that [E], [S], and [ES] are in equilibrium throughout the reaction while with the quasi-steady-state approach, we assume $d[ES]/dt = 0$. The resulting expression for reaction velocity, $V(= d[P]/dt$ or $-d[S]/dt)$, is still used widely today to determine rate constants for enzyme-catalyzed reactions that obey Michaelis-Menten, or saturation, kinetics:

$$V = \frac{V_{max} + [S]}{[S] + K_m} \tag{1}$$

where V_{max} is the limiting rate of the reaction when the enzyme is saturated with substrate, and K_m is the Michaelis-Menten constant, the identity of which is $K_m = k_{-1}/k_1$ for the case of rapid equilibrium, or $K_m = (k_{-1} + k_2)/k_1$ for the case of quasi-steady state. A graphical representation of Michaelis-Menten kinetics is shown in Fig. 1. The conversion rate of the enzyme–substrate intermediate into enzyme and product is governed by the turnover number (k_{cat}). When the reaction mechanism is that of Michaelis-Menten, and all the reaction steps are fast, k_{cat} is simply a first-order rate constant, and describes the maximum amount of substrate that can be converted to product per time ($k_{cat} = V_{max}/[E]$). The specificity constant (k_{cat}/K_m) is an apparent second order rate constant that describes the specificity of the enzyme for the substrate.

Michaelis-Menten parameters (e.g., K_m, k_{cat}, and k_{cat}/K_m) should only be confidently used when Michaelis-Menten kinetics are obeyed. When this is not the case, it is essential to derive the rate equation that reflects the more appropriate kinetic mechanism governing the reaction being studied. Even then, there may be complications if the temperature and pH are not constant, if the substrate is not fully soluble in an aqueous solution, or if substrate inhibition is occurring.[2–5]

Estimating Reaction Rate Parameters

In the common case of Michaelis-Menten kinetics, the relevant parameters are obtained by plotting kinetic data (i.e., V vs. [S]) according to Eq. 1, and applying a nonlinear parameter estimation method (e.g., gradient or derivative-free search methods). A number of other methods, fully developed in standard biochemistry texts, are widely used as well, each involving a linearization of the Michaelis-Menten equation. Among the most popular of these are the Lineweaver-Burke, Hanes-Woolf, and Eadie-Hofstee linearizations.

Encyclopedia of Agricultural, Food, and Biological Engineering
DOI: 10.1081/E-EAFE 120007200
Copyright © 2003 by Marcel Dekker, Inc. All rights reserved.

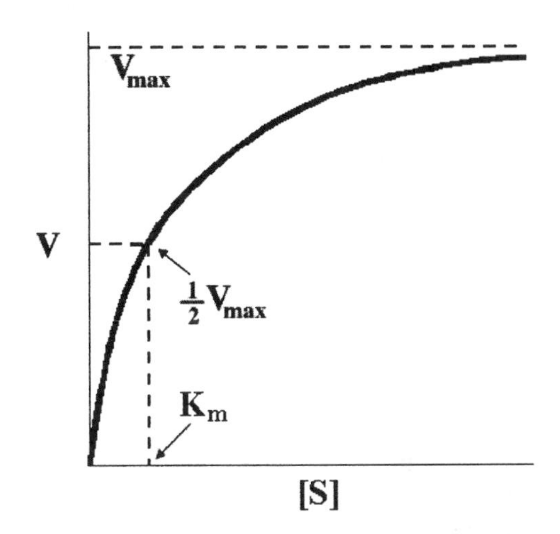

Fig. 1 The velocity (V) of an enzyme-catalyzed reaction displaying Michaelis-Menten kinetics is plotted as a function of substrate concentration [S]. The limiting rate (V_{max}) represents the theoretical maximum velocity of the reaction; K_m is equal to the substrate concentration at which $V = 0.5V_{max}$.

The ability of each linearization to accurately estimate the reaction parameters is affected by experimental error.[6] The inversion of the variables in the Lineweaver-Burke and Hanes-Woolf linearizations cause the smallest values of the dependent variable to be more heavily weighted in determining the placement of the fitted line. In addition, the independent variable of the Hanes-Woolf plot appears on both sides of the equation, producing a correlation effect.[7] There are statistical arguments against the use of the Eadie-Hofstee linearization as well. In this method, the dependent variable is plotted against itself, producing a correlation effect and introducing error on both the abscissa and ordinate. In fact, the method of least squares is not theoretically applicable to data linearized by the Eadie-Hofstee method because of the error associated with the abscissa.[6] Statistical problems associated with the linearizations can be avoided by using gradient and derivative-free search methods that minimize the sum of error-squared residuals of the nonlinear model (i.e., Eq. 1).

Considerations for Multisubstrate and Inhibition Kinetics

Biochemical reactions involving a single substrate reacting with an enzyme to form a single product are not common. Multisubstrate reactions predominate, and they can occur by several different kinetic mechanisms. In such cases, a mechanism capturing the essential features of the multisubstrate reaction, particularly those features not characteristic of Michaelis-Menten kinetics, must be postulated. Kinetic data can then be compared to the model evolving from that mechanism using steady-state methods (e.g., primary plots, product inhibition studies), as well as nonsteady-state methods (e.g., equilibrium isotope exchange, rapid reaction techniques).[3,5,8]

The kinetics of an enzymatic reaction can be altered by the reversible or irreversible binding of inhibitory compounds to the enzyme. There are three main types of inhibitors that reversibly bind to the enzyme: competitive, noncompetitive, and uncompetitive. A competitive inhibitor typically resembles the shape of the substrate, and binds to the enzyme's active site. Competitive inhibition can be overcome by increasing the concentration of the substrate. A noncompetitive inhibitor binds to the enzyme at sites other than the active site, and increasing the substrate concentration will not overcome the effect of the inhibitor. Another compound that reduces or blocks the binding of the noncompetitive inhibitor to the enzyme must be added in order to reduce the effect of the inhibitor. Uncompetitive inhibitors bind the enzyme–substrate complex only. Like reactions that involve multiple substrates, rate expressions for enzyme systems that include an inhibitor can be derived by postulating a mechanism that captures the essential features of the reaction. Rate expressions for the three inhibitor types discussed above are fully developed in standard texts of biochemistry and biochemical engineering.[1,2]

ENZYME INTERACTIONS WITH INSOLUBLE SUBSTRATES

Kinetic evaluation of systems involving immobilized enzymes or insoluble substrates can be complicated. In the case of an insoluble substrate, diffusion of enzyme into and product out of the substrate matrix is extremely important, as well as enzyme adsorption on the substrate surface. Further, a reaction may involve multiple enzymes, exhibiting synergistic or inhibitory effects on overall process efficiency.

In a variety of natural and practical circumstances, enzyme binding to the substrate surface is the controlling step of the reaction, and enzyme adsorption often serves as the starting point for model derivation.[9,10] However, many researchers continue to fit adsorption kinetic data to Langmuir and Langmuir-type equations, despite their substantial shortcomings in accurately describing enzyme behavior at interfaces, and their lack of providing any real utility in design. Although enzyme adsorption data generally fit the Langmuir equation quite well, there is no evidence supporting its use mechanistically. The Langmuir mechanism is predicated on the reversible binding of enzyme to form a uniform enzyme–substrate complex. The equation evolving from this simple mechanism is similar in form to that describing

Michaelis-Menten kinetics (Eq. 1), and relates adsorbed amount to free enzyme concentration in solution:

$$\Gamma = \frac{\Gamma_{\max} + [E]}{[E] + K} \qquad (2)$$

where Γ is the surface coverage of enzyme, $\Gamma_{\max}$ is the maximum surface coverage (corresponding to monolayer coverage), and K is the equilibrium constant governing the reversible formation of enzyme–substrate complex.

There is abundant evidence indicating that enzyme binding does not comply with these assumptions, resulting in data that deviates from model predictions.[7,9,11–13] Estimates of equilibrium constants derived from fitting data to the Langmuir equation should be considered as at best apparent, and at worst pseudo binding constants.[13] Past theoretical and experimental work with proteins at interfaces would however support the thought that a given enzyme can adsorb in different structural states, characterized by exhibiting different catalytic function, occupied areas, binding strengths, and propensities to undergo exchange events with other enzymes and surface active species. One might then best evaluate kinetic data with reference to a mechanism allowing enzyme molecules to adopt multiple states at the surface.[14] The mechanisms of Fig. 2 drawn to depict molecules adsorbing into one of two different states, either directly from solution (Fig. 2b), or through a surface-induced conversion (Fig. 2c) illustrate two of many alternative approaches to estimating adsorption kinetic rate constants, and examining the practical validity of assumptions. While such mechanisms are not comprehensive in a biophysical sense, they can certainly be constructed to embody the essential features of enzyme adsorption, in particular to account for multiple adsorption states and the relative tendencies among enzymes to adopt them.

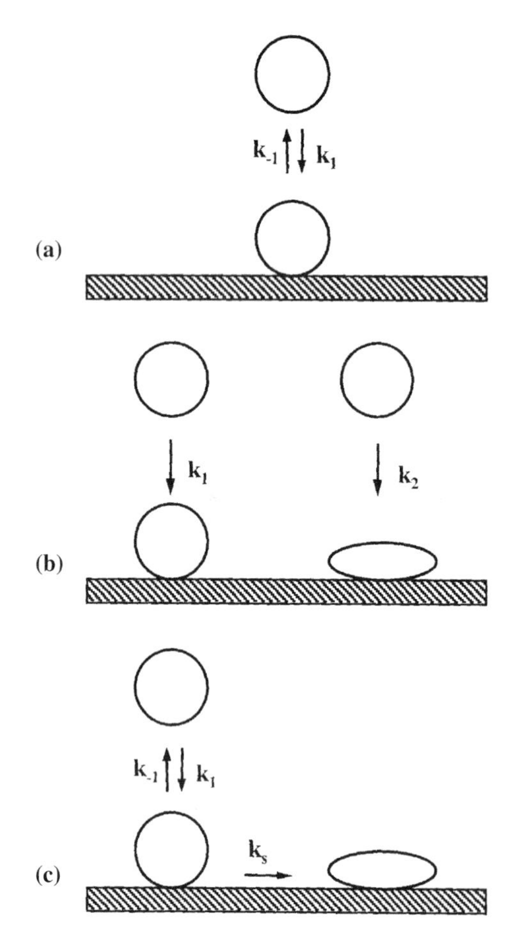

Fig. 2 Example mechanisms for enzyme binding, allowing for the existence of different binding states at the substrate surface: (a) single state, Langmuir adsorption; (b) adsorption into functionally dissimilar forms directly from solution; (c) reversible adsorption, followed by irreversible conversion of adsorbed enzyme to a functionally dissimilar form. Reversibility in these cases is defined with respect to dilution; i.e., exchange reactions can involve enzyme bound in either state.

METHODS USED TO INVESTIGATE SUBSTRATE KINETICS

There are a variety of instruments capable of collecting data for substrate kinetics. The most common methods involve photometric techniques (e.g., spectrophotometer, fluorometer). The change in substrate or product concentration can be monitored at the specific wavelength at which it absorbs light. The molar concentration (c) can be directly calculated from the absorbance (A) using the Beer-Lambert law ($A = \varepsilon cl$) where ε is the molar absorptivity coefficient and l is the path length of the cell.

Electrochemical methods can also be used to detect changes that occur during an enzyme-catalyzed reaction. For example, ion-selective electrodes relate a change in electrical potential to a change in the concentration of one of the reactants, while conductivity techniques measure the variation in electrical conductance with time for reactions involving a change in the number of charged molecules as substrates are converted to products. Assays for substrate kinetics can also be based on enthalpy (using a microcalorimeter to measure the heat gained or lost as the reaction proceeds), on radiochemical methods (such as a radioactively-labeled substrate that can be easily separated from the product before measurement), as well as on dry-reagent techniques (in which a reagent-containing test strip develops color proportional to the quantity of the specific compound present).

Rapid reaction techniques have been developed to better understand the kinetic mechanisms of enzyme–substrate reactions. These methods provide information about the intermediate compounds that form during the reaction, as well as data that can be used to calculate

S

individual rate constants. Examples of rapid reaction techniques include continuous flows systems (that pump enzyme and substrate together at a fixed speed, and then use the size of the tubing and the flow rate past a detector to monitor the reaction), stopped-flow techniques (where enzyme–substrate reactions take place in an observation chamber as a discrete process, rather than in continuous flow), and relaxation methods (where the combined enzyme and substrate are in equilibrium until a sudden temperature or pressure change is introduced to initiate a new equilibrium, which is then monitored). However, there are limitations to rapid reaction techniques. Reactions that proceed too quickly, (i.e., enzymes with turnover numbers greater than a thousand per second) cannot be analyzed in this way.[5,8]

REFERENCES

1. Lehninger, A.L.; Nelson, D.L.; Cox, M.M. *Principles of Biochemistry*, 2nd Ed.; Worth Publishers: New York, NY, 1993; 1013.
2. Shuler, M.L.; Kargi, F. *Bioprocess Engineering—Basic Concepts*; PTR Prentice Hall: Englewood Cliffs, NJ, 1992; 479.
3. Taylor, K.B. *Enzyme Kinetics and Mechanisms*; Kluwer Academic Publishers: Dordrecht, The Netherlands, 2002; 227.
4. Berg, O.G.; Jain, M.K. *Interfacial Enzyme Kinetics*; John Wiley & Sons, Ltd: West Sussex, England, 2002; 301.
5. Fersht, A. *Enzyme Structure and Mechanism*, 2nd Ed.; W.H. Freeman and Company: New York, NY, 1985; 475.
6. Dowd, J.E.; Riggs, D.S.J. A Comparison of Estimates of Michaelis-Menton Kinetic Constants from Various Linear Transformations. Biol Chem. **1965**, *240*, 863–869.
7. Bothwell, M.K.; Walker, L.P. Evaluation of Parameter Estimation Methods for Estimating Cellulase Binding Constants. Bioresour. Technol. **1995**, *53*, 21–29.
8. Bisswanger, H. *Enzyme Kinetics: Principles and Methods*; John Wiley & Sons: New York, NY, 2002; 260, 260.
9. Stahlberg, J.; Johansson, G.; Petterson, G. A New Model for Enzymatic Hydrolysis of Cellulose Based on the Two-Domain Structure of Cellobiohydrolase. Bio/Technol. **1991**, *9*, 286–290.
10. Nidetzky, B.; Steiner, W. A New Approach for Modeling Cellulase–Cellulose Adsorption and the Kinetics of the Enzymatic Hydrolysis of Microcrystalline Cellulose. Biotechnol. Bioeng. **1993**, *42*, 469–479.
11. Woodward, J.; Hayes, M.K.; Lee, N.E. Hydrolysis of Cellulose by Saturating and Non-saturating Concentrations of Cellulase: Implications for Synergism. Bio/Technol. **1988**, *6*, 301–304.
12. Bothwell, M.K.; Walker, L.P.; Wilson, D.B.; Irwin, D.C. Binding reversibility and surface exchange of *Thermomonospora fusca* E3 and E5, and *Trichoderma reesei* CBHI. Enzyme Microb. Technol. **1997**, *20*, 411–417.
13. Brash, J.L.; Horbett, T.A. Proteins at Interfaces—An Overview. In *Proteins at Interfaces II—Fundamentals and Applications*; Horbett, T.A., Brash, J.L., Eds.; Symp. Ser. 602; ACS: Washington D.C., 1995; 1–23.
14. McGuire, J.; Krisdhasima, V.; Wahlgren, M.; Arnebrant, T. Comparative Adsorption Studies with Synthetic, Structural Stability and Charge Mutants of Bacteriophage T4 Lysozyme. In *Proteins at Interfaces II—Fundamentals and Applications*; Horbett, T.A., Brash, J.L., Eds.; Symp. Ser. 602; ACS: Washington D.C., 1995; 52–65.

Subsurface Drainage Systems

Forrest T. Izuno
University of Minnesota, Waseca, Minnesota, U.S.A.

Raymond M. Garcia
LBFH Inc., Palm City, Florida, U.S.A.

INTRODUCTION

Subsurface drainage is the removal of excess water sourced from intentional or unintentional overirrigation, precipitation, seepage, or naturally high water tables from agricultural lands to achieve a suitably aerated root zone for crop growth, characteristically using manmade conduits buried beneath the soil surface. Essentially, water from the agricultural land above seeps into the drains and flows to buried collector conduits or open ditches on its way to storage or disposal water systems. Beginning with simple management techniques to provide an adequate root zone and field trafficability, basically involving installation and repairs, subsurface drainage system management has evolved into a complex science encompassing the desires to enhance crop nutrient uptake, control salinity, reduce erosion, maintain optimum water tables, and improve discharge water quality.

HISTORY

Drainage of land to enable agricultural production has its roots in ancient times, probably concurrent with the beginnings of agriculture. However, the first historical documentation of drainage as an art or science to improve agricultural lands dates back to the days of the Roman Empire.[1,2] From those days, until Sir James Graham introduced tile drains on his estate in Northumberland, England in 1810, drainage methods and philosophies changed very little. In 1835, a Scottish-born farmer, John Johnson, is credited with having started the modern era of drainage in the United States when he imported the patterns for molding clay tile from Scotland,[3–5] hand-molded the tiles, and installed them on his farm in Geneva, New York. By 1838, tile was being manufactured in the United States, and by 1880, 1140 tile factories, primarily in the Midwestern states of Illinois, Indiana, and Ohio were operating. As of 1982, approximately 110 million acres, representing over 20% of rural land in the United States, was being artificially drained.[6] Of that total, approximately 72% was cropland, much of it located in the North Central Region (Great Lakes and Cornbelt states), the Atlantic Coastal Plains (North and South Carolina and Georgia), the Mid-Atlantic States, Florida, Louisiana, and Texas.[3,7,8] In most of those areas, except areas such as south Florida where the topography causes surface drainage to prevail, subsurface drainage is the method of choice.

The first tile drains were made by hand. Rectangles of clay about one-half-inch thick were simply draped over a pole, a person's shinbone, or some other object to give the pieces a horseshoe shape. The shaped tiles were then air-dried and kiln-baked, resulting in arch-shaped pieces of pottery, which when placed open side down sequentially in a ditch and backfilled, resulted in an underground conduit for drainage.[2,8] Between 1848 and 1853, the first tile-making machine, based on extruding clay, was brought to the United States from England. This resulted in lowering the cost of tile for drainage, as well as the potential for experimenting with different shapes. Between 1860 and the turn of the century, concrete pipe was introduced where good clay was not available. Corrugated polyethylene plastic pipe was introduced in 1967. In addition to clay and concrete, perforated corrugated metal, bituminous-fiber, and rigid plastic tubing have all been installed for subsurface drainage conduits.[9] Perforated or slotted flexible corrugated plastic tubing is the material most often used.[10]

The impetus to dewater lands was solely to drain land on which to grow crops. Optimum drainage depths and spacings for specific crops under specific hydrologic conditions have been developed. Sand, gravel, and/or soil are placed around the drains to improve bearing strengths and to reduce soil inflow.[10–13] Layouts have been developed for different field needs. Mathematical equations were developed to simulate flow into the drains under saturated conditions.[11–17] Drainage system design procedures were developed, as were improved methods of installation. Numerous design, installation, and operating guidebooks, computer models, and references were written. It was demonstrated that subsurface drainage could help to mitigate salinity problems in irrigated areas.[11,18,19] Additionally, the appropriate application of

Encyclopedia of Agricultural, Food, and Biological Engineering
DOI: 10.1081/E-EAFE 120006924
Copyright © 2003 by Marcel Dekker, Inc. All rights reserved.

subsurface drainage could reduce soil erosion.[8,20] The loss of nutrients to ground or surface waters where they have the potential to become pollutants is also reduced by subsurface drainage.[7,8,21] Finally, subsurface drainage systems are applicable for water table control, which combines subsurface drainage, controlled drainage, and subsurface irrigation[6,22] to achieve the optimum aerated root zone by satisfying both irrigation and drainage needs while minimizing nutrient runoff.

DRAINAGE SYSTEM LAYOUTS

There are four basic patterns by which subsurface drains are laid out.[5,11] Collector pipes are laid out towards small, generally isolated areas in need of drainage (Fig. 1a). Pipe drains are then laid out strategically beneath the areas in need of drainage. These areas are usually slight depressions in the land or areas where ponding occurs due to the impervious soil substratum. Fig. 1b shows a typical herringbone pattern, appropriately named due to its resemblance to a fish bone structure. In this case, a large area is drained by lateral drains that empty into a central collector pipe. The pattern follows the surface path that drainage water would take on a concave field surface, emptying to the main collector pipe, which is laid out in the low line of the draw or swale. Fig. 1c shows a grid iron or parallel pattern. In this case, the lateral drains are laid out parallel to each other, but not necessarily perpendicular to the main collector conduit. The laterals drain to the main on only one side. This pattern is generally applicable for large flat or uniformly sloping areas. The cutoff or interceptor drain (Fig. 1d), laid out normal to the slope, intercepts flow from an area that is prone to lateral seepage. In this case, the subsurface drain keeps the water table from emerging above the land surface as it flows

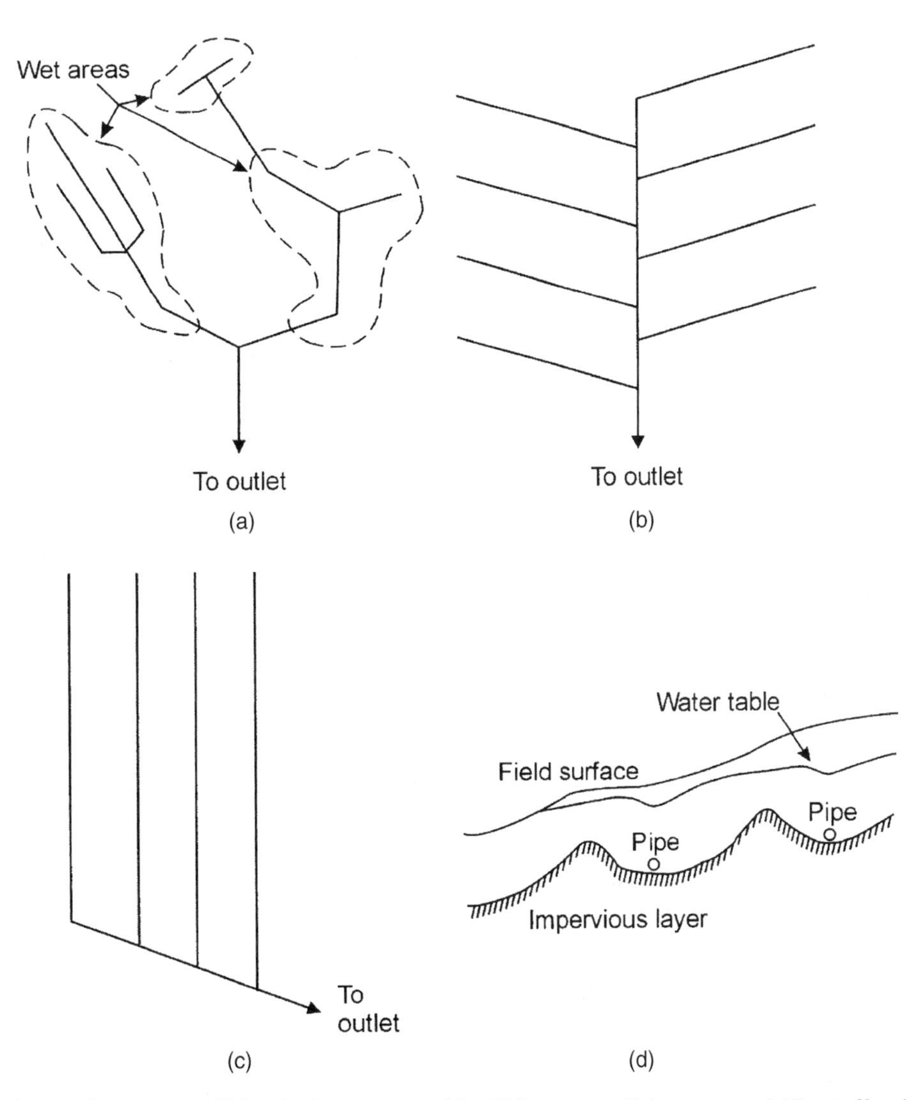

Fig. 1 (a) Natural or random pattern; (b) herringbone pattern; (c) grid iron or parallel pattern; and (d) cutoff or interceptor system.

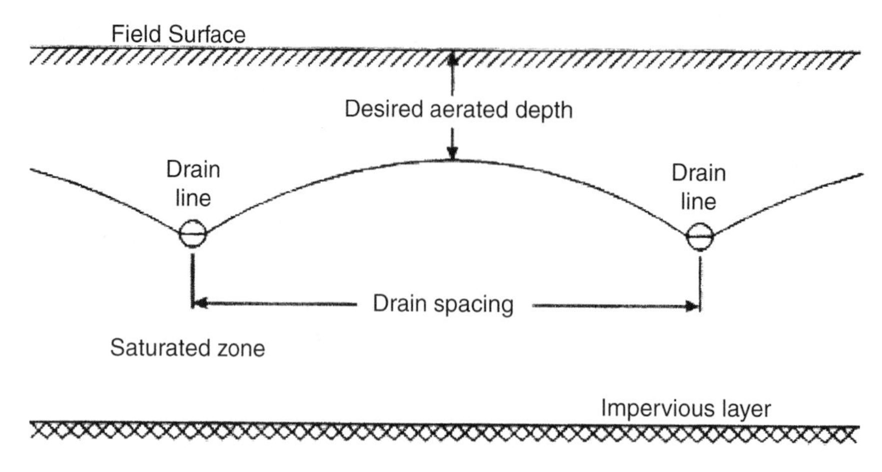

Fig. 2 Cross-section of typical drain system showing water table during drainage.

down a relatively steep hillside. In the absence of a drain, water would continue its flow path, exit the soil profile, and result in a wet area below.

DRAIN INSTALLATION

The effectiveness of subsurface drains is predicated on the fact that they are installed beneath the water table. Water, with the propensity to flow naturally from areas of positive pressure to zones of atmospheric or zero pressure, enters the drain and flows to open waterways or control structures. Subsurface drains are installed in two primary ways. The first, in its most basic form, requires the digging of trenches. Most pipe drains are laid in trenches prior to backfilling with soil, sand, gravel, and/or geotextiles. With corrugated flexible drain pipe installed with a drain plow, the grade is more difficult to control than with an open trench. These methods are detailed in Ref. 10. Drains are typically installed from 2.5-ft to 8.0-ft deep.[5] At least 2 ft of cover is required above the drains to keep field traffic and implements from damaging or collapsing the drains. Summarizing the installation depth issue, the drains need to be installed considering both depth and spacing. Basically, the vertical distance between the top of the capillary fringe above the water table and the soil surface, at the midpoint between drains (Fig. 2), represents the aerated root zone. This depth must satisfy the crop requirements. Drain depth and spacing being interrelated, the further apart the drain laterals are placed, the deeper they must be installed. Economics will play an important role in optimizing the installation, along with soil physical properties, constraints on the depth of the main, topography, depth to the impermeable layer, hydrology of the area, and the desired drainage rate.

A special type of subsurface drainage is the mole drain. This is simply a subsurface drain without the drain pipe.[2] Mole drainage is not very common. It has been used in the organic soils in Florida, in some mineral soil areas in the United States, and in England during World War II. Installation of mole drains requires an implement with a steel shank capable of ripping through the soil profile to a depth of between 2 ft and 3.5 ft.[11] At the toe of the shank is a steel cylinder, which forms a channel through the soil profile.

CONCLUSION

Subsurface drains play an important role in agricultural production in the United States. Their widespread use has allowed unfarmable land to be farmed and marginal land to be productive. Benefits other than simply draining the land are being realized, including an important role in ensuring the coexistence of agricultural and environmental concerns. The reader is directed to the internet (keyword search on subsurface drainage) to discover a vast amount of information regarding subsurface drainage history, current design and implementation specifications, and ongoing research.

REFERENCES

1. Luthin, J.N. Preface. In *Drainage of Agricultural Lands*; Luthin, J.N., Ed.; Monograph 7; American Society of Agronomy: Madison, Wisconsin, 1957; vii–x.
2. Beauchamp, K.H. A History of Drainage and Drainage Methods. In *Farm Drainage in the United States—History, Status, and Prospects*; Pavelis, G.A., Ed.; United States Department of Agriculture: Washington, D.C., 1987; 13–29.
3. Wooten, H.H.; Jones, L.A. The History of Our Drainage Enterprises. In *Water—The Yearbook of Agriculture*;

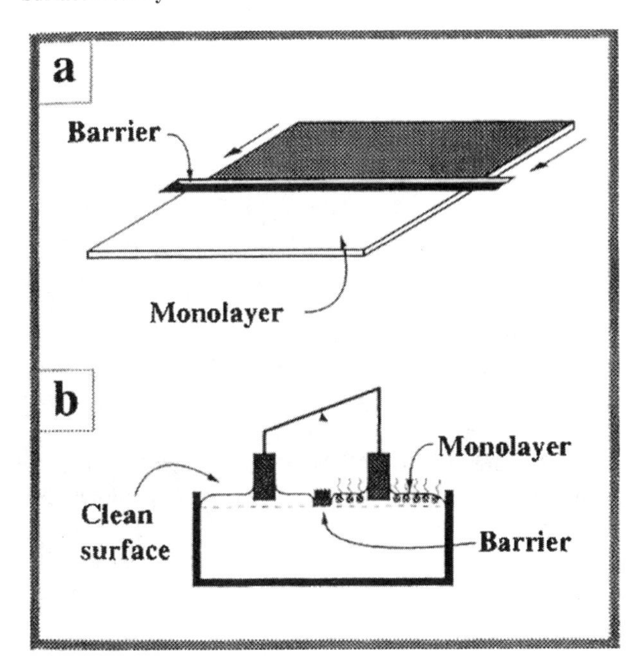

Fig. 2 Schematic illustration of a monolayer and a Wilhelmy-plate arrangement for surface tension measurement: (a) schematic illustration of a barrier delineating the area of a monolayer and (b) a Wilhelmy-plate arrangement for measuring the difference in γ on opposite side of barrier.

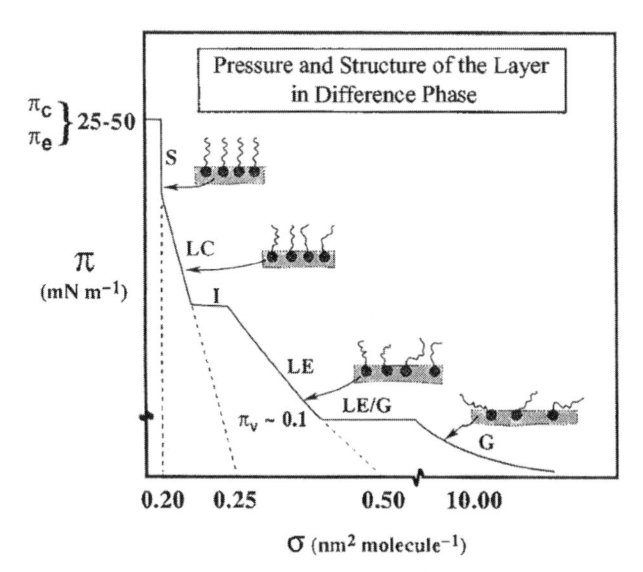

Fig. 3 Composite 2-D pressure π vs. area σ isotherm, which includes a wide assortment of monolayer phenomena. Note that the scale of the figure is not uniform so that all features may be included on one set of coordinates. The sketches of the surfactants show the orientations of the molecules in each phase at various stages of compression.

limiting area per molecule, which is found to be greater than the actual cross-section of the polar head but smaller than the area if the entire tail were free to rotate. As expected, in the LE state, there is lateral interaction between the tails of the neighboring molecules. If the area of the molecule is reduced further, there is a transition to a more compressed state known as liquid-condensed (LC) phase. LC phase will become a "solid-like" phase (S) if further compressed, where the molecules are closely packed. It is clear from the low compressibility of these states that the intermolecular forces are very strong.

Surface Equation of State for Monolayer

In the monolayer $\pi \sim \sigma$ isotherm, we can see that it will display gaseous behavior at low π or large σ. In this region, a 2-D equivalent to ideal-gas law applies:

$$\pi\sigma = k_b T \tag{19}$$

It is same as Eq. 16 for surface adsorption isotherm.

In this region, the nonpolar "tail" of molecule is expected to lay flat on the surface. That is, the molecule is expected to occupy an area of πl^2, where l is the length of nonpolar "tail." Just as van der Waals equation corrects ideal-gas law for nonideality, a 2-D Van der Waals equation can be derived to correct Eq. 19 for higher surface pressure.

$$\left(\pi + \frac{a}{\sigma^2}\right)(\sigma - b) = k_b T \tag{20}$$

where, a and b are the 2-D analogues of van der Waals constant.

REFERENCES

1. Darling, D.F.; Birkett, R.J. Food Colloids in Practice. In *Food Emulsions and Foams*; Dickinson, E., Ed.; Special Publication No. 58; Royal Society of Chemistry: Burlington House, London, 1986; 9.
2. Narsimhan, G. Emulsions. In *Physical Chemistry of Foods*; Schwartzberg, H.G., Hartel, R.W., Eds.; Marcel Dekker, INC: New York, 1992; 307–386.
3. Davies, J.T. *Proc. Int. Congr. Surface Activity*; Butterworth: London, 1957; Vol. 1.
4. Hiemenz, P.C.; Rajagopalan, R. *Principles of Colloid and Surface Chemistry*, 3rd Ed.; Dekker: New York, 1997; 300–333.
5. Feinerman, V.B.; Lucassen-Reynders, E.H.; Miller, R. Colloids and Surfaces A, 1999, *143*, 141–165.

Surface Drainage Systems

Lyman S. Willardson
Utah State University, Logan, Utah, U.S.A.

INTRODUCTION

Surface drainage, as an activity, is the orderly removal of excess water from the soil surface. The water itself can also be called surface drainage because it moves over the surface of the soil. The excess surface water can come from floods or from short high intensity rainfalls or from long duration low intensity rainfalls. Excess surface water can also result from the activities of man, such as urbanization and industrial development, where, under normal conditions, there would not be excess surface water.

Precipitation, in the form of rain or snow, is the primary source of runoff or surface drainage water. When rain falls on the surface of the soil, part or all of it is absorbed by the soil itself. If the rainfall intensity rate is less than the absorption rate of the soil, all of the rain will be absorbed and there will not be any surface runoff. The rate at which soil can absorb water depends on the soil type. Coarse sand and gravel soils can absorb water at a very high rate, so there is usually no surface drainage or runoff from such soils. Clay soils absorb water much more slowly than sand. The ability of soil to absorb water also changes over time. When the soil is dry, it can absorb water rapidly. As a soil becomes more wet, the rate of absorption decreases. The way a given soil absorbs water is called the infiltration characteristic of the soil. Fig. 1 shows how the infiltration rate of a Columbia Silt Loam soil changes over time. The infiltration rate of a soil is measured in $mm\,hr^{-1}$, which is the same as the way rainfall rates are measured. As time increases, the rate at which the soil can absorb water decreases. Whenever the rainfall rate is less than the infiltration rate of the soil, the rain will all be absorbed and there will be no runoff or surface drainage. In Fig. 1, if the rainstorm continued at a rate of $18\,mm\,hr^{-1}$ for a period of more than 30 min, surface runoff would begin to occur. Also, if the rainfall intensity was greater than $60\,mm\,hr^{-1}$, surface runoff would begin immediately, inasmuch as the rainfall rate is higher than the infiltration of the dry soil. Floods or surface drainage that occurs due to very high intensity, short duration rainfalls, are called flash floods. Flash floods are of short duration and high volume and are very dangerous because there is little fore-warning of their occurrence.

Large floods, such as those that occur along the Mississippi River and its tributaries in the central part of the United States are the result of slowly melting snow or of longer periods of rainfall that produce enough surface drainage water to exceed the natural carrying capacity of the river channels. With modern communications the progress of the excess surface drainage water along the river can be followed and people have time to raise levees or to find safety on higher ground until the flood, like a large flat wave or swell, passes.

NEED

It is important to understand the principle of surface drainage in order to avoid suffering damage that might be caused by excess water. One of the simplest examples of surface drainage occurs when rain falls on the roof of a house. The roof does not absorb more than a few drops of rain before runoff begins. The water flows down the slope of the roof and falls over the edge. If the water falls off the roof in an inconvenient location, rain gutters and downspouts are installed to move the water in an orderly manner to a place where it can be safely discharged. Sometimes the water is discharged into the street where it makes its way into a storm sewer system designed to safely dispose of the water in some natural channel. Other times, the water from the roof can be discharged onto a grassy area where it is absorbed by the soil. If the surface drainage water from the roof is properly handled, it does not cause problems. The roof area can be called a watershed and can be a valuable source of drinking water. In some areas, drainage water from roofs is stored in rain barrels. The owners of the rain barrels know from experience how many barrels can be filled by the surface drainage from the roof. Engineers can estimate surface drainage amounts from past rainfall and runoff measurements. Surface drainage from natural watersheds can be stored in large reservoirs for later uses.

HISTORY

Surface drainage began to be practiced when people came together to establish villages and communities. Houses

Encyclopedia of Agricultural, Food, and Biological Engineering
DOI: 10.1081/E-EAFE 120006923
Copyright © 2003 by Marcel Dekker, Inc. All rights reserved.

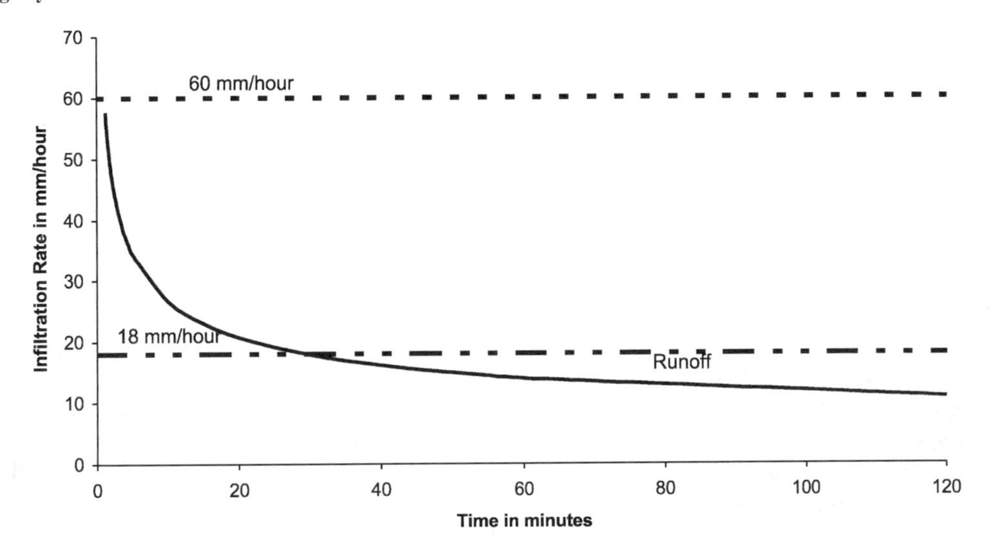

Fig. 1 Infiltration rate curve for a Columbia Silt Loan Soil.

were built that covered the soil, which had previously absorbed the rain that fell. The houses had roofs that produced runoff water that collected around the houses and in the streets. It became necessary to construct channels to carry the excess water away from the houses and dispose of it in a safe manner. People learned, by trial and error, where to put the channels and how large they should be. By removing the excess surface water, they also eliminated breeding places for mosquitoes and other insects. Surface drainage is very important to modern civilization as well.

Every modern structure, i.e., roads, bridges, dams, river channels, railroads, buildings, and cities, must take surface drainage into account when they are designed. When a dam is designed, it must be large enough and strong enough to store the water expected from the watershed. When a road or a railroad is built, passage for surface drainage water that will be produced by upslope areas, must be provided. If the structures are large and expensive, engineering studies must be made to determine the quantity of water that must be safely passed around or through the structure.

Beginning in the 1930s, the Soil Conservation Service[1] (SCS, now NRCS) of the U.S. Department of Agriculture began to make engineering studies of the relation between rainfall and surface drainage runoff. They considered the effect of rainfall intensity and duration for different statistical rainstorm recurrence intervals. They also classified the ability of different soils to absorb part of the expected water. As a result of this work, there are now engineering procedures that can be followed to design drainage channels that can safely move excess surface drainage water to a natural outlet. The same procedures are used by the U.S. Army Corps of Engineers to design flood control works that protect cities along the large rivers. Techniques have been developed to predict how often large floods will occur and how large the floods will be. This information is used to design flood control dams and levee systems. The U.S. Bureau of Reclamation[2] is concerned with both subsurface and surface drainage problems because they are related.

TYPES OF SURFACE DRAINAGE

Surface drainage can be divided into two types, rural and urban. Rainfall occurs in both kinds of environments and the excess water is handled in different ways.

Rural surface drainage usually takes the form of open channels constructed in the lowest parts of the landscape. Generally, the channels follow the natural drainage patterns of the area. Excess water collects in the lowest areas and flows downslope to a natural stream or river. Artificial channels are designed to carry the surface drainage water to an acceptable outlet without causing erosion of the surface soil or the channel. The volume of water to be carried by a channel must be determined and the channel must be designed to carry the required quantity of water without damaging the soil. Structures must be properly sized to pass the required flows. The infiltration characteristics of the soils are important in the design because the soils absorb significant amounts of water that does not have to be removed by the drainage channels. The quality of the water is usually good and so the water can be discharged into the nearest available stream. The design must also consider the possibility of downstream flooding because the water moves faster in the channels than is does when it moves naturally over the soil surface as a thin layer. The primary quality concern is the possible sediment content of the water.

Urban surface drainage has recently become very important because of the water quality and safety concerns of the public. As an urban area becomes more intensely developed, there is less exposed soil surface to absorb rainfall. Roofs and roads and parking lots essentially have a zero infiltration rate so that nearly all of the precipitation that falls on their surfaces immediately becomes surface drainage water. Water flows quickly over the smooth paved surfaces and can rapidly accumulate in volumes that could cause flood damage to businesses and homes. It has become necessary to require cities and businesses with large roofed areas and large parking lots to provide temporary storage facilities and controlled water release rates to avoid exceeding the capacity of the storm sewer system. Storm water (surface drainage water) in a city cannot be mixed with sanitary sewer water because it would overtax sewage treatment plant capacity. There is also an increasing problem with urban surface drainage water quality because of all the chemical and industrial by-products that are washed off roads and roofs by the rainfall. Urban surface drainage designers must consider the quality of the runoff water in seeking a suitable outlet for disposal of the water. In some cases, separate surface drainage water treatment facilities must be developed to treat the water before it can be safely discharged into a natural stream or river. Wetland technologies[3] are being developed to help solve the quality treatment problem of urban surface drainage water.

DESIGN CONSIDERATIONS

The three important factors to consider in the design of a surface drainage system are: 1) the volume of water to be removed; 2) the timing of the discharge; and 3) the quality of the water to be discharged into an appropriate outlet. The volume of water to be removed can be estimated from engineering information on rainfall duration, rainfall rate, infiltration rate, and recurrence interval. The timing of discharge can be computed from the hydraulic characteristics of the natural and artificial channels that will be carrying the water to the disposal point and from the capacity of the downstream channels. The quality of the water to be discharged must be known and controlled so that downstream users will not be adversely affected and so that the natural wildlife and aquatic environment in the receiving waters will not be adversely affected by the discharge. In the future, the runoff water quality may be the controlling criterion for the safe removal of surface drainage water in urban areas.

REFERENCES

1. SCS (National Resource Conservation Service). *Drainage of Agricultural Land. 1973*; Water Information Center: Port Washington, New York, 1973; 244–267.
2. U.S. Bureau of Reclamation. *Drainage Manual*; U.S. Department of the Interior: Washington, DC, 1993; 37–59.
3. Hammer, D.A. *Constructed Wetlands for Wastewater Treatment*; Lewis Publishers: Chelsea, Michigan, 1989.

Surface Fouling During Heating

Hongda Chen
United States Department of Agriculture (USDA),
Washington, District of Columbia, U.S.A.

S

INTRODUCTION

Fouling is defined as the accumulation of unwanted deposits on the surfaces of food processing equipment. The presence of these deposits represents a resistance to heat or mass transfer and therefore reduces the efficiency of the particular food process. The foulant may be constituents of the fluid food being processed, the products of chemical reactions occurred during the processes, particulate matter suspended in the fluid, and/or micro-organisms. Two different types, namely thermal fouling and membrane fouling, with distinctly different mechanisms and processes are most commonly seen in food industry.

The consequence is equally serious to food processing industry no matter what type of fouling is encountered. It increases capital cost due to necessary oversizing and special equipment design, energy consumption in operation to overcome increased pressure drop, maintenance cost for more frequent cleaning and treatment of cleaning effluents, and production losses during plant downtime for cleaning. The loss due to fouling and subsequent cleaning of heat exchangers was estimated between $4.2 and 10 billion per year for the entire U.S. industry, corresponding to 0.25% of the GNP.[1] Similar figures were reported in other industrialized countries.[2] Food industry is particularly vulnerable due to the easiness of fouling and high rate of deposition in liquid food processes. The deposits may adversely affect finished product quality and sterility, thus further reducing the profit of food industry.

The primary difference of thermal fouling from membrane fouling is the gradual reduction of heat transfer efficiency due to formation of deposit layer on heating surface. Commonly seen thermal processing equipment include various heat exchangers (plate, tubular, etc.), evaporators, and holding tanks. The following discussion is limited to thermal fouling. Discussion of membrane fouling may be found elsewhere.

FOULING OF HEAT TREATMENT EQUIPMENT

Formation

Several possible thermal fouling mechanisms are reported in literature for general heating equipment.[3]

1. Particulate deposition: Small particles agglomerate and bond at the surface.
2. Chemical reaction: The reactions among food constituents, which may occur either in the bulk fluid or on heating surface, result in polymers or large molecules.
3. Crystallization: Inorganic salts crystallize and orient into a coherent structure often referred to as scale.
4. Biofouling: Matrix of microbial cells and extracellular polymers adhere to the surface.
5. Corrosion: Chemical attack to the surface produces "new" chemical compounds.

In complex reality, fouling may be a combination of more than one of the above mechanisms. The particulate deposition frequently acts in conjunction with other mechanisms. For instance, the agglomeration of protein molecules due to intermolecular chemical reactions transfers through bulk fluid and boundary layer to the surface via particulate deposition. In biofouling, the micro-organisms responsible for biofilm formation must first encounter the surface before colonization can be formed. In crystallization, the crystal growth may depend on the transport of solid crystallites across the boundary layer to reach crystal seeds formed on the surface.

Chemical reaction fouling is associated with organic chemicals as opposed to reaction of metals with aggressive agents in corrosion fouling. The reactions are usually complex and may involve several chemical mechanisms. Furthermore, it is unlikely that the physical and chemical properties of the deposit due to the chemical reactions will be uniform throughout its thickness because the temperature profile of the deposit layer changes as it thickens. Dairy fluids have been most extensively studied and

Encyclopedia of Agricultural, Food, and Biological Engineering
DOI: 10.1081/E-EAFE 120007146
Copyright © 2003 by Marcel Dekker, Inc. All rights reserved.

documented in thermal fouling literature. A few other products, such as corn mill[4] and juice,[5] have also been investigated. The following discussion will focus on dairy fluids such as various milk and whey products, desserts, and cream.

Chemical Composition

Natural whey protein molecules have the tendency to adsorb onto stainless steel surface at room temperature.[6] It forms at most a bilayer, with the first layer of molecules irreversibly bound to the steel and cannot be removed by rinsing with distilled water.[7] Although this kind of fouling bears no consequence for processing concerning heat transfer and flow, it modifies the surface properties of steel, which becomes more adhesive to further deposition of proteins. Even at this thin layer of adsorption, it may also nurture an environment, friendly to microbial adhesion for biofilm, which is the fourth fouling mechanism discussed above.

It is generally agreed that the heat denaturation of whey proteins and the inverse solubility of calcium phosphate with temperature are the main causes of fouling in the thermal processing of dairy fluids.[2] Two types of deposits can be distinguished by their physical appearance and chemical composition.[8] Type A, sometimes referred to as "milk film," is wet, creamy white, and spongy. This happens while heating between 65 and 110°C. It contains about 60% proteins, with β-lactoglobulin being the richest species, and 30%–40% ash, on dry matter basis. Type A consists of two layers, a protein-rich layer on top of calcium-phosphate rich layer directly attached to the heating surface.[9] Type B deposition, on the other hand, occurs at temperature between 110 and 140°C. It contains much more ash (80%) and less protein (20%), on dry matter basis. It looks more compact, crystalline, and glassy.

Whey protein molecules unfold its compact spherical structure and become active when subjected to heating above 65°C. Intramolecular disulfide bridges are broken and free thiol- and SH-groups are exposed to allow intermolecular crosslinking to form covalently polymerized aggregates. The increase in molecular size facilitates agglomeration and deposition to the heating surface. The whey proteins in active state are ready to attach themselves to the already bound proteinaceous bilayer formed at room temperature. The kinetics of whey protein denaturation depends on types of the dairy products that differ in composition, pH, and heating treatment. Consequently, the fouling rate and extent vary according to products and processes.[2] The active whey proteins may act as bridge agent with κ-casein to bring casein micelles to the fouling layer.

Fouling due to mineral deposition in thermal processing of dairy fluids is attributed to the inverse solubility of calcium phosphate with temperature.[2] Calcium phosphate may deposit as such or in conjunction with proteins. The calcium phosphate deposited is a mixture of calcium phosphate dihydrate ($CaHPO_4{\cdot}2H_2O$) and octacalcium phosphate ($Ca_8H_2(PO_4)_6{\cdot}5H_2O$), which eventually under prolonged heating is transformed into hydroxy apatite [$Ca_5OH(PO_4)_3$], the least soluble calcium phosphate complex with a Ca/P ratio of 1.5.[2]

Small amounts of fat and lactose are entrapped in the deposits. However, they do not play important roles in fouling process.

Influence of Processing Parameters

Both protein denaturation and solubility of calcium phosphate are a function of temperature. The processing temperature governs the quantity and the composition of the deposits on heat exchanger surfaces. Any temperature increase, either at the wall or in the bulk, promotes fouling through the increased rate of the denaturation of milk serum proteins, especially β-lactoglobulin. Temperature increase also results in more precipitation of calcium phosphates. The precipitation of calcium phosphates and calcium ions on the surface of the depositing serum protein aggregates facilitates the fouling process. In the presence of calcium ions, the binding of proteins may occur through the formation of noncovalent bonds, thus, the specific alignment of two approaching protein molecules seeking disulfide linkage is no longer necessary. At temperatures higher than 110°C, the precipitation of calcium ions to the heating wall is at a much higher rate. Less calcium ions are available in bulk for the formation of more adhesive calcium-serum protein complex, therefore, less protein in the fouling layer.

Fluid dynamics, namely fluid velocity and Reynolds number, plays a critical role in deposit formation. Increase in velocity results in less fouling. Deposit formation from reconstituted whey protein concentrates reduces from Reynolds number 1800 (end of laminar flow region) to 9000 (end of the transition region) in a tubular heat exchanger.[10] It can be explained by an increase both in the wall shear stress and in the turbulence that is an indicator of mixing intensity in the bulk. The aggregation of activated protein molecules in the bulk reduces the probability of binding to the deposited proteins at high velocity. Plate heat exchangers are generally less prone to fouling than tubular heat exchangers because they are frequently operated in turbulent flow regime. Carefully designed corrugated channels in the plate heat exchangers enhance mixing intensity and thus reduce fouling deposit. Surface properties of different materials have been found to have no influence on fouling.[11]

The geometry of heating equipment may contribute to the fouling formation. Upon heating whey proteins, there is a critical period of time when the concentration of activated protein monomers is at its highest in the bulk fluid. The most severe fouling will occur if the activated proteins are allowed to access large surface area. A holding device with high volume to surface ratio is thus recommended to be installed in heating process of dairy fluids to minimize protein deposition because the deposition rate correlates with the concentration of activated whey protein molecules.[2] De Jong found that this approach reduced fouling by more than 50%.[12] Addition of citrate, pyrophosphate, and ethylenediamine tetraacetic acid (EDTA) to milk can reduce mineral deposit.

Mathematical Modeling of Fouling

During the last decade, considerable progress has been made in the modeling of fouling in heat exchangers in which dairy products are processed. Both classical chemical engineering approach and the kinetics approach through heat denaturation of β-lactoglobulin have contributed to a better understanding of the fouling mechanisms. A thorough review of the mathematical modeling of thermal fouling is not possible here. Only a brief review is presented below.

The first models for the thermal performance evolution with time during fouling of milk in a tubular heat exchanger were attempted by Fryer.[13] The decrease in the overall heat transfer coefficient, U, with time was described by a dimensionless fouling Biot number (Bi):

$$Bi = \frac{U_0 - U}{U} = R_f U_0 \tag{1}$$

where U ($\mathrm{W\,m^{-2}\,K^{-1}}$) is the overall heat transfer coefficient at any time t, U_0 the heat transfer coefficient of the clean surface, and R_f ($\mathrm{K\,m^2\,W^{-1}}$) the thermal fouling resistance. As fouling proceeds, the fouling resistance increases and the overall heat transfer coefficient decreases. In reality, R_f varies with processing time because the thickness, uniformity, and thermal conductivity of the fouling layer constantly change.

Fouling of milk-based fluids in a heating tube is considered a two-stage process.[13] During the induction period, it is controlled by the rate of salt deposition and protein adsorption. The surface is conditioned for further heavy deposition. The induction period ends when Bi is greater than 0.05, and the fouling period follows. The fouling stage is controlled by the reaction in the bulk fluid where materials produced can rapidly form surface deposit.

For the fouling in the postinduction period, the process is regarded as a balance between deposition, d, and removal, r, occurring at the solid–liquid interface. The following model results[13]

$$\frac{dBi}{dt} = \frac{k_d}{Re}\exp\left(\frac{-E}{Re}\frac{1+Bi}{T_w + T_b Bi}\right) - k_r Bi \tag{2}$$

where Re is dimensionless Reynolds number, T is temperature (K), and the subscripts w and b indicate wall and bulk, respectively. Empirical fitting gave $k_d = 4.85 \times 10^{13}\,\mathrm{sec^{-1}}$, $k_r = 1.3 \times 10^{-3}\,\mathrm{sec^{-1}}$, and the thermal activation energy, $E = 89 \pm 6\,\mathrm{kJ\,mol^{-1}}$. The model depicts that the initial fouling rate is thermally activated and inversely proportional to the fluid Reynolds number. Although this model is useful for predicting the thermal performance of heat exchangers, the link between the amount of deposit, the heat denaturation of β-lactoglobulin, calcium phosphate precipitation, and the fouling resistance, R_f, as a function of time could not be established.

Delplace, Leuliet, and Tissier[14] successfully obtained a good fit between an empirical correlation-based prediction and experimental data for the overall heat transfer coefficient in a plate heat exchanger as a function of time in the following general form.

$$U(t) = \frac{U_0 \lambda(t)}{U_0 \sigma t + \lambda(t)} \tag{3}$$

where the mean velocity of deposit layer growth, σ ($\mathrm{m\,sec^{-1}}$), is related to both the heat denaturation of β-lactoglobulin and the fouling behavior of a heat exchanger:

$$\frac{m_d}{S_u V} = \sigma \Delta C^{0.5} \tag{4}$$

The terms on the left-hand side of the equation represent the dry mass of deposition, m_d (kg), after heat processing a product of volume, V ($\mathrm{m^3}$), in a channel having a heat transfer area of S_u ($\mathrm{m^2}$). The terms on the right-hand side of the equation are the concentration difference of the native β-lactoglobulin between the inlet and the outlet of the heat processing channel (ΔC) ($\mathrm{kg\,m^{-3}}$) and the empirical constant, σ, which is $0.127\,\mathrm{m\,sec^{-1}}$ for straight corrugation plates and $0.06\,\mathrm{m\,sec^{-1}}$ for herringbone plates. Finally, the deposit thermal conductivity, $\lambda(t)$ ($\mathrm{W\,m^{-1}\,K^{-1}}$), as a function of time was established by Delplace, Leuliet, and Tissier[14] as follows:

$$\lambda(t) = 3.73e^{-2.5\times10^{-4}t} + 0.27 \tag{5}$$

where t is time (sec). Using this approach, the modeling of the overall heat transfer coefficient as a function of time from the induction phase to the postfouling period is possible.

Monitoring Thermal Fouling

The ability to accurately determine the extent of fouling is critical to a successful process, i.e., to decide an optimal cleaning schedule neither too frequent nor too long to assure the best operation and product quality. The reduction in heat transfer efficiency and the increase in the pressure drop are the two most obvious characteristics of thermal fouling. Therefore, different schemes can be developed based on these principles.

Theoretically, the overall heat transfer coefficient, U, and in turn the fouling thermal resistance, R_f, can be determined by measuring heat flux, q, and log mean temperature difference, $(\Delta T)_{lm}$, over a heat exchanger. This approach works well under well-defined and instrumented conditions in a research laboratory. However, many difficulties such as complex geometry, nonuniform heat transfer in different parts of the heat exchanger as a result of fouling, reliability of temperature sensors, and the number of sensors required in processing plants make this approach not practical. On the other hand, monitoring the pressure drop is rather easy with reasonable accuracy by using differential pressure sensors.

Cleaning

Timely cleaning is essential to return the heating equipment to its best operation condition. Cleaning of food processing equipment is discussed elsewhere in this Encyclopedia. For thermal fouling in dairy pasteurizers, cleaning with a dilute solution of EDTA and NaOH at 65–70°C was found to be the most effective. There is no real advantage of altering the sequence of alkaline and acid wash in a cleaning process. Use of enzymes is still too expensive.

CONCLUSION

The serious nature of fouling on operation efficiency as well as product quality and safety has been well recognized by food industry. Substantial progress has been made in understanding the fouling mechanisms, the effects of processing parameters, and chemistry of food being processed. Useful mathematical models have been developed to aid better design of heating equipment and processes. Certain guidelines are available to mitigate the fouling problem. Additional investigations are needed to fully reveal the fundamentals of fouling in various processes of a wide range of foods to minimize the impacts of fouling.

REFERENCES

1. Smith, S.A.; Dirks, Y.A. Costs of Heat Exchanger Fouling in the U.S. Industrial Sector. In *Industrial Heat Exchangers*; Hayes, A.J., Liang, W.W., Richlen, S.L., Tabb, E.S., Eds.; Am. Soc. Metals: Metals Park, OH, 1985; 339.
2. Visser, H. Fouling and Cleaning of Heat Treatment Equipment. *Bulletin of the International Dairy Federation*; IDH: Brussels, Belgium, 1997; Vol. 328, 1–44.
3. Bott, T.R. *Fouling of Heat Exchangers*; Elsevier: Amsterdam, The Netherlands, 1995.
4. Singh, V.; Panchal, C.B.; Eckhoff, S.R. Effect of Corn Oil on Thin Stillage Evaporators. Cereal Chem. **1999**, *76* (6), 846–849.
5. Watson, L.J.; Wright, P.G. Fouling Rates for Tubular Juice Heaters. In *Proceedings of the Conference of the Australian Society of Sugar Cane Technologists*, Bundaberg, Queensland, April 29–May 3, 1985; 201–208.
6. Roscoe, S.G.; Fuller, K.L.; Robitaille, G. An Electrochemical Study of the Effect of Temperature on Adsorption Behavior of β-Lactoglobulin. J. Colloid Interface Sci. **1993**, *160*, 243–251.
7. Arnebrant, T.; Ivarsson, B.; Larsson, K.; Nylander, T. Bilayer Formation at Adsorption of Proteins from Aqueous Solutions on Metal Surfaces. Prog. Colloid Polym. Sci. **1991**, *70*, 502–511.
8. Burton, H. Reviews of the Progress of Dairy Science, Section G. Deposits from Whole Milk in Heat Treatment Plants—A Review and Discussion. J. Dairy Res. **1968**, *35* (2), 317–330.
9. Tissier, J.P.; Lalande, M. Experimental Device and Methods for Studying Milk Deposit Formation on Heat Exchanger Surfaces. Biotechnol. Prog. **1986**, *2* (4), 218–229.
10. Belmar-Beiny, M.T.; Gotham, S.M.; Paterson, W.R.; Fryer, P.J. The Effect of Reynolds Number and Fluid Temperature in Whey Protein Fouling. J. Food Eng. **1993**, *19* (2), 119–139.
11. Sandu, C.; Lund, D. Fouling of Heating Surfaces—Chemical Reaction Fouling Due to Milk. In *Fouling and Cleaning in Food Processing*; Lund, D., Platt, E.A., Sandu, C., Eds.; University of Wisconsin: Madison, WI, 1985; 122–167.
12. De Jong, P.; Bournan, S.; Van der Linden, H.J.L.J. Fouling of Heat Treatment Equipment in Relation to the Denaturation of β-Lactoglobulin. J. Soc. Dairy Technol. **1992**, *45* (1), 3–8.
13. Fryer, P.J. Modeling Heat Exchanger Fouling. Ph.D. Thesis, Cambridge University, Cambridge, U.K., 1985.
14. Delplace, F.; Leuliet, J.C.; Tissier, J.P. Fouling Experiments of a Plate Heat Exchanger by Whey Proteins Solutions. Food Bioprod. Process.: Trans. Inst. Chem. Eng., Part C **1994**, *72* (C3), 163–169.

Surface Irrigation

C. Dean Yonts
University of Nebraska–Lincoln, Scottsbluff, Nebraska, U.S.A.

S

INTRODUCTION

Surface irrigation was likely first accomplished by placing a dam across a stream or river and flooding land nearby. As technology improved, uniform sloped canals were constructed, which allowed water to be moved further away from a river. Large dams were built which meant water could be stored from spring runoff for use during the entire growing season. The result was an abundance of land being irrigated by a system that relied on gravity to distribute water across a field using the soil surface as a means of conveyance. This is not to say surface irrigation only occurs with water from a river. Well water is also used to surface irrigate by using large pumps to bring the water to the soil surface. Regardless of the source of water, surface irrigation in the United States during 1999 accounted for 47% of the irrigated acres, nearly 30,000,000.[1]

SURFACE IRRIGATION METHODS

There are two general methods of surface irrigation, flooding where the entire land area to be irrigated is covered with water or furrow in which small ditches or corrugations are used to convey water across only a portion of the land surface. Flooding systems can be as simple as diverting water directly on to a field with no method of controlling water flow other than through the topography of the land. This method is commonly used to irrigate pastures or high mountain meadows where land grading is not economically feasible. When land grading is applied, topography is controlled which in turn allows better control of water flow.

Border Strip Flooding

Border strip flooding is a method of directing water flow between two borders or dikes that are constructed of soil. Water is forced to flow uniformly between borders in a thin layer across the entire area being irrigated. Leveling the land between the borders eliminates cross slope or the slope that would allow water to flow at an angle to the desired direction of flow. In practice, a series of border strips are usually constructed adjacent to each other. Land leveling to eliminate cross slope within a border strip is done independently of adjacent border strips as a method of eliminating most of the existing cross slope. Width of a border strip is therefore determined by the amount of cross slope that exists and the amount of water available for irrigation. More cross slope and less water means narrower strips. This method of surface irrigation is commonly used on alfalfa, small grains, and pasture.

Level Basin Flooding

Level basins are constructed by forming a dike of soil that surrounds a field or basin, which has been leveled meaning the basin has no slope in any direction. Because the basin is level, the water covers the entire soil surface with a predetermined depth of water and is allowed to infiltrate the soil. Water is diverted into the basin at relatively high flow rates to ensure that the basin is filled to the desired depth as quickly as possible. This method of irrigation is also adapted to row crop production by constructing small ditches and planting crop on the ridges that are formed between the ditches. Water is diverted into the level basin in much the same way only in this case water is ponded in the furrows and does not submerge the growing crop. For flood systems, the more control over the quantity of water applied and the uniformity of the depth of application, the higher the system application efficiency. Therefore, uncontrolled flood irrigation has a lower application efficiency than level basin irrigation.

Furrow

Furrow irrigation is used to irrigate most crops that are planted in rows and require cultivation to remove weed competition. Small channels or ditches are constructed between the planted rows and are used to convey water across the field being irrigated. Rather than water infiltrating the entire soil surface, as with flooding, with furrow irrigation water only infiltrates the soil through the furrow. As water infiltrates through the furrow, water is then moved within the soil both laterally and vertically to saturate the soil profile.

The process of infiltrating water through the furrow takes longer than if the same amount of water were applied by flooding. Because of the additional time needed to infiltrate water into the soil, water must be allowed to flow the entire length of the field for a period of time in order to apply the desired amount of water. This means a portion of the water

Encyclopedia of Agricultural, Food, and Biological Engineering
DOI: 10.1081/E-EAFE 120006934
Copyright © 2003 by Marcel Dekker, Inc. All rights reserved.

applied will run off the end of the field. Although water lost to runoff reduces application efficiency, with furrow irrigation it is necessary in order to apply water uniformly.

Water application uniformity is improved when water in the furrow has approximately the same amount of time to infiltrate the soil at the head and bottom ends of a field. The lower line in Fig. 1 shows the rate of water advance in a furrow to the bottom end of a field. Once the desired amount of water has been applied and water is shut off, water flow in the furrow does not stop instantaneously, but rather slowly stops flowing in the furrow. An example of time required for water to stop flowing in a furrow, referred to as recession time, can be seen as the upper line in Fig. 1. In Fig. 1, the difference between the advance time line and the recession time line is known as the opportunity time. It is called opportunity time because it is the time that water is flowing in the furrow and has the opportunity to infiltrate the soil. Although difficult, the goal of furrow irrigation is to have a consistent opportunity time along the entire length of the field. This is accomplished if the advance time curve and the recession time curve (Fig. 1) are parallel.

Little can be done to change rate of water recession once flow has been stopped. On the other hand, rate of advance can be made more equal to the rate of recession by increasing the rate of water advance. This can be accomplished by increasing furrow stream size. This results in more equal opportunity time but also increases the amount of runoff from the field. To maintain application efficiency in this situation, the runoff water must be recycled and reused for irrigation. Another method to reduce runoff is to cutback the furrow stream size once water reaches the bottom end of the field. However, this practice increases labor requirements.

A recent and popular method of making furrow advance time more equal to recession time is surge irrigation.[2] Surge irrigation is a method of automatically alternating or cycling water between two sets of irrigation furrows using expanding short time intervals of approximately

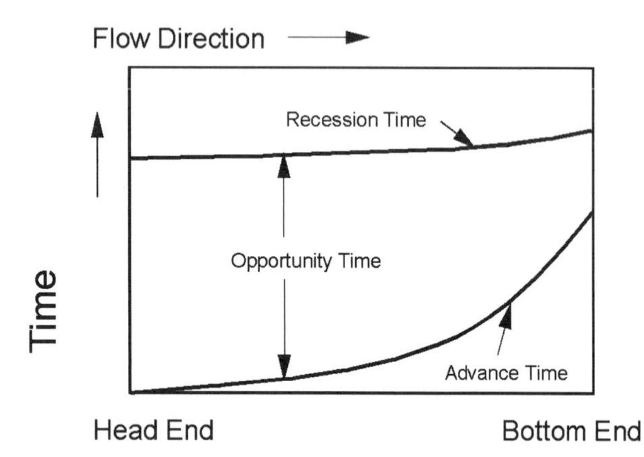

Fig. 1 Advance rate and recession rate determine opportunity time for water intake in a furrow.

15 min–120 min, Fig. 2. The process of cycling water between sets of furrows reduces infiltration rate and thus water advance is faster to the bottom end of the field. Fig. 3 shows the depth of water infiltration during four successive surge cycles. In this example, each surge cycle is intended to advance water approximately one-fourth the length of the field. Once water reaches the bottom end of the field, surge cycle times are reduced to complete the irrigation and reduce runoff. The surge infiltration line shows the final infiltration for surge irrigation. For comparison, Fig. 3 also shows the infiltration that might be expected for the example in Fig. 1 where water flowed into the furrow continuously. Surge irrigation can reduce infiltration at the top end of the field and get water to the bottom end of the field faster. This improves opportunity time and results in a more uniform water application.

Corrugation

Corrugation is another term often used to describe the construction of a small ditch or furrow. Corrugations however are generally used to irrigate close growing crops such as alfalfa or small grains where the crop grows both in and between the corrugations. Normally with this type of irrigation, the corrugations provide the primary mechanism for channeling the water. However, flooding between the corrugations usually occurs because plant material growing in the corrugations reduce the water carrying capacity.

SURFACE IRRIGATION IN THE FUTURE

Uniform application of water is very important. The soil profile where roots grow can only hold a given amount of water. This amount varies by soil type and is also dependent on the depth to which plants grow their roots. Applying more water than what the soil can hold results in water being moved below the zone where roots can utilize the water. If enough water is forced below the rooting depth, the water eventually replenishes the ground water with the risk of taking contaminants from the soil surface into the ground water.

There are relatively few new surface irrigation systems being installed in the United States. At one time, surface irrigation was considered low cost because power requirements for pumping were less when compared to pressurized sprinkler irrigation systems or not needed in the case of gravity systems. However, labor requirements are generally higher for surface irrigation systems and labor is no longer considered a minor cost. In addition to increasing costs for labor, the ability to find skilled labor

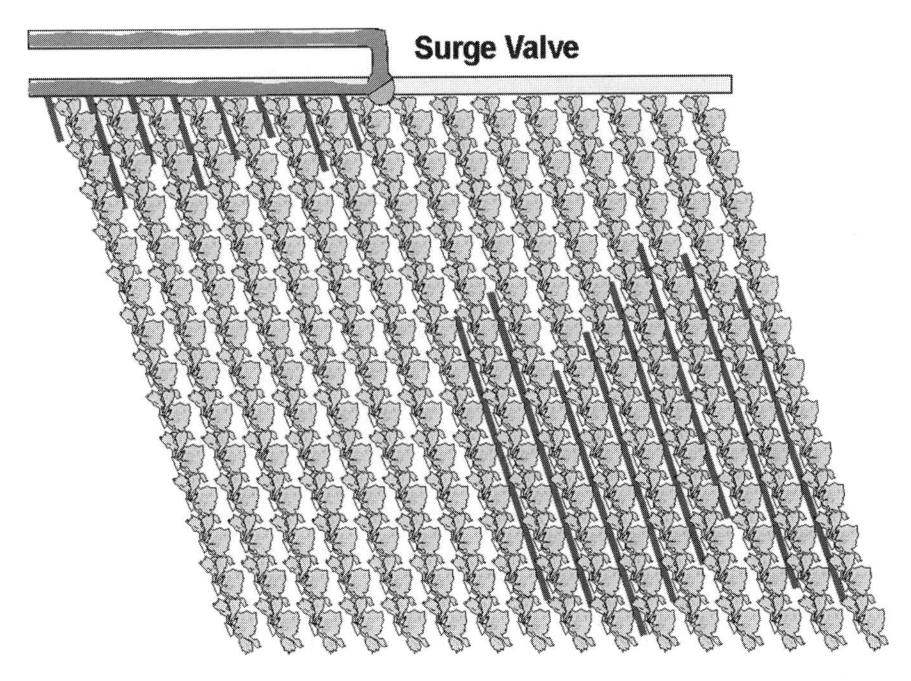

Fig. 2 Surge irrigation alternates water flow between two irrigation sets.

that can understand and are willing to operate surface irrigation systems is becoming even more difficult.

Application efficiency is also a significant issue for surface irrigation systems. As municipal, industrial, and environmental water needs increase, conversion to alternate more efficient irrigation systems occur. Because of producers converting to more efficient irrigation systems as a method to reduce water needs and cut labor costs, surface irrigation in the United States has dropped by over 6,000,000 acres from 1987 to 1999.[1,3] This reflects a 6% reduction in total surface irrigated acres over a 13-yr period. During this same time period, sprinkler irrigation in the United States increased by approximately 6,700,000 acres. This indicates total irrigated acres in the United States are increasing at a nominal rate while the biggest changes in the industry focuses on making improvements to existing surface irrigation systems.

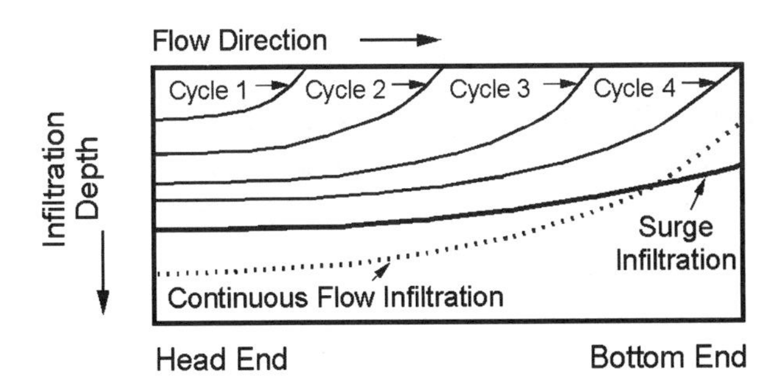

Fig. 3 Surge irrigation results in more uniform infiltration compared to continuous irrigation.

Conversion from surface irrigation systems will not and cannot occur overnight. Nor will all surface irrigation systems need to be converted. Many surface irrigation systems have been designed to use labor efficiently with acceptable application efficiencies. For example, there are a number of irrigation districts in the United States that use water from reservoirs and river systems. In many of these cases, the system wide efficiency is good because runoff water from surface irrigated fields is returned to the river for use downstream.

Water traditionally used for irrigation is increasingly being used for urban and industrial applications. Because of this, surface irrigation systems are continually being converted to alternative irrigation systems like sprinkler and microirrigation that apply water more uniformly and efficiently than surface irrigation. As the value of water used for irrigation continues to increase, conversion of surface irrigation will continue at a similar or perhaps slightly increased rate in future.

REFERENCES

1. 1999 Irrigation Survey. Irrig. J. **2000**, *50* (1), 16–31.
2. Stringham, G.E.; Keller, J. Surge Flow for Automatic Irrigation. Proceedings 1979 Irrigation and Drainage Specialty Conference, American Society of Civil Engineers, 1979.
3. 1996 Irrigation Survey. Irrig. J. **1997**, *47* (1), 27–42.

Surface Tension

Ganesan Narsimhan
Zebin Wang
Purdue University, West Lafayette, Indiana, U.S.A.

INTRODUCTION

Surface tension is a manifestation of imbalance of intermolecular forces experienced by a molecule at an interface (air–water or oil–water) compared to a molecule in the interior of a phase. This imbalance occurs because the surface molecule is not surrounded by like molecules. In order for the surface molecule to be at the interface, this imbalance has to be counterbalanced by a contractile force, which is referred to as the surface/interfacial tension. As one would expect, the surface tension would depend on the intermolecular interactions and would therefore be different for different materials. This force, denoted by γ, acts perpendicularly and inward from boundaries of the surfaces, tending to decrease the area of surfaces.

THERMODYNAMICS OF INTERFACES, SURFACE TENSION

Surface tension can be illustrated by a simple apparatus. Fig. 1 represents a loop of wire with one movable side on which a film could be formed by dipping the frame into a liquid. The surface tension will cause the frictionless slide wire to move in the direction of decrease film area unless an opposing force F is applied. The force evidently operates along the entire edge of the film and varies with the length of the slide wire. Therefore, the force per unit length of edge is the intrinsic property of the liquid surface. The force balance for the film wire is:

$$F = 2l\gamma \tag{1}$$

In the above equation, the factor, 2, appears as the film has two sides. From the above equation, the surface tension can be got by:

$$\gamma = F/2l \tag{2}$$

Eq. 2 also defines the units of surface tension to be those of force per length ($\mathrm{N\,m^{-1}}$ in SI or $\mathrm{dyn\,cm^{-1}}$ in cgs system).

If a force infinitesimally larger than the equilibrium force is applied to the slide wire, the wire will be displaced through a distance dx. The energy spent to increase the area of the film by the amount of dA is equal to $2l\,\mathrm{d}x$. Therefore, the work done on the system is given by

$$\text{Work} = F\,\mathrm{d}x = \gamma 2l\,\mathrm{d}x = \gamma\,\mathrm{d}A \tag{3}$$

This provides another definition of surface tension as the work required to increase the surface by unit area. According to this definition, the units of γ are energy per unit area ($\mathrm{J\,m^{-2}}$ in SI or $\mathrm{erg\,cm^{-2}}$ in cgs system).

Although the surface tension has two interpretations (force per unit length of boundary and energy per unit area of surface), the dimensions of the two definitions are equivalent. Energy/area will obtained if the numerator and denominator of force/length are multiplied by length.

In thermodynamics, surface tension has more strict definition as:

$$\gamma = \left(\frac{\partial F}{\partial A}\right)_{T,p} \tag{4}$$

Here F is free energy and A is area. For water at 20°C, γ is $7.28 \times 10^{-2}\,\mathrm{J\,m^{-2}}$; usually, it is written as $72.8\,\mathrm{mJ\,m^{-2}}$, which is numerically equal to the cgs value.

LAPLACE EQUATION

Consider an interface with R_1 and R_2 as the two principal radii of curvature (Fig. 2). Based on Eq. 4, one can obtain the pressure difference between the two sides of the interface as:

$$\Delta p = \gamma\left(\frac{1}{R_1} + \frac{1}{R_2}\right) \tag{5}$$

This equation is known as the Laplace equation, which states that the pressure inside a convex surface is greater than the outside pressure. For a spherical drop or a bubble, the excess pressure Δp inside the sphere is given by

$$\Delta p = \frac{2\gamma}{R} \tag{6}$$

R being the radius of the drop or bubble.

Encyclopedia of Agricultural, Food, and Biological Engineering
DOI: 10.1081/E-EAFE 120021381
Copyright © 2003 by Marcel Dekker, Inc. All rights reserved.

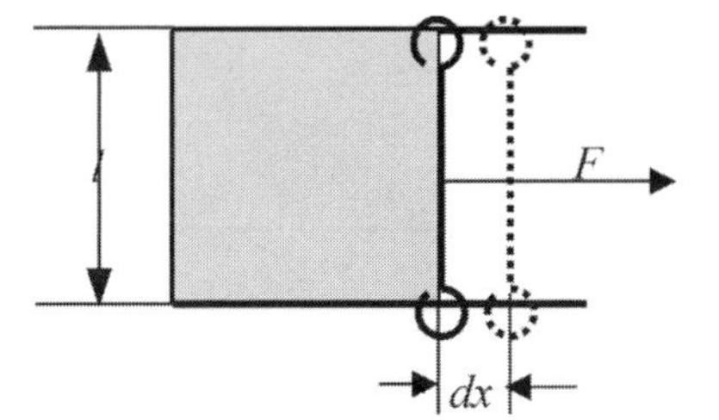

Fig. 1 Illustration of liquid film formation and surface tension.

METHODS OF MEASUREMENT

A brief description of different experimental methods of measuring surface tension is given below.

Capillary Rise

When a capillary is immersed inside a liquid, the liquid rises inside the capillary as shown in Fig. 3(a). The liquid surface inside the capillary has a curved meniscus. As a result, the pressure inside the meniscus is less than the atmospheric pressure. This pressure difference is the reason for the capillary rise. This difference in pressure is compensated by the hydrostatic head of the liquid column

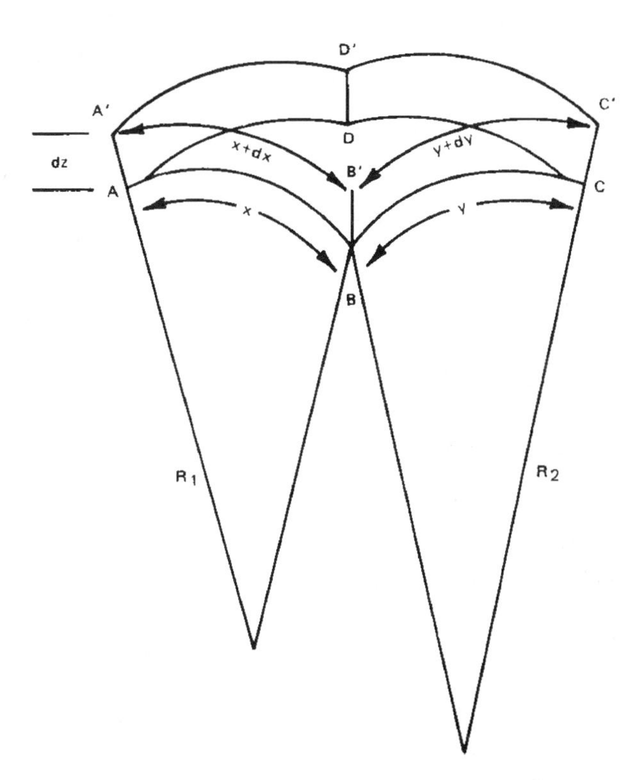

Fig. 2 Definition of coordinates describing the displacement of an element of curved surface ABCD and A′B′C′D′.

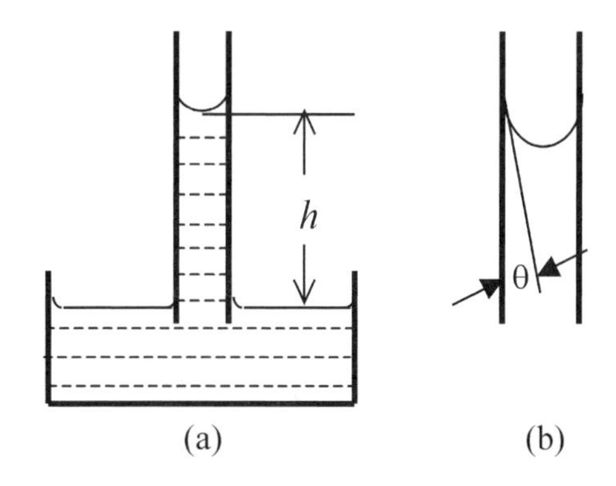

Fig. 3 (a) Capillary rise and (b) contact angle.

that rises inside the capillary. In other words,

$$\Delta p = \gamma\left(\frac{1}{R_1} + \frac{1}{R_2}\right) = \Delta\rho g h \tag{7}$$

where Δp is the pressure difference across the meniscus, R_1 and R_2 the two principal radii of curvature of the meniscus, h the height of liquid column inside the capillary, $\Delta\rho$ the density difference between the liquid and air, and g the acceleration due to gravity. If the liquid completely wets the wall of the capillary, the meniscus is hemispherical with the radius equal to the radius of the capillary. Under these conditions, we have

$$\frac{2\gamma}{R_{\text{cap}}} = \Delta\rho g h \tag{8}$$

There is an angle (Fig. 3(b)) between the tangent to the surface at the point of contact with surface and the surface itself. This angle, called as contact angle, is 0° for clean glass and most aqueous liquids and alcohol. But if the glass is dirty, it may be as high as 8°.[1] For nonzero contact angles, the above equation is modified as

$$\frac{2\gamma\cos\theta}{R_{\text{cap}}} = \Delta\rho g h \tag{9}$$

In Eq. 9, the weight of liquid above the bottom of meniscus is neglected. For more accurate result, it can be accounted for to give $(h + r/3)$ instead of h.[1]

Ring Method

The Du Nouy ring method[2,4] uses a vessel containing liquid and a ring to measure surface tension (Fig. 4). The vessel is capable of being moved upward and downward in a controlled manner. Initially, the vessel is positioned so that the ring is immerged into the liquid. Then the vessel is lowered slowly and the force exerted on the ring is recorded. When the ring is slightly above the liquid surface, some of the liquid "clings" to the ring and

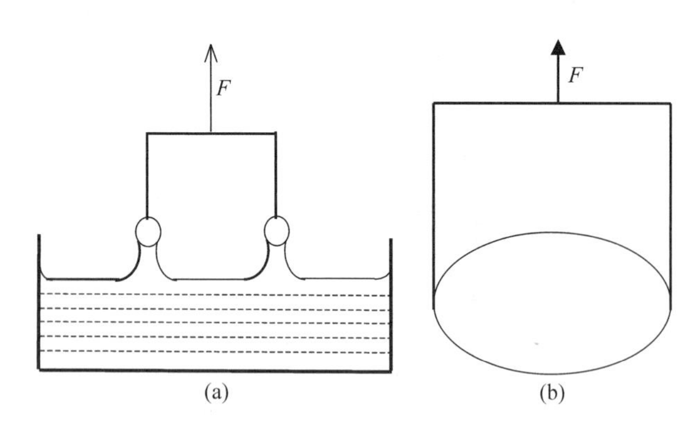

Fig. 4 Du Nouy ring method to measure surface tension.

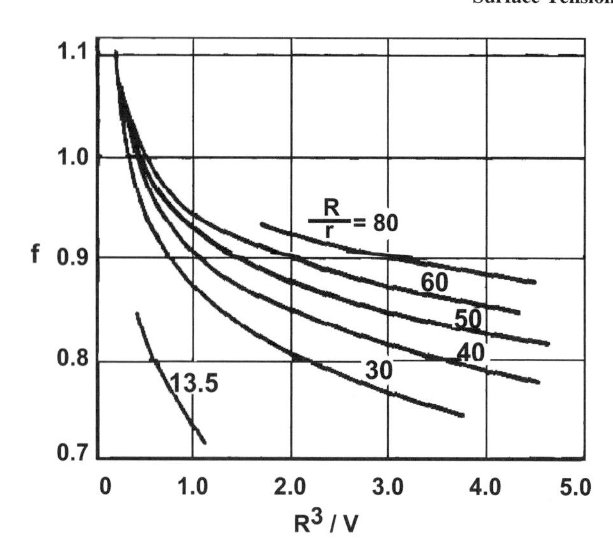

Fig. 5 Correction factor plots for the ring method.

forms a meniscus due to surface tension (see Fig. 4(a)). The vessel is lowered further to a position where the meniscus detaches from the ring. The force exerted on the ring at that point is approximately equal to the surface tension multiplied by the perimeter of the ring, i.e.,

$$F = 4\pi R\gamma \tag{10}$$

where R is the radius of ring. However, the surface tension does not act completely vertical and some of liquid clings to the ring even after the detachment. In practice, Eq. 10 is corrected as

$$F = 4\pi R\gamma\beta \tag{11}$$

where β, correction factor, is given by

$$\beta = f\left(\frac{R^3}{V}, \frac{R}{r}\right) \tag{12}$$

where V is the meniscus volume, R the radius of the ring, and r the radius of the wire. The values of correction factors[3] are summarized in Fig. 5. Values of β can also be calculated using semiempirical equations.[4] For accurate measurements, it is important that the edge of the ring be kept parallel to the surface of the liquid and that the contact angle is close to zero. Rings are usually made from platinum or platinum–iridium, which give contact angles very close to zero.

Wilhelmy Plate

The apparatus for Wilhelmy plate method[2,4] is similar to the ring method (Fig. 6). A rectangular plate is immersed in the sample. Some of the liquid "climbs" up the edges of the plate due to surface tension, thus forming the meniscus. The force recorded by the device is equal to the weight of the meniscus. This weight is balanced by the force due to surface tension. Equating the two forces, one obtains

$$F = 2(l + L)\gamma \cos\theta \tag{13}$$

where l and L are the length and thickness of the plate and θ the contact angle. Plates are often made from platinum or platinum–iridium to make the contact angle close to zero.

Drop Weight

Drop weight method is a very simple method. It measures the volume or mass of a droplet as it forms at the end of a capillary tube (Fig. 7). At the time of detachment from

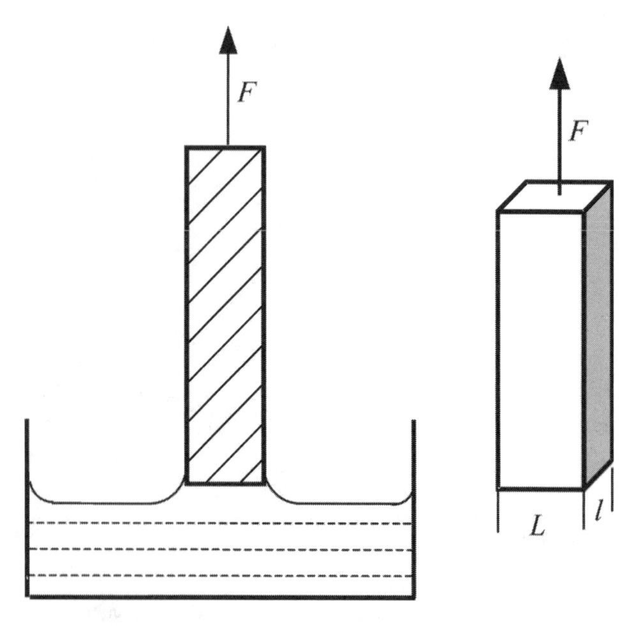

Fig. 6 Wilhelmy plate to measure surface tension.

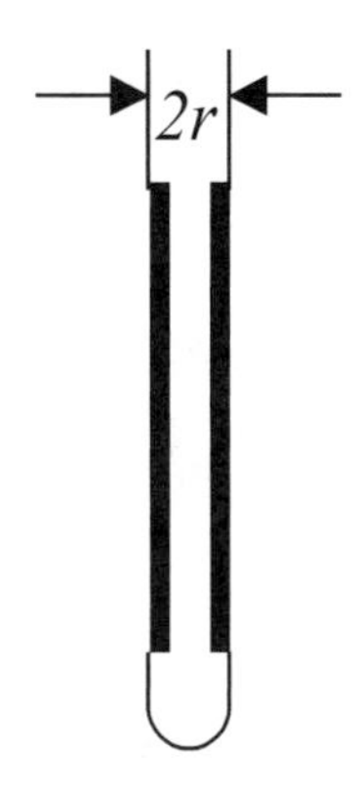

Fig. 7 Drop weight methods to measure surface tension.

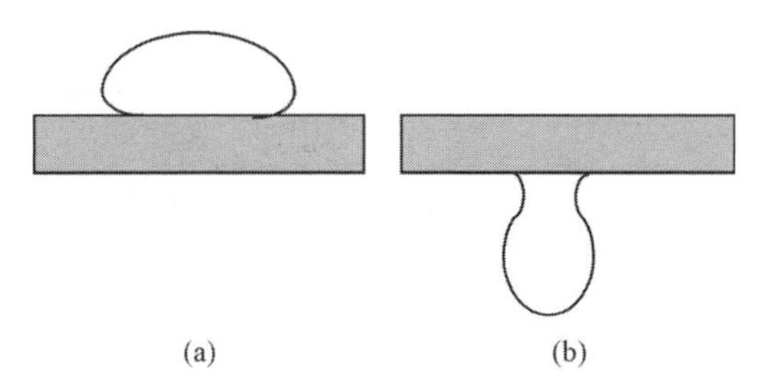

Fig. 8 (a) Sessile drop and (b) pendant drop to measure surface tension.

the capillary, the weight of the droplet is counterbalanced by the surface tension force. Equating the two, one obtains

$$\gamma = \frac{mg}{2\pi r} \tag{14}$$

where m is the mass of a drop, g the acceleration due to gravity, and r the radius of the capillary. If the liquid completely wets the capillary walls (which is usually the case), outer radius of the capillary is to be used. On the other hand, if the liquid does not wet the capillary walls, the inner radius of the capillary is used. In practice, the weight of the drop is less than the ideal weight because of the formation of smaller secondary droplets due to mechanical instability of the cylindrical neck connecting the primary droplet to the capillary tip. As a result, a correction needs to be applied to the above equation, i.e.,

$$\gamma = \frac{mg}{2\pi rf} \tag{15}$$

where the correction factor f is a function of $r/V^{1/3}$, V being the drop volume. The correction factor is given in Table 1.

The application of correction factor can be avoided by using a fluid with known surface tension.[1] If m_1 and m_2 are the drop masses of two fluids with surface tensions γ_1 and γ_2, respectively, then the ratio $m_1/m_2 = \gamma_1/\gamma_2$ will apply.

Table 1 Correction factor for the drop weight method

$r/V^{1/3}$	f	$r/V^{1/3}$	f	$r/V^{1/3}$	f
0.00	(1.0000)	0.75	0.6032	1.225	0.656
0.30	0.7256	0.80	0.6000	1.25	0.652
0.35	0.7011	0.85	0.5992	1.30	0.640
0.40	0.6828	0.90	0.5998	1.35	0.623
0.45	0.6669	0.95	0.6034	1.40	0.603
0.50	0.6515	1.00	0.6098	1.45	0.583
0.55	0.6362	1.05	0.6179	1.50	0.567
0.60	0.6250	1.10	0.6280	1.55	0.551
0.65	0.6171	1.15	0.6407	1.60	0.535
0.70	0.6093	1.20	0.6535		

Sessile and Pendent Drop

A sessile drop refers to a flattened drop (Fig. 8(a)) and a pendent drop refers to a hanging one (Fig. 8(b)). In the absence of gravity, a sessile drop will be spherical as this geometry results in a minimum surface area for the volume of the drop. Consequently, any departure from this spherical shape results in an increase in the surface energy of the drop. Gravity forces will tend to flatten the drop thus increasing its surface area. The equilibrium shape of the drop will depend on the balance of gravity and surface forces.

The shape of the drop can be predicted using Laplace equation. Consider the profile of a sessile drop[5] as shown in Fig. 9. Radii of curvature at the apex O of the drop is denoted by b.

Based on the pressure balance, we get,

$$\frac{1}{R_1/b} + \frac{\sin\phi}{x/b} = 2 + \frac{(\rho_A - \rho_B)gb^2}{\gamma} \times \frac{z}{b} = 2 + \beta\frac{z}{b} \tag{16}$$

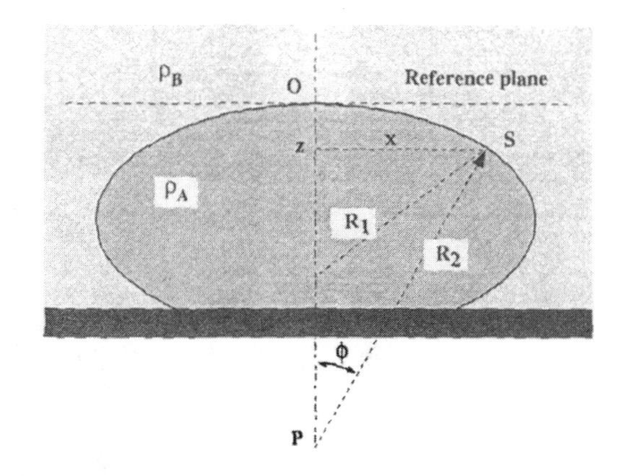

Fig. 9 Definition of coordinates for describing surfaces with an axis of symmetry (OP).

Table 2 x/b and z/b for $\beta = 25$ and $0° < \phi < 180°$

ϕ (°)	x/b	z/b	ϕ (°)	x/b	z/b
5	0.08521	0.00368	95	0.48092	0.28243
10	0.16035	0.01348	100	0.47934	0.29435
15	0.22230	0.02712	105	0.47682	0.30594
20	0.27250	0.04288	110	0.47345	0.31665
25	0.31333	0.05974	115	0.46931	0.32662
30	0.34684	0.07713	120	0.46452	0.33585
35	0.37455	0.09475	125	0.45911	0.34433
40	0.39755	0.11236	130	0.45319	0.35204
45	0.41666	0.12985	135	0.44682	0.35901
50	0.43249	0.14711	140	0.44008	0.36519
55	0.44551	0.16405	145	0.43302	0.37061
60	0.45609	0.18063	150	0.42571	0.37526
65	0.46451	0.19678	155	0.41823	0.37915
70	0.47101	0.21246	160	0.41062	0.38231
75	0.47579	0.22761	165	0.40296	0.38472
80	0.47905	0.24221	170	0.39528	0.38643
85	0.48089	0.25626	175	0.38766	0.38744
90	0.48148	0.26966	180	0.38014	0.38776

where the dimensionless group β is given by

$$\beta = \frac{(\rho_A - \rho_B) g b^2}{\gamma} \tag{17}$$

From geometry, the radius of curvature of drop profile along the plane of Fig. 9 is given by

$$\frac{1}{R_1} = \frac{d^2z/dx^2}{\left[1 + (dz/dx)^2\right]^{3/2}} \tag{18}$$

Eqs. 16 and 18 can be solved for different values of the dimensionless group β to obtain the drop shape (x as a function of z). The results are given in Table 2[5]. Usually, the drop profile is obtained by photography (or imaging) and the closest value of β that predicts the drop shape can then be obtained by matching the experimental and theoretical drop shapes. The surface tension γ can then be calculated using Eq. 17. Even though this method is tedious, it has an error of less than 0.1%.

REFERENCES

1. Lewis, M.J. *Physical Properties of Foods and Food Processing Systems*; Ellis Horwood: Chichester, 1987.
2. Hiemenz, P.C. *Principles of Colloid and Surface Chemistry*; Marcel Dekker: New York, 1986.
3. Adamson, A.W. *Physical Chemistry of Surfaces*, 4th Ed.; John Wiley & Sons, Inc.: New York, 1982, 38–49.
4. Couper, A. Surface Tension and Its Measurement. In *Physical Methods of Chemistry*; Rossiter, B.W., Baetzold, R.C., Eds.; John Wiley & Sons: New York, 1993; Vol. IXA.
5. Hiemenz, P.C.; Rajagopalan, R. *Principles of Colloid and Surface Chemistry*, 3rd Ed.; Marcel Dekker: New York, 1997, 279–283.

Temperature Measurement

S. Wang
Juming Tang
F. Younce
Washington State University, Pullman, Washington, U.S.A.

INTRODUCTION

Temperature is a principle parameter that needs to be monitored and controlled in most food processing operations such as heating, cooling, drying and storage. Temperature sensors have been developed based on different temperature-dependent physical phenomena including thermal expansion, thermoelectricity, electrical resistance, and thermal radiation.[1,2] Temperature sensors used in agricultural and food industries and research vary from simple liquid-in-glass thermometers to sophisticated and state-of-art thermal imaging. This article focuses only on temperature sensors that generate digital signals readily used for on-line monitoring and automatic temperature control purposes. The article focuses on the commonly used temperature sensors, principles, precision, response time, advantages, and limitations for agricultural and food engineering applications.

PRINCIPLES AND PROPERTIES

Resistance Thermometers

Resistance thermometers measure temperature of an object based on changes in electrical resistance of the sensing element. Two types of resistance thermometers are commonly used: thermistors and resistance temperature devices (RTD).

A thermistor uses a semiconductor as the sensing element. The electrical resistance of the semiconductor, made mostly from metallic oxides (e.g., NiO), decreases sharply as the temperature increases. Response follows the general relationship[3]:

$$R = R_0 e^{\beta(1/T - 1/T_0)} \tag{1}$$

where R is the electrical resistance (Ω) at temperature T (K), R_0 is the electrical resistance at a reference temperature, T_0. β is a constant for specific materials (K) and is in the order of 4000. The reference temperature, T_0, is generally taken as 298K (25°C).

The sensing elements of commercially available thermistors are protected by noncorrosive materials (e.g., epoxy, glass, stainless steel) and are in the form of beads, rods, probes, and disks. Sensors can be made as small as 0.1 mm in diameter to give short response time. Glass and stainless steel-sheathed probes between 2 mm and 3 mm are widely used. In measurement, a thermistor is connected to one leg of a Wheatstone bridge to yield a resolution as precise as 0.0005°C.[3] Accuracy of a thermistor sensor is often limited by the readout device with a resolution of only ± 0.1°C. Nonlinearity in electrical resistance output limits application of a thermistor to relatively narrow temperature ranges (e.g., within 100°C). Long-term output stability is another common problem of thermistor probes.

The RTD uses metal materials (e.g., platinum) as the sensing element. Electrical resistance in a metal changes in a predictable manner with temperature. A general relationship is used to convert electrical resistance R to temperature T:

$$R_T = R_0(1 + aT + bT^2 + cT^3 + \cdots) \tag{2}$$

where R_T is the electrical resistance at temperature T, R_0 is the electrical resistance at a reference temperature, usually 0°C, and a, b, and c are material constants. The number of terms used in Eq. 2 depends on the material used in the sensor, the temperature range to be measured, and the accuracy required. In many cases, only constant, a, is used to provide satisfactory accuracy over limited temperature ranges. Fig. 1 shows the temperature range and the output resistance ratio of the three commonly used metals: platinum, nickel, and copper. The RTD has a small positive temperature resistance coefficient of $0.0039\,\Omega\,\Omega^{-1}\,°C^{-1}$ for a platinum RTD at 25°C as compared with a high negative temperature resistance coefficient of about −0.045 for a thermistor with a β value of 4000.

The most widely used RTD is a platinum resistance sensor, which is measured to be 100 Ω and 139 Ω corresponding with 0 and 100°C, respectively. Lead wire resistance and changes in lead wire resistance with temperature may have significant effects on the resistance readings. In application, RTD resistance is transformed into voltage and measured in a bridge circuit connected by two, three, or four wires according to application objectives and the need to minimize the influence of lead

Encyclopedia of Agricultural, Food, and Biological Engineering
DOI: 10.1081/E-EAFE 120006895
Copyright © 2003 by Marcel Dekker, Inc. All rights reserved.

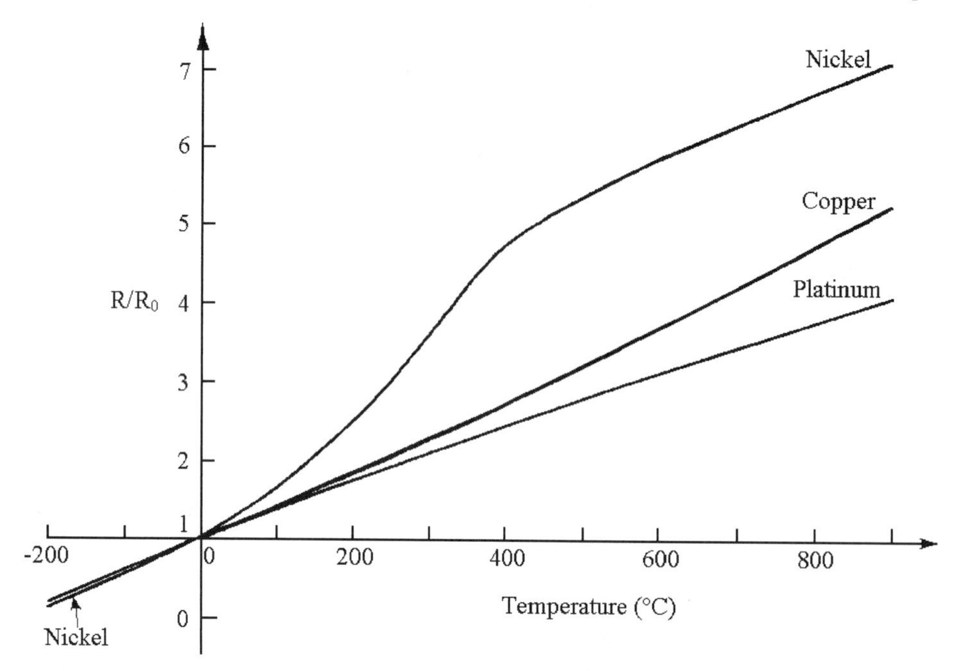

Fig. 1 Resistance–temperature relationship for platinum, nickel, and copper.

wire resistance. The RTD can be applied to measure solid, liquid, and gas substances at temperatures between −200 and 800°C. A major problem with RTD sensors is the relatively slow response time due to the protective sheath (ceramic, glass, or stainless steel). Sensor accuracy of the order of 0.2°C can be reached with careful calibration in specific temperature ranges.

Thermocouples

When two dissimilar metals are joined at two points (Fig. 2), a thermocouple circuit is formed. When these two junctions are at different temperatures, an electromotive force (emf) develops, causing an electric current to flow through the circuit. The magnitude of this emf is proportional to the temperature difference between the two junctions. This physical phenomenon, referred to as thermoelectricity, was discovered by T. J. Seebeck in 1821. The thermoelectricity is used for temperature measurement in thermocouple systems, in which one junction is maintained at a known reference temperature (T_{ref}) while the other is positioned for temperature measurement. The temperature to be measured (T) is converted from thermoelectric voltage V according to:

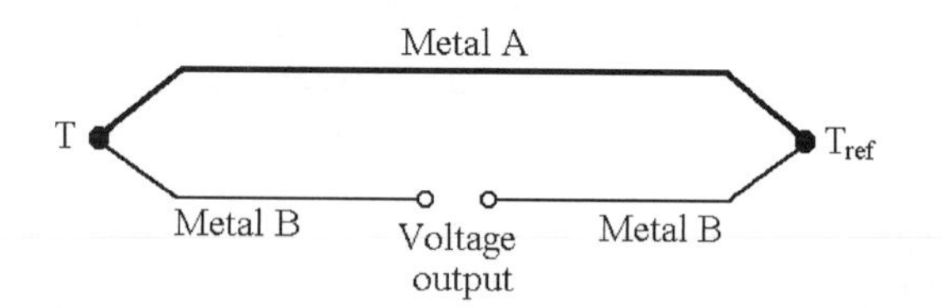

Fig. 2 Thermocouple circuit to measure temperature (T) with the cold junction (T_{ref}) from the output voltage (V).

$$T = T_{ref} + \frac{V}{\alpha} \tag{3}$$

where V is the measured voltage in the thermocouple circuit and α is the voltage/temperature proportionality constant (V °C^{-1}). Voltage can be measured with a potentiometer or a high-impedance solid-state digital voltmeter. The value of constant α depends on the type of wires used to make the thermocouple. Theoretically, for precision work the reference junctions should be kept in a triple-point-of water apparatus that maintains a fixed temperature of 0.01 ± 0.0005°C. In reality, the reference temperature, T_{ref}, can be room temperature, measured with a thermistor in most computer-based data acquisition systems.

Different pairs of metals are used for thermocouple types commercially labeled as types S (platinum/platinum–10% rhodium), R (platinum/platinum–13% rhodium), B (platinum–30% rhodium/platinum–6% rhodium), J (iron/constantan), T (copper/constantan), K (chrome/aluminum), and E (chromel/constantan). Typical characteristics of these thermocouples are shown in Fig. 3.[3] Thermocouples give relatively low voltage outputs (e.g., 0.04 mV °C^{-1} for type T). Type E thermocouples have the highest voltage output per degree among all commonly used thermocouple types, but type E

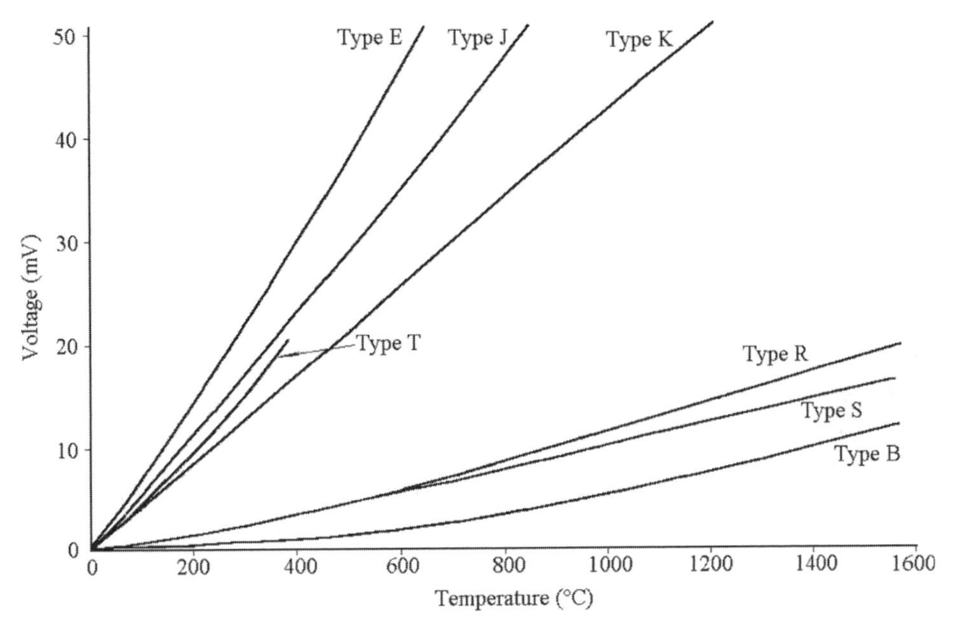

Fig. 3 Typical voltage–temperature relationship for common thermocouple materials when the reference temperature is 0°C.

thermocouples are not suited for reducing, sulfurous, vacuum, and low-oxygen environments. Types T and J are commonly used in industrial applications, because of their resistance to corrosion and harsh environments. For example, type J can be used between −150 and 760°C in both oxidizing and reducing environments, while type T can be used between −200 and 350°C. The upper temperature limit for type T is due to oxidation of copper at temperatures greater than 350°C.

The temperature-sensitive element in thermocouples can be made very small to provide a response time of only a few seconds.[4] Thermocouples can be used to detect the transient temperature in a solid, liquid, and gas, with an accuracy of ±0.25% of reading for types R and S, and ±0.5% reading for type T over a measurement range of 100°C. Generally, thermocouples cannot be used in electromagnetic fields, because the metallic elements may interact with the electromagnetic field.

Sonic Thermometer

The basic principle of a sonic thermometer is the measurement of the time of an ultrasound pulse traveling between two transducers (Fig. 4). It is well known that the speed of sound increases when the air moves in the same direction as the sound. This pair of transducers (a, b) act alternately as transmitters and receivers, exchanging high-frequency ultrasound pulses. This speed of sound in air is used for rapid calculation of air temperature. The output of the sonic temperature (T_s, °C) according to Kaimal and Gaynor[5] is:

$$T_s = \frac{L^2}{1612}\left(\frac{1}{t_1} + \frac{1}{t_2}\right)^2 - 273.15 \tag{4}$$

where L is the distance between the transducers (m), t_1 and t_2 are the travel times in each direction (sec). This sensor has a high sampling rate of up to 50 times per second. The measurement precision may reach 0.4% after careful calibration.

Radiation Thermometer

Radiation thermometers (often referred to radiometer, radiation pyrometer, or optic pyrometer) operate using electromagnetic radiation in the visible spectrum

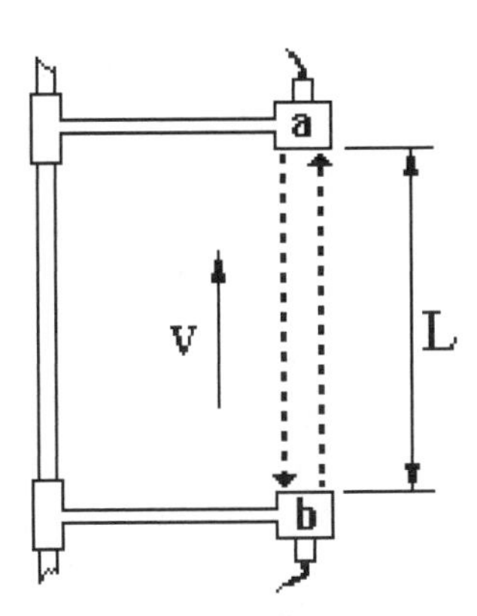

Fig. 4 Sonic anemometer (v, air speed; L, the distance between transducers a and b). (From Ref.[11].)

(0.3 μm–0.72 μm) and a portion of the infrared region (0.72 μm–40 μm). A radiation thermometer may use either a thermal detector or photon detectors as its sensing element. Thermal detectors are blackened elements designed to absorb incoming radiation at all wavelengths radiating from the measured object. Temperature rise of the detector caused by the absorbed radiation is measured by specially made RTDs, thermistors, or thermopiles (consisting of up to 30 thermocouples).

Another type of thermal detector is the pyroelectric detector that is based on pyroelectric crystals (e.g., lithium tantalate crystals). Pyroelectric thermal detectors have shorter response times than the thermal detectors using a RTD, thermistor, or thermopiles and are responsive to a wide radiation range between x-ray (0.001 μm) and far-infrared (1000 μm).

The radiation thermometers can measure the targeted surface temperature from a distance, so there is no contact between the thermometer and the object. As the infrared sensors measure an object temperature at a distance, there is no limitation on the thermal tolerance of materials used to make the sensor. The time constant of radiation thermometers is short, often in the range between several microseconds to a few seconds. Radiation thermometers, however, only read the temperature information at the surface of a relatively opaque material. It cannot be used to measure the transparent materials because the sensor will detect surface temperatures behind the object. Radiation thermometers require information on emissivity of the measured body surface and an internal or external reference temperature to provide accurate temperature information.

Fiber-Optic Thermometers

In radio frequency and microwave heating and drying applications, temperature sensors with metal components cannot be used directly because of electromagnetic noise and interaction between the metal parts and electromagnetic fields. For these applications, fiber-optic thermometers are often used.[6] Fiber-optic temperature sensors are developed from one of three methods: fluoroptic thermometry, Fabry-Perot interferometry, and absorption shift of semiconductor crystals.

The fundamental mechanisms behind fluoroptic thermometry rely on the use of phosphor (magnesium fluorogermanate activated with tetravalent manganese) as the sensing element. This material fluoresces in the deep red region when excited with ultraviolet or blueviolet radiation. The rate of decay of the afterglow of this material varies with temperature (e.g., 5 msec at −200°C, 0.5 msec at 450°C). Measuring the rate of afterglow decay allows an indirect determination of the temperature.

A Fabry-Perot interferometer (FPI) consists of two parallel reflective surfaces that form a cavity resonator. The space separating these two surfaces, also called the FPI cavity (1–2 wavelengths deep), varies with temperature. Changes in temperature result in the change of optical length (refractive index multiplied by cavity depth) of this resonator, even though the actual physical thickness of the film exhibits no measurable changes. Changes in the FPI cavity path, or in the optical path length of the resonators for light from a white-light source, are measured to determine the temperature of the sensing element.

A fiber-optic sensor based on absorption shift of semiconductor crystals relies on the temperature-dependent light absorption/transmission characteristics of gallium arsenide (GaAs). A unique feature of this crystal is that when temperature increases, the crystal transmission spectrum shifts to higher wavelengths (Fig. 5). The position of the absorption shift indicates the temperature measurement of the sensing element.[7]

Probe size of fiber-optic sensors is generally small (as small as 0.8 mm in diameter, Fisco Technologies, Inc., Quebec, Canada). Fiber-optic temperature sensors provide accuracy comparable to thermocouples. They generally have short response times (from 0.05 sec to 2 sec in liquid foods) and are well suited to relatively fast microwave or radio frequency heating, or in applications with strong influence of electromagnetic fields.

SENSOR CALIBRATION AND RESPONSE TIME

Calibration of a given temperature probe is important to ensure reliable measurement. In general, two calibration methods are used. One method is to expose the sensor to an established fixed-point environment, such as the triple-point of pure water and the freezing and boiling points of water. Another method is to compare readings with those of calibrated and traceable temperature sensors when they

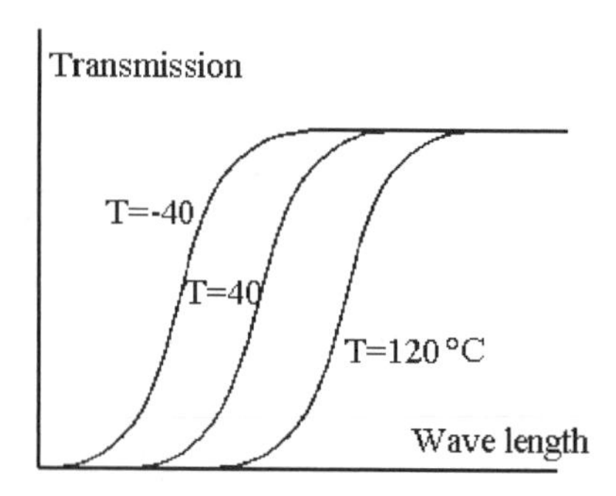

Fig. 5 Temperature-dependent light transmission of GaAs crystal as the sensing element of a fiber-optic thermometer.

are placed in the same thermal environment. Calibration procedures are standardized for specific applications.[8]

Besides accuracy, precision, and resistance to corrosion, an important consideration in selecting temperature sensors is the sensor's response time. Response time is a measure of how quickly a sensor follows rapid temperature changes. The smaller the response time, the more closely the sensor follows the temperature change of the measured medium. For a step-change in the medium temperature, the temperature of the sensor $T(t)$ changes with time t following the relationship:

$$\frac{T_m - T(t)}{T_m - T_0} = \exp(-t/\tau) \tag{5}$$

where constant τ is the sensor response time (also referred to as the time constant). τ is defined as the time for a sensor to reach 63.2% of a step-change in temperature ($T_m - T_0$) [Fig. 6(a)]. The response time can be estimated from the overall surface heat transfer coefficient, h (W m^{-2} °C^{-1}), and other physical properties:

$$\tau = \frac{\rho C_p V}{hA} \tag{6}$$

where A is the surface area of the sensor (m^2), V is the volume of the sensor (m^3), ρ is the density of the sensor (kg m^{-3}), and C_p is the specific heat of the sensor (J m^{-2} °C^{-1}). It is clear from Eq. 6 that reducing the volume, V, of the sensing element (including the protecting sheath) and increasing heat transfer coefficient, h, can reduce sensor response time. That is why temperature probes are often made small. For a same sensor, however, the response time in water can be up to 100 times shorter than in air because of the difference in the overall surface heat transfer coefficient h. For many applications in which air temperature is to be measured, an aspirated temperature sensor is used to increase surface heat transfer coefficient and reduce sensor response time.

Sensor response time can be used to assess measurement error in various applications. Fig. 6(b) demonstrates the thermal lag of a temperature sensor when exposed to a medium with a temperature ramp of dT/dt. At steady-state, the temperature lag is calculated by multiplying the thermal constant, τ, with the ramp rate, dT/dt. For example, when the medium temperature changes at 10°C min^{-1} and the time constant of the sensor is 0.5 min, the maximum thermal lag will be 5°C. Knowledge of the time constant of sensors is essential in designing effective dynamic temperature monitoring and control systems.

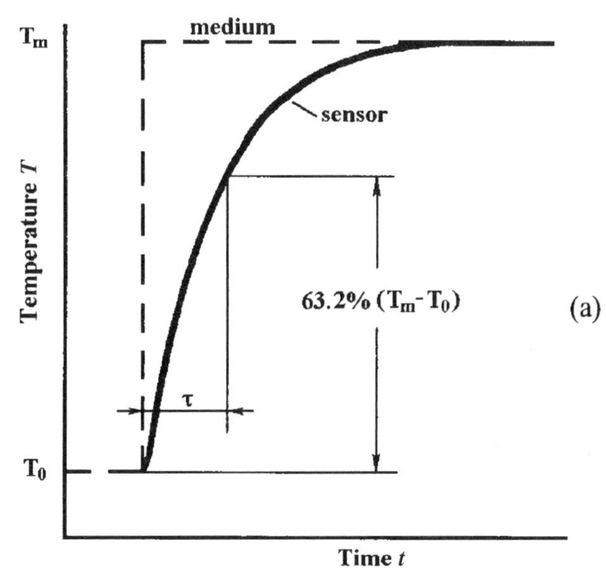

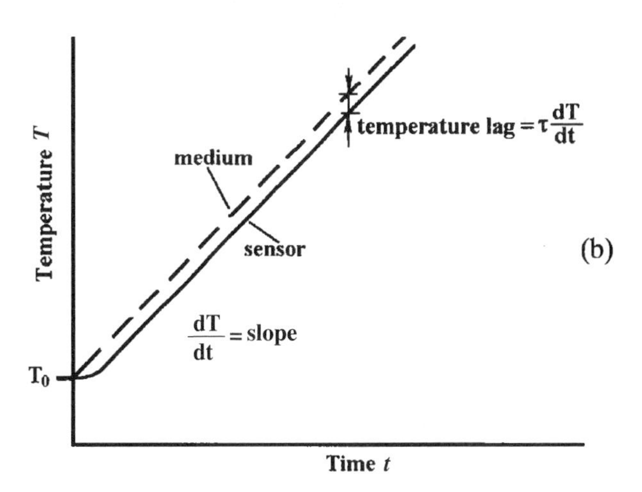

Fig. 6 (a) Response time, τ, defined under a step-change in medium temperature and (b) temperature lag of a sensor following a medium temperature ramp.

APPLICATION OF TEMPERATURE SENSORS

Temperature sensors are selected according to required accuracy, response time, initial investment, maintenance cost, ambient condition, and stability of calibration. Examples of typical applications of temperature sensors discussed above are summarized in Table 1. Thermistors are generally used for measuring temperature in a narrow range, because of their high precision (± 0.1°C) in small ranges.[2] Thermocouples are widely used because of acceptable precision, rapid response time, and low cost. RTDs, however, provide better accuracy and stability. The sonic thermometer is commonly used to measure air temperature in the turbulent environment of storage houses and bins. Radiation thermometers are especially suitable for measurement of moving objects or objects inside vacuum or pressure vessels. Radiation sensors respond very quickly, but are more costly than thermocouples or RTDs.

Table 1 Property and examples of typical applications of temperature sensors

Sensor	Signal output	Measurement range (°C)	Accuracy	Example of typical applications
Thermistor	Resistance	− 50 ~ 200	± 0.1%	Reference for data loggers
RTD	Resistance	− 200 ~ 850	± 0.2°C	Precise and stable measurements
Thermocouple	Voltage	− 250 ~ 400	± 0.5°C	Most food and agricultural applications
Sonic anemometer	Voltage	− 10 ~ 80	± 0.4%	Airflow temperature in storage and bins
Far-infrared thermometer	Voltage	~ 4000	< 1%	Surface temperature of foods in drying and baking
Fibre-optic thermometer	Voltage	− 50 ~ 250	± 0.5%	Microwave and radio frequency heating and drying

The sighting path and optical elements of the radiation detectors must be kept clean. Fiber-optic thermometers can be used in very strong electromagnetic fields.

Even for the same type of temperature probe, a large variety of styles is available for use in food and agricultural engineering applications. Precautions are needed in selecting appropriate probe style to match mechanical, chemical, and sanitation requirements. Some styles directly expose the sensing element to the environment while others are protected with a sheath or thermal well for durability and maintaining a seal. Heat sink paste should be used with thermal wells to increase heat transfer to the thermometer and avoid offsets and thermal time lags. Thermal wells and sheaths should be selected to protect the thermometer, but not add excessive mass and create thermal lags.

For measuring the temperature of fluid in a pipe, a probe in a sheath may by swaged into a bore-through style compression fitting. Care must be taken when selecting and installing exposed thermocouples or those in grounded metal sheaths. When an exposed or grounded style thermocouple contacts highly conductive materials such as metal piping or the food itself, "ground loops" may produce signal noise and inaccurate readings in the measuring system. In such cases, it is advisable to use an ungrounded or electrically isolated probe construction.

Government regulations must also be considered in selection of temperature probe type and documentation of probe calibration. For example, USDA regulations require that each retort system (for canning) be equipped with at least one mercury-in-glass thermometer having readable 0.5°C divisions as the reference for the process, even when other electronic methods are used for recording and process control.[9] Regulations may also require specific probe types for sanitation and clean-in-place (CIP) systems.

Special systems are available for heat penetration studies in cans and pouches. Generally, these special systems consist of the thermocouple probe and a receptacle, which provide means to penetrate the package, hold the probe in place, and maintain a seal (e.g., Ecklund Harrison Technologies of Fort Myers, Florida). Probes are available as flexible thermocouple wire, rigid plastic bodies, and as metal "needle" types. A needle type is mechanically stronger, but it can conduct heat along its length and lead to inaccurate results. For viscous food with low thermal conductivity, a plastic probe is preferred. Placement of a probe in the can requires consideration of can geometry and food material properties to ensure that the critical point (cold spot) is measured. For more viscous products heated by conduction, the critical point is generally the geometric center of the container. But for those less viscous, and therefore heated by internal convection, the critical point may be closer to the bottom of the container.[10]

Selection of temperature probes also includes selection of compatible signal conditioning equipment. Common bench-top display units or PC based equipment for the laboratory may be entirely unsuitable for use in processing. It is important to select probes styles that can be connected to suitable cables or enclosures for signal conditioners and transmitters that are protected from possible heat, moisture, chemical, and mechanical damages. Transmitters should be selected according to the type of temperature sensor, precision, range, and type of output required. Although standards such as 4 mA–20 mA d.c. analog signals are still commonly used, modern temperature sensors are now being interfaced to measurement and process control systems using highly sophisticated technologies such as Ethernet, Field Bus, and most recently, wireless networks.

REFERENCES

1. Doebelin, E.O. *Measurement Systems Application and Design*; Mcgraw-Hill Book Company: New York, 1983.
2. Magison, E.C. *Temperature Measurement in Industry*; Instrument Society of American: Durham, 1990.
3. Simpson, J.B.; Pettibone, C.A.; Kranzler, G. Temperature. In *Instrumentation and Measurement for Environmental Sciences*; Henry, Z.A., Zoerb, G.C., Birth, G.S., Eds.; American Society of Agricultural Engineers: St. Joseph, 1991; 601–617.

T

4. Baker, H.D.; Ryder, E.A.; Baker, N.H. *Temperature Measurement in Engineering*; John Wiley and Sons, Inc: New York, 1953.
5. Kaimal, J.C.; Gaynor, J.E. Another Look at Sonic Thermometry. Boundary-Layer Meteorol. **1991**, *56*, 401–410.
6. Kyuma, K.; Tai, S.; Sawada, T.; Nunoshita, M. Fiber-Optic Instrument for Temperature Measurement. IEEE J. Quantum Electron. **1982**, *18*, 676–679.
7. Belleville, C.; Duplain, G. White-Light Interferometric Multimode Fiber-Optic Strain Sensor. Optic Lett. **1993**, *18*, 78–80.
8. ASTM. *Standard Guide for the Use in the Establishment of Thermal Process for Food Packaged in Flexible Containers*; American Society for Testing Materials: Philadelphia, PA, 1988.
9. Gavin, A.; Weddig, L. *Canned Foods*; The Food Processors Institute: New York, 1999.
10. Alstrand, D.V.; Ecklund, O.F. The Mechanics and Interpretation of Heat Penetration Tests in Canned Foods. Food Technol. **1952**, *6* (5), 185–189.
11. Campbell, G.S.; Unsworth, M.H. An Inexpensive Sonic Anemometer for Eddy Correlation. J. Appl. Meteorol. **1979**, *18*, 1027–1077.

Therapeutics—Biomedical Biomaterials

Joel D. Bumgardner
Mississippi State University, Mississippi State, Mississippi, U.S.A.

INTRODUCTION

Biomaterials are materials used to direct, augment, or replace the function of damaged, diseased, or missing tissues. Gold, wood, glass, ivory, silk, gems, and tissues from animals have been recorded throughout much of history for the replacement of teeth, bones, eyes, to sew blood vessels together, and to replace cartilage. Table 1 lists a variety of metals/alloys, polymers, ceramics/glasses, and composites used in modern biomedical devices such as contact lenses, dialysis machines, sutures, catheters, pacemakers, stents, dental fillings, artificial hearts, and bone plates. These devices are often complex containing multiple components and materials. In 1998, the estimated worldwide medical device and supplies market was valued at more than $145 billion dollars.[1]

The term *biomaterial* is defined as any nonviable material used in a medical device intended to interact with biological systems.[2] Nonviable materials may be natural or synthetic in origin or a combination. Biomaterials are different from *biological materials* which are viable materials produced by a living system such as skin, bone, ligaments, wood, exoskeleton, etc.

To be successful in a biomedical application, a biomaterial must perform with an appropriate host response in a specific application, i.e., be biocompatible.[3] This definition indicates that the use of a material must be considered for a particular device for a particular application. For example, gold performs well as a dental filling material, but does not exhibit sufficient strength to be used as a hip implant, and is too dense and stiff to be used to replace a blood vessel or as a catheter. This article briefly introduces the reader to common biomaterials, their compositions, properties, and applications. A list of resources is provided at the end of the article for further information.

METALLIC BIOMATERIALS

Metals and alloys are widely used in orthopedics, dental/craniofacial, and cardiovascular applications including fracture fixation, total joint replacement, orthodontics, crowns, bridges, dentures, fillings, dental implants, stents, and wires. The most common metals and alloys used in orthopedics, craniofacial/dental, and cardiovascular applications are 316L stainless steel, cobalt–chromium (Co–Cr) alloys, titanium (Ti), and Ti-alloys. Gold (Au) and Au-alloys, Co–Cr and nickel–chromium (Ni–Cr) alloys, amalgams, and Ni–Ti alloys are widely used in dentistry. Ni–Ti alloys are also used in stents for cardiovascular and urological applications. These metals and alloys were selected in part for their mechanical properties, formability, and ability to resist corrosion. Corrosion resistance is particularly important since corrosion will lead to disintegration and loss of mechanical strength of the device and to the release of metallic corrosion products (e.g., Cr, Ni, Co, and Al ions) to the host which have the potential to cause metal hypersensitivity and other adverse tissue reactions.

The composition and mechanical properties of 316L stainless steel, Co–Cr, Ti, and Ti–aluminum (Al)–vanadium (V) alloys are given in Table 2. 316L stainless steel and Co–Cr contain Cr to provide a protective Cr oxide surface layer for corrosion resistance. The L designation of the 316 stainless steel indicates that the carbon (C) content of the alloy is limited to 0.03% to prevent formation of Cr-carbides which reduce corrosion resistance. An adherent Ti oxide surface film provides excellent corrosion resistance for Ti and its alloys and may be important in the integration of the implant device by bone cells and tissues.[4] In general, Ti and Ti alloys exhibit the lowest corrosion rates of the three alloys, followed by the Co–Cr alloys and then stainless steel.[5] The corrosion rate of 316L stainless steel though is still considered to be quite low.

The alloys exhibit a range of mechanical properties, depending upon processing and manufacturing, which are generally more than sufficient to resist physiological loads (Table 2). However, the 316L alloy is susceptible to attach by chlorides present in the body, particularly in small crevices and holes that occur for example between the screw head and sink hole of a fracture fixation plate or wires of a cardiovascular stent. The susceptibility to corrosion of 316L stainless steel has limited its current uses primarily to temporary devices such as fracture plates, screws (Fig. 1A), orthodontic wires and brackets, and medical/surgical instruments and tools.

The strength of the Co–Cr alloys is generally greater than stainless steel and Ti alloys. The American Society for Testing and Materials (ASTM) F562 Co–Cr alloy has highest strength values. The high strength of the Co–Cr alloys may be of concern since bone adjacent to a Co–Cr implant may be shielded from normal loading and be resorbed. The ASTM F75 Co–Cr alloy is used in

Encyclopedia of Agricultural, Food, and Biological Engineering
DOI: 10.1081/E-EAFE 120007222
Copyright © 2003 by Marcel Dekker, Inc. All rights reserved.

Table 1 Brief list of materials used in medical applications

Types of materials	Example applications
Metals	
Stainless steel, Co–Cr, titanium and titanium alloys, nickel–titanium, gold alloys, Ni–Cr alloys, and amalgams	Fracture fixation plates, screws, nails, joint replacements, orthodontic wires, stents, cases for pacemakers, supports for heart valves, dental implants, dental crowns, bridges, fillings, and inner ear bone replacements
Ceramics	
Carbon coatings, alumina oxides, zirconia, glass, glass ceramics, and calcium phosphates	Heart valves, dental implants, joint implants, coatings for dental and joint implants, bone defect filler, tissue scaffolds, drug delivery systems, and inner ear implants
Polymers	
Polyethylene, polyester, polytetraflouroethylene, PMMA, hydrogels, silicone rubber, PGA/PLA, collagen, cellulose, and chitosan	Joint replacement, vascular grafts, bone cement, contact and intraocular lenses, catheters, hand and toe joints, artificial tendon and ligament, reconstructive surgery, sutures, staples, tissue scaffolds, drug delivery systems, and hemostatic bandages

dentistry as the framework for removable dentures, in orthopedics for total joint replacements (Fig. 1B,i,iii and C,i) and as fracture fixation plates and screws. Castable Co–Cr alloys with compositions similar to ASTM F75 are also used in dental porcelain–metal restorations. The ASTM F562 Co–Cr alloy is used in orthopedics for total joint replacement devices. Co–Cr alloys with compositions similar to ASTM F562 may be used as orthodontic wires and brackets and as housings for pivoting valves or leaflets in artificial heart valves (Fig. 2).

Ti exists in four grades based on impurity content, especially oxygen and nitrogen, since they significantly

Table 2 Biomedical alloy composition and properties[a]

Alloy	Composition (wt%)	Yield strength[b] (MPa)	Ultimate tensile strength[c] (MPa)	Modulus (GPa)	% Elongation[f]
316L stainless steel (ASTM F138/139)	59.6–64.3Fe, 17–19Cr, 13–15Ni, 2.25–3Mo, 2(max) Mn, 0.03 (max)C	190–690	490–1,350	190	12–40
Cast[d] Co–Cr–Mo (ASTM F75)	63–68Co, 27–30Cr, 5–7Mo	450	655	210	8
Wrought[c] Co–Cr (ASTM F562)[e]	29–39Co, 19–21Cr, 33–37Ni, 9–10.5Mo	241–1,586	793–1,793		8–50
Ti (ASTM F67)[b]	98.9–99.5Ti, 0.18–0.4O, 0.03–0.05N, 0.2–0.5Fe	170–483	240–550	110	15–24
Ti–6Al–4V ELI (ASTM F136)	88.5–90.5Ti, 5.5–6.5Al, 3.5–4.5V	760–795	825–860	116	8–10

[a] Adapted from *1999 Annual Book of ASTM Standards, vol 13.01 Medical Devices; Emergency Medical Services* and *Biomaterials Science: An Introduction to Materials in Medicine, 1996.*

[b] Maximum percent composition of O, N, and Fe depend on grade of Ti.

[c] *Wrought* designation means implants are made by pressing, hammering, bending, and or cutting alloy into final shape.

[d] *Cast* designation means implants are made by melting the alloy and pouring into a final shape.

[e] Alloy is also known as MP35N (Standard Pressed Steel Co., now SPS Technologies, Inc., Jenkintown, PA, USA) in which the MP stands for multiphase, referring to the different metallographic phases present in the alloy microstructure and approximately 35 wt% Ni in alloy composition.

[f] Values depend upon the processing and manufacturing conditions. Values for Ti also depend on alloy grade (from *1999 Annual Book of ASTM Standards, vol. 13.01*, and *Biomaterials Science: An Introduction to Materials in Medicine*, 1996).

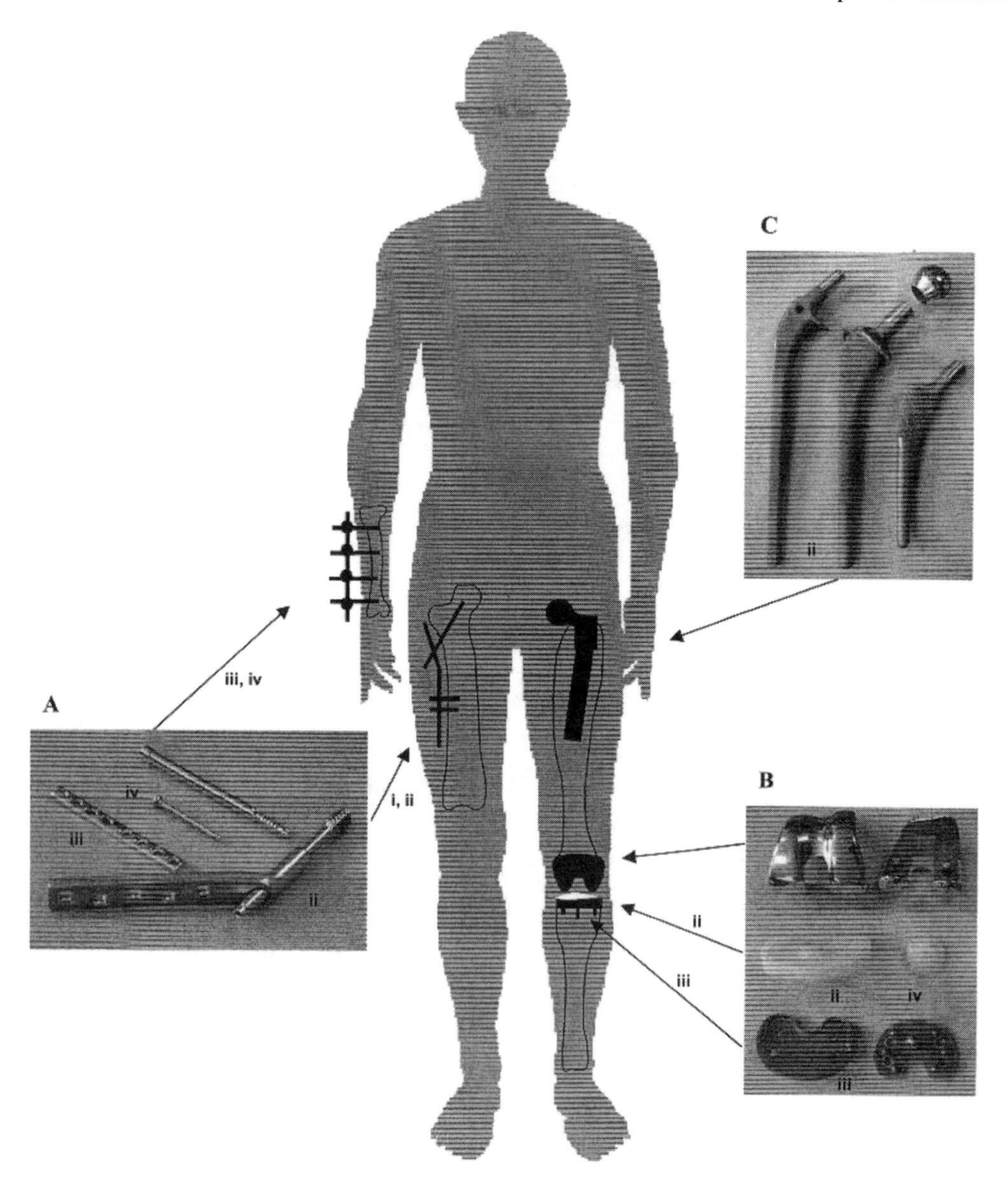

Fig. 1 Cartoon indicating location of orthopedic implants in host and images of the implant devices. (A) 316L stainless steel fracture fixation devices; i) femoral nail, ii) hip screw and plate, iii) bone plate, and iv) bone screw. (B) Components for a total knee replacement; i) Co–Cr femoral component, front and back view, ii) UHMWPE tibial bearing surface, iii) titanium alloy tibial plate, top and bottom view, and iv) UHMWPE patella. (C) Components of a femoral hip implant; i) Co–Cr femoral head and ii) titanium alloy femoral stem.

affect its mechanical properties. For example, increasing the oxygen content from approximately 0.2% (grade 1) to 0.4% (grade 4) increases its yield strength from 170 MPa to 485 MPa with a concomitant decrease in fatigue properties. Alloying Ti with Al and V increases the mechanical strength properties. While the strength properties of Ti and its alloys may not be a high as Co–Cr or 316L alloys (Table 2), it does have excellent strength to weight ratio, due to its lower density than the other two alloys. Ti exhibits poor shear strength and wear properties, which limits it application in fracture fixation plates and in articulating devices. Machined Ti and its alloys are widely used as the femoral stem component of total hips (Fig. 1C,ii), dental implants (Fig. 3), pace-maker casings, and heart valve housings. Cast and machined Ti and Ti alloys are finding applications for dental restorations and partial denture frameworks.

Ni–Ti alloys comprise a special class of alloys known as shape memory since they may be deformed and when heated, revert back to their original shape. These alloys are characterized by their high resiliency, limited formability, and thermal memory. Composition and thermal and

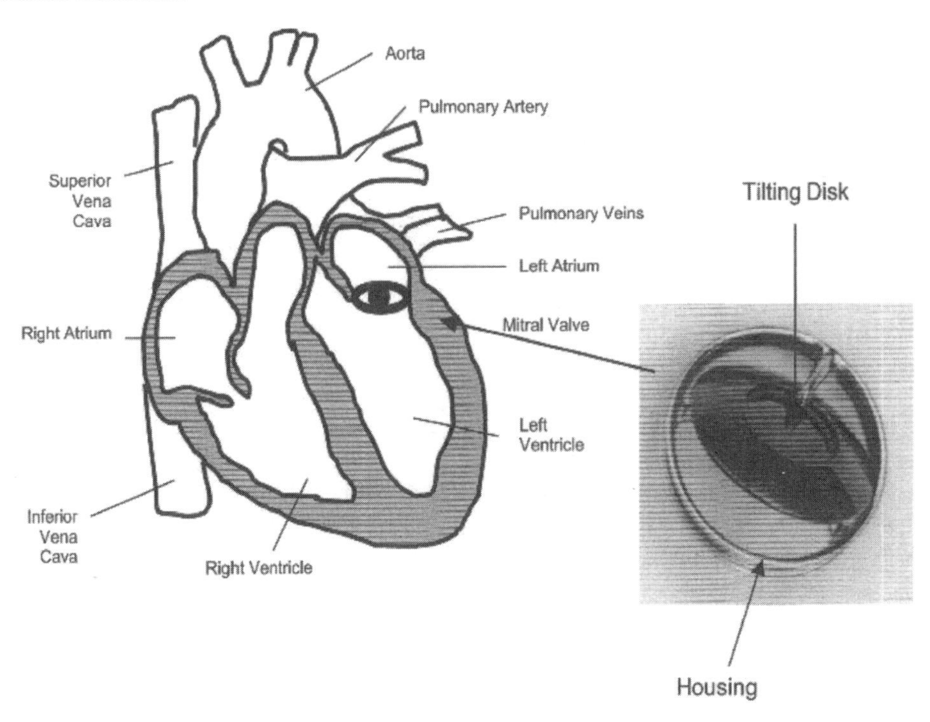

Fig. 2 Tilting pyrolytic carbon disk artificial heart valve in a Co–Cr metallic housing.

mechanical processing influence the temperature at which the alloys will return to their original shape. The most common shape memory alloy used in biomedical applications is Nitinol®, 55 wt% Ni + 45 wt% Ti. The high resiliency and low stiffness of the alloy provides high spring-back or super-elasticity which is important in orthodontics when teeth are poorly aligned. Ni–Ti alloys are also used as stents in cardiovascular and urological applications to hold open vessels.

Dental Alloys

Au, Au-based alloys, Co–Cr, and Ni–Cr alloys are widely used in dental applications as crowns, bridges, ceramic-to-metal restorations, and as frameworks for dentures. Au is used because of its durability, corrosion resistance, and ease of casting and forming. Dental Au alloys are primarily alloyed with silver (Ag) and copper (Cu) with variations in minor alloying elements like zinc

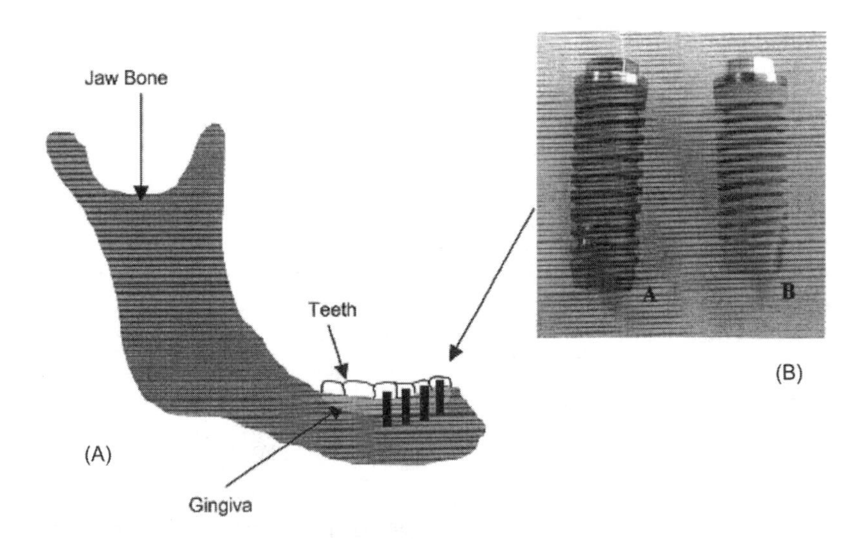

Fig. 3 Cartoon indicating the location of dental implants in the jaw. (A) Screw form implant made of titanium alloy and (B) implant with a HA (calcium-phosphate) coating.

(Zn), platinum (Pt), palladium (Pd), and indium (In). Au alloys retain their corrosion resistance as long as they contain 75 wt% or more of Au and other noble metals like Pt. High Au alloys (+ 83 wt% Au) are used to fill cavities which do not experience high stresses. The more highly alloyed materials with their increased strength and wear resistance are used as crowns, bridges, and dental frameworks. The Ni–Cr alloys also exhibit high strengths suitable for crowns, bridges, and dentures. These alloys are a highly heterogeneous group with wide variations in Cr and molybdenum (Mo) contents as well as secondary alloying elements such as niobium (Nb), gallium (Ga), beryllium (Be), Al, Fe, and others.[6] The Ni–Cr alloys rely on a protective Cr oxide layer for their corrosion resistance in the oral environment. The alloys have coefficients of thermal expansion similar to those of dental porcelains which reduces cracking of the aesthetic porcelain veneers during heating and cooling. Other dental alloy systems include the Ag–Pd, Pd–Ag, high Pd, and Cu-based alloys.

Amalgam is an alloy of mercury (Hg) with other metal(s). Dental amalgam is a specific family of alloys produced when liquid Hg is mixed (approximately in a 1:1 ratio) with solid alloy particles of 40%–70% Ag–22%–30% tin (Sn)–2%–40% Cu (element%). The Ag–Sn–Cu alloys, a.k.a. amalgam alloy, may also contain Zn, Pd, In, and selenium (Se) as secondary alloying elements. Dental amalgam is used to fill tooth cavities since when freshly mixed, it is very easy to deform and pack or condense into a cavity, it maintains anatomical form, has a reasonable resistance to fracture, and has a relatively long service life. However, the silver-gray color of dental amalgam has a poor aesthetic quality. Amalgams have been used in dentistry for approximately 175 yr. While case reports of allergic reactions to Hg in amalgams occur from time to time, there is no well-documented scientific evidence linking dental amalgams to disease.

BIOCERAMICS

The ceramics, glass, and glass–ceramic biomaterials are a heterogeneous group of biomaterials with applications as blood contacting devices like heart valves, aesthetic restorations in dentistry, articulating surfaces in total joint devices, coatings on devices for tissue in-growth, and in resorbable/degradable devices for filling tissue defects, and or delivery of drugs. The brittle nature and sensitivity to cracks and notches of the bioceramics generally limits their use under tensile conditions, but improvements in chemistry, composition, and processing continue to expand their use. Bioceramics may be divided into four groups based on their relative chemical activity: (1) inert, (2) nonresorbable or relatively inert, (3) bioactive, or surface reactive, and (4) biodegradable or resorbable.[7] The inert bioceramics include the graphite and glassy-carbons, and the nonresorbables include alumina and zirconia materials. The bioactive ceramics include the glass–ceramics and hydroxyapatite (HA), and the resorbable/degradable bioceramics include the calcium-phosphates (Ca-P) and Ca-carbonates.

Carbon-Based Bioceramics

Pyrolytic carbon is made through the decomposition of methane or propane at controlled temperatures and pressures in a fluidized bed using argon or helium to deposit the carbon atoms onto a substrate or as a single piece of material. Silicone carbide may be added to the reaction mixture to create a silicon-alloyed carbon with increased strength and wear resistance. Pyrolytic carbon is used predominantly in artificial heart valves and heart valve leaflets due to its high hemocompatibility and durability (Fig. 2). Hemocompatibility is due in part to an electronegative surface charge which minimizes protein and cell interactions to reduce hemolysis and thrombogenesis. The crystalline structure of the pyrolytic carbons is similar to the planar hexagonal arrays of graphite but with increased bonding between layers. The pyrolytic carbons have a high hardness which reduces wear, and may essentially undergo a near infinite number of cycles without failing under cardiovascular loading conditions. Diamond coated implant surfaces are currently under investigation but are not yet commercially available.

Nonresorbable Bioceramics

Nonresorbable or nearly inert bioceramics resist degradation and wear and have been used primarily as bone plates and screws, components of total joint replacement devices, dental implants, ventilation tubes, middle ear ossicles, and in the reconstruction of orbital rims. The two most common nonresorbable ceramics used in implants are alumina (Al_2O_3), and zirconia (ZrO_2). High density, high purity (> 97%) alumina is used in load-bearing hip devices and as dental implants due to its good biocompatibility and excellent tribological properties. Most alumina devices are fine grained (< 4 μm) polycrystalline materials formed by high pressing and sintering at 1600–1700°C. Some dental implants are made of a single crystal of alumina also known as sapphire. Mechanical properties are very sensitive to impurities and changes in grain size. When an alumina ball and socket in a hip prosthesis are highly polished and used together, the coefficient of friction may, eventually approach values of a normal joint. However, the very high modulus of alumina as compared to bone has raised concerns similar to the Co–Cr alloys over stress shielding and bone atrophy. Zirconia is produced by processes similar to alumina, but can be made with a much smaller grain size ($\sim 0.5\,\mu m$). This smaller grain size allows zirconia to be polished to a higher degree than alumina and, therefore, to exhibit better wear characteristics. Zirconia may be partially stabilized with 3%–9% magnesia (MgO) or yttria (Y_2O_3) to improve resistance to fracture. Zirconia has a lower modulus than alumina which, in addition to its low wear rates, is increasing its use as a bearing surface in total joint implants.

Bioactive Bioceramics

The bioactive glasses, ceramics, and glass–ceramics when implanted, develop with time an active carbonated Ca-P surface layer that is capable of bonding with tissues. This bonding allows the material to become incorporated into the tissues and withstand mechanical forces. The strength of the interfacial bond may even become greater than the strength of the bulk material or the tissue immediately adjacent to the material.

The basic composition of the bioactive glasses is 45 wt% SiO_2, 19.5 wt%–24.5 wt% Na_2O plus P_2O_5 with a 5:1 ratio of CaO to P_2O_5 often referred to as 45S5 bioglass. Variations include substituting 5%–15% B_2O_3 for SiO_2 and up to 12.25% CaF_2 for CaO. Glasses with lower than a 5:1 CaO to P_2O_5 ratio will not bond tissues. The brittle nature of bioglasses limits their use to low load applications like fillers for bone defects (e.g., alveolar ridge augmentation) maxillofacial reconstruction, percutaneous access, otolaryngological devices (Fig. 4), and as coatings for dental and maxillofacial prosthetics.

Ca-P ceramics have compositions similar to bone making them attractive for bonding implants to bone. The two most common forms are tri-Ca-P (TCP), $Ca_3(PO_4)_2$, and HA, $Ca_{10}(PO_4)_6(OH)_2$. Hydroxyapatite is very similar to the crystalline mineral phase of bone and teeth. The TCP has a much higher dissolution rate than the crystalline HA. Ceramic powders may be made by mixing molar ratios of Ca-nitrate and ammonium phosphate in an aqueous solution to form HA precipitate. Variations in temperature, pressure, and moisture during subsequent processing determines the relative amount of HA and TCP phases present in the final material and hence its resorption and strength properties. Porosity will increase dissolution and decrease strength, the substitution of carbonates for phosphates will also increase dissolution, and the substitution of fluorine ions for hydroxyl groups, known as fluorapatite and found in tooth enamel, will decrease dissolution. Due to limited strength in bulk form, their primary use has been as powders for bone filler, in low load

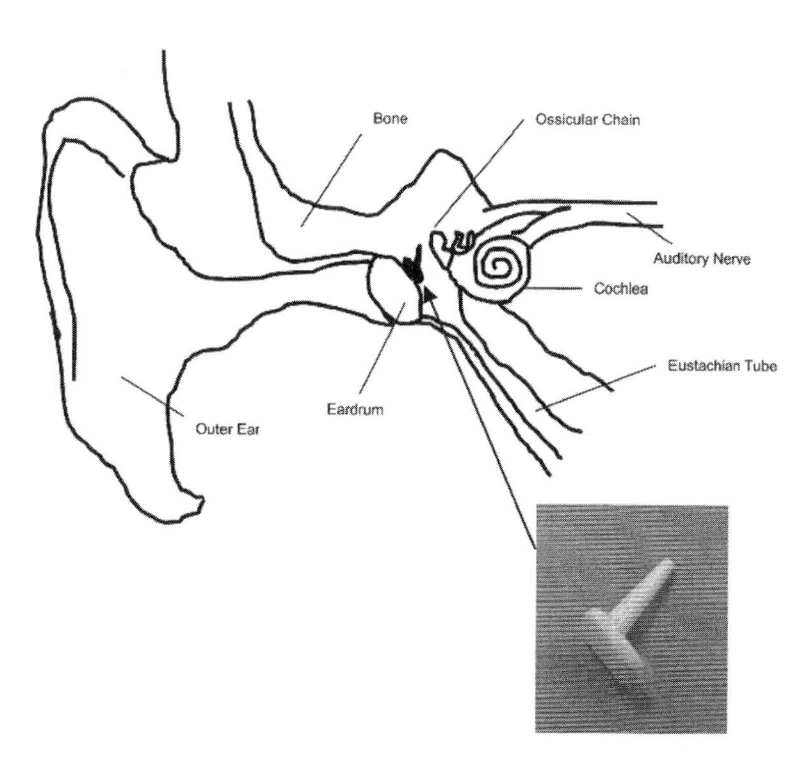

Fig. 4 Calcium fluoroaluminosilicate glass inner ear ossicle implant.

applications as in the middle ear and as coatings on alloys in joint and dental implants (Fig. 3B).

Degradable/Resorbable Bioceramics

Degradable/resorbable ceramics slowly degrade after implantation to be replaced with normal tissues. Most of the bioresorbable ceramics are Ca-P based and include TCP. Coralline is a Ca-carbonate materials derived from the exostructure of corals with a structural similarity to bone (genus *Porites*, and *Goniopora*). Coralline may also be transformed via a hydrothermal process to HA. The resorbable ceramics and coralline materials may be used as drug delivery devices (i.e., as material degrades, therapeutic agent is released), filling bone defects, and in repair and fusion of spinal vertebrae (Fig. 5).

POLYMERIC IMPLANT BIOMATERIALS

Natural and synthetic polymers are used in medical disposables, orthopedic, dental/craniofacial, and cardiovascular implants, engineered tissue scaffolds, drug delivery systems, resorbable structures, ophthalmological devices, and as dressings, sutures, and coatings (Fig. 6). Through careful control of polymerization reaction temperatures, pressures, formation times, and starting chemicals, a variety of polymers may be created with selected properties for specific applications. While the diversity of polymeric biomaterials is too large to cover in this brief section, a few of the commonly used polymers, their properties and applications are presented.

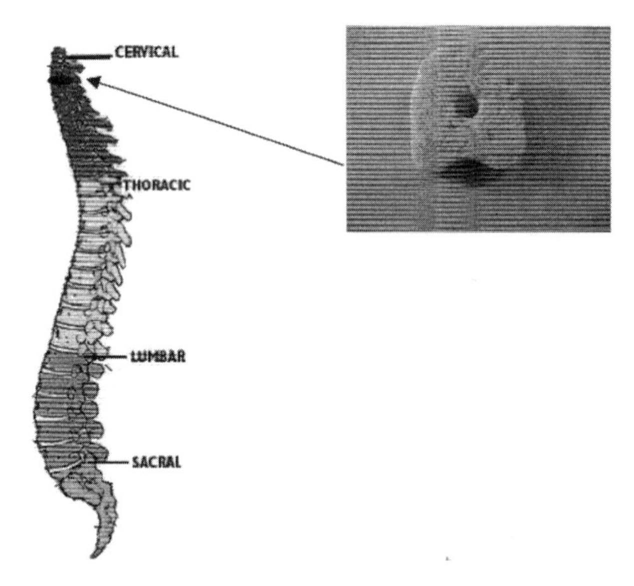

Fig. 5 Calcium carbonate (natural coral) used for cervical spine fusion.

Polyethylene

Only high density polyethylenes (HDPE) such as ultrahigh molecular weight polyethylene (UHMWPE) and HDPE with densities of $0.94\,g\,cm^{-3}$–$0.96\,g\,cm^{-3}$ are used in biomedical applications. Lower density forms cannot withstand sterilization treatments. High density polyethylene is used in pharmaceutical bottles, catheters, tubing and drains, nonwoven fabrics, and caps. Ultrahigh molecular weight polyethylene has a very high molecular weight ($MW > 2 \times 10^6\,g\,mol^{-1}$), exhibits high stiffness ($E = 2200\,MPa$) and hardness, and is therefore used in load bearing applications such as the acetabular cup in total hip implants, and as the tibial plateau and patellar surfaces in total knee replacements (Fig. 1B,ii,iv).

Polymethylmethacrylates

Polymethylmethacrylate (PMMA) has a range of medical applications including blood pumps, intravenous systems, membranes for blood dialyzers, contact lenses, intraocular lenses, dentures, maxillofacial prostheses, and as bone cement for total joint devices. Polymethylmethacrylate has excellent optical (92% transparency; refractive index = 1.49) and coloring properties, hardness, and stability, and is easily machined and molded. Substitution of a hydroxyethyl group for the methoxy group in the methacrylate monomer results in a hydrophilic polymer or hydrogel called polyhydroxyethyl methacrylate (poly-HEMA). Poly-HEMA may absorb more than 30% of its weight in water and is highly permeable to oxygen, making it popular for extended-wear soft contact lenses.

Polytetrafluoroethylene

Polytetrafluoroethylene (PTFE) or Teflon® (DuPont) is similar to polyethylene except all the hydrogen atoms are replaced by fluorine. The polymer is extremely stable, and resistant to thermal and chemical degradation. While it has a low tensile strength (14 MPa) and poor wear properties, it does have a very low coefficient of friction (0.1). Polytetrafluoroethylene may be expanded on microscopic scale into a microporous fabric (Gore-Tex) and used as a vascular graft and thermal insulator.

Polyesters

Several types of polyesters are used in medical devices such as polyethyleneterephthalate (PET), polyglycolic acid (PGA), and polylactic acid (PLA). Polyethyleneterephthalate is resistant to hydrolysis, and is hydrophobic. It is made into fabrics for vascular grafts (Fig. 7) and into fibers for sutures. Polyglycolic acid is subject to hydrolysis

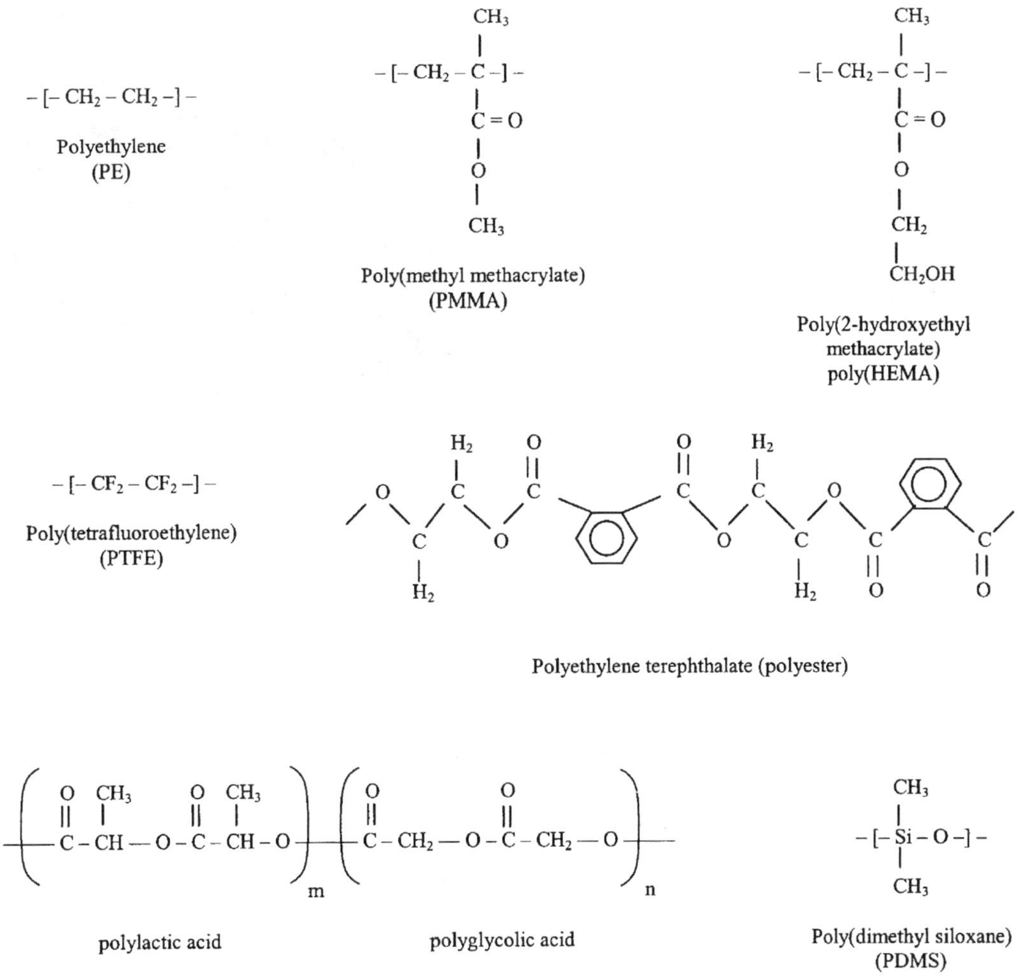

$-[-CH_2-CH_2-]-$

Polyethylene (PE)

$-[-CH_2-C(CH_3)(C{=}O\,O\,CH_3)-]-$

Poly(methyl methacrylate) (PMMA)

$-[-CH_2-C(CH_3)(C{=}O\,O\,CH_2\,CH_2OH)-]-$

Poly(2-hydroxyethyl methacrylate) poly(HEMA)

$-[-CF_2-CF_2-]-$

Poly(tetrafluoroethylene) (PTFE)

Polyethylene terephthalate (polyester)

$\left(\overset{O}{\overset{\|}{C}}-\overset{CH_3}{\overset{|}{CH}}-O-\overset{O}{\overset{\|}{C}}-\overset{CH_3}{\overset{|}{CH}}-O\right)_m \left(\overset{O}{\overset{\|}{C}}-CH_2-O-\overset{O}{\overset{\|}{C}}-CH_2-O\right)_n$

polylactic acid polyglycolic acid

$-[-Si(CH_3)_2-O-]-$

Poly(dimethyl siloxane) (PDMS)

Fig. 6 Common polymeric biomaterials.

and is used for absorbable sutures, staples, and bone pins. Copolymerization of PGA with PLA (PLA is hydrophobic) reduces the rate of hydrolysis of PGA. The PGA–PLA copolymers are used as absorbable sutures and staples (Fig. 8). The PGA–PLA materials are studied as resorbable fracture fixation plates since they will slowly absorb as the bone heals, thereby avoiding a second surgery. The materials may also be used in drug delivery systems for the slow controlled release of therapeutic compounds.

Silicone Rubber

Unlike the other polymer implant materials, poly-dimethyl siloxane or silicone rubber has a Si-O polymer backbone instead of a C backbone. The polymer is very elastic (percent elongation = 360–600), has low tensile strength (9.5 MPa), elastic modulus (350 MPa), and density ($1.12\,g\,cm^{-3}$–$1.23\,g\,cm^{-3}$), is highly oxygen permeable, and is stable. Low molecular weight ($750{,}000\,g\,mol^{-1}$) polymers are crosslinked to make higher molecular weight materials. Due to its high flexibility and stability, the silicone rubber has been used as finger and toe joints (Fig. 9), blood vessels, heart valves, breast implants, outer ears, chins, and

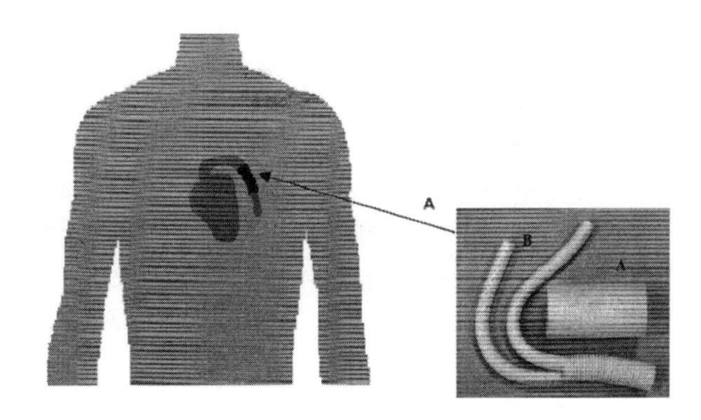

Fig. 7 Polyethyleneterephthalate vascular grafts. (A) Woven vascular graft and (B) knitted vascular graft.

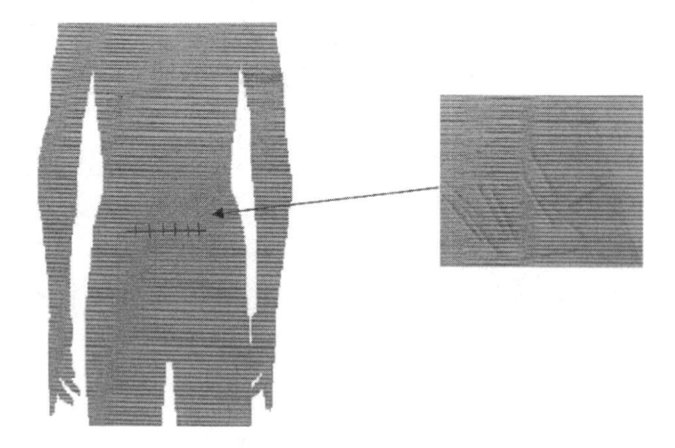

Fig. 8 Polyglycolic acid–polylactic acid copolymer staples for abdominal surgery.

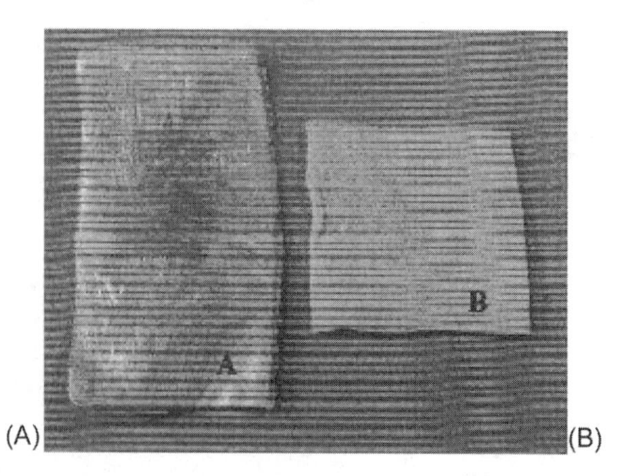

Fig. 10 Hemostatic sponges. (A) methylcellulose and (B) collagen.

nose implants. It has also been used for catheter and drainage tubes, and insulation for pacemaker leads.

Natural Polymers

Natural polymers, such as silk, cellulose (cotton), alginate, starch, collagen, and chitin are used in biomedical applications. These materials have similarities to components of host tissues which increase their biocompatibility, and their degradation products are nontoxic. However, protein components of these biopolymers can present significant immunogenic challenges. Silk and cellulose have been routinely used as sutures and hemostatic agents (Fig. 10). Collagen has been investigated and developed for sutures, hemostatic agents (Fig. 10), blood vessels, heart valves, ligaments/tendons, burn treatments, drug delivery systems, peripheral nerve regeneration, and intradermal/reconstructive and plastic surgery. Alginates and starch-based materials have been investigated for drug delivery systems, tissue scaffolds, and resorbable bone fixation devices. Chitin is the main structural fiber in the exoskeleton of arthropods and is a polysaccharide similar to cellulose. Its potential uses include resorbable sutures, hemostatic devices, blood vessels, tissue scaffolds, burn treatments, ligaments/tendons, contact lenses, and drug delivery systems.

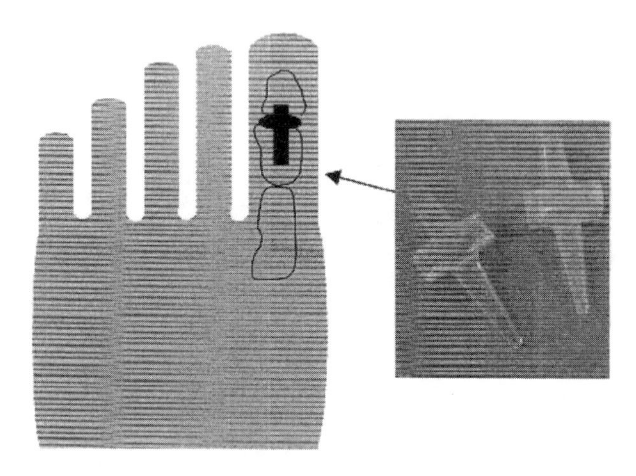

Fig. 9 Silicone rubber hinged toe joint implants.

CONCLUSION

This article has provided a very brief introduction to many of the common biomaterials used in medicine. While, many biomaterials have had a long and successful history of use in their particular applications, much research continues with the aim of developing new biomaterials and or improving current compositions and surface properties to improve implant function. Advances in nanotechnology, materials, bioprocessing/bioconversion, biology, medicine, and biotechnology are making possible the development of biomaterials that are capable of recruiting and interacting with cells and tissues to direct healing and or regeneration of damaged or diseased tissues and organs. Natural materials present special opportunities for continued development due to their similarity to host tissues and as a renewable and value-added resource. Biological engineers, with their interdisciplinary training in both biology and engineering, are especially well suited to effectively interact with scientists and clinicians and contribute to the advancement of biomaterials.

ACKNOWLEDGMENTS

The author wishes to express his gratitude to Ms. Jennifer Woodfield for her most patient and creative work with the figures and to Dr. Jerome A. Gilbert for his valuable comments and suggestions in preparing this manuscript.

REFERENCES

1. Schaaf, T.A. The Medical Device Market Today, Medical Device Link, http://www.devicelink.com, September 1–12, 1999 (accessed June 2001).
2. Williams, D.F., Ed. *Concise Encyclopedia of Medical and Dental Materials*, 1st Ed.; Pergamon Press: Oxford, 1990; 20.
3. Williams, D.F. Biomaterials and Biocompatibility—An Introduction. In *Fundamental Aspects of Biocompatibility*; Williams, D.F., Ed.; CRC Press: Boca Raton, FL, 1981; Vol. I, 2–5.
4. Kasemo, B.; Lausmaa, J. Biomaterials from a Surface Science Perspective. In *Surface Characterization of Biomaterials*; Ratner, B.D., Ed.; Elsevier: New York, 1988; 15–35.
5. Bundy, K.J. Corrosion and Other Electrochemical Aspects of Biomaterials. Crit. Rev. Biomed. Eng. **1994**, *22* (3/4), 139–251.
6. Bumgardner, J.D.; Lucas, L.C.; Johansson, B.I. In Vitro and In Vivo Evaluations of Base Metal Dental Casting Alloys. In *Encyclopedia of Biomaterials and Bioengineering*; Wise, D.L., Trantolo, D.J., Altobelli, D.E., Yaszemski, M.J., Gresser, J.D., Schwartz, E.R., Eds.; Marcel Dekker, Inc.: Chicago, IL, 1995; Vol. II, 1739–1763.
7. Hench, L.L. Ceramics, Glasses and Glass–Ceramics. In *Biomaterials Science—An Introduction to Materials in Medicine*; Ratner, B.D., Hoffman, A.S., Schoen, F.J., Lemons, J.E., Eds.; Academic Press: San Diego, CA, 1996; 73–84.

FURTHER READING

Annual Book of ASTM Standards vol 13.01: Medical Devices; Emergency Medical Services; American Society for Testing and Materials International: Conshohocken, PA, 2001.

Black, J. *Biological Performance of Materials: Fundamentals of Biocompatibility*; Marcel Dekker, Inc.: New York, 1992.

Bronzio, J.D., Ed. In *The Biomedical Engineering Handbook*; CRC Press: Boca Raton, FL, 1995.

Craig, R.G. *Restorative Dental Materials*, 9th Ed.; Mosby: St. Louis, MO, 1993.

Dumitriu, S., Ed.; In *Polymeric Biomaterials*, 2nd Ed.; Marcel Dekker Inc.: New York, 2002.

Greco, R.S. Ed. *Implantation Biology: The Host Response and Biomedical Devices*; CRC Press: Boca Raton, FL, 1994.

Johansson, B.I.; Lucas, L.C.; Bumgardner, J.D.; Metal Release from Casting Alloys for Fixed Prostheses. In *Encyclopedia of Biomaterials and Bioengineering*; Wise, D.L., Trantolo, D.J., Altobelli, D.E., Yaszemski, M.J., Gresser, J.D., Schwartz, E.R., Eds.; Marcel Dekker, Inc.: Chicago, IL, 1995; Vol. II, 1765–1783.

Ong, J.L.; Chan, D.C.N. Hydroxyapatite and Their Use as Coatings in Dental Implants: A Review. Crit. Rev. Biomed. Eng. **1999**, *28*, 667–707.

Park, J.B.; Lakes, R.S. *Biomaterials: An Introduction*, 2nd Ed.; Plenum Press: New York, 1992.

Petty, W. *Total Joint Replacement*; WB Saunders Company: Philadelphia, PA, 1991.

Ratner, B.D.; Hoffman, A.S.; Schoen, F.J.; Lemons, J.E., Eds.; In *Biomaterials Science—An Introduction to Materials in Medicine*; Academic Press: New York, 1996.

von Recum, A.F., Ed. In *Handbook of Biomaterials Evaluation: Scientific, Technical and Clinical Testing of Implant Materials*; Macmillan Publishing: New York, 1986.

Wise, D.L., Trantolo, D.J., Altobelli, D.E., Yaszemski, M.J., Gresser, J.D., Schwartz, E.R., Eds. In *Encyclopedia of Biomaterials and Bioengineering*; Marcel Dekker, Inc.: Chicago, IL, 1995.

Thermal Conductivity of Foods

Yonghee Choi
Kyungpook National University, Taegu, South Korea

Martin R. Okos
Purdue University, West Lafayette, Indiana, U.S.A.

INTRODUCTION

Knowledge of thermal conductivity of food substances is essential to researchers and designers for predicting the drying rate or temperature distribution within foods of various compositions when subjected to different drying, heating, and cooling conditions in the field of food engineering. This information is also necessary for the optimization design of heat transfer equipment, dehydrating and sterilizing apparatus. With the increasing amount of food preparation in industry and food institutions, the importance of such fundamental data also increases. General mathematical models to predict the thermal conductivity of food products based on temperature, composition, and structure of food would be valuable for engineers and scientists. Previous investigators have determined, to a limited extent, some of these values for tomato products and fruit juices. Riedel[1] measured thermal conductivity of fruit juices such as apple, grape, and pear of various water contents at temperatures of 20 and 80°C. Dickerson[2] reported the thermal properties such as thermal productivity, thermal diffusivity, density, and specific heat of some fruit juices. Sweet and Haugh,[3] using a probe method, measured thermal conductivities of cherry tomato and other fruits at room temperatures. Bhowmik and Hayakawa[4] measured thermal diffusivity and density of cherry tomato at 26°C.

Thermal property data have been collected by several investigators. Qashou, Vachon, and Touloukian[5] presented a preliminary compilation of experimental thermal conductivity data for foods and food products including fruit juices. Polley, Snyder, and Kotnour[6] compiled the thermal property data of various foods. Existing thermal properties of tomato juice products are primarily for room temperature, and very limited information is available on the effect of concentration and temperature. The thermal conductivity is affected by composition, density, and temperature.

Several researchers have developed mathematical models, which can be used to predict the thermal conductivities of food products. However, they are for specific foods and do not apply to all the physical situations. When this property is needed for various process conditions, the most efficient and practical way to obtain them is by models based on the process conditions. In general, composition and temperature are the main factors or process conditions affecting this property.

The overall objective of this study is to develop general models to predict the thermal conductivities of food products based on the weight fractions and the thermal conductivities of major pure components. The thermal conductivity values predicted by the proposed models will be compared to literature and experimental thermal conductivity data of foods. In order to accomplish the overall objective, it will be necessary:

1. To measure the thermal conductivity of major component of food products.
2. To determine the effects of ice during freezing processes on the thermal conductivities of food products.
3. To develop general mathematical models to predict the thermal conductivities of food products based on the thermal conductivities of each major component for the temperature range of -40–150°C.

METHODOLOGY

Basic Theory of Line Heat Source Probe

The line heat source probe was employed for the determination of thermal conductivity and thermal diffusivity simultaneously. This method has been used in recent years for the determination of thermal conductivity and thermal diffusivity of silicone rubber.[7] The thermal conductivity probe approximates a line heat source. The theory of this method has been reviewed by Van der Held and Drunen,[8] Hooper and Lepper,[9] Nix et al.,[7]

Encyclopedia of Agricultural, Food, and Biological Engineering
DOI: 10.1081/E-EAFE 120006988
Copyright © 2003 by Marcel Dekker, Inc. All rights reserved.

and Reidy and Rippen.[10] The theory is based on the fact that the temperature rise at a point close to a line heat source, in a semi-infinite solid, subjected a step change heat source, is a function of time, the thermal properties of the solid, and the source strength. The expression from which the thermal conductivity may be obtained is:

$$K = \frac{Q}{4\pi(T_2 - T_1)} \left| \begin{matrix} t_2 - t_0 \\ t_2 - t_0 \end{matrix} \right.$$

A time correction factor, b, was introduced to correct the effect of finite heater diameter and finite heat resistance between the heat source and the sample.[8] Details of theory are given in the paper by Nix et al.[7] and Baghe-Khandan.[11] After an initial transient period, a plot of temperature vs. the logarithm of time is linear until heat penetrates to the sample boundary, so that its slope, $Q/4\pi K$, is used for the calculation of thermal conductivity.

$$\frac{\Delta T}{\Delta} = b = Q/4\pi K$$

Experimental Equipment

The probe used in this study, as shown in Fig. 1, was developed by Baghe-Khandan[12] and consists of an insulated constantan heater wire (0.0076 cm in diameter) and an insulated chromel–constantan thermocouple wire (0.0076 cm in diameter) in a sewing needle (2.54 cm–3.81 cm in length and 0.0254 cm–0.127 cm in diameter). The wires were placed outside a sewing needle through the needle eye by gluing to the needle side with super glue and connected to their leads. The needles were fixed to a commercial thermocouple connector. The wires were covered with a small amount of epoxy glue for protection except the point of thermocouple junction in order to eliminate the error due to glue.

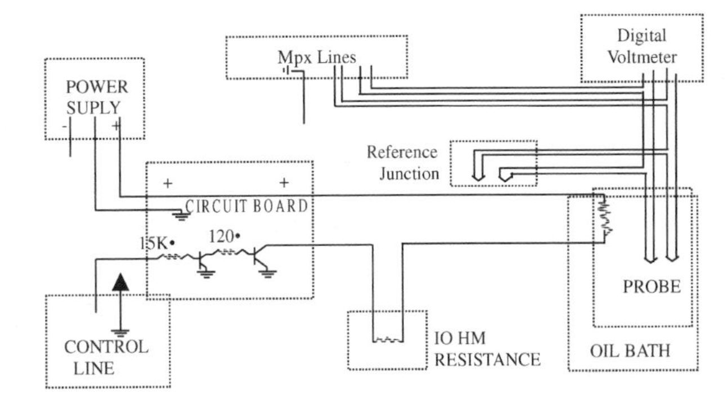

Fig. 2 The electrical circuit of thermal conductivity and thermal diffusivity measurement apparatus.

A d.c. power supply was used to provide power to the heater wire. A digital multimeter was used to check the current and voltage, measured to the nearest 0.1 mV during the data collection. The system was computerized to control the power (on or off) and collect all the data. The current to the probe heater wire was controlled by a switchboard activated by logic level signals. The signal from the probe thermocouple was amplified by a factor of 1000 and then sent to a 1-Hz low pass filter to eliminate the possible high frequency noise. The signal from the filter was transmitted to a solid state multiplexer, then to the analog-digital converter. The signals were read by a digital computer, which calculated thermal conductivity, thermal diffusivity, temperature, temperature rise, and test time. The electrical circuit of the apparatus is shown in Fig. 2.

To overcome the problem of water boiling at temperatures above 90°C, a sample holder, as shown in Fig. 3, was used. Brass cylinders having 1.905 cm inside diameter and 4.445 cm length were used in this experiment. A probe and lid was mounted to the pressure vessel after filling a vessel (16 mL) with samples and with

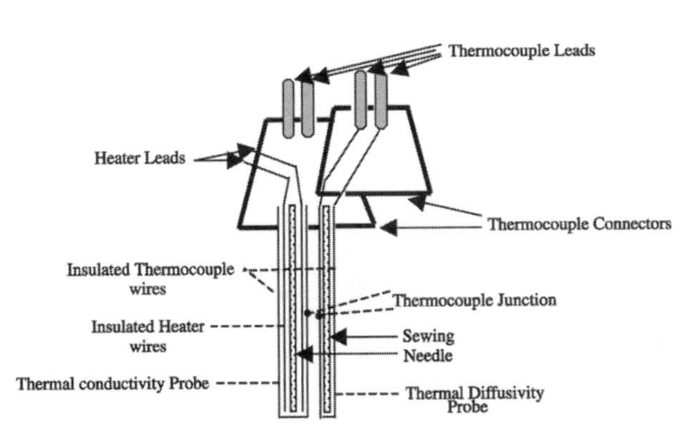

Fig. 1 A schematic of thermal conductivity and thermal diffusivity.

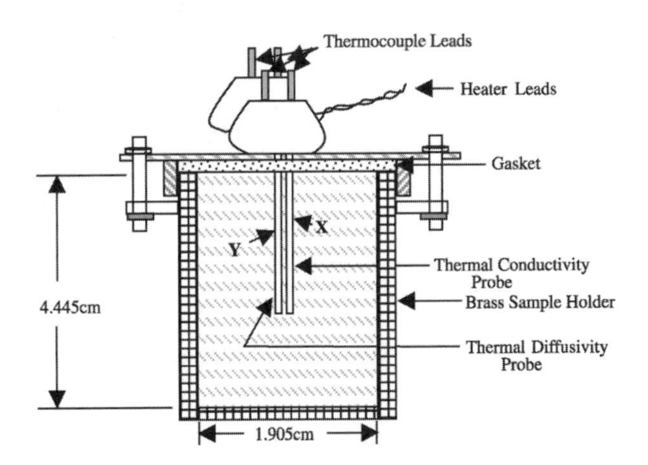

Fig. 3 A schematic of pressure sample holder.

0.7 gr glass wool to prevent convection. Epoxy glue was used to seal around the lid.

Sample Preparation

The major components of milk protein are casein and whey protein. The casein is present in the form of micelles made up of the various components of casein bounded together as calcium caseinate and complexed further with calcium, phosphate, magnesium, and citrate ions. Therefore, salt–casein serum was made by dialyzing skim milk with a salt solution using a 50,000 molecular weight cut-off membrane. It was dialyzed for 48 hr with the salt solution being changed every 6 hr. Then by placing it in a freeze dryer for 48 hr, salts–casein powder was obtained. Proteins remaining after the casein has been removed from skim milk are known as whey protein or milk serum proteins. Therefore, for whey protein powder, after precipitating skim milk by 0.5 N HCl at pH = 4.6 through the centrifuge at 2400 rpm for 15 min, the casein was removed. Then, it was dialyzed with a 3500 molecular weight cut-off membrane for 48 hr with distilled water being changed every 6 hr. Finally, it was placed in a freeze dryer for 48 hr, and whey protein powder was obtained. The schematic preparation process of casein–salts powder and whey protein powder is shown in Fig. 4. Meat protein was prepared by the similar method from ground beef. Egg albumin powder for egg while protein and gluten powder for plant protein were

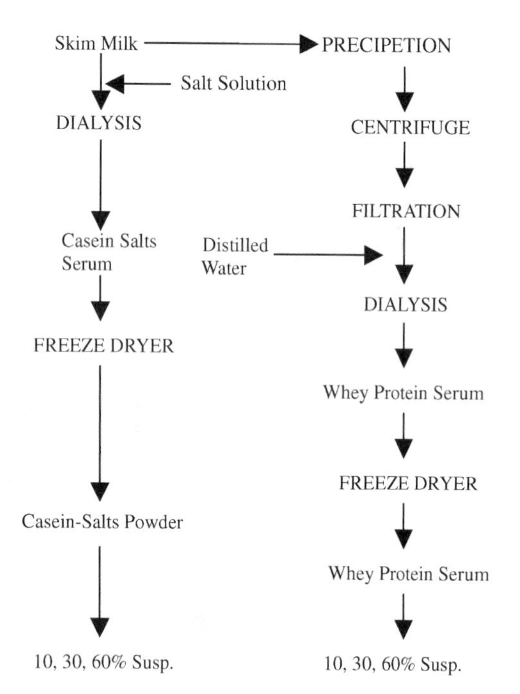

Fig. 4 Schematic process for casein-salts powder and whey protein powder.

Table 1 Quantities of chemicals for making milk salt solution

Chemicals	Quantity (g)
KH_2PO_4	1.580
K_3 citrate–H_2O	0.508
Na_3 citrate–$5H_2O$	2.120
K_2SO_4	0.180
$CaCl_2$–$2H_2O$	1.320
Mg_3 citrate–H_2O	0.502
K_3CO_3	0.300
KCl	1.078

purchased from Fisher Scientific Company and Sigma Chemical Company, respectively.

For a milk fat, butter oil was used for the measurement in this study. Butter oil is a refined product made by separating the milk fat from high fat cream. The product contains only small amounts of moisture and protein. The composition is 99.5% milk fat, 0.2% moisture, and 0.3% protein. Commercial oil products, such as corn oil for grain foods, vegetable oil for vegetable foods, and lard for meat foods, were used for the measurement of the properties of the fat component of food products.

Dextrose powder was purchased from Fisher Scientific Company. Lactose powder for milk carbohydrates was also purchased from Pfanstichl Laboratories, Inc. Pectin powders from by Sigma chemical Company and microcrystalline cellulose powder from FMC Corporation were used for the property measurements of fiber materials. Commercial pure cane sugar and corn starch powder were also used.

The salt components for all the food products are comprised of some of the chlorides, phosphates, citrates, and sulfates along with such elements as sodium, potassium, calcium, magnesium, and so on. The amounts

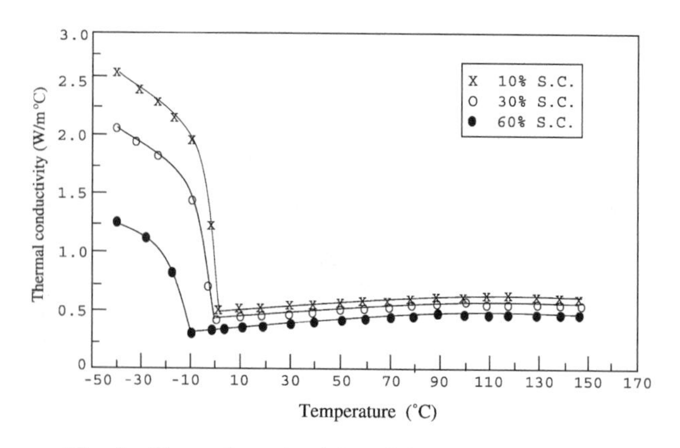

Fig. 5 Thermal conductivity of dextrose suspensions.

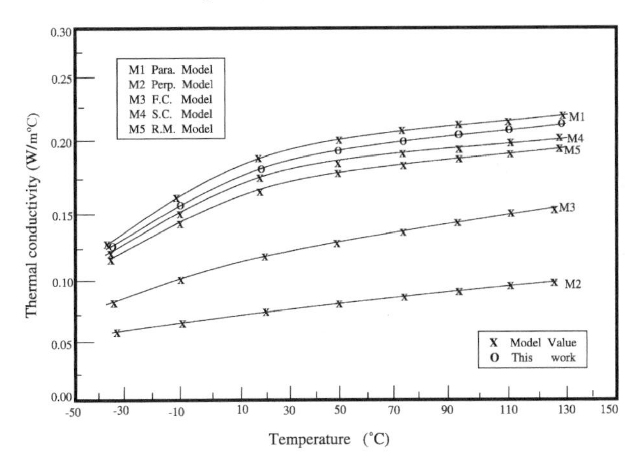

Fig. 6 Thermal conductivity of dextrose powder.

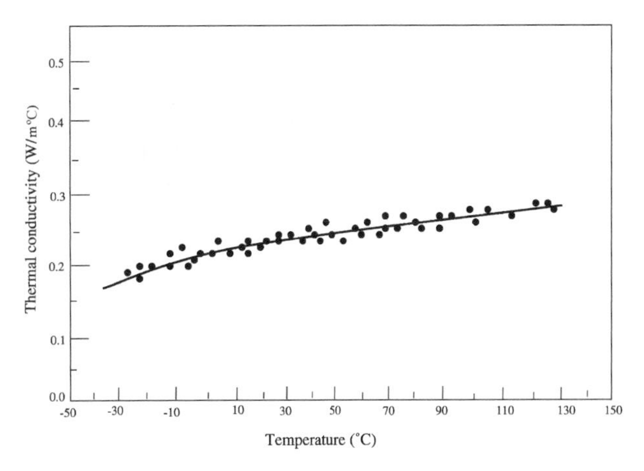

Fig. 7 Thermal conductivity of dextrose solids.

of each salt component are very slightly different in different kinds of food products. In addition, the percentage of salt components among food products is very small compared to the other major components of food products. Therefore, the milk salts, which have all the above salt components and can be easily prepared, were assumed as a basic salt for all kinds of food products. Jenness[13] reported on the preparation of a salt solution, which simulates milk salt solution, by using a dry blended mixture, as shown in Table 1.

Prediction of Unfrozen Water Fractions

Staph and Woolrich[14] proposed that a gradual depression of the freezing point in the unfrozen product fraction occurs throughout the freezing process. Based on the chemical potential of pure solute and pure liquid, the equation for freezing point depression can be derived:

$$\frac{\lambda}{R}\left[\frac{1}{T_0} - \frac{1}{T}\right] = \ln X_A$$

Table 2 Comparison of thermal conductivities of powder samples to the proposed models

Powder samples	Model no.	SE	SE (%)
Dextrose	M1	0.0087	4.70
	M2	0.1113	59.91
	M3	0.0674	36.30
	M4	0.0112	6.04
	M5	0.0191	10.34
Whey protein	M1	0.0093	6.21
	M2	0.0908	60.70
	M3	0.0574	38.33
	M4	0.0105	7.25
	M5	0.0213	14.26

Therefore, the unfrozen water fraction of a food system can be predicted at a given temperature below the initial freezing point.

Modeling

Based on the theoretical view of the rate of heat transfer to the material, the structural arrangement of the constituents should be considered in the model to predict the properties. The model proposed for parallel structural arrangement of two-component system:

$$K = K_s X_s^v + K_f X_f^v \quad (1)$$

For perpendicular structural arrangement:

$$K = \left[\frac{X_s^v}{K_s} + \frac{X_f^v}{K_f}\right]^{-1} \quad (2)$$

For fluid continuous system of material:

$$K = K_f\left[\left(1 - 2X_s^v\frac{1-\frac{K_s}{K_f}}{2+\frac{K_s}{K_f}}\right) \Big/ \left(1 + X_s^v\frac{1-\frac{K_s}{K_f}}{2+\frac{K_s}{K_f}}\right)\right] \quad (3)$$

Table 3 Thermal conductivity models of water and ice

	Property models	SE	SE (%)
Water	$K = 5.7109\times10^{-1} + 1.7625\times10^{-3}T - 6.7036\times10^{-6}T^2$	0.0028	0.45
Ice	$K = 2.2196 - 6.2489\times10^{-3}T + 1.0154\times10^{-4}T^2$	0.0078	0.79

Table 4 Thermal conductivity models of pure components of foods

Pure components	Thermal conductivity models ($W\,m^{-1}{}^{\circ}C^{-1}$)	SE	SE (%)
Albumin	$K = 1.8068 \times 10^{-1} + 1.1462 \times 10^{-3}T - 2.6888 \times 10^{-6}T^2$	0.0086	3.84
Casein	$K = 1.7138 \times 10^{-1} + 1.1234 \times 10^{-3}T - 2.4592 \times 10^{-6}T^2$	0.0066	2.98
Whey protein	$K = 1.8627 \times 10^{-1} + 1.2444 \times 10^{-3}T - 2.9499 \times 10^{-6}T^2$	0.0060	2.57
Meat protein	$K = 1.6266 \times 10^{-1} + 1.1726 \times 10^{-3}T - 2.3735 \times 10^{-6}T^2$	0.0129	5.49
Gluten	$K = 1.8671 \times 10^{-1} + 1.3229 \times 10^{-3}T - 3.4197 \times 10^{-6}T^2$	0.0108	4.58
Milk fat	$K = 1.7809 \times 10^{-1} - 2.4381 \times 10^{-4}T - 5.5169 \times 10^{-7}T^2$	0.0020	1.23
Vegetable oil	$K = 1.8224 \times 10^{-1} - 2.1949 \times 10^{-4}T - 7.3411 \times 10^{-7}T^2$	0.0041	2.39
Lard	$K = 1.8220 \times 10^{-1} - 2.0565 \times 10^{-4}T - 7.3267 \times 10^{-7}T^2$	0.0020	1.18
Corn oil	$K = 1.8109 \times 10^{-1} - 2.0145 \times 10^{-4}T - 7.8395 \times 10^{-7}T^2$	0.0035	2.15
Dextrose	$K = 2.1277 \times 10^{-1} + 1.2946 \times 10^{-3}T - 3.9135 \times 10^{-6}T^2$	0.0133	5.19
Lactose	$K = 1.9898 \times 10^{-1} + 1.4760 \times 10^{-3}T - 4.5666 \times 10^{-6}T^2$	0.0079	3.16
Sugar	$K = 2.0456 \times 10^{-1} + 1.3774 \times 10^{-3}T - 4.2079 \times 10^{-6}T^2$	0.0066	2.62
Starch	$K = 1.9001 \times 10^{-1} + 1.3698 \times 10^{-3}T - 4.4318 \times 10^{-6}T^2$	0.0125	5.33
Cellulose	$K = 1.7944 \times 10^{-1} + 1.3698 \times 10^{-3}T - 3.2086 \times 10^{-6}T^2$	0.0110	4.92
Pectin	$K = 1.8644 \times 10^{-1} + 1.2914 \times 10^{-3}T - 3.1286 \times 10^{-6}T^2$	0.0112	4.78
Milk salt	$K = 3.2962 \times 10^{-1} + 1.2914 \times 10^{-3}T - 2.9070 \times 10^{-6}T^2$	0.0083	2.15

For a solid continuous system of material:

$$K = K_s\left[3\frac{K_f}{K_s} + 2X_s^v\left(1\frac{K_f}{K_s}\right)\right] \Big/ \left[3 - X_s^v\left(1\frac{K_f}{K_s}\right)\right] \quad (4)$$

For a random mixture of the two phases:

$$K = \frac{1}{4}[3X_s^v - 1]K_s + (3X_f^v - 1)K_f + \left[(([3X_s^v - 1]K_s + [3X_f^v - 1]K_f)^2 + 8K_sK_f\right]^{1/2} \quad (5)$$

Another type of models based on a packed bed system of material have been proposed by Yagi and Kunii,[15] Kunii and Smith,[16] Okazaki, Ito, and Toei,[17] and Chen and Heldman.[18] Generally speaking, these models are complicated. A number of experiments and calculations are required to obtain the necessary input data. They are less applicable for practical applications, but may provide insights to basic researchers.

RESULTS AND DISCUSSION

Thermal Conductivity

Thermal Conductivities for pure component suspensions were measured at different concentrations such as 10%, 30%, 60%, then at temperature range of − 40–150°C. For dextrose, one of the samples, the experimental data is shown in Fig. 5. Thermal conductivities of sample suspensions at the temperature below initial freezing point were much higher than that at the temperature above initial freezing point. Fats are relatively poor conducts of heat. Thermal conductivity of milk fat in liquid state at the melting point of 34°C was 6.1% lower than that in solid state. Thermal conductivities of whey protein, milk fat, lactose, starch, and milk salts were measured in the test run of heating, cooling, reheating and recooling. The cycling results on milk fat, lactose and milk salts suspensions show that the heating and cooling did not have a significant effect on the thermal conductivity of samples. For whey protein suspension, thermal conductivity had a 4.9%–11.7% lower value in the recycling process below 70°C because of thermal denaturation. Thermal denaturation of protein, which usually occurs at 60–70°C, is a radical change in the protein structure. This change had a decreasing effect on the thermal conductivity. In the case of polysaccharides, starch was gelatinized at the temperature between 62 and 70°C. When starch was gelatinized, the crystalline region was disrupted and would gradually disappear. With this reason, thermal conductivity of corn starch suspension was 3.6%–10.8% lower in the recycling process below 60°C.

Table 5 Average molecular weights of pure components used to predict freezing point depression

Components	Molecular weight	Components	Molecular weight
Albumin	45,000	Dextrose	180
Whey protein	30,000	Lactose	342
Casein	1,000,000	Sugar	342
Meat Protein	350,000	Starch	30,0000
Gluten	60,000	Cellulose	30,0000
Milk Salt	158	Pectin	50,000

Table 6 Group models of major components of foods

Thermal property	Major component	Group models temperature function	SE	SE (%)
K (W m^{-1}°C^{-1})	Protein	$K = 1.7881 \times 10^{-1} + 1.1958 \times 10^{-3}T - 2.7178 \times 10^{-6}T^2$	0.012	5.91
	Fat	$K = 1.8071 \times 10^{-1} - 2.7604 \times 10^{-3}T - 1.7749 \times 10^{-7}T^2$	0.0032	1.95
	Carbohydrate	$K = 2.0141 \times 10^{-1} + 1.3874 \times 10^{-3}T - 4.3312 \times 10^{-6}T^2$	0.0134	6.42
	Fiber	$K = 1.8331 \times 10^{-1} + 1.2497 \times 10^{-3}T - 3.1683 \times 10^{-6}T^2$	0.0127	5.55
	Ash	$K = 3.2962 \times 10^{-1} + 1.4011 \times 10^{-3}T - 2.9069 \times 10^{-6}T^2$	0.0083	2.15

Prediction Model

The experimental thermal conductivity values of dextrose powder were compared with the five proposed model values at the temperature range of − 40–150°C, as shown in Fig. 6. It was found that the parallel model has a less error of 4.7% than the other models. The statistical comparison of the experimental thermal conductivity values of prepared powder sample to the five proposed model values are listed in Table 2. Based on these results and simplicity in models, the parallel model was proposed in this study for the prediction of properties of foods.

The thermal conductivity of a sample solid at a given temperature was determined by the following equation because a suspension was composed of a pure component and water.

$$K_s = \frac{K - K_w X_w^v}{X_s^v}$$

As the calculated thermal conductivities of pure component solids from three different solid content suspensions showed that they were not dependent linearly on temperature, a quadratic model was proposed for thermal conductivity of pure solids. Based on the theoretical view of the rate of heat transfer to the parallel structured arrangement of the constituents, it was found that the values of thermal conductivity of a material were expressed as the sum of the each property value proportional to the fraction of each component. The thermal conductivity of dextrose solids was plotted in Fig. 7.

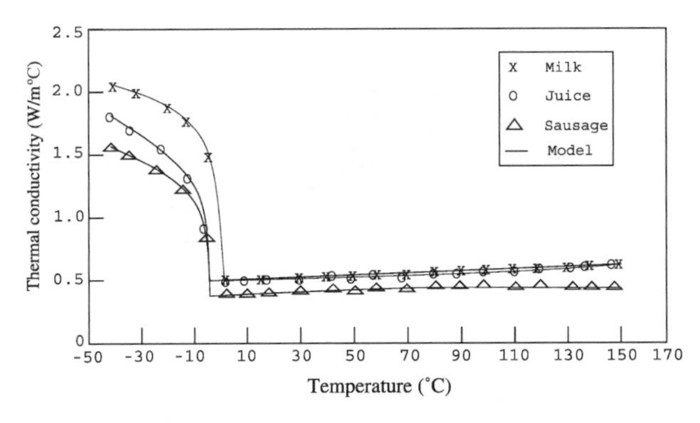

Fig. 8 Thermal conductivity of food product samples.

The coefficients in the proposed thermal conductivity model of each pure component solid at the temperature range of − 40–150°C were determined by the Optimization Computer Subroutine from the calculated values. Quadratic models for the thermal conductivities of liquid water and ice were developed, as shown in Table 3, and the obtained thermal conductivity models of pure components of foods are listed in Table 4.

For the frozen foods, the unfrozen water fractions at the different temperatures below the initial freezing point were determined from the product freezing point depression using the average molecular weights given in Table 5. For practical use, group models given in Table 6 for the thermal conductivities of major components, such as protein, fat, carbohydrate, fiber, and ash, are more applicable than the models of each pure component in predicting the thermal conductivities of food products, because the weight fractions of these major components in food samples are more conveniently obtainable factors than the weight fractions of each pure component within each group. Therefore, the group models were also developed within 6% error.

Comparison of Model Values

A comparison of the measured value of thermal conductivity for three foods, with using pure components

Table 7 Comparison between model and experimental values of thermal conductivity for food products

Thermal conductivity	Foods	SE	SE (%)
K	Evaporated milk	0.0289	3.52
	Concentrated orange juice	0.0227	2.94
	Bratwurst sausage	0.0291	4.54

was shown in Fig. 8. The thermal conductivity values by the proposed model of the food product samples were predicted within 4.6% error of the measured values as shown in Table 7.

REFERENCES

1. Riedel, L. Thermal Conductivity Measurement on Sugar Solution, Fruit Juice and Milk. Chem.-Ing.- Tech. **1949**, *21* (17), 340–341.
2. Dickerson, R.W. Thermal Properties of Foods. *The Freezing Preservation of Foods*, 4th Ed.; The AVI Publishing Company, Inc.: Westport, CT, 1968.
3. Sweet, V.E.; Haugh, C.G. A Thermal Conductivity Probe for Small Food Samples. Trans. ASAE **1974**, *17* (1), 56–58.
4. Bhowmik, S.R.; Hayakawa, K.I. A New Method for Determining the Apparent Thermal Diffusivity of Thermally Conductive Food. J. Food Sci. **1979**, *44* (2), 469–474.
5. Qashou, M.S.; Vachon, R.I.; Touloukian, Y.W. Thermal Conductivity of Foods. ASHRAE Semi-Annual Meeting, New Orleans, LA, 1972.
6. Polley, S.I.; Snyder, O.P.; Kotnour, P. A Compilation of Thermal Properties of Foods. Food Technol. **1980**, *34*, 76–94.
7. Nix, G.H.; Lowery, G.W.; Vachon, R.I.; Tauger, G.E. Direct Determination of Thermal Diffusivity and Conductivity with a Refined Line Source Technique. Prog. Aeronautics Astronautics **1967**, *20*, 865–878.
8. Van der Held, E.F.M.; Drunen, F.G. A Method of Measuring the Thermal Conductivity of Liquids. Physica **1949**, *15*, 865–881.
9. Hooper, F.C.; Lepper, F.R. Transient Heat Flow Apparatus for the Determination of Thermal Conductivities. ASHVE Trans. **1950**, *56*, 309–324.
10. Reidy, G.A.; Rippen, A.L. Methods for Determining Thermal Conductivity for Foods. Abstracts of Papers, ASAE Meeting, St. Joseph, MI, 1969; ASAE 69-383.
11. Baghe-Khandan, M.S. Experimental and Mathematical Analysis of Coking Effects on Thermal Conductivity of Beef. Ph.D. Thesis, Purdue University, West Lafayette, IN, 1978.
12. Baghe-Khandan, M.S.; Choi, Y.; Okos, M.R. Improve Line Heat Source Thermal Conductivity Probe. J. Food Sci. **1981**, *46* (5), 1432–1450.
13. Jenness, R.; Koops, J. Preparation and Properties of a Salt Solution Which Simulates Milk Ultrafiltrate. Neth. Milk Dairy J. **1962**, *16* (3), 153–164.
14. Staph, M.E.; Woolrich, W.R. Specific and Latent Heats of Foods in the Freezing Zone. Refrig. Eng. **1981**, *59*, 1086–1089.
15. Yagi, S.; Kunii, D. Studies on Effective Thermal Conductivities in Packed Beds. AICHE J. **1957**, *3* (3), 373–381.
16. Kunii, D.; Smith, J.M. Heat Transfer Characteristics of Porous Rocks. AICHE J. **1960**, *6* (1), 71–78.
17. Okazaki, M.; Ito, I.; Toei, R. Effective Thermal Conductivities of Wet Granular Materials. AICHE Symposium Series No. 163 **1977**, *73* (1), 164–176.
18. Chen, A.C.; Heldman, D.R. An Analysis of the Thermal Properties of Dry Food Powder in a Packed Bed. Trans. ASAE **1972**, *15* (5), 951–955.

Thermal Death Time

Vijay K. Juneja
Lihan Huang
United States Department of Agriculture (USDA), Wyndmoor, Pennsylvania, U.S.A.

T

INTRODUCTION

For centuries, humans have struggled to find suitable technologies for long-term preservation of perishable foods. For a long time, drying or dehydration, a technology learnt from nature, was probably the only method available. However, the quest for safe, nutritious, and organoleptic foods has never ended. Nicholas Appert, a Frenchman, deserves the most of the credit for canning. In 1810, Nicholas Appert for the first time demonstrated the possibility of preserving perishable foods in glass jars and bottles by heat. In the early days, however, canning or thermal processing was more an art than science. Modern canning did not come into birth until a century later in 1920s when Bigelow and Ball incorporated inactivation kinetics into thermal processing and established a scientific foundation for safe canning. This article introduces a fundamental parameter—thermal death time (TDT) in kinetic analysis of thermal inactivation and thermal process calculation.

BACKGROUND

The use of adequate heat treatment to destroy pathogenic and spoilage micro-organisms is the most effective food preservation process in use today and has been for decades. Heat treatment designed to achieve a specific lethality for food-borne pathogens is a critical control point and fundamentally important to assure the shelf life and microbiological safety of thermally processed foods. The use of relatively mild heat treatment (pasteurization) is widely accepted as an effective means for destroying all nonspore-forming pathogenic micro-organisms and significantly reducing the number of natural spoilage microflora, thereby extending the shelf life of such products. The heat is applied at very high temperatures (such as 121°C or 250°F) for a short time to render food free of viable micro-organisms that are of public health concern or capable of growing in the food at temperatures at which the food is likely to be stored under normal nonrefrigerated storage conditions. Specifically, the objective of sterilization is to reduce the probability of an organism's survival in a food to an acceptably low level.

DEFINITION

In early studies, the heat resistance of bacterial spores was quantified in order to calculate thermal processing times for canned foods and is the basis of all modem thermal process calculations. The basic concept associated with this technology was TDT and was defined in a classic paper by Bigelow and Esty[1] as "the length of time at different temperatures necessary to completely destroy a definite concentration of spores in a medium of known hydrogen concentration." According to Ball and Olson,[2] TDT is defined as the time needed under a constant temperature to completely destroy all micro-organisms in a population capable of causing food spoilage. In other words, it is the time point at which no microbial growth can be observed in the test media or food samples as tested using scientifically proven methods. It can be noted that the temperature is kept constant and the time required for killing all organisms is determined. Theoretically, TDT is completely different from the decimal reduction time or D value, which is the time needed to achieve a 90% or one log reduction in a cell population tested under a constant temperature. The higher the initial microbial population in a food, the longer the processing/heating time at a given temperature that is required to achieve a specific lethality of micro-organisms. Accordingly, the thermal process design is based on the expected microbial load in the raw product. For TDT to have meaning, it is necessary to specify number or concentration of organisms being heated, suspending liquid (chemical and physical properties), age of culture, medium in which the organisms are grown (chemical and physical properties), whether or not the organisms are subcultured after heating, manner of

Mention of brand or firm names does not constitute an endorsement by the U.S. Department of Agriculture over others of a similar nature not mentioned.

Encyclopedia of Agricultural, Food, and Biological Engineering
DOI: 10.1081/E-EAFE 120007094
Copyright © 2003 by Marcel Dekker, Inc. All rights reserved.

heating (whether agitated or not), container in which the heating is done and the heating medium.[2]

DETERMINATION OF TDT

The first step in calculating thermal processing time for canned foods involves determining the TDT of the critical organism being destroyed by the process. When test samples inoculated with a known initial concentration or number of a target micro-organism are subjected to heating under a constant temperature, samples with heating times less than the TDT will survive the treatment and recover if grown under suitable conditions, while all the samples with heating times greater than the TDT will show no signs of microbial growth. Obviously, TDT is temperature dependent. A TDT curve can be constructed by plotting log time of complete destruction against temperature. A TDT curve is the plot on a semilog paper depicting the best fit straight line above all positive points and below as many as possible negative points with time on the log scale and the heating temperature on the linear scale.[2] Fig. 1 is a typical TDT curve using the data listed in Table 1.[3] The straight line in Fig. 1 is obtained by linear regression of the log heating time (in bold type) and the heating temperature.

The log-linear TDT model assumes first-order kinetics, i.e., a constant microbial population is inactivated in each successive time period. This model has a long history of success, modeling the thermal inactivation of spores in retort processing. The approach assumes that all of the cells or spores in a population have the same heat resistance, and it is merely the chance occurrence of a quantum of heat energy impacting a heat sensitive target in a cell or spore that determines the death rate.

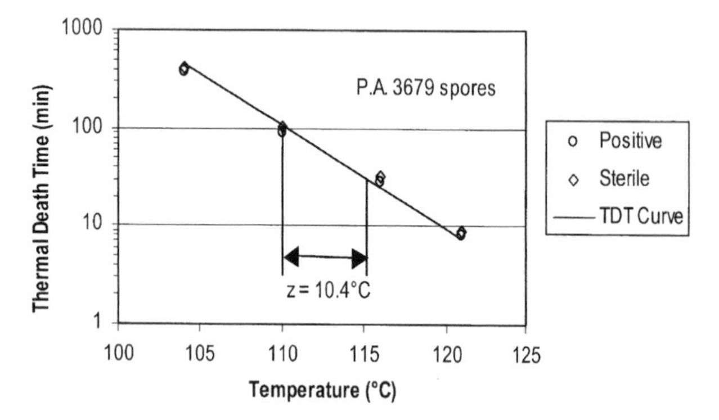

Fig. 1 Thermal destruction curve for PA 3679 spores suspended in pureed peas.

As the apparent or measured heat resistance of spores depends upon the heating menstruum, phosphate buffer at pH 7 containing spore suspension was used, as a suitable reference medium, in early studies to assess the heat resistance prior to determining the heat resistance in foods. Esty and Meyer[4] reported the following time/temperature combinations for sterilization of phosphate buffer (pH 7) supplemented with billions of *C. botulinum* spores: 120°C/4 min; 115°C/10 min; 110°C/33 min; 105°C/100 min; or 100°C/330 min. The authors also reported that: 1) the linear decline in the log number (by $>$ 10 log cycles) of survivors occurred with increased heating time; 2) heated surviving spores may germinate after a prolonged period; and 3) the heat resistance of the spores was higher in juices from some canned foods than in the reference phosphate buffer. This may be attributed to differences in composition (more solids in juices) among the substrates. In the following years, the concept of z-value and F-value associated with the thermal destruction of micro-organisms was introduced. The z-value is expressed as the °C or F required for the TDT curve to traverse one log cycle and is mathematically equal to the slope of the TDT curve. The z-value provides information on the relative resistance of an organism to different destructive temperatures. A value of $z = 8$°C implies that if the D-value at 68°C is 0.3 min, then at 60°C it will be 3 min, and at 52°C it will be 30 min. These time/temperature combinations are considered equivalent processes. The canning industry has adopted a D-value of 0.2 min at 121°C and a z-value in the order of 10°C (18°F) as a standard for calculating the required thermal process.[5] The F-value is used to express the time in minutes necessary to destroy a specific number of microbial spores or cells at a specified temperature. The minimum thermal process applied to commercial low acids canned food is 12*D*, based on the most heat resistant *C. botulinum* spores.[5] Accordingly, the F-value for *C. botulinum* in low acids canned food is 2.45 equivalent minutes at 121°C or 250°F (reference temperature) with a z-value of 10°C. This follows that processing for 2.45 min at 121°C should reduce the spores to one spore in 1 of 1 billion containers. The canning industry's adherence to the 12*D* concept has always guarded against this deadly pathogen and ensured safe products. It is worth mentioning that this canning industry process was developed after corrections for heating and cooling lags were incorporated in the *C. botulinum* heat resistance data collected in earlier studies.[4]

TDT data can be used to calculate the D-value of a micro-organism using the Halvorson-Ziegler equation,[3] which, in turn, can be used to calculate endpoint micro-organism concentration in the test samples.

T

Table 1 TDT determination

Temperature (°C)	Heating time (min)	Number of tubes tested	Number of positive tubes	Number of sterile tubes	*D* (min)
104	232.1	6	6	0	
	278.5	6	5	1	
	324.9	6	5	1	
	371.4[a]	6	3	3	
	417.8	6	0	6	71.6
110	77.5	6	6	0	
	90.4[a]	6	2	4	
	103.4	6	0	6	
	116.4	6	0	6	17.4
116	17.8	6	6	0	
	21.4	6	4	2	
	25.0	6	2	4	
	28.5[a]	6	1	5	
	32.0	6	0	0	5.50
121	5.0	6	6	0	
	6.0	6	4	2	
	7.0	6	4	2	
	8.0[a]	6	1	5	
	9.0	6	0	6	1.54

Data adapted from Ref. [3].
[a] t_c for each heating time.

The Halvorson-Ziegler equation is expressed as

$$B = 2.303 \log_{10} \frac{N_{Heated}}{N_{Sterile}} \tag{1}$$

$$D = \frac{t_c}{\log_{10} A - \log_{10} B} \tag{2}$$

where t_c is the net heating time after which all the samples are sterile in TDT study; A is the initial number of organisms; B is the final number of organisms; N_{heated} is the total number of samples heated in a TDT study under a constant temperature; and N_{sterile} is the number of sterile samples at t_c.

Existing methods for determining TDT include the tube, pouch (nylon), can, flask, thermoresistometer, and capillary tube methods.[3] The methodology for determining TDT involves sealing a sufficiently high number of bacterial cells in suspension in a tube/container, submerging in a water/oil bath, followed by maintaining at a particular temperature for given time periods after which the tubes are removed and cooled instantaneously. This is either followed by incubation of the heated samples in the same containers at temperatures and atmospheres conducive for the growth of the test organisms, or alternatively, plating the surviving spores/cells after heat exposure on a suitable growth medium before incubation. No growth indicates the lethal effects of heat. These traditional methods for determining TDT suffer from many disadvantages, including time consuming operations, appreciable heating and cooling lags, splashing of contents, flocculation, high initial cost, and hazards of contamination during subculturing.[3] Nevertheless, TDT determined by these methods has been instrumental for guaranteeing the safety of thermally processed food commodities for decades, and will continue to be a fundamental approach for developing and evaluating thermal processes in the future.

REFERENCES

1. Bigelow, W.D.; Esty, J.R. Thermal Death Point in Relation to Time of Typical Thermophilic Organisms. J. Infect. Dis. **1920**, *27*, 602–617.
2. Ball, C.O.; Olson, F.C.W. Bacteriology. *Sterilization in Food Technology*; McGraw-Hill Book Company, Inc.: New York, 1957; Chap. 4, 133–192.
3. Stumbo, C.R. Thermal Resistance of Bacteria. *Thermobacteriology in Food Processing*; Academic Press: New York, 1973; Chap. 7, 79–104.
4. Esty, J.R.; Meyer, K.F. The Heat Resistance of Spores of *Bacillus botulinus* and Allied Anaerobes. XI. J. Infect. Dis. **1922**, *31*, 650–663.
5. Houschild, A.H.W. *Clostridium botulinum*. In *Foodborne Bacterial Pathogens*; Doyle, M.P., Ed.; Marcel Dekker: New York, 1989; 111–189.

Thermal Diffusivity

Edgar G. Murakami
National Center for Food Safety and Technology/Food and Drug Administration, Summit-Argo, Illinois, U.S.A.

INTRODUCTION

Thermal diffusivity is a fundamental property of materials that indicates how fast heat propagates through a sample. It is used to calculate time–temperature distribution in materials undergoing heating or cooling. As many food processing unit operations involve heat transfer, knowledge of thermal diffusivity is required to optimize energy consumption, design equipment, maintain product quality, and to determine and control process parameters. Many food-processing systems require precise control of the process temperature to produce safe and high quality products. Thermally processed foods must maintain a product temperature which is sufficient to destroy microorganisms but does not significantly affect food quality. Quality attributes of foods such as color, vitamin, and texture are temperature dependent and degrade at high temperatures.

The time–temperature distribution in a product during processing is typically evaluated using the following equation:

$$\rho C_p \frac{\partial T}{\partial t} = k\left[\frac{\partial^2 T}{\partial x^2} + \frac{\partial^2 T}{\partial y^2} + \frac{\partial^2 T}{\partial z^2}\right] \qquad (1)$$

where C_p is the specific heat, kJ/kg-°K; ρ the density, kg m^{-3}; k the thermal conductivity, W m^{-1} K^{-1}; T the temperature, °C; t the time, sec; and x, y, z the dimensional position.

To simplify Eq. 1, the material properties are grouped together into the following equation:

$$\alpha = k/\rho C_p \qquad (2)$$

where α is the thermal diffusivity, m^2 sec^{-1}. Although thermal diffusivity is a convenient combination of three physical properties, it portrays a distinct characteristic of the thermal response to a transient heat transfer condition. In contrast, thermal conductivity is associated with steady-state heat transfer and specific heat with heat capacity.

The thermal diffusivity of food materials is influenced by composition and temperature. In foods, water and air are the two most influential of all the components due to their prevalence and extreme thermal property values.[8] For example at 0°C, the thermal conductivity of water, air, and the rest of food components (proteins, fats, carbohydrates, and ash) are approximately 0.6 W m^{-1} K^{-1}, 0.02 W m^{-1} K^{-1}, and 0.2 W m^{-1} K^{-1}, respectively.[8] The density values for air and water are also vastly different from the other food components. Thus, changes in the volume fractions of air and water can significantly alter the thermal diffusivity of food materials. For vegetables, every 1% increase in moisture content corresponds to a 1%–3% increase in thermal diffusivity.[20,28] For cassava, every 1% increase in moisture content corresponds to approximately a 4% increase in thermal diffusivity.[31] The effect of temperature on thermal diffusivity is minimal above freezing. In freezing temperatures, thermal diffusivity can increase several fold since the thermal diffusivity of ice is an order of magnitude higher than that of water.

Thermal diffusivity data are commonly used to determine the temperature profile during transient heat transfer situations. It is used to predict the location and magnitude of cold and hot spots during heating and cooling,[39] evaluate insulation properties of foams,[33] design process parameters, and measure flow rate. However, its application is not limited to processes involving heat transfer. Taking advantage of its strong correlation with moisture content, thermal diffusivity devices have been used to measure moisture content of soil[6,35] and drying rate.[39]

Thermal diffusivity data can be obtained from Eq. 2, published values, models, and direct measurement. Eq. 2 is commonly used because of its simplicity and theoretical consideration. However, this approach has a high degree of uncertainty due to the accumulation of measurement errors from the other properties. Data on thermal diffusivity of food materials can be obtained from the literature on fruits, vegetables, meats and fish[38]; fruits, juices, food components[8]; and dough and bakery products.[34] They can also be calculated from empirical models.[8,37,39] The application of data from published studies and empirical models should be limited to the materials and temperature range used to obtain the data. It is recommended to measure the thermal properties of specific food materials at intended process conditions.

Encyclopedia of Agricultural, Food, and Biological Engineering
DOI: 10.1081/E-EAFE 120006990
Copyright © 2003 by Marcel Dekker, Inc. All rights reserved.

MEASUREMENT TECHNIQUES

There is no universally accepted measurement technique for thermal diffusivity.[15,16,25–27] Selection of an appropriate technique depends on the desired accuracy, complexity of device, type and size of sample, and test conditions. For applications in foods, there are no commercially available instruments for measuring thermal diffusivity. Users still have the burden of making their own device.

The techniques for measuring thermal diffusivity can be grouped into three categories: no heat source, a constant heat source, and a modulated heat source. These techniques involve measuring heat flow through the samples, requiring either a heat source or a heat sink.

No Heat Source

The heat source/sink is located outside the analytical domain. Thermal diffusivity is evaluated only from temperature data. These techniques are popular among researchers due to their simplicity and low cost. However, they often require long test times (e.g., several hours) and a large temperature change in the product. To reduce the time and improve accuracy, heat transfer is enhanced by introducing a heat source inside or outside the sample.

The sample is shaped into a standard geometry, i.e., slab, cylinder, or sphere, and maintained in a constant temperature environment. Thermal diffusivity is calculated by fitting the temperature data to the analytical solution of Eq. 1 or estimated from time–temperature charts.[19] To simplify data analysis, various assumptions are used:

1. Constant temperature change. The temperature anywhere in the sample is assumed to change at a constant rate, eliminating the time variable in Eq. 1. This technique has been used on fruits[12]; orange juice[40]; and bulk foods.[22]
2. Constant slope of log of temperature change vs. time. This approach uses the cold spot temperature history taken during thermal processing. It was used on soursoup[21]; grains[39]; and various food materials.[32]

Constant Heat Source

These devices have a heater and several temperature sensors and they are usually designed by the researchers.

1. Line-heat source. By adding a second temperature sensor to the popular thermal conductivity probe, thermal conductivity and diffusivity can be measured simultaneously.[30] The resulting device is also called a dual probe since the second sensor is usually encased in a separate sheath. Thermal diffusivity is calculated by either fitting a nonlinear equation to the time–temperature data or from the time when the maximum temperature is reached.[6] This probe has been used on saturated soil and various liquids,[5,8] sandy soil,[6] and grains.[24] The apparatus is simple to make and operate. However, in dry bulk samples, (e.g., powders) temperature rise at the second sensor can be difficult to detect; there is also a problem with filling the gap between the heater and the second sensor with the sample. Watanabe[43] designed a single probe device by considering the heater wire as both a heated rod and a temperature sensor. Although the single probe is easy to use, it requires predrilling to make insertion possible in hard samples. Agrawal et al.[1] replaced the linear heat source with a nonintrusive plane heater for flat samples. This procedure requires density and specific heat values for simultaneous determination of thermal diffusivity and conductivity. Mathis Instruments (New Brunswick, Canada) makes a commercial version of a plane heater called the TC probe. The manufacturer states that the test time is as short as 1 sec and can be used on samples with thicknesses as small as 5 mm.
2. Bead thermistor. This technique consists of a spherical heat source that is also used as a temperature sensor. It has been used in biomaterials[4,42] and various food materials.[23]

Modulated Heat Source

The modulated heat source techniques involve either a periodic or pulse heating of one surface of a slab and monitoring the temperature rise on the other side. The energy source may be in the form of light (e.g., laser, high intensity lights, electron beam) or planar electrical heaters. Thermal diffusivity is calculated from data on temperature, phase shift of temperature, and heat emission. Although this technique is popular in engineering materials, it is not commonly used in food materials due to the complexity of data analysis, costly equipment, and high sample temperature. However, recent innovations in equipment design and experimental techniques have been developed that allow its application in food materials. Some of the modulated techniques published in the literature are:

1. Periodic hot wire. A platinum heater wire is heated with a sinusoidal electric current.[17] This technique is attractive to food engineers since the construction is similar to the thermal conductivity probe.

2. Flash technique. A short burst of radiant energy is applied on the surface of a thin disk and the temperature rise at the other side is measured. The test time can be as low as < 1 sec. It has been used in liquids, frozen and unfrozen food gels, and model food systems,[2,3,36] wood,[18] and beef, potatoes, and apples.[41] Holometrix Micromet (Bedford, MA) manufactures several designs of the Flash apparatus with accuracies of $\pm 3\%$ for a temperature range of 24–1100°C.
3. Dynamic Angstrom (DA) and photopyroelectric (PPE) methods. Samples are exposed to a fluctuating heat flux and calculations are based on the phase shift of the temperature.[16] The PPE is a modification of the DA method in which a laser is used as a heat source. The DA method has been used on polyurethane foams[33] and the PPE on various food materials,[9–11] fruit juices,[14] and on candy.[13]
4. Beam deflection. A light source heats the sample, and thermal emission from the samples causes the second light beam to bend (beans by Brown et al,[7] candy by Faviar et al.[13])
5. Thermo-acoustic technique. A metal sheet is attached to the back of a thin slab and a sinusoidal heat source is applied to the front. Thermal diffusivity is calculated from the phase difference between the source and temperature of the metal sheet.[29]

STATE-OF-THE ART

Developments of measurement techniques for thermal diffusivity are evolving. Most of the current techniques were developed before the 1970s and have been constantly refined to improve accuracy and allow measurements of smaller samples. The advent of computers has made it easier and more affordable to interface measurement devices to fast and reliable data acquisition and control systems. However, the search for nondestructive techniques that can be used on small and irregularly shaped food materials is continuing.

REFERENCES

1. Agrawal, R.; Saxena, N.S.; Mathew, G.; Thomas, S.; Sharma, K.B. Effective Thermal Conductivity of Three-Phase Styrene Butadiene Composites. J. Appl. Polym. Sci. **2000**, *76* (12), 1799–1803.
2. Andrieu, J.; Gonnet, E.; Laurent, M. Pulse Method Applied To Foodstuffs: Thermal Diffusivity Determination. In *Food Process Engineering and Process Applications*; LeMaguer, M., Jelen, P., Eds.; Elsevier Applied Sci.: NY, 1986; Vol. 1, 103–121.
3. Andrieu, J.; Laurent, M.; Puaux, J.; Oshita, S. Thermal Properties of Unfrozen and Frozen Food Gels Determined by an Automatic Flash Method. In *Engineering and Food*; Spiess, W., Schubert, H., Eds.; Elsevier Applied Science Pub.: NY, 1990; Vol. 1, 447–455.
4. Balasubramaniam, T.; Bowman, H. Thermal Conductivity and Thermal Diffusivity of Biomaterials: A Simultaneous Measurement Technique. Trans. ASME **1977**, *99* (3), 148–154.
5. Bilskie, J.R.; Horton, R.; Bristow, K.L. Test of a Dual-Probe: Heat-Pulse Method for Determining Thermal Properties of Porous Materials. Soil Sci. **1998**, *163* (5), 346–355.
6. Bristow, K.L. Measurement of Thermal Properties and Water Content of Unsaturated Sandy Soil Using Dual-Probe Heat-Pulse Probes. Agric. For. Meteorol. **1998**, *89* (2), 75–84.
7. Brown, S.M.; Bicanic, D.; Vanasselt, K. Photothermal Beam Deflection Measurements on Agricultural Produce. J. Food Eng. **1996**, *28* (2), 211–223.
8. Choi, Y.H. Food Thermal Property Prediction as Effected by Temperature and Composition. Ph.D. Thesis, Purdue University, W. Lafayette, IN, 1985.
9. Dadarlat, D.; Surducan, V.; Riezebos, K.J.; Bicanic, D. A New Photopyroelectric Cell for Thermal Characterization of Foodstuffs—Application to Sugar Systems. Instrum. Sci. Technol. **1998**, *26* (2–3), 125–131.
10. Dadarlat, D.; Gibkes, J.; Bicanic, D.; Pasca, A. Photopyroelectric Measurement of Thermal Parameters in Food Products. J. Food Eng. **1996**, *30* (1–2), 155–162.
11. Dadarlat, D.; Bicazan, M.; Frandas, A.; Morariu, V.; Pasca, A.; Jalink, H.; Bicanic, D. Photopyroelectric Measurements of Thermal Parameters in Margarines—Influence of Water Content. Instrum. Sci. Technol. **1997**, *25* (3), 235–243.
12. Dickerson, W. An Apparatus for Measurement of Thermal Diffusivity of Foods. Food Technol. **1965**, *19* (5), 198–204.
13. Favier, P.; Dadarlat, D.; Gibkes, J.; Vandenberg, C.; Bicanic, D. Thermal Diffusivity of a Hard Boiled Candy Obtained by Photothermal Beam Deflection and Standard Photopyroelectric Method. Instrum. Sci. Technol. **1998**, *26* (2–3), 113–124.
14. Frandas, A.; Bicanic, D. Thermal Properties of Fruit Juices as a Function of Concentration and Temperature Determined Using the Photopyroelectric (PPE) Method. J. Sci. Food Agric. **1999**, *79* (11), 1361–1366.
15. Gaffney, J.; Baird, C.; Eshleman, W. Review and Analysis of the Transient Method for Determining Thermal Diffusivity of Fruits and Vegetables. ASHRAE Trans. **1980**, *86* (2), 261–280.
16. Graebner, E. Measuring Thermal Conductivity and Diffusivity. In *Thermal Measurements in Electronics Cooling*; CRS Press Inc: FL, 1997; 243–271.
17. Griesinger, A.; Heidemann, W.; Hahne, E. Investigation on Measurement Accuracy of the Periodic Hot-Wire Method by Means of Numerical Temperature Field Calculations. Int. Commun. Heat Mass Transfer **1999**, *6* (4), 451–465.

18. Harada, T.; Hata, T.; Ishihara, S. Thermal Constants of Wood During the Heating Process Measured with the Laser Flash Method. J. Wood Sci. **1998**, *44* (6), 425–431.
19. Heldman, D.; Singh, R. *Food Process Engineering,* 2nd Ed.; AVI Publishing Co., Inc.: Westport, NY, 1981.
20. Jankowski, T.; Jankowski, S.; Koziol, K. Some Thermal Properties of Root Vegetables. Acta Aliment. Pol. **1981**, *7* (3/4), 137–146.
21. Jaramillo-Flores, E.; Hernandez-Sanchez, H. Thermal Diffusivity of Soursoup Pulp. J. Food Eng. **2000**, *46* (2), 139–143.
22. Kostaropoulos, A.E.; Saravacos, G.D. Thermal Diffusivity of Granular and Porous Foods at Low Moisture Content. J. Food Eng. **1997**, *33* (1–2), 101–109.
23. Kravets, R.; Larkin, J. Bead Thermistor for Determination of Thermal Properties in Foods. ASAE Paper, No. 86-6517, 1986.
24. Kusterman, M.; Scherer, R.; Kutzbach, H. Thermal Conductivity and Thermal Diffusivity of Shelled Corn and Grain. J. Food Process. Eng. **1981**, *4* (3), 137–153.
25. Maglic, K.; Cezairliyan, A.; Peletsky, V. *Compendium of Thermophysical Property Measurement Methods*; 1. Survey of Measurement Techniques; Plenum Press: NY, 1984; 299–453.
26. Mohsenin, N. *Thermal Properties of Foods and Agricultural Materials*; Gordon and Breach Sci. Publishers: NY, 1980; 104–111.
27. Nesvadba, P. Methods for the Measurement of Thermal Conductivity and Thermal Diffusivity of Foodstuffs. J. Food Eng. **1982**, *1*, 93–113.
28. Niesteruk, R. Changes of Thermal Properties of Fruits and Vegetables During Drying. Drying Technol. **1996**, *14* (2), 415–422.
29. Niskanen, K.; Simula, S. Thermal Diffusivity of Paper. Nord. Pulp Paper Res. J. **1999**, *14* (3), 236–242.
30. Nix, G.H.; Vachon, R.I.; Lowery, G.W.; McCurry, T.A. The Line Source Method: Procedure and Iteration Scheme for Combined Determination of Conductivity and Diffusivity. Proceedings of Eight Conference on Thermal Conductivity, Purdue Univ., Oct 7–10, 1968; Ho, C.Y., Taylor, R.E., Eds.; Plenum Press: NY, 1969; 999–1008.
31. Njie, D.N.; Rumsey, T.R.; Singh, R.P. Thermal Properties of Cassava, Yam and Plantain. J. Food Eng. **1998**, *37* (1), 63–76.
32. Poulsen, K.P. Thermal Diffusivity of Foods Measured by Simple Equipment. J. Food Eng. **1982**, *1*, 115–122.
33. Prociak, A.; Sterzynski, T.; Pielichowski, J. Thermal Diffusivity of Polyurethane Foams Measured by the Modified Angstrom Method. Polym. Eng. Sci. **1999**, *39* (9), 1689–1695.
34. Rask, C. Thermal Properties of Dough and Bakery Products: A Review of Published Data. J. Food Eng. **1989**, *9*, 167–193.
35. Ren, T.; Noborio, K.; Horton, R. Measuring Soil Water Content, Electrical Conductivity, and Thermal Properties with a Thermo-time Domain Reflectometry Probe. Soil Sci. Soc. Am. J. **1999**, *63* (3), 450–457.
36. Renaud, T.; Briery, P.; Andrieu, J.; Laurent, M. Thermal Properties of Model Foods in the Frozen State. J. Food Eng. **1992**, *15*, 83–97.
37. Riedel, L. Measurements of Thermal Diffusivity of Foodstuffs Rich in Water. Kaltetechnik–Klimatisierung **1969**, *21* (11), 315–319.
38. Singh, R.P. Thermal Diffusivity in Food Processing. Food Technol. **1982**, *36* (2), 87–91.
39. Tavman, S.; Tavman, I.H.; Evcin, S. Measurement of Thermal Diffusivity of Granular Food Materials. Int. Commun. Heat Mass Transfer **1997**, *24* (7), 945–953.
40. Telis-Romero, J.; Telis, V.R.N.; Gabas, A.L.; Yamashita, F. Thermophysical Properties of Brazilian Orange Juice as Affected by Temperature and Water Content. J. Food Eng., *inpress*.
41. Vacek, V. The Measurement of Thermal Diffusivity of Freeze Dried Foodstuffs (in Czech). Prum Potravin **1977**, *28*, 626–630.
42. Van Gelder, A.F.; Diehl, K.C. A Thermistor-Based Method for Measuring Thermal Conductivity and Diffusivity of Moist Food Materials at High Temperatures. In *Thermal Conductivity 23*, Proceedings of the Twenty-Third International Thermal Conductivity Conference, Nashville, TN, Nov 6–8, 1996; Wilkes, K.E. Dinwiddie, R.B. Graves, R.S., Eds.; 627–638.
43. Watanabe, H. Accurate and Simultaneous Measurements of the Thermal Conductivity and Thermal Diffusivity of Liquids Using the Transient Hot-Wire Method. Metrologia **1996**, *33* (5), 101–105.

Thermal Process Calculations

Pieter Verboven
Nico Scheerlinck
Bart M. Nicolaï
Katholieke Universiteit Leuven, Leuven, Belgium

INTRODUCTION

Thermal food processes are one of the few production processes in industry that rely on a mathematical model to ensure the safety of the process.[1] Mathematical process models and simulation software represent a powerful alternative to the traditional, time consuming temperature measurements and quantitative microbiological and food quality analyses. Such software links heat transfer models to predictive models for microbiological growth/destruction and quality kinetics. Heat transfer models allow the calculation of food center temperature during the subsequent stages of the production process. Consequently, the effect of thermal treatment on the microbiological and sensorial quality can be evaluated using well-established predictive methods that were outlined in preceding chapters.

Heat transfer models exist for conduction and convection heat transfer during food manufacture as well as for electric heating methods (ohmic and microwave heating). As these models can only be solved for very simple problems, numerical solution is usually mandatory. A wide variety of numerical techniques and corresponding software is now available to solve these models.

HEAT TRANSFER MODELS

Transient *heat conduction* in an isotropic object Ω with boundary Γ is governed by the Fourier equation[2]

$$\rho c \frac{\partial T}{\partial t} = \nabla k \nabla T + Q \quad \text{on } \Omega \tag{1}$$

with the initial and boundary conditions

$$T(x,y,z,t) = T_0(x,y,z) \quad \text{at } t = 0 \tag{2}$$

$$T(x,y,z,t) = f(x,y,z,t) \quad \text{on } \Gamma_\mathrm{d} \quad \text{(Dirichlet conditions)} \tag{3}$$

$$-k\frac{\partial}{\partial n_\perp}T = h(T_\infty - T) + \varepsilon\sigma(T_\infty^4 - T^4)$$

$$\text{on } \Gamma_c \text{ (general Neumann conditions)}$$

$$\text{or } -k\frac{\partial}{\partial n_\perp}T = q_\mathrm{c} \text{ on } \Gamma_c$$

$$\text{(Constant Flux conditions)} \tag{4}$$

where T is the temperature [°C], ρ the density [kg m^{-3}], c the heat capacity [J kg^{-1} °C^{-1}], k the thermal conductivity [W m^{-1} °C^{-1}], Q the volumetric heat generation [W m^{-3}], and t the time [sec]. Dirichlet boundary conditions correspond to the specification of a temperature on the boundary. The function $f(x,y,z,t)$ is then known, $n_\perp$ is the outward normal to the surface. Neumann conditions reflect convection and radiation at the boundaries with h the surface heat transfer coefficient [W m^{-2} °C^{-1}], T_∞ the ambient temperature, ε the emission coefficient, σ the Stefan-Boltzmann constant, and q_c a fixed heat flux. The thermal properties of food and the surface heat transfer coefficient can be found elsewhere in this encyclopedia.

When a pressure gradient is applied to a fluid or when the fluid is heated, convection occurs. To solve the *convection heating* problem, the conservation laws of mass and momentum for an incompressible fluid flow must be solved[3]

$$\nabla \cdot \mathbf{u} = 0 \tag{5}$$

$$\rho\frac{\partial \mathbf{u}}{\partial t} + \rho(\mathbf{u}\cdot\nabla)\mathbf{u} = -\nabla p + \nabla\cdot\boldsymbol{\tau} - \rho\mathbf{g}\beta(T - T_0) + \mathbf{B} \tag{6}$$

where $\mathbf{u}(u_i, u_j, u_k)$ is the velocity vector [m sec^{-1}] in the Cartesian coordinate system (x_i, x_j, x_k), p the pressure incorporating the hydrostatic pressure [Pa], $\mathbf{g}$ the gravity vector [m sec^{-2}], β the thermal expansion coefficient [K^{-1}], $\mathbf{B}$ the body forces [N m^{-3}] and $\boldsymbol{\tau}$ the viscous stress tensor [N m^{-2}], which for an incompressible Newtonian fluid equals

Encyclopedia of Agricultural, Food, and Biological Engineering
DOI: 10.1081/E-EAFE 120007107
Copyright © 2003 by Marcel Dekker, Inc. All rights reserved.

$$\tau_{ij} = \mu\left(\frac{\partial u_i}{\partial x_j} + \frac{\partial u_j}{\partial x_i}\right) - \frac{2}{3}(\nabla \cdot \mathbf{u})\delta_{ij} = \mu\left(\frac{\partial u_i}{\partial x_j} + \frac{\partial u_j}{\partial x_i}\right) \tag{7}$$

Non-Newtonian fluids and rheological food properties are discussed elsewhere in this encyclopedia. The temperature T_0 is a buoyancy reference temperature. The accompanying energy equation for the above system is

$$\rho c\frac{\partial T}{\partial t} + \rho c(\mathbf{u} \cdot \nabla)T = \nabla k \nabla T + Q \tag{8}$$

which equals Eq. 1 plus a convective contribution. Unless a very fine spatial and temporal resolution is employed, Eqs. 5–8 are not valid for turbulent flow. In that case, the equations have to be rewritten for the average flow, and additional closures are required to calculate the turbulence properties of the flow.[4] The boundary conditions for the above equations include Dirichlet or Neumann boundary conditions at the flow boundaries and wall boundary conditions (a wall function reflecting the behavior of the flow near the wall). Initial values must be provided for all variables. For a full account, the reader is refered to Hirsch[3] and Ferziger and Peric.[4]

In *ohmic and microwave heating*, energy is transferred from an electromagnetic source to the food, where it is dissipated into heat. In the electromagnetic spectrum, ohmic heating is situated in the 50/60 Hz range (domestic electric power) while the microwave region is 300 MHz–30 GHz. In food applications, the latter is restricted to the fixed frequencies of 2450 MHz (domestic) and 915 MHz (industrial). In ohmic heating, a voltage source V is placed in contact with food resulting in a current flow as dissolved ions are displaced. The electric field can be obtained from the Laplace equation, which requires a numerical solution in 2-D and 3-D heterogeneous food.[5] The volumetric ohmic heat dissipation Q [W m^{-3}] is derived from the computed electric field intensity ∇V as a result from Ohm's law:

$$Q = \sigma|\nabla V|^2 \tag{9}$$

where ∇V is the voltage gradient in food [V m^{-1}] and σ the conductivity [S m^{-1} = ohm^{-1} m^{-1}]. Foods containing sufficient ionic salts dissolved in free water are good conductors, as opposed to nonionized fluids (oils, fats) and nonmetallic solids (bone, crystalline structures). Conductivity typically increases with increasing temperature and increasing ionic salt content.[6]

Electromagnetic fields inside a microwave oven are generated from sources not in contact with the food. Microwave heating occurs because of dielectric heating, when water dipoles in the food attempt to align themselves with the alternating electromagnetic field. The electric and magnetic fields and the resulting energy storage and dissipation will change because of the presence of the dielectric material and the air in the cavity. The fields are time varying and 3-D and are described by Maxwell's equations, which must be solved numerically in complex 3-D cases.[7] When the electric field is known, the dissipated heat Q can be obtained from the power density [W m^{-3}]:

$$Q = \frac{1}{2}\omega\varepsilon_0\varepsilon''_{\text{eff}}|E|^2 \tag{10}$$

with $|E|$ the amplitudes of the electric field [V m^{-1}], ε_0 the absolute permittivity of free space [F m^{-1}], and ω the angular frequency [rad sec^{-1}]. The interaction between electromagnetic field and food is characterized by the effective loss factor $\varepsilon''_{\text{eff}}$ (incorporating both ionic conduction and dielectric relaxation) and the dielectric constant ε'. These material properties depend on moisture and salt content, temperature, and frequency.[7] The heat source terms in Eqs. 9 and 10 are incorporated into the appropriate heat transfer equation (Eq. 1 or Eq. 8) to obtain the food temperature distribution.

NUMERICAL SOLUTION

Eq. 1 can be solved analytically under a limited set of initial and boundary conditions for simple geometries only. Such solutions are given elsewhere in this encyclopedia. Due to significant advances in computer technology and numerical solution techniques, it is now possible to solve the heat transfer models for complex 3-D cases, which render the models excellent design tools for food processes.

The *finite difference method* (FDM) is the oldest discretization method for the numerical solution of differential equations and has already been described in 1768 by Euler. The method is based on the approximation of the derivatives in the governing equations by the ratio of two differences, i.e., using Taylor series expansions. For this purpose, the computational domain is subdivided in a regularly spaced grid of lines, which intersect at common nodal points (Fig. 1a and b). Subsequently, the space and time derivatives are replaced by finite differences in terms of nodal values of the functions. Algebraic equations are established for all nodes of the grid, which can be readily solved. The implementation of the FDM is straightforward for structured grids that are uniformly distributed on regular geometries even when higher order differencing schemes are used (Fig. 1a). On curved grid lines (Fig. 1b) in complex geometries, however, the procedure becomes a tedious task.[8]

The *finite volume method* (FVM) starts from the integral form of the governing equations. The computational

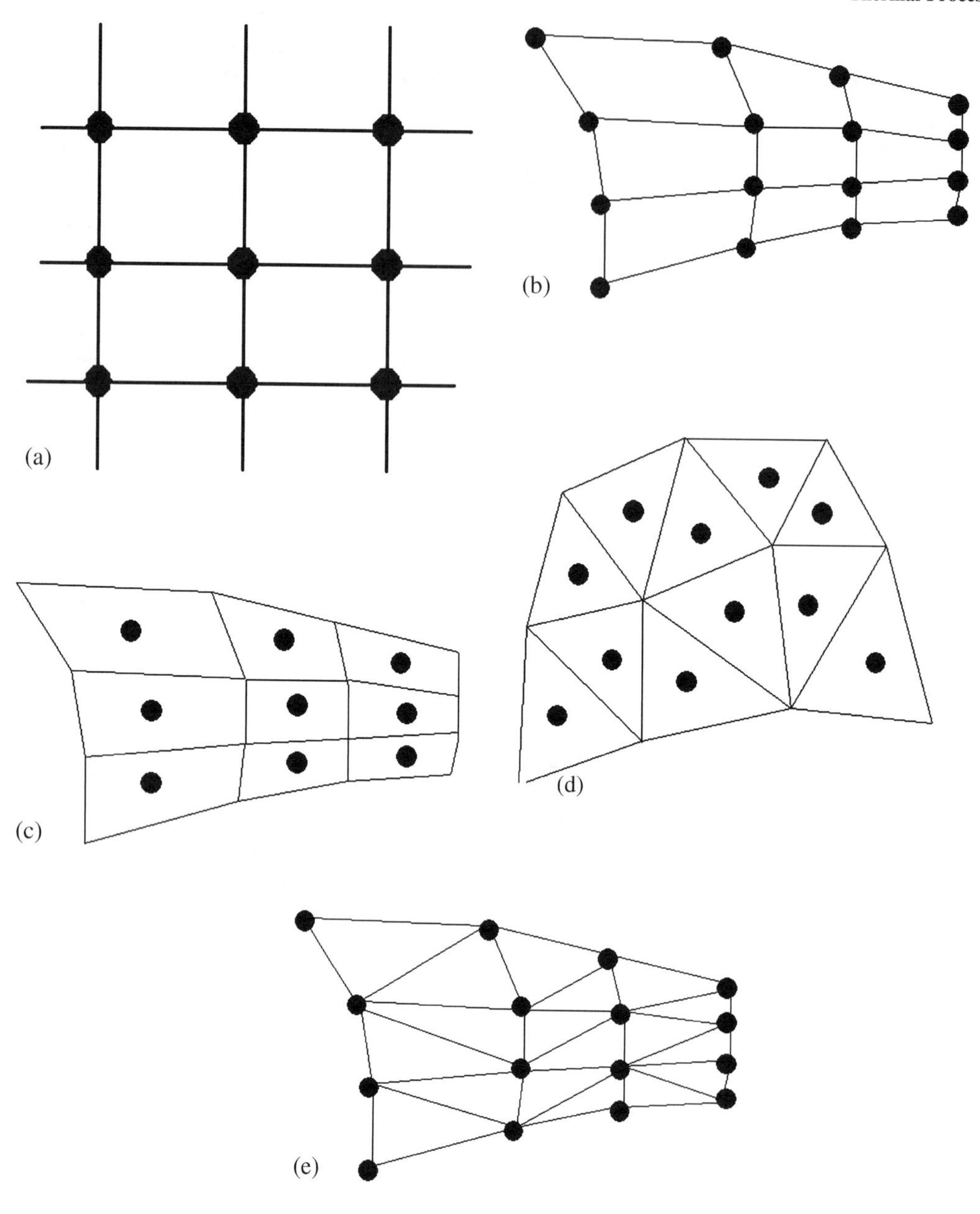

Fig. 1 Typical spatial grids used in discretization methods: (a) orthogonal structured FD grid, (b) nonorthogonal FD grid, (c) structured FV grid, (d) nonstructured FV grid, (e) 3-node triangular FE grid.

domain is subdivided into a number of interconnected but nonoverlapping subdomains called control volumes. Computational nodes are situated at the centroid of the control volumes (Fig. 1c). By application of the equations to the control volumes, the conservation form of the equations is transferred from the original infinitesimal scale to the discrete scale. Surface integrals and volume integrals are approximated in terms of values of the variables at the cell face and nodes, respectively. Cell face values are themselves expressed in terms of nodal values by means of interpolation. As a result, algebraic equations are obtained for all nodes, which can be solved by well-known solution methods. The FVM can be easily applied to any geometry, as the grid only defines the control volume boundaries. The combination of integration and interpolation, however, makes it difficult to apply higher order approximations on nonstructured grids. On nonstructured grids, control volumes cannot be reordered into a rectangular lattice (Fig. 1d). The main application area of FVM is Computational Fluid Dynamics (CFD).[4]

In the *finite element method* (FEM), a given computational domain is subdivided as a collection of finite elements, subdomains of variable size and shape, which are interconnected in a discrete number of nodes (Fig. 1e). Such elements are typically quadrilaterals or triangles in 2-D and tetrahedra or hexahedra in 3-D. The solution of the governing partial differential equation is approximated in each element by a low-order polynomial in such a way that it is defined uniquely in terms of the (approximate) solution at the nodes. The global approximate solution can then be written as a series of low-order piecewise polynomials with the coefficients of the series equal to the approximate solution at the nodes. Substitution of the approximate solution in the differential equation produces in general a nonzero residual. In the Galerkin FEM, the unknown coefficients of the low-order piecewise polynomials are then found by orthogonalization of this residual with respect to these polynomials. This results in a system of algebraic or ordinary differential equations, which can be solved using the well-known techniques. The advantage of FEM is its ability to deal with complex shapes in a straightforward way and the easy way to refine meshes simply by subdividing elements. The FEM originates from the mechanical engineering field.[9]

APPLICATIONS

There exists a wide range of applications for thermal process calculations in the food industry, from heating of foods (canning, frying, baking, ovens,...) to cooling (storage rooms, display cabinets, refrigerated trucks) and freezing. Thermal process calculations are used for design of processes and, more and more importantly, for optimization.

Design of Thermal Processing of Canned Liquid Food

Canned foods were one of the first applications of numerical thermal food process calculations, because of their economic importance and the large concern for food safety. In the case of sterilization/pasteurization of liquid foods in cylindrical cans, the main heating mode is convection of the liquid inside the container. The position and heating rate at the slowest heating point are by no means obvious or easily determinable. In such circumstances, analytical solutions of heat transfer models are not applicable, and one has to rely on numerical solutions to find the slowest heating location and the process value in the container. Such studies were first undertaken by Engelman and Sani[10] for pasteurization of beer in bottles and by Datta and Teixeira[11,12] for sterilization of canned foods. Their studies consisted of solving the full system of governing equations (Eqs. 5–8) using the FEM[10] and the FDM.[11,12] The FVM has been used at later stages to find similar results.[13] Fig. 2 shows the convection currents and temperature distribution in a heated can of water. Clearly, the slowest heating point is situated away from the centerline near the bottom of the cylinder. Procedures have been developed to obtain the sterilization value of the moving liquid elements by Datta[14] and Ghani et al.[13] The conventional procedure of obtaining sterilization based on temperature at the slowest heating zone was shown to be lower than the actual least sterilization of a real fluid element in the system.

Design of Heating Processes Involving Frying, Baking, Hot-Air, Infrared and Microwave Energy

Processes such as frying and baking are not pure heat transfer processes, although heating is applied as the main driving force. They also involve mass transfer, which makes the models outlined earlier significantly more

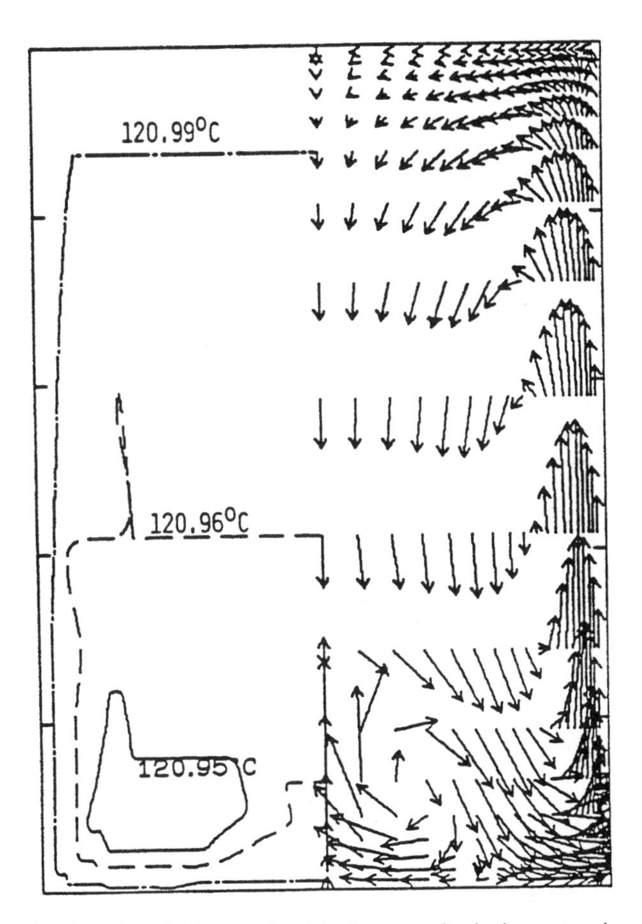

Fig. 2 Numerically calculated isotherms and velocity vectors in a cylindrical can after 30 min of heating at 121°C. (From Ref. 12.)

complex. The solution of coupled heat and mass transfer in these processes has recently received considerable attention. Such models can help to investigate uptake of fat, excessive water loss, browning, and nonuniform heating. Ni and Datta[15] used a multiphase porous model to compute heat and mass transfer inside potato slabs during baking. Singh and Vijayan[16] considered heat transfer with a moving crust–core interface during frying of foods.

In hot-air ovens, hot air is forced over foods. The spatial variation in air temperature, air velocity, and turbulence could be large and could lead to different heating rates of foods at different positions. Additionally, infrared energy can be applied, which will result in a different heating rate. Finally, foods do not heat uniformly by application of microwaves. Thermal calculations can help to investigate the temperature distribution inside the oven and food and improve the oven design. Zhang and Datta[17] modeled the combined heat transfer and electromagnetics in a domestic oven with foods. Datta and Ni[18] calculated temperature and moisture profiles in foods during combined infrared, hot-air, and microwave heating. It was shown that surface moisture can be reduced by application of infrared energy or hot air. Fahloul et al.[19] simulated the spatial distribution of temperature and moisture of biscuits and the temperature and humidity of the air inside a tunnel oven. Verboven et al.[20,21] used a model to calculate the airflow and the temperature distribution in a forced convection oven (Fig. 3). The predicted temperature response of polymer bricks subject to forced convection heating in the oven was compared to experiments. The cold and hot spots could be predicted by means of this model, which is a basis for design improvement.

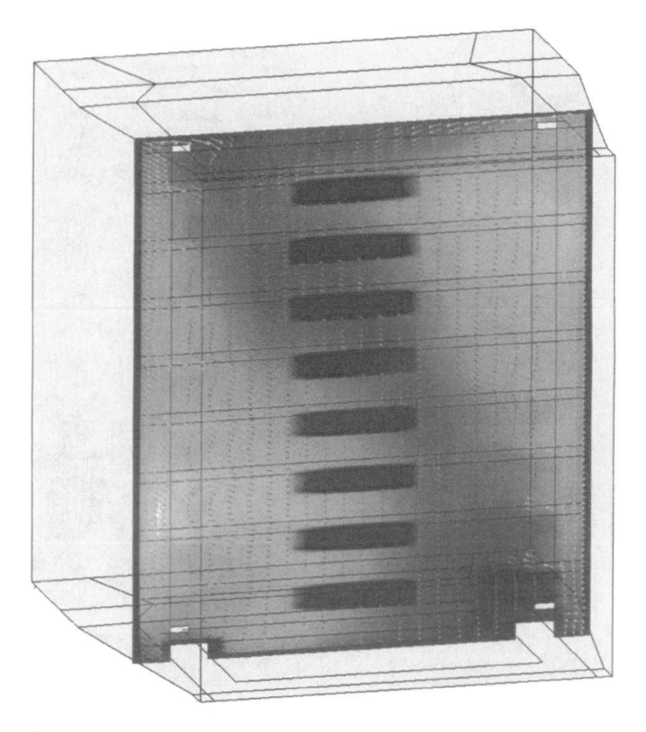

Fig. 3 CFD model predictions of the velocity and temperature field in a forced convection oven with eight rectangular food items. The wire frame is the internal geometry of the oven cavity. The arrows represent the velocity vectors (note the high velocities at the top of the cavity and through the orifices in both side walls). The shaded contours give the temperature field (dark shades are cold regions, light shades are hot regions). Hot air enters the cavity at the bottom left and top right of the cavity.

Design of Cooling and Freezing Processes

Frozen and chilled storage is important to extend the shelf life and preserve the quality and safety of foods. Thermal calculations can help to determine cooling and freezing times, to assess quality and safety, and to improve technologies and the design of equipment and installations. The use of 3-D transient numerical calculations of food temperature during cooling is becoming widespread (e.g., Nicolaï et al.,[22] Wang and Sun[23]). Scheerlinck et al.[24] developed a phase-change finite element model to predict the temperature of arbitrary 3-D food shapes during freezing and thawing. Based on thermal calculations, Kondjoyan and Daudin[25] assessed the effect of spatial heterogeneity of airflow conditions during chilling and storage of carcasses and meat products. Verboven et al.[26] calculated the sensitivity of the food temperature with respect to the air velocity and the turbulence intensity in different processes. The distribution of airflow, temperature, and humidity inside cool stores were modeled to assess the effect of cool room design and product stacking on temperature uniformity (Mirade, Kondjoyan, and Daudin,[27] Hoang et al.[28]). Similar studies were performed on chilled cabinets (Cortella[29]) and refrigerated trucks (Moureh, Menia, and Flick[30]). Models are being developed for heat and mass transfer inside bulks of food products (Xu, Burfoot, and Huxtable,[31] Hoang et al.[32]).

Optimal Control of Thermal Food Processes

Optimization of thermal processes is becoming increasingly important in the food industry. Traditional driving forces of profit and product safety have been joined by new forces such as the public awareness of product quality and environmental impact. The optimal control problem for a fixed terminal time is formulated as a mathematical objective function to be maximized or minimized (e.g., minimal energy consumption, maximal retention of a quality factor,...) by means of applying a time profile of certain control variables (e.g., the input power). Traditional optimization procedures no longer meet the requirements of convergence and computation speed when considering complex thermal processes, involving 3-D nonlinear heat transfer, microbial growth and inactivation, and quality degradation, i.e., when there are many nonlinear constraints. In such circumstances, optimal profiles of the control variables are calculated by means of advanced optimization algorithms.[33,34] The objective function is subject to the physics of the problem, described by mathematical models, such as heat transfer models (Eqs. 1–4 or Eqs. 5–8) coupled to microbial growth/inactivation and quality retention models, and several equality and inequality algebraic constraints (limits to the process control temperature, maximum product temperature, minimum desired lethality, minimum final retention of a quality factor).

The theoretical solution of this problem involves the application of Pontryagin's minimum principle.[34,35,36] Numerical solution approaches include methods that transform the original formulation into a constrained nonlinear program in which the control variables are parameterized by polynomials in time. Using these polynomials, the constraining model equations are solved using the FDM, FVM, or FEM and the objective function is calculated. The coefficients of these polynomials are adjusted at each iteration of an optimization algorithm. The algorithms are often gradient-based, i.e., they improve the coefficients using the calculated gradients of the objective function with respect to the coefficients in order to find a minimum, but stochastic methods have been proven to be very efficient as well.[33] This approach has been used for the optimization of thermal processing of conduction-heated canned food,[37] using a finite element formulation of the heat transfer model (Eqs. 1–4), and first order degradation kinetics for micro-organisms and quality factors and nutrients. The process temperature profile given in Fig. 4 is the optimal profile for a minimal process time, subject to specified limits of the process temperature, a specified minimal microbial inactivation, a specified maximum final product temperature, a specified minimum final average retention of a nutrient, and a specified minimum final surface retention of a quality factor. Compared to the optimal constant profile, process times were reduced by 20%–30%.

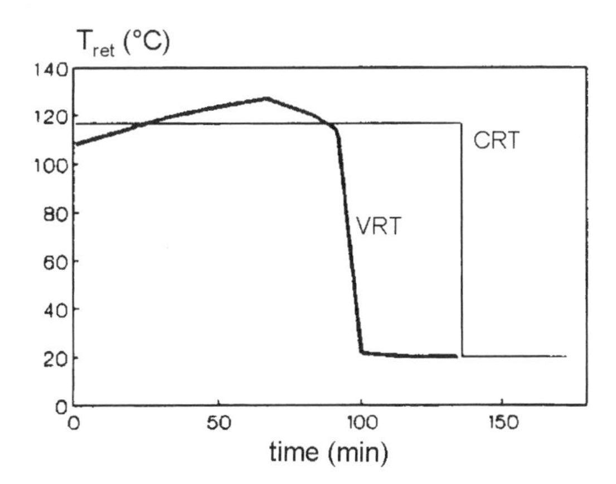

Fig. 4 Optimal (minimum process time) variable retort temperature profile (VRT) compared with optimal constant temperature process (CRT). (From Ref. 37.)

CONCLUSION

Models to describe thermal food processes were presented. It was shown that the models can be used to predict the temperature response of the food in many applications, which can be a basis for process design and optimization. In many studies, it was shown that the spatial and temporal variation of process temperature, humidity, and air velocity significantly affect food heating and have to be considered in the calculation. Heat transfer to and in foods is often strongly coupled to other mechanisms involving mass transfer and micromechanics. Processes involving airflow are complicated due to the effect of turbulence. Furthermore, the biological variability of thermophysical properties of foods is considerable. These factors need to be taken into account to improve the reliability of model predictions.

ACKNOWLEDGMENTS

Author Pieter Verboven is a Postdoctoral Researcher with the Flemish Fund for Scientific Research (F.W.O.-Vlaanderen). Author Nico Scheerlinck is a Postdoctoral Fellow with the Research Fund of the Katholieke Universiteit Leuven.

REFERENCES

1. Holdsworth, S.D. Optimization of Thermal Processing—A Review. J. Food Eng. **1985**, *4*, 89–116.
2. Incropera, F.P.; De Witt, D. Introduction to Conduction. In *Fundamentals of Heat and Mass Transfer*, 3rd Ed.; John Wiley and Sons: New York, 1990; 43–66.
3. Hirsch, C. The Mathematical Nature of the Flow Equations and Their Boundary Conditions. In *Numerical Computation of Internal and External Flows. Vol. 1: Fundamentals of Numerical Discretization*; John Wiley and Sons: Chichester, U.K., 1991; 133–152.
4. Ferziger, J.H.; Peric, M. Boundary Conditions for the Navier-Stokes Equations. In *Computational Methods for Fluid Dynamics*; Springer-Verlag: Heidelberg, Germany, 1996; 192–194.

5. Zhang, L.; Fryer, P.J. Models for the Electrical Heating of Solid–Liquid Food Mixtures. Chem. Eng. Sci. **1993**, *48* (4), 633–642.
6. Sastry, S.K. Advances in Ohmic Heating for Sterilization of Liquid Particle Mixtures. In *Advances in Food Engineering*; Singh, R.P., Wirakartakusumah, M.A., Eds.; CRC Press: Boca Raton, FL, 1992; 139–148.
7. Datta, A.K.; Anantheswaran, R.C. *Handbook of Microwave Technology for Food Applications*; Marcel Dekker: New York, 2001.
8. Özisik, M.N. *Finite Difference Methods in Heat Transfer*; CRC Press: Boca Raton, FL, 1994.
9. Zienkiewicz, O.C. *The Finite Element Method*, 3rd Ed.; McGraw-Hill Book Company: Maidenhead, U.K., 1977.
10. Engelamn, M.S.; Sani, R.L. Finite-Element Simulation of an In-Package Pasteurization Process. Numerical Heat Transfer **1983**, *6*, 41–54.
11. Datta, A.K.; Teixeira, A.A. Numerical Modelling of Natural Convection Heating in Canned Liquid Foods. Trans. ASAE **1987**, *30* (5), 1542–1551.
12. Datta, A.K.; Teixeira, A.A. Numerically Predicted Transient Temperature and Velocity Profiles During Natural Convection Heating of Canned Liquid Foods. J. Food Sci. **1988**, *53* (1), 191–195.
13. Ghani, A.G.A.; Farid, M.M.; Chen, X.D.; Richards, P. An Investigation of Deactivation of Bacteria in a Canned Liquid Food During Sterilization Using Computational Fluid Dynamics (CFD). J. Food Eng. **1999**, *42* (4), 207–214.
14. Datta, A.K. Mathematical Modelling of Biochemical Changes During Processing of Liquid Foods and Solutions. Biotechnol. Prog. **1991**, *7* (5), 397–402.
15. Ni, H.; Datta, A.K. Heat and Moisture Transfer in Baking of Potato Slabs. Drying Technol. **1999**, *17* (10), 2069–2092.
16. Singh, R.P.; Vijayan, J. Predictive Modeling in Food Process Design. Food Sci. Technol. Int. **1998**, *4* (5), 303–310.
17. Zhang, H.; Datta, A.K. Coupled Electromagnetic and Thermal Modeling of Microwave Oven Heating of Foods. J. Microwav. Power Electromagn. Energy **2000**, *35* (2), 71–85.
18. Datta, A.K.; Ni, H. Infrared and Hot-Air-Assisted Microwave Heating of Foods for Control of Surface Moisture. J. Food Eng. **2002**, *51* (4), 355–364.
19. Fahloul, D.; Trystram, G.; Duquenoy, A.; Barbotteau, I. Modelling Heat and Mass Transfer in Band Oven Biscuit Baking. Food Sci. Technol. **1994**, *27* (2), 119–124.
20. Verboven, P.; Scheerlinck, N.; De Baerdemaeker, J.; Nicolaï, B.M. Computational Fluid Dynamics Modelling and Validation of the Isothermal Airflow in a Forced Convection Oven. J. Food Eng. **2000**, *43*, 41–53.
21. Verboven, P.; Scheerlinck, N.; De Baerdemaeker, J.; Nicolaï, B.M. Computational Fluid Dynamics Modelling and Validation of the Temperature Distribution in a Forced Convection Oven. J. Food Eng. **2000**, *43*, 61–73.
22. Nicolaï, B.M.; Verlinden, B.; Beuselinck, A.; Jancsók, P.; Quenon, V.; Scheerlinck, N.; Verboven, P.; De Baerdemaeker, J. Propagation of Stochastic Temperature Fluctuations in Refrigerated Fruits. Int. J. Refrig. **1999**, *22*, 81–90.
23. Wang, L.J.; Sun, D.W. Modelling Three-Dimensional Transient Heat Transfer of Roasted Meat During Air Blast Cooling by the Finite Element Method. J. Food Eng. **2002**, *51* (4), 319–328.
24. Scheerlinck, N.; Verboven, P.; Fikiin, K.A.; De Baerdemaeker, J.; Nicolaï, B.M. Finite Element Computation of Unsteady Phase Change Heat Transfer During Freezing or Thawing of Foods by Using a Combined Enthalpy and Kirchhoff Transform Method. Trans. Am. Soc. Agric. Eng. **2001**, *44*, 429–438.
25. Kondjoyan, A.; Daudin, J.D. Optimisation of Air-Flow Conditions During the Chilling and Storage of Carcasses and Meat Products. J. Food Eng. **1997**, *34* (3), 243–258.
26. Verboven, P.; Scheerlinck, N.; De Baerdemaeker, J.; Nicolaï, B.M. Sensitivity of the Food Temperature with Respect to the Air Velocity and the Turbulence Kinetic Energy. J. Food Eng. **2000**, *48* (1), 53–60.
27. Mirade, P.S.; Kondjoyan, A.; Daudin, J.D. Three-Dimensional CFD Calculations for Designing Large Food Chillers. Comput. Electron. Agric. **2002**, *34* (1–3), 67–88.
28. Hoang, M.L.; Verboven, P.; De Baerdemaeker, J.; Nicolaï, B.M. Analysis of the Air Flow in a Cold Store by Means of Computational Fluid Dynamics. Int. J. Refrig. **2000**, *23*, 127–140.
29. Cortella, G. CFD-Aided Retail Cabinets Design. Comput. Electron. Agric. **2002**, *34* (1–3), 43–66.
30. Moureh, J.; Menia, N.; Flick, D. Numerical and Experimental Study of Airflow in a Typical Refrigerated Truck Configuration Loaded with Pallets. Comput. Electron. Agric. **2002**, *34* (1–3), 25–42.
31. Xu, Y.; Burfoot, D.; Huxtable, P. Improving the Quality of Stored Potatoes Using Computer Modelling. Comput. Electron. Agric. **2002**, *34* (1–3), 159–171.
32. Hoang, M.L.; Verboven, P.; Baelmans, M.; Nicolaï, B.M. Air Flow Effects on Heat and Mass Transfer During Cooling of Chicory Roots. In *Proceedings*, Annual Meeting of the ASAE, Chicago, Illinois, July 28–31; ASAE: St. Joseph, Michigan, 2002; Paper No. 026048, 16 pp.
33. Banga, J.R.; Irizarry-Rivera, R.; Seider, W.D. Stochastic Optimization for Optimal and Model-Predictive Control. Comput. Chem. Eng. **1998**, *22* (4–5), 603–612.
34. Stigter, J.D.; Scheerlinck, N.; Nicolaï B.M.; Van Impe, J.F. Optimal Heating Strategies for a Convection Oven. J. Food Eng. **2001**, *in press*.
35. Nadkarni, M.N.; Hatton, T.A. Optimal Nutrient Retention During the Thermal Processing of Conduction-Heated Canned Foods: Application of the Distributed Minimum Principle. J. Food Sci. **1985**, *50*, 1312–1321.
36. Saguy, I.; Karel, M. Optimal Retort Temperature Profile in Optimizing Thiamine Retention in Conductive-Type Heating of Canned Foods. J. Food Sci. **1979**, *44*, 1485–1492.
37. Banga, J.R.; Alonso, A.A.; Pérez-Martin, R.I.; Singh, R.P. Optimal Control of Heat and Mass Transfer in Food and Bioproducts Processing. Comput. Chem. Eng. **1994**, *18* (Suppl.), S699–S705.

Thermal Processing of Meat Products

Pie-Yi Wang
ConAgra Foods, Downers Grove, Illinois, U.S.A.

T

INTRODUCTION

Processed meats can be grouped into three categories: muscle products, coarse ground products, and emulsified products. In the processed meat industry, raw meats, either whole muscle or alternated meat particles, are restructured to the predetermined product characteristics. The major processing includes pickle injecting, massaging, blending, emulsifying, stuffing, and thermal processing. Processed meats are usually stuffed into casings, bags, or molds to form a desirable shape. After stuffing, meat products are subjected to thermal processing, which provides the product color, flavor, texture, and inhibits microbial growth. In the meat industry, the major heat processing equipment is the smokehouse. Inside the smokehouse, drying, smoking, cooking, and chilling processes occur. Simultaneous heat and mass transfer as well as microbial and biochemical reactions are induced.

SMOKEHOUSE

According to product handling methods, two types of smokehouses are available. One is the batch-type smokehouse in which products are handled manually. The drying, smoking, cooking, and prechilling processes occur in the same chamber. The other type is continuous smokehouse in which the products are hung on a chain and automatically go through drying, smoking, cooking, and chilling zones. The products out of batch-type smokehouses often require further chilling such as blast air or brine chilling. Most of continuous smokehouses are equipped with brine chilling. Products out of a continuous system can meet the temperature criteria of 40°F or lower. However, the air handling system is basically the same irrespective of the type of smokehouse. The fundamental design of smokehouse is illustrated in Fig. 1.[1] In a typical smokehouse, the main blower supplies the heated air to the air ducts located at both sides of the smokehouse. The air is then forced down along the walls through supply cones or slots that are attached to the bottom of air ducts. The airflow comes down from the walls, across the floor, up through the products, and returns to the return duct that is located at the middle of the smokehouse as illustrated in Fig. 2.[2] The returned airflow is reheated and recirculated back to the smokehouse. The heating source can be steam, gas fire, or electricity. As one of the major functions of smokehouse is drying, the returned airflow usually picks up substantial amount of moisture from the product. Part of the moist air is exhausted and replaced with fresh air. During the hot and humid summer season, a larger quantity of air has to be exhausted and replaced in order to maintain certain dry and wet bulb temperatures. In the cold and dry winter or during cooking cycle, steam is injected into the airflow before going to the smokehouse to satisfy the humidity setting as summer. The dry and wet bulb temperatures of return air are usually used as process control references. The conditioned air is then supplied by the main blower to the two sides of air duct inside the smokehouse. The point where the air streams coming from two sides meet is known as the "break point." The airflow at the break point is very turbulent and has the highest velocity in the smokehouse. Most smokehouses use the rotating dampers as shown in Fig. 2 to oscillate and gradually sweep airflow from one side to the other to ensure the uniformity of heat and mass transfer. The damper rotation rate is often set at 1 rpm. The rotating pattern of airflow is illustrated in Fig. 3.[2] In Fig. 3A, the rotating damper at the left side is closed and the right side is open. The airflow is forced from the right side to the left side of smokehouse. In Fig. 3B, both sides of damper are at 45° or have the same opening. The main airflow or break point is located at the middle of smokehouse. In Fig. 3C, the right side of damper is closed and the left side damper is open. The airflow reverses from the left to the right side of smokehouse. The rotating damper never completely closes the airflow from one side of air duct. The airflow from one side to the other is normally set at the ratio of 1:3, i.e., the open side air velocity is three times higher than the closed side air velocity. This ratio is adjusted according to the product height relative to the smokehouse height. The air circulation rate is normally greater than 10 air changes. One air change is equivalent to one smokehouse volume per minute.

Encyclopedia of Agricultural, Food, and Biological Engineering
DOI: 10.1081/E-EAFE 120007245
Copyright © 2003 by Marcel Dekker, Inc. All rights reserved.

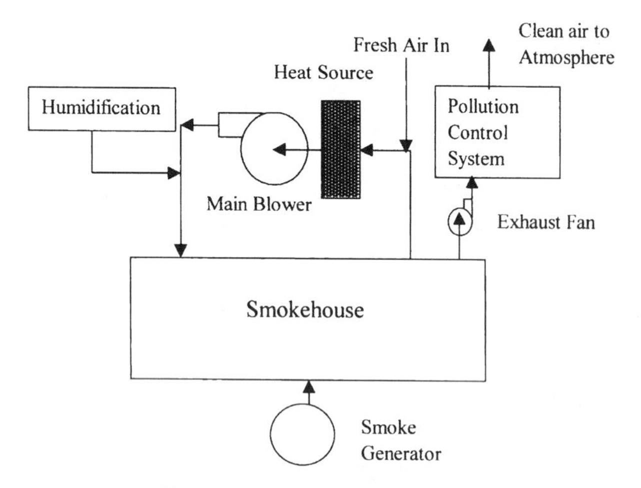

Fig. 1 Schematic of smokehouse.

DRYING

The heat processing schedule significantly affects the quality of processed meat.[3] As there are many kinds of quality attributes, a wide range of heat processing cycles has been employed in the meat industry. High humidity in a smokehouse can cause surface grease and poor product color. That is the reason why smoked meats must be predried before smoking. The variables that influence drying rate are dry and wet bulb temperatures as well as air velocity in smokehouse. Fig. 4A–C illustrates the effects of dry bulb temperature, air velocity, and relative humidity on the yield of frankfurters (25 mm × 140 mm). The yield is the most common term used in the industry to represent the weight change during drying. The original weight is 100%. The data in Fig. 4 show that the most critical variable to affect the drying process is air humidity. Air velocity higher than 900 ft min^{-1} (FPM) does not increase drying rate significantly as shown in Fig. 4B. The batch-type smokehouse usually starts at low temperature. If wet bulb temperature setting is higher than the house temperature, it will call an excessive amount of steam into the smokehouse. The excessive moisture then causes condensation on product surface and induces grease and color problems. In order to prevent excessive moisture at the beginning, it is common practice to set the wet bulb temperature at 0°F. Because of this setting, the exhaust and fresh air intake will run at maximum capacity disregarding the humidity condition in the smokehouse. There is a substantial variation in weather conditions year-round. Therefore, the exhaust and fresh air intake should run according to the dryness or humidity of intake air. Microprocessors have been used to control the smokehouse. The exhaust, intake air, and steam injection can be controlled and operated independently without interlocking problem. The cold startup causes excessive steam

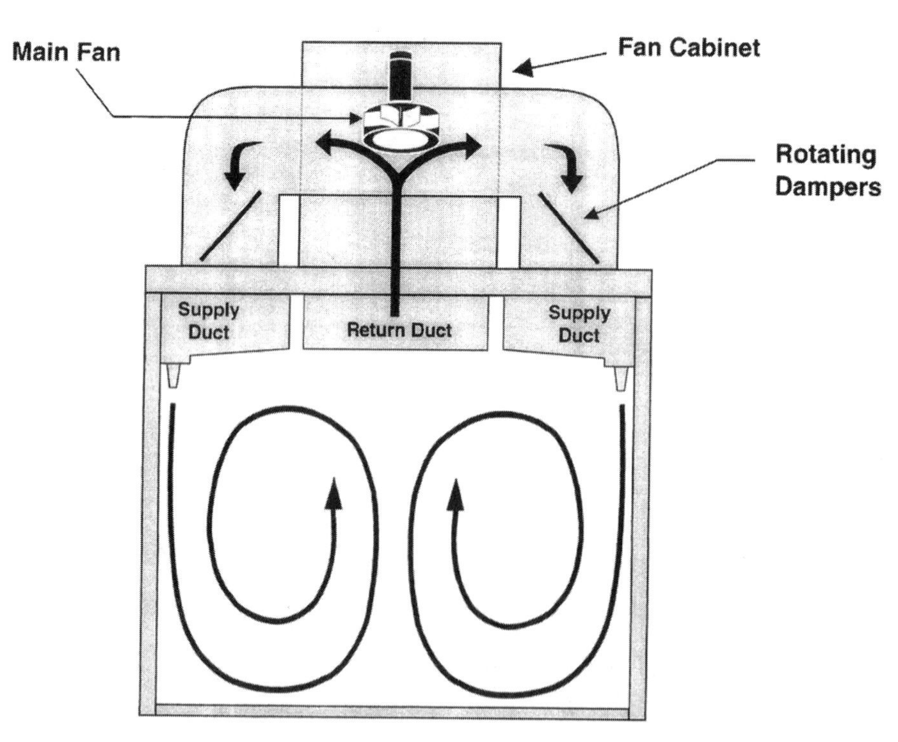

Fig. 2 Air flow in smokehouse. (Courtesy of Alkar Co.)

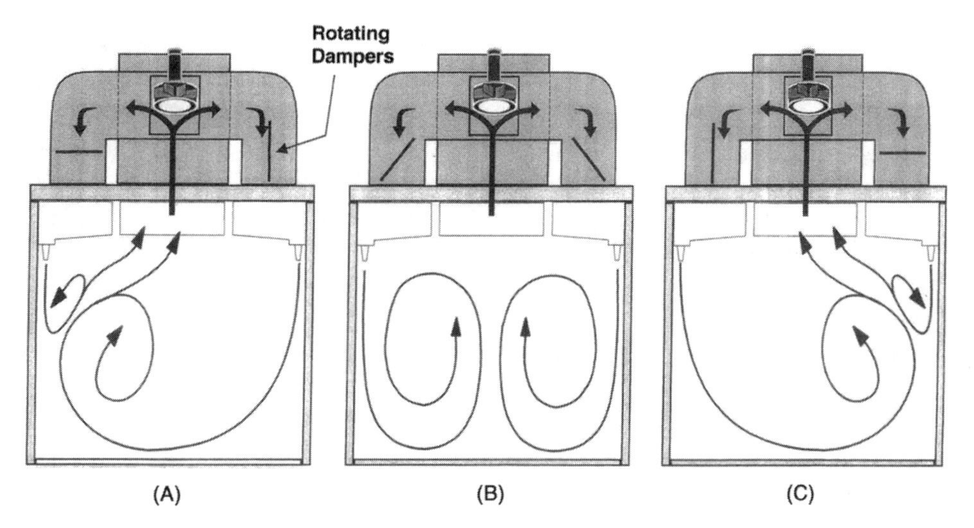

Fig. 3 Oscillating air flow in smokehouse. (Courtesy of Alkar Co.)

injection, which can be solved by setting steam injection off instead of setting wet bulb temperature at zero so that product can have more consistent quality all year-round and production can utilize less energy in winter.

SMOKING

The other function of a smokehouse is the smoking process. There are two different types of smoking processes: natural and liquid smoking. Natural smoke is produced from thermal decomposition of wood chips or sawdust and is also known as pyrolysis. Thermal decomposition of wood is induced by high temperature. Under normal smoking conditions, the smoking temperature ranges from 300 to 400°C. As long as wood chip or sawdust internal temperature reaches this range, the decomposition occurs and smoke is given off. This results in the generation of more than several hundreds of smoke compounds.[4] The effective smoke compounds are phenols, acids, and carbonyls that change with generation of temperature.[5] Higher thermal decomposition temperature will produce more effective smoke. As natural smoke introduced into the smokehouse cannot be completely absorbed by product, some smoke will be discharged to the atmosphere with the exhausted air that is necessary to control humidity in the smokehouse. A pollution control system is required to eliminate or reduce air pollution. In addition, a large quantity of tar produced during natural smoke generation often causes fire in smoke generator. It is difficult to maintain a consistent smoke quality and good operating conditions for the equipment. To improve smoking process, liquid smoke has hence emerged. Liquid smoke is generally produced from natural smoke by condensation or water scrubbing. The most prevalent method is a water scrubbing system. This method includes the smoldering of sawdust, controlling oxidation condition, and absorption of smoke in water. The smoke solution is recycled until a given smoke concentration is developed. Liquid smoke is often applied on product surface. The application methods include dipping, spraying, deluging, atomization, and regeneration. The most widely used methods are spraying, deluging, or atomization. Spraying or deluging is implemented immediately after the product is stuffed. Atomization employs high liquid and air pressure to form a cloud of liquid smoke to which products are then exposed. Atomization normally follows predrying process to assure better and more uniform color. For the last few years, liquid smoke has been improved to the point that the flavor profile is very similar to natural smoke.

COOKING

The processed meat after drying and smoking is cooked to a high internal temperature to assure sufficient microkill. The time–temperature relationship that is required for each product category is defined in FSIS Directive. Some processed meats do not require smoke flavor and color such as cook-in-bag products, which can be subjected to cooking only without drying and smoking. The most commonly used cooking medium is high humidity air, saturated steam, or hot water. In order to prevent smoke color from being washed out and to reduce product surface grease, the predried and smoked products are usually cooked with high humidity air. The relative humidity is normally no less than 50% in the cooking cycle.

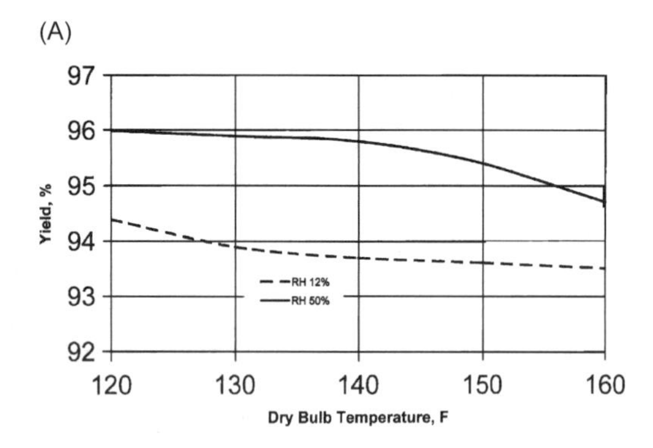

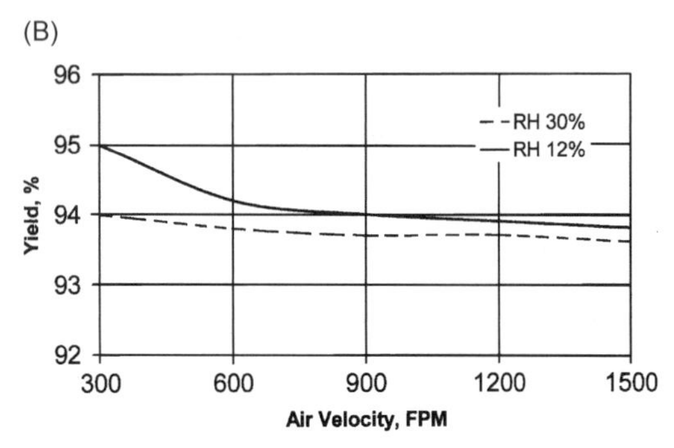

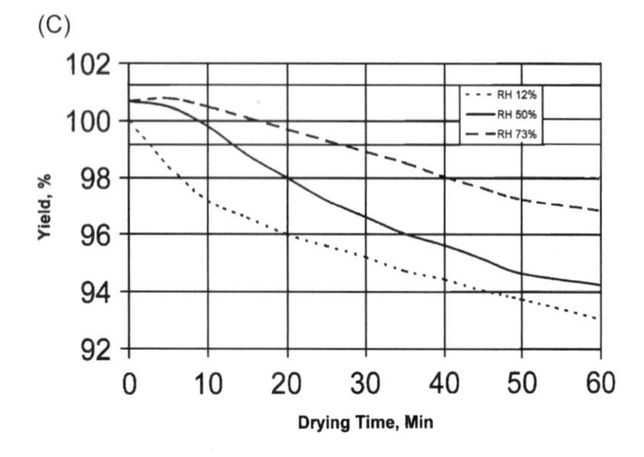

Fig. 4 (A) Effect of dry bulb temperature, frankfurter (25 mm × 127 mm), drying time 40 min; (B) effect of air velocity, frankfurter (25 mm × 127 mm), dry bulb 160°F, drying time 40 min; (C) effect of humidity, frankfurter (25 mm × 127 mm), dry bulb 120°F.

The critical process control point of humid air cooking is the wetness and cleanness of wet bulb sock. If wet bulb sock dries out or is covered with creosote from the smoking process, the wet bulb temperature will not reflect the true temperature. Then the product can be undercooked and overdried. The commonly used cooking medium is saturated steam. Steam cooking is condensation heat transfer that has a high heat transfer coefficient. The other widely employed cooking medium is hot water. Hot water also has a high heat transfer coefficient and is normally used to cook molded products. If water is circulated during cooking, there is no cooking rate difference between steam and hot water as illustrated in Fig. 5.[6] Because of multiquality attributes required for cooked meat, one constant cooking temperature may not produce a desirable product. Therefore, step changes in cooking temperature are often utilized to cook meat products.

COOLING

After drying, smoking, and cooking, products are chilled to less than 40°F to delay the onset of microbial growth. The time–temperature relationship required to chill a specific product is dictated by government regulations. Air cooling and brine chilling are the two most commonly used methods in the meat industry. Depending on the circumstances and product requirement, one method may be preferred over the other. However, there are several disadvantages with air cooling such as slow cooling rate, nonuniform cooling, and more moisture loss. As rapid cooling is needed, brine chilling is used in almost every continuous system, especially for products that require casing to be removed on line. Brine chilling can keep the casing moist and improves peelability substantially. In general practice, regular cold-water shower is employed before refrigerated brine chilling. The time ratio of water shower to brine chilling depends on the costs of water and refrigeration. Product shrink and flavor can be affected by brine chilling. When brine has higher salt concentration than that in the product, the moisture will migrate from product to the brine and salt will migrate from brine to the product. Higher shrink and salty flavor will occur. Therefore, salt concentration is critical in brine chilling process. The optimum operating conditions should maintain brine at the equilibrium condition with product.

CONCLUSION

In designing a thermal processing system to dry, smoke, cook, and chill meat products, the production rate and the product drying, heating, and cooling rates with a specific set of heat processing schedule should be determined. Then the required equipment capacity for air handling, heating, and cooling can be calculated by setting up material and energy balances. In meat industry, exhausting humid air and replacing with outside air is extensively used to control humidity in the smokehouse. In summer time, outside air is hot and humid, so some product may not dry properly. In this case, a mechanical

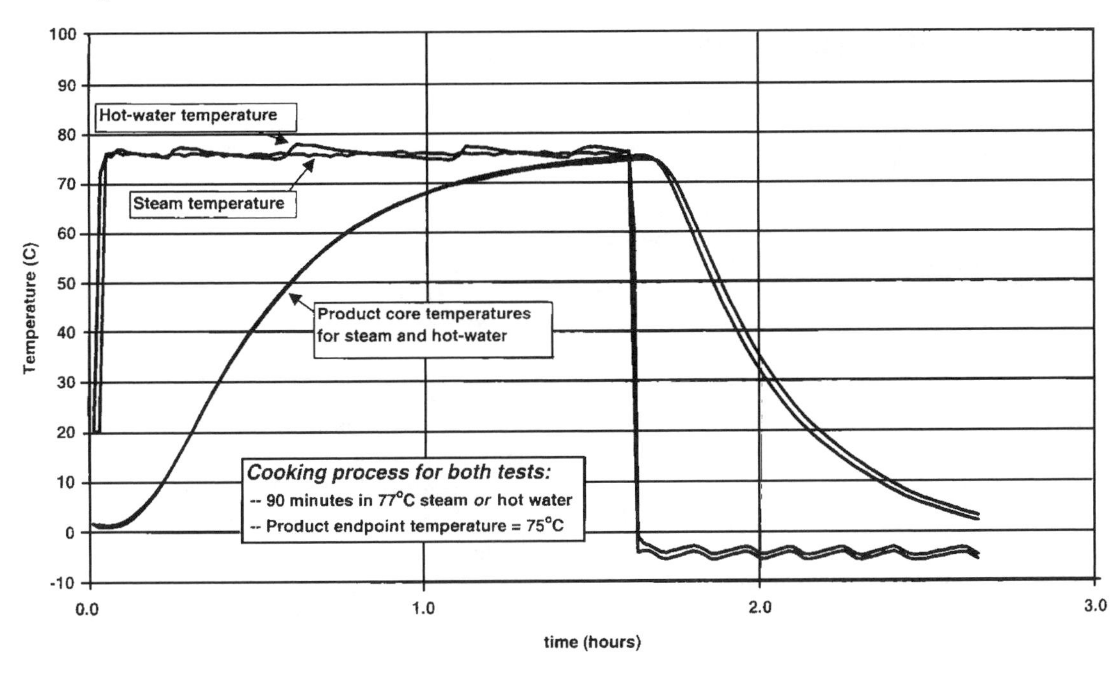

Fig. 5 Steam vs. hot-water cooking of sausage product stuffed in plastic casings (55 mm diameter). (Courtesy of Alkar Co.)

dehumidification system is required to achieve a proper drying. If natural smoking is employed, a pollution control unit should be built into the smokehouse exhaust system to prevent the pollutants discharge to the atmosphere.

REFERENCES

1. Wang, P.Y. Meat Processing Technology and Engineering. *Encyclopedia of Food Science and Technology*; John Wiley & Sons: New York, 1992; 1696–1704.
2. Hanson, R.E. Reducing Processing Variation in the Cooking and Smoking Process. Proceedings of the 50th Annual Reciprocal Meat Conference, 1997; 33–42.
3. Stech, I.; Osborne, W.R.; Mittal, G.S. Influence of Smokehouse Air Flow Patterns, Air Changes and Cook Cycle on Texture and Shrinkage of Wieners. J. Food Sci. **1988**, *53* (2), 421–424.
4. Maga, J.A. *Smoke in Food Processing*; CRC Press: Boca Raton, FL, 1988.
5. Simon, S.; Rydinski, A.A.; Tauber, F.W. Water-Filled Cellulose Casing as Model Absorbents for Wood Smokes. Food Technol. **1966**, *20*, 1494–1498.
6. Pulsfus, S. Steam Cooking Vs. Hot Water Cooking. Alkar Technical Report 2002, Lodi, WI.

Thermal Resistance Constant

Elton F. Morales-Blancas
Universidad Austral de Chile, Valdivia, Chile

J. Antonio Torres
Oregon State University, Corvallis, Oregon, U.S.A.

INTRODUCTION

The inactivation of micro-organisms by heat became a fundamental operation in food preservation during the 20th century and will remain so in the 21st century.[1] A critical aspect in thermal processing is the need to know the parameter quantifying the influence of lethal temperatures on the inactivation of microbial populations. This parameter is the thermal resistance constant or z-value.[2–10] It is important to note that the complete characterization of the impact of lethal temperature on microbial population requires reference to the decimal reduction time (D-value) and the thermal death time (TDT) (F-value) in addition to the z-value concept.[9,10] Therefore, this article will focus on the development of the z-value concept and its mathematical relationship with the D-value and F-value. The article concludes with the application and importance of thermal resistance parameters.

CONCEPT AND MATHEMATICAL REPRESENTATIONS

The temperature influence on microbial inactivation rates has been expressed in terms of the z-value coefficient beginning with the early years of thermobacteriology. Those beginning efforts include the quantitative measurements by Chick[11] on the thermal inactivation of vegetative cells, and by Esty and Meyer[12] on *Clostridium botulinum* spores, complemented with the development of the first graphical procedure based on "thermal death point" studies to represent on a semilog scale the resistance of bacteria to a lethal temperature.[13,14] The z-value concept was applied extensively in several primary reports during the development of the graphical and mathematical method for thermal process evaluation.[15–22] The general principles of thermal resistance of micro-organisms are well presented in the classical book by Ball and Olson,[2] in several textbooks and book chapters (e.g., Refs. [3]–[10]), and more recently in an electronic report prepared by the Institute of Food Technologists to the FDA/CFSAN.[10]

The z-value is a basic parameter in the evaluation scheme of microbial inactivation by heat. This term appeared in the first efforts to fit raw thermal death rate data to a temperature coefficient using a TDT curve, i.e., plotting on a semilog graph the TDT values as a function of lethal temperature.[15,16,18] Thus, originally z-values were calculated from values of TDT (F-values) defined as the "time necessary to inactivate all bacterial spores in a specific medium at a specified temperature (e.g., 250°F)."[14] For thermal resistance studies, the bacteriologist's definition of death is simple and practical, i.e., "a bacterium is dead when it has lost its ability to reproduce."[4,23]

The z-value, lately referred as *thermal resistance constant* and denoted by $z(T)$,[10] is a characteristic of a micro-organism and expresses the change in death rate with respect to a change in lethal temperature. Numerically it is equal to the number of degrees (in °F or °C) required to change by one logarithmic cycle (tenfold factor) the value of the *decimal reduction time* (D-value). Graphically, it can be determined by plotting the logarithm of D-value as a function of the lethal temperature, which since 1950 is frequently referred to as the TDT curve, instead of using the traditional TDT plot.[7,22,24–32] The difference between both graphs is trivial since F-values are multiples of D-values.[4] Sometimes the graph is also called the *thermal resistance*[3,8,9,33,34] or *thermal destruction* (TD) curve.[4,35–37]

Over the temperature range of concern in food sterilization technology, TDT curves approximate straight lines[4] (Fig. 1). Mathematically, z-value is equal to the negative reciprocal of the slope of the death rate curve and may be expressed as follows:

$$\frac{\log D_{T_1} - \log D_{T_2}}{T_1 - T_2} = -\frac{1}{z} \qquad (1)$$

Eq. 1 is used when z-value is calculated from D-values[22,31,38–44] and it is sometimes denoted by the z_D symbol. Also, the z-value can be determined from TDT

Encyclopedia of Agricultural, Food, and Biological Engineering
DOI: 10.1081/E-EAFE 120006980
Copyright © 2003 by Marcel Dekker, Inc. All rights reserved.

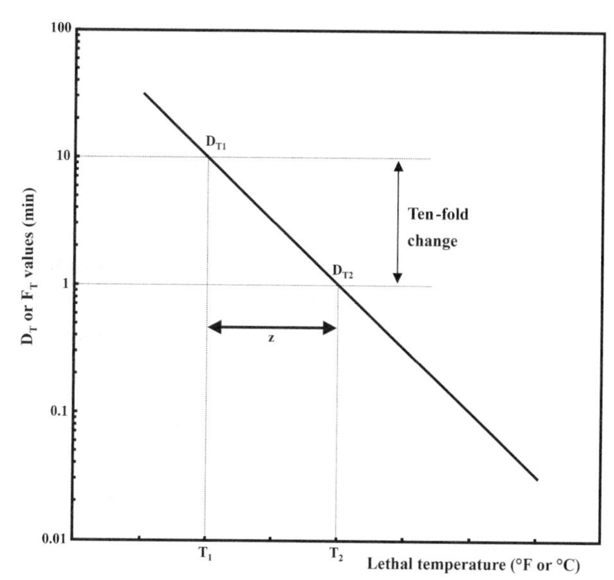

Fig. 1 Schematic representation of a TDT curve.

(F-value) data[8,18,42,45–47] and in this case the symbol z_F is sometimes used. Thus, the slope of the TDT curve may be expressed as follows:

$$\frac{\log F_{T_1} - \log F_{T_2}}{T_1 - T_2} = -\frac{1}{z} \quad (1a)$$

Eq. 1 is frequently rewritten in the following form[6,10]:

$$\log\left[\frac{D_T}{D_{T_{\mathrm{ref}}}}\right] = -\frac{(T - T_{\mathrm{ref}})}{z} \quad (2)$$

or rearranged as[6,8,48,49]:

$$D_T = D_{T_{\mathrm{ref}}} 10^{(T_{\mathrm{ref}} - T)/z} \quad (3)$$

Based on Eq. 1a, we obtain the equation of the TDT curve[6,8,48]:

$$\mathrm{TDT} = F_T = F_{T_{\mathrm{ref}}} 10^{(T_{\mathrm{ref}} - T)/z} \quad (3a)$$

The *decimal reduction time* term, using the abbreviation DRT instead of the symbol D used nowadays, was first introduced in 1943 by Katzin, Sandholzer, and Strong[50] in their quantitative determinations of microbial death rate by heat. At the same time, Ball[51] followed by Stumbo[19,20] used the symbol Z (Zeta) to represent the "slope value of the logarithmic survivor curve" and defined it as "the number of minutes required for the survivor curve to traverse one logarithmic cycle." To avoid confusion of the symbol Z with the z value, which represent the slope of the TDT curve, researchers in the field agreed to use the symbol D instead of Z.[22] The decimal reduction time (D-value) concept came into general use after 1950.[2,4,6]

The D-value is defined as "the time required to inactivate 90% of the spores or vegetative cells of a given organism when exposed to a constant lethal temperature in a given medium."[8,10,52] The expanded definition as "the time required for the concentration of any component, including micro-organisms, to change by a factor of ten at a fixed temperature"[48] allows its application to chemical factors following also first-order reaction kinetics.[6] Numerically, it is equal to the number of minutes required for the survivor or chemical concentration curve to cross a 1-log cycle. In the case of microbial inactivation, it can be determined graphically by plotting the number of surviving micro-organisms as a function of the time of heating at a constant lethal temperature using what is typically called a *survivor curve*[8,9] (Fig. 2). Mathematically, it is equal to the negative reciprocal of the slope of the survivor curve and can be calculated as follows:

$$\frac{\log N_{t_1} - \log N_{t_2}}{t_1 - t_2} = -\frac{1}{D_T} \quad (4)$$

Eq. 4 can be rearranged into the survivor curve equation to describe changes in microbial population as a function of heating time:

$$\log\frac{N_t}{N_0} = -\frac{t}{D_T} \quad (5)$$

As suggested by Pflug,[53,54] Eq. 5 is the empirical model expressing the F-value as multiple of D-values as follows:

$$F^z_{T_{\mathrm{ref}}} = D_{T_{\mathrm{ref}}}(\log N_0 - \log N_t) \quad (5a)$$

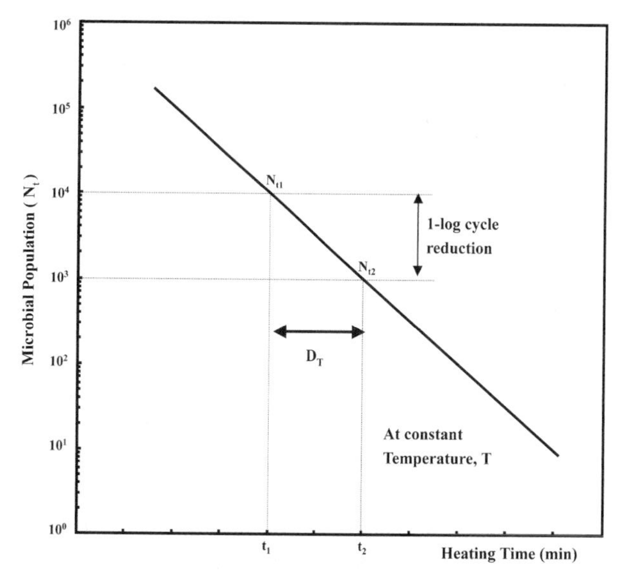

Fig. 2 Schematic representation of a survivor curve.

Eq. 5a is widely used by food microbiologist to determine the time required in a thermal process to achieve a desired "safe" final concentration of spores established from public health ("commercial sterility") or "economic spoilage" considerations.[6,48,55] Unfortunately, most authors do not use the superscript z to emphasize the type of micro-organism considered in this determination.

The parameters D_T and z are empirical coefficients determined from the kinetics for the reduction of a microbial population. Their definitions were developed on the basis that members of a homogeneous microbial population exposed to a lethal temperature are affected only by heat inactivation and that death rate is logarithmic in nature. The logarithmic death rate has been well substantiated in many independent studies[18,19,22,25,27,29,30,32–34,36–41,44,45,56–82] and for a given microbial population appears as a straight line on a semilog graph (Fig. 2). However, deviations that can be characterized as early shoulders and late tailings are sometimes observed in survivor curves.[56,83–93] Alternative mechanistic[94–101] and vitalistic[102–110] models have been proposed to account for these experimental deviations from the logarithmic death model. Future studies will be needed to demonstrate the advantages and then implement model alternatives to the D- and z-value concepts used today in thermal process calculations.

APPLICATION AND IMPORTANCE OF THERMAL RESISTANCE PARAMETERS

The traditional empirical approach, expressed in the definition of z- and D-values and the related mathematical models, has been widely accepted to describe the thermal inactivation rate of microbial populations and makes the calculation of thermal process equivalency relatively straightforward. This approach was developed as a pragmatic and practical tool for the commercial and industrial user in contrast to the more fundamental approach based on chemical reaction kinetics and the Arrhenius model. Food scientists and engineers have used this empirical approach to develop effective graphical and mathematical procedures for calculating and evaluating thermal processes for foods.[2,15–17,19,21,22,51,111]

The logarithmic nature of the survivor or destruction curve indicates that complete inactivation of microbial populations is impossible, since a decimal fraction of the population will survive even after an infinite number of D values. In practice, survivors are expressed using a probability approach; e.g., a surviving population of 10^{-n}/unit indicates one survivor in 10^n units subjected to a given heat treatment.[6,8] Large z values suggest that a given increase in lethal temperature of exposure for a microbial population results in a small change in D- or F-values. In general, z-values for quality factors are larger than those observed for microbial populations explaining the advantage of short processing times at high temperature. Finally, larger D- or F-values at a given temperature reflect the higher thermal resistance of a microbial population.[9]

As indicated by Pflug,[52–54] the safest way of designing heat treatments for sterilization purposes is to start from the most adverse and conservative situation. When establishing a heat process for low-acid foods, the manufacturer must always take into account public health (i.e., *C. botulinum*) considerations and the spoilage by mesophilic and thermophilic sporeforming organisms. Townsend, Esty, and Baselt[18] corrected the data of Esty and Meyer[12] with respect to lags in heating and cooling and obtained a $F_{250°F}$ of 2.45 min and a z-value of 18°C (actually 17.6°F) for *C. botulinum* spores. Afterwards, by applying the semilog model on Esty and Meyer's data the $D_{250°F}$-value was revised to 0.2 min.[53] These revised parameters were then used as the basis for selecting heat process schedules for preserving low-acid canned foods against botulism.[2,4,61] The procedures and criteria used to develop these schedules explain the remarkable record of safety achieved by the thermally processed food industry. The only known cases of underprocessing from commercial operations have resulted from the improper use of data, usually by inexpert personnel on food sterilization technology.[112] Errors in characterizing the heating of food, or in estimating the kinetics for the inactivation of micro-organisms, have caused deterioration of the food to such an extent as to render it unfit for consumption.[113,114]

Finally, in order to respond to the higher quality demands of modern consumers, the food industry needs the thermal resistance parameters (D- and z-values) for quality attributes, in addition to those for microbial factors. This will allow industry to establish the time/temperature relationship needed to provide a desired shelf life and product safety while avoiding overprocessing and thus unnecessary reductions in product quality.[9,115–119]

CONCLUSION

The thermal resistance constant (z-value) is a quantitative parameter utilized throughout the thermal processing literature to quantify the influence of lethal temperatures on homogeneous microbial populations. Its magnitude allows us to quantify the relationships between D- or F-values and lethal temperature. Although there is evidence of deviation from the logarithmic death model, at least for some microbial populations, a universally accepted alternative model has not been established. Consequently, the logarithmic death model, including the D-, z-, and F-value concepts, developed in the first 50 yr of the 20th century, is still widely used by the food industry and the authorities overseeing the correct application of food sterilization technologies.

NOMENCLATURE

D_T	Decimal reduction time, time required for 90% microbial inactivation at a constant lethal temperature *T*, dimension of time (e.g., min)
F_T	Time at a constant lethal temperature *T* required to inactivate a given percentage of a microbial population, dimension of time (e.g., min)
F_T^z	*F*-value specifying thermal resistance of a microbial population characterized by *z*-value
N_0	Initial microbial population
N_t	Microbial population at any time, *t*
t	Time (e.g., min)
T	Relative temperature (°C or °F)
T_{ref}	Reference temperature in the temperature range used to generate kinetic parameters (e.g., °C or °F)
z	Thermal resistance constant, a typical parameter for a micro-organism that measures the influence of temperature on microbial population inactivation rates; dimension of temperature degrees (e.g., °C or °F)
z_D	*z*-value determined from *D*-value data
z_F	*z*-value determined from *F*-value data
$z(T)$	Coefficient *z* indicating heat as the primary mode of microbial inactivation and used to distinguish it from $z(P)$ and $z(E)$ used in nonthermal processes such as high hydrostatic pressure and pulsed electric fields processing, respectively
Z	Zeta, used in the 1940s to represent the time (e.g., min) required for the survivor curve to traverse one logarithmic cycle and changed in the 1950s to the symbol *D*

ACKNOWLEDGMENTS

Author Elton F. Morales-Blancas acknowledges Comisión Nacional de Investigación Científica y Tecnológica of Chile (CONICYT—Project No. 1970303), Universidad Austral de Chile and Ministerio de Educación of Chile (MECESUP Program—Project No. AUS9908) for supporting a professional opportunity as an Oregon State University Faculty on courtesy appointment.

REFERENCES

1. Brody, A.L. Food Canning in the 21st Century. Food Technol. **2002**, *56* (3), 75–78.
2. Ball, C.O.; Olson, F.C.W. *Sterilization in Food Technology. Theory, Practice and Calculations*; McGraw-Hill Book Co.: New York, 1957.
3. Pflug, I.J.; Schmidt, C.F. Thermal Destruction of Microorganisms. In *Disinfections, Sterilization and Preservation*; Lawrence, C.A., Block, S.S., Eds.; Lea and Febiger: Philadelphia, 1968; 63–105.
4. Stumbo, C.R. *Thermobacteriology in Food Processing*, 2nd Ed.; Academic Press, Inc.: New York, 1973.
5. Casolari, A. Microbial Death. In *Physiological Models in Microbiology*; Bazin, M.J., Prosser, J.I., Eds.; CRC Press Inc.: Boca Ratón, FL, 1988; Vol. II, 1–44.
6. Toledo, R.T. *Fundamentals of Food Engineering*, 2nd Ed.; Van Nostrand Reinhold: New York, 1991.
7. Texeira, A. Thermal Process Calculations. In *Handbook of Food Engineering*; Heldman, D.R., Lund, D.B., Eds.; Marcel Dekker, Inc.: New York, 1992; 563–619.
8. Ramaswamy, H.S.; Singh, R.P. Sterilization Process Engineering. In *Handbook of Food Engineering Practice*; Valentas, K.J., Rotstein, E., Singh, R.P., Eds.; CRC Press LLC: Boca Raton, FL, 1997; 37–69.
9. Heldman, D.R.; Hartel, R.W. *Principles of Food Processing*; Chapman & Hall-International Thomson Publishing: New York, 1997.
10. http://vm.cfsan.fda.gov/(comm/ift-over.html (accessed Aug 2002).
11. Chick, H. The Process of Disinfection by Chemical Agencies and Hot Water. J. Hyg. **1910**, *10* (2), 237–286.
12. Esty, J.R.; Meyer, K.F. The Heat Resistance of the Spores of *Clostridium botulinum* and Allied Anaerobes. XI. J. Infect. Dis. **1922**, *31*, 650–663.
13. Bigelow, W.D.; Esty, J.R. The Thermal Death Point in Relation to Time of Typical Thermophilic Organisms. J. Infect. Dis. **1920**, *27*, 602–617.
14. Bigelow, W.D. The Logarithmic Nature of Thermal Death Time Curves. J. Infect. Dis. **1921**, *29* (5), 528–536.
15. Ball, C.O. *Thermal Process Time for Canned Foods*; Bull No. 37; Natl. Research Council: Washington, D.C., 1923; Vol. 7, Part 1.
16. Ball, C.O. Mathematical Solution of Problems on Thermal Processing of Canned Foods. Univ. Calif. (Berkeley) Publ. Public Health **1928**, *1* (2), 15–245.
17. Ball, C.O. Advancements in Sterilization Methods for Canned Foods. Food Res. **1938**, *3* (1/2), 13–55.
18. Townsend, C.T.; Esty, J.R.; Baselt, F.C. Heat-Resistance Studies on Spores of Putrefactive Anaerobes in Relation to Determination of Safe Processes for Canned Foods. Food Res. **1938**, *3* (3), 323–346.
19. Stumbo, C.R. Bacteriological Considerations Relating to Process Evaluation. Food Technol. **1948**, *2* (2), 115–132.

T

20. Stumbo, C.R. Further Considerations Relating to Evaluation of Thermal Processes for Foods. Food Technol. **1949**, *3* (4), 126–131.
21. Stumbo, C.R. New Procedures for Evaluating Thermal Processes for Foods in Cylindrical Containers. Food Technol. **1953**, *7* (8), 309–315.
22. Stumbo, C.R.; Murphy, J.R.; Cochran, J. Nature of Thermal Death Time Curves for P.A. 3679 and *Clostridium botulinum*. Food Technol. **1950**, *4* (8), 321–326.
23. Gombas, D.E. Bacterial Spore Resistance to Heat. Food Technol. **1983**, *37* (11), 105–110.
24. Baird-Parker, A.C.; Boothroyd, M.; Jones, E. The Effect of Water Activity on the Heat Resistance of Heat Sensitive and Heat Resistant Strains of Salmonellae. J. Appl. Bacteriol. **1970**, *33* (3), 515–522.
25. Odlaug, T.E.; Pflug, I.J. Thermal Destruction of *Clostridium botulinum* Spores Suspended in Tomato Juice in Aluminum Thermal Death Time Tubes. Appl. Environ. Microbiol. **1977**, *34* (1), 23–29.
26. Cameron, M.S.; Leonard, S.J.; Barret, E.L. Effect of Moderately Acidic pH on Heat Resistance of *Clostridium sporogenes* Spores in Phosphate Buffer and in Buffered Pea Puree. Appl. Environ. Microbiol. **1980**, *39* (5), 943–949.
27. Put, H.M.C.; De Jong, J. The Heat Resistance of Ascospores of Four Saccharomyces spp. Isolated from Spoiled Heat-Processed Soft Drinks and Fruit Products. J. Appl. Bacteriol. **1982**, *52* (2), 235–243.
28. Brown, K.L.; Ayres, C.A.; Gaze, J.E.; Newman, M.E. Thermal Destruction of Bacterial Spores Immobilized in Food/Alginate Particles. Food Microbiol. **1984**, *1* (3), 187–198.
29. Splittstoesser, D.F.; Leasor, S.B.; Swanson, K.M.J. Effect of Food Composition on the Heat Resistance of Yeast Ascospores. J. Food Sci. **1986**, *51* (5), 1265–1267.
30. Sumner, S.S.; Sandros, T.M.; Harmon, M.C.; Scott, V.N.; Bernard, D.T. Heat Resistance of *Salmonella typhimurium* and *Listeria monocytogenes* in Sucrose Solutions of Various Water Activities. J. Food Sci. **1991**, *56* (6), 1741–1743.
31. Török, T.; King, A.D., Jr. Thermal Inactivation Kinetics of Food-Borne Yeasts. J. Food Sci. **1991**, *56* (1), 6–9, 59.
32. Bremer, P.J.; Osborne, C.M.; Kemp, R.A.; Van Veghel, P.; Fletcher, G.C. Thermal Death Times of *Hafnia alvei* Cells in a Model Suspension and in Artificially Contaminated Hot-Smoked Kahawai (*Arripis trutta*). J. Food Protect. **1998**, *61* (8), 1047–1051.
33. Esselen, W.B.; Pflug, I.J. Thermal Resistance of Putrefactive Anaerobe No. 3679 Spores in Vegetables in the Temperature Range of 250–290°F. Food Technol. **1956**, *10* (11), 557–560.
34. Mulak, V.; Tailliez, R.; Eb, P.; Becel, P. Heat Resistance of Bacteria Isolated from Preparations Based on Seafood Products. J. Food Protect. **1995**, *58* (1), 49–53.
35. Löwik, J.A.M.; Anema, P.J. Effect of pH on the Heat Resistance of *Clostridium sporogenes* Spores in Minced Meat. J. Appl. Bacteriol. **1972**, *35* (1), 119–121.
36. Reichart, O. A New Experimental Method for the Determination of the Heat Destruction Parameters of Microorganisms. Acta Aliment. **1979**, *8* (2), 131–155.
37. Feeherry, F.E.; Munsey, D.T.; Rowley, D.B. Thermal Inactivation and Injury of *Bacillus stearothermophilus* Spores. Appl. Environ. Microbiol. **1987**, *53* (2), 365–370.
38. Xezones, J.; Hutchings, I.J. Thermal Resistance of *Clostridium botulinum* (62A) Spores as Affected by Fundamental Food Constituents. Food Technol. **1965**, *19* (6), 113–115.
39. Alderton, G.; Chen, J.K.; Ito, K.A. Effect of Lysozyme on the Recovery of Heated *Clostridium botulinum* Spores. Appl. Microbiol. **1974**, *27* (3), 613–615.
40. Perkins, W.E.; Ashton, D.H.; Evancho, G.M. Influence of the z Value of *Clostridium botulinum* on the Accuracy of Process Calculations. J. Food Sci. **1975**, *40* (6), 1189–1192.
41. Odlaug, T.E.; Pflug, I.J.; Kautter, D.A. Heat Resistance of *Clostridium botulinum* Type B Spores Grown from Isolates from Commercially Canned Mushrooms. J. Food Protect. **1978**, *41* (5), 351–353.
42. Bradshaw, J.G.; Peeler, J.T.; Corwin, J.J.; Hunt, J.M.; Twedt, R.M. Thermal Resistance of *Listeria monocytogenes* in Dairy Products. J. Food Protect. **1987**, *50* (7), 543–544, 556.
43. Foegeding, P.M.; Leasor, S.B. Heat Resistant and Growth of *Listeria monocytogenes* in Liquid Whole Egg. J. Food Protect. **1990**, *53* (1), 9–14.
44. Schuman, J.D.; Sheldon, B.W.; Foegeding, P.M. Thermal Resistance of *Aeromonas hydrophila* in Liquid Whole Egg. J. Food Protect. **1997**, *60* (3), 231–236.
45. Reed, J.M.; Bohrer, C.W.; Cameron, E.J. Spore Destruction Rate Studies on Organisms of Significance in the Processing of Canned Foods. Food Res. **1951**, *16* (5), 383–408.
46. Foegeding, P.M.; Stanley, N.W. *Listeria monocytogenes* F5069 Thermal Death Times in Liquid Whole Egg. J. Food Protect. **1990**, *53* (1), 6–8, 25.
47. Osborne, C.M.; Bremer, P.J. Application of the Bigelow (z-Value) Model and Histamine Detection to Determine the Time and Temperature Required to Eliminate *Morganella morganii* from Seafood. J. Food Protect. **2000**, *63* (2), 277–280.
48. Merson, R.L.; Singh, R.P.; Carroad, P.A. An Evaluation of Ball's Formula Method of Thermal Process Calculations. Food Technol. **1978**, *32* (3), 66–72, 75.
49. Ávila, I.M.L.B.; Silva, C.L.M. Methodologies to Optimize Thermal Processing Conditions: An Overview. In *Processing Foods Quality Optimization and Process Assessment*; Oliveira, F.A.R., Oliveira, J.C., Eds.; CRC Press Inc.: Boca Ratón, FL, 1999; 67–82.
50. Katzin, L.I.; Sandholzer, L.A.; Strong, M.E. Application of the Decimal Reduction Time Principle to a Study of the Resistance of Coliform Bacteria to Pasteurization. J. Bacteriol. **1943**, *45* (3), 265–272.
51. Ball, C.O. Short-Time Pasteurization of Milk. Ind. Eng. Chem. **1943**, *35* (1), 71–84.

T

52. Pflug, I.J. Calculating F_T-Values for Heat Preservation of Shelf-stable, Low-Acid Canned Foods Using the Straight-Line Semilogarithmic Model. J. Food Protect. **1987**, *50* (7), 608–615.
53. Pflug, I.J. Using the Straight-Line Semilogarithmic Microbial Destruction Model as an Engineering Design Model for Determining the F-Value for Heat Processes. J. Food Protect. **1987**, *50* (4), 342–346.
54. Pflug, I.J. Factors Important in Determining the Heat Process Value, F_T, for Low-Acid Canned Foods. J. Food Protect. **1987**, *50* (6), 528–533.
55. Stumbo, C.R.; Purohit, K.S.; Ramakrishnan, T.V. Thermal Process Lethality Guide for Low-Acid Foods in Metal Containers. J. Food Sci. **1975**, *40* (6), 1316–1323.
56. Shull, J.J.; Cargo, G.T.; Ernst, R.R. Kinetics of Heat Activation and Thermal Death of Bacterial Spores. Appl. Microbiol. **1963**, *11* (6), 485–487.
57. Koch, A.L. The Logarithm in Biology. II. Distributions Simulating the Log-Normal. J. Theor. Biol. **1969**, *23* (2), 251–268.
58. Alderton, G.; Chen, J.K.; Ito, K.A. Heat Resistance of the Chemical Resistance Forms of *Clostridium botulinum* 62A Spores over the Water Activity Range 0 to 0.9. Appl. Environ. Microbiol. **1980**, *40* (3), 511–515.
59. Davies, F.L.; Underwood, H.M.; Perkin, A.G.; Burton, H. Thermal Death Kinetics of *Bacillus stearothermophilus* Spores at Ultra High Temperatures. I. Laboratory Determination of Temperature Coefficients. J. Food Technol. **1977**, *12* (2), 115–129.
60. Perkin, A.G.; Burton, H.; Underwood, H.M.; Davies, F.L. Thermal Death Kinetics of *Bacillus stearothermophilus* Spores at Ultra High Temperatures. II. Effect of Heating Period on Experimental Results. J. Food Technol. **1977**, *12* (2), 131–148.
61. Pflug, I.J.; Odlaug, T.E. A Review of z and F Values Used to Ensure the Safety of Low-Acid Canned Food. Food Technol. **1978**, *32* (6), 63–70.
62. Beuchat, L.R.; Brackett, R.E.; Hao, D.Y.-Y.; Conner, D.E. Growth and Thermal Inactivation of *Listeria monocytogenes* in Cabbage and Cabbage Juice. Can. J. Microbiol. **1986**, *32* (10), 791–795.
63. Bucknavage, M.W.; Pierson, M.D.; Hackney, C.R.; Bishop, J.R. Thermal Inactivation of *Clostridium botulinum* Type E Spores in Oyster Homogenates at Minimal Processing Temperatures. J. Food Sci. **1990**, *55* (2), 372–373, 429.
64. Fain, A.R., Jr.; Line, J.E.; Moran, A.B.; Martin, L.M.; Lechowich, R.V.; Carosella, J.M.; Brown, W.L. Lethality of Heat to *Listeria monocytogenes* Scott A: D-Value and z-Value Determinations in Ground Beef and Turkey. J. Food Protect. **1991**, *54* (10), 756–761.
65. Line, J.E.; Fain, A.R., Jr.; Moran, A.B.; Martin, L.M.; Lechowich, R.V.; Carosella, J.M.; Brown, W.L. Lethality of Heat to *Escherichia coli* 0157:H7: *D*-Value and *z*-Value Determinations in Ground Beef. J. Food Protect. **1991**, *54* (10), 762–766.
66. Schoeni, J.L.; Brunner, K.; Doyle, M.P. Rates of Thermal Inactivation of *Listeria monocytogenes* in Beef and Fermented Beaker Sausage. J. Food Protect. **1991**, *54* (5), 334–337.
67. Condon, S.; Sala, F.J. Heat Resistance of *Bacillus subtilis* in Buffer and Foods of Different pH. J. Food Protect. **1992**, *55* (8), 605–608.
68. Kornacki, J.L.; Marth, E.H. Thermal Inactivation of *Salmonella senftenberg* and *Micrococcus freudenreichii* in Retentates from Ultrafiltered Milks. Lebensm.-Wiss. Technol. **1993**, *26* (1), 21–27.
69. Silla Santos, M.H.; Torres Zarzo, J.; Arranz Santamarta, A.; Peris Toran, M.J. Citric Acid Lowers Heat Resistance of *Clostridium sporogenes* PA 3679 in HTST White Asparagus Purée. Int. J. Food Sci. Technol. **1993**, *28* (6), 603–610.
70. Fernández, P.S.; Ocio, M.J.; Sánchez, T.; Martinez, A. Thermal Resistance of *Bacillus stearothermophilus* Spores Heated in Acidified Mushroom Extract. J. Food Protect. **1994**, *57* (1), 37–41.
71. Sörqvist, S. Heat Resistance of Different Serovars of *Listeria monocytogenes*. J. Appl. Bacteriol. **1994**, *76* (4), 383–388.
72. Ababouch, L.H.; Grimit, L.; Eddafry, R.; Busta, F.F. Thermal Inactivation Kinetics of *Bacillus subtilis* Spores Suspended in Buffer and in Oils. J. Appl. Bacteriol. **1995**, *78* (6), 669–676.
73. Palumbo, M.S.; Beers, S.M.; Bhaduri, S.; Palumbo, S.A. Thermal Resistance of Salmonella spp. and *Listeria monocytogenes* in Liquid Egg Yolk and Egg Yolk Products. J. Food Protect. **1995**, *58* (9), 960–966.
74. Haas, J.; Behsnilian, D.; Schubert, H. Determination of the Heat Resistance of Bacterial Spores by the Capillary Tube Method. II—Kinetic Parameters of *Bacillus stearothermophilus* Spores. Lebensm.-Wiss. Technol. **1996**, *29* (4), 209–303.
75. Kotrola, J.S.; Conner, D.E. Heat Inactivation of *Escherichia coli* O157:H7 in Turkey Meat as Affected by Sodium Chloride, Sodium Lactate, Polyphosphate, and Fact Content. J. Food Protect. **1997**, *60* (8), 898–902.
76. Orta-Ramirez, A.; Price, J.F.; Hsu, Y.-C.; Veeramuthu, G.J.; Cherry-Merritt, J.S.; Smith, D.M. Thermal Inactivation of *Escherichia coli* O157:H7, *Salmonella senftenberg*, and Enzymes with Potential as Time–Temperature Indicators in Ground Beef. J. Food Protect. **1997**, *60* (5), 471–475.
77. Veeramuthu, G.J.; Price, J.F.; Davis, C.E.; Booren, A.M.; Smith, D.M. Thermal Inactivation of *Escherichia coli* O157:H7, *Salmonella senftenberg*, and Enzymes with Potential as Time–Temperature Indicators in Ground Turkey Thigh Meat. J. Food Protect. **1998**, *61* (2), 171–175.
78. Annous, B.A.; Kozempel, M.F. Influence of Growth Medium on Thermal Resistance of *Pediococcus sp.* NRRL B-2354 (Formerly *Micrococcus freudenreichii*) in Liquid Foods. J. Food Protect. **1998**, *61* (5), 578–581.
79. Cho, H.-Y.; Yousef, A.E.; Sastry, S.K. Kinetics of Inactivation of *Bacillus subtilis* Spores by Continuous or Intermittent Ohmic and Conventional Heating. Biotechnol. Bioeng. **1999**, *62* (3), 368–372.

80. López-Malo, A.; Guerrero, S.; Alzamora, S.M. *Saccharomyces cerevisiae* Thermal Inactivation Kinetics Combined with Ultrasound. J. Food Protect. **1999**, *62* (10), 1215–1217.
81. Dock, L.L.; Floros, J.D.; Linton, R.H. Heat Inactivation of *Escherichia coli* O157:H7 in Apple Cider Containing Malic Acid, Sodium Benzoate, and Potassium Sorbate. J. Food Protect. **2000**, *63* (8), 1026–1031.
82. Murphy, R.Y.; Marks, B.P.; Johnson, E.R.; Johnson, M.G. Thermal Inactivation Kinetics of Salmonella and Listeria in Ground Chicken Breast Meat and Liquid Medium. J. Food Sci. **2000**, *65* (4), 706–710.
83. Stabel, J.R.; Steadham, E.M.; Bolin, C.A. Heat Inactivation of *Mycobacterium paratuberculosis* in Raw Milk: Are Current Pasteurization Conditions Effective? Appl. Environ. Microbiol. **1997**, *63* (12), 4975–4977.
84. Berry, M.R., Jr.; Bradshaw, J.G.; Kohnhorst, A.L. Heating Characteristics of Ravioli in Brine and in Tomato Sauce Processed in Agitating Retorts. J. Food Sci. **1985**, *50* (3), 815–822.
85. Rhan, O. The Non-logarithmic Order of Death of Some Bacteria. J. Gen. Physiol. **1930**, *13* (4), 395–407.
86. Withell, E.R. The Significance of the Variation on Shape of Time-Survivors Curves. J. Hyg. **1942**, *42* (2), 124–183.
87. Hansen, N.-H.; Riemann, H. Factors Affecting the Heat Resistance of Nonsporing Organisms. J. Appl. Bacteriol. **1963**, *26* (3), 314–333.
88. Moats, W.A. Kinetics of Thermal Death of Bacteria. J. Bacteriol. **1971**, *105* (1), 165–171.
89. Moats, W.A.; Dabbah, R.; Edwards, V.M. Interpretation on Nonlogarithmic Survivor Curves of Heated Bacteria. J. Food Sci. **1971**, *36* (3), 523–526.
90. Gould, G.W. Heat-Induced Injury and Inactivation. In *Mechanisms of Action of Food Preservation Procedures*; Gould, G.W., Ed.; Elsevier Science Publishers: London, 1989; 11–42.
91. Mackey, B.M.; Pritchet, C.; Norris, A.; Mead, G.C. Heat Resistance of Listeria: Strain Differences and Effects of Meat Type and Curing Salts. Lett. Appl. Microbiol. **1990**, *10* (6), 251–255.
92. Bhaduri, S.; Smith, P.W.; Palumbo, S.A.; Turner-Jones, C.O.; Smith, J.L.; Marmer, B.S.; Buchanan, R.L.; Zaika, L.L.; Williams, A.C. Thermal Destruction of *Listeria monocytogenes* in Liver Sausage Slurry. Food Microbiol. **1991**, *8* (1), 75–78.
93. Fujikawa, H.; Itoh, T. Tailing of Thermal Inactivation Curve of *Aspergillus niger* Spores. Appl. Environ. Microbiol. **1996**, *62* (10), 3745–3749.
94. Le Jean, G.; Abraham, G.; Debray, E.; Candau, Y.; Piar, G. Kinetics of Thermal Destruction of *Bacillus stearothermophilus* Spores Using a Two Reaction Model. Food Microbiol. **1994**, *11* (3), 229–241.
95. Abraham, G.; Debray, E.; Candau, Y.; Piar, G. Mathematical Model of Thermal Destruction of *Bacillus stearothermophilus* Spores. Appl. Environ. Microbiol. **1990**, *56* (10), 3073–3080.
96. Texeira, A.A.; Rodriguez, A.C. Microbial Population Dynamics in Bioprocess Sterilization. Enzyme Microbiol. Technol. **1990**, *12* (6), 469–473.
97. Sapru, V.; Texeira, A.A.; Smerage, G.H.; Lindsay, J.A. Predicting Thermophilic Spore Population Dynamics for UHT Sterilization Processes. J. Food Sci. **1992**, *57* (5), 1248–1252, 1257.
98. Sapru, V.; Smerage, G.H.; Texeira, A.A.; Lindsay, J.A. Comparison of Predictive Models for Bacterial Spore Population Resources to Sterilization Temperatures. J. Food Sci. **1993**, *58* (1), 223–228.
99. Rodriguez, A.C.; Smerage, G.H.; Texeira, A.A.; Busta, F.F. Kinetic Effects of Lethal Temperatures on Population Dynamics of Bacterial Spores. Trans. ASAE **1988**, *31* (5), 1594–1601, 1606.
100. Rodriguez, A.C.; Smerage, G.H.; Texeira, A.A.; Lindsay, J.A.; Busta, F.F. Population Model of Bacterial Spores for Validation of Dynamic Thermal Processes. J. Food Process. Eng. **1992**, *15* (1), 1–30.
101. Campanella, O.H.; Peleg, M. Theoretical Comparison of a New and the Traditional Method to Calculate *Clostridium botulinum* Survival During Thermal Inactivation. J. Sci. Food Agric. **2001**, *81* (11), 1069–1076.
102. Peleg, M.; Cole, M.B. Reinterpretation of Microbial Survival Curves. Crit. Rev. Food Sci. Nutr. **1998**, *38* (5), 353–380.
103. Cole, M.B.; Davies, K.W.; Munro, G.; Holyoak, C.D.; Kilsby, D.C. A Vitalistic Model to Describe the Thermal Inactivation of *Listeria monocytogenes*. J. Ind. Microbiol. **1993**, *12* (3/5), 232–239.
104. Little, C.L.; Adams, M.R.; Anderson, W.A.; Cole, M.B. Application of a Log-Logistic Model to Describe the Survival of *Yersinia enterocolitica* at Sub-optimal pH and Temperature. Int. J. Food Microbiol. **1994**, *22* (1), 63–71.
105. Ellison, A.; Anderson, W.A.; Cole, M.B.; Stewart, G.S.A.B. Modelling the Thermal Inactivation of *Salmonella typhimurium* Using Bioluminescence Data. Int. J. Food Microbiol. **1994**, *23* (3/4), 467–477.
106. Stephens, P.J.; Cole, M.B.; Jones, M.V. Effect of Heating Rate on the Thermal Inactivation of *Listeria monocytogenes*. J. Appl. Bacteriol. **1994**, *77* (6), 702–708.
107. Anderson, W.A.; McClure, P.J.; Baird-Parker, A.C.; Cole, M.B. The Application of a Log-Logistic Model to Describe the Thermal Inactivation of *Clostridium botulinum* 213B at Temperatures below 121.1°C. J. Appl. Bacteriol. **1996**, *80* (3), 283–290.
108. Kilsby, D.C.; Davies, K.W.; McClure, P.J.; Adair, C.; Anderson, W.A. Bacterial Thermal Death Kinetics Based on Probability Distributions: The Heat Destruction of *Clostridium botulinum* and *Salmonella bedford*. J. Food Protect. **2000**, *63* (9), 1179–1203.
109. van Boekel, M.A.J.S. On the Use of the Weibull Model to Describe Thermal Inactivation of Microbial Vegetative Cells. Int. J. Food Microbiol. **2002**, *74* (1), 139–159.
110. Peleg, M.; Engel, R.; Gonzalez-Martinez, C.; Corradine, M.G. Non-Arrhenius and Non-WLF Kinetics in Food Systems. J. Sci. Food Agric. **2002**, *82* (12), 1346–1355.

T

111. Bigelow, W.D.; Bohart, G.S.; Richardson, A.C.; Ball, C.O. *Heat Penetration in Processed Canned Foods*; Bull. No16-L, Natl. Canners' Assoc.: Washington, D.C., 1920.
112. Lund, D.B. Statistical Analysis of Thermal Process Calculations. Food Technol. **1978**, *32* (3), 76–78, 83.
113. Mulvaney, T.R.; Schaffner, R.M.; Miller, R.A.; Johnston, M.R. Regulatory Review of Scheduled Thermal Processes. Food Technol. **1978**, *32* (6), 73–76.
114. ICMSF. *Microorganisms in Foods 5. Characteristics of Microbial Pathogens*, 1st Ed.; Blackie Academic & Professional: London, 1996.
115. Lund, D.B. Design of Thermal Processes for Maximizing Nutrient Retention. Food Technol. **1977**, *31* (2), 71–78.
116. Holdsworth, S.D. Kinetic Data—What Is Available and What Is Necessary. In *Processing and Quality of Food: High Temperature Short Time (HTST) Processing*; Field, R.W., Howell, J.A., Eds.; Elsevier Applied Science: London, 1990; Vol. 1, 74–90.
117. Nasri, H.; Simpson, R.; Bouzas, J.; Torres, J.A. An Unsteady State to Determine Kinetic Parameters for Heat Inactivation of Quality Factors: Conduction-Heated Foods. J. Food Eng. **1993**, *19* (3), 291–301.
118. Villota, R.; Hawkes, J.G. Kinetics of Nutrient and Organoleptic Changes in Foods During Processing. In *Physical and Chemical Properties of Foods*; Okos, M.R., Ed.; American Society of Agricultural Engineering: St. Joseph, Michigan, 1986; 266–366.
119. Villota, R.; Hawkes, J.G. Reaction Kinetics in Food Systems. In *Handbook of Food Engineering*; Heldman, D.R., Lund, D.B., Eds.; Marcel Dekker, Inc.: New York, 1992; 39–144.

Thermal Resistance Parameters, Determination of

Elton F. Morales-Blancas
Universidad Austral de Chile, Valdivia, Chile

J. Antonio Torres
Oregon State University, Corvallis, Oregon, U.S.A.

INTRODUCTION

Thermal process calculations require an understanding of the thermal inactivation kinetics of micro-organisms, particularly the parameters describing the impact of lethal temperatures on the inactivation of a microbial population.[1–6] The published literature contains extensive data on the thermal resistance parameters of microbial populations to provide an initial estimate of the thermal process required for a given micro-organism and food product. In most situations, these initial estimates must be confirmed through evaluation under specific circumstances.[6] This article provides a brief overview and discussion of the methods used to determine basic kinetic parameters, namely the decimal reduction time (D-value) and the thermal resistance constant (z-value). Also included is a discussion on the limitations of calculated kinetic parameters found in the literature.

UNIVERSALITY OF THE LOGARITHMIC DEATH MODEL

Most researchers prefer, at least implicitly, the log-linear model to explain microbial death rate. The latter is referred as the "mechanistic theory"[7] and it assumes that "all cells or spores in a population have identical heat resistances, and it is merely the chance of a heat quantum impacting a heat-sensitive target in a cell or spore that determines death rate."[8–11] Deviations such as survivor curve shoulders and tailings[12–20] raise doubts about the universality of the logarithmic death model, and consequently the validity of the approaches described in the *Thermal Resistance Constant* article of this encyclopedia. Some researchers prefer to reject these deviations as experimental artifacts or explain them using several interpretations of the mechanism of heat inactivation or resistance.[9,21]

Survivor curve deviations characterized by an early shoulder have been successfully analyzed.[22–29] These researchers have developed more complex kinetic models of the mechanistic type considering the heat activation of dormant spores. The early shoulder is expressed as the result of these additional competing reactions obtaining a good fit between experimental and predicted data. Recently, a mechanistic model was proposed to predict survivor curve tails,[30] however, the mathematical procedure proposed needs additional experimental verifications.

An alternate microbial death rate model is the "vitalistic theory,"[7] which assumes that "individuals in a population do not have identical heat resistances and that these differences are permanent."[9,31] The main objection to this model is the relative scarcity of experimental evidence, compared to the number of logarithmic curves where tails or shoulders appear to be artifacts independent from the mechanism of inactivation.[9,21,32] However, some researchers have applied successfully the vitalistic theory to describe more accurately the death kinetics of a number of bacterial species in response to heat by fitting the log-logistic model.[33–39]

To obtain unquestionable survival curves, and therefore reliable models and kinetic parameters, researchers should follow three key recommendations[9,21]: 1) avoid the mere suspicion of an experimental/methodology artifact by choosing a reliable heating method and checking experimental procedures such as homogeneity of the treatment, treatment medium, clumping, protective effect, and enumeration of survivors among others; 2) study the survival curve over the highest possible number of decimal reductions; and 3) validate the data at low survivor concentrations by repeating these experiments.

Recent research[39–41] has emphasized the use of nonlog-linear models, including statistical interpretations, to describe the inactivation kinetics of microbial populations during preservation processes. However, currently there is insufficient information to support their general acceptance by researches, industry, and regulatory agencies. Thus, the logarithmic death model, including the D- and z-value concepts, is still universally used to

Encyclopedia of Agricultural, Food, and Biological Engineering
DOI: 10.1081/E-EAFE 120017721
Copyright © 2003 by Marcel Dekker, Inc. All rights reserved.

describe kinetics of thermal inactivation of microbial population when calculating sterility in thermal preservation processes.

METHODS OF ESTIMATION

The literature reports two methods for estimating microbial inactivation kinetic parameters, isothermal and nonisothermal heating procedures.[21,42–48] The traditional and most commonly used method for determining kinetic parameters under isothermal conditions for first-order kinetics is a two-step linear regression procedure.[1,49–55] Decimal reduction time (D_T) values are calculated first by linear regression of survivor data at different lethal temperatures and then the thermal resistance constant (z) is obtained by linear regression of these D_T-values. This procedure uses the following expressions for the decimal reduction time and the thermal resistance constant[47]:

$$\log N_t = \log N_0 - \frac{t}{D_T} \tag{1}$$

$$\log D_T = \log D_{T_{\mathrm{ref}}} - \frac{(T - T_{\mathrm{ref}})}{z} \tag{2}$$

Generally, the two-step method results in a relatively large confidence interval because of the small number of degrees of freedom.[56] Consequently, D_T-values predicted from the thermal death time curve also presents large confidence intervals.[47] Thus, the estimation of the kinetic parameters from the appropriate model requires careful attention to the statistical limits created by the experimental data. Statistical procedures to estimate the error in kinetic parameters must be followed with caution.[6,57]

In a study on the thermal inactivation kinetics of thiamine, Arabshahi and Lund[58] proposed a one-step nonlinear regression procedure. The objective was to increase the degrees of freedom and reduce the confidence interval for the kinetic parameters. The procedure uses the data as a whole and does not estimate parameters from values obtained from other statistical regressions. These authors and Cohen and Saguy[59] have indicated that the one-step method should be applied in all kinetic analysis. The advantages of the one-step nonlinear regression method were emphasized by van Boekel.[57] Under isothermal conditions, the z and $D_{T\mathrm{ref}}$ parameters can be estimated by a one-step nonlinear regression of the following form[46,47]:

$$\log N_t = \log N_0 - \frac{t}{D_{T_{\mathrm{ref}}} 10^{(T_{\mathrm{ref}} - T)/z}} \tag{3}$$

Periago et al.[47] determined the thermal inactivation parameters for *Bacillus stearothermophilus* spores using logarithms as in Eq. 3. The same authors concluded that although the two-step linear regression method has the advantage of being simple, the one-step nonlinear regression procedure is recommended when the confidence intervals of the kinetic parameters calculated by the two-step method are large making a comparison of these parameters statistically difficult.

On the other hand, the nonisothermal method for calculating basic kinetic parameters implies an iterative process. Initial values for D_T and z are assumed, and by using a nonlinear least squares program, the values of the D_T and z parameters are altered until the least sum of squares of the residuals is obtained.[60] Reichart[61] indicated that nonisothermal methods offer the advantage of determining the inactivation parameters under conditions in which temperature changes with time as it does during the actual sterilization process and there is also a reduction in the experimental effort. Lenz and Lund[62] indicated that a disadvantage of nonisothermal methods is their greater calculation complexity. However, an important advantage is the amount of information that can be obtained from a single experiment saving time, materials, and cost.[48] Periago et al.,[47] Fujikawa and Itoh,[46] and Leontidis et al.[48] calculated thermal resistance parameters (z and $D_{T\mathrm{ref}}$) under nonisothermal conditions following the guidelines of Tucker[63] for a numeric integration of the following expression:

$$\log \frac{N_t}{N_0} = -\int_{t_0}^{t} \frac{\mathrm{d}t}{D_{T_{\mathrm{ref}}} 10^{(T_{\mathrm{ref}} - T(t))/z}} \tag{4}$$

The traditional isothermal method is normally used to validate studies using a nonisothermal methodology as was done by Burton et al.,[64] Periago et al.,[47] and Leontidis et al.[48] to estimate basic kinetic parameters for thermal inactivation of *B. stearothermophilus* spores and Fujikawa and Itoh[46] for the thermal inactivation of mesophiles.

LIMITATIONS OF USE FOR THERMAL RESISTANCE PARAMETERS

Kinetic parameters for microbial populations exposed to thermal treatments, referred also as thermal resistance parameters, have been collected over several decades and the literature provides an impressive array of data for the development of thermal processes.[1,2,5,6,50,65,66] Unfortunately, many authors have failed to include information on the culture growth medium, composition of the recovery medium, menstruum in which the organisms were heated, pH, water activity (a_w), physiological age of the culture, and specific experimental conditions used such as the initial concentration of organisms, heating system tests, and most importantly the lethal temperature range employed.[1,9,49,66–70] Consequently, care should be taken when published *D*- and *z*-values are used to establish processes, compare the resistances of different microbial populations, or identify appropriate surrogate microorganisms.[6,65]

In this article, published data is briefly summarized (Table 1) showing the data ranges found in the literature. It is important to emphasize that data reported in the literature cannot be extrapolated and used to determine the *D*-value and the *z*-value in ranges where no kinetic data was ever published.[9,53,66,71,72] Experiments on the heat resistance of *Clostridium botulinum* and other microorganisms do not warrant extrapolations to *infinitum* of a survivor curve and do not ensure that the logarithmic model is always applicable. Reed et al.[73] pointed out that the majority of survivor curves are not exponential after the death of 99.99% of the spores, especially *C. botulinum* spores. This observation suggests that the tailing-off phenomenon cannot be disregarded as an experimental artifact[9] and heat-inactivation experiments must be extended to very low survival levels.[66] Casolari[74] indicated that the tailing-off phenomena might produce dramatic increases of the apparent *D*-value and, in turn, the *z*-value.

The literature lacks survival observations for 9 decimal reductions and it is extremely difficult to measure over 12 decimal reductions. Casolari[66] pointed out that a conservative criterion is to choose the highest *D*-value observed experimentally. Several researchers have emphasized that to determine a thermal process with reasonable accuracy it is necessary to know the initial microbial test concentration in addition to the *D*- and *z*-values obtained.[58,67,69] However, initial microbial and food component concentrations are not frequently available in most reports. Casolari[66] noted that the *D*- and *z*-values reported in the literature are correlated with larger *z*-values for large *D*-values. The same author remarked that high temperatures should be preferred for low-acid canned foods, provided they do not interfere with organoleptical, chemical, or technical factors, because the probability of deviations from exponential kinetics increases at lower temperatures. However, processing errors are more severe at high temperature suggesting that larger rather than smaller *z*-values should be used from the public health and quality preservation points of view.[50,69]

z- and *D*-values have also been used to express the temperature dependence of quality degradation by heat.[2,75–77] However, while most of the published data on food quality attributes have been presented as first-order reaction parameters, i.e., rate constant (k) and activation energy (E_a),[2,78–80] only a limited amount of the microbial inactivation data has been analyzed using these parameters. Published data on first-order kinetic parameters can be transformed into thermal resistance parameters using Eqs. 6 and 7 from the *Activation Energy in Thermal Process Calculations* article of this encyclopedia.

The data presented in Table 1 show that the quality attributes are more resistant (larger values of D and z constants) to thermal treatment than the microbial populations.[5,66] This observation is generally true for all quality attributes (vitamins, pigments, sensory quality among others) in most food products when compared to the typical microbial populations (spores and vegetative cells) of concern in foods preserved by thermal processing. The relationships presented suggest that the use of higher

Table 1 Range of thermal resistance parameters for microbial populations and food quality attributes

Constituent	pH	T_{ref} (°C)	$D_{T_{ref}}$ (min)	z (°C)
Spores				
C. botulinum type A and B, *B. stearothermophilus*, *C. sporogenes* (PA 3679)	> 4.5	121.1	0.1–5.0	7.7–12.2
B. coagulans	4.0–4.5	121.1	0.01–0.07	7.7–10
B. polymyxa, *B. macerans*, *C. pasteurianum*	4.0–4.5	100	0.1–0.5	6.7–8.9
Vegetative cells				
Mycobacterium tuberculosis, *Salmonella* spp., *Staphylococcus* spp., *Lactobacillus* spp., yeasts, and molds	< 4.0	82.2	0.0032–0.0095	4.4–7.0
Vitamin degradation				
Ascorbic acid, thiamine, vitamin B_6, vitamin A	—	121.1	43.5–921	17.8–50.7
Pigment degradation				
Chlorophyll, Anthocyanins	—	121.1	13.0–123	25.6–53.4
Sensory quality loss	—	121.1	5–500	25–44.4

(From Refs. [1], [2], [5], [75], [76].)

temperature and shorter time processes should favor the retention of nutritional and sensorial quality while achieving the desired level of microbial inactivation.

CONCLUSION

Although the literature shows alternatives to the log-linear microbial inactivation model for specific microbial populations, the kinetic parameters developed using the log-linear death model are still being used to develop thermal preservation processes ensuring safety, reducing spoilage, and minimizing food quality losses. Future studies are needed to demonstrate the advantages of alternatives to the log-linear death model currently used for thermal processing calculations. Several authors have demonstrated that alternative models can account for experimental deviations in survivor counts. The easy access to computational tools makes the evaluation of multiple model alternatives a practical choice. This will assist food processors achieve the main purpose of a thermal process, i.e., to ensure a safe and nutritious food supply with the optimum quality demanded by modern consumers.

NOMENCLATURE

D_T Decimal reduction time, time required for 90% microbial inactivation at a constant lethal temperature T, dimension of time (e.g., min)

E_a Activation energy constant, dimension of energy per mole (e.g., $J\,mol^{-1}$)

k_T First-order rate constant for microbial inactivation or quality attribute degradation at a constant temperature T, dimension of inverse units of time (e.g., min^{-1})

N_0 Initial microbial population or initial quality attribute

N_t Microbial population or quality attribute at any time, t

t Time (e.g., min)

T Relative temperature (°C or °F)

T_{ref} Reference temperature in the temperature range used to generate kinetic parameters (e.g., °C or °F)

z Thermal resistance constant, a typical parameter of each micro-organism that measures the influence of temperature on microbial population inactivation rates; dimension of temperature degrees (e.g., °C or °F)

ACKNOWLEDGMENTS

Author Elton F. Morales-Blancas acknowledges Comisión Nacional de Investigación Científica y Tecnológica of Chile (CONICYT—Project No. 1970303), Universidad Austral de Chile, and Ministerio de Educación of Chile (MECESUP Program—Project No. AUS9908) for supporting a professional opportunity as an Oregon State University Faculty on courtesy appointment.

REFERENCES

1. Stumbo, C.R. *Thermobacteriology in Food Processing*, 2nd Ed.; Academic Press, Inc.: New York, 1973.
2. Toledo, R.T. *Fundamentals of Food Engineering*, 2nd Ed.; Van Nostrand Reinhold: New York, 1991.
3. Teixeira, A. Thermal Process Calculations. In *Handbook of Food Engineering*; Heldman, D.R., Lund, D.B., Eds.; Marcel Dekker, Inc.: New York, 1992; 563–619.
4. Ramaswamy, H.S.; Singh, R.P. Sterilization Process Engineering. In *Handbook of Food Engineering Practice*; Valentas, K.J., Rotstein, E., Singh, R.P., Eds.; CRC Press LLC: Boca Raton, FL, 1997; 37–69.
5. Heldman, D.R.; Hartel, R.W. *Principles of Food Processing*; Chapman & Hall-International Thomson Publishing: New York, 1997.
6. http://vm.cfsan.fda.gov/~comm/ift-over.html (accessed Aug 2002).
7. Lee, R.E.; Gilbert, C.A. On the Application of the Mass Law to the Process of Disinfection—Being a Contribution to the "Mechanistic Theory" as Opposed to the "Vitalistic Theory". J. Phys. Chem. **1918**, *22* (5), 348–372.
8. Gombas, D.E. Bacterial Spore Resistance to Heat. Food Technol. **1983**, *37* (11), 105–110.
9. Cerf, O. Tailing of Survival Curves of Bacterial Spores: A Review. J. Appl. Bacteriol. **1977**, *42* (1), 1–19.
10. Charm, S.E. The Kinetics of Bacterial Inactivation by Heat. Food Technol. **1958**, *12* (1), 4–12.
11. McKee, S.; Gould, G.W. A Simple Mathematical Model of the Thermal Death of Microorganisms. Bull. Math. Biol. **1988**, *50* (5), 493–501.
12. Rhan, O. The Non-logarithmic Order of Death of Some Bacteria. J. Gen. Physiol. **1930**, *13* (4), 395–407.
13. Withell, E.R. The Significance of the Variation on Shape of Time-Survivors Curves. J. Hyg. **1942**, *42* (2), 124–183.
14. Hansen, N.-H.; Riemann, H. Factors Affecting the Heat Resistance of Nonsporing Organisms. J. Appl. Bacteriol. **1963**, *26* (3), 314–333.
15. Moats, W.A. Kinetics of Thermal Death of Bacteria. J. Bacteriol. **1971**, *105* (1), 165–171.
16. Moats, W.A.; Dabbah, R.; Edwards, V.M. Interpretation on Nonlogarithmic Survivor Curves of Heated Bacteria. J. Food Sci. **1971**, *36* (3), 523–526.
17. Gould, G.W. Heat-Induced Injury and Inactivation. In *Mechanisms of Action of Food Preservation Procedures*;

Gould, G.W., Ed.; Elsevier Science Publishers: London, 1989; 11–42.
18. Mackey, B.M.; Pritchet, C.; Norris, A.; Mead, G.C. Heat Resistance of Listeria: Strain Differences and Effects of Meat Type and Curing Salts. Lett. Appl. Microbiol. **1990**, *10* (6), 251–255.
19. Bhaduri, S.; Smith, P.W.; Palumbo, S.A.; Turner-Jones, C.O.; Smith, J.L.; Marmer, B.S.; Buchanan, R.L.; Zaika, L.L.; Williams, A.C. Thermal Destruction of *Listeria monocytogenes* in Liver Sausage Slurry. Food Microbiol. **1991**, *8* (1), 75–78.
20. Fujikawa, H.; Itoh, T. Tailing of Thermal Inactivation Curve of *Aspergillus niger* Spores. Appl. Environ. Microbiol. **1996**, *62* (10), 3745–3749.
21. Fujikawa, H.; Morozumi, S.; Smerage, G.H.; Teixeira, A.A. Comparison of Capillary and Test Tube Procedures for Analysis of Thermal Inactivation Kinetics of Mold Spores. J. Food Protect. **2000**, *63* (10), 1404–1409.
22. Shull, J.J.; Cargo, G.T.; Ernst, R.R. Kinetics of Heat Activation and Thermal Death of Bacterial Spores. Appl. Microbiol. **1963**, *11* (6), 485–487.
23. Le Jean, G.; Abraham, G.; Debray, E.; Candau, Y.; Piar, G. Kinetics of Thermal Destruction of *Bacillus stearothermophilus* Spores Using a Two Reaction Model. Food Microbiol. **1994**, *11* (3), 229–241.
24. Abraham, G.; Debray, E.; Candau, Y.; Piar, G. Mathematical Model of Thermal Destruction of *Bacillus stearothermophilus* Spores. Appl. Environ. Microbiol. **1990**, *56* (10), 3073–3080.
25. Teixeira, A.A.; Rodriguez, A.C. Microbial Population Dynamics in Bioprocess Sterilization. Enzyme Microb. Technol. **1990**, *12* (6), 469–473.
26. Sapru, V.; Teixeira, A.A.; Smerage, G.H.; Lindsay, J.A. Predicting Thermophilic Spore Population Dynamics for UHT Sterilization Processes. J. Food Sci. **1992**, *57* (5), 1248–1252, 1257.
27. Sapru, V.; Smerage, G.H.; Teixeira, A.A.; Lindsay, J.A. Comparison of Predictive Models for Bacterial Spore Population Resources to Sterilization Temperatures. J. Food Sci. **1993**, *58* (1), 223–228.
28. Rodriguez, A.C.; Smerage, G.H.; Teixeira, A.A.; Busta, F.F. Kinetic Effects of Lethal Temperatures on Population Dynamics of Bacterial Spores. Trans. ASAE **1988**, *31* (5), 1594–1601, 1606.
29. Rodriguez, A.C.; Smerage, G.H.; Teixeira, A.A.; Lindsay, J.A.; Busta, F.F. Population Model of Bacterial Spores for Validation of Dynamic Thermal Processes. J. Food Process Eng. **1992**, *15* (1), 1–30.
30. Campanella, O.H.; Peleg, M. Theoretical Comparison of a New and the Traditional Method to Calculate *Clostridium botulinum* Survival During Thermal Inactivation. J. Sci. Food Agric. **2001**, *81* (11), 1069–1076.
31. Peleg, M.; Cole, M.B. Reinterpretation of Microbial Survival Curves. Crit. Rev. Food Sci. Nutr. **1998**, *38* (5), 353–380.
32. Hugo, W.B. *Inhibition and Destruction of the Microbial Cell*; Academic Press: London, 1971.
33. Cole, M.B.; Davies, K.W.; Munro, G.; Holyoak, C.D.; Kilsby, D.C. A Vitalistic Model to Describe the Thermal Inactivation of *Listeria monocytogenes*. J. Ind. Microbiol. **1993**, *12* (3/5), 232–239.
34. Little, C.L.; Adams, M.R.; Anderson, W.A.; Cole, M.B. Application of a Log-Logistic Model to Describe the Survival of *Yersinia enterocolitica* at Sub-optimal pH and Temperature. Int. J. Food Microbiol. **1994**, *22* (1), 63–71.
35. Ellison, A.; Anderson, W.A.; Cole, M.B.; Stewart, G.S.A.B. Modelling the Thermal Inactivation of *Salmonella typhimurium* Using Bioluminescence Data. Int. J. Food Microbiol. **1994**, *23* (3/4), 467–477.
36. Stephens, P.J.; Cole, M.B.; Jones, M.V. Effect of Heating Rate on the Thermal Inactivation of *Listeria monocytogenes*. J. Appl. Bacteriol. **1994**, *77* (6), 702–708.
37. Anderson, W.A.; McClure, P.J.; Baird-Parker, A.C.; Cole, M.B. The Application of a Log-Logistic Model to Describe the Thermal Inactivation of *Clostridium botulinum* 213B at Temperatures Below 121.1°C. J. Appl. Bacteriol. **1996**, *80* (3), 283–290.
38. Kilsby, D.C.; Davies, K.W.; McClure, P.J.; Adair, C.; Anderson, W.A. Bacterial Thermal Death Kinetics Based on Probability Distributions: The Heat Destruction of *Clostridium botulinum* and *Salmonella bedford*. J. Food Protect. **2000**, *63* (9), 1179–1203.
39. Peleg, M.; Engel, R.; Gonzalez-Martinez, C.; Corradine, M.G. Non-Arrhenius and Non-WLF Kinetics in Food Systems. J. Sci. Food Agric. **2002**, *82* (12), 1346–1355.
40. Peleg, M. On Calculating Sterility in Thermal and Non-thermal Preservation Methods. Food Res. Int. **1999**, *32* (4), 271–278.
41. van Boekel, M.A.J.S. On the Use of the Weibull Model to Describe Thermal Inactivation of Microbial Vegetative Cells. Int. J. Food Microbiol. **2002**, *74* (1), 139–159.
42. Haas, J.; Behsnilian, D.; Schubert, H. Determination of the Heat Resistance of Bacterial Spores by the Capillary Tube Method. II—Kinetic Parameters of *Bacillus stearothermophilus* Spores. Lebensm.-Wiss. Technol. **1996**, *29* (4), 209–303.
43. Haas, J.; Behsnilian, D.; Schubert, H. Determination of the Heat Resistance of Bacterial Spores by the Capillary Tube Method. I. Calculation of Two Borderline Cases Describing Quasi-isothermal Conditions. Lebensm.-Wiss. Technol. **1996**, *29* (3), 197–202.
44. Rodrigo, C.; Rodrigo, M.; Alvarruiz, A.; Frígola, A. Thermal Inactivation at High Temperatures and Regeneration of Green Asparagus Peroxidase. J. Food Protect. **1996**, *59* (10), 1065–1071.
45. Welt, B.A.; Teixeira, A.A.; Balaban, M.O.; Smerage, G.H.; Hintinlang, D.E.; Smittle, B.J. Kinetic Parameter Estimation in Conduction Heating Foods Subjected to Dynamic Thermal Treatments. J. Food Sci. **1997**, *62* (3), 529–534, 538.
46. Fujikawa, H.; Itoh, T. Thermal Inactivation Analysis of Mesophiles Using the Arrhenius and *z*-Value Models. J. Food Protect. **1998**, *61* (7), 910–912.
47. Periago, P.M.; Leontidis, S.; Fernández, P.S.; Rodrigo, C.; Martínez, A. Kinetic Parameters of *Bacillus stearothermo-*

philus Spores Under Isothermal and Non-isothermal Heating Conditions. Food Sci. Technol. Int. **1998**, *4* (6), 443–447.
48. Leontidis, S.; Fernández, A.; Rodrigo, C.; Fernández, P.S.; Magraner, L.; Martínez, A. Thermal Inactivation Kinetics of *Bacillus stearothermophilus* Spores Using a Linear Temperature Program. J. Food Protect. **1999**, *62* (8), 958–961.
49. Lee, J.; Kaletunç, G. Calorimetric Determination of Inactivation Parameters of Microorganisms. J. Appl. Microbiol. **2002**, *93* (1), 178–189.
50. Pflug, I.J.; Odlaug, T.E. A Review of z and F Values Used to Ensure the Safety of Low-Acid Canned Food. Food Technol. **1978**, *32* (6), 63–70.
51. National Canners Association. *Laboratory Manual for Food Canners and Processors*; AVI Publishing Co.: Westport, CT, 1968; Vol. 1.
52. Mallidis, C.G.; Scholefield, J. Determination of the Heat Resistance of Spores Using a Solid Heating Block System. J. Appl. Bacteriol. **1985**, *59* (5), 407–411.
53. Gaze, J.E.; Brown, K.L. The Heat Resistance of Spores of *Clostridium botulinum* 213B over the Temperature Range 120° to 140°C. Int. J. Food Sci. Technol. **1988**, *23* (4), 373–378.
54. Harrison, M.A.; Huang, Y.-W. Thermal Death Times for *Listeria monocytogenes* (Scott A) in Crabmeat. J. Food Protect. **1990**, *53* (10), 878–880.
55. Schaffner, D.W.; Labuza, T.P. Predictive Microbiology: Where Are We, and Where Are We Going? Food Technol. **1997**, *51* (4), 95–99.
56. Lund, D.B. Considerations in Modeling Food Processes. Food Technol. **1983**, *37* (1), 92–94.
57. van Boekel, M.A.J.S. Statistical Aspects of Kinetic Modelling for Food Science Problems. J. Food Sci. **1996**, *61* (3), 477–485, 489.
58. Arabshahi, A.; Lund, D.B. Considerations in Calculating Kinetic Parameters from Experimental Data. J. Food Process Eng. **1985**, *7* (4), 239–251.
59. Cohen, E.; Saguy, I. Statistical Evaluation of Arrhenius Model and Its Applicability in Prediction of Food Quality Losses. J. Food Process Preserv. **1985**, *9* (4), 273–290.
60. Welt, B.A.; Teixeira, A.A.; Balaban, M.O.; Smerage, G.H.; Sage, D.S. Iterative Method for Kinetic Parameter Estimation from Dynamic Thermal Treatments. J. Food Sci. **1997**, *62* (1), 8–14.
61. Reichart, O. A New Experimental Method for the Determination of the Heat Destruction Parameters of Microorganisms. Acta Aliment. **1979**, *8* (2), 131–155.
62. Lenz, M.K.; Lund, D.B. Experimental Procedures for Determining Destruction Kinetics of Food Components. Food Technol. **1980**, *34* (2), 51–55.
63. Tucker, I. Nonisothermal Stability Testing. Pharm. Technol. **1985**, *6*, 68–78.
64. Burton, H.; Perkin, A.G.; Davies, F.L.; Underwood, H.M. Thermal Death Kinetics of *Bacillus stearothermophilus* Spores at Ultra High Temperatures. III. Relationship Between Data from Capillary Tube Experiments and from UHT Sterilizers. J. Food Technol. **1977**, *12* (2), 149–161.
65. ICMSF. *Microorganisms in Foods 5. Characteristics of Microbial Pathogens*, 1st Ed.; Blackie Academic & Professional: London, 1996.
66. Casolari, A. About Basic Parameters of Food Sterilization Technology. Food Microbiol. **1994**, *11* (1), 75–84.
67. Mulvaney, T.R.; Schaffner, R.M.; Miller, R.A.; Johnston, M.R. Regulatory Review of Scheduled Thermal Processes. Food Technol. **1978**, *32* (6), 73–76.
68. Ramaswamy, H.S.; Singh, R.P. Sterilization Process Engineering. In *Handbook of Food Engineering Practice*; Valentas, K.J., Rotstein, E., Singh, R.P., Eds.; CRC Press LLC: Boca Raton, FL, 1997; 37–69.
69. Pflug, I.J. Calculating F_T-Values for Heat Preservation of Shelf-Stable, Low-Acid Canned Foods Using the Straight-Line Semilogarithmic Model. J. Food Protect. **1987**, *50* (7), 608–615.
70. Gibson, B. The Effect of High Sugar Concentrations on the Heat Resistance of Vegetative Micro-organisms. J. Appl. Bacteriol. **1973**, *36* (3), 365–376.
71. Jones, M.C. The Temperature Dependence of the Lethal Rate in Sterilization Calculations. J. Food Technol. **1968**, *3* (1), 31–38.
72. Jonsson, U.; Snygg, B.G.; Härnulv, B.G.; Zachrisson, T. Testing Two Models for the Temperature Dependence of the Heat Inactivation Rate of *Bacillus stearothermophilus* Spores. J. Food Sci. **1977**, *42* (5), 1251–1252, 1263.
73. Reed, J.M.; Bohrer, C.W.; Cameron, E.J. Spore Destruction Rate Studies on Organisms of Significance in the Processing of Canned Foods. Food Res. **1951**, *16* (5), 383–408.
74. Casolari, A. Microbial Death. In *Physiological Models in Microbiology*; Bazin, M.J., Prosser, J.I., Eds.; CRC Press Inc.: Boca Raton, FL, 1988; Vol. II, 1–44.
75. Lund, D.B. Design of Thermal Processes for Maximizing Nutrient Retention. Food Technol. **1977**, *31* (2), 71–78.
76. Holdsworth, S.D. Kinetic Data—What Is Available and What Is Necessary. In *Processing and Quality of Food: High Temperature Short Time (HTST) Processing*; Field, R.W., Howell, J.A., Eds.; Elsevier Applied Science: London, 1990; Vol. 1, 74–90.
77. Nasri, H.; Simpson, R.; Bouzas, J.; Torres, J.A. An Unsteady State to Determine Kinetic Parameters for Heat Inactivation of Quality Factors: Conduction-Heated Foods. J. Food Eng. **1993**, *19* (3), 291–301.
78. Labuza, T.P. Enthalpy/Entropy Compensation in Food Reactions. Food Technol. **1980**, *34* (2), 67–77.
79. Villota, R.; Hawkes, J.G. Kinetics of Nutrient and Organoleptic Changes in Foods During Processing. In *Physical and Chemical Properties of Foods*; Okos, M.R., Ed.; American Society of Agricultural Engineering: St. Joseph, MI, 1986; 266–366.
80. Villota, R.; Hawkes, J.G. Reaction Kinetics in Food Systems. In *Handbook of Food Engineering*; Heldman, D.R., Lund, D.B., Eds.; Marcel Dekker, Inc.: New York, 1992; 39–144.

Thermodynamics of Food Freezing

Henry G. Schwartzberg
University of Massachusetts, Amherst, Massachusetts, U.S.A.

INTRODUCTION

Thermodynamics provides means for predicting and correlating: 1) equilibrium freezing points of foods; 2) food's equilibrium liquid water and ice contents as functions of temperature during freezing; 3) heat-removal requirements for freezing; and 4) effects of pressure and ice crystal size on freezing points.

At atmospheric pressure, the equilibrium freezing point of pure water, T_0, is 273.16K (0°C or 32°F). T_0 decreases by 0.0074K atm^{-1} as pressure increases.[1] Water often has to be subcooled before the ice crystal nuclei form. Once nucleation occurs, the temperature of the ice–water mix rises rapidly to T_0 and remains there as cooled liquid converts to ice. When only ice exists, its temperature drops below T_0 and gradually approaches that of the coolant.

ΔH_0, water's latent heat of freezing at T_0, is 333.6 kJ kg^{-1}(143.4 Btu lb^{-1}). C_w, liquid water's heat capacity, averages 4.187 kJ kg^{-1} K^{-1} or 1.00 Btu lb^{-1} °F^{-1}. At T_0, C_I, ice's heat capacity, is 2.093 kJ kg^{-1} K^{-1} or 0.500 Btu lb^{-1} °F^{-1}. C_I decreases roughly by 0.007 kJ kg^{-1} K^{-1} per K or 0.00093 Btu lb^{-1} °F^{-1} per °F as ice cools.

FREEZING POINT DEPRESSION

Dissolved solutes depress water's freezing point. As Fig. 1shows, freezing point depression increases as solute concentration increases.[2] As water in a solution freezes, solute concentration in the remaining solution increases except when solutes also solidify; and T, the equilibrium temperature of the ice–solution mix, decreases as ice formation continues. If the solution contains a single solute, cooling-induced ice formation causes T to drop until it reaches T_E, the eutectic temperature, where it remains while the remaining water and the solute solidify. After both completely solidify, continued cooling will cause T to drop further and gradually approach the temperature of the coolant.

Equilibrium freezing points of aqueous solutions are governed by the following equation[2]

$$\ln(a_w) = \ln(\gamma_w X_w) = -\frac{18.02\Delta H_{av}(T_0 - T)}{RT_0 T} \quad (1)$$

where a_w is the thermodynamic activity of water, γ_w the activity coefficient of water, and X_w the total mole fraction of water in the solution. T_0 and T are in °K. Water's molecular weight is 18.02. ΔH_{av} is the average latent heat of fusion between T_0 and T. The ideal gas law constant R is 8.314 kJ (kg mole)$^{-1}$ K^{-1}. The freezing point depression ($T_0 - T$) is the same in K and °C. γ_w is difficult to predict and correlate for foods. Because aqueous solutions contain both solvent water and water, which is bound to solute molecules and acts as part of those molecules,[3]

$$a_w \approx X_{we} = \frac{N_{wu}}{N_{wu} + N_s} = \frac{n_w - bn_s}{n_w - bn_s + En_s} \quad (2)$$

where X_{we} is the effective mole fraction of solvent water in the solution, N_{wu} the moles of solvent water, i.e., unbound water, N the moles of solute, n_w the total mass fraction of water in the solution, n_s the mass fraction of solute in the solution, b the mass of water bound per unit mass of solute, $E = 18.02/M_s$, and M_s the solute's molecular weight. Eq. 1 with X_{we}, given by Eq. 2, substituted for a_w will be called modified Eq. 1.

Food solutions usually contain many solutes. Unless a solute precipitates, the relative weight proportions of the dissolved solutes do not change during freezing. Therefore, constant effective E and b can usually be used for solute mixtures. Very good fits between modified Eq. 1 and experimental freezing point depressions vs. solute concentration data are obtained for single and mixed solutes when best-fit values of E and b are used. X_{we}, E, and b can similarly be determined for moist solid foods that contain insoluble solids and water sorbed by those solids in addition to solvent water and solute-bound water.

By using $2(X_{we} - 1)/(X_{we} + 1)$, the first term of a series expansion for $\ln(X_{we})$, in place of $\ln(X_{we})$ in modified Eq. 1 and noting that $\Delta H_{av}/(T_0 T) \approx \Delta H_0/T_0^2$, one obtains

$$\frac{En_s}{n_w - bn_s + 0.5En_s} = \frac{En_s}{n_w - Bn_s} \approx \frac{18.02\Delta H_0(T_0 - T)}{RT_0^2} \quad (3)$$

where $B = (b - 0.5E)$. Applying Eq. 3 when $n = n_{wo}$, the total weight fraction of water in the food prior to freezing, and $T = T_i$, its equilibrium initial freezing point, one

Encyclopedia of Agricultural, Food, and Biological Engineering
DOI: 10.1081/E-EAFE 120007023
Copyright © 2003 by Marcel Dekker, Inc. All rights reserved.

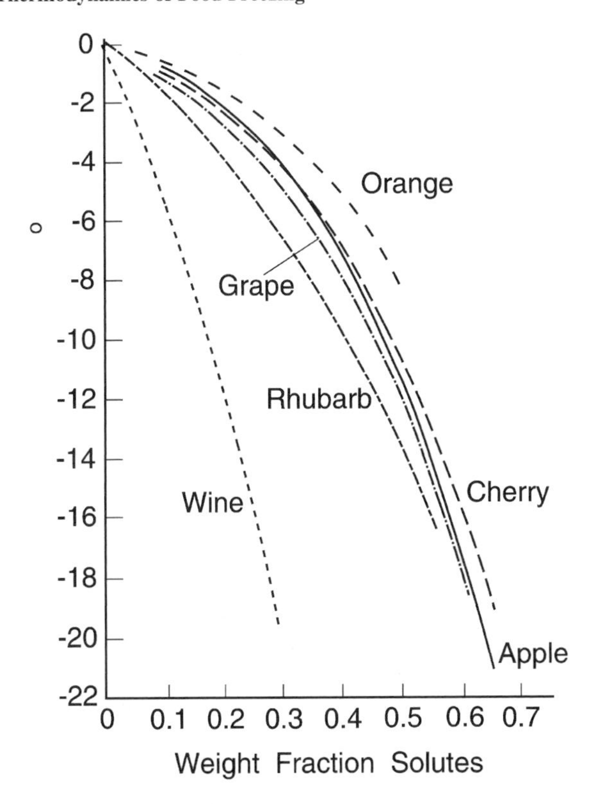

Fig. 1 Freezing point vs. solute weight fraction for wine and juices. (From Ref. 4, p. 137.)

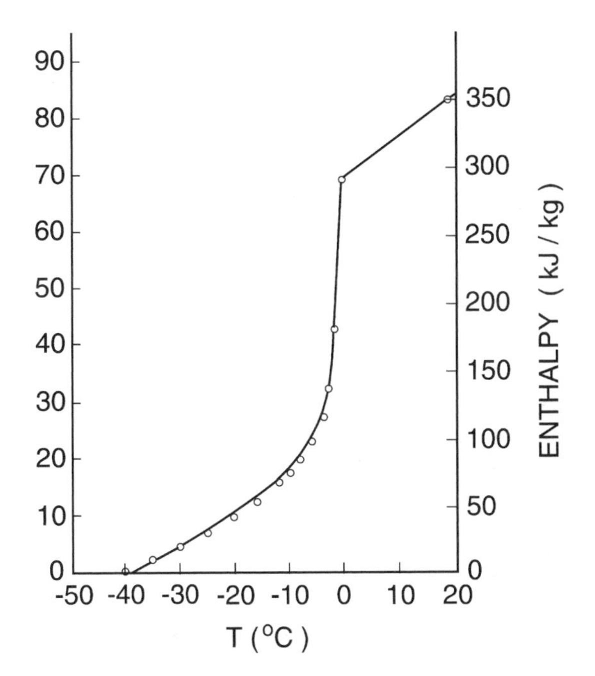

Fig. 2 Enthalpy vs. temperature for lean beef. Solid lines—data from Eqs. 6 and 7. O—calorimetric data by Reidel.[5]

obtains

$$\frac{En_s}{n_{wo} - Bn_s} \approx \frac{18.02\Delta H_0(T_0 - T_i)}{RT_0^2} \tag{4}$$

Dividing Eq. 4 by Eq. 3, one obtains

$$\frac{n_w - Bn_s}{n_{wo} - Bn_s} \approx \frac{T_0 - T_i}{T_0 - T} \tag{5}$$

where T is the current temperature during freezing and n_w the equilibrium weight fraction of unfrozen water in the food at T. The weight fraction of ice in the food is

$$n_I = n_{wo} - n_w \approx (n_{wo} - Bn_s)\left[\frac{T_i - T}{T_0 - T}\right] \tag{6}$$

Dimensionless Eqs. 5 and 6 can be used with temperatures in K, °C, or °F. The combined weight fraction of solutes and solids in the food is $n_s = (1 - n_{wo})$.

VOLUME CHANGES

The respective densities of liquid water and ice are $1000\,\mathrm{kg\,m^{-3}}$ and $917\,\mathrm{kg\,m^{-3}}$. Therefore as aqueous solutions freeze, expansion occurs. Freeze-concentrated solution may exude from spaces between the growing ice crystals. Gas-filled pores in foods may compact as ice forms. Therefore, extents of freezing-induced expansion in porous foods are difficult to predict.

ENTHALPIES

At temperatures $T < T_i$, enthalpies (heat contents) of foods undergoing freezing can be obtained by summing up the enthalpy contributions of the components involved, i.e.,

$$H = n_w H_w + n_I H_I + n_s H_s \tag{7}$$

where H, H_w, H_I, and H_s are the enthalpy per unit mass for the food, liquid water, ice, and the combined solids and solutes, respectively. The H's are measured with respect to a reference temperature T_R, usually 233.16K (− 40°C or − 40°F), where each H is zero. n_w at T can be obtained from modified Eq. 1, $n_I = n_{wo} - n_w$. If, instead, n_w and n_i are obtained from Eqs. 5 and 6, respectively, the relative error will be at most 0.4%. $H_w = \Delta H_0 + C_w(T - T_R)$, $H_I = C_I(T - T_R)$, and $H_s = C_s(T - T_R)$. C_s is the partial heat capacity for the combined solutes and solids in the food. Using Eqs. 5 and 6, substituting for H_s, H_w, H_I, n_s, n_w, and n_I, and rearranging, one obtains

$$H = (T - T_R)\left[C_f + \frac{(n_{wo} - Bn_s)\Delta H_0}{T_0 - T_R}\frac{T_0 - T_i}{T_0 - T}\right] \tag{8}$$

The heat capacity of the food in the completely frozen state is $C_f \approx (n_{wo} - bn_s)C_I + bn_sC_w + n_sC_s$. If H vs. T data obtained by calorimetry are available, C_f, B, and T, for use in Eq. 8, can be found by best-fit methods. Above T_i,

$$H = (T_i - T_R)\left[C_f + \frac{(n_{wo} - Bn_s)\Delta H_0}{T_0 - T_R}\right] + C_0(T - T_i) \quad (9)$$

C_0 is the heat capacity of the food in the thawed state. Fig. 2 depicts an H vs. T plot for beef.[5] Eqs. 5, 6, 8, and 9 are more accurate than similar equations derived earlier[4] where bn_s was used instead of Bn_s. The amount of heat removed per unit mass in cooling a food initially at temperature $T_1 > T_i$ to subfreezing temperature T_2 is $H_1 - H_2$, where H_1 is obtained by substituting T_1 for T in Eq. 9 and H_2 by substituting T_2 for T in Eq. 8.

Effective Heat Capacity

As freezing of foods takes place over a range of temperatures, the effective heat capacity C_e at T for $T < T_i$ is obtained by differentiating Eq. 6 with respect to T. $C_e = dH/dT$, yielding

$$C_e = C_f + \frac{\Delta H_0(n_{wo} - Bn_s)(T_0 - T_i)}{(T_0 - T)^2} \quad (10)$$

Eq. 10 is valid only at T below T_i. As T increases, C_e increases and peaks sharply at T_i, e.g., for lean beef it goes from 2.20 kJ kg^{-1} K^{-1} at 233.16K (−40°C) to 220.6 kJ kg^{-1} K^{-1} at T_i, i.e., 272.16K (−1.00°C). As soon as T is above T_i, Eq. 10 no longer applies and C_e immediately drops to C_0, e.g., 3.456 kJ kg^{-1} K^{-1} for lean beef.

THERMAL CONDUCTIVITY

The respective thermal conductivities of ice and water at 273.16K are $k_I = 2.25$ W m^{-1} K^{-1} and $k_w = 0.569$ W m^{-1} K^{-1}. Therefore, the thermal conductivity k of a food increases markedly as it freezes. Thermal conductivity is a transport property, not a thermodynamic property. Nevertheless, Eq. 9 correlates k for foods fairly well at $T < T_i$ when, as frequently occurs, dendritic (tree-like) ice crystals form during freezing.

$$k = k_f + (k_0 - k_f)\left[\frac{T_0 - T_i}{T_0 - T}\right] \quad (11)$$

k_0 and k_f are the thermal conductivities of food in the fully thawed and fully frozen states, respectively. At $T \geq T_i$, Eq. 11 no longer applies and $k = k_0$. Eq. 11 does not fully account for increases in k_I that occur as T decreases, but is accurate enough for most engineering purposes. Eq. 11 is not valid for materials that contain dispersed, rounded ice crystals instead of dendritic crystals. k, given by Eq. 11, and C_e from Eq. 10 or H from Eqs. 8 and 9 have been used in modeling how T varies with time and position during freezing and thawing.

DEVIATIONS FROM EQUILIBRIUM

During freezing, heat transfers from ice growth sites, and water diffuses through water-depleted regions to those sites. In cellular foods, membranes impede such diffusion. As freezing progresses and solute concentration in the residual solution increases, food solutions become very viscous. Because of high viscosity, water diffuses more slowly to ice growth sites. Solute concentrations at those sites rise above those in the rest of the residual fluid; and the temperature falls below the equilibrium temperature for the current average bulk solute concentration. Deviations from equilibrium are greatest when freezing is carried out rapidly and at T below −20°C. Therefore, extremely slow cooling has to be used in calorimetric measurements intended to determine equilibrium properties of freezing foods. Deviations from equilibrium often occur in industrial freezing processes. Thus, the equations presented earlier will be somewhat in error for such processes. As most heat removal and ice formation occurs just below T_i and well above −20°C, the error involved will be tolerable for most practical purposes.

EFFECTS OF CRYSTAL SIZE

Small crystals possess more surface energy per unit mass than larger ones. Therefore, equilibrium T for small ice crystals are slightly lower than that for larger crystals. The depression in T is inversely proportional to ice crystal diameter. For 1 μm crystals the depression is roughly 0.1°K, for 10 μm crystals roughly 0.0 1°K, and for 100 μm crystals roughly 0.001°K. Smaller crystals tend to melt and larger crystals tend to grow in foods containing ice crystals of mixed size. This effect, Ostwald ripening, can lead to graininess in ice cream. Ostwald ripening is used to produce large, readily separable ice crystals in modern freeze-concentration systems.[2]

REFERENCES

1. Hobbs, P.J. *Ice Physics*; Oxford University Press: Oxford, 1974; 346–364.
2. Schwartzberg, H.G. Food Freeze Concentration. In *Biotechnology and Freeze Concentration*; IFT Basic Symposium

Series, Schwartzberg, H.G., Rao, M.A., Eds.; Marcel Dekker, Inc.: New York, 1990; 127–202.
3. Riedel, L. On the Problem of Bound Water in Meat. Kaltetechnik **1961**, *13*, 41–43.
4. Schwartzberg, H.G. Effective Heat Capacities for the Freezing and Thawing of Foods. J. Food Sci. **1976**, *41*, 152–156.
5. Riedel, L. Calorimetric Investigation of the Meat Freezing Process. Kaltetechnik **1957**, *9*, 38–40.

Thermodynamics of Refrigerants

M. Shafiur Rahman
Shyam S. Sablani
Sultan Qaboos University, Muscat, Sultanate of Oman

INTRODUCTION

Quality and shelf-life of stored product are greatly influenced by processing and storage temperature. Lowering the temperature reduces the rates of reactions that cause quality deterioration. In earlier days, ice was used to obtain lower temperatures. Mechanical refrigeration systems are most commonly used today to achieve cooling. These systems transfer the heat from cooling chamber or medium to surrounding environment usually in the atmosphere. The transfer of heat is achieved by using a refrigerant through its phase-change cycle. A wide variety of refrigerants have been used in mechanical refrigeration system. Halocarbons, mostly chlorofluorocarbons (CFCs), have been extensively used as refrigerants due to their excellent thermodynamic properties. During early 1990s, there was an increasing concern about damaging effect of CFCs on protective ozone layer surrounding the planet earth. Use of alternate refrigerants for mechanical refrigeration system has been a topic of major interest among engineers and environmental scientists. This article provides brief historical background of refrigeration, refrigeration cycle, characteristics of refrigerants, and properties of ideal refrigerants.

HISTORICAL BACKGROUND OF REFRIGERATION

The year 1755 can be considered the starting point of the history of artificial refrigeration.[1] Starting at 1755, the prehistory of refrigeration began: humans then knew how to produce low temperatures from which he have benefited subsequently. During a span of 120 yr from 1755 to 1875, the first refrigerating apparatuses and machines were made and developed by several precursors. During the same period, the branches of physics were developed and organized. It was mainly after 1875 that refrigeration techniques began to benefit from thermodynamics. For the historian of refrigeration, the most important event was the construction of the first apparatus to make ice by vaporization of water at reduced pressure by William Cullen in 1755. In the first half of the 19th century three groups of fundamental important events occurred. The first one was the systematic work on the liquefaction of gases done by Faraday. The genesis of thermodynamics, which was originated by Carnot in 1824, finally realized between 1842 and 1852 as a result of great debates amongst Mayer, Joule, Clausius, and William Thompson. The invention of the refrigeration machine using compression of a liquefiable gas by Perkins in 1834, and of the air cycle machine in 1844 by Gorrie, but both of them developed significantly only after two decades of their invention.[1]

In the third quarter of the 19th century, the refrigeration machine found its technical personality and industrial identity. It was a period of intense creativity and four families of production of cold were created. The first was based on use of compression of liquefiable vapors. The vapors used were carbon dioxide, ammonia, sulfur dioxide, and methyl chloride. A number of other substances have been used as refrigerants since. Ferdinand Carre invented the absorption machine in 1859, and its use spread rapidly. The third type of machine was based on expansion of compressed air. It was developed into a commercial possibility by Kirk in 1862. The fourth type of machine using evaporation of water under reduced pressure was put into commercial use by Edmond Carre (brother of Ferdinand) in 1866.[1]

REFRIGERATION CYCLE

In a refrigeration system, heat is transferred from one environment to another at a higher temperature by evaporating a volatile fluid and recovery of the vapor for reuse is done by condensing or absorbing it. There are two types of refrigeration systems in use today: the compression or mechanical system, which is more popular, and the absorption system, which is used in specialized applications.[2]

Compression System

Major components of a simple compression refrigeration system are shown in Fig. 1. It consists of a closed system containing a fluid medium, known as a refrigerant that can be easily changed from its liquid state to vapor and back to liquid during its passage through the system, with different

Encyclopedia of Agricultural, Food, and Biological Engineering
DOI: 10.1081/E-EAFE 120007020
Copyright © 2003 by Marcel Dekker, Inc. All rights reserved.

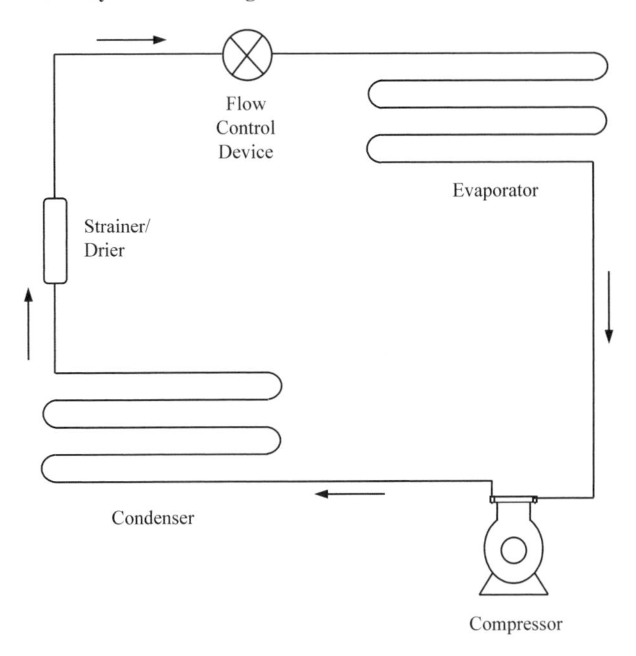

Fig. 1 Flow diagram of simple vapor compression refrigeration system showing the major components.

pressure zones. The major components of the simple vapor compression refrigeration system are evaporator, compressor, condenser, strainer/drier, and flow control or expansion valve. A liquid medium at higher pressure is allowed to flow through an expansion valve into a low-pressure zone, the evaporator, where it evaporates. This change needs energy, which is absorbed from the immediate surroundings. After expansion the vapor is continuously drawn into a pump, which compresses the vapor to a liquid state. During compression, the energy generated will heat the liquid. It is necessary to remove that heat by external cooling. In a domestic refrigerator it is accomplished by blowing air over the condenser, which is a coil of tube in rectangle behind the cabinet. In all but roof-top refrigerator installations, air cooling is augmented by fans. In large industrial systems the media condenser is usually cooled by water.[3] After cooling liquid refrigerant travels through the strainer/drier, which prevents plugging of the flow control device by trapping scale, dirt, and moisture. The flow of refrigerant into the evaporator is controlled by a pressure differential across a flow control device or expansion valve.[2,4]

Absorption System

The mechanical refrigeration cycle is based on the use of a compressor as a source of energy to transfer heat from one fluid to another. A different method of moving heat (or refrigerating) is the absorption refrigeration cycle. A simple absorption system is illustrated in Fig. 2. The absorption cycle accomplishes condensation of refrigerant vapor by using a secondary fluid to absorb it. In the absorption cycle, the compressor employed in the compression cycle is replaced by an absorber and a generator, which perform all the functions performed by the compressor in the compression cycle. In addition, whereas the energy input required by the compression cycle is supplied by the mechanical work of the compressor, the energy input in the absorption cycle is in the form of heat supplied directly to the generator. The source of the heat supplied to the generator is usually low-pressure steam or hot water. In smaller systems the heat is usually supplied by the combustion of an appropriate fuel, such as natural gas, propane, or kerosene, directly in the generator or by an electric resistance heater installed in the generator.[4]

High-pressure liquid refrigerant from the condenser passes into the evaporator through an expansion device that reduces the pressure of the refrigerant to a lower pressure. The liquid refrigerant evaporates by absorbing latent heat from the material being cooled, and the resulting low-pressure vapor then passes from the evaporator to the absorber. Absorbent solution is pumped from the absorber, which is on the low-pressure side of the system to the generator, which is on the high-pressure side of the system. Condensation of the refrigerant occurs in the absorber. In the generator, the refrigerant is separated from the absorbent by heating the solution and evaporating the refrigerant. The resulting high-pressure refrigerant vapor then passes to the condenser, where it is condensed by giving up latent heat to the condensing medium, which is the heat sink for the overall process. The absorbent solution left in the generator returns to the absorber through the return pipe.

The major advantage of absorption chillers is that they can utilize waste heat resulting in less energy expense. Another advantage is lower sound and vibration level compared to mechanical systems, making them ideally suited for installation in almost any part of a building or on the roof.

Pressure–Enthalpy Diagram

A good knowledge of compression cycles requires study of individual processes that make up the cycles and the relationship that exist among the several processes. This is usually accomplished using a pressure–enthalpy (p–h) diagram for the refrigerant as shown in Fig. 3. As the refrigerant is pumped through various components of a compression refrigeration system both pressure and enthalpy change. The state of the refrigerant can be represented as a point on the p-h chart. Saturated refrigerant enters the expansion device, at point d in the chart, as a liquid and passes through the device.

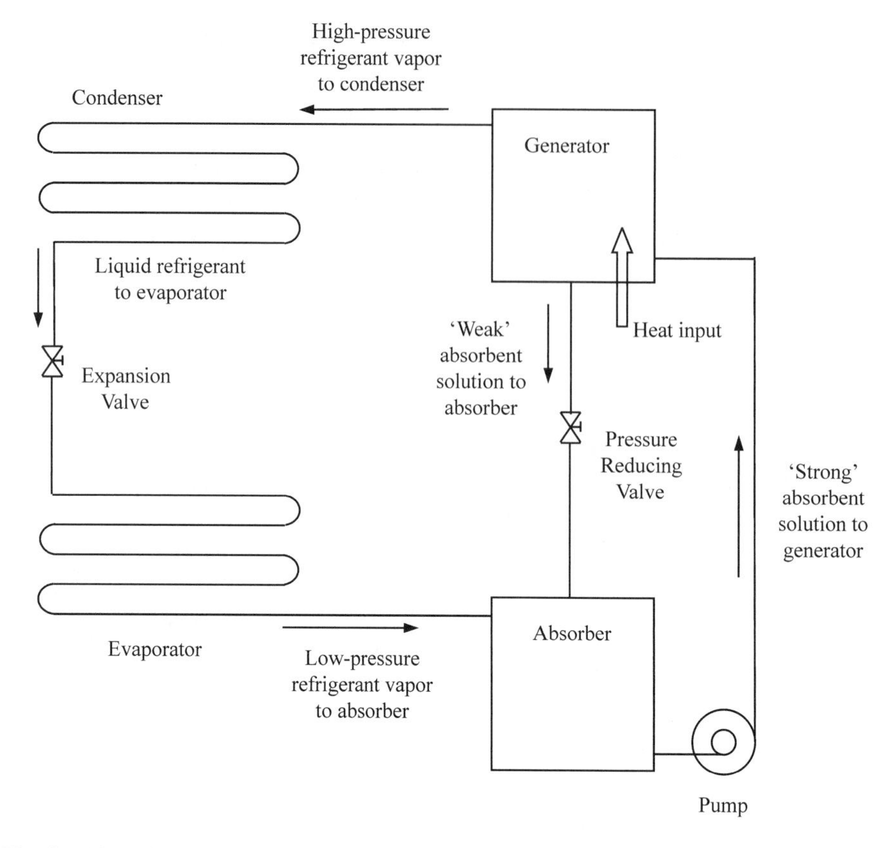

Fig. 2 Flow diagram of basic absorption refrigeration system showing the major components.

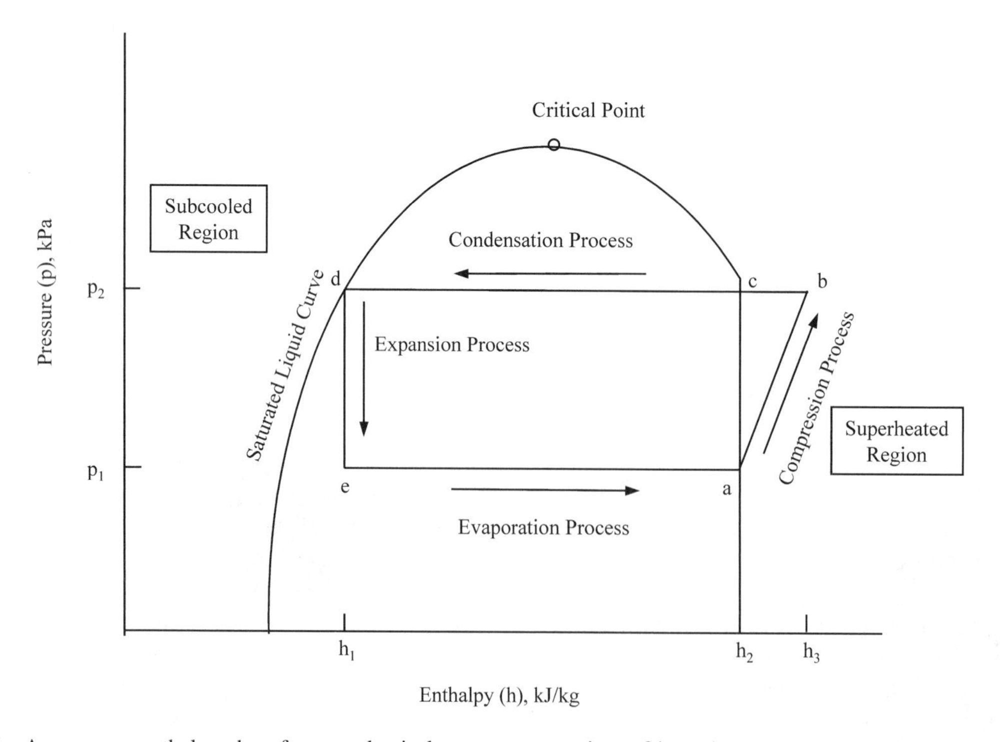

Fig. 3 A pressure–enthalpy chart for a mechanical vapor compression refrigeration cycle under saturated conditions.

The pressure drops from p_2 to p_1 (line d-e on the chart) while the enthalpy remains constant (h_1). Some flashing of the refrigerant occurs within the expansion valve resulting in a liquid/vapor mixture at point e. The vapor–liquid mixture absorbs heat as it is conveyed through the evaporator (line e-a); the enthalpy of the refrigerant increases from h_1 to h_2. Dry saturated vapor coming from evaporator enters the compressor, where state of the refrigerant is represented by point a. The refrigerant vapor at pressure p_1 and enthalpy h_2 is compressed at constant entropy (isoentropically) to pressure p_2. At the end of compression cycle the refrigerant vapor is superheated and its enthalpy increased to h_3. Then it enters the condenser where both superheat (b-c) and latent of condensation (c-d) are removed. The enthalpy of the refrigerant is decreased to H_1 while the pressure remains at p_2.[5]

Coefficient of Performance

In a refrigeration system heat is transferred from a low-temperature environment to the one that is at a higher temperature. The performance of a refrigeration system is measured by the ratio of the useful refrigeration effect obtained from the system to the work used. This ratio is called the coefficient of performance. Mathematically it can be expressed as:

$$\text{C.O.P.} = \frac{\text{Heat absorbed by the refrigerant}}{\text{Energy given to compressor}} = \frac{h_2 - h_1}{h_3 - h_2}$$

Cooling/Refrigeration load

The capacity of the refrigeration plant must be based on an accurate heat-load calculation. Several factors are considered in calculating the overall heat load, i.e., heat infiltration through insulation system (i.e., roof, wall, floor); heat given by circulating fan motors, forklift trucks, and lighting; heat given by men working; infiltration of outside air; precooling of product and heat of respiration; and defrost heat (Du Chattelier, 1995). A typical unit for cooling load used in commercial practice is "*ton of refrigeration.*" One ton of refrigeration is equivalent to the latent heat of fusion of one ton of ice 303, 852 kJ/24 hr (= 3.5168 kW). Prior to commencing with the heat-load calculation one must review the required room temperatures for the intended operations in the cooled space.

CHARACTERISTICS OF REFRIGERANTS

The refrigeration system could use refrigerants as shown in Table 1. No refrigerant has any major thermodynamic advantage, but the issues such as safety and environment friendliness could play an important role in selecting one. The ozone depletion issue and green house effect have resulted in rapid changes in the availability of refrigerants. Commercially available refrigerants can be split into two primary groups as: category 1 (CFCs, HCFCs, and the applicable newly developed replacement refrigerants), category 2 (ammonia refrigerant). The groups differ with respect to safety of personnel in the event of a leak from the refrigeration plant circulation.

The refrigeration system selected for a cooling system need to satisfy the following criteria: environmental issues, safety of personnel, installation considerations (like size of installation, applications to be satisfied

Table 1 List of some commercially used refrigerants

Refrigerant number	Chemical name	Chemical formula
Commonly used refrigerants		
R-11	Trichloromonofluoromethane	CCl_3F
R12	Dichlorodifluoromethane	CCl_2F_2
R13	Monochlorotrifluoromethane	$CClF_3$
R22	Monochlorodifluoromethane	$CHClF_2$
R30	Methylene chloride	CH_2Cl_2
R113	Trichlorotrifluoroethane	CCl_2FCClF_2
R114	Dichlorotetrafluoroethane	$CClF_2CClF_2$
New and potential refrigerants		
R134a	1,1,1,2-tetrafluoroethane	CF_3CH_2F
R500	A 73.8/26.2 azeotropic mixture of R12 and difluoroethane	CCl_2F_2/CH_3CHF_2
R502	A 48.8/51.2 azeotropic mixture of R12 and chloropentafluoroethane	$CHClF_2/CClF_2CF_3$
R503	A 40.1/59.9 azeotropic mixture of trifluoromethane and R13	$CHF_3/CClF_3$
R-717	Ammonia	NH_3

within the cold storage complex, i.e., chilling, freezing, long-term storage, conditioning of processing areas), capital cost, operating cost, flexibility/standby.[6] The general refrigerant property classes are: thermodynamic and heat transport properties and chemical and physiological properties.

Solubility of Refrigerants in Lubricants

All gases are soluble to some extent in mineral oils, and many of the refrigerant gases are highly soluble. The amount dissolved depends on the pressure, the temperature, and the chemical compatibility and the lubricant. Since refrigerants are much less viscous than oils, any appreciable amount dissolved in the oil causes a marked reduction in lubricant viscosity.[7] Two refrigerants usually regarded as poorly soluble in mineral oil are ammonia and carbon dioxide. The amount absorbed increases with increasing pressure and decreases with increasing temperature. In ammonia systems, where pressures are moderate, the 1% or less refrigerant that dissolves in the lubricant should have little effect on lubricant viscosity. In case of carbon dioxide, operating pressures are much higher and the quantity of gas dissolving in the lubricant may be enough to substantially reduce viscosity[7] and endanger adequate lubrication of the compressor.

Thermodynamic and Heat Transport Properties

These properties determine whether it is possible to construct an efficient and cost-effective refrigeration system with a selected refrigerant. The applicable relevant thermodynamic properties are: vapor pressure, boiling and condensing temperatures as a function of pressure, molecular mass, P–V relation, and enthalpy. These properties in turn determine the system displacement requirements and operation range in terms of pressure and temperature. The heat transfer properties, such as specific heat, thermal conductivity, and heat transfer coefficient with phase change relate the characteristics of the fluid, to condensation and evaporation. This in turn affects the selection of condensers and evaporators for the system. Thermodynamic properties of refrigerant are given in Table 2.

Chemical and Physiological Properties

These properties influence selection of refrigerants: effects on the external environment and suffocating, toxic and flammability and explosion hazards. The last three determine the safety standards required for systems operations and nearby personnel. In addition to the above, typical practical considerations, which have to be taken into account, are: price of refrigerant, water and oil solubility, and capability to detect leaks.

Table 2 Some physico-chemical properties of refrigerants

			Critical properties			Ideal gas state properties	
Refrigerants	Molecular weight	Normal boiling point (K)	P_c (Mpa)	T_c (K)	Δ_c (g cm^{-3})	h_0 (kJ kg^{-1})	s_0 (kJ kg^{-1}K^{-1})
R11	137.380	296.38	4.4026	471.15	0.5590	405.41	1.6937
R12	120.925	242.99	4.1290	384.95	0.5600	372.02	1.6995
R13	104.470	191.49	3.8770	301.88	0.5820	329.62	1.6876
R14	88.010	144.92	3.7500	227.51	0.6257	437.07	2.4363
R21	102.920	281.68	5.1812	451.48	0.5260	460.99	1.9232
R22	86.457	232.06	4.9900	369.33	0.5130	430.68	1.9866
R23	70.019	190.86	4.8162	299.01	0.5200	397.39	2.0675
R12B1	165.370	268.94	4.2500	426.88	0.6732	346.77	1.5417
R13B1	148.930	215.06	3.9611	340.15	0.7448	313.37	1.5232
R113	187.390	320.37	3.4100	487.25	0.5760	374.59	1.5520
R114	170.930	276.40	3.2570	418.83	0.5800	355.45	1.5586
R115	154.480	233.74	3.1600	353.10	0.6135	332.50	1.5554
RC318	200.040	266.86	2.7775	388.38	0.6200	335.50	1.5015
R500	99.300	239.43	4.4256	378.70	0.4970	409.35	1.8637
R502	111.600	227.53	4.0645	355.31	0.5610	373.11	1.7510
R503	87.280	185.06	4.3316	292.40	0.5605	344.60	1.8113

P_c = pressure at critical point; T_c = temperature at critical point; Δ_c = density at critical point; h_0 = enthalpy at standard state; s_0 = entropy at standard state.

PROPERTIES OF IDEAL REFRIGERANTS

Environmental Issues

The DuPont Company in the early 1930s developed a group of compounds (Freon) that became the dominating media for commercial refrigeration for more than 60 yr. Freon became a generic name for three media, which are also referred to as chlorofluorocarbons, CFCs (R12, R22, and R502). It was due to their useful physical properties such as low reactivity, appropriate boiling points, and heat capacities that they quickly found a wide range of applications as coolants in refrigerators and air conditioners, propellants for medical inhalers, industrial solvents and electronic cleaning agents.[8] They are notably stable/inert, nonflammable, noncorrosive, nonexplosive, and low in toxicity. Conventionally, unwanted CFCs were disposed of into the atmosphere because they are so stable. However, in the stratophere, UV radiation breaks the carbon–chlorine bonds, freeing chlorine radicals to take part in ozone-depleting reactions.

Clearly, to be a long-term CFC replacement a product needs to have no ozone depleting potential. Hydrocarbons and HFCs are oxidized more rapidly than CFCs and they have no effect on stratospheric ozone levels.[8]

New Refrigerants

New refrigerants need to have the right thermodynamic properties, be efficient, safe to use, and have minimal effects on the environment.[8] The main difficulty of using new refrigerants is the requirement of modifying or replacing existing equipment.

Regarding space cooling applications, the phase-out of CFCs affects R12 systems installed for chiller duties and R502 refrigeration plants used in freezer applications. The HCFC phase-out ultimately will affect R22, which has been traditionally used for mid-range temperature applications. Refrigerant R134A has been developed to replace R12. For R502 applications a number of blended refrigerants are available, marketed under various trade names. Currently these blends still incorporate R22 and as such should be seen as an interim solution. Alternatively R22 is being used to replace some of the traditional R502 applications, however, when used in this capacity liquid injection systems have to be incorporated in the system design to limit discharge temperatures to an acceptable level.

What are the alternatives? As a first step, HCFCs were selected as CFC alternatives because they have broadly similar chemical and physical characteristics but a much lower contribution to ozone depletion. It is obvious that CFC replacements based on HCFC compounds are only temporary solutions, likely to exist up to 2005 at the most. It is internationally recognized that the only current viable alternative, HCFC is only a short-term solution as they contain chlorine and need to be replaced soon. Instead, refrigerants with no influence on the ozone layer at all will be a long-term solution. Their chemical composition must thus be absolutely free from chlorine. Attention turned to nonchlorinated compounds, most notably the hydrofluorocarbons (HFCs). In addition, hydrocarbons, such as propane and butane, and ammonia have been suggested as CFC alternatives as none of them deplete ozone nor do they make significant contributions to global warming. However, these latter compounds suffer from certain drawbacks in terms of flammability and safety.[8]

A combination of mixtures may also be a possible alternative. From now on, mixtures will be much more common than before. The use of these mixtures will also require higher technical skill—for e.g., the HFC media operating with the new lubricating oils are more hygroscopic, which means that it is more difficult to keep moisture out of the systems. Another disadvantage of carbon-based refrigerants is that they contribute to global warming when discharged.

Ammonia is a traditional refrigerant, which has been in use for over 100 yr and which is acknowledged as having no ozone depleting characteristics or an adding to global warming, and as such is not subject to any proposed phase-out. But ammonia is poisonous and under certain circumstances—explosive. Even so, ammonia has kept its position as the foremost refrigerant in industrial applications, and continuous developments are being made related to safety regulations. To avoid risk, relevant codes and regulations need to be followed. Furthermore, its strong smell gives early leakage warning even at very low concentrations, unlike carbon-based refrigerants.

Regardless of which refrigerant is used, risk occurs in the event of uncontrolled leakage. The emphasis should thus be on the prevention of such leaks, which could be caused by inappropriate design, noncompliance with codes of construction, and human operating error. Alarms are set at 25 ppm for alerting engineers to a minor leak. The 500-ppm alarm will, apart from alerting to a major leak, stop the refrigeration plant and the machinery room ventilation fans. It will also activate the scrubber system.[9]

There is a trend towards low refrigerant charge indirect systems utilizing calcium chloride brine as a secondary refrigerant. In future carbon dioxide may be used instead of calcium chloride brine. However, such a system is still in its infancy.[9]

REFERENCES

1. Thevenot, R. *A History of Refrigeration Throughout the World*; (translated by Fidler, J.C.) International Institute of Refrigeration: Paris, 1979.
2. Langley, B. C. *Refrigeration and Air Conditioning*, 3rd Ed.; Prentice-Hall: New Jersey, 1986.
3. Nordmark, B. In the 'Cold Business'. Food Technol. **1999**, Apr*il*, 32–35.
4. Dossat, R.J. *Principles of Refrigeration*, 3rd Ed.; Prentice-Hall: New Jersey, 1991.
5. Singh, R.P.; Heldman, D.R. *Introduction to Food Engineering*; Academic Press: San Diego, CA, 1993.
6. Du Chattelier, J.L.J. Mechanical Refrigeration Systems in Cold Stores. Food Ind. S. Afri., January, 23–25, 1995.
7. ASHRAE, Lubricants in Refrigerant Systems. *Handbook Refrigeration Systems and Application*; American Society of Healing, Refrigerating, and Air-Conditioning Engineers: Atlanta, 1994; 7.1–7.7.
8. Campbell, N.; McCulloch, A. Coping Without the Common Coolant. Chem. Ind. **1999**, Apr*il*, 262–284.
9. Stera, A.C. Marine Refrigerated Transport: Recent Achievements and Outlook for the Future. Food Ind. S. Afr. **1996**, *May*,18–22.

FURTHER READING

Campbell, N.J.; McCulloch, A. Trans. IChemE **1998**, *76B*, 239–244.

Derwent, R.G.; Jenkin, M.E.; Saunders, S.M. Atmos. Environ. **1996**, *30*(2), 181–199.

Platzer, B.; Polt, A.; Maurer. *Thermophysical Properties of Refrigerants*; Springer-Verlag: Germany, 1986.

UNEP, *The Montreal Protocol on Substances That Deplete the Ozone Layer*; United Nations Environment Programme: Nairobi, 1997.

UNFCCC, Kyoto Protocol to the UN Framework Convention on Climate Change, Bonn, 1998.

Transducers

Yubin Lan
Fort Valley State University, Fort Valley, Georgia, U.S.A.

Yanbo Huang
Texas A&M University, College Station, Texas, U.S.A.

T

INTRODUCTION

For bioinstrumentation system, the physiological variable to be measured, the measurand, maybe the result of a molecular, cellular, or systemic event and can be mechanical, electrical, or chemical in nature. The transducer takes the output of the sensing element and transforms it into an electrical signal. The most common analog signal-conditioning elements are amplifiers, filters, rectifiers, triggers, comparators, and wave shapers.

Generally, a transducer is defined as a device that converts energy from some other form into electrical energy for the purposes of measurement or control. However, the most widely used definition for sensor is that which has been applied to electrical transducers by the Instrument Society of America[1]: "Transducer—A device which provides a usable output in response to a specified measurant." The measurand can be any physical, chemical, or biological property or condition to be measured. A usable output refers to optical, electronic, or mechanical signal. Following the advent of the microprocessor, a usable output has come to mean an electronic output signal. Sensors are critical components in all measurement and control systems of science and medicine and automated manufacturing and processing.[2]

Transducers may be classified as self-generating or externally powered. Self-generating transducers develop their own voltage or current and in the process, absorb all the energy needed from the measurand. Externally powered transducers, as the name implies, must have power supplied from an external source, though they may absorb some energy from the measurand.[3] In this article, only transducers used often in bioinstrumentation are introduced.

STRAIN GAGES

A strain gage is a resistive element that produces a change in its resistance proportional to an applied mechanical strain. A strain is a force applied in either compression (a push along the axis toward the center) or tension (a pull along the axis away from the center). The gages have been developed based on mechanical, optical, electrical, acoustical, and even pneumatic principles. A strain gage has several characteristics that should be considered in judging its adequacy for a particular application. These characteristics are as follows[4]:

1. The calibration constant for the gage should be stable; it should not vary with time, temperature, or other environmental factors.
2. The gage should be able to measure strains with an accuracy of $\pm 1\,\mu\text{m}\,\text{m}^{-1}$ over a large strain range ($\pm 10\%$).
3. The gage size (the gage length L and width W) should be small so that strain (a point quantity) is approximated with small error.
4. The response of the gage, largely controlled by its inertia, should be sufficient to permit recording of dynamic strains with frequency components exceeding 100 kHz.
5. The gage system should permit both on-location and remote readout.
6. The output from the gage during the readout period should be independent of temperature and other environmental parameters.
7. Both the gage and the associated auxiliary equipment should be low in cost to permit wide usage.
8. The gage system should be easy to install and operate.
9. The gage should exhibit a linear response to strain over a wide range.
10. The gage should be suitable for use as the sensing element in other transducer systems where an unknown quantity such as pressure is measured in terms of strain.

Although no single gage system can be considered optimum, the electrical resistance strain gage has nearly all the required characteristics listed above.[4]

Encyclopedia of Agricultural, Food, and Biological Engineering
DOI: 10.1081/E-EAFE 120007210
Copyright © 2003 by Marcel Dekker, Inc. All rights reserved.

INDUCTIVE TRANSDUCERS

Almost any electrical property that can be made to vary in a predictable manner under the influence of a physical stimulus may be used for transduction of that stimulus. Inductance, e.g., can be varied easily by physical movement of a permeable core with an inductor. Inductors, therefore, can be used to make transducers. There are, in fact, three basic forms of inductive transducers: *single coil*, *reactive Wheatstone bridge*, and *linear voltage differential transformer* (LVDT).[5]

QUARTZ PRESSURE SENSORS

Another modern form of sensor, especially in medical pressure measurements, is the quartz transducer. These devices are basically capacitive-based but are made differently than other capacitive transducers. The pressure sensor capsule of these devices is made of homogenous fused quartz. There are two capacitors in the capsule: a pressure capacitor and a reference capacitor. The capacitor plates are made of noble metals vacuum deposited onto their respective surfaces of the quartz capsule. These capacitors are connected in a radiometric series arrangement so that differences in dielectric properties of the quartz material are compensated. The capacitor can be connected in a capacitive bridge circuit, a mixed RC bridge circuit (both similar to Wheatstone bridges), or oscillator circuit. Advantages of the quartz transducer include very low (some sources claim zero) hysteresis, very low slippage of the metals and alloys with respect to the crystal, very low temperature sensitivity, good elastic properties, and ruggedness.

CAPACITIVE TRANSDUCERS

The capacitance between two parallel plates of area A separated by distance χ is

$$C = \varepsilon_0 \varepsilon_r A / \chi$$

where ε_0 is the dielectric constant of free space and ε_r is the relative dielectric constant of the insulator (1.0 for air). In principle, it is possible to monitor displacement by changing any of the three parameters: ε_r, A, or χ. However, the method that is easiest to implement and most commonly used is to change the separation between the plates.[6]

For bioinstrumentation application, compliant plastics of different dielectric constants may be placed between foil layers to form a capacitive mat to be placed on a bed. Patient movement generates charges, which is amplified and filtered to display respiratory movements from the lungs and ballistographic movements from the heart.[7]

TEMPERATURE TRANSDUCERS

The different sensors available for temperature measurement include resistance temperature detectors (RTDs), thermistors, expansion thermometers, integrated-circuit sensors, thermocouples, and pyrometers. Each type of sensor or instrument has advantages and disadvantages; selection of the proper sensor for a particular application is usually based on considerations of temperature range, *accuracy* requirements, environment, dynamic response requirements, and available instrumentation.

There are two types of common temperature transducers: thermocouples and thermistors. They are all commonly used in biomedical and biophysical research applications.

A thermocouple consists of two dissimilar conductors or semiconductors joined together at one end. Because the work functions of the two materials are different, a potential will be generated when this junction is heated. The potential is roughly linear to changes of temperature over a relatively wide range, although at the extreme limits of temperature for any given pair of materials, nonlinearity increases markedly.

RTDs and thermistors are usually employed when a high sensitivity is required. Because of the high-voltage output, higher accuracies can be achieved. The range of thermistors is limited and the output is extremely nonlinear.

OPTICAL RADIATION SENSORS

The intensity and frequency of optical radiation are parameters of growing interest and utility in consumer products, such as the video camera and security systems, and in optical communications system. The conversation of optical energy to electronic signals can be accomplished by several mechanisms; however, the most commonly used is the photogeneration of electrons in semiconductors. The most often used device is the PN junction photodiode. The construction of this device is very similar to that of the diodes used in electronic circuits as rectifiers. The diode is operated in reverse bias, where very little current normally flows. When light is incident on the structure and is absorbed in the semiconductor, energetic electrons are produced. These electrons flow in response to the electric field sustained internally across the junction, producing an externally measurable current. The current magnitude is proportional to the light intensity and also depends on the frequency of the light.

Sensing infrared radiation (IR) by means of photogeneration requires very small band-gap semiconductors, such as HgCdTe. Since electrons in semiconductors are also generated by thermal energy (e.g., thermistors), these IR sensors require cryogenic cooling in order to achieve reasonable sensitivity. Another means of sensing IR is to first convert the optical energy to heat and then measure the temperature change. Accurate and highly sensitive measurements require passive sensors, such as a thermopile or a pyroelectric sensor, as active devices will generate heat themselves.[2]

MICROSENSORS

Microsensors are sensors that are manufactured using integrated-circuit fabrication technologies and/or micromachining. Integrated circuits are fabricated using a series of process steps that are done in batch fashion, meaning that thousands of circuits are processed together at the same time in the same way. The patterns that define the components of the circuit are photolithographically transferred from a template to a semiconducting substrate using a photosensitive organic coating.[2]

Microsensors include thermal microsensors, radiation microsensors, mechanical microsensors, magnetic microsensors, (bio) chemical microsensors, and smart sensors. The recent advance in processor technology has led to a considerable demand for small sensors or microsensors that can fully exploit the benefits of integrated-circuit microtechnology. Integration of the sensor and part of the processor is often desirable as the characteristics of the sensor can be improved, e.g., by linearizing sensor output, compensating for temperature or humidity, and noise reduction.[8]

FIBER-OPTIC TRANSDUCERS

The role of optical fibers in different applications such as telecommunication system and biomedical instrumentation are well established. The recognition of their role in transducers, about two decades ago, led to an explosive number of applications in physical, chemical, and biochemical transducers.

Fiber optics is an efficient way of transmitting radiation from one point to another.[9] Fiber-optic sensors are replacing some conventional sensors for measuring a variety of electrical, electronic, mechanical, pneumatic, and hydraulic variables.[10,11] They are chemically inert and have freedom from electromagnetic interference. The fiber-optic sensors have been found in applications such as pesticide detection and protein and drug determination.

BIOSENSORS

Biosensors respond to biological measurands, which are biologically produced substances, such as antibodies, glucose, hormones, and enzymes. A biosensor can be defined as a device incorporating a biological sensing element to a transducer. A transducer converts change (physical or chemical) into a measurable signal, usually an electronic signal whose magnitude is proportional to the concentration of a specific chemical or set of chemicals. Biosensors are of special interest because of the very high selectivity of biological reactions and binding.

Biosensors can be further divided into metabolism biosensors, affinity biosensors, and recombinant biosensors depending on the sensing agent employed. For metabolism biosensors, sensing agents include isolated enzymes, combinations of enzymes and cofactors, and whole biological cells such as bacteria or algae. Biosensors based on isolated enzymes provide high selectivity for specific saccharides, alcohols, amino acids, organic acids such as lactate, and many more compounds. For affinity biosensors, monoclonal antibodies can be employed as sensing agents. The resulting "immunosensors" have the potential of providing a more rapid and simple-to-use alternative to immunoassays. For recombinant biosensors, the sensing agents are DNA probes.[12–14] Rapid detection techniques based on DNA probes for food-borne pathogens have been described by Bsat et al.[15]

REFERENCES

1. ANSI. *Electrical Transducer Nomenclature and Terminology*; ANSI Standard MC6.1, Instrument Society of America: Research Triangle Park, NC, 1975.
2. Dorf, R.C. *The Engineering Handbook*; CRC Press: Boca Raton, FL, 1996; 2298.
3. Coombs, C.F., Jr. *Electronic Instrument Handbook*, 3rd Ed.; McGraw-Hill: New York, 2000.
4. Dally, J.W.; Riley, W.F.; McConnell, K.G. *Instrumentation for Engineering Measurements*, 2nd Ed.; John Wiley & Sons: New York, 1993; 584.
5. Carr, J.J.; Brown, J.M. *Introduction to Biomedical Equipment Technology*; Prentice-Hall: Upper Saddle River, NJ, 1998; 703.
6. Webster, J.G. *Medical Instrumentation*; John Wiley & Sons, Inc.: New York, 1998; 961.
7. Alihanka, J.; Vaahtoranta, K.; Bjorkqvist, S.E. Apparatus in Medicine for the Monitoring and/or Recording of the Body Movements of a Person on a Bed, for Instance of a Patient. US Patent 4,320,766, 1982.
8. Gardner, J.W. *Microsensors: Principles and Applications*; John Wiley & Sons: New York, 1994; 331.

9. Epstein, M. Fiber Optics in Medicine. In *Encyclopedia of Medical Devices and Instrumentation*; Webster, J.G., Ed.; Wiley: New York, 1988; 1284–1302.
10. Sirohi, R.S.; Kothiyal, M.P. *Optical Components, Systems, and Measurement Techniques*; Marcel Dekker: New York, 1991.
11. Udd, E. *Fiber Optic Sensors: An Introduction for Engineers and Scientists*; Wiley: New York, 1991.
12. Kress-Rogers, E. *Handbook of Biosensors and Electronic Noses: Medicine, Food, and the Environment*; CRC Press: Boca Raton, FL, 1997; 695.
13. Wolbeis, O.S. *Fiber Optic Chemical Sensors and Biosensors*; CRC Press: Boca Raton, FL, 1991; Vols. I and II.
14. Kunnecke, W.; Mohns, J.; Rohm, I.; Bilitewski, U. Development of Screen-Printed Biosensors for Process Monitoring. In *Biosensors 94*; Turner, A.P.F., Karube, I., Heineman, W.R., Scheller, F., Eds.; Elsevier: Oxford, 1994.
15. Bsat, N.; Wiedemann, M.; Czajka, J.; Barany, F.; Batt, C.A. Food Safety Applications of Nucleic Acid-Based Assays. Food Technol. **1994**, *48* (6), 142.

Transient Heat Transfer Charts

Shri K. Sharma
International Food Network, Ithaca, New York, U.S.A.

T

INTRODUCTION

Heat transfer is one of the most important unit operations in processing foods. Almost every process requires heat transfer either as heat input or heat removal to alter the physical, chemical, and biological characteristics of the product. During storage of fruits, vegetables, meats, and dairy products, heat is removed to cool the product for preservation for a longer period of time. Heating involves destruction of pathogenic and other micro-organisms responsible for food spoilage, thus making food safe and stable for longer period of storage. Heat transfer is governed by certain physical laws enabling us to predict heating phenomenon and determining optimum operating conditions. When heating and cooling foods, transient or unsteady-state heat transfer takes place in the beginning of the process. The transient heating and cooling charts are simple and reasonably accurate to analyze this heat transfer phenomenon.

HEAT TRANSFER

Fourier's Law of Heat Conduction

Fourier's law of heat conduction states that if a temperature gradient exists across a material, heat will be transferred in the direction of decreasing temperature at a rate that is proportional to the temperature gradient ($\mathrm{d}T/\mathrm{d}X$) and the area (A) through which the heat is moving. The proportionality constant is characteristic of the particular material and is called the "thermal conductivity" of the material (Eq. 1).

$$Q = -kA\frac{\mathrm{d}T}{\mathrm{d}X} \qquad (1)$$

where Q is the rate of heat transfer, $\mathrm{J\,sec^{-1}}$ (W) or $\mathrm{Btu\,hr^{-1}}$, k the thermal conductivity of the material, $\mathrm{W\,m^{-1}\,K^{-1}}$, A the area of the conducting material perpendicular to the temperature gradient, $\mathrm{m^2}$, and $\mathrm{d}T/\mathrm{d}X$ the rate of temperature change per unit distance (the thermal gradient).

The minus sign in this equation indicates that heat moves from high temperature to low or down the temperature gradient.

For a finite thickness and steady-state conditions, Eq. 1 becomes:

$$Q = -kA\frac{\Delta T}{\Delta X} \qquad (2)$$

where ΔT is the temperature difference across the material and ΔX the thickness of the material.

Fourier's law can be expressed in terms of the ratio of a driving force to thermal resistance as shown in Eq. 3.

$$Q = -\frac{\Delta T}{\Delta X/kA} = -\frac{\Delta T}{R} \qquad (3)$$

where $R = \Delta X/kA$ and is the thermal resistance of the material.

Newton's Law of Heat Convection

When a fluid at one temperature comes in contact with a solid of a different temperature, a thermal boundary layer forms in the liquid. Usually, both a velocity gradient and a temperature gradient exist across this layer. Heat is transferred between the bulk of the liquid and the solid across this layer at a rate determined by the relationship (Eq. 4).

$$Q = -hA\Delta T \qquad (4)$$

where Q is the rate of heat transfer, W, A the area of contact between the liquid and solid, $\mathrm{m^2}$, ΔT the difference in temperature between the bulk of the liquid and the solid across the boundary layer, and h a proportionality constant called the "convective heat transfer coefficient" that depends on the nature of the system.

As with Fourier's Law, this equation can be expressed in terms of the ratio of a driving force to thermal resistance, thus

$$Q = \frac{\Delta T}{1/hA} = \frac{\Delta T}{R} \qquad (5)$$

where $R = 1/hA$ and is the thermal resistance of the boundary layer.

Encyclopedia of Agricultural, Food, and Biological Engineering
DOI: 10.1081/E-EAFE 120006997
Copyright © 2003 by Marcel Dekker, Inc. All rights reserved.

Overall Heat Transfer Coefficient

When heat must pass through several layers in series, the thermal resistances of the various layers are summed. Any number of conductive and convective layers may be summed in this way.

$$R = \sum_{i=1}^{n} R_i = R_1 + R_2 + R_3 + \cdots \tag{6}$$

Using total thermal resistance (R), we can express heat transfer through several layers in the form

$$Q = \frac{\Delta T}{R} \tag{7}$$

The overall heat transfer coefficient (U) is defined as

$$U = \frac{1}{RA} \tag{8}$$

Using this equation, we can express heat transfer through several layers in a form similar to the equations for Fourier's and Newton's laws.

$$Q = -UA\Delta T \tag{9}$$

TRANSIENT HEATING AND COOLING

When heating and cooling foods or any other material, transient or unsteady-state heat transfer takes place initially. The Schmidt plot and Gurney-Lurie type of charts are generally used to analyze the unsteady-state diffusion of temperature changes. Before explaining the mechanics of the heating and cooling curves, let us review unsteady-state heat transfer.

When a slab of food material is allowed to equilibrate to some initial temperature, say T_0, as shown in Fig. 1A, the temperature profile is horizontal, indicating that the temperature is uniform throughout the slab. Later, one side of the slab is exposed to a new temperature, T_1. For an instant, the temperature profile looks as shown in Fig. 1B,

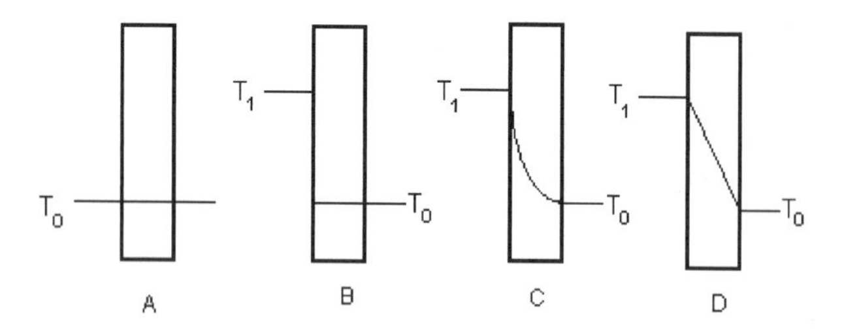

Fig. 1 Temperature profile during transient heating of an object.

with an abrupt temperature change at the surface. Soon, however, heat penetrates and the profile within the slab begins to assume the exponential shape as shown in Fig. 1C. This shape arises because heat reaches the near side of the slab faster than the farther side. This profile constantly changes during unsteady-state heat transfer. Eventually, if the external temperatures, T_0 and T_1, remain unchanged, the slab will warm until the temperature profile becomes a straight line as shown in Fig. 1D. This is the steady state and although heat will continue to flow through the slab, no further change in the temperature will occur within the slab and the profile will remain constant.

Schmidt Graphical Plot

The Schmidt graphical plot is a method for approximating the exponential profile at various times between the start of heating or cooling and the development of steady-state conditions. It is a simple graphical method to estimate the temperature at any point in the slab at any time, after heating or cooling begins. The distance and time variables' notations are used to describe the Schmidt plot to predict the heating or cooling time in an object.

Gurney-Lurie Charts

Gurney-Lurie charts are commonly used to estimate the time and temperature data during transient heat transfer. The rate of transient heating and cooling is affected by two factors:

- The rate of heat transfer between the medium and the object.
- The rate of heat transfer within the object.

In most cases, one or the other factor is limiting. We can determine which factor is limiting by computing the Biot number (N_{Bi}).

$$N_{Bi} = \frac{hx_1}{k} \tag{10}$$

where h is the convective heat transfer coefficient between the medium and the object, k the thermal conductivity of the object, and x_1 the characteristic dimension of the object = volume/area.

- When $N_{Bi} < 0.1$, we can assume negligible resistance within the object, and the rate of heating is limited by convective heat transfer.
- When $N_{Bi} > 40$, we can assume negligible resistance at the surface, and the rate of heating is limited by conductive heat transfer.
- When $0.1 < N_{Bi} < 40$, both factors are limiting.

T

Gurney-Lurie Charts for Infinite Plates, Cylinders, and Spheres

Variables

For situations where the Biot number is large, Gurney-Lurie type charts are used to estimate the rate of heat penetration into an object. In case of infinite plates, the objects resemble plate of infinite width and length compared to their thickness ($2x_1$). Heating or cooling takes place in x direction only, whereas y and z directions are infinite. In infinite cylinders, heat transfer occurs in the radial direction only. Gurney-Lurie type charts for infinite plates, cylinders, and spheres are shown in Figs. 2–4, respectively. This chart uses the following variables:

- x_1 = the half thickness of the plate, if heat is transferring through both surfaces and the full thickness, if heat is transferring through only one surface, and radius of cylinder and sphere.
- x = the distance between the center of the plate, cylinder, or sphere and the point under study.

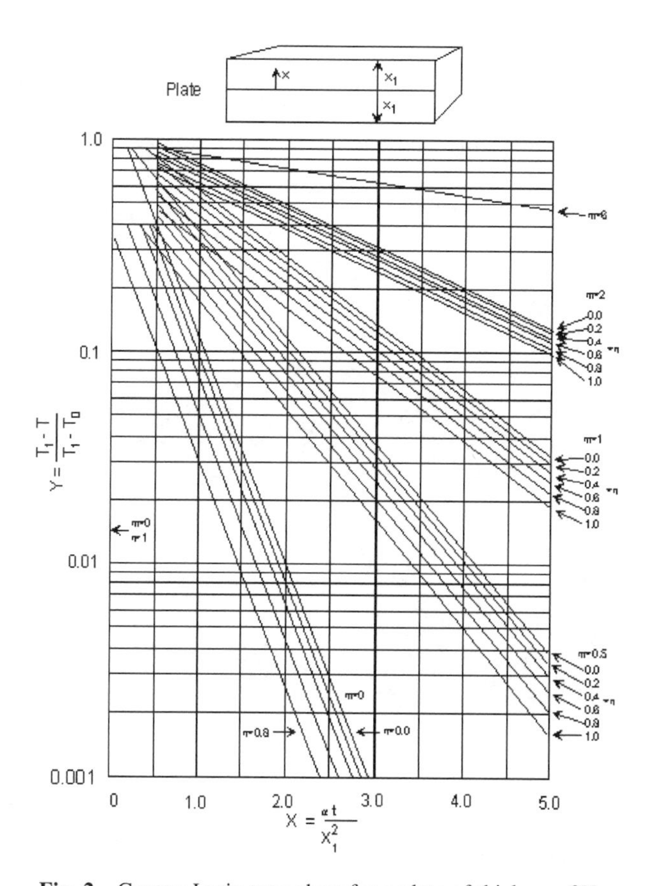

Fig. 2 Gurney-Lurie type chart for a plate of thickness $2X_1$.

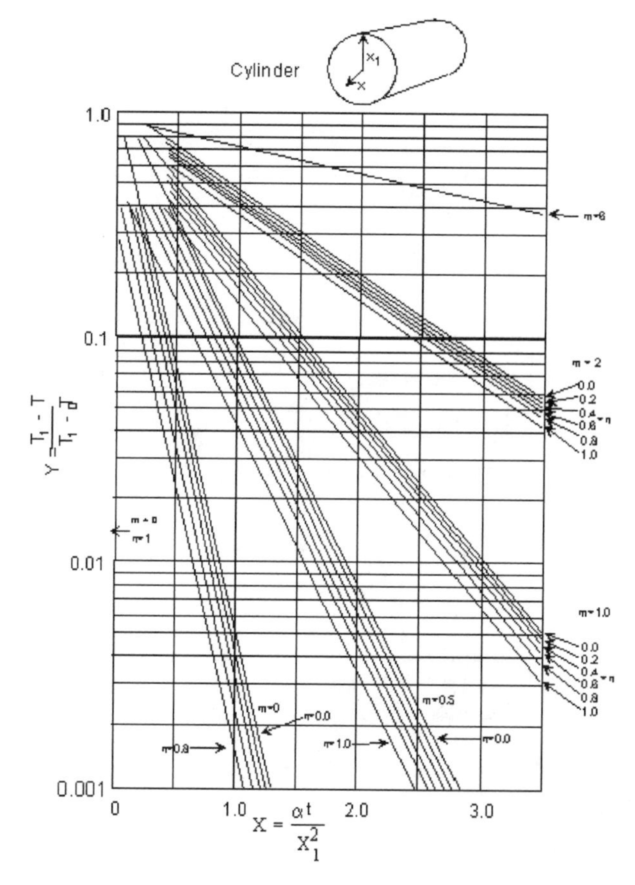

Fig. 3 Gurney-Lurie type chart for a cylinder of radius X_1.

- t = the time after the object is first immersed in the medium.
- T_1 = the temperature of the surrounding medium.
- T_0 = the initial temperature of the object, assumed to be the same throughout the object.
- T = the temperature at the point under study at time t.
- α = the thermal diffusivity coefficient of the object = $k/\rho C_p$.

Chart parameters

From these variables, the following parameters are computed before using the charts.

$$n = \frac{x}{x_1} \tag{11}$$

$$m = \frac{k}{hx_1} \tag{12}$$

In some cases, m value is close to the inverse of Biot number.

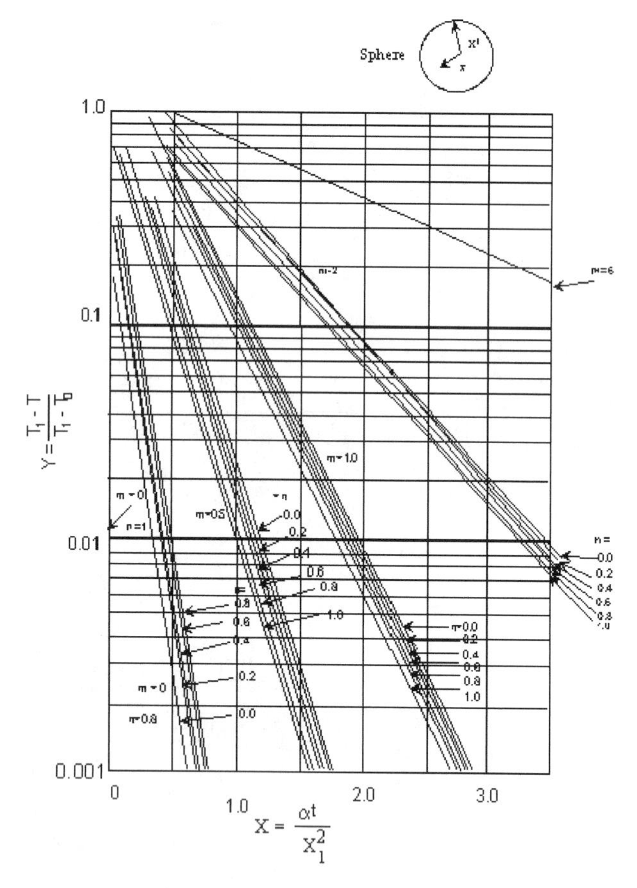

Fig. 4 Gurney-Lurie type chart for a sphere of radius X_1.

If time is specified and temperature is to be determined, compute the Fourier number (N_{FO}) or X as shown in Eq. 13.

$$N_{FO} = X = \frac{\alpha t}{x_1^2} \tag{13}$$

If temperature has been specified and time is to be determined, compute the dimensionless temperature (Y):

$$Y = \frac{T_1 - T}{T_1 - T_0} \tag{14}$$

Read the chart

If time has been specified, find X along the bottom of the chart (Figs. 2–4 for plates, cylinders, and spheres, respectively). Go up to the group of lines that match your computed value of m and find the specific line for your computed value of n. Read Y off the vertical axis. If temperature has been specified, reverse the process, starting with Y on the vertical axis and reading X from the horizontal axis.

Solve for the unknown

Solve for time (t) using Eq. 13 or temperature (T) using Eq. 14.

Transient Heat Transfer for Finite Object

The procedure described above is useful when the characteristic dimension of an object is small compared to the other dimensions, i.e., for plates that are wide compared to their thickness or cylinders that are long compared to their radius. Most objects are not this way. A can, e.g., has a length that is not much greater than its diameter. For such finite objects, the following procedure can be used to estimate the temperature after a specified period of time.

1. For a cylinder of radius x_1 and length $2y_1$, compute the Fourier number separately for each of the two dimension.

$$X_x = \frac{\alpha t}{x_1^2} \quad X_y = \frac{\alpha t}{y_1^2}$$

2. For a rectangular solid of dimensions $2x_1$ by $2y_1$ by $2z_1$, compute the Fourier number separately for each of the three dimensions.

$$X_x = \frac{\alpha t}{x_1^2} \quad X_y = \frac{\alpha t}{y_1^2} \quad X_z = \frac{\alpha t}{z_1^2}$$

3. Compute the appropriate m and n values for each Fourier number.
4. For each Fourier number, look up the corresponding Y value.
5. Compute the product of the Y values.
   - For a finite cylinder:

$$X_{xy} = Y_x Y_y \tag{15}$$

   - For a rectangular solid:

$$X_{xyz} = Y_x Y_y Y_z \tag{16}$$

6. Set the product equal to the dimensionless temperature and solve for T.

FURTHER READING

1. Geankoplis, C.J. Principles of Unsteady State Heat Transfer. *Transport Processes and Unit Operations*; PTR Prentice Hall: Englewood, NJ, 1993; 330–375.
2. Gebhart, B. Unsteady State Condition. *Heat Transfer*; McGraw Hill Inc.: New York, NY, 1971; 64–94.

3. Kakac, S.; Yener, Y. *Heat Conduction*; Taylor and Francis: Washington, D.C., 1993; 177–218.
4. Keith, F.; Black, W.Z. *Basic Heat Transfer*; Harper Row Publishers Inc.: New York, NY, 1980; 136–190.
5. Keith, F. *Principles of Heat Transfer*; International Text Book Company: Scranton, PA, 1966; 127–189.
6. Keith, F.; Bohn, M. *Principles of Heat Transfer*; West Publishing Company: St. Paul, MN, 1993; 77–154.
7. Sharma, S.K.; Mulvaney, S.; Rizvi, S.S.H. *Food Process Engineering: Theory and Laboratory*; John Wiley & Sons, Inc.: New York, NY, 2000; 68–95.

Tubular Heat Exchangers

Yasuyuki Sagara
The University of Tokyo, Tokyo, Japan

INTRODUCTION

In the modern industrialized food industries, it is common to find unit operations such as refrigeration, freezing, thermal sterilization, drying, and evaporation. These unit operations involve the heat transfer between a product and some heating or cooling medium, thus requiring heat exchangers. The variation in the viscosity and form of the food product requires the use of many different types of heat exchangers. In addition, the type of fluid flow determines the thermal performance and the pressure drop in the heat exchanger.

The effect of the condition of a fluid flow on the heat exchanger is discussed by introducing general laws of heat transfer such as conduction, convection, and radiation. This description also indicates how the turbulent flow affects the thermal performance of a heat exchanger.

Simple mathematical equations are presented that can allow prediction of the heat-transfer rate in heat exchangers. These equations provide us with useful tools to design and evaluate the performance of simple heat exchangers. Then, a good understanding of heat transfer under steady or unsteady state is required to design and operate the overall heating or cooling systems. As typical heat exchangers used for food applications, the performance characteristics and actual operations are shown for both plate and tubular heat exchangers.

OVERVIEW

Heat exchangers serve a straightforward purpose: controlling a system's or substance's temperature by adding or removing thermal energy. Although there are many different sizes, levels of sophistication, and types of heat exchangers, they all use a thermally conducting element, usually in the form of a tube or plate, to separate two fluids, such that one can transfer thermal energy to the other. Heating and cooling are the most common processes found in many industrial plants. Especially in the modern industrialized food industry, it is common to find unit operations such as refrigeration, freezing, thermal sterilization, drying, and evaporation in a heat exchanger. These unit operations involve transfer of heat between a product and some heating or cooling medium. Heating and cooling of food products is necessary for processes that result in preventing microbial and enzymatic degradation. In addition, desired sensorial properties are imparted to foods when they are heated or cooled. Table 1 indicates how machinery and equipment must be used for the transfer of heat in the many different food-processing operations.

However, capital investment, safety, and economics of processing are important operational factors to consider when purchasing or installing exchanger equipment. Of equal value in determining the type or model of equipment to install is the effect on the physical and chemical properties of the end products. Maintaining or improving nutritional value, aesthetics, safety of the products, and sensory attributes affect the marketability.

Paramount to the successful long-term operation of heat exchangers in meeting the above goals is the efficiency and ease of sanitizing the entire equipment. This is especially important for closed systems that cannot be dismantled while cleaning. The operation of cleaning in place (CEP) requires sanitary design that insures complete sanitation when cleaning liquids are circulated through the exchangers. Hence, engineering design is the controlling factor in insuring successful heat-transfer operations involving food processing.

FLUIDS FLOW IN A HEAT EXCHANGER

Inside a heat exchanger, the fluid flow is either turbulent or laminar. Turbulent flow produces better heat transfer because it mixes the fluid. Laminar-flow heat transfer relies entirely on the thermal conductivity of the fluid to transfer heat from inside a stream to a heat exchanger wall.

An exchanger's fluid flow can be determined from its Reynolds number (N_{Re}):

$$N_{\text{Re}} = \frac{\rho v D}{\mu}$$

where v is the flow velocity and D the diameter of the tube in which the fluid flows. The units cancel each other, making the Reynolds number dimensionless. If the Reynolds number is less than 2000, the fluid flow will be laminar; if the Reynolds number is greater than 6000, the fluid flow will fully turbulent. The transition region

Encyclopedia of Agricultural, Food, and Biological Engineering
DOI: 10.1081/E-EAFE 120007004
Copyright © 2003 by Marcel Dekker, Inc. All rights reserved.

Table 1 The state of a food product as related to classification of heat exchanger

Food form	Heat-transfer media	Classification	Example
Vapor	Liquid	Noncontact	Condensing vapors
Liquid	Vapor	Contact	Steam infusion, steam injection
		Noncontact	Heating by steam (condensing), refrigerant cooling
	Liquid	Noncontact	Transferring heat between liquids
	Solid	Contact	Melting ice
		Noncontact	Heating on metal hot plate in a container
Solid	Vapor	Noncontact	Extrusion (heating by steam jacket)
		Contact	Drying, cooking
	Liquid	Contact	Blanching, immersion freezing, deep-fat frying, poaching, steeping
		Noncontact	Extrusion (heating by liquid jacket)
	Solid	Contact	Dry ice cooling, freezing on plates
		Noncontact	Cooking on stove
Any food	None	Noncontact	Radiant heating, irradiation

between laminar and turbulent flow produces rapidly increasing thermal performances as the Reynolds number increases.

The type of flow determines how much pressure a fluid loses as it moves through a heat exchanger. This is important because higher-pressure drops require more pumping power. Although a manufacturer will normally determine the pressure drop, it is useful to predict the pressure drops that can occur with changing rates of flow. Laminar flow produces the smallest loss, which increases linearly with flow velocity. For example, double the flow velocity doubles the pressure loss. For Reynolds numbers beyond the laminar region, the pressure loss is a function of flow velocity raised to a power in the range of 1.6–2.0. In other words, doubling the flow could increase the pressure loss by a factor of four.

EXCHANGER EQUATION

Heat and mass transfer between fluids follows general laws of heat transfer and thermodynamics. Heat may be transferred by conduction, convection, or radiation. The heat-transfer rate (Q) of a given exchanger depends on its design and the properties of the two fluid streams. This characteristic can be defined as:

$$Q = UA\Delta T_{\text{log mean}}$$

where U is the overall heat-transfer coefficient or the ability to transfer heat between the fluid streams. The heat-transfer area of the heat exchanger or in other words, the total area of the wall that separates the two fluids, and $\Delta T_{\text{log mean}}$ the average effective temperature difference between the two fluid streams over the length of the heat exchanger.

A heat exchanger's performance is predicted by calculating the overall heat-transfer coefficient U and the area A. The inlet temperatures of the two streams can be measured, which leaves three unknowns; two exit temperatures and heat-transfer rate. These unknowns can be determined from three equations (the above using an arithmetic average for $\Delta T_{\text{log mean}}$ plus the heat-balance equation for each stream):

$$Q = UA\frac{(T_{\text{inhot}} - T_{\text{outcold}}) + (T_{\text{outhot}} - T_{\text{incold}})}{2}$$

$$= [mQC_p(T_{\text{out}} - T_{\text{in}})]_{\text{cold}} = [mQC_p(T_{\text{out}} - T_{\text{in}})]_{\text{hot}}$$

Solving these equations simultaneously usually requires iteration. In any case, a heat exchanger's manufacturer usually completes them.

HEAT EXCHANGER SYSTEMS AND PRINCIPLES OF OPERATION

Heat is transferred to or from a fluid in batch and continuous systems. Batch systems involve unsteady-state heat transfer whereby the food being heated or cooled begins at a given temperature and increases or decreases until the desired temperature is reached. The heat-transfer medium can vary in temperature (e.g., a hot surface or liquid that changes temperature as it gives or receives heat) or can be at steady state (e.g., condensing stream).

Batch steady-state systems involve a series of batch systems that give the overall effect of steady-state processing. For example, liquid-filled tanks in series can be stepwise heated by batch heating but can be connected so that the end result of the overall heating is a steady-state emission of constant-temperature liquid continuously flowing at the same rate as the cold liquid entering the first tank.

A true steady-state system is found in flowing liquids or viscous solids whereby mass flow rate, temperature, pressure, and physical properties of the fluid and the heat-transfer medium are constant at any given cross-section.

TYPES OF EXCHANGERS

There are two basic classifications of heat exchangers. One is the contact type in which there is direct physical contact between the fluid and the heating or cooling medium. The other is noncontact type in which the heat is transferred through a body that separates the fluid from heating or cooling source. Within these two categories, there are many proprietary designs and models of heat exchangers depending on the specific requirement for transferring heat to or from a given type of fluid. Here are a few of the most common changers.

Turbulent Heat Exchangers

The simplest continuous heat exchange occurs when two fluids of different temperatures are flowing through concentric pipes or tubes. The flow in steady-state heat exchangers can be either cocurrent (parallel) or countercurrent flow. During countercurrent flow, one stream (liquid or vapor) is introduced at the opposite end of the unit. By controlling the flow rates, it is possible to heat the cold liquid above the outlet temperature of the entering hot stream. Conversely, when two liquids are introduced at the same point, the stream on the surface can never leave at a temperature above that of the stream being cooled. This cocurrent system is normally less efficient than a countercurrent system because the temperature-difference driving force can become quite small as the temperatures of the two streams meet. However, there are circumstances whereby cocurrent flow can be used to insure that a heat-sensitive material does not rise above a certain temperature during processing. In the case of using stream to heat a flowing liquid, it is heated while the condensing steam is maintained at the saturation temperature of the stream. As in the case of steam kettle cooker, the most

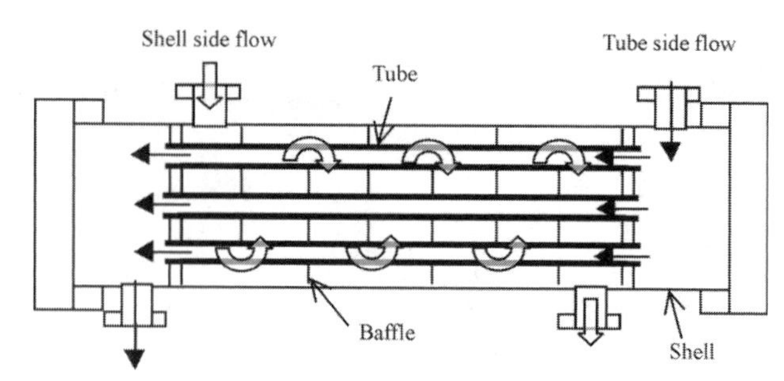

Fig. 1 Shell-and-tube heat exchanger.

common and efficient heating medium is condensing steam. Shell-and-tube heat exchangers are essentially improved tubular heat exchangers where a few to many tubes replace the single concentric inner tube, as shown in Fig. 1. The tube-side fluid passes axially through the inside of the tubes; the shell-side fluid passes over the outside of the tubes. Baffles, external and perpendicular to the tubes, direct the flow across the tubes and provide tube support. Tubesheets seal the end of the tubes, ensuring separation of the two streams. The process fluid is usually placed inside the tubes for ease of cleaning or to take advantage of the higher-pressure capability inside the tubes. The thermal performance of such an exchanger usually surpasses a coil type but is less than a palate type. Pressure capability of shell-and-tube exchangers is generally higher than a plate type but lower than a coil type.

FURTHER READING

Bartlett, D.A. The Fundamentals of Heat Exchangers. The Industrial Physicist. Am. Inst. Phys. **1996**, *2* (4), 18–21.

Singh, R.P. Heating and Cooling Processes for Foods. In *Handbook of Food Engineering*; Heldman, D.R., Lund, D.B., Eds.; Marcel Dekker, Inc.: New York, 1992.

Singh, R.P.; Heldman, D.R. *Introduction to Food Engineering*; Academic Press: Orlando, FL, 2001.

Toledo, R.T. *Introduction to Food Process Engineering*; Van Nostrand Reinhold: New York, 1991.

Tunnel Drying

N. Suzan Kincal
Middle East Technical University, Ankara, Turkey

T

INTRODUCTION

Tunnel drying is described with emphasis on typical food applications, equipment, and operating conditions. Information is provided on drying media, design considerations, and finally on costs and efficiency.

DESCRIPTION

The method of drying is suitable for particulate foods, foods that can be formed into reasonably sized particles to be placed over conveyor belts or trays on trucks, and foods like spaghetti that are placed on slowly traveling hangers. The belt, trucks, or hangers move progressively through the tunnel in contact with hot gases. Belt-type dryers are almost fully continuous in operation. When trucks or hangers are used, they either move at a constant speed through the drying tunnel, or occupy successive positions in the tunnel for given periods of time. As the product does not move with the container, the shape is well preserved.[1] The drying medium is indirectly heated air or diluted combustion products, partially purged and the remainder recirculated to maintain desired humidity levels and improve energy economy. The flow of the drying gases may be in the same direction as (cocurrent), in the opposite direction to (countercurrent) the material traveling through the tunnel, or a combination of the two flow schemes. One other flow regime is cross-flow, in which the gas is made to flow back and forth across the belt or the trucks in series. When the food to be dried has suitable properties, perforated-belts or trays along with suitable flow baffles are used with cross-flowing hot gases to provide through-circulation for improved heat and mass transfer rates.[2]

USES

The tunnel dryer is widely used in the food industry to dehydrate potatoes, vegetables, fruits, berries, pasta products, cereals, grains, starches, gelatins, yeast, and similar products. It is a versatile unit that can be easily adapted to various uses, except for the materials that cannot withstand long times of exposure to hot gases.[3] Sticky materials constitute another exception.[4]

MAJOR TYPES AND TYPICAL OPERATING CONDITIONS

Tunnel dryers are generally operated at atmospheric pressure, but can be placed under vacuum or pressure for specific purposes. The temperature of the drying medium may be as low as $-15°C$ in the case of vacuum dryers and may go up to 260°C for heat resistant foods.[3] Air recirculation is very commonly used, at a level of 60%–95%. Most tunnel dryers are of the continuous type with many variations, and some are operated batch-wise.

Batch Compartment Dryer

When the production rate is less than $500\,\text{tn}\,\text{yr}^{-1}$, the drying takes longer than 8 hr, or multiproduct operations are desired, a short tunnel with a few trucks is used batch-wise. The gas flow is parallel to the trays with a velocity of $1\,\text{m}\,\text{sec}^{-1}$–$10\,\text{m}\,\text{sec}^{-1}$, to eliminate stagnant air pockets and to improve the surface heat transfer.[6] Variable speed fans are used to provide higher gas velocity over the material during the early stages of drying. The trays or conveyors are loaded to a depth of 20 mm–100 mm, deep loadings reducing the labor but resulting in longer drying times as the falling rate drying time varies roughly with the square of the loading depth.[5]

Shallow loading results in faster drying, but care is needed to ensure depth uniformity and labor is increased. A clearance of not less than 40 mm is allowed between the material in one tray and the bottom of the tray immediately above. Vaporization rates of $0.2\,\text{kg}\,\text{hr}^{-1}\text{m}^{-2}$ – $2.0\,\text{kg}\,\text{hr}^{-1}\text{m}^{-2}$ based on exposed material surface is typical.

Through-Circulation Compartment Dryer

This is the perforated-tray, cross-flow version of the batch compartment dryer. The perforated-trays permit vapor escape through the bottom as well as the other surfaces. Rigid foods, resistant to attrition are dried in these units. Water vaporization rates are typically $1\,\text{kg}\,\text{hr}^{-1}\text{m}^{-2}$–$10\,\text{kg}\,\text{hr}^{-1}\text{m}^{-2}$ based on tray area.[5]

Encyclopedia of Agricultural, Food, and Biological Engineering
DOI: 10.1081/E-EAFE 120007092
Copyright © 2003 by Marcel Dekker, Inc. All rights reserved.

Continuous Tunnel Dryer

These dryers are characterized by continuous, nonmixed material flow through the tunnel, so that the residence time is uniform. The foods to be dried are conveyed on trays with solid or screen bottoms, placed on trucks, on hangers in cases like spaghetti, or moving belts traveling through the tunnel.

Truck conveying

When trays on trucks are used, one truck leaves from the discharge end as each new truck is fed into the inlet end. The trucks usually move on tracks or monorails, operated mechanically using chain drives. Air flow can be totally cocurrent, countercurrent, or a combination of both. Cocurrent flow is preferred with heat sensitive foods, so that the air at highest temperature is in contact with wet material fed. The air temperature, and therefore the heat transfer rate drop by the time the material is about to leave the tunnel. Countercurrent flow is most suitable from the point of view of energy economy, and is therefore generally preferred when heat sensitivity is not a problem. Another frequently used flow regime is cross-flow, with the air flowing through the trays in series. Reheat coils may be installed after each cross-flow pass to maintain constant temperature operation.[4] Large propeller type circulation fans are installed at each stage, and air may be introduced or exhausted at any desirable point. Tunnel dryers possess maximum flexibility through combinations of air flow and temperature staging.[6] The other operating conditions and evaporation rates are similar to those for batch compartment dryers.

Conveying on hangers

This is almost exclusive for spaghetti production. As the properties of the final product depend strongly on how well the drying has been carried out, the temperature, humidity, and the flow velocity of the drying medium is strictly adjusted and controlled in order to keep the drying rate moderate in all stages of drying, especially the final ones.

Belt conveying

The use of belt conveyors, or screen conveyors with suitable materials, bring about considerable savings on labor costs compared to the use of trays to be loaded manually, but require additional investment for automatic feeding and unloading devices. Generally, chain driven belts are used for foods.[7] Arrangements are often made to clean the conveyor by washing, brushing, and buffing on the return run. Typical conveyor widths range between 0.6 m and 3.6 m. Similar to the case of trucks, several hot gas flow regimes are possible, and cross-flow with intermittent heating is most common. It is usual to blow the hot gas upward in the feed end and down through the material on the delivery end. In the drying of cut vegetables, a belt made of fine mesh and shaped into an inclined trough is widely used.[4] While the physical characteristics of many foods are such as to permit through-circulation of air, many other products require preforming. The major preforming method used with food materials is extrusion, forcing of the product through a small diameter hole of constant area.[4] During this process, some of the water content is also squeezed out, leaving a material which is easier to dry. Filter cakes such as those in starch processing can often be preformed by scoring and breaking into small chunks. Proper preforming also helps the dusting problem.

Multiple belt tunnels

In order to present new drying surfaces, to agitate the drying foods, to take care of the shrinkage, and to combine the merits of cocurrent and countercurrent flow regimes; multiple belt conveyors placed at different levels can be used. The tunnel will need to be higher but shorter. The belts can be made to move at different speeds, and be loaded to different depths as the product properties change along the course of drying, providing higher efficiencies.[7]

DRYING MEDIUM

Diluted combustion products of lighter fuels, such as natural gas or LPG are extensively used as the drying medium in tunnel drying. Alternatively, ambient air can be heated to about 200°C using steam heated, extended surface coils. The latter is preferred when contact with combustion gases is to be avoided. A popular technique is to recycle the dryer exit gas, allowing only enough purge to maintain internal humidity.[5] In most cases, more gas is needed to transport heat than to purge vapor. The greater the gas velocity (over, through, or impinging upon the material), the greater the convection heat transfer coefficient.

DESIGN

Generally, the design of tunnel dryers are based on performance data or pilot tests[4,6] since accurate design methods are not available. Current publications, manufacturers of equipment, and research are possible sources. Methods for preliminary design and product surface temperature estimation as well as some performance data is given in Perry's Handbook.[6]

T

COST AND EFFICIENCY

The investment and operating cost for tunnel dryers are moderate, and depend strongly on the efficiency of the process. The efficiency varies between 20% and 70%, depending strongly on the recycle ratio, and also on whether diluted combustion products or indirectly heated air is used as the drying medium, the temperatures of the drying medium, feed, and product; the quality and thickness of insulation of the dryer, the type of product, and the quantity of moisture to be removed. The tunnel enclosure comprises insulated panels designed to limit exterior surface temperatures to less than 50°C for safe operation, which helps improve the energy economy. In general, continuous operation provides higher efficiencies.[3]

REFERENCES

1. Peck, R.E. Drying, Solids. In *Encyclopedia of Chemical Processing and Design*; McKetta, J.J., Cunningham, W.A., Eds.; Marcel Dekker, Inc.: New York, 1983; Vol. 17, 1–29.
2. Considine, D.M. Drying, Solids. In *Chemical and Processing Technology Encyclopedia*; McGraw Hill: New York, 1974; 374–375.
3. Hall, C.W.; Farral, A.W.; Rippen, A.L. Tunnel Dryer. *In Encyclopedia of Food Engineering*, 2nd Ed.; AVI Publishing Co.: Westport, CT, 1986; 805–807.
4. Okos, M.R.; Narasimhan, G.; Singh, R.K.; Weitnauer, A.C. Food Dehydration. In *Handbook of Food Engineering*; Heldman, D.R., Lund, D.B., Eds.; Marcel Dekker, Inc.: New York, 1992; 514–522.
5. McCormick, P.Y. Drying. In *Kirk—Othmer Encyclopedia of Chemical Technology*, 4th Ed.; Howe-Grant, M., Kroschwitz, J.I., Eds.; John Wiley and Sons, Inc.: New York, 1993; Vol. 8, 475–499.
6. Moyers, C.G.; Baldwin, G.W. Solids-Drying Equipment. In *Perry's Chemical Engineers' Handbook*, 7th Ed.; Perry, R.H., Green, D.W., Eds.; McGraw Hill: New York, 1997; 12.36–12.51.
7. Wilson, J.P. Drying Equipment. In *The Encyclopedia of Chemical Processing Equipment*; Mead, W.J., Ed.; Reinhold Publ. Co.: New York, 1964; 272–283.

Turbulent Flow

Jay Marks
Purdue University, West Lafayette, Indiana, U.S.A.

INTRODUCTION

Flow of fluids (liquids and gases) is a significant factor in a variety of operations in the area of food and agriculture. Prediction of the behavior of flowing fluids is important in determining flow rates and energy requirements in systems such as pumping, conveying, heating, cooling, spraying, etc. One common system of characterizing fluid flow is to divide the regime into two classifications, laminar and turbulent. These two regimes behave differently, and are handled differently for purposes of engineering calculations. Laminar flow (also called viscous flow or streamline flow) can be described and predicted by equations derived from momentum balances on the fluids in question. Turbulent flow, however, is much more complex. Comparison of the flow regimes, with particular attention to the turbulent regime, will help in understanding basic engineering calculations on fluid flow.

Laminar flow is described as the situation where fluid particles move in smooth paths, parallel to each other. The particles remain in the same location relative to the cross-section of the conduit as they move lengthwise through the conduit. Turbulent flow is described as the situation where fluid particles move in irregular paths as they progress through the conduit. The direction of flow and the velocity of each individual particle vary continually with time, resulting in swirling patterns or eddies. If the velocity of a particle of liquid would be measured at various times, the resultant plot would look similar to Fig. 1. Turbulence itself may be defined as the velocity fluctuations in the flow of a fluid over time and distance. These fluctuations may appear random. In theory, however, the fluctuations could be described mathematically if enough were known about the characteristics of the fluid and about the details of the physical containment of the fluid.

At the current state-of-the-art, even with many decades of study of the flow of fluids, the flow of fluids in turbulent mode is not completely predictable. Many of the engineering calculations dealing with turbulent flow of fluids still depend on correlations of experimental measurements.

It may be easier to evaluate turbulent fluid flow by considering initially laminar flow, and the factors that cause laminar flow to become turbulent.

REYNOLDS NUMBER

Various researchers in the mid-1800s investigated the different characteristics of fluid flow in conduits to determine which variables affected the flow patterns. The conclusions were that the diameter of the conduit, the velocity of the fluid, the density of the fluid, and the viscosity of the fluid were all significant. Dimensional analysis was used to derive a single expression that could be used to characterize fluid flow. The expression was:

$$\frac{DV\rho}{\mu} \tag{1}$$

where D is the diameter of the conduit (for noncircular conduits an equivalent value of D can be substituted), V is the average value of the velocity (volumetric flow rate divided by the cross-sectional area of the conduit), ρ is the fluid density, and μ is the fluid viscosity. If consistent units are selected for the variables in the term, they will cancel and the term is dimensionless. This is convenient, since the value of the expression is the same in the English and metric system if appropriate units are used. This expression is called the Reynolds number, in honor of the English engineer and physicist Osbourne Reynolds (1842–1912). The Reynolds number is commonly referenced as Re. Note that sets of consistent units for the equations in this article are listed in Table 1.

Laminar flow of fluids in a conduit of uniform cross-section is characterized by "layers" of fluid flowing past each other, each at a constant velocity. The velocity profile of the flow through a circular conduit can be calculated through conservation of momentum considerations, assuming there is no momentum transfer between the fluid layers. These calculations, verified by numerous empirical measurements, show that the velocity profile will be parabolic, measured across the diameter of the pipe, as shown in Fig. 2.

Encyclopedia of Agricultural, Food, and Biological Engineering
DOI: 10.1081/E-EAFE 120006956
Copyright © 2003 by Marcel Dekker, Inc. All rights reserved.

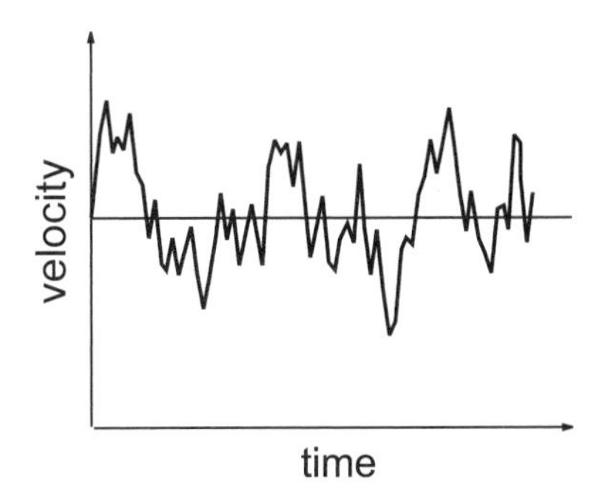

Fig. 1 Individual particle velocity vs. time.

Velocity is essentially zero at the wall, and twice the average velocity at the centerline.

Studies done over many years found that the Reynolds number serves as an indicator of whether the fluid flow is laminar or turbulent. At very low flow rates in a conduit, the flow regime will be in stable laminar flow. As the flow rate is increased and the Reynolds number approaches 2100, the flow will occasionally show short periods of turbulent flow. In other words, the laminar flow regime begins to show instability. This degree of instability increases with increase in flow rate. When the flow rate is such that the Reynolds number is approximately 4000, the fluid flow regime typically will be unstable, and the flow is considered turbulent. The Reynolds number range from approximately 2100 to approximately 4000 is called the critical range. (The flow regime in this range is transitioning from laminar to turbulent flow.) The beginning point of turbulent flow is not clearly defined due to the instability of the flow regime in this range. This instability means turbulence in the fluid (also called eddies), and can be triggered by small perturbations in the flow. These perturbations may be caused by fittings, changes in direction, rough spots in the conduit wall, etc. They may be even caused by vibrations due to machinery. The variety of possible perturbations leads to a range of Reynolds numbers where this change occurs. As the Reynolds number continues to increase, turbulence increases to a point considered "completely turbulent."

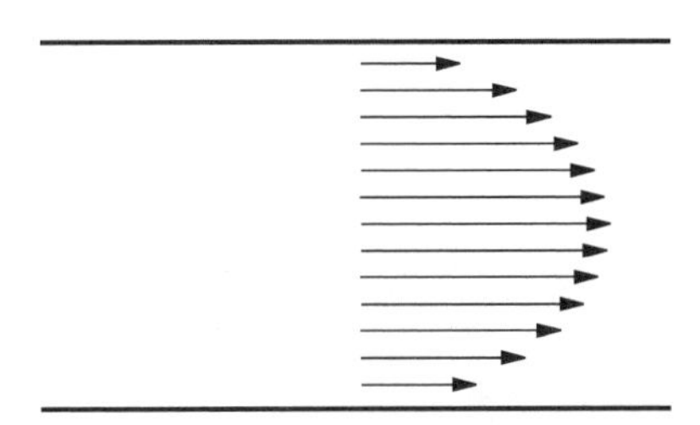

Fig. 2 Parabolic velocity profile.

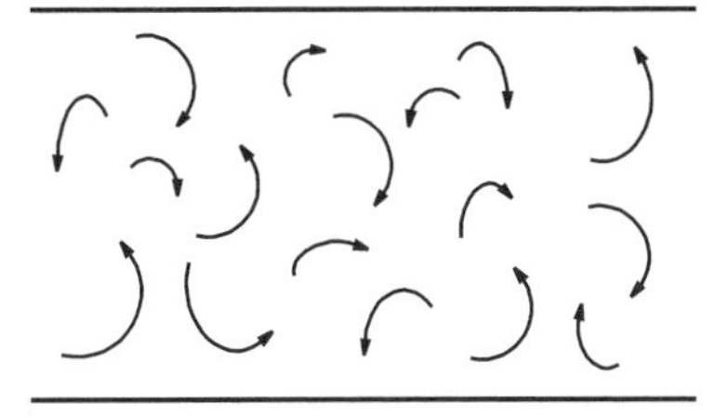

Fig. 3 Turbulent eddies.

Fig. 3 shows that at complete turbulence, eddy currents cause momentum transfer between nearly all layers of the fluid.

This means that, except adjacent to the wall, the average longitudinal velocity is the same throughout the conduit. This is shown in Fig. 4.

Because the velocities at nearly all points on the conduit diameter are the same, the fluid flows as though it were a solid "plug" of material. Hence the term "plug flow," sometimes used as a synonym for turbulent flow.

BERNOULLI EQUATION

Fluid flow through a conduit results in friction adjacent to the wall or between fluid particles. This friction converts pressure energy to thermal energy, causing a pressure drop in the direction of bulk flow. Prediction of the amount of this pressure drop is of considerable importance in the design of fluid handling equipment used for agricultural and food production. The equation commonly used for this prediction is the Bernoulli equation, also called the Mechanical Energy Balance equation:

$$z_1 \frac{g}{g_c} + \frac{P_1}{\rho} + \frac{V_1^2}{2\alpha g_c} + W = z_2 \frac{g}{g_c} + \frac{P_2}{\rho} + \frac{V_2^2}{2\alpha g_c} + F \qquad (2)$$

In this equation, z is the height, P is the pressure, ρ is the fluid density, g is the acceleration of gravity, g_c is the gravity constant, V is the average fluid velocity, W is the mechanical work introduced into the system (sometimes termed shaft work), and F is the friction due to flow. α is a variable with only two values. $\alpha = 0.5$ for

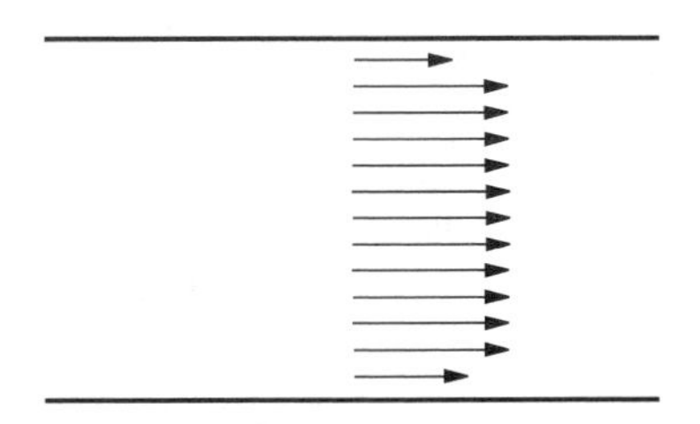

Fig. 4 Turbulent velocity profile.

laminar flow and $\alpha = 1.0$ for turbulent flow. The values of 0.5 for α arises from the fact that for a parabolic velocity distribution, the kinetic energy term based on the average velocity is twice that of a uniform velocity distribution. α, thus, adapts the equation to fit both laminar and turbulent flow regimes. Consistent units are necessary for the Bernoulli equation to balance. Table 1 lists sets of consistent units for the English system and the metric system.

If the units listed in Table 1 are used, the Reynolds number will be dimensionless, and each term in the mechanical energy balance equation will have the consistent units of ft $\mathrm{lb_f/lb_m}$ (English) or $\mathrm{N\,m\,kg^{-1}}$ (metric).

When the Bernoulli equation is used for flow regimes in the critical flow region, the pressure drop will be between that calculated for laminar or turbulent. For engineering calculations in the critical region, the pressure drop that will give the most conservative mechanical design is usually used. There are techniques for estimating pressure drop in the critical region, however.

The friction term in the Bernoulli equation can be calculated from the expression:

$$F = \frac{2fV^2L}{g_c D} \tag{3}$$

This equation is called the Fanning equation. In this equation, f is the Fanning friction factor, and L is the length of the conduit. As mentioned, flow characteristics for laminar flow regimes can be derived from theoretical considerations. This gives:

$$f = \frac{16}{Re} \tag{4}$$

For turbulent flow, the situation is more difficult. Theoretical solutions to many general cases of turbulent flow have not yet been developed, and the friction term in the Bernoulli equation is usually derived from empirical data correlated from many experiments.

FLUID FRICTION

Friction of fluid flow in conduits was found early in the 20th century to be related to the inside surface roughness of the conduit. Nikuradse[1] was able to demonstrate that the friction in a conduit was not due to the surface roughness of the inside wall of the conduit, but rather due to the roughness relative to the diameter of the conduit. This is termed *relative roughness*. He showed the friction of fluid flowing in a conduit in a turbulent regime correlated to a function the Reynolds number and the relative roughness:

$$f = \phi\left(\frac{DV\rho}{\mu}, \frac{\varepsilon}{D}\right) \tag{5}$$

ε is the roughness of the conduit wall, in length units. ε/D is thus dimensionless. Moody[2] then used the work of Nikuradse to develop what have become now as the Moody diagram. This is a very usable correlation of the empirical data on turbulent flow into a graphical form for determination of the Fanning friction factor. The Moody diagram is shown in Fig. 5. (Note that some authors use a Fanning friction factor with a value 4 times that in Fig. 5, and then adjust the equations utilizing f accordingly.) Table 2 gives experimentally determined values of ε for various common pipe materials.

In a real flow situation, there are additional components of F in Eq. 2. In addition to the friction due to the fluid flow through the conduit, there are additional friction components due to changes in direction, changes in diameter, and valves or fittings or other projections into the flow stream. Techniques to account for these components have been developed based on extensive experimental measurements. These are usually grouped into three categories: valves and fittings, sudden expansion, and sudden contraction. Valves and fittings are handled by converting them to an equivalent length of conduit and adding this length to the actual length of conduit. This is simplified by experimental findings that various sizes

Table 1 Consistent units

Variable	Description	English units	Metric units
D	Conduit diameter	ft	m
V	Fluid velocity	$\mathrm{ft\,sec^{-1}}$	$\mathrm{m\,sec^{-1}}$
ρ	Fluid density	$\mathrm{lb_m/ft^3}$	$\mathrm{kg\,m^{-3}}$
μ	Fluid viscosity	$\mathrm{lb_m/ft\,sec}$	$\mathrm{kg\,m^{-1}\,sec^{-1}}$
z	Elevation	ft	m
P	Pressure	$\mathrm{lb_f/ft^2}$	Pa ($=\mathrm{N\,m^{-2}}$)
g	Acceleration of gravity	$32.2\ \mathrm{ft\,sec^{-2}}$	$9.807\ \mathrm{m\,sec^{-2}}$
g_c	Gravitational constant	$32.2\ \mathrm{lb_m\,ft/lb_f\,sec^2}$	Not used
ε	Relative roughness	ft	mm

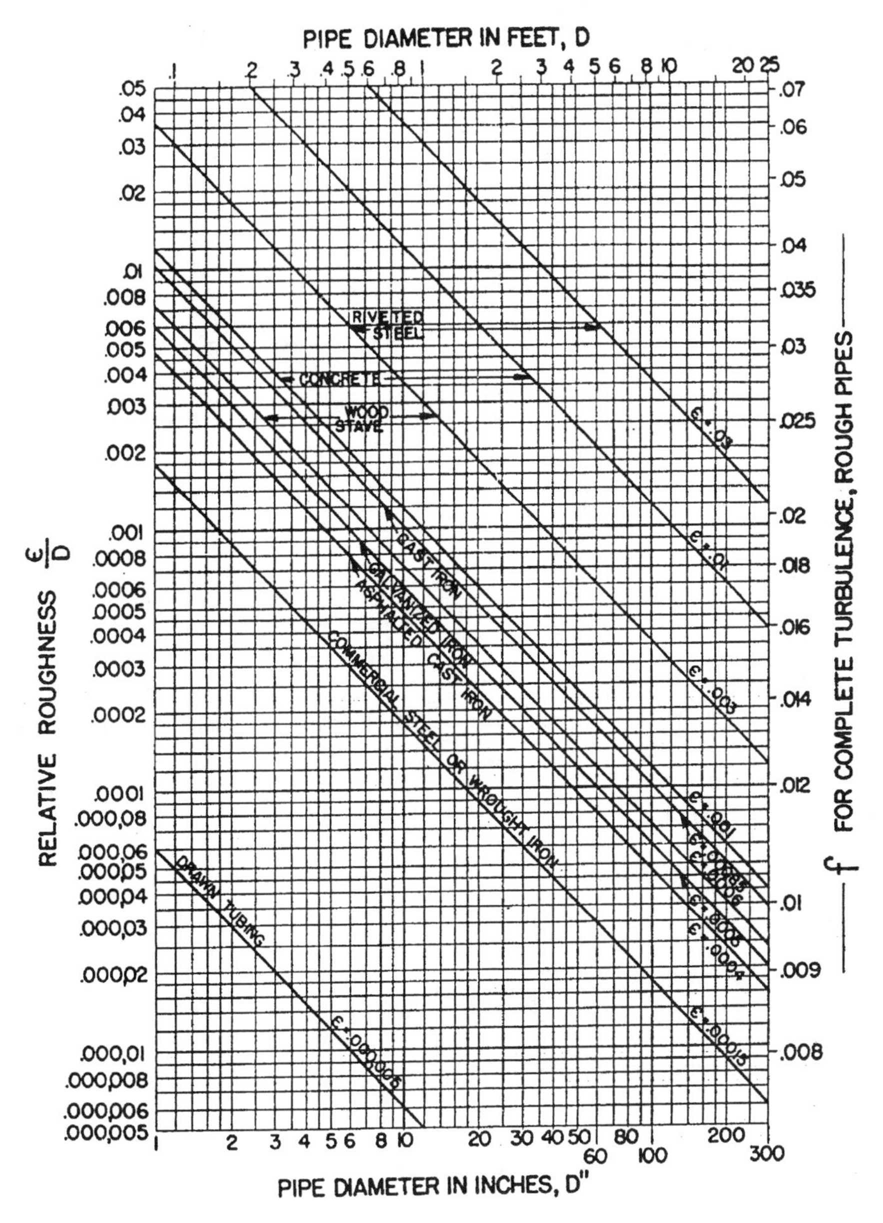

Fig. 5 Moody diagram. (From Ref. [1].)

of a single type of fitting (for example an elbow) have a variable of L_e/D that is approximately constant over a range of sizes. Tables of L_e/D values have been developed for a variety of valves and fittings, and a several of these are given in Table 3. The values of L_e/D are for liquids similar to water, and typically should be accurate to $\pm$ 30%. For gases similar to air, multiply the L_e/D values by 1.2.

Sudden contraction and sudden expansion (a change in pipe size, e.g., or entry or exit from a tank) are handled by calculating values of F for these occurrences and adding them to the value of F calculated for straight conduit flow. Eqs. 5 and 6 are used for this.

For sudden contraction:

$$F_c = \frac{K_c V_2^2}{2\alpha g_c} \tag{6}$$

The values for K_c vary with the ratio of the square of the diameters of the conduit segments. Values are given in Table 4.

For sudden expansion:

$$F_e = \frac{(V_1 - V_2)^2}{2\alpha g_c} \tag{7}$$

Table 2 Equivalent roughness values

Type of pipe or lining (when new)	ε mm	ε ft
Drawn tubing	0.0001	0.000005
Concrete	0.03–0.3	0.001–0.01
Cast iron		
Uncoated	0.015	0.0006
Asphalt coated	0.12	0.0004
Galvanized steel	0.013	0.0005
Glass	0 (smooth)	0 (smooth)
Plastic	0 (smooth)	0 (smooth)
Sanitary stainless	0 (smooth)	0 (smooth)
Welded steel	0.004	0.00015
Wrought iron	0.004	0.00015

Table 4 K_c values

D_2^2/D_1^2	K_c
0	0.5
0.1	0.46
0.2	0.42
0.3	0.38
0.4	0.34
0.5	0.30
0.6	0.26
0.7	0.22
0.8	0.15
0.9	0.075
1	0

Total friction in the mechanical energy balance Equation thus becomes:

$$F = F_{\text{total}} = F_{\text{pipe}} + F_c + F_e \quad (8)$$

OTHER TURBULENT SITUATIONS

Although flow of fluids in conduits is the fluid flow likely of most interest in agricultural, food, and biological engineering, it is helpful to have a brief overview of other turbulent flow situations. Consideration of flow past an infinite circular cylinder will give indications of the turbulent flow of fluids in such situations as flow past cooling or heating coils, turbulence in mixing, flow past bodies in a fluid stream, etc.

Table 3 Friction loss for turbulent flow through valves and fittings

Valve or fitting	L_e/D
45° Elbow, standard	16
45° elbow, long radius	9
90° elbow, standard	34
90° elbow, long radius	20
Tee, along run	18
Tee, entering run	45
Tee, entering branch	45
Coupling	Negligible
Union	Negligible
Gate valve, open	8
Globe valve, open	270
Diaphragm valve, open	100
Check valve, swing	90
Check valve, ball	3000

The following discussion applies to a constant flow past an infinite cylinder, with the flow perpendicular to the axis of the cylinder. The average flow velocity is measured for enough upstream of the cylinder that the flow diversion around the cylinder does not affect the velocity. Consider the flow vertical so that gravitational effects do not cause differing effects on either side of the cylinder. Thus, at low flow, the fluid layers separate smoothly and uniformly around the cylinder, coming together smoothly downstream of the cylinder.

As the flow rate is increased, symmetrical eddies form in the flow following the cylinder on either side of the cylinder centerline. This flow is still considered in the laminar range because the eddies are stable. As the flow rate approaches the critical Reynolds number, perturbations result instability in the eddies caused by the flow past the cylinder. At higher Reynolds numbers, these eddies begin to oscillate periodically, into a formation termed the Von Karmen vortex street. Continued increasing of the fluid velocity causes turbulence in the wake downstream some distance from the cylinder, breaking up the periodic oscillations. As flow increases, the turbulent region approaches the cylinder, finally becoming fully turbulent downstream of the cylinder. This sequence of development of the turbulent flow helps explain why speed of a mixer blade gives variable results. At low speed mixers causes "folding" of the liquid layers and gentle mixing. High shear mixers operate well into the turbulent region, causing rapid mixing (and likely break up of suspended particles) due to the turbulence causing high and erratic shear between fluid layers.

ADDITIONAL INFORMATION SOURCES

Eqs. 1–6 are useful in general fluid flow situations where a reasonably accurate prediction is needed and the characteristics of the fluids are simple or are not known in depth. These equations will handle the majority of flow calculations needed in practical applications. However, much has been done on the experimental and theoretical studies of fluid flow in laminar, critical, transition, and fully turbulent flow in conduits of various size, materials, and configurations. This research has resulted in a variety of correlations and empirical and theoretical equations of increasing complexity. These are of value to the engineer who is designing complex systems, or who is attempting to develop very precise fluid flow predictions. These equations usually require detailed information on the characteristics of the fluids involved, the specific configuration of the system, and the temperatures involved throughout the system. In addition, the computations are more complex. If such calculations are required, there are a variety of texts available, such as Brater,[3] Heldman,[4] Perry,[5] and Toledo.[6] In addition, Mathieu[7] gives a thorough and extensive coverage of turbulent flow concepts.

T

REFERENCES

1. Moody, L. Friction Factors for Pipe Flow. Trans. ASME **1944**, *66*, 671–684.
2. Nikuradse, J. *Stromnugsgesetze in Rohren*; VDI-Farschunghs, 1933.
3. Brater, E. *Handbook of Hydraulics*, 6th Ed.; McGraw-Hill: New York, 1976.
4. Heldman, D.; Singh, R. *Food Process Engineering*, 2nd Ed.; AVI Pub. Co. Inc.: Westport, CT, 1981.
5. Perry, R.; Green, D. *Perry's Chemical Engineers' Handbook*, 7th Ed.; McGraw-Hill: New York, 1997.
6. Toledo, R. *Fundamentals of Food Processing Engineering*, 2nd Ed.; Aspen Pub., Inc.: Gaithersburg, MD, 1999.
7. Mathieu, J.; Scott, J. *An Introduction to Turbulent Flow*; Cambridge Univ. Press: Cambridge, 2000.

Ultrafiltration

Gun Trägårdh
Lund University, Lund, Sweden

INTRODUCTION

Ultrafiltration (UF) is a pressure-driven operation in which porous membranes are used for fractionation of solutes, most often in aqueous solutions. It has several advantages, such as unique separation properties, concurrent fractionation and concentration, product quality improvement, low energy consumption, and sometimes, increased yield. Today, UF is an established technology in some areas of the food industry for purification and concentration of different types of liquid foods. The annual growth is considerable, and the number of applications are also likely to increase as new needs arise and equipment is developed.

In UF, components of a fluid are predominantly fractionated according to size and shape differences, even if other factors, e.g., electrical charge or type of active groups (hydrophilic, hydrophobic) in the molecules of the components, could play a role as well.[1] The membranes used in UF are characterized by pore diameters in the range 5 nm–50 nm, thus UF is very suitable for concentrating high molecular weight compounds, e.g., proteins, while low molecular weight compounds such as sugar and salts pass through the membrane. In this way, large molecules are concentrated and purified in one step. The driving force is a pressure gradient across the membrane, and the pressure on the feed side must be higher than the osmotic pressure of the components to be retained. As the osmotic pressure of macromolecules is quite low, the applied pressure is generally in the range of 0.1 MPa–1 MPa in UF.

TERMINOLOGY

The stream passing through the membrane is named permeate, the stream that is retained is named retentate. The capacity of a membrane is expressed as flux, the flow through the membrane per unit membrane area. Most often, volume fluxes are given ($\mathrm{L\,m^{-2}\,hr^{-1}}$). The retention of a component is defined as the ability of a membrane to hinder the component from passing through it. Factors, such as temperature, pressure, fluid velocity, pH, concentration effect flux and retention. Fouling, the deposition of suspended or dissolved substances on the membrane surface, at pore openings, or within the porous networks in the membrane leads to a decrease in flux and sometimes changes in the separation properties. Many factors effect fouling and it is important to consider the properties of both feed and membrane, as well as the process conditions, in order to minimize fouling.

The term membrane configuration describes the spatial arrangement of the membrane. For food processing, particularly in the dairy industry, the use of spiral wound membrane modules has increased rapidly and is now standard configuration whenever possible because of the low cost compared to other configurations. Plants with thousands of square meters of membrane area have been installed. For liquids containing suspended material, a tubular configuration is preferred.

DEVELOPMENT IN THE DAIRY INDUSTRY

In the mid-1960s, UF membranes with reasonable flux became available and the first applications were mentioned. The dairy industry was the first one to develop and use the technology, seeing the potential for recovery and purification of proteins from whey and to incorporate whey proteins into dairy products, thus increasing the yield. A major breakthrough was the commercialization of polysulfone UF membranes in the mid-1970s. This membrane material has much better resistance to pH, heat, and chemicals than the cellulose acetate membranes used earlier, making it possible to operate at higher temperatures and improve cleaning technology although the higher hydrophobicity meant more fouling problems. Technology for producing yogurt, various types of soft cheese, feta cheese, etc. from UF retentates was developed. In the 1980s, there was an expansion in this area, and technology for producing cheddar cheese from UF retentates was developed in Australia. In addition, the standardization of the protein content of cheese milk was developed. In the 1990s, the use of cheese milk standardization spread around the world, the standardization of protein for milk powder started, and market milk got increased attention.[2]

Encyclopedia of Agricultural, Food, and Biological Engineering
DOI: 10.1081/E-EAFE 120007074
Copyright © 2003 by Marcel Dekker, Inc. All rights reserved.

About 220,000 m^2 membrane area had been installed world wide as of 1997 for UF in the dairy industry, of which about 150,000 m^2 is for whey and 60,000 m^2 for milk treatment.[3] Over the last few years, there has been quite a large increase and by 1999, the installed area in the dairy industry was estimated to be about 300,000 m^2.[4] Some reasons for this increase are lower prices for membrane systems, increased processing capacity, development of new markets, and replacement of old equipment and technology.[3] A general trend in whey processing is that the operating temperature is lowered to 10°C because operating at this temperature overcomes problems of calcium phosphate precipitation occurring at higher temperatures.

The whey plants are used for production of whey protein concentrates (WPC) with varying protein contents, including very high ones for specific purposes. For milk, a major area is standardization of the milk protein content. The protein content of milk varies during the year and UF of the milk results in a more homogeneous and constant quality. In cheese making, the capacity of existing equipment can be doubled if the milk proteins are preconcentrated by UF. This, however, does not increase the yield substantially. UF is also used to obtain total solids content, equal or almost equal to the concentration in the final cheese. This technology is used for several types of cheese with moderate total solids content. So far, the use of UF for the production of semihard and hard cheese is very limited, and the production of hard cheese of cheddar type from UF retentate developed in Australia has stopped.[2] In many cases, the traditional process technology must be modified when UF concentrates are used instead of milk. It is, for instance, very important to carefully adjust the mineral balance in order to obtain the correct rheological properties and taste in the final product.

OTHER PROTEINACEOUS MATERIALS

There are also applications of purifying and concentrating proteins from sources other than milk. In the meat industry, animal blood is collected and the blood serum proteins are widely used in various food products in a concentrated form. These proteins are very heat sensitive and need gentle treatment to avoid coagulation. They can be successfully concentrated by UF, and commercial plants have been in operation for many years.[5,6]

Gelatin is one of the most important by-products from the meat processing industry and is widely used in food products as well as in the pharmaceutical and photographic industries. Different qualities are available. UF is commercially used for simultaneous concentration and desalting of gelatin solutions; sometimes, diafiltration is also used to obtain the right composition.

A third example of established applications in this field is the purification and concentration of enzymes, e.g., standardization of rennet used in cheese-making.

FRUIT JUICE AND WINE INDUSTRY

In juice processing, compounds such as pectins, cellulose, hemicellulose, starch, and proteins cause undesired turbidity during storage and the juice thus has to be clarified. Since the late 1970s, UF has been applied commercially for the clarification of different types of fruit juices. Most UF plants have been installed for apple juice clarification, but commercial systems are also in operation for other types of juices. There are several advantages of using UF over traditional clarification technology, such as increased yield (96%–98% recovery), elimination of filter aid and filter presses, better product quality (less haze), and reduced enzyme usage.[5]

UF is also an accepted technology within the wine industry. Several wineries are using UF to clarify, stabilize, and enhance their wines. UF is used to remove thermally unstable proteins, tannins, color components, yeast, carbohydrate gums, and oxidized phenolics.[6]

SUGAR INDUSTRY

Membrane applications in the sugar industry are still in its infancy, but progressing. For sweeteners (glucose, fructose), membranes are widely used to obtain higher purity, e.g., by removing suspended solids. Many thousands of square meter are reported to be installed worldwide. Concerning purification of sucrose from sugar cane and sugar beet processing, pilot plants have been installed. Potential applications are clarification of juice, syrups, etc. This involves more difficult separation problems and more fouling problems than with sweeteners.[7]

REFERENCES

1. The European Society of Membrane Science and Technology. *Nomenclature and Symbols in Membrane Science and Technology*; Koops, G.H., Ed.; CIP-Data Koninklijke Bibliotheek: Den Haag, The Netherlands, 1995.
2. Horton, B.S. Whatever Happened to the Ultrafiltration of Milk? Aust. J. Dairy Technol. **1997**, *52* (1), 47–49.

3. Timmer, J.M.K.; van der Horst, H.C. Whey Processing and Separation Technology: State-of-the-Art and New Developments. In *Whey*, Proceedings of the Second International Whey Conference, Chicago, Oct 17–29, 1997; International Dairy Federation: Brussels, Belgium, 1998.
4. Maubois, J.-L. Personal communication, 2001.
5. Cheryan, M., Ed. *Ultrafiltration and Microfiltration Handbook*; Technomic Publishing Company: Lancaster, PA, 1998.
6. Mohr, C.M.; Engelgau, D.E.; Leeper, S.A.; Charboneau, B.L. *Membrane Applications and Research in Food Processing*; Noyes Data Corporation: Park Ridge, NJ, 1989.
7. Hlavacek, M. Membrane Applications in the Sugar Industries. In *Applications of Membrane Technology in the Food and Dairy Industry*, Proceedings from Advanced Course on the Use of Membrane Technology in Environmental Applications, Oviedo, Spain, 1999; Coca, J., Luque, S., Eds.

UHT Processing—Impact on Dairy Product Quality

Arthur P. Hansen
North Carolina State University, Raleigh, North Carolina, U.S.A.

INTRODUCTION

Ultra high temperature (UHT) processing of fluid dairy products produces flavor compounds whose flavor profile changes over storage time. Milk lipids generate a variety of flavor compounds during UHT processing. These flavor compounds include: aldehydes, methyl ketones, lactones.[1] In addition, flavor compounds that contain sulfur are generated due to the formation of sulfhydryls resulting from the breaking of the cross linkage of these bonds in β-lactoglobulin.[2,3] This sulfur flavor is made up of several compounds, such as methyl sulfide, hydrogen sulfide, methyl mercaptan, and other sulfur containing compounds.[3,4] The flavor of UHT dairy products starts to change almost immediately after processing. This article describes multiple research projects conducted on UHT milk flavors as it changes in stored containers of milk and cream.

MATERIALS AND METHODS

The milk used in a series of studies from 1977 to 1984 was obtained from North Carolina State University Dairy. The microbial quality ranged from 50,000 to 300,000 CFU mL^{-1} of raw milk. Psychotrophs were less than 2000 CFU mL^{-1} and somatic cells were about 300,000 per mL. The milk received at the NCSU dairy plant was collected, processed, and stored under commercial conditions. The milk was processed at 300 L hr^{-1} using two processors: Cherry Burrel UHT steam injection system and Cherry Burrel Unitherm. The following products were processed: low fat milk (0.5% fat), whole fat milk (3.2% fat), and coffee cream (10% fat). Processing temperatures were 138, 143, and 149°C and hold times were 3.4 sec, 6.9 sec, and 20.3 sec. All products were packaged using an AB3-250ML Tetra Brik packaging machine. The packaged milks were placed in tray packs at 27 cartons to the pack and stored at 4, 20, 24, 30, or 40°C. Chemical changes were monitored in the stored products using pH, SH titer, YSI oxygen meter, gel electrophoresis, and gas liquid chromatography. Changes in sensory characteristics of the stored products were monitored using both trained and untrained taste panels. The milks were monitored for flavor and chemical changes over 1 yr.

RESULTS

Flavor changes in the milk at room temperature storage showed flavor loss each month that the product was stored. In milk, both saturated aldehyde and methyl ketone concentrations decreased over a 6 mo storage period. As the storage period increased, the concentration of the flavor compounds continued to decrease and the flavor of the milk became blander. The loss of aldehydes and ketones was similar in both processing profiles.[5]

When coffee cream was processed at the four processing conditions, the same phenomena occurred. The concentration of aldehydes decreased over the 12 mo storage period. The C_9 and C_{10} aldehyde concentrations decreased to about 20%–25%. The total carbonyl concentration decreased over the entire storage period. After the 12 mo period, the flavor compound concentration was about 20%–25% of the original resulting in a very bland coffee cream.[6] In observing the aroma and flavor compounds in UHT milk, dynamic changes occurred in the aroma profile, especially and immediately after processing. Approximately 14 peaks representing aroma compounds were detected by GC analysis immediately after processing.[7] The aroma profile of the flavor compounds decreased within four weeks with only half the peaks remaining.

After 8- and 12-weeks storage, the number of peaks decreased to 4 and the intensity of each of the peaks decreased in the headspace making the flavor of the milk less intense. Most of these flavors are cooked or sulfur type flavors, which oxidize out over storage and react with the low-density polyethylene (LDPE) in the package. The rate of change in the aroma profile was dependent upon the level of the O_2 in the package and by the thickness of the inner polyethylene layer of the package. These packages contained 1 ppm–2 ppm of O_2 and the polyethylene coating as the inner layer, causing fairly rapid loss of the aromatic compounds.[8] If the level of O_2 in the package was 8 ppm, such as indirect heated product, the loss of cooked flavor and aroma would have been faster than four weeks.

Low fat milk, whole milk, and cream were presented to trained and untrained taste panels for flavor evaluation for up to 1 yr storage. All three processing temperatures and holding times were tested by the panelists for off-flavor

Encyclopedia of Agricultural, Food, and Biological Engineering
DOI: 10.1081/E-EAFE 120007242
Copyright © 2003 by Marcel Dekker, Inc. All rights reserved.

and acceptability. Low fat milk, whole milk, and cream stored at 4°C showed flavor improvement, as the cooked flavor dissipated at the 4°C storage temperature. The whole milk and cream stored at 4°C received scores as high as the pasteurized control and higher as 4°C storage continued.[9]

The low fat milk, whole milk, and cream stored at room temperature showed improvement for at least 3 mo. As the cooked flavor dissipated from the milk at 3 mo–6 mo, the flavor scores leveled off and then started to decline. This decrease in flavor score was due to enzymatic activity in the milk. The milk and creams held for the shorter hold time of 3.4 sec declined much more quickly in flavor score due to higher levels of enzyme activity. The same milks held for 20.4 sec showed less tendency to decrease in flavor score. These more stable flavor scores were due to the lower degree of enzymatic activity in the milks and creams as compared with the short holding times. Enzymatic degradation caused higher levels of free fatty acids due to lipase activity in the UHT milk and cream.[10]

The whole milk samples exhibited similar changes to the low fat milk except the flavor tended to be more stable, and the degradation or development of off-flavors tended to be slower or harder to perceive. This flavor stability may be due to the masking effect of the milk fat. As the cooked flavor dissipated from the refrigerated UHT milk, the flavor scores continued to improve, as there were no enzymatic activity and/or chemical reactions occurring in the milk. The milks stored at room temperature showed flavor improvement as the cooked flavor dissipated.[11] This off-flavor development was due to Maillard reaction products and enzymatic activity in the UHT milk. The milks processed at 149°C and held for 3.4 sec did not last past 6 mo due to gelation of the milk. The panels were discontinued at this time due to off-flavor development and gelation of the milk. The milk held at 6.9 sec but processed at the same temperature did not start to develop off-flavors until 5 mo–6 mo as a greater amount of the enzymes were destroyed with the longer holding time causing less change in the flavor composition of the UHT milk. The product held for 6.9 sec also did not gel until the milk was almost 1-yr old. The milk held for 20.3 sec did not start to deteriorate until about 7 mo–8 mo at room temperature. This long shelf life may be due to greater enzyme inactivation at the longer hold time.[12] So the curves tended to flatten out and the flavor degradation was slower in the milk.

The coffee cream tended to exhibit similar data as the milk with cooked flavor dissipating in the refrigerated samples. As the storage time increased, the flavor scores improved and eventually the refrigerated cream was as good as the pasteurized control. The short hold tube of 3.4 sec gelled at 6 mo making it impossible to present to the taste panel, only the gel was weaker in cooked flavor than the milk. The 6.9 sec hold did better as it took almost a year before it gelled. The flavor scores improved as the cooked flavor dissipated and then flattened out and turned down as the off-flavors developed in the coffee cream. The 20.9 sec hold seemed to be more stable with flavor scores decreasing at 8 mo–9 mo storage. This cream lasted 2.5 yr before it gelled. The flavor panel rated the cream as unsaleable after 11 mo–12 mo storage.

High storage temperatures of 40°C tend to become unsaleable quickly due to enzymatic activity early on followed by the Maillard reaction. The shorter hold time, 3.4 sec, deteriorated quickly, the 6.9 sec hold lasted 4–6 weeks, and the 20.3 sec hold lasted 10–12 weeks in the best scenario with milk and cream. In every case, 40°C storage caused even the highest quality of UHT milk to breakdown. In areas that have summer temperatures of 35–40°C, shelf life of UHT milk may be drastically reduced. This reduction in shelf life occurs even with the highest quality milk due to the Maillard reaction.

The untrained taste panel data follows the trends established by the trained taste panel. For the untrained panel, 2.8 was the score for the control milk, UHT milk scored 2.5, and scores below 2.0 were unsaleable. In all cases, the UHT milk received a lower score than the fresh pasteurized milk. As the length of storage increased, the cooked flavor dissipated at refrigerated temperatures, and the flavor of the milk and cream samples continued to improve until the scores equaled the fresh pasteurized control. Those samples stored at 20 and 40°C tended to deteriorate at 6 mo and 9 mo, respectively, until they were equivalent to the trained taste panel data.[11]

The major flavor identified in the samples by the untrained taste panel was cooked. After the cooked flavor dissipated, the milk became blander. The flavor responses indicated by the panelists included flat, lacks flavor, and sweet or malty. In the shorter hold time, 3.4 sec, the flavor tended to deteriorate faster as off-flavors began to develop.

The longer hold times did not develop these off-flavors until 6 mo–9 mo and 7 mo–11 mo for 6.9 sec and 20.3 sec, respectively. The flavor defects found after 6 mo and 9 mo storage included chalky, coconut, musty, rancid, scorched, cheese whey, fruity, foreign, astringent, oxidized, oily, buttery, and soapy. These off-flavors represented 3.9% of the 3000 samples analyzed.[11] These percentages represent about 3000 different analyses with taste panelists and triplicate testing of all milk and cream. Data from the three different tank trucks of milk were averaged.

The loss of flavor compounds in UHT milk was continuous and the milk was in a state of dynamic change due to storage temperatures, enzymatic activity, and package interactions. The role of packaging becomes very important in measuring flavor loss. Polyethylene used in aseptic packaging was tested in model systems with dairy flavor compounds to determine the quality of flavor compounds lost to polyethylene film. As the chain length

of the aldehydes, methyl ketones, sulfur compounds, and esters increased, the compounds became more lipophilic and tended to bind to the LDPE. The LDPE acted like a scavenger absorbing the flavor compounds.

The aldehydes from C_{7-10} bound to LDPE 11%, 16%, 41%, and 63%, respectively. The methyl ketones showed less affinity for the LDPE with 1.5%, 3.0%, 9.0%, and 38% for the C_{7-10} methyl ketones. The esters (methyl caproate, methyl enathate, and methy caprylate) bound at 6.0%, 9.0%, and 41.0%. The sulfur compounds (methyl thiophene, methyl thioproponal, and benzyl methyl sulfide) bound at 8.4%, 12.8%, and 20.8%.[8]

O'Neil and Kinsella[13] reported the binding of nonanone to β-lactoglobulin in dairy systems. They attributed the binding of these lipid-like flavor compounds to hydrophobic patches on the protein. Upon heating in an UHT system, the protein's helical structure opened up allowing these lipid-like flavor compounds to bind to the hydrophobic patches causing a reduction in free flavor compounds in the milk system. The flavor loss dynamic was probably due to several reasons: binding to the LDPE and hydrophobic patches on the protein. Other changes could have been caused by the interaction of the flavor compounds with each other while being stored at room temperature or above or could have been lost due to oxidation reactions. Longer hold times cause greater enzymatic degradation, which tends to give more flavor stability to the products. On the front side of this scenario, as more cooked flavors were generated, they took longer to dissipate.

CONCLUSION

The processing of milk and cream using UHTs produces new groups of flavor compounds. These flavor compounds include aldehydes, methyl ketones, enals, dienals, lactones, methylsulfide, hydrogen sulfide, methyl mercaptan, and other sulfur containing compounds. As the dairy products are stored at room temperature in aseptic milk cartons, the flavor compounds bind with the polyethylene coating causing a reduction in flavor compounds each month. This causes the flavor intensity to decrease over time causing the milk to become blander in flavor until 6 mo–9 mo. At this time, new flavors can start to develop in the shorter hold time UHT milk (3.4 sec). The flavor defects found after 6 mo–9 mo storage included chalky, coconut, musty, rancid, scorched, cheese whey, fruity, foreign, astringent, oxidized, oily, buttery, and soapy. Longer hold times of 6.9 sec and 20.3 sec had fewer problems with gelation and off-flavor development due to less enzyme survival in the finished UHT milk or cream. Both the LDPE and β-lactoglobulin bind flavors in UHT dairy products causing flavor loss during storage at room temperature.

REFERENCES

1. McCarty, W.O.; Hansen, A.P. Effects of Ultra-high-temperature Steam Injection Processing on Composition of Carbonyls in Milk Fat. J. Dairy Sci. **1981**, *64*, 581.
2. Hansen, A.P.; Melo, T.S. Effects of Ultra-high-temperature Steam Injection upon Constituents of Skim Milk. J. Dairy Sci. **1977**, *60*, 1368.
3. Melo, T.S.; Hansen, A.P. Effects of Ultra-high-temperature Steam Injection on Model Systems of α-lactalbumin and β-Lactalglobulin. J. Dairy Sci. **1978**, *61*, 710.
4. Aboshama, K.; Hansen, A.P. Effects of Ultra-high-temperature Steam Injection Processing on Sulfur Containing Amino Acids in Milk. J. Dairy Sci. **1977**, *60*, 1374.
5. Earley, R.R.; Hansen, A.P. Effects of Process and Temperature During Storage of UHT Steam-Injected Milk. J. Dairy Sci. **1982**, *65*, 11.
6. Hutchins, R.K.; Hansen, A.P. The Effect of Various UHT Processing Parameters and Storage Conditions on the Saturated Aldehydes in Half-and-Half Cream. J. Food Prot. **1991**, *54*, 109–112.
7. Nash, J.B. The Effect of Processing and Storage on the Flavor and the Aromatic Compounds in Ultra-high-temperature Milk. M.S. Thesis, North Carolina State University, Raleigh, NC, 1984.
8. Arora, D.K.; Hansen, A.P.; Armagost, M.S. Sorption of Flavor Compounds by Polypropylene. *Food and Packaging Interactions II*; ACS Symposium Series No. 473; ACS: Washington, D.C., 1991; Chap. 17, 203–211.
9. Hansen, A.P.; Swartzel, K.R.; Giesbrecht, F.L.G. Effect of Temperature and Time of Processing and Storage on Consumer Acceptability of Ultra-high-temperature Steam Injected Whole Milk. J. Dairy Sci. **1980**, *63*, 187.
10. Hansen, A.P.; Swartzel, K.R.; Earley, R.R. Effect of UHT Processing and Storage on the Chemical and Physical Properties of UHT Milk. *Proceedings of Conference on UHT Processing and Aseptic Packing of Milk and Milk Products*, 1980; 153.
11. Hansen, A.P.; Swartzel, K.R. Taste Panel Testing of UHT Fluid Dairy Products. J. Food Qual. **1982**, *4*, 203.
12. Swartzel, K.R.; Hamaan, D.D.; Hansen, A.P. Rheological Behavior of Ultra-high-temperature Steam Injected Dairy Products in Aging. J. Food Process. Eng. **1980**, *3*, 143.
13. O'Neil, T.E.; Kinsella, J.E. Binding of Alkanone Flavors to β-Lactoglobulin: Effects of Conformational and Chemical Modification. J. Agric. Food Chem. **1987**, *35*, 770.

Unsteady-State Heat Transfer in Biological Systems

Fu-hung Hsieh
University of Missouri–Columbia, Columbia, Missouri, U.S.A.

INTRODUCTION

In an unsteady-state heat transfer process, the temperature at any location in the system changes with time. The transfer of heat in this process is unsteady or transient. Unsteady-state heat transfer is important because of the large number of heating and cooling problems occurring in biological systems. In food processing, perishable foods are heated by immersion in steam baths or chilled by immersion in cold water in order to increase their shelf life.[1] Skin burns and frostbite are caused by accidental heat and cold exposure. In hyperthermia and cryosurgery tissues are heated or cooled rapidly to achieve controlled destruction of the cancerous growth.[2] The regular day to night changes in air temperature and solar radiation cause bulk warming and cooling of ground.[3] In fact, any time a heat transfer processing operation is started, there will be a transient or unsteady-state heat transfer period before steady-state heat transfer conditions can be reached.

GENERAL GOVERNING EQUATION

As temperature (T) is a function of two independent variables, time (t) and location (x,y,z), the general governing equation for unsteady-state heat transfer is[4]:

$$\rho c_p \frac{\partial T}{\partial t} = \frac{\partial}{\partial x}\left(k\frac{\partial T}{\partial x}\right) + \frac{\partial}{\partial y}\left(k\frac{\partial T}{\partial y}\right) + \frac{\partial}{\partial z}\left(k\frac{\partial T}{\partial z}\right) + \dot{q} \qquad (1)$$

where ρ is the density, c_p the heat capacity, k the thermal conductivity, and $\dot{q}$ the rate of heat generation.

When the thermal conductivity is constant and the rate of heat generation is zero, Eq. 1 becomes[5]:

$$\frac{\partial T}{\partial t} = \alpha\left(\frac{\partial^2 T}{\partial x^2} + \frac{\partial^2 T}{\partial y^2} + \frac{\partial^2 T}{\partial z^2}\right) \qquad (2)$$

where $\alpha = \frac{k}{\rho c_p}$ is thermal diffusivity.

LUMPED CAPACITY METHOD

Consider a solid which is at a constant temperature T_0 at time $t = 0$ and is immersed in an environment held at constant temperature T_1. The temperature profile in the solid at time $t > 0$ will depend on the relative value of the convective heat transfer coefficient (h) at the surface of the solid (external surface resistance) and the thermal conductivity (k) of the solid (internal conductive resistance). If the internal conductive resistance to heat transfer is very small compared to the external surface resistance ($k \gg h$), the heat transfer inside the solid is instantaneous and the temperature within the solid is essentially uniform at any given time. Making an energy balance on the solid for a small time interval of time dt, the heat transfer from the environment to the solid must be equal to the change in internal energy of the solid[5]:

$$hA(T_1 - T)\,dt = c_p \rho V\,dT$$

where A is the surface area of the solid, T the average temperature of the solid at time t, and V the volume of the solid. Rearranging this equation and integrating between the limits of $T = T_0$ at time $t = 0$ and $T = T$ at time $t = t$,

$$\int_{T_0}^{T} \frac{dT}{T_1 - T} = \frac{hA}{c_p \rho V}\int_0^t dt$$

$$\frac{T - T_1}{T_0 - T_1} = \exp\left(-\frac{hA}{c_p \rho V}\right)t \qquad (3)$$

This equation shows the time–temperature history of the solid. The term $c_p \rho V$ is called the lumped thermal capacitance of the system. Note that negligible internal conduction resistance was assumed in the derivation of this equation. The parameter that compares the internal conductive resistance and external surface resistance is the Biot number,

$$N_{Bi} = \frac{hL}{k} \qquad (4)$$

where L is the characteristic dimension of the solid and is equal to V/A. The lumped capacity method can be used when $N_{Bi} < 0.1$.

For a sphere, $L = V/A = (4\pi R^3/3)/(4\pi R^2) = R/3$ (R is radius of sphere).
For a long cylinder, $L = V/A = (\pi R^2 L)/(2\pi RL) = R/2$ (R is radius of cylinder).

Encyclopedia of Agricultural, Food, and Biological Engineering
DOI: 10.1081/E-EAFE 120007238
Copyright © 2003 by Marcel Dekker, Inc. All rights reserved.

For a cube, $L = V/A = (a^3)/(6a^2) = a/6$ (a is side of cube).

SEMI-INFINITE SOLIDS

When a cold wave suddenly reduces the air temperature far below freezing, the depth in the soil of the earth at which freezing temperatures penetrate is important in agriculture and construction. This is the case where heat transfer occurs only in one direction. The general governing equation becomes[3]:

$$\frac{\partial T}{\partial t} = \alpha \frac{\partial^2 T}{\partial x^2} \tag{5}$$

Initially, the temperature of the solid is uniform at T_0. At time $t = 0$, the solid environment is suddenly brought at a constant temperature T_1. For time $t > 0$ the solid surface temperature T_S is different from the environment temperature T_1 due to the convective heat transfer coefficient at the surface. Eq. 5 has been solved for these conditions and the solution is[3,5]:

$$\frac{T - T_0}{T_1 - T_0} = \operatorname{erfc}\left(\frac{x}{2\sqrt{\alpha t}}\right) - \exp\left[\frac{h\sqrt{\alpha t}}{k}\left(\frac{x}{\sqrt{\alpha t}} + \frac{h\sqrt{\alpha t}}{k}\right)\right] \operatorname{erfc}\left(\frac{x}{2\sqrt{\alpha t}} + \frac{h\sqrt{\alpha t}}{k}\right) \tag{6}$$

where x is the distance into the solid from the surface and $\operatorname{erfc}(\eta) = 1 - \operatorname{erf}(\eta)$. The function, $\operatorname{erf}(\eta)$, is called the error function and is given by:

$$erf(\eta) = \frac{2}{\sqrt{\pi}} \int_0^{\eta} \exp(-\eta^2)\, d\eta$$

Eq. 6 holds only for short times. When the surface heat transfer coefficient is high ($h \gg 0$), the solid surface temperature is equal to the environment temperature ($T_S = T_1$). The solution then becomes[3]:

$$\frac{T - T_0}{T_1 - T_0} = \operatorname{erfc}\left(\frac{x}{2\sqrt{\alpha t}}\right) = 1 - \operatorname{erf}\left(\frac{x}{2\sqrt{\alpha t}}\right) \tag{7}$$

Both solutions 6 and 7 can also be evaluated using the transient heating and cooling charts presented in the next entry of this encyclopedia or found in other references.[5–7]

INFINITE SLAB

A common geometry that occurs in unsteady-state heat transfer problems is a flat plate or slab of thickness $2x_1$ in the x direction and having very large or infinite dimensions in the y and z directions. Under this condition, heat transfer occurs only in the x direction and the general governing equation is the same as Eq. 5. The slab has an initial uniform temperature T_0, and at time $t = 0$, the slab is exposed to an environment at temperature T_1 and unsteady-state heat conduction occurs. For time $t > 0$, the solid surface temperature T_S is different from the environment temperature T_1 due to the convective heat transfer coefficient at the surface.

The analytical solution of Eq. 5 is[4,6,8,11]:

$$\frac{T - T_1}{T_0 - T_1} = \sum_{n=1}^{\infty} \frac{2 \sin \lambda_n}{\lambda_n + \sin \lambda_n \cos \lambda_n} \cos\left(\lambda_n \frac{x}{x_1}\right) \exp(-\lambda_n^2 N_{Fo}) \tag{8}$$

where the eigenvalues λ_n are the roots of $\lambda_n \tan \lambda_n = N_{Bi}$ and $N_{Fo} = \frac{\alpha t}{x_1^2}$ is the Fourier number.

For higher values of h ($N_{Bi} > 100$), the slab surface temperature $T_S = T_1$ and the solution becomes[5]:

$$\frac{T - T_1}{T_0 - T_1} = \sum_{n=0}^{\infty} \frac{4(-1)^n}{(2n+1)\pi} cos \frac{(2n+1)\pi x}{2x_1} exp\left[-\left(\frac{2n+1}{2}\right)^2 \pi^2 N_{Fo}\right] \tag{9}$$

The numerical results of both Eqs. 8 and 9 can also be evaluated graphically using the transient heating and cooling charts presented in the next entry of this encyclopedia or found in other references.[5–9,11]

INFINITE CYLINDER

For unsteady-state heat transfer in a long cylinder, the conduction of heat occurs only in the radial direction since the conduction at the ends can be neglected or the ends are insulated. The general governing equation in cylindrical coordinates is[5]:

$$\frac{1}{\alpha}\frac{\partial T}{\partial t} = \frac{1}{r}\frac{\partial}{\partial r}\left(r\frac{\partial T}{\partial r}\right) = \frac{\partial^2 T}{\partial r^2} + \frac{1}{r}\frac{\partial T}{\partial r} \tag{10}$$

If the cylinder has an initial uniform temperature T_0, and at time $t = 0$, the cylinder is exposed to an environment at temperature T_1 and unsteady-state heat conduction occurs. For time $t > 0$, the cylinder surface temperature T_S is different from the environment temperature T_1 due to the presence of surface resistance. The solution of Eq. 9 is[4,6,10,11]:

$$\frac{T - T_1}{T_0 - T_1} = \sum_{n=1}^{\infty} \frac{2J_1(\lambda_n)}{\lambda_n[J_0^2\lambda_n + J_1^2\lambda_n]} J_0\left(\lambda_n \frac{r}{R}\right) \exp(-\lambda_n^2 N_{Fo}) \quad (11)$$

where R is the radius of cylinder and the eigenvalues λ_n are the roots of

$$\frac{J_0(\lambda_n)}{J_1(\lambda_n)} = \frac{\lambda_n}{N_{Bi}}$$

where J_0 and J_1 are the Bessel function of the first kind of order 0 and 1. Note the Fourier number is defined as $N_{Fo} = \frac{\alpha t}{R^2}$ and the Biot number $N_{Bi} = \frac{hR}{k}$.

For high values of h ($N_{Bi} > 100$), the cylinder surface temperature $T_S = T_1$ and the solution becomes[6]:

$$\frac{T - T_1}{T_0 - T_1} = 2\sum_{n=1}^{\infty} \frac{J_0\left(\lambda_n \frac{r}{R}\right)}{\lambda_n J_1(\lambda_n)} \exp(-\lambda_n^2 N_{Fo}) \quad (12)$$

where the eigenvalues λ_n are the roots to $J_0(\lambda_n) = 0$.

The numerical results of both Eqs. 11 and 12 can also be evaluated graphically using the transient heating and cooling charts presented in the next entry of this encyclopedia or found in other references.[5–9,11]

SPHERE

The unsteady-state heat transfer in a sphere also occurs in the radial direction only. The general governing equation in spherical coordinates for unsteady-state heat transfer is[5]:

$$\frac{1}{\alpha}\frac{\partial T}{\partial r} = \frac{1}{r^2}\frac{\partial}{\partial r}\left(r^2 \frac{\partial T}{\partial r}\right) = \frac{\partial^2 T}{\partial r^2} + \frac{2}{r}\frac{\partial T}{\partial r} \quad (13)$$

If the sphere is at a uniform temperature T_0 at $t = 0$ and is exposed to constant environment temperature T_1 at time $t > 0$, Eq. 13 can be solved analytically and the solution is[4,6,10,11]:

$$\frac{T - T_1}{T_0 - T_1} = \sum_{n=1}^{\infty} \frac{2(\sin\lambda_n - \lambda_n \cos\lambda_n)}{\lambda_n - \sin\lambda_n \cos\lambda_n} \frac{\sin\left(\lambda_n \frac{r}{R}\right)}{\lambda_n \frac{r}{R}} \exp(-\lambda_n^2 N_{Fo}) \quad (14)$$

where R is radius of sphere and the eigenvalues λ_n are the positive roots of $\tan\lambda_n = -\frac{\lambda_n}{N_{Bi} - 1}$.

For high values of h ($N_{Bi} > 100$), the cylinder surface temperature $T_S = T_1$ and the solution becomes[6]:

$$\frac{T - T_1}{T_0 - T_1} = \sum_{n=1}^{\infty} 2(-1)^{n+1} \frac{\sin\left(n\pi \frac{r}{R}\right)}{n\pi \frac{r}{R}} \exp[-(n\pi)^2 N_{Fo}] \quad (15)$$

The numerical results of both Eqs. 14 and 15 can also be evaluated graphically using the transient heating and cooling charts presented in the next entry of this encyclopedia or found in other references.[5–9,11]

FINITE SOLIDS

The unsteady-state heat transfer problems considered so far are limited to 1-D only. Many practical problems in biological systems frequently are involved with simultaneous unsteady-state conduction in two or three directions. These multidimensional unsteady-state heat transfer can be solved using the principle of superposition.[5,10] A rectangular block with dimensions of $2x_1$, $2y_1$, and $2z_1$ can be considered as intersections of three infinite slabs of half thickness x_1, y_1, and z_1 and the temperature at location (x,y,z), $T(x,y,z)$, for time $t > 0$ is given by[5,10]:

$$\frac{T(x,y,z) - T_1}{T_0 - T_1} = \left(\frac{T(x) - T_1}{T_0 - T_1}\right)\left(\frac{T(y) - T_1}{T_0 - T_1}\right) \times \left(\frac{T(z) - T_1}{T_0 - T_1}\right) \quad (16)$$

Similarly, a finite cylinder with radius of R and height of $2z_1$ can be considered as the product of the conduction through an infinite slab of half thickness z_1 and an infinite cylinder of radius R[5,10]:

$$\frac{T(r,z) - T_1}{T_0 - T_1} = \left(\frac{T(r) - T_1}{T_0 - T_1}\right)\left(\frac{T(z) - T_1}{T_0 - T_1}\right) \quad (17)$$

REFERENCES

1. Singh, R.P.; Heldman, D.R. *Introduction to Food Engineering*, 3rd Ed.; Academic Press: London, 2001; 659.
2. Berger, S.A.; Goldsmith, W.; Lewis, E.R. *Introduction to Bioengineering*; Oxford University Press: Oxford, 1996; 526.
3. Datta, A.K. *Biological and Bioenvironmental Heat and Mass Transfer*; Marcel Dekker, Inc.: New York, 2002; 383.
4. Carslaw, H.S.; Jaeger, J.C. *Conduction of Heat in Solids*, 2nd Ed.; Oxford University Press: London, 1959; 510.

5. Geankoplis, C.J. *Transport Processes and Unit Operations*, 3rd Ed.; Prentice Hall: Englewood Cliffs, New Jersey, 1993; 921.
6. Schneider, P.J. *Conduction Heat Transfer*; Addison-Wesley Publishing Company, Inc.: Cambridge, MA, 1955; 395.
7. Perry, R.H.; Green, D. *Perry's Chemical Engineers' Handbook*, 7th Ed.; McGraw-Hill Inc.: New York, 1997.
8. Middleman, S. *An Introduction to Mass and Heat Transfer*; John Wiley & Sons, Inc.: New York, 1997; 672.
9. Heisler, M.P. Temperature Charts for Induction and Constant Temperature Heating. Trans. ASME **1947**, *69*, 227–236.
10. Kakaç, S.; Yener, Y. *Heat Conduction*, 2nd Ed.; Hemisphere Publishing Corp.: Washington, D.C., 1985; 397.
11. Singh, R.P. Heating and Cooling Processes for Foods. In *Handbook of Food Engineering*; Heldman, D.R., Lund, D.B., Eds.; Marcel Dekker, Inc.: New York, 1992; 247–276.

Unsteady-State Heat Transfer in Foods

Vikram Ghosh
R. C. Anantheswaran
The Pennsylvania State University, University Park, Pennsylvania, U.S.A.

INTRODUCTION

Heat transfer is an important unit operation in many of the food manufacturing operations. Success of operations such as pasteurization and sterilization depends on the ability to predict and control the temperature within the food product. Heat transfer is also crucial in several other food processing operations such as drying, cooling, and freezing of foods. When any material is heated or cooled, it initially goes through an unsteady state heat transfer process before reaching a steady state heat transfer. During unsteady state heat transfer, the temperature at any given location is a function of time and location within the product, whereas during steady state transfer the temperature is independent of time.

The unsteady state heat transfer can be described by the following governing equation:

$$\frac{\partial T}{\partial t} = \alpha\left(\frac{\partial^2 T}{\partial x^2} + \frac{\partial^2 T}{\partial y^2} + \frac{\partial^2 T}{\partial z^2}\right) + \frac{q_{gen}}{\rho C_p} \tag{1}$$

where α is the thermal diffusivity of the food material ($\alpha = k/\rho C_p$), ρ, the density of the food material, k, the thermal conductivity of the food material, and C_p is the specific heat of the food material. When we are dealing with heat transfer involving exothermic or endothermic chemical reactions (such as during cooling of fresh produce item which involves heat generation due to respiration of fresh produce), a heat generation term (q_{gen} which is the heat generated per unit volume) needs to be included. In most of the food processing situations, the heat generation term is generally equal to zero and Eq. 1 reduces to:

$$\frac{\partial T}{\partial t} = \alpha\left(\frac{\partial^2 T}{\partial x^2} + \frac{\partial^2 T}{\partial y^2} + \frac{\partial^2 T}{\partial z^2}\right) \tag{2}$$

For solving problems with different geometries, Eq. 2 can also be written in spherical and cylindrical coordinates. The transformations of Eq. 2 in cylindrical and spherical coordinates are given in Eqs. 3 and 4, respectively.[1]

$$\frac{\partial T}{\partial t} = \alpha\left[\frac{1}{r}\frac{\partial}{\partial r}\left(r\frac{\partial T}{\partial r}\right) + \frac{1}{r^2}\frac{\partial^2 T}{\partial \theta^2} + \frac{\partial^2 T}{\partial z^2}\right] \tag{3}$$

$$\frac{\partial T}{\partial t} = \alpha\left[\frac{1}{r^2}\frac{\partial}{\partial r}\left(r^2\frac{\partial T}{\partial r}\right) + \frac{1}{r^2 \sin\theta}\frac{\partial}{\partial \theta}\left(\sin\theta\frac{\partial T}{\partial \theta}\right) + \frac{1}{r^2 \sin^2\theta}\frac{\partial^2 T}{\partial \phi^2}\right] \tag{4}$$

Eq. 2 has been solved for simple geometries and the solutions can be found in Ref. 2. The solution gives a temperature function that describes the temperature at different locations as a function of time during unsteady state heat transfer. Graphical techniques can also be employed to determine the transient time–temperature distribution with infinite slabs, infinite cylinders, and spheres as further discussed below (and in the article *Transient Heat Transfer Charts*). The two dimensionless numbers used in the graphical techniques are the Biot and the Fourier numbers.

BIOT NUMBER

The Biot number (Bi_0) is a dimensionless number and is described as:

$$Bi_0 = \frac{hx_1}{k} \tag{5}$$

where h is the convective heat transfer coefficient, x_1 is the characteristic dimension that represents the shortest distance from the surface and the center (x_1 is equal to radius/3 for a sphere, radius/2 for an infinitely long cylinder, and thickness/2 for an infinitely long square rod).[3] When a food product is heated or cooled, the heat transfer resistance is due to a combination of the heat transfer coefficient at the surface of the food (external heat transfer resistance) and/or due to the thermal conductivity of the food material (internal heat transfer resistance). The Biot number compares the relative values

Encyclopedia of Agricultural, Food, and Biological Engineering
DOI: 10.1081/E-EAFE 120006996
Copyright © 2003 by Marcel Dekker, Inc. All rights reserved.

of internal conduction resistance and surface convective resistance to heat transfer (more information on Biot number is available in the article *Biot Number*). When Biot number is less than 0.1, the internal heat transfer resistance is negligible and the external heat transfer resistance becomes the rate-limiting factor in determining the transient temperature distribution within the food product. When Biot number is greater than 40, the external heat transfer is negligible, and the internal heat transfer becomes the rate limiting factor and the rate of heat transfer is controlled by the conductivity of the food material. When the Biot number is between 0.1 and 40, there is finite internal and external resistance.

FOURIER NUMBER

The Fourier number (Fo) is also a dimensionless number and is described as:

$$Fo = \frac{\alpha t}{x_1^2} \tag{6}$$

where, t is the time, and x_1 is once again the characteristic dimension as described in Eq. 5. The Fourier number represents the ratio of the rate of heat conduction to the rate of heat storage.[4]

SYSTEM WITH NEGLIGIBLE INTERNAL RESISTANCE

This case usually occurs when the thermal conductivity of the material is very high as in the case of metals. Then the internal heat transfer resistance is negligible as compared to convective heat gain or loss at the surface of the material. An example of such a situation is the heat loss from steam pipe. The heat transfer under these circumstances can be described as:

$$hA(T_\infty - T)\mathrm{d}t = C_\mathrm{p}\rho V \mathrm{d}T \tag{7}$$

where T_∞ is the temperature of the fluid, V, the volume in m^3, A, the surface area of the object (m^2), T, the average temperature at time t, and ρ is the density of the object in $\mathrm{kg\,m^{-3}}$. Rearranging Eq. 7 and integrating between limits $t = 0$, when $T = T_0$ and $T = T$ at $t = t$,

$$\int_{T=T_0}^{T=T} \frac{\mathrm{d}T}{T_\infty - T} = \frac{hA}{C_\mathrm{p}\rho V}\int_{t=0}^{t=t} \mathrm{d}t \tag{8}$$

$$\frac{T - T_\infty}{T_0 - T_\infty} = \mathrm{e}^{-(hA/C_\mathrm{p}\rho V)t} \tag{9}$$

Eq. 9 can be used to describe the unsteady state heat transfer and the resulting time–temperature distribution within the bulk of the food material.

SYSTEMS WITH FINITE INTERNAL AND EXTERNAL RESISTANCE TO HEAT TRANSFER

Graphical solutions for the time–temperature distribution within infinite cylinder, infinite slab, and a sphere can be obtained from Heisler chart and/or Gurney and Lurie charts.[3] The Heisler charts give the temperature as a function of time at the geometric center of the solid, while the charts by Gurney and Lurie gives the temperatures as a function of time t at any location. The Heisler charts for different geometries are shown in Figs. 1–3. The charts use three dimensionless numbers, namely, the Fourier number (Fo), the reciprocal of the Biot number ($1/Bi_0$), and θ_0, as defined below[5]:

$$\theta_0 = \frac{T - T_\infty}{T_i - T_\infty} \tag{10}$$

where T is the temperature at the center or the mid-plane temperature of the material, T_i, the initial temperature, and T is the ambient temperature. θ_0 represents the unaccomplished temperature ratio within the food material.

For situations when there is negligible surface resistance to heat transfer, the charts can be used with $(1/Bi_0) = 0$. The overall approach to evaluating the time–temperature distribution involves the evaluation of the Biot number and Fourier numbers for the given

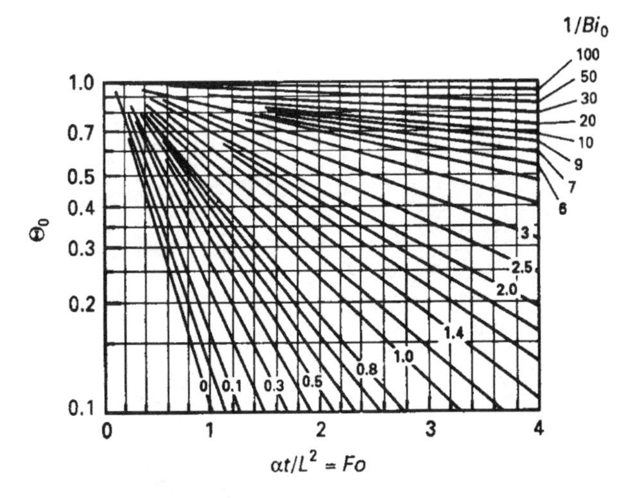

Fig. 1 Heisler chart for a slab of Thickness 2L subjected to convection at both boundary surfaces. (From Ref. 5.)

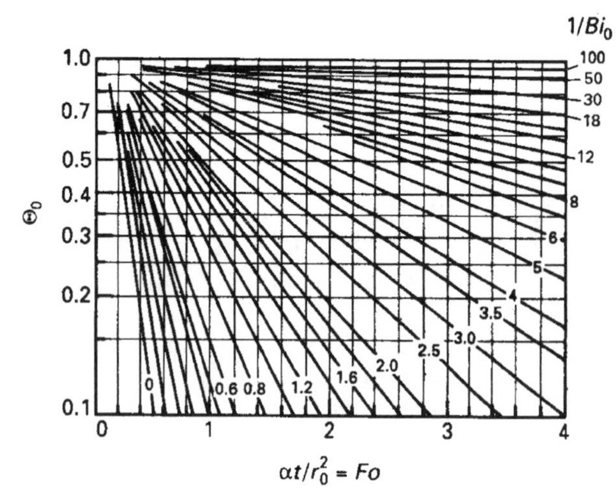

Fig. 2 Heisler chart for a solid cylinder of radius $r = r_0$, subjected to convection at boundary surface $r = r_0$. (From Ref. 5.)

situation, and then the temperature ratio is evaluated from the chart.

Finite cylinders and slabs can be described as an intersection of two infinite surfaces. This approach is also known as Newman's rule.[2] Eqs. 11 and 12 can be used to determine the time–temperature data from the data evaluated from infinite geometries.[4]

$$\left[\frac{T_\infty - T}{T_\infty - T_i}\right]_{\text{finite cylinder}} = \left[\frac{T_\infty - T}{T_\infty - T_i}\right]_{\text{infinite cylinder}} \times \left[\frac{T_\infty - T}{T_\infty - T_i}\right]_{\text{infinite slab}} \tag{11}$$

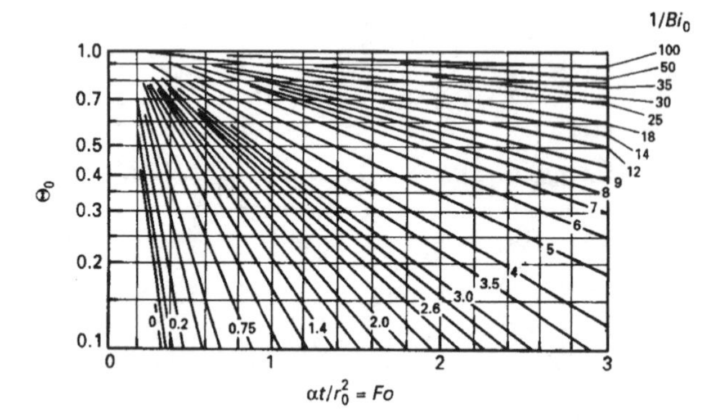

Fig. 3 Transient temperature chart for a solid sphere of radius $r = r_0$, subjected to convection at boundary surface $r = r_0$. (From Ref. 5.)

$$\left[\frac{T_\infty - T}{T_\infty - T_i}\right]_{\text{finite brick shape}} = \left[\frac{T_\infty - T}{T_\infty - T_i}\right]_{\text{infinite slab, width}} \times \left[\frac{T_\infty - T}{T_\infty - T_i}\right]_{\text{infinite slab, depth}} \times \left[\frac{T_\infty - T}{T_\infty - T_i}\right]_{\text{infinite slab, height}} \tag{12}$$

NUMERICAL METHODS

In most cases, the food material is not of regular shape and the boundary conditions vary with time. In these cases, neither the graphical method can be used nor is any mathematical solution possible. Such situations can be best handled by the use of numerical methods.

Let us consider the 1-D heat transfer problem as shown in Eq. 13.

$$\frac{\partial T}{\partial t} = \alpha\left(\frac{\partial^2 T}{\partial x^2}\right) \tag{13}$$

The partial derivative of Eq. 13 can be evaluated as shown in Eqs. 14 and 15.

$$\frac{\partial^2 T}{\partial x^2} \approx \frac{1}{(\Delta x)^2}(T_{i+1} - 2T_i + T_{i-1}) \tag{14}$$

and

$$\frac{\partial T}{\partial t} \approx \frac{1}{\Delta t}(T_i^{j+1} - T_i^j) \tag{15}$$

where the subscript i denotes position and superscript j denotes time. Eqs. 14 and 15 can be substituted in Eq. 13 and the time–temperature function can be solved for iteratively. This method can be extended to 2- and 3-D problems. A number of numerical methods have been suggested to solve a variety of problems and these methods can be found in Ref. 6. Numerical methods used in thermal processing of foods are given in the article *Thermal Process Calculations*.

REFERENCES

1. Bird, R.B.; Stewart, W.E.; Lightfoot, E.N. *Transport Phenomena*; John Wiley & Sons: New York, NY, 1960.
2. Carslaw, H.S.; Jaeger, J.C. *Conduction of Heat in Solids*; Clarendon Press: Oxford, 1959.
3. Geankoplis, C.J. *Transport Processes and Unit Operations*; Allyn and Bacon, Inc.: Newton, MA, 1983.
4. Singh, R.P.; Heldman, D.R. *Introduction to Food Engineering*; Academic press: San Diego, CA, 1993.
5. Thomas, L.C. *Heat Transfer*; Capstone Publishing Corporation: Tulsa, OK, 1999.
6. Holman, J.P. *Heat Transfer*; McGraw Hill Book Co.: Singapore, 1992.

Unsteady-State Mass Transfer in Biological Systems

U

Rakesh Ranjan
Applied Geoscience and Engineering, Reading, Pennsylvania, U.S.A.

Joseph Irudayaraj
The Pennsylvania State University, University Park, Pennsylvania, U.S.A.

INTRODUCTION

Industrial applications in the agricultural industry often require unit operations involving mass transfer. A number of food processing operations in drying, convective heating, infrared heating, packaging, extraction processes, osmotic processes, etc. involve mass transfer. The main focus of this article is to describe unsteady-state mass transfer with the help of governing equations and applications. Both steady-state and unsteady-state mass transfers are described and the coupled effect of mass and heat transfer is pointed out. Mass transfer is one of the fundamental processes in food during processing and/or storage.[1] Two terminologies may exist for mass transfer, mass or bulk flow of fluid from one location to the other due to convection flow or diffusion of constituents from one point to the other because of concentration gradients or a driving force. Unsteady state denotes a condition where the movement of mass occurs either due to convection or diffusion and changes with time.

Transient mass transfer in food related applications include diffusion of salt within a food matrix, diffusion of antimicrobial agents into a food system, diffusion of moisture in drying-related applications, migration of oils and fatty acids through package materials and within a composite food structure under favorable conditions, diffusion of volatile flavors through package materials and food system, mass transfer through polymeric materials, diffusion of nutrients and metabolites through porous media and redistribution of mass during storage. Selected applications of mass transfer in the food processing industry include food packaging[2,3] where the main concern is the diffusion of components into and out of the package environment, drying,[4,5] heat and mass transfer in composite food,[4,5] baking of bread,[6] and acidification of cut root vegetables,[7] to name a few.

Mass transfer is governed by a differential gradient in concentration of the migrating species across a boundary that contributes to the diffusion mass transport. The difference in the concentration of the migrating species provides the driving force for transfer of mass across a surface. There are two stages of mass transfer. The first stage is often denoted as the transient stage in which mass transfer is either time dependent, transient, or unsteady. The second stage occurs after some time when equilibrium has been established between the concentrations of the species across the surface and is called the steady state. In the rest of the article, the component that is being transferred across a boundary will be denoted as the migrating species and the boundary across which the transfer of species occurs will be denoted as the surface.

FUNDAMENTALS OF MASS TRANSFER

Mass transfer, by its very nature, is intimately involved with mixtures of chemical species[8] contributing to both mass diffusion occurring at the molecular level due to concentration gradient and bulk transport due to convection because of pressure gradients. If "i" is the species migrating across a surface and v_i the velocity of the migrating species and v the average velocity of the mixture, the relative velocity of the migrating species i in comparison to the whole mixture can then be denoted by $v - v_i$. The flux of the species i can then be given as $\rho_i v_i$, where ρ_i is the partial density of the migrating species i. The diffusion flux of the ith species relative to the mean flux is:

$$j_i = \rho_i(v_i - v) \tag{1}$$

where j_i represents the diffusion component of the migrating species i. The total mass flux of the ith species given as $n_i = \rho_i v_i$ is then given by

$$n_i = \rho_i v_i = \rho_i v + \rho_i(v_i - v) = \text{convection} + j_i \tag{2}$$

This expression comprises of two components, the convection component due to contribution from the mean flow and the diffusion component that occurs because of the relative velocity of the migration species with the mean flow by convection.

DOI: 10.1081/E-EAFE 120007190
Copyright © 2003 by Marcel Dekker, Inc. All rights reserved.

CONSERVATION EQUATIONS

A mass balance in a control volume (Fig. 1), fixed in space (R) through which there is a flux entering, a flux leaving, and a regeneration of mass $\dot{r}_i$ for species i inside the control volume (Fig. 1) has the form:

$$\frac{d}{dt}\int_R \rho_i\,\mathrm{d}R = -\int_S \vec{n}_i\cdot\mathrm{d}\vec{S} + \int_R \dot{r}_i\,\mathrm{d}R$$
$$= -\int_S \rho_i\vec{v}\cdot\mathrm{d}\vec{S} - \int_S \vec{j}_i\cdot\mathrm{d}\vec{S} + \int_R \dot{r}_i\,\mathrm{d}R \tag{3}$$

Here, ρ_i is the partial density of the species i in the mixture, $\mathrm{d}R$ a differential elemental volume inside the control volume, $d\vec{S}$ the product of the unit normal vector and $\mathrm{d}S$ (normal to the surface of the volume; expressing the area as a vector perpendicular to the surface). Regeneration of mass implies the generation of mass inside the control volume for each species in the mixture (for a general treatment) and hence it appears as a source term in the equation. Integration is performed over the volume because it is assumed that the regeneration of mass occurs over each elemental volume of the domain, and summation over the whole volume gives the total contribution due to mass regeneration. Using the mass conservation equation for each species i in R, the rate of increase of species i in $r =$ $-$ (rate of convection of species i out of r + rate of diffusion of species i out of r) + rate of creation of species i in r. The conservation of mass of all species flowing through a control volume in stationary space has been expressed mathematically in Eq. 3.

Using Gauss's theorem (for converting the surface integrals to volume integrals) the mass conservation equation can be expressed as:

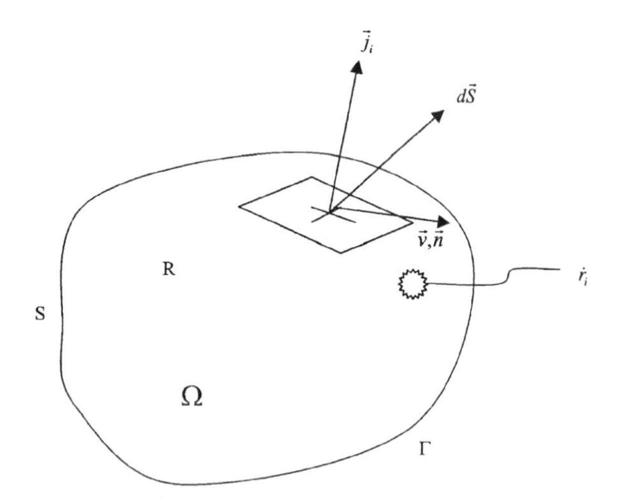

Fig. 1 Control volume in the mass diffusion field.

$$\frac{\partial\rho_i}{\partial t} + \nabla\cdot(\rho_i\vec{v}) = -\nabla\cdot\vec{j}_i + \dot{r}_i \tag{4}$$

The contribution of all the species to the transfer process can be written as:

$$\sum\nabla\vec{j}_i = 0 \tag{5}$$

$$\sum r_i = 0 \tag{6}$$

If the net mass created inside a control volume is equal to zero and the net diffusion flux is also zero, $\sum\nabla\vec{j}_i = \sum\rho_i(v_i - v) = 0$. Eq. 4 can be rewritten as:

$$\frac{\partial\rho}{\partial t} + \nabla\cdot(\rho\vec{v}) = 0 \tag{7}$$

Eq. 7 assumes that there is no regeneration of mass inside the control volume for the total mixture. If the mixture is assumed to be incompressible then $\nabla\cdot\vec{v} = 0$ and Eq. 4 can be reduced to:

$$\frac{\partial\rho_i}{\partial t} + \vec{v}\cdot\nabla\rho = -\nabla\cdot\vec{j}_i + \dot{r}_i \tag{8}$$

Eq. 7 for a stationary and nonreacting medium ($\vec{v} = 0$ *and* $\dot{r}_i = 0$) reduces to

$$\frac{\partial\rho_i}{\partial t} = -\nabla\cdot\vec{j}_i \tag{9}$$

Fick's law states that the diffusion component of mass transfer occurs because of the differential gradient in mass concentration. The transfer of mass is from a higher concentration region to a lower concentration region. It is expressed in mathematical form as

$$-\nabla\cdot\vec{j}_i = \rho D_{\mathrm{im}}\nabla m_i \tag{10}$$

where D_{im} is the mass diffusion coefficient for species i, the proportionality factor that correlates the flux of mass transfer to the actual gradient; m_i the mass fraction of the ith species in the mixture expressed by ρ_i/ρ, ρ the total density and ρ_i the density of the ith component. Using Eqs. 9 and 10 we have

$$\frac{\partial m_i}{\partial t} = D_i\nabla^2 m_i \tag{11}$$

Eq. 11 is the most frequently referenced unsteady-state mass transfer equation for a migrating species i without accounting for the chemical reaction inside the control volume and is based on the assumption that the mass transfer medium is stationary. It should be kept in mind that the applicability of this equation is limited to situations where the mass transport is purely diffusive

(i.e., the bulk transport by mean flow is zero). Eq. 11 is very similar in formulation to the heat conduction Fourier's equation. If the mass transfer is considered to occur across a convective boundary then the boundary condition for the surface Γ of the control volume for the ith component is given as:

$$-D\frac{\partial m}{\partial t} = \frac{k_c}{k_p}(m_e - m_s) \quad (12)$$

Here, k_c is the mass transfer coefficient, k_p the partition coefficient across the surface Γ, m_e the mass fraction of species i in the migrating mixture of the control volume, and m_s the mass fraction of the same species in the ambient surrounding. Mass transfer coefficient describes the transfer rate of mass from the convecting surface of any region. The higher is the mass transfer coefficient, the higher is the transfer of mass from the surface of the domain. Partition coefficient is found to be present particularly where there is a phase change of the mass across boundaries, as in polymer study applications. Phase change of a substance involves reaching equilibrium of the vapor pressure of the migrating species and the chemical potential of the same in the liquid phase; however, a numerical treatment of this is beyond the scope of this work. Partition coefficients across an interface for polymers and gels have been discussed in Refs. 3 and 4, respectively.

Eq. 11 can be further simplified to obtain the steady-state mass transfer equation where the mass transfer in the domain of interest is invariant with respect to time and is expressed as:

$$D_{\text{im}}\nabla^2 m_i = 0 \quad (13)$$

Here, mass transfer is purely diffusive in nature and the restrictions mentioned for Eq. 11 hold. Solution to Eq. 13 for a particular set of boundary conditions has been presented in "Application." Mass transfer for food materials seldom occurs in isolation; heat transfer and concurrent fluid flow is associated with it.[8] Mass transfer is often affected by heat transfer and the two phenomena are interdependent in many practical situations. It is thus instructive to view the coupled effects of heat and mass transfers, as this is realistic of most food-related drying applications. In addition, an understanding of coupled heat and mass transfer gives enough knowledge to grasp a single physical process such as mass or heat transfer.

COUPLED TRANSFER PHENOMENON

Different heat and mass transfer models have been proposed to model the coupling effects of this simultaneous process.[3,4,9,10] Absorption or evolution of water by solid results in evolution or absorption of heat. This heat diffuses through a solid causing a change in temperature, which affects the ability of solid to absorb or evolve water. Thus, the transfers of moisture and heat are coupled and have to be considered simultaneously. Coupled processes mentioned above are the most frequently found practical problems in mass transfer in drying applications. Luikov noted that there was a direct relationship between liquid and heat transfer in a capillary-porous body and proposed a system of governing equations to describe the coupled heat and mass transport process based on moisture potential and temperature gradients. Fortes and Okos[11,12] extended Luikov's equations and developed flux equations based on a phenomenological theory of irreversible thermodynamics and applied it to grain drying. They also concluded that the mass and heat transfers are coupled strongly in any drying application for a food material. Liquid transfers not only due to volumetric liquid concentration gradient but also due to temperature gradient; when liquid is present in the vapor form an additional transfer is also possible.

Luikov's two-way coupled equation deals with coupled heat and mass transfer while the three-way coupled equation describes the interdependence of heat, mass, and pressure transfer in a porous system. These two equations have been applied to food drying in 2-D[4] and 3-D.[5] The influence of temperature gradient as a moisture driving potential was investigated by Whitney and Poterfield[13] as a means of controlling the moisture gradient during the drying process. An application considering the shrinkage effects using coupled heat and mass transfer can be found in Browser and Wilhelm.[14] The same concept when applied in 2-D is called an adaptive mesh, where the solution domain (Ω) of the partial differential equation changes with time as does the moisture and temperature fluxes during transient heat and mass transfers. A more recent solution to the coupled heat and mass transfer problem can be found in Ref. 15.

SOLUTION TECHNIQUES

Analytical solutions to the above set of equations (transient mass transfer, Luikov coupled equations in heat and mass transfers for porous bodies) have been proposed.[16–19] However, analytical solutions are incapable of accommodating either nonlinear material properties, anisotropy, nonhomogeneity, or multidimensionality of the solution. Analytical solutions in any of the above situations become extremely complicated. Anisotropy, nonlinearity, and nonhomogeneity can be better handled by numerical techniques. Analytical solution for simple unsteady-mass transfer equations as in Eq. 11 for simple cases are provided,[20] and an excellent preliminary discussion of the topic can be found in Singh and Heldman.[1]

Numerical solution techniques of partial differential equations can in general be obtained by the finite difference, the finite volume (also called the control volume method), finite element methods, and the boundary element method. Finite difference method in essence is a Taylor's series expansion of the partial differential equation that incorporates the boundary conditions into the main formulation. The finite volume method (control volume method) involves the balancing of the fluxes on an arbitrary control volume and is similar to the finite difference method in formulation. The finite element method uses a weak formulation to develop the equations for an element followed by a numerical integration subroutine to calculate the element stiffness matrix.[21]

STEADY-STATE SOLUTION OF LAPLACE EQUATION EXAMPLE

The governing partial differential equation for steady-state mass transfer is governed by Eq. 13. We assume insulated boundary conditions on the three sides [(0,0) and (1,0); (0,0) and (0,1); and (0,1) and (1,1)] in the food model (Fig. 2) to have the form

$$-D\frac{\partial m}{\partial n} = 0 \tag{14}$$

As there is no mass transfer across the insulated surfaces, the convective mass transfer coefficients for these surfaces are zero. n is the surface normal vector that becomes x for the south and north boundaries and the dimension y for the west boundary. Eq. 14 upon integration for the surface is given as $m = c$ where c is a constant. Let us assume that the value of the constant is 0 (for demonstration purposes).

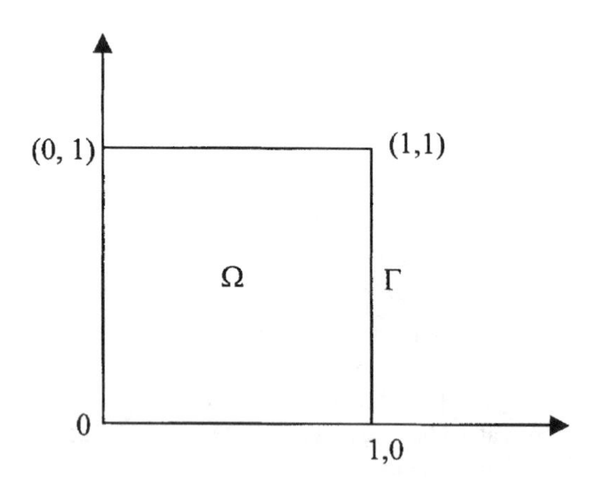

Fig. 2 Schematic of the food model.

If we assume that the boundary condition on the surface Γ (east boundary) to be

$$m(1, y) = \sin(\pi y) \tag{15}$$

then the problem has been defined completely in the domain with Eqs. 13–15 with the governing partial equation for mass transfer (Eq. 13) and the boundary conditions (Eqs. 14 and 15). The solution to the above set of equations was obtained on the domain shown in Fig. 2. Twenty-seven discretizations were considered each on x and y axis, respectively, and the [729 × 729] banded matrix was solved by the multigrid method. The solution to the system of Eqs. 13–15 by the finite difference method in the domain Ω (Fig. 2) is provided in Fig. 3.

UNSTEADY-STATE MASS TRANSFER WITH MASS REGENERATION

Let us assume that there is a regeneration of mass inside the domain (Fig. 2) and the governing partial differential equation for transient mass transfer is given by Eq. 16. The mass regeneration term appears in the right hand side of the equation and provides the driving force for mass transfer.

$$\frac{\partial m}{\partial t} - \left(\frac{\partial^2 m}{\partial x^2} + \frac{\partial^2 m}{\partial y^2}\right) = 1 \tag{16}$$

Insulated boundary conditions at the defined edges are

$$-D\frac{\partial m}{\partial n} = 0 \ \ [(0,0) \text{ through } (1,0)]$$

$$\text{and } [(0,0) \text{ through } (0,1)] \tag{17}$$

implying that there is no mass transfer across the insulated surfaces and hence the convective mass transfer coefficients at these surfaces are zero. If we assume that the boundary condition on the surface Γ to be (north and east boundaries)

$$m(1, y, t) = 0 \tag{18}$$

and

$$m(x, 1, t) = 0 \tag{19}$$

then the initial condition throughout the domain will be

$$m(x, y, 0) = 0 \text{ for all } (x, y) \text{ in } \Omega. \tag{20}$$

The solution to the system of Eqs. 16–19 by the finite element method in the domain Ω (Fig. 2) at the end of

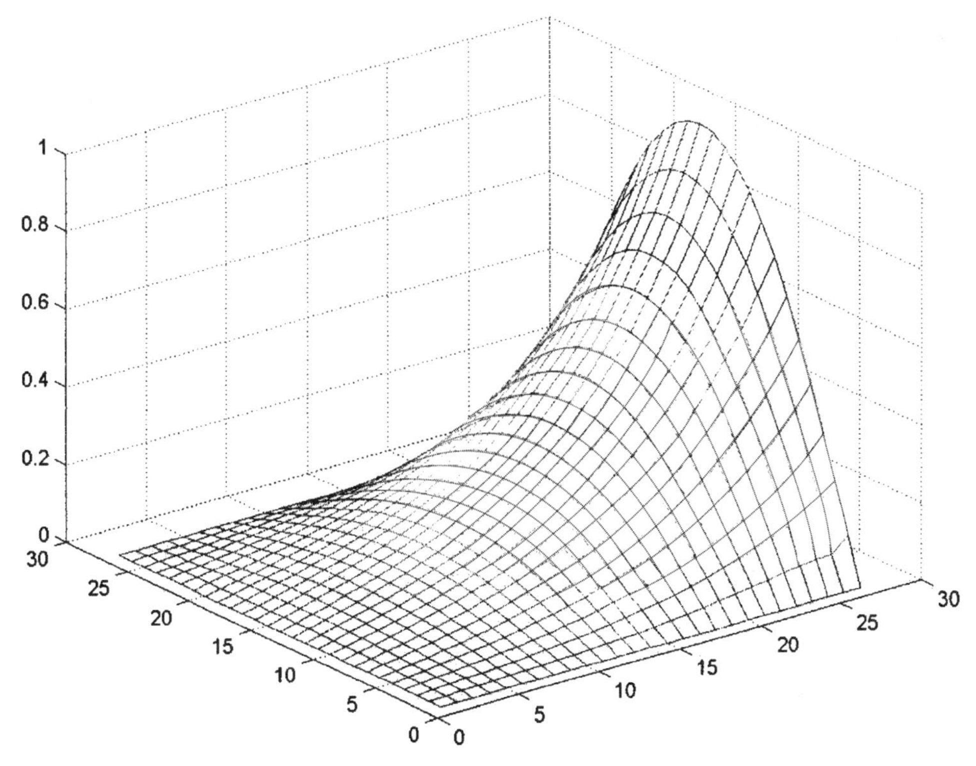

Fig. 3 Solution of the Laplace equation with given boundary conditions. (Courtesy of Ranjan, R., http://www.personal.psu.edu/users/r/x/rxr228/demonstration.html with source code, accessed September 2002.)

0.2 sec is given in Fig. 4, and at the end of 1 sec (steady state is reached) is presented in Fig. 5. The equation was solved with quadrilateral bilinear elements and Lagrange interpolation functions with iso-parametric formulation. Details on the finite element method can be found in Reddy[21] and solution to a coupled heat and mass transfer process can be found in Irudayaraj, Haghighi, and Stroshine.[22]

APPLICATIONS

The basic equation (Eq. 11) holds for cases where the mass transfer is purely diffusive in nature with no regeneration of mass inside the control volume. Eq. 11 models the mass transfer for a large number of problems such as the diffusion of water from the inside to the outside of foods when heated with convective air drying,[9,11] mass

Fig. 4 Moisture profile in a rectangular domain at the end of 0.2 sec.

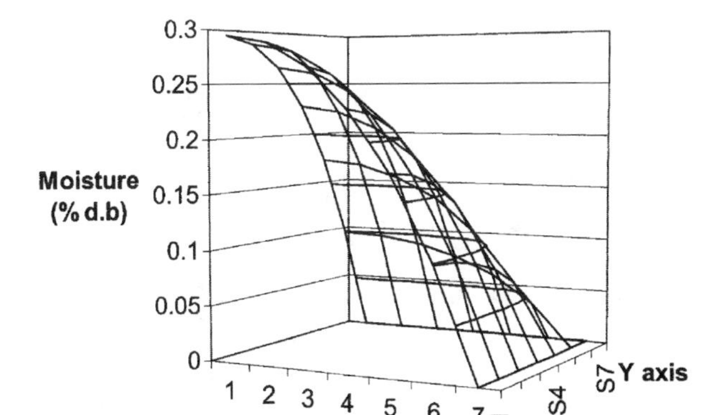

Fig. 5 Moisture profile in a rectangular domain at the end of 1.0 sec.

diffusion of pollutants into the food from packaging materials,[2] bread baking process[6] to list a few. The solution of the mass transfer equation (Eq. 11) needs the specification of the moisture diffusivity of the food sample. Moisture diffusivity of the food sample may not be a constant and may depend upon temperature, moisture, and the spatial variation. This makes the solution to the mass transfer equation nonlinear. The loss or gain of moisture inside food in turn leads to the increase or decrease in the area and the volume of the food sample being investigated. This implies that the solution domain of interest as represented by Eq. 11 also changes, and often there is a need to solve the equation using an adaptive mesh.[14,23] The liquid diffusion equation was solved using the finite element method for nonlinear water diffusion in rice.[23] A procedure for the estimation of the effective moisture diffusivities for bulk food grains has been presented by Thrope, Tapia, and Whitaker.[24] The rate of diffusion mass transfer depends to a great extent on mass diffusivity. Diffusivity values for gases diffusing within gases are in the order of $10^{-5}\,m^2\,sec^{-1}$, gases and liquids diffusing within liquids are in the order of $10^{-9}\,m^2\,sec^{-1}$ while gases, liquids, and solids diffusing within a solid have diffusivities in the order of 10^{-10}–$10^{-34}\,m^2\,sec^{-1}$. Mass diffusivity values of selected solute diffusion (e.g., sucrose diffusing through agar) or water diffusing through starch are also provided.[25]

Other applications of unsteady-state mass transfer include the porous media approach[26] and an extension to model unsteady-state moisture and temperature during baking of potato.[27] Extension of the convective drying process to model transient moisture migration during microwave[26,28] and infrared[29] heating has been successfully accomplished. Another approach includes the incorporation of chemical potential for moisture potential in Luikov's coupled equations to predict mass transfer in a composite food system.[30] As mathematical models and computational techniques advance, more complex equations could be solved. In the context of food systems, this will imply that the chemical and microbial kinetics could be combined with the mass transfer model to present an integrated approach to understand mass transfer in foods.

CONCLUSION

Mass transfer equations were derived for species conservation for a control volume with different components of migrating species. A steady-state mass transfer example for a simple mass diffusion problem was presented in two dimensions, along with an example of unsteady-state mass transfer with regeneration of mass inside a control volume. Unsteady-state mass transfer problem was also discussed in the context of temperature coupling. Literature indicates that unsteady-state mass transfer with coupling effects from heat transfer is the most frequently encountered problem in the various drying unit operations related to food. Future issues in mass transfer should address chemical and microbial kinetics as quality and safety is an integral part of the food system.

LIST OF SYMBOLS[a]

ρ	Density of the whole mixture ($kg\,m^{-3}$)
ρ_i	Partial density of the ith component of the migrating species ($kg\,m^{-3}$)
v_i	Velocity of the ith component of the migrating species ($m\,sec^{-1}$)
v	Mean velocity of bulk flow ($m\,sec^{-1}$)
j_i	Diffusion mass flux of the species i ($kg\,m^{-2}\,s^{-1}$)
t	Instant in time for unsteady-state mass transfer (sec)
R	Volume of the domain (m^3)
dS	Elemental area of the surface for an arbitrary control volume (m^2)
dR	Elemental volume of the control volume for deriving the conservation equations (m^3)
D_{im}	Mass diffusion coefficient of the species i diffusing in the mixture m (m^2/s)
m_i	Mass fraction of the ith species in the mixture (%)
k_c	Mass transfer coefficient ($kg\,m^{-2}\,sec^{-1}$)
k_p	Partition coefficient across the surface (dimensionless)

REFERENCES

1. Singh, R.P.; Heldman, D. *Introduction to Food Engineering*, 3rd Ed.; Academic Press: London, 2001; 500–528.
2. Laoubi, S.; Vergnaud, J.M. Theoretical Treatment of Pollutant Transfer in a Finite Volume of Food from a Polymer Packaging Made of Recycled Film and a Functional Barrier. Food Addit. Contam. **1996**, *13* (3), 293–306.
3. Hills, P.B.; Harrison, M. Two-Film Theory of Flavor Release from Solids. Int. J. Food Sci. Technol. **1995**, *30*, 425–436.
4. Irudayaraj, J.; Wu, Y. Numerical Modeling of Heat and Mass Transfer in Starch Systems. Trans. ASAE **1999**, *42* (2), 449–455.
5. Ranjan, R.; Irudayaraj, J.; Jun, S. A Three-Dimensional Control Volume Approach to Modeling Heat and Mass

[a]List of symbols for those units not mentioned in text for conciseness.

Transfer in Food Materials. Trans. ASAE **2001**, *44* (6), 1975–1982.
6. Zanoni, B.; Pierucci, S.; Peri, C. Study of Bread Baking Process II. Mathematical Modeling. J. Food Eng. **1994**, *23*, 321–336.
7. Azevedo, C.A.I.; Oleveria, A.R.F. A Model Food System for Mass Transfer in the Acidification of Cut Root Vegetables. Int. J. Food Sci. Technol. **1995**, *30*, 473–483.
8. Lienhard, J.H., IV. Lienhard, J.H., V. An Introduction to Mass Transfer. *A Heat Transfer Textbook*, 3rd Ed.; Phlogiston Press: Cambridge, 2001; 563–631.
9. Sokhansanj, S.; Bruce, D.M. A Conduction Model to Predict Grain Temperature in Grain Drying Simulation. Trans. ASAE **1987**, *30* (4), 1181–1184.
10. Luikov, A.V. *Heat and Mass Transfer in Capillary Porous Bodies*; Pergamon Press: Oxford, 1966.
11. Fortes, M.; Okos, M.R. A Nonequilibrium Thermodynamics Approach to Transport Phenomenon in Capillary Porous Media. Trans. ASAE **1981**, *24* (3), 756–760.
12. Fortes, M.; Okos, M.R. Non-equilibrium Thermodynamics Approach to Heat and Mass Transfer in Corn Kernels. Trans. ASAE **1981**, *24* (3), 761–769.
13. Whitney, J.D.; Poterfield, J.G. Moisture Movement in Porous, Hygroscopic Solid. Trans. ASAE **1968**, *11* (5), 716–719.
14. Browser, T.J.; Wilhelm, L.R. Modeling Simultaneous Shrinkage and Heat and Mass Transfer of a Thin, Nonporous Film During Drying. J. Food Sci. **1995**, *60* (4), 753–757.
15. Oliveira, L.S.; Haghighi, K. Conjugate Heat and Mass Transfer in Convective Drying of Multiparticle Systems. II. Soybean Drying. Drying Technol. **1998**, *16* (3/5), 463–483.
16. Crank, J. *Mathematics of Diffusion*, 2nd Ed.; Oxford University Press: London, 1979; 48–61.
17. Mikhailov, M.D. Exact Solution of Temperature and Moisture Distributions in a Porous Half Space with Moving Evaporation Front. Int. J. Heat Mass Transfer **1974**, *18*, 797–804.
18. Gupta, L.N. An Approximate Solution of the Generalized Stephan's Problem in a Porous Medium. Int. J. Heat Mass Transfer **1974**, *17*, 313–321.
19. Kumar, I.J. An Extended Variational Formulation of the Non-linear Heat and Mass Transfer in a Porous Medium. Int. J. Heat Mass Transfer **1971**, *14*, 1759–1770.
20. Trebal, R.E. *Mass Transfer Operations*, 2nd Ed.; McGraw-Hill: New York, 1968.
21. Reddy, J.N. *An Introduction to the Finite Element Method*, 2nd Ed.; McGraw-Hill: New York, 1993; 3–63.
22. Irudayaraj, J.; Haghighi, K.; Stroshine, R. Finite Element Analysis of Drying with Application to Cereal Grains. J. Agric. Eng. Res. **1992**, *53* (4), 209–229.
23. Zhang, T.Y.; Bakshi, A.S.; Gustafon, R.J.; Lund, D.B. Finite Element Analysis of Non linear Water Diffusion During Rice Soaking. J. Food Sci. **1984**, *49*, 246–250.
24. Thrope, G.R.; Tapia, O.A.J.; Whitaker, S. The Diffusion of Moisture in Food Grains—II. Estimation of Effective Diffusivity. J. Stored Prod. Res **1991**, *27*, 11–30.
25. Johnson, A.T. *Biological Process Engineering—An Analogical Approach to Fluid Flow, Heat Transfer, and Mass Transfer Applied to Biological Systems*, 1st Ed.; John Wiley and Sons, Inc.: New York, 1999; 494–520.
26. Datta, A.K.; Zhang, J. *Porous Media Approach to Heat and Mass Transfer in Solid Foods*; ASAE Paper No. 99-3068; ASAE: St. Joseph, MI, 1999.
27. Ni, H.; Datta, A.K. Heat and Moisture Transfer in Baking of Potato Slabs. Drying Technol. **1999**, *17* (10), 2069–2092.
28. Zhou, L.; Puri, V.M.; Anantheswaran, R.C.; Yeh, G. Finite Element Modeling of Heat and Mass Transfer in Food Materials During Microwave Heating: Model Development and Validation. J. Food Eng. **1995**, *25* (2), 509–529.
29. Ranjan, R. Mathematical Modeling of Convective and Infrared Drying of Foods Using the Control Volume Method. M.S. Thesis, Pennsylvania State University, University Park, 2001.
30. Sakai, N.; Hayakawa, K.I. Two-Dimensional Simultaneous Heat and Mass Transfer in Composite Food. J. Food Sci. **1992**, *57* (2), 475–480.

Unsteady-State Mass Transfer in Foods

J. Welti-Chanes
F. Vergara
D. Bermúdez
Universidad de las Américas–Puebla, Puebla, Mexico

H. Mújica-Paz
A. Valdez-Fragoso
Universidad Autónoma de Chihuahua, Chihuahua, Chih., Mexico

INTRODUCTION

There are many processes of industrial importance where food solids are subjected to batch processes and unsteady-state transfer conditions arise. In such unsteady-state situations, the food's concentration distribution varies with both time and position. Mass transfer under unsteady-state plays a key role in drying, lixiviation, infusion, osmotic dehydration, salting or desalting, and frying processes. The process of contaminant transfer from the packages into the food, and controlled release of active compounds can also be considered as unsteady-state systems. Among the components involved in these transfer situations are water, sugars, salt, oils, acids, flavors, oxygen, carbon dioxide, and contaminants.

This article is devoted to introduce the concept of unsteady-state regime and the mathematical models expressing the rate of unsteady-state diffusion. Methods of solution of such models are reviewed briefly and typical applications in several food processes are commented.

UNSTEADY-STATE DIFFUSION

Processes in which diffusion fluxes and concentrations are time-dependent are referred as unsteady-state processes.[1] To analyze the unsteady-state diffusion process, one-directional diffusion equation will be developed. Refer to Fig. 1 where mass is diffusing in the x direction in a cube of dimensions Δx, Δy, Δz, composed of a solid, stagnant gas, or stagnant liquid.

The molal diffusion flux of component A in the x direction ($N_{A,x}$) is related to the concentration gradient $\left(\frac{\partial C_A}{\partial x}\right)$ by Fick's first law

$$N_{A,x} = -D\frac{\partial C_A}{\partial x} \tag{1}$$

where D is the diffusion coefficient.

The mass balance of component A in terms of moles can be stated as

$$\text{rate of input} + \text{rate of production}$$
$$= \text{rate of output} + \text{rate of accumulation} \tag{2}$$

where

$$\text{rate of input} = N_{A,x}\big|_x = -D\frac{\partial C_A}{\partial x}\bigg|_x \tag{3}$$

rate of production $= 0$ (considering there is no chemical reaction in the process)

$$\text{rate of output} = N_{A,x}\big|_{x+\Delta x} = -D\frac{\partial C_A}{\partial x}\bigg|_{x+\Delta x} \tag{4}$$

$$\text{rate of accumulation} = \frac{\partial C_A}{\partial t}\Delta x\Delta y\Delta z \tag{5}$$

Substituting each term in Eq. 2, dividing by the volume ($\Delta x\Delta y\Delta z$), and rearranging,

$$-D\frac{\frac{\partial C_A}{\partial x}\big|_x - \frac{\partial C_A}{\partial x}\big|_{x+\Delta x}}{\Delta x} = \frac{\partial C_A}{\partial t} \tag{6}$$

When Δx tends to zero, Eq. 6 becomes

$$\frac{\partial C_A}{\partial t} = D\frac{\partial^2 C_A}{\partial x^2} \tag{7}$$

This equation relates the concentration C_A (kg mol m^{-3}) with position x and time t (sec), with a constant diffusion coefficient D (m^2 sec^{-1}), and it is known as the Second Law of Fick.[1] In the case of a concentration-dependent diffusion coefficient, Eq. 7 is expressed as

$$\frac{\partial C_A}{\partial t} = \frac{\partial}{\partial x}\left[D(c)\frac{\partial C_A}{\partial x}\right] \tag{8}$$

Encyclopedia of Agricultural, Food, and Biological Engineering
DOI: 10.1081/E-EAFE 120007038
Copyright © 2003 by Marcel Dekker, Inc. All rights reserved.

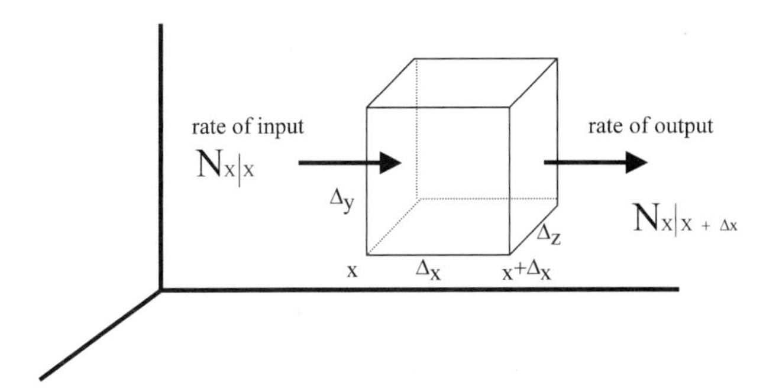

Fig. 1 Differential volume element (dV) in which 1D mass transfer is occurring by unsteady-state diffusion.

Eq. 7 can be written for an infinite cylinder as:

$$\frac{\partial C_A}{\partial t} = D\frac{1}{r}\frac{\partial}{\partial r}\left(r\frac{\partial C_A}{\partial r}\right) \quad (9)$$

where r is the radial coordinate. In this infinite body, mass transfer occurs mainly in a radial direction and not in an axial way.

It can also be written for a sphere as:

$$\frac{\partial C_A}{\partial t} = D\frac{1}{r^2}\frac{\partial}{\partial r}\left(r^2\frac{\partial C_A}{\partial r}\right) \quad (10)$$

Eq. 7 can also be extended to the 3D case for a slab, which is considered as a finite body:

$$\frac{\partial C_A}{\partial t} = D\left(\frac{\partial^2 C_A}{\partial x^2} + \frac{\partial^2 C_A}{\partial y^2} + \frac{\partial^2 C_A}{\partial z^2}\right) \quad (11)$$

Similar equations can be obtained for other geometries.[2,3]

Eqs. 7 and 9–11 are used to find the concentration of a solute as a function of time and position, and are mainly applicable to diffusion in solids and to limited situations in fluids. The analysis of unsteady-state systems, however, is frequently simplified by reducing the problem to consider only 1D diffusion.

SOLUTIONS OF THE FUNDAMENTAL EQUATIONS

The analysis of unsteady-state diffusion problems involves the solution of partial differential equations, which have more than one independent variable. The techniques of solution of Eq. 7 include analytical solutions (transformation of variables, separation of variables, or Laplace transforms), numerical, and graphical methods.[2]

Analytical Solutions

Many unsteady diffusion problems can be solved analytically, if partial differential equations are linear or have boundary conditions constant with time, by applying the separation of variables technique. To apply this technique, it is necessary to separate the partial differential equation into two ordinary differential equations and independently solve them. Then, the solution for the partial differential equation is a product solution of the form

$$C(t,x) = f(t)g(x) \quad (12)$$

where $C(t,x)$ is the concentration of a solute as a function of time and position represented as the product of two functions, one of time $f(t)$ and one of position $g(x)$.

The resulting ordinary differential equations may be integrated by using appropriate boundary conditions. Eqs. 7, 9, and 10 have been solved by the method of variable separation for an infinite slab, a cylinder, and a sphere,[2] considering average concentrations of the material that is diffused inside the body, a constant value of D along the process, and that the conditions outside the body (convective resistance negligible) do not affect the internal phenomenon:

1. Slab

$$\frac{C - C_0}{C_S - C_0} = 1 - \sum_{n=0}^{\infty}\frac{8}{(2n+1)^2\pi^2}\exp\left\{\frac{-(2n+1)^2\pi^2 Dt}{4L^2}\right\} \quad (13)$$

where C_0 is the initial uniform concentration (kg mol m^{-3}), C_S is the superficial concentration constant along time (kg mol m^{-3}), C is the average concentration (kg mol m^{-3}) at time t, D is the diffusion coefficient, n is the number of terms of the series, and L is half of the slab thickness.

2. Cylinder

$$\frac{C - C_0}{C_S - C_0} = 1 - \sum_{n=1}^{\infty}\frac{4}{r^2\alpha_n^2}\exp(-\alpha_n^2 Dt) \quad (14)$$

where r is the radius (m) and α_n are the positive roots of the first Bessel function of order zero.

3. Sphere

$$\frac{C - C_0}{C_S - C_0} = 1 - \frac{6}{\pi^2}\sum_{n=1}^{\infty}\frac{1}{n^2}\exp\left\{\frac{-Dn^2\pi^2 t}{r^2}\right\} \quad (15)$$

These series solutions are convenient to analyze long-time processes because they converge rapidly, thus in such cases the use of the first term of the series can be employed to model the diffusion process under unsteady-state.

If D varies with temperature, such variation can be modeled in many cases with the equation of Arrhenius.

Numerical Methods

The numerical solutions are applied when the initial concentration distribution is not uniform and the boundaries have different concentration levels, and to have an analysis of punctual values, not only of mean values. A commonly used numerical method is the finite-difference method.[3,4] The first step in this method is the finite-difference representation of the differential equation of diffusion in both position and time domains. To illustrate the method, the equation of one-directional diffusion in a flat plate will be considered.

The finite-difference form of Eq. 11 is then obtained by dividing the position (x) and time (t) domains into small steps Δx and Δt (Fig. 2). Thus, the derivative on the left can be expressed as a finite-difference at a point x

$$\frac{\partial C}{\partial t} = \frac{C_{x|t+\Delta t} - C_{x|t}}{\Delta t} \tag{16}$$

and the second derivative can be approximated as

$$\frac{\partial^2 C}{\partial x^2} = \frac{C_{t|x+\Delta x} - 2C_{t|x} + C_{t|x-\Delta x}}{(\Delta x)^2} \tag{17}$$

Upon substitution of Eqs. 16 and 17 in Eq. 7, the explicit finite-difference form of the 1D time-dependent diffusion equation is obtained:

$$\frac{C_{x|t+\Delta t} - C_{x|t}}{\Delta t} = D\frac{C_{t|x+\Delta x} - 2C_{t|x} + C_{t|x-\Delta x}}{(\Delta x)^2} \tag{18}$$

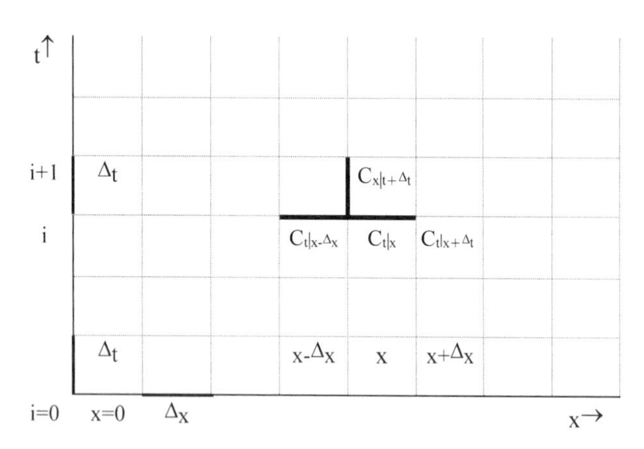

Fig. 2 Subdivision of the x–t domain into intervals of Δx and Δt for finite-difference representation of the 1D diffusion.

This equation can be rearranged in the form

$$C_{x|t+\Delta t} = r\left(C_{t|x+\Delta x} + C_{t|x-\Delta x}\right) + (1 - 2r)C_{t|x} \tag{19}$$

where $r = \frac{D\Delta t}{(\Delta x)^2}$.

According to Eq. 19, the concentration at location x and a new time $t + \Delta t$ can be evaluated from three concentrations that are known at the time t. Initial and boundary values along $t = 0$ become valuable to accomplish calculations. In order to obtain meaningful solutions, the term 1-2r should not be negative.[4]

Convective Mass Transfer

In many situations, a fluid flows over a solid body and convective mass transfer between the fluid and the solid surface takes place. If convective resistance at the surface is not negligible, the coefficient of mass transfer becomes important in the transport process.[3] Exact solution of mass transfer problems, where the primary mechanism is convection, is almost never feasible and resort to the concept of a mass transfer coefficient is necessary. A mass transfer coefficient k_c may be defined as

$$N_A = k_c(c_{L1} - c_{Li}) \tag{20}$$

where c_{L1} is the bulk fluid concentration, c_{Li} is the concentration in the fluid adjacent to the solid surface, and $(c_{L1} - c_{Li})$ is the partial concentration gradient across the fluid. Fig. 3 schematically shows this situation.

The equilibrium concentrations c_i and c_{Li} are related by a partition coefficient $K = c_{Li}/c_i$, which is a measure of the enrichment of the species, and can be less than, equal to, or greater than 1, according to the equilibrium conditions of the system.

Graphical Technique

For unsteady-state heat conduction problems in solids, solutions have been presented for simple geometries subjected to a given set of boundary and initial conditions. These solutions are presented in charts in terms of dimensionless ratios and can be translated directly to analogous problems of unsteady-state diffusion, allowing

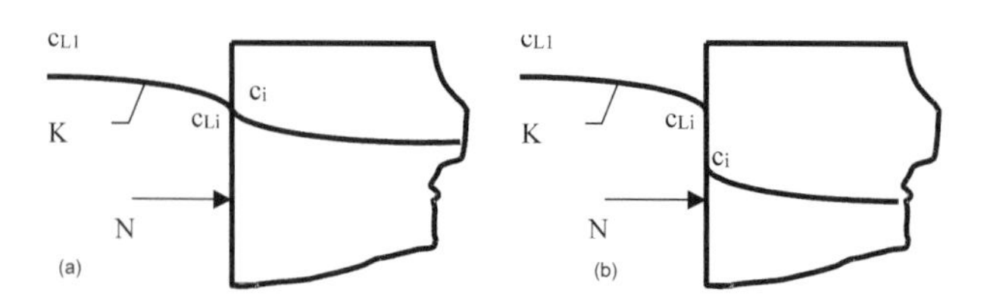

Fig. 3 Concentration gradients between a fluid and a solid phase, and partition coefficients ($K = C_{Li}/C_i$); (a) $K = 1.0$ and (b) $K > 1.0$.

Table 1 Relation between mass and heat transfer variables for unsteady-state

	Heat transfer	Mass transfer $K = 1$	Mass transfer $K \neq 1$
Y, unaccomplished change	$\frac{T-T_S}{T_0-T_S}$	$\frac{c-c_S}{c_0-c_S}$	$\frac{c-c_S/K}{c_0-c_S/K}$
X, relative time	$\frac{\alpha t}{x_1^2}$	$\frac{Dt}{x_1^2}$	$\frac{Dt}{x_1^2}$
m, relative resistance	$\frac{k}{hx_1}$	$\frac{D}{k_c x_1}$	$\frac{D}{Kk_c x_1}$
n, relative position	$\frac{x}{x_1}$	$\frac{x}{x_1}$	$\frac{x}{x_1}$

α is the thermal diffusivity, D is the diffusion coefficient, k is the thermal conductivity, k_c is the mass transfer coefficient, and x_1 is the characteristic dimension.
(Adapted from Refs. 3 and 5.)

the application of heat transfer process to solve mass diffusion problems. The ordinate of these charts represents the fraction of the unaccomplished change $\left(\frac{c-c_S}{c_0-c_S}\right)$ and the abscissa the relative time $\left(\frac{Dt}{x_1^2}\right)$. These charts, their physical significance, and use are presented in several books.[3,5] Table 1 summarizes the correspondence between variables for unsteady heat and mass transfer (molecular diffusion).

APPLICATIONS

The use of the unsteady-state Fickian equation has proved to be adequate when applied to model diffusion of water and solutes in the osmotic dehydration of fruits and vegetables.[6,7] In addition, the diffusion of salt into blocks of cheese has been modeled by the unsteady-state diffusion theory. The drying process may be characterized by the moisture diffusion as the controlling phenomena and experimental results have been successfully interpreted by using Fick's second law.[8] The release rate of an active compound has been modeled considering the active agent dissolved in the matrix at or below the saturation concentration, which is immersed in an well-agitated and infinite medium.[9] In lixiviation calculations for solids of common sample geometries, the problem can be solved with the use of graphical methods for transient diffusion. The permeation process of gases and flavors through a packaging film can be described by unsteady-state diffusion and the time-lag permeation technique allows simultaneous determination of the three main quantities characterizing mass transfer: the diffusion coefficient (D), the gas or vapor solubility (S), and the permeability ($P = DS$).

REFERENCES

1. Treybal, R.E. *Mass Transfer Operations*, 3rd Ed.; McGraw-Hill Book Co. International Edition: New York, 1981.
2. Crank, J. *The Mathematics of Diffusion*, 2nd Ed.; Oxford University Press: London, 1975.
3. Geankoplis, Ch.J. *Transport Processes and Unit Operations*, 3rd Ed.; Prentice-Hall International, Inc.: London, 1993.
4. Smith, G.D. *Numerical Solution of Partial Differential Equations*, Oxford University Press, Ely House: London, 1971.
5. Welty, J.R.; Wicks, Ch.E.; Wilson, R.E. *Fundamentals of Momentum, Heat and Mass Transfer*, 3rd Ed.; John Wiley & Sons: New York, 1984.
6. Mauro, M.A.; Menegalli, F.C. Evaluation of Diffusion Coefficients in Osmotic Concentration of Bananas (*Musa cavendish* Lambert). Int. J. Food Sci. Technol. **1995**, *30*, 199–213.
7. Wang, W.C.; Sastry, S.K. Salt Diffusion into Vegetable Tissue as a Pretreatment for Ohmic Heating: Determination of Parameters and Mathematical Model Verification. J. Food Eng. **1993**, *20*, 311–323.
8. Vergara, F.; Amézaga, E.; Bárcenas, M.E.; Welti, J. Analysis of the Drying Processes of Osmotically Dehydrated Apple Using the Characteristic Curve Model. Drying Technol. **1997**, *15* (3–4), 949–963.
9. Pothakamury, U.R.; Barbosa-Cánovas, G.V. Fundamental Aspects of Controlled Release in Foods. Trends Food Sci. Technol. **1995**, *6*, 397–406.

Vacuum Cooling

Banu F. Ozen
Purdue University, West Lafayette, Indiana, U.S.A.

Rakesh K. Singh
The University of Georgia, Athens, Georgia, U.S.A.

INTRODUCTION

Vacuum cooling is proven to be an effective precooling method for porous and large surface to mass ratio products. Energy required for cooling is provided by the product itself with the evaporation of water during application of vacuum on the product. Vacuum cooling provides a uniform, fast, and economical cooling compared to conventional cooling methods. Although it might cause water loss from the product during cooling, vacuum cooling found applications in food industry to cool specific products such as lettuce, bakery products, and some meat products.

PRINCIPLES AND EQUIPMENT

Vacuum cooling is based on the principle that the boiling point of water changes with pressure. As the pressure decreases, boiling point temperature also drops as illustrated in Fig. 1. Vacuum cooling is achieved by evaporation of water from the product at very low air pressure, and energy required to evaporate the water (latent heat of vaporization) is provided by the heat available in the product itself. This allows uniform cooling of even tightly packed produce.[1] Quantity of heat given off by the product, indicated by θ (kcal hr^{-1}), during cooling could be expressed by[2]:

$$\theta = \frac{W \times C \times (T_2 - T_1)}{t} \quad (1)$$

where W is the weight of the product (kg), C is the specific heat of the product (kcal $\mathrm{kg}^{-1}\mathrm{C}^{-1}$), T_1 is the product's entering temperature (C), T_2 is the product's leaving temperature (C), and t is the time of cooling (hr).

Vacuum cooling system consists of mainly vacuum chamber, vacuum pump, and vapor compressing unit (Fig. 2). Product is stored in the chamber during the cooling process. Vacuum pumps such as mechanical rotary pumps or steam jets with barometric condensers are used to evacuate the vacuum from the chamber. In mechanical systems, evaporated water is passed over refrigerated coils and condenses on the cold surfaces. A steam augmenter is appended to the steam jet and condenser systems to increase the temperature of vapor; therefore, the need for the refrigeration in condenser is eliminated[3] (Fig. 3). Reducing the pressure as quickly as possible to the point at which the produce temperature coincides with saturation temperature of water (flash point) is important from quality and economical point of views.[4] Capacity of the pump is one of the determining factors for the time needed to reduce the pressure to the flash point. It is necessary to remove the water vapor formed during the cooling to prevent the saturation of air. Pumping rate of the vacuum pump, S, is given by[2]:

$$S = \frac{V}{t} \ln \frac{P_1}{P_2} \quad (2)$$

where V is the volume of the vacuum tank (m^3), t is the time desired to reach flash point (hr), P_1 is the initial pressure (atmospheric pressure) (mbar), P_2 is the pressure equivalent to flash point (mbar).

Vacuum cooling is the most applicable to products that easily release water and have high surface to mass ratios such as leafy vegetables.[5] Vacuum cooling is a batch process, and faster than conventional cooling methods.[6,7] The ratio between the mass of the produce and its evaporation area determine the speed and the effectiveness of the cooling process.[8] Temperature could be precisely controlled during cooling by adjusting the vacuum level.[1] Capital cost of vacuum cooling installations is high compared to other methods of cooling used for the same purposes, and operation cost depends on the size of the installation and throughput.[1,4]

APPLICATIONS

Fruits and Vegetables

Field heat causes rapid deterioration of some horticultural products after harvest. It is difficult to reduce the core temperature of leafy vegetables with conventional cooling methods due to overlapping of leaves on each other and

Encyclopedia of Agricultural, Food, and Biological Engineering
DOI: 10.1081/E-EAFE 120007016
Copyright © 2003 by Marcel Dekker, Inc. All rights reserved.

V

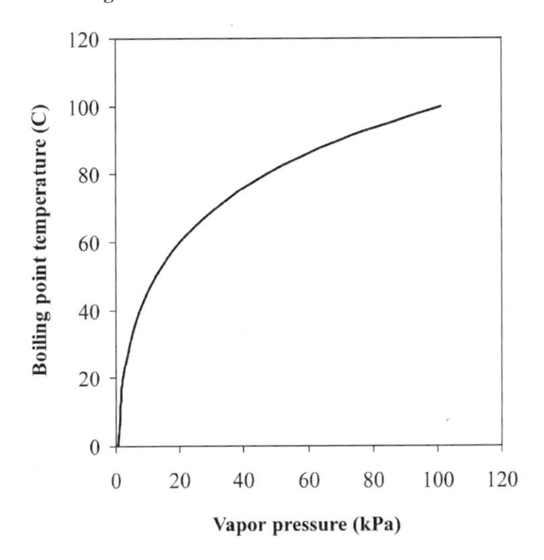

Fig. 1 Boiling point of water as a function of pressure.

high surface/volume ratio.[9] Vacuum cooling provides rapid reduction in the temperature of these types of products compared to other conventional methods of cooling. Placing vegetables or fruits in cold storage without precooling results in significant moisture losses because of the large vapor pressure difference caused by the temperature difference between the product and the storage room.[9] Vacuum cooling has become the standard method for cooling leafy vegetables, particularly lettuce, in many European countries and in the United States before storage. Due to their high water content (approximately 90%) and porous structure, which would allow water to escape easily, mushrooms are also suitable for vacuum cooling.[1,10] Vacuum precooling improves the quality of mushrooms compared to products not cooled before storage. However, browning becomes worse for slightly deteriorated mushrooms after vacuum cooling; therefore, only good quality mushrooms benefit from this type of cooling.[11] Cooling of fruits and vegetables such as eggplant, cucumber, carrot, pepper, and cauliflower by vacuum has also been reported.[6,12] Although vacuum cooling of nonleafy vegetables takes longer compared to leafy vegetables, it is still shorter than conventional cooling methods.

As vacuum cooling is based on the evaporation of water from the product, the moisture loss in vacuum cooled products is higher compared to conventional methods of cooling (Fig. 4). Increase in surface area also results in an increase in weight loss. For products having high moisture content, every 5°C drop results in about 1% weight loss during vacuum cooling.[13] Despite the moisture loss, wilting of leafy vegetables after vacuum cooling is generally not a big concern due to uniform removal of moisture.[2] Also, both instrumental and visual ratings did not indicate any significant color change compared to conventional cooling methods.[5,14,15] Prewetting the product before cooling could reduce the weight loss and cooling times for some vegetables[16] (Fig. 5).

Vacuum cooling could be applied to products either before wrapping in an unperforated plastic material or after packaging in perforated bags. Vacuum cooling and

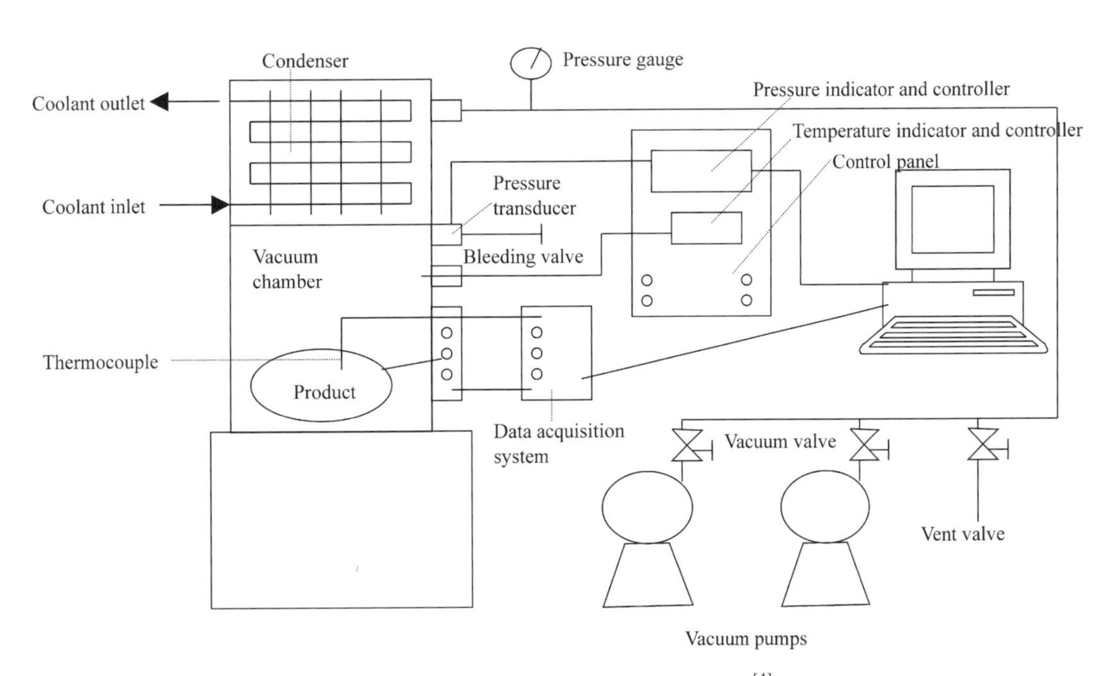

Fig. 2 A laboratory scale vacuum cooler.[4]

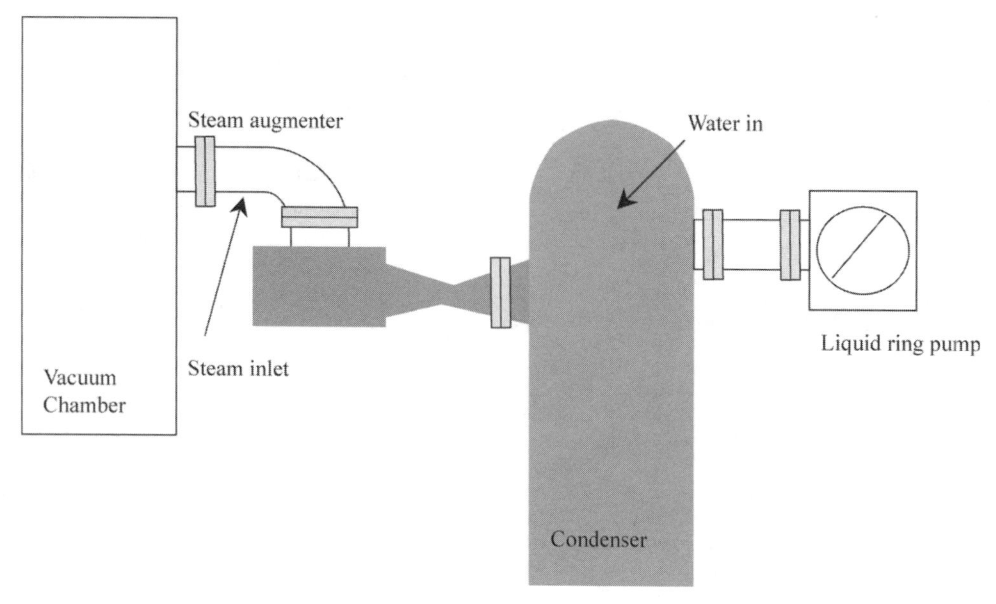

Fig. 3 Vacuum cooling system with steam augmenter.[3]

storage in the package could have significant effect in prevention of weight loss.[9] Polypropylene and perforated polyvinyl chloride and polyethylene bags are commonly used in packaging of lettuce.[17] However, some packaging materials such as unperforated stretch films prevents proper cooling of the product since they present a severe barrier to moisture vaporization.[18]

Meat and Meat Products

As for the fruits and vegetables, vacuum cooling provides rapid cooling of cooked meat products compared to other conventional cooling methods[7,19] (Fig. 6). Guidelines established in the United States and European countries require fast cooling of cooked meat products due to the concerns in growth of bacteria derived from heat resistant

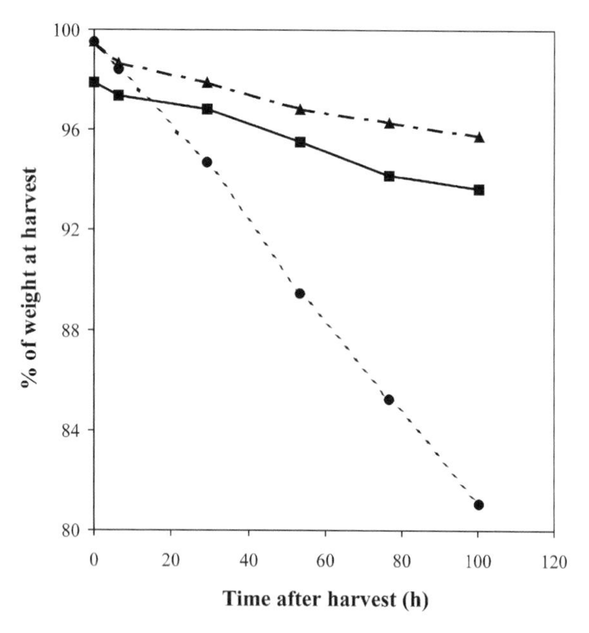

Fig. 4 Weight loss of mushrooms during storage; —●— stored at 18°C; –■– vacuum cooled and stored at 5°C; –▲– conventionally cooled and stored at 5°C.[10]

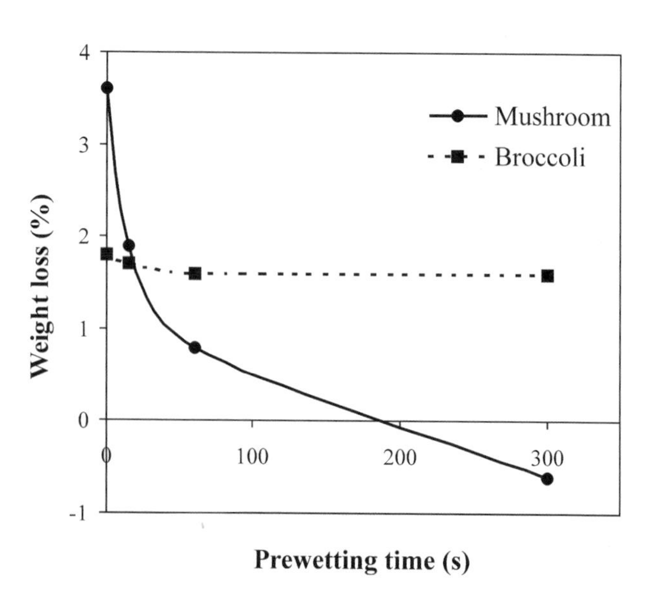

Fig. 5 Effect of prewetting time on weight loss of mushroom and broccoli during vacuum cooling. (Data from Ref. [15].)

spores. American Meat Institute recommends that cooling should begin within 90 min of cooking, and cooling times required to reduce the temperature from 48 to 12.7°C should not exceed 6 hr with cooling continuing to 4.4°C.[20] As the thermal conductivity of meat is low, conventional cooling systems such as air blast and water immersion that depend on the conduction of heat inside the product do not meet the guidelines for cooked meat cooling.[7,19,21] On the other hand, vacuum cooling can achieve the time and temperature requirements for cooling the cooked meats stated in regulations. Increased porosity of the meat samples that develops during vacuum cooling significantly increases the cooling rates. Preparation of samples such as whether the meat is minced or whole muscle, packaged in casings or netting, is a significant factor in the development of porosity.[22] As a result of high moisture loss and increased porosity, vacuum cooled meat samples have lower thermal conductivity, specific heat capacity, and thermal diffusivity than samples cooled with conventional methods.[23] A model developed using computational fluid dynamics successfully simulated the heat and mass/vapor transfer within the cooked meat during vacuum cooling.[24] This model was also very effective in predicting temperature profiles and weight loss simultaneously.

Weight loss is also a concern for vacuum cooled meat products. Although the chill loss is higher and total product yield is lower for vacuum cooled meat products compared to conventional cooling methods, improvements in percent yield and mass loss could be obtained by regulation of evacuation rate.[25] The efficiency of vacuum cooling depends on the surface area to mass ratio; therefore, its applicability to large meat products such as beef and ham joints is limited.[26] Although the results of instrumental analysis revealed that vacuum cooled meat had a darker color and tougher texture due to water loss and compression of muscles, no negative impact on sensory attributes was detected by panelists. In addition, rapid decrease in temperature during vacuum cooling and lower water activity of cooled meat results in lower total viable count.[7] Injection of meat with a brine solution could improve its quality attributes such as texture and color. However, cooling times may increase and undesirable flavor problems may arise as a result of brine injection.[26]

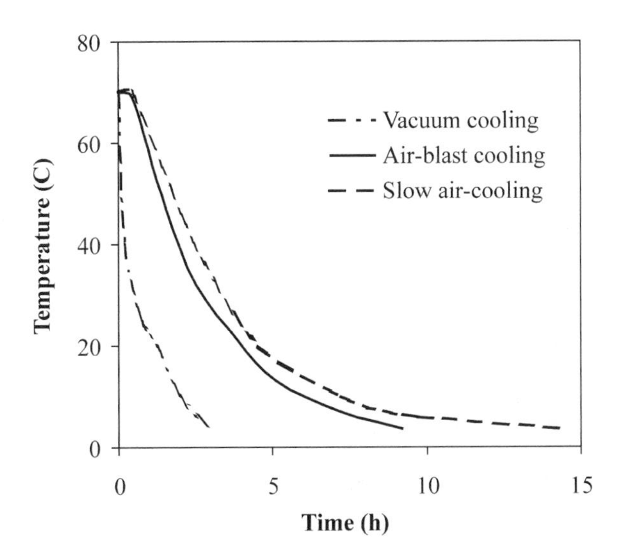

Fig. 6 Cooling curves of cooked beef cooled by different methods.[7]

Bakery Products

Bakery products are another food group that might benefit from vacuum cooling. As conventional vacuum cooling was not very successful due to the excessive moisture loss and structural changes on the crust of the product, modulated vacuum cooling systems (MVC) were developed to eliminate the undesirable effects of vacuum cooling on bakery products. In MVC systems, speed at which the water vapor generated is modulated rather than applying a continuous vacuum on the product.[4] As the last step of baking in which stabilization takes place by completion of gelatinization of starch and coagulation of proteins could be carried out during cooling at very high speed, MVC accelerates the baking + cooling process considerably.[27] MVC also extends the shelf life of breads in terms of crust characteristics, and prevents the collapse of the crust resulting from pulling action of contracting inner core. As cooling takes place at a high speed, migration of fillings such as jams or meat sauces in pastries to outer structure is also prevented.[26] However, one adverse effect of vacuum cooling is the reduction of the aroma substances of the bakery products.[28]

Ready-to-Eat Meals

Vacuum cooling also has applications in the cooling of ready-to-eat meals and sauces used in these types of products. Vacuum cooling provides much faster reduction in temperature and uniform cooling of these products compared to air blast cooling. In addition, since both cooking and cooling could be performed in the same unit, delays in transferring the product to a separate unit is eliminated.[4] However, high vacuum applied to cooked product may cause splash of the sauces to the processing vessel; therefore, extensive cleaning of the equipment is necessary to prevent the microbial growth. There are also safety concerns due to operation of these systems over a wide range of positive and negative pressures.

CONCLUSION

Vacuum cooling is a very effective cooling method for large surface area to volume products that have free water. Although it causes water loss during cooling, vacuum cooling is a fast and economical method that could be used especially as a precooling technique to reduce the postharvest deterioration of vegetables. Recent research also showed that this type of cooling could be applicable to meat and bakery products and ready-to-eat meals.

REFERENCES

1. Noble, R. A Review of Vacuum Cooling of Mushrooms. Mushroom J. **1985**, *149*, 168–170.
2. Haas, E.; Gur, G. Factors Affecting the Cooling Rates of Lettuce in Vacuum Cooling Installations. Int. J. Refrig. **1987**, *10*, 82–86.
3. Fish, A.R. Vacuum Cooling. Proceedings of Annual Meeting of the American Society of Bakery Engineers, Chicago, IL, 1980; American Society of Bakery Engineers: Wilmette, IL, 1980.
4. McDonald, K.; Sun, D.W. Vacuum Cooling Technology for the Food Processing Industry: A Review. J. Food Eng. **2000**, *45*, 55–65.
5. DeEll, J.R.; Vigneault, C.; Favre, F.; Rennie, T.J.; Khanizadeh, S. Vacuum Cooling and Storage Temperature Influence the Quality of Stored Mung Bean Sprouts. Hortic. Sci. **2000**, *35* (5), 891–893.
6. Sun, D.W. Comparison of Rapid Vacuum Cooling of Leafy and Non-leafy Vegetables. Annual ASAE International Meeting, Toronto, Canada, July 18–21, 1999; American Society of Agricultural Engineers: St. Joseph, MI, 1999; ASAE Paper No. 99-6117.
7. McDonald, K.; Sun, D.W.; Kenny, T. Comparison of the Quality of Cooked Beef Products Cooled by Vacuum Cooling and Conventional Cooling. Lebensm.-Wiss. Technol. **2000**, *33*, 21–29.
8. Sun, D.W.; Wang, L. Vacuum Cooling. In *Advances in Food Refrigeration*; Sun, D.W., Ed.; Leatherhead Publ.: Surrey, U.K., 2001; 264–304.
9. Turk, R.; Celik, E. The Effects of Vacuum Cooling on the Quality Criteria of Some Vegetables. Actae Hortic. **1994**, *368*, 825–829.
10. Burton, K.S.; Frost, C.E.; Atkey, P.T. Effect of Vacuum Cooling on Mushroom Browning. Int. J. Food Sci. Technol. **1987**, *22*, 599–606.
11. Gormley, T.R. Vacuum Cooling and Mushroom Whiteness. Mushroom J. **1975**, *27*, 84, 86.
12. Hayakawa, A.; Kawano, S.; Iwamoto, M.; Onodera, T. Vacuum Cooling Characteristics of Fruit Vegetables and Root Vegetables. Rep. Natl. Food Res. Inst. **1983**, *43*, 109–115.
13. Barnard, N. Some Experiments in Vacuum Cooling. Mushroom J. **1974**, *14*, 48–51.
14. Kim, B.S.; Kim, D.C.; Lee, S.E.; Nahmgoong, B.; Choi, M.J.; Jeong, M.C. Freshness Prolongation of Crisphead Lettuce by Vacuum Cooling. Agric. Chem. Biotechnol. **1995**, *38*, 239–247.
15. Nahmgung, B.; Kim, B.S.; Kim, O.Q.; Chung, J.W.; Kim, D.C. Influence of Vacuum Cooling on Browning, PPO and Free Amino Acid of Shiitake Mushroom. Agric. Chem. Biotechnol. **1995**, *38*, 345–352.
16. Sun, D.W. Effect of Pre-wetting on Weight Loss and Cooling Time of Vegetables During Vacuum Cooling. Annual ASAE International Meeting, Toronto, Canada, July 18–21, 1999; American Society of Agricultural Engineers: St. Joseph, MI, 1999; ASAE Paper No. 99-6119.
17. Martinez, J.A.; Artes, F. Effect of Packaging Treatments and Vacuum-Cooling on Quality of Winter Harvested Iceberg Lettuce. Food Res. Int. **1999**, *32*, 621–627.
18. Cheyney, C.C.; Kasmire, R.F.; Morris, L.L. Vacuum Cooling Wrapped Lettuce. Calif. Agric. **1979**, *33*, 18–19.
19. Burfoot, D.; Self, K.P.; Hudson, W.R.; Wilkins, T.J.; James, S.J. Effect of Cooking and Cooling Method on the Processing Times, Mass Losses and Bacterial Condition of Large Meat Joints. Int. J. Food Sci. Technol. **1990**, *25*, 657–667.
20. Code of Federal Regulations. Animals and Animal Products. *Requirements for the Production of Cooked Beef, Roast Beef and Cooked Corn Beef*; 9 CFR; US Department of Agriculture: Washington, D.C., 1998; 270–271.
21. Sun, D.W.; Wang, L. Heat Transfer Characteristics of Cooked Meats Using Different Cooling Methods. Int. J. Refrig. **2000**, *23*, 508–516.
22. McDonald, K.; Sun, D.W. The Formation of Pores and Their Effects in a Cooked Beef Product on the Efficiency of Vacuum Cooling. J. Food Eng. **2001**, *47*, 175–183.
23. McDonald, K.; Sun, D.W.; Lyng, J.G. Effect of Vacuum Cooling on Thermophysical Properties of a Cooked Beef Product. J. Food Eng. **2002**, *52*, 167–176.
24. Hu, Z.; Sun, D.W. Simulation of Heat and Mass Transfer for Vacuum Cooling of Cooked Meats by Using Computational Fluid Dynamics Code. Proceedings of the Eight International Congress on Engineering and Food, Puebla, Mexico, 2000; Technomic Publ.: Lancaster, PA, 2000.
25. McDonald, K.; Sun, D.W. Effect of Evacuation Rate on the Vacuum Cooling Process of a Cooked Beef Product. J. Food Eng. **2001**, *48*, 195–202.
26. McDonald, K.; Sun, D.W.; Kenny, T. The Effect of Injection Level on the Quality of a Rapid Vacuum Cooled Cooked Beef Product. J. Food Eng. **2001**, *47*, 139–147.
27. Bradshaw, W. Modulated Vacuum Cooling for Bakery Products. Baker's Dig. **1976**, *50*, 26–31.
28. Kratochvil, J. Effect of Vacuum Cooling on Bread Aroma. Proceedings of the Fifth Symposium on Aroma Substances in Foods, Prague, Czech Republic, 1981.

Vegetable Production Machine Design

V

James L. Glancey
University of Delaware, Newark, Delaware, U.S.A.

INTRODUCTION

Vegetable production continues to be a significant component of the agricultural crop sector in several areas of the United States. In 2001, the USDA reported that 6.4 million acres of vegetables were harvested with a farm value of 14.8 billion dollars.[1] Fresh market production exceeded 2 million acres with a value of 9 billion dollars, and vegetables for processing were grown on 1.3 million acres with a value of 1.3 billion dollars. The importance of these commodities to the US population has grown as per capita consumption of fruits and vegetables increased 12% in the 1990s to 454 lbs annually.

Methods for producing vegetable crops vary greatly depending on the crop and region. For some crops like sweet corn, production equipment requirements are similar to grain crops. However, most vegetable crops require production techniques that necessitate the use of special equipment. This article outlines the basic cultural practices and equipment designs used in the production of vegetables.

SOIL PREPARATION

Generally, good vegetable production is achieved on well drained soils with good soil structure, fertility, organic mater, and pH. Proper seedbed preparation is an essential part of commercial vegetable production. Objectives of soil preparation include developing proper soil structure for good seed-soil contact, controlling weeds, managing organic matter, incorporate soil amendments, improving water infiltration, reducing or eliminating the effects of compaction, and establishing a soil surface appropriate for a specific production method or crop.

Tillage

The production of most vegetable crops utilizes intense tillage practices to insure good soil condition at planting. Adequate soil preparation is required to promote uniform germination and maturity, and eliminate the compaction that often occurs during the harvest of vegetables. In addition, the relatively small size of many vegetable seeds requires a soil structure with small soil particles to promote good seed-soil contact.

Commercial vegetable production has traditionally relied on conventional tillage practices to achieve the soil conditions necessary for crop growth. Moldboard plowing is still utilized to develop and maintain good soil structure while burying weed seeds and organic residues that can harbor insects and diseases. Because the selection of labeled herbicides is limited for many vegetable crops, adequate weed control can only be achieved through primary tillage, subsequent cultivation, and crop rotation. Chisel plowing is used for crops such as sweet corn that have several commercially available pesticides, and may not be affected by crop residues. Secondary tillage may utilize a disc harrow after plowing; however, field cultivators have become an effective alternative. Reduced compaction and improved leveling of the soil surface are some of the reasons field cultivators have grown in popularity.

Final tillage operations can serve two primary purposes in vegetable production. The first objective is firming the soil prior to planting. This not only promotes good seed-soil contact, but also helps insure accurate and consistent seed depth during planting. A second objective in final tillage may be to form the soil into the configuration required for a particular cultural practice. As illustrated in Fig. 1, wide beds can be formed using PTO-powered bed shapers for crops including tomatoes and melons. These beds promote soil warming, facilitate furrow irrigation, and allow crops including lettuce, spinach, onions, and cucumbers to be planted in multiple rows on a bed. Other bed shapers are available for narrow or single row beds, and may not be PTO-powered.

Reduced or conservation tillage practices have been used for some vegetables including sweet corn, tomatoes, snap beans, lima beans, and pumpkins. Unlike agronomic crops, input costs for most vegetables are high and the resulting savings from reduced tillage may be small compared to total production costs. In addition, risks associated with conservation tillage include compaction, low soil temperatures, additional insect, disease or weed pressures, and poor drainage. Together, these factors have limited the adoption of reduced and no-till tillage practices.

Encyclopedia of Agricultural, Food, and Biological Engineering
DOI: 10.1081/E-EAFE 120006863
Copyright © 2003 by Marcel Dekker, Inc. All rights reserved.

Fig. 1 Four-row PTO-powered bed shaper. (Courtesy of Johnson Manufacturing, Woodland, California.)

Deep tillage is a practice that has been utilized in the production of several vegetable crops, especially where mechanical harvesters are used. Wet soil conditions at harvest coupled with the large mass of some harvesters often result in severe, deep compaction. Subsoilers have traditionally been used to eliminate compacted zones, however, a more recent practice has been toward rippers configured with a variety of shanks and chisel points. Among the more common configurations are straight shanks with winged point geometries that promote significant subsurface soil fracture with minimal surface disturbance.

Plasticulture

The use of plastic mulches to promote rapid and early plant development by increasing soil temperatures has been used to grow vegetables since the 1960s. A plasticulture system typically consists of a polyethylene plastic film used as a ground cover or mulch, a raised bed, and drip irrigation. Clear film, often used in early sweet corn production, provides excellent soil warming but no weed control; black film not only promotes elevated soil temperatures but also reduces weed pressure and conserves soil moisture for a variety of crops including watermelons, tomatoes, cucumbers, and peppers. Recent advances in plastic technology have provided wavelength selective films that allow light penetration but still provide good weed control.

Equipment for plasticulture production is available in a wide variety of configurations. Fig. 2 illustrates a one-row black plastic layer that forms the soil bed, dispenses the plastic, and covers the edges of the plastic with soil. Some plastic layers are also equipped with features that place drip tubing under the plastic. Although some biodegradable plastics have been developed, most plastic is pickup at the end of the growing season; some equipment has been developed and is commercially available for gathering plastic.

In addition to bed covers, plastic is also used to create tunnels over early seeded crops. The clear tunnels promote in-the-row soil warming and some moisture conservation, and are removed when the ambient temperature is high enough to facilitate plant growth. Fig. 3 illustrates a typical one-row clear plastic tunnel layer used for sweet corn. Hoops or ribs are used to form the plastic into a tunnel and coulters direct soil onto the edges of the plastic to secure the tunnel.

PLANT ESTABLISHMENT

A vegetable crop can be established using two methods. *Direct seeding* is most frequently used for a variety of vegetable crops in which seeds are sown directly into the soil. Variations of this technique include the seeding of primed seeds, pregerminated seeds, and seeds that are encapsulated in a gel. Alternatively, crops can be *transplanted* using machines that allow plants established in a greenhouse or warmer region to be placed in the soil. Although transplanting is usually more labor intensive than direct seeding, it promotes earlier harvest and provides a means of field production for crops that are best germinated in a greenhouse. Crops that are typically transplanted include broccoli, cabbage, lettuce, and peppers, however, for small scale production, almost any vegetable crop can be transplanted.

Seeds and Seeding Requirements

Precise row and plant spacing, good seed-soil contact, and accurate and consistent depth of seed placement are among

Fig. 2 Single bed plastic layer used in plasticulture production. (Courtesy of the Mechanical Transplanter Co., Holland, Michigan.)

the most important factors in establishing good vegetable crops. Crops that are properly planted are more likely to produce uniform fruit and higher yields at a lower cost.

For the purposes of planter design, the most relevant classification of vegetable crops is based on seed size. Typical seed sizes vary greatly for vegetables; sweet corn averages 50 seeds per 10 g sample, while celery may have as many as 25,000 seeds in a sample of the same mass. As a result, equipment to meter seeds and establish good seed-soil contact are typically designed for crops with similar seeds and seeding requirements. A summary of average seed sizes for most vegetable crops is provided in Ref. 2; typical seeding requirements (plant and row spacing) for vegetable crops are provided in Ref. 3.

Fig. 3 Single-row tunnel layer used for some early seeded crops. (Courtesy of the Mechanical Transplanter Co., Holland, Michigan.)

In order to insure good germination, seeds must be properly handled and treated. Generally, corn, pea, and bean seeds are most susceptible to mechanical damage, and need to be handled carefully. Seed treatment methods are used to kill organisms or pathogens within or attached to the seed. Once planted, treatments protect the seed from rot and fungi, or can be used as inoculums to promote nodule formulation on legume roots. Seed treatments can be applied dry prior to planting as a power or slurry.

Direct Seeding

Planters for direct seeding vegetables are generally designed for specific vegetable crops. Exceptions are sweet corn in which a conventional field corn planter with finger (plateless) or plate-type metering is used, peas in which a grain drill is used for seeding, and beans (snap, lima) in which bulk metering cups are used with a conventional row crop planter. Despite these examples, most direct seeded vegetable crops require precise seed metering and placement in the soil. Precision seeding, defined as the precise placement of seeds at the correct in-row spacing and depth, is commonly used as a standard practice in most production systems. Although planters for precision seeding are expensive, the efficient use of seed, the improvement in uniformity and germination, and the elimination of subsequent population thinning operations usually make this planting method economically viable.

There are three basic types of precision planter metering designs that are currently used for vegetable planting, each with advantages and disadvantages.

Plate: A rotating horizontal plate with notches is used to singulate and move seeds to a drop tube. Most planters are ground drive and correct in-row seed spacing is achieved by changing the drive ratio between the plate and drive wheel. These planters are best suited for round seeds that singulate well. With the correct plate selection, they can be used for large and medium seed sizes.

Belt: Holes in a conveyor belt capture and move seeds to a drop tube. Most planters are ground drive and correct in-row seed spacing is achieved by changing the drive ratio between the plate and drive wheel. Seed singulation is best with spherical seeds.

Vacuum: A rotating vertical plate with equally spaced through-holes near the outer circumference and a vacuum applied on the back side of the plate are used to singulate individual seeds (Fig. 4). Both the drive ratio between the plate and the drive wheel as well as the number of holes in the plate can be used to change the in-row seed spacing. Additionally, plates with different hole sizes can be used to accommodate a wide variety of seed sizes. With the correct selection of plate hole size, this metering

Fig. 4 Vacuum metering unit. (Courtesy of Monosem, Inc., Lenexa, Kansas.)

system design is less sensitive than other systems to the shape of the seed.

Several studies have been conducted to evaluate the metering performance of precision planters. Generally, proper configuration of the metering unit coupled with good quality, uniform seed provide acceptable performance. However, comparisons of vacuum and belt-type metering characteristics for nonuniform seeds found surprising results.[4] Neither of the belt configurations used in the belt-type unit nor the vacuum unit demonstrated good performance for graded and ungraded turnip seeds; both skips and doubles were excessively high. Additional work with belt-type planters found significant variations in performance among six identical metering units.[5] Seed delivery was found to vary as much as 116% between metering units operated under the same test conditions.

As shown in Fig. 5, planter frame designs are similar to other conventional planters. Accommodations for starter fertilizer are common; some planters also provide hoppers and metering units for fungicides or other seed treatments to be applied in-row. Individual planter units can be moved laterally on the toolbar to provide different row spacing; some designs can be configured with row spacings as narrow as 20 cm for planting multiple rows on raised beds. Furrow openers and press wheels are similar to conventional planters. For some crop like sweet corn and pumpkins, no till coulters are available.

Potato planters differ from other vegetable planters because they must handle pieces of cut potatoes. Traditionally, metering has been achieved with either pick- or cup-type conveyor mechanisms. Pick-type metering units utilize metal spears mounted on a conveyor chain; singulation is achieved as the spears pierce and then transport individual potato pieces to a drop tube where a knock-off mechanism removes the pieces from the conveyor. The cup-type metering unit uses a similar principle, however, a conveyor with cups is used to capture and transport potato pieces. More recently, vacuum type metering systems have been commercially introduced that use a concept similar to the illustration in Fig. 4.

Direct seeding for plasticulture production systems is accomplished with a punch planter. As illustrated in Fig. 6, a typical punch planter consists of a rotating drum with hollow punches geared to the seed metering unit.[6] As the drum rotates, the punches puncture holes in the plastic mulch, and also create cavities in the soil to accommodate seed placement. Once the seed is dropped, the punch is removed and a press wheel is used to firm the soil around the seed.

Transplanters

Vegetable transplants are available either with bare roots, or contained in a growing tray with the growing media (peat) still attached to the roots. The handling and subsequent planting of each of these two types of transplants are fundamentally different, and transplanting machines are typically classified as either bare root or cell type. The most common designs today still require manual placement of the transplant into the mechanism that delivers to transplant to the desired location in the soil. As

Fig. 5 Six-row vacuum-type vegetable planter. (Courtesy of Monosem, Inc., Lenexa, Kansas.)

Fig. 6 One-row punch planter used for planting in plastic mulch. (Courtesy of Ferris Farms, New Wilmington, Pennsylvania.)

illustrated in Fig. 7, personnel typically ride on the planter and are responsible for transferring transplants to the delivery system for one or sometimes two rows.

The delivery mechanism is typically ground driven and can consist of a rotating carousel or drum in which an operator loads transplants. In-row spacing is adjusted by changing the drive ratio between the ground drive wheel and carousel or drum. During operation, once the carousel or drum rotates to the proper location, the transplant is ejected and placed in a furrow or opening created by a shoe. Wheels are pressed firmly in the soil around the transplant once it is placed in the soil. Typically, fertilizers and pesticides are not applied with the transplanter.

Multiple row transplanters are common for large scale production. Planter units are mounted on a tool bar that allows for adjustable row spacings. Automatic or

Fig. 7 Four-row vegetable transplanter. (Courtesy of the Mechanical Transplanter Co., Holland, Michigan.)

self-feeding transplanters have been studied[7] and developed experimentally[8] but have limited commercial availability and acceptance.

CHEMICAL APPLICATION

Agricultural chemicals are an integral and essential component of commercial vegetable production. They are applied to plants and the soil as a means to control weeds, diseases, and insects. In addition, chemicals are used as soil amendments to improve fertility and adjust pH. In vegetable production, chemicals may be applied prior to, during, or after planting, and may be applied as solid, liquid, or gas. As a result, the choice of application equipment is a function of both the timing and form of the application.

The most common applicator used in vegetable production is the boom-type sprayer. A typical applicator, shown in Fig. 8, can be used to broadcast or band fertilizers, pesticides, or mixtures of chemicals. Smaller boom sprayers can be three-point-hitch mounted or towed by a tractor; larger sprayers are typically self-propelled. Generally, boom sprayers are equipped with low-pressure systems for vegetables; nozzles type and system pressure can be changed to provide sufficient atomization for the chemical being applied. One advantage of this type of applicator is the ability to tank-mix fertilizers and pesticides. This practice, common in vegetable production, can save time, reduce costs, and enhance the action of some herbicides.

Extremely low thresholds for disease and insect damage on most vegetable crops necessitate pesticide postplanting applications during the growing season.

Fig. 8 High clearance sprayer for pre- and postemergence pesticide applications. (Courtesy of Ag-Chem Equipment Co., Inc., Jackson, Minnesota.)

As a result, high clearance self-propelled sprayers have been developed (Fig. 8). As illustrated in Fig. 9, hoods can also be incorporated to isolate vegetable plants from the pesticide being applied. Typically, hooded sprayers are used in vegetable production for herbicide application to control weeds between crop rows.

Within the last 20 yr, electrostatic sprayers have become commercially available, and have been used in production. For vegetables, these sprayers have usually been configured as boom sprayers and improve chemical deposition, especially under the leaves where many insects reside. In addition, the potential for reduced spray drift provides a larger window of opportunity for insecticide application, which is critical for the control of insect damage in vegetables.

Aerial applicators have several advantages in vegetable production including the ability to apply pesticides when ground equipment can no longer drive through a crop. Aerial applications of insecticides are common, especially on crops like peas and cucumbers that cannot accommodate wheel traffic later in the growing season. Airplanes are used for most aerial applications, however, helicopters have been used in some instances. Although this method provides rapid and timely coverage, cost per hectare compared to ground applicators is higher, and the increased potential for drift can limit use.

Subsurface application of chemicals is a practice used for both fertilizers and pesticides. Anhydrous ammonia is occasionally used as a preplant nitrogen source for some vegetables. In soils that are infested with soil-borne pathogens or nematodes, the subsurface application of a soil fumigant can be used to reduce pest populations. Prior to a fumigant treatment, fields must be relatively free of debris and the soil must be sufficiently loose to allow for penetration of the gaseous fumigant. Most fumigants and other low-pressure liquids should be injected to a depth of at least 15 cm; after application, the soil should be rolled or dragged to reduce volatilization losses.[9] Some fumigants like methyl bromide require the use of a plastic film cover over the soil to prevent losses.

Electronic controllers for chemical applicators have become an integral part of most dry and liquid applicator configurations. Monitoring and control of chemical application information coupled with geo-referenced location data provide the ability to record as-applied spatial maps of pesticides and fertilizers. In vegetable production, this provides managers with convenient and accurate documentation of chemical application information for nutrient and pesticide management plans. This is especially important in developing crop rotation plans since several vegetables are sensitive to pesticide carry-over effects.

Fig. 9 Fully mounted hooded sprayer. (Courtesy of Redball, LLC, Benson, Minnesota.)

POSTPLANTING OPERATIONS

Vegetable production frequently requires operations subsequent to planting to control pests and insure fruit uniformity and quality. In many cases, limited commercially available herbicides also necessitate cultivation for controlling weeds.

Mechanical Cultivation

The primary objective of cultivation is to reduce weed pressure to acceptable levels. This not only reduces competition for nutrients and light, but also improves harvested product quality by reducing foreign material in some mechanically harvested crops. However, excessive cultivation may promote soil moisture loss and impede effective fruit recovery during vegetable machine harvest because of unlevel soil conditions.[10,11] In vegetable production, a number of different cultivation methods and configurations are available, each with advantages and disadvantages.

Cultivator Designs

Types of mechanical cultivators include rigid, spring tooth, finger-wheel, rotary hoe, and rolling basket. Most individual tools or gangs of tools are mounted on a ridge tool bar that allows variable tool spacing to accommodate different crop row spacings. Depth of cultivation is controlled by gauge wheels, however, some fully mounted cultivators may also rely on the tractor three-point-hitch to assist in controlling depth.

Fig. 10 illustrates a typical spring tooth type cultivator with S-tine shanks. With this type of design, the tine is allowed to vibrate as it moves though the soil, thus promoting soil fracture and weed disturbance. Several types of tools have been used for vegetable cultivation. The most common type is a sweep (Fig. 11), which can be configured for significant soil disturbance, or to sever weed roots under the surface of the soil at lower speeds. A finger-wheel cultivator has also been shown to be effective

Fig. 10 S-tine cultivator. (Courtesy of Kelly Manufacturing Co., Tifton, Georgia.)

Fig. 11 Sweep-type tool. (Courtesy of Kelly Manufacturing Co., Tifton, Georgia.)

in vegetable production, even for small crops. In one study, at operating speeds of approximately $8\,km\,hr^{-1}$, tests in mustard, kale, and spinach crops found the finger-wheels were effective in weed control and did not induce crop damage or reduce yields.[12]

Rotary Tillers

Rotary tillers are PTO-powered, and consist of several sections of rotating L-shaped blades. As shown in Fig. 12, shrouding around the rotating tools promotes soil pulverization and prevents soil from being thrown out of the row. Although these cultivators require additional power and are heavier than conventional cultivators, rotary tillers are very effective in weed control. In addition, they provide excellent leveling of soil between rows, and are sometimes used for machine harvested vegetables like pickling cucumbers and lima beans that can be difficult to harvest if unlevel ground surfaces or furrows are present between crop rows.

Other Postplanting Practices

Flame cultivation has been used to control weeds in some culture systems. Typically, heating chambers located between crop rows are mounted on a three-point-hitch, and burning propane or other fuel is used to heat the weed plants sufficiently to kill the plants. Guards are used to insulate the vegetable plants from the heating chamber. Although this technique provides an alternative to herbicides, commercial adoption is limited because throughput rates are low and the amount of fuel required for large plantings is costly.[13]

Thinning is occasionally used to remove plants and achieve desired plant populations. Although the advent of precision planting has reduced the need for this practice, mechanical thinning may still be performed early in the growing season. One method to eliminate unwanted plants is cross-blocking; cultivator shanks mounted on a toolbar are operated perpendicular to the row. Another technique consists of a blade that oscillates laterally as it is moved down the row. Each of these methods is nonselective; selective techniques are inherently more complex and are still being developed.

Biological control of pests is a method that has been studied for several decades. Research has demonstrated that deploying parasitic or predatorily insects (or their eggs) that feed on harmful pests can be an effective alternative to chemical. Mechanical systems to transport and distribute beneficial insects to standing vegetable

Fig. 12 Two-row PTO-powered rotary cultivator. (Courtesy of Ford Distributing, Marysville, Ohio.)

crops have been designed and tested; performance has been positive both in terms of uniformity of application and efficacy of the organisms.[14] Some "biosprayers" are commercially available; however, biological control can be expensive due to the cost associated with the large quantities of insects required for adequate pest control.

NEW TECHNOLOGIES

Several innovations have recently been introduced that have the potential to significantly enhance several aspects of vegetable production. The objectives with these new approaches include improving performance of existing equipment, reducing labor requirements, providing more efficient use of pesticides, and reducing the potential environmental impact of production.

The study of site-specific management of chemicals has been under investigation for several years. Because of the intense use of fertilizers and pesticides, coupled with the need for crop uniformity, this approach has the potential to significantly improve several vegetable crops. Using Differential Global Positioning Systems (DGPS), spatial data has been collected regarding nutrient, pH, weed, and pests. Variable rate applicators equipped with DGPS, have then been used to apply fertilizers, lime, and pesticides only where they are needed. Although the performance of the equipment has generally been acceptable, the economic and environmental consequences of this approach in vegetable production are not well understood and need additional research.[15]

One of the most promising new technologies for vegetable production is guidance systems. Tractors equipped with DGPS capable of real-time kinematic (RTK) positioning have been used for several vegetable field operations.[16] Although the costs are relatively high for this equipment, the benefits, especially for vegetable production, are significant. Precision steering not only reduces the variation in row spacing between adjacent planter spacing, but also provides exact position information in which the row location is recorded. As a result, postplanting operations can utilize the row position information for machine guidance purposes. For example, once a crop has been planted with a precision guidance system, cultivation can be performed in close proximity to the crop without damaging roots; subsurface irrigation tubing location is known and can be avoided. In addition, some operations can be performed at a higher velocity, and in the dark. Specific production systems that utilize these technologies are currently under development.[17]

Machine vision systems have been developed and used to enhance performance of several vegetable operations. Cameras have been used to identify tomato plants from weeds for the purposes of weed control.[18] Robotic systems have been tested for cultivation in which a vision system is used to identify in-row weeds. Once a weed is located, a microprocessor based controller selectively applies herbicide using a precision chemical metering system.[19] Although not commercially available, these prototype spraying systems have been tested in commercial settings, and can operate at speeds of $1.2\,\text{km}\,\text{hr}^{-1}$. Other similar systems have visually identified the crop row and controlled the lateral position of a cultivator to provide precise cultivation in close proximity to the crop row.

REFERENCES

1. Anonymous, *Vegetables at a Glance: Area, Production, Value, Unit Value, Trade and per Capita Use*; Vegetables and Melons Briefing Room, Economic Research Service, U.S. Department of Agriculture: Washington, D.C., 2002; 1, http://www.ers.usda.gov/briefing/vegetables/vegpdf/VegAtAGlance.pdf (accessed Sept 2002)
2. Nonnecke, I.L. Classification of Vegetables. *Vegetable Production*; AVI Publishing: Wesport, CT, 1989; 29.
3. Fordham, R.; Biggs, A.G. Plant Establishment. *Principles of Vegetable Crop Production*; Collins Professional and Technical Books: London, 1985; 91–92.
4. Parish, R.L.; Bracy, R.P. Metering Nonuniform Vegetable Seed. Hortic.Technol. **1998**, *8* (1), 69–71.
5. Parish, R.L.; McCoy, J.E. Inconsistency of Metering with a Precision Vegetable Seeder. J. Vegetable Crop Prod. **1999**, *4* (2), 3–7.
6. Shaw, L.N.; Kromer, K.H. A Punch Planter for Vegetables on Plastic Mulch Covered Beds. In *Agricultural Engineering, Volume 1, Land and Water: Proceedings of 11th International Congress on Agricultural Engineering Use*; Dodd, V., Grace, P., Eds.; CIGR: Balkema, Rotterdam, 1989; 1711–1714.
7. Munilla, R.D.; Shaw, L.N. An Analysis of the Dynamics of High-Speed Vegetable Transplanting. Acta Hortic. (ISHS) **1987**, *198*, 305–317.
8. Suggs, C.W.; Peel, H.B.; Seaboch, T.R.; Eddington, D.L. Self Feeding Transplanters. In *Agricultural Engineering, Volume 1, Land and Water: Proceedings of 11th International Congress on Agricultural Engineering Use*; Dodd, V., Grace, P., Eds.; CIGR: Balkema, Rotterdam, 1989; 1715–1721.
9. Kepner, R.A.; Bainer, R.; Barger, E.L. Applying Fertilizers and Granular Pesticides. *Principles of Farm Machinery*, 3rd Ed.; AVI Publishing: Wesport, CT, 1977; 274–275.
10. Glancey, J.L.; Kee, W.E.; Wootten, T.L. Machine Harvest of Lima Beans for Processing. J. Vegetable Prod. **1996**, *3* (1), 59–68.
11. Glancey, J.L.; Kee, W.E.; Wootten, T.L.; Dukes, M.D.; Postles, B.C. Field Losses for Mechanically Harvested Green Peas for Processing. J. Vegetable Prod. **1996**, *2* (1), 61–81.
12. Parish, R.L.; Bracy, R.P.; Porter, W.C. Finger-Wheel Cultivators for High-Speed Cultivation of Vegetable Crops. J. Vegetable Crop Prod. **1996**, *2* (1), 3–11.
13. Parish, R.L.; Porter, W.C.; Vidrine, P.R. Flame Cultivation as a Complement to Mechanical and Herbicidal Control. J. Vegetable Crop Prod. **1997**, *3* (2), 65–83.
14. Gardner, J.; Giles, D.K. Mechanical Distribution of Chrysoperla rufilabris and Trichogramma pretiuosum: Survival and Uniformity of Discharge After Spray Dispersal in Aqueous Suspension. Biol. Control **1997**, *8* (1), 138–142.
15. Roberson, G.T. Precision Agriculture Technology for Horticultural Crop Production. Hortic. Technol. **2000**, *10* (3), 448–451.
16. Upadhydya, S.K.; Ehsani, M.R.; Mattson, M. An Ultra-precise, GPS Based Planter for Site-Specific Cultivation and Plant-Specific Chemical Application. In *Proceedings of the 5th International Conference on Precision Agriculture and Other Precision Resources Management*, Bloomington, MN, July 16–19, 2000.
17. Rosa, U.A.; Upadhyaya, S.K.; Koller, M.; Josiah, M.; Pettygrove, S. Precision Farming in a Tomato Production System. In *Proceedings of the 5th International Conference on Precision Agriculture and Other Precision Resources Management*, Bloomington, MN, July 16–19, 2000.
18. Tian, L.; Slaughter, D.C.; Norris, R. Outdoor Field Machine Vision Identification of Tomato Seedlings for Automated Weed Control. Trans. ASAE **1997**, *40* (6), 1761–1768.
19. Lee, W.S.; Slaughter, D.C.; Giles, D.K. Robotic Weed Control System for Tomatoes. Precision Agric. **1999**, *1*, 95–113.

Vegetable Storage Systems

Kenneth Hellevang
North Dakota State University, Fargo, North Dakota, U.S.A.

INTRODUCTION

Providing the appropriate environment that will maintain vegetable quality is the goal in vegetable storage. The physiology of the vegetable during the storage period must be understood to provide the correct environment. The required environment must be specific for the vegetable being stored. Generally, the environmental system will need to provide a different environment for the different phases of the storage period. For example, the vegetable may need to be cooled or warmed when it first enters the storage, then there may be a curing period, followed by a cooling, storage, and reconditioning phase. Vegetables are considered as living organisms, so the heat of respiration and the removal of respiration by products must be considered in the design of the storage environmental system. Providing the appropriate environment requires applying principles of physical science for designing the storage facility and environmental control system. The primary focus of this article is on potato storage, but also includes information on onion and carrot storage.

POTATO STORAGE

Storage is an integral part of potato production and marketing. Effective potato storage requires a properly designed, constructed, and managed facility.

Storage Conditions

The potato tuber in storage is a living organism. It needs oxygen for respiration, and generates heat, carbon dioxide, and water. The condition of the potatoes going into storage determines to a large extent how well they will store and what their condition will be while coming out of storage. All storage can do is help maintain quality. Quality does not improve during storage.

Moisture loss from potatoes is weight loss and is comparable to a reduction in yield. However, a greater consideration may be the quality factor. Shrink affects the appearance and firmness of the tubers. In general, if weight losses are kept below 5%, tubers will have a smooth skin and be sufficiently firm for all uses. The tubers will feel increasingly soft as weight losses increase. At 8%–9% weight loss, the skin will be wrinkled and the potatoes are unsuitable for sale for most uses. Ventilation of air at a relative humidity above 90% is recommended.

In general, temperature recommendations for long-term storage are: table and seed potatoes, 38–40°F; French fries, 40–50°F; and chip potatoes 45–55°F. At a temperature of about 45°F and lower, reducing sugars normally start forming in the potato tuber. The actual temperatures, where reducing sugars begin to form, depend on variety and growing conditions. Although reducing sugars seldom build to the point of causing appreciable sweetness in table potatoes, they are very important in the quality of chipping and French fry potatoes. Reducing sugars in chipping potatoes result in an undesirable dark color in the fried chips. Reducing sugars do not normally occur at temperatures exceeding 50°F. Growers can recondition potatoes that have been stored below 50°F by holding them for a period of time at temperatures around 60°F. During reconditioning, the reducing sugars turn back to starches. However, the quality of reconditioned potatoes seldom is as good as the quality of potatoes that have been stored constantly at temperatures over 50°F. Reconditioned potatoes need to be used soon after reconditioning. Higher temperatures increase the respiration rate of potatoes and increase the likelihood of sprouting.

The greatest weight loss from potatoes normally occurs during the first two to three weeks of storage. During this period, high respiration rates, high moisture loss, and high heat production occurs. To minimize the amount of weight loss or shrink during early storage, proper suberization or wound healing must occur. Conditions for good suberization are 50–60°F temperatures and relative humidities above 90%.

A supply of oxygen during suberization is critical. Because of the high respiration rate, a carbon dioxide buildup may occur, which will adversely affect potato quality and storability. Occasional air circulation with some air exchange with outside is recommended to supply oxygen during suberization.

General Storage Layout

Three types of storage layouts commonly used are: 1) single bin (Fig. 1); 2) door-per-bin (Fig. 2); and 3) cross-alley (Fig. 3). There are many variations in these

Encyclopedia of Agricultural, Food, and Biological Engineering
DOI: 10.1081/E-EAFE 120006902
Copyright © 2003 by Marcel Dekker, Inc. All rights reserved.

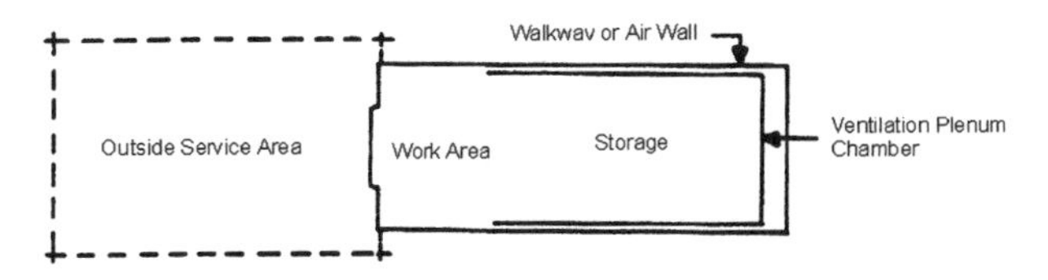

Fig. 1 Single bin.

basic layouts of storage buildings. Common bin dimensions in multiple bin storages range from 16 ft to 24 ft wide, 80 ft to 120 ft long, and 18 ft to 20 ft high. Bin capacities range from 10,000 cwt to 20,000 cwt. The type and size of equipment for handling potatoes will be a major factor in the door size as well as the space needed for maneuvering inside and outside of the storage. When planning a new storage facility, plan for expansion and leave plenty of maneuvering room.

Structural Requirements

Basic structural requirements for a potato storage include: 1) wall strength to resist the pressure of the potatoes, (A professional should design potato storage structures. see Ref. 1); 2) floor strength to support loads due to truck and equipment traffic; 3) insulation to reduce or prevent condensation; and 4) vapor retarders to protect the insulation and structural framework and reduce moisture loss from the storage.

Insulation and Vapor Retarders

Excellent insulation and vapor retarders are required in potato storages, to maintain the high humidity needed in a storage. Insulation is required to reduce heat loss and to stop or reduce condensation on walls and ceilings.

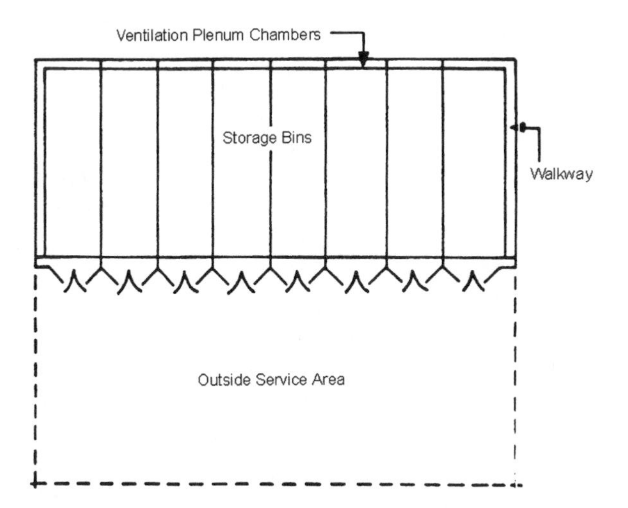

Fig. 2 Door-per-bin.

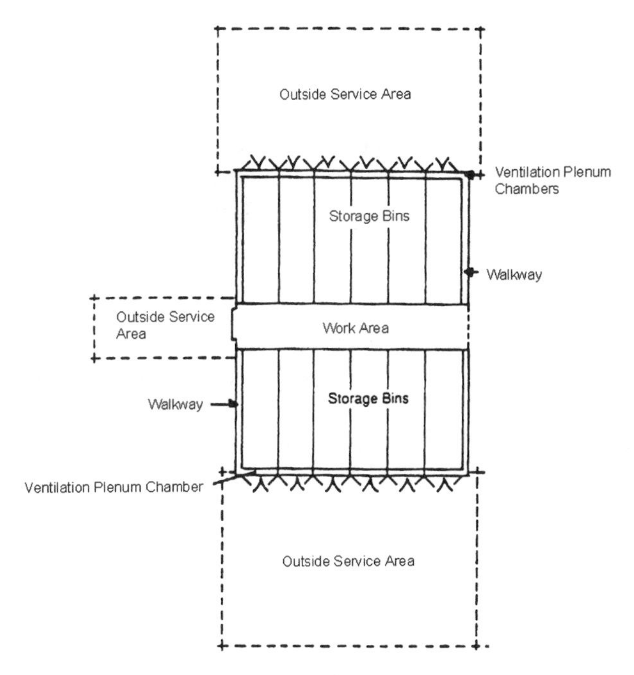

Fig. 3 Cross-alley storage.

Condensation forms on surfaces because the surface temperature is colder than the air dew point. Condensation is an indication of inadequate insulation, not excess humidity. Condensation can be controlled by reducing the relative humidity, but this leads to increased potato shrink during storage. Relative humidity should not be reduced below 90% or shrink may become a problem. An absolute minimum insulation requirement is an R-20 rating. At this insulation level, condensation would still be expected when the outdoor temperature is below − 20°F with an inside temperature of 50°F and 90% relative humidity.

In warm storages (above 50°F), a minimum R-value of 35 is recommended for ceilings and R-value of 25 for exterior walls. This equals 10 in. of fiberglass in the ceiling and 8 in. in the walls. No potatoes should be against the exterior walls since the wall surface will be cold and there will likely be condensation on it. There is commonly an air space between the exterior wall and the potato wall. In cold storages (below 45°F), a less insulation may be used as temperature differentials between inside and outside are not as great, but this limits the use of the storage to cold storage. An R-value of 30 is recommended for ceilings and R-value of 20 for nonpotato contact walls. This equals 8 in. of fiberglass insulation in the ceiling and 6 in. in the wall. Many types of insulation can be used.

Vapor retarders are the most important part of a potato storage structure for protecting the structure and are crucial for maintaining the insulation value. Moisture moves through all types of insulation and all insulations absorb moisture. This drastically reduces the insulating ability of the insulation. Moisture not only reduces insulation value but also will cause rotting or rusting and structural deterioration. Building failure from structural deterioration has occurred in less than 10 yr due to improper use of vapor retarders. It is imperative that a continuous vapor retarder be placed between the potatoes and the insulation. All seams in the vapor retarder must be rolled and taped with an appropriate tape or sealed in some other manner that will limit vapor permeability at the seams. The connection between the vapor retarder and structural members such as the sill should be sealed with a sealant. All holes must also be sealed. The term vapor retarder is used since most products allow water vapor to pass through them. The term vapor barrier has been used in years past, but a product can be defined as a vapor barrier and still allow water vapor to pass through, depending on application.

Vapor retarders are rated according to a "perm" rating. For most construction, anything with a perm rating of less than 1 is considered a vapor retarder. However, in potato storage a retarder should be used with a perm rating considerably less than 1, preferably less than 0.10.

Table 1 lists the permeability of some materials. The permeability can be estimated for different thicknesses of some of the materials by dividing the listed permeability by the desired thickness. For example, the permeability of urethane insulation is about 1 perm for a 1-in. thickness. The expected permeability for a 4-in. thickness would be about 0.25 perms.

Because moisture problems often occur in potato storage structures, an air space between the insulation and roof sheathing is required. Water vapor that has passed through the vapor retarder must be removed by ventilation. Provide at lease 1 ft^2 of open attic vent area for each 200 ft^2 of attic floor area. Place one-half of this vent area at the ridge and one-fourth of the vent area at each eave. In flat or near flat roof construction, leave at least 4 in. of air space between the insulation and roof sheathing. Continuous vents at both ends of the roof are considered a minimum with these flatter roofs. A small amount of air infiltration inside the exterior wall sheathing should be considered to remove the moisture that will move into the walls.

Table 1 Permeability of some materials

Material	Permeability (perms)
Built-up roofing	0
1/4 in. Exterior plywood	0.7
1/4 in. Interior plywood	1.9
Extruded polystyrene—1 in. thick	1.2
Polystyrene-bead board—1 in. thick	2.0–5.8
Expanded polyurethane—1 in. thick	0.4–1.6
Aluminum foil (1 mil)	0
Polyethylene film (2.0 mil)	0.16
Polyethylene film (4.0 mil)	0.08
Polyethylene film (6.0 mil)	0.06
Paint—2 coats	
Asphalt on wood	0.3–0.5
Paint—3 coats	
Exterior lead-oil on wood	0.3–1.0
Latex	5.5–11

Permeability: 1 perm = 1 gr of water per hour per ft^2 per in. of mercury pressure difference.

Ventilation Systems

Ventilation systems that move air through the potatoes are recommended because control of the potato environment may be achieved. The airflow required to keep potatoes in good condition will vary depending on climate and potato type. From 1.00 cfm (cubic feet per minute) to 1.50 cfm of air per cwt (hundredweight) of potatoes is recommended for warm storages and from 0.7 cfm to 1.0 cfm per cwt for cold storages. To determine the cwt of potatoes in storage, first find the volume occupied by the potato pile in cubic feet and multiply by 0.42. For example, $24\,\text{ft wide} \times 62\,\text{ft long} \times 16\,\text{ft high} \times 0.42 = 10,000\,\text{cwt}$. This quantity of potatoes would typically require a minimum of 10,000 cfm of ventilation air.

Maximum airflow usually is not needed throughout the entire storage season. However, it should be available for rapid cooling and if deterioration of the potatoes "rots" begins to develop in the potato pile. Make provisions to reduce the total quantity of air flow when the maximum is not needed. An interval timer is frequently used to control the amount of total daily airflow. In addition, reduced airflow can be accomplished using a two-speed fan or more than one fan per plenum chamber. Unused fans should be covered to prevent backflow, if more than one fan is installed in a plenum chamber. Airflow is also varied using a variable speed fan.

In multibin storages, air may be diverted from several bins to one in an emergency. To do this, block the ducts in one or two adjacent bins and force the air into the bin in trouble. Under these circumstances, more than 2 cfm per cwt can be provided to a given bin. Plywood sheets will work for blocking ducts.

Fan Selection

Once the required ventilation air is determined, a fan must be selected. The fan selected must have the required air capacity at the static pressure or backpressure generated by the potatoes and the duct and damper system. Normally, choose fans that deliver the required air at about 0.75 in.–1.0 in. of static pressure. So called "basket fans" will not work. They deliver large amounts of air at minimal pressures, but virtually no air at pressures caused by the potatoes. Normally, "axial flow" panel fans are used for potato storages.

The Complete System

Fig. 4 shows a complete ventilation system including damper, fan, plenum chamber, and ducts. Building the plenum chamber, damper controls, and mounting the fan all in one location makes for more efficient operation of the storage. The fan discharges to a plenum chamber, which reduces the air velocity and allows for more even air distribution to the ducts.

The damper controls the amount of incoming and exhaust air for the potato storage. With the damper in a closed position, all the air coming off the top of the potato pile will be returned to the fan and recirculated through the pile. With the damper turned 90° to the full open position, all the air coming from the top of the pile will be exhausted and outside air will be pulled through the fan. Provide about 1 ft^2 of cross-sectional area per 1500 cfm in both the inlet and exhaust openings. The bin exhaust openings should be as high as possible, preferably at ceiling level. This helps keep the air moving along the ceiling and will help reduce condensation during cold weather. The damper may be built in one piece as shown in Figs. 4 and 5, or may be composed of two or more doors (Fig. 6). Airflow can also be directed through louvers (Fig. 7).

Controls for the ventilation system may be either manual or automatic. In a manual control system, ropes and pulleys commonly control the damper or doors, and the fan is switched manually and/or by a thermostat. The simplest automatic control consists of a low temperature cut-off thermostat on the fan, which prevents air that is too cold from being drawn into the storage. The thermostat uses a sensor located in the main distribution duct. A differential thermostat controlling the damper, doors, or louvers and a low temperature cut-off on the fan are other options for automatic control. Control panels are commonly used today that control louvers and humidification. The automatic systems generally do a better job because they make adjustments to the system continually.

Always use a low temperature cut-off on the fan in case a louver or damper does not operate. If you use automatic motors on the louvers or dampers, provide a manual override and built-in motor protection since frost may accumulate on dampers and louvers. If these louvers should freeze open or closed, the controls will be unable to do their job and problems can result.

The duct or air distribution system is the final step in delivering air to the potatoes. Ducts must be large enough to carry the desired airflow without creating excessive static pressure. High static pressures will reduce the fan output. Duct cross-sectional area of 1 ft^2 is recommended per 1200 cfm of air delivered from the fan, which gives an air velocity of 1200 ft min^{-1}. Duct velocity should not exceed 1500 ft min^{-1} unless the ventilation system has been designed by an engineer competent in potato storage ventilation design. Ducts are commonly placed no farther than 10 ft–12 ft apart or 60%–70% of the potato depth. Ducts may be circular, triangular, square, or rectangular.

Under-floor ducts of 2 ft wide are commonly covered with 3-in. bridge plank cover boards. Triangular ducts are usually built from 2-in. nominal structural lumber laid flat or plywood with structural framing.

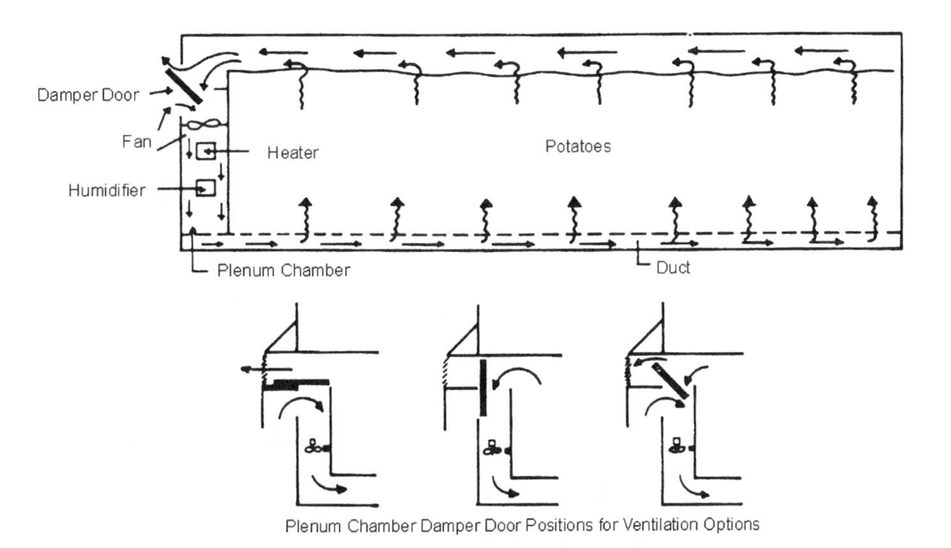

Fig. 4 Ventilation system components.

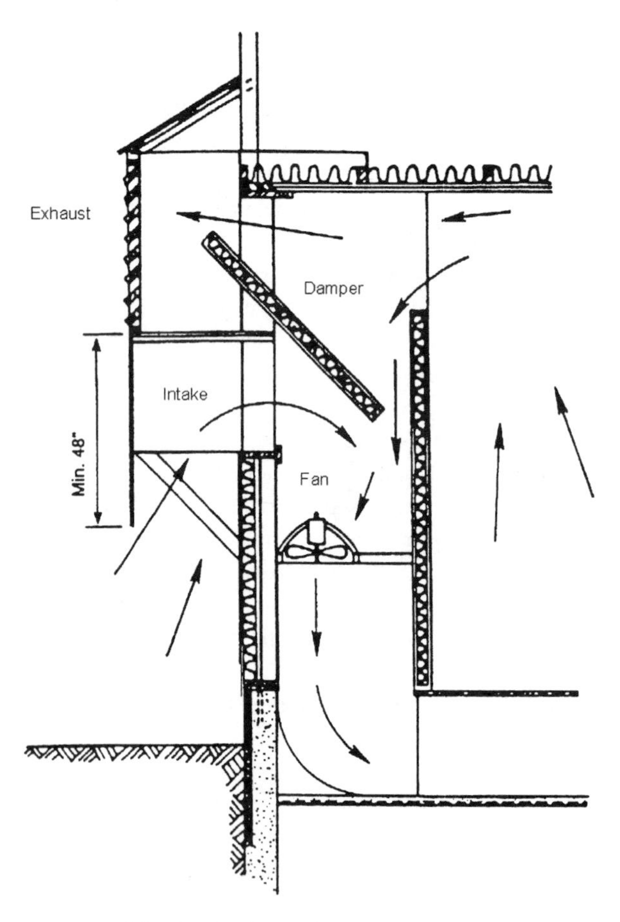

Fig. 5 Fan house for damper system.

The effective air discharge area from the duct affects the distribution of the air. Excess open area results in poor distribution and not enough open area results in excess duct air pressure. Research results show that an air duct should have a total effective discharge area equal to about 90% of the cross-sectional area of the duct. As potatoes cover about 70% of the opening area facing the potatoes, a discharge area of 2.7 times the cross-sectional area should be provided if potatoes are sitting on the opening. For example, a 2-ft wide by 2.5-ft deep under-floor trench duct would require 13.5 ft^2 of duct discharge area. This requires a 19-in. long slot, approximately 1.25 in. wide for each foot of duct length in an 80-ft long duct (Fig. 8).

Culvert-type ducts and the openings from those ducts need to be designed for the potatoes to receive proper airflow. As the holes in a culvert are normally in the recessed part of the corrugation, it is assumed that none of the holes is blocked by potatoes. The total cross-sectional area of all the outlet holes in a culvert should be 90% of the duct cross-sectional area. If the air discharge opening is recessed in a floor or leaner duct, this will perform as a guarded slot, so the amount of discharge area is 90% of the duct cross-sectional area.

Duct cross-sectional area may be reduced as the distance from the fan increases in order to reduce material costs and to maintain a more uniform air velocity. Adequate cross-sectional area must always be provided, so duct velocities do not exceed the design velocity. In addition, the duct cross-sectional area should not be reduced in a step more than one-third of the preceding cross-sectional area. For example, if the duct area is 3.5 ft^2 before the first step, the duct area should not be less than 2.3 ft^2 of cross-sectional area after the step.

Heat Requirements

Under ordinary operating conditions, even during cold weather, potatoes will give off enough heat through respiration to keep a well-insulated storage warm if it is full of potatoes. However, outside air that is brought into the storage, either by damper leakage or with ventilation, may need to be heated, especially when storing potatoes at warmer temperatures. If a storage problem such as "leak" or "wet rot" develops, the relative humidity must be reduced to control the problem. In this case, heat must be available to warm incoming air. Recommendations are for 8 Btu–10 Btu per hour per cwt of potatoes in storage for an LP gas or fuel oil hater. In a well-insulated storage, 5 Btu per cwt per hour output is satisfactory. Use only an electric ignition furnace, since pilot lights tend to be blown out. In addition, exhaust the combustion gases outdoors. An electrically heated storage requires about 1.5 kW (kilowatts) per 1000 cwt of potatoes. Thus, a 10,000 cwt bin would require a 15 kW electric heater.

Humidification

Relative humidities of 90%–95% are required to keep potato shrinkage to a minimum since most shrinkage results from water evaporating from the potatoes. This water evaporation increases rapidly as the relative humidity of the air decreases.

There are two basic types of humidifiers. The most common types draw heat for evaporation from the air. The centrifugal atomizer shown in Fig. 9 is an example. Water fed onto a rotating disc slides to the edge, where it is thrown off by centrifugal force onto a fine screen or ring of teeth. This breaks the moisture into a very fine mist. A fan mounted behind the atomizer unit delivers an air stream, which mixes with the mist and evaporates it. In another type of evaporative humidifier, the air passes through a moist pad or a thin spray of water. With the pad type, water is recirculated over the pad to keep it wet. Another type of evaporative humidifier uses a pneumatic system in which

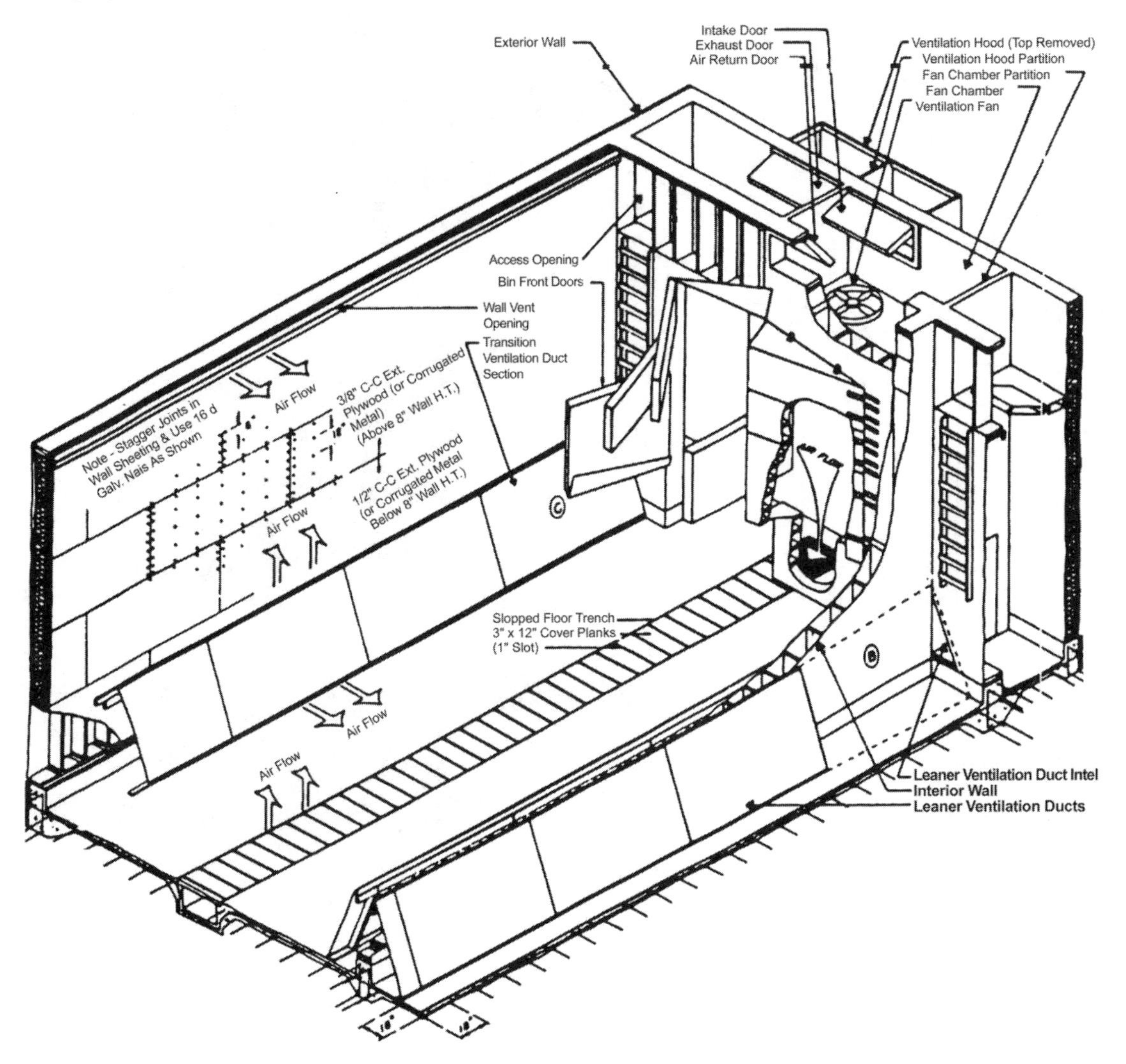

Fig. 6 Bin ventilation diagram.

air breaks up the water to form a fine spray mist. In steam-type humidifiers, heat helps evaporate the water. These are generally high-capacity units.

A humidifying system consists of both a humidifier and a humidistat, which senses the relative humidity in the air after humidification. Humidistats require frequent calibration when used, where the relative humidity is over 80%. Use a sling psychrometer to calibrate the humidistat. Humidistats used in potato storage must be cleaned periodically and maintained according to recommended procedures for use in a dirty, high humidity environment.

A humidifier is normally placed in the plenum chamber in the airstream of the circulation fan. In this location, all of the air, whether it is recirculated or fresh incoming air, will pass over the humidifier. Some free water accumulation in the immediate area of the humidifier will result from incomplete evaporation. Anticipate this and take measures to protect any equipment or structural components in the immediate area. Floor drainage must be designed into the building so water does not flow under the potato pile.

Humidification rates are usually given in gallons of water per hour per 1000 cfm of outside air being brought into the storage. A capacity of 1 ga–3 ga per hour per 1000 cfm is typical depending on the climate.

In many areas, the water has a high dissolved mineral content. These high mineral waters, when used in humidifiers, cause problems of mineral build-up on screens, on teeth, in orifices, and in solenoid valves. Clean frequently under these conditions to keep the humidifiers operating properly, or condition the water being used.

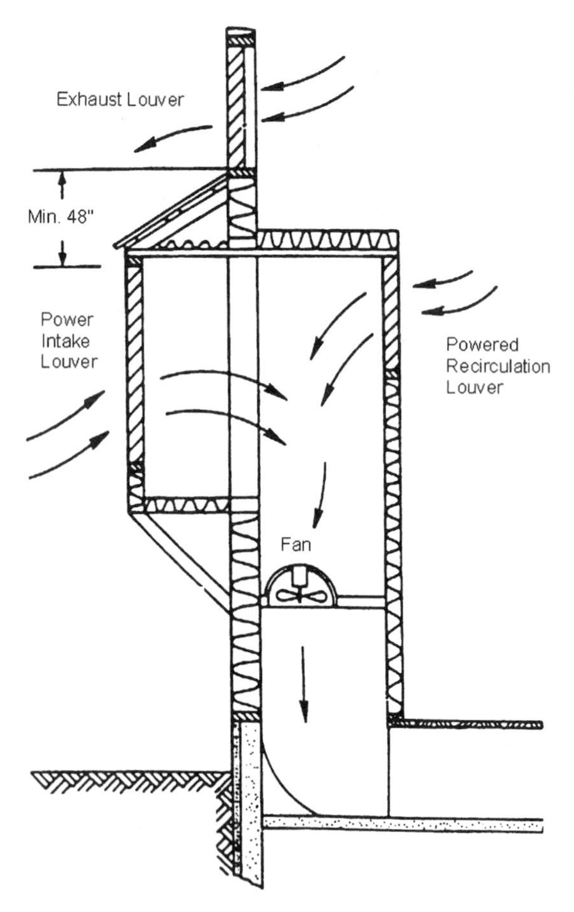

Fig. 7 Fan house for louver system.

Fig. 9 A centrifugal humidifier.

Refrigeration

Refrigeration is used for storing seed and table potatoes, where cold temperatures are required, when outside temperatures do not permit ambient cooling, and for long-term storage of processing potatoes. Using refrigeration permits storage into June and July and even later is possible.

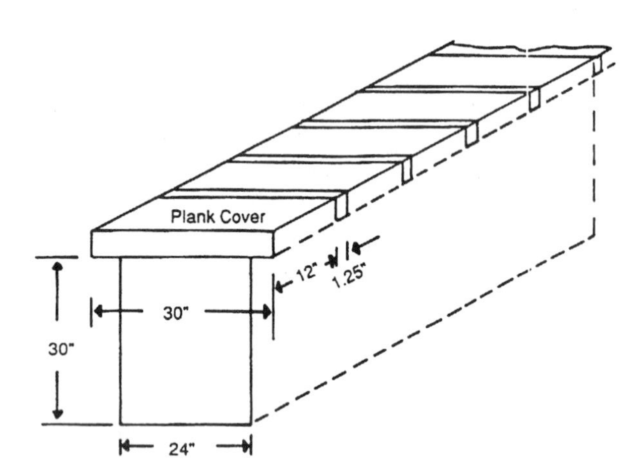

Fig. 8 Under-floor duct with exhaust openings sized for 80-ft length.

Applying refrigeration to potato storages is different from most other refrigeration applications. Because of the extremely high relative humidities required in potato storage, a large temperature difference between evaporator and storage air cannot be used. A large temperature difference would dehumidify the air. Therefore, an extra large evaporator with a high airflow across the coil must be utilized.

The size of cooling equipment required depends upon the total cooling load. Experience in the Red River Valley of North Dakota and Minnesota has shown cooling capacities of 0.5 tn–0.7 tn per 1000 cwt of potatoes are normally adequate. This is equivalent to 6000 Btu–8400 Btu per hour cooling capacity per 1000 cwt of potatoes. In a 10,000 cwt bin, a 5-tn–7-tn refrigeration unit utilizing a 5 hp–7.5 hp compressor unit would be required.

Storage Management

Check temperatures and relative humidities daily at several locations in the storage every day. Check the relative humidity with a sling psychrometer, because a sling psychrometer is the only instrument, which retains accuracy for long periods at high relative humidities.

Check the ventilation equipment closely during operation. Never trust automatic controls completely; they are not foolproof. Dampers may freeze in place in cold weather, heaters may quit working, or other equipment may fail. A good storage operator keeps the equipment in good condition and checks often to see that it is operating properly.

Environmental control will vary for different potato uses and at different stages during the storage period. Limited fan operation is required during suberization and the holding period. More fan operation will be required during cooling or reconditioning the potatoes. Maximum airflow will be required if storage problems develop. Storage temperatures and rates of potato temperature changes will vary according to the use of the potato. Processing potatoes are cooled at a rate of about 0.5°F day^{-1} while seed and table potatoes are normally cooled faster.

ONION STORAGE

The bulk density of onions is approximately 40 lb ft^{-3}. Bulk piles of onions should not exceed 10 ft in height if resistance to airflow through the pile is to be maintained at a reasonable level. Onions in bulk storage exert lateral pressures on walls that restrain them. Lateral pressures for onions should be approximately equal to those for potatoes because the bulk densities and angles of repose are nearly equal for the two products.

Depending upon their initial conditions, onions may need to go through several phases in storage including drying, curing, cooling, holding, and conditioning. Very specific environmental conditions are required for each phase of the storage. The components of the environmental systems are similar to that used in potato storage systems, but the required environment is much different.

CARROT STORAGE

Physiologically mature carrots store longer than immature carrots. Carrots should be harvested in cool weather from dry fields for optimum storage. The most critical factor affecting storage life is the time it takes after harvest to get the internal temperature of the carrot to about 32°F. Carrots have been stored up to 9 mo at a relative humidity maintained near the saturation point (98%–100% relative humidity) at a temperature of 32–34°F. A more common duration is about 4 mo–6 mo in bulk storages.

A ventilation rate of 40 cfm tn^{-1}–60 cfm tn^{-1} is recommended for carrots. The recommended height to pile carrots is about 10 ft–14 ft. The bulk density of carrots will range from 27 lb ft^{-3}–34 lb ft^{-3}.

REFERENCE

1. ASAE. Loads Exerted by Irish Potatoes in Shallow Bulk Storage Structures. *ASAE Standards 2000*; ANSI/ASAE EP446.2 DEC95; ASAE: St. Joseph, MI.

Water Quality in Biological Systems

Steven G. Hall
Caye Drapcho
Louisiana State University, Baton Rouge, Louisiana, U.S.A.

INTRODUCTION

Water quality refers to the physical, chemical, and biological characteristics of water. Specifically, harmful effects on humans, other organisms, or ecosystems imply poor water quality, which may derive from microorganisms, heavy metals, organic pesticides, nutrients, and physical properties of water. Biological aspects of water quality include considerations of the water environment for aquatic organisms, ecosystem, and health aspects in municipal, agricultural, and industrial systems.

Water's (H_2O) polar structure and hydrogen bonding impart unique properties of density, specific heat, and surface tension that greatly impact aquatic ecosystems. Water expands in the solid phase, forming ice which floats (the maximum density of water occurs at 3.94°C) and insulates, protecting aquatic organisms in cold conditions. Water also has a high specific heat value (4.217 kJ/kg K at 0°C), reducing temperature fluctuations in and near water, enhancing life. Its surface tension, due to hydrogen bonds, allows economical oxygen transfer via natural or mechanical means in aquaculture and waste treatment applications.

Water is ubiquitous, being held in oceans ($\sim 10^{21}$ kg), groundwater ($\sim 10^{19}$ kg), glaciers ($\sim 10^{19}$ kg $\sim 1\%$), surface water ($\sim 10^{16}$ kg, $< 0.01\%$), and atmospheric water ($\sim 10^{16}$ kg).[1] Thus, while salt water is plentiful, freshwater is relatively rare and is mostly in groundwater and glaciers. As a result, good management of rivers, lakes, and aquifers is an important aspect of biological engineering.

Water can be characterized based on physical, chemical, and biological characteristics. Physical characteristics include solids content, as well as temperature, color, and odor. Chemical characteristics include dissolved and suspended materials. Organic compounds include biodegradable compounds quantified as chemical oxygen demand (COD) or biochemical oxygen demand (BOD) and organic compounds such as aromatic and chlorinated hydrocarbons and pesticide.[2] Inorganic compounds include nitrogen and phosphorus, which as nutrients for plant and algal growth play a large role in eutrophication of saltwater and freshwater environments. Metals include chromium, lead, and mercury, and other priority pollutants.[3] Biological characteristics include bacteria, viruses, protozoa, and helminths of pathogenic or nonpathogenic nature. Many parameters are measured in concentration, for example, percent or parts per million (ppm). Table 1 presents some important parameters, typical biological effects, typical sources, and recommended or measured concentrations.[2,4,5] Van Der Leeden et al.,[6] offer further information (Table 1).

CAUSES OF WATER QUALITY DEGRADATION

Water may dissolve or carry contaminants, derived from natural or anthropogenic sources. Natural contamination in surface and groundwater includes constituents from rock and soil such as sodium, magnesium, calcium, iron, and even selenium or arsenic.

Anthropogenic sources of water contamination include point-source and nonpoint source discharges (Fig. 1), and atmospheric deposition. Point source discharges may be industrial, agricultural, or municipal, and are characterized by quantity and quality characteristics. Municipal wastewater is comprised of domestic and industrial wastewater and stormwater runoff. Nonpoint source pollutants are classified into categories based on source, including urban runoff, agriculture, silviculture, construction, resource extraction, and septic tank discharge (Fig. 1).

EFFECTS OF WATER QUALITY DEGRADATION

Water quality is critical for aquatic organisms. For example, temperature has significant effects on respiration, food consumption and conversion, growth and reproduction. Excess nutrients, DO, pH, pathogens, and other contaminants are critically important. For aerobic organisms, DO is often the limiting water quality parameter.[7–9] Contaminants can have direct toxicity or indirect impact such as excess nutrients producing eutrophication and increased swings in available DO.[10]

High levels of contaminants can be toxic to humans and animals.[3,11] Drinking water should be low in pathogenic organisms and toxic compounds. Salinity can be a factor in

Encyclopedia of Agricultural, Food, and Biological Engineering
DOI: 10.1081/E-EAFE 120007219
Copyright © 2003 by Marcel Dekker, Inc. All rights reserved.

W

Table 1 Physical, chemical, organic, and inorganic components impact water quality. Shown are selected water quality parameters, sample sources, biological effects, and recommended or measured concentrations

Parameter	Effects	Typical sources	Concentrations
Physical properties			
Temperature	Influences DO	Solar radiation; wind	Depends on ecosystem
Solids	Affects clarity, health	Agriculture; industry	80 ppm TSS/400 TDS[a]
Chemical properties			
DC	Needed for respiration	Aeration; plants; wind	> 5 ppm[a]
pH	Activity affects chemical processes	Basic or acid inputs; CO_2 concentration	6.5–8[a]
Organic components	Increase oxygen demand of water	Agricultural, urban wastewater	< 30 mg/L BOD[a]
Total coliform bacteria	Indicator of fecal bacterial presence	Human and animal wastes	10^5-10^6/ml[b]
Gierdia cysts	Giardiasis/diarrhea		10^1-10^2/ml[b]
Inorganic components			
Nitrate	Toxic at high range	Agriculture; Municipal waste	0–3 ppm[a]; 20–40 ppm[c]
Calcium	Component of bone; water hardener	Minerals	4–160 ppm[a]; 6–16 ppm[c]
Chlorine	Toxic in high doses	Water treatment	< 0.003 ppm[a]
Zinc	Toxic in high doses	Groundwater/industrial input	< 0.005 ppm[a]
Salinity	Osmotic balance	Irrigation water, saltwater intrusion	< 5%[a]

[a]Recommended water quality standards for aquaculture.[4]
[b]Typical mineral increase from domestic water.[2]
[c]Representative numbers in domestic wastewater.[17]

water quality, especially in coastal and arid areas. In addition to individual organisms, ecosystems can also be impacted by water quality. Eutrophication may occur due to excess nitrogen, phosphorous, or other nutrients, often due to agricultural runoff, and may cause shifts in species present and reduce biodiversity.

POLITICAL AND REGULATORY CONCERNS

Due to the importance of water quality on human and ecosystem health and productivity, policies and regulations have been developed. Water quality and quantity are concerns in arid areas, while drinking and recreational

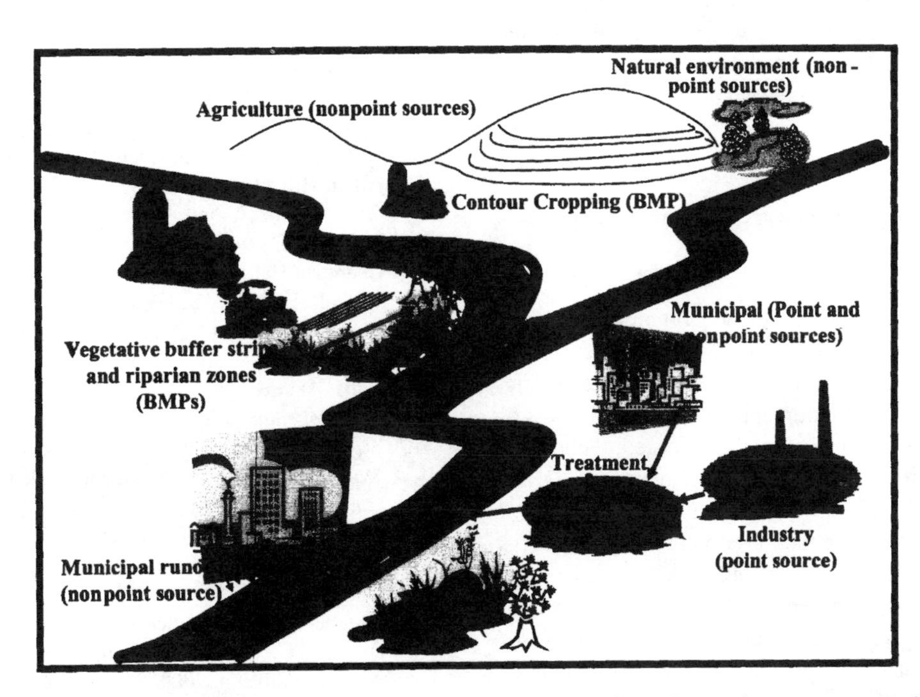

Fig. 1 Agriculture, industry, cities, and the natural environment all contribute to point and nonpoint source water pollution. Treatment is common for point sources such as municipal and industrial wastes, while BMPs such as contour cropping and buffer zones can reduce nonpoint source runoff.

quality is a concern nationwide. Historically, the work of Pasteur as well as others[12–15] pointed toward the need to address water quality issues, including microbial pathogens. Legal precedents date back several decades. The Federal Water Pollution Control Act of 1972 established basic water quality goals and policies for the United States. The Clean Water Act of 1987 aimed to restore and maintain the chemical, physical, and biological integrity of the nation's waters and established and National Pollutant Discharge Elimination System (NPDES) for permitting of wastewater point source discharges.[2] The NPDES permits typically address effluent limitations (discharge standards) for pollutants, based on control technologies, required removal efficiencies, or water quality standards. The NPDES permits are required for industrial stormwater discharges, wastewater discharges from municipal wastewater treatment plants, and from confined animal feeding operations (CAFOs) that confine and provide feed for animals rather than relying on grazing.

Some policies, whether regulatory or voluntary, incorporate biological and engineering knowledge and practices. Total maximum daily loads (TMDLs) for example, limit contaminants that may be discharged legally, while best management practices (BMPs) focus on practical methods that will help reduce contaminant loading (Fig. 1) of ground and surface waters.[10] Watersheds that drain land areas may cross political boundaries and force consideration of policies that accommodate multistate or international issues as policies and practices may vary.[16,17] Water bodies serve both as water source and pollution outlet. Policies developed due to social and biological realities may both drive and be influenced by engineering and science.

MEASUREMENT AND CONTROL METHODS FOR WATER QUALITY

Analytical techniques for in situ or laboratory measurement or organic and inorganic quality parameters include gravimetric, titrimetric, colorometric, spectrophotometric, potentiometric, and chromatographic methods. Analysis of microbial populations include culture-based techniques (plate counts and multiple tube fermentations), DNA-based techniques, and antibody resistance techniques.[18] Ecosystem impacts can be measured via biological assays of zooplankton, fish, and insect populations (see Ref. [17]).

Technologies and policies have been codeveloped for water quality. In cases where the complete banning of toxic chemicals has not been feasible, control methods have included municipal and industrial water treatment, agricultural BMPs and other methodologies (Fig. 1). Wastewater treatment methods for point sources, as well as management techniques for nonpoint sources are discussed in Refs. [2,5,6].

CONCLUSION

Water quality engineering requires location specific, biological viewpoints, consideration of microbial interactions, watershed, regional and global pollutant transport. Traditional physico-chemical techniques plus biofiltration, bioremediation, and preventive methods such as BMPs are used. Recognition of the nonlinear, time varying, and stochastic nature of biological systems are leading engineers to new biological methods relevant to aquatic, human, and ecosystem biology to measure and improve water quality. Engineers increasingly must consider regulatory implications and rely on biological knowledge to design more sustainable water quality management systems.[16]

ACKNOWLEDGMENT

This publication has been approved by the LSU AgCenter as manuscript number 02-22-0721.

REFERENCES

1. Freeze, R.A.; Cherry, J.A. *Groundwater*; Prentice Hall: Englewood Cliffs, NJ, 1979; 604.
2. Liu, D.H.F.; Liptak, B.G., Eds. *Environmental Engineers' Handbook*; Lewis Publishers: Boca Raton, 1997; 1431.
3. Allen, H.E.; Garrison, A.W.; Luther, G.W., III. *Metals in Surface Waters*; Ann Arbor Press: Ann Arbor MI, 1998.
4. Lawson, T. *Fundamentals of Aquacultural Engineering*; Chapman Hall: New York, 1994.
5. Metcalf, Eddy Inc. In *Wastewater Engineering, Treatment, Disposal and Reuse*; Tchobanoglous, G., Burton, F.L., Eds.; Irwin McGraw-Hill: Boston, 1991.
6. Van Der Leeden, F.; Troise, F.; Todd, D. *The Water Encyclopedia*; Lewis Publishers: Boca Raton, 1990.
7. Boyd, C.E.; Tucker, C.S. *Pond Aquaculture Water Quality Management*; Kluwer Academic Publishers: Boston, 1998.
8. Culberson, S.D.; Piedrahita, R.H. Aquaculture Pond Ecosystem Model: Temperature and Dissolved Oxygen Prediction—Mechanism and Application. Ecol. Modelling **1996**, *89*, 231–258.
9. Summerfelt, S.T.; Vinci, B.J.; Piedrahita, T.H. Oxygenation and Carbon Dioxide Control in Water Reuse Systems. Aquacultural Eng. **2000**, *22*, 87–108.
10. Blake, J.D.; Magette, W., Eds. *National Livestock, Poultry, and Aquaculture Waste Management*; American Society of Agricultural Engineers: St. Joseph, MI, 1992.

11. Aindur, M.O.; Doull, J.; Klaassen, C.D. *Casarett and Doull's Toxicology, The Basic Science of Poisons, 4th Ed.*; McGraw Hill: New York, 1991.
12. Theriault, E.J. *The Dissolved Oxygen Demand of Polluted Waters*; Public Health Bulletin #173; US Public Health Service: Washington D.C., 1927.
13. Thomas, H.A. Pollution Load Capacity of Streams. Water Sewage Works **1948**, *95* (11), 409.
14. Monod, J. The Growth of Bacterial Cultures. In *Annual Review of Microbiology*; Clifton, C.E., Raffel, S., Barker, H.A., Eds.; 1949; Vol. 3, 371–394.
15. McCarty, P.L. Energetics and Bacterial Growth. In *Organic Compounds in Aquatic Environments*; Faust, S.C., Hunter, J.V., Eds.; Marcel Dekker: New York, 1971; 495–512.
16. Bennett, L.L. The Integration of Water Quality into Transboundary Allocation Agreements. Lessons from the Southwestern United States. Agric. Econ. **2000**, *24* (1), 113–125.
17. Metcalfe, J.L. Biological Water Quality Assessment of Running Waters Based on Macroinvertebrate Communities: History and Present Status in Europe. Environ. Pollut. **1989**, *60* (1/2), 101–139.
18. Clesceri, L.S.; Greenberg, A.E.; Eaton, A.D., Eds. *Standard Methods for the Examination of Water and Wastewater*, 20th Ed.; American Public Health Association, American Water Works Association, Water Environment Federation, 1998.

Water Quality in Precision Agriculture

U. Sunday Tim
Iowa State University, Ames, Iowa, U.S.A.

INTRODUCTION

Rapid technological change has always been the hallmark of American agriculture. Growing competitive pressure originating from international agriculture markets, improving yields of crops, and controlling or alleviating environmental degradation have encouraged farmers to seek and adopt new production technologies and sustainable management practices. The adoption of space-based technologies [e.g., global positioning systems (GPS)], in conjunction with new demands from a diverse set of environmental interest groups, are setting the broad context for more changes to U.S. agriculture. Along with changes in the structure of production agriculture, increased contractual arrangements, new and far-reaching government policies and regulations, and intensification of production systems are also altering U.S. agriculture.

In response, farmers are seeking competitive ways of producing food and fiber. With the rapid advances in GPS, geospatial technologies, and variable rate application systems, today's agriculture is entering a new era of precision agriculture. Also called precision farming, site-specific crop production, or site-specific management, precision agriculture refers to a suite of technologies and implements that allow farmers to apply more precise amount of inputs based on the within-field spatial and biophysical variability. The goals of precision agriculture include increased yields, reduced production costs, and enhanced environmental quality. Precision agriculture has also been shown to have the potential for reducing the need for environmental regulation by documenting and monitoring commercial fertilizer and pesticide use by identifying environmentally sensitive areas where reduced inputs may be appropriate. The ability to characterize and respond to variability in productivity within a field, maximize net economic returns, reduce wastes, and minimize impact on the environment constitute some of the objectives of precision agriculture. This section provides an overview of precision agriculture and explores its implications for water quality, particularly issues related to the water quality effects of variable rate application of commercial fertilizer and pesticides. An attempt is made to synthesize the important aspects of precision agriculture while guiding the reader to the growing volume of literature on the subject. Readers seeking more detailed information are referred to numerous publications and reviews.[1–3]

AGRICULTURE AND WATER QUALITY

The use of commercial fertilizer and animal manure in agriculture has increased the availability of essential nutrients needed for plant growth and thus contributed to increases in total yields and crop quality. However, overapplication of fertilizer and manure may cause soil and water quality degradation as well as human health problems. For example, excessive delivery of nutrients, primarily nitrogen (N) and phosphorus (P), to surface waters has been linked to the eutrophication of aquatic and estuarine systems. In particular, agricultural sources have been estimated to contribute about 65% of the nitrogen loads entering the Gulf of Mexico from the Upper Mississippi River Basin.[4] In addition, recent research has found that 40% of major estuaries in the United States exhibit highly eutrophic conditions from nutrient enrichment. In terms of human health, nitrate-N in excess of $10\,mg\,L^{-1}$ has been shown to cause methemoglobinemia in infants and may be a risk to humans due to nitrosamine formation.

Concerns about agriculture's dependence on chemicals have also extended to the environmental and health impacts of pesticides. The U.S. Environmental Protection Agency (EPA) estimated that pesticide use in the United States was over 4.6 billion pounds of active ingredients in 1997, the most recent year for which detailed estimates are available. During the same year, an estimated 568 million pounds of herbicide were used in agriculture, commerce, and homes, lawns, and gardens. Insecticide and fungicide applications consisted of 168 lb and 165 lb, respectively.[5] In general, pesticide use in agriculture has: 1) increased crop yield; 2) increased farm efficiency; 3) improved food quality; 4) improved soil conservation; and 5) ensured a stable, predictive food supply.[6] However, several studies have reported the presence of pesticide residues in food and water resource systems. Results from large-scale stream water quality sampling conducted under the National Water Quality Assessment by the U.S. Geological Survey from 1992 to 1996 indicate that over 95% of the samples collected from rivers and streams contained at

Encyclopedia of Agricultural, Food, and Biological Engineering
DOI: 10.1081/E-EAFE 120007158
Copyright © 2003 by Marcel Dekker, Inc. All rights reserved.

least one pesticide.[7] The National Survey of Pesticides in Drinking Water Wells, which the EPA conducted between 1988 and 1990, also found 10% of the community water wells and 4% of the rural domestic wells to contain residue of at least one pesticide.[8] Chronic exposure to agricultural pesticides has been associated with many adverse human health outcomes including acute poisoning, cancer, non-Hodgkin's lymphoma, and neurological and developmental malignancies.[9] In addition to the troubling but unsettled link between pesticide exposure and a variety of illnesses and diseases, the possibility that man-made chemicals, including pesticides, may disrupt the functioning of both wildlife and human endocrine systems is a growing concern. The endocrine system regulates many critical biological processes, including growth, development, and fertility.[10]

Additional evidence of agriculture's impact on the environment can be found in the EPA's most recently published National Water Quality Inventory. This report documents, at the national and state level, the degree of surface water impairment and the primary causes and sources of that impairment. The report notes that over 20,000 water bodies across the United States do not meet their designated beneficial uses, and 218 million Americans reside within 10 mi of a polluted water body.[11] Controlling nonpoint pollution from U.S. cropland is considered to be the most significant water quality need. According to the EPA, water quality impairment from cropland sediment alone has been estimated to cause from $2 to $8 billion a year. Pimentel et al.[12] estimate the groundwater and public health costs of pesticide pollution annually at $1800 million and $787 million, respectively.

Ecological and human health implications of using agricultural chemicals in large amounts are central targets of environmental or legislative reform. Farmers throughout the United States are confronting new legislative and business demands brought upon by changing markets and economic pressures. Furthermore, market-based global competition in agricultural products is challenging economic viability and sustainability of traditional production systems, necessitating development of new farming systems and practices. While changes in U.S. agriculture are intended to create price stability and maximize global competitiveness and free-market performance, the pathway for implementing them has not been void of challenges and problems. To effectively anticipate change in this increasingly dynamic environment, many U.S. farmers are adopting precision agriculture practices to improve the linkage between on-farm production, profitability, and good environmental stewardship. The enthusiasm for precision agriculture can be traced to its ability to improve farm profitability, enhance global competitiveness, and protect or improve environmental quality.

OVERVIEW OF PRECISION AGRICULTURE

The term "precision agriculture" encompasses those agricultural production practices that use rapidly advancing geospatial and production technologies to: 1) tailor crop production inputs to achieve desired outcomes such as economic profitability and reduced environmental impacts and 2) monitor these outcomes at fine spatial and temporal scales. The National Research Council defines precision agriculture as "a management strategy that uses information technologies to bring data from multiple sources to bear on decisions associated with crop production."[1] Wolf and Buttel[13] define precision agriculture as a "production and planning tool supporting more refined application of inputs and better investment decisions" and a "coordinating technology providing a digital interface between the farm field and other stages of production, upstream and downstream." Common to these definitions is the notion that precision agriculture uses information technologies and intelligent farm implements to tailor production inputs to specific areas in a field. The size of the area typically depends on the degree of spatial variability in the field, the tools available to document this variability, and the farmer's implementation strategy. By matching management practices to specific locations in each field, farmers can potentially reduce the cost of inputs (e.g., fertilizer, pesticides, lime, or irrigation water), increase overall crop yields, reduce production risks, and alleviate potential environmental impacts. Rather than uniformly applying crop production inputs to the entire field, farmers are able to integrate GPS with a variable rate application technology (VRT) to apply the appropriate amount of inputs needed to alleviate production problems in specific areas of the field.

Precision agriculture is typically not a single technology, rather, a suite of information and geospatial technologies uniquely integrated to enhance decision-making in agriculture. Components of a typical integrated system include equipment for locating a position in a field (e.g., GPS); equipment for sampling soils and plant tissue; implements for harvesting and monitoring grain yield (e.g., yield monitor); a software system for managing, manipulating, analyzing, and displaying farm data [e.g., geographic information systems (GIS)]; environmental sensing systems and sensors for monitoring plant growth, plant health, and soil physical and chemical properties; and variable rate application systems and input application controls. These technological components are uniquely combined to enable the farmer to: 1) capture data at an appropriate scale; 2) add value to the data; 3) implement an effective management response at an appropriate scale; and 4) monitor the results.[1,14,15]

Precision agriculture focuses on minimizing the loss of pollutants and sediment from agricultural lands because of edge-of-field runoff and leaching from the root zone. Once nutrients and pesticides are applied to the soil, their movement is largely controlled by the amount of soil and water and must therefore be managed through structural and conservation practices. For example, effective nutrient management abates nutrient movement by minimizing the amount of nutrients available for export. This can be realized by applying the right amount of nutrients, at the right time, at the right place, and in the right way. The technological component of precision agriculture that facilitates efficient management of production inputs is VRT. Existing VRT systems have the ability to spatially vary the application rate of farm inputs using specialized controllers that determine the flow rates of crop protection chemicals, nutrients, seed, and irrigation water in response to the desired change in local application rates at specific locations in the field. The systems for variable rate application include soil and plant sensors and environmental sensing systems that are responsive to climate and production.[16–18]

VRT involves the application of crop production inputs on a spatially variable basis to match the conditions in each part of the field. Advantages include higher average yields of crops, lower farm input costs, enhanced food safety and security, and improved environmental quality from reduced chemical use and losses.[2] The hypothesis supporting VRT is that the collection of detailed data on field variability and the use of this data to avoid excess fertilizer and chemical application should decrease the potential for runoff and leaching losses of the chemicals.[1] However, in the absence of spatial variability and soil heterogeneity, variable rate application of chemical inputs serves no specific environmental and water quality purpose. Therefore, it is important to characterize what spatial configuration of terrain, soils, climate, cropping, and management practices exhibit sufficient variability to make VRT economically profitable and environmentally beneficial.

PRECISION AGRICULTURE AND WATER QUALITY

The many water quality problems associated with production agriculture are due to excessive application of chemicals or the application of chemicals at the wrong place in the field. To mitigate these problems, management practices [collectively called best management practices (BMPs)] and new farming systems have been developed that can sustain crop yield, while reducing loss of chemicals to vital water resources. For example, by implementing conservation practices (e.g., no till), farmers can reduce pollutant loss by decreasing soil erosion and reducing the delivery of chemicals to surface waters through runoff and sediment. Through integrated nutrient management, they can reduce the nutrients available for transport in sediment, runoff, and leaching. Finally, by implementing integrated pest management practices such as biological pest control, field scouting, and using pesticides with environmentally friendly properties, farmers are reducing pesticide residues in soil and water. By adopting VRT, farmers can define homogeneous spatial units and management zones for fertility management and target pesticide application to problem areas in the field.[14,19] Taken together, these practices have reduced the off-site effects of production agriculture.

In this section, the limited number of studies available on the water quality impacts of site-specific nutrient management, focusing on N and herbicide losses from conventional and variable rate application of the chemicals, are reviewed. This is not an exhaustive review because numerous ongoing projects hold promise to provide new insights into chemical losses under site-specific crop production and provide data upon which to judge the efficacy of precision agriculture. Notable examples of ongoing projects include the North Central Regional Project (NCR-180) titled "Site-specific Management," the Southern Regional Project (S-283) titled "Develop and Assess Precision Farming Technology and its Economic and Environmental Impacts," the Multi-regional, Site-specific Soybean Systems Research, and many others.

Precision Agriculture Effects on Nutrient Losses

In contrast to studies that examine natural resource/landscape variability (e.g., soil, climate, and terrain), the effects of conservation tillage on nutrient losses, and/or the agronomic and economic impacts of precision agriculture, very few studies have examined the water quality implications of site-specific or variable rate application of N and P. Most of the previous studies focused on the site-specific or variable rate application of N for improved crop yields and on evaluating N balances in agricultural soils.[20] Pierce and Nowak[2] identified conditions in which variable rate N application provides water quality benefits. These include conditions where: 1) terrain variability regulates water and N availability; 2) excess N influences crop quality; 3) spatial variability in crop yield is high; and 4) net N mineralization is high and covaries with soil characteristics and landscape position in the field. Larson et al.[21] provided a comprehensive review of the implications of site-specific management on nonpoint pollution.

Much of the previous research on precision agriculture focused on characterizing the spatial variability of soil's physical and chemical properties and their impact on plant

growth. The premise behind this research was that as soils vary in their ability to supply plant nutrients and crops vary in their demand for nutrients, opportunities exist to match plant nutrient demands with soil fertility tests. For example, Walters and Goesch[22] examined the spatial and temporal variation in soil nitrate-N to determine the variable rate application of N to a cornfield. Nolin et al.[23] evaluated the within-field spatial variability of P to improve fertilization efficiency and reduce losses to surface water. Algerbo and Thylen[24] measured the effects of variable nitrate-N on yield and crop quality. Mallarino et al.[25] evaluated the response of corn and soybean crops to uniform and variable rate P fertilization. Examining data on soluble P and sediment-bound P collected from many field experiments, Larson et al.[21] concluded: "When good management practices, such as amounts, timing, and placement of fertilizers and tillage systems are applied on a variable-rate basis, movement of P off the land can be minimized." Many other studies have used yield maps, soil fertility tests, and soil fertility factors to develop site-specific nutrient management strategies.[26,27]

Several case studies exist on the agronomic and economic implications of precision agriculture, particularly the effects of precision nutrient management on yields and farm income.[28] Only a limited number of long-term field studies have been conducted to document the effects of variable rate application of N on water quality. Using data collected from an irrigated cornfield in Nebraska, Ferguson et al.[29] evaluated a VRT approach to site-specific nutrient management and its potential to reduce nitrate-N loss to groundwater. They observed no significant differences between nitrate-N concentrations under variable rate N application compared to conventional uniform application. Kitchen et al.[30] compared corn yield and residual soil nitrate-N under conventional uniform application of N fertilizer with variable rate N application. In general, because the amount of N required under variable rate application is usually lower than that for conventional uniform application, the potential for N losses in runoff and leaching is smaller. Overall, regardless of the application technique (conventional uniform rate or variable rate), the amount of N unavailable to the growing crop, due to position in the field, is highly susceptible to leaching and losses in runoff and sediment.

Precision Agriculture Effects on Pesticide Losses

Site-specific management of herbicides, insecticides, and fungicides has gained considerable support because of perceived environmental quality benefits. Motivation for the variable rate application of pesticides is due to improved understanding of weed biology and insect ecology, prerequisites for effective weed management and insect control strategies. In weed management, e.g., it has been well documented that weeds are highly aggregated in most production fields.[31,32] Johnson, Cardina, and Mortensen[31] provide a review of the literature on site-specific management of weed in agricultural systems, while Fleischer, Weiss, and Smilowitz[33] reviewed studies of the spatial distribution of insects in agricultural crops. Thus, common approaches for controlling and reducing weed pressure have involved applying herbicides to areas of the field where the weed pressure is heavy or patch spraying areas where weeds are randomly distributed and varying herbicide application practices (e.g., timing, rate, and method of application) according to soil physical, chemical, and biological properties as well as weed characteristics (e.g., growth stage, species, and density).[2,34] The first approach requires prior knowledge of historical weed distribution, while the second approach requires accurate documentation of the spatial and temporal variation in those soil properties that influence herbicide efficacy.[35,36] The overall benefits are reduced herbicide use, reduced impact on nontarget organisms, and improved water quality and public health.

Because the effectiveness of variable rate herbicide application depends on the ability to document the spatial distribution of weeds in the field, as well as those soil properties that impact herbicide efficacy, research activities have focused on topics such as weed sensing, mapping, and modeling; patch spraying practices and technologies; and sprayer equipment performance and accuracy. For example, weed population in a field has been identified using farmer survey,[37] aerial photography,[38] or a GPS-based method to locate weed patches in the field.[39] Mortensen, Dieleman, and Johnson[32] discussed the spatial variation of weeds and the management of weed problems in a field. Williams, Mortensen, and Doran[40] assessed weed and crop interactions for integrated weed management in corn. Regardless of the method used to document the spatial and temporal distributions of weeds in a field, the environmental benefits of VRT accrue from applying herbicides only to those areas of the field where weed pressure is significant and adopting integrated pest management strategies.

Only a few studies have reported the soil and water quality benefits of variable rate herbicide management.[1,21,41] Blumhorst, Weber, and Swain[42] evaluated the efficacy of selected herbicides on the physical and chemical properties of certain soils; they concluded that herbicide activity was highly correlated to soil organic matter content. Qui, Shearer, and Watkins[43] developed site-specific application strategies for herbicides based on soil properties, weed competition, and yield potential.

MODELING WATER QUALITY EFFECTS OF PRECISION AGRICULTURE

Field monitoring is one approach to improve our understanding of the relationship between agricultural management and water quality. According to the Intergovernmental Task Force on Water Quality Monitoring,[44] water quality monitoring is "an integrated activity for evaluating the physical, chemical, and biological character of water in relation to human health, ecological conditions, and designated water uses." It includes monitoring of rivers, lakes, reservoirs, estuaries, coastal waters, atmospheric precipitation, and general water. Monitoring provides insight into water quality conditions and is integral to agricultural management and watershed planning. Monitoring programs also provide data for the verification, calibration, and validation of models. However, the development of effective monitoring programs, even at the field scale, is an extremely difficult task. Furthermore, the financial ramifications associated with long-term field monitoring are increasing dramatically, making computer simulation modeling a cost-effective alternative. Simulation models provide advanced analytical tools to explore and predict the interrelationship among the hydrologic, agronomic, ecological, and socioeconomic components of agro-ecosystems. The collective experience from the water quality modeling community is that, despite the limitations of existing simulation models, they are indispensable in ecological research, technology transfer, and management decision-making. A wide array of process-based, field-scale models has been used to evaluate water quality implications of precision agriculture. These models include Chemicals, Runoff, and Erosion from Agricultural Management Systems (CREAMS), Erosion Productivity Impact Calculator (EPIC), Groundwater Loading Effects of Agricultural Management Systems (GLEAMS), Leach Estimation and Chemistry Model (LEACHM), Pesticide Root Zone Model (PRZM), and Root Zone Water Quality Model (RZWQM). Detailed information on these models and their applications in agriculture can be found in the literature.[45,46]

A number of modeling experiments have evaluated the water quality implications of precision agriculture at the plot, field, and whole farm levels. For example, Watkins, Lu, and Huang[47] used the EPIC model[48] to predict crop yields and N losses for different management units within a 63-ha field under conventional and variable rate N fertilization and water application. They concluded that variable rate N application did not significantly reduce N losses compared to conventional N application. Using monitoring data from a tile-drained field in southern Ontario, Rolloff, MacDonald, and Couturier[49] assessed the EPIC model's ability to simulate potential effects of N fertilization; they reported negligible effects of varying N fertilization rates on stream N loads. Delgado[50] used the Nitrogen Leaching Economic Analysis Package (NLEAP) to evaluate the effects of soil type on residual nitrate-N and assess the model's ability to simulate N processes in fields under precision agriculture management. Differences in residual soil nitrate-N were observed for the different soil types at the field site. Larson et al.[21] used the LEACHM to compare chemical losses under conventional, uniform rate management with site-specific management on two Minnesota fields. They observed some reduction in the amount of N leached. De Koeijer and Oomen[51] used a simulation modeling approach to evaluate relationships among yield, N applied, and N leached from a field in the Netherlands; they reported no significant reduction in average N applied or N leached below the root zone under conventional and site-specific N management. Wang[52] developed a problem-solving environment that integrated ArcView GIS, the SPLUS statistical analysis program, and the RZWQM to enhance simulation of the interrelated agronomic and environmental impacts of precision agriculture. Bakhsh[53] used field data with RZWQM to evaluate the effects of N fertilizer and swine manure application rates on nitrate-N losses with subsurface drainage water and crop yields. Other field-scale simulation modeling of soil and water quality impacts of precision agriculture have been reported.[54,55] In summary, results from these models to assess water quality improvements from implementation of precision agriculture are mixed.

In general, simulation models allow estimation of crop yield, chemical fate, and transport for a wide range of climate, management, and soil conditions. By running simulation models, scientists are able to diagnose crop production and environmental conditions under diverse landscapes and production practices. However, a number of issues are not well addressed in the current modeling of the water quality impacts of precision agriculture. For example, most of the existing models do not adequately consider spatial variable properties of the agricultural landscape or the heterogeneity in field and transport processes. To apply these models, GIS and techniques of spatial statistical analysis are required. Furthermore, a gap remains between the current understanding of processes under site-specific management, the ability to incorporate this knowledge into existing models, and the data needed to establish the reliability of the models. Finally, current models need to be made user friendly, easy-to-use, interoperable, and suitable for making routine tactical and strategic crop production decisions.

FUTURE CHALLENGES AND RESEARCH NEEDS

That American farmers and ranchers are currently under severe economic pressure is no secret. In response, they are continually searching for new techniques and practices to reduce input costs while maintaining productivity and enhancing profitability. Concomitantly, their general concern for environmental quality and good natural resource stewardship is increasing. Therefore, the demand for reliable information on alternative farming practices, such as precision agriculture, has increased significantly, creating exciting new challenges and research opportunities. Some areas where basic and applied research is needed include: 1) evaluating precision agriculture systems performance, including research on agronomic and socioeconomic impacts; 2) conducting on-farm research that identifies unique combinations of precision technologies and conservation management practices that provide a desirable level of economic benefits; 3) evaluating the environmental effects of precision agriculture, including field monitoring research to document impacts of variable rate application of N, P, and synthetic herbicides on water quality; 4) documenting the cumulative effects of precision agriculture practices at the watershed level to enhance implementation of programs such as the total maximum daily load (TMDL); 5) developing environmental sensing systems and sensors to reduce the cost of characterizing field/soil spatial variability; 6) designing and prototyping decision support systems that enhance holistic assessments of production risks and trade-off frontiers associated with the adoption of precision agriculture; and 7) funding of social, equity, and behavioral research to explore institutions and public policies that enhance or impede adoption of precision agriculture practices.[56,57]

Since its debut in the mid-1980s, the concept of precision agriculture has intrigued many farmers and raised concerns in others. Researchers at land grant institutions have initiated new projects; extension personnel have engaged farmers, crop advisors, and the entire agribusiness community in dialogue on issues related to precision agriculture; and the agribusiness industry has actively evaluated production technologies and implements. However, many social, equity, and behavioral challenges remain. For example, how much do income level and profit margin influence the choice of precision agriculture technologies and practices? Is good environmental stewardship alone sufficient to facilitate the adoption of precision agriculture? How important are age and education, compared with experience, in the adoption of precision agriculture management systems? How do regional differences in soil, terrain, and climate impact the adoption of precision agriculture practices? The future directions of precision agriculture will depend upon the ability of researchers to address these research challenges and to provide answers to questions related to the environmental benefits of this emerging crop production practice.

From a purely technical perspective, there are two major challenges. First, there is the issue of scale of field monitoring and modeling experiments. To date, most studies on the water quality impacts of precision agriculture have concentrated on the field and farm levels and none at the watershed scale. According to the 1997 report by the National Research Council, the "spatial patterns of farming activity within the watershed may have more of an impact on environmental quality than do improvements in environmental management within farm fields." The second issue relates to modeling philosophy and modeling uncertainty. Lumped, deterministic, field-scale simulation models have now become acceptable analytical tools for predicting fate and transport of chemicals in agricultural landscapes. With the reduced funding for on-site field monitoring, many researchers anticipate an even greater role of models in agricultural management decision-making. To improve the capability of models, there is a critical need to: 1) create a collaborative environment between model developers and field scientists to ensure that appropriate experimental data is collected for model evaluation and application; 2) develop procedures for characterizing errors and uncertainties in models and their effects on production management decisions; 3) develop problem-solving environments that link field-scale biophysical models with those that describe economic, geochemistry, meteorology, and biology of agricultural systems; and 4) create authentic modular interfaces that enhance the ease-of-use of models. The future of precision agriculture will depend upon collaboration between the private sector and scientists at land grant universities and colleges to address existing research issues and challenges.

CONCLUSIONS

Compared to other agricultural management practices, precision agriculture is considered a revolutionary practice that can add fundamentally new elements of value to agriculture's traditional method of producing food, feed, and fiber. The adoption of precision agriculture practices and technologies is proceeding at an accelerated rate, and farmers and ranchers are beginning to exploit its full potential to enhance profitability and productivity. The vast amount of site-specific data generated in precision agriculture has helped producers and growers make more technically defensible and economically sound management decisions. Indeed, precision agriculture has evolved from a collection of partially compatible technologies (e.g., the use of GPS or differential GPS with yield monitors and combines to document grain yield) into a sophisticated management system that accommodates diverse biophysical settings, cropping practices, and landscapes that are characteristic of American agriculture.

In this article, the concepts of precision agriculture and its potential water quality implications were examined. Future challenges and research needs were also identified. Compared to the agronomic and economic aspects of precision agriculture, long-term studies documenting the effects of precision agriculture on water quality are limited. Research studies at the watershed level are nonexistent. From existing evidence, assembled from actual field data or from model simulations, it is not clear what the potential water quality benefits of precision agriculture are. It is clear, however, that both surface runoff losses and subsurface leaching of agricultural chemicals can be reduced under precision agriculture compared to traditional farming practices.

Farmers, ranchers, resource agencies, watershed stakeholders, and regulators critically need data upon which to judge the long-term water quality benefits of precision agriculture.

REFERENCES

1. National Research Council. *Precision Agriculture in the 21st Century: Geospatial Information Technologies in Crop Management*; National Academy of Sciences: Washington, D.C., 1997.
2. Pierce, F.J.; Nowak, P. Aspects of Precision Agriculture. Adv. Agron. **1999**, *67*, 1–85.
3. Pierce, F.J.; Sadler, E.J., Eds. *The State of Site-Specific Management for Agriculture*; ASA, CSSA, and SSSA: Madison, WI, 1997.
4. Goolsby, D.A.; Coupe, R.C.; Markovechick, D.J. *Distribution of Selected Herbicides and Nitrate in the Mississippi River and Its Tributaries. April Through June, 1991*, U.S. Geological Survey Water Resources Investigation Report 91-163, 1991.
5. U.S. Environmental Protection Agency. *National Water Quality Inventory: 1998 Report to Congress*, EPA 841-R-00-008; Office of Waters, U.S. Environmental Protection Agency: Washington, D.C., 2000.
6. National Research Council. *Clean Coastal Waters: Understanding and Reducing the Effects of Nutrient Pollution*; National Academy Press: Washington, D.C., 2000.
7. U.S. Geological Survey (USGS). *The Quality of Our Nation's Waters: Nitrates and Pesticides*, Circular 1225; U.S. Geological Survey: Reston, VA, 1999.
8. U.S. Environmental Protection Agency. *Another Look: National Survey of Pesticides in Drinking Water Wells*, Phase II Report, EPA/579-09-91/020; U.S. Environmental Protection Agency: Washington, D.C., 1992.
9. Blair, A.; Zahm, S.H. Agricultural Exposures and Cancer. Environ. Health Perspect. **1995**, *103* (Suppl 8), 205–208.
10. Eubanks, M.W. Hormones and Health. Environ. Health Perspect. **1997**, *105*, 482–487.
11. U.S. Environmental Protection Agency. *Atlas of America's Polluted Waters*, EPA/840-B-00/002; Office of Waters, U.S. Environmental Protection Agency: Washington, D.C., 2000.
12. Pimentel, D.; Acquay, H.; Biltonen, M.; Rice, P.; Silva, M.; Nelson, J.; Lipner, V.; Giordano, S.; Horowitz, A.; D'Amore, M. Assessment of Environmental and Economic Impacts of Pesticide Use. In *The Pesticide Question: Environment, Economics, and Ethics*; Pimentel, D., Lehman, H., Eds.; Chapman and Hall: New York, 1993; 49–84.
13. Wolf, S.A.; Buttel, F.H. The Political Economy of Precision Farming. Am. J. Agric. Econ. **1996**, *78*, 1269–1274.
14. Carr, P.M.; Carlson, G.R.; Jacobsen, J.S.; Nielsen, G.A.; Skogley, E.O. Farming Soils, Not Fields: A Strategy for Increasing Fertilizer Profitability. J. Prod. Agric. **1991**, *4*, 57–61.
15. Lowenberg-DeBoer, J.; Swinton, S.M. Economics of Site-Specific Management in Agronomic Crops. In *The State of Site-Specific Management for Agriculture*; Pierce, F.J., Sadler, E.J., Eds.; ASA, CSSA, and SSSA: Madison, WI, 1997; 369–396.
16. Stone, M.L.; Solie, J.B.; Raun, W.R.; Whitney, R.W.; Taylor, S.L.; Ringer, R.D. Use of Spectral Radiance for Correcting In-Season Fertilizer Nitrogen Deficiencies in Winter Wheat. Trans. ASAE **1996**, *39*, 1623–1631.
17. Sudduth, K.A.; Hummel, J.W.; Birrell, S.J. Sensors for Site-Specific Management. In *The State of Site-Specific Management for Agriculture*; Pierce, F.J., Sadler, E.J., Eds.; ASA, CSSA, and SSSA: Madison, WI, 1997; 69–79.
18. Colburn, J.W. Soil Doctor Multi-parameter, Real-Time Soil Sensor and Concurrent Input Control System. In *Proceedings of the Fourth International Conference on Precision Agriculture*, St. Paul, MN, July 19–22, 1998; Robert, P.C., Rust, R.H., Larson, W.E., Eds.; ASA, CSSA, and SSSA: Madison, WI, 1999; 1011–1022.
19. Fiez, T.E.; Miller, B.C.; Pan, W.L. Assessment of Spatially Variable Nitrogen Fertilizer Management in Winter Wheat. J. Prod. Agric. **1994**, *7*, 17–18, 86–93.
20. Sawyer, J.E. Concepts of Variable Rate Technology with Considerations for Fertilizer Application. J. Prod. Agric. **1994**, *7*, 195–201.
21. Larson, W.E.; Lamb, J.A.; Khakural, B.R.; Fergeson, R.B.; Rehm, G.W. Potential of Site-Specific Management for Nonpoint Environmental Protection. In *The State of Site-Specific Management for Agriculture*; Pierce, F.J., Sadler, E.J., Eds.; ASA, CSSA, and SSSA: Madison, WI, 1997; 337–367.
22. Walters, D.T.; Goesch, J.E. Temporal and Spatial Variation in Soil Nitrate Acquisition by Maize as Influenced by Nitrate Depth Distribution. In *Proceedings of the Fourth International Conference on Precision Agriculture*, St. Paul, MN, July 19–22, 1998; Robert, P.C., Rust, R.H.,

Larson, W.E., Eds.; ASA, CSSA, and SSSA: Madison, WI, 1999; 43–54.

23. Nolin, M.C.; Simard, R.R.; Cambouris, A.N.; Beauchemin, S. Spatial Variability of Phosphorus Status and Sorption Characteristics in Clay Soils of the St. Lawrence Lowlands (Quebec). In *Proceedings of the Fourth International Conference on Precision Agriculture*, St. Paul, MN, July 19–22, 1998; Robert, P.C., Rust, R.H., Larson, W.E., Eds.; ASA, CSSA, and SSSA: Madison, WI, 1999; 395–406.
24. Algerbo, P.A.; Thylen, L. Variable Nitrogen Application: Effects on Crop Yield and Quality. In *Proceedings of the Fourth International Conference on Precision Agriculture*, St. Paul, MN, July 19–22, 1998; Robert, P.C., Rust, R.H., Larson, W.E., Eds.; ASA, CSSA, and SSSA: Madison, WI, 1999; 709–717.
25. Mallarino, A.P.; Wittry, D.J.; Dousa, D.; Hinz, P.N. Variable-Rate Phosphorus Fertilization: On-Farm Research Methods and Evaluation for Corn and Soybean. In *Proceedings of the Fourth International Conference on Precision Agriculture*, St. Paul, MN, July 19–22, 1998; Robert, P.C., Rust, R.H., Larson, W.E., Eds.; ASA, CSSA, and SSSA: Madison, WI, 1999; 687–696.
26. Vetsch, J.A.; Malzer, G.L.; Robert, P.C.; Huggins, D.R. Nitrogen Specific Management by Soil Condition: Managing Fertilizer Nitrogen in Corn. In *Proceedings of the Second International Conference on Site Specific Management for Agricultural Systems*, Bloomington/Minneapolis, MN, March 27–30, 1994; Robert, P.C., Rust, R.H., Larson, W.E., Eds.; ASA, CSSA, and SSSA: Madison, WI, 1995; 465–473.
27. Wibawa, W.D.; Dludlu, D.L.; Swenson, L.J.; Hopkins, D.G.; Dahnke, W.C. Variable Fertilizer Application Based on Yield Goal, Soil Fertility, and Soil Map Unit. J. Prod. Agric. **1993**, *6*, 255–261.
28. Malzer, G.L. Crop Yield Variability and Potential Profitability of Site-Specific N Management. Better Crops Plant Food **1996**, *3*, 6–8.
29. Ferguson, R.B.; Hergert, G.W.; Schepers, J.S.; Gotway, C.A.; Cahoon, J.E.; Peterson, T.A. Site-Specific Nitrogen Management of Irrigated Maize: Yield and Soil Residual Nitrate Effects. Soil Sci. Soc. Am. J. **2002**, *66*, 544–553.
30. Kitchen, N.R.; Hughes, D.F.; Suduth, K.A.; Birrell, S.J. Comparison of Variable Rate to Single Rate Nitrogen Fertilizer Application: Corn Production and Residual Soil NO_3-N. In *Proceedings of the Second International Conference on Site Specific Management for Agricultural Systems*, Bloomington/Minneapolis, MN, March 27–30, 1994; Robert, P.C., Rust, R.H., Larson, W.E., Eds.; ASA, CSSA, and SSSA: Madison, WI, 1995; 427–441.
31. Johnson, G.A.; Cardina, J.; Mortensen, D.A. Site-Specific Weed Management: Current and Future Directions. In *The State of Site-Specific Management for Agriculture*; Pierce, F.J., Sadler, E.J., Eds.; ASA, CSSA, and SSSA: Madison, WI, 1997; 101–130.
32. Mortensen, D.A.; Dieleman, J.A.; Johnson, G.A. Weed Spatial Variation and Weed Management. In *Integrated Weed and Soil Management*; Hatfield, J.L., Buhler, D.D., Stewart, B.A., Eds.; Sleeping Bear Press: Chelsea, MI, 1998; 293–309.
33. Fleischer, S.J.; Weiss, R.; Smilowitz, Z. Spatial Variation in Insect Population and Site-Specific Integrated Pest Management. In *The State of Site-Specific Management for Agriculture*; Pierce, F.J., Sadler, E.J., Eds.; ASA, CSSA, and SSSA: Madison, WI, 1997; 101–130.
34. Stafford, J.V.; Miller, P.C.H. Spatially Selective Application of Herbicides to Cereal Crops. Comput. Electron. Agric. **1993**, *9*, 217–229.
35. Cardina, J.; Sparrow, D.H.; McCoy, E.L. Analysis of Spatial Distribution of Common Lampsquarters (Chenopodium Album) in No-Till Soybeans (Glycine Max). Weed Sci. **1995**, *43*, 258–268.
36. Brown, R.B.; Steckler, J.P. Prescription Maps for Spatially Variable Herbicide Application in No-Till Corn. Trans. ASAE **1995**, *38*, 1659–1666.
37. Stafford, J.V.; Ambler, B.; Lark, R.M.; Catt, J. Mapping and Interpreting the Yield Variation in Cereal Crops. Comput. Electron. Agric. **1996**, *14*, 101–119.
38. Stafford, J.V.; Benlloch, J.V. Machine Assisted Detection of Weeds and Weed Patches. In *Proceedings of the First European Conference on Precision Agriculture*, September 7–10, 1997; Stafford, J.V., Ed.; BIOS Scientific Publ. Ltd: Oxford, U.K., 1997; 511–518.
39. Colliver, C.T.; Maxwell, B.D.; Tyler, D.A.; Robert, D.W.; Long, D.S. Geo-referencing Wild Oats Infestations in Small Grains: Accuracy and Efficiency of Three Weed Survey Techniques. In *Proceedings of the Third International Conference on Precision Agriculture*, June 22–23, 1996; Robert, P.C., Rust, R.H., Larson, W.E., Eds.; ASA, CSSA, and SSSA: Madison, WI, 1997.
40. Williams, M.M.; Mortensen, D.A.; Doran, J.W. Assessment of Weed and Crop Fitness in Cover Crop Residues for Integrated Weed Management. Weed Sci. **1998**, *46*, 595–603.
41. Webber, J.B.; Tucker, M.R.; Isaac, R.A. Making Herbicide Recommendations Based on Soil Tests. Weed Technol. **1987**, *1*, 41–45.
42. Blumhorst, M.R.; Weber, J.B.; Swain, L.R. Efficacy of Selected Herbicide as Influenced by Soil Properties. Weed Technol. **1990**, *4*, 279–283.
43. Qui, W.; Shearer, S.A.; Watkins, G.A. *Modeling of Variable Rate Herbicide Application Using GIS*; Paper No. 94-3522; American Society of Agricultural Engineers: St. Joseph, MI, 1994.
44. Intergovernmental Task Force on Water Quality Monitoring (ITFM). *Ambient Water Quality Monitoring in the U.S. First Year Review. Evaluations and Recommendations*; U.S. Geological Survey: Reston, VA, 1992.
45. Ahuja, L.R.; Ma, L.; Howell, T.A. *Agricultural Systems Models in Field Research and Technology Transfer*; CRC Press: Boca Raton, FL, 2002.
46. Singh, V.P.; Woolhiser, D.A. Mathematical Modeling of Watershed Hydrology. J. Hydrol. Eng. **2002**, *7*, 270–292.
47. Watkins, B.; Lu, Y.C.; Huang, W.Y. Economic Returns and Environmental Impacts of Variable Rate Nitrogen Fertilizer and Water Application. In *Proceedings of the Fourth*

International Conference on Precision Agriculture, St. Paul, MN, July 19–22, 1998; Robert, P.C., Rust, R.H., Larson, W.E., Eds.; ASA, CSSA, and SSSA: Madison, WI, 1999; 1667–1679.

48. Mitchell, G.; Griggs, R.H.; Benson, V.; Williams, J. EPIC User's Guide Version 5300. The EPIC Model. Environmental Policy Integrated Climate, Formerly Erosion Productivity Impact Calculator; USDA-ARS, Grassland Soil and Water Research Lab: Temple, TX.
49. Rolloff, G.; MacDonald, K.B.; Couturier, A.R. Potential Effects of Nitrogen Fertilization Scenarios in Small Watersheds in Southern Ontario. In *Proceedings of the Fourth International Conference on Precision Agriculture*, St. Paul, MN, July 19–22, 1998; Robert, P.C., Rust, R.H., Larson, W.E., Eds.; ASA, CSSA, and SSSA: Madison, WI, 1999; 1395–1407.
50. Delgado, J.A. NLEAP Simulation of Soil Type Effects on Residual Nitrate Nitrogen and Potential Use of Precision Agriculture. In *Proceedings of the Fourth International Conference on Precision Agriculture*, St. Paul, MN, July 19–22, 1998; Robert, P.C., Rust, R.H., Larson, W.E., Eds.; ASA, CSSA, and SSSA: Madison, WI, 1999; 1367–1378.
51. De Koeijer, T.J.; Oomen, G.J.M. Environmental and Economic Effects of Site-Specific and Weather Adapted Nitrogen Fertilization for a Dutch Field Crop Rotation. In *Proceedings of the First European Conference on Precision Agriculture*, September 7–10, 1997; Stafford, J.V., Ed.; BIOS Scientific Publ. Ltd: Oxford, U.K., 1997; 379–386.
52. Wang, X. Integrated Spatial Decision Support System for Precision Agriculture. Ph.D. Dissertation, Iowa State University, Ames, IA, 1999.
53. Bakhsh, A. Use of Site-Specific Farming Systems and Computer Simulation Models for Agricultural Productivity and Environmental Quality. Ph.D. Dissertation, Iowa State University, Ames, IA, 1999.
54. Paz, J.O.; Batchelor, W.D.; Babcock, B.A.; Colvin, T.S.; Logsdon, S.D.; Kaspar, T.C.; Karlen, D.L. Model-Based Technique to Determine Variable Rate Nitrogen for Corn. Agric. Syst. **1999**, *61*, 69–75.
55. Verhagen, A.; Booltink, H.W.G.; Bouma, J. Site-Specific Management: Balancing Production and Environmental Requirements at Farm Level. Agric. Syst. **1995**, *49*, 369–384.
56. Robert, P.C. Precision Agriculture: Research Needs and Status in the USA. In *Precision Agriculture '99: Proceedings of the 2nd European Conference on Precision Agriculture*, July 11–15, 1999; Stafford, J.V., Ed.; BIOS Scientific Publ. Ltd: Oxford, U.K., 1999; 19–33.
57. Fixen, P.E. Research Needs for Site Specific Nutrient Management to Benefit Agriculture. Better Crops **1998**, *82*, 16–18.

Water Table Management

Robert O. Evans
North Carolina State University, Raleigh, North Carolina, U.S.A.

INTRODUCTION

Excessive soilwater is a major concern of soils with seasonally shallow water tables. Smedema[1] estimates that excess water poses a significant cropping limitation on about one-third of the rainfed cropland globally. Drainage is the practice of removing excess water from land. Roughly 400 million ha of the world's rainfed cropland require artificial drainage improvement for efficient crop production of which about 25%–30% (100 to 150 million ha) have been adequately drained to date. The primary goal of agricultural drainage in humid regions is to lower the water content of the root zone to facilitate seedbed preparation and planting (often referred to as trafficability criteria[2]) and to provide adequate soil aeration following excessive rainfall so as to minimize crop stress and yield reductions caused by anaerobiosis.[3] Anaerobiosis describes a low-oxygen growing environment that impairs the growth rate of plants, which require oxygen for respiration. Drainage and anaerobiosis relationships were recently summarized by Evans and Fausey[4] and Evans.[5]

DRAINAGE BENEFITS AND IMPACTS

Several techniques are available to improve drainage and reduce crop stress related to excess water within the root zone. These include both surface practices[6] and subsurface practices.[7–9] Historically, drainage systems were designed, installed, and managed to minimize the trafficability and anaerobiosis problems caused by excess water in the root zone. From a management standpoint, this is referred to as conventional drainage. Conventional drainage systems in the United States have generally been designed with the capacity to lower the water table sufficiently to satisfy extreme crop drainage requirements in about 4 yr out of 5 yr.[10]

For many crops, there is an optimum planting date.[11] Yields decline when planting is delayed past that date. In humid climates, the most extreme drainage requirements typically occur during the wetter time of the year (early spring in most years). During this period, water loss by evaporation is fairly low so excess water must be removed by the drainage system in order to lower the water table and accommodate planting by the optimum date. While wetness is the major concern, soil moisture conditions under variable rainfall can fluctuate greatly. As a result, crops may periodically suffer from drought stresses even on traditionally shallow water table soils. Intensive drainage systems are often necessary to provide trafficability during extreme wet periods but these conventional drainage systems tend to remove more water than necessary during drier periods, a condition referred to as temporary overdrainage.[12] These drought conditions may reduce yields substantially in some years and it is common for both wet and drought stresses to occur at different times during the same cropping season.

Many artificially drained soils are adjacent to environmentally sensitive and ecologically important surface water resources. These natural streams and surface water bodies often serve as the primary outlet for artificial drainage systems. In many surface water bodies, nutrient levels, particularly nitrogen and phosphorus, have become high enough that a very delicate balance exists between desirable flora and undesirable species such as blue–green algae.[13] Water bodies receiving excessive nutrient loads are susceptible to blue–green algae blooms, which unchecked can alter the aquatic food chains. The algae blooms are unsightly and may pose problems such as toxicity, bad taste, and/or odor to recreational users of the water. These blooms can also consume much of the dissolved oxygen, leaving the water anoxic (deprived of oxygen). Anoxic conditions are usually stressful and sometimes fatal to fish. Drainage water from agricultural cropland is one source of nitrogen and phosphorus that can contribute to surface water nutrient enrichment.

NEED FOR WATER TABLE MANAGEMENT

In many locations, conventional drainage methods have been associated with problems of overdrainage and surface water pollution caused by excessive transport of fertilizer nutrients in drainage waters. In response to these problems, there has been a general transition in drainage philosophy from conventional drainage practices to those related to water table management. Water table management provides drainage during wet periods but also incorporates measures to manage the water level at the drainage outlet, making it possible to reduce overdrainage.

DOI: 10.1081/E-EAFE 120007236
Copyright © 2003 by Marcel Dekker, Inc. All rights reserved.

The process of managing or controlling the amount of drainage that occurs is referred to as controlled drainage. While conventional drainage systems are sometimes prone to overdrainage, the reduced drainage volume realized with water table management often results in a reduction in the nutrient load being discharged with the drainage water.[14,15] Although recent growth in the use of controlled drainage has been to conserve water and enhance drainage water quality,[16] controlled drainage has been used historically to reduce subsidence in drained organic soils.[17] This application continues in places such as the Everglades Agricultural Area in Florida, the Wester Johor area in Malaysia, and many other locations around the world.[18]

During dry periods, the controlled drainage process can be reversed and the drainage system used to supply water to the root zone, a process referred to as subirrigation.[19] Subirrigation is not a new process given that it has been practiced in scattered locations for nearly a century.[20,21] Early applications were on very permeable organic or sandy soils. Skaggs, Kriz, and Bernal,[19] Skaggs,[22] and Doty, Currin, and McLin[23] showed that subirrigation could be applied on finer textured soils by proper design of the system to match soil and site conditions. Management practices that involve conventional drainage, controlled drainage, and/or subirrigation are referred to collectively as water table management. The goal of water table management is to manage the level of the shallow water table in order to minimize wet stress, reduce over drainage and subsequent off-site transport of agricultural chemicals, conserve water, and provide for more effective utilization of rainfall. From these perspectives, water table management applications have been confined predominately to humid areas.

WATER TABLE MANAGEMENT PRACTICES

Water table management involves combinations of surface drainage, subsurface drainage, controlled drainage, and/or subirrigation. Surface drainage refers to the removal of excess water from a field by water movement across the soil surface to an open ditch or other drainage outlet. Surface drainage is achieved by a system of open ditches typically installed at 100 m–200 m intervals. Ditches are parallel to each other and oriented in the general direction of the prevailing slope. To encourage surface runoff and reduce surface ponding, fields are often graded and sometimes crowned into a turtleback shape. Field ditches are typically 0.5 m–1.5 m deep and discharge to collector canals typically laid out on 1 km or 2 km grids. Irregularly spaced ditches are often used to drain depressional areas. Bedded rows parallel to the field ditches are sometimes used to elevate the seedbed and help reduce water stress when plants are small. Surface drainage systems influence the position of the water table by minimizing surface ponding and by reducing the volume of water that infiltrates. Once elevated, the water table is lowered primarily by evapotranspiration (ET) rather than removal of water from the soil profile by direct drainage. The effectiveness of surface drainage is controlled by the relative smoothness of the soil surface (which affects surface depressional storage) and the slope of the land surface (which affects how quickly excess water drains from the field).

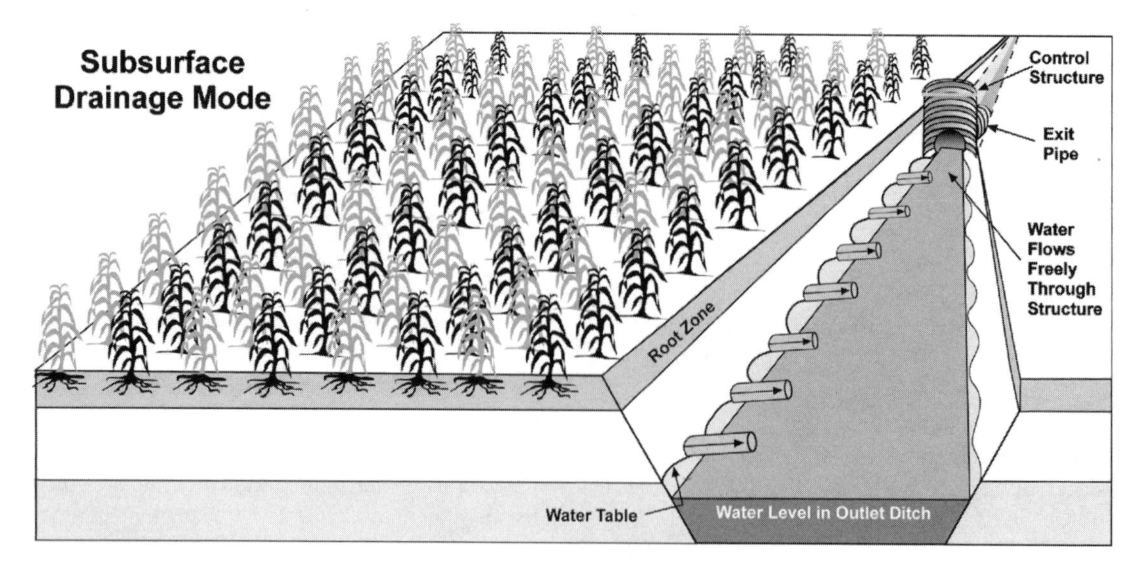

Fig. 1 Schematic of the typical conventional (free flowing outlet) subsurface drainage operational mode. Subsurface drainage occurs as long as the water table is at a higher elevation than the drain pipe or ditch bottom.

Subsurface drainage is achieved by a system of buried clay tile, perforated plastic tubing, or unlined mole drains (Fig. 1). In general, 100 mm–150 mm diameter tile or tubing is buried 1 m–1.5 m deep at intervals of 10 m–50 m. The subsurface drainage pipes generally outlet to an open canal or stream. Subsurface drainage systems lower the water table following heavy rainfall. In many cases, crop protection is also provided as a result of the water table having been lowered by the drainage system in advance of the rainfall. In this latter case, subsurface drainage has increased soil storage, which prevents the water table from rising into the root zone during the rainfall event. Primary factors that affect the rate of subsurface drainage include drain spacing and depth and transmissivity of the soil profile.[24] As a general rule, subsurface drainage systems have been sized to remove about 1.25 cm day^{-1} (drainage coefficient), which translate to roughly 12 in. of drawdown[25] if the water table is initially at the soil surface. The actual drawdown per unit of drainage coefficient depends on the drainable porosity of the soil. For a given drainage coefficient, the water table drawdown is less for course-textured soils with high drainable porosity and greater in fine-textured soil with low drainable porosity. Overdrainage becomes problematic in cases where the drainage intensity is high due to deep drain depths and/or high permeability. In many cases, adequate crop protection from excess soilwater stresses is provided as long as the water table is not closer than 0.3 m from the soil surface for more than 1 day.[10,26,27] Depending on soil texture, optimum soilwater conditions occur for water tables between 0.3 m and 1.0 m.[28] In the conventional subsurface drainage mode, outlet pipes are free flowing so drainage continues as long as the water table is above the elevation of the pipes. When pipes are less than 1 m deep, there is little risk of overdraining the soil profile, but if pipes are placed greater than 1 m, particularly in soils with high transmissivity, significant soilwater will continue to drain from the soil profile well after the water table has receded below the crop protection level of 0.3 m. Under these drainage conditions, controlled drainage is the most effective water table management strategy to employ.

Controlled drainage involves the use of some type of adjustable, flow-retarding structure placed in the drainage outlet that allows the operator to manually establish a desired water level within the outlet (Fig. 2). Many types of structures can be used depending on the layout of the drainage system. Where drain tubing or field ditches outlet directly to an open channel such as a canal or stream, the system is referred to as an open system. Water control structures for open systems may range from simple, stop-log, weir-type structures often referred to as flashboard risers (Fig. 3)[29] to automated inflatable dam-type structures.[12] Where drain tubes outlet to main drains rather than open channels, the system is referred to as a closed system.[30] Several tubing manufacturers have designed and marketed barrel-type structures for use in closed drainage systems, which function as a weir in the main drain line allowing the water level to be controlled.

When operated in the controlled drainage mode, drainage occurs as long as the water table in the field is at a higher elevation than the weir elevation at the control structure. Once the water table in the field drops below the weir setting, drainage stops; however, the water table will continue to recede as the crop removes water by ET. Once the field water table drops below the water level in

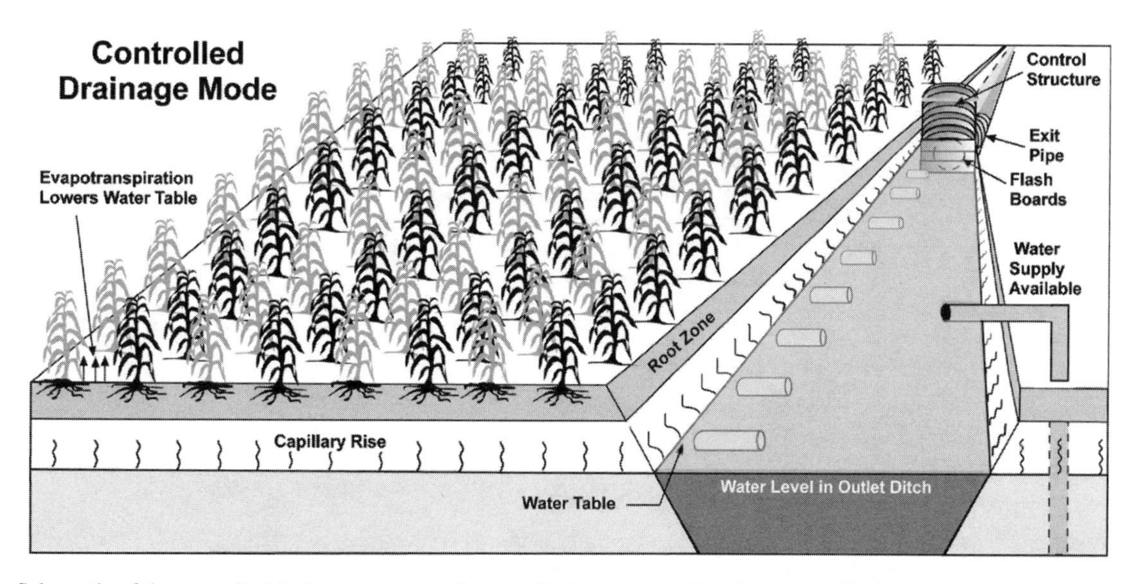

Fig. 2 Schematic of the controlled drainage operational mode. Drainage stops when the water table drops to the same level as the top of the control structure (weir). The water table may continue to drop due to ET.

Fig. 3 Flashboard riser type water control structure used to manage the outlet water level in an open ditch system.

the outlet, the process may reverse whereby water stored in the outlet ditch passively flows back through the drains into the soil profile. The amount of water stored in the outlet depends on the dimensions of the outlet. Large canals may supply the equivalent of 5 mm–10 mm of water while tubing outlets store very little water. In either case, water stored in the soil profile that would otherwise drain is typically of greater magnitude than the amount of water stored in the outlet. In the controlled drainage mode, the water level in the outlet typically fluctuates several times between the weir setting and the bottom of the outlet (Fig. 4). Controlled drainage is most effective where drought conditions are intermittent and of short duration. For a single event, controlled drainage may retain up to 25 mm of water in the soil profile that would otherwise drain from the system. The water saved could delay drought stress for a period of 3 day–7 day depending on ET. Over the course of a growing season, controlled drainage systems may conserve up to 75 mm of water that would otherwise be lost from the soil through over-drainage.[31] Actual storage depends on the drainage intensity, drainage system layout, soil drainable porosity, and water requirements of the crop. Compared to conventional drainage, operation of the system in the controlled drainage mode typically results in less subsurface drainage accompanied by a modest increase in surface drainage with a net reduction in total annual drainage of about 10% depending on rainfall.[14]

Subirrigation involves the introduction of supplemental water to the system that subsequently moves into the field through the underground tubing or field ditches to satisfy the crop ET requirements (Fig. 5). The field water level is influenced by the elevation of the control weir, pumping strategy, drain system intensity, transmissivity of the soil, and rate that water is removed by the crop. The height of the water table needed to avoid drought stress is a function of the soil and crop but normally will be between 0.5 m and 1 m from the soil surface. Subirrigation provides water to the root zone from the underlying water table through the process of capillary rise. The pumping strategy and optimum water table control level also depends on many factors including crop, stage of growth, soil physical properties, system design, and prevailing weather conditions.[32] Water may be pumped continuously or cycled by automatic control. The water level in the outlet can be maintained relatively constant or allowed to cycle between the weir control level and elevation of the drain tubing. While in the subirrigation mode, the water level in the outlet should not be allowed to drop below the top

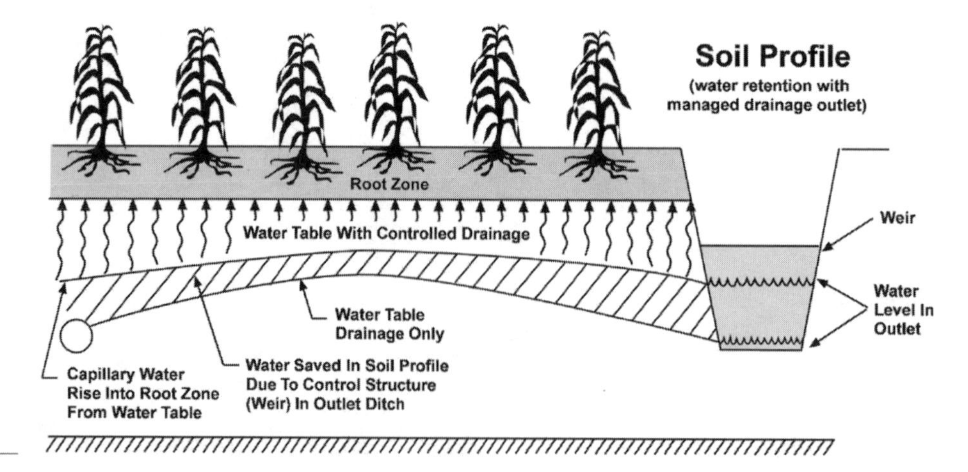

Fig. 4 Water level fluctuation with a controlled drainage system. Controlled drainage can be practiced with a subsurface drain tube system (represented by the circle on the left) or an open ditch system. The cross-hatched area represents the amount of water saved during one cycle. Once the water table drops below the weir, it does not rise again until the next rainfall event large enough to cause percolation below the root zone.

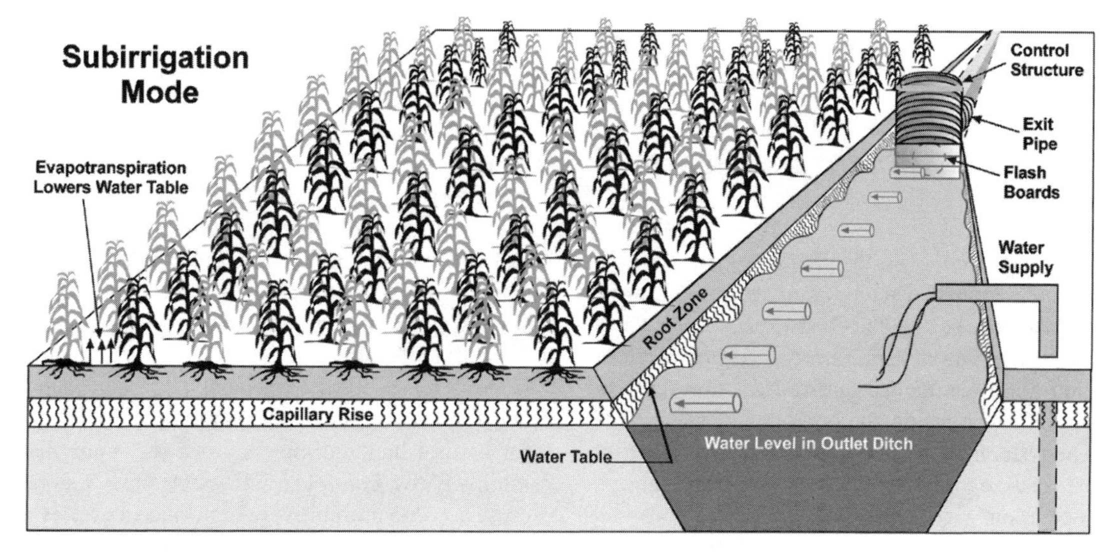

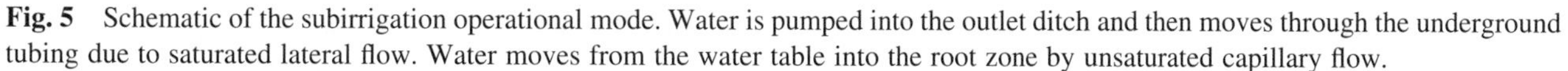

Fig. 5 Schematic of the subirrigation operational mode. Water is pumped into the outlet ditch and then moves through the underground tubing due to saturated lateral flow. Water moves from the water table into the root zone by unsaturated capillary flow.

of the drain tubing in order to prevent the introduction of floating debris and sediment into the drain lines that could clog the drains. Maintaining a relatively constant water level in the outlet simplifies system management and can be achieved with simple float switches to regulate the "on" and "off" cycling of the pump. However, this approach tends to maintain low soilwater storage. As a result, much of the rainfall that occurs during the irrigation season simply drains from the profile resulting in increased irrigation to supply most of the crop water requirements. A cyclic operation of the pump that allows the water level to fluctuate between the weir control elevation and the top of the drains results in better rainfall utilization and reduced irrigation pumping requirements (Fig. 6). With this approach, once the water table is raised to the desired level, the pump is shut off. The water table is then allowed to recede due to ET until it drops to a lower predefined allowable limit. Pumping is then resumed and the water table is again raised to the predefined upper level and the cycle repeats. This approach requires more intensive management and greater system soilwater transmission capacity in order to avoid drought stress that may develop when the water table is near the lower set point. As the water table is maintained at a higher elevation during both

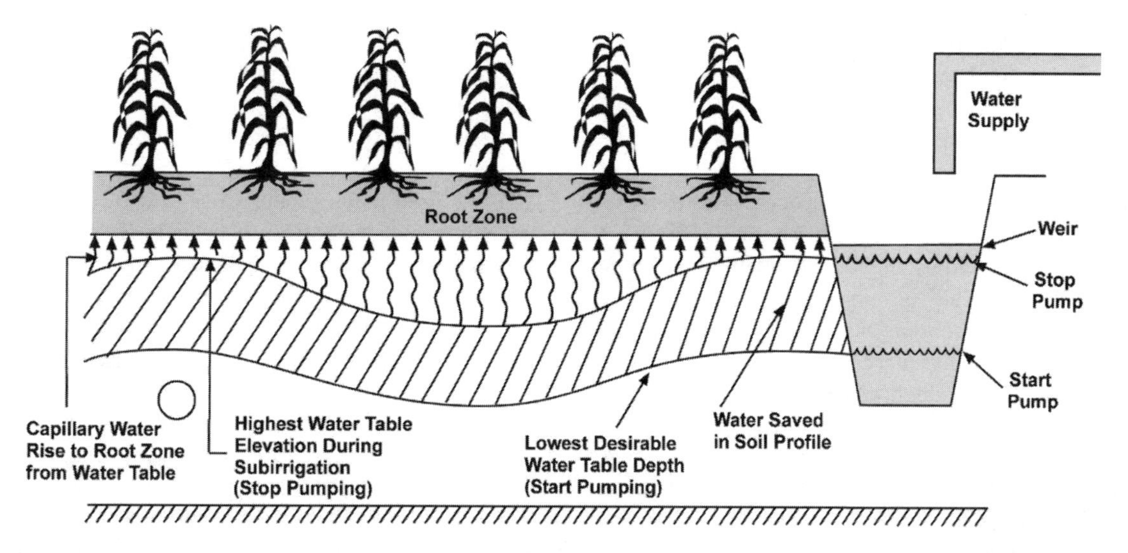

Fig. 6 Cyclic method of water table control during subirrigation. Subirrigation can be practiced with a subsurface drain tube system (represented by the circle on the left) or an open ditch system. Pumping occurs until the water table rises to the "stop pump" level, then the water table is allowed to recede due to ET until the "start pump" level is reached. Pumping resumes and the water level again rises to the "stop pump" level.

controlled drainage and subirrigation modes of operation, good surface drainage becomes very important.[24,33]

WATER TABLE MANAGEMENT DESIGN AND APPLICATION

Water table management gives the operator greater control over the drainage system whereby the water table can be adequately lowered below the root zone during wet periods (conventional drainage), raised during dry periods (subirrigation), and maintained during transition (controlled drainage). Proper design and operation of water table management systems have supported crop yield increases of 5%–30% relative to yields under conventional drainage modes of operation.[12,15,34–42]

It has long been recognized[43] that the design of drainage and water table management systems should be tailored to the soils, crops, and climatological conditions that prevail in the area. Over the past 30 yr, researchers have developed improved methods for designing and operating water table management systems.[30,32,44–50] Computer simulation models such as DRAINMOD,[51] SWARTRE,[52] and WATRCOM[53] provide objective methods for relating water management system design to soil properties and climatological conditions. Simplified methods for estimating drain spacings for drainage and subirrigation have also been derived.[24,25,30,48] Management strategies to provide efficient utilization of water[12,31] and improved drainage water quality[15,54–57] have been developed.

There are two approaches to water table management: on-farm field scale systems and watershed or hydrologic unit scale systems. Most water table management installations have occurred at the farm scale[29,58] although a few watershed scale projects have been implemented.[35,59] The advantage of field scale systems is that water table levels within individual fields can be managed to satisfy the unique requirements of the crop in each field. Management decisions become the responsibility of each landowner individually because water level adjustments in one field usually do not affect up or down stream fields. Field scale systems typically require more control structures, which translate into higher system costs and management requirements. Detailed site evaluation procedures and design guidelines for field scale water table management systems have been developed.[25,29,48,50,60] Key parameters to consider include the need for drainage improvement and existing drainage intensity; soil slope, permeability, and location of restrictive soil horizons; and availability of supplemental water for subirrigation.

Controlled drainage at the watershed scale is a promising strategy in those cases where deep drainage outlets have been constructed to provide drainage for the lowest elevations in the watershed.[12,59] With suitable soils and site conditions, Doty and associates showed that watershed scale channel control could conserve soilwater and provide a water supply for irrigation, thus reducing the demands on off-site water supplies. Crop yields were higher during each year of their 4-yr study due to higher stream water levels upstream of the water control structure and due to sprinkler irrigation that utilized water supplied solely from the controlled main channel. As with field scale systems, topography is a critical determinant of the feasibility of watershed level control. Where the watershed slope is relatively flat (channel slope less than 0.2%[56]), the water table can be maintained within a suitable range for several hundred meters along the main channel with relatively few control structures. At greater slopes, system costs become prohibitively high since a greater number of water control structures are required to maintain segments of uniform water level and also because larger-sized, more expensive structures are required for the larger channels.[29]

SUMMARY

The technical feasibility of water table management at both a farm field scale and watershed scale is fairly well documented. Water table management can increase crop yields, reduce overdrainage, reduce the transport of fertilizer nutrients and other potential pollutants, improve water use efficiency, and at the watershed scale, reduce demands on other water sources within the watershed to facilitate irrigation. The magnitude of the benefits varies among fields and watersheds as well as from year to year. The success of water table management at any scale is influenced by soils, crops, topography, seasonal rainfall, hydraulic properties within the controlled area, and overall management of the system. While research over the past 30 yr has lead to significant improvements in design and operational methods, there still remains a need to improve and fine tune management strategies to optimize the net benefits of these practices, especially for watershed scale systems. Apart from the technical challenge underlying the management of the outlet water level to satisfy multiple objectives, there are several legal and institutional barriers that must also be addressed. In the United States, drainage district laws in some states prohibit the placement of structures that retard flow in drainage district channels. Where multiple landowners are involved, questions arise regarding who has the responsibility and authority to establish water control levels in the channel and who regulates individual landowner rights to withdraw water. For multiple purpose channels, there is concern about the impact of drainage control on downstream users and the impact of water control structures on the migration

of anadromous fish. Until these issues are resolved, there continues to be a need for research and development efforts that evaluate and enhance the sustainability of water table management systems.

REFERENCES

1. Smedema, L.K.; Abdel-Dayem, S.; Ochs, W.J. Drainage and Agricultural Development. Irrig. Drain. Syst. **2000**, *14*, 223–235.
2. Reeve, R.C.; Fausey, N.R. Drainage and Timeliness of Farming Operations. In *Drainage for Agriculture*; van Schilfgaarde, J., Ed.; Agron. Monogr. 17; ASA: Madison, WI, 1974; 55–66.
3. Fausey, N.R.; Doering, E.J.; Palmer, M.L. Purposes and Benefits of Drainage. In *Farm Drainage in the United States: History, Status, and Prospects*; Pavelis, G.A., Ed.; Economics Research Service, U.S. Department of Agriculture. Misc. Publication No. 1455, 1987; 170.
4. Evans, R.O.; Fausey, N.R. Effects of Inadequate Drainage on Crop Growth and Yield. In *Agricultural Drainage*; Skaggs, R.W., van Schilfgaarde, J., Eds.; Agron. Monogr. 38; ASA, CSSA and SSSA: Madison, WI, 1999; 13–54.
5. Evans, R.O. Drainage and Anaerobiosis. In *Encyclopedia of Soil Science*; Lal, R., Ed.; Marcel Dekker, Inc.: New York, 2002; 364–369.
6. Carter, C.E. Surface Drainage. In *Agricultural Drainage*; Skaggs, R.W., van Schilfgaarde, J., Eds.; Agron. Monogr. 38; ASA, CSSA and SSSA: Madison, WI, 1999; 1023–1050.
7. Schwab, G.O.; Fouss, J.L. Drainage Materials. In *Agricultural Drainage*; Skaggs, R.W., van Schilfgaarde, J., Eds.; Agron. Monogr. 38; ASA, CSSA and SSSA: Madison, WI, 1999; 911–926.
8. Broughton, R.S.; Fouss, J.L. Subsurface Drainage Installation Machinery and Methods. In *Agricultural Drainage*; Skaggs, R.W., van Schilfgaarde, J., Eds.; Agron. Monogr. 38; ASA, CSSA and SSSA: Madison, WI, 1999; 967–1004.
9. Spoor, G.; Leeds-Harrison, P. Nature of Heavy Soils and Potential Drainage Problems. In *Agricultural Drainage*; Skaggs, R.W., van Schilfgaarde, J., Eds.; Agron. Monogr. 38; ASA, CSSA and SSSA: Madison, WI, 1999; 1051–1182.
10. Soil Conservation Service (SCS). Subsurface Drainage. *National Engineers Handbook*; USDA Soil Conservation Service, 1971; Chap. 4, 4–43.
11. Seymour, R.M.; Skaggs, R.W.; Evans, R.O. Corn Yield Response to Plant Date. Trans. ASAE **1992**, *35* (3), 865–870.
12. Doty, C.W.; Parsons, J.E.; Tabrizi, A.N.; Skaggs, R.W.; Badr, A.W. Deep Ditch Overdrainage Affects Water Table Depth and Crop Yield. *Proceedings of the Specialty Conference, Environmentally Sound Water and Soil Management*; ASCE, 1982; 113–121.
13. Paerl, H.W. Dynamics of Blue–Green Algal (*Microcystis aeruginosa*) Blooms in the Lower Neuse River, North Carolina: Causative Factors and Potential Controls, Report No. 229; North Carolina Water Research Institute: Raleigh, NC, 1987; 164.
14. Evans, R.O.; Skaggs, R.W.; Gilliam, J.W. Controlled Versus Conventional Drainage Effects on Water Quality. J. Irrig. Drain. Eng. ASCE **1995**, *121* (4), 271–276.
15. Zucker, L.A., Brown, L.C., Eds. *Agricultural Drainage: Water Quality Impacts and Subsurface Drainage Studies in the Midwest*; Ohio State University Extension Bulletin 871; The Ohio State University: Columbus, OH, 1998; 40.
16. Evans, R.O. Controlled Drainage. In *Encyclopedia of Water Science*; Stewart, B.A., Howell, T., Eds.; Marcel Dekker, Inc.: New York, 2003; *in press*.
17. Stevens, J.C. Drainage of Peat and Muck Lands. *Water: The 1955 Yearbook of Agriculture*; U.S. Gov. Print. Office: Washington, DC, 1955; 539–557.
18. Skaggs, R.W. Water Table Management: Subirrigation and Controlled Drainage. In *Agricultural Drainage*; Skaggs, R.W., van Schilfgaarde, J., Eds.; Agron. Monogr. 38; ASA, CSSA and SSSA: Madison, WI, 1999; 695–718.
19. Skaggs, R.W.; Kriz, G.J.; Bernal, R. Irrigation Through Subsurface Drains. J. Irrig. Drain. Eng. ASCE **1972**, *98* (IR2), 363–373.
20. Spencer, A.P. Subirrigation. Fla Agric. Ext. Bull. 99 **1938**.
21. Clinton, F.M. Invisible Irrigation on Egin Bench. Reclam. Era **1948**, *34*, 182–184.
22. Skaggs, R.W. Water Table Movement During Subirrigation. Trans. ASAE **1973**, *16* (5), 988–993.
23. Doty, C.W.; Currin, T.S.; McLin, R.E. Controlled Subsurface Drainage for Coastal Plain Soils. J. Soil Water Conserv. **1975**, *30*, 82–84.
24. Skaggs, R.W.; Tabrizi, A.N. Design Drainage Rates for Estimating Drain Spacing in North Carolina. Trans. ASAE **1986**, *29* (6), 1631–1640.
25. Doty, C.W.; Evans, R.O.; Gibson, H.J.; Hinson, R.D.; Williams, W.B. *Agricultural Water Table Management: A Guide for Eastern North Carolina*; N.C. Agricultural Extension Service, N.C. Agricultural Research Service, USDA Agricultural Research Service, and USDA Soil Conservation Service: Raleigh, 1986; 205.
26. Purvis, A.C.; Williamson, R.E. Effects of Flooding and Gaseous Composition of the Root Environment on Growth of Corn. Agron. J. **1972**, *64*, 674–678.
27. Jackson, M.B. Rapid Injury to Peas by Soil Waterlogging. J. Sci. Food Agric. **1979**, *30*, 143–152.
28. Visser, W.C. De Landbouwwaterhuishouding in Nederland. Comm. Onderz. Landb. Waterhuish. Ned. TNO. Rapport nr. 1. 231 (as cited by Wesseling, J. Crop Growth and Wet Soils. In *Drainage for Agriculture*; van Schilfgaarde, J., Ed.; ASA Monograph No. 17; Madison, WI, 1974; Chap. 2.); 1958.
29. Evans, R.O.; Parsons, J.E.; Stone, K.; Wells, W.B. Water Table Management on a Watershed Scale. J. Soil Water Conserv. **1992**, *47* (1), 58–64.
30. Belcher, H.W.; Merva, G.E.; Shayya, W.H. SI-DESIGN—A Simulation Model to Assist with the Design of Subirrigation Systems. In *Water Management in the Next Century*. Trans. Workshop Subsurface Drainage Models,

15th Int. Congress ICID, The Hague; Lorre, E., Ed.; CEMAGREF-DICOVA: Antony, France, 1993; 295–308.
31. Evans, R.O.; Skaggs, R.W. *Operating Controlled Drainage and Subirrigation Systems*; North Carolina Agricultural Extension Service Bulletin AG-356: Raleigh, NC, 1985; 10.
32. Fouss, J.L.; Evans, R.O.; Thomas, D.L.; Belcher, H.W. Operation of Controlled Drainage and Subirrigation Facilities for Water Table Management. In *Agricultural Drainage*; Skaggs, R.W., van Schilfgaarde, J., Eds.; Agron. Monogr. 38; ASA, CSSA and SSSA: Madison, WI, 1999; 743–763.
33. Skaggs, R.W. Effect of Surface Drainage on Water Table Response to Rainfall. Trans. ASAE **1974**, *17*, 406–411.
34. Carter, C.E.; Fouss, J.L.; McDaniel, V. Water Management Increases Sugarcane Yields. Trans. ASAE **1988**, *31*, 503–507.
35. Parsons, J.E.; Evans, R.O.; et al. Stream Water Level Control for Irrigation Supplies. In *Managment of Farm Irrigation Systems*; Hoffman, G.L. et al., Ed.; ASAE: St. Joseph, MI, 1990; 971–981.
36. Cooper, R.L.; Fausey, N.R.; Streeter, J.G. Yield Potential of Soybean Under a Subirrigation/Drainage Water Management System. Agron. J. **1991**, *83*, 884–887.
37. Busscher, W.J.; Sadler, E.J.; Wright, F.S. Soil and Crop Management aspects of Water Table Control Practices. J. Soil Water Conserv. **1992**, *47*, 71–74.
38. Camp, C.R.; Thomas, W.M.; Doty, C.W. Drainage and Irrigation Effects on Cotton Production. Trans. ASAE **1994**, *37* (3), 823–830.
39. Carter, C.E.; Camp, C.R. Drain Spacing Effects on Water Table Control and Cane Sugar Yields. Trans. ASAE **1994**, *37* (5), 1509–1513.
40. Broughton, R.S. Economic Production and Environmental Impacts of Subirrigation and Controlled Drainage. In *Subirrigation and Controlled Drainage*; Belcher, H.W., D'Intri, F.M., Eds.; Lewis Publishing: Boca Raton, FL, 1995; 183–191.
41. Fausey, N.R.; Cooper, R.L. Subirrigation Response of Soybeans Grown with High Yield Potential Management. In *Subirrigation and Controlled Drainage*; Belcher, H.W., D'Intri, F.M., Eds.; Lewis Publishing: Boca Raton, FL, 1995; 225–230.
42. Madramootoo, C.A.; Broughton, S.R.; Dodds, G.T. Water Table Management for Soybean Production on a Sandy Loam Soil. Can. Agric. Eng. **1995**, *37*, 1–7.
43. van Schilfgaarde, J. Transient Design of Drainage Systems. J. Irrig. Drain. ASCE **1965**, *91* (IR3), 9–22.
44. Skaggs, R.W. Evaluation of Drainage-Water Table Control Systems Using a Water Management Model. In Proceedings 3rd National Drainage Symposium, Chicago, IL, Hoffman, G.J. Ed.; ASAE, 1976; 61–88.
45. Skaggs, R.W. Water Movement Factors Important to the Design and Operation of Subirrigation Systems. Trans. ASAE **1981**, *24*, 1553–1561.
46. Smith, M.C.; Skaggs, R.W.; Parsons, J.E. Subirrigation System Control for Water Use Efficiency. Trans. ASAE **1985**, *28*, 489–496.
47. Fouss, J.L. Simulated Feedback-Operation of Controlled-Drainage/Subirrigation Systems. Trans. ASAE **1985**, *28*, 839–847.
48. Evans, R.O.; Skaggs, R.W. Design Guidelines for Water Table Management Systems on Coastal Plain Soils. Appl. Eng. Agric. ASAE **1989**, *5* (4), 539–548.
49. Fouss, J.L.; Rogers, J.S. Drain Outlet Water Level Control: A Simulation Model. In Proceedings of the 6th International Drainage Symposium, Nashville, TN, Fouss, J.L., Ed.; ASAE, 1992; 46–61.
50. Fouss, J.L.; Evans, R.O.; Belcher, H.W. Design of Controlled Drainage and Subirrigation Facilities for Water Table Management. In *Agricultural Drainage*; Skaggs, R.W., van Schilfgaarde, J., Eds.; Agron. Monogr. 38; ASA, CSSA and SSSA: Madison, WI, 1999; 719–742.
51. Skaggs, R.W. A Water Management Model for Shallow Water Table Soils. Technical Report No. 134; Water Resources Research Institute of the University of North Carolina: Raleigh, NC, 1978; 178.
52. Feddes, R.A.; Kowalik, P.J.; Zaradny, H. Simulation of Field Water Use and Crop Yield. *Simulation Monographs*; PUDOC: Wageningen, The Netherlands, 1978; 189.
53. Parsons, J.E.; Skaggs, R.W.; Doty, C.W. Development and Testing of a Water Management Simulation Model (WATRCOM): Development. Trans. ASAE **1991**, *34*, 120–128.
54. Gilliam, J.W.; Skaggs, R.W. Use of Drainage Control to Minimize Potential Detrimental Effects of Improved Drainage Systems. In: Proceedings of the Specialty Conference, Development and Management Aspects of Irrigation and Drainage Systems. J. Irrig. Drain. Div. ASCE **1985**, 352–362.
55. Evans, R.O.; Gilliam, J.W.; Skaggs, R.W. *Controlled Drainage Management Guidelines for Improving Drainage Water Quality*; N.C. Cooperative Extension Service Bulletin AG-443: Raleigh, NC, 1991; 15.
56. Gilliam, J.W.; Osmond, D.L.; Evans, R.O. *Selected Agricultural Best Management Practices to Control Nitrogen in the Neuse River basin*; North Carolina Agricultural Research Service Technical Bulletin 311; North Carolina State University: Raleigh, NC, 1997; 53.
57. Gilliam, J.W.; Baker, J.L.; Reddy, K.R. Water Quality Effects of Drainge in Humid Regions. In *Agricultural Drainage*; Skaggs, R.W., van Schilfgaarde, J., Eds.; Agron. Monogr. 38; ASA, CSSA and SSSA: Madison, WI, 1999; 801–830.
58. Shirmohammadi, A.; Camp, C.R.; Thomas, D.L. Water Table Management for Field-Sized Areas in the Atlantic Coastal Plain. J. Soil Water Conserv. **1992**, *47* (1), 52–57.
59. Doty, C.W.; Parsons, J.E.; Badr, A.W.; Tabrizi, A.N.; Skaggs, R.W. Water Table Control for Water Resource Projects on Sandy Soils. J. Soil Water Conserv. **1985**, *40* (4), 360–364.
60. Hancor, Inc. *Subsurface Irrigation and Drainage Systems Contractor Design Manual*; Hancor, INC.: Findlay, OH, 1985.

Water Vapor Properties

Graham Thorpe
Victoria University of Technology, Melbourne, Victoria, Australia

INTRODUCTION

The textures and shelf lives of foods are profoundly affected by their moisture contents. For example, if cookies are too moist they cease to crumble, and bread may become so dry that it is inedible. When certain foods such as corn and peanuts are too moist, they are likely to become infested with molds. These molds can give rise to mycotoxins that are poisonous to humans and animals in concentrations of parts per billion, and they are responsible for liver cancer being endemic in some countries.

The moisture contents of foods and the vapor pressures of moisture vapor in equilibrium with them are intimately related. The properties of water vapor are important because they can be used to infer the moisture content of foods, and they can be used to estimate how quickly foods dry or become moist. Knowledge of the thermal properties of water vapor can be used to calculate how much energy must be used to dry foods. This is essential in many of the calculations used in food-processing operations.

SATURATION VAPOR PRESSURE

Water consists of molecules that are composed of hydrogen and oxygen atoms. The molecules are in constant motion, and there are attractive forces between them. If a quantity of liquid water in a closed container has a free surface that is in contact with air say, some of the molecules have so much kinetic energy that they can escape the attractive pull of other water molecules. As a result, they escape into the vapor region in which the intermolecular attractive forces are negligibly small. A dynamic equilibrium is established between the liquid water and its vapor when the number of molecules leaving the liquid in a given time equals the number that enters the liquid. At this point of equilibrium, the pressure that the water molecules exert on the walls of the container and the liquid water is known as the saturation vapor pressure. In the case of a plane or flat surface of water, this pressure is a function only of temperature. One of the most accurate expressions that relate the saturation vapor pressure, p_s (Pa), and temperature, T (°C), is the Goff–Gratch[1] equation, namely

$$p_s = e_{st} 10^z$$

in which

$$z = a\left(\frac{T_s}{T + 273.15} - 1\right) + b\log_{10}\left(\frac{T_s}{T + 273.15}\right) + c\left(10^{d\left(1 - \frac{T+273.15}{T_s}\right)} - 1\right) + f\left(10^{h\left(\frac{T_s}{T+273.15}\right)} - 1\right)$$

where $a = -7.90298$; $b = 5.02808$; $c = -1.3816 \times 10^{-7}$; $d = 11.344$; $f = 8.1328 \times 10^{-3}$; $g = -3.49149$; $e_{st} = 101,324.6$; and $T_s = 373.16$.

A simpler expression for the relationship between the vapor pressure of water and temperature has been proffered by Hunter,[2] namely

$$p_s = \frac{6 \times 10^{25}}{(T + 273.15)^5} \exp\left(\frac{-6800}{T + 273.15}\right)$$

which is accurate to within $\pm$ 0.3% in the range $0°C \leq T \leq 60°C$.

SPECIFIC HEATS

The two principal specific heats of a material are those that are obtained when the material is held at either constant pressure or at constant volume during a heating or cooling process. In the case of water vapor, the former always has a higher value than the latter because when water vapor expands against its surroundings at a constant pressure it must perform mechanical work. Energy is required to perform this work. The specific heat of water vapor, $C_p \, \mathrm{kJ\,kg^{-1}\,K^{-1}}$, is a function of absolute temperature, T_{abs}, and it can be calculated using the expression adapted from that given by Çengel and Boles,[3] i.e.,

$$C_p = a + bT_{abs} + cT_{abs}^2 + dT_{abs}^3$$

in which $a = 1.7896$, $b = 0.1067 \times 10^{-3}$, $c = 0.5856 \times 10^{-6}$, and $d = -0.1995 \times 10^{-9}$. This equation has an average error of 0.24% and a maximum error of 0.53% in the temperature range $273\mathrm{K} \leq T_{abs} \leq 1800\mathrm{K}$. Under the conditions normally encountered when processing food

Encyclopedia of Agricultural, Food, and Biological Engineering
DOI: 10.1081/E-EAFE 120007049
Copyright © 2003 by Marcel Dekker, Inc. All rights reserved.

and agricultural produce, the specific heats at constant pressure, C_p, and constant volume, C_v, are related by the expression derived for ideal gases

$$C_p = C_v + R$$

in which R is the specific gas constant of water vapor, $0.461\,\mathrm{kJ\,kg^{-1}\,{}^\circ C^{-1}}$. The ratio of specific heat at constant pressure, C_p, to the specific gas constant, R, is given[4] by the equation

$$\frac{C_p}{R} = f(T_{abs})$$

where

$$f(T_{abs}) = a + bT_{abs} + cT_{abs}^2 + dT_{abs}^3 + eT_{abs}^4$$

in which $a = 4.070$, $b = -1.108 \times 10^{-3}$, $c = 4.152 \times 10^{-6}$, $d = -2.964 \times 10^{-9}$, $e = 0.807 \times 10^{-12}$, and T_{abs} is the absolute temperature. This equation is accurate in the range $300\mathrm{K} \leq T_{abs} \leq 1000\mathrm{K}$. It is now possible to calculate the specific heat at constant volume by means of the equation

$$C_v = R\big(f(T_{abs}) - 1\big)$$

LATENT HEAT OF VAPORIZATION

The latent heat of vaporization, $h_v\,\mathrm{kJ\,kg^{-1}}$, of water is expressed by the following function of temperature, T °C

$$h_v = 2501.3 - 2.301T - 0.00142T^2$$

This expression is accurate to $\pm 0.03\%$ in the range $0°\mathrm{C} \leq T \leq 100°\mathrm{C}$.

SPECIFIC ENTHALPY OF WATER VAPOR

The specific enthalpy, h, of water vapor can be estimated using the above relationships for the specific heat, $C_p\,\mathrm{kJ\,(kg\,K)^{-1}}$, at constant pressure and latent heat of vaporization. Hence, at an absolute temperature T_{abs}K

$$h = 2004.8 + aT_{abs} + (b/2)T_{abs}^2 + (c/3)T_{abs}^3 + (d/4)T_{abs}^4$$

in which the constants a, b, c, and d are those used in the expression for specific heat, C_p, at constant pressure given above.

THERMAL CONDUCTIVITY

The thermal conductivity of moisture vapor is a function of temperature and pressure. At a pressure of 100 kPa, approximately one standard atmosphere, Sengers and Watson[5] report the following values of the thermal conductivity, $k\,\mathrm{mW\,(m\,K)^{-1}}$, of water vapor:

Temperature (K)	300	400	500	600
Thermal conductivity ($\mathrm{mW\,m^{-1}\,K^{-1}}$)	18.7	27.1	35.7	47.1

A discussion of the variation of the thermal conductivity of gases with temperature and pressure is given by Bird, Stewart, and Lightfoot.[6]

VISCOSITY

Like thermal conductivity, the viscosity of water vapor is a function of temperature and pressure. At a pressure of 100 kPa, Sengers and Watson[5] report the following values of the viscosity, $\mu\,\mathrm{Pa\,s}$, of water vapor:

Temperature (K)	300	400	500	600
Viscosity ($\mathrm{Pa\,s} \times 10^6$)	10.0	13.3	17.3	21.4

Bird, Stewart, and Lightfoot[7] present methods of calculating the viscosities of gases under a range of temperatures and pressures.

MASS DIFFUSIVITY

In many foods and agricultural products, moisture diffuses along water vapor concentration gradients. For example, it has been shown[8,9] that moisture migration through bulks of cereal grains, such as wheat, is

Table 1 Values of mass diffusivity of water vapor through air

	Mass diffusivity ($\mathrm{m^2\,sec^{-1}}$) ($\times 10^5$)	
Temperature (K)	**Kusuda**	**Marrero and Mason**
293.15	2.49	2.42
373.15	3.96	3.99
473.15	6.17	6.38
573.15	8.76	8.73
673.15	11.65	11.35

(From Refs. 10 and 11.)

Table 2 The permittivity of water vapor at a pressure of 1 atm

Permittivity	Temperature (°C)	Permittivity	Temperature (°C)
1.00007	0	1.00144	60
1.00012	10	1.00213	70
1.00022	20	1.00305	80
1.00037	30	1.00428	90
1.00060	40	1.00587	100
1.00095	50		

controlled by the rate at which moisture vapor diffuses through the intergranular pores between the grain kernels. The moisture does not diffuse along a straight path but it primarily follows the tortuosities of the intergranular spaces. In this case and in many others that involve the estimation of mass transfer in the drying of foods, it is essential to know the mass diffusivity, D_v m^2 sec^{-1}, of moisture vapor through air. Kusuda[10] presents the following relationship between the diffusivity of moisture vapor in air and absolute temperature, T_{abs} K

$$D_v = 9.1 \times 10^{-9} T^{2.5} / (T + 245.18)$$

Values of the mass diffusivities of water vapor through air estimated by the above equation and those presented by Ref. 11 are shown in Table 1.

PERMITTIVITY

Permittivities of water vapor at atmospheric pressure and over the temperature range 0–100°C, given by Birnbaum and Chatterjee,[12] are presented in Table 2.

REFERENCES

1. Goff, J.A.; Gratch, S. Low-Pressure Properties of Water from −160F to 212F. Trans Am. Soc. Heat Vent. Eng. **1946**, *52*, 95–121.

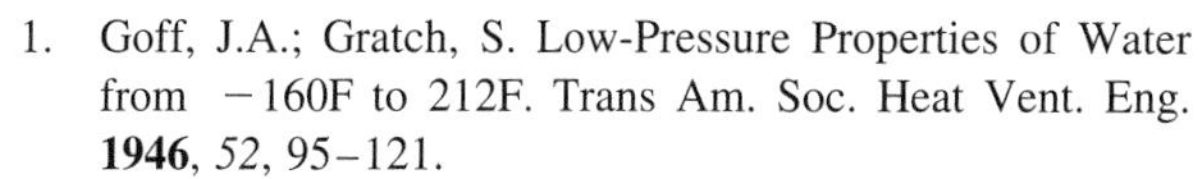

2. Hunter, A.J. An Isostere Equation for Some Common Seeds. J. Agric. Eng. Res. **1987**, *24*, 219–232.
3. Çengel, Y.A.; Boles, M.A. *Thermodynamics—An Engineering Approach*, 3rd Ed.; WCB/McGraw-Hill: Boston, 1998; 901.
4. Wark, K.; Richards, D.E. *Thermodynamics*, 6th Ed.; WCB/McGraw-Hill: Boston, 1999; 1031.
5. Sengers, J.V.; Watson, J.T.R. Improved International Formulations for the Viscosity and Thermal Conductivity of Water Substance. J. Phys. Chem. Ref. Data **1986**, *15*, 1291–1314.
6. Bird, R.B.; Stewart, W.E.; Lightfoot, E.N. Thermal Conductivity and the Mechanism of Energy Transport. In *Transport Phenomena*; John Wiley and Sons: New York; Chap. 8.
7. Bird, R.B.; Stewart, W.E.; Lightfoot, E.N. Viscosity and the Mechanism of Momentum Transport. In *Transport Phenomena*; John Wiley and Sons: New York; Chap. 1.
8. Thorpe, G.R. Moisture Diffusion Through Bulk Grain. J. Stored Prod. Res. **1981**, *17*, 39–42.
9. Thorpe, G.R.; Ochoa, J.A.; Whitaker, S. The Diffusion of Moisture in Food Grains. I. The Development of a Mass Transfer Equation. J. Stored Prod. Res. **1991**, *27*, 1–9.
10. Kusuda, T. Calculation of the Temperature of a Flat-Plate Wet Surface Under Adiabatic Conditions with Respect to the Lewis Relation in Humidity and Moisture. In *Measurement and Control in Science and Industry*; Wexler, A., Ed.; Reinhold: New York, 1985; Vol. 1.
11. Marrero, T.R.; Mason, E.A. Gaseous Diffusion Coefficients. J. Phys. Chem. Ref. Data **1972**, *1*, 3–118.
12. Birnbaum, G.; Chatterjee, S.K. The Dielectric Constant of Water in the Microwave Region. J. Appl. Phys. **1952**, *23*, 220–223.

Wet Bulb Temperature

Curtis L. Weller
Alejandro Amézquita
University of Nebraska–Lincoln, Lincoln, Nebraska, U.S.A.

INTRODUCTION

In order to understand the concept of wet bulb temperature for air–water vapor mixtures, it is necessary to understand two processes for saturating any gas with water. One of the processes can be visualized through use of the following experiment. In a well-insulated chamber (Fig. 1), an entering gas contacts a spray of recirculating liquid water. The gas leaving the chamber is at a higher humidity and lower temperature than the gas that enters. The evaporation of water into the gas results in saturation of the gas by converting part of the enthalpy (sensible heat) of the entering gas into latent heat for vaporizing water. Exchange of heat between the gas and the water with no loss across the chamber walls is defined as adiabatic saturation.

The temperature of the water being recirculated in the first process reaches a steady-state temperature called adiabatic saturation temperature (T_{sat}). This T_{sat} is attained when a large amount of water continuously contacts the entering gas in an adiabatic chamber. The enthalpy balance over the chamber (Fig. 1) can be expressed as follows:

$$\frac{W - W_{sat}}{T - T_{sat}} = -\frac{C_s}{\lambda_{as}} = \frac{(1.005 + 1.884W)}{\lambda_{as}} \quad (1)$$

where T and W are gas temperature (°C) and absolute humidity (kg water per kg dry gas) at entering conditions, respectively; W_{sat} is absolute humidity (kg water per kg dry gas) at saturation conditions; C_s is humid heat or specific heat of the entering gas–water vapor mixture ($\mathrm{kJ\,kg^{-1}\,°C^{-1}}$); and λ_{as} is latent heat of vaporization of water ($\mathrm{kJ\,kg^{-1}}$).

In the other process, when a small amount of water is exposed to a continuous stream of entering gas under adiabatic conditions, the water temperature decreases to a steady-state nonequilibrium temperature. As the amount of liquid is small, the temperature and humidity of the gas are not modified, as is the case in the first process, adiabatic saturation. Rather, the cooling effect of evaporating water into gas decreases the temperature of the remaining water to what is known as the wet bulb temperature (T_{wet}). This temperature of the water also describes the wet bulb temperature of the gas.

WET BULB TEMPERATURES

The adiabatic saturation temperature, T_{sat}, is sometimes referred to as the thermodynamic wet bulb temperature whereas the wet bulb temperature, T_{wet}, is known as the psychrometric wet bulb temperature. For air–water vapor mixtures, the difference between T_{sat} and T_{wet} is negligible when temperatures of interest are low enough (<50°C) to form dilute air–water vapor mixtures. For other gas–vapor mixtures, this difference is usually not negligible.

Thermodynamic Wet Bulb Temperature

For air–water vapor mixtures, T_{sat} is defined as the temperature at which water, by evaporating into entering air at a given dry bulb temperature (T_{dry}) and W, adiabatically brings the air to saturation (100% relative humidity) at the same temperature as the water, T_w, with no change in barometric pressure. This can be easily observed by revisiting the experiment described in Fig. 1. During the adiabatic saturation process, heat leaves the air, cooling it and allows water to evaporate into the air. Saturated air leaves the adiabatic chamber at a temperature equal to that of the recirculating water ($T_{sat} = T_w$).

Psychrometric Wet Bulb Temperature

A liquid-filled thermometer with its bulb covered with a wet wick or cloth may be used for the determination of a T_{wet} as depicted in Fig. 2. The wick is exposed to a flowing stream of unsaturated air with T_{dry} and W at high velocity ($> 5\,\mathrm{m\,sec^{-1}}$). If the air is unsaturated, some liquid evaporates from the wick into the air stream as the vapor pressure of the saturated wick is higher than that of the unsaturated air.

The evaporation process requires latent heat that comes from sensible heat within the liquid in the wick and causes the temperature of the covered bulb to decrease. As the temperature of the wick decreases, sensible heat is transferred by convection from the air stream and tends to raise the temperature of the wick. A steady state is reached when the heat flow from the air stream to the wet wick is

Encyclopedia of Agricultural, Food, and Biological Engineering
DOI: 10.1081/E-EAFE 120007051
Copyright © 2003 by Marcel Dekker, Inc. All rights reserved.

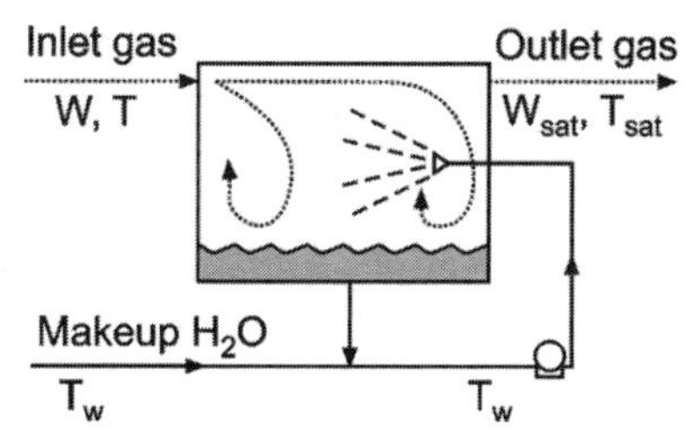

Fig. 1 Adiabatic saturation of a gas in an insulated chamber.

equal to the latent heat of vaporization required to evaporate the liquid from the wick. Therefore, at steady state, the net heat flow to the wick is zero and the temperature of the wick and bulb is constant. This equilibrium temperature is the T_{wet}.

Thermodynamic Vs. Psychrometric Wet Bulb Temperature in Air–Water Vapor Mixtures

The process described above for the determination of T_{wet} is not one of adiabatic saturation. However, this process involves simultaneous heat and mass transfer from the wet bulb and is described by the Lewis relation. The relation states that the ratio of the heat transfer coefficient to the mass transfer coefficient is proportional to the specific heat of the air at constant pressure. For air–water vapor mixtures at low mass transfer rates, this relation approaches unity and is approximated as follows:

$$\frac{h_a}{k_a C_a} = 1 \tag{2}$$

where h_a is heat transfer coefficient at the interface of water and air; k_a is gas-phase mass transfer coefficient of water vapor in air; and C_a is specific heat of air at constant pressure.

Consequently, the agreement between T_{wet} and T_{sat} is a direct result of the nearness of the Lewis relation to unity

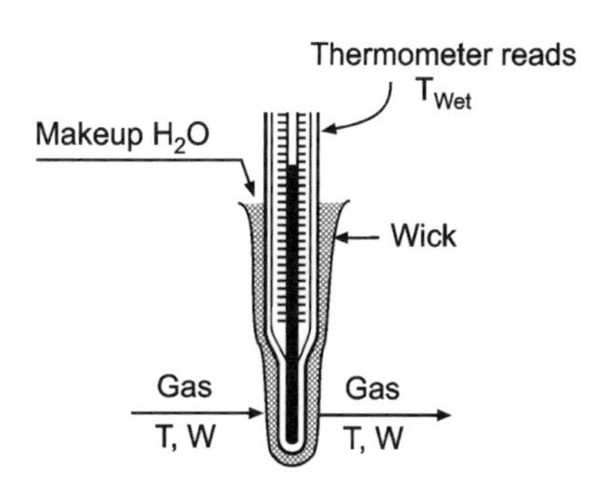

Fig. 2 Determination of wet bulb temperature for a gas.

for air and water vapor. Hence, only small corrections must be applied to T_{wet} readings to obtain T_{sat}.

Heat and Mass Transfer Balances

A heat balance on the wick of Fig. 2 used to determine T_{wet} is as follows:

$$q_t = q_s + N_v \lambda_{wb} + N_a \lambda_a \tag{3}$$

where q_t is total heat transfer flux; q_s is sensible heat transfer flux; N_v is water vapor mass flux; λ_{wb} is latent heat of vaporization for water vapor; N_a is air mass flux; and λ_a is latent heat of vaporization for air. In equilibrium, q_t and N_a are zero, and the quantities q_s and N_v can be expressed as follows:

$$q_s = h_a(T - T_{wet}) \tag{4}$$

$$N_v = kM_a p_a(W - W_{wet}) \tag{5}$$

where h_a and k are convective heat transfer coefficient of air to wick and molar transfer coefficient of the water vapor, respectively; W_{wet} is absolute humidity of saturated air; M_a is molecular weight of air; p_a is partial pressure of the air; and the other terms defined elsewhere. Eq. 3 can now be rearranged and expressed as:

$$\frac{W - W_{wet}}{T - T_{wet}} = -\frac{h_a/kM_a p_a}{\lambda_{wb}} = -\frac{h_a/k_a}{\lambda_{wb}} \tag{6}$$

The ratio h_a/k_a, called the psychrometric ratio, varies from 0.96 to 1.005 for air–water vapor mixtures. Its value is close to the value of C_s in Eq. 1 and C_a in Eq. 2, and assuming λ_{wb} is equal to λ_{as} for the condition when W_{wet} equals W_{sat}, Eq. 6 can then be expressed as:

$$\frac{W - W_{wet}}{T - T_{wet}} = -\frac{C_s}{\lambda_{wb}} \tag{7}$$

Eq. 7 gives the slope of the wet bulb lines in a psychrometric chart. The approximation of Eq. 6 with Eq. 7 is only true for air–water vapor mixtures, where the psychrometric ratio is close to unity. For other gas–vapor systems, the psychrometric ratio varies substantially and Eq. 7 is not useful to approximate Eq. 6.

Measurement of Wet Bulb Temperature

A simple yet effective method for measuring T_{wet} uses a liquid-filled (e.g., mercury or aqueous ethanol) thermometer that has its bulb covered with a wet wick or porous cotton cloth. The thermometer is exposed to a stream of air moving rapidly as illustrated in Fig. 3. If

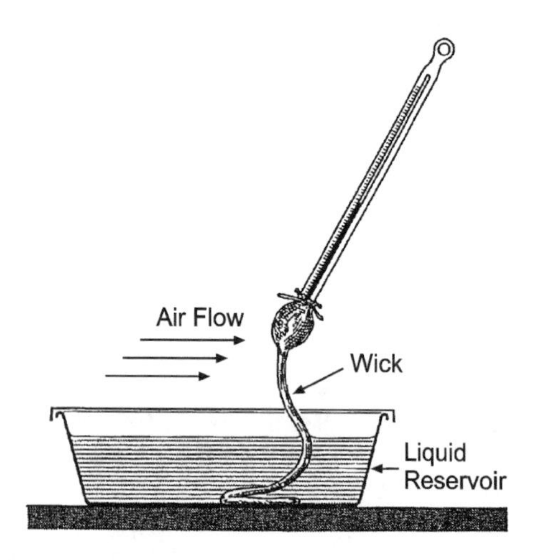

Fig. 3 Sketch of a wet bulb thermometer. (From Bennett and Myers, 1982. Copyright© 1982, 1974, 1962 by McGraw-Hill, Inc.)

the air is unsaturated, the temperature reading will reach a value lower than the T_{dry} of the air.

A second and more practical option is to use a psychrometer, which evaluates both T_{dry} and T_{wet} simultaneously. The psychrometer consists of two identical thermometers. The dry bulb thermometer has its bulb directly exposed to air and senses the temperature of ambient air. The wet bulb thermometer has a close-fitting porous cotton cloth over its bulb that is kept moist by a wick connecting it to a reservoir of distilled water. Unless the air is saturated, water will evaporate from the cloth when exposed to the air stream, which cools the wet bulb and hence registers a lower temperature than the dry bulb thermometer.

The most common and simplest version of this apparatus is called the sling psychrometer (Fig. 4A) in which two thermometers are mounted in a sling that is whirled rapidly by hand over one's head. The wick on the wet bulb thermometer is wetted with distilled water before whirling and remains damp for several minutes. During this time, the psychrometer is whirled for several minutes until at least two consecutive readings of the same value are recorded. Another variation of this apparatus is the Assmann psychrometer (Fig. 4B) in which the air is moved over the thermometer bulbs by forced ventilation. Air velocity must be at least 5 m sec^{-1}.

Variations in other psychrometers are based on using other techniques to measure T_{wet} or to calculate it from dry bulb temperature and relative humidity measurements. Those directly measuring temperature typically use thermistors and thermocouples. The accuracy with which T_{wet} can be determined depends on the accuracy of the instrumentation. For psychrometers calculating T_{wet} from dry bulb and relative humidity measurements, the typical technique for determining relative humidity involves correlating capacitance (thermoset and thermoplastic polymer sensors), resistance (thermoplastic and AlO_3 sensors), or conductivity (ceramic LiCl film sensors) measurements with moisture in air around the sensor.

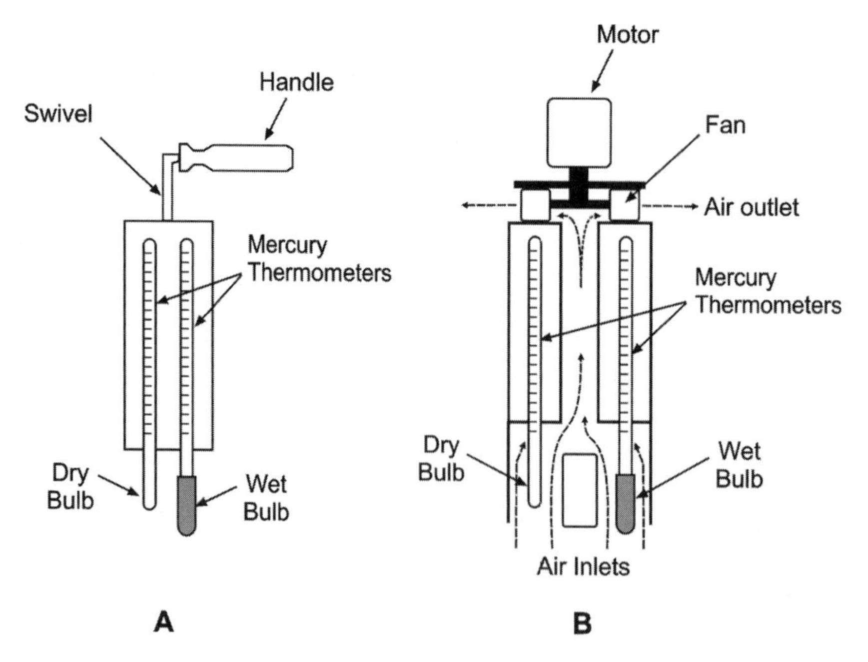

Fig. 4 Schematic diagrams of (A) the sling and (B) Assmann psychrometers. (From White and Ross, 1991. Copyright© 1991 by American Society of Agricultural Engineers.)

CONCLUSION

T_{wet} is one of the properties of an air–water vapor mixture that can be easily determined. It is of importance because T_{wet} and T_{dry} can specify the state of any mixture. State properties such as relative humidity, dew point, enthalpy, and absolute humidity can be readily evaluated from a psychrometric chart or program using the T_{dry} and T_{wet}. Measurement of T_{wet} does not have to be either costly or time consuming.

FURTHER READING

American Society of Heating, Refrigerating and Air-Conditioning Engineers, Inc. *2001 ASHRAE Handbook of Fundamentals*; ASHRAE, Inc.: Atlanta, GA, 2001; 5.9, 6.9–6.10.

Barbosa-Cánovas, G.V.; Vega-Mercado, H. *Dehydration of Foods*; Chapman and Hall: New York, 1996; 9–27.

Bennett, C.O.; Myers, J.E. *Momentum, Heat and Mass Transfer*, 3rd Ed.; McGraw-Hill Book Co.: New York, 1982; 638–642.

Geankoplis, C.J. *Transport Processes and Unit Operations*, 3rd Ed.; Prentice-Hall, Inc.: Englewood Cliffs, NJ, 1993; 525–532.

Himmelblau, D.M. *Basic Principles and Calculations in Chemical Engineering*, 6th Ed.; Prentice-Hall, Inc.: Upper Saddle River, NJ, 1996; 487–501.

Mujumdar, A.S.; Menon, A.S. Drying of Solids: Principles, Classification and Selection of Dryers. In *Handbook of Industrial Drying*, 2nd Ed.; Mujumdar, A.S., Ed.; Marcel Dekker, Inc.: New York, 1995; 1–14.

White, G.M.; Ross, I.J. Humidity. In *Instrumentation and Measurement for Environmental Sciences*, 3rd Ed.; Henry, Z.A., Zoerb, G.C., Birth, G.S., Eds.; American Society of Agricultural Engineers: St. Joseph, MI, 1991; 8-01–8-13.

WLF Equation

M. Erhan Yildiz
Jozef L. Kokini
Rutgers, The State University of New Jersey, New Brunswick, New Jersey, U.S.A.

INTRODUCTION

The WLF equation is an empirical equation that describes the strong temperature dependence of various changes. It is applicable to a temperature range between T_g and $T_g + 100°C$. The WLF equation has found extensive application in synthetic polymer research. There is a growing interest in the application of the WLF equation to materials of biological origin. The temperature dependence of several physical/chemical phenomena that take place during processing and storage has been studied using the WLF equation. Generally, so-called "universal constants" have been used. However, studies on biomaterials have shown that the constants are material specific and should be obtained experimentally. The effects of moisture on the WLF constants have been studied. The WLF equation was modified to describe the effect of moisture on the material specific constants.

WLF EQUATION

The WLF (Williams–Lendel–Ferry) equation[1,2] is an empirical equation that has been used to describe the universal non-Arrhenius effect of temperature on viscous flow,[3–6] viscoelastic behavior,[3,7–13] current conductivity and dielectric properties,[3,14] nuclear magnetic resonance response, dynamic light scattering, and molecular relaxation processes of polymers/copolymers and supercooled liquids.[2,14–16] In addition, lifetime predictions in polymer aging,[17] the effect of cross-linking,[18] composition and plasticizers[5,19] on viscoelastic properties of polymers, the temperature dependence of tracer diffusion coefficients,[4,20–22] and the effect of molecular weight[4–6] on viscosity—based on free volume considerations—were determined using the WLF equation. Several experimental methods including the tensile, dynamic mechanical, creep and stress relaxation experiments, and time–temperature superposition principle have been used to describe the time, temperature, and frequency dependence of material properties with the aid of the WLF equation.[2,4,8,9,23–26]

The general form of the WLF equation is given as[1,2,7]:

$$\log a_T = \left(- \frac{C_1(T - T_s)}{C_2 + (T - T_s)} \right) \tag{1}$$

where a_T is the reduced variables shift factor, C_1 and C_2 are constants, T is the temperature, and T_s is a reference temperature. The form of the above equation is independent of the choice of the reference temperature.[2]

The physical interpretation of the WLF equation can be based on the observation that viscosity is closely connected with molecular mobility, which is related to free volume.[1,7] The knowledge of the free volume–temperature relationship of a polymeric material is useful in understanding free volume-based principles. The volumetric behavior of a supercooled liquid as a function of temperature is shown in Fig. 1. It is generally accepted that the volume of a liquid is composed of occupied volume (V_o) and free volume (V_f).[2,7] The specific volume (V) defined as the total volume per gram, is the sum of occupied volume and free volume. Occupied volume includes not only the van der Waals radii volume of the molecules but also the volume associated with the vibrational motions. Occupied volume, which can be estimated only indirectly, increases as temperature increases with a constant expansion coefficient.[2,23] This indicates that the actual dimensions of the molecules are a weak function of temperature.[27]

Many of the properties of polymeric liquids demonstrate the presence of a substantial portion of free volume. Free volume may be present as holes of the order of molecular (monomeric) dimensions or smaller voids associated with packing irregularities. The free volume per gram, V_f, is a useful semiquantitative concept despite being poorly defined.[2,7]

Relationship Between Free Volume and the WLF Equation

Flow is a form of molecular motion that requires a critical amount of free volume. Doolittle[28] used the free volume dependence of the molecular mobility to express viscosity of an ordinary liquid as a function of free volume.[2,7,25] The work of Doolittle[28] on the viscosity of nonassociated

Encyclopedia of Agricultural, Food, and Biological Engineering
DOI: 10.1081/E-EAFE 120006982
Copyright © 2003 by Marcel Dekker, Inc. All rights reserved.

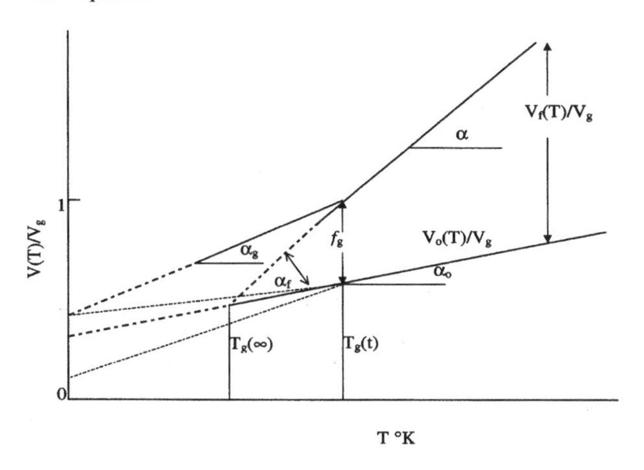

Fig. 1 Volumetric behavior of a supercooled liquid. (Re-drawn from Ref. [2].)

pure liquids such as *n*-alkanes resulted in the following relationship:

$$\ln \eta = \ln A + B\left(\frac{V_0}{V_f}\right) \tag{2}$$

where V_0 is the difference between the specific volume and the free volume, A and B are constants.

Eq. 2 is based on molecular transport in a liquid consisting of hard spheres. It suggests that the rate of molecular rearrangements and transport phenomena, such as diffusion and viscosity, which depend on molecular rearrangements, are dictated by available free volume.[2,7] In the derivation of the WLF equation, the fractional free volume, f, is accepted as the part of the total volume accessible to a kinetic process of interest.[7] Williams, Landel, and Ferry[1] suggested that $f = V_f/V_o$ for small V_f, whereby the Doolittle equation can be written as follows:

$$\ln \eta = \ln A + \left(\frac{B}{f}\right) \tag{3}$$

Fractional free volume, f, is generally assumed[2,7,25] to be a linear function of temperature. The following equation can be used to represent f at any temperature above T_g:

$$f = f_g + \alpha_f(T - T_g) \tag{4}$$

where f_g is the fractional free volume at the glass transition temperature and α_f is the thermal expansion coefficient.

When T_g was used as the reference temperature, after algebraic manipulations one can obtain:

$$\log a_T = -\frac{B}{2.303 f_g}\left(\frac{(T - T_g)}{f_g/\alpha_f + (T - T_g)}\right) \tag{5}$$

Eq. 5 indicates that a shift in the log time scale will produce the same change in molecular motion as the indicated nonlinear change in temperature. It is obvious from Eq. 5 that as the difference between experimental temperature and reference temperature is increased, the polymer will have higher mobility and shorter relaxation times. Finally, the WLF equation with the so-called "universal constants" can be given in the following form[1]:

$$\log a_T = \left(-\frac{C_1^g(T - T_g)}{C_2^g + (T - T_g)}\right) \tag{6}$$

where $C_1^g = 17.44$ and $C_2^g = 51.6$ are universal constants.

The parameters in Eqs. 5 and 6 are related to each other:

$$\frac{B}{2.303 f_g} = C_1^g \quad \text{and} \quad \frac{f_g}{\alpha_f} = C_2^g \tag{7}$$

It is generally suggested that B takes the value of unity.[2,23] As a result, $f_g = 0.025$, i.e., the fractional free volume of many glass forming liquids is 2.5% at T_g and $\alpha_f = 4.8 \times 10^{-4}\ \text{deg}^{-1}$.[2,23,27] Constant universal values indicate that T_g is an iso-free volume state; however, f_g has inherent time-scale dependence on T_g. The thermal expansion coefficient shows variations for different polymers. Certain polymers whose shear viscoelastic behaviors reveal distinct contributions from chain backbone and side chain motions tend to have low α_f values.[2] Others have noted that C_1^g is more constant than C_2^g.[15] The magnitude of C_2^g, and in particular C_2/T_g carries information on the character of the polymer. Angell[15] suggested that C_1^g should be 16 for relaxation times and 17 for the viscosity of systems that obey the WLF equation over the whole possible relaxation time range with a single parameter set.

Different systems can be compared using the glass transition temperature (T_g) as the reference temperature where segmental mobility ceases.[23] However, T_g is a kinetic phenomenon, which can be influenced by traces of impurities, thermal history, and several other factors.[2,7,25] Due to the difficulties associated with obtaining viscoelastic data at the vicinity of T_g, the reference temperature was always selected in the middle of the softening range to avoid complication of selecting T_g.[1,24] An interesting finding was that the two temperatures were related; $T_s - T_g = 50 \pm 4°\text{C}$.[1,24]

When a wide range of polymers was considered, C_1^g values ranged between 15 K and 26 K and C_2^g took values between 20 K and 130 K.[14] The compilation of available WLF constants for synthetic polymers is given by Ngai and Plazek.[14] Actual variation from one polymer to another is too high and therefore, universal values should be used as a last resort when data is not available.[2] This statement has been repeated as a word of caution by several researchers including the original authors.[1,2] Same concern specifically focusing on biological systems has been repeated and researchers were cautioned about the difference between Eqs. 1 and 6.[29] Eq. 6 with fixed universal coefficients can result in erroneous results. There have been fundamental studies that clearly showed the inapplicability of the WLF equation to biological systems with universal constants.[30–32] Several other researchers showed that when the WLF equation is applied to various processes such as the temperature dependence of non-enzymatic browning rates, the best fit of the data yielded constants different from the universal values.[33,34]

This significant difference observed in biological materials can be attributed to the complexity of the biosystems. Biopolymers are much more complex than synthetic polymers. They have large polydispersities, many different chemical groups, including 20 amino acids for proteins, various carbohydrate units, different charges on the polymer surfaces, large degrees of branching, and branches of different sizes. Therefore, C_1 and C_2 should be treated as material-specific coefficients rather than universal values.

Time–Temperature Superposition Principle

Time–temperature superposition principle indicates that for viscoelastic materials, time (or frequency) and temperature are equivalent to the extent that data at one temperature can be superimposed upon data at another temperature by shifting the curves along the log time axis.[2,7,25]

In the transition zone from glass to rubber, the dependence of viscoelastic functions on the time, temperature, and frequency is the strongest.[2] In this zone, viscoelastic spectra will be dictated by the retardation and relaxation times. The retardation and relaxation times strongly depend on temperature. Higher relaxation times indicate no configurational changes taking place within the period of deformation, whereas lower relaxation times indicate that all configurational modes of motion within entanglement coupling points can freely occur.

The analysis of viscoelastic functions as affected by time and temperature was carried out using the reduced variables.[2] Reduced variables help to separate the two principle variables of time and temperature on which the viscoelastic properties depend. Thus, we can express the properties in terms of single functions (time and temperature) whose form can be determined experimentally. The reduced variables concept is based on the time–temperature superposition principle.[2]

Using the time–temperature superposition principle, time and frequency dependent modulus and compliance functions can be shifted to obtain unique reduced curves with an extended time or frequency range as shown in Fig. 2.[25] Extended time dependence of modulus at extremely long times given by the master curve may be physically questionable, yet reduced variables and related functions are very important in predicting changes that occur in small and long temperature/times that are experimentally inaccessible.[2]

Horizontal shifting of the curves compensates for a change in the time scale brought about by changing temperature.[25] This is illustrated as follows:

$$G(T_1, t) = G(T_2, t/a_T) \tag{8}$$

where G represents the viscoelastic property under consideration (e.g., modulus).

Modulus is defined per unit cross-sectional area. As the volume of polymer is a function of temperature, when volume changes, the modulus will vary with the amount of matter contained in unit value. By using density, we can make the corresponding correction to account for the mass change per unit volume due to the temperature change[14,23,25]:

$$G(T_1, t)/\rho(T_1)T_1 = G(T_2, t/a_T)/\rho(T_2)T_2 \tag{9}$$

Division by temperature corrects the changes due to the temperature dependence of the modulus, while division by density corrects the changing number of chains per unit volume due to temperature change. However, due to slow temperature variation, this term is generally ignored.[2,14]

Applications of the WLF Equation

The WLF equation describes the kinetic nature of the glass transition and has been shown to be applicable to any glass-forming polymer, oligomer, or monomer. Different approaches used to obtain the WLF constants are summarized by Ferry.[2] WLF parameters, as is usually

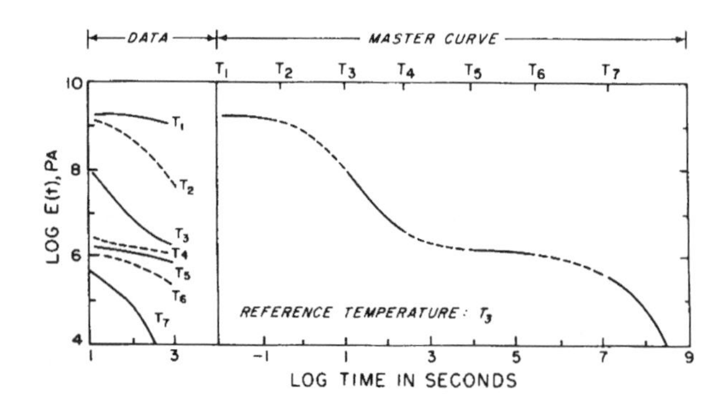

Fig. 2 Stress relaxation master curve obtained by horizontal shifting of stress relaxation curves (on the left) at different temperatures. (Adapted from Ref. [25].)

done in food literature, could also be obtained fitting the equation to available data (usually reaction rates).[35]

The WLF equation applies to the temperature range between T_g and $T_g + 100°C$ for amorphous polymers and between T_g and T_m for partially crystalline polymers.[2,7,14] However, there is evidence indicating different temperature dependencies for different modes of chain motions. Factors such as crystallinity, nonhomogeneity, and temperature induced interactions may also change the temperature dependence. Therefore, it is important to know from which region C_1 and C_2 were determined.[14,36]

The power of the WLF lies in its generality. In the original development, no particular chemical structure was assumed other than a linear amorphous polymer above T_g. However, there are certain challenges in the application of the WLF equation to systems with certain characteristics. The applicability of the WLF equation to polymer blends is one of the complex fields. Amorphous polymers and their random copolymers exhibit a single T_g. Phase separated blends possess multiple and relatively sharp transitions for each phase.[10] When there is increased molecular mixing or compatibility, then the T_g of each component can shift, broaden, or even merge to a certain extend depending on both the compatibility and the relative component proportions.[10] Existence of multiple T_g and/or broad glass transition range complicates the application of the WLF equation to such systems.

The time–temperature equivalence principle has reported to fail for multiphase blends.[2] However, for phase-separated systems, the WLF shift factors have been shown to hold around the T_g of each component.[10] The WLF equation has been used for several miscible blends.[14] However, its success for miscible polymers is yet to be clarified due to variable reports.[10] Akay and Rollins[10] studied the glass transition behavior of interpenetrating networks (IPNs) of polyurethane and poly(methyl methacrylate). The viscoelastic properties were studied using the WLF equation. The reference temperatures were selected to provide best fits to the experimental shift factors. They reported much larger fractional free volumes for IPNS than the individual components. The WLF equation fitted the data reasonably well over wide glass transition regions of the IPNs.

Application of the WLF Equation to Biological Systems

The WLF equation is a useful tool in the prediction of temperature-induced physical/chemical changes in foods as a function of specific processing and storage conditions.

The changes in the viscoelastic properties of food polymers as a function of time and temperature above the T_g have significant effects on the functional properties of foods. Several physical and chemical changes and deteriorative processes in foods are molecular mobility and diffusion-controlled and they are related to the T_g. For example, the changes in textural characteristics and physicochemical properties of foods around the T_g, which for many cereal foods is near their typical storage temperature, can be predicted if WLF constants are known.[37] Predicting changes in viscoelastic properties and designing foods with well-defined, reproducible properties can be related to the behavior and acceptability of food materials.

The WLF equation utilizes a specific reference temperature, and the ability to reduce the experimental values to a reference temperature such as T_g has great potential because it enables a fundamental comparison that uses material specific temperature. For a low moisture systems T_g and for frozen systems ($W > W'_g$) T'_g could be selected as a reference temperature.[1,9,37–39] However, the selection of a reference temperature in the softening zone may give better results.[1,2]

Readers should be aware that there is a disagreement between scientists on assigning a single temperature to the glass transition. It is important to note that several groups including ours are using T_g to represent the glass transition with the understanding that glass transition takes place over a temperature range that varies from system to system.

The WLF equation has been successfully applied to a variety of food systems from simple sugars to complex systems that are heterogeneous mixtures including multi-domains of linear/branched structures. The application of the WLF equation to food systems has led to an ability to describe the temperature dependence of a number of phenomena, including viscosity of concentrated solutions of mixed sugars,[40,41] amorphous solids,[38] supercooled fructose and glucose,[42] amorphous isomalt[43] and elongational viscosity of processed cheese[44]; perceived iciness of ice cream and frozen desserts[37,39]; the collapse of dried fish hydrolizate[45,46] and sugar glasses[47,48]; the crystallization and crystallization times of amorphous carbohydrates above their T_g[49–51] and the effect of collapse and crystallization on retention and release of encapsulated oils[52,53]; the rates of nonenzymatic browning reactions in skim milk, various different model food systems and dehydrated vegetables[33–35,54–58]; (re)crystallization[37,59] and enzyme hydrolysis in frozen foods[60] and ascorbic acid degradation in frozen starch hydrolyzates[61]; thermal inactivation times of bacterial spores[62] and lag times of psychrotrophic bacteria[63]; diffusion in food polymers[48,64]; viscoelastic properties of hydrated elastin rubbers,[22] concentrated, thermoset, aqueous gels of whey protein,[65] 7S and 11S soy protein isolates,[30] gelatinized starch,[31] pectin-co-solute,[66] various sugar/gum mixtures including glucose, sucrose, galactomannan, κ-carrageenan, and gellan gum,[67–69] and soy flour.[32]

Several physical and chemical changes in foods, whether deteriorative or not, depend on temperature. Even though the WLF equation has found wide application avenues, it has been constantly compared to Arrhenius model that is used to describe various kinetic phenomena as a function of temperature. Temperature ranges indicating the applicability of different models are shown in Fig. 3.[70] At high enough temperatures the WFL equation becomes inapplicable and the relaxation process is governed by more specific features.[2] Relaxation processes are generally described by Arrhenius kinetics above $T_g + 100°C$ or melting point, T_m. The Arrhenius type approach is also associated with glassy kinetics. When the temperature is lowered below T_m, more precisely between T_m and T_g, non-Arrhenius behavior with much stronger temperature dependence is observed. The WLF equation defines the kinetics of molecular-level relaxation processes that will take place in the rubbery state within the experimental time scale in terms of an exponential, but non-Arrhenius function of T-T_g.

The studies that compared both models revealed variations in different systems. Sa and Sereno[34] showed that the nonenzymatic browning reaction rates in dehydrated onions were described better with the WLF equation than Arrhenius model in a temperature range from 15 to 45°C. This behavior was consistently observed at a_w values of 0.33 ($T_g = -11°C$), 0.44 ($T_g = -21.1°C$), and 0.55 ($T_g = -26.1°C$). Sapode et al.[71] tested several methods including Arrhenius model and reported that the WLF equation was the most suitable for the viscosity–temperature relationship of Australian honeys. On the other hand, there are other reports showing that the reaction rates of anthocyanin degradation in strawberries and ascorbic acid degradation in frozen model systems were described by the WLF and Arrhenius models equally well.[34,61]

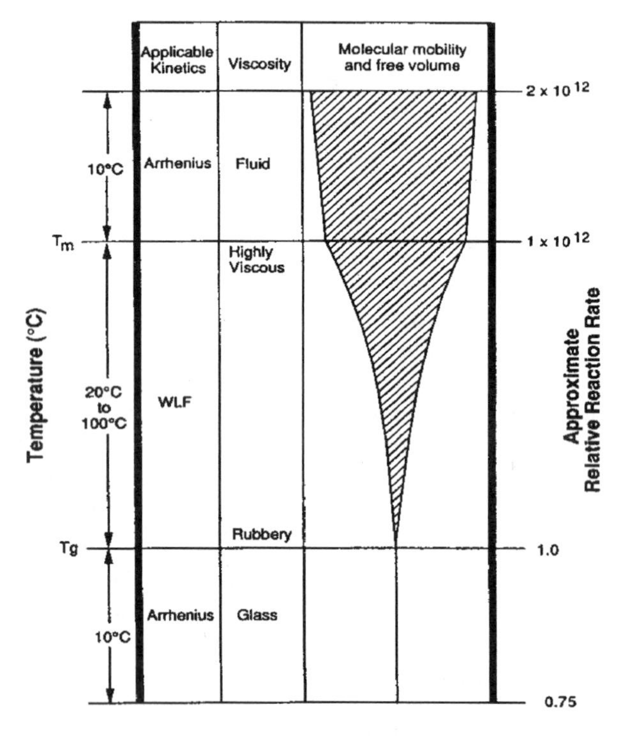

Fig. 3 Schematic interrelations among temperature, appropriate type of kinetics, viscosity, molecular mobility, free volume, and relative rates of diffusion-dependents events. (Reprinted from Ref. [70].)

The applicability of the WLF equation in the transition zone is based on fundamental principles. Ability of the WLF equation to define the temperature dependence of activation energy at the vicinity of the glass transition is one of these. When Arrhenius type dependence is extrapolated from rubbery to glassy state, there is usually a break in the curve and activation energies are generally underestimated. In the WLF consideration, activation energy is not constant as a function of temperature, but varies as temperature is changed. Therefore the plot of a viscoelastic property or relaxation time under consideration vs. $1/T$ exhibits a nonlinear form in the temperature range where the WLF equation is applicable.

Arrhenius model has been used to describe temperature effects on several physical/chemical changes in food systems with a great success. However, it has to be noted that the available data is mostly in a relatively narrow temperature range and generally well above the T_g of the systems studied. For a conclusive comparison of the applicability of Arrhenius model and the WLF equation to physical/chemical changes, experimental data over a wide temperature range ($T \leq T_g \leq T_g + 100°C$) should be obtained. This information can enable researchers to better evaluate the curvature in the plots of studied changes (e.g., viscosity) vs. temperature.

In many cases, the WLF equation was applied with universal constants. Several biological systems showed good agreement between the WLF equation and the experimental data when universal constants were used. For example, the temperature dependence of aqueous fructose:sucrose blend was described well using universal constants as shown in Fig. 4. The WLF equation suggests strong temperature dependence for molecular mobility such as diffusion,[20,48] viscosity,[40,42,72] and the rates of processes such as crystallization[49–51,53] over a 20°C interval above T_g, which is characteristic of WLF behavior in the rubbery fluid range.[37] Levi and Karel[48] showed that the rate of diffusion of propanol in low moisture sucrose rubbers increases by five orders of magnitude from T_g to $T_g + 20°C$. Slade and Levine[37] noted that viscosity data such as that of concentrated sugar solutions[40,42,73]

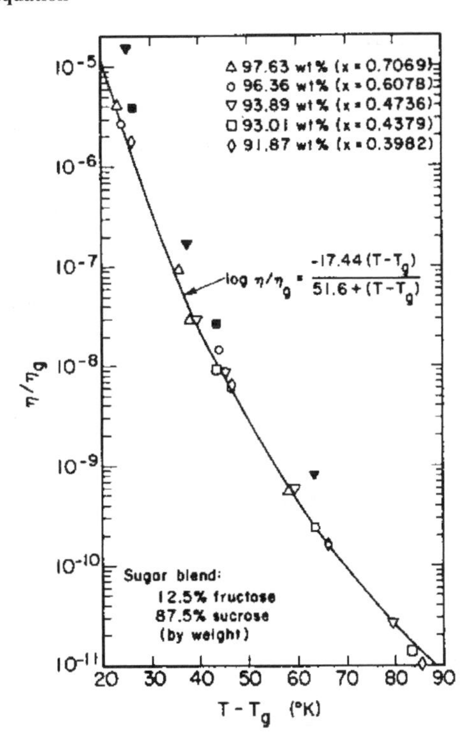

Fig. 4 Temperature dependence of viscosity of aqueous solutions of a 12.5:87.5 (w/w) fructose:sucrose blend. (Reprinted with permission from Ref. [40].)

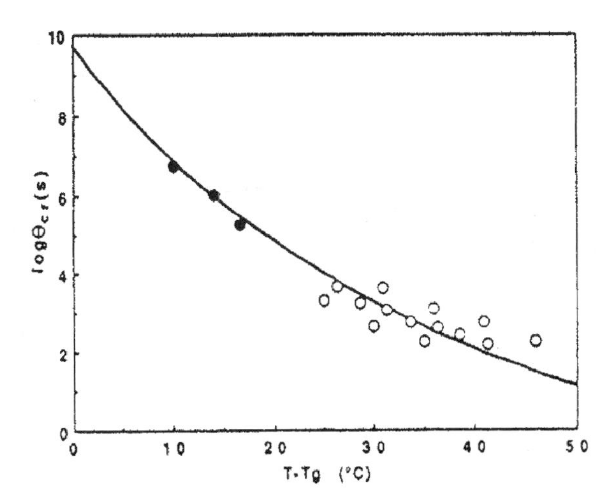

Fig. 5 Crystallization times (θ_{cr}) for amorphous lactose were predicted well with the WLF equation. (Adapted from Ref. [51].)

showed five orders of magnitude change, over 20°C interval near T_g, which is in good agreement with the WLF equation using universal constants. In these experiments viscosity at T_g was selected to be $\eta_g = 10^{12}$ Pa sec. Despite the lack of experimental evidence for the viscosity data at T_g, the T_g values predicted using the universal constants and $\eta_g = 10^{12}$ Pa sec, were in good agreement with the independent experimental measurements of T_g.[42,73] The crystallization of sucrose or lactose in completely amorphous powders at low moisture contents was also shown to change by five orders of magnitude over a 20°C interval above T_g as shown in Fig. 5.[49–53]

The WLF equation with universal constants was reported to fail for several other systems. The solubility of proteins in frozen cod and decrease of ascorbic acid in frozen peas deviated from the WLF kinetics with universal constants.[74] Upon evaluating the available data, Simatos, Blond, and Le Meste[74] concluded that the reaction rate constants in the temperature range of $T_g' < T < T_g = +40$°C showed a weaker temperature dependence than that predicted by the WLF equation with the universal constants. These deviations can be as a result of ice melting and dilution of reactants that masks accelerated reaction rates, ceased diffusivity limitations, and effect of other components on reaction kinetics.[61,74] Apparent failure of the WLF equation is merely the indication of the inapplicability of universal constants to the system under consideration rather than the WLF equation itself. Indeed, many researchers showed that the WLF equation, with values of the coefficients obtained using time–temperature-superposition principle or dictated by experimental data, correctly described the kinetics of several different phenomena.[29,30,32,33,37,66,75,76] The WLF constants obtained in these studies were different than the universal values. Lillie and Gosline[76] showed that over a range of water contents for rubbery elastin the average WLF coefficients were $C_1 = 9.7$ and $C_2 = 28.8$. Cocero and Kokini[75] noted that viscosity of glutenin (40% moisture w/w) could be described with the WLF equation using $C_1 = 10.8$ and $C_2 = 40.2$. The WLF constants for the browning reaction rates in onion at different water activities varied between 15.49–30.04 and 28.94–10.8 for C_1 and C_2, respectively.[34] The WLF constants for anthocyanin degradation rates in dehydrated strawberries varied with water activity and ranged from 53.2 to 124 for C_1 and from 20.8 to 733 for C_2.[34] The material specific WLF constants were reported for several foodstuff including Australian honeys, $C_1 = 13.7–21.1$ and $C_2 = 55.9–118.7$[71] and dehydrated vegetables, $C_1 = 7.2–25.1$ and $C_2 = 50–130$.[33]

Time–temperature superposition data has been applied to obtain viscoelastic properties of biopolymers such as soy protein isolates 7S and 11S,[30] soy flour films,[32] gelatinized starch,[31] and several high sugar–polysaccharide systems[66,67] over extended time or frequency ranges. Morales-Diaz and Kokini[30] showed that the WLF constants for 7S and 11S soy protein isolates were $C_1 = 31 \pm 3$, $C_2 = 207 \pm 23$ and $C_1 = 67 \pm 4$, $C_2 = 482 \pm 19$, respectively. These differ remarkably from the universal constants. Stress relaxation data and the resulting master curve of 7S-soy protein isolate

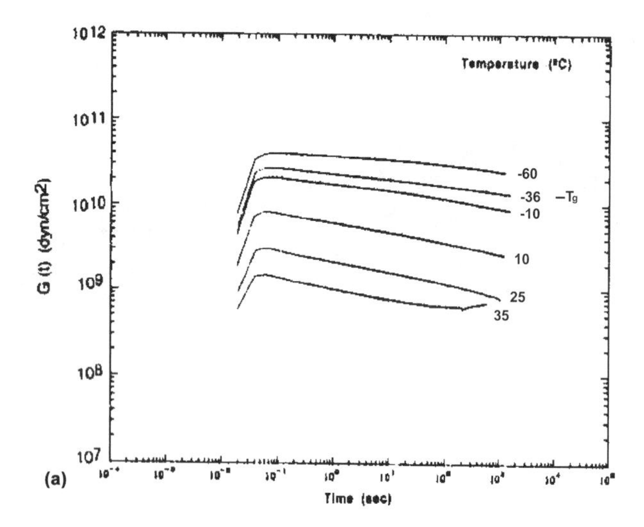

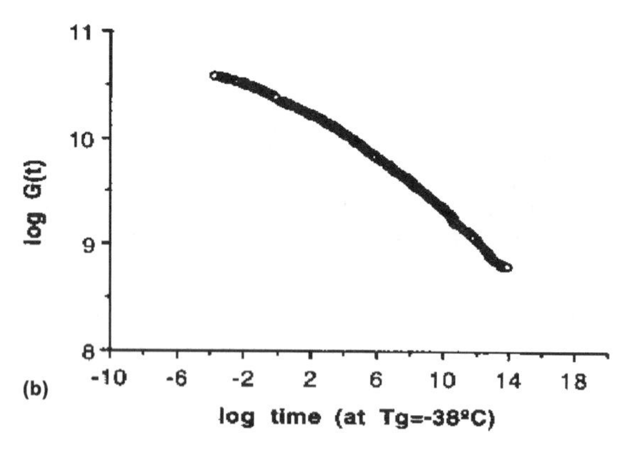

Fig. 6 Stress relaxation curves (a) of 7S soy flour isolate and resulting master curve (b) obtained from time–temperature superposition principle. (Reprinted from Ref. [30].)

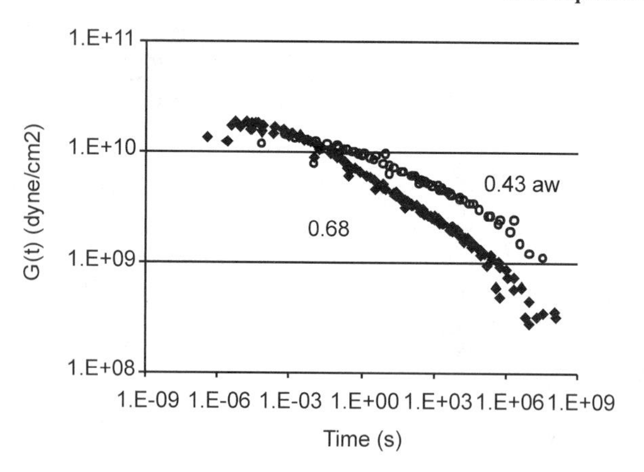

Fig. 7 Stress relaxation master curves for soy flour films at different water activities. (From Ref. [32].)

(20.8% moisture content) are shown in Fig. 6a, b. $T_g = -38°C$ was selected as the reference temperature to construct the master curve. The relaxation modulus shown in Fig. 6b is equivalent to the behavior that would have been obtained if stress relaxation experiments were carried out at the reference temperature (T_g) and enough time was available to conduct the experiment.[30] Kasapsis et al.[66] applied the time temperature superposition principle to pectin-co-solute mixture systems and by reducing the mechanical response to an iso-free volume temperature showed that temperature dependence can be explained by a single curve. Kasapsis, Sablani, and Biliaderis[31] also used a modified form of the WLF equation to account for the moisture loss in starch matrices at temperatures higher than 100°C and reported that $C_1 = 11.6$ and $C_2 = 2.1$.

The free volume in the temperature range between T_g and $T_g + 100°C$ controls the effect of temperature as well as molecular weight and plasticizers on the relaxation process. Therefore, factors that affect the available free volume and molecular mobility will have an impact on viscoelastic properties of materials under consideration. When a plasticizer is added to an undiluted polymer, the relaxation times decrease rapidly and the magnitude of the effect can be successfully interpreted in terms of additional free volume introduced by the plasticizer.[2,19]

Water is one of the most important plasticizers for food polymers. The effect of moisture content on the network structure of food polymers using the WLF equation was studied using the time–temperature superposition principle.[32] The stress relaxation master curves for several water activities (a_w) between 0.43 and 0.75 (7%–16% moisture content w/w) are given in Fig. 7. The experimental temperature closest to the average T_g of the material at a given moisture content was selected as the reference temperature. Soy flour films showed imperfect master curves, which have not been reported in the literature before. This is probably due to the complex nature of the material. The WLF constants in a_w range of 0.43–0.75 ranged between 17–27 for C_1 and 116–164 for C_2.[32] It was shown that when the mechanical response of soy flour films in a given moisture content range was reduced to a common temperature (20°C), the temperature dependence can be expressed by a single curve indicating that a change in moisture content did not effect the relaxation behavior of soy flour films. Shift factors as a function of temperature for four different a_w reduced to 20°C are given in Fig. 8.

Yildiz and Kokini[32] proposed the following equations that can be used to obtain the WLF constants at different moisture contents for soy flour films when the T'_g(K) at that moisture content is known:

$$C'_1 = \frac{2761}{430 - T'_g} \tag{10}$$

$$C'_2 = 430 - T'_g \tag{11}$$

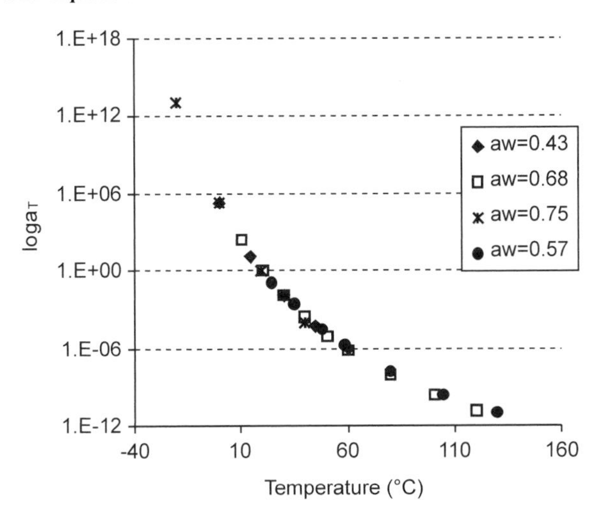

Fig. 8 Temperature dependence of the shift factor, for a fixed reference temperature of 20°C (293 K) (From Ref. [32].)

Similar sensitivity to water was observed for oriental lacquer films.[19]

In the temperature range where the WLF equation applies, $T - T_g$ (or $T - T'_g$ for frozen systems) and T_m/T_g are of key importance.[69] Slade and Levine[37] noted that in $\log a_T$ vs. $(T - T_g)$ plots, the ratio of T_m/T_g can be used to represent relative mobilities for polymers with a common value of T_g that have different crystallizing properties. Such plots are also used to differentiate kinetic properties and suitability of the WLF equation in the temperature region of interest. These plots are similar to the strong and fragile classification[72] of glass forming liquids.[37] High T_m/T_g values result in relatively weak temperature dependence, which are identified as strong liquid.[77] On the other hand, low T_m/T_g values correspond to fragile liquid classification with very strong temperature dependence. Angell[77] also noted that glassy water near (−137°C)[78] behaves like a very strong liquid, whereas, glucose, fructose, maltose, and sucrose behave somewhat fragile. Another interesting feature of the WLF equation is the similarity between WLF kinetics and Michaelis-Menten kinetics, which may enhance the practical utility of the WLF equation.[37]

CONCLUSION

The WLF equation is a useful tool to describe the strong temperature dependence of various processes in the glass transition region. The viscoelastic properties of biopolymers as well as the kinetics of several mobility-limited changes (e.g., nonenzymatic browning) in food systems have been studied using the WLF equation. Food polymers constitute major components of many food products and dictate their functional properties. The determination of the viscoelastic properties of biopolymers is essential in designing foods with well-defined reproducible properties for food processing, storage, and consumption.

Water is a very significant component of foods. It was shown that water activity and moisture content play important role in the viscoelastic properties of food materials. Time temperature superposition principle and certain modifications can be used to apply the WLF equation to determine moisture dependence of viscoelastic properties.

In many applications, the universal WLF constants are used to describe the effect of temperature on viscous flow, viscoelastic response, and reaction rates of both biomaterials and synthetic polymers. However, for food systems, the WLF constants have been shown to be material-specific properties that may significantly differ from the universal constants. Therefore, the use of the universal values for such food polymers should be avoided, and material-specific WLF constants should be determined experimentally.

REFERENCES

1. Williams, M.L.; Landel, R.F.; Ferry, J.D. The Temperature Dependence of Relaxation Mechanisms in Amorphous Polymers and Other Glass-Forming Liquids. J. Am. Chem. Soc. **1955**, *77*, 3701–3706.
2. Ferry, J.D. *Viscoelastic Properties of Polymers*, 3rd Ed.; John Wiley and Sons: New York, NY, 1980.
3. Koike, T. Relationship Between Melt Viscosity and Dielectric Relaxation Time for a Series of Epoxide Oligomers. J. Appl. Polym. Sci. **1993**, *47*, 387–394.
4. Gell, C.B.; Graessley, W.W.; Fetters, L. Viscoelastic and Self-diffusion in Melts of Entangled Linear Polymers. J. Polym. Sci., Part B: Polym. Phys. **1997**, *35*, 1933–1942.
5. Marchionni, G.; Ajroldi, G.; Cinquina, P.; Tampellini, E.; Pezzin, G. Physical Properties of Perfluoropolyethers: Dependence on Composition and Molecular Weight. Polym. Eng. Sci. **1990**, *30* (14), 829–834.
6. O'connor, K.M.; Scholsky, K.M. Free Volume Effects on the Melt Viscosity of Low Molecular Weight Poly(methyl methacrylate). Polymer **1989**, *30*, 461–466.
7. Sperling, L.H. *Introduction to Physical Polymer Science*; John Wiley & Sons: New York, NY, 1986.
8. Dean, D.; Husband, M.; Trimmer, M. Time –Temperature-Dependent Behavior of a Substituted Poly(paraphenylene): Tensile, Creep, and Dynamic Mechanical Properties in the Glassy State. J. Polym. Sci. Part B: Polym. Phys. **1998**, *70*, 2971–2979.
9. Fetters, L.J.; Graessley, W.W.; Kiss, A.D. Viscoelastic Properties of Polyisobutylene Melts. Macromolecules **1991**, *24*, 3136–3141.
10. Akay, M.; Rollins, S.N. Transition Broadening and WLF Relationship in Polyurethane/Poly(methyl methacrylate) Interpenetrating Polymer Networks. Polymer **1993**, *34* (5), 967–971.

11. Lobbrecht, A.; Friedrich, C.; Sernetz, F.G.; Mulhaupt, R. Viscoelastic Properties of Poly(ethylene-co-styrene) Copolymers. J. Appl. Polym. Sci. **1997**, *65*, 209–215.
12. Simon, S.L.; Plazek, D.J.; Sobieski, J.W.; McGregor, E.T. Physical Aging of a Polyetherimide: Volume Recovery and Its Comparison to Creep and Enthalpy Measurements. J. Polym. Sci. B: Polym. Phys. **1996**, *35*, 929–936.
13. Han, Y.; Yang, Y.; Li, B.; Feng, Z. Viscoelastic Properties of Phenolphthalein Poly(ether ketone). J. Appl. Polym. Sci. **1995**, *56*, 1349–1353.
14. Ngai, K.L.; Plazek, D.J. Temperature Dependencies of Viscoelastic Response of Polymer Systems. In *Physical Properties of Polymers Handbook*; Mark, J.E., Ed.; AIP Press: Woodbury, NY, 1996; 341–362.
15. Angell, C.A. Why $C_1 = 16 - 17$ in the WLF Equation Is Physical- and the Fragility of Polymers. Polymer **1997**, *38* (26), 6261–6266.
16. Read, B.E. Mechanical Relaxation in Isotactic Polypropylene. Polymer **1989**, *30*, 1439–1445.
17. Huy, M.L.; Evrard, G. Methodologies for Lifetime Predictions of Rubber Using Arrhenius and WLF Models. Die Angew. Makromol. Chem. **1998**, *261/262*, 135–142.
18. Ogata, M.; Kinjo, N.; Kawata, T. Effects of Crosslinking on Physical Properties of Phenol–Formaldehyde Novolac Cured Epoxy Resins. J. Appl. Polym. Sci. **1993**, *48*, 583–601.
19. Ogawa, T.; Inoue, A.; Osawa, S. Effect of Water on Viscoelastic Properties of Oriental Lacquer Film. J. Appl. Polym. Sci. **1998**, *69*, 315–321.
20. Ehlich, D.; Sillescu, H. Tracer Diffusion at the Glass Transition. Macromolecules **1990**, *23*, 1600–1610.
21. Kim, H.; Waldow, D.A.; Han, C.C.; Tran-Con, Q.; Yamamoto, M. Temperature Dependence of Probe Diffusion in Bulk Polymer Matrices. Polym. Com. **1991**, *32* (4), 108–112.
22. Gosline, J.M. Structure and Mechanical Properties of Rubber-like Proteins in Animals'. Rubber Chem. Technol. **1987**, *60*, 417–438.
23. Ward, I.M. *Mechanical Properties of Solid Polymers*; Wiley-Interscience: New York, NY, 1971.
24. Ferry, J.D. Temperature Dependence of Viscoelastic Properties: The Fitzgerald Apparatus and the WLF Equation. J. Polym. Sci. Part B: Polym. Phys. **1999**, *37* (7), 621–622.
25. Aklonis, J.J.; MacKnight, W.J. *Introduction to Polymer Viscoelasticity*, 2nd Ed.; John Wiley and Sons: New York, NY, 1983.
26. Shutilin, Y.F. Use of the Williams–Landel–Ferry and Arrhenius Equations in Describing the Relaxational Properties of Polymers and Polymer Homologues. Vysokomol. Soyed. **1991**, A*33* (1), 120–127.
27. Hamed, G. Free Volume Theory and the WLF Equation. Elastomerics **1988**, *1*, 14–17.
28. Doolittle, A.K. Studies in Newtonian flow. II. The Dependence of the Viscosity of Liquids on Free-Space. J. Appl. Phys. **1951**, *22* (12), 1471–1475.
29. Peleg, M. On the Use of the WLF Model in Polymers and Foods. Crit. Rev. Food Sci. Nutr. **1992**, *32*, 59–66.
30. Morales-Diaz, A.; Kokini, J.L. Understanding Phase Transitions and Chemical Complexing Reactions in 7S and 11S Soy Protein Fractions. In *Phase/State Transitions in Foods*; Rao, M.A., Hartel, R.W., Eds.; Marcel Dekker: New York, NY, 1998; 273–313.
31. Kasapsis, S.; Sablani, S.S.; Biliaderis, C.G. Dynamic Oscillation Measurements of Starch Networks at Temperatures Above 100°C. Carbohydr. Res. **2000**, *329*, 179–187.
32. Yildiz, M.E.; Kokini, J.L. Determination of WLF Constants for a Food Polymer System: Effect of Water Activity and Moisture Content. In Review. J. Rheol. **2001**, *45*, 903–912.
33. Buera, M.P.; Karel, M. Application of the WLF Equation to Describe the Combined Effects of Moisture and Temperature on Nonenzymatic Browning Rates in Food Systems. J. Food Process. Preserv. **1993**, *17*, 31–45.
34. Sa, M.M.; Sereno, A.M. The Kinetics of Browning Measured During the Storage of Onion and Strawberry. Int. J. Food Sci. Technol. **1999**, *34*, 343–349.
35. Nelson, K.A.; Labuza, T.P. Water Activity and Food Polymer Science: Implications of State on Arrhenius and WLF Models in Predicting Shelf Life. J. Food Eng. **1994**, *22*, 271–289.
36. Plazek, D.J.; Chelko, A.J. Temperature Dependence of the Steady State Recoverable Compliance of Amorphous Polymers. Polymer **1977**, *18*, 15–18.
37. Slade, L.; Levine, H. Glass Transitions and Water–Food Structure Interactions. Adv. Food Nutr. Res. **1995**, *38*, 103–269.
38. Levine, H.; Slade, L. A Polymer Physico-chemical Approach to the Study of Commercial Starch Hydrolysis Products (SHPS). Carbohydr. Polym. **1986**, *6*, 213–244.
39. Levine, H.; Slade, L. A Food Polymer Science Approach to Cryostabilization Technology. Comments Agric. Food Chem. **1989**, *1*, 315–396.
40. Soesanto, T.; Williams, M.C. Volumetric Interpretation of Viscosity for Concentrated and Diluted Sugar Solutions. J. Phys. Chem. **1981**, *85*, 3338–3341.
41. Maltini, E.; Anese, M. Evaluation of Viscosities of Amorphous Phases in Partially Frozen Systems by WLF Kinetics and Glass Transition Temperatures. Food Res. Int. **1995**, *28* (4), 367–372.
42. Ollet, A.L.; Parker, P. The Viscosity of Supercooled Fructose and Its Glass Transition Temperature. J. Texture Stud. **1990**, *21*, 355–362.
43. Raudonus, J.; Bernard, J.; JanBen, H.; Kowalczyk, J.; Carle, R. Effect of Oligomeric and Polymeric Additives on Glass Transition, Viscosity and Crystallization of Amorphous Isomalt. Food Res. Int. **2000**, *33*, 41–51.
44. Campella, O.H.; Popplewell, L.M.; Rosenau, J.R.; Peleg, M. Elongational Viscosity Measurements of Melting American Process Cheese. J. Food Sci. **1987**, *52*, 1249–1251.
45. Aguilera, J.M.; Levi, G.; Karel, M. Effect of Water Content on the Glass Transition and Caking of Fish Protein Hydrolyzates. Biotechnol. Prog. **1993**, *9*, 651–654.
46. Aguilera, J.M.; del Valle, J.M.; Karel, M. Caking Phenomena in Amorphous Powders. Trends Food Sci. Technol. **1995**, *6*, 149–155.

47. Levi, G.; Karel, M. Volumetric Shrinkage (Collapse) in Freeze-Dried Carbohydrates Above Their Glass Transition Temperature. Food Res. Int. **1995**, *28* (2), 145–151.
48. Levi, G.; Karel, M. The Effect of Phase Transitions on Release of *n*-Propanol Entrapped in Carbohydrate Glasses. J. Food Eng. **1995**, *24*, 1–13.
49. Roos, Y.H.; Karel, M. Plasticizing Effect of Water on Thermal Behavior and Crystallization of Amorphous Food Materials. J. Food Sci. **1991**, *56*, 38–43.
50. Roos, Y.H.; Karel, M. Phase Transitions of Amorphous Sucrose and Frozen Sucrose Solutions. J. Food Sci. **1991**, *56* (1), 266–267.
51. Roos, Y.H.; Karel, M. Crystallization of Amorphous Lactose. J. Food Sci. **1992**, *57* (3), 775–777.
52. Labrousse, S.; Roos, Y.; Karel, M. Collapse and Crystallization in Amorphous Matrices with Encapsulated Compounds'. Sci. Aliment. **1992**, *12*, 757–769.
53. Shimada, Y.; Roos, Y.; Karel, M. Oxidation of Methyl Linoleate Encapsulated in Amorphous Lactose-Based Food Model. J. Agric. Food Chem. **1991**, *39*, 637–641.
54. Franzen, K.; Singh, R.K.; Okos, M.R. Kinetics of Nonenzymatic Browning in Dried Skim Milk. J. Food Eng. **1990**, *11*, 225–239.
55. Buera, M.P.; Karel, M. Effect of Physical Changes on the Rates of Nonenzymatic Browning and Related Reactions. Food Chem. **1995**, *52*, 167–173.
56. Roos, Y.H.; Himberg, M.J. Nonenzymatic Browning Behavior, as Related to Glass Transition, of a Food Model at Chilling Temperatures. J. Agric. Food Chem. **1994**, *42*, 893–898.
57. Karel, M.; Buera, P.; Roos, Y. Effect of Glass Transitions on Processing and Storage. In *The Glassy State in Foods*, 1st Ed.; Blanshard, J.M.V., Lillford, P.J., Eds.; Nottingham University Press: Loughborough, 1993; 13–35.
58. Karmas, R.; Buera, M.P.; Karel, M. Effect of Glass Transition on Rates of Nonenzymatic Browning in Food Systems. J. Agric. Food Chem. **1992**, *40*, 873–879.
59. Slade, L.; Levine, H. Beyond Water Activity: Recent Advances on an Alternative Approach to the Assessment of Food Quality and Safety. CRC Crit. Rev. Food Sci. Nutr. **1991**, *30*, 115–360.
60. Kerr, W.L.; Miang, H.L.; Reid, S.D.; Chen, H. Chemical Reaction Kinetics in Relation to Glass Transition Temperatures in Frozen Food Polymer Solutions. J. Sci. Food Agric. **1993**, *61*, 51–56.
61. Biliaderis, C.G.; Swan, R.S.; Arvanitoyannis, I. Physicochemical Properties of Commercial Starch Hydrolyzates in the Frozen State. Food Chem. **1999**, *64* (4), 537–546.
62. Sapru, V.; Labuza, T.P. Glassy State in Bacterial Spores Predicted by Polymer Glass-Transition Theory. J. Food Sci. **1993**, *58* (2), 445–448.
63. Schaffner, D.W. The Application of the WLF Equation to Predict Lag Time as a Function of Temperature for Three Psychrotropic Bacteria. Int. J. Food Microbiol. **1995**, *27*, 107–115.
64. Roos, Y.H. Glass Transition-Related Physicochemical Changes in Foods. Food Technol. **1995**, *10*, 97–102.
65. Katsuta, K.; Miura, M.; Nishimura, A. Kinetic Treatment for Rheological Properties and Effects of Saccharides on Retrogradation of Rice Starch Gels. Food Hydrocoll. **1992**, *6*, 187–198.
66. Kasapsis, S.; Al-Alawi, A.; Guizani, N.; Khan, A.J.; Mitchell, J.R. Viscoelastic Properties of Pectin-Co-solute Mixtures at Iso-free-volume States. Carbohydr. Res. **2000**, *329*, 399–407.
67. Kasapsis, S.; Al-Marhoobi, I.M.A.; Khan, A.J. Viscous Solutions, Networks and the Glass Transition in High Sugar Galactomannan and κ-Carrageenan Mixtures. Int. J. Biol. Macromol. **2000**, *27*, 13–20.
68. Sworn, G.; Kasapsis, S. The Use of Arrhenius and WLF Kinetics to Rationalise the Mechanical Spectrum in High Sugar Gellan Systems. Carbohydr. Res. **1998**, *309*, 353–361.
69. Kasapsis, S. Advanced Topics in the Application of the WLF/Free Volume Theory to High Sugar/Biopolymer Mixtures: A Review. Food Hydrocoll. **2001**, *15*, 631–641.
70. Fennema, O.R. Water and Ice. In *Food Chemistry*, 3rd Ed.; Fennema, O.R., Ed.; Marcel Dekker: New York, NY, 1996; 17–95.
71. Sopade, P.A.; Halley, P.; Bhandari, B.; D'Arcy, B.; Doebler, C.; Caffin, N. Application of the Williams–Landel–Ferry Model to the Viscosity–Temperature Relationship of Australian Honeys. J. Food Eng. **2003**, *56*, 67–75.
72. Bellows, R.J.; King, C.J. Product Collapse During Freeze Drying of Liquid Foods. AIChE Symp. Ser. **1973**, *69* (132), 33–41.
73. Noel, T.R.; Ring, S.G.; Whittam, M.A. Kinetic Aspects of the Glass-Transition Behavior of Maltose–Water Mixtures. Carbohydr. Res. **1991**, *212*, 109–117.
74. Simatos, D.; Blond, G.; Le Meste, M. Relation Between Glass Transition and Stability of a Frozen Product. Cryo-Lett. **1989**, *10*, 77–84.
75. Cocero, A.M.; Kokini, J.L. Prediction of Temperature Dependence of the Apparent Shear Viscosity of 40% Moisture Glutenin Using Arrhenius and WLF Equations. Inst. Food Techol. Annu. Meet., New Orleans, LA, 1992.
76. Lillie, M.A.; Gosline, J.M. The Effects of Hydration on the Dynamic Mechanical Properties of Elastin. Biopolymers **1990**, *29*, 1147–1160.
77. Angell, C.A.; Bressel, R.D.; Green, J.L.; Kanno, H.; Oguni, M.; Sare, E.J. Liquid Fragility and the Glass Transition in Water and Aqueous Solutions. In *ISOPW-V-Water Foods: Fundamental Aspects and Their Significance in Relation to Processing of Foods*, 1st Ed.; Fito, P., Mulet, A., McKenna, B., Eds.; Elsevier Applied Science: London, 1994; 115–143.
78. Johari, G.P.; Hallbrucker, A.; Mayer, E. The Glass–Liquid Transition of Hypersequenced Water. Nature **1987**, *330*, 552–553.

Index

Brief Contents

FIG. 325. AN INLAID STANDARD OF UR. (*Second Section.*)

By permission of the University Museum.

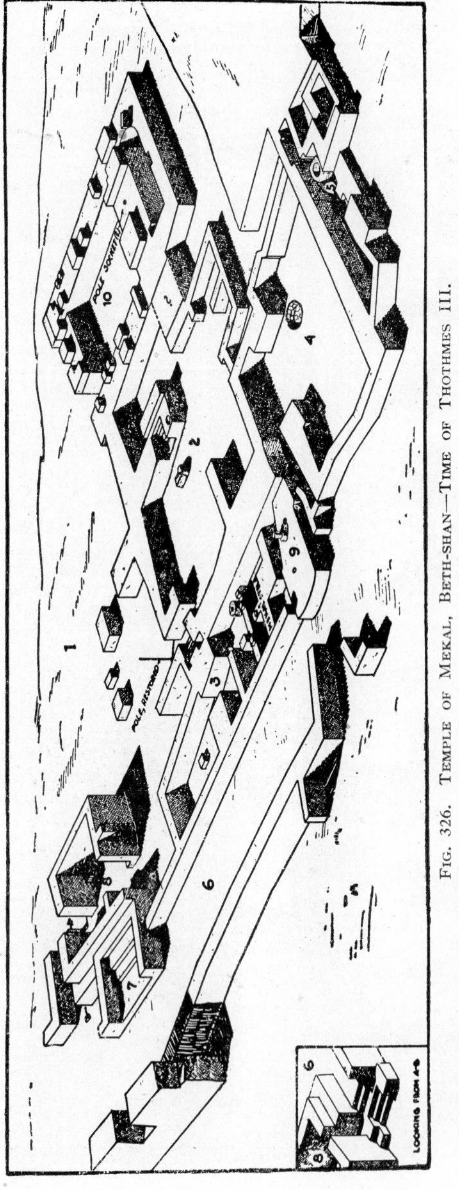

FIG. 326. TEMPLE OF MEKAL, BETH-SHAN—TIME OF THOTHMES III.

By permission of the University Museum.

FIG. 327. TEMPLE OF AMENOPHIS III, BETH-SHAN (RESTORED).
By permission of the Palestine Exploration Fund.

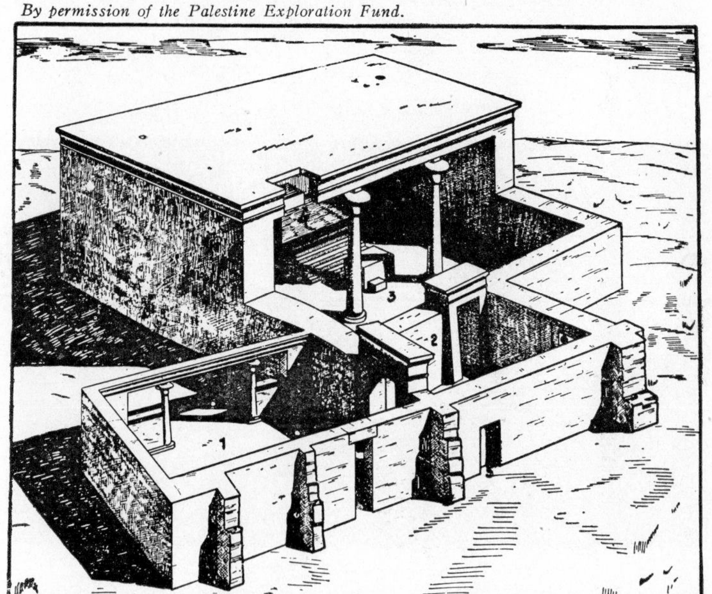

FIG. 328. TEMPLE OF SETI I, BETH-SHAN (RESTORED).
By permission of Palestine Exploration Fund.

FIG. 329. TEMPLE OF RAMSES II, BETH-SHAN (RESTORED. PART REMOVED TO SHOW INTERIOR).

By permission of Palestine Exploration Fund.

FIG. 330. SOLOMON'S STABLES, MEGIDDO (RESTORED).

By permission of the Oriental Institute of the University of Chicago.

FIG. 331. ARCHITECTURAL RESTORATION OF A STONE AGE TEMPLE AT TEPE GAWRA, STRATUM IX

By permission of the American Schools of Oriental Research

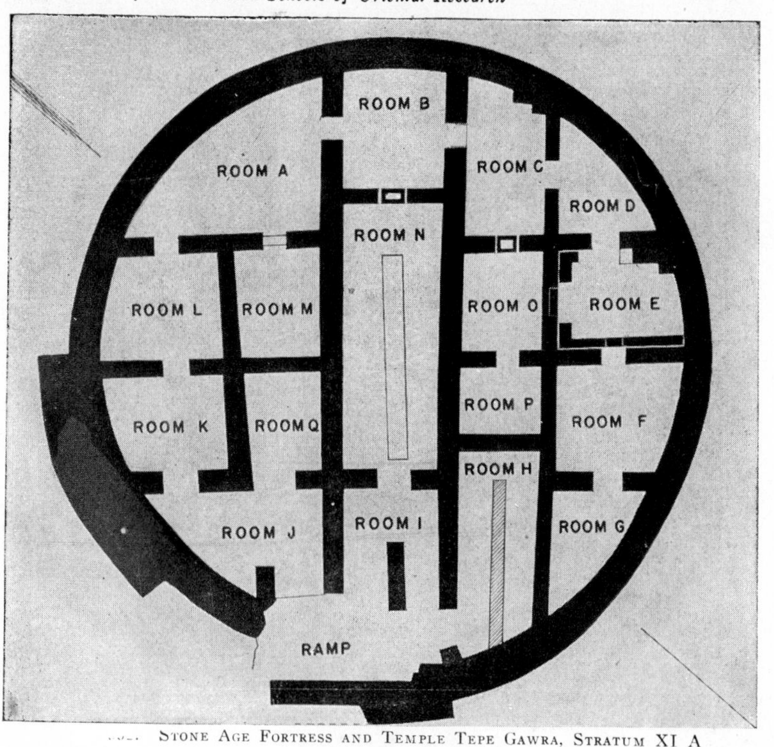

STONE AGE FORTRESS AND TEMPLE TEPE GAWRA, STRATUM XI A

By permission of the American Schools of Oriental Research

FIG. 333. ENTRANCE TO LARGE COPPER MINE AT UMM EL-AMAD, IN THE ARABAH. PROBABLY WORKED BY SOLOMON

By permission of the London Illustrated News

Copy photograph from *Oriental Institute Communications*, No. 16, p. 39.

FIG. 334. A CHILD'S TOY OF POTTERY FROM TELL ASMAR—2800–2600 B.C. A BROKEN EYELET FOR TYING A PULL CORD IS ON ITS BREAST

By permission of the Oriental Institute of the University of Chicago

FIG. 335. CYLINDER SEAL OF INDIAN-MAKE FOUND AT TELL ASMAR IN A HOUSE EARLIER THAN 2600 B.C.

By permission of the Oriental Institute of the University of Chicago

FIG. 336. STAMP SEALS FROM INDIA, SHOWING ELEPHANT AND RHINOCEROS SIMILAR TO THOSE FROM TELL ASMAR

By permission of the American Schools of Oriental Research

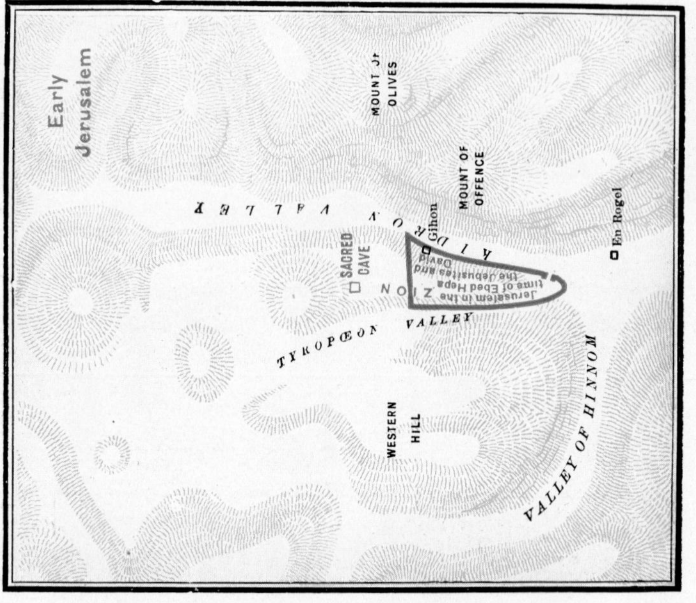
Early Jerusalem
MOUNT Jr OLIVES
MOUNT OF OFFENCE
KIDRON VALLEY
Gihon
En Rogel
SACRED CAVE
ZION
Jerusalem in the time of Ebed Hepa the Jebusites and David
TYROPŒON VALLEY
WESTERN HILL
VALLEY OF HINNOM

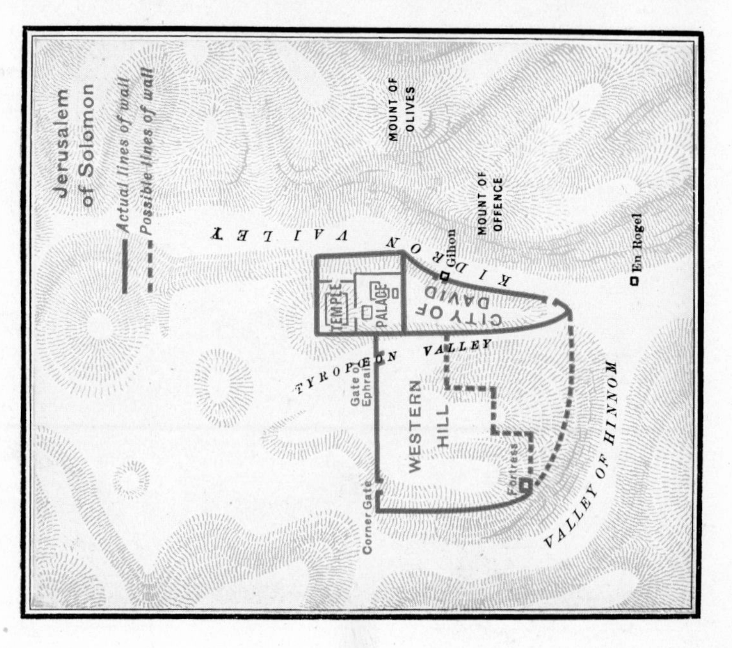
Jerusalem of Solomon
Actual lines of wall
Possible lines of wall
MOUNT OF OLIVES
MOUNT OF OFFENCE
KIDRON VALLEY
Gihon
En Rogel
TEMPLE
PALACE
CITY OF DAVID
TYROPŒON VALLEY
Gate of Ephraim
WESTERN HILL
Fortress
Corner Gate
VALLEY OF HINNOM

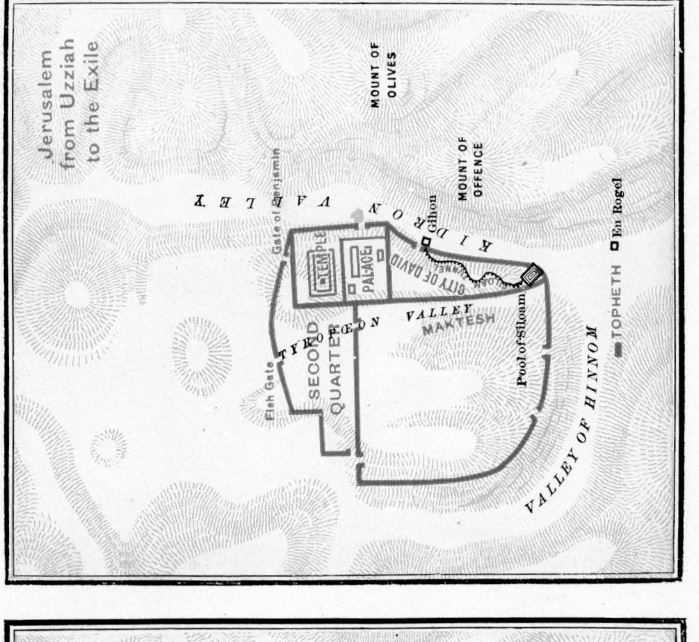
Jerusalem from Uzziah to the Exile
MOUNT OF OLIVES
MOUNT OF OFFENCE
KIDRON VALLEY
Gate of Benjamin
TEMPLE
PALACE
CITY OF DAVID
Gihon
SECOND QUARTER
Fish Gate
TYROPŒON VALLEY
MAKTESH
Pool of Siloam
VALLEY OF HINNOM
TOPHETH
En Rogel

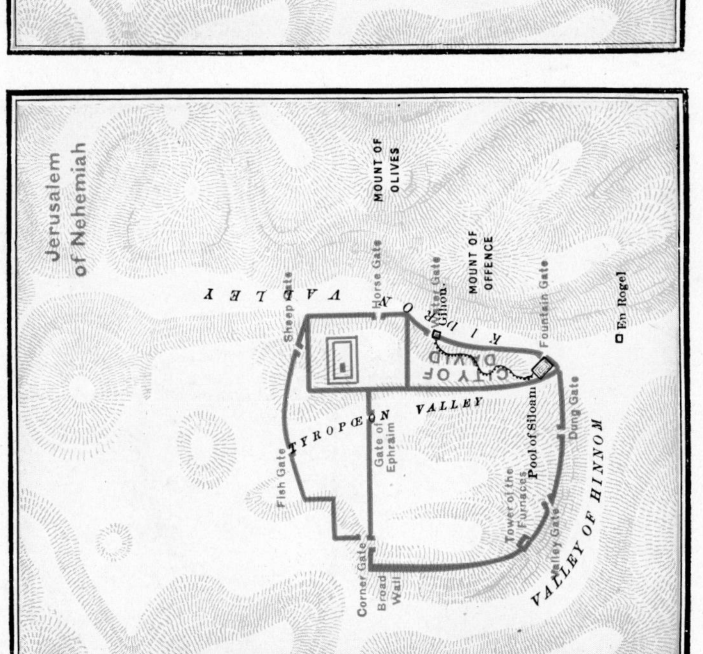
Jerusalem of Nehemiah
MOUNT OF OLIVES
MOUNT OF OFFENCE
KIDRON VALLEY
Sheep Gate
Horse Gate
Water Gate
Gihon
Fountain Gate
CITY OF DAVID
TYROPŒON VALLEY
Fish Gate
Gate of Ephraim
Corner Gate
Broad Wall
Tower of the Furnaces
Pool of Siloam
Dung Gate
Valley Gate
VALLEY OF HINNOM
En Rogel

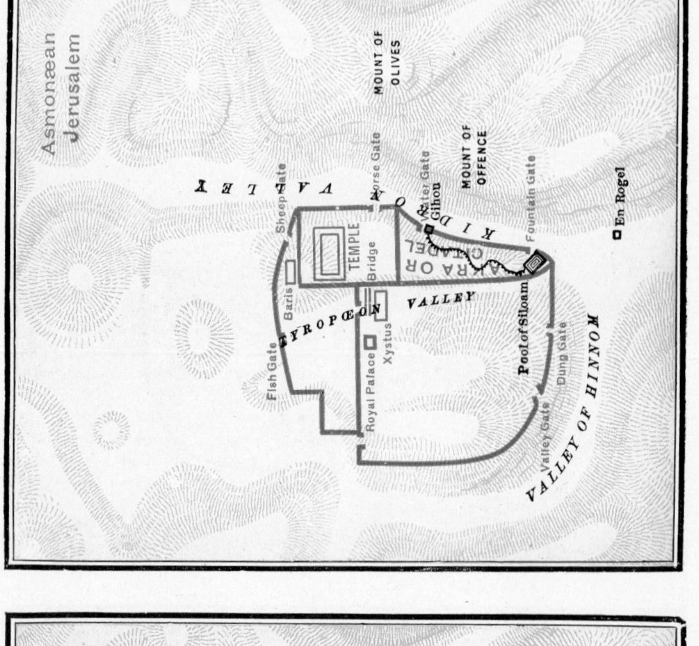
Asmonæan Jerusalem
MOUNT OF OLIVES
MOUNT OF OFFENCE
VALLEY
KIDRON
Sheep Gate
Horse Gate
Water Gate
Gihon
Fountain Gate
En Rogel
TEMPLE
Bridge
AKRA OR CITADEL
Baris
TYROPŒON
VALLEY
Pool of Siloam
Fish Gate
Royal Palace
Xystus
Dung Gate
Valley Gate
VALLEY OF HINNOM

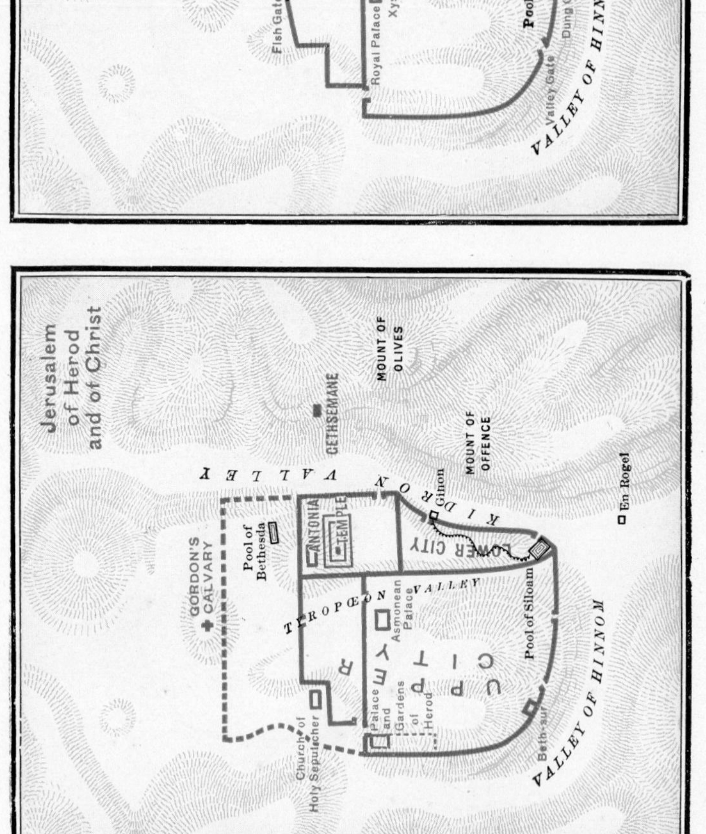
Jerusalem of Herod and of Christ
MOUNT OF OLIVES
GETHSEMANE
MOUNT OF OFFENCE
VALLEY
KIDRON
Gihon
En Rogel
GORDON'S CALVARY
Pool of Bethesda
ANTONIA
TEMPLE
LOWER CITY
TYROPŒON VALLEY
Asmonean Palace
UPPER CITY
Pool of Siloam
Palace and Gardens of Herod
Church of Holy Sepulcher
Beth-sur
VALLEY OF HINNOM